P9-CKX-060

Riegel's

Handbook of Industrial Chemistry

Riegel's

Handbook of Industrial Chemistry

EIGHTH EDITION

Edited by

James A. Kent

Dean of the College of Engineering and Science
University of Detroit
Detroit, Michigan

VNR VAN NOSTRAND REINHOLD COMPANY
NEW YORK CINCINNATI TORONTO LONDON MELBOURNE

Library of Congress Catalog Card Number: 82-4806
ISBN: 0-442-20164-8

Manufactured in the United States of America

Published by Van Nostrand Reinhold Company Inc.
135 West 50th Street, New York, N.Y. 10020

Van Nostrand Reinhold Publishing
1410 Birchmount Road
Scarborough, Ontario MIP 2E7, Canada

Van Nostrand Reinhold
480 Latrobe Street
Melbourne, Victoria 3000, Australia

Van Nostrand Reinhold Company Limited
Molly Millars Lane
Wokingham, Berkshire, England

15 14 13 12 11 10 9 8 7 6 5 4 3 2 1

Library of Congress Cataloging in Publication Data
Riegel, Emil Raymond, 1882–1963.
Riegel's Handbook of industrial chemistry.

Includes bibliographical references and index.
1. Chemistry, Technical. I. Kent, James Albert, 1922- II. Title. III. Title: Handbook of industrial chemistry.
TP145.R54 1982 660 82-4806
ISBN 0-442-20164-8 AACR2

To my Wife
ANITA

Preface

As always, the aim of this book is to present an up-to-date account of the many facets of as broad a cross section of the chemical process industry as is reasonable in a single volume of this size. Since the last edition, which was published in 1974, the industry has continued to demonstrate that is is still in a period of dynamic change and that while some segments are more mature than others, none can be described as static. Indeed, developments in some areas of the industry can only be described as revolutionary and exciting. This edition brings together the contributions of thirty-four experts from industry, academe, and government. Many of these are probably known to the reader for their numerous contributions to scholarly literature. Through their excellent work, contents of the chapters from the previous edition have at the least been updated and revised. Several of the chapters, which now have new authors, have been completely redone, using new approaches and different perspectives.

The contributors have done an outstanding job, and credit for the quality of the individual chapters goes to them. Errors of omission or duplication, and shortcomings in organization of the material are mine.

Grateful acknowledgment is made to the editors of technical magazines and publishing houses for permission to reproduce illustrations and other material and to the many industrial concerns which contributed drawings and photographs.

Comments and criticism by readers will be welcome.

JAMES A. KENT
Birmingham, Michigan

Contents

Riegel's

Handbook of Industrial Chemistry

1

Economic Aspects of the Chemical Industry

F. E. Bailey, Jr.*

Within the formal departments of science at the traditional university, chemistry has grown to have a unique status because of its close correspondence with an industry, the chemical industry, and a branch of engineering - chemical engineering. There is no biology industry, though drugs, pharmaceuticals, and agriculture are closely related disciplines. While there is no physics industry, there are the power generation and electronics industries. But, connected with chemistry, there is an industry. This unusual correspondence probably came about because in chemistry one makes things–chemicals–and the science and the use of chemicals more or less grew up together during the past century.

Since there is a chemical industry, which serves a major part of all industrialized economies, providing in the end synthetic drugs, fertilizers, clothing, building materials, paints, elastomers, etc., there is also a subject, "chemical economics," and it is this subject, the economics of the chemical industry, which is the concern of this chapter.

*Union Carbide Corporation, South Charleston, West Virginia 25303.

DEFINITION OF THE CHEMICAL INDUSTRY

Early in this century, the chemical industry was considered to have two parts: the manufacture of inorganic chemicals and the manufacture of organic chemicals. Today, the Standard Industrial Classification (SIC Index) of the United States Bureau of the Census defines "Chemical and Allied Products" as comprising three general classes of products: "(1) basic chemicals such as acids, alkalis, salts, and organic chemicals; (2) chemicals to be used in further manufacture such as synthetic fibers, plastics materials, dry colors, and pigments; and (3) finished chemical products to be used for ultimate consumption as drugs, cosmetics, and soaps, or to be used as materials or supplies in other industries such as paints, fertilizers, and explosives."[1] An even broader description that is often considered is "chemical process industries," major segments of which include: chemical and allied products and petrochemicals; paper and pulp; petroleum refining; rubber and plastics; and stone, clay and glass products.

THE PLACE OF THE CHEMICAL INDUSTRY IN THE ECONOMY

The total value of manufacturer's sales and shipments in the United States in 1979 was about \$1.7 trillion. In comparison with this total, the value of sales and shipments of the chemical and allied products industry was about \$149 billion. For perspective, these chemical sales were equal to 60 percent of the total value of food and food products and about two-and-a-half times that of paper product shipments and sales and about the same sales value as that of all primary metals. In Table 1.1, these chemical sales to selected markets are indicated for 1971[1]. It is interesting to note that "Chemicals and Allied Products" is the industry's own best customer, reflecting the sale of reactive chemical intermediates used in the manufacture of more complex chemical products.

To further gauge the place of chemicals in the economy of the United States, comparisons in growth can be made with the *gross national product* (GNP). The gross national product, an index of the size of the economy, is the sum for any one year of a nation's output in terms of expenditures for goods and services by consumers, government, business, and foreign interests; that is, it is the total of personal consumption, government purchases, gross private domestic investment, and net export of goods and services. For 1979, the gross national product of the United States was over two-and-a-quarter trillion dollars (Fig. 1.1). In terms of current dollars (reflecting both real growth and inflation) the United States' GNP doubled between 1950 and 1960 (\$286 billion to \$506 billion) and essentially

TABLE 1.1 Chemical Industry Sales to Selected Markets in the United States, 1971

Market	*Percent of Sales*
Chemical and Allied Products	46%
Textile Mill Products (Including apparel industry)	16
Rubber and Plastic Products	11
Food and Farming	6
Paper and Allied Products	4
Petroleum Industry	3

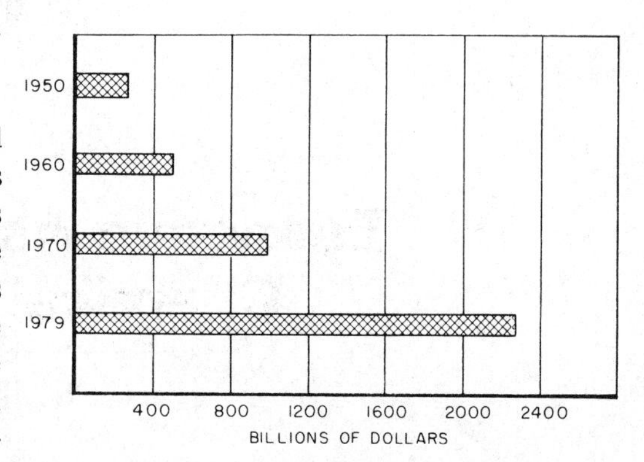

Fig. 1.1. The gross national product (GNP) of the United States in current dollars for the years 1950–1979.

doubled again by 1970 (\$982 billion). These figures indicate average growth rates for the GNP of about 7.2 percent per year of which about four percent represents real growth in output of goods and services and the remainder, inflation factors.

In these terms, the chemical and allied products industry kept pace with the general economy. Sales in 1950 amounted to about \$15.4 billion and in 1960, \$26.3 billion. In 1970, sales amounted to about \$50 billion - essentially doubling in dollar value each decade, producing an industry annual growth rate of 7–8 percent per year.

In the decade of the 1970's, however, inflation became an enormous factor in the economy. In 1979, the GNP of the United States in current dollars was \$2369 billion. While a number of factors have been identified as the basis for inflation, the principal factors in the chemical industry have been the cost of energy (power and heat) and the cost of raw materials (petroleum and natural gas). It is instructive, then, to put the GNP in terms of constant dollars, so as to be an index of the "volume" of goods and services produced. In Figure 1.2, the GNP is expressed in terms of 1972 dollars in which case the GNP in 1970 is \$1075.3 billion and in 1979, \$1431.6 billion. In terms of either of these indices, GNP in current or constant dollars, the chemical industry grew in the 1970's at a pace faster than that of the overall economy. Chemical and allied products sales in 1979 in current dollars were about \$149 billion (Figure 1.3).

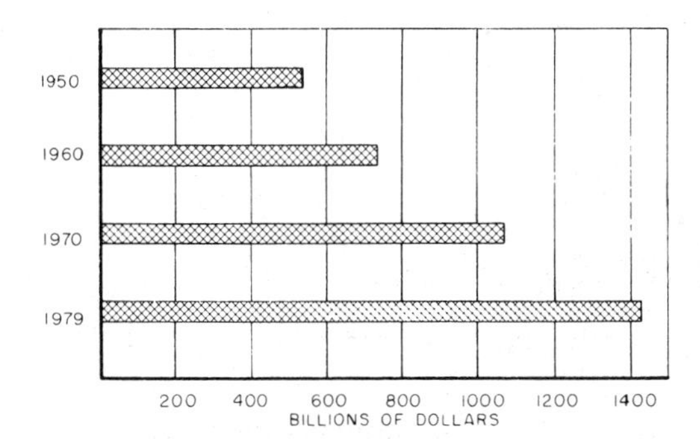

Fig. 1.2. The gross national product (GNP) of the United States for the years 1950–1979 in constant dollars (1972).

These average growth rates for an entire industry cover a broad spectrum of rates of growth for various product classes. The historic growth of chemicals has been characterized by rapid development of new products which in time achieve the status of high volume, bulk-shipment products having an established place in the economy with correspondingly slower overall growth. Meanwhile, new materials are introduced which grow at a more rapid rate. The net result is an industry growing with the economy and an aggregate growth somewhat faster than GNP.

A spectrum of the chemical industry is examined in more detail in Table 1.2. The industry growth rate in the 1970's in value added was about 12 percent per year. The highest growth rates in value added tend to be in high technology areas with new products (Table 1.2). Agricultural chemicals is an area of both increased demand and high technology, reflecting the leadership of the United States in agricultural technology and production. On the other hand, synthetic rubber in the 1970's experienced a relatively low growth rate, reflecting the importance of the automobile tire (decline in sales growth) in this market as well as rapid inflation in raw material costs.

TABLE 1.2 1972–1977 Growth Rates in Value Added in the United States for Selected Parts of the Chemical Process Industries

Agricultural Chemicals	24.0%/year
Surface Active Agents	17.9
Plastics Materials	17.2
Adhesives and Sealants	14.5
Industrial Gases	12.4
Toilet Preparations	12.2
Chemical and Allied Products	12.0
Paints	11.8
Pharmaceutical Preparations (Human and veterinary)	8.9
Organic, Non-cellulosic fibers	8.0
Soaps and other Detergents	7.8
Cellulosic Fibers	6.7
Synthetic Rubber	3.2

While growth in total dollar volume of the chemicals produced has kept pace or exceeded the growth of the total United States' economy, the actual production of chemicals has grown even faster. In terms of physical volume—the tons produced annually—the chemical process industries have grown faster than the industrial average represented by the Federal Reserve Board index of industrial production (1967 = 100). (Fig. 1-4.) Relative to 1967, the industrial production index in 1979 was 152.1, while that for chemical and allied products was 210.4 and plastic products was 270.5. These industry indices cover a broad spectrum of product classes (Table 1.3), ranging in growth from the huge increases in plastics, synthetic fibers,

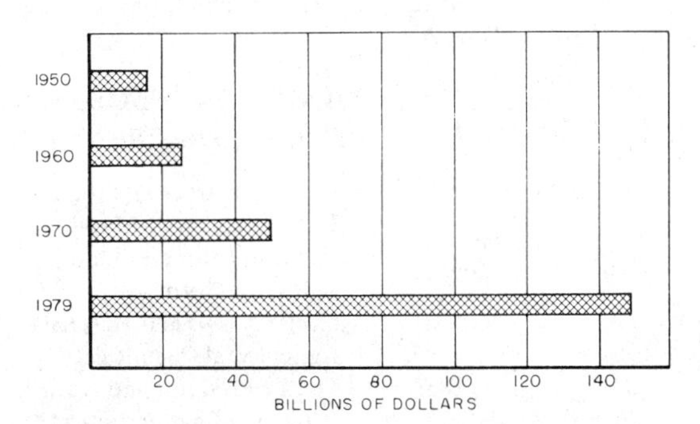

Fig. 1.3. Chemical and allied product sales in the United States for the years 1950–1979.

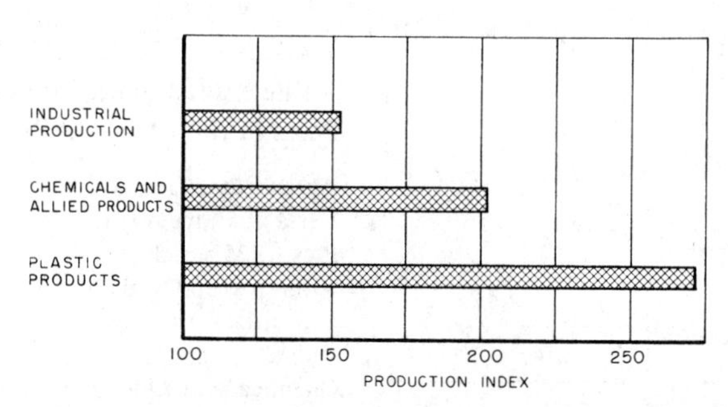

Fig. 1.4. United States Federal Reserve production indices for 1979 for industrial production, chemical and allied products and plastics products (1967 = 100).

and agricultural chemicals to the modest growth in elastomers and paints.

If the amount in tons produced by the chemical industry has grown faster than average industrial production, while the total dollar sales have grown at a rate closer to that of the economy, the relative unit price of chemical products might be expected to have been less subject to inflation. In Table 1.4, the producer price indices in the United States for a number of industries are given for 1978 relative to 1967. Relative to the economy, the price indices for chemicals products have held about even, reflecting primarily increases in raw material costs. The two most disruptive factors in the economy of the United States in the 1970's, the cost of energy and the cost of petroleum and natural gas, have been offset in the chemical industry to a considerable degree through chemical process improvements and energy conservation. The chemical industry actually consumed three percent less total energy in operations in 1978 than in 1972[2]. (Fig. 1.5.)

TABLE 1.3 Production Indices for Selected Parts of the Chemical Process Industries (1967 = 100)

Year	*All Chemicals*	*Basic Inorganics*	*Basic Organics*
1969	118	98	125
1972	144	107	155
1976	171	124	190
1979	210	134	233
Year	*Synthetic Fibers*	*Synthetic Elastomers*	*Plastics Materials*
1969	140	118	149
1972	189	127	217
1976	214	119	280
1979	271	131	416
Year	*Soaps and Toiletries*	*Paints*	*Agricultural Chemicals*
1969	111	112	106
1972	130	118	109
1976	143	123	185
1979	172	126	222

CHARACTERISTICS OF THE CHEMICAL INDUSTRY

Investment Trends

The chemical industry tends to be a high investment business. Capital spending by the chemical and allied products industry in the United States has been a sizable percentage of that spent for all manufacturing, some ten percent (Table 1.5). The amount spent for all chemical process industries has been, of course, even larger. For perspective, annual expenditures for new plant and equipment in the United States for chemical and allied products industry in recent years has averaged about two-and-one-

TABLE 1.4 Producer Price Indices–United States, 1978 (1967 = 100)

Refined Petroleum Products	321
Lumber and Wood Products	276
Metal and Metal Products	227
Industrial Chemicals	226
Plastic Resins and Materials	200
Agricultural Chemicals	198
Textile Products and Apparel	160
Drugs and Pharmaceuticals	148

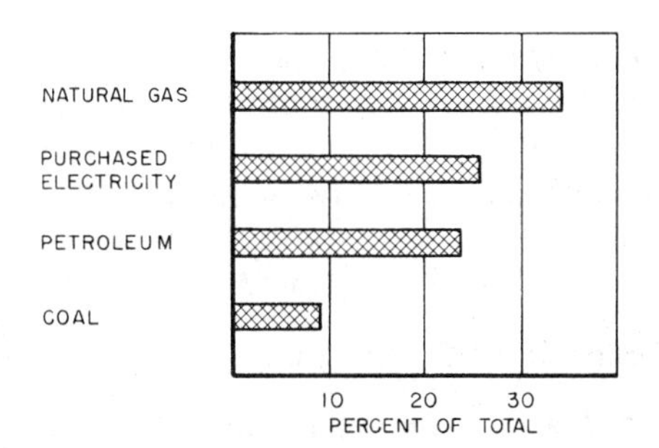

Fig. 1.5. Sources of energy used in chemical industry in the United States in 1978 as a percent of the total used by the industry.

half times that for iron and steel and about half of the invested in the petroleum industry. For the past decade a significant part of these capital investments have been made in pollution control projects.

Much of the capital investment in the chemical industry is spent for facilities to produce major chemicals (Table 1.6) in truly enormous quantities. The volume produced has been reflected in the size of plants being built to achieve the required economies of scale. That such economies are achieved is seen in the more modest increases in the chemical producers' price indices relative to the inflation levels in the general economy. (Economy of scale refers to the relative cost of building a larger plant. A rule of thumb is that the relative cost of building a smaller or a larger plant is the ratio of the productivities of the two plants being considered, raised to the 0.6 power. In other words, the unit cost of producing a chemical decreases markedly as the size of the plant producing it is increased, providing that the plant can be operated near capacity.)

TABLE 1.5 Capital Spending by Industries in the United States (Billion of Dollars), 1979

All United States Corporations	180
All Manufacturing	80
Chemical Process Industries	30
Chemical and Allied Products	8
Industrial Chemicals	6.5

Capital Spending by Industries in the United States on Pollution Control Projects (Billions of Dollars), 1979

All Manufacturing	4.0
Chemical Process Industries	2.3
Chemical and Allied Products	.5

TABLE 1.6 Production of Selected Chemicals and Plastics in the United States, 1979 (Billions of Pounds)

Organic Chemicals	*Billion Pounds*
Ethylene	29.2
Propylene	14.3
Styrene (monomer)	7.5
Phenol (synthetic)	3.0
Inorganic Chemicals	
Chlorine	24.2
Ammonia	36.2
Sulfuric Acid	84.0
Plastics	
Low density polyethylene	8.8
High density polyethylene	5.0
Polypropylene	3.8
Polystyrene (all types)	6.2
Poly (vinyl chloride)	6.1
Phenolics	1.8
Synthetic Rubber	2.5

Today, a typical base petrochemicals plant will consume the equivalent of 30,000 barrels per day of naphtha to produce about one billion pounds of ethylene a year, plus 2.5 billion pounds of coproducts. To be economically feasible, for example, vinyl chloride and styrene monomer plants must be scaled in the billion pound-a-year range.

Along with these very large plants and the associated enormous investment, most of the chemical industry is characterized by high investment versus low labor components in cost of manufacture. The National Industrial Conference Board statistics list the chemical industry as one of the highest in terms of capital investment per production worker. The investment per worker in a base petrochemicals olefins plant may well exceed a quarter of a million dollars. Once again, however, such an index covers a spectrum of operations, and for a profitable chemical specialties manufacturer the investment may be of the order of 25,000 dollars per worker. Employment in selected parts of the chemical industry is given in Table 1.7.

TABLE 1.7 Employment in Selected Parts of the Chemical Industry in 1979

Chemical Industry	*Thousands of Workers*
Chemical and Allied Products	1100
Plastics Materials	220
Drugs	190
Inorganic Chemicals	170
Organic Chemicals	170
Petroleum Products	200
Rubber and Plastic Products	770

Commercial Development and Competition Factors

In an earlier period in the development of the chemical industry, chemical companies were generally production oriented, exploiting a process to produce a chemical and then selling it into rapidly expanding markets. Plant sizes and investments required to participate were small fractions of those of today. Raw materials were often purchased to produce chemical intermediates for sale. Smaller plants operating in smaller manufacturing complexes did not present the obvious problems of environmental pollution about which we have become more aware during the past decade. A new investment in chemical production today includes a significant proportion to abate and control environmental intrusion.

As the industry has grown, there has been a strong tendency toward integration both forward and back. Petroleum producers have found opportunities based on their raw materials position to move into chemical manufacturing. Chemical companies, on the other hand, have moved to assure their access to low-cost raw materials. Similarly, producers of plastic materials have moved forward to produce fabricated products, such as films, fibers, and consumer items, while fabricators have installed equipment to handle and formulate the plastic materials in order to provide a supply at the lowest possible cost.

With ever higher investment and increasing cross-industry competition, increasingly greater sophistication has been required of marketing analysis and selection of investment. The enormity of the investment required today to participate successfully does not permit multiple approaches for the private investor. Consequently, a high degree of market orientation tends to predominate the chemical industry, as well as increasingly targeted research and development programs.

Technological Orientation

The chemical industry is a high technology industry, albeit now more marketing oriented and competitive than in its earlier period of development. This orientation is seen in the number of scientists and engineers employed in Research and Development relative to other industries (Table 1.8). In general, the chemical industry is among the largest employers of scientists and engineers, and it invests a sizable percentage of the total United States business investment in R&D (Table 1.9).

The contemporary scientist and engineer engaged in research and development in the chemical industry represents individually a high investment occupation. Since the mid-1950's, chemistry has become increasingly an instrumental science. The instruments now routinely used are both highly sophisticated and costly. A major research project would not be undertaken without access to a variety of spectrophotometers, spectrometers, and chromatographs, as well as the needed physical chemical instruments for structure determinations and reaction kinetics. Pilot plants are highly automated and instrumented. Both the basic researcher and pilot plant engineer require access to computer facilities. In 1978, the average cost to maintain an operating R&D scientist or engineer in the chemical industry was $74,000 per year. Impressive as these statistics may be in representing the business investment in chemicals R&D in the United States, R&D

TABLE 1.8 Scientists and Engineers in Research and Development in the United States, 1979 (Thousands)

Aircraft and Missiles	86
Electrical Equipment and Communications	90
Machinery	64
Chemical and Allied Products	52
Motor Vehicles and other Transportation	34
Petroleum Refining	17

TABLE 1.9 United States' Business Investment in Research and Development, 1979

	Billions of Dollars
All Business	37.7
Chemical and Allied Products	4.1
Industrial Chemicals	2.1

spending in the chemicals industry as a percent of sales has declined from about four percent in 1970 to about 2.5 percent in 1980 (Fig. 1.6). This decline has become a concern in that reinvestment in R&D, particularly in West Germany and Japan, has remained at a higher level as measured by this index.

Obsolescence and Dependence on Research

The high technology level that characterizes the chemical industry, and which is reflected in heavy investments in R&D, generally concerns discovery and development of new products and improvement in the manufacture of known products. The first area is the more conspicuous: the pharmaceutical for a specific disease; the narrow spectrum, transient pesticide; the new, super-performance, composite system for an internal combustion engine; or the new composite sails on the 1980 entries in the America's Cup races. The second area, however, makes viable the circumstances outlined

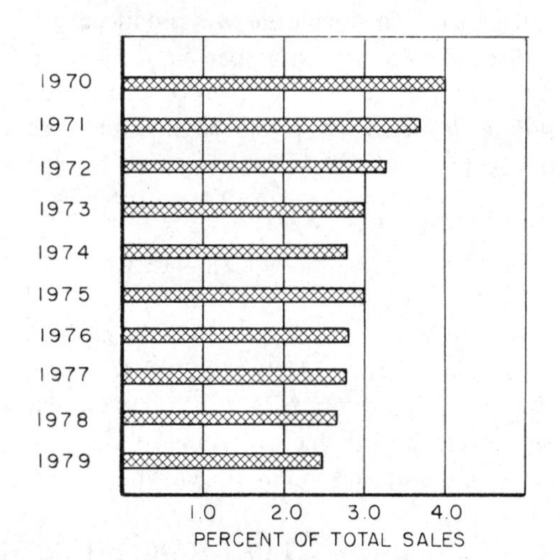

Fig. 1.6. R and D spending by the 14 largest chemical companies in the United States as a percent of total sales for the years 1970–1979.

earlier under which increasing investments can be made to produce larger quantities of materials. The development of a new, lower cost process for a commercial product can permit development of a profitable opportunity, or spell disaster for a company with existing investment in a now obsolete plant. Major reductions in manufacturing cost can be achieved, for example, by reducing the number of reaction steps required, changing to a lower cost or more available raw material, or eliminating coproducts or costly separations. The ability of a process scheme to contain or avoid a pollutant can be a deciding factor in continuance of a manufacturing operation. Examples of the above factors will make clear the economic consequences.

Acetic acid production in the United States has increased eightfold in the last 30 years. From the 1930's, acetic acid was produced by a three-step synthesis from ethylene: acid hydrolysis to ethanol, then catalytic dehydrogenation to acetaldehyde, then direct liquid-phase oxidation to acetic acid and acetic anhydride as coproducts.

$$CH_2{=}CH_2 \xrightarrow[H_2O]{H_2SO_4} C_2H_5OH \xrightarrow[Cu/Cr]{}$$

$$CH_3CHO \xrightarrow[\substack{Co\\(liq)}]{O} (CH_3CO)_2O + CH_3COOH$$

In the 40's, a major process change was introduced–direct oxidation of butane to acetic acid and coproducts (such as methylethylketone).

$$C_4H_{10} \xrightarrow{O_2} CH_3COOH + CH_3{-}\overset{\overset{\displaystyle O}{\|}}{C}{-}C_2H_5 + \text{etc.}$$

Fewer steps in synthesis were reflected in lower cost and investment. In 1969, another advance was announced, synthesis of acetic acid from methanol and carbon monoxide with essentially no coproducts[3,4].

$$CH_3OH + CO \xrightarrow[Rh]{I^-} CH_3C(=O)OH$$

Lack of coproducts reduces costs and investment in distillation and other separations systems, very attractive process features in an industry where the principally accepted measure of business quality is return on investment.

Acetic anhydride is needed as a process intermediate in acetylations. To obtain acetic anhydride from acetic acid, acetic acid is first pyrolized to ketene which then reacts with recovered acetic to yield the anhydride.

$$CH_3C(=O)OH \xrightarrow[\Delta]{} CH_2{=}C{=}O$$

$$CH_2{=}C{=}O + CH_3C(=O)OH \longrightarrow (CH_3C{=}O)_2O$$

In 1980, the Tennessee Eastman unit of Eastman Kodak announced that it would begin construction of a plant to make acetic anhydride from coal[5,6]. This decision reflected the changing of the raw materials base of much of the chemical industry due to such factors as the rising cost of natural gas and of petroleum and the large coal reserves of the United States.

In the new process, Eastman will make synthesis gas (carbon monoxide and hydrogen) from coal and then from synthesis gas, methanol. Up to now, methanol has been chiefly produced from natural gas methane.

$$CO + 2H_2 \longrightarrow CH_3OH$$

Methanol can then react with acetic acid to make methyl acetate

$$CH_3OH + CH_3C(=O)OH \longrightarrow CH_3{-}O{-}\overset{O}{\overset{\|}{C}}{-}CH_3$$

Acetic anhydride is then obtained from the catalytic carbonylation of methyl acetate with carbon monoxide[4]

$$CH_3{-}O{-}\overset{O}{\overset{\|}{C}}{-}CH_3 + CO \longrightarrow (CH_3C{=}O)_2O$$

The attractiveness of this process is twofold: (1) the raw materials base of synthesis gas from coal and (2) the avoidance of the energy consuming manufacture of ketene by pyrolyzing acetic acid.

The increase in the production of vinyl chloride, the principal monomer for poly(vinyl chloride) plastics, which are used in vinyl flooring, phonograph records, shower curtains, rain coats, car seat upholstery, house siding, pipe, and so on, is even more spectacular. Production in the United States has increased from 250 million pounds in 1950 (when it was declared by many industry economic forecasters to be a mature chemical commodity) to over one billion pounds in 1960, to about 3.5 billion pounds in 1970, and to over seven billion pounds in 1980.

During the early development period of vinyl resins in the 1930's, vinyl chloride was produced via catalytic addition of hydrogen chloride to acetylene.

$$CH{\equiv}CH + HCl \xrightarrow[HgCl]{} CH_2{=}CHCl$$

Later, a so-called "balanced" process was introduced in which, by addition of chlorine to ethylene, ethylene dichloride was produced. Ethylene dichloride could then be cracked to vinyl chloride and HCl, with the hydrogen chloride recycled to produce vinyl chloride from acetylene

$$CH_2{=}CH_2 + Cl_2 \longrightarrow CH_2Cl{-}CH_2Cl \xrightarrow[\Delta]{} CH_2{=}CHCl + HCl$$

At this point, vinyl chloride was being produced from chlorine, acetylene, and ethylene. More recently, catalytic oxychlorination has been developed in which vinyl chloride is produced from ethylene and hydrogen chloride.[7]

$$CH_2{=}CH_2 + HCl \xrightarrow[Cu]{[O]} CH_2{=}CHCl + H_2O$$

The hydrogen chloride can be obtained via

cracking of ethylene dichloride. The oxychlorination process freed vinyl chloride from the economics of the more costly raw material, acetylene (Deliberate acetylene manufacture is energy intensive. Less expensive byproduct acetylene from gas cracking has not been available in sufficient supply for the large, near billion pound-per-year vinyl chloride units).

During the long period of development of poly(vinyl chloride) into one of the major plastics materials, several basic processes for making PVC evolved. In all of these processes, vinyl chloride was handled as a liquid under pressure. Other than the relative ease with which vinyl chloride could be polymerized by free radical initiators, the monomer, vinyl chloride, was regarded as an innocuous, relatively inert chemical. A number of producers of PVC resins were caught by total surprise in the 1970's when it was found that long-term (20 years) exposure to vinyl chloride monomer could cause rare forms of tumors[8].

During the 1960's, vinyl chloride sold in the United States for five to six cents per pound. In the presence of traces of air (oxygen), vinyl chloride could form low concentrations of peroxide which could collect in compressors and on occasion decompose to blow out compressor seals. Rather than recover and compress the inexpensive monomer for recycle from stripping and drying operations at the end of the polymerization process, some manufacturers vented it to the atmosphere. After discovery that vinyl chloride was a carcinogen, venting was not permissible. Containment and recovery were mandatory. Some older processes and manufacturing facilities could not be economically modified to incorporate monomer containment, so operations were discontinued. This case is but one example of the impact of the need for environmental controls on manufacturing processes and operations.

Propylene oxide is another basic chemical, used in manufacturing intermediates for urethane foams (used in cushioning and insulation) and brake and hydraulic fluids. The volume of propylene oxide produced has increased from 310 million pounds in 1960 to two-and-a-quarter billion pounds in 1979. The classical industrial synthesis has been the reaction of chlorine with propylene to produce the chlorohydrin followed by dehydrochlorination with caustic to produce the epoxide, propylene oxide, plus salt. In this case, both the chlorine and the caustic used to effect this synthesis are discarded as a valueless salt byproduct.

$$CH_3CH{=}CH_2 + Cl_2 + H_2O \longrightarrow CH_3CH(OH)CH_2Cl$$

$$CH_3CH(OH)CH_2Cl \xrightarrow[\text{Caustic}]{} CH_3\overset{O}{CH{-}CH_2} + \text{Salt}$$

A more economic process has been commercialized. In one version, a hydroperoxide is produced by catalytic air oxidation of a hydrocarbon such as ethylbenzene. Reaction of this hydroperoxide with propylene yields propylene oxide plus a coproduct. This peroxidation can be carried out with other agents to give different coproducts such as t-butanol or benzoic acid [9,10].

$$C_6H_5{-}C_2H_5 \xrightarrow[V]{[O]} C_6H_5{-}C_2H_4OOH$$

$$C_6H_5{-}C_2H_4OOH + CH_3CH{=}CH_2 \longrightarrow CH_3\overset{O}{CH{-}CH_2} + C_6H_5{-}C_2H_4OH$$

When the economics are balanced, a significant cost reduction is achieved by eliminating the coproduct, salt, which is of low value and presents a disposal problem. Further, a process can be designed to produce a coproduct which can be used or sold as a chemical intermediate.

If a company is in the business of making and selling products such as acetic acid, vinyl chloride, propylene oxide, or other chemicals and has plans to stay in business and to expand its facilities to serve growing markets, it must have at least economically competitive processes. Today, this means being competitive with new processes in the United States, Western Europe, Japan, and the Soviet Union—for the chemical industry is a world-wide industry. Further, the processes which are operative must be environmentally compatible–all toxic or carcinogenic by-products or waste must be contained and disposed of harmlessly. Even a relatively innocuous by-product such as salt must be disposed of so as not to intrude on the environment.

The profound effect of environmental concerns on the manufacture of a chemical is reflected in the history of aerosol pressurized products. These products are familiar to the consumer as the aerosol spray can containing hair spray, deodorant, desert cream topping, or insect spray (Table 1.10). The market for these products grew enormously in the 1960's, with rapid consumer acceptance of these convenience, packaged products[11,12].

The aerosol spray product is a pressurized formulation with a propellant gas. During the rapid growth of these products, the major propellant gases were chlorofluorocarbons. Then in 1973, uncontrolled release of chlorofluorocarbons into the atmosphere was linked to possible depletion of the ozone layer in the Earth's atmosphere. Since stratospheric ozone provides significant protection at the Earth's surface from ultraviolet radiation from the Sun, depletion of the ozone layer could be forecast to lead to skin cancers, reduced seafood and grain crops, and alteration of the carbon dioxide level in the atmosphere.

The most easily controlled source of chlorofluorocarbon discharge into the environment–aerosol propellants–was removed from the marketplace. These products have been removed from the marketplace and replaced by formulations in which the propellant gas is

TABLE 1.10 Aerosol, Pressurized Product Sales in the United States in 1978

Use	*Percent of Sales*
Personal Products and Toiletries	33%
Household Products	28
Coatings and Finishes	14
Automotive	7
Food Products	6
Insect Sprays	6
Industrialized Products	5
Animal Products	1
Miscellaneous	0.5

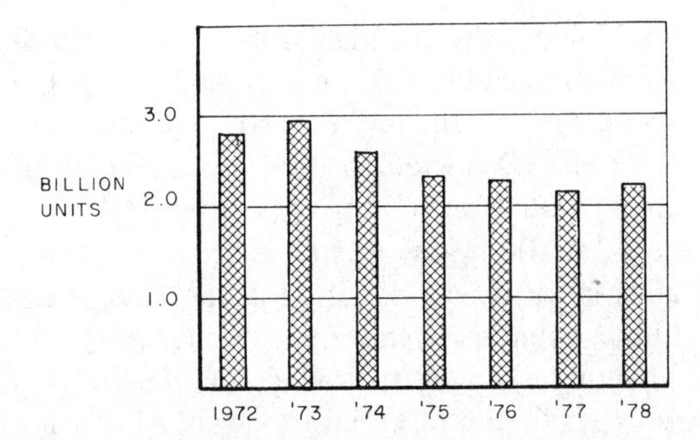

Fig. 1.7. Billion units of aerosol-pressurized products sold in the United States for the years 1972–1978. The year 1973 marked the announcement of the potential effect of chlorofluorocarbons on the stratospheric ozone layer.

hydrocarbon or carbon dioxide. The consumers, however, have registered their concern by switching away altogether from aerosol products. This consumer concern was noted in the rapid decline is sales of these products from 1973 until 1977 (Fig. 1.7). Since then, consumers have begun to return to the convenience products as an understanding of the new propellant systems has been accepted. Meanwhile, one of the fastest growing chemical product markets of the 1960's, the fluorocarbon propellant gases, has abruptly ended. However, typical of such situations in the chemical industry, replacement systems have been rapidly developed.

WORLD-WIDE CHEMICAL INDUSTRY

The major chemical producers in the United States have developed very significant overseas business through trade and investment (Table 1.11). In companies such as DuPont, Union Carbide, and Monsanto, foreign assets and sales amount to about 25 to 30 percent of total business, while Dow Chemical derives an even higher proportion from overseas investment[2]. The major chemical companies truly do business on a world-wide basis, as with the established and familiar pattern in the petroleum industry. As United States producers have expanded abroad, competitors based in the large economies of Western Europe and Japan have developed comparably in size with those in the United States. These chemical companies compete in exports to developing areas

TABLE 1.11 1979 Overseas Business of United States Chemical Producers

Producer	*Foreign Assets as a Percent of Total Assets*	*Foreign Sales as a Percent of Total Sales*
DuPont	26.0%	31.7%
Dow Chemical	52.6	50.5
Union Carbide	28.6	30.3
Monsanto	30.7	33.1
Celanese	19.9	21.1
W. R. Grace	23.1	29.9
Allied Chemical	25.8	25.5

and in United States domestic markets by both trade and investment.

For the overall economy of the United States, the United States chemical exports are a very important factor. This importance is emphasized if the United States balance of trade is examined (Table 1.12). In 1969, the United States enjoyed a favorable balance of trade of over one billion dollars. If just the balance of trade in chemical products were counted, the 1969 "chemical balance of trade" was over two billion dollars. At that time the preponderance of chemical trade was with Western Europe.

In the decade from 1969 to 1979, the dollar value of United States trade has increased fivefold; however, principally due to oil imports and a fifteenfold increase in the price of imported oil, the United States' trade balance in

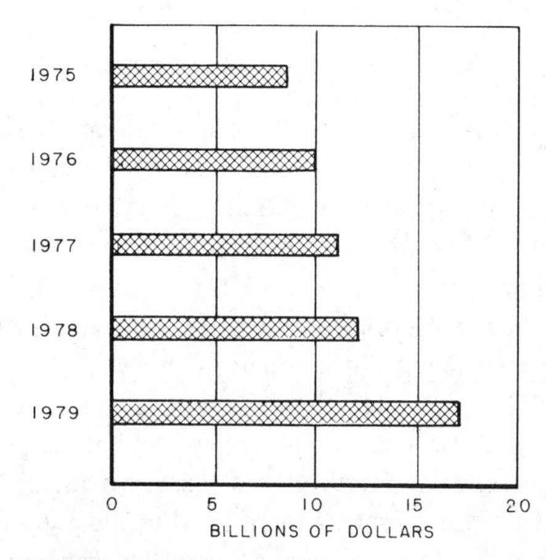

Fig. 1.8. The total chemical exports from the United States in billions of dollars for the years 1975–1979.

TABLE 1.12 United States Balance of Trade and Major Chemical Trading Areas (Millions of Dollars)

	1979	*1969*
Total Chemical Exports	$17,306	$3,383
Total Chemical Imports	7,485	1,232
Chemical Trade Balance	+9,821	+2,151
All Export Trade	$181,639	$37,314
All Import Trade	206,327	36,052
Official Balance of Trade	-24,688	+1,262
Chemical Trade	*Exports to (1979)*	*Imports from (1979)*
Western Europe	$5,585	$3,449
Canada	1,991	2,362
Japan	1,652	616
Latin America	3,814	405
All Other Areas	4,264	653

TABLE 1.13 World's Largest Chemical Companies

Company	*Country*		*Total Sales Millions of Dollars in 1969*	*1979*
Hoechst	Germany		$2,550	$15,870
Bayer	Germany		2,550	15,079
BASF	Germany		2,430	15,018
DuPont	United States		3,655	12,572
ICI	United Kingdom		3,250	11,389
Dow Chemical	United States		1,876	9,255
Union Carbide	United States		2,933	9,177
Montedison Group	Italy	(Montecatini)	2,620	8,224
Rhone-Poulenc	France		1,840	7,940
Monsanto	United States		1,939	6,193
Exxon (Chemical Sales Only)	United States		1,004	5,807
Celanese	United States		1,250	3,010

1979 was a negative 25 billion dollars. The "chemical balance of trade," however, was very favorable, nearly ten billion dollars (Fig. 1.8). While Western Europe remains the largest trading partner of the United States, trade within the Americas has enormously increased, as has trade with other areas of the world.

While the United States is the world's leading chemical producer–and many of the largest chemical manufacturers are located principally in the United States–there are European-based chemical concerns that are of comparable size in sales and resources (Table 1.13). The three giants of West German chemicals, Hoechst, Bayer, and Badische Anilin & Soda Fabrik (BASF) have grown enormously and have been notably successful in international trade. All three operate in the United States through subsidiaries.

In recent years, the Japanese chemical industry has also grown to rank with the world's giants (Table 1.14). The industry development in Japan has been carefully guided by the Ministry of International Trade and Industry. Capital demands have been met by cooperation within the industry through joint ventures involving the most basic processes to produce, for example, methanol, olefins, and ammonia. The basic plants in Japan are "world-scale," that is, approximately a billion pounds of annual capacity. The Japanese chemical industry is active both in export and also investment abroad as a means of introducing new technology on the world market.

TABLE 1.14 Major Japanese Chemical Producers

Company	*1979 Sales (Millions of Dollars)*
Mitsubishi Chemical	2,967
Sumitomo Chemical	2,716
Asahi Chemical	2,359
Toray Industries	2,094
Mitsui Toatsu	1,767
Teijin Ltd.	1,756
Showa Denko	1,706
Mitsubishi Petrochemical	1,558
Dainippon Ink and Chemical	1,416
Sekisui Chemical	1,300

The trend of "foreign" expansion in the chemical industry on a world-wide basis by companies based in Japan, Europe, and the United States can be expected to continue. The chemical giants are multinational in seeking raw materials and markets for the increasingly high technology products of the chemical industry. Operating from the strong investment and technological base in the United States, the chemical industry in the United States in one of the country's major assets in world trade.

Acknowledgement: Many references and data for the discussion of the economic aspects of the chemical industry were obtained through Dr. N. D. Holden and the Union Carbide Corporation Corporate Planning Department.

REFERENCES

1. *Chemical Economics Handbook*, Stanford Research Institute, Menlo Park, California.
2. "Facts and Figures," *Chemical and Engineering News*, June 9, 1980.
3. Belgian Patent 713,296 (to Monsanto).
4. Parshall, G. W., "Organometallic Chemistry in Homogeneous Catalysis," *Science*, **208** 1221 (1980).
5. *Chemical Week*, **126** No. 3, p. 40 (1980).
6. *Chemical and Engineering News*, **58**, No. 2, p. 6 (1980).
7. British Patents 1,027,277 (to PPG) and 1,016, 094 (to Toyo Soda).
8. *Chemical Week*, **125** No. 6, p. 22 (1979).
9. French Patent 1,460,575 (to Halcon).
10. *Chemical Week*, **127** No. 14, p. 47 (1980).
11. *Chemical and Engineering News*, **58** No. 1 p. 6 (1980).
12. *Chemical Week*, **124** No. 1, p. 32 (1979).

2

Industrial Wastewater Technology

William J. Lacy*

INTRODUCTION

Industries use huge quantities of the nation's waters and are the major factor in the continuing rise in water pollution. In the U.S., they utilize over 50 trillion gallons of water and, prior to discharge, treat more than 30 trillion gallons. In terms of a single pollution parameter, BOD, the waste generated by industries is equivalent to that generated by the population of over 360 million people. Even more undesirable than the BOD loads of industrial effluents are the enormous quantities of mineral and chemical wastes from factories which steadily become more complex and varied. They include metals such as iron, chromium, nickel, and copper; salts such as compounds of sodium, calcium, and magnesium; acids such as sulfuric and hydrochloric; petroleum wastes and brines; phenols; cyanides; ammonia; toluene; blast furnace wastes; greases; many varieties of suspended and dissolved solids; and numerous other waste compounds. These wastes degrade the quality of receiving waters by imparting tastes, odors and color; and through excess mineralization, salinity, hardness, and corrosion. Some are toxic to plant and animal life.

The variety and complexity of inorganic and organic components contained in industrial effluents present a serious liquid wastewater treatment control problem in that the pollution and toxicity effects of these constituents are of greater significance than those found in domestic wastewaters.

Conventional wastewater treatment technology, which is often barely adequate for existing waste types, offers even less promise of providing the type and degree of treatment which will be required in the near future. Therefore, industrial pollution-control technology must be developed to achieve effective and economical control of pollution from such varied industries as those producing metals and metal products, chemicals and allied products, paper and allied products, petroleum and coal products, food and kindred products, textiles and leather goods.

TREATMENT LEVELS

There are at present two basic methods of treating wastes. They are called *primary* and *secondary*. In primary treatment, solids are

*Director, Water, Wastewater and Hazardous Materials, R&D Division, Environmental Protection Agency, Washington, D.C.

allowed to settle and are removed from the water. Secondary treatment, a further step in purifying wastewater, uses biological processes.

Primary Treatment

As wastewater enters a plant for primary treatment, it flows through a screen which removes large floating objects such as rags and sticks that may clog pumps and pipes. The screens vary from coarse to fine—from those consisting of parallel steel or iron bars with openings of about half an inch or more to screens with much smaller openings.

Screens are generally placed in a chamber or channel in a position inclined with respect to the flow of the sewage to make cleaning easier. The debris caught on the upstream surface of the screen can be raked off manually or mechanically. Some plants use a device known as a comminutor which combines the functions of a screen and a grinder. These devices catch and then cut or shred the heavy solid material. In this method the pulverized material remains in the sewage flow to be removed later in a settling tank.

After the wastewater has been screened, it passes into what is called a grit chamber where sand, grit, cinders, and small stones are allowed to settle to the bottom. The unwanted grit or gravel from this process is usually disposed of by filling land near a treatment plant.

In some plants, another screen is placed after the grit chamber to remove any further material that might damage equipment or interfere with later processes.

With the screening completed and the grit removed, the wastewater still contains suspended solids. These are minute particles of matter that can be removed from the sewage by treatment in a sedimentation tank. When the speed of the flow of wastewater through one of these tanks is reduced, the suspended solids will gradually sink to the bottom. This mass of solids is called raw sludge.

Various methods have been devised for removing sludge from the sedimentation tanks. In older plants it was removed by hand. After a tank had been in service for several days or weeks, the sewage flow was diverted to another tank. The sludge in the bottom of the out-of-service tank was pushed or flushed with water to a nearby pit and then removed for further treatment or disposal.

Almost all plants built within the past 30 years have included mechanical means for removing the sludge from sedimentation tanks. In some plants it is removed continuously, while in others it is at intervals. To complete the primary treatment, the sludge-free effluent is chlorinated to kill harmful bacteria and then discharged into a stream or river. The chlorination also helps to reduce odors.

Although 10 percent of the municipalities in the United States give only primary treatment to the sewage, this process by itself is considered entirely inadequate for most needs. Municipalities and industry, faced with increased amounts of wastes and wastes that are more difficult to remove from water, have turned to secondary and even advanced waste treatment.

Secondary Treatment

Secondary treatment removes about 90 percent of the organic matter in wastewater by making use of the bacteria it contains. The two principal processes for secondary treatment are *trickling filters* and the *activated-sludge* process. The effluent from the sedimentation tank in the primary stage of treatment flows or is pumped to a facility using one or the other of these processes.

Trickling Filter. A trickling filter is simply a bed of stones from three- to ten-feet deep through which the wastewater passes. Bacteria gather and multiply on these stones until they can consume most of the organic matter in the wastewater. The cleaner water trickles out through pipes in the bottom of the filter for further treatment. The wastewater is applied to the bed of stones in two principal ways. One method consists of distributing the effluent intermittently through a network of pipes laid on or beneath the surface of the stones. Attached to these pipes are smaller, vertical pipes which spray the effluent over the stones. Another much-used method consists of a vertical pipe in the center of the filter con-

A. DIFFUSED AERATION

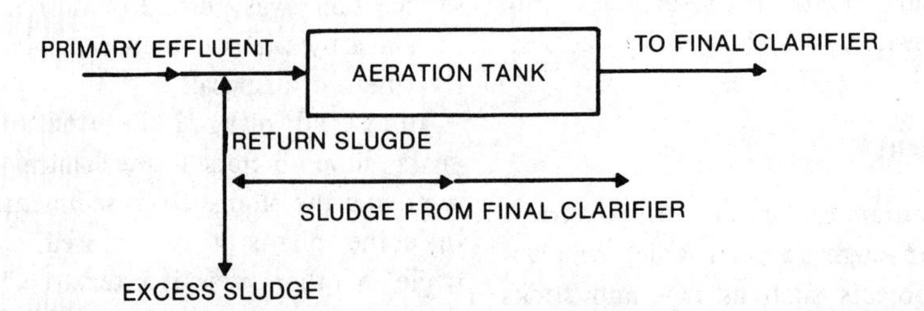

B. MECHANICAL AERATION SEE DIFFUSED AERATION FOR TYPICAL FLOW DIAGRAM.

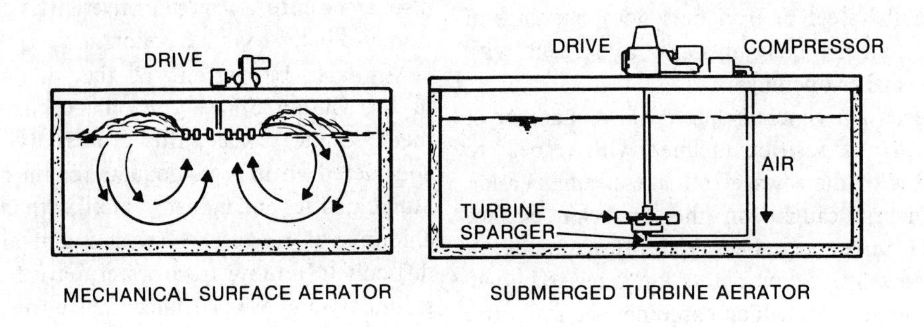

C. MODIFIED AND HIGH RATE AERATION

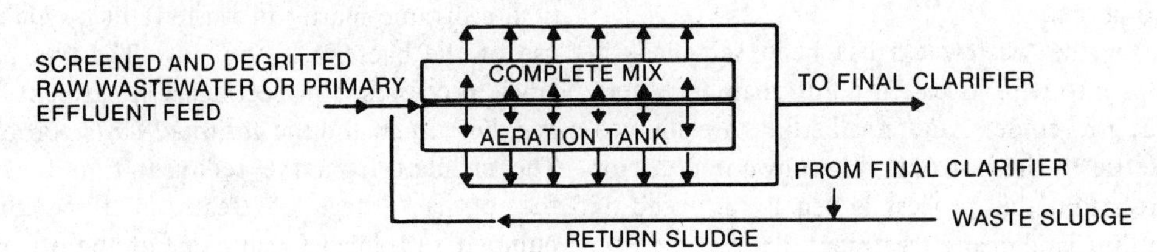

Fig. 2.1. Types of secondary biological treatment methods. (*Courtesy EPA*)

nected to rotating horizontal pipes which spray the effluent continuously upon the stones.

Activated-Sludge Process. The trend today is toward the use of the activated-sludge process instead of trickling filters. The former process speeds up the work of the bacteria by bringing air and sludge heavily laden with bacteria into close contact with the wastewater.

In the activated-sludge process, the wastewater from the sedimentation tank in primary treatment is pumped to an aeration tank where it is mixed with air and sludge loaded with bacteria and allowed to remain for several hours. During this time, the bacteria break down the organic matter. From the aeration tank, the wastewater, now called mixed liquor, flows to another sedimentation tank to remove the solids. Chlorination of the effluent completes the basic secondary treatment. The sludge, now activated with additional millions of bacteria and other tiny organisms, can be used again by returning it to an aeration tank for mixing with new wastewater and ample amounts of air.

The activated-sludge process, like most other techniques, has advantages and limitations. The size of the units needed is small so that they require comparatively little land space. Also, the process is free of flies and odors. But it is more costly to operate than the trickling filter, and it sometimes loses its effectiveness when faced with difficult industrial wastes.

A. PURE OXYGEN (COVERED).

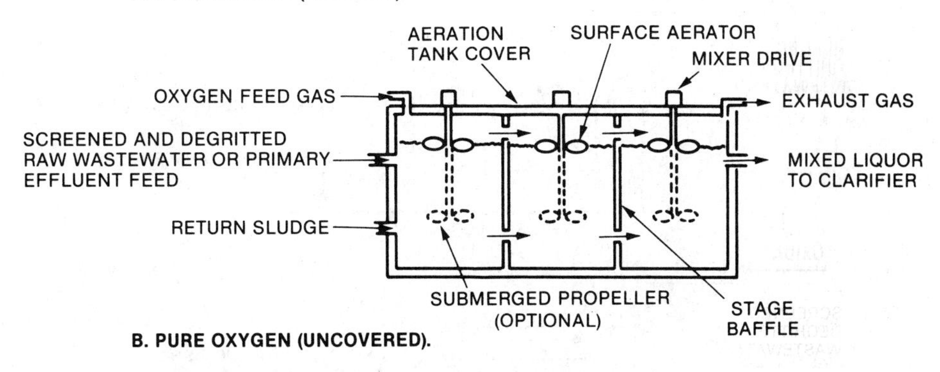

B. PURE OXYGEN (UNCOVERED).

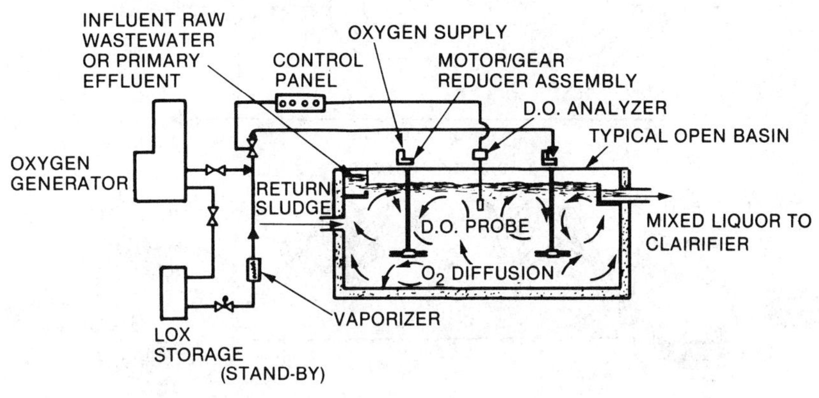

C. CONTACT STABILIZATION.

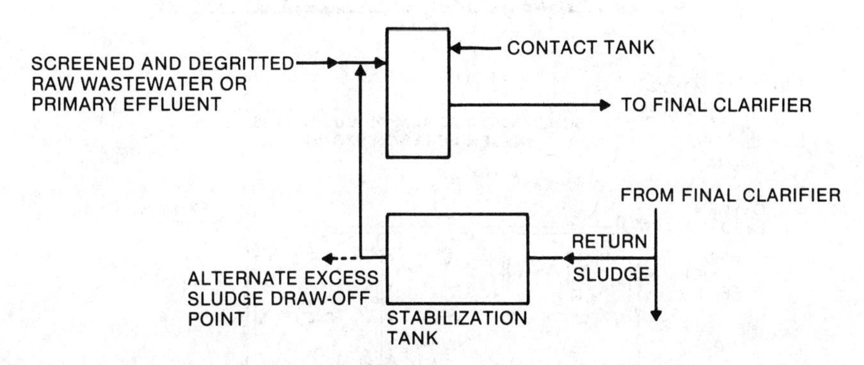

Fig. 2.2. Oxygen and contact stabilization processes for wastewater treatment. (*Courtesy EPA*)

An adequate supply of oxygen is necessary for the activated-sludge process to be effective. Air is mixed with wastewater and biologically active sludge in the aeration tanks by three different methods. The first, mechanical aeration, is accomplished by drawing the wastewater from the bottom of the tank and spraying it over the surface, thus causing the wastewater to absorb large amounts of oxygen from the atmosphere. In the second method, large amounts of air under pressure are piped down into the wastewater and forced out through openings in the pipe. The third method is a combination of mechanical aeration and the forced-air method.

Tertiary Treatment

Tertiary treatment is used when the waste stream must meet strict requirements governing

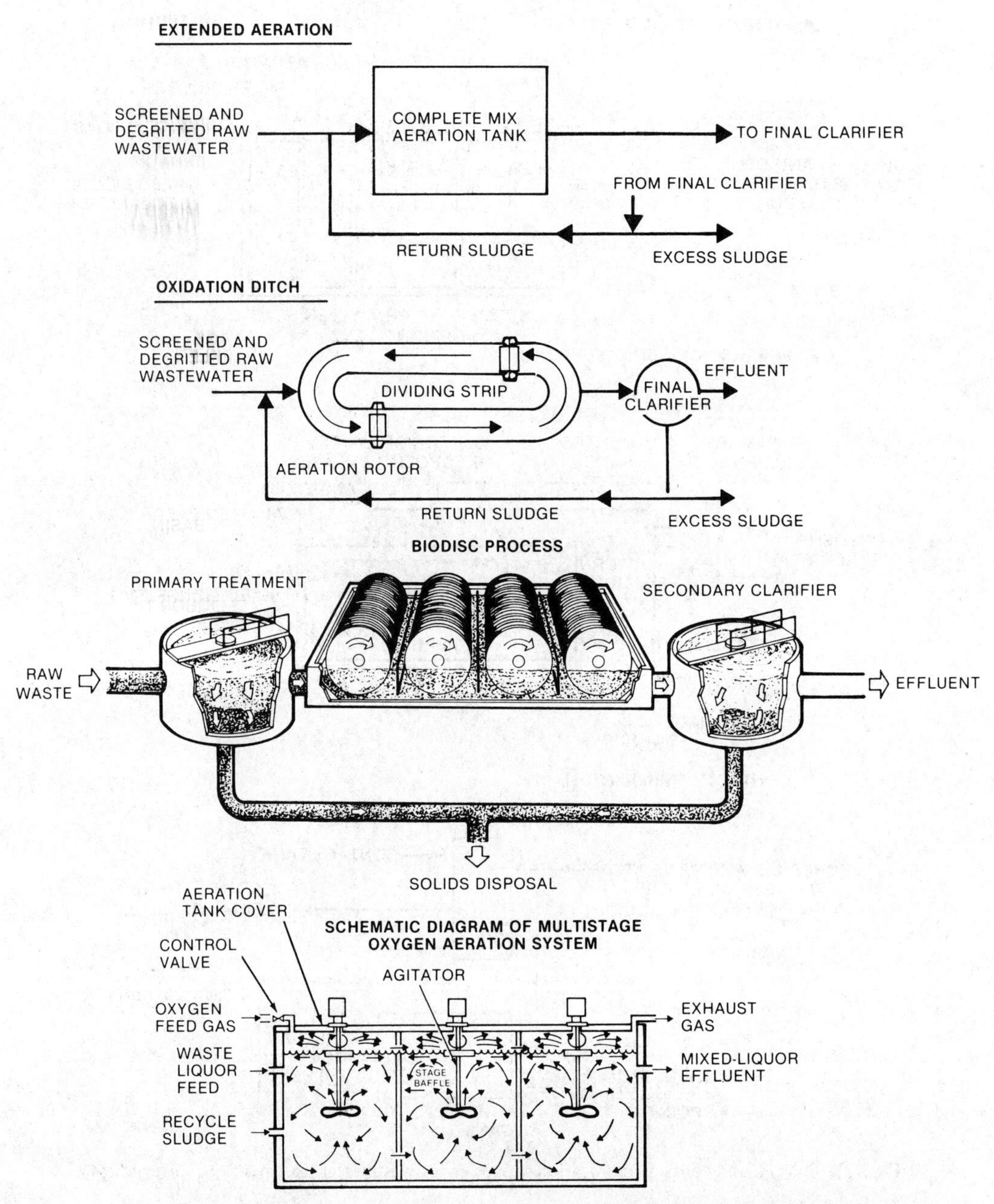

Fig. 2.3. Other types of aeration and biological treatment processes. (*Courtesy EPA*)

recreational bodies of water, or must approach drinking-water standards. This may require one or several of the following processes: slow filtration; rapid filtration with activated carbon; adsorption by activated carbon; application of ozone; high-rate chlorination or use of other oxidizing chemical; or lagooning.

At each plant the question may arise as to what degree of treatment is actually required. Water quality criteria imposed by different waste-stream discharges may vary widely. Even within the same state, or for a particular river basin, different limits for each of the contaminants may be set for the section of the river under consideration.

LAGOONS AND SEPTIC TANKS

There are many well-populated areas in the United States that are not served by any sewer systems or waste-treatment plants. Lagoons and septic tanks are the usual alternatives in such situations.

Lagoons or, as they are sometimes called, stabilization or oxidation ponds, also have several advantages when used correctly. They can give sewage primary and secondary treatment or they can be used to supplement other processes. A lagoon is a scientifically constructed pond, usually three- to five-feet deep, in which sunlight, algae, and oxygen interact to restore water to a quality equal to or better than effluent from secondary treatment. Changes in the weather affect how well a lagoon will break down the sewage. When used with other waste-treatment processes lagoons can be very effective.

TYPES OF INDUSTRIAL WASTES

The great variety of industrial pollutants argues against an attempt to catalog them, but for the purposes of general description we may recognize at least five distinct categories of wastes from industrial sources:

1. Oxygen-demanding materials.
2. Settleable and suspended solids.
3. Many materials which impart acidity or alkalinity.
4. Heat.
5. Toxic compounds.

Within these general categories of pollutants are included most of the possible sources of recognized water-quality problems. Only bacterial and viral presences fall outside of the group of polluting effects to which industry contributes materially. (Although these are not entirely out of the range of parameters to be included in the polluting activities of industry: meat packing plants, and other food processors in lesser measure, contribute to the presence and viability of water-borne bacteria.) Manufacturing must stand near the top, and it very probably leads, in any list of potential sources of water pollution.

And for any possible strategy of water-pollution control, industrial wastes are of critical importance. Over the last ten years, the amount and composition of industrial output has been such that for every incremental pound of BOD that has been generated directly by population increase, twenty more have been generated by increased industrial output. Increased per capita production is the essence of improvement in the standard of living, and the production of wastes is an inescapable concommitant of the production of goods. So as population and per capita production continue to advance, we can anticipate a continuing and unavoidable advance in the volume of wastes to be managed.

Fortunately, industry has added rapidly to its inventory of waste-treatment facilities over the last decade and a half, and it appears that provision for waste treatment is routinely designed into new plants and plant additions. While additional treatment of municipal wastes had an unquestionable influence on the nation's ability to contain the level of waste discharges, the preponderance of industrial wastes and their more rapid rate of increase would have made containment impossible if incremental industrial-waste-treatment effectiveness had not occurred at more advanced rates than incremental industrial production.

Strategies for Management of Industrial Wastes

Unlike the public sector, where only variations on a single theme of waste treatment are possible, industrial pollutants can be managed through at least four distinct strategies. (1) The obvious procedure is the installation of industrial-waste-treatment plants. (2) A second, and increasingly prevalent procedure is to discharge industrial wastes to public systems for treatment. Industry is estimated to account for half of the current BOD loading to metropolitan area waste-treatment plants. (3) Process modification and changed product formulations are probably the most effective—as well as the most efficient—means of reducing wastes. The outstanding example is the shift of the pulp and paper industry from the sulfite pulping process to the sulfate process, calculated to have been responsible for a greater reduction in the aggregate level of BOD than the combination of all the waste-treatment plants in the U.S. More recent examples include the rising prevalence of cooling-water recycle, development of biodegradable detergents, and substitution of reclaimable hydrochloric acid for sulfuric acid in metal-pickling liquors. (4) In the absence of alternative control procedures, industry has on occasion abandoned a product line or production procedure. At the present time, phosphorus-based detergents, DDT, mercury-battery production of chlorine and alkali, and the Solvay process for production of soda ash all seem likely candidates for abandonment.

QUANTITY OF INDUSTRIAL WASTES

Table 2.1 shows forecasts of water use and wastewater discharge for several categories of industry for the years 1975, 1985, and 2000.

Industrial wastes differ markedly in chemical composition, physical characteristics, strength, and toxicity from wastes found in normal domestic sewage. Every conceivable toxicant and pollutant of organic and inorganic nature can be found in industrial wastewaters (see Table 2.2). Thus, the BOD_5 or solids content often are not adequate indicators of the quality of industrial effluents. For example, industrial wastes frequently contain persistent organics which resist the secondary treatment procedures applied normally to domestic sewage. In addition, some industrial effluents require that specific organic compounds be stabilized or that trace elements be removed as part of the treatment process.

It is therefore necessary to characterize each industrial wastewater to permit comparative

TABLE 2.1 Department of Commerce Forecasts of Manufacturing Water Use Trillions Gallons Per Year (10^{12} Gallons) for 1975, 1985, and 2000

	1975	*1985*	*2000*
Primary Metals–SIC 33			
Gross Water Use	10.2	12.4	15.4
Intake	6.9	2.1	1.3
Discharge	6.1	1.2	.26
Chemical–SIC 28			
Gross Water Use	14.6	27.3	52.8
Intake	7.1	2.1	1.9
Discharge	6.6	1.3	.42
Paper, Pulp and Board–SIC 26			
Gross Water Use	9.5	14.4	23.0
Intake	3.3	2.2	2.0
Discharge	2.9	1.5	.41
Petroleum–SIC 29			
Gross Water Use	8.6	10.0	14.5
Intake	1.7	.53	.44
Discharge	1.5	.30	.09
Food and Kindred Products–SIC 20			
Gross Water Use	1.6	2.0	2.7
Intake	1.0	.54	.43
Discharge	0.9	.35	.14
All Manufacturing Water Use			
Gross Water Use	49.9	73.8	121.0
Intake	22.5	8.6	7.2
Discharge	20.3	3.2	5.4

pollutional assessments to be made for individual industries as well as industry groups. Characterization will permit classifying the components of industrial wastewaters into as few as four basic classes of pollutants to more readily collate pollution statistics and to evaluate the economics of methods of treatment, as well as to project least-cost methods. Proposed generalized basic classification parameters are BOD, COD, SS, and TDS into which all known pollutants can be classed. Also required is the establishment of a relative-pollution comparative index for all significant pollutants. This index, in combination with the known characteristics and volume of a wastewater, will determine the relative gross-pollution severity of all industrial wastes and establish a basis for comparing the severity of pollution from different industries.

Table 2.3 shows the pollution abatement capital expenditures and operating cost by form of abatement and major industry group for 1978 and 1977 as compiled by the U.S. Department of Commerce. Table 2.4 presents a list of industries which figure most significantly in water pollution, arranged according to their Standard Industrial Classification. It is anticipated that treatment technology for similar types of wastewater will be interchangeable between industries.

A wide spectrum of technology is available for controlling and treating industrial water pollution. Some of the more important unit operations and unit processes are given in Table 2.5.

Alternatives in wastewater treatment are shown in the sketch below:

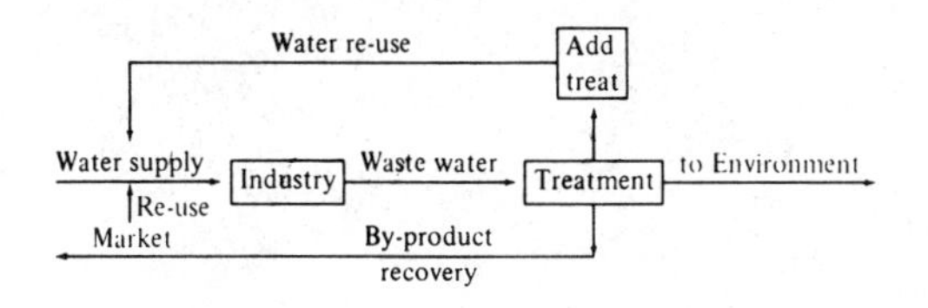

The alternatives shown above primarily consist of:

a. Wastewater treatment (as required to abate pollution to meet water quality standards.
   (1) For discharge to environment (to meet necessary water quality criteria).
   (2) For reuse (to meet industrial water-quality demands to conserve water and offset cost of treatment).
b. In-plant measures (to reduce pollutants and water discharge).
   (1) Operational (housekeeping techniques and manufacturing procedures).
   (2) Design (to permit reuse, to reduce wastewater generation).
c. Residue treatment.
   (1) By-product recovery (to reduce gross disposal, utilizes values).
d. Combined methods.
   (1) Joint treatment (to utilize scale factors, off-peak capacity, synergistic effects).
   (2) Other (combined a-b-c methods as appropriate).

The alternatives best suited for implementation will depend on many factors and local conditions. For nonprogressive industries, inplant measures should be explored for potential application. For industries which have demonstrated effective treatment methods, lower cost alternatives of treatment stressing reuse and by-product recovery should be given consideration.

INDUSTRIAL WATER REUSE

Waste products must be considered as an integral part of any manufacturing process and the cost of treatment of industrial wastes must be included in the pricing of the product. Waste-disposal operations normally result in a net cost to the industry producing the waste. However, by-product recovery and utilization techniques can reduce the cost of treatment and frequently prove to be less expensive than other methods of disposal.

Recycled water ultimately may be the most valuable product due to supply shortages, increasing water-supply costs, increasing water-treatment costs, and mounting municipal sewerage charges. The recovery of product fines, useable water, and thermal energy are key methods of reducing over-all waste treatment costs and must always be considered.

Frequently, waste streams can be eliminated or significantly reduced by process modifications. One notable example is the application of save-rinse and spray-rinse tanks in plating

TABLE 2.2 Wastewater Characteristics and Pollutants of Selected Industry

Liquid Waste Characteristic	*Domestic*	*Meat Products*	*Canned and Frozen Foods*	*Sugar*	*Textile Mill Products*	*Paper and Allied Products*
Unit volume	x	x	x	x	x	x
pH		x		x	x	x
Acidity				x	x	
Akalinity		x		x	x	x
Color					x	x
Odor					x	
Total solids		x		x	x	x
Suspended solids	x	x	x	x	x	x
Temperature		x				x
BOD_5/BOD ultimate	x	x	x	x	x	x
COD	x					
Oil and grease		x		x	x	x
Detergents						
(Surfacants)	x				x	x
Chloride		x				x
Heavy metals						
Cadmium						
Chromium					x	
Copper					x	
Iron						
Lead						
Manganese						
Nickel						
Zinc					x	
Nitrogen						
Ammonia		x		x	x	x
Nitrate					x	
Nitrite					x	
Organic		x		x		x
Total		x			x	x
Phosphorus		x				
Phenols					x	
Sulfide					x	x
Turbidity	x				x	
Sulfate					x	x
Thiosulfate					x	
Mercaptans						x
Lignins						x
Sulfur						x
Phosphates						x
Potassium						x
Calcium						x
Polysaccharides						x
Tannin						x
Sodium						x
Fluorides						
Silica						
Toxicity					x	
Magnesium						
Ammonia						
Cyanide						
Thiocyanate						
Ferrous iron						
Sulfite						x
Aluminum						
Mercury						x

[a]Source: *The Cost of Clean Water*, Volume II, FWPCA, U.S. Department of the Interior,

Basic Chemicals	*Fibers Plastics and Rubbers*	*Fertilizer*	*Petroleum Refining*	*Leather Tanning and Finishing*	*Steel Rolling and Finishing*	*Primary Aluminum*	*Motor Vehicles and Parts*
x	x	x	x	x	x	x	x
x	x	x	x	x	x	x	x
x		x	x	x	x	x	x
x			x	x	x		x
x	x		x				x
x	x		x				x
x	x	x		x	x	x	x
x	x		x	x	x		x
x	x	x			x		x
x	x		x	x	x		x
x	x		x		x		x
x	x		x		x		x
x	x						x
x	x		x	x	x		x
x							x
x							x
x							x
x			x		x		x
x							x
x			x				x
x							x
x							x
x		x	x	x	x		
x							
x							
x			x	x			
x		x		x			
x		x					
x	x		x		x		x
x			x	x	x		
x			x				
x	x		x	x			
x							
x			x				
x							
x		x					x
x			x				
x			x				
x							
				x			
x			x	x			
x		x	x		x	x	
x		x					x
x	x						
			x				
x		x	x		x		
x				x	x		x
x					x		
x					x		
x							
x							
x		x	x				

U.S. Government Printing Office, Washington, D.C., April 1, 1968.

TABLE 2.3 Pollution Abatement Capital Expenditures and Operating Costs, by Form of Abatement and Major Industry Group: 1978 and 1977*
(Value figures in millions of dollars)

			Pollution abatement capital expenditures (PACE)				*Pollution abatement gross annual costs (GAC) including payments to governmental units*				*Percent change*		*Standard error of estimates (Percent)*	
SIC code	*Industry*		*Total*	*Air*	*Water*	*Solid waste*	*Total*	*Air*	*Water*	*Solid waste*	*PACE*	*GAC*	*PACE*	*GAC*
	All industries, total[1]	1978	3,226	1,813	1,222	191	6,224	2,481	2,496	1,246	−7	15	1	1
		1977	3,484	1,652	1,674	158	5,425	2,240	2,203	981	(X)	(X)	1	1
20	Food and kindred products	1978	187	69	103	15	409	71	232	106	2	15	9	4
		1977	184	68	104	13	357	56	211	89	(X)	(X)	4	3
21	Tobacco products	1978	13	6	5	2	14	6	5	3	160	27	1	3
		1977	5	2	1	2	11	5	4	3	(X)	(X)	1	3
22	Textile mill products	1978	58	42	14	2	93	19	53	21	57	24	11	8
		1977	37	21	15	1	75	11	46	18	(X)	(X)	7	4
24	Lumber and wood products	1978	82	45	25	12	85	27	19	39	37	16	1	1
		1977	60	34	19	7	73	20	19	34	(X)	(X)	10	5
25	Furniture and fixtures	1978	10	8	1	1	26	9	4	13	−17	8	13	8
		1977	12	10	1	1	24	10	4	10	(X)	(X)	12	7
26	Paper and allied products	1978	325	115	181	29	595	149	345	100	−24	12	1	1
		1977	427	134	262	32	529	134	309	86	(X)	(X)	2	2
27	Printing and publishing	1978	12	7	4	1	40	8	11	21	50	33	11	8
		1977	8	3	3	2	30	7	6	17	(X)	(X)	10	7
28	Chemicals and allied products	1978	815	365	378	72	1,474	399	786	289	−17	19	3	1
		1977	983	340	593	50	1,238	336	685	218	(X)	(X)	1	1
29	Petroleum and coal products	1978	418	318	100	8	995	635	303	57	13	5	2	2
		1977	369	168	196	5	948	601	289	57	(X)	(X)	1	1
30	Rubber, miscellaneous plastics products	1978	24	17	5	2	81	17	23	42	−35	9	52	20
		1977	37	17	14	5	74	20	19	35	(X)	(X)	6	3
31	Leather and leather products	1978	7	1	5	1	17	1	10	6	−36	13	46	14
		1977	11	2	9	(Z)	15	1	9	5	(X)	(X)	48	16
32	Stone, clay, glass products	1978	121	90	27	4	246	162	33	52	−9	14	10	2
		1977	133	86	39	9	215	142	28	44	(X)	(X)	17	4
33	Primary metal industries	1978	788	560	218	9	1,302	772	333	197	−10	16	1	1
		1977	875	616	250	8	1,122	722	268	132	(X)	(X)	2	1
34	Fabricated metal products	1978	56	30	23	2	186	50	73	63	−27	25	23	11
		1977	77	33	39	5	149	40	66	43	(X)	(X)	7	3

35	Machinery, except electrical	1978	77	37	27	13	152	38	52	61	-14	13	6	5
		1977	90	42	42	6	135	33	50	53	(X)	(X)	3	2
36	Electric, electronic equipment	1978	68	32	33	4	144	30	69	45	6	13	44	15
		1977	64	23	36	5	128	28	63	37	(X)	(X)	4	2
37	Transportation equipment	1978	139	70	58	11	287	77	112	98	67	23	14	2
		1977	83	37	39	6	234	61	97	76	(X)	(X)	1	1
38	Instruments, related products	1978	17	7	10	1	56	7	27	21	-29	19	2	1
		1977	24	15	9	1	47	9	23	15	(X)	(X)	4	2
39	Miscellaneous manufacturing industries	1978	9	2	5	2	22	4	6	12	29	16	28	5
		1977	7	3	4	(Z)	19	5	5	8	(X)	(X)	28	5

Note: Totals may not agree precisely with detail because of independent rounding.
(X) Not applicable.
(Z) Represents less than $500,000.
[1]Major industry group 23, apparel and other textile products, was not included in this survey and therefore, is excluded from the U.S. level.
*U.S. Department of Commerce, Bureau of Census, Industry Division.

TABLE 2.4 Standard Industrial Classification of Industries of Significance for Water Pollution

CODE	
20	FOOD AND KINDRED PRODUCTS
201	Meat products
2011	Meat slaughtering plants
2013	Meat processing plants
2015	Poultry dressing plants
202	Dairy products
2021	Creamery butter
2022	Natural and process cheese
2023	Condensed and evaporated milk
2026	Fluid milk
203	Canned and frozen foods
2033	Canned fruits and vegetables
2034	Dehydrated food products
2035	Pickled foods, sauces, salad dressings
2037	Frozen fruits and vegetables
204	Grain mill products
2041	Flour and other grain mill products
2043	Cereal preparations
2046	Wet corn milling
205	Bakery products
206	Sugar
207	Candy and related products
208	Beverage industries
2082	Malt liquors
2084	Wines and brandy
2085	Distilled liquors
2086	Soft drinks
209	Miscellaneous foods and kindred products
2091	Cottonseed oil mills
2092	Soybean oil mills
2094	Animal and marine fats and oils
2096	Shortening and cooking oils
22	TEXTILE MILL PRODUCTS
2211	Weaving mills, cotton
2221	Weaving mills, synthetic
2231	Weaving, finishing mills, wool
225	Knitting mills
226	Textile finishing, except wool
228	Yarn and thread mills
229	Miscellaneous textile goods
24	LUMBER AND WOOD PRODUCTS
2421	Sawmills and planning mills
2432	Veneer and plywood plants
2491	Wood preserving
26	PAPER AND ALLIED PRODUCTS
2611	Pulp mills
2621	Paper mills, except building
2631	Paperboard mills
264	Paper and paperboard products
265	Paperboard containers and boxes
2661	Building paper and building board mills
28	CHEMICALS AND ALLIED PRODUCTS
281	Basic chemicals
2812	Alkalies and chlorine
2818	Organic chemicals, n.e.c.
2819	Inorganic chemicals, n.e.c.
282	Fibers, plastics, and rubbers
2821	Plastics materials and resins
2823	Cellulosic man-made fibers
2824	Organic fibers, noncellulosic
283	Drugs
284	Cleaning and toilet goods
2851	Paints and allied products
2861	Gum and wood chemicals
287	Agricultural chemicals
289	Miscellaneous chemical products
29	PETROLEUM AND COAL PRODUCTS
2911	Petroleum refining
295	Paving and roofing materials
30	RUBBER AND PLASTICS PRODUCTS, n.e.c.
3069	Rubber products, n.e.c.
3079	Plastics products, n.e.c.
31	LEATHER AND LEATHER PRODUCTS
3111	Leather tanning and finishing
32	STONE, CLAY, AND GLASS PRODUCTS
3211	Flat glass
3241	Cement, hydraulic
325	Structural clay products
326	Pottery and related products
327	Concrete and plaster products
3281	Cut stone and stone products
329	Nonmetallic mineral products
33	PRIMARY METAL INDUSTRIES
331	Steel rolling and finishing mills
332	Iron and steel foundries
333	Primary nonferrous metal
3341	Secondary nonferrous metals

lines. This measure brings about a substantial reduction in waste volume as well as a net reduction in metal dragout.

Industries in general are becoming more aware of the need for overall pollution control and product (or by-product) recovery. This is not only because pollution affects the environment but also because it affects the general public, who are the customers. In addition, industries, too, depend upon our nation's rivers and streams for suitable water for their manufacturing processes.

Table 2.6 shows the treatment levels suitable for industrial wastewater reuse and typical resulting costs.

Pretreatment

Most states or municipalities will require pretreatment to comply with the effluent limitations in NPDES permits. Pollutants which would interfere with or pass through the publicly-owned treatment works, (POTW), resulting in a violation of any of these NPDES permit requirements, must be pretreated or rejected from the system. Pretreatment is most commonly required for incompatible pollutants to prevent interference with treatment processes or then passing through to receiving waters. Where design capacity is not available, pretreatment for compatible pollutants may be necessary to comply with NPDES permit effluent limitations.

Pretreatment of incompatible wastes offers several operational advantages to POTW's. One significant advantage to the municipality is the specialized treatment that each wastewater contribution receives, as well as the advantage that the potential for plant upset is greatly reduced by pretreatment.

Joint Treatment

Joint treatment is an alternative that can be advantageous to both the POTW and the industry. Treatment of industrial wastewaters is incidental to a POTW's primary function of treating domestic sewage. Where the industrial contribution constitutes a significant portion of the total flow and substantially alters the concentration of pollutants normally contained in domestic sewage, the public agency may resort to the joint treatment approach. Then the industry or industries contributing the pollutants are made partners in the design and construction of the system, and the treatment works are designed to specifically remove the industrial pollutants. Both capital costs and operating costs are allocated to the industry and the public agency according to an agreement arrived at through negotiation, or as required by Federal regulations, if construction grant funds are involved.

Joint treatment of industrial wastewaters with municipal domestic sewage offers these advantages:

- Savings in capital and operating expenses due to the economics of large scale treatment facilities
- Increased flow which can result in reduced ratios of peak to average flows
- More efficient use of land resources, particularly in cases where available land for treatment facilities is scarce
- Improved operation (larger plants are potentially better operated than smaller plants)
- Increased number of treatment modules with resultant gains in reliability and flexibility
- More efficient disposal of sludges resulting from treatment of wastewaters containing pollutants susceptible to treatment in POTW's
- Utilization of the nutrients available in domestic wastes for biological treatment of industrial wastes which are nutrient deficient

Possible disadvantages of joint treatment are as follows:

- Where the pollutants are different from those usually treated in a POTW, a design to treat the combined industrial/domestic waste stream for these pollutants may not be cost effective.
- Joint treatment by definition implies that the POTW was designed so as not to be interfered with by industrial wastes. However, where this requires design modifications ordinarily not required for domestic wastes, joint treatment may not be cost effective.
- If joint treatment results in sludge disposal or utilization problems, it may not be acceptable.
- Some costs for the construction of joint treatment works which are solely to treat industrial pollutants are not eligible for Federal construction grants.

Sludge Disposal

The ultimate disposal of sludges produced by either pretreatment or joint treatment operations is an important factor to consider. The POTW must be aware of the effect on environmental problems that may result from sludge disposal. A comparison of solid waste estimates for manufacturing industries is given in Table 2.7. Pretreatment facilities normally remove incompatible pollutants that may be deposited in the POTW sludge. This can be an advantage in terms of the environmental problems connected with the ultimate disposal of sludge. Incompatible pollutants in sludges can cause problems in most disposal techniques utilized, including incineration, landfills, ocean dumping,

TABLE 2.5 Unit Operations and Processes Applicable to Treatment and Control of Industrial Water Pollution

	Dissolved BOD Removed	*Suspended and Colloidal Solids Removed*	*Dissolved Refractory Organics Removal*	*Dissolved Inorganics Removal*	*Dissolved Nutrient Removal*	*Microorganisms Removal*	*Concentrate Removal*
Biological processes							
Activated sludge	x	x	–	–	–	x	–
Anaerobic digestion	x	x	–	–	–	–	x
Bio-filters	x	–	–	–	x	–	–
Biomass treatment (algae harvesting)	x	–	–	–	x	–	x
Biological PO_4 removal	x	–	–	–	x	–	–
Extended aeration							
Bio-denitrification	L[a]	–	L	–	x	–	–
Bio-nitrification	x	x	–	–	–	–	–
Pasveer oxidation ditch	x	x	–	–	–	x	–
Chemical processes							
Chemical oxidation							
Catalytic oxidation	x	x	x	–	–	x	–
Chlorination	x	x	x	–	–	x	–
Ozonation	L	–	x	–	–	x	–
Wet oxidation	x	x	x	–	–	x	–
Chemical precipitation	–	–	x	x	x	–	–
Chemical reduction	–	–	–	–	x	–	–
Coagulation							
Inorganic chemicals	x	x	–	–	x	x	–
Polyelectrolytes	x	x	–	–	–	x	–
Disinfection	–	–	–	–	–	x	–
Electrolytic processes							
Electrodialysis	–	–	–	x	x	x	–
Electrolysis	–	–	x	x	–	–	–
Extractions							
Ion exchange	–	–	–	x	x	–	–
Liquid-liquid (solvent)	–	–	x	x	–	–	–
Incineration							
Fluidized-bed	x	x	x	–	–	x	x
Physical processes							
Carbon adsorption							
Granular activated	x	–	x	–	–	–	–
Powdered	x	x	x	–	–	x	–
Distillation	x	x	x	x	x	x	–

Filtration							
Coal filtration	L	x	–	–	–	x	–
Diatomaceous-earth filtration	–	x	–	–	–	x	–
Dual-media filtration	–	x	–	–	–	x	–
Micro-screening	–	x	–	–	–	x	–
Sand filtration	–	x	–	–	–	x	–
Flocculation-sedimentation	–	x	–	–	–	x	–
Foam separation	–	x	x	x	–	–	–
Freezing	x	–	x	x	x	–	–
Gas hydration	x	–	x	x	x	–	–
Reverse osmosis	x	x	x	x	x	x	–
Stripping (air or steam)	x	x	x	–	x	–	–

[a]Under specific conditions there will be limited effectiveness.

TABLE 2.6 The Treatment Levels Suitable for Industrial Wastewater Reuse and the Resulting Typical Costs (Including sludge processing) are:

Unit Costs of Wastewater Treatment[1]
(Million gallons per day)

Treatment Level	*Treatment System*	*Unit Cost, ¢/1000 gal* 1 mgd	10 mgd	50 mgd
1a	Activated-sludge	100.0	47.4	36.3
1b	Trickling filter	104.7	52.6	39.5
1c	Rotating biological contactors	107.2	65.1	55.8
2a	2-State nitrification	130.7	58.8	44.7
2b	Rotating biological contactors	142.3	98.2	85.0
2c	Extended aeration	43.2	28.4	–
3a	Nitrification-denitrification	150.2	73.2	56.0
3b	Selective ion exchange	189.3	89.9	65.5
4	Filtration of secondary effluent	132.8	55.1	42.7
5a	Alum added to aeration basin	172.8	70.3	57.4
5b	Ferric chloride added to primary	156.4	77.5	57.8
5c	Tertiary lime treatment	196.1	81.7	59.3
6a	Tertiary lime, nitrified effluent	217.0	83.0	60.6
6b	Tertiary lime plus ion exchange	224.4	108.9	77.2
7	Carbon adsorption, filtered secondary effluent	176.5	72.7	57.2
8	Carbon, tertiary lime effluent	238.7	99.4	74.2
9	Carbon, tertiary lime, nitrified effluent	263.8	100.6	75.5
10	Carbon, tertiary lime, ion exchange	288.0	126.5	91.9
11	Reverse osmosis of AWT effluent	481.3	214.6	171.8
12a	Physical-chemical system, lime	250.4	103.3	81.2
12b	Physical-chemical system, ferric chloride	255.6	115.0	86.4
13a	Irrigation	79.5	61.5	53.5
13b	Infiltration-percolation	38.1	19.7	15.8
13c	Overland flow	52.4	35.4	30.1

Note: Certain levels of treatment are additive; for example Level 5c is Level 1a followed by tertiary lime treatment.
[1] "Water Reuse and Recycling, Volume 2–Evolution of Treatment Technology" CWC Engineers for OWRT. PB-80 131469.

and land spreading. Consequently, the removal of incompatible pollutants at their source by pretreatment is advantageous.

However, incompatible pollutants removed by pretreatment still require an ultimate disposition. The sludge produced by pretreatment operations may be a source for by-product recovery or recycle. When this is not economically or technically feasible, disposal of sludge is necessary. Although the sludges produced by industrial pretreatment may not technically be under municipal regulatory control, the impact on other environmental areas should be noted.

Pollutants that Interfere with POTW

Interference can be defined as any situation where the addition of a new pollutant or pollutants inhibits the functioning of an existing waste treatment facility or unit. Interference can be caused by a variety of chemical, physical, and biological phenomena. It can be caused not only by materials which inhibit biological sewage treatment processes, but also by substances which cause problems in sewage collection systems, sludge disposal or utilization methods, water reuse, land application of wastewater, or other operations. Interference with sludge disposal or utilization and reuse of wastewater is caused primarily by incompatible pollutants which become concentrated in sludge or through reuse techniques.

Following primary treatment such as chemical coagulation and filtration, consideration should be given to employing activated carbon adsorption to avoid interference with existing physical-chemical sewage treatment systems. The characteristics of the activated carbon adsorption

TABLE 2.7 Comparison of Solid Waste Estimates for Manufacturing Industries

SIC No.	Industry	Waste in Millions of Pounds per Year ‡ (Total U.S.) Combustion Engineering Study *	California Study †
19	Ordnance and accessories	711	NA
20	Food and kindred products (excludes meat packing)	14,260 (or, 7.1 × 10^6 tons)	43,880 (or, 21.9 × 10^6 ton)
21	Tobacco manufacture	813	–
22	Textile mill products	2,158	340
23	Apparel and textile products	719	650
24	Lumber and wood products	76,107	137,660
25	Furniture and fixtures	3,877	430
26	Paper and allied products	10,189	2,550
27	Printing and publishing	15,221	1,030
28	Chemicals and allied products	6,048	5,050
29	Petroleum and coal products	1,148	5,800
30	Rubber and plastics products	4,927	2,690
31	Leather and leather products	6,325	–
32	Stone, clay and glass products	4,915	2,830
33	Primary metal industries	3,503	9,800
34	Fabricated metal products	7,660	3,300
35	Machinery, except electrical		8,400
36	Electrical equipment and supplies	3,047	4,640
37	Transportation equipment	3,479	3,080
38	Instruments and related products	1,665	–
39	Miscellaneous manufacturing	1,696	–

**Technical-Economic Study of Solid Waste Disposal Needs and Practices, Volume II, Industrial Inventory*, Combustion Engineering, Inc., for the U.S. Department of Health, Education and Welfare, Public Health Service, Clearinghouse for Federal Scientific and Technical Information, Pub. 1886, Report SW-7c, Pub. PB-187-712, 1969.

†*California Solid Waste Planning Study*, California Department of Public Health, 1969.

‡Divide pounds/year by 2.2 to obtain kilograms/year.

process are such that few substances are left behind which can cause interference. Although many facilities of this type are currently being designed and constructed, operating data from full-scale physical-chemical plants are limited.

Materials Which Inhibit Biological Treatment Processes

Three basic categories of treatment processes have been delineated, including aerobic processes, anaerobic processes, and nitrification. The aerobic processes have been subdivided into activated-sludge and trickling filter operations, while anaerobic processes consist of sludge digestion. Although many currently operating biological treatment plants utilize the trickling filter process, relatively little data is available on pollutant interferences. Information on aerobic biological treatment presented is predominantly concerned with the activated-sludge process.

Interfering substances are generally categorized as inorganics and organics. Acidity, alkalinity, pH, ammonia, transition metals, metals, sulfate and sulfide comprise the major components of the inorganic category. Organic substances include alcohols, amines, chlorinated hydrocarbons, pesticides and herbicides, phenol, surfactants, and miscellaneous organic chemicals.

A precipitated pollutant has relatively little inhibitory impact compared to soluble components. Settleable solids are usually removed in primary treatment and therefore seldom reach the potentially inhibited biological-unit process. Because the insoluble fraction of a pollutant parameter will have relatively little effect on

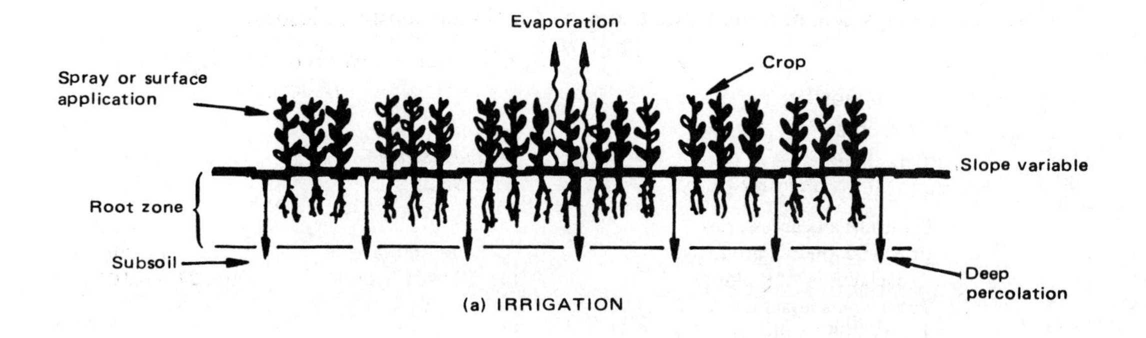

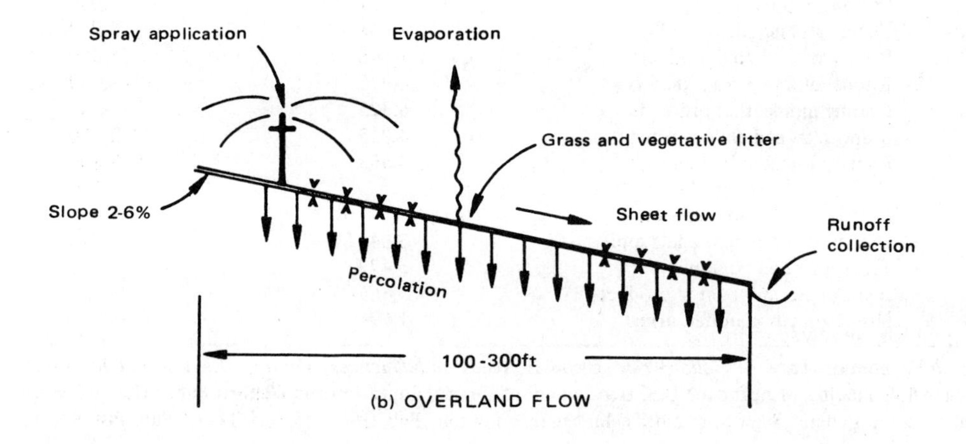

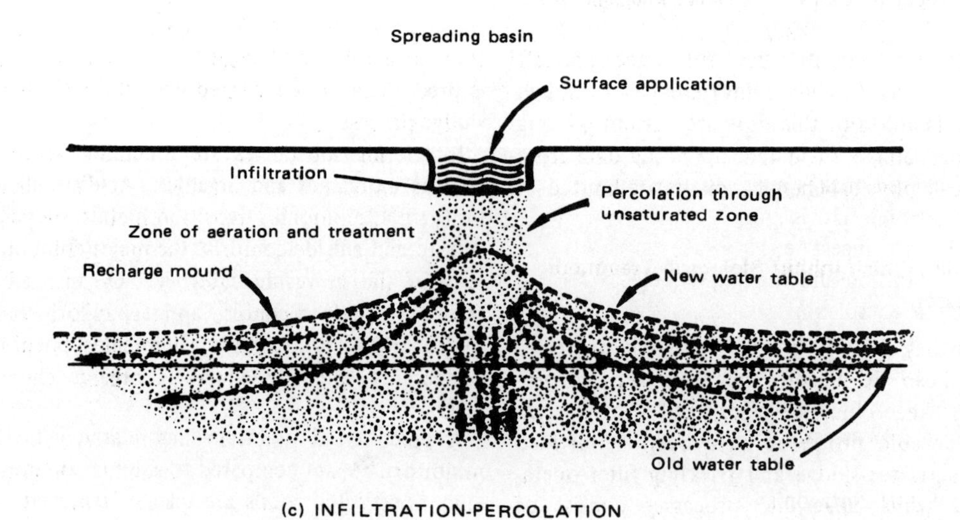

Fig. 2.4. Three basic approaches to land treatment systems. (*Courtesy EPA*)

biological-treatment processes, most investigators have conducted their tests using only dissolved pollutants.

Whether or not a substance is inhibitory depends on factors such as concentration and the presence of other chemicals which have synergistic or antagonistic effects. Some substances, such as mercury, even when present in wastewater at a very low concentration, can disrupt one or more functions of a biological treatment system. Others, such as chloride ion, are inhibitory only at relatively high concentrations. Special phenomena that may alter the inhibitory nature of a substance are outlined below.

Synergism. Synergism is characterized as an increase in the inhibiting effect of one substance by the presence of another. Synergism, and its opposite–antagonism–is most prevalent in situations involving combinations of transitions metals or heavy metals. The inhibitory effects are also enhanced by acidity.

The synergistic effects of such metals in combination with acidity can be understood from the chemistry of these metals. In the usual pH range of sewage influents (somewhat basic), heavy metals tend to be insolubilized by hydrolysis. They precipitate or adsorb on solids, and they interact with polyelectrolytes or various chemical species containing anionic functional groups. Acidity suppresses hydrolysis, keeping the metal ions in solution, and the hydrogen ions compete with the metal ions for adsorption on solids or anionic functional groups in solution.

The synergism of cyanide or other complexing substances makes metal wastes more easily biodegradable. It is possible for the microorganism to ingest excessive levels of complexed metal ion and then to destroy by assimilating the complexing substance shielding the microorganism from the metal ion with the subsequent release of an excessive level of the metal within the organism, upsetting its biological life processess.

Antagonism. Antagonism is the opposite of synergism; it is characterized as a decrease in the inhibitory effect of one substance by the presence of another. Antagonistic effects occur with the combination of metallic and anionic pollutants. Chelating agents, such as EDTA (disodium salt of ethylenediamine tetra-acetic acid) and HEDTA (disodium salt of hydroxyethylenediamene triacetic acid) exhibit antagonistic properties with metals. Chelating agents are used in culturing microorganisms to regulate the level of metals needed to grow bacterial cultures. Bacteria survive in culture solutions containing concentrations of these metals high into the inhibitory range, when chelating agents are present.

Ambiguity regarding the inhibitory effects of metals on sludge digestion exists in the literature. The ambiguity may be explained by the antagonistic effect of the sulfide normally present in a digester. Sulfide ion precipitates metals, removing them from solution thus eliminating their inhibitory effect. This has been used effectively in restoring upset digesters to operation. Additionally, other ions such as hydroxide, chromate, ferr-cyanide, phosphate, carbonate and arsenate will tend to precipitate with metals, thereby reducing the inhibitory effect.

Acclimation. Significant biological factors must be considered to understand the inhibitory effect of pollutants. In the activated-sludge process, a healthy biomass contains a broad distribution of microorganisms, including many species of bacteria and protozoa. In the absence of sufficient food, reproduction slows and the microorganism devour each other. Whenever the environment changes due to the introduction or omission of a given pollutant, the opportunities for reproduction and growth of different species change.

The history of an activated-sludge biomass affects the way in which it will respond to a new pollutant. When a new pollutant is introduced, species which cannot tolerate this substance fail to reproduce and grow and tend to die off while more tolerant species consume the food supply and grow and reproduce.

When a biomass becomes accustomed to the presence of a normally inhibitory concentration of a substance, it can be characterized as acclimated to that pollutant. But sludge digestion and nitrification do not have the same flexibility of adaptation to changing environmental conditions as do other biological processes. Both nitrification and sludge digestion are biological processes that rely on particular strains of bacteria. Sludge digestion proceeds in two steps, using two specific bacterial strains to achieve digestion. Nitrification is also limited to particular bacterial types. When adverse conditions are encountered in these processes, there is no possibility of another organism taking over for the affected strain of bacteria. Neither nitrification nor sludge digestion are readily acclimated to a new pollutant and may be easily upset.

A POTW may contain any of thousands of inorganic compounds present as major or minor constitutents. Most inorganic substances dissolved in wastewater are present in ionic form and can reduce the number of parameters of interest to a few cations and anions and nonionics.

The major cations consist of the ammonium ion and various metal ions. A few metal ions, sodium, calcium and magnesium, are prevalent in wastewaters but are not inhibitory except at very high concentrations.

IRON AND STEEL

The iron and steel industry encompasses a variety of processes for transformation of iron ore into fabricated iron and steel products. In addition to the manufacture of steel products, most large steel mills operate by-product coke plants, producing metallurgical coke and coke by-products. The industry is divided into five segments: 1) ore preparation, 2) coke production, 3) coke by-products recovery, 4) pig iron production, 5) steel manufacturing.

The processes involved in ore preparation are iron ore beneficiation–including mining, upgrading, and concentration operations–and agglomeration or preparation of the ore for charging into a blast furnace.

In the coking segment, mined metallurgical coal is prepared for charging into coke ovens, coked (nondestructive distillation) and quenched. Coke-oven gas, a by-product from coking, is treated for by-product recovery and also used as a fuel. Crude tar, ammonia, light oil, phenol and other by-products of coking are further processed, depending on plant design and on markets for specific products.

Pig iron production involves production of pig iron from iron ore, coke, and limestone in a blast furnace.

The steel manufacturing segment primarily involves production of steel from pig iron and scrap in electric, open hearth or basic oxygen furnaces and finishing operations in which raw steel is shaped, rolled, drawn, coated or otherwise treated to produce sheets, strips, plates, pipe, wire, or other forms of steel products.

Many steel plants generally do not incorporate all combinations and variations of the operations described in this document. The industry encompasses a variety of plants ranging from small to very large and from older, marginally operating facilities built early in the century to more efficient modern facilities built or upgraded in recent years.

Major portions of iron ore come from the Lake Superior region. Other areas of iron ore production are California, Utah, Wyoming, Texas, Missouri, Alabama, Pennsylvania, and New York. Combined output of these areas makes the U.S. the world's third largest producer of iron ore.

Most of the major steel companies own or control domestic mines that supply at least part of their ore needs. These companies also have invested substantially in iron mines in Canada, Venezuela, Chile, Brazil, Liberia and Australia. The major companies producing iron ore in Canada are owned or controlled principally by U.S. interests. It is estimated that captive mines furnish about 85 percent of the ore used by the U.S. domestic iron and steel industry.

Trends in iron ore mining are significant. Taconite pellets are replacing iron ores, huge open-pit mines are replacing underground mines, and mining is changing from a seasonal to a year-round basis. Changes in steel manufacturing operations are equally significant. Oxygen furnaces, and to a lesser degree electric furnaces, are rapidly replacing the open hearth, while continuous casting eliminates some of the conventional steel mill operations such as production of ingots and the use of slabbing, blooming, or billet mills.

Significant waterborne wastewaters result from all steel mill manufacturing operations. These wastes are principally suspended solids, oils, waste acids, ammonia, cyanides, phenols, chlorides, fluorides, sulfides, heavy metals, and heated discharges.

Basic processing operations include by-product coke manufacturing; sintering operations; blast furnaces; steel making (electric furnaces, basic oxygen furnaces, and open hearth furnaces); vacuum degassing operations; continuous casting; rolling mill operations; and finishing operations.

Sources of Pollution

Each of these basic operations in iron and steel production contains a large complexity of pollutant discharge into the environment. For the coking operation, wastes are emitted from the waste ammonia liquor, still wastes, final cooler wastes, and light oil recovery wastes. The blast furnace, with pollutants similar to those from the coking operation, has its main aqueous waste resulting from gas cleaning with wet washers. The actual steel-making operation also generates liquid wastes from the air pollution control equipment, such as: sparking boxes, spray chambers, and venturi scrubbers. The degassing operation has liquid wastes from the barometric-condenser cooling water. Cooling waste is discharged from the casting operations, and sintering operations associated with dust control produce another stream of wastewater. The rolling mill operation has wastewater from the scale, lubricating oils, spent pickling operations, and pickling rinse waters.

Treatment Technology

Treatment technology differs for each of the various unit operations. The coking operation uses several different technologies. For example, for the waste ammonia liquor, the coking operation employs, anhydrous recovery, bio-oxidation of cyanides and phenols, ammonia denitrification, and possibly total incineration. A closed-loop operation is achievable for the final cooler water with a minimum blowdown. Recycle is available for the light oil recovery wastes. Cooling-water wastes from the vacuum degassing operation are treated by sedimentation, filtration, cooling, and recycle. In the continuous-casting process for pig iron, the spray cooling waters are recycled after sedimentation of scale pit, followed by filtration or cooling. In newer plants, the sintering phase employs dry-dust collection and gas-clean equipment, whereas the wet systems in new plants incorporate a complete recycle consisting of a thickener (polyelectrolyte addition), vacuum filter, and clarifier. The recycle blowdown consists of, aeration for oil removal, lime precipitation, fluoride neutralization, and final thickeners.

In rolling-mill operations, one finds crude settling chambers being used for scale and oil recovery and lime neutralization, and evaporation-condensation used for treatment of strong pickling wastes. Recovery of oily wastewater is accomplished by primary belt skimming, chemical coagulation, magnetic agglomeration, and deep filtration. The main waste-treatment operations employed in the finishing operations include chrome reduction, cyanide oxidation, sedimentation, and metal sludge filtration.

The oxidation of cyanides to cyanates (CNO-) or carbon dioxide (CO_2) by various oxidizers (alkaline chlorination or hypochlorite) has been accomplished for this industrial liquid waste. Bio-oxidation is effective but generally susceptible to shock loads and requires a long residence time. Cyanide removal by ion exchange column is effective but expensive.

The phenolic wastes can also be removed by either liquid extraction or vapor recirculation, but these methods are not very economical. The bio-oxidation of phenols is being widely used and is economical.

Effective methods of ammonia recovery and removal require the use of an ammonia still and gas-liquid adsorption. Biological nitrification-denitrification of ammonia is not currently feasible because of inherent system instabilities.

Disposal of Pickling Solutions

The simplest method is neutralization, but this unfortunately leaves a great deal of dissolved solids. Ion exhange removal has been demonstrated for sulfuric (H_2SO_4) and hydrochloric (HCl) acid. Direct crystallization of either

$FeSO_4$ from sulfuric or $FeCl_2$ from hydrochloric liquors is possible. The acid is regenerated in the process and the entire procedure looks promising. Finally, spent pickling solution separation by reverse osmosis is now only a matter of speculation.

METAL FINISHING*

Operations

Electrocleaning. Electrocleaning is a process used as a final preparation of parts after heavy soil has been removed by any of the cleaning processes previously mentioned above. The operation takes place in a heavy-duty alkaline solution and the work may be anodic, cathodic or alternating between each condition. As in all the electrochemical processes that will be described, the electrical source will be direct current at relatively low voltages and high amperages.

The function of electrocleaning is to remove remaining soil and to make the surface chemically active.

Anodizing. Anodizing is a process used to produce an oxide on a metal in a controlled manner. Its purpose is to give the base metal a corrosion resistant surface with some capability for resisting abrasion. It will also present a surface capable of accepting other finishes and, particularly in the case of aluminum, for accepting dyes.

While some limited anodizing is performed on zinc, by far the greatest amount is done on aluminum.

Electroplating. Electroplating is the electrodeposition of an adherent metallic coating upon an electrode, which is the workpiece, for the purpose of obtaining a surface with properties or dimensions different from those of the basic metal. These properties may include improvement of appearance, corrosion protection, wear resistance, and so forth.

The operation takes place in aqueous solutions containing the metal ion to be plated. The work piece is cathodic and in most instances the metal ion is constantly replenished from an anode containing the metal. A notable exception is chromium where the anode is insoluble and metal ions are replenished by additions of chromic acid.

Electroplating must be preceded by cleaning and activating operations, and a typical sequence would be as follows:

a) Vapor degrease or soak clean in an emulsion or detergent cleaner
b) Spray clean in a detergent cleaner
c) Electroclean in an alkaline cleaner
d) Sulfuric acid dip
e) Electroplate
f) Electroclean
g) Sulfuric acid dip
h) Second electroplate

Rinsing would follow each process stage except (a).

There may be a single electroplate only, and it may or may not be followed by a chemical dip such as with sodium dichromate or an oxidizing agent. The former most likely would be on zinc or cadmium and the latter on copper or brass.

Electropolishing. This process, which is the reverse of plating, removes metal from the working surface. The effect is to smooth and brighten the surface, and in some applications it is used for deburring.

Baths contain a variety of chemicals, depending on the metal being polished and the desired effect.

Electrocoating. Electrocoating, or electrophoresis, is a process in which an organic coating is deposited electrolytically. When direct current is applied in a bath where the work is cathodic, there is a migration of pigment and resin particles to the part.

The excess paint is washed from the surface of the part as in other chemical treatments. Paints used include 8-10% solids, surfactants and possibly organic solvents. The pH in the bath is between 7 and 10.

Electroforming. Electroforming is a process using plating solutions to deposit a layer of

*From "Review of the Canadian Metal Finishing Industry," Environmental Protection Services of Canada, EPS 3-WP-75-2, Mar., 1975.

metal of such thickness as to have structural strength of its own. It is used for paint masks and other applications where a configuration must be closely followed.

In-Plant Water Pollution Control

Water Uses. The function of water rinse in metal finishing is to remove from the suface of the workpiece dissolved and adhering materials that might be carried from one bath to another of which would stain the surface of the finished product.

Water rinsing is used after alkaline cleaners, acid pickling, and etching, and after each process bath that has a subsequent non-compatible process operation. It is also used to dilute dumps and to wash down floors.

Evaporators. Evaporators have been in use in the metal finishing industry for many years; in the larger plating plants they are used for recovery of chromium, nickel, and cyanides. The minimum initial capital cost and stream requirements prevent their initial capital cost and stream requirements prevent their economic use in smaller plants, but where applicable they have recovered their initial investment in a rather short time.

As with all recovery systems where the chemicals are returned to the bath, there is a built-up of impurities and these are usually removed by filtration and/or iron exchange. It is possible to form a closed-loop system in which the chemicals are continuously returned to the bath and the condensate returned to the rinse tanks.

Reverse Osmosis. Reverse osmosis accomplishes much the same result as distillation in that it separates an effluent into a concentrated solution of the dragged-out chemicals and nearly pure water. The process reverses the normal osmotic pressures by applying a considerably greater pressure to the concentrated solution, thus forcing the water into the diluted solution, leaving behind most of the dissolved salts. This is accomplished by creating a pressure differential across a membrane using pumps producing 400 to 800 PSIG. Maintenance problems are normally associated with such pumps.

Most of the work on reverse osmosis has been done in the last ten years, and there are only a few installations in operation. However, industry sees the applications of this method as having good potential. Increasing effort is being made in the United States and Europe to improve the quality of membranes, develop new ones, and simplify production units. The ability to concentrate dilute solutions and return water suitable for rinsing makes this equipment a likely component of future installations that are required to meet more stringent discharge regulations.

Cellulose acetate is the membrane most commonly used; its use is limited to pH ranges between 2.5 and 7. Its temperature limits are between 65 and 90 degrees F., and it will not resist strong oxidizing agents. Further practical parameters are that it be used only for rinse waters, that flow rates should not exceed 5 GPM, and that there be no oil or heavy particles. Other membranes being studied include polypropylene, polyurethane, and the polyamides. These latter are being used for cyanide recovery since they can handle pH up to 12.

Since each of the three most likely types of equipment for recovery (evaporators, ion exchangers, and reverse osmosis units) have their limitations, it is possible that combinations of them would permit a company to maximize their advantages while minimizing their disadvantages. Closed-loop requirements will certainly employ these systems in the next few years, and one of the probable combinations will use reverse osmosis to remove the bulk of the contaminant, followed by ion exchange to remove the remainder. The backwash of the ion exchanger would then also be treated.

External Controls

Cyanide Destruction. By far the most universally used system for the destruction of cyanide is that using chlorine gas or sodium hypochlorite in an alkaline solution.

The reaction is believed to take place in three steps:

1) $NaCN + Cl_2 \longrightarrow CNCl + NaCl$

2) $CNCl + 2NaOH \longrightarrow NaCNO + NaCl + H_2O$

3) $2NaCNO + 3Cl_2 + 4NaOH \longrightarrow$

$$N_2 + 2CO_2 + 6NaCl + 2H_2O$$

The first reaction is almost instantaneous, but the second reaction takes a minimum of 5 minutes at the optimum pH of 11.5 and much longer at a pH below 9. The third reaction is longer again, about 30 minutes at its optimum pH of 8.5.

In practice it is wise to maintain a high pH during reaction (1), not only to speed the reaction but also to eliminate any possibility of evolution of the toxic gas cyanogen chloride (CNCl).

The amount of chlorine required per pound of cyanide as CN is 4.25 to 4.5 pounds for the complete reaction through (3). If destruction to the cyanate form only is required, the chlorine demand would be about 1.25 pounds. Additional quantities of chlorine will be required if there are any other oxidizable compounds present. A copper solution containing Rochelle salts will demand more chlorine as will any organic materials present. The consumption of sodium hydroxide for the complete reaction is approximately 4.25 pounds.

The presence of nickel or iron, and to a lesser extent copper, will further complicate the reaction by forming complexes that take time to break down. These create so much difficulty that every effort is made to avoid or minimize their presence in the cyanide stream.

The selection of the specific chemicals to be used is dependent on local cost, availability, storage arrangements, control methods and personnel.

Chlorine gas is the cheapest source of chlorine in large quantity in most locations but presents hazards in storage and in use that some companies prefer not to cope with. Calcium hypochlorite is often preferred in batch treatments, but the most frequently used source of chlorine in medium-sized plants is sodium hypochlorite.

A process that may have application in some circumstances has been on the market for a little over a year. It converts the cyanide to the cyanate and precipitates the metal oxide from zinc and cadmium solutions. Work is being done to extend the application to copper solutions. The active reagents for this process are formalin and hydrogen peroxide. The process becomes proprietary by the addition of stabilizers and compounds which help the reaction and the settling of the precipitate.

This method lends itself to small plant applications where metal levels of 1-2 mg/1 and cyanate are acceptable in the final effluent. However, it is a requirement that there by sufficient counterflow rinsing to achieve maximum levels of cyanide in the rinse water. While it is simpler to use this system in a batch treatment plant, continuous flow treatment is possible.

Chromium Reduction and Removal. Most of the chromium in metal finishing plant effluents will be in the hexavalent form originating from rinses from chromium plating, chromic acid cleaning and etching baths, or chromate bright dips. There will be, in addition, concentrated dumps from the cleaning, etching, and bright dip baths.

The most frequently used method for removing chromium from the discharge waters requires the reduction of hexavalent chromium to the trivalent state and subsequent precipitation, usually as the hydroxide.

The choice of reduction agent is dependent on cost, availability, and convenience. The chemicals most frequently used are sulphur dioxide, sodium bisulfite, sodium metabisulfite, and furrous sulfate. The latter is most often used when it is available as a by-product to sulfuric acid descaling within the same plant.

The chemical reaction involved with sulfur dioxide can be shown as follows:

$$3SO_2 + 2H_2CrO_4 \longrightarrow Cr_2(SO_4)_3 + 2H_2O$$

Using sodium metabisulfite with sulfuric acid, the reaction is:

$4H_2CrO_4 + 3Na_2S_2O_5 + 3H_2SO_4 \longrightarrow$

$$Na_2SO_4 + 2Cr_2(SO_4)_3$$

In one of the systems which will be discussed later, where treatment is in line with the plating process, a second alkaline reduction is added using sodium hydrosulfite or hydrazine. This latter reaction may be shown as follows:

$4H_2CrO_4 + 3N_2H_4 \xrightarrow{Na_2CO_3}$

$$4Cr(OH)_3 + 3N_2 + 4H_2O$$

The acid reaction occurs very rapidly at a pH

of 2-3. If the chromium bearing stream is separated from any alkaline streams it will have a pH close to 3 in most plants. When the pH must be lowered sulfuric acid is used.

Removal of Other Metals. The final removal of all metals usually found in metal finishing operations is most frequently carried out by precipitation of the metal as the hydroxide.

The pH of acid solutions is raised toward neutrality and the pH of alkaline solutions is lowered. Metal salts from simple inorganic compounds will tend to become insoluble in the neutral pH range, but not all metals will precipitate just on neutralization, and not all metals will precipitate at the same point and to the same extent.

ORGANIC CHEMICALS; PETROLEUM REFINING

In the petrochemical and petroleum refining industries, most of the pollution comes from such process operations as the following: crude oil processing; steam distillation; steam stripping; product rinses; barometric condensers; lube oil manufacturing; petrochemical manufacturing; boiler and cooling tower blowdowns; contaminated area drainage runoff; special additives used in water treatment; oil storage and transfers; vessel clean-outs; unrecovered reaction products (by-products and coproducts); process leaks, spills, and purges; and inorganic chemical production of acids, halogens, and nitrogenous salts. Wastewaters may contain various salts, acids and alkalies, ammonia, sulfides, solids, mixtures of organics of varying biodegradability, phenols and other taste- and odor-producing chemicals, and heavy and light oils. They typically have a BOD and COD range of 100–10,000 mg/l. and 200–15,000 mg/l., respectively, with an average of 1150 mg/l. BOD and 3100 mg/l. COD. Table 2.8 lists some typical wastewater characteristics from these industries.

These wastewaters are relatively nonbiodegradable and have relatively high concentrations of pollutants, particularly compared to domestic sewage. Figure 2.5 breaks down different types of petrochemical wastewater: type (a) represents wastewater similar to domestic waste; type (b) represents a semibiodegradable waste;

TABLE 2.8 Refinery and Petrochemical Wastewater Characteristics

Petrochemical principal products	*Waste flow, mgd*	*BOD, mg/l.*	*COD, mg/l.*	*SS, mg/l.*
Phenol, ethylene	2.0	300	1 200	300
Acrylonitrile	0.302	–	1 200	239
Fatty acids, esters, glycerol	0.10	10 000	14 000	–
Azo and anthraquinone dyes	0.94	352	1 760	152
Ethylene, alcohols, phenol	5.9	1 700	3 600	610
Acrylonitrile, acetonitrile, hydrogen cyanide	3.9	390	830	106
Butadiene, alkalate, MEK, styrene, maleic anhydride	2.0	1 870	–	10
Butadiene, maleic acid, fumaric acid, tetrahydrophthalic anhydride	3.605	959	1 525	–
Phenols	0.215	6 600	13 200	–
Acids, formaldehyde, acetone, methanol, ketones, nitric acid, nylon salt, vinyl acetate, acetaldehyde	3.46	530	10 130	160
Isocyanates, polyols, urethane foam	0.57	421	1 200	50
Acetaldehyde	1.15	20 000	50 000	200
Ethylene, propylene, butadiene, alpha olefins, polyethylenes	0.750	155	380	120
Butylene isomers, butadiene, maleic anhydride, fumaric acid, tetrahydrophthalic anhydride, alkalate, aldehydes, alcohol	1.50	1 960	2 980	–
Refinery–Class A	0.22	20	120	40
Refinery–Class B	0.99	250	750	180
Refinery–Class C	2.98	300	1 080	240
Refinery–Class D	4.35	160	1 080	130
Refinery–Class E	7.93	200	520	90

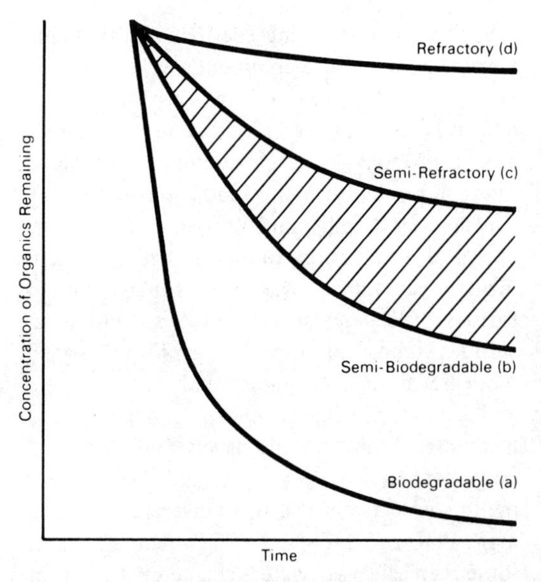

Fig. 2.5. Types of petrochemical wastewater and their relative biodegradability. *(des Rosiers, P.E. and Rey, G., The EPA Program for Environmental Control in the United States and Puerto Rican Petrochemical Industry, 1972.)*

type (c) requires a long period of "seed" acclimation before assimilation of organics proceeds; and type (d) represents a refractory waste with essentially no biochemical actions. Thus, figure 2.5 shows that petrochemical wastewaters generally fall into the range between a type (b) and type (c) waste.

Several theories are proposed for this behavior: (1) reaction rates of individual components (organics) may differ markedly; (2) the bio-mass present may not utilize some of the constituents due to the absence of the proper enzymes (catabolic and biosynthetic enzymes and permeases must all be present in proper concentrations); and (3) there may exist materials causing inhibition or repression (interference in the natural metabolic pathway). An additional factor, and one that may be the most significant, is the formation of intermediates and end-products of metabolic pathways that diffuse out of or are released by cell lysis. These products may profoundly affect the biodegradability of the waste. Extracellular quantities of these intermediates and metabolites often are indistinguishable from the original waste components and due to the nonspecificity of either the BOD or COD analysis, such a change in concentration cannot be monitored directly. Therefore, a waste which is biodegradable may appear to be "hard" because of intermediates which have passed into the medium during the given detention time.

Table 2.9 contains a list of 34 organic compounds found to be biorefractory.

In addition to wastewaters, all industries have inherent pollution control concerns with process sludges, waste chemicals, and thermal discharges in the form of cooling waters. The question of

TABLE 2.9 Refractory Industrial Wastes

Nitrobenzene[a]	Bromobenzene
Trichloroethane	Dichlorobenzene
Tetrachloroethylene	Bromochlorobenzene
Chloroethyl ether	Ethylbenzene[a]
Chloromethyl ethyl ether	Chloroform
Chloropyridine	Styrene[a]
Chloronitrobenzene	Isopropylbenzene
Dichloroethyl ether	Butylbenzene
Benzene[a]	Dibromobenzene
Toluene[a]	Isocyanic acid[a]
Camphor	Methylchloride[a]
Veratrole (1,2-dimethoxy benzene)	Bromophenylphenyl ether
Guaiacol (methoxy phenol)	Dinitrotoluene[a]
Borneol (bornyl alcohol)	Methylbiphenyl
Isoborneol	Acetone
Ethylene dichloride[a]	2-Ethylhexanol
Chlorobenzene	2-Benzothiozole

[a]These compounds also have been found to impart taste and odor to drinking water supplies in trace amounts.

toxic, hazardous, and/or taste, odor, and tainting constituents in discharges of any sort must also be confronted.

Warm water discharge can be controlled by use of cooling towers or ponds with provisions for the cooled water to be reused. There are indications that industry is well aware of the benefits of recycling the cooling towers blowdowns. One such example concerns a Class "E" refinery where the cooling towers are used to reduce the concentration or organics and to concentrate the undesirable precipitates. Thus, all wastewaters originating from plant operations and storm water runoff are either recycled into the process stream or reused or recycled in the cooling towers.

If wastewater is reused to its maximum potential, it could provide industry with a viable technique for pollution control, which could conceivably be the least expensive alternative for fulfilling future regulatory requirements. Indeed, it is conceivable that in the extreme case there would be no discharge of wastewater. Add to this the water conservation aspect and the potential savings resulting from reduced effluent monitoring, and the total recycle concept has even greater appeal.

The largest amounts of the waste load entering our waterways are derived from the more complex types of refinery operations. It should be evident that more than half of each of the five important water pollution loads originate from Class D and E refineries.

The more complex the refinery, the greater is the need for higher efficiency pollution control systems in order to achieve the same water discharge load per unit of capacity as for less complex refineries. These data also indicate that the more complex refineries have larger capacities and are the larger point sources of waste loads.

Table 2.10 shows treatment control efficiencies achievable through biological systems and activated carbon adsorption, as compared to the overall industry average being achieved at this time. It illustrates that the refining industry could substantially improve its pollution control by applying either well-demonstrated technology (biological systems) or newer and emerging technology, such as activated carbon adsorption.

Water Pollution Control Technology

Table 2.11 illustrates the technological areas that are, in general, pertinent to effective pollution control for the organic chemicals and petroleum refining industries. The specific solutions available to each plant will be determined by several factors—and economics will often be the most important of these. By and large, one must recognize the need to detoxify and/or degrade toxic wastes, control discharges of nutrients and salts (possibly through by-product recovery techniques), and reduce inadvertent high strength wastewater releases to quantities that produce little or no ecological effects on receiving streams. Where the receiving waters are too sensitive for any permissible waste discharge, the development and implementation of "no discharge" systems of pollution control may be required as the means to

TABLE 2.10 Waste Treatment Effectiveness in Petroleum Refining Industry

Water pollution parameter	*Overall API separator effluent, lb × 10^6/yr*[a]	*Overall final effluent discharges, lb × 10^6/yr*[a]	*Current treatment effectiveness, %*	*Potential percent effectiveness possible with*		*Concentration, mg/l.*
				Established practice[b]	*Current technology*[c]	
BOD	413	292	30	77-94	97	15
COD	952	910	4	44-75	89	80-115
Oil	175	131	25	26-63	76	5
Phenol	33	20	39	97-99.5	99.97	0.1
Suspended solids	233	183	22	15-75	78	2.2 × BOD_5

[a]Derived from data in 1967 Domestic Refinery Effluent Profile (American Petroleum Institute).
[b]Current established practice, e.g., activated sludge (EPA-API).
[c]Technologically feasible but economics are refined, e.g., activated sludge followed by carbon adsorption.

TABLE 2.11 Pollution Control Technology Status in Petrochemical and Petroleum Refining Industries

Technology	*Research*	*Development*	*Demonstration*	*Applications and objectives*
Biological oxidation	Some	Some	Yes	(1) Multistage Biox systems (2) To achieve and maintain 90–95% BOD_{μ} removal for single-stage systems. (3) Need pre-and post-treatment process/ operations to improve performance and reliability of both single- and multistage biox
Sludge disposal	Yes	Yes	Yes	(1) Solids disposal (a) Land assimulation (b) Incineration (fuel value) (c) Catalylic oxidation (d) Solvent extraction (e) Wet oxidation and pyrolysis
Advanced treatment	Yes	Yes	Yes	(1) Biox alternatives (2) To Achieve and maintain 95–99% pollutant reductions (3) By- and co-product recovery (4) Refractory organic removal (5) Biox supplement
Closed loop systems	Yes	Yes	Yes	(1) No discharge (closed cycled systems) (2) Cost of water (3) Water conservation and reuse (4) Containment of trace toxic chemicals
Comprehensive approach	–	Yes	Yes	(1) Total environmental effects (2) Basin planning (3) Joint treatment (4) Synergistic benefits

meet local or natural environmental standards (present and in the future).

The wastewater treatment methods presently employed by these industries may be classified as follows: (a) disposal by dilution; (b) process improvements and modification, including waste stream segregation for specialized treatment prior to final treatment; (c) bio-treatment consisting of both single and multistage aerobic and anaerobic systems in series or parallel operation; (d) physical treatment methods; (e) chemical treatment methods; (f) joint industrial/municipal treatment; and (g) limited water recycle and reuse.

Table 2.12 lists the various industrial organic chemical processes as waste sources.

PRIMARY ALUMINUM INDUSTRY*

The primary aluminum industry consists of processing bauxite ore to produce alumina, and occasionally aluminum hydroxide, and processing alumina to produce aluminum. The manufacture of other aluminum chemical compounds is not considered a part of this industry since only two companies that produce primary aluminum also produce a few aluminum based chemicals.

The two main industry segments consist of processing bauxite to alumina and then processing alumina to aluminum. They are not usually performed at the same site since they are distinct and separate processes. A third industry segment, electrode preparation, is accomplished at the aluminum production site. Production of aluminum from recycled scrap aluminum and the production of many other aluminum-based chemicals are not considered parts of this industry. These industries utilize different raw materials and include mainly companies and/or divisions which do not produce aluminum from alumina.

Approximately 7.6 billion kg (8.4 million tons) of alumina were produced in this country

*EPA 600/2-77-023 y, Feb. 1977.

TABLE 2.12 Industrial Organic Chemical Processes as Waste Sources

Process	*Source*	*Pollutants*
Alkylation:		
Ethylbenzene		Tar, hydrochloric acid, caustic soda, fuel oil
Cyanide production	Water slops	Hydrogen cyanide, unreacted soluble hydrocarbons
Dehydrogenation:		
Butadiene product from n-butane and butylene	Quench waters	Residue gas, tars, oils, soluble hydrocarbons
Ketone production	Distillation slops	Hydrocarbon polymers, chlorinated hydrocarbons, glycerol, sodium chloride
Styrene from ethylbenzene	Catalyst	Spent catalyst (Fe, Mg, K, Cu, Cr, Zn)
	Condensates from spray tower	Aromatic hydrocarbons, including styrene, ethyl benzene, and toluene, tars
Extraction and purification:		
Isobutylene	Acid and caustic wastes	Sulfuric acid, C_4 hydrocarbon, caustic soda
Butylene	Solvent and caustic wash	Acetone, oils, C_4 hydrocarbon, caustic soda, sulfuric acid
Styrene	Still bottoms	Heavy tars
Butadiene	Solvent	Cuprous ammonium acetate, C_4 hydrocarbons, oils
Extractive distillation	Solvent	Furfural, C_4 hydrocarbons
Halogenation (principally chlorination)		
Addition to olefins	Separator	Spent caustic
Substitution	HCl absorber, scrubber	Chlorine, hydrogen chloride, spent caustic, hydrocarbon isomers and chlorinated products, oils
	Dehydrohalogenation	Dilute salt solution
Hypochlorination	Hydrolysis	Calcium chloride, soluble organics, tars
Hydrochlorination	Surge tank	Tars, spent catalyst, alkyl halides
Hydrocarboxylation (OXO Process)	Still slops	Soluble hydrocarbons, aldehydes
Hydrocyanation (for acrylonitrile, adipic acid, etc.)	Process effluents	Cyanides, organic and inorganic
Isomerization in general	Process wastes	Hydrocarbons; aliphatic, aromatic, and derivative tars
Nitration		
Paraffins		Byproduct aldehydes, ketones, acids, alcohols, olefins, carbon dioxide
Aromatics		Sulfuric acid, nitric acid, aromatics
Oxidation		
Ethylene oxide and glycol mfg.	Process slops	Calcium chloride, spent lime, hydrocarbon polymers, ethylene oxide, glycols, dichloride
Aldehydes, alcohols, and acids from hydrocarbons	Process slops	Acetone, formaldehyde, acetaldehyde, methanol, higher alcohols, organic acids

TABLE 2.12 (Continued)

Process	*Source*	*Pollutants*
Acids, anhydrides from aromatic oxidation	Condensates	Anhydrides, aromatics, acids
	Still slops	Pitch
Phenol, acetone from aromatic oxidation	Decanter	Formic acid, hydrocarbons
Sulfation of olefins		Alcohols, polymerized hydrocarbons, sodium sulfate, ethers
Sulfonation of aromatics	Caustic wash	Spent caustic
Utilities	Boiler blow-down	Phosphates, lignins, heat, total dissolved solids, tannins
	Cooling system blow-down	Chromates, phosphates, algicides, heat
	Water treatment	Calcium and magnesium chlorides, sulfates, carbonates

in 1972, including 0.27 billion kg (0.3 million tons) in the Virgin Islands, from processing about 15.4 billion kg (17 million tons) of bauxite. Approximately 11 percent of this bauxite was mined in this country, and the balance was imported mainly from Jamaica and Surinam. The primary aluminum industry utilized about 94 percent of this alumina to make aluminum. Bauxite is processed to alumina by five companies at nine sites (one in the Virgin Islands). Plants range in capacity from 0.32 to 1.22 billion kg (0.350 to 1.380 million tons) per year. Limited growth of bauxite processing plants in this country is expected.

Aluminum is produced by 13 companies at 32 locations but three companies produce most (87 percent) of the alumina and 65 percent of the aluminum in the U.S.

The basic process used for producing raw aluminum metal is all plants in the U.S. consists of the electrolytic dissociation of alumina (Al_2O_3) dissolved in a molten bath of cryolite (Na_3AlF_6). Oxygen released in this process reacts with the carbon anode to form carbon dioxide and some carbon monoxide. The molten aluminum settles to the bottom of the cell, directly above the cathode. The three types of cells or pots used to accomplish this reaction are the prebake cell, the vertical, and the horizontal Stud Soderberg designs. These cells all operate under the same principle, but they differ in anode configuration. The prebake design is used in about 70 percent of total production.

Individual aluminum plants range in size from 31.7 to 254 million kg (35,000 to 280,000 tons) per year capacity. Total capacity is about 4.445 billion kg (4.9 million tons) per year. The industry operates at about 85 percent of capacity and employs approximately 146,600 people under Standard Industrial Classification Code 3334.

Aluminum plants are generally not located near large population centers and are always located near sources of relatively low-cost electrical power. Approximately 70 percent of the aluminum manufacturing capacity is located in areas with a population density of less then 80 people per square mile. The environmental effects from aluminum production on livestock and vegetation are very significant due to the potentially high fluoride emissions. Alumina plants are located near mining areas in Arkansas and in port cities along the Gulf of Mexico and tend to be in more populated areas.

Aluminum production is projected to grow at a rate of about 5.2 percent per year until 1980, based on a demand growth of 7.5 percent. The excess in demand over growth will be made up by an increased operating ratio, by the gradual disposal of the GSA stockpile, and by construction of new plants.

The promulgation of emission standards by the EPA is not expected to limit industry growth significantly. Increased electrical power costs will, however, cause secondary (recycled) aluminum to compete more economically with primary aluminum for some markets.

A number of aluminum reduction plants produce their own electricity by fuel combustion and/or hydroelectrical power.

Raw Materials

Bauxite, of which approximately 87 percent is imported, is the raw material for alumina production. This material is composed mainly of metallic oxides with aluminum oxide comprising from 20 to 60 percent of the ore as mined. Domestic ores tend to be leaner in alumina content. Heavy metals are not usually found in this ore. The major environmental problems

TABLE 2.13 Raw Materials Required to Produce Aluminum*

Material	*Amount, % by weight of aluminum products*
Sulfur	1-5
Alumina (Al_2O_3)	190
Cryolite (Na_3AlF_6)	3-5
Aluminum fluoride (AlF_3)	3-5
Fluorspar (CaF_2)	0.3
Anode:	
Petroleum Coke	49 (Prebake) 45 (Soderberg)
Pitch Binder	12.5 (Prebake) 16.7 (Soderberg)
Cathode (carbon)	2

*EPA-600/2-77-023Y, Feb., 1977.

associated with bauxite mining and shipping are caused by fugitive dust and water runoff.

Alumina and carbon (petroleum coke) are the major constituents used for making aluminum as shown in Table 2.13. Calcined petroleum coke, pitch, and some anthracite are used to make anodes and cathodes for the aluminum reduction process.

Products

Alumina and aluminum metal are the primary products of the industry. Much smaller amounts of aluminum hydroxide, sodium aluminate, and aluminum fluoride are also made by a few plants. The alumina is used largely in the aluminum smelting industry, and the aluminum is used for manufactured products.

Environmental Impact

Waste sludge (red mud) is the primary pollution problem associated with bauxite processing. For each kilogram of alumina product, two kilograms of bauxite must be processed. Atmospheric emissions of particulate are caused by raw material and product handling, by calcining the hydrated alumina in a rotary furnace, and by sintering of the mud when required.

Atmospheric emissions of coke dust from crushing and screening operations occur in the electrode manufacturing process. Volatile hydrocarbons from the paste and binder mixing operations also occur. When prebaked electrodes are manufactured, atmospheric emissions of fluorides, hydrocarbons, and sulfur oxides also occur. The fluoride compounds come from the recycled anode scrap material from the aluminum smelting operation. Emissions of SO_x amount to between 0.35 to 1.0 percent and emissions of total fluorides between 0.15 and 0.75 percent of aluminum produced.

Primary aluminum reduction mainly causes atmospheric emissions of particulate and gaseous fluorides. Emissions vary with the type of reduction cell and with the type of hooding and control device utilized. Approximately 20 to 30 grams of fluoride are evolved per kilogram of aluminum (40 to 60 pounds/ton), with the gaseous fluorides accounting for up to about 60 percent of the total emissions. On the average, about 25 to 30 percent of this emission escapes from the cell to the atmosphere. Total particulate (including solid fluoride) emissions from the cell have not been well quantified, but amount to approximately 45 grams/kg of product (90 pounds per ton).

An overall control efficiency of 70 to 80 percent of fluoride emissions is achieved by most plants. A primary control system vents the cell, and the pollutants captured in this system are reduced by about 95+ percent with wet scrubbers, electrostatic precipitators, or alumina coated fabric filters. The emissions that escape from the cell are vented through the cell building roof. A number of plants are now venting the entire cell-room gases through low pressure drop-baffled spray systems to reduce emissions.

Recovery of fluoride compounds from dry sorption systems employing fabric filters or from the scrubber sludges is practiced at some plants.

Sources of wastewater from primary aluminum reduction include wet scrubbing, boiler blow-down, and cooling water. The reported volume of solid wastes resulting from water treatment in a number of plants was 15 to 30 kg per metric ton of aluminum produced. These solid wastes are composed for cryolite, carbon, and calcium fluoride sludge. Spent carbon cathode pot linings are another source of solid waste. The estimated annual volume of such wastes produced is 1200 cubic meters (about one acre foot). The EPA Office of Solid Waste Management Programs is currently sponsoring an investigation of the wastes and disposal technologies for the primary metals industry which includes aluminum production.

PLASTICS AND RESINS INDUSTRY*

The plastics and resins industry includes operations which convert monomer or chemical intermediate materials obtained from the basic petrochemicals industry and the organic chemicals industry into resinous polymer products. Figure 2.5 is a simplified diagram showing the interrelation of industries closely related to the

*Based on EPA Report EPA-600/2-77-023j, Feb., 1977.

TABLE 2.14 Major Raw Materials for the Plastics and Resins Industry

Raw Materials	*1972 Consumption*
acrylates and methacrylates, monomers	358.4 Gg (789.5 × 10^6 lb)
acrylonitrile	124.4 Gg (274.1 × 10^6 lb)
alcohols (except ethyl)	1.86 hm^3 (490.4 × 10^6 gal)
carbon black	Not reported
cellulose acetate	Not reported
extender oils (petroleum derived)	Not reported
formaldehyde (37%)	618.4 Gg (1362.2 × 10^6 lb)
glycerin	11.8 Gg (26.1 × 10^6 lb)
liquid refinery and petroleum gases	
butadiene	156.4 Gg (344.6 × 10^6 lb)
ethylene	1.978 Tg (4,356.7 × 10^6 lb)
other (isoprene, propylene, isobutylene, etc.)	530.5 Gg (1168.5 × 10^6 lb)
melamine	23.0 Gg (50.7 × 10^6 lb)
phenol	285.7 Gg (629.2 × 10^6 lb)
phthalic anhydride	103.1 Gg (227.2 × 10^6 lb)
plasticizers	93.5 Gg (205.9 × 10^6 lb)
rubber processing chemicals (accelerators, antioxidants, blowing agents, inhibitors, peptizers, etc.)	Not reported
soap and detergents	18.1 Gg (39.9 × 10^6 lb)
sodium hydroxide	3.54 Gg (3.9 × 10^6 tons)
styrene	1.417 Tg (3122.2 × 10^6 lb)
sulfuric acid (100%)	81 Gg (90 × 10^3 tons)
thermoplastic resins	242.8 Gg (534.9 × 10^6 lb)
thermosetting resins	17.2 Gg (37.9 × 10^6 lb)
urea	157.5 Gg (346.9 × 10^6 lb)
vinyl acetate monomer	193.1 Gg (425.3 × 10^6 lb)
vinyl chloride monomer	1.176 Tg (2591.4 × 10^6 lb)
woodpulp (excluding wood flour)	40.0 Gg (44.1 × 10^3 tons)

plastics and resins industry. Table 2.14 is a listing of the major raw materials used in the industry. Approximately three hundred producers accounted for the 1974 production of 22 billion pounds of plastics and resins. A detailed treatment of this industry is provided in Chapter 10.

Environmental Impact

Liquid waste streams, generally aqueous, are encountered throughout the industry. Much of the wastewater originates from processing in which the process streams come into direct contact with water. Wastewater may also be formed during the course of a chemical reaction; it may arise from cleaning process vessels, area housekeeping, utility boiler and cooling water blow-down, and other sources such as laboratories. The contaminants encountered in the wastewater include organic reactants, monomers, oligomers, polymers, and salts.

The extensive heating and cooling requirements necessitate elaborate cooling tower, refrigeration, and steam generation facilities. The wastewater streams from these points are generally combined to be sent to the water treatment plants. Cooling tower and boiler blow-down streams may contain toxic anti-corrosion chemicals such as chromium compounds and anti-fouling agents.

SYNTHETIC FIBER INDUSTRY*

For purposes of this section, synthetic fibers are defined as noncellulosic fibers of synthetic origin. The activities of this industry start with

*Based on EPA Report EPA 600/2-77-023k, Feb., 1977.

TABLE 2.15 Raw Material Consumption for Production of Synthetic Fibers by Companies in SIC Code 2824 in 1971

Raw Material	*Consumption, Gg*
Acrylonitrile	290
Acrylates and methacrylates	12
Caprolactam	347
Glycols	518

Source: U. S. Bureau of Census. Census of Manufacturers, 1972. Industries Series: Plastics Materials, Synthetic Rubber, and Man-made Fibers. MC72(2)-28B. Washington, D.C., GPO, 1974.

a synthetic long-chain polymer and end with the formation of a marketable filament or threadlike material. Thus, as it is defined, the synthetic fiber industry employs 97,000 people in 149 plants, with most of the production coming from a few large plants having capacities of from 110 to 330 million pounds per year.

The primary raw material used in the fibers industry is bulk polymer obtained either directly from the polymerization process or indirectly in the form of dried polymer chips. Table 2.15 gives 1974 industry consumption of major raw materials. Many additives are blended with the polymer before actual fiber production including delustrants, pigments, dyeing assistants, dye receptors, optical brightners, antioxidants, light stabilizers, and heat stabilizers. Materials added to the fiber to enhance product utility include lubricating agents, bacteriostats, humectants, antistatic agents, and others. Both organic solvents and aqueous solutions of inorganic salts are used in some processes.

Environmental Impact

Relatively little data were available on the environmental impact of the synthetic fibers industry. It can be assumed that the industry produces all three types of waste (gaseous, solid, and liquid) in varying degrees. Aqueous emissions appear to represent the largest potential source of pollution.

In general, the polymer raw materials are not toxic or otherwise hazardous unless heated to temperatures at which decomposition can occur. Emissions from the fiber industry usually arise from mechanical treatment of the polymer or are associated with solvents, additives, lubricants, or finishes used in processing. Companies which use integrated polymerization spinning systems produce waste which contains unreacted monomer.

The major sources of gaseous emissions are from cooling chambers, conditioning chambers, solvent removal chambers, and from solvent make-up. The use of hot solvents in several processes results in the entrainment of solvent by the fibers. Solvent may be subsequently emitted as vapor during processes such as drawing or heat-setting.

Particulate emissions of the solid polymer are possible from most of the processes but primarily from the cutting, winding, crimping, and baling process steps. Other solid wastes result from disposal of substandard material, filter solids, and water treatment sludge. Some four percent of fiber produced is of substandard quality. This waste fiber may be buried, incinerated, reprocessed, or sold depending on supply and demand.

Liquid emissions generally are termed "spin-finish wastes." Wastes included in this category are water used for purging the spinning baths and washing the filaments, lubricants used in finish applications, and solvent wastes from dry and wet spinning. Other liquid effluents arise from wash water in filtration steps, solvent spills, and drawing baths. Periodic cleaning of process equipment also contributes significantly to the total wastewater load. Sanitary wastes resulting from the large number of people employed at fiber plants are a significant portion of the total effluent load. Cooling water blowdown also contributes to the liquid effluent from melt spinning.

Wastewater emissions from some operations have been classified according to waste load and treatability. This information is summarized in Table 2.16. Analysis of samples from a settling pond at an acrylic fibers production facility indicated the presence of acrylonitrile (100 mg/l), 2,3-dibromo-l-propanol (0.5 mgl/l), an isomer of dibromopropene, and 2,4-dimethyldiphenylsulfone.

TABLE 2.16 Summary of Wastewater Data for Selected Fibers

Fiber	*Wastewater Loading (m^3/kkg)*	*Raw Waste Loads (kg/kkg)* BOD_5	*COD*	*SS*
Nylon	1.3–30.9	0.1–60	0.2–90	0.1–6
Olefin[1]	8.3–14.2	0.4–1.1	1.8–2.6	0.2–2.2
Spandex	–	20[2]	40[2]	–

[1]polypropylene
[2]estimated

SYNTHETIC RUBBER*

The synthetic rubber industry as described in this section comprises companies which produce a synthetic vulcanizable elastomer by polymerization or copolymerization of monomers derived from petroleum or natural gas. The 1976 Directory of Chemical Producers indicates that there are 111 facilities for producing synthetic elastomers, and it is estimated that over 4.2 billion pounds of synthetic rubber were produced in 1976.

Ethylene, propylene, isobutylene, methane, benzene, and butylene are the basic petrochemical feedstocks used to produce monomers for use in the synthetic rubber industry. These feedstocks are used directly as monomers or as feedstocks to produce other monomers such as butadiene, styrene, chloroprene, acrylonitrile, and isoprene. The tire and rubber industries consume 90 to 95 percent of all the carbon black produced in the United States. About 90 percent of the synthetic rubber produced in the U.S. is extended with hydrocarbon oils, and extenders such as carbon black or oil can comprise as much as 50 percent of the rubber produced. Other materials used directly in elastomer production include antioxidants, catalysts, initiators, reaction terminators, soaps, modifiers, and other additives.

Environmental Impact

The synthetic rubber industry produces all three types of waste (gaseous, solid, and liquid) in varying degrees. The largest source of pollution is aqueous emissions. Parameters of interest in characterizing liquid waste streams include biological oxygen demand (BOD), chemical oxygen demand (COD), total suspended solids (TSS), and total dissolved solids (TDS). The petrochemical feedstocks used as raw materials may be emitted in low concentrations in wastewater, as fugitive gaseous emissions, or in solid residues. Other materials which may be emitted during the manufacturing process include all of the catalysts, initiators, modifiers, etc. which are added in small quantities in various processes. Table 2.17 lists some organic chemicals which were identified in aqueous waste streams from some synthetic rubber plants.

Solid wastes are present primarily as suspended solids (SS) in plant waste streams. Rubber solids collect on much of the process equipment and are either removed by hand or washed with water and then added to the general plant waste stream. Water effluents can be categorized into three types of wastes. They are: 1) utility wastes such as cooling tower blow-down, 2) process waste streams such as the decant water from solvent separators, and 3) equipment and area washdowns. Some plants purchase their steam while others generate it on site. Process waste streams and equipment cleanup wastes are usually combined to give a single plant effluent stream.

*Based on EPA Report EPA 600/2-77-023i, Feb., 1977.

PULP AND PAPER*

Pulp- and paper-making processes, until the last few years, changed relatively little since the industry began centuries ago. They generally cause considerable pollution, especially during pulpmaking, although intensive pollution con-

*Based on an article by Bette, John L. and Glen A. Allard, EPS, Department of the Environment, Canada Industry and Environment, UNEP, Vol. 2, J/F/M (1979).

TABLE 2.17 Organic Compounds Identified in Effluents From Synthetic Rubber Plants

Compound	*Waste Stream*	*Concentration (mg/l)*
2-Benzothiazole	Latex accelerators and thickeners–plant's holding pond	0.16
2-Benzothiazole	Synthetic rubber plant's aerated lagoon.	NA
n-Butylisothiocyanate	Latex accelerators and thickeners–plant's holding pond	0.1
bis-(2-Chloroethoxy) methane	Synthetic rubber plant's treated waste	140
bis-2-Chloroethyl ether	Synthetic rubber plant's treated waste	0.16
Dibutylamine	Latex accelerators and thickeners–plant's raw effluent	<1
N,N-Diethylformamide	Latex accelerators and thickeners–plant's raw effluent	<1
Diethylphthalate	Synthetic rubber plant's settling pond	NA
Dimethylphthalate	Synthetic rubber plant's settling pond	NA
p-Dithiane	Synthetic rubber plant's treated waste	0.12
Ethyl isothiocyanate	Latex accelerators and thickeners–plant's raw effluent	<1.5
Furfural	Synthetic rubber plant's settling pond	0.002
2-Mercaptobenzothiazole	Synthetic rubber plant's aerated lagoon	NA
1-Methyl naphthalene	Synthetic rubber plant's settling pond	0.002
Pentachlorophenol	Latex accelerators and thickeners–plant's holding pond.	0.4
Styrene	Synthetic rubber plant's settling pond	0.003
2,2'-Thiodiethanol (Thiodiglycol)	Synthetic rubber plant's treated waste	2(E)
Triethylurea	Latex accelerators and thickeners–plant's raw effluent	6.4

Notes: NA - not available
E - Estimated

Source: Webb, Ronald G., et al. Current Practice in GC-MS Analysis of Organics in Water. EPA-R2-73-277. 16020 GHP. Athens, Ga., EPA, Southeast Environmental Research Lab., 1973.

trol efforts are bearing fruit. This industry uses large amounts of water, several billion gallons per year, which, while it is being used and before it is discharged into the hydrographic system, is loaded with pollutants. Most of these can be characterized as suspended solids (organic and inorganic) and BOD. Other water pollutants must also be taken into account for their possible toxic effects on fish and the biota. These pollutants may come from the raw materials or from different stages in the production process. They include resin acids, organic sulfides, and traces of heavy metals. Finally, mention should be made of the color of the effluent, caused mainly by lignin from the wood.

Pulp and paper manufacture are described in considerable detail in Chapter 15, "Industrial Chemistry of Wood."

Pollution Control

The levels of discharge and their composition vary from process to process; however, both the BOD and SS material can be effectively reduced by conventional wastewater treatment technology.

Suspended solids are traditionally removed by

sedimentation with or without flocculants. Sedimentation is normally accomplished with mechanical clarifiers, although some installations utilize sedimentation basins. Large land requirements,, inefficient settling, and difficulties encountered in cleaning the ponds have made the clarifier the more predominant method.

Wastewater treatment for the reduction of BOD_5 is traditionally carried out in biological treatment systems, either aerated stabilization basins with approximately five days retention of wastewater or by activated-sludge systems. Various high-rate activated-sludge systems have been tried by the pulp and paper industry. Some of these systems have been found to be susceptible to shock loads and others incapable of breaking down some of the longer chain fatty and resin acids that are known to be toxic to fish.

Currently, the emphasis is shifting dramatically from the control of the conventional pollutants of suspended solids and BOD to the control of toxic pollutants such as specific organic constituents. This change results from the success of industry in dealing internally with BOD and TSS and the realization that the most cost-effective next step in controlling pollution is to deal with specific contaminants within the process rather than implementing high-cost technology for questionable pollution reductions. Concerns about specific contaminants will result in continuing efforts to find improved approaches and in some instances conversion to new processes.

Internal Pollution Abatement

Internal pollution abatement measures are aimed at reducing pollutant discharges at their orgin—which in many cases result in the recovery of chemical products, the conservation of heat, and a reduction in water consumption. In addition to general good housekeeping practices to reduce spills and eliminate the unnecessary use of fresh water, a number of significant techniques are mentioned below.

Dry Debarking. The use of dry debarking (with its inherent reduction in wastewater load and its production of dry bark fuel, which is more readily burned to recover energy) is becoming more common for the production of both mechanical and chemical pulps. Most new debarking installations will be of the dry type.

Pulp Washing. The use of four-stage washing with the resultant reduction in carry-over of cooking chemicals to bleach plants is becoming more prevalent. Many mills are finding that four-stage washing is justified from an economic viewpoint when pulp quality energy savings and reduced effluent treatment costs are taken into consideration.

In addition to the multistage vacuum drum washers that are being used more commonly, a belt washing system which offers the potential to reduce capital costs and operate with lower energy requirements has been installed at Thorold, Ontario, Canada; Salem, Oregon, United States; and Amayaski, Japan. The belt washer is reported to save on space requirements and eliminates the interstage repulping associated with drum washers.

Chlorine Dioxide Substitution. There has been an increasing trend to substitute chlorine dioxide for chlorine in the first stage of Kraft bleaching. The use of chlorine dioxide changes the chemical reaction from an additive to an oxidative chemical process which reduces the production of toxic chlorinated organic compounds. A prerequisite of chlorine dioxide substitution is a high washing efficiency to reduce the bleach chemical consumption in the first stage, as well as the production of extraneous, environmentally obnoxious products.

Condensate Recycle. The cleanest condensates from the evaporators are commonly utilized in causticizing plants. Foul condensates from the evaporators and digesters need to be treated before they can be reused. This is most commonly done by steam stripping. The stripped condensates are recycled to the brown stock washers, the smelt-dissolving tank scrubber, the lime-kiln scrubber, or to other miscellaneous processes. Condensate stripping may be necessary to reduce fish tainting and to produce clean condensates that may be reused within the mill—and without creating an air pollution problem due to the emission of reduced sulfur compounds. Condensate stripping involves high operating costs because five percent steam must be supplied to remove the reduced sulfur compounds and at least ten percent to remove compounds such as methanol and turpentine. New plants can be designed with an appropriate intergrated heat-recovery system to reduce operating costs. To prevent excessive contamination of condensates, large vapor entrainment areas are now incorporated into evaporator designs.

Odor Control. The use of low-odor recovery boilers or, alternatively, black-liquor oxidation has resulted in greatly reduced sulfur emissions from Kraft mill recovery boilers. It is expected that new plants will use one of these systems. Incineration of noncondensable odor-containing gases in lime kilns has also become quite a common practice. A number of mills utilize small incinerators for noncondensable gases to eliminate any emissions during lime kiln shutdowns. To smooth out flows to the lime kiln or incinerator, it is common practice to install digester vapor surge tanks.

Major Process Modification

As previously indicated, major process changes have been developed over the last ten years with appropriate attention to the resultant pollution load. In fact, in some instances, the process changes were instituted with environmental protection as the prime objective. The following are some of the recent process changes.

Rapson-Reeve Effluent Free Kraft Mill. In the Rapson-Reeve process, the total bleach plant effluent is used to wash the unbleached pulp and to prepare the digester cooking-liquor. This eliminates the need for costly external wastewater treatment. The organic wastes formed in the bleaching process enter the black-liquor system where they are concentrated and burned. The sodium chloride formed from the spent bleaching chemicals is removed in a unique salt-recovery process (SRP) plant by evaporating fresh white liquor and crystallizing the crude NaCl (salt). The salt is recovered and used to produce more bleaching chemicals in a R-3 chlorine dioxide generator. The Great Lakes Paper Company at Thunder Bay, Ontario, Canada, started operation of a new 550-tons-per-day Kraft mill in March, 1977, incorporating such a closed-cycle system that sends bleach plant effluent through the brown stock system to the chemical recovery boiler. The mill, as expected, has encountered some initial difficulties with the closed concept but system modifications should overcome these problems.

Oxygen Bleaching. The commercial use of oxygen bleaching has grown markedly in the 1970's. There are now ten commercial plants using this process that are situated in all the major forest-producing areas of the world. The process consists of replacing in part, or completely, the first two stages of conventional bleaching (chlorination and extraction) with gaseous oxygen in an alkaline medium such as sodium hydroxide. From a pollution perspective, the obvious advantage is reduced water usage when performed at high consistencies and the absence of chlorides in the effluent. Approximately twenty to thirty percent of the chlorine in a conventional bleach sequence is discharged as organochloride compounds. As the oxygen stage effluent contains no chlorides, it can be recycled to the Kraft recovery boiler, resulting in a reduction of from thirty to fifty percent of BOD_5, COD, and color as compared to the effluent of a conventional bleach sequence.

FOOD PROCESSING

The food processing industries are extraordinarily diverse and include the following common unit operations:

Material handling. Conveying, elevating, pumping, packing, and shipping.

Separating. Centrifuging, draining, evacuating, filtering, percolating, fitting, pressing, skimming, sorting, and trimming. (Drying, screening, sifting, and washing also fall into this category.)

TABLE 2.18 Estimated Quantities of Wastewater Volume and Raw Loads from Major Food Industries in the United States, 1974*

Industry	*Wastewater Volume (10^9 liters)*	*BOD_5 (10^6 kg)*	*Suspended Solids (10^6 kg)*
Meat	307	331	NA
Dairy	212	182	91
CPFV	492	91	68
Sugar	341	68	455
Malt beverage	151	205	82
Seafood	49	45	43
Grain milling (wet corn)	0.04	47	24
Malt industry	19	12	2.27
Flavoring extracts, syrups	NA	11	0.91
Distilled liquor	6.74	4.68	4.68
Soft drink	18	4.05	–
Coffee	1.89	1.8	0.45

Notes: NA = not available.
*EPA 600/2-79-009, January, 1979.

Heat exchanging. Chilling, freezing, and refrigerating; heating, cooking, broiling, roasting, baking, and smoking.

Mixing. Agitating, beating, blending, diffusing, dispersing, emulsifying, homogenizing, kneading, stirring, whipping, and working.

Peeling and size reduction. Peeling, breaking, chipping, chopping, crushing, cutting, grinding, milling, maturating, pulverizing, refining (as by punching or rolling), shredding, slicing, and spraying.

Forming. Casting, extruding, flaking, molding, pelletizing, rolling, shaping, stamping and die casting.

Coating. Dipping, enrobing, glazing, icing and panning.

Decorating. Embossing, imprinting, sugaring, topping.

Controlling. Controlling air humidity, temperature, pressure and velocity; inspecting, measuring, tempering, weighing.

Packaging. Capping, closing, filling, labeling, wrapping.

Storing. Piling, stacking, warehousing.

The estimated volumes of wastewater and raw loads from major food industries in the United States for 1974 are given in Table 2.18. It can be seen from this table that the canned, preserved, and frozen fruits and vegetables industry (CPFV) produces the largest volume of wastewater in the food industries and that the meat industry and malt beverage industry produce the major portion of the BOD load. It is worth noting here that while sugar refining is still the major source of suspended solids load in the food industries, the volume of wastewater, BOD, and suspended solids for that industry have been substantially reduced from earlier levels through improvements in in-plant practices, process water reuse, and sugar recovery.

Waterborne Toxic Pollutants

The possible waterborne toxic pollutants in the food industry may include pesticides that are removed from the surface of commodities during washing operations, solvents used for extraction, PAH* contained in wastewaters from washing of smokehouses, disinfecting chemicals and fungicides applied to commodities or used in the washing of vessels and other equipment, and naturally occurring phenolic compounds

*polyaromatic hydrocarbons

present in specific commodities. The most pervasive waterborne sources of toxics in the food industry are the washing operations (and perhaps certain other wastewaters such as blanching water and flume water). Certain commodities are stored after harvesting in order to allow actual processing to proceed over a period of several months or until the next harvest. In such a case, pesticides are required for the storage of certain commodities.

Canned, Frozen and Preserved Fruits and Vegetables (CPFV)

The CPFV industry is an excellent example of a complex industry that is a major source of water pollutants. The industry basically extends the shelf life of raw commodities through the use of various preservation methods such as canning, freezing, dehydrating, and brining. It is difficult to give a general description of the wastewater problems of the CPFV industry because there is a great diversity in geographical location, size of plant, and number of products processed.

Whereas other industries such as the organic chemicals industry may have a wider range of products and a comparable number of plants, the CPFV industry is unique in the seasonal variation of the waste loads and the need to meet high sanitation standards in compliance with state and federal laws.

The wastewater pollutants generated by the CPFV industry result predominantly from the unit operations of transport, peeling, washing, and blanching. Table 2.19 shows the sources and estimated average volumes of wastewaters from processing steps in the canning of fruits. The volumes given are those that result if only fresh water is used in all operations and no attempt is made to reuse water. The volumes may be greatly reduced by recycling; for example, can cooling is now accomplished at some plants using recirculated cooling water systems.

TABLE 2.19 Sources and Estimated Volumes of Wastewaters from Processing Steps in Canning of Fruits (highly variable by commodity)*

	Waste Flow (liters)		*Percent of*
Operation	*per hour*	*per ton*	*total flow*
Peeling	4,542	48	2
Spray washing	41,635	385	17
Sorting, slicing, etc.	11,355	120	5
Exhausting of cans	4,542	48	2
Processing	2,271	24	1
Cooling of cans	90,840	945	37
Plant cleanup	79,485	840	33
Box washing	7,149	70	3
Total	241,819	2,480	100

*EPA 600/2-79-009, January, 1979.

The industry has no general rules for wastewater treatment. Single-commodity plants (many very large) such as corn processors, citrus processors, pickle producers, potato processors, and sauerkraut producers are more likely to have on-site treatment with subsequent discharge to a stream than are multi-commodity plants. Tomato processors and apple processors are two notable exceptions, with wastes from more than 70% of the tomato products production capacity going to municipal treatment systems and wastes from more than 90% of apple products production capacity going to municipal or on-land systems. One important difference between large corn and potato processing plants and tomato processing plants is the length of the processing period or "campaign." Citrus and potato processing operations may run for 9 months of the year while tomato processing operations run for only 2 to 3 months. The shorter the season the less likely a company will be to invest large sums of money for pollution control equipment that will sit idle for 75% of the year. More than two-thirds of the CPFV industry's wastewater is treated in municipal treatment systems or by on-land treatment.

It was previously mentioned that water reuse and recylcing can significantly reduce the volume of wastewater in the CPFV industry. However, extensive water reuse or recycling in any food industry does raise questions concerning public health. Public health issues may arise if water that has been used to transport wastes (equipment washing) or wash commodities is to be used in areas where product contact may occur. Although cascading water reuse from a process demanding high water quality to a process demanding lower water quality may be

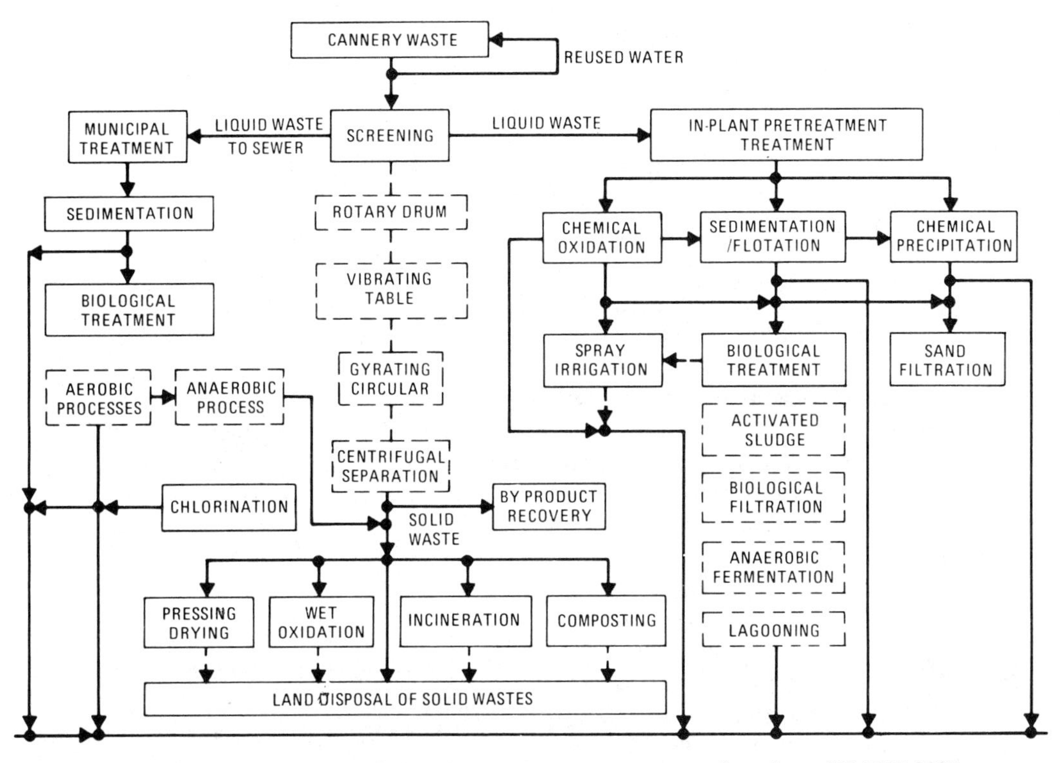

Fig. 2.6. Canned and frozen fruits and vegetables waste treatment flow chart—SIC 2033.2037.

acceptable, complete reuse by recycling to all areas of the plant may be limited because of pesticide buildup or the potential for pathogenic organism transport.

A flow chart for treatment of wastewater from canned and frozen fruits and vegetables is shown in Figure 2.6.

Seafood Processing

The canned and preserved seafood industry can be classified into several categories, some of which are as follows: farm-raised catfish, conventional blue crab, mechanized blue crab, Alaskan crab, dungeness and tanner crab, shrimp, tuna, salmon, sardine, oyster, and clam.

Treatment of wastewaters from seafood processing has been almost nonexistent because the processing plants are typically located on the seacoast, many in remote areas. Since the passage of PL92-500, increasing concern about the condition of the environment has stimulated activity in the application of existing waste treatment technologies to the seafood industry. For example, the tuna processing industry has done considerable work on by-product recovery, and some other segments of the seafood industry screen their wastes for recovery of solids for use in meal and feeds. Further research is needed on in-plant water use as well as end-of-pipe treatment.

Sugar Processing

The sugar processing industry encompasses operations that convert raw sugar cane to raw sugar, raw sugar to refined sugar, and raw sugar beets to refined sugar. Table 2.20 shows the annual discharge of wastewater volume, BOD_5, and suspended solids from the sugar processing industry in 1974.

As shown above, raw cane sugar represents the major source of water pollution in the sugar processing industry (with 56% of the wastewater volume 90% of the BOD_5 load, and 98.5% of the suspended solids). Beet sugar generates a significant portion of wastewater volume, but a less significant portion of BOD_5 and suspended solids loads. The water pollution problem from

TABLE 2.20 Annual Wastewater Discharge of Sugar Processing Industry, 1974

	Total sugar Processing Industry	*Beet sugar*	*Cane sugar refining*	*Raw sugar cane*
Wastewater volume (10^9 liters)	341	132	18.9	189
BOD_5 (10^6 kg)	68.2	4.55	2.28	61.4
Suspended solids (10^6 kg)	455	4.55	2.28	448

*EPA 600/2-79-009, January, 1979.

cane sugar refining is small in the sugar processing industry.

About 65% of the sugar processing industry uses lagoon treatment for wastewater. About 30% of the industry uses the land disposal system, and only 5% of the industry discharges to municipal sewage treatment plants.

Lagoon systems are effective in removing all the floating and suspended solids. Although BOD reduction on a percentage basis is good, the final effluent BOD is often too high to be in compliance with stream standards. Effluent with low BOD can be attained only by maintaining long retention periods, which requires large land areas. There is much interest in developing systems that require less land area.

Suspended solids loads in wastewaters from sugar industries are the highest among the food industries. More research is needed for improving harvesting equipment to reduce the amounts of soil, leaves, and stems in the raw cane sugar, since these materials require the use of excessive volumes of water to clean the product. Improvements are also needed in the equipment used for dry separation of the unwanted material from the sugar-bearing material.

Meat Industries (Red Meat and Poultry)

Red meats and poultry processing both require facilities for receiving; killing; removal of hide, hair, or feathers; eviscerating and trimming; cooling; and packing. Further processing of the meat is more extensive with swine and beef than with poultry.

The quantity and characteristics of the wastewater from this industry are influenced by plant size, species slaughtered, amount of on-site processing, extent of wastewater segregation, and recycling and reuse. The reported ranges in water use, BOD, and suspended solids for wastewaters from red meat packing and processing are quite large, as shown below.

Flow, liter/head	575-685
BOD, mg/liter	320-5440
BOD, kg/1000 kg live weight	1.9-27.6
Suspended solids, mg/liter	240-7220
Suspended solids, kg/1000 kg live weight	1.2-53.8
Total volatile solids, kg/1000 kg live weight	3.1-56.4

Meat industry wastewaters contain carbon, nitrogen, and phosphorus in proportions such that aerobic or anaerobic biological treatment is readily accomplished without the addition of nutrients. The treatment of wastewater may be handled in a municipal sewage treatment plant. In many communities, the only pretreatments required prior to discharge of wastewater to the public sewer are treatment by screening to remove manure solids and gravity separation or air flotation to reduce grease to 400 mg/liter. Between 50% and 70% of the total discharges from the meat processing industry are to municipal sewer systems.

During the 1950s and early 1960s, the meat industry abandoned many old packing plants in larger cities and built new slaughter or packing plants in small communities near the source of animal production. Increased mechanization in processing meat products has tended to reduce wastewater flow per unit of production. Location of new plants in small communities has in many cases required the installation of secondary treatment facilities on site, such as trickling filters, activated sludge, anaerobic digestors, and anaerobic contact processes.

TERMINOLOGY

Many of the more common terms encountered in wastewater-treatment technology are given below to assist the reader in developing his vocabulary in this field and in understanding what follows.

Abatement. The method of reducing the degree or intensity of pollution, also the use of such a method.

Absorption. The penetration of a substance into or through another. For example, in air pollution control, absorption in a liquid from which it can then be extracted.

Acclimation. The physiological and behavioral adjustments of an organism to changes in its immediate environment.

Acclimatization. The acclimation or adaptation of a particular species over several generations to a marked change in the environment.

Activated carbon. A highly adsorbent form of carbon, used to remove odors and toxic substances from gaseous emissions. In advanced waste treatment, activiated carbon is used to remove dissolved organic matter from wastewater.

Activated sludge. Sludge that has been aerated and subject to bacterial action, used to remove organic matter from sewage.

Activated-sludge process. The process of using biologically active sewage sludge to hasten breakdown of organic matter in raw sewage during secondary waste treatment.

Acute toxicity. Any poisonous effect produced within a short period of time, usually from 24-96 hours, resulting in severe biological harm and often death.

Adaption. A change in structure or habit of an organism that allows it to exist in a changed environment.

Adsorption. The adhesion of a substance to the surface of a solid or liquid. Adsorption is often used to extract pollutants by causing them to be attached to such adsorbents as activated carbon or silica gel. Hydrophobic, or water-repulsing adsorbents, are used to extract oil from water-ways in oil spills.

Advanced waste treatment. Wastewater treatment beyond the secondary or biological stage that includes removal of nutrients such as phosphorus and nitrogen and removal of a high percentage of suspended solids. Advanced waste treatment, known as tertiary treatment, is the "polishing stage" of wastewater treatment and produces a high quality effluent.

Aeration. The process of being supplied or impregnated with air. Aeration is used in wastewater treatment to foster biological and chemical purification.

Aerobic. This refers to life or processes that can occur only in the presence of oxygen.

Agricultural pollution. The liquid and solid wastes from all types of farming, including runoff from pesticides, fertilizers, and feedlots; water and wind erotion and dust from plowing; animal manure; carcasses; and crop residues and debris. It has been estimated that agricultural pollution in the U.S. has amounted to more than $2\frac{1}{2}$ billion tons per year.

Anaerobic. Refers to life or processes that occur in the absence of oxygen.

Anti-degradation clause. A provision in air quality and water quality laws that prohibits deterioration of air or water quality in areas where the pollution levels are presently below those allowed.

Aquaculture project. A controlled discharge of pollutants to enhance growth or propagation of harvestable freshwater, estuarine, or marine plant or animal species.

Aquifer. An underground bed or stratum of gravel or porous stone that contains water.

Aquatic plants. Plants that grow in water, either floating on the surface, growing up from the bottom of the body of water, or growing under the surface of the water.

Assimilation. Conversion or incorporation of absorbed nutrients into protoplasm. Also refers to the ability of a body of water to purify itself of organic pollution.

Autotrophic. Self-nourishing; denoting those organism capable of constructing organic matter from inorganic substances.

Bacteria. Single-celled microorganisms that

lack chlorophyll. Some bacteria are capable of causing human, animal, or plant diseases; others are essential in pollution control because they break down organic matter in the air and in the water.

Baghouse. An air pollution abatement device used to trap particulates by filtering gas streams through large fabric bags, usually made of glass fibers.

Baling. A means of reducing the volume of solid waste by compaction.

Bar screen. In wastewater treatment, a screen that removes large floating and suspended solids.

Benthic region. The bottom of a body of water. This region supports the benthos, a type of life that not only lives upon, but contributes to the character of the bottom.

Benthos. The plant and animal life whose habitat is the bottom of a sea, lake, or river.

Bioassay. The employment of living organisms to determine the biological effect of some substance, factor, or condition.

Biochemical oxygen demand (BOD). A measure of the amount of oxygen consumed in the biological processes that break down organic matter in water. Large amounts of organic waste use up large amounts of dissolved oxygen, thus the greater the degree of pollution, the greater the BOD.

Biodegradable. The process of decomposing quickly as a result of the action of microorganisms.

Biological oxidation. The process by which bacterial and other microorganisms feed on complex organic materials and decompose them. Self-purification of waterways, activated sludge, and trickling filter wastewater treatment processes depend on this principle. The process is also called biochemical oxidation.

BOD_5. The amount of dissolved oxygen consumed in five days by biological processes breaking down organic matter in an effluent. (See biochemical oxygen demand.)

Boom. A floating device that is used to contain oil on a body of water.

Channelization. The straightening and deepening of streams to permit water to move faster, to reduce flooding, or to drain marshy acreage for farming. However, channelization reduces the organic waste assimilation capacity of the stream and may disturb fish breeding and destroy the stream's natural beauty.

Chemical oxygen demand (COD). A measure of the amount of oxygen required to oxidize organic and oxidizable inorganic compounds in water. The COD test, like the BOD test, is used to determine the degree of pollution in an effluent.

Chlorination. The application of chlorine to drinking water, sewage or industrial waste for disinfection or oxidation of undesirable compounds.

Chlorinator. A device for adding a chlorine-containing gas or liquid to drinking or wastewater.

Chlorine-contact chamber. A chamber in a waste treatment plant in which effluent is disinfected by chlorine before it is discharged to the receiving waters.

Clarification. In wastewater treatment, the removal of turbidity and suspended solids by settling, often aided by centrifugal action and chemically induced coagulation.

Clarifier. In wastewater treatment, a settling tank which mechanically removes settleable solids from wastes.

Coagulation. The clumping of particles in order to settle out impurities; often induced by chemicals such as lime or alum.

Coliform index. An index of the purity of water based on a count of its coliform bacteria.

Coliform organism. Any of a number of organisms common to the intestinal tract of man and animals whose presence in wastewater is an indicator of pollution and of potentially dangerous bacterial contamination.

Combined sewers. A sewerage system that carries both sanitary sewage and storm-water runoff. During dry weather, combined sewers carry all wastewater to the treatment plant. During a storm, only part of the flow in intercepted, because of plant overloading; the remainder goes untreated to the receiving stream.

Combustion. Burning. Technically, a rapid oxidation accompanied by the release of energy in the form of heat and light. It is one of the three basic contributing factors causing air pollution; the others are attrition and vaporization.

Comminution. Mechanical shredding or pulverizing of waste; a process that converts it into a homogeneous and more manageable material. Used in solid waste management and in the primary stage of wastewater treatment.

Comminutor. A device that grinds solids to make them easier to treat.

Compaction. Reducing the bulk of solid waste by rolling and tamping.

Compost. Relatively stable decomposed organic material.

Composting. A controlled process of degrading organic matter by microorganisms. (1) mechanical–a method in which the compost is continuously and mechanically mixed and aerated. (2) ventilated cell–compost is mixed and aerated by being dropped through a vertical series of ventilated cells. (3) windrow–an open-air method in which compostable material is placed in windrows, piles, or ventilated bins or pits and occasionally turned or mixed. The process may be anaerobic or aerobic.

Decomposition. Reduction of the net energy level and a change in chemical composition of organic matter brought about by the actions of aerobic or anaerobic microorganisms.

Diatomaceous earth (Diatomite). A fine siliceous material resembling chalk used in wastewater treatment plants to filter sewage effluent in order to remove solids. May also be used as inactive ingredient in pesticide formulations paplied as dust or powder.

Digester. In a wastewater treatment plant, a closed tank that decreases the volume of solids and stabilizes raw sludge by bacterial action.

Digestion. The biochemical decomposition of organic matter. Digestion of sewage sludge takes place in tanks where the sludge decomposes, resulting in partial gasification, liquefaction, and mineralization of pollutants.

Dissolved oxygen (DO). The oxygen dissolved in water or sewage. Adequately dissolved oxygen is necessary for the life of fish and other aquatic organisms and for the prevention of offensive odors. Low dissolved oxygen concentrations generally are due to discharge of excessive organic solids having high BOD, the result of inadequate waste treatment.

Dissolved solids. The total amount of dissolved material, organic and inorganic, contained in water or wastes. Excessive dissolved solids make water unpalatable for drinking and unsuitable for industrial uses.

Effluent. A discharge of pollutants into the environment, partially or completely treated or in its natural stage. Generally used in regard to discharges into waters.

Electrodialysis. A process that uses electrical current and an arrangement of permeable membranes to separate soluble minerals from water. Often used to desalinize salt or brackish water.

Enrichment. The addition of nitrogen, phosphorus, and carbon compounds or other nutrients into a lake or other waterway in order to greatly increase the growth potential for algae and other aquatic plants. Most frequently, enrichment results from the inflow of sewage effluent or from agricultural runoff.

Environment. The sum of all external conditions and influences affecting the life, development and, ultimately, the survival of an organism.

Environmental impact statement. A document prepared by a Federal agency on the environmental impact of its proposals for legislation and other major actions significantly affecting the quality of the human environment. Environmental impact statements are used as tools for decision making and are required by the National Environmental Policy Act.

Eutrophication. The normally slow aging

process by which a lake evolves into a bog or marsh and ultimately assumes a completely terrestrial state and disappears. During eutrophication the lake becomes so rich in nutritive compounds, especially nitrogen and phosphorus, that algae and other microscopic plant life become superabundant, thereby "choking" the lake and eventually causing it to dry up. Eutrophication may be accelerated by many human activites.

Eutrophic lakes. Shallow lakes, weed-choked at the edges and very rich in nutrients. The water is characterized by large amounts of algae, low water transparency, low dissolved oxygen and high BOD.

Evaporation ponds. Shallow, artificial ponds where sewage sludge is pumped, permitted to dry, and either removed or buried by more sludge.

Fecal coliform bacteria. A group of organisms common to the intestinal tracts of man and of animals. The presence of fecal coliform bacteria in water is an indicator of pollution and of potentially dangerous bacteria contamination.

Filtration. In wastewater treatment, the mechanical process that removes particulate matter by separating water from solid material, usually by passing it through sand.

Floc. A clump of solids formed in sewage by biological or chemical action.

Flocculation. In wastewater treatment, the process of separating suspended solids by chemical creation of clumps or flocs.

Flume. A channel, either natural or man-made, which carries water.

Fly ash. All solids, including ash, charred paper, cinders, dust, soot or other partially incinerated matter, that are carried in a gas stream.

Groundwater. The supply of freshwater under the earth's surface in an aquifer or soil that forms the natural reservoir for man's use.

Groundwater runoff. Groundwater that is discharged into a stream channel as spring or seepage water.

Habitat. The sum total of environmental conditions of a specific place that is occupied by an organism, a population, or a community.

Heavy metals. Metallic elements with high molecular weights, generally toxic in low concentrations to plant and animal life. Such metals are often residual in the environment and exhibit biological accumulation. Examples include mercury, chromium, cadmium, arsenic, and lead.

Infiltration/inflow. Total quantity of water entering a sewer system. Infiltration means entry through such sources as defective pipes, pipe joints, connections, or manhole walls. inflow signifies discharge into the sewer system through service connections from such sources as area of foundation drainage, springs and swamps, storm waters, street wash waters, or sewers.

Interceptor sewers. Sewers used to collect the flows from main and trunk sewers and carry them to a central point for treatment and discharge. In a combined sewer system, where street runoff from rains is allowed to enter the system along with sewage, interceptor sewers allow some of the sewage to flow untreated directly into the receiving stream, to prevent the plant from being overloaded.

Lagoon. In wastewater treatment, a shallow pond–usually man-made–where sunlight, bacterial action, and oxygen interact to restore wastewater to a reasonable state of purity.

Leachate. Liquid that has percolated through solid waste or other mediums and has extracted dissolved or suspended materials from it.

Leaching. The process by which soluble materials in the soil, such as nutrients, pesticide chemicals, or contaminants, are washed into a lower layer of soil or are dissolved and carried away by water.

NTA. Nitrilotriacetic acid, a compound once used to replace phosphates in detergents.

Nutrients. Elements or compounds essential as raw materials for organism growth and development; for example, carbon, oxygen, nitrogen, and phosphorus.

Oligotrophic Lakes. Deep lakes that have a

low supply of nutrients and thus contain little organic matter. Such lakes are characterized by high water transparency and high dissolved oxygen.

Outfall. The mouth of a sewer, drain, or conduit where an effluent is discharged into the receiving waters.

Oxidant. Any oxygen-containing substance that reacts chemically in the air to produce new substances. Oxidants are the primary contributors to photochemical smog.

Oxidation. A chemical reaction in which oxygen unites or combines with other elements. Organic matter is oxidized by the action of aerobic bacteria; thus oxidation is used in wastewater treatment to break down organic wastes.

Oxidation pond. A man-made lake or pond in which organic wastes are destroyed by bacterial action. Often oxygen is bubbled through the pond to speed the process.

Packed tower. An air pollution control device in which polluted air is forced upward through a tower packed with crushed rock or wood chips while a liquid is sprayed downward on the packing material. The pollutants in the air stream either dissolve or chemically react with the liquid.

Pathogenic. Causing or capable of causing disease.

PCBs. Polychlorinated biphenyls, a group of organic compounds used in manufacture of plastics. In the environment, PCBs exhibit many of the same characteristics as DDT and may, therefore, be confused with that pesticide. PCBs are highly toxic to aquatic life; they persist in the environment for long periods of time, and they are biologically accumulative.

Percolation. Downward flow or infiltration of water through the pores or spaces of a rock or soil.

Polyelectrolytes. Synthetic chemicals used to speed flocculation of solids in sewage.

Pretreatment. In wastewater treatment, any process used to reduce pollution load before the wastewater is introduced into a main sewer system or before it is delivered to a treatment plant for substantial reduction of the pollution load.

Primary treatment. The first stage in wastewater treatment in which substantially all floating or settleable solids are mechanically removed by screening and sedimentation.

Resource recovery. The process of obtaining materials or energy, particulary from solid waste.

Runoff. The portion of rainfall, melted snow, or irrigation water that flows across ground surface and eventually is returned to streams. Runoff can pick up pollutants from the air or the land and carry them to the receiving waters.

Sanitation. The control of all the factors in man's physical environment that exercise or can exercise a deleterious effect on his physical development, health, and survival.

Sanitary landfilling. An engineered method of solid-waste disposal on land in a manner that protects the environment; waste is spread in thin layers, compacted to the smallest practical volume and covered with soil at the end of each working day.

Screening. The removal of relatively coarse floating and suspended solids by straining through racks or screens.

Scrubber. An air pollution control device that uses a liquid spray to remove pollutants from a gas stream by absorption or chemical reaction. Scrubbers also reduce the temperature of the emission.

Secondary treatment. Wastewater treatment, beyond the primary stage, in which bacteria consume the organic parts of the wastes. This biochemical action is accomplished by use of trickling filters or the activated-sludge process. Effective secondary treatment removes virtually all floating and settleable solids and approximately 90 percent of both BOD_5 and suspended solids. Customarily, disinfection by chlorination is the final stage of the secondary treatment process.

Sedimentation tanks. In wastewater treatment, tanks where the solids are allowed to settle or to float as scum. Scum is skimmed off; settled solids are pumped to incinerators, digesters, filters, or other means of disposal.

Septic tank. An underground tank used for the deposition of domestic wastes. Bacteria in the wastes decompose the organic matter, and the sludge settles to the bottom. The effluent then flows through drains into the ground. Sludge is pumped out at regular intervals.

Settleable solids. Bits of debris and fine matter heavy enough to settle out of wastewater.

Settling tank. In wastewater treatment, a tank or basin in which settleable solids are removed by gravity.

Sewage. The total of organic waste and wastewater generated by residential and commercial establishments.

Sewerage. The entire system of sewage collection, treatment, and disposal. Also applies to all effluent carried by sewers whether it is sanitary sewage, industrial wastes, or storm water runoff.

Sludge. The construction of solids removed from sewage during wastewater treatment. Sludge disposal is then handled by incineration, dumping, or burial.

Solid waste. Useless, unwanted, or discarded material with insufficient liquid content to be free flowing. (Also see *Waste.*) (1) Agricultural–solid waste that results from the raising and slaughtering of animals, as well as the processing of animal products and orchard and field crops. (2) Commercial–waste generated by stores, offices, and other activities that do not actually turn out a product. (3) Industrial–waste that results from industrial processes and manufacturing. (4) Institutional–waste originating from educational, health care, and research facilities. (5) Municipal–residential and commercial solid waste generated within a community. (6) Pesticidal–the residue from the manufacturing, handling, or use of chemicals intended for killing plant and animal pests. (7) Residential–waste that normally originates in a residential environment. Sometimes called domestic solid waste.

Sorption. A term including both adsorption and absorption. Sorption is basic to many processes used to remove gaseous and particulate pollutants from an emission and to clean up oil spills.

Stabilization. The process of converting active organic matter in sewage sludge or solid wastes into inert, harmless material.

Suspended solids (SS). Small particles of solid pollutants in sewage that contribute to turbidity and that resist separation by conventional means. The examination of suspended solids and the BOD test constitute the two main determinations for water quality performed at wastewater treatment facilities.

Synergism. The cooperative action of separate substances so that the total effect is greater than the sum of the effects of the substances acting independently.

Tertiary treatment. Wastewater treatment beyond the secondary, or biological stage that includes removal of nutrients such as phosphorus and nitrogen, and a high percentage of suspended solids. Tertiary treatment, also known as advanced waste treatment, produces a high quality effluent.

Thermal pollution. Degradation of water quality by the introduction of a heated effluent. Primarily a result of the discharge of cooling waters from industrial processes, particularly from electrical power generation. Even small deviations from normal water temperatures can affect aquatic life. Thermal pollution usually can be controlled by cooling towers.

Toxic pollutants. A combination of pollutants including disease-carrying agents which, after discharge and upon exposure, ingestion, inhalation, or assimilation into any organism can cause death or disease, mutations, deformities, or malfunctions in such organisms or their offspring.

Trace metals. Metals found in small quantities or traces, usually due to their insolubility.

Trickling filters. A device for the biological or secondary treatment of wastewater consisting of a bed of rocks or stones that support bacterial growth. Sewage is trickled over the bed, enabling the bacteria to break down organic wastes.

Urban runoff. Storm water from city streets and gutters that usually contains a great deal of litter and organic and bacterial wastes.

Waste. (Also see *Solid waste.*) (1) bulky waste–items whose large size precludes or complicates their handling by normal collection, processing, or disposal methods. (2) construction and demolition waste–building materials and rubble resulting from construction, remodeling, repair, or demolition operations. (3) hazardous waste–wastes that require special handling to avoid illness or injury to persons or damage to properly. (4) special waste–those waste that require extraordinary management. (5) wood pulp waste–wood or paper fiber residue resulting from a manufacturing process. (6) yard waste–plant clippings, prunings, and other discarded material from yards and gardens (also known as yard rubbish).

Water pollution. The addition of sewage, industrial wastes, or other harmful or objectionable material to receiving water to concentrations or in sufficient quantities to result in measurable degradation of water quality.

Water quality criteria. The levels of pollutants that affect the suitability of water for a given use. Generally, water use classification includes: public water supply, recreation, propagation of fish and other aquatic life, agricultural use, and industrial use.

Water quality standard. A plan for water quality management containing four major elements: (1) The use (recreation, drinking water, fish and wildlife propagation, industrial, or agricultural) to be made of the water, (2) criteria to protect those uses, (3) implementioned plans (for needed industrial-municipal waste treatment improvements), and (4) enforcement plans and an anti-degradation statement to protect existing high-quality waters.

REFERENCES

1. "Summary Report on the Effects of Heavy Metals on the Biological Treatment Processes," Barth, E. F., et al., *Journal of the Water Pollution Control Federation*, Vol. **37**, No. 1, (January, 1965) p, 86.
2. "Review of Literature on Toxic Materials Affecting Sewage Treatment Processes, Streams, and BOD Determinations," Rudolfs W., et al., *Sewage and Industrial Wastes*, Vol. **22**, No. 9, (September, 1959) p. 1157.
3. "Zinc in Relation to Activated Sludge and Anaerobic Digestion Processes," McDermott, Gerald N., et al., *Proceedings of the 17th Industrial Waste Conference*, Purdue University, p. 461 (1962).
4. "The Effects of Industrial Wastes on Sewage Treatment," Masselli, Joseph W., et al., *Report prepared by New England Interstate Water Pollution Control Commission*, (June, 1965).
5. *Environmental Effect of Photoprocessing Chemicals*, Vol. **1**, Report by the National Association of Photographic Manufacturers, Inc., 200 Mamaroneck Ave., Harrison, N.Y. 10528 (1974).
6. *Environmental Effect of Photoprocessing Chemicals*, Vol. **II**, Report by the National Association of Photographic Manufacturers, Inc., 600 Mamaroneck Ave., Harrison, N.Y. 10528, 324 pp. (1974).
7. *Fate of Benzidine in the Aquatic Environment*: *A Scoping Study*, U.S. EPA Contract #68-01-2226, January, 1974.
8. "Anaerobic Processes–Literature Review," Ghosh, S., *Journal of the Water Pollution Control Federation*, Vol. **44**, No. 6, p. 948 (June 1972).
9. "Anaerobic Processes," Pohland, F. G. and S. J. Kang, *Journal of the Water Pollution Control Federation*, Vol. **43**, No. 6, p. 1129 (June 1971).
10. "Digestion Fundamentals Applied to Digester Recovery–Two Case Studies," Dague, Richard R., et al., *Journal of the Water Pollution Control Federation*, Vol. **42**, No. 9, p. 1666 (September 1970).
11. "The Effects of Heavy Metals and Toxic Organics on Activated Sludge," Goss, Thomas A., *Masters Thesis*, Univ. of Pittsburgh (1969).
12. "Effects of Copper on Aerobic Biological Sewage Treatment," McDermott, Gerald N., et al., *Journal of the Water Pollution Control Federation*, Vol. **35**, No. 2, p. 227 (February 1963).
13. *A Handbook on the Effects of Toxic and Hazardous Materials on Secondary Biological Treatment Processes, A Literature Review*, Environmental Quality Systems, Inc., Rockville, Maryland, (prepared for the Allegheny County Sanitary Authority and the EPA, Sept. 1973, unpublished).
14. "Effects of Metallic Ions on Biological Waste Treatment Processes," Reid, George W. et al., *Water and Sewage Works*, Vol. **115**, No. 7, p. 320, (July 1968).
15. "The Influence of Trivalent Chromium on the Biological Treatment of Domestic Sewage," Bailey, D. A., et al., *Water Pollution Control*, Vol. **69**, No. 2, p. 100, (1970).

16. "The Effect of Phenols and Hetero-Cyclic Bases on Nitrification in Activated Sludges," Stafford, D. A., *Journal of Applied Bacteriology*, Vol. 37, p. 75, (1974).
17. "Accumulation of Methanogenic Substrates in CCl_4 Inhibited Anaerobic Sewage Sludge Digester Cultures," Sykes, Robert and E. J. Kirsch, *Water Research*, Vol. 6, p. 41, (1972).
18. "The Toxicity of Cadmium to Anaerobic Digestion: Its Modification by Inorganic Anions," Mosey, F. E., *Water Pollution Control*, Vol. 70, p. 584 (1971).
19. "Inhibition of Anaerobic Digestion of Sewage Sludge by Chlorinated Hydrocarbons," Swanwick, J. D. and Margaret Foulkes, *Water Pollution Control*, Vol. 70, p. 58 (1971).
20. *The Impact of Oily Materials on Activated Sludge Systems*, Environmental Protection Agency, NTIS #PB 212-422, EPA #12050 DSH (March, 1971).
21. "Effect of Toxic Wastes of Treatment Processes and Watercourses," Jackson, S. and V. M. Brown, *Water Pollution Control*, Vol. 69, p. 292 (1970).
22. "The Effect of Chloroform in Sewage on the Production of Gas from Laboratory Digesters," Stickley, D. P., *Water Pollution Control*, Vol. 69, p. 585 (1970).
23. "Effect of Boron on Anaerobic Digestion," Banerji, S. K. and P. R. Parikh, *Proceedings of the 4th Mid-Atlantic Industrial Waste Conference* (1970).
24. *Correlation of Advanced Wastewater Treatment and Ground Water Recharge*, Beckman, W. J. and R. J. Advendt, (prepared for U.S. Environmental Protection Agency, Office of Water Program Operations).

Additional References

1. Isherwood, W. R., and McVaugh, J., "Wastewater Treatment and Reuse in an Independent Rendering Company." *Proc. 9th Nat'l Symp. Food Processing Waste*, 321 EPA 600/2-78-188, Cincinnati, Ohio (1978).
2. Shubin, R. M., et al., "Relation of the Electroflotation of Fat to the pH of Wastewater," *Myasn. Ind.* SSSR (USSR), 1, 18 (1978); Chem. Abs., 89, 48389a (1978).
3. Ramirez, E. R., and Clemens, O. A., "Physiochemical Treatment of Rendering Wastewater by Electrocoagulation." *Proc. 9th Nat'l Symp. Food Processing Wastes*, 265, EPA 600/2-78-188, Cincinnati, Ohio (1978).
4. Dencker, D. L., et al., "Expanded Secondary Waste Treatment Makes Zero Discharge to Streams." *Food Processing*, 39, 10, 52 (1978).
5. Szabo, A. J., et al., "Dissolved Air Flotation Treatment of Gulf Shrimp Cannery Wastewater." *Proc. 9th Nat'l Symp. Food Processing Wastes*, 221, EPA 600/2-78-188, Cincinnati, Ohio (1978).
6. Ketchum, L. H., Jr., "Overland Flow Treatment of Duck Processing Wastewater in a Cold Climate." *Proc. 9th Natl. Symp. Food Processing Wastes*, 77, EPA 600/2-78-188, Cincinnati, Ohio (1978).
7. Miller, B. F., et al., "Using Microwaves to Reduce Pollution from Scalding Chickens." *Proc. 9th Nat'l. Symp. Food Processing Wastes*, 173, EPA 600-2-78-188, Cincinnati, Ohio (1978).
8. Rogers, C. J., "Recycling of Water in Poultry Processing Plants." EPA 600/2-78-039, Cincinnati, Ohio (1978).
9. Hamza, A., et al., "Potential for Water Reuse in an Egyptian Poultry Processing Plant." *Jour. Food Sci.*, 43, 1153 (1978).
10. "Calendar of EPA Regulations Under Consideration," *Pollution Eng.*, 10, 12, 61 (1978).
11. Ferguson, J. R., "Impact of EPA Regulations on the Steel Industry," *Pollution Eng.*, 10, 8.23 (1978).
12. Prober, R., et al., "Ozone-Ultraviolet Treatment of Coke Oven and Blast Furnace Effluents for Destruction of Ferricyanides," *Proc. 32nd Ind. Waste Conf.*, Purdue Univ. W. Lafayette, Ind., 17 (1977).
13. Holladay, D. W. et al., "Biodegradation of Phenolic Waste Liquors in Stirred-tank, Packed-bed, and Fluidized-bed Bioreactors," *Jour. Water Poll. Control Fed.*, 50, 2573 (1978).
14. Brower, G. R., et al., "Control of Water Problems in a Steel Mill's Blast Furnace Gas Wash Recirculation System," *Proc. 32nd Ind. Waste Conf.*, Purdue Univ., W. Lafayette, Ind., 549 (1977).
15. Beall, J. F., and McGathen, R., "Guidelines for Wastewater Treatment I–How to Minimize Wastewater," *Metal Finish.*, 75, 9, 13 (1977); Metals Abs., 11 58-0007 (1978).
16. Yeats, A. R., "Ion Exchange Selectively Removes Heavy Metals from Mixed Plating Wastes," *Proc. 32nd Ind. Waste Conf.*, Purdue Univ., W. Lafayette, Ind. 467 (1977).
17. Grutsch, J. F., "Wastewater Treatment: The Electrical Connection," *Environ. Sci. and Technol.*, 12, 9, 1022 (1978).
18. Steiner, J. L., et al., "Pollution Control Practices: Air Flotation Treatment of Refinery Wastewater," *Chem. Eng. Progr.*, 74, 12, 39 (1978).
19. Genkin, V. E., and Belevtsev, A. N., "Electro-chemical Treatment of Industrial Wastewaters," *Pro. Joint US/USSR Symp. on Advanced Equip and Facilities for Wastewater Trt.*, U.S. EPA, Cincinnati, Ohio (1978).
20. Burbank, N. C., Jr., et al., "A Comparison of Ozone and Hydrogen Peroxide in the Emergency Oxidation of a Refinery Wastewater." *Proc. 32nd Ind. Waste Conf.*, Purdue Univ., (Ann Arbor Science Publishers, Ann Arbor, Mich.) 1977, 402 (1978).

21. "Development Document for Proposed Existing Source Pretreatment Standards for the Electroplating Point Source Category," Environmental Protection Agency, EPA-440/1-78-085 (Feb. 1978).
22. Marino, M., "Achieving Industrial Wastewater Reuse by Application of the Best Available Technology System," *Metal Finishing*, 77, 1, 46 (1978).
23. "Recovery Pays," *Plating and Surface Finishing*, **65**, 2, 45 (1978).
24. Lancy, L. E., and Steward, F. A., "Disposal of Metal Finishing Sludges–The Segregated Landfill Concept," *Plating and Surface Finishing*, **65**, 2, 14 (1978).
25. Cairns, J. et al., "Developing an On-Site Continuous Biological Monitoring System for the Chemical Industry," *Proc. 5th WWEMA Ind. Poll. Conf.*, 285, WWEMA, McLean, Virginia (1978).
26. Nebel, C., et al., "Ozone Oxidation of Phenolic Effluents," *Proc. 31st Ind. Waste Conf.*, Purdue Univ., (Ann Arbor Science Publishers, Inc. Ann Arbor, Mich.) 1976, 940, (1977).
27. Patterson, J. W., et al., "Pollution Abatement in the Military Explosive Industry," *Proc. 31 Ind. Waste Conf.*, Purdue Univ., (Ann Arbor Science Publishers, Inc., Ann Arbor, Mich.) 1976, 385 (1977).
28. Pallanich, P. J. "Pure Oxygen Treatment of Pesticide Plant Wastewater," *Chem. Eng. Progr.* **74**, 4, 79 (1978).
29. Campbell, H. J., et al., "Evaluation of Chromium Removal from a Highly Variable Wastewater Stream," *Proc. 32nd Ind. Waste Conf.*, Purdue Univ., (Ann Arbor Science Publishers, Inc., Ann Arbor, Mich.) 1977, 102 (1978).

3

Coal Technology

James E. Funk*

INTRODUCTION

Coal, and the technology by which coal is mined and utilized, is an important part of the U.S. energy picture today, and recent events indicate an increasing importance in the future. Coal represents over 90 percent of the U.S. proven reserves of fossil fuels, and it is again being considered as a source of synthesis gas for the petrochemical industry, as well as a source of electric power production and process heat generation. Table 3.1, a recent estimate of U.S. energy resources and reserves, shows the vast quantity of coal reserves relative to the quantities of other fossil fuels in the U.S.

Thirty-one percent of the world's coal reserves are in the United States. The USSR follows with 23 percent; Western Europe has 18 percent, and the People's Republic of China has 15 percent. The remaining reserves lie in other countries and areas of the world. Exxon has estimated that the United States has about 250 billion recoverable tons of coal, distributed as shown in Table 3.2. Eastern U.S. coals are generally bituminous with a heating value of 10,000 to 13,000 Btu/lb. Bituminous coals comprise roughly one-half of the total U.S. reserves.

The western and southwestern U.S. coals are mainly subbituminous, with a heating value of roughly 8,000 Btu/lb., and lignite, with a heating value of 6,000 Btu/lb.

Our 250-billion-ton coal reserve represents an enormous energy base. In terms of heating value, it is equivalent to approximately one trillion barrels of oil, almost 25 times larger than the U.S. reserves of crude oil and natural gas liquids combined. Although there are recoverable reserves sufficient to take care of our foreseeable needs centuries into the future, of greatest importance is that coal can provide the energy for our immediate needs during the next 40 or 50 years. This will enable us to continue research and development of other alternatives and to bring them on-line as they become economically feasible.

Figure 3.1 shows the sources of U.S. energy

*Institute for Mining and Minerals Research, University of Kentucky, Lexington, Kentucky. Contributions to this chapter by the staff of the Institute for Mining and Minerals Research are gratefully acknowledged.

TABLE 3.1 U. S. Energy Resources and Reserves

	Quads (10^{15} Btu)	
	Estimated Resources*	*Proven* Reserves*
Crude Oil	290–2,150	154
Natural Gas	325–1,220	210
Unconventional Gas		
Coal Seams	410	–
Tight Sands	–	220
Devonian Shale	–	17
Geopressured Brine	3,310–98,600	–
Tar Sands	150–215	12
Heavy Oil	320	10
Coal	26,100–98,600	6960
Peat	168–1,450	–
Shale Oil	5,950–34,800	430

**Estimated resources* are defined as concentrations of naturally occurring solid, liquid, or gaseous materials in or on the earth's crust in such form that economic extraction is currently or potentially feasible. Estimates of resources are professional judgements based on a variety of geological data, exploration records, production histories, and various assumptions. As a subset of estimated resources, *proven reserves* are defined as those quantities which geological and engineering data demonstrate, with reasonable certainty, to be recoverable in future years from known deposits under existing economic and operating conditions.
Source: Hill, R. F.,"Synthetic Fuels Summary" ESCOE (Engineering Societies Commission on Energy, Inc.), Washington, D.C., FE-2468-82, August, 1980.

from 1947 to the present, along with a projection of sources to 1990. The energy consumption pattern in the U.S. does not reflect the character of our resources. Energy consumption by source and sector for 1979 is shown in Table 3.3. The numbers are in Quads, or 10^{15} Btu. The important thing to notice is that 47 percent of our energy comes from petroleum and 26 percent from natural gas. Coal provides only 19 percent. Clearly we are using the most of what we have the least. Over 24 percent of the energy we used in 1979 was imported, causing a huge deficit in our balance of payments and real problems for our domestic economy.

TABLE 3.2 Geographical Distribution of U.S. Recoverable Coal Reserves (Source: Exxon Corp.)

	Billion Tons
Appalachian Basin	50
Eastern Interior Basin	38
Gulf Coast	13
Northern Rocky Mountains	97
Southern Rocky Mountains	34
West Coast and Alaska	18
	250

Our energy consumption has not always followed this pattern. There was a time when coal was the major supplier of U.S. energy needs. In the early 1920's, shortly after World War I, coal was providing over 75 percent of the energy used in the U.S. The use of oil and natural gas expanded rapidly during the 1930's and 40's; and by the end of World War II, oil and gas had surpassed coal as the principal source of energy for the U.S.

The dominant influence in the rapid expansion of the use of gas and oil was government policy which controlled prices of these fuels at low levels and encouraged switching to these fuel forms. Construction of major pipelines from the oil fields to the East Coast brought a cleaner and cheaper fuel to large metropolitan markets. The railroads, once major users of coal, changed to diesel locomotives. More recently the EPA required conversion of many electric generating plants from coal to oil and gas. The promise of nuclear power and strong

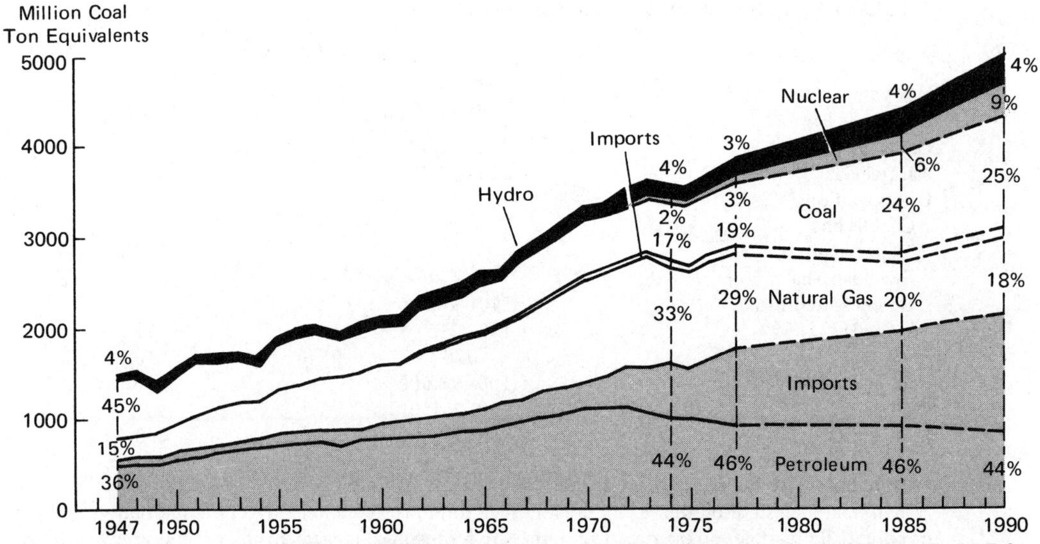

Fig. 3.1. U.S. energy supply. *(The President's Commission on Coal, Coal Data Book, Washington, D.C., Feb. 1980.)*
Note: Percent figures represent percent shares of total supply.
Source: See accompanying table.

efforts by government, through the Atomic Energy Commission, to promote nuclear power also diverted attention from coal as a major energy source. It is now clear, however, that coal will play a major role in the U.S. energy picture during at least the next three to four decades.

There are no shortages of predictions and projections that coal production must increase substantially in the next 10 to 20 years in order to meet U.S. energy needs. In 1979, the U.S. produced 770 million tons of coal. Studies by various federal agencies (such as the National Energy Plan, Project Independence, and Department of Commerce) and independent technical groups (CONAES) project coal production of 1,200-1,300 million tons per year by 1990 and 1,700-1,900 million tons per year by 2000. Exxon recently published its *Energy Outlook for 1980 - 2000*, which calls for coal production of almost 2,300 million tons per year by the year 2000. This projection has coal supplying 33 percent of our energy needs in 2000 and shows natural gas dropping from 26 percent to 17 percent and petroleum dropping from 46 percent to 33 percent. The compound annual growth rates in the use of coal are 4.2 percent from 1980-90, and 3.8 percent from 1990-2000. The use of coal in traditional markets will have to be expanded if we are to realize these growth rates, and coal will have to be moved into those sectors of our energy economy now being served by petroleum and natural gas. The development of a synfuels industry, using coal as a feedstock will facilitate such a move.

COAL RESERVES, PRODUCTION AND CONSUMPTION

Estimated remaining coal resources of the United States are shown in Table 3.4. North Dakota has the largest amount of lignite; Montana has the most subbituminous coal, while Illinois has the largest tonnage of bituminous coal. The demonstrated reserve base of U.S. coals by rank is shown in Table 3.5. Tables 3.6 and 3.7 show the reserve base by mining method – underground and surface.

The leading producing states are shown in Table 3.8. Kentucky has the largest production, followed by West Virginia and Pennsylvania. Yearly domestic consumption of bituminous coal and lignite is shown in Table 3.9. Since

TABLE 3.3 1979 U.S. Energy Consumption by Sector and Source (in Quads, 10^{15} Btu)

Sector	*Source* Coal	*Natural Gas*	*Petroleum*	*Hydro*	*Nuclear*	*Total*
Residential & Commercial	.24	8.0	6.66	–	–	14.90
Industrial	3.56	7.70	7.95	.04	–	19.25
Transportation	–	.53	19.18	–	–	19.71
Electric Power	11.41	3.61	3.26	3.12	2.77	24.17
Total	15.21	19.84	37.05	3.16	2.77	78.03
					Other	.09
						78.12
					Rounding	.07
					Total	78.19

Sector	*Quads*
Residential & Commerical	29.28
Industrial	29.15
Transportation	19.76
Total	78.19

Source	*Quads*
Coal	15.21
Natural Gas	19.84
Petroleum	37.05
Hydro	3.16
Nuclear	2.77
Coke Import	.07
Other	.09
Total	78.19

1 Quad/year = 0.47 million bbls/day Crude Oil Equivalent @ 5.8×10^6 Btu/bbl
= 43.5 million tons/year Coal Equivalent @ 11,500 Btu/lb

Source: Monthly Energy Review, USDOE, Energy Information Administration, DOE/EIA-0035/02(80), Feb., 1980.

1954 the electric utilities have been the single largest consumer. The sharp decline in retail sales of coal is mainly due to the change from coal to natural gas or fuel oil for home heating.

ORIGIN AND CLASSIFICATION OF COAL

Coal originated from the remains of trees, bushes, ferns, mosses, vines, and other forms of plant life that flourished in huge swamps and bogs many millions of years ago, during prolonged periods of humid, tropical climate and abundant rainfall. The precursor of coal was peat, which was formed by bacterial and chemical action on heavy accumulations of plant debris comprising bark, leaves, seeds, spores, various secretions, and the remains of trunks and branches.

Exclusion of oxygen from the peat by overlying water retarded its rate of decay, and accumulations of more plant debris and sediments compressed and solidified it, resulting in the beginning of coalification. The extent to which coalification progressed, which determines the *rank* of the coal, depended largely on the temperatures existing during this process. When the prevailing temperature corresponded to a depth of only a few hundred feet below the surface, coalification was relatively slow and resulted in low-rank coal, such as lignite and subbituminous coal. At increased depths, higher temperatures in the organic masses caused increases in rank, producing bituminous and anthracitic coal. Intrusions of igneous rocks sometimes accelerated these increases. The significant chemical changes that occur during coalification, from lignite to anthracite, are the decrease in hydrogen and oxygen content and the increase in carbon content. Some recognized authorities now believe that pressures on the coal deposits due to overburden affected the physical-structural development but had little influence on chemical coalification.

TABLE 3.4 Estimated Remaining Coal Resources of the United States*
(In millions (10^6) of short tons. Estimates include beds of bituminous coal and anthracite generally 14 in. or more thick, and beds of subbituminous coal and lignite generally $2\frac{1}{2}$ ft. or more thick, to overburden depths of 3,000 and 6,000 ft. Figures are for resources in the ground.)

	Overburden 0-3,000 feet							Overburden 3,000-6,000 feet	Overburden 0-6,000 feet
	Remaining identified resources, Jan. 1, 1974								
State	*Bituminous coal*	*Sub-bituminous coal*	*Lignite*	*Anthracite and semi anthracite*	*Total*	*Estimated hypothetical resources in unmapped and un-explored areas*[1]	*Estimated total identified and hypothetical resources remaining in the ground*	*Estimated additional hypothetical resources in deeper structural basins*[1]	*Estimated total identified and hypothetical resources remaining in the ground*
Alabama	13,262	0	2,000	0	15,262	20,000	35,262	6,000	41,262
Alaska	19,413	110,666	(2)	(3)	130,079	130,000	260,079	5,000	265,079
Arizona	[4]21,234	(4)	0	0	21,234	0	21,234	0	21,234
Arkansas	1,638	0	350	428	2,416	[5]4,000	6,416	0	6,416
Colorado	109,117	19,733	20	78	128,948	161,272	290,220	143,991	434,211
Georgia	24	0	0	0	24	60	84	0	84
Illinois	146,001	0	0	0	146,001	100,000	246,001	0	246,001
Indiana	32,868	0	0	0	32,868	22,000	54,868	0	54,868
Iowa	6,505	0	0	0	6,505	14,000	20,505	0	20,505
Kansas	18,668	0	(6)	0	18,668	4,000	22,668	0	22,668
Kentucky:									
Eastern	28,226	0	0	0	28,226	24,000	52,226	0	52,226
Western	36,120	0	0	0	36,120	28,000	64,120	0	64,120
Maryland	1,152	0	0	0	1,152	400	1,552	0	1,552
Michigan	205	0	0	0	205	500	705	0	705
Missouri	31,184	0	0	0	31,184	17,489	48,673	0	48,673
Montana	2,299	176,819	112,521	0	291,639	180,000	471,639	0	471,639
New Mexico	10,748	50,639	0	4	61,391	[7]65,556	126,947	74,000	200,947
North Carolina	110	0	0	0	110	20	130	5	135
North Dakota	0	0	350,602	0	350,602	180,000	530,602	0	530,602
Ohio	41,166	0	0	0	41,166	6,152	47,318	0	47,318
Oklahoma	7,117	0	(6)	0	7,117	15,000	22,117	[8]5,000	27,117
Oregon	50	284	0	0	334	100	434	0	434

Pennsylvania	63,940	0	0	18,812	82,752	[9]4,000	86,752	[10]3,600	90,352
South Dakota	0	0	2,185	0	2,185	1,000	3,185	0	3,185
Tennessee	2,530	0	0	0	2,530	2,000	4,530	0	4,530
Texas	6,048	0	10,293	0	16,341	[11]112,100	128,441	*(11)*	128,441
Utah	[12]23,186	173	0	0	23,359	[13]22,000	45,359	35,000	80,359
Virginia	9,216	0	0	335	9,551	5,000	14,551	100	14,651
Washington	1,867	4,180	117	5	6,169	30,000	36,169	15,000	51,169
West Virginia	100,150	0	0	0	100,150	0	100,150	0	100,150
Wyoming	12,703	123,240	*(2)*	0	135,943	700,000	835,943	100,000	935,943
Other States[14]	610	[15]32	[16]46	0	688	1,000	1,688	0	1,688
Total	747,357	485,766	478,134	19,662	1,730,919	1,849,649	3,580,568	387,696	3,968,264

[1]Source of estimates: Alabama, W.C. Culbertson: Arkansas, B.R. Haley: Colorado, Holt (1975): Illinois, M.E. Hopkins and J.A. Simon: Indiana, C.E. Wier: Iowa, E.R. Landis: Kentucky, K.J. Englund: Missouri, Robertson (1971, 1973): Montana, R.E. Matson: New Mexico, Fassett and Hinds (1971): Oklahoma, S.A. Friedman: Oregon, R.S. Mason: Pennsylvania, anthracite, Arndt and others (1968): Pennsylvania, bituminous, W.E. Edmunds: Tennessee, E.T. Luther: Texas lignite, Kaiser (1974): Virginia, K.J. Englund: Utah, H.H. Doelling: Washington, H.M. Beikman: Wyoming, N.M. Denson, G.B. Glass, W.R. Keefer, and E.M. Schell: remaining states, by the author. [2]Small resources of lignite included under subbituminous coal. [3]Small resources of anthracite in the Bering River field believed to be too badly crushed and faulted to be economically recoverable (Barnes, 1951). [4]All tonnage is in the Black Mesa field. Some coal in the Dakota Formation is near the rank boundary between bituminous and subbituminous coal. Does not include small resources of thin and impure coal in the Deer Creek and Pinedale fields. [5]Lignite. [6]Small resources of lignite in western Kansas and western Oklahoma in beds generally less than 30 in. thick. [7]After Fassett and Hinds (1971), who reported 85,222 million tons "inferred by zone" to an overburden depth of 3,000 ft. in the Fruitland Formation of the San Juan basin. Their figure has been reduced by 19,666 million tons as reported by Read and others (1950) for coal in all categories also to an overburden depth of 3,000 ft. in the Fruitland Formation of the San Juan basin. The figure of Read and others was based on measured surface sections and is included in the identified tonnage recorded in table 2. [8]Includes 100 billion tons inferred below 3,000 ft. [9]Bituminous coal. [10]Anthracite. [11]Lignite, overburden 200-5,000 ft.; identified and hypothetical resources undifferentiated. All beds assumed to be 2 ft. thick, although many are thicker. [12]Excludes coal in beds less than 4 ft. thick. [13]Includes coal in beds 14 in. or more thick, of which 15,000 million tons is in beds 4 ft. or more thick. [14]California, Idaho, Nebraska and Nevada. [15]California and Idaho. [16]California, Idaho, Louisiana, and Mississippi.

*Source: U.S. Geological Survey (Taken from 1980 Keystone Coal Industry Manual.)

TABLE 3.5 Demonstrated Reserve Base[1] of Coals in the United States on January 1, 1976, According to Rank[2]

(million short tons)

State	*Anthracite*	*Bituminous*	*Subbituminous*	*Lignite*	*Total*
Alabama	–	2,008.7	–	1,083.0	3,091.7
Alaska	–	697.5	5,446.6	14.0	6,158.2
Arizona	–	325.5	–	–	325.5
Arkansas	96.4	270.1	–	25.7	392.2
Colorado	25.5	9,144.0	4,121.3	2,965.7	16,256.4
Georgia	–	0.9	–	–	0.9
Idaho	–	4.4	–	–	4.4
Illinois	–	67,969.3	–	–	67,969.3
Indiana	–	10,714.4	–	–	10,714.4
Iowa	–	2,202.2	–	–	2,202.2
Kansas	–	998.2	–	–	998.2
Kentucky, East	–	13,540.1	–	–	13,540.1
Kentucky, West	–	12,460.8	–	–	12,460.8
Louisiana	–	–	–	(3)	(3)
Maryland	–	1,048.3	–	–	1,048.3
Michigan	–	126.8	–	–	126.8
Missouri	–	5,014.0	–	–	5,014.0
Montana	–	1,385.4	103,416.7	15,766.8	120,568.9
New Mexico	2.3	1,859.9	2,735.8	–	4,598.0
North Carolina	–	31.7	–	–	31.7
North Dakota	–	–	–	10,145.3	10,145.3
Ohio	–	19,230.2	–	–	19,230.2
Oklahoma	–	1,618.0	–	–	1,618.0
Oregon	–	(3)	17.5	–	17.5
Pennsylvania	7,109.4	23,727.7	–	–	30,837.1
South Dakota	–	–	–	426.1	426.1
Tennessee	–	965.1	–	–	965.1
Texas	–	–	–	3,181.9	3,181.9
Utah	–	6,551.7	1.1	–	6,552.8
Virginia	137.5	4,165.5	–	–	4,302.9
Washington	–	255.3	1,316.7	8.1	1,580.1
West Virginia	–	38,606.5	–	–	38,606.5
Wyoming	–	4,002.5	51,369.4	–	55,371.9
Total	7,371.1	228,924.6	168,425.0	33,616.6	438,337.3

[1] Includes measured and indicated resource categories as defined by the USBM and USGS and represents 100% of the coal in place.
[2] Data may not add to totals shown due to rounding.
[3] Quantity undetermined (basic resource data do not provide the detail required for delineation of reserve base).
Source: U.S.B.M. (taken from 1980 Keystone Coal Industry Manual).

Coal is a heterogeneous substance, being grossly characterized by layers or bands having glossy or dull appearances. The glossy layers are composed mainly of *vitrinite*, which was formed from woody parts of plants: trunks, limbs, branches, twigs, and roots. The dull bands consist of finely divided material, formed from leaves, pollen, spores, seeds, and resins (collectively termed *exinite*), in addition to fine vitrinite and more coalified *micrinite* and *fusinite*. Besides these banded coals, there are two types of dull, nonbanded coals-*cannels* which are rich in spores, and *bogheads* which contain abundant remains of algae.

Coals are classified by rank and/or type. Tables 3.10 and 3.11 give a classification of coals in the United States according to rank, whereas Table 3.12 classifies coal on the basis of type.

The main rank classes seen in Table 3.10 are

TABLE 3.6 Demonstrated Reserve Base[1] of Coals in the United States on January 1, 1976 Potentially Minable by Underground Methods[2]

(million short tons)

State	*Anthracite*	*Bituminous*	*Subbituminous*	*Lignite*	*Total*
Alabama	–	1,724.2	–	–	1,724.2
Alaska	–	617.0	4,805.9	–	5,423.0
Arkansas	88.6	163.1	–	–	251.7
Colorado	25.5	8,467.9	3,972.1	–	12,465.4
Georgia	–	0.5	–	–	0.5
Illinois	–	53,128.1	–	–	53,128.1
Indiana	–	8,939.8	–	–	8,939.8
Idaho	–	4.4	–	–	4.4
Iowa	–	1,736.8	–	–	1,736.8
Kentucky, East	–	9,072.5	–	–	9,072.5
Kentucky, West	–	8,510.4	–	–	8,510.4
Louisiana	–	–	–	–	–
Maryland	–	913.8	–	–	913.8
Michigan	–	125.2	–	–	125.2
Missouri	–	1,418.0	–	–	1,418.0
Montana	–	1,385.4	69,573.5	–	70,958.9
New Mexico	2.3	1,258.8	889.0	–	2,150.1
North Carolina	–	31.3	–	–	31.3
North Dakota	–	–	–	–	–
Ohio	–	13,090.5	–	–	13,090.5
Oklahoma	–	1,192.9	–	–	1,192.9
Oregon	–	(3)	14.5	–	14.5
Pennsylvania	6,966.8	22,335.9	–	–	29,302.7
South Dakota	–	–	–	–	–
Tennessee	–	627.2	–	–	627.2
Texas	–	–	–	–	–
Utah	–	6,283.8	1.1	–	6,284.9
Virginia	137.5	3,277.0	–	–	3,414.5
Washington	–	255.3	835.3	–	1,090.6
West Virginia	–	33,457.4	–	–	33,457.4
Wyoming	–	4,002.5	27,644.8	–	31,647.2
Total	7,220.7	182,019.6	107,736.1	–	296,976.3

[1] Includes measured and indicated resource categories as defined by the USBM and USGS and represents 100% of the coal in place.
[2] Data may not add to totals shown due to rounding.
[3] Quantity undetermined (basic resource data do not provide the detail required for delineation of reserve base).
Source: U.S.B.M. (From 1980 Keystone Coal Industry Manual.)

grouped according to certain chemical and physical properties. It will be noted that the factors determining rank, or degree of coalification, are moisture, volatile matter (material that is volatilized when the coal is heated in the absence of air at a certain temperature and for a certain length of time), fixed carbon (the residue after the loss of volatile matter), heating value, caking, and weathering properties.

A more commonly used rank indicator for coals higher in rank than lignite is the reflectance of vitrinite which is measured by comparing the intensity of a beam of light incident on a polished surface of vitrinite with the light reflected directly back from that surface. Table 3.11 gives a classification of coal according to vitrinite reflectance under oil.

An international classification of coals by type was recently developed by the Coal Committee Economic Commission for Europe, to eliminate confusion in evaluating coals shipped in international trade. Table 3.12 shows this system. It classifies high rank coals according to their volatile-matter content,

TABLE 3.7 Demonstrated Reserve Base[1] of Coals in the United States on January 1, 1976 Potentially Minable by Surface Methods[2]

(million short tons)

State	*Anthracite*	*Bituminous*	*Subbituminous*	*Lignite*	*Total*
Alabama	–	284.4	–	1,083.0	1,367.4
Alaska	–	80.5	640.7	14.0	735.2
Arizona	–	325.5	–	–	325.5
Arkansas	7.8	107.0	–	25.7	140.5
Colorado	–	676.2	149.2	2,965.7	3,791.0
Georgia	–	0.4	–	–	0.4
Illinois	–	14,841.2	–	–	14,841.2
Indiana	–	1,774.5	–	–	1,774.5
Iowa	–	465.4	–	–	465.4
Kansas	–	998.2	–	–	998.2
Kentucky, East	–	4,467.6	–	–	4,467.6
Kentucky, West	–	3,950.4	–	–	3,950.4
Louisiana	–	–	–	(*3*)	(*3*)
Maryland	–	134.5	–	–	134.5
Michigan	–	1.6	–	–	1.6
Missouri	–	3,596.0	–	–	3,596.0
Montana	–	–	33,843.2	15,766.8	49,610.1
New Mexico	–	601.1	1,846.8	–	2,447.9
North Carolina	–	0.4	–	–	0.4
North Dakota	–	–	–	10,145.3	10,145.3
Ohio	–	6,139.8	–	–	6,139.8
Oklahoma	–	425.2	–	–	425.2
Oregon	–	–	2.9	–	2.9
Pennsylvania	142.7	1,391.8	–	–	1,534.4
South Dakota	–	–	–	426.1	426.1
Tennessee	–	337.9	–	–	337.9
Texas	–	–	–	3,181.9	3,181.9
Utah	–	267.9	–	–	267.9
Virginia	–	888.5	–	–	888.5
Washington	–	–	481.5	8.1	489.5
West Virginia	–	5,149.1	–	–	5,149.1
Wyoming	–	–	23,724.7	–	23,724.7
Total	150.5	46,905.0	60,688.9	33,616.6	141,361.0

[1] Includes measured and indicated resource categories as defined by the USBM and USGS and represents 100% of the coal in place.
[2] Data may not add to totals shown due to rounding.
[3] Quantity undetermined (basic resource data do not provide the detail required for delineation of reserve base).
Source: U.S.B.M. (From 1980 Keystone Coal Industry Manual.)

calculated on a dry, ash-free (d.a.f.) basis. Since volatile matter is not an entirely suitable parameter for coals containing more than 33 percent volatile matter, the calorific value on a moist, ash-free basis is included as a parameter for such coals. The resulting nine classes of coal based on volatile-matter content and calorific value, are grouped according to their caking properties when the coal is heated rapidly, employing either the free-swelling or the Roga test. These groups are further subgrouped according to coking properties, using either the Audibert-Arnu or the Gray-King test. A three-figured code number is used to classify a coal. The first figure indicates the class of the coal, the second figure the group, and the third

TABLE 3.8 Coal Production by States

(1,000 Net Tons)

State	*1974*	*1975*	*1976*	*1977*	*1978p*	*% Change 78/77*
Alabama	19,824	22,644	21,537	21,545	20,553	−4.6
Alaska	700	766	706	705	731	+3.7
Arizona	6,448	6,986	10,420	11,059	9,054	−18.1
Arkansas	455	488	534	563	519	−7.8
Colorado	6,896	8,219	9,437	11,989	13,814	+15.2
Georgia	–	74	186	226	113	−50.0
Illinois	58,215	59,537	58,239	53,493	48,600	−9.2
Indiana	23,726	25,124	25,369	27,797	24,182	−13.0
Iowa	590	622	616	513	450	−12.3
Kansas	718	479	590	897	1,226	+36.7
Kentucky	137,197	143,613	143,972	146,262	135,689	−7.2
Maryland	2,337	2,606	2,830	3,036	2,998	−1.3
Missouri	4,623	5,638	6,075	6,366	5,665	−11.0
Montana	14,106	22,054	26,231	27,226	26,600	−2.3
New Mexico	9,392	8,785	9,760	11,083	12,632	+14.0
North Dakota	7,463	8,515	11,102	12,028	14,028	+16.6
Ohio	45,409	46,770	46,582	47,918	41,237	−13.9
Oklahoma	2,356	2,872	3,635	5,978	6,070	+1.5
Pennsylvania	80,462	84,137	85,777	84,639	81,477	−3.7
Tennessee	7,541	8,206	9,283	9,433	10,032	+6.4
Texas	7,684	11,002	14,063	15,865	20,020	+26.2
Utah	5,858	6,961	7,967	8,581	9,141	+6.5
Virginia	34,326	35,510	39,996	37,624	31,946	−15.1
Washington	3,913	3,743	4,109	5,057	4,708	−6.9
West Virginia	102,462	109,283	108,834	95,433	85,314	−10.6
Wyoming	20,703	23,804	30,836	46,028	58,328	+26.7
Total U. S.	603,406	648,438	678,685	691,344	665,127	−3.8

pPreliminary
Source: 1980 Keystone Coal Industry Manual

figure the subgroup. The details of its use and the test methods on which it is based have been described.[1]

COAL COMPOSITION

Coal is defined as an organic rock composed chemically of carbon, hydrogen, oxygen, nitrogen, sulfur, and mineral matter. These components can vary considerably both chemically and structurally, depending on the rank of the coal and even within coals of the same rank. Of these parameters, sulfur and mineral matter exhibit the greatest variation for coals of the same rank, while for coals of different ranks the greatest characteristic variations will occur in the carbon and hydrogen values.

There are two primary types of coal analysis—proximate and ultimate. In addition, there are a number of miscellaneous analyses. Since many of these tests are empirical, requiring strict adherence to specified conditions, it is essential that authoritative standard methods be established. In the United States, Technical Committee D-05 of the American Society for Testing and Materials (ASTM) develops and sponsors these methods. International standards are developed and sponsored by Technical Committee 27, Solid Mineral Fuels, of the International Organization for Standardization (ISO/TC27).

Proximate analysis includes the determination of moisture, ash, volatile matter, and then fixed carbon by difference. These values, in addition

TABLE 3.9 U.S. Consumption of Coal* by End-Use Sector, 1947-1977 and Projected**

(million tons)

	ELECTRIC UTILITIES		*COKE PLANTS*		*GENERAL INDUSTRY AND OTHER†*		*RETAIL‡*		
	Million Tons	*Percent of Total*	*Million Tons*	*Percent of Total*	*Million Tons*	*Percent of Total*	*Million Tons*	*Percent of Total*	*TOTAL*
1947	86	16	105	19	258	47	97	18	546
1948	96	18	107	21	230	44	87	17	520
1949	81	18	91	20	185	42	88	20	445
1950	88	19	104	23	178	39	84	19	454
1951	102	22	113	24	179	38	75	16	469
1952	103	25	98	23	151	36	67	16	419
1953	112	26	113	26	142	34	60	14	427
1954	115	32	85	23	111	31	52	14	363
1955	141	33	107	25	122	29	53	13	423
1956	155	36	106	25	123	28	49	11	433
1957	157	38	108	26	113	27	36	9	414
1958	153	41	76	21	102	28	36	10	367
1959	166	45	79	22	92	25	29	8	366
1960	174	46	81	21	95	25	30	8	380
1961	179	48	74	20	93	25	28	7	374
1962	191	49	74	19	95	25	28	7	388
1963	209	51	78	19	99	24	23	6	409
1964	223	52	89	21	100	23	19	4	431
1965	243	53	95	21	102	22	19	4	459
1966	264	54	96	20	106	22	20	4	486
1967	272	57	92	19	99	21	17	3	480
1968	295	59	91	18	98	20	15	3	499
1969	308	61	93	18	91	18	15	3	507
1970	319	62	96	19	89	17	12	2	516
1971	326	66	83	17	75	15	11	2	495
1972	349	67	87	17	72	14	9	2	517
1973	387	70	94	17	67	12	8	1	556
1974	390	70	90	16	64	12	9	2	553
1975	403	72	83	15	63	12	7	1	556
1976	447	75	84	14	61	10	7	1	599
1977	475	77	77	12	61	10	7	1	620
1985	746	79	97	10	106	11	(a)	–	949
1990	912	80	101	9	131	11	(a)	–	1144

*Bituminous coal and lignite.

**DOE mid-level projection assumes medium economic growth and medium availability of energy with constant oil prices.

†Includes manufacturing and mining industries, steel and rolling mills, cement mills, and railroads.

‡Coal dealers.

(a)Projected retail consumption is included in "General Industry and Other" projection total.

Source: Energy Information Administration: *Annual Report to Congress*, Vols, II & III, 1977.

TABLE 3.10 Classification of Coals by Rank[a]

Class	Group	Fixed Carbon Limits, per cent (Dry, Mineral-Matter-Free Basis)		Volatile Matter Limits, per cent (Dry, Mineral-Matter-Free Basis)		Calorific Value Limits, Btu per pound (Moist,[b] Mineral-Matter-Free Basis)		Agglomerating Character
		Equal or Greater Than	Less Than	Greater Than	Equal or Less Than	Equal or Greater Than	Less Than	
I. Anthracitic	1. Meta-anthracite	98	-	-	2	-	-	-
	2. Anthracite	92	98	2	8	-	-	-
	3. Semianthracite	86	92	8	14	-	-	Nonagglomerating[c]
II. Bituminous	1. Low volatile bituminous coal	78	86	14	22	-	-	Commonly agglomerating[e]
	2. Medium volatile bituminous coal	69	78	22	31	-	-	
	3. High volatile *A* bituminous coal	-	69	31	-	14,000[d]	-	
	4. High volatile *B* bituminous coal	-	-	-	-	13,000[d]	14,000	
	5. High volatile *C* bituminous coal	-	-	-	-	11,500	13,000	
						10,500	11,500	Agglomerating
III. Subbituminous	1. Subbituminous *A* coal	-	-	-	-	10,500	11,500	Nonagglomerating
	2. Subbituminous *B* coal	-	-	-	-	9,500	10,500	-
	3. Subbituminous *C* coal	-	-	-	-	8,300	9,500	-
IV. Lignitic	1. Lignite *A*	-	-	-	-	6,300	8,300	-
	2. Lignite *B*	-	-	-	-	-	6,300	-

[a]This classification does not include a few coals, principally nonbanded varieties, which have unusual physical and chemical properties and which come within the limits of fixed carbon or calorific value of the high-volatile bituminous and subbituminous ranks. All of these coals either contain less than 48 per cent dry, mineral-matter-free fixed carbon or have more than 15,500 moist, mineral-matter-free British thermal units per pound.

[b]Moist refers to coal containing its natural inherent moisture but not including visible water on the surface of the coal.

[c]If agglomerating, classify in low-volatile group of the bituminous class.

[d]Coals having 69 per cent or more fixed carbon on the dry, mineral-matter-free basis shall be classified according to fixed carbon, regardless of calorific value.

[e]It is recognized that there may be nonagglomerating varieties in these groups of the bituminous class, and there are notable exceptions in high volatile *C* bituminous group.

TABLE 3.11 Classification of Coals by Vitrinite Reflectance (Oil Reflectance Limits of ASTM Coal Rank Classes)

Rank	*Maximum Reflectance, %*	*Maximum Reflectance, %*†	*Random Reflectance, %**
Subbituminous	-0.47		
High Volatile Bituminous	C 0.47-0.57 B 0.57-0.71 A 0.71-1.10	-1.03	0.50-1.12
Medium Volatile Bituminous	1.10-1.50	1.03-(1.35-1.40)	1.12-1.51
Low Volatile Bituminous	1.50-2.05	(1.35-1.40)-	1.51-1.92
Semianthracite	2.05-3.00 (approx.)		1.92-2.50
Anthracite	>3.00 (approx.)		>2.50

†Procedure of Bethlehem Steel Corporation using reactive vitrinite reflectance.
*After McCartney and Teichmüller (1972, 1973).
Source: Davis, DOE FE-2030-TR10, 1978

TABLE 3.12 International Classification of Coals by type. (Courtesy Bureau of Mines, U.S. Department of the Interior.)

GROUPS (determined by caking properties)			CODE NUMBERS										SUBGROUPS (determined by coking properties)		
GROUP NUMBER	ALTERNATIVE GROUP PARAMETERS: Free-swelling index (crucible-swelling number)	ALTERNATIVE GROUP PARAMETERS: Roga index	The first figure of the code number indicates the class of the coal, determined by volatile-matter content up to 33% V. M. and by calorific parameter above 33% V. M. The second figure indicates the group of coal, determined by caking properties. The third figure indicates the subgroup, determined by coking properties.										SUBGROUP NUMBER	ALTERNATIVE SUBGROUP PARAMETERS: Dilatometer	ALTERNATIVE SUBGROUP PARAMETERS: Gray-King
3	>4	>45					435	535	635				5	>140	>G_8
						334	434	534	634				4	>50-140	G_5-G_8
						333	433	533	633	733			3	>0-50	G_1-G_4
						332 a / 332 b	432	532	632	732	832		2	≤ 0	E-G
2	$2\frac{1}{2}$-4	>20-45				323	423	523	623	723	823		3	>0-50	G_1-G_4
						322	422	522	622	722	822		2	≤ 0	E-G
						321	421	521	621	721	821		1	Contraction only	B-D
1	1-2	>5-20			212	312	412	512	612	712	812		2	≤ 0	E-G
					211	311	411	511	611	711	811		1	Contraction only	B-D
0	0-$\frac{1}{2}$	0-5		100 (A, B)	200	300	400	500	600	700	800	900	0	Nonsoftening	A
CLASS NUMBER →			0	1	2	3	4	5	6	7	8	9	As an indication, the following classes have an approximate volatile-matter content of: Class 6 33-41% volatile matter; 7 33-44% " "; 8 35-50% " "; 9 42-50% " "		
CLASS PARAMETERS	Volatile matter (dry, ash-free) →		0-3	>3-10 (A: >3-6.5; B: >6.5-10)	>10-14	>14-20	>20-28	>28-33	>33	>33	>33	>33			
	Calorific parameter [a] →		–	–	–	–	–	–	>13,950	>12,960-13,950	>10,980-12,960	>10,260-10,980			
CLASSES (Determined by volatile matter up to 33% V. M. and by calorific parameter above 33% V. M.)															

Note: (i) Where the ash content of coal is too high to allow classification according to the present systems, it must be reduced by laboratory float-and-sink method (or any other appropriate means). The specific gravity selected for flotation should allow a maximum yield of coal with 5 to 10 percent of ash.
(ii) 332a . . . >14-16% V. M.
332b . . . >16-20% V. M.

[a] Gross calorific value on moist, ash-free basis (30 C., 96% relative humidity) B. t. u. / lb.

to total sulfur and heating values*, are usually sufficient to provide the information necessary for comfort and process heating, steam generation, and coking considerations.

The *ultimate analysis* which includes the determination of carbon, hydrogen, nitrogen, and total sulfur, in addition to the moisture and ash values—with oxygen being calculated by difference—is important for the calculation of material balances in thermal and chemical processes which use coal as a feedstock.

The procedures which comprise the miscellaneous analysis group are the greatest in number and the most varied in scope. Included are not only standard methods such as heating value, free-swelling index, and ash analysis but also those procedures used to determine trace elements, mineral phases, surface parameters, conversion potential, and so forth. While, in the near future, the purchase of coal for combustion will continue to be based primarily on moisture, ash, heating value, and sulfur values, the acquisition of vast quantities of coal for conversion processes will be dictated by parameters such as conversion potential, trace element content, mineral phases, and surface characteristics.

The proximate and ultimate analyses are described in ASTM[2] and Bureau of Mines publications[3] as are a number of other miscellaneous analyses. Many of the procedures, however, are not standard methods and must be acquired from one or more of the many publications dealing with the specific area of interest.

Figure 3.2 shows the proximate analyses of coals selected to represent the various ranks or types of coal reported on the ash-free basis. Analyses listed by coal bed, county, and state can be found in a number of publications such as Bureau of Mines Bulletin #446 and Illinois State Geological Survey Circular #499. Furthermore, a large amount of information is being acquired in the Pennsylvania State University data base and will also provide an excellent data source.

COAL MINING

Coal mines fall into two general classifications: surface and underground. Production by type of mining is shown in Figure 3.3; a projection made by Exxon of total coal production is shown in Figure 3.4.

Surface Mining

Surface mining techniques are used when the overlying rock, or overburden, is thin enough to make the mining economical. Also, coal near the surface cannot be mined underground in a safe manner. Surface mining is done by a number of methods: contour mining, moutaintop removal, area mining, and auger mining.

Contour mining is used in mountainous terrain where the slope of the surface will permit only a narrow bench to be cut around the side of a hill and then backfilled immediately after the removal of the coal. It enables small deposits to be mined, and it is the only method that can be used on slopes of 15 degrees or greater.

Mountaintop removal has been used in the Appalachian coal fields. In this method of mining, all the overburden on top of the mountain is removed. After the coal is recovered and the surface reclaimed, flat land remains which has many uses in areas which otherwise consist of steep slopes.

---

*Heating value, usually reported as Btu per pound, is the *high-heat value* (HHV), also referred to as the *gross heating value*. This quantity is defined as the heat produced by combustion of unit quantity, at constant volume, under specified conditions such that the products are ash, sulfur dioxide and carbon dioxide gases, nitrogen and liquid water. It is often helpful to correct the heating value for differences between laboratory conditions and actual conditions occurring in combustion of net heating values. The *low-heat value* (LHV) is calculated from the HHV by deducting 1030 Btu's for each pound of water originally present or formed during combustion and is defined as the heat produced by combustion of unit quantity at constant atmospheric pressure, under conditions leaving all water in the products as vapor.

The high-heating value can also be reasonably calculated from a number of formulas using determined values for carbon (C), hydrogen (H), oxygen (O), sulfur (S) and additional data as deemed necessary. While the Dulong formula, using only C, H, O, and S values, has been the most generally used in the United States, several others are available which may better fit the wide range of raw materials and products of modern technology.

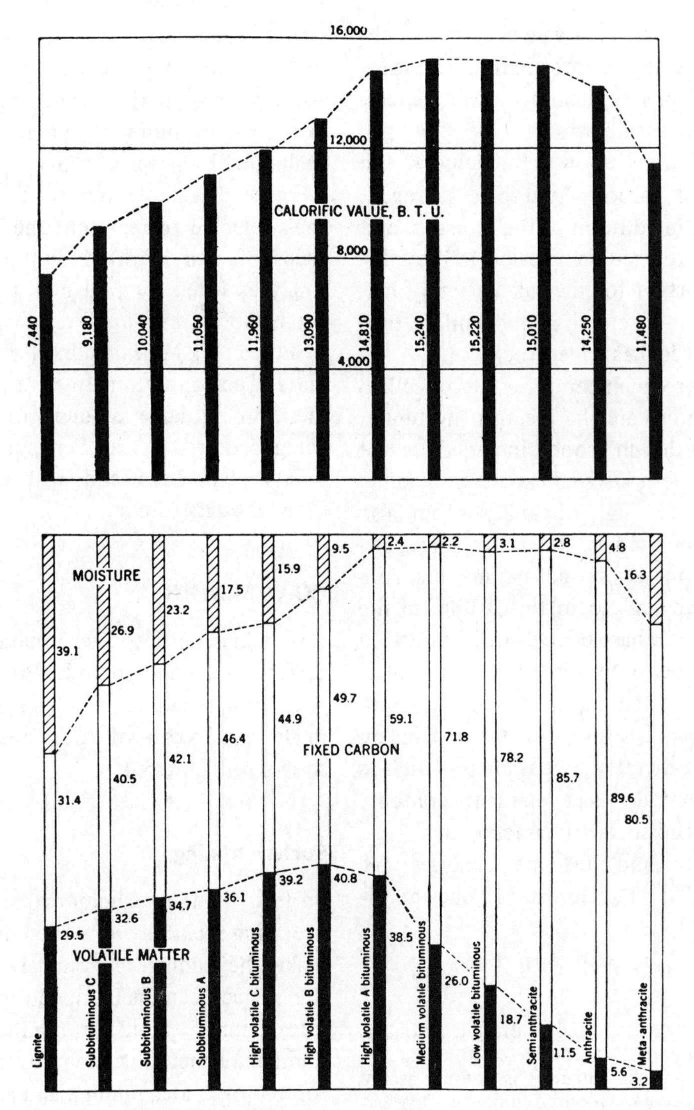

Fig. 3.2. Analysis of United States coals selected to represent the various ranks. (*Courtesy Bureau of Mines, U.S. Department of the Interior.*)

Area Mining is used in the flat or gently rolling lands of the midwest and west where large and efficient equipment can be used. Larger areas of disturbed lands are generally produced by this method, but reclamation is simpler.

Auger mining is a supplementary method used to reach coal in stripped areas where the overburden has become too thick to economically remove. Large augers are operated from the floor of the surface mines and bore horizontally into the coal face to produce some reserves not otherwise mineable.

Underground Mining

Underground mining techniques are used to reach coal located below too much overburden for surface mining. There are three types of underground mines, each defined by the method used to reach and remove the coal (Figure 3.5).

A *drift mine* is one which enters a coal seam exposed at the surface on the side of a hill or mountain; the mine follows the coal horizontally.

A shaft mine is one where a vertical shaft is

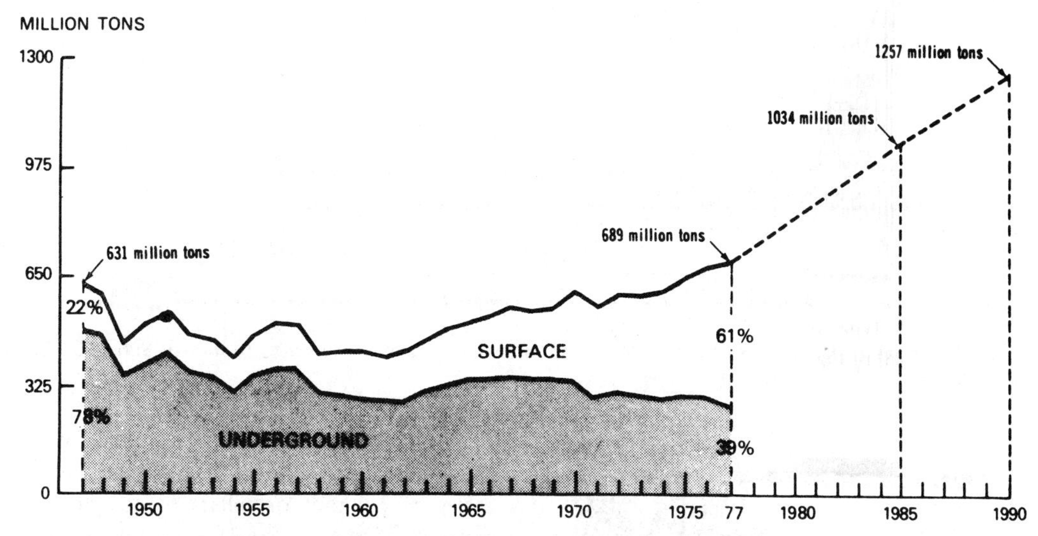

Fig. 3.3. U.S. coal production by methods of mining. (*The President's Commission on Coal, Coal Data Book, Washington, D.C., Feb. 1980.*)
Note: Percent figures represent shares of total production.
Source: See accompanying table.

dug through the rock to reach the coal which may be at great depths below the surface. The coal is then mined by horizontal entries into the seam with the resulting coal hoisted to the surface through the vertical shaft.

A *slope mine* is one where an inclined tunnel is driven through the rock to the coal, with the mined coal removed by conveyors or track haulage. This type of operation is sometimes matched with a shaft mine.

Two general mining systems are in use in underground mining: room-and-pillar and longwall.

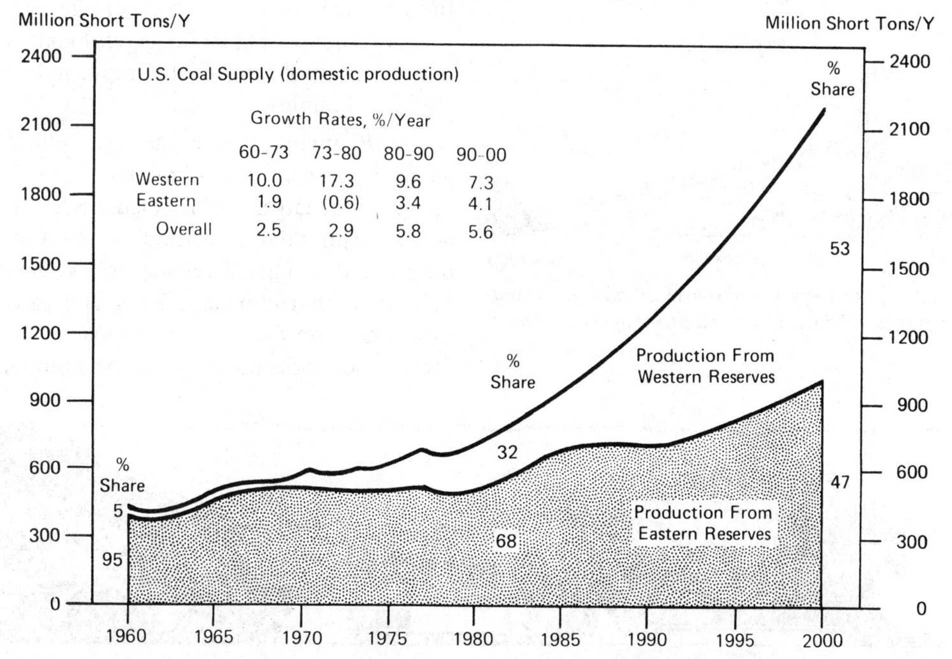

	60-73	73-80	80-90	90-00
Western	10.0	17.3	9.6	7.3
Eastern	1.9	(0.6)	3.4	4.1
Overall	2.5	2.9	5.8	5.6

Fig. 3.4. U.S. coal supply–domestic production. (*Exxon Company, Energy Outlook 1980-2000, Houston, Texas, Dec., 1979.*)

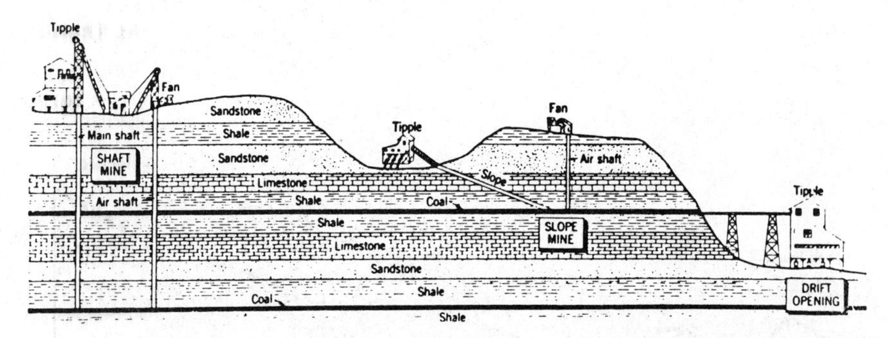

Fig. 3.5. Three types of entrances to underground mines–shaft, slope, and drift. (*Courtesy Bureau of Mines, U.S. Department of the Interior.*)

Fig. 3.6. Basic steps in conventional mining. (*Courtesy Bureau of Mines, U.S. Department of the Interior.*)

Room-and-pillar mining removes coal from a series of rooms in the coal bed and leaves a series of columns or pillars of coal for the support of the roof. This system recovers about 50% of the coal and leaves an area much like a checkerboard. It is used in areas where the overlying rock or "roof" has geologic characteristics which will offer the possibility of good roof support. This system was used in the old mines where the coal was hand dug. Two methods in use today for extracting the coal from the seam are the conventional method where the coal is under-cut and blasted free (Figure 3.6) and the continuous method where a machine moves along the coal face to extract the coal instead of blasting it loose (Figure 3.7). Roof control is the major problem of the room-and-pillar method of mining.

Longwall mining uses a machine which is pulled back and forth across the face of the coal seam in larger rooms (Figure 3.8). It is not a major mining method in the United States at this time.. Coal recovery is greater than in room-and-pillar mining and can be used where roof conditions are fair to poor. Strong roof rock, however, can be a problem.

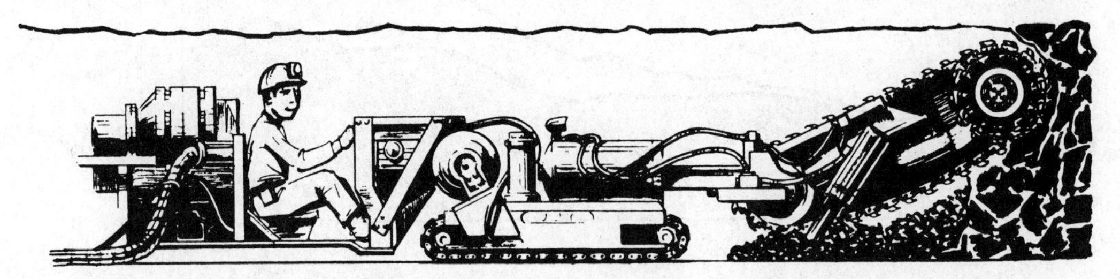

Fig. 3.7. Continuous mining machine. (*Courtesy Bureau of Mines, U.S. Department of the Interior.*)

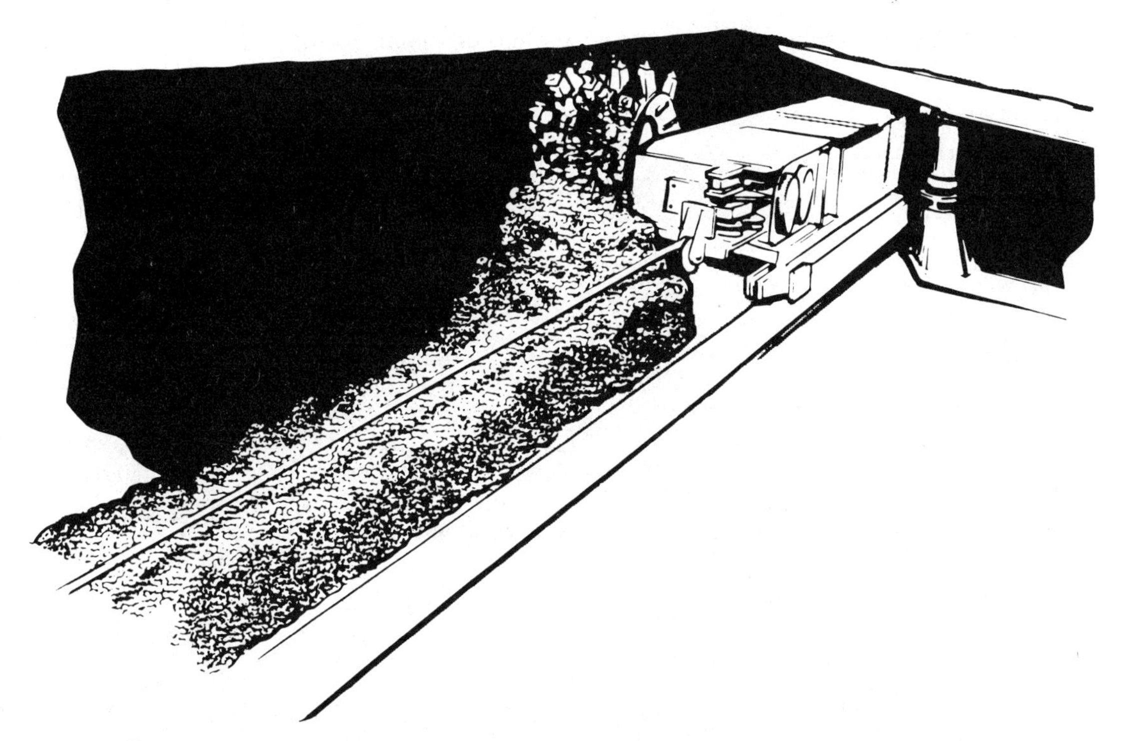

Fig. 3.8. Longwall mining machine. (*Courtesy Bureau of Mines, U.S. Department of the Interior.*)

The seam should be over 42" thick to accommodate this type of operation and the large coal cutter or plow that it uses. A large reserve is necessary.

A modification of this method using a continuous mining machine on faces up to 150 ft. long is under development and is known as *shortwall* mining. It uses the room support system of self-advancing chocks developed for longwall operations.

Productivity trends by method of mining are shown in Figure 3.9. The drop in underground mining activity beginning in 1969 is attributable at least in part, and perhaps entirely to, the passage of the Coal Mining Health and Safety Act of 1969. Figure 3.10 shows recent employment trends in the U.S. coal industry. The number of people working has been increasing since 1970.

COAL PREPARATION

Coal preparation is a collective term for the physical and mechanical processes applied to coal to make it suitable for a particular use. Some of these processes are breaking and crushing, screening, wet and dry concentrating, and dewatering. This section will discuss briefly only some wet methods of concentration to reduce the ash and sulfur content of coal. All aspects of coal preparation are described comprehensively in a recent book.[4]

Coal comes from the mine in a wide range of sizes and contains rock slate, pyrites, and other impurities. Wet or dry concentration, hereafter called *cleaning*, is used to remove the impurities, thereby upgrading the coal for markets. Many users have rather rigid specifications concerning size and the ash and sulfur content. For example, coal for making metallurgical coke is specified as 80 percent through a $\frac{1}{8}$-in screen, 6 ± 1 percent ash, and not more than 1.25 percent sulfur.

The percentage of total coal production cleaned between 1940 and 1975 is illustrated in Table 3.13. These statistics show a peak in cleaning production in the late 60's. The decline in percent of total production cleaned in the late 60's and early 70's reflects an increase in production of western U.S. lignites which are difficult to clean. The decline in tonnage

PRODUCTIVITY TRENDS BY METHOD OF MINING, 1948-1978

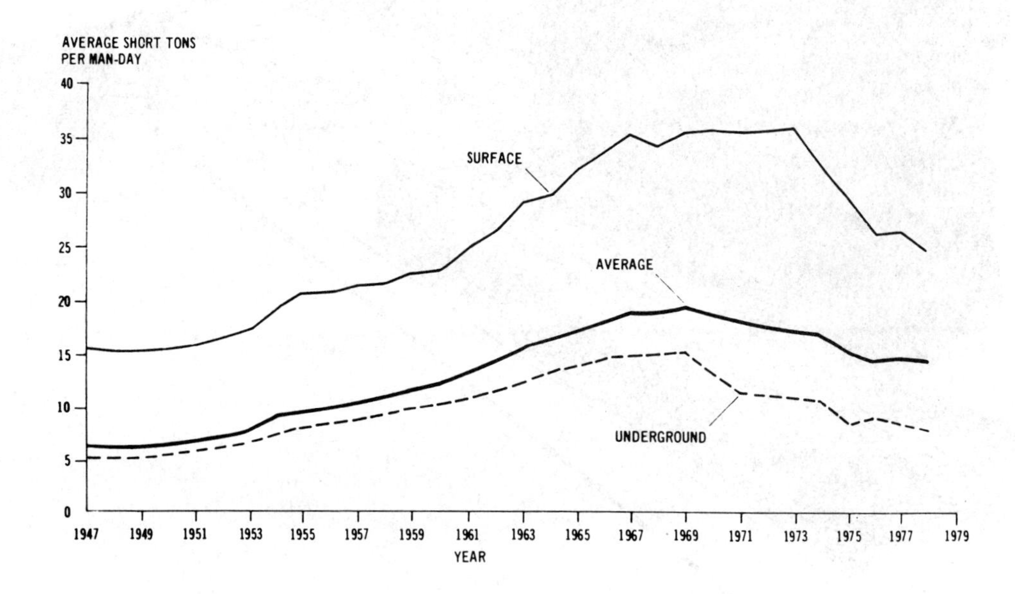

1948-1978	Annual increase rate of	(+) 2.8%
1960-1978	Annual increase rate of	(+) 0.9%
Since 1969	Productivity decreasing at annual rate of	(-) 4.1%

Source: Energy Information Administration, Annual Report to Congress, Vol. II, 1978.

Fig. 3.9. Productivity trends by method of mining, 1948-1978. (*Energy Information Administration, Annual Report to Congress, Vol. II, 1978.*)

cleaned in the early 70's is probably a reflection of production of coals capable of combustion directly with stack gas scrubbing systems. However, recent studies have shown considerable economic advantage of combining flue gas desulfurization with coal preparation. As a result of these advantages, coupled with the need to use run of mine coals with higher ash and sulfur, will encourage a reversal in the late 70's and 80's for increasing coal cleaning.

For discussion purposes, it is convenient to divide cleaning into two parts: *coarse-coal cleaning* and *fine-coal cleaning*, the dividing coal size being arbitrarily $\frac{3}{8}$ in. Cleaning methods and devices are designed to separate coal from impurities by the differences in specific gravity, coal being less dense than its impurities. The exception is *flotation*, which depends on the differences in the surface characteristics of coal and its associated impurites and is mostly used for −28 Mesh coals.

Coarse-Coal Cleaning

The most popular concentrating processes in this country are *jigging* and *dense-medium washing*.

Jigging. Separation of coal from its impurities by jigging consists of inducing pulsations in water in which the raw coal is suspended. An open-top rectangular box having a perforated bottom, or screen plate, is used. The pulsations are induced through the screen plate by reciprocating members, causing alternate expansion and compaction of the coal-water mixture, re-

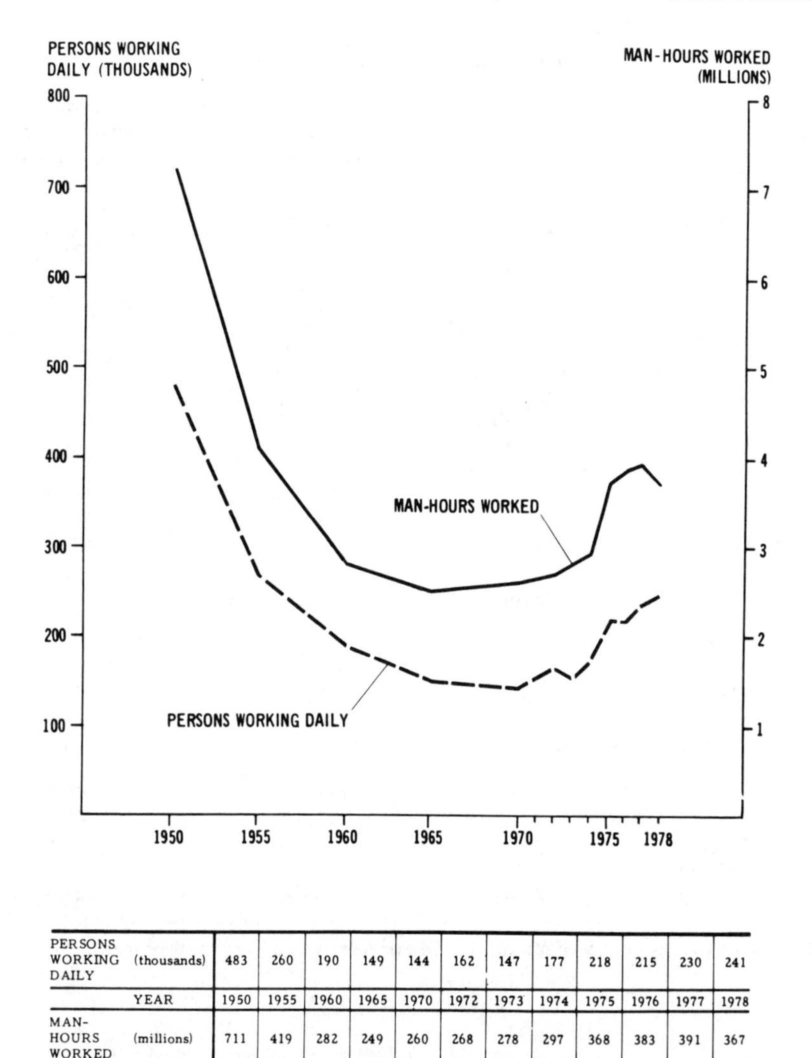

PERSONS WORKING DAILY	(thousands)	483	260	190	149	144	162	147	177	218	215	230	241
	YEAR	1950	1955	1960	1965	1970	1972	1973	1974	1975	1976	1977	1978
MAN-HOURS WORKED	(millions)	711	419	282	249	260	268	278	297	368	383	391	367

Source: Mine Safety and Health Administration.

Fig. 3.10. U.S. coal industry employment, 1950–1978. (*The President's Commission on Coal, Coal Data Book, Washington, D.C. Feb., 1980.*)

sulting in stratification of the coal according to the specific gravity of the particles. The lighter coal particles rise to the top, overflow the end of the jig, and are removed as clean product. The denser impurities settle on the screen plate and are withdrawn as *refuse*. Most jigs are of the Baum type, in which the pulsations are generated by air pressure. Plungers and diaphragms are also used as pulsers.

Dense-medium Washing. Coal is cleaned by immersing it in a fluid having a density between that of the coal and its impurities. Most dense-medium washers use a suspension of sand or magnetite in water to achieve the desired density. The modern highly automated dense-medium plants are capable of high throughput and sharp separation of the raw coal into a clean product and refuse.

Fine-Coal Cleaning

The principal fine-coal washers are the *wet concentrating table, dense-medium cyclone, hydro-*

TABLE 3.13 Mechanically Cleaned Bituminous Coal (1000 Tons)

Year	*By Wet Methods*	*By Pneumatic Methods*	*Total Mechanically Cleaned*	*Percent of Total Production Mechanically Cleaned*
1940	87,290	14,980	102,270	22.2
1945	130,470	17,416	147,886	25.6
1950	183,170	15,529	198,699	38.5
1955	252,420	20,295	272,715	58.7
1956	268,054	24,311	292,365	58.4
1957	279,259	24,768	304,027	61.7
1958	240,153	18,882	259,035	63.1
1959	251,538	18,249	269,787	65.5
1960	355,030	18,139	273,169	65.7
1961	247,020	17,691	264,711	65.7
1962	252,929	18,704	271,633	64.3
1963	269,527	19,935	289,462	63.1
1964	288,803	21,400	310,203	63.7
1965	306,872	25,384	332,256	64.9
1966	316,421	24,205	340,626	63.8
1967	328,135	21,268	349,402	63.2
1968	324,123	16,804	340,923	62.5
1969	315,596	19,163	334,761	59.7
1970	305,594	17,855	323,452	53.6
1971	256,892	14,506	271,401	49.1
1972	281,119	11,710	292,829	49.2
1973	278,413	10,505	288,918	48.8
1974	257,592	7,557	265,150	43.9
1975	260,289	6,704	266,993	41.2

Source: Minerals Yearbook, Bureau of Mines, U.S. Department of the Interior, Washington, D.C.

cyclone, launder, feldspar jig, hydrotator, and *froth-flotation cell.* Of these, the concentrating table is the most popular, especially for cleaning bituminous coal. Cyclones are becoming increasingly important for washing both anthracite and bituminous coal, while launders and feldspar jigs are used to a limited extent for cleaning bituminous coal. The hydrotator is used extensively for anthracite.

Wet Concentrating Table. This is essentially a continuous mechanized form of the classical miner's pan. It consists of a transversely and longitudinally-tilted rectangular or rhombohedral table. A one-quarter size version of a commercial table is shown in Fig. 3.11. The diagonal ripples seen through the flowing coal-water mixture are caused by riffles in the deck. These riffles are directly responsible for the capacity of the table and represented a major improvement in tabling when it was first used to clean coal in 1980.

The upper side of the deck contains a feedbox at the corner near the head motion, and a wash-water distribution box. The deck is vibrated longitudinally with a low forward stroke and a rapid return. Normal decks are 8 feet wide and 16 feet long, and will clean up to 12 tons per hour per deck.

Dense-medium Cyclone. Figure 3.12 shows the essential features of a dense-medium cyclone. The mixture of dense medium and raw coal enters tangentially near the top of the cylindrical section, producing free-vortex flow. The refuse moves downward along the wall and is discharged through the underflow orifice. The cleaned coal moves radially toward the cyclone axis, passes through the vortex finder

Fig. 3.11. A wet-concentrating table for cleaning fine coal. (*Courtesy Bureau of Mines, U.S. Department of the Interior.*)

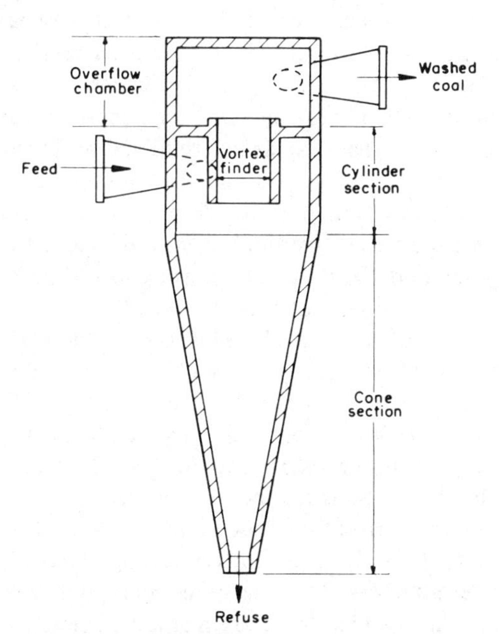

Fig. 3.12. Dense-medium cyclone for cleaning coal. (*Courtesy Bureau of Mines, U.S. Department of the Interior.*)

to the overflow chamber, and is discharged from a tangential outlet.

The dense medium used exclusively is pulverized magnetite suspended in water. The magnetite is recovered magnetically for re-use.

Cyclones are available in several sizes, 20-inch and 24-inch diameters being most common. A 20-inch unit will clean about 50 tons of raw coal per hour, and the 24-inch one about 75 tons per hour.

Hydrocyclone. The hydrocyclone, as the name implies, is a cyclone cleaning device that uses only water as a medium. Its design differs from that of the conventional dense-medium cyclone by having a greater cone angle and a longer vortex finder. Hydrocyclones currently are used to clean coal 0.025 inch and smaller, but can be used for coal as coarse as $\frac{1}{4}$ inch.

Froth Flotation. In the froth-flotation process, fine coal slurry, to which a small quantity of flotation agents is added, is processed through a flotation cell. In the cell, finely disseminated air bubbles pass upward through the slurry. The hydrophobic coal particles attach themselves to the surface of the air bubbles and rise to the top to form a froth. The hydrophilic impurities sink in the cell and are removed as refuse.

There are three general classes of flotation agents; *frothers, collectors or promoters, and modifying agents*. The frother promotes stable froths. Examples are amyl and butyl alcohols and methylisobutyl carbinol (MIBC). The collector or promoter improves adherence of the coal particles to the air bubbles. Some frothers, e.g., MIBC, also function to a degree as collectors. Modifying agents have several functions, such as depressing flotation of unwanted material, altering the surface of the particles to aid the attachment of air bubbles, and providing the desired acidity or alkalinity of the flotation pulp.

The froth flotation process is used to clean minus 35-mesh coal.

Other Cleaning Devices. Because of their limited use in the United States, the launder, feldspar jig, and hydrotator are not described in this section, and the reader is referred to an earlier reference[4] for complete information.

RECENT DEVELOPMENTS IN COAL PREPARATION

Coal preparation is changing rapidly in the United States. Table 3.14 shows a change in emphasis away from predominantly using jigs to using dense-medium processes, as well as an increase in fine-coal cleaning. In addition, there has been a decrease in thermally dried coal as a result of more stringent particulate emissions regulations. However, fluidized-bed driers have maintained their share of the market. The major emphasis in coal preparation is the reduction of sulfur and the maximizing of Btu recovery. The need for sulfur reduction results primarily from the enforcement of SO_X emission regulations and the high cost of stack-gas scrubbing systems. The need for higher Btu recovery in coal preparation results from the increased value of coal in the market place. New developments in coal preparation have recently been implemented to better meet these requirements. These advances include fine coal cleaning, fine coal dewatering, and closed water circuits. Much of the information presented in this section is abstracted from articles by Killmeyer[6] and Cavallaro and Deurbrock[7].

Fine-Coal Cleaning

The trend toward fine-size coal cleaning in order to remove more sulfur and recover more Btu's has spurred the development of many new techniques and circuits in washing, dewatering, sizing, and closing water circuits. The Batac jig, developed in Germany, has shown increasing popularity in the U.S. and has been installed in at least three U.S. plants to process $\frac{1}{2}$ inch by 0 raw coal. It generally follows the Baum jig, a coarse coal cleaning device, which cleans the plus $\frac{1}{2}$ inch raw coal. In the Batac jig, air pulsations are produced in multiple chambers directly beneath the bed screen. The pulses in each chamber are controlled independently and precisely. The jig provides significantly higher throughput in the same space as concentrating tables, and it cleans with a sharpness of separation similar to that of concentrating tables for the same size range.

An interesting new fine-size coal cleaning circuit has been installed in a Pennsylvania preparation plant by Roberts & Schaefer Co. Originally, the plant used air tables, but a dense-medium cyclone has been incorporated to reclean the second and third refuse draws from the air table. The first draw is essentially pure refuse with no misplaced float material. With the second and third draws wide open, the amount of sink material reporting to the clean coal is reduced. At these table settings, about 20 percent of the feed is sent to the cyclone to maximize Btu recovery, while upgrading the final product. The cyclone produces a clean coal with 5.6 percent ash, which is then combined with the clean coal from the air table. The net result is a 5 percent increase in yield and a 20 percent decrease in ash content. The air table itself is presently receiving considerable attention from western coal producers who are looking for a dry method of cleaning coal in the water-scarce western states.

Another fine-coal cleaning technology which

TABLE 3.14 Mechanical Cleaning and Thermal Drying of Bituminous Coal and Lignite, by Type of Equipment, (Thousand short tons)

	1968	*1969*	*1970*	*1971*	*1972*	*1973*	*1974*	*1975*
Cleaners:								
Wet methods:								
Jigs	159,028	155,027	140,457	115,407	127,591	132,655	129,302	124,317
Concentrating tables	47,268	45,328	44,058	35,656	40,257	34,935	28,869	28,682
Classifiers	4,871	3,401	3,593	2,071	2,980	3,297	2,698	6,176
Launders	4,498	4,644	5,199	4,896	5,467	5,121	3,577	2,664
Dense-medium processes:								
Magnetite	70,633	71,701	76,590	70,262	74,667	74,605	68,749	72,448
Sand	27,027	24,023	23,290	17,802	15,273	12,617	12,427	13,533
Calcium chloride	1,839	1,911	1,714	1,702	2,081	981	1,107	951
Total	99,497	97,636	101,593	89,764	92,021	88,203	82,283	96,931
Flotation	8,961	9,560	10,694	9,098	12,802	14,201	10,863	11,519
Total, wet methods	324,123	315,596	305,594	256,892	281,119	278,413	257,592	260,289
Pneumatic methods	16,804	19,163	17,855	14,506	11,710	10,505	7,557	6,704
Grand total	340,923	334,761	323,452	271,401	292,829	288,918	265,150	266,993
Dryers:								
Fluidized-bed	38,153	40,639	42,595	32,564	34,118	30,907	24,616	25,866
Multilouver	13,831	10,067	8,554	6,337	2,861	1,616	3,006	1,969
Rotary	1,995	1,204	876	835	6,924	5,519	697	794
Screen	5,897	5,100	4,806	1,176	2,776	2,484	1,960	2,798
Suspension or flash	9,202	7,381	5,410	5,432	6,098	5,575	5,766	4,184
Vertical tray and cascade	4,225	2,691	1,924	1,761	459	100	–	70
Total	73,303	67,082	64,165	48,105	53,235	46,202	36,045	35,681

Source: Minerals Yearbook, Bureau of Mines, U.S. Department of the Interior, Wash., D.C.

is receiving attention is high-gradient magnetic separation. High-gradient magnetic separation (HGMS) uses large-capacity magnetic devices as a practical means to separate small, weakly-magnetic particles, predominantly pyrite, from the coal. This technology, which is being utilized commercially to purify kaolin clay, is being investigated to determine its potential for application in the coal preparation industry, and attempts are being made to establish the technical feasibility of removing a substantial fraction of the inorganic sulfur from dry coal powders by HGMS.

The present methods of physical coal preparation in relation to sulfur reduction apply only to the partial removal of pyritic sulfur. As these physical methods do not remove organically bound sulfur, they can be applied effectively only to selected coals. However, the Bureau of Mines is presently engaged in a research program to develop chemical treatments specifically aimed at organic sulfur removal. The chemical oxidation of the organic sulfur to a form which could be removed more readily appears to have potential, and it is being investigated actively.

It is envisioned that the optimum desulfurization of coal will include both physical and chemical treatment. The raw coal will be crushed fine enough to maximize pyrite release and then be physically treated using wet processes to remove the liberated pyrite. The fine-size wet coal product will then be treated by a chemical process that will remove sufficient organic and remaining pyritic sulfur to provide an environmentally acceptable final product.

Fine-Coal Drying. One of the biggest problems in processing fine-size coal is dewatering and sizing. The severe blinding tendency of minus 28-mesh coal slurries on conventional screens makes separations extremely inefficient and unreliable. A new screening device has been developed by Derrick Manufacturing Corp. which purports to have solved the blinding problem. It consists of a "sandwich screen" concept, which binds two fine screen surfaces, one on top of the other. The action of the two screens against each other, due to a special high-frequency low-amplitude vibrating motor, prevents near-size particles from becoming lodged in the top screen. The openings in the bottom screen are slightly larger than those in the top to insure that particles do not become trapped between screens. It is claimed that these screens can make sharp separations down to 400 mesh. The final section of the screen may be a slotted natural rubber surface fitted with a patented vacuum arrangement to help remove additional surface moisture.

Closed Water Circuits. In addition to the enforcement of air emission standards, stream pollution and refuse disposal laws recently have become more stringent. Thus, new preparation plants employ some type of closed water circuit with the gravitational clarifier as the primary piece of equipment in the system. Two new thickener designs, the Lamella and the Enviro-Clear, have gained interest for their possible application at plant sites where land area is at a premium. These units have been installed at older plants that had space limitations and were expanding their plant capacity or eliminating waste ponds.

Both compact thickeners may require as little as one-tenth of the space occupied by a conventional circular tank thickener. Their overflow clarities are competitive with those of conventional thickeners at 500 ppm or less of solids; underflow concentrations are slightly higher at 30 to 40 percent solids.

Each thickener operates on a different design principle. The Lamella has the same settling area for a given flow rate as a conventional thickener by virtue of a series of plates. The settling of solids on these plates is much like that of a stack of conventional thickeners, but the plates are inclined to allow continuous removal of the solids. Lamellas have feed-loading rates of 0.5 to 0.6 gpm/ft^2, and use flocculant dosages of about 2 ppm. These figures are in line with those of the large circular tank thickeners. At present, the largest Lamella will handle 1,500 gpm.

The Enviro-Clear has a much higher feed-loading rate, 3 to 4 gpm/ft^2, because it does not have the settling area of the Lamella. It is a bottom-fed unit that actually uses the settled solids as a filtering medium for the water as it flows upward to the overflow. There are only two zones in this thickener, clarified liquid and compressed solids. Feed fluctuations can be absorbed by allowing the sludge bed level to rise or fall without affecting overflow clarity.

Both units are more sensitive to feed variations, and they lack large storage capacity, which is the great advantage of conventional thickeners.

COAL CLEANING AND AIR POLLUTION CONTROL

Until recently, coal cleaning was performed mainly to reduce the amount of ash and any concomitant reduction of sulfur was coincidental. However, owing to national efforts to reduce the amount of sulfur oxides discharged to the atmosphere, most of which come from coal-fired electric utility plants, the removal of sulfur from steam coal is now an end in itself.

Sulfur occurs in coal in three forms: organic, pyritic (FeS_2) and sulfate. Sulfur as sulfate occurs up to only a few tenths of a percent; the bulk occurs in the organic and pyritic forms. Since organic sulfur is part of the coal molecule, no method has been proven for removing it economically, although research is presently in progress on chemical methods for doing so. Consequently, the amount of organic sulfur represents the lowest limit to which a coal can be mechanically freed of sulfur.

Since pyrite is nearly four times as dense as coal, it would seem possible to separate all of it from the coal substance. In practice, however, the amount of pyritic sulfur that can be removed varies considerably from one coal to another, depending on the way the pyrite is distributed throughout the coal.

Where most of the pyrite can be removed, mechanical coal preparation is an economically attractive process. Since pyritic sulfur comprises about one-half of the sulfur in coal, one of the more serious air pollution problems would be substantially reduced in such cases.

In one study[8], the economic potential of coal preparation combined with stack-gas scrubbing was evaluated (although the concept of physical coal cleaning combined with flue-gas desulfurization is not new). In this study, actual coal use areas, coal source areas, and the most probable coalbed sources are defined and used to make an economic evaluation of physical coal cleaning. Then the cost of a new utility plant that can remove SO_2 (exclusively) by stack-gas scrubbing to meet the new source emission standard was evaluated. This was followed by a similar evaluation of the combined use of physical coal cleaning plus stack-gas scrubbing to attain the same SO_2 emission level.

Such case studies indicate that for many situations the economic advantage of a combined approach is quite significant. This advantage depends on:

(1) The availability of coals capable of significant reductions in ash and sulfur by physical upgrading with minimal Btu loss;

(2) A clean-coal sulfur level that is compatible with significantly less than full-scale scrubbing requirements. Even so, the range of variables is such that each coal source-user combination must be individually assessed.

COAL UTILIZATION

Combustion

The greatest percentage of the coal consumed in this country is burned in boilers to generate steam for turboelectric plants in the electric utility industry. For example, of the 620 million tons consumed in 1977, the utilities used 475 million tons, or 77 percent of the total.

Remarkable advances have been made in the energy efficiency of electric utility plants because of improvements in the steam-generating equipment and turbines. The average heat rate of 1937 was 1.4 pounds of coal per kilowatt-hour, and 0.8 pounds in 1968. Some plants recently placed in operation have heat rates of about 0.7 pound of coal per kilowatt-hour, which corresponds to an overall thermal efficiency of 39 percent, based on coal having an as-fired heating value of 12,500 Btu per pound.

Further improvements in thermal efficiency may be achieved with a new system of power generation that is now being developed. Known as MHD (magnetohydrodynamics), this system is expected to have a thermal efficiency of between 50 and 55 percent if coal is used as a fuel. The principle of MHD is to burn the fuel under conditions that will give a combustion gas temperature of about 4500°F, add an easily ionizable salt like potassium carbonate (seeding) so that the gas will have a relatively high concentration of free ions and electrons, and then pass this conducting gas through a magnetic field and draw off electric energy through electrodes placed in the path of the gas. By analogy, the hot conducting gas moving through a magnetic field behaves like the armature in a conventional turbogenerator. A coal-fired MHD generator would discharge less sulfur and nitrogen oxides to the atmosphere than a conventional plant of equivalent size because the seeding salt would absorb these oxides, which would be recovered during regeneration of the seed.

In the early days, boilers were rated by horsepower. One rated horsepower was defined as equivalent to 10 square feet of boiler heating surface. A developed horsepower was arbitrarily set as equal to the evaporation of 34.5 pounds of water per hour at 212°F. Later, an equivalent heat unit of 33,480 Btu per hour was adopted. From these definitions, R_b, percent boiler rating, is obtained:

$$R_b = \frac{w(h - h_w) \times 100}{3348 \times S}$$

where w = steam flow, lb/hr
h = enthalpy of steam at boiler or superheater outlet, Btu/lb
h_w = enthalpy of water at boiler or economizer inlet, Btu/lb
S = boiler heating surfaces, sq ft

Percent boiler rating is still used for small boilers and is a relative measure of "size" where furnace-wall cooling and superheating surfaces are absent or relatively small compared with boiler convection surface.

Utility and industrial boilers have a relatively large amount of heat-transfer surface other than boiler convection surface, and boiler horsepower and boiler rating are not suitable means of designating boiler output. Almost invariably output is now given in pounds per hour of steam flow at maximum continuous load, or in energy equivalent megawatts (6000 lb/hr steam per megawatt). Peak or overloads are given as percentages of full load.

Combustion Equipment. Industrial and electric utility boilers are fired with either stokers, pulverized-coal-fired burners, or cyclone burners, the choice depending on the kind of coal and the amount of steam needed. Table 3.15 gives the kinds of coal that can be burned with the various kinds of firing. Although good results will usually be obtained with the respective coal and firing equipment, this table should be used only as a rough guideline, and final equipment selection should be based on a sound engineering analysis.

(1) *Stoker Firing.* Table 3.16 gives the approximate range of capacity for each kind of stoker. The principles of stokers, showing the relative movement of fuel and air, are illustrated in Fig. 3.13. Both fuel and air have the same direction in retort stokers; this is called underfeed burning. The fuel moves across the air direction in chain- or traveling-grate stokers; this is called crossfeed burning. The spreader stoker approximates overfeed burning, the in-

TABLE 3.15 Burning Equipment for Various Coals

	Stokers				
Fuel	*Under-feed*	*Traveling or chain grate*	*Spreader*	*Pulverized-coal burner*	*Cyclone burner*
Anthracite		x		x	
Bituminous:					
17–25% volatile	x		x	x	x
25–35% volatile				x	x
strongly coking	x	x	x	x	x
weakly coking		x	x	x	x
Lignite		x	x	x	x

TABLE 3.16 Approximate Range of Capacity of Stokers

Type	*Steam, M lb/hr*	*Grate Heat Release, M Btu/hr sq ft (max.)*
Single retort	5–50	200
Multiple retort	40–300	300
Traveling or chain grate	10–300	300
Spreader	10–300	1,000

coming fuel moving toward the air. Except for certain types of coal gasifiers, in which lump coal moves downward toward a grate against air (or oxygen and steam) coming through the grate, no conventional combustion system operates purely in the overfeed mode.

(2) *Pulverized-Coal Firing and Cyclone Firing.* Electric utility and large industrial plants favor pulverized-coal-fired and cyclone-fired furnaces because of their inherent flexibility regarding the kind and quality of coal, their comparatively good availability, their quick response to load changes, and their extremely high steam-generating capacity. Recent units generate as high as 9 million pounds of steam per hour at 3500 psig and 1000°F.

The burner and furnace configurations for the main types of pulverized-coal firing (often called suspension firing), and cyclone firing, are shown in Fig. 3.14. There are some design variations in vertical, impact, and horizontal suspension firing, but these schematic drawings serve to illustrate the principles.

The first suspension-fired furnace in this country was designed like the one shown for

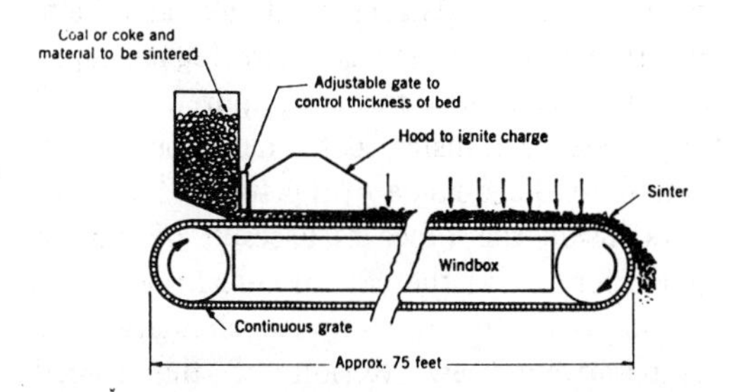

Fig. 3.13. Dwight Lloyd stoker for manufacturing sintered products. (*Courtesy Bureau of Mines, U.S. Department of the Interior.*)

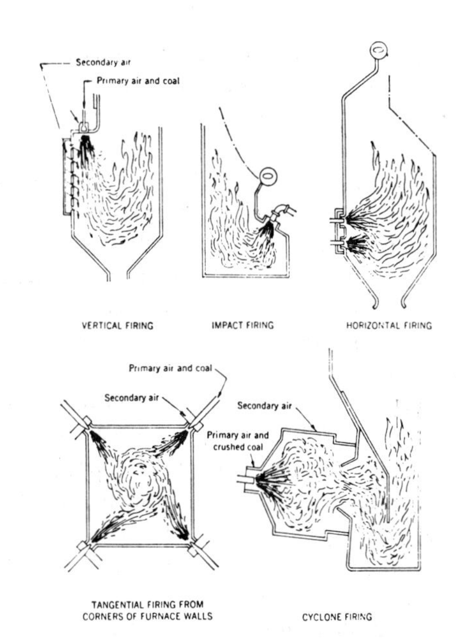

Fig. 3.14. Methods of firing pulverized and crushed coal. (*Courtesy Bureau of Mines, U.S. Department of the Interior.*)

vertical firing in Fig. 3.14. Pulverized coal (about 70 percent through a 200-mesh screen) is transported to the burner with primary air, the amount of this air being about 20 percent of that needed for complete combustion. The balance of the air, known as secondary air, is admitted through openings in the furnace wall. Because a large percentage of the total combustion air is withheld from the fuel stream until it projects well down into the furnace, ignition stability is good.[10] This type of firing is well suited for coals that are difficult to ignite, such as those with less than 15 per cent volatile matter. Although no longer used in central-station power plants, this design, with delayed admission of secondary air, may find favor again if low-volatile chars from various coal-conversion processes are burned for heat and power.

The other types of suspension firing use burners in which the primary air and coal, and the secondary air are mixed just before or immediately after entering the furnace. With tangential firing the burners are arranged in vertical banks at each corner of a square, or nearly square, furnace and directed toward an imaginary circle in the center of the furnace. This produces a vortex with its axis on the vertical center line. The burners consist of an arrangement of slots one above the other admitting through alternate slots, primary air-fuel mixture and secondary air. The burners can be tilted upward or downward 30 degrees from the horizontal plane, enabling the operator to control superheat and to permit selective utilization of furnace heat-absorbing surfaces. Basically, the turbulence needed for mixing the fuel and air is generated in the furnace instead of in the burners.

In cyclone firing, the coal is not pulverized as for suspension firing but is crushed to 4-mesh size and admitted tangentially with primary air to a water-cooled cylindrical chamber called a cyclone furnace. The finer particles burn in suspension, and the coarser ones are thrown by centrifugal force to the furnace wall. The wall, having a sticky coating of molten slag, retains the coal until it burns. The secondary air, which is admitted tangentially along the top of the furnace, completes the combustion of the coarse particles. Slag drains continuously into the main boiler furnace.

(3) *Fluidized-Bed Combustion.* Recently, interest has been growing in another mode for burning coal—the fluidized bed. Although still in the experimental stage, there are clearly seen potential advantages to this burning mode. a) The fuel-bed temperature is low, about 1550°F, which means less formation of nitrogen oxides, and retention of some of the sulfur in the ash of certain coals. (Adding dolomite or limestone to the fuel bed greatly improves sulfur retention.) Equally important is less volatilization of sodium and potassium in the coal; consequently, there are less deposits on and corrosion of furnace, superheater, and reheater tubes. b) Heat-transfer rates from the fluidized bed to immersed heat-transfer surfaces are relatively high, as much as 100 Btu per hr per sq ft per degree Fahrenheit.

It is believed that successful application of fluidized-bed boilers to electric-utility stations would reduce capital costs about 10 percent compared with conventional plants of the same capacity. The principle of the fluidized-bed combustor is illustrated schematically in Fig. 3.15.

There is a commerical stoker that burns low-grade coals in a fluidized mode, the *Ignifluid Stoker*. The Ignifluid combustion system was first installed in France, and numerous industrial plants in Europe have since used it.

Basically, the Ignifluid system is a combustion chamber with a narrow chain grate inclined from the front to the rear of the chamber about 15 degrees. Naturally formed fuel banks between the front and sides of the chamber and the stoker surround the fluidized bed and protect the lower parts of the wall against excessive heat and adhering clinkers. The fuel banks have no part in the combustion because no air passes through them, but their surface is continually renewed by incandescent particles thrown up from the fluidized bed. Since some of the fuel is burned in suspension and the rest on the grate, this system differs from the fluidized-bed system described previously, in which the entire coal charge is fluidized. The Ignifluid system is not adaptable in its present form to steam-generating tubes within the bed, which would be necessary to achieve the high-steaming capacities of modern electric utility plants.

Boiler Types. There are various kinds of industrial and utility boilers, broadly classified as fire-tube and water-tube. In the former, the hot combustion gases pass through tubes, and heat is transferred to water outside the tubes. The most common and least expensive boiler of this type is the horizontal return tubular (HRT) boiler. However, because of the design and construction of fire-tube boilers, there is a definite limitation to their size and the pressure which they can tolerate.

Water-tube boilers may be broadly classified as straight-tube and bent-tube types, the latter having several variations in design and being preferred for applications where higher capacities and steam pressures are required. In both types, heat is transferred by radiation or convection to the outside of the tubes, and water flows inside the tubes as a result of thermal circulation, or in the case of certain bent-tube boilers, as the result of forced circulation. A comparatively new version of the forced-circulation, bent-tube boiler for central-station power plants is the "once-through" type. The

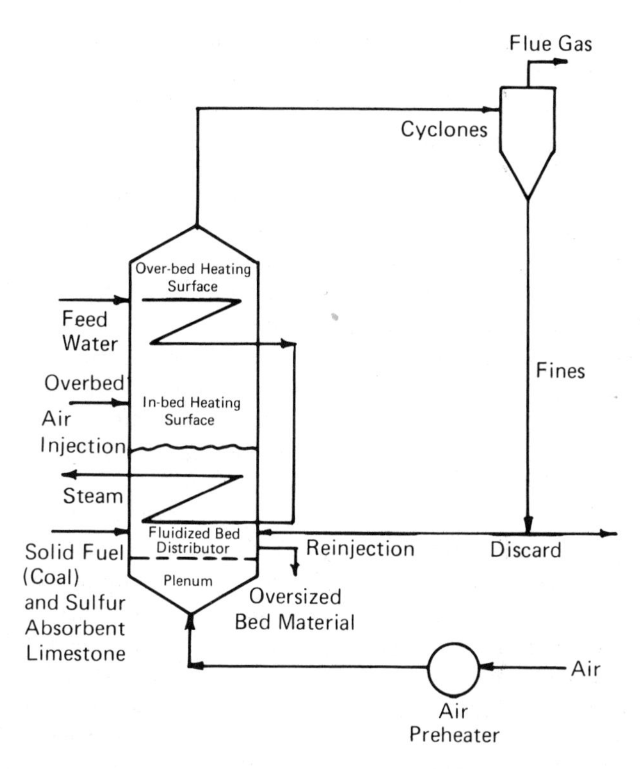

Fig. 3.15. Basic components of a fluidized-bed combustion boiler. *(Shang, J. Y., Department of Energy, Morgantown, W. Va.)*

feed-water passes progressively through the heating, evaporation, and superheater sections; no drum is used for separating the steam from the boiler water as in other boilers, consequently, the ratio of water circulated in the boiler to steam generated is unity. The data in Table 3.17 approximate the range of steam capacities and pressures for the principal types of boilers.

Combustion Calculations. The amount of air or oxygen just sufficient to burn the carbon, net hydrogen, and sulfur to carbon dioxide is the theoretical or stoichiometric amount of air or oxygen. The general expression for combustion of a fuel is

$$C_mH_n + [(4m + n)/4]O_2 = mCO_2 + (n/2)H_2O$$

m and n being the number of atoms of carbon and net hydrogen, respectively, in the fuel. For example, one mole (an amount equal to the

TABLE 3.17 Approximate Range of Capacities of Various Types of Industrial and Utility Boilers

Type	*Capacity, lb steam/hr*	*Maximum Design Pressure, psig*
Fire tube (HRT)	1,000–15,000	250
Water tube		
straight	15,000–150,000	2000
bent, 3-drum, low-head	1,000–35,000	400
2-drum, vertical	1,000–350,000	1000
electric utility	up to 9,000,000	3500

molecular weight of the fuel) of methane (CH_4) requires 2 moles of oxygen for complete combustion to 1 mole of carbon dioxide and 2 moles of water. If air is used, each mole of oxygen is accompanied by 3.76 moles of nitrogen.

The theoretical weight or volume of oxygen or air to burn a given weight of fuel is of primary interest in designing combustion equipment. The *volume of theoretical oxygen* to burn any fuel can be calculated from the ultimate analysis of the fuel as follows:

$$359\left(\frac{C}{12}+\frac{H_2}{4}-\frac{O_2}{32}+\frac{S}{32}\right) = \text{cu ft of oxygen/lb of fuel}$$

where C, H_2, O_2, and S are the fractional weights of these elements in 1 lb of fuel. The weight of oxygen in pounds is obtained by multiplying the cubic feet of oxygen by 0.0891.

The *volume of theoretical air* is obtained by using a coefficient of 1710 instead of 359 in the equation above.

More than the theoretical amount of air is always necessary in practice to achieve complete combustion. The *excess air* can be calculated by the equation:

$$A_{xs} = 100\ [(A/A_t - 1)]$$

in which A_{xs} is the percentage of excess air, A_t is the theoretical air in lb/lb, and A is the air actually used in lb/lb. If one wishes to know the percentage of excess air, and only the flue gas composition is known, it can be calculated from

$$A_{xs} = 100\ [O_2/(0.266N_2 - O_2)]$$

where O_2 and N_2 are the percentages by volume of these components in the dry flue gas. This equation is applicable only when the nitrogen in the fuel is negligible and there are no combustible gases, such as carbon monoxide or hydrogen, in the flue gas.

Coal Gasification

Currently in this country there is an urgent need to supplement the U.S. reserves of natural gas. Also, there is a serious need for sulfur-free gas from coal to replace raw coal as a boiler fuel, thus abating atmospheric pollution by sulfur oxides. Processes based on coal gasification or hydrogasification are currently being investigated in the United States to satisfy these needs.

Coal gasification has been discussed recently by many authors, including Bodle and Schora[11], Miller[12], Considine[13], Baughman[14], and Hill[15,16]. The topic will be covered only briefly here.

Gasification as used here means reacting coal with air, oxygen, steam, carbon dioxide, hydrogen, or a mixture of these, in a manner that yields a gaseous product suitable for use either as a source of energy, or as a raw material for the synthesis of industrial chemicals, liquid fuels, or other gaseous fuels. Some of the important synthesis products are synthetic pipeline gas (methane), hydrogen, ammonia, alcohols, olefins, and distillable oils.

The reaction of coal with air and steam produces a gas (low-Btu, or producer gas) with a heating value of about 150 Btu per standard cubic foot. This gas, after cleaning, can be used directly as a fuel for steam or power generation. If oxygen, rather than air, is used with steam, the product gas is not diluted with nitrogen and has a heating value of up to 500 Btu per cubic foot, depending upon the equipment, coal, and operating conditions employed. This medium-Btu, or synthesis gas, can be combusted directly as a fuel or can be upgraded into a variety of synthetic products. Figure 3.16 illustrates the various processing steps required to produce synthetic gas from coal. By making suitable adjustments to the carbon monoxide to hydrogen ratio in the shift converter, gas mixtures suitable for conversion to methane, methanol, alcohols, distillable oils and hydrogen can be produced. Still in the pilot stage is hydrogasification, which involves reacting coal with hydrogen at high pressure and temperature to obtain a methane-rich gas suitable for subsequent processing to synthetic pipeline gas.

Gasifiers are the mechanical devices used to contact the coal with the gasifying medium. These contactors may be divided into three main categories:

1. *Fixed-Bed Gasification*. Lump coal is sup-

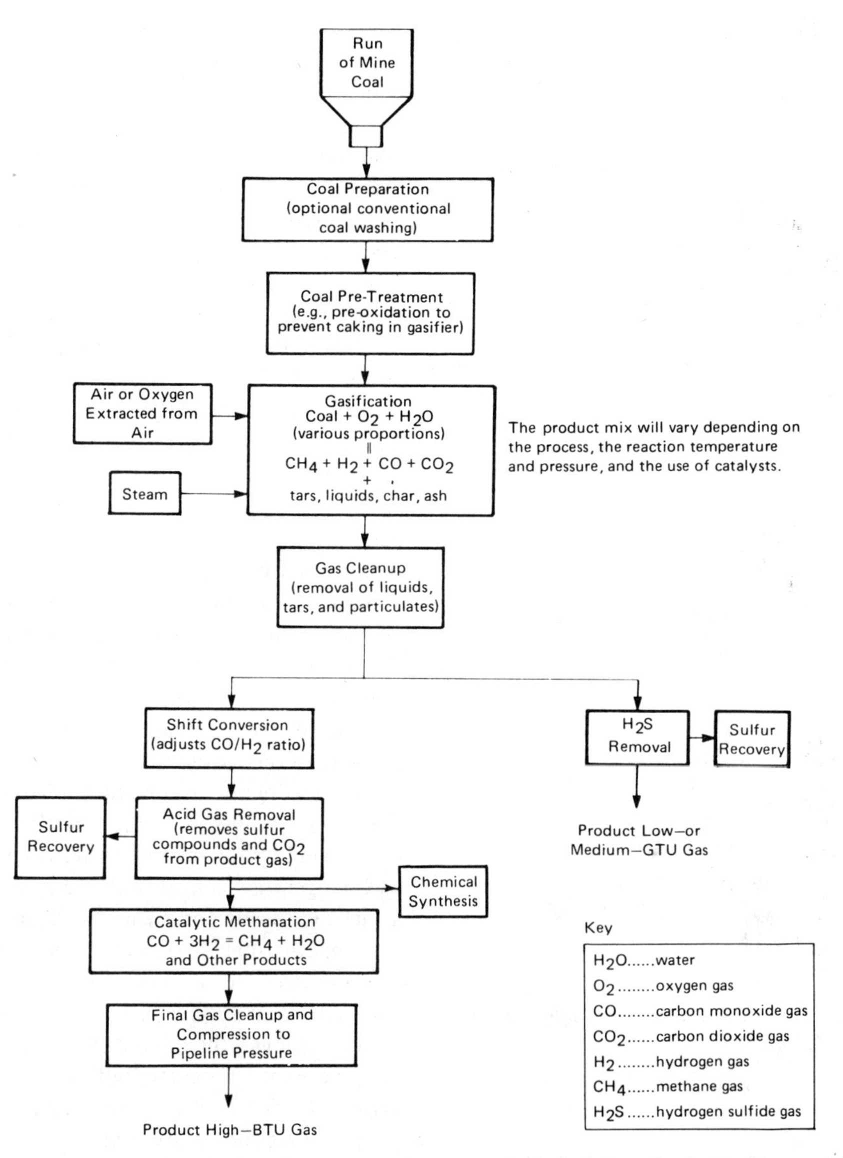

Fig. 3.16. Coal gasification. (*The President's Commission on Coal, Coal Data Book, Washington, D.C., Feb., 1980.*)

ported on a grate or by other means and the flow of gas and coal may be countercurrent or co-current. The former is the most common process. (Figure 3.17).

2. *Fluidized-Bed Gasification*. Crushed or fine coal is fluidized by the gasifying medium, giving an expanded fuel bed that can best be visualized as boiling or jiggling. (Figure 3.18).
3. *Suspension or Entrained-Flow Gasification*. Fine coal is suspended in the gasifying medium, the particles moving with the gas, either linearly or in a vortex pattern. (Figure 3.19).

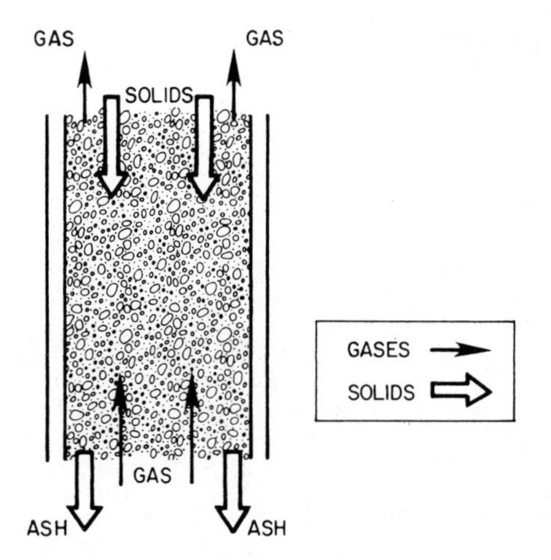

Fig. 3.17. Fixed-bed gasifier. *(Bodle, W. W. and Shora, F. C.*[11])

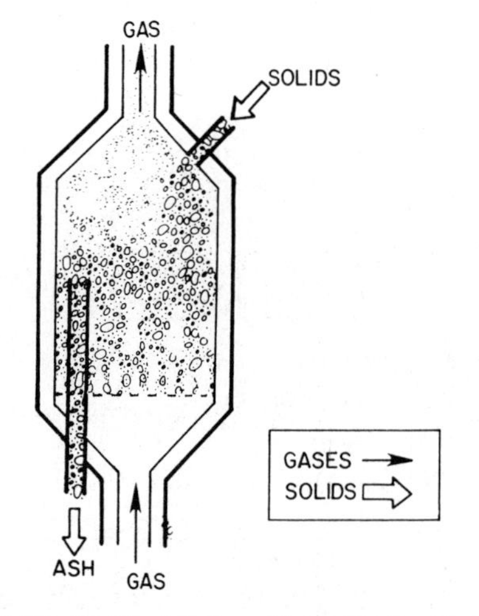

Fig. 3.18. Fluidized-bed gasifier. *(Bodle, W. W. and Schora, F. C.*[11])

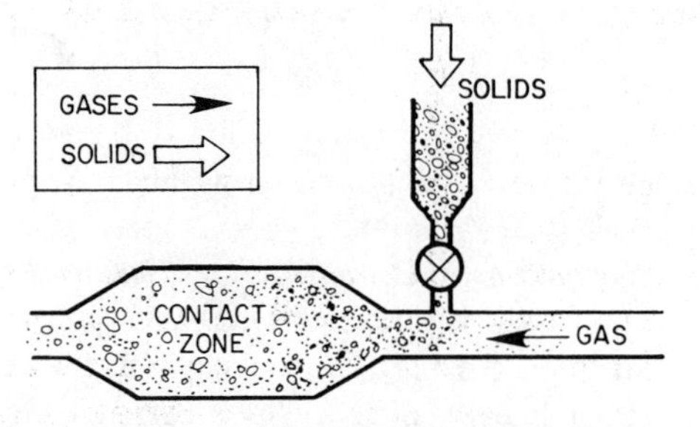

Fig. 3.19. Entrained-flow gasifier. *(Bodle, W. W. and Schora, F. C.*[11])

Rotary kilns and molten salt baths are also being considered as contactors for coal gasification.

Fixed-Bed Gasifiers. Fixed bed gasifiers consist of a descending bed of coal contacted countercurrently by a flow of gases moving upward through the interstices between the coal lumps. As coal moves downward through the bed it is subjected sequentially to drying, devolatilization, gasification, combustion, and ash cooling.

In the drying zone at the top of the bed, water is evaporated by contact of the coal with hot gases flowing from lower portions of the gasifier. As the dry coal descends further down the bed, its temperature rises until at about 600°F, devolatilization occurs, liberating tars, oils, and gases. The devolatilized coal, or char, is further heated as it descends through the bed, and is gasified by reaction with steam, carbon dioxide and hydrogen to produce a mixture of carbon monoxide, hydrogen, and methane, along with unreacted steam and carbon dioxide. The rates of these reactions and the final composition of the gases produced are strongly influenced by the gasifier temperatures and pressure of operation.

The maximum temperature reached within the gasifier occurs in the combustion zone and is regulated by the relative proportions of steam and oxygen (air) added to the vessel. If dry ash is desired, sufficient steam is added to offset the exothermic oxidation reactions with endothermic steam-carbon reactions and to maintain temperatures in the combustion zone safely below the ash-fusion temperature. Slagging gasifiers operate with correspondingly higher oxidant/steam ratios and maximum temperatures and remove ash in a molten state.

Fixed-bed gasifiers require sized coal for proper operation; typically coal between $\frac{1}{8}$ inch to 2 inches in diameter is used. The countercurrent flow of gases and solids leads to high thermal efficiencies with exit gas temperatures usually in the range of 500 to 1000°F. The raw product gas contains dust as well as the tars and oils produced during gasification, and these materials must be removed prior to downstream processing.

Fixed-bed gasifiers can be operated at

TABLE 3.18 Characteristics of Fixed-Bed Gasifiers

	OPERATING PRESSURE	*OXYGEN OR AIR*	*AGGLOMERATION PREVENTION*	*STATUS*
		DRY ASH, SINGLE STAGE		
GEGAS	TO 500 PSIG	AIR	STIRRER PADDLES	5-FT DIAMETER TEST UNIT
LURGI	TO 450 PSI	OXYGEN OR AIR	ROTATING BLADES	COMMERCIAL
MERC	TO 105 PSI	AIR	SPIRALING STIRRER	42-IN. DIAMETER TEST UNIT
RILEY-MORGAN	40 IN. H_2O	AIR	AGITATOR IN ROTATING BED	10.5-FT TEST UNIT
WELLMAN-GALUSHA	10 IN. H_2O	AIR	SPIRALING ARMS	COMMERCIAL
WILPUTTE	ATM	AIR	ROTATING ARM	COMMERCIAL
		DRY ASH, 2-STAGE		
ATC/WELLMAN (Incandescent)	ATM	AIR	NONE	COMMERCIAL
FW/STOIC	ATM		NONE	COMMERCIAL
RUHR-100	1500 PSI	OXYGEN	STIRRER BLADES	3-7 TON/HR TEST UNIT
WOODALL-DUCKHAM	40 IN. H_2O	AIR OR OXYGEN	NONE	COMMERCIAL
		SLAGGING		
BGC/LURGI	TO 400 PSI	OXYGEN	STIRRER	6-FT DIAMETER TEST UNIT
GFERC	TO 400 PSI	OXYGEN	STIRRER	16-IN. DIAMETER TEST UNIT

Source: Bodle, W. W. and Schora, F. C., "Coal Gasification Technology Overview," *Symposium Papers: Advances in Coal Utilization Technology*, Institute of Gas Technology, May 14–18, Louisville, Ky. pp. 11–34 (1979).

atmospheric pressure or can be pressurized: with either air or oxygen as the oxidant; with either dry ash or molten ash removal; with or without stirrers to prevent agglomeration; and either as single-stage or two-stage gasifiers. Two stage gasifiers have the advantage of enabling a tar-free gas to be removed, thereby reducing the size of the equipment required for tar and oil recovery. Tables 3.18, 3.19 and 3.20 summarize the characteristics and status of several fixed-bed gasifiers available commercially or under development. Figure 3.20 illustrates the mechanical configuration for the Lurgi fixed-bed gasifier. The unit is commercially available.

Fluidized-Bed Gasifiers. Fluidized-bed gasifiers consist of coal particles suspended in an upward flow of reactant gas. Reactant gases are introduced through a distributor at the bottom of the bed at velocities sufficient to suspend the

TABLE 3.19 Commercial Fixed-Bed Gasifiers

		UNITS	
	NUMBER OF INSTALLATIONS	*NUMBER*	*LARGEST CAPACITY, TONS/DAY COAL*
LURGI	16	102	800
WELLMAN-GALUSHA	270	550	100
WILPUTTE	20	47	60
ATC/WELLMAN (Incandescent)	20	30	103
FW/STOIC	3	3	108
WOODALL DUCKHAM	41	111	108
	370	843	

Source: Bodle, W. W. and Schora, F. C., "Coal Gasification Technology Overview," *Symposium Papers: Advances in Coal Utilization Technology*, Institute of Gas Technology, May 14–18, 1979, Louisville, Ky. pp. 11–34.

TABLE 3.20 Recent Commercial Gasification Installations

	OWNER	*LOCATION*	*YEAR OPERATION*
WELLMAN-GALUSHA	CAN DO INDUSTRIAL PARK	HAZELTON, PA	IN CONSTRUCTION
	DOUGLAS ENERGY CENTER	PIKEVILLE, KY	IN CONSTRUCTION
	HOWMET ALUMINUM CO.	LANCASTER, PA	IN CONSTRUCTION
	CHEMICAL EXCHANGE CO.	HOUSTON, TX	IN CONSTRUCTION
	U.S. BUREAU OF MINES	TWIN CITIES, MN	1978
	GLEN-GERY CORP.	YORK, PA	1977
FW/STOIC	UNIVERSITY OF MINNESOTA	DULUTH, MN	1978
ATC/WELLMAN	CATERPILLAR TRACTOR CO.	YORK, PA	1978

Source: Bodle, W. W. and Schora, F. C., "Coal Gasification Technology Overview," *Symposium Papers: Advances in Coal Utilization Technology*, Institute of Gas Technology, May 14-18, 1979, Louisville, Ky. pp. 11-34.

incoming particles, but at velocities less than those which would cause the particles to be blown out of the vessel. The result is a bed of highly mixed solids in intimate contact with the gas phase. The biolent agitation leads to very uniform temperatures throughout the bed and the smaller particle diameters and high degree of mixing lead to reaction rates which are generally higher than those encountered in fixed-bed gasifiers.

Single-stage, fluidized-bed gasifiers do not generally achieve the high thermal efficiencies associated with fixed-bed gasifiers. However, a higher degree of countercurrent operation can be achieved by operating such systems in stages, and several gasifiers presently under development employ this principle. A staged operation not infrequently simulates the major zones in a fixed-bed gasifier, with a low-temperature stage employed for drying and devolatilization, an

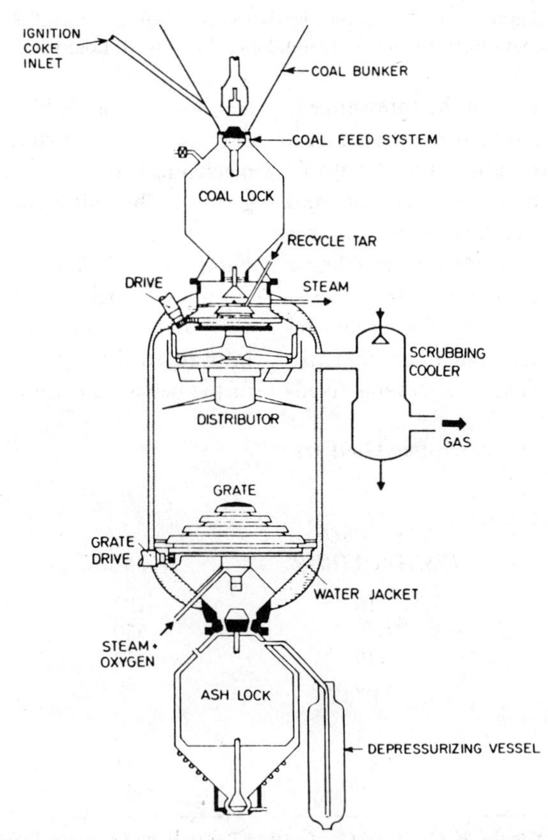

Fig. 3.20. Typical Lurgi gasifier. (*"Comparative Study of Coal Gasification Processes–Koppers–Totzek, Lurgi, and Winkler," Koppers Co., Pittsburgh, Pa.*)

intermediate-temperature stage used for gasification and a high-temperature stage to accomplish combustion. In some designs, each stage is a separate vessel, although it is not uncommon to combine several stages in a single vessel.

Exit gas temperatures for fluidized-bed gasifiers are usually higher than those for fixed-bed gasifiers, and as a result the product gas contains much lower levels of tars and oils produced during devolatilization. Particle attrition in the gasifier, however, generally leads to higher levels of dust carryover. Particles and ash are removed in fluidized-bed systems as dry solids, and coal pretreating or mechanical modifications to the gasifier are required to handle strongly caking coals.

Fluidized-bed gasifiers can be operated at atmospheric pressure or pressurized. Fluidizing gases can be mixtures of steam with either air or oxygen and, in some cases, mixtures of hydrogen with other gases. They may be operated either as single or as multiple-stage units. Generally, special provisions are required in order to handle strongly caking coals. Table 3.21 summarizes the characteristics of several fluidized-bed gasifiers presently under development, as well as of the Winkler gasifier which is commercially available. There are 70 Winkler gasifiers at 24 installations worldwide, and the largest unit processes 650 tons of coal per day. The Winkler gasifier is illustrated in Figure 3.21.

Entrained-Flow Gasifiers. Entrained-flow gasifiers consist of finely ground coal entrained in a flow of reactant gases. Contact between the solids and gases is cocurrent with little or no back-mixing occurring. Entrained-flow gasifiers may be either single-stage or two-stage.

In single-stage operation, high temperatures, 2200-3200°F, are used to completely gasify the coal in mixtures of steam and oxygen (air). Due to the high temperatures employed, gasification rates are considerably higher than those achieved with either fixed-bed or fluidized-bed gasifiers; however, heat economy is generally lower. The lack of back-mixing encountered with cocurrent flow and the rapid reaction rates enable these gasifiers to handle all coals, including those which are strongly caking, without pretreatment. The high temperatures of operation produce a gas devoid of both methane and volatile tars and oils. Dust loadings, though, are greater than those encountered with either of the two other gasifier types.

Two-stage, entrained-flow gasifiers are those in which the incoming coal is first entrained with reactant gases to produce gas; the resultant char is further gasified in a second stage, which may or may not be entrained. As is the case with fluidized-bed gasifiers, staged operation enables better overall thermal efficiencies to be achieved without sacrificing higher gasifier throughputs, as the more reactive incoming coal can be gasified at lower temperatures than the less reactive chars. Entrained-flow gasifiers can be operated at atmospheric pressure or pressurized, and ash may be removed either dry or molten.

Table 3.22 summarizes the characteristics of several entrained-flow gasifiers presently under development as well as those of the Koppers-Totzek gasifier which is commercially available. There are 53 Koppers-Totzek gasifiers in 31

TABLE 3.21 Characteristics of Fluidized-Bed Gasifiers

	DEVELOPER	*OXYGEN OR AIR*	*PRESSURE*
WINKLER	DAVY POWERGAS	OXYGEN OR AIR	ATM
RHEINBRAUN	RHEINISCHE BRAUNKOHLENWERKE	OXYGEN	150 PSI
CO_2 ACCEPTOR	CONOCO DEVELOPMENT	AIR	150 PSI
HYGAS	IGT/GRI/DOE	OXYGEN OR AIR	1200 PSI
SYNTHANE	DOE/PETC	OXYGEN	1000 PSI
WESTINGHOUSE	WESTINGHOUSE ELECTRIC/DOE	AIR	225 PSI
U-GAS	IGT/DOE	OXYGEN OR AIR	350 PSI
COGAS	COGAS DEVELOPMENT CO.	AIR	10 PSIG
CATALYTIC	EXXON	NONE	500 PSI

Source: Bodle, W. W. and Schora, F. C., "Coal Gasification Technology Overview," *Symposium Papers: Advances in Coal Utilization Technology*, Institute of Gas Technology, May 14-18, 1979, Louisville, Ky., pp. 11-34.

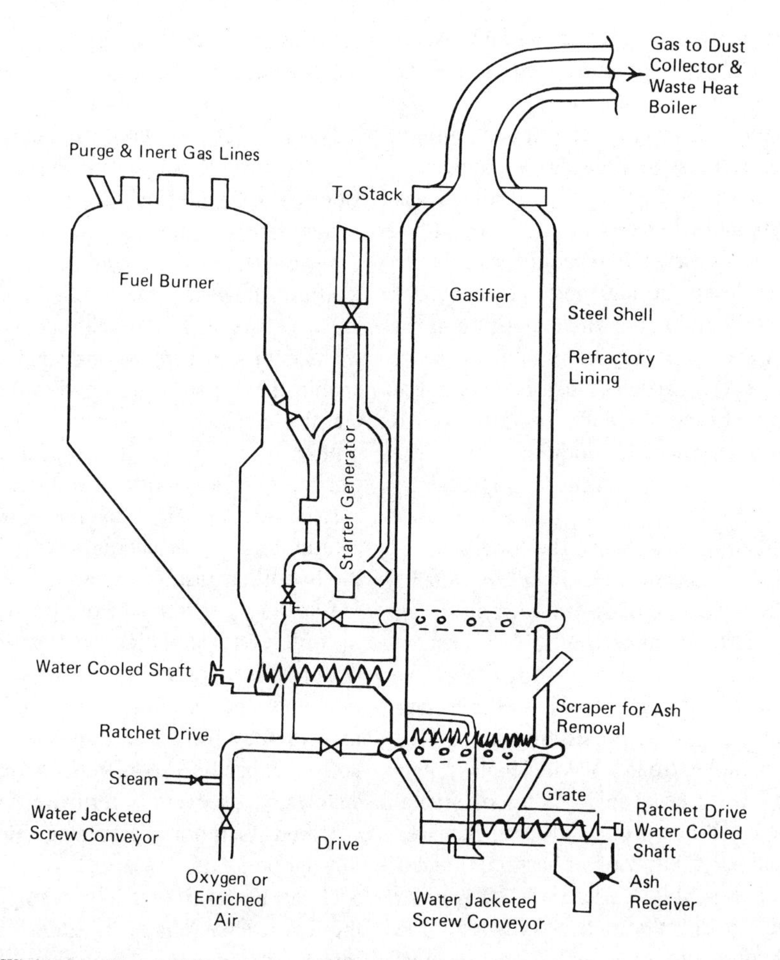

Fig. 3.21. A Winkler gasifier. ("*Comparative Study of Coal Gasification Processes–Koppers–Totzek, Lurgi, and Winkler,*" *Koppers Co., Pittsburgh, Pa.*)

installations worldwide; the largest unit processes 672 tons of coal per day.

In late 1980, Koppers Company, Inc. announced a new entrained-bed process designated as the *KBW Coal Gasification Process.* The process is based on the partial oxidation of carbonaceous material in the presence of oxygen and steam. In the case of coal feedstocks, carbon conversion is said to be a function only of the reactivity of the coal and approaches 100 percent for lignites. It is claimed that at the high operating temperatures, contaminants such as tars, phenols, cyanides, or ammonia are minimized or eliminated.

In the KBW process, (Figure 3.22), powdered coal (minus 200 mesh—"face powder" consistency) is entrained with steam and oxygen and, under slight pressure, is fed into the gasifier where gasification takes place at high temperatures. The gas stream leaves the gasifier at about 2800°F. The slagged ash drops into a water seal from which it is removed by a continuous scraper conveyor. The high temperature reportedly results in production of a gas mixture rich in the desirable constituents of CO and H_2, with no methane and only limited amounts of CO_2. Characteristically, coal yields a gas of 50-55 percent CO and 30-35 percent H_2, on a dry basis. The exit gas is cooled to about 400°F in a waste heat boiler and cleaned to remove particulate matter.

Molten Bath Gasifiers. Molten bath gasifiers are those in which the reactants, coal, steam, and oxygen (air) are introduced below the surface of a molten liquid and unreacted coal and ash are removed from the surface. Gasifiers employing molten salt, molten iron, and molten ash are presently under development. The molten iron system will operate at near atmo-

TABLE 3.22 Characteristics of Entrained-Flow Gasifiers

DEVELOPER	AIR OR OXYGEN: 1ST STAGE	AIR OR OXYGEN: 2ND STAGE	2ND STAGE CONTACTING	OPERATING PRESSURE	STATUS
		SINGLE-STAGE GASIFICATION			
BELL HMF	AIR			TO 225 PSI	0.5 TON/HR TEST UNIT
KOPPERS-TOTZEK	OXYGEN AND STEAM			10–12 PSIG	COMMERCIAL
KOPPERS KBW	OXYGEN AND STEAM			NOT DISCLOSED	PILOT PLANT
MOUNTAIN FUEL	OXYGEN AND STEAM			TO 150 PSIG	0.5 TON/DAY TEST UNIT
SHELL-KOPPERS	OXYGEN OR AIR			TO 450 PSI	6-TON/DAY PILOT PLANT
TEXACO	OXYGEN OR AIR			TO 1200 PSI	24-TON/DAY PILOT PLANT
		2-STAGE GASIFICATION			
BI-GAS	SYN GAS AND STEAM	OXYGEN AND STEAM	ENTRAINED	TO 1500 PSI	5-TON/HR PILOT PLANT
C-E	COMBUSTION GAS AND AIR	AIR	ENTRAINED	ATM	5-TON/DAY PDU
FOSTER WHEELER	SYN GAS AND STEAM	AIR AND STEAM	ENTRAINED	ATM	PILOT PLANT UNDER CONSTRUCTION
PEATGAS	SYNGAS	OXYGEN AND STEAM	FLUID BED	TO 500 PSI	5–10 LB/HR PDU
ROCKWELL INTERNATIONAL	HYDROGEN	NOT KNOWN	NOT KNOWN	TO 1500 PSI	0.25 TON/HR TEST UNIT

Source: Bodle, W. W. and Schora, F. C., "Coal Gasification Technology Overview," *Symposium Papers*: *Advances in Coal Utilization Technology*, Institute of Gas Technology, May 14–18, 1979, Louisville, Ky., pp. 11–34.

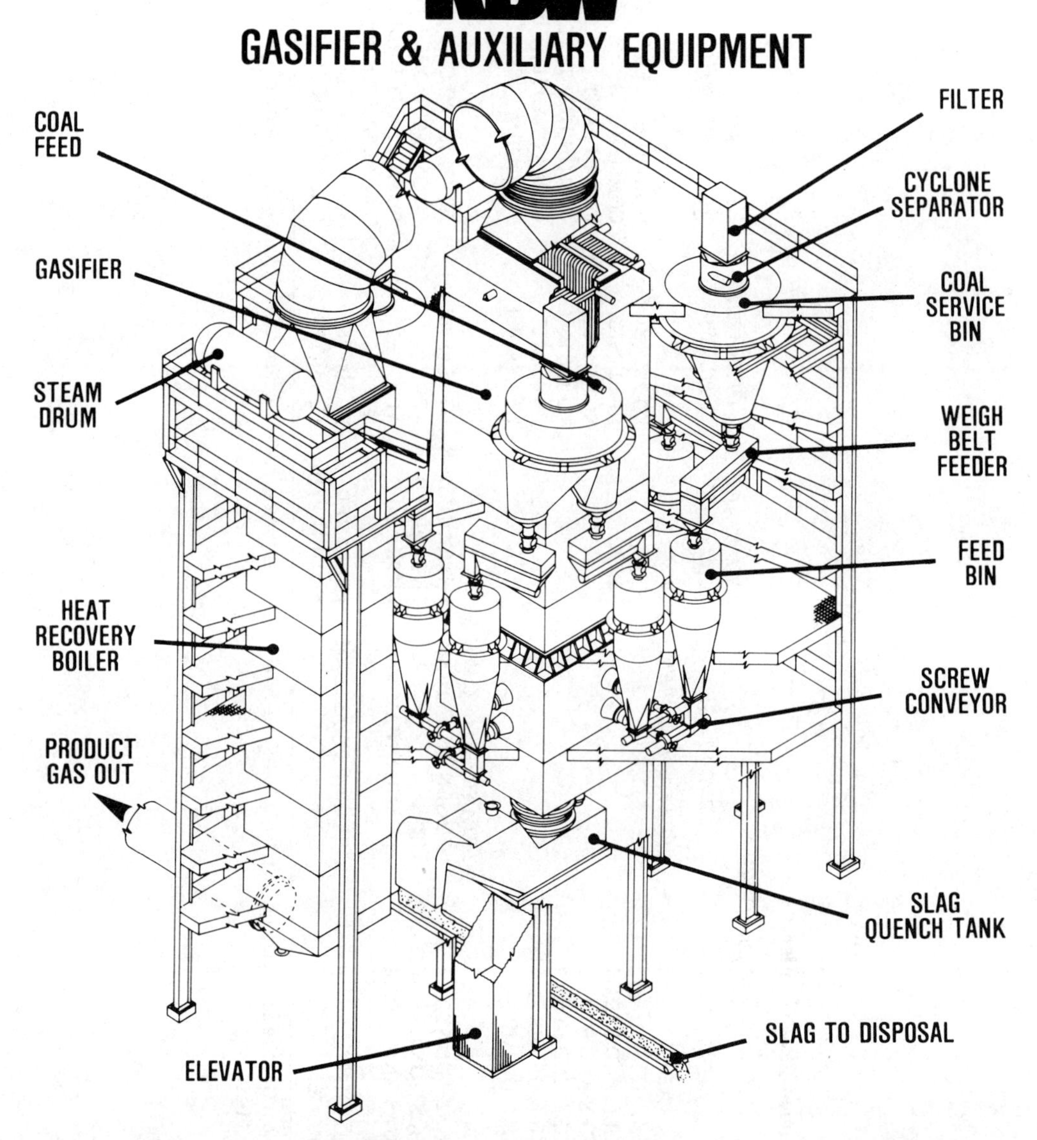

Fig. 3.22. Koppers Totzek gasifier. (*"Comparative Study of Coal Gasification Processes–Koppers–Totzek, Lurgi, and Winkler," Koppers Co., Pittsburgh, Pa.*)

spheric pressure while the others are being developed for high pressure (30-35 atm.) operation.

Rotary-Kiln Gasifiers. The rotary-kiln gasifier being developed by the Allis-Chalmers Corporation consists of a tumbling bed of solids moving countercurrently to the flow of hot gases. Reaction zones in the kiln parallel those in a fixed-bed gasifier with the coal sequentially dried, devolatilized, gasified, and combusted. Steam and air are injected into the kiln through ports located beneath the tumbling solids. While rotary kilns are commercially available, the gasifier is still in the development stage with a 600 ton/day demonstration plant being contemplated.

Low-Btu Gasification for Power Generation

Several studies have indicated that low-Btu gasifiers coupled with gas turbine/steam turbine combined cycles represent one of the most promising technologies for future coal-based power generation systems. These technologies offer the potential for high electrical conversion efficiencies (perhaps up to 50%) and economies of scale which can be achieved in sizes considerably below 1000 megawatts.

High reliability with low capital and operating costs, along with minimal environmental impacts, are prerequisites for gasification systems proposed for utility applications. Operating costs can be minimized with a gasifer which is capable of operating on run-of-mine coals, including caking coals as well as fines. The gasifier should achieve high thermal efficiencies and high carbon utilizations, and it should operate with air rather than oxygen to avoid the expenses of an oxygen plant. The unit should be capable of high coal throughputs, but should be constructed and operated in such a way that maintenance expenses are minimized.

Environmental impacts are minimized if the gasifier is designed and operated in such a way that tars are not produced, or, if they are produced, that they can be consumed in the system. Overall, the gasifier should use all of the water produced in the system, or be a net consumer of water so as to avoid liquid effluent disposal problems. Ash leaving the system should be in a form which is benign for disposal.

Table 3.23 summarizes the features generally characterizing fixed-bed, fluidized-bed, and entrained-bed gasifiers, as well as advantages and disadvantages inherent in each when being considered for gasification-combined-cycle power generation applications. No particular gasifier type is uniquely suited to these applications. To date, there is only one integrated gasification-combined-cycle plant in the world—an air-blown, pressurized Lurgi gasifier in Lunen, West Germany, on the STEAG utility system. In the United States, fixed-bed, fluidized-bed, entrained-bed, and rotary kiln gasifiers are being developed for combined-cycle applications. Each of these systems incorporates design features and special provisions which make it more versatile and adaptable to power generation needs.

General Electric Fixed-Bed Gasifier. General Electric, at its Corporate Research and Development Center in Schenectady, N.Y., is operating a one-ton-per-hour, air-blown, advanced fixed-bed gasifier. (See Figure 3.23.) The unit operates at 300 psig and is equipped with a mechanical agitator to enable caking coals to be gasified.

The fact that the volume of coal in the gasifier at any time is large compared to the steam and air injection rates allows considerable flexibility in handling feed interruptions. The pilot gasifier has run for as long as twenty minutes without coal feed, with the result that although the bed level was lowered but operability was maintained. Similarly, the gasifier has operated for as long as twenty seconds without steam addition and without slagging.

Westinghouse Fluidized-Bed Gasifier. Westinghouse, at its Advanced Coal Conversion Department in Madison, Pa., is operating a 15-ton-per-day air (or oxygen) blown, fluidizied-bed gasifier. A schematic of the unit is shown in Figure 3.24. Unlike most fluidized-bed gasifiers, the Westinghouse system can accept caking coals without pretreatment and is able to achieve high carbon utilization with very little carbon lost to the outgoing ash.

Operation of the gasifier has been described as follows:

> The fresh, unpretreated coal is fed to the gasifier along its center line, where it is combusted in a stream of oxygen (or air) through the central feed tube. Steam fed with the oxygen (or air) and in the grid zone of the gasifier reacts with the coal and char to form hydrogen and carbon monoxide. As the bed of char circulates through the jet, the carbon in the char is consumed by combustion and gasification, leaving particles that are rich in ash.
>
> The ash-rich particles contain mineral compounds and eutectics that melt at temperatures of 1000°F to 2000°F. These liquid phases within the char particles extrude through the pores to the surface of the char, where they stick to other liquid droplets on adjacent particles. In this way, ash agglomerates form that are larger and denser than the particles of char in the bed. The agglomerates defluidize, migrate to the annulus around the feed tube, and are continuously removed by

TABLE 3.23 Features of Fixed-Bed, Fluidized-Bed, and Entained-Bed Gasifiers

	FIXED BED	*FLUIDIZED BED*	*ENTRAINED BED*
PARAMETERS	• GRAVITATING BED OF COAL ASH • MECHANICAL GRATES/ DISTRIBUTORS • DISCRETE ZONES –PREHEATING-DRYING DEVOLATIZATION –GASIFICATION –COMBUSTION • TEMPERATURE GRADIENT • SLOW PROCESS RESPONSE	• FLUIDIZED BEDS ARRANGED IN ONE OR MORE ZONES • UNIFORM TEMPERATURE AND COMPOSITIONS THROUGHOUT EACH FLUIDIZED ZONE • MODERATE PROCESS RESPONSE	• UP FLOW OR DOWN FLOW SUSPENSION GASIFICATIONS • HIGH TEMPERATURE–HIGH RATE PROCESS • FAST PROCESS RESPONSE
ADVANTAGES	• HIGH CARBON CONVERSION EFFICIENCY • LOW ASH CARRYOVER • LOW TEMPERATURE OPERATION • LOWEST AIR/OXYGEN REQUIREMENT	• HIGH DEGREE OF PROCESS UNIFORMITY • EXCELLENT SOLIDS/GAS CONTACT • LOWER RESIDENCE TIME THAN FIXED BED GASIFIER • HIGHER COAL THROUGHPUT PER UNIT VOLUME OF REACTOR	• HANDLES ALL TYPES OF COAL–NO PRETREATMENT • LOW STEAM CONSUMPTION • EXCELLENT SOLIDS/GAS CONTACT • NO TAR FORMATION • NO PHENOL FORMATION • ABILITY TO SLAG ASH • HIGH CAPACITY PER UNIT VOLUME OF REACTOR • PRODUCES INERT SLAGGED ASH
DISADVANTAGES	• SIZED COAL REQUIRED • COAL FINES MUST BE BRIQUETTED • LOW CAPACITY • LOW OFFGAS TEMPERATURE	• SIZED COAL REQUIRED • DRY COAL REQUIRED FOR FEEDING • REQUIRES COMPLICATED GAS DISTRIBUTOR	• REQUIRES FINELY CRUSHED COAL 70% < 200 MESH • SMALL SURGE CAPACITY REQUIRING CLOSE CONTROL

• PRODUCES TARS AND HEAVIER HYDROCARBONS • HIGH STEAM CONSUMPTION • PRODUCES PHENOLS • USE OF CAKING COALS NOT COMMERCIALLY PROVEN	• CAKING COALS REQUIRE PRETREATMENT • HIGH CARBON LOSS WITH ASH • FLUIDIZATION REQUIREMENT SENSITIVE TO FUEL CHARACTERISTICS

Source: Miller C. L., "The U.S. Coal Gasification Program: Progress and Projects," *Mech. Eng. Journal*, August, 1980, pp, 34–40.

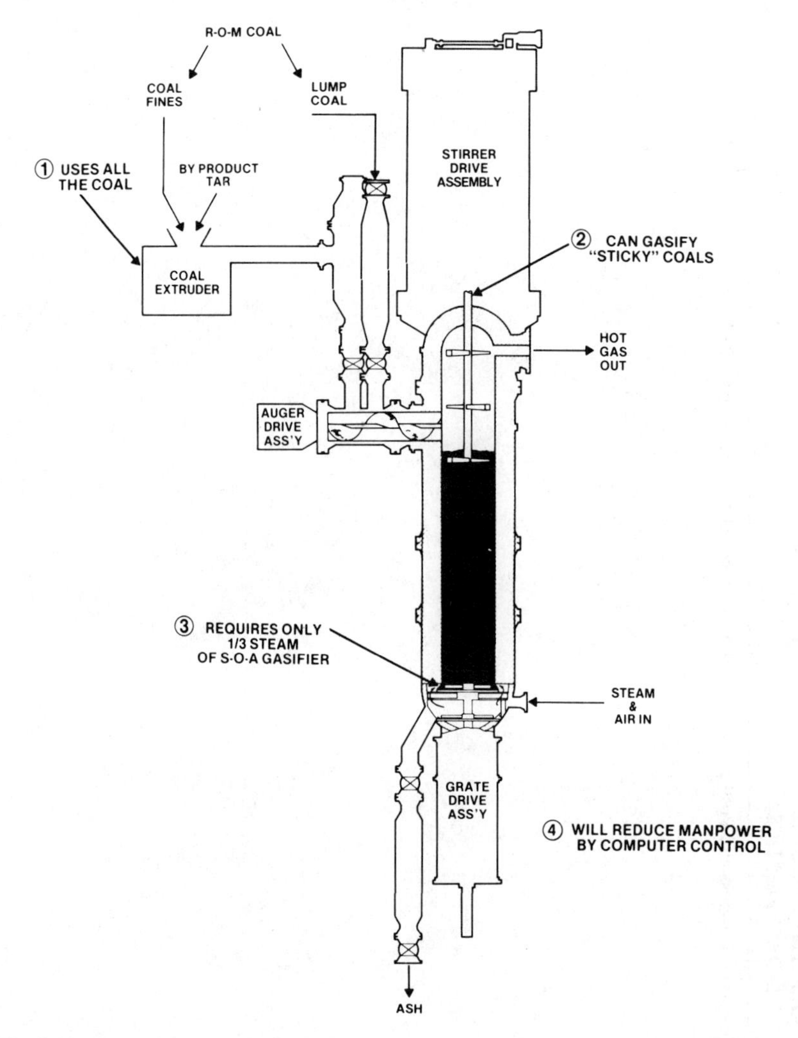

Fig. 3.23. General Electric GEGAS–D advanced gasifier. (*Courtesy General Electric Co.*)

a rotary feeder to lock hoppers. Recycled product gas or steam is used to partially fluidize the ash and cool it as it is withdrawn.

The raw product gas, containing methane, hydrogen, carbon monoxide, carbon dioxide, and gaseous impurities (but no tars or other hydrocarbons) exits the reactor at 1800°F. A refractory-lined cyclone is used to remove char particles from the raw gas before it is quench-cooled in a quench scrubber that also removes most of the remaining particulate matter. The char fines collected in the cyclone are pneumatically transported to lock hoppers from which they are reinjected into the gasifier along with the fresh coal. All of the fines collected and recycled are consumed by the combustion, gasification, and agglomeration processes within the reactor. A flow chart for this process is shown in Figure 3.25.

Combustion Engineering Entrained-Bed Gasifier. A conceptual diagram of a 100 ton/hour air-blown entrained-bed gasifier based on a process developed by Combustion Engineering is shown in Figure 3.26.[24] In this system coal is pulverized and transported pneumatically to the gasifier. There, it is partially combusted in a deficiency of air, and converted to a product gas stream containing char and H_2S. Coal ash is fused and tapped from the bottom of the gasifier as molten slag. Product gas leaving the top of the gasifier is cooled and ducted to a particulate removal system, where the char is removed and cycled back to the gasifier for

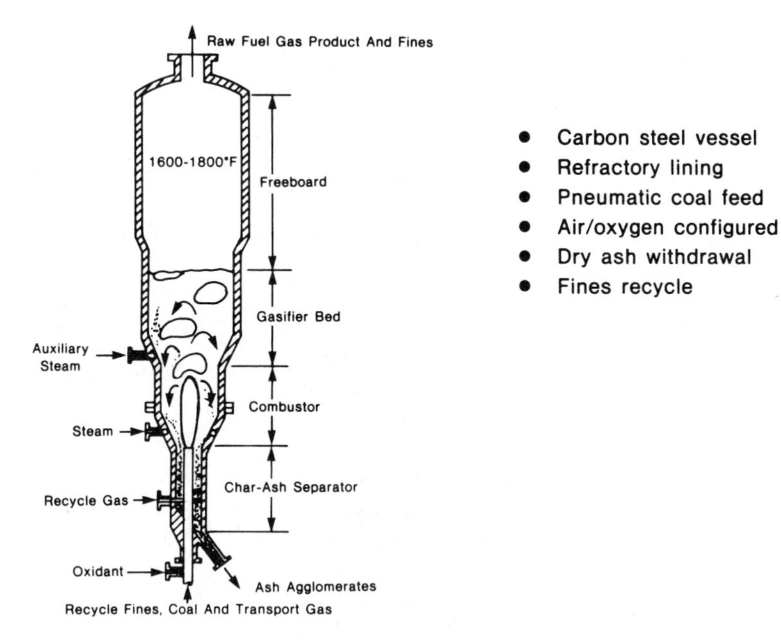

Fig. 3.24. Westinghouse pressurized fluidized-bed gasifier. (*Courtesy Westinghouse Electric Corporation.*)

consumption. The cooled char-free gas then moves to a Stretford absorber, where the H_2S is removed and converted to elemental sulfur. The cleaned gas stream, with a heating value of about 110 Btu/scf is suitable for fueling steam generators, gas turbines, or industrial process equipment–without the need for stack gas cleanup systems.

The gasifier itself consists of two stages, a combustor and reductor. About one-third of total gasifier coal is fed to the combustor, where the coal is burned with air nearly to completion to provide gas temperatures of about 3,200°F. This hot gas stream passes upward to the inlet of the reductor zone, where the remaining two-thirds of the total coal is

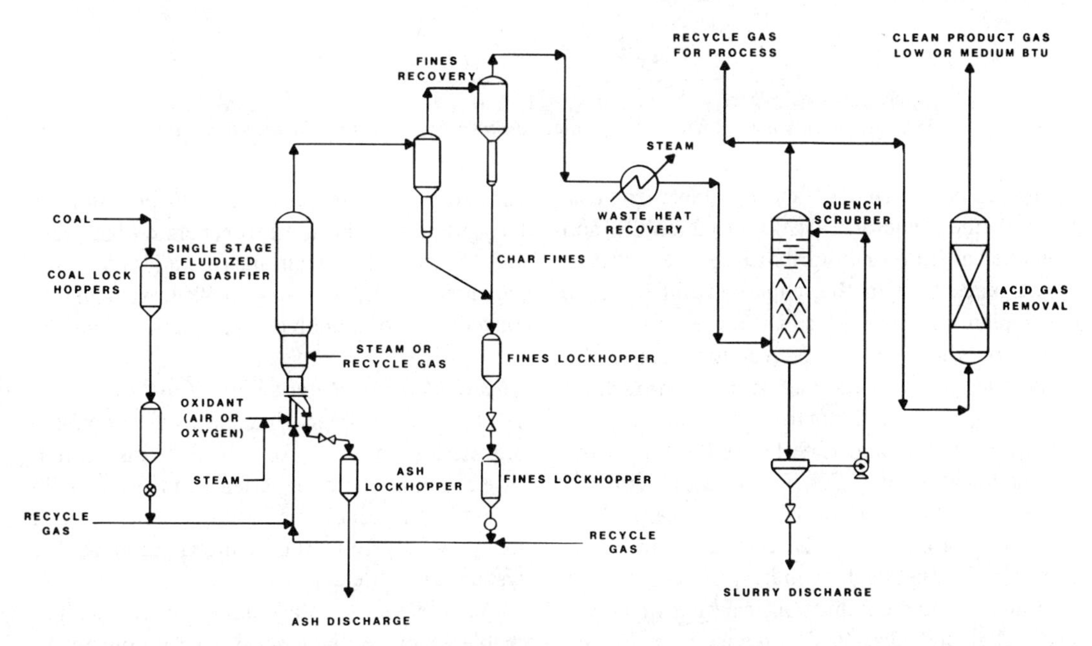

Fig. 3.25. Westinghouse pressurized fluidized-bed gasification system. (*Courtesy Westinghouse Electric Corporation.*)

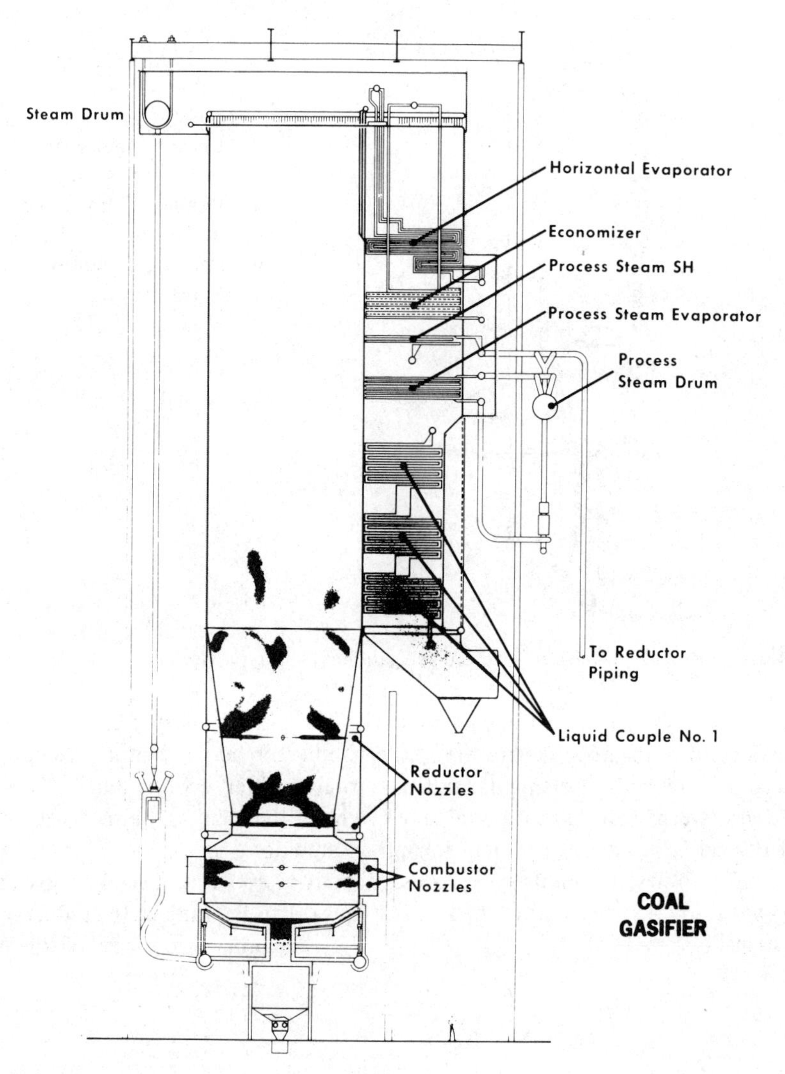

Fig. 3.26. Combustion Engineering gasifier. (*Courtesy Combustion Engineering, Inc.*)

injected with a large deficiency of air. The coal fed to the reductor is devolatilized rapidly, and gasified as it flows upward through the reductor. Residence time in the reductor is sufficient to complete the desired gasification reactions.

Since the gasification reactions are endothermic, gas temperatures decrease throughout the height of the reductor. The product gas, containing char and H_2S, leaves the top of the reductor at about 1,700°F, above the temperatures at which tars or oils are formed. Char removed from the gas is recycled to the combustor. The high combustor temperature promotes char combustion, slagging of ash in the coal and char, and provides heat for the endothermic reactions in the reductor.

In commercial applications, the vessel heat absorption may be used to generate high pressure steam. This steam may be added to that generated in the gas-using facility, particularly conventional or combined-cycle power plants.

Allis-Chalmers Rotary-Kiln Gasifier. Since 1971, Allis-Chalmers has been developing a proprietary process for low-Btu gasification based upon their experience in rotary-kiln technology. Since the gasifier was developed solely with private funds, information on the system is proprietary.

The use of a rotary kiln for gasification enables a proven large-scale reactor concept to be employed, thereby reducing the risk associated with scale-up. The rotary bed provides the ability to process caking coals without pretreatment and permits the unit to process fines. The large coal inventory in the system provides for safety and variable load response, leading to high turndown ratios (10:1).

The unit will handle coal sized 4″ X 0″ and is capable of operating at moderate pressures (50–100 psig). Allis-Chalmers is planning to build a gasification plant to supply gas to Illinois Power's 50 Mw Wood River Station No. 3. Construction of the system is expected to be completed by 1983.

Retrofitting Burners for Low-Btu Gas. Before the widespread use of natural gas and the enactment of environmental regulations, several types of fuel gases were produced from coal and used for firing a wide variety of industrial and heating processes. However, in the past 35 years industry has relied on the use of clean-burning natural gas as a primary gaseous fuel. Today, natural gas supplies to industry may be interrupted and many industries are considering converting to fuel gas produced from coal. Because existing burners are designed to burn natural gas, a significant factor in the cost of converting a particular process to coal gas is the extent to which existing burner design must be altered or retrofitted for the new fuel.

To evaluate the convertibility of the burners, it is necessary to know certain properties of the fuel gas being considered, such as those shown in Table 3.24. Important characteristics include:

- Adiabatic flame temperature
- Flame length
- Flame stability
- Flame emissivity

For industrial purposes, gases containing hydrocarbons may be burned by premixing them with a certain amount of air. In the design or redesign of such a burner, the first consideration is the volume of gas required and the next is the flame characteristic. If a sharp, hot flame is necessary, then a high primary air to gas ratio is necessary. The minimum and maximum gas pressures available from a given process are also important.

A burner consists of three major parts:

1. the orifice used for measuring;
2. an injection tube used for mixing;
3. the parts from which the air-gas mixture is discharged.

In changing from one fuel to another, the orifice size would have to be enlarged, for example, if going from natural gas to low-Btu gas. This sizing of the orifice may be accomplished using the standard sharp-edged orificed equation from any handbook. This will determine the volume of fuel to the combustion process. The air-gas ratio which is determined in relation to pressure drop in the injection tubes will fix the primary air-gas ratio. If this air-gas ratio is too low, the flame length will increase and impingement may result in carbon monoxide production. A suitable practice for manufactured gas is to use as primary air thirty to forty percent of the total air required for combustion, with some flexibility being provided in the burner to handle changes in composition. If the primary air-gas ratio is too high, the probability of blow-offs and strike backs increases.

In the burner, the port area, spacing, and distribution of secondary air are important design parameters. The individual port design is

TABLE 3.24 Typical Fuel Gas Properties Composition, Vol. %

Gasifier	N_2	CO_2	CO	H_2	CH_4	C_3H_8	*HHV Btu/scf*	*Adiabatic Flame Temperature °F*
Lurgi (O_2)	1	29	19	40	9	0	285	3156
Winkler (O_2)	1	20	32	41	3	0	269	3327
Koppers-Totzek (O_2)	1	9	52	34	0	0	284	3578
Wellman-Galusha (Air)	46	7	26	14	3	0	160	2948
Natural Gas	0.6	0.9	0	0	91.5	1.3	1066	3358
Propane	0	0	0	0	0	98.6	2521	–

Source: Institute for Mining and Minerals Research, University of Kentucky.

a factor in flame stability. Energy fluxes above 20,000 Btu/sq. in. may result in loss of the flame due to blow-off. For more details, the U.S. Bureau of Standards should be consulted.

PETROCHEMICAL FEEDSTOCKS

Feedstocks for the petrochemical industries are now produced mainly from crude oil and natural gas. About 90% of all organic chemicals are manufactured from only six feedstocks: synthesis gas, ethylene, propylene, butadiene, benzene, and p-xylene.[17] Synthesis gas accounts for 56% of these six feedstocks. Synthesis gas, a term used to describe various mixtures of carbon monoxide and hydrogen, is now produced by steam reforming of natural gas or light oils and is used in the manufacture of ammonia, methanol and other chemicals. However, coal can also be gasified to produce a synthesis gas.

There are many references available which discuss and evaluate processes for the production of chemicals from coal.[18,19,20,21] A typical flow sheet for a plant, taken from O'Hara,[18] is shown in Figure 3.27. This plant uses hydroliquefaction, pyrolysis and gasification and produces a product slate as shown in Figure 3.28. The acronym POGO refers to Power Oil Gas Other, a project recently completed by the Ralph M. Parsons, Co. for the U.S. Department of Energy.

The future of coal as a chemical feedstock depends on the relative economics of synthesis gas production from coal versus production from natural gas and crude oil. Some projections can be found which show coal to be competitive in 1980, while others show that coal will not be competitive until after the end of this century. Even though the future oil and natural gas situation is not clear, some companies are proceeding to build plants which use coal as feedstock. Tennessee Eastman,[22] for example, plans to use 6,000 tons of coal per day to manufacturer acetic anhydride by 1983.

Indirect Coal Liquefaction

In indirect coal liquefaction, a synthesis gas—primarily a mixture of carbon monoxide and hydrogen—is first produced in a suitable gasifier by reactions with oxygen and steam. This gas is then treated to remove sulfur compounds, and the ratio of CO to H_2 is fixed by means of the watergas shift reaction. The purified and composition-adjusted synthesis gas is then reacted to produce the desired liquid products. Three near term applications are now under consideration—synthesis of methanol, catalytic conversion of methanol to gasoline, and liquid hydrocarbons via Fischer-Tropsch synthesis.

Fischer-Tropsch. Fischer-Tropsch synthesis is the best known indirect liquefaction method. It was developed in 1925 by the German chemists Franz Fischer and Hans Tropsch. In this process, synthesis gas is reacted over an iron or cobalt catalyst at 350 psi and 330°C to produce largely aliphatic hydrocarbons, including gasoline, jet fuel, diesel oil, middle distilates, heavy oil, and waxes. The Germans built several plants utilizing this technology during World War II to provide nearly 15,000 barrels per day of military fuels. During the 1950's a 50 barrels-per-day pilot plant was operated by the U.S. Bureau of Mines in Louisiana, Missouri. A 7,000-barrels-per-day commercial plant, using a natural gas feedstock, was also operated in the 1950's in Brownsville, Texas.

The South Africans have applied the Fischer-Tropsch technology in their SASOL facility. The Sasol (South African Coal, Oil and Gas Corporation) installation is the largest synfuel plant based on coal operating anywhere. It converts an inferior high-ash (35 percent) coal into a broad spectrum of products. These may include ethylene, propylene, butylene, gasoline, and long straight-chain alcohols and hydrocarbons. The Sasol plant has been operating for about 25 years. During that time efficiency has increased and much has been learned about control of products. Sasol is a profitable operation, while selling gasoline at the refinery gate for about 50¢ a gallon (14¢ a liter) in 1979.

South Africa has substantial deposits of coal but no oil, and it obtains 75 percent of its energy from coal. The plant at Sasolburg is located on a huge coal field. Sasol I, the existing synthetics plant, is part of a versatile chemical complex that supplies much of South Africa's needs for materials such as nitrogen fertilizers, plastics, and pipeline gas. Consump-

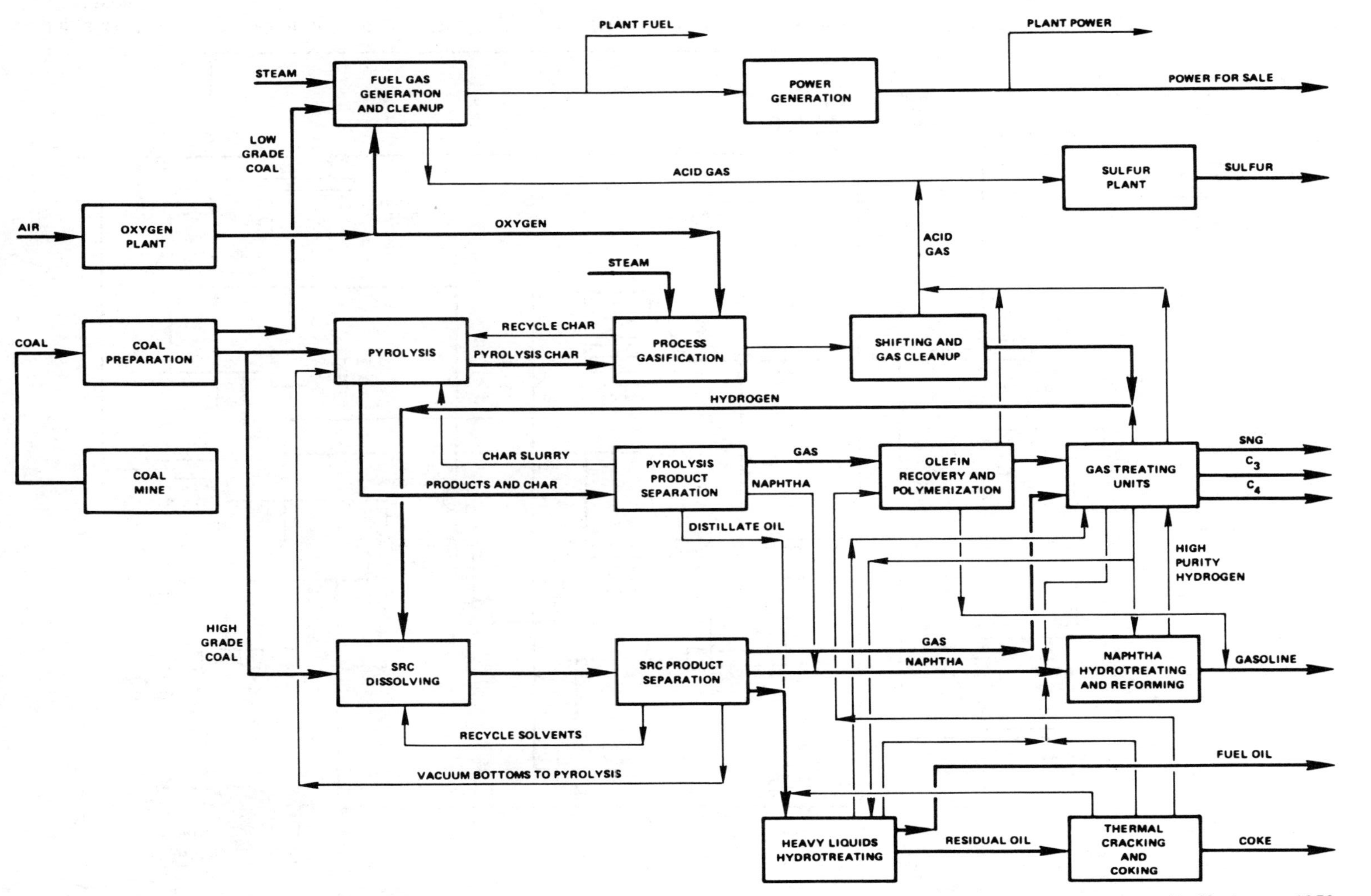

Fig. 3.27. Simplified block diagram of the POGO complex. *(O'Hara, J. B., et al, "Project POGO-A Coal Refinery," Chem. Engr. Prog., pp 49–63, August, 1978. By permission.)*

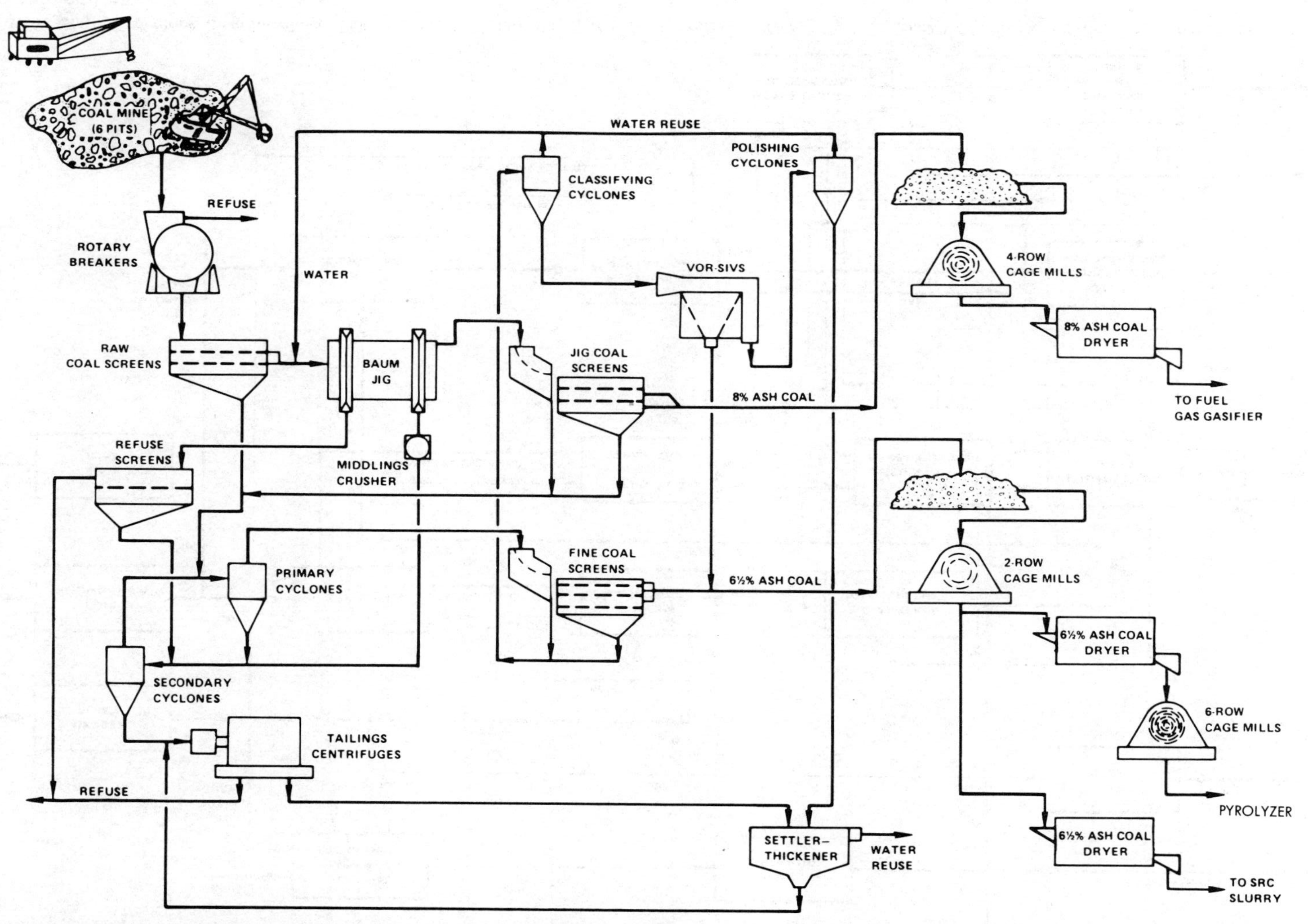

Fig. 3.28. Projected POGO plant yields. *(O'Hara, J. B. et al, "Project POGO–A Coal Refinery," Chem. Engr. Prog., pp. 49–63, August, 1978. By permission.)*

tion of coal at Sasol is 5 million tons a year. Coal is gasified in Lurgi reactors with steam and oxygen at total pressures of about 20 atmospheres. Principal products are H_2, CO, CO_2, and CH_4. The key components are H_2 and CO. Their ratio can be adjusted by varying the amounts of H_2O and O_2 used in the gasification. Constituents present in addition to the major gaseous products include NH_3, H_2S, and other sulfur-containing substances. To purify the raw products the gas is cooled, condensing out water and phenolic-type substances, and the remaining gas is passed through three absorption trains employing very cold methanol. This removes essentially everything except H_2, CO, and CH_4. Sulfur abundance in the purified gas is less than 1×10^{-8}. This guarantees relative freedom from sulfide poisoning of catalysts and a low sulfur content in the final products. A flow sheet is shown in Figure 3.29.

In October 1973, a large-scale plant, SASOL II, designed principally to produce gasoline, was authorized by South Africa. This plant should begin operating in 1984. In 1978, another major synthetic fuel plant was started. These two new plants are designed for zero emissions of liquid waste and will produce over 100,000 barrels of coal-derived liquids per day. The total cost of these projects is over $7 billion, a huge investment for such a small country.

Methanol. Methanol is a water-soluble, low-molecular-weight alcohol which may be of increasing importance as a low-sulfur fuel, a chemical feedstock, and perhaps as an intermediate for the production of gasoline. The synthesis of methanol is accomplished by the catalytic conversion of synthesis gas containing two moles of hydrogen for each mole of carbon monoxide. Methanol synthesis is widely practiced in industry on a commercial scale and will not be discussed further here.

Methanol to Gasoline–The Mobile Process. Mobil Research and Development Corporation, under a contract with the U.S. Department of Energy, has developed a process which catalytically dehydrates and polymerizes methanol to produce a high octane unleaded gasoline. The catalyst is one of a new family of synthetic zeolites designated ZSM-5 by Mobil. These new zeolites have a unique channel structure, different from previously known wide-pore (9–10 Angstrom units in diameter) and other narrow-pore (~5 Angstrom units) zeolites. One of these new zeolites converts methanol into a mixture of hydrocarbons corresponding to high quality gasoline. The new catalyst is said to have the ability to operate for long periods of time with negligible coke formation. The mass balance shows 0.44 lb. of gasoline and 0.56 lb. of water from each pound of methanol and the heating value of the gasoline produced is 95% of that in the methanol feed. A schematic diagram of Mobil's four-barrel-per-day pilot plant is shown in Figure 3.30. Crude methanol is vaporized at the bottom of the reactor, and passes through the dense fluidized catalyst bed at 775°F and 25 psig. The methanol is converted to hydrocarbons and water. The catalyst is removed from reaction products in a disengager section at the top of the reactor, the reactor effluent is condensed, and the water and hydrocarbon products are separated. To make additional gasoline, propene and butenes can be alkylated with isobutane by conventional petroleum technology. The reactor is 25 feet long and four inches in diameter.

Portions of thc catalyst powder are periodically removed from the reactor and regenerated with air. The regenerated catalyst is transferred back to the reactor. Small amounts of carbon monoxide, carbon dioxide, and coke are formed as by-products.

Table 3.25 shows a comparison of two coal-to-gasoline plants under consideration for construction in the Western U.S. One uses the Mobile methanol conversion process (M-Gasoline) and the other uses the Fischer-Tropsch synthesis. The gasoline yield is substantially higher in the M-gasoline process and the overall process thermal efficiency, based on higher heating values, is four percentage points higher.

Synthetic Pipeline Gas

Synthesis gas, mostly carbon monoxide and hydrogen, has a heating value which is much too low for direct injection into the natural gas network. Methane may be produced from synthesis gas by adjusting the ratio of carbon monoxide to hydrogen and then causing the

Sasol's Two Main Coal-Into-Oil Processes

Coal gasification and purification flow diagram

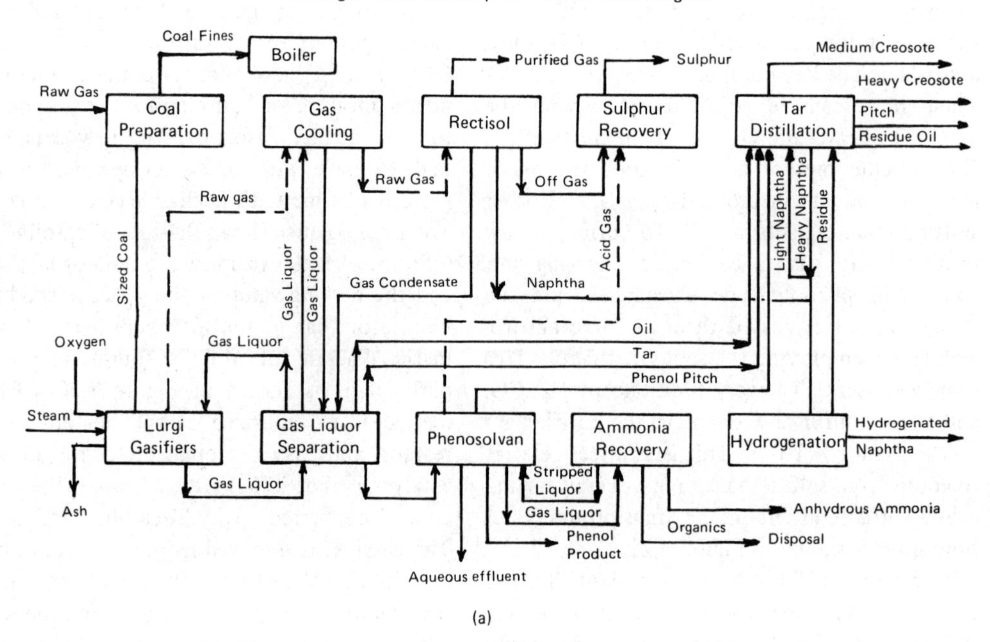

Fischer-Tropsch Synthesis Flow Diagram

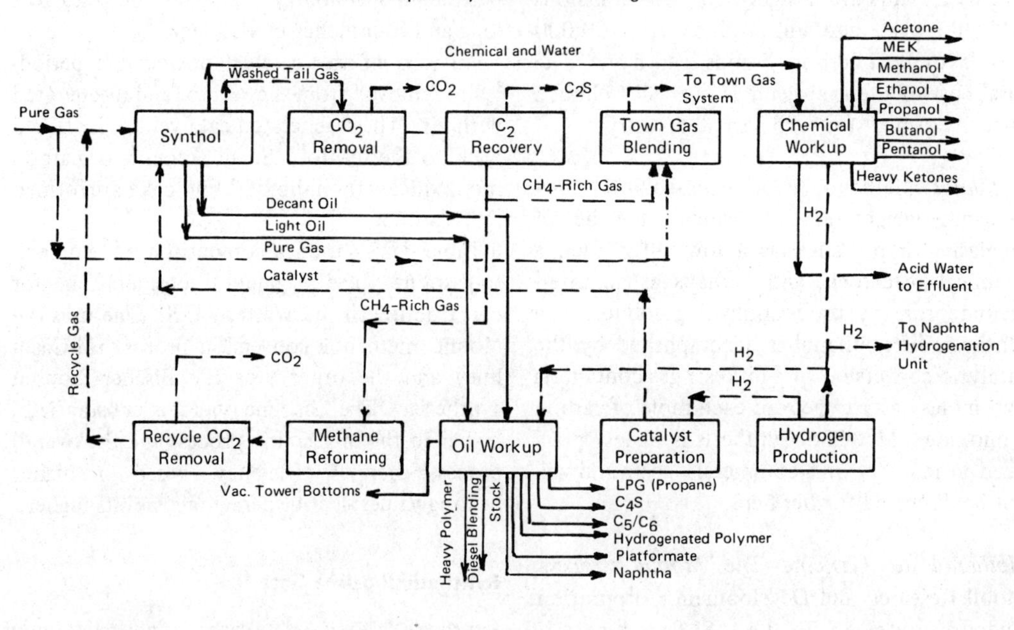

Fig. 3.29. Sasol's two main coal-into-oil processes. (*Encyclopedia of Chemical Processing Design, pp. 310, 320, Marcel Dekker, Inc. 1979. By permission.*)

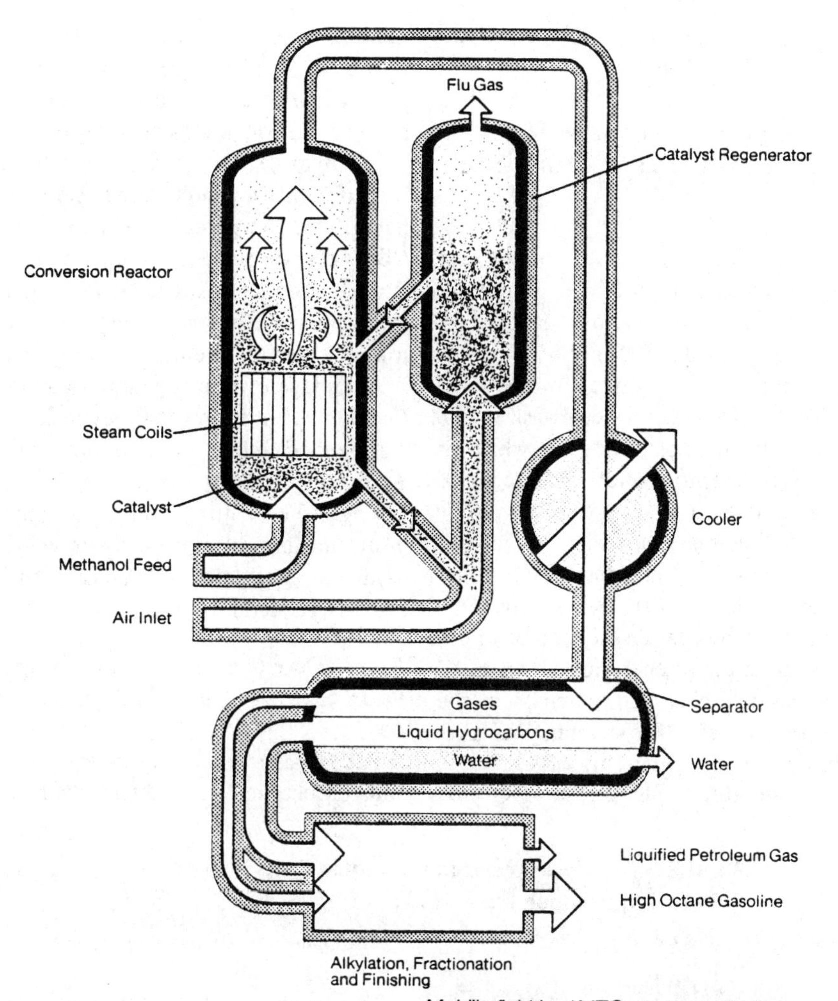

Fig. 3.30. Mobil's *fluid-bed* MTG process, which requires additional scale-up studies, uses a unique zeolite catalyst in converting methanol directly to high octane, unleaded gasoline. (*Courtesy the Pace Company, Denver, Colorado.*)

mixture to undergo a highly exothermic catalytic reaction.

Production of synthetic pipeline gas from coal gasified in a Lurgi gasifier has been studied extensively. In such a plant, the gases from the gasifier are cleaned to remove dust, tars, oils, and sulfur-bearing materials. The resulting mixture is shifted to the appropriate carbon dioxide–hydrogen ratio and the methanation reaction then produces methane, carbon dioxide, and water. After purification and compression, the methane is ready to be added to the natural gas pipeline. Such a plant is now under construction near Beulah, North Dakato, by American Natural Resources Company. Called the Great Plains Gasification Associates Coal Gasification Project, it is designed to produce 125×10^6 SCFD of pipeline quality gas from North Dakota lignite, utilizing Lurgi technology. The project is sponsored by five of the largest pipeline companies in the U.S.

A number of other pipeline quality gas processes are, or have been, under development, including the following:

CO_2 Acceptor	Bigas	COED/COGAS
Hygas	Synthane	

These processes are described in detail elsewhere[13,14,16] and will be discussed only briefly here.

CO_2 *Acceptor*. The CO_2 Acceptor process was designed to accomplish two important functions within the same reactor: coal gasification and carbon dioxide (CO_2) removal. The commercial process uses three fluidized-bed reactors. The gasifier achieves coal gasification and the combination of the CO_2 which is formed with an acceptor solid. The acceptor solid is regenerated in a second vessel. The acceptor is the lime material which reacts with CO_2 in the reactor and incidentally provides reaction heat which contributes to the heat input required for the coal gasification reaction. Heat for the acceptor regeneration is generated by burning residual char withdrawn from the top of the gasifier bed. The combustion takes place within the acceptor regenerator.

Lignite and subbituminous coals are preferred feeds because of their high reactivity. With these, the gasification temperature is sufficiently low to avoid ash fusion and consequent solid deposition and particle agglomeration.

The hydrogen-rich gas mixture from the gasifier is purified, methanated, compressed, and dehydrated to produce pipeline quality gas. The CO_2 Acceptor process is unique among the high-Btu gas processes as it does not require a separate acid gas removal system. It also has the lowest water requirement. It has the particular disadvantage of poor operability on feeds other than lignites or subbituminous coals.

A second mode of operation for the CO_2 Acceptor process produces syngas by not going through the methanation step. Syngas is a useful medium-Btu intermediate which may be used to synthesize methanol, other hydrocarbons, or alcohols.

Hygas. Development of the Hygas process began in 1954. Pilot plant operation began in 1972.

The Hygas reactor is a single vessel containing four separate, vertically-stacked contacting

TABLE 3.25 Coal Conversion to Liquid Fuels by Two Indirect Liquefaction Processes

Input	*M-Gasoline*	*Fischer-Tropsch*
Coal (Millions Tons/Day)	27.3	27.8
Water (Gallons/Minute)	6,300	6,600
Product		
Gasoline (Barrels/Day)	22,045	13,580
Synthetic Natural Gas, (Million Cubic Ft/Day)	148.5	173.3
LPG (C_3) (Barrels/Day)	2,555	1,107
Butanes (Barrels/Day)	2,205	146
Diesel Fuel (Barrels/Day)	–	2,307
Fuel Oil (Barrels/Day)	–	622
Alcohol (Thousand Pounds/Day)	–	510
FOE* (Barrels/Day)	45,550	44,950
Other		
Power, MW Electrical	5.31	3.31
Coal Fines (Thousand Tons/Day)	1.2	0
Ammonia (Tons/Day)	103	103
Sulfur (Tons/Day)	61	61
Efficiency (HHV), %	62	58

*All products converted to fuel oil equivalent 6.0 MM Btu per barrel.
Source: U.S. Department of Energy.

zones. Coal is pumped into the reactor as a slurry containing up to 50 weight percent solids in recycle oil. The slurry is dried in the top zone with hot gases rising from the lower reactor zones. The dried coal is then gravity fed down through the remaining reactor zones, countercurrent to the ascending reacting gas stream. There are two hydrogasification zones. The first, with a short residence time (1-10 seconds), low temperature (650-760°C), and dilute-phase transport reactor, is followed by a second with a longer residence time (20-40 mintues), higher temperature (870-925°C), and dense-phase fluidized-bed reactor. Sixty to seventy percent of the total product methane is produced in these two hydrogasification reactors. In the bottom zone the remaining residual char is converted to high hydrogen-content synthesis gas by dense-phase, fluidized-bed steam-oxygen gasification at temperatures of 950-1010°C and 10-20 minutes residence time. This hot synthesis gas is the hydrogen source for the two upper hydrogasification zones. Ash is withdrawn from the bottom zone, water-quenched, and removed from the process as a water slurry.

The raw gas mixture from the reactor is scrubbed, then passed through an acid-gas removal system. A final methanation process converts most of the unreacted hydrogen and carbon monoxide present to methane.

A pretreatment step is required for caking coals which would otherwise agglomerate in the reactor. The pretreatment consists of mild air oxidation which destroys the coals' agglomerating properties but also reduces product yield.

Bigas. Development of the Bigas process began in late 1963 on the hypothesis that a high methane yield would result from reactions of coal with steam at high temperature and pressure.

Coal is fed into the first gasifier stage as a water slurry. Steam is injected separately, combines with the coal, and the mixture is contacted by hot gas from the oxygen-fed lower stage, causing convertion to methane, carbon monoxide, carbon dioxide, hydrogen, and char. This mixture rises through the gasifier and is discharged from the gasifier to a water quench. The gas is scrubbed further, then treated in a shift reactor which combines carbon monoxide and steam to form more hydrogen. The gas is then passed through acid-gas scrubbers to remove hydrogen sulfide and carbon dioxide. This is followed by a methanation reactor which combines most of the carbon monoxide and remaining hydrogen and forms more methane, raising the heating value of the product gas significantly.

This reactor system is designed to handle all types of coals without pretreatment. Unfortunately there are materials problems due to the high operating temperature and the slagging section is particularly prone to flow interruptions due to solidification of the molten ash. Also, solids circulations throughout the system pose a difficult problem. The process has never operated satisfactorily.

Synthane. Under development since the early 1960's, the Synthane process uses sized coal which is fed by a weigh-bit lock-hopper system into the fluidized-bed pretreater where it is dried before flowing to a single-stage fluidized-bed reactor. Steam and oxygen are injected below the fluidized bed with gasification occuring within the fluidized bed. Unreacted char flows downward into another bed fluidized and cooled by steam. Char is removed from the pressurized system as a water slurry.

Product gases are scrubbed prior to shift conversion which gives the proper carbon monoxide-to-hydrogen ratio for methanation.

Large quantities of char which can be used as a boiler fuel are produced. However, this solid fuel is undesirable compared to fluid fuels.

COED/COGAS. The COED/COGAS process involves the flow of coal through four fluidized-bed pyrolysis steps, each operating at a higher temperature than the preceding one. Temperatures are selected to prevent agglomeration and depend on the coal being processed. Typical temperatures are 600-650°F for the first stage, 800-850°F for the second, 1,000°F for the third, and 1,600°F for the fourth. Hot gases from the fourth stage, in which some of the char is burned with oxygen, flow countercurrent to the solids and provide the fluidizing medium for the intermediate stages. Fines from the first and second stages are removed in cyclones and the gases are then quenched to remove oils. The char is desulfurized with hydrogen and then gasified in the COGAS process.

Two synthetic pipeline gas processes are being developed as part of the U.S. Department of Energy's High-Btu Coal Gasification Demonstration Program—one by the CONOCO Coal Development Company and the other by the Illinois Coal Gasification Group (ICGG). The CONOCO project uses a British Gas/Lurgi slagging gasifier followed by shift, cooling and cleaning, methanation, compression, and drying. The demonstration plant is designed to produce 19×10^6 SCFD of pipeline gas from 1,800 tons of coal per day (1,000 tons per day to the gasifier). The commercial plant would produce 250×10^6 SCFD.

The ICGG project employs the COED/COGAS technology and the demonstration plant is designed to produce 24×10^6 SCFD of gas and 1,400 barrels per day of fuel oil from 2,330 tons of coal per day.

The U.S. Department of Energy is funding the design phase for both of these projects and hopes to choose one for construction during 1980. Plans call for the completion of construction of the demonstration plant in 1983.

Direct Coal Liquefaction

Liquids may be produced from coal by pyrolysis, direct catalytic liquefaction, and indirect liquefaction. In pryrolysis, coal is contacted with a carrier gas at 900 - 1200°F and pressures up to 1000 psig. Gases, liquids, and char are the products, and the char usually accounts for more than 50 percent of the feed coal. The liquid yield is about one barrel per ton of the feed coal. The U.S. Deparment of Energy recently entered into a three-year program with Occidental Research to develop a flash hydropyrolysis process which is an outgrowth of modifications made to earlier pyrolysis work. A flow diagram for this process, which shows promise of increasing the quantity and quality of oil yields, is given in Figure 3.31. Hot dried coal is conveyed into an entrained-bed reactor by recycle gas, while hot recycle char increases the temperature to about 1800°F. Short reactor residence, or contact time, and quenching in a reducing medium characterize this process.

Several processes for direct liquefaction are undergoing extensive development at this time: pilot plants sized for 200 - 600 tons of coal per day are now operating for the H-Coal and the Exxon Donor Solvent (EDS) processes and demonstration plants sized for 6000 tons of coal per day are being designed for the Solvent Refined Coal processes (SRC I and II). Detailed descriptions of these processes may be found in references 13, 14 and 23. A brief description of each is given below.

H-Coal Process. The original H-Coal process was developed on a bench scale by Hydrocarbon Research, Inc. and Cities Service Oil Company. This work, conducted in 1964, was an extension of the ebullated-bed processing technology originally used to convert heavy oil residues from petroleum into lighter fractions (H-Oil Process). Under sponsorship by the Office of Coal Research, HRI constructed a process development unit (PDU) capable of treating about three tons of coal per day, yielding in excess of three barrels of crude liquids per ton of coal. The process has operated on eastern U.S. bituminous coals, western U.S. subbituminous coals and lignite, and Australian brown coal.

Based on the data acquired from the bench-scale unit and the PDU, design and engineering began in 1973 on a 600-ton-per-day pilot plant. The plant is located adjacent to the Ashland Oil, Inc. refinery in Catlettsburg, Kentucky. The location will allow much of the utility system to be shared with the refinery. The plant was completed early in 1980, and the first runs on coal were made in April of that year. The plant was constructed under the joint sponsorship of the U.S. Department of Energy, the Electric Power Research Institute, Ashland Synthetic Fuels, Inc., Conoco Coal Development Company, Mobil Oil Corporation, Standard Oil Company (Indiana), Ruhrkohle AG of West Germany, and the Commonwealth of Kentucky. Those same groups are sponsoring continuing laboratory studies.

The H-Coal process converts coal to hydrocarbon liquids by hydrogenation with a cobalt-molybdenum catalyst in an ebullated-bed reactor. Depending on the operating conditions,

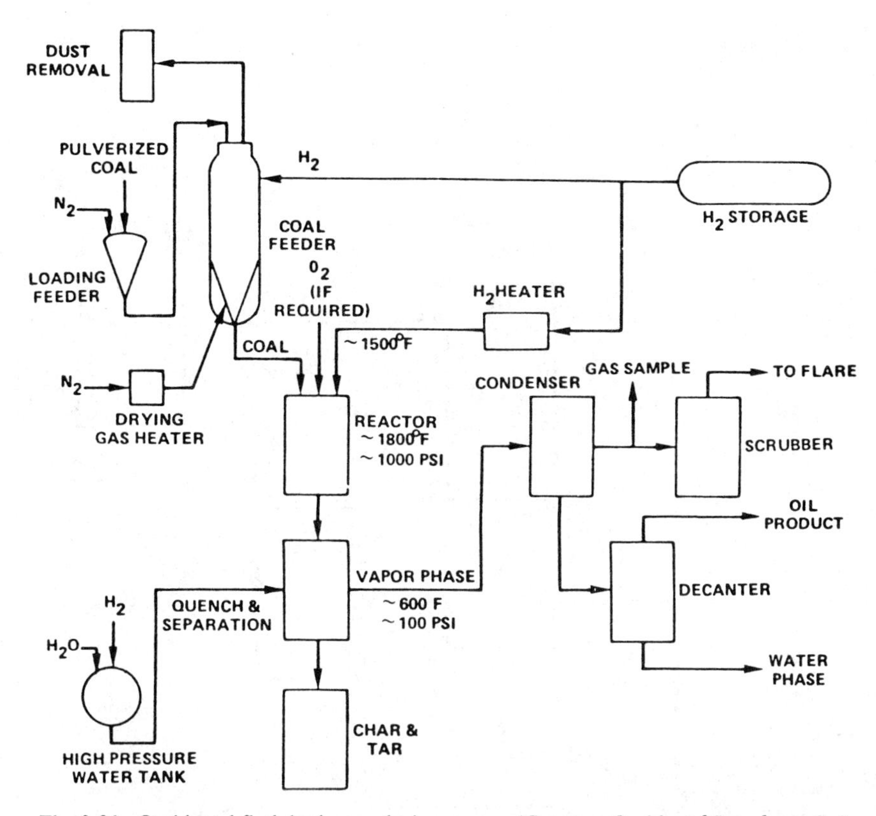

Fig. 3.31. Occidental flash hydropyrolysis process. (*Courtesy Occidental Petroleum Co.*)

the liquid product may range from a heavy boiler fuel to a synthetic crude product. A schematic of the process is shown in Figure 3.32.

Coal is first crushed, dried, slurried with recycle oil, and pumped to a pressure of approximately 230 atmospheres. Compressed hydrogen is then added to the slurry and the mixture is preheated. The material is then charged continuously to the bottom of the ebullated-bed reactor. The upward flow of slurry through the reactor maintains the catalyst in a fluidized state. Catalyst activity is maintained by the withdrawal of spent catalyst. Reactor temperature is controlled by adjusting the preheater outlet temperature. Typically, the temperature of the slurry entering the reactor is 850°F.

The vapor product from the reactor is cooled and the heavier components are collected as a liquid. The liquid stream from the primary condenser is fed to an atmospheric pressure distillation unit, where it is distilled to produce a light distillate and a heavy distillate product. The solid-liquid product stream from the reactor, which contains unconverted coal, ash and oil is fed to a flash separator. The overhead products are treated in the distillation unit and the bottoms product is further treated with a hydroclone, a liquid-solid separator, and a vacuum-distillation unit.

The operating conditions of the H-Coal process can be altered to produce various types of primary product. To produce a synthetic crude product, relatively high temperatures and hydrogen partial pressures are used and the solids-liquid separation can be conducted by vacuum distillation. The production of gas and low-sulfur residual oil requires lower temperatures and pressures. Less hydrogen is consumed during the production of residual oil. The H-Coal process requires between 14,000 and 20,000 standard cubic feed of hydrogen for each ton of coal processed, depending on the type of fuel produced. About three barrels of coal-derived liquids are obtained per ton of coal feed.

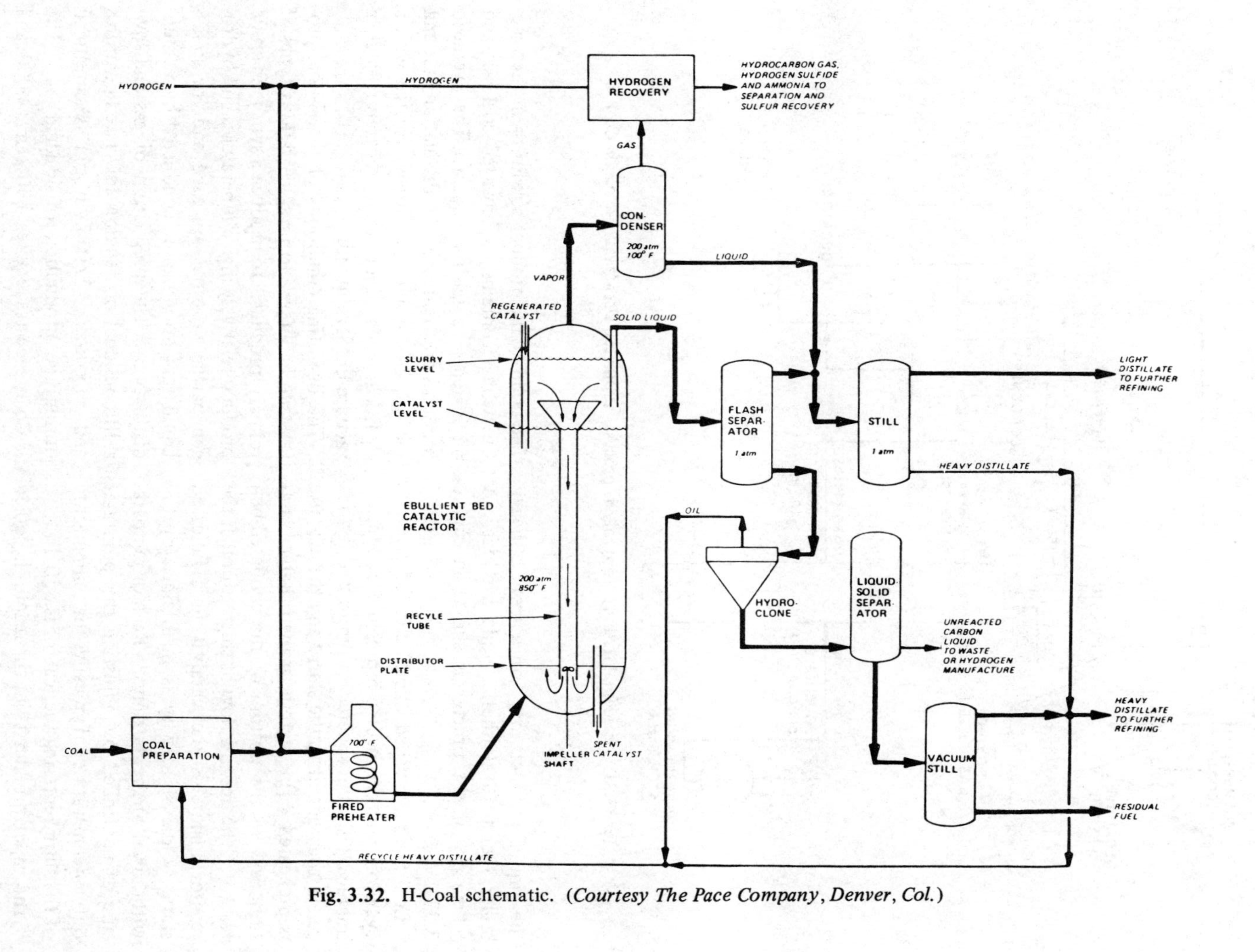

Fig. 3.32. H-Coal schematic. *(Courtesy The Pace Company, Denver, Col.)*

Exxon Donor Solvent Process. Initial research on the Exxon Donor Solvent (EDS) process extended from 1966 to 1973. During that time, extensive research was conducted in experimental equipment ranging in complexity from 100-cubic centimeter batch units to a continuous $\frac{1}{2}$-ton-per-day integrated pilot plant. Various reactor configurations, separation techniques, solvents and process variables were studied. The result of these studies was a non-catalytic, tubular, plug-flow liquefaction reactor that circumvented the need for ash-tolerant catalysts and mechanical separation devices.

The essential features of the EDS process are shown in Figure 3.33. Crushed coal is liquefied in a non-catalytic tubular reactor in the presence of molecular hydrogen and the hydrogen-donor solvent. The liquefaction reactor operates at 800–880°F and 1500–2000 psig. The hydrogen-donor solvent is a material with a 400–850°F boiling range. The solvent is a catalytically-hydrogenated-recycle stream, fractionated from the middle boiling range of the liquid product. After hydrogenation, the solvent is mixed with fresh coal-feed and pumped through a preheat furnace into the liquefaction reactor. Slurry leaving the liquefaction reactor is separated by distillation into gas, naphtha, distillates, and a vacuum bottoms slurry. The vacuum bottoms slurry is coked to produce additional liquids.

The critical processing steps are essentially adaptations of Exxon's petroleum refinery technology. Distinguishing features are the decoupled configuration of the liquefaction and catalytic-hydrogenation sections and the use of vacuum distillation for solids/liquid separation. The catalyst does not contact coal minerals or high-boiling liquids, thereby leading to longer catalyst life at high activity. Use of hydrogenated rather than unhydrogenated recycle solvent produces an improvement in process operability, particularly in downstream processing vessels. Also, hydrogenated solvent produces higher distillate product yields than unhydrogenated solvent. The use of mechanical separation devices for solids/liquids separation is avoided.

The EDS process produces liquids suitable for motor gasoline, blending stocks, low-sulfur fuel oil, and utility fuel. From 2.5 to 3 barrels of liquid can be produced from each ton of coal.

Solvent Refined Coal (SRC I and II) Processes. The development of the Solvent Refined Coal (SRC) process is being conducted in pilot plants at Fort Lewis, Washington, and Wilsonville, Alabama. Pittsburg & Midway Coal Mining Company has the responsibility for the operation of the 50-ton-per-day plant at Fort Lewis, which is funded by the U.S. Department of Energy. The operation of the six-ton-per-day

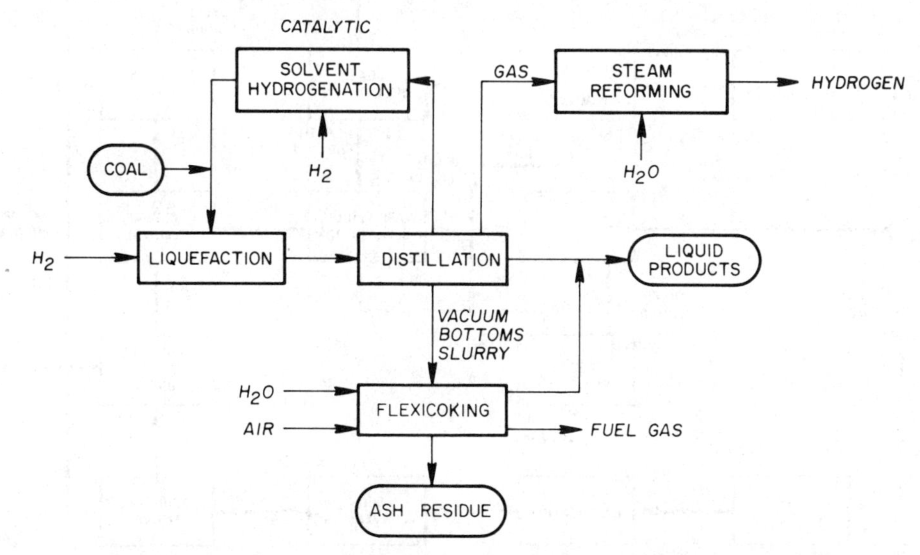

Fig. 3.33. Donor solvent process. (*Courtesy The Pace Company, Denver, Col.*)

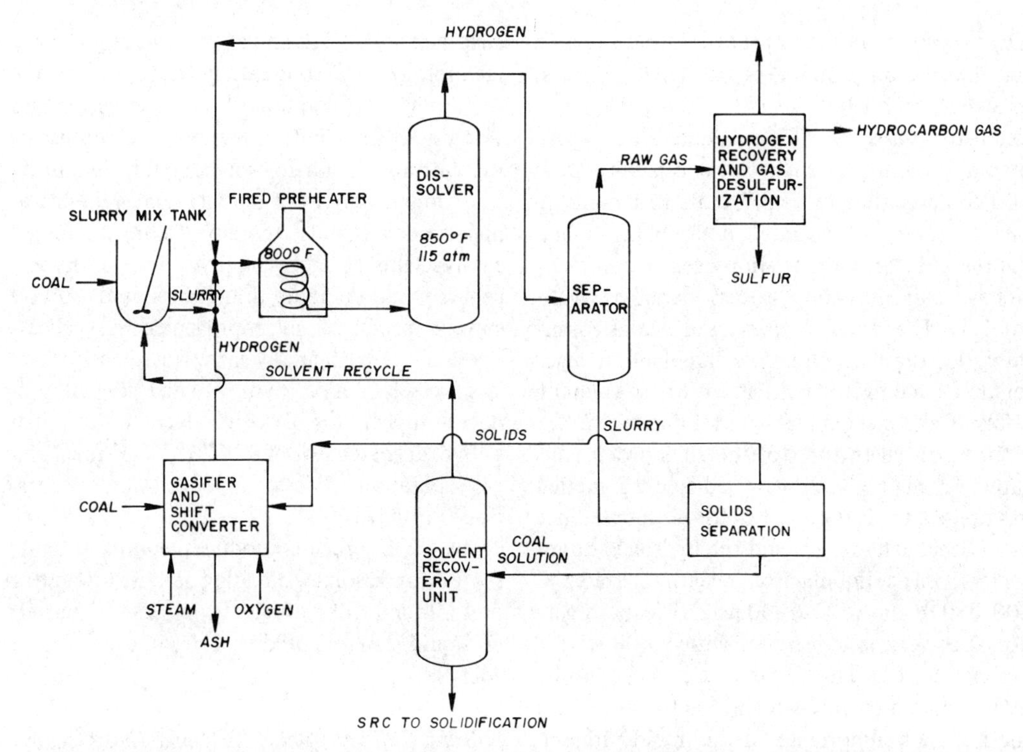

Fig. 3.34. SRC-I process for solvent refined coal. (*Courtesy The Pace Company, Denver, Colorado.*)

plant at Wilsonville is jointly sponsored by the Department of Energy and the Electric Power Research Institute, and Southern Company Services, Inc.

Initial investigation of the feasibility of a coal de-ashing process was performed by Spencer Chemical in 1962, under a contract to the U.S. Office of Coal Research. The process was successfully demonstrated in a 50-pound-per-hour continuous-flow unit in 1965. In 1969, Stearns-Roger Corporation completed the design of the 50-ton-per-day plant for Pittsburg &

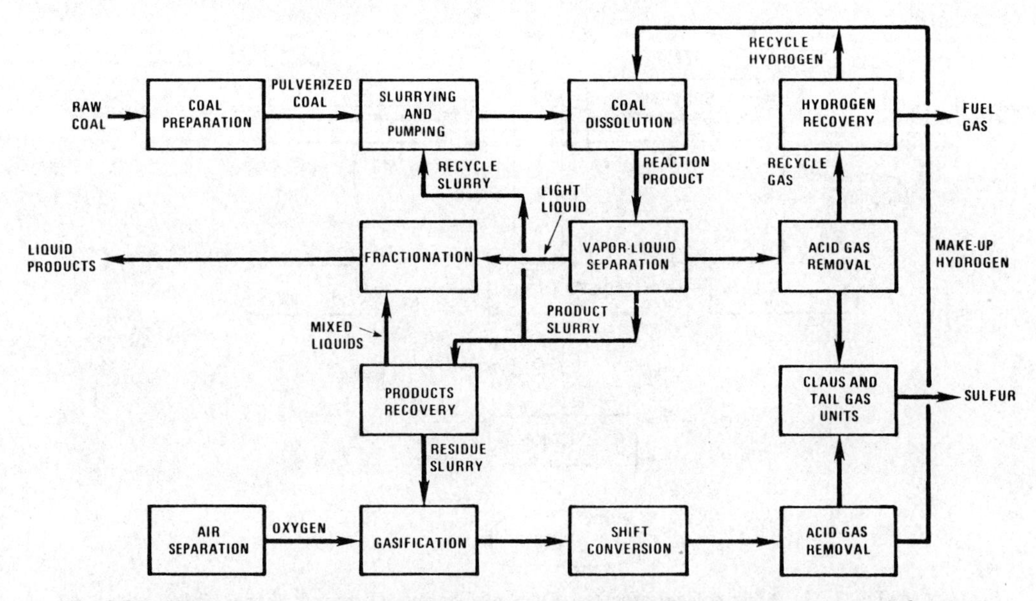

Fig. 3.35. SRC-II process for solvent refined coal. (*Courtesy The Pace Company, Denver, Colorado.*)

TABLE 3.26 Typical Properties of SRC-I Product (Coal Feed: Western Kentucky Bituminous)

Carbon, wt. %	88.0
Hydrogen, wt. %	5.9
Nitrogen, wt. %	2.2
Sulfur, wt. %	0.7
Oxygen, wt. % by difference	3.1
Ash, wt. %	0.2
Forms of Sulfur	
Sulfate, wt. %	–
Pyritic, wt. %	–
Organic, wt. %	0.7
Heating Value, Btu/lb	16,250
Fusion Point, °F (gradient bar)	350
Density, gm/cc	1.2

Source: Baughman, G.L., *Synthetic Fuels Data Handbook*, Second Ed, Cameron Engineers, Inc., Denver, Colorado, 1978, pp. 177–237 (By permission)

Midway Coal Mining Company, and Rust Engineering Company was responsible for the construction of the plant which became operational in October 1974. Initial tests studied the effects of different coal feed rates and dissolver temperatures. By late 1976, 3,000 tons of SRC had been produced for use in combustion testing by Southern Services Company's Georgia Power Company. In 1977, the plant was modified, and operations were conducted in the SRC-II mode. In 1977–78, almost 6,000 barrels of liquid SRC fuel were produced. A portion of this fuel was burned in a successful test conducted by Consolidated Edison in New York during the summer of 1978.

Catalytic Inc. designed, built, and is operating the six-ton-per-day pilot plant for EPRI and Southern Company Services in Wilsonville. The pilot plant began operation in January 1974 and has since operated with both bituminous and subbituminous coals. In 1979, the International Coal Refining Company, a joint venture of Air Products and Chemicals, Inc. and Wheelabrator-Frye, Inc. was formed to design, construct and operate a 6,000-tons-of-coal-per-day SRC-I demonstration plant located near Newman, a small town in Western Kentucky. Plans call for this plant to begin operation in 1984.

The SRC I process converts high-sulfur, high-ash coals to a low-sulfur, low-ash solid fuel. The SRC-II process results in a liquid rather than a solid, by recycling the product slurry as solvent, thereby increasing the conversion of the coal to lower molecular weight fuels. A schematic of the SRC-I process is shown in Figure 3.34 and a schematic of the SRC II process is shown in Figure 3.35.

In the SRC process, the coal is first pulverized to less than $\frac{1}{8}$ inch and mixed with process-derived solvent. The solvent-to-coal ratio is about 1.5 for SRC I and 2.0 for SRC II. The slurry, at 450°F and 2,000 psig, is combined with hydrogen and is pumped to a preheater to increase the temperature to 700–750°F, and is then fed into a dissolver. In this unit, which operates at a temperature of 840–870°F, the coal is hydrogenated, and thereby depolymerized, leading to an overall decrease in product molecular weight and dissolution of the coal. The solvent is also hydrocracked in the dissolver unit, yielding lower molecular weight hydrocarbons ranging from light oil to methane. The hot effluent from the dissolver is treated in a series of high pressure separators where

TABLE 3.27 Typical Properties of SRC Fuels Using Recycle SRC Process

	Solid Fuel	*Distillate Fuel*
Gravity: ° API	–18.3	5.0
Approximate Boiling Range: °F	800+	400–800
Fusion Point: °F	350	–
Flash Point: °F	–	168
Viscosity: SUS at 100°F	–	50
Sulfur*	0.8	0.3
Nitrogen*	2.0	0.9
Heating Value: Btu/lb.	16,000	17,300

*Assuming Western Kentucky coal feed with 4% Sulfur and 2% Nitrogen

Source: Baughman, G.L., *Synthetic Fuels Data Handbook*, Second Ed., Cameron Engineers, Inc., Denver, Colorado, 1978, pp. 177–237 (By permission)

TABLE 3.28 Coal Conversion Process Summary

Process	*Developer*	*Primary Product*	*Secondary Product*	*Feed Coal Type/Size*	*Year Begun*	*Conf. Index*
SRC - I	Southern Co. Services + EPRI + DOE + Gulf	Solid Boiler Fuel	Naphtha	All types	1962	B-3
SRC - II	Gulf + DOE	Liquid Boiler Fuel	Gas LPG Naphtha	All types with ash restrictions	1976	B-4
EDS	Exxon	Liquid Boiler Fuel	LPG Naphtha Gas	All acceptable	1966	C-3
H-Coal Fuel Oil	HRI	Fuel Oil	Naphtha Gas			C-2
H-Coal Syncrude		Syncrude				C-2
Fischer-Tropsch	Standard Technology used over 50 years	Range of Hydrocarbons	LPG Alcohols No. 2 Oil Fuel Oil Gas	All coal is gasified	Before 1930	A-2
M-Gasoline	Mobil for methanol to Gasoline Conversion	Premium Gasoline	LPG	Any gasifier to methanol process may be used		C-3
Methanol		Methanol	–	Depends on gasifier	–	A-2
CO_2 Acceptor	Conoco Coal Development Co.	High-Btu Gas Low-Btu Gas	None	Lignite or Sub-Bituminous /8×100 mesh	1968	B-2
Hygas	IGT	High-Btu Gas	Naphtha	All with pretreat	1954	C-3
Bigas	Bituminous Coal Research	High-Btu Gas	None	All	1963	C-4
Synthane	BuMines PERC Lummus	High-Btu Gas	Char	All/20 to max 20%–200	1961	C-3
CE	Combustion Engineering	Low-Btu Gas Electric Power	None	All	1974	C-4
Westing-house	Westinghouse	Electric Power	None	All	1972	C-4
		Medium-Btu Gas	None			C-4
Lurgi	Lurgi	High-Btu Gas	Tar, Oils Char Naphtha Coal Fines	Non-to low caking	Before 1930	A-2

Source: Rogers, K.A. and Hill, R.F., "Coal Conversion Comparisons," ESCOE, The Engineering Societies Commission on Energy, Inc., Washington, D.C., FE-2468-82, Aug. 1980.

TABLE 3.29 Process Confidence Index

Process Development	*Economic Reliability*
D–Exploratory state - not beyond simple bench tests	4–Screening estimate, very approximate
C–Development stage - operated on small integrated scale only	3–Incomplete definition for estimates used
B–Pre-commercial - successful pilot plant operation	2–Firm basis for values developed
A–Complete - process demonstrated sufficiently to insure commercial success	1–Values considered to be satisfactory for commercial venture

Source: Rogers, K.A. and Hill, R.F., "Coal Conversion Comparisons," ESCOE, The Engineering Societies Commission on Energy, Inc., Washington, D.C., FE-2468-82, Aug. 1980.

gases and light hydrocarbons are removed as vapor, leaving a heavy liquid slurry stream suitable as solvent. The vapor stream is then cooled to condense liquid hydrocarbons which are separated from the gases.

The liquid slurry is subjected to a mineral separation step where the undissolved solids are removed. Process solvent is recovered from the coal solution by distillation and recycled to slurry the coal feed. The residue which remains is the solvent-refined coal product.

In the SRC II process, the slurry stream from the dissolver is split into two portions, one of which is recycled to provide solvent for coal-slurry mixing, while the other is fractionated to recover the primary products of the process. The light hydrocarbon liquid stream from the dissolver is also fractionated. The products from the fractionation system are naphtha, low-sulfur fuel oil, and a vacuum residue consisting of heavy oil, ash, and undissolved organic material from the coal. The gases from the dissolver are treated to remove hydrogen sulfide and carbon dioxide. A cryogenic separation unit yields liquid petroleum gases and pipeline gas for sale. Unreacted hydrogen is recovered and recycled to the process. Recently, tests have been conducted in which the process features of the SRC II process (recycle) are used to produce a combination of solid and liquid products.

Typical properties of solid SRC-I product are shown in Table 3.26. Table 3.27 shows the properties of both the solid and liquid products which can be produced by using the recycle SRC process. Depending on the severity of the recycle process, a wide combination of the solid and liquid products can be produced.

Coal Conversion Process Comparisons

The state of development of various coal conversion processes and the dearth of commercial operating data make it very difficult to make meaningful comparisons of these processes. The Engineering Societies Commission on Energy, Inc. (ESCOE), under contract to the U.S. Department of Energy, recently undertook the task of making such comparisons.[15,16]

Table 3.28 shows the developer, products, coal-feed type, year started, and a confidence index for various coal conversion processes. The process confidence index is a measure of the state of development of the particular process and the reliability of available economic estimates. Table 3.29 is a key to the meaning of the confidence index.

Table 3.30 shows the products and associated costs for the various processes for a dry coal feed rate of 25,000 tons per day. The higher heating value (HHV) of the coal is 11,200 Btu/lb dry coal; the cost is $30 per ton; and the plant operating factor is 0.9 (330 operating days per year). The product and energy cost figures are for utility-type financing. The results of any economic analysis depend strongly on the assumptions made concerning efficiency and plant capital costs. A more complete description of the economic comparison and the assumptions may be found in the work of Rogers and Hill.[16]

TABLE 3.30 Products and Associated Costs for Various Coal Conversion Processes

Process	*Product*	*Quantity Per Day*	*Product Cost*	*Energy Cost $/10^6 Btu*
SRC-I	SRC Solid	10,880 Tons	106.72 $/Ton	3.38
	Fuel Oil	7,500 Bbl	23.53 $/Bbl	
SRC-II	LPG	5,500 Bbl	19.23 $/Bbl	3.62
	Naphtha	10,700 Bbl	23.84 $/Bbl	
	Fuel Oil	45,300 Bbl	19.72 $/Bbl	
	Gas	23.1 MMSCF	5.59 $/$10^6$ Btu	
EDS	Propane	3,270 Bbl	18.66 $/Bbl	3.96
	Butane	3,500 Bbl	20.22 $/Bbl	
	Naphtha	19,900 Bbl	23.03 $/Bbl	
	Fuel Oil	28,700 Bbl	19.05 $/Bbl	
	C2 - Gas	41.9 MMSCF	5.04 $/$10^6$ Btu	
H-Coal Fuel Oil	Naphtha	18,200 Bbl	21.70 $/Bbl	3.30
	Fuel Oil	42,200 Bbl	17.96 $/Bbl	
	Gas	19.7 MMSCF	5.09 $/$10^6$ Btu	
H-Coal Syncrude	Naphtha	31,900 Bbl	20.51 $/Bbl	3.58
	Fuel Oil	24,300 Bbl	16.97 $/Bbl	
	Gas	56.3 MMSCF	4.81 $/$10^6$ Btu	
Fischer-Tropsch	Gasoline	18,200 Bbl	24.84 $/Bbl	4.99
	LPG	18,800 Bbl	18.99 $/Bbl	
	No. 2 Oil	1,200 Bbl	23.54 $/Bbl	
	Fuel Oil	2,000 Bbl	19.47 $/Bbl	
	Med Btu Gas	90.7 MMSCF	5.52 $/$10^6$ Btu	
	C2 - Gas	37.2 MMSCF	5.52 $/$10^6$ Btu	
M-Gasoline	Gasoline	52,700 Bbl	24.55 $/Bbl	4.84
	LPG	7,300 Bbl	16.89 $/Bbl	
Methanol	Methyl Fuel	113,400 Bbl	11.33 $/Bbl	4.37
	Methanol	8,400 Bbl	12.26 $/Bbl	
HYGAS	SNG	333.8 MMSCF	3.51 $/$10^6$ Btu	3.45
	Naphtha	6,800 Bbl	14.97 $/Bbl	
Synthane	SNG	350.0 MMSCF	3.58 $/$10^6$ Btu	3.46
	CHAR	1,344 Tons	26.92 $/Ton	
CO_2 Acceptor	SNG	397.9 MMSCF	3.55 $/$10^6$ Btu	3.55
LURGI	SNG	326.4 MMSCF	3.95 $/$10^6$ Btu	3.89
	Tar Oil	4,500 Bbl	16.79 $/Bbl	
	Naphtha	2,200 Bbl	16.84 $/Bbl	
BIGAS	SNG	388.2 MMSCF	3.67 $/$10^6$ Btu	3.62
CO_2 Acceptor	SYNGAS	1143 MMSCF	2.79 $/$10^6$ Btu	2.79
Westinghouse Syngas	SYNGAS	1272 MMSCF	2.42 $/$10^6$ Btu	2.42
Westinghouse	Electric Power	2625 MW	6.14 $/$10^6$ Btu	6.12
			2.1 ¢/kWh	
C-E	Electric Power	2838 MW	6.08 $/$10^6$ Btu	6.08
			2.1 ¢/kWh	

Source: Rogers, K.A., and Hill, R.F., "Coal Conversion Comparisons," ESCOE, The Engineering Societies Commission on Energy, Inc., Washington, D.C., FE 2468-82, Aug., 1980.

REFERENCES

1. Ode, W.H., and Frederic, W.H., "The International Systems of Hard-Coal Classification and Their Application to American Coals," *Bureau of Mines Report of Investigations*, 5435, 1958.
2. American Society for Testing and Materials, *Coal and Coke*, Part 8 of 1970 Book of ASTM Standards, Philadelphia, Pa., 1970.
3. "Methods of Analyzing Coal and Coke," *U.S. Bureau Mines Bull.*, No. 638, 1967.
4. Leonard, J.W., Ed., *Coal Preparation*, American Inst. of Mining, Metallurgical, and Petroleum Engineers, Inc., New York, 4th Edition, 1979.

5. Bureau of Mines, U.S. Department of the Interior, *Minerals Yearbook*, Washington, D.C.
6. Killmeyer, R.P., Jr., "State-of-the-Art of Coal Cleaning and Recent Developments in Equipment and Circuits," Third Kentucky Coal Refuse Disposal and Utilization Seminar, Pineville, Ky., May 11-12, 1977.
7. Cavallaro, J.A. and Deurbrock, A.W., "An Overview of Coal Preparation" in *Coal Desulfurization-Chemical and Physical Methods*, T.D. Wheelock, Ed., ACS Symposium Series No. 64, American Chemical Society, Washington, D.C. 1977.
8. Hoffman, L., Anesco, S.J., Holt, E.C., Jr., "Engineering Economic Analyses of Coal Preparation with SO_2 Cleanup Processes for Keeping Higher Sulfur Coals in the Energy Market," NCA/BCR Coal Conference and Expo III, Louisville, Ky., October 19-21, 1976.
9. McConnell, J.F., "Homer City Coal Cleaning Demonstration," 4th National Conference on Energy and Environment, (EPA).
10. "Combustion Engineering," Combustion Engineering, Inc., New York, 1966.
11. Bodle, W.W. and Schora, F.C., "Coal Gasification Technology Overview," *Symposium Papers: Advances in Coal Utilization Technology*, Institute of Gas Technology, May 14-18, 1979, Louisville, Kentucky, pp. 11-34.
12. Miller, C.L., "The U.S. Coal Gasification Program: Progress and Projects," *Mechanical Engineering Journal*, August 1980, pp. 34-40.
13. Considine, D.M., *Energy Technology Handbook*, McGraw-Hill, New York, 1977, pp. 1-188 - 1-285.
14. Baughman, G.L., *Synthetic Fuels Data Handbook*, second edition, Cameron Engineers, Inc., Denver, Colorado, 1978, pp. 177-237.
15. Hill, R.F., "Synthetic Fuels Summary," ESCOE, The Engineering Societies Commission on Energy Inc., Washington, D.C., FE-2468-82, August 1980.
16. Rogers, K.A. and Hill, R.F., "Coal Conversion Comparisons," ESCOE, The Engineering Societies Commission on Energy Inc., Washington, D.C., FE-2468-51, July 1979.
17. McKelvey, K.N., "Alternative Raw Materials for Organic Chemical Feedstocks." *Chem. Eng. Prog.*, pp. 45-48, March, 1979.
18. O'Hara, J.B. et al, "Project POGO-A Coal Refinery," *Chem. Eng. Prog.*, pp. 49-63, August, 1978.
19. O'Hara, J.B. et al, "Petrochemical Feedstocks from Coal," *Chem. Eng. Prog.*, pp. 64-72, June, 1977.
20. *Conference on Chemical Feedstocks Alternatives*, American Institute of Chemical Engineers and National Science Foundation, Houston, Texas, October 2-5, 1977.
21. Leonard, J.P. and Frank, M.C., "The Coal Based Chemical Complex," *Chem. Eng. Prog.*, pp. 68-72, June, 1979.
22. "Chemicals from Coal: Tennessee Eastman Makes the Move," *Chem. Eng. Prog.*, p. 89, March, 1980.
23. "Advances in Coal Utilization Technology," Symposium Papers, sponsored by Institute of Gas Technology, May 14-18, 1979, Louisville, Kentucky.
24. *Handbook of Gasifiers and Gas Treatment Systems*, DRAVO Corp., Pittsburgh, Pa., Contract No. E(49-18) 1772, FE-1772-11, National Technical Information Service, U.S. Department of Commerce, February, 1976.
25. Salvador, L.A., Rath, L.K., Carrera, J.P. and Vidit, E.J., "The Westinghouse Coal Gasification Process," International Gas Research Conference, June, 1980.

Additional References

1. Anderson, L.L. and Tillman, D.A., *Synthetic Fuels From Coal*, New York, 1979.
2. Hoffman, E.J., *Coal Conversion*, Energon Co., Laramie, Wyoming, 1978.
3. Howard-Smith, I. and Werner, G.J., "Coal Conversion Technology," *Chemical Technology Review*, #66, Noyes Data Corporation, Park Ridge, N.J., 1976.
4. Cassidy, S.M., *Elements of Practical Coal Mining*, Port City Press, Inc., Baltimore, Maryland, 1973.
5. Loftness, R.L., *Energy Handbook*, Van Nostrand Reinhold Co., New York, 1978.
6. Lom, W.L. and Williams, A.F., *Substitute Natural Gas, Manufacture and Properties*, Wiley, New York, 1976.

4

Sulfuric Acid and Sulfur

James R. West* and Gordon M. Smith**

SULFURIC ACID

Sulfuric acid, a strong acid, is an oily, viscous, water-white, nonvolatile liquid. It absorbs water from the atmosphere. A drop of it on the skin causes a severe burn. It is made in large volumes by the fertilizer and chemical industries. It is used as a solvent, a dehydrating agent, a reagent in chemical reactions or processes, an acid, a catalyst, an absorbent, and in many other applications. The concentrated acid is usually stored in steel tanks. The dilute acid may be stored in lead-lined or plastic tanks. It is used in very dilute concentrations and as strong fuming acid. It is often recovered and reused. After use in some phases of the explosives, petroleum, and dye industries it is often recovered in a form unsuitable for reuse in that industry but suitable for use in another industry; however, in some processes it may be reconcentrated or regenerated for reuse in the same process system. It is a versatile, useful acid and has been called the "work horse" of the chemical industry.

Uses of Sulfuric Acid

Sulfuric acid is one of the most widely used of all manufactured chemicals, and its rate of production has long been a reliable index of the total chemical production and the industrial activity of a nation. For 1979, the average per capita consumption of sulfur in the United States was 135 pounds, most of which went into the manufacture of sulfuric acid. In the world, the per capita consumption in 1979 was about 27 pounds.

The sulfuric acid consuming industries in the United States are listed in Table 4.1. Of these, the largest consumer is the fertilizer industry which treats phosphate rock with sulfuric acid to produce superphosphate (a mixture of monocalcium phosphate and calcium sulfate) or crude ("wet process") phosphoric acid. There is hardly an article of commerce which has not come into contact with sulfuric acid at one time or another during its manufacture or in the manufacture of its components. The

*Retired. Formerly Vice President, Research, Engineering and Construction, Texasgulf Inc.
**Texasgulf Inc.

TABLE 4.1 Sulfuric Acid-Consuming Industries in the United States (Millions of Short Tons, 100% H_2SO_4)

Consuming Industries	*1970*		*1975*		*1978*	
	Tons	*%*	*Tons*	*%*	*Tons*	*%*
Agriculture						
Phosphatic	14.3	49	20.0	61	26.0	63
Ammonium Sulfate	1.6	5	1.6	5	1.6	4
Total	15.9	54	21.6	66	27.6	67
Other Industries						
Chemicals	3.5	12	2.3	7	3.2	8
Petroleum Refining	2.2	8	2.0	6	2.2	5
Iron and Steel	0.5	2	0.3	1	0.4	1
Other Metals	1.2	4	1.8	6	2.2	5
Uranium/Vanadium Ore Processing	0.3	1	0.4	1	0.8	2
Paints and Pigments	1.7	6	1.1	3	0.9	2
Rayon and Cellulose Film	0.7	2	0.3	1	0.5	1
Miscellaneous	3.5	11	2.7	9	3.7	9
Total	13.6	46	10.9	34	13.9	33
GRAND TOTAL	29.5	100	32.5	100	41.5	100

Source: Marketing Research Department estimates, Texasgulf Inc.

consumption of acid in the various industries is undergoing constant change. Economic progress demands that manufacturers strive to decrease the consumption of acid per unit product manufactured. However, progress also is continually turning up new uses for sulfuric acid.

Kinds of Acid

Sulfuric acid is marketed in the United States as a large tonnage product. It is made in numerous grades and strengths, and shipments are made in both packaged containers and in bulk. It is produced in grades of exacting purity for use in storage batteries and for the rayon, dye

TABLE 4.2[a] Acid Strengths and End Uses

Percent H_2SO_4	*°Bé*	*Percent Oleum (Percent Free SO_3)*	*Specific Gravity*	*Uses*
35.67%	30.8°	–	1.27	Storage batteries
62.18–69.65	50–55	–	–	Normal superphosphate and other fertilizers
77.67	60.0	–	1.7059	Normal superphosphate and other fertilizers; isopropyl and sec-butyl alcohols
80.00	61.3	–	1.7323	Copper leaching
93.19	66.0	–	1.8354	Phosphoric acid, titanium dioxide
98.99	–	–	–	Phosphoric acid, alkylation, boric acid
104.50	–	20%	1.9056	
106.75	–	30	1.9412	
109.00	–	40	1.9737	Caprolactam, nitrations and sulfonations,
111.25	–	50	1.9900	dehydration, blending with weaker acids
113.50	–	60	1.9919	
114.63	–	65	1.9842	

[a]Sulfuric Acid, Chemical Economics Handbook, SRI, Menlo Park, California, August, 1979.

and pharmaceutical industries. It is produced to less exacting purity specifications for use in the steel, heavy chemicals, and fertilizer industries. Originally, sulfuric acid was marketed in four grades known as: chamber acid, 50° Beaumé (Bé); tower acid, 60° Bé; oil of vitriol, 66° Bé; and fuming acid. Table 4.2 lists typical acid strengths produced and some of the end uses for these strengths. Other concentrations may also be used for many of the applications indicated. Sulfuric acids are commonly specified as commercial, electrolyte (high purity for batteries), or C.P. (reagent) grades.

THE MANUFACTURE OF SULFURIC ACID

History

Sulfuric acid is formed in nature by the oxidation and chemical decomposition of naturally occurring sulfur and sulfur-containing compounds. It is formed by the weathering of coal brasses or iron disulfide, discarded on refuse dumps at coal mines. It is formed in the atmosphere by the oxidation of sulfur dioxide emitted from the combustion of coal, oil, and other substances. It is also formed by chemical decomposition resulting from geological changes.

In ancient times, sulfuric acid was probably made by distilling niter (potassium nitrate) and green vitriol (ferrous sulfate heptahydrate). Weathered iron pyrites were usually the source of the green vitriol. About 1740, the acid was made in England by burning sulfur in the presence of saltpeter (potassium nitrate) in a glass balloon flask. The vapors united with water to form acid which condensed on the walls of the flask. In 1746, the glass balloon flask was replaced by a large lead-lined box or chamber, giving rise to the name "chamber process." In 1827 Gay Lussac, and in 1859 Glover, changed the circulation of gases in the plant by adding towers which are now known as Gay-Lussac and Glover towers. These permit the recovery from the exit gases of nitrogen oxides which are essential to the economic production of chamber acid. Today, most sulfuric acid made in the United States is produced by the contact process which is based on scientific technology developed about 1900 and thereafter.

BASF built a successfully operating contact plant in the United States in 1898. In 1900, General Chemical erected a pyrite-burning contact plant in the United States using the Herreshoff furnace.

Development of the Sulfuric Acid Industry in the United States

The manufacture of sulfuric acid has been a basic industry in the United States for many years. It has been made by two well-established methods: the chamber process and the contact process. Initially, the production of acid in the United States was concentrated on the eastern seaboard. After the Civil War, the industry spread to the West and into the South. Between 1899 and 1904, the number of acid manufacturers increased rapidly in Ohio and Illinois; while just before the turn of the century, there was great manufacturing activity in the South as a result of the discovery of phosphate deposits in South Carolina and Florida and the development of the phosphate fertilizer industry.

Over the years, the number of contact plants has increased while the number of chamber plants has decreased. During the mid 1940's, the number of chamber plants was approximately equal to the number of contact plants. Today, the contact plants have replaced the chamber plants, and the trend is toward contact plants of larger and larger capacity. Contact plants producing 2400 tons of acid a day in a single train are not unusual. Texasgulf is building a single-train contact plant with a rated capacity of 3100 tons per day.

Users of sulfuric acid confronted with the task of disposing of waste or spent acid find it advantageous to arrange with an independent producer to exchange the waste acid for fresh acid. Methods have been developed which permit such producers to reprocess the waste acid and obtain a product of virgin quality. Also they can operate centrally located plants of large size which can produce acid at a much lower cost than can be done in the small plant needed by the average consumer. The end use of sulfuric acid, more than any other factor, determines the location of sulfuric acid plants. However, sulfuric acid plants using metallurgi-

TABLE 4.3 New Sulfuric Acid Production in the United States (Thousands of Short Tons, 100% H_2SO_4)

	Number of Establishments			*Sulfuric Acid Production*		
	1970	*1975*	*1978*	*1970*	*1975*	*1978*
Alabama	7	4	3	378.1	291.0	(*2*)
Arizona	4	6	6	338.1	1,088.5	1,340.7
California	12	10	10	1,474.3	1,232.4	1,088.9
Florida	16	14	17	8,088.2	9,735.2	14,839.1
Georgia	10	6	4	530.5	414.4	495.4
Illinois	12	10	9	1,126.4	922.6	990.0
Louisiana	9	8	10	2,891.0	4,260.1	5,622.6
Maryland & Delaware	6	3	2	810.7	476.9	(2)
New Jersey	9	10	10	1,422.9	1,222.8	1,036.2
New York	2	2	2	(*2*)	(*2*)	(*2*)
Ohio	8	7	8	641.6	597.7	550.0
Pennsylvania	9	7	7	832.3	475.4	577.3
Texas	13	16	14	2,389.4	2,187.3(*3*)	2,931.7(*3*)
Other	70	58	56	7,336.4	8,280.3	9,961.5
Total	187	161	158	28,259.9	31,184.6	39,433.4
Chamber Plants	23	7	3	320.5	96.8	(*4*)
Contact Plants	165(*1*)	155(*1*)	155	27,939.4	31,087.8	39,433.4

1. One establishment in Georgia reports production by both processes.
2. Data included in total.
3. Includes Oklahoma.
4. Chamber plants constitute less than 1% of total production.
Source: Current Industrial Reports, Series M28A and M28B, U.S. Department of Commerce, Bureau of the Census.

Fig. 4.1 Metallurgical gas sulfuric acid plant. (*Courtesy Texasgulf Inc.*)

cal gas as the source of sulfur are usually located near the smelters producing such gas. Data on the production of acid in the United States are listed in Table 4.3.

The Chamber Process for Making Sulfuric Acid

Once of great importance in the United States, chamber process plants have become almost extinct. For a discussion of the chamber process, the reader should consult the texts of A. M. Fairlie and of W. W. Duecker and James R. West.

The Contact Process

The basic features of the contact process for making sulfuric acid, as practiced today, were described in a patent issued in England in 1831. It disclosed that if sulfur dioxide, mixed with oxygen or air, is passed over heated platinum the sulfur dioxide is rapidly converted to sulfur trioxide, which can be dissolved in water to make sulfuric acid. The practical application of this disclosure, however, was delayed. An understanding of the complex reactions occurring in the gas phase over the catalyst required the development of that branch of physical chemistry known as chemical kinetics, and also the development of that branch of engineering known as chemical engineering. A demand for acid stronger than that which could be produced readily by the chamber process stimulated this development. The eventual success of the contact process for making sulfuric acid led to the development of other catalytic processes used for making many of the synthetic chemicals known today.

The heart of the contact sulfuric acid plant is the converter in which sulfur dioxide is converted catalytically to sulfur trioxide. Over the course of the years, a variety of catalysts have been used, including platinum and the oxides of iron, chromium, copper, manganese, titanium, vanadium, and other metals. The first catalyst used was platinum. It proved to be extremely sensitive to poisons such as arsenic compounds present in small amounts in some sources of sulfur dioxide. The successful development of the contact process depended in part on the recognition of the existence of catalytic poisons, and in devising methods for their removal. Platinum and iron catalysts were the main catalysts used prior to World War I. At present, vanadium catalysts in various forms, combined with promotors, are generally used.

A number of different plant designs have been developed for the efficient production of sulfuric acid by the contact process. In the United States, these are often referred to by the name of the builder or designer, e.g., a Lurgi, a Monsanto, or a Davy McKee plant. In Europe and other parts of the world, Lurgi Gesselschaft für Chemie and Hüttenwesen mbH, Chemiebau Dr. A. Zieren GmbH & Co. KG, and Simon-Carves Chemical Engineering Ltd. are noted for designing and building contact sulfuric acid plants. Contact plants are also classified according to the material used in the production of the sulfur dioxide charged to the plant, e.g., sulfur, hydrogen sulfide, gypsum, iron pyrites, smelter gas, or spent and sludge acids.

The principal steps in a contact plant in which sulfur is burned are shown in a simplified diagram in Fig. 4.2. These are: (1) the production of sulfur dioxide with dry air; (2) cooling the gas; (3) conversion or oxidation of the sulfur dioxide to sulfur trioxide; (4) cooling the sulfur trioxide gas; and (5) absorption of the sulfur trioxide in sulfuric acid.

If sulfur is the raw material, it is burned in a sulfur burner in the presence of dried air to yield a gas containing 8 to 11 percent sulfur dioxide and 8 to 13 percent oxygen. The gas produced by burning sulfur usually is clean and requires little further treatment; it is cooled in a waste-heat boiler to about 760° to 840°F before it enters the converter. If the raw material is a sulfide ore, the equipment may include roasters, sintering machines, copper converters, reverberatory furnaces, flash smelters, and so forth. The gas from roasters usually contains from 7 to 14 percent sulfur dioxide, while that from sintering machines seldom contains more than 8 percent sulfur dioxide. Gases from metallurgical operations are contaminated with fumes, water and other vapors, and solid impurities which must be removed in cyclone dust collectors, electrostatic dust and mist precipitators, or scrubbing towers. The gases must then

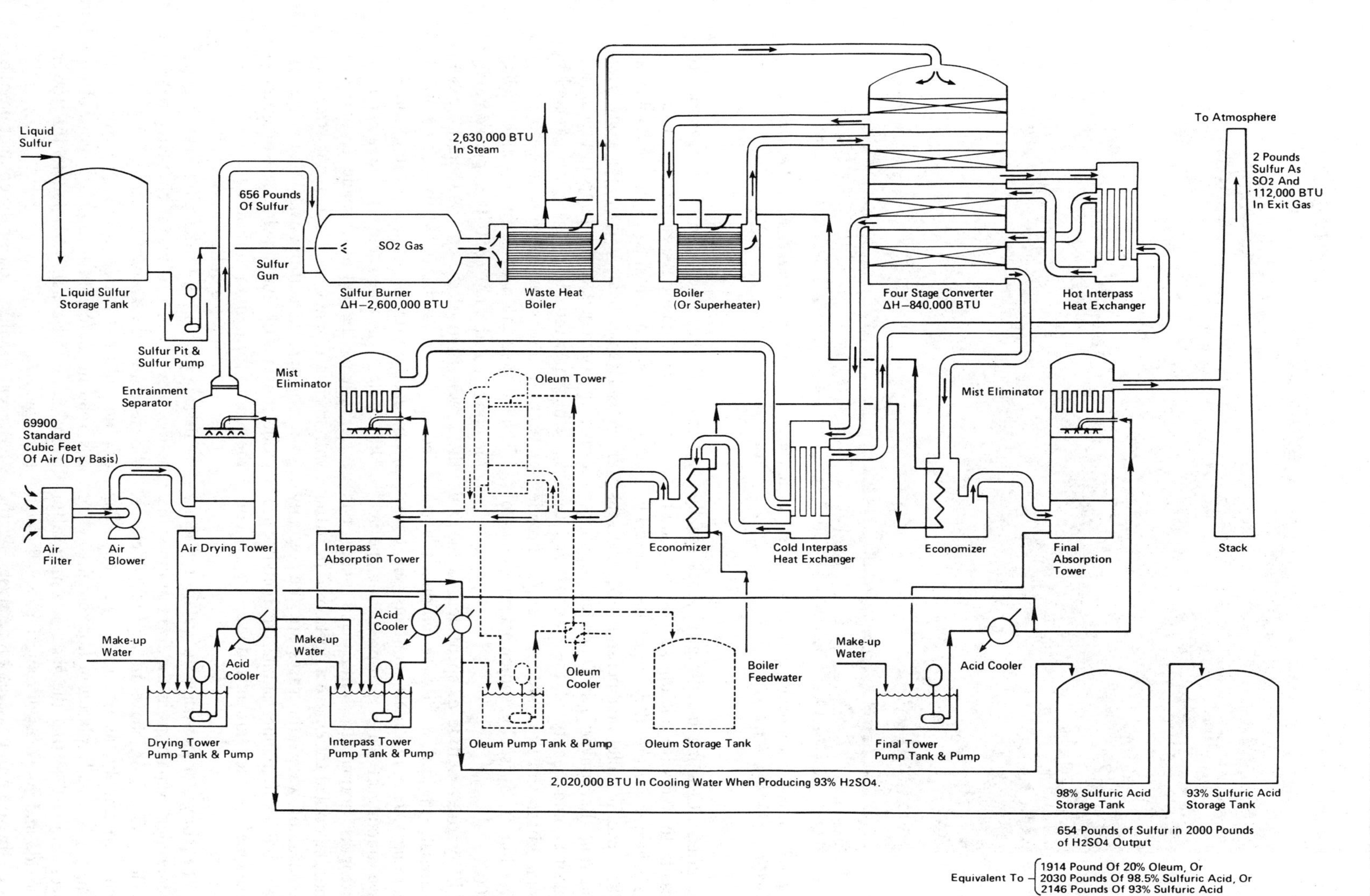

Fig. 4.2 Schematic flow diagram of a double absorption contact process for sulfuric acid manufacture.

Fig. 4.3 Sulfur-burning double absorption contact sulfuric acid (2 trains) of Texasgulf Inc. at Lee Creek, N.C. (*Courtesy Texasgulf Inc.*)

be dried. Cleaning metallurgical gases usually results in excessive cooling so that they must be reheated before they enter the converter. The specific inlet temperature of the gas entering the converter is dependent upon the quantity and quality of the catalyst, and the composition and flow rate of the sulfur dioxide gas, but it is usually somewhat in excess of 800°F.

If hydrogen sulfide or spent acid is the raw material, the sulfur dioxide produced when burning such materials contains an excessive amount of water for producing the acid strengths normal to contact-acid plants and must be cooled to remove a portion of it. After removal of water, the gases are then usually reheated in a manner similar to metallurgical gases before they are introduced to the converter.

The converter contains the catalyst, usually pentavalent vanadium in the form of individual pellets, similar in shape to aspirin tablets or, more commonly, as extruded cylinders. The catalyst is usually placed in horizontal trays or beds arranged so that the gas containing sulfur dioxide and an excess of oxygen passes through two, three, or four stages of catalyst. As the gas passes through the converter, approximately 95 to 98 percent of the sulfur dioxide is converted to sulfur trioxide, with the evolution of considerable heat. Maximum conversion cannot be obtained if the temperature in any stage becomes too high. Therefore, gas coolers are employed between converter stages.

The concentration of sulfur trioxide leaving the converter at 800 to 850°F is approximately the same as that of the entering sulfur dioxide. The converter gas is cooled to 450 to 500°F in an economizer or tubular heat exchanger. The cooled gas then enters the absorption tower where the sulfur trioxide is absorbed with high efficiency in a circulating stream of 98 to 99 percent sulfuric acid. The sulfur trioxide unites with the small excess of water in the acid to form more sulfuric acid.

Oleum, or fuming sulfuric acid, is produced by passing part or all of the cooled gas through a tower over which oleum is circulated before the gas enters the absorption tower. A portion of the sulfur trioxide in the gas stream is absorbed in the oleum to produce more oleum from the sulfuric acid also fed to this tower (see Fig. 4.2).

A recent modification of the contact process,

which was developed in the 1960s to reduce the emission of sulfur dioxide in the exit gases, includes two stages of absorption. Because of chemical equilibria conditions it is impossible to obtain conversion efficiencies much higher than 98 percent in plants using only one stage of absorption. The reversible reaction

$$SO_2 + \frac{1}{2}\ O_2 \rightleftharpoons SO_3$$

will, however, proceed further in the righthand direction if the sulfur trioxide is removed. The modified process takes advantage of this fact and overall conversion efficiencies of SO_2 to SO_3 exceeding 99.7 percent are achieved with sulfur burning plants. Stringent government requirements regulating the allowable amount of sulfur dioxide emissions from new sulfuric acid plants in the United States, Canada, and many other countries necessitate the use of a double-absorption plant, or other means such as a scrubber, to reduce the sulfur dioxide released to the atmosphere in the exhaust gases. Federal regulations in the United States require that no more than four pounds of sulfur dioxide may be emitted in the stack gas per ton of sulfuric acid produced for all new sulfur-burning sulfuric acid plants. This is equivalent to 99.7% efficiency. Although there is increased yield of sulfuric acid obtained from the double-absorption plant, the incremental production (on the order of 1.8%) does not usually offset the increased cost (approximately 10%) over that of a single-absorption plant.

The flow sheet of a double-absorption, sulfur-burning contact plant is shown in Fig. 4.2. The flow sheet is very similar to that of a single-absorption plant, with the exception that an additional absorption step is added after the second or third catalyst bed to remove the sulfur trioxide formed before the sulfur dioxide containing gas is passed over the remaining catalyst. Additional heat exchangers, pumps, and related equipment are required in addition to the second absorption tower.

The gases leaving the first absorbing tower, instead of discharging directly to the atmosphere as they would in a single absorption type plant, pass through a heat exchanger in which they are reheated to about 800°F before reentering the converter. They then pass through additional catalyst, are cooled, and flow through a second absorbing tower and then to the atmosphere. The overall conversion of SO_2 may exceed 99.5 percent. The second conversion step may be made with one or two trays of catalyst contained in the same vessel used in the first conversion step. The heat exchanger may be substituted for the economizer in the first stage of the flow sheet, and the economizer may be used to cool the gases in the last stage. Many variations are possible.

SULFUR

The 1970 and 1975 consumptions of sulfur in the Western Bloc Nations are given in Table 4.4, together with estimates for 1978. It is seen that of the total sulfur consumed in 1970, about 87 percent went to the manufacture of sulfuric acid, while in 1975 about 89 percent went to the same end. There are, however, many other important uses of sulfur. Nonacid consumption in the United States for 1970, 1975, and 1978 (estimated) is shown in Table 4.5.

Table 4.6 shows the total sulfur consumption in the United States for 1970, 1975, and 1978. Between 1970 and 1978, the total sulfur consumption increased by about 37 percent. The amount of total sulfur consumed in the manufacture of sulfuric acid remained essentially constant at approximately 90 percent of the total.

Table 4.7 gives the production of sulfuric acid in the United States in 1970, 1975, and 1978 by source of sulfur. The amount of elemental sulfur used went from about 77 percent in 1970 to about 80 percent in 1978. In 1978 approximately 45 percent more elemental sulfur was burned to make sulfuric acid than in 1970. Table 4.7 illustrates the dominance of elemental sulfur for the production of sulfuric acid in the United States.

Sources

Iron pyrites at one time were the main sulfur-containing material used in the manufacture of sulfuric acid. With the development of the Frasch process for producing elemental sulfur,

TABLE 4.4 Western Bloc Nations Consumption of Sulfur (Millions of Long Tons)

	1970		*1975*		*1978*	
	Amount	*% of Total*	*Amount*	*% of Total*	*Amount*	*% of Total*
Agricultural						
Phosphatic	9.7	35	13.0	43	15.5	43
Ammonium Sulfate	2.0	7	1.6	5	1.4	4
Total Agricultural	11.7	42	14.6	48	16.9	47
Industrial Acid						
TiO_2	1.6	6	1.6	5	1.6	4
Textiles	1.4	5	1.6	5	1.8	5
Metallurgy	1.4	5	1.6	5	2.0	6
Others	8.1	29	7.7	26	9.4	26
Total Industrial Acid	12.5	45	12.5	41	14.8	41
Industrial Nonacid						
Pulp	1.3	5	1.0	3	1.1	3
Liquid SO_2	0.1	1	0.1	1	0.3	1
CS_2	0.8	3	0.9	3	0.7	2
Refined Sulfur	0.6	2	0.7	2	0.7	2
Others	0.7	3	0.7	2	1.3	4
Total Industrial Nonacid	3.5	13	3.4	11	4.1	12
TOTAL INDUSTRIAL	16.0	58	15.9	52	18.9	53
TOTAL SULFUR	27.7	100	30.5	100	35.8	100

Source: Marketing Research Department estimates, Texasgulf Inc.

elemental sulfur replaced iron pyrites as the raw material for making sulfuric acid; this was done in many places for several reasons. Frasch sulfur is exceptionally pure, burns in relatively simple equipment to yield a gas containing a high concentration of sulfur dioxide at a constant rate, eliminates the need for gas cleaning, reduces capital cost, creates less pollution, and leaves no residue requiring disposal. However, sour gas sulfur and refinery gas sulfur, most always even more pure than Frasch sulfur, have supplemented or replaced Frasch sulfur in many parts of the world. Sulfur dioxide from metallurgical smelting processes also is used in manufacturing sulfuric acid, often for prevention of atmospheric pollution.

Some have predicted that sulfur dioxide in combustion gases would become a substantial source of sulfur dioxide for making sulfuric acid. Yet in the United States, the "throwaway" system represents over 90% of the currently installed flue-gas desulfurization capacity. So the amount of sulfur dioxide in combustion gases forecast to be made into sulfuric acid has not thus far become a reality.

In 1953, sulfur mined by the Frasch process

TABLE 4.5 Nonacid Consumption of Sulfur in the United States (Thousands of Long Tons)

USE	*1970*	*1975*	*1978*
Carbon disulfide	265	150	160
Pulp and paper	425	300	240
Rubber	60	65	75
Sulfur dioxide	50	65	70
Agriculture	80	90	160
Phosphorus pentasulfide	35	40	50
Other	185	290	345
TOTAL	1,100	1,000	1,100

Source: Marketing Research Department estimates, Texasgulf Inc.

TABLE 4.6 Total Sulfur Consumption in the United States (Millions of Long Tons)

	1970		*1975*		*1978*	
	Tons	*%*	*Tons*	*%*	*Tons*	*%*
Acid	8.2	88	9.7	91	11.4	91
Nonacid	1.1	12	1.0	9	1.1	9
TOTAL	9.3	100	10.7	100	12.5	100

Source: Marketing Research Department estimates, Texasgulf Inc.

TABLE 4.7 U.S. Sulfuric Acid Production by Sulfur Source (Thousands of Short Tons, 100% H_2SO_4)

	From New Sulfur											
			Smelter Gases									
	Elemental Sulfur	H_2S	*Copper*	*Zinc*	*Lead*	*Total*	*Pyrites*	*Total*	*Sludge Acids*	*Total New Acid*	*Fortified Acid*	*Total*
1978	32,863	300	2,740	825	155	3,720	950	37,833	1,600	39,433	1,654	41,087
1975	25,940	400	1,785	710	130	2,625	630	29,595	1,590	31,185	1,175	32,360
1970	22,630	675	750	920	170	1,840	1,250	26,395	1,865	28,260	1,265	29,525

Source: Current Industrial Reports, Series M28A and M28B, U.S. Department of Commerce, Bureau of the Census; Mineral Industry Surveys, "Sulfur," U.S. Department of the Interior, Bureau of Mines; "Sulfur" Chemical Economics Handbook; SRI, December, 1979, Marketing Research Department estimates, Texasgulf Inc.

in the United States dominated the world sulfur market. By 1958, however, Mexico had become well established as a major supplier of Frasch sulfur. Other present suppliers of Frasch sulfur include Poland and Iraq. By 1963, both Canada and France became important suppliers of brimstone recovered from natural gas. This world-wide production is now augmented by recovered sulfur from oil and gas fields in the Middle East. With the decline of the Frasch sulfur industry in the United States, Canada and Poland emerged as the main suppliers of brimstone on the world markets by 1977. Future sources of sulfur include additional recovery from hydrocarbons in the Middle East and by-product recovery from energy developments in coal, oil shale, and tar sands, as well as recovery for pollution control from nonelemental sources.

Sulfur is shipped on world markets both as a solid and a liquid. In 1958, Texasgulf began the important trend towards the transportation of sulfur in liquid form. However, the emphasis on environmental control has led to forming the solid sulfur for shipment, a departure from the bulk, run-of-mine shipments of the past. There exist a number of forming processes which produce granules, prills, agglomerates, or slates.

Because of the vast resources of sulfur in the earth, there will never be a sulfur shortage in the long term. However, producing sulfur from most of the above sources in the future will probably bring higher costs to the product.

REFERENCES

1. Allied Chemical Corp., "Sulfuric Acid," 1978.
2. American Petroleum Institute, *API Toxicological Review*: *Sulfuric Acid*, 2nd Ed., New York, 1963.
3. Anderson, J. C., *Sulphur*, No. 140, 29-31 (Jan/Feb 1979).
4. ASTM, Standard Methods for Analysis of Sulfuric Acid, Designation E-233-1968.
5. "Atmospheric Emissions from Sulfuric Acid Manufacturing Processes," Public Health Service Publication No. 999-AP-13, Cincinnati, U.S. Dept. of Health, Education, and Welfare, 1965.
6. Bauer, R. A., and Vidon, B. P., *Chem. Eng. Progr.*, **74**, No. 9, 68, 69 (Sept. 1978).
7. Bhattacharjya, S. K., *Chem. Age India*, **21**, No. 6, 585-590 (1970).
8. Bischoff, W. F., and Habib, Y., *Chem. Eng. Progr.*, **71**, No. 5, 59, 60 (May 1975).
9. Boase, J. L., and Duckworth, R. A., *Chem. Process Eng.*, Part 1, 58-67 (Feb. 1963); Part 2, 132-137 (Mar. 1963).
10. Brink, J. A., Jr., Burggrabe, W. F., and Greenwell, L. E., *Chem. Eng. Progr.*, **64**, No. 11, 82-86 (Nov. 1968).
11. The British Sulphur Corp. Ltd., *Sulphur*, No. 136, 45-48, 67, 68 (May/June 1978).
12. The British Sulphur Corp. Ltd., *Sulphur*, No. 136, 54-57, 60, 61 (May/June 1978).
13. The British Sulphur Corp. Ltd., *Sulphur*, No. 136, 22-26 (May/June 1978).
14. The British Sulphur Corp. Ltd., *Sulphur*, No. 136, 28-33 (May/June 1978).
15. The British Sulphur Corp. Ltd., *Sulphur*, No. 140, 44-45 (Jan/Feb 1979).
16. The British Sulphur Corp. Ltd., *Sulphur*, No. 141, 19, 20, 22 (March/April 1979).
17. The British Sulphur Corp. Ltd., *Sulphur*, No. 141, 23, 26, 27 (March/April 1979).
18. The British Sulphur Corp. Ltd., *Sulphur*, No. 141, 29-31 (March/April 1979).
19. The British Sulphur Corp. Ltd., *Sulphur*, No. 141, 32-37 (March/April 1979).
20. The British Sulphur Corp. Ltd., *Sulphur*, No. 142, 30-39 (May/June 1979).
21. Browder, T. J., "Latest Developments and Modern Technology in the Sulfuric Industry," Preprint 2a, American Institute of Chemical Engineers, Dec. 1970.
22. Burkhardt, D. B., *Chem. Eng. Progr.*, **64**, No. 11, 66-70 (Nov. 1968).
23. Cameron, G. M., Nolan, P. D., and Shaw, K. R., *Chem. Eng. Progr.*, **74**, No. 9, 47-50 (Sept. 1978).
24. Chari, K. S., *Chem. Process Eng.* (Bombay), **3**, No. 2, 35-36 (Feb. 1969).
25. *Chem. Eng.*, **70**, No. 8, 106 (Apr. 15, 1963).
26. *Chem. Eng. News*, **42**, No. 51, 42-43 (Dec. 21, 1964).
27. *Chem. Week*, **102**, No. 19, 55-56 (May 11, 1968).
28. Clayton, Jr., C. H., and Johnson, T. E., *Chem. Eng. Progr.*, **74**, No. 9, 54-57 (Sept. 1978).
29. Collins, J. J., Fornoff, L. L., Manchanda, K. D., Miller, W. C., and Lavell, D. C., *Chem. Eng. Progr.*, **70**, No. 6, 58-62 (June 1974).
30. Connor, J. M., *Chem. Eng. Progr.*, **64**, No. 11, 59-65 (Nov. 1968).
31. Davies, D. S., Jiminez, R. A., and Lemke, P. A., *Chem. Eng. Progr.*, **70**, No. 6, 68, 69 (June 1974).
32. DeWolf, P., and Larison, E. L., *American Sulphuric Acid Practice*, 1st Ed. New York, McGraw-Hill, 1921.

33. Donovan, J. R., Palermo, J. S., and Smith, R. M., *Chem. Eng. Progr.*, **74**, No. 9, 51-54 (Sept. 1978).
34. Donovan, J. R., and Stuber, P. J., *J. Metals*, **19**, No. 11, 45-50 (Nov. 1967).
35. Donovan, J. R. and Stuber, P. J., "The Technology and Economics of Interpass Absorption Sulfuric Acid Plants," presented Dec. 1968 at Annual Meeting, American Institute of Chemical Engineers.
36. Doyle, K., and Deszynski, A., *Sulphur*, No. 139, 26-31 (Nov/Dec 1978).
37. Duecker, W. W., and West, J. R., Eds., *The Manufacture of Sulfuric Acid*, New York, Van Nostrand Reinhold, 1959.
38. Duros, D. R., and Kennedy, E. D., *Chem. Eng. Progr.*, **74**, No. 9 70-77 (Sept. 1978).
39. DuPont, "Sulfuric Acid."
40. Eckert, Jr., G. F., *Eng. Mining J.* **180**, No. 3, 115-118 (March, 1979).
41. Epstein, M., Sybert, L., Wang, S. C., Leive, C. C., and Princiotta, F. T., *Chem. Eng. Progr.*, **70**, No. 6, 53, 54 (June 1974).
42. Fairlie, A. M., *Sulfuric Acid Manufacture*, New York, Van Nostrand Reinhold, 1936.
43. Farkas, M. D., and Dukes, R. R., *Chem. Eng. Progr.*, **64** No. 11, 54-58 (Nov. 1968).
44. Fasullo, O. T., *Sulfuric Acid Use and Handling*, New York, McGraw-Hill, 1965.
45. Gall, R. L., and Piasecki, E. J., *Chem. Eng. Progr.*, **71**, No. 5, 72-76 (May 1975).
46. Gamse, R. N., and Speyer, J., *Chem. Eng. Progr.*, **70**, No. 6, 45-48 (June 1974).
47. Hensinger, C. E., et al., *Chem. Eng.*, **75**, No. 12, 70-72 (June 3, 1968).
48. Hiroshi, K., *Chem. Economy Eng. Rev.*, 29-31 (Dec. 1969).
49. *J. Air Poll. Control Assoc.* **13**, No. 10, 499-502 (Oct. 1963).
50. Johnstone, H. F., and Moll, A. J., *Ind. Eng. Chem.* **52**, No. 10, 861-863 (Oct. 1960).
51. Kappanna, A. N., and Chaudhuri, B. P., *Indian J. Appl. Chem.*, **26**, No. 4, 91-96 (1963).
52. Kennedy, J. G., *Metal Finishing J.*, 374-376 (Nov. 1967).
53. Kerr, J. R., et al., "The Sulfur and Sulfuric Acid Industry of Eastern United States," U.S. Bureau Mines, Inform. Circ., No. 8255, 1965.
54. Koehler, G. R., *Chem. Eng. Progr.*, **70**, No. 6, 63-65 (June 1974).
55. Kronseder, J. G., *Chem. Eng. Progr.*, **64**, No. 11, 71-74 (Nov. 1968).
56. Labine, R. A., *Chem. Eng.*, **67**, No. 1, 80-83 (1960).
57. LaMantia, C. R., Lunt, R. R., and Shah, I. S., *Chem. Eng. Progr.*, **70**, No. 6, 66, 67 (June 1974).
58. Lunge, G., *The Manufacture of Sulphuric Acid and Alkali*, Vol. 1 (3 parts), New York, Van Nostrand Reinhold, 1913.
59. Mandelik, B. G., and Pierson, C. U., *Chem. Eng. Progr.*, **64**, No. 11, 75-81 (Nov. 1968).
60. Marshall, V. C., *Brit. Chem. Eng.*, **6**, No. 12, 841-850 (Dec. 1961).
61. Martin, G., and Foucar, J. L., *Sulphuric Acid and Sulphur Products*, New York, Appleton-Century-Crofts, 1916.
62. "Methods of Test for Sulfuric Acid," B.S. 3903, London, British Standards Institution, 1965.
63. Meyer, B., *Sulfur, Energy and Environment*, New York, Elsevier Scientific Publishing Co., 1977.
64. Miles, F. D., *The Manufacture of Sulphuric Acid (Contact Process)*, New York, Van Nostrand Reinhold, 1925.
65. Miller, W. E., *Chem. Eng. Progr.*, **70**, No. 6, 49-52 (June 1974).
66. Monsanto Co., "Monsanto Designed Sulfuric Acid Plants," 1944.
67. Moore, H. C., *Sulphuric Acid Tables; Hand Book*, 2nd Ed., rev., Chicago, Hillson and Etten, 1926.
68. Newby, R. A., Keairns, D. L., and Vidt, E. J., *Chem. Eng. Progr.*, **71**, No. 5, 77-79 (May 1975).
69. *Oilweek*, **15**, No. 9, 23 (Apr. 13, 1964).
70. Olin Mathieson Chemical Corp., "Sulfuric Acid," 1958.
71. Parkes, J. W., *The Concentration of Sulphuric Acid*, New York, Van Nostrand Reinhold, 1924.
72. Parrish, P., and Snelling, F. C., *Sulfuric Acid Concentration*, New York, Van Nostrand Reinhold, 1925.
73. Rampacek, C., "Sulfuric Acid from Sulfur Dioxide by Autoxidation in Mechanical Cells," U.S. Bur. Mines, Rept Invest., No. 6236, 1963.
74. Rosenberg, H. S., Engdahl, R. B., Oxley, J. H., and Genco, J. M., *Chem. Eng. Progr.*, **71**, No. 5, 66-71 (May 1975).
75. Sander, U., and Daradimos, G., *Chem. Eng. Progr.*, **74**, No. 9, 57-67 (Sept. 1978).
76. Slin'ko, M. G., and Beskov, V. S., *Intern. Chem. Eng.*, **2**, No. 3, 388-393 (Jul. 1962).
77. Smith, G. M., and Mantius, E., *Chem. Eng. Progr.*, **74**, No. 9, 78-83 (Sept. 1978).
78. Snowden, P. N., and Ryan, M. A., *J. Inst. Fuel*, **42**, No. 34, 188-189 (May 1969).
79. Strom, S. S., and Downs, W., *Chem. Eng. Progr.*, **70**, No. 6, 55-57 (June 1974).
80. "Sulfuric Acid," U.S. Dept. of Commerce, Bureau of Census, Current Industrial Reports, Series M28A-13, Supplement I, issued annually.
81. Sullivan, T. J., *Sulphuric Acid Handbook*, 1st Ed., New York, McGraw-Hill, 1918.
82. Tamaki, A., *Chem. Eng. Progr.*, **71**, No. 5, 55-58 (May 1975).
83. Tennessee Corp., "Sulfuric Acid," 1962.

84. Texas Gulf Sulphur Co., "Sulphur Manual," Section VIII, Sulphuric Acid, 1965.
85. U.S. Industrial Chemicals Co., "Sulfuric Acid and Oleum Technical Data," New York, 1978.
86. Vasan, S., *Chem. Eng. Progr.*, 71, No. 5, 61-65 (May 1975).
87. Waeser, B., *Die Schwefelsäurefabrikation*, Braunschweig, Fridr. Vieweg and Sohn, 1961.
88. Walitt, A., "A Process for the Manufacture of Sulfuric and Nitric Acids from Waste Flue Gases," presented Dec. 1970, at Second International Clean Air Congress of the International Union of Air Pollution Prevention Assoc.
89. Warren, I. H., *Australian J. Appl. Sci.*, 7, 346-358 (1956).
90. Wells, A. E., and Fogg, D. E., "The Manufacture of Sulphuric Acid," U.S. Bur. Mines, Bull., No. 184, 1920.
91. Wheelcock, T. D., and Boylan, D. R., *Chem. Eng. Progr.*, 64, No. 11, 87-92 (Nov. 1968).
92. Wyld, W., *Raw Materials for the Manufacture of Sulphuric Acid and the Manufacture of Sulphur Dioxide*, New York, Van Nostrand Reinhold, 1923.
93. Wyld, W., *The Manufacture of Sulphuric Acid (Chamber Process)*, New York, Van Nostrand Reinhold, 1924.
94. Zwemer, J. H., and Dean, C. M., *Chem. Eng. Progr.*, 56, No. 2, 39-41 (Feb. 1960).

5

Synthetic Nitrogen Products

Ralph V. Green*
Hosum Li**

NITROGEN FIXATION AND AMMONIA SYNTHESIS

As long ago as 1780, Cavendish caused atmospheric nitrogen and oxygen to combine by means of an electric spark. The first practical large-scale manufacture of a nitrogen compound from atmospheric nitrogen was that of Birkeland and Eyde, at Nottoden, Norway, early in the century. In this process, air is passed at a rapid rate through an arc spread out to form a flame. Earlier, Bradley and Lovejoy had used the arc method in 1902 at Niagara Falls, but the attempt failed because the area of the arc flame was too small and the gases were not removed from the reaction chamber fast enough. The Norwegian process benefited from the faults demonstrated in this installation.

The manufacture of synthetic ammonia, tried a little later, first succeeded with the Haber process in which a mixture of nitrogen and hydrogen is passed at a moderately high temperature and under pressure over a contact catalyst, with a partial conversion of the elemental gases to ammonia. Several modifications for making ammonia from these elements have been so successful that this process is now more important than all other synthetic processes combined. Ammonia, some of its salts, and its derivatives are valuable fertilizers. Moreover, if nitric acid is called for, ammonia may be oxidized with atmospheric oxygen, aided by a contact catalyst, so that the synthetic ammonia process may also produce from atmospheric nitrogen, in an indirect way, what the arc process furnishes directly. There is a continuing need for ammonia, ammonium nitrate, and urea as fertilizer materials.

An entirely different process for the fixation of atmospheric nitrogen is the calcium cyanamide process, which depends on the fact that metallic carbides, particularly calcium carbide, readily absorb nitrogen to form the solid cyanamide. This substance is a fertilizer. By further treatment it may be transformed into cyanide; by another, into ammonia; but this ammonia is more costly than direct synthetic ammonia. The process was developed by Frank and Caro in Germany during the years 1895 to 1897, and has been introduced in many countries, includ-

*Retired E. I. DuPont de Nemours & Co.
**Area Supervisor (DuPont) 1980

ing the United States and Canada since that time.

Nitrogen can also be fixed by the high-temperature contacting of nitrogen with oxygen in the air. Experimental work on such a process has been undertaken by several investigators and results have been reported in the literature. The concentration of nitric oxides in the product gas has been too low to make the process commercially feasible. If this deficiency could be rectified, the process might become commercially attractive; however, there would still remain engineering problems associated with the design of the high temperature equipment.

This high-temperature nitrogen process operates in the range of 2,000°C, hence the materials of construction are critical. The feed to the process is raised to the reaction temperature by means of a pebble preheater, which is heated by burning a fuel such as natural gas or fuel oil. Pebble attrition is a serious problem in the operation of the preheater.

Nitrogen may be passed over metals at suitable temperatures to form nitrides which, when treated with steam, will yield ammonia. The best known process embodying this principle is the Serpek process, introduced in France but not in America, for the manufacture of aluminum nitride.

Nitrogen Fixation

New processes have been proposed for nitrogen fixation, but none of these have approached a commerically attractive stage.[1,2,3] One of these processes[1] is based upon reactions of nitrogen obtained from air in a nuclear reactor where the energy density corresponds to a temperature of 8000°K. This is high enough to break the nitrogen bonds and react the nitrogen and oxygen. Separation of the active particles for recycle and the NO_2 from the oxygen and nitrogen would be required. Also, radioactivity of the product and impurities must be overcome. One gram of fully enriched uranium is reported to produce two tons of nitric acid.

In another recent process[2] nitrogen is fixed by using a homogeneous molecular catalyst which allows a tremendous reduction in the 2000 psig synthesis pressure that is normally used. Molecular-nitrogen complexes are formed—in one case, containing rhodium, and in another iridium. These react with an acid azide giving complexes in which the nitrogen shares two loosely bound electrons in the metal. The iridium complex contains two triphenyl phosphate groups, chloride, and a carbonyl group.

Research work in Russia and the United States is based on the use of dicyclopentadienyltitanium dichloride in the presence of ethyl magnesium bromide.[2,3] This system is based on the fact that enzymes are able to fix nitrogen at room temperature and atmospheric pressure. The fixation of nitrogen occurs by allowing a mixture of diethyltitanium dichloride [$(C_2H_5)_2TiCl_2$] and ethylmagnesium bromide (C_2H_5MgBr) in ether to react in a nitrogen atmosphere at room temperature and 150 atm pressure for 7 to 31 hours. It is reported that 0.8 to 0.9 mole of NH_3 is formed per mole of titanium dichloride.

A unique method for fixation of nitrogen is given in a French Patent.[4] In this process, purified hydrogen and nitrogen are passed into an electrolyte solution of an alkali at a concentration of 5 to 15 percent at 20 to 50 atm. A catalyst consisting of tablets of powdered nickel promoted with cobalt is used.

The liquid electrolyte is removed from the reactor and the dissolved ammonia removed at lower pressure by distillation. The electrolyte is returned to the reactor where more ammonia is formed. A temperature of 50 to 80°C is used. Yields of 4.2 percent from the gases per pass are reported. The process claims to reduce the plant investment and reduce compression costs.

Research continues on the fixation of nitrogen[5]. It has been shown that low energy ultraviolet photons can fix nitrogen at room temperature under oxidizing conditions on the surface of a titanium dioxide-containing material. Nitric oxide is formed.

The various processes with actual or potential value for the fixation of atmospheric nitrogen may be grouped together as follows:

Arc Processes. Air is passed at a rapid rate through a broad or long electric arc to give nitrogen oxides. This was the earliest method for fixation of nitrogen, and it is reported that the process is still used in Norway or Sweden.

Other sources of high temperature are used such as preheating with natural gas and air, or by use of nuclear energy.

Direct Synthetic Processes. Nitrogen is combined with hydrogen to produce ammonia by contacting the two gases over a catalyst at elevated pressure and temperature. The ammonia can then be oxidized with air to produce nitric acid which will combine with additional ammonia to produce ammonium nitrate.

The great expenditure of electric energy required by the arc process is avoided; however, energy requirements for compression of the hydrogen-nitrogen mixture are large. In addition, a large expenditure of energy is necessary to produce the hydrogen from natural gas and water, or from water by electrolysis.

Cyanamide Process. This process requires fairly pure nitrogen as well as calcium carbide, a product of the electric furnace ($CaC_2 + N_2 \xrightarrow{1000^\circ C} CaNCN + C$). Calcium cyanamide may be used as a fertilizer, or ammonia may be produced by steaming in autoclaves ($CaCN_2 + 3H_2O \longrightarrow CaCO_3 + 2NH_3$). However, the latter is rarely done.

Cyanide Process. Nitrogen is passed into a vessel containing an alkali and coal. The products are cyanides and ferrocyanides. This process has no commercial importance at present.

Nitride Process. Nitrogen reacts with certain metals to produce nitrides, such as aluminum nitride, which in turn, when treated with water, yield ammonia.

Fixation with Complexes. Nitrogen is reacted with acid azides in the presence of rhodium or iridium catalysts. Also, organic complex compounds have been shown to fix nitrogen in the presence of enzymes. At present, these have no commercial importance.

Fixation of atmospheric nitrogen has been a difficult task because elemental nitrogen is comparatively unreactive. It combines with few other elements, and then usually only under drastic conditions (elevated temperatures).

Nitrogen Consumption

During the period 1955-1965, world consumption of nitrogen as a fertilizer increased from 8 million tons to 16.3 million tons. From 1965 to 1975, consumption increased to 39 million tons. By 1985 the consumption is expected to be 67 million tons. See reference 6.

World capacity[6] for production of nitrogen as a fertilizer from 1975 to the present has increased, and is expected to continue to increase because of the new installations in the U.S.S.R., Asia, and Eastern Europe. A recent disclosure[7] has shown that China is building six, 1000 T/D plants, India sixteen plants with a total capacity of 13,000 T/D, and the U.S.S.R. sixteen plants with a total capacity of 22,600 T/D.

Up until 1972, nitrogen fertilizer consumption in North America exceeded that in Asia. However, by 1985, consumption in Asia is expected to be 1.4 times that in North America and 1.6 times that in the U.S.S.R.

The per capita consumption of nitrogen in the U.S.A. is about 80 lb./yr, whereas it is only 50 lb/yr in Europe and 8 lb/yr in China. With the population explosion in the Far East, a total of 30 or more plants will be needed by 1985, if the per capita consumption approaches that in the U.S.A. Hence, there is potentially a great need for new construction and engineering services to meet the expansion of the industry.

Nitrogen capacity in various areas of the world for 1975 are given in Table 5.1.

Materials consumed as fertilizer in the United States are given in Table 5.2.

Ammonia, including direct application and solutions, shows the greatest tonnage increase. Urea continues to be of interest because of its large nitrogen content, i.e., 46.7% N vs. 35% N for ammonium nitrate. Ammonium sulfate

TABLE 5.1 Capacities for Fertilizer Nitrogen–1975[6]

Area	*1000 M Tons N*
Asia	15,093
North America	14,850
Western Europe	14,360
U.S.S.R.	10,364
Eastern Europe	8,385
Latin America	2,672
Africa	832
Oceania	446
TOTAL	67,002

TABLE 5.2 Selected Nitrogen Fertilizer Materials[6] (Millions of Short Tons)

Material	*1965*	*1970*	*1977*
Ammonia	8.9	13.8	17.6
Ammonium Sulfate	0.71	0.60	0.47
Ammonium Nitrate	4.1	5.4	5.7
Urea	1.0	2.6	4.2

consumption continues to decrease because it is a by-product and contains only 21.2% nitrogen.

Ammonia

Since 1963, there has been a revolution in the ammonia-manufacturing business. The advent of the large single-train plants has resulted in a large increase in production capacity, the shutdown of a number of smaller plants, and a reduction in manufacturing cost. The total capacity of twenty of the largest U.S. ammonia producers in 1980 was estimated at 24,107 M tons per year. Annual capacity of these producers ranged from a high of 2160 M tons per year to a low of 450 M tons per year.[8]

Prior to World War II, the production capacity for ammonia was relatively unchanged. During the war the need for explosives caused an increase in the production of ammonia for nitric acid manufacture.

After the war, available ammonia capacity was directed to the manufacture of fertilizers. Accordingly, there was a rapid increase in fertilizer consumption. The advantages of fertilizer were therefore emphasized, and productive capacity increased by leaps and bounds. From 1940 to 1950, the number of ammonia plants doubled; then from 1950 to 1960 the number more than doubled again. It was in the early sixties that the single-train concept was developed and many smaller plants were slated for shutdown. However, capacity tripled in the period from about 1958 to 1968. Yet a large increase in capacity is still needed throughout the world to meet the current and even increasing food needs of an expanding population.

Ammonia processes currently in use are the following:

BASF	Kellogg
Braun	Lummus
Chemico	Mont Cenis
Chemo Project	OSAG-Linz, Austria
Fauser Montecatin	Pritchard
Foster-Wheeler Casale	SBA
Grande Paroisse	Topsoe
ICI	TVA
Japan Consulting Inst.	Uhde

A promoted iron catalyst is used for all processes with the exception of Mont Cenis. Pressures employed vary from 100 to 900 atmospheres and temperatures from 400° to 650°C. In general, gas recirculation in the synthesis section is used to give good utilization of hydrogen and nitrogen since only 9 to 30 percent conversion is obtained per pass over the catalyst. The synthesis gas source varies depending on the location. In some instances

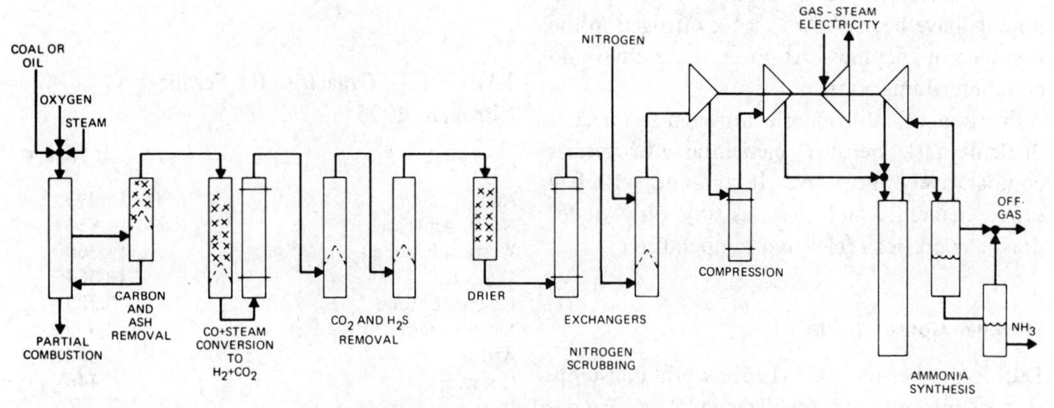

Fig. 5.1 Ammonia from coal or oil.

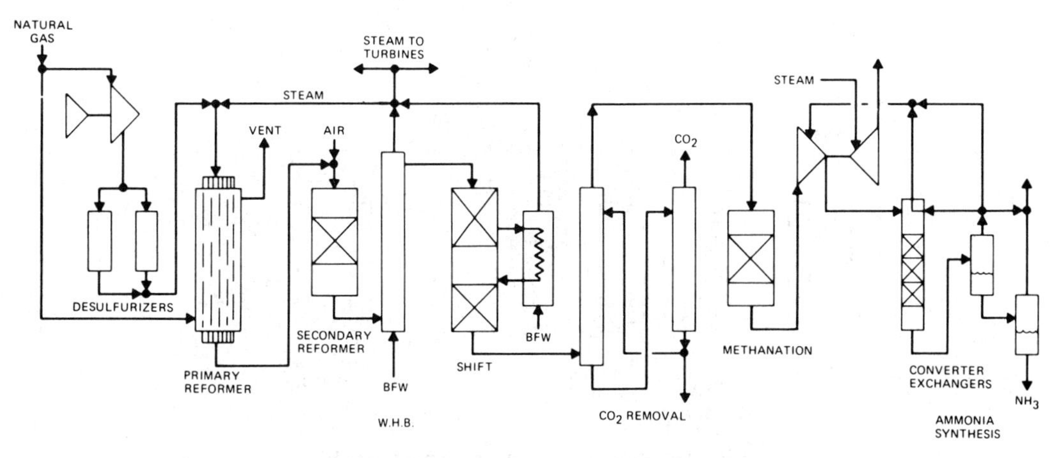

Fig. 5.2 Ammonia synthesis via natural gas reforming.

electrolytic hydrogen is used; in others, hydrogen from petroleum products or natural gas is used.

Processes[10,11] for producing ammonia from several raw materials are shown in Figs. 5.1, 5.2, 5.3, and 5.4.

Fig. 5.1 illustrates a process that employs coal or oil as a raw material. The process includes coal handling, preparation, and pulverization, partial oxidation of coal or oil to synthesis gas, followed by carbon and ash removal, carbon monoxide conversion, carbon dioxide and H_2S removal, low temperature scrubbing with liquid nitrogen, and compression for ammonia synthesis. Air separation and sulfur recovery are steps unique to coal or heavy oil feed stocks.

With natural gas as the raw material, the plants depend on steam reforming to produce hydrogen and carbon monoxide. Air is bled into a secondary reformer to supply nitrogen. The carbon monoxide is then reacted with steam to produce hydrogen and carbon dioxide; the latter is removed and the last traces of carbon monoxide and carbon dioxide converted to methane, which is not a poison in the ammonia synthesis step. The purified hydrogen and nitrogen are compressed and fed to the ammonia synthesis loop.

Recent innovations[12,13,14] have been made to improve the energy efficiency of the natural gas fired plants.

These include:

improved heat recovery
process air preheat
combustion air preheat

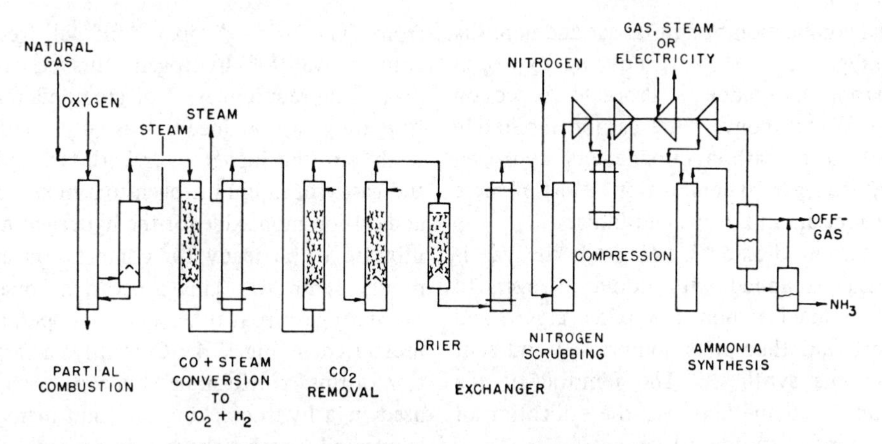

Fig. 5.3 Ammonia synthesis via partial combustion of natural gas.

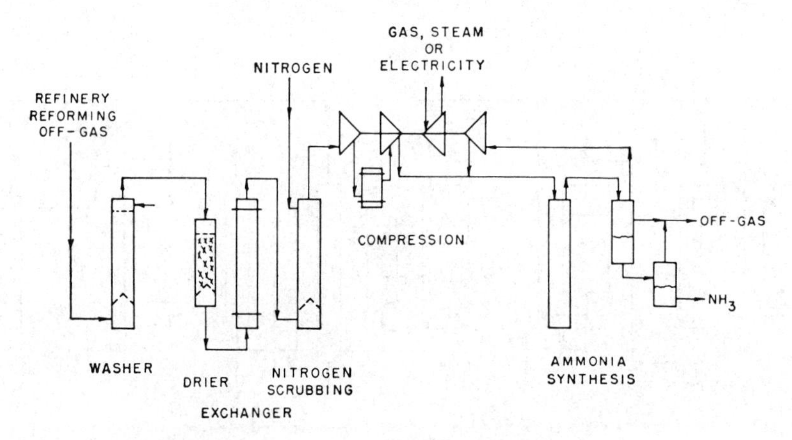

Fig. 5.4 Ammonia synthesis from refinery H_2.

improved CO_2 removal
low synthesis pressure
horizontal ammonia converter
systems low pressure drop
power generation.

Site facilities and economics will dictate in most cases where these new innovations are applicable.

For the Kellog-designed plants, the range of feed and utilities required are as follows:

Natural Gas (feed and fuel)	31.73-28.5	MMBtu/short ton
Demineralized Boiler Feed Water	520-616	gal/short ton
Cooling Water Circulation	68,730-45,920	gal/short ton
Power	1.46--	KWH/short ton
Steam Export (credit)	--1.33	M lbs/short ton

Partial combustion of natural gas and naphtha with oxygen is used for producing hydrogen and carbon monoxide[15]. Processes based on this initial step convert the carbon monoxide with steam to carbon dioxide and hydrogen directly after partial combustion. Most of these installations are in Europe and India.

The carbon dioxide is removed; the gas is dried and scrubbed with liquid nitrogen to remove residual carbon monoxide, argon, and methane, and then it is compressed and sent to ammonia synthesis. The removal of the argon and methane improves the utilization of hydrogen in the synthesis loop.

Partial oxidation has the advantage that several different feed stocks such as natural gas, naphtha, fuel oil, and petroleum fractions can be used. In addition, the H_2/CO ratios of the synthesis gas can be made more favorable for end products such as methanol, oxo alcohols, and acetic acid (pure CO), if these alternatives are desired.

With the advent of "platforming" of petroleum products to give unsaturated compounds for high test gasoline, large volumes of hydrogen off-gas have been produced[16]. The off-gas from "platforming" operations will frequently contain over 90% hydrogen. In such cases, the processing requirements for ammonia manufacture are greatly reduced. It is only necessary to wash out the higher hydrocarbons, and scrub the gas with liquid nitrogen to remove methane and carbon monoxide for the hydrogen-nitrogen mixture to be ready for compression and ammonia synthesis. This process, or one based on hydrogen from other cracking operations, is illustrated in Fig. 5.4. Currently, a large portion of this "platforming" hydrogen is generally used in a hydrocracking unit and many petroleum producers are short of hydrogen.

For an ammonia plant at a new location

TABLE 5.3 Energy Reserves—Proved and Currently Recoverable[17]

	QUADS(A)	
Material	*U.S.A.*	*World*
Natural Gas	213	2290–2680
Natural Gas Liquids	24	250–290
Crude Oil	171	309–380
Shale and Tar Sands	444	1540
Coal	4796	1740–1870
Uranium	20700	–

(A)One Quad is 10^{15}–One Quadrillion-Btu.

where by-product hydrogen is not available, selection of the most economical process will depend to a large extent on the raw material used. The data given in Table 5.3 indicate the relative availability of various energy sources.

At current consumption rates of natural gas in the United States, the proved reserves will be ample for only about another ten years, but estimated total remaining recoverable reserves of 776 to 1195 quads will be ample for about 35 to 55 years. However, the type of feed stock used for ammonia will no doubt be based upon the process that is developed for its utilization and upon regulations governing the end use of available energy sources.

Natural gas is currently the preferred source of hydrogen and energy. Coal is a good source of energy, but a poor source of hydrogen. When coal is used, the hydrogen must be obtained from water. Nuclear fission will supply abundant energy; but the source of hydrogen that fission will utilize is yet to be determined.

Feed Stock Purification

Since most of the ammonia plants today are based on the reforming of natural gas with steam over a nickel catalyst to produce hydrogen, it is imperative that the sulfur content of the feed stock be reduced to as low a level as possible. Natural gas for transportation by pipeline is usually desulfurized to minimize pipeline corrosion. Hence, by the time it reaches an ammonia producer, the only sulfur present is that used to odorize the gas. On the other hand, naphtha which is used as a raw material may require more drastic treatment to remove sulfur.

For natural gas purification, the usual treatment is to pass the gas over specially prepared activated carbon to remove the sulfur. The treatment takes place at ambient temperature and at the pressure level used in the reformer or natural gas line. The higher the pressure, the more effectively the carbon is used.

The activated carbon may be impregnated with iron or copper to enhance the removal of sulfur from the gas. About 100,000 to 200,000 cu ft of natural gas can be treated per cu ft of carbon. Then the carbon must be regenerated with high temperature steam at low pressure. The temperature of regeneration must exceed the boiling point of the adsorbed compound with the highest boiling point. Usually a temperature of 250 to 300°C is satisfactory. A two-bed system is normally used in the activated carbon process—one is in operation while the other is being regenerated.

For desulfurization of naphtha, a more complicated process is required. Hydrodesulfurization is quite often used. About 0.5 mole of H_2 is mixed with 1 mole of vaporized naphtha or 250 SCF per barrel depending upon the sulfur and olefin content. The mixture is preheated to 600°F. It is then passed over a cobalt-molybdenum catalyst where the olefins are hydrogenated to hydrocarbons and the sulfur compounds are converted by the hydrogen to H_2S. The gas is then passed over a sulfur adsorbent such as iron or zinc oxide. It may or may not be necessary to condense the naphtha. This will depend upon the amount of hydrogen used and the need to remove it from the naphtha.

All olefins are hydrogenated to paraffin hydrocarbons in this process. This is advantageous since olefins will otherwise readily crack to carbon in the next step (reforming) of the process. United States Patent 3,106,457[18] gives conditions under which olefins will crack to carbon. The minimum steam to carbon ratio reported to avoid carbon is given as $(Ri)_o = (0.244p + 8.15 \times 10.^{0.0052M})$ where $(Ri)_o$ is the minimum steam to carbon ratio (moles per mole organic carbon atom), p is the operating pressure in psia, and M is the molecular weight of the olefin.

For preparation of coal, after tramp iron and wood are removed, it must be crushed, sized, pulverized, and in some cases slurried with water for injection into the reactors. It may also be profitable to use some applicable method for the type of coal available to remove sulfur-bearing gangue.

Coal selection is important for proper gasifier operation. For slagging, liquid-bottom gasifiers, the melting point of the ash will be critical since a high melting point may result in a frozen slag-tap opening. In other cases, with fixed-bed gasifiers, caking coals may result in agglomeration of ash and coal resulting in blow holes which will allow oxygen and steam to break through into the product gas. High ash content will also result in poor yields and increased costs for ash handling. High sulfur coals may overload the gas clean-up section of the plant if it has not been designed so as to avoid this problem.

Metals content of the coal ash may also be critical since they can form deposits on waste heat boiler tubes or water wall tubes of the gasifiers and decrease tube life.

Hydrogen Production

A list of processes for producing hydrogen is given in Table 5.4, along with the hydrogen to carbon monoxide ratio of the synthesis gas produced. When synthesis gas is made by the gasification of coke with steam and oxygen, the hydrogen to carbon monoxide ratio is 0.6; with the intermittent gasification of coke with air and steam the synthesis gas has a ratio of 0.9. By using excess steam with the reforming of methane, the hydrogen to carbon monoxide ratio can be increased to as high as 5. Thus by selection of the proper process, the equipment requirements for conversion of carbon monoxide to hydrogen with steam can be reduced.

TABLE 5.4 Synthesis Gas H_2/CO Ratio

Method of Manufacture	*H_2/CO Ratio*
Oxygen–coke–steam	0.6
Air–coke–steam	0.9
Oxygen–coal–steam	1.0
Oxygen–fuel oil–steam	1.0
Propane–steam	1.33
Methane–oxygen	1.7
Methane–oxygen–steam	2.3
Methane–steam	3.0–5.0

Very often, the raw material that is available such as natural gas, coal, or oil, will dictate the type of equipment or process employed. Table 5.5 gives raw material requirements for various processes in the production of synthesis gas.

Selection of a process for hydrogen manufacture depends not only on the raw material and its cost, but also on the scale of operation, the purity of the synthesis gas produced, the pressure level of the natural gas, and other carbon monoxide and H_2 requirements.

Today hydrogen is manufactured by four principal processes: (1) steam reforming of natural gas; (2) partial combustion of natural gas or oil with pure oxygen; (3) recovery of hydrogen from petroleum refinery gases or other cracking operations; and (4) gasification of coal or coke with air or oxygen and steam. Small amounts of hydrogen are also manufactured by electrolysis. The four processes are discussed below.

Reforming. The reforming reaction is as follows:

$$CH_4 + H_2O \longrightarrow 3H_2 + CO.$$

The reaction is endothermic and requires the input of a large amount of heat. The heating-gas requirement is about 80 percent of the process-gas requirement. The reaction is carried out at temperatures in the range of 800°C and at pressures ranging from atmospheric up to 500 psig.

Other reactions which proceed at the same time as the reforming reaction are:

$$CO + H_2O \rightleftharpoons CO_2 + H_2$$

$$CH_4 \rightleftharpoons C + 2H_2$$

$$2CO \rightleftharpoons C + CO_2$$

The equilibrium composition dependends upon the steam to gas ratio entering the reactor, the temperature, the pressure, and the quantity of inerts in the reaction mixture. Hougen and Watson[19] give a general equation which simplifies to the following for reforming of CH_4 with steam:

$$K = \frac{(N_{CO})(N_{H_2})^3}{(N_{CH_4})(N_{H_2O})}(Kv)\left(\frac{\pi}{N}\right)^2$$

where K is the equilibrium constant; N_{CO}, N_{H_2}, N_{CH_4}, N_{H_2O} are the moles of each species present; Kv is the fugacity factor; π is the total pressure; and N is the total number of moles present. The equilibrium constant for the methane reforming reaction is as follows:

Temperature, °F	*Equilibrium Constant*
600	2.186×10^{-7}
800	2.659×10^{-4}
1,000	4.900×10^{-2}
1,200	2.679
1,400	0.6343×10^{2}
1,600	8.166×10^{2}
1,800	6.755×10^{3}

The final composition will also satisfy the carbon monoxide-shift reaction. The equilibrium constant for this reaction is:

Temperature, °F	*Equilibrium Constant*
600	31.4
800	9.03
1,000	3.75
1,200	1.97
1,400	1.20
1,600	0.819
1,800	0.604

To avoid carbon formation as indicated by the foregoing reactions, the steam-to-gas ratio must be maintained high enough to favor the reforming and shift reactions in preference to those forming carbon.

TABLE 5.5 Synthesis Gas Raw Material Requirements

Method of Manufacture	*Requirement per 1,000 cu ft of ($CO + H_2$)*
Air-coke-steam	40 lb Coke
	80 lb Steam
	2,200 cu ft Air
Oxygen-coke-steam	30 lb Coke
	300 cu ft Oxygen
	34 lb Steam
Coal partial combustion	37 lb Coal
	350 cu ft Oxygen
	37 lb Steam
Natural gas partial combustion	360 cu ft Natural gas
	260 cu ft Oxygen
Natural gas reforming	285 cu ft Natural gas
	230 cu ft Natural gas for fuel
	60 lb Steam
Propane reforming	3 gal Propane
	1.67 gal Distillate for fuel
	380 lb Steam[a]
Naphtha reforming	14.7 lb Naphtha
	5.8 lb Naphtha for fuel
	105 lb Steam
Fuel oil partial combustion	20.5 lb Oil
	320 cu ft Oxygen
Electrolysis	163 Kw-h
Steam-iron	290 lb Steam
	0.3 lb Ore
	206 cu ft Fuel gas
	65 lb Coke
Coke-oven gas	353 lb Coal[b]
"Platforming" off-gas	84 gal Feed stock

[a]Includes CO shift to CO_2 and H_2 with steam.
[b]Excludes CH_4 + CO in coke-oven gas which can be converted to H_2.

A catalyst is required for this reaction. Requirements are: high activity, long life, good physical strength, and low cost. Such a catalyst can be prepared by mixing nickel oxide powder, alumina, and calcium aluminate cement, forming the mixture into the desired shapes (for example, $\frac{5}{8} \times \frac{5}{8} \times \frac{1}{4}$ inch Raschig rings) and then calcining the catalyst at a temperature of 1200 to 2000°F.

Excellent properties are also obtained by impregnating a support material, such as alumina that has been fired at 1500°C, with a nickel salt solution, and then firing the catalyst.[20]

Although silica and silica-bearing materials are very useful in making catalyst supports, their use is prohibited because the silica volatilizes and migrates from the hotter zone to lower temperature zones downstream. Usually it deposits on the waste heat boiler tubes after the secondary reformer.

Normally, the nickel oxide is reduced to nickel and water by the hydrogen that is produced in the operation. However, it has been shown[21] that in some cases, the reduced nickel can be reoxidized by high temperature steam to the nickel oxide when large amounts of steam and small amounts of H_2 are present.

$$Ni + H_2O \rightleftharpoons NiO + H_2$$

U.S. Patent 3,827,987 claims that addition of 0.5% to 10% by volume of H_2 in the natural gas feed will keep the nickel in the reduced state, thus making it more active. Also, the hydrogen will retard the formation of nickel sulfide, thus avoiding poisoning of the catalyst.

$$NiS + H_2 \rightleftharpoons Ni + H_2O$$

Poisons[20] for the nickel catalyst are: sulfur, arsenic, chlorides or other halogens, phosphates, and copper or lead. A 15% nickel catalyst is poisoned at 775°C when it contains 0.005 percent sulfur. This is equivalent to reaction of all the nickel on the surface of the crystallites 1 micron in diameter. For lower operating temperatures the amount required for poisoning is still lower. When using naphtha as a feed stock, 0.5 ppm of sulfur (w/w) in the naphtha is the maximum allowed concentration for operation at 775°C.

Carbon can also reduce the effectiveness of the catalyst. When conditions are favorable for the reactions, carbon deposits as follows:

$$CH_4 \longrightarrow C + 2H_2$$

$$2CO \longrightarrow C + CO_2$$

Thus, where there is insufficient steam, carbon will be deposited on the catalyst, reducing its effectiveness. Carbon will gasify with steam at 800°C or higher in the reformer.

$$C + H_2O \longrightarrow CO + H_2$$

Operation and changes in operation should always be made to insure sufficient steam over the catalyst; thus, if the feed rate is to be increased, the steam rate should be increased first. Conversely, if the feed rate is to be decreased, the natural gas rate should be decreased first.

A reforming reaction furnace usually consists of a number of 2 to 8 inch centrifugally cast chrome-nickel tubes (25% Cr, 20% Ni, 0.35–0.45% C) (Grade HK-40) about 10 to 40 feet long, mounted vertically in a refractory-lined furnace. The process gas and steam are fed downward over the catalyst and removed from the bottom of the tubes. The tubes are heated by burners located strategically in the sides, top, or bottom of the furnace. Close control of the heat is essential to ensure uniform temperature distribution throughout the tubes.

To save natural gas for ammonia manufacture, some plants have converted from natural gas firing of the reformer to fuel oil firing. This requires a change in the burners with addition of steam to atomize the oil. Careful selection of the oil is necessary to avoid damage to the tubes. Fuel oils contain various amounts of vanadium, sodium, and sulfur which will form deposits and cause pitting of tubes.

During change from atmospheric temperature to reaction temperature (800°C), expansion of the catalyst tubes is considerable; thus provision must be made for this growth. Usually, the tubes are suspended by springs, or counterweights are attached to the tops. Feed materials are introduced through flexible tubing.

At the high operating temperature in the reformer, severe tube damage can result[22] from carry-over of boiler feed-water salts. The carry-over of boiler feed-water salts can cause sulfidation and molten salt attack on the tubes, manifold, and transfer lines. Carbonization of methane and higher hydrocarbons will carburize most metals rapidly. Also, trace amounts of lead in the alloy will promote catastrophic oxidation. and burner fuels may contain sodium, vanadium, and sulfur compounds which will attack the outer surface of the tubes.

Some of the high-temperature transfer lines are protected by castable or firebrick linings, with the outer retaining piping water-cooled. The inner refractory liners are also protected by shrouds of Type 304 or 321 stainless steel.

Preheated steam and natural gas are mixed in the ratio of 3 to 5 prior to introducing these into the catalyst tubes. Usually, the flue gases are sufficient to provide the preheat.

Steam to carbon ratios of 4 to 8 are claimed to be advantageous; such ratios result in less fouling of waste-heat boilers and give better reformer-catalyst and tube life.[23] Tube life is in the range of 6 to 12 years.

The hydrogen to carbon monoxide ratio for the product gas is usually 4 to 5; however, this can be increased if additional steam is used in the reaction. This will reduce the demand in the carbon monoxide-shift converters which follow the secondary reformer. Steam is sometimes introduced after the reformer before the gas is fed to the carbon monoxide-shift converter.

In the single-train ammonia plants, the natural gas is reformed in two steps. In the first step, the reaction takes place in the primary reformer in tubes suspended in a refractory-lined furnace. The large endothermic heat is supplied by burning natural gas with air in the furnace. Heat flux in the tubes will be as high as 35,000 Btu/hr/sq ft. The methane leakage is about 10 percent in the effluent dry gas, or about 60 to 65 percent of the feed methane is converted to synthesis gas.

For ammonia manufacture, it is necessary to introduce nitrogen into the system. This is done as indicated in Fig. 5.2 by mixing air with the primary reformer gas in a refractory-lined vessel which contains additional reforming catalyst. The temperature at the point of mixing will reach 2,300°F because of the fast reaction of the oxygen in the air. But, then the endothermic reforming reaction takes place and the gases cool to about 1,800°F. At this temperature, the methane concentration has been reduced to about 0.3 percent. The concentration is dependent upon the pressure, temperature, and the quantity of nitrogen and steam present.

The catalyst used in this reforming step does not need to be as active as that in the primary reformer. Hence, the usual nickel concentration is about 15 percent compared with 25 percent in the primary-reformer catalyst.

Mixing of the air and primary reformer gas is critical and specially designed burners are used.

The hot gases from this section of the process provide sufficient heat to generate the large amount of the steam necessary to supply the process and drive the compressors downstream in the plant.

Partial Combustion. By burning natural gas with a limited quantity of oxygen, a synthesis gas containing approximately 2 moles of hydrogen for each mole of carbon monoxide can be produced.[24]

$$CH_4 + \tfrac{1}{2}O_2 \longrightarrow CO + 2H_2$$

The reaction is exothermic and must be carried out in a refractory-lined or water-cooled vessel. High temperatures favor the formation of carbon monoxide and hydrogen.[25] Pressure will hinder the reaction; however, pressures up to 1,500 psi can be used and still produce a gas of low methane content. If steam is introduced into the reaction, the hydrogen to carbon monoxide ratio can be increased to over 2. Because of the high temperatures involved, e.g., 1,400°C, special care must be observed in processing the gas after it leaves the reactor.

Preheating the natural gas and oxygen will reduce the requirements for oxygen and natural gas, since some of the natural gas must be burnt to carbon dioxide in order to raise the products to their proper temperature. The partial combustion reaction itself is not sufficiently exothermic to raise the products to the desired temperature.

The partial combustion of fuel oil or crude oil is practiced in many areas where natural gas is not available. When using crude oil, care

TABLE 5.6 Product Gas Compositions from Various Feeds by Partial Combustion (percent by volume dry)

	Feedstocks					
Gas Components	*Nat. Gas*	*Light Naphtha*	*Heavy Fuel Oil*	*Vac. Resid.*	*Propane Asphalt*	*Coal*
Hydrogen	61.80	52.09	46.74	44.80	43.69	34.39
Carbon Monoxide	33.75	42.59	48.14	49.52	50.09	44.22
Carbon Dioxide	3.10	4.88	3.80	4.16	4.45	18.64
Methane	1.00	0.30	0.30	0.30	0.30	0.38
N_2 + A	0.27	0.13	0.23	0.18	0.30	0.68
H_2S	–	0.01	0.76	1.00	1.12	1.46
COS	–	–	0.03	0.04	0.05	0.10
NH_3	–	–	–	–	–	0.13
H_2/CO mol/mol	1.83	1.22	0.97	0.90	0.87	0.78

must be taken to provide for removal of sulfur compounds and ash-containing materials in the crude oil. These have been known to damage the refractory and, of course, to introduce sulfur into the product gas. In the case of crude-oil or fuel-oil partial combustion, the carbon content is much greater than in the case of natural gas, and special design considerations are necessary to produce a satisfactory gas.

Gas compositions from partial combustion of various fuels are given in Table 5.6.

Nitrogen to produce H_2/N_2 ratio is usually introduced later in the processing sequence.

Coal and Coke Gasification. An important process for the gasification of coal and coke is the Lurgi process, used in South Africa at the "coal to oil" plant; it is also used in Europe. This process requires the use of coke or a non-caking coal and oxygen; the product gas is usually high in methane. The reaction is carried out with a fixed bed of burning coal or coke into which steam and oxygen are introduced. Oxygen required is 0.11 SCF O_2/SCF of synthesis gas when using lignite and 0.21 SCF O_2/SCF of synthesis gas when using anthracite.[27] This and other gasification processes are described in Chapter 3, Coal Technology.

There are several processes for purifying synthesis gas made from coal, coke, or oil. The impurities usually present are H_2S, CS_2, COS, ungasified carbon, and residual ash. Other factors which influence the selection of a purification process are the hydrogen to carbon monoxide ratio, and the carbon dioxide content.

If the gas contains large amounts of sulfur, it is quite often the practice to remove this sulfur before the carbon monoxide-shift operation. There are a number of sulfur removal processes, namely, amine scrubbing, hot potassium carbonate scrubbing, and the Rectisol process. Ash and carbon removal are usually carried out before the gas is treated for sulfur or carbon dioxide removal and before the shift operation.

If a gas does not contain large quantities of sulfur and ash, the carbon monoxide-shift reaction is carried out directly after the synthesis operation. In some instances, it is also desirable to remove the carbon dioxide before the shift operation, since it has an adverse effect on the carbon monoxide-shift equilibrium. Carbon dioxide may be removed by scrubbing with water, monoethanol-amine (MEA), or hot potassium carbonate. The preferred process depends to some extent on the local economics and the particular conditions under which the synthesis gas is produced.

Hydrogen from Petroleum Gases. Petroleum gases from platforming operations, hydroformers, and butadiene plants may be processes to recover H_2.[28] The gas from the platformer is particularly rich in H_2, containing as much as 90 to 95 percent. This gas is usually purified by low temperature fractionation or washing with liquid nitrogen.

Electrolysis. In some areas hydrogen is obtained by electrolysis, but normally this is not economical. Electric power must be available at a very low cost such as that from hydroelectric power.[29] In some cases, by-product hydrogen from the electrolysis of brine has been used to manufacture ammonia.

Carbon Monoxide Shift

Conversion of carbon monoxide to carbon dioxide and hydrogen with steam is necessary to make economical use of the raw synthesis gases produced by the foregoing processes. In this operation, the following reaction is promoted:

$$CO + H_2O \rightleftharpoons CO_2 + H_2$$

The reaction is exothermic; the equilibrium is not affected by pressure, but high temperatures are unfavorable for complete conversion.

The equilibrium constant for the reaction is as follows:

Temperature, °F	*Equilibrium Constant*
400	207
450	119
500	72.8
550	46.7
600	31.4
650	22.0
700	15.9

In recent years the shift operation has been carried out in two steps, one at a high temperature of 350 to 450°C, and one at a low temperature of 200 to 300°C. The high-temperature shift reaction is carried out over an iron-chromium catalyst which is not particularly sensitive to sulfur, although high concentrations have been found to be detrimental. At the usual effluent temperature, the dry gas will contain about 3 percent carbon monoxide.

To operate the low-temperature shift reaction, it is necessary to remove heat between the two catalyst beds. The low-temperature shift catalyst contains copper and zinc and is very sensitive to sulfur and halogens.[30] Therefore a guard catalyst is frequently used to protect the shift catalyst.

High steam-to-gas ratios favor conversion of the carbon monoxide to hydrogen and a low temperature rise, resulting in a more favorable equilibrium. The normal carbon monoxide leakage is about 0.3 to 0.5 percent at an exit temperature of 240°C

The low-temperature shift catalyst can be purchased in a reduced state. If it is not, it must be reduced before placing it in operation. The cupric oxide must be reduced to metallic copper.

Experience in operation of the large single-train ammonia plants has shown that performance of the low-temperature shift catalyst is critical to good operating continuity. The use of a small guard vessel ahead of the main low-temperature shift catalyst bed is valuable.[31] The following operating practices have been found to be helpful.[31]

1. Dry reduction of catalyst using natural gas or nitrogen as the carrier instead of steam.
2. The use of a low-temperature shift catalyst instead of zinc oxide as a guard catalyst.
3. Elimination of known sources of poisons.
4. A separate guard vessel.
5. Regular replacement of guard catalyst.
6. Dry inlet temperature control in place of quench.
7. Avoidance of approach to the dew point of less than 22°C.

Carbon Dioxide Removal

Following the shift conversion of CO with steam to CO_2 and H_2, the bulk of the CO_2 must be removed from the hydrogen and nitrogen. There are numerous processes for this operation and the selection of the most favorable one depends upon local circumstances.[32] Carbon dioxide can be removed by scrubbing the gas under pressure with various solvents. Processes for CO_2 removal are:[9]

Hot Potassium Carbonate: Benfield or Catacarb
Vetrocoke (arsenical)
Monoethanol Amine
Triethanolamine
Water
Sulfinol–Sulfolane ($C_4H_8SO_2$); (diisopropanol amine)
Selexol–Dimethyl ether of polyethylene glycol
Rectisol–Refrigerated Methanol
Fluor Solvent–Propylene Carbonate

Alkazid–Potassium-N-methyl-aminopropionate
Purisol–N-methyl-pyrrolidone (NMP)

Water Scrubbing. This system, once used widely, has not remained competitive. However, it appears that such a system is currently being offered by Friedrich Uhde GMBH in Germany.[32] A large quantity of water is required to dissolve the carbon dioxide and special design is required to avoid excessive loss of nitrogen and hydrogen. Power recovery is important, which is achieved by use of properly designed turbines.

Hot Potassium Carbonate.[34] In this system, an aqueous solution containing 40 percent K_2CO_3 is circulated at 118°C in the absorber. The CO_2 picked up is released in the stripper at atmospheric pressure by heating the solution to 230°C. An inhibitor is generally needed to reduce corrosion. Stainless steel equipment is used in many places. This system is characterized by its low steam requirements and its compatibility with the catalysts.

The system may be modified by adding materials which enhance the solubility of CO_2 and hence result in more complete recovery. Alkanolamines and As_2O_3 are typical additives.

Monethanolamine (MEA). A 15 to 30 percent solution of MEA in water is used to absorb the CO_2 under pressure. The solution is then regenerated by heating in a stripper in which the carbon dioxide is released. This system is characterized by good absorption properties at low pressure.

A corrosion inhibitor is also used with this process. With the Amine Guard[35] inhibitor, the MEA concentration in the circulating solution can be increased to 30% from a normal 20%. Hence, the circulating rate can be decreased by 33% and the heat load is decreased 43%. Stainless steel is used in critical areas, where hot carbon dioxide-laden solutions are present. Fig. 5.5 gives a picture of a conventional ammonia plant designed by Kellogg with natural gas feed and "Amine Guard-MEA CO_2 Removal."

It is reported that antifoam agents, e.g. certain silicones, are used in some cases with this process.

Sulfinol.[36] The sulfinol process is a development by the Shell Development Corp., in Emoryville, California. The solution used consists of sulfolane ($C_4H_8SO_2$), a ring compound, disisopropanolamine, and water. This process is characterized by its high CO_2 retention under pressure, and by the fact that steam requirements for regeneration are small. A sidestream regenerator is required to remove by-products which build up in the system. See U.S. Patent 3,347,621.[37]

Mild steel is suitable in a number of places; however, stainless is preferred where CO_2 concentrations are high.

Fig. 5.5 Large ammonia plant. *(Courtesy Pullman-Kellogg.)*

The process must be carefully engineered to protect the downstream nickel catalysts from the sulfur which may carry over from the absorber. Likewise, the CO_2 that is produced must be protected from entrainment of the solvent and contamination with sulfur.

Propylene Carbonate.[38,39] This process was developed primarily for the removal of CO_2 from natural gas. Two 1200 T/D ammonia plants use the Fluor Solvent, propylene carbonate. Regeneration of the solvent is accomplished by pressure release rather than by heating. The heat saved can be used in other areas of the plant, such as in ammonia recovery in the synthesis loop.

The absorber operates at a relatively low temperature, made possible by the refrigerating effect of the released CO_2. Mild steel is suitable for construction in most areas of the system. If the CO_2 is needed for urea manufacture, it can be released at 100–200 psi.

Refrigerated Acetone.[40] Refrigerated acetone has been studied for CO_2 removal. It is not widely used.

Rectisol-Refrigerated Methanol.[41,42] This system is in use at the South African coal to oil plant. The methanol is cooled to $-50°C$ at which temperature the vapor pressure is less than 1 mm Hg. The Rectisol process is used chiefly for the following applications:

1. "Complete and simultaneous removal of CO_2, H_2S and organic sulfur, NH_3; hydrogen cyanide, gumformers, higher hydrocarbons and other impurities from coal gasification crude gas to produce gas for synthesis or pipeline transmission."
2. "Removal H_2S, COS, and CO_2 from reformed gas, in particular from gas produced by partial oxidation of hydrocarbons, to yield synthesis gas."
3. "Integration of gas purification with low temperature plants (liquefaction and fractionation) for the complete removal of moderate contents of acidic components."

The system can be designed to produce a stream rich in H_2S for processing in a Claus unit to recover sulfur and meet emission standards.

Giammarco-Ventrocoke.[43] This process is based upon the absorptive power of an arsenic-activated potassium carbonate solution. Both H_2S and CO_2 are removed. The CO_2 level can be reduced to about 0.05 percent, and CO_2 with a purity of 99 percent can be produced.

$$CO_2 + K_2CO_3 + H_2O \rightleftharpoons 2KHCO_3$$
$$6CO_2 + 2K_3AsO_3 + 3H_2O \rightleftharpoons 6KHCO_3 + As_2O_3$$

The solution is reported to be noncorrosive to steel.

Final Purification

Before the ammonia synthesis gas ($3H_2$ and $1N_2$) is admitted to the ammonia synthesis step, essentially all of the caron monoxide and carbon dioxide must be removed since they are temporary poisons for the ammonia catalyst. Also, the presence of carbon dioxide in the make-up gas causes difficulty in the synthesis loop because it forms carbamates and ammonium carbonate which are solids that can damage compressors and valves.

Methanation. This is the most popular process for final purification or removal of the last traces of carbon monoxide and carbon dioxide. This process is the reverse of the methane reforming process.

$$CO + 3H_2 \rightleftharpoons CH_4 + H_2O$$
$$CO_2 + 4H_2 \rightleftharpoons CH_4 + 2H_2O$$

A large excess of hydrogen present and a lower temperature favor the complete removal of the oxides of carbon. Usually, the remaining oxides are less than 10 ppm.

In this process, the synthesis gas, which has been scrubbed to remove the bulk of the CO_2, is preheated to about 600°F and then passed over a nickel catalyst (NiO and Al_2O_3). Ruthenium is also effective as a catalyst. The space velocity used depends upon the pressure of the process. At 450 psig the space velocity (standard cu ft of gas per hour per cu ft of catalyst) is about 5,000 hr^{-1}. At lower pressures a space velocity of 2,000 hr^{-1} is used.

The catalyst life is normally one to three years. Sulfur and halogens are poisons and shorten the life.

Equilibrium constants[44] for the two reactions are:

T°C	$\dfrac{(P_{CH_4})(P_{H_2O})}{(P_{CO})(P_{H_2})^3}$	$\dfrac{(P_{CH_4})(P_{H_2O})^2}{(P_{CO_2})(P_{H_2})^4}$
200	0.21547×10^{12}	0.94748×10^{9}
260	0.45626×10^{9}	0.62706×10^{7}
320	0.32635×10^{7}	0.11001×10^{6}
380	0.56011×10^{5}	0.38882×10^{4}
440	0.18550×10^{4}	0.23282×10^{3}
500	0.10219×10^{3}	0.20997×10^{2}

Both reactions are exothermic; the heats of reaction are ($\Delta H_{25°C}$) -49.27 and -39.44 K cal/mole of carbon monoxide and carbon dioxide, respectively.

Nitrogen Wash Operation. The last traces of carbon monoxide are removed by liquid nitrogen scrubbing in some plants, generally those using a partial oxidation process for generation of the synthesis gas. The nitrogen is obtained from the air-separation plant used for oxygen production. In the nitrogen-wash operation, the gas must be thoroughly dried and the carbon dioxide removed before the gas is cooled to the point where it can be scrubbed with liquid nitrogen. Otherwise, the moisture and carbon dioxide would freeze out in the equipment, causing a number of problems. Hence, the gas must be passed through silica gel driers and a caustic wash to remove the water and carbon dioxide. At his point, the gas can safely be cooled to liquid-nitrogen temperature and scrubbed to remove the carbon monoxide, methane, and argon.

The overhead gases from the wash tower usually contain about 90 percent hydrogen and 10 percent nitrogen, thus requiring the addition of more nitrogen to give the required ratio for ammonia synthesis. The gas so produced is suitable for ammonia synthesis.

Copper Ammonium Carbonate Scrubbing. In some instances, it is preferable to use copper scrubbing to remove the last traces of carbon monoxide in the ammonia synthesis gas. This process is usually carried out at pressures in the range of 1000 to 1500 psi, one of the intermediate pressures of the high pressure compressors. The copper liquor is circulated counter-current to the synthesis gas, picking up the carbon monoxide as a copper ammonium carbonate-carbon monoxide complex and producing a relatively carbon monoxide-free gas. Methane and argon, however, are not removed; these pass on with the hydrogen to the ammonia synthesis units.

The copper liquor is regenerated by reducing the pressure on the liquid, which allows the carbon monoxide complex to decompose. The carbon monoxide is separated from the copper liquor, and the liquid is compressed and recirculated to the scrubbing operation.

Selective Oxidation of Carbon Monoxide. This process[45] is based on the fact that over a platinum catalyst at 250 to 320°F, carbon monoxide will be selectively oxidized to carbon dioxide. Excess oxygen is converted to water by hydrogen.

$$CO + \tfrac{1}{2}O_2 \rightarrow CO_2$$

$$H_2 + \tfrac{1}{2}O_2 \rightarrow H_2O$$

The disadvantage of the process is that a second scrubbing step is necessary to remove the CO_2 formed by oxidation of the carbon monoxide.

Cryogenic Purifier.[46] In the C.F. Braun process for ammonia manufacture, excess air is added in the second reformer. The excess nitrogen thus introduced must be removed prior to the synthesis step to avoid excessive loss of hydrogen and excessive compression costs. At the same time that the excess nitrogen is condensed and removed, the remaining traces of carbon monoxide, the methane, and most of the argon are removed, leaving a gas comparable to that produced by the nitrogen wash operation.

U.S. Patent 3,584,998 claims the primary reformer can be eliminated by adding more excess air, preheating the air, natural gas, and steam and carrying out the entire reforming step in a single-bed vessel. In this process modification, the feed to the purifier would contain 60 to 70% hydrogen, 30 to 40% nitrogen, 2 to 3% methane, and about 0.5% argon.

Compression

With the advent of large-capacity plants (over 600 T/D), it has been possible to design centrifugal compressors for the services required. The pressure level has also dropped to about 2,000 psi to favor compressors which had been successfully used by industry. However, more and more machines are now being built for higher pressures, up to 5,000 psig. One of the advantages of the centrifugal compressors is that the need for lubricating oil is minimal, for this material causes difficulty in the synthesis loop.

The centrifugal compressors are driven by steam turbines. High-pressure, high-temperature steam is generated in the plant using the heat in the process gas leaving the secondary reformer and the flue gases from the primary reformer.

Design of the compressors is critical from the stand-point of efficiency and economy. About 25 percent of the downtime of large plants results from compressor troubles. As aid to compressor operation, vibration sensors have been found to pinpoint sources of difficulty.[47]

Nonideality of the gases must be considered when working at high pressures. Compressibility factors for a mixture of 3 hydrogen and 1 nitrogen are available.[48]

About 20 percent of the energy used in ammonia manufacture goes into compression (synthesis gas, air, and refrigeration). A study[49] of new alternatives suggests that gas turbines and more efficient steam systems (higher temperatures and reheat) should be considered.

For smaller plants, under 600 T/D, reciprocating compressors are still used. These often contain several services on a common drive shaft: air compression, natural gas compression, synthesis gas compression and recycle, and ammonia compression for refrigeration.

Ammonia Synthesis

Ammonia synthesis is carried out at pressures ranging from 2,000 to 10,000 psi. The preferred pressure depends largely on the quality of the synthesis gas and certain other conditions, such as production requirements per converter.

Quartulli, et al,[50] published a detailed study of the factors to be considered in selecting the proper synthesis pressure. Reaction equlibrium, reaction rate, and condensation of ammonia are all favored by high pressure. Mechanical problems and cost considerations are also considered. In general, the study concluded that operation at 2100 psi is more economical than that at 3,300, 4,700, or 6,100 psi. The following were some of the factors considered: catalyst volume, recycle flow, horsepower for make-up-gas compression, recycle-gas flow, refrigeration, equipment cost, and equipment reliability.

The ammonia reaction is $N_2 + 3H_2 \longrightarrow 2NH_3$; it is exothermic. See Reference (51) for a discussion of equilibrium values for the reaction.

Lower temperatures favor higher equilibrium concentrations of ammonia; however, the rate of reaction at low temperature is so slow that the process is uneconomical at temperatures below about 400°C. Usually a conversion of about 15 to 30 percent per pass is obtained when the temperature is in the range of 400 to 600°C.

The heat of reaction for the synthesis of ammonia is given by the following equation:[51]

$$\Delta H \text{ (cal/mole)} = -\left[0.54526 + \frac{840.609}{T} + \frac{459.734 \times 10^6}{T^3}\right] p - 5.34685T - (0.2525 \times 10^{-3})T^2 + (1.69167 \times 10^{-6})T^3 - 9157.09$$

where T is °K and p is in atmospheres.

Reaction Rate. The rate of reaction over the synthesis catalyst dependends upon the type of catalyst, rate constants for synthesis and decomposition, and the partial pressures of nitrogen, hydrogen, and ammonia. Temkin and Pyzhev[51] propose the following equation:

$$W = K_1 P_{N_2}\left(\frac{(P_{H_2}^3)}{(P_{NH_3}^2)}\right)^{\alpha} - K_2\left(\frac{(P_{NH_3}^2)}{(P_{H_2}^3)}\right)^{\beta}$$

where W is the reaction rate; K_1 and K_2 are the rate constants for synthesis and decomposition, respectively; P_{H_2}, P_{N_2}, and P_{NH_3}, are partial pressures; α, β are constants. The constant α has been found to be in the range of 0.5 to 0.75 and β, 0.4 to 0.3; α, $\beta = 1.0$. Tempkin, and others, established this equation on the basis

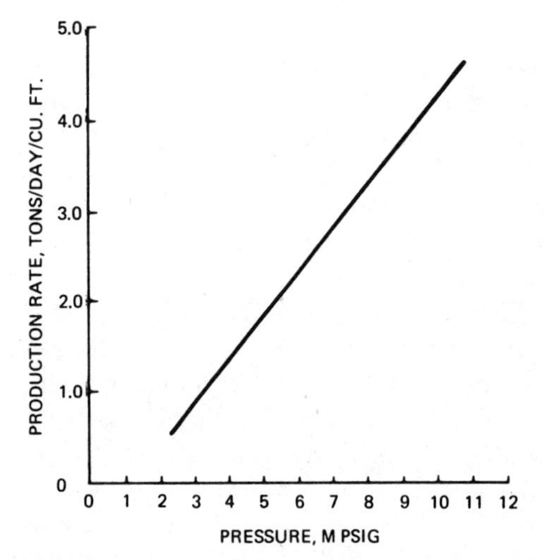

Fig. 5.6 Ammonia production rate as a function of pressure.

that the rate determining steps are the chemisorption of nitrogen in a diatomic form and the combustion of a chemisorbed nitrogen molecule with a hydrogen molecule.

At high pressure and with some ammonia in the gas, the effect is to have only one rate-determining step.[52] Figure 5.6 gives some indication of the production rate as a function of pressure. Most systems are designed for 2,000 to 5,000 psi.

The gas composition for ammonia synthesis is also critical, relative to the cost of ammonia. High concentrations of inert materials such as argon or methane will build up in the circulating system and produce a low partial pressure of hydrogen and nitrogen resulting in low conversions to ammonia. In such cases, it is necessary to purge these materials from the system and maintain the proper ratio of hydrogen to nitrogen. Also, an excess of either one of these reactants will act as an inert material and lower the effective pressure of the controlling component. Again, it would be necessary to purge excessively to restore the desired degree of conversion.

In design of the equipment, it is necessary to take into account the solubility of the various gases in the synthesis loop. The solubilities[53] of the gases in cc of gas at normal pressure and temperature per gram of liquid ammonia per atm partial pressure are:

Gas	*5°C*	*30°C*
Hydrogen	0.095	0.112
Nitrogen	0.117	0.126
Argon	0.151	0.161
Methane	0.270	0.392

It should be noted that all four gases are more soluble at the higher temperature.

The optimum concentration of inerts will depend on the temperature and pressure of the operation and the concentration of inerts in the make-up gas. The solubilities of argon, methane, hydrogen, and nitrogen in liquid ammonia determine the amounts of these gases which are removed from the system with the liquid ammonia. The designer has some control over this purge. The conditions of synthesis can be selected to give optimum purge rates once solubility, make-up gas composition, and degree of conversion are determined.

The production rate of ammonia is dependent upon the recycle rate for the gases in the loop, the degree of conversion attained in the catalyst and the efficiency with which the ammonia is removed. Table 5.7 gives some ranges for the saturated-vapor ammonia concentration in a $3H_2 : 1N_2$ gas mixture over liquid ammonia.

TABLE 5.7 Ammonia Concentration Over Liquid Ammonia

Pressure (Atm)	*Temperature (°C)*	*% NH_3 in Gas*
50	-20	5.7
	0	10.0
	20	19.0
	30	25.5
100	-20	3.25
	0	5.9
	20	11.0
	30	15
200	-20	2.05
	0	3.95
	20	7.5
	30	10.04
300	-20	1.62
	0	3.25
	20	6.4
	30	8.8
700	-20	1.19
	0	2.35
	20	4.90
	30	6.80

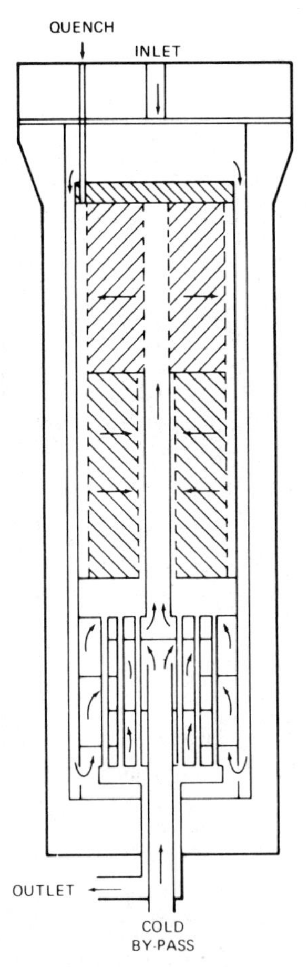

Fig. 5.7 Topsoe radial-flow converter. *("Nitrogen," Sept. 1964, The British Sulfur Corporation.)*

Ammonia liquid viscosity[64] may be calculated from the following equation:

$$\mu = \mathrm{b}\mathrm{T}^{m}$$

μ = Viscosity in centipoises, cP
b = 6.301×10^8 Correlation constant
$m = -3.921$
T = Temperature, °K
μ at $-50°$ C is 0.317 cP.

Converter Details. Two relatively new converter arrangements are shown in Figs. 5.7 and 5.8. These both contain an internal heat exchanger, and the bypass inlet is for temperature control purposes. As in most converters, the cold inlet gas passes between the catalyst basket and the shell to prevent overheating of the shell and hydrogen embrittlement.

In Fig. 5.7, a radial-flow converter is shown. This is a design by Topsoe. The main feature of this converter is the provision of a much greater cross section to the flow of gas, thereby reducing pressure drop. This low pressure drop makes possible the use of smaller catalyst particles with some increase in activity per unit volume.

Kellogg has introduced the horizontal converter[54], as shown in Fig. 5.8. This arrangement gives a large cross-sectional area for the flow of gas, while still using a shell that permits full closure. A small catalyst size is thus permitted, with an accompanying reduction in volume.

In addition to the care which must be exercised with respect to materials of construction, it is also necessary to design for heat removal. This is usually accomplished by internal heat

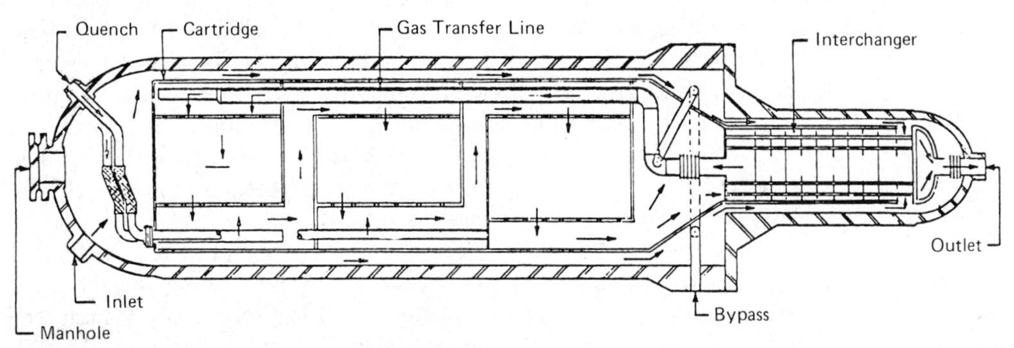

Fig. 5.8 Horizontal ammonia converter, Kellogg design. *(Hydrocarbon Processing, pp 115–122, Dec. 1978, copyright 1978 by the Gulf Publishing Company.)*

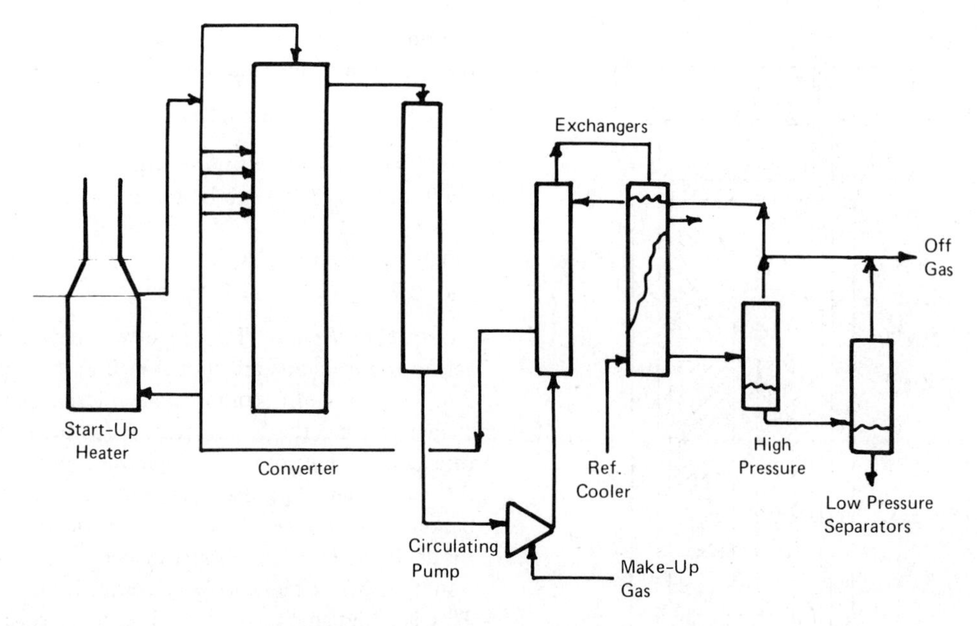

Fig. 5.9 Circulating system for ammonia synthesis.

exchangers in the converter. The incoming gas is heated by the exit gas, which is thereby cooled. A temperature rise of about 150°C is used in the design of a converter.

One of the current circulating systems for ammonia synthesis is shown in Fig. 5.9. Control of the gas composition is important in the synthesis loop. Means of achieving this is revealed in U.S. Patent 2,894,821.[55] The hydrogen to nitrogen ratio must be maintained close to 3 and the concentration of inert materials, methane and argon, must be limited to levels which will permit a satisfactory partial pressure of the H_2 and N_2. In the process described in the patent, the circulating gas in the synthesis loop is analyzed for hydrogen and nitrogen and the inlet air flow to the secondary reformer is adjusted to maintain the ratio of hydrogen to nitrogen at 3.0.

Control of many modern ammonia plants is now accomplished via gas analyzers and on-line computers to adjust conditions for optimum performance. An increased production rate of 25 T/D has been claimed by computer control[61]; in addition, there has been no change in feed gas consumption. Control strategy includes adjustments to the H_2/N_2 ratio, purge rate, and reformer furnace control. Heat recovery, too, during regeneration of catalysts is included.

One company offers a "Total Sensor-based Computer-Control System" which includes control of:

- Steam/gas ratio
- Reformer temperature
- Air/gas ratio in reformer
- H_2/N_2 Ratio
- Ammonia Converter Catalyst Temperature
- Ammonia Converter Pressure

Other instrument companies offer computer control systems including automatic analysis of the various gas streams involved.

Catalysts[56] for ammonia synthesis were originally made by melting relatively pure iron in an electric furnace. The oxide was mixed with promoters such as potassium, calcium, magnesium or aluminum. This process has been superceded by mixing natural or synthetic magnetite and promoters, and melting the mixture in an electric furnace.

The cooled mass is broken up, sized, and in some cases pre-reduced before charging to the converter. Particle size is 2 to 10 mm, while the bulk density is in the range of 2.5 to 2.9 Kg/liter. A typical chemical composition is as follows:

68.4% total Fe, 3.16% Al_2O_3, 0.56 to MgO, 0.50% Si O_2, 3.54% Ca O, 0.58 % K_2O.

The catalyst must be reduced to the metallic constituents before it is placed in operation. (See page 173 of reference 20.) For a multi-bed quench converter, the synthesis gas is circulated in the loop at operating pressure, heated at 50°C per hour to 340°C and then at 10°C per hour until the desired temperature is reached. The water in the exit gas should be about 5000–10,000 ppm. As ammonia is made, it will reduce the heat input that is required at the start. When temperatures rise of their own accord, the circulation rate can be increased. Temperatures should be held below 500°C.

A highly active catalyst has been described in U.S. Patent 3,992,328. The reduced iron catalyst is treated with an aqueous solution of cerium nitrate and re-reduced after packing in the converter. Exit ammonia concentrations from a laboratory test unit are about 3 to 4 percentage points higher in the exit gas for the Lummus catalyst.

Although no known use is made of a synthesis system modified by an electrical potential[57] on the catalyst, the process is claimed to increase the conversion efficiency by 38.6 percent. A static negative charge of about 6,500 volts was used on the catalyst.

Innovations for the ammonia synthesis loop include the recovery of argon and hydrogen, with recycle of the hydrogen, from the synthesis-loop purge gas. With a system designed for argon recovery[62], streams rich in methane and hydrogen can be obtained. The purge gas is cooled, washed with liquid nitrogen, and then the liquid nitrogen stream is fractionated in a two-column train to give liquid argon and methane. A stream of H_2 and N_2 is recycled to the loop.

With a system designed by Petrocarbon Developments, Inc.[63], the purge gas is first treated with aqueous ammonia to recover the ammonia as anhydrous ammonia, then given a final cleanup with molecular sieves before treating in a cyrogenic system which concentrates the H_2 to 91%, with 8% N_2 and less than 1% argon and 1% CH_4. The recycle fuel-gas stream contains 24% CH_4, 28% H_2, 38% N_2 and 10% argon.

Environmental control is essential to meet E.P.A. demands. It has been estimated that for a plant consuming 37. million Btu per ton of ammonia produced, environmental control will require 0.7 million Btu, and environmental costs will be 1.1% of the operating costs. Discharge streams which require processing to eliminate undesirable contaminants include: (1) the regeneration acid and base for boiler feed water treatment units, (2) the regeneration steam for natural gas desulfurization catalysts, (3) the low temperature shift condensate (ammonia, amines, and methanol), and (4) CO_2 removal process regeneration effluents.

The ammonia process can be combined with a urea process and some operating and investment efficiencies realized.[58] The synthesis gas generated by conventional means is mixed with liquid ammonia before the carbon dioxide is removed. This mixture of ammonia and carbon dioxide is synthesized to urea, and the off-gas from this operation is used for ammonia synthesis. The carbon dioxide-removal step is thereby eliminated and the compression of carbon dioxide for urea is avoided.

Uses of Ammonia

The uses for ammonia are given in Tables 5.8 and 5.9. By far the greatest consumption is in the manufacture of fertilizers or as a direct application for fertilizer.

Storage and Transport

Ammonia is usually stored in large 20,000 ton atmospheric-pressure tanks at a temperature of -28° F. With this system, the ammonia vapor-

TABLE 5.8 Uses of Ammonia[a]

	1,000 Short Tons
Ammonium nitrate	2,900
Direct fertilizer application	3,300
Solid and liquid fertilizers	830
Urea	1,700
Ammonium sulfate	590
Nitric acid, industrial	380
Ammonium phosphates	2,000
Others	1,500
Total	13,200

[a]*Chem. Week* (September 11, 1965)[59]

TABLE 5.9 Industrial Uses of Ammonia[a]

Industry	*Usage*
Explosives	Nitrates, dynamite, azides
Plastics	Nitrocellulose, urea-formaldehyde, melamine
Metallurgy	Bright annealing of steel, dry reducing gas
Pulp and Paper	Ammonium bisulfite, melamine
Rubber	Aniline, acrylonitrile, polyurethanes, chemical blowing agents (for foam rubber)
Textiles	Nylons, acrylonitrile, terephthalates
Foods	Amino acids, sodium nitrite, and nitric oxides
Drugs	Vitamins, nitrofurans
Miscellaneous	Refrigerant, detergents, insectides, nitroparaffins, hydrazine

[a]Based on data in *Chem. Eng.*, 62, 11, 280–282 (1955).

ized by heat-leak must be compressed and condensed, and returned to the system.

Some of the low-pressure tanks are built with a double wall; this adds some protection from a leak and also provides a means for retaining insulation.

One unique storage system consists of an underground cavern mined from a strata of limestone.[60]

Normally, ammonia is transported by barge, ship, or tank cars to terminals located at strategic points in the consuming agricultural areas. Some ammonia is transported by interstate pipelines.

Of current world ammonia capacity—about 77 million short tons of nitrogen equivalent—64% is based on natural gas, 13% on naphtha, and 12% on coal or coke, with the remaining 11% equally divided among other feedstocks. Within the next 30 years, coal will probably be replacing a major portion of the natural gas based plants. For instance, with coal at $25/ton and natural gas at $3/Mcf, the ammonia selling price is a break-even. Fuel oil at $18 per barrel gives the same ammonia selling price.

TVA has an experimental coal-based plant approved for operation in 1980, and the Department of Energy has approved two coal-gasification projects in conjunction with the W. R. Grace Company and Ebasco Services and the other with Air Products and Chemicals.[27]

NITRIC ACID

Nitric acid has been known since the 13th century. Glauber devised its synthesis from strong sulfuric acid and sodium nitrate; however, it was Lavoisier who showed that nitric acid contained oxygen; and Cavendish showed that it could be made from moist air by an electric spark.

In the oldest methods used, Chile saltpeter was reacted with concentrated sulfuric acid in heated cast iron retorts; the evolved nitric acid vapors were condensed and collected in stoneware vessels.

Today, nitric acid is made by oxidation of ammonia with air over a precious-metal catalyst at 800 to 950°C and at atmospheric or pressures of 120 psig. The overall reaction is:

$$NH_3 + 2O_2 \longrightarrow HNO_3 + H_2O + \text{98.7 Kcal evolved}$$

The concentration of nitric acid is usually about 60% with conventional equipment. If higher concentrations are desired, special equipment or processes are required.

Nitric acid can also be produced by the high temperature combination of nitrogen with oxygen in the air, and by the use of radiation.[65] One method which has received considerable attention is the so-called Wisconsin Process. Although a 40-ton per day plant was built, it could not compete economically with the conventional ammonia-oxidation route.

Nitric acid is generally a light-amber liquid; however, the pure acid is colorless, strongly hydroscopic, and corrosive. It is a strong oxidizing acid. It boils at 78.2°C, freezes at -47°C, and forms a constant boiling mixture with water (68% nitric acid by weight) which boils at approximately 122°C at atmospheric pressure. White fuming nitric acid usually contains 90 to 99 percent by weight HNO_3, from 0 to 2 percent by weight NO_2, and up to 10 percent by weight water. Red fuming acid usually contains about 70 to 90 percent by weight HNO_3, 2 to 25 percent by weight NO_2, and up to 10 percent by weight water.

Chemistry

The oxidation of ammonia catalytically is an extremely rapid heterogeneous reaction. The

TABLE 5.10 Equilibrium Constants for Reactions of Ammonia Oxidation with Air[67]

Equilibrium Temp. °F	*K_p, atm, for Formation of: (Equation No.)* N_2 (1)	NO_2 (4)	NO (3)	NO_2 (4)	H_2 (5)
440	7.07×10^{34}	4.44×10^{28}	1.13×10^{26}	1.43×10^{28}	3.33
800	2.61×10^{25}	2.69×10^{20}	2.11×10^{19}	5.04×10^{19}	1.11×10^{2}
1160	1.49×10^{20}	7.36×10^{15}	3.8×10^{15}	9.94×10^{14}	8.84×10^{2}
1520	6.71×10^{16}	9.12×10^{12}	1.54×10^{13}	–	3.21×10^{3}
1860	3.18×10^{14}	8.85×10^{10}	3.36×10^{11}	–	8.14×10^{3}

platinum or platinum-rhodium catalyst in the form of a fine gauze is very selective and very active under optimum conditions. Side reactions can occur resulting in a loss of product.

When ammonia is oxidized in air over a catalyst the following reactions take place:

$$NH_3 + \tfrac{3}{4} O_2 \longrightarrow \tfrac{1}{2} N_2 + \tfrac{3}{2} H_2O \quad (1)$$

$$NH_3 + O_2 \longrightarrow \tfrac{1}{2} N_2O + \tfrac{3}{2} H_2O \quad (2)$$

$$NH_3 + \tfrac{5}{4} O_2 \longrightarrow NO + \tfrac{3}{2} H_2O \quad (3)$$

$$NH_3 + \tfrac{7}{4} O_2 \longrightarrow NO_2 + \tfrac{3}{2} H_2O \quad (4)$$

$$NH_3 \longrightarrow \tfrac{1}{2} N_2 + \tfrac{3}{2} H_2 \quad (5)$$

The equilibrium constants for these reactions are given in Table 5.10.

Equation 3, for the formation of NO can be promoted by the selection of the right conditions and the proper catalyst. Nitric oxide yield increases as temperature is increased. At atmospheric pressure, with 80 mesh gauze - 0.003 inch wire (8% rhodium-platinum), 1650°F, and 100 pounds of NH_3 per troy oz. of catalyst per day, the conversion efficiency is about 99%.[67]

As pressure for the platinum-rhodium gauze process increases, efficiencies for conversion of NH_3 to NO fall off[68]; at one atmosphere - 97-98%; 4 atmospheres - 95-96% and at 10 atmospheres 42-93%. Increased pressure does, however, increase the amount of ammonia that can be oxidized per unit of catalyst.

Space velocity should be maintained high in order to avoid reaction of ammonia in the feed gas on the reactor walls ($2\ NH_3 + \tfrac{3}{2}O_2 \longrightarrow N_2 + 3\ H_2O$). Low temperature of the reactor walls minimizes this reaction.

Also, contact of NO_2 with NH_3 must be kept as low as possible ($3NO_2 + 4NH_3 \longrightarrow \tfrac{7}{2} N_2 + 6H_2O$); hence, the catalyst bed must be shallow, and the velocity high enough to avoid back mixing.

The effect of reactor residence time on nitric oxide yield is indicated by the following[67]:

Reactor Residence Time, Seconds	*Nitric Oxide Yield, percent NH_3 Oxidized*
0.28	82.1
0.11	85.7
0.061	90.2
0.023	91.8

The concentration of ammonia in the feed gas is kept below 14% because of the lower spontaneous inflammability of the mixture.

The desired reactions taking place in the process are:

$$NH_3 + \tfrac{5}{4} O_2 \longrightarrow NO + \tfrac{3}{2} H_2O + 69.9 \text{ Kcal evolved/g mol}$$

$$\tfrac{3}{2} NO + \tfrac{3}{4} O_2 \longrightarrow \tfrac{3}{2} NO_2 + 13.6 \text{ Kcal evolved/g mol}$$

$$\tfrac{3}{2} NO_2 + \tfrac{1}{2} H_2O \longrightarrow HNO_3 + \tfrac{1}{2} NO + 8.35 \text{ Kcal evolved/g mol}$$

The oxidation of ammonia to NO takes place primarily over the gauze catalyst, while the oxidation of the NO to NO_2 takes place in the absorber of the process. Equilibrium constants for the reaction

$$NO + \tfrac{1}{2} O_2 \longrightarrow NO_2$$

are as follows:

Temp. °F	*Equilibrium Constant,* Kp; atm$^{-0.5}$*
35	9.42×10^{6}
80	1.19×10^{6}
170	4.61×10^{4}
260	3.95×10^{3}
350	7.24×10^{2}
530	35.2

*See Reference 67

Note the negative temperature coefficient of the reaction. Low temperature favors the nitrogen oxide oxidation reaction as well as the absorption.

Processes

The first processes for oxidation of ammonia operated at atmospheric pressure. Over the years both high pressure and dual pressure processes have been used. Early processes used are given in Table 5.11.

Today the dual pressure process is widely used. In this process, the oxidation of ammonia is carried out at 72.5 psi and the absorption of the oxidation products at 160 psi. The low pressure for oxidation promotes high efficiencies of conversion and avoids high losses of the precious metal catalyst. Fig. 5.10 shows a flow chart for a dual-pressure plant using the Grande Paroisse process. Consumption of ammonia, catalyst, and energy for this process is estimated as follows:

Material Consumed	*Per Ton of 100% HNO_3*
Ammonia	280 Kg
Platinum	90 mg
Electric Power	9 Kwh
Cooling Water	140 m^3
Steam, excess	600 Kg

In the Grande Paroisse dual-pressure process, the ammonia is vaporized, preheated, and filtered before mixing with the air which is filtered and compressed to 6 to 35 atm. Part of the air goes to the bleacher. The mixture of air and ammonia passes down over the platinum-rhodium catalyst. The converter effluent is cooled in a tail-gas heater exchanger and other exchangers before entering the NO compressor where the pressure is raised to 8 to 13 atm. Both the air compressor and NO compressor are driven by a steam turbine.

The compressed converter gas is cooled in a series of exchangers and enters an absorption tower where water is added. The tail gas from the absorber is warmed through three exchangers and then enters a power recovery-expansion turbine. The tower effluent from the bottom contains the nitric acid and enters a bleacher at the top, flowing down countercurrent to a stream of air. The air which leaves at the top of the bleacher combines with the converter effluent and is recycled to the absorption tower.

The tail gas from the power recovery turbine contains less than 150 ppm nitrogen oxides.

In the first stage of a nitric acid process where ammonia and air are mixed, it is important to hold the proper ratio of constituents. Usually a mixture of about 10 percent ammonia and 90 percent air is employed. At this point, it is also important to avoid the decomposition of ammonia on the converter walls. Aluminum is superior to mild steel in this respect. At 350°C the rate of decomposition is 300 times faster on mild steel than on aluminum. Ammonia is decomposed 70 times faster on stainless steel than on aluminum.[69] However, by keeping the preheat temperature low, satisfactory yields can be obtained with mild steel. The temperature level used should be below 200°C.

In the reaction which takes place over the catalyst bed, the rate-limiting step is the diffusion of ammonia to the catalyst surfaces.[69] With a high gas-rate through the gauze, some of the ammonia will not reach the catalyst surfaces. It reacts instead with the nitric oxide product to give elemental nitrogen

$$4NH_3 + 6NO \longrightarrow 5N_2 + 6H_2O$$

TABLE 5.11 Early Processes for Nitric Acid Manufacture

Process	*Temperature °C*	*Pressure*	*Acid Strength %*
Low Pressure	800	Atmos.	50–52
Medium Pressure	850	40 psi	60
High Pressure (DuPont)	950	120 psi	60
Pintsch. Bamag	850	Atmos.	98–99

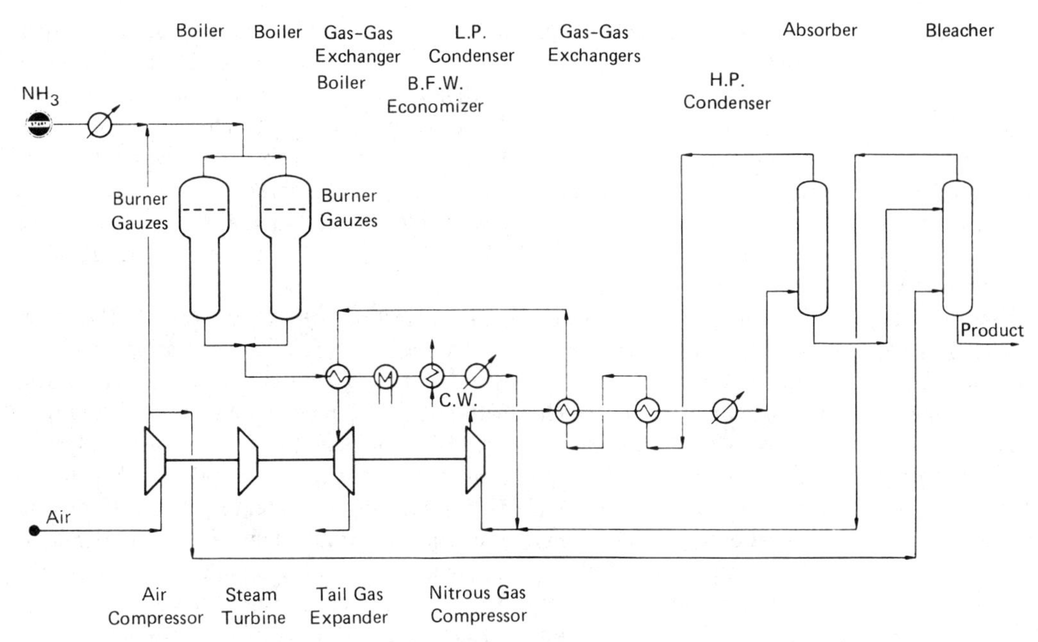

Fig. 5.10 Dual pressure nitric acid process—Grand Pariosse. (*Hydrocarbon Processing*, **58**, *No. 11, 1979.*)

The linear flow of gas across the catalyst is therefore important and will depend on the arrangement of the catalyst. Linear flow rates of 1 to 3 feet per second have given high efficiency. These rates are for gas at standard temperature and pressure. Without selection of proper flow rates or, conversely, without selection of the proper converter size for a given production, inefficient operation may result not only from too high a velocity with the aforementioned yield loss, but also from too low a velocity which can produce gas channels along the walls of the converter.

The catalyst generally used for ammonia oxidation is a platinum-rhodium alloy containing about 10 percent rhodium. A gauze made of wire 0.003 inch in diameter with about 80 meshes to the inch is used. During the life of the catalyst, it becomes polycrystalline in appearance. There is also some weight loss, the effect of which is more pronounced at the higher converter temperatures employed in the pressure process. With the atmospheric processes at lower temperature the platinum loss is about 45 mg per metric ton of acid produced. Recovery steps are required for operation at 120-pound pressure. This usually involves periodic cleaning of the converter and heat exchange equipment, and the use of a filter in the gas stream beyond these pieces of equipment.

A summary of catalyst developments is given in reference 74. The materials examined as a catalyst are platinum, platinum-iridium alloys, and platinum-rhodium (various compositions, see Fig. 5.11). Other alloys of platinum were

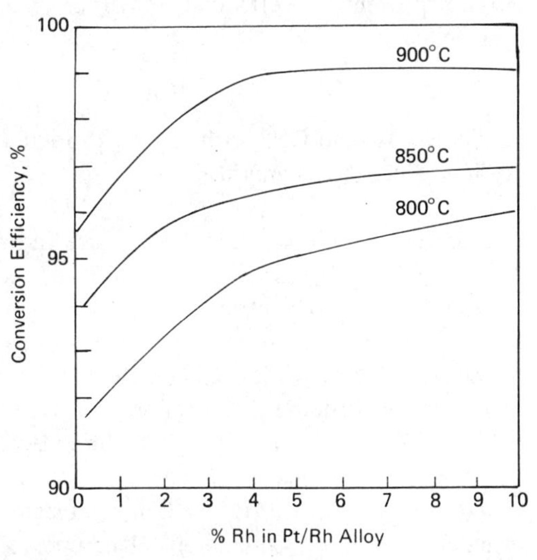

Fig. 5.11 Effect of rhodium content on catalyst efficiency.[74]

examined, such as those containing palladium, ruthenium, rhenium, and tungsten. The addition of these latter materials to the platinum-rhodium alloy gave inferior catalysts. Loss of activity results from contaminates such as iron, calcium, MoS_2 (lubricants), and phosphorous in the gas streams.

Several systems to recover catalyst which would otherwise be lost have been investigated. These include filters of glass wool, asbestos fibers, Raschig rings, and marble chips (3 mm to 5 mm dia.). Shortcomings of these recovery systems have been overcome by a process which uses gold-palladium gauzes. The gauze is placed as close to the production catalyst as possible, the idea being to recover the platinum-rhodium catalyst while it is in the vapor form. Palladium, by itself, is reported to be the most effective "getter"; but when used alone, the palladium becomes too brittle. Thus, a 20% gold-palladium alloy is generally used. The total recovery, S, for a number of gauzes can be estimated by the following:

$$S = 1 - \left(\frac{a}{100}\right)^n \times 100$$

a = % recovery by the first gauze
n = number of gauzes

Normally a recovery of 55-70% is economical.

Base metal catalysts have been investigated in fixed bed and fluidized bed systems but these have not proven to be effective. Materials tried are as follows:

$Fe_2O_3-Mn_2O_3-Bi_2O_3$ (Braun oxide)

$CoO-Bi_2O_3$; and CoO with Al_2O_3, thorium, cerium, zinc, and cadmium.

Adsorption of nitric oxide takes place in a sieve or bubble-cap tower into which air is added to oxidize the nitric oxide to nitrogen dioxide. The nitrogen dioxide must be absorbed in water to liberate nitric acid and nitric oxide. There are two equilibria as follows: $N_2O_4 = 2NO_2 = 2NO + O_2$. The first of these two equilibria is established relatively quickly; the second more slowly. At 150°C the nitrogen tetroxide is almost completely decomposed into nitrogen dioxide. At 500°C the equilibria lies at 25 percent nitrogen dioxide and 75 percent nitric oxide. The formation of nitrogen dioxide from nitric oxide and oxygen is relatively slow at atmospheric pressure, but proceeds quite rapidly under higher pressure. Because this reaction has a negative temperature coefficient, the absorber is operated at the lowest temperature obtainable with available cooling water, in the temperature range of 10 to 40°C.

Concentration of Nitric Acid. The most widely used method of concentrating nitric acid consists of mixing the 60 percent nitric acid with strong (93 percent) sulfuric acid, and then passing the mixture through a distillation system from which concentrated (95 to 98 percent) nitric acid and denitrated, residual sulfuric acid containing approximately 70 percent H_2SO_4 is obtained. The dilute, residual sulfuric acid may be reconcentrated for further use.

One process which employs magnesium nitrate as a dehydrating agent[70] is illustrated in Fig. 5.12. An aqueous solution containing 72 percent magnesium nitrate is fed to the middle of the tower which also receives the 60 percent nitric acid feed. Pure nitric acid of 95 to 98 percent strength is taken overhead. A solution containing about 55 percent magnesium nitrate is removed from the base. After reconcentration to 72 percent, the magnesium nitrate is recirculated to the nitric acid tower.

A Pintsch Bamag process[71] concentrates nitric acid by using the difference in the composition of the nitric acid-water constant boiling mix-

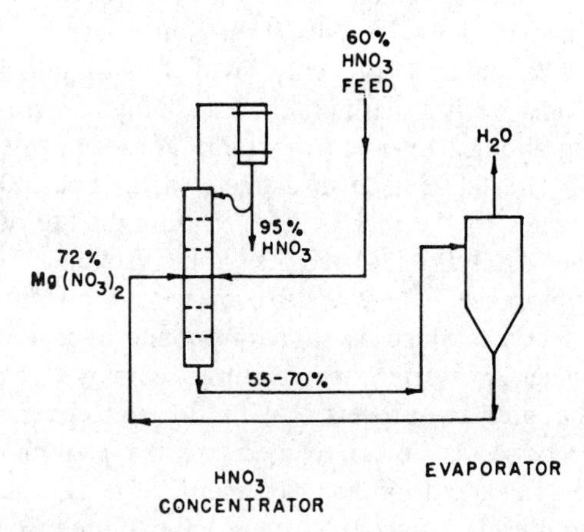

Fig. 5.12 Magnesium nitrate process for nitric acid concentration. (*Hydrocarbon Processing*, **58**, *No. 11, 209, Nov. 1979.*)

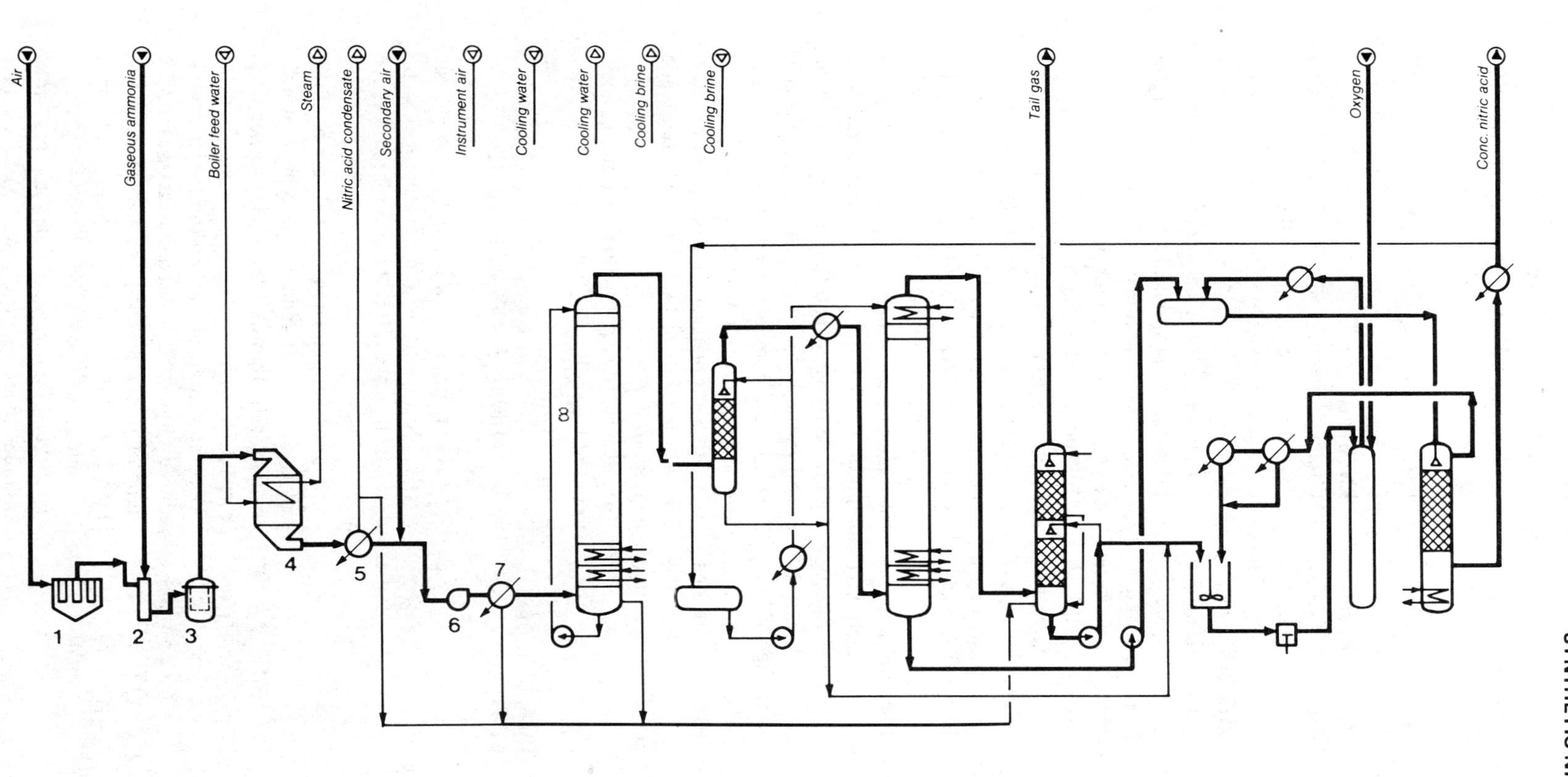

1. Air filter
2. Ammonia-air-mixer
3. Mixed gas filter
4. Burner with waste-heat boiler
5. Gas cooler I
6. NO blower
7. Gas cooler II
8. Oxidation column
9. Circulating acid tank
10. Acid cooler
11. Final oxidator
12. Brine gas cooler
13. Absorption column
14. Tail gas scrubber
15. Head tank for raw acid
16. Raw acid cooler
17. Liquefier
18. Agitator vessel for raw mixture
19. Reactor vessel
20. Bleaching column
21. Final acid cooler

Fig. 5.13 Concentrated nitric acid by Uhde. (*Courtesy of Uhde.*)

tures at different pressures. A two column distillation system is employed.

Uhde has developed a process for producing highly concentrated nitric acid (98-99%) without using dehydrating agents. The NO from the ammonia burner is converted to NO_2 by means of highly concentrated HNO_3; the NO_2 is cooled with brine, and after removal of the residual water, it is physically absorbed in chilled, highly concentrated nitric acid; then it is separated from the concentrated nitric acid in a distillation column, and then liquefied by refrigeration whereby the NO_2 is dimerized to N_2O_4. The liquid N_2O_4 is mixed with dilute nitric acid and fed to the reactor by a high pressure centrifugal pump. Gaseous oxygen is likewise fed to the reactor where the $N_2O_4/HNO_3/H_2O$ mixture reacts at a pressure of about 720 psi to form highly concentrated nitric acid.

The acid leaving the reactor contains about 20% dissolved N_2O_4 which is separated in a distillation column and returned to the reactor cycle. The product is bleached and sent to storage, with a portion recycled to the absorber cycle. (See Fig. 5.13.) Fig. 5.14 gives a view of a Uhde plant for production of concentrated nitric acid.

Typical consumption figures for a medium-sized plant (150 T/D), per ton of HNO_3 (100%) are as follows:

NH_3	282 Kg
O_2	125 m^3 @ NPT
Electricity	285 KWHr
Cooling Water	200 m^3
Excess Steam	0.6 t
NO_x in Waste Gas	200 ppm

Fig. 5.14 Plant using Uhde process for direct production of concentrated nitric acid. (*Courtesy of Uhde.*)

Purification. Chemically pure nitric acid is required for some applications, and as a reagent for analytical measurements. The commercial grade may contain impurities, such as arsenic, sulfur, and phosphorous.

In one purification process (U.S. Patent 3,202,481) potassium or sodium permanganate is used to oxidize the impurities. Other oxidizing agents can also be used, such as persulfates or chlorates. Molybdenum trioxide is used to form a complex with the oxidized phosphorous impurities. The nitric acid is contacted with the oxidizing agent and the molybdenum compound at the boiling point of the acid; the acid is continuously removed by distillation. The oxidized impurities remain in the residue. As a result, the purified acid will contain less than 0.5 part per billion by weight of arsenic and phosphorous.

Stabilizers. Over a period of time, concentrated nitric acids tend to decompose according to the equation

$$4HNO_3 \longrightarrow 4NO_2 + 2H_2O + O_2$$

As a result, pressures will build up in storage vessels. Nitric acid is also very corrosive. So some stabilizers and/or corrosion inhibitors are used.

Corrosion of aluminum by red fuming nitric acid is reduced by adding 4 percent by weight of hydrogen fluoride, 52% HF. (See U.S. Patent 2,760,845.)

Decomposition of the concentrated acid is reduced by such substances as quaternary ammonium compounds, organic sulfones, inorganic persulfates, and organic sulfonium compounds.[72]

Pollution Abatement. The tail gas from ammonia-oxidation processes may contain sufficient nitrogen oxides to result in serious atmospheric pollution (smog) or in serious corrosion of power-recovery equipment, if such is used, unless some means for removal of the oxide is employed. The oxides are usually reduced with natural gas, hydrogen, or a carbon monoxide-containing gas, or they can be absorbed by alkaline liquors (soda, calcium, or magnesium) or ammonium sulfite-bisulfite.[72]

As an example of the reduction procedure, a gas containing the materials listed below was treated with 100 scf of natural gas and 1000 scf of air over a nickel catalyst on alumina at 90 psig, 1150°F (1600°F exit), and with a space velocity of 20,000 cu. ft. of exit gas per cu. ft. of catalyst.

Gas	*%*
N_2	97.2
O_2	2.0
H_2O	0.6
$NO + NO_2$	0.2

The effluent gas contained about 75 ppm $NO-NO_2$ mixture; 0.1% CO: 0.3% H_2; and 1.5% CO_2.

With methane as the reducing gas, materials which are particularly effective as catalysts are rhodium, palladium, platinum, and ruthenium. Alumina or Torvex®, a ceramic honeycomb material, can be used as a base.

These catalysts are usually used at about 0.5 percent by weight on Al_2O_3. With these materials, the initial reaction temperature is from 700 to 1000°F depending upon the catalyst and oxygen concentration.

Harding and others[73] developed an absorption-desorption process using zeolites. It is stated that this process will result in the recovery of the oxides and add up to 4 or 5 tons per day of capacity to a 300 T/D plant. Exit NO_x concentrations of 0.001 percent were obtained for three hours and 0.02 percent after three hours, when feeding 2.4 scfm of tail gas (0.18 percent nitric oxides) over 4.9 pounds of zeolite. The columns were regenerated with steam at 300-350°F, or by hot air.

Production. In 1977, production of nitric acid in the United States was 7,950,734 short tons. Major producers are: DuPont, Allied Chemical, Monsanto, Hercules, and Gulf. In other parts of the world, Uhde has built plants with capacities up to 1800 T/D (Norsk Hydro, Oslo-Norway), and Grande Pariosse has built a plant with capacity up to 1250 T/D (Sluiskil, Holland). A Du Pont plant in Texas was built in 1977 for a capacity of 907 T/D.

Uses of Nitric Acid

The primary use of nitric acid is in the manufacture of ammonium nitrate for the fertilizer industry. Approximately 65 percent of the total production goes into fertilizers. Solution Fertilizers, many of which contain nitrates, are growing in importance. Nitric acid is used in some cases for the acidulation of phosphate rock to produce mixed fertilizers. (See Chapter 8.)

About 25 percent of the nitric acid produced is consumed in industrial explosives. (Chapter 19.) Approximately 1500 thousand tons per year of nitric acid is used for explosives which, in turn, are used in coal mining, quarrying, metal and nonmetal mining, heavy construction, and petroleum exploration. Fertilizer grade ammonium nitrate may also be used as an explosive or blasting agent. It is mixed with oil and set off by priming with a high explosive.

Nitric acid has a number of other industrial applications. It is used in pickling stainless steel, in steel refining, and in the manufacture of dyes, plastics, and synthetic fibers. Most of the methods used for the recovery of uranium, such as ion exchange and solvent extraction, use nitric acid. Thus with the expansion of the nuclear program nitric acid consumption may increase.

In 1976, it was reported[57] that 75 percent of nitric acid manufactured in the U.S. went into the manfuacture of fertilizers, while about 15 percent was used in the manufacture of explosives. The remaining 10 percent went into a number of miscellaneous products.

AMMONIUM NITRATE

Total ammonium nitrate production has about leveled off in the United States at 7,200,000 short tons since 1973.[6] There has been a decline in the U.S.A. consumption of ammonium nitrate as a direct application fertilizer material over these same years. However, the greater use of liquid solution fertilizers containing ammonium nitrate have helped to maintain the consumption. This usage is attributed to the greater ease of application for the liquids.

Large amounts of ammonium nitrate are still consumed for explosives and in the manufacture of nitrous oxide.

There are five different crystalline forms of ammonium nitrate.[67] One is stable below -18°C, another at -18 to 32°C, a third at 32 to 84°C, a fourth at 84 to 125°C, and a fifth at 125 to 169.6°C the melting point. The molecular weight is 80.05, the nitrogen content 35%, and the solubility in water, 118.3 grams per 100 grams at 0°C and 576 grams per 100 grams at 80°C. It decomposes on heating to N_2O and water; however, great

$$NH_4NO_3 \longrightarrow N_2O_{(g)} + 2H_2O_{(g)} + 6720 \text{ cal/gram mole (evolved)}$$

care must be exercised to avoid local overheating which could cause an explosion.

It is important to mention the hazards involved in handling ammonium nitrate and the precautions that should be taken. Large masses of concentrated solutions at high temperatures should be avoided; in particular, avoid contamination with decomposition catalysts such as chlorides and organic materials. Keep the melt above a pH of 4.5, and keep the neutralizer well ventilated.

The Texas City disaster was dramatic proof that organic material and ammonium nitrate decompose with explosive violence. Following a fire under confinement conditions, an explosion occurred which took almost 600 lives, injured 3,500, and destroyed 33 million dollars worth of property. Ammonium nitrate solutions also explode with disastrous results. Remember–above 250°F ammonium nitrate must be considered an explosive hazard.

Stafford, Samuels, and Croysdale[75] give a comprehensive review of ammonium nitrate plant safety. They discuss how coating ammonium nitrate affects its explosibility. For instance, 2.5 percent by weight of diatomaceous earth applied as a coating on ammonium nitrate reduces its detonation sensitivity, and 2.5 percent by weight of finely divided sulfur increases the detonation sensitivity. At a 5 percent concentration, urea decreases the thermal stability of ammonium nitrate.

Some precautions that have been suggested in the manufacture of ammonium nitrate are as follows:[76] (1) operation should be conducted below 250°F; (2) when the pH is less than 4.0 for a prolonged period of time, type 304L stainless steel may be used; and (3) when the pH is over 4.0 for solids and over 6.0 for solutions and the temperature is less than 240°F, aluminum may be used.

The process for manufacturing ammonium nitrate consists of neutralizing ammonia with nitric acid under controlled conditions.

$$HNO_{3(l)} + NH_{3(g)} \longrightarrow NH_4NO_{3(ag)} + 26000 \text{ cal/gram mole evolved}$$

Water is evaporated, and the anhydrous, or nearly anhydrous, solution is prilled or spherodized to a solid product. Fig. 5.15 shows the C&I/Girdler, Inc. process with either the prilling tower or the spherodizer. Anhydrous ammonia and 50 to 60 percent nitric acid are fed to the neutralizer where the water is evaporated by the heat of reaction (550 to 620 Btu/lb of ammonium nitrate, depending on the acid strength). Final concentration is done in a falling film evaporator in which a stream of air helps to carry off the water vapor. The air stream from the evaporator is mixed with that from the spherodizer for clean-up in a wet scrubber.

A C&I/Girdler additive or nucleating agent for the prilling process is added to the melt before prilling or spherodizing. This produces unstressed crystals and stabilizes the prills against temperature cycling through crystal transition phases. It reduces prill breakage and the amount of coating agent required. The product from the prill tower or spherodizer is sent to a final cooler.

In the spherodizer granulation process, the molten ammonium nitrate is sprayed onto the rolling bed of solid particles in the drum, forming granules. These are cooled and screened. Advantages of spherodizing include: a variation in product size, atmospheric pollution from the prill tower is eliminated, and the product has a

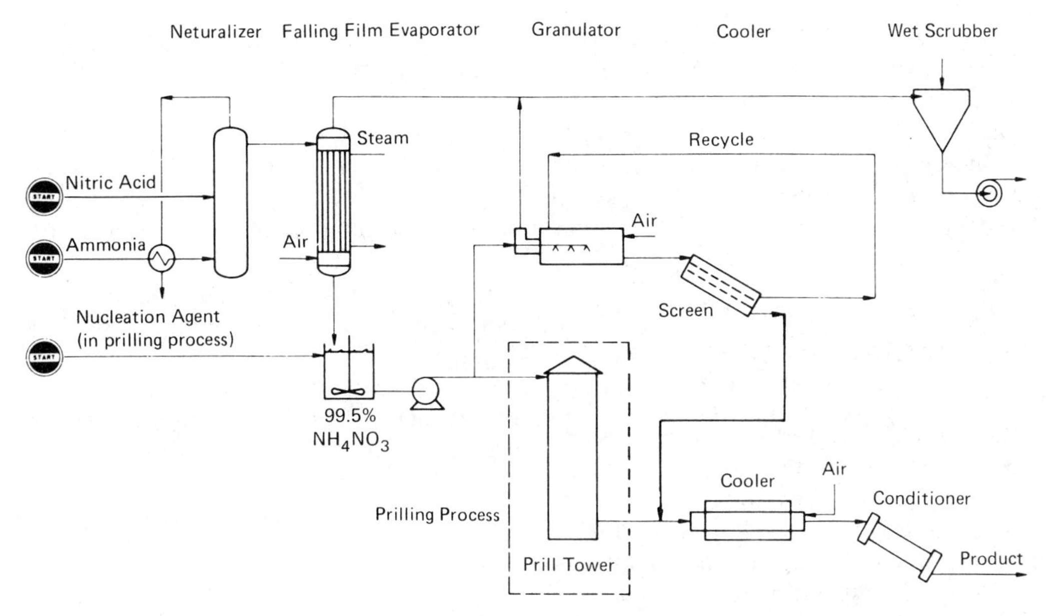

Fig. 5.15 Ammonium nitrate manufacture per C&I Girdler, Inc. (*Hydrocarbon Processing*, **58**, *No. 11, 1979, copyright 1979 by the Gulf Publishing Co.*)

greater hardness. A yield of over 99.5% from ammonia and nitric acid is claimed.

To prill ammonium nitrate, a solution is evaporated in a vacuum or falling-film evaporator until it is almost free of water. The concentration of ammonium nitrate will vary from 95 to 99.95 percent. The prilling tower itself is about 50 to 100 feet tall, depending on the process being used. The concentrated ammonium nitrate solution is fed through sprayformers which break up the stream into small droplets that solidify during free fall in the tower. In some instances limestone is mixed with the ammonium nitrate before prilling; in other cases, the prills are coated with limestone dust. A mixture of limestone and ammonium nitrate is known as "Nitro chalk" in the fertilizer industry. Plants that use the 95 percent ammonium nitrate feed to the prilling tower usually require a drying step before the material can be bagged.

UREA

Urea was first synthesized by Wohler in 1828 from ammonia and cyanic acid. It is a colorless crystalline material, soluble in water and in alcohol, but not in ether. Urea contains 46 percent nitrogen, the most of any ordinarily solid fertilizer material. Ammonium nitrate contains 34 percent nitrogen. Since urea is converted into ammonia and then into nitrates in the soil, it makes a very concentrated form of nitrogen fertilizer. It is also used in resin and plastic manufacture and in the synthesis of organic materials.

The production of urea increased almost threefold from 1965 to 1975,[6] but excluding a peak production of 5MM tons in 1977, the production has been relatively constant at 4.5MM tons. Capacity in 1978 and 1979 reached a new high of 7.3MM tons.[77]

Urea can be produced by the hydrolysis of cyanamide (melting point 44°C) according to the equation, $CNNH_2 + H_2O \longrightarrow CO(NH_2)_2$. At high pressure, 33 to 53 atm, urea can also be formed by heating ammonium carbonate. Equilibrium is obtained at 130 to 150°C with a 30 to 45 percent yield. A process has been described by Raymond Fanz and Fred Applegath of Lion Oil Company which produces urea from ammonia, carbon monoxide, and sulfur (dissolved in methanol).[78] The process operates at 100°C and 20 atm and produces H_2S along with the urea. The H_2S is oxidized to sulfur for recycle.

The current method of manufacture is to combine ammonia with carbon dioxide under pressure to form ammonium carbamate, which is then decomposed into urea and water. The unreacted carbon dioxide and ammonia are recovered and recycled to the synthesis operation

$$2NH_3 + CO_2 \rightleftharpoons NH_2COONH_4 \quad \text{(Exothermic)}$$

$$NH_2COONH_4 + \text{heat} \rightleftharpoons CO(NH_2)_2 + H_2O \quad \text{(Endothermic)}$$

The main differences between the various urea processes are in the methods used to handle the converter effluent, to decompose the carbamate and carbonate, to recover the urea, and to recover the unreacted ammonia and carbon dioxide for recycle with a minimum expenditure of energy and a maximum recovery of heat. In some processes, a liquid is used to recycle a solution of carbamate, carbon dioxide, ammonia, and water. In others, the amount of water recycled is minimized, and only carbon dioxide and ammonia are recycled. In the older plants, or once-through plants, the off-gases are used as feed to ammonium nitrate or ammonium sulfate plants.

The *Du Pont* process represents a combination of a number of total recycle and partial recycle processes. It was one of the first processes developed. Unconverted carbon dioxide and ammonia are recovered and recycled as a water solution.[79] The converter, lined with silver, operates at approximately 400 atm and at 200°C. Conversion is approximately 70 percent. In partial recycle processes, only a portion of the unreacted ammonia is recycled, the rest is used to produce by-product ammonium sulfate or ammonium nitrate.

In the *Pechiney* process, the reaction takes place in an oil medium which is later used to return the unconverted ammonia and carbon dioxide to the converter. The converter is lined with lead and operates at approximately 200 atm and at 180°C. Conversion is approximately 50 percent.

The *Stamicarbon* process is based upon the use of carbon dioxide to strip ammonia from the reactor effluent counter-currently. As the carbon dioxide removes the ammonia from the solution, the carbamate decomposes leaving a minimum amount in the effluent. The solution leaves the stripper at about 150–180°C. The stripped gases flow to the reactor along with ammonia in an amount equivalent to the amount of carbon dioxide added for stripping. Most of the off-gas from the reactor is condensed, and inert gases are bled from the system before the condensate is returned to the base of the reactor. This process is shown in Fig. 5.16.

For a Stamicarbon total recycle plant with CO_2 stripping to produce fertilizer grade prills, typical consumptions per metric ton of product are:

	Biuret Content Of 0.8%	*Biuret Content Of 0.2.-0.25%*
Kgs. CO_2	755	755
Kgs. NH_3	570	570
Kwh power	115	135
Kgs. Steam (25 atmg)	900	1000
m^3 Cooling Water (Δ T=11°C)	65	65
Export Steam, Sat. 3 atm	100–300	350–400

The *Allied Chemical-C.P.I.* process, which was developed by T. O. Wentworth (U.S. Patent 3,107,149), is based on recycle of NH_3 which is separated from the reactor effluent gases by scrubbing with monoethanolamine. The carbon dioxide is absorbed in the amine, and the gaseous ammonia is recycled to the reactor. (See Fig. 5.17.) The reactor effluent is first decomposed; then the urea solution is removed and concentrated. A specially designed reactor lined with zirconium is needed. Carbon dioxide conversion is relatively high—80 to 85 percent per pass.[80]

The *Chemico "Thermo-Urea"* process[81] employs centrifugal compressors to recycle the carbon dioxide, ammonia, and water vapor to the reactor. The reactor effluent is passed into a separator from which the hot gases pass to the compressor, and the liquid effluent enters a carbamate decomposer where the remaining gases are removed and sent to the compressor. The liquid effluent from the first decomposer enters a low-pressure decomposer which removes the last traces of carbon dioxide, ammonia, and inert gas. The ammonia and carbon

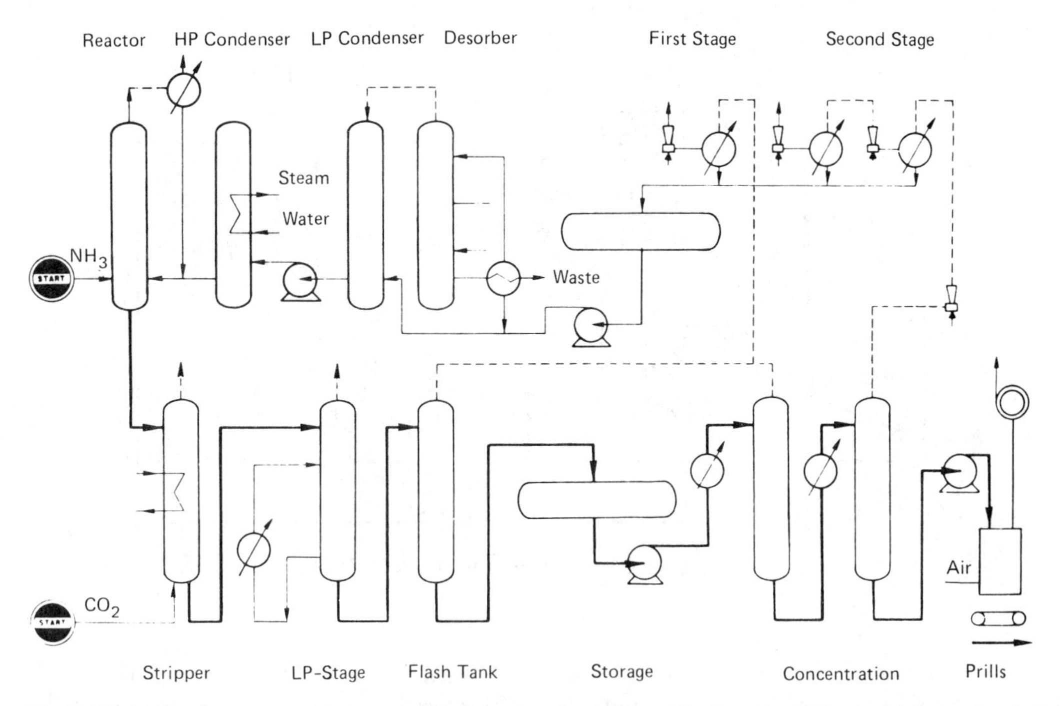

Fig. 5.16 Stamicarbon process for urea. (*Hydrocarbon Processing,* **58,** *No. 11, 1979, copyright by the Gulf Publishing Company.*

dioxide are absorbed in water for recycle to the compressor after vaporization; and the inert gas is removed from the top of the absorber. Aqueous urea is removed from the low pressure decomposer (See Fig. 5.18.) An improved process is described in U.S. Patent 3,301,897.

Mitsui Toatsu Chemicals, Inc. and *Snamprogetti* offer plants with capacities up to 1800 metric tons per day. A variety of process modifications are available with the Mitsui Toatsu design by J. F. Prichard and Company. The latest modifications incorporate energy-

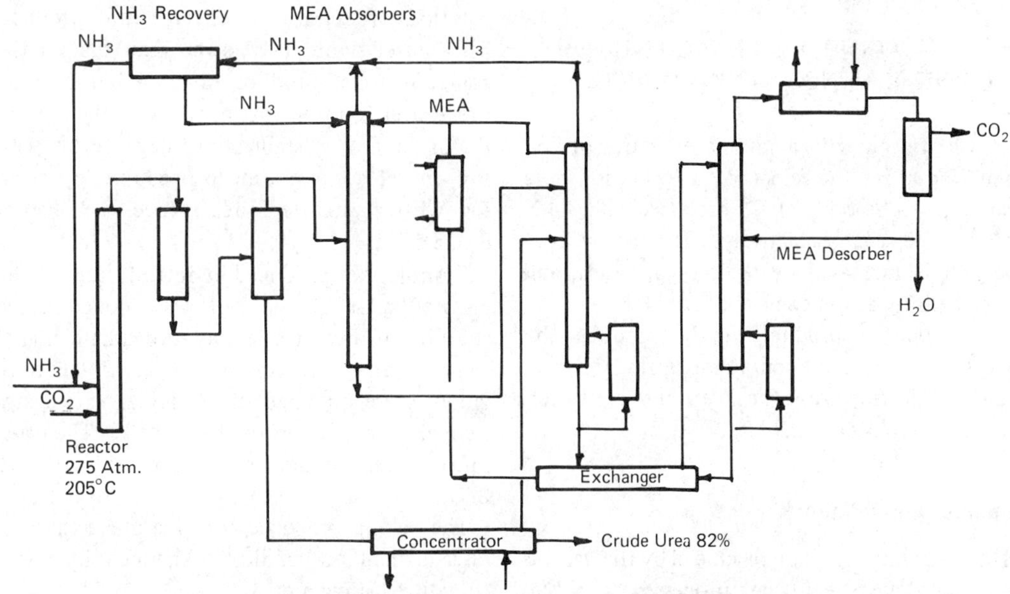

Fig. 5.17 Allied Chemical-C.P.I. process for urea.[80, 87, 88]

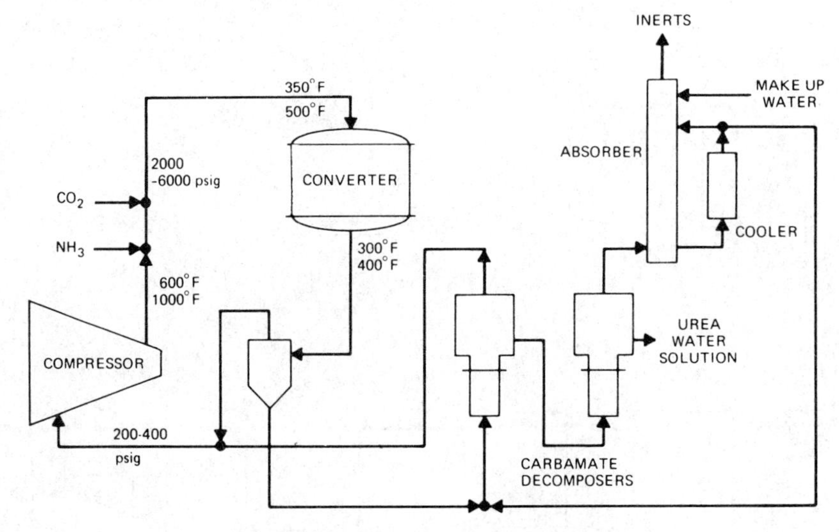

Fig. 5.18 Chemico thermo-urea process.[81]

saving developments. Power requirements are reported to be as low as 52 to 66 Kwh per metric ton.[82]

To avoid a yield loss and minimize product contamination, it is important to minimize biuret formation. The concentration of biuret in urea used for foliar feeding must be low so as to avoid mottling of the leaves. Consequently, a foliar grade of urea is produced containing less than 0.035 percent biuret. Its formation is as follows:

$$CO(NH_2)_2 \rightleftarrows HNCO + NH_3$$
$$CO(NH_2)_2 + HNCO \rightleftarrows NH_2CONHCONH_2$$

At 140°C and with ammonia pressures of 760 and 70 mm Hg, the velocities of formation were reported to be 0.9×10^{-4} mole/sec. and 3.4×10^{-4} mole/sec., respectively. The rate of conversion is increased by basic compounds, and decreased by acidic compounds.[83]

A low-biuret product can be produced by recycling the biuret-containing effluents from the purification zone back to the converter. See U.S. Patent 3,232,984.

Product Formulation

The urea-water solution produced by the various processes must be further processed to obtain the final products, in the form of crystals, flakes, prills, or as granular material. By far the largest amount of urea is produced as a fertilizer in the form of prills, which are produced by removing water from the solution, and dropping the molten urea down through spray formers in a tower countercurrent to a stream of air. The droplets of molten urea, which solidify before hitting the base of the tower, are cooled, and then bagged or sold in bulk.

Crystallization of urea from the urea-water solution produces a relatively pure product. The biuret content is lower than that in the standard prilled product, since the biuret in the urea-water solution is removed with the mother liquor in the crystallization step. Prills with low biuret content can be produced by starting with crystallized urea. (See U.S. Patent 3,025,571.)

Granules are produced by introducing into an agitated mass of dry particles a concentrated solution of urea at a temperature a little higher than its set point. Agitation is continued until granules are obtained; then they are dried at a temperature of no more than 120°C. The usual ratio of urea solution to dry particles is 0.4:1 to 0.7:1.

The caking properties of urea make the product difficult to handle.[80] The technique used to reduce caking are:

1. Coating.
2. Mixing with additives.
3. Thermal treatment.

Additives such as ammonium sulfate, acetylene-diurea, magnesium carbonate, and substituted ureas, e.g., p-toluidine, are used.

Coatings which are reported to be effective are sugar, formaldehyde, and acetaldehyde.

In the thermal treatment, the prills are passed through a zone of temperature higher than the melting point of urea.[80]

Anhydrous urea can be produced directly, according to J. S. Mackay,[84] by synthesizing the urea at a temperature above its melting point and simultaneously keeping the partial pressure of the water below the condensation point. A pressure above 1000 psig and a temperature above 200°C are used. The preferred conditions are 2000 psig and 225°C. Nitrogen, but preferably ammonia, can be used to reduce the partial pressure of the water vapor.

$$10NH_3 + CO_2 \longrightarrow (NH_2)_2CO + H_2O + 8NH_3$$

Uses of Urea

Industrially, when combined with formaldehyde, urea gives resins that can be machined. Urea is used as an ingredient in softeners for cellulose, cellophane, and wood. It is also used to retard end-checking of boards during drying. Added to glue, gelatin, or starch, it reduces the viscosity and permits the use of higher concentrations of the active material. Urea is used in the petroleum industry to separate straight-chain hydrocarbons by forming crystalline complexes between the urea and a hydrocarbon with a long straight carbon chain. Urea is used in the preparation of barbituric acid, caffein, ethylurea, hydrazine, melamine, guanidine, and sulfamic acid. It is also used in the manufacture of medicinals, in some cosmetic applications and deodorants, and in skin creams to improve the texture. With hydrogen peroxide, urea forms a crystalline additive $CO(NH_2)_2 \cdot H_2O_2$ which is employed as a disinfectant and oxidizing agent.

Urea-formaldehyde, urea-furfural-formaldehyde, and urea-resorcinal mixtures form thermosetting resins which are used in molding applications such as radio and TV cabinets, in the treatment of paper and fabrics, and as adhesives for bonding wood. The resins are nearly colorless but opaque, and may be pigmented as desired. They improve crush and crease resistance in textiles, and impart wet strength to paper and paper coatings. A small amount of acid, or acid generating catalysts, may be added to make the resin set.

However, the principal outlet for urea is in the fertilizer field, not only as a solid fertilizer but also as a liquid in combination with ammonia and ammonium nitrate. The commercial product is known as UAN. A continued growth is expected since the solution can be handled as if it were water, in contrast to that required for anhydrous ammonia.

In 1977, 2,684,095 tons urea were produced in the United States as a solid fertilizer, and 1,473,299 tons in the form of solutions.

In the world market, urea is the leading nitrogen fertilizer material.[85] High nitrogen analysis saves transportation costs; also, urea can be used on almost any crop, and production costs have been lowered by the large single-train ammonia and urea plants. In 1978, urea surpassed the use of ammonium nitrate in the U.S.A. All indications are that the U.S.A. fertilizer market is moving toward the greater use of fluids. When combined with superphosphate, urea produces a fertilizer known as "Phosphazote." A urea salt with calcium nitrate is also recommended as a fertilizer. In addition, urea can be combined with formaldehyde to produce a slowly soluble compound which will supply nitrogen to the soil during the entire growing season.

Urea has found acceptance as a protein substitute for ruminant animals (cattle, goats, and sheep). In this application, it can efficiently replace about one-third of the natural protein in ruminant feeds, but care in formulating must be exercised to optimize the concentration. Urea is toxic in large doses. It is reported in a recent patent[86] that urea-hydrocarbon solids adducts are more palatable than urea alone, making the product more readily accepted by the animals. The hydrocarbons used are alkanes, such as n-hexane to n-tetradecane and olefins, such as n-hexene and n-octadecene. The following table shows the advantage of an urea adduct over urea.

Feed Additive	*Nitrogen Retained (grams/day)*
Urea	1.73
n-tetradecane-urea adduct	2.16
1-tetradecene-urea adduct	2.22

Melamine

Melamine or cyanuramide, $C_3N_3(NH_2)_3$, a monoclinic compound, decomposes at its melting point of 360°C. Its molecular weight is 126.13. It forms white monoclinic crystals and has a bulk density of 20.6 lb cu ft, a nitrogen content of 66.6 percent, and forms a solution in formalin with a color of 15 APHA. It sublimes at its boiling point. It is nonflammable and is only mildly toxic.

```
          N
        //  \
  H2N—C      C—NH2
       |      ||
       N      N
        \\  /
          C
          |
         NH2
```

Melamine is produced by heating urea and passing the resulting mixture of isocyanic acid and ammonia over a suitable catalyst, quenching the products with water or aqueous mother liquor, and working them up by filtration, centrifuging, or crystallization.[89]

The reaction takes place in two stages:

$$CO(NH_2)_2 \longrightarrow HNCO + NH_3 \qquad (1)$$

This reaction is endothermic requiring 800 kilocalories per kg of urea or 21,800 kilocalories per lb mole of urea.

$$6HNCO \longrightarrow C_3N_3(NH_2)_3 + 3CO_2 \qquad (2)$$

This reaction is exothermic, liberating 1100 kcal per kg of melamine or 63,000 kcal/lb mole of melamine. The overall reaction is endothermic.

The net yield is 84 percent of theoretical; but if the mother liquor can be recycled to a urea plant a better yield is possible.

The O.S.W. process is shown in Fig. 5.19.[89] Urea, along with recycle ammonia, is fed into a fluidized-bed reactor containing inert solids, where the temperature is quickly raised to 350°C within a few seconds. The resulting mixture of isocyanic acid and ammonia is then fed into a reactor containing a solid catalyst, such as silica gel, alumina, or boron phosphate. Products from the isocyanic acid reactor are quenched with water, the resulting slurry is centrifuged, and the mother liquor is sent to recovery or a urea plant.

Off-gases from the quench system contain ammonia, carbon dioxide, and water vapor.

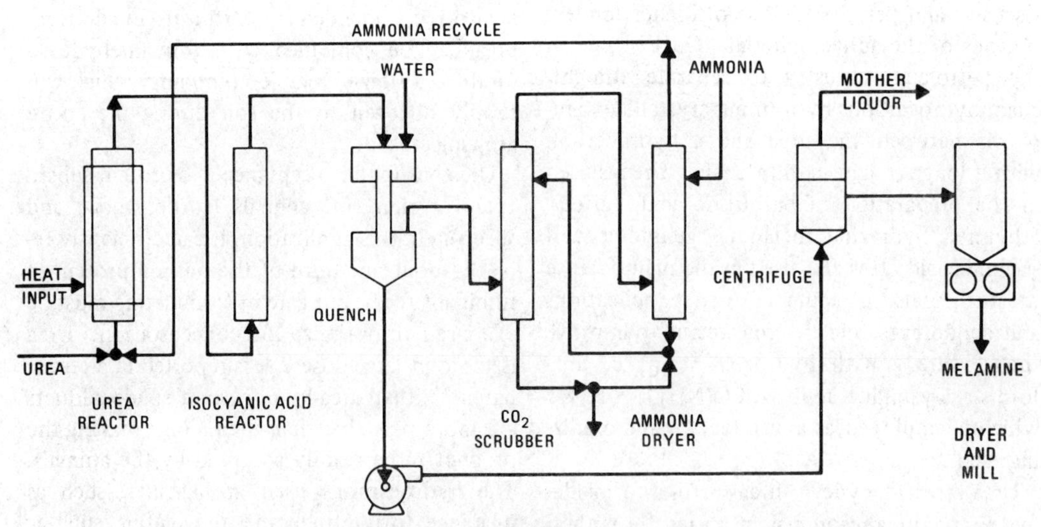

Fig. 5.19 Melamine synthesis—O.S.W.[89]

The CO_2 is removed as ammonium carbonate, and the ammonia is returned to the urea reactor.

The design of the urea reactor is critical. The wall temperature must not be too high or corrosion is a problem.[90] This can be avoided by using stationary layers of catalyst along the inner wall, held in place by corrosion-resistant flights secured to the periphery at suitable intervals. With this method, the reactor wall can be made of steel containing 0.16 percent carbon, 0.35 percent silicone, 0.40 to 2.00 percent manganese, not more than 0.5 percent phosphorous, and not more than 0.05 percent sulfur.

The formation of melamine from isocyanic acid takes place at 400 to 430°C. The heat of reaction is absorbed as sensible heat by raising the entering vapors from 320 to 430°C. Melam ($C_6H_9N_{11}$) and melem ($C_6H_6N_{10}$), condensation products of melamine, are by-products that form by the splitting off ammonia when the temperature is too high. Melamine will also hydrolyze if allowed to remain too long with water at high temperature, the resultant products being ammeline $(CH)_3(NH_2)_2OH$ and ammelide $(CH)_3(NH_2)(OH)_2$.

Melamine is milled to produce a fine particle size, 5-10 microns.

Methods of purifying crude melamine are being studied[91], since recrystallization is required to obtain a pure product. In the reference given, crude melamine is dissolved in aqueous ammonia from which it is crystallized. The obtainable yields are reported to be 10% higher than for conventional processes.

Melamine is used for the manufacture of molding compounds, laminates, coatings, paper and textile-treating resins, and adhesives. Producers of melamine are:

Allied Chemical Corp.
American Cyanamid Co.
Melamine Chemicals, Inc.
First Mississippi Corp.

ALIPHATIC AMINES

Amines are derivatives of ammonia in which the hydrogen atoms in the ammonia have been displaced by monovalent hydrocarbon radicals. Depending on the number of hydrogen atoms displaced, amines are classified as primary (RNH_2), secondary (R_2NH) and tertiary (R_3N). According to the type of hydrocarbon radicals that substitute the hydrogen atoms, amines are further categorized as aliphatic (saturated or unsaturated), aromatic, heterocyclic, alicyclic or any combination of these. *Amines* are distinguished from *imines* in which two hydrogen atoms in an ammonia molecule are displaced by bivalent hydrocarbon radicals (R=NH); and also distinguished from *nitriles* or *cyanides* in which all the hydrogen atoms in ammonia are displaced by a trivalent hydrocarbon radical ($RC\equiv N$).

Although the amine family consists of a multitude of compounds, the scope of this section is confined to the two commercial aliphatic amines produced in large volume:

- Methylamines
- Ethylenediamine

In general, most aliphatic amines are toxic and corrosive to skin. Low molecular weight amines are volatile and may cause bronchitis and even suffocation if inhaled. Full protective equipment should be worn when handling the material. The first aid treatment upon superficial contact with amines is to flush the exposed area with large quantities of water.

Methylamines

Early commercial use of methylamines was in the leather industry as a depilatory agent. Since the 1950's, the amines have been developed into major industrial chemicals, with end uses spanning from military rocket fuels to building blocks for pharmaceuticals, solvents, and agricultural chemicals. The demand for methylamines is expected to exceed 250 million pounds per year in the 1980's.

Methylamines are colorless liquids and are volatile under atmospheric conditions. The amines have threshold odor limits less than 10 ppm. At low concentration, they smell fishy. At high concentration, they bear the odor of ammonia. Nevertheless, smelling is not a reliable means to detect methylamines because the nasal membrane can be rapidly desensitized. The "MSA" Universal Tester with Detector Tube 92115 can be used to detect less than 10 ppm of the amines in air. Methylamines are more basic than ammonia. They are very soluble in water, alcohols, ethers and most polar solvents. At low temperatures, aqueous methylamines form crystalline hydrates: $CH_3NH_2 \cdot 3H_2O$, $(CH_3)_2NH \cdot 7H_2O$ and $(CH_3)_3N \cdot 10H_2O$. In the presence of ammonia, methylamines react explosively with

TABLE 5.12 Physical Properties of Anhydrous Methylamines

	Anhydrous Methylamine		
	MMA	*DMA*	*TMA*
Chemical formula	CH_3NH_2	$(CH_3)_2NH$	$(CH_3)_3N$
Molecular weight	31.06	45.08	59.11
Freezing point, °C (in air at 1 atm)	–93.5	–92.2	–117.3
Boiling point, °C	–6.3	6.9	2.9
Vapor density at 1 atm			
(25 C, 77 F) g/liter	1.3	1.9	2.5
air = 1	1.1	1.6	2.0
Liquid density at sat'n pressure			
(25 C, 77 F), g/mL (Mg/m^3)	0.6562	0.6496	0.6270
lb/gal	5.48	5.42	5.23
Refractive index at sat'n pressure, n_D^{25}	1.3491	1.3566	1.3443
Liquid viscosity (25 C, 77 F), cP(mPa · s)	–	0.190	0.175
Critical temperature, C	156.9	164.5	160.1
Critical pressure, atm	73.6	52.4	40.2
Enthalpy of formation, ΔH_f° (298.15 K, 25 C), kcal/mol			
–gas	–5.49	–4.41	–5.81
–liquid	–11.3	–10.5	–11.0
–solution (1000 moles H_2O)	–16.78	–17.3	–18.6
Vapor pressure, psia			
25°C	50	30	32
100°C	400	230	205
Heat capacity, C_p° (298.15 K, 25 C), cal/C · mol			
–gas	12.7	16.9	21.9
–liquid	–	32.9	32.31
Entropy, S° (298.15 K, 25 C) cal/C · mol			
–gas	58.15	65.24	68.6
–liquid	35.90	43.58	49.8
Heat of fusion (mp),			
cal/g	47.20	31.50	26.46
Btu/lb	84.96	56.70	47.63
Heat of vaporization (bp),			
cal/g	198.6	140.4	92.7
Btu/lb	357.5	252.7	166.9
Gibbs energy of solution, ΔG_f° (molality = 1; 25 C), kcal/mol	4.94	13.83	22.22
Entropy of solution, S° (molality = 1; 25 C), cal/C · mol	29.5	31.8	31.9
Autoignition temperature, C	430	400	190
F	806	752	374
Flammable limits, vol %			
–lower	4.9	2.8	2.0
–upper	20.7	14.4	11.6

TABLE 5.13 Physical Properties of Aqueous Methylamines

Property	*40% MMA*	*40% DMA*	*60% DMA*	*25% TMA*
Boiling point, C	48	54	36	43
F	118	129	97	109
Freezing point, C	−38	−37	−74.5	6
F	−36	−35	−103	43
Density (liquid), at 25 C (77 F)				
g/mL (Mg/m^3)	0.897	0.892	0.829	0.930
lb/gal	7.49	7.44	6.92	7.76
Vapor pressure, at 25 C (77 F)				
psia	5.8	4.2	9.7	6.6
mm Hg	300	215	500	340
kPa	40	29	67	45
Flash point, Closed cup,				
(ASTM D-92), C	−12	−18	−52	6
F	10	−1	−61	42

mercury. Hence, instruments containing mercury should not be used on methylamines. The amines are also flammable in air. Their flammability ranges as well as some of their important physical properties are listed in Tables 5.12 and 5.13.

Manufacture. Numerous chemistry routes are known to be feasible for the synthesis of methylamines. Among them, at least three routes are reported to exhibit significant yield of methylamines worthy of industrial attention:

- Alkylation of ammonia by methanol or methyl ether.
- Reaction of carbon monoxide, ammonia and hydrogen[138]
- Hydrogenation of hydrogen cyanide[139]

As of the end of the 1970's, practically all the commerical methylamine processes are based on the technology of methanol alkylation of ammonia.

1. Alkylation of Ammonia by Methanol.[157] The alkylation reactions of ammonia were originally discovered by Sabatier and Maihle in the 1900's for the synthesis of higher amines. The chemistry had been adopted for methylamines in the 1930's where mono-, di- and tri-methylamines (MMA, DMA & TMA) were co-produced in the following vapor phase reactions in the presence of a dehydrating catalyst:

(I) $NH_3 + CH_3OH \xrightleftharpoons{Cat.} CH_3NH_2 + H_2O +$ 5.358 Kcal/mol

(II) $CH_3NH_2 + CH_3OH \xrightleftharpoons{Cat.} (CH_3)_2NH +$ $H_2O +$ 9.598 Kcal/mol

(III) $(CH_3)_2NH + CH_3OH \xrightleftharpoons{Cat.} (CH_3)_3N +$ $H_2O +$ 14.098 Kcal/mol

(IV) $NH_3 + (CH_3)_3N \xrightleftharpoons{Cat.} CH_3NH_2 +$ $(CH_3)_2NH -$ 8.74 Kcal/mol

(V) $NH_3 + (CH_3)_2NH \xrightleftharpoons{Cat.} 2CH_3NH_2 -$ 4.24 Kcal/mol

(VI) $CH_3NH_2 + (CH_3)_3N \xrightleftharpoons{Cat.} 2(CH_3)_2N -$ 4.50 Kcal/mol

The reaction mechanism and kinetics were studied by Novella and Una[140]. While it was not possible to develop theoretical kinetic equations based on Langmuir-Hinshelwood theory, the empirical kinetic expressions deduced from their experiments using aluminum silicate catalyst are in the form of:

$$\frac{M}{A} = C_1 \exp\,[C_2 T + C_3 X]$$

for each of the amines

where M = gm of catalyst
A = mole MeOH fed per hour
T = reaction temperature
X = MeOH conversion
C_1, C_2, C_3 = Coefficients corresponding to the ratio of MeOH and ammonia fed

Reactions (I) to (III) are known as the alkylation reactions. They are exothermic and highly irreversible except for reaction (III)[141]. Reactions (IV) to (VI) are known as disproportionation reactions which are reversible and are endothermic if the reactions proceed from left to right. The alkylation reactions dictate the rate of consumption of methanol, and are somewhat faster than the disproportionation rates which govern the selectivity of the three amines. However, the relative rate of alkylation and disproportionation depends on the type of catalyst and may be changed if a new catalyst is developed.

Thorium oxide[140] was the original catalyst but was soon replaced by alumina and alumina silica catalyst[142] in the early 1930's because of their catalytic activity and selectivity. The improvement in the catalytic activity reduced the required reaction residence time and thus decreased the investment in the reactor. The catalyst selectivity determines the distribution of the three amines in the reactor effluent gases. Numerous patents have been issued based on different recipes for preparing the catalyst for improvement in selectivity[143,144,145,140].

Conventional catalysts consist of about 86% silica and 14% alumina and yield predominantly TMA which is of least commercial demand. The undesired TMA is separated and recycled to the reactor for disporportionation into other amines. U.S. patent 3,387,032 claimed that by adding 0.05 to 0.95% silver phosphate, rhenium heptasulfide, molysulfide or cobalt sulfide to the conventional catalyst and with proper steam treatment, the selectivity of the three amines is shifted in favor of mono- and di- at the expense of tri-. The relative increases of MMA and DMA in the reactor effluent are 67% and 98% respectively. The corresponding decreases in TMA is 63%.

Japan patent 1623('52) also claimed improvement in selectivity in favor of MMA and DMA with a MgO/SiO_2 catalyst (1:1 ratio). The most recent development in catalytic technology is covered in West Germany patent DT-2916-060 for the use of zeolite to produce predominantly MMA and DMA. Although the cited patents, as well as many others, have claimed improvement in the catalysis for the synthesis of methylamines, no publication is available to account for how many of these inventions are actually used industrially or to what extent.

The selectivity is also dependent on the operating parameters[141] of the reaction, such as the feed composition, the temperature and the residence time. Since all the reactions are equilmolar, pressure is not a direct operating variable, although increasing pressure will also increase the residence time for a given reactor volume. Serban[146,147,158] used the nitrogen/carbon (N/C*) mole ratio to correlate the effects of feed composition on selectivity. As expected, the MMA content in the reactor effluent increases with respect to increasing the N/C as the ammonia disproportionates TMA into lower amines. DMA, however, exhibits a maximum at a N/C of approximately 1, beyond which MMA begins to dominate. TMA, of course, decreases as the N/C increases.

Although the increase of the N/C can induce a more favorable selectivity, the benefit is counter-balanced by the recovery cost of unconverted ammonia. In addition to its dependency on the N/C, TMA content is reported to be reduced as water is fed to the reactor (i.e., reversibility of Reaction III). The reaction temperature has a twofold effect on selectivity. Increasing the temperature will increase the reaction rate of the disproportionation reactions. Furthermore, higher temperature favors the equilibrium distribution of MMA and DMA in reactions (IV) to (VI). The published temperature dependency of the equilibrium constants is:

$$\frac{[MMA]\,[DMA]}{[NH_3]\,[TMA]} = \text{Exp}\ \frac{1725.1}{T} + 1.673 = K(\text{IV})$$

$$\frac{[MMA]^2}{[NH_3]\,[DMA]} = \text{Exp}\ \frac{-1028.4}{T} + 0.877 = K(\text{IV})$$

$$\frac{[DMA]^2}{[MMA]\,[TMA]} = \frac{K(\text{IV})}{K(\text{V})}$$

The increase in reaction residence time also reduces the TMA content in the reactor efflu-

*N/C was calculated as the mole of nitrogen atoms at the feed divided by the mole of carbon atoms.

ent. Since the disproportionation reactions are the rate-limiting steps, increase in residence time will favor the disproportion of TMA into lower amines.

In addition to the six primary reactions discussed above, formaldehyde, N,N,N',N',-tetra methyldiaminomethane and other trace impurities are known to be generated by side-reactions. Although the formation of by-products does not constitute a major yield loss, they present operating problems in the separation of amines. U.S. patent[181] 3,720,715 showed that the injection of hydrogen into the reaction can suppress the side-reactions.

A typical methylamines process is divided into three sections: Synthesis, Refining, and Waste Recovery.

In the synthesis step (Figure 5.20), the pressure is operated at about 70 ~ 500 psig. Methanol, ammonia and the recycled amines which are in excess of market demand are vaporized in a vaporizer. The mixture is then superheated to about 300° ~ 400°C before entering the reactor. The feed temperature is controlled to maintain the MeOH conversion close to completion. The feed temperature is increased as the throughput in the reactor increases and as the catalyst deactivates during its life. The reactor is typically an adiabatic fixed-bed catalytic reactor where the residence time is about 10 to 40 seconds.[145] The heat of reaction is recovered by preheating the reactant mixture in a superheater.

The temperature rise in the reactor depends on the MeOH conversion and the selectivity. The reactor exit temperature is typically at about 400° to 500°C. The reactor effluent gases are condensed and stored for refining. Small amounts of noncondensible CO, H_2, CH_4 are vented from the condenser. The yield of the reactions is over 90% for both ammonia and methanol. While the MeOH conversion is controlled at close to 100%, the ammonia conversion is not a control variable and is dependent on the N/C ratio. The ammonia conversion is typically at about 10 to 40%. The crude product in a conventional process consists of:

Product	*Wt. %*
NH_3	20 ~ 50
MMA	7 ~ 12
DMA	10 ~ 22
TMA	10 ~ 30
MeOH	0 ~ 3
H_2O	10 ~ 25

The challenge in the near term technological improvement in the synthesis step is to reduce the throughput per product poundage by reducing the amines recycle. The incentives here are lower capitol investment as well as lower energy consumption.

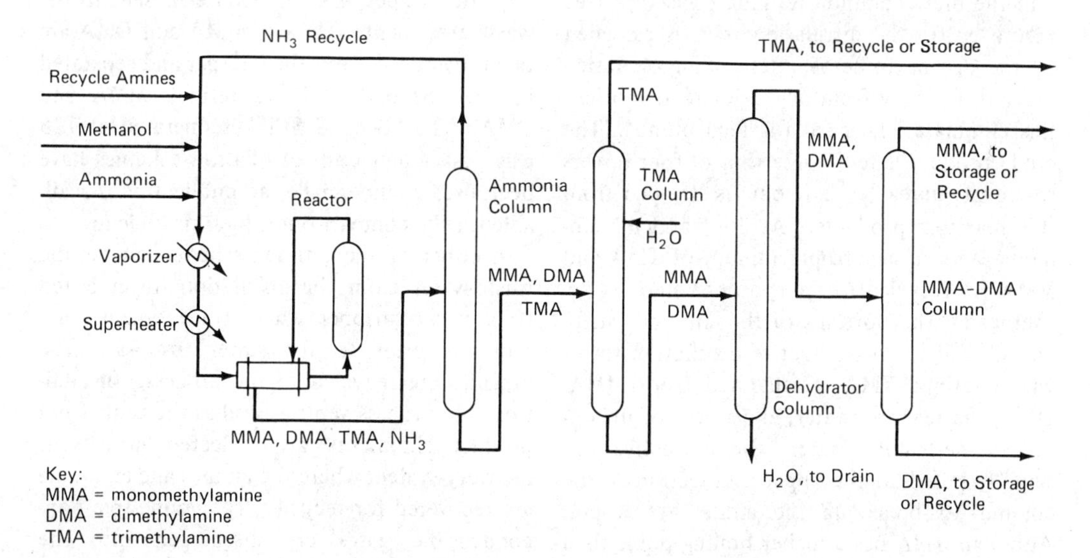

Fig. 5.20 Schematic of methylamines synthesis and refining process. *(Adapted from Chemical Engineering, Oct. 29,* **70** 1962.)

In the refining step, the amines are separated from the unreacted raw materials as well as from the reaction water by a series of distillations. Since the amines have small differential relative volatility and are affinitive to water, the fractionations of the amines are very energy consuming. Reportedly, it takes about 10 to 30 pounds of steam to produce one pound of amines. The energy efficiency of the process is dependent on the design of the equipment—which is also a trade-off between operating cost and capitol investment. Most of the distillation towers are operated under a pressure of between 25 to 300 psig. There are at least three reasons for distilling under pressure. First, increasing the pressure also increases the condensing temperature of the volatile amines and ammonia in the distillation overhead and thus eliminates the use of refrigerating condensers. Second, operating under pressure increases the throughput capacity of a given tower diameter. Third, some of the ammonia-amines azeotropes disappear at elevated pressure. Azeotropic mixtures[139] of ammonia-TMA, TMA-MMA and TMA-DMA were reported in the literatures. Extractive distillation[148,149,150] is also a viable means to overcome the difficulties of separating the amines. The extractive solvents are selected for their different solubilities of the three amines. Two alternative distillation sequences have been published.

In the more common refining process[151] (see Figure 5.20), the distillation train is cascaded by the driving force of a descending operating pressure for flow from one column to another; this eliminates the need for feed pumps. The crude product is fed to the first of four towers where the unreacted ammonia is stripped from the reaction products. As low boilers, ammonia and an azeotropic amount of TMA and MMA are recycled to the synthesis step via the overhead. The bottoms of the ammonia stripper are fed to the extractive distillation tower where refined TMA is separated from MMA, DMA, the reaction water, and traces of unconverted methanol. Water, the extractive solvent[148], is fed to the upper mid-section of the column to break up the amine azeotropes. Although TMA has a higher boiling point than MMA and DMA in anhydrous form, water reverses the volatility of the three amines due to the higher solubilities of MMA and DMA. TMA, after being dehydrated in the top section of the column, is either shipped as product or recycled to the synthesis step. The tails of the extractive distillation tower are fed to the dehydrator where MMA and DMA are taken overhead, and the reaction water and traces of unreacted MeOH are sent to waste treatment. The final refining step separates MMA from DMA. In Figure 5.21, an alternative refining scheme[152] for methylamines is shown.

Five distillation towers are used. The crude product is first fed to a low-pressure stripper at about 60 psig. The reaction water and traces of unreacted MeOH are stripped from the crude product as bottoms and are sent to waste treatment. The alkaline compounds are taken overhead from the low-pressure stripper and are pumped to the "First Still" for ammonia removal. The "First Still" is operated at 250 psig where ammonia and an azeotropic amount of TMA (~7%) are taken overhead for recycle to the synthesis step. The ammonia-free amines mixture is fed from the bottoms of the "First Still" to the extractive distillation tower similar to the one described previously. The tower is operated at 175 psig where TMA is purified as an overhead product. The bottoms are a 25–35% aqueous solution of MMA and DMA. The aqueous solution is dehydrated in the subsequent stripper where the water is sent to the waste treatment. The dry MMA and DMA are taken overhead from the stripper and separated in the "Second Still" as refined MMA and DMA. The "Second Still" is operated at 125 psig. Although both distillation schemes have been used commercially, no publication is available for the comparison of their proficiency.

In either of the refining schemes above, the waste-water from the distillation tower is fed to a waste stripper where trace organics are removed prior to the sewer disposal. The organics are recycled to the process. In addition, the process vents as well as the vents from product shipments are collected in a vent recovery system where the amines and ammonia are recovered for recycle. The amine-free noncondensible gases are discharged into the atmosphere.

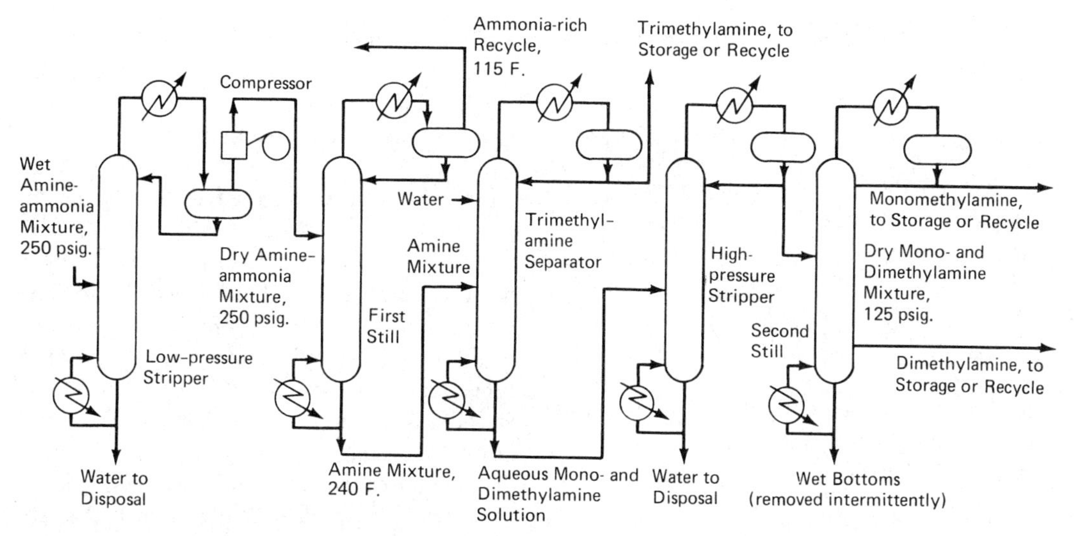

Fig. 5.21 Alternative distillation scheme for refining methylamines. (*Chem. Engr., 101-102, Aug. 21 1961.*)

Other Routes. U.S. patent 3,444,203 described the synthesis reactions of methylamines via carbon monoxide, ammonia, and hydrogen. The $CO/NH_3/H_2$ route has the advantage of backward integration to cheaper raw materials and also eliminates the expenditure for manufacturing methanol. Although the yields are reported to be in the range of 90 plus percent, the current state of the art calls for high pressure equipment (eg. 5000 psig) and requires an inhibitively high capital investment.

The reduction of hydrogen cyanide produces predominantly MMA. The chemistry is described in DBP[153] 1,019,313 and others. The present technology is not economically attractive except for backward integration to more elementary raw materials using HCN as an intermediate.

The reduction of formamide to methylamines was also published in French Patent[154] 762194.

Economics and Marketing.[155,156] Methylamines are diversified in their end uses. Monomethylamine is used as a raw material for insecticides, explosives (monomethylamine nitrate as sensitizer in Tovex®), surfactants (N-Oleytamine, $HOC_2H_4SO_3Na$), rocket fuels, photographic developer, and others. Dimethylamine is consumed in the manufacture of industrial solvents (Dimethylformamide & Dimethylacetamide), water treatment agents, pesticides, rubber accelerators, rocket fuel, and additives in petrochemicals. Trimethylamine has as its major outlet the manufacture of choline chloride as animal feed supplement and gastro intestinal medicines.

The distribution of the end uses for methylamines at the end of the 1970's is shown below:

End Uses

MMA	
Explosives	35%
Insecticides	38%
Surfactants	8%
Others	19%

DMA	
Water treatment	13%
Pesticides	14%
Rubber accelerator	15%
DMF & DMAC	40%
Other	18%

TMA	
Choline chloride	90%
Other	10%

The average historic growth in the 1970's was 7.7% per year. The growth rate in the 1980's is expected to be about 6% per year as some of the end uses become matured. A review of the historic annual production is shown in Figure 5.22.

Methylamines are marketed as anhydrous grade; 25% and 40-60% aqueous MMA solutions; 25% and 40% aqueous DMA solutions; and 25% and 40-60% aqueous TMA solutions. Since the early 1950's, the price has risen from 13 cents per pound to $33\frac{1}{2}$ cents per pound of anhydrous grade and to 44 cents per pound of solution in 1979. The major U.S. Producers are listed below:

Producer	*Capacity in million pounds per year as of the end of 1979*
DU PONT	185
AIR PRODUCTS	100
IMC	25
GAF	10
TOTAL	320

Air Products has announced a major expansion to 175 million annual pounds to be started up in the early 1980's.

Ethylenediamine.

Ethylenediamine is a slightly viscous liquid which bears a strong odor of ammonia. In most conventional processes, ethylenediamine (EDA) is co-produced with diethylenetriamine (DETA), triethylenetetramine (TETA), tetraethylene pentamine (TEPA) and pentaethylenehexamine (PEHA). Some of their physical properties are outlined below:

	EDA	*DETA*	*TETA*	*TEPA*	*PEHA*
Molecular Formula	$NH_2(CH_2)_2NH_2$	$NH_2CH_2CH_2NH$ | CH_2CH_2NH	$NH_2(CH_2CH_2NH)_2$ | $CH_2CH_2NH_2$	$NH_2(CH_2CH_2NH)_3$ | $CH_2CH_2NH_2$	$NH_2(CH_2CH_2NH)_4$ | $CH_2CH_2NH_2$
Molecular Wt.	60.11	103.17	146.24	189.31	232.36
Normal Boiling Pt.°C	117°	207°	277°	decomposes @ 340°	*
Melting Pt.°C	11°	-39°	-35°	-30°	-26°
Density, lb/gal	7.67	7.96	8.19	8.33	8.39
Flash Point (Open Cup)°C	38°	102°	143°	185°	185°

Ethylenediamine is soluble in water, ethanol, acetone, ether and benzene. The threshold odor limit for EDA is 10 ppm.

Commercial ethylenediamine is produced by three different chemistry routes:

- Ethylene dichloride and ammonia
- Ethylene glycol (or Ethylene Oxide) and ammonia
- Ethanolamine and ammonia

Most existing processes use the ethylene dichloride route. The ethylene oxide route is employed by Union Carbide in their new facilities in Taft, Louisiana which are scheduled to start up in the early 1980's. Both the ethylene dichloride and ethylene oxide routes produce other higher amines (eg. DETA, TETA, TEPA, and PEHA) as well. The ethanolamine route, however, produces predominantly ethylenediamine. Reportedly, BASF was planning to construct a new plant in Belgium in the late 1970's using a process based on the ethanolamine technology.

Ethylene Dichloride[162] *and Ammonia.* This process is typically operated in an ammonia rich condition. The primary reaction is:

$$ClCH_2CH_2Cl + 2NH_3 \longrightarrow NH_2CH_2CH_2NH_2 + 2HCl$$

Other by-products are diethylenetriamine, triethylenetetramine, tetraethylenepentamine, pentaethylenehexamine and polyethylene polyamines. Low reaction temperature and pressure yield a more favorable selectivity. However, it results in a low conversion rate. High tempera-

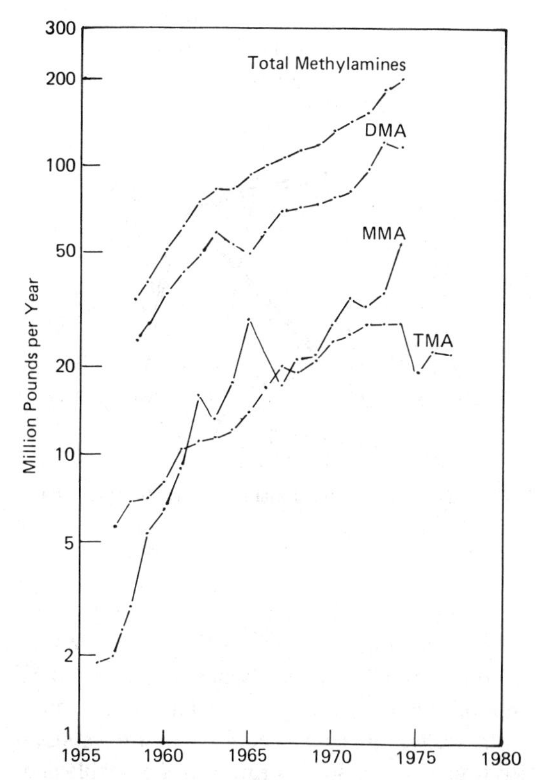

Fig. 5.22 Historical annual production of methylamines.[155, 156]

ture and pressure, on the other hand, increase the conversion rate, but become less selective for EDA. The reaction system is noncatalytic. In a typical EDA process, ethylene dichloride reacts with excess aqueous ammonia at about 100 ~ 200°C and at 10 to 20 atm. The crude product is an aqueous solution of ammonia, ammonium chloride, ethylenediamine hydrochloride and higher amine hydrochlorides. Caustic is added to neutralize the crude product and free the ethylenediamine, the higher amines, and the ammonia from the amine hydrochloride and ammonium chloride. The crude product is refined by fractionations where the unreacted ammonia is recycled and the amines are purified. The EDA yield of the process is estimated to be between 45 to 70%.

Ethylene Glycol and Ammonia. This reaction takes place over a hydrogenation catalyst such as nickel or copper supported on alumina, titania or similar metal oxide. The primary reaction is:

$$\underset{\text{OH}}{\text{CH}_2}-\underset{\text{OH}}{\text{CH}_2} + 2NH_3 \xrightarrow{\text{cat}} NH_2CH_2CH_2NH_2 + 2H_2O$$

with coproducts DETA, TETA, TEPA and PEHA produced simultaneously. The ammonia to ethylene glycol ratio[163] ranges from 20 ~ 30:1 (mole basis) with water present to the extent of at least 30% by weight of the glycol. Hydrogen is required to activate the catalyst at about 0.2% by weight of the glycol. The liquid phase reaction temperature is 220 to 270°C at 3000 to 6000 psig. The crude product is refined by fractionations.

Ethanolamine and Ammonia. The reaction for this process is as follows:

$$NH_2CH_2CH_2OH + NH_3 \longrightarrow NH_2CH_2CH_2NH_2 + H_2O$$

Ethylenediamine is the major product.

Economics and marketing. The fastest growing end use of ethylenediamine is in the manufacture of the chelating agent EDTA. EDTA, ethylenediaminetetraacetate, is the product of the reaction of EDA with chloroacetic acid or sodium cyanide/formaldehyde. The average growth rate in the late 1970's was about 4 to 5%.

The second major use of EDA is as an intermediate in the production of carbamate fungicides. However, the carbamate fungicide markets have appeared to reach maturity.

Sharing approximately the same proportion of EDA markets is the manufacture of aminoethanolamines. A small proportion of EDA is used to manufacture dimethylolethylene urea resins for permanent-press and easy-care fabric finishes in the textile industry.

Up until recently, the export market has occupied a substantial share of EDA production. Nevertheless, the export business is likely to be curtailed in the 1980's because of the new manufacturing facilities in Europe.

The distribution of EDA among end uses is shown below. The average growth rate through the early 1980's is expected to be about 2%. The demand for EDA in 1978 and 1979 was 65 and 66 million pounds per year, respectively. The demand is likely to rise to 72 million pounds per year by 1983.

Uses of EDA	*Percent*
Dimethyloethylene urea resins	4
Aminoethylethanolamines	15
Carbamate fungicides	15
Chelating agent	18
Miscellaneous	24
Export	24

The pricing history of EDA is shown in Figure 5.23. The major U.S. producers are:

	*Current Capacity**	*Projected capacity**	
	1979	*1980*	*1982*
Union Carbide	63	133	133
Dow	30	40	60
TOTAL	93	173	193

*Millions of pounds per year

HEXAMINE

Hexamine, $(CH_2)_6N_4$, is also known as hexamethylenetetramine, aminoform, or formin. It was first prepared in 1859 by Butlerov of Russia. Hexamine has a three-dimensional molecular structure with the carbon atoms occupying the peaks of an octahedron and the nitrogen atoms occupying the peak of a tetradron which circumscribes the octahedron.

Hexamine is a colorless solid at temperatures below 285°C. It has a crystalline structure of rhombic dodecahedrons. It does not melt, but it sublimes at 285° ~ 295°C. Hexamine is odorless with a sweet metallic taste. It is nonpoisonous although it is known to cause skin rashes on some people who are exposed to it in vapor or solid form. It is flammable and burns in a bluish-yellow flame.

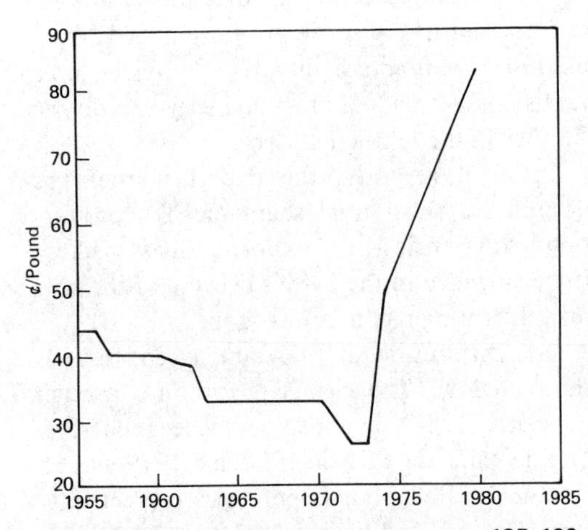

Fig. 5.23 Pricing history of ethylenediamine.[165, 166, 167]

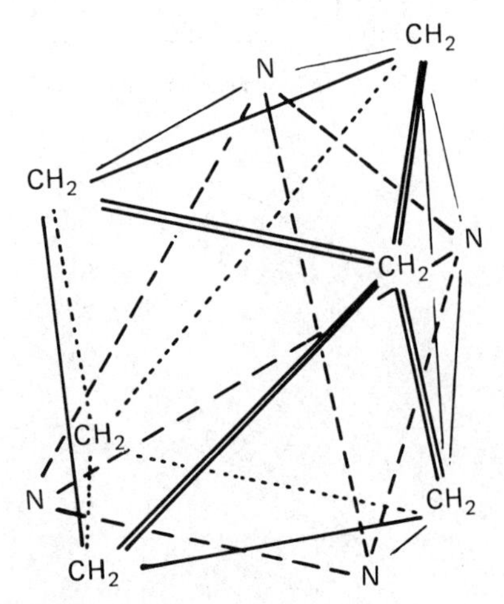

Fig. 5.24 Three dimensional schematic of hexamine.

Thermal decomposition of hexamine produces ammonia, hydrogen cyanide, methane, nitrogen, hydrogen, and residue oils. It also hydrolyzes into ammonia and formaldehyde in acidic solutions. Hexamine reacts with hydrogen peroxide in acid solution to form hexamethylenetriperoxide diamine (HMTP) which is an explosive. When nitrated, hexamine is converted into cyclotrimethylene-trinitramine which is also known as hexahydro-1,3,5-trinitro-s-triazine, cyclonite, hexagen, or RDX. It is an ingredient of the so called "blockbuster" bombs which were listed among the most powerful explosive devices in World War II.

The aqueous solution of hexamine exhibits the phenomenon of inverse solubility. At 25°C, water can dissolve up to 0.867 gm hexamine/cc whereas at 50°C, the solubility is decreased to 0.807 gm/cc. Hexamine in aqueous solution demonstrates the properties of a weak mono base with pH ranging from 8 to 9. Hydrate $(CH_2)_6N_4 \cdot 6H_2O$ of hexamine can be crystallized from aqueous solution at temperatures below 14°C. Some physical properties of hexamine are listed below:

Molecular Weight	140.19
Sublimation Temperature	285°–295°C
Density @ –5°C	1.331 gm/cc
Specific heat	36.4 cal/°C
Heat of Formation @ 25°C	28.8 Kcal/mole
Heat of Combustion @ 25°C	1003 Kcal/mole
Free Energy	102.7 Kcal/mole
Entropy	37.048 cal/°C
Solubility @ Room Temperature,	
Chloroform	0.134 gm/cc
Methanol	0.073 gm/cc
Ethanol	0.029 gm/cc
CCl_4, acetone, benzene, ether	<1%

Manufacture

Hexamine is manufactured by the liquid phase reaction[128] of:

$$4NH_3 + 6HCHO \longrightarrow (CH_2)_6N_4 + 6H_2O + 55 \text{ Kcal/mole Hexamine}$$

The reaction is controlled at pH 7~8 because formic acid and carbon dioxide are formed under acidic conditions. The reaction mechanism is studied by Baur and Ruetschi[119,126] who reported a one-step trimerization reaction in which the kinetic is first order in ammonia and second order in formaldehyde. Boyd and Winkler[120,127], however, have proposed a more complex mechanism. The exact mechanism may well be dependent on whether the reaction is under ammonia rich or formaldehyde rich conditions.

Early commercial production of hexamine used a batch reaction of an aqueous solution of formaldehyde and ammonia at ~30% and ~20% concentration respectively. The reaction temperature was controlled at ~20°C via a reactor cooler to remove 55 Kcal of heat of reaction per mole of hexamine produced. Due to the low reaction temperature and the relatively high exotherm, the temperature control was difficult as well as expensive because of the large cooling surface required. The hexamine concentration in the crude product was about 15% by weight. The aqueous hexamine solution was concentrated and crystallized under 40 mm Hg of vacuum. Approximately 5500 Btu per pound of hexamine produced were required to evaporate 5.7 pounds of water per pound of hexamine in the crude product solution. Ammonia was added during the evaporation step to avoid hydrolysis of hexamine. The batch cycle ranged from eight hours to over a day, depending on the design criteria.

In the 1940's and 50's, the development work was aimed at speeding up the synthesis rate via continuous processes as well as steam conservation via increasing the hexamine concentration in the crude product.

Fig. 5.25 is the schematic representation of a continuous hexamine process. Here, anhydrous ammonia is used to reduce the amount of water entering the reactor and thus reducing the heat requirement in the evaporation step. Industrial grade formaldehyde is usually available in 30~40% aqueous solution. The reacting solu-

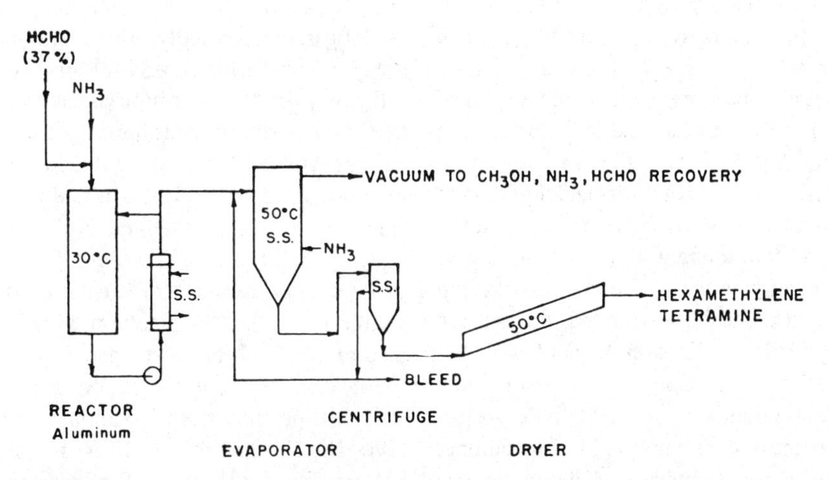

Fig. 5.25 Manufacture of hexamethylenetetramine. (*Adapted from Petrol. Ref.*, **37**, *No. 9, 351-353 1958.*)

tion is controlled at 30° to 50°C by being circulated through an external water-cooled heat exchanger. The residence time in the reactor is in the range of 15~30 minutes. Some processes use gaseous ammonia instead of liquid ammonia as raw material. In that case, the reactor cooler must be capable of accomodating the 60% additional heat generated from the heat of solution of the gaseous ammonia. If gaseous ammonia is used, the heat load in the reactor cooler is approximately 88 Kcal per mole of hexamine vs 55 Kcal/mole if liquid ammonia is used. To avoid local hot spot and hammering effect of mixing gaseous ammonia and the formaldehyde, a stoicheometric amount of the reactants is added to a large circulating stream of hexamine solution.

Although the use of liquid ammonia has the advantage of causing a smaller heat load in the reactor cooler, it has the disadvantage of product contamination caused by impurities in the liquid ammonia. But whether the ammonia is introduced to the reactor in liquid or gas, the use of anhydrous ammonia reduces the water content in the crude product solution by 34% and increases the hexamine concentration to 21%.

The reactor effluent is fed to a vacuum evaporator where the product is concentrated and crystallized. Approximately 3700 Btu per pound of hexamine produced is required to evaporate 3.8 pounds of water per pound of hexamine in the crude product solution. The slurry is centrifuged and the crystals are washed and discharged to a dryer. The mother liquor is recycled to the reactor system; however, a small bleed is necessary in order to avoid the buildup of impurities in the system. A small amount of ammonia is continuously added during the evaporation step to reduce the decomposition of the hexamine. If the formaldehyde feed contains methanol, it will be removed from the evaporator system along with the ammonia and water. The temperature in the final drying step is held to below 50°C to avoid yield losses due to decomposition. The overall yield is in the range of 95 to 96% based on formaldehyde; that is, 3.64 pounds of formaldehyde are required to manufacture one pound of hexamine. The reactor is usually made of stainless steel or aluminum; whereas the reactor cooler and the evaporator are made of stainless steel.

In the early 1950's, the Meissner process[119 120] (East Germany) was introduced, using the heat of reaction and the heat generated from the mixing of ammonia and formaldehyde in the reactor to vaporize the water from the reaction. The reactants are introduced to the reactor as gaseous ammonia and aqueous formaldehyde. The reactor temperature is controlled at 50°~70°C by regulating the total pressure at which the reaction mixture is allowed to boil, or by varying the partial pressure of inert gases. The amount of water vaporized corresponds with the heat generated in the reactor. Reportedly, the hexamine solution can be concentrated to 25~30% in a reactor which has the configuration of a continuous vacuum distillation column. If additional heat is added to the reactor, hexamine can be crystallized in the reactor and, thereby, eliminating the need for a crystallizer. To crystallize the hexamine, a total of 3700 Btu per pound of product is required for removing the water of reaction and the water in the formaldehyde solution (3.8 lb H_2O/lb hexamine). But, the heat of reaction and of the mixing of ammonia and formaldehyde supplies approximately 1100 Btu per pound of hexamine (ie., 30% of the heat required). Therefore, a net heat input of 2600 Btu to the reactor per pound of hexamine is required if the product is to be crystallized in the reactor. The overhead vapor from the reactor is condensed and the methanol and other volatile impurities are removed from the system as off-gas. Reportedly, the reaction goes to 98%[121] completion based on both reactants.

Ideally, the Meissner process can manufacture hexamine without heat input if both formaldehyde and ammonia are introduced into the reactor as gases. Unfortunately, industrial transactions of formaldehyde are in the form of aqueous solution. However, if the Meissner process is integrated[120] with the formaldehyde synthesis process, the reactants to the hexamine reactor can both be in the gaseous state. Here, the heat generated in the reactor is increased by 1160 Btu per pound of hexamine to a total of 2290 Btu/lb due to the heat of solution of gaseous formaldehyde. The use of gaseous formaldehyde also reduces the amount of water to be vaporized from the reactor to 0.77 lb/lb hexamine, which needs 750 Btu/lb of hexamine to vaporize. The overall energy balance shows an excess of 1540 Btu/lb of hexamine produced. Furthermore, the integrated process reduces the manufacturing cost of formaldehyde because the formaldehyde gas can be fed directly into the hexamine reactor from the formaldehyde reactor without separation from other components. Unconverted methanol and other volatile impurities from the formaldehyde reactor leave the hexamine system via the overhead vapor and may be condensed for recovery. The reactor off-gas consists of 18~20% hydrogen and may be used as fuel. The slurry from the reactor is centrifuged to separate the crystals for subsequent washing and drying. Approximately one liter of water per Kg of hexamine is used. The wash solution is recycled to the reactor where the water is vaporized. A yield of 98%[120] was reported in semi-work studies.

In 1970, Tenneco, Inc. patented a hexamine process[123] (U.S. 3,538,199) in which the hexamine is separated from the reaction mixture via a spray dryer instead of an evaporator. In this process, gaseous ammonia is reacted with a 30~50% aqueous formaldehyde solution in approximately 0.67 NH_3/formaldehyde (mole) ratio. The formaldehyde solution is prepared from deionized water and thus is low in cation content. The reaction mixture is maintained at 50°~90°C and at a pH of 7.5~8. The reaction residence time ranges from 2 to 5 minutes. The aqueous reaction solution (25~35 wt % of hexamine) is fed continuously into a spray dryer in which it is atomized and is contacted with a stream of air or inert gas preheated to 200°~400°C. The hexamine in the spray is solidified and dehydrated to 0.1~0.2 wt % of water. The exit temperature of the inert gas is maintained below the decomposition temperature of the product, eg. at 100°~120°C. The typical residence time of the drying step is 0.5~5 seconds. The hexamine particles are then separated from the inert gas and subjected to 1000 to 2500 psi pressure in a pair of cast steel rolls where sheets of densified hexamine are formed. In the final step, the sheets are subdivided into dust-free granules of sizes ranging from 100~300 microns.

Economics and Marketing

Military use of hexamine in the manufacture of cyclonite accounted for most of its rapid growth in the United States in the second half of the 1960's. When nitrated into cyclotrimethylene-trinitramine, hexamine is converted into an ingredient for the production of the powerful cyclonite explosives known as the "blockbuster" bombs in World War II. The military consumption of hexamine has been severely curtailed since the mid-1970's and is expected to be low in military demand unless there is an outbreak of warfare involving the United States. The principal uses of hexamine in the peacetime economy are:

- Curing agent in thermosetting resin production—It serves as a methylating agent in the curing of phenol-formaldehyde resins.
- Chelating agent—The production of nitrilotriacetic acid and its salt is the second largest peacetime consumption of hexamine.
- Accelerator—In the rubber industry it is used to prevent vulcanized rubber from blocking.

Other uses for hexamine are in pharmaceuticals, as an inhibitor of corrosion caused by strong mineral acids, for shrink-proofing agents in the textile industry, and as an agent to enhance color fastness and elasticity to cellulose fibers. It also has potential for end use as a fungicide in the citrus fruit industry. Hexamine produces methylene groups without forming water, as is the case with paraformaldehyde. The growth rate between 1979 and 1983 is expected to be 2-3% per year. The U.S. manufacturing capacity[124] was 149 million pounds per year as of January 1979. The major U.S. Manufacturers are:

Manufacturer	*Capacity In Million PPY*
Borden Inc.	30
W.R. Grace & Company	30
Occidental Petroleum Corp.	28
Plastics Eng. Company	8
Tenneco, Inc.	22
Wright Chemical Corp.	31
Total	149

Chemical Economics Handbook 658.5033X

HYDRAZINE

Hydrazine, $H_2N—NH_2$, is the simplest diamine. Anhydrous hydrazine was first prepared by Lobry de Bruyn in 1894. In 1938, Fairmont Chemicals began the first production of hydrazine in the United States. The quantity produced was small and the product was mainly for captive use in boiler-water treatment. In World War II, hydrazine became important to Germany as a component for the fuel of their missiles. In 1954, Olin Chemicals built the first U.S. commercial N_2H_4 plant of about 4 million pound per year capacity. Hydrazine is mainly consumed as high energy rocket fuel, oxygen scavenger for boiler-water treatment, blowing agents for polymer industries, agricultural chemicals, and pharmaceuticals. In addition, it has a promising future as a fuel for fuel cells. Because of its dual active nitrogen sites, hydrazine and its derivatives are versatile tools for building heterocycle compounds, although the relatively high prices may affect its growth in new end uses in this area.

Hydrazine is a clear, hygroscopic fuming liquid which bears the odor of ammonia. It is a mild base and is highly polar. It is miscible in polar solvents such as water, ammonia, amines, and alcohols. Its high dielectric constant accounts for its unusually high boiling point (@ 113.5°C) and makes it a good solvent for ionic reactions. In liquid phase, hydrazine is known to be associated although it is not associated in vapor phase. It is believed to be in gauche form and exhibits dipole moment. It forms hydrogen bonds with water. It forms an azeotropic mixture with water at 68% hydrazine with a boiling point of 120.5°C. Hydrazine Hydrate (H_2NNH_3OH) is obtained by fractional distillation from water. The monohydrate has a boiling point of 119°C, a melting point of -40°C, and a density of 1.03 gm/ml at 21°C. The physical properties of hydrazine are tabulated in Table 5.14.

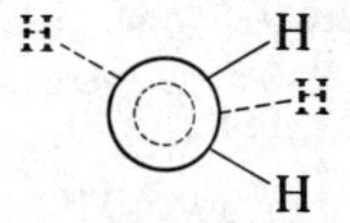

TABLE 5.14 Physical Properties of Hydrazine

Molecular Weight:	Anhydrous	32
	Hydrate	50
Boiling Point (°C)		113.5
Melting Point (°C)		1.4
Specific Gravity (gm/ml)	@ 0°C	1.025
	@ 15°C	1.014
	@ 25°C	1.004
	@ 50°C	0.982
Critical Temperature (°C)		380
Critical Pressure (atm)		145
Vapor Pressure (mm Hg)	@ 25°C	14
	@ 31°C	20
	@ 36°C	100
Viscosity (cp)	@ 5°C	1.2
	@ 25°C	0.9
Basicity, pKa		8.07
Heat of Vaporization (Kcal/mole)		9.6
Heat of Solution (Kcal/mole)	@ 25°C	-3.9
Heat Capacity (cal/mol)	@ 25°C	23.6
Heat of Combustion (Kcal/mole)		-146.6
Heat of Formulation (Kcal/mole)	Liquid	12
	Gas	23
Dielectric Constant	@ 25°C	52
Flash Point (°C)		52
Explosive Limits in Air by Vol %		4.7 - 100

Hydrazine is a strong reducing agent. When catalyzed by metal, it decomposes into nitrogen, ammonia, and hydrogen:

$$3NH_2NH_2 \longrightarrow 4NH_3 + N_2$$

$$NH_2NH_2 \longrightarrow 2H_2 + N_2$$

The four hydrogen atoms in hydrazine molecules can be substituted to form monosubstituted $RNHNH_2$, unsymmetrical disubstituted R_2NNH_2, symmetrical disubstituted RNHNHR, trisubstituted R_2NNHR, and tetrasubstituted R_2NNR_2. Alkylhydrazines, such as monoalkylhydrazine and dialkylhydrazine, resemble hydrazine in their properties. They are weak bases, strong reducing agents, and water soluble. In fact, monomethyl and unsymmetrical dimethyl hydrazines are listed with hydrazine as some of the highest performing high energy fuels; tri and tetra-alkyl hydrazines are relatively weak reducing agents.

Hydrazine vapor is flammable in air in the range of 4.7 to 100 vol %. Liquid hydrazine burns like gasoline when ignited freely in air, although it burns fiercely and at an elevated temperature. An aqueous solution below 40% is not ignitable. Small spillage of hydrazine is best handled by deluging with large quantity of water.

Hydrazine can be fatally toxic via injection, ingestion, inhalation of vapors, or contact with skin. Exposure to hydrazine causes irritation to eyes, nose and throat, and leads to dizziness and nausea. The symptoms may appear gradually over several hours of exposure. Although hydrazine has a relatively low threshold odor limit (70 to 80 ppm), smelling is not a reliable detection for the presence of hydrazine because the membranes of the nose may be desensitized rapidly. The threshold value adopted by the American Conference of Governmental hygienists is 1 ppm. For less than ten minutes of exposure, the maximum tolerable concentration in air is recommended to be 10 ppm.

Manufacture

Three routes for the synthesis of hydrazine are of industrial significance. The Raschig (Olin) process and the Hofmann (Urea) process are the conventional routes to the manufacture of hydrazine. The third is a novel process recently developed by Produits Chimiques Ugine Kuhlmann of France. The Raschig process is presently the principal process for large scale manufacture of hydrazine. Ammonia and sodium hypochlorite are the raw materials with chloramine as an intermediate. The Hofmann process substitutes urea for ammonia as raw material and thus eliminates the need for ammonia recovery facilities. Thus, although the raw material cost in the urea process is somewhat higher than that of the Raschig process, it has the advantage of lower investment cost. Both the Raschig and Hofmann processes produce sodium chloride as by-product and, therefore, require investment in facilities, as well as energy, for the crystallization of NaCl.

The Ugine Kuhlmann process uses hydrogen peroxide instead of sodium hypochlorite as raw material and eliminates the capital and energy costs for crystallizing the by-product NaCl produced in the conventional processes. The capital as well as manufacturing costs with the Ugine Kuhlmann process are estimated to be reduced by 10%.

Raschig and Olin Processes.[135] The Raschig process was originally discovered in 1907 and subsequently modified into the Olin process. The chemistry takes place in liquid phase and is divided into three steps:

$$(1)\ NaOH + Cl_2 \longrightarrow NaOCl + HCl$$

$$(2)\ NH_3 + NaOCl \longrightarrow NH_2Cl + NaOH$$

$$(3)\ NH_2Cl + NH_3 + NaOH \longrightarrow N_2H_4 + NaCl + H_2O$$

The side reactions are:

$$(4)\ 2NH_3 + 3NH_2Cl \xrightarrow{Cu^{++}} 3NH_4Cl + N_2$$

$$(5)\ 2NH_2Cl + N_2H \longrightarrow 2NH_4Cl + N_2$$

Protein like materials such as albumin and glue are used as inhibitors by forming complexes with the metal ions which would otherwise promote the side reactions.

In Reaction 1, the formation of sodium hypochlorite is spontaneous with in situ mixing of chlorine and aqueous caustic solution. In Reaction 2, the reaction rate for chloramine

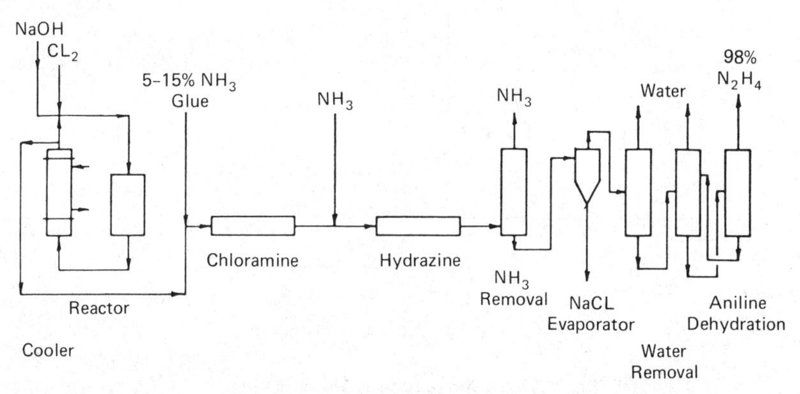

Fig. 5.26 Hydrazine manufacture. (*Adapted from Chilton, C. H., Chem. Eng.,* **65**, *No. 14, 123, July 14, 1968.*)

formation in Reaction 3 is rapid relative to the formation of hydrazine in Reaction 3. Since the decomposition rates are relatively insensitive to temperature, Reaction 3 is operated at 130°C to speed up the rate determining step. Excess ammonia is used to minimize the hydrazine-chloramine decomposition.

Fig. 5.26 is the process schematic of the Raschig process. Sodium hypochlorite is produced by continuous passage of chlorine into aqueous sodium hydroxide (30%) in a cooler-reactor system.

The temperature is maintained below 30°C in order to prevent the formation of sodium chlorate. For the same reason, sodium hydroxide is kept below 1 gm/liter. Excess caustic will also overload the subsequent crystallization step. The inhibitor for side reactions such as glue is added to the solution until the mixture is viscous. In the chloramine reactor, dilute ammonia solution (5 to 15%) is added to the hypochlorite solution at a ratio of 3:1. Excess anhydrous ammonia is then added to the chloramine solution in the hydrazine reactor at a ratio of 25:1 at 130°C. Synthesis efficiency favors a dilute system although the increase in operating cost due to the low concentrate may ultimately become inhibiting.

In the first step of refining, the unreacted ammonia is removed from the reactor effluent in the overhead of the ammonia stripper. The tails are fed into an evaporator where sodium chloride is concentrated and removed. The vapor from the evaporator is dehydrated in the three subsequent columns. In the first dehydrator, water is removed until the $N_2H_4-H_2O$ mixture approaches the azeotropic concentration of about 65% hydrazine. In the second dehydrator, aniline is used as the extracting agent to break the azeotrope, and the water is removed in the overhead with the aniline, which is subsequently recovered in a decanter for recycle. The final column removes aniline from the hydrazine, producing 98% purity in N_2H_4.

Vacuum distillation is used to reduce decomposition. However, the leakage of air into the process may result in decomposition of hydrazine into nitric oxide, ammonia, and water. In some cases, anhydrous hydrazine is obtained by dehydrating with 50% caustic.

The overall yield based on chlorine is 65%. The combined yield for Reactions 1 and 2 is about 95% while the hydrazine step has a yield of 70%.

The Raschig process can also be extended to react amines with chloramine to produce monosubstituted, or unsymmetrical disubstituted hydrazines, for example:

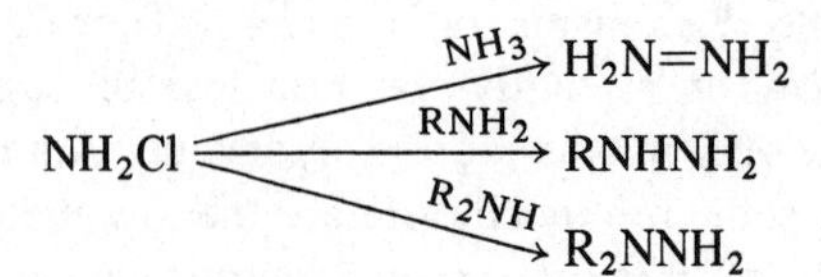

Hofmann (Urea) Process. The overall reaction proceeds in liquid phase as:

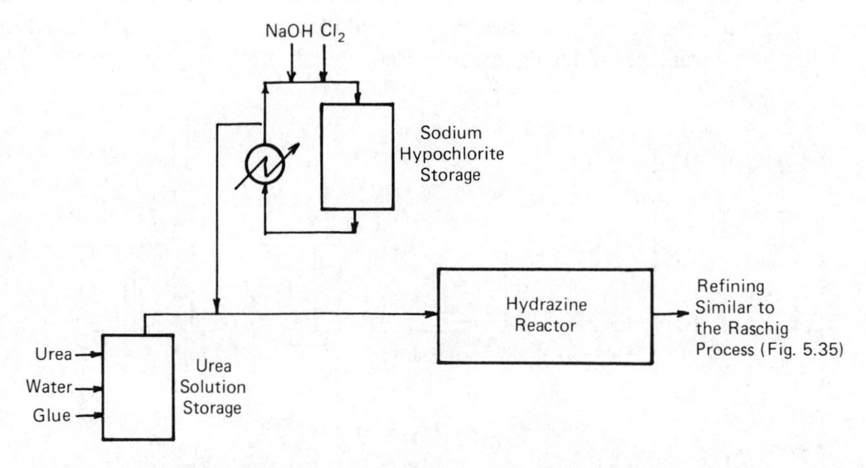

Fig. 5.27 Hoffman process for hydrazine.

$$NH_2\overset{\overset{O}{\|}}{C}NH_2 + NaOCl + 2NaOH \longrightarrow$$
$$N_2H_4 + NaCl + Na_2CO_3 + H_2O$$

As in the Raschig process, aqueous caustic is chlorinated into sodium hypochlorite solution. (Figure 5.27.) The urea solution is prepared by dissolving urea in water with the addition of steam to override the endothermic dissolution. The temperature is maintained at about 5°C for the 43% urea solution. Glue is added as an inhibitor to the side reaction at a ratio of 0.5 gm: liter solution. The urea and hypochlorite solutions are added to the hydrazine reactor at a ratio of 1 : 4. The reaction temperature is allowed to rise to 100°C. The crude product contains approximately 35 gm of N_2H_4/liter and can be refined in the same procedure as the Raschig process.

Ugine Kulmann Process. The technology for this process was developed by Produits Chemiques Ugine Kulmann of France, based on the ammonia, hydrogen peroxide, and ketone route to hydrazine.[130,131,132,133,134] The reaction sequence is believed to be:

$$CH_3\overset{\overset{O}{\|}}{C}CH_2CH_3 + NH_3 \longrightarrow CH_3\overset{\overset{NH_2}{|}}{\underset{\underset{OH}{|}}{C}}CH_2CH_3 \longrightarrow$$
$$CH_3\overset{\overset{NH}{\|}}{C}CH_2CH_3 + H_2O \quad (1)$$

$$CH_3\overset{\overset{NH}{\|}}{C}CH_2CH_3 + H_2O_2 \longrightarrow$$
$$CH_3\overset{NH-O}{\overset{\diagdown\ \diagup}{C}}CH_2CH_3 + H_2O \quad (2)$$

$$CH_3\overset{NH-O}{\overset{\diagdown\ \diagup}{C}}CH_2CH_3 + NH_3 \longrightarrow$$
$$CH_3\overset{\overset{NNH_2}{\|}}{C}CH_2CH_3 + H_2O \quad (3)$$

$$CH_3\overset{\overset{NNH_2}{\|}}{C}CH_2CH_3 + CH_3\overset{\overset{O}{\|}}{C}CH_2CH_3 \longrightarrow$$
$$(CH_3)(CH_3CH_2)C{=}N{-}N{=}C(CH_3)(CH_3CH_2) + H_2O \quad (4)$$

$$(CH_3)(CH_3CH_2)C{=}N{-}N{=}C(CH_3)(CH_2CH_3) + 2H_2O \longrightarrow$$
$$2CH_3\overset{\overset{O}{\|}}{C}CH_2CH_3 + N_2H_4 \quad (5)$$

The net reaction is $2NH_3 + H_2O_2 \longrightarrow N_2H_4 + 2H_2O$ with the ketone being recycled.

The schematic of the process is shown in Fig. 5.28. The manufacturing procedure is divided into five major steps: namely, the synthesis of azine, the removal of low boilers, high boilers separation, hydrolysis of azine followed by the dehydration of hydrazine. The synthesis of azine requires a residence time of seven hours. Ammonia, hydrogen peroxide and methylethyl ketone (MEK) are fed into the reactors along with methanol and acetonitrile at about 50°C. Small amounts of disodium salt of ethylenediamine tetracetic acid are added to the reactants. In the low-boiler removal step, the unreacted ammonia, MeOH, MEK, and acetonitrile are separated and recycled to the reactors. The high-boilers solution contains water, MEK-azine, and acetamide. The acetamide is formed in the reaction of

$$2NH_3 + H_2O_2 + CH_3\overset{\overset{O}{\|}}{C}CH_2CH_3 + CH_3CN \longrightarrow$$
$$(CH_3)(CH_3CH_2)C{=}N{-}N{=}C(CH_3)(CH_2CH_3)$$
$$+ CH_3\overset{\overset{O}{\|}}{C}NH_2 + 3H_2O$$

After being separated from the solution as a high boiler, the acetamide is dehydrated into acetonitrile at high temperature by a catalyst of orthophosphoric acid impregnated alumina. The acetonitrile is recycled to the reactors. In the hydrolysis step, the aqueous MEK-azine solution is partially decomposed into an aqueous solution of MEK and hydrazine.

MEK is separated and recycled to the synthesis step. The hydrazine is fractionated from the unhydrolyzed aqueous azine solution. The azine solution is reworked through the hydrolysis step. The hydrazine is marketed as hydrate or dehydrated by aniline.

The estimated yield is 42 mole percent in ammonia and 73 mole percent in hydrogen peroxide.

Other Technologies.[136] There are several other processes which are not used commercially. These include the following:

• The Bergbau or Bayer Process is somewhat similar to the Ugine Kuhlmann technology where a ketazine intermediate is formed. The reaction routes are:

$$NH_3 + Cl_2 + CH_3\overset{\overset{O}{\|}}{C}CH_2CH_3 \longrightarrow CH_3-\underset{\diagdown\,\diagup}{\overset{HN-NH}{C}}-CH_2CH_3 \xrightarrow[H_2O]{H^+} NH_2\overset{+}{N}H_3 \quad \text{or} \quad \xrightarrow{H_2O} NH_2NH_2 \cdot H_2O$$

$$\downarrow \text{MEK},\ H^+$$

$$NH_3 + Cl_2 + CH_3\overset{\overset{O}{\|}}{C}CH_2CH_3 \longrightarrow \begin{matrix} CH_3 \\ CH_3CH_2 \end{matrix}\!\!>C=N-N=C<\!\!\begin{matrix} CH_3 \\ CH_2CH_3 \end{matrix} \xrightarrow{H^+} NH_2\overset{+}{N}H_3 \quad \text{or} \quad \xrightarrow{H_2O} NH_2NH_2 \cdot H_2O$$

This process is not yet been used on a commercial scale.

• Direct synthesis of hydrazine is reported via passage of ammonia through an extended glow discharge

$$2NH_3 \rightleftharpoons N_2H_4 + H_2$$

Highly turbulent flow and effective cooling are required. Allyl compounds are used to increase the yield.

• Anhydrous hydrazine is also formed by the reaction of chlorine and ammonia:

$$Cl_2 + 4NH_3 \longrightarrow N_2H_4 + 2NH_4Cl$$

The reaction sequence is postulated to be

$$Cl_2 + 2NH_3 \longrightarrow NH_2Cl + NH_4Cl \qquad (1)$$

$$NH_2Cl + NH_4Cl + 2NH_3 \longrightarrow N_2H_4 + 2NH_4Cl \qquad (2)$$

High temperature and pressure increase the reaction rate. A large ammonia to chlorine ratio (50:1 to 350:1) is required to obtain an acceptable yield.

Economics and Marketing

The versatile end uses of hydrazine reflect its attractive properties as a strong reducing agent, a reactive building block for other chemicals, and a source of powerful energy. The consumption of hydrazine as high energy fuel was the dominating market through the 1960's. In the 1970's, the end uses in the manufacture of agricultural chemicals became the major customer for hydrazine and is expected to remain as the fastest growing market in the 1980's. In spite of its versatile chemical properties, the relatively high price of hydrazine may be a hindrance to its growth. The end use pattern in the mid 1960's is compared with the pattern in late 1970's below:

	Mid-1960's Basis = 15 million ppy consumed (pure Hydrazine)	*Late-1970's Basis = 20 million ppy consumed (pure Hydrazine)*
High energy fuels	73%	20%
Agricultural chemicals	11%	35%
Blowing agents	8%	25%
Boiler water treatment	3%	10%
Pharmaceuticals	3%	
Miscellaneous	2%	10%

The projection of the overall growth rate in the 1980's is forecasted to be in the proximity of 7%, since the markets in blowing agents and boiler-water treatment appear to have matured.

Hydrazine is marketed as anhydrous hydrazine, hydrazine monohydrate (64% hydrazine), and 54.5% aqueous solution (85% hydrazine monohydrate). In 1979, the U.S. capacity was estimated to be 49 million ppy. The current major U.S. producers are:

Producer	*Estimated 1979 Capacity (million pounds per year)*
Fairmont Chemicals	3
Mobay Chemicals	22
Oline Chemicals	21
Uniroyal	3
TOTAL	49

Agricultural Chemicals. The fastest growing

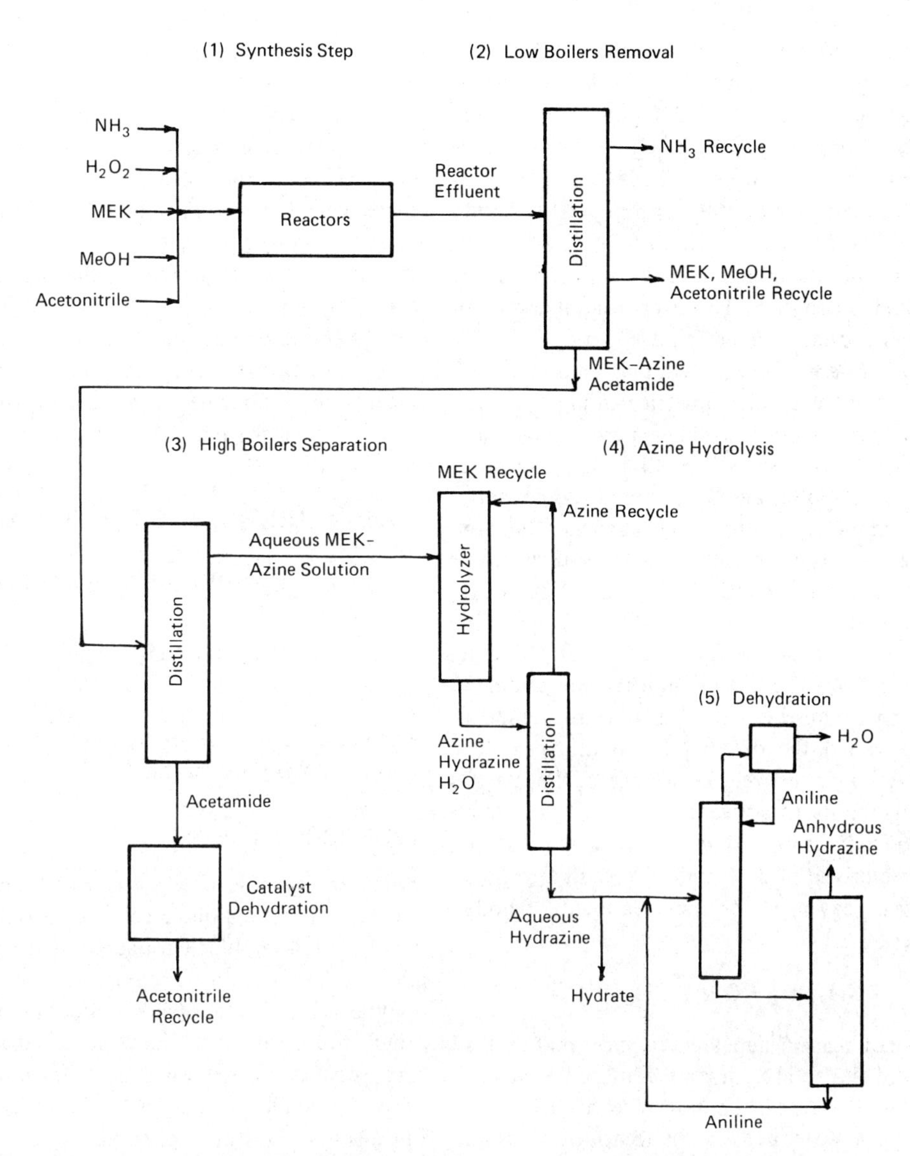

Fig. 5.28 Schematic of Ugine Kuhlman process for hydrazine.

application of hydrazine is in the manufacture of herbicides, pesticides, fungicides, and plant growth regulators. In the late-1970's, this application accounted for 35% of the total consumption of hydrazine. For example, as a plant-growth regulator, maleic hydrazine interferes with the metabolism of the growth tips of some root crops and grasses, and thus temporarily inhibits their growth. It is used to prevent the sprouting of potatoes and onions in storage, to control the growth rate of grass so as to reduce the frequency of mowing in public parks, and to combat the growth of suckers on tobacco plants. It also has many other usages.

The N-N linkage of some of the hydrazine derivatives adds to their attractiveness as agricultural chemicals because of their self-degradation mechanism. Thiadiazole is an example of this unique property where the biological activeness of the compound is complemented by its self-destruct mechanism.

Blowing Agents. In the late-1970's, 25% of the hydrazine produced was converted to blowing agents such as 2,2′-Azoisobutyronitrile, benzene sulfonyl hydrazide, and azodicarbonamide for the manufacture of foam plastics. These compounds* decompose at elevated temperature (e.g., >100°C) to yield large quantities of nitrogen (e.g., >100 ml/gm) together with some other compounds. Azodicarbonamide has the advantage of decomposing into nontoxic compounds in addition to the desired nitrogen. However, the market for blowing agents seemed to have matured.

High Energy Fuels. Hydrazine, monomethylhydrazine, and unsymmetrical dimethylhydrazine are classified as high performing liquid rocket fuels because of their high specific/density impulses and spontaneous combustion. The mono- and di-methylhydrazines have an advantage over hydrazine due to their lower freezing points (<50°C vs 2°C). The rocket fuel markets constituted 20% of hydrazine consumption in the late-1970's. The fluctuation in the U.S. defense budgets and funding in the space missile programs has from time to time varied the demand for hydrazine. The safety concerns over the storage of the fuel has hampered its attractiveness.

Boiler Water Treatment. The strong reducing characteristic of hydrazine makes hydrazine an oxygen scavenger for the treatment of boiler water.

$$NH_2NH_2 + O_2 \longrightarrow 2H_2O + N_2$$

This application appeared to have matured to 10% of the total hydrazine demand in the late-1970's. It is by far the most effective treatment for boiler water because the treatment does not produce solid residues. Recently Olin advertised for improvements in speeding up the reaction even at low temperature by catalyzing the solutions. Hydrazine also helps to form an adherent layer of magnetite, Fe_3O_4, over the iron surfaces of the boilers and thereby reduces corrosion. However, in spite of its efficiency as an oxygen scavenger, the high price of hydrazine has limited its growth in the boiler-water treatment market.

Miscellaneous. In addition to its versatile chemical nature for the manufacture of pharmaceuticals, hydrazine has a high potential for growth as fuel for fuel cells because of its reactivity:

$$\text{Anode: } NH_2NH_2 + 4OH^- \longrightarrow N_2 + 4H_2O + 4e \quad E^\circ = 1.16V$$

$$\text{Cathode: } O_2 + 2H_2O + 4e \longrightarrow 4OH^- \quad E^\circ = 0.4V$$

$$\text{Overall: } NH_2NH_2 + O_2 \longrightarrow N_2 + 2H_2O \quad E^\circ = 1.56\ V$$

Again, the high price of hydrazine has been a hurdle for this potential application.

HYDROGEN CYANIDE

Hydrogen cyanide (HCN) is also known as hydrocyanic acid, prussic acid, or formonitrile. It is the building block for adiponitrile, methyl methacrylate, cyanuric chloride, amino carbonylic acid, sodium cyanide, methionine, and other intermediates. These intermediates, in turn, serve as raw materials for the manufacture of synthetic fibers, plastics, herbicides, chelatting agents, dyestuffs, pharmaceuticals, explosives, surfactants, and metal treatment agents. A detailed discussion of the major end uses of hydrogen cyanide is presented later in this section. U.S. consumption of hydrogen cyanide grew at an average rate of 6% in the 1970's and is expected to maintain the same rate at least through the first half of the 1980's.

Hydrogen cyanide is an extremely toxic chemical in both vapor and liquid states. It is colorless with a fragrance similar to bitter almonds. The threshold odor limit for most people is <5 ppm. It is highly soluble in water, alcohols, and ethers. The acidity of hydrogen cyanide is weak, with a dissociation constant at the range of neutral amino acids. It polymerizes

*2,2'-Azoisobutyronitrile

$$(CH_3)_2C(CN)-N=N-C(CN)(CH_3)_2$$

Benzensulfonyl Hydrazide

$$NH_2NHSO_2-C_6H_4-O-C_6H_4-SO_2NHNH_2$$

Azodicarbonamide

$$NH_2\overset{O}{\overset{\|}{C}}N=N\overset{O}{\overset{\|}{C}}NH_2$$

under basic conditions such as in the presence of ammonia and water and produces "azulmic" acids. The polymerization can become explosively violent if confined. The change of appearance from colorless to yellowish brown is the first symptom of polymerization, and disposition by combustion is considered safe provided that ventilation is adequate. Alternatively, hydrogen cyanide solution can be detoxified by first reacting with excess caustic to form sodium cyanide and subsequently reacting with ferrous sulfate solution in 1:1 weight ratio. Ferro-cyanide, the resulting product, is relatively nontoxic. Instruments for detecting and alarming of minute amounts of the poisonous hydrogen cyanide are readily available in the market. Test paper for detecting HCN can also be prepared by moistening filter paper with 1:1 mixture of cupric acetate solution (3 gm/1) and benzidine acetate solution (60 vol %). The test paper will turn blue if HCN content is greater than 20 ppm. The physical properties of hydrogen cyanide are tabulated in Table 5.15.

TABLE 5.15 Physical Properties of Hydrogen Cyanide

Molecular Formula	HC≡N	
Molecular Weight	27.03	
Melting Point	-14°C	
Boiling Point (1 atm)	26°C	
Vapor Pressure -10°C	160 mm Hg	
0°C	264 mm Hg	
10°C	415 mm Hg	
20°C	625 mm Hg	
Critical Point: Temperature	183.5°C	
Pressure	50 atm	
Density: Gas @ 31°C	0.765 lb/ft^3	
Liquid @ 18°C	43.5 lb/ft^3	
Aqueous Solution	*(% HCN)*	*(lb/ft^3)*
	10	61.4
	20	59.8
	60	51.7
Surface Tension @ 20°C	19.68	dyn/cm
Viscosity @ 20°C	0.201	cp
Heat of formation @ 18°C, 1 atm =		
vapor	-30.7 Kcal/mol	
liquid	-24.0 Kcal/mol	
Heat of combustion	159.4	Kcal/mole
Heat of polymerization	10.2	Kcal/mole
Latant heat	6.027	Kcal/mole
Dissociation constant	7.2×10^{-6}	
Conductivity, Ω^{-1} cm^{-1}	3.3×10^{-6}	
Flamability in air, low	5.6%	
high	40%	

Manufacture

Hydrogen cyanide can be manufactured from almost any form of hydrogen, carbon, and nitrogen if sufficient energy is provided. However, only five processes are presently of economic significance. The Andrussaw process is currently the principal HCN manufacturing process in the world. The Degussa process is a modified version of the Andrussaw technology, suitable for applications where savings from improved yield and selectivity can justify a complex reaction system. The Shawinagan process is of interest to locations where electricity is inexpensive. The Formamide process is useful for sites where methane is unavailable. The Sohio acrylonitrile process produces HCN as a by-product.

Andrussaw Process. This process was developed by Dr. L. Andrussaw of Argentina in the early 1930's. The process (Fig. 5.29) has been the major source for large scale HCN manufacture during the last two decades. The reaction involves

$$NH_3 + CH_4 + \tfrac{3}{2} O_2 \xrightarrow[Pt]{1100°C} HCN + 3H_2O + 144 \text{ Kcal/mole}$$

the reaction mechanism[115] is believed to be

$$\text{Step 1: } NH_3 + O_2 \longrightarrow NH_3O_2 \longrightarrow HNO + H_2O$$

$$\text{Step 2: } HNO + CH_4 \longrightarrow HNCH_2 + H_2O \longrightarrow HCN + H_2 + H_2O$$

Although nitric oxide and ammonia (Step 2) react at a high rate above 850°C, a reaction temperature at about 1100°C is necessary to promote the formation of nitric oxide (Step 1). A Platinum/Rhodium alloy is employed as catalyst for the reaction at a ratio between 8:2 to 9:1. Platinum alone will volitilize under the reaction conditions. When alloyed with Rhodium, the catalyst life can range from 4,000 to as long as 10,000 hours. The catalyst bed is

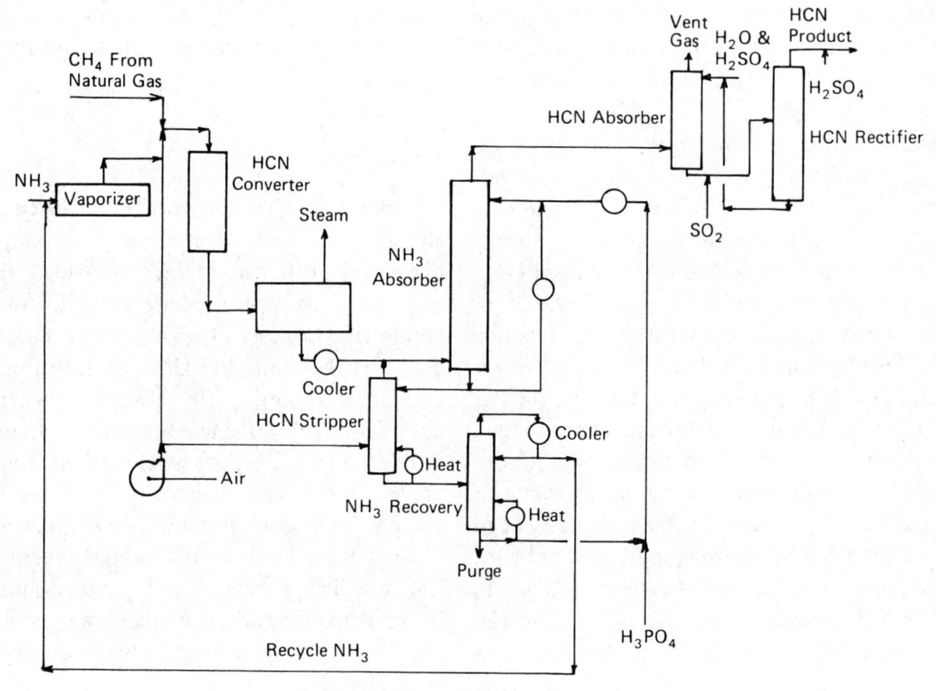

Fig. 5.29 Andrussaw process for HCN.

usually made of several layers of 80 mesh gauze of the noble metal wire (3μ). The spent catalyst is mostly recoverable.

The reaction takes place under fuel rich condition. Typical feed composition to the converter is 11~12% ammonia, 12~13% methane and 74~78% air on a volumetric basis. The control of feed composition is essential to guard against explosions as well as excessive yield loss. The reaction is usually initiated by preheating a section of the catalyst bed with electricity or torch. Once lit off, the reaction is self-sustained by the exothermic heat. The performance of a typical Andrussaw reaction system is:

	Conversion (mole %)	*Yield (mole %)*	*Yield (lb/lb HCN)*
CH_4	60~80%	60~70%	0.90~1.05
NH_3	60~70%	60~90%	0.70~1.05

The crude product at the exit of the converter consists of:

Product	*mole %*	*Wt.%*
HCN	6~12	~8
NH_3	1.5~3	~2
H_2	6~8	~1
N_2	56~57	~66
H_2O	23~24	~19
CO	3~4	~4
CO_2	0.3	–
CH_4	0.1	–

The heat of reaction is recovered in the waste heat boiler to generate 2~6 pounds of steam per pound of HCN produced. The crude product is then quenched to below 350°C to avoid cracking the HCN. The unreacted ammonia is removed from the crude in an absorption system. The ammonia-free crude gas is then passed through the HCN absorber where the inert gases are vented. Water is used as the absorbent with H_2SO_4 added to inhibit the polymerization of HCN. In the final step of the refining process, the bottoms of the HCN absorber are dehydrated in the HCN rectifier and concentrated to >99% hydrogen cyanide with ~0.1% H_2SO_4 added as a polymerization inhibitor.

Several systems have been patented for removal of the unreacted ammonia from the crude product gas. The early Andrussaw processes used dilute sulfuric acid (~6%) as absorbent and produced ammonium sulfate as a by-product. However, the high cost for purifying the by-product has exceeded 4¢ lb since the mid-70's and made it increasingly difficult to compete in the already over-supplied fertilizer market. Unless there are special circumstances, present economics and environmental considerations favor the use of a recycle system to recover the unconverted ammonia. The typical ammonia recovery systems use mono-ammonium phosphate at ~70°C as the absorbent to react with the unconverted ammonia in the crude product gas to form di-ammonium phosphate. The di-ammonium phosphate solution then undergoes a thermal reversal process to liberate ammonia for recycle. The regenerated mono-ammonium phosphate solution is then reused in the absorption system. Polyhydroxyboric acid complex is an alternative absorbent for ammonia recovery which works on the same principles.

Some Andrussaw processes are equipped with filters and scrubbing systems to ensure the quality of the raw material. Phosphorous compounds in the feed are known to deactivate the catalyst. Also, unsaturated hydrocarbons will crack to coke over the catalyst screen. Reportedly, iron rust poisons the catalyst. A dual system for filtering ammonia and air separately is recommended because acidic compounds in air are known to react with ammonia to form soots and cause plugging in the feed system.

Since the converter is operated near flame conditions, the linear velocity of the reactant should be higher than 14 ft/sec to avoid back flame. Reportedly, the reaction residence time of 10^{-8} [97,98] second is adequate. The Andrussaw process has the advantage of a simple reaction system in comparison with the Degussa and Shawinagan processes. However, the yield and conversion of both methane and ammonia in the Andrussaw process are lowest among the major commercial processes. Thus, the rising cost of raw materials is expected to hamper the economic attractiveness of the Andrussaw process. The low HCN concentration in the crude product gas (6~7%) requires a higher equipment size and energy consumption per unit production of HCN as compared with the Degussa and Shawinagan processes.

In 1963, Jenks and Shepard of DuPont[100] reported that, by preheating the reactants to 505°C (versus 95°C conventional), the HCN concentration in the crude can be increased to 12% with close to 90% yield in ammonia. Stanford Research Institute estimated[99] that, for a 31 million ppy HCN plant, the production cost can be reduced by 0.3¢/lb HCN as a result of preheating the feed streams. The additional total fixed investment is estimated to be $300,000 (CE cost index=180) above the conventional Andrussaw process. The estimate is based on installed cost for two preheaters, one for ammonia and the other for methane and air mixture. In spite of the improvement, the HCN concentration in the crude is still substantially below the Degussa and Shawinagan processes.

Degussa Process[116]. Also known as the BMA process, this route to HCN was developed by Degussa of Germany in 1949. The reaction involves:

$$CH_4 + NH_3 + 60\ \text{Kcal/mole} \xrightarrow{1200^\circ C \sim 1300^\circ C} HCN + 2H_2$$

A platinum, aluminum, and ruthenium alloy is used as a catalyst at a ratio of 7.5 : 4:1 (wt.).

Except for the reactor system and the elimination of the waste heat boiler, the process is similar to the Andrussaw process. The endothermic reaction takes place inside a gas fired tubular reactor in which the tubes of sintered alumina are coated internally with a thin (15μ) layer of catalyst. The catalyst life is typically longer than 6500 hr. Reportedly[98], a Degussa reactor of 3300 pph HCN capacity is comprised of 8 furnaces with 13 reactor tubes fired to 1200°~1300°C. The energy requirement of the reaction system is in the range of 14~15 Btu/lb HCN. The performance of the Degussa reaction system is reported to be:

	Conversion	*Yield*	
	(mole %)	*(mole %)*	*(lb/lb HCN)*
CH_4	90~91%	90~93%	0.64~0.66
NH_3	83~85%	84~91%	0.69~0.75

The crude product at the exit of the reactor contains:

Product	*mole %*
HCN	20~23
NH_3	2~3
CH_4	1~2.4
N_2	0.5~1
H_2	70~72

Since the HCN concentration in the crude product is as much as twice that of the Andrussaw process, the quenching of the crude is even more critical in order to avoid cracking and hydrolysis of hydrogen cyanide. The unreacted ammonia is removed from the crude for recycle or disposal. Because of the high ammonia conversion relative to the Andrussaw process, the choice between recycle or disposal of the ammonia depends on each application. Similar to the Andrussaw process, the HCN is separated from the crude via absorption into water with H_2SO_4 added as inhibitor against polymerization. The vent gas from the HCN absorber consists of:

Product	*Vol %*
Hydrogen	96%
CH_4	2.5%
N_2 & O_2	1.5%

The relatively pure hydrogen by-product is either purified for sales or is used as fuel gas in the furnace. The generation rate of hydrogen is in the range of 0.19 to 0.23 lb H_2/lb HCN.

The higher conversion and yield of the Degussa process has the advantage over the Andrussaw process in lower production cost and small equipment size except for the reactor system. Unfortunately, the long residence time required for the reaction makes the process less attractive for large scale production. The complexity of the reactor system and the fragility of the reactor tubes are additional drawbacks.

Improvement of the heat transfer mechanism in the reaction system is an area of interest. The conventional Degussa process requires heat transfer across the nonconductive reactor tubes. U.S. Patent 1,190,922[101] introduced the use of a "Pebble" heat exchanger where a circulating stream of ceramic pebbles is used as the heat transfer media. The pebbles are heated in a combustion vessel and then pumped or fluidized to another vessel where they heat the reactants. Other mechanisms, such as rotary preheaters, have also been studied. However, no publication on commercial use of either mechanism in the Degussa process is available.

The Degussa process appears to be of interest for small scale production of HCN where CH_4 is expensive.[99]

Shawinagan Process. This process, also known as the Fluohmic process, was developed in the early 1960's by Shawinagan Chemicals (now a division of Gulf Oil Canada Ltd.). The reaction involves:

$$3NH_3 + \underset{\text{(eg. propane)}}{\text{Hydrocarbon}} + 151\ \text{Kcal/mole} \xrightarrow{1370^\circ C} 3HCN + 7H_2$$

No catalyst is required for the reaction. The heat input to the endothermic reaction is furnished by an electrically heated fluidized-bed of coke particles. Reportedly, the electricity consumption is near 3 Kwh[99]/lb HCN produced which is roughly six times that of the Andrussaw process. The performance of the reaction system is reported to be:

	Conversion	*Yield*	
	(mole %)	*(mole %)*	*(lb/lb HCN)*
C_3H_8	88%	88 ~ 90%	0.60 ~ 0.62
NH_3	86%	86 ~ 90%	0.70 ~ 0.73

The crude product gas at the exit of the reactor consists of:

Product	*mole %*
HCN	25
NH_3	0.05
H_2	72
N_2	3

The HCN is separated from the crude for refining and the hydrogen is vented from the system as by-product together with the small amount of nitrogen and unreacted ammonia.

Approximately 0.2[102] lb of H_2 is generated per lb of HCN produced. The by-product hydrogen stream is sufficiently pure for direct recycle to manufacture ammonia if the two manufacturing facilities are ideally arranged within pipe line distances. Alternatively, the hydrogen can be marketed as fuel.

The Shawinagan process has the advantage of high HCN concentration in the crude as well as high conversion of ammonia and propane. The low ammonia concentration in the crude eliminates the need for recovery facilities. Since H_2O is absent from the system, the formation of the undesirable "azulmic" acids is precluded. However, because of the high electricity consumption, the process is only attractive to locations where electricity cost is extremely low. Even with moderately low-cost electricity such as on the Gulf Coast, the HCN production cost would still be 3.7¢/lb* higher than with the Andrussaw process. Reportedly, the Shawinagan process can be used with heavy or light naphtha and is of interest to areas where methane is unavailable.

The reported construction cost for a 9 million ppy licensed Fluohmic process plant in Spain was $2.6[99] million in 1974.

Formamide Process. Developed independently by DuPont and BSAF, this process was used commercially in Japan and Western Europe. Since the 1950's, it has been largely replaced by the Andrussaw process. The reaction involves the following steps:

(1) Methyl formate synthesis

$$CO + CH_3OH \xrightarrow[100^\circ C]{150 \sim 3500\ psi} HC(=O)OCH_3$$

(2) Formamide Synthesis

$$HC(=O)OCH_3 + NH_3 \xrightarrow[40^\circ C]{200\ psi} HC(=O)NH_2 + CH_3OH$$

(3) HCN Formation[103,104]

$$HC(=O)NH_2 + 20.4\ Kcal/mole \xrightarrow[400^\circ C\ Vacuum]{Acidic\ Fe_2O_3} HCN + H_2O$$

The conversion of ammonia is 60 ~ 85%. The overall performance of the reaction system starting from the synthesis of methyl formate is:

Product	Yield (mole %)	Yield (lb/lb HCN)
NH_3	79 ~ 90%	0.7 ~ 0.8
CO	52 ~ 94%	1.1 ~ 2.0
CH_3OH	–	0.2 ~ 0.3

The crude product consists of 60 ~ 70 vol % of HCN. The process is of interest for the in situ manufacture of small amounts of hydrogen cyanide. Since the transport of formamide is easier and safer than methane, ammonia, or HCN, the process is useful for areas where methane and ammonia are unavailable. The conversion of foramide to HCN (Reaction 3) is reported to be in the range of 92 ~ 95 mole %.

Sohio Acrylonitrile Process. This process produces hydrogen cyanide and acetonitrile as by-products of the acrylonitrile production. It was developed in the late 1950's by Standard Oil Company of Ohio. Ironically, the manufacture of acrylonitrile used to be the largest consumer of HCN until the Sohio process became the principal source of acrylonitrile in 1960's. The by-product from the Sohio process was estimated to have supplied ~20% of the HCN demand in the late 1970's. The chemistry involves:

Primary Reaction:

$$CH_2{=}CHCH_3 + NH_3 + \tfrac{3}{2}O_2 \xrightarrow{Cat} \underset{\text{acrylonitrile}}{CH_2{=}CHCN} + 3H_2O$$

Side Reactions:

$$CH_2{=}CHCH_3 + \tfrac{3}{2}NH_3 + \tfrac{3}{2}O_2 \xrightarrow{Cat} \tfrac{3}{2}\underset{\text{acetonitrile}}{CH_3CN} + 3H_2O$$

$$CH_2{=}CHCH_3 + 3NH_3 + 3O_2 \xrightarrow{Cat} \underset{\text{Hydrogen Cyanide}}{3HCN} + 6H_2O$$

The Conventional Sohio process produces 0.1 lb HCN/lb acrylonitrile or 0.10 ~ 0.13 lb HCN/lb propylene.

Several new catalysts have recently been

*Assumed 1.5¢/Kwh

patented to boost acrylonitrile yield by curtailing the by-product formation. It is claimed that the new catalysts can produce on an average of 0.06 lb HCN/lb acrylonitrile or ~0.07 lb HCN/lb propylene. Since acrylonitrile has a higher commercial value than hydrogen cyanide, the availability of HCN via the Sohio route depends mainly on the influx of the demand for acrylic fibers.

Other Routes to HCN. The Andrussaw and Degussa processes are difficult to surpass on the basis of raw material cost. Methane is by far the most economical source of carbon except for coal. Ammonia is the cheapest source of nitrogen unless there is a technological breakthrough in the use of elementary nitrogen. A brief description of several processes using coal or nitrogen is presented below. However, these processes have yet to demonstrate commercial significance at the present.

1. $N_2 + H_2 + C$[117] $\xrightarrow[2000^\circ C]{\text{electric arc}}$ HCN where the electric arc was generated by two carbon electrodes. However, only low conversion has been reported.
2. $2CH_4 + N_2 \xrightarrow{2200^\circ C} 2HCN + 3H_2$ is described in U.S. Patent 2,776,872.[105]
3. Ammoniated Coal + NH_3 $\xrightarrow{1000^\circ \sim 1300^\circ C}$ $HCN + H_2$ developed by U.S. Bureau of Mines[106]. The conversion of ammonia was close to 100%. The yields obtained were: 0.9 ~ 1.4 lb HCN/lb NH_3 and 0.04 ~ 0.20 lb HCN/lb coal. The crude product composition was 11% in HCN. However, the data reported were obtained under a partial helium atmosphere and performance under realistic conditions may be different. The process is of interest in coal mining states such as West Virginia where the cost of carbon from coal is only $\frac{1}{3}$ that of methane. More work is required to commercialize the technology.
4. HCN from coking[99] of coal is reported to be 0.7 lb HCN/ton of coal. However the low concentration makes the recovery uneconomical.
5. High temperature plasma synthesis of HCN was reported by Fidler of Czech in Chem. Prum.,[107]

In special circumstances where the availability of hydrocarbons is a logistical constraint, several alternative processes can be considered, especially for small-scale production of HCN:

1. $CH_3OH + O_2 + NH_3 \longrightarrow HCN + 3H_2O$ has been reviewed by Stanford Research Institute. Early patents reported only 3 mole % of HCN in the crude product. However, a recent patent showed improvement of the process to approach 70% of the Andrussaw Process.[109]
2. toluene ($C_6H_5CH_3$) $+ NH_3 \xrightarrow[N_2,\ 1000^\circ C]{\text{Pt, P} + O_2 \text{ Cat}}$ benzene + HCN + $\frac{5}{2}$ H_2, where the yields obtained were 100% conversion of toluene to 100 mole % yield of benzene and 77% yield of HCN.[110]
3. $2NaCN + H_2SO_4 \longrightarrow Na_2SO_4 + 2HCN$. Although the reaction is simple, the disposal of sodium sulfate may be a drawback.

In addition, Japanese Patent 13459[108] reported noncatalytic oxidation of $2CH_4 + NH_3 + \frac{1}{2}O_2 \longrightarrow HCN + 5H_2 + CO$. The significance of this process is currently being investigated.

Economics and Marketing

The historic growth and outlook of HCN consumption in relation to its major end uses are shown in Fig. 5.30. The capacity in 1979 was reported to be 1.27 million ppy. Current business forecasts projected 70% utilization of capacity by 1980. Major U.S. producers of HCN and the capacities[111] of their manufacturing facilities are summarized in Table 5.16.

End Uses.[112] Hydrogen cyanide is the main raw material used in the manufacture of:

- Adiponitrile by reacting with butadine. The end use of adiponitrile is mainly for conversion to hexamethylene diamine for manufacture of Nylon 6/12 and Nylon 6/6.

$$(NCCH_2CH_2CH_2CH_2CN)$$

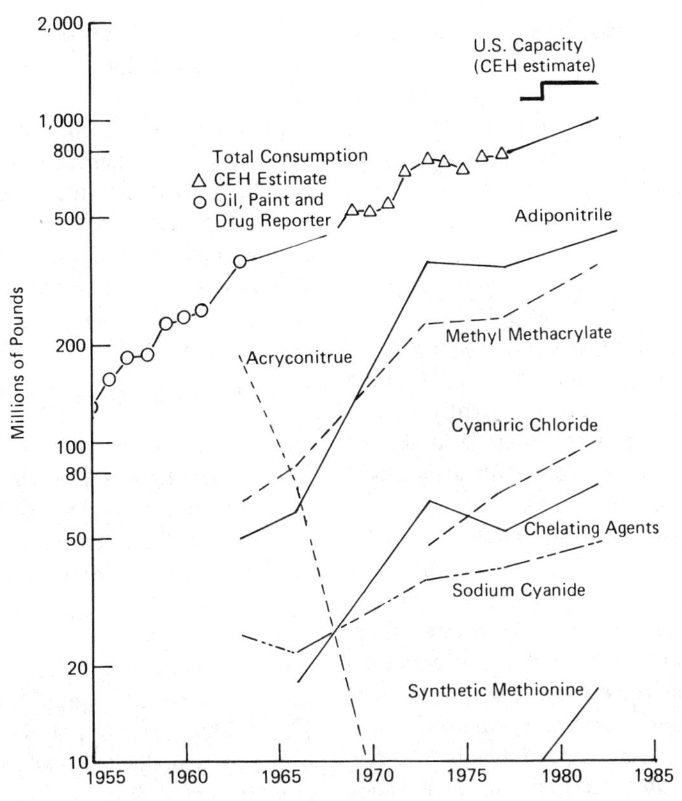

Fig. 5.30 Hydrogen cyanide capacity and consumption.

TABLE 5.16 U.S. Producers of Hydrogen Cyanide

Company	*Annual Capacity As of January 1, 1979 (Millions of Pounds)*
American Cyanamid Company	32
Ciba-Geigy Corporation	90
Degussa Corporation	53
Dow Chemical U.S.A.	20
E.I. DuPont de Nemours & Co. Inc.	653
Hercules Incorporated	3
Monsanto Company	178
Rohm and Haas Company	200
The Standard Oil Company	40
TOTAL	1,269

- Methyl methacrylate by reacting with acetone, sulfuric acid and methanol. The markets are mainly in castings and moldings resins.

$$CH_2{=}C(CH_3){-}C({=}O){-}OCH_3$$

- Cyanuric chloride by reaction with chlorine via cyanogen chloride intermediate. Cyanuric chloride is used as an intermediate for herbicides, explosives, pharmaceuticals, surfactants, and dyestuffs.

$$C_3N_3Cl_3 \text{ (2,4,6-trichloro-1,3,5-triazine ring: Cl–C=N–C(Cl)=N–C(Cl)=N)}$$

- Chelating agents by reacting with formaldehyde and amines to form amino carbonylic acid and subsequently saponified to produce salt of the acid. The common acids are NTA, EDTA, etc. with major markets in the manufacture of soaps, water treatment agents, and cleaning agents.

- Sodium Cyanide (NaCN) by reaction with caustic. The markets are in electroplating, metal treatment, pharmaceuticals, dye stuffs, and plastic industries.
- Methionine for poultry feeds.

ANILINE

Aniline or amino benzene, $C_6H_5NH_2$, is a colorless oil with a boiling point of 184.4°C and a freezing point of -6.1°C. It is slightly soluble in cold water and infinitely soluble in alcohol and ether. It is highly toxic with a threshold limit value of 5 ppm by volume.

Aniline is produced from nitrobenzene by hydrogenation of the amino group to an amine group. A catalyst is used; metals of Groups I, V, VI, and VII, as well as iron, copper, tungsten, molybdenum, nickel and cobalt are effective. Impurities which may need to be considered are: aminophenols, cyclohexylamine, phenylene diamine, and diphenylamine.

The reaction can be carried out in the vapor phase using a fluidized bed with a molar ratio of three hydrogen to one nitrobenzene, at a temperature of 250 to 300°C and at a pressure of 10 to 25 atm. A linear flow rate of 0.1 to 5.0 ft/sec. is used.

Reduction of nitrobenzene can also be carried out in the liquid phase with H_2S, although it is believed that this is not a commercial process at present. A fixed bed of catalyst is used in the Lonza process[82], which has been installed at the First Chemical Corp. plant in Pascagoula, Miss. Plant design rate is 150,000,000 lb/yr. Yield is about 85%, or 1.35 pounds of nitrobenzene per pound of aniline.

Nitrobenzene is produced by the reaction of nitric acid on benzene at 50°C in the presence of sulfuric acid. The reaction is postulated to proceed as follows:

$$HONO_2 + H^{\oplus} \longrightarrow H_2ONO_2^{\oplus}$$

$$H_2ONO_2^{\oplus} \rightleftarrows H_2O + NO_2^{\oplus} \text{ (nitronium ion)}$$

$$C_6H_6 + NO_2^{\oplus} \longrightarrow [C_6H_6 \cdot NO_2]^{\oplus} \longrightarrow C_6H_5NO_2 + H^{\oplus}$$

In concentrated sulfuric acid, HNO_3 is completely converted to $NO_2^{\oplus}$, thus favoring the nitration reaction.

Impurities in the nitrobenzene result from impurities in the benzene. Compounds which may be present are: nitrocresols, nitrous acid, nitrogen oxides, dinitrobenzene, nitrotoluene, carbolic acid, and nitrophenols.

Aminolysis of chlorobenzene will also produce aniline, although this is probably not used commercially.

$$C_6H_5Cl + 2NH_3 \longrightarrow C_6H_5NH_2 + NH_4Cl$$

For the aminolysis of chlorobenzene, three moles of ammonia are usually used at an operating temperature of 180 to 220°C and a pressure greater than the vapor pressure of the reactants. This reaction is carried out in the liquid phase with copper compounds as catalysts.

The process used by Scientific Design Co., Inc. for aminolysis of phenol[82] is shown in Fig. 5.31. This mixed-vapor feed of phenol and ammonia is preheated and passed over a fixed catalyst bed. A unique HALCON-developed catalyst is used, probably based on alumina. The reactor effluent is partially condensed, then sent to an ammonia recovery still. The ammonia is then recycled. The ammonia recovery still bottoms are dried by distillation and the aniline is removed overhead in a purification still. Nearly quantitative yields are said to be obtained and the aniline has a high purity. An azeotrope from the purification still is recycled to the reactor, and a long catalyst life is claimed. A 30,000 metric ton/year plant has been in satisfactory operation for Mitsui Petro-chemical Industries, Ltd., Japan. This process is favorable where low cost phenol is available and when high purity aniline is desired. Capital costs are low also because the nitration of ben-

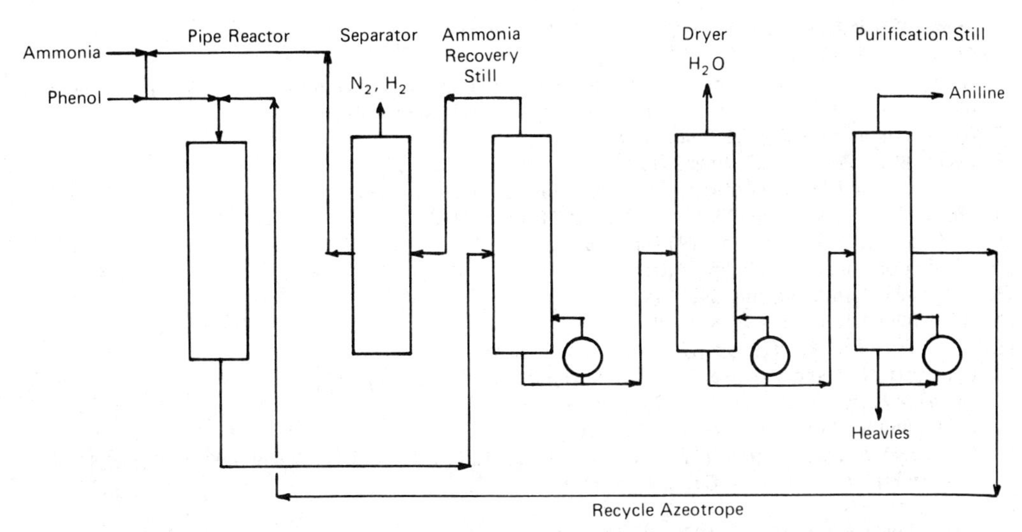

Fig. 5.31 Aniline by aminolysis of phenol. (*Hydrocarbon Processing*, **58**, *No. 11, 137 1979, copyright 1979 by the Gulf Publishing Company.*)

zene is avoided. Waste disposal problems are minimal.

Aniline is used for the production of rubber chemicals, isocyanates, dyestuffs, pharmaceuticals, pesticides, inks, synthetic fibers, and photographic film. U.S. Producers of aniline are:

American Cyanamid Co.
E. I. Du Pont de Nemours and Co.
Mobay Chemical Corp.
Rubicon Chemicals, Inc.
Uniroyal, Inc.
First Mississippi Corp.

Demand in 1975 was 412 million pounds; in 1978, it was 605 million pounds; and for 1979, it is expected to be 630 million pounds.

OTHER COMPOUNDS

Additional nitrogen compounds of commercial importance are:

Hexamethylenediamine, (Ref. 94, 162, 168), which is used primarily in the manufacture of nylon.

Ethanolamines (Ref. 159, 150), which are used for manufacture of detergents and as scrubbing agents for removal of acidic compounds from gases.

Acrylonitrile, (Ref. 82, 92, 93, 95, U.S. Patent 3,499,025), which is used in the manufacture of acrylic and modacrylic fibers, plastics and resins, and nitrile rubbers or elastomers.

Dimethylformamide, (Ref. 169, 170, 171, 172, 173, 174, 175), which is a versatile and powerful solvent for organic and inorganic compounds, and which is a vital reaction medium for ionic and non-ionic compounds.

Dimethylacetamide, (Ref. 177, 180) which is an important industrial solvent for polyacrylonitrile, vinyl resins, cellulose derivatives, styrene polymers, and linear polyesters. It is also used as a paint remover.

Isocyanates, (Ref. 89, 95, 96) which are important for the production of polyurethane foams, resins, and rubbers.

REFERENCES

1. *Chem. Eng. News* 37, No. 32, 46–47 (Aug. 10, 1959).
2. *Chem. Week.*, **100**, No. 3, 64 (Jan. 21, 1967).
3. *Akad. Noak S.S.S.R. Seriya Khimicheskaya Izvestiya*, No. 9, 1728–1729, (1964).

4. French Patent 1,391,595, *Chem. Eng.* **65**, 58 (July 28, 1958).
5. *Chem. Eng. News*, p. 16, (Aug. 27, 1979).
6. Harris, Gene T. and Harre, Fowin A. "World Fertilizer Situation And Outlook, 1978–85– and Trends 1979" *International Fertilizer Development Center - TVA*, Muscle Shoals, Ala.
7. *Hydrocarbon Processing*, Section 2, Oct. 1979.
8. *Chem. Week*, Pages 22–23, June 1, 1977.
9. Slack A. V. and James, G. Russell, *Ammonia*, Part III, p. 291, p. 387.
10. *Chem. Eng. Progress*, Vol. **70**, No. 10, Oct. 1974, Pages 21-35.
11. *Hydrocarbon Processing*, Nov. 1977, pp. 361-366.
12. Private Communication, Pullman Kellogg Co., Oct. 8, 1979.
13. U.S. Patent 4,162,290, July 24, 1979.
14. U.S. Patent 4,153,673, May 8, 1979.
15. *Chem. Eng.*, Oct. 8, 1979, pp. 57–59.
16. Updegraff, N. C. and Mayland, B. J. Petrol, *Refines* **33**, No. 12, 156–159, (Dec. 1954).
17. *Chem. and Eng. News*, Aug. 27, 1979, p.24.
18. U.S. Patent 3,106,457, Lockerbie, T.E. et.al.
19. Hougen, O. A. et.al., *Chemical Process Principles* Part II, p. 1107, New York, John Wiley & Sons, 1959.
20. *Catalyst Handbook*, Katalco Corp., Oak Brook, Ill., p. 79.
21. U.S. Patent 3,827,987.
22. *Chem. Engineering*, Dec. 18, 1978, pp. 115–128.
23. Synthetic Fuels Symposium, A.C.S. Sept. 12–15, 1955.
24. *Chem. Engineering*, Oct. 8, 1979, pp. 57–59.
25. "Selected Values of Properties of Hydrocarbons", Am. Petrol. Inst. Research, Project 44, Nat. Bureau of Standards.
26. *Hydrocarbon Processing*, Sept. 1979, p. 192.
27. *Chemical Engineering*, Jan. 30, 1978, pp. 69–71.
28. *Petrol. Processing* (Sept. 1956).
29. "Abundant Nuclear Energy," U.S. Atomic Energy Commission, CFSTI, May 1969.
30. Lombard, J. F., *Hydrocarbon Process*, **48**, No. 8, p. 111 (Aug. 1969).
31. Lundberg, W. C., *Chemical Engineering Progress*, June 1979, p. 81.
32. Swaim, C. D., *Hydrocarbon Process*, **49**, No. 3, 127–30 (Mar. 1970).
33. "Water Scrubbing for CO_2 Removal," Fredrich Uhde-Gmbh, BCE Processes in Europe, Nov. 1969.
34. *Chemical Engineering Process* Vol. **69**, No. 2, Feb 1973, pp. 67–70.
35. "Amine Guard Systems in Hydrogen Production," AICHE - 86th National Meeting, Houston, Texas, April 1–5, 1979.
36. U.S. Patent 3,352,631.
37. U.S. Patent 3,347,621.
38. *Chemical Engineering*, Dec. 3, 1979, pp. 88–89.
39. *Hydrocarbon Process.*, April 1978, pp. 145–151.
40. U.S. Patent 2,880,591.
41. U.S. Patent 2,863,527.
42. Private Communication, Lurgi-Gesellschaft.
43. *Chem. Eng.* 67, No. 19, pp. 166–9 (Sept. 19, 1960).
44. *Catalyst Handbook* Springer-Verlag, New York, Inc. p. 119.
45. "Removing Carbon Monoxide from Ammonia Synthesis Gas," *Ind. Eng. Chem.* **53**, No. 8, 645 (Aug. 1961).
46. Grotz, B. J., *Hydrocarbon Process* **46**, No. 4, 197 (1967).
47. "A Practical Vibration Primer," *Hydrocarbon Process*, Nov. 1975, p. 251.
48. Perry, J. H. *Chemical Engineers Handbook*, McGraw Hill, New York.
49. "Compression Systems for Ammonia Plants," *Chem. Eng. Progress*, July 1979, p. 88.
50. "Best Pressure for Ammonia Plants," *Hydrocarbon Process*, Vol. **47**, No. 11, 1968, pp. 153-161.
51. Nielsen Andres, *An Investigation of Promoted Iron Catalysts for the Synthesis of Ammonia*, 3rd Ed., Jul. Gjellerup Forlag, Copenhagen; 1956.
52. "Catalyst Reviews," 1–25, (April 9, 1970).
53. Ulmanns, *Encyklopadie de Technischen Chemie* Munchen, Berlin, Urban and Schwarzenberg, 1953.
54. *Hydrocarbon Processing*, Dec. 1978, pp. 115–122.
55. U.S. Patent 2,894,821.
56. Slack, A. V. and James, G. Russell *Ammonia*, Part III, pp. 123–153.
57. NIOSH Criteria for Recommended Standards, Occupational Exposure to Nitric Acid, 1976).
58. U.S. Patent 3,310,376.

59. *Chem. Week.* 97, No. 11 (Sept. 11, 1965).
60. "Mined Underground Storage," *C.E.P.* Vol. **11** Safety in Air and Ammonia Plants.
61. *Chemical Processing*, July 1979, p. 140.
62. *Chem. Engineering*, July 16, 1979, p. 63.
63. *Chem. Engineering*, Oct. 10, 1977, p. 90.
64. "Viscosity-Temperature Correlations for Liquids," *Chemical Engineering*, July 16, 1979, p. 83.
65. *Chem. Eng. News*, **36**, No. 38, 62–64, Sept. 22, 1958.
66. *Chem. Eng. Progress*, **49**, 349 (1953).
67. "Fertilizer Nitrogen," Sauchelli, A.C.S. Monograph Series 161.
68. Thompson, R. *The Modern Inorganic Chemical Industry*, The Chemical Society-Burlington House, London-1977.
69. Spratt, D. A. Proc. Fertilizer Soc. of England (Mar. 25, 1958).
70. *Chem. Trade Journal*, **75** (July 11, 1958).
71. *Chem. Eng.*, **67**, No. 8, 94 (1960).
72. Powell, R. "Nitric Acid Technology-Recent Developments" Noyes Development Corp., 1969.
73. *Chem. Eng. News*, **44**, No. 27, 13 (July 4, 1966).
74. "Catalytic Processes in Nitric Acid Manufacture," Fertilizer Society Proceedings; London, 1978.
75. Stafford, J. D., Samuels, W. E., and Croysdale, L. G. "Ammonium Nitrate Plant Safety," Am. Inst. of Chem. Eng., Sept. 27–29, 1965.
76. *Chem. Week*, **101**, No. 2, 75 (July 8, 1967).
77. *Chem. Eng. News*, Jan. 22, 1979.
78. *Chem. Eng.*, **67**, No. 25, 71 (1960).
79. *Chem. Week*, **75**, No. 22, 90 (1954).
80. "Urea Process Technology," p. 139, Park Ridge, N.J. Noyes Development Corp.
81. U.S. Patent 3,200,148.
82. *Hydrocarbon Processing*, Petrochemical Handbook Issue, Nov. 1979.
83. *Nitrogen*, No. 15, p. 37 (Jan. 1962).
84. U.S. Patent 2,527,315.
85. Harri, Edwin A., "The Outlook for Nitrogen Fertilizers," TVA, Muscle Shoals, Ala. Forest Fertilizer Conference–Union, Washington 9/25–27, 1979.
86. U.S. Patent 3,502,478.
87. U.S. Patent 3,107,149.
88. U.S. Patent 3,236,888.
89. *Hydrocarbon Processing*, **48**, No. 11, 209 (Nov. 1969).
90. Schwartzmann, M., U.S. Patent 3,498,982 (to B.A.S.F.).
91. Takakuwa, Y., et al., "Development of a Process for the Purification of Crude Melamine Manufactured from Urea," *Int. Chem. Engineering*, **Vol. 19**, No. 4, Oct. 1979, p. 624.
92. *Chemical Engineering*, **62**, pp. 288–291, Sept. 1955.
93. *Hydrocarbon Processing*, p. 124, Nov. 1979.
94. Sittig, M., "Amines, Nitriles, and Isocyanates, Processes and Products," Park Ridge, N.J., Noyes Development Corp. 1969.
95. *Chem. Eng.*, March 14, 1977.
96. *Chem. Eng.*, Sept. 24, 1979, p. 80.
97. Andrussaw, L., *Angew. Chem.* 48,593 (1935).
98. Kirk & Othmer, Vol. **6**, pp. 577.
99. SRI Report X-3-2, pp. 10.
100. U.S. Patent 3,104,940.
101. U.S. Patent 1,190,922.
102. Hahn, A. V. G., *The Petrochemical Industry: Market & Economics*, pp. 154, McGraw-Hill.
103. German Patent 944,547.
104. German Patent 1,209,501.
105. U.S. Patent 2,776,872.
106. U.S. Patent 3,501,267.
107. Fidler, *Chem. Prum.*, 68, Series 18, issue 2, pp. 94–9.
108. Japanese Patent 13459.
109. Sedriks, Walter, "Hydrogen Cyanide from Methanol," SRI publication.
110. U.S. Patent 3,501,267.
111. *Chemical Economics Handbook* 664.5020 E.
112. W. H. Ton, *Chem. Eng.*, **62**, No. 10, 188 (1955).
113. *Chem Profile*, "Hydrogen Cyanide," Schnell Publishing Co., April 1, 1977.

114. *Chem. Eng.* **66**, No. 9,289 (1955).
115. *Refining Engineer*, C-22-C-27 (Feb. 1959).
116. *Hydrocarbon Processes* **46**, No. 11, 189 (Nov. 67).
117. *Ind. Eng. Chem.*, **52**, No. 7, (Jul. 1960).
118. Chang, S. S. and Westrum, E. F. Jr., *J. Phy. Chem.* 64-1547-1551 (1960).
119. Astle, M. J., *Industrial Organic Nitrogen Compounds*, pp. 12–13 (1961).
120. Meissner, F. Schwiedessen, E. and Othmer, D. F., *Ind. Eng. Chem.* **46**, 724 (1954).
121. U.S. Patent Appl. 271,628 (1952).
122. U.S. Patent Appl. 366,064 (1953).
123. U.S. Patent 3,538,199.
124. Chemical Economics Handbook 658,5033X.
125. *Chem. Profile*, April 1, 1969, Schnell Publishing Co.
126. *Helv Chem Acta*, **24**, 754 (1951).
127. *Can J. Res.* **25**, Sec. B.
128. *Petrol Refiner* **37**, No. 9, 351–353 (1958).
129. Kirk & Othmer, "Chemical Technologies," Vol. **11**, pp. 183.
130. W. German Ausleg. 2,143,516 (Jan. 18, 1973)–to Ugine Kuhlmann.
131. Belg. Patent 766,845 (May 7, 1971) to Ugine Kuhlmann.
132. Netherlands Appl. 72,08669 (Dec. 28, 1972) to Ugine Kuhlmann.
133. W. German Offen 2,210,790 (Oct. 5, 1972) to Ugine Kuhlmann.
134. W. German Offen 2,403,810 (Aug. 29, 1974) to Ugine Kuhlmann.
135. *Chem. Eng.* **65**, No. 14, pp. 120–123 (July 14, 1958).
136. Powell, R., "Hydrazine Manufacturing Processes," Park Ridge, N.J., Noyes Development Corp. 1968.
137. Du Pont Methylamines Product Bulletin.
138. U.S. Patent 344,203.
139. Ullmanns, *Encyklopadie de Technischen Chemie*, 3rd Ed., Munich-Berlin: Urban & Schwarzenberg (1960) Vol. **12**, pp. 421–425.
140. *An. Quim.* 1969, **65** (7–8), pp. 699–708 (Spain).
141. Egly, R. S. and Smith, E. F., *CEP* **44**, No. 5, Trans. Am. Inst. Chem. Engrs pp. 387–98 (1948).
142. U.S Patent 1,799,722.
143. E. German Patent 108,275.
144. Japan patent 1623 ('52).
145. U.S. Patent 3,387,032.
146. Serban, S. *Rev. Chim* (Bucharest) **14** (8), pp. 451–4 (1963).
147. Issoire, J. & Long, C. V., Bull. Soc. Chem. France, 1960, 2004–12.
148. U.S. Patent 2,119,474.
149. French Patent 956085/87.
150. U.S. Patent 2,547,064.
151. *Chem. Eng.*, Oct. 29, 1962, pp. 72.
152. *Chem. Eng.*, Aug. 21, 1961, pp. 100–102.
153. DBP 1,019,313 (Old German Patent-1957).
154. French Patent 762,194.
155. Chemical Profile, July 1, 1979, Schnell Publising Co., Inc.
156. *Chemical Economics Handbook*, 611,5030 Schnell Publishing Co. Inc.
157. *Hydrocarbon Processes* **44**, No. 11, p. 241 (Nov. 1965).
158. U.S. Patent 2,153,405.
159. *Hydrocarbon Processing*, 48, No. 11, p. 175 (Nov. 1969).
160. Moran & Lowenhein, *Faith, Keyes & Clark's Industrial Chemicals*, 4th Ed. pp. 341.
161. *Chemical Profile*, Oct. 1, 1979, Schnell Publishing Co. Inc.
162. Sitig, M., "Amines, Nitriles & Isocyanate Processes and Products," Park Ridge, N.J., Noyes Development Corp. 1969.
163. U.S. Patent 3,137,730.
164. *Chemical Profile*, July 1, 1979, Schnell Publ. Co. Inc.
165. *Chemical Economics Handbook* 648.5054.
166. *Chemical Economics Handbook* 651,5032.
167. Moran & Lowenhein, *Faith, Keyes & Clark's Industrial Chemicals*, 4th Ed., pp. 386.
168. U.S. Patent 3,017,331.
169. Morrison & Boyd, *Organic Chemistry*, 3rd Ed. pp. 31.
170. DuPont Dimethylformamide Product Bulletin.
171. Astile, M. J., *Industrial Organic Nitrogen Compounds*, Van Nostrand Reinhold (1961), pp. 260.

172. Winteler, H, Beler, A., and Guyer, A., *Helv. Chem. Acta*, 37,2370 (1954).
173. Shulman & Nolstad, *Ind. Eng. Chem.*, **42**, 1058 (1950).
174. French Patent 1,029,151.
175. U.S. Patent 2,677,706.
176. U.S. Patent 4,042,621 (1977).
177. DuPont Dimethylacetamide Product Bulletin.
178. Ruhoff & Reid, *J. A. Ch. Soc.*, **59**, 401 (1937).
179. Mitchell & Reid, *J. A. Ch. Soc.* **53**, 1879 (1931).
180. Ratchford & Fisher, *J. Org. Ch.* **15**, 317 (1950).
181. U.S. Patent 3,720,715.

6

Salt, Chlor-Alkali and Related Heavy Chemicals

James J. Leddy*

SODIUM CHLORIDE

Common salt, sodium chloride, occurs in nature in almost unlimited quantities. It is a direct source of chlorine, caustic soda (sodium hydroxide), sodium chlorate, synthetic soda ash (sodium carbonate), sodium metal, and sodium sulfate. Indirectly, it is also the source of hydrochloric acid and a host of sodium salts. It has an imposing list of uses, placing it among the more important substances in the economic world. It is used to preserve meat, fish, and hides; it is a necessary component of the animal diet; it is used in refrigeration systems; and large quantities are used for ice control on high ways in colder climates. Salt is used by the soap maker to separate the soap from glycerine and lye, and by the dye manufacturer to precipitate his products. In addition, salt is used extensively for the regeneration of water-softening resins.

Salt is mined as a solid in shaft mines with depths of 500–2000 ft., in Michigan, Ohio, New York, Kansas, Louisiana, and Texas. Run-of-the-mine salt contains 98–99% NaCl. Over twenty-five percent of the world's salt is produced in the United States.

More frequently, salt is obtained by solution mining. Typically, this is achieved by drilling a well into the salt formation, installing concentric piping into the well, pumping water in one pipe, and retrieving nearly saturated brine from the other pipe. The brine concentration is controlled by the rate of pumping, and is kept slightly undersaturated to avoid salting up the brine lines at the wellhead. The brine is purified by chemical treatment, settling, and filtration. Such artificial brines permit a cheaper operating cost, and are well adapted to the manufacture of synthetic soda ash, chlor-alkali products, and table salt. Very extensive production of artificial brines is carried out along the U.S. Gulf Coast, principally in Texas and Louisiana, where huge domelike deposits of rock salt exist in readily accessible areas. Some of these domes are four miles in diameter and over eight miles deep,[1] containing over 100 billion tons of salt each. Crude oil and processed hydrocarbons are conveniently stored in the caverns produced in these domes by solution mining of the salt. The hydrocarbons float on the brine within the cavern, and are readily

*Dow Chemical Co., Freeport, Texas

TABLE 6.1[3] 1978 Uses of Salt (Millions of Short Tons)

Chlor-Alkali	22.9
Highway Ice Control	8.8
Food Processing	3.3
Synthetic Soda Ash	2.8
Feed and Mixes	2.1
Other Chemicals	2.0
Industrial	1.2
Misc.	4.0

recovered by simply pumping brine back into the well. The Gulf Coast, with its salt and hydrocarbon deposits, deep water ports, and ready sources of energy is ideally suited for the chemical manufacture of salt derivatives, especially chlorinated hydrocarbons.[2]

Artificial brines are also extensively produced from stratified salt deposits in Michigan, Ohio, New York, and West Virginia, and Ontario, Canada. Much of the early chemical industry in North America was concentrated in these areas because of the availability of these brines and the need for salt in the production of soda ash, chlorine, and caustic. The stratified salt exists in layers 20 to 200 feet thick, associated with anhydrite ($CaSO_4$). The nature of these deposits is such that it leads to higher calcium levels in the brine when recovered by solution mining compared with the brine obtained from salt domes. In the vicinity of many of these stratified salt deposits, there also exist naturally occurring brines. These natural brines consist mainly of $CaCl_2$, with lesser amounts of $MgCl_2$, KCl, NaCl, a few thousand ppm of bromide, and up to a few hundred ppm of iodide. These natural brines are mined principally for their bromine content.

Salt is also derived from seawater in those places in the world, such as California, where solar evaporation is sufficient to concentrate the seawater in large basins until some of the salt deposits. The salt-based chemical industry in Asia and the West Coast of the U.S. derives its salt from solar evaporation, which accounts for about half the total world production.

TABLE 6.2[3] U.S. Salt Production in 1978 (Millions of Short Tons)

Evaporated Salt	6 ($40/ton)
Rock Salt	15 ($12/ton)
In Brine	26 ($ 6/ton)
Total	47

Salt and other chemicals are also derived from naturally occurring brines, such as brine from Searles Lake in California, which typically contains 16.5% NaCl, 6.8% Na_2SO_4, 4.8% KCl, 4.8% Na_2CO_3, and 1.5% $Na_2B_4O_7$. Another source is the Great Salt Lake, which has a 27% salinity, of which 80% is NaCl.

At Searles Lake, the brine is concentrated in a multi-effect evaporator, from which NaCl is recovered. The process liquor is fractionated further to yield a KCl-borax concentrate and burkeite, a double salt of Na_2CO_3 and Na_2SO_4, which is processed further to yield soda ash and salt cake (Na_2SO_4) as products.

SODA ASH

Soda ash, sodium carbonate, is derived commercially from three sources; in order of decreasing importance they are: (1) naturally occurring trona ore; (2) the Solvay ammonia-soda process; and (3) naturally occurring alkaline brines. There has been a dramatic shift away from the synthetic, or Solvay process ash over the last decade, as is shown in Fig. 6.1.

The Solvay process like its predecessor, the Leblanc process, is on the decline because of economic factors. It is still a major factor in ash production in Europe, but may decline there as well. Increasing costs of production and environmental factors are the key issues in the many closings of synthetic ash plants in the last several years. However, a complete world-wide take-over by natural ash is doubtful

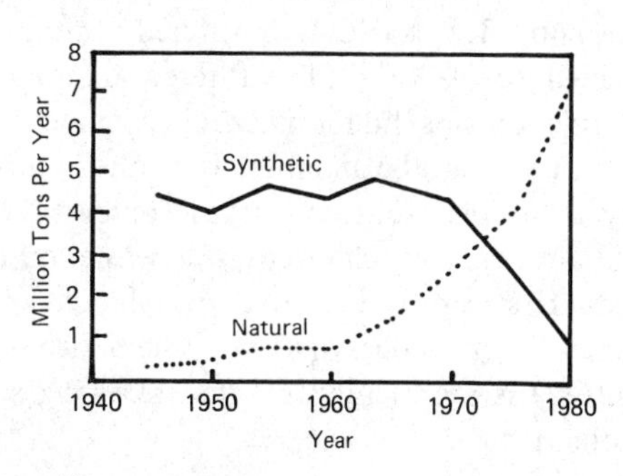

Fig. 6.1[3] U.S. production of synthetic vs. Natural Soda Ash.

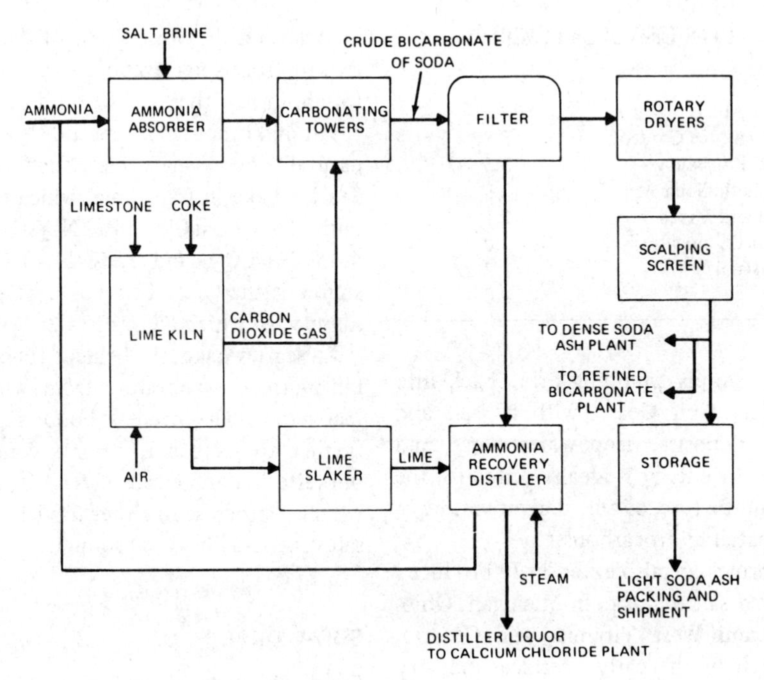

Fig. 6.2 Simplified diagrammatic flow sheet for the Solvay ammonia soda process.

because of two factors: (1) the limited amount of natural ash compared with the widespread availability of salt and limestone (the essential ingredients of the synthetic ash process); and (2) the locations of natural ash deposits relative to the locations of the ash consumers. Most of the U.S. natural ash is derived from the area of Green River, Wyoming. Significant amounts are also recovered from the alkaline brines of Searles Lake, California.

The synthetic ash process, or ammonia-soda process, was perfected by Ernest Solvay in 1865. The process is based on the precipitation of $NaHCO_3$ when an ammoniated solution of salt is carbonated with CO_2 from a coke-fired lime kiln. The $NaHCO_3$ is filtered, dried, and calcined to Na_2CO_3. The filtered ammonium chloride process liquor is made alkaline with slaked lime and the ammonia is distilled out for recycle to the front end of the process. The resultant calcium chloride is a waste or by-product stream. The net overall chemical change in the process (shown schematically in Fig. 6.2) is represented by the following stoichiometry:

$$CaCO_3 \text{ (limestone)} + C \text{ (coke)} + O_2 \text{ (air)} + 2NaCl \text{ (brine)} \longrightarrow Na_2CO_3 + CO_2 + CaCl_2 \text{ (brine)}$$

The process requires a large amount of fuel to calcine both the limestone and sodium bicarbonate, and to generate steam for ammonia recovery. For the reaction proper, no fuel is required. In fact, large volumes of cooling water are required to remove the heat generated by the absorption and reaction of ammonia and carbon dioxide. The process has an imperfection in that an undesirable solution of calcium chloride is also produced.

Prior to being fed to the process, the NaCl brine must be purified to remove calcium and magnesium ions so that they will not be precipitated when the brine is carbonated, producing objectionable scale on the equipment surfaces, as well as contaminating the product. A solution of soda ash and caustic is added to the brine to precipitate the calcium as $CaCO_3$, and the magnesium as $Mg(OH)_2$. These are flocculated, removed in a settler, and discarded. The purified brine is fed to the ammonia absorber in the process.

Lime for recovering ammonia in the process and CO_2 for reacting with the ammoniated brine are produced by calcining the best available limestone with foundry coke in vertical shaft kilns. This type of equipment is preferred for producing the maximum yield of active lime and the maximum concentration of CO_2 in the kiln gas. The dry lime, drawn from the vertical kiln, is cooled by the entering air, which is thereby preheated. Likewise, for maximum fuel economy, the exiting gas preheats the incoming limestone and fuel. The gas composition should exceed 40% CO_2 with only fractional percentages of CO and O_2.

The dry lime and hot water are fed to a rotating cylindrical slaker to produce milk of lime, which is pumped to the distillers for the recovery of ammonia.

The Solvay process recycles large quantities of NH_3, and it is necessary to minimize losses in scrubbing the various NH_3-containing gas streams. Thus, the incoming purified brine is used to wash the ammonia-bearing air which is pulled through the bicarbonate cake on the vacuum filters. This is accomplished in a packed absorber. The brine then passes through a second packed section where it absorbs the NH_3 in the gases from the carbonating towers. The brine then flows to the main NH_3-absorber, which is also a packed tower. The brine is circulated through water-cooled heat exchangers to remove the heat of absorption. Ammonia, carbon dioxide and small amounts of H_2S released in the distiller are absorbed in the packed tower. In addition, a small amount of make-up ammonia is added. A typical analysis of the cooled, ammoniated brine is:

Temperature	38°C
NH_3	90 g/1
CO_2	40 g/1
NaCl	260 g/1
H_2S	0.1 g/1

The total heat removed from the ammonia absorption is 1.25 million Btu per net ton of soda ash produced.

Next, the ammoniated brine is carbonated to a point just short of crystallization. Then the brine is given a final carbonation and cooled to produce the crude $NaHCO_3$ crystals. The equipment used in the carbonation step consists of groups of five identical towers having alternate rings and discs in the upper section to assure mixing of the falling liquor with the rising gas stream without becoming plugged by the crystallizing solid phase. The lower section of each tower consists of a series of heat-exchanger bundles alternating with rings and discs.

The ammoniated brine is passed downward through one of the group of five columns which has become fouled with sodium bicarbonate after four days' operation as a crystallizing unit. The 40% CO_2 gas from the kilns is pumped into the bottom of the column to provide agitation and heat in order to dissolve the crystalline scale and bring the liquor to a composition just short of crystallization. The liquor is adjusted to the desired temperature by passage through a heat exchanger in preparation for feeding to the crystallizing towers. The liquor is then fed into the top of each of the other four columns in the group.

A mixed gas of 60-75% CO_2 derived from mixing the 40% CO_2 from the kilns and the 90% CO_2 from the calcination of bicarbonate, is fed to the bottom of these crystallizing units. Absorption of CO_2 in the highly alkaline ammoniated brine results in crystallization of crude sodium bicarbonate. Due to the heat evolved in the absorption and neutralization of the carbonic acid gas, and from crystallization of the sodium bicarbonate, the temperature of the liquor in the column rises from 38°C to a maximum of about 62-64°C. In normal operation, the temperature of the discharge slurry is maintained at about 27°C by automatic adjustment of the water flow through the cooling tubes. A tower such as that shown in Fig. 6.3 has a capacity for producing 50 tons of finished soda per day.

The heat removed from the carbonator is about 260,000 Btu per net ton of ash produced, while that removed from the crystallizing units is 1.25 million Btu per ton of product.

It is noteworthy that the presence of sulfide in the feed liquor to the towers serves to maintain a protective film of iron sulfide on the cast iron equipment, which minimizes the contamination of product crystals from iron corrosion products.

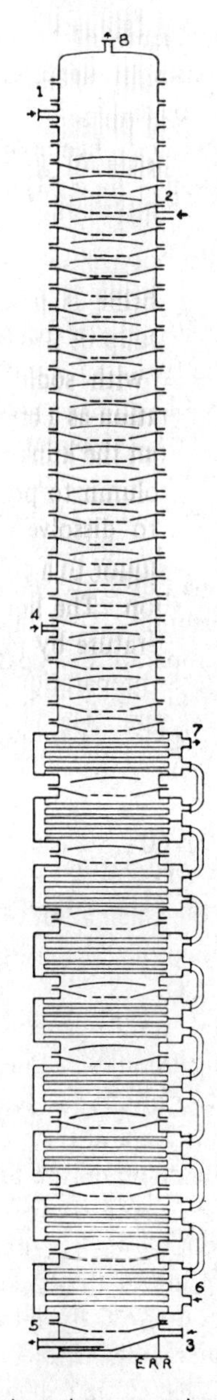

Fig. 6.3 A carbonating tower in the ammonia soda process (Solvay); it is 69 feet high and 6 feet in diameter: (1) entry for ammoniated brine, used when the tower is being cleaned; (2) entry for the ammoniated brine for the regular bicarbonate precipitation; (3) and (4) carbon dioxide entries; (5) outlet for bicarbonate slurries; (6) cooling water inlet; (7) cooling water outlet; (8) escape for uncondensed gases. (*Modeled after Kirschner.*)

Slurry drawn from the crystallizing columns is filtered in rotary vacuum filters where the $NaHCO_3$ crystals are water washed. The filtered liquor then flows to the distillation unit for recovery of ammonia. The crude bicarbonate filter cake contains 3% NH_4HCO_3, 12–15% H_2O and 80–85% $NaHCO_3$.

Next, the liquor is pumped to the NH_3-still preheater. At this point, the sulfide solution required for corrosion protection is added. The preheated process stream enters a stripper where excess CO_2 is removed prior to treating with lime and distilling out the NH_3. The hot liquor from the stripper flows to an agitated vessel into which is also added the milk of lime to release NH_3:

$$Ca(OH)_2 + 2NH_4Cl \longrightarrow CaCl_2 + 2NH_3 + 2H_2O$$

The lime-treated solution is then fed to the top of the bubble-cap distillation unit. Steam is injected at the bottom, stripping out the ammonia down to a residual level of only 0.001%.

The crude $NaHCO_3$ is calcinated in dryers which are constructed with rotating seals and gas-tight feed and discharge mechanisms. This is to ensure the production of CO_2 which is undiluted with air:

$$2NaHCO_3 \longrightarrow Na_2CO_3 + H_2O + CO_2$$

The heat requirement is about 2 million Btu per ton of soda ash produced. Product from the dryers is cooled for shipment, or converted to other products. A typical analysis of good commercial light soda ash made by this process is:

Na_2CO_3	99.70%
NaCl	0.12%
H_2O	0.12%
Fe	15 ppm
Ca and Mg	75 ppm

The waste liquor from the process may be evaporated to produce by-product $CaCl_2$ and NaCl. However, the major portion is clarified and pumped into water courses having natural flow sufficient to provide necessary dilution for disposal.

Production of soda ash from trona ore mined

in the Green River area of Wyoming is rapidly expanding. Two basic processes are used involving dry mining of the ore from a depth of about 1500 ft.; attention is also being given to the development of a third process based on solution mining of the ore. In the older dry-mining process, the ore is crushed and dissolved at the surface. The solution is then purified by settling and filtration, followed by evaporative crystallization to form sesquicarbonate: $Na_2CO_3 \cdot NaHCO_3 \cdot 2H_2O$. The crystals are centrifuged off and calcined to ash in steam-tube dryers. In a more recent process the crushed ore is calcined immediately, then dissolved, purified and evaporatively crystallized to sodium carbonate monohydrate, which is centrifuged and dried to product ash.

Relatively small quantities of soda ash are produced from alkaline brines at Searles Lake, California, by a process of fractional crystallization which also produces other sodium and potassium salts. Table 6.3 shows the current distribution of soda ash in the U.S.

The 1980 contract price of soda ash at plants in Wyoming is $76–86/ton.

SODA ASH-RELATED PRODUCTS

Most manufacturers convert a large portion of the light ash produced by the Solvay process to a form known as dense ash, because this is the form preferred by the glass industry, which is the major consumer of soda ash. Dense ash has almost twice the bulk density of light ash, 65 lbs/cu. ft. versus 35 lbs/cu. ft., and consists of much coarser granules, approximating the sand used in the glass batch, thus giving uniform mixing. Dense ash is manufactured by hydrating light ash to the monohydrate and then calcinating back to anhydrous ash.

TABLE 6.3[3] Percentage Distribution of U.S. Soda Ash

Consuming Industry	%
Glass	55
Chemical Mfg.	25
Pulp and Paper	5
Alkaline Cleaners	5
Water Treatment	3
Exports and Misc.	7

TABLE 6.4[3] Uses for Sodium Bicarbonate

Industry	%
Food (Baking Powder)	40
Chemicals	20
Pharmaceuticals	10
Fire Extinguishers	10
Soaps and Detergents	5
Leather, Textiles Paper, and Misc.	15

Some light ash manufacturers convert a portion of their production to refined sodium bicarbonate, or baking soda. This is done by carbonating a purified soda ash solution, centrifuging the bicarb crystals, and drying them at a low temperature to avoid reversion to the carbonate. U.S. capacity in 1980 for bicarb is 425,000 tons per year, with estimated demand at only 270,000 tons.[4] Table 6.4 gives a breakdown of uses for $NaHCO_3$.

By-product $CaCl_2$ and NaCl are produced from the waste liquor in the ammonia-soda process. Clarified waste is fed to multiple-effect evaporators. During evaporation, NaCl precipitates out. It is filtered, washed, and dried for sale. The remaining $CaCl_2$ liquor is either sold directly or concentrated further and sold in flake form. Major uses are for road ice and dust control, and refrigeration brines. In 1978, 600,000 tons were sold as solid and 200,000 tons as liquid, including the $CaCl_2$ produced from naturally occurring brines,[3] which now accounts for more production of $CaCl_2$ than Solvay-derived $CaCl_2$.

SODIUM SULFATE

Production capacity for sodium sulfate, or salt cake, has remained fairly constant over the last decade, with the number of producing plants continuting to decline. Over half of the U.S. supply of Na_2SO_4 is produced from natural brines. Most of the remainder is by-product derived from the production of viscose rayon, sodium dichromate, and sulfonation process phenol. Salt cake produced in the manufacture of HCl from salt by the action of sulfuric acid (Mannheim process), or SO_2 and air (Hargreaves process), is on the decline, there being now only two Mannheim plants and one Hargreaves plant

TABLE 6.5[3] 1978 Sodium Sulfate Production by Process

Process	*Thousands of tons*	*%*
Natural Brine	635	53
Viscose Rayon	194	16
Mannheim, Hargreaves	118	10
Dichromate, phenol and other	248	21

in the U.S. Table 6.5 lists the 1978 U.S. production of salt cake. In addition, a small amount of Glauber's salt ($Na_2SO_4 \cdot 10H_2O$) is still produced for sale, but it affects very little in the world of sodium sulfate.

Salt cake is a term which now refers generally to sodium sulfate, but originally meant sulfate produced in the Mannheim furnace, which yields a product containing less than 99% Na_2SO_4 compared to the much purer anhydrous Na_2SO_4 derived from natural brines.

In the latter process (shown schematically in Fig. 6.4) the natural brine, containing about 10% Na_2SO_4, is saturated with NaCl by pumping it into a salt deposit. The concentrated brine is then pumped from the salt well through a refrigerated heat-exchanger, where it is cooled to around -10°C, and then into a crystallizer where quite pure crystals of Glauber's salt are formed. This solid hydrate of Na_2SO_4 is then melted and the water removed by evaporation using a submerged burner as the heat source. The wet Na_2SO_4 is dried further in a rotary kiln, producing anhydrous Na_2SO_4. Over ten tons of natural brine and 1.6 million Btu of heat energy are required to produce one ton of product. The most significant production location is in Texas, where the sulfate-containing brine is conveniently located near deposits of both domed salt and natural gas.

A similar process is practiced at Searles Lake. There the brine is first carbonated and chilled to remove sodium carbonate and borax. Further chilling crystallizes Na_2SO_4 as the hydrate, Glauber's salt, and some remaining borax. The coarse crystals of Glauber's salt are separated from the fine crystals of borax in a hydraulic classifier. The sulfate fraction is then filtered, washed, dried, and evaporated to produce anhydrous Na_2SO_4.

In another process at Searles Lake, the brine is first evaporated to produce NaCl and a double salt of Na_2CO_3 and Na_2SO_4. The two solids are separated in a hydraulic classifier, and the sulfate fraction is redissolved and recrystallized to Glauber's salt. Anhydrous Na_2SO_4 is recovered by mixing the Glauber's salt with sodium chloride brine. This dissolves the Glauber's salt and allows anhydrous Na_2SO_4 to precipitate because of its lowered solubility in the presence of sodium chloride.

In the Mannheim process, NaCl and 100% H_2SO_4 in amounts equivalent to complete conversion to Na_2SO_4 are fed to a circular muffle furnace made of cast iron. The furnace is equipped with a shaft, which penetrates from below and carries four arms, each of which is fitted with two cast-iron plows. The shaft

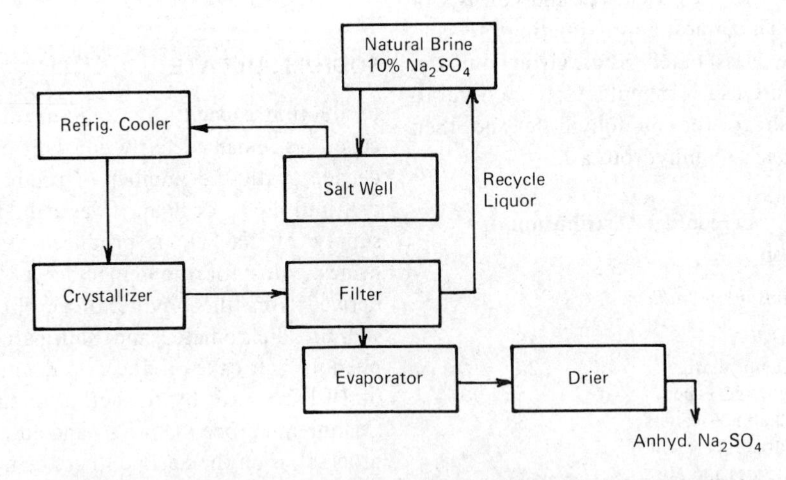

Fig. 6.4 Simplified flow sheet for the production of Na_2SO_4 from natural brine.

rotates at 0.5 rpm, slowly plowing the heated mixture to the circumference where the burned cake discharges through a chute. The furnace charge is heated to about 840°C, just below the fusion temperature of the salt cake. HCl is liberated, cooled, and absorbed in water to produce 32% HCl for sale. The salt cake is really a by-product of the operation.

A substantial amount of by-product Na_2SO_4 is also produced in a variety of processes involving the use of H_2SO_4 to react with a sodium salt, or to neutralize free caustic soda. Chief among these are the viscose rayon process and the manufacture of sodium dichromate. In the viscose process, Na_2SO_4 is formed by the reaction between H_2SO_4 and cellulose xanthate plus free caustic in the rayon fiber spinning bath. Spin-bath liquor is evaporated to crystallize Glauber's salt, which is centrifuged off, melted, and then evaporated to anhydrous Na_2SO_4. This is a significant source of sulfate, since 1.1 lb of Na_2SO_4 are produced per pound of rayon fiber spun.

In the manufacture of sodium dichromate, by-product Na_2SO_4 crystallizes directly as the anhydrous material when H_2SO_4 is added to a boiling solution of sodium chromate:

$$2Na_2CrO_4 + H_2SO_4 + H_2O \longrightarrow Na_2Cr_2O_7 \cdot 2H_2O + Na_2SO_4$$

Decreased uses of chrome by the auto and small appliance manufacturers, and mandated recycling of chromium wastes in the electroplating industry have led to a decline of Na_2SO_4 produced from this source.

Table 6.6[3] shows the U.S. distribution for consumption of Na_2SO_4. The price of salt cake is $46/ton, and for the higher quality detergent and rayon grades $60-70/ton.

The use of Na_2SO_4 in the manufacture of kraft paper is giving way somewhat to the use of sodium sulfide and sodium hydrosulfide, which are becoming more readily available as by-products of the petroleum industry, derived from the caustic scrubbing of sulfide-containing hydrocarbon gases. The sulfides are often preferred by the paper pulp mills because the sulfate which is used must be reduced to the sulfide in the pulping process.

TABLE 6.6 U.S. Distribution of Na_2SO_4.

Industry	*Thousands of tons*	*%*
Kraft pulp	580	45
Detergents	520	40
Glass	150	11
Misc.	50	4

SODIUM SULFIDES

Sodium sulfide (Na_2S) and sodium hydrosulfide, sometimes referred to as sodium sulfhydrate, (NaHS), share the same derivation: caustic soda and H_2S. Earlier processes based on reduction of salt cake with coal have largely been supplanted. In 1979, combined U.S. production was over 300,000 tons.[3] The source of H_2S used in the manufacture of these chemicals is essentially all by-product. Controlled caustic scrubbing of H_2S forms a solution of NaHS:

$$NaOH + H_2S \longrightarrow NaHS + H_2O$$

The NaHS solution is filtered to remove the sulfides of heavy metals, such as Fe, Hg, Ni, Mn and Cu. The clear filtrate may be sold as a 44-46% solution of NaHS or evaporated in stainless steel equipment to crystallize a solid hydrate containing 70-72% NaHS, which is sold as a flake product.

NaHS is easily converted to Na_2S by further reaction with caustic:

$$NaHS + NaOH \longrightarrow Na_2S + H_2O$$

By using a NaHS solution of the proper concentration with a flake caustic, a hydrate product containing 60-62% Na_2S is obtained. This is sold directly as a flake product, or fused solid in drums. When high quality H_2S and NaOH are used, the product Na_2S is suitable for use in dyes, photography, rayon and leather manufacture.

Lower quality sulfides are obtained by using H_2S produced as a by-product in the manufactue of CS_2 from methane (or other low molecular weight hydrocarbons) and sulfur:

$$CH_4 + 4S \longrightarrow CS_2 + 2H_2S$$

The gas mixture from this catalytic reaction is cooled and scrubbed with caustic. The sulfides so produced contain small amounts of mercap-

TABLE 6.7 Uses for Na_2S and NaHS

Industry	*% Total Market* Na_2S	*NaHS*
Pulp and paper	5	50
Dyestuffs	18	15
Rayon and film	13	12
Metals and minerals	12	–
Leather	20	9
Export and miscellaneous	32	14

tans (e.g., CH_3SH), which lend a very objectionable odor to the product.

Sulfides are also obtained as by-products in the manufacture of $BaCO_3$ from the barite ore, $BaSO_4$, by roasting with coal, water leaching, and treatment with soda ash:

$$BaSO_4 + 4C \longrightarrow BaS + 4CO$$

$$BaS + Na_2CO_3 \longrightarrow Na_2S + BaCO_3$$

Table 6.7 shows the distribution for consumption of Na_2S and NaHS by industry use. They are both extensively used in a variety of industries, and the use of NaHS in paper pulping is growing. The 1980 carload price for Na_2S flake in drums is \$275/ton. NaHS similarly sold for \$320/ton flake, and \$305/ton for tank cars of 45% liquid (100% basis).

SODIUM THIOSULFATE

Ninety percent of sodium thiosulfate (hypo) is used in photography because of its ability to dissolve water-insoluble silver salts. However, because substantial market erosion has occurred concurrently with the growth in use of videotape, the market for sodium thiosulfate continues a steady decline from 40,000 tons per year thirty years ago to well less than half that today.

Most current production is derived as by-product from the manufacture of sulfur dyes and Na_2S.

$$Na_2S + SO_2 + H_2O \longrightarrow Na_2SO_3 + H_2S$$

$$Na_2SO_3 + S \longrightarrow Na_2S_2O_3$$

$$2Na_2S + Na_2CO_3 + 4SO_2 \longrightarrow 3Na_2S_2O_3 + CO_2$$

In producing sulfur dyes, by-product leach liquor contains sodium thiosulfate, which is simply derived by evaporation and crystallization. Much of the thiosulfate is sold as the pentahydrate, $Na_2S_2O_3 \cdot 5H_2O$. Stainless steel equipment is used in processing thiosulfate.

Of less importance today is the older soda ash-sulfur process. Ash is dissolved in hot water and the solution is pumped to SO_2 absorption towers. The resulting $NaHSO_3$ liquor is then heated with powdered sulfur in an agitated stainless steel digestion tank at elevated temperature. The product solution is filtered, evaporated, and crystallized to yield $Na_2S_2O_3 \cdot 5H_2O$ product, which is centrifuged, washed, dried, and packaged in air-tight containers (to discourage efflorescence).

SODIUM BISULFITE

Most of the sodium bisulfite of commerce is really the anhydride, $Na_2S_2O_5$, sodium metabisulfite or sodium pyrosulfite. The usual assay is 98% $Na_2S_2O_5$, 1.5% Na_2SO_3, and 0.5% Na_2SO_4. Annual U.S. production in 1980 was estimated to be 450,000 tons at a worth of \$0.15-0.20/lb.[5] Methods of manufacture are all variations of the same theme:

$$Na_2CO_3 + 2SO_2 \longrightarrow Na_2S_2O_5 + CO_2$$

In one variation, SO_2 is sparged into a stainless steel absorber through which is passed a solution of Na_2CO_3. Formation of product crystals from the saturated solution is achieved by lowering the temperature. The crystals are centrifuged off and dried rapidly in a flash drier in order to avoid air oxidation to sulfate.

The anhydrous product is used principally in dye manufacture and fixing, in the leather industry, in the photographic industry as an antichlor, and in the manufacture of hydrosulfite.

SODIUM HYPOSULFITE (HYDROSULFITE)

Not to be confused with "hypo" (which is the term used in photography to refer to thiosulfate), sodium hyposulfite (or hydrosulfite) $Na_2S_2O_4$, is a powerful reducing agent which is used principally for the reduction of vat dyes. Current price of this chemical is \$0.62/lb.

There were approximately 62,000 tons produced in 1979.

There are two principal methods of manufacture. One involves reduction of $NaHSO_3$, or SO_2, with zinc dust. The other reduces $NaHSO_3$ with sodium amalgam from an electrolytic mercury chlorine cell.

There are two variations of the zinc method. In the first, a solution of sodium hydrogen sulfite is reduced with zinc dust in the presence of excess SO_2:

$$2NaHSO_3 + Zn + SO_2 \longrightarrow Na_2S_2O_4 + ZnSO_3 + H_2O$$

Milk of lime is added to neutralize remaining sulfurous acid. $CaSO_3$ and $ZnSO_3$ are removed by filtration. Common salt is added to the remaining solution to salt out $Na_2S_2O_4 \cdot 2H_2O$. The suspension of crystals is heated to 60°C to dehydrate the product. Water is extracted with alcohol and the product crystals are vacuum dried. The crystals are stable only when completely dry.

The second method based on zinc consists of treating an aqueous suspension of zinc dust with SO_2 at 80°C.

$$Zn + 2SO_2 \longrightarrow ZnS_2O_4$$

Treatment of this solution with soda ash precipitates a basic zinc carbonate, which is filtered off. The dry crystalline product $Na_2S_2O_4$ is obtained by salting out and drying as in the first method.

The amalgam process uses a dilute sodium amalgam (0.5% Na) derived from the cathode reaction of a mercury chlorine cell.[6] The amalgam is admixed with a solution of $NaHSO_3$ whose pH is controlled in the range of 5 to 7.

$$4NaHSO_3 + 2NaHg \longrightarrow Na_2S_2O_4 + 2Na_2SO_3 + 2H_2O + Hg$$

The mercury is returned to the mercury cell to form more amalgam, and the Na_2SO_3 is allowed to react with aqueous SO_2 to form more $NaHSO_3$.

Eighty percent of all $Na_2S_2O_4$ is used in dyeing fibers (60% for cotton and 20% for other fibers and blends).

SODIUM PHOSPHATES

Twenty percent of the phosphorous produced in the U.S. is utilized in the manufacture of 1.4 million tons per year of a family of sodium phosphate products, chief among which is sodium tripolyphosphate. In turn, 85% of the total sodium phosphates produced goes into detergent manufacture, with 7% used as food additives and in baking powders. This major family of heavy chemicals turns over several hundred million dollars per year. Basically, production is derived from reaction between soda ash and phosphoric acid.[7] By far, the major source of the acid is furnace grade acid, because of its high purity as compared with the less pure wet process acid.

Since such a high percentage of these products goes into detergent manufacture, the business, though good, is somewhat precarious in that phosphate pollution of waterways could be mandated against *if* a suitably innocuous and economical substitute could be found for effective detergent use.

The largest volume of the sodium phosphates is produced as the tripolyphosphate, $Na_5P_3O_{10}$. It is produced by reaction between a hot solution of Na_2CO_3 and 60% H_3PO_4, maintaining a molar ratio of sodium to phosphorous of 1.67 in order to achieve the desired stoichiometry. The solution is purged with steam to evolve all the CO_2. It is then filtered, spray dried, and granulated for shipment. By varying the ash to acid ratio, a variety of products can be produced.

This process forms the basis for an entire family of products, fundamental to which, in addition to the tripolyphosphate, are the production of the di- and trisodium salts. These may be produced in the same process by crystallizing out and recovering $Na_2HPO_4 \cdot 12H_2O$, which is dried to Na_2HPO_4, and then further processing the liquor by adding strong caustic, evaporating, and crystallizing out $Na_3PO_4 \cdot 12H_2O$.

The mono-, di-, and pyrophosphates are used in foods, the latter two in baking powders. The monosodium salt is produced from Na_2HPO_4 by further processing with H_3PO_4, evaporation, and crystallization. Prolonged heating of the monosodium derivative at 230°C produces the pyrophosphate, $Na_2H_2P_2O_7$. Further, heating the monosalt to the point of fusion produces hexametaphosphate, a glassy substance of uncertain stoichiometry, which is used in water treating.

SODIUM SILICATE

The sodium silicate family is analogous to the phosphates in that a whole series of derivatives are produced in a reaction between Na_2CO_3 and silica by varying the ratio of soda ash to silica. The products are often referred to by the ratio of SiO_2 to Na_2O in a given composition. U.S. Production is almost a million tons per year, with a value of about $250 million.[5] Table 6.8 gives a typical use pattern for this heavy chemical group.

Sodium silicates are made in batches by fusing sand and soda ash in a furnace at about 1300°C, in whatever ratio is necessary to achieve the desired properties of the final compound. In general:

$$Na_2CO_3 + nSiO_2 \longrightarrow Na_2O(SiO_2)_n + CO_2$$

The product is called *water glass*, because, when solid, it actually is a glass. But, unlike lime-soda glass, (ordinary window glass), it is soluble in water. The process is carried out in large tank furnaces similar to window glass furnaces. The materials are introduced in batches at intervals, but the products may be drawn off continuously if desired. A mixture of salt cake and coal may replace a portion of the soda ash.

As the melt leaves the furnace, a stream of cold water shatters it into fragments. These are dissolved with super-heated steam in tall, narrow steel cylinders with false bottoms,[8] and the product liquor is clarified.[9] Sodium silicates are sold as solutions which vary from the most viscous, 69°Bé, to the thinner solutions down to 22°Bé, suitable for use in paints. A dry powder is also marketed. This is made by forcing the thick liquor through a very fine opening into a chamber which is swept with a rapid current of cold air, which carries off the moisture.[10] Because the solid silicates are hygroscopic, they are often blended with Na_2SO_4 to prevent caking. Silicate solutions are also processed further with mineral acids to manufacture hydrated silica, which yield silica gel upon being dried. Silica gel is extensively used in air-drying equipment and in the manufacture of catalysts.

TABLE 6.8 Uses of Sodium Silicates

Industry	*Thousands of tons*	*%*
Catalysts and gels	293	35
Soaps and detergents	151	18
Boxboard adhesives	84	10
Pigments	126	15
Water treatment	50	6
Other	136	16

CHLOR-ALKALI (CHLORINE AND CAUSTIC SODA)

Ranking among the higher volume chemicals produced in the world are chlorine and caustic soda (sodium hydroxide). The chlor-alkali value to the U.S. economy in 1980 was $5 billion. Table 6.9 lists world production figures for both products,[11] by the countries in which they were produced.

Over a third of the total world chlor-alkali capacity exists in the U.S. in about 70 plants, by far the larger of which are concentrated in the Gulf Coast area of Louisiana and Texas.[12] This concentration derives from the combined availability in that region of salt, hydrocarbons (fuel and organic products based on chlorine and caustic), and deep water ports for trans-

TABLE 6.9 1979 Chlor-Alkali Production

	Thousands of short tons	
	Caustic	*Chlorine*
U.S.A.	12,300	12,100
Canada	1,250	1,130
Britain	1,100	1,000
W. Germany	3,210	3,300
France	1,390	1,370
Italy	1,400	1,340
U.S.S.R.	3,700	3,550
Japan	3,250	3,100
Africa	135	130
Australia	210	200
Argentina	180	175
Brazil	300	350
Mexico	315	300

TABLE 6.10 Market Distribution of U.S. Chlorine

Market	%
Vinyl chloride	17
Propylene oxide	10
Methylene chloride	3
Chlorinated solvents	14
Inorganic chemicals	11
Water treatment	5
Pulp and paper	13
Fluorocarbons	8
Organic chemicals	10
Miscellaneous	9

portation. The intercoastal canal, the Mississippi and Ohio Rivers, and their connecting waterways are also well suited to heavy barge traffic from the Gulf to much of the heartland.

Table 6.10 gives the distribution of U.S. chlorine by industry use. Table 6.11 similarly lists the market distribution for caustic.[13]

The heart of the chlor-alkali process is the electrolytic cell in which saturated, purified NaCl brine is electrochemically decomposed. There are basically two types of cell:[14,15] diaphragm and mercury. The diaphragm cell derives its name from the separator, or diaphragm, which is a mat of asbestos fiber vacuum-deposited directly on a woven-wire, mild steel member. The steel mesh serves doubly as a support for the diaphragm and as the cathode, or negative electrode, of the cell. The diaphragm separates the anode and cathode compartments of the cell to discourage direct mixing of the chlorine and caustic, and also to prevent chlorine and hydrogen from mixing, which can be a dangerous situation since they are capable of reacting violently with one another under certain conditions. A head of brine is maintained higher on the anode (the positive electrode, at which chlorine is produced) side of the diaphragm than the cell liquor level on the cathodic side. Thus brine slowly percolates through the diaphragm and into the cathode chamber of the cell where water is decomposed to form hydrogen gas and hydroxide ion. The effluent cell liquor contains about 12% NaOH, 14% NaCl, small amounts of chlorate, hypochlorite, sulfate, and carbonate—with the remainder being water. Since the usual caustic product for sale is a 50% NaOH solution, this cell effluent requires evaporation and separation of the salt.

TABLE 6-11 Market Distribution of U.S. Caustic

Market	%
Organic chemicals	34
Inorganic chemicals	9
Petroleum	5
Food	1
Pulp and paper	15
Soap and detergent	6
Exports	12
Alumina	5
Textiles	3
Rayon	3
Miscellaneous	7

The second type of cell differs from the diaphragm cell in that the cathode is a thin film of mercury, with no separator between it and the anode. While chlorine is produced at the anode in both types of cell, the cathode reactions differ markedly. Instead of water decomposing at the cathode of the mercury cell, as it does in the diaphragm cell, sodium metal is deposited in the mercury to form a dilute (0.5%) amalgam. This amalgam flows out of the cell into a decomposer, in which water and the amalgam react directly to form a 50% NaOH solution of high purity, there being no evaporation required. The denuded mercury is returned to the cell.

Another difference between the two cell types is the brine flow. In the diaphragm cell about 50% of the NaCl is decomposed, the remainder flows out of the cell in the weak caustic effluent. In the mercury cell, only a small fraction of the salt is decomposed per pass through the cell. It is immediately dechlorinated, resaturated with salt and returned to the cell.

At first glance, it would appear that the mercury cell process is far superior to the diaphragm cell process. In many ways it is. However, the mercury cell requires much greater input of electrical energy than does the diaphragm cell. In certain areas, especially where electrical energy is cheap and where small-capacity, easily run plants are required, the mercury cell is quite economical. This is especially so if fuel costs are high, which discourages the plant designer from installing evaporators which require burning fuel just to make steam to run the evaporators. In larger installations where it makes more sense to install a powerhouse at the site, the generation of power co-generates lower pressure steam which may be used quite economically to drive a multiple-effect evaporator.

There is a third type of cell which has some operating features reminiscent of both the diaphragm and mercury cell. In some respects the new cell is like a diaphragm cell in that it has a separator. However, the similarity ends there. The separator in the newer type cell is a perfluorinated ion-exchange membrane essentially through which only hydrated sodium ions migrate under the influence of the electrical field.[16,17] Since there is no fluid flow through the membrane, the brine in the anode compartment becomes depleted in salt. As it does so, it is removed, dechlorinated, resaturated, and returned to the cell, just as in the case of the mercury cell. Also, very high quality caustic is made in this cell in concentrations ranging from 20-40%. These solutions are more economically evaporated than is the diaphragm cell effluent, because they are more concentrated and contain no salt to speak of.[18] This technology is just emerging, but will in time largely displace the other cells.

A most significant advance in cell anode technology is being used to great advantage in all three cell types to conserve energy, reduce impurities, and lengthen cell life.[19,20] Prior to this development the cell anodes were made out of graphite. The new anodes are made out of titanium, which has excellent corrosion resistance in hot, chlorinated brine. However, the very oxide film which lends corrosion resistance is not an electrical conductor. By applying electroactive coatings, largely composed of mixed oxides of ruthenium and titanium, to the surface of the titanium it remains conductive. These anodes give a vast improvement in performance over graphite for two reasons: (1) they do not wear away like graphite, which, as it becomes increasingly thinner, widens the electrode gap in the cell, requiring higher and higher voltages to operate the cell; (2) the electroactive coating is actually a catalyst which allows chlorine to be formed on the anode surface with essentially zero activation energy. Both properties contribute to much reduced power consumption in running the cells. In the case of the mercury cell, there is the added advantage that the dimensional stability of the new anode precludes the continual adjustment of electrode spacing, which is constantly necessary with graphite electrodes.

The ion-exchange membrane and dimensionally stable anode developments are impacting greatly on chlor-alkali operations, though the former is in a much younger stage of development. Their impact has been accelerated by both environmental and energy cost considerations. Cell capital cost is skyrocketing because of the expense of these new components; however, their usage does lower operating costs. In practice, a balance between capital costs and operating costs is sought in order to achieve an economic optimum for a given plant site.

The chemistry at the anode is identical for all three cells:

$$2Cl^- \longrightarrow 2e^- + Cl_2$$

The cathode reactions are also the same for the diaphragm and membrane cells:

$$2H_2O + 4e^- \longrightarrow H_2 + 2OH^-$$

In the mercury cell, however, the cathode reaction is quite different, there being no hydrogen formed in the cell:

$$Na^+ + e^- \xrightarrow{(Hg)} Na \text{ (amalgam)}$$

The amalgam may be used as a chemical reductant (e.g. to reduce $NaHSO_3$ to the hydrosulfite,[6] or to hydrodimerize acrylonitrile to adiponitrile[21]), or decomposed with water to make caustic as is generally done.

Environmental pressures on mercury discharges have dealt a blow to the practice of mercury cell technology. As a result, there have been many plant closings. An excellent job has been done in achieving substantial reductions of Hg losses, and this technology will continue to be practiced. However, most recent emphasis is on the diaphragm cell and ion-exchange membrane technology. As the latter matures, further inroads will be made against the mercury cell. This is already happening in Japan.

The theoretical decomposition potential, E°, of a cell (i.e., the voltage below which it is impossible to form any products at all) is strictly dependent on the nature of the electrode reactions. It is related to the Gibbs free energy of the net overall chemical change occurring in the cell: $\Delta G = -nFE^\circ$, where n is the number of electrons involved in the electrode reaction, and F is the Faraday constant.[22] E° is the thermodynamic potential, to which the *IR* drops (voltage necessary to drive the current through the electrical resistances in the cell), and the overpotential (voltage associated with the kinetics of the reactions) must be added to get the actual cell voltage:

$$\text{Cell Voltage} = \text{Thermodynamic Potential} + \text{IR Drop} + \text{Overvoltage}$$

Using a modern diaphragm cell running at 1500 amps/m^2 as an example, the percentages of the total cell voltage attributable to these thrcc components are: 65% thermodynamic, 25% IR, and 10% overvoltage. Cell design and development programs continue to search for better ways to reduce energy consumption. Rapidly rising power costs have made power consumption the primary criterion in evaluating cell performance.[23] Thus, higher cell capital becomes justifiable on the basis of performance. But as cell capital increases, it tends to force operation at increased current density in order to produce more pounds per cell per unit of time. In a practical sense, this translates to a compromise between cell cost and current density in order to minimize production cost. Fig. 6.5 is a typical current-voltage curve for a diaphragm cell. The current density simply refers to total current divided by the nominal electrode area of either the mode or cathode, often expressed as thousands of amperes (KA) per square meter. Fig. 6.5 also shows the electrical energy consumption in kilowatt hours per 2,000 lbs. of chlorine (KWH/T Cl_2). The minimum energy requirement, unattainable in practice, is 1543 KWH/T Cl_2. This corresponds to 100% current efficiency (685.8 KAH/T Cl_2) and 100% voltage efficiency (i.e., no IR drops and zero over-voltage at both electrodes, so that the voltage is the thermodynamic potential, or 2.17 volts). The energy efficiency of a typical diaphragm cell is easily expressed as:

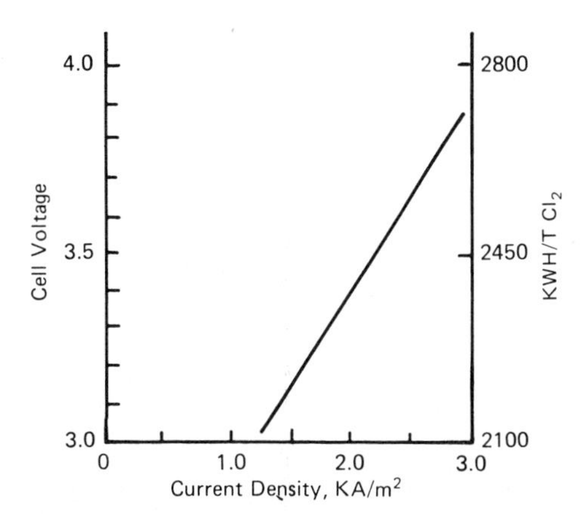

Fig. 6.5 Current-Voltage Curve for a typical diaphragm chlor-alkali plant.

$$\frac{(2.17) \times (\%\ \text{Current Efficiency})}{(\text{Cell Voltage})}$$

For example, if a cell running at 80,000 amperes at a current density of 2.2 KA/m^2, and a current efficiency of 96%, requires a cell voltage of 3.40, the overall energy efficiency is:

$$\frac{2.17 \times 96}{3.40} = 61.3\%$$

Much of the inefficiency results in heat generation; a cell will heat itself almost to boiling when fed brine at about 40°C. The self-heating tends to lower the cell voltage because the electrical resistance of the cell decreases with increasing temperature. Since modern rectification equipment (transformers, controllers, and rectifiers) are very efficient, the conversion of AC electricity into electrochemical products is essentially dependent on the cell efficiency. Typically, AC-KWH/lb. product = 1.03 × (DC-KWH/lb. product).

Figure 6.6 is a schematic of a commercial diaphragm cell, typically operated in the range of 75-150,000 amperes. Many such cells are connected in series as shown in Figure 6.7, so that the current leaving at the anode of one cell

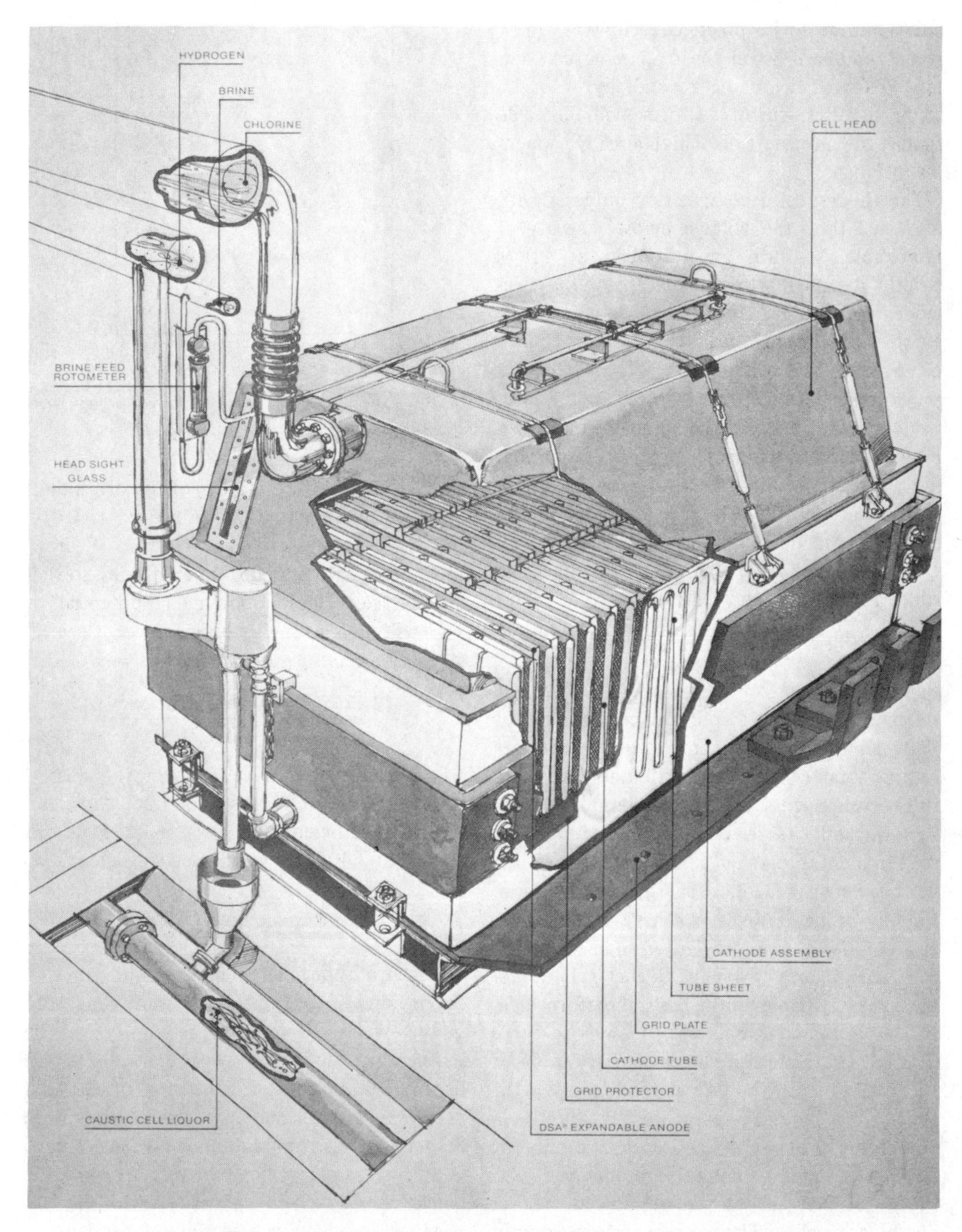

Fig. 6.6 Diamond Shamrock diaphragm cell. (*Courtesy Diamond Shamrock Corp.*)

enters the cathode of the next cell, with the anode terminal cell connected to the positive bus of the rectifier, and the cathode terminal cell connected to the negative bus of the rectifier.

With allowance for inefficiencies, each 30 KA increment of current which passes through the series circuit of cells produces 2000 lbs. of chlorine, 2100 lbs. of caustic, and 57 lbs. of hydrogen, per cell, per 24 hours of operation.

If a production plant is required to produce 900 tons of chlorine per day using rectification

Fig. 6.7 A series of Diamond Shamrock Cells. *(Courtesy of Diamond Shamrock Corp.)*.

equipment capable of sustained operation at 90,000 amperes, the plant must operate a minimum of 300 cells. If the cell voltage is 3.4 volts at 90 KA, rectification equivalent to 1020 volts is required. In practice, the logistics of production requirements, plant reliability, and maintenance outages will dictate the size (i.e. number of cells and nominal amperage) of each circuit. In the example above, the plant would use a minimum of two circuits, with two nominal 90 KA, 600 volt rectifiers.

Figure 6.8 is a simplified flow sheet for a typical diaphragm cell chlor-alkali plant. The process begins with brine treating, in which the well brine is saturated with return salt from the caustic evaporators, and then treated with hypochlorite to destroy ammonia and amines, and also treated with carb-caustic liquor to remove calcium and magnesium impurities.

The removal of trace ammonia-nitrogen impurities is essential to avoid buildup of NCl_3 in the chlorine purification and liquefaction stages. The NCl_3 is unstable in concentrations above a few thousand ppm, and can decompose violently. The Ca and Mg impurities are removed to reduce plugging of the cell diaphragms, and to lengthen the time between rediaphragming. This was typically done every few months, but has now been extended to over a year, principally by replacement of the graphite anodes with the newer metal anodes. In the older cells, the slowly oxidizing graphite formed a sludge which tended to plug the diaphragm.

The hydrogen from the cells was formerly discarded, (vented to the atmosphere), unless it was used in some of the larger plants to make NH_3 and other chemicals. It takes a large plant to produce a useful amount of H_2, since the cell produces only 0.028 lbs. H_2 per lb. Cl_2. Also, the fuel value of H_2 is only a third that of methane. But with our energy crisis, the fuel value of the hydrogen is now sufficient to recover it for boiler fuel. This requires cool-

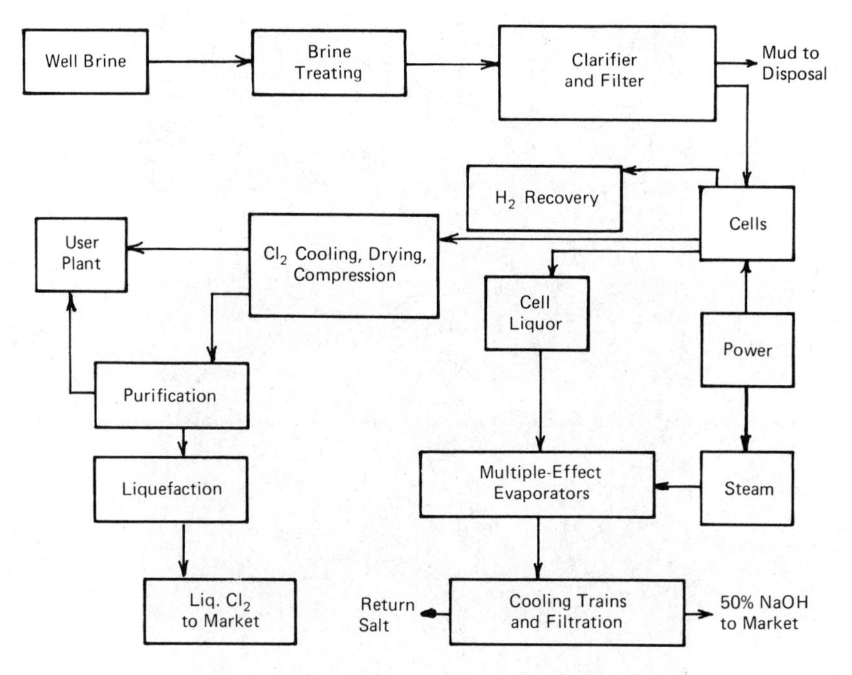

Fig. 6.8 Simplified flow sheet for chlor-alkali Plant.

ing to remove water, scrubbing to remove traces of chlorine, caustic and salt, and then compressing it sufficiently to satisfy the feed requirements of the burner. Of increasing importance is the burning of the hydrogen with the dilute chlorine stream known as tail gas from the liquid chlorine plant.

The chlorine from the cell is saturated with water at the cell temperature, and contains lesser amounts of NCl_3, organic chlorides, and salt, from brine entrainment. The gas is cooled to knock out the water. This chlorinated water is returned to the brine plant. The cooled gas is then dried by passage through towers of sulfuric acid. Strong (98%) acid is fed to one end of the drying train at a rate equivalent to the removal of weak (65%) acid from the opposite end of the drying train. The waste acid is often used to neutralize alkaline wastes.

The moisture content of the chlorine at this stage is usually sufficiently low to allow the use of steel in the remainder of the process. The gas is compressed, usually to whatever pressure is required by the user plant, or to sufficient pressure to liquefy it, the exact pressure dependent upon the refrigeration capacity and liquid chlorine storage requirements of a given plant. Figure 6.9 is the liquefaction curve for chlorine.[26]

In northern climates, where ambient temperatures can fall below those required to liquefy chlorine at rather modest pressures, it is necessary to provide means for maintaining the temperatures in chlorine-gas distribution lines above the liquefaction point. This can be done by tracing or wrapping the lines with heating tape, using steam lines, and insulating the pipes. These precautions are especially important in

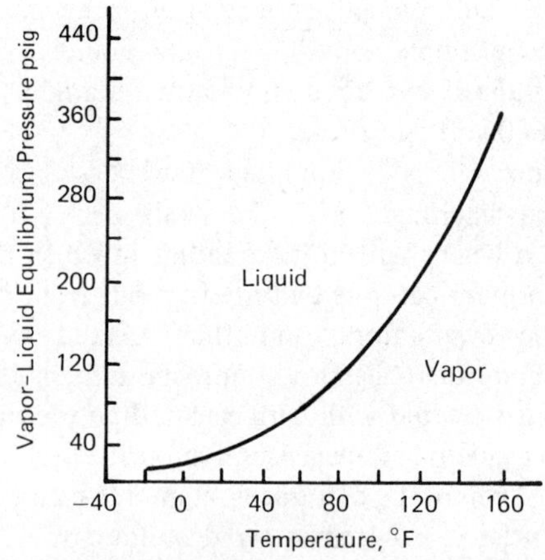

Fig. 6.9 Liquefaction curve for chlorine.

organic process plants, which can run without incident while using gaseous chlorine but which could explode if fed liquid chlorine.

Organic impurities and trace amounts of NCl_3 may be removed from gaseous chlorine in a simple extraction in which liquid chlorine is used as the extractant. The liquid bottoms flow from the extractor to a still in which chlorine is flashed off to recycle and the organic fraction (largely low molecular weight organic chlorides) is recovered for disposal by combustion to CO_2 and HCl, which is recovered for use.

There are many options in the liquefaction of chlorine, depending upon the storage requirements for the liquid and the system used to handle the tail gas. Low pressure storage is recommended because of the reduced hazard in the event of a spill. This, however, requires lower storage temperatures. Low temperatures also reduce the amount of chlorine in the tail gas from liquefaction. Yet lower temperature may not be necessary if a plant also deliberately produces bleach or HCl from the tail gas.

The source of the tail gas goes back to the cells. At the anode, some water is oxidized to produce small amounts of oxygen. To this amount is added more oxygen from air which may leak into the chlorine header from outside, since the header is operated under a slight vacuum to prevent gas from escaping to the atmosphere. The cell gas typically contains about 97% Cl_2. Since the remaining 3% is noncondensable under the conditions in the liquid chlorine plant, it remains gaseous at the tail end of the liquefaction process, saturated with chlorine in an amount determined by the vapor pressure of chlorine at the temperature and pressure in that stage of the process. When air is deliberately added to the system to reduce the hydrogen content of tail gas to below 5%, the tail gas often increases in volume. With hydrogen above that level, there is great risk of explosion. The hydrogen comes from the cathode compartment of the cell, which is operated under a slight pressure to avoid sucking air into the hydrogen collection system. Ordinarily, this leads to quite small amounts of hydrogen in the cell chlorine gas. But when diaphragm problems arise, the level can rise to undesirable values. In good plant practice, this is monitored continuously.

The cell liquor flowing from the cathode compartment of each cell is collected for evaporation to 50% product caustic. The cell liquor contains 10-12% NaOH, 14-16% NaCl, 0.1-0.5% Na_2SO_4, 0.02-0.04% Na_2CO_3, and 0.01-0.2% $NaClO_3$. The evaporation is carried out in reverse fed nickel-lined multi-effect systems (i.e. feed liquor enters the opposite end of train with respect to the primary steam entry). With rapidly rising fuel costs, it has become economically justifiable in the construction of new plants to install quadruple-effect evaporators. Previously triple-effect systems were installed. Most caustic evaporation is still done in triple-effect.

Because of the high salt content of the feed liquor, the evaporation system is really an evaporative-crystallizer. Salt is removed from each effect in the evaporation train. Hydroclones, centrifuges, settlers, and filters are extensively used to remove the salt as efficiently as possible and return it to the brine plant with as little caustic loss as possible. In this regard, good plant practice strikes an economic balance between caustic loss versus the addition of wash water to the system, which requires additional heat input.

The sulfate content of caustic evaporator feed can cause a problem during evaporation. When the caustic concentration reaches about the 32% level, the triple salt $NaCl \cdot Na_2SO_4 \cdot NaOH$ crystallizes out along with NaCl crystals. To avoid sulfate build up in the system which would lead to excessive caustic loss, the sulfate must be purged from the system. If recycled with the salt, its concentration will rise to unacceptable levels. Various schemes are used, but all plants purge sulfate at some point.

The hot 50% caustic from the evaporator train is cooled and filtered to remove salt down to the solubility limit. Typical analysis of product caustic, on a 50% NaOH basis (actual caustic content varies from 48.5 to 50.5%) is: 1.0% NaCl, 0.15% Na_2CO_3, 0.025% Na_2SO_4, 0.15% $NaClO_3$, 5 ppm Fe, and a maximum of 30 ppm combined heavy metals (Sb, As, Bi, Cd, Cu, Pb, Hg, Ag, and Sn). Most often the billing weight of a shipment of caustic is based on its Na_2O content, equivalent to 76% Na_2O:

$$\frac{(\text{lbs solution}) \times (\%\ Na_2O)}{76} = \text{net lbs NaOH}$$

(Na_2O basis)

Actually, pure NaOH contains 77.5% Na_2O. The value of 76% is derived from tradition, based on the fact that many decades ago the purest form of commercially available caustic contained 76% Na_2O. In isolated cases, where the market is far removed from the source, the caustic is further concentrated and shipped hot as 73% NaOH. This saves on freight costs, but the solution must be kept hot because it freezes at about 160°F. Similarly, 50% caustic must be maintained above 60°F.

In selected markets, such as the rayon industry, low salt levels are required (typically 100 ppm). Mercury cell and membrane cell caustic are ideally suited for these markets. Diaphragm cell caustic is purified further in a few installations, but primarily to remove $NaClO_3$, which is corrosive to the equipment used to manufacture anhydrous caustic. Chlorate alone may be removed by reaction with a reducing agent, such as sugar, which is injected into the hot caustic feed to the anhydrous concentrator system. This has the disadvantage of increasing both the salt and carbonate levels in the caustic.

To remove chlorate and other salts, the 50% caustic may be extracted with liquid ammonia in a pressurized system.[27] The ammonia fraction is then processed through a stripper to remove and recycle the ammonia. The alkaline stripper bottoms are useful in neutralizing acidic waste streams. The purified caustic is evaporated further, and then fed to an anhydrous concentrator, typically an Inconel falling film evaporator, heated with molten salt. The anhydrous caustic, containing at least 97.5% NaOH, is marketed as one solid mass in drums, as flake caustic, or more desirably as beads or prills. The latter are marketed in bulk or bags.

Since the binary system H_2O–NaOH contains several hydrates which may be conveniently crystallized, another possibility exists for purifying caustic. Of special interest are the dihydrate, $NaOH \cdot 2H_2O$, which melts at about 55°F, and $NaOH \cdot 3.5\ H_2O$, which melts at 60°F. The impurities concentrate in the mother liquor; purified caustic is then produced by melting the crystals. The most recent attention on this method of purification has been in Japan.[28] These processes are more difficult than they appear, because of the high viscosity of concentrated caustic solutions near their freezing points, and the resultant challenges of both adequately transfering heat and separating the crystals from the mother liquor. But with the advent of the membrane cell, there may be more development activity in this area.

In the handling and storage of 50% caustic, it is important to limit the amount of iron pick-up. The iron spec is 5 ppm on a solution basis. Stainless steel offers little advantage over mild steel with respect to iron contamination. Furthermore, hot, salt-containing caustic leads to severe stress corrosion cracking of stainless steel. Carbon steel storage tanks are often lined with an epoxyphenolic or neoprene latex coating. Iron pick-up from transfer piping increases with increasing temperature and increasing linear velocity of pumping. Unnecessary recirculation or pumping through steel lines should be avoided.[29] Iron, copper, and nickel impurities are especially undesirable for the bleach manufacturer.[30]

About 75% of the caustic produced is concentrated. The remainder is used directly as alkaline cell liquor—as, for example, in the conversion of propylene to propylene oxide by the chlorhydrin process. Similarly, there is some chlorine produced by methods which do not produce caustic, as shown in Table 6.12. Fused chloride salt electrolysis produces chlorine in the manufacture of magnesium metal by the Dow process, and of sodium metal in the Downs cell. The only other process of note is the Kel-Chlor process.[31] This process converts by-product HCl to chlorine by oxidation with NO_2 through the intermediates NOCl and $HNSO_5$. There is one 500 ton/day plant operating in

TABLE 6.12 U.S. Chlorine Production by Process

Process	*Tons Per Day*	*%*
Diaphragm	32,000	78
Mercury	7,800	19
Other	1,200	3

the U.S. An earlier variation which employed KCl and HNO_3 feed to produce Cl_2 and KNO_3 has been shut down.

A small (200,000 tons Cl_2 per year), but important, variation of NaCl electrolysis substitutes KCl as the feed. Both mercury cells and diaphragm cells are used, producing chlorine and KOH (caustic potash). The KOH is concentrated for sale as a 45% solution or in solid form, containing 88-92% KOH. A big use for KOH is in the manufacture of liquid soaps and detergents. The analogous sodium soaps and detergents are generally solids.

Table 6.13 gives the 1980 U.S. prices for chlorine and for caustic soda in its various forms.[5] These are typical values. For significant volumes, much movement of chlor-alkali goods is contracted below market price.

The entire chlor-alkali world must constantly perform a balancing act. The electrolysis of salt brine yields almost equal amounts of chlorine and caustic. It is rare that the growth or use-rates for the two products are equal. Furthermore, the chlor-alkali balance problem extends to related products such as soda ash, lime, salt cake, and hydrochloric acid, even to the extent of influencing (and being influenced by) the current economics of chlorinated hydrocarbons.

For example, in the 1960's the demand for chlorine was so much greater than for caustic that concerted efforts had to be made to move caustic into new markets, such as the glass industry. Another effect of the same economic climate was a glut of HCl from the tremendous growth in vinyl chloride and a host of other chlorinated organic derivatives. Necessity being the mother of invention, this situation led to the development of the oxy-chlorination process for the production of vinyl chloride, resulting in no net production of HCl; to a route for producing methyl chloride using HCl and methanol; to the Kel-Chlor process for production of chlorine without caustic; and to the recycle of by-product HCl to chlorine by electrolysis.[32,33,34] There was a host of developmental activity both in the U.S. and Europe to convert caustic to soda ash and to improve the old Deacon process, which is a non-electrolytic, catalytic conversion of HCl to chlorine.[35] In another case, a large consumer of sulfuric acid, the steel industry, was converted to the use of HCl for pickling steel. Huge amounts of caustic cell effluent were used to replace lime in a variety of applications.

By 1973 the picture was changing. Environmental pressures on fluorocarbon derivatives (derived from chlorinated compounds, and creating by-product HCl), and on some of the chlorinated solvents, impacted negatively on chlorine growth. As the 1970's began to spell stagflation (a stagnant economy with rapidly increasing inflation), the automotive and construction industries were brought to their knees, which meant increasingly poor markets for vinyl chloride, thereby further tightening the availability of caustic. At the same time, the use of caustic for flooding crude oil formations to enhance recovery began to show promise of consuming tremendous tonnages. Caustic remains tight in 1980, which translates to high HCl demand and increased consumption of lime. Consideration is being given to resurrecting Mannheim operations to produce HCl, and the old lime-soda process for the conversion of lime and soda ash into caustic. Until the early 1940's, this process was the major source of caustic. While it is doubtful that such processes will make a comeback, the scenario serves to illustrate the interesting situations which arise in trying to maintain chlor-alkali balance in the economy, and how far-reaching the balance problem extends.

TABLE 6.13 1980 Value of Chlor-Alkali Products

Product	*$/Ton (100% Basis)*
Liquid chlorine	145
50% Caustic	190
50% NaOH (Rayon)	205
73% Caustic	245
Solid Caustic	350
Flake Caustic	350
Bead Caustic	350

HYDROCHLORIC ACID

As shown in Table 6.14, the predominant source of HCl is as by-product from the chlorination of hydrocarbons, including the incineration of chlorinated organic wastes in thermal oxidizers.[36] HCl from the salt cake process has

TABLE 6.14 1980 Sources of HCl

Source	*Thousands of Tons (100% Basis)*	*%*
Salt cake	96	3
HCl burner	280	9
By-product and other	2700	88

declined, while the installation of HCl burners to convert chlorine plant tail gas to HCl has increased.

Hydrogen chloride is marketed both as anhydrous HCl and as hydrochloric acid, usually a 32% solution. The absorption of HCl in water customarily is achieved in a falling film absorber, followed by a packed tails tower which vents inerts to the atmosphere. The weak acid from the secondary absorber is fed to the primary absorber. The degree of absorption of the HCl is dependent on the concentration of inerts in the gas stream. The heat of absorption of HCl in water is about 700 Btu per lb of HCl, which requires that the primary absorber be water cooled.

Acid made by burning hydrogen in chlorine tail-gas is quite pure and requires no further treatment. By-product acid requires purification in packed scrubbers in which the scrubbing medium is the hydrocarbon which is circulated in the scrubbing tower before being sent to the chlorinator. A refrigerated cooling system is frequently employed to maintain low temperature in the circulating scrubber liquor in order to maximize absorption efficiency.

Current prices for HCl are $250/ton for the anhydrous, and $200/ton (100% basis) for high quality 32% acid. Upward pressure is being felt on production and prices due to the reduced availability of by-product acid from fluorocarbon manufacture, as well as the conversion of methyl chloride manufacture from methane and Cl_2 to methanol and HCl. Also, the conversion of steel pickling from sulfuric to hydrochloric acid, and increasing use of HCl in oil well acidification are impacting considerably on the HCl economy.

BROMINE AND BRINE CHEMICALS

Previously derived from seawater and brines, bromine is now largely derived from naturally occuring brines. Seawater contains only about 70 ppm, requiring the processing of about 4 million gallons for every ton of bromine produced. Though some seawater operations remain in Europe, naturally-occurring brines are now the principal source. Bromide levels in these brines are typically 2000 ppm in Michigan, 5000 ppm in Arkansas, and 5000 ppm in the Dead Sea. The oil-field brines in Arkansas are considerably warmer and more concentrated than the Michigan brines. Present U.S. production is principally in the area of Magnolia, Arkansas.

In 1970, 73% of bromine production went into ethylene dibromide (EDB), which enjoyed many good years as a lead scavenger in leaded gasoline. With the U.S. Government mandate on car exhaust emissions, including the intended demise of leaded gasoline, the EDB business is collapsing in this country. Due mainly to exports, EDB still accounts for a respectable amount of bromine production, being some 40% of the total in 1980. Ten percent of total bromine is marketed as CH_3Br, a fumigant, while sundry organic compounds and high density brine fluids account for 25% and 20%, respectively, of total bromine production. Both of the latter represent relatively new uses for bromine. Major bromo-organics are tetrabromobisphenol A, decabromodiphenyl oxide, hexabromocyclododecane, and pentabromochlorocyclohexane. These are blended with a variety of polymeric materials to modify the properties of the finished article.

A rising success story is the sudden prominence of bromine chemicals used for their high solution densities as completion fluids in oil wells the world around. As drillers go deeper and deeper in their search for crude oil, they have increasing need for high density fluids. $CaBr_2$ solution, the mixed $CaBr_2-ZnBr_2$ solutions are becoming widely used in that application. These are readily produced by reaction between the oxides and HBr, which, in turn, is produced by burning hydrogen in bromine. Also, pure KBr for photographic use is made from K_2CO_3 and HBr. Total bromine volume in 1980 was about 200,000 tons for all uses.

The bromine process is simple. The feed brine is heated by cross-exchanging with spent brine and fed to a steaming-out tower. Chlorine and steam are admitted to the tower, and the liberated bromine is condensed, distilled, and dried. Recovery of bromine is 95% based on the bromide content of the feed brine. The spent brine is neutralized and pumped back into subterranean strata.

In Michigan, this debrominated brine (containing 10-20% $CaCl_2$ and 3-5% $MgCl_2$) is processed further to produce calcium and magnesium brine chemicals. For example, the brine is treated with slaked lime to precipitate $Mg(OH)_2$, referred to as mag hydrate, which is settled and removed in a clarifier. The remaining brine is evaporated in a multi-effect system to recover $CaCl_2$ as a flake product containing about 76% $CaCl_2$ (essentially $CaCl_2 \cdot 2H_2O$), or concentrated further to almost anhydrous $CaCl_2$ (95%), sold in pellet form. $CaCl_2-MgCl_2$ liquor is often used for road dust control. The amount of $CaCl_2$ produced is dependent on the market for $Mg(OH)_2$, which is principally used for the manufacture of MgO refractories and pharmaceuticals.

By-product $CaCl_2$ from the Solvay soda ash process has been largely replaced by $CaCl_2$ from brine due to the many Solvay plant closings. Production of $CaCl_2$ is about 700,000 tons per year. $CaCl_2$ liquor (40%) sells for $40/ton, while the 95% pelletized product sells for $185/ton in bags, and $140/ton in bulk. Principal uses of $CaCl_2$ are: dust control 30%, deicing roads 25%, industrial use (e.g. refrigeration) 20%, concrete additive 10%, crude oil production 10%.

BLEACHES

Since the fledgling days of industrial chemical practice, the bleaching properties of chlorine have been in demand. It is noteworthy that the beginnings of our industry, some two hundred years ago, centered upon materials and methods of bleach manufacture. Much of modern chemical engineering practice derives from the struggles and developments of this heritage, particularly in the closing years of the last century, when the electrolytic production of chlorine and caustic began to take hold. Rampant spread of the great killer diseases such as cholera and typhoid were shown to be easily eliminated by the treatment of wastewater and drinking water with small amounts of bleach. Textile manufacturers and the paper industry had established the value of bleach in their operation early on.

Prior to the development of the liquid chlorine industry, the only means of transporting chlorine was as bleach. Common practice in the early years was to absorb the chlorine in hydrated lime, $Ca(OH)_2$, to form chloride of lime, $CaOCl_2$. Easily transported as a solid, this chemical contains the equivalent of 35% Cl_2. When dissolved in water it forms equal moles of $CaCl_2$ and $Ca(OCl)_2$:

$$2CaOCl_2 \text{ (dissolved)} \longrightarrow CaCl_2 + Ca(OCl)_2$$

It is not the same as calcium hypochlorite, which contains over 99% available chlorine. The term "available chlorine" refers to the oxidizing power of a bleach equivalent to that of free chlorine, as determined by the standard acidic-KI-thiosulfate titration. The most practical route to achieve a high available-chlorine content in a lime-based bleach involves chlorinating a slurry of lime in caustic,[6] followed by cooling to $-10°F$. Crystals are centrifuged off and added to a slurry of chlorinated lime containing $CaCl_2$ in an amount equivalent to the NaOCl content of the original crystals. Warming this solution produces crystals of $Ca(OCl)_2 \cdot 2H_2O$, containing around 3% lime as an impurity. This material is granulated, dried, and packaged for sale. It has an available chlorine content over 70%.

Another approach uses chlorine monoxide and water to form a solution of HOCl, which is neutralized with a lime slurry, spray-dried, and granulated to yield a product containing 70% available chlorine.[30]

The lime-based products have the disadvantage of containing insoluble components, which lead to the necessity for settling and sludge disposal in some applications. This is especially objectionable in laundry and dishwashing uses. A granular bleach based on LiOCl[30] is quite useful in these applications since it is completely soluble. It is prepared as a 35% solution of LiOCl by mixing solutions of LiCl and NaOCl, from which NaCl precipitates. The solution is evaporated, and the solid (containing LiOCl and NaCl) is dried and formulated with salt cake to a white, free-flowing, granular product which retains its bleaching power (35% available chlorine) reasonably well during storage.

In the bleaching of fibers for the manufacture of paper and textiles, sodium chlorite ($NaClO_2$) and/or chlorine dioxide (ClO_2) are often used because of their ability to achieve excellent whiteness without as much fiber degradation as occurs in chlorine or hypochlorite bleaching. For example, in the production of kraft paper, known for its strength, the product is brown when conventional bleach is used, because bleaching to whiteness would destroy the strength. By using chlorite, a high strength, white product is obtained.

$NaClO_2$ is manufactured[37] from ClO_2 (derived, in turn, from sodium chlorate), hydrogen peroxide, and caustic:

$$2ClO_2 + H_2O_2 + 2NaOH \longrightarrow 2NaClO_2 + 2H_2O + O_2$$

It is a powerful but stable oxidizing agent, the commercial form of which contains 125% available chlorine.

$NaClO_2$ is used in some applications as a source of ClO_2. Chlorite and chlorine react in water to produce ClO_2 and salt:

$$2NaClO_2 + Cl_2 \longrightarrow 2ClO_2 + 2NaCl$$

The use of chlorine dioxide in scrubbing foul air from fat rendering, in fish processing plants, and in treating cannery water for recycle (eliminating a source of pollution) has solved problems for which chlorine by itself was ineffective becasue of its own odor and taste.[30]

Chlorine dioxide is a very unstable molecule. It cannot be economically transported, but is easily produced *in situ.* The rapidly growing use of ClO_2 in the pulp and paper industry has led to the rapid growth of $NaClO_3$ production in recent years. Sodium chlorate, $NaClO_3$, is readily produced in solution form by the electrolysis of NaCl brine in a cell which is very similar to a diaphragm chlor-alkali cell, except it has no diaphragm. By allowing the chlorine and caustic produced in the cell to react immediately, and by providing an additional vessel where the cell liquor can be kept hot, with sufficient residence time all the hypochlorite disproportionates to chloride and chlorate:

$$3NaOCl \longrightarrow NaClO_3 + 2NaCl$$

Coated titanium metal anodes and steel cathodes are used. Small amounts of dichromate are added to the cell liquor to passivate the steel so it doesn't corrode. The cell liquor contains about 50% $NaClO_3$, which may be used as-is, or evaporated to 75% to drop out salt, filtered, and cooled to crystallize out $NaClO_3$. Current production efficiency is typically 85%. Over 75% of the sodium chlorate produced is used in pulp bleaching. Its bulk price is $350/ton.

Akin to $NaClO_3$ is the similar manufacture of bromates, starting with NaBr, or KBr, cell feed. Bromates are also produced by bromination of the respective carbonates. The bromate business is extremely small by comparison to other oxidants. But bromates are important in the bread-making industry for maturing flour and in conditioning dough. They are also used in permanent wave lotions and in the manufacture of dyes. The bromates are specialty chemicals, selling for $1.25/lb.

Treatment of a solution of $NaClO_3$ and NaCl with acid (H_2SO_4 or HCl) produces ClO_2 and Cl_2 which is immediately absorbed for use in bleaching. The ClO_2 generator must be operated with care to avoid ClO_2 concentrations above 10%, which can lead to explosion from self-decomposition. The molar ratio of ClO_2 to Cl_2 formed in the generator is 2:1. If this level of Cl_2 is objectionable, the gas mixture from the generator, which is mostly air (for safety), is passed through a tower through which chilled water is circulated, which dissolves the ClO_2, but only 25% of the Cl_2. The remaining Cl_2 is scrubbed with alkali. This raises the molar ratio to $8ClO_2 : 1Cl_2$.

An important consideration in the storage and handling of any bleachant, but especially dry bleaches, is that they are reactive chemicals of substantial oxidizing power. Solid $NaClO_3$, $NaClO_2$, or $NaBrO_3$ in contact with organic materials can be easily detonated. Heat and/or moisture lead to the exothermic decomposition of hypochlorites. Contact with reducing agents or combustibles can be disastrous. It is to be noted that not only are hypochlorites oxidants, but they yield oxygen on decomposition:

$$2OCl^- \longrightarrow O_2 + 2Cl^-$$

Because of its ease of manufacture, cost, handling convenience, and wide acceptance for a variety of applications, sodium hypochlorite bleach enjoys widespread uses, most well known of which is as ordinary household bleach. It is produced mostly by reaction of liquid chlorine with 20% NaOH. Soda bleaches vary in their NaOCl content up to about 15%. The higher the NaOCl content, the higher the required excess of caustic to maintain stability. For this reason, household bleach contains 5.5% NaOCl. In the trade it is common to speak of the available chlorine level of soda bleach as the "trade percent." This is related to the NaOCl concentration, expressed in grams per liter, as:

$$(\text{Trade \% available } Cl_2) = (\text{GPL NaOCl}) \times 0.0953$$

$$(\text{Weight \% available } Cl_2) = (\text{Trade \%})/(\text{Specific gravity})$$

Household bleach has a wt.% available Cl_2 of 5.2, and a trade % of 5.6, which is the percent of available chlorine per volume of solution.

REFERENCES

1. Kauffman, D. W., *Sodium Chloride*, New York, Van Nostrand Reinhold, 1960.
2. Leddy, J. J., *J. Chem. Ed.* 47, p. 386 (1970).
3. *Chemical Economics Handbook*, Stanford Research Institute, Menlo Park, Calif. (1979).
4. *Chemical Week* (McGraw-Hill) June 4, 1980.
5. *Chemical Marketing Reporter*, Schnell Publishing Co., New York, Dec. 1980.
6. Sconce, J., "Chlorine," ACS Monograph No. 154, 180 (1962).
7. Van Wazer, J. R., *Phosphorous and Its. Compounds*, Interscience, New York (1958).
8. U.S. Patent 1,385,595.
9. U.S. Patent 1,132,640.
10. German Patent 249,222.
11. *International Electrochemical Progress*, International Electrochemical Institute, p. 3 (February, 1980).
12. Leddy, J. J., *J. Chem. Ed.* 57 640 (1980).
13. Leddy, J. J., et al., *Kirk-Othmer Encyclopedia of Chemical Technology*, John Wiley (1978).
14. Mantell, C. L., *Electrochemical Engineering*, 4th ed., McGraw-Hill, Inc. (1960).
15. Kuhn, A., *Industrial Electrochemical Processes*, Elsevier Publishing Co. (1971).
16. U.S. Patent 3,718,627.
17. Grot, E. G., *Chem.-Ing. Tech.* **44**, p. 167 (1972).
18. Seko, M., New York A.C.S. Meeting, April (1976).
19. Beer, H., *Chem. Tech.* 9, p. 150 (1979).
20. U.S. Patent 3,632,498.
21. German Patent 71937.
22. Latimer, W. M., *Oxidation Potentials* Prentice-Hall (1952).
23. Dotson, R. L., *Chem. Eng.*, p. 106, July 17 (1978).
24. *Chemical Engineering*, p. 84, July 27 (1970).
25. Hine, F., et al., *Int'l. Chem. Eng.* 17, No. 1, p. 1 (1977).
26. Kapoor, R. and J. J. Martin, "Thermodynamic Properties of Chlorine," Eng. Res. Inst., Univ. of Mich. (1957).
27. *Chem. Met. Eng.* **51**, p. 119 (1944).
28. U.S. Patent 3,983,215.
29. Considine, D. M., *Chem. Proc. Tech. Encyclopedia*, McGraw-Hill, p. 233 (1974).
30. White, G. C., *Handbook of Chlorination*, Van Nostrand Reinhold, (1972).
31. Van Dýk, C. P., *Chem. Eng. Progress* 69, 57 (1973).
32. U.S. Patent 3,635,804.
33. Grosselfinger, F. B., *Chem. Eng.*, Sept. 14 (1964).
34. U.S. Patent 3,242,065.
35. U.S. Patent 3,210,158.
36. U.S. Patent 3,989,807.
37. Sneed, M. C., *Comprehensive Inorganic Chem.*, Van Nostrand (1954).

7

Phosphorus and Phosphates

J. H. McLellan*

PHOSPHORUS

Phosphorus is the twelfth most abundant element in the earth's crust and is an essential constituent of all living tissue. While a large number of minerals are known that contain phosphorus, the only ones of significant economic value are the several varieties of apatite. In the form of the mineral fluorapatite, $Ca_5(PO_4)_3F$, phosphorus is found in many granitic and metamorphic rocks. The principal mineral constituent of animal bone and teeth is hydroxyapatite, $Ca_5(PO_4)_3OH$. Phosphorus forms one of the important biochemical cycles in the sea and passes up the food chain from smaller to larger animals. The cycle is closed by bacterial degradation of bone, teeth, and other tissue.

In fertilizer technology, phosphorus content (P) is expressed in terms of the anhydride, P_2O_5. For phosphate ores, the phosphorus content often is expressed as equivalent tricalcium phosphate, $Ca_3(PO_4)_2$, also known as bone phosphate of lime (BPL).

$$P = 0.4364 \times P_2O_5$$

$$BPL = 2.1852 \times P_2O_5$$

PHOSPHATE ORES

The phosphate of commerce is usually spoken of as phosphate rock, even though it is often in an unconsolidated form such as sand, fine gravel, or clay-like lumps. Phosphate deposits are quite common, but are unevenly distributed. They differ widely in age, being found throughout the geological column.

Crystalline apatite is widespread in igneous and metamorphic rocks but is seldom found in commercial amounts. An exception is the Khibiny complex in the Kola Peninsula of Russia which supplies much of that country's needs. There, the apatite is associated with nepheline-syenite rocks, and the ore is beneficiated to produce an exceptionally high grade concentrate. Apatite is also found in commercial amounts associated with ring structures of alkaline rocks known as carbonatites in several parts of the world. Mines have operated for a number of years at Jacupiranga and other localities in Brazil. A similar deposit is operated

*Austin, Texas

on a large scale at Phalaborwa in the Republic of South Africa. Extensive deposits of apatite associated with carbonatites near Kapuskasing and elsewhere in Ontario are known, but have not been developed. However, in the overall picture crystalline apatite is a relatively unimportant raw material for the phosphate industries.

Most phosphates of commercial importance are derived from ores which formed as marine sediments and which are commonly referred to as phosphorites. The particles of phosphate in phosphorites usually occur intermixed with limestone, quartz sand, and clay minerals and the mined material must therefore be processed in a beneficiation plant. Individual grains, pellets, concretions, or masses of phosphate mineral are themselves impure, with a composition corresponding only approximately to fluorapatite. The mineral is often referred to as carbonate-apatite, with an apparent substitution of part of the $(PO_4)^{\equiv}$ by $(CO_3)^{=}$ and $(F)^{-}$. Fluorapatite has a CaO/P_2O_5 weight ratio of 1.31, whereas most phosphate concentrates show a higher ratio, ranging from 1.40 to 1.70. A high ratio is undesirable, since a greater amount of sulfuric acid is required to produce a given quantity of phosphoric acid.

The world's reserves of phosphate are extensive. Over 70% are located in northern and northwestern Africa, mostly in Morocco. In recent years the phosphate rock of international commerce has come principally from the United States, Morocco, and Russia. Current world consumption is about 130 million metric tons. Large deposits are known in other areas of Africa, and in South America and Australia. Pacific Ocean islands have been major producers in the past, but these reserves are being exhausted.

United States Deposits

Marine phosphate deposits are found widely in the United States, but only five districts have been important to commerce: South Carolina, Tennessee, Florida, areas in the western United States, and North Carolina. The Florida and Carolina deposits are found in a long narrow belt of phosphatic sediments of Tertiary age that stretches from central Florida to Virginia. Phosphate mining started in the United States in the vicinity of Charleston, South Carolina, in 1867, and it was in this area that the phosphate fertilizer industry got its start. The ore consisted of thin layers of pebbles, sand, and fossil bone, and was mined with hand shovels and wheelbarrows. Production in the district declined after the turn of the century as the Florida deposits became more important.

Tennessee Deposits. Phosphate deposits were discovered in central Tennessee in the 1890's, and production has been continuous since that time. Geologically different deposits known as blue, white, and brown rock were exploited, but the brown rock deposits have been of chief importance. Tennessee brown rock is an impure phosphatic clay which formed by the weathering of Ordovician age limestones. The ore is washed and dried and is siliceous in nature. It is used to prepare electric furnace burden for the manufacture of elemental phosphorus. Most of the operating mines are in Maury and Williamson Counties. However, as shown in Table 7.1 the production of Tennessee brown rock is declining.

Florida Deposits. Hard rock phosphate was first exploited in Alachua and Levy Counties, in Florida in the 1870's. There has been no significant production of this material for many years, although extensive reserves still remain.

Pebble phosphate deposits were discovered a little later in Polk County, Florida, and these are the principal source of current production. Initially, phosphatic sand and gravel bars were mined along stream channels. The product was collected and washed by screening and was known as river pebble. Later, it was found that layers of phosphatic material (called matrix) were present under relatively thin overburden in the divide areas between the streams. This product is called land pebble.

Land pebble occurs in horizontal layers and lenses which range in thickness from a few to 40 feet. The range of thickness of the overburden is similar. The matrix is a variable mixture of phosphate sand, phosphate pebbles, quartz sand, and clay. The principal ore horizon is called the Bone Valley Formation and is Pliocene in age. It overlies the Hawthorne

TABLE 7.1 U.S. Production of Marketable Phosphate Rock (Thousand Metric Tons; Million $)

	1970		*1975*		*1978*	
	Production	*Value*	*Production*	*Value*	*Production*	*Value*
Florida/N.C.	28,375	$159	36,922	$1,000	42,649	$783
Tennessee	2,869	15	2,078	29	1,671	14
Western U.S.	3,899	29	5,285	93	5,031	96
Total U.S.	35,143	$203	44,285	$1,122	49,351	$893

Source: *Mineral Industry Surveys*, U.S. Dept. of Interior, Bureau of Mines.

Formation of Miocene age which is a limestone containing phosphatic grains and pellets. The Bone Valley Formation is considered to be a residue resulting from extensive submarine weathering and reworking of sandy phases of the Hawthorne Formation.

Until recently all of the phosphate from the central Florida district was mined in Hillsborough and Polk Counties. Now, new mines are being opened to the south in Manatee, Hardee, and DeSoto Counties. Here, the overburden and ore layers are thicker and the ore is of lower grade.

Phosphate also has been mined in northern Florida near White Springs since 1965. These deposits are similar to those of central Florida and are mined and processed in a similar manner.

Western States. The largest phosphate reserves in the United States are found in the Phosphoria Formation of Permian age in Idaho, Montana, Utah, and Wyoming. Mining was started in 1906 and has increased steadily, with most of the production from mines in southeastern Idaho. Whereas the phosphates of Florida can be readily moved to prime agricultural areas and can be shipped overseas through the nearby port facilities at Tampa, the western phosphates are in rugged mountainous country, far removed from markets. Therefore, the chief use of these western ores has been in making elemental phosphorus by the electric furnace process.

The best ore occurs as a phosphatic shale in the Meade Park member of the Phosphoria Formation. Some weathered outcrops yield material that can be used directly. Other ores require treatment by washing, flotation, and calcination before use in making furnace burden or for acidulation.

North Carolina. The latest phosphate district to be developed in the United States is the North Carolina field, in Beaufort County, about 60 miles west of Cape Hatteras. Production started there in 1966. The phosphate occurs as a dark brown sand in the Pungo River Formation, of middle Miocene age. High grade sand layers up to 40 feet in thickness are present, along with thin limestone layers and indurated low grade sand layers. Overburden consisting of quartz sand and silt, fossil shell beds, and clay is about 90 feet thick in the initial mining area and is up to 200 feet thick in other parts of the deposit.

Mining of Phosphates

In 1980, mining in the four phosphate districts in the United States was done at a rate in excess of 45 million metric tons of beneficiated product per year. Several times this amount of ore had to be excavated and moved to the beneficiation plants and additional large amounts of overburden had to be handled. About 85% of the production was from the central and northern Florida area. Production statistics for U.S. phosphate rock are shown in Table 7.1.

In Tennessee, some blue rock has been mined underground, but the brown rock is produced from scattered open cuts, using small draglines and trucks.

In the western states, the productive layers are highly folded competent beds and have been mined by both underground and open pit methods. Bulldozers, scrapers, and draglines are used in the open pit mines.

Originally, phosphate deposits in central Florida were mined from open cuts by hand labor and mule drawn scrapers. Later, mining was done hydraulically, and with steam shovels. Mining on a very large scale was then made possible by the development of huge walking electric draglines, with buckets of 30-40 cubic yard capacity. Even larger machines are now being introduced.

Mining in North Carolina is based upon the open pit techniques developed and used in Florida, but includes several modifications required by the greater overburden and greater thickness of ore, and by the coastal setting of the newer district. Here, the second growth pine timber is cut and the land is cleared. Then a suction dredge with a rotary cutter is moved in by having the dredge cut its own channel. The dredge excavates a block of about 100 acres to a depth of 40 feet, pumping the material up to two miles away to reclaim and level a previously mined area. After the dredge moves out, the water is pumped from the dredge basin, and it is allowed to dry for several months.

The second stage of mining in the North Carolina field is completed by using large electric draglines, with buckets ranging in size from 50 to 72 cubic yards, suspended from 300 ft long booms (see Figure 7.1). Working from the dry basin floor, the dragline makes a long cut about 150 ft wide and several thousand ft long and casts the remaining 50 ft of overburden to a row in the previously mined cut. The ore layer is now exposed, and the phosphate sands are removed and stacked in piles behind the machine. Here a smaller dragline feeds the ore into a shallow sump where it is jetted with high pressure water nozzles to form a slurry. The slurry, containing about 35% solids by weight, is pumped to the beneficiation plant through 18-inch diameter pipelines laid on the ground. Large centrifugal pumps are used and several booster pumps may be required to move the ore two or three miles. All of the equipment is mounted on skids and is frequently moved.

The North Carolina phosphate beds overlay an important aquifer, the Castle Hayne limestone, which furnishes water to communities and farms over a large area. The water level in the aquifer is depressed in the vicinity of the open pit by pumping from a battery of deep wells. The water is collected and used in beneficiating the ore and in associated fertilizer plant operations.

Fig. 7.1. Large electric dragline mining phosphate in North Carolina. (*Courtesy Texasgulf, Inc.*)

The use of dredges for prestripping is very helpful in reclaiming mined out areas. The dredge spoil can be moved long distances, to wherever needed, to fill in low areas and to provide level surfaces for agricultural use or reforesting.

Mining operations in Florida have been simpler because the ground is more elevated and not swampy, and the ore is not as deep. Draglines alone have been used for mining, with a hydraulic system for transporting ore. As the mining area moves southward in central Florida and as land reclamation requirements become more exacting, the use of dredges for prestripping may prove beneficial.

Beneficiation of Phosphates

In Tennessee and the western states, suitable grades of phosphate are made by selective mining, washing, screening, and drying. Much of this rock is used for the electric furnace process and lower P_2O_5 grades are acceptable compared to the grades required for acidulation.

In Florida and North Carolina, careful sizing and flotation are the principal beneficiation steps. In some cases, clay is a problem and the ore must be intensely agitated or scrubbed to disperse the clay masses and liberate the phosphate sands and gravels.

Individual particles of Florida phosphate, because of their geological history, vary widely in grade, size, porosity, and impurities. North Carolina phosphate, on the other hand, is very uniform in physical and chemical properties. For years, the Florida plants produced phosphate rock varying in grade from about 66 BPL to 78 BPL (30.2-35.7% P_2O_5). Most of the high grade material has been mined, and some rock is now being produced at grades lower than 66 BPL. Double-floated North Carolina phosphate typically is very uniform at 66.5% BPL, increased to about 72 BPL upon calcination.

In Florida, pebble rock is produced by washing and screening to obtain a size range of about $-\frac{3}{4}''$ X +14 mesh. The -14 mesh X +35 mesh particles are often separated in hydroclassifiers and sent to a separate coarse-rougher flotation circuit or to an agglomeration circuit. In the latter case, the phosphate particles are treated with a reagent which causes the particles to form agglomerates. The agglomerates are then separated from quartz sand particles in spiral classifiers, vibrating tables, or flooded belts.

The -35 mesh X +150 mesh size material is concentrated by two stages of froth flotation. In the first stage, the feed is treated with a fatty acid reagent which causes the phosphate particles to float and the silica sand to drop to the bottom of the cell. The froth product is acidized to destroy the fatty acid film and is then sent to a second flotation circuit, with an amine reagent. In this stage, the residual silica sand is the froth product, and this is discarded. The sink material is the final product, and it is dewatered and moved on belt conveyors to wet storage piles. A typical flowsheet for a Florida concentrator with pebble rock and froth flotation products is shown in Figure 7.2.

In North Carolina, the concentrator flow sheet is quite simple. Only 3-4% of the ore is coarser than 14 mesh and this is separated by screens. This coarse ore is low in grade and contains limestone particles, so it is discarded. The -14 X +150 mesh fraction is concentrated in one or two stages of froth flotation. An agglomeration circuit is not used. A single stage of fatty acid flotation is used to produce concentrate for local consumption in making wet process phosphoric acid. The two-stage flotation process is used to make concentrate for shipment away from the area.

In Florida, the phosphate is dried in a fluid bed or in rotary units to about 1% moisture, before shipping. Some of the phosphate used locally to make phosphoric acid is not dried. It is ground in wet-circuit ball mills and fed to the acid plants as a slurry in order to avoid the dust problems of dry grinding.

North Carolina phosphate contains organic matter to the extent of about 1%, expressed as carbon. To destroy the organic matter and to remove CO_2, most of the production is calcined in multistage fluid-bed units at a temperature of 1500°F. The organic matter supplies part of the fuel requirement, and the rest is supplied as heavy fuel oil, injected in the high temperature compartment of the calciners. Typical

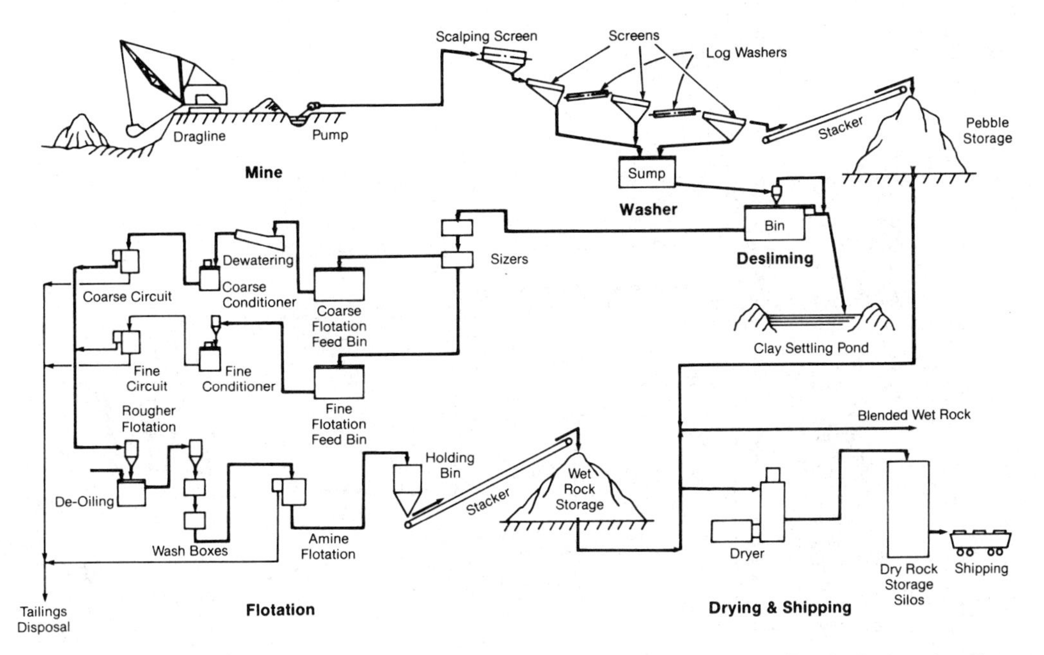

Fig. 7.2. Typical flow diagram, Florida phosphate rock concentration. (*Courtesy Jacobs Engineering Group, Inc.*)

analyses for double-floated North Carolina concentrate are given in Table 7.2.

ELEMENTAL PHOSPHORUS

Phosphate rock is converted into usable chemicals by two methods. In one, the rock is charged to an electric furnace with silica and coke to produce elemental phosphorus. The phosphorus is then converted into phosphoric acid and other compounds. In the other, the phosphate rock is pulverized and reacted with sulfuric acid to form dilute, impure phosphoric acid. The acid is concentrated in evaporators and used to make fertilizers. This is known as the wet process method.

TABLE 7.2 Typical Analyses, North Carolina Phosphate Concentrate

	Weight Percentage	
Component	*Dry Concentrate*	*Calcined Concentrate*
Moisture, as H_2O	1.3	0.1
BPL	66.5	71.9
P_2O_5	30.4	32.9
CaO	49.0	52.8
Acid Insoluble	2.0	1.6
Fe_2O_3	0.7	0.8
Al_2O_3	0.4	0.5
F	3.7	4.0
MgO	0.5	0.6
C, total	3.0	0.7
CO_3, as C	1.4	0.6
Na_2O	1.0	1.0
S, total	1.0	1.2
CaO/P_2O_5 ratio	1.6	1.6

(Courtesy Texasgulf, Inc.)

The furnace or thermal process is shown in Figure 7.3. The approximate reaction is:

$$2Ca_3(PO_4)_2 + 6SiO_2 + 10C \longrightarrow 6CaSiO_3 + P_4 + 10CO$$

The charge or furnace burden must have certain physical and chemical properties. The particles should be dry durable masses in the $\frac{1}{4}$ to $\frac{3}{4}$ inch size range that won't break up or segregate in the furnace. Phosphatic shale is siliceous and is a suitable material, but usually it must be sintered or briquetted before use. Quartz gravel is a good source of supplementary silica and metallurgical coke is the form of carbon used.

The energy required is about 12,000 KWH per ton of phosphorus produced. The most modern furnaces are quite large, with ratings in the range of 60,000 KVA.

The furnace is a water-cooled circular steel

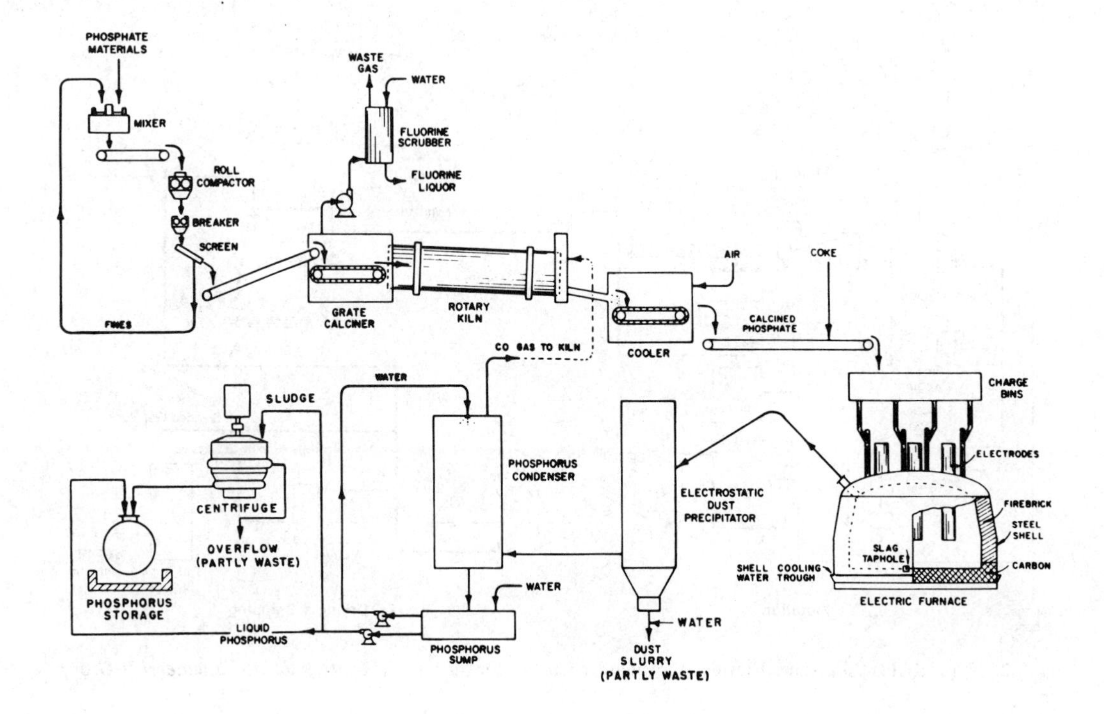

Fig. 7.3. Electric furnace process for production of elemental phosphorus. (*Courtesy Tennessee Valley Authority.*)

shell lined with refractories. The bottom or hearth is made of thick blocks of carbon. Three large carbon electrodes are mounted through the roof of the furnace and these are slowly lowered into the charge as the bottom ends are consumed. A three-phase electrical system is used, and the electrodes must be precisely positioned to balance the phases and maintain a high power factor.

The phosphorus leaves the furnace as a vapor. After passing through an electrostatic precipitator, it is condensed by direct contact with water. The liquid phosphorus is then filtered or centrifuged to remove dust particles and is shipped in tankcars. CO discharged from the furnace is used as fuel in the feed preparation operations or is burned in a flare.

A calcium silicate slag is tapped from the bottom of the furnace at intervals, and at some locations this material has been sold as an aggregate. Any iron present in the furnace burden combines with phosphorus to form ferrophosphorus and this is tapped off separately.

In 1980, there were about eight plants in operation, some with multiple furnaces. No new electric furnaces have been built in the United States since 1968; moreover, several old units have been closed down during this period. New emission standards and the rapidly rising cost of electrical power make it unlikely that new furnaces will be built in the near future.

Furnace Phosphoric Acid

Phosphoric acid of high purity is made by burning phosphorus with air and hydrating the resulting P_2O_5 with water, according to the reaction:

$$P_4 + 5O_2 + 6H_2O \longrightarrow 4H_3PO_4$$

Equipment for this operation has been constructed both from carbon blocks and stainless steel. The water quantity is controlled to give product acid corresponding to 75% or 85% H_3PO_4. If food grade acid is specified, some additional chemical treatment of the acid may be required. For example, slight traces of arsenic are removed by treatment with H_2S.

If even less water is used for hydration, a product known as polyphosphoric acid or superphosphoric acid results. Ordinary phosphoric

TABLE 7.3 U.S. Phosphoric Acid Production[1,2] (Thousand Metric Tons P_2O_5)

	1970	*1975*	*1978*
Wet Process:			
Ordinary	N.A.	5,745	7,628
Super	N.A.	505	421
Total	4,211	6,250	8,049
Furnace Acid Total	944	709	628

Sources: [1] "Inorganic Chemicals M28A", U.S. Dept. of Commerce, Bureau of Census
[2] "Inorganic Fertilizer Materials and Related Products M28B", Current Industrial Reports, U.S. Dept. of Commerce, Bureau of Census.

acid is a solution of the monomer, H_3PO_4, in water, and is called orthophosphoric acid. If a molecule of water is removed between two orthophosphoric molecules the dimer, pyrophosphoric acid, $H_4P_2O_7$, is formed. Similarly, the trimer and higher polymers can be made. Superphosphoric acid is an equilibrium mixture of orthophosphoric acid and its polymers. Thermal superphosphoric acid corresponding to about 105% H_3PO_4 was developed by TVA and has some very interesting properties for making stable solution fertilizers. However, thermal phosphoric acid has become too expensive for widespread use in the fertilizer industry. Furnace phosphoric acid production in the United States is shown in Table 7.3.

Industrial Phosphates

About one-fourth of the 410,000 metric tons of phosphorus currently produced in the United States is consumed to make compounds such as phosphorus pentoxide, phosphoric trichloride, and phosphorus pentasulfide which find use for the preparation of drying agents, plasticizers, oil additives, fire retardants, and insecticides. Most of the phosphorus, however, is converted into orthophosphoric acid, some of which is used in soft drinks, candy, baked goods, and various other food products. Furnace grade acid finds wide use in metal treating methods. About 75% of all furnace acid is converted into sodium and potassium phosphates.

The cheapest and most important salts of phosphoric acid are the sodium salts, made by reacting the acid with sodium carbonate or sodium hydroxide. Sodium phosphates may be classified in a general way as (1) orthophosphates, (2) crystalline condensed phosphates, and (3) glassy condensed phosphates.

Three sodium orthophosphates can be prepared, depending on whether one, two, or three hydrogen atoms are replaced by sodium. Monosodium phosphate is formed in the reaction

$$2H_3PO_4 + Na_2CO_3 \longrightarrow 2NaH_2PO_4 + H_2O + CO_2$$

Sodium carbonate is also used to make disodium phosphate, Na_2HPO_4, but sodium hydroxide must be used to replace the third hydrogen in trisodium phosphate, Na_3PO_4. The orthophosphates have a wide range of uses in industry. Sodium phosphates are made abroad in large quantities from wet process phosphoric acid by stepwise neutralization. Impurities must be purged from the solutions and the salts purified by recrystallization.

Condensed phosphates are made by eliminating water from orthophosphates. The most important compound is sodium tripolyphosphate, made according to the following reaction:

$$2Na_2HPO_4 + NaH_2PO_4 \xrightarrow{-H_2O} Na_5P_3O_{10}$$

The most important use of sodium tripolyphosphate is as a builder of detergents. However, legislative restrictions on the use of phosphorus compounds in household detergents have caused a large decline in the amount of this salt made in the last few years.

Glassy condensed phosphates are represented by sodium hexametaphosphate, in which the O:P ratio is 3:1. There can be considerable variation in the Na_2O to P_2O_5 ratio.

The principal use of the condensed phosphates is to sequester metallic ions in water. They form water-soluble complexes with the metals and prevent metallic compounds from precipitating to cause discoloration, scale, and sludges.

Wet Process Phosphoric Acid

Most phosphate rock is converted into useful products by the acidulation or wet process. In this process, the phosphate rock is pulverized to a fine powder and reacted with a mineral acid, usually sulfuric acid. The reaction mixture is a slurry of dilute, impure phosphoric acid and calcium sulfate crystals. The solids are separated and washed on a vacuum filter and the acid, containing about 30% P_2O_5, is usually concentrated to about 54% P_2O_5 (75% H_3PO_4). Fluorine in the phosphate rock, as well as impurities such as compounds of iron, aluminum, and magnesium also react with the acid and are carried over into the crude phosphoric acid. If phosphate rock is considered, for the sake of simplicity, to be tricalcium phosphate, the overall reaction is as follows:

$$Ca_3(PO_4)_2 + 3H_2SO_4 \longrightarrow 3CaSO_4 + 2H_3PO_4$$

When there is a high sulfate ion concentration in the liquid phase, the phosphate rock particles become coated with calcium sulfate layers and the reaction stops. To overcome this difficulty, the phosphate rock is added to a large volume of phosphoric acid. The rock dissolves, forming a dilute solution of monocalcium phosphate in phosphoric acid. Simultaneously, sulfuric acid is added to maintain a low concentration of sulfate ions in the phosphoric acid. Under these conditions, calcium sulfate forms by crystallizing on the surfaces of existing calcium sulfate crystals, which grow larger. In the typical reaction system the addition of rock, sulfuric acid, and water is controlled to give phosphoric acid strength of about 30% P_2O_5, sulfate ion concentration in the liquid phase of about 2-3%, and a solids concentration in the slurry of about 35%. The solids consist almost entirely of calcium sulfate crystals.

Calcium sulfate can be present in several degrees of hydration, depending on temperatures and acid concentrations. Most wet process phosphoric acid is made at a strength of 30% P_2O_5 or less and at a reaction temperature of 175°F or less. Under these conditions, the stable form of the crystallized calcium sulfate is the dihydrate, $CaSO_4 \cdot 2H_2O$, or gypsum. At higher temperatures and concentrations, the calcium sulfate crystallizes as the hemihydrate, $CaSO_4 \cdot \frac{1}{2} H_2O$, or plaster of paris. Under even higher temperatures, the stable crystal form is $CaSO_4$, or anhydrite.

Since the solids must be separated from the phosphoric acid, it is important that the calcium sulfate crystals be as large and regular as possible. Gypsum is by far the easiest form of calcium sulfate to separate, and therefore almost all wet process phosphoric acid is made under conditions that yield this form. Relatively weak phosphoric acid is made and a large amount of heat must be removed from the reaction zone to obtain a gypsum slurry that can be filtered readily.

Modern phosphoric acid units are large plants, some of them being able to produce over 1000 tons per day of P_2O_5 in a single train. Figure 7.4 shows such a plant. The wet process phosphoric acid plant typically has four sections: (a) rock grinding; (b) reaction; (c) filtration; and (d) evaporation.

Rock Grinding. Most phosphate rock needs to be pulverized before it is used. Roller ring mills once were popular, but as plants became larger, a change was made to air-swept ball mills. There are dust problems associated with dry grinding, and the trend today is toward the wet grinding of phosphate rock in ball mill circuits. The fineness of grind depends on the reactivity of the rock, but usually is 60% minus 200 mesh or finer. Some flotation concentrates with high reactivity, such as North Carolina phosphate, are used without being ground.

Reaction. The reaction or attack section consists of one or more agitated compartments or tanks having a total volume which can provide an average retention time for the slurry of from 4-8 hours. The tanks are covered and fumes are drawn off through a scrubbing system. Pulverized rock, sulfuric acid, and water are fed into different points in the circuit. Water is brought in with dilute wash acids from the filtration section, and in some cases water is introduced by using it to dilute the sulfuric acid.

The reactions generate a large amount of heat which has to be continuously removed in order to keep the slurry at 175°F or lower and to have the calcium sulfate crystallize in the gypsum form. Generally this is done by pumping large volumes of slurry to an elevated flash cooler. The cooled slurry returns to the first tank or compartment and recirculates through the tanks. High slurry recirculation rates are

Fig. 7.4. A large U.S. wet process phosphoric acid plant. (*Courtesy Texasgulf, Inc.*)

characteristic of many phosphoric acid plant designs. Alternatively the slurry can be cooled by blowing large volumes of air under the surface of the slurry. Some plants are abandoning this method because of the problems in scrubbing the air to meet emission standards.

The rock feed to the circuit is fixed, and samples of slurry are analyzed at regular intervals to determine the amount of sulfate in solution, the P_2O_5 concentration in the liquid, and the percentage of solids in the slurry. The most important control is to adjust the sulfuric acid addition rate in order to keep the sulfate concentration within narrow limits. A flow sheet for an attack section using multiple compartments is illustrated in Figure 7.5.

Filtration. The development of very large filters has made it possible to process over 1000 tons per day of P_2O_5 in a single line of equipment. The most popular filter in use today is the Bird-Prayon tilting pan filter. Another large filter being used for phosphoric acid is the Ucego horizontal filter. One of the first filters used for phosphoric acid was the belt filter, but the early designs were small units. There are now three or four types of very large belt filters which are being installed in new phosphoric acid plants. All of these filters operate under vacuum, which is maintained by water-sealed vacuum pumps. The slurry is fed to the filter and the strong filtrate is the product acid, usually 28-30% P_2O_5. The filter cake then passes through several zones where it is washed in stages by counter-flowing wash water. These wash waters containing the P_2O_5 removed from the filter cake are pumped to the reaction system to furnish the water needed by the process. The gypsum cake usually is discharged from the filter into a repulping tank from which it is pumped to a disposal area for impoundment. It may also be carried away on a belt conveyor. Figure 7.6 is a filtration flow diagram.

Evaporation. Very little phosphoric acid is used at filter strength. Most of it is concentrated to about 54% P_2O_5 for shipment to other fertilizer plants. Often 54% P_2O_5 acid is blended with 30% P_2O_5 filter acid to give a strength of 40% P_2O_5, used to make solid fertilizers such as diammonium phosphate or triple superphosphate. Forced circulation evaporators are used, heated by low pressure steam. The evaporator loop consists of an axial flow

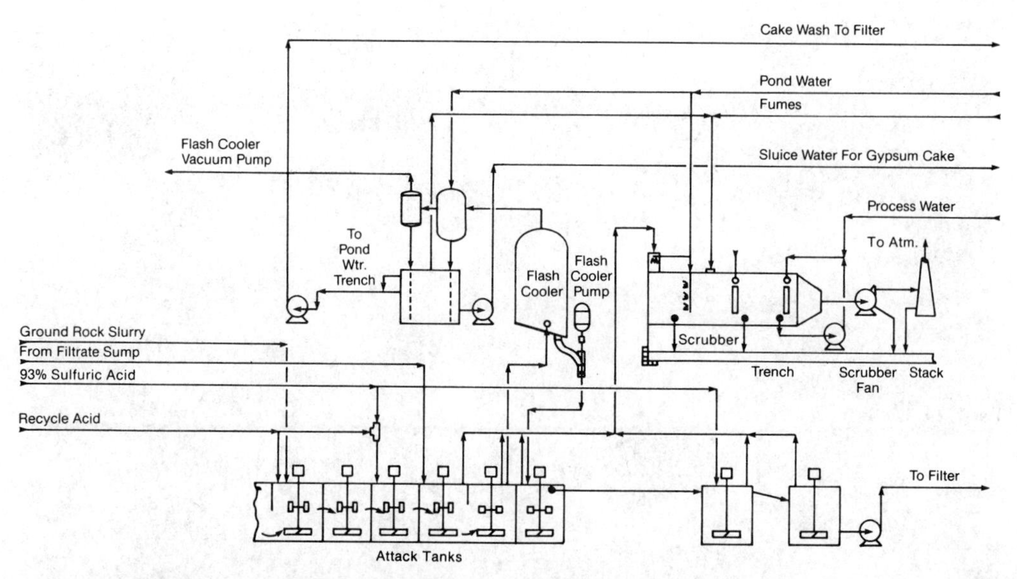

Fig. 7.5. Flow diagram of attack or reaction section of a wet process phosphoric acid plant. (*Prayon process courtesy Davy McKee Corp.*)

circulating pump, a vertical heater containing tubes made from carbon, and a rubber-lined flash chamber. Vacuum is maintained by a barometric condenser and steam ejectors. Single, double, and triple stage evaporators are in service. Figure 7.7 shows the equipment arrangement for phosphoric acid evaporation. Typical concentrated wet process phosphoric acid composition is given in Table 7.4.

Much work has been done abroad to develop acidulation processes that use nitric acid or hydrochloric acid. Calcium nitrate and calcium chloride are both soluble in dilute phosphoric acid, and therefore difficult to separate. Numer-

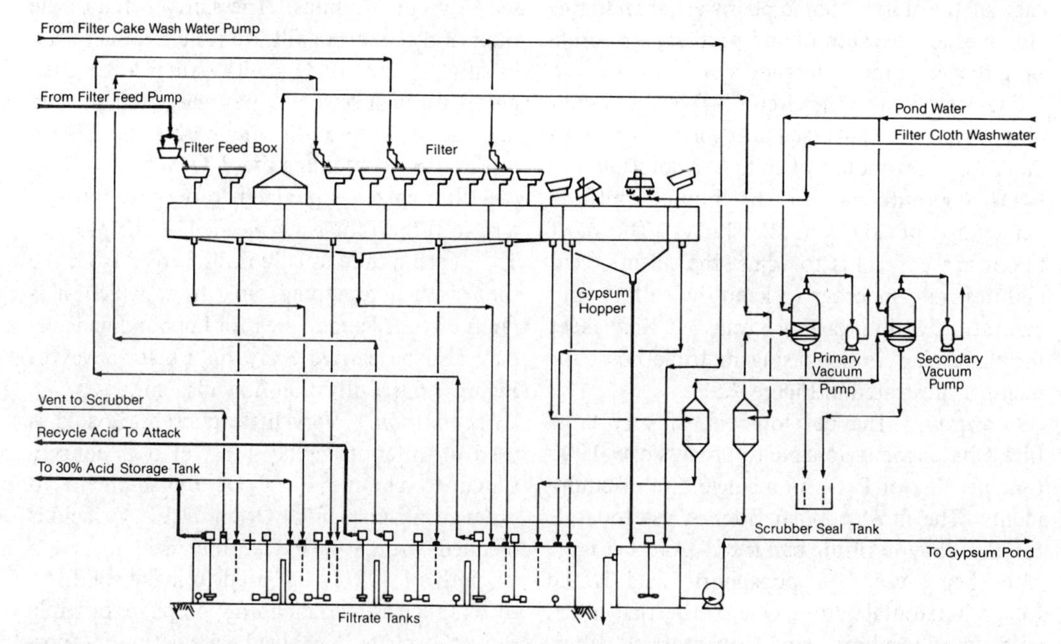

Fig. 7.6. Flow diagram of filtration section of a wet process phosphoric acid plant. (*Prayon process courtesy Davy McKee Corp.*)

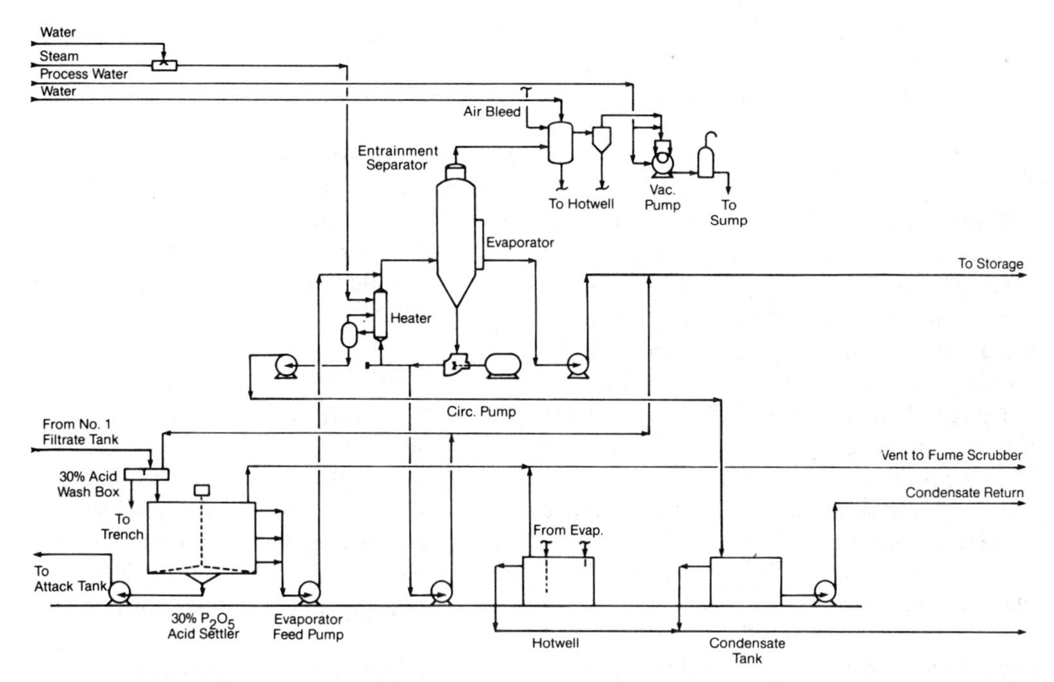

Fig. 7.7. Flow diagram of evaporation section of a wet process phosphoric acid plant. *(Courtesy Davy McKee Corp.)*

ous nitrophosphate plants have been built in Europe but in recent years many have been shut down in favor of sulfuric acid acidulation. Most nitrophosphate fertilizers have low water solubility and are not suited to American farming practices.

Superphosphoric Acid. Solution fertilizers have become very popular in the United States.

TABLE 7.4 Typical Analyses, Wet Process Phosphoric Acid Made From North Carolina Calcined Concentrate

	Weight Percentage	
Component	*Concentrated Acid*	*Superphosphoric Acid*
P_2O_5, total	53.0	*69.5
Solids	0.2	
Free water	22.0	
Fe_2O_3	1.5	2.0
Al_2O_3	0.7	1.0
F	0.6	0.3
MgO	1.1	1.3
SO_4	2.7	3.7
CaO	0.1	0.2
Sp.g, at 75°F	1.68	2.0

*About 36% of the total P_2O_5 is present as polyphosphates

(Courtesy Texasgulf, Inc.)

The principal source of P_2O_5 for these fertilizers is wet process superphosphoric acid containing about 70% P_2O_5. 35% or more of the P_2O_5 is present in the polyphosphoric form. When this acid is ammoniated and diluted, the iron, aluminum, and magnesium compounds naturally present remain in solution, sequestered by the polyphosphates. Clear solutions result and there is no clogging of the sprays used for applying the fertilizer. When ordinary phosphoric acid is neutralized with ammonia, heavy sludges form and the resulting solution is difficult to store and apply.

Superphosphoric acid is made by the additional concentration of clarified phosphoric acid in vacuum evaporators of the falling-film or forced-circulation type. High pressure steam or Dowtherm vapor is used for heating. Corrosion is a problem, so the equipment is made from high alloy stainless steels. The acid is shipped in special insulated tank cars to the solution fertilizer plants which are located close to the farm areas they serve. Organic matter contributes to sludge problems in making solution fertilizers; therefore, calcination of the phosphate rock used for making the acid is advantageous.

Additional details of the manufacture of superphosphoric acid, and of the triple superphosphate, are to be found in Chapter 8.

Wet Process Acid By-Products

Phosphogypsum. Between 5 and 6 tons of gypsum on a dry basis are made for each ton of P_2O_5 produced in a wet processs phosphoric plant. This material usually is disposed of as waste, impounding it in old mine pits, or stacking it in huge piles, or in some cases discharging it into very large rivers. Most of the gypsum produced in California is sold to farmers for control of salt build-up in irrigated soils; a small quantity is sold to peanut farmers in North Carolina and Virginia.

Much attention has been given to using phosphogypsum in place of natural gypsum to make plaster, wallboard, and Portland cement. Phosphogypsum is quite impure, containing sand, fluosilicates, and P_2O_5. The cost of removing impurities and drying the phosphogypsum is too great for it to be competitive with mined gypsum. The hemihydrate-dihydrate phosphoric acid process used widely in Japan yields a good quality gypsum but the operation is too expensive to be an economical source of gypsum under U.S. conditions.

Another research goal has been to recover SO_2 from phosphogypsum for recycling to the sulfuric acid plants. Several small plants that can do this have been built abroad. The phosphogypsum is washed, dried, mixed with clay, and calcined at a high temperature to yield SO_2 and cement clinker. About one ton of cement is obtained per ton of sulfuric acid. Energy costs are very high and prospects for economical SO_2 recovery are very poor.

Fluorine Recovery. Phosphate rock contains 3-4% fluorine, and some of this is recovered as a by-product in manufacturing wet process phosphoric acid. During acidulation, the fluorine is released as hydrofluoric acid, HF, and this reacts with the silica present as an impurity in the rock to form fluosilicic acid, H_2SiF_6. Some of the fluorine is lost with the gypsum as sodium or potassium fluosilicates and some remains dissolved in the filter acid. When the acid is concentrated much of the fluorine in the feed is boiled off, appearing as HF and silicon tetrafluoride, SiF_4, in the vapors.

Fluorine is recovered at the evaporator station by scrubbing the vapors leaving the flash chamber. The vapors pass through an entrainment separator to remove fine droplets of phosphoric acid and then into a spray tower where the vapors are scrubbed with a weak solution of fluosilicic acid according to the reaction

$$2HF + SiF_4 \xrightarrow{H_2SiF_6} H_2SiF_6$$

Part of the circulating solution is continuously withdrawn as a 20-25% aqueous solution of H_2SiF_6. The solution is shipped in rubber-lined tank cars and is used for fluoridation of drinking water and the preparation of fluosilicates. These salts find use in ceramics, pesticides, wood preservatives, and concrete hardeners.

Uranium Recovery. Uranium occurs in many phosphate rocks in the range of 0.005-0.03% of U_3O_8. This is too low in grade to economically process the rock for uranium alone. However, in making wet process acid, most of the uranium is put into solution and can be extracted from the filter strength acid. A number of uranium extraction units have been placed into service in U.S. phosphate fertilizer plants in the last few years.

Typically, wet process phosphoric acid made from Florida rock contains about 1.0 lb of U_3O_8 per ton of P_2O_5. It is claimed that as much as 90% of this can be extracted. The uranium is present in solution in the acid in an oxidized state, U(VI). Research done at the Oak Ridge National Laboratory resulted in a reductive stripping process based on using a mixture of di (2-ethylhexyl) phosphoric acid and trioctyl phosphine oxide (DEPA-TOPO) dissolved in kerosene as the extractant.

Acid from the filter is first treated to remove fine suspended solids and organic matter and then is contacted with the extractant in mixer settlers. The aqueous phase or raffinate is sent to a plant section to remove residual traces of kerosene and then is returned to the phosphoric acid plant for concentration. If the kerosene is not completely removed the rubber linings in the evaporators may be damaged.

The organic or solvent phase containing the uranium is treated with a reducing agent and the U(IV) is stripped from the solvent by a suitable agent, such as a small amount of concentrated phosphoric acid. Thus, the uranium can be transferred back and forth from phosphoric acid to solvent by controlling its state of oxidation.

Two extraction steps usually are employed and then the uranium is precipitated from the second step solvent as ammonium uranyl carbonate, using ammonium carbonate solution. This material is dried and calcined to U_3O_8. A second process developed by the same laboratory uses octylphenyl phosphoric acid (OPPA) in kerosene as the extractant. This reagent extracts U(IV) from wet process phosphoric acid and is useful when dealing with phosphoric acid in a reduced state, such as that made from calcined phosphate rock. Such acid is easier to process since it is substantially free of organic matter.

Some western U.S. rock contains uranium at about the same concentration as Florida rock. North Carolina phosphate contains only about 40% as much uranium. Since the world reserves of phosphate rock are very large, the amount of uranium in these deposits is also large. However, the extraction plants are expensive to build and operate and can only be justified when uranium prices are high.

Animal Feed Supplement. Calcium phosphates for use in animal and poultry feeds are made from both furnace and wet process phosphoric acids. Dicalcium phosphate, $CaHPO_4$, containing 18.5% P, and monocalcium phosphate, $Ca(H_2PO_4)_2$, containing 21.0% P, are made in large tonnages. Both grades are prepared by reacting phosphoric acid with pulverized limestone in a pug mixer. The limestone must be quite pure and the phosphoric acid must have a low fluorine content. If 54% P_2O_5 wet process phosphoric acid is used, it is defluorinated first by adding diatomaceous earth and then sparging the acid with steam. A preferred method is to use wet process superphosphoric acid, which has a low fluorine content. The superphosphoric acid is hydrolysed by diluting with water and heating. The pug mixer product is a fine granule, -12 mesh, which is dried and shipped in bulk to feed-mixing plants.

Clean Phosphoric Acid

High power costs and environmental problems have discouraged the installation of new phosphorus furnaces in the United States. There is considerable interest in processes for cleaning or purifying wet process phosphoric acid so that it can be substituted for furnace acid in such uses as metal treating and the preparation of sodium phosphates. Phosphoric acid cleaning plants using solvent extraction are operating successfully in Europe and Japan.

Wet process phosphoric acid made from calcined rock is preferred because organic matter in the acid is very troublesome. Acid at a concentration of 35-50% P_2O_5 is treated with various reagents and adsorbents and refiltered. The acid is then fed to a column or a battery of mixer-settlers and extracted with a solvent such as butyl alcohol or tributyl phosphate. Generally about three fourths of the phosphoric acid transfers to the organic phase, leaving the impurities in the raffinate. The raffinate is sent to a fertilizer unit to recover its P_2O_5. The yield of cleaned acid can be increased by adding another mineral acid such as sulfuric acid or hydrochloric acid to the extraction step.

After washing, the phosphoric acid is stripped from the solvent with water and the solvent is returned to the extraction section. The phosphoric acid is now quite dilute and still contains small amounts of impurities. The acid is then concentrated; the impurities are removed by steam stripping and the addition of reagents and adsorbents followed by filtration. The exact details of the process vary depending upon the impurities present in the feed acid and the solvent used.

It is unlikely that clean phosphoric acid can compete successfully with food grade furnace acid in the United States. However, it should be quite competitive for most industrial uses. A cleaning plant is quite expensive, but new electric furnaces are even more costly. It is reasonable to expect that future needs for industrial phosphates will be met in part by cleaned wet-process phosphoric acid.

REFERENCES

1. Hird, John M., "Overburden Stripping–Combination Use of Dredge and Dragline," *Mining Engineering*, Vol. **32**, 1980, pages 311-314.
2. Gurr, T.M., "Geology of U.S. Phosphate Deposits," *Mining Engineering*, Vol. **31**, 1979, pages 682-691.
3. Slack, A.V., Editor, *Phosphoric Acid*, Parts I and II, Marcell Dekker, Inc., New York, 1968, 1159 pages.
4. The British Sulphur Corporation Limited, *World Survey of Phosphate Deposits*, Third Edition, London, 1971.

8

Phosphate Fertilizers; Potassium Salts; NPK and Natural Organic Fertilizers

Ronald D. Young*

PHOSPHATE FERTILIZERS

Early History of Fertilizers

Through the centuries of ancient and medieval times, men have been deeply interested in improving crop yields through the addition of various mineral or organic substances. Up until the last two or three hundred years, however, the approach to the subject was highly empirical; only by accident or by trial and error was it found that applications to the soil of various organic wastes or naturally occurring mineral substances such as manure, ground bones, wood ashes, saltpeter, and gypsum, dramatically improved plant growth. But the results were not predictable; a treatment that benefited one field might have no effect—or even an adverse effect—on another.

As more and more chemical elements were identified, scientists became interested in determining the amount and relative importance of various elements in plants. The German scientist Liebig, who stressed the value of elements derived from the soil in plant nutrition and the necessity of replacing those elements to maintain soil fertility, may be credited with laying the foundation for the modern fertilizer industry. He recognized the value of nitrogen but believed that plants could get their nitrogen from the air. He also envisioned a fertilizer industry with nutrients such as phosphate, lime, magnesia, and potash prepared in chemical factories. In 1840, Liebig published results of his work and recommended treatment of bones with sulfuric acid to make the phosphate more readily available to plants. This was actually the start of chemically processed fertilizers.[1]

Ordinary Superphosphate (OSP)

In 1842, Lawes in England followed up on the pioneering work of Liebig and received a patent on the use of sulfuric acid and organic phosphate material to produce "superphosphate." An industry grew slowly, and in 1862 about 150,000 tons of what later became referred to as ordinary or normal superphosphate (16 to 18% P_2O_5) was produced in England. By 1870, there were 70 ordinary superphosphate

*Division of Chemical Development Tennessee Valley Authority Muscle Shoals, Alabama 35660

plants in the United Kingdom and 7 in the Charleston, South Carolina, area of the United States.[2]

In 1888, the commercial shipment of phosphate rock from Florida was initiated and a major industry in that state followed. Ordinary superphosphate (OSP) was produced in local plants from phosphate rock shipped from the mining area.[3]

OSP is the simplest and oldest of manufactured phosphate fertilizers—being produced since 1842. For this purpose, pulverized phosphate rock is treated with sulfuric acid in a comparatively simple plant to produce a product containing about 20 percent P_2O_5. Products in the earlier years had a lower analysis, ranging from 14 to 18 percent P_2O_5. OSP can be used for direct application, bulk blending, or in production of granular NPK fertilizers. If the product is to be used for direct application or bulk blending, it should be granulated.

The main equipment for manufacturing OSP consists of a mixer, a den, and if granulation is used, some suitable type of equipment for granulation with steam or water. The simple TVA cone mixer is used widely for mixing of phosphate rock and sulfuric acid. The dens can be either of the batch or continuous type. The slat-conveyor continous den, commonly referred to as a Broadfield den, is used in many continuous systems.[4] A diagram of a typical cone mixer and continuous den (slat-conveyor) is shown in Figure 8.1.

There have been various types of continuous mixers, but the lowest cost and simplest is the TVA cone mixer that has no moving parts. (See Figure 8.2.) Mixing is accomplished by the swirling action of the acid. Short, single-shaft or double-shaft pug mills are also used for mixing.

The reaction of phosphate rock with sulfuric acid to produce OSP can be expressed in chemical stoichiometric terms. However, in general practice the proportioning is usually based on a simpler relationship of about 0.6 pound of sulfuric acid (100% H_2SO_4 basis) per pound of phosphate rock (30-32% P_2O_5). The phosphate rock usually is pulverized to about 90 percent −100 mesh and 70 percent −200 mesh.

Gases that are released while the superphosphate is solidifying (setting) cause the mass in the den to become porous and friable so that it can then be cut and handled readily. OSP made from typical rock will "set" in 40 to 50

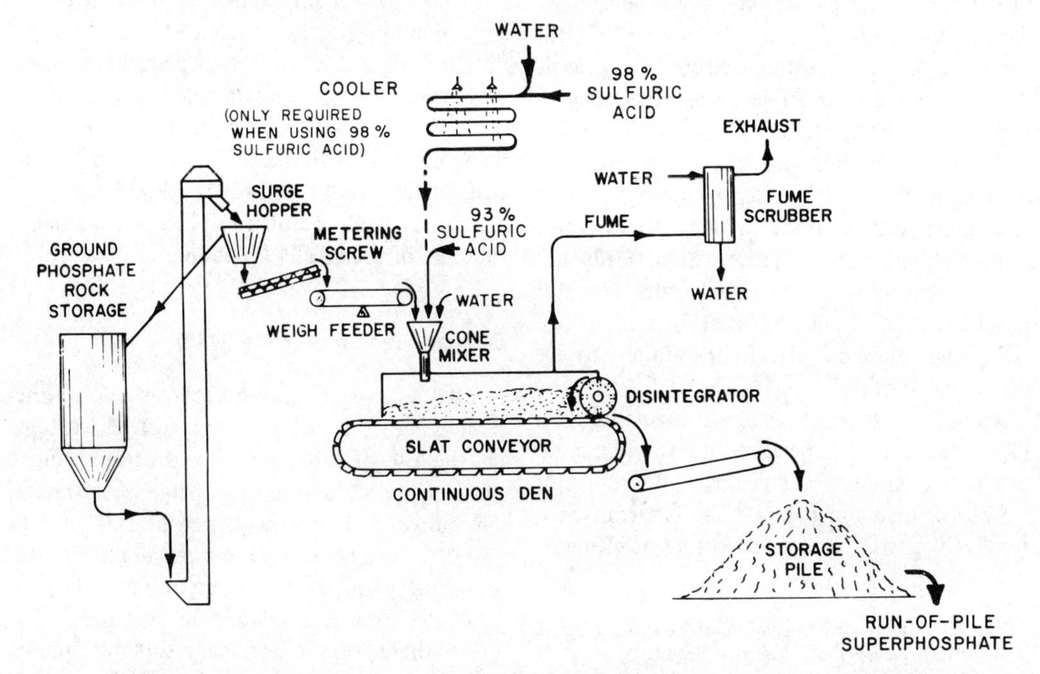

Fig. 8.1. Continuous process for the manufacture of normal superphosphate. (*Courtesy TVA.*)

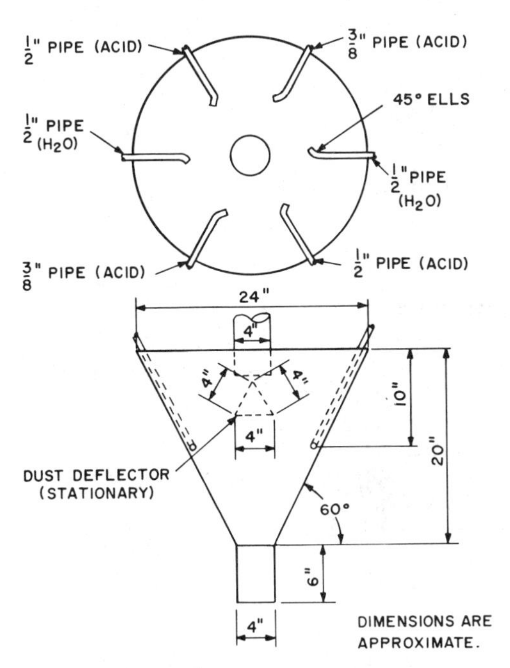

Fig. 8.2. Typical cone mixer for production of 25 to 30 tons per hour of normal superphosphate. (*Courtesy TVA.*)

minutes in a continuous den, whereas the set time in a batch den requires on the order of $1\frac{1}{2}$ to 2 hours.

The superphosphate is usually held in storage piles (cured) for 4 to 6 weeks in order to obtain better handling properties and to allow the chemical reactions to continue. The usual grade of OSP made from Florida rock is 20 percent available P_2O_5; an analysis is shown below.[5]

Analysis, % by Weight

P_2O_5								
Total	*Available*	*H_2O Soluble*	*CaO*	*Free Acid*	*SO_4*	*F*	*R_2O_3*	*MgO*
20.2	19.8	18.0	28.1	3.7	29.7	1.6	1.6	0.15

Because of the low phosphate analysis of OSP (20% P_2O_5), economics favor shipping the phosphate rock (32% P_2O_5) to local plants where the superphosphate is produced, and there usually used in formulations for granular NPK fertilizers. The maximum amount of OSP possible is used in formulations for grades in which it can be used, since it is the lowest cost form of P_2O_5. OSP ordinarily will readily absorb about 6 pounds of ammonia per 20 pounds of P_2O_5 during the ammoniation-granulation process.

OSP is granulated in large quantities as it is discharged from the den. Pan granulators and rotary drums are used. Some OSP is granulated in western Europe by use of steam and water in a rotary drum.

Triple Superphosphate (TSP)

TSP is made by acidulation of phosphate rock with phosphoric acid, using equipment and processes similar to that for OSP. TSP (45–46% P_2O_5) did not appear on the scene in any appreciable quantity until wet-process phosphoric acid (see wet-process phosphoric acid in Chapter 7) was produced commercially. TVA initiated production of similar concentrated superphosphate in the late 1930's by using electric-furnace phosphoric acid. Widespread agronomic testing, and market development through the use of large tonnages by fertilizer manufacturers in TVA demonstration programs, led to the rapid acceptance of this much higher analysis phosphate intermediate. Producers of phosphate rock in the United States and other countries moved into production of wet-process acid and TSP. Logistics favored production of the higher analysis TSP (46% P_2O_5 versus 32% P_2O_5 for phosphate rock) near the source of the phosphate rock, and then shipping this intermediate to mixed-fertilizer plants near the markets.

The TVA cone mixer has been almost universally used in production of nongranular TSP. Since the "set time" for TSP is only 14 to 20 minutes, as compared with 40 to 50 minutes for OSP, a simpler, cupped conveyor belt is

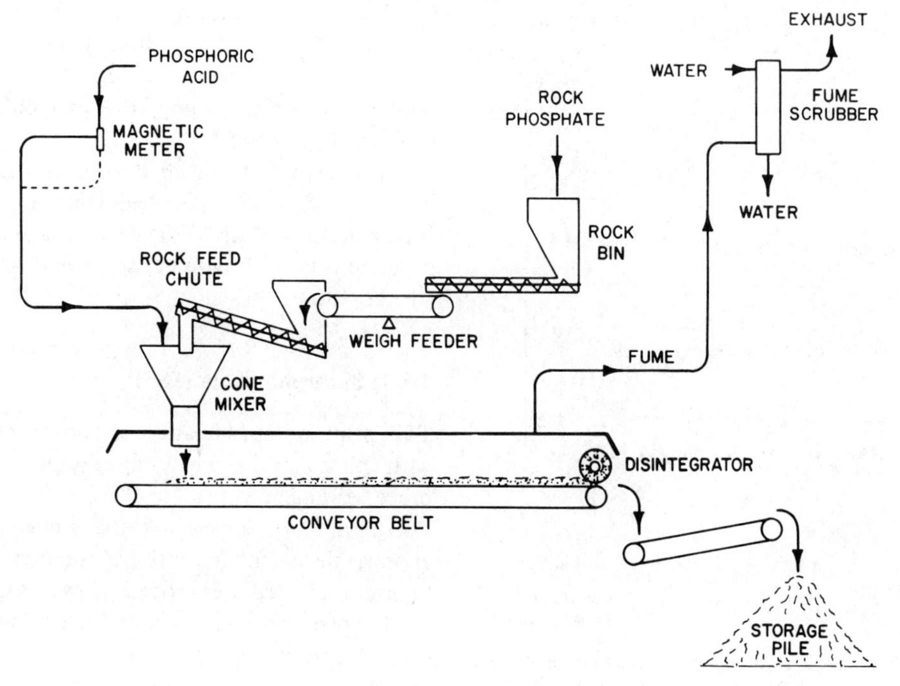

Fig. 8.3. Continuous system for triple superphosphate production. (*Courtesy TVA.*)

usually used to hold the acidulate until the TSP solidifies, instead of the slat-type den used for OSP. With a belt about 5 feet wide and 100 feet long, the production rate is usually 40 to 50 tons per hour. A diagram of the cone mixer and "wet-belt" system is shown in Figure 8.3.[6]

Proportioning for TSP is typically 2.4 to 2.5 pounds of P_2O_5 from acid for each pound of P_2O_5 from rock. The TSP is usually cured 4 to 6 weeks prior to shipment or use at the site. Typical chemical analysis of TSP made with Florida rock is shown below.

Analysis, % by Weight								
P_2O_5								
Total	*Available*	*H_2O Soluble*	*Free Acid*	*CaO*	*R_2O_3*	*MgO*	*F*	*H_2O*
46.9	46.3	42.0	3.4	19.3	3.1	0.5	2.7	4.5

Use of TSP in granular NPK fertilizer formulations, together with or in place of OSP, allowed production of higher analysis grades of granular NPK fertilizers, such as 13-13-13 instead of 10-10-10. The TSP can be readily ammoniated to about 3.5 pounds of ammonia per 20 pounds of P_2O_5 during the granulation process.

Granular TSP. TSP is produced in large quantities in granular form for use in direct application; it is also used in bulk blends. In some processes, cured TSP is granulated in a rotary drum or pan granulator, using steam and water to promote granulation. In Australia and the United Kingdom, the superphosphate usually is granulated in a pan or drum as it comes from the den.

A slurry-type process (as shown in Figure 8.4) is used in the United States and other countries. Pulverized phosphate rock is treated with wet-process phosphoric acid in a two-stage reaction system, and the slurry is sprayed into a pug mill or rotary drum for layering on recycle at a ratio of 10 to 12 pounds of recycle per pound of

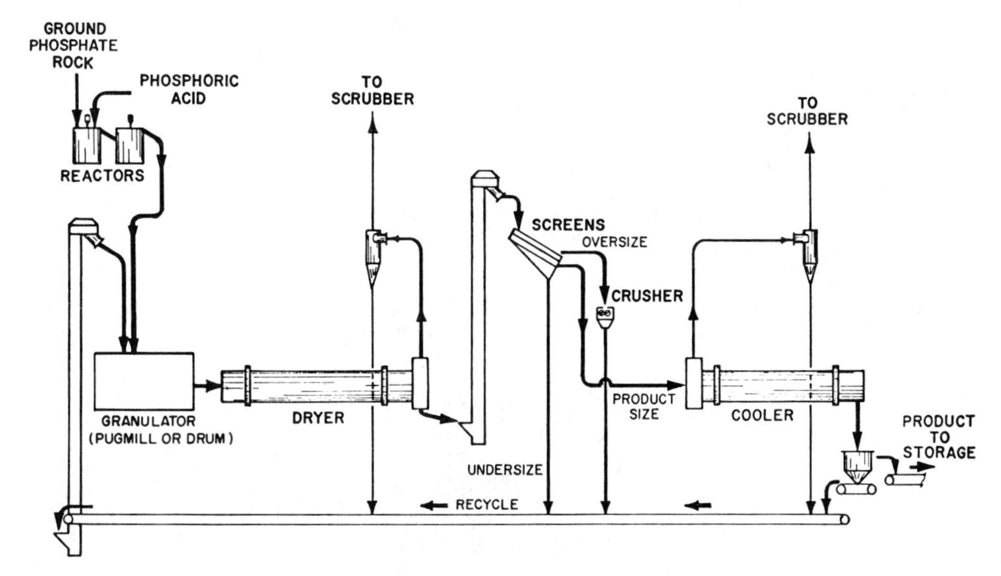

Fig. 8.4. Process for granular triple superphosphate. (*Courtesy TVA.*)

product. Product granules (45% P_2O_5) are quite spherical and dense. The lower grade results from the need to decrease the acidulation ratio from the usual 2.45 pounds of acid P_2O_5 per pound of rock P_2O_5 to about 2.25 in order to control free-acid content and to minimize resultant prolonged stickiness. For production rates higher than about 25 tons per hour, a rotary drum is usually used instead of a pug mill.[6]

Most of the granular TSP in the United States is used in bulk blending, as discussed later.

Ammonium Phosphates

Although ammonium phosphates did not come on the scene in significant quantities until the early 1960's, they have become the leading form of phosphate in the United States and in the world. Almost all new phosphate fertilizer complexes built in recent years and those planned and being built in the 1980's are for the production of ammonium phosphate as the major product.

Crystalline diammonium phosphate (DAP) was produced by TVA starting in the late 1940's, and demonstration programs showed it to be a very good high-analysis fertilizer. Smaller amounts were produced by others as a by-product. When the simple and dependable TVA process for granular DAP (18-46-0) was developed in 1960-1, it was rapidly adopted by the industry.[7] Many granular DAP plants have production capacities of about 50 tons per hour, with a few as high as 70 to 100 tons per hour.

Granular DAP. A flow diagram of a typical granular DAP production unit of the TVA type that has become standard in the industry is shown in Figure 8.5. Wet-process phosphoric acid of about 40 percent P_2O_5 content (or a mixture of 54% P_2O_5 acid and acid from the scrubbing circuit of 28 to 30% P_2O_5 content) is fed to a preneutralizer. Anhydrous ammonia is sparged through open-end pipes, which project through the walls of the tank, in order to neutralize the acid to an $NH_3{:}H_3PO_4$ mole ratio of about 1.4. This is in a range of maximum solubility of ammonium phosphate as shown in the solubiltiy curve of Figure 8.6. The heat of reaction evaporates considerable water, and the water content of the slurry is 16 to 20 percent.

The slurry is pumped at a controlled rate and distributed on the bed in a rotary-drum, TVA-type ammoniator-granulator. The most commonly used metering system for preneutralized slurry is a variable-speed centrifugal pump with automatic control signal from a magnetic flow-

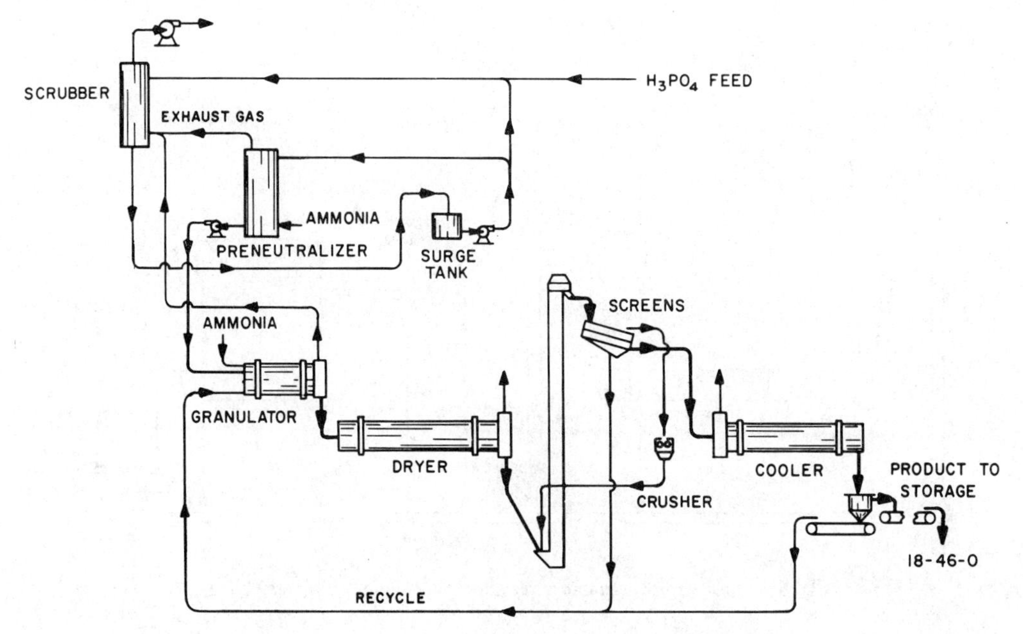

Fig. 8.5. TVA process for granular diammonium phosphate.

meter. Some plants have had success with a magnetic flowmeter and automatic control valve of a special ball type.

Ammonia is sparged beneath the bed in the rotary drum to ammoniate the slurry further to near DAP; the usual finishing $NH_3:H_3PO_4$ mole ratio is 1.85 to 1.94. Ammonia evolved from the preneutralizer and granulator is recovered in acid of about 30 percent P_2O_5 content in a scrubbing circuit. Material recycled at a rate of 5 to 7 pounds per pound of product is the primary control of granulation.

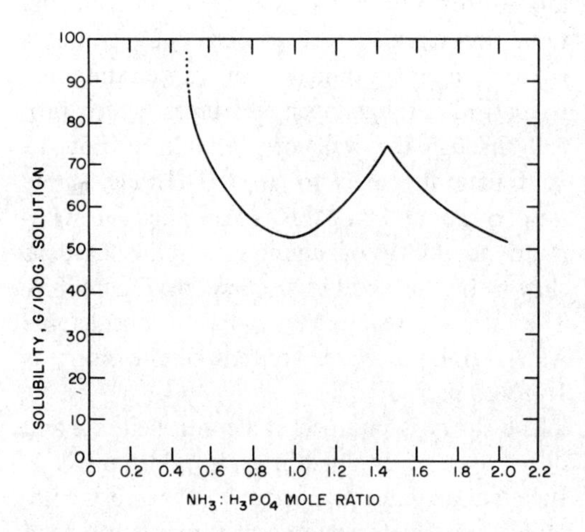

Fig. 8.6. Effect of NH_3/H_3PO_4 mole ratio on solubility in the ammonium phosphate system.

Discharge from the granulator is dried with moderate heat to 180° to 190°F product temperature. Most plants screen hot, and cool only the product fraction, since the material is not very sticky. Rotary coolers or the compact and very efficient fluidized-bed type coolers are used. The product, with a moisture content of 1.5 to 2 percent, does not require a conditioner. It has excellent storage and handling properties, in bags or in bulk.

Construction materials other than mild steel are required only for the acid lines, the preneutralizer, the slurry handling system, and the scrubbing circuits. Type 316L stainless steel or rubber- and brick-lined mild steel is used for the preneutralizer. Type 316L stainless is also used for the slurry pumps and piping. Fiberglass-reinforced polyester plastic and high-density polyvinyl chloride are sometimes used for wet process acid pipes and for scrubbers that may also be constructed of rubber-lined mild steel.[7]

Granular Monoammonium Phosphate (MAP). Granular DAP and the grade 18-46-0 have become household terms in the world fertilizer industry. But substantial interest in granular

MAP has developed, particularly where soils are mainly alkaline, as in Canada and Pakistan. Also, where the primary interest is in producing and shipping phosphate, the practical 11-52-0 to 10-54-0 grades are attractive in that they can provide higher phosphate payload.

TVA developed two comparatively minor modifications of the granular DAP process to allow production of MAP.[8] In one method, the acid is ammoniated to an $NH_3:H_3PO_4$ mole ratio of only about 0.6 in the preneutralizer and then to about 1.0 in the granulator drum. In another procedure that has been adopted by industry, acid in the preneutralizer is ammoniated to a $NH_3:H_3PO_4$ ratio of about 1.4, as it is with DAP; additional wet-process acid is distributed onto the bed in the granulator to adjust back to the MAP mole ratio of about 1.0 (see Fig. 8.6). The remainder of the process with either modification is the same as for DAP, but higher drying temperature can be used to increase the production rate for MAP.

Nongranular MAP. Starting about 1968, simple processes were developed for production of nongranular (sometimes called powdered) MAP. The main processes were developed by Fisons and Scottish Agricultural Industries (SAI) in the United Kingdom, Swift in the United States, and Nissan in Japan. TVA has done some pilot-plant work in the production and use of this intermediate product.[51] A number of these comparatively low-cost units have been built commercially, including plants in the United Kingdom, the Netherlands, Japan, Australia, Spain, United States, Brazil, and Iran. This intermediate usually is shipped to other plants where it is then used in production of NPK fertilizers. However, the nongranular MAP has not attained the popularity that had been predicted in the mid-1970's.

U.S. and World Production of OSP, TSP, MAP, and DAP

The production of TSP in the United States and the world overtook OSP in the early 1960's to become the leading phosphate fertilizer for a short period. The spectacular growth in the popularity of ammonium phosphates (DAP and MAP) led to their production surpassing that of OSP and TSP in 1967 in the United States. By about 1975, the ammonium phosphates were leading in the world.

Figure 8.7 shows the annual United States production of these main phosphate fertilizers from 1960 through 1978.[10] OSP production declined gradually from about 1.3 million tons of P_2O_5 in 1960 to only about 0.3 million tons in 1978. TSP production increased from about 1 million tons of P_2O_5 in 1960 to about 1.8 million tons in 1966 and was essentially steady from 1966 onward. Production of ammonium phosphates in the United States skyrocketed from only 0.3 million tons of P_2O_5 in 1961 to about 5 million tons of P_2O_5 in 1979. Growth in production of ammonium phosphates continued very sharply from 1974 through 1978.

World production of all phosphate fertilizers totaled 27.3 million metric tons of P_2O_5 in 1977. This represented a substantial increase from 18.5 million metric tons in 1969. World production of phosphate fertilizers has been projected to reach 33.3 million metric tons of P_2O_5 in 1980 and 38.9 million metric tons of P_2O_5 by 1985. In 1977, North America produced 29 percent of the world's total phosphate fertilizers, and western Europe produced 20 percent.[11]

Nitric Phosphates

Fertilizers that are referred to as nitric phosphate or nitrophosphate are produced by acidulation of phosphate rock with nitric acid or with mixtures of nitric and sulfuric or phosphoric acids. There are a variety of processes and equipment that have been used in Europe since the late 1930's.[12, 13] There are also a number of plants in Central and South America, and in Asia. In 1980, there are only three nitric phosphate plants in the United States, and these have been modified substantially to use higher proportions of phosphoric acid and other materials.

The primary advantage of nitric phosphate processes is that no sulfur or less sulfur is required as compared with ammonium phosphates; this is particularly important during a shortage of sulfur, or where sulfur must be shipped long distances over land.

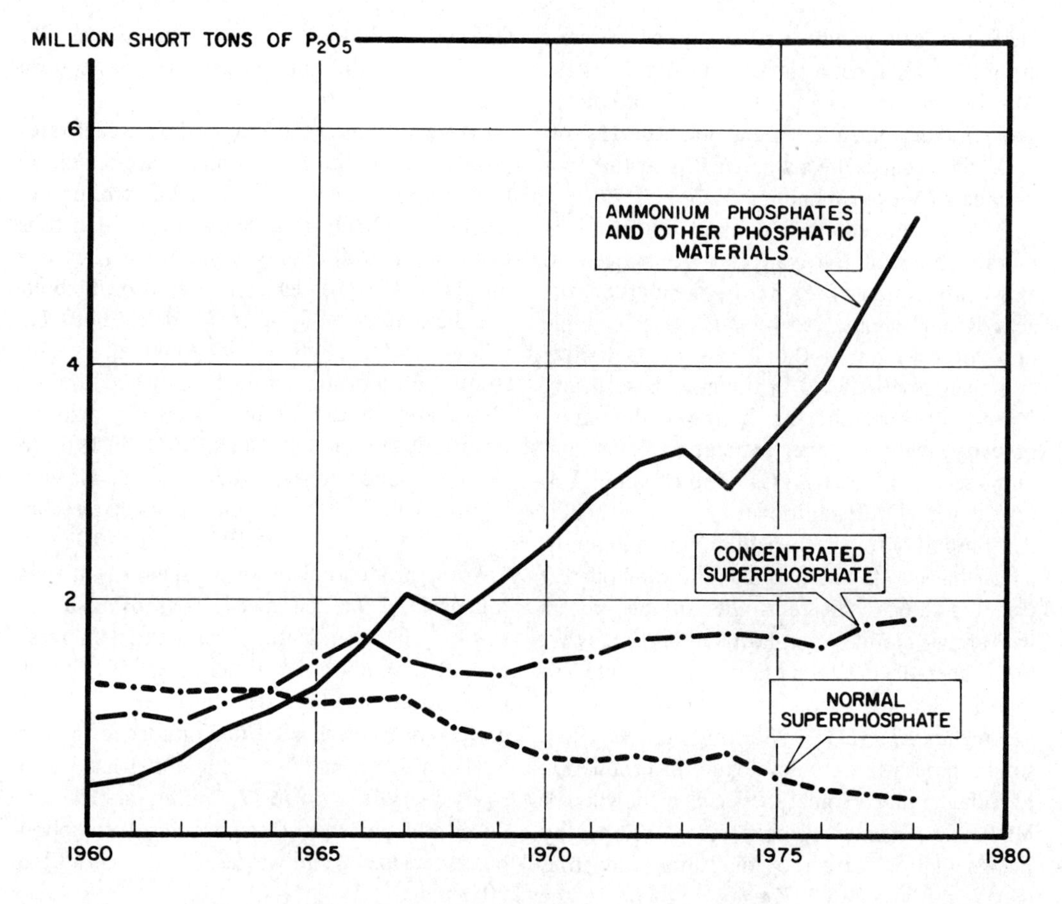

Production of Phosphate Fertilizer Materials–United States

Year	Superphosphate		Multinutrient Materials			Total
	Normal	Concentrated	Ammonium Phosphates	Other	Total	
	(thousand short tons of P_2O_5)					
1960	1,270	986	269	131	400	2,656
1961	1,247	1,024	370	102	472	2,743
1962	1,213	960	536	113	649	2,822
1963	1,227	1,113	...	...	891	3,231
1964	1,206	1,225	...	...	1,034	3,465
1965	1,113	1,466	1,081	172	1,253	3,832
1966	1,138	1,696	1,376	239	1,615	4,449
1967	1,184	1,481	1,747	284	2,031	4,696
1968	914	1,389	1,633	215	1,848	4,151
1969	807	1,354	1,844	288	2,132	4,293
1970	670	1,474	2,092	361	2,453	4,597
1971	626	1,513	2,395	468	2,863	4,992
1972	677	1,659	2,577	570	3,147	5,483
1973	620	1,693	2,919	347	3,266	5,578
1974	698	1,719	2,654	296	2,950	5,367
1975	484	1,678	3,193	218	3,411	5,573
1976	383	1,595	3,614	232	3,846	5,824
1977	340	1,791	4,325	243	4,568	6,699
1978p.	291	1,820	4,875	356	5,231	7,342

Source: USDC, *Inorganic Fertilizer Materials and Related Products*, Series M28B-13, annual reports.

Fig. 8.7. Production of phosphate fertilizer materials–United States.

One process modification uses only nitric acid for acidulation; the extraction slurry is cooled to crystallize calcium nitrate that is removed by centrifugation. This process is referred to as the Odda process. As such, the calcium nitrate is either sold as a fertilizer or converted to co-product ammonium nitrate. The various process modifications utilize a rotary drum, pug mill, or spray drum (Spherodizer) for granulation.[12] NPK nitric phosphate grades have been produced by a prilling process in a few European plants.

In the earlier years, a main disadvantage of nitric phosphate processes was the low water solubility of phosphate in the products. Use of supplemental phosphoric acid, or "deep cooling" by refrigeration to remove a higher proportion of calcium nitrate in the Odda-type processes, now allows water solubility of 60 percent or higher. Popular grades of nitric phosphates include 14-14-14, 22-11-11, 20-20-0, and 26-13-0.

Growth in production of nitric phosphates has been slow since the popular processes for granular DAP and MAP started their rapid growth in the early 1960's throughout the world. World capacity for nitric phosphate production was about 3.2 million metric tons of P_2O_5 in 1978. Actual likely production was about 2.5 million tons of P_2O_5.[14]

Potassium Phosphates

Potassium phosphates and processes for their production have been studied for a long time, but commercial production has been very limited and primarily experimental. Some early TVA studies were with wet-process orthophosphoric acid in reaction with potassium chloride. The reaction mixture was then heated to produce potassium metaphosphate. Grade of the product was about 0-54-37. Major problems were corrosion due to hydrochloric acid evolved and poor market prospects for the by-product hydrochloric acid.

Pilot-plant studies of several schemes for production of potassium phosphate fertilizers were conducted during the 1970's by major organizations in the United States, Ireland, Scotland, Israel, and Spain. Product grades included 9-48-16, 0-60-20, 8-56-18, and 0-47-31. One process was tested in a retrofitted system on a semicommercial scale, but work was discontinued after very limited operation. Potassium phosphates appear to have potential but have never really approached commercialization because of economic and marketing restraints and technological problems.[15]

Other Phosphates

Phosphate Rock for Direct Application. Direct application of finely pulverized phosphate rock has been utilized from near the beginning of fertilization practice. This unprocessed phosphate is, of course, the lowest cost material, but its agronomic effectiveness depends upon the chemical and mineralogical nature of the rock, the pH and other characteristics of the soil, and the crop produced. TVA researchers have characterized phosphate rock from a large number of sources as to their reactivity.[16] Phosphate rock from North Carolina (United States) and Gafsa (Tunisia) are at the top and about equal in reactivity and suitability for direct application. Other phosphate rocks that are not so reactive but are reasonably effective and marketed for this purpose include some types from Morocco, Israel, and a few other locations.

In 1974, a world total of about 3.8 million metric tons of phosphate rock was used for direct application. The United States used only about 100,000 tons for direct application that year. Interest in use of this lowest cost form of phosphate appears to be growing. Tests with North Carolina rock in the Southeastern United States showed promise in the late 1970's.

Partially Processed Phosphates. Aluminum phosphates from Christmas Island are calcined at 930° to 1100°F and sold under the trade name Calciphos for direct application. Similar material from Senegal is sold under the name Phosphal. TVA has granulated finely pulverized North Carolina phosphate rock with concentrated urea solution and partial (25%) acidulation with sulfuric acid in a pilot-plant pan granulator. The same types of products also were prepared with Christmas Island Calciphos. These fertilizers are being tested agronomically.

Studies of granulation with other fertilizer salts also were conducted. The premise is that the soluble binding material would allow the granules to disintegrate in the soil moisture and yield the original finely divided phosphate. The partially acidulated rock and the material granulated with urea solution would provide a portion of the phosphate in a readily soluble form to promote early plant growth. The International Fertilizer Development Center at Muscle Shoals, Alabama, is continuing developmental work on partially processed phosphate rock directed toward utilization of material in developing countries.

Partially acidulated phosphate rock is produced and marketed in substantial quantities in Germany, Finland, and Brazil. Finely pulverized (90% -300 mesh) phosphate rock is granulated commercially in France by using moisture and a binder.

Rhenania Phosphate. Kalichemie in Germany has produced a calcium-sodium phosphate ($CaNaPO_4$), known as Rhenania phosphate, fertilizer for several years. Substantial tonnages have been used in Germany and exported to other European countries, to Africa, and to South America. The general formula for the reaction is:

$$\underset{\text{(phosphate rock)}}{Ca_{10}F_2(PO_4)_6} + 2SiO_2 + 4Na_2CO_3 \longrightarrow \underset{\text{(Rhenania phosphate)}}{6CaNaPO_4} + 2Ca_2SiO_4 + 2NaF + 4CO_2$$

The feed materials–Kola apatite, silica sand, and sodium carbonate–are proportioned, mixed, and then calcined in a rotary kiln at a maximum temperature of about 2300°F. The discharged clinker is cooled, finely pulverized, and then granulated in a drum by addition of water and potato starch as a binder. The granules are dried and screened to -6, +12 mesh. Magnesium and/or boron are added to some products. Grade of the straight Rhenania phosphate is 28 to 30 percent P_2O_5. Granular mixed grades include 0-15-25 and 0-12-20 + 5MgO + 0.15B.

Basic Slag. Basic slag, a by-product of the steel industry, has been widely used in large tonnages as a fertilizer in Europe and South America and to a lesser extent in Asia and North America. This material, which is also referred to as Thomas slag, has a P_2O_5 content of about 17 percent. World production of basic slag totaled about 1.4 million tons of P_2O_5 in 1962. It stayed at that level through 1971 but decreased to about 1.1 million tons of P_2O_5 in 1977.

Fused Calcium Magnesium Phosphate. Fused calcium magnesium phosphate is produced by adding a magnesium source (usually olivine or serpentine) and phosphate rock to an electric furnace. The resultant product, containing about 20 percent P_2O_5 and 15 percent MgO, is pulverized to increase its availability to plants. Some 900,000 tons of this material was produced in Japan and Taiwan in 1977.

Investment Requirements for Phosphate Fertilizer Production Facilities

A major cost item in production of fertilizers, that overshadows the cost of energy and manpower, is the very large investment required for production facilities. As shown below, escalation of capital costs for facilities has been very steep during the 1970's, and there is no sign of any appreciable easing as we enter the 1980's.

C&E News Construction Cost Index for U.S. Plant Facilities

1970 = 125.7
1979 = 229.7
1980 = 250.0

Countries that have indigenous phosphate rock suitable for development must invest a large amount of capital just to produce phosphate rock suitable for use in production of phosphoric acid and phosphate fertilizers. Facilities for mining 2.5 to 3 million metric tons of rock per year, facilities for beneficiation, and systems for storage and handling, would cost 200 to 250 million dollars in the United States. The investment likely would be about one-third higher in developing countries.

The investment in a major complex for production of WPA, TSP, and ammonium phosphates is also very high. Estimates on a mid-1979 U.S. basis indicate that a complex to produce wet-process phosphoric acid (600 tpd P_2O_5), diammonium phosphate (900 tpd DAP), and triple superphosphate (650 tpd TSP) would require an investment of about 150 million dollars, battery limits basis. At a new site, the total investment including auxiliary, support, and service facilities would total 200 to 250 million dollars.

Environmental Controls in the Phosphate Fertilizer Industry

Environmental problems in the phosphate fertilizer industry go all the way back to the source of materials necessary for the industry. The problems in adequate control of dust and noise during mining of phosphate rock, maintenance and control of slimes and tailings ponds resulting from beneficiation, and restoration of mined-out areas are being dealt with effectively in the United States and other major mining areas. There have been some serious incidents, though, including broken dikes that caused major spills of tailings into streams.

In the processes for production of major phosphate intermediates, the main problems are collection of particulates and control of fluorine from the plant environment and from stack emissions. Noise control in operating areas also is being dealt with. Although the imposed standards are quite rigid, they are being complied with by applying practical engineering judgment, good housekeeping and maintenance, and when necessary, by large investments in well-engineered plant equipment for dust collection and scrubbing. Where controls are very rigid, phosphate plants are moving to essentially zero discharge of contaminants into the plant wastewater. Anyone visiting the major phosphate production centers in the United States and other countries for the first time in several years can readily see the dramatic improvements.

Plant management has been diligent in making needed investments and in insisting upon better operating and maintenance practices that decrease environmental problems at the source. Many of the large production complexes now recover fluorine in salable form for use by the aluminum or other industries.[17,18] Fluosilicic acid of about 25 percent concentration is recovered in a number of plants (including OSP) and marketed for use in water fluoridation.

With the escalated price for uranium in the late 1970's, its recovery from wet-process phosphoric acid proliferated. Semiworks systems are in operation, and commercial units are being installed in the United States and Canada. A total of 19 such units was reported in 1980. Plans for recovery of this increasingly valuable by-product is spreading to other parts of the world.[19,20] Present uranium recovery processes however, are not very effective when used in processing the more concentrated acid (40% P_2O_5), produced by hemihydrate processes. But TVA is conducting research work on modifications of present recovery processes in efforts to obtain satisfactory recovery.

The Energy Situation in the Fertilizer Industry

The depletion of energy resources and the increasing cost of energy pose serious problems for the fertilizer industry. Energy dependency is laced throughout the industry, from processing of basic raw materials to production, transport, and application of finished fertilizers. Considerable quantities of energy are required in mining phosphate rock, sulfur, and potash; for refining these primary fertilizer raw materials; and for transporting them to points of use.

Wet grinding of phosphate rock and feeding it in that form to wet-process acid plants are being carried out in several plants as mentioned earlier.[21] This saves fuel for drying the rock and eliminates a major source of dust. Other advantages of wet grinding include economical outdoor storage of unground rock, easier unloading of hopper cars, and in some cases, more consistent filtration in production of wet-process acid. On the negative side, accurate feeding of rock slurry can pose a problem, and rock with high clay content is difficult to process by wet grinding.

In plants producing intermediates or finished fertilizers, the main energy requirements are for process steam and for fuel for drying the fertilizers. Electrical energy for driving the large equipment also is a significant cost component. In large phosphate fertilizer complexes, the evolved process heat is used effectively to produce much of the steam required at the complex. One large source of this evolved energy is from the sulfuric acid plants used to produce the acid needed in manufacture of wet-process phosphoric acid. Prior to the early 1970's, costs of steam, natural gas, and fuel oil were so low that this cost component in production of fertilizers was not very significant. The "energy crunch" starting in 1972-73 changed all of this. Allocation of natural gas and fuel oil resulted in curtailed operation, and the severalfold increases in price led to serious planning and action in conservation of energy.

Energy requirements for production of some main fertilizer intermediates are shown below.[22]

Product	*Million Btu per Ton of Product*
Wet-process phosphoric acid	5.4
Triple superphosphate (TSP)	8.8[a]
Diammonium phosphate (DAP)	13.6[a]
Ammonia	34.5
Urea	22.5[a]

[a]Includes energy for production of the ammonia and/or wet-process acid used as intermediates.

Transport of fertilizer raw materials, intermediates, and finished products requires a substantial amount of energy. Comparative energy requirements per metric ton-mile for various modes of transport are tabulated below.

Type of Transport	*Btu/Metric Ton-Mile*
Pipeline	495
Rail	740
Water	750
Truck	2,650

Plastic film bags are a petrochemical-based product and therefore are a comparatively high consumer of petroleum. The heavy trend toward bulk handling, transport, and application of granular and fluid fertilizers results in savings of this energy component. Energy as fuel for operation of fertilizer-application vehicles is not a very important component.

Important energy-saving measures are growing, like wet grinding of phosphate rock for use in production of wet-process phosphoric acid, and use of pipe reactors and melt granulation (described under production of granular mixed fertilizers) to eliminate or greatly decrease the drying of ammonium phosphate and NPK products. Further innovations are likely as the cost of energy continues to increase and as more engineering effort is directed toward conservation.

POTASSIUM SALTS

Potassium salts are essential to plant growth and are needed particularly for the formation of seeds. The main salt, potassium chloride, is generally referred to in the fertilizer trade as potash.

Soluble Potassium Salts

European Deposits. The first deposit of soluble potassium salts was discovered in the area around Stassfurt, Germany, and supplied nearly all of the world's requirements until 1914. After World War I, deposits in Alsace were developed and became another important source. The deposits near Stassfurt are located about 1,000 feet below ground level and consist of three distinct layers. The upper layer is principally carnallite ($MgCl_2 \cdot KCl \cdot 6H_2O$), while polyhalite ($2CaSO_4 \cdot MgSO_4 \cdot K_2SO_4 \cdot 2H_2O$) lies below the carnallite, and kainite ($MgSO_4 \cdot KCl \cdot 3H_2O$) lies below the polyhalite. The Alsatian deposits lie about 1,500 feet below the surface in two layers, both of which arc principally sylvanite ($KCl \cdot NaCl$). Both the Stassfurt and Alsace deposits are recovered by crystallization from solution. Part of the material is purified to 98 percent KCl (61% K_2O) and some is purified at lower concentrations.[23]

U.S. Deposits. U.S. potash production was started in the Carlsbad, New Mexico, area in 1931. This was the first production of sizable tonnages in the United States. The minerals are found at depths between 800 and 1,800 feet below ground level. Deeper deposits occur in nearby counties in Texas. Many layers exist, with a total cumulative thickness of about 36 feet. The potassium minerals exist as sylvanite, kainite, and polyhalite. Potassium chloride is recovered by crystallization, and the production of some potassium sulfate provides a valuable by-product, because when applied to the soil, it supplies potassium without the introduction of chloride. The deposits near Carlsbad are the largest within the continental United States.

Potassium Salts from Brines. A unique approach to production of potassium salts from brines is carried out at Searles Lake, located near the California-Nevada border. The composition of this lake brine is given below.

Lake Brine Component	*% by Weight*
NaCl	16.35
Na_2SO_4	6.96
KCl	4.75
Na_2CO_3	4.74
$Na_2B_4O_7$	1.51
Na_3PO_4	0.155
NaBr	0.109
Miscellaneous	0.076
TOTAL solids	34.65

Potassium chloride, borax, sodium carbonate, sodium sulfate, lithium phosphate, sodium phosphate, potassium sulfate, boric acid, and bromine are produced from the brine. Solar evaporation is now used to the extent that is practical.

Canadian Deposits. Work toward mining potash ore in Canada was begun during the mid-1950's. The deposit, some 300 miles long and 10 miles across, is found largely at 3,000 to 3,500 feet below the surface, but in some cases it extends to a depth of 5,000 feet. Difficulties were anticipated, partially because of the depth, but mostly because of water-logged sands existing below the 1,250-foot level. The sands, known as the Blairmore water sands, are 200 to 300 feet in depth. They presented a major obstacle because of their fluid consistency and the troublesome water that flowed into the mine below that level. The first mining operation was carried out in 1956, but during the same year seepage of water into the shaft made shutdown necessary for grouting and shaft repair. It was not until 1967, that the mine was back in production. In the meantime, another mine began to operate, and by 1967 four mines producing potassium chloride were in operation.[24]

The number of mines in Canada increased to 10 by 1979, strengthening that country's second-place position behind the U.S.S.R. in world production of potash.

Figure 8.8 shows the surface installation at one of the mines. The shaft house is in the background in the photograph; processing and storage buildings are to its left. Figure 8.9 shows the movement of ore in the mine to the bottom of the shaft. The first picture gives some idea of the size of the operation.

Another company started a mine in 1957; but because of problems in the Blairmore water sands, production was not achieved until 1962. Water was retained outside the mine by installing a series of cast iron rings to seal off the Blairmore. Other operations used a technique of freezing the walls of the shaft until a casing could be installed.

Another operation in the area consists of solution mining. In this case a shaft is not needed because a solution is pumped down to the potash ore where it becomes more concentrated in potassium chloride and is then returned to the surface for refining. The solution is recycled, but details of the process have not been released.[24]

About 10 years of development of novel mining techniques and large capital expenditures were required for the Canadian projects to reach fruition. About two-thirds of the potash used in the United States is imported from Canada.

World production of potash totaled 26.4 million tons of K_2O in 1978. This was up from about 17 million tons in 1970 and 12 million in 1965. A breakdown of 1978 production by countries is listed below.[25]

Country	*1978 Production Millions of Tons, K_2O*
U.S.S.R.	8.9
Canada	6.1
East Germany	3.3
West Germany	2.5
United States	2.3
France	1.8
Israel	0.7
Spain	0.6
Italy	0.15
United Kingdom	0.14

U.S. production of potash was 2.3 million tons of K_2O in 1978; consumption was 6.8

Fig. 8.8. Surface structures at a Canadian potash mine. The shaft house is in the background and processing and storage buildings are to its left. (*Courtesy Potash Co. of America.*)

Fig. 8.9. Movement of ore in the mine to the shaft at a Canadian potash mine. (*Courtesy Potash Company of America.*)

million tons of K_2O. Prices of commercial products ranged from $90 to $120 per ton and total commercial value of U.S. production was $200 million in 1979.[26]

About 95 percent of the total potash production in North America in 1979 was KCl, and 95 percent of the total was used as fertilizer. Potash is added during the production of finished fertilizers of granular, bulk blended, and fluid types. There are no major problems in the addition of potash salts except solubility, which is the controlling factor in clear liquid fertilizers and which establishes maximum grades.[27]

A large part of the U.S. production of potash is in granular form. The need for granular potash developed when the practice of bulk blending of granular N, P, and K materials started to flourish in the United States in the late 1960's. (See. "Preparation of NPK Fertilizers" in this chapter.)

Potassium Nitrate

Potassium nitrate has been produced historically either through the reaction of potassium chloride and sodium nitrate or reclaimed from natural deposits. The chemical reaction is:

$$NaNO_3 + KCl \longrightarrow KNO_3 + NaCl$$

The sodium chloride is crystallized and filtered from the hot solution. The potassium nitrate remaining in the dilute solution is crystallized and dried. Since the raw material costs are high, the process has limited commercial value.

A process for production of potassium nitrate from potassium chloride and nitric acid, after several years of development, was placed in commercial production in the United States in 1963. The overall process is represented by the following equation:

$$2KCl + 2HNO_3 + \tfrac{1}{2}O_2 \longrightarrow 2KNO_3 + Cl_2 + H_2O$$

It is rather complicated but the key to its success is the oxidation of nitrosyl chloride according to the reaction:

$$NOCl + 2HNO_3 \longrightarrow 3NO_2 + \tfrac{1}{2}Cl_2 + H_2O$$

The nitrogen dioxide product is converted to nitric acid for recycle in the process. The corrosive nature of various intermediates requires the use of costly materials such as Inconel, stainless steel, and titanium. The process has been described in detail.[28]

The major agronomic interest in the more expensive potassium nitrate and potassium sulfate is for use on crops such as tobacco, where chloride should be avoided.

Some producers of chlorine and caustic use potassium chloride in part of their cells to produce potassium hydroxide, which has various industrial uses. The high solubility of potassium hydroxide makes it particularly well-suited for use in liquid fertilizers, but cost greatly limits this usage.

PREPARATION OF FINISHED (NPK) FERTILIZERS

History

The fertilizer industry in the United States began during the middle years of the 19th century when Justus von Liebig (1803-1873) in Germany was teaching the dependence of vegetation on the mineral components of the soil. The first American edition of Liebig's book, entitled *Organic Chemistry in Its Application to Agriculture and Physiology*, appeared in 1841. Materials such as animal manures, wood ashes, bones, fish, guano, wool waste, chalk, and marl had been recognized as aids to plant growth in earlier times. Now they began to take on added significance because of the newly acquired knowledge that their agricultural value was due to their content of nitrogen, phosphorus, potassium, calcium, and other chemical elements.

In the United States, Phillip S. and William H. Chappell of Baltimore, Maryland, obtained the first U.S. patent for mixed fertilizer on May 27, 1849. Baltimore rapidly became a manufacturing center for superphosphate fertilizers, and the industry soon spread to other principal cities of the Atlantic seaboard.

In 1856, Samuel W. Johnson (1830-1909) returned to the United States from Europe, where he had studied under Liebig. He contributed greatly to the technology and use of fertiliziers in the United States, and he was one of the foremost leaders in the establishment of state experimental stations and the testing of the fertilizer industry's products.

Phosphate rock deposits were opened in South Carolina in 1867, Florida in 1887, Tennessee in 1894, Idaho and Wyoming in 1906, Montanna in 1921, and North Carolina in 1965. The first imports of potassium products from Germany, referred to in the trade as "potash," began in 1870. The last great barriers to our domestic sufficiency in the basic fertilizer elements were overcome with successful operation of the first U.S. plant for the direct synthesis of nitrogen from the atmosphere at Syracuse, New York, in 1921 and the opening of deposits of potassium ores at Carlsbad, New Mexico, in 1931.[29]

The average total plant food content (N, P_2O_5, and K_2O) of mixed fertilizers rose from 13 percent in 1880 to 15 percent in 1910, and to 20 percent in 1940. The total nutrient content averaged 40 percent in 1970 and about 43.5 percent in 1979.[30]

U.S. consumption of N, P_2O_5, and K_2O in all fertilizers totaled 1.8 million tons in 1940. This total reached 16 million tons in 1970 and 22.3 million tons in 1979—more than a twelvefold increase during the past 40 years. Consumption of the individual nutrients in 1979 were:

Nutrients	*Million Tons of Nutrient*
N	10.6
P_2O_5	5.5
K_2O	6.2

Plant nutrient consumption data for the United States from 1965 to 1979 are shown in Figure 8.10.

Following the advent of synthetic nitrogen production in 1921, high-analysis compounds largely replaced the low-analysis organic materials, such as cottonseed meal, fish scrap, etc., previously used. By 1935, ammoniation of superphosphate fertilizers with low-pressure nitrogen solutions was well established, solid urea was introduced, and granulation of superphosphate to improve its physical condition became a commercial accomplishment.[31] During World War II, ammonium nitrate came on the market in substantial quantities. In 1945, the introduction of the TVA cone mixer for

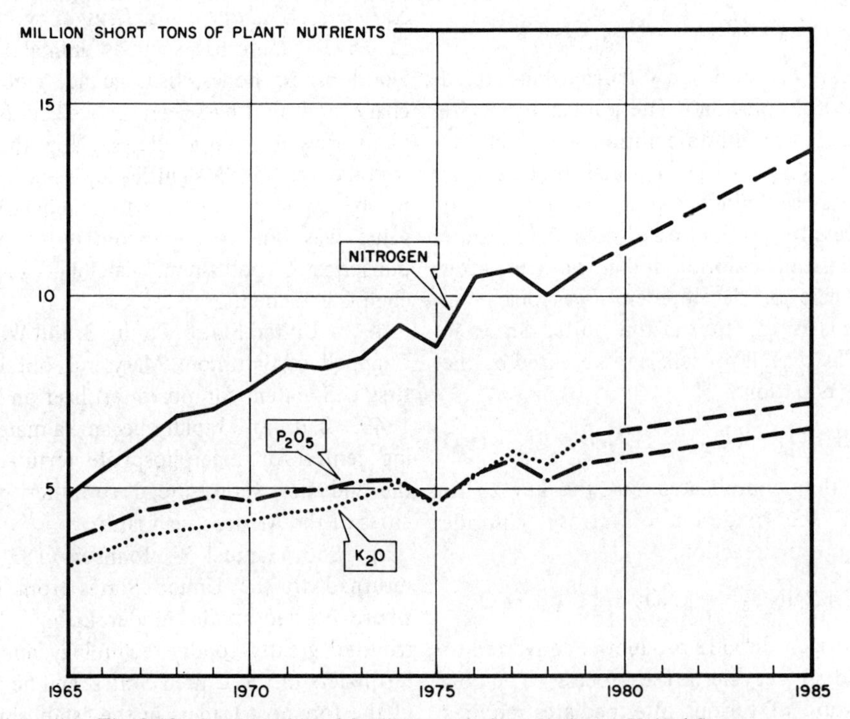

Fig. 8.10. Consumption of fertilizers and plant nutrients—United States.

the continuous acidulation of phosphate rock signaled a rapid rise in the production of TSP.

Since 1950, there has been a dramatic rise in the production of granular fertilizers, spurred largely by the TVA ammoniation-granulation process. As discussed in the section on phosphate fertilizers, ammonium phosphates (MAP and DAP) were the star performers starting in 1961. Their production dominated in the United States by 1970, and in the world by 1977.

Plant Nutrient Elements

Sixteen chemical elements are known to be essential to the normal development and growth of plants. Their percentage in the dry matter of the plant is known to vary widely with different species of plants, and with different soil and climate conditions under which plants are grown. However, the order of magnitude expressed as the average percentage of each element present in the dry weight of the whole plant may be represented roughly as follows:[32]

Element	*Amount in Whole Plant % (Dry Weight)*
Oxygen	45
Carbon	44
Hydrogen	6
Nitrogen	2
Phosphorus	0.5
Potassium	1.0
Calcium	0.6
Magnesium	0.3
Sulfur	0.4
Boron	0.005
Chlorine	0.015
Copper	0.001
Iron	0.020
Maganese	0.050
Molybdenum	0.0001
Zinc	0.0100
TOTAL	99.9011

Oxygen, carbon, and hydrogen make up the plant structure and comprise about 95 percent of the elemental content of the plant. The plant obtains them from the atmosphere and water. The remaining 13 essential elements in plants are supplied in large part by soils and fertilizers. They comprise only about 4.9 percent of the elemental content of the plant.

Fertilizer Production Patterns

The fertilizer industry in the United States and world has evolved and developed markedly during its 140-year history since Liebig. Its greatest development has occurred since about 1945, with production spreading to many developing countries. The U.S. fertilizer industry is organized basically into separate segments for production and marketing of nitrogen, phosphate, or potash intermediates and products. There is some overlapping, but the large basic complexes produce predominately N, P, or K products. The finished NP or NPK fertilizers usually are prepared in small- or medium-sized plants near the markets. Ammonium phosphates are mainly produced near the phosphate rock mines.

Nitrogen fertilizer production complexes are located near natural gas sources. Many ammonia plants are sited at wellheads; others are located on natural gas pipelines. Nearness to markets is another important consideration, as are transportation alternatives. Locations on waterways facilitate domestic shipments by low-cost barge and export shipments via oceangoing vessels.

Large amounts of ammonia are shipped by rail and truck. Substantial quantities move by pipeline from plants in gas-producing areas to major fertilizer markets. Ammonia is shipped in refrigerated barges; upon arrival, the liquid ammonia often is stored at atmospheric pressure in insulated and refrigerated tanks.

Ammonia is often dissolved in water to make nitrogen solutions, which are low or nonpressure materials that are shipped by truck, rail, barge, and pipeline. Other major sources of nitrogen are prilled or granular ammonium nitrate and urea. Urea is shipped by truck, rail, or barge. Ammonium nitrate, because of the hazards associated with it, usually moves by rail or truck.

The major U.S. phosphate production centers are located near the phosphate rock mines in Florida, North Carolina, and western areas of Idaho, Utah, and Wyoming. There is a substantial economic advantage for upgrading phosphate rock (28-32% P_2O_5) to wet-process acid (52-54% P_2O_5), TSP (44-46% P_2O_5), DAP (18-46-0), and MAP (11-52-0) for long-distance shipment to regional or local plants that produce finished granular or fluid fertilizers. Major phosphate fertilizer complexes in Florida and North Carolina are favorably located for export of phosphate rock, wet-process acid, TSP, and DAP.[2]

Most of the potash used in the United States was mined near Carlsbad, New Mexico, until the early 1960's, when the deposits in Saskatchewan, Canada, came into production. In 1978, the Canadian mines supplied about two-thirds of the potash consumed in the United States. Major U.S. potash corporations built mining and processing operations there, but several of these have been taken over by the Government of Saskatchewan.

Most finished fertilizers in the United States are processed in regional or local plants comparatively near the markets. Production and marketing techniques for the three main systems—granulation, bulk blending, and fluid fertilizer production—are described later.

Most finished fertilizers in Europe, Asia, and Latin America are produced at major production complexes, located mainly to facilitate receipt of raw materials and shipment of products to the markets. Locations that allow shipment by water are highly preferred. Major raw materials and intermediates—such as phosphate rock, phosphoric acid, and ammonia—are often shipped to these complexes from domestic or outside sources.

The 51 million tons of fertilizers consumed in the United States in 1979 represented 22.3 million tons of nutrients ($N + P_2O_5 + K_2O$). This was up from 40 million tons of fertilizers consumed in 1970 that represented 16 million tons of the primary nutrients.[33]

Direct Application. Fertilizer materials applied directly in the United States in 1979 totaled 25.3 million tons; this was about one-half of all fertilizers used. The remaining half was used in preparation of mixed (NPK) fertilizers.

The main nitrogen materials used for direct application and their percentage of total N applied directly in 1979 are shown below.

Product	*Percentage of Total N Applied Directly*
Anhydrous and aqua ammonia	54
Nitrogen solutions	21
Urea	13
Ammonium nitrate	12

Only about 15 percent of all phosphate fertilizer was used for direct application in 1979. Ordinary and triple superphosphate accounted for 69 percent, and ammonium phosphates for 20 percent of the phosphate (P_2O_5) applied directly.[33]

Systems Used in the United States for Preparation of Finished (NPK) Fertilizers

About 61 percent of the total world fertilizers in 1979 were mixed (NPK) types. In the United States, about 53 percent of all fertilizers were mixed—granular and fluid. As we reach the 1980's, the production pattern for finished fertilizers in the United States has evolved to three main systems. These are:

- Granulation
- Bulk blending
- Fluid fertilizer (liquids and suspensions) production

The proportion of finished NPK fertilizers produced in the United States by ammoniation-granulation processes reached a maximum of about 80 percent during the early 1960's. Since that time, there has been a decreasing trend in granulation as the production of bulk blends and fluid fertilizers continues to increase.

The practice of physically mixing granular N, P, and K intermediates (bulk blending) started in the North Central States in the mid-1950's. Growth was spectacular, and more than 5,000 blending units were in operation by 1970; about 5,500 are believed to be in operation in 1980.

Production of liquid fertilizers started in California in 1923. Growth was slow for the next 25 years and had reached only 22,000 tons in that state in 1953. In 1955, 147 companies were reported to be manufacturing liquids. Growth was rapid in the 1960's, the number of plants increasing from 355 in 1959 to more than 1,700 in 1968. The number of plants increased spectactularly to about 2,800 in 1971.[34]

By the late 1970's, a general stabilization in the production-marketing system indicated production of about 40 percent of U.S. mixed fertilizers each by ammoniation-granulation and bulk blending, and about 20 percent as fluids.

Production of Granular Mixed Fertilizers

The earliest mixed fertilizers in the United States were based on waste materials from the meat-processing industry. In 1913, about 42 percent of the nitrogen in U.S. fertilizers was supplied by these wastes; in 1979, the proportion was less than 1 percent. The first mineral-mixed fertilizers were in powdered form and caked badly. The practice of ammoniating superphosphate during production of these fertilizers became popular in the 1930's. Analysis was low, such as 6-8-4 and 2-10-7 (N-P_2O_5-K_2O). The practice of producing fertilizers in granular form, which was already popular in Europe, did not develop to any appreciable extent in the United States until the dependable and comparatively simple TVA ammoniation-granulation process was developed and demonstrated in 1952. This versatile process became widely used, and more than 60 percent of U.S. fertilizers were prepared this way during the 1960's.[35]

Materials Used. The use of N, P_2O_5, and K_2O intermediates in local plants for production of solid mixed fertilizers started in the United States and Europe with use of potash, calcium cyanamide, sodium nitrate, guano or other solid nitrogen materials, and OSP to produce low-grade pulverized mixtures. Ammoniating solutions came on the scene later and gave more versatility in fixing lower cost nitrogen. Ammonium sulfate became available as supplemental nitrogen, mainly as by-product material. With the advent of granulation in the United States in the early 1950's, the use of intermediates became more important. Granulation had started as early as the mid-1930's in Europe, and by 1950, it was quite popular in some countries. In the United States, anhydrous ammonia became more economical and practical to transport. Nitrogen solutions containing free ammonia were used to ammoniate locally produced OSP (20% P_2O_5) and shipped-in TSP (46% P_2O_5) that became an important intermediate for upgrading the phosphate content. Higher analysis granular products with better handling, storage, and application properties were produced. Sulfuric acid was used to promote granulation and allow greater fixing of ammonia, and later wet-process phosphoric acid (53-54% P_2O_5) came into use in moderate amounts as an intermediate to provide higher analysis grades with higher water solubility. Secondary nutrients (S and Mg) and micronutrients (Zn, Cu, B, Mo, Fe, Mn, etc.) were added as needed.

Equipment and Operation Technique. In the United States, the TVA ammoniation-granulation process and equipment (rotary cylinder or drum) that originated in 1953 were widely adopted. Within a year, about 50 plants of this type were in operation. By 1962, there were 164 plants in the United States known to be using this process, and 200 or more local and regional granulation plants of this type were estimated to be in operation by the mid-1960's.[36] The rotary drum-type granulation equipment is by far the most widely used because of its versatility in combining the mixing, ammoniation, and granulation in a single unit. Other types of granulators including pug mills, spray drums, and inclined pan granulators are used to a lesser extent.

The ammonium phosphate "boom" in the middle 1960's and the growth of bulk blending and liquid fertilizer production slowed the growth in local granulation plants and they declined in relative importance. However, the approximately 125 NPK granulation plants remaining in 1980 continue to have an effective place in the production and marketing system in the United States. They are estimated to have produced nearly 40 percent of all finished (NPK) fertilizers in 1979. This system has several advantages, including those listed below.

- A variety of NPK grades of various ratios can be produced with economical formulations. Secondary and micronutrients can be readily added to produce special grades.

- Recycle ratios are low, and grades can be changed more easily than in large complex plants.
- Investment in basic production facilities is comparatively low, and fewer costly auxiliary and support facilities are required than for a large complex.
- Operation can be seasonal if desired with use of a small crew of local operating personnel that do not require a high level of training.
- Ownership can be local with obvious advantages in marketing of products due to personal acquaintance with customers and their preferences.

Some typical formulations for several grades of granular fertilizers that are produced in NPK granulation plants are shown in Table 8.1.[29]

A schematic diagram of a typical local granulation plant of the type used in the United States is shown in Figure 8.11.

Such plants include storage bins for several solid intermediate materials and tanks for ammonia, nitrogen solution, and acids. In most plants, the solid materials are moved from ground-level bins by front-end loader and elevated to a cluster hopper. The components used are batch weighed in the proportions required in the formulation and fed collectively at a fairly steady rate to the ammoniator-granulator. Some plants use individual feeders for the solid components. Ammonia or nitrogen solution and sulfuric acid are sparged beneath the bed in the rotary drum. When phosphoric acid is used, it is sprayed onto the bed in the granulator. Magnetic flowmeters and automatic rotameters are commonly used for controlling the rate of feed of the liquids. The formulations are devised to provide the proper balance of heat, moisture, and total fluid phase in order to allow good granulation. They are balanced to ammoniate properly and to produce the desired grade.

Material from the granulator flows into a direct fuel-fired rotary dryer and then into a rotary countercurrent cooler. The cooled material is screened to separate the desired product size (usually about -6 +12 mesh), oversize is crushed, and undersized fines are recycled to the granulator. The product may be treated with a surface conditioner to minimize caking. Products usually are stored in bulk prior to bagging and shipment.[37] There are some variations in equipment and operating technique, such as elimination of the dryer by using a pipe reactor for melting granulation as described later. NPK granulation plants usually produce from 15 to 30 tons per hour. They are comparatively simple and flexible in operation, and allow the use of a variety of materials in formulations. Starting up and shutting down operations are not particularly difficult, and grades can be changed frequently without great difficulty. In the early years of local granulation-plant practice in the United States, essentially all of the output was bagged for shipment. After bulk blending became popular, bulk handling of granular fertilizers directly from the granulation plant in spreader trucks and "tote boxes" became fairly important. By the

TABLE 8.1 Some Typical Formulations for Grades of Granular Fertilizers.

Materials Used	*Grade*				
	10-10-10	*12-12-12*	*5-20-20*	*6-24-24*	*10-20-10*
	(Kg/metric ton)				
Ammonia	–	–	63	25	–
N solution, 44.8% N[a]	180	204	–	48	277
Ammonium sulfate, 20.5% N	98	147	–	–	–
Diammonium phosphate (18-46-0)	–	–	–	106	–
Single superphosphate, 20%	510	254	158	–	280
Triple superphosphate, 46%	–	156	377	425	324
Potassium chloride, 60%, K_2O	167	200	334	400	167
Sulfuric acid, 93% H_2SO_4	49	70	64	–	38
Filler or conditioner	36	–	–	–	–
Steam	–	–	–	75	–

[a]25% NH_3, 69% ammonium nitrate, 6% H_2O.

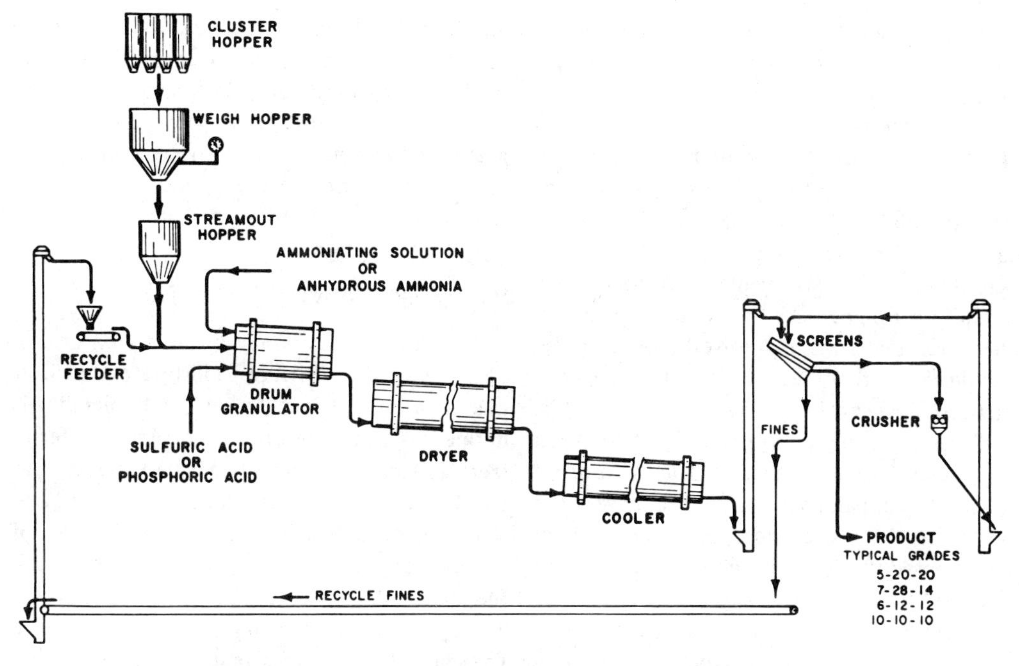

Fig. 8.11. Typical granulation plant–United States. (*Courtesy TVA.*)

mid-1970's, this practice had become quite common, and by the late 1970's a number of plants were moving 50 to 70 percent or more of their production in bulk. This practice is an effort to incorporate some of the advantages and economy of bulk blending in moving granular-mixed fertilizers to customers.

Environmental Problems. Perhaps the greatest problem facing local granulation plants in the 1970's was meeting required regulations to avoid atmospheric and stream pollution. The earlier plants were usually located in rural or suburban areas, but neighborhoods eventually grew up nearby. It became more important to control and remove dust and fumes from stack effluents, and the need to avoid dumping liquid wastes into streams increased. With continuing intense interest in environmental factors, good control of all aqueous effluents and stack emissions became essential. Addition of scrubbers to existing plants is costly and difficult; return of liquid effluents to the process is also difficult, and technical personnel often are not available to fully plan for and deal with these problems.

Environmental problems resulted in some plants being shut down and production consolidated at other more favorable locations. It was found that some changes in formulations and in operating techniques could greatly decrease dust and fume problems. TVA and industry showed in early tests that use of wet-process phosphoric acid instead of sulfuric acid greatly decreased fumes and dust. Fuming is less of a problem because ammonium chloride aerosol fume is not formed with this acid. When sulfuric acid is used, it reacts with potassium chloride in the granulator to produce the troublesome fumes. A more effective and lower cost scheme is melt granulation by use of a pipe reactor for ammoniation of phosphoric and sulfuric acids (discussed later).

Engineering firms and equipment manufacturers have devised simpler and more effective dust and fume control equipment; these joint efforts have made it less difficult to deal with environmental problems than had been visualized in the early 1970's. The development of bag filters and their operating technique, which allowed their use for very effective collection of dust in fertilizer plants, was a major accomplishment. What had been visualized in 1970 by many as an impossible or impractical task has resulted in many plants fully meeting the standards for aqueous effluents and stack emissions. Good housekeeping, improved maintenance, and tighter operating discipline have facilitated effective environmental controls.

Local Granulation Practices in Other Countries. In continental Europe, less use is made of local granulation plants than in the United States, although several plants of this general type that are at least regional in nature are utilized in France, Germany, and Spain. In other countries, such as the Netherlands, Italy, and Belgium, the local plant concept is not practiced and several grades of finished NPK fertilizers of complex types are produced in larger centralized plants. Nitric phosphate plants of comparatively large size have been used rather extensively in Europe, and ammonium phosphate-based systems came into use in the 1970's.

In the United Kingdom, there is substantial practice of the regional or local granulation-plant concept, although there are large major producers. There is considerable practice of blending granular intermediates in preparation of final varieties of grades in large granulation plants. Equipment is provided for blending immediately before bagging or bulk loading. Blending in granulation plants is said to comprise 60 to 70 percent of production output in some plants in the United Kingdom.[38]

Production of granular compound fertilizers is practiced quite widely in several Central and South American countries, such as Colombia, Ecuador, Peru, Brazil, Venezuela, and Costa Rica. Several years ago, the Eirich pan granulator was introduced, and this system proliferated for mechanical granulation with moisture. In programs to modernize existing plants, as well as in new plants, rotary drum-type granulators are being installed. In larger complex plants, the spray-drum granulator (Spherodizer) is used quite extensively, but this type of granulator is not being installed in new facilities. Earlier plants were of the nitric phosphate-type, but since about 1965 ammonium phosphate-based fertilizers have predominated.

Production in Asia is primarily of complex NPK fertilizers at large production complexes. Ammonium phosphate-based grades predominate and have become more prevalent in Japan since about 1970. Considerable amounts of ammonium chloride are used in finished granular fertilizer formulations in Japan, and proportions of organic slow-release nitrogen fertilizers are added during granulation to provide controlled-release properties.

BULK BLENDING

History and Growth

A blended fertilizer may be described as one consisting of a mechanical mixture of fertilizer materials. As referred to here, bulk blending is defined as a mixing process for granular fertilizers (simple or binary) in small plants that usually receive their intermediates in bulk and that are located very near the point of use of the blended products. The practice of bulk blending of granular fertilizers has grown rapidly in the United States, particularly since the early 1960's. This practice started in Illinois in 1947, growth was slow, reaching only 200 units in the United States by the mid-1950's. The real explosion came in the 1960's when the number reached 1,700 by 1964 and nearly doubled to 3,100 by 1966. In 1976, around 5,600 bulk-blending units produced about 10.7 million tons of products. The number of bulk-blending units in operation in 1980 is believed to have stabilized at 5,200 to 5,500; they now account for preparation of about 40 percent of all finished (NPK) fertilizers.[39]

Key factors in the rapid growth of bulk blending were convenience in providing a variety of plant food ratios or grades (or prescription amounts of N, P, and K), economy due to handling in bulk, and services provided by the blender. The blender is the retail dealer in the stream-lined bulk-blending marketing system, and he usually transports the fertilizer to the farms and applies it when needed. Typical blending plants produce and market only 2,000 to 6,000 tons per year on a seasonal basis. There are a few plants with a much larger annual production. Capital investment in blending units is comparatively low ($250,000 or less), but most blenders have about an equal investment in application equipment.[40] The typical blender is equipped with high-speed application equipment that has high-flotation tires. Usually, spinner-type applicators are used.[41] Considerable care is required during handling and application to prevent segregation of ingredients which can result in a nonuniform application of the blend.[42] Moreover, nonuniform application of blends has been especially troublesome when herbicides are sprayed onto and applied with the blends. Usually some nonuniform weed kill occurs in the field. New types of application and handling equipment, plus improved operating procedures, should help to eliminate this problem.

Blending plants operate on a quite seasonal basis, and usually the operators have supplemental business enterprises or they provide additional farm services. Many people have attributed the success of bulk blending in the United States largely to the additional services usually provided by the blender. These include a range of activities from soil sampling to establish a basis for fertilizer needs to the service of applying the fertilizers to the soil. Farmers generally are willing to pay for the application service since there are many other demands for available farm labor at the time.

Intermediates Used in Bulk Blending

The granular or prilled intermediates most commonly used in bulk-blending plants in the United States are urea, ammonium nitrate, diammonium phosphate (18-46-0), triple superphosphate, and potassium chloride. Lesser amounts of ammonium sulfate, granular ordinary superphosphate, and potassium sulfate are used. Operating experience and TVA studies have pointed out some combinations of materials that are incompatible and should be avoided.[44] These include urea with unammoniated superphosphate and, of course, urea with ammonium nitrate. Secondary and micronutrients can be suitably provided in bulk blends, but there are some problems.[44a] Equipment used for transport of the intermediates in bulk must keep the material dry and avoid spillage.

Equipment and Operating Technique

There are many varied arrangements of equipment in the thousands of plants throughout the United States. The most popular type of blending plant layout is shown in Figure 8.12. This type of plant has a horizontal rotary mixer at ground level. In a variation of this general type, the rotary mixer is in an elevated position that allows discharge directly into spreader trucks.

The key equipment is prefabricated by a number of equipment manufacturers in a variety of types. The total system is quite simple as indicated by the diagram in Figure 8.13. Typically, the blending operation involves receipt of the intermediate granular components in bulk by railroad car or truck, storage of the intermediates separately in bins, weighing of the granular components in desired proportions, and mixing to obtain essential uniformity. The weighing and mixing operations are combined in some types of equipment such as the ribbon mixer mounted on weigh scales. The most popular type of rotary mixer usually receives preweighed batches of the granular intermediate components. Weigh hoppers are used in some plants while other plants weigh front-end loaders with the scoop load of material. Some plants simply use volumetric measurement of components. The mixing or blending step usually requires only 1 to 2 minutes, so high hourly rates of production are achieved.

The entire cycle of weighing, mixing, and discharging may be automated.[43] The weight of each component of the blend is punched on a card which is fed into a control mechanism that is set to repeat the cycle for the desired number of batches. Once started by a push button, operation continues until the entire consignment has been prepared. Timing of the various operations is overlapped, and the output, even with a 1-ton mixer, may be 15 to 20 tons per hour.

Rotary mixers are available with capacity of 1 to 6 tons per batch and ribbon blenders with 1 to 4 tons, so high hourly capacity of 15 to 30 tons per hour can be obtained with simple but ingenious arrangements.

Problems in Bulk Blending

Bulk blending is a simple type of practice and can provide uniform mixtures in almost any

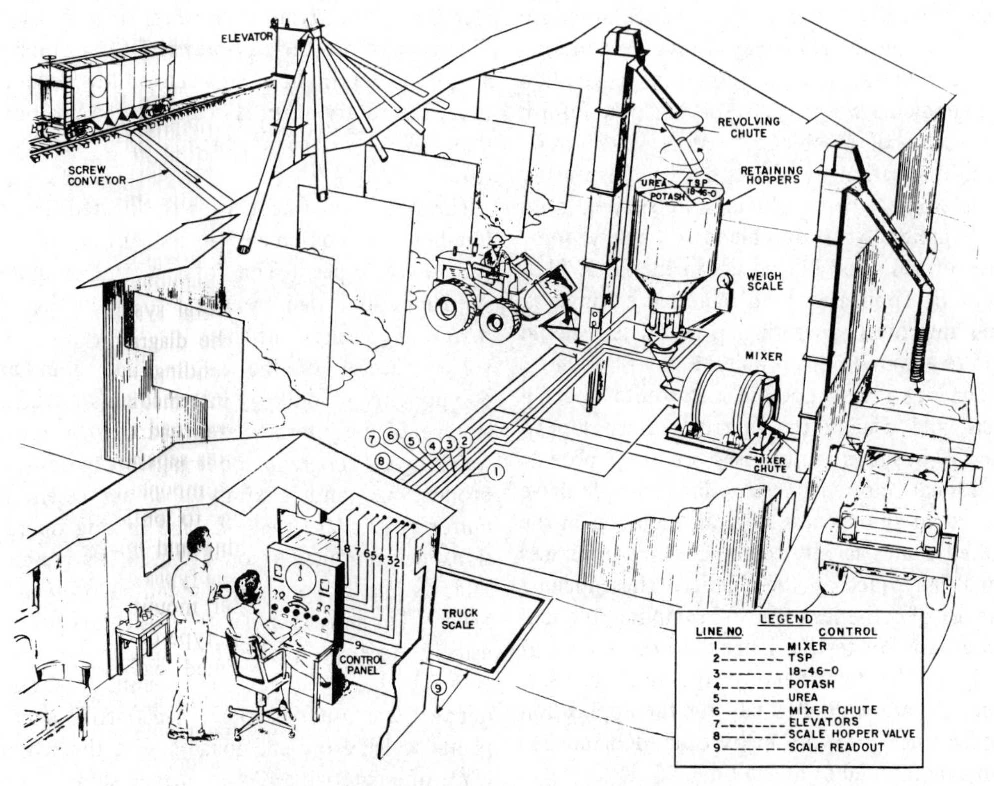

Fig. 8.12. Bulk-blending plant layout. (*Courtesy TVA.*)

desired proportions. Blends can be prepared to provide a given weight of N, P, and K rather than specific grades. However, unless proper materials are used and certain practices are followed, segregation can occur and non-uniform blends will then be delivered and applied to the farm. The factors involved and the precautions that are necessary have been thoroughly described by Hoffmeister et al., of TVA.[44] The main requirements are the use

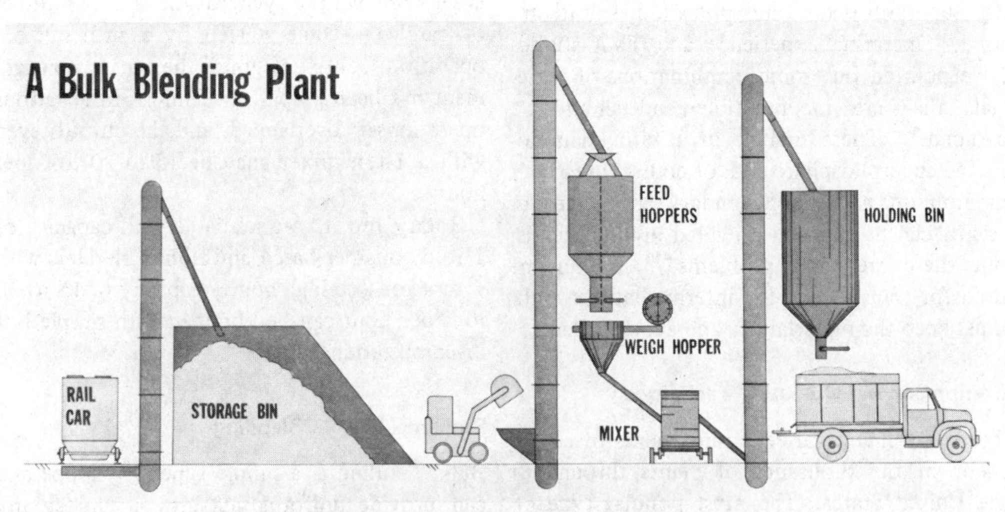

Fig. 8.13. Bulk blending–a simple concept. (*Courtesy TVA.*)

of granular materials with a well-matched range of particle size and the handling of the mixture after blending in ways that will minimize segregation. It is quite easy to obtain a uniform blend by use of proper materials, but maintaining this uniformity on the way to the farm and onto the soil involves precautions. Shape and density of the granules are much less important factors than is matching the particle size for obtaining uniformity in blending. Handling procedures that may cause segregation include coning (as occurs when the blend is allowed to drop onto sloping piles), vibration in hauling vehicles, and ballistic action imparted by some types of spreaders.

Practice in Other Countries

Although bulk blending has increased spectacularly in the United States and now accounts for a large part of the total finished fertilizers, this practice has not been very widely adopted in other countries. Canada uses this system quite extensively. But there is limited practice of this concept in the United Kingdom, as well as in Brazil and other South American countries. Several blending plants were built in the Caribbean area. In Korea, ammonium phosphate and potash are cogranulated and additional nitrogen is provided by blending with prilled urea.

In one bulk-blending operation in England, an annual output of about 15,000 tons was reached in the second year of operation at a single well-located plant. Their goal of 50,000 tons was reached in the fifth year. The single location at seaside, with shipment of finished blends over a 90- to 100-mile radius, was decided upon after a logistical study of three alternative locations.

It is doubtful that the combination of factors that led to the popularity and rapid growth of bulk blending in the United States will be repeated to this extent in other countries. However, this system with its ability to concentrate the chemical processing in the hands of prime producers, and to simplify the final fertilizer preparation, will certainly be given consideration. The blending operation using intermediates eliminates one step in the marketing system, and the expense of bagging usually is avoided.

FLUID FERTILIZERS

Fluid fertilizers are of three types–liquids, suspensions, and slurries. Properly prepared, *liquids are clear solutions* that remain essentially free of any precipitates for extended periods. The liquids usually have a brown or green color derived from the wet-process acid used in their preparation. Clear liquid fertilizers of comparatively high analysis and high quality are dependent on ammonium polyphosphates in the intermediates used in their preparation. TVA researchers, starting in the late 1940's, pioneered in the development of polyphosphates and processes for preparation of liquid fertilizers of high quality.

Suspension fertilizers have substantial proportions of plant nutrients suspended in a finely divided state in the liquid. About two percent of a gelling type of clay is added to enhance suspending properties. *Slurry fertilizers do not have extended suspending properties.* They are not nearly so popular as suspensions, since they should be used as soon as is practical after preparation. Slurries usually are prepared by acidulation of phosphate rock with nitric and sulfuric or phosphoric acids, followed by ammoniation of the slurry.

Growth and Popularity of Fluid Fertilizers.

Fluid fertilizer use in the United States was only 27,000 tons in 1954. This increased to nearly 15 million tons of such fertilizers by 1979. This tonnage included anhydrous and aqua ammonia and nitrogen solutions applied directly, and accounted for nearly one-third of total fertilizers used.

Fluid-mixed fertilizers accounted for 18 percent of the U.S. total in 1979, representing a continuing steady increase from 12 percent in 1970.[45] The main fluid-fertilizer activity is in the United States, but there is substantial production in France and a few plants in Canada, Mexico, Belgium, Brazil, the Soviet Union, South Africa, and the United Kingdom.

The market continues to expand for fluids because of increased practice of direct application of ammonia and nitrogen solutions, and the continuing growth in popularity of suspensions.

The dominant features of the fluid production and marketing system are the simplicity of mixing facilities and the ease of handling. Also, fluid fertilizers can be applied uniformly at rates greater than an acre per minute with standard high flotation-type equipment. As farm mechanization increased, fluids became more popular. Uniform application, ease of adding pesticides and micronutrients, greater homogeneity, fewer dust problems, and adaptability to irrigation systems are some of the incentives for greater use of fluids. Custom application by dealers or farmer application with dealer-owned equipment constitute additional advantages.[46]

Fluid fertilizers supplied about 60 percent of the total nitrogen and 12 percent of the total P_2O_5 applied in 1978. There were about 300 fluid-fertilizer plants in the United States in 1950 and 750 in 1960. The number increased by a "quantum jump" to about 2,700 in 1970.[34] Estimates indicated that about 3,500 plants were operating in 1978.

Clear Liquid Fertilizers

TVA pioneered the development of liquid and suspension fertilizers. Development of polyphosphate liquid intermediates (11-33-0 and later 11-37-0) in the early 1960's by TVA, led to clear liquid grades of higher analysis with good storage properties. The polyphosphates sequester (hold in solution) the plant food elements and impurity contents for extended periods of storage. The base liquids are shipped to local plants and used to produce NPK grades such as 8-8-8, 8-16-8, and 7-21-7.[47] In 1971, TVA developed a very simple process for production of high-polyphosphate liquids. A pipe reactor is used in which ammonia reacts with superphosphoric acid of low (20 to 30%) polyphosphate content almost instantaneously. The heat of reaction drives off most of the water and forms a polyphosphate melt that is processed to 10-34-0 or 11-37-0 grade liquids of high quality with about 75 percent of the P_2O_5 as polyphosphate. Investment is quite low.[48,49] The polyphosphate liquid is often shipped to local plants for use in preparation of finished NPK liquid grades.

Materials, Equipment, and Processes. Ammonium polyphosphate liquid (11-37-0), made by ammoniation of electric-furnace superphosphoric acid (76-78% P_2O_5), was developed by TVA, and this intermediate was provided to the liquid fertilizer industry until about 1974. After that, the main step was wet-process superphosphoric acid (WPSA), which was also developed by TVA. Until 1971, the WPSA was produced at a concentration of 73 to 74 percent P_2O_5 to allow the preparation of mixed liquids with about 50 percent of the P_2O_5 as polyphosphate. This acid was ammoniated in a tank-type reactor in a "hot mix" type of operation.

After rapid development in pilot-plant studies, TVA introduced the simple pipe-reactor system for production of high-polyphosphate liquid fertilizers in 1971. Broad and intense interest by the U.S. liquid fertilizer industry developed quickly, and 135 commercial units of this type were documented in 1979. The pipe reactor is about the same as that used in melt granulation as described previously. However, wet-process superphosphoric acid (68-70% P_2O_5) of low polyphosphate content (20-35%) is used, and the absence of free water to be evaporated results in a much higher temperature of 650°-700°F. At this temperature, additional polyphosphate is rapidly formed so that 75-80 percent of the P_2O_5 in the melt is polyphosphate.

A typical commercial plant that uses the pipe reactor is shown in Figure 8.14. The mix tank is usually about 5 to 6 feet in diameter and 6 to 12 feet high. The inverted "U"-type pipe reactor used in this system is made of stainless steel. It usually has an overall length of about 20 feet. Most reactors are made of 4-inch standard pipe and tee section.

Superphosphoric acid is fed through one arm and gaseous ammonia through the other. Melt from the reactor is discharged into the mix tank where it is quenched with recirculating 10-34-0 product and reacted with additional ammonia so that the pH of the product is in the range of 5.8 to 6.2. Exit gases from the mix tank are scrubbed with cooled product and the water needed in the process. The temperature of the liquid in the mix tank is kept at 180°-200°F

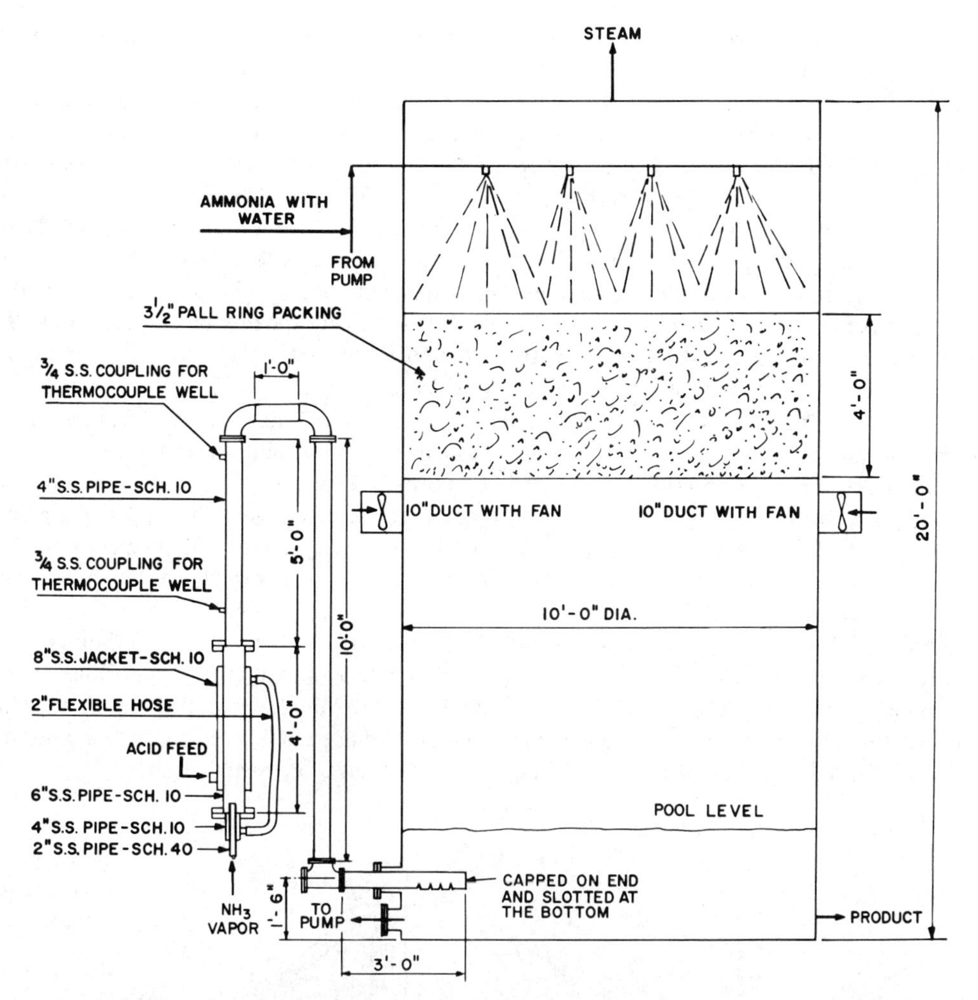

Fig. 8.14. TVA pipe-reactor system for liquid fertilizers of high polyphosphate content. (*Courtesy TVA.*)

by regulating the quantity of cooled product returned to the mix tank. Essentially, no hydrolysis of the polyphosphates occurs because of the short retention time in the mix tank. The hot liquid is used to vaporize the liquid ammonia used in the process, thereby conserving energy.

After the liquid is partially cooled in the vaporizer, it is finally cooled in a conventional evaporative-type cooler that is often mounted above the mix tank. Product from the cooler is pumped to storage. The high polyphosphate content of the liquids greatly facilitates sequestration of Fe, Al, and Mg impurities; and these liquids and NPK finished liquid products have good storage properties for extended periods.

The 10-34-0 or 11-37-0 clear liquid is blended in cold mix-type local plants with potash, urea-ammonium nitrate (UAN) solution, and sometimes other components, to prepare a variety of NPK grades. Some popular clear liquid grades include 8-8-8, 7-21-7, and 8-16-8. The comparatively low potash solubility is the limiting factor on concentration.

Suspension Fertilizers

TVA also developed and demonstrated suspension fertilizers which have substantial portions of small crystalline plant nutrients suspended in the fluid. About 2 percent of gelling clay is used as a suspending agent. Suspensions allow production of high-analysis grades with about twice the concentration found in clear liquid grades and near that of most granular NPK mixtures. They also provide a means of suspending secondary nutrients and micronutrients so that they can be applied uniformly. Probably the most important reason for the increased popularity of fluid mixtures and nonpressure nitrogen solutions is their use as carriers of herbicides. Experience has shown this is a way to accomplish uniform weed and pest kill. It also helps avoid soil compaction, because one pass across the field is eliminated.[50]

An orthophosphate suspension intermediate (13-38-0) was developed by TVA in the mid-1970's and is being used in demonstration programs in industry plants. A companion urea-ammonium nitrate (UAN) suspension with N content of 30 or 31 percent was developed by TVA and introduced to industry in 1977. A fluid clay suspension that contains 25 percent clay and 20 percent urea (9% N) also has been introduced by TVA. This pregelled source of clay is very convenient for use by producers of NPK suspensions to supply the 2 percent needed as the suspending agent. Equipment that would be needed for handling, feeding (metering), and gelling of dry clay is expensive and requires considerable attention.[51]

Since about 1975, there has been an upswing in use of solid intermediates in production of NPK suspensions. The main solid materials used are urea, potash, granular DAP (18-46-0), granular MAP (10-54-0), and nongranular MAP (11-52-0). There is considerable use of 11-55-0 granular ammonium phosphate (15% of the P_2O_5 as polyphosphate) introduced by TVA. Ammonia and wet-process phosphoric acid are optional materials.

A typical local mix plant for preparation of suspensions from solid materials and ammonia is shown in Figure 8.15. The main features of the plant are the large recirculating pump and the high-intensity mixer. Many plants of this type have annual production greater than 3,000 tons of NPK suspensions–such as 11-33-0, 20-10-10, and 7-21-21–and their hourly production rate is about 20 tons per hour. Cost studies show that suspensions made from solid materials can be produced about as economically as bulk blends of comparable grades.

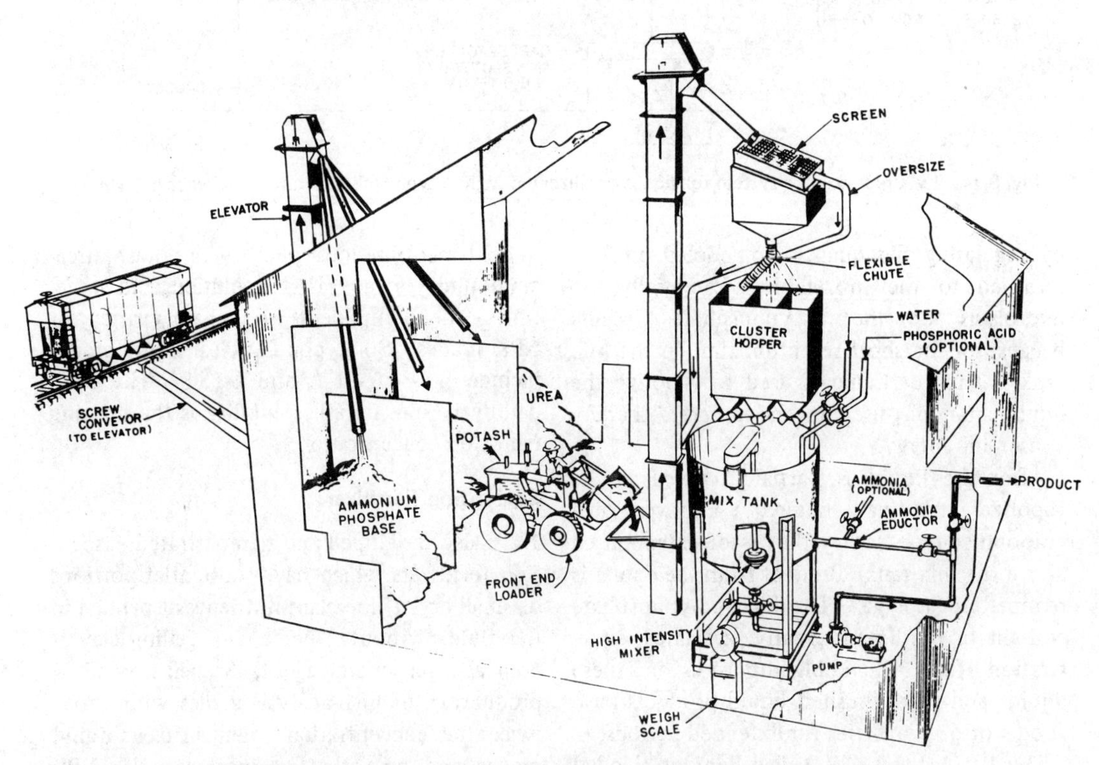

Fig. 8.15. Suspension fertilizer system using solid feeds. (*Courtesy TVA.*)

Suspension fertilizers have continued to increase in popularity in the United States. This has been particularly true in the southeastern states where the use of solids in preparation of suspensions is popular. It has been estimated that about 2 million tons of suspension mixtures (NPK) was produced in 1976. Suspension fertilizers are popular because of:

- Relatively high analysis–about twice that of clear liquid fertilizers
- Ease of handling when compared to solid mixed fertilizers
- Convenience of incorporating micronutrients, pesticides, and herbicides
- Fewer problems with air and stream pollution from fluid fertilizer plants

REFERENCES

1. Hignett, T. P., "History of World Fertilisers and Manufacturing Processes," *Indian Chemical Manufacturer* 9, No. 4, 13–17 (1971).
2. Young, Ronald D., and Achorn, Frank P., "Trends in U.S. Fertilizer Technology," *TVA Bulletin Y-133*, 5–7, Tennessee Valley Authority, Muscle Shoals, Alabama (Aug. 1978).
3. Young, R. D., "Phosphate Fertilizers, How Manufactured–Superphosphates, Phosphoric Acid, Ammonium Phosphates," *Fertilizer-Farm Chemical Workshops*, 74–92, University of Georgia (Nov. 1970).
4. *The Role of Phosphorus in Agriculture*, Am. Soc. Agron., Crop Science Soc. Am., and Soil. Science Soc. Am., Madison, Wisconsin, 198–199 (1980).
5. Ibid., p. 199.
6. Ibid., p. 201.
7. Young, R. D., Hicks, G. C., and Davis, C. H., "TVA Process for Production of Granular Diammonium Phosphate," *J. Agr. Food Chem.* **10**, 442–7 (Nov. 1962).
8. Young, R. D., and Hicks, G. C., "Production of Monoammonium Phosphate in a TVA-Type Ammonium Phosphate Granulation System," *Com. Fert.* **114**, No. 2, 26–7 (Feb. 1967).
9. Blouin, G. M., "Urea Costs–Granulation vs. Prilling," *TVA Bulletin Y-92*, Tennessee Valley Authority, Muscle Shoals, Alabama (April 1975).
10. Bridges, J. Darwin, "Fertilizer Trends 1979," *TVA Bulletin Y-150*, p. 23, Tennessee Valley Authority, Muscle Shoals, Alabama (Jan, 1980).
11. Harris, Gene T. and Harre, Edwin A., "World Fertilizer Situation and Outlook 1978–85," 8–13, International Fertilizer Development Center, Muscle Shoals, Alabama (March 1979).
12. Hignett, T. P., "Nitrophosphate Processes Advantages and Disadvantages," *Proc. 15th Annual Meeting Fert. Ind. Round Table 1965*, 92–5.
13. Hignett, T. P., and Newman, E. L., *Liquid Fertilizer Round-up, Proc.*, National Fertilizer Solutions Association, St. Louis, Missouri, 32–3 (July 1967).
14. Harris, Gene T., and Harre, Edwin A., op.cit., p. 8.
15. *The Role of Phosphorus in Agriculture*, Am. Soc. Agron., Crop Science Soc. Am., and Soil Science Soc. Am., Madison, Wisconsin, pp. 215–217 (1980).
16. Lehr, James R., and McClellan, Guerry H., "Phosphate Rocks: Important Factors in Their Economic and Technical Evaluation," *CENTO Symposium on the Mining and Beneficiation of Fertilizer Minerals*, 194–242 (Nov. 19–24, 1973).
17. Heil, F. G., Young, R. D., and Stumpe, J. J., "Pilot-Plant Study of an Ammonium Fluoride Process for Recovery of Fluorine from Superphosphate," *J. Agr. Food Chem.* 9, 457–9 (Nov./Dec. 1961).
18. English, M., "Fluorine Recovery from Phosphatic Fertilizer Manufacture," *Chem. and Proc. Eng.* 48, No. 12, 43–7 (Dec. 1967).
19. "Uranium Recovery from Phosphoric Acid–The Gardinier-Uranium Pechiney Ugine Kuhlmann Process," *Phosphorus and Potassium* No. 99, 31–3 (Jan./Feb. 1979).
20. *Chem. Week* **125**, No. 15, 42 (Oct. 10, 1979).
21. Houghtaling, Samuel V., "Wet Grinding of Phosphate Rock Holds Down Dollars, Dust, and Fuel," *Engineering and Mining Journal* 176, No. 1, 94–96 (Jan. 1975).
22. Blouin, G. M., "Effects of Increased Energy Costs on Fertilizer Production Costs and Technology," *TVA Bulletin Y-84*, Tennessee Valley Authority, Muscle Shoals, Alabama (Nov. 1974).
23. Adams, Samuel S., *The Role of Potassium in Agriculture*, Chapter 1, American Society of Agronomy, 1968.
24. Kapusta, E. C., *The Role of Potassium in Agriculture*, Chapter 2, American Society of Agronomy, 1968.
25. "Potash, Europe Increases Production in 1978," *Phosphorus and Potassium* no. 99, p. 7, (Jan/Feb 1979).
26. *Chem. Eng. News* 58, No. 11, 13 (March 17, 1980).
27. Young, Ronald D., *The Role of Potassium in Agriculture*, Chapter 3, American Society of Agronomy, 1968.

28. Spealman, Max L., "New Route to Chlorine and Saltpeter," *Chem. Eng.* 72, No. 23, 198–200 (Nov. 8, 1965).
29. Young, R. D., "Production of Compound Fertilizers from Intermediates in Local Plants," op.cit., p. 14.
30. Bridges, J. Darwin, op.cit., p. 7.
31. Mackall, J. N., and Shoeld, Mark, "Granulating Phosphate Fertilizers," *Chem. and Met. Eng.* 47, 102–5 (1940).
32. *Farm Chemicals 1970 Handbook*, **134**, Meister Publishing Company, Willoughby, Ohio.
33. Bridges, J. Darwin, op.cit., pp. 6–8.
34. Hignett, Travis P., "Liquid Fertilizer Production and Distribution," *TVA Circular Z-27*, 7, Tennessee Valley Authority, Muscle Shoals, Alabama (Oct. 1971).
35. Yates, L. D., Nielsson, F. T., and Hicks, G. C., "TVA Continuous Ammoniator for Superphosphates and Fertilizer Mixtures," Part I *Farm Chemicals* **117**, 38, 41, 43, 45, 47, 48 (July 1954). Part II, Ibid. **117**, 34, 36–8, 40–1 (August 1954).
36. Young, R. D. "Production of Compound Fertilizers from Intermediates in Local Plants," *TVA Bulletin Z-30*, 12–20, Tennessee Valley Authority, Muscle Shoals, Alabama (Sept. 1971).
37. Achorn, Frank P., and Lewis, J. S., "Producing Granular Fertilizers," *TVA Circular Z-18*, Tennessee Valley Authority, Muscle Shoals, Alabama (May 1970).
38. Young, R. D., "Production of Compound Fertilizers from Intermediates in Local Plants," op.cit., pp. 21–27.
39. Ibid., pp. 27–33.
40. Hignett, Travis P., *Bulk Blending of Fertilizers: Practices and Problems*, The Fertilizer Society, Proceedings No. 87, London (March 25, 1965).
41. *Application of Granular Fertilizers*, Circular Z-12, Tennessee Valley Authority, Muscle Shoals, Alabama (Sept. 1970).
42. Hoffmeister, George, Watkins, S. C., and Silverberg, Julius, "Bulk Blending of Fertilizer Material: Effect of Size, Shape, and Density on Segregation," *J. Agr. Food Chem.* **12**, 64–9 (Jan./Feb. 1964).
43. Young, R. D., "Production of Compound Fertilizers from Intermediates in Local Plants," op.cit., pp. 30–32.
44. Hoffmeister, George, "Compatibility of Raw Materials in Blended Fertilizers–Segregation of Raw Materials," *Proc. 12th Annual Meeting Fert. Ind. Round Table 1962*, 83–8.
44a. Young, Ronald D., "Providing Micronutrients in Bulk-Blended, Granular, Liquid, and Suspension Fertilizer," *Commercial Fertilizers* **118**, No. 1, 21–4 (Jan. 1969).
45. Bridges, J. Darwin, op.cit., p. 9.
46. Smith, Yates C., and Culp, John E., "Service Sells Fluid Fertilizers in Texas," *Fert. Soln.* **22**, No. 6, 86, 88, 90, 92, 94 (Nov./Dec. 1978).
47. Huffman, E. O., and Newman, E. L., "Polyphosphates are Revolutionizing Fertilizers, Part IV Behavior and Outlook," *Farm Chemicals* **133**, No. 2, 28–30, 32 (Feb. 1970).
48. Meline, R. S., Lee, R. G., and Scott, W. C., "Use of a Pipe Reactor in Production of Liquid Fertilizers with Very High Polyphosphate Content," *Fert. Soln.* **16**, No. 2, 32–45 (March/April 1972).
49. Achorn, F. P., Kimbrough, H. L., and Meyers, F. J., "Latest Developments in Commercial Use of the Pipe Reactor Process," *Fert. Soln.* **18**, No. 4, 8–9, 12, 14, 16, 20–1 (July/Aug. 1974).
50. Achorn, Frank P., and Kimbrough, Homer L., "New Processes and Products for the Fluid Fertilizer Industry–Phosphates," *Proceedings 1978 Round-Up*, National Fertilizer Solutions Association, Kansas City, Missouri, 23–31 (July 20–21, 1978).
51. Scott, W. C., Burns, M. R., and Watson, J. R., "New Developments Streamline Suspension Fertilizer Production," *Agrichemical Age*, **22**, no. 2, pp. 10, 12, 14 (March 1978).

9

Rubber

E. E. Schroeder*

INTRODUCTION

The term rubber was derived from the use of a resinous material as an eraser, an application popularized about 1770 by the renowned chemist Joseph Priestley. Today, rubber not only suggests the resinous substance from a tree or a synthetic polymer with elastic properties, but also represents the generic term for products such as tires, hose, and belting.

Rubber is one of the most essential of the world's raw materials. The strategic position of rubber in our civilization is evident from the effect that wars have had on the supply of rubber and from the way in which countries have reacted to find substitutes when the supply of natural rubber was threatened.

During World War I, Germany produced 2500 tons of methyl rubber from 2,3-dimethylbutadiene, which they synthesized with considerable difficulty. This was part of a program to make them independent of curtailed supplies of natural rubber. Methyl rubber was better suited for making hard rubber, and its production was stopped as natural rubber became available at the end of the war.

After World War I, research in the synthetic rubber field continued in Germany, and production of the numbered "Bunas" was started. This name was developed from *Bu*tadiene *Na*trium, since butadiene was polymerized by sodium in the German process.

Three types, buna 32, 85, and 115, were made in this series and used for special applications. Buna 32 was a softener used in rubber products, as well as hard rubbers. Buna 85 and 115 saw some use in conventional rubber applications.

Emulsion polymerization entered the picture in Germany in 1927, first with the polymerization of butadiene, and then with its copolymerization with styrene and with acrylonitrile. By the beginning of World War II, these bunas (identified by letters S, SS, N) had achieved commercial importance, and a number of types were produced as rubber substitutes when natural rubber supplies were cut off.[31]

The first synthetic rubbers to be commercially available in the United States were "Thiokol"

*The Firestone tire and Rubber Co.—Presently with Shintech, Inc., Freeport, Texas.

(The author wishes to acknowledge the use of material from an earlier edition prepared by Dr. R.L. Bebb.)

TABLE 9.1 Production Capacity[a] of U.S. Synthetic Rubber

Type Code	*Description*	*Capacity Metric Tons*
SBR	Emulsion Styrene/Butadiene Rubber	1,261,979
HS/B	Emulsion High Styrene&Butadiene Latex	345,100
BR	Butadiene Rubber–Solution	415,072
SSBR	Solution Styrene/Butadiene Rubber	209,100
IIR	Butyl or Isobutylene–Isoprene Rubber	227,500
IR	Isoprene Rubber	60,000
EPDM	Ethylene Propylene Terpolymer	223,110
CR	Neoprene or Polychloroprene	209,995
NBR	Acrylonitrile-Butadiene or Nitrile Rubber	92,000

[a]The Rubber Industry Statistical Report by Clayton F. Ruebensaal (Copyright 1979), International Institute of Synthetic Rubber Producers, Inc. 2077 South Gessner, Suite 133, Houston, Texas 77063.

(1930) and "Neoprene" (1931), or "Duprene" as it was first called. Both of these are still being produced commercially because they have special properties that are not matched by other synthetics or natural rubber. Prior to the beginning of the World War II, a number of U.S. laboratories had investigated the German products and had initiated programs of their own in order to develop practical synthetic rubbers both for tire applications and for specialty uses. By the date of Pearl Harbor, December 7, 1941, Firestone had over 100,000 tires tested on the road. Other companies also had active programs.

The capacity of the synthetic rubber plants in the United States alone (Table 9.1) and the spectrum of properties offered to the manufacturers of rubber goods has led to a usage pattern (Table 9.2) heavily weighted in favor of the synthetic rubbers. While similar synthetic rubbers are being produced throughout the world, a higher proportion of natural rubbers is being used outside the United States. (Table 9.3)

The United States synthetic rubber industry was a very effective stabilizer for the price of natural rubber. After the start of the Korean War, the price of natural rubber rose from 15 to 73.4 cents in February 1951. Synthetic rubber replaced an increasing amount of the natural rubber then in use and the price of natural rubber was forced down as a result. As natural rubber came down to 20 cents in 1954, some consumers changed from synthetic rubber to the natural product, but with each rise in the price of natural rubber, synthetic rubber became more established and a swing back to natural rubber was resisted. Synthetic rubber now has the advantage of having many suppliers who are located in the United States and who produce a uniform product at competitive prices.

As the United States entered into World War II, many companies entered into research, development, and production contracts with the government. In the period from December, 1941, through April, 1955, these companies conducted, under government sponsorship, closely coordinated programs in which back-

TABLE 9.2 Rubber Consumption in U.S., 1978 (Metric Tons)

Total New Rubber	3,201,053
Natural	764,654
Dry	708,753
Latex	55,901
Total Synthetic	2,436,399
% of Total New Rubber	76.11%
SBR	1,417,168
Dry	1,259,697
Latex	257,471
Polybutadiene	406,314
Butyl	147,886
Nitrile	73,940
Dry	62,202
Latex	11,738
Polyisoprene	75,804
Ethylene-Propylene	140,025
Polychloroprene	120,947
All Other	54,315
*Reclaimed	118,732

*Not included in Total New Rubber

RMA Statistical Report, Rubber Manufacturers Association, 1901 Pennsylvania Avenue, N.W., Washington, D.C. 20006.

TABLE 9.3 1978 World Production and Consumption of Rubber

	Production (MT)	*Consumption (MT)*
Natural Rubber	3,715,000	3,715,000
Synthetic Rubber	8,855,000	8,760,000
		Percentage Consumption of Synthetic Rubber
U.S.A.		76.6 (est.)
Total, World		70.2 (est.)

[1] Rubber Statistical Bulletin, Volume 34, No. 4 (January 1980).

ground and experience were exchanged on the production of synthetic rubber. This close cooperation, especially at the beginning of the program, enabled a rapid buildup of the synthetic rubber industry.

The U.S. Government's synthetic rubber program provided that the plants should ultimately be sold. With the crisis over, the government sold the plants in 1955 to private companies who, in turn, were required to maintain the plants so that GR-S type synthetic rubber could be produced in an emergency. Production capacity for both monomers and rubber has increased considerably under private operation. Today, there are a number of synthetic rubber producers who were not active in the government program, and the types of synthetic rubber have been considerably expanded, especially by the new stereospecific solution rubbers, which will be discussed later in this chapter.

The following are some important dates in the commercial history of synthetic rubber:

- 1914–18 Methyl rubber produced in Germany.
- 1930 Thiokol, an organic polysulfide rubber resistant to oils and solvents, introduced in the U.S.
- 1931 Neoprene (originally DuPrene), a polymer of 2-chlorobutadiene-1,3, production started.
- 1933 Buna S, a butadiene-styrene copolymer, made in emulsion in Germany.
- 1936 Perbunans, or Buna N, specialty oil-resistant rubbers from butadiene and acrylonitrile, manufactured in Germany.
- 1939 Mercaptan-persulfate emulsion recipes patented. French Patent 843,903 (July 12, 1939) served as the basis for the butadiene-styrene "Mutual" hot-polymerization recipe later adopted as standard for hot GR-S production in the U.S.
- 1940 Butyl rubber, a copolymer of isobutylene and isoprene characterized by a very low permeability to air and especially suited to inner tubes, production started.
- 1942 GR-S (hot rubber), a copolymer of butadiene and styrene, produced in the U.S.
- 1944 Silicone elastomers, characterized by retention of elastomeric properties over a wide range of temperatures introduced.
- 1946 Polyurethane rubbers, prepared by diisocyanate-coupling of dihydroxy compounds, introduced in Germany.
- 1947 Cold rubber, a copolymer of butadiene and styrene emulsion polymerized at 41°F, produced in the U.S.
- 1954–55 Synthetic natural rubber, prepared by the polymerization of isoprene, announced.
- 1955 Government-owned synthetic rubber plants purchased by private industry; expansion of production capacity begun.

This chapter will include a brief review of the raw materials used in the synthetic rubber industry, a description of both the emulsion and nonaqueous routes for manufacturing synthetic rubbers, the production of natural rubber and natural latex, a review of the compounding of rubbers, and the evolution of thermoplastic elastomers.

SYNTHETIC RUBBERS BY THE EMULSION PROCESS

Butadiene-Styene Copolymers

The flow sheet shown in Fig. 9.1 represents the route used for making the synthetic rubber which is produced in a larger volume than any other elastomer in the world. Butadiene ($CH_2{=}CH{-}CH{=}CH_2$) and styrene ($C_6H_5{-}CH{=}CH_2$) are mixed with an appropriate emulsifying solution, catalyst, and modifying agent (mercaptan), in an agitated pressure vessel with suitable temperature control until the desired conversion of the monomers (butadiene and styrene) is achieved. The polymerization is then stopped with a suitable chemical, the latex is stripped of unreacted monomer, antioxidant is added, and the rubber is isolated from the latex by coagulation with salt (NaCl), salt-acid, or aluminum sulfate solution.

Polybutadiene rubber can be made in the same way by omitting the styrene; oil-resistant

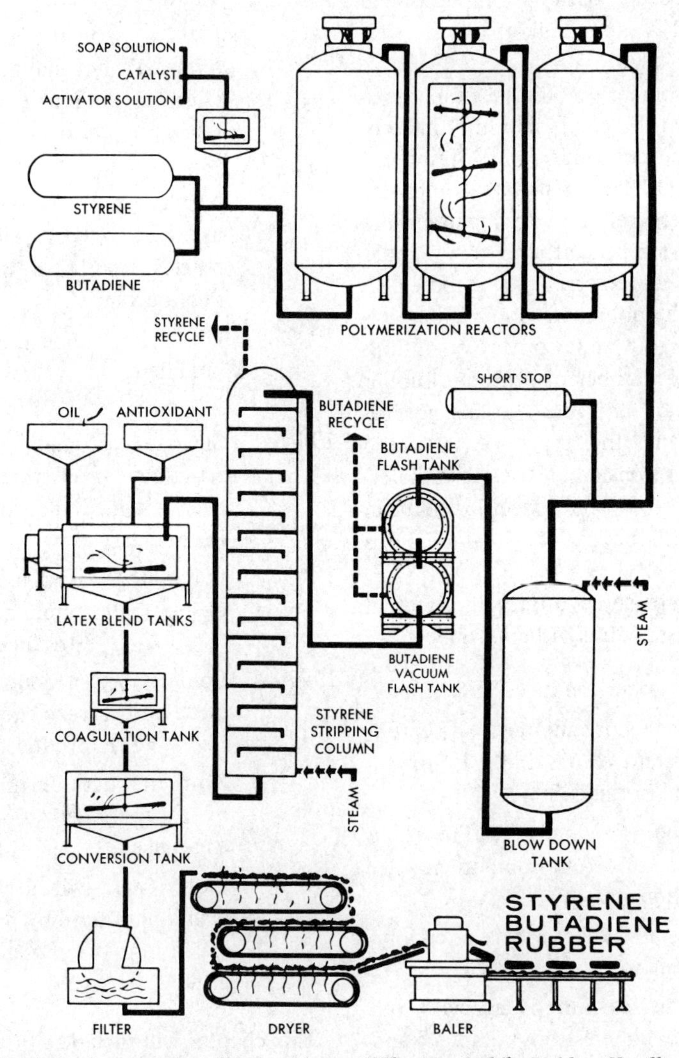

Fig. 9.1. The production of styrene-butadiene rubber. (*"The Vanderbilt Rubber Handbook,"* *R. O. Babbit, ed.; by permission of the R. T. Vanderbilt Co.*)

rubbers result from substituting acrylonitrile for the styrene. The formula adopted March 26, 1942, by the Technical Advisory Committee of Rubber Reserve, for the preparation of "hot" rubber, has remained the basis for its commercial production, with only minor modifications:[18]

	Parts Per 100 Monomer
Butadiene	75.0
Styrene	25.0
Water	180.0
Soap	5.0
"Lorol" mercaptan (n-$C_{12}H_{25}SH$)	0.50
Potassium persulfate	0.30
Polymerization temperature	50°C
Time	12 hours
Conversion	75%

The object of the polymerization is to form rubbery polymers of butadiene and styrene of a desired molecular weight. Polymerization is initiated by potassium persulfate and the molecular weight is regulated by a mercaptan. The role of the mercaptan in this polymerization is particularly important because its type and its concentration regulate the molecular weight of the polymer and determine its processing characteristics.

Chemical reactions occurring in hot (50°C) emulsion polymerization are given below:

A review of the mechanism of emulsion polymerization may be found in References 5, 22, 61 and 64.

The polymerization reactors are designed to withstand pressures over 100 psig. The temperature of the reactants is adjusted to the desired level by controlling the temperature in the reactor jacket or in the internal coils.

The reactors vary in size from 3700 gallons, in the early plants, to over 5000 gallons. They are either constructed of stainless steel or of iron clad with stainless steel; they are jacketed and agitated. Different types of agitators have been used.

The selection depends considerably on the type of mercaptan used, since the efficiency of the mercaptan action is affected by the type and rate of agitation. In hot polymerization, the pressure within the reactor may reach 70 psig, dropping to about 40 psig at approximately 70% conversion of the monomer. The reactors are usually connected in series and may be equipped for either batch or continuous operation.[46] The extent of polymerization is monitored by measurement of the solids content in the aqueous emulsion.

During polymerization, the hydrocarbons are emulsified in the soap solution; a portion is dissolved in the soap micelles where polymerization is initiated. As polymerization proceeds, the soap micelle loses importance and the balance of the reaction takes place in the

(1) $K_2S_2O_8 + C_{12}H_{25}SH \longrightarrow C_{12}H_{25}S^{\bullet}$ — Generation of free radicals

Potassium persulfate; Dodecyl mercaptan; Mercaptyl free radical

(2) $C_{12}H_{25}S^{\bullet} + CH_2{=}CHCH{=}CH_2 \longrightarrow C_{12}H_{25}SCH_2CH{=}CHCH_2^{\bullet}$ — Initiation

Butadiene; Monomer radical

(3) $C_{12}H_{25}SCH_2CH{=}CHCH_2^{\bullet} + n(CH_2{=}CHCH{=}CH_2) \longrightarrow C_{12}H_{25}S(CH_2CH{=}CHCH_2)_{n+1}^{\bullet}$ — Propagation

Butadiene; Homopolymer radical

and/or

$m(C_6H_5CH{=}CH_2) \longrightarrow C_{12}H_{25}S(CH_2CH{=}CHCH_2)(CH_2\underset{\displaystyle C_6H_5}{\underset{|}{C}H})_m^{\bullet}$

Styrene; Copolymer radical

(4) $C_{12}H_{25}S(CH_2CH{=}CHCH_2)_{n+1}^{\bullet} + C_{12}H_{25}SH \longrightarrow C_{12}H_{25}S(CH_2CH{=}CHCH_2)_{n+1}H + C_{12}H_{25}S^{\bullet}$ — Termination and chain transfer

Polymer

polymer-monomer particles.[25] Butadiene and styrene do not copolymerize at the same rate. In a 75/25 charge, the initial polymer contains 17.2% styrene, whereas at 80% conversion the polymer contains 21.2% styrene and at total conversion, 25.0%.

After approximately 12 hours, the polymerization is terminated at about 70% conversion, since higher conversions lead to polymers with inferior properties, presumably because of crosslinking to form highly branched structures. Termination is affected by the addition of a "shortstop" such as a hydroquinine which destroys the free radicals and arrests further polymerization. At this stage, the rubber is contained in a stable milky suspension known as latex. Before being used as such, or coagulated to isolate the rubber, the latex must be stripped of unreacted monomers. This is done in two steps; butadiene is removed by exposing the latex in flash tanks to atmospheric and reduced pressure, while styrene is removed by steam stripping in a vacuum distillation column. Before the latex is coagulated, an antioxidant must be added to prevent deterioration of the rubber during drying, storage, and processing. The latex is then coagulated with an acidified brine solution, or by means of an aluminum sulfate solution. One practice involves creaming the latex with concentrated brine and adding it to an agitated vessel along with dilute acid until the particle size of the crumb reaches the desired value. The fatty acids are released from thc soap on the surface of the copolymer particles and large porous crumb aggregates are formed. The crumbs pass over a vibrating screen or a dewatering press to remove the serum.[10] The rubber is reslurried, washed, and filtered. It then passes over a continuous belt dryer where the moisture content is reduced to less than 0.75%. The dried rubber is pressed into 75 pound bales for shipment.

The finished butadiene-styrene rubber contains fatty acids, rosin acids, antioxidants, moisture and some inorganic materials, mainly sodium chloride.

Since termination of the government program, many types of hot rubber have been introduced to the trade for different uses. These represented variations in butadiene-styrene ratios, changes in plasticity, changes in conversion, and changes in antioxidant and in methods of coagulation. Important objectives have been improvements in properties, processing characteristics, and color.

Cold Rubber

Throughout the government program, there was a persistent belief that a better rubber could be made if lower polymerization temperatures were used. Efforts to activate the hot recipe at lower temperatures were not particularly successful and the polymerization time was long at temperatures lower than 50°C (122°F). In the early part of the program, a number of revised recipes were proposed, but these did not reduce the polymerization temperature substantially below 30°C. There was no major improvement in the quality of this rubber over that of a normal hot control.

The successful development of ways to accelerate polymerization at low temperatures centered around the "Redox" system which was developed independently in Germany, Great Britain, and the United States. The Redox system involves the presence of an oxidizing agent, usually a peroxide or hydroperoxide, a reducing agent, and a soluble salt of a metal capable of existing in several states of oxidation.

Chemical reactions in cold (5°C) emulsion polymerization area shown on page 287.

At the end of the war, it was learned that the Germans had developed a system using benzoyl peroxide as a oxidizing agent and sugar as the reducing agent. Good rubbers were produced from these systems in the United States, but polymerization characteristics were considered unsatisfactory. The German process was therefore modified in the United States by the introduction of hydroperoxides and by the replacement of sugar with amines. Since synthetic rubber plants are now privately owned, polymerization recipes are no longer published. It is felt, however, that the "cold" recipe still remains a variation of the SFS type developed toward the end of the government program. This was such a dependable recipe in 1955 that

(1) $ROOH + Fe^{++} \longrightarrow RO^{\bullet} + Fe^{+++}$ — Generation of free radicals

Diisopropylbenzene hydroperoxide; Ferrous ion; Peroxyl free radical; Ferric ion

(2) $RO^{\bullet} + CH_2{=}CHCH{=}CH_2 \longrightarrow ROCH_2CH{=}CHCH_2^{\bullet}$ — Initiation

Butadiene; Monomer radical

(3) $ROCH_2CH{=}CHCH_2^{\bullet}$ +

$n(CH_2{=}CHCH{=}CH_2) \longrightarrow RO(CH_2CH{=}CHCH_2)^{\bullet}_{n+1}$ — Propagation

Butadiene; Homopolymer radical

and/or

$m(C_6H_5CH{=}CH_2) \longrightarrow RO(CH_2CH{=}CHCH_2)(CH_2CHC_6H_5)^{\bullet}_m$

Styrene; Copolymer radical

(4) $RO(CH_2CH{=}CHCH_2)^{\bullet}_{n+1} + RSH \longrightarrow RO(CH_2CH{=}CHCH_2)_{n+1}H$ — Termination and chain Transfer

Polymer + $RS^{\bullet}$ mercaptan radical

it continues to be in use with, at most, small modification:

	Parts per 100 Monomer
Butadiene	71.5
Styrene	28.5
Water	200.0
Mixed tert.-mercaptans	0.125–0.15
Potassium fatty acid soap	4.7
"Daxad-11"	0.1
KCl	0.5
$FeSO_4\ 7H_2O$	0.004
Sodium formaldehyde sulfoxylate (SFS)	0.0228
Ethylene diamine tetraacetic acid (Sequestrene AA)	0.0246
NaOH	0.0024
Diisopropylbenzene hydroperoxide	0.03–0.10
Sodium dimethyl dithiocarbamate (SDD) Stopping Agent	0.10

With the introduction of cold polymerization, it was found that the physical properties of synthetic polymers and the wear characteristics of tires made from them were superior to those of hot rubbers.[57] As a consequence, the synthetic rubber program moved increasingly toward cold rubbers. Variations of the polymerization recipe resulted in smoother polymerization rates, better reaction times and more reproducible performance. The production equipment used for cold polymerization is similar to that used for hot rubber manufacture, except provision is made for cooling the reactors with a refrigerant in jackets or in coils within the reactor.

Developments in SBR production have resulted in separate grades for specific uses.[4] High molecular weight elastomers have been modified with the addition of up to 50 phr of petroleum base oils to permit easy factory processing. Light colored rubbers have been developed for non-tire applications by using lighter colored emulsifiers, antioxidants and extending oils. Carbon black masterbatch types have been developed for manufacturers having limited mixing capacity or who wish to avoid handling loose carbon black. The following numbering system was developed under the government synthetic rubber program to classify polymers; it is still in use except for newer products:

Series	
1000	hot polymers
1500	cold polymers
1600	cold black masterbatch with 14 or less phr oil
1700	cold oil masterbatch
1800	cold oil-black masterbatch with more than 14 phr oil
1900	miscellaneous dry polymer masterbatches
2000	hot latices
2100	cold latices

Butadiene-Acrylonitrile Copolymers

Butadiene-acrylonitrile copolymers are produced in much the same way as butadiene-styrene rubber. Because this rubber was not made under government operation, there are many trade names for its various types. Each supplier offers a range of plasticities and acrylonitrile contents, thereby achieving considerable variation in oil resistance. Some are stabilized with discoloring anti-oxidants; others contain light-stable materials for use in special applications.

$$n\mathrm{CH_2{=}CH(CN)} + m\mathrm{CH_2{=}CH{-}CH{=}CH_2} \longrightarrow$$

Acrylonitrile   Butadiene

$$(-\mathrm{CH_2{-}CH{=}CH{-}CH_2{-}CH_2{-}CH(CN){-}})_{n+m}$$

Butadiene-acrylonitrile polymer

The outstanding characteristics of the acrylonitrile rubbers are their resistance to oils, fats and solvents in general, and their performance at low and high temperatures (-70 to 300°F).

Since these rubbers are unsaturated, as are the butadiene-styrene rubbers, they may be vulcanized with sulfur in the classical compounding routes. Latices of butadiene-acrylonitrile copolymers may be used for making foam rubber products, as well as special purpose papers. An extensive review on nitrile elastomers is given in reference 19.

Neoprene[9, 20, 64]

The polymerization of chloroprene to neoprene has been described by Walker and Mochel,[48] and resembles the preparation of butadiene-styrene rubbers.

	Parts per 100 Monomer	
Chloroprene	100	
N-Wood rosin	4.0	Dissolved in
Sulfur	0.6	the monomer
Water	150	
Sodium hydroxide	0.8	
Sodium salt of naphthalene sulfunic acid formaldehyde condensation product	0.7	
Potassium persulfate	0.2-1.0	

Change in the specific gravity of the contents of the reactor during polymerization is large enough for the conversion to be followed by measuring that propery. At the desired conversion, tetramethylthiuram disulfide is introduced to stop polymerization; the latex is then ready for removal of unreacted chloroprene and coagulation. The stripping is done by vacuum steam distillation.

One method for isolating neoprene from the latex involves freezing.[69] Acetic acid is added to the alkaline latex just short of coagulation. The sensitized latex is then passed over the surface of a large brine-cooled drum where a sheet is frozen and rubber is coagulated. The rubber is removed from the drum, washed with water, passed through squeeze rolls, and dried in air at 120°C. Then the dried film is gathered into ropes and cut into short lengths for bagging and use. The finished rubber has a specific gravity of 1.23.

There are a number of different types of neoprene polymers available. These are sulfur-modified polymers (GN, GNR, GRT), which can be cured by the addition of metal oxides alone. The unmodified types (W and WRT) require accelerators as well as the metal oxides. For vulcanization, the neoprene rubbers are usually compounded with light-calcined magnesia and zinc oxide; cross-linking occurs at the 1,2-addition sections. The cured strength may reach 4000 psi with an elongation of 900% and a 600% modulus value near 1000 psi.

Neoprene is both stable to oxidation and flame resistant; it swells only moderately in oils and chemicals and has good retention of properties in the swollen state. It has been used in wire insulation, cable jackets, gaskets for aliphatic liquids, belts for power transmission, and conveyor belt covers, especially where oil and heat are encountered. Since it is available as a latex, there are many applications where its unusual properties are an advantage.

Latex[6,27,67]

A latex is a stable dispersion of a polymeric substance in an aqueous medium. The dispersed component of the two-member system thus consists of essentially spherical particles of varying size suspended in a continuous phase containing suitable emulsifiers, or stabilizers.

Natural latex came into commercial use in the early 1920's, and by the beginning of World War II had become an important raw material. It was, therefore, logical that as the natural latex supply became limited, synthetic rubber latex should be substituted as far as possible.

In the commercial manufacture of synthetic rubbers where water and a dispersing system are charged into the reactor, the process also forms a latex from which the rubber is isolated. A disadvantage of latices made in early polymerization recipes was the very small particle size of the dispersed phase. As these latices made by "hot" recipes were concentrated, the viscosity increased so rapidly that 40% solids was the upper limit. The low-temperature "cold" polymerization recipes with proper selection of dispersing agents led to 62-65% solids latex (Type GR-S 2105).

The total-solids percentage to which a latex can be concentrated is limited by the packing effect of the particles themselves; thus, a latex of very small particle size is limited to 40% solids. Some improvement can be made by adding electrolytes, such as trisodium phosphate, to the charge; but a greater improvement results from increasing the particle-diameter so that an optimum of 74% of the rubber occurs as large particles. Several methods have been used commercially for increasing this diameter by destabilizing the particles: (1) careful additions of hydrocarbon; (2) freezing the latex under rigid pH control; (3) passing the latex through a tight colloid mill (pressure process); and (4) addition of a polyether glycol (chemical process). The agglomeration may be effected in the presence of other latices such as polystyrene.

Latices have been prepared to a more limited extent by dissolving rubbers in suitable solvents, dispersing the resulting solutions, and subsequently recovering the solvent. The "solution" rubbers described later are well suited to this technique, and latices of high *cis*-polyisoprene have been offered commercially. This process permits the latex to be tailored for particular applications by changing the type and level of the stabilizing system. Since one or more additional steps are involved in manufacturing these latices, they may be more expensive to produce than the latices produced by emulsion polymerization.

Synthetic latices are widely used in foam rubbers, paper-coating processes, saturation of wet and dry web in paper manufacture, carpet backings, adhesives, tire-cord treatments and "latex" paints.

NONAQUEOUS METHODS OF POLYMERIZATION

Nonaqueous processes for making synthetic rubbers have a commercial history dating back to the German numbered Buna rubbers, the Russian polybutadienes, polyisobutylene (Vistanex) in Germany, butyl rubber in the United States, Thiokol (U.S.), and silicone rubbers (U.S.). These processes cover the spectrum extending from those methods which produce an insoluble polymer which separates during polymerization (butyl rubber) to the numbered Bunas which remain in solution throughout.

Reaction conditions for the nonaqueous polymerizations are quite different from those in emulsion. In these processes, the main considerations are the solubility of the catalyst, its sensitivity to impurities, and the viscosity of the polymer solution (or cement) in the reactor system. The concentration of impurities must be monitored and controlled more carefully than in the emulsion processes, since even a few parts per million of water, combined sulfur, or carbonyls, as well as other reactive materials can have pronounced effect on the polymerization. Because viscosity of the cement increases exponentially with molecular weight and concentration, the control of reaction conditions can be difficult. Yet, the nonaqueous routes offer the greatest promise for new products and are becoming more important than the emulsion process previously described.

Butyl Rubber[64]

Butyl rubber is a copolymer of isobutylene with about 3% isoprene polymerized in a solvent at temperatures near 150°F. The catalyst, aluminum chloride, is added as a dilute solution in methyl chloride. As polymerization proceeds, the polymer precipitates and the resultant slurry overflows into an agitated hot-water tank. Unreacted monomer flashes off and is recovered by drying and condensation. An antioxidant is added to prevent deterioration during drying and storage. The crumb is passed through a tunnel dryer at 200 to 350°F to remove most of the water; then it is fed into an extruder and onto rubber mills. It is removed continuously to the suitable packaging machine.

Butyl rubber resembles natural rubber in appearance, but it cannot be made into hard rubber products because of extremely low (3%) unsaturation. Since butyl rubber is the least permeable to gases of all rubbers, much of it is consumed in the manufacture of inner tubes for automobile tires. It is also of considerable value in making air bags for curing tires where repeated exposure to high temperatures would greatly shorten the life of normal rubbers.

High *Cis*-Polyisoprene (Synthetic Natural Rubber)

In preparing butadiene-styrene polymers during World War II, it was recognized that the products were not true duplicates of natural rubbers. Although Hevea is composed of isoprene units, an emulsion polyisoprene does not duplicate the high gum tensile strength of natural rubber nor many of its other properties. Furthermore, emulsion diene polymers are composed primarily of *trans*-diene units, whereas natural rubber is largely *cis*-1,4-polyisoprene. In the latter case, the regular arrangement of molecules permits natural rubber to crystallize during extension,[16,50] butadiene-styrene polymers show no comparable tendency to crystallize.

However, a new era in synthetic rubber began in the mid-1950's when Firestone, Goodrich, and Goodyear each revealed it had made a rubber from isoprene that matched natural rubber, molecule for molecule. The discovery of how to duplicate in a reactor what nature does in the Hevea brasiliensis tree marked the beginning of the so-called stereo rubbers.

In November 1955, The Firestone Tire & Rubber Company announced the successful synthesis of a *cis*-1,4-polyisoprene supported by tire-test results.[2,3] This polymer, known as "Coral" rubber,[58] was prepared by the polymerization of isoprene catalyzed by powdered lithium. The catalyst was made by melting the metal in petroleum jelly and agitating the product in a special high speed stirrer to produce a fine dispersion of lithium having a mean diameter of 20 microns.

A reactor was charged with 100 parts of pure dry isoprene and 0.1 part lithium. After an induction period at 30 to 40°C, the charge thickened, and it became solid toward the end of the polymerization. The catalyst was decomposed with isopropyl alcohol, and the rubber was stabilized with a suitable antioxidant. The polymer was washed, then dried at 50°C.

Lithium was reported to be specific in its effect; other alkali metals gave mixtures of *cis*- and *trans*-polymers.

In November 1955, B. F. Goodrich Company reviewed for the American Chemical Society the structure and properties of *cis*-polyisoprene polymerized according to information supplied by Dr. Karl Ziegler. From the infrared absorption spectra, x-ray diffraction patterns, and second-order transition temperatures, the synthetic polymer was considered substantially the same as natural rubber.

Polyisoprenes require antioxidants as do the emulsion butadiene-styrene copolymers. Since fatty acid is not present in the final polymer, it is necessary to compensate by adjustments in the compounding recipe. Physical tests in the laboratory for pure gum, and body and tread stocks indicate that the polymer and Hevea are virtually identical.[26]

According to a Belgian patent[23] issued to Goodrich-Gulf Chemicals Company, *cis*-1,4-polyisoprene and mixtures of *cis*- with *trans*-1,4-polyisoprene may be prepared by controlling the catalyst, which consists of a reaction product of an alkylaluminum with titanium tetrachloride in a hydrocarbon medium. A *cis*-polymer is obtained if the mole ratio of Ti to Al is 1; if the mole ratio is 1.5 to 3, a *trans*-polymer results.

In 1955, the Goodyear Tire & Rubber Company disclosed the preparation of *cis*-1,4-polyisoprene[12] in which triethylaluminum was used as a catalyst together with a cocatalyst. The infrared spectrum was comparable to that of Hevea rubber.

A description of the pilot plant for the preparation of *cis*-polyisoprene has been given by the Goodyear Tire & Rubber Company.[65] Extremely pure isoprene and a hydrocarbon solvent are used in this process. Special precautions are required, since oxygen and certain unsaturated compounds are severe catalyst poisons. Both isoprene and the solvent are distilled, mixed, and passed through a silica gel or alumina dryer. They are pumped through the dehydrating bed into the reactor where the catalyst is added. The temperature is controlled with water or brine in the reactor jacket at a polymerization temperature of 50°C. In this system, lower temperatures give higher molecular weight polymers. At about 7 percent solids, the contents of the reactor become viscous and temperature-control is difficult. The final solids content of the cement is reported as being about 25 percent. At this stage, the cement is extremely viscous.

At the end of polymerization, the cement is pumped to a tank where the catalysts can be deactivated and the necessary antioxidant added. The cement is then heated and transferred to an extruder dryer where the solvent is vaporized for recovery and reuse. The product, as extruded, contains less than 1% volatile material.

The catalyst for the polymerization consists of two parts. One, triisobutylaluminum, is spontaneously flammable and must be handled carefully. It should be mixed with the correct proportion of the second component in storage cylinders and the mixture then be pressured into the reactor.

Although the discovery of methods to synthesize a synthetic polyisoprene was a major scientific breakthrough, commercial production has been limited by the availability of isoprene at competitive costs. Additional information on polyisoprene is available in Reference 55.

Solution Polybutadiene

The catalysts for producing polybutadiene have been adapted from the organolithium compounds[21] and Ziegler-type systems[17] which were originally developed for polyisoprene.

Whereas *cis*-polyisoprene described previously was developed in an effort to synthesize natural rubber, polybutadiene took a place in the synthetic rubber field because of the unique character it imparts when blended with natural rubber, emulsion SBR, polyisoprene, and solution butadiene-styrene copolymers.

Compounds containing polybutadiene show high rebound and low heat build-up, characteristics which make this product valuable for truck and bus tire manufacture,[11] and equally valuable in the production of passenger car tires.

High 1,4-Polybutadiene.[21] While the polymerization of butadiene catalyzed by alkyllithium compounds in ethers had been known since 1934, it was polymerization in the absence of ethers that gave high 1,4-configuration in polybutadiene and polyisoprene.

The polymerizations are run in very pure, dry hexane or heptane in the presence of an organoalkali catalyst, butyllithium being used most frequently. Pure butadiene may be added in a batchwise operation or in a continuous process involving one or more stirred pressure vessels. Heat is removed from the jackets. In general, the polybutadienes are similar regardless of the alkyl- or aryllithium selected. The linear polymeryllithium remains active at the end of the polymerization unless the charge is overheated. These are known as "living polymers," for they can continue polymerizing if more monomer is added. If a second monomer is charged to a living-polymer solution, the block polymer that results may have unusual properties that have special applications.

The molecular weight distribution from a single polybutadiene preparation is very narrow and the solutions are quite viscous.

High cis-Polybutadiene.[17] The Ziegler-Natta Catalysts[32,42,71] were reaction products of alkylaluminum compounds with $TiCl_4$ or $TiCl_3$ at the start. They were extended to salts of other transition metals, with Ti, V, Mo, Co and Ni being the most important. Commercial practice is directed toward the reaction products of $TiCl_4$ and TiI_4 with alkylaluminum compounds, alkylaluminum halides with cobalt compounds, and aluminum trihalides with cobalt compounds.

Solution Butadiene-Styrene Rubbers

Sodium catalysts for the copolymerization of butadiene and styrene were developed in the mid-1940's by A. A. Morton and termed "Alfin" catalysts. These catalysts were reaction products of sodium, isopropyl alcohol and olefins (Alfin).[38] However, the extremely high molecular weight of the polymers produced with these catalysts made processing into finished articles very difficult.

Alkyllithium catalysts previously described for polybutadiene production are also used extensively in the manufacture of butadiene-styrene copolymers. Figures 9.2 and 9.3 show that styrene has much less tendency to enter the copolymerization in the early stages than does butadiene. As the butadiene supply is depleted in the neighborhood of 80 percent conversion, styrene starts entering very rapidly. This difference in reactivity is somewhat typical of the catalysts used in solution processes. Whereas copolymer elastomers are readily prepared with free radical initiators, ionic copolymerizations tend to be more discriminating.

Random solution butadiene-styrene copolymers may be prepared by careful regulation of the monomer mixture present at any specific time, or by the addition of randomizing agents such as alkali alkoxides or tetramethylethylenediamine (see Fig. 9.3). These polymers are being marketed under the trade name Stereon®

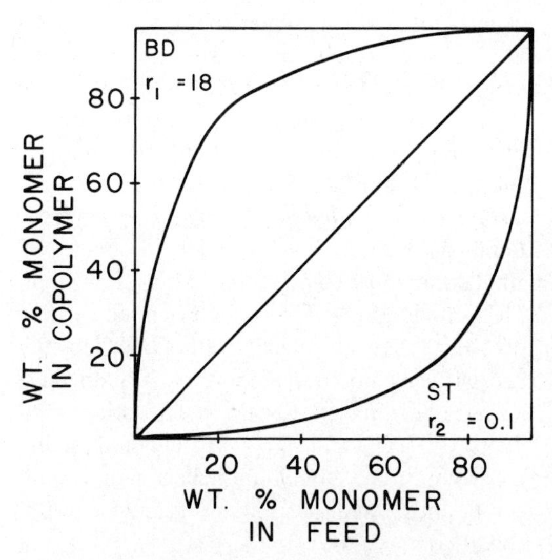

Fig. 9.2. Copolymerization of butadiene and styrene in benzene with butyllithium. (*"Polymer Chemistry of Synthetic Elastomers," J. P. Kennedy and E. G. Tornquist, eds., Interscience, 1969; by permission of John Wiley and Sons, Inc.*)

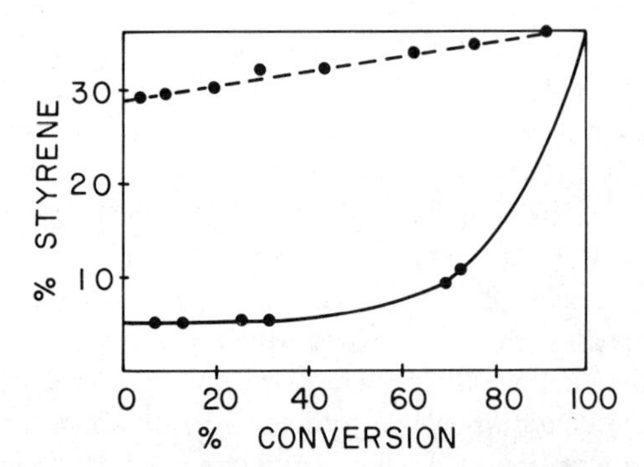

Fig. 9.3. Butadiene/styrene 65/35 copolymerization at 90°F;–butyllithium initiator, no modifier;–butyllithium/TMEDA, 1/2. (*R. L. Bebb and A. E. Oberster, "Angew. Makromol. Chem.*, **16-17** (*1971*))

by The Firestone Tire & Rubber Co., and Solprene® by the Phillips Petroleum Co.

Recently, attention has been turning to the use of "living" systems in making block polymers, which will be described in greater detail in the later section on thermoplastic elastomers.

Polyolefin Rubbers

The transition-metal catalyst systems have been extended from low-pressure polymerization of ethylene to mixtures of ethylene and propylene, and interesting rubbers have resulted.[41] These have required specific catalyst systems, usually involving mixtures of alkyl aluminum compounds with derivatives of vanadium (VCl_4, VCl_3, $VOCl_3$, V triacetate) or titanium ($TiCl_4$, $TiCl_3$), and again, hydrocarbon solvents.

Ethylene-propylene rubbers (EPR) have been offered commercially in the range 55-60% ethylene content. These contain no unsaturation and must, therefore, be cured with peroxide or radiation, or grafted with unsaturated diacids. Therefore, terpolymers (EPDM or EPT) have been made containing 2.5-4% of a nonconjugated diene (dicyclopentadiene, 1-5-hexadiene, ethylenenorborene, methylenenorbornene). These can be vulcanized either with sulfur systems, with peroxides, or by radiation.

The EPR and EPDM vulcanizates have good low-temperature properties and excellent resistance to oxygen and ozone. They were first used in mechanical rubber goods, but are expanding into the tire field, especially in sidewalls, as polymer variations and compounding techniques are making them more adaptable.

Condensation Rubbers[39]

The urethane rubbers constitute examples of condensation polymers, wherein monomers combine with the elimination of simple molecules, such as water or ammonia. Occasionally, as in the case with urethane rubber, nothing is eliminated. Here isocyanates react with dihydroxy alcohols:

$$RNCO + HO{-}R'{-}OH \longrightarrow RNHCOO{-}R'{-}OCONHR$$

There are many possibilities for varying the nature of the polymer by selecting different diisocyanates, as well as different dialcohols such as polyether glycols, branched glycols, and polyester glycols for the starting materials.[54] Either millable or castable polymers may then be mixed with suitable crosslinking agents and cured to usable rubbers.[40]

Urethane rubbers have high tensile strengths and resilience; they are generally resistant to oils, ozone, and oxygen.

Silicone Rubbers

Silicone rubbers differ from those previously described in that their chains consist of alternate atoms of silicon and oxygen, with no carbon atoms. Although work on the preparation of the silicone polymers was started in the Corning Research Laboratories, the General Electric Company initiated a program at about the same time, and in 1945, both companies announced the development of silicone rubber. Most of the silicone rubbers are derived from dimethyldichlorosilane, but variations include a partial substitution of other groups, such as the phenyl for the methyl radical, and the preparation of polymers containing vinyl or allyl radicals.

Dimethyldichlorosilane (boiling point 70°C) is prepared by passing methyl chloride over powdered silicon with copper catalysts at 275 to 375°F. The general reaction is:

$$2CH_3Cl + Si \xrightarrow[cat]{} (CH_3)_2SiCl_2$$

The conversion of dimethyldichlorosilane into the polymer results from the addition of water and subsequent hydrolysis in the presence of small proportions of iron chloride, sulfuric acid, or sodium hydroxide. These catalysts must be washed out of the polymer. Low polymeric materials are removed by distillation.

$$R_2SiCl_2 + 2H_2O \longrightarrow R_2Si(OH)_2 + 2HCl$$

$$nR_2Si(OH)_2 \longrightarrow HO(SiR_2O)_nH + (n-1)H_2O$$

Silicone rubbers are useful over a remarkably wide range of temperatures, extending from -130 to 550°F. Although their tensile strengths are low, other properties compensate. One advantage is the fact that the rubbers are white and can be used for the preparation of light-colored stocks. They are not attacked by ozone, and they have good electrical properties. Swelling in oils is relatively low, and they are affected by very few chemicals.

Silicone rubbers are generally vulcanized by the addition of a peroxide, such as di-tert-butyl peroxide or dicumyl peroxide. Oxides of certain metals, such as lead and zinc, accelerate vulcanization; silica, titania, ferric oxide, and alumina are good fillers.[33,51] Additional background on silicone elastomers is given in Reference 62.

NATURAL RUBBER

Rubber is found in varying amounts in many plants throughout the world. Although other sources have been used in time of war, only one is commercially important at present. This is the *Hevea brasiliensis*, a tree native to Brazil, but now grown on plantations throughout the tropics.

Rubber occurs in Hevea as a milky latex. The dry product is obtained by a process known as coagulation, during which the latex is destabilized by the addition of acids or salts.

Hevea brasiliensis was found growing wild in the tropics of Brazil and was taken into the Far East by the British. Vast natural rubber plantations have grown up from this beginning (see Table 9.4). Stock from carefully selected trees giving high yields were grafted onto ordinary seedlings to produce a family of descendants from a single tree known as a "clone," and production was improved from an average of

TABLE 9.4 Production of Natural Rubber 1978[a]

	Metric Tons
Asia and Oceania	
Malaysia	1,606,500
Indonesia	902,500
Thailand	466,968
Sri Lanka	155,662
Vietnam	40,000
Cambodia	18,000
India	132,991
Burma	20,000
China	35,000
Philippines	63,000
Other	4,750
Africa	
Liberia	78,500
Nigeria	57,500
Ghana	5,000
Zaire	25,500
Cameroon	17,171
Ivory Coast	17,526
Other	1,250
Latin America	
Brazil	23,708
Other	19,000
Total	3,715,000

[a]Taken from Rubber Statistical Bulletin 34, No. 4 (Jan. 1980).

250 to 500 pounds per acre to more than 2000 pounds per acre. Recent developments include the yield stimulants such as ethylene gas, which is the active ingredient in Ethephone and Ethad (commercial stimulants).

As new seedlings develop in the plantation nursery, they are arranged in regular plantings around a collecting house. After the trees are six years old, a program is established for tapping them and collecting the latex. In *Hevea brasiliensis*, the latex occurs in tiny ducts or tubes found under the bark and just outside the green cambium, or growing layer. Each morning a diagonal cut, just deep enough to sever the latex vessels, is made with a special knife. The liquid is collected in a small cup at the bottom of the diagonal cut. When the cut is first made, a small amount of preservative is placed in the cup to prevent coagulation. During the several hours before the latex stops flowing, a tree will yield about 100 cubic centimeters of normal latex (30 to 40% solids). This is strained in the collecting station to remove dirt and bark, and treated with more preservative. Then it is transferred to a central factory where it is centrifuged or coagulated.

Because latex is very susceptible to bacterial action, an adequate preservative must be added to protect it from the time it leaves the tree until it is used. Dilute ammonia is commonly used despite its volatility. It does not kill bacteria but reduces their growth. Some plantations use a small amount (0.15-0130%) of formaldehyde to sterilize the latex, following it up with ammonia before shipment. Still other producers use Santobrite (sodium pentachlorophenate) at approximately 0.3% based on the latex, along with 0.1% ammonia, to produce a latex with especially good keeping qualities.[66] With stricter environmental regulations and the concern for worker exposure to ammonia, recent developments have led to using a composition of low ammonia, tetramethylthiuram disulfide, and zinc oxide as preservatives.

Natural Latex

About 10% of the natural rubber is sold as a latex concentrate. The field latex, as received at the central processing stations, is concentrated to over 60% solids to reduce ocean freight and to facilitate processing. About 90% of this is concentrated by centrifuging. Most of the balance is concentrated by the creaming method, with a little being concentrated by evaporation. Electrodecantation has been used for concentration, but only to a limited extent.

The *centrifuge* method depends on the difference in specific gravities between the rubber and the serum; rubber has a specific gravity of 0.91, and the serum, 1.02. In this case, a special centrifuge separates a cream from a low solids skim containing about 10% solids. The cream fraction is further stabilized by the addition of preservatives and adjusted to the correct solids level for shipment. Special concentrates for applications requiring low water absorption and high dielectric properties have also been developed by diluting the recentrifuging the cream.

In the *creaming* process, a small quantity of a gum, such as ammonium alginate, gum tragacanth or Irish moss, is added to produce a reversible agglomeration of the rubber particles. With an increased size and slower Brownian movement, the particles cream, and a high solids fraction can be removed from the top. Creamed latices, like centrifuged latices, lose the major water soluble impurities with removal of the serum. These latices are, therefore, of particular value where a low level of impurities is desirable.

Concentration by *evaporation* requires the addition of stabilizers, alkalis, and soap to the latex. Concentration is effected in a rotating drum in which a smaller rolling drum furnishes additional evaporation and agitates the latex. This route differs from the two described previously in that all ingredients in the original latex, plus any additives introduced for stability, remain in the finished product. The nonrubber ingredients in the evaporated latex may amount to 6 to 7%. Since latex concentrated by this method can reach as much as 75% solids, it is of use in special applications.

In the *electrodecantation* method,[56] latex is added to a rectangular tank with an electrode at each end and having many grooves in its sides, about 1 centimeter apart, in which sheets of cellophane are placed. When an electric current is applied, particles build up on the cellophane and float to the top as a cream. Fresh latex is added continuously at about the mid-point of the tank to displace the cream from the top of the cell. Latex of 60 to 62% concentration has been produced by this process.

Dry Rubber

Most field latex is converted into "ribbed smoke" (RSS). Field latex is diluted to about 15%, and coagulated by the addition of dilute formic or acetic acid. After 1 to 18 hours, the fine particles of rubber agglomerate to large masses. These are transferred to washing and dewatering mills. In the manufacture of "pale crepe" rubber, the material is washed thoroughly before it is dried in a hot oven. In the preparation of "smoked sheets," the freshly coagulated rubber is not washed but dewatered in mills with even-speed rolls; the wet sheets are then dried in wood smoke. The slow-smoking process produces a brown rubber which resists deterioration by mold and bacteria. In addition to pale crepe and smoked sheets, plantations produce a number of different grades of rubber including bark, earth scrap, and factory salvage.

The following table, based on an analysis of 35 samples of smoked sheets and 102 samples of pale crepe, gives some idea of the impurities present in both types of rubber.[15]

	Smoked Sheet (Average) %	*Pale Crepe (Average)* %
Moisture	0.61	0.42
Acetone extract	2.89	2.88
Protein (N × 6.25)	2.82	2.82
Ash	0.38	0.30
Rubber hydrocarbon (by difference)	93.30	93.58

The acetone-extract fraction contains fatty acids, sterols, and esters. The fatty acids have an important effect on vulcanization; the sterols and esters are believed[52] to contain the natural antioxidants which protect the rubber during processing and storage. The protein fraction has an important effect on the vulcanization rate of the natural rubber. If the coagulum deteriorates during drying or if putrefaction has occurred prior to coagulation, cure rates will vary from lot to lot. The ash content in natural rubber is generally not important unless it is found to contain copper and manganese. These particular elements catalyze oxidation of the rubber and subsequent deterioration. Since these elements are concentrated in the bark of the tree, it is important to remove the bark from the latex as soon as possible.

Grades of natural rubber have been established by the Rubber Manufacturers' Association[53,66] in an effort to classify the various rubbers produced through the world. In 1965, Technically Specified Rubbers (TSR) came onto the market to assure high standards of quality and provide a baled product competitive with the synthetic rubbers in handling, storage, and processing.

Recently there has been renewed interest in guayule, a shrub which grows in the arid regions of Mexico and in the Southwest U.S., as a

TABLE 9.5 Physical Properties and Relative Performance for 15 Types of Elastomers

COMPOUND SELECTION AND SERVICE GUIDE

			1	2	3	4	5	6
Base Polymer — Common Name			NATURAL RUBBER	SYNTHETIC RUBBER	GRS	BUTYL	BUTADIENE	E P RUBBER
Chemical Name			Polyisoprene	Synthetic Polyisoprene	Styrene Butadiene	Isobutylene Isoprene	Polybutadiene	Ethylene Propylene
SAE J200. — ASTM D-2000 Classification			AA	AA	AA, BA	AA, BA	AA	CA
ASTM D-735, SAE J-14: MIL-R-3065 (MIL-STD-417)			R(N)	R(S)	R(S)	R(S)	R(S)	R(S)
ASTM Designation (D 1418)			NR	IR	SBR	IIR	BR	EPDM, EPR
WEIGHT BASE ELASTOMER		LB/CU IN.	0.033	0.033	0.034	0.033	0.034	0.031
		SPEC. GR.	0.93	0.93	0.94	0.92	0.94	0.86
PHYSICAL PROPERTIES FOR ELASTOMER COMPOUNDS	DUROMETER, RANGE		30-100	40-80	40-100	30-100	45-80	30-90
	RESILIENCE		OUTSTANDING	OUTSTANDING	GOOD	FAIR	OUTSTANDING	GOOD
	TENSILE STRENGTH, PSI (REINFORCED)		4000 +	2000-3000	2000 +	2000 +	2500	2000-3000
	ELONGATION, % (REINFORCED)		500	300-700	450	300-800	450	500
	DRIFT, ROOM TEMP.		EXCELLENT	GOOD	EXCELLENT	FAIR	GOOD	FAIR
	COMPRESSION SET		GOOD	FAIR	GOOD	FAIR	FAIR	FAIR
	ELECTRICAL RESISTIVITY		EXCELLENT	EXCELLENT	EXCELLENT	EXCELLENT	EXCELLENT	EXCELLENT
	IMPERMEABILITY, GAS		GOOD	GOOD	FAIR	OUTSTANDING	GOOD	GOOD
RESISTANCE PROPERTIES — MECHANICAL	RESISTANCE TO	IMPACT	EXCELLENT	EXCELLENT	EXCELLENT	GOOD	GOOD	GOOD
		ABRASION	EXCELLENT	EXCELLENT	EXCELLENT	GOOD	EXCELLENT	GOOD
		TEAR	EXCELLENT	GOOD	FAIR	GOOD	GOOD	POOR
		CUT GROWTH	EXCELLENT	EXCELLENT	GOOD	EXCELLENT	FAIR	GOOD
RESISTANCE PROPERTIES — TEMPERATURE	TENSILE STRENGTH: PSI, AT	250 F	1800	1800	1200	1000	1200	2000
		400 F	125	125	170	350	170	400
	ELONGATION, %, AT	250 F	500	500	250	250	250	300-500
		400 F	80	80	60	80	60	0-120
	DRIFT AT 212 F		GOOD	GOOD	GOOD	FAIR	GOOD	FAIR
	HEAT AGING AT 212 F		GOOD	FAIR	GOOD	EXCELLENT	FAIR	EXCELLENT
	FLAME RESISTANCE		POOR	POOR	POOR	POOR	POOR	POOR
	LOW TEMPERATURE	STIFFENING, F	20 TO 50	20 TO 50	0 TO 50	−10 TO −40	−30 TO −60	20 TO 50
		BRITTLE POINT, F	80	80	80	−80	−100	90
RESISTANCE PROPERTIES — ENVIRONMENTAL	WEATHER		FAIR	FAIR	FAIR	EXCELLENT	FAIR	EXCELLENT
	OXIDATION		GOOD	EXCELLENT	GOOD	EXCELLENT	EXCELLENT	GOOD
	OZONE		POOR	POOR	POOR	EXCELLENT	POOR	EXCELLENT
	RADIATION		FAIR TO GOOD	FAIR TO GOOD	GOOD	POOR	POOR	POOR
	WATER		EXCELLENT	EXCELLENT	EXCELLENT	EXCELLENT	EXCELLENT	GOOD TO EXCELLENT
	ACID		FAIR TO GOOD	FAIR TO GOOD	FAIR TO GOOD	EXCELLENT	FAIR TO GOOD	GOOD TO EXCELLENT
	ALKALI		FAIR TO GOOD	FAIR TO GOOD	FAIR TO GOOD	EXCELLENT	FAIR TO GOOD	GOOD TO EXCELLENT
	GASOLINE, KEROSENE, ETC. (ALIPHATIC HYDROCARBONS)		POOR	POOR	POOR	POOR	POOR	POOR
	BENZOL, TOLUOL, ETC. (AROMATIC HYDROCARBONS)		POOR	POOR	POOR	FAIR TO GOOD	POOR	FAIR
	DEGREASER SOLVENTS (HALOGENATED HYDROCARBONS)		POOR	POOR	POOR	POOR	POOR	POOR
	ALCOHOL		GOOD	GOOD	FAIR	EXCELLENT	FAIR TO GOOD	POOR
	SYNTHETIC LUBRICANT (DIESTER)		POOR TO FAIR	POOR TO FAIR	POOR	FAIR	POOR TO FAIR	POOR TO FAIR
	HYDRAULIC FLUIDS	SILICATES	POOR	POOR	FAIR TO POOR	FAIR	POOR	FAIR TO GOOD
		PHOSPHATES	POOR TO FAIR	POOR TO FAIR	POOR	GOOD	POOR TO FAIR	GOOD TO EXCELLENT
SUBJECTIVE PROPERTIES	TASTE		FAIR TO GOOD	FAIR TO GOOD	FAIR TO GOOD	FAIR TO GOOD	FAIR TO GOOD	GOOD
	ODOR		FAIR TO GOOD	GOOD	GOOD	GOOD	GOOD	GOOD
	NONSTAINING		POOR TO GOOD	EXCELLENT	POOR TO GOOD	GOOD	GOOD	GOOD
BONDING TO RIGID MATERIALS			EXCELLENT	EXCELLENT	EXCELLENT	AIR TO EXCELLENT	EXCELLENT	POOR

By permission of the Stalwart Rubber Co.

potential source of natural rubber. Proponents of guayule suggest it could be grown on land of little economic value and provide the U.S. with a domestic source of natural rubber to decrease dependence on imported petroleum and natural rubber. Reserach programs are directed to improve yield and quality to make guayule economically attractive.[13]

A comparison between the properties of natural rubber with the various synthetics is provided in Table 9.5.

POLYMER STRUCTURE AND ELASTICITY

A rubber-like solid is unique in that its physical properties resemble those of solids, liquids, and gases in different respects. It is solid-like in that rubber has dimensional stability. It behaves like a liquid because its coefficient of thermal expansion and isothermal compressibility are of the same order as those of liquids. It resembles gases in the sense that stresses in deformed rubbers increase with increased temperature, much as the pressure in compressed gases increases with increased temperature. Thus, the characterization of a synthetic rubber is much more complex than the characterization of a simple organic compound, and a highly specialized science has developed for this field.[34]

Microstructure, sequence distribution, molecular weight, and glass transition temperature are frequently used to characterize a rubber and

7 NEOPRENE	8 NITRILE	9 THIOKOL	10 URETHANE	11 SILICONE	12 FLUOROSILICONE	13 HYPALON	14 ACRYLIC	15 FLUOROCARBON
Chloroprene	Butadiene Acrylonitrile	Polysulfide	Polyester/Polyether Urethane	Polysiloxane		Chlorosulfonated Polyethylene	Polyacrylate	Fluorinated Hydrocarbon
BC, BE	BF, BG, BK	AK	BG	FC, FE, GE	FK	CE	DF, DH	HK
SC	SB, SA	SA	SB	TA	TA, TB	SC	TB	TB
CR	NBR	PTR	AU, EU	PSi, PVSi, Si, VSi	FVS	CSM	ACM, ANM	FPM
0.044	0.036	0.048	0.039	0.036	0.051	0.040	0.040	0.05-0.07
1.23	1.00	1.34	1.05	0.95	1.4	1.10	1.10	1.4-1.95
40-95	20-90	20-80	55-100	25-90	40-80	50-95	40-90	65-90
EXCELLENT	GOOD	FAIR	GOOD TO EXCELLENT	POOR TO EXCELLENT	GOOD	GOOD	GOOD	FAIR
3000+	1000-3500	500-1500	4000-8000	600-1500	1300	1500-2500	1700	1500-3000
650-850	400-600	200-550	250-800	90-900	500	250-500	450	100-450
FAIR TO GOOD	GOOD	POOR	GOOD TO EXCELLENT	FAIR TO EXCELLENT	EXCELLENT	FAIR	FAIR	GOOD
FAIR TO GOOD	GOOD	POOR TO FAIR	EXCELLENT	GOOD TO EXCELLENT	GOOD TO EXCELLENT	FAIR TO GOOD	FAIR TO GOOD	GOOD TO EXCELLENT
FAIR	POOR	FAIR	GOOD	EXCELLENT	OUTSTANDING	GOOD	FAIR	GOOD
GOOD	GOOD	EXCELLENT	GOOD	FAIR	GOOD	EXCELLENT	GOOD	EXCELLENT
GOOD	FAIR	POOR	EXCELLENT	POOR TO GOOD	FAIR	FAIR TO GOOD	POOR	POOR TO GOOD
GOOD TO EXCELLENT	EXCELLENT	POOR TO FAIR	EXCELLENT	POOR TO EXCELLENT	FAIR	GOOD	FAIR TO GOOD	GOOD
GOOD	GOOD	POOR TO FAIR	EXCELLENT	POOR TO GOOD	FAIR	FAIR TO GOOD	FAIR	POOR TO GOOD
GOOD	GOOD	POOR	FAIR TO EXCELLENT	POOR TO GOOD	FAIR	FAIR TO GOOD	GOOD	POOR TO GOOD
1500	700	700	1800	550	EXCELLENT	500	1300	300-800
180	130	UNDER 25	200	450	GOOD	200	225	150-300
350	120	140	300	200	EXCELLENT	60	400	100-350
0-100	20	UNDER 25	140	100	GOOD	20	150	50-160
FAIR TO GOOD	EXCELLENT	POOR	EXCELLENT	EXCELLENT	EXCELLENT	FAIR	FAIR	GOOD TO EXCELLENT
GOOD	GOOD	GOOD	FAIR TO GOOD	EXCELLENT	OUTSTANDING	EXCELLENT	EXCELLENT	OUTSTANDING
GOOD	POOR TO FAIR	POOR	POOR TO FAIR	FAIR TO GOOD	OUTSTANDING	GOOD	POOR TO FAIR	EXCELLENT
+10 TO 20	+30 TO 20	10 TO 45	10 TO 30	65 TO 180	70	30 TO 50	+35 TO +10	+20 TO 30
45	65	60	60	90 TO -180	85	60	10	+10 TO 60
EXCELLENT	GOOD	EXCELLENT	EXCELLENT	EXCELLENT	OUTSTANDING	EXCELLENT	EXCELLENT	EXCELLENT
GOOD	FAIR TO GOOD	EXCELLENT	EXCELLENT	EXCELLENT	OUTSTANDING	EXCELLENT	EXCELLENT	OUTSTANDING
EXCELLENT	POOR	EXCELLENT	EXCELLENT	EXCELLENT	OUTSTANDING	OUTSTANDING	EXCELLENT	OUTSTANDING
FAIR TO GOOD	FAIR TO GOOD	FAIR TO GOOD	GOOD	FAIR TO EXCELLENT	GOOD TO EXCELLENT	FAIR TO GOOD	POOR TO GOOD	FAIR TO GOOD
GOOD	EXCELLENT	GOOD	GOOD	EXCELLENT	OUTSTANDING	GOOD	FAIR	GOOD
GOOD	GOOD	FAIR	POOR TO FAIR	POOR TO GOOD	OUTSTANDING	EXCELLENT	FAIR	GOOD TO EXCELLENT
GOOD	FAIR TO GOOD	GOOD	POOR TO FAIR	POOR TO FAIR	OUTSTANDING	EXCELLENT	POOR	POOR TO GOOD
GOOD	EXCELLENT	EXCELLENT	EXCELLENT	POOR TO FAIR	OUTSTANDING	FAIR	EXCELLENT	EXCELLENT
POOR	GOOD	EXCELLENT	POOR TO FAIR	POOR TO FAIR	EXCELLENT	POOR TO FAIR	POOR	EXCELLENT
POOR	POOR	FAIR TO GOOD	FAIR TO GOOD	POOR TO GOOD	EXCELLENT	POOR TO FAIR	POOR	GOOD
FAIR	EXCELLENT	GOOD	GOOD	GOOD	GOOD TO EXCELLENT	GOOD	POOR	EXCELLENT
POOR	FAIR TO GOOD	GOOD	POOR	POOR TO FAIR	OUTSTANDING	POOR	GOOD	FAIR TO GOOD
GOOD	FAIR	POOR TO GOOD		POOR	OUTSTANDING	GOOD	GOOD	GOOD
POOR	POOR	POOR TO FAIR	POOR	GOOD	POOR	POOR TO FAIR	POOR	POOR
FAIR TO GOOD	FAIR TO GOOD	POOR TO FAIR	GOOD	GOOD TO EXCELLENT	EXCELLENT	FAIR TO GOOD	FAIR TO GOOD	FAIR TO GOOD
FAIR TO GOOD	GOOD	POOR	GOOD	GOOD TO EXCELLENT	EXCELLENT	GOOD	FAIR TO GOOD	GOOD
GOOD TO EXCELLENT	POOR TO GOOD	POOR TO FAIR	GOOD	OUTSTANDING	OUTSTANDING	EXCELLENT	GOOD	POOR TO GOOD
GOOD TO EXCELLENT	GOOD TO EXCELLENT	FAIR TO GOOD	FAIR TO GOOD	FAIR TO EXCELLENT	GOOD TO EXCELLENT	FAIR TO GOOD	GOOD	POOR TO GOOD

predict the usefulness of a polymer for a particular application.

Microstructure

Microstructure refers to the way in which monomer units are ordered along a polymer chain and includes the geometric formation in which they are assembled. In diene polymers, the isoprene units found in natural rubbers and the butadiene and isoprene units found in polybutadiene and polyisoprene synthetic rubbers are arranged in various combinations of *cis* 1,4 and *trans*-1,4 additions, as well as 1,2 structure, and in the case of isoprene, additions between the 3,4 positions. Natural rubber of the Hevea type is essentially all *cis* structure; the hard, more resilient plastic counterparts (balata and gutta percha) are primarily *trans*-polyisoprene structures. (See Table 9.6 and Fig. 9.4)

Sequence Distribution

When two or more monomers are charged into a polymerization system, copolymers may result. Depending on the relative reactivities of the monomers, the second or other monomers may enter the polymer chain at a different rate than the first monomer. Copolymerizations in synthetic rubber production generally result in two types of structures: a) Amorphous polymers with relatively uniform distribution

TABLE 9.6 Microstructures of Synthetic Rubbers

	Percent			
	Cis-1,4	*Trans-1,4*	*1,2*	*3,4*
Emulsion rubbers				
SBR "hot"[a]	18	65	17	–
SBR "cold"[a]	12	72	16	–
Solution rubbers				
Polybutadiene				
Butyllithium catalyst	34	54	12	–
Ziegler cobalt system	98	–	2	–
Ziegler nickel system	95	4	1	–
Alfin catalyst	5	75	20	–
Polyisoprene				
Butyllithium catalyst	94	–	–	6
system	99	–	–	1

[a]Structure calculated on butadiene content only.

of the monomer units along the chain, i.e. ABABAABABB, or b) Block copolymers of varying lengths, i.e. AAAAAABBBBBB.

The greater the size of these blocks, the more the rubber is affected by the characteristics of the homopolymer of the second monomer. Thus, as the polystyrene block size increases on butadiene-styrene copolymers, the elastomers begin to show the hardness of polystyrene at room temperature and its tendency to soften with increasing temperature. Nuclear magnetic resonance, infrared, glass transition temperature, rheological behavior, solubility properties and chemical reactivity are frequently used to establish sequence distribution.

Glass Transition Temperature

Polymeric materials can be divided into three classes: (a) Elastomers, having a glass transition temperature (Tg) below room temperature; (b) Rigid polymers such as polystyrene having Tg

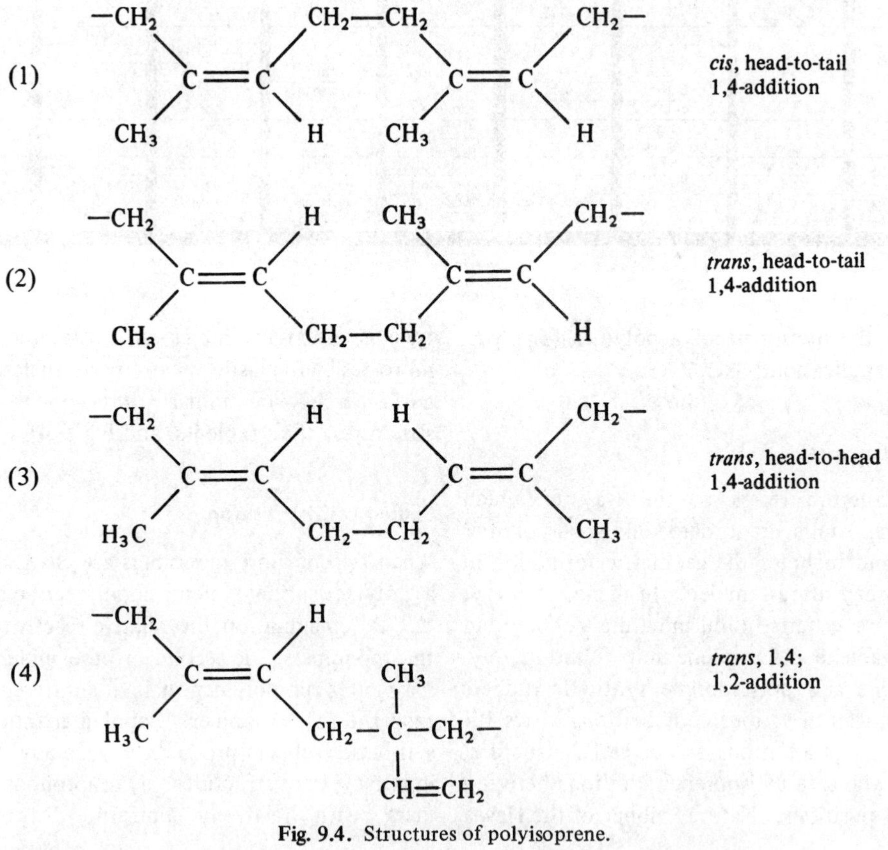

(1) $-CH_2$, CH_3 / C=C / H, CH_2—CH_2 / C=C / CH_3, H, CH_2- — *cis*, head-to-tail 1,4-addition

(2) $-CH_2$, CH_3 / C=C / H, CH_2—CH_2 / C=C / CH_3, CH_2-, H — *trans*, head-to-tail 1,4-addition

(3) $-CH_2$, H_3C / C=C / H, CH_2—CH_2 / C=C / H, CH_2-, CH_3 — *trans*, head-to-head 1,4-addition

(4) $-CH_2$, H_3C / C=C / H, $CH_2-C(CH_3)(CH{=}CH_2)-CH_2-$ — *trans*, 1,4; 1,2-addition

Fig. 9.4. Structures of polyisoprene.

above room temperature; (c) Partially crystalline polymers having Tg either below or above room temperature (e.g. nylon, polyethylene, polypropylene). Thus the Tg is the one characterization of a polymer which tells the most about the mechanical properties. In elastomers, the Tg not only defines the service temperature, but also influences the performance; in tires, for instance, the Tg of a rubber is a dominant factor in determining the traction and treadwear which can be achieved in a tire.[30]

Molecular Weight and Microstructure

The molecular weight of a polymer represents an average expression of the various chain lengths in a particular sample. There are now four ways of expressing the average molecular weight:[8, 21]

1. $M_{\overline{v}}$, viscosity average molecular weight based on viscometric techniques.
2. $M_{\overline{n}}$, number average molecular weight based on osmometer determinations.
3. $M_{\overline{w}}$, weight average molecular weight based on light scattering.
4. $M_{\overline{k}}$, kinetic molecular weight based on polymer yield divided by the moles of initiator used in the charge.

In industrial applications of polymers, the molecular weight specification generally represents a compromise between the practical aspects of production and the theoretical performance which could be achieved with a polymer of infinite molecular weight. Of special interest in this regard is the molecular weight distribution (MWD) of a rubber as it relates to processing, physical properties, and polymerization conditions. Recent advances in gel permeation chromatography (GPC)[37] have combined the GPC with multi-detectors and microprocessors to provide not only the MWD but also the composition of the various weight fractions.

THERMOPLASTIC ELASTOMERS[36,45]

Thermoplastic elastomers are a new class of block copolymer elastomers with "thermally reversible crosslinks." The chief benefit of these new polymers is that they can be shaped into useful articles using high speed thermoplastic processing techniques, such as injection molding. No vulcanization is required and any scrap can be reprocessed. Yet they are truly elastic, having a high coefficient of friction, and they retain a high degree of flexibility at low temperatures.

The molecular chain of thermoplastic elastomers generally consists of "hard" (A) or "soft" (B) segments in a polymer having an ABA structure. At service temperatures, the A block is a thermoplastic polymer well below its Tg, while B is an elastomeric polymer well above its Tg. The dominant soft "B" segments are soft, flexible, and elastomeric (Figure 9.5). At ordinary temperatures, the hard segments associate to function as pseudo crosslinks. As the temperature in increased, the attractive fources (whether due to crystallization, hydrogen bonding or ionic forces) between the hard segments decrease and the polymer can be processed as a conventional thermoplastic.

Butadiene/Styrene Thermoplastic Elastomers

Thermoplastic elastomers have been prepared by grafting & poly-blending; but by far, the most important method involves "living" lithium polymerization systems of butadiene and styrene. Potential polymerization techniques include:

1. Sequential addition of monomers
   RLi + styrene monomer (S) ⟶ RSLi
   RSLi + butadiene monomer (B) ⟶ RSBLi
   RSBLi + styrene monomer (S) ⟶ RSBSLi
2. Dilithium initiators
   LiRLi + butadiene monomer (B) ⟶ LiBRBLi
   LiBRBLi + styrene monomer (S) ⟶ LiSBRBSLi
3. Coupling of living diblock polymers
   $2SBLi + RCL_2 \longrightarrow SBRBS + {}_2LiCl$

Polymer properties can be varied by altering the molecular weight of the elastomeric B block

Fig. 9.5. Model of the ABA block polymer molecular structure. (*Block Copolymers, "Network Characteristics of Thermoplastic Elastomers," E. T. Bishop, S. Davison, J. Moacanin, G. Holden and N. W. Tschoegl (1969). By permission of Interscience Publishers, Div. of John Wiley and Sons, Inc.*)

and the plastic S block. The effect of block molecular weight on the properties of these thermoplastic elastomers is illustrated in Figure 9.6.

Thermoplastic elastomeric can be compounded to modify properties or reduce costs. Fillers are used to increase modulus, tear strength, hardness, and flex life. No reinforcement is necessary since this is provided by the domain structure of the polymer. Calcium carbonates and clay increase hardness while melt flow is generally reduced by mineral fillers.

Oils are added to reduce cost or increase melt flow and flex life. However, extender oils tend to reduce tensile strength, hardness, and abra-

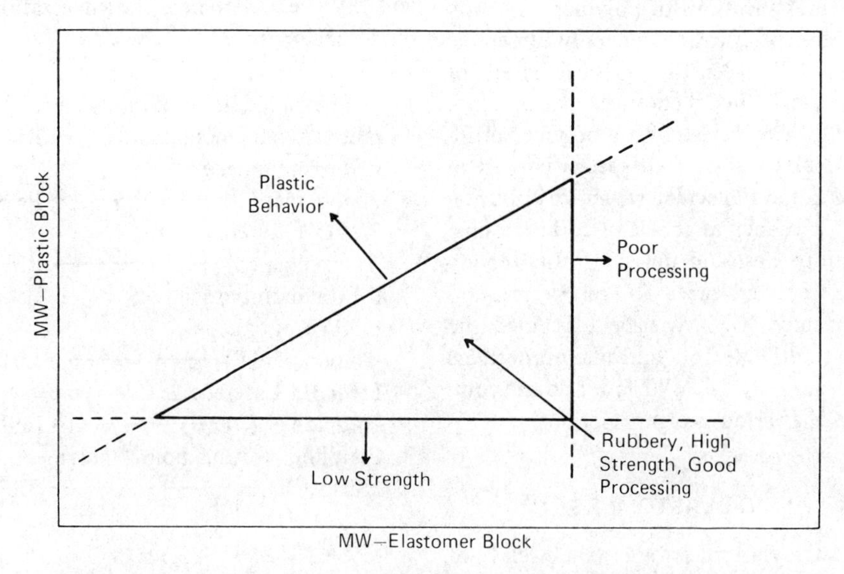

Fig. 9.6. Effect of block molecular weight on properties of thermoplastic elastomers. (*Block Copolymers, "Thermoplastic Elastomers," G. Holden, E. T. Bishop, N. R. Legge, J. Moacanin and H. W. Tschoegl (1969); by permission of Interscience Publishers, Division of John Wiley and Sons, Inc.*)

sion resistance. Since domain formation is the key to the tensile properties of thermoplastic elastomers, an extender oil which softens the domain will reduce the physical properties of the compound. Thus, naphthenic and paraffinic oils have less effect on tensile strength and hardness than aromatic oils in styrene-butadiene thermoplastic elastomers. Resins are frequently added to thermoplastic elastomers to modify the hardness, increase melt flow, or improve adhesion. The effect of the resins on physical properties depends on their characteristics and whether they associate with the elastomeric or plastic phase.

The physical properties of styrene block copolymers are comparable to those of conventional compounded natural rubber and random SBR polymers. Tensile strengths over 4000 psi with elongations of over 1000% have been reported.

The largest use of styrene block copolymers is in direct molding and unit sole footwear. The ease of processing, plus the ability to recycle any scrap material, represents a major advantage over conventially-cured elastomers. Other applications include adhesives, sporting goods, and tubing. These polymers can also be used to modify the properties of thermoplastics to increase toughness and impact resistance.

Olefinic Thermoplastic Elastomers

Olefinic thermoplastic elastomers are copolymers of propylene and ethylene. Crystalline homopolymer segments form the domain structure and random ethylene-propylene segments form the elastomeric phase. In general, the olefinic elastomers require little compounding.

One of the principal uses of the olefinic thermoplastic elastomers is in flexible automobile parts designed to withstand low velocity impact. Typical parts include filler panels, bumper guards, and fender extensions. Other applications of these polymers include sporting goods, wire and cable insulation, hose, and tubing.

RUBBER TECHNOLOGY

With the exception of the newly developed thermoplastic elastomers, the commercial application of either raw natural or raw synthetic rubber is very limited. Since most raw rubbers are plastic and soluble, their uses are restricted to adhesives and sealants, for example, friction tape and electrical tape.

When rubber is mixed with sulfur and heated, vulcanization occurs. The rubber changes from a plastic material to a strong elastic substance which is tack-free, abrasion resistant and no longer readily soluble in common solvents.

For Hevea and the butadiene-based polymers, sulfur is the normal vulcanizing or curing agent; however, a curing material with a high sulfur content such as an organic polysulfide (tetramethy-thiuram disulfide, alkylphenol disulfides, and aliphatic polysulfides), may be substituted. Another class of curing agent is found among the organic peroxides such as di-tert-butyl and dicumyl peroxides.

Some synthetic rubbers, such as neoprene and copolymers containing methacrylic acid, possess functional groups through which cross-linking may occur without a sulfur cure. In the case of neoprene, zinc oxide and magnesium oxide are the normal curing agents; with butyl rubber, alkylphenolformaldehyde resins may serve the purpose. Otherwise, the cure depends on reaction with sulfur or sulfur-containing agents.

The type and concentration of the various ingredients to be added in the development of a rubber stock depend on the properties desired in the finished product. The purpose of the stock will determine the preferred kind of rubber and compounding recipe.

The most frequently used curing system is based on sulfur. Although it may be used alone, vulcanization may be shortened by adding accelerators. These are usually selected from the aldehyde amines, guanidines, thiazoles, or the ultra-accelerators ordinarily derived from dithiocarbamic acid. The usual concentration of these materials is in the range of 0.1 to 1.5 parts per 100 parts of rubber.

In the development of a usable compounding recipe, activators such as zinc oxide, and fatty acids such as stearic acid to solubilize the zinc, are often added. The concentration of these materials ranges from 1 to 5% and 0.5 to 4%, respectively. The compounded rubber ordinarily contains an antioxidant to improve aging.

This is often a secondary aromatic amine or a substituted phenol depending on whether the rubber is to be used in a dark or light-colored application, respectively.

Because "pure gum" products containing the ingredients listed above are not suitable for applications such as tires, reinforcing agents must be added to the stock. Selected for economy and for development of optimum performance, these are normally carbon blacks, inorganic fillers, or reinforcing resins. The many different types of carbon black available for the manufacture of block stocks vary in particle size and in their effect on the properties of rubber stocks. Proper choice of a black or reinforcing pigment is possible only after careful study of available materials. (For a detailed review of compounding ingredients and vulcanization, see Reference 1.)

Stock Preparation

The operations involved in preparing stocks of Hevea or butadiene-styrene rubbers may involve some of the following: breakdown in an oven, a plasticator, or on open-roll mills (in the case of natural rubber); batch mixing in a mill or a Banbury; warm-up in a mill or Banbury if the stock has been allowed to cool during storage; calendering for preparation of sheet skim coating, or skimming onto fabric to produce stocks or plies of the desired dimensions; extrusion, for treads, onto wires, or over hose carcasses; building into green tires, shoes, and hose. (See Figure 9.7 for typical flow diagram for the manufacture of various rubber goods.)

Natural rubber differs from the synthetic butadiene-styrene types in its greater tendency to soften during milling. For this reason, the synthetic rubbers are manufactured to a low plasticity suitable for subsequent processing in the factory. The natural product is prepared by mill mastication at as low a temperature as is practical, normally around 200°F, but at least below 270°F; in a plasticator the rubber may be heated as high as 350°F for breakdown.

The rubber mills used in preparing stocks consist of two parallel steel rolls which vary in size from 6- to 10-inch models in the laboratory to 84- and 120-inch mills in the factory. The selection of surfaces, roll speed, and ratio of speed of the two rolls depends on the particular types of rubber being handled. The Banbury mixer is an enclosed machine containing two water-cooled rotors operating in a water-cooled chamber. The Banbury must be of such a size that the batch will completely fill the mixing chamber. Advances in the design of the Banbury mixer now permit a mixing time of 8 minutes or less.

After being mixed in the Banburys, the stocks are fed to a three- or four-roll calendar used for three types of work: frictioning, skim coating and sheeting. Fabric is frictioned when it is passed between rolls running at different speeds, but the speed of both rolls is the same so that a film of rubber is pressed into the fabric. Fabric used for manufacturing tire plies is usually dipped in a bath containing either a latex-casein compound or a latex-resorcinol-formaldehyde mixture to improve adhesion between the rubber and the cords. It is then dried and fed into the calendar.

Automobile Tires and Tubes

One of the most important applications of all rubbers is tires. Because of the increased demand for speed and endurance in modern vehicles, it has been necessary to design rubber compounds and build tires with these factors in mind. Tires must be designed to withstand heavy loads and high speeds for long periods; the body of the tires must be able to withstand severe shocks and must not show excessive heat build-up.

For years, tires have been built on a flat drum. Successive plies of fabric, which have been friction-coated and cut on a bias, are applied on the drum in an alternating fashion. Circles of wire, known as "beads", which will reinforce the tire where it touches the rim, have been prepared previously from a number of wires embedded in rubber. The beads are attached at the end of the fabric on the drum. The beads may be further protected by "chafer strips." The center of the tire is protected by "breaker strips" topped by a slab of highly abrasion-resistant stock comprising the tread. The last part added to the tire is a sidewall,

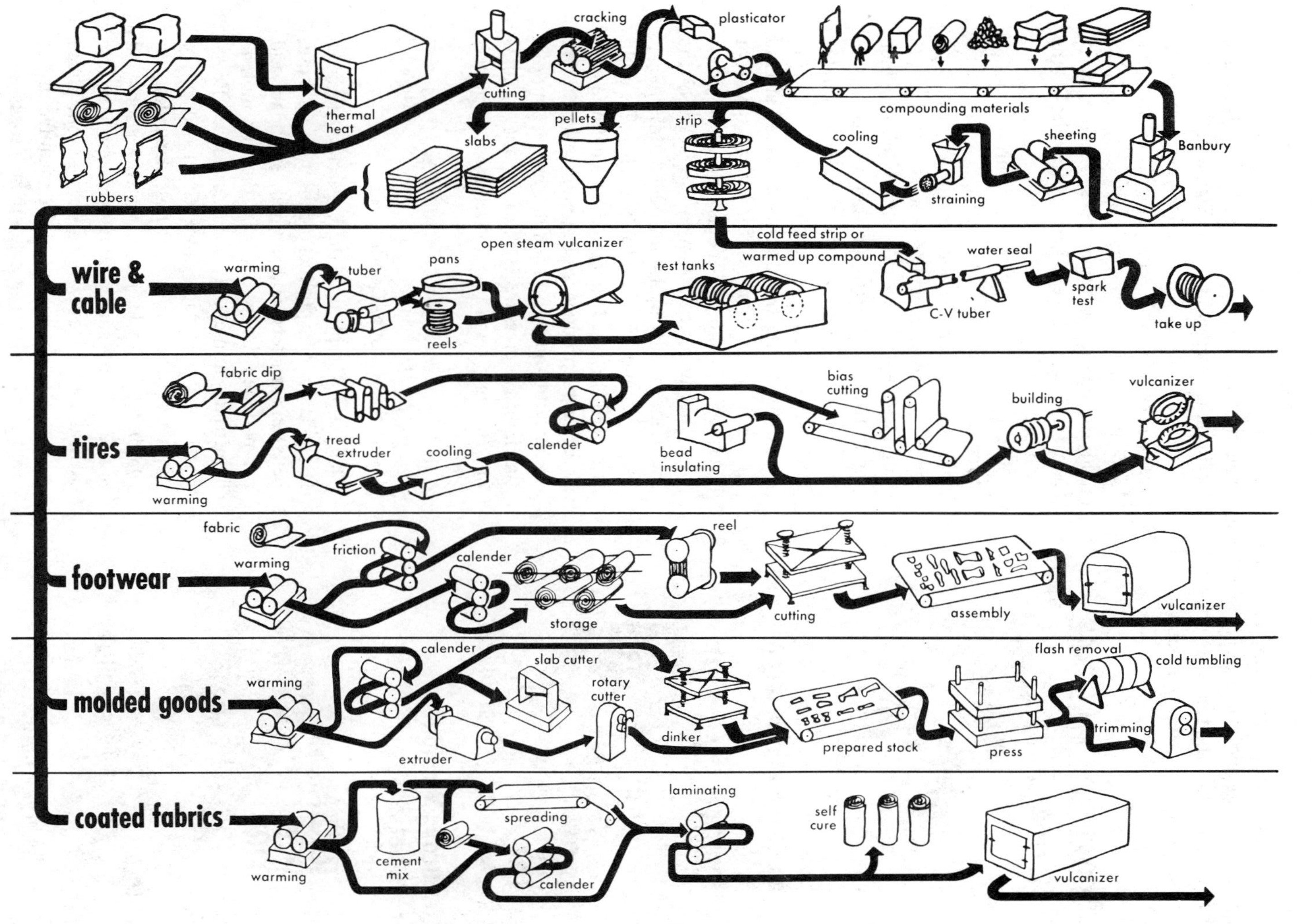

Fig. 9.7. Typical flow diagram for rubber goods manufacture. ("*The Vanderbilt Rubber Handbook*," *R. O. Babbit, ed.; by permission of the R. T. Vanderbilt Co.*)

compounded for white or black and for maximum resistance to ozone deterioration and curb chafing.

Radial tires involve different construction techniques in that the plies run at right angles to the center line of the tread. In conventional bias tires, they run at approximately a 35 degree angle. A belt of tire cord applied circumferentially acts as a tread stabilizer.

CONVENTIONAL TIRE
(2 or 4 plies)

Body ply
cords run at bias angle

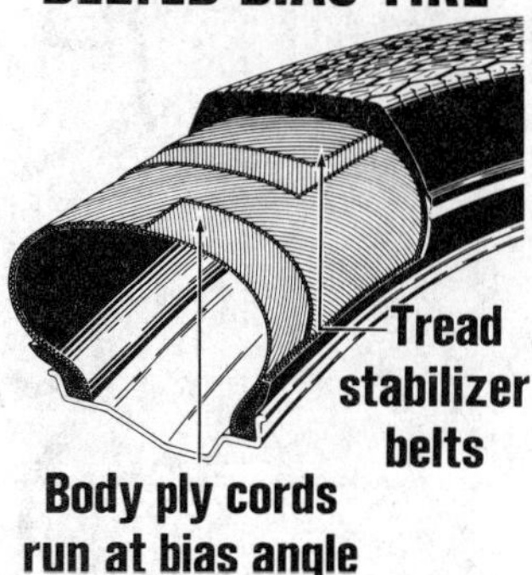

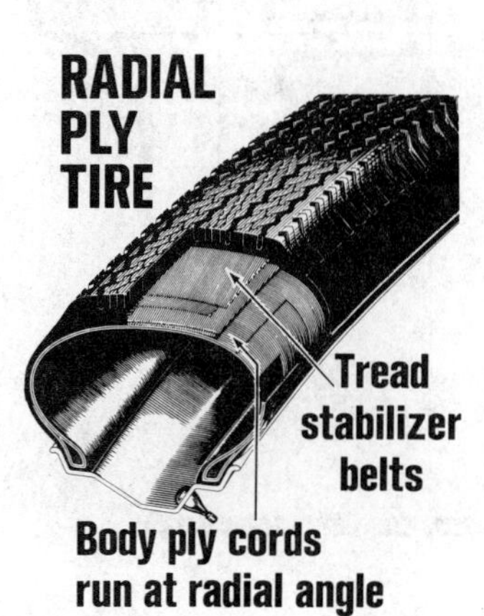

Fig. 9.8. Variations of tire construction. (*Courtesy The Firestone Tire and Rubber Co.*)

The belted-bias tire is a type of hybrid tire in which the body plies are positioned as in the bias tires while outer tread stabilizer ply is used as in radial tires as shown in Figure 9.8.

The rayon tire cord of a few years ago has received considerable competition from nylon and polyester cords, especially for bias plies and for the radial under plies. Rayon, glass, and steel cords have taken active roles in the belt plies.

Because of the high temperatures at which tires must operate on passenger cars, trucks and buses, considerable attention has been given to polymers which generate less heat, as well as to replacements for cotton fabric, which tends to lose strength and fail at higher temperatures. Rayon and nylon cords are finding application for high-speed operation.

Tires from the flat drum are shaped around an air bag in a mold for vulcanization or "curing." In modern molds, the green tire is cured by means of steam in the jacket of the mold; steam and air are introduced into the air bag to expand the tire into the mold. The mold is heated for a specified time at a definite temperature to produce an optimum state of cure throughout the tire. Then, the heat is shut off, the internal pressure is released, the mold is opened, and the tire is removed for final inspection.

Mechanical Rubber Goods

This term is applied to the large class of manufactured rubber products that includes hose, belts and gaskets, and molded, extruded, and compounded items used in the automotive, industrial, household and appliance fields. These may be made from either natural or synthetic rubber, or mixtures, depending on whether the stocks are expected to show oil resistance or some particular characteristic such as resistance to aging or abrasion. Since price is often important in the manufacture of mechanical rubber goods, these stocks may include more filler than those used for tires.

In applications such as automotive transmission seals, pump impellers, O-rings, wire insulation and hose, resistance to extremes in temperature, chemicals, oils, and atmospheric conditions is frequently required. Neoprene, butadiene-acrylonitrile and various specialty elastomers are generally used in these applications.

Hard rubbers may be made from natural or synthetic rubber by increasing the amount of sulfur to 30 or 35 parts per 100 of polymer. Hard rubber stocks may contain extenders such as hard rubber dust, coal dust, carbon black, or inorganic pigments.

RAW MATERIALS FOR SYNTHETIC RUBBER

Butadiene

Butadiene ($CH_2{=}CH{-}CH{=}CH_2$) is the primary raw material for the synthetic rubber industry. Its commercial production has involved the following methods:

1. Thermal cracking of hydrocarbons.
2. Catalytic or oxidative dehydrogenation of butylenes.
3. Catalytic dehydrogenation of butane.
4. Catalytic treatment of ethyl aclohol and ethanol-acetaldehyde mixtures.
5. Dehydrochlorination of chlorinated butanes.
6. Aldol condensation of acetaldehyde followed by conversion of aldol to butadiene.
7. Reaction of acetylene and formaldehyde to form 2 butyne -1, 3 diol followed by hydrogenation and dehydration.

Thermal Cracking of Hydrocarbons. Of the various routes for manufacturing butadiene, the thermal cracking of hydrocarbons has become the dominant source of butadiene throughout the world. The use of naphtha as the basic feedstock for petrochemical production, particularly ethylene and propylene, has resulted in large quantities of coproduct butadiene.[47] This is especially true in Western Europe and Japan which have been totally dependent on the coproduct from steam cracking for over ten years.

Typical yields of the various products from a steam cracker are shown in Table 9.7. It is apparent that with lighter feeds, for example, ethane, there is a very high yield of ethylene, while the heavier feeds produce increasing quantities of other materials such as coproduct butadiene. The feedstock type not only affects the cost of the primary product, ethylene, but also has an effect on product slate, utilities consumption, and plant investment.[70]

Butadiene dimerizes to vinylcyclohexene at a rate such that at ambient temperature, industry specifications of 0.2% dimer are exceeded in only two or three weeks. At -5°C, dimerization, in essence, stops, but the cost of storage and refrigeration is substantial. Therefore, butadiene is generally consumed as fast as it is produced. Because the cracking of hydrocarbons depends on the markets for the primary products such as ethylene and propylene, the economics dictate that coproduct butadiene be sold at prices which insure that on-purpose facilities limit output to the differential

TABLE 9.7 Typical Steam Cracker Yields–Basis: Feed = 100

	Feedstock			
	Ethane	*Propane*	*Naphtha*	*Gas Oil*
Products				
Ethylene	80	45	29.7	25.7
Propylene	1.4	14.5	14.1	13.3
Butadiene	0.0	2.7	4.5	4.2
Aromatics	0.0	3.4	12.3	11.0
Subtotal Yield Primary Products	81.4	65.6	60.6	54.2
Gasoline and C_4's	5.0	6.2	19.7	14.1
Fuel Oil and Gas	13.6	28.2	19.7	31.7
TOTAL YIELD	100	100	100	100

Reference: Hydrocarbon Processing, April 1977, Pages 145–148.

TABLE 9.8 U.S.A. Butadiene Production (Millions of pounds)

	n-butane	*n-butene*	*Coproduct From Ethylene Operations*	*Total*	*% Coproduct*
1970	1225	950	925	3100	30
1975	950	550	1100	2600	43
1978	350	1050	2100	3500	60

between current requirements and the availability of coproducts. This has resulted in exports of coproduct butadiene from Europe and Japan to the U.S.A. and the gradual shutdown of on-purpose facilities (Tables 9.8 and 9.9).

Catalytic Dehydrogenation of N-Butylenes. Various processes have been used to dehydrogenate n-butylenes into butadiene. However, as the cost of energy and hydrocarbons increase, the oxydehydrogenation process of Petro-Tex has generally been selected because of its energy efficiency and yield.

In this process, air, steam, and normal butanes with less than 1% isobutylene are passed over a fixed bed catalyst at a temperature of about 1100°F. Acetylenes and oxygenated by-products are formed which must be removed in subsequent purification steps. Yields approaching 90% are possible. Frequently, coproduct extraction processes are combined with the oxydehydrogenation process to utilize the butylenes present in the C_4 streams. A complete description of the oxydehydrogenation process is provided in Reference 63.

Butane Dehydrogenation. A direct route for producing butadiene from butane (Houdry process) is based on a regenerative operation[68] in which a chromia-alumina catalyst is employed at low temperature and pressure. The system is operated on a cyclic basis. First, the feed stream is cracked for a short period (10 minutes); then the catalyst chamber is evacuated and given an air blow to bring the catalyst back to operating temperatures; finally, the chamber is evacuated to complete the cycle. As the conversion of butane to butadiene is only about 55%, butane dehydrogenation plants have generally been shut down due to their relatively high energy and raw materials consumption.

TABLE 9.9 Butadiene Production (1000 tons)

	Western Hemisphere	*Western Europe*	*Others*	*Total*
1975	1424	1145	578	3147
1976	1695	1421	673	3789
1977	1698	1543	707	3948
1978	1845	1590	791	4226

Reference: European Chemical News, June 4, 1979, pages 14–15.

Butadiene From Ethanol. An ethanol-acetaldehyde mixture is fed over copper-containing catalysts or silica gel containing tantalum oxide in the Ostromislenski process.[14,28,29,49,59] Part of the ethanol is converted to acetaldehyde by dehydrogenation over a copper catalyst at temperatures above 200°C and atmospheric pressure. Ethanol and acetaldehyde are then mixed at about 3 to 1 mole ratios. The catalyst gradually becomes covered with carbon and must be burned off periodically to restore its activity.

The products of this cracking reaction are simpler than those derived from the petroleum cracking processes. They consist primarily of ethylene, propylene, butadiene, unchanged starting materials, and some oxygenated by-products. The butadiene fraction containing 96 percent butadiene is refined to extractive distillation by treatment with B,B'-dichloroethyl ether ("Chlorex"). The resulting purity is at least 98.5 percent.

Alcohol-based butadiene plants were a major source of butadiene in the U.S.A. during World War II, and more recently in India and Brazil. Generally, it is much cheaper to produce butadiene from petroleum. However, with the recent interest in using renewable resources, it is possible that some petrochemicals, for example, butadiene, will again be produced from ethanol.

Isoprene

Isoprene is one of the most common structures in nature and is present not only in natural rubber but also in camphors, terpenes, and various other plant and animal tissues. Isoprene, however, is much more difficult to produce than butadiene. Stereopolymerization systems have now increased the demand for "synthetic natural rubber" and commercial plants have been built using a number of routes from different starting materials.

Recovery from C_5 Streams. This was the easiest route used to obtain commercial quantities of isoprene, and it remains the preferred route for a new producer. The cracking of naphtha or gas oil not only yields 15 to 20 pounds of coproduct butadiene per 100 pounds of ethylene, but also 2 to 3 pounds of coproduct isoprene. A typical analysis for a C_5 stream is shown in Table 9.10. The C_5 stream can either be hydrogenated and blended into gasoline, or the isoprene can be recovered by solvent extraction. The purification process resembles that used for butadiene, but with a greater limitation on the extraction solvent (acetonitrile, dimethyl formamide, or N-methyl pyrrolidine).

Dehydrogenation of Isoamylenes. This process is effected in the same manner as the catalytic dehydrogenation of n-butylenes, which was described in the section on manufacture of butadiene. The economics of this route depend on having access to a favorably priced feed stream. The value of this feedstock varies with changes in the production of gasoline. Isoamylenes can be used for alkylation or for direct blending in gasoline itself. Its availability may be further limited by the current trend toward low-lead gasolines.

TABLE 9.10 C_5 Stream from Naphtha Cracker

Component	*Amount %*
Isoprene	15
Pentadienes	11
Cyclo pentadienes	15
Pentanes	37
Pentenes	19
Other	3
	100

Isobutylene and Formaldehyde. The reaction product of these compounds has been explored by many workers, including some in France and Russia.

$$(CH_3)_2C{=}CH_2 + 2HCHO \longrightarrow$$

$$(CH_3)_2\overset{}{C}\langle\begin{matrix}CH_2{-}CH_2\\ O{-}CH_2\end{matrix}\rangle O$$

4,4 dimethyl-1,3-dioxane

$$\downarrow$$

$$CH_2{=}\overset{\displaystyle CH_3}{\overset{|}{C}}{-}CH{=}CH_2 + HCHO + H_2O$$

Depending on the available feedstocks, a crude isobutylene stream may be used and formaldehyde may be manufactured from methanol. Aqueous formaldehyde and isobutylene are reacted at low pressures at about 95°C. Unreacted components are flashed off, and the dimethyldioxane is recovered by distillation from a first bottoms stream known as Residols I. The catalytic decomposition of dimethyldioxane occurs at atmospheric pressure and at a temperature below 400°C. Both fixed and moving catalyst beds have been used. In the final purification, a second bottoms stream is separated.

The economics of this process again depend on the prices and availability of feedstocks, as well as the value of the recovered Residol streams. While these can be burned, they are being considered as a source of chemicals, thus ultimately reducing the isoprene cost.

Acetone and Acetylene. The reaction products of these compounds is developed by SNAM in Italy.[32]

$$CH_3{-}CO{-}CH_3 + CH{\equiv}CH \longrightarrow CH_3{-}\underset{\displaystyle OH}{\overset{\displaystyle CH_3}{\overset{|}{\underset{|}{C}}}}{-}C{\equiv}CH \tag{1}$$

2-methyl-3-butyn-2-ol

$$CH_3{-}\underset{\displaystyle OH}{\overset{\displaystyle CH_3}{\overset{|}{\underset{|}{C}}}}{-}C{\equiv}CH + H_2 \longrightarrow CH_3{-}\underset{\displaystyle OH}{\overset{\displaystyle CH_3}{\overset{|}{\underset{|}{C}}}}{-}CH{=}CH_2 \tag{2}$$

2-methyl-3-butene-2-ol

$$CH_3{-}\underset{\displaystyle OH}{\overset{\displaystyle CH_3}{\overset{|}{\underset{|}{C}}}}{-}CH{=}CH_2 \longrightarrow CH_2{=}\overset{\displaystyle CH_3}{\overset{|}{C}}{-}CH{=}CH_2 + H_2O \tag{3}$$

In step (1) acetone and an excess of acetylene react in liquid ammonia over a suitable catalyst (aqueous KOH). The yield at 10-40°C and 285 psi is high. The catalyst is neutralized at the end of the reaction, and the unreacted acetylene is recycled.

Two distillation steps are used. First, the unreacted acetone is separated and then the methybutynol; the catalyst and heavier components are left in the bottoms. Partial hydrogenation is effected at 30-80°C and 70-140 psi hydrogen over a supported palladium catalyst. This is separated from the product stream by a centrifuge and re-used. The methylbutynol is fed, as an azeotrope with water, over high purity aluminas at atmospheric pressure and 250-300°C. After a water wash the product is separated by distillation. The economics of this route obviously depend on having a source of low-cost acetylene. With the construction of large-scale coal gasification plants, it is possible that acetylene will be available for isoprene production.

Chloroprene[20]

The manufacture of chloroprene

$$(CH_2{=}CCl{-}CH{=}CH_2)$$

starts with the conversion of acetylene to monovinylacetylene[44] by polymerizing the acetylene in the presence of an aqueous catalyst solution of cuprous chloride and ammonium chloride. The off-gas is cooled to -70°C to condense the vinylacetylene ($CH{\equiv}C{-}CH{=}CH_2$, boiling point 5°C) while the excess acetylene is recycled. The condensate is distilled in a column, pure vinylacetylene comprising the overhead while the bottoms contain divinylacetylene. Monovinylacetylene is then converted to chloroprene by reaction with hydrogen chloride in the presence of a cuprous chloride solution.[9]

REFERENCES

1. Alliger, G. and Sjothun, I. J., *Vulcanization of Elastomers*, New York, Van Nostrand Reinhold 1964.
2. Alliger, G., Willis, J. M., Smith, W. A., and Allen, J. J., *Mech. Eng.* 1098-1102 (1956).
3. Alliger, G., Willis, J. M., Smith, W. A., and Allen, J. J., *Rubber World*, **134**, 549-59 (1956).
4. Bauer, R. G., *The Vanderbilt Rubber Handbook*, 51-77, R. T. Vanderbilt Co. (1978).
5. Blackley, D. C., *Emulsion Polymerization*, John Wiley & Sons, 1975.
6. Blackely, D. C., *High Polymer Latices*, Vols. I and II, New York, Palmerton, 1966.
7. Blow, C. M., *Rubber Technology and Manufacture*, pg. 29, Butterworth & Co., 1971.
8. Bovey, F. A., *Polymer Conformation and Configuration*, New York, Academic Press, 1964.
9. Carothers, W. H., Williams, I., Collins, A. M., and Kirby, J. E., *J. Am. Chem. Soc.*, **53**, 4203-25 (1931).

10. *Chemical Engineering*, Reprint No. 94, pg. 43, McGraw-Hill.
11. *Chem. Eng. News*, **37**, 23 (1959).
12. *Chem. Eng. News*, **33**, 4518 (1955).
13. *Chem. Eng. News*, Economics Improving for Guayule Rubber, pg. 10, Vol. **56**, No. 35, Aug. 28, 1978.
14. Corson, B. B., Jones, H. E., Welling, C. E., Hinckley, J. A., and Stahly, E. E., *Ind. Eng. Chem.*, **42**, 359–73 (1950).
15. Davis, C. C., Ed., *Chemistry and Technology of Rubber*, New York, Van Nostrand Reinhold, 1937.
16. D'Ianni, J. D., *Eng. Chem.*, **40**, 253–6 (1948).
17. DiGiacomo, A. A., Maerker, T. B., and School, J. W., *Chem. Eng. Progr.* **57**, 35 (1961).
18. Dunbrook, R. F., *India Rubber World*, **117**, 203–7 (1947).
19. Dunn, J. R., Coulthand, D. C., and Pfisterer, K. A., *Rubber Chem. and Technol.* **51**, 389 (1978).
20. DuPont, "The Neoprenes."
21. Forman, L. E., Chap. 6 in *Polymer Chemistry of Synthetic Elastomers*, J. P. Kennedy and E. G. M. Tornquist, Eds., Part ii, 491–596, John Wiley & Sons, 1969.
22. Gardon, J. L., "Mechanism of Emulsion Polymerization," *Rubber Chem. Technol.* **43**, 74–94 (1970).
23. Belgian Patent 543, 292 (1956; to Goodrich-Gulf).
24. Happel, J., Cornell, P. W., Eastman, D. B., Fowle, M. J., Porter, C. A., and Schutte, A. H., *Trans Am. Inst. Chem. Eng.*, **42**, 189–214, 1001–7 (1946).
25. Harkins, W. D., *J. Am. Chem. Soc.*, **69**, 1428–44 (1947).
26. Horne, S. E., Jr., Kiehl, J. P., Shipman, J. J., Folt, V. L., Gibbs, C. F., Willson, E. A., Newton, E. B., and Reinhart, M. A., *Ind. Eng. Chem.*, 48, 784–91 (1956).
27. Howland, L. H., and Brown, R. A., *Rubber Chem. Technol.* **34**, 1501 (1961).
28. Jones, H. E., Stahly, E. E., and Corson, B. B., *J. Am. Chem. Soc.*, **71**, 1822–8 (1949).
29. Kampmeyer, P. M., and Stahly, E. E., *Ind. Eng. Chem.*, **41**, 550–5 (1949).
30. Kienle, R. N., Dizon, E. S., *Rubber Chem. Technol.* **44**, No. 5, 1971.
31. Logemann, H. and Pampus, G., *Kautschuk Gummi-Kunststoffe*, **23**, 479–486, (1970).
32. Malde, M. de, DiCio, A., and Mauri, M. M., *Hydrocarbon Process Petrol. Refiner* **43**, 149 (1964); *Chem. Eng.* **71**, No. 20, 78 (1964); and *Chem. Eng.* **78** (Sept. 28, 1969).
33. McGregor, R. R., *Silicones and Their Uses*, New York, McGraw-Hill, 1954.
34. Miller, M. L., *The Structure of Polymers*, New York, Van Nostrand Reinhold, 1966.
35. Mitchel, S., Hall, W. F., DeWames, R. E., *Rubber Chem. Technol.* **45**, No. 3, 1972.
36. Moacanin, J., Holden, G., Tschoegl, N. W., *Block Copolymers*, John Wiley & Sons, 1969.
37. Moore, J. E., *J. Polymer Sci.* **A2**, 835 (1964); Gamble, L. W., Westerman, L. and Knipp, E. A., *Rubber Chem. Technol.*, **38**, No. 4, 823 (1965); Cazes, J., *J. Chem. Educ.*, **43**, No. 7, A-567 and No. 8, A-625 (1966).
38. Morton, A. A., "Alfin Catalysts" in *Encyclopedia of Polymer Science and Technology*, Vol. **1**, 629-238, New York, John Wiley & Sons, 1964.
39. Morton, M., *Rubber Technology*, pg. 440–458, Van Nostrand Reinhold, 1973.
40. Myer, D. A., Chapter 10 in Reference 1.
41. Natta, G., Crespi, G., Valvassouri, A. and Sartori, J., "Polyolefin Elastomers," *Rubber Chem. Technol.*, **36**, 1583 (1963).
42. Natta, G., and Porri, L., in *Polymer Chemistry of Synthetic Elastomers*, J. P. Kennedy and E. G. M. Tornquist, Eds., Part II, Chapter 7, New York, John Wiley & Sons, 1969.
43. Nielson, L. E., *Rubber & Plastics News*, Nov. 26, 1979.
44. Nieuwland, J. A., Calcott, W. S., Downing, F. B., and Carter, A. S., *J. Am. Chem. Soc.* **53**, 4197–4202 (1931).
45. The Plastics and Rubber Inst., Conference on "The Technology of Plastics and Rubber Interface," Brussels, 1976.
46. Poehlein, G. W. and Dougherty, D. J., *Rubber Chem. Technol.* **50**, No. 3, pg. 601, (1977).
47. Ponder, T. C., *Hydrocarbon Processing*, Vol. **55**, No. 10, pg. 119–121, Oct. 1976.
48. "Proceedings Second Rubber Technical Conference," 69–78 (1948).
49. Quattlebaum, W. M., Toussaint, W. J., Jr., and Dunn, J. T., *J. Am. Chem. Soc.*, **69**, 593–9 (1947).
50. Richardson, W. S., and Sacher, A., *J. Polymer Sci.*, **10**, 353–70 (1953).
51. Rochow, E. G., *Introduction to the Chemistry of Silicones*, 2nd Ed., New York, John Wiley & Sons, 1951.
52. *Rubber Chem. Technol.*, 7, 633 (1934).
53. Rubber Manufacturers Assoc., Inc., New York, "Type Description and Packing Specifications for Natural Rubber," Revised, Dec. 1954.
54. Saunders, J. H., and Frisch, K. C., *High Polymers*, Vol. **XVI**, New York, John Wiley & Sons, 1962.
55. Schoenberg, E., Marsh, H. A., Walters, S. J. and Saltman, W. M., *Rubber Chem. Technol.*, **52**, No. 3, pg. 526–605, 1979.

56. British Patent 459,972 (Jan. 19, 1937) ("Semperit" Oesterreichisch-Amerikanische Gummiwerke A. B., to Metallgesellschaft A.-G.).
57. Sjothun, I. J., *Rubber Age*, 74 (1953).
58. Stavely, F. W., et al, *Ind. Eng. Chem.*, 48 778-83 (1956).
59. Toussaint, W. J., Dunn, J. T., and Jackson, D. R., *Ind. Eng. Chem.*, 39, 120 (1947).
60. Treloar, L. R., *Rubber Chem. Technol.*, 47, No. 3, 1974.
61. Ugelstad, J. and Hansen, F., *Rubber Chem. Technol.*, **50**, No. 3, pg. 639 (1977).
62. Warrick, E. F., Pierce, O. R., Polmanter, K. E., *Rubber Chem. Technol.*, **52**, No. 3, 1979.
63. Welch, M., Croce, L. J., Christman, H. F., *Hydrocarbon Processing*, Vol. **57**, No. 11, pg. 131-136, Nov. 1978.
64. Whitby, G. S., Davis, C. C., and Dunbrook, R. F., eds., *Synthetic Rubber*, New York, John Wiley & Sons, 1954.
65. Winchester, C. T., *Ind. Eng. Chem.*, **51**, 19 (1959).
66. Winspear, G. G., *Vanderbilt Latex Handbook*, R. T. Vanderbilt Co., 1954.
67. Winspear, George, G. and Watermann, R. R., in *Introduction to Rubber Technology*, M. Morton, Ed., New York, Van Nostrand Reinhold, 1959.
68. Womeldorph, D. E., Stevenson, D. H., and Friedman, L., *Am. Petrol. Inst.* (May 14, 1958).
69. Youker, M. A., *Chem. Eng. Progr.*, 43, No. 8, 391 (1947).
70. Zdonik, S. B., Bassler, E. J., Hallee, L. P., *Hydrocarbon Processing*, pg. 73-81, Vol. **53**, No. 2, Feb. 1974.
71. Ziegler, K., Dersch, F., and Wollthau, H. *Ann. Chem.*, **511**, 13 (1934); Ziegler K. and Jakob, L., *Ann. Chem.*, **511**, 45 (1934); Ziegler, K., Jakob, L., Wollthau, H., and Wenz, A., *Ann. Chem.*, **511**, 64 (1934).

10

Synthetic Plastics

Robert W. Jones* and Robert H. M. Simon*

INTRODUCTION

The word "plastic" was originally used as an adjective to denote a degree of mobility or formability. In the 1909 edition of "Webster's International Dictionary," the noun was not listed. Shortly thereafter, with the introduction of "Bakelite" by Dr. Baekeland, the word was often used as a noun, most frequently referring to "Bakelite," (phenol-formaldehyde resins) "Celluloid," (cellulose nitrate) and casein plastics.

The American Society for Testing Materials (D 1695-77) has defined a plastic as "a material that contains as an essential ingredient an organic substance of large molecular weight, is solid in its finished state, and, at some stage in its manufacture or in its processing into finished articles, can be shaped by flow." According to this definition, synthetic fibers, all rubbers, and even bread doughs are plastics, but glass is not. Those who insist that glass is a plastic omit "organic substance" as part of the definition. In this chapter, synthetic fibers, regenerated cellulose, e.g., rayon, (Chapter 11), rubber (Chapter 9), glass, those materials used exclusively in surface coatings (Chapter 22), and, of course, bread dough will not be considered.

The word "resin" is an old one derived from the Latin "resina" and the Greek "rhetine." Originally it referred to the natural exudates (or their fossil remains) of vegetable origin. The ancient Egyptians used such materials to help preserve their mummies.[1] Frankincense and myrrh, the Wise Men's gifts to the infant Jesus, are both natural resins. There are many such natural products: accroide, congo, rosin, copal, dammar, sandarac, elemi, kauri, manila, mastic, batu, pontianak, and shellac. Today they are used principally in surface coatings or as binders and adhesives. When identified as "synthetic" (the adjective is frequently omitted for brevity), the current meaning of "resin" in the plastics industry is "that base substance of high molecular weight" before it has been mixed with colorants, fillers, plasticizers,†

*Monsanto Plastics and Resins Co., Springfield, Mass.

†A plasticizer is a substance oridinarily (though not necessarily) of lower molecular weight that makes the plastic more flexible (lower elastic modulus); hopefully it should improve impact strength without markedly increasing creep or lowering ultimate strength.

lubricants and/or stabilizers to make a finished commercial plastic molding powder.† Similarly, in the surface-coatings industry, resin refers to the base binding material of high molecular weight before it has been formulated into a paint, varnish, or enamel.

The commercial history of synthetic plastics begins with the development of a practical molding process for cellulose nitrate (Celluloid), in 1869, by John Wesley Hyatt and his brother Isaiah, who were seeking a substitute for ivory. In about 1899, Adolf Spitteler treated casein with formaldehyde to make commercial plastic materials. In 1909, Leo Baekeland developed the first practical process for making moldings from phenol-formaldehyde resins. The organic and physical chemistry of these inventions remained an art for a long time, awaiting two concepts. One, championed by Hermann Staudinger and a few others in the 1920's, is that plastics owe their most significant properties to the extremely large size of their "giant" molecules. The other, first clearly implied by the experiments of Wallace Hume Carothers in 1929, is that the reaction mechanisms and thermodynamic equilibria normally associated with reactive organic groups are almost the same for groups attached to large molecules.

With these concepts as guideposts, many thousands of polymers have since been made, or rediscovered. Less than fifty basic types have attained commercial success.

VIEWPOINT OF THE CONSUMER

The ultimate judgment about the utility of plastics is made by the consuming public; and judging by the response to the use of plastics in a variety of areas, the consumer is beginning to realize that, properly applied, plastics are superior to the materials formerly used and not just a "cheap substitute," unfortunately still a common image. The superior stain, abrasion, and mar resistance of laminated melamine (e.g., Formica®) table and counter tops is well known. The superiority of plastic laminates (with fiber glass) for boats is also well known. (However, note the use of the term "fiber glass boat hulls" rather than "plastic boat hulls" due to the early poor image of the word "plastic.") Automobile, furniture, airplane, and bus upholstery of vinyl sheeting is superior to previous materials in many respects; replacement slip covers for automobiles are nearly obsolete due to the superior wear resistance of vinyl sheeting. The use of plastics in football helmets, in golf club shafting, in fishing rods, and in shampoo bottles is beginning to destroy the myth that all plastics are brittle. Developers of synthetic leathers were careful in avoiding the use of the word "plastic", although that is what the materials are. The industry must constantly guard against misapplication which leads to the image of a "cheap substitute."

The typical consumer is unaware of the extensive use of plastics where their superior price-performance dictates their use, e.g., insulation for nearly all wiring, drain and vent plumbing, printed circuits, automobile instrument panels and front grills—now chrome plated, in windshields to make safety glass, agricultural piping, greenhouses, and mulching equipment.

Frequently, plastics will give a necessary balance of properties which cannot be matched by other materials, e.g., color, light weight, warmth of touch, low thermal and electrical conductivity, and resistance to biological and environmental degradation. Ablating shields for space vehicles and coatings for the leading edges of supersonic aircraft wings are two exotic applications.

Some of the other major areas of plastics application include: packaging, housings (e.g., for appliances, tools, TV's, radios, and telephones), furniture, automotive equipment, electrical and thermal insulation, toys, building products, boats and recreational equipment, and piping. New and novel uses include: synthetic printing paper—principally in Japan; artificial organs and tissues, and prosthetic devices in medicine; unbreakable glazing and "safety-glass" laminates; soft drink and beer bottles; synthetic leather for shoes; chrome-plated plastic parts for superior wear and corrosion resistance.

†The term "powder" is a misnomer. Most commercial molding powder is in the form of granules or pellets which provide high bulk density, good flow, no dusting, and other desirable handling characteristics.

VIEWPOINT OF THE INDUSTRIAL DESIGNER

In plastics, the industrial designer has found a versatile raw material. Plastics have a tremendous advantage over natural products in that they can be tailor-made for a specific use. Plastics may be classified in many ways. The industrial designer uses conventional physical properties to characterize plastics, e.g., density, tensile strength, impact strength, modulus of elasticity, and creep rate. These, however, are not enough. Price, fabricating costs, electrical characteristics, thermal conductivity, colorability, other appearance factors, and chemical resistance frequently determine the selection of a particular plastic in preference to another plastic or other materials.

For information about specific properties of various plastics, consult Table 10.1 and Selected References PP-1 through PP-5.

Testing plastics is a highly specialized art, which has been developed in the U.S. primarily through the efforts of the members of the American Society for Testing Materials (ASTM). Plastics are usually excellent electrical insulators, but their electrical properties must be measured over a wide range of electrical and climatic conditions. Few plastics obey Hooke's Law; many types creep even at very low temperatures and stresses. Specialized stress-strain testing machinery has been developed which permits extensive variations in temperature and rate of loading. Since coloring versatility and simplicity are prime factors in many plastics markets, tests for color reproducibility and stability are often essential. (See Selected References T-1 through T-4 for testing procedures for plastics.)

VIEWPOINT OF THE FABRICATOR

Not only must the fabricator keep in mind the industrial designer and consumer, he must also consider the types of equipment required and the behavior of the plastics in relation to this equipment. Plastics are divided into thermosetting and thermoplastic materials. Thermoplastics are materials which become fluid upon heating above a certain temperature, frequently called the "heat distortion temperature." Upon cooling they set to an elastic solid. The process can be repeated many times since no primary chemical bonds are normally made or broken. Thermosetting materials, on the other hand, may normally be heated to the fluid state only once. The chemical structure of a thermosetting plastic is altered by heat, and crosslinked products are formed which cannot be resoftened. Thermosetting plastics are not normally used in injection molding machines and extruders since these machines contain spaces where hot molten plastic may remain for indefinite lengths of time, thus causing varying levels of cure and possible set-up in the machine. Chemically, the molecules of a thermosetting resin cross-link* irreversibly through primary chemical bonds to form one large molecule. It may seem strange to call a washing machine agitator a single molecule, but this is the modern view of highly crosslinked structures.†

Compression and transfer molding are the two main methods used to produce molded parts from thermosetting plastics[2]; however, injection molding is under development and may become important in the future.[3,4] In compression or transfer molding, the thermosetting molding compound is heated to about 150°C so as to soften it sufficiently so that it will flow into the mold cavity. The soft plastic is held under a pressure often greater than 140 kg per cm^2 long enough for the material to polymerize or crosslink, resulting in a hard, rigid product. A simple compression molding machine is driven by a vertical hydraulic ram press as shown in Fig. 10.1.

Thermoplastics may be molded in a compression molding machine or in the faster, more versatile, and economical injection molding machine (Fig. 10.2). The concept of maintaining a stable thermoplastic material in a fluid state and squirting it under pressure into a

*A term more inclusive than thermosetting resin is crosslinked resin, applied equally well to resins such as polyurethane, some types of which form crosslinked structures at room temperature.

†There are undoubtedly many individual molecules of varying sizes trapped in the over-all network, but it is probable that any macroscopic region of the agitator is connected through primary chemical bonds to all other regions of the agitator.

TABLE 10.1 Basic Commercial Plastics

Descriptive Name	*Chemical Structure of Monomers*	*Outstanding and/or Typical Properties or Uses*[a]	*Thermosetting, Thermoplastic, or Casting*	*Commercial Methods of Manufacture*
ADDITION POLYMERS				
Polyethylene	$H_2C{=}CH_2$	Excellent electrical properties. Good impact strength. Excellent colorability but translucent in thick sections. Good chemical resistance. Used in film, moldings, wire insulation, pipe, paper coating, and flexible bottles. Available in flexible and semirigid forms only unless heavily filled. Film widely used in packaging.	Thermoplastic	(a) High pressure; dense gas with free radical catalysts. (b) Medium pressure, solvent-nonsolvent; transition-metal oxide catalysts. (c) Low- and medium-pressure, solvent-nonsolvent, Ziegler catalysts.
Polypropylene	$H_3C(H)C{=}CH_2$	Lowest density of any plastic. Fair to good impact. Fair rigidity and dimensional stability. Excellent colorability. Translucent in thick sections. Good chemical resistance Properties vary widely with degree of crystallinity.	Thermoplastic	Solvent-nonsolvent; Ziegler catalysts.
Polyvinyl chloride and copolymers with vinyl acetate and vinylidine chloride ("Vinylite")	$H_2C{=}CHCl$	Good electrical properties and flame resistance with proper plasticizers. Rigid as polymerized–made flexible with plasticizers. Good impact strength especially in polyblends with elastomers. Good chemical resistance. Requires heat stabilizers. Many diverse uses.	Thermoplastic	(a) Suspension. (b) Solvent-nonsolvent. (c) Emulsion, used primarily for surface coatings, textile and paper treating, and "Plastisols."
Polyvinylidene chloride, always as a copolymer with vinyl chloride	$H_2C{=}CCl_2$	Very low moisture vapor transmission. Good chemical resistance. Self-extinguishing flame resistance. Used in pipe linings, film.	Thermoplastic	(a) Suspension. (b) Emulsion.
Polystyrene; also copolymerized with α-methyl styrene and vinyl toluenes to improve heat resistance	$H_2C{=}CH(C_6H_5)$	Excellent color, transparency, rigidity, dimensional stability, electrical properties, molding speeds. Used principally in moldings; also in phonograph records and rigid foams.	Thermoplastic	(a) Mass (polymer soluble in monomer). (b) Suspension. (c) Emulsion, used principally for surface coatings and polishes.

Polystyrene-based graft and/or polyblend polymers with styrene/butadiene copolymer synthetic rubbers	$(H)(H)C{=}C(H)(C_6H_5)$ $H_2C{=}CH{-}CH{=}CH_2$	Excellent rigidity, dimensional stability, molding speeds. Good impact strength. Good colorability in opaques. Used extensively in moldings and for extruded sheet which is then vacuum formed. Packaging is large market.	Thermoplastic	(a) Mass. (b) Emulsion. (c) Suspension.
Styrene/acrylonitrile copolymer grafted or polyblended to butadiene/acrylonitrile rubber (ABS)	$(H)(H)C{=}C(H)(C_6H_5)$ $H_2C{=}CH{-}CH{=}CH_2$ $CH_2{=}CHCN$	Outstanding impact. Good chemical resistance. Fair weatherability. Fair colorability in opaques. Used in pipe, moldings, extruded sheet.	Thermoplastic	(a) Emulsion. (b) Suspension. (c) Mass.
Styrene/acrylonitrile copolymer	$(H)(H)C{=}C(H)(C_6H_5)$ $CH_2{=}CHCN$	Better chemical resistance, crazing resistance, weatherability, impact strength than polystyrene. Good colorability including transparents. Used in moldings, film.	Thermoplastic	(a) Solution. (b) Mass to low conversion followed by devolatilization.
Polymethylmethacrylate ("Plexiglass")	$(H)(H)C{=}C(CH_3){-}C({=}O){-}O{-}CH_3$	Excellent color, transparency, rigidity, dimensional stability, outdoor stability. Good impact strength. Used in moldings and as heavy sheeting. High impact graft copolymers available.	Thermoplastic and casting	(a) Suspension. (b) Casting between glass for transparent sheeting.
Polyesters (Combination of condensation and addition polymers.)	$HC{=}CH$ ring with $C({=}O){-}O{-}C({=}O)$ (maleic anhydride) $(H)(H)C{=}C(H)(C_6H_5)$ HOROH	Used with glass cloth or fibers to achieve very high tensile and flexural strengths approaching those of steel. Used to make laminates in aircraft, boats, furniture. Some potting of electrical components. Some compression and low-pressure molding.	Thermosetting and casting	See text.

TABLE 10.1 (Continued)

Descriptive Name	*Chemical Structure of Monomers*	*Outstanding and/or Typical Properties or Uses*[a]	*Thermosetting, Thermoplastic, or Casting*	*Commercial Methods of Manufacture*
Polyvinyl butyral (Chemically modified addition polymer)	$H_2C{=}CH{-}O{-}C({=}O){-}CH_3$ $CH_3CH_2CH_2CH{=}O$	Primarily used for safety glass interlayer. Has excellent adhesion to glass. Excellent ultraviolet light resistance in absence of air.	Thermoplastic	See text.
Polychlorotrifluoroethylene ("Kel-F")	$ClFC{=}CF_2$	Excellent resistance to chemicals. Used as gasketing, packing, and corrosion resistant coatings.	Thermoplastic	(a) Emulsion. (b) Solvent-nonsolvent.
Polytetrafluoroethylene ("Teflon")	$F_2C{=}CF_2$	Excellent resistance to chemicals. Used as gasketing, packing, and corrosion resistant coatings. Attacked only by alkali metals and hot strong bases.	Fuse slowly under high pressure and 450–500°F	Emulsion.
CONDENSATION POLYMERS				
Phenol-formaldehyde ("Bakelite")	C_6H_5OH (phenol) $H_2C{=}O$	Good electrical properties with proper formulation. High heat resistance. Good to excellent impact and tensile strengths. Very slow to essentially no burning rate depending on filler. Acid resistant. Used for moldings, particularly for industrial and electrical applications where its poor colorability is not important. Used extensively for decorative laminate backing.	Thermosetting and casting	See text.
Nylon 66	$HOC(=O)(CH_2)_4C(=O)OH$ $NH_2(CH_2)_6NH_2$	Excellent wear resistance, tensile and impact strengths. Low coefficient of friction. High heat distortion temperature. Self-extinguishing to fire. Used for gears, rolls, and other moving parts.	Thermoplastic	See Chapter 11.

ε-Caprolactam (nylon 6)	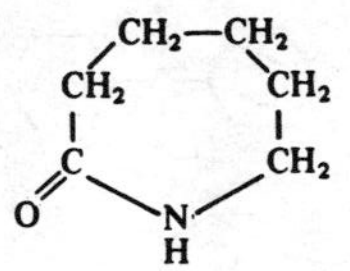 cyclic: CH_2—CH_2—CH_2—CH_2—CH_2—N(H)—C(=O)	Properties very similar to nylon 66.	Thermo-plastic	See Chapter 11.
Melamine-formaldehyde	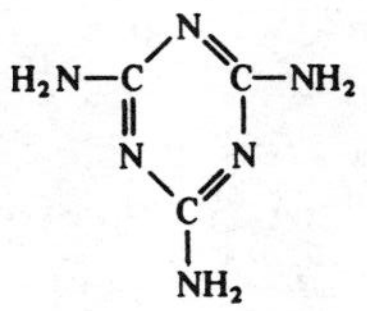 H_2N—C(=N—C(NH_2)=N—C(NH_2)=N) (melamine ring); $CH_2{=}O$	Excellent hardness abrasion resistance, flame resistance, heat resistance. Good colorability in opaques, stain resistance, electrical properties. Used in laminates. Used with alkyd surface coatings and for adhesives.	Thermo-setting	See text.
Urea-formaldehyde	$H_2N\overset{O}{\overset{\Vert}{C}}NH_2$ $CH_2{=}O$	Similar to melamine but poorer alkali resistance, crazing resistance, heat resistance, dimensional stability. Also poorer wear and hardness. Used in adhesives.	Thermo-setting	See text.
Epoxy	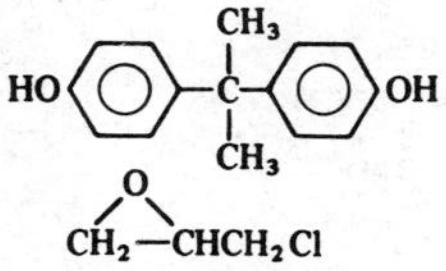 HO—C_6H_4—C(CH_3)$_2$—C_6H_4—OH; $\overset{O}{CH_2—CH}CH_2Cl$; amines, organic anhydrides (see text for other monomers)	Outstanding adhesion to metals. Good chemical resistance. Fair to excellent impact in flexible formulations. Glass laminates have excellent tensile and flexural strengths plus good chemical and outdoor resistance if properly formulated. Widely used in surface coatings and adhesives.	Thermo-setting and casting	See text.
Ethylene glycol-terephthalic acid ("Mylar")	$HOCH_2CH_2OH$; 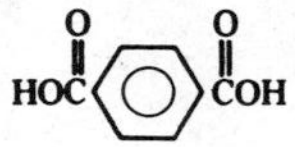 HO$\overset{O}{\overset{\Vert}{C}}$—$C_6H_4$—$\overset{O}{\overset{\Vert}{C}}$OH	Outstanding toughness and tear strength in films. Also widely used in fibers. (See Chapter 11.)	Thermo-plastic	

TABLE 10.1 (Continued)

Descriptive Name	*Chemical Structure of Monomers*	*Outstanding and/or Typical Properties or Uses*[a]	*Thermosetting, Thermoplastic, or Casting*	*Commercial Methods of Manufacture*
Silicones	$(CH_3)_2SiCl_2$; $(C_6H_5)_2SiCl_2$ (see text)	Outstanding heat resistance. Unusual solubilities give anti-foam and water-repellent properties. Used for insulating varnish on electric motors, for high temperature rubbers, greases, and lubricants, for high temperature glass laminating, to increase adhesion of polyesters to glass fibers, and a multitude of other specialty uses most of which are not "plastic" applications in the strict sense.	Thermosetting in plastic applications	See text.
Polyurethanes	$CH_3C_6H_3(N{=}C{=}O)_2$; HOC(=O)RC(=O)OH (see text); HOROH	Best properties as rubbers with excellent wear resistance. Widely used as flexible foams; also for rigid foams and surface coating applications.	Thermosetting	See text.
Polyoxymethylene	$H_2C{=}O$	Excellent abrasion resistance. Tough. Excellent colorability in opaques only. Good chemical resistance. Excellent dimensional stability.	Thermoplastics	Ionic solvent-nonsolvent (probably a fast chain reaction mechanism like vinyl polymers).
Polyoxetanes ("Penton")	$(ClCH_2)_2{-}C{-}CH_2$ / $H_2C{-}O$ (ring) (see text)	Best chemical resistance of easily molded rigid polymers. Self-extinguishing to fire.	Thermoplastic	Ionic solvent-nonsolvent.
Polycarbonates	$C_6H_5{-}O{-}C(=O){-}O{-}C_6H_5$ or $Cl{-}C(=O){-}Cl$; $HO{-}C_6H_4{-}C(CH_3)_2{-}C_6H_4{-}OH$	Excellent impact, dimensional stability at high temperatures, colorability. Good electrical properties. Self-extinguishing.	Thermoplastic	See text.

Polyphenylene oxides	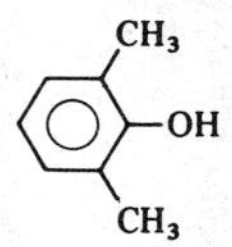	Very high heat distortion resistance for a thermoplastic; nondripping and self-extinguishing in flammability tests.	Thermoplastic	See text.
Polysulfone	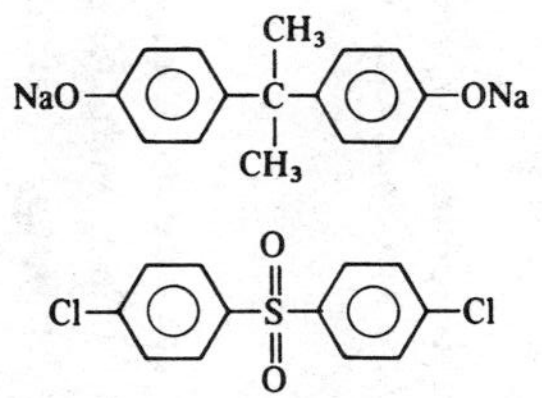	Very high heat distortion resistance for a thermoplastic; nondripping and self-extinguishing in flammability tests.	Thermoplastic	See text.
CELLULOSIC	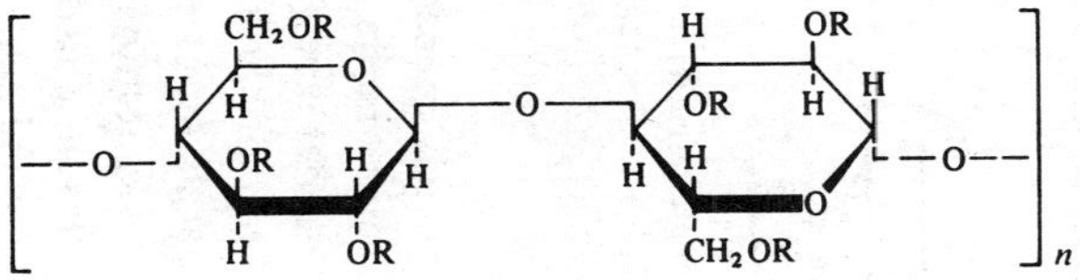			
Cellulose acetate	$n = 50\text{-}100$ $R = -C(=O)-CH_3$ (2.2-2.3 R's per ring)	Good impact strength. Good transparency. Excellent colorability. Fair outdoor stability. Slow to self-extinguishing burning rate. Requires plasticizers to form. Used for photographic film and moldings.	Thermoplastic	See text.
Cellulose nitrate	$n = 250$ $R = NO_2$ (1.9-20 R's per ring)	Outstanding impact strength. Uses as plastic declining.	See text	See text.
Cellulose acetate butyrate	$R = -C(=O)-CH_3$ (1 per ring) $R' = -C(=O)CH_2CH_2CH_3$ (1.7 per ring)	Similar to cellulose acetate with better dimensional stability and weatherability. Used for pipe, telephone housings, moldings.	Thermoplastic	See text.
Ethyl cellulose	$n = 250$ $R = -CH_2CH_3$	Outstanding impact of all cellulosics, otherwise similar.	Thermoplastic	See text.

[a]Within each class of polymer the properties can be widely varied depending on the plasticizer, copolymer type and the ratio of filler and stabilizer in the basic formulation, particularly in condensation polymers. The properties given are only the outstanding and characteristic ones.

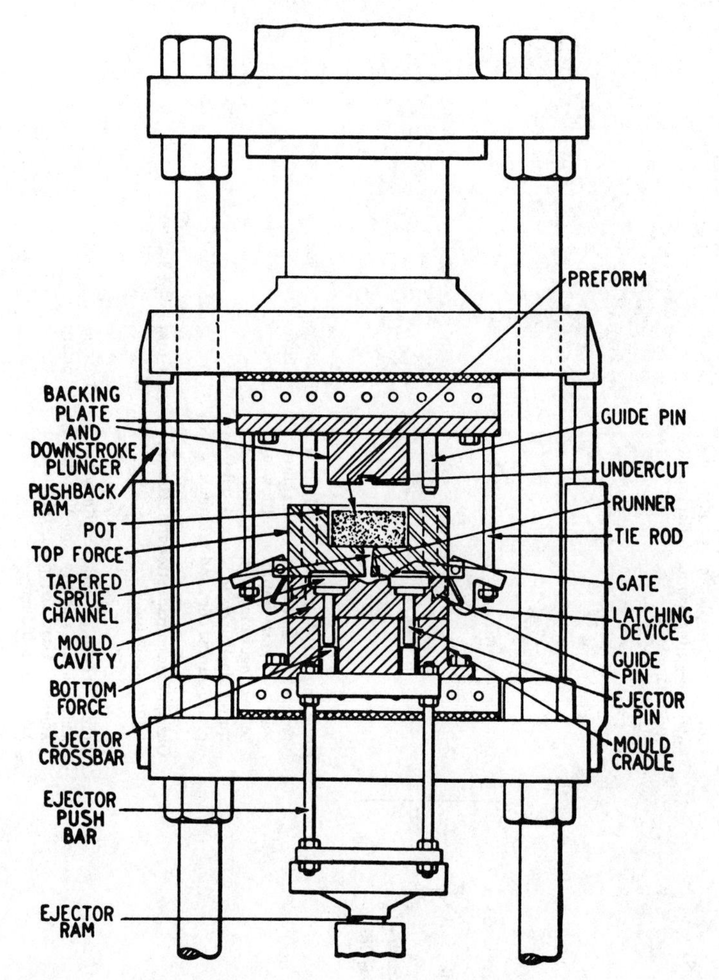

Fig. 10.1 Transfer mold for thermosetting plastics which permits heating of preform pellets during curing time of piece in mold cavity. (*Redfarn, C. A., "A Guide to Plastics," Cliffe and Sons, Ltd., London, 1958; by permission.*)

cooler mold was first developed in the middle 1920's by Dr. Arthur Eichengum and the German firm of Eckert and Ziegler. It was not until the late 1930's that successful automatic machines were used in the United States. Once a thermally stable material became commercially available, development of the injection molding machine followed logically. The first such commercial plastic was cellulose acetate, introduced in 1933. Injection molding of a thermoplastic is a physical process and is based on the ability of the thermoplastic to reversibly soften with heat and harden on cooling. No chemical reaction takes place as in the case of thermosetting resins. Heating and cooling may be repeated many times if desired. Polymer beads or pellets are heated in a cylinder or, in more modern machines, with the aid of an extruder screw until molten. The melt is pushed out of the cylinder through a nozzle by either a piston or the extruder screw, now acting as a piston, into a relatively cool, closed mold. Here it solidifies to form a part shaped like the reverse image of the mold. A schematic of a screw injection molding machine is shown in Fig. 10.2.

Reaction injection molding, i.e., RIM, is a relatively new, rapidly expanding, fabricating method for producing molded parts of large size[6,7,8,9,10]. Polyurethanes, using isocyanates plus a polyol, are the principal materials, although other experimental polymers are

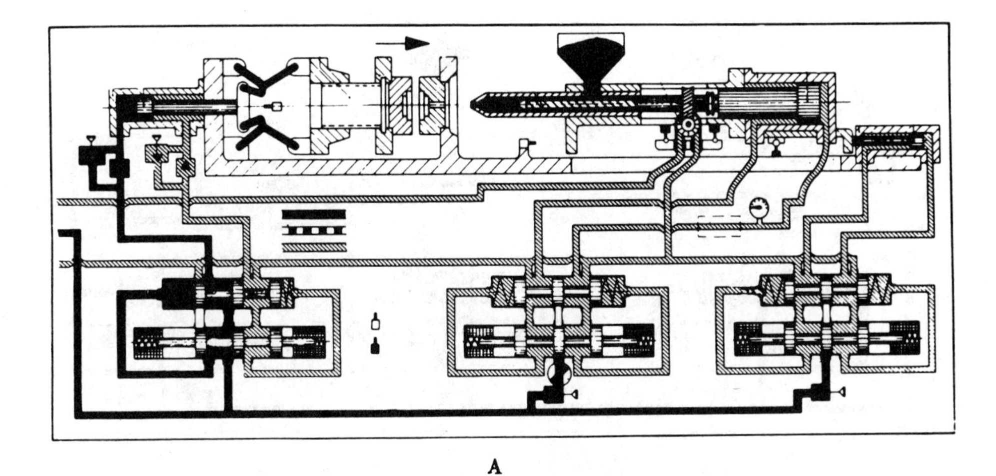

A

B

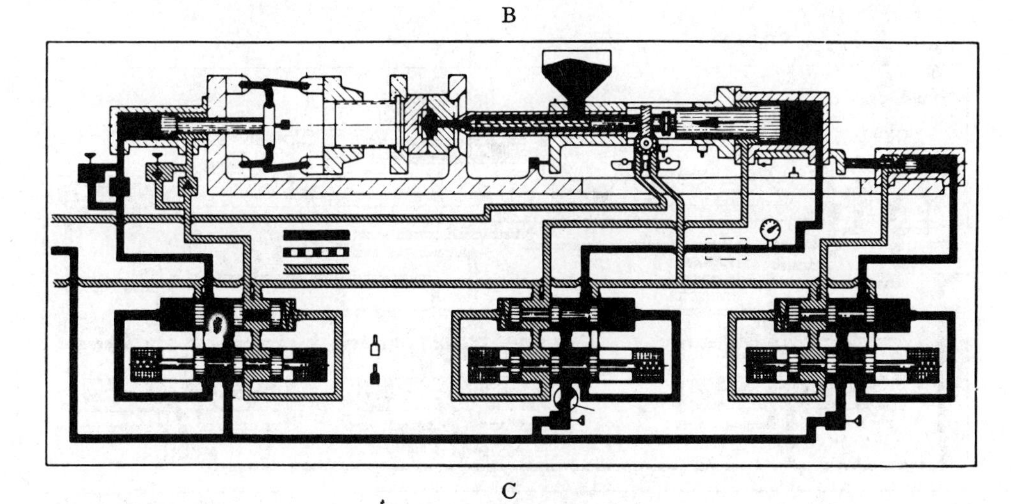

C

Fig. 10.2 Operational details of in-line reciprocating screw injection molder. *A.* When the mold starts to close, the screw has just finished charging the front end of the cylinder. *B.* With the mold closed, the heating cylinder then moves toward the sprue bushing and is ready to commence the injection cycle. *C.* The hydraulic cylinder then forces the screw forward under the injection head and fills the cavity. *D.* Cooling and charging cycle: the mold is filled and, after holding time, the screw starts to rotate and charge the front end of the cylinder. *E.* The screw is still charging the front end of the cylinder, and the cylinder carriage starts to move away from the sprue bushing. *F.* The finished cycle: mold opens and molded part is ejected with the sprue bushing. *(Du Bois, J. H., and John, F. W., "Plastics," Reinhold, 1967; by permission of Van Nostrand Reinhold Co.)*

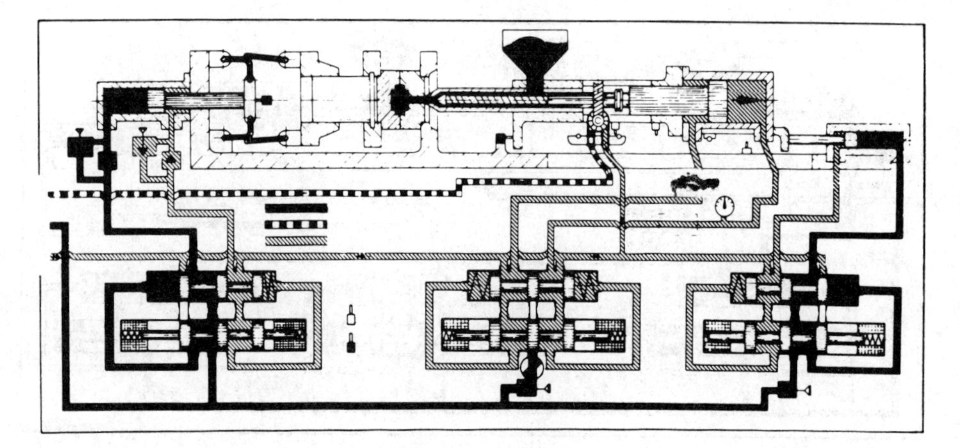

D

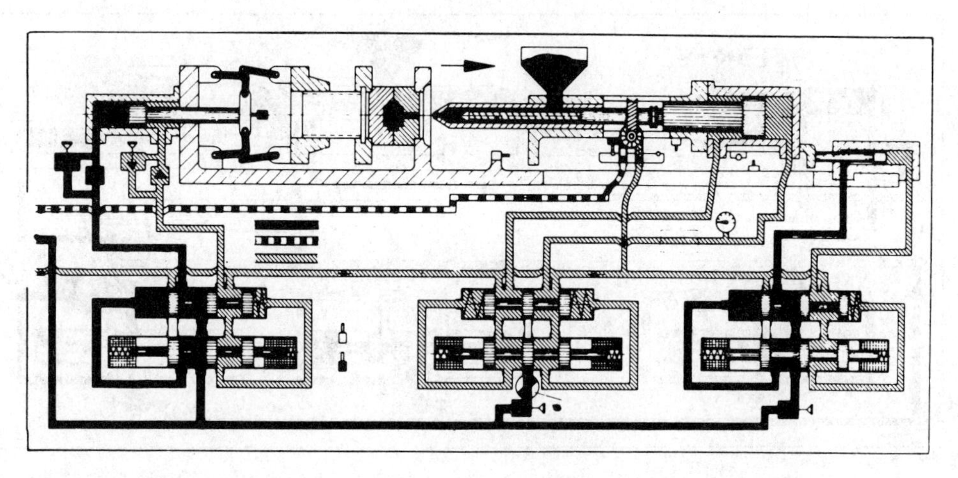

E

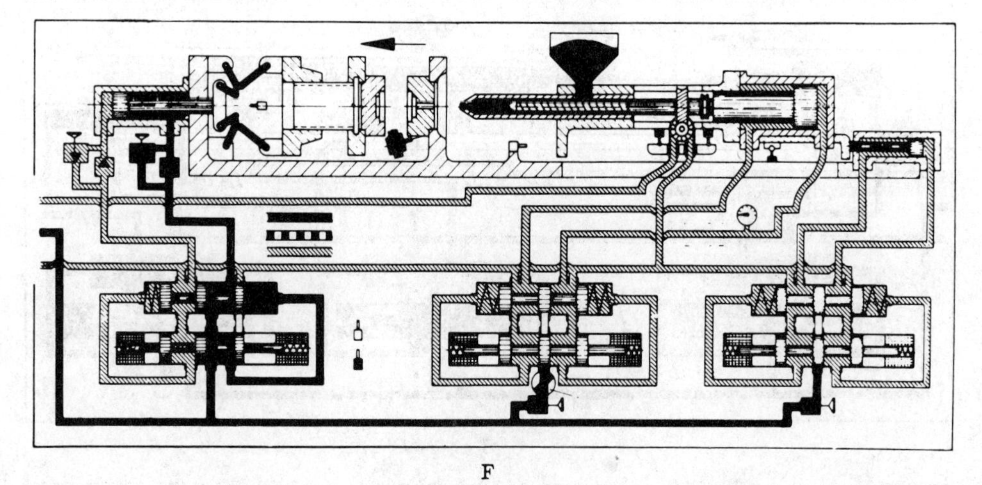

F

Fig. 10.2 (Continued)

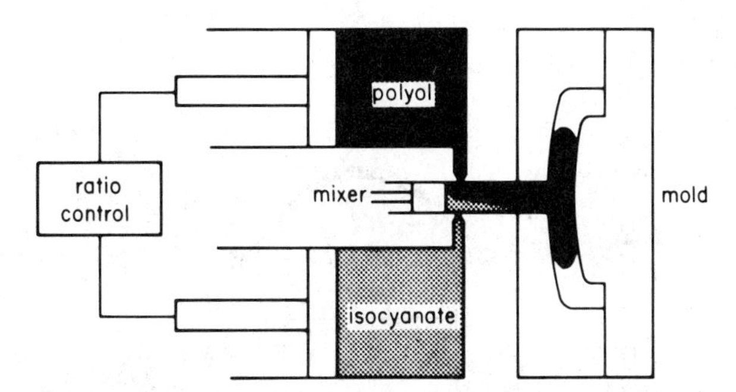

Fig. 10.3 Schematic of the reaction injection molding process. (*Lee, J. L. et al., Poly. Engr. and Sci.,* 20, *no. 13, 868, (1980); Soc. of Plastics Engrs.*)

under development. In the RIM process, the low viscosity, liquid monomers are mixed rapidly by jet impingement and/or mechanical methods just prior to introduction to the mold cavity. (See Figure 10.3.) The monomers must have very rapid reaction rates–such as the polyurethanes which react in a very few seconds–in order (1) to achieve rapid molding cycles, (2) to prevent flashing through mold contact surfaces and, also, (3) to prevent penetration of the thin liquid into microscopic "holes" in the mold surface which can cause severe adhesion (mold release) problems. The automotive industry, particularly, is rapidly expanding applications for this process using formulations of both rigid and flexible polyurethanes. Fillers are also being incorporated in the monomers to lower cost and improve physical properties, e.g., rigidity and heat distortion.

If sheets, rods, tubes, or profiles of various lengths are desired, thermoplastics can be conveniently extruded (Fig. 10.4). Rigid sheets of rubber-modified polystyrene and styrene acrylonitrile copolymers are frequently vacuum formed into complex shapes, e.g., containers, refrigerator liners, boats (Fig. 10.5). Many thermoplastics are calendered into films or sheets (Fig. 10.6).

Blow molding (Fig. 10.7) is a very popular technique for fabricating hollow containers (bottles) from a wide variety of polymers, e.g., rigid and flexible polyvinylchloride, polyethylene, polypropylene, rubber-modified poly-

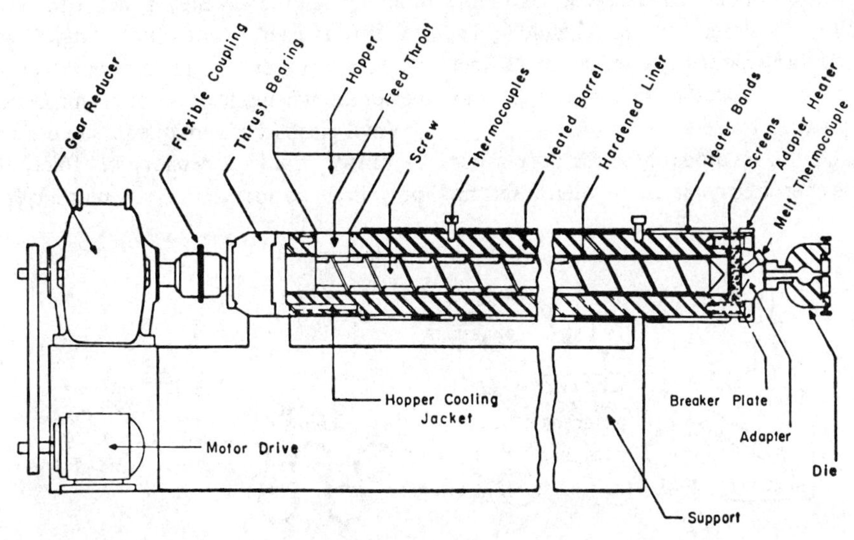

Fig. 10.4 Elements of an extruder. (*Bernhardt, E. C., "Processing of Thermoplastic Materials," copyright by the Society of Plastics Engineers, Inc. 1959. Van Nostrand Reinhold, New York.*)

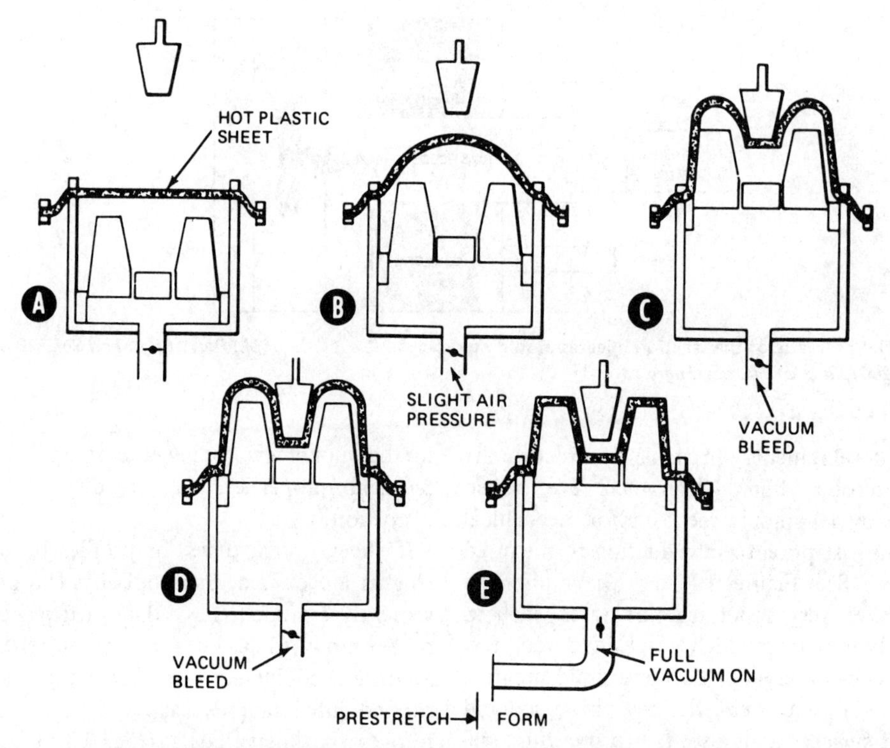

Fig. 10.5 Vacuum-forming using combination or air-slip-plug assist. (*Basdekis, C. H., "ABS Plastics," Reinhold, 1964; by permission Van Nostrand Reinhold.*)

styrene, and rubber-modified styrene-acrylonitrile copolymers (ABS).

Most polyethylene film is made by a blow extrusion process (Fig. 10.8) where air is maintained in a "balloon" formed at the exit of a circular extrusion die. The top (cool exit end) of the balloon is closed by a set of pinch rolls. Film, particularly polyethylene, up to 40 feet wide is routinely manufactured in tremendous volume by such a process.

There are a few methods of fabricating plastics in which either thermoplastic or thermosetting base resins may be used. Laminates are made by pressing–under heat and pressure–layers of paper, cloth, glass fiber, or glass cloth which have been impregnated with a liquid resin. Thermosetting resins are the only types used commercially since they make a harder, more solvent resistant, and more rigid laminate. Kitchen counter and table tops are one of the better known applications of such materials.

Vinyl plastisol is a unique plastic consisting of a highly fluid dispersion of finely divided polyvinylchloride resin in a plasticizer. Mild

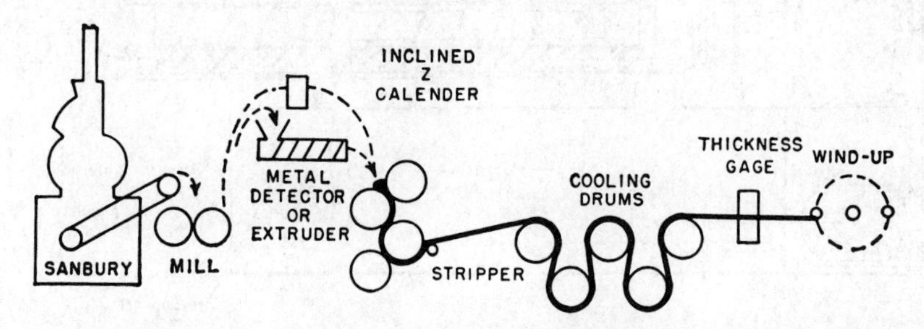

Fig. 10.6 Parts of calendar layout for plastic film production. (*Bernhardt, E. C., "Processing of Thermoplastic Materials," Van Nostrand Reinhold, copyright by the Society of Plastics Engineers, Inc., 1959; by permission.*)

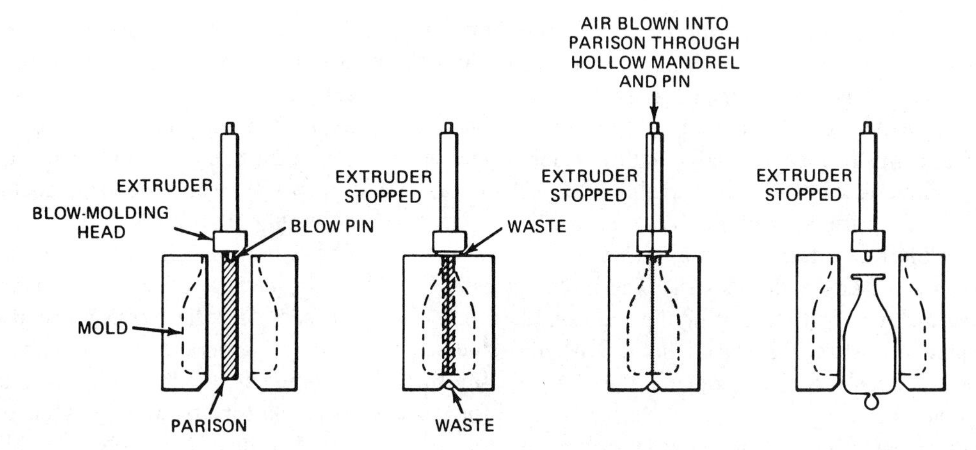

Fig. 10.7 Simple intermittent extrusion blow molding of hollow articles. *A.* Complete extrusion of parison. *B.* Close mold. *D.* Eject. (*"Encyclopedia of Polymer Science and Technology,"* vol. 9, *H. F. Mark and W. G. Gaylord, eds., Interscience, 1968; by permission of John Wiley & Sons, Inc.*)

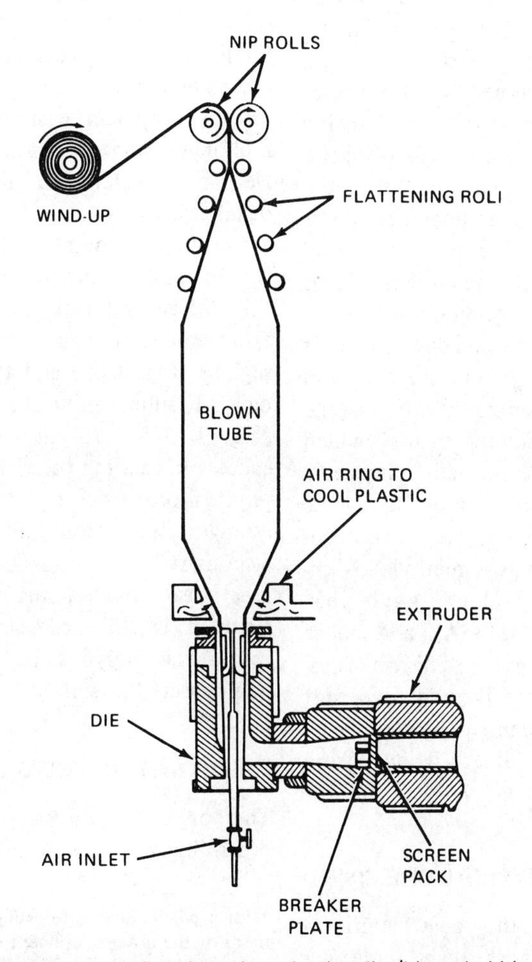

Fig. 10.8 Formation of sheet by extrusion through a circular die (blown-bubble extrusion). Ingredients are resins, stabilizers, pigments and other additives. (*"Encyclopedia of Polymer Science and Technology,"* vol. 6, *H. G. Mark and W. G. Gaylord, eds., Interscience Publishers, 1968; by permission of John Wiley & Sons, Inc.*)

heat causes the resin and plasticizer to fuse into a homogeneous solid. Vinyl plastisols–and also other thermoplastics, such as polyethylene, polypropylene, rubber-modified polystyrene–are frequently formed into large hollow shapes by a technique known as rotational molding or casting. The plastic is normally added to the mold while it is rotated about at least two axes to evenly distribute the plastic. The mold is heated either prior to or during the addition of the plastic. After the plastic has fused, the mold is cooled and opened, and the part is removed.

Plastic foams made from a wide variety of polymers have a wide range of applications and are made by a variety of methods depending upon the polymer and the application. Polystyrene beads containing some pentane blowing agent are prefoamed with high pressure steam and then formed into a wide variety of shapes from drinking cups to fish-net floats and insulated paneling. Rigid urethane foams provide the insulating material for most home refrigerators. They are formed in place by filling the enclosed space between the inner and outer shells (both of which may be vacuum formed plastic sheets) with the urethane monomers plus a small amount of water to generate the "pneumatogen" carbon dioxide. The above materials are rapidly mixed in an injection nozzle or gun to prevent premature foaming and/or setting. Other techniques use "blowing agents" which will decompose into gases–usually nitrogen–upon heating. Commercial blowing agents include azobisformamide (ABFA), azobisisobutyronitrile (AIBN), N,N′-dimethyl-N,N′-dinitrosoterephthalamide (DNTA), 4,4′-oxybisbenzenesulfonylhydrazide (DBSA), and, more recently, several sulfonyl semicarbazides. Recent methods have been devised to permit injection molding of foamed articles, e.g., furniture pieces resembling wood from rubber-modified polystyrene.

VIEWPOINT OF THE PHYSICAL CHEMIST

To the physical chemist the most significant characteristic of plastics is high molecular weight. Man has always used giant molecules–or high polymers–to feed, clothe, and house himself. Wood is a high polymer, so are cotton, silk, wool, and meat. The realization that plastics are high polymers wherein the atoms are joined by primary chemical bonds, and vice versa, is a surprisingly recent concept championed by pioneers like Hermann Staudinger and Theodor Svedberg, both of whom received the Nobel Prize for their work characterizing "makromolekules."* To the physical chemist the separation of polymers into plastics, resins for surface coatings, synthetic fibers, rubbers (or more recently, elastomers) destroys some of the unity in the subject of high polymers. To a physical chemist, a rubber is a polymer with certain rheological properties which can be fairly well explained by a cross-linked network present in a model of a polymer above its second-order ("melting") transition point. Similarly, the physical chemist can explain and correlate many of the properties which make certain polymers desirable for fibers, films, moldings, surface coatings, and other uses. He can measure, with great precision in some instances, the size and size-distribution of the molecules, the number and size of branches in the chain, the randomness of the copolymers, the degree and type of stereospecificity, the arrangement and patterns of the crystalline regions (Fig. 10.9) and the complex morphology of rubber-reinforced rigid "polyblends" (Fig. 10.10). He has achieved considerable success in relating these measurements to gross physical properties, e.g., tensile strength, impact strength, elongation, and solubility. However, although the physical chemist has explained a great deal, much remains to be done in this field; it is still necessary to synthesize and test a new polymer before its properties can be precisely known.

VIEWPOINT OF THE ORGANIC CHEMIST

The organic chemist, usually interested in describing chemical reactions and the resulting

*For a particularly interesting history of the development of the modern concept of the structure of organic polymers, see P. J. Flory, "Principles of Polymer Chemistry," pages 3-27.

Fig. 10.9 Spherulites of crystalline polystyrene growing in a melt. Electron micrograph. *(Courtesy G. C. Claver, Monsanto Co.)*

products, ignored much of polymer chemistry for a long time. The traditional organic chemist of the nineteenth century regarded the uncrystallizable "tars" that formed from certain reactions as unsuccessful experiments, since the objective was the preparation of crystals whose physical chemistry was better understood. The progeny of the polymeric "gunks" which the nineteenth and early twentieth century chemist threw away in disgust, today, constitute the largest tonnage market for synthetic organic materials.

Now with better methods of characterizing the molecular structure of polymers, the modern organic chemist has developed well-understood syntheses for many polymers and logical mechanisms which explain vast bodies of data. The inventor of nylon, Wallace Hume Carothers,[11] beginning in 1929, showed that the reaction mechanisms and the thermodynamic equilibria ordinarily associated with reactive organic groups were not altered merely because these groups were attached to large rather than small molecules. This vital concept removed the mystery from polymer synthesis and funneled the vast knowledge of organic chemistry into the polymer field, resulting, in the early 1930's, in the dramatic development of new polymers.

Carothers defined two types of polymers: (1) *Addition polymers* are those like polyethylene in which the monomer units, on polymerizing do not completely rupture a chemical bond but, in the usual case, utilize the π-bonding electrons in a double bond to form a new bond between the monomers. Most addition polymers are obtained from monomers containing carbon-carbon double bonds, e.g., vinyl monomers ($H_2C{=}CHR$), vinylidene monomers ($H_2C{=}CR_2$), and diene monomers ($H_2C{=}CR{-}CH{=}CH_2$), but some commercial addition polymers are also made from formaldehyde ($CH_2{=}O$), and ethylenic

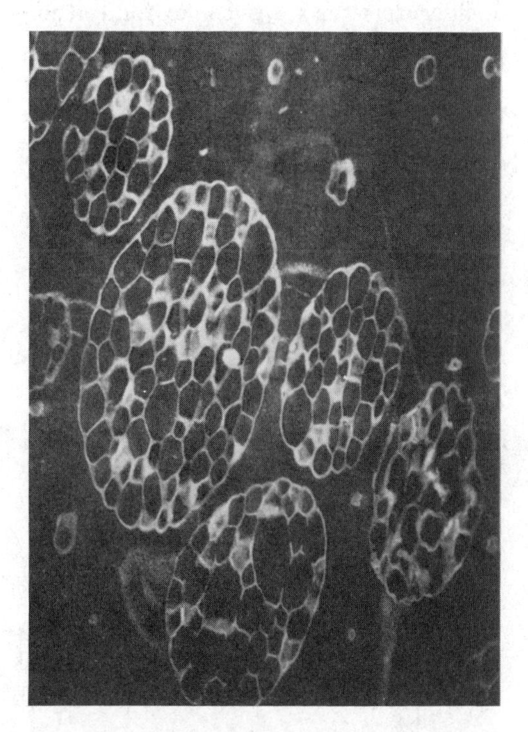

Fig. 10.10 Complex "honeycomb structure of rubber-modified ("impact") polystyrene. Rubber is white, polystyrene is dark. Electron photomicrographs of thin (ca. 0.1 micron) sections stained with OsO_4 per Kato[10a] *(Courtesy G. C. Claver, Monsanto Co.)*

monomers ($CF_2{=}CF_2$); (2) *Condensation polymers* are those wherein chemical bonds between atoms completely rupture during polymerization, causing formation of low molecular weight fragments, frequently water. By using pure materials and forcing the reactions to completion through removal of the low molecular weight fragment, many varieties of high polymers can be made from multifunctional organic molecules, e.g., synthetic fibers (Chapter 11) prepared from organic acids and amines (nylon) and from organic acids and alcohols (polyesters such as "Dacron" and "Mylar"), as well as alkyd paints (Chapter 22), from organic aldehydes and phenols or amines. Catalysts vary, depending on the nature of the functional organic group. Usually Lewis acids or bases are required.

Some reactions do not fit neatly into either the addition- or condensation-polymer classification, but are, nonetheless, usually considered condensation polymers, e.g., reactions between a diisocyanate and a glycol to form a polyurethane,

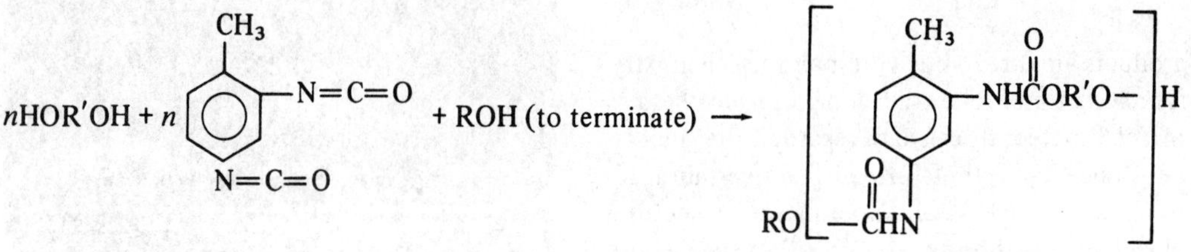

$$n\text{HOR}'\text{OH} + n\ \text{C}_6\text{H}_3(\text{CH}_3)(\text{N{=}C{=}O})_2 + \text{ROH (to terminate)} \longrightarrow \text{RO}\left[-\overset{\overset{\text{O}}{\|}}{\text{C}}\text{HN}-\text{C}_6\text{H}_3(\text{CH}_3)-\text{NH}\overset{\overset{\text{O}}{\|}}{\text{C}}\text{OR}'\text{O}-\right]\text{H}$$

or ring-opening reactions such as the formation of nylon 6 from caprolactam,

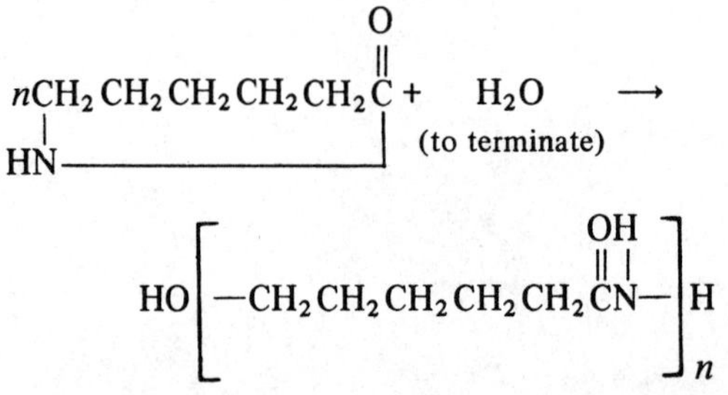

$$n\underbrace{\text{CH}_2\text{CH}_2\text{CH}_2\text{CH}_2\text{CH}_2\overset{\overset{\text{O}}{\|}}{\text{C}}}_{\text{HN}\text{———}} + \underset{\text{(to terminate)}}{\text{H}_2\text{O}} \longrightarrow \text{HO}\left[-\text{CH}_2\text{CH}_2\text{CH}_2\text{CH}_2\text{CH}_2\overset{\overset{\text{O}}{\|}}{\text{C}}\overset{\overset{\text{H}}{|}}{\text{N}}-\right]_n\text{H}$$

In addition to the large number of vinyl and vinylidene monomers which form addition polymers, and the even larger number of difunctional and multifunctional organic chemicals which form condensation polymers, the combination of various addition monomers into addition copolymers, and mixtures of condensation reactants form literally millions of combinations. Also, by adding monomers to previously formed polymers, e.g., adding styrene to a synthetic rubber, graft polymers* with unusual properties can be formed; this is done commercially to produce vinyl chloride copolymers, styrene-acrylonitrile-butadiene copolymers (ABS), polystyrenes, and polymethylmethylacrylates possessing high impact strength.

*Graft polymers are a special type of copolymer and resemble block polymers, e.g.:

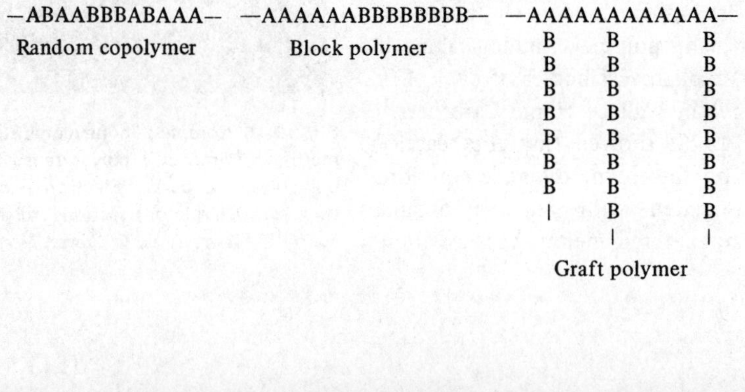

```
—ABAABBBABAAA—   —AAAAAABBBBBBBB—   —AAAAAAAAAAAA—
Random copolymer    Block polymer       B     B     B
                                        B     B     B
                                        B     B     B
                                        B     B     B
                                        B     B     B
                                        B     B     B
                                        B     B     B
                                        |     B     B
                                              |     |
                                          Graft polymer
```

Also, variations in the molecular weight, molecular-weight distribution, amount of chain branching and crosslinking, molding or forming conditions, types and amounts of plasticizers, stabilizers, colorants, and fillers further increases the possibilities for various material specifications to astronomical levels. Table 10.1 gives a very brief summary of the principal plastics and the monomers from which they are made.

Addition polymerization may be initiated by free radicals or ions. At the present time, free-radical polymerization is commercially the most important. However, the discovery that polymer properties can be greatly influenced by their degree of stereoregularity is currently creating great interest in ionic polymerization. Free radicals can be generated by the action of ultraviolet light, X-rays, and high-energy particles. They are commercially produced for industrial polymerizations by heat, peroxides, oxygen, and diazo compounds. The polymerization propagates itself by a chain-reaction mechanism; i.e., when a free radical and a monomer molecule react, a new free radical is formed. The reacting monomer unit captures the odd electron and becomes an initiating free radical itself. By such a mechanism, polymer molecules with thousands of monomeric units are generated in a few seconds or less; the monomers are normally linked together in a head-to-tail fashion.

The chain reaction may be stopped by the combination of two growing chains or by an initiating radical colliding with the end of a growing chain rather than with a monomer molecule (this is unlikely), in which both ends of the polymer chains will contain initiator fragments. Chain transfer agents used to control molecular weight, e.g., mercaptans and halogenated aliphatic hydrocarbons, may also terminate a growing chain by splitting into two free-radical fragments, one terminating the growing chain and the other behaving as a free-radical initiator. Disproportionation, i.e., the rearrangement of the polymeric free radical into a more stable configuration, usually with the formation of a double bond, is also possible. In addition, chain terminators or inhibitors, such as quinones, oxygen, sulfur, and amines, may stop a growing chain by forming stable adducts with the initiator and/or the polymeric free radical. Early chemists were plagued with the question of what stops the chain; this is still a difficult problem to solve quantitatively for any particular polymerization scheme.

Ionic catalysis of vinyl and diene monomers is a very old art. Matthews[12] in England and Harris[13] in Germany discovered simultaneously in 1910 that metallic sodium would polymerize butadiene. Yet little commericalization was achieved until very recently. Then around 1952, Karl Ziegler, who had been working in the field of metal-organic compounds for nearly 30 years, discovered that aluminum trialkyls complexed with titanium tetrachloride would polymerize ethylene to high molecular weight at low pressure.[14,15,16,17] At about the same time, John Hogan and Robert Banks[18] of Phillips Petroleum discovered that certain hexavalent chromium oxides on silica or aluminum gel would polymerize ethylene to commercial polymers. Shortly thereafter G. Natta and coworkers in Italy discovered that a modified "Ziegler catalyst" would cause "stereospecific" propagation* of the growing chain for α-olefins. Other soluble ionic catalysts, particularly lithium alkyls and cobaltous alkyls, have been found to influence chain configuration, particularly in dienes.

Frequently, polymer molecules possessing a regular structure crystallize readily, producing unusual properties, i.e., they are higher melting, more rigid, stronger, and less soluble than their noncrystalline stereo-irregular counterparts.

This field excited great commercial and scientific interest. To date *cis*-polypropylene, *cis*-polyisoprene, and *cis*-polybutadiene have been produced commercially. But crystalline polymers of styrene, methyl methacrylate, 3-methyl-1-butene, and several methylpentenes and methylhexenes have all been made only in the laboratory.

In contrast to the general nature of catalysis in addition polymerization, condensation polymerization uses the acid or base catalyst appropri-

*Natta's modified Ziegler catalyst forbids *trans* addition of an incoming α-olefin molecule; hence all *cis*-polymer is produced.

ate for the organic groups involved. Some pairs or groups require no catalyst at all. No chain reaction mechanism occurs, and the molecules gradually increase in size as the polymerization proceeds. If trifunctional or higher functional molecules are present, cross-linking will occur, leading first to gelation and eventually to complete hardening.

VIEWPOINT OF THE CHEMICAL ENGINEER

The plastics industry provides a fascinating field for the chemical engineer. He can investigate polymerization kinetics, reactor design, process control, the rheology of polymer melts, slurries, solutions or latices, the unique problems of mixing or compounding various polymer blends, and the problems of fabricating the polymer into the final product. In addition, the entire range of transport phenomena (mass, heat, and momentum transfer) are involved in the engineering of polymer plants. The objective of the chemical engineer is the commercial development of a process which will produce the highest quality polymer for a particular use at the lowest possible cost within proper safety and environmental constraints.

Polymerization mechanisms and kinetics are central to the design of the optimum process for the optimum product. For example, the process by which unsaturated monomers are converted to polymers of high molecular weight through addition polymerization is a chain reaction. In free-radical polymerization in the presence of an initiator, the first step is the decomposition of the initiator I (e.g., benzoyl peroxide) to yield two free radicals ($I \xrightarrow{kd} 2R^{\cdot}$). (The $^{\cdot}$ symbolizes the odd electron of the free radical.) Then a chain is initiated by the addition of a monomer molecule M to the primary radical $R^{\cdot}$ ($R^{\cdot} + M \xrightarrow{ki} M_1^{\cdot}$). The polymer molecule grows with successive addition of monomer molecules:

$$M_1^{\cdot} + M \xrightarrow{kp} M_2^{\cdot}$$

$$M_2^{\cdot} + M \xrightarrow{kp} M_3^{\cdot}$$

or in general:

$$M_x^{\cdot} + M \xrightarrow{kp} M_{x+1}$$

Normally, the propagation reaction above proceeds extremely rapidly. Thousands of monomer units add in fractions of a second. Finally the growth of the polymer molecule may be terminated by its combination with another free radical, for example, by coupling

$$M_x^{\cdot} + M_y^{\cdot} \xrightarrow{k_t} M_{x+y}$$

Alternatively, the chain may be interrupted by chain transfer agents, which terminate the chain but initiate a new chain simultaneously. Such agents permit control of the molecular weight to nearly any degree. Control of molecular weight distribution is also possible, but very narrow distributions can only be obtained by using "living," growing anionic chains. The above simplified kinetic scheme for addition (vinyl) polymers illustrates the complexity of most polymerizations; condensation polymerizations, "living" ionic chain-growth reactions, and polymerizations of olefins using the heterogeneous solid catalysts exhibit different but equally or more complex kinetics.

The chemical engineer in the plastics industry has a few unique problems which he does not have to the same degree in other branches of his industry. These problems stem from the fact that polymers cannot be purified economically, i.e., the separating operations of distillation, extraction, adsorption, etc., are not commercially practical methods for isolating good polymeric molecules from undesirable ones. Ordinarily, all polymeric molecules produced by a particular polymerization scheme will end up in the final product sent to the customer. This means that any changes in the polymerization process will be reflected in the properties of the final product. The process engineer in the plastics industry must be completely aware of polymerization mechanisms, kinetics, and quality evaluation schemes. In addition, he must appreciate that tests for "quality" cannot be simple or definitive; variations in the process due to scale-up or modifications made for economic gain may alter "quality" irrevocably in subtle ways not always measurable by simple routine control tests.

VIEWPOINT OF THE ECONOMIST

The plastics industry began a period of rapid growth in the late 1930's, which continued into the mid-seventies. (See Figure 10.11). In 1974, both prices and supplies of essential raw materials were severely disrupted by the OPEC oil cartel. Prices for plastics, which had been steadily decreasing as technology improved and as sales volume increased, were forced to double or triple from their early 70's lows in order to accommodate the nearly ten fold increase in the cost of crude oil. (See Table 10.2.)

Plastics–as a material only–contain less than 1% by weight of the total consumption of oil and gas in the United States for energy purposes. Also, the total amount of energy required to fabricate a finished part is generally less for plastics than that required for many other materials, e.g., aluminum, steel and glass. Therefore, the energy crunch per se is not expected to materially slow the future rate of plastics growth compared to the total economy. (Whether the energy problem will continue to slow the rate of growth of the total economy–as it has since 1974–is still a matter of debate and concern among economists.) Figure 10.11 shows that sales of plastic materials have indeed begun to continue their upward trend after suffering a severe disruption in 1975-77. Because of the energy problem, the light weight of plastics will offer incentives for even greater utilization in the future in all transportation fields.

MANUFACTURING PROCESSES

The manufacturing processes for polymers are nearly as diverse as the number of polymers, but it is possible to classify many of them as follows:

A. Addition polymerization processes.
   1. Mass
   2. Emulsion
   3. Suspension
   4. Solvent
   5. Solvent-nonsolvent
B. Condensation polymerization processes.
C. Ring opening and other "addition" polymerization processes in which the polymer grows in a manner other than by a chain reaction.

TABLE 10.2 Sales in 1979 of Synthetic Plastics and Resin Materials in the United States

Material	*Sales, Millions of lbs.*	*Approx. Avg. $ Value/Pound, 1980*	*Approx. Avg. $ Value/Pound, 1971*
Polyethylenes	11,368	0.42	0.15
Polyvinylchlorides	5,560	0.34	0.13
Polystyrenes	3,658	0.47	0.14
Polypropylenes	3,526	0.42	0.20
Polyurethanes	1,732	0.74	n.a.
Phenolics	1,408	0.51	0.20
ABS & P(S/AN)	1,292	0.68	0.35
Urea and Melamines	1,164	0.67	0.20
Polyesters, Thermoset	1,120	0.52	0.30
Polyesters, Thermoplastic (non-textile use)	816	1.15	n.a.
Acrylics	532	0.68	0.45
Epoxies	310	0.93	0.50
Nylons (non-textile use)	280	1.34	0.80
Polycarbonates	218	1.21	0.75
Cellulosics	154	0.97	0.48
Polyphenylene Oxides	136	1.05–1.50*	0.70–1.20*
Polyacetals	96	1.12	0.60
Others	1,592	–	–
TOTAL	35,180		

*PPO is normally sold blended with rubber-modified polystyrene; the grades with less PPO are lower in cost and in heat distortion temperature; the higher cost grades will be fire retardant.

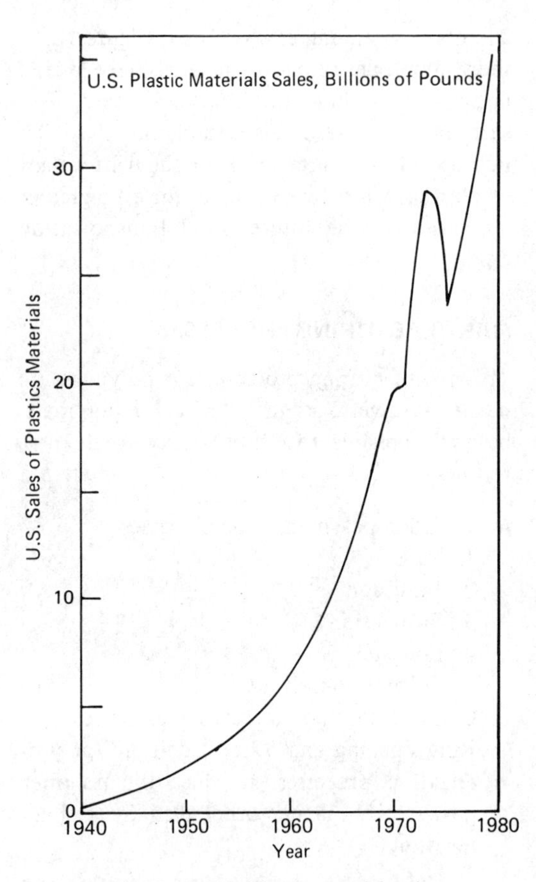

Fig. 10.11 U.S. Plastic materials sales.

Addition Polymerization

Addition polymerization has evolved along many paths, as engineers have tried to solve the major problem of removing the large heat of polymerization, e.g., 12-26 kg-cal/mole of vinyl monomer.*[20] For polyethylene this amounts to more than 800 calories per gram or a temperature rise of over 1000°C under adiabatic reaction conditions. Such temperature changes are intolerable, causing complete breakdown. For example, carbonization of ethylene occurs under high pressure at about 300-350°C. The thermal conductivity of such organic materials is very low—less than in most common insulating materials—and thus magnifies the already difficult heat transfer problem. Unusual methods are required to solve this problem which is peculiar to the polymer industry.

Commercial reaction temperatures range from -90°C for isobutylene, to 180-270°C for high pressure polyethylene. For polymers that dissolve in their monomers, extreme viscosities are reached at low conversions, e.g., 100,000-200,000 centipoise is the viscosity of 30-percent converted styrene syrups at the polymerization temperature of 85-90°C. For polymers which will precipitate from their monomers (e.g., polyvinylchloride, polyacrylonitrile), slurries or pastes of low mobility result at 25-35 percent conversion.

Mass Polymerization. The term mass or bulk polymerization, as used here, refers only to monomers which dissolve their polymers (styrene, methyl methacrylate) and not those from which the polymer precipitates (e.g., vinyl chloride, vinylidene chloride, acrylonitrile). Mass polymerization is the method frequently used in the laboratory to study a new monomer and its copolymers; no extra variables are introduced and the heat removal problem is trivial provided thin (10-25 mm) glass tubing is used. However, scaling-up such a process in a practical manner presents tremendous problems. Only polystyrene and some of its copolymers are commercially manufactured today by a pure mass polymerization process.**

In the middle 1930's, Badische Anilin in Germany developed a continuous process for manufacturing polystyrene which may still be used in at least Germany and Italy.[21] The process consists of polymerizing styrene monomer to 30-33 percent in stirred jacketed kettles at about 90°C and then feeding this viscous syrup to the top of a vertical jacketed tower, about 30 inches in diameter by 19 feet long, containing a few temperature regulating coils. This tower is operated at atmospheric pressure

*The difference in energy between a "standard" carbon-carbon double bond and two single bonds is about 16 kg-cal. Variations from this level are due to substituents.

**Methyl methacrylate and "allyl" casting-resin syrups are cast and polymerized in thin sections by a pure mass process, but these processes, though practiced commercially on a small scale for speciality items, will not produce molding powders economically. Methyl methacrylate molding powder is made via a suspension process.

and is not quite full; thus temperatures are prevented from rising much above the boiling point of styrene monomer (145°C) until high conversions are attained, at which point temperatures rise to about 200°C. Molten polystyrene is pumped from the bottom of the tower at a rate of about 900 lb/hr by means of screw extruders. The molten strands from the extruder are cooled and then cut into transparent granules or pellets.

In the United States, the degree of temperature control attained in the German tower process has not been adequate to achieve optimum quality and versatility. Equipment has been devised[23,24,25,26,27,28] which can control the reaction temperatures more effectively to higher conversions than in the German tower process. Most of these processes use slowly revolving screws or paddles which gently agitate the viscous polymerizing syrup past tubes filled with circulating oil, thereby aiding heat transfer. A small amount of ethylbenzene may also be added to aid viscosity reduction and heat transfer. To reach very low monomer contents, devolatilization is carried out under high vacuum (29 in. of Hg) using vented twin-screw extruders or similar devices (see Fig. 10.12).

Methyl methacrylate is mass polymerized to make sheets using plate glass as the mold material.[31] Also, certain allyl resins containing some trifunctional molecules for crosslinking are mass polymerized to encase or "pot" electrical components. The volume manufactured by these latter processes is very small.

Emulsion Polymerization. Emulsion polymerization is a technique by which an addition polymer or copolymer is produced in a two-phase system. Its application requires the emulsification of the monomer in a medium, usually water, through the use of emulsifiers, such as soaps, alkyl sulfates, and alkyl sulfonates. These are supplied in addition to the other ingredients that go into most polymeriza-

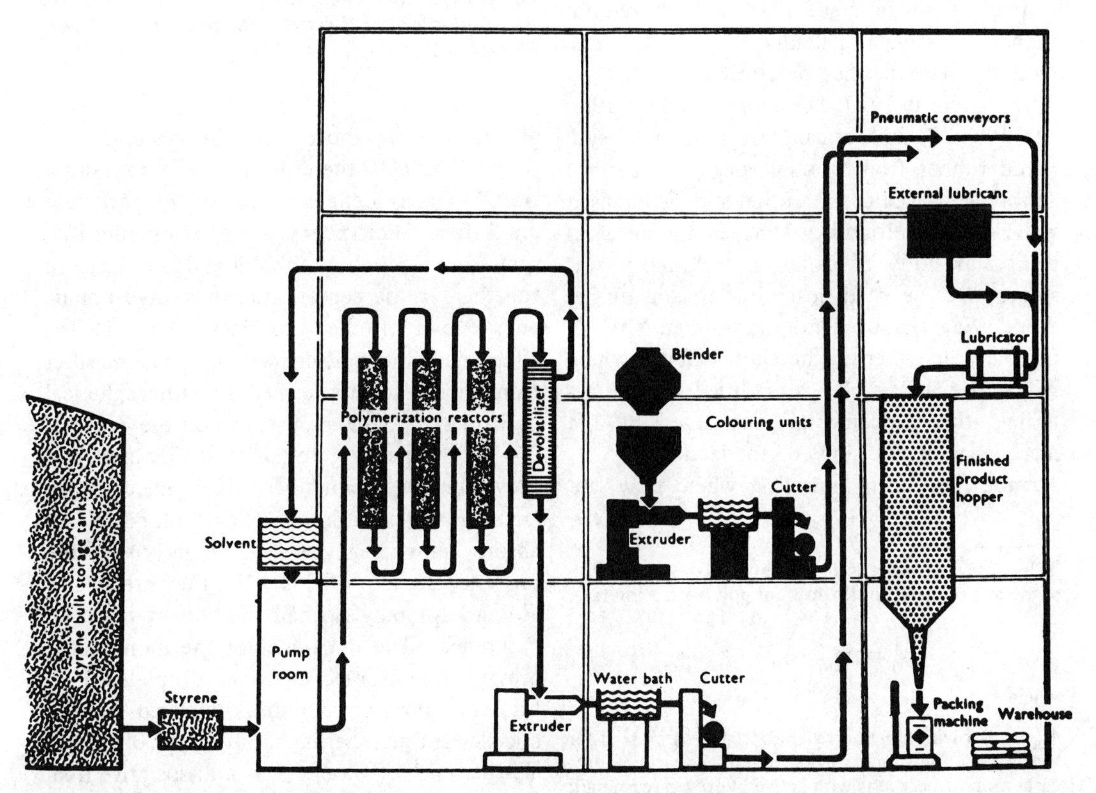

Fig. 10.12 Styrene polymerization, continuous mass process of Dystrene Ltd. (*Reproduced with permission from British Plastics,* 30, *26* (*Jan. 1957*))

tions, such as the initiator and chain transfer agents.* The use of this simple, additional reagent, the emulsifier, leads to several important characteristics which make the emulsion process unique.[32]

Three of the most important of these are:

1. The ability to form a polymer of high molecular weight at a very high rate of polymerization.
2. The maintenance of low viscosity throughout the reaction.
3. The relative ease of heat transfer.

The product of an emulsion-polymerization reactor is a latex, which is simply a stable colloidal suspension of polymer particles in a continuous phase, generally water. The small polymer spheres are much smaller than the initial monomer droplets and range in size between 0.01 and 1.0 micron in diameter (Fig. 10.13).

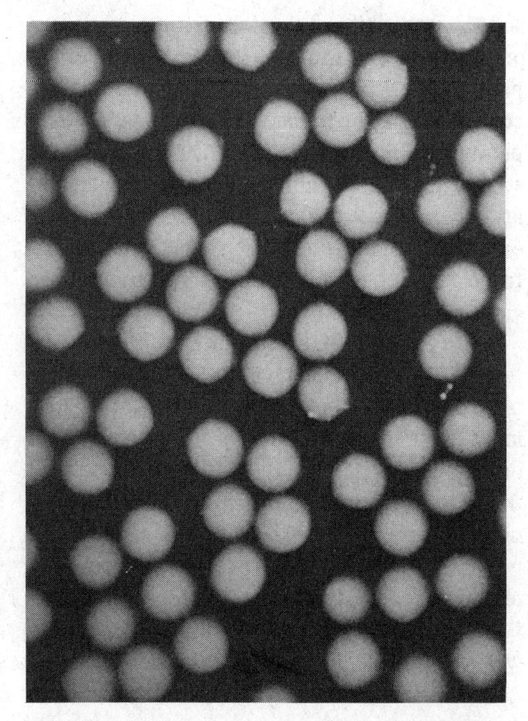

Fig. 10.13 Electron micrograph (10,000 X mag.) of polyvinyl chloride latex particles with unusually small distribution of particle size. *(Courtesy G. C. Claver, Monsanto Co.)*

Emulsion polymerization was first attempted with natural gums and other hydrophilic protective colloids in an effort to duplicate natural rubber. Though patented in 1913, these methods were neither practical nor reproducible. Then in 1927, Dinsmore,[33] and Luther and Heuck[34] in Germany tried to make synthetic rubber from normal soaps and surface active agents. Later on, Luther and Heuck used water-soluble initiators. Much of the development and study of emulsion technology was spurred by the need, just prior to and during World War II, for synthetic rubber, first in Germany and then in the United States where W. D. Harkins[35,36,37] is creditied with establishing the fundamental postulates of the mechanism of emulsion polymerization.

Emulsifiers form micelles when they are added to water above a certain concentration, which is called "the critical micelle concentration." This is a characteristic of the particular emulsifier. Micelles are aggregates of emulsifier molecules whose hydrophobic ends are clustered together at the center and whose hydrophilic ends extend into the water phase. The micelles, which have the ionic character of the emulsifier from which they are formed, are either spherical or lamellar[38] with dimensions on the order of 25-50 Å, depending on the emulsifier used, and they number around 10^{17}-10^{18} micelles per cm^3 of water. The micelles "solubilize" or absorb monomer, swelling slightly in the process. Because of their size, they are able to account for only a small fraction of the total monomer. The remainder of the monomer is present as emulsified monomer droplets 10-30 thousand angstroms in size, stabilized like the micelles and numbering about 10^{10}-10^{11} particles per cm^3 of water.[37] If a dissociative free-radical initiator is added to the aqueous phase, the free radicals generated by thermal decom-

---

*Chain transfer agents lower molecular weight with slight or no change in the over-all polymerization rate, i.e.,

$$M_n^{\cdot} + RH \longrightarrow M_nH + R^{\cdot}$$

where

$M_n^{\cdot}$ is a growing polymer chain; RH is a chain transfer agent, e.g., mercaptan, halogenated aliphatic hydrocarbons, hydrocarbons with active hydrogen (cumene); $R^{\cdot}$ is a new radical ready to begin generation of a new polymer chain.

position of the initiator diffuse through the water to the micelles where they initiate a polymer chain. The micelle now becomes a polymer particle. Harkins established that the micelles are the source of polymer particles and that these particles are then the locus of polymerization. Termination of a growing polymer chain occurs when a new free radical enters the particle and terminates the chain. Harkins showed that.[36]

1. There are more polymer particles produced than there are emulsified monomer droplets initially.

2. The sizes of the polymer particles are several orders of magnitude smaller than the emulsified monomer droplets.

3. The emulsified monomer droplets, if tested for polymer formation during the reaction, show only insignificant amounts of polymer.

Harkins postulated that the emulsified monomer droplets must serve as a reservoir of monomer supplying the growing particles. A pictorial representation of the emulsion polymerization system very early in the reaction is shown in Fig. 10.14.[32] A typical emulsion polymerization recipe would be as follows (in parts by weight):

Styrene	100
Sodium oleate	6
Potasium persulfate	0.3
Water	150

The reaction temperature is 70°C.

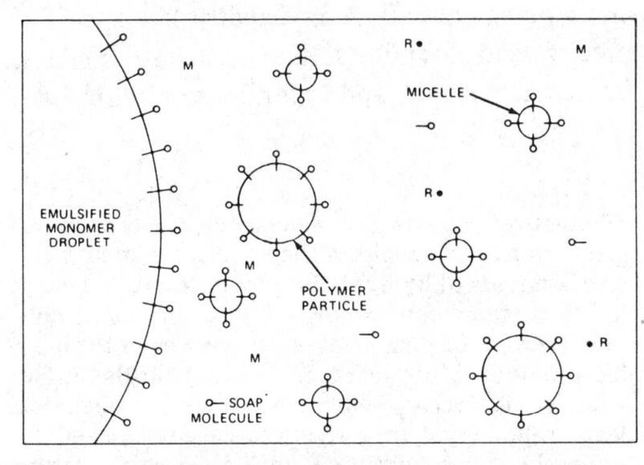

Fig. 10.14 Schematic representation of emulsion polymerization: M = solubilized monomer; R· = catalyst, monomer, or low molecular weight polymer radical. (*Courtesy D. C. Sundberg.*[32])

The kinetics of emulsion polymerization is considerably more complex than that of mass polymerization. The initiator and emulsifier concentrations, the concentration of monomer present, the number of particles formed and their size, and the effect of temperature on the decomposition of the initiator and the propagation reaction of the polymer chain must all be considered. Extremely important work providing a quantitative understanding of the emulsion-polymerization mechanism has been carried out by Smith and Ewart,[40] and more recently, by Gardon.[41] Sundberg[32] has provided an excellent review of the current quantitative understanding of emulsion polymerization and has developed a set of mathematical models based on Harkins' postulates which provide a means of predicting the physical characteristics of the latex as well as the chemical characteristics of the polymer.

Modern emulsion polymerization solves the major problem of heat transfer admirably. Rapid polymerization rates may be obtained at precisely defined low temperatures while high molecular weights are maintained, particularly when "redox" catalyst systems are used.[146] Also, the use of water-soluble catalysts frequently promotes stability of the latex particles which are much smaller than those obtained by dispersion of a polymer through intensive mixing. (Fig. 10.14) Continuous polymerization is possible in tubular reactors, in towers, or in a series of agitated reactors.

In spite of its many advantages, however, emulsion polymerization has not been used extensively to make solid plastics.* Since it is commercially impractical to remove the emul-

*Latices commercially processed to plastics include: polyvinyl chloride plastisol resin (produced by spray drying the latex); "Kralastic," "Cycolac," and "Lustran" (styrene-acrylonitrile graft polyblends with synthetic rubbers);[42] "Teflon" (polytetrafluoroethylene); and possibly, "Kel-F" (polytrifluorochloroethylene). Many other latices are produced in large volume to be sold as such for use in formulations for surface coatings, such as textile finishes, adhesives, paper binders and coatings, waxes and polishes.

It is important to distinguish between plastics, rubbers, resins for surface coatings, and synthetic fibers; the latter three industries utilize emulsion technology extensively.

sifying agents completely, emulsion-produced polymers are poorer in haze, color, heat stability, and electrical properties than products made by other processes. Also, some water-soluble initiators such as potassium persulfate make comparatively unstable chain-end groups for at least vinyl chloride and vinylidene chloride polymers. In addition, the cost of obtaining a dry, pelleted polymer from a latex is considerably higher than for mass or suspension polymer. First, the emulsion must be coagulated. This may be accomplished in many ways, such as adding acid or other electrolytes, vigorous mixing, ultrasonics, freezing, or forcing the latex through jets. The coagulated crumb will ordinarily contain most of the water originally present in the latex, e.g., from 40-65 percent water. "Synerizing," i.e., heating near the softening point, the coagulated crumb can squeeze out some of the water inclusions.[43] This water is not on the surface but must slowly diffuse out through the polymer. Drying is therefore slow and expensive. After drying, the crumb (or sometimes powder) must be densified into pellets. Drum drying and spray drying (commercially used for polyvinyl chloride plastisols) dry without requiring coagulation, yet both of these methods are comparatively expensive.

Emulsion polymerization equipment is almost always glass-lined because plating or scaling of polymer on metals is frequently very severe. In the laboratory, tumbling 6-ounce soft-drink bottles may be used for carrying out the polymerizations at pressures up to 150 psig.†

Suspension Polymerization. For a long time suspension and emulsion polymerization were considered synonymous. Then Mark and Hohenstein[44] pointed out the characteristics which today serve to separate the two processes: emulsion polymerization utilizes comparatively low molecular weight substances of high surface activity as emulsifiers, i.e., substances which markedly lower the interfacial tension; suspension polymerization, on the other hand, requires finely divided solids or a protective colloid,* which, though they preferentially locate themselves at the interface, do not greatly lower the surface tension. In the emulsion process, initiation begins in the continuous water phase; in suspension polymerization the catalyst is soluble in the monomer phase and initiation begins there. The most apparent physical effect resulting from these essential differences is the much larger size of the polymer particle produced by suspension polymerization, e.g., the diameters of suspended particles are usually between 50 and 2,000 microns (about 0.002-0.08 inch), while those of most emulsion particles are between 0.01 and 1.0 micron (Fig. 10.15). The large suspension particles are easily separated from the water phase by filtration or centrifugation, whereas the smaller emulsion particles are separated with greater difficulty. Also, the importance of agitation differs in the two systems. In emulsion polymerization, a certain minimum level of agitation is required to break up the pure monomer phase, but after this minimum level is attained agitation may vary greatly up to the point where it causes coagulation of the emulsion particles; nor does agitation ordinarily affect particle size or particle-size distribution as it does in suspension polymerization.

Many types of finely divided solids, frequently combined with protective colloids or very small quantities of low-molecular weight surface active agents have been used as suspension agents. These include commercial tricalcium phosphate (hydroxy apatite),[45] specially precipitated barium sulfate, calcium oxalate, bentonite clay, and aluminum hydroxide.

†Premission should be obtained from the appropriate company for such use, and the bottles should be kept segregated.

*Protective colloids are water-soluble substances of great molecular weight. They usually contain sufficient amounts of hydrophobic groups to cause them to locate preferentially at the oil-water interface; however, because of their great molecular weight the interfacial tension is not markedly lowered but the surface viscosity is greatly enhanced, slowing coalescence. Water-soluble polymers that have been utilized for suspension polymerization include gelatin, natural water-soluble gums (usually polysaccharides), polyvinyl pyrrolidone, polyvinyl alcohol, sulfonated polystyrene, polyacrylic acid and its salts, and polymethacrylic acid and its salts.

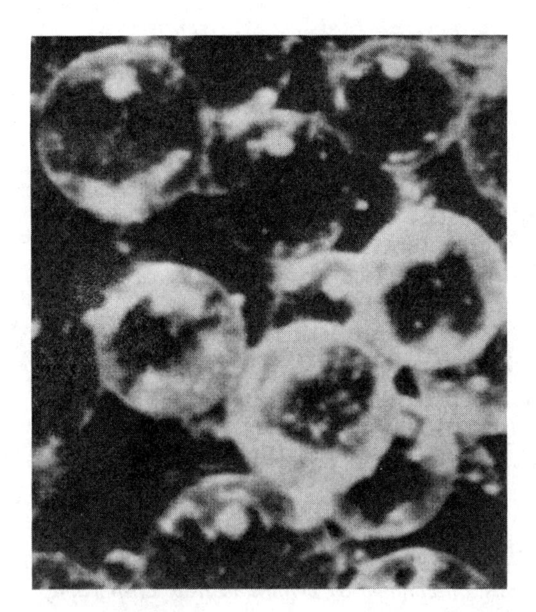

Fig. 10.15 Polystyrene beads. 20 X mag. *(Courtesy of Monsanto Co.)*

Water-soluble high polymers are also effective, e.g., polyvinyl alcohol, carboxymethylcellulose, hydroxyethylcellulose, polyacrylic and polymethacrylic acid, and copolymers of acrylic and methacrylic acid and their esters.

The use of an oil-soluble catalyst gives reaction kinetics and mechanisms essentially identical to those of mass polymerization rather than of emulsion polymerization. Therefore, reaction rates are slower in suspension polymerizations than in emulsion polymerizations aiming for the same molecular weight product.

Suspension polymerization is of great importance in the plastics industry. It not only solves the heat-transfer problem but also permits simple and complete separation of the polymer from water and suspending agents. Haze, water absorption, color, and electrical properties of a polymer made by a good suspension process will approach those of a polymer made by a mass process.

A typical suspension process would be carried out as follows:

> Into a glass-lined, stirred reaction vessel are charged 200 parts freshly distilled water and 0.1 part of a water soluble interpolymer of acrylic acid and 2-ethylhexyl acrylate. The atmosphere in the reaction vessel is swept free of oxygen with nitrogen and then 100 parts of styrene and 0.1 part benzoyl peroxide are charged to the reaction vessel. The reaction mixture is heated and stirred for 6 hours at 90°C and then for 8 hours at 130°C. The polymer is obtained at close to 100% conversion, in the form of small spherical beads of an average diameter of about 3 mm, suspended in the water as a slurry.[46]

The slurry is transferred to a slurry hold tank, the polymer is separated by centrifugation, washed, and dried. (Fig. 10.16). From the dryer the beads are mixed with colorants, stabilizers, plasticizers (if polyvinyl chloride), etc; melted in a Banbury mixer or an extruder; extruded, cooled, and finally chopped into pellets for bagging as a molding powder.

Plastic molding powders that are commercially made via a suspension process include:

1. Polymethylmethacrylate.[47]
2. Polystyrene and rubber-modified polystyrene.[48]
3. Polyvinylidene chloride-vinylchloride copolymer.
4. Polyvinyl chloride and copolymers.
5. Polyvinyl acetate.
6. Poly(styrene-acrylonitrile) copolymer.
7. Rubber-modified styrene-acrylonitrile copolymer, i.e., ABS.[42]

As far as is known, commercial suspension processes are all batch, but several designs for continuous suspension polymerization have been patented.[49,50] A semicontinuous suspension polymerization process is employed by Wacker-Chemie[51] and Tenneco for the production of PVC paste resin. The reactors may be glass-lined or of stainless steel and up to 16,000 gallons in volume. The difficulties of commercially obtaining a "clean" bead with no contamination or agglomeration are treated by Tromsdorff.[52] In vinyl chloride polymerization, a further difficulty is that of obtaining a porous bead that will soak up plasticizer rapidly and uniformly in subsequent processing (Fig. 10.17). The commercial art of suspension polymerization is complicated and many details have not been disclosed. The complete suspending-agent recipe is the most important part

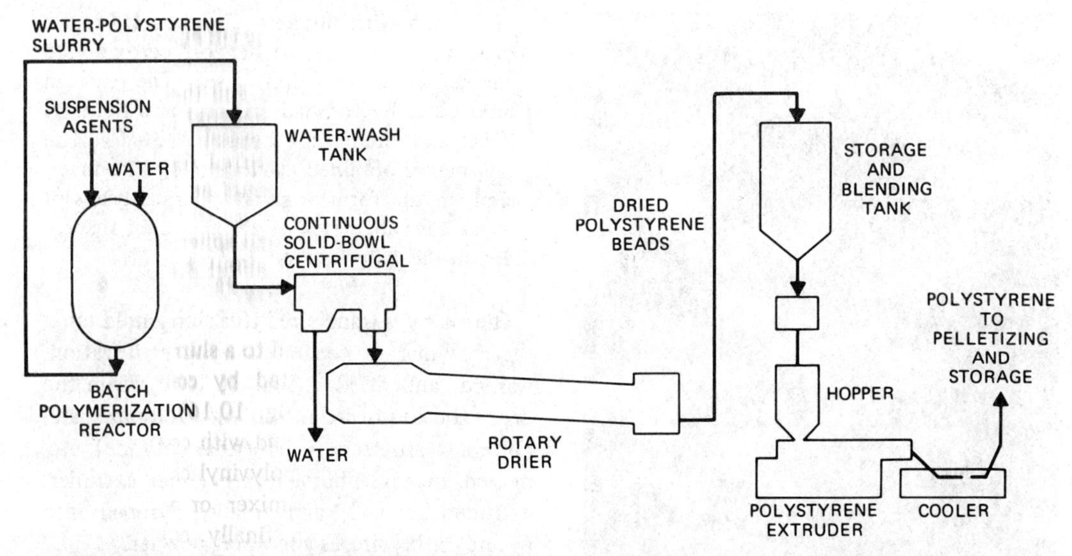

Fig. 10.16 Flow sheet for polystyrene by suspension process. *(Reprinted by permission from Chem. Eng., 65, 100 (Dec. 1958); by McGraw-Hill, Inc.)*

Fig. 10.17 Polyvinyl chloride suspension particles showing porosity. Taken with a scanning (surface reflecting) electron microscope. 150 X mag. *(Courtesy G. C. Claver, Monsanto Co.)*

of the process, although agitation also plays a vital role.

Solution Polymerization. Addition-type polymerization carried out in a solvent which will dissolve the monomer, the polymer, and the polymerization catalyst, is termed solution polymerization. Processes for polyolefins which use a solvent insoluble solid catalyst are–by this definition–not considered truly solution polymerization since their kinetics are substantially different. Polymerization of pure monomers that dissolve their polymer is solution polymerization provided that the polymerization is halted at a low enough conversion to maintain a reasonable level of fluidity. The eventual high-viscosity levels of complete bulk polymerization are avoided in solution polymerization, and heat transfer is greatly facilitated. By proper adjustment of the reaction temperature and pressure, reflux of monomer or solvent may remove heat. (Because of the high molecular weight of polymers, boiling point elevation is insignificant until high-percent conversions by weight are reached.)

Solution polymerization will generally slow down the reaction rate and lower the molecular weight. Depending on the polymer, such effects may be commercially desirable or undesirable. In some instances the rate at which monomers will graft onto preformed polymer molecules can be lessened; this leads to lower levels of chain branching and crosslinking.

Solution polymerization, however, has been used sparingly in the plastics industry, primarily because most of the polymers marketed to date have been successfully polymerized by the more economical mass and suspension processes. The cost of solvent removal and recovery may be appreciable. The development of machinery that can successfully and economically devolatilize a solvent from a polymer is a specialized art. Although vacuum drum driers will do the job, they are extremely costly for the rates attainable. Multiple-vented single or twin screw extruders have been used commercially; pressures below 30 torr are frequently required.[24] Another commercially used technique involves strands or sheets falling through a vacuum chamber.[29] Production rates are slow because of the slow diffusion rates in polymers.

A solution process for making styrene-acrylonitrile copolymers and other copolymers of styrene[53,54,55] which is probably being used for the commercial production, has been disclosed. (Styrene-alpha-methyl styrene copolymers are probably also being made commercially by a similar process, but here the solvent is merely the monomer.) For styrene-acrylonitrile copolymers, ethylbenzene or toluene at 30 percent concentration by weight are the preferred solvents; they ensure a favorable balance between the reaction rate and chain transfer: i.e., reaction rates and molecular weights are both lowered to commercially desirable levels. The process is a continuous one with the reactant (70/30 styrene/acrylonitrile) and solvent being fed to a recycling loop of a 20 percent by weight polymer solution maintained at 150°C by a heat exchanger. The reaction rate under these conditions is 27 percent/hour. A form of twin-screw extruder called the "Plastruder"[56] recovers the solvent and the unreacted monomer for recycle to the reactor.

Solution polymerizations of the addition type using ionic catalysis have been studied in the laboratory without notable commercial success as yet, except in the rubber industry where nearly all plant expansions are currently using this route (see Chapter 9). Cationic catalysts, e.g., $AlCl_3$, HF, are used extensively to polymerize olefins to a very low degree in the petroleum industry (Chapter 14). Also, many tons of low-molecular weight petroleum resins and asphaltic polymers are produced using cationic catalysts, e.g., concentrated sulfuric acid, aluminum halides, ferric chloride, and boron halides.[57] To describe all these resins and their manufacture is beyond the scope of this chapter.

Solvent-Nonsolvent Polymerization (Precipitation Polymerization). Some polymers are insoluble in their monomers, e.g., polymers of vinyl chloride, vinylidene chloride, acrylonitrile, chlorotrifluoroethylene, and ethylene (at least at lower temperatures and pressures). If polymerization is initiated in the pure monomers, a polymeric precipitate forms. Sometimes other solvents are added to increase or decrease solubility of the monomer in the solid polymer. Surprisingly, with free radical catalysts, molecular weights are frequently much higher in such systems than in nearly equivalent homogeneous systems. The reason for this molecular weight effect is probably similar to the explanation given for the higher molecular weight of emulsion polymers, i.e., initiation occurs primarily in the continuous phase so that each precipitated particle is likely to have only one growing radical, to which fresh monomer can readily diffuse while other polymeric radicals cannot. Termination by combination of two polymeric radicals, therefore, is infrequent. And because of the relative immobility of the growing chains, even if two precipitated particles should fuse, the polymeric radicals will probably not combine.

Chain termination mechanisms have not been explored in detail for any particular system, but chain transfer to monomer, to the polymeric chain, to the solvent, or to a chain transfer agent is a possibility as is disproportionation or termination by catalysts, inhibitors, or retarders. If chain transfer to polymer takes place as it does with polyvinyl chloride, branched structures develop.

Solvent-nonsolvent polymerization was the method selected for preparing the vinyl chloride-vinyl acetate copolymers ("Vinylites") first introduced commercially in 1935. The monomer itself is the principal nonsolvent. At about 25-30 percent conversion, the polymer adsorbs most of the monomer and heat transfer is substantially lowered. The unreacted vinyl chloride monomer, which is a gas at normal temperature and pressure, must be recovered by flash drying or other methods. A French process[58,59] uses reflux in a ribbon-agitated batch reactor of special design to achieve sufficient heat transfer to permit conversions of about 80-85 percent.

Polyoxymethylene (polymerized formaldehyde) has been developed commercially using basic ionic catalysts, such as quarternary ammonium bases, trialkyl amines of high molecular weight, and trialkyl phosphines.[60,61] Thoroughly dried hydrocarbon solvents which do not dissolve the polymer are used, e.g., hexane, cyclohexane, and toluene. Pure formaldehyde completely free of water and acids must be used to obtain a polymer of high molecular weight with reasonably narrow molecular weight distribution and good thermal stability. After polymerization is completed, the polymer is stabilized by acetylating the end groups at high temperatures, 700°F, with acetic anhydride in the presence of pyridine or an alkali acetate as a catalyst.[62] Additional stability can be incorporated by copolymerizing ethylene oxide with the formaldehyde.

A continuous low-temperature (32°F) solvent-nonsolvent process has been described[63,64,65] for polytrifluorochloroethylene ("Kel-F"). The reactive, free-radical catalyst, bis-trichloroacetyl peroxide, dissolved in trichlorofluoromethane, is added continuously to a stirred, jacketed, stainless steel reactor along with the monomer. The polymer is separated from the slurry containing monomer and trichlorofluoromethane by means of a continuous filter or centrifuge, and it is then dried further in a vacuum drier to remove all the remaining monomer. The unreacted monomer and diluent are recycled to the reactor with a portion being sent to a repurification distillation column.

Low-pressure polymerization of ethylene and propylene with heterogeneous catalysts could be described as a solvent-nonsolvent type of polymerization; however, the insoluble solid catalysts used here make the reaction kinetics and purification methods substantially different. These processes will be described later in a separate section.

Condensation Polymerization

Generalizations concerning processes and equipment for the various commercial condensation polymerizations would be meaningless. These subjects will be treated in the following sections where each type of condensation polymer is discussed separately.

Phenol-Formaldehyde Resins. It is not surprising that the first completely synthetic plastic was made from the common and highly reactive multifunctional chemicals, phenol and formaldehyde. As early as 1870, Baeyer, in Germany, observed the reactions between phenol and formaldehyde and characterized the phenol alcohols which were formed. Around the turn of the century Delair, Smith, and Leback attempted to utilize the resinous reaction products as shellac substitutes. Beginning in the early 1900's Baekeland discovered that useful moldings could be made if the final stages of the reaction were carried out under heat and pressure, preferably in the presence of a suitable filler, e.g., sawdust, wood flour, and cotton cloth. Since that time, phenolic resins have become the "workhorse" of the plastics industry. They are used in electrical components, insulating varnishes, industrial laminates, binders, etc.

The chemistry of this complex reaction is, even today, not completely understood. When crosslinked structures of large size are formed, isolation and characterization of these structures is very difficult. Much investigation has been done, however, and a great deal is known—so much, in fact, that the brief description given below suffers from oversimplification.

The first series of reactions is the formation of phenol alcohols under the influence of either acids or bases.

(1) 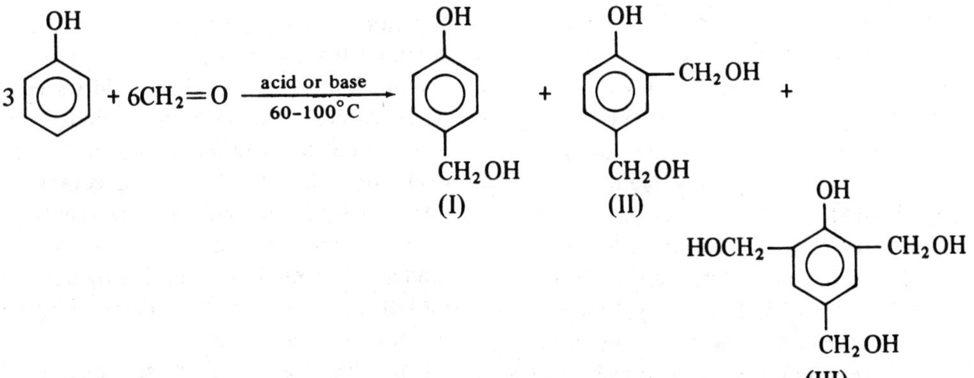 3 [phenol: OH on benzene ring] + $6CH_2{=}O$ $\xrightarrow[60\text{–}100^\circ C]{\text{acid or base}}$ [OH, CH_2OH on ring] (I) + [OH, CH_2OH, CH_2OH on ring] (II) + [OH, $HOCH_2$, CH_2OH, CH_2OH on ring] (III)

Next, these phenol alcohols condense in a complex manner. Using acid catalysts the following reactions are favored.

(2) 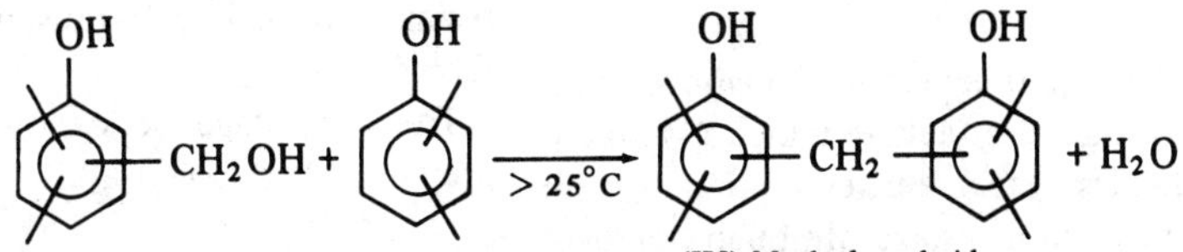 [OH ring]$-CH_2OH$ + [OH ring] $\xrightarrow[> 25^\circ C]{}$ [OH ring]$-CH_2-$[OH ring] + H_2O

(IV) Methylene bridge

(3) 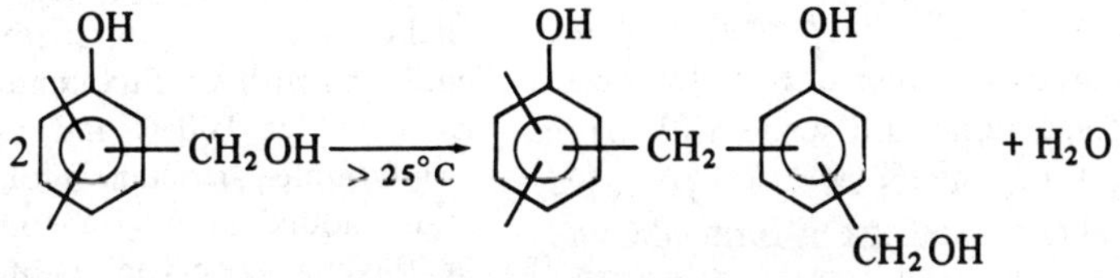 2 [OH ring]$-CH_2OH$ $\xrightarrow[> 25^\circ C]{}$ [OH ring]$-CH_2-$[OH ring, CH_2OH] + H_2O

(V) Methylene bridge

(4) 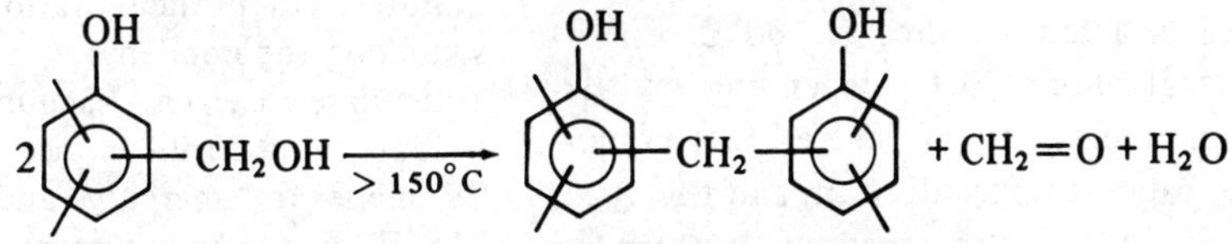 2 [OH ring]$-CH_2OH$ $\xrightarrow[> 150^\circ C]{}$ [OH ring]$-CH_2-$[OH ring] + $CH_2{=}O$ + H_2O

(VI) Methylene bridge

(5) 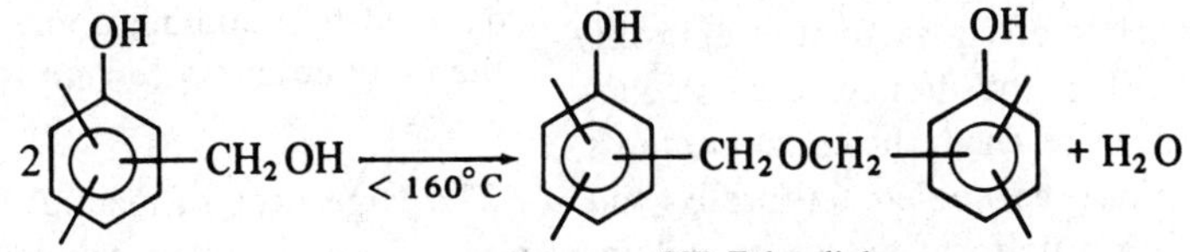 2 [OH ring]$-CH_2OH$ $\xrightarrow[< 160^\circ C]{}$ [OH ring]$-CH_2OCH_2-$[OH ring] + H_2O

(VI) Ether linkage

Reactions (2) and (3) with the benzene nucleus proceed quickly with acid catalysts even at temperatures below the point where phenol alcohols form rapidly. Hence, under acid-catalyzed conditions unreacted phenol alcohols cannot ordinarily be isolated, but are intermediates. Reaction (5), though catalyzed by weakly acidic conditions, is slower than Reactions (2) and (3) and occurs, therefore, at neutral pH's and/or when all of the *ortho* and *para* positions on the phenol nuclei are occupied by either alcoholic groups or unreactive ones. At high temperatures (around 160°C) the ether linkage becomes unstable, losing formaldehyde and reverting to a methylene bridge (Reaction (4)).

With basic catalysts, Reaction (5) is slow at the high pH's used commercially. The condensation reactions (2), (3), and (4) also take place, but at a much slower rate than under acid catalysis. These reactions are catalyzed by a surprisingly low concentration of strong base; and further amounts of base do not increase their rate. Since the formation of phenol alcohols (Reaction (1)) is more strongly catalyzed by bases, pure or partially condensed alcohols can be isolated.

Commercial equipment for the above reactions usually includes jacketed anchor-agitated steel or stainless steel kettles equipped with vacuum reflux condensers, although scraped-wall heat exchangers might be used for a continuous process.

Two major types of commercial molding powders are based on either one- or two-stage resins. One-stage resins are made with basic catalysts, e.g., 1 to 2 parts calcium hydroxide to 100 parts phenol, and use a formaldehyde phenol mole ratio between 1.1 and 1.5. Formalin or paraformaldehyde* may be used to supply the formaldehyde. Phenol alcohols may be formed in a few minutes at 100°C or in a few hours at about 70°C, depending on the heat-exchange capacity of the reaction equipment, the catalyst concentration, and the type of resin desired. After most of the formaldehyde has combined with the phenol, the water is removed and condensation is continued under 27-29 inches of Hg vacuum at approximately 75°C. When the desired level of condensation has been reached, the entire contents of the kettle must be cooled uniformly and rapidly to ensure a uniform degree of polymerization. One commercial method involves dumping the entire contents onto a large steel cooling floor through which cold water flows. After cooling, the resin is ground to a fine powder and blended about half and half with wood flour or other filler and colorants plus small quantities of mold release agents and/or cure accelerators (magnesium oxide, calcium oxide). This blend is then densified and fused on hot mill rolls, cooled, and ground into molding powder (Fig. 10.18).

One-stage resins, though still manufactured, are not as important for molding powders as two-stage resins (novalaks) which usually have superior flow characteristics. The polymerization equipment used for two-stage resins is essentially the same as that used for one-stage resins, although acid-resistant kettles are preferred.

The first stage of a two-stage resin may be prepared as follows:

> One hundred parts of phenol and 0.5 part of concentrated H_2SO_4 are charged into an anchor-agitated stainless-steel reaction vessel equipped with a reflux condenser. The charge is heated at 100°C and 69 parts (0.8 mole CH_2O/mole of phenol) of 37 percent formalin added at a rate compatible with the heat-exchange capacity of the reflux condenser. After all the formalin has been added, the charge is refluxed for an additional 30 minutes, then dehydration is begun by switching the condensate flow from reflux to a distillate receiver. Vacuum is applied as the boiling point begins to rise until a vacuum of 28 inches of mercury and temperature of 215°F are reached, corresponding to approximately 4 percent free phenol in the final resin. The charge is neutralized with lime and dumped to a cooling floor. The solid resin is then ground into a fine powder.

The large heat of reaction*,[66] and the rapid rate of condensation under acid conditions

*In the past, formalin (a 37 percent solution of formaldehyde in water) was almost always used, but with the advent of cheap solid paraformaldehyde (a solid polymer of formaldehyde) from natural gas, the latter is frequently used to increase capacity and save on subsequent dehydrating costs.

*For the complete condensation of formaldehyde to form a methylene-bridge structure plus liquid water, about 21 kcal/mole of formaldehyde are released. This can be split into 4.1 kcal/mole for the formation of the phenol alcohol and 16.9 kcal/mole for the formation of the methylene bridges [reaction (2)].

makes the above procedure desirable for safety reasons. The resulting resin is thermoplastic and is termed a novalak. To make a thermosetting molding powder, hexamethylenetetramine† crystals (about 10 to 15 parts/-100 parts of novalak resin) are blended with wood flour, colorants, mold release agents, and cure accelerators using a fifty/fifty ratio of total resin including "hexa" to total fillers. After blending, the resin is fused on hot mill rolls, cooled, and ground into a molding powder in a manner similar to the method used for one-stage resins.

To explain the final cure of one- and two-stage resins, a highly reactive quinone methide is postulated as an intermediate which combines to form complex condensed rings of red or brown color. These quinone methides may also explain the reaction of phenolic resins with the double bonds in rosin, drying oils, rubber, and other addition-type monomers. Addition polymerization with rosin and drying oils has been used extensively to increase the oil solubility of phenol-resin varnishes.

To explain the reaction of hexa, the following principal reactions are postulated:

$$(6)\quad 2\,C_6H_5OH + \tfrac{1}{3}\,(\text{hexa}) \longrightarrow HOC_6H_4\text{-}CH_2NHCH_2\text{-}C_6H_4OH\ (\text{VII}) + \tfrac{1}{3}NH_3$$

$$(7)\quad (\text{VII}) \xrightarrow{\text{heat}} HOC_6H_4\text{-}CH{=}N{-}CH_2\text{-}C_6H_4OH\ (\text{VIII})\ \text{yellow} + \tfrac{1}{2}H_2\uparrow$$

$$(8)\quad 2\,C_6H_5OH + \tfrac{1}{6}\,(\text{hexa}) \longrightarrow HOC_6H_4\text{-}CH_2\text{-}C_6H_4OH\ (\text{IV}) + \tfrac{2}{3}NH_3\uparrow$$

Many more reactions during curing have been proposed, and undoubtedly many of them do occur to some extent. Several authors[67,68,69,70,71] have given excellent summaries of experimental evidence for the above and other possible reactions.

Aminoplast Resins. Urea was known to react with formaldehyde as early as 1884, but commercial molding powders were not developed until 1926. As in the case of phenolic resins, commercial development preceded an understanding of the basic chemistry involved. Again, this is primarily due to the difficulty of characterizing complicated crosslinked structures. The following equations oversimplify a very complex picture:[73,74,75,76,77]

$$(1)\quad CH_2{=}O + H_2NC({=}O)NH_2 \rightleftarrows H_2NC({=}O)NHCH_2OH\ \ (\text{I})$$

$$\updownarrow + CH_2{=}O$$

$$HOCH_2NHC({=}O)NHCH_2OH\ \ (\text{II})$$

$$\updownarrow + CH_2{=}O$$

$$(HOCH_2)_2NC({=}O)NHCH_2OH\ \ (\text{III})$$

†Commonly called "hexa," it is the nearly instantaneous reaction product of ammonia and formaldhyde.

$$(CH_2)_6N_4$$

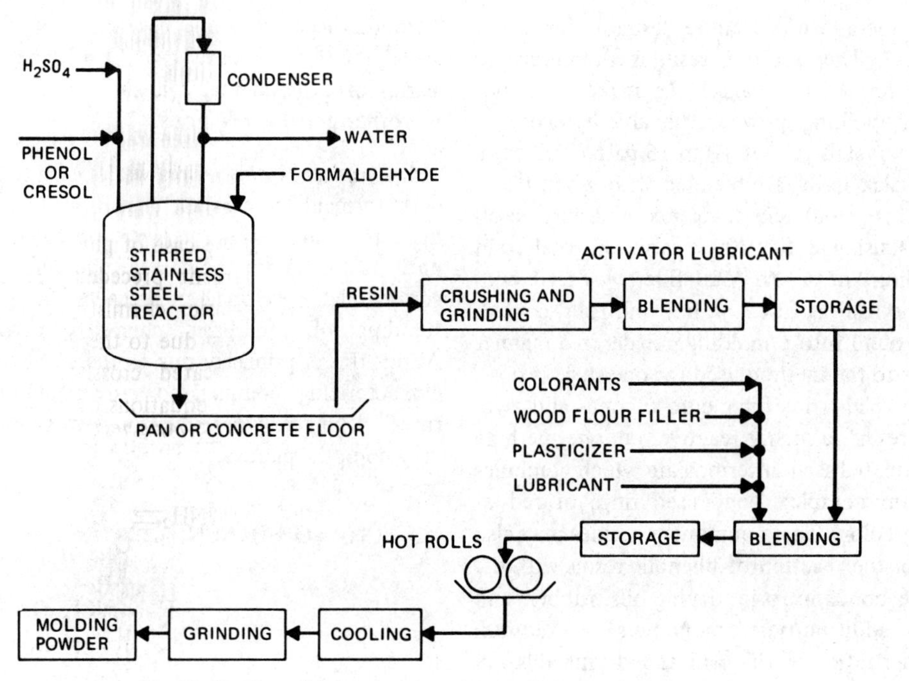

Fig. 10.18 Flow diagram for phenol-formaldehyde molding powder.

Reaction (1) is catalyzed by base or acid and is an equilibrium type. Under commercial conditions very little trimethylolurea is formed. If acid catalysts are used the methylolureas cannot be isolated since further condensation occurs as follows:

$$(2)\quad \left[-\overset{\overset{\large O}{\|}}{C}\overset{R'}{N}CH_2OH + H_2N\overset{\overset{\large O}{\|}}{C}\overset{R'}{N}-\right] \longrightarrow \left[-\overset{\overset{\large O}{\|}}{C}\overset{R'}{N}CH_2\overset{H}{N}\overset{\overset{\large O}{\|}}{C}\overset{R'}{N}-\right] + H_2O \quad \text{(IV)}$$

$$(3)\quad \left[-\overset{\overset{\large O}{\|}}{C}\overset{R'}{N}CH_2OH + H\overset{R}{N}\overset{\overset{\large O}{\|}}{C}\overset{R'}{N}-\right] \longrightarrow \left[-\overset{\overset{\large O}{\|}}{C}\overset{R'}{N}CH_2\overset{R}{N}\overset{\overset{\large O}{\|}}{C}\overset{R}{N}-\right] + H_2O$$

$$(4)\quad 2\left[-\overset{\overset{\large O}{\|}}{C}\overset{R'}{N}CH_2OH\right] \longrightarrow \left[-\overset{\overset{\large O}{\|}}{C}\overset{R'}{N}CH_2OCH_2\overset{R'}{N}\overset{\overset{\large O}{\|}}{C}-\right] + H_2O \quad \text{(V)}$$

$$\text{where } R = -CH_2OH \text{ or } -CH_2\overset{R'}{N}\underset{\underset{\large O}{\|}}{C}\overset{R'}{N}R$$

$$R' = R \text{ or } H$$

With equal concentrations of possible reaction sites, Reaction (2) with an unsubstituted amide hydrogen to form methylene bridges (IV) and Reaction (4) to form ether linkages (V) are favored over Reaction (3) with a substituted amide hydrogen to form methylene bridges. Under basic conditions, Reactions (3) and (4) are not favored and Reaction (2) will proceed very slowly.

At the higher temperatures and acid conditions which exist during molding, the ether linkages are unstable, breaking to form methylene bridges and formaldehyde. For this reason, low formaldehyde/urea ratios of about 1.5 are used in molding powder or laminating resins to minimize ether formation and subsequent shrinkage and cracking during molding or after long use.

Melamine was an expensive laboratory curiosity until about 1939. Low cost commercial manufacture was then stimulated when the outstanding properties of melamine resins were discovered (Table 10.1).

Melamine has the following structure. The possibility of many resonating configurations account for its excellent thermal stability.

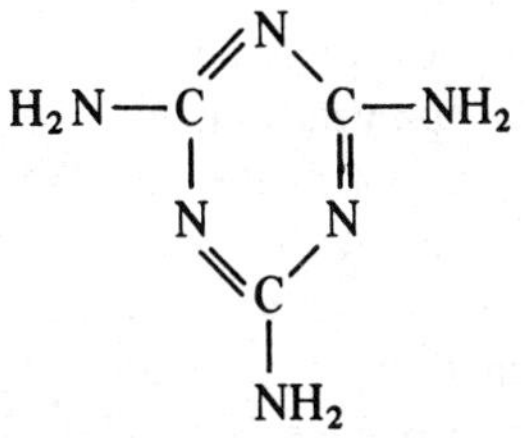

```
          N
        //  \
H2N—C        C—NH2
      |       ||
      N       N
        \\  /
          C
          |
         NH2
```

The chemistry of the condensation of melamine with formaldehyde is analogous to that of urea, with the following exceptions:

1. Six reactive hydrogen atoms exist, three of which are the highly reactive unsubstituted amine* type.

2. The reactivity of these hydrogens is greater than that for urea, and methylene bridges are readily formed under basic conditions.

3. Since the melamine molecule contains three highly reactive sites it can more readily form crosslinked structures than can urea with only two highly reactive sites, e.g., formaldehyde/melamine ratios for molding and laminating resin are usually 2.5 compared to 1.5 for urea.

4. The methylolmelamines are much more resistant to dissociation into formaldehyde and melamine. The hexamethylol compound is readily formed, and the di- and trimethylol compounds do not dissociate rapidly.

The equipment and processing steps for the preparation of melamine and urea resins are very similar to those described earlier for phenolic resins. Corrosion resistant equipment is used to retain the excellent color inherent in these resins.[78] Although water may be removed by simple vacuum distillation as it is for phenolic resins, spray drying is frequently the method employed. If the laminates or molding powders are made at the resin manufacturing site, impregnating the laminating paper or the molding powder filler with the wet resin is the most economical method, followed by drying at atmospheric pressure in tunnel or tray driers. The dried impregnated sheets or molding powders are then completely cured under heat and pressure as are phenolic molding powders. The spray-dried, low molecular weight resins used for laminates are redissolved in water and alcohol to make an impregnating solution. Fillers are ordinarily highly refined α-cellulose and cotton fibers which retain the excellent color inherent in the base resins. Paper with a naturally high wet strength is used for laminates.

The adhesive, textile, paper-treating, and surface-coating industries comprise a slightly larger market for aminoplast resins than do the molding powder and laminating markets. To ensure compatibility with surface coatings, butyl ethers are usually formed with butanol, using higher formaldehyde ratios to increase ether formation. Other nitrogen-containing chemicals used in aminoplast formulations include dicyandiamide substituted melamines, cyclic ethylene urea, and thiourea.

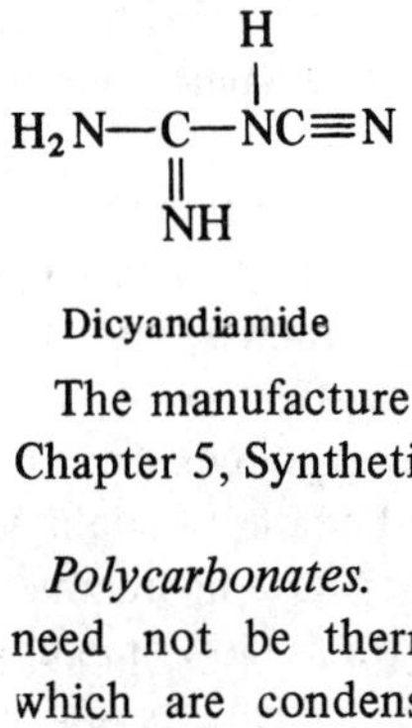

```
          H                 H2C——CH2
          |                  |     |
H2N—C—NC≡N                  HN     NH
    ||                        \   /
    NH                          C
                                ||
                                O
```

Dicyandiamide 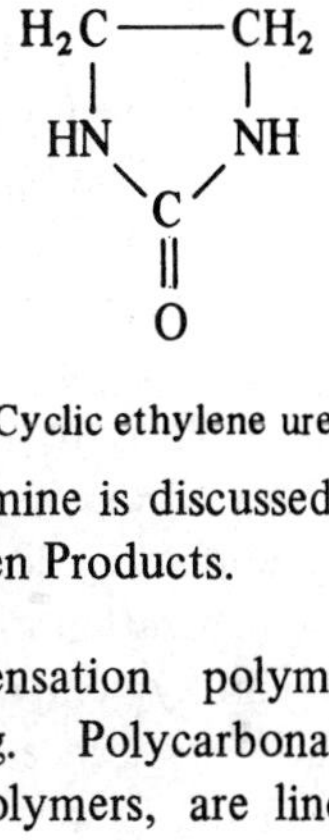   Cyclic ethylene urea

The manufacture of melamine is discussed in Chapter 5, Synthetic Nitrogen Products.

Polycarbonates. Condensation polymers need not be thermosetting. Polycarbonates which are condensation polymers, are linear thermoplastic polyesters of carbonic acid with aliphatic or aromatic hydroxy compounds. They may be represented by the general structure.[79,22,80,81,82]

$$\mathrm{H{+}OR{-}O{-}\overset{O}{\overset{\|}{C}}{+}_{n}OROH}$$

where R is normally the hydrocarbon radical in bisphenol A

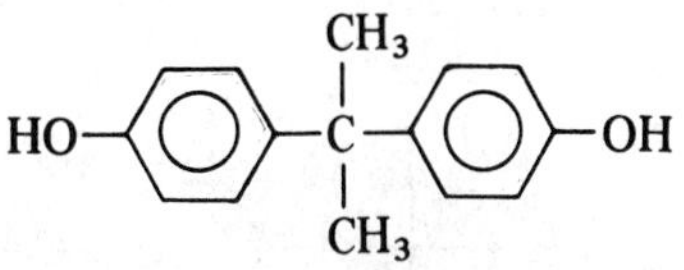

```
              CH3
              |
HO—⟨◯⟩—C—⟨◯⟩—OH
              |
              CH3
```

Though Einhorn[83] first reported preparing linear polyesters of carbonic acid in 1898,

*The $-NH_2$ groups in melamine behave even more like amide groups than do the $-NH_2$ groups in urea.

polycarbonates are relatively new polymers industrially, with the first commercially important patents being issued in 1956[84] and 1959.[85] The most readily available monomer is bisphenol-A and the polycarbonate that derives from this monomer has been found to have the best balance of properties. Hence, when full-scale commercial production began in Germany in 1959 and in the U.S. in 1960, it was bisphenol-A polycarbonate that was produced. The polymer found quick applications in cast-film form as electrical foil and as a base for photographic film, and in pellet form as an injection molding and extrusion compound for use in parts for electrical and electronic components.[22]

Polycarbonates can be broadly divided into aliphatic polycarbonates, aliphatic-aromatic polycarbonates, and aromatic polycarbonates. Bisphenol-A polycarbonate is an aromatic polycarbonate. As free carbonic acid, H_2CO_3, cannot be used directly but only as a derivative, a classical esterification reaction where the acid is heated with an aliphatic or aromatic alcohol is not possible.

Aliphatic polycarbonates can be prepared by the reaction of aliphatic dihydroxy compounds with phosgene, bis-chloroformates, or bis-chlorocarbonates by transesterification of esters of aliphatic dihydroxy compounds, or by polymerization of cyclic carbonates of aliphatic dihydroxy compounds. They have no commercial significance due to their low melting points (below 120°C), their high solubility, their hydrophilic nature, and their low thermal stability.

Aliphatic-aromatic polycarbonates are prepared by polycondensation of the bisalkyl or bisaryl carbonate of p-xylene glycol in the presence of titanium catalysts, such as tetrabutyl titanate. These have not found any practical application either.

Aromatic polycarbonates, of which bisphenol-A polycarbonates are the most important, are the polycarbonates which have found wide commercial application. They are prepared by phosgenation of aromatic dihydroxy componds in the presence of pyridine, by interfacial polycondensation between aromatic dhydroxy compounds in aqueous alkaline solutions with phosgene or bischlorocarbonic acid esters of aromatic dihydroxy compounds in the presence of inert solvents, or by transesterification.

The transesterification reaction of bisphenol-A and diphenyl carbonate is shown in the following equation:

$$2(C_6H_5O)_2CO + HOC_6H_4C(CH_3)_2C_6H_4OH \longrightarrow C_6H_5OCO\text{+}C_6H_4C(CH_3)_2C_6H_4OCO\text{+}{-}C_6H_5 + 2C_6H_5OH$$

The reaction is run in a well stirred reactor between 180 and 300°C and 1 to 30 mm Hg pressure. The polymer molecular weight is limited by the high melt viscosity which is characteristic of polycarbonates due to the relative inflexibility of the polymer chains (e.g., 5×10^5 poise at 240°C for a 30,000 osmotic molecular weight).

A plant for the preparation of bisphenol-A polycarbonate by phosgenation in the presence of pyridine has been in operation since 1960.[86] The reaction is:

$$x\,HO-C_6H_4-C(CH_3)_2-C_6H_4-OH + x\,CO\,Cl_2 \xrightarrow{\text{pyridine}} H\left[O-C_6H_4-C(CH_3)_2-C_6H_4-O-\overset{O}{\overset{\|}{C}}\right]_x Cl + (2x-1)\,HCl\ (\text{as pyridine salt})$$

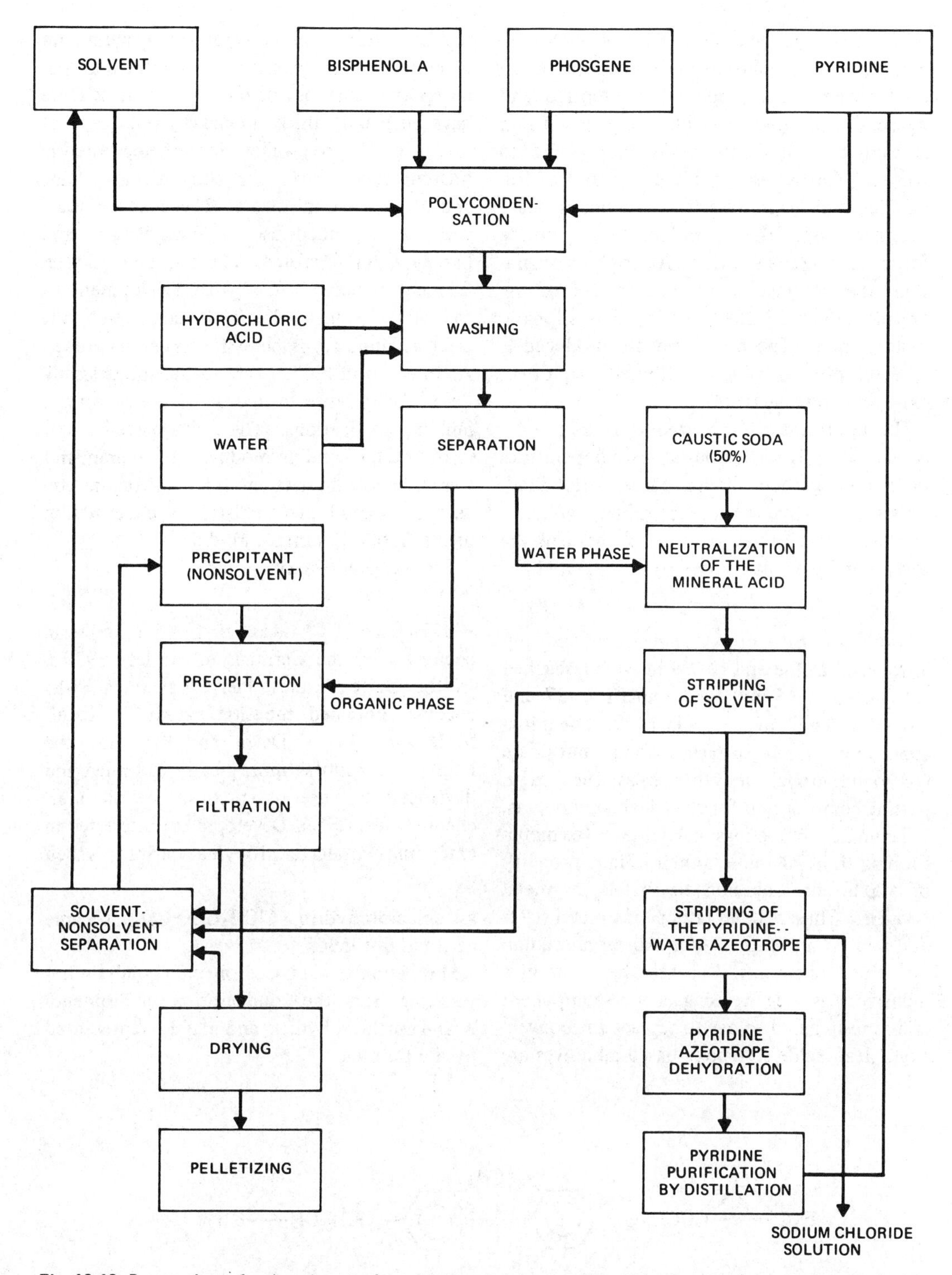

Fig. 10.19 Preparation of polycarbonate from bisphenol A by reaction with phosgene in the presence of pyridine. (*"Encyclopedia of Polymer Science and Technology," vol.* 10, *H. G. Mark and N. G. Gaylord, eds., Interscience Publishers, 1968*; *by permission of John Wiley & Sons, Inc.*)

A simplified process flow sheet is shown in Fig. 10.19. A solution of bisphenol-A in an inert solvent, such as methylene chloride, and pyridine is phosgenated. The polymer solution is washed with dilute hydrochloric acid to convert the excess pyridine to the hydrochloride which dissolves in the aqueous phase together with the pyridine hydrochloride formed during the reaction. After phase separation, the polymer is precipitated out of the organic solvent by the addition of an aliphatic hydrocarbon. The polycarbonate produced is a white powder which is filtered off, dried, extruded, and pelletized.

The advantage of the process is that polycondensation is carried out at low temperatures in a homogeneous liquid phase. Its disadvantages are that it uses pyridine, which is expensive and causes odor problems, and the solvent and precipitant have to be separated.

Epoxy Resins. Epoxy resins, commercially introduced at the end of World War II, reached a capacity of 35 million pounds in 1957 and around 310 million pounds in 1979. The initial application was in surface coatings, but there was rapid growth in other areas, such as in potting compositions for electrical components, in laminates, in castings, particularly for metal-forming tools, and due to their unique property of exceptional adhesion to metals, as metal bonders. Thus, epoxies rather than metal rivets are used to apply the outer skin of aluminum on high-speed aircraft. Recently the automotive industry has begun using a one-component epoxy resin-based solder to replace the conventional lead solders. Epoxy-based adhesives are replacing brazing methods for joining aluminum to copper in low-pressure refrigerator coils. It is reported that out of one million units there have been only thirty failures. As a result of this success, epoxy adhesives are being tested in high-pressure refrigeration coils, dehumidifiers, and room air-conditioners. The next few years will see expanded use of epoxy resins and hardeners in reinforced plastics, especially in filament winding. With the development of specially designed electrical-grade resins, the electrical industry is increasing its use of epoxies. Architects and builders are seeing the possibilities in epoxy resins for industrial and decorative building applications. The resins played a part in the Lunar Landing Module test program and were used as adhesives and fillers in conjunction with tapes and honeycombs in the construction of the Apollo Command Module.[87]

The reactive "epoxy" group

$$-OCH_2\overset{\overbrace{\quad O \quad}}{CH-CH_2}$$

* was first used in resin chemistry by the Germans in the late 1930's. In the United States, Pierre Castan, a Swiss chemist, obtained the first patent.[88] Later, S. O. Greenlee of Devoe and Raynolds, the large paint manufacturer, greatly extended and developed the use of epoxies.[89] Still later, chemists from Shell Development, interested in extending markets for propylene derivatives such as epichlorohydrin, $ClCH_2\overset{\overbrace{\quad O \quad}}{CH-CH_2}$, became active in this field.[90]

Most epoxies are based on the glycidyl ether resulting from the condensation of bisphenol A and epichlorohydrin, and may be represented by the formulas:

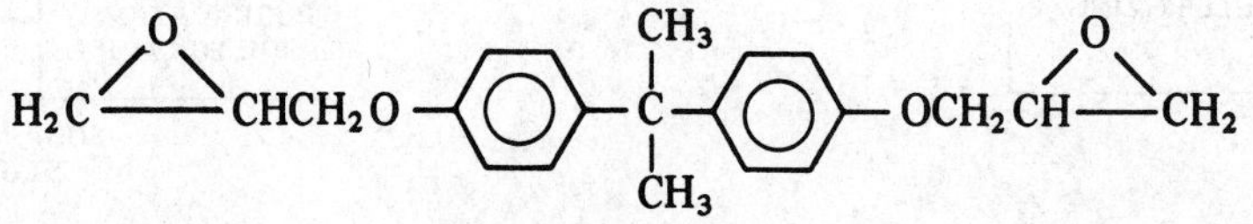

$$H_2\overset{\overbrace{\quad O \quad}}{C-CHCH_2}O-C_6H_4-\underset{CH_3}{\overset{CH_3}{\underset{|}{\overset{|}{C}}}}-C_6H_4-OCH_2\overset{\overbrace{\quad O \quad}}{CH-CH_2} \qquad \text{(I)}$$

*Glycidyl ether is a more correct term since epoxy groups without oxygen on the a carbon atom do not have the same reaction characteristics as glycidyl ether groups. However, in resin chemistry the term epoxy is synonymous with glycidyl ether and will be so used here.

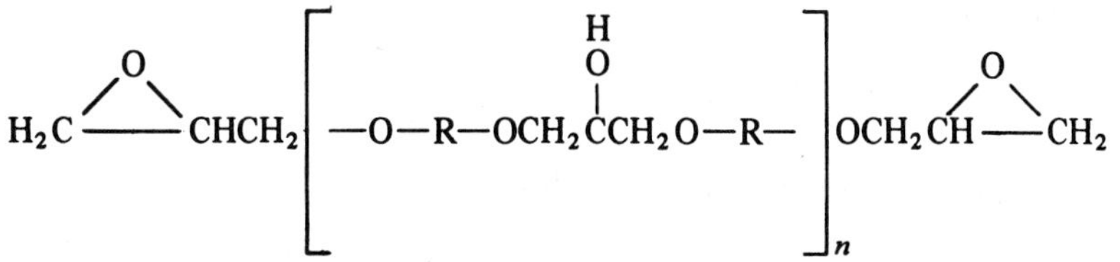

$$H_2C\overset{O}{\frown}CHCH_2\left[-O-R-OCH_2\underset{}{\overset{\overset{H}{O}}{C}}HCH_2O-R-\right]_n OCH_2CH\overset{O}{\frown}CH_2 \qquad (II)$$

where, for most commercial resins, $n = 1$ to 8, and

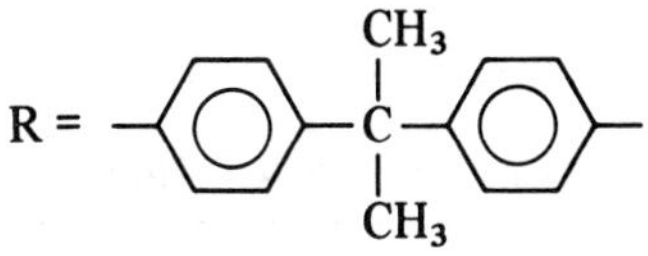

$$R = -C_6H_4-\underset{CH_3}{\overset{CH_3}{C}}-C_6H_4-$$

The reaction scheme by which these bis-glycidyl ethers are produced is discussed in greater detail by Skeist,[91] Lee and Neville,[92] and Coderre,[93] a brief summary follows:

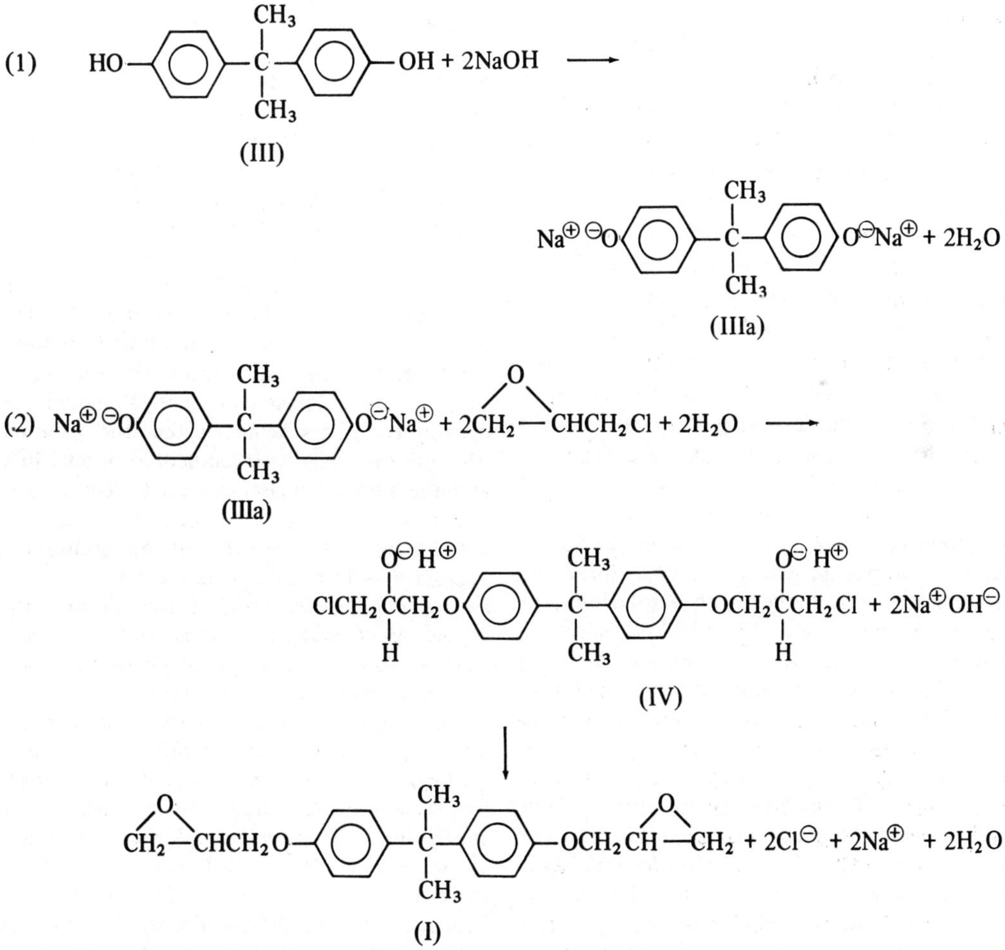

(1)

$$HO-C_6H_4-\underset{CH_3}{\overset{CH_3}{C}}-C_6H_4-OH + 2NaOH \longrightarrow$$
(III)

$$Na^{\oplus}\,{}^{\ominus}O-C_6H_4-\underset{CH_3}{\overset{CH_3}{C}}-C_6H_4-O^{\ominus}Na^{\oplus} + 2H_2O$$
(IIIa)

(2)

$$Na^{\oplus}\,{}^{\ominus}O-C_6H_4-\underset{CH_3}{\overset{CH_3}{C}}-C_6H_4-O^{\ominus}Na^{\oplus} + 2CH_2\overset{O}{\frown}CHCH_2Cl + 2H_2O \longrightarrow$$
(IIIa)

$$ClCH_2\underset{H}{\overset{O^{\ominus}H^{\oplus}}{C}}CH_2O-C_6H_4-\underset{CH_3}{\overset{CH_3}{C}}-C_6H_4-OCH_2\underset{H}{\overset{O^{\ominus}H^{\oplus}}{C}}CH_2Cl + 2Na^{\oplus}OH^{\ominus}$$
(IV)

$$\downarrow$$

$$CH_2\overset{O}{\frown}CHCH_2O-C_6H_4-\underset{CH_3}{\overset{CH_3}{C}}-C_6H_4-OCH_2CH\overset{O}{\frown}CH_2 + 2Cl^{\ominus} + 2Na^{\oplus} + 2H_2O$$
(I)

(3) $Na^{\oplus} + Cl^{\ominus} \longrightarrow NaCl\downarrow$

(4) 2(IIIa) + (I) + $2H_2O$ ⟶

$$Na^{\oplus}\ {}^{\ominus}OR{-}OCH_2\underset{|}{\overset{OH}{C}}HCH_2O{-}R{-}OCH_2\underset{|}{\overset{OH}{C}}HCH_2ORO^{\ominus}Na^{\oplus} + 2NaOH$$

(V)

$$R = -C_6H_4-C(CH_3)_2-C_6H_4-$$

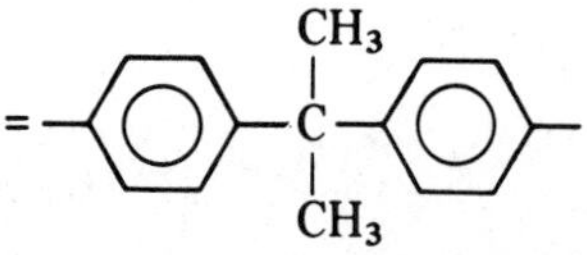

where

(5) (V) + $\overset{O}{CH_2-CHCH_2Cl}$ → higher molecular weight analogs of (IV) and (I).

Reaction (1) in the above requires a small amount of water (about 1 percent) to permit formation of the ionized species (IIIa). Reaction (2) to form the chlorohydrin (IV) is slow compared to the formation of the epoxy from the chlorohydrin and to the formation of salt [Reaction (3)]. If an excess of base or water is present, the epichlorohydrin and the epoxy group in the resin (I) will undergo reaction to the glycol. The rate of addition of base must be adjusted to the rate at which Reaction (2) occurs if optimum basicity is to be maintained. Also, to make the monomeric bisepoxy (I), an excess of epichlorohydrin will favor Reaction (2) over Reaction (4). The above principles are embodied in the following preparation of a nearly pure monomeric bisepoxy resin (I).

A reaction vessel fitted with a heater, stirrer, thermometer, and distilling head having a separator providing return to the reactor of the lower layer was charged with a solution containing a mole ratio of epichlorohydrin to bisphenol of 10:1. The solution was heated to about 100°C and maintained at this temperature during addition of 1.90 moles of sodium hydroxide per mole of the bisphenol, the caustic being introduced as a 40 percent aqueous solution. Water and epichlorohydrin distilled from the reaction mixture was condensed in the head, and only the epichlorohydrin layer returned to the reaction mixture. The rate of addition of caustic and rate of distilling kept the temperature at about 100°C so that the reaction mixture contained about 1.5 percent water, the addition taking approximately 2 hours. Upon completing the caustic addition the bulk of the unreacted epichlorohydrin was distilled from the reaction mixture followed by application of vacuum to a pressure of 1 mm Hg and 160°C to remove residual epichlorohydrin. The residue consisting of ether product and salt was cooled and to it was added an equal weight based on the ether of methyl isobutyl ketone along with three times this weight of water. The mixture was agitated at about 25°C and then allowed to separate into two phases. The brine phase containing about 9.5 percent salt was removed and discarded. The organic phase with ether product containing about 1.0 percent chlorine was then contacted with an equal weight of 5 percent aqueous sodium hydroxide solution, and the mixture was agitated for an hour at about 80°C. This quantity of excess caustic amounted to about 8.9 times that needed to react with the organically bound chlorine [Product IV in the preceeding reaction scheme] in the ether product. The mixture was next cooled to about 50°C and the aqueous phase separated. The organic phase was then agitated with about half an equal weight of 2 percent aqueous solution of sodium dihydrogen phosphate at about 25°C to neutralize any residual sodium hydroxide. After separation of phases the methyl isobutyl ketone was distilled from the organic phase, first up to a temperature of 160°C under atmospheric pressure, then down to a pressure of about 1 mm Hg at the same temperature. The resulting diglycidyl ether of bisphenol [Product I above] was a pale yellow liquid which analyzed as containing 0.25 percent chlorine and 0.521 epoxy equivalents per 100 grams, and had a molecular weight of 355. The product had high reactivity with added bisphenol, giving 100 percent reaction when heated at 190°C for 6 hours with an added 35.6 percent bisphenol.[94]

An alternate method of removing salt from the bisepoxy monomer involves filtering it, then washing the filter cake with isopropanol to recover the adhering resin, and finally distilling the isopropanol from the filter cake wash.

If higher molecular weight solid epoxy ethers are desired, as is the case in surface coatings, lower mole ratios of epichlorohydrin to bisphenol A, e.g., 1.2 to 2.0, are used. Also, less caustic catalyst is required per mole of bisphenol A since this is partially consumed by the condensation reaction (4) which does not destroy the caustic catalyst through formation of NaCl. For these higher molecular weight resins, salt is ordinarily removed by stirring the resins with water that is above the melting point of the resin, then decanting off the salt layer. Washing is repeated with hot water until the resin reaches a neutral pH.

Equipment for preparing base epoxy resin is similar to that used for phenolics and aminoplasts, i.e., batch operations are performed in an anchor-agitated kettle which has a vacuum reflux condenser. The high temperatures (400°F) required for alkyds are not necessary, but corrosion-resistant material is usually used to maintain a good level of color.

Besides bisphenol A, other polyhydroxyl molecules are used for commercial base resins, e.g., glycerol and novalac (phenol/formaldehyde) resins; however, at present bisphenol A is by far the principal polyol. Other epoxy materials which have been used in specific formulations include butadiene dioxide, the dioxide of vinyl cyclohexene, styrene oxide, and an epoxidized Diels-Alder condensation product (see below) of butadiene and crotonaldehyde.

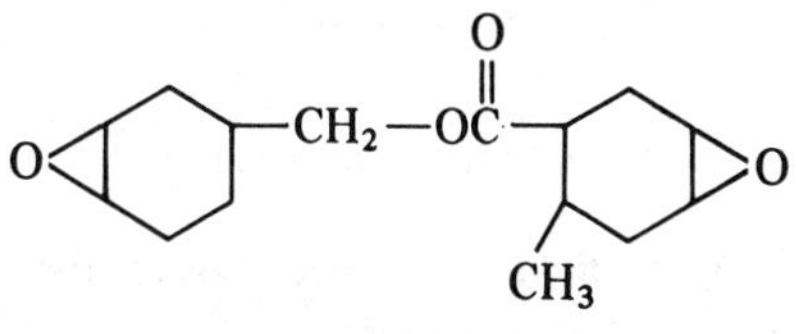

Condensation product

Epoxy groups may be introduced into olefins by means of the peracetic acid-epoxidation process,[75] valuable in manufacturing epoxy plasticizers from unsaturated natural oils (e.g., soybean oil). The epoxy group acts as a heat stabilizer for vinyl chloride resins and also promotes resin compatability:

The base bisepoxy resins are very stable, requiring catalysts and/or coreactants, e.g., amines, organic acid anhydrides, boron trifluoride-tertiary amine complexes, to effect cure.

In the curing reactions no volatile products are evolved; thus shrinkage strains, which cause problems in other thermosetting condensations where volatile products are evolved during the final curing operations, are minimized. The lack of volatiles during cure and the inherent stability of most of the bonds in the cured epoxy, plus the high softening point imparted by the aromatic groups account for many of the desirable properties of epoxy materials.* The ester bond formed with acid anhydride catalysts is the weakest with respect to caustic or water hydrolysis while the heat distortion temperature of aliphatic amine-hardened resins is frequently lower than acid-hardened resins, not because of bond rupture but because of the flexibility of the aliphatic crosslinkages.

The secondary amine first formed when using amine coreactants is even more reactive than the original primary amine, giving an autocatalytic effect to the reaction rate even if temperatures are held constant. In addition, the reaction produces a large amount of heat, on the order of 20-25 kg-cal/mole of condensing epoxide group. For liquid resins with a large number of epoxy groups per unit weight, adiabatic temperature rises of 350°F or more are easily possible in either thick laminates or potting molds. Many amines, particularly those aliphatic amines partially reacted with ethylene oxide or bisphenol A to decrease the dermatitis frequently caused by amines, will react with epoxy groups even at room temperature, giving handling times (i.e., "pot lives") sometimes as

*Although no volatile products result, the catalysts themselves are sometimes volatile. At the high temperatures reached in the exotherm with amine coreactants, many of the amines are quite volatile, causing toxicity and odor as well as foaming. Also, BF_3 is a toxic gas and poses similar problems.

short as 15 minutes. To solve the short pot-life problem, automatic two-fluid metering devices have been developed. Aliphatic amine hardeners used commercially include ethylene diamine, diethylenetriamine, triethylene tetramine, their partially hydroxyethylated or cyanoethylated counterparts, and their glycidyl ether derivatives. Commercially important aromatic amines are *m*-phenylenediamine, 4,4′-methylenedianiline, and 4,4′-diaminodiphenyl sulfone. The aromatic groups increase the softening temperature of the cured resin.

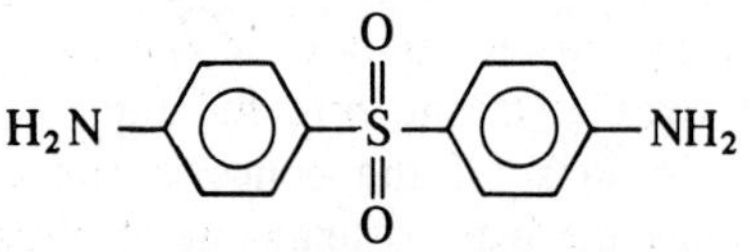

4,4′-Diaminodiphenyl sulfone

The polymerization catalyzed by tertiary amines is believed to be a true chain reaction type of addition polymerization, but the length of the chain is probably very low. The reaction is sensitive, and the presence of alcohols such as the secondary alcohol in the epoxy-base resins containing more than one bisphenol A molecule, or water, though accelerating the reaction, will cause chain transfer: the molecular weight is thereby drastically lowered, and the properties of the cured resin are significantly altered. One commercial tertiary amine, 2,4,6-tris(dimethylaminomethyl)phenol, is commonly used, however. In addition, this polymerization reaction undoubtedly occurs with the tertiary amines that are formed by epoxide condensation with primary and secondary amine-curing agents, thus accounting for greater than stoichiometric yields from many amines, particularly piperidine.

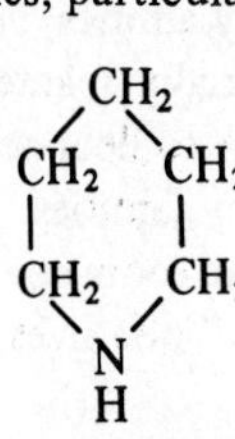

Piperidine

The amine hardeners discussed so far give resins with high softening temperatures but poor impact resistance. To improve impact strength and flexibility, multifunctional amines having long flexible aliphatic chains between the amino groups are used, e.g., hexamethylenediamine and mixed diamines derived from fatty acids. Similar products are the so-called polyamides ("Versamids"), the reaction products of aliphatic polyamines, e.g., diethylenetetramine and the diacid obtained via a Diels-Alder dimerization of unsaturated fatty acids (e.g., linoleic isomers). Another means of imparting flexibility to the cured resin is to include a Thiokol "liquid polymer" which has a structure similar to $HS(CH_2CH_2OCH_2CH_2SS)_nCH_2CH_2OCH_2CH_2SH$. In the presence of amine hardeners, the mercaptan groups will react with epoxy groups incorporating the flexible "rubber" chain in the molecule.

Acid anhydrides react with the epoxy group to form ester linkages. These reactions are only mildly exothermic and do not cause the exotherm problems associated with amine hardeners. Also, the acidic conditions catalyze formation of ethers between the secondary alcohol in (V) and other epoxy groups; therefore less than stoichiometric quantities of acids are frequently used to achieve cure. Ether formation, is it believed, does not occur under the basic conditions existing in amine curing. Pot lives of acid anhydride-catalyzed formulations are very long at room temperatures, but high-temperature treatment or baking ovens are necessary to effect cure (300–400°F *vs.* 150–250°F for amine catalysts). Commercial acid anhydrides are classified as solids, liquids, or chlorinated derivatives. The solids include phthalic anhydride, hexahydrophthalic anhydride, and pyromellitic dianhydride.

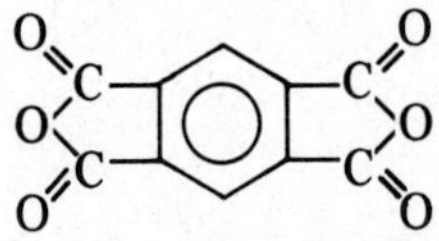

Pyromellitic dianhydride

The liquids include dodecenylsuccinic anhydride and a methylated maleic adduct of phthalic anhydride ("Methyl Maleic Anhydride"). Flame resistance is given by chlorinated anhydrides, such as dichloromaleic anhydride and hexachloroendomethylenetetrahydrophthalic anhdride.

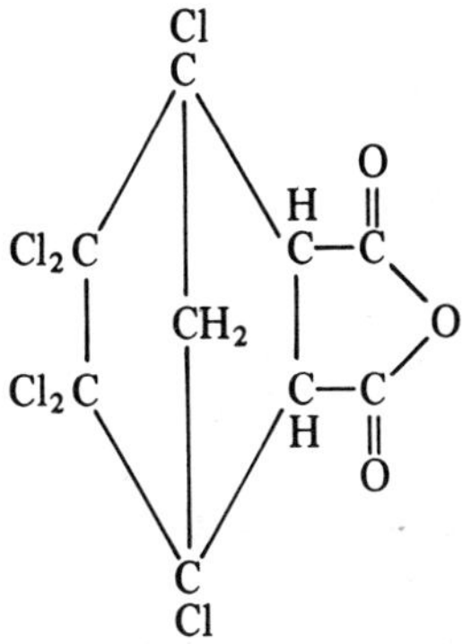

Hexachloroendomethylenetetrahydrophthalic anhydride

The diversity of chemicals used during curing and the complexity of the curing reactions explain why those who use epoxies (in contrast to many other plastics) frequently have an independent chemical laboratory directed by a competent industrial chemist or chemical engineer.

Unsaturated Polyesters. In the plastics industry the term *polyesters* has generally had a far narrower connotation than is implied chemically. In the plastics field a polyester is ordinarily the base resin consisting of a liquid unsaturated polyester plus a vinyl-type monomer; this liquid mixture is capable of reacting to form infusible crosslinked solids under the influence of catalysts and/or heat.

One of the largest uses of polyesters is in fibers for textiles and automobile tires. However, this use will not be discussed in this chapter. In the "plastics" industry, unsaturated polyesters, also called thermosetting polyester resins, are used primarily in rigid laminates, moldings, or castings and are almost always reinforced with glass cloth or glass fibers. They were developed during World War II and were first used very successfully in self-sealing gas tanks containing a rubber liner. When pierced by a bullet, metal tanks splay or "flower" and prevent the rubber tank liner from swelling shut and closing the hole. Polyester laminates do not splay and therefore allow the rubber liner to close the hole. When reinforced with glass, their high flexural strength is outstanding, approaching that of metals. The major markets are the automotive and marine industries. Parts now being molded include fender extensions, rocker panels, and roofs. In housing, the use of polyesters in modular bathrooms has become significant. Approximately 1120 million pounds of polyester resins were produced in the United States in 1970.

Maleic anhydride is the principal unsaturated dibasic acid used in the polyester, although fumaric acid is also used in limited amounts. Uniquely among ordinary unsaturated mono-

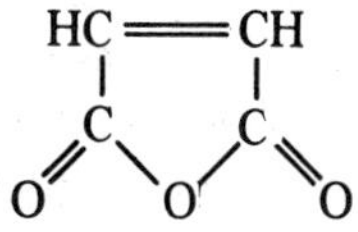

Maleic anhydride

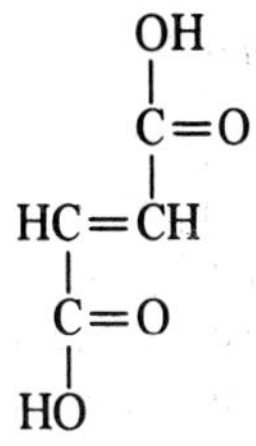

Fumaric acid

mers, the double bond in maleic anhydride, its acid, its esters, and similar α,β-carboxyl-substituted olefins will not undergo homogeneous polymerization even at high temperatures but will copolymerize rapidly with a wide variety of vinyl monomers at even faster rates than these monomers will homopolymerize. It is because of this peculiar property of maleic anhydride that polyesters are prepared by first making (at the necessarily high temperatures) a linear low-viscosity gel-free polyester containing several double bonds per molecule and then mixing this unsaturated polyester with a vinyl type monomer, usually styrene, which, under the influence of catalysts and/or heat, will crosslink the polyester molecules to form a rigid infusible polymer.

The difunctional alcohol ethylene glycol is frequently used for the coreactant, but it is supplemented with propylene glycol, diethylene glycol, or dipropylene glycol to decrease the tendency for the liquid resin to crystallize and to increase the flexibility of the cured resin.

To promote compatibility with the styrene monomer used to crosslink the polyester, phthalic anhydride is incorporated into the polyester backbone in mole ratios with maleic anhydride of from 1:1.5 to 1:1. Besides giving compatibility with styrene, phthalic anhydride imparts some flexibility to the cured resin and lowers the cost. Adipic acid, because of its long flexible aliphatic carbon chain, is used to promote a high degree of flexibility.

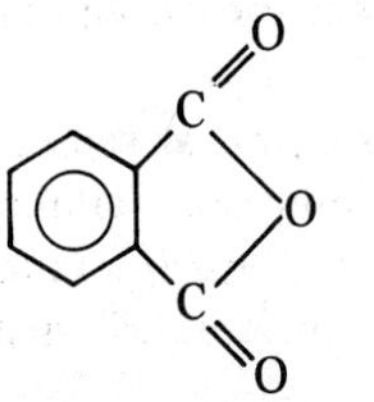

Phthalic anhydride

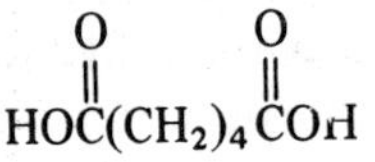

$$HO\overset{O}{\overset{\|}{C}}(CH_2)_4\overset{O}{\overset{\|}{C}}OH$$

Adipic acid

Styrene is by far the most common crosslinking agent and is usually added to the base polyester by the manufacturer of the polyester. To ensure that most of the maleic unsaturated groups are reacted and optimum strength is achieved, an excess of styrene must be used (30-40 percent is the average) because the styrene monomer will polymerize not only with an active maleic radical but also with an active styrene radical; maleic radicals, however, will react only with styrene monomer. Low temperatures favor the maleic-styrene reaction while high temperatures favor the styrene-styrene reaction, accounting for the advantages frequently found in initial low temperature cures.

Besides styrene, many other vinyl monomers may be used to develop special properties, e.g., triallyl cyanurate to promote heat resistance, diallyl phthalate to reduce volatility during cure, acrylic acid ester to promote flexibility via internal plasticization and to impart improved weatherability.

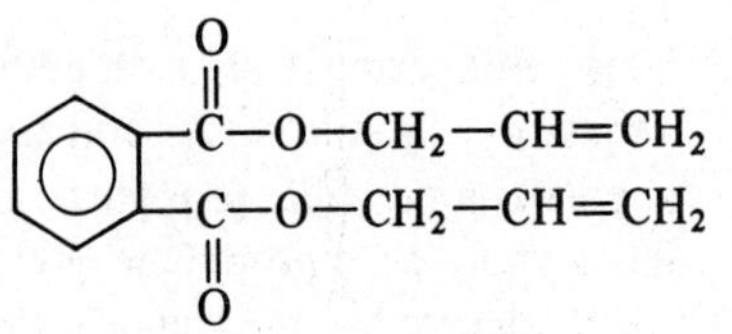

$$C_6H_4(\overset{O}{\overset{\|}{C}}-O-CH_2-CH=CH_2)_2$$

Diallyl phthalate

$$CH_2=\overset{H}{C}-\overset{O}{\overset{\|}{C}}-OR$$

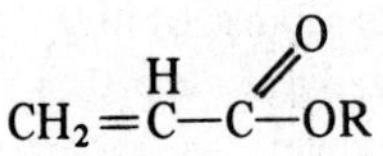

Acrylic acid ester

In addition to the difunctional acids and alcohols, monofunctional acids and alcohols may be added in small amounts to limit polyester molecular weight. Allyl alcohol has been used to give additional unsaturation for subsequent crosslinking.

The properties of polyesters depend on the nature of the reactants. High levels of crosslinking and high percentages of aromatic rings promote hardness, rigidity, strength, and brittleness. Lower levels of crosslinking (less maleic anhydride) and long flexible aliphatic chains impart flexibility and some impact strength. Larger polyester molecules promote some strength although this effect has a definite upper limit which is reached at a moderate molecular weight.

The above brief picture illustrates how complicated and varied the formulation of polyesters can be. As an example[95] the preparation of a particular polyester might be carried out in a jacketed stainless steel reaction vessel equipped with an agitator, and a reflux condenser followed by a total condenser. The charge is 5 moles of maleic anhydride, 3 moles of phthalic anhydride, 4 moles of ethylene glycol, and 4 moles of diethylene glycol. The temperature is raised to 375°C until an acid number* of 60-65 is obtained, while maintaining an inert atmosphere. The inert purge gas (CO_2 or N_2) is introduced through a submerged sparger. In addition to excluding oxygen, which discolors the resin, the purge gas aids in removing the water formed during the condensation reaction. The temperature is then raised and held at 440°C until the acid number is 45-50 or until a Gardner-Holt† viscosity of N to Q is

*Acid number is the number of milligrams of potassium hydroxide necessary to neutralize the free acid in one gram of sample. Since each free carboxyl group represents, on the average, one molecule, the number average molecular weight of the resin, Mn, is given by the following:[96]

$$Mn = \frac{1000 \times 56}{\text{Acid Number}}$$

†Gardner-Holt viscosities are widely used in the surface-coating industry and are obtained by comparing the time it takes for a standard bubble of air to rise in a sample tube with the time it takes for a similar bubble to rise in a set of standard tubes identified by letters. N to Q Gardner-Holt corresponds roughly to 3.40 to 4.35 poise at 77°F.

reached when 100 parts of polymer are combined with 50 parts of styrene. When the reaction is completed (about 5 hours total time), the reaction mass is cooled to about 150°C and mixed with 6.5 moles of styrene containing *p-t*-butylcatechol inhibitor (0.02 percent of final mixture); the temperature during mixing is maintained at about 40°C. The mixture is cooled to 20°C and pumped to drums or tank cars.

Many catalysts, most of which are free radical generators, have been proposed to cure the resins. Benzoyl peroxide, mixed with 50 percent dibutyl phthalate to facilitate mixing with the resin, is most frequently used for hot cures (those above 50°C) at levels of 1 to 2 percent. Although methyl ethyl ketone peroxide, as a 60 percent solution in dimethyl phthalate, is sometimes used for lower temperature cures, particularly when combined with paint "driers" such as cobalt naphthenate. Sometimes combinations of these lower temperature catalysts with a much higher temperature catalyst, e.g., di-*t*-butyl peroxide (activity beginning at about 115°C) or dicumyl peroxide (activity beginning at about 105°C), may be used for composite cures leading to improved properties.

$$(CH_3)_3—C—O—O—C—(CH_3)_3$$

Di-*t*-butyl peroxide

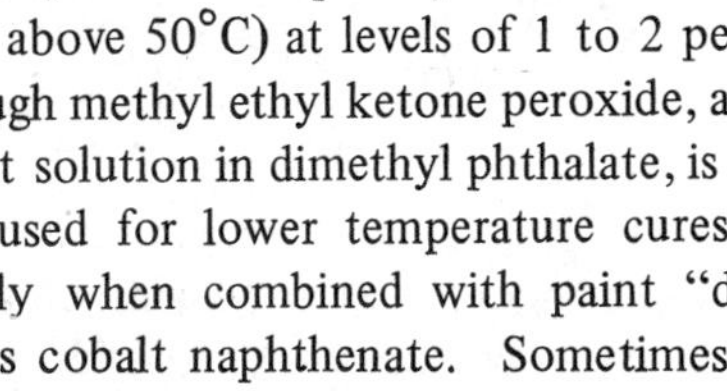

$$C_6H_5—C(CH_3)_2—O—O—C(CH_3)_2—C_6H_5$$

Dicumyl peroxide

Promoters (amines) are sometimes used to lower the temperatures at which the catalysts become effective. Many other catalyst systems are used as well, for example, the ozonides in styrenated polymers, and organic azo compounds and azines as crosslinking catalysts.

In addition to the action of free radicals, some cure is possible by further condensing the polyester itself, using acids or bases which will catalyze esterification, e.g., calcium hydroxide, barium hydroxide, *p*-toluenesulfonic acid.

The properties, such as filament size, type of weave, etc., of the laminating material (almost always glass fibers or glass cloth) are as important to the final properties of the cured laminate or casting as the resin itself.[97,98] To ensure a good bond between resin and glass, the starch lubricants, necessary during weaving, must be burned off; adhesive coatings are also normally used.

Polyesters, thermoplastics. In the textile industry (see Chapter 11), polyester fibers made from ethylene glycol and terephthalic acid (PET) are the largest volume synthetic fiber. Until recently, nontextile usage was limited to films (Mylar®). Soft-drink bottles are now also being manufactured in large volume. Also, very recently, grades of polyester have been made which will crystallize with sufficient rapidity to give satisfactory injection moldings. However, most moldings are made using butylene glycol which imparts sufficient mobility to the polymer chain to cause more rapid crystallization during rapid injection molding cycles. Usage of these materials has grown rapidly with the recent soft-drink bottle development plus the introduction of injection moldable types; usage in 1979 was 816 million pounds in the U.S.

Polyphenylene Oxide Resins. In 1959, Hay and co-workers[99,100] reported that certain 2,6-di-substituted phenols could be oxidatively polymerized using cuprous ion as a catalyst to give aromatic polyethers (PPO)

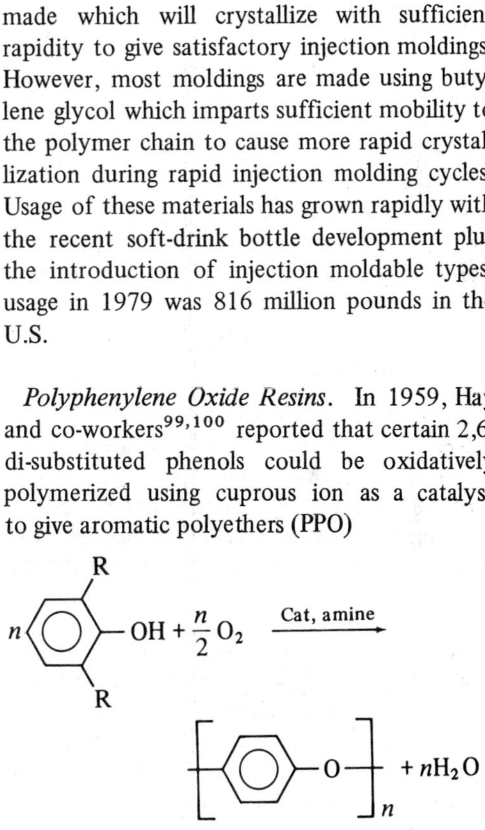

$$n\ \text{(2,6-R}_2\text{C}_6\text{H}_3\text{)}—OH + \frac{n}{2}O_2 \xrightarrow{\text{Cat, amine}} [—C_6H_4—O—]_n + nH_2O$$

In commercial PPO, R is a methyl group. The product has an exceptionally high softening point, about 210°C, higher than that of most other commercially available thermoplastics. (See polysulfones in the following pages.) It is hard, tough, modestly transparent and self-extinguishing. As a pure material, it is very difficult to mold, having a very high viscosity at temperatures below its decompositon temperature. However, it was discovered that polystyrene forms a true blend at the molecular level giving a great degree of plasticization and making fabrication commercially feasible. However, the polystyrene lowers the heat distortion temperature and lessens the self-extinguishing character unless other fire retardant additives are used. If rubber-modified polystyrenes are used, toughness is not harmed. Nearly all commercial products use substantial (up to 50 percent) amounts of rubber-modified polystyrene, thereby lowering the cost as well as improving the ease of fabrication.

A typical laboratory polymerization process has been described:

> Charge 200 ml of nitrobenzene, 70 ml of pyridine, 1 gm of cuprous chloride to a vigorously agitated glass flask. Oxygen is bubbled in at 300 ml/min followed by 15 g of 2,6-dimethylphenol. The reaction is continued for about 30 minutes maintaining the temperature at about 30°C. After dilution of the polymeric solution with 100 ml of chloroform, the solution is precipitated in 1.1 liters of methanol containing 3 ml of concentrated hydrochloric acid. Filtering, methanol washing and/or redissolving in chloroform followed by reprecipitation in methanol are used to separate the pyridine and catalyst. The yield was about 91%.

The details of commercial practice are not known.[100] General Electric is the sole manufacturer, using Noryl® as their trademark.

Polysulfone Resins. In the early 1960's, chemists at Union Carbide developed a new type of polymer called "polysulfones" containing both ether linkages and sulfoxide linkages between phenylene groups. They are prepared by reacting bisphenol A with di-*p*-chlorophenyl sulfoxide in the presence of exactly two moles of sodium hydroxide per mole of bisphenol A. A dipolar aprotic solvent, completely free of water, e.g., dimethyl sulfoxide, is required.

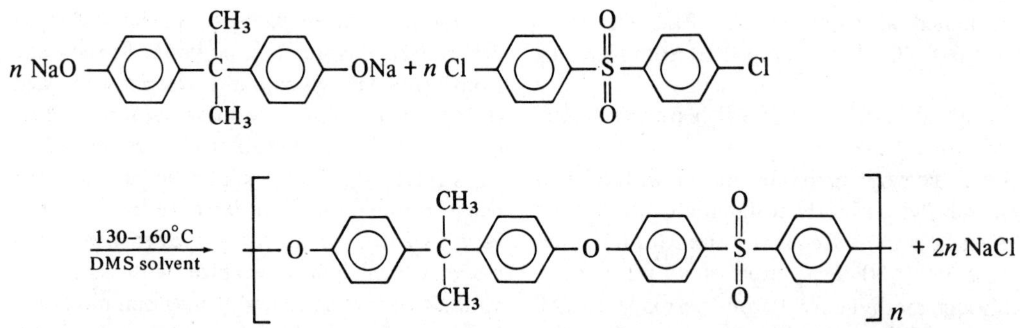

$$n\,NaO-C_6H_4-C(CH_3)_2-C_6H_4-ONa + n\,Cl-C_6H_4-SO_2-C_6H_4-Cl \xrightarrow[\text{DMS solvent}]{130\text{–}160^\circ C} \left[-O-C_6H_4-C(CH_3)_2-C_6H_4-O-C_6H_4-SO_2-C_6H_4-\right]_n + 2n\,NaCl$$

Molecular-weight control is achieved by the addition of a monofunctional chain terminator, e.g., sodium phenate or methyl chloride.

The structure of these materials is similar to the pure polyphenylene oxides (PPO); hence, their properties are similar. The color and transparency is somewhat inferior to PPO, but the heat resistance is claimed to be slightly superior.[101]

Polychloromethylether (a polyoxetane). Hercules Chemical Co. has commercialized a poly(3,3-bis(chloromethyl)oxetane) under the trade name of "Penton."

$$H-\left[-O-CH_2-\underset{CH_2}{\overset{CH_2Cl}{C}}-CH_2-\right]-OH$$

The oxetane monomer is synthesized from pentaerythritol by esterification with acetic acid (or anhydride) followed by chlorination with dry hydrochloric acid at 200°C in the presence of zinc chloride catalyst. Treatment with sodium hydroxide cyclizes the trichloroacetate to the oxetane monomer.

$$(HOCH_2)_4C + 4CH_3COOH \longrightarrow C(CH_2OCCH_3)_4 + 4H_2O \quad (C{=}O)$$

$$C(CH_2O\overset{O}{\overset{\|}{C}}CH_3)_4 + 3HCl \xrightarrow[ZnCl_2]{200^\circ C}$$

$$(ClCH_2)_3C(CH_2O\overset{\overset{O}{\|}}{C}CH_3) + 3CH_3COOH$$

$$(ClCH_2)_3C(CH_2O\overset{\overset{O}{\|}}{C}CH_3) + 2NaOH \xrightarrow[25°C]{H_2O}$$

$$ClCH_2-\underset{\underset{CH_2-O}{|}}{\overset{\overset{CH_2Cl}{|}}{C}}-CH_2 + NaCl + NaO\overset{\overset{O}{\|}}{C}CH_3 + H_2O$$

To polymerize the oxetane monomer, a rapid continuous mass process has been developed using a special trialkylaluminum catalyst (plus promoters); the heat of reaction raises the melt to the exit temperature of about 200°C. To control temperature, viscosity, and molecular weight, catalyst levels are kept very low and conversions are about 80 percent.[102]

Other oxetanes can be polymerized, but this bischloromethyl derivative is the only one commercialized to date.

The principal uses are for fabricating corrosion-resistant equipment, particularly where the Penton is bonded to metals. Such composite fabrication is simpler for Penton than for the fluorinated polymers, Teflon or Kel-F.

POLYOLEFIN RESINS

Polyolefins, particularly polyethylene and polypropylene, are by far the largest selling class of synthetic plastics. Domestic sales volume in 1979 was 12.8 billion pounds of polyethylene and 3.8 billion pounds of polypropylene.[103] They have enjoyed rapid sales growth; since 1971, sales have averaged about 9%/year for polyethylene and 15%/year for polypropylene. By contrast, 1979 domestic sales volume of polyvinyl chloride, the next largest selling resin group, was about 6.1 billion pounds. Polyolefin sales in 1980 amounted to about 84% of nominal installed plant capacity.[104]

The huge polyolefins sales reflect their broad useful property spectrum and ease of fabrication, combined with a favorable cost structure. The latter results from a relatively favorable raw material supply and from highly evolved continuous polymerization processes. Major markets include blow-molded bottles, extruded pipe, films, coatings, fibers and filaments, and automotive applications.

Polyethylene

There are dozens of grades of polyethylene, each intended for a different set of end uses. However, two principal categories have been generally recognized: low-density polyethylene (LDPE) and high-density polyethylene (HDPE).

The density of LDPE varies roughly from .91 to .94 g/cm^3, while HDPE varies from .95 to .97 g/cm^3. LDPE is generally characterized by molecular chains with longer and more frequent branching. It also has lower crystallinity than HDPE. As polyethylene density increases, impact strength and tensile elongation at fail tend to drop off, while tensile strength, hardness, flexural modulus, and heat distortion temperature increase.

High Pressure Polyethylene Processes. LDPE was the first of these polyolefin resins produced commercially, having been developed in England during the 1930's.[105] Until the late 1970's, it was produced almost exclusively via free radical initiators in continuous processes operating at 15,000 to 50,000 lb./in^2 and temperatures of 150 to 250°C. Without the very high pressures, the polymers formed would have excessively low molecular weight. Molecular weight is largely determined by the rate of chain growth divided by the rate of chain termination. In the case of polyethylene free radicals, the chain termination step is very rapid. To obtain sufficiently high molecular weight, the relative chain growth rate must be increased. This calls for high temperatures and high ethylene concentrations, the combination of which results in the need for very high pressure.

Polymerization is initiated by molecular oxygen or other free radical initiators such as peroxides and azo compounds, preferably those with a 1-minute half-life in the temperature range of 110–210°C.[106] The high temperatures and the highly reactive nature of the polyethylene free radical result in both intramolecular and intermolecular chain transfer reactions. Short chain branches are produced by the former during propagation by a "backbiting" mechanism where the end of the chain coils backward, extracting a hydrogen from the fifth carbon back in the chain. Chain propagation then continues from this relocated radical site. Intermolecular chain transfer is largely carried out to adjust molecular weight and generally involves the addition of small amounts of compounds such as branched alkanes, olefins, as well as hydrogen. Chain transfer to butane, for example, is illustrated below.

$$R{-}(CH_2CH_2)_n{-}CH_2CH_2\cdot + CH_3CH_2CH_2CH_3 \longrightarrow R{-}(CH_2CH_2)_n{-}CH_2CH_3 + CH_3\dot{C}HCH_2CH_3$$

The secondary carbons of the butane chain transfer agent are more subject to hydrogen abstraction. An isobutyl free radical is therefore formed from which chain propagation then starts. Chain transfer can also occur with other polymer molecules, resulting in long-chain branching. Termination of the growing chains can occur by combination of two free radicals or by a disproportionation reaction. With disproportionation there is a transfer of a hydrogen radical between the two reacting long-chain free radicals producing two polymer molecules: one with a saturated end group and the other with an unsaturated end group.

The initiator is metered into the compressed ethylene stream just before it enters the reactor, which may be either a tubular type or an autoclave. In the former case, the feed stream containing initiator is preheated to 100-200°C where polymerization begins. The tubular reactor consists of jacketed heavy-walled pipe, up to a mile in length, and designed with carefully configured bends to maintain desired flow patterns. Ethylene polymerization is highly exothermic (25.4 kcal/mole at 25°C.), and a large fraction of the generated heat is removed via heat transfer oil or water circulating in the jacket. The heat which is not removed via the jacket raises the temperature of the reactant stream which could reach 300°C or more. In some processes, additional ethylene and initiator may be injected at points along the reactor length to improve reaction rates. The reactor may also be pressure-pulsed at approximately 1-minute intervals to help remove wall deposits. The reactor discharge will contain from 8 to 20% polymer, depending to a degree on the extent of the multiple additions.

Autoclaves are continuous flow, stirred tank vessels. Depending on design, a broad range of backmixing can be achieved, with some approaching the near-plug flow of tubular reactors. Although heat transfer surfaces may be provided to remove a portion of the heat of reaction, these reactors are frequently operated almost adiabatically; the heat generated being nearly balanced by the sensible heat required to bring the cool feed stream to reactor temperature. The ranges of average reactor residence time and conversions are 25 to 40 seconds and 10 to 18% respectively.[107] The reactant mixture is viscous, and effective agitation is required to promote temperature uniformity, and to minimize dead spots. Mixing may be varied to provide different residence time distributions, permitting a degree of control of polymer properties principally via the molecular weight distribution.

Termination Reactions

$$2R{-}CH_2{-}CH_2\cdot \xrightarrow{\text{Combination}} R{-}CH_2{-}CH_2{-}CH_2{-}CH_2{-}R$$

$$2R{-}CH_2{-}CH_2\cdot \xrightarrow{\text{Disproportionation}} R{-}CH_2{-}CH_3 + R{-}CH{=}CH_2$$

High Pressure Process Details. A flow diagram of a commercial high pressure polyethylene process is shown in Fig. 10.20. Ethylene, along with chain transfer agents, enters the process via a primary compressor (not shown) which performs the first step in building up the pressure of the reactants. The stream then passes to a large compressor which raises the stream pressure to the 15,000-50,000 lb/in^2 range. Higher pressures will generally result in higher product density, principally because the "backbiting" reaction referred to earlier becomes less favored and fewer long chain branches are formed. However, since the chain propagation rate tends to increase with higher pressure, molecular weight will also increase.

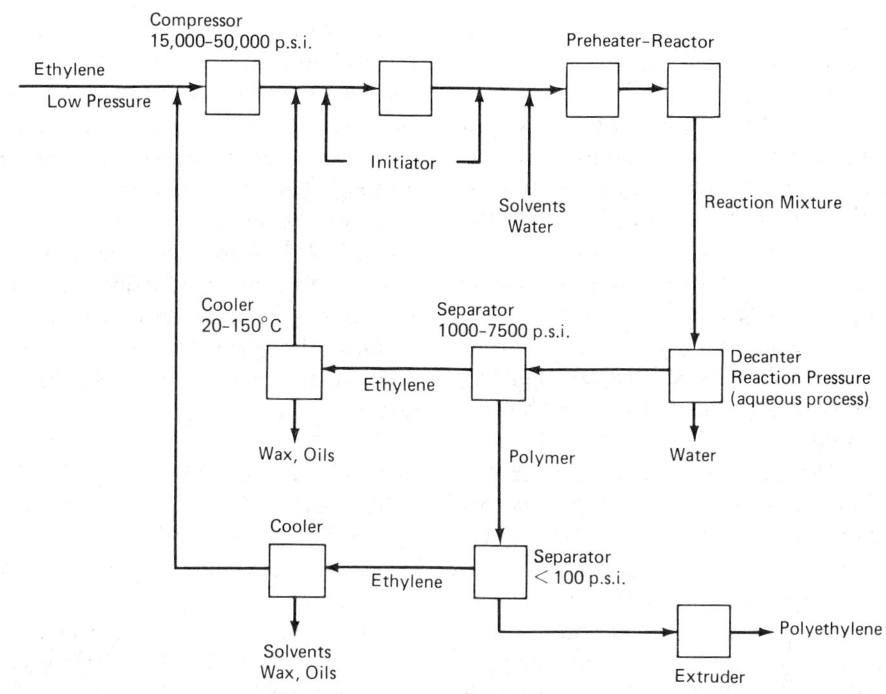

Fig. 10.20 Generalized flow diagram for high-pressure polyethylene processes. (*"Crystalline Olefin Polymers," part 1, R. A. V. Raff and K. W. Doaks, eds., Interscience Publishers, 1965; by permission of John Wiley & Sons, Inc.*)

The discharge from the reactor passes through separators (generally two in series) to remove the polymer from other components. Although a single separator could be used, lowering the pressure directly to atmospheric, this would be very wasteful of compressor energy, a major item of production cost. By putting in a separator stage operating at 1000–7500 lb/in^2, considerable savings in recompression costs of the recycled ethylene are possible. If the pressure is too high in the first separation, considerable low molecular weight polyethylene will be carried over, dissolved in the recycle ethylene, causing recompression problems. Molten polyethylene is drawn off from the final low-pressure separator via gear pumps, screw pumps, or extruders. Further blending of the base resin with lubricants, stabilizers, colorants, etc. may then be made using extruders or Banbury mixers followed by pelletization.

The very high pressure requirements of these processes have resulted in many challenging design problems involving materials of construction, seals, pumps, and reactors. Safety barriers must be provided. Explosive decomposition of ethylene is thermodynamically favored and may be encountered at reactor conditions. Safety disc and vent lines must be designed to handle the resultant pressure surge. Hot spots and locally high initiator concentrations are among the conditions which promote the problem. These and others can be minimized via proper design and operating procedures.

Low Pressure Polyethylene Processes. Starting around the mid-1950's, ethylene polymerization processes were developed which were not based on free radical initiators and could be generated at far lower pressures: usually 450 lbs/in^2 or less. Initially, these "low pressure" processes produced HDPE because little or no branching occurred along the polymer chains. During the 1970's, however, these processes evolved to the point where it was possible to make a broad range of product densities down to those corresponding to LDPE.

The low pressure processes employ heterogeneous catalyst systems of which there are two principal classes: a) transition metal oxide catalysts and b) Ziegler-Natta catalysts.

The metal oxide catalysts are usually based on chromium or molybdenum with silica or silica-alumina supports. These were first commercialized for polyethylene by Phillips Petroleum following the discovery by Hogan in 1951 of the chromium oxide catalyst system.[108,109] There is a broad patent literature covering the preparation and use of these catalysts. Chromium-based catalysts can be prepared by impregnating the support with an aqueous solution of CrO_3.[110] Following drying, the finally divided catalyst is normally activated by fluidizing in air for several hours at 500 to 1000°C. It is then stored in dry air or nitrogen. The chromium content of the catalyst is generally in the range of 1 to 5% and predominantly in the highest oxidation state (hexavalent). Supported molybdenum oxide catalysts have been used in a process developed by the Standard Oil Company of Indiana.[107]

Ziegler-Natta catalysts were first developed during the early 1950's through the discoveries of K. Ziegler in Germany[111] and G. Natta[112,113] in Italy. This work led to the award of the Nobel Prize in chemistry to these men in 1963. These catalysts are formed by the combination of a reducible compound of the transition elements of Group IV to VII with a reducing organometallic compound of metals of Groups I and III. Typically, combinations of aluminum alkyls and aluminum alkyl halides are contacted with the halides of Ti (III) and Ti (IV) to form the catalysts. (The titanium halide component is often referred to as the catalyst and the aluminum compound as the cocatalyst.) Compositional variables, the purity of the inert diluent in which the ingredients are mixed, mixing temperature, and agitation intensity are among the parameters which must be controlled with exceptional care during catalyst preparation to achieve desired operating conditions, polymer properties, and yields. Vanadium halides are also used to form Ziegler-Natta catalysts.

With both the transition metal oxide and the Ziegler-Natta catalysts, extreme care must be exercised in manufacturing operations to remove or exclude O_2, water, polar materials, and other impurities which could alter their activity. Although both catalyst types have been intensively studied and an extensive patent literature exists,[107,114] there is not yet complete agreement on their mode of action, and a considerable amount of proprietary art exists with their commercial preparation and use.

Transition metal oxide and Ziegler-Natta catalysts share three features:

a) Solid surface with the ability to adsorb monomer
b) Transition metals that can exist in several valance states, easily converted from one to another
c) Ability to form organometallic compounds by reaction with other organometallics or with monomer

Operating temperatures and pressures for processes using Ziegler-Natta catalysts are generally lower than those employing transition metal oxides. With the latter, however, chromium-based catalysts permit lower pressures than those based on molybdenum.[107]

During the 1970's, other chromium compounds, such as bis-triphenylsilyl chromate and chromocene deposited on high surface area silica, were disclosed by Union Carbide as high productivity ethylene polymerization catalysts.[115,116]

In the case of Ziegler-Natta catalysts, the polymerization is believed to be initiated at active centers on the surface of the finely divided transition metal halide. One of many proposed mechanisms where titanium is used postulates the following steps when a dissolved ethylene molecule encounters an active site of octahedral structure.[117] R may be an ethyl group–C_2H_5, or an alkyl chain–$C_{2n}H_{4n+1}$.

$$\mathrm{Cl{-}Ti(R)(Cl)(Cl)(Cl){-}\!\!-} + CH_2{=}CH_2 \xrightarrow{I} \mathrm{(R)(Cl)(Cl)(Cl)Ti{-}\!\!-} \cdots \underset{CH_2}{\overset{CH_2}{\|}} \xrightarrow{II} \mathrm{Cl{-}Ti(R)(Cl)(Cl){-}\!\!-}\underset{CH_2}{\overset{CH_2}{\|}} \xrightarrow{III} \mathrm{Cl{-}Ti(CH_2CH_2R)(Cl)(Cl)(Cl){-}\!\!-}$$

In Step I, the ethylene molecule is coordinated into the vacant position in the octahedral structure by means of π-bonds. When this π-bond forms, the transitional Ti-C bond (Step II) beccomes more susceptible to cleavage (Step III). Following Step III, there is once again a vacant octahedral position which permits another ethylene molecule to react and add to the R-tail already formed. In principle, monomer addition can continue until a polar compound, either a contaminant or purposely added, reacts irreversibly at the active octahedral site. Another mechanism[114] postulates that the monomer unit is inserted into alkyl bridges linking titanium and aluminum atoms in a bimetallic electron-deficient complex.

The kinetic schemes may be represented with the simplifying assumption that monomer is added to an organometallic compound with an active center. Chain growth then proceeds via addition of monomer units via an anionic mechanism. Thus:

Initiation

$$\overset{(+)}{\text{Cat}} \cdot \overset{(-)}{\text{R}} + CH_2{=}CH_2 \longrightarrow \overset{(+)}{\text{Cat}} \cdot \overset{(-)}{CH_2}{-}CH_2{-}R$$

Propagation

$$\overset{(+)}{\text{Cat}} \cdot \overset{(-)}{CH_2}CH_2(CH_2{-}CH_2)_n{-}R + CH_2{=}CH_2 \longrightarrow \overset{(+)}{\text{Cat}} \cdot \overset{(-)}{CH_2}CH_2(CH_2{-}CH_2)_{n+1}{-}R$$

Several chain termination reactions can occur besides the type cited above where the active site is destroyed. These include:

Transfer with Hydrogen

$$\overset{(+)}{\text{Cat}} \cdot \overset{(-)}{CH_2}{-}CH_2{-}R + H_2 \longrightarrow \overset{(+)}{\text{Cat}} \cdot \overset{(-)}{H} + CH_3{-}CH_2{-}R$$

Hydride Ion Transfer

$$\overset{(+)}{\text{Cat}} \cdot \overset{(-)}{CH_2}{-}CH_2{-}R \longrightarrow \overset{(+)}{\text{Cat}} \cdot \overset{(-)}{H} + CH_2{=}CH{-}R$$

$$\overset{(+)}{\text{Cat}} \cdot \overset{(-)}{H} + CH_2{=}CH_2 \longrightarrow \overset{(+)}{\text{Cat}} \cdot \overset{(-)}{CH_2}{-}CH_3$$

Transfer by Metal Alkyls

$$\overset{(+)}{\text{Cat}} \cdot \overset{(-)}{CH_2}CH_2{-}R + ZnR'_2 \longrightarrow \overset{(+)}{\text{Cat}} \cdot \overset{(-)}{R'} + R'ZnCH_2CH_2{-}R$$

During the 1970's, Ziegler-Natta catalysts supported on carriers such as magnesium compounds were introduced[118] which had far higher productivity than those used previously. In conventional Ziegler-Natta catalysts, a large fraction of the potentially active polymerization sites remain unused since they are buried deep within the particles. When placed on a proper support, however, the metal atoms are isolated, providing more available sites for polymerization. As a result of these and other improvements, polymer yields per unit weight of catalyst have been increased to the point where the costly step of removing catalyst or catalyst residues from the product is generally no longer necessary. Other developments have significantly improved catalyst activity, and selectivity, as well as control of molecular weight parameters and polymer chain branching.

In low pressure polyethylene processes, it is common practice to add small quantities (generally below 10%) of olefins as comonomers, especially to improve environmental-stress crack resistance and impact strength. Butene-1 and hexene-1 are frequently used.[107] Short side chains form which result in improved low temperature toughness, while product density and stiffness are reduced.

A large number of low pressure polyethylene processes have been commercialized. However, for purposes of illustration we will discuss two: the Phillips Particle Form Process and the Union Carbide Unipol Process.

Phillips Particle-Form Process. This process produces polyethylene as a slurry in a hydrocarbon diluent. It is a second generation process; the earlier version, operating at higher temperature, formed polyethylene as a solution in a hydrocarbon solvent. A simplified schematic of the slurry process is shown in Fig. 10.21.

The catalyst is a chromium oxide type supported on silica gel, similar to that described earlier, and suspended in the purified n-pentane diluent. The streams of catalyst suspension, purified ethylene, and comonomer (typically butene-1) are fed to the loop reactor, operating typically at around 90-100°C and pressures up

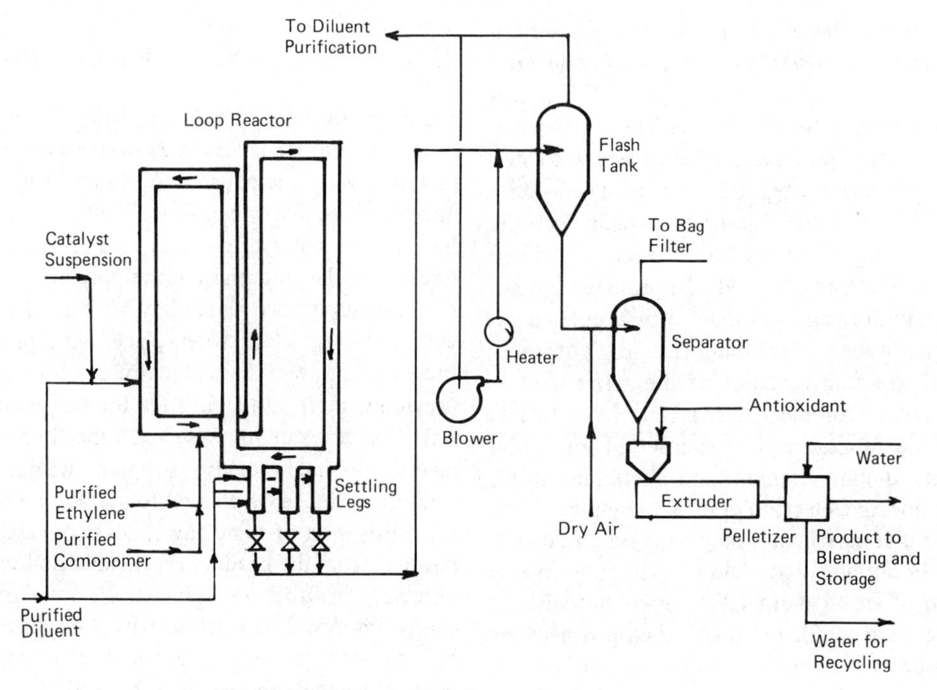

Fig. 10.21 Phillips particle form polyethylene process.

to 425-450 psig.[107] The reactor impeller develops a rapid circulation of the resulting slurry of polymer and catalyst in the monomer-diluent solution. The reaction temperature plays an important role in controlling polymer molecular weight, and pressure is set to insure sufficient dissolving of ethylene in the liquid phase at the slurry temperature.

The reactor is jacketed for circulating coolant liquid to remove the heat of polymerization generated. Very high space-time yields for this design have been claimed.[119] Tubing diameters of 20 inches are reported for commercial units[107,120,121] with vertical sections each about 50 feet long. The equivalent total tubing length is estimated to be up to 300 ft. with an effective volume of about 600 ft.^{3} Linear velocities of 10 to 30 ft./sec. are recommended so as to prevent polymer solids from depositing on the reactor walls. Little polymer dissolves in the diluent, and with the polyethylene solids in the reactor slurry normally maintained at 18 to 25%, the slurry viscosity is kept low. The flow regime is therefore turbulent which results in good heat transfer.

Screw conveyors are used along with the settling legs shown to withdraw concentrated slurries containing 50 to 80% solids from the reactor.[107] These techniques minimize polymer plugging in the exit lines and reduce the amount of monomer-diluent solution withdrawn. If necessary, fresh diluent can be added as shown to the concentrated slurry; this promotes its discharge to the flash tank which is maintained at 1 psig. to boil off the pentane. Heat for the flashing is supplied by recirculating and heating a portion of the discharged pentane vapor and mixing this vapor with the inlet stream to the flash tank. The main stream of pentane vapor is recycled to the diluent purification area (not shown) where it is compressed and distilled to remove the small amount of very low molecular weight polymers formed in the reactor.

The polyethylene solids discharged pneumatically from the flash tank pass to a separator and then to an extruder along with an antioxidant stream. The molten extrudate, with the antioxidant thoroughly blended, then passes to a pelletizer, from which the product is then pneumatically conveyed to silos for further blending and storage prior to shipment.

Union Carbide "Unipol" Process. This gas phase process was first commercialized in 1978 following a development period starting in the 1960's.[122] Since no diluent is used, the costly steps for purifying and recycling this component (as illustrated with the Phillips process discussed above) are eliminated, further simplifying the overall process. The process is capable of providing a broad range of product densities via use of comonomers, although it was first commercialized to produce LDPE. Since the backbone of the polymer chain is linear, compared to the highly branched conventional LDPE, the material has become known as LLDPE (linear low density polyethylene). This resin has better strength and toughness than conventional LDPE, especially in film applications. Other producers such as Dow Chemical and Exxon Chemical have since commercialized their own LLDPE processes.

A key to the Unipol process is a proprietary catalyst (possibly related to Carbide's high-yield chromium-based catalysts discussed earlier) which allows reactor operation at 100°C and pressures in the range of 100-300 lbs/in.^{2} These relatively mild conditions, compared to the older LDPE processes, have led to significant reductions in plant capital. Energy consumption, moreover, was reduced by 75% and operating costs by 50%. A flowchart for the process is shown in Fig. 10.22.

A stream of ethylene and a comonomer (such as butene-1, pentene-1 or hexene-1) is fed continuously to a fluid bed reactor. The monomers are pretreated to remove impurities which could poison the catalyst (which is added separately). The comonomer, generally 4 to 10% of the feed, is added in order to lower the polymer density by introducing short branches on the chain. The fluid bed in the reactor is made up of granular polyethylene, formed by the polymerization. The ethylene-comonomer gas stream circulates up through the fluid bed and exits the reactor through an enlarged top section designed to disengage most of the fine particles. The gas then passes to a cycle compressor and external cooler to remove the exothermic heat of polymerization before returning to the reactor. The average polymer residence time in the fluid bed is 3 to 5 hours. The exiting particles have an average size of about 1 mm. Since the fluid bed is continuously growing, the product is removed intermittently through a gaslock chamber so as to maintain an approximately constant bed volume. The small amount of residual monomer which

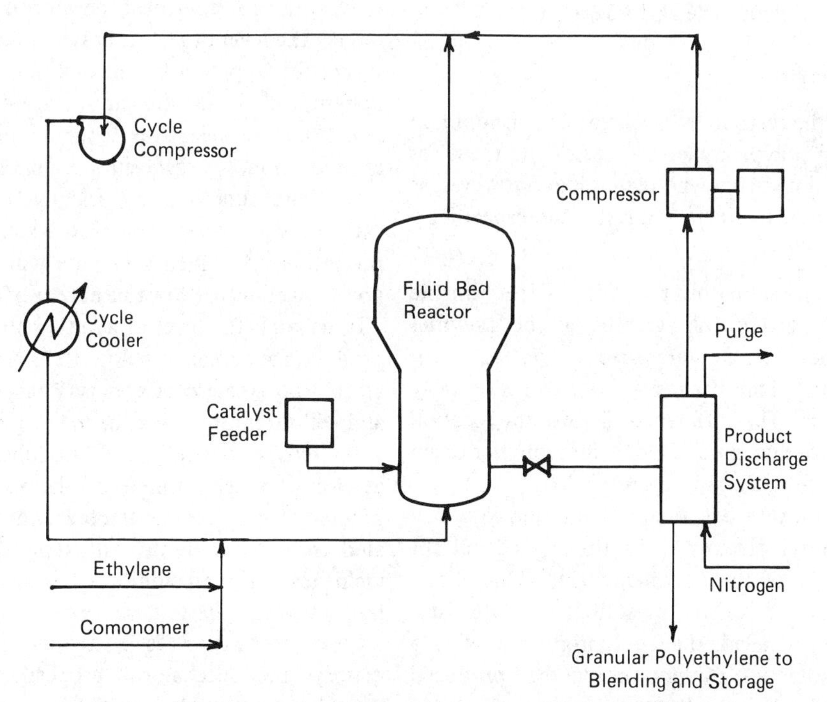

Fig. 10.22 Union Carbide's "Unipol" process for LLDPE.[119]

accompanies the granular free-flowing product into this chamber is easily purged so that the product can be safely conveyed by air. Catalyst productivity is so high that it is not considered necessary to remove catalyst residues from the product. Similarly, the level of low molecular weight impurities is too small to warrant removal. One or more additives, such as stabilizers, antislipping agents, antioxidants, etc., may be blended into the granular product before it is stored or shipped. There is no need to pelletize the product as practiced with the older processes. Product properties are set primarily by catalyst composition. Reactor conditions are reportedly unchanged, regardless of the product being made.

Bedsides development of the highly productive catalyst, it was necessary to devise methods to avoid particle agglomeration in the fluidized bed, and to achieve a desirable particle size, shape, and bulk density. The solution to these problems involved producing a catalyst with uniform activity (to avoid excessively hot particle surface temperatures that would lead to agglomeration), and defining reactor conditions which would assure uniform heat removal and minimize production of oligomers (which also promote particle agglomeration.)

Polypropylene

As shall be seen, there is a great deal in common between polypropylene processes and those for HDPE. Polypropylene technology, however, is complicated by the fact that the stereospecificity of the repeating unit, $-CH_2-\overset{\overset{\displaystyle CH_3}{|}}{C}H-$, in the polymer chain will greatly affect polymer properties. Three stereospecific configurations were first identified by Natta in the early 1950's.[114] The differences involve the location of the methyl groups along the polymer chain and are illustrated in Figure 10.23.

Isotactic refers to the configuration in which the methyl groups, $-R$, are aligned on the same side of the polymer chain. This allows the polymer to be crystalline. Natta discovered in 1954 that polypropylene with a high isotactic content could be produced using the Ziegler-Natta catalysts discussed earlier. Commercial polypropylene is based on the isotactic configuration.

In the *syndiotactic* arrangement, the methyl groups are aligned alternately on opposite sides of the chain. Syndiotactic polypropylene has been prepared in the laboratory,[123] but no commercial interest in it has yet developed.

In the third configuration shown in Figure 10.23, the methyl groups are randomly arranged along the chain. This random arrangement, called *atactic*, prevents crystallization, and the polymer has a low melting point as well as poor solvent resistance. For these reasons there is only limited commercial use for pure atactic polypropylene. Polymerizations based on free radical initiators are not suitable for making polypropylene commercially since the product is atactic and of low molecular weight.

Polypropylene processes are generally based on Ziegler-Natta catalysts. As with the HDPE processes discussed earlier, much effort has been devoted to developing catalysts with higher polymer productivity, improved selectivity, higher polyrates, better control of molecular parameters, and improved product form (e.g. better particle size and uniformity, higher bulk density, minimal fines.) In addition, catalysts have been developed which can reduce the level of atactic polymer formed. One successful approach[124] involves a third component called an adjuvant or stereoregulator. It is usually an electron donor compound such as an amine, organosulfide, phosphine, or ether, and functions by selectively poisoning catalyst sites responsible for atactic polymer formation.[125] Since adjuvants will also lower polyrates their use must be carefully optimized.

In general, the level of atactic polymer in the product is maintained below 10%. As indicated, atactic polymer interferes with crystallization and will therefore lower the melting point along with tensile strength and modulus. Excess atactic polymer formed in the reactor must generally be solvent extracted and then separated from the solvent. This represents both a yield loss and an additional processing step, both of which add to costs.

Copolymerization is a favored method to modify the mechanical properties of polypropylene. Up to about 4% of ethylene is

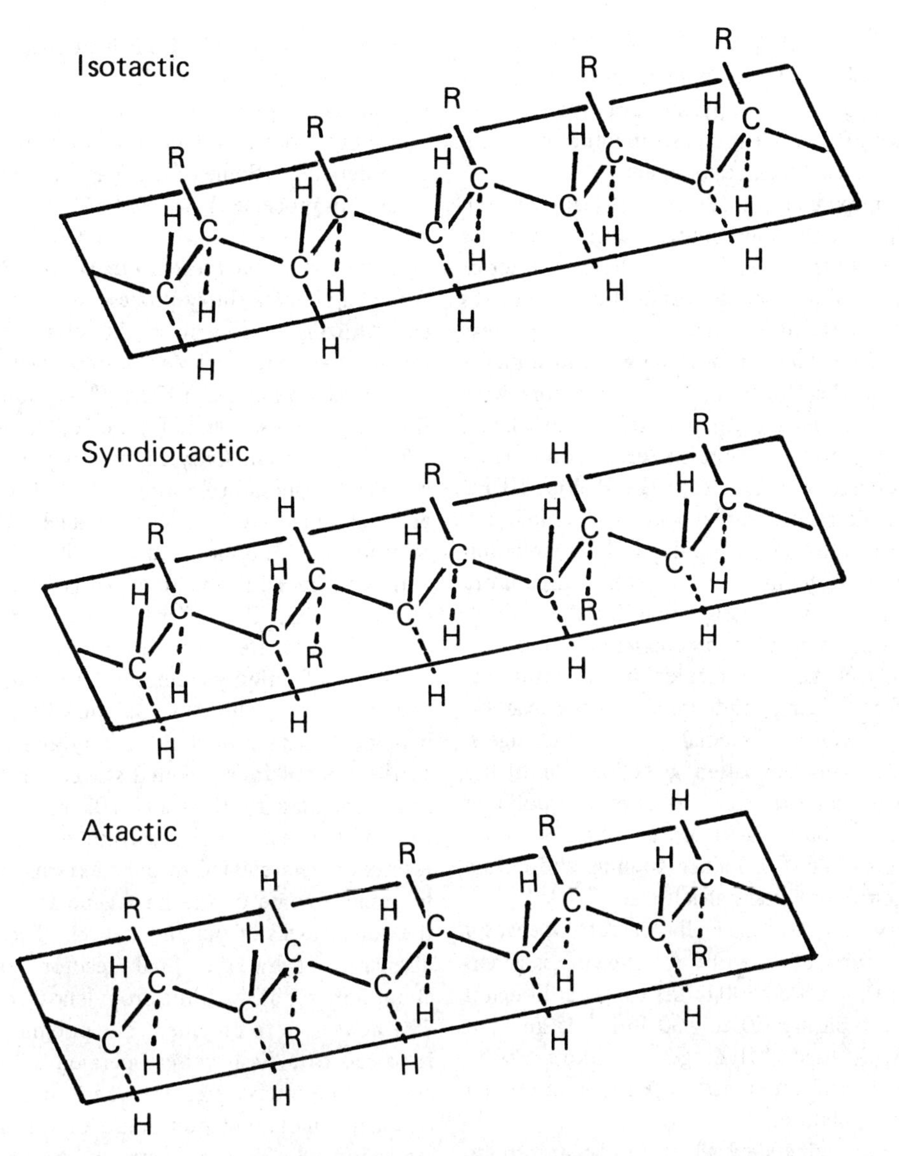

Fig. 10.23 Polypropylene configurations.

frequently added to form a random copolymer. The ethylene in the chain tends to reduce crystallinity, resulting in improved elongation and toughness. Although atactic polypropylene in the product can have a similar effect, copolymerization accomplishes it more efficiently and with fewer process difficulties.

For applications requiring higher impact strength, especially at lower temperature, further increases of the random copolymer ethylene level become unsuitable because other key properties, such as stiffness and melting point, are too seriously degraded. Here, so-called block copolymers are used instead. In a reacting mixture with Ziegler-Natta catalysts. there normally co-exist growing chains, "dead" polymer, and inactive catalyst sites. If propylene is replaced in the mixture by a second monomer such as ethylene, the growing chains then add ethylene, forming a block copolymer. Along with this block copolymer are the previously formed "dead" polypropylene homopolymer chains. If ethylene is added to propylene in the mixture, a copolymer is formed, consisting of a block of largely isotactic polypropylene bonded to a block of random ethylene-propylene copolymer. As much as 30% ethylene can be incorporated with propylene via block copolymers.

Polypropylene is made commercially by two principal continuous routes: slurry processes and gas phase processes. Each has a number of variants. For example, some slurry processes employ inert diluents such as heptane; whereas with others, liquid propylene without diluents serves as the slurry medium. Reactors with slurry processes are generally stirred autoclaves, frequently four or more in series. Loop reactors, similar in principle to the Phillips HDPE reactor described earlier, are also employed. As with ethylene, propylene polymerization is very exothermic and reactor temperature control is a prime design consideration. Cooling can be carried out by coolant circulation in reactor jackets, monomer or diluent refluxing, cold feed streams, and circulating the reactant slurry through external heat exchangers. Reactor walls are often glass-lined or highly polished stainless steel to suppress polymer fouling.[107] Slurry agitation is critical to insure good heat transfer, reduce fouling, and obtain satisfactory polymer particle size.

Slurry processes generally operate at reactor temperatures of 40 to 85°C. Pressure is generally in the 300- to 500-lb/in^2 range, although it can be typically 50 to 100 lb/in^2 if an inert diluent is used. Hydrogen is almost always added as a chain transfer agent for molecular weight regulation.

The processing steps which can occur in polypropylene manufacture are as follows:

a) Catalyst preparation. (Sometimes completed in situ in the reactor.)
b) Polymerization.
c) Recovery and recycling of unreacted propylene and any comonomers.
d) Recovery and recycling of diluent if one is used.
e) Destruction and separation of the catalyst from the final polymer. (This can be avoided if catalyst productivity is high enough.)
f) Removal of atactic (amorphous) and low molecular weight fractions in the polypropylene formed. (Part or all of this step can sometimes be eliminated.)
g) Finishing operations which include product drying, additives blending, extrusion, and pelletizing. (Some of these are eliminated in certain processes.)

Mitsui Petrochemical-Montecatini Process. An example of a slurry process for polypropylene utilizing a diluent is the one based on Mitsui Petrochemical-Montedison technology developed during the 1970's.[126] A simplified flow diagram is shown in Figure 10.24.

The high productivity catalyst is prepared in batches. Magnesium chloride and ethyl benzoate are ballmilled to produce a finely divided mixture in 6.5:1 molar ratio. This is treated with 15 times its weight of $TiCl_4$ in heptane diluent at 130°C. After further washings in diluent, the treated particles are mixed with a solution of triethylaluminum to form the catalyst slurry. This slurry, along with purified heptane diluent, propylene, and hydrogen is fed to the first of four jacketed stirred reactors in series operated at 40°C and 103 psia. A portion of the reacting slurry from the first two reactors is recirculated through external coolers to remove some of the heat generated during the early stages of polymerization. The slurry discharged from the final reactor contains about 35% solids, with the propylene feed about 79% converted to polymer. The polymer slurry is passed to a flash vessel operated at slightly above atmospheric pressure where most of the unreacted monomer and some of the diluent are vaporized. This vapor stream is compressed and cooled in the monomer recovery section of the plant and then recycled. The slurry from the flash vessel is centrifuged to recover a polymeric wet cake which is dried in a closed-circuit two-stage fluid bed drier that reduces diluent content to less than 0.1%. The dried product is pneumatically conveyed to the product finishing and storage area where it is compounded, pelletized, and stored for packaging or bulk transport. Vapor evaporated from the wet cake is recovered from the recirculating inert gas and sent to the diluent recovery section.

The mother liquor from the centrifuge

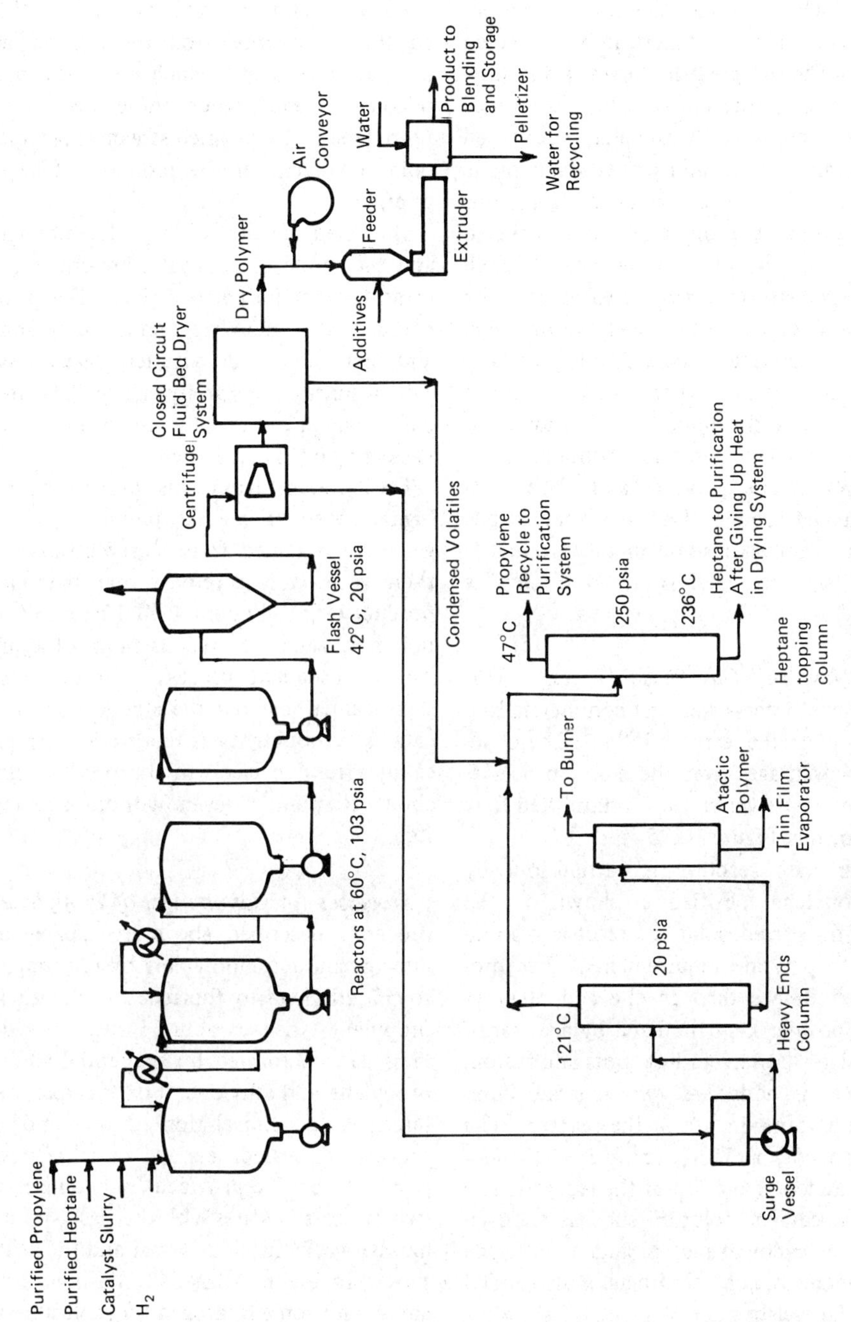

Fig. 10.24 Polypropylene via Mitsui Petrochemical-Montedison technology.

contains diluent contaminated with atactic polymer, catalyst components, propylene, and low volatility by-products. This mixture passes to the heavy ends distillation column, the bottoms of which contain the atactic polymer fraction. The atactic polymer can be recovered via the thin film evaporator shown in the flow diagram. The overhead from the heavy ends column is combined with the condensed volatiles from the driers and sent to the heptane topping column. The overhead from this column is primarily propylene and is recycled to the monomer purification system. A small portion of this stream is purged to prevent the accumulation of impurities such as propane. The hot bottoms stream passes through the drier system which utilizes its thermal energy, and it is then recycled to the diluent purification area.

The polypropylene product contains only about 0.16% atactic polymer and about 160 ppm of catalyst residue. For most applications this level of catalyst residue in the product is acceptable, so no catalyst removal step is employed.

BASF "Novolen" Gas Phase Process. This polypropylene process was first commercialized in Germany in the early 1970's. There are economic advantages over the slurry processes since it requires fewer steps. A simplified flow sheet is shown in Figure 10.25.[127,128,129]

The gas phase reaction is carried out by adding propylene, purified as shown, to the bottom of the stirred, solid bed reactor at a rate greater than polyrate requirements. The propylene feed flashes through the bed of polypropylene powder kept fluidized by a stirrer of special design[130] to avoid hot spots and fusion. The catalyst is added as two separate components and forms in situ in the reactor. The first component, a $TiCl_3$ complex in cyclohexane, is added at the top of the reactor. The second, a cocatalyst solution such as triethylaluminum in cyclohexane, is added with the inlet propylene stream. Hydrogen is also added for molecular weight control.

Reactor residence time is 5 to 6 hours with the reactor operated 60 to 80% full. Temperature control is primarily accomplished by flashing of the propylene feed.

Polypropylene powder exiting the reactor is stripped of its volatile constituents in the cyclone and polymer stripper. Propylene vapor escaping the agitated solids bed in the reactor is condensed and recycled directly to the feed stream. Exit vapors from the cyclone pass to a distillation system which removes propane, oligomers, and other impurities from the propylene. The purified stream is recycled to a point upstream of the monomer feed stream drier.

The dried powder is fluffy. Reactor agitation and the mode of catalyst component addition greatly affect its particle size. The powder is blended with stabilizers and other additives, and then fed to an extruder-pelletizer system. The resulting polypropylene pellets are then conveyed pneumatically to storage silos for packaging or bulk shipment.

The product from this process contains a higher level of atactic polymer than those generally obtained from slurry processes since there is no step to remove that material. The product originally contained 15 to 25% atactic polymer, which limited its range of applicability. Subsequent process improvements have significantly lowered the atactic fraction. The catalyst productivity is reportedly so high that catalyst residue levels in the product are only about 20 ppm,[126] even without any removal step.

Processes for Copolymers of Propylene. The processes described above are capable of producing random copolymers by charging, say, 3 to 5% ethylene to the reactors along with the propylene. As was pointed out, block copolymers can be formed by sequential addition of propylene and ethylene. For example, with the Mitsui Petrochemical-Montedison homopolymer process discussed earlier, block copolymer capability can be provided by adding a train of two reactors in series with the flash vessel in the line between the flash vessel and the centrifuge shown in Figure 10.24.[126] Purified ethylene, along with some hydrogen for molecular weight control, are charged to these two additional reactors. There are also some modifications to the catalyst for these block copolymer processes.

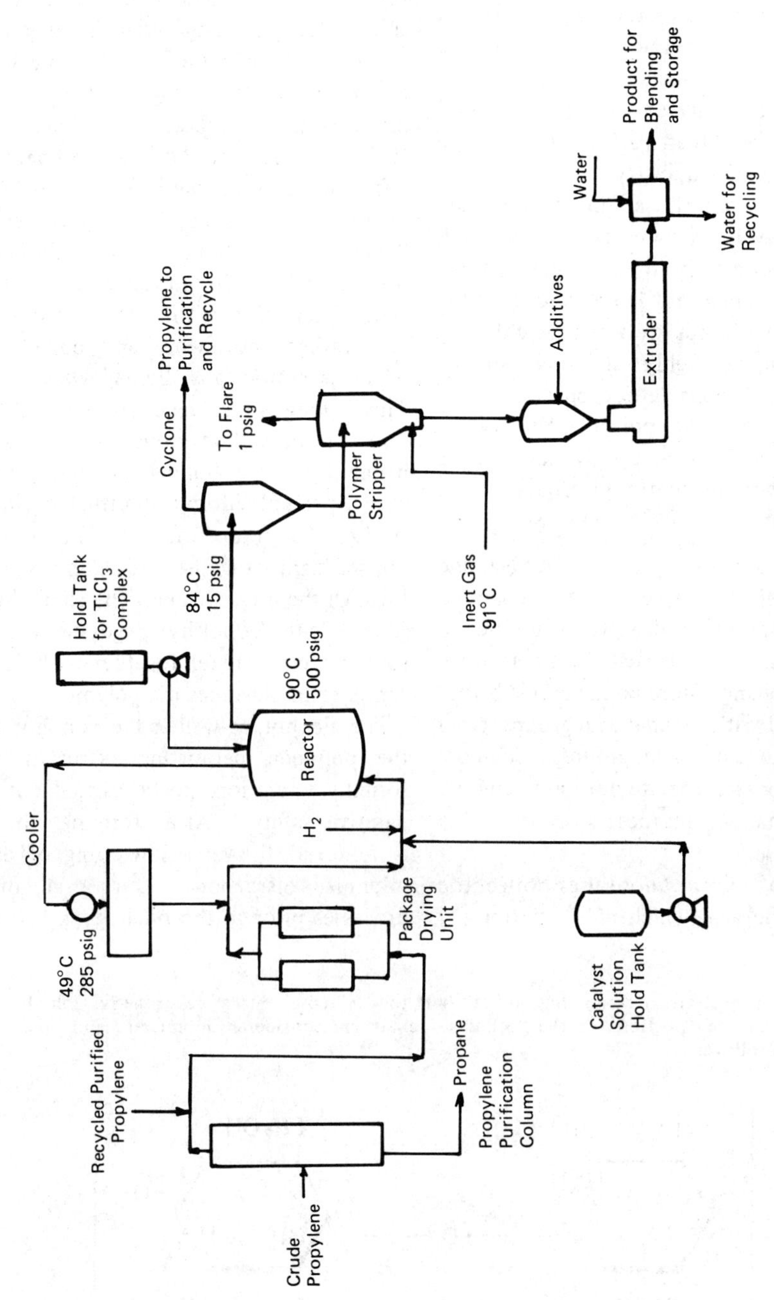

Fig. 10.25 BASF gas phase process for polypropylene.

CELLULOSE DERIVATIVES

Cellulose is the most common naturally occurring high polymer of organic chemical nature. The cell walls of nearly all plants consist of cellulose.* Throughout the world, over a trillion pounds of cellulose are converted annually to paper products, regenerated cellulose (rayon and cellophane), and cellulose-based plastics, with less than 0.1 percent being utilized in the plastics industry.

Although cellulose is a linear molecule, heating it below its charring temperature will not soften it sufficiently to permit plastic flow. The cause of this behavior is believed to be the strong hydrogen bonds between neighboring molecules forming highly stable crystalline structures with regular lattice spacings. To solubilize or fluidize the molecule, the crystalline structure is broken up by substituting various groups at the hydroxyl groups of the cellulose. The xanthate group, $-\mathrm{O}\overset{\mathrm{S}}{\overset{\|}{\mathrm{C}}}\mathrm{SNa}$, and the cuprammonium complex are used commercially to temporarily dissolve cellulose in caustic solutions for eventual regeneration to rayon or cellophane. More permanent substitution is achieved with organic acid groups, ether groups, and inorganic acid groups. Cellulose acetate, propionate, acetate-butyrate, and nitrate are commercial plastics, as is the ethyl ether of cellulose.

Preparation of *nitrocellulose* uses both nitric acid and sulfuric acid which aids nitration (1) by forming cellulose sulfate which causes the fibers to swell rapidly, and (2) by complexing with the water formed, thus forcing nitration to a high level. Since only about 1.9-2.0 of the three hydroxyls available in a "glucose" ring are substituted for plastic applications, more water is used in the nitrating acid than would be used in the preparation of gun-cotton. A suitable nitrating acid for plastics consists of 61 percent H_2SO_4, 21 percent HNO_3 and 18 percent H_2O. After nitration at low temperatures (75-85°F) in order to minimize acid cleavage of the "glucose" rings which lowers the molecular weight and impairs impact strength, the spent acid is centrifuged off; the centrifuged fibers are washed thoroughly and boiled in water which is periodically drained off and replaced. This boiling, which takes from 10 to 60 hours, removes the sulfate groups that cause product instability. After boiling, the fibers are bleached with 1 percent chlorine and treated with sodium sulfite. To remove the water the wet fibers are pressed tightly together and alcohol is percolated through the pressed cake. The alcohol-wet cake (30 percent alcohol by weight) is then charged to a powerful mixer together with 26 percent camphor to plasticize the polymer.

The alcohol, as well as the camphor, fluidizes the polymer, permitting extrusion or other forming operations to be carried out in comparative safety. After forming, the alcohol is removed by very slow aging or drying to minimize distortion. During all the above processes prior to the final aging, it is essential

*Native cellulose may contain some rings other than the "glucose" (or more properly, anhydro-β-D-glucopyranose) ring, but such portions of the "cellulose" chain are considered anomalous, and "pure" cellulose usually means the following:

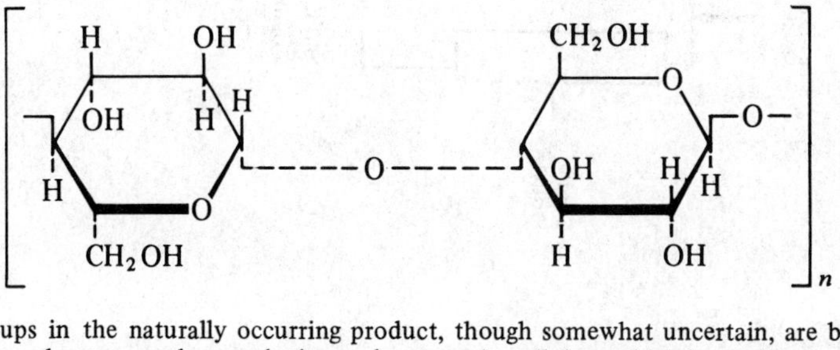

H OH CH₂OH O OH H H O OH H H H O CH₂OH H OH n

where the end groups in the naturally occurring product, though somewhat uncertain, are believed to consist equally of reducing end groups and nonreducing end groups; in cellulose which has undergone hydrolytic chain cleavage to lower the molecular weight, these end groups are sometimes acidic in nature, presumably because of oxidation.

that the nitrated cellulose be kept wet with water or alcohol since serious fires can easily be started by slight frictional heating. Because of the necessity for final aging and because of great sensitivity to heat degradation, nitrocellulose is not a thermoplastic in the ordinary sense; however, neither is it a thermosetting plastic.

To make *cellulose acetate* (or propionate or butyrate), the purified alpha-cellulose pulp [usually obtained by the sulfite process from wood (Chapter 15)] is shredded and mixed with acetic acid to wet and swell the fibers. This mixture is then fed to the acetylation reactor where sulfuric acid catalyst (1 percent cellulose), methylene chloride solvent (or more acetic acid as solvent), and acetic anhydride are added. Acetylation is done at moderate temperatures, e.g., 90-120°F, to minimize acid cleavage. Following acetylation, wet acetic acid is added to remove the sulfate ester groups and to deesterify the cellulose to the desired level. During the acetylation process, the fibers dissolve in the solvents. This requires precipitation in water and subsequent milling to make a flake and to squeeze out most of the acid and water. The flake is then washed with water in a countercurrent band filter; centrifuging, drying, and blending follow (Fig. 10.26). To make a molding powder, the flake is rapidly cold-mixed with a plasticizer but not homogeneously blended. The mix is then fed to a set of hot mill rolls where colorants are added; the plasticizer is then homogeneously mixed. The mill-rolled sheet is diced to molding powder pellets. Alternatively, the flake and the plasticizer containing dispersed colorants are fed continuously to a heated extruder which homogenizes the mixture and forms strands which are subsequently cooled and chopped into molding powder. To make the mixed acetate-butyrate* or acetate-propionate, butyric acid or propionic acid is added to the acetylating mixture.

Ethyl cellulose for plastic purposes usually contains 2.3-2.5 ether groups per "glucose" ring. It is made by reacting cellulose with ethyl chloride under alkaline conditions in the presence of water. By-products are ethyl alcohol and ethyl ether formed via hydrolysis of the ethyl chloride.

Following the reaction, the reaction mass is mixed and simultaneously precipitated and atomized to fine granules; a special two-fluid spray nozzle is used in which the "gaseous" fluid stream is steam.

The polymer spray is contained in a precipitating tank where the volatile benzene, ethyl chloride, ethyl ether, and ethyl alcohol are steam distilled off. After precipitation, the polymer is centrifuged, washed with hot water, and bleached with dilute sodium hypochlorite (pH 10.5-11.3 at 175°F); another centrifuging and washing follow. Stabilization (deashing) is achieved with a dilute acetic acid rinse (pH 4-4.2) and a final centrifuging and water rinsing; this is followed by drying in a countercurrent rotary steam tube or a concurrent rotary hot air drier. The dried crumb is mixed

*Usually about 1.7 butyryl and 1.0 acetyl groups are substituted per "glucose" ring.

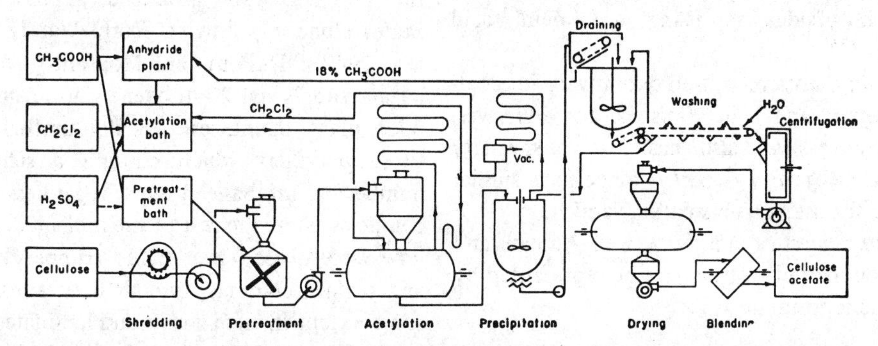

Fig. 10.26 Manufacture of cellulose acetate by the methylene dichloride method. *(After Ranby, B. G. and Rydholm, S. A., "Cellulose and Cellulose Derivatives," in "Polymer Processes," C. E. Schildknecht, ed., Interscience Publishers, 1956; by permission John Wiley & Sons, Inc.)*

with colorants, stabilizers, and plasticizers; then it is extruded to strands which are chopped into molding-powder pellets.

The above is but a brief description of the cellulosic plastics; Ott, Spurlin, Graffin, Bikules, and Segal,[147] and others[131,132,133,134,135] give more complete accounts.

OTHER PLASTICS

Besides the plastics already discussed, there are several others that are used either in small amounts as plastics or extensively for applications other than as plastics as narrowly defined in this chapter.

Polyvinyl butyral is used for the interlayer of safety glass. Although for this purpose it is the best plastic on the market, it has few other uses. Its formula may be represented by:

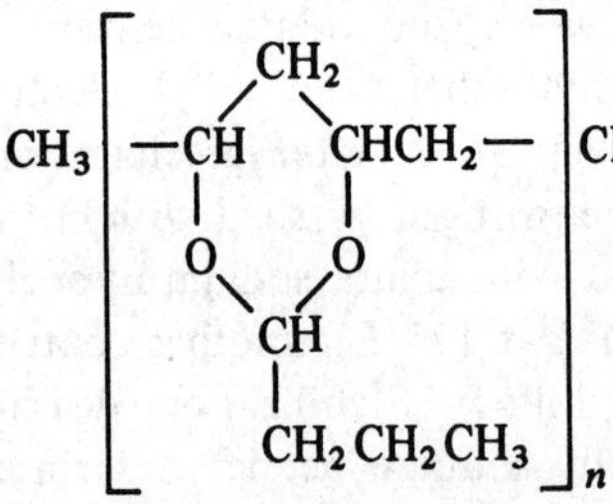

$$CH_3\left[\begin{array}{c} -CH \overset{CH_2}{\frown} CHCH_2- \\ | \qquad\quad | \\ O \qquad O \\ \diagdown CH \diagup \\ | \\ CH_2CH_2CH_3 \end{array}\right]_n CH_2OH$$

although some acetyl groups and alcohol groups are also present. It is made by polymerizing vinyl acetate ($CH_2{=}CHO\overset{O}{\overset{\|}{C}}CH_3$) either in suspension or solution with free-radical catalysts, hydrolysis of the resulting polymer with sulfuric acid in an alcoholic solution to yield polyvinyl alcohol, and then acetal formation with butyraldehyde using a mineral acid catalyst.[136]

Polyvinyl acetate is used extensively for coatings, textile and paper finishes, and adhesives. *Polyvinyl alcohol* also has a large specialty market. *Polyvinyl formal* is made by a similar process for use as an insulating varnish.

Silicone plastics are very well known and much publicized. They may be represented by the generic formula

$$R'\left[\begin{array}{c} R \\ | \\ -SiO- \\ | \\ R \end{array}\right]_n R'$$

where R is usually $-CH_3$ or $-C_6H_5$ (phenyl). They are produced from the corresponding diorganosilicon chlorides (R_2SiCl_2) plus water, and acid or base condensation catalysts.[137,138] A very small amount of the triorgano-substituted silicon chloride will terminate the chain and control molecular weight. Branching, and eventual crosslinking may be achieved by using a monoorganosilicon chloride ($RSiCl_3$) or by adding oxidizing agents (e.g., benzoyl peroxide) to oxidize the methyl groups as in the vulcanization of silicone rubbers. Allyl groups have been utilized to permit vinyl-type crosslinking to occur with free radical-generating catalysts.

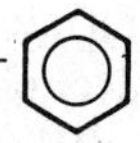

Depending on the conditions, oils, heat-hardenable resins, or rubbers may be formed. Silicones are used commercially for potting compounds, surface-coating enamels (particularly in electrical applications), stable rubbers, lubricating oils, and greases which have extraordinarily low variations in viscosity with temperature changes. Specialty uses include mold release agents, antifoam agents, water repellents, polyester-glass bonding agents, and more recently, due to their inert character, surgical uses, especially in cosmetic surgery. Their use in special heat-stable moldings, castings, and laminates is minimal.

Polyurethanes were manufactured in Germany during World War II when the Bayer laboratories developed their use in rigid foams, adhesives, and coatings. Polyurethanes have developed rapidly in the United States. Since their discovery in 1937, the application of polyurethanes has grown until in 1968, 500 million pounds of the polymer were produced in the United States alone. Following World War II, in the late 1940's, DuPont and Monsanto Company began supplying 2,4-tolylene diisocyanate in pilot plant quantities.[139] Polyurethanes are those polymers which contain a significant number of urethane groups, regardless of the composition of the rest of the molecule, and are now largely used in flexible (rubbery) foams and surface coatings, though they are used to some extent in rigid foams and laminates. Two methods are used for producing foam.[140,141] (1) The "one-shot" process involves simultaneous mixing of all reactants, catalyst, and additives. The reactions begin at once; foaming is completed in a couple of minutes, but several hours may be required to complete the cure. (2) In the "prepolymer" method the isocyanate and a diol are reacted to produce a prepolymer. This product is foamed later by reaction with water. The molecular weight is increased during foaming. Typical reactions for the second method are:

$$(1)\quad 2R(C{=}N{=}O)_2 + HOR'OH \longrightarrow$$

$$O{=}C{=}N{-}R{-}\overset{H}{N}{-}\overset{\overset{O}{\|}}{C}{-}O{-}R'{-}O{-}\overset{\overset{O}{\|}}{C}{-}\overset{H}{N}{-}R{-}N{=}C{=}O \quad \text{“Prepolymer”}$$

$$(2)\quad n(\text{Prepolymer}) + nH_2O \longrightarrow \left[{-}\overset{H}{N}\overset{\overset{O}{\|}}{C}\overset{H}{N}{-}R{-}\overset{H}{N}\overset{\overset{O}{\|}}{C}O{-}R'{-}O\overset{\overset{O}{\|}}{C}\overset{H}{N}{-}R{-} \right]_n + nCO_2$$

If only the diol is used, no foam is produced and the process is called reaction injection molding (RIM). (See previous section "Viewpoint of the Fabricator.")

A number of new *heat-resistant polymers* were developed during the 1960's and 1970's. Heat resistance, or thermal stability, is a measure of the ability of a material to maintain mechanical properties such as strength, toughness, and elasticity at a given temperature.[142] The "use temperature" of many polymers is limited by changes in their physical character at high temperatures, that is they soften or melt, rather than by changes in their chemical character. A typical linear polymer such as polystyrene has a maximum use-temperature of about 80-100°C. The melting point of a polymer may be increased by the introduction of polar substituents, such as halogen atoms and nitrile groups–polytetrafluoroethylene, for example, has a maximum use-temperature of about 290-310°C; or hydrogen-bonding groups such as amides.

In addition to physical factors, there are chemical factors which determine the thermal stability of a polymer. Polymers with high heat resistance must have bonds of high dissociation energy as well as structural features which do not allow degradation by low-energy processes. Another chemical factor affecting heat stability is the formation of crosslinked structures under the influence of heat with resultant changes in the physical properties of the polymer. Furthermore, for a thermally stable polymer to be useful it must also be chemically inert to oxygen, moisture, acids, bases, or other substances to which it might be exposed.

The synthesis of polymers containing aromatic units in the polymer-chain backbone was prompted by the known high bond energies, low degree of reactivity, and the rigidity of aromatic structures. These are the most successful of the new heat-resistant polymers. One of the first developed by Marvel[143,144] is *poly*-p-*phenylene* which is an insoluble, infusible material with good thermal stability. *Polythiazoles*, among the first fully aromatic condensation polymers to be made, are composed of alternating phenyl and thiazole rings. Though they are quite stable thermally, they cannot be produced at a sufficiently high molecular weight to form films. *Polyimides* are extremely heat stable, in addition to being flame-, radiation-, and oxidation-resistant; they also possess excellent electrical characteristics. For example, they have a thermal stability of over a year at 275°C.[142] Due to this unique combination of properties, polyimides, e.g., Skybond (Monsanto Co.) or Kaplon (DuPont), are used in many areas. For example, polyimides reinforced with glass or boron, or glass-based polyimide honeycombs are used as structural components in aircraft. Polyimide-graphite composities show promise in the construction of fan blades in jet-engine aircraft because of their high heat resistance and the weight savings over titanium blades. Polyimide adhesives are used for metal surfaces, such as titanium and aluminum. They are also used in the production of printed circuit boards with high-temperature operating capabilities.

Other polymers with superior thermal stability

are the *polybenzimidazoles* and derivatives, *polytetrazopyrenes*, *polyquinoxalines*, and the various *ladder* and *spiro polymers.* Spiro polymers are non-crosslinked structures in which two molecular strands are joined to each other. These polymers could be soluble and fusible, and would be expected to have better thermal properties than the simple stranded type, as at least two bonds must be broken for a reduction in molecular weight to take place. Ladder polymers are still in the developmental stage and techniques to make structures with higher molecular weights and no breaks in the ladder are being investigated.[145] Recently Marvel and co-workers have developed a pyrrolone-type ladder polymer which is thermally stable above 500°C.

Nylon is used extensively as a plastic as well as a fiber. Over 280 million pounds were used in plastic applications in 1979. Gears, bearings, and other unlubricated machine parts are particularly appropriate applications. Its manufacture is discussed in Chapter 11.

PLASTICIZERS

Besides the large industry engaged in manufacturing base resins, a sizable synthetic organic chemical industry is based on plasticizers for these resins. These materials impart flexibility and formability to many polymers, particularly those based on vinyl chloride and cellulose. Chemically they are frequently aliphatic esters (usually C_4 or higher) of dibasic acids, e.g., phthalic, adipic, azelaic, and sebacic acids. Epoxidized unsaturated natural esters (soybean oil) and similar unsaturated esters are increasing in importance. Phosphate esters, polyester resins, and aromatic hydrocarbons of high molecular weight are also common.

The list of commercial polymeric materials is still growing despite the fact that each new product must compete with the properties and prices of existing polymers. The perfect polymer for all plastic applications is an impossibility. A plastic that is resistant to high temperatures yet is heat-sealable and readily molded at low temperatures, that has excellent outdoor stability but is transparent, that has excellent hardness and abrasion resistance yet is strong and tough, that may be either rigid or flexible as desired, that is an excellent electrical insulator but also dissipates static charge at a high rate, and that is resistant to all chemicals yet can form solution coatings–such a plastic exists only in salesmen's dreams and chemist's nightmares.

REFERENCES

1. Mantell, C. L., Kopf, C. W., Curtis, J. L., and Rodgers, E. M., *The Technology of Natural Resins*, New York, John Wiley & Sons, 1942.
2. Hull, J. L., in *Encyclopedia of Polymer Science and Technology*, Vol. IX, pp. 1-47, New York, John Wiley & Sons.
3. Kestler, J., *Mod. Plastics*, Aug., 1969, pp. 58-59.
4. Wright, V., *Modern Plastics*, Oct., 1969, pp. 54-58.
5. Sloane, D. J., "Injection Molding Machinery," in *Machinery and Equipment for Rubber and Plastics*, Seaman and Merril, Eds., Vol. I, New York, Rubber World, 1952.
6. Macosko, S. W., et al, SPE 38th ANTEC, May 5-8, 1980, p. 423. (References by this author et al at end of this paper.)
7. McBrayer, R. L., Carver, T. G., SPE National Tech. Conf., Nov. 6-8, 1979, p. 127.
8. Seel, K., Klier, L., SPE National Tech. Conf., Nov. 6-8, 1979, p. 117.
9. Suh, N. P., Okine, R. K., SPE 37th ANTEC, May 7-10, 1979, p. 21.
10. Manzione, L. T., SPE RETEC, Feb. 25-27, 1980, p. 105.

10a. Kato, K., *Polymer Eng. Sci.*, 7, No. 38 (1967).

11. Carothers, W. H., *Collected Papers*, H. Mark and G. S. Whitby, Eds., New York, John Wiley & Sons, 1940.
12. Matthews, F. E., and Strange, E. H., German Patent 249,868.
13. Harris, C., *Ann. Chem.*, 383,213 (1911).
14. Doak, K. W., and Schrage, A., "Commercial Polymerization and Copolymerization Processes," in *Crystalline Olefin Polymers*, R. A. V. Raff and K. W. Doak, Eds., Part I, Chapter 8, New York, John Wiley & Sons, 1964.
15. Ziegler, K., and Gellert, H. G., U.S. Patents 2,699,457 (1955); 2,695,327.
16. Ziegler, K., Holzkamp, E., Breil, H., and Martin, H., *Angew. Chem.*, 67,541 (1955).

17. Ziegler, K., Martin, H., *Makromol. Chem.*, 18-19, 186 (1956).
18. Hogan J. P., and Banks, R. L., U.S. Patent 2,825,721 (March, 1958; to Phillips Petroleum).
19. Natta, G., and Danusso, F., Eds., *Stereoregular Polymers and Stereospecific Polymerizations*, 2 Volumes, Elmsford, N.Y., Pergamon Press, 1967.
20. Roberts, D. E., *J. Res. Nat. Bu. St.*, **44**, 221 (1950).
21. Boundy, R. H., and Boyer, R. F., *Styrene*, New York, Van Nostrand Reinhold, 1952.
22. Christopher, W. F., and Fox, D. W., *Polycarbonates*, New York, Van Nostrand Reinhold, 1962.
23. Allen, I., *et al.*, U.S. Patent 2,496,653 (Feb., 1952; to Union Carbide).
24. Allen, I., *et al.*, U.S. Patent 2,614,910 (Oct., 1952; to Union Carbide).
25. Amos, J. L., *et al.*, U.S. Patents 2,494,924; 2,530,409; 2,714,101; 2,941,985; 3,058,965 (to Dow Chemical).
26. McDonald, D. L., Colomer, Coultor, K. E., and McCurdy, J. L., U.S. Patent 2,727,884 (to Dow Chemical).
27. Ruffing, N. R., U.S. Patent 3,243,481 (to Dow Chemical).
28. Stober, K. E., *et al.*, U.S. Patent 2,530,409 (May, 1948; to Dow Chemical).
29. Amos, J. L., *et al.*, U.S. Patent 2,849,430 (to Dow Chemical).
30. Charlesworth, R. K., Murdock, S. A., Shaw, K. G., U.S. Patent 3,201,365 (to Dow Chemical).
31. Schildknecht, C. E., *Vinyl and Related Polymers*, pp. 197-203, New York, John Wiley & Sons, 1952.
32. Sundberg, D. C., Ph.D. Thesis, Univ. of Delaware, 1970.
33. Dinsmore, R. P., U.S. Patent 1,732,795 (Oct. 22, 1929; filed Sept. 13, 1927; to Goodyear).
34. Luther, M., and Heuck, C., U.S. Patent 1,860,681 (filed July 17, 1928; issued June 21, 1932; filed in Germany Jan. 8, 1927; to I. G. Farben).
35. Harkins, W. D., *Physical Chemistry of Surface Films*, p. 332, New York, Van Nostrand Reinhold, 1952.
36. *Ibid*, p. 339.
37. Harkins, W. D., *J. Chem. Phys.*, **13**, 47, 381 (1945); and *J. Am. Chem. Soc.*, **69**, 1928 (1947).
38. Klevins, H. B., *Chem. Revs.*, **47**, 1 (1950).
39. Gemens, H., *Advan. Polymer Sci.*, **1**, 2349 (1959).
40. Smith, W. V., and Ewart, R. H., *J. Chem. Phys.*, **16**, 592 (1948).
41. Gardon, J. L., *J. Polymer Sci.*, **A-1**, 6, 623, 643, 665, 687 (1968).
42. Basdekis, C. H., *ABS Plastics*, New York, Van Nostrand Reinhold, 1964.
43. Simon, R. H. M., and Oster, B., U.S. Patent 3,345,430 (Oct. 1967; to Monsanto).
44. Mark, H., and Hohenstein, W. P., *J. Polymer Sci.*, **1**, 127 (1946).
45. Grim, J. M., U.S. Patent 2,715,118 (Aug., 1955; to Koppers).
46. Ott, J. B., U.S. Patents 2,862,912 and 3,051,682 (to Monsanto).
47. Horn, M. B., *Acrylic Resins*, Van Nostrand Reinhold, 1960.
48. Teach, W. C., Kiessling, G. C., *Polystyrene*, Van Nostrand Reinhold, 1960.
49. German Patent 1,125,175 (1962; to B.A.S.F.).
50. Shanta, P. L., U.S. Patent 2,694,700 (1954).
51. U.S. Patent 2,981,722 (1961; to Wacker-Chemie).
52. Tromsdorf, E., "Polymerizations in Suspension," in *Polymer Processes*, Schildknecht, C. E., Ed., Chapter 3, New York, John Wiley and Sons, 1956.
53. Hanson, A. W., U.S. Patents 2,488,198 (1949; to Dow Chemical), and 2,769,804 (Nov. 1956; to Dow Chemical).
54. Hanson, A. W., and Zimmerman, R. L., *Chem. Eng. News*, 77 (1957).
55. Jones, C., Harris, B., and Ingley, F. L., U.S. Patent 2,739,142 (March, 1956; to Dow Chemical).
56. Hanson, A. W., Heston, A. L., and Buecken, H. E., U.S. Patent 2,519,834 (1950; to Dow Chemical).
57. Ellis, C., *The Chemistry of Synthetic Resins*, pp. 123-141, 201, 231, New York, Van Nostrand Reinhold, 1935.
58. Baeyaert, A. E. M., U.S. Patent 2,715,117 (Aug., 1955); U.S. Patent 2,856,272 (Oct., 1958; to Saint-Gobain).
59. Thomas, J., *Hydrocarbon Process*, **47**, No. 11, 192 (Nov., 1968).
60. Akin, R. B., *Acetal Resins*, New York, Van Nostrand Reinhold Co., 1962.
61. Schneider, A. K., McDonald, R. N., Cairns, T. L., *et al.*, U.S. Patent 2,768,994 (Oct., 1956); 2,775,570 (Dec., 1956); 2,795,571 (June, 1957); 2,828,286 (March, 1958); 2,828,287 (March, 1958; to DuPont).
62. British Patent 770,717 (April, 1955; to DuPont).
63. Miller, W. T., U.S. Patent 2,579, 437 (Dec., 1951; to M. W. Kellogg).
64. Rearick, J. S., U.S. Patent 2,600,804 (June, 1952; to M. W. Kellogg).
65. Wrightson, J. M., U.S. Patent 2,600,821 (June, 1952; to M. W. Kellogg Co.).
66. Greenwalt, C. H., U.S. Patent 2,388,138 (Oct., 1945; to DuPont).
67. Carswell, T. S., *Phenoplast S*, New York, John Wiley & Sons, 1947.
68. Gould, D. F., *Phenolic Resins*, Van Nostrand Reinhold Co., 1959.
69. Martin, R. W., *The Chemistry of Phenolic Resins*, New York, John Wiley & Sons, 1956.

70. Megson, J. I. L., *Phenolic Resin Chemistry*, New York, Academic Press, 1958.
71. Robitschek, P., and Lewin, A., *Phenolic Resins*, London, Iliffe and Sons, 1950.
72. Zavistsas, A. A., Beaulieu, R. D., Leblanc, J. R., *J. Polymer Sci.*, A-1, 6 No. 9, 2541 (1968).
73. Blais, J. F., *Amino Resins*, New York, Van Nostrand Reinhold, 1959.
74. DeJong, J. I., *Rec. Trav. Chim.*, **71**, 643 (1952); **72**, 88, 653, 1027 (1953); **73**, 139 (1953).
75. Ilicoto, *Chimi. Ind.* (Milan), **34**, 688 (1952).
76. Vale, C. P., *Aminoplastics*, London, Cleaver-Hume Press, 1950.
77. Walker, J. F., *Formaldehyde*, pp. 281-317, New York, Van Nostrand Reinhold, 1953.
78. Perry, E., and Reese, F. E., *Ind. Eng. Chem.*, **40**, 2039 (1948).
79. Bottenbruch, L., in *Encyclopedia of Polymer Science and Technology*, Vol. **X**, p. 710, New York, John Wiley & Sons, 1969.
80. Schnell, H., *Angew. Chem.*, **68**, 633 (1956).
81. Schnell, H., *Chemistry and Physics of Polycarbonates*, New York, John Wiley & Sons, 1964.
82. Thompson, R. J., and Goldblum, K. B., *Mod. Plastics*, **35**, 131 (1958).
83. Einhorn, A., *Ann. Chem.* **300**, 135 (1898).
84. Schnell, H., Bollenbruch, L., and Krim, H., Belgian Patent 532,543 (1956; to Fabenfabriken Bayer).
85. Fox, D. W., Australian Patent 221,192 (1959; to General Electric).
86. *Chem. Eng.*, **38**, 124 (Nov. 14, 1960).
87. Hartman, S. J., *Plastics World*, **27**, No. 8, 42 (1969).
88. Castan, Pierre, *et al.*, U.S. Patents 2,324,483 (July, 1943); 2,444,333; 2,458,796; 2,637,715 (to Ciba).
89. Greenlee, S. O., *et al.*, U.S. Patents 2,456,408 (Dec., 1948); 2,503,726; 2,510,885-6; 2,511,913; 2,512,996; 2,521,911-12; 2,528,359-60; 2,538,072; 2,558,949; 2,581,464; 2,582,985; 2,589,245; 2,592,560; 2,615,007-8; 2,694,694; 2,698,315; 2,712,000; 2,717,885 (to Devoe and Raynolds).
90. Werner, E. G., Wiles, Q. T., *et al.*, U.S. Patents 2,467,171 (April, 1949); 2,528,932-3-4; 2,541,027; 2,575,558; 2,640,037; 2,642,412; 2,682,515; 2,735,329; 2,716,009; 2,752,269; 2,840,511; 2,841,595 (to Shell Development).
91. Skeist, I., and Somerville, G. R., *Epoxy Resins*, New York, Van Nostrand Reinhold, 1958.
92. Lee, H., and Neville, K., *Epoxy Resins*, New York, McGraw-Hill, 1957.
93. Coderre, R. A., "Epoxy Resins," in *Encyclopedia of Chemical Technology*, R. E. Kirk and D. F. Othmer, Eds., Vol. I (Supplement), pp. 312-329, New York, John Wiley & Sons, 1957.
94. Pezzoglia, P., U.S. Patent 2,841,595 (July, 1958; to Shell Development).
95. Bjorksten Research Laboratories, Inc., *Polyesters and Their Applications*, New York, Van Nostrand Reinhold, 1956.
96. Flory, P. J., *Principles of Polymer Chemistry*, Ithaca, Cornell University Press, 1953.
97. Lawrence, J. R., *Polyester Resins*, New York, Van Nostrand Reinhold, 1960.
98. Sonneborn, R. H., *Fiberglass Reinforced Plastics*, New York Van Nostrand Reinhold, 1954.
99. Hay, A. S., Blanchard, H. S., Endres, E. F., and Estance, J. W., *J. Am. Chem. Soc.*, **81**, 6335 (1959).
100. Hay, A. S., *et al.*, in *Encyclopedia of Polymer Science and Technology*, Vol. X, p. 94, New York, John Wiley & Sons, 1969.
101. Johnson, R. N., in *Encyclopedia of Polymer Science and Technology*, Vol. XI, pp. 447-463, New York, John Wiley & Sons, 1969.
102. Chopey, N. P., *Chem. Eng.*, **68**, No. 2, 112 (1961).
103. *Modern Plastics*, January 1980, pp. 72-76.
104. *Ibid*, p. 106.
105. Swallow, J. C., *The History of Polythene*, Renfrew and Morgan, Eds., Chap. I, London, Iliffe and Sons, 1957.
106. British Patent 921,542 (March 20, 1963; to Monsanto).
107. Albright, L. F., *Processes for Major Addition-Type Plastics and Their Monomers*, Chaps. 3 and 4, New York, McGraw-Hill, 1974.
108. Hogan, J. P., and Banks, R. L., U.S. Patent 2,825,721 (to Phillips Petroleum) March, 1958.
109. Clark, A., *Ind. Eng. Chem.*, **59**, 29 (1967).
110. Hogan, J. P., *ACS Polymer Preprints*, **10**, No. 1, pp. 240-247, April, 1969.
111. Ziegler, K., *et al.*, *Angew. Chem.* **67**, 426 (1955).
112. Natta, G., *J. Polym. Sci.* **16**, 143 (1955).
113. Natta, G., *Chem. Ind.* **47**, 1520 (1957).
114. Natta, G., and Danusso, F., Eds., *Stereoregular Polymers and Stereospecific Polymerizations*, Vols. 1 and 2, Pergamon Press, New York, 1967.
115. Johnson, R. N. (to Union Carbide), U.S. Pat. 3,687,920 (Aug. 29, 1972).
116. Karol, F. J., *J. Polymer Sci.* **10**, Part A-1, 2609-37 (1972).
117. Cossee, P., *J. Catalysis* **3**, 80 (1964).
118. Delbouille *et al.*, German Pat. 2,146,688 (to Solvay), April 27, 1972.

119. Short, J. N., *Ind. Research and Development*, Sept., 1980, pp. 109–112.
120. Norwood, D. D., U.S. Patent 3,248,179 (to Phillips Petroleum), April 26, 1966.
121. Marwil, S. J. *et al.*, U.S. Patent 3,374,211 (to Phillips Petroleum), March 19, 1968.
122. *Chemical Engineering*, Dec. 3, 1979, pp. 80–85.
123. Natta, G. *et al.*, *J. Am. Chem. Soc.* 84, 1488 (1962).
124. Keii, T., *Kinetics of Ziegler-Natta Polymerization*, Chapman & Hall, Ltd., London, 1972, p. 105.
125. Buls, V. W., and Higgins, T. L., *J. Polymer Sci.* (A-1) 8, 1025 (1970).
126. Fong, W. S. *et al.*, "Polypropylene," Process Economics Program Report No. 128, SRI International, Menlo Park, Cal., April, 1980. A private report.
127. "BASF Novel Process for Polypropylene," *Oil and Gas Journal*, **68** (47), p. 64 (1970).
128. Trieschmann, H., *et al.*, U.S. Patent 3,652,527 (to BASF), March 28, 1972.
129. British Patent 1,372,717 (to BASF) 1974.
130. Wisseroth, K., *et al.*, U.S. Patent 3,545,729 (to BASF), Dec. 8, 1970.
131. Bracken, W. O., "Cellulose Plastics," in *Encyclopedia of Chemical Technology*, R. E. Kirk and D. F. Othmer, Vol. **III**, pp. 391–411, New York, John Wiley & Sons, 1949.
132. Hill, R. O., "Cellulose Esters, Organic," in *Encyclopedia of Polymer Science and Technology*, Vol. **III**, p. 307, New York, John Wiley & Sons, 1965.
133. Miles, F. D., *Cellulose Nitrate*, New York, John Wiley & Sons, 1955.
134. Paist, W. D., *Cellulosics*, New York, Van Nostrand Reinhold, 1958.
135. Ranby, B. G., and Rydholm, S. A., "Cellulose and Cellulose Derivatives," in *Polymer Processes*, Schildknecht, C. E. Ed., pp. 351–428, New York, John Wiley & Sons, 1956.
136. *Chem. Eng.*, **61**, 346 (Feb., 1954).
137. Lichtenwalner, H. K., Sprung, M. N., "Silicones," in *Encyclopedia of Polymer Science and Technology*, Vol. **XII**, p. 464, New York, John Wiley & Sons, 1970.
138. Meals, R., in *Encyclopeida of Chemical Technology*, R. E. Kirk and D. F. Othmer, Eds., 2nd Ed., Vol. **XVIII**, p. 221, New York, John Wiley and Sons, 1969; (Meals, R. N., and Lewis, F. M., *Silicones*, New York, Van Nostrand Reinhold, 1959.)
139. Pigott, K. A., in *Encyclopeida of Polymer Science and Technology*, Vol. **XI**, p. 506, New York, John Wiley & Sons, 1969.
140. Dombrow, B. A., *Polyurethanes*, Van Nostrand Reinhold, 1965.
141. Saunders, J. H., Frisch, K. C., *Polyurethanes Part I: Chemistry*; *Part II: Technology*, New York, John Wiley & Sons, 1962, 1964.
142. Mulvancy, J. G., "Heat-Resistant Polymers," in *Encyclopedia of Polymer Science*, Vol. **VII**, p. 478, New York, John Wiley & Sons, 1967.
143. Frey, D. A., Hasegawa, M., and Marvel, C. S., *J. Polymer Sci.*, A, 1, 2067 (1963).
144. Marvel, C. S., and Hartzell, G. E., Jr., *J. Am. Chem. Soc.*, 81, 488 (1959).
145. *Chem. Eng. News*, 48, No. 34, 36 (1970).
146. Bovey, F. A., Kolthaff, I. M., Modulin, A. I., and Mechon, E. J., *Emulsion Polymerization*, pp. 71–93, New York, John Wiley & Sons, 1955.
147. Ott, Spurlin, Graffin, Bikules, and Segal, "Cellulose and Cellulose Derivatives," vol. **5**, pts. 1, 2, 3, and 4 of *High Polymers*, Interscience, 1954, 1955, 1971.

Selected Special References

Properties, Practical

PP-1 ASTM Standards, Part 26, "Plastics-Specifications; Methods of Testing Pipe, Film, Reinforced and Cellular Plastics," reissued annually, Philadelphia, American Society for Testing and Materials.
PP-2 Baer, E., Ed., *Engineering Design for Plastics*, New York, Van Nostrand Reinhold, 1964.
PP-3 Brydson, J. A., *Plastic Materials*, New York, Van Nostrand Reinhold, 1966.
PP-4 Lever, A. E., Ed., *The Plastics Manual*, London, Scientific Press, 1966.
PP-5 *Modern Plastics Encyclopedia*, New York, McGraw-Hill, Annually.

Testing of Polymers

T-1 ASTM Standards, Part 27, "Plastics–General Methods of Testing, Nomenclature," reissued annually, Philadelphia, American Society for Testing and Materials.
T-2 Level, A. E., and Rhys, J. A., *The Properties and Testing of Plastic Material*, 3rd ed., Temple Press, 1968.
T-3 Schmitz, J. V. and Brown, W. E., *Testing of Polymers*, 4 volumes, New York, John Wiley and Sons, 1965, 1966, 1967, 1969.
T-4 Schmitz, J. V., *Bibliography of Polymer Testing, Processing, and Applications*, Plastics Institute of America and Society for Plastics Engineers, 1965.
(See also PP-1.)

11

Man-Made Textile Fibers

Colin L. Browne* and Robert W. Work**

TEXTILE BACKGROUND

The first conversion of naturally occurring fibers into threads strong enough to be looped into snares, knotted to form nets, or woven into fabrics is lost in prehistory. Unlike stone weapons, such threads, cords, and fabrics—being organic in nature—have in most part disappeared, although in some dry caves traces remain. There is ample evidence to indicate that spindles used to assist in the twisting of fibers together had been developed long before the dawn of recorded history. In that spinning process, fibers such as wool were drawn out of a loose mass, perhaps held in a distaff, and made parallel by human fingers. (A servant girl so spins in Giotto's, *The Annunciation to Anne*, c. 1306 A.D., Arena Chapel, Padua, Italy.)[1] A rod (spindle), hooked to the lengthening thread, was rotated so that the fibers while so held were twisted together to form additional thread. The finished length was then wound by hand around the spindle, which in becoming the core on which the finished product was accumulated, served the dual role of twisting and storing, and in so doing, established a principle still in use today. (Even now, a "spindle" is 14,400 yards of coarse linen thread.) Thus, the formation of any thread-like structure became known as spinning, and it followed that a spider spins a web, a silkworm spins a cocoon, and man-made fibers are spun by extrusion, although no rotation is involved.

It is not surprising that words from this ancient craft still carry specialized meanings within the textile industry and have entered everyday parlance, quite often with very different meanings. Explanations are in order for some of those that will be used in the pages that follow.

For example, as already indicated, "spinning" describes either the twisting of a bundle of essentially parallel short pieces of wool, cotton, or precut man-made fibers into thread or the extrusion of continuous long lengths of man-made fibers. In the former case, the short lengths are known as "staple" fibers and the resulting product is a "spun yarn," the long lengths are called "continuous filament yarn,"

*Celanese Fibers Co., Charlotte, N.C.
**North Carolina State University, Raleigh, N.C.

or merely "filament yarn." Neither is called a "thread," for in a textile industry that term is reserved for sewing thread and rubber or metallic threads. Although to the layman "yarn" connotes a material used in hand knitting, the term will be used in the textile sense hereinafter.

Before discussing man-made fibers, it is necessary to define some of the terms used.* The "denier" of a fiber or a yarn defines its linear density, i.e., the weight in grams of a 9000-meter length of the material at standard conditions of 70°F and 65 percent relative humidity. Although denier is actually a measure of linear density, in the textile industry the word connotes the size of the filament or yarn. Fibers usually range from 1 to 15 denier, yarns from 15 to 1650. Single fibers, usually 15 denier or larger, used singly, are termed "monofils." The cross-sectional area of fibers of identical deniers will be inversely related to their densities which range from 0.92 for polypropylene to 2.54 for glass. Since by definition denier is measured at standard conditions, it describes the amount of "bone-dry" materials plus the moisture regain, which ranges from zero for glass and polypropylene to 13 percent for rayon. It should be mentioned that some years ago scientific organizations throughout the world accepted the word "tex," this being the weight of one kilometer of the material, as a more useful term than denier. "Tex" is an accepted adjunct to the SI or "International System of Units," but it has received only limited acceptance in commerce, while the SI systems of units is being employed increasingly in scientific organizations. Furthermore, the sizes of cotton, wool, and worsted yarns, and yarns containing man-made fibers but produced by the traditional cotton, wool, or worsted systems are still expressed in the inverse-count system used by each for centuries past.

The "breaking tenacity" or more commonly, "tenacity," is the breaking strength of a fiber or a yarn expressed in force per unit denier,

*Each year the ASTM publishes in its "Book of Standards" the most recent and accepted definitions and test methods used in the textile and fiber industries.

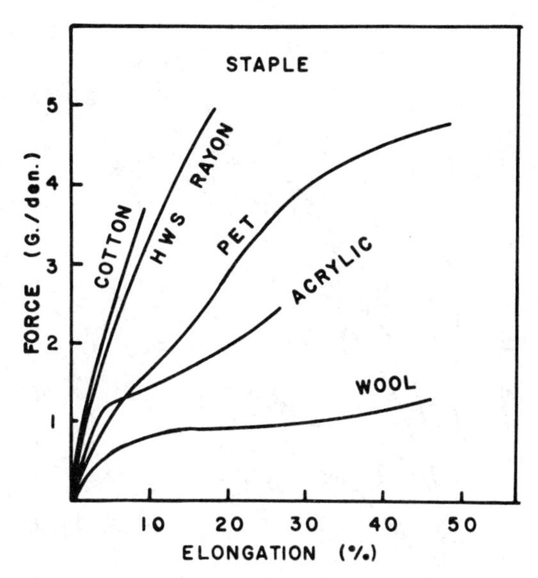

Fig. 11.1 Force-elongation relationships of natural and man-made staple fibers at standard conditions of 70° F and 65% relative humidity.

i.e., in grams per denier, calculated from the denier of the original unstretched specimen. "Breaking length" expresses the theoretical length of yarn which would break under its own weight and is used mostly in Europe. "Elongation" means "breaking elongation" and is expressed in units of length calculated as a percentage of the original specimen length.

Typical force-elongation curves of some man-made and natural staple fibers, and textile-type man-made fibers are shown in Fig. 11.1 and 11.2.

HISTORY

The earliest expression of the concept that fibers could be made by man, as well as in nature, is often attributed to Hooke. In 1664, he remarked that not only should it be possible to make a silk-like fiber, but also that such a fiber would be of value in the marketplace. Yet almost two hundred years elapsed before this concept was realized by Andemars, who drew fibers from a solution of cellulose nitrate containing some rubber; and only after considerable development and the passage of a generation did Chardonnet's patent (1885) open the door for commercialization of man-made fibers. During

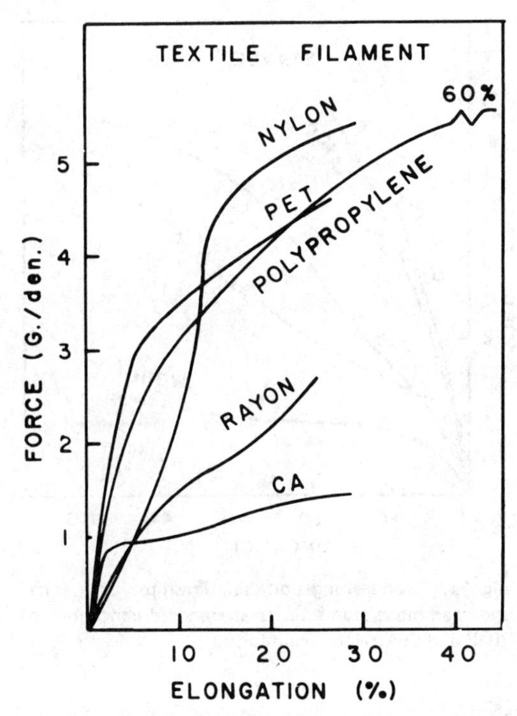

Fig. 11.2 Force-elongation relationships of man-made textile continuous filament yarns at standard conditions of 70°F and 65% relative humidity.

the last years of the nineteenth century and the beginning of the twentieth, progress was so rapid that production of the fiber known then as artificial silk, and since named rayon, increased from several thousands of pounds in 1891 to over 2 million pounds in 1910. Although in the United States the production of rayon had risen to about 150 million pounds a year by the time the revised first edition of this book was published (1933), its future was still so unpredictable that the following statement was made: "the artificial silk industry may therefore be regarded as supplementing worm silk production rather than rivaling it." That same edition also mentioned that worm-silk importation into the United States in 1931 totaled almost 90 million pounds. It is with this background in mind that the development and production of man-made textile fibers should be examined.

At first, man attempted to duplicate the fibers found in nature. Materials were developed which differed from the natural fibers but which were based, nevertheless, on such natural polymers as cellulose and proteins. The former was the basis of the two cellulosics that became successful: rayon and cellulose acetate. The second period, initiated by a technological breakthrough, was marked by the work of W. H. Carothers.[2] Fibers were described as being composed of high-molecular weight linear polymers, and the first of these, nylon 66, was synthesized and produced in large volumes. It was quickly followed by nylon 6, polyester, and acrylic fibers, and polyolefins shortly thereafter. Glass, although for unknown reasons not termed as "man-made" in the textile industry, joined this group of large production items.

At about the time that the previous edition of this book was being prepared, it was predicted that the man-made fiber industry was entering a much changed period.[3] Essentially, it was said that no new, large-volume, high-profit fibers would be developed. Instead, the existing ones would become "commodities," with all the economic impact thereby implied. It will be seen in the text to follow that the prognostication has been borne out. No major chemical engineering processes have been added, and previously described ones remain virtually unchanged. Volumes have increased enormously for nylon and polyester, but even some large producers have been forced out of competition. Small volume items and their manufacturers have fared even worse. Research activity has been reduced and centered on fiber modifications (for example, see a complilation of recent patents), and many such variants are now successfully marketed.[4]

VOLUME OF PRODUCTION*

Figure 11.3 compares population growth with the production of man-made fibers and the mill consumption of natural fibers in the United States. In this connection it should be mentioned that the per capita consumption of all fibers, starting at a level in the 1920's of about 30 pounds, rose to approximately 40 pounds following World War II, and reached 59 pounds during 1972. Since then it has topped this amount only twice and currently (1981) has dropped to 53.3 lb/capita. But clearly overshadowing the increases resulting from popula-

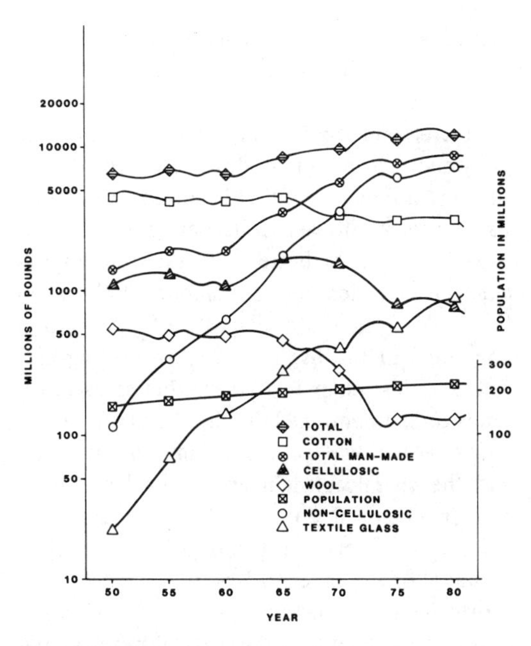

Fig. 11.3 Comparative growth of population and fiber consumption in U.S., 1950–1080.

tion growth and a higher standard of living are the volumes produced of first the cellulosic fibers and then the noncellulosic, or completely synthetic fibers. As may be seen from Fig. 11.3, the rise of the noncellulosics had great impact on the cellulosics. The combination of the two has reduced wool to a quite secondary position while at the same time deposing "King Cotton."

Perhaps the most dramatic changes have taken place in the use of the materials required in the manufacture of tire cords. Originally made from cotton, rayon took a commanding position during World War II. But as late as 1951 cotton comprised about 40 percent of the total output of tire cords of approximately half a billion pounds, and nylon was at a negligible level of 4 million pounds. By 1960, cotton had all but disappeared; nylon represented about 37 percent of the total (on a weight basis), even though only about 0.8 pounds of nylon is needed to replace 1.0 pounds of rayon. Whereas rayon for several years had dominated the so-called original-equipment tire market and nylon had held a corresponding position for replacement tires, more recently glass and polyester have made heavy inroads into both—especially in belted constructions. The situation continued to change in favor of noncellulosic man-made fiber usage in tires so that by 1972, rayon was down to 14 percent, nylon up to 42 percent, polyester up to 32 percent, glass up to 7 percent, and steel at 5 percent, all on a weight basis. In the late seventies, tire markets were dominated by wholly man-made fibers with polyester holding over 90 percent of the passenger car original equipment market and nylon commanding over 90 percent of the truck original equipment market. This division of markets is a direct result of the performance characteristics of the two fibers. Polyester-containing tires are free of "flat spotting" or cold-morning thump and so are preferred in passenger cars for their smooth ride. On the other hand, nylon-containing tires are tougher and more durable and so are the choice for trucks and off-road vehicles.

The production of man-made fibers throughout the world has developed in a manner that rather parallels the situation in the United States, as may be seen by reference to Fig. 11.4. There are some expected differences, and obviously the data for the world usage are strongly influenced by the large components attributable to the United States. The output of the cellulosics has leveled off, but the expansion of the noncellulosics has continued un-

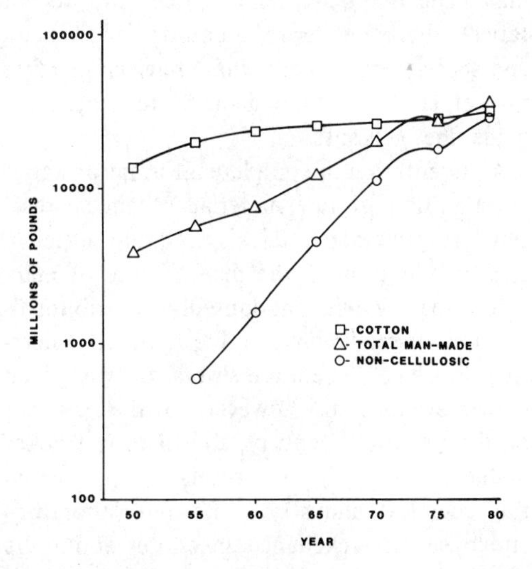

Fig. 11.4 Production of fibers in the world, 1950–1978.

abated. The use, or at least the recorded use, of the natural fibers, cotton and wool, rose rapidly in the 1950's, as the world economy rebounded at the conclusion of World War II. Since then (1960-1970) a modest increase has continued, being essentially parallel with the growth of world population. But, when compared with population trends, it appears that the great demand area has been for the man-made textile fibers. Much of this increase has resulted from an improved standard of living and the absence of major wars.

A detailed economic examination of the prices of fibers and the changes that have taken place during the last half century would show two rather vivid occurences. The first of these is the rapid decrease in the prices of the newer fibers as they became established, followed by a leveling out and stabilization. The second is the relative stability of prices of the man-made fibers on short-term and even long-term bases as compared with fluctuations in the prices for the natural fibers where governmentally imposed stability has not been in effect. Data are not presented in this text, but suffice it to say that in the first half of the twentieth century there was a truism in the textile industry of the effect that the man who made or lost money for the company was the one who was responsible for buying cotton and wool "futures." But it should also be emphasized that list prices of man-made fibers are ceiling prices and do not reflect the short-term discounts, allowances, and special arrangements that are given in a free marketplace when the demand for any man-made fiber gets soft.

A presentation of complete information about consumption of raw materials, chemical reactions, reagents and catalysts used, and efficiencies of operation in the manufacture of man-made fibers would undoubtedly contribute to a better understanding of the industrial chemistry involved. There are several factors which have prevented this, however. In the first half of the twentieth century, the historical belief in the efficacy of trade secrets still permeated the chemical industry. Retired laboratory employees still exchange anecdotes about the use in the rayon industry of thermometers calibrated in arbitrary units and of tire manufacturers whose rubber compounds might contain three portions of the same chemical taken from three differently coded bins. Even with the increased mobility of technical and scientific personnel during and following World War II, the idea still pervaded that if nothing other than patents were allowed to become public knowledge, so much the better. But the situation has changed considerably since about 1960, as can be noted from the availability of information contained in the list of suggested reading that follows this chapter. Nevertheless, secrecy tends to be maintained despite the fact that key employees move from company to company, and the chemical engineering knowledge available in chemical concerns that produce large volumes of fibers permits an almost complete appraisal of a competitor's activities.

In general, in the early period of production of a fiber, the cost of the original raw material may have had very little bearing on the selling price of the final fiber. A most important factor is the action of the producer's competitors and the conditions of the market and the demand that can be developed. But the complexity of the processes involved in conversion determines the base cost of the fiber at the point of manufacture. As a process becomes older, research reduces this complexity; with simplification, there may be rapid drops in plant cost. If demand remains high, such reductions will not be expected to be reflected in selling prices; rather, profits are high. As more producers enter the field in order to share in those profits, output capacity surpasses demand, and in accordance with classical economic theory, major selling price reductions result. This was happening, in general, in the 1960s and for cellulose based man-made fibers in the 1970s. But beginning in 1973, the cost of petroleum-based products started to rise steeply and erratically. This rise was based not on economic considerations alone but on political considerations among the oil producing and exporting countries (OPEC) as well. Further upward pressure on man-made fiber prices has resulted from governmental limitations placed on chemical usage and exposure and on amounts of chemicals that can be discharged into the air and water. To meet these limitations, the man-made fiber industry has made large infusions of capital. In some instances, the expenditure could not be justified and plant capacity was shut down permanently. This was particularly true in the case of filament rayon. In recent years, the factors of rapidly rising raw material/energy prices and the costs of meeting environmental regulations have not allowed the prices of man-made fibers to fall as production experience is gained and technological advances introduced. Instead, selling prices are continually adjusted upward in an effort to pass along unavoidable cost increases so as to maintain profitability.

The great importance of man-made fibers in the chemical industry and in the over-all economy of the United States becomes apparent when the volume of production of these materials is considered and compared with the market value of even the least expensive of the raw materials used by them. The amounts of oil and natural gas consumed by the man-made fiber industry represent around one percent of national annual usage. Of this amount, about one-half is used to produce raw materials from petrochemicals while the other half is used for energy to convert trees to wood pulp for cellulose-based fibers and to convert the wood pulp and petrochemical derived raw materials to fibers.

RAYON

Chemical Manufacture

Rayon, the first of the man-made fibers produced in large volume, depends on the natural polymer cellulose for its raw material. In the early days the main source of raw material was cotton linters, the short fibers left on the cotton seed after the lint was removed for direct use in textile yarns. This was natural since linters, a relatively pure source of cellulose, were readily available and inexpensive, and the noncellulosic portions could be removed without excessivly drastic treatment. The combination of improved technologies for purifying cellulose derived from wood, and the shortage of cotton linters during World War II when they were used for cellulose nitrate, resulted in the virtual elimination of cotton linters as a raw material for rayon.

The manufacture of rayon starts with the cellulose raw material, i.e., wood pulp. (A flow diagram for the complete process is shown in Fig. 11.5.) This is received by the rayon manufacturer in sheet form and in appearance is not unlike a thick unbacked blotter. In the manufacture of this pulp, impurities are removed, special care being given to traces of metallic elements, such as manganese and iron, which effect the manufacturing process and the quality of the final product. Since the production of the wood pulp from wood involves drastic chemical action at elevated temperatures, it is not surprising that degradation occurs which substantially reduces the originally very high molecular weight of the cellulose. Some of this breakdown is essential, but it also results in a wide distribution of molecular weights. That portion which is not soluble in 17 to 18 percent aqueous caustic is known as alpha cellulose. The lower-molecular weight beta and gamma fractions are soluble and are lost in the first step of the manufacture of rayon. The composition of the pulp is therefore aimed at high alpha content. A typical economic trade-off is involved. The pulp producers can secure an alpha content up to 96 percent or even higher by means of a cold caustic extraction or, on the other hand, the rayon manufacturer can use a less expensive, lower alpha-content pulp and expect to secure a lower yield. It goes without saying that the sellers have numerous grades available to meet the specific process needs and end-product requirements of each of the buyers.

In the manufacture of rayon it is the usual practice to begin "blending" at the first step, which involves steeping the pulp. Further blending proceeds throughout succeeding steps. The warehouse supply of pulp consists of numerous shipments from the pulp source or, in some cases, sources. In making up the batches for the conventional process, a few sheets are taken from each of several shipments, the ratio being so predetermined that all units of the stock are used, and every exhausted shipment is replaced by a new one. This serves two purposes. It prevents a slight variation in a

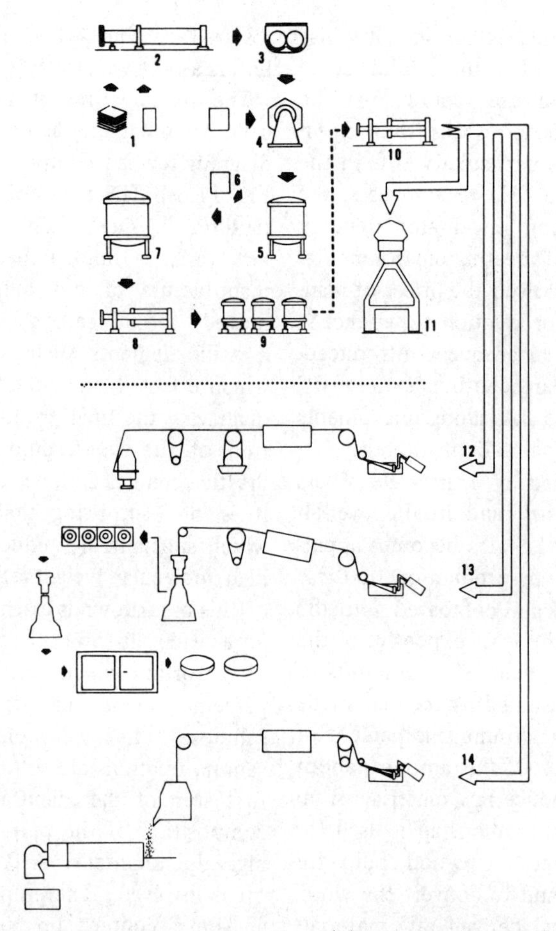

Fig. 11.5 Flow diagram for manufacture of viscose rayon: (1) cellulose sheets and caustic soda; (2) steeping press; (3) shredder; (4) xanthating churn; (5) dissolver; (6) caustic supply; (7) ripener; (8) filtration; (9) deaeration; (10) filtration; (11) continuous process; (12) tire cord; (13) pot spinning; (14) staple spinning.

single pulp lot from unduly affecting any given volume of production, and it provides a moving average so that changes in time are reduced to a minimum.

The cellulose sheets are loaded vertically, but loosely, into a combination steeping bath and press (Fig. 11.6), which is slowly filled with a solution of 17 to 20 percent caustic, and in which they remain for about 1 hour. In the steeping, the alpha cellulose is converted into alkali or "soda" cellulose; at the same time and, as already mentioned, the caustic solution removes the beta and gamma celluloses (also called hemicelluloses) which are too low in molecular weight to be used in the final rayon fiber. The exact chemical composition of the soda cellulose is not known, but there is evidence that one mole of NaOH is associated with two anhydro-glucose units in the polymer chain.

$$(C_6H_{10}O_5)_n \text{ (cellulose)} + 18\% \text{ aqueous NaOH} \longrightarrow [(C_6H_{10}O_5)_2 \cdot NaOH]_n \text{ (swollen, insoluble, soda cellulose I)} + \text{soluble soda cellulose from } \beta \text{ and } \gamma \text{ celluloses}$$

The excess caustic solution is drained off for reuse. Additional amounts are removed by forcing the sheets between press plates with a hydraulic ram. Even after this operation the sheets remain in a highly swollen condition and retain from 2.7 to 3.0 parts of the alkali solution. The spent steeping solution squeezed out

Fig. 11.6 Steeping of cellulose in the manufacture of viscose rayon. *(Courtesy Avtex Fibers, Inc.)*

of the pulp is pumped to dialyzers which separate and recover the caustic from the organic material.

As might be expected, attempts have been made to convert this batch operation to a contincuous process. In order to do so, the pulp in roll form can be fed through the caustic bath with the excess solution being removed by vacuum or other means. Another approach is to shred the sheet so that upon immersion of the free fibers in caustic a slurry is formed. This provides a rapid and complete contact between the reactants and a correspondingly accelerated formation of the soda cellulose. But, as is the case with every continuous process, there is the added burden of securing a flow of raw materials of identical properties and the need for exact control of processing conditions.

Following removal of the solution, the sheets of soda cellulose in the batch process are discharged into a shredder. Blending is again accomplished by mixing the charges from two or more steeping presses in a single shredder where the already soft sheets are torn into small pieces or "crumbs"; cooling is provided to prevent thermal degradation. Shredding is controlled to produce crumbs that are open and fluffy, and that will allow air to penetrate the mass readily; this is essential in aging.

Soda cellulose is aged by holding it at a constant temperature in perforated containers so that air can contact all the material. The oxygen in the air produces uniform aging accompanied by a reduction in molecular weight and an increase in the number of carboxyl groups present. The target of aging is an average molecular weight high enough to produce satisfactory strength in the final rayon but low enough that the viscosity of the spinning solution will not be excessively high at the desired concentration and not require reduction of the concentration to an uneconomically low point. The aging proceeds for periods of up to two or three days, although the tendency is to speed up the operation by using higher temperatures and traces of metal ions to catalyze the reaction. A combination of experience and constant quality-control testing guarantees that the material will reach the correct point for conversion to cellulose xanthate.

Cellulose xanthate, or more exactly, sodium cellulose xanthate, is obtained by mixing the aged soda cellulose with carbon disulfide in a vapor-tight xanthating churn. Based upon weight of cellulose, the amount of carbon disulfide used will be in the range of 30 percent for regular rayon to 50-60 percent for improved and modified varieties.

$$[(C_6H_{10}O_5)_2 \cdot NaOH]_n + CS_2 \text{ (30-60\% based on weight of cellulose in soda cellulose)} \longrightarrow (C_6H_{10}O_5)_n[C_6H_7O_2(OH)_x(O{-}\underset{\underset{S}{\|}}{C}{-}S \cdot Na^+)_{3-x}]_m \text{ or,}$$

$$\text{for simplicity, (cellulose}{-}O{-}\underset{\underset{S}{\|}}{C}{-}SNa)$$

The xanthate is soluble in a dilute solution of sodium hydroxide–a characteristic discovered by Cross and Bevan in 1892–and this property makes the spinning of rayon possible. It is a yellow solid; when dissolved in a dilute solution of alkali, it becomes a viscous, honey-colored liquid, hence the word "viscose." At this stage the viscous solution may contain about 7.25 percent cellulose as xanthate in about 6.5 percent solution of sodium hydroxide, although concentrations of both of these vary, depending on what end products are desired. The solution is ready for mixing with other batches to promote uniformity, to be followed by filtration, ripening, deaeration, and spinning. Filtration takes place in plate and frame filter presses, usually in several stages so that filter dressings of decreasing pore size may be used to secure a balance of through-put and step-wise particle and gel removal.

Such an operation is a straightforward one for "bright" rayon, but only in the days of "artificial silk" did the shiny fiber alone satisfy the market. After a few years, a dull appearing fiber was also demanded. At first, fine droplets of oil in the filaments were used to produce dullness until it was discovered that titanium dioxide pigment having a particle size smaller than one micron in diameter was even more satisfactory. It has since become the universal delustrant for all man-made fibers. With the use of pigments of any type, problems of dispersion and agglomerate-formation must be faced. The usual practice has been to add this pigment when mixing the cellulose xanthate into the dilute solution of caustic. However, it must be realized that not only bright and dull, but also semidull fiber is needed, and the demand for each of these three kinds of rayon varies from time to time. It should be mentioned that the producer may add a few parts per million of a tracer element to the solution so that the product can be identified in the event of later complaints.

When the "dope dyeing" of rayon (the use of colored pigments in the fiber) became important, it was no longer possible in all cases to make the final spinning solution during the process of dissolving the xanthate. ("Dope dyeing" will be discussed in greater detail under another heading; it is sufficient to point out here that a large volume of this type of rayon is manufactured and is of considerable importance.) Cleaning or flushing equipment, made necessary by change-overs, involves additional costs. Thus, development of a process which would inject colored pigments as close to the spinning as is possible is desirable. Moreover, in recent years, chemicals other than pigments have had to be added in order to obtain a rayon different from the simple material which was the standard product in the early days of the industry. In order to produce rayon of different physical properties, the rate of precipitation and regeneration of the cellulose is modified by the addition of various chemicals to the spinning bath.

From the standpoint of chemical processing,

it is obvious that pigments and other chemicals may be added when the sodium cellulose xanthate is dissolved in dilute caustic solution, or at any time up to the moment before it enters the spinnerette prior to being extruded. To keep the operations as flexible as possible, the additives should be injected at the last possible moment so that when a change-over is desired there will be a minimum amount of equipment to be cleaned. On the other hand, the farther along in the operation that additives are placed into the stream the greater the problem of obtaining uniformity in an extremely viscous medium, and the greater the difficulty in maintaining exact control of proportions before the viscous solution is passed forward and spun. Furthermore, all insoluble additives must be of extremely small particle size, and all injected slurries must be freed of agglomerates by prefiltration; if not, the viscous solution containing the additives must be filtered. Each manufacturer of viscose rayon develops his own particular conditions for making additions, depending on a multitude of factors, not the least of which is the existing investment in equipment. All manufacturers must face the universal necessity of filtering the solution with or without pigments or other additives, so that all impurities and agglomerates which might block the holes in the spinnerette are removed.

Although it was known in the years following the discovery by Cross and Bevan that a viscose type of solution could be used in the preparation of regenerated cellulose, the conversion of this solution into useful fibers was not possible until the discovery that the solution required aging until "ripe." Ripening is the first part of the actual chemical decomposition of cellulose xanthate which, if allowed to proceed unhampered, would result in gelation of the viscose solution.

$$\text{Cellulose}-\text{O}-\underset{\underset{\text{S}}{\|}}{\text{C}}-\text{SNa} \overset{H_2O}{\rightleftarrows} \text{Cellulose}-\text{O}\underset{\underset{\text{S}}{\|}}{\text{C}}-\text{SH} + \text{NaOH} \overset{H_2O}{\rightleftarrows}$$

$$\text{Cellulose} + \text{HO}\underset{\underset{\text{S}}{\|}}{\text{C}}-\text{SH} \longrightarrow \text{Cellulose} + CS_2 + H_2O$$

Experience has taught the manufacturer the correct time and conditions for his aging operation, but the requirement of aging itself demands that the entire process be so planned that the viscose solution will arrive at the spinnerette possessing, as nearly as possible, the optimum degree of ripeness so as to produce fibers having the desired characteristics. This degree of ripeness is determined by an empirical test made periodically. It is a measurement of the resistance of the solution to precipitation of the soda cellulose when a salt solution is titrated into it. Thus it is known as the "salt index" or "Hottenroth number" after its originator.

An additional step in the over-all ripening operation involves the removal by vacuum of residual, dissolved and mechanically-held air. More recent practice is to use a vacuum on a moving thin film rather than on the bulk solution in a tank.

It should be mentioned that so inevitable is decomposition of cellulose xanthate and consequent gelation of the contents of pipes and tanks, that all viscose rayon plants must be prepared to pump in-process viscose solutions to a waste receiver, purging the entire system with dilute caustic solution in the event of a long delay in spinning.

Wet Spinning

Spinning a viscose solution into rayon fibers (wet spinning) is the oldest of the three common ways of making man-made fibers. In this method, the polymer is dissolved in an appropriate solvent, and this solution is forced through fine holes in the face of the spinnerette which is submerged in a bath of such composition that the polymer precipitates. The pressure necessary for this extrusion is supplied by a gear pump which also acts as a metering device; the solution is moved through a final or "candle" filter before it emerges from the holes of the spinnerette. There is immediate contact between these tiny streams and the liquid or "wet" bath. As the bath solution makes contact with the material extruded from the holes, chemical or physical changes take place. These changes, whether of lesser or greater complexity, convert the solution of high-molecular weight linear polymer first to a gel structure and then to a fiber. It is an interesting fact that the spinning of viscose rayon, with all of the ramifications made possible by variations in the composition of the solution and the precipitating bath, as well as in the operating conditions, presents the chemist and chemical engineer with both the oldest and the most complex wet-spinning process.

The formation of rayon fibers from viscose solution is far from being simple, either from the physical or chemical standpoint. The spinning bath usually contains 1 to 5 percent zinc sulfate and 7 to 10 percent sulfuric acid, as well as a surface-active agent, without which minute deposits will form around the holes in the spinnerette. Sodium sulfate (15 to 22 percent) is present, formed by the reactions, and as sulfuric acid is depleted and sodium sulfate concentration builds up, an appropriate replenishment of the acid is required. There is a coagulation of the organic material as the sulfuric acid in the spinning bath neutralizes the sodium hydroxide in the viscose solution; at the same time, chemical decomposition of the sodium cellulose xanthate takes place to regenerate the cellulose. If zinc ions are present, which is the usual situation in the production of the improved types of rayon, an interchange takes place so that the zinc cellulose xanthate which precipitates less quickly, becomes an intermediate. Chemical additives are usually present to repress hydrogen-ion action. The gel-like structure, the first state through which the material passes, is not capable of supporting itself outside the spinning bath. As it travels through the bath, however, it quickly becomes transformed into a fiber that can be drawn from the spinning bath and that can support itself in subsequent operation (Fig. 11.7). The reactions between the bath and the fiber which is forming are paramount in determining the characteristics of the final product; it is for this reason that additives (previously mentioned), as well as zinc ions, may be used to control both the rate of coagulation and regeneration. In this manner, the arrangement of the cellulose molecules may be controlled to produce the structural order desired. One important outcome of this flexibility has been the development of "high wet strength, high modulus" rayons, commercialized under a number of trade names. Unlike the older "regular" rayon, the rupture tenacity of which drops to about 1.5 g/den. when wetted, these go only to about 3 g/den. (see Figures 11.1 and 11.2). As the generic description implies, their properties tend in the direction of those of cotton, and are usable as a replacement for cotton, especially in blends.

(Rapid reaction)

$$2\ \text{Cellulose}-O-\underset{\underset{S}{\|}}{C}-SNa + H_2SO_4 \longrightarrow \text{Cellulose} + Na_2SO_4 + CS_2$$

(Slow reaction)

$$2\ \text{Cellulose}-O-\underset{\underset{S}{\|}}{C}-SNa + ZnSO_4 \longrightarrow (\text{Cellulose}\ O-\underset{\underset{S}{\|}}{C}-S)_2\ Zn + Na_2SO_4$$

$$(\text{Cellulose}\ O-\underset{\underset{S}{\|}}{C}-S)_2\ Zn + H_2SO_4 \longrightarrow \text{Cellulose} + ZnSO_4 + CS_2$$

Because of hydraulic drag, stretching occurs in the bath and also in a separate step after the yarn leaves the bath. In both cases, the linear molecules of cellulose are oriented from random positions to positions more parallel to the fiber axis. If a rayon tire cord is to be the final product, the fibers must be severely stretched to produce a very high orientation of the molecules, this being the basis of the tire cord's high strength and ability to resist stretching without which growth of the tire body would occur. For regular textile fibers such high strengths are not desired, and the spinning and stretching operations are controlled to produce rayon of lower strength and greater stretchability under stress.

In order to stretch the yarn uniformly during the manufacturing process, two sets of paired

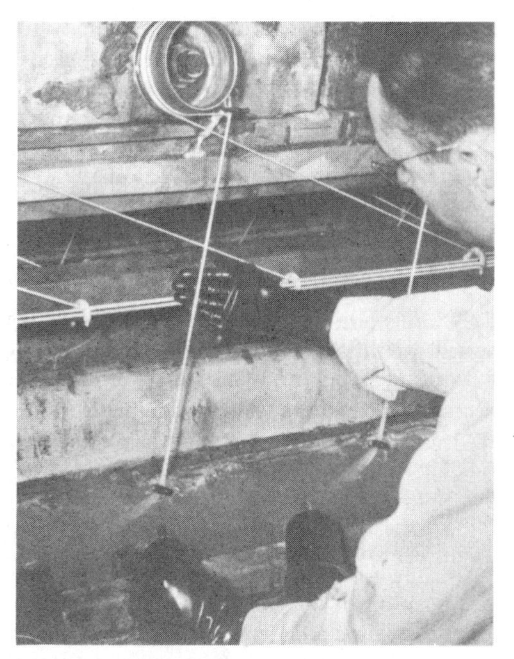

Fig. 11.7 Spinning of viscose rayon. *(Courtesy of Avtex Fibers, Inc.)*

rollers or "godets" are employed, each of the two sets operating at different rotational speeds. The yarn is passed around the first set of godets several times to prevent slippage and is supplied to the stretching area at a constant speed. A second set of godets moves it forward at a more rapid rate, also without slippage. Stretching may range from a few to 100 or more percent. Spinning speeds are on the order of 100 meters per minute, but these may vary with both the size of the yarn and the process used.

Spinning conditions, composition of the spinning bath, and additives to the viscose solution determine the physical characteristics of the rayon, its breaking strength and elongation, modulus, ability to resist swelling in water, and its characteristics in the wet state as compared with those of the dry material. Not only must the chemical composition of the spinning bath be carefully controlled but the temperature must also be regulated at a selected point, somewhere in the range of 40° to 60°C, to ensure those precipitation and regeneration conditions essential to the manufacture of any particular viscose rayon having the properties needed for a selected end use.

After the precipitating bath has reacted chemically with the viscose solution to regenerate the cellulose and raw rayon fiber has been formed, the subsequent steps must be controlled so that differences in treatment are minimized; otherwise such sensitive properties as "dye acceptance" will be affected and the appearance of the final product will vary. It is not the purpose of this description of processes to include information on dyeing and finishing of fabrics used for garments and in households. But it is well to emphasize an old adage at this point, as true today as ever. In essence, fibers that cannot be supplied to the public in a full range of colors and shades are fated to fit only limited specialty or industrial uses. The manufacturers of rayon learned early how to meet these requirements.

Minute traces of suspended sulfur resulting from the chemical decomposition of cellulose xanthate must be removed by washing with a solution of sodium sulfide. It is expedient to bleach the newly formed fibers with hypochlorite to improve their whiteness; an "antichlor" follows. The chemicals originally present and those used to purify the fibers must be removed by washing. As a final step, a small amount of lubricant must be placed on the filaments to reduce friction in subsequent textile operations.

There are several different processes used for the steps involved in spinning and purifying continuous filament rayon. One of the most common involves the formation of packages of yarn, each weighing several pounds, for separate treatment. After it has been passed upward out of the spinning bath and stretched to the desired degree, the yarn is fed downward vertically into a rapidly rotating can-like container called a spinning pot or "Topham" box (after the man who invented it in 1900). It is thrown outward to the wall of the pot by centrifugal force and gradually builds up like a cake with excess water being removed by the same centrifugal force. This cake is firm, although it must be handled with care, and is sufficiently permeable to aqueous solutions to permit purification.

In another method of package-spinning, the yarn is wound onto a mandrel from the side at a uniform peripheral speed. With this process, the yarn may be purified and dried in the package thus formed. In any of these systems, the spinning and stretching, as well as subsequent steps, may involve separate baths.

The continuous process for spinning and purifying textile-grade rayon yarn merits particular mention from the standpoint of industrial chemistry since it is rather an axiom that a continuous process is to be preferred over a batch or discontinuous operation. This method employs "advancing rolls" or godets which make it possible for the yarn to dwell for a sufficient length of time on each pair, thus allowing the several chemical operations to take place in a relatively small area. Their operation depends on the geometry existing when the shafts of a pair of adjacent cylindrical rolls are oriented slightly askew. Yarn led onto the end of one of these and then around the pair will progress toward the other end of the set with every pass, the rate of traversing, and therefore the number of wraps, being determined by the degree of skewness.

The production of rayon to be converted to staple fiber is also amenable to line operation. Here the spinnerette has many thousands of holes and a correspondingly large number of filaments are formed in the precipitation bath. The resulting tow is then stretched to the desired degree and immediately cut in the wet and unpurified condition. The mass of short lengths can be belt conveyed through the usual chemical treatments after which it is washed and dried. It is fluffed to prevent matting and packaged for shipment in large cases.

Nitro, Cuprammonium, and Cellulose Acetate Processes for Rayon

Nitrocellulose. Although nitrocellulose was spun and regenerated into cellulose in the early days of the rayon industry, it is a good many years since this process has been used commercially, and today it is only of historical interest.

Cuprammonium Cellulose. Cellulose forms a soluble complex with copper salts and ammonia. Thus, when cellulose is added to an ammonical solution of copper sulfate which also contains sodium hydroxide, it dissolves to form a viscous blue solution and in this form it is known as cuprammonium cellulose. When used in determining the viscosity of cellulose as a measure of its molecular weight, it has been called "Schweizer solution." The principles on which the chemical and spinning steps of this process are based are the same as those for the viscose process. Cellulose is dissolved, in this case in a solution containing ammonia, copper sulfate, and sodium hydroxide. Unlike the viscose solution, the cuprammonium solution need not be aged and will not precipitate spontaneously on standing except after long periods. It is, however, sensitive to light and oxygen. It is spun into water and given an acid wash to remove the last traces of ammonia and copper ions. Although this rayon was never manufactured in volume even approaching that achieved by the viscose process, the smaller individual filaments inherent to it made it useful in certain specialty markets. It is no longer manufactured in the United States but continues to be made abroad.

Cellulose Acetate. The conversion of cellulose to cellulose acetate is still another method of making cellulose soluble, and therefore, spinnable. (The manufacture of cellulose acetate will be described in a later section.) A solution of the acetate in acetone may be wet spun under conditions that cause it to coagulate and precipitate in typical fiber form and then it may be saponified back to cellulose in batch operation or continuous process. On the other hand, cellulose acetate fibers, dry spun in a manner that will be described in a later section, can be stretched in the presence of steam, which plastifies them, followed by saponification in a caustic solution containing a relatively large amount of sodium acetate to prevent disorientation from taking place.

Regenerated cellulose yarn of extremely high strength but low breaking elongation was produced in small volume by these methods under the trade name Fortisan.®* but its production was discontinued in 1970.

---

*Registered trade-mark, Celanese Corporation.

Textile Operations

After the filament rayon fiber has been spun and chemically purified, much of it passes through what are known as "textile operations" before it is ready to be knitted or woven. Since these steps of twisting and packaging or beaming are common to the manufacture of all man-made fibers, it is advisable to review briefly the background and processes.

Rayon, the first man-made fiber, not only had to compete in an established field, but also had to break into a conservative industry. Silk was the only continuous filament yarn, and products made from it were expensive and had high prestige. This offered a tempting market for rayon. Thus, the new product entered as a competitor to silk and, as already noted, became known as "artificial silk." Under the circumstances it was necessary for rayon to adapt itself to the then existing processing operations and technologies.

When making yarn, it was customary to twist several of the silk filaments together in order to secure a yarn of desired size and strength and prevent the breaking of a single filament from pushing back to form a fuzz ball. Since rayon was weaker than silk and the individual filaments were smaller, it required the same amount or even more twisting. In pot spinning, the twists were two to three per inch at the start; in package spinning, none. Experience proved that both the degree of twisting and the size of the package desired by the weaver or knitter varied according to the final application.

These operations could have been carried out in the same plant where the yarn was spun, but the existence of silk "throwsters" (from the Anglo-Saxon "*thrāwan*," to twist or revolve) made this unnecessary. As the rayon industry developed, the amount of yarn twisted in the producing plant or sent forward to throwsters was the result of many factors. Over the years the trend has been to use less twist and to place, instead, several thousand parallel ends directly on a "beam," to form a package weighing as much as 300 to 400 pounds. Such a beam is shipped directly to a weaving or knitting mill. The advent of stronger rayons, as well as other strong fibers, and the diminishing market for crepe fabrics which required highly twisted yarns, accelerated the trend away from twisting.

In all twisting and packaging operations the yarn makes contact with guide surfaces and tensioning devices, often at very high speeds. To reduce friction it is necessary to add a lubricant as a protective coating for the filaments. This is generally true of all man-made fibers, and it is customary to apply the lubricant or "spinning finish" or "spinning lubricant" as early in the manufacturing process as possible. In the case of those materials which develop static charges in passing over surfaces this lubricant must also provide antistatic characteristics.

It is difficult to overstate the importance of fiber lubricants to the successful utilization of man-made fibers. There are few problems which can be more damaging to a fiber handling operation than a lubricant upset. A separate chapter could be written on lubricant usage but some of the more important factors will be mentioned here. Obviously, lubricants must change friction between the fiber and various surfaces to allow movement without excessive damage to the fiber or the surface contacted. The surface could be any one of a variety of metal or ceramic surfaces which are contacted over a wide range of relative velocities. Another important surface interface is that with other fibers. This is critical in staple fiber processing and in maintaining yarn package integrity. The lubricant composition must be stable under a variety of storage conditions without decomposing or migrating within the package or being lost from the fiber surface by adsorption within the fiber. It is essential to those handling the fiber that the fiber finish be nontoxic and nondermatitic. It must also be compatible with other materials added during textile processes such as the protective size coat added to warp yarns before weaving or the wax often applied to yarns before circular knitting. Possible metal corrosion must be evaluated for each lubricant composition. Finally, after having performed its function, the fiber finish or lubricant must be completely scoured from the fabric to permit uniform adsorption of dyes and fabric finishes. The application of sewing lubricants to fabrics to be cut and sewn is yet another topic.

Spun Yarn. After rayon became established in the textile industry where it could be used as a silk-like fiber, and its selling price was greatly reduced, other markets were developed. The cotton, wool, worsted, and linen systems of converting short discontinuous fibers to yarns were well established, and their products were universally accepted. Here again it was necessary to make rayon fit in with the requirements of available and historically acceptable operations. The first of these was that it be cut into lengths the same as those found in cotton and wool. The viscose rayon process was and is eminently suited to the production of tows containing thousands of filaments. The pressure required to force the solution through the holes is so low that neither thick metal sections nor reinforcement of the surface is necessary to prevent bulging and large spinerettes containing several thousand holes can be used. Furthermore, the spinning bath succeeds in making contact with all the filaments uniformly. As a result, the spinning of viscose rayon tow is very similar in principle to the production of the smaller continuous filament yarns.

Since both cotton and wool possess distortions from a straight rod-like structure, machinery for their processing was designed to operate best with such crimped fibers. It was necessary for rayon staple to possess similar lengths and crimpiness in order to be adapted to existing equipment. The crimp, that is, several deformations from straightness per inch, may be produced in rayon either "chemically" or mechanically. In the former case, the precipitation-stretching step in spinning is carried out so that the skin and core of the individual filaments are radially non-uniform and constantly changing over very short lengths along the filaments. Since the skin and core differ in sensitivity to moisture, a situation occurs which is not unlike the thermal effect on a bimetallic strip, with consequent distortion of the filaments. Mechanical crimp is produced by feeding the tow between two wheels, which in turn force it into a chamber already tightly filled with tow. As soon as the tow leaves the nip of the feeding wheels it is forced against the compacted material ahead of it, and the straight filaments collapse immediately. As the mass of material is pressed forward it becomes tightly compacted in the distorted condition and remains that way until it escapes through a pressure loaded door at the opposite end of the chamber.

CELLULOSE ACETATE

Historical

Cellulose acetate was known as a chemical compound long before its potential use as a plastic or fiber-forming material was recognized. The presence of hydroxyl groups had made it possible to prepare cellulose esters from various organic acids, since cellulose consists of a long molecular chain of beta-anhydro-glucose units, each of which carries three hydroxyl groups—one primary, the other two secondary. The formula for cellulose (already noted) is $[C_6H_7O_2(OH)_3]_n$; when this is fully esterified, a triester results. It was learned quite early that whereas cellulose triacetate is soluble only in chlorinated solvents, a product obtained by partial hydrolysis of the triester to a "secondary" ester (having about 2.35 to 2.40 acetyl groups per anhydro-glucose unit) was easily soluble in acetone containing a small amount of water. Many other cellulose esters have been prepared, but only the acetate has been commercialized successfully as a man-made fiber. Propionates and butyrates, and mixed esters of one or both with acetate, have applications as plastics.

Cellulose nitrate, a well-known material, was used to "dope" the fabric wings of fighter planes in World War I to render them taut and impermeable to air. With the advent of tracer bullets these aircraft were dubbed "flaming coffins," and the need for expanded production of the less flammable secondary cellulose acetate became imperative. Thus at the end of the war the brothers Camille and Henri Dreyfus found themselves in possession of a large factory and a product having no apparent usefulness. Necessity being the mother of invention, they not only developed a method for spinning cellulose acetate into a fiber, they also sponsored the research which resulted in new dyestuffs to color it and thus make it a broadly

Fig. 11.8 A modern integrated plant designed for the continuous flow of materials from raw chemicals to the shipment of final textile yarns. *(Courtesy Celanese Fibers Co.)*

salable textile product. Their story is often told because in the minds of many scientists their contributions to fiber science seem to have concluded an era. They were among the last inventors who built and controlled industries based upon their inventions and who reaped huge financial rewards therefrom.

Manufacture of Secondary Cellulose Acetate

Cellulose acetate was originally made from purified cotton linters, but this raw material has been entirely replaced by wood pulp. The other raw materials used are acetic acid and acetic anhydride, which for many years have been produced from natural gas by the petrochemical industry.

The manufacture (See Fig. 11.9) of cellulose acetate is a batch operation. There has been mention in the patent literature of a continuous process, but its utilization as a production process has not been announced. The "charge" of cellulose is of the order of 800 to 1500 pounds. It is pretreated with about one-third its weight of acetic acid and a very necessary amount of moisture, about 6 percent of its weight. If it is too dry at the time of use, more water must be added to the acetic acid. A small amount of sulfuric acid may be used to assist in swelling the cellulose and make it "accessible" to the esterifying mixture.

Although there has been much discussion of the chemistry of cellulose acetylation, it is now generally agreed that the sulfuric acid is not a "catalyst" in the normal sense of the word, but rather that it reacts with the cellulose to form a sulfo ester. The acetic anhydride is the reactant which provides the acetate groups for esterification. The acetylation mixture consists of the output from the acetic anhydride recovery unit, being about 60 percent acetic acid and 40 percent acetic anhydride, in an amount 5 to 10 percent above the stoichiometric requirement,

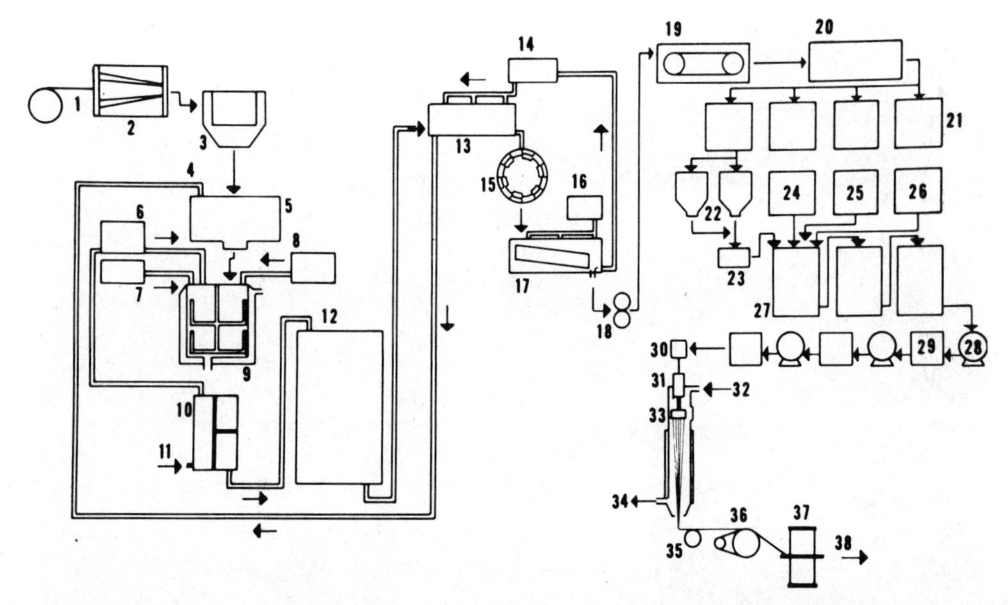

Fig. 11.9 Flow diagram for manufacture of cellulose acetate yarn: (1) wood pulp; (2) attrition mill; (3) cyclone; (4) 35% acetic acid; (5) pretreater; (6) magnesium acetate solution; (7) precooled acetylation mix; (8) sulfuric acid; (9) acetylator; (10) ripener; (11) steam; (12) blender; (13) precipitator; (14) dilute acetic acid; (15) hammer mill; (16) water; (17) rotary screen washer; (18) squeeze rolls; (19) drying oven; (20) blender; (21) storage bins; (22) silos; (23) weight bins; (24) acetone; (25) wood pulp; (26) pigment; (27) mixers; (28) hold tanks; (29) filter press; (30) pump; (31) filter; (32) air; (33) jet; (34) acetone recovery; (35) oiling wheel; (36) feed roll; (37) bobbin; (38) inspection.

to which has been added 10 to 14 percent sulfuric acid based on the weight of cellulose used. The reaction is exothermic and requires that the heat be dissipated.

In preparing for acetylation, the liquid reactants are cooled to a point where the acetic acid crystallizes, the heat of crystallization being removed by an appropriate cooling system. When a temperature of about 0°C is reached, the slush of acetic acid crystals in the acetic anhydride-sulfuric acid mixture is pumped to the acetylizer, a brine-cooled mixer of heavy construction. This may be either a unit equipped with sigma blades on horizontal axes or a tank carrying a vertically-mounted stirrer. The pretreated cellulose is dropped in from the pretreating unit located above. The reaction is highly exothermic, and at the start large amounts of heat are produced. As the temperature of the reaction mixture rises to the melting point of the acetic acid (16.6°C) its large heat of fusion (45.91 cal/gram) prevents a dangerous rise in temperature which would degrade the molecular weight of the cellulose chain. As the

Fig. 11.10 Process vessel for acetylation of cellulose. (*Courtesy Celanese Fibers Co.*)

reaction proceeds, brine in the jacket of the acetylizer provides additional cooling.

$$\text{Cellulose} + (CH_3CO)_2O + H_2SO_4 \text{ (10–15\% based on weight of cellulose)} \xrightarrow[\text{Anhydrous}]{CH_3COOH} [C_6H_7O_2(OSO_3H)_{0.2}(CH_3COO)_{2.8}]_n$$

The reaction product is soluble in the acetylation mixture; as it is formed and dissolved, new surfaces of the cellulose are presented to the reagents. One variation of this procedure uses methylene chloride, rather than an excess of acetic acid in the reaction mixture. This chemical is used both to dissipate the heat by refluxing (boiling point, 41.2°C) and to dissolve the cellulose ester as it is formed. As the reaction proceeds, the temperature is allowed to rise. Since the cellulose is a natural product obtained from many sources it varies slightly in composition and the end of the reaction cannot be predicted exactly; the disappearance of fibers as determined by microscopic examination is therefore the usual means of following its progress.

During the acetylation operation, a certain amount of chain fission is allowed to take place in the cellulose molecule. This is to ensure that the viscosity of the cellulose acetate spinning solution will be as low as possible for ease of handling but high enough to produce fibers with the preferred physical characteristics. The temperature of the reaction controls the rates of both acetylation and molecular-weight degradation. By carefully regulating the temperature, acetylation can be carried to completion in less time than is needed to reduce the molecular weight to the desired value. The final adjustment to this target can then be easily made by continued stirring at a constant temperature.

The next step in the manufacture of cellulose acetate is "ripening," whose object is to convert the triester, this is, the "primary" cellulose acetate, to a "secondary" acetate having an average of about 2.35 to 2.40 acetyl and no sulfo groups (if any sulfuric acid is used in pretreatment) per anhydro-glucose unit. While the cellulose sulfo-acetate is still in the acetylizer, sufficient water is added to react with the excess anhydride and start the hydrolysis of the ester. Usually the water is used as a solution of sodium or magnesium acetate which increases the pH and promotes hydrolysis. In the early years of the industry, ripening was allowed to proceed slowly at room temperature for a period of over 24 hours. Currently the procedure is to raise the temperature to about 70 to 80°C, by direct injection of steam, to speed the reaction. Although this introduces additional water, the total amount is far below the point where insolubility occurs; with constant stirring, hydrolysis, first of the sulfo and then of the acetyl groups, is relatively homogeneous. Hydrolysis is continued until the desired acetyl content is obtained. When this value is reached, an aqueous solution of magnesium or sodium acetate is added to cool the batch and stop the hydrolysis. It is then ready for precipitation. For example,

$$[C_6H_7O_2(OSO_3H)_{0.2}(CH_3COO)_{2.8}]_n + (CH_3COO)_2Mg \xrightarrow[]{\text{aqueous conc. } CH_3COOH} [C_6H_7O_2(OH)_{0.65}(CH_3COO)_{2.35}]_n + MgSO_4$$

The solution is carried to the verge of precipitation by adding dilute acetic acid. It is then flooded with more dilute acetic acid and mixed vigorously, so that the cellulose acetate comes out as a "flake" rather than a gelatinous mass or fine powder. The flake is then washed by standard countercurrent methods to remove the last traces of acid, and dried in a suitable dryer.

Manufacture of Cellulose Triacetate

When completely acetylated cellulose, rather than the secondary ester, is the desired product, the reaction must be carried out with perchloric acid rather than sulfuric acid as the catalyst. In the presence of 1 percent perchloric acid, a mixture of acetic acid and acetic anhydride converts a previously "pretreated" cellulose to triacetate without changing the morphology of the fibers. If methylene chloride rather than an excess of acetic acid is present in the acetylation mixture, a solution is obtained. However, it is not imperative to secure an ester with no hydroxyl groups whatsoever in order to obtain a fiber which will behave in essentially the same way as the triester. It is possible to use about 1 percent sulfuric acid instead of perchloric acid. When the sulfoacetate obtained from such a reaction is hydrolyzed with the objective of removing only the sulfo-ester groups, the resulting product has about 2.94 acetyl groups per anhydroglucose unit instead of the theoretical 3.00 of the triacetate. The preparation, hydrolysis, precipitation, and washing of "triacetate" are in all other respects similar to the corresponding steps in the manufacture of the more common secondary acetate.

Acid Recovery. In the manufacture of every pound of cellulose acetate about 4 pounds of acetic acid are produced in 30 to 35 percent aqueous solution. Needless to say, every attempt is made to move the more dilute solutions countercurrent to the main product so as to raise the concentration of the material to be recovered. For example, as already mentioned, when the acid dope is diluted and precipitation is produced by flooding, previously obtained dilute acid rather than water is used.

The accumulated acid contains a small amount of suspended fines and some dissolved cellulose esters of low acetyl value and molecular weight. To remove the suspended material the acid is passed slowly through settling tanks. Then it is mixed with organic solvents, for example, a mixture of ethyl acetate and benzene, which concentrates the acid in an organic layer which is then decanted. Distillation separates the acid from the solvent and concentrates the acid to a glacial grade needed for conversion to the anhydride.

To produce the acetic anhydride, the acid is dehydrated to ketene and reacted with acetic acid using a phosphate catalyst at 500°C or higher in a tubular furnace.

$$CH_3COOH \xrightarrow[\text{catalyst}]{\text{heat}} H_2O + CH_2{=}C{=}O$$

$$CH_2{=}C{=}O + CH_3COOH \longrightarrow (CH_3CO)_2O$$

Fig. 11.11 Recovery of acetic acid. *(Courtesy of Celanese Fibers Co.)*

Both the dehydration and the reaction with acetic acid may be carried out in separate steps; or they may be allowed to take place with the products quenched so that the water produced does not react with the anhydride. In the latter operation, the mixture of unreacted acid, water, and anhydride is fed directly to a still which yields dilute acetic acid overhead and an anhydride-acetic acid mixture at the bottom. Conditions are controlled in such a way that the raffinate is about 40 percent anhydride and 60 percent acetic acid. As already mentioned, this is the desired ratio for the reaction mixture used for acetylation of cellulose.

Blending of Flake. As in the manufacture of viscose, the products of batch operations are blended to promote uniformity in the manufacture of cellulose acetate, since it has not been found possible to manufacture a consistently uniform product. Although a blend of different celluloses is selected in the beginning, the pretreatment, acetylation, and ripening are batch operations with little or no mixing. Before precipitation, a holding tank provides an opportunity for mixing; then precipitation, washing, and drying–all continuous–promote uniformity. The dried cellulose acetate flake moves to holding bins for analysis; the moisture content, acetyl value, and viscosity being especially important. The results of the analyses determine what further blending is necessary to obtain a uniform product; rather elaborate facilities are used to select, weigh, and mix the flake. After blending and mixing portions of selected batches, the lot is air-conveyed to large storage bins or "silos" which are filled from the center of the top and emptied from the center of the bottom, thus bringing about further mixing.

Spinning Cellulose Acetate

Although cellulose acetate was originally dissolved in acetone containing a small amount of water in large rotating drum-like mixers, this has long since been replaced by a continuous operation. The acetone is metered into a vertical tank equipped with a stirrer, and the cellulose acetate flake and filter aid are weighed in an automatic hopper; all operations are controlled by proportioning methods common to the chemical industry. The ratio of materials is about 25 percent cellulose acetate, 4 percent water, less than 1 percent ground wood pulp as a filter aid, and the remainder acetone. The mixture moves forward through two or three stages at the rate at which it is used, the hold-time being determined by experience. After dissolution is completed, filtration is carried out in batteries of plate and frame filter presses in three or even four stages. The filter medium of the first stages may be any of several highly open-type pulps; the latter are often the blotter-type cellulose sheets so that the passage of the "dope" is through presses of decreasing porosity. The wood-pulp filter aid is removed early in the process and by building up in the presses, presents new filter surfaces.

Much of the cellulose acetate is delustered by the addition of titanium dioxide pigment, as with viscose rayon. Similarly, from the standpoint of thorough dispersing and mixing, it is desirable to add the pigment as early as possible in the dissolving and mixing operation. But this creates cleaning problems if the equipment is to be used alternately for bright and dull dope. Generally speaking, the ratio of demand for these two products is fairly constant and it is preferable to use separate systems to produce them. Each manufacturer must make compromises according to his particular needs.

Between each (and after the last) filtration, the dope goes to storage tanks which serve to remove bubbles; in this case, a vacuum is not necessary. From the final storage tank it is pumped into a header located at the top of each spinning machine; then it is directed to a series of metering gear pumps, one for each spinnerette, which in some sectors of the acetate industry is traditionally called a "jet." Since the holes in the cellulose acetate-spinning spinnerette are smaller (0.03 to 0.05 mm) than those in the corresponding viscose devices, great care must be taken with the final filtration. An additional filter for the removal of any small particles that may have passed through the large filters is placed in the fixture, sometimes called the "candle," to which the spinnerette asembly is fastened. A final filter is placed in the spinnerette-assembly unit over the top of the spinnerette itself.

The operating unit for spinning cellulose acetate illustrates one of the three common spinning methods, that is "dry" spinning. Having come after the development of viscose rayon manufacture, which had been named "wet" spinning, the absence of water and its replacement by heated air was logic enough for the use of the contrasting word. The dope is heated (in some cases above the boiling point of acetone, 56.5°C) to lower its viscosity and thus reduce the pressure required to extrude it, and to supply some of the heat needed for evaporating the acetone solvent. Heat may be produced in a thermostatically controlled preheater or supplied from the heated cabinet itself.

The spinnerette is stainless steel, and because the filaments must be heated and prevented from sticking together, and because space must be allowed for the escape of acetone vapor, the holes must be kept farther apart than those of the spinnerettes used for wet spinning. As the hot solution of cellulose acetate in acetone emerges downward into the spinning cabinet,

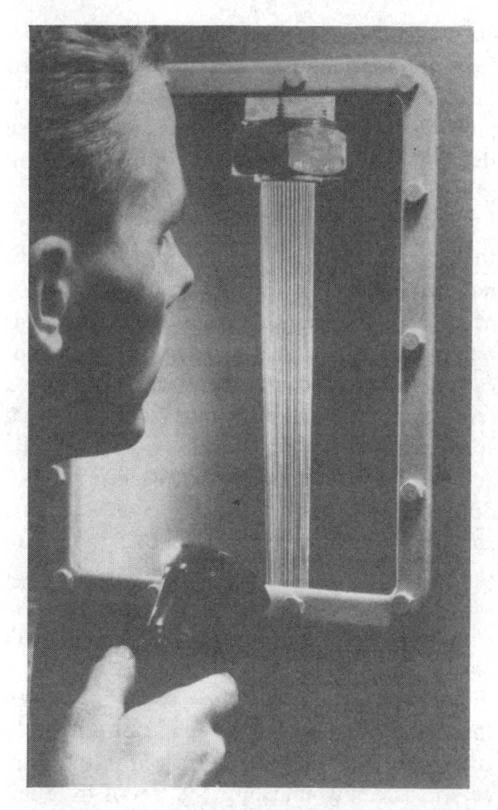

Fig. 11.12 Dry spinning of cellulose acetate. (*Courtesy Tennessee Eastman Co.*)

an instantaneous loss of acetone takes place from the surface of the filaments which tend to form a solid skin over the still liquid or plastic interior. A current of air, either in the direction the filaments are moving, or countercurrent, heats the filaments, and as the acetone is diffused from the center through the more solid skin, each filament collapses to form the indented cross-sectional shape typical of cellulose acetate. The heated air removes the vaporized acetone. Each manufacturer uses a preferred updraft, downdraft, or mixed-draft operation, as his needs dictate.

The cabinet through which the yarn passes vertically downward must be long enough to allow sufficient acetone to diffuse outward and evaporate from the surfaces of the filaments so that the latter will not stick to the first surface contacted nor fuse to each other. The temperature of the air in the cabinet, the rate of flow, the length of the cabinet, the size and number of filaments, and the rate of travel are all interrelated in the spinning process. Since it is desirable to increase spinning speeds to the limit of the equipment, the tendency has been to construct longer spinning cabinets as each new plant is built. Present spinning speeds are of the order of 600 or more meters per minute, measured as the yarn emerges from the cabinet.

Other dry-spinning operations follow essentially the same pattern. For example, the dry spinning of cellulose triacetate is identical to the process used for the older and more familiar secondary acetate except that the acetone solvent is replaced by a chlorinated hydrocarbon such as methylene chloride. The solubility is improved by the addition of a small amount of methanol (5 to 15 percent).

The acetate or triacetate yarn may emerge from the cabinet through an opening on the side near the bottom after passing over a deflecting pin or it may be brought out of the bottom directly below the spinnerette. In either case, it makes contact with an applicator which provides the lubricant required to reduce both friction and static formation in subsequent operations. With its surface lubricated, the yarn passes around a "feed" roll which determines the rate of withdrawal from the spinning cabinet, and then to any of several desired packaging devices.

Unlike the packaging of rayon yarn, cellulose acetates are either "ring" spun or wound into a package called a "disc," "zero twist," or "cam wound." In the ring-spun package, the yarn carries a slight twist of less than one turn per inch, but it requires a relatively expensive bobbin. Since the trend is toward less twisting, much acetate yarn is "beamed" in the producers plant after little or no twisting, the heavy beams being shipped directly to knitters or weavers.

Filament yarns are twisted for two reasons. One is to supply certain aesthetic characteristics such as touch, drapability, and elasticity. The other more fundamental reason is to provide physical integrity to the filament bundle so that it can be warped, woven, and knit without excessive breakage or fraying of individual filaments.

Fig. 11.13 Beaming cellulose acetate yarn from a reel holding about 800 packages of yarn. *(Courtesy Tennessee Eastman Co.)*

The yarn just mentioned as having no twist imparted before beaming may have been subjected to intermingling just prior to wind-up after extrusion. In the intermingling process, yarn with no twist, and usually under low tension, is passed through a zone where it is impinged upon by a jet stream of compressed air. This causes the filaments to interlace or intermingle with each other and they then can become metastable in this configuration when tension is reapplied. In this condition, the yarn has the integrity of twisted yarn and will pass through several textile processing steps without difficulty but with each handling, some of the intermingling is worked out. This type of yarn is also described as twist substitute yarn. The duPont process referred to by their trademark Rotoset® was the first example of this type of yarn treatment.

Solvent Recovery. The air containing the acetone vapor is drawn out of the spinning cabinet and passed through beds of activated carbon which sorb the organic solvent. The acetone is recovered by steaming and then separating it from the water by distillation. Efficiency of recovery is above 95 percent, and about 3 pounds of acetone are recovered per pound of cellulose acetate yarn. The recovery of methylene chloride and methanol from the manufacture of cellulose triacetate follows the same general procedure but the combination of chlorinated hydrocarbon, steam, and activated charcoal tend to produce traces of corrosive products which must be guarded against.

Dope-Dyed Yarn. As with viscose rayon, colored pigments or dyestuffs may be added to the spinning solution so that the yarn will be colored as it is produced, thus eliminating the necessity of dyeing the final fabric. Although methods of dope dyeing may differ in detail, depending on the fiber and the process used, certain essentials are the same throughout. As mentioned earlier, even when using titanium dioxide, a compromise must be made on the basis of two competing needs. Complete mixing, uniformity, and filtration require that the addition be made early in the operation; minimal cleaning problems during change-overs require just the opposite. This problem is much more difficult to solve when colored pigments are involved since the demands for colors are determined by everchanging style trends. A manufacturer may plan a long run of a given color, arranging the changes from shade to shade so that the differences in successive runs will be negligible. This might, for example, be through a series of blues ranging from a pastel to navy; but such is not always possible. There exist two solutions to the problem. If a manufacturer must produce a multitude of colors in relatively small amounts, it is desirable to premix individual batches of spinning dope. Each batch should be pretested on a small scale, even sampled and submitted to the customer, to ensure that the desired color will be acceptable when it is produced. Facilities must be provided to allow each batch of colored dope to be cut into the system very close to the spinning operation in order to minimize pipe cleaning. Permanent piping must be flushed with solvent or the new batch of colored dope; some of the equipment may be disassembled for mechanical cleaning after each change of color.

Another method of producing spun-dyed yarn involves using a group of "master" dopes of such color versatility that when they are injected by appropriate proportioning pumps into a mixer located near the spinning operation they will produce the final desired color. The advantages of such an operation are obvious; the disadvantage lies in the textile industry's demand for an infinite number of colors. No small group of known pigments will produce final colors of every desired variety.

PROTEINS

As previously mentioned, the use of naturally existing polymers to produce fibers has had a long history. In the case of cellulose the results were fabulously successful. An initial investment of $930,000 produced net profits of $354,000,000 in 24 years for one rayon company.[5] On the other hand, another family of natural polymers—proteins—has thus far resulted in failure or at best very limited production.

These regenerated proteins are obtained from milk (casein), soya beans, corn, and peanuts. More or less complex chemical separation and purification processes are required to isolate them from the parent materials. They may be dissolved in aqueous solutions of caustic, and wet spun to form fibers which usually require further chemical treatment as, for example, with formaldehyde. This reduces the tendency to swell or dissolve in subsequent wet-processing operations or final end uses. These fibers are characterized by a wool-like feel, low strength, and ease of dyeing. Nevertheless, for economic and other reasons they have not been able to compete successfully with either wool (after which they were modeled) or other man-made fibers.

Based on their later publications, Courtaulds, Ltd. apparently gave careful study in the 1940s to the synthesis of polypeptide fibers, perhaps targeting at the properties of the major ampullate fibers of orb-web-building spiders. These have a rupture tenacity of 70-90 g/tex and elongations of 25-35%,[6] a combination unreached by any other fiber, natural or man-made.

NYLON

Historical

Nylon was the first direct product of the technological breakthrough achieved by W. H. Carothers of E. I. du Pont de Nemours & Co. Until he began his classic research on high polymers, the manufacture of man-made fibers was based almost completely on natural linear polymers. Such materials included rayon, cellulose acetate, and the proteins. His research showed that chemicals of low molecular weight could be reacted to form polymers of high molecular weight. By selecting reactants which produce molecules having great length in comparison with their cross-section, that is, linear molecules, fiber-forming polymers are obtained.

With this discovery the man-made fiber industry entered a new and dramatic era.

Manufacture

Nylon 66. The word "nylon" was established as a generic name for polyamides, one class of the new high-molecular weight linear polymers. The first of these, and the one still produced in the largest volume, is nylon 66 or polyhexamethylene adipamide. Numbers are used with the word nylon to indicate the number of carbon atoms in the constituents, in this case hexamethylenediamine and adipic acid.

To emphasize the fact that it does not depend on a naturally occurring polymer as a source of raw material, nylon has often been called a "truly synthetic fiber." To start the synthesis, benzene, as a by-product from the coking of coal, may be hydrogenated to cyclohexane,

$$C_6H_6 + 3H_2 \xrightarrow{\text{catalyst}} C_6H_{12}$$

or the cyclohexane may be obtained from petroleum. The next step is oxidation to a cyclohexanol-cyclohexanone mixture by means of air.

$$x\ C_6H_{12} + O_2\ (\text{air}) \xrightarrow{\text{catalyst}} y\ C_6H_{11}OH + z\ C_6H_{10}O$$

In turn, this mixture is oxidized by nitric acid to adipic acid.

$$C_6H_{11}OH + C_6H_{10}O + HNO_3 \xrightarrow{\text{catalyst}} (CH_2)_4(COOH)_2$$

Adipic acid so obtained is both a reactant for the production of nylon and the raw-material source for hexamethylenediamine, the other reactant. The adipic acid is first converted to adiponitrile by ammonolysis and then to hexamethylenediamine by hydrogenation.

$$(CH_2)_4(COOH)_2 + 2NH_3 \xrightarrow{\text{catalyst}} (CH_2)_4(CN)_2 + 4H_2O$$

$$(CH_2)_4(CN)_2 + 4H_2 \xrightarrow{\text{catalyst}} (CH_2)_6(NH_2)_2$$

Another approach is through the series of compounds furfural, furane, cyclotetramethylene oxide, 1,4-dichlorobutane, and adiponitrile. The furfural is obtained from oat hulls and corn cobs.

$$\text{Furfural (furan-2-CHO)} \xrightarrow[\text{catalyst}]{\text{Steam}} \text{Furane } (C_4H_4O) \xrightarrow[\text{catalyst}]{H_2} \text{Cyclotetramethylene oxide } (C_4H_8O) \xrightarrow[\text{catalyst}]{HCl} Cl(CH_2)_4Cl \xrightarrow{NaCN} NC(CH_2)_4CN$$

Or 1,4-butadiene, obtained from petroleum, may be used as a starting raw material to make the adiponitrile via 1,4-dichloro-2-butene and 1,4-dicyano-2-butene.

$$CH_2{=}CHCH{=}CH_2 \longrightarrow ClCH_2CH{=}CHCH_2Cl \xrightarrow[\text{catalyst}]{HCN} NCCH_2CH{=}CHCH_2CN \xrightarrow[\text{catalyst}]{H_2} NC(CH_2)_4CN$$

When hexamethylenediamine and adipic acid are mixed in solution in a one to one molar ratio, the so-called "nylon salt" hexamethylenediammoniumadipate, the direct progenitor of the polymer, is precipitated. After purification, this nylon salt is polymerized to obtain a material of the desired molecular weight. It is heated to about 280°C under vacuum while being stirred in an autoclave for 2 to 3 hours; a shorter holding period follows; and the process is finished off at 300°C. The molecular weight must be raised to a level high enough to provide a fiberforming material, yet no higher. If it is too high, the corresponding viscosity in the sub-

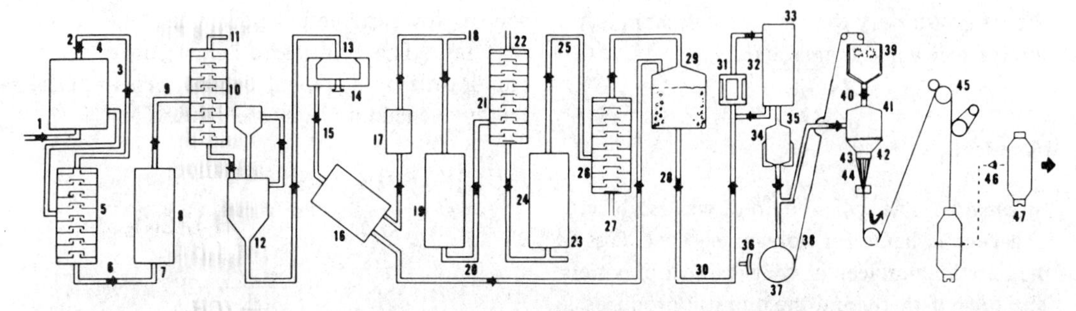

Fig. 11.14 Flow diagram for the manufacture of nylon 66 yarn: (1) air; (2) cyclohexane from petroleum; (3) reactor; (4) recycle cyclohexane; (5) still; (6) cyclohexanol-cyclohexanone; (7) nitric acid; (8) converter; (9) adipic acid solution; (10) still; (11) impurities; (12) crystallizer; (13) centrifuge; (14) impurities; (15) adipic acid crystals; (16) dryer; (17) vaporizer; (18) ammonia; (19) converter; (20) crude adiponitrile; (21) still; (22) impurities; (23) hydrogen; (24) converter; (25) crude diamine; (26) still; (27) impurities; (28) nylon salt solution; (29) reactor; (30) stabilizer; (31) calandria; (32) evaporator; (33) excess water; (34) autoclave; (35) delustrant; (36) water sprays; (37) casting wheel; (38) polymer ribbon; (39) grinder; (40) polymer flake; (41) spinning machine; (42) heating cells; (43) spinnerette; (44) air; (45) drawtwisting; (46) inspection; (47) nylon bobbin. (Note: Wherever the demand for liquid polymer at a spinnerette is large, as for example in the spinning of tire yarn, it is pumped directly from the autoclave.

sequent spinning operation will require extremely high temperatures and pressures to make it flow. Accordingly, a small amount of acetic acid is added to terminate the long-chain molecules by reaction with the end amino groups.

The polymerized product is an extremely insoluble material and must be melt spun (see below). Therefore, should a delustered fiber be desired it is necessary to add the titanium dioxide pigment to the polymerization batch before that reaction occurs. For ease of handling, the batch of nylon polymer may be extruded from the autoclave to form a thin ribbon which is easily broken down into chips after rapid cooling. But, whenever possible, the liquid polymer is pumped directly to the spinning operation.

Nylon 6. Nylon 6 is made from caprolactam and is known as Perlon® in Germany where it was originally developed by Dr. Paul Schlack.[4] Its production has reached a very large volume in the United States in recent years.

Like nylon 66, nylon 6 uses benzene as raw material, which is converted through previously mentioned steps to cyclohexanone. This compound is in turn converted to the corresponding oxime by reaction with hydroxylamine, and cyclohexanone oxime is made into caprolactam by the Beckmann rearrangement.

$$\text{(cyclohexanone)}\ C_6H_{10}C{=}O + H_2NOH \longrightarrow C_6H_{10}C{=}NOH + H_2O$$

$$C_6H_{10}C{=}NOH \xrightarrow{H_2SO_4} \overline{CH_2(CH_2)_4C{=}O}\ \text{(ring closed through NH)}$$

After purification, the lactam is polymerized by heating it at elevated temperatures in an inert atmosphere. During self-condensation, the ring structure of the lactam is opened so that the monomer acts as an epsilon-aminocaproic acid radical. Unlike nylon 66, polymerization of caprolactam is reversible; the polymer remains in equilibrium with a small amount of monomer. As with nylon 66, nylon 6 is extruded in ribbon-like form, quenched,

Fig. 11.15 Aerial view of a nylon filament yarn plant: shipping docks, textile spinning area, and plant office buildings (left center), chemical intermediates (right center); and the technical center (left foreground). (*Courtesy Monsanto Co.*)

and broken into flakes for subsequent spinning, or the molten polymer is pumped directly to the spinning equipment.

Melt Spinning

Because of its extremely low solubility in low-boiling and inexpensive organic solvents, nylon 66 required a new technique for converting the solid polymer into fibers, hence the development of "melt" spinning, the third basic method for manufacturing man-made fibers. The following description refers essentially to nylon 66 since it was the first to use the method, but the process applies, in general, to all melt-spun man-made fibers.

In the original production of nylon fiber by melt spinning, the chips of predried polymer were fed from a chamber onto a melting grid whose holes were so small that only passage of molten polymer was possible. Both solid and liquid were prevented from contacting oxygen. The polymer melted in contact with the hot grid and dripped into a pool where it became the supply for the spinning itself. This melting operation has been entirely replaced by delivery of the molten polymer pumped directly from the polymerization stage or by "screw" melting. In the latter process, the solid polymer in chip form is fed into an extrusion-type screw contained in a heated tube. The depth and helix angle of the grooves are engineered in such a way that melting takes place in the rear section, and the molten polymer is moved forward under increasing pressure to a uniformly heated chamber preceding the metering pump.

Whatever means is used to secure the molten polymer, it is moved forward to a gear-type pump which provides both high pressure and a constant rate of flow to the final filter and spinnerette. The filter consists of several metal screens of increasing fineness or graded sand arranged in such a way that the finest sand is at the bottom. After being filtered, the molten polymer at a pressure of several thousand pounds per square inch is extruded through the small holes in the heavily constructed spinnerette. It is necessary to maintain the temperature of the pool, pump, filter, spinnerette assembly, and spinnerette at about 20–30°C above the melting point of the nylon, which is about 264°C for nylon 66 and 228°C for nylon 6.

When the extruded fibers emerge from the spinnerette face into the relatively cool quench chamber where a cross current of air is provided, rapid solidification takes place. The solid filaments travel downward to cool, and an antistatic lubricant is applied before they make contact with the wind-up rolls to prevent static formation and to reduce friction in subsequent textile operations. Great care must be used in conveying the freshly spun yarn from the spinning chamber to the yarn package if it is to be "drawn" in a separate operation. (The combination of spinning and drawing, known as "spin-drawing" will be described later.)

Cold Drawing

It was learned early that the fibers made from nylon 66 could be extended to about four times their original length with very little effort, but that thereafter a marked resistance to extension took place. It was discovered that during this high extension the entire length of fiber under stress did not extend uniformly. Rather, a "necking down" occurred at one or more points and when the entire length under tension had passed through this phenomenon, a high-strength fiber was obtained. It was also found that when more than one necking down was allowed to take place in a given length of fiber, a discontinuity occurred at the point where the two came together. Accordingly, the drawing operation is aimed at forcing the drawing to occur at a single point as the yarn advances from the supply to the take-up package.

Cold drawing consists essentially of removing the yarn from the package prepared in the melt-spinning operation and feeding it forward at a uniformly controlled rate under low tension. It is passed around a pulley or roller which determines the supply rate and prevents slippage; for nylon 66 it is then wrapped several times around a stationary snubbing pin. From there it goes to a second roller which rotates faster than the supply roller to produce the desired amount of stretch, usually about 400 percent. The necking down occurs at the snubbing pin. In the case of nylon 6, drawing may be effected satisfactorily without passing the yarn around such a snubbing pin.

The long molecules of the nylon 66 or 6 polymer, which are randomly positioned in the molten polymer, when extruded from the spinnerette tend to form "crystalline" areas of molecular dimensions as the polymers solidify in the form of freshly spun fibers. In the drawing operation, both these more ordered portions as well as the amorphous areas tend to become oriented so that the lengthwise dimensions of the molecules become parallel to the long axis of the fiber and additional intermolecular hydrogen bonding is facilitated. It is this orientation which converts the fiber having low resistance to stress into one of high strength.

By controlling the amount of drawing as well as the conditions under which this operation takes place, it is possible to vary the amount of orientation. A minimal amount is preferable in the manufacture of yarns intended for textile applications wherein elongation of considerable magnitude and low modulus or stiffness is required rather than high strength. On the other hand, strength and high modulus are at a premium when fibers are to be used in products such as tire cords. High resistance to elongation is imperative if the tire is not to "grow" under conditions of use. In this connection it should be noted that nylon tire cord that has been produced by twisting the original tire yarn and plying the ends of these twisted yarns together is hot stretched just before use at the tire plant to increase strength and reduce even further the tendency to elongate under tension.

The separate operations of spinning and drawing nylon presented a challenge whose object was combination of these two into a single continuous step. But the problem was obvious, for the operating speeds of the two separate steps had already been pushed as high as was thought to be possible. How then would it be possible to combine them into a continuous spin-draw, wherein a stretching of about 400 percent could take place? The answer appears to lie in the manner in which the cooling air is used and in the development of improved high-speed winding devices. By first cooling the emerging fibers by a cocurrent flow of air and then cooling it further by a countercurrent flow, the vertical length of the cooling columns can be kept within reason. In-line drawing may occur in one or two stages and relaxation may be induced if needed. The final yarn is said to be packaged at speeds of 6000 meters per minute.

ARAMIDS

Historical

In 1973, the Federal Trade Commission recognized aramids as a distinct generic classification of man-made fibers. They are defined as "a long-chain synthetic polyamide in which *at least* 85% of the amide linkages are attached directly to two aromatic rings." This distinguishes aramids from nylon, which was redefined as a polyamide with *less than* 85% of the amide linkages attached to two aromatic rings. Aramid fiber was first produced in the United States in the early sixties by E. I. du Pont de Nemours & Co. This fiber is now known under their trade mark Nomex®. A second fiber in this classification, Kevlar®, was introduced by du Pont in the early seventies.

Manufacture/Properties

The synthesis of the polymers on which these two wholly aromatic polyamides are based is different from the routes used to produce either of the aliphatic polyamides, nylon 66 or nylon 6. Rather, a reaction sequence based on the Schotten-Baumann procedure is used in which an amine and an acyl chloride are reacted in the presence of an acid acceptor to give an amide. In the case of Nomex®, the reactants are m-phenylene diamine and isophthaloyl chloride to give poly-m-phenylene-isophthalamide (MPD-I). This polymer cannot be melted without decomposition, so the preferred fiber formation route is dry spinning. Patent literature suggests that it is spun from a solvent system composed of dimethyl-formamide and lithium chloride. Final properties are achieved by stretching in steam after washing in order to remove residual solvent.

MPD-I

The physical and chemical properties of this fiber are not remarkably different from other strong polyamides. It does, however, have excellent heat resistance which makes it particularly suited for use in protective clothing and in specified industrial end uses. Military flight suits, fire-fighter uniforms, and hot gas filtration are few among many possible applications.

The other commercially available fiber in this category, Kevlar®, is significantly different from Nomex® in both method of extrusion and fiber properties. It is based on poly-para-phenylene terephthalamide (PPD-T), which results from the condensation of para-phenylene diamine and terephtaloyl chloride. In contrast to MPD-I, which is based on meta-substituted aromatic nuclei, PPD-T contains only aromatic nuclei linked through para-substitution positions.

PPD-T

This results in a polymer which has the ability to form optically anisotropic or ordered solutions and which can be spun into fibers with very high strength and very high modulus, surpassing even inorganic fibers. It is believed that the para-linked polymer molecules resemble rigid rod-like structures which with the aid of the regularly spaced polar polyamide linkages form highly compact oriented fiber microstructures. There now exists extensively published scientific literature listing other fully aromatic para-linked structures. It is apparent that several conditions must be met for an anisotropic solution to form:

1. A rigid polymer chain,
2. A medium giving the necessary solvent/polymer interaction,
3. The required solute concentration,
4. The appropriate temperature.

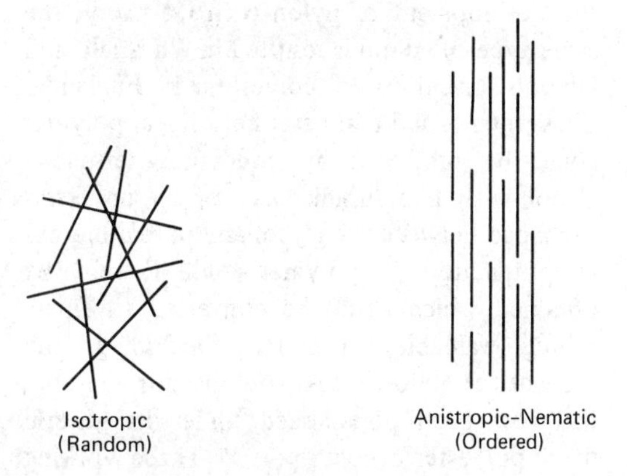

It is believed that PPD-T fiber is spun from a solution in concentrated sulfuric acid. To achieve the highest degree of molecular orientation, the fibers may be spun by a hybrid process referred to as "dry jet/wet spinning." In this process, the extrusion jet is placed several centimeters above the coagulation bath, the nascent fibers descend into the liquid, pass under a guide and proceed in the bath while undergoing stretch; then they are withdrawn from the bath and wound up. A subsequent washing step may be required to remove residual acid solvent.

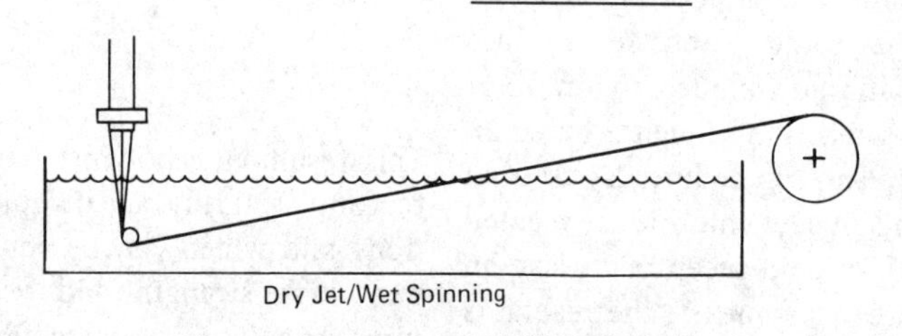
Dry Jet/Wet Spinning

Kevlar® is reported to have about twice the breaking strength (~20 g/d) of high tenacity nylon and polyester, but its most outstanding physical property is its high stiffness (~1000 g/d) which can range up to an order of magnitude greater than standard polyester. This property has led to volume usage of the fiber as reinforcement in composite materials such as the belt in radial tires and aerospace structures.

POLYESTERS

Historical

Just as the original work of Carothers regarding polyamides inspired research which resulted in the development of nylon 6 in Germany, the same type of stimulus resulted in Whinfield and Dixon's invention of polyesters in England.[5] These men found that a synthetic linear polymer could be produced by condensing ethylene glycol with terephthalic acid or by an ester-exchange between the glycol and pure dimethyl terephthalate. A polymer could thereby be obtained which could be converted to fibers having valuable properties, including the absence of color. Like nylon, this material has now been popularized under its generic name polyester or just "poly." Those working with it commonly refer to it as PET. It first appeared under the trade name Terylene® (Imperial Chemical Industries, Ltd.) in England, and was first commercialized in the United States as Dacron® (E. I. du Pont de Nemours & Co.). By 1972 it was being sold under about 100 additional trade names throughout the world.

Manufacture

When the development of polyethylene terephthalate (PET) occurred, ethylene glycol was already being produced in large amounts from ethylene, a by-product of petroleum cracking, by oxidation of ethylene to ethylene oxide and subsequent hydration to ethylene glycol, which, in a noncatalytic process, uses high pressures and temperatures in the presence of excess water.

$$CH_2{=}CH_2 + O_2 \rightarrow \underset{\text{(epoxide, O bridging)}}{CH_2{-}CH_2} \xrightarrow{H_2O} HOCH_2CH_2OH$$

On the other hand, although *o*-phthalic acid, or rather its anhydride, had long been produced in enormous amounts for use in the manufacture of alkyd resins, the *para* derivative was less well known and not available on a large scale. The synthesis is a straightforward one, however, from *p*-xylene which is oxidized to terephthalic acid, either by means of nitric acid in the older process or by air (catalyzed) in the newer one. As already mentioned, in the early years this was then converted to the easily purified dimethyl ester in order to obtain a colorless polymer adequate for the manufacture of commercially acceptable fibers.

Several other methods were developed for producing the desired dimethyl terephthalate. The Witten (Hercules) process goes from *p*-xylene to toluic acid by oxidation of one methyl group on the ring, following which the carboxyl group is esterified with methanol. This process is then repeated with the second methyl group to secure the dimethyl ester of terephthalic acid.

$$CH_3C_6H_4CH_3 \xrightarrow[190°C]{O_2} HOOCC_6H_4CH_3 \xrightarrow[150°C]{CH_3OH} CH_3OOCC_6H_4CH_3 \xrightarrow{\text{Same two steps}} CH_3OOCC_6H_4COOCH_3$$

Either phthalic anhydride or toluene, both in ample supply as raw materials, may be used in the Henkel processes. Use of phthalic anhydride depends upon dry isomerization of the potassium salt of the *ortho* derivative to the *para* from at about 430°C and 20 atm pressure; or toluene is oxidized to benzoic acid, whose potassium salt can be converted to benzene and the potassium salt of terephthalic acid by disproportionation.

The first step in the reaction of dimethyl terephthalate and ethylene glycol is transesterification to form bis(*p*-hydroxyethyl) terephthalate (bis-HET) and eliminate methanol.

$$CH_3OOCC_6H_4COOCH_3 + 2HOCH_2CH_2OH \xrightarrow{\sim 200°C} HOCH_2CH_2OOCC_6H_4COOCH_2CH_2OH + 2CH_3OH\uparrow$$

This product is then polymerized in the presence of a catalyst to a low molecular weight, and the by-product glycol is eliminated. In a second stage, at a temperature of about 275°C and under a high vacuum, the molecular weight is raised to secure the melt viscosity desired for the particular material involved. Like nylon, this final material may be extruded, cooled, and cut into chips for storage and remelting, or it may be pumped directly to the spinning machines.

From the beginning, it was obvious that it would be a considerable advancement in industrial chemistry, to say nothing of cost reduction, if the process could be simplified by making it unnecessary to go through the dimethyl derivative to secure a product of adequate purity. This was accomplished in the early 1960's when methods of purifying the crude terephthalic acid were developed, and conditions and catalysts were found which made possible the continuous production of a color-free polymer. It is said that the selection of the catalyst is especially aimed at the prevention of ether linkages in the polymer chain due to intracondensation of the glycol end groups.

Two additional rather similar routes are also known. Both depend upon the reaction between ethylene oxide, rather than ethylene glycol, and terephthalic acid to form the bis-HET monomer already mentioned. The difference between the two methods lies in the point where purification is done; in one case, the crude terephthalic acid is purified, in the other, the bis-HET monomer. In both cases this monomer is polymerized by known procedures to form a fiber-grade polyester. The titanium dioxide delustrant is added, as might be expected, early in the polymerizing process.

Only one other polyester has reached long term commercialization. It is now produced in limited volume as Kodel 200® by Tennessee Eastman Co. and is considered to be 1-4 cyclo hexylene dimethylene terephthalate. The glycol which is used instead of ethylene glycol in this process, exists in two isomeric forms, one melting at 43°C and the other at 67°C. This makes their separation possible by crystallization so as to secure the desired ratio of the two forms for conversion to the polymer.

This ratio determines the melting point of the polymer, this being a most important property for a material which is to be melt spun. The polymer from the 100%-*cis* form melts at 275°C and from the 100%-*trans* form at 318°C. Indications are that the commercial product is about 30/70 *cis-trans.*

Polyesters are melt spun in equipment essentially the same as that used for nylon, which has already been described. Wherever the volume is large and the stability of demand is adequate, the molten polymer is pumped directly from the final polymerization stage to the melt-spinning machine. The molten polymer is both metered and moved forward at high pressure by a gear-type pump through filters to the spinnerette which contains capillaries of about 9 mils (230 microns) diameter. Great care is used to eliminate moisture and oxygen from the chips if these are used, and from the spinning chamber. When the polyester fibers are destined to become staple, the output of a number of spinnerettes are combined to form a tow which can be further processed as a unit. Continuous filament yarn is packaged for drawing. Spin-drawing has become common.

Drawing

Unlike nylon, which contains a fairly high amount of crystalline component, polyethylene terephthalate fibers are essentially amorphous as spun. In order to secure a usable textile yarn or staple fiber, this product must be drawn under conditions that will result in an increase in both orientation and crystallinity. This is done by drawing at a temperature above the glass transition point, tg, which is about 80°C. Conditions of rate and temperature must be selected so that the amorphous areas are oriented and crystallization will take place as the temperature of the drawn fibers drops to room temperature. An appropriate contact-type hot plate or other device is used and about 300 to 400 percent extension is effected. As with nylon, the conditions of draw, especially the amount, determine the force-elongation properties of the product. Industrial-type yarns, such as those intended to be used as tire cord, are more highly drawn.

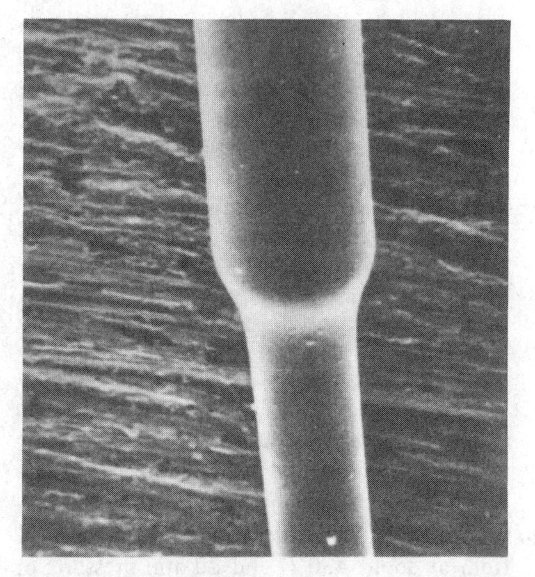

Fig. 11.16 Drawnecking-polyester single filament. *(Courtesy E. I. du Pont de Nemours and Co.)*

Heat Setting

The ability of textile fibers to be "set" is not characteristic of man-made fibers alone. Aided in many cases by the presence of starch, cotton fabrics have been ironed to a smooth and wrinkle-free condition; also, the sharp crease in wool trousers has been commonplace for

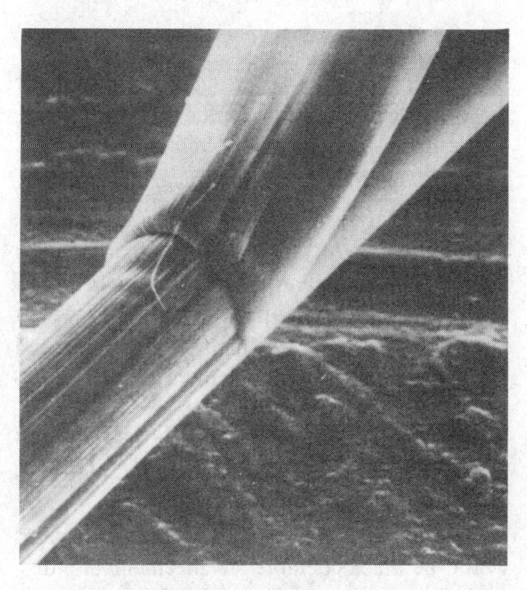

Fig. 11.17 Skin-peeling polyester filament showing fibrillar structure. *(Courtesy E. I. duPont de Nemours and Co.)*

generations. In other words, these fabrics were exposed to moisture at elevated temperatures while being held or pressed into desired geometrical configurations and then allowed to cool before being released from constraint. Such fabrics tend to remain unchanged while cool and dry, even though the fibers from which they are formed carry internal stresses. But reversion takes place upon washing or exposure to high relative humidity.

With the development of nylon, and especially polyesters, a durable kind of setting has become possible. When fabrics made from these fibers are shaped and then exposed to elevated temperatures either in the dry condition or, in the case of nylon particularly, in the presence of water vapor, thermoplastic relaxation of induced stresses in the fiber takes place and configurations at the molecular level adjust to a new and lower energy level. This depends on not only the temperature used but also the duration of the exposure. Thus a few seconds at 230°C will produce the same results as exposure for a considerably longer period at a temperature 50 to 75 degrees lower. The permanency of the setting, that is, the ability of a fabric or garment to return to its original configuration after temporary distortion even while exposed to moisture and raised temperatures, is a function of the severity of the heat setting.

It is this property of polyamides and polyesters that has been the main factor contributing to "ease of care" and the "wash and wear" characteristics of garments made from these polymers. In turn, these garments have revolutionized both the textile and apparel industries.

Textured Yarns

Fundamentally, the manufacture of "textured" yarns is closely related to the heat setting of fabrics, the difference being that the individual filaments or bundle of filaments in textured yarns are distorted from an essentially straight rod-like form and then heat set. In some instances, the fibers are distorted in a more or less random way; at other times a regular pattern is desired.

The first commercially successful textured yarn was produced by highly twisting nylon-66, heat setting it as a full package of yarn, and then untwisting it through zero and applying a small amount of twist in the opposite direction. This process changed an individual filament from unitary close-packed structure to one which was voluminous due to mutual interference. The technique of heat setting the twisted yarn as a batch-unit operation has now been displaced by a continuous one, using what is known as a "false twisting" process. This is based upon the principle that if a length of yarn in prevented from rotating at both ends but is rotated on its axis at its center point, the resulting two sections will contain "Z" and "S" twists in equal amounts. When this occurs with a moving yarn, any element in it will first receive a twist in one direction but after passing the false twisting point must revert to zero twist. If it is then made to pass over a hot plate while in the twisted state and is heat set in that configuration, even after returning to the untwisted condition, the individual filaments will tend to remain distorted when lengthwise stress is released. Because of the low mass and diameters of textile yarns or monofilaments, it is possible to false twist them at extremely high rotational speeds. Yarn forward speeds of about 1000 m/min. are currently obtainable by passing the yarn between, and in contact with, belts opposably driven. (When attempts are made to secure higher rates, problems of twist control develop.) The same technique is now more commonly applied to undrawn or partially drawn yarn at the spin-drawing machine. The resulting yarn may be heat set as part of the same continuous operation by passing it through a second heater under conditions of overfeed or little or no tension in order to secure both thermally stable geometric configurations in the individual distorted filaments that comprise the yarn and the degree of "stretchiness" and bulk desired in the final product.

Since these yarns are being made in one less step and also within the plants spinning the parent product, this latest development may be said to constitute another advance in the industrial chemistry of man-made-textile products. This draw-texturing appears to be especially applicable to polyester yarns intended for fabrics known as "double knits" and "textured wovens."

Yarn can be forced forward by means of "nip" rolls, although this may seem to be quite contrary to the old adage that one cannot push on an end of string. When this is done so that the yarn is jammed into a receiver already full of the preceding material, it collapses with sharp bends between very short lengths of straight sections. In this condition heat is applied to set it. In practice, the mass of such yarn is pushed through a heated tube until it escapes at the exit past a spring loaded gate. During this passage it is heat set in a highly crimped configuration. It is cooled before being straightened and wound onto a package. In another continuous process, the yarn or monofilament is pulled under tension over a hot sharp edge so that it is bent beyond its elastic limit and heat set in that condition. The result is not unlike that produced by drawing a human hair over the thumb nail.

When such yarns are knitted or woven into fabric, the filaments tend to return to the configurations in which they were originally heat set. Contraction takes place in the direction of the yarn axis and this in turn converts the smooth flat fabric into a "stretch" fabric and gives the surface a textured appearance. These fabrics or the garments made from them, whatever the process used to produce the yarns, may be heat-treated to secure stability in a desired geometric configuration. A degree of stretch may be retained, or a flat and stable textured surface may be produced. There are a number of variations of the texturing process, these, combined with the many possibilities of heat-setting, impart considerable versatility to the final product. The growth in the use of these products in the 1960's is well known. Carpeting also provides a significant market since texture is one of the most important characteristics of soft floor coverings. Such products have been important to the successful use and expanded development of nylon and polyester yarns in recent years.

Recent Developments

The composition and properties of wholly aromatic polyamides or aramids were discussed in a previous section. When both the diacid and diamine components are *para*-substituted aromatic compounds, the resulting polymer is capable of forming lyotropic liquid-crystalline solutions. These solutions can be dry or wet spun into fibers with unusually high tensile strength and tensile modulus. When a similar strategy is tried to make polyester fiber from a homopolymer of a para-substituted aromatic diacid and a *para*-substituted aromatic diol, only infusible and intractable materials are obtained. A solution to this problem has been found in the development of polyester copolymers which give thermotropic liquid-crystalline melts over a useful temperature range and have viscosities suitable for melt extrusion into fibers or films having high levels of orientation. Spin line stretch factors of the order of several hundred are used to achieve orientation, and physical properties are developed further by heat treatment at temperatures approaching melting conditions.

The first fibers from a thermotropic liquid-crystalline melt whose properties were reported were spun from a copolyester of *para*-hydroxy benzoic acid (PHB) and poly-ethyleneterephthalate (PET) by workers at Tennessee Eastman Co. The preparation of the copolymer proceeds in two stages. First, *para*-acetoxybenzoic acid is reacted with poly-ethyleneterephthalate in an acidolysis step to give a copolyester prepolymer which in the second step is further condensed to a higher degree of polymerization suitable for fiber formation.

$$\underset{\text{PHB}}{CH_3\overset{O}{\overset{\|}{C}}-O-C_6H_4-\overset{O}{\overset{\|}{C}}-OH} + \underset{\text{PET}}{\left[\overset{O}{\overset{\|}{C}}-C_6H_4-\overset{O}{\overset{\|}{C}}-OCH_2CH_2O\right]_n} \xrightarrow[\text{2) condensation}]{\text{1) acidolysis}}$$

$$CH_3\overset{O}{\overset{\|}{C}}-OH + \left[\overset{O}{\overset{\|}{C}}-C_6H_4-O\right]_a - - \left[\overset{O}{\overset{\|}{C}}-C_6H_4-\overset{O}{\overset{\|}{C}}\right]_b - - - \left[OCH_2CH_2O\right]_b - -$$

When the mol% of PHB in the copolymer exceeds about 30–40%, a liquid-crystalline melt is obtained. Up to about 60 mol%, order in the melt increases and melt viscosity decreases. Compositions containing about 60 mol% PHB can be melt spun into fibers using standard extrusion techniques. It is this unusual combination of properties that makes this class of materials valuable for the formation of high-strength fibers and plastics.

Among melt-spun fibers, those based on thermotropic liquid-crystalline melts have the highest strength and rigidity reported to date and appear comparable to polyamides spun from lyotropic liquid-crystalline solutions. This was a very active field of research in the seventies, and many compositions giving anisotropic melts based on a wide variety of comonomers have been reported. Obviously, these compositions must contain three components at minimum, but many have four or five components. Some frequently-used constituents, in addition to those mentioned above, are 2,6-naphthalene-dicarboxylic acid, hydroquinone, 4,4′-biphenol, iso-phthalic acid, and 4,4′-dihydroxy-diphenyl ether.

ACRYLICS

Polymer Manufacture

Acrylic (commonly not designated as polyacrylic) fibers are spun from polymers that are made from monomers of which a minimum is 85 percent acrylonitrile. This compound may be made from hydrogen cyanide and ethylene oxide through the intermediate ethylene cyanohydrin,

$$\overset{O}{\widehat{CH_2CH_2}} + HCN \longrightarrow HOCH_2CH_2CN \xrightarrow[-H_2O]{\text{catalyst}} CH_2{=}CHCN$$

It may also be made directly from acetylene and hydrogen cyanide,

$$CH{\equiv}CH + HCN \longrightarrow CH_2{=}CHCN$$

But the reaction which is currently preferred uses propylene, ammonia, and air,

$$3CH_2{=}CHCH_3 + 3NH_3 + 7O_2 \text{ (air)} \xrightarrow[<500^\circ C]{\text{catalyst}} CH_2{=}CHCN + 2CO + CO_2 + CH_3CN + HCN + 10H_2O$$

Pure acrylonitrile may polymerize at room temperature to polyacrylonitrile (PAN), a compound which, unlike polyamides and polyesters, does not melt at elevated temperatures but only softens and finally discolors and decomposes. Nor is it soluble in inexpensive low-boiling organic solvents. Since fibers made from it resist the dyeing operations commonly used in the textile industry, the usual practice is to modify it by copolymerization with other monomers, e.g., vinyl acetate, vinyl chloride, styrene, isobutylene, acrylic esters, acrylamide, or vinyl pyridine in amounts up to 15 percent of the total weight (beyond which the final product may not be termed an acrylic fiber). The choice of modifier depends on the characteristics which a given manufacturer considers important in a fiber, the availability and cost of the raw materials in his particular area of production, and the patent situation.

In copolymerizating acrylonitrile with another monomer, conditions must be controlled in such a way that the reaction produces a polymer having the desired chain construction and length. Generally the reaction takes place in the presence of substances capable of producing free radicals. In addition, certain trace metals which have been found to increase reaction rates offer a means of controlling chain length. When polymerization is carried out in solution, after an induction period, the reaction is rapid and liberates a considerable amount of heat. Furthermore, since the polymer is not soluble in the monomer, a thick paste is formed. These facts limit the usefulness of such a process. Carrying out the polymerization in the presence of a large amount of water is a convenient method and the one most gener-

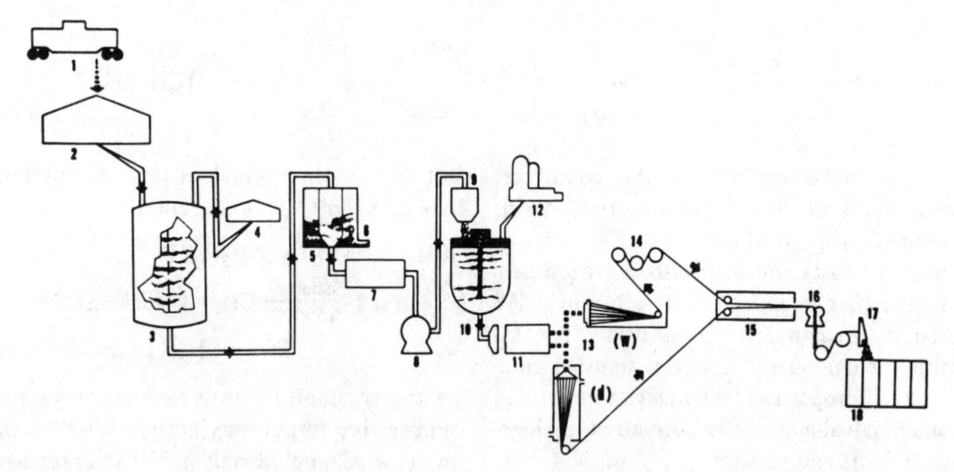

Fig. 11.18 Flow diagram for the manufacture of acrylic fiber: (1) acrylonitrile; (2) tank farm; (3) polymerizer; (4) comonomer and catalyst; (5) centrifuge; (6) waste liquid; (7) dried polymer; (8) grinding; (9) polymer storage; (10) dissolver; (11) filter; (12) solvent plant; (13) spinnerette; (13w) wet spinning; (13d) dry spinning; (14) roller drier; (15) additional treatment; (16) crimper; (17) cutter; (18) acrylic fiber bale.

ally used. In this case the polymer forms a slurry and the water provides a means for removing the heat from the site of the reaction. Moreover, most of the common redox-catalyst systems are water-soluble. Polymerization may be carried out batchwise or by a continuous process.

In the standard batch method, the monomers and catalyst solutions are fed slowly into an agitated vessel containing a quantity of water. The heat of reaction is removed either by circulating cooling water through the jacket surrounding the vessel or by operating the reaction mixture at reflux temperature and eliminating the heat through the condenser water. The monomer and catalyst feeds are stopped when the desired amounts have been added and polymerization is allowed to continue until there is only a small amount of monomer remaining in the reaction mixture. Then the slurry is dumped from the reaction vessel, filtered, washed, and dried.

If a continuous process is used, e.g., moving batch or pipeline, the raw materials are metered into one end and polymerization proceeds as the mixture moves forward. The resulting polymer slurry emerges from the other end. This method requires exceedingly careful control of feed ratios and temperatures to maintain consistent polymer quality.

In the continuous-overflow method, rather than stopping the monomer and catalyst feed when the reaction vessel is full, the slurry is simply allowed to overflow; the solids are removed by filtration, washed, and dried. The filtrate contains a certain amount of unreacted monomer, and this is recovered by steam distillation after the catalyst has been destroyed to prevent further polymerization. The dried polymer is the raw material from which fibers are spun.

Spinning

As already mentioned, pure polyacrylonitrile softens at elevated temperatures and thermal decomposition starts before the molten state is reached. The same is true of the copolymers commonly used to produce fibers. Accordingly, melt spinning is impossible; spinning must be done from a solution of the polymer. Both dry and wet spinning are carried out in current commercial operations.

The operations which are used to either wet- or dry-spin acrylics are essentially the same as those already described for rayon and acetate, respectively. The polymer must be completely dissolved in solvent and the solution filtered to remove any impurities which would cause spinnerette blockage. Because acrylic polymers are not soluble in common nonpolar solvents, polar substances such as dimentyl formamide, dimethyl acetamide, or solutions of inorganic salts such as zinc chloride-calcium chloride mixtures are required. Only wet spinning is possible with the latter. Dimethyl formamide boils at 152.8°C and exerts a vapor pressure of 3.7 mm of Hg at 25°C compared with acetone (used in dry spinning of cellulose acetate) which has a vapor pressure of 228.2 mm of Hg at 25°C. It follows that, unlike acetone which requires an activated-carbon system for recovery, dimethyl formamide may be condensed directly from the air stream used to evaporate the solvent from the forming fiber.

Acrylics, like rayon, require stretching either during spinning or immediately after fiber formation, or both, followed by a relaxing treatment in order to obtain the desired characteristics of the modulus, rupture tenacity, and rupture elongation. These same properties are influenced by spinning speeds, and the temperature of the drying air, if dry spun, or temperature and composition of the bath, if wet spun. The multitude of combinations made possible by the use of various comonomers and the flexibility of the fiber-forming operations furnish the several manufacturers with versatility and the users with a variety of acrylic fibers.

Acrylic fibers possess a property that made it possible for them in the late 1950's and early 1960's to find immediate, even spectacular, acceptance in the knitted sweater field, until then dominated by wool. When acrylic fibers, normally in the form of a heavy tow, are hot-stretched (for example, by being drawn over a hot plate and then cooled under tension) they are converted to a labile state. Upon immersion in hot water such fibers will contract considerably but not to their prior unstretched length. In practice this characteristic is used to produce a bulky yarn resembling the woolen yarns long accepted for use in sweaters. The process is described briefly below.

The stretched labile fibers are further cold-stretched to the breaking point so that non-uniform lengths are produced similar to the lengths found in wool. These are crimped and then mixed with thermally stable acrylic fibers which have been stretched and relaxed and have about the same length and degree of crimp. The blend is converted to a spun yarn by the same process used in making woolen yarns, and in turn this yarn is knitted into sweaters and other similar products. When such garments

Fig. 11.19 Wet spinning of acrylic tow. (*Courtesy Monsanto Co.*)

are dyed in hot water, the labile fibers, intimately blended with stable ones, contract lengthwise individually. In the process segments of the stable units tend to be carried along physically by entrapment and friction. But since each such fiber does not change its overall length, only the yarn as a whole decreases in length. Lateral displacement of the large volume of stable fibers results in the formation of a more voluminous structure known as a "bulky" or "hi-bulk" yarn.

Bicomponent or Conjugate Spun Fibers

As will be shown, it should be theoretically possible to make any of the common man-made fibers in bicomponent forms. But acrylics have received the most attention for quite good reasons. Their general characteristics have tended to make them competitive with wool. This means that they should be processable upon machinery developed for handling wool, as well as being capable of being accepted into markets previously dominated by wool. It follows that since the natural fiber possesses crimp which produces the cohesion that determines its behavior in processing and in part its appearance and "hand" in usage, a similar crimp was desired for acrylics. Unlike the polyamides and polyesters, heat setting the crimp is not possible for very fundamental reasons.

The principle which is the basis for bicomponent fibers is usually likened to that which underlies the bicomponent metal strips often used in temperature controllers. With the latter, differential-thermal expansion of the two joined components results in a bending of the thermal element. With fibers, moisture is usually the agent that acts upon the two side-by-side portions. Differential swelling or shrinkage causes the fiber to be brought into a crimped, or preferably, a spirally distorted condition. Incidentally, the side-by-side structure occurs naturally in wool.

The combination of small size and large number of holes in a spinnerette might lead to the conclusion that it would be almost impossible to design a spinnerette assembly which would bring two streams of polymer or polymer solutions together at each such hole and extrude them side by side to form a single filament. Such designs have, in fact, been made. But solutions of fiber-forming polymers fortunately possess properties which encourage laminar flow and thus make other approaches possible. This phenomenon was remarked upon earlier in connection with dope dyeing; when a suspension of a colored pigment is injected into a dope stream, a considerable problem must be overcome in order to secure adequate mixing so as to secure "dope dyed" fibers of a uniform color. Thus, it was known that when two streams of essentially the same solution of a fiber-forming polymer are brought together, side-by-side, and moved forward down a pipe or channel by the same amount of pressure behind each, virtually no mixing takes place. By bringing these streams to each spinnerette hole in such an individual side-by-side arrangement and using appropriate mechanical separators, the extruded filament from each hole will have a bicomponent structure. In addition to producing fibers in which the two components form a bilateral symmetrical structure, ingenious arrangement of predividers of the two streams can produce from the full complement of holes in a single spinnerette a selected group of fibers wherein the amount and position of each of the two components is randomly distributed throughout their cross sections. It follows that curls of uniform or random geometry may be produced to meet required needs.

VINYL AND MODACRYLIC FIBERS

Vinyls

When nylon-66 was developed, it was described as being "synthetic" or "fully-synthetic" in order to differentiate it from rayon and acetate. This was no small act of courage, since the word "synthetic," in that period just following the repeal of Prohibition in the United States, was often associated in the public mind with the least palatable kind of alcoholic beverages. In due time, what is known in the advertising business as "puffing" led it to be known as the "first fully synthetic fiber," which was an anachronism. It so happens that fibers based upon polyvinyl chloride predated it by several years.

About 1931, the production of fibers from polyvinyl chloride (PVC) was accomplished by dry spinning from a solution in cyclohexanone. But by chlorinating the polymer, it was possible to secure solubility in acetone which has the advantage of possessing a boiling point about 100°C lower than cyclohexanone. Several million pounds per year of this fiber were produced in Germany during World War II to relieve the shortages of other materials. Unfortunately, PVC begins to soften at about 65°C and in the fibrous state shrinks disastrously upon heating. Because of its low softening point, it cannot be dyed at the temperatures commonly used for this purpose and, furthermore, it resists dyeing.

Modifications of PVC have been produced by copolymerization with other monomers. The first successful one consisted of 90 percent vinyl chloride copolymerized with 10 percent vinyl acetate. It was dry spun from acetone and given the trade name Vinyon by its producer, Union Carbide Corporation. (In 1960 vinyon was accepted as a generic name for fibers containing not less than 85 percent vinyl chloride.) It has never been produced in large volume; it is used for heat-sealable compositions.

A copolymer of vinyl chloride with vinylidene chloride was used for a number of years to produce melt-spun, heavy monofilaments. These found use in heavy fabrics where the chemical inertness of the polymer was needed, in outdoor furniture and in upholstery for seats in public-transportation vehicles.

Another vinyl-based fiber was developed in Japan but has not been produced or used in the United States. As such, it illustrates the importance both of relative availability of raw materials and differences in markets, in the success of a chemical product. Acetylene made from calcium carbide is converted to vinyl acetate which, following polymerization, is saponified to polyvinyl alcohol.

$$CH{\equiv}CH + CH_3C(=O){-}OH \longrightarrow CH_2{=}CHOC(=O)CH_3 \xrightarrow[\text{catalyst}]{\text{heat}}$$

$$\text{+}CH_2CH(OC(=O)CH_3)\text{+}_n \xrightarrow{H_2O} \text{+}CH_2CH(OH)\text{+}_n + CH_3COOH$$

This is soluble in hot water, and the solution is wet spun into a coagulating bath consisting of a concentrated solution of sodium sulfate. The fibers are heat treated to provide temporary stabilization so that they may be converted to the formal derivative by treatment with an aqueous solution of formaldehyde and sulfuric acid. This final product resists hydrolysis up to the boiling point of water. It seems reasonable to assume that it contains hemi-acetal groups and some unreacted hydroxyls on the polymer chain as well as crosslinking acetyl groups between the adjacent molecules.

Under the trade name Kuralon®*, it achieved a production level of about 180 million pounds in 1970 but has dropped to 16 million of continuous filament and 87 million of staple in 1980. The former is mainly used in industrial rubber products; the latter is used for the uniforms of Japanese school children, nonwoven and coated fabrics, and filters. It is thought to be made in even larger volumes in the People's Republic of China.

Modacrylics

In the United States, the modification of PVC has moved in the direction of copolymerizing vinyl chloride with acrylonitrile, or perhaps it should be said that PAN has been modified by copolymerizing the acrylonitrile with chlorine-containing vinyl compounds. In any case, two modacrylic fibers are now produced in the United States, a modacrylic being defined as one containing at least 35 percent but not over 85 percent acrylonitrile. Verel.®** is said to be a 50-50 copolymer of vinylidene chloride and acrylonitrile with perhaps a third com-

*Registered trademark, Kuraray Co., Ltd.
**Registered trademark Tennessee Eastman Co.

ponent graft-polymerized onto the primary material to secure dyeability.

The method used to produce Verel has not been revealed. Since it is being manufactured by a company experienced in and equipped for dry spinning cellulose acetate, it seems reasonable to suspect that this is the method used for this polymer.

Modacrylic fibers require after-stretching and heat stabilization in order to develop the necessary properties. It seems probable that the stretching is of the order of 900–1300 percent, and that, in a separate operation, shrinkage of about 15 to 25 percent is allowed during the time the fibers are heat-stabilized.

The modacrylic fibers, like vinyon, and unlike the acrylic fibers, have not become general-purpose fibers. They can be dyed satisfactorily and, thus, are acceptable in many normal textile products. But their nonflammability tends to place them in uses where such a property is important, even vital. Blended with other fibers, they are used in carpets; but their largest market is in deep-pile products, such as "fake-furs," or in doll hair where a fire hazard cannot be tolerated.

ELASTOMERIC FIBERS

The well-known elastic properties of natural rubber early led to processes for preparing it in forms which could be incorporated into fabrics used for garments. One such process uses standard rubber technology. A raw rubber of high quality is compounded with sulfur and other necessary chemicals, calendered as a uniform thin sheet onto a large metal drum, and vulcanized under water. The resulting skin is then spirally cut into strips which may be as narrow as they are thick, for example, 0.010 in. by 0.010 in. square in cross section. These strips are desulfurized, washed, dried, and packaged. Larger cross sections are easier to make. This product, coming out of the rubber rather than the textile industry, is known as a thread.

Another method produces a monofilament known as a latex thread. As the name would indicate, rubber latex is the raw material, and since extrusion through small holes is required, the purity of the material must be of a high order. With proper stabilization, the latex solution may be shipped from the rubber plantation to the plant where it is compounded with sulfur and other chemicals needed for curing, as well as with pigments, antioxidants, and similar additives. This is followed by "precuring" in order to convert the latex to a form which will coagulate upon extrusion into a precipitating bath of dilute acetic acid and form a filament having sufficient strength for subsequent operations. It passes out of the bath and is washed, dried, vulcanized in one or two stages, and packaged.

The rubber threads manufactured by either process can be used as such in combination with normal nonelastomeric yarns in fabrics made by weaving or knitting. But most of these, especially those made by a latex process, are first covered by a spiral winding of natural or man-made yarns. Often two layers are applied in opposite directions to minimize the effects of torque. Such coverings have two purposes. The first is to replace the less desirable "feel" of rubber on human skin by that of the more acceptable "hard" fiber. The second concerns the engineering of desired properties into the product to be woven or knitted into fabric. As an elastomeric material begins to recover from a state of high elongation, it supplies a high stress, but as it approaches its original unstretched condition the stress drops to a very low order. When wound in an elongated state with a yarn having high initial modulus and strength, the elastomeric component cannot retract completely because its lateral expansion is limited and jamming of the winding yarn occurs. Thus the combination of such materials can be made to provide stretch and recovery characteristics needed for a broad spectrum of applications.

But the traditional elastomeric threads have been subject to certain inherent limitations. The presence of unreacted double bonds leads to a sensitivity to oxidation, especially when exposed to the ultraviolet radiation of direct sunlight. There is also low resistance to laundry and household bleaches and dry cleaning fluids.

During recent years, elastomeric yarns or threads have been used in garments intended to restrain the female figure in a form approaching the ideal of youth and at the same time give the illusion of the garments' nonexistence. Such garments must be thin and highly effective per unit of weight. The materials from which they are composed must be compatible with these requirements.

Thus, it is not unexpected that the producers of man-made fibers, already eminently successful in meeting the needs of the market place, should look to the field of elastomeric fibers for new possibilities. Given the limitations of rubbers, both natural and synthetic, as well as the relationships between molecular structure and behavior of fiber-forming linear polymers, the scientists faced new challenges.

As an over-simplification, it can be said that within limits a rubber-like material can be stretched relatively easily but reaches a state where crystallization tends to occur. The structure produced in this manner resists further extension, and the modulus rises sharply. In contrast with the conditions that occur when the man-made fibers discussed earlier in this chapter are drawn to form fibers of stable geometry in the crystalline and orientated states, the crystalline state of the elastomeric fibers is labile unless the temperature is lowered materially. This elastomeric aspect is reasonably straightforward; synthetic rubbers have been around for a long time. But to improve on the chemical sensitivity of rubber, new approaches were necessary. The solution was found in linear polymers containing "soft" sections connected by "hard" components.

The soft, flexible, and low-melting part is commonly an aliphatic polyether or a polyester with hydroxyl end groups and a degree of polymerization of 10 to 15. The hard portion is derived from an aromatic diisocyanate supplied in an amount that will react with both end groups of the polyether or polyester to form urethane groups. The product, an intermediate known as a prepolymer, is a thick liquid composed essentially of molecules carrying active isocyanate groups at each end. For example:

$$\overset{\displaystyle H}{\diagup}\!\!\mathrm{HO\!\left(RO\right)_{\!n}} + 2\,\mathrm{O{=}CNR'NC{=}O} \longrightarrow$$

$$\mathrm{O{=}CNR'N{=}\underset{\displaystyle |\atop OH}{C}O\!\left(RO\right)_{\!n}\!-\!\underset{\displaystyle |\atop OH}{C}{=}NR'NC{=}O}$$

where $\left(RO\right)$ is an aliphatic polyether chain, R′ is one of several commonly available ring structures, and $n \sim 10$.

The elastomeric polymer is secured by "capping" or "extending" the prepolymer through its reaction with short-chain glycols or diamines, thus completing the formation of hard groups between soft, flexible chains. The conversion of these polymers into usable fibers may be accomplished by wet-, dry-, or melt-spinning operations, depending on the polymer. Additives to impart color or improve resistance to ultraviolet radiation and oxidation may be incorporated in the spinning solutions or in the melts.

The development of elastomeric fibers has resulted in a variant of wet spinning called "reaction" or "chemical" spinning. In point of fact, rayon, the first wet-spun material, might properly be said to be produced by "reaction wet spinning" or "chemical wet spinning" since complex chemical reactions have always been involved in that operation. Be that as it may, it has been found that the prepolymer of an elastomeric fiber may be extruded into a bath containing a highly reactive diamine so that the chemical conservation from liquid to solid occurs there.

The elastomeric fibers produced in this fashion are based upon segmented polyurethanes and by definition are known generically as spandex yarns. Each manufacturer uses a trade name for the usual commercial reasons. Perhaps the most noteworthy aspect from the standpoint of industrial chemistry is the multitude of options available to the manufacturer through the ingenious use of various chemicals for soft segments, hard units, chain extenders, and conditions of chemical reaction, followed by numerous possibilities for extrusion and after-treatments. However, at this time virtually all of the companies which entered the market with new products and enthusiasm in the middle 1960s have since dropped out of it.

POLYOLEFIN FIBERS

Although polyethylene was considered as a source of useful fibers at an early data, its low melting point (110-120°C) as well as other limitations precluded active development during the period when production of other fibers based upon the petrochemical industry expanded enormously. The higher melting point of high-density polyethylene gave some promise, but this was overshadowed by the introduction of polypropylene around 1958-59. Great expectations were held out for the latter as a quick contender with the polyamides and polyesters, already successful, as well as the acrylics then coming into the fiber field in volume. Several facts were thought to be in favor of polypropylene. The first was raw material costs—a few cents per pound; the second was the high level of sophistication in the spinning and processing of fibers and the presumption that this would readily lead to development of means of converting the polymer to fibers; and finally, there was a belief that the American consumer would be ready to accept, and perhaps even demand, something new and different. However, the limitations found in polypropylene fibers, such as extremely poor dyeability and low heat stability, combined with the lower prices and increased versatility offered by the already established fibers dashed the hopes for quick success. Nevertheless polypropylene has found an increasingly important place, and its properties have led to new techniques of manufacture and specialized uses.

Early in the manufacture of polyolefins, a concept was developed for dry spinning directly from the solution obtained in the polymerization operation. Had it been feasible it would have been the realization of a chemical engineer's dream; the gaseous olefin fed into one end of the equipment and the packaged fiber, ready for shipment to a textile mill, coming out the other end. But it did not turn out that way, and today melt spinning is the accepted technique for the production of monofils and multifilament yarns. To this usual method there has been added a fibrillation or "split film" procedure, and, as might be expected from a film-forming polymer, splitting of thin sheets is also possible.

The polyolefins are completely resistant to bacterial attack, are chemically inert, and are unaffected by water. Monofilaments can be produced that possess high strength, low elongation under stress and dimensional stability at normal atmospheric temperatures. Polyolefin monofilaments have found broad application in cordage and fishing nets (which float), and when woven into fabrics they are used for outdoor furniture, tarpaulins, and similar applications. Large filament denier staple is widely used in "indoor-outdoor" carpets.

The granules of polypropylene are prepared for usage by tumble mixing them with the selected pigments (often premixed and suspended in polyethylene by the supplier to obtain adequate dispersion), anti-oxidant, and ultra-violet light inhibitor. This combination may then be put through a screw type extruder, quenched in water as "spaghetti," dried, and cut into short lengths for feeding to the melt-spinning machine. Because of high melt viscosity, polypropylene must be extruded at 100–150°C above its melting point. The usual metering pumps follow the heated melt screws, and screen filtration units are placed above the spinnerette. The filaments may be extruded into water for quenching the heat removal. The materials thus produced must be after-drawn in a heated condition to several times their original length, this being determined by the molecular weight and the properties desired in the end product. Stabilization follows in a heat-setting operation at constant length or with a limited shrinkage being allowed.

The production of multifilament yarns follows along the lines commonly used in melt-spinning processes, with air quenching replacing the water quenching mentioned above. If anything, a greater drawdown, that is, a higher ratio of the velocity of solid fiber wound up to the average velocity of liquid polymer in the spinnerette capillary, is used in this spinning operation than is commonly used for other melt-spun fibers. This elongating operation does not preclude the need for an afterdrawing step to produce the orientation necessary to achieve the desired physical properties. As with monofilaments, hot drawing is required. Where the fibers are to be used as multifilament yarns a higher amount of stretch is used than when staple fibers are to be the end product. Higher strength and lower elongation are normally required of multifilament yarns as compared with staple fibers.

Fig. 11.20 The melt spinning of polypropylene, showing a heavy yarn emerging from the cooling column, passing in contact with the lubricant applicator, and over a pair of draw-off godets that determine its speed, and finally to a tension-controlled packaging device. *(Courtesy Hercules, Inc.)*

The production of "spunbonded" materials from a number of fiber-forming polymers has been especially applied to polypropylene (Typar®* to manufacture a sheeting which is used as the primary backing in tufted carpets. In this case the strength-contributing fibers are spun and drawn in the continuous operation. The oriented fibers are randomly laid as a web and bonded by fusing under heat and pressure at selected points.

As already noted, since polyolefins are converted in great volume to thin films, it is logical to slit these into narrow strips for uses where they can compete with the more conventional fibers. But the polyolefins have also made possible the "split-film" method of producing fibers. This is due to their ability to be cast into films which, when stretched, become highly crystalline and oriented in the direction of stretch. The very low strength in the direction perpendicular to the axis of orientation causes the film to split and fibrillate. The resulting web-like structure, comprising interattached fibrils of high longitudinal strength may be converted by twisting into yarn-like structures or cut into staple fibers.

*Registered trademark, E. I. du Pont de Nemours & Company.

POLYTETRAFLUOROETHYLENE

Teflon®* is unique in that it is not soluble in any known solvent and will not melt below about 400°C, at which point it is not stable enough to allow spinning. Such a combination would seem to pose an impossible problem. Research into the fundamental characteristics of the polymer, however, revealed that the submicroscopic particles precipitated from the polymerization reaction were about 100 times as long as they were thick. If a thin stream of polytetrafluoroethylene suspension is extruded,

*Registered trademark, E. I. du Pont de Nemours & Co.

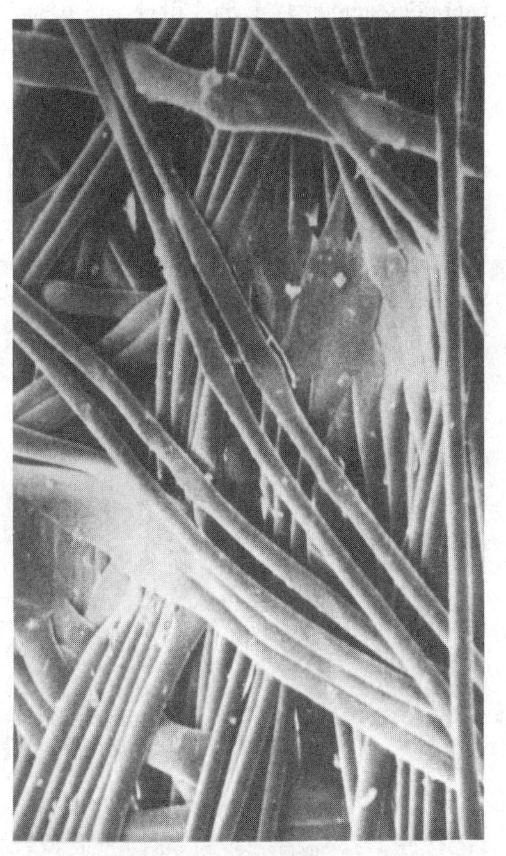

Fig. 11.21 Spunbonded polypropylene shows embossing which helps bind the structure. *(Courtesy E. I. du Pont de Nemours and Co.)*

when the dispersion is broken the particles line up to form a discrete filament which has sufficient strength to allow its transportation to a sintering operation. This fiber has adequate tenacity (1.5 grams per denier) for ordinary textile handling, knitting, and weaving.

The chemical inertness and thermal stability of this material is so great that in spite of its extremely high price ($15-$30 per pound, depending upon filament and yarn size, and considerably higher for specialty items in 1980) it is used in chemical operations where drastic conditions exist and no other organic fiber is suitable.

GLASS AND CARBON FIBERS

Among the man-made inorganic fibers, by far glass is produced by the largest volume. There has been a rapid increase in the use of textile grades of glass fibers. Outside the textile field, enormous amounts of glass fibers are used in air filters, thermal insulation (glass wool), and for the reinforcement of plastics.

Glass possesses obvious and well-known characteristics which have largely determined the methods used to form it into large objects. It flows readily when molten and can be drawn into filaments, whose extreme fineness appears to be limited only by the drawing speed. The method used in producing textile-grade glass fibers follows this principle.

In the commercial operation the molten glass, produced either directly or by remelting of marbles, is held at a uniform temperature in a vessel, whose bottom carries a bushing containing small uniform holes. The molten glass flows through these as tiny streams which are attenuated into filaments at speeds of the order of 3,000 meters per minute, coated with a lubricant, gathered into groups to form yarns, and wound up. For a particular glass viscosity, the size of the individual filaments is determined by the combination of hole-size and speed of attenuation.

Because of the inherently high modulus of glass, very fine filaments are required in order to approach the required properties of textile materials. Thus, the diameter of glass filaments falls in the range of 3.8 to 7.6 microns whereas the average diameter of the finest organic fibers is about twice as large. The fiber- and yarn-numbering system is based on nomenclature used in the glass industry and differs from the traditional systems accepted in the textile and organic-fiber industries.

The method of manufacture of glass staple fibers differs from those used to produce the corresponding organic materials, all of which are based upon cutting the continuous-filament product. Air jets, directed in the same line of flow as the emerging streams of glass, attenuate them and break the solid glass into the lengths desired for further processing. These are

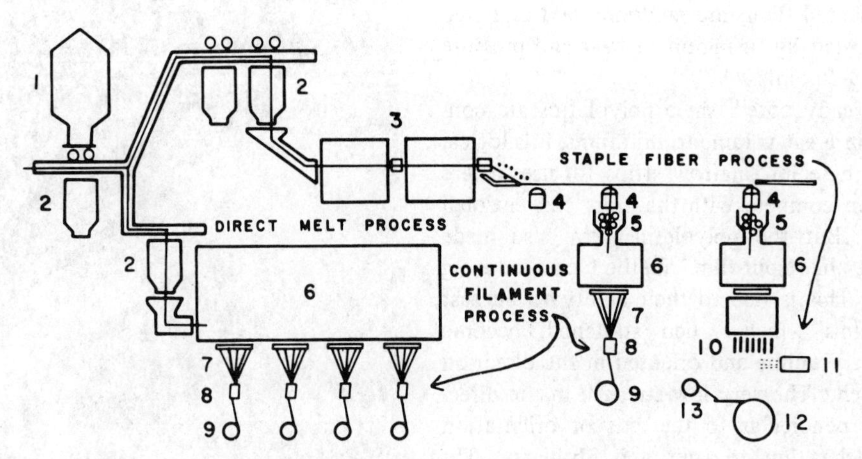

Fig. 11.22 Flow diagram for the manufacture of textile glass fiber: (1) glass batch; (2) batch cans; (3) marble-forming; (4) cullet cans; (5) marbles; (6) melting furnaces; (7) filament yarn formation; (8) gathering and sizing; (9) yarn packaging; (10) air jets; (11) lubricant spray; (12) collection for staple fibers; (13) staple fiber packaging. (*Courtesy Owens-Corning Fiberglass Corp.*)

gathered on an appropriate vacuum drum and delivered as slivers or a matte. To produce fibers which may be coarser and considerably less uniform to be used for the production of filters, paper, or thermal insulation, large streams of molten glass are cross blown by blasts of hot air, steam, or burning gas.

As might be expected from the nature of glass, the conversion of glass fibers into the final products has required the development of new lubricants, finishes, and processing techniques. For example, because glass fabrics cannot be dyed directly or printed with the colors demanded for acceptance as draperies, the colorant must be applied to a resin coating. But before applying the coating it is necessary to remove the lubricant which was placed on the fibers to permit their conversion into a fabric. This is done by burning. The elevated temperature resulting from this operation also relaxes the internal strains developed in the glass fibers during the steps of the textile operations and sets the yarns in the required geometry. The fabric is then resin treated, cured, and dyed or printed.

Another inherent property of glass is the tendency of unprotected fiber surfaces to abrade each other to destruction under the action of very little mechanical working. Its poor adhesion to rubber and the inadequacy of then existing bonding agents, when it was first considered for rubber reinforcing purposes, frustrated attempts by manufacturers to take advantage of the very high tensile strength, completely elastic behavior, high modulus, and lack of moisture sensitivity of glass fibers. But it has been possible to modify the fiber surfaces so that satisfactory adhesion is achievable, and the impregnant can be applied in such a way that fiber to fiber contact is prevented. With adhesion problems solved, glass in cord forms has found large markets in belt-type tire construction and in all kinds of power transmitting rubber belts. It is estimated that glass fibers used as reinforcing agents are approaching a volume of a billion pounds a year in the U.S.

Following World War II the development of jet aircraft and rockets brought about demands for fibers having thermal resistance, strength, and modulus far beyond what could be obtained in existing organic fibers. Much of this need was for reinforcing materials which would be embedded in matrices of one type or another. As a result, techniques have been developed for preparing fibers from a good many metals and refractory inorganic compounds. Although these materials are essential for certain uses, the volume of production is still low and the prices correspondingly high (as much as $1,000 per pound).

Carbon and graphite fibers are made from rayon and acrylic progenitors by driving off virtually all of the hydrogen and oxygen contained in them. The principal is essentially the same as that which brought about the formation of coal or, citing a more recent and dramatic example, the conversion of the original wooden beams of buildings in Herculaneum, buried by a flow of mud from Vesuvius in 67 A.D., to what appears to be charcoal. In the present commercial process the starting material is selected so as to produce a final product of the desired size and morphology, since the brittleness precludes usual textile operations. It is heated at 200–400°C to cause thermal decomposition but not combustion. Heating follows in an inert atmosphere at temperatures as high as 1,700°C, with conditions chosen so as to secure desired end product properties in the resulting "carbon" fibers ($C \sim 97\%$). If "graphite" ($C > 99.6\%C$) is wanted, temperatures as high as 3000°C are used, as well as stretching. Thus millennia in nature are shortened to hours or less by industrial chemistry. Carbon fibers are currently priced at about $15 per pound and graphite at perhaps twice that figure. Production in the U.S. is well over a million pounds a year.

HIGH-TEMPERATURE-RESISTANT FIBERS[9]

At present only one high-temperature-resistant fiber is produced in what may be termed commercial volume. The need for such fibers has arisen from both the general economy and the space program where the usual characteristics of organic-based fibers are desired but the high temperature resistance of inorganic fibers is required. Thus, they are expected to retain their structural integrity at temperatures of 300°C and above for considerable periods of time, but otherwise their properties should resemble those of the more common man-made textile fibers.

The one fiber referred to above was announced in early 1962 as HT-1 nylon, but it is presently sold as Nomex®*. It is generally accepted in the industry as being formed by solution polymerization of m-phenylenediamine and isophthalic acid chloride. Since it does not melt but rather decomposes at temperatures well above 300°C, it must be spun from solution. It appears that a dry-spinning technique is used. Polyamides of this type must be hot drawn and then relaxed in order for them to have the desired dimensional stability when later used as elevated temperatures. Variants of the original polymer have been developed. The total U.S. production is of the order of 15 million pounds per year, and the price is about $6/lb.

Poly-2-2′-(*m*-phenylene)-5,5′-bibenzimidazole, commonly designated as PBI, was developed under the aegis of the Air Force Materials Laboratory in cooperation with the Celanese Corp. It is a condensation polymer obtained from the reaction of 3,3′-diaminobenzidine and diphenylisophthalate in a nitrogen atmosphere at temperatures which may reach 450°C in the final stages. The polymer is dissolved at a high temperature under nitrogen pressure in dimethylacetamide to which a small amount of lithium chloride is added to increase the stability of the solution. It is then dry spun into an atmosphere of heated nitrogen (about 200°C), from which the solvent is recovered, stretched slightly in steam, washed to remove the lithium chloride and the last traces of solvent, dried, and packaged. Drawing and relaxing are done in an inert atmosphere, as might be expected, since temperatures as high as 250°C or higher are used.

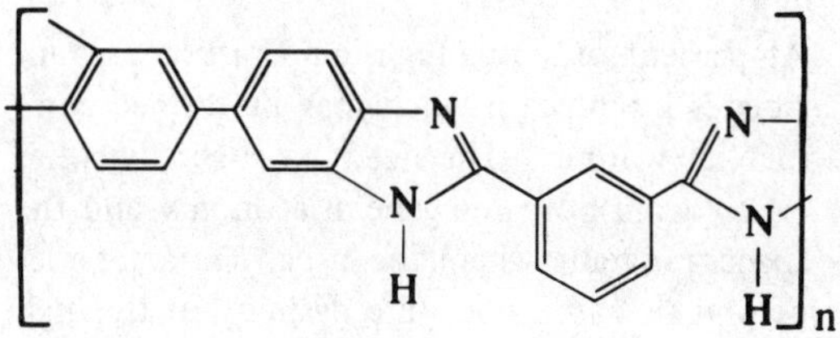

The final yarn is golden yellow, and because this color appears to be an intrinsic property of the polymer, it may have a built-in limitation as far as the civilian market is concerned. But since it is capable of retaining about one half of its original strength (~5 g/den) upon exposure to air for 18 hours at 350°C or one hour at 425°C, it would appear to have considerable promise in a limited field. Celanese has announced the intention to investigate commercial development of this fiber.

*Registered trademark, E. I. de Pont de Nemours & Co.

FIBER VARIANTS

Introduction

In a previous section, data and plots were given showing the rapid rise in consumption and production of man-made fibers at the expense of natural fibers. The principal reason for this has been the wide range of man-made fiber variants which can be produced from a single fiber forming polymer. The wide range of polymers each with their particular properties adds yet another dimension. This is not to say that there is only one cotton, wool, silk, or asbestos. There are many varieties of natural fibers, but their supply is limited by natural factors such as climate and genetics. The relative availabilities of man-made fiber types can be altered by controlled chemical-process changes, while the amount and quality of a desired cotton type that can be grown is determined in large part by climatic conditions which man has not yet learned to control. Another factor that has aided the growth of man-made fibers is their consistent quality and properties. Again the grade and quality of natural fibers are subject to the vagaries of nature.

For the purposes of this discussion, fiber variants will be divided into two types: chemical and physical. Chemical variants will be those involving a small but significant change in composition, whereas physical changes will be those involving a change in either the dimensions of the fiber or its stress/strain or stability features. The definitions of the two variants could also be based on modification of esthetics or modification of functionality.

Physical Variants

Most man-made fibers are available as staple, tow, and filament. Natural fibers are available only in the characteristic forms in which they occur, with filament silk and cotton staple as examples. All man-made fibers are formed initially as filament yarns. The german adjective *endlos* which translates literally as "endless" is very descriptive in that filament yarns are continuous strands consisting of one or more members that for most practical purposes are infinite in length. Fine filament yarns (40 to 100 denier) are used in producing lightweight apparel fabrics while coarse filament yarns (800 to 1200 denier) are found as reinforcement in tires or conveyor belts. These examples are chosen as extremes to show the range or applicability of man-made fibers and represent only a hint of the actual range of end uses.

When many filament yarns are collected into a bundle immediately after formation, the resulting structure is called a tow. Tows may range from 10,000 to over a million in total denier. In the next step, tow is crimped by the process previously described which imparts what is usually a saw-tooth appearance along the length of all the filaments. For some end uses, the crimped tow is the product which is shipped out by the fiber maker. One example already discussed is that of acrylic tows which are converted to staple as part of the spinning of yarns with a wool-like character. Another example is the use of cellulose acetate tows to form cigarette filters or the ink reservoirs for marking pens. In this case, the compact tow bundle is first treated to separate the individual filaments giving a voluminous structure which is then gathered into a continuous rod, wrapped in paper, and cut into appropriate lengths.

In the fiber-making plant, tow is cut into short lengths or staple which range in length from $\frac{1}{16}''$ to $6''$. The end-use product determines the staple length used. The very short staples are used in making flocked structures or in the manufacture of papers containing blends of natural cellulose and man-made fiber. The longest lengths are used in spinning heavy yarns that are used in carpets or for cordage. Most staple produced in $1\frac{1}{2}''$ to $3''$ in length and is used to form blends with cotton, rayon, or wool in the yarns that are used in standard apparel fabrics. The staple length of the man-made fiber is chosen to match that of the other blend component; otherwise, uneven yarn of poor quality will result.

The size of a man-made fiber can be altered by changing the size of the hole through which it is extruded and maintaining a constant take-up speed. Expressed in denier (weight in grams of 9000 meters of fiber), commercial fibers range from about 1.25 to 25 denier. This corresponds to average diameters of about 5 to 50 microns. The very large fibers would be used to make doll's hair or wigs. The bulk of man-made fiber staples are made in the 1.5 to 6.0 denier range which corresponds to cotton blending fiber at the low denier end and coarse wool fiber at the high end. Staple and filament yarns used in carpets are in the 12 to 16 denier range, while industrial filament yarns such as tire cord will be about 6 denier per filament. The size of a fiber is a determinant of its stiffness which in turn influences the draping quality and surface feel of a fabric made from it.

Usually fibers are extruded through circular jet holes, but it is possible to use non-circular holes in the jets which results in a wide variety of cross-sectional shapes being available. In the cases of fibers spun from solutions, as in dry or wet spinning, a majority of the mass exiting the jet hole is removed during fiber formation. For, example, cellulose acetate fiber is made from 25% solution of cellulose acetate in acetone containing a small amount of water. After the fiber leaves the jet face, solvent begins evaporating and as a result the fiber cross section decreases in area. The final result is a fiber of roughly circular cross section but with a serrated edge and much smaller in area than the parent jet hole. When cellulose acetate is extruded through a triangular jet hole, the end result is a fiber of "Y" cross section which results from shrinkage down from the original triangular shape.

In the melt-spinning process, there is no solvent loss to influence cross-sectional shape formation. In the case of a triangular jet hole in melt spinning, the molten fiber leaves the jet face with a triangular cross section but immediately tends to return to a circular cross section driven by surface tension forces. It is necessary, therefore, to quench or freeze the fiber as soon as possible to maintain cross-sectional definition. Some loss of definition is unavoidable during the subsequent drawing step which almost always follows extrusion in a melt spinning process.

Fibers of non-circular cross section can modify and change both esthetic and functional qualities in textile structures. The triangular cross section is typical in those regards. Its shape makes it stiffer than circular fiber of the same cross-sectional area. In a fabric, this results in less drapability and a crisper surface feel. The flat surfaces will reflect light in a different way and can create desirable lustrous effects. These optical effects are subject to many subtle influences having to do with the sizes of the reflecting surfaces and the amount of internal reflection that takes place. Triangular or "Y" cross-sectional fibers have greater specific surface area (square meters per gram) than their circular counterparts. This accounts for the use of non-circular fibers in aerosol filtration where surface area is a major factor in efficiency. Yarns made with triangular cross-sectional fiber are more voluminous than those from round cross-sectional fibers. This means that when made into fabrics of equal weight, the variant cross-sectional fabric will transmit less light and be less permeable to air.

When spinning blended staple yarns, maximum strength is obtained if the blend components have similar load/elongation characteristics. The stiffness of polyester staple fibers can be varied mainly by changing the draw ratio used to orient the fiber after extrusion. At standard draw ratio, the fiber obtained is most suitable for blending with standard rayon, acrylic, or wool fibers. At a higher draw ratio, the staple produced is stiffer and is similar to cotton in load/elongation properties. In most industrial uses, fiber and yarns are used under conditions where they bear a load while hot. Examples of this are tires and power transmission belts; it is important that they not grow or stretch significantly under these conditions. Accordingly, industrial yarns are drawn to the highest extent to take all the possible stretch out of them during the manufacturing process. Relatively, these yarns have high strength, high stiffness, and low elongation to break.

While in most instances it is desirable to have fibers that are dimensionally stable, in some structures it is advantageous to use mixtures of fibers that are stable with fibers that will shrink in heat or steam. The yarn bulking that occurs with blend yarns of high- and low-shrinkage acrylic staples has already been described in the section on that fiber. Using the same principle, felt-like structures can be made by heat treatment of non-woven battings containing low- and high-shrink staple. High-shrinkage potential is usually built into a fiber by stretching it and failing to give it a stabilization treatment.

Chemical Variants

The use of titanium dioxide as pigment in the delustering or dulling of man-made fibers was discussed in the section on viscose rayon. The addition of these pigment particles influences the processing and performance of the fiber, as well as changing its appearance. Because of the whiteness of the delustered fiber, it requires more dyestuff to reach a given shade than in the case of a bright fiber. The sliding friction of a delustered fiber is lower because the pigment particles protruding from the fiber surface reduce the contact area between a fiber and a guide, for example. By the same token, some pigments can accelerate the wear of contacted surfaces. It is said that the orientation or drawing of delustered fibers proceeds more smoothly because the pigment particles act as nucleation sites where molecular motion is initiated. Unless specially treated, the surface of anatase—one of the crystal forms of titanium dioxide—can accelerate the ultraviolet light degradation of acetate or nylon fibers. It is postulated that the crystal surface catalyzes the formation of peroxides from the water and oxygen under the influence of ultraviolet light and that peroxides are the active species in the resulting polymer degradation. For this reason, rutile—the other common crystal form of titanium dioxide—is used to deluster fibers when improved sunlight resistance is needed.

The degree of polymerization or molecular weight of the polymers on which man-made fibers are based can be controlled as part of one of the early steps in the process. The polymer molecular weight chosen for a fiber has a strong influence on process economics, ease of conversion to fiber, and end use performance. Standard fibers are based on the best balance of these factors. However, in fabrics that are open in texture and made from standard polyester staple spun yarns of low twist, a condition known as "pilling" will develop as a result of wearing. These pills are made up of fiber ends which have worked loose from the yarn bundles as a result of surface rubbing and have wrapped around themselves. In the case of fabrics from natural fibers, which are generally less wear resistant than man-made fibers, these pills or fiber bundles will be lost by attrition with continuing wear. Since the wear resistance of a man-made fiber is related directly to its molecular weight in a general way, the pilling tendency of a polyester staple can be reduced by lowering its degree of polymerization. This compromises the tensile characteristics of the fiber only to a small extent while all the other desirable properties such as minimum care characteristics are essentially unaffected. When the molecular weight of a polyester or nylon polymer is increased above the standard, the fibers which can be produced will have increased tensile properties and fatigue resistance. The filament polyester and nylon yarns used in end products such as tires and conveyor belts are based on such polymers.

The technology for dope dyeing or mass coloration of fiber as part of the fiber manufacturing operation was described in the sections on viscose rayon and cellulose acetate. The use of this technology has decreased largely due to the problems of profitably managing the required inventory of colors in a rapidly changing fashion market. Mass coloration is only used extensively if the fiber cannot be dyed by any other means. Polypropylene is a good example in this instance. A hypothetical example of a market where mass coloration could succeed would be in military uniforms where the color selection is very limited and fastness requirements are high. However, if one regards "white" as a color one finds that a substantial portion of the polyester staple fiber produced for blending with cotton or rayon contains an optical brightener or fluorescing agent. This is required to overcome the yellowing tendency of polyester in use as a result of the absorption of hydrophobic soils. The cotton or rayon fibers in a blend are continually rewhitened by the fluorescing agents added to laundry compounds for that purpose, but these agents are without effect on the polyester blend component. Since the polyester component is usually at least half of the blend, the spun-in optical brightener it contains is vital for the maintenance of overall whiteness.

In the cases of polyester, nylon, and acrylic fibers, their manufacturers have developed fiber variants with a wide range of dyeing characteristics. These are referred to as dye variant fibers. Polyester fibers are usually dyed with what are described as disperse dyes. These dyes are only slightly soluble in boiling water and are used in the form of dispersions. The dissolved dye is absorbed from the dyebath into the polyester fiber by a process of solid solutioning. The dissolved dye concentration is then replenished by some of the dispersed dye passing into solution. The overall rate of dyeing is determined by the molecular structure of the fiber through which the dye must diffuse. Generally, the more the molecules have been organized by draw orientation and crystallization, the slower is the dyeing rate. The ability of the molecules to organize can be reduced by polymerizing a small amount of a foreign dibasic acid or glycol into the structure. Usually, 5-10 mol% is sufficient to break up the regularity and prevent optimum structure development. Adipic acid or isophthalic acid and polyethylene glycols are some of the comonomers used frequently to form fast- or deep-dyeing polyesters. These fibers are more economical to dye or print since special dyebath additives, high dyeing temperatures, and high-pressure steam-print fixation are usually not required.

Polyester fibers can be given an additional mechanism for dyeing if an ionic comonomer is added during polymerization. A common additive is an alkali metal salt of dimethyl–5-sulfo–isophthalate which gives sulfonic (anionic) groups as part of the polymer structure. These groups allow the fiber to absorb basic (cationic) dyes by a specific ionic mechanism. The amount of cationic dye which can be absorbed by the fiber is stoichiometrically related to the number of anionic sites present in the fiber. This is quite distinct from the general solid solutioning observed with disperse dyes and polyester. A cationic dyeable polyester is useful in two main ways in fabric coloration. First, cationic dyes give brighter, clearer shades than disperse dyes, and this can be important during fashion cycles for both solid-dyed shades and prints. Second, arrangements of unmodified and cationic dyeable polyesters in fabrics or yarns can be dyed in the piece to a variety of color/white combinations by selection of dyestuffs. This is more economical than dyeing fibers or yarns to different colors in separate procedures and then combining them into yarns and fabrics.

When made by the typical process, nylon 6,6 can be dyed either with disperse dyes or with acid (anionic) dyes. The dyeing of nylon with disperse dyes proceeds by essentially the same steps as described above in the case of polyester except that nylon absorbs dye much more readily. The ability of nylon to absorb acid (anionic) dyes is the result of a significant number of accessible free amine (cationic) end groups being present in the nylon polymer. The dyeing of nylon with acid dyes is analagous to the dyeing of wool or to the dyeing of modified polyester with basic dyes except that the polarities of the interacting groups are reversed in the latter case.

For best fastness to light and washing, nylon 6,6 is dyed with acid dyes, and nylon dye variants are based therefore on manipulation of the level of acid dye uptake. By adding a monobasic acid such as acetic acid to the reaction mix near the end of the polymerization process, the amine end groups arc converted to amide groups which have no affinity for acid dyes under normal dyeing conditions. This technique creates light acid dyeing or acid reserve dye variants which can have some capacity to absorb basic dyes on carboxylic acid end groups. Nylon dye variants which have increased acid dye uptake can be made by using a slight excess of diamine in the polymerization. In this way, there are no free carboxylic acid end groups. Nylon dye variants have found the greatest acceptance in floor coverings where attractive patterns can be piece dyed using controlled dyebath conditions and selected acid dyestuffs.

Acrylic fibers are dyed most frequently with basic dyes. This is made possible by copolymerizing acrylonitrile with an acidic monomer such as styrene-para-sulfonic acid. Acid dyeing acrylic fiber can be made by using a basic comonomer such as a vinyl pyridine or a vinyl pyrrolidone.

Fibers not having inherent flame resistant properties can often be given this property by incorporation of an additive. This may be done by copolymerization into the polymer, reaction with the polymer after polymerization, use of a polymeric additive, or use of a monomeric additive. These additives usually contain bromine, nitrogen, or phosphorous or a combination of these. Great care must be used in choosing the additive and its level of addition in order to prevent loss of other desirable fiber properties and to avoid possible harmful effects to wearers of garments or handlers of yarns and fabrics.

A wide variety of special durable surface treatments have been used on man-made fibers with differing degrees of technical and commercial success. These include treatments for soil resistance, antistatic electricity, and promotion of wearer comfort through moisture wicking. Fiber finishes for promoting adhesion have been successful in the cases of polyester tire cord to rubber and glass fiber to polyester resin.

REFERENCES

1. *Seven Centuries of Art*, p. 12, New York, Time-Life Books, 1970.
2. Mark, H., and Whitby, G. S., *Collected Papers of W. H. Carothers*, New York, John Wiley & Sons, 1940.
3. Hindle, W. H., "Fibers Outlook," Chemical Manufacturers Research Association Paper No. 760, May 1972.
4. Robinson, J. S., ed., *Fiber-Forming Polymers: Recent Advances*, Park Ridge, New Jersey: Noyes Data Corp., 1980.
5. Markam, J. W., *Competition in the Rayon Industry*, p. 16, Cambridge, Mass., Harvard University Press, 1952.
6. Work, R. W., "Dimensions, Birefringences, and Force-Elongation Behavior of Major and Minor Ampullate Silk Fibers from Orb-Web-Spinning Spiders–The Effects of Wetting on These Properties," *Textile Res. J.*, 47, 650–662 (1977).
7. German Patent 748,253.

8. British Patent 578,079.
9. Preston, J., and Economy, J., eds., *High Temperature and Flame-Resistant Fibers*, New York: John Wiley & Sons, 1973.

SUGGESTED FURTHER READING

The reader is referred to the two encyclopedias listed below for additional information. They contain enormous quantities of information on man-made fibers as well as comprehensive bibliographies.

1. *Kirk-Othmer Encyclopedia of Chemical Technology*, 2nd ed., Interscience Publishers, New York. (Twenty one volumes and a supplement, 3rd ed., to date, sixteen volumes.)
2. *Encyclopedia of Polymer Science and Technology*, Interscience Publishers, New York. (Sixteen volumes.)

The following books contain broad discussions of man-made textile fibers in their many ramifications.

1. Morton, W. E. and Hearle, J. W. S. *Physical Properties of Textile Fibres*, Manchester, The Textile Institute, London, Butterworths, 1962.
2. Hearle, J. W. S. and Peters, R. H., Ed., *Fibre Structure*, Manchester, The Textile Institute, London, Butterworths, 1963.
3. Moncrieff, R. W., *Man-Made Fibres*, 6th Ed., New York, John Wiley and Sons, 1975.
4. Peters, R. H., *Textile Chemistry*; Vol. **I**, "The Chemistry of Fibers," and Vol. **II**, "Impurities in Fibers; Purification of Fibers," New York, Elsevier, 1963 and 1967.
5. Mark, H. F., Atlas, S. M., and Cernia, E., Eds., *Man-Made Fibers*; *Science and Technology*, Vol. I, Vol. II, and Vol. III, New York, John Wiley and Sons, 1967, 1968, and 1968.
6. Ciferri, A. and Ward, I. M., Eds., *Ultra-High Modulus Polymers*, London, Applied Science Publishers, 1979.

12

Animal and Vegetable Oils, Fats, and Waxes

Glenn Fuller*

INTRODUCTION

Oils, fats, and waxes belong to the class of compounds called lipids, which are insoluble in water, but soluble in ether and other organic solvents. Lipids have long been considered important as one of the three major classes of food nutrients. They differ from the other two, proteins and carbohydrates, in that they provide more than twice the energy per unit weight when metabolized to carbon dioxide and water. Hence, they were thought to be primarily compounds for storage of energy in plants and animals. Lipids provide over 40 percent of the calories in the U.S. diet.[1] More recently, lipids have been shown to be extremely important in the organization of the living cell. Various metabolic functions are carried out in isolated parts of cells, and cell membranes which separate and isolate the cell organelles consist of complex lipids and proteins. Although the functions of lipids in cell membranes are not well understood, the selective permeability of lipid-membrane constituents certainly plays a role in the isolation of cell functions.

Oils and fats are usually distinguished from one another by their melting points and, to a large extent, by their sources. Oils are liquid at ambient temperatures while fats are solid or semisolid. The fats are usually of animal origin while oils are extracted from plant tissue, fish, or marine animals. The prinicipal components of both fats and oils are glycerol esters of fatty acids. These esters can be considered to be formed by combination of fatty acids and alcohols with loss of one molecule of water for each ester linkage formed. Since glycerine (glycerol) is a trihydric alcohol, one molecule can combine with three molecules of fatty acid to form a triglyceride, as illustrated:

$$\begin{array}{l} CH_2OH \\ | \\ CHOH \quad + \; 3C_{17}H_{35}COOH \longrightarrow \\ | \\ CH_2OH \end{array}$$

Glycerol Stearic acid

*Western Regional Research Center, Agricultural Research Service, U.S. Department of Agriculture, Albany, California.

$$\begin{array}{l} CH_2OOCC_{17}H_{35} \\ | \\ CHOOCC_{17}H_{35} + 3H_2O \\ | \\ CH_2OOCC_{17}H_{35} \end{array}$$

Glycerol tristearate
(tristearin)

Fats and oils as obtained from animal or vegetable tissues are predominantly triglycerides with traces of mono- and diglycerides, free fatty acids, sterols, phospholipids and other minor constituents. Procedures for refining and processing crude oils are designed to remove most of the trace constituents.

Waxes differ from fats and oils in that a mono- or dihydric alcohol has replaced glycerol in combination with a fatty acid. Waxes may be of either animal or vegetable origin and wax alcohols may be either aliphatic or alicyclic. In mammals, significant amounts of cholesterol esters of fatty acids are found, particularly in the adrenals, the liver, and blood plasma.[2] Cholesterol esters are of interest in nutrition because they may affect blood cholesterol levels, which may be important in development of atherosclerosis.

In addition to cholesterol (illustrated), other sterols are found in minor amounts in fats, either as esters or in uncombined form.[3] Cholesterol is present only in trace

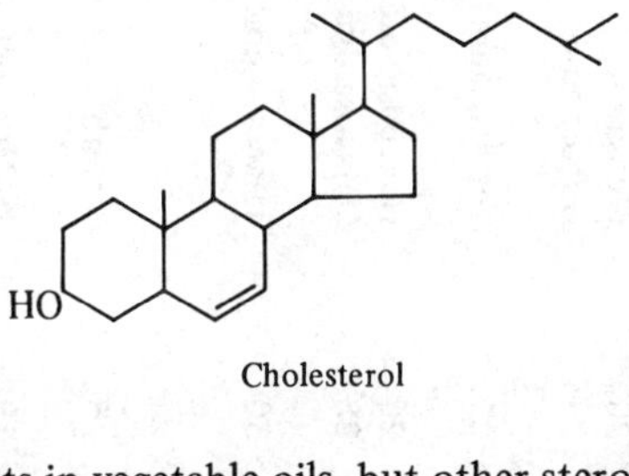

Cholesterol

amounts in vegetable oils, but other sterols such as stigmasterol, lanosterol, β-sitosterol, are commonly present in vegetable oils. Vegetable oils may also contain xanthophylls and carotenes (terpenoid compounds) which impart color to the oil, sometimes desirable from the standpoint of appearance, but usually removed in processing. β-Carotene is a precursor of Vitamin A.

One important group of compounds which are similar in structure to the triglycerides is the phospholipids. Three important phosphatidic esters are phosphatidylcholine, phosphatidylserine, and phosphatidylethanolamine. Phosphatidylinositol is also often present with one or more sugar moieties

$$\begin{array}{l} RCOOCH_2 \\ | \\ RCOOCH \\ | \\ CH_2OP(=O)(OH){-}OY \end{array}$$

$Y = -CH_2CH_2\overset{\oplus}{N}Me_3\overset{-}{O}H$	3-phosphatidylcholine
$Y = -CH_2CH_2NH_2$	3-phosphatidylethanolamine
$Y = -CH_2CH(CO_2H){-}NH_2$	3-phosphatidylserine

attached. Biologically, the phosphatides are important, but most processing of oils for food requires their early removal. This "degumming" process yields "lecithin" from soybean oil, a product which has some commercial value because of its surface-active properties.

FATTY ACIDS

Fatty acids are obtained from triglycerides by the process of fat splitting. The physical and chemical properties of fats and oils are determined essentially by the fatty acid composition of their triglycerides. Most of the common commercial fats and oils are principally mixed triglycerides of five common fatty acids; palmitic, stearic, oleic, linoleic and linolenic acids. All are carboxylic acids with unbranched chains and, except for palmitic acid, they all have eighteen carbon atoms. Their structures and properties, along with those of some of the many less common fatty acids, are shown in Table 12.1. The acids in this table are only representative of the many such acids identified as metabolities of living organisms, but they illustrate some rules which generally apply in their chemistry.

TABLE 12.1 Some Important Fatty Acids

No. of Carbon Atoms	*Systematic Name*	*Common Name*	*Melting Point, °C*	*Boiling Point, °C/mm*	*Methyl Ester (BP°C/mm)*	*Common Sources*
			Unbranched Saturated Fatty Acids			
3	Propanoic	Propionic	−20.8	141	80	Bacterial fermentations
4	Butanoic	Butyric	−5.3	164	103	Milk fats
5	Pentanoic	Valeric	−33.8	186	126.5	Bacterial fermentations
6	Hexanoic	Caproic	−3.2	206	151	Milk fats and some seed oils
8	Octanoic	Caprylic	16.5	240	195	Milk fats and *Palmae* seed oils
10	Decanoic	Capric	31.6	271	228	Sheep and goats' milk palm seed oils, sperm head oil
12	Dodecanoic	Lauric	44.8	130/1	262	Coconut oil
14	Tetradecanoic	Myristic	54.4	149/1	114/1	Palm and coconut oils
16	Hexadecanoic	Palmitic	62.9	167/1	136/1	Palm oil
18	Octadecanoic	Stearic	70.1	184/1	156/1	Animal fats
20	Eicosanoic	Arachidic	76.1	204/1	188/2	Some animal fats
22	Docosanoic	Behenic	80.0	–	206/2	Various seed oils incl. peanut oil
24	Tetracosanoic	Lignoceric	84.2	–	222/2	Minor amounts in some seed oils
26	Hexacosanoic	Cerotic	87.8	–	237/2	Plant waxes
28	Octacosanoic	Montanic	90.9	–	–	Beeswax and other waxes
30	Triacontanoic	Mellissic	93.6	–	–	Beeswax and other waxes
			Branched Saturated Fatty Acids			
5	3-Methylbutanoic	Isovaleric	−51.0	177	–	Dolphin and porpoise fats
19	10-Methylstearic	Tuberculostearic	11	–	–	Lipid of turbercle bacillus
			Monoenoic Fatty Acids			
10	9-Decenoic	Caproleic	–	142–8/15	115–6/12	Milk fats

14	9-Tetradecenoic	Myristoleic	–	–	108–109	Some feed fats--milk fats
16	9-Hexadecenoic	Palmitoleic	–0.5 to 0.5	–	134–5/1	Many fats and marine oils
17	9-Heptadecenoic	–	14.0–14.1	–	–	*Candida tropicallis* yeast
18	6-Octadecenoic	Petroselinic	30–33	208–10/10	–	Parsley seed oil
18	9-Octadecenoic	Oleic	16.3	153/0.1	152.5/1	Almost all fats and oils
22	12-Docosenoic	Erucic	33.5	241–3/5	169–170/1	Rapeseed oil
			Dienoic Fatty Acids			
10	2,4-Decadienoic	Stillingic	–	–	–	Stillingia oil
18	9,12-Octadecadienoic	Linoleic	–5	202/1.4	149.5/1	Many vegetable oils
22	13,16-Docasadienoic	–	–	–	–	Rapeseed oil
			Trienoic Fatty Acids			
16	6,10,14-Hexadecatrienoic	Hiragonic	–	180–190/15	–	Sardine oil
18	6,9,12-Octadecatrienoic	–	–	–	–	Seeds of the evening primrose family
18	9,11,13-Octadecatrienoic	Eleostearic	–	–	–	Tung oil
18	9,12,15-Octadecatrienoic	Linolenic	–11	157–158/0.001–0.002	184/4	Linseed oil
20	5,8,11-Eicosatrienoic	–	–	–	–	Brain phosphatides
20	8,11,14-Eicosatrienoic	–	–	–	–	Shark liver oil
			Fatty Acids of More Unusual Structure			
18	12-Hydroxy-9-octadecenoic	Ricinoleic	–	–	–	Castor oil
20	14-Hydroxy-11-eicosenoic	Lesquerolic	–	–	–	*Lesquerella* seed oil
18	12,13-Epoxy-9-octadecenoic	Vernolic	–	–	–	Some *Compositae* seeds
18	8,9-Methylene-8-heptadecenoic	Malvalic	–	–	–	*Malvaceae* seeds
19	9,10-Methylene-9-octadecenoic	Sterculic	–	–	–	*Sterculiaceae* seeds
18	13-(2-Cyclopentenyl) tridecanoic	Chaulmoogric	–	–	–	*Chaulmoogra* oil

The common fatty acids have unbranched chains with even numbers of carbon atoms. Some with odd carbon numbers are found in trace amounts in most fats, while branched fatty acids are produced by bacteria. Heptadecenoic acid (17 carbons) is a major component of the lipids of *Candida tropicallis* yeast.[4] Unsaturated acids usually are in the *cis* configuration; that is, the hydrogens at each end of the double bond are on the same side. This preference for *cis*-fatty acids occurs in almost

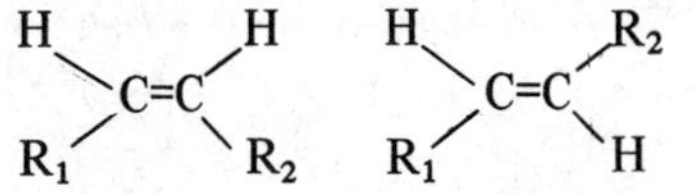

$$\underset{Cis}{\begin{matrix} H & & H \\ & C{=}C & \\ R_1 & & R_2 \end{matrix}} \qquad \underset{Trans}{\begin{matrix} H & & R_2 \\ & C{=}C & \\ R_1 & & H \end{matrix}}$$

all living organisms, even though the *trans*-configuration is thermodynamically more stable. Also, in the common polyunsaturated acids (more than one double bond), the unsaturation is "methylene-interrupted" with the structure $-CH{=}CH{-}CH_2{-}CH{=}CH{-}$, although conjugated double bonds are thermodynamically preferred. When bonds are conjugated, however, as in eleostearic acid, they may occur as a mixture of *cis-trans* or all *trans* isomers.

The physical properties of fatty acids show regular patterns characteristic of their chemistry. Melting points of saturated acids rise with molecular weight but show an alternating pattern, with those acids that have odd numbers of carbon atoms having lower melting points than the even-carbon acids that have one less carbon. Double bonds, particularly those with the *cis* configuration, produce a kink in the regular alternation of the hydrocarbon chain, making it more difficult to fit into a regular crystal structure. Hence, saturated C_{18} stearic acid melts at 70.1°C, while monoenoic C_{18} oleic acid melts at 16.3°C and diunsaturated C_{18} linoleic acid melts at -11°C. Boiling points of acids and their esters rise smoothly with molecular weight and are not greatly affected by unsaturation, so separation by distillation of acids having the same number of carbons is not practicable. Branching of the carbon chain usually lowers both boiling point and melting point, while polar groups such as hydroxyl functions increase the boiling point and the viscosity while lowering the melting point.

Isolation, identification, and quantitative analysis of fatty acids have been greatly facilitated by development of improved chromatographic techniques. Triglycerides are not distillable but they can be readily converted to methyl or ethyl esters by alcohol interchange. The esters are readily separated by gas chromatography. High performance liquid chromatography and thin-layer techniques may also be used for separation and identification. Triglycerides may even be partially separated according to the number of carbon atoms and double bonds by high-temperature gas chromatographic techniques.[5]

As components of animal fats and vegetable oils, fatty acids represent a significant part of the human diet. Among the unbranched fatty acids, the specific mixture of acids consumed is unimportant from the standpoint of caloric content. However, fatty acids, such as linoleic acid (18:2) and arachidonic acid (20:4) are considered essential in small amounts,[6] since they are precursors to prostaglandins, compounds of some biological significance. It is also recommended by some groups of medical researchers that a reasonably high polyunsaturate to saturate ratio be maintained to help prevent heart disease. Branched, cyclic, or substituted fatty acids are reported to have deleterious effects in the diet. The cathartic effect of ricinoleic acid in castor oil has long been well-known.

TRIGLYCERIDES

Each molecule of triglyceride contains three fatty acid moieties and, in general, high-melting fatty acids produce high melting glyceride esters. However, in natural fats properties are influenced by the number and distribution of molecular species. Consider a mixture of triglycerides containing only two fatty acids, A and B. The following sequences are possible, since the middle 2-position of glycerol differs from the 1- and 3-positions. AAA, AAB, ABA, ABB, BAB, BBB. These species may all occur; distribution may be random, but biosynthesis is usually directed, so that a given triglyceride often appears in higher concentration than would be expected from strictly statistical considerations. Natural fats contain many more than two fatty acids, so the distributional possibilities are numerous. A result of this multiplicity of molecular species is that fats do not possess sharp melting points; instead, they freeze or melt over a wide temperature range. The first molecules to form solid crystals during cooling are those with the highest saturated fatty acid content. The crystallized fat may be removed by filtration to leave an oil with higher unsaturated content. Such a process, called "winterization," is used in making salad oils which can remain liquid under refrigeration.[7]

There are a few exceptions to the rule that fats behave as mixtures of many types of molecules. Castor oil contains 90 percent ricinoleic acid and thus approximates triricinolein (glycerol triricinoleate) in its physical and chemical behavior. Cocoa butter, which is used in confectionery coatings, melts in the rather narrow range of 89-95°F. Its major ingredient is 2-oleopalmitostearin

$$\begin{array}{l} CH_2OOCC_{17}H_{35} \\ | \\ CHOOC(CH_2)_7CH{=}CH(CH_2)_7CH_3 \\ | \\ CH_2OOCC_{15}H_{31} \end{array}$$

2-Oleopalmitostearin

Since cocoa butter has become increasingly expensive in recent years, considerable research has been directed toward production of other confectionery coating materials with a narrow melting range slightly lower than body temperature. The approaches to this problem involve hydrogenation, fractional crystallization, rearrangement of fatty acids and proper blending of the products.[8]

Although a wide variety of molecular species exists in natural fats, the oils from many plant seed sources show little variation in fatty acid composition within a given species, especially if the plants are grown under similar climatic conditions. Interesting genetic and environmental variations do occur, however. Ordinary safflower oil usually contains about 75 percent linoleic acid and 12-15 percent oleic acid. Another variety of safflower, reported by Knowles[9], reverses the ratio of these two fatty acids. Sunflower oil exhibits considerable variation of fatty acid composition with climate. Sunflower seeds grown in the Southern United States have a much lower linoleic to oleic acid ratio than seeds grown in Minnesota and North Dakota.[10] The composition of depot fats of nonruminant animals is largely a function of the fats ingested. If a high percentage of unsaturated oils is eaten, depot fats will contain more unsaturated fatty acids than if saturated fats predominate in the diet. Ruminant animals show more consistency in fatty acid composition, since bacteria in the rumen reduce unsaturated acids to more saturated ones. Experimental feedings of encapsulated polyunsaturated fats have prevented this reduction in the rumen of cattle, producing beef of higher than normal unsaturated fat content.[11]

EDIBLE VEGETABLE OILS

Vegetable oils, after they are extracted and processed, go to a large number of end uses, both edible and industrial. The edible end products come from a variety of sources, depending on the desired physical properties and on the economics of production of the starting materials. End uses include the following:

Shortening. This product is a liquid or solid fat added to baking dough to prevent excessive cohesion of wheat gluten from the flour. The strands of gluten are "shortened" or modified so that baked products remain tender.[12] Shortenings may be made from lard or from partially hydrogenated oils.

Margarine. Originally, margarines were hydrogenated oils of fairly constant composition sold as substitutes for butter. Margarines sold today are formulated in ways to give edible spreads with high or low polyunsaturate content, varying physical properties, and a number of compositions, even including "low-calorie" margarines that have a high water content. Formulation of margarines has become a very important part of edible oil technology.[13]

Salad oils. Salad oils may be blends of oils from different sources, or they may be derived from a single source. They are usually unhydrogenated or lightly hydrogenated winterized oils which have been refined and deodorized. An exception is olive oil, which is prized for the distinctive flavor of the "virgin" pressed oil.

Frying oils. Frying oils are refined oils which have usually been hydrogenated to increase their oxidative stability at high temperatures.

Confectionery fats and hard butters. These products are usually specific oils which have been hydrogenated, then interesterified, rearranged and blended to confer the desirable properties described above.

Surface active agents. This class of oil-derived products includes mono- and di-glycerides, as well as esters of sugars, sorbitol, and other polyols.[14] Characteristically, the hydrocarbon chains of the fatty acids provide hydrophobic portions of the molecule while residual hydroxyls of polyols or sugars provide a hydrophilic moiety. They are used to stabilize emulsions in foods, provide thickening, modify bread dough, and do a host of other functions.

Table 12.2 shows the amounts of some principal fat and oil products consumed in the United States in selected recent years.

The chemical and physical properties of the oils affect processing and formulation of products. It is therefore important to consider the fatty acid composition of these oils. A shorthand nomenclature can be used for the fatty acids, in which the first number represents the total number of carbon atoms and the second number indicates the unsaturation. Hence, oleic acid with 18 carbon atoms and one double bond is represented by 18:1, whereas stearic acid is 18:0 and linoleic acid is 18:2. This nomenclature is used in Table 12.3 which shows representative compositions for various fats and oils.

An indication of the total unsaturation of oils is the iodine value. (See the section on Analytical Procedures in this chapter.). High iodine values indicate high polyunsaturation. Consumption patterns of oils and fats in the United States are outlined in Table 12.4.

Soybean Oil

Soybean production in the United States was over 1.8 million bushels in 1978. Soybean oil is by far the most plentiful oil in the United States and in world trade. There is also significant production of soybeans and oil for export in Brazil and Argentina. U.S. exports of oil were approximately 2.2 billion pounds in 1978. Soybean oil is highly polyunsaturated with iodine value ranges of 120–141. Soybean oil which has not been hydrogenated tends to oxidize and produce reverted flavors, often described as "painty" or "grassy", attributed to the relatively high linolenic acid content (8 percent). Consequently, the oil is not commonly used for cooking. However, it can be partially hydrogenated to a product of greater stability, and in this form it is a major component of salad oils (after winterizing), shortenings, and margarines. The meal remaining after extraction of the oil is particularly valuable as a component of animal feeds since it contains approximately 44 percent protein. Protein isolates and concentrates for food use are also prepared from soybean meal.

TABLE 12.2 Civilian Consumption of Edible Fat and Oil Products in the United States

	1970		*1975*		*1978 (preliminary)*	
	Total Mil. lb.	*Per Capita lb.*	*Total Mil. lb.*	*Per Capita lb.*	*Total Mil. lb.*	*Per Capita lb.*
Butter	1,061	5.3	1,012	4.8	968	4.5
Margarine	2,223	11.0	2,375	11.2	2,494	11.4
Shortening and frying fat	3,496	17.3	3,661	17.3	3,972	18.2
Salad and cooking oil	3,125	15.5	3,856	18.2	4,484	20.5
Other edible products	480	2.4	436	2.1	454	2.1

TABLE 12.3 Fatty Acid Composition of Some Fats and Oils[a]

Source	Fatty Acids, %						
	12:0	*14:0*	*16:0*	*18:0*	*18:1*	*18:2*	*Others*
Soybean Oil	–	0.1	10.5	3.2	22.3	54.5	9.4[b]
Cottonseed Oil	–	1.0	25.0	2.8	17.1	52.7	1.4
Peanut Oil	–	–	11.0	2.3	51.0	30.9	4.8
Sunflower Oil[c]	–	–	7.0	3.3	17.7	72.0	–
Safflower Oil	–	0.1	6.7	2.7	12.9	77.5	0.1
Safflower Oil (High Oleic)	–	–	5.4	1.7	80.7	12.2	–
Olive Oil	–	–	6.9	2.3	84.4	4.6	1.8
Coconut Oil	48.2	16.6	8.9	3.8	5.0	2.5	15.9
Rapeseed Oil	–	0.1	4.0	1.3	17.4	12.7	64.5[d]
Rapeseed Oil (Low Erucic)	–	–	5.2	1.9	59.8	20.4	12.7[e]
Palm Oil	0.1	1.2	46.8	3.8	37.6	10.0	0.5
Palm Kernel Oil	50.9	18.4	8.7	1.9	14.6	1.2	4.3
Corn Oil	–	–	11.5	2.2	26.6	58.7	1.0
Lard	–	1.0	27.0	14.0	43.0	10.0	5.0
Tallow	–	3.0	30.0	19.0	44.0	–	4.0
Linseed Oil	–	–	5.0	5.0	20.0	17.0	53.0[f]

[a]The composition is somewhat variable. These are average values from recent years' crops.
[b]Soybean oil contains approximately 8 percent linolenic (18:3) acid.
[c]Grown in colder climates. In the Southern United States sunflower oil may contain more than 50 percent oleic acid and as little as 35 percent linoleic acid.
[d]Older varieties of rapeseed oil contain 45–46 percent erucic (22:1) acid.
[e]Low erucic acid rapeseed oil contains *ca.* 10.8 percent linolenic (18:3) acid and only traces of erucic (22:1) acid.
[f]Linseed oil contains approximately 53 percent linolenic (18:3) acid.

TABLE 12.4 U.S. Domestic Disappearance of Food Fats and Oils[a]

	Million Pounds			
	Year			
Item	*1970*	*1973*	*1976*	*1978[b]*
Butter	1074	964	941	1013
Lard	1645	1150	814	935
Edible Tallow	518	500	534	869
Coconut Oil	644	539	1075	945
Corn Oil	445	450	581	613
Cottonseed Oil	890	991	532	619
Olive Oil	67	49	56	53
Palm Oil	182	294	611	277
Palm Kernel Oil	94	120	138	138
Peanut Oil	193	150	265	120
Safflower Oil	100	98	70	80
Soybean Oil	6253	7255	7454	8867
Sunflower Oil	–	83	26	166

[a]Source: Fats and Oils Situation, February 1978, FOS290
[b]Preliminary

Cottonseed Oil

Cottonseed oil is important in the United States, although it is a by-product from manufacture of cotton fiber. Its iodine value is approximately 100–110, so it is considered to be of medium polyunsaturated content. Most cottonseed oil is winterized and used for salad oil. Another important use of unhydrogenated cottonseed oil is in frying potato chips and other snack foods. Since cottonseed oil contains no linolenic acid it is not readily subject to flavor reversion. Hydrogenated cottonseed oil is very important in the manufacture of shortening. In 1978, over 1.3 billion pounds of this oil was produced, about half of which was exported.

Corn Oil

Corn oil is also a by-product, coming from corn starch production. The seed germ is the source of the oil, which is quite stable, though it has a relatively high iodine value of about 125. Much corn oil is used in soft margarines, and the liquid oil is used extensively in salad oil. Estimated production in 1978 was 700 million pounds.

Peanut Oil

Peanut oil is also a by-product oil, generally coming from low-grade nuts. It has a high oleic acid content and an iodine value of 84-100. Unhydrogenated peanut oil is an excellent cooking oil, used by manufacturers of potato chips and other snack foods because of its good flavor. Domestic production of peanut oil has varied between 140 million and 494 million pounds in recent years.

Palm Oil and Palm Kernel Oil

Palm oil is extracted from the fruit of the oil palm tree, while palm kernel oil comes from the seed. In composition, these oils differ significantly from one another. Palm oil is high in palmitic (16:0) acid while palm kernel oil, like coconut oil, has a high content of lauric (12:0) and myristic (14:0) acids. During the mid 1970's, palm oil was imported extensively into the United States as increased production in Malaysia, Indonesia, and the Ivory Coast coincided with scarity and high prices for other oils. However, as the price of palm oil rose from 21 cents to 30 cents per pound, imports dropped from 933 million pounds in 1975-76 to 362 million pounds in 1976-77, and this drop has continued. Palm oil, with its new plantings and high production per acre will remain a significant factor in the world oil picture. In the United States, it will be used in cooking oils and shortening and confectionery fats, expecially at times when there is a price advantage over soybean oil. Palm kernel oil has been a relatively minor import compared to coconut oil, its primary competitor.

Sunflower Oil

Sunflower seed and oil are seeing large increases in production in the United States in recent years. Sunflowers grow well in cool climates; consequently, this native American plant is a major oilseed in countries such as the USSR. While consumption has been relatively low in the U.S., there have been considerable increases in production of sunflower seed and oil, especially in Minnesota, North Dakota, South Dakota, and Texas. The bulk of U.S. sunflower seed is exported to Western Europe, Eastern Europe, Mexico, and Venezuela.[15] As noted previously, the composition of sunflower oil changes when the seed is grown in warmer climates. Northern sunflower contains about 68-72 percent linoleic acid and 14 percent oleic acid, while southern sunflower oils may have over 50 percent oleic acid and as little as 35 percent linoleic.[16] Respective iodine values are 130 for northern sunflower oils and 106 for those grown in southern states.

Olive Oil

Olive oil is desired for its flavor; hence the best oil, called virgin oil, is pressed from the fruit pulp of the olive and not refined. Oil extracted with solvent, refined, and deodorized is called "pure" olive oil and is quite bland. Most olive oil sold in the United States is imported, though small amounts may be produced in California. Olive oil is expensive, and is therefore a specialty item in the United States, used primarily for its flavor. It has been blended with other oils for use in salad dressings. Olive oil is low in polyunsaturates but high in oleic acid, with iodine values ranging from 80 to 90.

Safflower Oil

Safflower is an ancient crop, grown in the Middle East since prehistoric times, but only produced in quantity since the early 1950's. It has the highest linoleic acid content of any commercial oil, but it contains no linolenic acid. It is therefore not subject to the severe flavor reversion problems of soybean oil. With increasing demand for polyunsaturated oils in the diet, safflower has found uses in soft margarines, salad oils, dietetic imitation ice creams, etc. Because of its high polyunsaturation, it is not used in batch frying operations where it is heated and cooled repeatedly, since it tends to produce off-flavors and polymeric products, at high temperatures in air. Safflower grows well in hot dry climates, for example, California and Arizona, where soybeans and other oilseeds are not well adapted. Economics of production have been marginal, so production has not increased in recent years.

A high-oleic variety of safflower oil has also been produced in recent years. In composition, it is quite similar to olive oil, with a low polyunsaturated content and high oxidative stability at frying temperatures. Regular safflower oil contains 75-80 percent linoleic acid and 13 percent oleic acid, while in the new variety these values are reversed.

Coconut Oil

Coconut oil differs from the oils grown in temperate climates because it has high percentages of saturated fatty acids with less than eighteen carbon atoms. Its composition gives it a high thermal and oxidative stability, so it is used in many products such as filled milks where oxidation of less saturated fats is likely to produce off-flavors. Coconut oil, however, is prone to hydrolytic rancidity since it contains significant amounts of caprylic (8:0) and capric (10:0) acids which when converted from their glycerides to free acids have strong, unpleasant odors and flavors. Enzymatic hydrolysis of the oil before refining often causes refining problems because of high free fatty acid content.

Rapeseed Oil

Commercially grown rapeseed oil has changed drastically in composition in the past decade. Prior to the 1970's, rapeseed oil contained large percentages of erucic (22:1) acid. Animal feeding studies showed that high consumption of erucic acid correlated with the presence of heart lesions. Luckily, plant breeders were able to develop the current varieties in which oleic acid is the predominant fatty acid and in which erucic acid concentration is nearly zero. Rape plants are adapted to cold climates, so rapeseed oil has been an important oil in Canada and Northern Europe. Mustard seed and a newer oilseed, crambe, also produce oils with high erucic acid content and composition similar to the older varieties of rapeseed oil. Before refining, these oils contain sulfur compounds which impart a strong cabbage-like odor. Refining the oil removes these compounds but glucosinolates and other sulfur compounds which are somewhat toxic cause problems with the use of the by-product meal for feed.

INDUSTRIAL OILS

The oils described above are used primarily for edible purposes, although all of them find industrial utilization as sources of fatty acids, soaps, detergents, biostatic agents, and resins. Soybean and safflower oils, being semidrying oils, are used in paints and coatings. Some oils, however, are produced almost exclusively as industrial raw materials. The trend of such use has been downward, because of competitive technology based on petroleum-derived products, although recent increases in the price of petroleum could reverse the trend.

Linseed Oil

Linseed oil comes from varieties of the flax plant which are grown for oil production. It contains over 50 percent linolenic acid, more than any other common oil. The unique structure of linolenic acid makes it susceptible to oxidation, a reaction which is responsible for the drying action of polyunsaturated oils.[17] Initially, drying oils such as linseed oil react to form hydroperoxides, but these, being thermally unstable, break down rapidly to undergo a series of complex reactions, including a free-radical polymerization process. Such polymerization takes place in oil-based paints and varnishes; it is often accelerated by adding to coating formulations catalysts such as cobalt naphthenate or other transition metal compounds which increase the rate of peroxide decomposition. Linseed oil disappearance in the U.S. during 1978-79 was 227 million pounds, the highest level since 1973-74.[15]

Castor Oil

Castor oil is an interesting raw material for chemical synthesis in that it has two points vulnerable to chemical attack, the double bond and the hydroxyl group of ricinoleic acid, which comprises over 90 percent of the fatty acids in the oil. Physically, castor oil is characterized by a high viscosity resulting from hydrogen bonding of its hydroxyl groups. It is a raw material for paints, coatings, lubricants, cosmetics, and a variety of other products including the chemicals, undecylenic acid and sebacic acid. Castor oil has been grown in the southern states from time to time, but our current annual consumption of 100 to 150 million pounds comes almost entirely from Brazil. The castor plant (*Ricinus communis*) produces an extremely toxic protein, ricin, several potent allergenic materials, and an alkaloid, ricinine, which all appear in castor pomace. Thus, the residual meal cannot be used as an animal feed without extensive further treatment, a fact which has been a deterrent to castor production in this country.[18]

Tall Oil

Tall oil is a by-product of paper manufacture. The noncellulosic portion of wood pulp is dissolved in a sodium hydroxide-sodium sulfide mixture which forms a black liquor. This liquor is neutralized with acid, blown with steam and skimmed to recover the tall oil which is a mixture of fatty acids and resin acids. About 70 pounds of tall oil are produced per ton of pulp. Tall oil is used to manufacture soap, paper sizing, surface coatings, printing inks, putties, and a variety of other end products.

Jojoba Oil

Jojoba oil (pronounced *ho hô ba*) is found in the seed of the Jojoba plant (*Simmondsia chinensis*), a plant which grows wild in drier climates of southwestern North America. The oil is not yet sold in high volume, but it has excited considerable interest, to the extent that many nurseries in the Southwest and in the Western States are selling plants. The oil is a liquid wax (mostly $C_{20\text{-}22}$ monounsaturated acid esters of $C_{20\text{-}22}$ monounsaturated fatty alcohols) which is chemically almost equivalent to sperm whale oil.[19] Whale oil was used for industrial products such as lubricant additives, leather lubricants, high quality waxes, and cosmetics, until whaling was banned by most western governments. Jojoba oil promises to fill the gap left by the unavailability of sperm whale oil.

FATS OF ANIMAL ORIGIN

Lard

Lard is the fat rendered from the fatty tissues of the pig. For many years it was the major shortening product, but it has been supplanted to a large extent by hydrogenated vegetable oils. A large quantity of lard is still used in commercial frying operations. Although it has low content of unsaturated acids (iodine value 55-60), it is often hardened further by hydrogenation and blended into shortenings. Domestic disappearance of lard has declined in recent years and presently amounts to only about 800 million pounds.

Tallow

Tallow is rendered from the fatty tissues of cattle in a manner similar to that used in the production of lard. After rendering, it is usually deodorized and used in shortenings and bakery margarines. It is also an industrial raw material, being utilized for manufacture of soap and fatty acids, particularly industrial oleic acid. Fractionation methods have been developed with tallow to yield fractions suitable for both food and industrial uses.[20]

Marine Oils

Marine oils are processed from menhaden, herring, pilchard, sardines, tuna, and anchovies.[21] In 1976, the production of all American fish oils was 204 million pounds, about 90% of which was menhaden oil. These oils are not used extensively in the United States for edible purposes, but they are hydrogenated and used for shortening in Canada and Europe. Fish oils are highly unsaturated, with fatty acids often containing five or six methylene-interrupted double bonds. Such polyunsaturation makes them highly susceptible to oxidation and formation of fishy flavors and odors.

PROCESSING AND REFINING FATS AND OILS

Extraction

The purpose of oil extraction is to obtain the maximum amount of good quality oil while getting maximum value from the residual press cake or meal. Oil is extracted from oilseeds by pressing, solvent extraction, or a combination of both. When brought in from the field, seeds must be cleaned and brought to a moisture content which gives minimum danger of microbial spoilage and spontaneous combustion during storage. After storage and shipping, seeds are brought into the oil mill for processing. The first step in extraction is preparation of the seed, which includes cracking, decortication, (removal of hulls), and reduction of particle size. The seeds are cracked in rollers and decorticated, if necessary, by a combination of air classification and screening or, in a few cases, by flotation. The seeds are reduced in particle size by grinding or rolling; then they are steam cooked and flaked. The heat treatment coagulates the proteins and adjusts the moisture content so that extraction is accomplished most easily.

Cooked seeds are then conveyed to screw presses for continuous pressing. A screw press consists of a worm shaft which revolves within a "cage." The bars of the cage are spaced so that oil escapes but press cake does not. Most screw presses have a feed gear and a choke gear.[22] The press cake emerging from a screw press will contain 8 to 14 percent residual oil if solvent extraction is to follow, or 2 to 8 percent if screw pressing is the only extractive process. Solvent extraction can be done in a number of systems, but it is accomplished by a countercurrent procedure in which intimate contact is achieved between seeds and solvent. The apparatus may be the basket type arranged

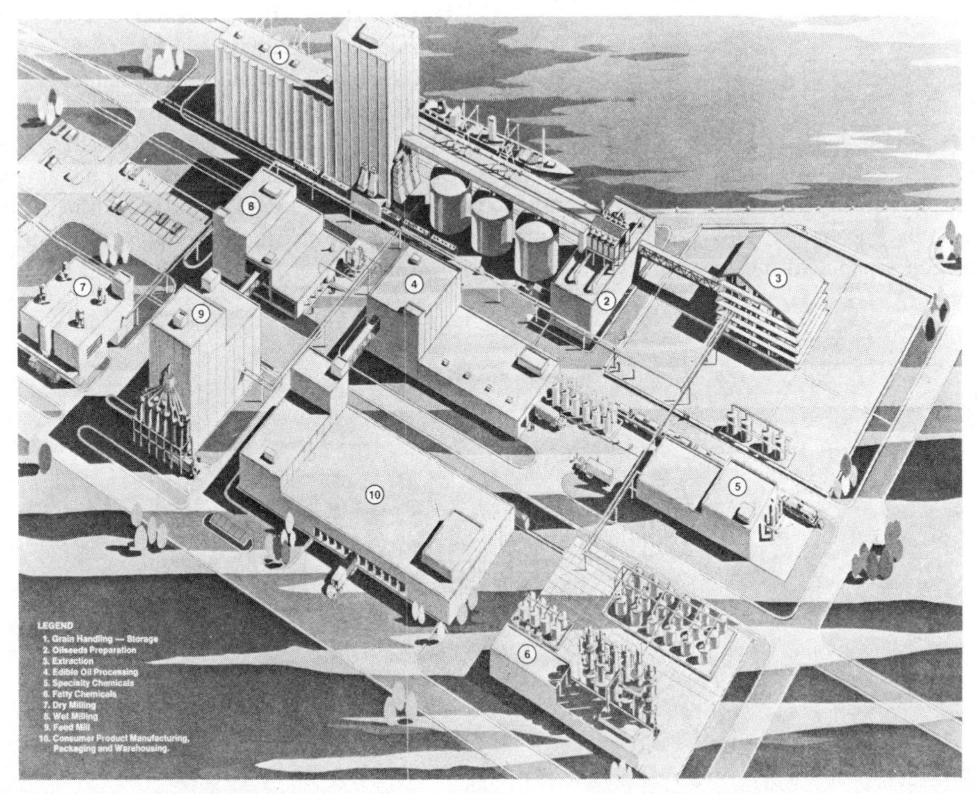

Fig. 12.1 Drawing of oilseed and protein plant. *(By permission of Davy-McKee Corporation, Chicago, Illinois.)*

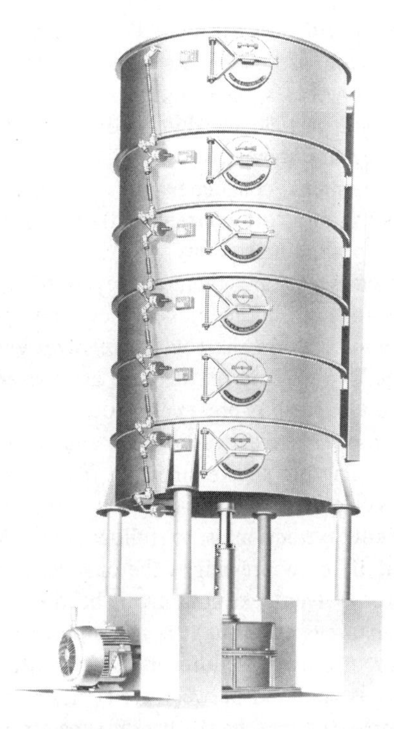

Fig. 12.2 Stacked cooker for oilseeds. (*By permission of Anderson International Corporation, Strongsville, Ohio.*)

vertically, the rotary-cell type in which containers rotate horizontally, or the full-loop type.[23] The common solvent for edible oils is hexane, or actually a hexane-containing naphtha boiling in the range 146–156°F. Solvent extraction equipment must be totally enclosed and thoroughly protected because of solvent flammability. There has been interest in extraction with nonflammable solvents, especially trichloroethylene, but the advantages of using a nonflammable solvent have been more than offset by the toxicity problems generated by use of chlorinated solvents. After extraction maximum solvent recovery is necessary for economical operation. Solvent is recovered by distillation from the miscella (mixed solvent and oil) and is also removed from the oilseed pomace by steam stripping. Recovered solvent is reused for extraction while the extract oil is usually combined with the prepress oil for refining, except for cases such as olive oil in which the virgin press oil has special flavor qualities. Desolventizing of the meal requires careful handling of time, temperature, and moisture so that protein quality is maintained. To produce protein isolates for food use, there must be as little denaturation as possible. At least two companies have introduced new flash desolventizing systems to achieve good solvent removal without denaturation.

Marine oils are often solvent extracted from fish meal; but animal fats such as lard or tallow, as well as some marine oils, are "rendered" by application of dry or moist heat. Dry rendering in steam jacketed tanks followed by pressing is satisfactory for inedible fats, but edible animal fats are produced by wet rendering, using steam at 40 to 60 psi. In the latter process 99.5 percent or more of the fat is recovered.

Fig. 12.3 Prepress expeller. (*By permission of Anderson International Corporation, Strongsville, Ohio.*)

Refining

Edible oils have a number of impurities, many of which are considered undesirable. It is particularly important to remove the phosphatides and free fatty acids to obtain a clear oil; degumming and alkali refining are the processes for removing these components along with some other undesirable substances such as phenolic materials and gossypol (found in cottonseed). The amount of free fatty acids is first determined by analysis of the crude oil. If there are significant amounts of phosphatides, the oil is degummed by contacting with an appropriate amount of water, steam, or phosphoric acid. The phosphatides which precipitate are recovered and sold as "lecithin," which has emulsifying properties useful in many food products. The most common refining procedure is to treat the oil with an appropriate amount of caustic soda solution based on free fatty acid content plus an amount empirically determined. The quantity, concentration, time of contact, and temperature of the caustic soda is very important to minimize losses. Reaction of caustic soda with free fatty acids produces soapstock which can be continuously removed by centrifugation. If the oil is not previously degummed, phosphatides are also removed during alkali refining, since hydration makes them insoluble in oil. After separation of soapstock, oil may be water-washed and again centrifuged to bring soap concentration to a satisfactory low level. In the United States, refining of vegetable oils is usually done in a continuous process, though it may be done batchwise. Good alkali refining reduces the free fatty acid content of oils to 0.01 to 0.03 percent. Residual soap content should be 50 ppm or less for good bleaching efficiency. The soapstock from alkali refining is treated with acid to yield a mixture of free acids along with some neutral oil entrained with the soap. The "split" soapstock is usually sold for use as a component of animal feeds.

In alkali refining, oils with a high free fatty acid content such as palm oil, palm kernel oil, coconut oil, and tallow, there are often large

Fig. 12.4 Continuous centrifugal separators used in refining and washing oils. *(By permission of Sullivan Systems, Inc., a subsidiary of Alfa-Laval, Inc., Tiburon, California.)*

refining losses because considerable neutral oil is entrained with soapstock. For such oils, steam refining–a process of steam distillation of free fatty acid from the neutral oil under high vacuum–is coming into increasing use. Steam refining may be employed to bring free fatty acids to a low value before alkali refining or it may be used to bring free fatty acids to <0.05 percent without alkali refining at all.[24] Usually, bleaching is accomplished prior to steam refining, while it is done afterward in alkali refining.

Other refining methods, such as acid refining are used for specialized purposes. Some solvent-extracted oils may be miscella-refined with alkali before the solvent is removed.[25] Better contact with alkali solution and improved separation of soapstock are among the claimed advantages.

Bleaching

Some pigmented materials such as chlorophyll or carotenoids, which are not eliminated by alkali refining can be removed by adsorption processes. Activated carbon has been used for bleaching, but bleaching clays or "earths" are less expensive and quite effective. Some oils may be decolorized by using an unactivated bentonite clay but most are bleached more effectively with acid-activated bentonites containing a high proportion of montmorillonite.[26] The amounts of activated clay used depend upon the type of oil and its prior handling; for vegetable oils 0.25-2.0 percent by weight quantities are common. Bleaching may be done at atmospheric pressure at 200-250°F or under a vacuum of at least 28 inches at 160-180°F. Palm oil is an exception and requires temperatures above 300°F, even under vacuum. The oil is agitated with clay at the proper temperature for 10 to 20 minutes. The clay is then removed by filtration. Many refineries discard the spent clay without recovering the retained oil, although this practice results in a loss of oil almost equal to the weight of clay used. Increasingly, efforts are made to recover this oil although it may be relatively low in quality. Bleaching is often done as a batch process, but continuous bleaching equipment is commercially available.

Hydrogenation

Edible oils are hydrogenated to produce hard or plastic fats or to enhance the oxidative stability of the oils by lowering their content of polyunsaturated fatty acids. Hydrogenation is the single most important chemical process in the vegetable oil industry.[27] In the hydrogenation process, one or more carbon-carbon double bonds in a triglyceride molecule are converted to single bonds by addition of hydrogen as illustrated below:

$$CH_3(CH_2)_7CH{=}CH(CH_2)_7COOR \xrightarrow[\text{Catalyst}]{H_2} CH_3(CH_2)_{16}COOR$$

The hydrogenation reaction requires hydrogen at moderate pressures (2 to 3 atmospheres) and a catalyst, which is normally finely divided nickel. Such catalysts may be prepared by reduction of nickel salts adsorbed on an inert support of high surface area, or they may be Raney nickel-type catalysts, prepared by dissolving aluminum from an aluminum-nickel alloy with strong alkali. Variables affecting hydrogenation include the catalyst, temperature, hydrogen pressure, and amount of agitation. If only partial hydrogenation is desired, the selectivity of the hydrogenation system is important. Consider for example the molecule of oleolinoleolinolenin illustrated below:

$$\begin{array}{ll}
CH_2OOC(CH_2)_7CH\overset{a}{=}CHCH_2CH\overset{b}{=}CHCH_2CH\overset{c}{=}CHCH_2CH_3 & \text{(linolenic)} \\
| & \\
CHOOC(CH_2)_7CH\overset{d}{=}CHCH_2CH\overset{e}{=}CH(CH_2)_4CH_3 & \text{(linoleic)} \\
| & \\
CH_2OOC(CH_2)_7CH\overset{f}{=}CH(CH_2)_7CH_3 & \text{(oleic)}
\end{array}$$

If a light, partial hydrogenation is desired to increase oxidative stability, it would be preferable to first reduce all the double bonds labeled *c* in the linolenic acid moiety. Because no catalyst is entirely selective, any of the other bonds in the triglyceride may be reduced first. Reduction of one of the bonds labeled *a* or *b* would produce an "isolinoleic" acid from the linolenic acid portion of the molecule. Reduction of bonds *d* or *e* would produce isooleic or oleic acid from the linoleic acid, while hydrogenation of bond *f* forms a stearic acid moiety from oleic acid. Not only is no catalyst completely selective, but hydrogenation catalysts may also cause isomerization of an unsaturated bond from the *cis* to a *trans* configuration, or the migration of double bonds up or down the fatty acid chain. There is a considerable amount of research relating to selectivity of hydrogenation systems.[28] For example, some copper catalysts are reported to be considerably more selective than the commonly used nickel catalysts.[29] Sulfur-poisoned nickel catalysts may be used to produce maximum amounts of *trans*-acid moieties and a higher melting point.

Hydrogenation is a very versatile reaction, allowing considerable variability in the properties of plastic fats formed, which is an important factor in the preparation of final products such as margarines and shortenings. When very light hydrogenation is used to produce a liquid oil of relatively high stability, some of the fat molecules may become quite saturated. The more saturated "stearine" can be removed by winterization and filtration.

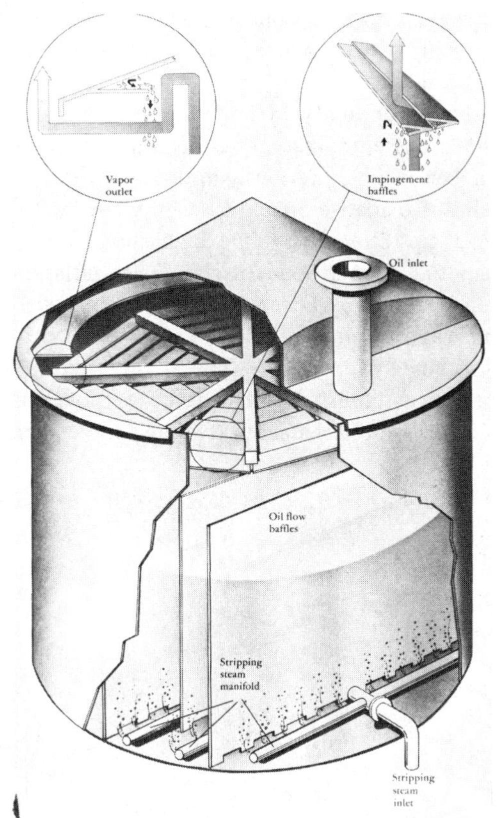

Fig. 12.5 Cutaway view of Sullivan deodorizer tray. *(By permission of Sullivan Systems, Inc., a subsidiary of Alfa Laval, Inc., Tiburon, California.)*

Deodorization

Deodorization is the last major processing step in handling refined vegetable oils.[30] Untreated, bleached, and hydrogenated oils all have minor amounts of free fatty acids, aldehydes, ketones, and other compounds which impart undesirable odors. Such oils are deodorized by steam stripping in stainless-steel vessels under vacuum at elevated temperatures. The process may be either batch, semicontinuous or continuous, but to insure good deodorization the oil must be heated several hours at approximately 6 mm pressure with sparge steam supplied at about 3 pounds per 100 pounds of oil per hour. The cooling process is very important afterward, since oxidation may take place rapidly at elevated temperatures. Deodorized oil should contain no more than 0.02–0.05 percent free fatty acids. Some plants combine steam refining (see above) with the steam deodorization. In this combination process, free fatty acids in the feed stock may be as high as 5 or 6 percent. A critical part of the process of steam refining is recovery of the fatty acids from the sparge stream.[31]

Protection from Autoxidation

It has become common practice in the edible oil industry to construct reaction vessels of stainless steel since iron is a prooxidant at elevated temperatures. Trace amounts of copper are even stronger oxidation catalysts so copper piping and copper or brass valves are especially to be avoided in handling oils for edible use.

The susceptibility of polyunsaturated oils to deleterious autoxidation has been mentioned a number of times previously in this chapter. The point of oxidative attack is a methylene hydrogen at a position next to the double bond, rather than the double bond, itself. The initiation step is attack by a free radical and removal of the weakly held methylene hydrogen to produce an allylic radical. The allylic radical is resonance stabilized and behaves as though it has free radical character at either end of the three-carbon system. This allylic radical reacts very quickly with oxygen to produce a mixture of peroxy radicals which can continue the autoxidation chain by removing an allylic hydrogen atom from another molecule of unsaturated fatty acid. Autoxidation continues in this manner until two free radicals react to disproportionate or dimerize or until a free radical intermediate reacts with some other substrate to form a stable radical incapable of initiating another oxidation chain. Trace metals, especially copper, catalyze autoxidation by reacting with hydroperoxides to create new free radicals and initiate new chain reactions. Rancid odors and polymeric products result from thermal decomposition of hydroperoxides.

Deodorized oils are very low in peroxides, but they must be handled carefully to insure that peroxide buildup does not occur. During the cooling step after deodorization, citric acid is often introduced as a metal sequestrant to effectively inactivate trace metal ions. In addi-

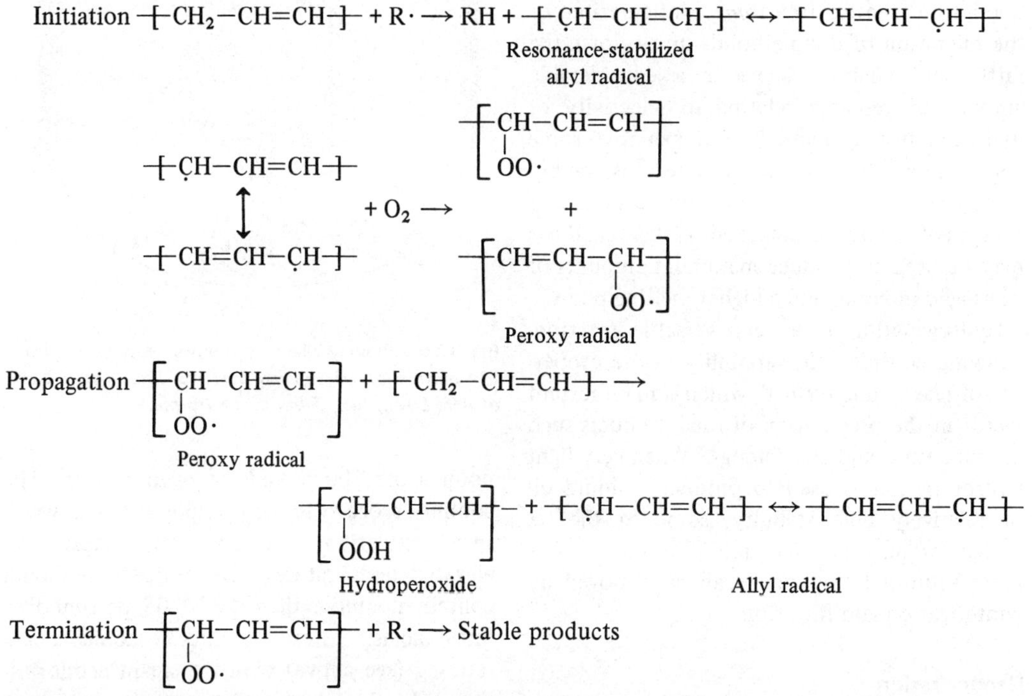

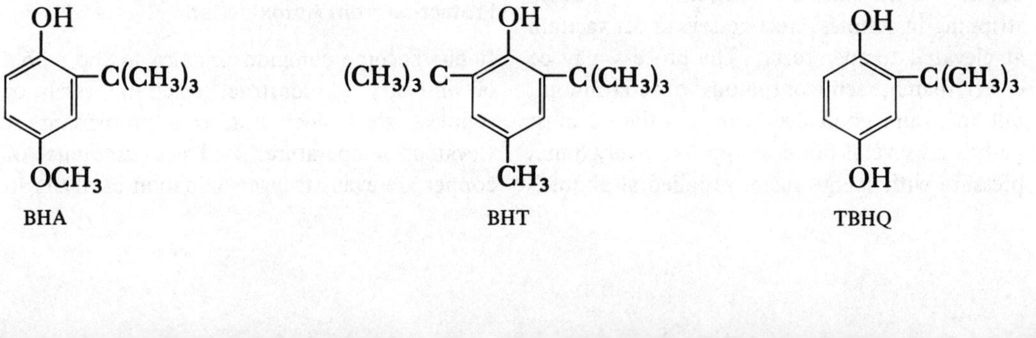

Initiation $-[CH_2-CH=CH]- + R\cdot \rightarrow RH + -[\dot{C}H-CH=CH]- \leftrightarrow -[CH=CH-\dot{C}H]-$

Resonance-stabilized allyl radical

$$\begin{matrix} -[\dot{C}H-CH=CH]- \\ \updownarrow \\ -[CH=CH-\dot{C}H]- \end{matrix} + O_2 \rightarrow \begin{matrix} -[CH(OO\cdot)-CH=CH]- \\ + \\ -[CH=CH-CH(OO\cdot)]- \end{matrix}$$

Peroxy radical

Propagation $-[CH(OO\cdot)-CH=CH]- + -[CH_2-CH=CH]- \rightarrow$

Peroxy radical

$-[CH(OOH)-CH=CH]- + -[\dot{C}H-CH=CH]- \leftrightarrow -[CH=CH-\dot{C}H]-$

Hydroperoxide — Allyl radical

Termination $-[CH(OO\cdot)-CH=CH]- + R\cdot \rightarrow$ Stable products

BHA: 2-tert-butyl-4-methoxyphenol (OH, $C(CH_3)_3$, OCH_3)

BHT: OH, $(CH_3)_3C$, $C(CH_3)_3$, CH_3

TBHQ: OH, $C(CH_3)_3$, OH

BHA BHT TBHQ

tion, small amounts of antioxidants are added, which act as free radical traps to terminate oxidation chains. Some compounds commonly added as antioxidants include BHT (butylated hydroxytoluene), BHA (butylated hydroxyanisole) and TBHQ (tertiary butylhydroquinone). Whether or not antioxidants are incorporated, many manufacturers of vegetable oils also introduce nitrogen into the space above oil packaged in bottles. Additional protection is often afforded by use of amber bottles which help prevent intiation of auto-oxidation by sunlight.

Fat Splitting

The other processes described in this section have been concerned with obtaining high quality liquid triglyceride oils, primarily for edible use. Industrially, fatty acids are an important raw material which are obtained by reaction of water or steam with triglycerides to produce fatty acids and glycerol (33).

$$\begin{array}{lcll} CH_2OOCR_1 & & CH_2OH & R_1COOH \\ | & & | & \\ CHOOCR_2 & + 3H_2O \longrightarrow & CHOH & + R_2COOH \\ | & & | & \\ CH_2OOCR_3 & & CH_2OH & R_3COOH \\ \text{Triglyceride} & & \text{Glycerol} & \text{Fatty Acids} \end{array}$$

Industrially, the process may be carried out with or without catalyst and often requires an emulsifying agent. The reaction occurs stepwise with formation of diglyceride and monoglyceride as intermediates. The first step is the most difficult since intimate contact with water is necessary. Intermediate mono- and diglycerides help to solubilize and emulsify the water in the fat phase. Atmospheric or low-pressure processes make use of sulfuric acid or zinc oxide catalyst with alkylbenzenesulfonic acids as emulsifiers. Higher pressure reactions require steam pressures of 700–725 psig and temperatures of 260°C and may be accomplished without catalyst. High pressure splitting is not satisfactory for highly polyunsaturated oils or castor oil since their fatty acids are too easily isomerized or dehydrated.

After splitting, various procedures for separation of the fatty acids may be employed. These are usually crystallization steps to separate more saturated from less saturated fatty acids. Processes include "panning and pressing," involving melting and cooling to a specified temperature until crystallization occurs, then pressing the liquid away from the crystals, or a variety of solvent recrystallization techniques. Solvents used for separation of fatty acids include liquid propane, methanol, acetone, and other ketones.[34]

ANALYTICAL PROCEDURES

Analytical methods for fats and oils have been published by the American Oil Chemists' Society.[35] There are a large number of procedures, though many of these have value only for testing properties which have special applications. It is useful, though, to describe a few of these tests because of their general applicability and wide use. AOCS method numbers are listed along with the method name.

Iodine Value (Cd 1-25)

Iodine value is defined as the number of centigrams of iodine absorbed by one gram of fat. Thus, it is a measure of unsaturation in fats and oils. In practice, it is determined by using iodine chloride or iodine bromide as a reagent, rather than free iodine. Even with these more reactive compounds, absorption by double bonds is not usually complete, so iodine value must be regarded as an empirical test which is of value to the processor. Since gas chromatography is now used widely, it is the practice in many research laboratories to first determine the composition of the fatty acids and then calculate the iodine value from that composition. Some iodine values of free fatty acids are presented in Table 12.5.

Saponification Value (Cd 3-25)

The saponification value is the weight in milligrams of potassium hydroxide needed to completely saponify one gram of fat. It can give an indication of the average molecular weight of the fatty acids in a fat, but this test presents less useful information than the fatty acid composition determined by gas chromatography.

TABLE 12.5 Iodine Values of Unsaturated Fatty Acids

Fatty Acid	*No. of C Atoms*	*No. of Double Bonds*	*Iodine Value*
Decenoic	10	1	149.1
Dodecenoic	12	1	128.0
Tetradecenoic	14	1	112.1
Hexadecenoic	16	1	99.78
Oleic	18	1	89.87
Linoleic	18	2	181.04
Linolenic	18	3	273.52
Eicosenoic	20	1	81.75
Erucic	22	1	74.98

Melting Point (Cc 1-25, Cc 2-38)

When applied to fats, the term "melting point" has little significance, since melting usually occurs over a wide range of temperature. Some useful information may be gained, though, if melting points are determined by rigidly reproducible methods. A number of different methods are used, including the FAC melting point, the Wiley melting point, the congeal point, and the titer (all described in Reference 35). In addition to these values, the Solid Fat Index (SFI) is often used for specifying oil properties. The SFI is related approximately to the percentage of solids in a fat at a given temperature. When determined at a number of specified temperatures, it can be especially useful to margarine manufacturers or other processors who need to control the characteristics of their manufactured products by blending. Other methods such as nuclear magnetic resonance and the Mettler dropping-point method are more reproducible than simple melting point techniques.

Peroxide Value (Cd 8-53)

Oxidation of fats is desirable in drying oils, but undesirable in most other products. The first step in oxidation is the formation of hydroperoxides. These hydroperoxides have no odors or flavors, but they readily decompose to produce aldehydes, hydrocarbons, ketones and other volatile products which are characteristic of oxidative rancidity. Peroxides react with potassium iodide to liberate free iodine which may then be titrated. The peroxide value is expressed as milliequivalents of iodine formed per kilogram of fat. The peroxide value is used as an indicator of oxidative rancidity. Organoleptic evaluation by a trained odor panel can often detect rancidity at peroxide levels as low as 1 meq/kg.

Active Oxygen Method (AOM) (Cd 12-57)

This method was developed to predict the shelf-stability of fats. It is useful with natural animal fats, but hydrogenated fats and oils which often are stabilized by the addition of antioxidants do not correlate well with AOM values. In this test, oils are heated at 97.8°C while air is blown through the sample. The AOM value is reported as the number of hours to reach a peroxide value of 100 meq/kg. A number of other tests have been devised to predict oil stability, but none of them have yet shown accurate correlations for all types of oils.

Free Fatty Acid (FFA) (Ca 52-40)

Free fatty acids result from hydrolysis of fats. Oilseeds contain lipase enzymes which catalyze such hydrolysis when seeds are damaged or become moist and begin to germinate. Lipases are also found in animal tissues. Consequently, if seeds are improperly stored, or if animal tissues are allowed to stand for a while before rendering, the FFA values will be high. Deodorization lowers the FFA to 0.05 percent or less. To determine FFA, the sample is titrated with sodium hydroxide. It is calculated as free oleic acid and reported as a percentage.

Smoke Point (Cc 92-48)

Smoke point is defined as the temperature at which a fat will produce continuous wisps of smoke. Impurities such as free fatty acids lower the smoke point drastically. A one percent FFA content may lower the smoke point of a shortening from 425° to 320°F.

Color (Cc 132-43 and Cc 136-45)

With a few exceptions, light color is desirable in oils. Color is best measured by recording the visible absorption spectrum. However, there are empirical methods which are quite satisfactory for quality control. For this purpose oils may be compared in a tube of standard depth with the color of glass standards.

TRENDS IN FATS AND OILS

This chapter has dealt with oil processing as it is at the beginning of the 1980's. However, there are many influences, both in raw materials and manufacturing processes, which will cause changes in the next few years. One factor which will greatly affect the world-wide availability, the price, and the use of oils is the increased production of palm oil. Malaysian palm yields are in the range of 3000 pounds of oil per acre. Once acreage is stabilized, total production of this oil will not vary greatly from year to year because the productive life of palm trees is at least 20 years. Use of palm oil in the United States will depend on its price relative to other oils. A situation like that in the mid 1970's when it was considerably less expensive than soybean oil will cause its use to rise. Sunflower seed oil seems to have found an important place in this country and there is a trend toward its increased production and use.

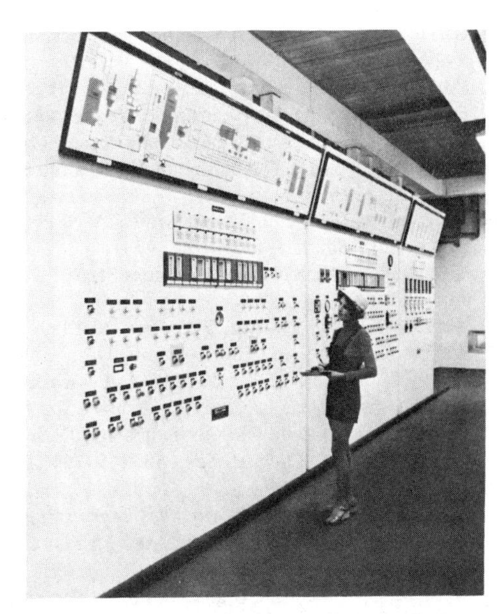

Fig. 12.6 Central control panel for a modern oil processing plant. *(By permission of Sullivan Systems, Inc., a subsidiary of Alfa-Laval, Inc., Tiburon, California.)*

The most important factor impacting on processing of fats and oils in recent years has been the greatly increased cost of energy. New oilseed plants are being built with a view toward energy conservation, effectively saving energy by using waste heat from one process to warm the input streams for another process. The microprocessor revolution shows promise for continuous data analysis and better control of unit processes. In the next few years continuous processing will probably continue to gain at the expense of batch processes, and use of physical refining (steam refining) will probably increase. Control of organic effluents will continue to be of concern as EPA regulations as well as those developed by local jurisdictions are increasingly promulgated and enforced.

REFERENCES

1. National Academy of Sciences, *Recommended Dietary Allowances*, 8th ed., Washington, D.C., 1974.
2. Gunstone, F. D., *An Introduction to the Biochemistry of the Fatty Acids and Their Glycerides*, 2nd ed., pp. 2–4, London, Chapman and Hall, 1967.
3. Weihrauch, J. L. and Gardner, J. M., "Sterol Content of Foods of Plant Origin," *J. Am. Dietetic Assn.*, **73**, 39–47 (1978).
4. Bauchart, D. and Aurousseau, B., "Preparation of Heptadecenoic Acid from *Candida Tropicallis* yeast," *J. Am. Oil Chemists Soc.*, **57**, 121–123 (1980).
5. Litchfield, C., Harlow, R. D., and Reiser, R., "Quantitative Gas Liquid Chromatography of Triglycerides," *J. Am. Oil Chemists Soc.*, **42**, 849–857 (1965).
6. Mathias, M. M. and Dupont, J., "The Relationship of Dietary Fats to Prostaglandin Biosynthesis," *Lipids*, **14**, 247–252 (1979).
7. Kreulen, H. P., "Fractionation and Winterization of Edible Fats and Oils," *J. Am. Oil Chemists Soc.*, **53**, 393–396 (1976).
8. Babayan, V. K., Hard Butters and Confectionery Coatings, *J. Am. Oil Chemists Soc.*, **55**, 845–848 (1978).
9. Knowles, P. F., "Variability in Oleic and Linoleic Acid Contents of Safflower Oil," *Economic Botany*, **19**, 53–62 (1965).

10. Gandy, D. E., 'Sunflower Situation in Russia and the United States," *J. Am. Oil Chemists Soc.*, **55**, 597A–598A (1978).
11. Edmondson, L. F., Yoncoskie, R. A., Rainey, N. H., Douglas, F. W. and Bitman, J., "Feeding Encapsulated Oils to Increase the Polyunsaturation in Milk and Meat Fat," *J. Am. Oil Chemists Soc.*, **51**, 72–76 (1974).
12. Thomas, A. E , "Shortening Formulation and Control," *J. Am. Oil Chemists Soc.*, **55**, 830–833 (1978).
13. Wiedermann, L. H., "Margarine and Margarine Oil Formulation and Control," *J. Am. Oil Chemists Soc.*, **55**, 823–829 (1978).
14. Feuge, R. O., White, J. L. and Brown, M., "Preparation of Fatty Acid Esters of Polyol Glucosides," *J. Am. Oil Chemists Soc.*, **55**, 699–702 (1978).
15. U.S. Dept. of Agriculture, Economics Statistics and Cooperatives Service, *Fats and Oils Situation*, FOS–296, July 1979.
16. Weiss, T. J., *Food Oils and Their Uses*, AVI Publishing Co., Westport, CT, 1970. pp. 32-33.
17. Wexler, H., "Polymerization of Drying Oils," *Chem. Revs.*, **64**, 591–611 (1964).
18. Fuller, G., Walker, H. G., Mottola, A. C., Kuzmicky, D. D., Kohler, G. O. and Vohra, P., "Potential for Detoxified Castor Meal," *J. Am. Oil Chemists Soc.*, **48**, 616–618 (1971).
19. Miwa, T. K., "Jojoba Oil Wax Esters and Derived Fatty Acids and Alcohols: Gas Chromatographic Analyses," *J. Am. Oil. Chemists Soc.*, **48**, 259–264 (1971).
20. Luddy, F. E., Hampson, J. W., Herb, S. F., and Rothbart, H. L., "Development of Edible Tallow Fractions for Specialty Fat Uses," *J. Am. Oil Chemists Soc.*, **50**, 240–244 (1973).
21. Stansby, M. E., "Development of the Fish Oil Industry in the United States," *J. Am. Oil Chemists Soc.*, **55**, 238–243 (1978).
22. Ward, J. A., "Processing High Oil Content Seeds in Continuous Screw Presses," *J. Am. Oil Chemists Soc.*, **53**, 261–264 (1976).
23. Becker, W., "Solvent Extraction of Soybeans," *J. Am. Oil Chemists Soc.*, **55**, 754–761 (1978).
24. Sullivan, F. E., "Steam Refining," *J. Am. Oil Chemists Soc.*, **53**, 358–360 (1976).
25. Cavanagh, G. C., "New Integrated Refining Process for Edible Oils," *J. Am. Oil Chemists Soc.*, **33**, 528–531 (1956).
26. Richardson, L. L., "Use of Bleaching Clays in Processing Edible Oils," *J. Am. Oil Chemists Soc.*, **55**, 777–780 (1978).
27. Allen, R. R., "Principles and Catalysts for Hydrogenation of Fats and Oils," *J. Am. Oil Chemists Soc.*, **55**, 792–795 (1978).
28. Scholfield, C. R., Butterfield, R. O. and Dutton, H. J., "Calculation of Catalyst Selectivity in Vegetable Oil Hydrogenation," *J. Am. Oil Chemists Soc.*, **56**, 664–667 (1979).
29. Johansson, L. E. and Lundin, S. T., "Copper Catalysts in the Selective Hydrogenation of Soybean and Rapeseed Oils: I. The Activity of Copper Chromite Catalyst," *J. Am. Oil Chemists Soc.*, **56**, 974–980 (1979).
30. Zehnder, C. T., "Deodorization 1975," *J. Am. Oil Chemists Soc.*, **53**, 364–369 (1976).
31. Gavin, A. M., "Edible Oil Deodorization," *J. Am. Oil Chemists Soc.*, **55**, 783–791 (1978).
32. Sherwin, E. R., "Oxidation and Antioxidants in Fat and Oil Processing," *J. Am. Oil Chemists Soc.*, **55**, 809–814 (1978).
33. Sonntag, N.O.V., "Fat Splitting," *J. Am. Oil Chemists Soc.*, **56**, 729A–732A (1979).
34. Zilch, K. T., "Separation of Fatty Acids," *J. Am. Oil Chemists Soc.*, **56**, 739A–742A (1979).
35. R. O. Walker, ed., *Official and Tentative Methods*, 3rd ed. (Revised annually) American Oil Chemists Soc., Champaign, Ill., 1978.

SELECTED REFERENCES

(For further reading on selected topics)

1. "AOCS Short Course on Industrial Fatty Acids," *J. Am. Oil Chemists Soc.*, **56** (11), November 1979.
2. Eckey, E. W., *Vegetable Fats and Oils*, New York, Van Nostrand Reinhold, 1954.
3. Emken, E. A. and Dutton, H. J., eds., "Geometrical and Positional Fatty Isomers," American Oil Chemists Soc. 1978.
4. *Fats and Oils Situation*, Economics, Statistics and Cooperatives Service, USDA, Washington, D.C. (Published Periodically).
5. Kuksis, A., ed., *Handbook of Lipid Research*, Vol. 1, "Fatty Acids and Glycerides," New York, Plenum Press, 1978.
6. Kunau, W-H, and Holman, R. T., eds., "Polyunsaturated Fatty Acids," American Oil Chemists Society, Champaign, Illinois, 1977.
7. Perkins, E. G., ed., "Analysis of Lipids and Lipoproteins," American Oil Chemists Society, Champaign, Illinois, 1975.

8. Proceedings, "Processing and Quality Control in Edible Fats and Oils," *J. Am. Oil Chemists Soc.*, 55 (11), November 1978.
9. Proceedings, "World Conference on Oilseed and Vegetable Oil Technology," *J. Am. Oil Chemists Soc.*, 53 (6), June 1976.
10. Proceedings, "World Conference on Vegetable Food Proteins," *J. Am. Oil Chemists Soc.*, **56**, (3), March 1979.
11. Pryde, E. H., ed., "Fatty Acids," American Oil Chemists Society, Champaign, Illinois, 1979.
12. Swern, Daniel, ed., *Bailey's Industrial Oil and Fat Products*, 4th ed., Vol. 1, New York, John Wiley and Sons, 1979.

13

Soap and Synthetic Detergents

Gene Feierstein* and William Morgenthaler**

SOAP

Introduction

Soap is believed to be one of the oldest chemical materials known to man obtained by reacting two substances to get a product with social significance. The word soap is derived from the latin "sapo" first used by Pliny The Elder, about A.D. 75. Although Pliny is credited[1] with the first written reference to soap, its use is believed to have begun long before recorded history. Soap "per se" was probably never actually discovered but evolved rather from various crude mixtures of alkali and fatty materials.

Over time it was learned that soap was not a mixture of alkali and fat but indeed resulted from a chemical reaction, later called saponification, and thus soap making changed from an art to an industry. Indeed soap remained the principal cleaning product, or surface active agent, well into the twentieth century. In the 1930's, synthetic detergents were developed and dramatically reduced world dependence on soap for cleaning.

It is interesting to note that the basic batchwise process for soapmaking remained practically unchanged for approximately 2000 years. It was not until the late 1930's that continuous soapmaking processes were developed and installed in large-scale manufacturing plants. Ironically, this timing coincides with the early stages of the tremendous growth of the synthetic detergent products. Nonetheless, a significant market remains for soap-based products for both consumer cleaning, primarily bar soap, as well as for industrial use.

Chemistry

Soap is the sodium or potassium salt of a long chain monocarboxylic (fatty) acid. It is made by the action of a hot caustic solution on tallows, greases, and fatty oils with the simultaneous formation of glycerine which at one time was wasted or left in the soap, as it still is in certain cases. However, most soapmakers collect the glycerin or glycerol as it is a valuable by-product with many important end uses.

*Elegene Consultants, Clearwater, Fla.
**Monsanto Industrial Chemicals Co., St. Louis Mo.

The reaction, starting with a typical fatty material (glyceride), is called saponification. It is composed of two separate and distinct steps, i.e., the controlling reaction to split, or hydrolyze, the fat and then after separation from the glycerine, a simple neutralization of the fatty acids generated with caustic soda.

$$\underset{\text{Glyceryl Tristearate}}{(C_{17}H_{35}COO)_3C_3H_5} + 3H_2O \longrightarrow \underset{\text{Stearic Acid}}{3C_{17}H_{35}COOH} + \underset{\text{Glycerin}}{C_3H_5(OH)_3}$$

$$\underset{\text{Stearic Acid}}{C_{17}H_{35}COOH} + \underset{\text{Caustic Soda}}{NaOH} \longrightarrow \underset{\text{Sodium Stearate (Soap)}}{C_{17}H_{35}COONa} + H_2O$$

Of course, if a fatty acid is available as a starting material, soap can be made simply by neutralizing (saponifying) with caustic soda as shown in the second step above.

Fats and Oils Used in Soap Manufacture

Although synthetic fatty acids can be used as a feedstock, the industry relies on naturally occurring fats and oils as the principal raw material for soapmaking. These natural products are triglycerides with three fatty acid groups randomly esterified with glycerol. Each fat suitable for soapmaking contains a number of long chain fatty acids groups with an even number of carbon atoms, ranging generally from C_{12} (lauric acid) to C_{18} (stearic acid) in the saturated members, as well as unsaturated fatty acids with the same chain lengths. The component fatty acids in a number of naturally occurring oils are listed in Table 13.1. The properties of the resulting soap are determined by the amounts and compositions of the component fatty acids in the starting fat mixture.

It has been shown,[2,3] using soaps made experimentally from single chain length fatty acids, that the important properties of the soap are highly dependent upon the length of the fatty acid chain and the degree of unsaturation. The principal considerations for selecting fatty materials as feedstock are the provision of a fat mixture containing saturated and unsaturated, and long- and short-chain fatty acids in suitable proportions to yield the desired qualities of stability, hardness, solubility, ease of lathering, etc., in the finished product, and sufficient refining and bleaching of the fat charge to ensure a good appearance. Due to the balance of physical and chemical properties regarded as highly desirable for sodium soaps, the choice of fatty acids which may be regarded as suitable feedstock is rather limited. Specifically, they consist of the saturated and unsaturated fatty acids with 12 to 18 carbons in the chain. In general, fatty acids with chain lengths less than 12 carbons are not preferred since their soaps have poor surface activity and skin irritation tendencies. The upper limit of chain length is 18, beyond which the resulting soaps are too insoluble, and both poor surface activity and poor sudsing is observed. Unsaturation in the fatty materials must also be carefully considered as fatty acid mixtures with a high degree of unsaturation yield soaps which tend to be soft, susceptible to oxidation and have poorer surface activity.

One method for the selection of a fat charge is the empirical INS method described by Webb.[4] The INS factor of a fat is defined as the saponification number minus the iodine number. For mixtures, the overall INS factor is derived from a weighted average of the factors of the individual fats. Choosing the fatty materials for soap production therefore is a chore unto itself when one considers both the chemical property requirements coupled with the commercial factors of price and availability. The most commonly used mixture of fats and oils used for soapmaking are tallow and coconut oil, in about a 3:1 proportion, although several other animal fats and vegetable oils described below are also useful raw materials. The major fatty acids found in beef tallow and coconut oil are shown in Table 13.1.

Note that tallow contains approximately 96% C_{16}-C_{18} fatty acids which is balanced in soap with coconut oil containing nearly 50% C_{12} (lauric) fatty acid and 20% C_{14} (myristic) fatty acid.

Tallow is well established as the basic component of soap made in the United States and is

TABLE 13.1 Average Fatty Acid Composition and Constants of Various Fats and Oils

		Animal Fats		*Vegetable Oils*			
	Chemical Formula	*Tallow*	*Lard*	*Coconut*	*Palm Kernel*	*Palm*	*Castor*
Saturated Acids							
Caproic	$C_6H_{12}O_2$			0.2	Trace		
Caprylic	$C_8H_{16}O_2$			8.0	3.0		
Capric	$C_{10}H_{20}O_2$			7.0	6.0		
Lauric	$C_{12}H_{24}O_2$			48.0	50.0		
Myristic	$C_{14}H_{28}O_2$	2.0	1.0	17.5	15.0	1.0	
Palmitic	$C_{16}H_{32}O_2$	30.0	26.0	8.8	7.5	42.5	
Stearic	$C_{18}H_{36}O_2$	21.0	11.0	2.0	1.5	4.0	2.0
Arachidic	$C_{20}H_{40}O_2$						
Behenic	$C_{22}H_{44}O_2$						
Lignoceric	$C_{24}H_{48}O_2$					Trace	
Unsaturated Acids							
Myristoleic	$C_{14}H_{26}O_2$						
Palmitoleic	$C_{16}H_{30}O_2$						
Oleic	$C_{18}H_{34}O_2$	45.0	58.0	6.0	16.0	43.0	8.6
Linoleic	$C_{18}H_{32}O_2$	2.0	3.5	2.5	1.0	9.5	3.5
Linolenic	$C_{18}H_{30}O_2$						
Elaedostearic	$C_{18}H_{30}O_2$						
Ricinoleic	$C_{18}H_{34}O_3$						85.9
C_{20} Unsaturated	–						
C_{22} Unsaturated	–						
Constants							
Saponification Value		196–200	195–200	251–264	240–250	196–206	175–183
Iodine Number		35–44	50–69	8–10	16–23	48–58	82–86
Titer-°C		37–46°	36–43°	20–23°	20–23°	38–47°	–

the principal animal fat in soapmaking. It is obtained as a by-product of the meat processing industry by rendering the body fat from cattle and sheep. Tallows from different sources may vary considerably in color (both as received and after bleaching), titer (solidification point of the fatty acids), free fatty acid content, saponification value (alkali required for saponification), and iodine value (measure of unsaturation). Beef or mutton fat with a titer of 40°C or higher is generally classed as tallow. Tallow has been used as the only fat in the manufacture of soap chips for use by commercial laundries, which employ relatively higher wash temperatures and higher alkalinity. However, in the majority of soap products, as mentioned previously, tallow is mixed with vegetable oils such as coconut to improve the solubility and lathering properties.

Grease and lard (hog fat) are the other animal fats sometimes used as soap raw materials. Both are less desirable than tallow as they must undergo various treatments for upgrading, including bleaching and partial hydrogenation to reduce unsaturation, prior to use in the soapmaking process.

Coconut oil is the most important vegetable oil used in soapmaking due primarily to its high lauric acid content. It is obtained by crushing and extracting the dried fruit, copra, of the coconut palm tree. Coconut oil is not ordinarily used alone except in the manufacture of special soaps for use with very hard water or with salt-water, in which lathering ability and stability to electrolytes are all-important. Coconut oil soap is typically very white, extremely resistant to oxidation and firm in consistency. Use of this oil in high-grade milled toilet bars contributes to the polished appearance desired in finished bars.

Other vegetable oils of lesser importance as fatty-acid sources for soapmaking include both solid and liquid materials such as palm oil, palm kernel oil, castor oil, olive oil, corn oil,

Edible Oils						*Marine Oils*		
Cottonseed	*Corn*	*Peanut*	*Soybean*	*Sunflower*	*Olive*	*Whale*	*Menhaden*	*Sardine*
0.5					Trace	8.0	7.0	5.0
21.0	7.5	7.0	6.5	3.5	9.0	11.0	16.0	14.0
2.0	3.5	5.0	4.5	3.0	2.3	2.5	1.0	3.0
Trace	0.5	4.0	0.7	0.6	0.2			
	0.2	3.0	Trace	0.4				
						1.5	Trace	Trace
						17.0	17.0	12.0
33.0	46.3	60.0	33.5	34.0	82.5	34.0	27.0	10.0
43.5	42.0	21.0	52.5	58.5	6.0	9.0	Trace	15.0
			2.3			Trace		
						5.0	20.0	22.0
						12.0	12.0	19.0
192-200	188-193	185-192	189-194	189-194	185-200	185-195	189-193	189-193
100-115	116-130	83-95	124-148	120-136	74-94	110-136	148-185	170-190
32-38°	18-20°	28-32°	20-21°	17-20°	18-25°	22-24°	31-33°	28-34°

and peanut oil. Generally, the oils from the edible category are the bottoms or "foots" remaining after the edible product is separated. Certain of these oils find use due to special properties, such as castor oil to make transparent soap, or to eliminate a by-product problem as in using the "foots" from edible oils to make lower quality industrial soaps.

Marine oils, such as whale oil and various fish oils, have found use in soapmaking as have some non-fatty materials such as rosin, tall oil, and certain naphthenic acids. Once again, when using the raw materials of lesser importance, consideration is given to pretreatment requirements, cost, and special properties of the final products.

In summary, the variety of fatty and oily feedstock for soapmaking is quite large in spite of the chain length restrictions, although as previously stated, tallow and coconut oil are the key raw materials for the industry at present.

Inorganic Raw Materials

Two products dominate the inorganic raw materials for soapmaking, viz., sodium hydroxide (caustic soda) and sodium chloride (salt). The latter is a key material in the processing of soap, although the quantity found in the finished product is nil.

Caustic soda is the principal inorganic ingredient used in the saponification process. Sodium carbonate (soda ash) is not suitable for saponifying glycerides, but is an inexpensive source of alkali for saponifying fatty acids. Potassium hydroxide (caustic potash) is employed almost exclusively in making soft soaps (liquids or pastes) since the potassium soaps are more water soluble than the sodium soaps. Blends of caustic soda and caustic potash are sometimes used to achieve special properties in the final soap. Potassium carbonate, like soda ash, may be used to saponify the fatty acids but is not useful for glyceride saponification.

Batch Manufacturing Methods

Prior to about 1940, kettle soap manufacture was the basic operation upon which the entire soap industry was built. In some parts of the world, this method remains as the primary soapmaking technique. While the chemical reactions are straightforward, practical soapmaking borders at time on an art (particularly in batch preparation) due to the extraordinary complex physical nature of soap and its aqueous systems. After saponification, the soap must be carried through a series of phase changes for the removal of impurities, the recovery of glycerine, and reduction of the moisture content to a relatively low level. Details on the physical chemistry that occurs in soapmaking are provided by McBain.[5] Consulting his phase diagram for the ternary system (sodium stearate–sodium chloride–water) permits one to follow the steps in the full-boiled kettle process. It is generally agreed that commercial soaps behave essentially like salts of single fatty acids, like stearic, so that McBain's phase diagram will depict very closely the equilibria involved in commercial soap production.

The complete series of operations in the production of an ordinary full-boiled or "settled" soap can be detailed as follows. Each of the successive operations performed on a single batch of material in the soap kettle are termed "changes."

1. Reaction of the glycerides with caustic soda until saponification is essentially complete.
2. Graining out of the soap from solution with salt in two or more stages for recovery of the glycerine produced by the reaction.
3. Boiling of the material with an excess of alkali to complete saponification, followed by graining out with alkali, and
4. Separation of the batch into immiscible phases of neat soap and nigre, which is called the "fitting" operation.

The final product, "neat soap," is made up of approximately 65% sodium soap, 35% water, and traces of glycerine, alkali, salt, etc. This is the feedstock from the kettle process which leads to toilet bars, chips, flakes, granules, and soap powders.

Full-Boiled Kettle Process. In the boiled process, a batch of 300,000 lbs. of soap, for example, is made in a steel kettle 28 ft. in diameter and 33 ft. deep, with a slightly conical bottom (see Fig. 13.1). A solution of caustic soda testing 18-20°C Bé (12.6-14.4 percent NaOH) is run into the kettle, and the melted fats, greases, or oils are then pumped in (a taber pump is suitable). The amount of caustic is regulated so that there is just enough to combine with all the fatty acids liberated. Heat is supplied by direct steam entering through a perforated coil laid on the bottom of the kettle. There is no stirrer, but agitation is provided by a direct steam jet entering at the base of a central pipe. The kettle is kept boiling until saponification is essentially complete; this requires about 4 hours. Salt (NaCl) is then

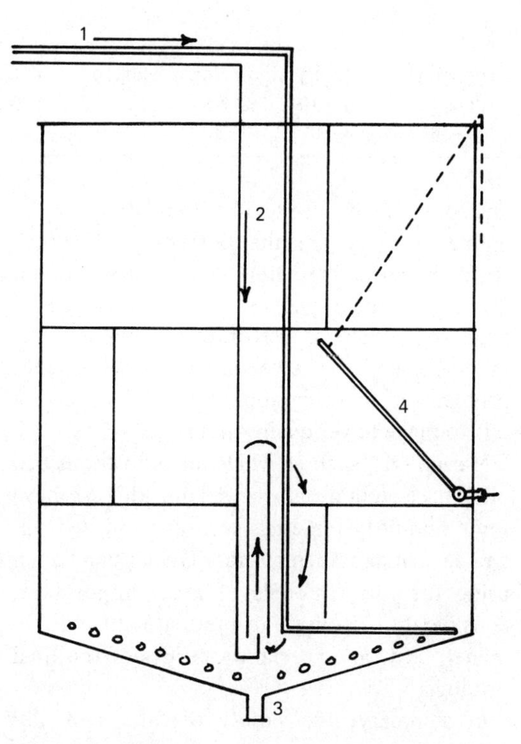

Fig. 13.1 Kettle for soap by the boiled process. (1) steam inlet for the perforated steam coil; (2) steam for agitation; (3) run-off for lye liquors and glycerine liquors; (4) the soap is pumped out through swing pipe.

shoveled in and allowed to dissolve, and the boiling is continued until the soap has separated, forming the upper layer. The lower layer contains glycerin (4 percent) and salt, and is drawn off at the bottom of the kettle. The whole operation which was just described is termed the saponification change, and requires about 8 hours. The salt used is chiefly rock salt; most of it is recovered and used again.

On the second day, water, and some caustic is run in and boiled with the soap; any glycerin caught in the soap is dissolved, and the solution, a lower layer again, is run off at the bottom and combined with the first glycerin water.

On the third day, a 10° Bé fresh lye (6.5 percent NaOH) is run into the kettle and boiled with the soap. Any glyceride which escaped the first treatment is saponified; any uncombined free acid is neutralized. The soap, which is not soluble in the alkaline liquor, acquires a grainy structure. This is called the strengthening change. After settling, the lye is run off and used in a new batch.

On the fourth day, the soap is boiled with water, which is chiefly incorporated in the soap. (Some salt is added at this point.) By this treatment, the melted soap acquires a smooth, glossy appearance. On settling, three layers are formed; the upper layer, the melted soap; the middle layer, or nigre, dark in color, consisting of a mechanical mixture of soap in a soap solution and impurities; and a very small lower layer containing some alkali. The melted soap is pumped away by means of a swing pipe without removing the nigre; the latter may remain in the tank and be worked into the next batch; the smaller lower layer is wasted. This operation is the finishing change, and lasts several days because the settling must be very thorough. Approximately one week is customary for the entire cycle of operations.

The melted "neat" soap is pumped to dryers, crutchers, or storing frames; it contains 30-35 percent water. One pound of fat makes about 1.4 lbs. of kettle soap; the factor varies with different raw materials.

Other Batch Processes. The cold and semiboiled processes are the simplest of the batch techniques, both requiring a minimum of equipment. Due to their simplicity, they do not permit recovery of the glycerin by-product and as the raw materials are not purified or the impurities separated during the process, the product is generally inferior to soap made by the time-consuming full-boiled method. Cold-made and semiboiled soaps are also frequently made from fatty acids.

Cold-process soapmaking is conducted by melting the fat charge in a vessel equipped with a mechanical stirrer and adding the calculated amount of caustic soda with vigorous agitation. A reasonably thick emulsion is obtained from the two immiscible liquids (glycerides and caustic soda solution) which is run into frames. Saponification is actually completed over several days of storage at ambient temperature, after which the cooled, solidified soap can be removed from the frames and cut into cakes.

The semiboiled process is similar to the cold process except more heat is applied to speed the saponification reaction. Typically, this procedure is used to produce soaps containing filler ingredients such as sodium silicate or abrasives. The fillers are mixed with the saponified soap until well dispersed in the kettle or mixing vessel after which the mass is poured into frames for cooling and solidification.

Batchwise production of soap requires a number of storage vessels to maintain a steady supply of neat soap for finishing operations that are usually continuous in large-scale operations. Work toward reducing the time and energy requirements of standard soap boiling gradually resulted in the development of sophisticated continuous saponification processes. The kettle soap process has been nearly abandoned for the large scale commercial production of soap in favor of the continuous equipment. Figure 13.2 shows a pilot plant setup that could be run by both batch or continuous modes to produce neat soap.

It is necessary to process the fats prior to kettle boiling, the kettle charge being adjustable as desired. Similar adjustments may be made in the stock delivered to the continuous fat splitter where the fatty acids are separated from the glycerin. The fatty acid is passed through the vaporizer where tars and decomposition

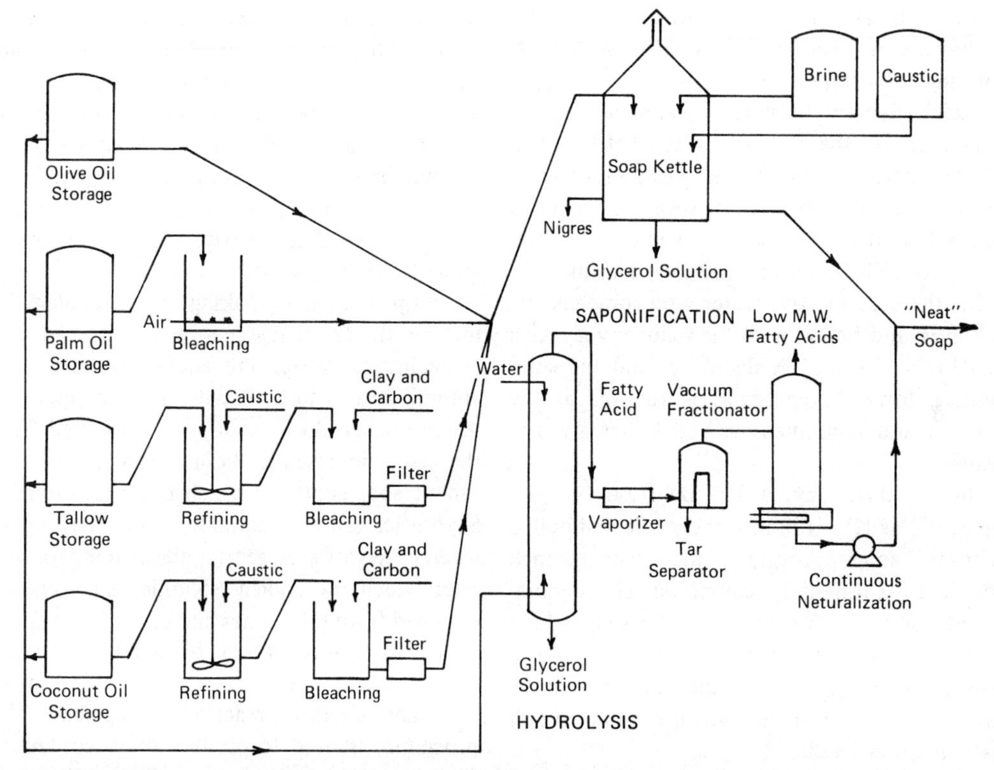

Fig. 13.2 Flowsheet for neat soap manufacture. (*Ind. Eng. Chem.*, **49**, *no. 3, 338* (*March 1957*); *copyright 1957 by the American Chemical Society and used by permission of the copyright owner.*)

products are removed; the water-white fatty acids pass either directly to the continuous Stratco contactor for continuous neutralization, or they may be vacuum fractionated to remove low molecular weight or special acids, and thence to the continuous neutralizer.

Continuous Soapmaking

Continuous processes can be grouped into two categories. The first group, including the DeLaval, Sharples, Mechaniche Moderne (Monsavon) and The Mazzoni SCN, are based on the continuous saponification of fats, followed by continuous washing and fitting, with removal of lye and nigre separations. The second group is based on continuous splitting, distillation, and neutralization of fatty acids. Examples of this latter group include the Mills, the Mazzoni SA, and the Armour processes.

Continuous Saponification. The Monsavon Process[6,7] is one of the earliest continuous processes for saponification, washing, and recovery of glycerol. The first step in this process employs a colloid mill to prepare a hot emulsion of aqueous caustic in the fats. The emulsion is fed through a reaction tube where additional saponification occurs and is finally fed into a soap kettle equipped with an agitator to complete the reaction. The crude neat soap is pumped from the bottom of the kettle up through a washing tower made up of several sections and counter-current to a flow of hot caustic solution. Each section is equipped with agitation to insure intimate mixing of the soap and caustic. At the top of the tower the washed soap is continuously removed and further mixed with water, brine, or caustic soda and pumped to a settling tank. The Monsavon Process produces an excellent quality neat soap of 60–65% solids in 24 hours.

The Sharples Process[9,10] is based on centrifugal separation of lyes, nigre, and glycerine. This system follows the same pattern as the traditional kettle process i.e. saponification, washing, and fitting. However, all steps are accomplished rapidly and continuously by separating soap and lyes and neat soap and nigre with the aid of high speed centrifuges. Four stages are used in this process. Two stages are used to complete the saponification reaction, a third to wash the reaction mass and the last fitting stage to facilitate the separation of neat soap and nigre. The entire process requires less than 2 hours to go from the fat feedstock to a high quality, bright and clean neat soap.

The DeLaval Centripure Process[7,8,24] for the continuous saponification of fat is similar to both the Sharples and the Monsavon methods. In the first stage of this three-stage process, the fatty oils and caustic are fed countercurrent to a recycle stream of preformed soap, the soap acting as both an emulsifier and a catalyst to effect nearly instantaneous saponification. The reaction mass then goes to the second, washing, stage to remove impurities and glycerine by centrifuge. The last stage is a fitting stage to separate neat soap and nigre with the former removed for processing to the final product form and the latter recycled to the washing section. Soap made by this technique is of excellent color.

The Mazzoni SCN Process,[7] which is one of several Mazzoni systems for processing soaps and detergents, is an automated process for the continuous saponification of fats, including the washing of the soap and the recovery of glycerol. In this process, proportioned amounts of fats, caustic soda, and brine solution are metered into a four-stage reaction autoclave, where heat, pressure, and recirculation contribute to the saponification reaction going to 99.5% completion. An amount of the reaction mixture equivalent to the combined feeds is continuously cooled and passed into a static separator for phase separation and removal of spent lye. The soap is washed with countercurrent lyes in several stages of mixers plus static separators. In the last stage, the soap is washed with fresh brine and separated by centrifugation into soap and lye phases. Both the Mazzoni and the Monsavon processes utilize a unique proportioning pump which is actually eight separate pumps in one unit to add the various raw materials to the processing equipment.

Continuous Fat Splitting and Neutralization. One of the prime advantages of the process involving splitting before saponification is its great flexibility. Fats, greases, and oils difficult or impossible to bleach satisfactorily can be split and distilled to yield light-colored fatty acids. Additionally, the producer is not limited to a product having the composition of neat soap but may directly produce a soap of substantially lower moisture content. Another advantage is the flexibility in choice of neutralizing cation, as either sodium or potassium ion is easily used.

The first continuous saponification process, involving fat splitting as the first step, to be operated on a large scale is that patented by Mills[11,12,13] for the Procter and Gamble Company. The process is operated by mixing a zinc oxide catalyst into the blended fat feedstock and reacting countercurrently with water in a 65-foot tall stainless-steel hydrolyzing tower at high temperature (ca. 500°F) and pressure (600–700 psi). A continuous stream of crude fatty acids is withdrawn from the top and crude glycerol is taken out the bottom. The fatty acids are purified by distillation, under vacuum, which takes the place of washing and the separation of neat soap and nigre for the removal of color bodies and other impurities. Neutralization is accomplished by continuous high-speed mixing with caustic soda solution and salt to produce neat soap ready for further processing to the desired final form. High grade toilet soaps equal in quality to those made by the best kettle-soap processes are made by the Mills process.

The Armour Process[14,15] represents another variation of a continuous fat splitting, distillation, and neutralization procedure. Tallow and coconut oil are not preblended in this process but instead are split separately without use of a catalyst in a Colgate-Emery fat splitter. The fatty acids are distilled in units similar to traditional fractionating stills, mixed to give the appropriate ratio of tallow fatty acids to coconut fatty acids and then neutralized in a DeLaval Centripure reactor. The resultant neat soap is pumped directly to a holding tank to await further processing.

Another process for continuous fat splitting

is available from the Mazzoni Company. It is a high-pressure process called the Mazzoni SA system. It can be coupled with other available Mazzoni systems to recover glycerine from sweet water (Mazzoni CGS), for fatty-acid distillation (Mazzoni DAG) and continuous fatty-acid neutralization, with either caustic soda (Mazzoni SC) or sodium carbonate (Mazzoni SCC).

Other neutralization systems currently available include the DeLaval Centripure system which is operated via viscosity control and a Meccaniche Moderne system which uses pH control similar to the Mazzoni process.

Soap Finishing Operations

Neat soap produced by either the kettle or any of the continuous saponification procedures contains 30-35% water. Conversion of the neat soap from these processes or from neutralized fatty acid processes requires a number of steps to obtain useful end products. The simplest method of converting neat soap or other hot liquid soap to a solid form suitable for forming into bars is known as framing. This consists simply of running the liquid soap into portable frames and allowing it to solidify in the form of large cakes. Up to 7 days is required for the soap to solidify in this fashion. More rapid processing is accomplished by drying the neat soap to 10-15% water for bar soap production or 5-10% for flake production in various types of drying equipment.

Drying. Three types of drying equipment have been routinely employed to make the different soap products, i.e. cabinet (conveyor or flake) drying, flash drying, and vacuum drying units. Cabinet drying is accomplished by first solidifying the neat soap on a chilling roll. The soap ribbons that come off the roll fall onto a wire mesh conveyor which passes back and forth through a chamber or cabinet where the residence time and the temperature of the drying air is adjusted to give the desired degree of drying; usually, 13-14% moisture is chosen for a product that is to be milled. In flash drying, the neat soap is superheated under pressure in a heat exchanger and then released through an orifice into a vented storage tank.

In the late 1940's, vacuum spray dryers made their appearance in the soapmaking industry and have become the most popular drying units. Vacuum drying is accomplished by spraying the hot neat soap onto the cold inner wall of a vacuum chamber and mechanically scraping the product off the walls. Vacuum dryers are capable of producing toilet soap base, soap-synthetic base and industrial soap, as well as filled or unfilled laundry soap.

Soap chips, flakes, or pellets obtained from the drying step are used as feedstock for various end products, including toilet bars, laundry soaps, and spray-dried laundry products.

Toilet Soap Bars. High-quality toilet soap bars are usually milled during the process. The first step in the production of milled soaps consists of mixing weighed batches of soap (10-15% water content) from the drying process, along with dyes, perfumes, titanium dioxide (whitening agent), and other additives in heavy-duty horizontal mixers called amalgamators. This coarse mixture is fed into the mills which consist of a series of rolls through which the soap mass is passed in the form of a thin sheet. The individual rolls rotate at different speeds so that the soap sheet, in passing between adjacent rolls, is not only compressed but is also subjected to an intensive shearing action. The soap ribbons scraped from the last roll are very uniform in composition.

The soap ribbons pass into the next piece of equipment in the process called a "plodder." The plodder compresses the milled soap into a dense mass which is forced through a plate with multiple perforations and then extruded through a die in the end or "nose" of the unit in the form of a continuous bar or log. Milling and plodding of the soap into bars accomplishes several objectives which cannot be obtained in frame-dried soap. First it changes the crystalline phase of the soap to the more desirable β-phase; second it reduces the moisture content below 15%; and lastly it thoroughly homogenizes the low concentration minor additives in the bar.

The continuous log of soap from the plodder is then cut into cakes or blanks with a wire cutter, passed through a conditioning tunnel, stamped into finished bars in automatic presses, fitted with dies bearing appropriate design or name features, wrapped, and finally packed into shipping cartons.

While the foregoing discussion is applicable for producing a standard toilet soap bar, other types or categories of soap, such as floating, transparent, marbelized, etc. have special processing and equipment requirements. Complete soapmaking plants from raw fats to packaged bars are available for purchase. One of the more popular systems in the United States was developed by the Mazzoni Company which finishes the soap via their LTC bar soap system. Jungerman[16], Spitz[17], and Herrick[18] have reviewed recent advances in soapmaking and bar-soap technology.

Types of soaps that can be produced by the milled method include deodorant soaps containing bacteriostatic additives, super-fatted soaps containing 2-6% excess of unsaponified oil, floating soap with air beat into the soap to reduce its density, and combination bars composed of mixtures of soaps and synthetic detergents or other major components.

Other Soap Forms. Soap flakes are produced from passing mixer product through a series of finishing rolls. The rolls are adjusted to a close tolerance which gives a thin, shiny film of soap. The flakes are scored by rotating cutters and stripped from the final roll with a stationary knife.

Spray-dried soap powders are typically formulated products containing various additives such as builders, optical brighteners, dyes, etc. The formulation ingredients are homogenized in a mixing vessel called a crutcher. Crutcher slurry is dried in a spray tower as described fully in this chapter. The primary advantage of a spray dried soap is its ability to dissolve rapidly in water.

Figure 13.3 depicts the three basic methods of producing "neat" soap from fats and the various finishing operations employed to obtain the final soap products.

FATTY ACIDS

Free fatty acids are not only a raw material for soap, but are becoming an ever-increasing

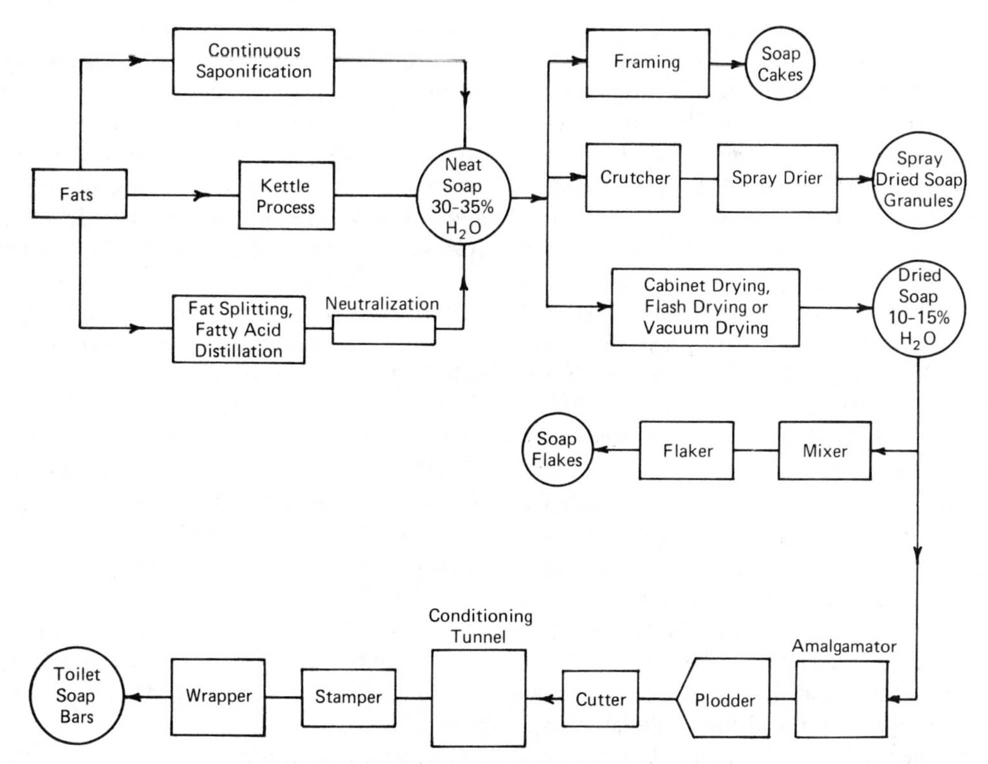

Fig. 13.3 Soap product flow chart.

source of raw materials for other chemical products. Although fatty acids are synthesized commercially, most of the commercial fatty acids are produced by the hydrolysis or splitting of natually occurring fats and oils.

Hydrolysis of the natural triglyceride-starting materials follows the reaction previously shown, but repeated below:

$$\begin{array}{l} CH_2-OOCR \\ | \\ CH-OOCR \\ | \\ CH_2-OOCR \\ \text{triglyceride} \\ \text{(fat)} \end{array} + 3H_2O \xrightarrow[\text{catalyst}]{\text{heat, pressure}} \underset{\substack{\text{fatty}\\\text{acid}}}{3\ RCO_2H} + \begin{array}{l} CH_2-OH \\ | \\ CH-OH \\ | \\ CH_2-OH \\ \text{glycerol} \end{array}$$

The R group on the triglyceride can contain varying amounts of unsaturation. When low-grade fats and oils are used as feed to fatty acid production, prior treatment to remove impurities is frequently required.

Manufacturing Methods

Over the years, fatty acids have been produced by four basic processes, i.e. saponification of fats followed by acidulation, the Twitchell Process, batch autoclave splitting, and continuous high-pressure, high-temperature hydrolysis. While the saponification approach is the earliest known process, fatty acids did not become commercially viable as a raw material until the development of the Twitchell Process around 1900.

Twitchell Process. This process[19] is a batchwise acid hydrolysis of the fats, at atmospheric pressure, in the presence of a catalyst called Twitchell's reagent. The initial Twitchell's reagent was benzenestearosulfonic acid, although later sulfonated-petroleum products were commonly used. After removal of impurities, the fats are mixed with water, the catalyst (0.75-1.25%), a small amount of sulfuric acid (0.5-1.0%) and boiled with live steam for 1-2 days. Sweet water is removed, replaced with fresh water and sulfuric acid, and then another boiling cycle is conducted. After 2-4 boiling cycles, the hydrolysis reaction approaches completion and the crude fatty acids are drawn off for purification. The effectiveness of the Twitchell reagent in fat splitting is generally attributed to both emulsifying and catalytic properties.

While the Twitchell Process is the simplest to run, it has several drawbacks compared to the more modern techniques. These include feedstock preparation, the need for acid-resistant equipment, long batch times, high steam consumption, and the emission of noxious, acidic fumes. Additionally, the fatty acids produced by this process are somewhat darker than those by alternate processes and the by-product glycol is more difficult to purify due to the presence of acid and catalyst in the sweet water.

Batch Autoclave Splitting. This process is conducted with or without a catalyst in a copper or stainless steel pressure vessel. The charge is made up of fats, water and, if desired, a catalyst—typically 1-2% of an oxide of barium, calcium, magnesium, or zinc.

Steam is then admitted to achieve the desired pressure of 75-150 psi and a temperature of 300-350°F. Hydrolysis is complete in 5-10 hours, after which the contents of the autoclave are blown into a separating tank where the glycerol is withdrawn from the fatty acid. An acid wash is essential to remove the catalyst, which is in combination with the fatty acid. Noncatalytic splitting requires a high-pressure autoclave, the temperature being increased to 450°F and the pressure to 425-450 psi. The reaction time is lowered considerably in this manner, requiring only 2-3 hrs. to effect 95-98% hydrolysis. This is possible because the reaction has a positive temperature coefficient and because the solubility of water in oil increases with temperature. This process generally produces lighter colored fatty acids than the Twitchell Process, and in considerably shorter times. However, it is run on a batch basis and thus is not well suited for large scale production.

Continuous Splitting. The process which

currently dominates fatty acid production from glyceride feedstocks is continuous, countercurrent, high-temperature, high-pressure splitting. The development of this technology was pioneered by workers at Colgate-Palmolive-Peet, Procter and Gamble, and Emery Industries during the 1930's and 1940's.[20] Some details on the Proctor and Gamble (Mills Process) are provided earlier in this chapter.

In the continuous process, the time involved is very short (2 hours or less), compared with 5-10 hours for the batch process and 1-2 days for the Twitchell Process. In the operation of a typical continuous, countercurrent fat hydrolysis plant, the feed is first pretreated to remove major impurities as in the Twitchell Process. The treated fats are deaerated under a vacuum to prevent darkening by oxidation during splitting, and charged at a controlled rate to the bottom of the hydrolyzing tower through a sparge ring which breaks the fat into droplets. These towers, constructed of fatty acid resistant, high-pressure, type 316 stainless steel, are generally 50-70 feet tall with diameters around 3 feet. The oil droplets in the bottom contacting section rise because of their lower density through the aqueous glycerol in the sweet water accumulating section at the bottom of the column. This effectively extracts entrained and dissolved fatty acid and oil from the countercurrent flow of sweet water. At the same time, deaerated and demineralized water is sparged to the top contacting section where it sweeps out the glycerine dissolved in the fatty phase and carries it down the tower.

The upward flowing fat-rich stream and the downward flowing water-rich stream from their respective contacting sections join in the reaction zone about midway in the column. Here they are brought to reaction (hydrolysis) temperature by direct injection of high-pressure steam and the final phases of splitting occur. Temperatures of 240-250°C and pressures in the 600-700 psi range are typically used in the reaction zone (figure 13.4).

The fatty acids are removed from the top of the tower to a decanter where the entrained water is separated. Simultaneously, the sweet water is withdrawn from the bottom of the tower and pumped to tanks to await further processing to commercial glycerol.

Hydrolysis of fats via continuous methods is capable of producing very high-quality fatty acid with 97-99% splitting efficiency and yields of glycerol in the 10-25% range. The cost of the fatty acid produced in a countercurrent

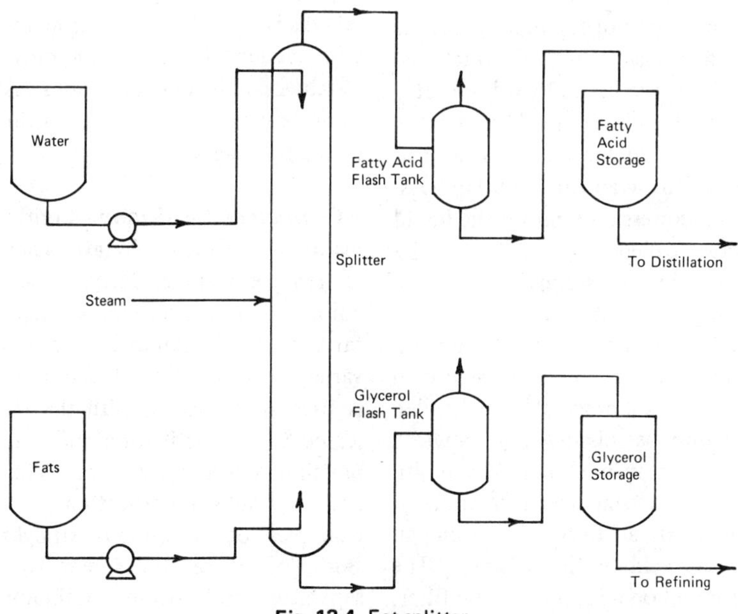

Fig. 13.4 Fat splitter.

tower is usually less than the cost of the original fat because the glycerol credit pays for the labor, overhead, and energy required.

Purification of Fatty Acids

Although the crude mixtures of fatty acids resulting from any of the methods described above may be used as they are, usually a separation into more useful components is made. The composition of the fatty acids from the splitter depends upon the fat or oil from which they were derived. Those most routinely used for the production of fatty acids include beef tallow, coconut oil, palm oil, fish oil, and the foots of the commonly-used vegetable oils such as cottonseed and soybean. Rarely will the composition of the fatty acid mixture be directly suitable for a specific end use. Therefore fractionation is used almost universally to prepare fatty acid products more desirable for specific end uses than the mixtures obtained from the continuous splitter. Fractionation according to the degree of unsaturation present is usually accomplished by a crystallization process, whereas fractionation according to molecular weight is done by distillation. In either case, the separations are usually aimed at providing feedstock for a specific end product rather than producing "pure" fatty acis.

Probably the most used of the older processes is "panning and pressing". This fractional crystallization process is limited to those fatty acid mixtures which solidify readily, such as tallow fatty acid. The molten fatty acid is run into pans, chilled, wrapped in burlap bags, and pressed. This expression removes the liquid unsaturated acids, called red oil, leaving the solid stearic acid. Red oil is composed mainly of oleic acid, and is sold in that form.

A more recently utilized method of purification is fractional crystallization of the acid with solvents. In the Emersol process[21] the distilled fatty acid mixture is dissolved in methyl alcohol and the solution is chilled in a multitubular crystallizer, causing the formation of crystals such as stearic acid crystals, leaving an acid such as oleic acid in the solvent. The crystals are separated on a rotary vacuum filter, washed with cold alcohol while still on the filter, and then melted and passed to a still where the solvent is driven off. The pure fatty acid is left as the finished product. Similar treatment yields a finished oleic acid.

An alternate process for separating fatty acids into liquid and solid fractions without solvents was developed by Henkel[22] and licensed by Lurgi. The solid-liquid fatty acid mixture is pumped through a scraper-cooler to precipitate the stearic portion (stearic/palmitic acids). The cooled mixture is then passed through a hydrophilizer, a mixer where a concentrated solution of a surfactant (an alkylbenzene sulfonate or an alkyl sulfate) is used to wet and suspend the stearin crystals. Then these are separated from the olein or liquid fatty acids in a conventional solid-bowl centrifuge. The olein phase is drained, washed, and vacuum dried. The stearic acid portion is heated to the acid-melting point to separate the water, then the acid is washed and dried. The separated water phase is recycled for further use with the wetting agent.

Stearic acid may be cast into cakes or fed to a flaker. Powdered stearic acid may be made by atomizing the molten acid at the top of a spray tower.

Distillation. The above methods separate fatty acids of different degrees of saturation. Separation of fatty acids of different chain lengths is typically accomplished via distillation, with vacuum distillation the most widely used. Batchwise and semicontinuous distillation have been largely replaced by continuous straight distillation units.

Continuous Distillation. Continuous vacuum distillation of the crude fatty acids is conducted to remove residual fatty glycerides, improve color, and improve odor so that the resultant fatty acids are suitable for soapmaking and a variety of other uses. Steam as well as vacuum is used in all cases, with the steam being injected to aid distillation and minimize decomposition of the fatty acids. The Lurgi design and the Mills modification of the Lurgi are examples of continuous straight distillation. Both of these types employ entrainment separators, and operate continuously with constant feed and constant removal of distillate and residue. They differ in that the Lurgi is a single-pot steam distillation system while the Mills unit uses flash (dry) distillation and two separate still pots, one smaller pot enclosed within the primary still.

Fractionating Distillation. Column designs for fractionating stills are based on pioneering work by Potts and co-workers. The first fractionating still for use with fatty acids was put in operation by Armour and Company. This type of still is used to remove an "odor cut" for soapmaking and to separate lauric acid from coconut oil; it also can be used to prepare a linoleic-acid concentrate from cottonseed oil fatty acids by fractionation removal of the palmitic acid. Bubble cap or sieve tray columns are employed when it is desired to remove a small amount of low boiling material from the fatty acids, either to improve odor or to obtain short-chain fatty acids for other applications.

Stills of both the continuous and fractionating types are offered commercially by a number of manufacturers, including Wurster and Sanger, Foster-Wheeler, Lurgi, and Mazzoni.

Hydrogenation of distilled fatty acids is often desirable to reduce the amount of unsaturation. Improvement of color and/or odor stability of soaps made from distilled fatty acids is often accomplished by partial selective hydrogenation at a point prior to saponification (neutralization) to soap. The hydrogenation of fatty acids is generally similar to the hydrogenation of fats and oils to produce hard fats, although certain differences are noteworthy. For example, corrosion-resistant convertors are mandatory to handle the fatty acids; processing temperatures are generally lower; the catalyst needs frequent changeout; and higher pressures are necessary to accelerate the reaction.

Fatty Acids From Tall Oil

Tall oil is a by-product from sulfate paper-pulp mills, accumulated as a result of the digestion operation. It is a nonglyceride oil obtained from the pine wood used in the papermaking process. The fats, fatty acids, and rosin acids in the wood are saponified during the process, skimmed off in solid form and converted into tall oil by acidulation with sulfuric acid.

Crude tall oil is a mixture of rosin acids, unsaturated fatty acids, and varying amounts of unsaponifiable material. Although crude tall oil from a given source can be relatively constant in composition, a wide fluctuation in the three basic components is evident from surveying commercial materials.

A variety of techniques may be used for refining and modifying tall oil, but fractional distillation is the method used commercially for the production of tall oil fatty acids. Three or more fractions can be generated by distillation, i.e., a nonvolatile residue of tall oil pitch, a volatile fraction rich in fatty acids and lean in rosin acids, and a less volatile fraction rich in rosin acids and lean in fatty acids.

Commercial tall oil fatty acids usually contain a minimum of 90% fatty acids with the principal components (40-50% each) being oleic and linoleic acids. Tall oil fatty acids have many applications, including use as a chemical intermediate, in paints and other protective coatings, in soaps and detergents, and in ore flotation.

Uses For Fatty Acids

A fatty acid molecule is multifunctional as an intermediate to a variety of end products, as it can be modified at the terminal carboxyl group, at the alpha carbon atom, or at a site of unsaturation. Various reactions at these sites such as esterification, polymerization, oxidation hydrogenolysis, salt formation, α-sulfonation, and preparation of amides are all practiced industrially. The two major uses for fatty acids and their derivatives are in plasticizers and surface active agents. Other important applications include use of lubricants, as paint and protective-coating ingredients, in rubber compounding, and a host of lesser examples.

Fat-Based Surface Active Agents. Fatty acids and/or derivatives are found in all four categories of surfactants i.e. anionic, cationic, nonionic, and amphoteric. As described by Jungermann[23], some of the more important anionic members include the carboxylic acid salts (soaps), sulfated alcohols, and sulfated ethoxylated alcohols. Toilet soap bars, heavy duty laundry detergents, and cosmetic products such as shampoos are major end products containing these anionic surfactants.

The two major applications of fatty-acid-based cationic surfactants are as germicides and fabric softeners. Varying the organic groups on quarternary ammonium salts gives rise to a multitude of properties and products in this group of surfactants.

The most important among the nonionic surfactants are the polyoxyethylene and polyoxpropylene derivatives of fatty alcohols, amids, and acids. Other examples of nonionic surfactants are glycol, glycerol, and sugar esters and fatty alkanolamides. These surfactants are used in a variety of consumer cleaning products, cosmetics, for emulsifiers in agricultural chemical formulations, and textile applications to name but a few of the many uses.

The last category, amphoteric surfactants, also contains examples of fat-based members. Fatty amine oxides and fatty imidazoline derivatives both are well represented in many consumer oriented products. Certain of the imidazoline derivatives have found application in mild, nonirritating shampoos while others are fabric softeners. The amine oxides are also used in shampoos as well as hand dishwashing liquid detergents.

While the old standby, soap, still constitutes a significant use of fat-based surfactants, it is easy to see that their use has become much more widespread indeed; the mixing of animal fats and wood ashes has progressed a long way since Pliny's description of *sapo*.

SYNTHETIC DETERGENTS

Synthetic detergents were initially developed as soap substitutes in an economy which was running short of edible fats and oils. These compounds were made resistant to deleterious and insoluble hard-water salt formation, and were a marked improvement in wetting, cleansing, and surfactancy in general. The term "synthetic detergents" has been shortened to "syndets" to describe the detergent compositions comprising the synthetic active ingredient along with other detergent additives. "Surface active agent" has been shortened to "surfactant" to describe the surface active principal or active ingredient (AI).

Even though surfactants have entered the toilet soap field, soap is still an excellent medium for combined usage, and fatty anionics are increasingly being used with it. While soap production in the United States has been reduced in proportion to syndets, increases in the population and general usage have resulted in an increase in soap production over the years. (See Table 13.2., p. 466.) Other advantages of soap are: (1) it is among the least toxic of all surfactants; (2) it is biodegradable and does not cause stream pollution; (3) it can be recovered where large amounts are used in an immediate area; (4) it makes antibacterial agents effective; and (5) it does not require added soil suspending agents as do the syndets.

The disadvantages of soap are pretty well recognized, not least among which is the use of fats and oils in competition with potential food uses.

Increase in liquid syndets has been a major continuing change. These liquid products were initially designed particularly for dishwashing and light-duty fabric cleansing or for medium-duty hard-surface cleaning. Fully automatic washers and washer-drier combinations have triggered the demand for these easily dispensed liquid products. At least half the total production of surfactants has applications in the household cleaning field, with smaller though appreciable tonnages used otherwise.

Surfactants may arbitrarily be subdivided into four categories, depending on ionic activity. These are: anionic, cationic, ampholytic, and nonionic.

Hydrophil-Hydrophobe Balance

Surface active agents are, as the term indicates, active at surfaces by preferential orientation of the molecule. This suggests that some built-in characteristic appears to contribute to, or control, molecular activity. Because surfactants are effective in either aqueous or nonaqueous systems, depending on their solubility characteristics, the molecule may be tailored for either system. Starting with alkylbenzenes as an example, they are slightly surface active in nonaqueous media but insoluble in water, hence ineffective. By sulfonating an alkylbenzene, for example, dodecylbenzene (DDB), a single SO_3Na group provides high water solubility and excellent surfactant characteristics in water, but the compound is then essentially insoluble in petroleum solvents. If DDB is di-or trisulfonated, the compound becomes more water soluble and, in effect loses much of its surface activity, thereby approaching a simple electrolyte such as sodium sulfate or, by analogy, a simple benzenesulfonate. In neither case is there sufficient hydrophobe influence to increase preferential orientation, the hydrophil balance having been exceeded. However, the DDB monosulfonate, highly water soluble and an excellent surfactant for an aqueous system, may be rendered hydrophobic and useful in nonaqueous systems by neutralizing the SO_3H group with a long-chain amine to render the molecule water insoluble. It is also possible, in the case of a nonaqueous surfactant with a single SO_3Na group, to increase the C_{12} side chain to approach C_{18} or higher; the same effect can be obtained as that resulting from neutralization of the shorter alkylbenzene sulfonate with an amine.

For nonionic systems the same hydrophil-hydrophobe balance exists, except that in place of the SO_3 or SO_4 water-solubilizing groups which form ionized aqueous solutions, nonionics depend on a multiplicity of oxygen groups or linkages which can unite with water by means of hydrogen bonds, thus inducing water solubility. Nonionics of the ethylene oxide-adduct type, therefore, introduce an extra dimension over ionics, since not only the hydrophobe may be varied but the hydrophil as well. This extra-dimensional feature may possibly account for some of the increasing usage of this class of compound.

In general, an optimum hydrophil-hydrophobe balance exists for a specific application, and for the class of compounds used for a given application. This optimum composition is generally arrived at by evaluating the hydrophil-hydrophobe characteristics of the specific purpose, and in at least one instance (emulsions), physico-chemical measurements have been used to predict the most effective surfactant or combination for the particular purpose.

In studying the literature, one might be led to believe that the various compounds mentioned, such as lauryl sulfate, sodium stearate, dodecylbenzenesulfonate, octylphenyl nonaethylene glycol ether, and the like, are pure compounds. The compounds used commercially are mixtures, lauryl sulfate being a generic term for a mixture of sulfates whose largest fraction is derived from the C_{12} alcohol; the remainder comes from higher and lower alcohols, the amount depending on the sharpness of the original alcohol distillation cut. The same is true for alkylbenzene derivatives and octylphenol compounds, while soaps are mixtures of the various fatty acids natural to fats and oils. A further example of mixtures is that of ethylene oxide adducts. The nonaethylene glycol ether designation suggests that this compound is the main constituent, but ethylene oxide adducts are manufactured on the basis of weight addition; the ethylene oxide adds to individual hydrophobe molecules in a manner which can give a normal (Poisson) distribution of adducts, the largest proportion being represented by a 9-molar adduct. Both lower and higher adducts are also present. It might seem that a competitive edge could be gained by supplying a highly purified compound, but this is not necessarily true. Most surfactants are used for many different purposes having many varied requirements; mixtures frequently permit usage where the pure compound might be less effective. Long experience with soaps has shown that except for very specific purposes, pure soaps are not competitive with properly chosen soap mixtures. This experience carries over to other surfactants.

Anionic Surfactants

Alkylaryl Sulfonates. Alkylbenzene sulfonates have been the "workhorse" of the detergent industry. They account for approximately 50% of the total synthetic anionic detergent volume used in liquid and spray-dried formulations. Three basic grades of alkylate (alkylbenzene) are manufactured with mo-

TABLE 13.2 U.S. Consumption of Soap and Synthetic Detergents[a] (millions of pounds)

	Soap	*Synthetic detergents*	*Total*
1940	2306	30	3236
1950	2882	1443	4325
1960	1230	3940	5170
1970	1050	5650	6700
1980*	1300	6400	7700

[a]CEH Marketing Report, Speciality Chemicals, Chemical Economics Handbook–SRI International, August, 1981.
*Estimated.

lecular weight ranges of approximately 235, 245, and 260. In general, the C_{10-12} range alkylates are used for light-duty liquid systems, and the C_{12-14} range alkylates for heavy-duty liquid and spray-dried detergents. The alkylate is sulfonated and neutralized, primarily to a sodium salt, prior to use.

Alkylbenzene production, prior to 1965, was synthesized from petroleum tetrapropylene reacted with an aluminum chloride or hydrogen fluoride catalyst and benzene. The resultant alkylate was a "hard" branched chain compound which was considered to be slowly biodegradable. A straight chain alkylate, termed LAB (linear alkyl benzene), has been produced since 1965. Extensive research[25,26] has demonstrated biodegradation effectiveness in sewage treatment plants in excess of 90–95%.

There are three basic processes for the manufacture of linear alkylbenzene. One route is by *partial dehydrogenation* of paraffins. This is followed by alkylation of benzene with a mixed olefin/paraffin feedstock, using a liquid hydrogen fluoride catalyst. A second route is via *partial chlorination* of paraffins. The chloroparaffin/paraffin feedstock is alkylated with benzene in the presence of an aluminum chloride catalyst. The third process uses *partial chlorination*, but includes a dehydrochlorination to olefin step prior to alkylation with aluminum chloride or hydrogen fluoride.

Products with low 2-phenyl isomer content (13-28%) are produced by the hydrogen fluoride alkylation of internal olefins from paraffin dehydrogenation. High 2-phenyl products (25-35%) are produced from aluminum chloride alkylation of chloroparaffins or olefins.

Sulfonation. For detergent use, the alkylate must be sulfonated to an acid form and then neutralized with a base, such as sodium hydroxide. The major use of the neutralized slurry is in the production of spray-dried detergents and both light- and heavy-duty liquid detergents. The slurry can also be drum dried to a powder or flake, or spray dried to light density granules. End use for the dried forms is primarily for institutional applications and car washing compounds. Sulfonation adds a hydrophilic group (SO_3H) to the hydrophobic alkylate to form the surfactant molecule. The resultant reaction is quite exothermic and almost instantaneous. In order to prevent decomposition and maintain optimum color, an efficient heat removal system is necessary.

Two chemicals are routinely used for sulfonation, i.e., oleum and sulfur trioxide, with the latter gaining increased popularity in recent years. Oleum sulfonation requires relatively inexpensive equipment and can be accomplished by either batch or continuous processes. However, the major disadvantages arc the SO_3H cost/lb. (as H_2SO_4), the need to dispose of the spent acid stream, and potential corrosion problems due to the dilute sulfuric acid generated. This process normally yields an 88-91% purity sulfonic acid with the remainder consisting of approximately 6-10% H_2SO_4, 0.5-1.5% water, and 0.5-1.0% unsulfonated oils. Reactions involved in oleum sulfonation are as follows:

Sulfonation

$$R{-}C_6H_5 + SO_3 \cdot H_2SO_4 \rightarrow R{-}C_6H_4{-}SO_3H + H_2SO_4$$

Alkyl benzene + Oleum (20–25%) → Whole sulfonic acid + 96% sulfuric acid

Dilution

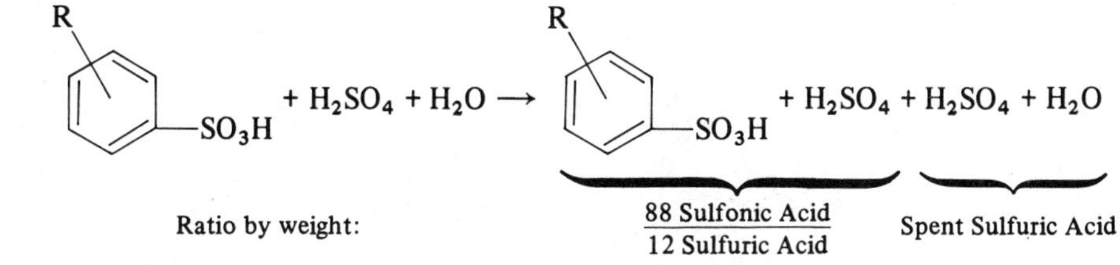

$$R\text{-}C_6H_4\text{-}SO_3H + H_2SO_4 + H_2O \longrightarrow R\text{-}C_6H_4\text{-}SO_3H + H_2SO_4 + H_2SO_4 + H_2O$$

Ratio by weight: 88 Sulfonic Acid / 12 Sulfuric Acid; Spent Sulfuric Acid

Neutralization

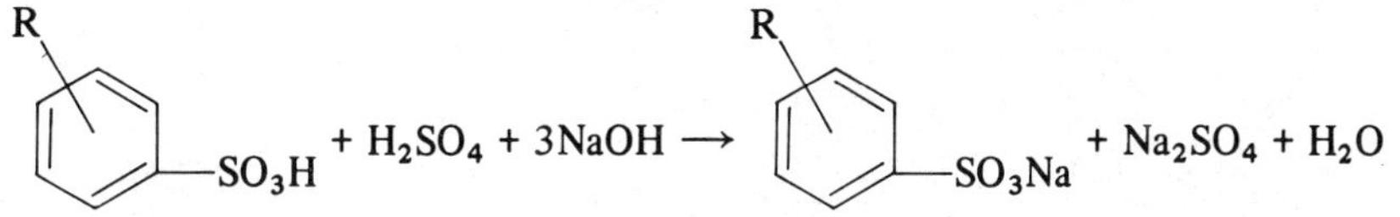

$$R\text{-}C_6H_4\text{-}SO_3H + H_2SO_4 + 3NaOH \longrightarrow R\text{-}C_6H_4\text{-}SO_3Na + Na_2SO_4 + H_2O$$

Sodium Alkyl benzene Sulfonate

The air-SO_3 process normally produces a 95-98% purity sulfonic acid with a 96-98% sulfonate to sulfate ratio. Although initial costs are much higher than for an oleum process, relative SO_3H costs are significantly lower.

The chemical reactions involved for SO_3 sulfonation are as follows:

Sulfonation

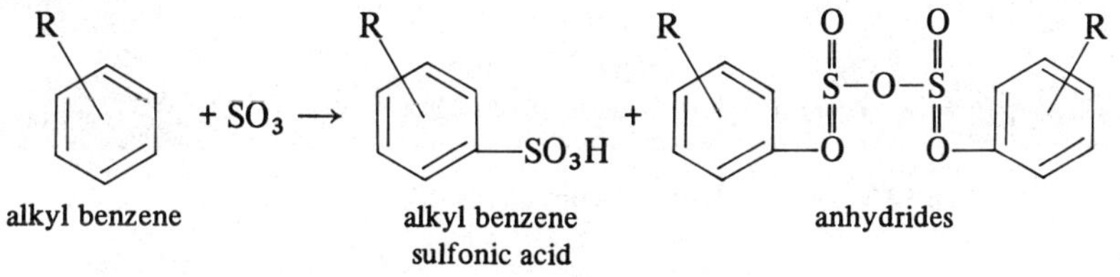

$$R\text{-}C_6H_5 + SO_3 \longrightarrow R\text{-}C_6H_4\text{-}SO_3H + R\text{-}C_6H_4\text{-}O\text{-}SO_2\text{-}O\text{-}SO_2\text{-}O\text{-}C_6H_4\text{-}R$$

alkyl benzene; alkyl benzene sulfonic acid; anhydrides

Hydrolysis

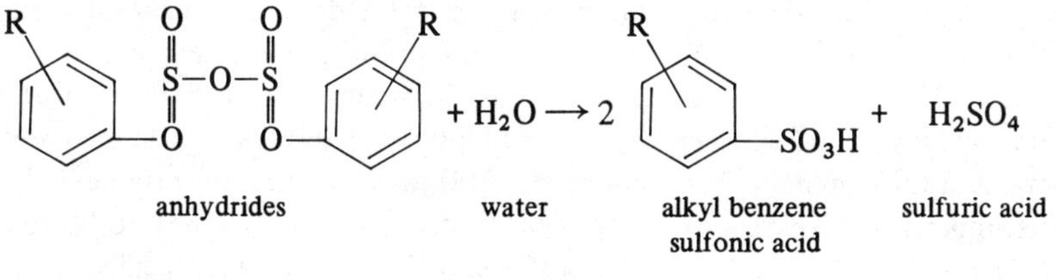

$$R\text{-}C_6H_4\text{-}O\text{-}SO_2\text{-}O\text{-}SO_2\text{-}O\text{-}C_6H_4\text{-}R + H_2O \longrightarrow 2\ R\text{-}C_6H_4\text{-}SO_3H + H_2SO_4$$

anhydrides; water; alkyl benzene sulfonic acid; sulfuric acid

Neutralization

$$R\text{-}C_6H_4\text{-}SO_3H + H_2SO_4 + NaOH \longrightarrow R\text{-}C_6H_4\text{-}SO_3Na + Na_2SO_4 + H_2O$$

alkyl benzene sulfonic acid; sulfuric acid; caustic soda; sodium alkyl benzene sulfonate; sodium sulfate; water

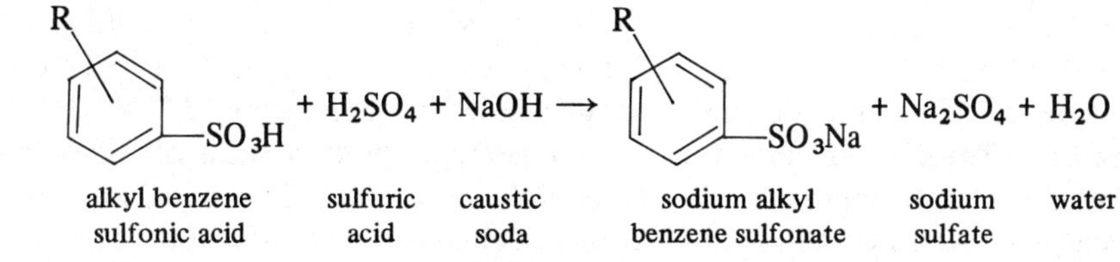

A typical continuous oleum sulfonation plant is shown in Figure 13.5. This process is divided into four zones: sulfonation, digestion or aging, dilution, and phase separation. Inline mixing of the alkylate and oleum is done in the sulfonation stage. The exothermic reaction is controlled by a heat exchanger. Particular attention to temperatures, acid strength, reaction time, and raw material ratio are imperative to sulfonation quality and acid color.

The product leaving the sulfonation zone is aged or digested 15-30 minutes. At this point, it is a mixture of sulfonic acid and highly con-

Fig. 13.5 Typical oleum sulfonation plant.

centrated sulfuric acid. After digestion, the mixture is diluted with water. The decanter zone permits separation of the sulfonic acid and the spent acid phase. The reaction mass, after decanting, contains two layers. The lower spent acid layer contains approximately 75-80% sulfuric acid. The upper layer consists of approximately 87-90% sulfonic acid, 1% free oils, and 7-12% sulfuric acid. Oleum sulfonation is preferred in some cases because the reaction is easier to control.

A typical continuous air-SO_3 sulfonation unit is shown in Figure 13.6. The SO_3 is maintained in vapor form and is diluted to approximately 3% in air, maintaining an SO_3/alkylate ratio of approximately 1.03/1.0. The Chemithon[27,28] and Ballestra[29] units have unique reactor designs to provide intimate mixing of the alkylate and air-SO_3 streams. The reactor temperature is controlled between 110-176°F, depending on production rates desired and product quality. Digestion and hydrolysis temperatures are maintained between 110-125°F. During hydrolysis, approximately two parts of water is added per 100 parts of reaction product. This step decomposes acid anhydrides to sulfonic and sulfuric acid. The sulfonic acid from the unit is approximately 95-98% pure, with a 1-2% free oil content and 1-3% sulfuric acid content. A typical flow sheet for alkylbenzene sulfonation is shown in Figure 13.7.

Both processes produce acids of comparable color. A Klett color value of 40 to 140 is normally obtained after sulfonation, dependent upon the alkylate molecular weight and control of process variables.

A novel venturi reactor has also been patented[30] by Chemithon Corporation. The

Fig. 13.6 Typical SO_3 sulfonation plant.

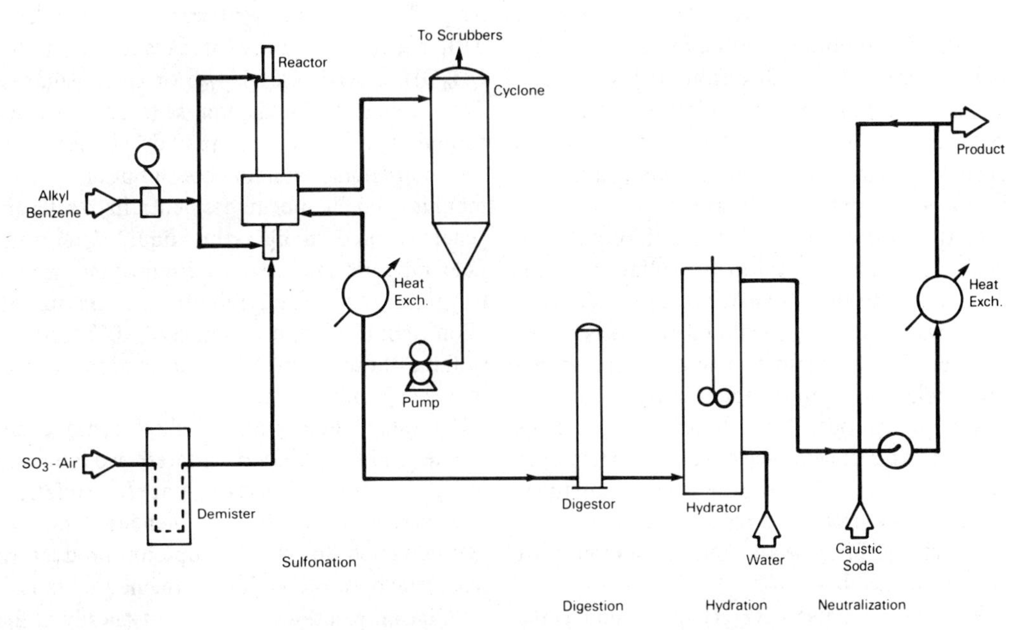

Fig. 13.7 Alkyl benzene sulfonation schematic flow diagram.

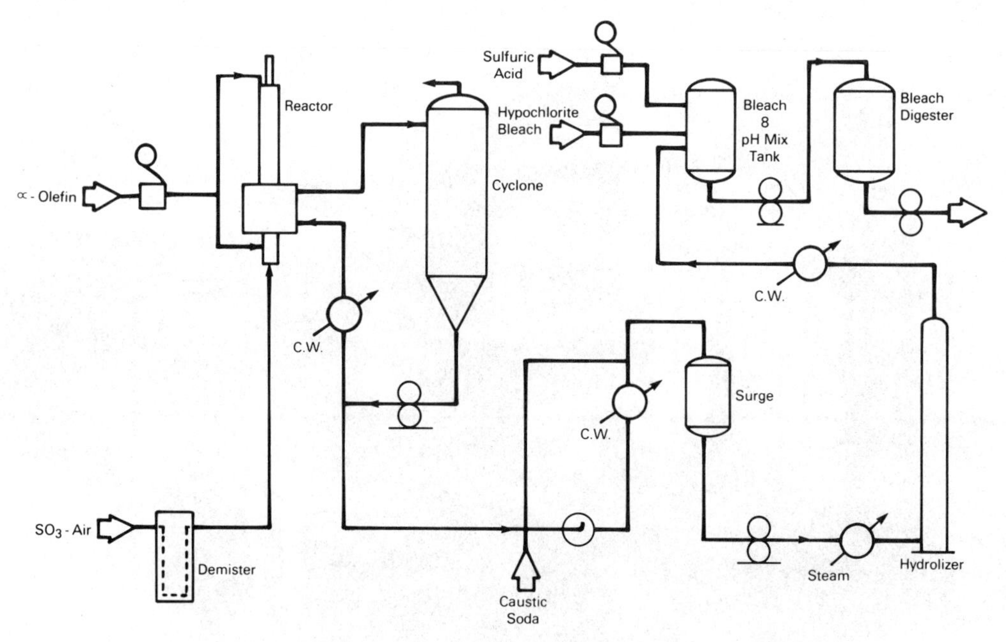

Fig. 13.8 Alpha-olefin sulfonation schematic flow diagram.

organic reactant is sulfonated by injecting it into a stream of gas comprising sulfur trioxide at a venturi. The reaction mixture is quenched and recycled downstream until sulfonation is complete.

For sulfation of fatty alcohols or ethoxylated fatty alcohols, the same reactor is used, but cooling water and sulfation temperatures are adjusted to an optimum value for the particular raw material. The digestion and hydration steps are not required in sulfation, and these systems are by-passed. The alcohol/sulfuric acids are invariably neutralized immediately, as reversion takes place in storage.

In the case of fatty alcohols, chlorosulfonic acid[31] can also be used as the sulfation agent. This is due to the simplicity of processing and the advantage of light-colored products. The reaction is stoichiometric and is irreversible after being taken to completion by the loss of hydrogen chloride. Some corrosion problems may be experienced due to chloride salts in the final product. Hydrotropes are sometimes added to the process to reduce the viscosity of the acid and improve the end use solubility of the neutralized base.

Sulfamic acid (NH_2SO_3H) has found some use in commercial sulfation. This acid provides simultaneous sulfation and neutralization to the ammonium salt.

Typical processes for SO_3 sulfation and neutralization are shown in Figure 13.8 (alpha olefins) and Figure 13.9 (fatty alcohols).

Neutralization of sulfonic acid from either an oleum or SO_3 sulfonation process is similar. The sulfonic acid can be neutralized with aqueous solutions of caustic; i.e., KOH, NH_4OH, NaOH, $Ca(OH_2)_2$, or alkanolamines. The sodium or potassium salts are used for formulating liquid and spray-dried detergents for household laundry consumption. Ammonium and alkanolamides neutralized salts are generally used in light-duty liquid detergents. Calcium salts are used as emulsifying agents. In general, neutralized salts from oleum sulfonation contain approximately 10% sodium sulfate; those from SO_3 sulfonation contain only 2-3% sodium sulfate.

For mixed detergents of alkyl benzene sulfonates and fatty alcohol sulfates, two loops are used, one for sulfonation, one for sulfation. The respective acids are blended prior to neutralization to yield a superior product to one made by postblending of the neutral salts.

Sulfation reactions are shown typically in the following equations:

$$C_{12}H_{23}OH + SO_3 \longrightarrow C_{12}H_{23} - O - SO_3H$$

Lauryl alcohol → Lauryl alcohol sulfuric acid

$$C_{12}H_{23}OSO_3H + NaOH \longrightarrow C_{12}H_{23}OSO_3Na + H_2O$$

Sodium lauryl sulfate

$$R - (OC_2H_4)nOH + SO_3 \longrightarrow R(OC_2H_4)n\ OSO_3H$$

Ethoxylated Fatty Alcohol → Ethoxylated fatty alcohol sulfuric acid

$$R(OC_2H_4)n\ OSO_3H + NaOH \longrightarrow R(OC_2H_4)n\ OSO_3Na$$

Sodium ethoxylated fatty alcohol sulfate

Sulfonic acid can also be processed to a dry powdered form by stoichiometric neutralization with a base salt, such as sodium carbonate. The neutralized mixture is slightly moist at this stage, and requires blending with other detergent additives to complete its transition to a dry free-flowing powdered detergent. The Ballestra process[32] mixes the acid and salts together, while the Stauffer process[33] meters the acid and caustic streams into a tumbling product bed. The A. G. Hoechst process[34] degasses the sulfonic acid prior to neutralization. Removal of SO_2 or SO_3 gases, and improved temperature control, produces a neutralized paste of excellent color quality.

Cationic Surfactants

A U.S. International Trade Commission Report shows that the production of cationic surfactants was in excess of 500M pounds in 1978.[35] These type products tend to be specialty items and are tailored for diverse uses; i.e., germicides, textile applications, sensitizers, flotation agents, corrosion inhibitors, and fabric softeners. The most recent growth has been in their use as fabric softeners for home laundry detergent. Since they are antistatic agents, they perform well in removing the static charge associated with synthetic fabrics.

These surfactants have at least one hydrophobic group attached directly to a positively charged nitrogen molecule. Cationics surfactants may be formed from nitriles, amines, amide-linked amines, or quaternary nitrogen bases. Significant literature reviews[36] cover this subject more thoroughly. Additional information has also been edited extensively by Jungermann.[37]

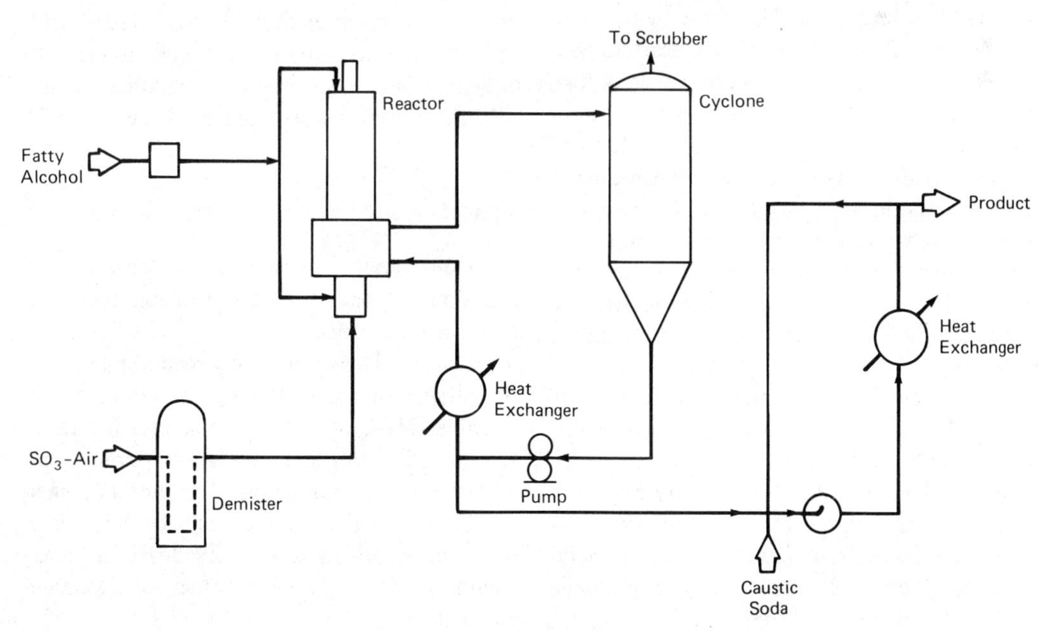

Fig. 13.9 Fatty alcohol sulfation schematic flow diagram.

Nonionic Surfactants

Nonionic surfactants are increasingly popular active ingredients for heavy-duty laundry detergent formulations. The majority of nonionic surfactants are polyoxyethylene or polyoxypropylene derivatives of alkylphenol, fatty acids, alcohols, and amides. It has been estimated that approximately one ethylene oxide unit is required to solubilize each methylene unit. Thus, by changing the ratio of ethylene oxide to fatty acid, the nature of the surfactant can be modified from a water soluble to oil soluble product. This versatility makes applications possible in numerous areas, i.e., detergents, agricultural products, metal protection, emulsions, polymerizations, textile applications, deinking, and penetrants. The reaction of ethylene oxide with fatty acids was described initially in 1928.[38] Shick[39] and Jungermann[40] have edited very thorough volumes concerning nonionic surfactants. In processing, the fatty acid and catalyst are charged to a reactor and purged with an inert gas such as nitrogen. The partial pressure is adjusted so that the explosive limits of the gas mixture in the reactor during and after addition of ethylene oxide are not exceeded.[41,42] The mass is heated to reaction temperature (115-200°C) and ethylene oxide added until the desired pressure is reached. A reaction[43] is indicated when a rise in temperature occurs. Cooling water is necessary to control the rate of reaction. If there is overcooling, the reaction can stop and lead to absorption of ethylene oxide. This can lead to a buildup of ethylene oxide in the reactor, which can cause a sudden reaction to take place. Reactor temperature and pressure are controlled until the desired quantity of ethylene oxide has been added. The ethylene oxide addition is then discontinued, the batch cooled, and the catalyst removed by filtration.

Other examples of nonionics are ethylene oxide adducts, glycols, glycerols, sugar esters, alkanolamides, and anhydrohexitols.

Polyethylene glycol esters of fatty acids are produced by reaction with ethylene oxide or by esterification of the fatty acid with polyethylene glycols. The reaction with ethylene oxide is described as:

$$RCOOH + nC_2H_4O \longrightarrow RCOO(C_2H_4O)_nH$$

Oxyethylation of fatty acid ester is used for production of other surfactants, such as sorbitol esters. Sorbitol, when reacted with fatty acid, yields an anhydro-hexitol ester.

Lauric, palmitic, stearic, and oleic acid are also ethoxylated and widely used as detergents and emulsifiers.

Amphoteric Surfactants

These surfactants contain both cationic and anionic groups. The products are relatively expensive due to the raw materials involved and processing costs. Typical uses include applications in shampoos, bubble baths, and other toiletries and cosmetics. A typical product is shown below[44]:

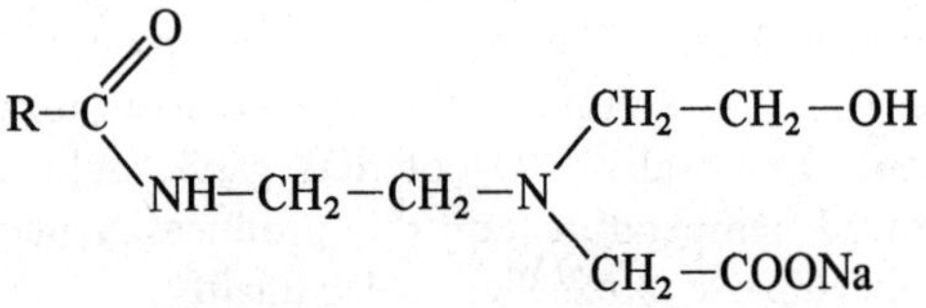

$$R-C(=O)-NH-CH_2-CH_2-N(CH_2-CH_2-OH)(CH_2-COONa)$$

N-β-hydroxyethyl-N-β-carboxymethyl fatty acid amidoethylamine, sodium salt.

Other examples of fatty amine oxides include dimethyl carboxymethyl coconut fatty acid, propylamide ammonium betaine, alkylamido betaine, and cocoamidopropylamine oxide. Additional varieties are cited by McCutcheon.[45]

Alkylolamides

A wide variety of amine condensates can function as foam boosters, wetting agents in aqueous-nonaqueous systems, emulsifiers, and dispersants. These are mainly reaction products of diethanolamine (DEA) or monoethanolamide (MEA) with fatty acids at a 2:1 ratio. Excess DEA converts both the amino and amido ester to an active 2:1 product, N,N-bis(2-hydroxyethyl) lauramide. In detergents, alkylamides are used at 1-2% levels as foam stabilizers. The general formula for alkylanolamide is:

$$\mathrm{R{-}\overset{\overset{\large O}{\|}}{C}{-}\underset{\underset{\large R^1}{|}}{N}{-}\overset{\overset{\large R^2}{|}}{\underset{\underset{\large R^2}{|}}{C}}{-}\overset{\overset{\large R^3}{|}}{\underset{\underset{\large R^3}{|}}{OH}}}$$

R^1 = alkyl; R^2 = hydrogen, alkyl, or hydroxyalkyl; and R^{2-3} = hydrogen or alkyl.

Detergent Builders and Additives

Present detergent products, whether in liquid or powdered form, are complex mixtures of several ingredients. The formulations consist primarily of surfactants, builders, and other additives designed to maximize performance for the consumer while maintaining reasonable raw material and manufacturing costs. Builders are typically added to a detergent formulation to extend or improve its cleaning performance across a wide range of use conditions. The combination of builders and surfactants exhibit a synergistic effect which boosts the detergency or cleaning efficacy of the mixture as compared with an equal amount of either compound alone. The major properties and characteristics desirable for a compound to be considered a detergent builder include:

1. The ability to control water hardness and other metal ions.
2. A contribution to final product alkalinity.
3. Buffer capacity in the proper pH range.
4. Deflocculation ability.
5. Compatibility with other formulation ingredients and detergent additives.
6. Consumer safety.
7. Environmental acceptability
8. Processability.
9. Adequate cost/performance.

In today's detergent products, a variety of builder materials, either alone or in combination, are used in formulating, with each compound contributing its own set of unique properties to enhance the final product's total performance. Over the past ten years combinations of new builders and those of proven performance have been examined in detergent products to meet changing consumer needs and demands, fabric changes, environmental considerations, and increased energy saving consciousness.

Phosphates. Tetrasodium pyrophosphate ($Na_4P_2O_7$-TSPP) became commercially available in the mid-30's and soon gained acceptance as the builder choice for use in soap-formulated products. The combination of TSPP and sodium carbonate was found to be more effective than either ingredient used separately. By the mid-40's, sodium tripolyphosphate ($Na_5P_3O_{10}$-STP) was commercially available and became the product of choice due to superior sequestering and detergent processing characteristics. Sequestration is the reaction by which a heavy metal cation combines with an anion to form a soluble complex. The sequestration of Ca^{++} and Mg^{++} ions leads to softened water, and is the most important function of any detergent builder.

Sodium tripolyphosphate is the major builder ingredient in heavy-duty laundry detergents, automatic dishwashing compounds, and industrial and institutional cleaners. In liquid laundry detergent, tetrapotassium pyrophosphate ($K_4P_2O_7$-TKPP) has been preferred due to superior solubility characteristics.

Other phosphate builder salts include sodium trimetaphosphate, trisodium and tripotassium phosphates, and sodium acid pyrophosphate. Each has a unique property whose utility is dependent upon the particular detergent system or end-use application envisioned. "Phosphorus and Its Compounds"[46] presents a complete assessment of phosphates and their role in detergents.

Silicates. Silicates, both sodium and potassium salts or solutions, have valued functional characteristics such as emulsification, buffering, deflocculation, and antiredeposition ability. An additional unique property of silicates is to provide corrosion protection to metal parts in washing appliances as well as protect the surface of china patterns and metal utensils in automatic dishwashers.

Silicates are manufactured in either liquid, crystalline, or powdered forms. They also vary in degree of alkalinity. The alkalinity of the silicate contributes to the concentration of hydroxide ions necessary to high-performance detergent products. The ability of the silicate to maintain pH (buffering capacity) in the presence of acid soils enhances the sequestration ability of the builder system in the formulation. The silica/alkali ratios of the silicate are selected by the formulator to meet specific product requirements. A sampling of the numerous grades commercially available and typical uses are as follows:

Type Salt	*Ratio SiO_2/Na_2O*	*Physical Form*	*End Use*
Sodium	3.22:1.0	Syrupy liquid	Laundry detergents, automatic dishwashing compounds
Sodium	2.4:1.0	Viscous liquid	Same as above plus liquid detergents
Sodium	2.0:1.0	Alkaline syrupy liquid	Industrial products
Sodium	2.0:1.0	Powder	Automatic dishwashing compounds–dry mix products
Sodium-meta	1.0:1.0	Crystalline, pentahydrate	Industrial–Institutional–dry mix products
Sodium-meta		Crystalline, anhydrous	Same as above
Sodium-sesqui	2.0:3.0	Crystalline	Same as above
Sodium-ortho	1.0:2.0	Crystalline	Same as above

When using liquid silicates, it is essential that the product with the proper ratio be selected for various process and performance parameters. For example, liquid silicates containing a high SiO_2/Na_2O ratio are susceptible to forming insolubles if overdried. In agglomeration processes, the use of lower SiO_2/Na_2O ratio liquid silicates requires additional mixing, aging, and drying to obtain a crisp, free-flowing product.

Carbonates. In certain areas of the United States, phosphate salts have been legislatively banned from use in laundry detergent products. In these areas, sodium carbonate (Na_2CO_3) has replaced the phosphate as the builder or hardness (ion) control agent. Sodium carbonate softens water by precipitating the hardness minerals. It also is one of the most economical sources of alkalinity for detergents and provides some degree of soil dispersion and suspending action. It is commonly used in laundry detergents, automatic dishwash compounds, hard surface cleaners, and presoak formulations.

Sodium Sulfate. Although considered a builder by some manufacturers, sodium sulfate (Na_2SO_4) contributes little to detergent performance. It is commercially available from natural sources and as a by-product from rayon processing. In dry mix and agglomerated type products, sodium sulfate is used because of its flow characteristics. In spray-dried products, sodium sulfate acts as an inert "filler" and aids in density control as well as contributing to the crisp granulated characteristics of the spray-dried detergent beads.

Sodium Citrate. Sodium citrate ($Na_3C_6H_5O_7$) has "builder" characteristics due to its ability to sequester water hardness ions and deflocculate soils.[47] Its principal current use is in non-phosphate liquid laundry detergents, but it can be found to a limited extent in dry powders.

Zeolites. Within the past three years, zeolites have gained attention and are being used as a builder for powdered laundry detergents.[48] Zeolites are crystalline hydrated aluminosilicates, of Group I and II elements. These are sodium, calcium, magnesium, potassium, strontium and barium salts which appear in natural or synthesized forms. One of the zeolites of particular interest to the detergent industry is a synthetic form, Type 4A. The empirical formula is $Na_2O \cdot Al_2O_3 \cdot 2SiO_2 \cdot 4.5H_2O$.

Zeolites are not water soluble, but reduce calcium water hardness by an ion exchange mechanism. The calcium ion passes through the pore openings in the particles and exchanges with the sodium ion. Zeolites, however, cannot remove magnesium ions to any great extent, due to the layer radius of the magnesium ions. Due to this limitation, zeolites are normally used in conjunction with other builders which have the ability to sequester magnesium and other ions present in the wash water.

Sodium Chloride. Sodium chloride (NaCl) is still used by some formulators in detergent manufacture. Its main function is as an inert "filler" or diluent. In spray-drying applications, it is used to control slurry viscosity and the density of the spray-dried bead or granules.

Although formulated to some extent in automatic dishwashing compounds, its use is not recommended.[49] When washing aluminumware, the aluminum oxide coating breaks down when exposed to sodium chloride. This causes corrosion in the form of pitting.

Nitrilotriacetic Acid. One of the first compounds developed as a replacement for sodium tripolyphosphate was nitrilotriacetic acid. The sodium salt of this compound

$$(N(CH_2COONa)_3 \cdot H_2O)$$

has been shown to have excellent sequestration and/or chelating ability. This product was quickly accepted by the detergent industry as an alternate builder, but its usage was terminated in the U.S. in December, 1970. Three major detergent manufacturers voluntarily agreed to discontinue its usage in the United States due to preliminary evidence indicating that the compound may or could be teratogenic. Subsequent assessment (May, 1980) by the EPA, culminating nearly 10 years of extensive safety studies, indicated that risk to persons exposed to NTA from detergents would be extremely low. The EPA stated, however, that NTA should not be used in such products as shampoos, hand dishwashing liquids, or foods.

Although NTA has not been used in U.S. household laundry products since the voluntary ban, it has been used as a detergent builder in Canada since 1970. The product is normally used in conjunction with phosphates to enhance the sequestration characteristics of the formulated detergent products and can readily be spray dried.[50]

Ethylenediaminetetracetic Acid. EDTA is a compound which has excellent sequestrating or chelating effectiveness. Its structure is as follows:

$$(HOOCCH_2)_2N-C_2H_4-N(CH_2COOH)_2$$

The sodium salt of this compound has function advantages in liquid detergent systems, but cost performance limits its usage beyond a role as a stabilizer.

Trisodium Oxapropane Tricarboxylate (Builder M). This compound was developed by the Monsanto Company[51] in the early-1970's as a replacement and/or partial substitute for phosphates in detergent formulating. The product's primary function is the sequestration of divalent metal ions, particularly calcium and magnesium, in aqueous solutions. It also acts as a buffering agent and exhibits deflocculating properties. This compound is trisodium 2-oxa-1,1,3-propane tricarboxylate salt. The product was indicated to have application in both liquid and dry laundry detergent formulations, although it was never commercialized.

Carboxymethyl Oxysuccinate (CMOS). Lever Brothers Company developed this compound, to be used as an adjuvant or replacement for phosphate salts in detergent formulations.[52] It is not a commercially available product at this time. CMOS is trisodium 2-oxa-1,3,3,4 butanetricarboxylate. It can be isolated in an anhydrous form, or as a mono, tetra, or pentahydrate. This product is similar to Builder M and is also readily degradable.[53]

Detergent Additives

In addition to detergent builders and surfactants, there are also numerous additives which provide specific functions in detergents. A few of the most widely used materials are described below.

Sodium Carboxymethylcellulose (NaCMC). The surfactants employed in synthetic detergents are very effective in removing soil and stains from the surface of the washed fabric.

Under heavy soil loading, however, there is a tendency for these soils to redeposit on the clothes. To reduce soil redeposition, NaCMC is added to the detergent system. Although several theories have been advanced, a true understanding of its redeposition characteristics is not known.

Optical Brighteners. Although the surfactant system in detergents cleans clothes effectively, white fabric tends to "grey" or lose its white appearance after numerous washes. Optical brighteners are added to detergent formulations to visually improve the "whiteness" or brightness of the fabric.

All major detergent products, liquids and powders, contain one or more fluorescent whitening agents. Substantive to most fabrics, these materials absorb ultraviolet radiation and convert it to a visible blue-white reflectance range. This characteristic has a tendency to mask the natural yellowing or greying of the fabric. Most brightener systems are more effective on cotton than synthetic fabrics. Usage range varies from 0.05% to 0.2% in the detergent formula.

Hydrotropes. Hydrotropes are used primarily in liquid detergents to increase the solubility of less soluble ingredients. They are also used on occasion as viscosity modifiers in the slurry phase of manufacturing spray-dried detergent powders. Reducing the high viscosity of the thick detergent slurry improves its processing characteristics, reduces product bulk, and insures crisp, free-flowing granules.

The most common hydrotropes are the ammonium, potassium, or sodium salts of toluene, xylene, or cumene sulfonates.

Enzymes. Proteolytic and amylolytic enzymes have been designed for use as adjuvants in detergent systems to remove stains. These complex proteinaceous molecules act as catalysts and tend to break down particular soils and stains to a form more readily removed from fabrics. Basic use of these ingredients are in presoak detergent systems, where sufficient time is available for the catalytic process to react on the stain. Enzymes are not compatible with liquid chlorine bleaches.

The bacteria from which the enzymes are formed are cultured in a protein media. The bacteria attack the protein and secrete proteolytic enzymes. After fermentation, the "beer" containing the enzymes is filtered from the grain media. The resultant liquor is concentrated via evaporation, and the enzymes are precipitated from the concentrate with a solvent. The solvent solution is removed from the enzymes through centrifugation, and the enzymes are then dissolved in a water media, filtered for spore removal, and spray dried to a fine powder. The powdered enzymes can be admixed with other salts and either prilled, compacted, encapsulated via a water-soluble wax solution, or agglomerated to granular spheres or beads for postmixing with detergent powders.[54,55,56]

Pearlescent-opacifying Agents. These compounds are used basically in liquid detergent systems to produce a specific esthetic effect. Pearlescent compounds (bismuth, titanium-dioxide coated mica) are normally insoluble in water or alcohol, but are dispersed in emulsion-type systems.

Opacifying agents (e.g., water-soluble salts of hydrolyzed copolymers of styrene and maleic anhydride) are used to reduce translucence, modify the viscosity characteristics of the liquid, and provide a cream-like texture or pearlescent effect to the product.

Anticaking Agents. Due to the reduced phosphate levels and use of nonphosphate builders in some powdered laundry detergents, the flow characteristics of these products have changed. Products made with increased anionic active levels, plus increased sodium carbonate usage, tend to exhibit compaction caking in the detergent carton. Anticaking agents have been added to the formula (generally admixed) to reduce this tendency, especially in warm humid climates. A wide range of products are available for this purpose, including sodium benzoate,[57] tricalcium phosphate, colloidal aluminum oxide, silicon dioxide, magnesium silicate, calcium stearate, sodium sulfosuccinate, and microcrystalline cellulose.

Spray Drying

A major portion of laundry detergents today are spray dried. In this process, the detergent raw materials are mixed with water to a thick paste or slurry, atomized into sperhical droplets, and dried to a crisp, free-flowing granular product. The art of spray drying was first patented in 1883 but was not practiced to any extent until the 1930's.

With the advent of synthetic detergents in the 1940's, spray drying became the dominant method of detergent processing. The advantages of spray drying are numerous; i.e., product density can be varied from 0.20 to 0.80 gm/cc, solubility of the detergent granules is significantly improved, production rates per man hour are increased, multiple formulations can be prepared using the same equipment, and the excellent free-flowing and solubility characteristics are advantageous to consumer usage. The major disadvantages of spray drying are the initial capital investment required, the relative inflexibility to vary product sizing and density in some units, and the energy intensive nature of this process.

Spray drying is divided into several process steps, i.e., sclcction of formulations, slurry preparation, atomization, drying, conditioning of product, packaging, waste reclamation, and pollution control.

In today's environment, detergents are manufactured for use in various areas of the country where there may or may not be legislation controlling the use of the phosphates. Typical formulations are as follows:

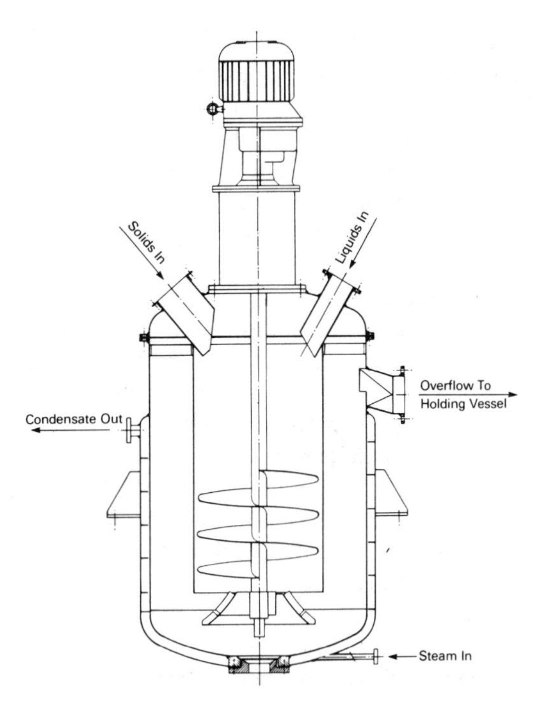

Fig. 13.10 Continuous mixer.

spray drying. A typical continuous mixer is shown in Figure 13.10. By proper maintenance of the order of addition for raw materials, slurry solids level, slurry temperatures, and close attention to viscosity and aeration trends, the desired final product characteristics of the detergent can be controlled. In essence, the spray tower itself is simply a contained heat source. If the same quality and quantity of product is introduced to the spray tower, it will dry or remove the moisture at a uniform rate. It is only when the aforementioned vari-

Ingredients	*Phosphate Built Detergent % in Formula*	*Non-Phosphate Built Detergent % in Formula*
Sodium Tripolyphosphate	20–35	0
Sodium Carbonate	5–10	10–40
Sodium Alkyl benzene Sulfonate	8–17	10–20
Sodium Silicate	6–12	5–20
Sodium Sulfate	10–30	10–30
Sodium CMC, Optical Brightener	1–2	1–2
Water	5–10	5–10

Slurry preparation of the detergent formulation is considered by the majority of manufacturers to be the single most important factor in ables are not controlled that problems arise in controlling the density, moisture level, aging, and/or packaging characteristics of the product.

For example, if the slurry solids level is not constant (±2%), the evaporative load, production rate, particle sizing, and moisture levels of the spray-dried granules will vary. Compensating for nonuniform feed to the spray tower is at best difficult due to the interaction of the controllable variables such as tower temperature, pump pressure, etc. Although some minor tower adjustments are necessary at times to control process conditions, the major controlling factor for routine operations is the preparation and uniformity of the detergent slurry.

Slurries are prepared via batch, semicontinuous, and continuous crutching units. A continuous slurry preparation system is shown in Figure 13.11. Solids levels are normally maintained as high as possible (64-82%) to reduce the sensible heat loading and maximize tower through-put. With the introduction of "soft" (biodegradable) alkyl benzene, the sodium tripolyphosphate used in processing slowly changed from low temperature rise (T.R.) grades to high T.R. grades. Low T.R. sodium tripolyphosphate tends to form more "lumps" and "grit" during hydration from the anhydrous to hexahydrate phase in the mixing stage. It is advantageous to maximize sodium tripolyphosphate hexahydrate formation ($Na_5P_3O_{10} \cdot 6H_2O$), as this water is "bound" and will be retained in the final product. The bound water acts as an inert filler as well as contributing to crisp, free-flowing characteristics during conveying, storage, and packaging.

After the detergent raw materials are mixed, the slurry is charged to a homogenizer and any large lumps or gritty particles are desized and screened to prevent plugging of the spray nozzle. The slurry is then charged to a booster pump, deaerator, and to a high pressure pump (300-1200 psig) for atomization. Most detergent slurries are atomized into spherical droplets via high pressure nozzles. The nozzles produce a hollow conical pattern with a spray angle of 45-90°.

The slurry emerges from the nozzle as a film. The centrifugal force developed shears the film into droplets. This action is caused by specially grooved cores inserted in the nozzle ahead of the point of discharge of the liquid mass. A series of spray nozzles (2 to 25) are placed near the top of the spray tower and angled so one edge of the desired spray pattern is vertical to the side walls of the spray tower. Pump pressures from 300 to 1200 psig may be used to atomize the slurry. Some spray towers use two or three rings of nozzles, depending upon the formulation and drying capacity of the unit. Production rates of detergent spray towers range from 5,000 to 80,000 lbs./hr. In most cases, the rates are controlled by the slurry feed facilities, solids level, and drying capacity.

Spray towers vary in size from approximately

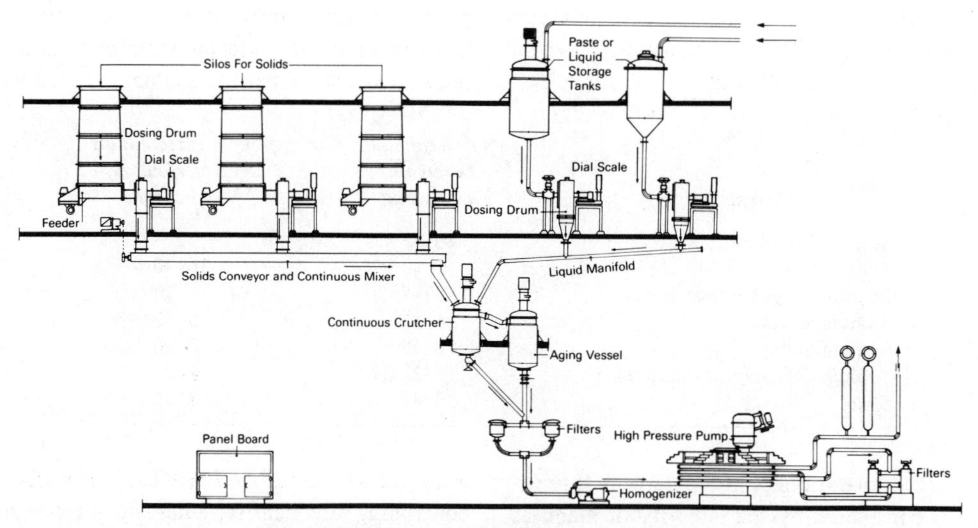

Fig. 13.11 Continuous slurry preparation system.

10 to 35 feet in diameter, and range from 40 tc 200 feet in height. A majority of detergent spray towers use counter-current air flow patterns, although some are co-current in design. Some have the capability of either mode of operation as described in Figure 13.12. In general, a tower of counter current design will produce a more dense particle. In this system, the drying gas stream (250–270°C) is uniformly dispersed into the lower section of the tower through a series of plenum chambers or baffles equally spaced around the periphery of the unit.

The hot gasses are cooled as they are exhausted from the top of the tower by the atomized droplets being dried upon contact with the hot air stream. Due to the relatively low temperatures at this point (110-140°C), the particles do not expand to their full capacity as the water is evaporated from the spheres. The resultant bead, therefore, normally has a thick shell, and subsequent increased bulk density. If a similar formulation was dried in a cocurrent spray tower, the atomized particle would be subjected to a much faster drying rate. The initial higher inlet air temperature (250-370°C) would increase the boiling rate and the elasticity of the particle would be expanded, resulting in a thin-walled bead of lighter density. Figure 13.13 represents typical hollow spheres produced by spray drying.

In most spray towers, the air flow and temperature are adjusted at the inlet ports to maintain balanced air/temperature patterns. When the atomized slurry is sprayed into the drying chamber, the air flow is disturbed and becomes very turbulent. This can cause "hot" and "cold" zones in the tower and affect the drying characteristics of the atomized particles if not controlled. For example, it is possible to have both over- and underdried particles discharged at the same time to the product conveyor. Adjustments in the air pattern are necessary to prevent this occurrence. Another effect of this turbulence, as well as the velocity of the atomized particles from the spray nozzles, is that a small portion of the product is impinged upon the tower walls. This "ring" of product, approximately 10–20 feet below the spray nozzles, tends to build up in size and eventually crack or scale off the walls. Sometimes this "ring" will build up to such an extent that several tons of dried product will break loose from the walls and lodge in the product discharge chute at the base of the tower. To prevent this occurrence, the tower walls are cleaned on a nonroutine basis by knocking off ring deposits with air and/or water lances. Some manufacturers also employ a cleaning ring which travels up and down the straight wall section of the spray tower to reduce such build-ups.

This wall residual can be reclaimed for processing by recycling a certain portion of it in dry or wet forms to subsequent slurry mixes. The quantity of scrap added is normally less than 10% of the total batch. Fines from the dust collectors or cyclones can also be added to the virgin mixes in a similar fashion.

After drying in the spray tower, the detergent granules are conveyed to storage silos prior to carton packaging. In some processes, storage of the product is preceeded by fluid bed conditioning. The product from the tower is very warm (75–120°C) and contains free moisture on the surface (1-3%). Depending on the formulation's ingredients, some products will set up to a cake if no effort is made to reduce the product's temperature. In other cases, the free moisture on the surface of the particles will migrate to a "bound-moisture" phase. This also will cause excessive caking and storage or handling problems.

Either before or after conditioning, perfume and other heat-sensitive ingredients (enzymes, surfactants) may be added to the product. This is generally accomplished by metering the additives with the detergent beads in a rotary drum. Post additions are normally no more than 2–15% of the total detergent product. An excellent review of major spray drying patents is provided by Sittig.[58]

Agglomeration Processing

Agglomeration is the technique of formulating a mixture of granular and/or dry powdered nonuniform raw materials into an attractive spherical granular form with enhanced esthetic and flow characteristics. The majority of

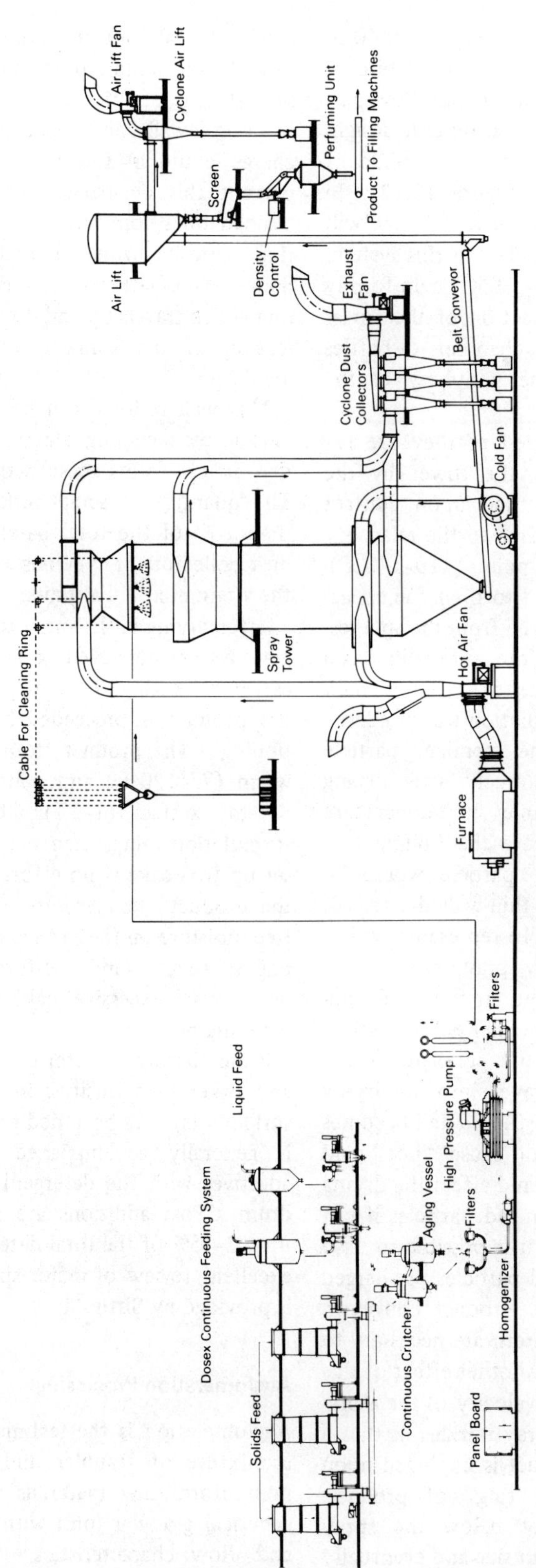

Fig. 13.12 Spray drying plant. (7 p)

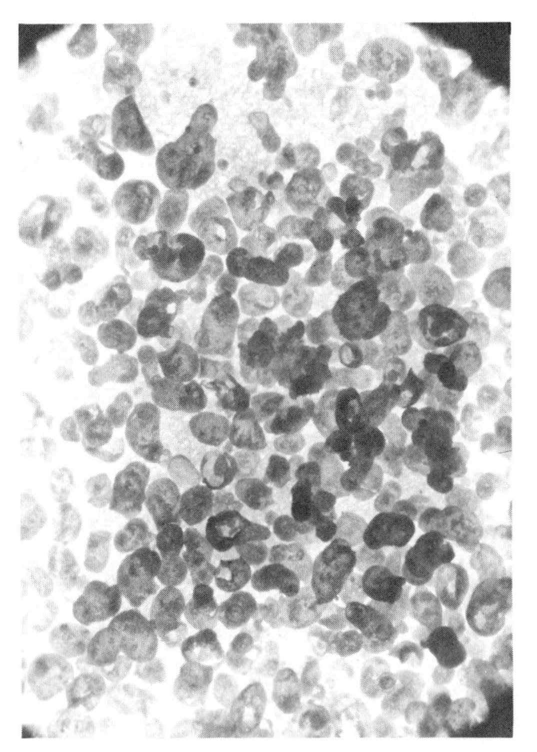

Fig. 13.13 Spray dried detergent bead. (Magnification = 15X).

automatic dishwashing compounds (ADWC's) manufactured today are processed in this manner. With energy costs becoming increasingly important, continued developments are predicted that will broaden the use of this process for heavy-duty laundry detergents. There are numerous types of equipment which are suitable for agglomeration, including horizontal and vertical mixers, orbital screw units, rotary drums, zig-zag mixers, and pan agglomerators. The advantages of agglomeration are basically centered around low capital costs (as compared to spray towers) and energy saving. A disadvantage of agglomeration processing is that the products manufactured in the different processing units vary in particle size uniformity and the bulk density is normally limited to a range of 0.45 gms/cc to 1.0 gm/cc. An additional disadvantage concerns the type of raw materials used in formulating. For example, the quantity of liquids that can be absorbed into the formulation is limited by the absorption characteristics of the dry raw materials.

The typical process steps for formulating a laundry detergent or ADWC product must be predetermined in the laboratory prior to continuous plant processing. In general, the dry raw materials in the plant are preblended or metered into the blender at a predetermined rate. The surfactant portion of the formula is metered onto the dry raw materials. The dry product bed is rolling and/or tumbling to expose a fresh surface area for surfactant absorption. The surfactant should be atomized to provide uniform distribution to the product bed, but a distribution "header" or spray bar may be preferred in some processes. Liquid sodium silicate is normally used to bind and assist in agglomerating the particles. In a rotary agglomerator-conditioner, (see Figure 13.14), the silicate is sprayed onto the product bed simultaneously with the surfactant. At this stage, the product is agglomerated, and physically has a wet or pasty texture. It can easily be compacted into large balls if the quantity of liquids, flow patterns, and drum speed are not controlled.

After agglomeration, the product can be desized to eliminate any lumps which may have formed during processing. The material is then charged to a drying or conditioning drum. The excess moisture is removed from the product and it can then be stored for packaging. Agglomerated formulations sometimes require

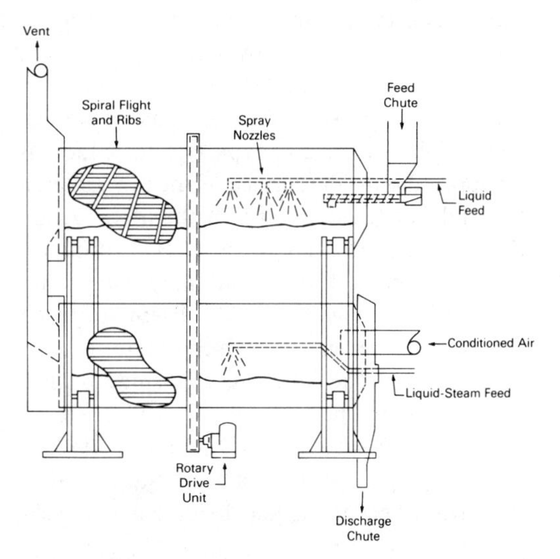

Fig. 13.14 Continuous rotary agglomerator-conditioner.

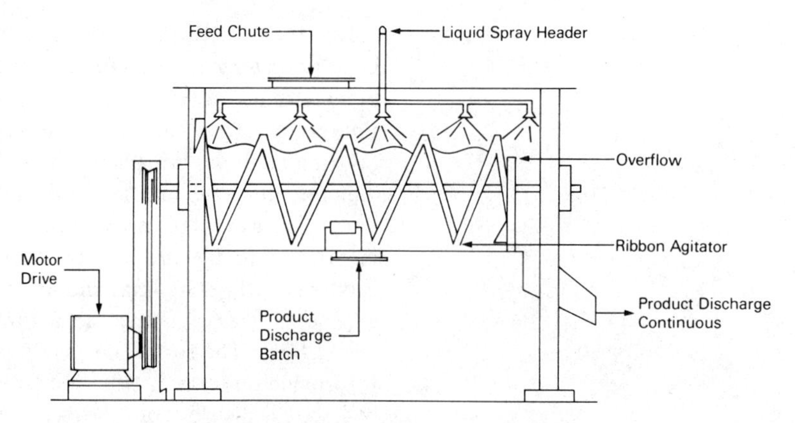

Fig. 13.15 Batch or continuous ribbon mixer.

additional aging. If the formula contains a large quantity of sodium tripolyphosphate (30–55%) and liquid silicate (20–28% as-is basis), additional time is required for the water in the silicate to migrate to the sodium tripolyphosphate and form the hexahydrate. With this type formula, the dried product is sometimes aged from six to twenty-four hours in tote bins to permit the caking associated with the hydration phase to take place. These totes are then "dumped," and the product screened and/or lightly milled prior to further silo storage and packaging. The hydration rate can be accelerated if the product is aged on a conveyor belt or in a rotary drum of limited bed depth for one-fourth to one hour. Other types of agglomeration equipment offer a wide variety of processes as potential alternates to spray drying.

Ribbon mixers, as shown in Figure 13.15, are used primarily for "dry mix" formulating. They are comprised of a "U" shaped trough and a ribbon or paddle-type agitator. Liquids are sprayed onto the surface of the raw material bed and slowly dispersed into the mix. Operation of the mixer can be batch or continuous. Orbital or conical type stationary mixers are also available (Figure 13.16). These are fitted with screw-type agitators which rotate through a static bed. The liquid feed is sprayed onto the surface of the mix and slowly dispersed throughout the total mass of raw materials. Stationary spray systems are generally used in instances where the liquid loading is relatively small.

Rotary mixers are manufactured in a wide variety of sizes and shapes. There are drum type (Figure 13.17), cone or double cone (Figure 13.18), zig-zag configuration (Figure 13.19), vertical in-line mixers (Figure 13.20), and rotary discs or pans (Figure 13.21). The drum-type units generally contain baffles, which "roll" or "lift" the product for uniform dispersion of the liquid feed. The cone or double done type unit is very versatile. Combinations of raw materials can be simultaneously added to the rotary drum in addition to feeding one or more liquid feed streams. Due to its length, it is possible to add and absorb one liquid preferentially, prior to additional liquids

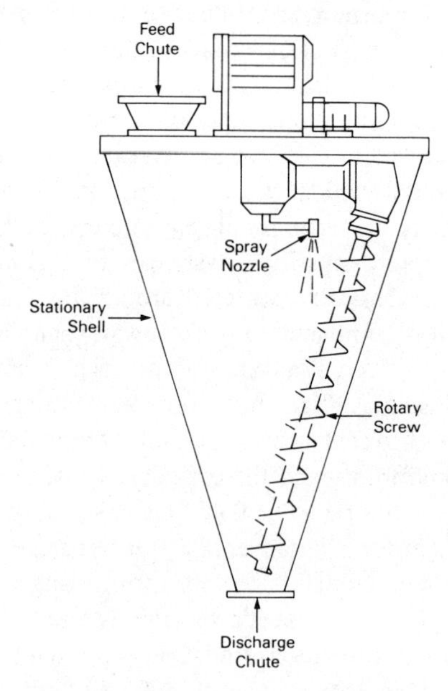

Fig. 13.16 Batch orbital screw mixer.

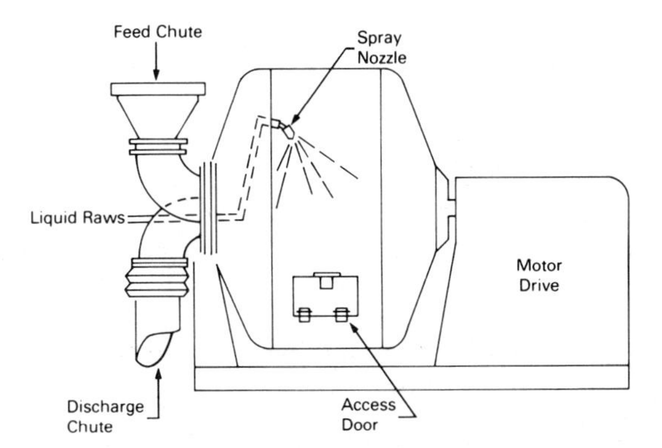

Fig. 13.17 Continuous rotary mixer-agglomerator.

or dry raws. The dry and liquid raws are thoroughly mixed in the stationary zone and receive additional blending and agglomeration in the rotary shell or cone. The vertical mixer utilizes high speed agitation to intimately mix and uniformly blend liquids and dry raws prior to discharge. Contact time is reduced to less than five seconds.

All of these units can be operated in continuous fashion. The rotary mixers, however, exhibit improved agglomeration, increased liquid feed dispersion levels, and optimized product uniformity. In addition to agglomeration of detergents and automatic dishwashing compounds, these units can be used for hydration of sodium tripolyphosphates, dry neutralization of sulfonic acids, post addition of raw materials and fines to spray-dried products, and partial encapsulation.

Unique Processes for Light Density Product

The "fluff" process,[59] developed by Monsanto Company, has the flexibility to produce granulated detergent products within a density range of 0.25 to 0.9 gms/cc. Sodium trimetaphosphate ($Na_3P_3O_9$) is used as the builder salt in the formulation. The phosphate is mixed into a slurry (45-75% solids) containing the other detergent ingredients. After heating the slurry to approximately 60-70°C, a sufficient quantity of caustic solution is added to convert the trimetaphosphate to sodium tripolyphosphate hexahydrate. During conversion of the phosphate salt, an exothermic reaction takes place which causes the slurry to expand to several times its original volume. This reaction produces light fluffy, semi-dry granules. The granules are further dried to remove free

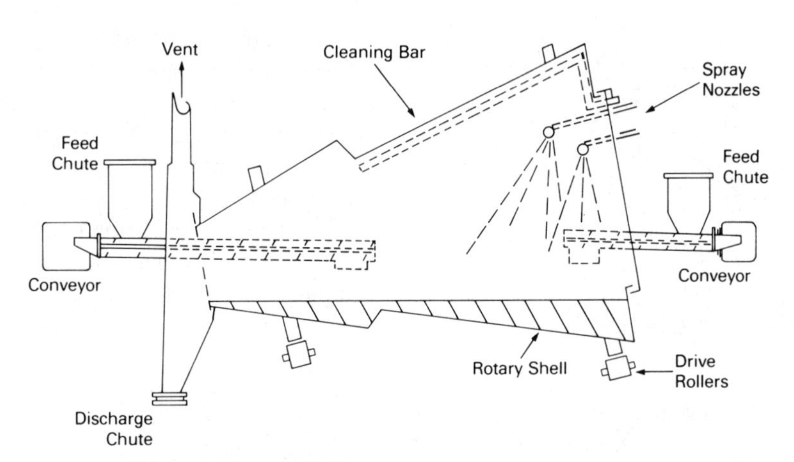

Fig. 13.18 Continuous two-stage double cone agglomerator.

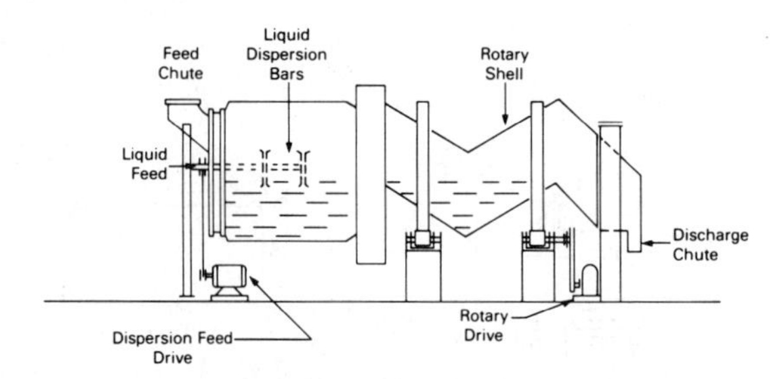

Fig. 13.19 Continuous blender-agglomerator.

moisture prior to packaging. Product density can be varied dependent upon the formulation ingredients, slurry solids levels, and process temperatures.

Under controlled conditions, sodium trimetaphosphate can be processed in spray-drying equipment. This procedure assures conversion of the phosphate salts to the hexahydrate form and maximizes the moisture content available from the phosphate salts. The resultant advantages include less severe product caking during silo storage, packaging, and product shelf life.

Other processes for the manfuacture of light density detergents have been patented by the Colgate-Palmolive Company.[60,61] These methods involve heating a detergent slurry to approximately 50°C to improve fluidity and flowability. Then a substantially nonreactive compound is added which releases carbon dioxide when heated further. An oxygen-liberating compound can also be employed. A sufficient quantity is used to form a paste capable of retaining small oxygen bubbles. When heated, the oxygen is liberated and expands the paste at least two times the original volume.

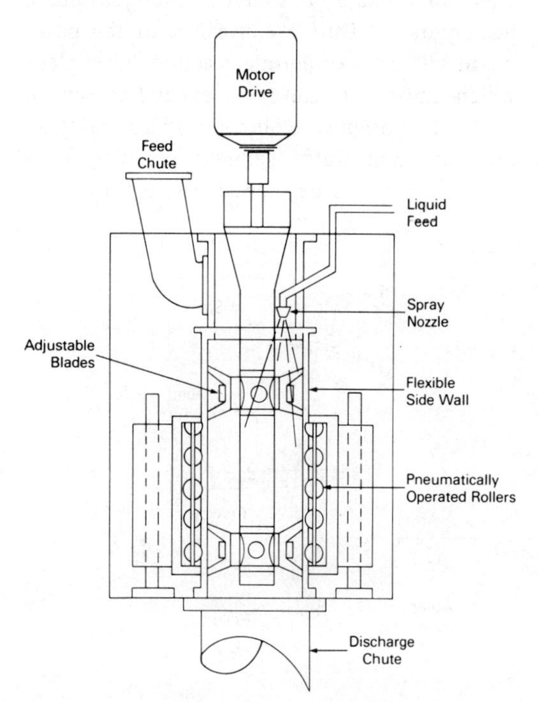

Fig. 13.20 Continuous in-line vertical agglomerator.

Liquid Detergent Processing

Liquid heavy-duty laundry products account for approximately 18% of the total heavy-duty laundry detergent market. The largest growth for heavy-duty liquids started in 1974 as a result of legislation banning phosphates. Unbuilt liquids, based on nonionic or nonionic/anionic active combinations, increased in volume due to improved performance of these type of active systems in the absence of phosphates. This is particlarly true in hard water areas.

These products are approximately 50-60% water, the remainder being a combination of surfactants, builders, foam regulators, solubilizers, antiredeposition agents, whitening and blueing agents, corrosion inhibitors, colorants, and perfume. Several formulations also contain antistatic and softening ingredients.

Liquid systems require careful blending of raw materials to produce a stable product. Special attention is necessary for the following items:

Viscosity. The product must be pourable and maintain the same viscosity characteristics from

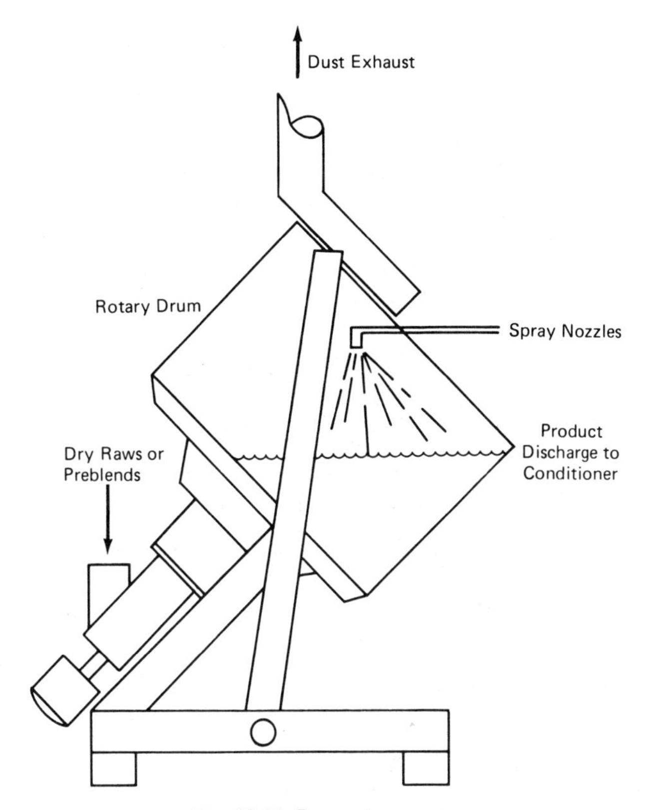

Fig. 13.21 Pan agglomerator.

batch to batch. This can be controlled by proper solvent or hydrotrope selection.

Clear-Cloud Point. The composition should have sufficient solubility to prevent "hazing" when subjected to storage in cool or cold environments.

Freeze-Thaw Stability. The formula must be compounded to prevent phase separation or solidification at freezing temperatures.

Light-duty liquid detergents are generally formulated with LAS (linear alkylaryl sulfonate). Heavy-duty liquid detergents generally use a combination of several actives; i.e., LAS, sodium alcohol ether sulfate, alcohol ethoxylates, and olefin sulfonates.

Manufacture of light- and heavy-duty formulations requires relatively basic equipment for blending of the various ingredients. The raw materials are generally added batchwise to a large mixing vessel and fed directly from the mixer to the filling lines. The filling line is a composite of stations, where the bottles are volumetrically filled, capped, weighed, labelled, and packaged into cartons. Due to the differences in physical and functional characteristics of various products, it is essential to thoroughly clean the system between production of various products.

Detergent Trends

To date, no new detergent builders have been developed which have the functional characteristics and cost advantages of phosphates. A variety of builders (previously discussed) are presently being formulated in nonphosphate legislated detergent products, but these salts are not as effective as phosphates. This performance differential is being compensated, to some extent, by the uses of increased synthetic surfactant levels in both liquid and powdered products. Zeolite builders have become a factor in formulating low phosphate and phosphate-free detergent formulations.

Future energy considerations may also hasten the change of detergent manufacturers from spray-dried products to agglomerated laundry detergents. This will dictate a denser product with improved solubility characteristics. Dense products will permit reduced use levels, for example $\frac{1}{4}$ to $\frac{1}{2}$ cup per load for laundry detergent, and necessitate new packaging concepts. Continued emphasis and growth will also be experienced in the use of liquid detergent formulations. It is projected that as the family unit decreases in size through the year 2000, washing appliances also will be of lower "load" capacity. In addition, it is probable that the water consumption of appliances and wash water temperatures will also be reduced. These changes will present a significant challenge to the detergent industry during the next decade.

REFERENCES

1. H. R. Galleymore, *Soap and Chem. Spec.*, **39**, No. 12, 83 (1963).
2. J. W. McBain and W. W. Lee, *Oil Soap*, **20**, 17 (1943).
3. J. W. McBain, R. D. Vold, and K. Gardiner, *Oil Soap*, **20**, 221 (1943).
4. E. T. Webb, *Modern Soap and Glycerin Manufacture*, Davis Brothers, London, 1927.
5. J. W. McBain and W. W. Lee, *Ind. Eng. Chem.*, **35**, 917 (1943).
6. F. Lachampt and R. Perron, in *The Chemistry of Fats and Other Lipids*, Vol. **5**, Pergamon Press, Inc., New York, 1958, p. 31.
7. A. L. Schulerud, *JAOCS*, **40**, No. 10, 609 (1963).
8. F. T. E. Palmquist and F. E. Sullivan, *JAOCS*, **36**, 173 (1959).
9. L. D. Jones, *JAOCS*, **35**, 630 (1958).
10. A. T. Scott (to Sharpless Corp.), U.S. Patents 2,300,749-51 (1942); 2,336,893 (1943).
11. V. Mills (to Procter and Gamble Co.), U.S. Patents 2,156,863 (1939); 2,159,397 (1939); 2,274,801-2 (1942).
12. Anon., *Soap, Perf., Cosmet.*, **20**, 1090 (1947).
13. G. W. McBride, *Chem. Eng.*, **54**, 94 (1947).
14. Anon., *Soap Chem. Spec.*, **40**, 67 (1964).
15. H. W. Ladyn, *Chem. Eng.*, **71**, 106 (1964).
16. E. Jungermann, *JAOCS*, **50**, 475 (1973).
17. L. Spitz, *JAOCS*, **45**, 423 (1967) and *JAOCS*, **50**, 151 (1978).
18. A. B. Herrick, *JAOCS*, **55**, 147 (1978).
19. E. Twitchell, U.S. Patent 601,603 (1898).
20. A. B. Herrick and E. Jungermann, *JAOCS*, **40**, 615 (1963).
21. Anon., *Ind. Eng. Chem.*, **39**, 126 (1947).
22. W. Stein and Co-workers (to Henkel & Cie, G.m.b.H.), U.S. Patent 2,800,493 (1957).
23. E. Jungermann, *JAOCS*, **56**, 827A (1979).
24. G. R. Platt, *Soap, Perf Cosmet.*, **43**, 233 (1970).
25. R. D. Swisher, "Surfactant Biodegradation," *Surfactant Science Series*, Vol. **3**, Marcel Dekker, New York (1970).
26. E. A. Matzner, M. M. Crutchfield, R. P. Langguth, and R. D. Swisher, *Tenside*, Vol. **10**, Nos. 3 and 5, p. 119-25 and 239-45 (1973).
27. R. J. Brooks, et al., U.S. Patent 3,257,715, June 21, 1966.
28. R. J. Brooks, et al., U.S. Patent 3,259,645, July 5, 1966.
29. M. Ballestra, U.S. Patent 3,884,643, May 20, 1975.
30. R. J. Brooks, et al., U.S. Patent 4,113,438, September 12, 1978.
31. P. O. Shull, et al., U.S. Patent 3,277,145, October 4, 1966.
32. M. Ballestra, U.S. Patent 3,180,699, April 27, 1965.
33. C. A. Sumner, U.S. Patent 3,597,361, August 3, 1971.
34. R. Frank, et al., U.S. Patent 3,867,316, February 18, 1975.
35. R. Williams, *Surfactants-1980*, Technical and Economic Services, Inc.
36. A. M. Swartz and J. W. Perry, *Surface Active Agents*, Vol. **1**, John Wiley and Sons, New York, (1949).
37. E. Jungermann, *Cationic Surfactants*, Marcel Dekker, Inc., New York (1970).
38. French Patent 664,261.
39. M. J. Shick, *Nonionic Surfactants*, Vol. **I**, Marcel Dekker, Inc., New York (1967).
40. E. Jungermann, *Cationic Surfactants*, Chapter 9, Marcel Dekker, Inc., New York (1967).
41. Dow Chemical Company, "Alkylene Oxides," *Technical Bulletin 125-276-60*, Midland, Michigan (1960).
42. L. G. Hess and V. V. Tilton, *Ind. Eng. Chem.*, **42**, 1251 (1950).

43. W. B. Satkowski, S. K. Huang, and R. L. Liss, *Nonionic Surfactants*, Vol. 1, Chapter 5, Marcel Dekker, New York (1967).
44. Emergy Industries, Inc., "Rewopon® AM-21," *Product Bulletin*, January, 1979.
45. McCutcheon's, *Detergents and Emulsifiers*, North American Edition, MC Publishing Company, New Jersey (1979).
46. VanWazer, *Phosphorus And Its Compounds*, Vol. II, John Wiley and Sons, New York (1961).
47. P. A. Brochert and R. A. Marino, *Soap/Cosmetics/Chem. Specialities*, Vol. I, January, 1980.
48. A. C. Savitsky, "Type A Zeolites as a Laundry Builder," *Household and Personal Products Ind.*, Vol. 14, No. 3, p. 52 (March, 1977).
49. M. E. Sorgenfrei, "Increased Use of STPP," *Soap/Cosmetics/Chem. Specialties*, Vol. 3, No. 2, p. 60 (February, 1980).
50. H. E. Feierstein, et al., U.S. Patent 3,717,589, February 20, 1973.
51. M. M. Crutchfield, et al., U.S. Patent 3,865,755, February, 1975.
52. M. E. Tuvell, "Sodium CMOS Properties and Performance," *Household and Personal Products Ind.*, Vol. 15, No. 4, p. 45 (April, 1978).
53. E. J. Singer, "CMOS Evaluation of Safety for the Environment," *Household and Personal Products Ind.*, Vol. 15, No. 8, p. 56 (August, 1978).
54. S. G. Clark et al., U.S. Patent 3,573,170, March 30, 1971.
55. H. E. Feierstein et al., U.S. Patent 3,594,325, July 20, 1971.
56. S. G. Clark et al., U.S. Patent 3,706,674, December 19, 1972.
57. J. A. Sagel et al., U.S. Patent 3,932,316, January 13, 1976.
58. M. Sittig, "Detergent Mfg. Including Zeolite Builders," *Chem. Tech. Review*, No. 128, Noyes Data Corporation, New Jersey (1979).
59. H. E. Feierstein et al., U.S. Patent 3,390,093, June 25, 1968.
60. B. B. Dugan et al., U.S. Patent 3,177,147, April 6, 1965.
61. L. Habicht et al., U.S. Patent 3,202,613, August 24, 1965.

General References

1. Soap and Synthetic Detergents by J. C. Harris; Chapter 13 in *Riegel's Handbook of Industrial Chemistry*, 7th edition, J. A. Kent, Editor.
2. *Bailey's Industrial Oil and Fat Products*, 3rd edition, D. Swern, Editor, Interscience Publishers, New York (1964).
3. E. G. Thomssen and J. W. McCutcheon, *Soaps and Detergents*, MacNair-Dorland Co., New York, 1949.

14

Petroleum and Its Products

H. L. Hoffman*

THE NATURE OF PETROLEUM

Petroleum is a mixture of hydrocarbons—chemical combinations of hydrogen and carbon. When burned completely, the hydrocarbons should yield only water (H_2O) and carbon dioxide (CO_2). When the burning is incomplete, carbon monoxide (CO) and various oxygenated hydrocarbons are formed. Since most burning uses air, nitrogen compounds also exist. In addition, there are other elements associated with the hydrocarbons in petroleum such as sulfur, nickel, and vanadium, just to name a few.

Petroleum is found normally at great depth underground or below seabeds. It can exist as a gas, liquid, solid, or a combination of these three states. Drilling is used to reach the gaseous and liquid deposits of petroleum. Then they are brought to the surface through pipe. The gas usually flows under its own pressure. The liquid may flow from its own pressure or be forced to the surface by submerged pumps. Solid or semisolid petroleum is brought to the surface in a number of ways: by digging with conventional mining techniques, by gasifying or liquefying with high temperature steam, or by burning a portion of the material in the ground so that the remainder can flow to the surface.

Natural gas is the gaseous form of petroleum. It is mostly the single-carbon molecule, methane (CH_4). When natural gas is associated with liquid petroleum underground, the methane will come to the surface in admixture with some heavier hydrocarbons. The gas is then said to be a wet gas. These heavier hydrocarbons are isolated and purified in natural gas processing plants. The operation yields ethane (petrochemical feed), propane (LPG), butane (refinery blending stock) and hydrocarbon liquids (natural gas condensate).

When the underground natural gas is associated with solid hydrocarbons like tar or coal, the methane will have little other hydrocarbons. Then the gas is said to be a dry gas.

Crude oil is the common name given to the liquid form of petroleum. In some writings, one will see reference to "petroleum and natural gas" suggesting petroleum and crude oil

*Hydrocarbon Processing, Houston, Texas.

are used as synonymous terms. Some crude oils have such great density that they are referred to as heavy oils and tars.

Tar sands are small particles of sandstone surrounded by an organic material called bitumen. The bitumen is so highly viscous and clings so tenaciously to the sandstone that it is easy to think of the mixture as a solid form of petroleum. Yet it is a mixture of high-density liquid on a supporting solid.

Oil shales are real petroleum solids. The curious thing about oil shales is that they do not contain petroleum crude oil. Instead, they contain an organic material called kerogen. The kerogen can be heated to yield a liquid substance called shale oil, which in turn can be refined into more conventional petroleum products.

Many products derived from petroleum are partly the consequence of the vast collection of hydrocarbons occurring in petroleum's natural state. A far more important factor is the ability of the hydrocarbon processing industries to transpose laboratory discoveries into large scale commercial operations. Thus, the petroleum industry is an interesting study in applied organic chemistry and physical property manipulation.

Most of this discussion will concern the processing of petroleum crude oil, the most widely used form of petroleum resources. Natural gas processing will come in briefly at a few points. And since most of the world's petroleum is consumed as energy fuels, it is appropriate to begin with a brief review of the world's total energy situation.

Largest Energy Supplier

In an earlier chapter entitled Coal Technology, the point is made that coal offers a much more abundant primary source of energy than does petroleum. This is certainly true, but another fact remains: the world presently gets most of its energy from crude oil and natural gas. Petroleum is the major source of fuel used in transportation, manufacturing, and home heating.

Primary energy sources are defined as those coming from natural raw materials. The primary energy sources for the world during 1979 are reported[1] in Fig. 14.1. Oil and gas together furnished 65% of the total world energy usage for that year.

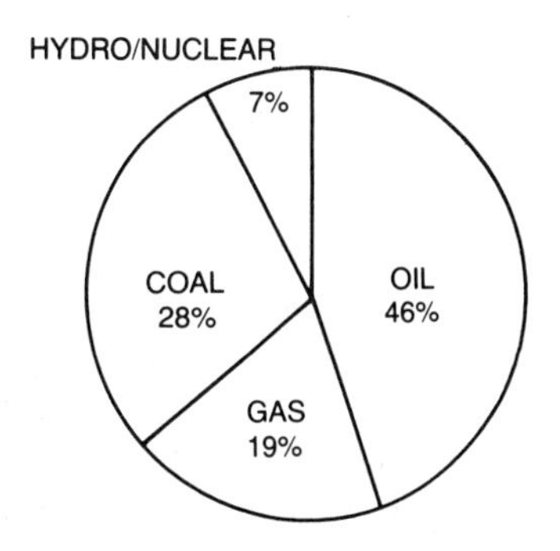

Fig. 14.1 World primary energy consumption, 1979. World total equals 140×10^6 daily barrels oil equivalent.

Note electricity is missing from this representation. Electricity is a secondary energy source because it is generated by consuming some of the other natural resources shown in Fig. 14.1. Thus, electricity should not appear in an energy balance unless the fuel from which it is generated is omitted from the other totals. Too often, this correction is not made and an inflated energy supply results.

Proven crude oil reserves determined for several earlier years[2] are shown in Table 14.1.

TABLE 14.1 Proven Crude Oil Reserves Compared With Production Rates (10^9 barrels)

Year	*Proven reserves, first of yr*[2]	*Annual production*[3]	*Ratio, reserves divided by production, years*
1935	23.4	1.65	14.2
1940	34.9	2.15	16.2
1945	51.2	2.59	19.8
1950	76.4	3.80	20.1
1955	153.7	5.63	27.3
1960	255.8	7.67	33.4
1965	341.7	11.06	30.9
1970	516.7	16.69	31.0
1975	569.1	19.50	29.2
1980	625.8	22*	28.4*

*Author's estimate

Also included in this table are the annual production rates[3] and the ratio of the reserves to annual production.

The ratio of reserves to production is a way of gauging natural reserves in terms of the years the proven reserves would sustain current production rates. Yet, if this measurement is taken literally, one might wonder how the petroleum industry grew and prospered during periods when there were less than 20 years of reserves. An explanation lies in the fact that the exploration and production segments of the petroleum industry seek and find oil sources to replace depleting ones. Note, for example, how the ratio jumped during the period between 1950 and 1955 as a result of the big oil reserves discovered in Middle Eastern countries.

What is disturbing at present is the fact that new oil sources are not being found fast enough to keep up with rapidly growing consumption rates. Furthermore, when oil is found, it is generally at greater depths or under seabeds farther from shore than was the case before. Reports proliferate now regarding the ultimate reserves likely to be found in the world.

Ultimate recoverable petroleum reserves have been estimated with wide variations, the consensus being between 1.5 to 2.5 trillion barrels (10^{12} bbl). Most estimators go to great length to explain the basis of their estimates. By comparison, proven reserves are in the neighborhood of 600 billion barrels. Proven reserves are generally taken to mean: "the oil remaining in the ground which geological and engineering information indicate with reasonable certainty to be recoverable in the future from known reservoirs under existing economic and operating conditions."[2]

Alternate feedstocks for refineries are being sought from coal, tar sands, and shale oil. Synthetic crude oil will be made from these raw materials so that conventional refining units can continue to be used to make consumer products.

In the meantime, top priority will be given to use crude oil to make liquid transportation fuels (because of their convenience) and petrochemical materials (because of their diversity of uses). It's a good bet that many of the gross conversion methods now applied to crude oil processing will be replaced in the future by more specific conversion. It is in this transition that knowledge of organic chemistry guides the development of new refining processes.

From Well to Refinery

A country-by-country listing[1] of crude oil production and consumption shows the importance of petroleum movement around the world. Production and consumption rates of crude oil for various countries are shown in Table 14.2. The movement of crude oil by seas from countries with high production rates to countries with high consumption rates is depicted in Fig. 14.2. The width of the lines represents the volume; the lines themselves show the origin and destination of oil flow, but not necessarily the specific routes followed.

The growth of world refining capacity attempts to keep up with the growing demand for petroleum products. A measure of this growth[1] is shown in Fig. 14.3. The upper curve shows total refining capacity, while the lower curve shows the amount of crude oil run through the refineries.

One might wonder why refining capacity continued to surge ahead when crude throughput took a dip in the period from 1974 through 1976. For one thing, the amount of crude oil available for processing is subject to the whims of international trade. Since so much crude oil comes from some countries (notably, Middle Eastern countries) to be refined in other countries, international relations between countries

TABLE 14.2 Distribution of Production and Consumption, 1979[1] (10^6 barrels daily)

Country/Area	*Produced*	*Consumed*
United States	10.2	17.9
Canada	1.8	1.9
Other W. Hemisphere	5.5	4.4
W. Europe	2.4	14.9
Africa	6.7	1.3
Middle East	21.8	1.5
Asia	2.4	3.1
Japan	0.0	5.5
Australasia	0.5	0.8
USSR, E. Europe, China	14.4	12.8
World total	65.7	64.1

Fig. 14.2 Major oil movement by sea, 1979. (*"BP Statistical Review of the World Oil Industry, 1979", The British Petroleum Co. Ltd.*)

is a strong factor that determines how much crude oil feedstock is available.

Another factor in refining growth is the time required to construct processing units. In highly industrialized countries like the United States, Japan, and Western European countries there are mounting restrictions on new refinery sites. Thus, a decision to build a refinery and the actual completion of that refining capacity will take several years in order to fulfill local and governmental requirements. Then one to four years of actual construction activities are required before a new refinery will start processing feedstocks.

80
60
40
20
0
'65 '70 '75 '80
CAPACITY
THRUPUT
10^6 DAILY BARRELS

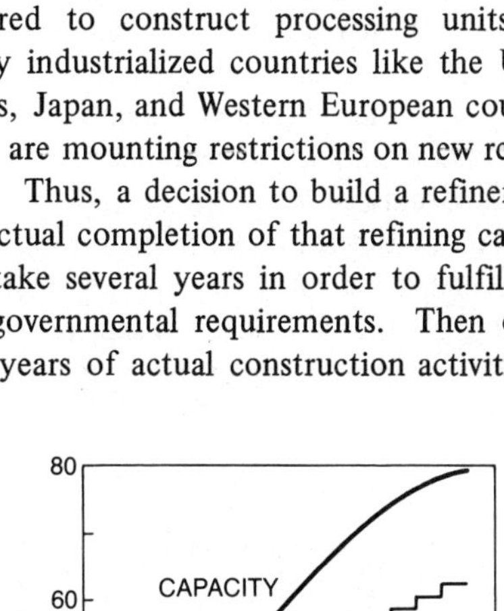

Fig. 14.3 World refining capacity and crude oil throughputs.

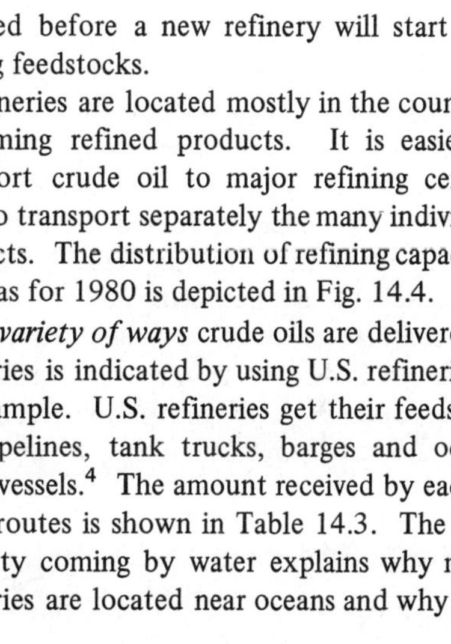

Refineries are located mostly in the countries consuming refined products. It is easier to transport crude oil to major refining centers than to transport separately the many individual products. The distribution of refining capacity[1] by areas for 1980 is depicted in Fig. 14.4.

The variety of ways crude oils are delivered to refineries is indicated by using U.S. refineries as an example. U.S. refineries get their feedstock via pipelines, tank trucks, barges and ocean-going vessels.[4] The amount received by each of these routes is shown in Table 14.3. The large quantity coming by water explains why many refineries are located near oceans and why they

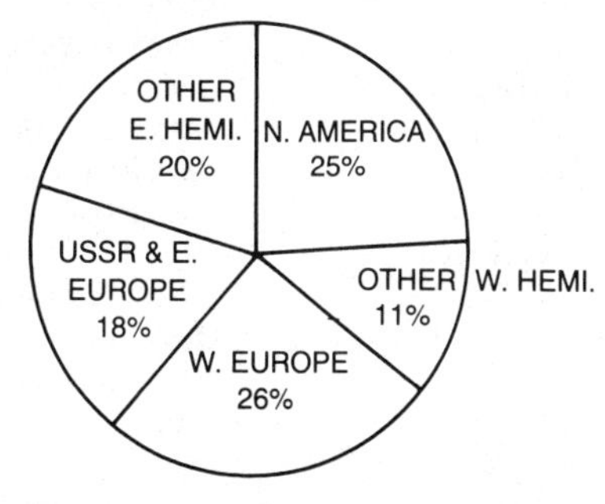

WORLD TOTAL = 80.37 × 10^6 DAILY BARRELS

Fig. 14.4 World refining capacity in early 1980.

TABLE 14.3 Method of Transportation for Crude Oil Received by U.S. Refineries, 1979[4]

Transportation method	*Vol %*
Domestic crude oil	
Pipelines	42.9
Tank cars and trucks	2.1
Tankers and barges	11.0
Subtotal	56.0
Foreign crude oil	
Pipelines	12.7
Tankers and barges	31.3
Subtotal	44.0
Total receipts	100.0

own or lease such a large fleet of barges and ocean-going vessels.

PRODUCT NAMES

The distinction between refined products and petrochemicals is often a subtle one. In general, when the product is a fraction from crude oil that includes a fairly large group of hydrocarbons, the fraction is classified as a refined product. Examples of refined products are: gasoline, diesel fuel, heating oils, lubricants, waxes, asphalts, and petroleum coke.

By contrast, when the product from crude oil is limited to only one or two specific hydrocarbons of fairly high purity, the fraction is called a petrochemical. Examples of petrochemicals are: ethylene, propylene, benzene, toluene, and styrene—to name only a few.

There are many more identifiable petrochemical products than there are refined products. There are many specific hydrocarbons that can be derived from petroleum. However, these hydrocarbons lose individual identity when they are grouped into a refined product.

Refined Products

Most refined products at the consumer level are blends of several refinery streams. Product specifications determine which streams are suitable for a specific blend. Part of the difficulty of learning about refining lies in the industry's use of stream names that are different from the names of the consumer products.

Consider the listing in Table 14.4. The names in the last column should be familiar because they are used at the consumer level. Yet within a refinery, these products will be blended from portions of crude oil fractions having the names shown in the first column. To make matters worse, specifications and statistics for the industry are often reported under yet another set of names—those shown in the middle column of Table 14.4.

Gasoline at the consumer level, for example, may be called benzol or petrol, depending on the country where it is sold. In the early stages of crude oil processing, most gasoline com-

TABLE 14.4 Several Names for the Same Material

Crude oil cuts	*Refinery blends*	*Consumer products*
Gases	Still Gases	Fuel gas
	Propane/Butane	Liquefied petroleum gas (LPG)
Light/Heavy	Motor Fuel	Gasoline
naphtha	Aviation turbine, Jet-B	Jet fuel (naphtha type)
Kerosine	Aviation turbine, Jet-A	Jet fuel (kerosine type)
	No. 1 Fuel oil	Kerosine (range oil)
Light gas oil	Diesel	Auto and tractor diesel
	No. 2 fuel oil	Home heating oil
Heavy gas oil	No. 4 fuel oil	Commercial heating oil
	No. 5 fuel oil	Industrial heating oil
	Bright stock	Lubricants
Residuals	No. 6 fuel oil	Bunker C oil
	Heavy residual	Asphalt
	Coke	Coke

ponents are called naphthas. Kerosine, is another example. It may be called coal oil to denote that it replaces stove oil (or range oil) once derived from coal. Kerosine's historical significance was first as an illuminating oil for lamps that once burned sperm oil taken from whales. But today, kerosine fractions go mostly into transportation fuels such as jet fuel and high quality No. 1 heating oil.

Product Specifications

Product application and customer acceptance set detailed specifications for various product properties. In the United States, the American Society for Testing and Materials (ASTM) and the American Petroleum Institute (API) are recognized for establishing specifications on both products and methods for testing. Other countries have similar referee organizations. For example, in the United Kingdom, it is the Institute of Petroleum (IP). In West Germany, it is Deutsches Institute suer Normung (DIN). In Japan, it is the Ministry of International Trade and Industry (MITI).

Boiling range is the major distinction among refined products, and many other properties are related directly to the products in these boiling ranges. A summary of ASTM specifications for fuel boiling ranges[5] is given in Table 14.5.

Boiling range also is used to identify individual refinery streams—as an example will show in a later section concerning crude oil distillation. The temperature that separates one fraction from an adjacent fraction will differ from refinery to refinery. Factors influencing the choice of cut point temperatures includes the following: type of crude oil feed, kind and size of downstream processes, and relative market demand among products.

Other specifications can involve either physical or chemical properties. Generally these specifications are stated as minimum or maximum quantities. Once a product qualifies to be in a certain group, it may receive a premium price by virtue of exceeding minimum specifications or by being below maximum specifications. Yet all too often, the only advantage for being better than specifications is an increase in the volume of sales in a competitive market.

The evolution of product specifications will, at times, appear sadly behind recent developments in more sophisticated analytical techniques. Certainly the ultimate specifications should be based on how well a product performs in use. Yet the industry has grown comfortable with certain comparisons, and these standards are retained for easier comparison with earlier products. Thus, it is not uncommon to find petroleum products sold under an array of tests and specifications—some seemingly measuring similar properties.

It is behind the scenes that sophisticated analytical techniques prove their worth. These techniques are used to identify specific hydrocarbons responsible for one property or another. Then suitable refining processes are devised to accomplish a desired chemical reaction that will increase the production of specific types of hydrocarbons.

In the discussion on refining schemes, major specifications will be identified for each product category. It will be left to the reader to remember that a wide variety of other specifications also must be met.

Product Yields

As changes occur in relative demand for refined products, refiners turn their attention to ways that will alter internal refinery streams. The big problem here is that the increase in volume of one fraction of crude oil will deprive some other product of that same fraction. This point is often overlooked when the question arises: "How much of a specific product can a refinery make?" Such a question should always be followed by a second question: "What other products will be penalized?"

Envision, for example, what would happen if the refining industry were to make all the gasoline it possibly could with today's present technology. The result would be to rob many other petroleum products. A vehicle which needs gasoline for fuel also needs such products as industrial fuels to fabricate the vehicle, lubricants for the engine's operation, asphalt for roads upon which the vehicle is to move,

TABLE 14.5 Major Petroleum Products and Their Specified Boiling Range[5]

Product designation	*ASTM designation*	*Specified temp. for vol % distilled at 1 atm., °F* 10%	50%	90%
Liquefied petroleum gas (LPG)	D 1835			
Commercial propane		–[a]		–[b]
Commercial butane		–[a]		–[c]
Aviation gasoline (Avgas)	D 910	158 max	221 max	275 max[d]
Automotive gasoline	D 439			
Volatility class A		158 max	170–250	374 max[e]
Volatility class B		149 max	170–245	374 max[e]
Volatility class C		140 max	170–240	365 max[e]
Volatility class D		131 max	170–235	365 max[e]
Volatility class E		122 max	170–230	365 max[e]
Aviation turbine fuel	D 1655			
Jet A or A-1		400 max		–[f]
Jet B		–[g]	370 max	470 max
Diesel fuel oil	D 975			
Grade 1-D				550 max
Grade 2-D				540–640
Grade 4-D		–	not specified	–
Gas turbine fuel oil	D 2880			
No. 0-GT		–	not specified	–
No. 1-GT				550 max
No. 2-GT				540–640
No. 3-GT		–	not specified	–
No. 4-GT		–	not specified	–
Fuel oil	D 396			
Grade No. 1		420 max		550 max
Grade No. 2		–[h]		540–640
Grade No. 4		–	not specified	–
Grade No. 5		–	not specified	–
Grade No. 6		–	not specified	–

[a]vapor pressure specified instead of front end distillation
[b]95% point, –37°F max
[c]95% point, 36°F max
[d]final point, 338°F max
[e]final point, all classes, 437°F max
[f]final point, 572°F max
[g]20% point, 290°F max
[h]flash point specified instead of front end distillation

and petrochemical plastics and fibers for the vehicle's interior. Until adequate substitutes are found for these other petroleum products, it would be unwise to make only one product, even though sufficient technology may exist to offer this option.

This is not to say that substitutes will not be found nor that these substitutes will not be better than petroleum products. In fact, many forecasts suggest that petroleum will ultimately be allocated only to transportation fuels and petrochemical feedstocks. It appears that these uses are the most suitable options for petroleum crude oil.

In the United States, the relative portions of refined products[4] made from crude oil are shown in Table 14.6. This distribution of products is the result of a long standing trend to convert the heavier less valuable fractions into lighter more valuable fractions. The ways this can be done will be discussed in the section on refinery schemes.

TABLE 14.6 Product Yields from U.S. Refineries, 1979 Preliminary[4]

Product	*Vol % of refinery input*
Still gas	3.8
Ethane/ethylene	0.1
Liquefied gas	2.3
Gasoline	42.4
Jet fuel	6.9
Kerosine	1.3
Special naphtha	0.7
Petrochemical feed	4.8
Distillates	21.7
Lubricants	1.3
Waxes	0.1
Coke	2.6
Asphalt	3.1
Road oil	0.1
Residuals	11.6
Miscellaneous	0.8
Total	103.6*

*Includes 3.6 vol % gain because most products are of less density than original feedstock.

Petrochemicals

The portion of crude oil going to petrochemicals may appear small compared to fuels, but the variety of petrochemicals is huge. The listing in Table 14.7 will give some idea of the range of petrochemical applications. Many of these products are described in the Chapter 25, *Synthetic Organic Chemicals.* A few will be included here as they come into competition with the manufacture of fuels.

Despite their variety, all commercially manufactured petrochemicals account for the consumption of only a small part of the total crude oil processed. In the United States, where petrochemicals have grown swiftly, the total petrochemical output at the beginning of 1980 was little more than $6\frac{1}{2}$ volume percent (vol %) of all petroleum feedstocks. An estimated $2\frac{1}{2}$ vol % went to energy of conversion leaving 4 vol % represented as petrochemical products. Even so, this quantity has been sufficient for petrochemical-based materials to replace many products once made from such raw materials as coal, lumber, metal ores, and so forth.

REFINING SCHEMES

A refinery is a massive network of vessels, equipment, and pipes. The total scheme can be divided into a number of unit processes. In the discussion to follow, only major flow streams will be shown, and each unit will be depicted by a single block on a simplified flow diagram. Details will be discussed later.

Refined products establish the order in which each refining unit will be introduced. Only one or two key product specifications are used to explain the purpose of each unit. Nevertheless,

TABLE 14.7 Petrochemical Applications

Absorbents	De-emulsifiers	Hair conditioners	Pipe
Activators	Desiccants	Heat transfer fluids	Plasticizers
Adhesives	Detergents	Herbicides	Preservatives
Adsorbents	Drugs	Hoses	Refrigerants
Analgesics	Drying oils	Humectants	Resins
Anesthetics	Dyes	Inks	Rigid foams
Antifreezes	Elastomers	Insecticides	Rust inhib.
Antiknocks	Emulsifiers	Insulations	Safety glass
Beltings	Explosives	Lacquers	Scavengers
Biocides	Fertilizers	Laxatives	Stabilizers
Bleaches	Fibers	Odorants	Soldering flux
Catalysts	Films	Oxidation inhib.	Solvents
Chelating agents	Finish removers	Packagings	Surfactants
Cleaners	Fire-proofers	Paints	Sweeteners
Coatings	Flavors	Paper sizings	Synthetic rubber
Containers	Food supplements	Perfumes	Textile sizings
Corrosion inhib.	Fumigants	Pesticides	Tire cord
Cosmetics	Fungacides	Pharmaceuticals	
Cushions	Gaskets	Photographic chem.	

the reader is reminded that the choice from among several types of units and the size of these units are complicated economic decisions. The trade-off among product types, quantity, and quality will be mentioned to the extent that they influence the choice of one kind of process unit over another.

Feedstock Identification

Each refinery has its own range of preferred crude oil feedstock for which a desired distribution of products is obtained. The crude oil usually is identified by its source country, underground reservoir, or some distinguishing physical or chemical property. The three most frequently specified properties are density, chemical characterization, and sulfur content.

API gravity is a contrived measure of density.[6] The relation of API gravity to specific gravity is given by the following:

$$°\text{API} = \frac{141.5}{sp\ gr} - 131.5$$

where *sp gr* is the specific gravity, or the ratio of the weight of a given volume of oil to the weight of the same volume of water at a standard temperature, usually 60°F.

An oil with a density the same as that of water, or with a specific gravity of 1.0, would then be 10°API oil. Oils with higher than 10°API gravity are lighter than water. Since lighter crude oil fractions are usually more valuable, a crude oil with a higher °API gravity will bring a premium price in the market place.

Heavier crude oils are getting renewed attention as supplies of lighter crude oil dwindle. In 1967, the U.S. Bureau of Mines (now part of the U.S. Department of Energy) defined heavy crudes as those of 25°API or less. More recently, the American Petroleum Institute proposed to use 20°API or less as the distinction for heavy crude oils.

A characterization factor was introduced by Watson and Nelson[7] to use as an index of the chemical character of a crude oil or its fractions. The Watson characterization factor is defined as follows:

$$\text{Watson } K = (T_B)^{1/3}/sp\ gr$$

where T_B is the absolute boiling point in degrees Rankin and *sp gr* is specific gravity compared to water at 60°F. For a wide boiling range material like crude oil, the boiling point is taken as an average of the five temperatures at which 10%, 30%, 50%, 70% and 90% is vaporized.

A highly paraffinic crude oil might have a characterization factor as high as 13 while a highly naphthenic crude oil could be as low as about 10.5. Highly paraffinic crude oils can also contain heavy waxes which makes it difficult for the oil to flow. Thus, another test for paraffin content is to measure how cold a crude oil can be before it fails to flow under specific test conditions. The higher the pour point temperature, the greater the paraffin content for a given boiling range.

Sour and sweet are terms referring to a crude oil's approximate sulfur content. In early days, these terms designated smell. A crude oil with a high sulfur content usually contains hydrogen sulfide—the gas associated with rotten eggs. Then the crude oil was called sour. Without this disagreeable odor, the crude oil was judged sweet. Today, the distinction between sour and sweet is based on total sulfur content. A sour crude oil is one with more than 0.5 wt % sulfur, whereas a sweet crude oil has 0.5 wt % or less sulfur. It has been estimated that 58% of U.S. crude oil reserves are sour. More importantly, an estimated 81% of world crude oil reserves are sour.[8]

ASTM distillation is a test prescribed by the American Society for Testing and Materials to measure the volume percent distilled at various temperatures.[5] The results are often reported the other way around: the temperatures at which given volume percents vaporize.[9] These data indicate the quantity of conventional boiling range products occurring naturally in the crude oil. Analytical tests on each fraction indicate the kind of processing that may be needed to make specification products. A plot of boiling point, sulfur content, and API gravity for fractions of Light Arabian crude oil are shown in Fig. 14.5. This crude oil is among the ones most traded in the international crude oil market.

In effect, Fig. 14.5 shows that the material in

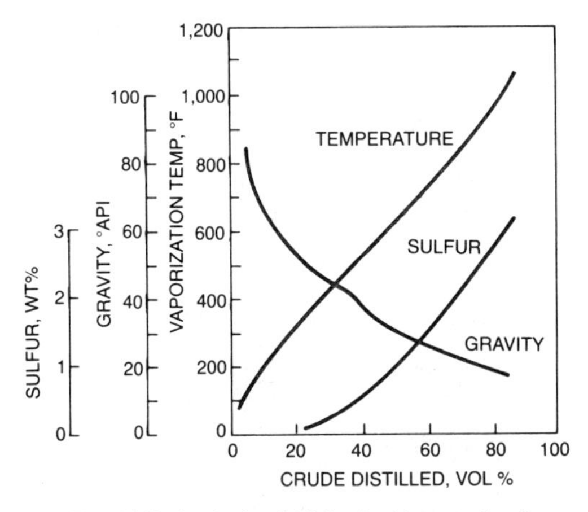

Fig. 14.5 Analysis of Light Arabian crude oil.

the mid-volume range of Light Arabian crude oil has a boiling point of approximately 600°F, a liquid density of approximately 30°API and an approximate sulfur content of 1.0 wt %. These data are an average of eight samples of Light Arabian crude oil. More precise values would be obtained on a specific crude oil if the data were to be used in design work.

Since a refinery stream spans a fairly wide boiling range, the crude oil analysis data would be accumulated throughout that range to give fraction properties. The intent here is to show an example of the relation between volume distilled, boiling point, liquid density and sulfur content.

Crude Oil Pretreatment

Crude oil comes from the ground admixed with a variety of substances: gases, water, and dirt (minerals). The technical literature devoted to petroleum refining often omits crude oil cleanup steps. It is likely presumed that the reader wishing to compare refining schemes will understand that the crude has already been through these cleanup steps. Yet cleanup is important if the crude oil is to be transported effectively and to be processed without causing fouling and corrosion. Cleanup takes place in two ways: field separation and crude desalting.

Field separation is the first attempt to remove the gases, water, and dirt that accompany crude oil coming from the ground. As the term implies, field separation is located in the field near the site of the oil wells. The field separator is often no more than a large vessel which gives a quieting zone to permit gravity separation of three phases: gases, crude oil, and water (with entrained dirt).

The crude oil is lighter than water but heavier than the gases. Therefore, crude oil appears within the field separator as a middle layer. The water is withdrawn from the bottom to be disposed of at the well site. The gases are withdrawn from the top to be piped to a natural gas processing plant or are pumped back into the oil well to maintain well pressure. The crude oil from the middle layer is pumped to a refinery or to storage awaiting transportation by other means.

Crude desalting is a water-washing operation performed at the refinery site to get additional crude oil cleanup.[10] The crude oil coming from field separators will continue to have some water and dirt entrained with it. Water washing removes much of the water-soluble minerals and entrained solids.

If these crude oil contaminants were not removed, they would cause operating problems during refinery processing. The solids (dirt and silt) would plug equipment. Some of the solids, being minerals, would dissociate at high temperature and corrode equipment. Still others would deactivate catalysts used in some refining processes.

Crude Oil Fractions

The importance of boiling range for petroleum products already has been discussed in connection with the earlier Table 14.5. The simplest form of refining would isolate crude oil into fractions having boiling ranges that would coincide with the temperature ranges for consumer products. Some treating steps might be added to remove or alter undesirable components, and a very small quantity of various chemical additives would be included to enhance final properties.

Crude oil distillation separates the desalted crude oil into fractions of differing boiling ranges. Instead of trying to match final product boiling ranges at this point, the fractions

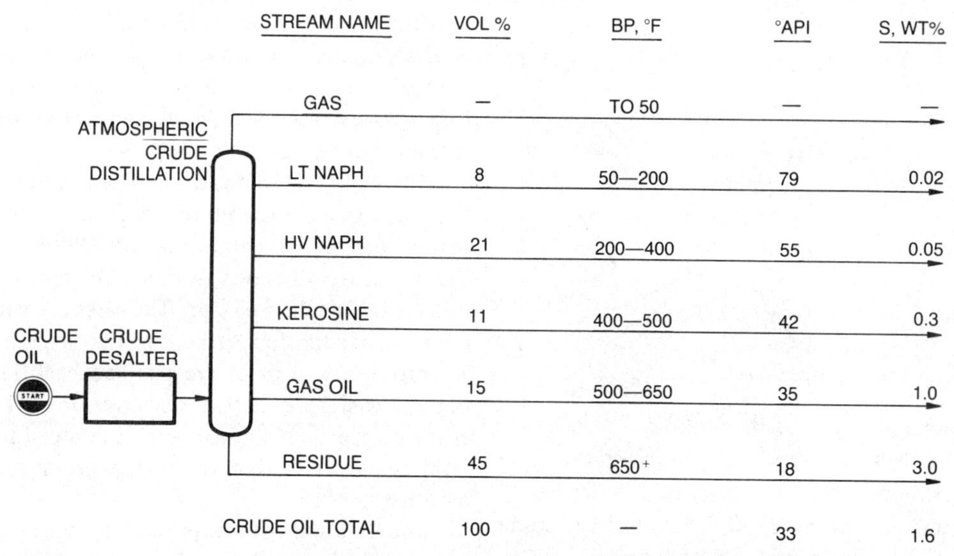

Fig. 14.6 Separating desalted crude oil into fractions.

are defined by the number and kind of downstream processes.

The desalting and distillation units are depicted in Fig. 14.6 to show the usual fractions coming from crude oil distillation units. The discussion in the following paragraphs shows the relationships between some finished products and downstream processing steps.

GASOLINE

The light and heavy naphtha fractions from crude oil distillation are ultimately combined to make gasoline. The two streams are isolated early in the refining scheme so that each can be refined separately for optimum blending in order to achieve required specifications–of which only volatility, sulfur content, and octane number will be discussed.

Volatility

A gasoline's boiling range is important during its aspiration into the combustion chamber of a gasoline-powered engine. Vapor pressure, a function of the fuel's boiling range, is also important. Boiling range and vapor pressure are lumped into one concept, volatility.[11]

Lighter components in the gasoline blend are established as a compromise between two extremes: enough light components are needed to give adequate vaporization of the fuel-air mixture for easy engine starting in cold weather, but too much of the light components can cause the fuel to vaporize within the fuel pump and result in vapor lock.

Heavier components are a trade-off between fuel volume and combustion chamber deposits. Heavier components extend the yield of gasoline that can be made from a given volume of crude oil. But heavier components also contribute to combustion chamber deposits and spark plug fouling. Thus, an upper limit is set on gasoline's boiling range to give a clean-burning fuel.

Sulfur Content

Sulfur compounds are corrosive and foul smelling. When burned in an engine, these compounds result in sulfur dioxide exhaust. Should the engine be equipped with a catalytic muffler, as is the case for many modern automobile engines, the sulfur is exhausted from the muffler as sulfur trioxide, or sulfuric acid mist.

Caustic wash or some other enhanced solvent washing technique is usually sufficient to remove sulfur from light naphtha. The sulfur compounds in light naphtha are mercaptans and organic sulfides that are removed readily by these washing processes.

Heavy naphtha is harder to desulfurize. The

sulfur compounds are in greater concentration and are of more complicated molecular structure. A more severe desulfurization method is needed to break these structures and release the sulfur. One such process is hydrotreating.

Hydrotreating is a catalytic process that converts sulfur-containing hydrocarbons into low-sulfur liquids and hydrogen sulfide.[12] The process is operated under a hydrogen-rich blanket at elevated temperature and pressure. A separate supply of hydrogen is needed to compensate for the amount of hydrogen required to occupy the vacant hydrocarbon site once held by the sulfur. Also, hydrogen is consumed to convert the sulfur to hydrogen sulfide gas.

Nitrogen and oxygen compounds are also dissociated by hydrotreating. The beauty of the process is that molecules are split at the points where these contaminants are attached. For nitrogen and oxygen compounds, the products of hydrotreating are ammonia and water, respectively. Thus, the contaminants will appear in the offgases and are easily removed by conventional gas treating processes.

Octane Number

Another condition to keep gasoline engines running smoothly is that the fuel-air mixture start burning at a precise time in the combustion cycle. An electrical spark starts the ignition. The remainder of the fuel-air mix should be consumed by a flame front moving out from the initial spark.

Under some conditions, a portion of the fuel-air mixture will ignite spontaneously instead of waiting for the flame front from the carefully timed spark. The extra pressure pulses resulting from spontaneous combustion are usually audible above the normal sounds of a running engine and give rise to the phenomenon called "knock." Some special attributes of the knocking phenomenon are called pinging and rumble. All of these forms of knock are undesirable because they waste some of the available power of an otherwise smooth-running engine.

Octane number is a measure of a fuel's ability to avoid knocking. The octane number of a gasoline is determined in a special single cylinder engine where various combustion conditions can be controlled.[5] The test engine is adjusted to give trace knock from the fuel to be rated. Then various mixtures of iso-octane (2,2,4-trimethyl pentane) and normal heptane are used to find the ratio of the two reference fuels that will give the same intensity of knock as that from the unknown fuel. Defining iso-octane as 100 octane number and normal heptane as 0 octane number, the volumetric percentage of iso-octane in heptane that matches knock from the unknown fuel is reported as the octane number of the fuel. For example, 90 vol % iso-octane and 10 vol % normal heptane establishes a 90 octane number reference fuel.

Two kinds of octane number ratings are specified, although other methods are often used for engine and fuel development. Both methods use the same reference fuels and essentially the same test engine. Engine operating conditions are the difference. In one, called the *Research method*, the spark advance is fixed, the air inlet temperature is 125°F, and engine speed in 600 rpm. The other, called the *Motor method*, uses variable spark timing, a higher mixture temperature (300°F) and a faster engine speed (900 rpm).

The more severe conditions of the Motor method have a greater influence on commercial blends than they do on the reference fuels. Thus, a Motor octane number of a commercial blend tends to be lower than the Research octane number. Recently, it has become the practice to label gasoline with an arithmetic average of both ratings, abbreviated (R+M)/2.

Catalytic reforming is the principal process for improving the octane number of a naphtha for gasoline blending.[10] The process gets its name from its ability to re-form or re-shape the molecular structure of a feedstock. The transformation that accounts for the improvement in octane number is the conversion of paraffins and naphthenes to aromatics. The aromatics have better octane numbers than their paraffin or naphthene homologs. The greater octane number increase for the heavier molecules explains why catalytic reforming is usually applied to the heavy naphtha fractions.

Catalysts for reforming typically contain

platinum or a mixture of platinum and other metal promoters on a silica-alumina support. Only a small concentration of platinum is used, averaging about 0.4 wt %. The need to sustain catalyst activity and the expense of the platinum make it common practice to pretreat the reformer's feedstock to remove catalyst poisons.

Hydrotreating, already discussed, is an effective process to pretreat reforming feedstocks. The two processes go together well for another reason. The reformer is a net producer of hydrogen by virtue of its cyclization and dehydrogenation reactions. Thus, the reformer can supply the hydrogen needed by the hydrotreating reactions. A rough rule-of-thumb is that a catalytic reformer produces 800 -1,200 scf of hydrogen per barrel of feed, while the hydrotreater consumes about 100-200 scf/bbl for naphtha treating. The excess hydrogen is available for hydrotreating other fractions in separate hydrotreaters.[13]

DISTILLATES

Jet fuel, kerosine (range oil), No. 1 fuel oil, No. 2 fuel oil, and diesel fuel are all popular distillate products coming from 400°F to 600°F fractions of crude oil. One grade of jet fuel uses the heavy naphtha fraction, but the kerosine fraction supplies the more popular heavier grade of jet fuel, with smaller amounts sold as burner fuel (range oil) or No. 1 heating oil.

Some heating oil (generally No. 2 heating oil) and diesel fuel are very similar and are sometimes substitutes for each other. The home heating oil is intended to be burned within a furnace for space heating. The diesel fuel is intended for compression-ignition engines.

Hydrotreating improves the properties of all these distillate products. The process not only reduces the sulfur content of the distillates to a low level but also hydrogenates unsaturated hydrocarbons so that they will not contribute to smoke and particulate emissions—whether the fuel is burned in a furnace or used in an engine.

RESIDUALS

Crude oil is seldom distilled at temperatures above about 650°F. At higher temperatures, coke will form and plug the lower section of the crude oil distillation tower. Therefore, the portion with a boiling point above 650°F is not vaporized—or at least not with the processing units introduced so far. This residual liquid is disposed of as industrial fuel oils, road oils, etc. The residual is sometimes called reduced crude because the lighter fractions have been removed.

PRODUCING MORE LIGHT PRODUCTS

The refining scheme evolved to this point is shown in Fig. 14.7. It is typical of a low investment refinery designed to make products of modern quality. Yet the relative amounts of products are dictated by the boiling range of the crude oil feed. For Light Arabian crude oil reported earlier (see Fig. 14.5), all distillate fuel oils and lighter products (those boiling below 650°F) would comprise only about 55 vol % of the crude oil feed rate.

For industrialized areas where the principal demand is for transportation fuels or high quality heating oils, a refining scheme of the type shown in Fig. 14.7 would need to dispose of almost half of the crude oil as low quality, less desirable, residual products. Moreover, the price obtained for these residual products is not only much lower than revenues from lighter products but also lower than the cost of the original crude oil. Thus, there are economic incentives to convert much of the residual portions into lighter products of suitable properties.

Relative volumes of petroleum product deliveries in the United States and the portions existing in Light Arabian crude oil are compared by boiling ranges in Fig. 14.8. Note that 80-85 vol % of all U.S. petroleum products are lighter than the boiling temperature of 650°F compared to the 55 vol % existing in the crude oil. Furthermore, half of all U.S. products are gasoline and lighter distillates (boiling temperatures less than 400°F) compared to 29 vol % in the crude oil.

This comparison can appear unfair since an array of products obtained easily from one crude oil would be difficult to obtain from another crude oil feed. Total product deliveries in the United States come from a variety of

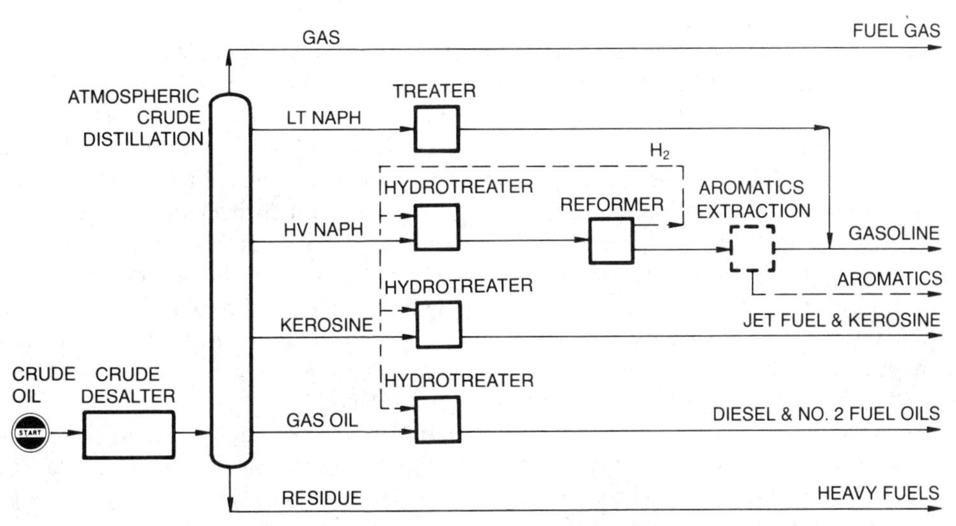

Fig. 14.7 Low investment route to modern products.

different crude oils, processed in a variety of different refining schemes. But the comparison serves to emphasize a long term trend in the refining industry—to convert heavy, less desirable fractions into lighter, more valuable products. The comparison also lays the foundation for the next group of processes to be discussed. The choice and arrangement of processes hereafter are intended to depict a breadth of refining technology, rather than suggest a commercial scheme for handling the example crude oil.

Cracking

These processes cause hydrocarbon molecules to break apart into two or more smaller molecules. Thermal cracking uses high temperature (above 650°F) and long residence time to accomplish the molecular split. Catalytic cracking accomplishes the split much faster and at lower temperatures because of the presence of a cracking catalyst.

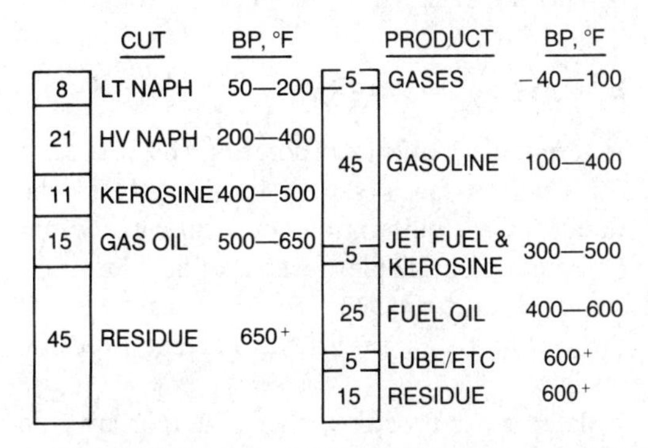

Fig. 14.8 Light Arabian crude oil compared to 1979 U.S. deliveries.

Catalytic cracking involves not only some of the biggest units, with their large catalyst reactor-separators and regenerators, but it is also among the more profitable operations with its effective conversion of heavy feeds to light products. Gasoline from catalytic cracking has a higher octane number than thermally-cracked gasoline. Yields include less gas and coke than thermal cracking; i.e., more useful liquid products are made. The distribution of products between gasoline and heating oils can be varied by different choices for catalysts and varying operating conditions.

The best feeds for catalytic crackers are determined by a number of factors. The feed should be heavy enough to justify conversion. This usually sets a lower boiling point of about 650°F. The feed should not be so heavy that it contains undue amounts of metal-bearing compounds nor carbon-forming materials. Either of these substances is more prevalent in heavier fractions and can cause the catalyst to lose activity more quickly.

Visbreaking is basically a mild, once-through thermal-cracking process. It is used to get just sufficient cracking of resid so that fuel oil specifications can be made. Although some gasoline and light distillates are made, this is not the purpose of the visbreaker.

Coking is another matter. It is a severe form of thermal cracking in which coke formation is tolerated to get additional lighter liquids from the heavier, dirtier fractions of crude oil. Here, the metals that would otherwise foul a catalytic process are layed down with the coke. The coke settles out in large coke drums that are removed from service frequently (about one a day) to have the coke cut out with high-pressure water lances. To make the process continuous, multiple coke drums are used so that some drums can be onstream while others are being unloaded.

Hydrocracking achieves cracking with a rugged catalyst to withstand resid contaminants and with a hydrogen atmosphere to minimize coking. Hydrocracking combines hydrotreating and catalytic-cracking goals, but a hydrocracker is much more expensive than either of the other two. The pressure is so high (up to 3,000 psi) that very thick walled vessels must be used for reactors (up to 9 inches thick). The products from a hydrocracker will be clean (desulfurized, denitrified, and demetalized) and will contain isomerized hydrocarbons in greater amount than in conventional catalytic cracking. A significant part of the expense of operating a hydrocaracker is for the hydrogen that it consumes.

Vacuum Distillation

Among the greater variety of products made from crude oil, some of the products (lubricating oils for example) have boiling ranges that exceed 650°F–the general vicinity where cracking would occur in atmospheric distillation. Thus, by operating a second distillation unit under vacuum, the heavier parts of the crude oil can continue to be divided into specific products. Furthermore, some of the fractions distilled from vacuum units are better than atmospheric residue for cracking because the metal bearing compounds and carbon forming materials are more highly concentrated in the vacuum residue.

Reconstituting Gases

Cracking processes to convert heavy liquids to lighter liquids also make gases. Another way to make more liquid products is to combine gaseous hydrocarbons. A few small molecules of a gas can be combined to make one bigger molecule with fairly specific properties. Here, a gas separation unit is added to the refinery scheme to isolate the individual types of gases. When catalytic cracking is also part of the refining scheme, there will be a greater supply of olefins–ethylene, propylene and butylene. Two routes for reconstituting these gaseous olefins into gasoline blending stocks are described below.

Polymerization ties two or more olefins together to make polymer gasoline. The double bond in only one olefin is changed to a single bond during each link between two olefins. This means the product will still have a double bond. For gasoline, these polymer stocks are good for blending because olefins tend to have higher octane numbers than their paraffin homolog.

However, the olefinic nature of polymer gasoline can also be a drawback. During long storage in warmer climates, the olefins can continue to link up to form bigger molecules of gum and sludge. This effect, though, is seldom important when the gasoline goes through ordinary distribution systems.

Alkylation combines an olefin and iso-butane when gasoline is desired. The product is mostly isomers. If the olefin were butylene, the product would contain a high concentration of 2,2,4-trimethyl pentane. The reader is reminded that this is the standard compound that defines 100 on the octane number scale. Alkylates are high quality gasoline-blending compounds, having good stability as well as high octane numbers.

A MODERN REFINERY

A refining scheme incorporating the processes discussed so far is shown in Fig. 14.9. The variations are quite numerious, though. Types of crude oil available, local product demands, and competitive quality goals are just a few of the factors that are weighed to decide a specific scheme.

Many other processes play an important role in the final scheme. A partial list of these other processes would have the following goals:

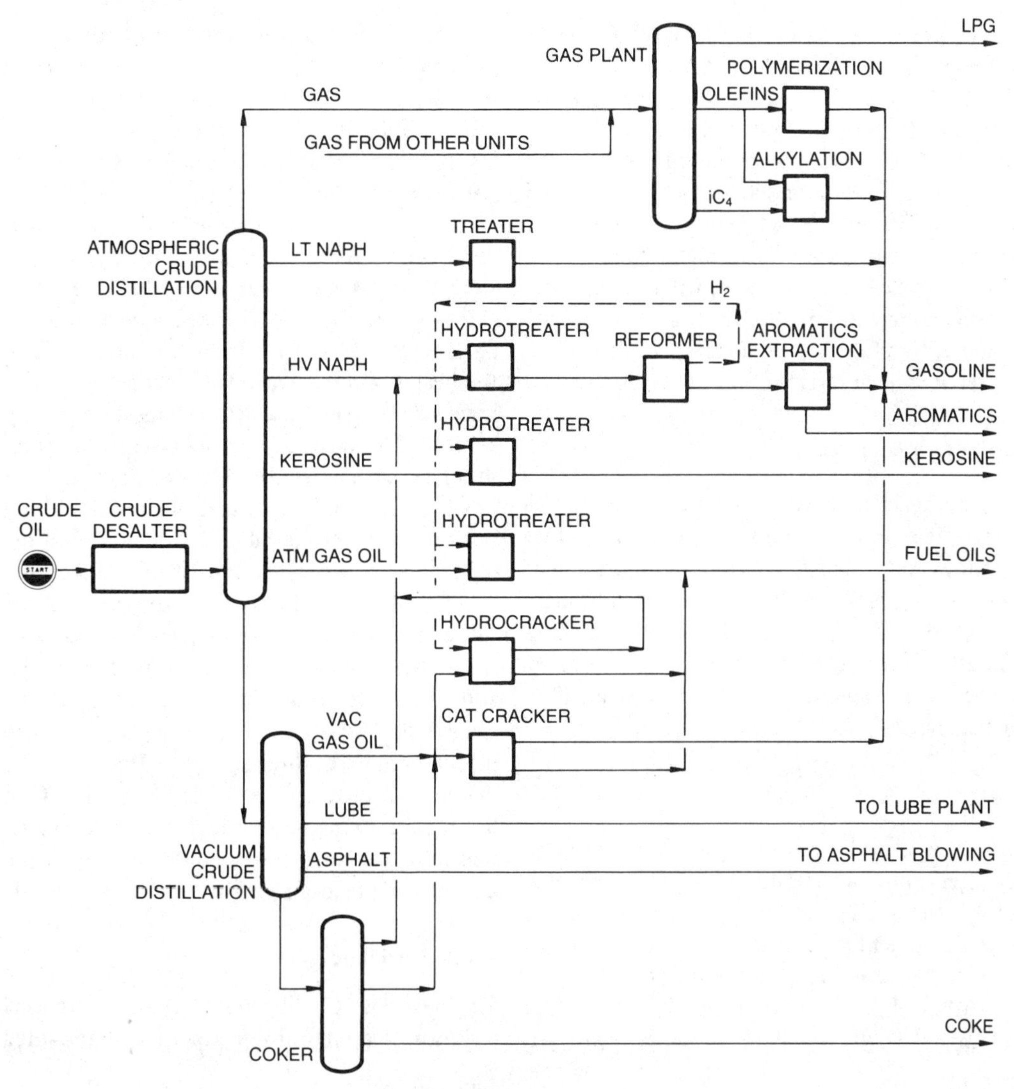

Fig. 14.9 High conversion refinery.

dewaxing lubricating oils, deoiling waxes, deasphalting heavy fractions, manufacturing specific compounds for gasoline blending (alcohols, ethers, etc.), and isolating specific fractions for use as petrochemical feedstocks.

Petrochemicals

It has already been mentioned that petrochemicals account for only a little more than $6\frac{1}{2}$ vol % of all petroleum feedstocks. Earlier, Table 14.7 gave the vast array of the applications of these petrochemicals. Olefins and aromatics make up a big part of the total.

Ethylene is one of the most important olefins. It is usually made by cracking gases–ethane, propane, butane or a mixture of these as might exist in a refinery's offgases. When gas feedstock is scarce or expensive, naphthas and even whole crude oil have been used in specially designed ethylene crackers. The heavier feeds also give significant quantities of higher molecular weight olefins and aromatics.

Aromatics, as has been pointed out, are in high concentration in the product from a catalytic reformer. When aromatics are needed for petrochemical manufacture, they are extracted from the reformer's product using solvents such as glycols (the Udex process, for example) and sulfolane, to name two popular ones.

The mixed aromatics are called BTX as an abbreviation for benzene, toluene and xylene. The first two are isolated by distillation and the isomers of the third are separated by partial crystallization. Benzene is the starting material for styrene, phenol, and a number of fibers and plastics. Toluene is used to make a number of chemicals, but most of it is blended into gasoline. Xylene use depends on the isomer, *para*-xylene going into polyester and *ortho*-xylene going into phthalic anhydride. Both are involved in a wide variety of consumer products.

PROCESS DETAILS

So far, refining units have been described as they relate to other units and to final product specifications. Now, typical flow diagrams of some major processes will be presented to highlight individual features. In many cases, the specific design shown herein is an arbitrary choice from among several equally qualified designers.

Crude Desalting

Basically a water-washing process, the crude desalter must accomplish intimate mixing between the crude oil and water, then separate them sufficiently so that water will not enter subsequent crude-oil distillation heaters.

A typical flow diagram is shown in Fig. 14.10. The unrefined crude oil is heated to 100–300°F for suitable fluid properties. The operating pressure is 40 psig or more. Elevated temperatures reduce oil viscosity for better mixing, and elevated pressure suppresses vaporization. The wash water can be added either before or after heating.

Mixing between the water and crude oil is assured by passing the mixture through a throttling valve or emulsifier orifice. Trace quantities of caustic, acid, or other chemicals are sometimes added to promote treating. Then the water-in-oil emulsion is introduced into a high voltage electrostatic field inside a gravity settler. The electrostatic field helps the water droplets to agglomerate for easier settling.

Salts, minerals, and other water-soluble impurities in the crude oil are carried off with the water discharged from the settler. Clean desalted crude oil flows from the top of the settler and is ready for subsequent refining.

Additional stages can be used in series to get additional reduction in salt content of the crude oil. Two stages are typical, but some installations use three stages. The increased investment cost for multiple stages is offset by reduced corrosion, plugging, and catalyst poisoning in downstream equipment by virtue of lower salt content.

Crude Distillation

Single or multiple distillation columns are used to separate crude oil into fractions determined

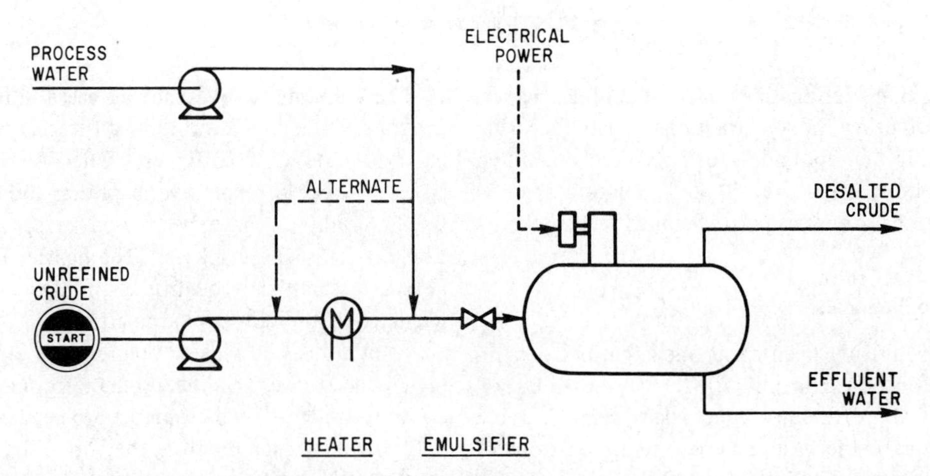

Fig. 14.10 Crude desalting. (*Hydrocarbon Processing,* **59**, *no. 9, 93–220, Sept. 1980; copyright 1980 by Gulf Publishing Co. and used by permission of the copyright owner.*)

by their boiling range. Common identification of these fractions was discussed in connection with Fig. 14.6, but these should only be considered as a guide since a variety of refining schemes call for altering the type of separation made at this point.

A typical flow diagram of a three-stage crude distillation system[14] is shown in Fig. 14.11. The crude oil is heated by exchange with various hot products coming from the system before it passes through a fired heater. The temperature of the crude oil entering the first column is 600-700°F, or high enough to vaporize the heavy gas oil and all lighter fractions. The first column is depicted as having a larger diameter in the lower section because the quantity of vapors and liquids passing through this section require more cross-sectional area to avoid high pressure drop across individual contacting trays or to prevent high-velocity vapors from blowing the liquid up the column rather than giving good mixing as the liquids follow their normal path down the column. The final design depends upon the quantities of individual fractions and permitted tower loadings.

Since light products must pass from the feed point up to their respective drawoff point, any intermediate stream will contain some of these lighter materials. Stream stripping (note the group of steam strippers beside the first column in Fig. 14.11) is a way to reintroduce these light materials back into the tower to continue their passage up through the column.

The next two fractionating columns of Fig. 14.11 are operated under vacuum. Steam jet ejectors are used to create the vacuum so that the absolute pressure can be as low as 30–40 mm Hg. The vacuum permits hydrocarbons to be vaporized at temperatures much below their normal boiling points. Thus, fractions with boiling points above 650°F can be separated with vacuum distillation without causing thermal cracking. Steam is often added to vacuum units to reduce the partial pressure of the hydrocarbons even further. If the steam is added ahead of the furnace associated with the vacuum columns, fluid velocity through the furnace tubes is increased and coke formation is minimized.

Lately, a popular addition to a crude distillation system has been a preflash column ahead of the two stages shown in Fig. 14.11. The preflash tower strips out the lighter portions of a crude oil before the remainder enters the atmospheric column. It is the lighter portions that set the vapor loading in the atmospheric column which, in turn, determines the diameter of the upper section of the column.

Incidentally, total refining capacity of a facility is reported in terms of its crude-oil handling capacity. Thus, the size of the first distillation column, whether a preflash or an atmospheric distillation column, sets the reported size of the entire refinery. Ratings in barrels per stream day (bpsd) will be greater than barrels per calendar day (bpcd). Processing units must be shut down on occasion for maintenance, repairs, and equipment replacement. The ratio of operating days to total days (or bpcd divided by bpsd) is called an "onstream factor" or "operating factor." The ratio will be expressed either as a percent or a decimal. For example, if a refinery unit undergoes one shutdown period of one month during a three year duration, its operating factor is (36-1)/36, or 0.972, or 97.2%.

Outside the United States, refining capacity is given normally in metric tons per year. Precise conversion from one unit of measure to the other depends upon the specific gravity of the crude oil, but the approximate relation is one barrel per day equals 50 tons per year.

Hydrotreating

This is a catalytic hydrogenation process that reduces the concentration of sulfur, nitrogen, oxygen, metals, and other contaminants in a hydrocarbon feed. In more severe forms, hydrotreating saturates olefins and aromatics.

A typical flow diagram is shown in Fig. 14.12. The feed is pumped to operating pressure and mixed with a hydrogen-rich gas, either before or after being heated to the proper reactor inlet temperature. The heated mixture passes through a fixed bed of catalyst where exothermic hydrogenation reactions occur. The effluent from the reactor is then cooled and sent through two separation stages. In the first,

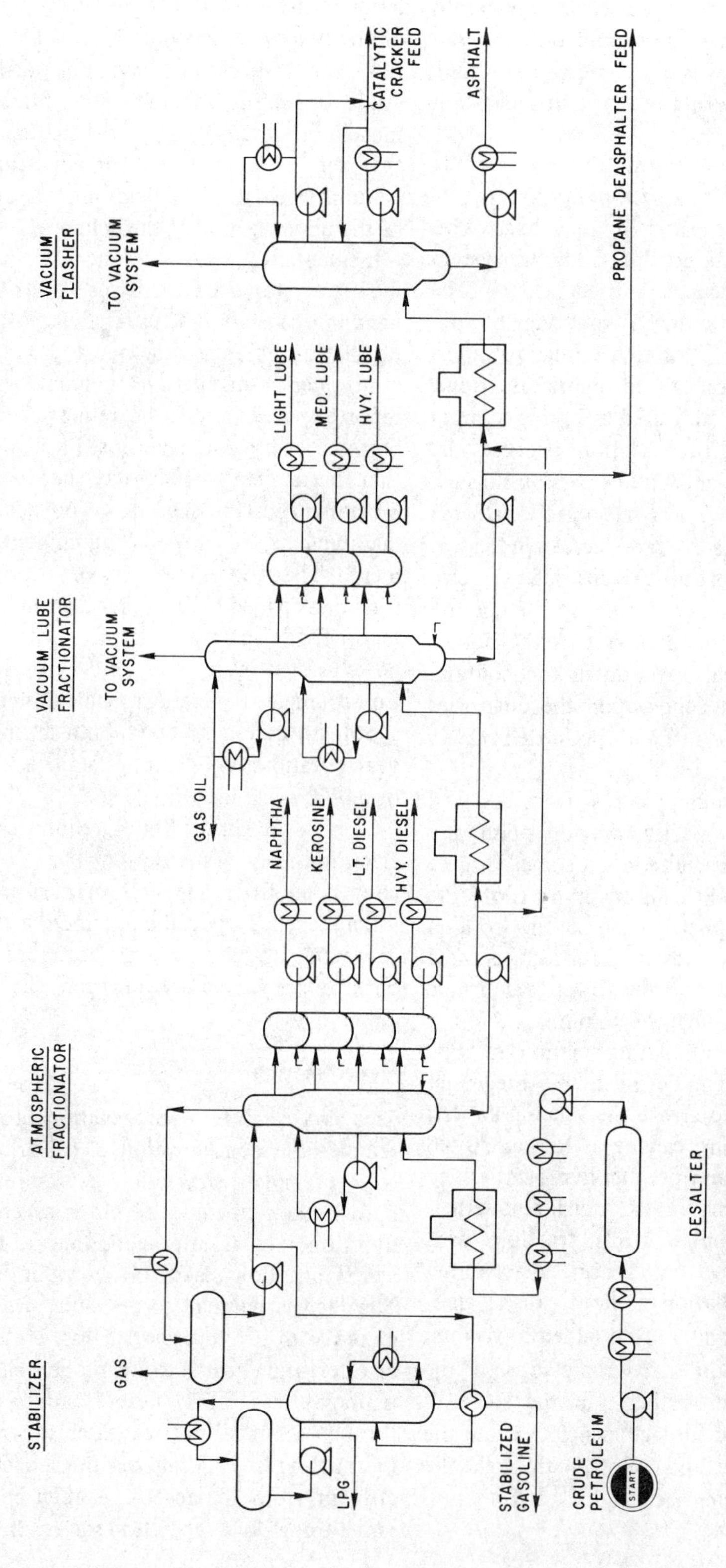

Fig. 14.11 Crude distillation—Foster Wheeler Corp. *(Hydrocarbon Processing,* 53, *no. 9, 106, Sept. 1974; copyright 1974 by Gulf Publishing Co. and used by permission of the copyright owner.)*

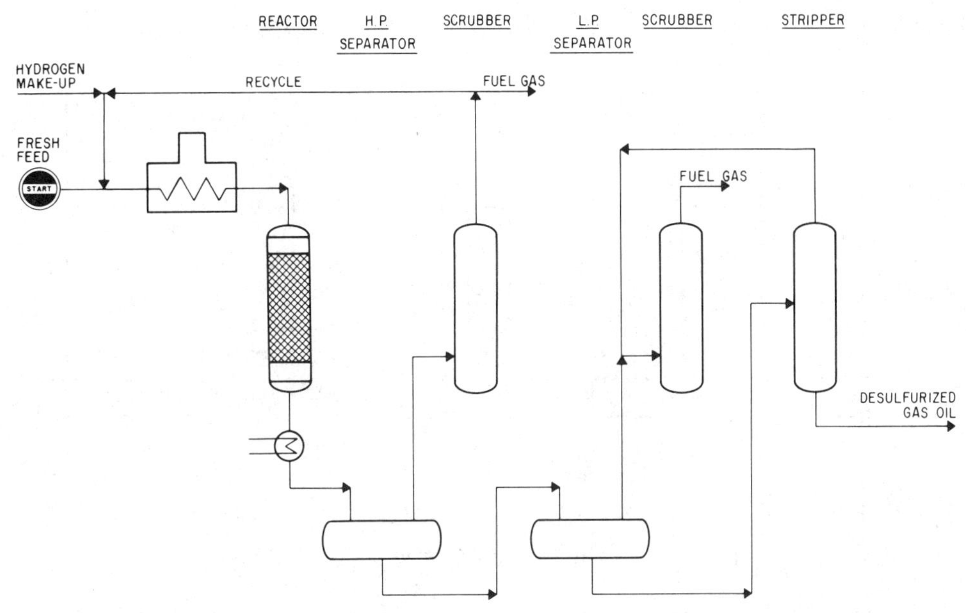

Fig. 14.12 Gulfining—Gulf R&D Co. and Houdry Division of Air Products and Chemicals, Inc. (*Hydrocarbon Processing,* **59**, *no. 9, 93-220, Sept. 1980; copyright 1980 by Gulf Publishing Co. and used by permission of the copyright owner.*)

the high-pressure separator, unreacted hydrogen is taken overhead to be scrubbed for hydrogen sulfide removal; the cleaned hydrogen is then recycled. In the second, the lower-pressure separator takes off the remaining gases and light hydrocarbons from the liquid product. If the feed is a wide-boiling range material from which several blending stocks are to be made, the second separator can be a distillation column for removing gases and light hydrocarbons, with sidestreams used for each of the liquid products.

The feed for hydrotreating can be a variety of different boiling range materials extending from light naphtha to vacuum residues. Generally, each fraction is treated separately to permit optimum conditions—the higher boiling materials requiring more severe treating conditions. For example, naphtha hydrotreating can be carried out at 200-500 psia and at 500-650°F, with a hydrogen consumption of 10-50 scf/bbl of feed. On the other hand, a residue hydrotreating process can operate at 1000-2000 psia and at 650-800°F, with a hydrogen consumption of 600-1200 scf/bbl.[13] Nevertheless, hydrotreating is such a desirable cleanup step that it can justify its own hydrogen manufacturing facilities, although the hydrogen-rich stream obtained as a by-product from catalytic reforming usually is sufficient for most operations.

Catalyst formulations constitute a significant difference among hydrotreating processes. Each catalyst is designed to be better suited to one type of feed or one type of treating goal.[15] When hydrotreating is done for sulfur removal, the process is called hydrodesulfurization and the catalyst generally is cobalt and molybdenum oxides on alumina. A catalyst of nickel-molybdenum compounds on alumina can be used for denitrogenation and cracked-stock saturation.

Catalytic Reforming

Some confusion comes from the literature when the term "naphtha reforming" is used to designate processes to make synthesis gas—a mixture containing predominantly carbon monoxide and hydrogen. However, naphtha reforming has another meaning and is the one intended here—production of an aromatic-rich liquid for use in gasoline blending.

A typical flow diagram is shown in Fig. 14.13.

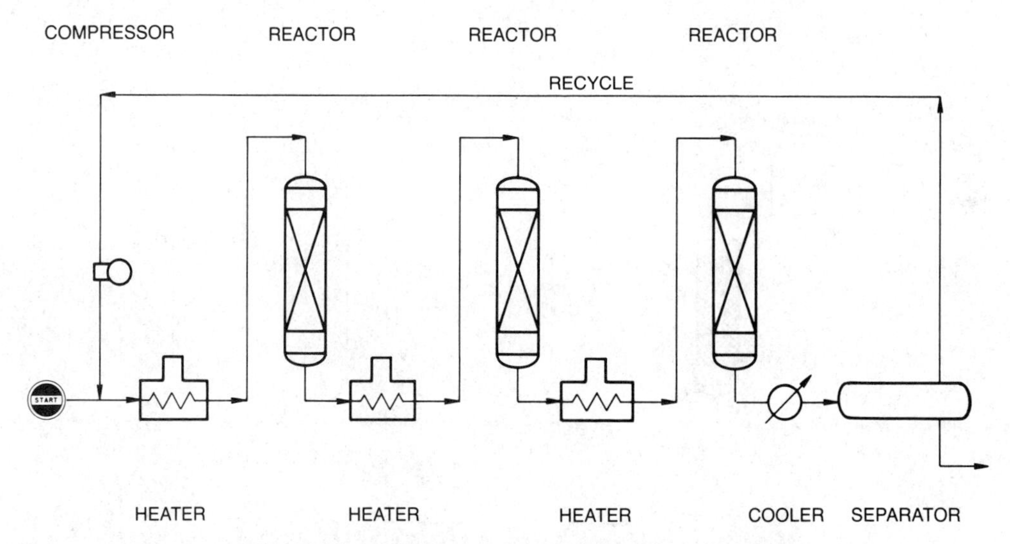

Fig. 14.13 Catalytic reforming, generalized flow diagram.

The feed is pumped to operating pressure and mixed with a hydrogen-rich gas before heating to reaction temperatures. Actually, hydrogen is a by-product of the dehydrogenation and cyclization reactions, but by sustaining a hydrogen atmosphere, cracking and coke formation are minimized.

The feed for catalytic reforming is mostly in the boiling range of gasoline to start with. The intent is to convert the paraffin and naphthene portions to aromatics. As an example, a 180-310°F fraction of Light Arabian crude oil was reported to have 8 vol% aromatics before catalytic reforming, but was 68 vol% aromatics afterwards. The feed paraffin content (69 vol%) was reduced to less than half, and the feed naphthene content (23 vol%) was almost completely absent in the product.[16]

The extent of octane number change with changes in molecular configuration is shown in Table 14.8 where normal paraffins and naphthenes are compared with their aromatic homologs.

If the naphthenes are condensed (multi-rings or indanes) they tend to deactivate the reforming catalyst quickly. Control of the end point of the feed will exclude these deactivating compounds.

TABLE 14.8 Aromatics Have Higher Octane Numbers[6]

Hydrocarbon homologs		*Octane number, clear* Motor	*Octane number, clear* Research
C_7 hydrocarbons			
n-paraffin	C_7H_{16} (n-heptane)	0.0	0.0
naphthene	C_7H_{14} (cycloheptane)	40.2	38.8
	" (methylcyclohexane)	71.1	74.8
aromatic	C_7H_8 (toluene)	103.5	120.1
C_8 hydrocarbons			
n-paraffin	C_8H_{18} (n-octane)	-15[a]	-19[a]
naphthene	C_8H_{16} (cyclooctane)	58.2	71.0
	" (ethylcyclohexane)	40.8	45.6
aromatic	C_8H_{10} (ethylbenzene)	97.9	107.4
	" (o-xylene)	100.0	120[a]
	" (m-xylene)	115.0	117.5
	" (p-xylene)	109.6	116.4

[a]Blending value at 20 vol % in 60 octane number reference fuel.

TABLE 14.9 Favored Operating Conditions for Desired Reaction Rates[17]

Feed	*Reaction*	*Product*	*Desired rate*	*To get desired rate* *Press*	*Temp.*
Paraffins	Isomerization	Iso-paraffins	Inc.	Inc.	Inc.
	Dehydrocyclization	Naphthenes	Inc.	Dec.	Inc.
	Hydrocracking	Lower mol. wt.	Dec.	Dec.	Dec.
Naphthenes	Dehydrogenation	Aromatics	Inc.	Dec.	Inc.
	Isomerization	Iso-paraffins	Inc.	Inc.	Inc.
	Hydrocracking	Lower mol. wt.	Dec.	Dec.	Inc.
Aromatics	Hydrodealkylation	Lower mol. wt.	Dec.	Dec.	Dec.

Catalysts which promote reforming reactions can give side-reactions. Isomerization is acceptable but hydrocracking gives unwanted saturates and gases. Therefore, higher operating pressures are used to suppress hydrocracking. This remedy has disadvantages. Higher pressures also suppress reforming reactions too, although to a lesser extent. Generally, a compromise is made between desired reforming and undesired hydrocracking. The effects of operating conditions on competing reactions[17] are shown in Table 14.9.

In the late 1960s, it was discovered that the addition of certain promoters, such as rhenium, germanium, or tin, to the platinum-containing catalyst would reduce cracking and coke formation. The resulting catalysts are referred to as bimetallic catalysts. These newer catalysts permit the process to enjoy the better reforming conditions of lower pressure without being unduly penalized by hydrocracking. Earlier pressures of 500 psig are now down to 150 psig.

Operating temperatures are important, too. The reactions are endothermic. Best yields would come from isothermal reaction zones, but this is difficult to achieve. Instead, the reaction beds are separated into a number of adiabatic zones operating at 500–1,000°F with heaters between stages to supply the necessary heat of reaction and hold the overall train near a constant temperature. Three or four reactor zones are commonly used when it is desired to have a product with high octane numbers.

In the recent push to make gasoline with high octane numbers but without the use of antiknock additives, high severity catalytic reforming is the prime route. The big disadvantage is a yield loss. Newer catalysts make the loss less dramatic, but the penalty remains[18] as can be seen from Fig. 14.14.

Catalytic Cracking

A typical diagram of a fluid catalytic cracker is shown in Fig. 14.15. The unit is characterized by two huge vessels, one to react the feed with hot catalyst and the other to regenerate the spent catalyst by burning off carbon with air. The activity of the newer molecular-sieve catalysts is so great that the contact time between feed and catalyst is reduced drasti-

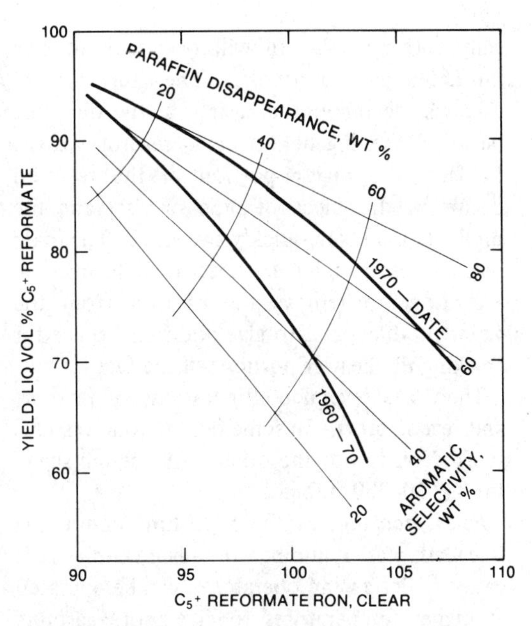

Fig. 14.14 Better octane numbers, less yield (Kuwait naphtha). *(McDonald, G. W. G., Hydrocarbon Processing, **56**, no. 6, 147–150, June 1977; copyright 1977 by Gulf Publishing Co. and used by permission of the copyright owner.)*

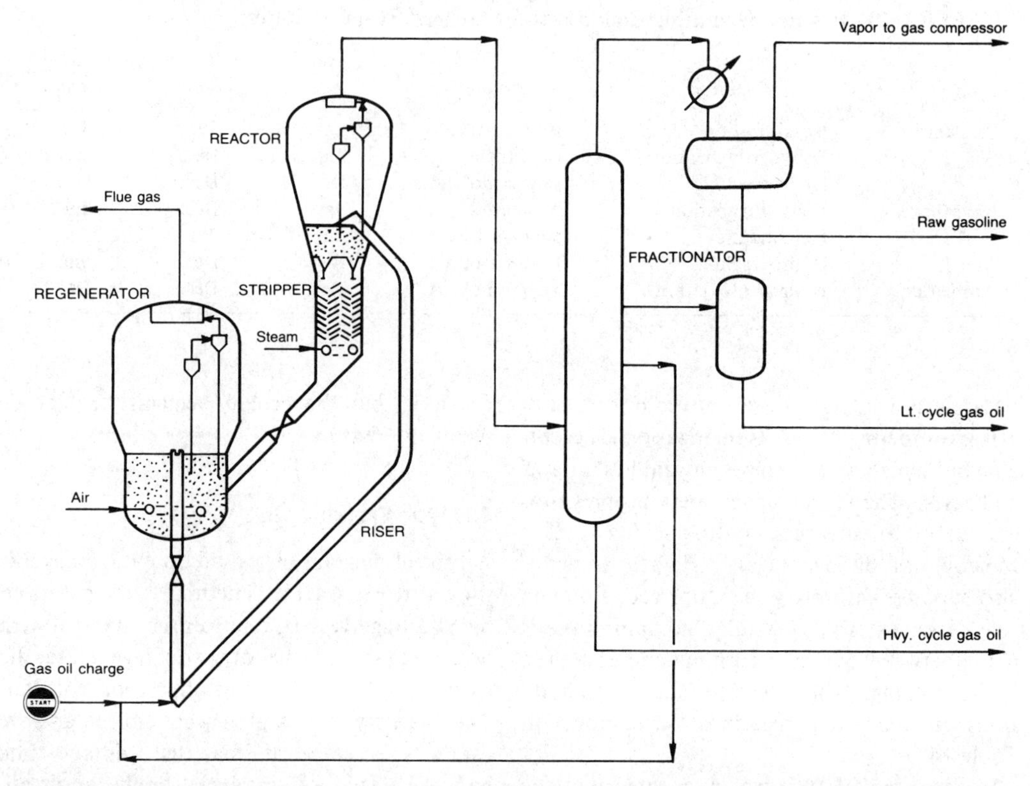

Fig. 14.15 Fluid catalytic cracking—Texaco Development Corp. (*Hydrocarbon Processing,* **59**, *No. 9, 93-220, Sept. 1980; copyright 1980 by Gulf Publishing Co. and used by permission of the copyright owner.*)

cally. If not, the oil will overcrack to give unwanted gases and coke. The short contact time is accomplished using a transfer line between the regenerator and reactor vessels. In fact, the major portion of the reaction occurs in this piece of pipe or riser and the products are taken quickly overhead. The main reactor vessels then are used to hold cyclone separators to remove the catalyst from the vapor products and to give additional space for cracking the heavier portions of the feed.

There are several configurations of reactors and regenerators. In some designs, one vessel is stacked on top of the other. All are big structures (150-200 ft. high).

Riser cracking, as the short-time contacting is called, has a number of advantages. It is easier to design and operate. It can be operated at higher temperatures to give more gasoline olefins. It minimizes the destruction of any aromatics formed during cracking. The net effect can be the production of gasoline having octane numbers 2 or 3 numbers higher than earlier designs would give.

Better regeneration of the spent catalyst is obtained by operating at higher temperatures (1,300-1,400°F).[19] The coke that is deposited on the catalyst is more completely burned away by higher temperature air blowing. The newer catalysts are rugged enough to withstand the extra heat, and newer metallurgy gives the regenerator vessel the strength it needs at higher temperatures.

Heavier feedstocks can be put into catalytic crackers. The nickel, vanadium, and iron in these heavier fractions do not deactivate the catalysts as fast as they once did because passivators are available now to add to the catalysts.[20] The extra sulfur that comes with heavier feeds can be prevented from exhausting into the atmosphere during regeneration because of catalysts that hold on to the sulfur compounds until the catalysts get into the reactor.[21] Then the sulfur compounds are cracked to light gases and leave the unit with the cracked products. Ordinary gas treating methods are used to capture the hydrogen sulfide coming from the sulfur in the feedstock.

Coking

Coking is an extreme form of thermal cracking. The process converts residual materials that might not easily be converted by the more popular catalytic cracking process. Coking is also a less expensive process for getting more light stocks from residual fractions. In the coking process, the coke is considered a by-product that is tolerated in the interest of more complete conversion of residues to lighter liquids.

A typical flow diagram of a delayed coker is shown in Fig. 14.16. There are several possible configurations, but in this one the feed goes directly into the product fractionator in order to pick up heavier products to be recycled to the cracking operation. The term "delayed coker" signifies that the heat of cracking is added by the furnace and the cracking occurs during the longer residence time in the following coke drums. Furnace outlet temperatures are in the range of 900 to 950°F while the coke drum pressures are in the range of 20 to 60 psig.

The coke accumulates in the coke drum and the remaining products go overhead as vapors to be fractionated into various products. In this case, the products are gas, naphtha, light gas oil, heavy gas oil, and coke. When a coke drum is to be emptied, a large drilling structure mounted on top of the drum is used to make a center hole in the coke formation. The drill is equipped with high-pressure water jets (3,000 psig or more) to cut the coke from the drum so that it can fall out a bottom hatch into a coke pit. From there, belt conveyors and bucket cranes move the coke to storage or to market.[22]

Fluid coking is a proprietary name given to a different type of coking process in which the coke is suspended as particles in fluids flowing from a reactor to a heater and back again. When part of the coke is gasified, the process is called *Flexicoking*. Both Fluid Coking and Flexicoking are proprietary processes of Exxon Research and Engineering Co.

A flow diagram for Flexicoking is shown in Fig. 14.17. The first two vessels are typical of Fluid coking in which part of the coke is burned in the heater in order to have hot coke nuclei to contact the feed in the reactor vessel. The cracked products are quenched in an overhead scrubber where entrained coke is returned to the reactor. Coke from the reactor circulates to the heater where it is devolatized to yield a light hydrocarbon gas and residual coke. A sidestream of coke is circulated to the gasifier

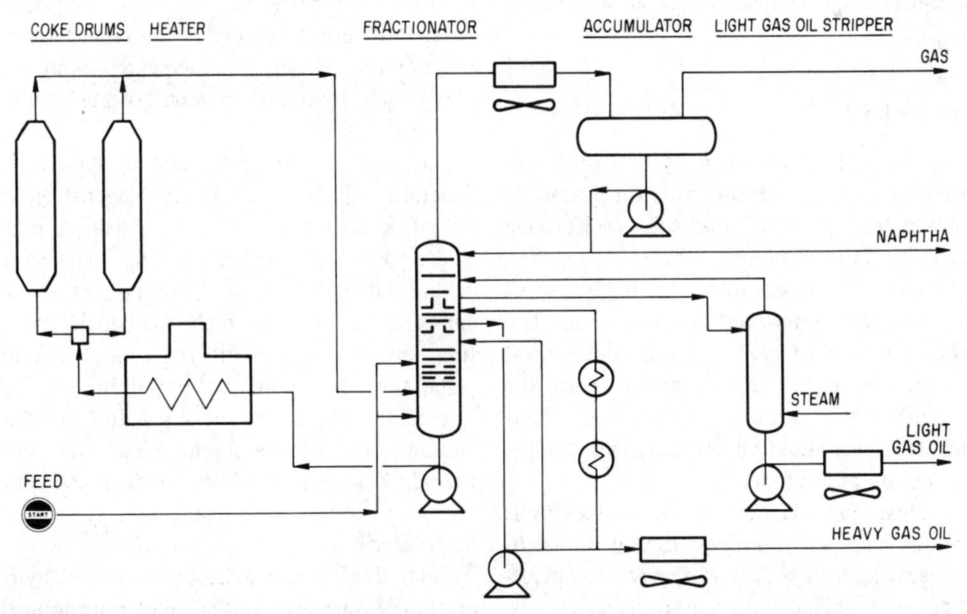

Fig. 14.16 Delayed coking—Foster Wheeler Energy Corp. (*Hydrocarbon Processing,* **59**, *no. 9, 93-220, Sept. 1980; copyright 1980 by Gulf Publishing Co. and used by permission of the copyright owner.*)

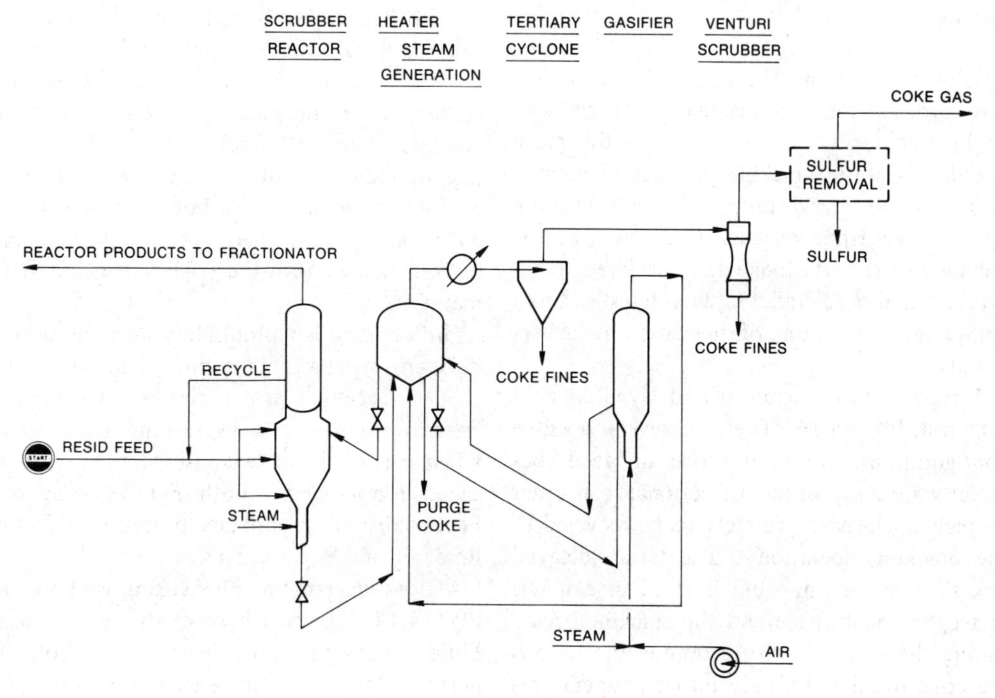

Fig. 14.17 Flexicoking—Exxon Research and Engineering Co. (*Hydrocarbon Processing,* **59**, *no. 9, 93-220, Sept. 1980; copyright 1980 by Gulf Publishing Co. and used by permission of the copyright owner.*)

where, for most feedstocks, 95% or more of the gross coke product from the reactor is gasified at elevated temperature with steam and air. Sulfur that enters the unit with the feedstock eventually becomes hydrogen sulfide exiting the gasifier and is recovered by a sulfur removal step.

Hydrocracking

Before the late 1960s, most hydrogen used in processing crude oil was for pretreating catalytic reformer feed naphtha and for desulfurizing middle-distillate products. Soon thereafter, requirements to lower sulfur content in most fuels became an important consideration. The heavier fractions of crude oil were the hardest to treat. Moreover, these fractions were the ones offering additional sources of light products. This situation set the stage for the introduction of hydrocracking.

A typical flow diagram for hydrocracking is shown in Fig. 14.18. Process flow is similar to hydrotreating in that feed is pumped to operating pressure, mixed with a hydrogen-rich gas, heated, passed through a catalytic reactor, and distributed among various fractions. Yet the hydrocracking process is unlike hydrotreating in several important ways. Operating pressures are very high, 2,000-3,000 psia. Hydrogen consumption also is high, 1,200-1,600 scf of hydrogen per barrel of feed depending on the extent of the cracking.[13] In fact, it is not uncommon to see hydrocrackers built with their own hydrogen manufacturing facilities nearby.

The catalysts for hydrocracking have a dual function. They give both hydrogenation and dehydrogenation reactions and have a highly acidic support to foster cracking. The hydrogenation-dehydrogenation component of the catalysts are metals such as cobalt, nickel, tungsten, vanadium, molybdenum, platinum, palladium, or a combination of these metals. The acidic support can be silica-alumina, silica-zirconia, silica-magnesia, alumina-boria, silica-titania, acid-treated clays, acidic-metal phosphates, or alumina, to name some given in the literature.[23]

Great flexibility is attributed to most hydrocracking processes. Under mild conditions, the process can function as a hydrotreater. Under

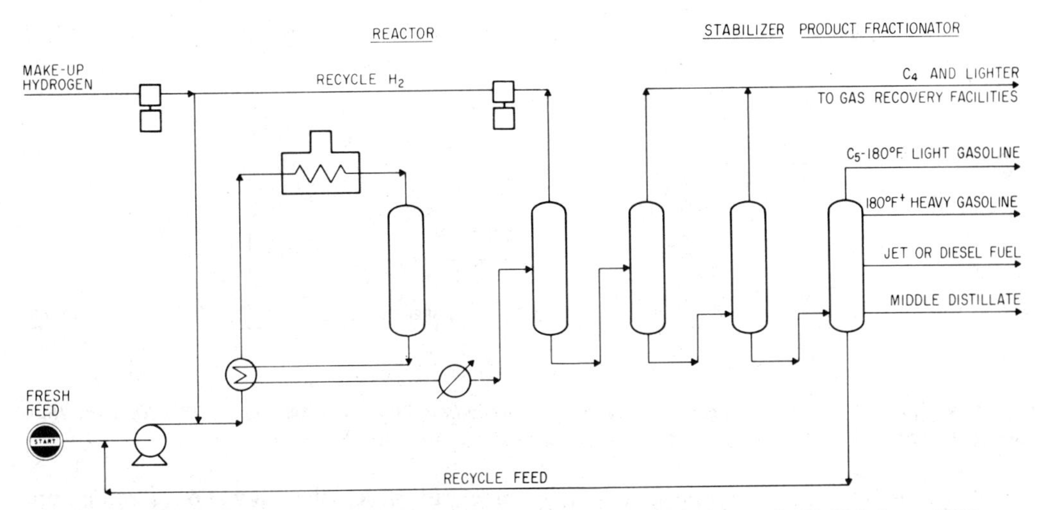

Fig. 14.18 Isocracking–Chevron Research Co. (*Hydrocarbon Processing,* **59,** *no. 9, 93–220, Sept. 1980; copyright 1980 by Gulf Publishing Co. and used by permission of the copyright owner.*)

more severe conditions of cracking, the process produces a varying ratio of motor fuels and middle distillates, depending on the feedstock and operating variables. Even greater flexibility is possible for the process during design stages when it can be tailored to change naphthas into liquefied petroleum gases or convert heavy residues into lighter products.

Because the hydrocracker is viewed as both a cracker and a treater, it can appear in refining process schemes in a number of different places. As a cracker, it is used to convert feeds that are too heavy or too contaminant-laden to go to catalytic cracking. As a treater, it is used to handle heating-oil fractions that need to be saturated to give good burning quality. But it is the trend to heavier feeds and lighter high-quality fuels that causes hydrocracking to offer advantages to future refining, even though the hydrocracking units are much more expensive to build and to operate.

The principle of an ebulliating catalyst bed is embodied in some proprietary designs, in contrast with the fixed-catalyst beds used in other versions of hydrocracking. The H-Oil process of Hydrocarbon Research, Inc. and the LC-Fining process jointly licensed by C-E Lummus and Cities Service Research and Development Co. are examples of hydrocracking processes that use a mixed-reaction bed instead of a fixed-bed of catalyst.

Polymerization

This process usually is associated with the manufacture of plastic films and fibers from light hydrocarbon olefins with products like polyethylene and polypropylene. As a gasoline manufacturing process, the polymerization of light olefins emphasizes a combination of only two or three molecules so that the resulting liquid will be in the gasoline boiling range.

For early polymerization units, the catalyst was phosphoric acid on a quartz or kieselguhr support. Many of these units were shut down when the demand for gasoline with increased octane numbers prompted the diversion of the olefin feeds to alkylation units that gave higher octane number products. Yet some refinery balances have more propylene than alkylation can handle, so a newer version of polymerization was introduced.[24] It is the Dimersol process of the Institut Francais du Petrole for which the flow diagram is shown in Fig. 14.19.

The Dimersol process uses a soluble catalytic complex injected into the feed before it enters the reactor. The heat of reaction is taken away by circulating a portion of the bottoms back to the reactor after passing through a cooling water exchanger. The product goes through a neutralizing system that uses caustic to destroy the catalyst so that the resulting polymer is clean and stable. Typical octane number

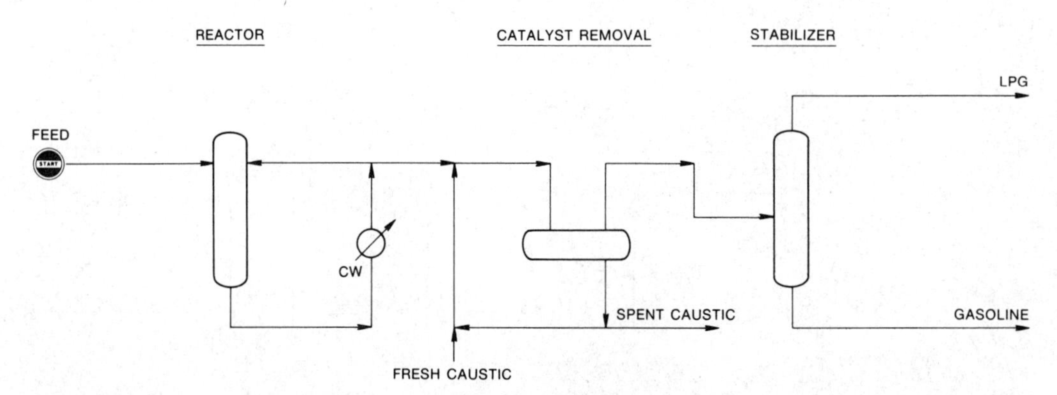

Fig. 14.19 Dimersol–Institut Francais du Petrole. *(Hydrocarbon Processing,* **59**, *no. 9, 93–220, Sept. 1980; copyright 1980 by Gulf Publishing Co. and used by permission of the copyright owner.)*

ratings for the product are 81 Motor and 96.5 Research, unleaded.

Alkylation

This is another process that increases the total yield of gasoline by combining some of the gaseous light hydrocarbons to form bigger molecules boiling in the gasoline range. Alkylation combines isobutane with a light olefin, typically propylene and butylene. A flow diagram for an alkylation unit using hydrofluoric acid as a catalyst is shown in Fig. 14.20.

Common catalysts for gasoline alkylation are hydrofluoric acid or sulfuric acid. The reaction is favored by higher temperatures, but competing reactions among the olefins to give polymers prevent high quality yields. Thus, alkylation usually is carried out at low temperatures in order to make the alkylation reaction predominate over the polymerization reactions. Temperatures for hydrofluoric acid catalyzed reactions are approximately 100°F and for sulfuric acid they are approximately 50°F. Since the sulfuric-acid catalyzed reactions are carried out at below normal atmospheric temperatures, refrigeration facilities are included.

Alkylate product has a high concentration of 2,2,4-trimethyl pentane, the standard for the 100 rating of the octane number scale. Other compounds in the alkylate are higher or lower in octane number, but the lower octane number materials predominate so that alkylate has a Research octane number in the range of 92 to 99. Developments are underway to slant the reactions in favor of the higher octane materials.[25] Random samples of alkylate quality reported in the literature[26] are summarized in Table 14.10.

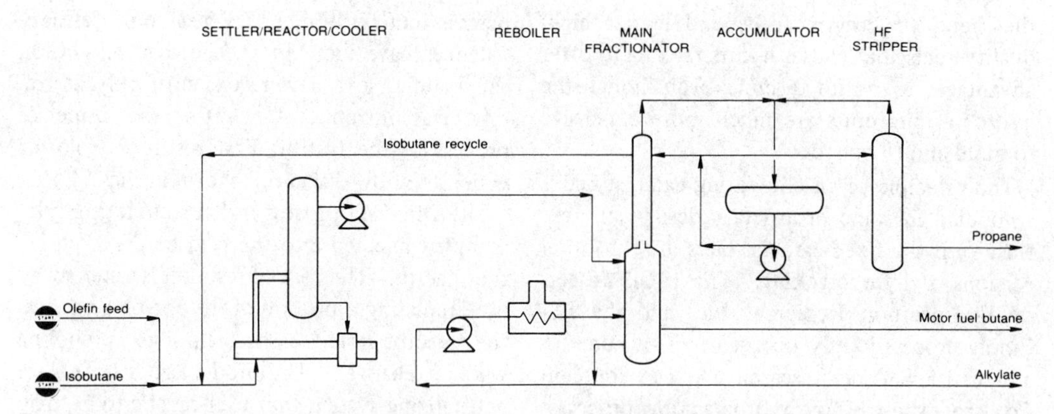

Fig. 14.20 HF alkylation–Phillips Petroleum Co. *(Hydrocarbon Processing,* **59**, *no. 9, 93-220, Sept. 1980; copyright 1980 by Gulf Publishing Co. and used by permission of the copyright owner.)*

TABLE 14.10 Typical Alkylate Octane Numbers[25]

	Feed olefin			
	C_2	C_3	C_3+C_4	C_4
Research octane number, clear	101.5	90.5	93	96.5
Motor octane number, clear	93	89	91	95.5

Ethylene

The evolution of cracking to make ethylene has progressed along two lines. In one, the ethylene is a by-product of fuel manufacturing, with the feedstock sometimes being a less desirable fuel material and at other times being a heart cut from some very desirable fuel material like naphtha. In the other line of progression, ethylene is pursued as a growing business of its own, with heavier by-product liquids being treated for use as gasoline blending stocks.

A popular starting material for ethylene cracking is ethane or propane. Some forecasts[27] suggest that these light hydrocarbon feeds may not be available in the growing volumes needed to keep up with a predicted 4.5-5% per year growth in ethylene demand. Thus many recent ethylene-cracking processes are tailored to handle heavier feedstocks. A flow diagram of an ethylene cracker[28] is shown in Fig. 14.21.

The feedstock is preheated and mixed with steam to be cracked in a tubular pyrolysis furnace. The products leave the furnace at 1,400-1,600°F and are rapidly quenched in an exchanger and sent to a gasoline fractionator where heavy fractions are removed. The gaseous products go to a quench tower where direct water quench stops any further reaction. The flow diagram shows that the remainder of the process is intended to get light gas separation of the product.

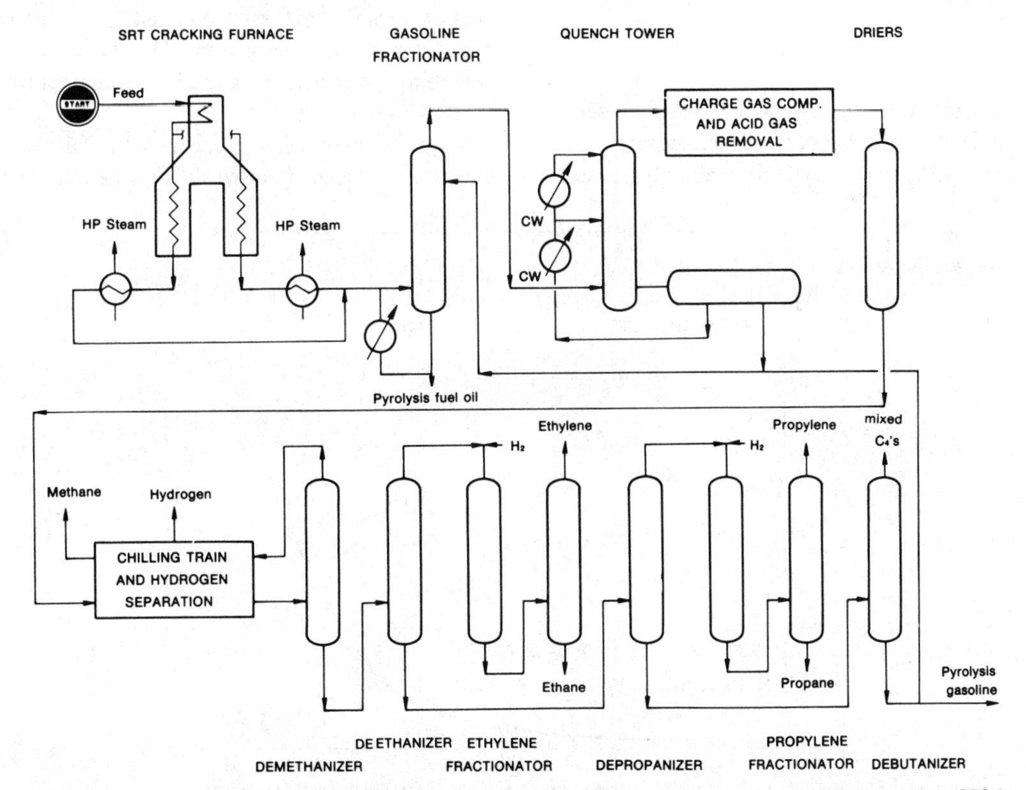

Fig. 14.21 Ethylene—C-E Lummus. *(Hydrocarbon Processing,* **58**, *no. 11, 160, Nov. 1979; copyright 1979 by Gulf Publishing Co. and used by permission of the copyright owner.)*

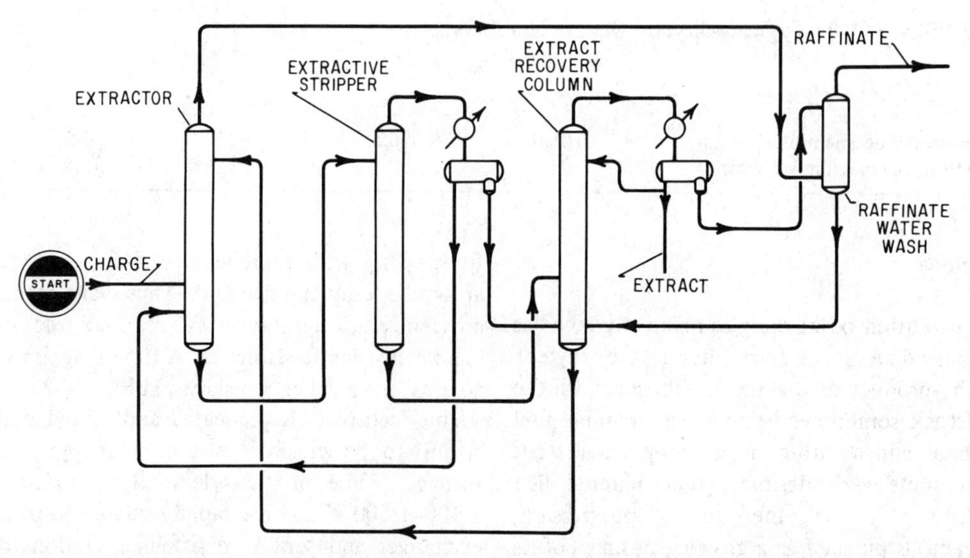

Fig. 14.22 Sulfone extraction—UOP Process Division of UOP, Inc. for process originally developed by Royal Dutch/Shell group. (*Hydrocarbon Processing,* **59**, *no. 9, 93–220, Sept. 1980; copyright 1980 by Gulf Publishing Co. and used by permission of the copyright owner.*)

The main derivatives of ethylene are polyvinyl chloride (PVC), ethylene glycol, and polyethylene.

Aromatics

Benzene, toluene, xylene, and ethylbenzene are made mostly from catalytic reforming of naphthas with units like those already discussed. As a gross mixture, these aromatics are the backbone of gasoline blending for high octane numbers. However, there are many chemicals derived from these same aromatics.[29] Thus many aromatic petrochemicals have their beginning by selective extraction from naphtha or gas-oil reformate.

A typical extraction process is the Sulfolane extraction process[10] shown in Fig. 14.22. The

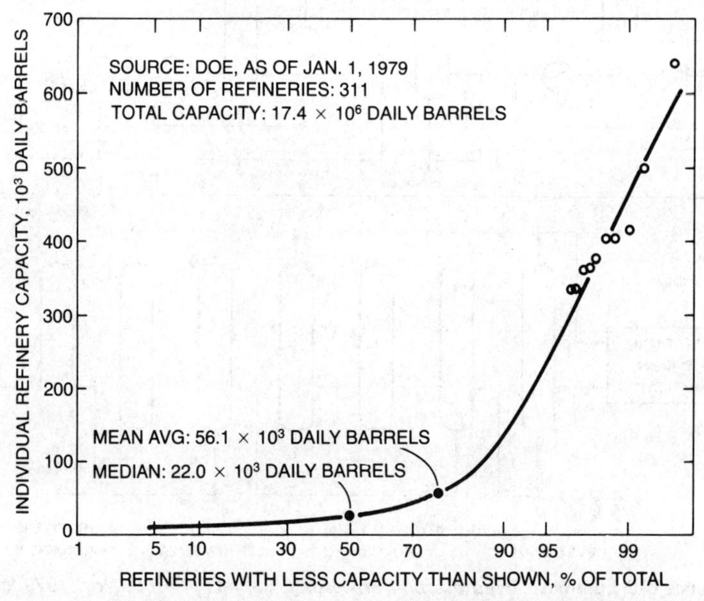

Fig. 14.23 U.S. Refinery size distribution as of Jan. 1, 1979. (*U.S. Department of Energy.*)

TABLE 14.11. Distribution of Unit Cost

Process unit	*Relative capacity, vol %*	*Unit cost, $/bpd*	*Cost as part of total refinery $/bpd*	*%*
Crude distillation, atm.	100	200	200	8
Crude distillation, vac.	35	230	80	3
Catalytic cracking, fresh feed	28	1,000	280	11
Hydrocracking	4	1,500	60	2
Coking	6	900	54	2
Thermal cracking	3	500	15	1
Catalytic reforming	21	800	168	7
Alkylation, product basis	5	1,500	75	3
Hydrotreating	46	300	138	5
Others & offsites	100	1,430	1,430	57
Total refinery	100	2,500	2,500	100

Basis: Total U.S. refining capacity, first of 1979. Cost: 1979 basis.

reformate is fed to a contactor for the countercurrent extraction of aromatic components with sulfolane solvent. The solvent, rich in aromatics, goes to a stripper and recovery column where the aromatics are separated from the solvent. Aromatic recoveries and purities are varied by choosing suitable operating parameters.

REFINERY SIZE AND COST

The few, truly huge refineries give the impression that most refineries are big. The bigger world refineries have a capacity of more than 600,000 bpd and growing as of the first of the 80's.[30] Yet one should not be misled by a so-called "average" refinery designed for 100,000 bpd. Actually, the mean average size of the 311 refineries in the United States operating as of the first of 1979 was 56,080 bpd, computed from U.S. Department of Energy data.[31] Moreover, half of these refineries are only about 22,000 bpd (the median) or less. A distribution plot for U.S. refinery sizes is shown in Fig. 14.23.

Costs are harder to nail down. Some reasonable estimates are given in Table 14.11. The relative capacity for each type of unit was fixed to match the relative capacity in existence in the U.S. at the first of 1979. The costs are also on a 1979 basis, although inflation has been changing these values rapidly.

An important item to notice is that more than half of the cost of a refinery is for materials and equipment other than those directly associated with specific petroleum processing. Some of these other items include storage, utilities, and environmental control systems. Land costs are excluded.

It should not be inferred that future construction will have the same cost distribution as shown in Table 14.11. The cost distribution in Table 14.11 is based on existing unit capacity. Future product demands will likely slant future construction in favor of one or another process with different relative unit costs based on future technology.

REFERENCES

1. The British Petroleum Co., Ltd., *BP Statistical Review of the World Oil Industry 1979*, published annually.
2. "International Outlook Issue," *World Oil*, Aug. 15, annually.
3. U.S. Dept. of the Interior, *Minerals Yearbook, Metals, Minerals and Fuels*, annually.
4. U.S. Department of Energy, "Petroleum Statement, Monthly," DOE/EIA-0109(yr/month).
5. American Society for Testing and Materials, *1980 Annual Book of ASTM Standards*, Parts 23 and 24.
6. American Petroleum Institute, *Technical Data Book--Petroleum Refining*, 3rd Ed., 1977.
7. Watson, K. M. and Nelson, E. F., "Improved Methods for Approximating Critical and Thermal Properties of Petroleum Fractions," *Industrial and Engineering Chemistry*, Vol. 25 (1933), p. 880.

8. Hoffman, H. L., "Sour Crude Limits Refining Output," *Hydrocarbon Processing*, Vol. 52, No. 9 (Sept. 1973), pp. 107–110.
9. Ferrero, E. P. and Nichols, D. T., "Analyses of 169 Crude Oils From 122 Foreign Oilfields," U.S. Department of the Interior, Bureau of Mines, Information Circular 8542, 1972.
10. "1980 Refining Process Handbook," *Hydrocarbon Processing*, Vol. 59, No. 9 (Sept. 1980), pp. 93–220.
11. Unzelman, G. H. and Forster, E. J., "How to Blend for Volatility," *Petroleum Refiner*, Vol. 39, No. 9 (Sept. 1960), pp. 109–140.
12. Kay, H., "What Hydrogen Treating Can Do," *Petroleum Refiner*, Vol. 35, No. 9 (Sept. 1956), pp. 306–318.
13. Corneil, H. G. and Heinzelmann, F. J., "Hydrogen for Future Refining," *Hydrocarbon Processing*, Vol. 59, No. 8 (Aug. 1980), pp. 85–90.
14. Foster Wheeler Corp., "Crude Distillation," *Hydrocarbon Processing*, Vol. 53, No. 9 (Sept. 1974), p. 106.
15. Kellett, T. F. et al. "How to Select Hydrotreating Catalyst," *Hydrocarbon Processing*, Vol. 59, No. 5 (May 1980), pp. 139–142.
16. Hughes, T. R. et al, "To Save Energy When Reforming," *Hydrocarbon Processing*, Vol. 55, No. 5 (May 1976), pp. 75–80.
17. Jenkins, J. H. and Stephens, J. W., "Kinetics of Cat Reforming," *Hydrocarbon Processing*, Vol. 59, No. 11 (Nov. 1980), pp. 163–167.
18. McDonald, G. W. G., "To Judge Reformer Performance," *Hydrocarbon Processing*, Vol. 56, No. 6 (June 1977), pp. 147–150.
19. Hartzell, F. D. and Chester, A. W., "FCCU Gets a Catalyst Promoter," *Hydrocarbon Processing*, Vol. 58, No. 7 (July 1979), pp. 137–140.
20. Dale, G. H. and McKay, D. L., 'Passivate Metals in FCC Feeds," *Hydrocarbon Processing*, Vol. 56, No. 9 (Sept. 1977), pp. 97–102.
21. Magee, J. S. et al, "A Look at FCC Catalyst Advances," *Hydrocarbon Processing*, Vol. 58, No. 9 (Sept. 1979), pp. 123–130.
22. Howell, R. C. and Kerr, R. C., "Moving Coke? What to Expect," *Hydrocarbon Processing*, Vol. 60, No. 3 (March 1981), pp. 107–111.
23. Sullivan, R. F. and Meyer, J. A., "Catalyst Effects on Yields and Product Properties in Hydrocracking," presented before American Chemical Society, Philadelphia meeting, April 6–11, 1975.
24. Benedek, W. J. and Mauleon, J-L., "How First Dimersol Is Working," *Hydrocarbon Processing*, Vol. 59, No. 5 (May 1980), pp. 143–149.
25. Heck, R. M. et al, "Better Use of Butenes for High-Octane Gasoline," *Hydrocarbon Processing*, Vol. 59, No. 4 (April 1980), pp. 185–191.
26. Logan, R. S. and Banks, R. L., "Disproportionate Propylene To Make More and Better Alkylate," *Hydrocarbon Processing*, Vol. 47, No. 6 (June 1968), pp. 135–138.
27. Klein, R. A., "Olefins Shift to Heavy Liquids," *Hydrocarbon Processing*, Vol. 59, No. 10 (Oct. 1980), pp. 113–115.
28. C-E Lummus, "Ethylene," *Hydrocarbon Processing*, Vol. 58, No. 11 (Nov. 1979), p. 160.
29. Dosher, J. R., "Toluene: Octanes or Chemicals?" *Hydrocarbon Processing*, Vol. 58, No. 5 (May 1979), pp. 123–126.
30. "World Refineries, 1980," *International Petroleum Times*, Vol. 84, No. 2119 (March 15, 1980), pp. 13–32.
31. U.S. Department of Energy, Energy Information Administration, "Petroleum Refineries in the United States and U.S. Territories, January 1, 1979," DOE/EIA-0111/79, released June 28, 1979.

15

Industrial Chemistry of Wood

Edwin C. Jahn* and Roger W. Strauss**

Wood was first used by man in his distant prehistoric antiquity. Certainly, the use of wood goes back to his most primitive culture. Wood has been employed by man for hundreds of millenia in two general ways, viz., (1) as found in nature and reshaped for specific uses, and (2) as a raw material to make something else, such as heat. Today, this ever-renewable resource from the forest provides the raw material for these same two basic classes of use, though on a greatly expanded level.

Wood remains today the world's most widely used industrial raw material. It is the source not only of lumber and wood products but also provides fiber for most of the world's paper, packaging materials, and rayon. A growing array of industrial chemicals is produced from wood, and wood is increasingly used for industrial and home fuel.

Forest lands account for one-third of the total land area of the United States. These lands are the source of the raw material for the fifth largest industrial complex in the United States, namely, the forest-products industries which provide a wide range of wood, fiber, and chemical commodities. These forest lands are also a source of many other values, often harder to define economically, such as grazing, water supply, wildlife and fishes, recreation, and aesthetics.

The earth's forest resources depend, either directly or indirectly upon photosynthesis. This basic process of life performed by green plants can be represented in simple form as:

$$H_2O + CO_2 \xrightarrow[\text{solar energy}]{\text{chlorophyll}} \text{organic compounds} + O_2$$

In addition to C, H and O, plants also incorporate N and S into organic compounds via light-dependent reactions. Thus, the basic processes of photosynthesis have determined and sustained life as we know it, producing organic materials and oxygen.

The photosynthetic process continuously stores solar energy in plant tissues in the amount[1] of about 2×10^{11} tons of carbon annually, with an energy content of 3×10^{21} J.

*State University of New York College of Environmental Science and Forestry, Syracuse, N.Y.
**Bowater Incorporated, Old Greenwich, Connecticut.

This is about 10 times the world's annual energy use and 200 times our food energy consumption. The photosynthetic process operates at only about an average of 0.2-0.3 percent efficiency on the earth's land surface and up to 0.5 percent on agricultural land. These efficiencies represent stored energy and not the initial conversion energies usually quoted in other energy systems. It is apparent that very small increases in the efficiency of conversions to stored energy could greatly increase the amount of plant material formed.

It is important that we examine how photosynthesis fits into the biosphere and how we may improve the utilization of biological solar energy conversion as a future source of raw materials via both traditional, as well as non-conventional mechanisms. Among these, according to D. O. Hall[1], are increasing the efficiency of photosynthetic conversion, increased fixation of CO_2 into C_4 compounds, control of photorespiration, genetic engineering, plant selection, regulation and selection of products, H_2 and H_2O_2 production and other techniques. All atmospheric CO_2 and O_2 are continually being recycled by plants every 300 and 2,000 years, respectively, and all H_2O every 2 million years. This recycling phenomenon is most important, and any interference with it by pollution could have serious effects on life. With care on our part, plants can continue indefinitely to supply renewable quantities of food, structural materials, fiber, fuel, and chemicals.

Trees, being long-lived and possessing a great volume of wood, serve as a major renewable resource and act as a storage reservoir until they are harvested. In other words, wood and bark can be held in storage on the stump, whereas biomass from annual plants must be collected and stored each year.

However, the U.S. forests are being put under increasing pressures due to a combination of causes, among which are increase in per capita demand for paper and wood products, population growth, and changes in patterns of land use due to urbanization and increased leisure time.

A few examples[2,3] are sufficient to point up dramatically these increasing pressures on our forest resources. Per capita consumption of paper products has increased from 255 pounds in 1940 to 604 pounds in 1979. Demand for plywood and other wood products is also increasing. At the same time, there is a constant reduction in timber-growing land due to shifting land use, such as for right-of-way, urban and industrial expansion, parks and recreation areas, and protection of scenic and aesthetic values. A net nationwide reduction of timber lands on the order of 30 million acres or nearly 6 percent of the 1968 total is considered possible by the end of the century.[3] For these combined reasons, the demand for wood may exceed the supply by the year 2000.

Besides human demands on our forests, there is the loss of trees to the relentless attack of diseases and insects upon our forests and ornamental trees. William E. Waters and Robert W. Brandt[4] of the U.S. Forest Service point out that 2.4 billion cu ft of timber are killed each year by insects and diseases. This does not include the losses of ornamental trees. Present disease and insect-pest control methods help stem the attack, but constant research in this technology is essential. A substantial cutting back on these losses would go a long way to extend our timber resources.[5]

Present practices in the harvesting of timber and the manufacturing of forest products leave large quantities of wood (estimated as high as 40 percent) that are not utilized. These include (1) the non- or under-utilized trees (non-commercial timber, such as salvable dead trees, defective trees, and non-merchantable species), with a combined inventory of about 956 million dry m. tons,[6] (2) non-utilized removals (for thinning, land clearing, etc), amounting to over 19 million dry m. tons annually,[6] (3) residues from harvesting amounting to 100 million dry tons per year,[7] (4) manufacturing residues, estimated at 102 million tons annually of which 20 million tons are presently unused,[7] and (5) post-consumer residues of paper and wood which make up about half of the total municipal trash or about 62-75 million tons per year.[8] In addition, many organic chemicals are not recovered from various stages of pulping processes, but are burned to provide energy and to recover the inorganic chemicals required for the pulping operations.

Progress is being made in utilizing these great quantities of unused wood materials, especially the manufacturing (mill) residues, of which about 80 percent are now used for fuel and for pulp. Chemical utilization of these materials has always been attractive to the chemist and the conservationist as a means for getting greater value from the forest.

The recent end of the era of cheap fossil hydrocarbons, together with growing public realization that petroleum is a finite resource, has stimulated serious review and consideration of the developments of alternative resources for energy and for organic chemicals. For the latter, forests are a major alternative. Our forests offer opportunity for both energy and chemicals because of their high productivity and sustained yield, if properly managed. For these reasons, much attention is now being given to the examination of our forests as a source of energy, through direct combustion and through conversion to synthesis gas, charcoal and methanol, and as a source of chemicals such as sugars, alcohols, phenols, and others for industrial use, including feedstocks for plastics.

Though the pulp and paper industry is a major industry, the other wood-chemical industries are small by comparison. Some have gone into decline, such as wood distillation, due to competition from synthetic processes, and others are comparatively new, such as chemically-modified wood. The chemical wood industries also differ in other ways. Some, such as charcoal burning, which is centuries old, are still primitive in many areas. Others, such as paper, are modern and aggressive. The products which are treated in this chapter include pulp and paper, fiber boards, paper-base laminates, products derived from pyrolysis and chemical degradation, wood sugars and alcohol, tree exudates, extractives, modified wood, wood-plastic combinations, and composite boards.

CHEMICAL NATURE OF WOOD

Wood is a supporting and conducting tissue for the tree. To serve these functions, about 90 percent of wood tissue is composed of strong, relatively thick-walled long cells. These, when separated from each other, are fine fibers very suitable for paper making.

Chemically, the cell-wall tissue of wood is a complex mixture of polymers. These polymers fall into two groups, the polysaccharides and lignin. The polysaccharides of wood are collectively known as *Holocellulose*, meaning total cellulosic carbohydrates. The holocellulose accounts for about 70 to 80 percent of the extractive-free woody tissues, with lignin making up the remainder.

The holocellulose is composed of cellulose and a mixture of other polysaccharides, collectively termed *hemicelluloses.* Cellulose is a high-molecular weight linear polymer of condensed glucose units. Hemicellulose is a mixture of lower D.P. linear and branched condensed polymers of pentose and hexose sugars (xylose, arabinose, glucose, mannose and galactose), with derivatives of xylan and galactoglucomannan as the most prevalent components. Cellulose makes up the main framework of the cell walls of the wood fibers. It is highly resistant chemically, whereas the hemicelluloses have relatively low resistance to acids and alkalis.

Lignin serves as the adhesive material of wood, cementing the fibers and other cells together to form the firm anatomical structure of wood. Lignin is a complex polymer of condensed phenylpropane units (see Fig. 15.1). It is susceptible to degradation and dissolution by strong alkalis at elevated temperatures, by acid sulfite solutions at elevated temperatures, and by oxidizing agents. Thus, lignin can be removed from the wood, leaving the separated cellulosic fibers in the form of a pulp.

In addition to the cell-wall substance, wood also contains extraneous materials present in the cavities of the cells. In some woods, these are present in considerable amounts and are commercially important. The volatile oils and resins of the southern pines are an example. The extraneous materials are numerous and cover a wide range of chemically different materials. Most of these may be separated from the wood by steam distillation and solvent extraction as indicated below.

Steam distillation—terpene hydrocarbons, esters, acids, alcohols, aldehydes, aliphatic hydrocarbons.

Ether extraction—fats, fatty acids, resins, resin acids, phytosterols, waxes, nonvolatile hydrocarbons, and the above volatile com-

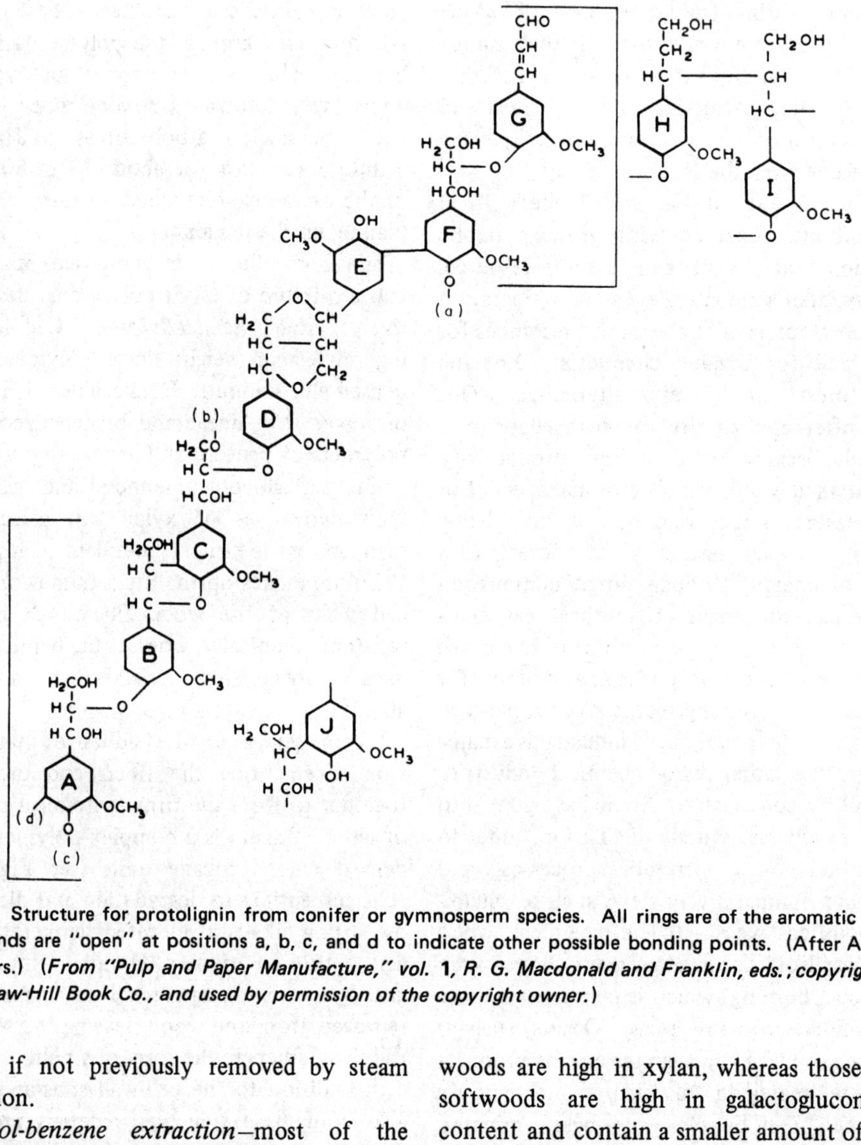

Fig. 15.1 Structure for protolignin from conifer or gymnosperm species. All rings are of the aromatic benzene type; bonds are "open" at positions a, b, c, and d to indicate other possible bonding points. (After Adler and coworkers.) (*From "Pulp and Paper Manufacture," vol. 1, R. G. Macdonald and Franklin, eds.; copyright 1969 by McGraw-Hill Book Co., and used by permission of the copyright owner.*)

pounds if not previously removed by steam distillation.

Alcohol-benzene extraction—most of the ether-soluble materials plus phlobaphenes, coloring matter, and some tannin.

Alcohol extraction—tannin and most of the above organic material, except some resins.

Water extraction—sugars, cyclitols, starch, gums, mucilages, pectins, galactans, and some inorganic salts, tannins, and pigments.

The chemical composition of wood varies between species. The hardwoods and softwoods of the temperate zone show consistent differences. The hardwoods have less lignin and more hemicelluloses than the softwoods. Furthermore, the hemicelluloses of the hardwoods are high in xylan, whereas those of the softwoods are high in galactoglucomannan content and contain a smaller amount of xylan. The chemical composition of a few North American woods is shown in Table 15.1.

MANUFACTURE OF PULP AND PAPER

The importance of the pulp and paper industry to the American economy is exemplified by the growth rate in the use of paper and paper products. New uses are continuously being found for paper, and these developments together with the rising standard of living have resulted in a constant increase in the per capita consumption of paper. As shown in Table 15.2 the paper industry in the 19th century,

TABLE 15.1 Chemical Analysis of Some North American Woods

	Hardwoods				*Softwoods*			
	Aspen	*White Birch*	*Beech*	*Basswood*	*White Spruce*	*Balsam Fir*	*Jack Pine*	*Eastern Hemlock*
Lignin[a]	20.9	19.0	22.1	19.8	28.6	30.0	28.6	32.5
Cellulose[a]	42.7	43.1	44.0	45.8	42.5	42.2	42.1	41.6
Acetyl-4-methyl-glucuronoxylan[a]	30.8	34.7	30.3	31.5				
Arabino-4-methyl-glucuronoxylan[a]					9.3	8.4	12.0	7.2
Glucomannan[a]	5.2	3.0	3.2	2.3				
Acetylgalacto-glucomannan[a]					19.2	19.1	17.1	18.5
Ash[a]	0.4	0.2	0.4	0.6	0.4	0.3	0.2	0.2
Solubility[b]								
Hot water	2.8	2.7	1.5	2.4	2.2	3.6	3.7	3.4
Ethyl ether	1.9	2.4	0.7	2.1	2.1	1.8	4.3	0.7

[a]Values based on oven-dry extractive-free (alcohol-benzene) wood. Analyses by Dr. Tore E. Timell.
[b]Values based on unextracted wood (oven-dry basis).

dependent on rags as the source of fiber, was stagnant until the middle part of the century when processes were invented for the production of fiber from wood. Since that time, the industry has shown a constant growth. Table 15.2 also shows the breakdown of paper consumption by various grades.

In 1978, the United States produced a total of 61,500,000 tons of paper and paperboard in hundreds of separate mill locations. This paper and board was made from 45,000,000 tons of pulp, produced primarily from wood in over 300 pulp mills, and 15,000,000 tons of recycled waste paper.

Table 15.3 lists the leading states for both pulp and paper production in 1968. Since only a small amount of paper is made from rags or other fiber sources (agricultural residues), it is obvious that most wood pulp is produced from those areas of the country that are heavily forested.

An examination of Table 15.3 shows that, with the exception of Maine and Wisconsin, pulp production is concentrated in the southern and northwestern sections of the United States. While a high percentage of pulp is converted into paper or board at the same plant site, a significant portion (classified as "market pulp") is sold in bales to mills in other sections of the country for subsequent manufacture into paper. Thus, while New York and Michigan are relatively important in paper making, they rank

TABLE 15.2 Paper Consumption in the United States

Paper Consumption by Year		*Paper Consumption by Grades (1969)*	
Year	*Consumption (lb/capita)*	*Grade*	*Consumption (lb/capita)*
1810	1	Newsprint	95
1869	20	Book papers	70
1899	58	Writing papers	30
1919	119	Packaging	60
1929	220	Tissue and sanitary	35
1948	356	Paper board	250
1956	434	Construction	40
1969	576		
1978	604		

TABLE 15.3 Paper and Pulp Production by States for 1968

	Paper Production		*Pulp Production*	
State	*Tons*	*Ranking*	*Tons*	*Ranking*
Georgia	4,035,000	(1)	4,829,000	(1)
Wisconsin	3,027,000	(2)	1,444,000	(10)
Alabama	3,002,000	(3)	3,134,000	(4)
Louisiana	2,621,000	(4)	2,433,000	(6)
Washington	2,594,000	(5)	3,883,000	(2)
Oregon	2,532,000	(6)	2,503,000	(5)
New York	2,463,000	(7)	520,000	(22)
Michigan	2,227,000	(8)	575,000	(20)
Florida	2,123,000	(9)	3,165,000	(3)
Maine	2,106,363	(10)	2,173,000	(7)
South Carolina	1,826,000	(14)	2,013,000	(8)
Virginia	2,013,000	(12)	1,857,065	(9)

quite low as pulp producers. However, both New York and Michigan are large users of waste paper which is reprocessed into usable fiber. With the current emphasis upon recycling, it is probable that the use of waste paper will be greatly accelerated in the near future.

Raw materials for the pulp and paper industry can be classified as fibrous and nonfibrous. Wood accounts for over 95 percent of the fibrous raw material (other than waste paper) in the United States. Cotton and linen rags, cotton linters, cereal straws, esparto, hemp, jute, flax, bagasse, and bamboo are also used and in some countries are the major source of paper-making fiber.

Wood is converted into pulp by mechanical, chemical, or semichemical processes. Sulfite, sulfate (kraft), and soda are the common chemical processes while neutral sulfite is the principal semichemical process. Coniferous wood species (softwoods) are the most desirable, but the deciduous, broad-leaved species (hardwoods) have gained rapidly in their usage and constituted about 25 percent of the pulpwood used in 1969. Table 15.4 lists the production of pulp by the various processes.

Nonfibrous raw materials include the chemicals used for the preparation of pulping liquors and bleaching solutions and the various additions to the fiber during the papermaking process. For pulping and bleaching, these raw materials include sulfur, lime, limestone, caustic soda, salt cake, soda ash, hydrogen peroxide, chlorine, sodium chlorate, and magnesium hydroxide. For papermaking they include rosin, starch, alum, kaolin clay, titanium dioxide, dyestuffs, and numerous other specialty chemicals.

Wood Preparation

The bark of trees contains relatively little fiber and much strongly colored nonfibrous material; what fiber there is, is of poor quality. The nonfibrous material will usually appear as dark colored dirt specks in the finished paper. Therefore, for all but low-grade pulps, bark should be removed as much as possible, and this must be very thorough in the case of groundwood and sulfite pulps if the finished paper is to appear clean.

Debarking is usually done in a drum barker where bark is removed by the rubbing action of logs against each other in a large rotating drum (see Fig. 15.2). Hydraulic barkers using high-pressure water jets are excellent for large logs and are common on the West Coast. Mechanical knife barkers are becoming more common and

TABLE 15.4 Woodpulp Production by Grade 1978

Grade	*Tons*
Dissolving and special alpha grades	1,415,000
Sulfite paper grades	1,760,000
Sulfate paper grades	35,750,000
Groundwood and TMP	4,400,000
Semichemical	4,000,000

Fig. 15.2 Wood being barked by two 12-ft by 45-ft barking drums. (*Courtesy Chicago Bridge and Iron Co.*)

are used extensively in smaller operations because of their lower capital cost. Also they have found widespread use in sawmills to debark logs prior to sawing so that the wood wastes can be used to produce pulp.

Wood cut in the spring of the year during the active growing season is very easy to peel. Much of the spruce and fir cut in the North is still hand peeled during this season and usually represents the optimum in bark removal.

The standard log length used in the Northeast is 48 inches while 63 inches is common in the South. Wood is generally measured by log volume, a standard cord being considered as containing 128 cubic feet. Large timber on the West Coast is generally measured in board feet of solid volume. Measurement and purchase of wood on a weight basis is practiced and has the advantage of being more directly related to fiber content.

The recent growth in the use of wood residues has been phenomenal. By barking the sawlogs, the slabs, edgings, and other trimmings that were formerly burned can now be used to make pulp. Table 15.5 shows the relative sources of wood in the United States.

Thus, almost 28 percent of the wood used by the pulp industry in 1970 could be classified as waste wood. Several mills have been built that use no logs whatsoever but depend on residuals from satellite sawmill operations. Special sawmilling equipment has been developed to produce sawdust of a proper size so that it can also be used. These residuals are usually purchased in units of 2400 pounds of dry wood.

Wood used in producing groundwood or mechanical pulp requires no further preparation after debarking, but that used in the other chemical processes must first be chipped into small pieces averaging $\frac{1}{2}$-1 in. in length and about $\frac{1}{8}$-$\frac{1}{4}$ in. in thickness.

Chipping is accomplished with a machine consisting of a rotating disc with knives mounted radially in slots in the face of the disc (Fig. 15.3). Modern chippers have up to twleve knives; the ends of the logs are fed against the disc at about a 45° angle. Each knife cuts a layer of wood equal in thickness to the distance the knife protrudes. In chipping, it is necessary

TABLE 15.5 Sources of Wood for Pulp in the United States (1978 - Equivalent Cords of 128 Cubic Feet)*

Source	*Cords*
Pulpwood	47,000,000
Residues	30,000,000

*(75% Softwood - 25% Hardwood)

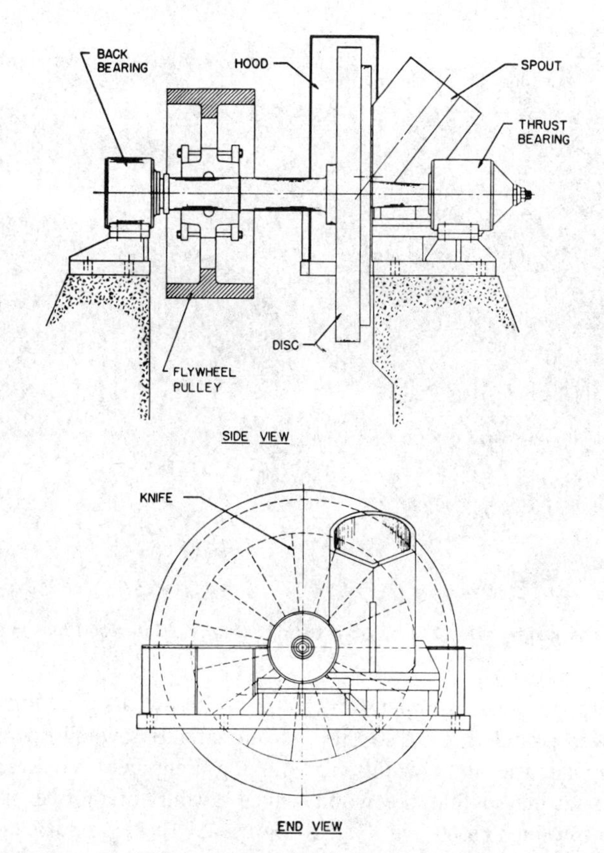

Fig. 15.3 Sketch of modern wood chipper. (*Courtesy Carthage Machine Co.*)

only to cut the wood across the grain since the layers of wood cut on the chipper immediately break up along the grain due to the forces exerted by the chipper. A typical chipper using pulpwood up to 20 in. in diameter would have a disc about eight feet in diameter rotating at 600 rpm and would handle 15-25 cords per hour. Whole-log chippers on the West Coast have been built with discs up to 14 ft in diameter that will accept logs up to 36-40 in. in diameter and 30 ft long. Chip size is not uniform and screens are necessary to separate the oversize chips and sawdust from the acceptable chips.

Mechanical Pulping

Mechanical pulping, as the term implies, does not involve a chemical process. However, it is one of the more important methods of making pulp, such as newsprint, which consists of about 80 percent mechanical (or groundwood) pulp.

Groundwood pulp is made by forcing the whole log against the face of a cylindrical abrasive stone rotating at relatively high speeds (Fig. 15.4). The logs are positioned so that their axes are parallel to the axis of the rotating stone. Sufficient water must be added to the stone to serve as a coolant and carry the pulp away.

At one time, natural sandstone was used for the grindstone but modern stones are either silicon carbide or aluminum oxide grits in a vitrified clay binder. Thus the characteristics of the stone can be varied to produce pulps "tailor-made" to fit their desired end use. Groundwood pulps for roofing or flooring felts must be extremely coarse and free draining, so they require a stone with large grits, whereas newsprint pulps are very fine and require the use of small grits.

Pulp characteristics can also be varied by changing the stone surface pattern, the stone speed, the pressure of the logs against the stone,

Fig. 15.4 Large two-pocket grinder for producing groundwood. (*Courtesy Koehring-Waterous Ltd.*)

and the temperature of the ground-pulp slurry. Generally a coarser and more freely draining pulp is obtained with a coarse surface pattern and high speed, pressure, and temperature. Type and condition of the wood are also factors, but groundwood pulps are usually made from the coniferous or long-fibered species since the deciduous or short-fibered species give very weak pulps.

Many designs of machines are used for grinding wood. The pocket grinder (Great Northern type) is usually equipped with two pockets for holding the wood and utilizing hydraulic pistons to force the wood against the stone. At the end of the cycle, the piston is retracted and a new charge of wood is placed in the pocket. Continuous or magazine type grinders use large chains to force the wood against the stone thus giving a more uniform grinding action. There are several advantages to each type and both are in common use. Modern grinders use up to 10,000 hp per stone and can produce up to 120 tons of pulp per day.

A recent development has been the production of groundwood or mechanical pulp from chips by using "refiners" (Fig. 15.5). The chips are fed between rotating discs containing steel plates with fine teeth. The rolling and cutting action of the plates as the chip moves from the center to the periphery of the discs reduces the chips to pulp. The original concept used chips with no pretreatment but the quality of the pulp was only marginally better. A new variation called TMP (thermo-mechanical pulping) was developed and has found widespread usage due to the greatly superior pulp strength. The chips are steamed at 40 psi for 2-4 minutes and then refined in a refiner that has been modified to operate under pressurized conditions. The use of small amounts of chemical in the pretreatment steaming stage will produce pulps of even greater strength. These pulps are sufficiently improved over stone groundwood that they can be used to partially substitute for chemical pulps. Newsprint consisting of 100% TMP is now being produced commercially.

Groundwood or mechanical pulp is low in strength compared to the chemical pulps. It is composed of a mixture of individual fibers, broken fibers, fines, and bundles of fibers. Papers made from groundwood also lose strength and turn yellow with time. Thus, groundwood pulps are used only in relatively impermanent papers such as for newsprint, catalogues, magazines, and paperboard. Groundwood papers have excellent printing qualities because of high bulk, smoothness, resiliency, and good ink absorption. Newsprint contains

Fig. 15.5 Single rotating disc refiner. (*Courtesy Sprout, Waldron and Co.*)

about 80 percent groundwood and the other papers mentioned about 30-70 percent with the remainder being a chemical pulp for greater strength. Groundwood is the cheapest pulp made and also utilizes the entire wood giving essentially 100 percent yield.

Chemical Pulping

Lignin is the noncarbohydrate portion of extractive-free wood and, as shown in Table 15.1, accounts for 20-30 percent of the weight of wood. In manufacturing highly purified pulps of high whiteness, it is necessary to remove all of the lignin with minimum loss or degradation of the carbohydrate cell wall. Usually this is done in two steps, involving pulping to liberate the individual fibers and then bleaching these fibers to the desired whiteness. Unbleached pulps are usually tan to dark brown in color and are used in that form for grocery bags, wrapping paper, and corrugated containers.

Lignin is not a specific chemical entity but is rather a very complex, heterogeneous, three-dimensional polymer consisting largely of phenylpropane units joined together by various ether and carbon-carbon linkages. It cannot be isolated from wood in its entirety without extensive alteration of its structure and requires rather drastic conditions for its removal, which in turn may cause it to undergo further condensation reactions. About 70 percent of the lignin exists in the intercellular regions, and the rest is intimately associated with the carbohydrates in the cell wall. Thus lignin should be considered more as a *class* of related materials since its composition is not the same in all living plants. For example, significant differences exist between softwood and hardwood lignin, so it is probably correct only to refer to it in relation to its plant source and the method of isolation. Figure 15.1 shows a proposed structure that accounts for many of the reactions and chemical properties of lignin, but it should not be construed as a definite chemical formula.

The objective of chemical pulping is to solubilize and remove the lignin portion of wood, leaving the industrial fiber composed essentially of pure carbohydrate material. While many variations are used throughout the world, the most convenient classification of pulping methods is by whether they are acidic or alkaline. Each has its own specific advantages and disadvantages, but as can be seen in Table 15.4

the alkaline (sulfate) process accounts for over 90 percent of all chemical pulp produced in the United States. All processes use aqueous systems under heat and pressure.

The sulfite process uses a cooking liquor of sulfurous acid and a salt of this acid. While calcium was the most widely used base at one time, it has been supplanted by sodium, magnesium, and ammonia.

The sulfate process uses a mixture of sodium hydroxide and sodium sulfide as the active chemical. The term sulfate process is misleading, but it is called that because it uses sodium sulfate as the make-up chemical. The term kraft is also used to describe this process and is derived from the Swedish or German word for strength since it does produce the strongest pulp. Historically, sodium hydroxide alone (soda process) was first used as the alkaline pulping agent, but very few mills are still in operation since the pulp is weak and inferior to sulfate pulps.

Alkaline Processes. The pulping (cooking) process traditionally was performed on a batch basis in a large pressure vessel called a digester. Conditions used will vary depending upon the type of wood being pulped and the quality of end product desired. Typical conditions for kraft cooking are listed in Table 15.6.

Digesters are cylindrical in shape with a dome at the top and a cone at the bottom. Ranging in size up to 40 ft high and 20 ft in diameter, the largest will hold about 7,000 cu ft of wood chips (about 35 tons) for each charge. The chips are admitted through a large valve at the top, and at the end of the cook they are blown from the bottom through a valve to a large blow tank. During the cook the liquor is heated by circulation through a steam heat exchanger, which also avoids the dilution of the cooking liquor that would occur from heating by direct injection of steam.

TABLE 15.6 Typical Sulfate Pulping Conditions

Pressure	100-110 psig
Temperature	170-175°C
Time	2-3 hrs
Alkali charge	15-25% of weight of wood (calculated as Na_2O but consisting of approximately $5NaOH + 2\ Na_2S$).

Liquor to wood ratio is 4 to 1 (by weight).

In recent years the development of the continuous digester (Fig. 15.6) has been a very important factor, especially in the kraft industry. Chips are admitted continuously at the top through a special high-pressure feeder, and the cooked pulp is withdrawn continuously from the bottom through a special blow unit. Recent installations range in size up to 150 ft high and are capable of producing about 1000 tons of pulp per day in one unit. Cooking liquors and conditions are approximately the same as for the batch digesters. These units offer both good economics in the production of pulp and a quality advantage compared to the batch digester. However, since capital investment is somewhat higher, both systems are being installed in new mills.

Due to the high alkali charge, the chemicals must be recovered and re-used. This also alleviates pollution problems since the yield of pulp is only about 45 percent of the original wood weight and these organic residues must be eliminated. After being cooked in the digester, the pulp is washed in a countercurrent rotary vacuum washer system using three or four stages. The pulp is then ready for bleaching or for use in papers such as grocery bags where the brown color is not objectionable.

The separated liquor is very dark and is known as "black liquor." It is concentrated in multieffect evaporators to 60-65 percent solids. At this concentration the quantity of dissolved organic compounds from the wood (lignin and carbohydrates) is sufficient to allow the liquor to be burned in the recovery furnace.

By controlling the amount of excess air admitted to the furnace and the temperatures, the organics in the liquor can be burned. The inorganics collect on the bottom of the furnace as a molten smelt of Na_2CO_3 and Na_2S. Sodium sulfate is added to the liquor as make-up and is reduced to Na_2S by carbon. After dissolving in water, this mixture (called "green liquor") is reacted with slaked lime.

$$Na_2CO_3 + Ca(OH)_2 \longrightarrow 2NaOH + CaCO_3$$

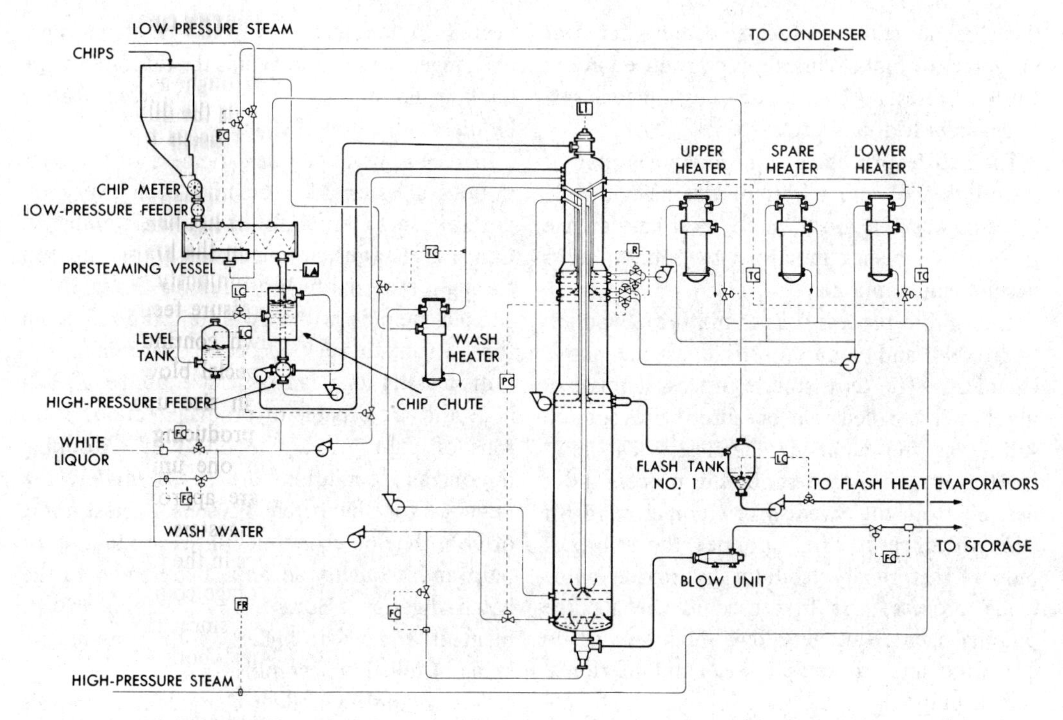

Fig. 15.6 Typical Kamyr continuous cooking system for sulfate pulp. (*Courtesy Kamyr, Inc.*)

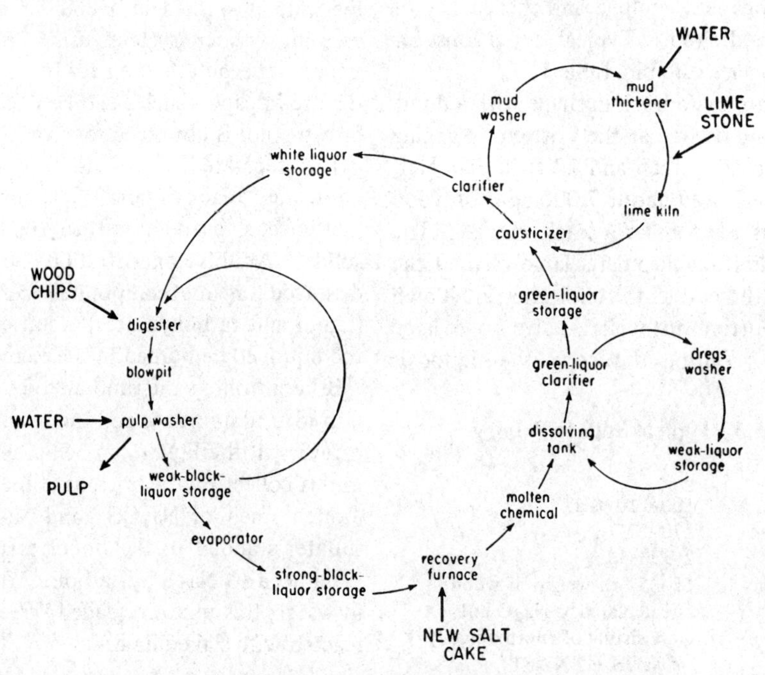

Fig. 15.7 Diagram showing cyclic nature of the Kraft recovery process. (*"Pulp and Paper," 2nd ed., Casey, J. P., Interscience Publishers, New York, 1967; by permission of John Wiley and Sons, Inc.*)

Since the Na_2S does not react with the lime the resultant mixture of NaOH and Na_2S (called "white liquor") can be reused to pulp more wood. The $CaCO_3$ sludge is filtered off, burned in a lime kiln and reused. Thus, the chemical system is a closed one as shown in Fig. 15.7 and this minimizes costs and pollution.

The kraft process has had a serious problem with air pollution due to the production of hydrogen sulfide, mercaptans, and other vile-smelling sulfur compounds. In recent years, the use of various techniques such as black-liquor oxidation, improved evaporators and furnaces, and control of emissions has greatly improved this situation. However, older mills are being forced to expend large amounts of money to bring their operations up to the new standards.

Practically any kind of wood can be pulped by the kraft process, and since it produces the strongest pulps with good economies, it has grown to be the dominant process in the world. While the odor problem is very severe, it does not appear as though this process will be supplanted in the near future; instead, it will be improved and modified. When pulping resin-rich woods such as southern pine, the kraft process yields turpentine and tall oil as valuable by-products. The steam generated in the recovery furnace is almost enough to make the pulp mill self-sufficient.

A recent discovery that small amounts of anthraquinone (0.05-0.25% on wood) added to alkali liquors can enhance delignification and produce pulp only slightly inferior to kraft has been of great interest. Similar compounds can act in catalytic amounts and much research into the mechanism is being done. The price of anthraquinone is high, and since it is consumed when the liquor is burned in the recovery cycle, this process has not yet been adopted. This development of a non-sulfur process is of tremendous technical and economic significance since it is a method which could eliminate the odor problem with kraft.

Sulfite Process. Lignin will react with the bisulfite ion (HSO_3^-) under acidic conditions to form lignosulfonates that are soluble in water. For many years this was the preferred process since it produced pulps of light color that could easily be bleached, it used cheap chemicals in fairly limited amounts so that no recovery was necessary, and it was a relatively simple process to operate.

While production of sulfite pulp has remained relatively constant for the last thirty years, the rapid growth of kraft pulping has reduced its share to less than 10 percent of the chemical pulp produced. There are several reasons for this, but the primary ones are the inability to cook resinous woods such as pine, problems in producing strong pulps from hardwoods, and recently, of the greatest importance, the lack of cheap and simple recovery systems such as in the kraft process to reduce water pollution problems. However, this process produces pulps with special qualities such as high-alpha-cellulose grades for rayon so that it will continue to be used.

Initially calcium was the preferred base since it was cheap and convenient to use. However, since no recovery system is available, most calcium-base mills have either ceased operation or have converted to sodium, magnesium, or ammonia for which recovery systems are available.

Regardless of the base used, the initial step is the burning of sulfur to produce sulfur dioxide (SO_2). The air supply to the burner must be carefully controlled, since too much air will enhance the formation of sulfur trioxide (SO_3) and subsequent production of sulfuric acid (H_2SO_4) which is very undesirable. The gas must also be cooled quickly from 1000°C, leaving the burner at below 400°C, also to minimize formation of SO_3. After cooling to 20-30°C, the SO_2 gas must be absorbed in water and reacted with the proper base to form the cooking liquor.

For calcium-base liquor, the gas was passed through towers packed with limestone with water flowing down through the tower. Because of the limited solubility of calcium bisulfite [$Ca(HSO_3)_2$], the pH of the liquor was very low (about 2) and free sulfurous acid was present. This is usually called the acid sulfite process. As mentioned before, calcium-base mills have essentially disappeared in the United States.

The so-called soluble bases are now used with

each having certain advantages. Since solutions of sodium, magnesium, and ammonium bisulfite are all soluble at pH 4.5, the current practice is to pulp at the higher pH and is usually called bisulfite pulping. Extremely long cooking times (7-10 hrs) are necessary with acid sulfite, whereas 4-5 hrs are sufficient with bisulfite.

Sodium base is the easiest to prepare (Na_2CO_3 or NaOH is usually used as the make-up chemical) and gives the highest quality pulp. However, while recovery processes are available, they are complicated and expensive. Magnesium base [from $Mg(OH)_2$] is somewhat more difficult to handle but good recovery systems are available and a majority of the sulfite pulp is now produced from this base. Ammonium base (from NH_4OH) has been used in the past. While the ammonia cannot be recovered, the liquor can be evaporated and burned without leaving any solid residue, thus reducing water pollution. As long as aqueous ammonia remains low in price, this process will be attractive since the SO_2 can be recovered from the waste gases by passing them through a wet scrubber flooded with fresh ammonium hydroxide.

Batch digesters are usually used in the sulfite process. Cooking temperatures are lower (140-150°C) and times are longer. Pulp yields are about the same as in the kraft process.

Spruce and fir are the preferred species for cooking by the sulfite process since they produce relatively stong, light-colored pulps. About 20 percent of newsprint consists of this type of pulp that has not been bleached. Thus the sulfite industry is concentrated in Canada, northern United States, and the Pacific Coast where the supplies of spruce and fir are greatest and the largest quantities of newsprint are produced.

A large amount of research has been done on developing products from the waste sulfite liquor and some success has been achieved. Vanillin, alcohol, and torula yeast can be produced as by-products, while the lignosulfonates are used as viscosity modifiers in drilling muds and similar uses. However, the majority of waste liquor is burned to recover the cooking chemicals and the heat values.

Other Pulping Processes. Various combinations of chemical and mechanical treatment have been used to produce pulps with specific properties. Mild chemical treatments to give partial delignification and softening are followed by mechanical means to complete fiber separation.

The neutral sulfite semichemical (NSSC) process is one in which wood chips, usually from hardwoods, are cooked with Na_2SO_3 liquor buffered with either $NaHCO_3$, Na_2CO_3, or NaOH to maintain a slightly alkaline pH during the cook. Unbleached pulp from hardwoods cooked to a yield of about 75 percent is widely used for corrugating medium. While bleachable pulps can be produced by this process, they require large quantities of bleaching chemicals and the waste liquors are difficult to recover. Currently many NSSC mills are located adjacent to kraft mills, and the liquors can be treated in the same furnace. Thus the waste liquor from the NSSC mill becomes the make-up chemical for the kraft mill, solving the waste problem. NSSC hardwood pulp is the premier pulp for corrugating medium and cannot be matched by any other process.

Chemi-mechanical pulps are usually produced by soaking the chips in solutions of NaOH or Na_2SO_3, and then refining them in disc refiners to produce a groundwood-type pulp. Chemical consumption is very low and yields are usually 85-95 percent. Currently much research is being done to improve the qualities of the pulp produced by this method.

Screening and Cleaning of Wood Pulp

The desired pulp fibers are usually between one and three mm in length with a diameter about one-hundredth as large. Any bundles of fibers or other impurities would show up as defects in the finished paper and must be screened out. Wood knots are usually difficult to pulp and must be removed.

Screening is usually a two-stage process with the coarse material being removed by screens with relatively large perforations ($\frac{1}{4}$ to $\frac{3}{8}$ in.). Additional fine screening is done with screens using very small (0.008 to 0.014 in.) slots to

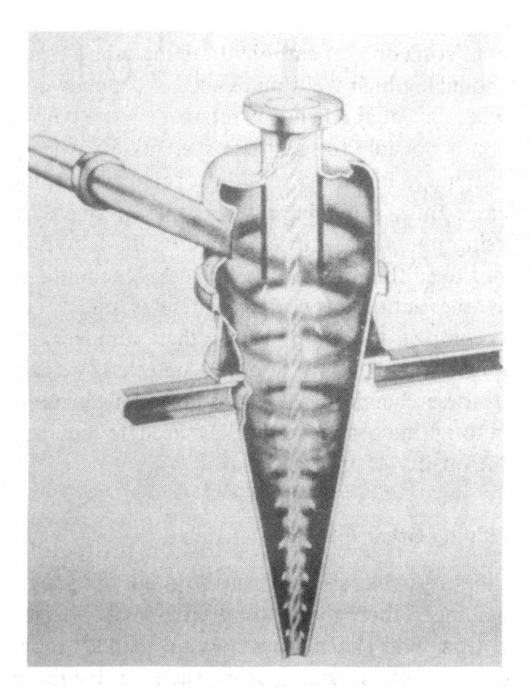

Fig. 15.8 Centrifugal cleaner for removal of dirt. (*Courtesy Bauer Brothers Co.*)

insure removal of oversized impurities. Screen size openings will depend on the species of wood being processed and the desired quality of end product. Because of the tendency of the fibers to agglomerate when suspended in water, it is customary to screen at very low solids (consistencies) of about $\frac{1}{2}$ percent fiber and $99\frac{1}{2}$ percent water.

To meet the ever increasing demands for cleaner pulps, the centrifugal cyclone cleaner (Fig. 15.8) has come into almost universal use. The screened pulp is pumped through these units at low consistencies and high velocities. The fiber slurry enters the cone tangentially at the top and a free vortex is formed with the velocity of the flow greatly increased as the diameter of the conical section is reduced. Heavier particles of sand, scale, or other dirt are forced to the outside of the cleaner and are discharged from the bottom tip through a small orifice. Due to the velocity gradients existing in the cone, the longer fibers (75-95 percent) are carried into the ascending center column and are discharged through the larger accept nozzle at the top. In a properly designed and operated unit, a shape separation is also made so that round particles, even though of the same specific gravity as the good fibers, will also be discharged as rejects through the bottom orifice. In this way, small pieces of bark are also removed. To reduce the quantity of rejects to an acceptable level, they in turn are processed through a second, third, or even a fourth stage of cleaners, thus holding the final loss of pulp to about $\frac{1}{4}$ to 1 percent of the feed depending upon quality demands and the dirt level of the incoming pulp.

Bleaching of Wood Pulp

The color of unbleached pulp ranges from the cream or tan of the sulfite process to the dark brown of the sulfate (kraft) process. While about 75-90 percent of the lignin has been removed by the pulping process, the remainder, along with other colored degradation products, must be removed by bleaching.

While it is possible to improve the brightness (whiteness) of the pulp in one stage, achievement of high brightness on an economic basis requires the use of several stages. Current practice uses combinations of chlorination with elemental chlorine (C), alkaline extraction with sodium hydroxide (E), and various oxidative stages using sodium or calcium hypochlorite (H), chlorine dioxide (D), or hydrogen peroxide (P). The pulp is washed between each stage to remove solubilized impurities. Many combinations are possible and each mill selects the sequence that fits its requirements the best. Those most commonly in use are:

CEH	CED
CEHD	CEDED
CEHDP	CEHDED

The greater the number of stages the higher the quality of the final pulp but at increased cost. Sulfite pulps are much easier to bleach and usually use only the 3- or 4-stage sequences, whereas kraft pulps require additional stages. Typical conditions for bleaching a kraft pulp would be as given in Table 15.7.

Chlorine and caustic are purchased, but

TABLE 15.7 Typical Conditions for Bleaching Kraft Pulp

Stage	*% Chemical on Pulp*	*Time (min)*	*Temperature (°F)*
Chlorination	5-6	30	70
Extraction	$2\frac{1}{2}$-3	60	140
Hypochlorite	1-2	90	105
Chlorine dioxide	$\frac{1}{2}$-1	240	160

chlorine dioxide must be generated at the site using sodium chlorate as the basic chemical.

Effluents from bleach plants are a source of great concern in regard to pollution. Large quantities of water are discharged (typically 10-20 thousand gal per ton) which are high in color, especially the water coming from the chlorination and extraction stages. Two experimental bleach plants are now operating with a new sequence that should alleviate this problem, and as soon as these prove commercially feasible it is anticipated that a significant change in bleaching practice will take place.

These new developments involve the replacement of the chlorination and extraction stages with a single stage involving gaseous oxygen and sodium hydroxide. The pulp is dewatered to about 15-25 percent solids, treated with about 4-6 percent NaOH, and passed through a reactor in a fluffed condition using oxygen gas at about 150 psi. The pulp consumes about 1-2 percent oxygen in about 15 mins and leaves in a semi-bleached condition. After washing, the use of conventional chlorine dioxide stages will produce the desired level of whiteness. The advantage of this process is that the effluent contains no chlorides and can be used as process water in the pulp mill. Any sodium and organics in the effluent will then enter the kraft recovery cycle and serve as make-up, thus reducing the pollution load. Currently the main disadvantage is the obvious one of introducing a new process involving rather unusual pieces of equipment; also, there is some degradation of the cellulosic fiber since oxygen will attack cellulose as well as lignin. However, it is expected that these problems will not prove insurmountable, and a rapid introduction of this process is anticipated.

These bleaching sequences are designed to remove lignin, yielding a highly purified fiber consisting only of carbohydrate material. When producing high-yield pulps such as groundwood where all of the lignin is retained in the pulp it is not possible to use these systems. However, extremely high brightness is not required in this case; thus, some improvement is attained by using one stage with either peroxide or hydrosulfite (dithionite). No yield loss is encountered as the action of both of these is merely to decolorize the pulp rather than remove any impurities. Usually about $\frac{1}{2}$ percent of either of these chemicals will give a noticeable increase in brightness and are widely used to upgrade the quality of groundwood.

Pulping Other Fibrous Materials

Fibrous materials that can be used for papermaking other than wood are waste paper, cotton and linen rags, waste manila rope, cotton linters, and grasses such as bamboo, esparto, bagasse (sugar cane residue), cereal straw, flax, jute, and hemp. As improved technology is developed, these other sources of fibers (particularly bagasse) will play a more important part in those countries where wood is in short supply. However, the United States, Canada, Scandinavia, and Soviet Russia will continue to depend on wood for all but a very small fraction of their needs. While the use of tropical hardwoods has been limited, it is anticipated that countries such as Brazil will assume major importance as pulp suppliers in the near future.

Currently about 25 percent of the paper produced in the United States is made from waste paper. The majority of this recycled paper (about 80 percent) is used "as is" without attempting to remove ink, dyes, or pigments from the paper. The resultant pulp is of rather poor quality and color and is used primarily as filler stock in paperboard. The higher quality waste paper is treated with sodium hydroxide and steam, and then bleached to produce a high-quality pulp that is used to replace virgin pulp from wood. Due to the pressure of municipalities to reduce their solid waste disposal, it is anticipated that the amount of waste paper used will increase in the next few years. Table 15.8 shows the types and quanti-

TABLE 15.8 Consumption Distribution of Selected Paperstocks (as in early 1970s)

Grade	*Paperstock (tons)*	*Consumption (%)*
Old newpapers	2,350,000	20.6
Old containers	3,500,000	30.7
Mixed waste	3,350,000	29.4
Special grades	2,200,000	19.3
TOTAL	11,400,000	100.0

Estimated Production of Selected Secondary Fiber Products

Product	*Production from Paperstock, Tons*	*Proportion of Total U.S. Consumption of the Grade Indicated, %*
Paper (from deinked waste)		
Newsprint	320,000	4.0
Book, uncoated	400,000	15.4
Printing, coated	300,000	9.1
Writing	300,000	10.7
TOTAL	1,320,000	
Paperboard (from waste used as is)		
Folding and Set-Up Boxboard	3,100,000	66.0
Other Boxboard	1,600,000	41.0
Linerboard	300,000	3.9
Corrugating Medium	800,000	18.2
Other Containerboard	1,500,000	55.6
TOTAL	7,300,000	
Building products	1,700,000	38.7

In the case of paper the amount made from waste paper is 4.3% of the total U.S. consumption of paper, whereas for paperboard it is 31% of the total.

ties of waste used, as well as the quantities of paper and board made from that waste.

Stock Preparation

Stock preparation in a paper mill includes all intermediate operations between preparation of the pulp and the final papermaking process. It can be subdivided into (1) preparation of the "furnish" and (2) "beating" or "refining." Furnish is the term used to describe the water slurry of fiber and other chemicals which goes to the paper machine. Beating or refining refers to the mechanical treatment given to the furnish to develop the strength properties of the pulp and impart the proper characteristics to the finished paper.

Cellulosic fibers are unique in that when suspended in water they will bond to each other very strongly as the water is removed by filtration and drying, without the necessity of an additional adhesive. This is due to the large number of hydrogen bonds which form between thc surfaces of fibers that are in close contact as the water is removed. This bonding is reversible and accounts for the well-known fact that paper loses most of its strength when wet. If paper is suspended in water and agitated, it will separate into the individual fibers which allows the easy reuse of waste paper or the processing-waste from the paper mill itself.

In order to enhance the bonding capability of the fibers, it is necessary to mechanically beat or refine them in equipment such as beaters, jordans, or disc refiners. This treatment of the pulp slurry at about 3-6 percent consistency is done by passing the pulp between the two rotating surfaces of the refiner. These surfaces contain metal bars and operate at very close clearances. As the fibers pass between the bars they are made more flexible and a larger surface for bonding is developed by the mechanical action.

This refining brings about fundamental changes in the pulp fibers and increases the degree of interfiber bonding in the final sheet of paper. Thus the final properties of the paper can be significantly changed by varying the degree and type of refining. As additional refining is performed, properties such as tensile strength, fold, and density are increased while tear resistance, opacity, thickness, and dimensional stability are decreased. Thus, the proper refining conditions must be selected to bring out the desired properties without detracting too much from other properties.

The furnish of a paper machine varies widely depending on the grade of paper being made. Newsprint usually consists of about 80 percent groundwood and 20 percent chemical fiber (sulfite or semibleached sulfate). Bag papers and linerboard are usually 100 percent unbleached softwood kraft, although obviously these could be made from bleached fibers if white containers are desired. Printing papers are made from bleached pulps and contain both hardwoods and softwoods. By selecting the proper pulps and refining conditions, a wide variety of paper qualities can be achieved.

However, the paper industry is a large user of chemicals, as it has been found that relatively small quantities of additives can materially change the properties of paper. Use of 1-2 percent rosin size and 2-3 percent alum $[Al_2(SO_4)_3]$ will greatly increase the resistance of paper to penetration by water or ink. Pigments such as kaolin clay, calcium carbonate, and titanium dioxide are added in amounts up to 15% to increase opacity and give a better printing surface. Organic dyes and colored pigments are added to produce the highly colored papers used in business and printing papers. Other additives such as wet-strength resins, retention aids, and starch can be used to give particular properties that are needed. Thus, in order to produce the wide variety of grades of paper now available the papermaker selects the proper pulps, refining conditions, and additives, and then combines the pulp and additives before sending them on to the paper machine for the final step in the process.

Papermaking Process

Some paper mills are not integrated with pulp mills, and it is necessary for these mills to use

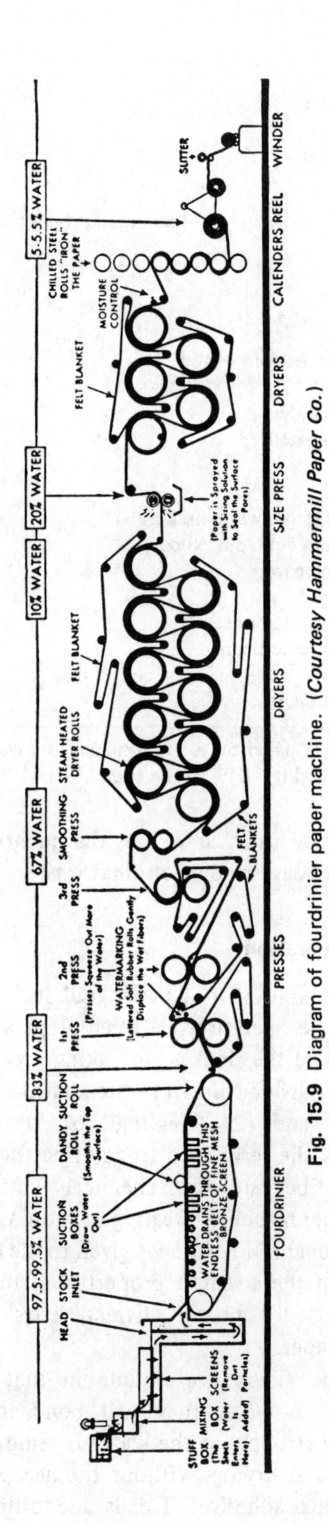

Fig. 15.9 Diagram of fourdrinier paper machine. (*Courtesy Hammermill Paper Co.*)

dried, baled pulp manufactured at a separate location. Many mills making limited quantities of highly specialized papers fall into this category since it allows maximum flexibility in selecting the optimum pulps for a particular paper grade. However, the papermaking process is the same regardless of the source of pulp.

After the furnish has been prepared with the proper refining treatment and additives, it is stored in the machine chest and then fed continuously into the paper machine system. A refiner or jordan is placed in this line to give the paper-machine operators the opportunity to make small adjustments in the quality of the furnish as needed to give the desired paper properties. Screens and centrifugal cleaners are also included to insure high quality paper.

The papermaking process is essentially a system whereby the pulp is diluted to a very low consistency (about $\frac{1}{2}$ percent) and continuously formed into a sheet of paper at high speeds, and then the water is removed by filtration, pressing, and drying. The basic units of the fourdrinier paper machine are diagramed in Fig. 15.9; a picture is shown in Fig. 15.10.

The section of the paper machine where the paper is formed is referred to as the "wet end." The fourdrinier machine is characterized by a headbox which allows the diluted stock to flow through an orifice (slice) onto the flat moving wire. This is actually an endless wire belt which returns on the underside of the machine thus allowing the process to proceed continuously. As a low headbox consistency (about $\frac{1}{2}$ percent) is necessary for good formation, the volumes of water handled are very large (about 10,000 gal/min for a machine producing 300 tons per day). Much of the water is removed through the wire by the action of the table rolls and foils that support the wires in the forming area immediately following the headbox. At this point the stock consistency has been raised to about 2 percent and the stock no longer drains freely. By passing over suction boxes operating at fairly high vacuum (6-8 in. Hg), the consistency is raised to about 15-20 percent. A dandy roll (covered with woven wire) rotating on top of the wet paper is used to improve the formation and can impart a watermark if it contains the proper pattern. The suction roll after the suction boxes has several functions: it removes additional water; it serves as the driving roll for the wire, and it serves as the point at which the paper separates from the wire and passes into the press section while the wire returns to pick up additional pulp at the headbox.

Modern fourdrinier machines are available up to a width of 360 inches and can operate at speeds up to 3,000 ft/min. Newsprint machines

Fig. 15.10 Modern high-speed paper machine. *(Courtesy Beloit Corp.)*

are usually the widest and fastest, while those making heavier grades such as linerboard run somewhat more slowly. In the United States, there are several machines capable of producing over 1,000 tons per day of linerboard. More specialized grades such as bond and printing papers are usually produced at a lower speed on narrower machines, and 150-250 tons per day would be considered a high output. Many specialized grades such as filter paper and tracing paper are produced on very small, slow machines producing only a few tons per day. Machines making tissue paper for sanitary purposes use modifications of the standard fourdrinier to produce tissue at speeds of up to 5,000 ft/min. Because of the lightweight of this paper, it is necessary to make many modifications in the equipment shown in Fig. 15.9.

There are many new methods of forming a sheet of paper being developed. The most common concept is to introduce the pulp slurry between two converging wires and remove the water from both sides. This is called "twin-wire" forming, and many variations are now in operation. This technique has many advantages and may eventually replace the fourdrinier.

A second method used to form paper is the cylinder machine. Actually these machines are used primarily to form the multi-ply board used in packaging such as cereal boxes. The cylinder wet end consists of one or more (up to eight) cylinder vats, each of which forms a separate wet web of fibers. Each vat contains a woven wire-covered cylinder rotating in the diluted pulp slurry. The liquid head on the outside of the cylinder is greater than that on the inside resulting in a flow of water through the wire and out of the vat. The pulp mat that is formed on the face of the cylinder is removed by an endless woolen felt which moves in contact with the cylinder by means of a rubber roll riding on top of the felt. With machines having more than one cylinder, the same felt moves from one cylinder to the next and the wet sheets from each cylinder are laminated to each other on the bottom side of the felt. Thus, very heavy papers or boards can be fabricated by multivat machines. Since each vat can be supplied with a different type of pulp, it is possible to make boards with a white surface of high-quality pulp and a center of low-cost pulp from waste newspaper or other cheap grades. Due to hydrodynamic problems, cylinder-machine speeds are limited to about 250 feet per minute and widths of about 150 in. However, because of the heavyweight board produced and the widespread use of cheap waste paper for most of the furnish, the cylinder machine is widely used. New forming units looking like miniature fourdrinier units (Ultra-former, Inverformer) have been developed and are rapidly replacing the old-fashioned cylinder vat since their speed is not as limited.

From the wet end of the machine, the wet sheet is conveyed by woolen felts through a series of roll-type presses for further water removal, increasing the consistency to about 35 percent. The sheet is then threaded through the dryer section consisting of a long series of steam-heated cast iron cylinders which reduce the moisture content to approximately 5 percent which is about the equilibrium moisture content for cellulosic fibers at 40-50 percent relative humidity. Tissue machines use one large dryer (called a Yankee dryer) ranging from 8 to 18 feet in diameter. Because of the light weight of the tissue paper, it can be dried at high speeds on a single dryer.

After drying, the paper is compacted and smoothed by passing through a calender stack consisting of a vertical row of highly polished cast iron rolls. The paper is then wound into rolls on the reel, and these rolls are then rewound on a winder into shipping rolls or sheeted and trimmed to the desired size. Figure 15.11 shows the large reels at the dry end of a large fourdrinier machine.

The quality of many papers is improved by a surface treatment of the paper. A size press about two-thirds of the way along the dryer section can apply a solution of starch to improve surface bonding. More sophisticated coating equipment can apply a layer of pigments and binders to give the desirable properties associated with printing papers such as those used in high quality magazines and advertising.

Paper is converted into its end product by many different methods. Some papers are sheeted, shaped, or fastened into final forms while others require more elaborate processing.

Fig. 15.11 Dry end of large fourdrinier machine. *(Courtesy St. Regis Paper Co.)*

Corrugated boxboard is made by gluing sheets of linerboard to each side of a fluted sheet of corrugating medium. Papers for packaging may be laminated to polyethylene film, aluminum foil, or coated with waxes and hot-melt resins. The printing and bag- and box-making industries depend on the production of the many mills which produce the several hundred grades of paper used in the United States, and each user may require special paper characteristics to match his process.

BOARD AND STRUCTURAL MATERIALS

Board, sheets, panels, and other structural materials are manufactured from wood fibers and various other vegetable fibers, from wood particles, and from paper. The industries making these products are not generally classified as chemical industries; nevertheless, they are closely related to chemical industry. Fiberboard manufacture is similar to papermaking; particle boards and paper laminates involve the use of synthetic resins and, therefore, chemical technology.

Fiberboard

Fiberboard is the term for rigid or semirigid sheet materials of widely varying densities and thicknesses manufactured from wood or other vegetable fibers. The board is formed by the felting of the fibers from a water slurry or an air suspension to produce a mat. Bonding agents may be incorporated to increase the strength, and other materials may be added to give special properties, such as resistance to moisture, fire, or decay.

Fiberboards are manufactured primarily for panels, insulation, and cover materials in buildings and other structures where flat sheets of moderate strength and/or insulating capacity are required. They are also used as components in doors, cupboards, cabinets, furniture, and millwork.

Classification of fiberboards is best done on the basis of density, since there is a great deal of overlap when classifying by use only. Table 15.9 shows the density classification of fiberboards as well as some of their major uses.

The production of fiberboards goes back to 1898 when the first plant was built in Great Britain. However, large-scale production, mainly of insulation board, developed in the United States between the two world wars. The United States is still the largest producing country and accounts for about a third of the world output. Fiberboard production figures for recent years are shown in Table 15.10.

TABLE 15.9 Classification and Uses of Fiberboards*

Fiberboards	*Density Classification* g/cm^3	*lb/cu ft*	*Major Uses*
Noncompressed (insulation board) semirigid insulation	0.02-0.15	1.25-9.5	Heat insulation as blankets and batts; industrial cushioning
Rigid insulation board (includes wallboard and softboard)	0.15-0.40	9.5-25.0	Heat and sound insulation as sheathing, interior panelling, base for plaster or siding, thick laminated sheets for structural decking, cores for doors and partitions, acoustical ceilings.
Compressed Intermediate or medium density fiberboard (includes laminated paperboards and homogeneous boards)	0.40-0.80	25-50	Structural use and heat insulation as sheathing base for plaster and siding, interior panelling, containers, underflooring
Hardboard	0.80-1,20	50-75	Panelling, counter tops, components in doors, cabinets, cupboards, furniture, containers, and millwork, concrete forms, flooring
Densified hardboard (superhardboard)	1.20-1.45	75-90	Electrical instrument panels, templets, jigs, die stock

*From information in "Fibreboard and Particle Board," Food and Agriculture Organization of the United Nations, Rome, 1958.

There has been a much more rapid increase in the production of compressed fiberboards (hardboards) than noncompressed fiberboards (insulation board) during recent years.

Wood is the principal raw material for the manufacture of fiberboards. The species used are numerous, including both softwoods (coniferous) and hardwoods (broad-leaved) and vary from region to region. The wood may come from harvesting of commercial timber and pulp species, as well as from species not commonly used for lumber or pulp, from cull timber, from logging and forest management residues, and from industrial wood residues.

TABLE 15.10 Fiberboard and Particle Board Production[a]

		Figures in 1000 metric tons		
Hard pressed board (Compressed fiberboard)	1966	1970	1974	1978
U.S.A.	980	1397	1889	2028
World	4177	6140	8055	8652
Structural and insulating board (Noncompressed fiberboard)				
U.S.A.	1130	4424	4699	5142
World	2113	8067	8501	9247
		Figures in 1000 meters^3		
Particle board				
U.S.A.	1155	3127	5476	6460
World	6237	19170	31747	38589

[a]*Source*: "Yearbook of Forest Products," Food and Agriculture Organization of the United Nations, Rome, Italy.

Other fiber raw materials for fiberboard manufacture are bagasse (sugar cane residue after sugar extraction) and waste paper. Only minor amounts of other plant fibers are used.

Wood handling and preparation for fiberboard manufacture is much the same as described for pulp and paper. Wood is debarked and chipped with the same type of equipment. If the chips are to be first extracted for resins or tannin, then cylinder or drum-type chippers may be used instead of disc chippers.

Fibers for fiberboard are coarser and less refined chemically than those used for paper. Processes are used which bring about fiber separation with minimum loss in chemical components and in maximum yield. The pulping processes used are generally the following: (1) mechanical, (2) thermal-mechanical, (3) semichemical, and (4) explosion methods.

Mechanical Pulping. The mechanical pulping process is the same as that described for making paper pulp. Stones of coarse grit are used to give somewhat coarser fiber of higher freeness than the usual groundwood for papermaking. Freeness is a measure of the ease of drainage of water from the pulp. A fast drainage rate is required since thick mat is produced in forming the wet sheet and this must drain rapidly to maintain an economical rate of production.

A very coarse shredded wood fiber is made by a shredder, consisting of two cylinders to which are attached numerous small pointed hammers, which swing freely as the cylinders are rotated. These hammers "comb" or shred wet green wood, such as aspen, into stiff coarse bundles of fibers, producing a bulky pulp. This type of equipment is not widely used, and most mechanical pulp is made on conventional stone grinders.

Untreated green or water-soaked wood chips may also be directly refined in disc mills of either the single-rotary or double-rotary type similar to those used for making refiner groundwood. This gives a coarse pulp acceptable for insulation board.

Thermal-Mechanical Pulping. Generally, chips are given a steaming or other heat-treatment prior to or during defibering in a disc mill. Steaming or heating in hot water softens the wood so that upon grinding a pulp is produced with fewer broken fibers and with less coarse fiber bundles. Steaming is generally preferred and is carried out in a digester under a variety of conditions of time and temperature. A typical steaming period is about 30 min at 75 psi. If iron digesters are used, a small amount of alkali may be added to the chips to prevent corrosion by the organic acids produced by the hydrolytic action on the wood. The steamed chips are defibered in a disc-type attrition mill having two discs made of special alloys, one or both rotating, similar to those used for refiner groundwood.

The pulp may or may not be screened, as necessary. It is usually given some further refining to give maximum strength. Sizing and other additives are introduced and the pulp suspension is delivered to the wet-sheet-forming machine.

A special continuous thermal-mechanical process has been developed whereby the wood chips are steamed and ground while at elevated temperatures and pressures. The feature of the process is the combination of steaming and defibering in one unit in a continuous operation. The entire operation is carried out under pressure and no cooling takes place prior to defibering. Wood chips are continuously introduced by a plunger feed mechanism into a preheater where they are heated to 170-190°C by steam at 100-165 psi. Passage through the preheater takes 20-60 seconds, after which the hot chips are fed by a screw directly to the single rotating disc refiner, where they are ground while at the above temperature and pressure conditions. At these conditions the lignin, which is concentrated in the intercellular regions (middle lamella) of the wood, becomes somewhat thermoplastic, permitting easier separation of the fibers. The fibrous material is exhausted to the atmosphere through relief valves. Due to the very short steaming time it is claimed that little hydrolysis takes place so that there is little loss in wood substance, the yields being 90-93 percent. Additional refining is necessary for the preparation of insulation board stock, and slight refining may be desirable for hardboard stock, especially to break down slivers.

Semichemical Pulping. In some cases wood or other fibrous raw material may be given a mild chemical pretreatment prior to mechanical defibering. The processes are similar to those described for making paper pulps by semichemical methods. Generally, the conditions of treatment are somewhat milder than for paper pulp in order to get maximum yields. The chemical treatments usually involve cooking with neutral sulfite, caustic soda, or lime solutions. Yields from wood chips generally are above 80 percent.

Explosion Process. A unique process for defibering wood was developed by W. H. Mason. Wood chips, about $\frac{3}{4}$ in. long, prepared in conventional chippers and screened, are subjected to high pressure, in a cylinder, commonly called a gun, about 2 feet by 6 feet in size, and ejected through a quick-opening valve. The elevated temperature softens the chips and, upon ejection, they explode into a fluffy mass of fibers and fiber bundles. The process involves thermal plasticization of the lignin, partial hydrolysis, and disintegration by the sudden expansion of the steam within the chip.

About 260 lb of wood chips are fed into the cylinder and steamed to 600 psi for 30-60 secs. Then the pressure is quickly raised to 1000 psi (about 285°C) and held only about 5 seconds before suddenly releasing the charge into a cyclone. The time of treatment at this high pressure and temperature is critical and depends upon the species of wood, chip size, moisture content, and the quality of product desired. The steam is condensed in the cyclone and the exploded fiber falls into a stock chest where it is mixed with water and pumped through washers, refiners, and screens.

The high temperature to which the chips have been subjected causes appreciable hydrolysis of the hemicelluloses, resulting in a somewhat lower yield of pulp than obtained by mechanical or thermal-mechanical pulping. The hydrolysis results in a final board product with an enriched lignin content; about 38 percent compared to about 26 percent in the original softwood. The lignin content of the pulp can be varied by controlling the steaming process.

Board Forming. Pulp prepared by any of the above processes may be used for making insulation board. Mechanical (groundwood) pulp was the first type of pulp used in large-scale production of insulation boards, and is still being used in many plants. Pulps from other sources, such as disc mills, may be admixed with it. Groundwood pulp is not considered satisfactory for hardboard, and most hardboard is made from pulp prepared by the explosion process or by defibering with disc refiners.

Board making is basically similar to papermaking and involves refining, screening, mixing of additives, sheet forming, and drying operations. Pressing is also required for hardboards. The pulp is refined and screened prior to sheet formation.

Sizing agents in amounts up to one percent of the fiber are added to the pulp in mixing chests. Paraffin wax emulsion is commonly used for all types of baord. For insulation boards, rosin, cumarone resin, and asphalt are also used. Often a mixture of rosin and paraffin emulsion is used, with 10-25 percent rosin in the mixture. For hardboards, paraffin wax is the most common sizing agent, though tall oil derivatives and phenol-formaldehyde resins are also used. The sizing agent is precipitated on the fibers by alum, with careful control of pH; the latter may be between 4.0 and 6.5, according to the conditions.

The strength properties of a fiberboard depend mainly upon the felting characteristics of the individual fibers and to a lesser degree upon their interfiber bonding. The felting or forming process is usually done from a water suspension of the fiber at a consistency of around one percent. This is the *wet-felting* process. A relatively new *air-felting* or dry-forming process has been developed and is being used in a few American plants for hard-board manufacture.

The wet-felting process is generally carried out in a manner similar to papermaking, i.e., in a continuous operation on a fourdrinier machine or on a cylinder machine. The machines move more slowly than in the case of papermaking (5-45 ft/min on the fourdriner) and a coarser mesh wire is employed. In the cylinder-machine method, a single large vacuum cylinder, 8-14 ft in diameter, or two cylinders counterrotating and forming a two-ply sheet are most commonly used. Further water removal is effected by suction boxes and press rolls. The wet sheet is cut to length and conveyed on rollers from the press sections to a tunnel type drier. Then the dried sheets are cut into desired sizes.

A third type of wet-felting is a discontinuous method, known as the *deckle-box* method. The deckle box consists of a bottomless frame which can be raised or lowered onto a wire screen. A measured quantity of stock sufficient to form one sheet is pumped into the deckle box and vacuum applied to the lower side of the screen. After most of the water has drained off, pressure is applied from the top to express more water and compact the sheet, reducing its thickness. The deckle-frame is then raised and the sheet conveyed to the driers.

For the recent air-felting process, defibering is usually done in disc mills with control of the moisture content to give the minimum amount possible, consistent with good defibering conditions. Additional moisture may be removed by preheating the air that conveys the fiber from the refiners to the cyclone. The fiber may be further dried in a tunnel or other type of drier. Fines are removed either by air classification or screens after the drier. Wax for sizing is introduced either with the chips or added as a spray before or after passing the disc mills (about 2.5 percent of the weight of the fiber). Sometimes 0.5–5 percent of phenolic resin is added, depending upon the quality of board desired. The fiber-blend is fed to a moving screen by a metering unit through a combined air and mechanical action. The fibers felt as they fall on the screen and the fiber mat thus formed is precompressed between belts and/or rollers. If the board is to be wet pressed, water is added by spraying.

After the felting or sheet-forming operation, the subsequent operations differ for insulation board and hardboard (see Fig. 15.12). For insulation board, the sheets are dried without further compression, whereas for hardboard, the sheets are either pressed and dried simultaneously (wet-pressing) or are first dried and then pressed (dry-pressing). Air-felted sheets are pressed directly after forming.

Drying and Pressing. The wet-felted sheets for insulation board or for dry-pressed hardboard, containing 50–80 percent water, may be dried by any of three methods, namely (1) tunnel kilns using racks or carts to support the sheets, (2) steam-platen dryers, and (3) continuous roller driers of single or multideck arrangement. Most widely used is the continuous roller-type multideck drier, which has an average length of 150 to 300 ft but may be more than 600 ft long. An average drier will have 8 decks and be 12 ft wide.

The pressing conditions greatly affect the board properties. The conditions of time, temperature, pressure, and moisture content will depend upon the fiber in the board and the product desired. In wet pressing, a typical cycle has 3 phases and lasts from 6–15 minutes. The first is a short high-pressure stage (up to 710 psi) to remove most of the free water and

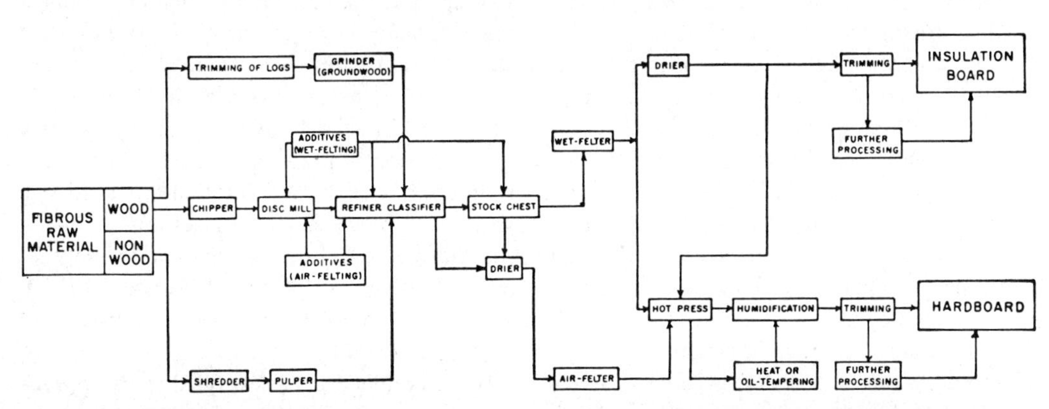

Fig. 15.12 Schematic outline for insulation board and hardboard manufacture. (*From "Fiberboard and Particle Board," Food and Agricultural Organization of the United Nations, Rome, 1958.*)

bring the board to the desired thickness; the second step serves to remove water vapor and requires most of the time; the third stage is a final short period at high pressure to effect a final "cure" or bonding by plastic flow of the lignin. To secure this fiber-to-fiber bond, a temperature of 185°C must be attained and temperatures up to 210°C may be used to increase production rate. In the dry-pressing process the cycles are shorter ($1\frac{1}{2}$-$3\frac{1}{2}$ min) and the temperatures and pressures are usually higher.

Conditioning. After hardboard has been hot-pressed or has been heat-treated or oil-tempered, its moisture content is well below what it will reach in equilibrium with the atmosphere in normal use. Such very dry boards will change dimensions upon picking up moisture and may warp. It is important, therefore, to humidify the boards under controlled conditions before packaging. The desired equilibrium moisture content (E.M.C.) reached will vary from 5 to 12 percent, depending upon the nature of the board and the general humidity conditions in the region of use.

Most humidifying is done in the chambers or tunnels, kept at 80-85 percent R.H. and 38-50°C. A lesser-used system of conditioning is by water spraying and dipping followed by standing to allow uniform absorption.

Special Treatments. Hardboards are often given a special treatment to improve strength and resistance to moisture. *Heat treatment* is a method which has come into wide use. The boards, which are kept apart to permit hot air circulation, are heated in chambers by either batch, continuous, or progressive systems. Typical heating conditions are 5 hr at 155-160°C. The heating increases the strength (except impact), sometimes as much as 25 percent, and the water resistance is improved. This operation may replace sizing wholly or in part. Some exothermic reaction takes place in the board and the heat developed must be removed by the hot circulating air to prevent burning. Probably some chemical condensations occur in the wood fiber, producing an internal resin system, and there is the possibility of some cross-linking of large molecules.

Some hardboard is *oil-tempered.* A drying oil, such as linseed, tung, perilla, soya, or tall oil, or an alkyd resin, is impregnated into the board, by passing the hot-pressed board through a hot oil bath. About 4-8 percent of the oil is absorbed. The board is then heated in a kiln with circulating air at 160-170°C for 6-9 hr. This treatment hardens the oil as well as bringing about chemical reactions in the fiber and results in greater strength and moisture resistance.

Various additives may be incorporated into insulation boards and hardboards, or they may be surface-treated to bring about resistance to decay, insects, and fire. Pentachlorophenol, copper pentachlorophenate, and several arsenicals are commonly used for preservative treatments. The sodium salt of pentachlorophenol is added before sizing and is precipitated onto the fiber along with the size by the alum. Arsenic trioxide is usually added at the headbox of the board-forming machine. Special fire-retarding-paint coatings are sometimes used to give resistance to the spread of flame.

Particle Board

Particle boards are composed of discrete particles of wood bonded together by a synthetic resin adhesive, most commonly urea-formaldehyde or phenol-formladehyde. The material is consolidated and the resin cured under heat and pressure. The strength of the product depends mainly upon the adhesive and not upon fiber felting as in the case of fiberboards, although the size and shape of the particles influence strength properties. They may be fine slivers, coarse slivers, planar shavings, shreds, or flakes. They are divided into two main groups, namely (1) hammer-mill produced particles (slivers and splinters from solid wood residues, feather-like wisps to block-shaped pieces from planer shavings) and (2) cutter-type particles sometimes called "engineered" particles (flakes and shreds). The various steps in particle-bond manufacture are illustrated in Fig. 15.13.

Hammer-milled particles usually vary appreciably in size. Dry raw material produces greater amounts of fines than green wood. Cutting

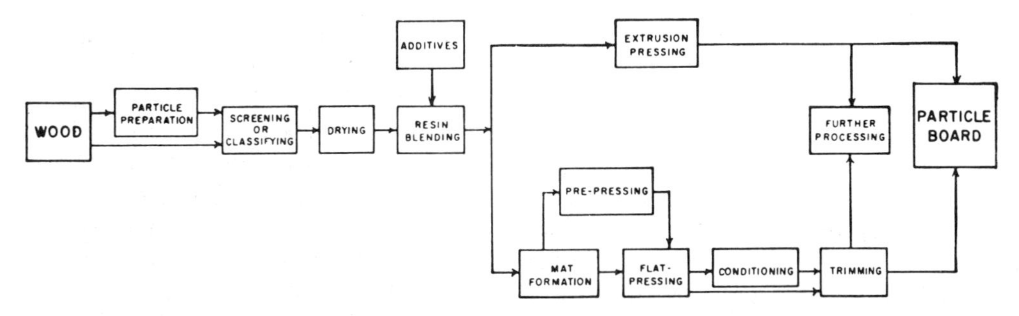

Fig. 15.13 Schematic outline for particle-board manufacture. *(From "Fiberboard and Particle Board," Food and Agriculture Organization of the United Nations, Rome, 1958.)*

machines (either cylinder-type or rotating-disc-type) give more uniform particles, with the length dimension in the direction of the grain of the wood. The thickness, size, and shape of particles influence the strength of the board. Boards made from sawdust have the lowest strength properties, hammer-milled particles give boards of intermediate strengths and solid wood cut to flakes gives boards of highest strengths.

Particle boards may be made in a wide range of densities. Low-density or insulating types are a comparatively recent development in Central Europe, whereas the high-density-hardboard types are an American development. Most particle-board production is in the middle-density range.

Particle boards are most commonly used as core stock for veneer in furniture and in doors, as interior panels for walls and ceilings, a sub-flooring, as sheathing and siding, and as components in interior millwork. The dense types are used in the same way as fiberboard hardboard, described in the previous section. Both dense particle boards are hardboards, after receiving a surface coating, may be printed with decorative designs.

Particle board production has increased rapidly, both in the United States and worldwide, in recent years. During the period 1966–78, the increase was about six-fold (see Table 15.10).

Paper-Base Laminates

Paper-base laminates are panels or other laminated assemblies composed of many plies of resin-treated paper molded together under high temperatures and pressure to produce rigid structures which no longer have the characteristics of paper. These products are widely used in the electrical and machine industries for insulators, gears, pulleys, and a multitude of machine parts. They possess high impact strength and toughness, good electrical insulation, high dimensional stability, and are not subject to corrosion; they also have a dampening effect on sound, eliminating rattle and drumming in steel cars and machinery. Furthermore, they can easily be manipulated into complex shapes and can be drilled, turned, and sawed. These properties make these products of great industrial value. They are also used in making trays, light flooring panels, table and counter tops, and many other products employing panels.

Paper-base laminates fall into four major classes, namely (1) mechanical or structural, (2) electrical, (3) punching, and (4) decorative. Phenolic resins are especially suitable and mostly used where mechanical strength and resistance to heat, water, and electricity are required. For punching-grade laminates, phenolic resins are specially modified with plasticizers or drying oils to yield laminates having good plasticity and elasticity. For electrical grades the phenolics are generally catalyzed with ammonia, amines, or less conductive catalysts in place of the stronger alkaline catalysts otherwise used.

For decorative laminates, urea and melamine resins are the principal resins used. The melamines are used where translucent, light-colored products with good heat and water resistance are required. Polyesters and some melamines are used for low-pressure laminating, enabling the continuous production of counter and table tops by passing the assembly through a set of rolls and then through a heating chamber.

There are two broad general methods for introducing resins into papers for making paper-base laminates; namely (1) by the beater addition process and (2) by impregnation of the paper sheet. In the beater addition process the resin is added to the pulp in the beater and then precipitated on the fiber by alum or an acid. The resin adheres to the pulp fibers, which are then formed into a sheet. This type of paper-plastic combination is commonly termed *resin-filled paper.* These resin-filled papers may be made into flat or shaped preforms. The former are flat sheets; the latter are vacuum-felted to a shape closely conforming to that of the final molded product. Shaped preforms are used for deep forms requiring high strength contours. Little flow is required on molding, hence the paper sheet does not break and greater strength is thereby obtained. Stretchable cross-creped paper for postformable laminates is a new development.

The more common method of application of the resin is by impregnation of the wet or dry paper sheet (web) with a solution of resin. Such papers are termed *resin-impregnated papers.*

Water-soluble and alcohol-soluble types of phenolic resins are both used for paper-base laminates. The former tend to give more brittle but more dimensionally stable products than the latter. Phenolic resins are mostly applied by impregnation of the wet or dry paper sheet producing *resin-impregnated papers*, although some progress has been made in the slush stock addition procedures to produce phenolic *resin-filled papers.*

In the wet-web process for impregnating paper for laminates, the wet sheet of paper on the paper machine is carried through a resin bath while the sheet is supported by a wire. The sheet contains up to 65 percent water, and the amount of resin taken up depends upon the moisture content of the paper, solids content of the resin solution, viscosity of the resin solution, temperature of the bath, machine speed, and pressure of the squeeze-rolls. Only water-soluble or water-dispersible resins can be used in the wet-web process.

Generally, however, phenolic impregnations of paper for laminates and other purposes demanding a high resin content are done with dry paper on off-machine equipment. The dry paper sheet is continuously passed through a resin solution, generally an alcoholic solution and then moves under and between two metering rolls, after which the paper is dried to remove solvent and to complete the condensation of the thermosetting phenolic resin.

The addition of phenolic resins in the form of emulsions to the slush stock by the beater-addition process has received considerable study in recent years. Papers containing 45 to 55 percent resin and even as high 65 percent have been prepared. The resin is precipitated by adding alum to the beater. Papers prepared in this way are highly plastic and are suitable for low-pressure molding. The phenolic resins can also be used in combination with elastomers, such as GR-S types, neoprene, hycar, and vinyl polymers, to produce laminates of high impact strength and greater elongation under tension.

Lignin is also commercially used as a resin for paper laminates. Its cheapness and availability as a by-product in pulp manufacture make it attractive. One of the most successful products is made with lignin recovered from the spent liquor of the soda pulping process. The lignin may be added to the pulp suspension in solution and precipitated directly among the fibers or added in a pre-precipitated form.

Properties of paper-base laminates depend both upon the resin and the paper. In general, the final product has the characteristics of the resin used, provided over 30 percent is resin. The paper acts as a structural reinforcer, greatly enhancing toughness, and tensile, flexural, and impact strength.

Electrical properties of the laminate depend both upon the paper and the resin, though the amount of resin absorbed is the most important factor affecting electrical insulating properties. Papers of low-power factor, good dielectric strength, and high dielectric constant are necessary for electrical grades of laminates. Other desired paper properties for good paper-base laminates are uniformity, cleanliness (freedom from slime, dirt, and fiber bundles), low finish, neutral pH, freedom from chemicals (bleach residues, etc.), low and uniform moisture content (under 4 percent), and uniform absorbency. High absorbency is desired when high electrical resistance and minimum water absorption are required, since greater amounts of resin are taken up by high absorbency papers. For high impact strength-punching grades of laminates, a low absorbency paper is necessary to reduce the amount of resin absorbed.

Substantial amounts of polymer resins are used in the production of paper laminates. During 1969 and 1970, an annual average of over 115 million pounds of phenolics and over 50 million pounds of urea and melamine resins were consumed in making paper laminates.

Polymer-Modified Papers, Overlays, and Honeycomb Cores

In addition to paper-base laminates, polymers are combined in many ways with paper to develop new properties or to modify or enhance certain properties. Paper is commonly combined with polymers, such as synthetic resins, elastomers, and plastics in general, to produce products which may be classified as follows: (1) wet strength paper, (2) special purpose papers containing large amounts of resins or elastomers (chemical resistant papers, sandpaper backings, gaskets, imitation leather, shoe parts, wood overlay and honeycomb core papers, air and oil filters, battery separators, etc.), (3) plastic coated papers, and (4) paper-base laminates. The use of plastics in combination with paper has grown enormously during the past two decades. For example, during 1969 and 1970 an annual average of 66 million lbs or urea and melamine resins were used for coating and treating paper and 326 million lbs of polyethylene were consumed in coating paper and paperboard. Many other polymers were also used in treating and coating papers; these included polyesters, styrene, polyvinyl chloride, their copolymers, and others. Polymer-modified papers are thus very important in our present-day economy.

The paper-base laminates have already been discussed. Two other polymer-treated papers used in construction and industrial materials will be briefly mentioned here, namely, overlay papers for lumber and plywood, and paper honeycomb cores. For the many other applications of polymer-modified papers the reader is referred elsewhere.

Overlaid lumber is a composite of lumber and phenolic resin-treated kraft paper. Similar paper overlays are applied to veneer, plywood, hard fiberboard and particle board. Paper honeycomb cores are also made from phenolic resin-treated kraft paper, which is formed into a honeycomb of different geometrical designs, such as figure eight or hexagonal, in special machines. Simpler types are made from resin impregnated corrugated sheets which can be assembled in several ways. This material permits construction of sandwich panels of light weight and high strength. Kraft papers for wood overlays or for honeycomb cores are treated with phenolic resin in the same ways as described above for paper-base laminates.

Paper overlays have three basic uses: masking, decoration, and structures. Masking overlays are used to cover minor defects and provide a more uniform paintable surface. Such overlays contain 20 to 25 percent phenolic resin based on the weight of the paper. For structural purposes, one or more sheets of paper may be used in the laminate. High-density and medium-density types are produced for plywood. The high-density type contains not less than 40 percent of a thermosetting resin, phenolic or melamine; it has a hard smooth surface not requiring further finishing and may be used for exterior service. The curing of the resin is completed at the same time the paper is bonded to the wood material in a hot press or, in the case of plywood, at the same time the veneer is assembled into plywood. The overlay sheet swells and shrinks much less than wood and thus exerts a resistance to the dimensional changes of the wood and may reduce lateral swelling by as much as 40 percent. Overlay papers also upgrade the appearance of low grade lumber, increase strength properties, improve finishability, and increase resistance to weathering.

Decorative surfaces are obtained by applying a top sheet of white paper on which is a printed design. This is covered with a clear coating of melamine or vinyl resin. Or, a thin transparent (when cured) paper impregnated with melamine resin may be applied to a decorative veneer, providing a permanent protective finish.

For honeycomb cores, either water- or alcohol-soluble phenolic resins may be used. Many types of facings may be glued to the honeycomb cores, viz.: veneers, plywood, hardboards, asbestos board, aluminum, stainless steel, and paper-plastic laminates. Thin sheet material may be used due to the almost continuous support of the core.

The honeycomb sandwich possesses great strength in relation to its weight. It may carry loads as much as 25 tons per sq ft. Strength and weight vary with weight of the paper, quantity of resin impregnated, and honeycomb design.

Honeycomb sandwich construction has many uses, such as in airplanes, cargo containers, truck and trailer bodies, railway passenger cars, cabins, barns, airplane hangars, house floors, walls, roofs, doors, and in a variety of other products. Besides combining strength with lightness, honeycomb sandwich material has high rigidity, good insulation properties, resistance to fungus and pests, and durability to temperature extremes.

NONWOVENS

Nonwovens are products made from fibers held together by a suitable adhesive. They are intermediate between true (woven) textiles and true (cellulose) papers and substitute for woven textiles in many uses, such as draperies.

The following processes are currently recognized:[9]

1. Carded (oriented), adhesive bonded.
2. Air-laid (random), adhesive bonded.
3. Water-laid (paper process), adhesive bonded.
4. Spun-bonded continuous filaments of synthetics, selfbonded (heat) or adhesive bonded.
5. Combinations of textile scrim and paper, adhesive bonded.

Currently, the textile-oriented processes (carded and air-laid) account for over 80 percent of production, whereas the paper-oriented processes account for about 15 percent of the total. The products made on the paper machine are stiffer, but they are less costly and can be made at relatively high speeds.

The typical water-laid furnish is composed of wood-pulp fibers and up to 25 percent of synthetic or rayon fibers. After forming, the nonwovens are treated with an adhesive of suitable properties for the end use. The main purpose of the adhesive is to impart strength. It may be a latex, a water soluble polymer, or a monomer or low-molecular weight product that polymerizes to an insoluble resin upon drying or curing. Some latices may be added to the wet sheet (wet-end addition) before the sheet is dried. Other additives such as fire retardants may be added with the adhesive or applied by a separate treatment.

WOOD-PLASTICS COMBINATIONS AND MODIFIED WOOD

As in the case of paper, the use of synthetic resins in the wood industry has advanced with the developments in polymer chemistry. Polymers permit greater and better use of wood raw material, including waste and low-grade wood. They also improve strength and appearance, and decrease dimensional change, thereby improving the competitive position of wood in relation to other materials.

The combintions of polymer resins with wood in the manufacture of particle board, as well as the application of polymer-treated papers as overlays for lumber and plywood and as honeycomb cores for sandwich construction have been discussed above. Brief consideration will be given to the "modified" woods.

Modified wood refers to wood that has been subjected to chemical or heat treatment, with or without pressure, to bring about changes in its properties. Much research has been done, especially by the U.S. Forest Products Laboratory, on the modification of wood to overcome its dimensional instability and to impart other properties.

Reduction in hygroscopicity and improved dimensional stability of wood have been experimentally developed by the acetylation of wood and by heat stabilization. Heat stabilization is brought about by heating wood, preferably under the surface of molten metal or a fused salt to exclude oxygen; this procedure gives good temperature control and a minimum of loss in strength. Temperatures varying from

150 to 320°C have been used, with the time necessary to give a certain dimensional stability being reduced by half for each 10°C increase in temperature. Wood treated in this way has been termed *staybwood.*

When wood is heated, it exhibits a certain degree of thermoplasticity, as has been described in the discussion of fiberboards. Therefore, if heated wood is compressed (165-175°C at 100-140 kg/cm^2), the internal stresses due to compression are relieved by the plastic flow of the wood. Such a compressed wood, termed *staypak*, has improved resistance to moisture and to shrinking and swelling and improved strength properties. The mechanism of the stabilization of hot-compressed wood is probably due to the plastic flow of the lignin component of the wood.

The most effective means for modifying wood and developing improved properties is to produce a wood-plastics composite by forming a synthetic polymer within the interior of the wood. Two methods have been developed, one by impregnating the wood with a water solution of phenol-formaldehyde and the other by using liquid vinyl monomers.

To make a phenolic-wood composite, a water solution of low-molecular weight phenol-formaldehyde is impregnated into thin veneers or strips of wood under pressure. They are then cured slowly in a kiln to remove the water and set the resin, i.e., to polymerize it to a hard thermoset resin. This material, termed *impreg*, contains resin bonded to the internal capillary surfaces within the cell walls of the wood and, due to the volume of the resin, this keeps the wood in a partly swollen condition. Therefore, the wood retains a volume nearly corresponding to its water-swollen dimensions and goes through a much smaller volume change than untreated wood when it is immersed in water.

Wood impregnated with phenol-formaldehyde resins and dried, but not cured, has greater plasticity than untreated wood and, therefore, can be compressed at considerably lower pressures than untreated wood. For example, impregnated spruce, aspen, and cotton-wood dried to 6 percent moisture can be compressed at 149°C to one-half their original thickness at a pressure of 17.6 kg/cm.[2] In contrast untreated woods of the same species and pressed under the same conditions will compress only 5-10 percent.

There are two types of compressed resin-treated wood (*compreg*) being produced. One is a high resin-content (25-30 percent) product of very high dimensional stability and suitable for electrical insulators in high tension lines, knife handles, and most important of all, for tooling jigs, and forming dies. Compreg dies are lower in cost, easier to repair, and harder to scratch than metal dies.

The other type of compreg contains 5-10 percent resin and has higher shock resistance and greater toughness than the high-resin-content material.

More recently, composites of wood with vinyl polymers have been developed. Since the vinyl polymers are clear, colorless thermoplastic materials they do not significantly discolor the wood, so its natural beauty is retained, whereas the phenolic resins darken the wood. Like the phenolic resins, the vinyl polymers also give improved mechanical properties and dimensional stability to the wood. The mechanisms of bringing about the property changes, however, are different. The phenolic resins are introduced in an aqueous solution, which swells the wood and at the same time penetrates within the cell walls. On drying and curing, a high degree of permanent dimensional stability is thus achieved. The vinyl polymers, however, are introduced only into the cell lumens and other voids in the wood. They, therefore, act as a bulking agent, resisting and slowing down dimensional change when the material is subjected to water or high humidity.

A variety of vinyl monomers such as methyl methacrylate and styrene, may be used. Completely filling the cell lumens and other voids (the "full-cell process") is easily accomplished by first subjecting the wood to a partial vacuum (about 0.3 in. of Hg), then covering the wood with the monomer and soaking for 2 to 6 hrs, depending upon the species of wood and its dimensions. Some penetration of the monomer into the cell walls may also be obtained by using a diffusion process, such as a solvent-exchange method.

Polymerization of the vinyl monomer in the wood may be done by either of two processes: (1) radiation and (2) free radical catalysts with heat. Some of the original research on the radiation process was done by Kenaga *et al.*[10] of the Dow Chemical Co. Solvent-monomer mixtures were used to swell the wood before irradiation, thus holding the wood in its expanded state. If the wood can be anchored permanently in its maximum water-swollen state, then the antishrink efficiency should be 100 percent. Kent and co-workers[11,12] at the University of West Virginia, during 1961-68, carried out an extensive study of radiation for the production of wood-plastic composites. Some of this technology is now in commercial practice. Gamma radiation is used for in-depth polymerization of vinyl monomers in wood, while beta radiation is used for polymerization of surface coatings on wood.

The catalyst-heat process was first mentioned[13] in a paper in 1936 describing the use of methacrylate resins by the Du Pont Company. This process has been intensively studied and further developed by Meyer and coworkers[14,15] at the State University of New York College of Forestry in Syracuse (1965-1971).

The polymerization of the vinyl monomers in both processes depends upon the same mechanism, namely initiation by free radicals. In the radiation process, the gamma rays passing through the monomer and the woody tissue create a large number of excited and ionized molecules, many of which break into fragments, namely organic free radicals ($R^{\cdot}$). These act as the initiator for the polymerization of an unsaturated monomer:

$$R^{\cdot} + CH_2{=}CH_2 \longrightarrow R{-}CH_2{-}CH_2^{\cdot}$$

$$R{-}CH_2{-}CH_2^{\cdot} + {}_nCH_2{=}CH_2 \longrightarrow R(CH_2{-}CH_2)_n{-}CH_2{-}CH_2^{\cdot}$$

$$2R(CH_2{-}CH_2)_n{-}CH_2{-}CH_2^{\cdot} \longrightarrow R(CH_2{-}CH_2)_{2n+2}R$$

Alternatively, the free radicals may be formed by thermal decomposition of compounds involving a weak bond. Commercially, the catalyst 2,2′-azobisisobutyronitrile is now most widely used, since it forms free radicals at a lower temperature than benzoyl peroxide.

$$CH_3{-}\underset{CN}{\overset{CH_3}{\overset{|}{\underset{|}{C}}}}{-}N{=}N{-}\underset{CN}{\overset{CH_3}{\overset{|}{\underset{|}{C}}}}{-}CH_3 \xrightarrow{\Delta} 2CH_3{-}\underset{CN}{\overset{CH_3}{\overset{|}{\underset{|}{C}}}}{-}N{-} \longrightarrow 2CH_3{-}\underset{CN}{\overset{CH_3}{\overset{|}{\underset{|}{C^{\cdot}}}}} + N_2$$

If the end use of the wood-polymer composite requires an abrasive (sanding) or cutting process that brings about high temperatures, the thermoplastic polymer will melt, causing machining difficulties. To prevent such melting, a cross-linking substance such as diethylene glycol dimethacrylate is added to the monomer before impregnation into the wood (about 5 percent of the volume of the monomer).[16] This brings about cross-linking of the long unbranched vinyl polymers, resulting in a substance which is not thermoplastic. The process is akin to the curing of rubber.

Recent work by Rowell and associates[17,18] has shown that the hydroxyl groups of wood may be grafted to form epoxide and urethane derivatives. In the first case, wood hydroxyls react with alkylene oxides forming an epoxide molecule with stable ether bonds:

$$\text{Wood}{-}OH + CH_3{-}CH_2{-}\overbrace{CH{-}CH_2}^{O} \longrightarrow \text{Wood}{-}O{-}CH_2{-}\underset{HO}{\underset{|}{C}H}{-}CH_2{-}CH_3$$

In the second case, methyl isocyanate reacts to form a urethane, which is very stable to acid and base hydrolysis:

$$\text{Wood}{-}OH + CH_3{-}N{=}C{=}O \xrightarrow[\text{pressure}]{\Delta} \text{Wood}{-}O{-}\overset{\overset{O}{\|}}{C}{-}NH{-}CH_3$$

The reaction takes place largely in the wood cell walls and does not fill the lumens. These polymer grafts in wood offer future potential for improving such wood properties as dimensional stability and decay resistance and for enhanced properties for special applications.

Both the radiation and heat-catalyst processes are used industrially. One of the major uses of the wood-vinyl polymer composite is for parquet flooring. It requires no finishing or waxing and is very easy to maintain. Other uses include bowling alley flooring, knife and brush handles, golf club heads, billiard cues, archery bows, shuttle cocks and kicker sticks, and tamping sticks for dynamite.

THE FORMING OF WOOD

In order to bend, shape, or otherwise form wood, it is necessary to induce temporary plasticization. The steaming of wood has been an art for generations. As pointed out earlier, wood is somewhat plastic in the presence of heat and water; thus, by steaming it, this property can be taken advantage of in the bending of wood into various shapes for furniture and other uses.

A new process for the temporary plasticization of wood has been developed by Schuerch[19,20] using anhydrous ammonia, either in the liquid or gaseous phase. Anhydrous ammonia causes swelling of both of the major polymer systems of wood: the lignin and the polysaccharides. Ammonia wets lignin and causes a lower tack temperature than water. Ammonia also enters the crystal lattice of cellulose, causing it to swell and to form the crystal structure of an ammonia-cellulose compound. Upon removal of the ammonia the ammonia-cellulose reverts to cellulose in a more distorted crystalline form.

The forming of wood by temporary plasticization with anhydrous ammonia takes advantage of these principles. Wood that has been immersed in liquid ammonia or treated with gaseous ammonia under pressure until the cell walls have been penetrated, becomes pliable and flexible. In this plasticized condition it can be readily and easily shaped and formed by hand or mechanically. The ammonia readily vaporizes and evaporates from the wood, whereby the wood regains its normal stiffness, but retains the new form into which it has been shaped. The wood can be distorted into complex shapes without spring-back. For exact control of shape, restraint is needed for a short period, until the greater part of the ammonia has vaporized, but thereafter, there is no tendency for the wood to spring back to its original form.

The process is still developmental, but particleboard, hardboard, and veneers, as well as wood strips may be formed by the ammonia process. Other possible uses, such as embossing, densification, and improved machining of wood may potentially be taken advantage of through the temporary modification of wood by anhydrous ammonia. Treating plants have been developed on a pilot-plant scale, whereby temperatures can be controlled between −35 and +35°C.

PRESERVATIVE AND FIRE-RETARDANT TREATMENT OF WOOD

Wood Preservation

Wood, a natural plant tissue, is subject to attack by fungi, insects, and marine borers. Some species of wood are more resistant to decay than others, as, for example, the heartwood of cedars, cypress, and redwood, due to the presence of natural toxic substances among the extractable components. Most woods, however, are rapidly attacked when used in contact with soil or water, or when exposed to high relative humidities without adequate air circulation. Wood for such service conditions requires chemical treatment with toxic chemicals, collectively termed *wood preservatives*. The service life of wood may be increased from 5-15 fold, depending upon the conditions of preservative treatment and the nature of the service.

The preservative treatment of wood is the second largest chemical wood-processing industry; pulp and paper manufacture is the most important. Since 1950, the average annual volume of wood treated has been 250 to 286 million cu ft (see Table 15.11). The more important types of wood products treated are also shown in Table 15.11.

Preservative Chemicals. Toxic chemicals used for the preservation of wood may be classified as follows:

1. Organic liquids of low volatility and limited water solubility

coal-tar creosote
creosote-coal tar solutions
creosote-petroleum solutions
other creosotes

2. Chemicals dissolved in organic solvents, usually hydrocarbons
   chlorinated phenols (principally pentachlorophenol)
   copper naphthenate
   solubilized copper 8-quinolinolate
3. Water-soluble inorganic salts
   acid copper chromate
   ammoniacal copper arsenite
   chromated copper arsenate
   chromated zinc chloride
   fluor chrome arsenate phenol

Cresote from coal tar is the most widely used wood preservative because (1) it is highly toxic to wood-destroying organisms; (2) it has a high degree of permanence due to its relative insolubility in water and its low volatility, (3) it is easy to apply and to obtain deep penetration; and (4) it is relatively cheap and widely available.

For general outdoor service in structural timbers, poles, posts, piling, mine props, and for marine uses, coal-tar creosote is the best and most important preservative. Because of its odor, dark color, and the fact that creosote-treated wood usually cannot be painted, creosote is unsuitable for finished lumber and for interior use.

Coal-tar creosote (see Chapter 8) is a mixture of aromatic hydrocarbons containing appreciable amounts of tar acids and bases (up to about 5 percent of each) and has a boiling range between 200 and 355°C. The important hydrocarbons present include fluorene, anthracene, phenanthrene and some naphthalene. The tar acids are mainly phenols, cresols, xylenols, and naphthols; the tar bases consist of pyridines, quinolines, and acridines.

Often coal tar or petroleum oil is mixed with coal-tar creosote, in amounts up to 50 percent, as a means of lowering preservative costs. Since coal tar and petroleum have low toxicity, their mixtures with creosote are less toxic than creosote alone.

A number of phenols, especially chlorinated phenols, and certain metal-organic compounds, such as copper naphthenate and phenyl mercury oleate, are effective preservatives. Pentachlorophenol and copper naphthenate are most commonly used, and they are carried into the wood in 1 to 5 percent solutions in petroleum oil. Pentachlorophenol is colorless, and can be applied in clear volatile mineral oils to millwork and window-sash which require a clean, nonswelling, and paintable treatment.

Inorganic salts are employed in preservative treatment where the wood will not be in contact with the ground or water, such as for indoor use or where the treated wood requires painting. They are also satisfactory for outdoor use in relatively dry regions.

Preservation Processes. The methods for applying preservatives to wood are classified as follows:

1. Nonpressure processes

TABLE 15.11 Wood Products Treated with Preservatives (Thousands of Cubic Feet)

Products	*1950*	*1967*	*1976*	*1977*
Crossties and switch ties	127,465	88,635	101,048	100,443
Poles	77,622	84,322	53,143	53,063
Lumber and timbers	32,821	62,241	67,123	61,734
Piling	12,523	16,627	8,478	11,346
Fence posts	12,269	21,018	13,769	10,735
Cross arms	2,493	4,605	4,628	1,347
Plywood			2,848	2,406
All other	4,792	8,910	6,194	9,213
TOTAL	269,985	286,358	257,231	250,287

Sources: "Statistical abstract of the United States," U.S. Dept. of Commerce, Bureau of the Census; and Proc. of the American Wood Preservers' Association.

Surface (superficial) applications by brushing, spraying, or dipping
Soaking, steeping, and diffusion processes
Thermal process
Vacuum Processes
Miscellaneous processes

2. Pressure processes
   Full-cell process (Bethell)
   Empty-cell processes (Rueping and Lowry)

Brush and spray treatments usually give only limited protection because the penetration or depth of capillary absorption is slight. Dip treatments give slightly better protection. Organic chemicals dissolved in clear petroleum solvents are often applied to window sash and similar products by a dip treatment of 1 to 3 minutes.

Cold soaking of seasoned wood in low-viscosity preservative oils for several hours or days and the steeping of green or seasoned wood in water-borne preservatives for several days are methods sometimes employed for posts, lumber, and timbers on a limited basis. The diffusion process employs water-borne preservatives that will diffuse out of the treating solution into the water in green or wet wood.

The most effective of the nonpressure processes is the thermal method of applying coal-tar creosote or other oil-soluble preservatives, such as pentachlorophenol solution. The wood is heated in the preservative liquid in an open tank for several hours, after which it is quickly submerged in cold preservative in which it is allowed to remain several hours. This is accomplished either by transferring the wood at the proper time from the hot tank to the cold tank, or by draining the hot preservative and quickly refilling the tank with cooler preservative. During the hot treatment, the air in the wood expands and some is expelled. Heating also lowers the viscosity of the preservative so that there is better penetration. When the cooling takes place, the remaining air in the wood contracts, creating a partial vacuum which draws the preservative into the wood. For coal-tar creosote, the hot bath is at 210-235°F and the cold bath at about 100°F. This temperature is required to keep the preservative fluid.

The hot- and cold-bath process is widely used for treating poles and, to a lesser extent, for fence posts, lumber, and timbers. The results obtained by this process are the most effective of the common nonpressure processes and most nearly approach those obtained by the pressure processes.

The vacuum processes involve subjecting the wood to a vacuum to draw out part of the air. The wood may be either subjected to a vacuum alone or to steaming and a vacuum before being submerged in a cold preservative. These methods are used to a limited extent in the treatment of lumber, timber, and millwork.

Commercial treatment of wood is most commonly done by one of the pressure processes, since they give deeper penetrations and more positive results than any of the nonpressure methods. The wood, on steel cars, is run into a long horizontal cylinder, which is closed and filled with preservative. Pressure is applied, forcing the preservative into the wood.

There are two types of pressure treatment, the full-cell and the empty-cell. The full-cell process seeks to fill the cell lumens of the wood with the preservative liquid, giving retention of a maximum quantity of preservative. The empty-cell process seeks deep penetration with a relatively low net retention of preservative by forcing out the bulk liquid in the wood cells, leaving the internal capillary structure coated with preservative.

In the full-cell process, the wood in the cylinder is first subjected to a vacuum of not less than 22 inches of mercury for 15-60 minutes, to remove as much air as possible from the wood. The cylinder is then filled with hot treating liquid without admitting air. The maximum temperature for creosote and its solutions is 210°F and for water-borne preservatives it is 120 to 150°F, depending upon the preservative. The liquid is then placed under a pressure of 125-200 psi and the temperature and pressure are maintained for the desired length of time, usually several hours. After drawing the liquid from the cylinder, a short vacuum is applied to free the charge of surface-dripping preservative.

In the empty-cell process, the preservative liquid is forced under pressure into the wood containing either its normal air content (Lowry process), or an excess of air, by first subjecting the wood to air pressure before applying the preservative under pressure (Rueping process). In the former case, the preservative is put in the cylinder containing the wood at atmospheric pressure, and, in the latter case, under air pressure of 25-100 psi. After the wood has been subjected to the hot preservative (about 190-200°F) under pressure (100-200 psi in Lowry process at 150-200 psi in the Rueping process) and the pressure has been released, the back pressure of the compressed air in the wood forces out the free liquid from the wood. As much as 20-60 percent of the injected preservative may be recovered, yet good depth of penetration of preservative is achieved.

Preservative Retention. Retention of preservative is generally specified in terms of the weight of preservative per cubic foot of wood, based on the total weight of preservative retained and the total volume of wood treated in a charge. Penetration and retention vary widely between different species of wood, as well as with woods of the same species grown in different areas. In most species, heartwood is much more difficult to penetrate than sapwood. Also, within each annual growth ring there is variability in penetration, the latewood generally being more easily treated than the earlywood.

The American Wood-Preservers' Association Standards specify methods of analysis to determine penetration and retention. They also specify minimum retention amounts for different preservatives according to the commodity, the species, the pretreatment of the wood, such as kiln drying, and the end use of the commodity. Heavier retention is required for products in contact with the ground (poles, timbers, etc.) or with marine waters (piles, timbers, etc.). Unprotected wood in contact with the ground is subject to severe attack by fungi and insects and, in contact with sea water, it is quickly destroyed by marine borers. For wood products to be used in contact with the ground or marine waters, creosote is the major preservative employed, since it can be readily impregnated to give high retention and good protection and is not leached out by water.

Fire-Retardant Treatments

Protection of wood against fire may be accomplished by application of certain chemicals. Because of the cost, commercial treatments are limited to materials used in localities where fire building codes require fire-retardation treatment. Fire-retardant chemicals are water soluble, which limits the use of treated wood to places where it is not subjected to leaching.

Two types of treatment are used for improving the fire resistance of wood; namely (1) impregnation of the wood with fire-retardant chemicals and (2) coating the surface of the wood with an oxygen-excluding envelope. Among the most commonly used chemicals for impregnation treatments are diammonium phosphate, ammonium sulfate, borax, boric acid, and zinc chloride. These compounds have different characteristics with respect to fire resistance. Ammonium phosphate, for example, is effective in checking both flaming and glowing; borax is good in checking flaming but is not a satisfactory glow retardant. Boric acid is excellent in stopping glow but not so effective in retarding flaming. Because of these different characteristics, mixtures of chemicals are usually employed in treating formulations.

The American Wood-Preservers' Association Standards specify the following four types of fire-retardant formulations.

Type A

Chromated zinc chloride—a mixture of sodium dichromate and zinc chloride having the composition: hexavalent chromium as CrO_3, 20% and zinc as ZnO, 80%.

Type B	
Chromated zinc chloride (as above)	80%
Ammonium sulfate	10%
Boric acid	10%

Type C	
Diammonium phosphate	10%
Ammonium sulfate	60%
Sodium tetraborate, anhydrous	10%
Boric acid	20%

Type D	
Zinc chloride	35%
Ammonium sulfate	35%
Boric acid	25%
Sodium dichromate	5%

Minimum and maximum limits of variation in the percentage of each component in the above formulations are specified in the standards.

The impregnation methods are similar to those employed for the preservative treatment of wood by water-borne salts using pressure processes. The maximum temperature of the solution must not exceed 140°F for formulations Types A, B, and D, and must not be over 160°F for Type C. Subsequent to treatment, the wood must be dried to remove the water solvent to a moisture content of 19 percent or less. For most uses, the wood is kiln dried to a moisture content of under 10 percent.

Larger amounts of chemical must be deposited in the wood for effective fire protection than is necessary for the water-borne chemicals used for decay prevention. Whereas retentions from 0.22 to 1.00 lb per cu ft of wood for the water-soluble toxic salts are specified according to commodity standards in order to give good protection against decay and insects, as much as 5 to 6 pounds of some fire-retardants may be required for a high degree of effectiveness against fire. Usually smaller amounts will give a good degree of protection. For example, formulation Type B when impregnated in amounts of 1.5 to 3 lb per cu ft of wood provides combined protection against fire, decay, and insects. The effectiveness of a fire-retardant treatment depends upon the performance rating of the treated material when tested in accordance with ASTM E84 (no greater flame spread than 25).

There are various explanations for the fire-retardant effect of the chemicals impregnated into wood for this purpose. There is probably no single mechanism but rather a combination of mechanisms which are operative, including (1) the fusing of the chemical within the wood at high temperatures to form thin noncombustible films which exclude oxygen; (2) the evolution of noncombustible gases in some cases; and (3) the catalytic promotion of charcoal formation, instead of volatile combustible gases, with the added benefit of increased thermal insulation.

Surface coatings for fire-retardation are less effective than impregnation of chemicals into the wood. Formulations are used containing either ammonium phosphate, borax, or sodium silicate, together with other constituents to provide good bonding to the wood. Paints containing substantial amounts (30 to 50 percent) of these fire-resistant chemicals provide moderate protection against fire.

WOOD HYDROLYSIS

Hydrolysis is the selective conversion (saccharification) of wood or other plant tissue into their constituent sugars. The cellulose in plant tissue is converted into glucose, while the hemicellulose yields mainly mannose from softwoods and xylose from hardwoods. Historically, hydrolysis has been done by the use of mineral acids. The first producing factory was built in the U.S.A. during World War I. Further developments in Germany led to factories in Germany and Switzerland during World War II. However, they proved to be not competitive during peace-time conditions when cheaper sources of sugars from agricultural products and of ethanol from petroleum and natural gas are available. Yet saccharification of wood and agricultural residues has continued to expand in the U.S.S.R.[21]

The availability of enormous quantities of unused wood residues from logging, thinnings and removals, sawmilling, and woodworking operations, estimated during the 1970's at 140 million tons annually, has stimulated much attention to the production of sugars and ethyl alcohol from this raw material. But problems were encountered in developing commercial-acid hydrolysis processes due to the serious corrosive action of the acid and the difficulty in removing the sugars from the scene of the reaction before they were subjected to too much degradation. As a result, a technique known as enzymatic hydrolysis has received considerable study during recent years as an alternative to mineral-acid hydrolysis for producing sugar from cellulosic materials. (This process will be discussed later in this chapter.)

Cellulose, the major component of wood, gives about 90 percent yield of pure glucose under laboratory conditions of hydrolysis, according to the following equation:

$$(C_6H_{10}O_5)_n + nH_2O \xrightarrow{acid} nC_6H_{12}O_6$$

where n is in the range of 10,000-15,000. The hemicellulose fraction gives a mixture of sugars, viz., xylose, arabinose, mannose, galactose, and glucose. Glucose, galactose, and mannose are yeast-fermentable sugars, whereas the pentoses (xylose and arabinose) are nonfermentable. The potential total reducing sugar yield from wood averages 65-70 percent, whereas the fermentable sugar yield is about 50 percent for the hardwoods and 58 percent for the softwoods. The lower quantity of fermentable sugar from the hardwoods is due to their higher content of pentosans, compared to the coniferous woods.

Hemicelluloses hydrolyze much more easily and rapidly than cellulose. Temperatures and acid concentrations that hydrolyze the cellulose to glucose in a matter of a few hours readily convert much of the hemicellulose into simple sugars in minutes or even seconds. Under industrial conditions of hydrolysis, the sugars formed undergo decomposition, with the pentoses decomposing more rapidly than the hexoses. Thus, the conditions of hydrolysis cause variations in the ratio and yields of the various sugars due to (1) their different rates of formation by hydrolysis and (2) their different rates of decomposition. Table 15.12 shows the relative rates of decomposition of the sugars obtained by the hydrolysis of wood.

TABLE 15.12 Decomposition of Wood Sugars in 0.8 Per Cent Sulfuric Acid at 180°C

Sugar	*First-Order Reaction Constant K (min⁻¹)*	*Half-life (min)*
Glucose	0.0242	28.6
Galactose	0.0263	26.4
Mannose	0.0358	19.4
Arabinose	0.0421	16.4
Xylose	0.0721	9.6

Sources: J. F. Saeman, *Ind. Eng. Chem.*, 37, 43 (1945).

The polysaccharides of wood (holocellulose) may be hydrolyzed by two general methods: (1) by strong acids, such as 70-72 percent sulfuric acid or 40-45 percent hydrochloric acid; or (2) by dilute acids, such as 0.5-2.0 percent sulfuric acid. The hydrolysis by strong acids is constant, proceeds at a first-order reaction and is independent of the degree of polymerization. The reaction may be represented as follows:

$$\text{Holocellulose} \xrightarrow[\text{acid}]{\text{strong}} \left.\begin{array}{l}\text{Swollen Cellulose}\\ \text{Soluble Pentosans}\end{array}\right\} \longrightarrow \text{Soluble Polysaccharides} \xrightarrow{\text{dilute acid}} \text{Simple Sugars}$$

In dilute-acid hydrolysis, the reactions are heterogeneous and more complex because no swelling and solubilizing of the cellulose occurs. Cleavage of the insoluble cellulose results directly in low-molecular weight oligosaccharides (intermediate products), which are rapidly converted to simple sugars, as indicated below:

$$\text{Holocellulose} \xrightarrow[\text{dilute acid}]{(1)} \begin{array}{l}\text{Insoluble "stable" cellulose}\\ \text{Soluble hemicellulose intermediates}\\ \text{Pentose sugars}\end{array}$$

$$\text{Insoluble "stable" cellulose} \xrightarrow{(2)} \text{Oligosaccharides (cellulose intermediates)} \xrightarrow{(3)} \begin{array}{l}\text{Hexose sugars}\\ \text{Pentose sugars}\end{array}$$

Reaction (1) is rapid and occurs under mild conditions, hydrolyzing mainly the hemicelluloses. Reaction (2) is slow, proceeds as a first-order reaction and is the limiting reaction in this process. Reaction (3) is rapid.

Based on the above methods, two classes of industrial processes have been developed, namely. the Bergius-Rheinau process, based on the use of concentrated hydrochloric acid at ordinary temperatures, and the Scholler-Tornesch process, in which very dilute sulfuric acid is used at temperatures of 170 to 180°C (338 to 356°F). The latter method in an improved form is known as the Madison process, based on work done at the United States Forest Products Laboratory in Madison, Wis. A number of modifications have been developed, including four in Japan.

Bergius-Rheinau Process

Hydrochloric acid of about 40-45 percent (by weight) is produced by reinforcing recovered, weaker acid with hydrogen chloride from salt-sulfuric acid retorts, or by burning chlorine with illuminating gas.

Wood chips are air dried, then charged into a tile-lined reactor and extracted countercurrently by the acid. The fresh strong acid enters that part of the battery of diffusers or reactors which contains the most nearly exhausted wood and is pumped through the next following containers until it is nearly saturated with the carbohydrates dissolved from the wood. Part of this solution is mixed, under slight cooling, with fresh wood and the mixture charged into the head container of the battery. After filling this and allowing a few hours for reaction time, an amount of solution is forced from this container through the pressure of the incoming acid, which corresponds to the yield from one charge.

The drawn-off solution contains hydrochloric acid and carbohydrates in about equal parts, and at a concentration of about 25 percent (by weight) each. It is concentrated in stoneware tubes under vacuum. The distillate, containing about 80 percent of the hydrochloric acid, with minor proportions of acetic acid and furfural is reused after fortification. The concentrated sugar solution is dried to a powder in a spray-drier, where is also loses most of the remaining acid.

The dry, somewhat acid, powder contains the carbohydrates in the form of intermediate polymers (oligosaccharides). They are water-soluble and must undergo further hydrolysis to obtain simple sugars, either for fermentation or cystallization.

This is done by dissolving the oligosaccharides, diluting the solution to approximately 20 percent sugar concentration, and heating it for 2 hr in the presence of 2 percent acid at 125°C. Part of the glucose can be crystallized from the neutralized and reconcentrated solution, while the mother liquors are fermented to alcohol or used for growing yeast. A diagrammatic flow chart for the Bergius process is shown in Fig. 15.14.

When hardwoods are to be used, it is necessary to remove a part of the hemicellulose first by prehydrolysis. This has been carried out on a large scale with straw, a substance chemically similar to hardwood, by heating it in a 8 to 1 liquid to solid ratio, with 0.5 percent sulfuric acid for 2 to 3 hr at 130°C (266°F). Without prehydrolysis, hardwoods and straws form slimy materials, probably because of their high hemicellulose content, which prevent the flow of the hydrolyzing acid.

In the Bergius-Rheinau process the concentrated hydrochloric acid employed requires

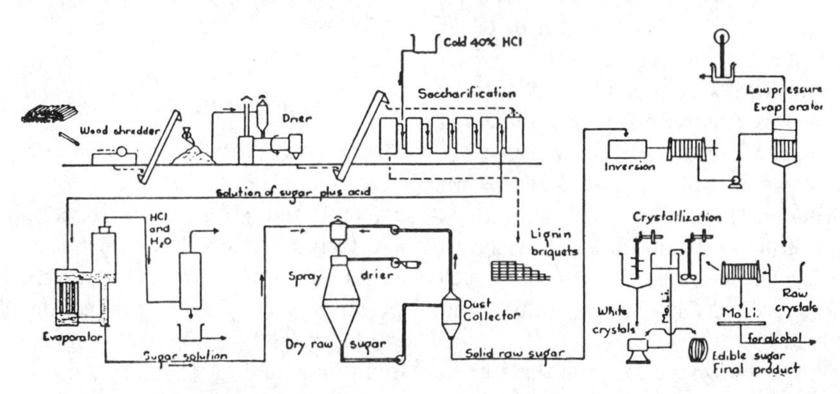

Fig. 15.14 Diagrammatic flow sheet for the Bergius wood hydrolysis process.

dried wood, and recovery of the acid is essential. The process gives high yields of sugars (60 to 65 percent) at high concentrations. The intermediate sugars first obtained, however, call for an extra processing step to reduce them to monomers, before fermentation or crystallization.

Madison Process with Continuous Percolation

In the Madison process, dilute sulfuric acid with an average concentration of 0.5 percent is pressed through wood in the form of sawdust and shavings. Regular flow of the acid and of the resulting sugar solution is one of the two principal requirements; the other is a lignin residue which can be discharged from the pressure vessels without manual labor. Both depend upon careful charging of the wood, which should not contain too many very fine particles, and upon maintaining a pressure differential of not more than 5 to 6 lbs sq in. between top and bottom of the digester.

The digesters or percolators are pressure hydrolyzing vessels, commonly employing a pressure of 150 lb/sq in., and having a capacity of 2000 cu ft each. In the original Scholler plants in Germany, the digesters were lined with lead and acid-proof brick; in the Madison process a lining of "Everdur" metal was found to give sufficient protection.

The wood, about 15 tons, is pressed down with steam, and then heated by direct steam, after which the acid is introduced. The practice in Germany was to bring the dilute acid into the digester in several batches, with rest periods of about 30 min, heating the wood to temperatures of 130°C (266°F) at first, then to 180°C (356°F), while keeping the temperature of the entering acid 10 to 20°C lower. A total of about 14 hr was required to exhaust the wood, yielding about 50 lb of carbohydrates for 100 lb of dry wood substance. In the Madison process, continuous flow of the acid, and correspondingly, of the sugar solution, is provided—in other words, continuous percolation. The cycle is thereby reduced to 6 hr, and the yields are increased somewhat.

The lignin is blown out of the digester by opening the specially constructed bottom valves while the vessel is still under pressure.

The sugar solutions usually contain about 5 percent carbohydrates and 0.5 percent sulfuric acid. The solutions, still under pressure (150 lb/sq in.), are flash-evaporated to 35 lb/sq in., neutralized with lime at that pressure, and filtered. Calcium sulfate is much less soluble at the elevated temperature corresponding to the pressure than it is at 100°C. This is a fortunate circumstance, for it must be removed to an extent sufficient to avoid difficulties caused by the formation of incrustations in the subsequent alcohol distillation. The filtered solution is cooled by further flash evaporation and heat exchange with water to fermentation temperature.

Sugar yields from coniferous woods (softwoods) are about 50 percent at an average concentration of 5 percent. When fermented, the average ethyl alcohol yield per ton of dry woods is 50–60 gal and sometimes higher.

The dilute sulfuric acid employed in the Madison process gives lower yields (49 to 55 percent) of sugars than the Bergius-Rheinau process. and only very dilute solutions are obtained directly. Recovery of heat is easier in the Madison process and the acid need not be recovered.

The dilute-acid hydrolysis method is presently preferred for production on a commercial scale, although the yield of glucose is only about 50 percent based on the cellulose. Nearly quantitative yields can be obtained by strong-acid hydrolysis, but this is offset by higher capital costs for corrosion resistant equipment and higher operating costs for acid recovery plus acid losses.

Other Developments

The Russians have energetically pursued the development of wood-hydrolysis technology; emphasis today is on the production of yeast and furfural. The latter, which is obtained as a by-product during wood hydrolysis, is used as an intermediate in the production of organic chemicals and is converted mostly to tetrahydrofuran and furfuryl alcohol.

Two Japanese processes have been developed[22] using concentrated (above 62%) sulfuric acid. In the Nihon-Mokuzai-Kagaku process, the acid is neutralized by lime and the gypsum may be used for making gypsum board and for other purposes. In the Hokkaido process, the acid is recovered by dialysis. Two other Japanese developments[22] are the Udic-Rheinau process, using 41% hydrochloric acid and the Noguchi-Chisso process using anhydrous hydrochloric acid. All four of these Japanese processes operate at atmospheric pressure and at temperatures below 100°C. All involve a prehydrolysis step to remove the hemicellulose since the main product is crystalline glucose.

More recently, a mechanochemical process, termed the Riga Hydrolysis Method[21] has been developed in Russia. In this process small amounts of concentrated sulfuric acid are mixed with lignocellulose material and subjected to mechanical treatment by feeding through a screw conveyor-press, a roller press, or other device. The acid consumption is reduced to 10-20% of the initial raw material weight, thereby eliminating the necessity of recovery.

Other recent proposals[23] include the dissolving of cellulose from lignocellulose by concentrated sulfuric acid, followed by precipitation and hydrolysis of the amorphous cellulose[24] and the use of acid catalysts in a mixture of water and organic solvents, such as lower alcohols and ketones.[25] The latter process is claimed to dissolve all the carbohydrate polymers in the wood, as well as the lignin if the hydrolysis is carried to completion.

Enzymatic Hydrolysis

Significant work by the U.S. Army Laboratory[26,27] at Natick, Massachusetts, has led to the development of very strong cellulase enzymes, such as from the tropical fungus, *Tricoderma viride* and its mutants, termed *Tricoderma reesei.* This process continues to be improved and holds promise as a noncorrosive, atmospheric-pressure method for converting cellulosic materials, especially cellulose wastes, into glucose. A 30-fold increase in the activity and yields of the cellulase enzyme has been achieved in less than a decade, including increasing the susceptibility of the substrate to hydrolysis.[28] Recent work indicates that two strains of thermophylic actinomyces can assimilate cellulose fiber at a rate faster than *Tricoderma viride.* However, like the latter, the rate and extent of saccharification are limited by substrate crystallinity and enzyme in activation. Cellulase is a mixture of several enzymes acting together, and all the components are required together and serve different functions.

The less the crystallinity of the cellulose, the more easily it is hydrolyzed by enzymes. Crystallinity of cellulose also impedes acid hydrolysis as well as enzyme hydrolysis. Moreover, the presence of lignin can completely prevent enzymatic hydrolysis by acting as a physical barrier.

Through enzymatic hydrolysis, it is theoretically possible to obtain a 100% yield of glucose, but the process still has many obstacles to overcome. The reaction is slow, the enzyme is expensive, its recycling is critical, and the process can be inhibited by lignin. Since most plant material, such as wood, is highly lignified, an inexpensive pretreatment process must be developed in order for the enzymatic hydrolysis process to become industrially usable on wood.

Prehydrolysis of Wood for Pulp Production

When wood is pulped by the sulfate process, a large part of the hemicelluloses are converted to sugars and organic degradation products, using up pulping chemicals. The spent sulfate liquor is burned to recover pulping chemicals, and the sugars and their decomposition products contribute to the calorific value of the total liquor, though only to the extent of 18 percent.

A process has been developed whereby the easily hydrolyzable sugars are removed from the wood by prehydrolysis and subsequently used for the production of fodder yeast or as a source of sugar for fermentation products. Hydrolysis of pine, fir, or spruce for 15 min at 165-175°C, using 0.3 percent sulfuric acid removes about 16 percent of the wood. The sugars contain about 50 percent hexoses (mainly glucose) and 50 percent pentoses, and have been found to have a yeast yield of about 5 percent of the weight of the wood.

The residual wood is pulped by the sulfate process and gives about 40 percent yield of high alpha-content pulp for dissolving purposes (rayon, film, plastics).

Fermentation of Sulfite Waste Liquor

The sulfurous acid used in the sulfite pulping liquor causes hydrolysis of the more easily hydrolyzable components of wood, especially the pentosans in the hemicellulose. About 35 percent of the pontentially fermentable sugars in the wood are hydrolyzed. However, most of these are decomposed during the long pulping procedure so that only one-fourth to one-third, including much of the more resistant hexoses, remains in the waste sulfite liquor. If these are fermented by yeast about 12.5 gal of 95 percent alcohol per ton of wood may be produced.

A large number of plants in Europe and a few in North America have been constructed to utilize the sugar hydrolyzate in sulfite waste liquors. The procedure as carried out in one American operation is as follows:

The liquor is recovered from the digester by discharging it in such a manner that as much sugar as possible is removed with a minimum of dilution by washing. Free sulfur dioxide is removed and recovered by blowing steam through the solution, which decreases the acidity from a pH of 2.2 to about 3.9. The liquors are cooled by a vacuum flash and neutralized by lime to a pH of about 4.2. A small amount of inorganic nitrogen is added for yeast growth and about 1 percent by volume of yeast is added continuously. Fermentation is carried out in a series of tanks, the solution flowing from one to the other with agitation to keep the yeast in suspension. The yeast is recovered by centrifuging and mixed with the new sugar solution entering the fermenter. About 30 hr is required for fermentation. The alcohol content of the fermented liquor is about 1 percent by volume. The ethyl alcohol is recovered in stainless steel stills. Methanol and other alcohols are obtained in small amounts as by-products.

Vanillin

Vanillin is not a product of hydrolysis or of fermentation; in fact, it does not originate from the holocellulose, but rather from the lignin portion of the wood. The major organic material in sulfite waste liquor (spent liquor, after pulping wood by the sulfite process) is the lignin dissolved from the wood as lignosulfonic acid. Alkaline degradation of this lignin product produces vanillin, the same substance which occurs naturally in the vanilla bean.

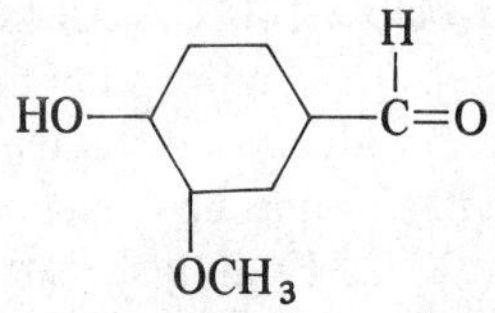

H
HO— —C=O
OCH_3

Vanillin

With the Howard process, 5-10 percent of vanillin is produced, based on the lignin in the waste sulfite liquor. Some vanillin is produced from sulfite waste liquor in both the United States and Canada.

ENERGY FROM WOOD

The average stored energy content of wood is 8600 Btu/lb. Variations in heat values for wood as a function of species occurs within a relatively narrow range of 8000 to 10,000 Btu/lb. Bark values are slightly higher, about 10,000 Btu/lb.

The greater the oxygen content of a carbon compound or carbonaceous substance the lower the heat of combustion. Lignin with an elemental analysis of $C_{10}H_{11}O_2$ compared to $C_6H_{10}O_5$ for cellulose, has a lower degree of oxidation and a higher heat of combustion that cellulose. The fossil fuels which have very low or no oxygen content have much higher energy contents, for example, 12,000-13,500 Btu/lb. for coal, 18,000-19,900 for oil and 18,550 for natural gas.

Wood and other biomass materials are hygroscopic and retain substantial amounts of moisture (15-60%). Hence, some of the thermal energy of the wood is lost in vaporizing this water to steam during combustion. In addition, wood has a lower density than coal, so that more volume of material must be gathered and transported to provide the same heating value. Wood has a negligible sulfur content, less than 0.1% compared to much larger amounts in coal (avg. 2%) and petroleum. Sulfur from coal and petroleum creates serious environmental problems by causing acid rain. Also, ash from coal is a disposal problem and creates unhealthy fly ash. The ash content of wood is less than 0.5%. Thus, the combustion of wood creates no significant environmental problem; wood is a clean, attractive fuel.

Surprisingly, fuel is still the major use of wood on a world-wide basis, amounting to an estimated 1.2×10^9 m^3/y compared to 0.8×10^9 m^3/y for lumber and other solid wood products, 0.34×10^9 m^3/y for pulp, and 0.2×10^9 m^3/y for other uses.[29] Thus, on a global basis, nearly 50 percent of the wood harvest is used directly for fuel. In many nonindustrialized countries, fuel amounts to 80-90 percent of total wood consumption, compared with an average of 7 percent for industrialized regions.

The present energy crisis brought on by accelerating costs and impending shortages of petroleum have rekindled interest in the biomass as a source of energy. In the U.S.A., wood today supplies only 2.5-3.0 percent of the total energy requirement. Most of this is used by the forest industry itself, which is now seeking complete energy independence. However, interest in biomass as a fuel is showing up in many areas. Manufacturing plants, public utilities, public institutions, and cities have suddenly become interested in wood, wood wastes, crop residues, municipal wastes (garbage), and manure as a source of fuel. Also, there has been an increase in the demand for wood stoves for home heating. It is likely that wood will play an increasing role in the future energy supply of the U.S.A. A major and, in the long run, overwhelming advantage of wood is that it is constantly renewable and can be developed in great quantities; whereas over the long-term, the fossil fuels will diminish in their availability.

Energy from wood and other biomass materials may be produced by several methods, namely:

1. direct combustion → heat
2. gasification → low Btu gas and/or methanol
3. pyrolysis → low Btu gas + charcoal
4. liquifaction → oil + hydrocarbon fuels
5. anaerobic fermentation → methane
6. hydrolysis → sugars $\xrightarrow{\text{fermentation}}$ ethanol

Direct combustion is the major technology presently used. Conversion to other energy forms, such as low Btu gas, and methanol are receiving increasing study.

Most wood and wood-derived materials (spent pulping liquors) that are used for energy are consumed by the forest products industry itself. The source of this fuel is almost entirely in the wood processing and manufacturing operations, termed "manufacturing residuals" or "mill residues." Forest residues from logging operations and the noncommercial trees in the forest are beginning to be used as the conditions become economical to do so. Energy plantations are farther down the road.

In 1977, wood wastes supplied the forest products industry with 45% of its energy requirements or 1.2 quads out of the total of 2.7 needed.[7] Of this amount, 0.7 quads were supplied to the pulp and paper industry by organic materials in its spent liquors and 0.5 quads were wood and bark residues from the mill. The spent pulping liquor contains 40-50% of the wood as dissolved lignin and a large variety of other organic compounds. The spent liquor is almost totally used, after evaporation to about 65% solids, to produce steam and to recover the inorganic chemicals used in the pulping operation. The Tomlinson furnace is almost universally used and is the most efficient system for combined energy and chemical recovery yet devised.

The mixture of wood and bark residues burned directly are collectively termed "hog fuel." Over the past decade, there has been a trend to channel more and more of the cleaner, drier and larger-sized mill residues into raw materials for products. The coarse wood residues (slabs, trim, edgings, etc.) are chipped and sent to pulp mills. Also, considerable sawdust is now being pulped instead of being burned for fuel. Some of the residues, including shavings are also used for particle board. This means that hog fuel contains the less desirable, dirtier, and wetter forms of waste.

Combustion Technology

The hog fuel boiler represents the conventional technology of using wood for energy to produce process steam. Modern wood-fuel boilers have traveling grates and use mechanical draft fans. Many are of gigantic size, handling as much as 500,000 lbs. of wood per hour. Spreader stokes distribute the fuel evenly over the large grate areas to ensure efficient combustion. The ash can be removed continuously with traveling grates. Modern instrumentation meters air flow, controls fuel-to-air ratios, and meters combustion efficiency and stack emissions. With these developments, the energy recovery and environmental acceptability of wood burning systems have been improved in the forest products industry.

The deterioration of hog fuel quality referred to above makes it more difficult to obtain efficient combustion, high reliability, and low stack emissions simultaneously. To overcome this, the fuel can be processed before combustion to remove dirt and moisture, and produce clean burning, efficient fuel. This can be done by drying, screening, grinding, washing and pelletizing or densification of the hog fuel in order to remove noncombustible dirt and moisture, reduce the size of large and slow burning material, and agglomerate small and fine material. Size uniformity is important for ease in transportation, storage and fuel feeding. Dirt and moisture reduction reduces transportation costs, promotes efficient combustion and minimizes air emissions. Another approach for using low quality hog fuel is the development of improved combustors. Examples are (1) the fluidized-bed combustors which obtain excellent wood combustion at relatively high efficiencies with low quality, nonuniform fuels and (2) the pyrolytic burner which has very low stack emissions with relatively high thermal efficiencies and wood combustion rates.

Present day wood-fired boiler systems are complicated and cost considerably more than a comparable petroleum fired installation. According to Jamison,[7] "As much as 25 percent of the capital cost is in the fuel handling equipment and another 20 percent is in the air pollution control system. Because of the high capital costs and the lower thermal efficiencies of burning wood compared to oil (68% vs. 82%), the success of the wood-fired systems depends on the low cost of the wood fuel supplies."

The North American forest products industry has been successful in developing useful technologies for recovering energy from wood residues as process heat or steam, which are cost effective. New technologies promise even greater economic benefits.

Cogeneration Technology

Cogeneration is the concurrent generation of electricity and the use of exhaust heat, usually in the form of process steam, for manufacturing operations. This is done by burning fuel (in this case, wood) to make high-pressure steam, 600–1200 lb./sq. in., passing this steam through a back-pressure or extraction turbine to drive a generator, and then using the steam exhausted from the turbine at lower pressures, 50–300 lb./sq.in., for process heat. This technology gets full use of the energy contained in the fuel. Wood at 55% moisture will generate power at about 60% efficiency.

The forest products industry is a major user of cogeneration technology, since it requires large quantities of process steam as well as electricity. It produces about 50% of its electricity needs in this way. The pulp and paper industry alone is the largest producer of energy by cogeneration of any U.S. industry. Electricity self-sufficiency is likely to increase to 80–90% for forest industries in the future through cogeneration. However, there is a minimum size of plant for economic power generation. Steam usage should be more than 70,000–120,000 lb./hr., equivalent to 3–5 MW of back-pressure, in order for a plant to economically employ cogeneration systems.

Pyrolysis

When wood is heated in the absence of air or with only limited amounts of air, thermal degradation takes place. This begins at about 100°C and increases with rising temperature. At about 270°C exothermic reactions set in causing a rise in temperature (usually held at 400-500°C) bringing about complete carbonization. The products are charcoal, condensible liquids, and noncondensible gases. The condensible liquids separate into aqueous (pyroligneous acid) and oil and tar fractions. The charcoal, the gases (low Btu gas) and the oil fractions all may be used as fuels.

The average composition of the wood gas is 29-33% carbon monoxide, 3.5-18% methane, 1-3% higher hydrocarbons, and 1-3% hydrogen. The hydrogen content of the gas increases with increasing temperature of pyrolysis. The wood gas has a fuel value of 300 Btu/cu. ft.

The yields of the different materials obtained by pyrolysis vary with the species of woods used and the type of equipment and system employed. Manipulation of three variables, mainly temperature, heating rate, and gas residence time, can greatly alter the relative proportions of gas, liquid, and char produced. For example, the oil and tar yields can be varied between 1 and 40% and the char between 40 and 10% or less. Low temperatures favor liquids and char; low heating rates favor gas and char; and short gas residence favors liquids. Conversely, high temperatures favor gas, high heating rates favor liquids, and long gas residence times favor gas. Thus, the various product fractions can be preferentially manipulated by proper combinations of these variables.[28]

Charcoal Production. Charcoal is the only significant product now produced in North America by pyrolysis (carbonization) of wood. Production has been increasing from a low figure of 232,000 tons in 1958. The reversal since 1958 of the previous downward trend in charcoal production is due to the great demand for charcoal for outdoor cooking. Today, most charcoal is crushed or ground and compressed into briquettes, a form especially suitable for home and recreational use. For these purposes, charcoal is a very satisfactory smokeless and low sulfur fuel.

The trend in the charcoal industry is towards greater use of wood residues from other forest industry operations. This requires mechanized continuous methods of production. Although most charcoal is still produced by conventional kilns, the trend persists towards both verticle gas-recycle retorts operating batch-wise and larger continuous conversion systems using saw mill residues or prepared wood fines. The more modern retorts recycle the wood gases and valatile products back through the combustion system, thereby diminishing the need for outside sources of heat and decreasing the exhaustion of organic materials to the atmosphere.

The yields of high-grade charcoal are 30 to 38% of the dry weight of the wood used. The labor requirements are low and, with good wood-handling facilities, including a mechanical conveyor and lift truck, charcoal can be manufactured in the more modern retorts at the rate of 1 ton per 4 man-hours of labor, compared to 11 man-hours of labor to produce 1 ton of charcoal in a typical kiln.

Low Btu Gas and Oil. Recently there has been much research activity to produce low Btu gas and oil from wood for energy. The Georgia Institute of Technology and the Tech-Air Corporation have intensively studied pyrolysis of biomass materials and have built several pilot plants.[30] There pyrolytic process (GT/T-A) has been applied to forestry residues. This plant differs from the older wood-distillation plants, such as the Badger-Stafford retorts used in the 1930's, in several respects, such as smaller-sized equipment to process 7 dry tons per hour compared to 3-4 tons per hour, use of wood dried to 7% moisture instead of to less than 0.5%, and use of a small quantity of air inside this retort (approx. 0.25 lb. per lb. feed) to sustain the reaction instead of gases in order to heat it to 1000°F. The on-line time of the rated capacity for the GT/T-A system is claimed to be better than 90% as compared to 67% for the Badger-Stafford units.

For the GT/T-A system the wood material is hogged so that the maximum size is not more than about 1 inch in any dimension. Sawdust may be used directly. The relatively uniform material is conveyed to a dryer where it is dried to about 7% water content. The dryer is heated by a portion of the wood gas from the reactor, and in the event that this should not be sufficient, the oil produced during pyrolysis can be used as back-up fuel.

The dried wood is fed into the top of the reactor through an air-lock and moves downward by gravity. A sensing device measures the bed height and controls the input of the wood. The temperature increases from 350-500°F at the top of the bed to 1000-1700°F in the pyrolysis zone. The gases move upward through the bed of descending wood and leave the top at 350-500°F. The reactor is operated at slightly under atmospheric pressure by an induced draft fan. The charcoal passes through an outfeed device at the bottom of the reactor into a sealed chamber. Here it is cooled by a water spray and then conveyed to a storage bin. The rate of charcoal discharge controls the rate of passage of material through the unit.

The gases from the reactor pass into a scrubber where they are sprayed with cooled pyrolytic oil, which removes particulates and cools the gas stream to 180-200°F, causing condensation of the condensible organic substances into an oil mixture. The oil is filtered, the filter cake returned to the reactor and the filtered oil pumped to a tank, cooled, and then recirculated through the scrubber. Excess oil is pumped to a storage tank.

The gases from the scrubber-condenser contain the non-condensible gases, low boiling organic vapors and water vapor. A portion of the gases are used to heat the wood dryer and the remainder are available as a fuel for other purposes, but it is desirable that they be used near the pyrolysis plant.

Operating conditions can be controlled to vary the distribution of the energy among the products: gas, oil, and charcoal. Thus, high yields of charcoal can be obtained with correspondingly reduced yields of oil and gas and vice-versa, or the unit may be operated as a gasifier with only 3.8% yield of charcoal. The off-gases in the latter case can be burned directly in a gas-fired boiler or the condensible oils removed first and used separately as fuel. The total energy recovery in the form of wood gas, oil, and charcoal is about 95% of the energy in the dry wood used in the process.

Gasification. Gasification is the thermal degradation of wood or other carbonaceous material in the presence of controlled amounts of oxidizing agents, e.g. air or pure oxygen. Gasification is carried out at higher temperatures than in the pyrolysis process, up to around 1000°C. Hence, the reaction rates are very fast, making equipment design critical. Thermal efficiency for conversion of wood to gases is 60-80%, which compares favorably with coal. Wood gasification offers several advantages over coal[28], namely, (1) much lower oxygen requirements, (2) practically no steam requirements, (3) lower costs for changing H_2/CO ratios which are already higher in wood gas and (4) no or very little desulfurization costs. Coal has an advantage in that larger plants can be built for coal gasification than is normally the case for wood, due to its procurement advantages.

As noted above, the GT/T-A pyrolysis system can also be used as a gasifier by increasing the amount of air to the reactor and screening the charcoal and recycling the coarser particles to the incoming feed material. In this way, most of the wood is converted to gas and oil. If all the off-gases from the converter are conducted directly into a boiler as a hot gaseous fuel, then the GT/T-A process is essentially a gasifier. This requires that the gases be utilized as a fuel close to the pyrolysis plant. The use of air in the reactor results in the presence of nitrogen in the gas, causing a reduction in fuel value from 360-420 Btu/cu. ft. (without air) to 140-200 Btu/cu. ft. (with air).

Because of the energy crisis, several gasifiers have been designed to handle municipal refuse, wood wastes, and other biomass materials. Basically, gasifiers fall into two types: (1) those that use air and (2) those using oxygen. An example of the air-system is the Moore-Canada gasifier. The wood residues are carried on a moving bed through stages of drying, reduction, and char oxidation. The ash is discharged in granular form. The maximum temperature in the reaction zone is about 1222°C. The hydrogen content of the crude gas is increased from 8-10% up to 18-22% by adding steam to the air intake.

The Union Carbide Company Purox gasifier is an example of the oxygen-system. This unit also uses a moving-bed reactor. Pure oxygen is the oxidizing agent to convert the char into CO and CO_2. Molten ash leaves the bottom at about 1670°C.

If air is used, the crude gases contain about 46% nitrogen which must be removed by cryogenic means; but if oxygen is used, it must be first separated from air into oxygen and nitrogen by a cryogenic system. The crude gases also contain an oil and tar fraction, about 2% of the wood (dry basis). Gasifiers designed for wood operate at atmospheric pressure in contrast to coal gasifiers which operate at pressures up to 400 lbs./sq. in.

The composition of the wood gas varies according to the technology used. If the limited oxygen required is supplied by air, a typical gas might contain 10-18% H_2, 22-30% CO, 6-9% CO_2, 45-50% N_2, and 3-5% hydrocarbons and have a heating value of about 1,700 K cal/m^3 (180 Btu/cu. ft.). If pure oxygen is supplied to the process, the gas might contain 24-26% H_2, 40% CO, 10% hydrocarbons and 23-25% CO_2 and have a heating value of 2,900 K cal/m^3 (350 Btu/cu. ft.).

The gas may be used directly as a fuel in a variety of ways; for example, supplying boiler energy in an industrial plant at the rate of 2.5×10^8 K cal/hr from 1460 tons (oven dry basis) of wood and/or bark per day; to small retrofitted furnaces in single-family houses, producing up to 2.5×10^4 K cal/hr from about 136 kg/day of dry wood.

Since wood gas from gasifiers approximates synthesis gas, it can be enriched with hydrogen for conversion of its CO and H_2 to methanol, a suitable liquid form of energy. The technology of gasification is under active development[30,31] in both equipment and process design to reduce capital costs, and in the chemistry of the process to improve yields and obtain more favorable gas ratios. The latter includes the use of catalysts to enhance the process and to promote the formation of specific products, such as methane or ethylene for increased thermal efficiency or for feedstocks for chemical synthesis.

Methanol Production. Methanol is produced by reacting synthesis gas, i.e., two volumes of H_2 and one volume of CO in a catalytic converter at pressures of 1500-4000 lbs/sq. in. Presently, 99% of the methanol produced in the U.S. is derived from natural gas or petroleum (see Chapt. 25). Methanol may also be produced from wood gas. Hence, wood could be a future additional raw material for making methanol, especially for use as an additive to gasoline for internal combustion engines.

If wood gas is used that is produced in a gasifier using air as the oxidizing agent, the nitrogen (about 40%), as well as the CO_2, must be removed. The CO_2 is removed by passing the gas through hot potassium carbonate followed by scrubbing with mono-ethanolamine. The nitrogen, hydrocarbons, residual CO_2 and water are removed by freezing them out in a cryogenic system. By means of an iron catalyst, a "shift conversion" of a portion of the CO is carried out with steam in order to reach the required 2 parts of H_2 to 1 part of CO as follows:

$$CO + H_2O \xrightarrow{\text{catal.}} H_2 + CO_2$$

$$\Delta h = -590 \text{ Btu/lb. CO}$$

The CO_2 produced in the shift conversion is removed by hot potassium carbonate. The pressurized gas is then led to the methanol reactor. Two different catalyst systems may be used: (1) zinc-chromium catalyst requiring gas pressures of 2000-4000 lb./sq. in. or (2) copper catalyst system at 1000-2000 lb/sq. in. About 95% of the gas is converted to methanol by this exothermic reaction:

$$2H_2 + CO \xrightarrow{\text{catal.}} CH_3OH$$

$$\Delta h = -1200 \text{ Btu/lb. CO}$$

The yield of methanol with present technology is about 390 l per ton of oven dry wood. Conversion of wood to methanol is less efficient than producing methanol from natural gas or coal; about 38% compared to about 60%, based on the heat value of methanol as a percent of the total energy input into the plant.

Raw materials cost is the most significant operating cost. Based on 1975 cost figures[31], the production cost for methanol from wood is estimated to be almost twice that produced from natural gas. Improving the efficiency of the gasifier reactor to increase the quantities of CO and H_2 produced from wood would enhance the process.

WOOD DISTILLATION CHEMICALS

The yield of charcoal by pyrolysis represents only about one-third of the weight of the wood, the remainder being accounted for as gases and vapors. The first commercial recovery of byproducts from the gases and vapors was undertaken by James Ward at North Adams, Massachusetts, in 1830.[32] The gases and vapors were cooled and the condensible portions converted to liquors. From the aqueous portion, known as pyroligneous acid, acetic acid and its salts were recovered. This was the beginning of the hardwood distillation industry. Later, methanol was also recovered from the pyroligneous acid, and the tars were separated and used for fuel, or were fractionated into creosote oil, soluble tars, and pitch.

With the development of low-cost continuous processes for the synthetic production of methanol, acetic acid, and acetone, the hardwood-distillation industry went into a decline after World War II. However, the demand for charcoal has held firm and even increased. As a result, improved methods for charcoal production have been developed and the recent trend is toward continuous carbonization. Thus, the carbonization of wood has gone full cycle, beginning with the production of only charcoal, to the production of industrial organic chemicals with charcoal playing a secondary role, and finally back to the production mainly of charcoal.

The products obtained by the distillation of hardwoods differ from those of the softwoods. This is primarily because only resinous softwoods can be profitably distilled, the resin giving rise to turpentine, pine oil, and rosin oil; products which are equally if not more important than the charcoal.

Hardwood Distillation

The once-flourishing hardwood-distillation industry has now ceased to exist in the United States due to the development of cheaper means of producing the chemical products from petroleum by synthetic methods. Therefore, this topic will be given only brief mention. The reader interested in further details is referred to the 1962 edition of *Riegel's Industrial Chemistry*, Chapter 15.

Two methods of hardwood distillation were developed: (1) the externally heated process and (2) the internally heated process.

The Externally Heated-Oven Process. An outgrowth of the kiln method for producing charcoal was the hand-loaded iron retort developed in 1850. However, the large-scale development of the chemical-wood industry did not begin until 1875, when the large carloaded ovens were employed.

The externally heated ovens (*retorts*) used steel cars or buggies. About 2.5 cords of wood in 4-foot lengths were piled on each car. The common 10-cord oven held 4 cars. The oven was made of steel, enclosed in a brick chamber with a space around the sides and top of the oven for circulation of the fire gases. Heat was applied by burning natural gas, oil, or coal. Two openings in the rear wall allowed the volatile products to pass out.

The cycle was 24 hrs. During the first few hours, the heating was rapid in order to reach the distillation temperature. Water came over first. An exothermic reaction took place next, after which the outside heat had to be decreased. The vapors passed out to the condenser where they formed the liquid condensate, the pyroligneous acid. The uncondensed gas was piped to the boiler house. After about 10 hrs, the flames in the burners were raised again, but not so high as at first. After 22 hrs the distillation was over; all burners were turned off, and the retort was allowed to cool for 2 hrs.

The buggies were placed in air-tight cooling chambers for two to three days, usually with a quenching spray of water after the first day. The charge shrunk considerably during the distillation, but the charcoal was obtained in the form of rather large pieces, with very little dust.

The Internally Heated-Retort Process. Several processes were developed whereby the retort was heated internally by hot circulating gases. A vertical retort was employed and filled with wood blocks or chunks. Oxygen-free gas, resulting from carbonization of wood in neighboring retorts, was freed from condensible gases in condensers and scrubbers and part of it was used to heat the remaining gas in a furnace, usually to 500–600°C. These hot gases were then circulated through the wood charge. As carbonization occured, the vapors were passed to condensers and the charcoal dropped to a cooling chamber at the base of the retort.

Several variations of this process were developed, some of which were continous instead of batch systems. In the continuous system, exothermic reaction was initiated by hot gases and then this reaction continued near the center of the retort as the wood in small sizes passed downward. The rising hot gases heated the incoming wood, and the charcoal dropped to a cooling chamber at the bottom (such as in Badger-Stafford, Lambiotte, and Pieters processes).

Chemical Recovery. The gases and vapors passing out of the oven or retort were cooled, the condensed pyroligneous acid was stored, and the gas was scrubbed and blown to the power house where it was burned under boilers, except the part used to bring a fresh retort to the proper temperature.

The pyroligneous acid was separated mechanically from the settled tar. Modern plants used a continuous refining process to fractionate the pyroligneous acid into water, tar, wood oils, and pure chemicals.

Destructive Distillation of Softwoods

Like the hardwood-distillation industry, the destructive distillation of softwoods is no longer practiced. Its major products, turpentine and pine oil, can be obtained in better quality and at lower cost by the extraction (so-called "steam distillation") process, or from the gum obtained by tapping southern pines, and as by-products of the sulfate pulp industry (see section in this chapter on Naval Stores).

Softwoods have less acetyl content than hardwoods and, therefore, yield less acetic acid upon distillation. The methanol yield is also much lower than from hardwoods. Recovery of these and their related chemicals was not practical in softwood distillation. Only resinous softwoods, especially longleaf and slash pine, were distilled because of the value of the products obtained from their resin content. Old stumpwood from logged-off areas and pitchy portions of fallen trees were the preferred material. This is because the sapwood, which is low in resin content had decayed away, leaving primarily the resin-rich heartwood. Some timber and sawmill wastes were also used, but generally only material which contained 20 percent or more resin.

Both the older small, hand-filled retorts of 1- to 2-cord-capacity and the larger car-loaded ovens up to 10-cord-capacity, similar to those used in hardwood distillation, were used for the destructive distillation of resinous softwoods. Also some retorts were made of concrete and internally heated by flues of large iron pipe. In general, the distillation procedure was similar to that for hardwoods.

Whether the crude oils were collected from the retort or oven as a whole, or in fractions, they were redistilled for separation into primary products. Copper stills, known as pinetar stills, were commonly used, and were provided with both steam coils and steam jets for steam distillation. The products from the pinetar still were usually light and heavy pine oils, pitch, and a composite of several light solvent oils. Further fractional distillation of the solvent oils yielded turpentine, dipentene, pine oil, and small amounts of other hydrocarbons. The yields vary greatly according to the resin content of the wood and the operating conditions. On the average, the yields per ton of southern pine stumpwood and "lightwood,"* are:

Total oils	35–40 gal
Wood turpentine	4–6 gal
Tar	20–30 gal
Charcoal	350–400 lb

*Resinous portions of wood, mainly heartwood, after decay of the sapwood; also termed "fatwood."

Products of Wood Distillation

Charcoal consists mainly of carbon, together with incompletely decomposed organic material and adsorbed chemicals. The amount of these secondary materials (containing hydrogen and oxygen) associated with the carbon decreases rapidly with increase in the distillation temperature as shown in Table 15.13. Commercial charcoal corresponds to about the 400 to 500°C product and has a volatile content (organic residues) of 15-25 percent. The average yield is 37-40 percent.

Charcoal is used mainly as a fuel. The use of charcoal in metallurgy has largely given way to coke. However, charcoal is an important material for the chemical industry, for the manufacture of items such as calcium carbide, sodium and potassium cyanide, carbon disulfide, magnesium chloride, hydrochloric acid, carbon monoxide, electrodes, fireworks, black powder, catalysts, pharmaceuticals, glass, resin moldings, rubber, brake linings, gas cylinder absorbent, paint pigment, and, in the form of activated carbon, as an adsorptive agent for the purification of gases and liquids. Miscellaneous uses include nursery mulch, crayons, and poultry and stock feeds.

The pyroligneous acid is the dilute aqueous condensate obtained by cooling the vapors from the retort or oven. It contains acetic acid, methanol, acetone, and minor quantities of numerous other organic compounds, of which more than 30 have been identified. Formerly hardwood distillation was the exclusive source of acetic acid, methanol, and acetone, and these were considered the primary products of the process. The average yields based on the dry weight of the wood are: acetic acid, 4-4.5 percent; methanol, 1-2 percent; and acetone, 0.5 percent.

The wood tars were largely used for fuel at the plant. There was a small demand for some of the tar components, however. The wood tars are of two types, namely (1) the soluble tars and (2) the settled tars. The soluble tars are those in the pyroligneous acid solution and they are separated as tars in the refining process. The settled tars are insoluble in, and heavier than, the aqueous pyroligneous acid and they are mechanically separated from it. The settled tars can be fractionated into (1) light oils with boiling points up to 200°C, specific gravities less than 1.0, and containing aldehydes, ketones, acids, esters, and steam-volatile phenols, and (2) heavy oils which boil above 200°C, have specific gravities greater than 1.0, and contain many phenolic components, and (3) pitch. Maple beech, and birch give total tar yields of 10-12 percent and oaks give 5-9 percent tar.

The heavy-oil fraction contains phenols, especially cresols, and is known as *wood-tar cresote*. It was used as a preservative for timbers, as a disinfectant, and for staining. Another important product was medicinal beechwood creosote, which was used as a disinfectant. The light-oil fractions were used as solvents, and the pitch was used for waterproofing and insulating agents.

The noncondensible gases produced during the distillation of hardwoods vary widely in amount and composition with the distilling conditions. The average range of composition of the gas is: 50-60 percent carbon dioxide; 28-33 percent carbon monoxide; 3.5-15 per-

TABLE 15.13 Composition and Amount of Charcoal Produced at Different Maximum Temperatures

Distillation Temperature	*Composition of Charcoal*			*Charcoal Yield on Dry Wt. of Wood*
	Carbon	*Hydrogen*	*Oxygen*	
°C	%	%	%	%
250	70.6	5.2	24.2	65.2
300	73.2	4.9	21.9	51.4
400	77.7	4.5	18.1	40.6
500	89.2	3.1	6.7	31.0
600	92.2	2.6	5.2	29.1
1000	96.6	0.5	2.9	26.8

cent methane; 1-3 percent higher hydrocarbons; and 1-3 percent hydrogen. The hydrogen content of the gas increases with increasing distillation temperature. The gas mixture has a heating value of about 300 Btu per cu ft. The gas was normally used as a fuel in the distillation plant and as a heat-conveyor in the internal gas-heated processes.

The major products obtained by the destructive distillation of southern pine wood were turpentine, dipentene, and pine oil. Wood turpentine contains a large amount of alpha-pinene and also a large amount of dipentene, differing in these respects from gum turpentine. The general properties of the two turpentines and their uses, however, are similar. Likewise the properties and uses of destructively distilled dipentene and pine oil are similar to those of the extracted or steam-distilled products.

Some of the common uses of the less important products of the destructive distillation of resinous pines were:

Tar and tar oils	Cordage, rubber, oakum, fish nets, tarpaulins, paper, soaps, insecticides, roofing cements, and paints.
Pyroligneous acid	Limited use in meat smoking, leather tanning, and as weed killer.
Charcoal	Similar to those for hardwood charcoal.

Pyrolysis of wood for producing industrial chemicals is now being reconsidered in the light of increasing oil costs. Modern technology could improve the yields of products and their separation into phenols and other desired chemicals. Goldstein[33] has reviewed the technical possibility of producing plastics, synthetic fibers, and rubber from wood instead of from petroleum and natural gas. He estimated that 95% of U.S. production of these materials could theoretically be produced from chemical feedstocks obtained from wood by pyrolysis, hydrolysis, and other processes. This would require 60 million tons of wood compared to the present 100 million tons required for pulp production.

NAVAL STORES

The naval stores industry in the United States began in the very early Colonial days, when wooden vessels used tar and pitch from the crude gum or oleoresin collected from the wounds of living pine trees. The demand for tar and pitch from crude gum is now of minor importance.

The industry is centered in the southeastern United States and is confined to the longleaf and slash pine areas. There is also a small, but locally important, naval stores-producing area in the Landes region of southwestern France, based on the maritime pine.

There are three routes by which naval stores are produced. The oldest method is by tapping living trees to cause a flow of oleoresin. The second method is removal of naval stores by solvent extraction. The latter process has now replaced steam distillation as a means of recovering turpentine. In the U.S., the latest and now the most important route is by the sulfate pulping of pines, during which turpentine and tall oil are recovered as by-products of kraft pulp manufacture. A fourth process, no longer used in the U.S., is recovery of turpentine and pine oils by the destructive distillation of pine wood.

World-wide, about 60% of naval stores are produced by tapping living trees, whereas gum naval stores account for only 4% of U.S. production. The sulfate process is now the major process in the U.S.

Turpentine is a volatile oil consisting primarily of terpene hydrocarbons, having the empirical formula $C_{10}H_{16}$. These 26 atoms can have many different arrangements. Only six are present in appreciable amounts in commercial turpentines, namely; alpha-pinene (b.p. 156°C), beta-pinene (b.p. 164°C), camphene (b.p. 159°C), Δ^3-carene (b.p. 170°C), dipentene (b.p. 176°C), and terpinoline (b.p. 188°C). The molecular configurations of some of these are shown in Fig. 15.15.

Gum and sulfate turpentines have similar compositions. Gum turpentine contains 60-65% α-pinene, 25-35% β-pinene and 5-8% other terpenes, compared to 60-70% α-pinene, 20-25% β-pinene and 6-12% other terpenes for sulfate turpentine. Wood turpentine, which has 75-80% α-pinene, has no or very little β-pinene (0-2%), and also contains 4-8% camphene and 15-20% other terpenes.

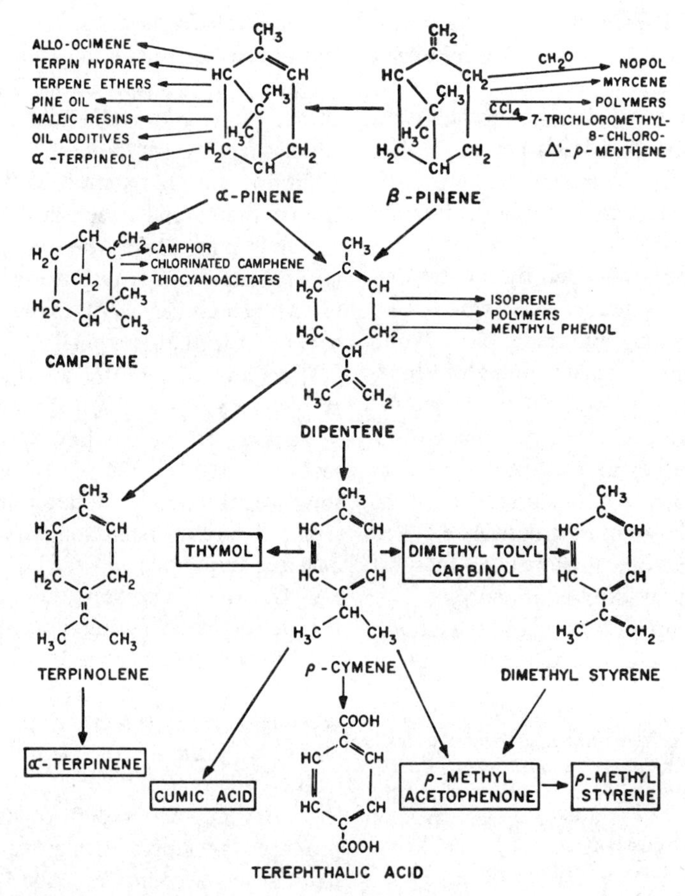

Fig. 15.15 Some reactions of alpha-pinene and beta-pinene. *(L. A. Goldblatt, "Yearbook of Agriculture," U.S. Department of Agriculture, 1950–1951.)*

Rosin, the other major naval stores product, is a brittle solid which softens at 80°C. Chemically it is composed of about 90 percent resin acids and 10 percent neutral matter. The resin acids are mainly *l*-abietic acid and its isomers, $C_{20}H_{30}O_2$. These are tricyclic monocarboxylic acids and are diterpenes.

Rosin is graded and sold on the basis of color, the color grades ranging from pale yellow to dark red (almost black). The color is due almost entirely to iron contamination and oxidation products. Fresh oleoresin, as it exudes from the tree, will yield a rosin that is nearly colorless. Color-bodies are removed by selective solvents and selective absorption from a 10–15 percent gasoline solution passed through beds of diatomaceous earth. About 70 percent of the world's rosin is produced in the United States.

Naval stores also include important fatty acids, as well as sterols and other products.

Gum Naval Stores

The crude gum or oleoresin is caused by flow from healthy trees by exposing the sapwood. The lower part of the tree is faced, i.e., a section of bark is removed, giving a flat wood surface for the gutters which are inserted into a slanting cut made by a special axe. The gutters conduct the gum to a container which can hold 1–2 qt of gum. At the top of the exposed face, a new V-shaped strip of bark is removed about every two weeks.

The operations of inserting gutters, hanging cups, and cutting the first bark is preferably done in December or January, since early facing stimulates early season gum flow. The gum continues to flow until November, with the height of the season being from March to September.

The collected gum is distilled from a copper still; turpentine and water pass over, and the rosin is left in the still. The remaining molten rosin, plus impurities, is passed through a series of strainers and cotton batting to remove dirt particles. The liquid rosin is then run into tank cars, drums, or multiwall paper bags for shipment. Annual production figures for gum turpentine and rosin are shown in Table 15.14.

Increases in yield of naval stores is brought about by chemical treatment of the exposed wood, especially with paraquat herbicides (dipyridyl compounds). This treatment stimulates extensive oleoresin formation and diffusion into the wood, extending to the pith of the tree and several feet above the treatment level. As much as 40% oleoresin content in the wood has been produced. Such treatment could double naval stores production, both for gum and sulfate processes. It also has the potential of providing a new type of wood naval stores by solvent extraction prior to sulfate pulping, or a combination of both methods.[34]

Extraction Process for Wood Naval Stores

"Wood" naval stores are produced by solvent extraction of resin-rich wood from old southern pine stumps and roots. The depletion of these stumps from the large trees of virgin forests, as well as high labor costs, have brought about a major decline in the production of wood naval stores by this process.

In modern practice, all the resin products are removed from the shredded wood by solvent extraction. The solvent retained by the extracted wood chips is recovered by steaming. Extraction is carried out with naphtha (b.p. 90-115°C fraction). Multiple extraction is carried out in a series of vertical extractors in a

TABLE 15.14 Naval Stores Production for U.S.A.

Rosin–1000 Drums of 520 lb net

Year	*Gum*	*Wood (Extracted)*	*Tall Oil*	*Total*
1955	528	1342	None	1870
1958	400	1196	269	1865
1963	459	1098	528	2085
1969	119	830	793	1742

Turpentine–1000 Barrels of 50 gal

Year	*Gum*	*Wood (Extracted)*	*Wood (Destr. Dist.)*	*Sulfate*	*Total*
1955	176	208	2.41	232	618
1960	119	163	none	323	605
1966	84	155	none	427	666
1970	35	103	none	480	618

Pine Oil and Other Terpenes–1000 Barrels of 50 gal

Year	*Pine Oil*	*Other Terpenes*
1955	189	108
1965	234	125
1970	278	100

Source: Naval Stores Annual Report, U.S. Dept. Agr. (1958, 1955); Agricultural Statistics, U.S. Dept. Agr. (1970); Statistical Abstracts of U.S., U.S. Dept. Commerce, Bureau of Census, 91st Ed., Washington (1970).

countercurrent manner, whereby fresh solvent is used for the final extraction of a charge.

The solution from the extractors is vacuum distilled and the solvent recovered. The remaining terpene oils are fractionally redistilled under vacuum and recovered as turpentine, dipentene, and pine oil. The nonvolatile rosin is of dark color and is upgraded by clarification methods, such as selective absorption of its solution (bed-filtering). Annual production figures of the products are shown in Table 15.14 during the period when gum and wood products were declining and tall oil and sulfate products were increasing.

Sulfate Naval Stores

Sulfate turpentine is obtained as a by-product during the kraft pulping of pine woods. Vapors are periodically released from the top of the digesters; these are condensed, and the oily turpentine layer is separated and purified by fractional distillation and treatment with chemicals to remove traces of sulfur compounds. Sulfate turpentine is very similar to gum turpentine obtained from the oleo-resin of the tree and contains 60-70% of α-pinene, 20-25% of β-pine and 6-12% other pinenes. Sulfate turpentine from western North American woods contains appreciable amounts of Δ^3-carene, which is used as a solvent. In the U.S.A., about 80% of the annual production of turpentine is by the sulfate pulping of southern pines (approximately 125 million 1.).

The spent black liquor from the sulfate pulping of pines contains the less volatile products of the wood resin in the form of sodium salts or soaps. The liquor is first concentrated in multiple-effect evaporators and then the concentrate is sent to settling tanks. The soaps rise to the surface and are skimmed off and then are acidified by sulfurous or sulfuric acid. The crude tall oil rises to the top and is mechanically separated. Crude tall oil from southern pines contains 40-60% resin acids and 40-55% fatty acids with 5-10% neutral substances. These are separated by fractional distillation under vacuum.

One m. ton of crude tall oil yields about 350 kg. of rosin, 300 kg. of fatty acids, and 300 kg. of head and pitch fractions. For each m. ton of pulp produced, northern pines yield about 50 kg. of tall oil and the southern pines about 125 kg. U.S.A. capacity for fractional distillation of tall oil is nearly 1 million m. tons per year.

The rosin component of tall oil is mostly made up of resin acids, where are diterpene derivatives. The major compounds (figures indicate averages) are as follow:

1. abietic type–abietic (32%), neoabietic (4%), palustric (10%), and dehydroabietic (30%) acids.
2. pimaric type–pimaric (4%), isopimaric (10%), and sandara copimaric (small amounts) acids.

The fatty acids from tall oil have the following components: oleic (50%), linoleic (35%), conjugated linoleic (8%), stearic (2%), palmitic (1%), and others (4%). From tall oil heads, a fraction is produced that is composed predominantly of saturated acids, containing 55% palmitic acid. Oleic acid is distilled in grades that are 99.5% pure.

The neutral or "unsaponifiable" materials present in tall oil include anhydrides, phenolics, diterpene aldehydes and alcohols, stilbenes, and sterols. In the neutral fraction of southern pine tall oil soap, 80 compounds have been identified.[35] These include 25.1% sistosterol and a total of 32.4% sterols. The sistosterol content of crude tall oil is 2 to 3% and is the main component of the neutral fraction.

Sulfate naval stores production has increased dramatically in the U.S. and now accounts for 80% of total turpentine production and about 60% of total naval stores production (combined rosin and turpentine).

Uses of Naval Stores Products

The naval stores products have a wide range of usefulness from ordinary household commodities to complex industrial uses, as outlined below.

Turpentine. Turpentine is used directly as a solvent, thinner, or additive for paints, varnishes, enamels, waxes, polishes, disinfectants, soaps, pharmaceuticals, wood stains, sealing wax, inks, crayons, and as a general solvent. The chemistry of its monoterpenes offers many possibilities for conversion to other substances as illustrated in Fig. 15.15. There is increasing use of turpentine to produce fine chemicals for flavors and fragrances.[34] A large use of turpentine is for conversion by mineral acids to synthetic pine oil. It is also a raw material for making terpin hydrate, resins, camphene, insecticides, and other useful commodities. These uses are outlined as follows:

- Solvents for paints, etc. (11%).
- Synthetic pine oil (48%), used for mineral flotation, textile processing, solvents, odorants, bacteriacides, and conversion to other chemicals such as terpin hydrate, etc.
- Polyterpene resins (600–1500 mol. wt.) (16%), used for adhesives, pressure-sensitive sizes (dry cleaning, paper, chewing gum).
- Camphene → toxophene insecticides (16%).
- Flavor and fragrance essential oils (9%).

Dipentene. Dipentene is present in the higher boiling fractions of wood turpentine. It is used in paints and varnishes and as a penetrating and softening agent in rubber reclamation.

Pine Oil. Pine oil obtained from wood naval stores has similar uses to the synthetic pine oil made from turpentine, as outlined above.

Rosin. Rosin is used mainly in some modified form. Because the abietic-type acids in rosin each contain a carboxyl group and double bonds, they are reactive and can be used to produce salts, soaps, esters, amines, anides, nitriles, and Diels-Alder adducts; and can be isomerized, disproportionated, hydrogenated, dimerized, and polymerized. When destructively distilled, rosin produces a viscous liquid, termed rosin oil, used in lubricating greases.

The paper industry uses large amounts of the sodium salt of rosin as paper size, which accounts for the greatest single use of rosin. The synthetic rubber industry is the second most important user of rosin. In making styrene-butadiene rubber, disproportionated rosin soaps are used alone or in combination with fatty acid soaps as emulsifiers in the polymerization process. Disproportionation decreases the number of double bonds in the abietic acid of the rosin, making a more stable material.

The adhesives industry is the third most important market for rosin. Rosin, modified rosins, and rosin derivatives are used in several types of adhesives, including the pressure-sensitive, hot-melt, and elastomer-based latices, and solvent rubber cements.

Protective coatings are the fourth major user of rosin, either directly or in a modified or derivative form. Varnishes and alkyds are the most common types of protective coatings using rosin. Rosin is combined with a heat-reactive phenol-formaldehyde resin to produce a widely used varnish. Printing inks also use substantial amounts of rosin.

The above uses are summarized as follows:

- Paper size (33%).
- Chemical intermediates and rubber (42%).
- Resins and ester gums (18%).
- Coatings (3.9%).
- Other uses (3.1%).

Fatty Acids. Of the total fatty acids produced annually in the U.S., amounting to more than 450,000 tons, 35% comes from tall oil.[36] The solvent extraction of pine wood yields about 1% fatty acids and their esters. Unlike it is for resin acids and terpenes, the yield is not increased, however, by paraquat (dipyridyl herbicides) treatment. Hence, the kraft-pulping industry will continue to be the major source of fatty acids from wood. The major uses of the fatty acids are:

- Intermediate chemicals (43.8%).
- Protective coatings (28.6%).
- Soaps and detergents (11.1%).
- Flotation (3.4%).
- Other uses (16.5%).

A future product from the neutral fraction of tall oil may be sistosterol. This chemical has potential use in the synthesis of cortisones and other steroids and hormones by fermentation processes.

TANNIN AND OTHER EXTRACTIVES

The tissues of wood, bark, and the leaves of trees contain a great variety of chemical substances of considerable scientific interest and some of practical value. Turpentine, pine oil, and rosin from the resins of pines are the most important commercial extractives from American woods.

Tannin is a commercially important substance that can be extracted from the wood, bark, or leaves of certain trees and other plants. Tannins are complex dark-colored polyhydroxy phenolic compounds, related to catechol or pyrogallol, and vary in composition from species to species. They have the important property of combining with the proteins of skins to produce leather.

For many years, most of the leather in the United States was tanned with domestic tannins from hemlock and oak bark and from chestnut wood. Today only a small amount of tannin comes from these and other domestic sources. The most important source of vegetable tannin today is the wood of the quebracho tree which grows mainly in Paraguay and Argentina. The tannin content of this tree and a few other sources of vegetable tannin are shown in Table 15.15.

TABLE 15.15 Tannin Content of Some Plant Materials

Plant Material	*Percent Tannin*
Domestic sources	
Eastern hemlock bark	9–13
Western hemlock bark	10–20
Tanbark oak	15–16
Chestnut oak	10–14
Black oak	8–12
Chestnut wood	4–15
Sumac leaves	25–32
Foreign sources	
Quebracho heartwood	20–30
Mangrove bark	15–42
Wattle (*acacia* bark)	15–50
Myrobalan nuts	30–40
Sicilian sumac leaves	25–30

The wood or bark for tannin production is reduced to chips and shreds by passing the material through hoggers or hammer mills. It is then extracted with warm water in diffusion batteries. The dilute solutions are evaporated to the desired concentration. Loss of solubility of the tannin can be counteracted by treatment of the concentrate with sodium sulfite.

ESSENTIAL OILS

Various essential oils are obtained by steam distillation of wood chips, barks, or leaves of trees. Chemically they are related to turpentine, being mainly terpene hydrocarbons. In the United States the following oils are produced; cedar-wood oil from the wood of red cedar; conifer-leaf oils from the needles and twigs of spruce, hemlock, cedar, and pine; sassafras oil from roots, wood, and bark of the sassafras tree; and sweet-birch oil from the bark and twigs of birch (*Betula lenta*).

MEDICINALS

The bark of the cascara tree of the northwestern region of the United States yields cascara, a laxative used in medicine. Several hundred tons of bark are harvested annually.

The red gum tree of the southern United States exudes a yellowish balsamic liquid or gum from wounds. This is known as storax. It is produced by removing a section of bark and incising the wood much in the same manner as that used for the production of naval stores gum described above. Storax is used in medicinal and pharmaceutical preparations, such as adhesives and salves. It is also used as an incense, in perfuming powders and soaps, and for flavoring tobacco.

REFERENCES

1. Hall, D. O., "Solar Energy Use Through Biology–Past and Future," in *Future Sources of Organic Raw Materials–CHEMRAWN 1*, L. E. St. Pierre and G. R. Brown, eds., pp. 579–599, Pergamon Press, New York/Toronto, 1979.
2. U.S. Forest Service, "Timber Trends in the United States," Forest Service Report No. 17, U.S. Dept. of Agriculture, Washington, D.C. Feb. 1965.

3. "The United States Timber Supply," National Policy Conference, Arden House, Harriman, N.Y. Convened by State University of New York College of Forestry, Nov. 23-24, 1970.
4. Waters, W. E. and Brandt, R. W., *American Forests* (Sept. 1970).
5. Anon., "One Third of the Nation's Land," A Report to the President and to the Congress by the Public Land Law Review Commission, Washington, D.C., June, 1970.
6. U.S. Forest Service, "The Outlook for Timber in the United States," U.S.D.A., Forest Service Rpt. No. 20, 367 pp., 1973.
7. Jamison, R. L., "Wood Fuel Use in the Forest Products Industry," in *Progress in Biomass Conversion*, Vol. 1, K. Sarkanen and D. A. Tillman, eds., pp. 27-52, Academic Press, New York, 1979.
8. Abert, J. G. and H. Alter, "Recovering Energy from Municipal Waste," in *Process in Biomass Conversion*, Vol. 1, K. V. Sarkanen and D. A. Tillman, eds., Academic Press, New York, 1979.
9. Maxwell, C. S., *Tappi*, **54**, 1932-1934 (1971).
10. Kenaga, D. L., Fennessey, J. P., and Stannett, V. I., *Forest Prod. J.*, **12**, No. 4, 16-168 (1962).
11. Kent, J. A., Winston, A., and Boyle, W., "Preparation of Wood-Plastics Combinations Using Gamma Radiation," ORO-612, Office of Technical Services, Dept. of Commerce, Washington, D.C., Sept. 1, 1963.
12. Kent, J. A., Winston, A., Boyle, W., and Taylor, G. B., "Preparation of Wood-Plastics Combinations Using Gamma Radiation to Induce Polymerization (Copolymers, Additives and Kinetics)," Interim Report, U.S. Atomic Energy Comm., ORO-2945-7, 1967.
13. Anon., *Ind. Eng. Chem.*, 28, 1160-1163 (1936).
14. Meyer, J. A., *Forest Prod. J.*, **15**, No. 9, 362-364 (1965).
15. Meyer, J. A. and Loos, W. E., *Forest Prod. J.*, **19**, No. 2, 32-38 (1969).
16. Meyer, J. A., *Forest Prod. J.*, **18**, No. 5, 89 (1968).
17. Rowell, R. M. and W. D. Ellis, "Chemical Modifications of Wood: Reaction of Methyl Isocyanate with Southern Pine," *Wood Science*, **12**, No. 1, 52-58, 1979.
18. Rowell, R. M., R. Moisuk and J. A. Meyer, "Wood-Polymer Composites: Cell Wall Grafting with Alkaline Oxides Followed by Lumen Treatments with Methyl Methacrylate," U.S. Forest Products Lab Report, Madison, Wisc., Jan. 1979.
19. Schuerch, C., *Forest Prod. J.*, **14**, No. 9, 377-381 (1964).
20. Schuerch, C., U.S. Patent 3,282,313 (Nov. 1, 1966).
21. Karlivan, V. P., "New Aspects of the Production of Chemicals from Biomass," in *Future Sources of Organic Raw Materials–CHEMRAWN 1*, L. E. St. Pierre and G. R. Brown, eds., pp. 483-494, Pergamon Press, New York/Toronto, 1979.
22. Locke, E. G. and Garnum, E., "FAO Technical Panel on Wood Chemistry: Working Party on Wood Hydrolysis," *Forest Prod. J.*, **11**, No. 8, 380-382 (1961).
23. Goldstein, I. S., "Hydrolysis of Wood," Proc. of Tech. Assoc. Pulp and Paper Ind., 1980 Annual Meeting, Atlanta, Ga.
24. Tsav, G. T., M. Ladisch, C. Ladisch, T. A. Hsu, B. Dale and T. Chan, "Fermentation Substrates from Cellulosic Materials," in *Annual Reports in Fermentation Processes*, Vol. 2, D. Perlman, ed., Academic Press, New York, 1979.
25. Chang, P. C. and L. Pasgnes, "Comparative Dissolution Rates of Carbohydrates and Lignin During Aqueous Acidified Organosolve Saccharification of Aspen and Douglas Fir Heartwood," TAPPI Forest Biology/Wood Chemistry Conference, Madison, Wisc., June 20-22, 1977.
26. Gaden, E. L., M. H. Mandels, E. T. Reese and L. A. Spano, "Enzymatic Conversion of Cellulosic Materials: Technology and Application," Biotech, and Bioeng. Symp., No. 6, 316 pp., Interscience, New York, 1976.
27. Mandels, M. H., S. Doroal and J. Madeiros, "Saccharification of Cellulose with *Trichoderma* Cellulase," Proc. Second Annual Fuels from Biomass Symp., pp. 627-684, Rensselaer Polytechnic Institute, Troy, N.Y., June 20-22, 1978.
28. Goldstein, I. S., "New Technology for New Uses of Wood," Proc. of Tech. Assoc. of Pulp and Paper Industry, 1980 Annual Meeting, Atlanta, Ga.
29. Kringstad, K., "The Challenge of Lignin," in *Future Sources of Organic Raw Materials–CHEMRAWN 1*, L. E. St. Pierre and G. R. Brown, eds., pp. 627-636, Pergamon Press, New York/Toronto, 1979.
30. Knight, J. A., "Pyrolysis of Wood Residues with a Vertical Bed Reactor," in *Process in Biomass Conversion*, Vol. 1, K. V. Sarkanen and D. A. Tillman, eds., Academic Press, New York, 1979.
31. Raphael Katzen Associates, "Report on Chemicals from Wood Waste," U.S.D.A., For. Prod. Lab., Madison, Wisc., 1975.
32. Brown, N. C., *Forest Products*, 197, New York, John Wiley & Sons, 1950.
33. Goldstein, I. S., "Potential for Converting Wood Into Plastics," *Science*, **189**, 847-852, Sept. 12, 1975.
34. Zinkel, D. F., "Naval Stores: Silvichemicals from Pine," *J. Appl. Polymer Sci.*, Appl. Polymers Symp. No. 28, 309-327, 1975.
35. Connor, A. H. and J. W. Rowe, *J. Am. Oil Chem. Soc.*, **52**(9): 334-338, 1975.

36. Herrick, F. W. and H. L. Hergert, "Utilization of Chemicals from Wood: Retrospect and Prospect" in *Recent Advances in Phytochemistry*, Vol. 11, F. A. Loewus and V. C. Runeckles, eds., pp. 443-515, Plenum Press, New York, 1977.

Selected References

Books

American Wood Preservers' Association, *American Wood Preservers' Standards*, Washington, D.C., 1971.

Britt, K. W., *Handbook of Pulp and Paper Technology*, 2nd ed., Van Nostrand Reinhold, New York, 1970.

Browning, B. L., *Methods of Wood Chemistry*, Vols. I and II, John Wiley and Sons, New York, 1967.

Casey, J. P., *Pulp and Paper*, 2nd ed., Vols. I, II and III, John Wiley and Sons, New York, 1960 (to be revised 1981).

Goldstein, I. S., eds., *Wood Technology: Chemical Aspects*, ACS Symposium Series No. 43, American Chemical Society, Washington, D.C., 1977.

Hillis, W. E., ed., *Wood Extractives*, Academic Press, New York, 1962.

Hunt, G. M. and G. A. Garratt, *Wood Preservation*, McGraw-Hill, New York, 1953.

Loewus, F. A. and V. C. Runeckles, eds., *Recent Advances in Phytochemistry*, Vol. 11, Plenum Press, New York, 1977.

MacDonald, R. G., ed., *Pulp and Paper Manufacture*, 2nd ed., Vols. I, II and III, McGraw-Hill, New York, 1969.

MITRE Corporation/METREK Div., *Silvicultural Biomass Farms*, Vols. I-VI, MITRE Technical Report No. 7347, May 1977.

Raphael Katzen Associates, *Report on Chemicals from Wood Waste*, U.S.D.A., For. Prod. Lab, Madison, Wisc., 1975.

Rydholm, S. A., *Pulping Processes*, John Wiley and Sons, New York, 1965.

Sarkanen, K. V. and Ludwig, C. H., eds., *Lignins*, John Wiley and Sons, New York, 1971.

Sarkanen, K. V. and D. A. Tillman, eds., *Progress in Biomass Conversion*, Vol. 1, Academic Press, New York, 1979.

Shafizadeh, F., K. V. Sarkanen and D. A. Tillman, eds., *Thermal Uses and Properties of Carbohydrates and Litnins*, Academic Press, New York, 1976.

St. Pierre, L. E. and G. R. Brown, eds., *Future Sources of Organic Raw Materials–CHEMRAWN 1*, Pergamon Press, New York, 1979.

Timell, T. E., ed., "Proc. of the Eighth Cellulose Conference," Applied Polymer Symposia No. 28, Vols. I, II and III, John Wiley and Sons, New York, 1976.

U.S. Department of Agriculture, "Crops in Peace and War," *The Yearbook of Agriculture*, Washington, D.C., 1950-51.

Wenzel, H. F. J., *The Chemical Technology of Wood*, Academic Press, New York, 1970.

Articles and Chapters

Canadian Wood Council, "Wood: Fire Behavior and Fire Retardant Treatment–A Review of the Literature," Ottawa, Nov. 1966.

Davidson, R. W. and W. G. Baumgardt, "Plasticizing Wood with Ammonia–A Progress Report," *Forest Prod. J.*, 20, No. 3, 19-24, 1970.

Jahn, E. C. and V. Stannett, "Polymer Modified Papers," in *Modern Materials*, Vol. II, H. H. Hausner, ed., Academic Press, New York, 1960.

Lawrence, R. V., "Naval Stores Products from Southern Pines," *Forest Prod. J.*, 19, No. 9, 87-92, 1969.

Meyer, J. A., "Wood Polymer Composites and Their Industrial Applications," in *Wood Technology: Chemical Aspects*, I. S. Goldstein, ed., ACS Symposium Series No. 43, American Chemical Society, Washington, D.C., 1977.

MITRE Corporation/METREK Division, Appendix on "Hydrolysis of Wood," pp. B-1-B-13, in *Silvicultural Biomass Farms*, Vol. V, Technical Report No. 7347, May 1977.

Rowell, R. M., "Chemical Modification of Wood: Advantages and Disadvantages," Am. Wood Preservers' Assoc. Proc. 71, 41-51, 1975.

Siau, J. F., J. A. Meyer and C. Skaar, "Wood-Polymer Combinations Using Radiation Techniques," *Forest Prod. J.*, 15, 426-435, 1965.

16

Sugar and Other Sweetners

Charles B. Broeg* and Raymond D. Moroz**

INTRODUCTION

Sugar and starch are among those organic "chemicals" found so abundantly in nature that no serious efforts have been made to synthesize them commercially from coal (or petroleum), air, and water. Both are available at such concentrations in some plants that sizable industries have resulted from growing those plants and extracting carbohydrates therefrom.

The primary use of sugar is in the manufacture of food or as a food in itself. Where used for such purposes, most of it is highly refined or purified, but considerable quantities are consumed in some areas as a crude product. Sugar is used to a limited extent in the production of other chemicals such as sucrose esters and in the form of by-product molasses as a substrate for fermentation processes. Currently, the juice of sugar-producing plants is also utilized for fermentation.[1]

Starch is extensively used in the textile and paper industries, but food is also a major outlet, especially if one considers flours that are primarily starch as a crude form of starch. However, a very important outlet for starch is its conversion to dextrose and glucose syrups for use in the food industry. Recent technological developments have greatly magnified the importance of starch as a source of sugars.[2] Those developments are based on the enzymatic conversion of glucose to fructose. The following discussion explores briefly the sources of various sugars and the processes by which they are converted into commercial products.

SUGAR

History

The ancestry of the sugar cane and its use as a food has been traced to the island of New Guinea. (However, a very interesting legend related by J. A. C. Hugill in Chapter 1 of *Sugar: Sucrie and Technology*[3] associates sugar cane with the origin of the human race.) Around 8000 B.C., the plant started on its migration from New Guinea to the many areas of southeastern Asia, Indonesia, the Philippines, Malay,

*Vice President Research and Technical Services Revere Sugar Corporation.

**Assistant Technical Director Revere Sugar Corporation.

Indochina, and eastern India, with man probably as its main dispersal agent.[4] It is in Bengalese India that sugar cane was first cultivated as a field crop and that the juice was manufactured into various solid forms. A general knowledge of sugar was prevalent throughout India by 400 B.C. By the tenth century A.D., sugar cultivation and manufacturing had become important industries in Persia and Egypt. The early Islamic movement spread the knowledge of the sugar industry throughout the Mediterranean area. On the second voyage of Columbus to America in 1493, sugar cane was introduced in Santo Domingo. It spread rapidly through the West Indies and Central America. Cortez brought the cane to Mexico, and Pizarro introduced it in Peru. By 1600, the sugar industry was the largest in tropical America.[5]

The modern sugar industry dates from the end of the 18th century when steam replaced animal energy and made possible the development of larger and more efficient production units. The vacuum pan appeared in 1813, bag filters in 1824, multiple-effect evaporators in 1846, filter presses in 1850, centrifugals in 1867, driers in 1878, and packaging machines in 1891.[6]

Cultivation of the sugar beet plant and the manufacture of sugar from the beet developed in the industrial nations of Europe within the last two centuries. In 1747, a German chemist, Marggraf, established that sugar from beets was the same as sugar from cane. His pupil, Achard, in 1799, demonstrated that sugar can be commercially prepared from beets. During the Napoleonic wars, a short-lived beet sugar industry was established in France. It was slowly revived 20 years after Waterloo and spread to most European countries.[7] Today, sugar from beets accounts for about 41 percent of the world's supply.

Cane Sugar

Agriculture. The sugar cane is a large perennial tropical grass belonging to the genus *Saccharum.* There are three basic species, *S. officinarum*, *S. robustum*, and *S. spontaneum*, and a large number of varieties. Sugar cane is propagated commercially by cuttings, each cutting consisting of portions of the cane plant having two or more buds. The buds sprout into shoots from which several other shoots arise below the soil level to form a clump of stalks or a "stool." From 12 to 20 months are required for crops maturity from new plantings and about 12 months for ratoon crops (that is, cane stalks arising from stools which have been previously harvested). Most of the field operations have been mechanized, but planting is still done by hand in some cane-producing areas. Fields are replanted after 2 to 5 "cuttings" have been made from the original plantings. As cane grows, its foliage is largely limited to the upper one third of the plant (the lower leaves die off from shading as the stalk grows taller). The cane stalk is round and from a few to more than ten feet long when mature. It is covered with a hard rind which is light brown or green, yellowish green, or purple in color, depending on the variety. The stalk consists of a series of joints or internodes separated by nodes. The rind and the nodes are of a woody nature, while the internode is soft pith. The internode contains the greater part of the juice containing sucrose, and the nodes contain "eyes" or buds which "sprout" when planted.[8]

Harvesting. The sugar cane is still cut by hand with machete-type knives in many producing areas. The canes are cut at ground level and, at the same time, leaves and tops of the stalk are removed. In areas where labor is scarce or expensive, machine harvesting has come into widespread use. Some harvesters cut, chop, and load the cane into transporting vehicles in one operation at rates of 30 to 45 tons per hour. Other harvesters cut, bundle, and dump cane in rows for pick-up by other machines. Transportation of field cane to the mill is accomplished by railcars, trucks, trailers, and carts.

Preparation of Cane for Milling. Mechanically harvested cane must be washed before milling in order to eliminate soil, rocks, and field trash. In some instances, "dry cleaning" precedes washing. Washing systems range from a simple spraying with warm water on the carrier or table to an elaborate system consisting of conveyors with water jets, stripping rolls, and baths for removal of stones.

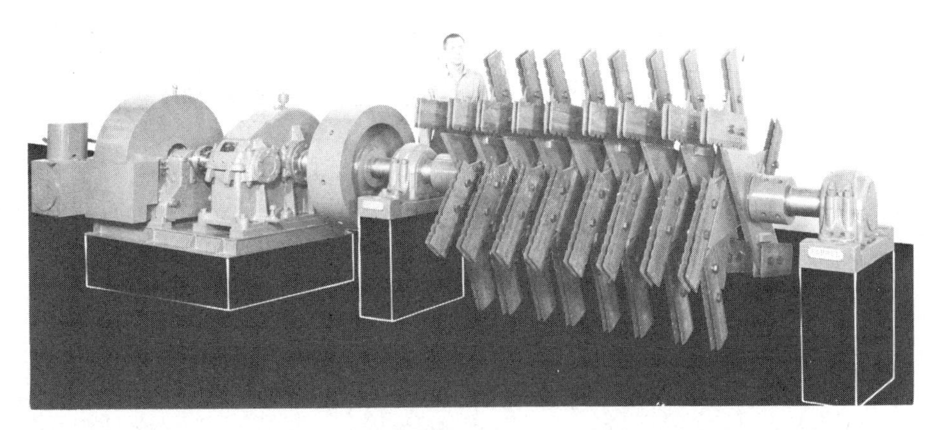

Fig. 16.1 Farrel Type K-4 cane knife to handle 4500 tons of cane per 24-hour day. (*Courtesy Farrel Co., Division of USM Corp.*)

Juice Extraction. The cane is first prepared for grinding by one or more of the following operations: (1) chopping into smaller pieces with one or two sets of rotating knives (400–600 rpm; see Fig. 16.1); or (2) disintegrating the cane into finer sizes by a crusher and/or shredder (Fig. 16.2). The crusher is usually a two-roller mill of the Krajewski or Fulton types.

Recently, a new machine has been developed which combines the action of the chopper and the shredder into a single unit. It is commonly called a Unigrator or the "Ducasse Shredder."[9,10] The advantages gained have been a greater increase in juice extraction by the mills, an increase in the grinding rate, and in some cases, a decrease in power consumption by the milling operation.

Fig. 16.2 Fulton crusher roll. (*Courtesy Fulton Iron Works Co., St. Louis, Mo.*)

Radically new approaches to the preparation of sugar cane for the extraction of sugar have been under development for more than a decade.[11] The basic concept consists of chopping cane stalks into short segments (8–12 inches long) and feeding the segments into a machine (Figures 16.3 and 16.4) which separates the cane into three major fractions: (1) epidermis, (2) rind, and (3) pith. This approach to the preparation of cane for the extraction of sugar has several potential advantages: (1) reduced power requirements, (2) simplification of juice extraction processes, and (3) the recovery of not less than three by-products having greater utility than bagasse.[11] Development of this process has advanced to a pilot plant capable of handling as much as 20 tons of cane per hour.

The crushed and shredded cane passes then through a series of three horizontal rollers (mill) arranged in a triangular pattern with the top roll rotating counterclockwise (Fig. 16.5), and the bottom two rollers clockwise. A series of three-roller mills, numbering 3 to 7, is called a tandem (Fig. 16.6). The pressure on the top roll is regulated by hydraulic rams and averages about 500 tons. Below each mill there is a juice pan into which expressed juice flows. The crusher and first mill extract 60–70 percent of the cane juice and the remaining mills take out 22–24 percent.

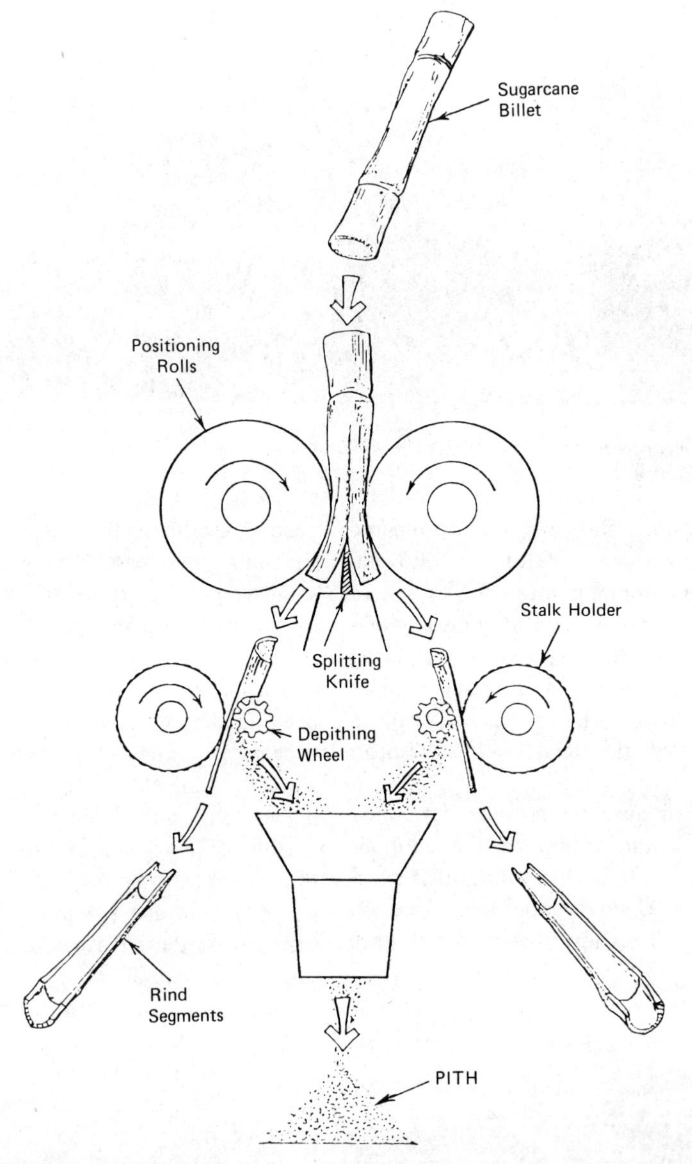

Fig. 16.3 Tilby separator process. (*Courtesy Ander-Cane, Inc., Naples, Fla.*)

When the fiber content of the bagasse reaches about 50 percent, extraction by conventional milling approaches zero. Since the juice that remains with the fiber contains the same proportion of sucrose as the original juice of the cane, yield of sugar would be substantially lowered if extraction were terminated at that point. Consequently, a process called "compound imbibition" is used to reduce the sucrose in the fiber by repeated dilution and milling. In a 5-mill tandem, water is added to the 4th mill and the expressed juice from that mill is brought back to the 2nd mill. The expressed juice from the 3rd mill is recirculated to the 1st mill and the 5th mill to the 3rd mill. In this way, the juice in the bagasse is always diluted ahead of each mill. The amount of imbibition applied at each mill is approximately equal to the amount of water applied to the 4th mill or to the penultimate mill in a differently numbered tandem.[12]

Before the expressed juice goes to clarification, it is screened through perforated metal screens with openings of about 1 mm in diam-

Fig. 16.4 Lignex sugar cane separator, full frontal working view. *(Courtesy Lignex Products Group, Hawker Siddeley Canada Ltd., Vancouver, B.C.)*

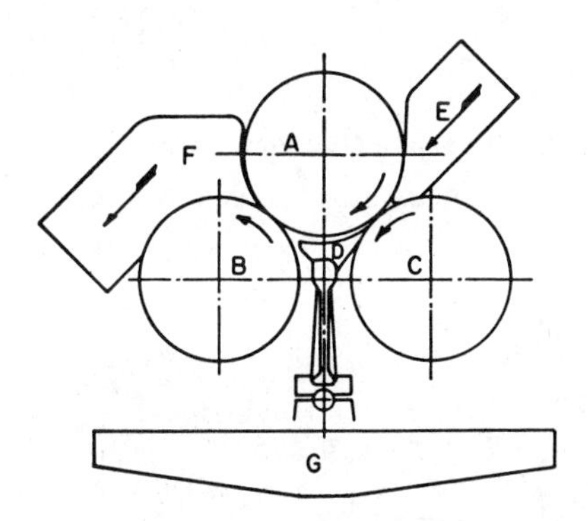

A - Top-roller.
B - Bagasse roller.
C - Feed roller.
D - Trash plate.
E - Feed of cane.
F - Discharge of crushed cane.
G - Juice-collecting trough.

Fig. 16.5 A cane mill with its three crushing rolls and with a typical turn-plate arrangement. A hydraulic piston maintains pressure on A, while B and C are fixed. The rolls are under pressure of 75–80 tons per foot of roll width.

eter. Additional screening may be provided by stationary or vibrating metal cloths.

Diffusion. Diffusion is used to extract juice from the cells in the sugar cane. The diffusers used for this have capacities that range from 1000 to 6000 tons of cane per day. They are normally used in conjunction with part of a milling tandem, the crusher and first mill. Preparation of the cane prior to its entry into the diffuser is essential for good extraction. In the diffuser, the crushed cane is countercurrently washed with imbibition water at temperatures ranging from 50 to 75°C. The last stage of a diffuser, the bagasse exit, receives water which gradually increases in sugar content as it proceeds to the first stage. Most of the sugar is extracted in the first four or five stages by simple displacement of sucrose from the open ruptured cane cells. In later stages, diffusion appears to take place in unruptured cane cells. One or two mills are used to express water from the bagasse after discharge from the diffuser. The crusher and one mill extract 70 percent of the juice from the cane and the diffuser removes 30 percent of the expected yield. Sucrose extraction from a mill-diffusion system averages 97 percent compared to 90–94 percent with a straight milling system.[13]

Since 1962, several diffusers have been installed in cane mills throughout the world. Cane diffusers are now manufactured by a number of companies including Braunschweigische Maschinenbauanstalt, CF&I Engineers Inc., De Danske Sukkerfabrikker, De Smet, S. A., Sucatlan Engineering, and Suchem Inc. of Puerto Rico.

Juice Purification. The juice from the milling station is an acidic, opaque, greenish liquid containing soluble and insoluble impurities such as soil, protein, fats, waxes, gums, fine bagasse and coloring matter in addition to soluble salts. The process designed to remove as much of these impurities as possible employs lime and heat and is called clarification or defecation. In simple lime defecation, milk of lime is added to the cold juice in amounts (about one pound of calcium oxide per ton of cane) sufficient to bring the pH to the range of 7.5–8.5. The limed juice is pumped through heaters in which the juice is heated to temperatures between 90

Fig. 16.6 Farrel four-unit milling train. (*Courtesy Farrel Co., Division of USM Corp., Ansonia, Conn.*)

Fig. 16.7 Continuous ring type (Saturne) total maceration sugarcane diffuser. (*Courtesy Sucatlan Engineering, Paris.*)

Fig. 16.8 Open top view of DDS double cylindrical tilted cane diffuser. *(Courtesy De Danske Sukker-fabrikker, Copenhagen.)*

and 115°C. There are many modifications to this basic process involving different sequences of lime and heat. For example, fractional liming and double heating involves liming the cold juice to pH 6.4, heating, liming to pH 7.6, and heating again before sending the juice to the clarifiers.

The combination of lime and heat forms a flocculent-type precipitate with various components in the juice. These consist mostly of insoluble lime salts, coagulated protein, and entrapped colloidal and suspended matter. The precipitate is removed by sedimentation or settling in continuous closed-tray clarifiers, i.e., Rapi-Dorr, Graver, BMA, and Bach Poly-Cell. The juice leaving the clarifier is a clear brownish liquid.

The flocculent precipitate or "muds" which settles on the clarifier trays contains about 5 percent solid matter. Sugar is recovered from the muds by means of rotary vacuum filters equipped with perforated metallic screen cloth. The turbid filtrate is returned to the clarification system and the press cake is discarded or sent to the fields as fertilizer.

Recent innovations in the filtration of muds

Fig. 16.9 Oliver Campbell vacuum mud filter applied to cane juice muds. *(Courtesy Dorr-Oliver, Inc., Stamford, Conn.)*

include the use of tighter filter media. A clearer filtrate is obtained making it possible to send it to the evaporators without reprocessing. The Eimcobelt rotary vacuum filter uses a tight filter cloth which is continuously removed from the drum and washed with hot water. Another type is the direct filtration of muds with horizontal leaf filters (Fas Flo Filters). One process uses polyelectrolytes to increase the filterability of the muds through cloth filters (Rapid-Floc system).

Good clarification depends upon the formation of a stable flocculent precipitate which settles rapidly. Juices low in phosphates and other minerals, such as magnesium salts, tend to be refractory. Fine soil carried in from the field may also interfere with clarification. Polyelectrolytes are being used to improve the coagulation and the settling rate of the precipitate during clarification.[14,15]

Evaporation. The clarified juice (about 85 percent water) is pumped to evaporators where it is concentrated to a clear heavy syrup containing about 65 percent solids. Evaporation is carried out in "multiple effect" evaporators in order to achieve maximum steam economy. Each "effect" is arranged in series and operated so that each succeeding one operates under a higher vacuum (lower pressure). This arrangement allows the juice to be drawn from one vessel to the next and permits the juice to boil at a low temperature. The concentrated juice (syrup) is removed from the last effect by pump. Triple, quadruple, and quintuple effect evaporators are used, but the quadruple effect is most common. In a four (quadruple) effect, one pound of steam evaporates four pounds of water. Occasionally, surfactants are added to the juice to improve the rate of evaporation.[6]

Crystallization. The syrup from the evaporator is pumped to a "vacuum pan" in which it is evaporated to supersaturation in order to cause sugar to crystallize. The vacuum pan is a single-effect evaporator designed to handle viscous materials. It is a vertical cylinder with its bottom designed to allow easy removal of the crystallized mass. The heating elements used in vacuum pans are either short, large-diameter vertical tubes (calandria pans) or coils around the inner surface of the truncated cone of the pan (coil pans). A typical vacuum pan with a "catchall" or entrainment separator for separating syrup from vapors measures approximately 25 feet in height and 15 feet in diameter. The working capacity is about 1000 cubic feet of massecuite (mixture of crystals and syrup or mother liquor). The shape of the pan and the positioning of the heating elements within the pan are important design factors in maintaining good circulation of the massecuite. For example, floating calandrias (calandria not attached to shell of pan), horizontal pans, and pans having mechanical circulators[16] are used.

Increasing use is made of continuous vacuum-pan systems introduced several years ago (Fives Lille-Cail). The continuous vacuum pan is a horizontal cylinder with compartments in its lower part. The pan is provided with an additional evaporator called the concentrator, where the density of the syrup is raised to 78-80° Brix.* Seed is added in the first compartment and the resulting massecuite moves progressively through the compartments of the pan. Additional syrup is added to each compartment to control the fluidity of the massecuite. Approximately 20 percent of the syrup is introduced at the concentrator and 80 percent of the syrup is fed to the various compartments through special feed headers. The rate of massecuite discharge is controlled by the level in the pan.[17]

The crystallization of sucrose in vacuum pans is called "sugar boiling" and each boiling is termed a "strike." Since a single crystallization does not recover all the sucrose from the syrup, mother liquor from a "strike" is recycled for recovery of additional sugar. There may be as few as two and as many as four reboilings to recover the maximum amount of sugar. A three-boiling system is the most popular. In order to distinguish among the boilings, a letter is assigned to each boiling and its products, i.e., the first boiling (syrup massecuite) would be given the letter *A* and its products after centri-

*The Brix scale is a density scale for sugar (sucrose) solutions. The degrees Brix are numerically equal to the percentage of sucrose in the solution. The term "Brix solids" refers to the solid content as determined by the Brix hydrometer.

Fig. 16.10 Top view of open Stork continuous crystallizer under construction. *(Courtesy Stork-Werkspoor Sugar, Hengelo, The Netherlands.)*

fugation, *A* sugar and *A* molasses; the second and third boilings would be assigned the letters *B* and *C*, respectively.

The boiling systems are based on "apparent purity," the amount of sucrose remaining in solution, and are usually expressed as the ratio of the polarization value to the total solids as measured by the Brix hydrometer. In a three-boiling system, the purity of the *A* massecuite is set between 80-85 by blending syrup with *A* molasses, the purity of the *B* massecuite is set between 70-75 by blending with syrup and *A* molasses, and the *C* massecuite is set between 55-60 by blending with syrup and *B* molasses. Other variations are possible.[6]

The *A* and *B* massecuites, after being discharged from a vacuum pan, are sent to centrifugals for separation. The *A* and *B* sugars are combined and become commercial raw sugar, the principal product of a cane sugar factory.

The *C* massecuite is a low-purity, highly viscous material which is not immediately sent to the centrifugals because of the large amount of recoverable sucrose remaining in solution. Instead, it is placed in crystallizers, U-shaped or horizontal containers equipped with coils attached to a hollow rotating shaft through which water circulates. The massecuite remains in the crystallizer from one to four days to allow additional crystallization to take place. Centrifugation of the *C* massecuite yields a final molasses (blackstrap) and a *C* sugar which is used for seeding the *A* and *B* boilings.

Centrifugation. Massecuite from the vacuum pans is sent to centrifugal machines wherein the crystals are separated from the mother liquor. A centrifugal consists of a cylindrical perforated basket lined with a screen of perforated sheet metal. The basket, enclosed in a metal casing, is mounted on a vertical shaft which rotates the basket imparting centrifugal force to the massecuite.

In a batch centrifugal, hot massecuite is fed into the basket through a short chute from the holding vessel. As the basket rotates, the massecuite forms a vertical layer on the screen lining. When the machine reaches operating speed (1000-1800 rpm) the syrup flows through the perforated lining and basket and is removed through an outlet at the bottom of the casing. The sugar on the lining is washed with a spray of water to decrease the amount of molasses adhering to the crystals. The basket continues rotating until the sugar is fairly dry, at which time the machine is switched off and brakes are applied. The sugar is discharged by a manually or automatically operated plough.

In continuous centrifugals, the machines do not stop but continue in motion while receiving fresh supplies of massecuite. The rate of feed must be carefully regulated to obtain optimum separation of molasses from crystals.[18]

Packaging and Warehousing. The packaging of raw sugar in jute bags has diminished to a very low level in most areas of the world. In fact, all of the raw sugar produced in Louisiana, Florida, Hawaii, Australia, and Puerto Rico is shipped in bulk. Where raw sugar is still bagged, bag weights vary from 140-150 lbs. in the Philippines to 250 lbs. in the West Indies. Bulk raw sugar is brought from the factories to seaport terminals by means of dump trucks, rail-

Fig. 16.11 Silver continuous centrifuges. (*Courtesy CF & I Engineers, Inc., Denver, Colorado.*)

road cars, or barges. These terminals, with capacities of 75,000 tons or more, are warehouses which were originally intended for bagged sugar, but now are specially designed structures for bulk handling. Unloading and loading of ships is done by bucket cranes or by gantries with traveling cranes.[12]

SUGAR REFINING

Affination and Melting

The first step in refining of raw sugar is called affination and consists of removing the film of molasses (in which a large portion of the impurities are contained) from the surface of the raw sugar crystal. The process involves the mixing of raw sugar with a saturated sugar syrup (72-75° Brix) at about 120°F in a U-shaped trough called a mingler. (Saturated syrup is used to avoid dissolving the raw sugar.) The mingler has a rotating agitator to maintain maximum contact between the sugar and syrup. The mixture is centrifuged to separate the crystals from the syrup and the crystals are washed with hot water. The "washed" sugar is "melted" or dissolved in water to a density of 55-60° Brix. Potable water, steam condensate, and "sweet waters" are used for dissolution. The liquor from the melter is screened to remove nonsugar materials such as sand, stones, wood, cane fibers, and lint. Screening is done by vibrating screens (Tyler-Hummer), cyclonic separator (Dorrclone), or centrifugal screens (DSM).[12, 15]

Purification

A number of processes are available for purifying the liquor from the melter prior to its crystallization into granulated sugar. The two combinations most commonly used are:

1. Clarification–Filtration–Decolorization
2. Clarification–Filtration–Decolorization–ion exchange

Clarification. One of three types of chemical treatment is used as the first step in the purification process. These are liming, phosphatation, and carbonation.

Liming is the simplest of the chemical treatments. Process liquor is "limed" to a neutral pH using milk of lime and heated. A slurry of diatomaceous earth is added and the treated liquor is pumped through a filter press, such as those discussed in the following section.

When phosphatation is used, the screened liquor is heated to 60-70°C and mixed with phosphoric acid (0.005 to 0.025 percent P_2O_5 on solids). The mixture is immediately limed to pH 7.0-8.0, aerated with compressed air, and sent to a clarifier (a rectangular tank equipped with heating coils). The liquor enters the clarifier at one end and is heated to 88°C while flowing to the outlet at the opposite end. The heated liquor releases air which rises carrying the flocculent calcium phosphate precipitate to the surface. A blanket of this precipitate (scum) forms at the surface and is skimmed from the liquor surface by moving paddles. Clarifiers differ in shape and design; some of the best known are Williamson, Jacobs, Buckley-Dunton, and Sveen-Pederson.

In carbonation, process liquor heated to 60-80°C is limed to about pH 10 (0.03 to 0.8 percent CaO on solids), gassed with carbon dioxide, heated to 85°C, and regassed until the pH drops to between 8.4-9.0. Carbonation, because of the two-stage gassing, requires two clarifiers, primary and secondary. Washed flue gases are the source of carbon dioxide. The calcium carbonate precipitate that forms entraps colored matter, colloidals, and some inorganic compounds.[19]

A novel addition to phosphate clarification has been the "Talofloc" process for the cane refining industry and the "Talodura" process for the raw cane factory. The "Talofloc" process consists of the addition of two chemicals at specific points in the clarification station. First, a cationic surfactant called "Talofloc" is added to the unclarified melt liquor in concentrations ranging from 0.02 to 0.07 percent on solids. This step is followed by the addition of lime and phosphoric acid and the incorporation of air. Just before the liquor enters the clarifying chamber a polyacrylamide flocculant (Taloflote) is added which induces the aggregation of the calcium phosphate precipitate, making a faster and more complete separation of the floc from the liquor. The Talofloc added prior to the chemical treatment causes at least a 50 percent increase in color removal when used in conjunction with Taloflote. The desugarization of the floc (scum) is best accomplished in a specially designed two-stage or three-stage countercurrent flotation system with further addition of Taloflote. For plants that are expanding or replacing their old clarifiers with the Talofloc process, special clarifiers designed to operate at maximum efficiency with Taloflote and Talofloc are recommended. One such unit with a capacity of 40 tons per hour is described in the literature.[20]

Filtration. The liquor from a phosphatation or a carbonation clarifier contains small amounts of finely dispersed particulate matter which require filtration for removal. The filtration process is similar for both types of clarified liquor.

A precoat of a filter aid (diatomaceous earth or perlite) is first deposited on the filter surfaces of the press. (Additional filter aid is added to phosphate clarified liquors to improve press runs.) Liquor at 160-185°F is fed to the press at a pressure of about 60 psig until the flow rate drops below a pre-determined level ending the filtration cycle. The calcium carbonate particles in carbonated liquor act as a fairly good filter aid as long as the particles are developed to optimum size during carbonation. Large quantities of sugar polysaccharides (starch) in the melt liquor will restrict the growth of calcium carbonate requiring the use of diatomaceous earth. A filter cycle may last from 2 to 12 hours depending upon the quality of the feed liquor.

Three types of filter presses are generally used: all are leaf presses:

1. Horizontal body with vertical leaf filter, i.e., Sweetland, Vallez, Auto-Filters and Industrial.
2. Vertical tank with vertical leaf filters; i.e., Niagara, Angola and Pronto.
3. Vertical tank with horizontal plates, i.e. Sparkler and Fas-Flo.

The filter leaves are dressed with cloth made of cotton or synthetic fibers (nylon, dacron, etc.) or with a wire screen of 60-80 mesh size.

The sugar remaining in the filter cake is recovered by washing the cake in place (sweetening off) with hot water, or the cake is sluiced off the leaves with hot water and flushed to a holding tank for refiltration. The filtered sweet water is generally added back to the melter to dissolve washed raw sugar.[12,15]

Decolorization. Filtered clarified liquor is a clear, dark brown liquid having a solids content between 55-65° Brix, a pH of 6.8-7.2, and temperatures between 65 and 85°C. If pulverized activated carbon is used as a decolorizing agent, it is usually added prior to filtration. Otherwise, the liquor is subjected to one or more decolorizing adsorbents, for example, bone char, granular carbon, and ion exchange resins. Bone char removes colorants, colloidal matter, and some inorganic substances while granular carbon is only a decolorizer. Ion exchange resins adsorb color and change the composition of the ash.

Bone char and granular carbon are generally used in fixed beds or in cylindrical columns 20-25 ft. high and about 10 ft. in diameter. More recent systems percolate liquor upward or downward through a stationary bed of adsorbent or by countercurrent flow of liquor and adsorbent (CAP or continuous adsorption process). (Figure 16.17)

Liquor flow through bone char is about 1500 gallons per hour over a period of 30 to 60 hours. For granular carbon, the flow rate is 3000 gallons per hour for 20 to 30 days.

After the decolorizing cycle is completed, the adsorbent is sweetened off by displacing the liquor with water and washed. It is transferred to regenerating equipment consisting of dryers, kilns, and coolers. Bone char is regenerated at 540°C in a controlled amount of air. Granular carbon is revivified at 950°C in a limited-oxygen and steam atmosphere. The kilns can be either the retort type or multiple hearths in columns (Herreshoff). After regeneration, the adsorbent is returned to the system for a new decolorizing cycle.

Ion exchange resins are used in columns 8-10 ft. high and 6-10 ft. in diameter holding between 100 and 300 cu. ft. of resin. The resin-bed depth is only 2-4 ft. Flow rates are rapid (3000 to 4500 gallons per hour) and the cycle is short (8-16 hrs.). Regeneration is accomplished in the column with a 10-15 percent salt solution at 140-160°F. The chloride form of a strong anionic resin decolorizes the liquor, and the sodium form of a strong cation resin softens the liquor. The sweet waters from the various adsorbent columns are recovered by returning them to the melter.[12,21]

Crystallization

Decolorized liquor is a pale yellow liquid with a solids content of 55-65° Brix. In some refineries, this liquor goes directly to a vacuum pan for crystallization. However, most refineries (both cane and beet) pre-evaporate the liquor in a multiple-effect evaporator to a solids content of about 72° Brix.

There are four stages in the process for crystallization of sucrose: (1) seeding or graining; (2) establishing the seed; (3) growth of crystals; and (4) concentration.

A sufficient quantity of evaporated liquor is drawn into the pan to cover the heating elements. Water is evaporated from the syrup until its supersaturation approaches 1.25. At this point, the steam pressure is lowered and seed crystals are added. The seed is finely pulverized sugar dispersed in isopropyl alcohol or sugar liquor. This method is called shock seeding because addition of the seed induces an immediate formation of crystal nuclei throughout the supersaturated syrup. The nuclei are "grown" to a predetermined size or grain. Once the grain is established, the crystals are "grown" to size by maintaining supersaturation between 1.25 and 1.40 through control of steam pressure, vacuum, and the feed rate of the evaporated liquor. Adequate circulation during crystal growth is important.

When the volume of the massecuite reaches the maximum working capacity of the pan, the syrup feed is shut off and evaporation is allowed to proceed until a thick massecuite is formed. When the massecuite concentration is considered "right," steam and vacuum are shut off and the contents dropped into a holding tank equipped with agitators where it is kept in motion until discharged into centrifuges.

Instrumentation is used extensively in sugar

boiling to control the progress of crystallization. Some of the principles used to provide control of vacuum pans for boiling a strike of sugar are:

1. Boiling point rise (BPR)–thermometers are used to measure the temperature of the massecuite and its vapors (difference between these two temperatures is the observed (BPR).
2. Electrical conductivity of the massecuite–based on the principle that conductivity is inversely proportional to the viscosity of the solution which in turn has a similar relationship to the water content and thus the degree of supersaturation.
3. Fluidity of the massecuite–an ammeter is used to measure the current used by the motor of a mechanical circulator; changes in current indicate changes in the fluidity of the massecuite.
4. Soluble solids–measured by refractometer attached to the pan (pan refractometer).

Several systems employing instrumentation for automatic control of pan boilings by supersaturation have been used successfully. However, controls are empirical in nature and are based on the conversion of the various techniques and arts of a "sugar boiler" into mechanical operations which lend themselves to automation.

The boiling system of a refinery is straightforward; the first strike is boiled from evaporated liquor, the second strike is boiled from the runoff syrup of the first strike, continuing on to three or four strikes. The runoff of the last strike can be used in a variety of ways: (1) as syrup for affination; (2) as syrup for remelt; (3) as syrup for producing a soft brown sugar; or (4) it can be reprocessed and reboiled.[12,16]

An innovative approach to the conversion of two-batch vacuum pans into a continuous crystallizing system was reported at the 1979 meeting of Sugar Industry Technologists.[22]

Remelt Sugars

The sugar contained in affination syrup must be recovered for economy. Recovery is accomplished by crystallizing the sugar in a vacuum pan. The resulting sugar is "raw" in composition and is returned to the refinery at the affination station. The residual syrup from the remelt station is known as refiners' blackstrap.

Centrifuging

Refined sugar crystals are recovered from the mother liquor by centrifuging the massecuite in

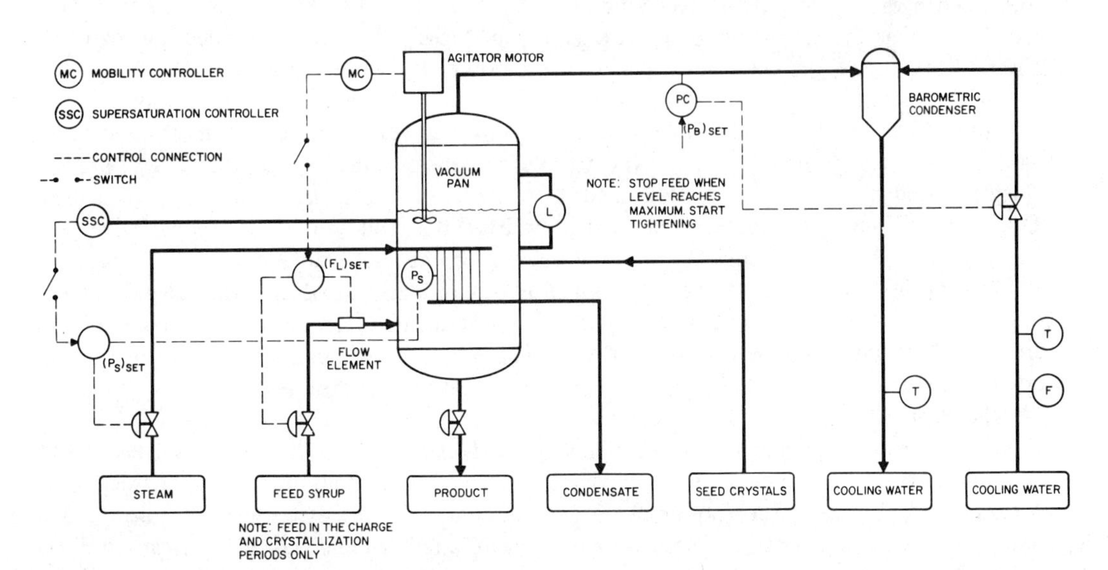

Fig. 16.12 Schematic drawing of vacuum pans and controls. *(Courtesy the Foxboro Co., Foxboro, Mass. Reproduced from Sugar y Azucar, Oct., 1970.)*

equipment similar to that used for affining raw sugar. However, at this point, the crystals are washed with a greater quantity of hot water. The washed crystals are discharged into a holding bin supplying a dryer.

Drying

Sugar from the centrifugals contains from 1 to 2 percent moisture and is too wet to be placed in storage or packages. The wet sugar is fed to drying equipment called granulators, which are usually rotating horizontal drums, 15 to 35 feet long and 6–7 feet in diameter, inclined slightly so as to be discharged by gravity. Heated air is blown through the dryer concurrent with the flow of the sugar. Two types commonly used are the Roto-Louvre granulator and the Standard-Hershey granulator. The Roto-Louvre is a single rotating drum in which hot air from louvres in the wall dries a moving bed of sugar. The Hershey granulator consists of a rotary dryer followed by a second unit in which the sugar is cooled to 45–55°C after leaving the dryer at 52–55°C.

Packaging and Storing of Refined Granulated Sugar

In the last three decades, there have been significant changes in the methods of handling refined sugar as it leaves the dryer. Bulk deliveries of refined granulated sugar to customers has made it necessary for the refiner to store finished product in bulk rather than in bags. Granulated sugar remains free flowing for a longer period of time if it is "conditioned." "Conditioning" reduces handling, packaging, and storage problems resulting from caking. Conditioning involves several factors, including control of the moisture content, the temperature, and the grain size as well as good inventory management.[23]

As the sugar leaves the dryer, it is cooled in one of a variety of equipment types including the Holoflite screw conveyor, thermal disc processor, or a Buttner turbo tray dryer. The cooled refined sugar is classified according to crystal size by a screening or sieving operation. It is stored in large bins or silos provided with an atmosphere of controlled temperature and humidity. Circulation of dehumidified air through or above the sugar pile and mechanical movement from silo to silo are also practiced to prevent caking.

Shipment of conditioned bulk sugar is made in specially designed rail cars (100,000 to 180,000 pounds capacity) or in hopper trucks (40,000 to 60,000 pounds capacity).

Conditioned sugar is also packaged in multiwall paper bags of 2-, 5- and 10-pound capacity. These sizes are overwrapped in 60 pound bales for shipment. Larger bags (100, 50 and 25 pounds) are also packed.

Granulated sugar is available in several sizes according to the needs of customers. These range from very large to very small crystals.

Direct-Consumption Sugar ("Plantation White")

Some cane sugar mills produce white sugar, usually by sulfitation or carbonation.

Sulfitation. The expressed juice from the mills is heated to 75°C, clarified with lime and sulfur dioxide, and in some cases filtered and then evaporated. After the syrup is treated with sulfur dioxide again, it goes through a three- or four-boiling system. The *A* and *B* sugars are mixed with a high-purity syrup, then centrifuged, dried, and screened for size and distributed as white sugar.

A recent improvement in the manufacture of plantation white sugar has been the application of a process called 'Talodura." The process consists of using a combination of sulfite clarification followed by subsequent clarification of the evaporated liquor with lime and phosphoric acid to which a flocculent is added. Improvements were made in the color of the mill white sugar, boiling-house recovery and chemical costs per ton of sugar.[24]

Carbonation. In this process, the mixed juices from the mills are heated, clarified with lime and evaporated to about 35° Brix. This syrup is relimed and treated with carbon dioxide, filtered, recarbonated, reheated and refiltered. After carbonation, the syrup is given a double sulfur dioxide treatment and filtered. The resulting syrup is subjected to a three- or four-boiling system with the *A* and *B* sugars used as the refined product.

Both processes are subject to a number of modifications, depending on the equipment of the mill and the quality of the sugar desired.[19]

Other Sugars

The cane and beet sugar industries manufacture several types of sugar in addition to granulated sugar. Among these are liquid sugars, brown or soft sugars, pulverized sugar, and special agglomerated sugars.

Liquid Sugars (sugars dissolved in water). Liquid sugars are economically important largely because a number of food manufacturers use sugar in the form of a syrup, and also because of the efficiency and ease of handling of a liquid product. Liquid sugars are produced by one of two methods: (1) dissolution or melting of refined sugar in water; or (2) further purification of in-process liquors by ion exchange treatment to remove minerals and further decolorization using bone char, and pulverized or granular carbon. Liquid sugar has other advantages over granulated in that it can be delivered as sucrose, a mixture of sucrose and invert sugar, invert sugar, or as a blend of those sugars with various glucose syrups from starch hydrolysis and dextrose from the same source. Sucrose liquid sugars are usually distributed at a 67 percent sugar concentration, while invert sugar and mixtures of invert sugar and sucrose are distributed at concentrations of 72-77 percent sugar. Invert sugar is an equimolecular mixture of dextrose and levulose resulting from the hydrolysis of sucrose.

Brown or Soft Sugars. These are specialty "dry" products ranging in color from light to dark brown. In addition to sucrose, soft sugars contain varying amounts of water, invert sugar, and nonsugars. These sugars have a characteristic flavor. In some cases, soft sugars are "boiled" from a low-purity process liquor to obtain the desired color, flavor, and composition. In other instances, refined granulated sugar is "painted" with an impure syrup or molasses to produce a product of similar appearance and characteristics.

Pulverized Sugars. These are manufactured by milling granulated sugar to the desired size. In most instances, pulverized sugars are mixed with small amounts of dried starch or tricalcium phosphate to prevent caking. A free-flowing agglomerated pulverized sugar without anticaking agents has been manufactured.[20]

Agglomerated Sugars. Several kinds are manufactured. Since soft sugars are subject to severe caking, one sugar refiner has resolved the problem by converting soft sugar into an agglomerated free-flowing dry product. Several companies manufacture "dry fondant" by agglomerating mixtures of invert sugar and very fine (less than 40 μ) sucrose particles.

Amorphous Sugar. Amorphous sugar is a microcrystalline sugar agglomerated during crystallization similar to dry fondants without having invert or reducing sugars added to it. Its polarization ranges from 99.1 to 99.2 with an invert content from 0.01 to 0.15 percent, a color of 20 to 40 m.a.u., and a size in the 40-80 mesh range.[25]

BEET SUGAR

Agriculture

The sugar beet, *Beta vulgaris*, is a temperate-zone plant grown largely in the Northern Hemisphere. A long growing season is required, and in the United States it is one of the first crops planted and the last harvested. The seed is planted in rows and after the plants have emerged, they are "thinned" to permit better development of a fewer number of beets. Unlike sugar cane, which cannot be cultivated after the cane has reached a certain height, sugar beets require a considerable amount of cultivation. Both cultivating and harvesting operations have been largely mechanized in the United States.

After plants have reached maturity in late fall (October and November), they are harvested by machines which remove the top growth of leaves, lift the roots from the ground, and deliver them to a holding bin or a truck. Because the harvesting season is much shorter than the processing season, beets are stored in piles at the factory or at outlying points near transportation. Frequently, storage piles are ventilated to lower the temperature of the beets, thus reducing sugar loss due to respiration during storage.[7,26]

Washing

Beets are transferred from storage into the factory in water flumes. These flumes lead directly into a rock-catcher which allows rocks to settle out and then on into a trash-catcher and a rotary washer.

Slicing

The washed beets are sliced into thin V-shaped cossettes by means of specially shaped knives set in frames mounted around the periphery of a rotating drum. Good removal of rocks and trash is essential to reasonable life for the knives.

Diffusion

The cossettes are weighed and transferred continuously into a diffuser where water passes countercurrently to the movement of the cossettes, and by dialysis, sugar and some of the nonsugars of the beet are extracted.

Both continuous and batch diffusers are used, but batch processes are largely being displaced. Continuous diffusers come in a variety of forms and shapes but all of them employ the same principle, i.e., movement of juice countercurrently to the movement of the cossettes. In batch diffusers, the cossettes are held stationary and the diffusion juice is moved from cell to cell. A few of the well-known types of continuous diffusers are the Silver chain or scroll type, RT, BMA, DDS, DeSmet, Buckau-Wolf, and Olier.

The RT-type diffuser is a large revolving drum with an internal helix which separates the drum into moving compartments. As the drum revolves, the cossettes travel by the action of the moving helix from one end of the drum to the other end while the juice moves in the opposite direction.

In a Silver chain-type continuous diffuser, the cossettes move on a drag chain through a series of U-shaped cells about 2 ft. wide and 12 ft. deep.

The BMA diffuser is a cylindrical tower with a conveyor mechanism attached to a central rotating shaft. Guide plates on the shaft direct the cossettes upward, while the juice exits through screens at the bottom.

Historically, a batch-type diffuser, the Roberts battery, was widely used. It consisted of 12 to 14 identical columns connected in series so that water pumped into the end cell (tail) passed through each cell to the cossette entrance cell (head). Normal circulation of the juices within the column was from top to bottom. During operation of the Roberts battery, three cells were not on stream, one being washed out, another being filled with cossettes, and the third cell being emptied. The cossettes were distributed in the cell by hanging chains.

The temperature of the juices during diffusion ranges from 60-85°C. Antifoaming agents are used to control foam in continuous diffusers and bactericides are added for microbiological control.[7]

Juice Purification

The diffusion juice is a dark turbid liquid containing 10 to 15 percent sucrose and 1 to 3 percent nonsugars including proteins, nitrogenous bases, amino acids, amides, inorganic material, and pectinous matter. These impurities are removed by a series of processes using lime and carbon dioxide (carbonation), sulfur dioxide (sulfitation) and crystallization. In recent years, there has been an increasing use of granular carbon and ion exchange resins in those plants where refined granulated sugar is made.

Carbonation. In this process, the juice is heated to 80-90°C, limed, and then sent to a carbonator for gassing with carbon dioxide. The resulting mixture containing insoluble lime salts, chiefly calcium carbonate, is pumped to

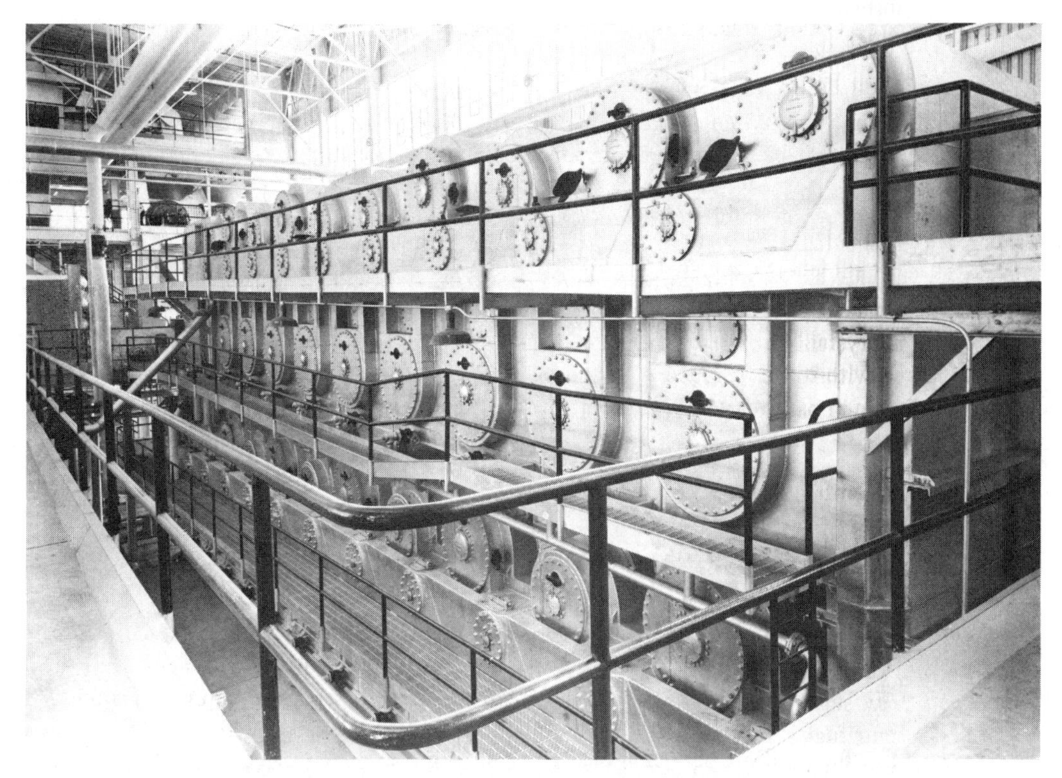

Fig. 16.13 Silver chain-type diffuser. *(Courtesy CF & I Engineers, Inc., Denver, Col.)*

subsiders (thickeners, clarifiers) to remove the insolubles by settling. From the subsider, the partially clarified filtrate is recarbonated and the residual lime is precipitated. This treatment is followed by press filtration. The sludge from the subsiders is filtered on rotary-drum filters, and the sugar-laden filtrate is returned to the first carbonation step. The carbonation process can be either continuous or batch.

One continuous system for carbonation widely used in the United States is the Dorr process. Heated juice is treated with milk of lime or lime saccharate in a cylindrical column (primary tank) where carbon dioxide gas is added. The contents of the secondary tank are continuously pumped into the bottom of the primary tank. The finished first-carbonated juice flows from an overflow line in the discharge of the recirculation pump to a subsider (Dorr multifeed thickener) in which the insoluble matter is separated from the juice. The thickener is a cylindrical tank consisting of a number of shallow compartments with trays and rotating paddles which continuously remove the settled sludge. The filtrate is reheated (90-95°C) and sent to another carbonation tank. The effluent from the second carbonation tank is pressure-filtered on a leaf press such as the Kelly, Sweetland, Vallez, or Angola. The filtered liquor is treated with a small amount of sulfur dioxide in a tower, heated, and refiltered.

Pre-liming. The European beet industry prefers a pre-liming system in which diffusion juice is sent through a series of tanks where lime is added to various levels of alkalinity. The limed juice is then carbonated and generally follows a purification pattern similar to the Dorr process.

Ion exchange resins have been used in some plants to decolorize, deionize, and to delime the juices. However, their use is the exception rather than usual practice.

Evaporation and Standard Liquor

The thin juice (about 15 percent solids) discharged from the filter presses after clarification is evaporated to thick juice (50-65 percent solids) in multiple-effect evaporators. If decolorizing adsorbents are used, they are usually added to the thick juice. Granular carbon and pulverized carbon have been used for this purpose. Thick juice treated with pulverized carbon usually must be double-filtered to remove all of the carbon.

Low raw sugars (sugars with purities below that of refined) are added to the thick juice in the melters to make standard liquor from which white sugar is crystallized. Standard liquor is usually filtered with diatomaceous earth before going to the vacuum pan.

Crystallization, Centrifuging, and Drying

Crystallization practices in United States beet sugar factories are similar to those in a cane sugar refinery and result in a white granulated sugar comparable in quality to those of refined cane sugar. However, some European factories may make "raw sugar" as a separate product. White-sugar centrifuge stations as well as drying operations are comparable to the same operations in a cane sugar refinery.

Sugar Recovery from Molasses

Beet sugar molasses from the normal operations is usually high in sucrose. Recovery of part of this sugar utilizes the "Steffan's Process" in which sucrose is separated by means of a lime salt known as calcium saccharate. In this process, molasses is diluted to 6-7 percent sucrose and cooled to 18°C. Finely pulverized lime is added with agitation to form a precipitate known as cold saccharate cake. This precipitate (containing about 90 percent of the sucrose) is filtered. The filtrate is heated, during which another precipitation occurs and this too is filtered. Cold saccharate precipitate is removed by rotary vacuum filters, and hot saccharate, by thickeners. Both precipitates are mixed with sweet water and returned to the first carbonation where lime saccharate is decomposed into $CaCO_3$ and sucrose.

A second molasses-desugaring process is known as the "barium process." Sucrose is precipitated in the form of barium saccharate. Inasmuch as soluble barium solids tend to be toxic, a major process requirement is that all barium be removed before the recovered sucrose is returned to process. Furthermore, economics dictate that the barium be recovered for re-use.

A third process which has been developed recently for the recovery of sugar from beet molasses is ion exclusion. A strong cationic resin, crosslinked with no more than 4-5 percent of divinyl benzene, in the monovalent salt form and whose size ranges between 20-40 U.S. mesh, is fed in a vertical column to alternate cycles of molasses solution (35-45 Brix) and water. Controlling the dimensions of the column is important. Height to diameter ratio should not be less than 4 to 1, and preferably around 20 to 1; control of flow rate (0.4 to 0.6 gal./sq.ft./min.), the temperature (70-80°C) and clarity of the feed solution are also equally important. Sugar and nonsugar components of the molasses are separated chromatographically. Four fractions are obtained: the first fraction contains most of the inorganic substances, colorants, and organic nonsugars; the second fraction is similar to the composition of the feed but contains about 25 percent more water; the third fraction is a high purity sucrose solution between 10-20 Brix; the fourth fraction is an extremely dilute solution of sucrose and invert sugar (2-5 Brix). The sucrose fraction either goes to pan crystallization or is further purified with activated carbon and ion exchange resins and sold as liquid sugar. The first or waste fraction is concentrated to 70 Brix and marketed as "fodder molasses." The resin, after a number of cycles, has to be regenerated with salt because of the accumulation of calcium and magnesium ions. In some cases, the feed molasses is first softened prior to ion exclusion treatment.[27,28]

Sugar cane molasses can be subjected to the same basic ion exclusion process with some modification: (1) the molasses must be clarified to remove as much suspended matter as possible, and (2) prior softening is necessary to extend operating cycles. Since cane molasses contains substantial quantities of reducing sugar, the resulting fractions consist of nonsugars, a mixture similar in composition to the feed, a mixture of sucrose and reducing sugars rich in sucrose, and a mixture of reducing sugars and sucrose.[29,30]

Sugar Recovery from Syrup

In 1962, a beet sugar company in the United States built a tank system for storing thick juice produced in excess of its crystallizing capacity. This storage system served two purposes: (1) it enabled the "beet end" of the factory to operate at full capacity even though the "sugar end" could not handle the output; and (2) it provided raw material for the sugar end to operate at full capacity when the beet end was in trouble.

The system proved such a success that other beet factories have adopted the practice. Furthermore, the improvements made in the storage systems have made it possible to extend the operating time of the sugar end far beyond the termination of the slicing campaign.[31]

Storage and Packaging

In the past 30 years, both beet sugar factories and cane sugar refineries have gone extensively to bulk storage of finished products. Beet sugar

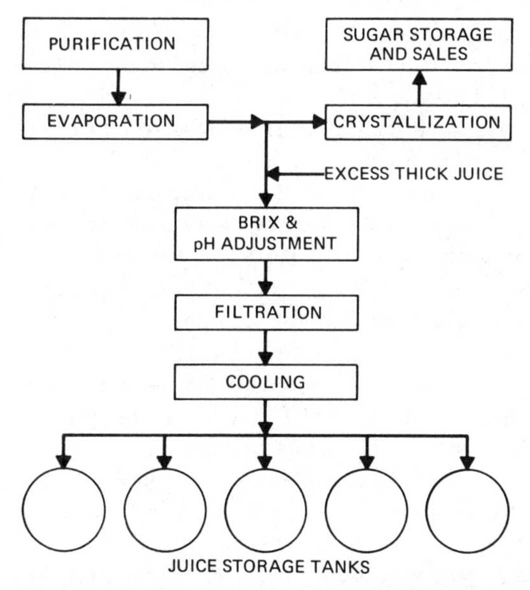

Fig. 16.14 Schematic flow diagram showing flow of thick juice to storage. (*Courtesy The Amalgamated Sugar Co., Ogden, Utah.*)

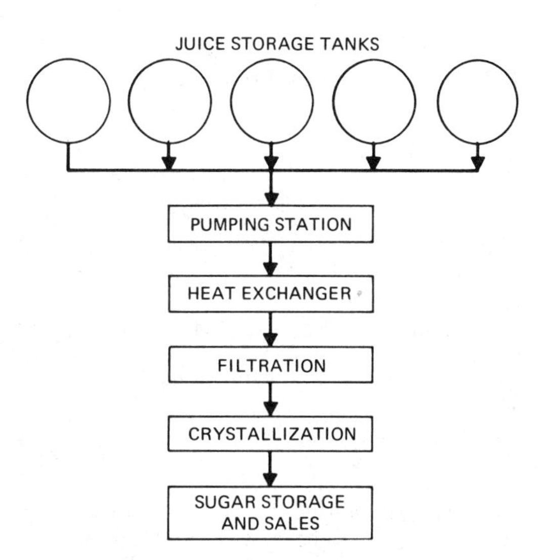

Fig. 16.15 Flow diagram showing flow of thick juice from storage through processing. (Courtesy The Amalgamated Sugar Co., Ogden, Utah.)

bulk storage is usually more extensive because of the need to hold sugar to meet customer needs from one processing season to the next. The daily capacities of cane sugar refineries are such, however, that it would be uneconomic to store more than a few day's output.

Prior to bulk storage, refined beet sugar was stored in bags (mostly 100 lbs.). Such storage was acceptable as long as large customers purchased sugar in bags. However, a shift in the consumption of sugar from the home to food processors, and the development of bulk transportation facilities made bulk storage a natural consequence. Thus, beet sugar is packed in the usual customer-size packages and 100-pound bags, and it is also shipped in bulk containers ranging in size from "tote bins" (2,000 lbs.) to rail cars holding several tons.

Cane and Beet Sugar Production

Cane sugar is primarily a product of tropical countries, whereas beet sugar is produced in temperate climates. However, both are produced in a few countries, among which are the United States of America, China, Afghanistan, Spain, Iran and Pakistan.[32] Total annual production of centrifugal or crystalline sugar is shown in Table 16.1, together with average

TABLE 16.1 Centrifugal[a] Sugar Production (Raw Value)[b] by Continent, 1970/71 - 1978/79[33] (In Units of One-Million Metric Tons)

	1970/71 thru 1974/75 Average	*1975/76*	*1976/77*	*1977/78*	*1978/79*
North America	17.1	19.2	18.8	19.2	19.3
South America	10.8	11.3	12.7	13.8	12.5
Europe (West)	20.8	22.6	23.7	26.9	27.0
Europe (East)	4.6	4.9	5.2	5.6	5.6
Africa	5.3	5.5	6.1	6.2	6.4
Asia	14.5	17.4	19.2	20.4	19.9
Oceania	3.1	3.3	3.7	3.7	3.3
U.S.S.R. (Europe & Asia)	8.1	7.7	7.3	8.8	9.0
WORLD TOTAL	75.1	81.7	86.3	92.5	90.9

[a]Noncentrifugal sugars are usually consumed in the area of origin. Estimated production since 1970 has ranged from 10.4 to 11.9 million metric tons (1979/80).
[b]Raw value is a term defined mathematically to reduce sugars of differing sucrose contents (from raw sugar containing varying amounts of sucrose to refined sugar containing approximately 100 percent sucrose) to an equivalent basis in pounds.

annual production for representative periods over the past 10 years.

Cane sugar production currently amounts to approximately 60 percent of the total of world sugar production. The largest five producers are the U.S.S.R., Brazil, India, Cuba, and the United States of America, in that order.[33]

The United States produced less than 6 million short tons, raw value, of sugar in 1977/78.[34] About half of this amount was cane sugar produced in Hawaii, Louisiana, Florida, and Puerto Rico. The remainder was beet sugar produced in all the states west of the Mississippi River except Arkansas, Missouri, Oklahoma, and Nevada, and, in addition, in the states of Michigan and Ohio. In spite of the fact that the United States is the fifth largest producer of sugar in the world, a substantial portion of its needs must be met by imports from other sugar-producing countries. Most of the imported sugar is produced from cane.

Sugar Consumption and Usage

Per capita consumption of sugar varies widely from country to country. While Europe, the Americas, and Oceania consume more than 40 kg per capita, the average for the world is only slightly more than 21 kg.[35] The rate of increase in sugar consumption is slowing apparently in relation to population growth and perhaps as a consequence of consumption saturation in some countries.[35]

Sugar usage patterns, particularly in developed countries, have not changed markedly in the past several years. In the United States, approximately two-thirds of sugar deliveries are used for the manufacture of food and beverages and the remaining one-third is used by institutions, (see Table 16.2). Nonfood uses continue to be a very small portion of the total. Among the food uses, only the beverage industry has shown an increase in sugar usage in the 1972-1978 period.

Nonfood uses of sugar are relatively small in comparison with its food use. However, sugar, because of its availability in large quantities at very high purities, has always been viewed as an attractive raw material for organic chemical synthesis. Research in this area has been aggressively pursued for more than 25 years[36], and a great number of products have resulted from those efforts although few of them have been commercialized for economic reasons or because of the complexity of the chemical synthesis. Among the derivatives produced in the laboratory over the years are ethers, fatty and other esters, acetals, reduction products, and halogen, sulfur, and metal derivatives, as well as others.[37] Potential applications include products for use such as surfactants, surface coatings, food and feed additives, polymers,

TABLE 16.2 U.S. Sugar Deliveries to Industrial and Nonindustrial Users, Calender Years, 1972–78

Type of buyer	*1972*	*1973*	*1974*	*1975*	*1976*	*1977*	*1978*
	Calendar year, 1,000 short tons, refined sugar						
Industrial users:							
Food use							
Bakery and cereal products	1,449	1,454	1,443	1,241	1,313	1,386	1,308
Confectionery products	1,057	1,035	1,018	795	915	951	912
Processed foods	987	1,025	949	743	737	746	690
Dairy products	599	595	570	511	553	567	550
Other	508	502	514	486	518	542	413
Total	4,600	3,611	4,494	3,776	4,036	4,192	3,873
Beverage use	2,437	2,469	2,350	2,074	2,252	2,454	2,588
Total industrial users	7,037	7,080	6,844	5,850	6,288	6,646	6,461
Nonindustrial users:							
Institutions							
Eating and drinking	85	94	91	72	65	69	104
Other[1]	88	106	121	85	133	148	125
Total institutions	173	200	212	157	198	217	229
Wholesale and retail							
Wholesalers, jobbers, and sugar dealers	2,103	2,064	2,002	1,919	2,143	2,095	1,996
Retail grocers, chain stores, and supermarkets	1,316	1,316	1,353	1,261	1,314	1,288	1,180
Total wholesale and retail	3,419	3,380	3,355	3,180	3,457	3,383	3,176
Minus consumer size packages[2]	2,557	2,530	2,581	2,409	2,439	2,460	2,267
Redistributed to industrial and other users[3]	862	850	774	771	1,018	923	909
Total nonindustrial user	3,592	3,580	3,567	3,337	3,655	3,600	3,405
Total food use	10,629	10,660	10,411	9,187	9,943	10,246	9,866
Nonfood use[4]	91	111	128	86	101	105	161
Total food and non-food use	10,720	10,771	10,539	9,273	10,044	10,351	10,027

[1] Includes deliveries to government agencies and the military. [2] Less than 50 pounds. [3] Includes some deliveries to eating and drinking places and institutions. [4] Used largely for pharmaceuticals and some tobacco.

Source: Fruit and Vegetable Division, AMS, USDA.

textile chemicals, pharmaceuticals, and pesticides.[38]

The reduction products, sorbitol and mannitol, have been manufactured from sugar and sugar products. Sugar's use for that purpose usually depends upon its being competitively priced with dextrose or glucose syrups which are the most commonly used raw materials. A review of the potential applications of sucrose derivatives in other areas is contained in *Sucrochemistry*, ACS Symposium Series 41, 1977, American Chemical Society, Washington, D.C.

Fatty acid esters of sucrose have been manufactured in the United States and some foreign countries on a commercial basis for many years.[36] Plants for a solventless process for manufacturing esters are in various stages of design and construction. An almost ironical laboratory development, casting chlorosucrose derivatives in the role of a noncaloric sweetener, has received widespread interest in recent years.[39,40]

SUGARS DERIVED FROM STARCH

Starch Hydrolysis

Undried starch, usually from wet milling, is pumped as a slurry to the conversion plant

TABLE 16.3 Carbohydrate Composition of Glucose Syrups (Saccharide as a % of Total Carbohydrates)

DE	*Mono*	*Di*	*Tri*	*Tetra*	*Penta*	*Hexa*	*Hepta*	*Higher*
15	3.7	4.4	4.4	4.5	4.3	3.3	3.0	72.4
35	13.4	11.3	10.0	9.1	7.8	6.5	5.5	36.4
45	21.0	14.9	12.2	10.1	8.4	6.5	5.6	21.3
55	30.8	18.1	13.2	9.5	7.2	5.1	4.2	11.9
65	42.5	20.9	12.7	7.5	5.1	3.6	2.2	5.5

where it undergoes one or more hydrolytic processes to yield mixtures of various carbohydrates in the form of syrups or as crystalline dextrose. The kind and amount of the various carbohydrates obtained depend upon the type of hydrolysis system used (acid, acid-enzyme, or enzyme-enzyme), the extent to which the hydrolytic reaction is allowed to proceed, and the type of enzyme used. The fact that most starches consist of two different kinds of polymers (amylose and amylopectin) also has an effect on the nature of the products obtained.

The extent to which starch is converted into simpler carbohydrates is indicated by the "dextrose equivalent" (D.E.), a measure of the reducing sugar content calculated as dextrose and expressed as a percentage of the dry substances.[41] Hydrolyzates having dextrose equivalents ranging from 5 to 100 are produced. Those having a very low dextrose equivalent are frequently referred to as "dextrins." They are produced by minimal acid hydrolysis or roasting. Starch-hydrolyzate syrups are commonly produced as "low," "regular," "intermediate,' "high," or "extra-high" conversion products, as more or less standard products. Table 16.3 shows the composition of some of the various syrups.[41] In addition to carbohydrates, the syrups contain minerals (largely sodium chloride) and nitrogenous substances.

Acid Hydrolysis. A starch slurry containing 35-45 percent solids is acidified with hydrochloric acid to about pH 1.8-1.9. The suspension is pumped into an autoclave (convertor) where live steam is gradually admitted to a pressure of 30-45 psi. The conversion time largely determines the D.E. of the hydrolyzate, i.e., eight minutes may produce 42 D.E. syrup, or ten minutes may produce 55 D.E.[42] Converted liquors are neutralized with sodium carbonate to a pH of 5-7, coagulating insoluble protein, fats, and colloidal matter. The scum is removed by centrifuge.

The dark-colored clarified liquor is pressure filtered and then concentrated to 60 percent solids in multiple-effect evaporators. The concentrated liquor is decolorized with granular carbon in columns of 12 ft. diameter and 30 ft. height in a countercurrent manner; i.e., liquor flows upward in the columns, while a portion of the carbon is removed from the bottom periodically. It has been reported that carbon is used at the rate of 2.5 percent of dry solid processed and that 5 percent of carbon is lost

Fig. 16.16 Merco centrifuge used for starch separation in a wet Milling Process.

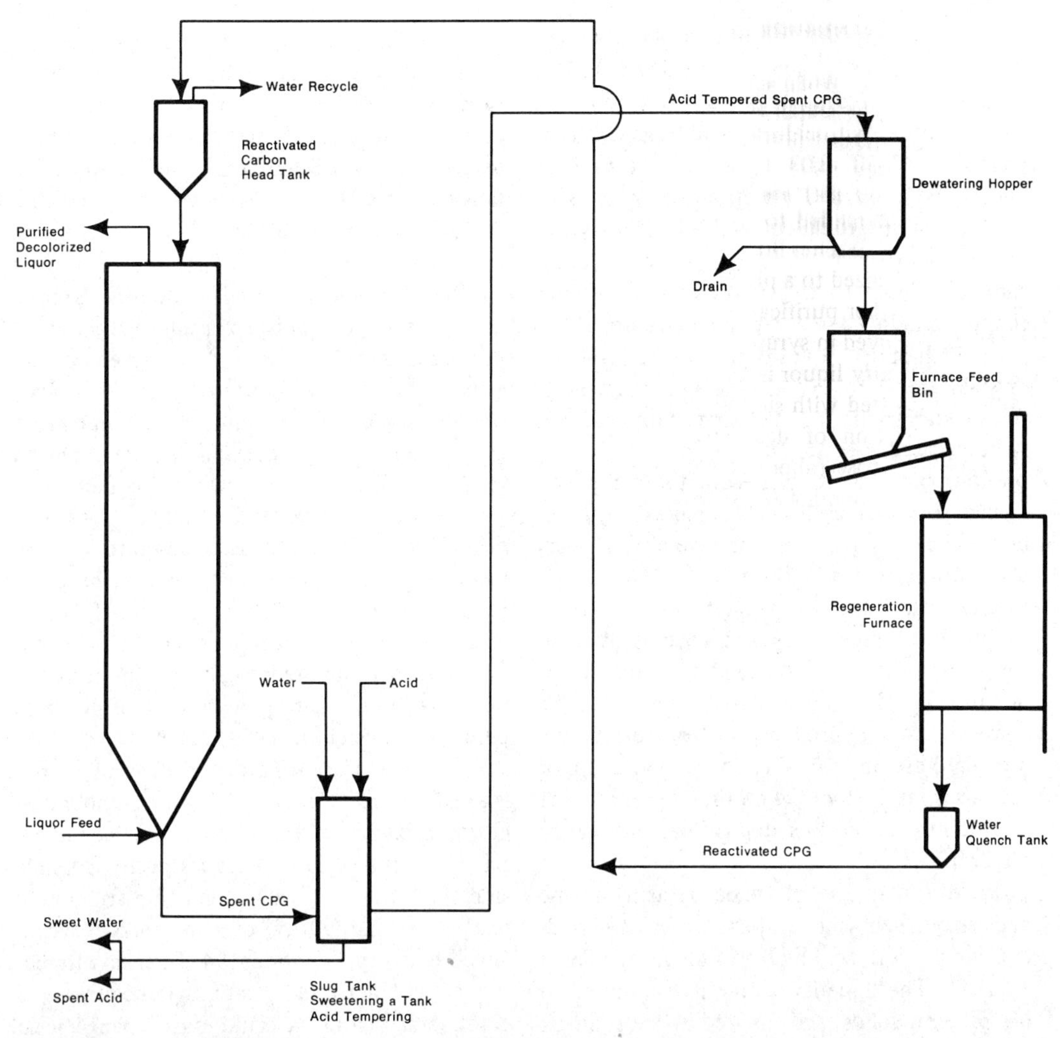

Fig. 16.17 Flow diagram of the Pittsburgh moving bed system for glucose-dextrose. (*Courtesy Activated Carbon Div., Calgon Corp., Pittsburgh.*)

during revivification.[43] "Low ash" syrups are usually deionized with ion exchange resins. Finally, the processed liquor is evaporated to a final solids content of 75 to 85 percent in a single-effect evaporator.[44]

Acid-Enzyme Hydrolysis. Starch is first liquified and hydrolyzed to specific dextrose equivalents with hydrochloric acid. After evaporation to 60 percent solids, a saccharifying enzyme (fungal amylases) is added to continue hydrolysis to the desired level. By choosing two or more types of enzymes (such as alpha-amylases, beta-amylases, or glucoamylases) and adjusting the initial acid hydrolysis, syrups with different ratios of dextrose, maltose, and higher saccharides can be obtained.[45]

Enzyme-Enzyme Hydrolysis. Enzyme-enzyme conversion employs an enzyme for starch liquefaction in place of acid. Subsequent hydrolysis is by enzymes as above. The choice of hydrolytic system depends upon economics and the kind of end product desired. Enzymes are usually inactivated by heating to 75-80°C.[45]

Conversion to Dextrose

The manufacture of dextrose requires conditions different from those used for the manufacture of syrups. When acid hydrolysis is used, the density of the starch slurry is lower (15 to 20 percent). Hydrochloric acid is added to a concentration of 0.03 N, and higher steam pressures (40-50 psi) are used. The time of hydrolysis is extended to about 30 minutes or until the purity reaches 90-91 D.E. The hydrolyzate is neutralized to a pH of 4-5 with sodium carbonate. Other purification steps are similar to those employed in syrup manufacture.

The high-purity liquor is pumped to insulated crystallizers fitted with slowly moving agitators for crystallization of dextrose monohydrate. A heavy seed base (about 25 percent) from a previous batch is mingled with the syrup and cooled to about 38°C. The seeded liquor is held at this temperature for several days until about 60 percent has crystallized. The mixture is then centrifuged to separate crystals from the mother liquor. The wet sugar is dried in rotary dryers or recrystallized into anhydrous dextrose. The monohydrate may also be converted to anhydrous dextrose by drying the monohydrate in hot air. A second crop of crystals is taken from the mother liquor and the run-off syrup from this step is final molasses or "hydrol."[44,46]

Dextrose may also be made by acid-enzyme and enzyme-enzyme systems. In the acid-enzyme process, acid hydrolysis is stopped near 18 D.E. The liquefied slurry is evaporated to 60 percent solids and treated with a fungal glucoamylase until the D.E. is above 90. With enzyme-enzyme processes, a 30 percent starch slurry is liquefied with a heat-stable bacterial amylase followed by treatment with fungal glucoamylase. The liquor is purified in the same manner as others.[45]

In addition to syrups, several solid products are manufactured, including several kinds of dried sugars. "Crude sugar" is a solidified liquor containing large amounts of dextrose and substantial quantities of nondextrose solids as well. A solid product of recent development, "noncrystalline" or "enzyme-converted" dextrose, is highly refined and contains about 92 percent dextrose.

High Fructose Syrup

The rapid development of enzyme technology has made available in commercial quantities a new sweetener derived from starch. This new product which is generally known as "High Fructose Corn Syrup" (HFCS) contains 30 percent fructose, 35 percent glucose, and 6 percent higher saccharides. The fructose fraction is obtained by the isomerization of glucose through the catalytic action of the enzyme, glucose isomerase.

Glucose isomerase can be derived from a variety of microorganisms; *Bacillus coagulans*,[47] *Actinoplanes missourensis*,[48] and several organisms from the *Streptomyces genus.*[49] Since glucose isomerase is an intracellular enzyme, it is commonly used today in an immobilized form fixed on an inert carrier, geometrically shaped for use in vertical columns. The high degree of efficiency attained through the technological progress that enzyme suppliers have made in a relatively short span of time is illustrated by comparing the enzyme cost of 75 cents per hundredweight of HFCS in the mid 70's to the approximate cost of 40 cents per hundredweight in 1979.[50]

The basic features of the isomerization process is a series of reactors containing the immobilized enzyme fixed in a packed bed. The feed material is of the highest practical purity possible since enzymatic activity and life is closely related to the purity of the glucose syrup. Glucose syrup of about 94 D.E. is filtered, treated with activated carbon to remove residual color, and deionized with ion exchange resin to lower the ash content, particularly the calcium ions. The dissolved oxygen is reduced by flash evaporation, which also concentrates the feed stock to 40-45 percent glucose and raises the temperature to 60-65°C. At this point, prior to entering the reactors, enzyme activators are added, particularly magnesium ions and, in some cases, hydrogen sulfite ions. The pH is also adjusted to about 8. Isomerized liquor is removed from the process when the equivalent of 42 percent fructose on a dry basis is reached. The isomerized syrup is further treated with activated carbon and ion exchange resin and finally concentrated to 71 percent solids.[43,47,51]

Recently, a "second generation" of HFCS has become commercially available containing 55 percent fructose on a dry basis. The higher fructose content is achieved by subjecting the

42 percent fructose syrup to a chromatographic separation of fructose from glucose by means of strong cationic resins of low cross-linkage in the calcium form. The fructose is preferentially held by the resin and is later eluted with water.[43,52] Recently, a process has been described which uses a nonresin absorbent, a rigid, nonswelling inorganic product. The process employs a unique cycling device to remove from, or feed to, various parts of the chromatographic column liquids of different composition. The product stream has a fructose content of 95 percent on a dry basis.[53]

Fructose

Fructose, also known as levulose and fruit sugar, has been known as a naturally occurring sugar found in several fruits as well as honey for many years, but it has become commercially significant only in the past several years.[54] Fructose, as a laboratory chemical, has been prepared from inulin and invert sugar,[55] but its commercialization was based on its extraction from hydrolized sucrose solutions or invert sugar syrup (crystalline fructose).

Fructose has had wider usage in Europe where it is manufactured by several companies. Those processes have been based on the hydrolysis of sucrose followed by chromatographic or ion exclusion separation with subsequent crystallization of fructose after its separation from glucose and sucrose. The development of processes for isomerizing glucose into fructose, together with commercialization of chromatographic separation of mixtures of sugars has led to the use of high fructose corn syrup as a raw material for the manufacture of pure fructose. The technology has been reviewed by M. Seidman.[56]

The role of fructose as a food and in the food processing industry in the United States is not clear at this time. While its desirable physical and chemical characteristics are well documented,[54] its basic attractiveness at this time appears to be related to its relative sweetness when compared with sucrose.[57] In theory, substitution of fructose for sucrose in foods (especially soft drinks) would permit a reduction in the caloric content of a food because of the greater relative sweetening power of fructose. However, experimental studies have demonstrated that fructose sweetness varies with the conditions under which it is used.[58] A theoretical basis explaining this variability in sweetness has been set forth by Shallenberger.[59]

BY-PRODUCTS

Molasses

Molasses, a residual mother liquor from which little or no additional sugars can be recovered economically, is a by-product common to the cane and beet sugar industries, and also to the dextrose industry. Each type of manufacturer has designated this liquid by-product with its own peculiar name: the molasses from cane sugar production is most commonly called "blackstrap," that from beet is simply "beet molasses," while that from starch hydrolysis is known in the United States as "hydrol."

The molasses from each source differs considerably, as is shown in Table 16.4.

Blackstrap contains a significant quantity of both sucrose and reducing sugars. A substantial amount of reducing sugar is produced during processing of cane juice into sugar and molasses. Beet molasses contains primarily sucrose and little or no reducing sugar because of highly alkaline processing conditions which destroy reducing sugars. Hydrol, of course, contains no sucrose. Higher saccharides result from incomplete hydrolysis or from the polymerization of "sugar" units during processing.

World production of blackstrap and beet molasses closely parallels sugar production.

TABLE 16.4 Analysis of Molasses from Various Sources

Constituent	*Blackstrap*[60] *(Dry Basis)*	*Beet Molasses*[61] *(Dry Basis)*	*Hydrol*[18] *(Wet Basis)*
Sucrose	37.4	63.0	
Reducing sugars	32.7	0.7	55.0
Higher saccharides		1.5	
Ash	13.5	12.0	7.2
Nitrogen	0.4	2.0	0.07

TABLE 16.5 Cane and Beet Molasses Production 1970/71–1979/80

(Millions of Metric Tons)

Year Average	*Production*
1970/71–1974/75	25.5
1975/76	27.2
1976/77	29.5
1977/78	33.3
1978/79	33.4
1979/80	31.8

Total world production is shown in Table 16.5.[33]

Published information on world production of hydrol is not readily available. However, production in the United States amounts to more than 20 million gallons annually.[62]

The major outlet for all molasses at the present time is in feed for cattle or other animals. Fermentation into rum, potable ethanol, citric acid, and vinegar continues to be a market for considerable quantities of molasses. Large quantities are used for growing yeasts. These uses have diminished in relation to feed use over the years. Beet molasses, because of its high content of nitrogeneous substances, is valued for the production of citric acid by fermentation.

The rapid rise in petroleum prices has greatly increased interest in converting molasses and cane juice into absolute ethanol as partial or total substitution for gasoline in automotive vehicles.[63,64,65]

Bagasse and Beet Pulp

An important by-product from a cane sugar mill is bagasse, the fibrous portion of cane from which juice is extracted. Bagasse, at the time it is discharged from the milling train, contains solid matter (short fibers and the spongy tissue of the pith) and 50 percent by weight of water. Since the average fuel value of ash-free dry bagasse is 8300 Btu/lb,[65] it is primarily used as fuel for the generation of steam in sugar factories. However, when there is an excess of bagasse available or alternate fuels are plentiful and reasonably priced, the bagasse is utilized as a raw material for the pulp, paper, paperboard, and wallboard industries, especially in wood-poor areas of the world. Other successful uses of bagasse include its conversion into furfural and a filler for explosives. Today, bagasse is receiving a great amount of attention as a source of sugar and fuel.[66]

Beet pulp is the structural portion of beet roots which remains after spent beets are discharged from a diffuser. Since it is highly hydrated at that time, the wet pulp is dewatered in pulp presses. It is then discharged into wet silos and may be sold as wet pulp for animal feed. However, it is more often dried or mixed with molasses and dried. In both cases, it is used in the manufacture of animal feeds.

Other By-Products

When grains such as corn, sorghum, and wheat are used as sources of starch, a number of important by-products are obtained. These include steep water, corn oil, gluten, and hulls. Most of these are used as ingredients in mixed feeds, but corn oil is widely used as a food product. By-products from the other commercial sources of starch are not as important as those from grains, but they usually find their way into feeds if they are salvaged.

TRADE IN SUGAR AND OTHER SWEETENERS

Many countries, especially the tropical, cane-sugar producing ones, produce more sugar than is consumed locally. A number of beet-sugar producing countries, particularly in Europe, have become self-sufficient or nearly so in recent years. However, some European countries continue to import raw sugar to be refined. Altogether, 25.1 million metric tons, raw value, of centrifugal sugar were exported in 1978. The Soviet Union, United States, United Kingdom, Japan, China (mainland), and Canada are the largest importers, while Cuba, France, Australia, Brazil, Philippine Islands, and Thailand are among the largest exporters.[67]

Trade in sugar throughout much of the world was influenced by agreements and special trading arrangements. For instance, the United States regulated sugar production, distribution, and importation by means of the Sugar Act of 1948, as amended.[68] However, many of the regional agreements and the United States Sugar Act have been supplanted by the "International Sugar Agreement" which became effective in 1977.[69] The current agreement differs markedly from its predecessors[70] in that it is supported by 58 signatory countries, including the United States.[67]

Other sweeteners such as dextrose, glucose syrups, honey, maple sugar and syrup, and edible molasses have been available in the United States for many years. However, only dextrose and glucose syrup have been produced and distributed in substantial quantities in the United States (and many other countries, particularly western Europe) as alternative or supplemental sweeteners.

Per capita consumption of dextrose in the United States has been relatively constant at 4 to 5 pounds for many years.[71] The growth in the per capita consumption of starch-derived sweeteners prior to 1970 occurred in glucose syrup (a family of hydrolyzed syrups manufactured from starch), but that growth was limited to less than two pounds per year.[71] The increased use in the starch-derived sweeteners had no apparent effect on the consumption of cane and beet sugar which had remained more or less constant at 97-99 pounds per capita.[71]

The introduction in the late sixties of the starch-derived product commonly referred to as HFS (high fructose syrup) introduced a period of substantial change in the sweetener industry. That change is still in progress largely because of the utilization of new technology as well as

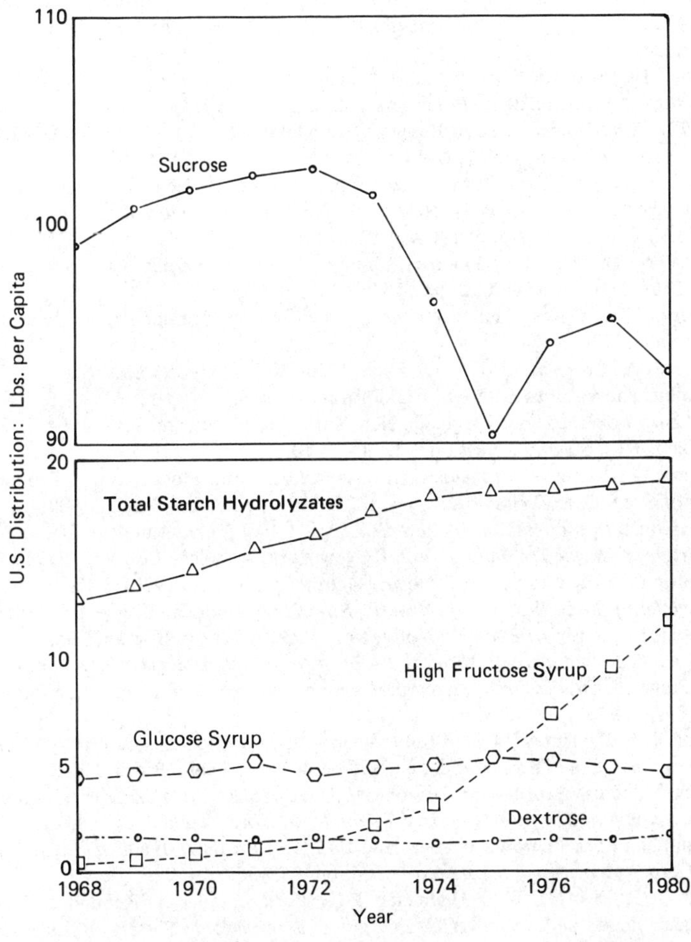

Fig. 16.18 Trends in distribution of sugars in the United States.

economic factors. The trend of the changes introduced by this product (now a family of products) is shown in Figure 16.18. HFS consumption has increased from less than one pound per capita in 1968 to more than fifteen pounds in 1979. HFS is for all intents and purposes a replacement for invert sugar (derived from sucrose) for many of the industrial food uses formerly supplied by invert sugar syrups. HFS also displaces sucrose syrups in some applications. Those trends are apparent from Figure 16.18 which shows steady increases in HFS consumption from 1973 to the present, and a decline in sugar consumption during the same period. However, the declining consumption of sugar is the result of several factors, only some of which are related to the availability of HFS.

HFS technology and products have spread to other countries, particularly Canada, Japan, and several countries of western Europe. The pace of commercial development in those other areas does not appear to be as rapid as that in the United States where a combination of plentiful supplies of starch at reasonable prices and an existing industry based on starch hydrolysis appear to catalyze the development. A review of current trends and competitive factors was presented at the XVII Congress of International Society of Sugar Cane Technologists.[50]

REFERENCES

1. P. M. A. M. Chenu, "Alcohol Manufacture in a Sugar Factory," *XVI Congress of International Society of Sugar Cane Technologists*, pp. 3241-3250, Sao Paulo, Brazil, 1978.
2. G. Tegge, "Development and Prospects of Sweeteners Obtained by Glucose Isomerization," *International Sweetener Report*, No. 1, pp. 3-8, December 1978, Edited by H. Ahlfeld, F. O. Licht GmbH, Ratzeburg, Fed. Rep. Germany.
3. J. A. C. Hugill, "History of the Sugar Industry," *Sugar: Science and Technology*, pp. 15-37, G. G. Birch and K. J. Parker, editors, Applied Science Publishers Ltd., London, 1979.
4. G. P. Meade, *The Sugar Molecule*, Sugar Research Foundation, N.Y., **11**, No. 1, pp. 11-19, 1959-60.
5. N. Deerr, *The History of Sugar*, Vol. 1, London, Chapman and Hall, 1949.
6. E. Hugot and G. H. Jenkins, *Hand Book of Cane Sugar Engineering*, New York, American Elsevier, 1960.
7. R. A. McGinnis, *Beet Sugar Technology*, New York, Van Nostrand Reinhold, 1951.
8. N. Deerr, *Cane Sugar*, London, Norman Rodger, 1921.
9. V. Ducasse, "Heavy Duty Chopper-Fiberiser," *XVI Congress of International Society of Sugar Cane Technologists*, pp. 1568-1601, Sao Paulo, Brazil, 1978.
10. L. B. Rodriguez, E. G. Garcia, and A. C. Alba, "New Concept in Cane Preparation," pp. 55-59, *Sugar y Azucar*, November 1977.
11. Battelle Staff, "Sugar Crops as a Source of Fuels," Vol. II, pp. 51-65, Report to U.S. Dept. of Energy by Battelle Columbus Laboratories, Columbus, Ohio, August 1978.
12. G. P. Meade, *Cane Sugar Hand Book*, 9th ed., New York, John Wiley and Sons, 1964.
13. "BMA Cane Diffusion," *Sugar y Azucar*, p. 22, July 1970.
14. E. Whayman and O. L. Crees, "Mechanistic Studies of Cane Mud Flocculation," *XVI Congress of International Society of Sugar Cane Technologists*, pp. 1175-1182, Sao Paulo, Brazil, 1978.
15. V. Baikow, *Manufacture and Refining of Raw Cane Sugar*, New York, American Elsevier, 1967.
16. P. Honig, *Principles of Sugar Technology*, Vol. II, New York, American Elsevier, 1959.
17. B. Silver, "Continuous Vacuum System," *Sugar y Azucar*, p. 36, June 1970.
18. P. Honig, *Principles of Sugar Technology*, Vol. III, New York, American Elseiver, 1963.
19. P. Honig, *Principles of Sugar Technology*, Vol. I, New York, American Elsevier, 1953.
20. J. T. Rundell, C. V. Rich and A. E. Norcott, "A New Automatic Defecation-Decolourization Station Using the Talofloc Process," *Publication of Technical Papers and Proceedings of Sugar Industry Technologists*, Vol. **32**, pp. 58-70, 1973.
21. F. X. Pollio and F. X. McGarvey, "Use of Ion Exchange in Cane Sugar Processing," *XVI Congress of International Society of Sugar Cane Technologists*, pp. 2729-2741, Sao Paulo, Brazil, 1978.
22. M. Dmitrovsky, "Sucrose Process for Evaporative Continuous Crystallization," *Publication of Technical Papers and Proceedings of The Sugar Industry Technologists, Inc.*, Vol. **38**, pp. 291-302, 1979.
23. T. Rodgers and C. Lewis, "Drying of Sugar and Its Effect on Bulk Handling," 15th Tech. Conf. Report, British Sugar Corp., *International Sugar Journal*, **64**, pp. 359-362, 1962.
24. M. C. Bennett, J. R. Elvin, H. W. B. Heineman, P. R. Pottage and J. T. Rundell, "Syrup Clarification for Improving Sugar Quality and Yield," *XVI Congress of International Society of Sugar Cane Technologists*, pp. 2797-2809, Sao Paulo, Brazil, 1978.

25. F. M. D. Leao, "Production of Amorphous Refined Sugar in Brazil," *XVI Congress of International Society of Sugar Cane Technologists*, pp. 1246-1254, Sao Paulo, Brazil, 1978.
26. W. G. Bickert, F. W. Bakker-Arkema and S. T. Dexter, "Refrigerated Air Cooling of Sugar Beets," *Journal of the American Society of Beet Sugar Technologists*, **14**, No. 7, pp. 547-554.
27. H. G. Schneider and J. Mikule, "Recovery of Sugar from Beet Molasses by the P. & L. Exclusion Process," *International Sugar Journal*, Vol. 77, pp. 259-264, 294-298, 1975.
28. M. Munir, "Molasses Sugar Recovery by Liquid Distribution Chromatography," *International Sugar Journal*, Vol. **78**, pp. 100-106, 1976.
29. H. Hongisto and H. Heikkila, "Desugarization of Cane Molasses by the Finnsugar Chromatographic Separation Process," *XVI Congress of International Society of Sugar Cane Technologists*, pp. 3031-3037, Sao Paulo, Brazil, 1978.
30. R. M. Rapaport, A. Monti, R. D. Moroz, and C. B. Broeg, "Process for Recovering Useful Products from Carbohydrate-Containing Materials," U.S. Patent No. 4,101,338, July 18, 1978.
31. "Thick Juice Storage for Off-Season Processing," *Beet Sugar*, **57**, pp. 46-49, January 1962.
32. F. O. Licht, *International Sugar Reports*, **102**, No. 20, pp. 1-4, 1970.
33. *Sugar: Foreign Agriculture Circular*, FS 1-80, U.S. Dept. of Agriculture, Washington, D.C., January 1980.
34. "Sugar and Sweetener Report," SSR Vol. **4**, No. 5, U.S. Dept. of Agriculture, Washington, D.C., May, 1979.
35. G. B. Hagelbert and H. Ahlfeld, "The Sugar Industry in a Changing Economic Climate," *XVII Congress of International Society of Sugar Cane Technologists*, Manila, 1980.
36. J. L. Hickson, "The Potential for Industrial Uses of Sucrose," *Sugar: Science and Technology*, pp. 151-180, G. G. Birch and K. J. Parker, editors, Applied Science Publishers Ltd., London, 1979.
37. V. Kollonitsch, *Sucrose Chemicals*, pp. 17-105, The International Sugar Research Foundation, Inc., 1970.
38. V. Kollonitsch, *Sucrose Chemicals*, pp. 173-224, The International Sugar Research Foundation, Inc., 1970.
39. A. J. Vlitos, *Sucrochemistry–New Products From Sugar*, World Sugar Research Organization Symposium in Caracas, Venezuela, March 1979, published by World Sugar Research Organization, London.
40. R. Khan, "Advances in Sucrose Chemistry," *Sugar: Science and Technology*, pp. 181-210, G. G. Birch and K. J. Parker, editors, Applied Science Publishers Ltd., London, 1979.
41. *Corn Syrups and Sugars*, Corn Industries Research Foundation, Washington, D.C., 1965.
42. J. A. Kooreman, "Physical and Chemical Characteristics of Enzyme Converted Syrup," *Manufacturing Confectioner*, pp. 35-90, June 1955.
43. R. V. MacAllister, *Advances in Carbohydrate Chemistry and Biochemistry*, Vol. **36**, R. S. Tipson and D. Horton, editors, New York, Academic Press, 1979.
44. R. W. Kerr, *Chemistry and Industry of Starch*, New York, Academic Press, 1944.
45. G. Reed, *Enzymes in Food Processing*, New York, Academic Press, 1966.
46. G. R. Dean and J. B. Gottfried, *Advances in Carbohydrate Chemistry*, Vol. **5**, pp. 127-137, New York, Academic Press, 1950.
47. B. Vabo, "Enzyme Systems for Production of HFCS Syrups," presented at International Biochemical Symposium, Stone & Webster Engineering Corp., Chicago, Illinois, October 1976.
48. J. V. Hupkes, "Practical Conditions for the Use of Gist-Brocades' Immobilized Glucose Isomerase: Maxazyme® GI-IMMOB," presented at International Biochemical Symposium, Toronto, October 1977.
49. J. M. Newton and E. K. Wardrip, "High-Fructose Corn Syrup," *Symposium: Sweeteners*, G. E. Inglett, editor, Westport, Connecticut, The Avi Publishing Company, Inc., 1974.
50. D. E. Nordlund, "High Fructose Syrups: The Competition to Sugar," *XVII Congress of International Society of Sugar Cane Technologists*, Manila, 1980.
51. M. Hoare, "Implications of Immobilized Glucose Isomerase Specifications for the Economics of High Fructose Glucose Syrup Production," *International Sweetener Report* No. 1, pp. 23-32, December 1978, Edited by H. Ahlfeld, F. O. Licht GmbH, Ratzeburg, Fed. Rep. Germany.
52. H. W. Keller and A. C. Reents, "Production of Very Enriched Fructose Corn Syrup (VEFCS) with the Techni-Sweet(TM) Systems," presented at International Biochemical Symposium, Toronto, Canada, October 1977.
53. D. B. Broughton, H. J. Bieser, R. C. Berg, and E. D. Connell, "High Purity Fructose via Continuous Adsorptive Separation," *La Sucrerie Belge*, Vol. **96**, pp. 155-162, May 1977.
54. T. E. Doty and E. Vanninen, "Crystalline Fructose: Use as a Food Ingredient Expected to Increase," *Food Technology*, pp. 34-38, November 1975.
55. F. J. Bates and associates, *Polarimetry, Saccharimetry and the Sugars*, Circular C440, pp. 398-403, National Bureau of Standards, U.S. Dept. of Agriculture, Washington, D.C., 1942.
56. M. Seidman, "New Technological Developments in D-Fructose Production," *Developments in Food Carbohydrate-1*, pp. 19-42, G. G. Birch and R. S. Shallenberger, editors, London, Applied Science Publishers, 1977.
57. W. E.Harris, J. A. Rubino and J. W. McNutt, *Food Processing Industry*, pp. 28-29, February 1978.

58. A. V. Cardello, D. Hunt, and B. Mann, *Journal of Food Science*, Vol. **44**, pp. 748-751, 1979.
59. R. S. Shallenberger, "Predicting Sweetness from Chemical Structure and Knowledge of Chemoreception," *Food Technology*, pp. 65-66, January 1980.
60. W. W. Binkley and M. L. Wolfram, "Composition of Cane Juice and Final Molasses," *Advances in Carbohydrate Chemistry*, Vol. 8, New York, Academic Press, 1953.
61. J. B. Stark and R. M. McCready, "The Relation of Beet Molasses Composition to True Purity," *Journal of American Society of Beet Sugar Technologists*, **15**, pp. 61-72, 1968.
62. "Molasses Market News Annual Summary," U.S. Dept. of Agriculture, Washington, D.C., 1969.
63. E. S. Lipinsky, R. A. Nathan, W. J. Sheppard, T. A. McClure, W. T. Lawhon and J. L. Itis, "Systems Study of Fuels from Sugarcane, Sweet Sorghum and Sugar Beets," Vol. 1: Comprehensive Evaluation, Report to U.S. Energy Research and Development Administration by Battelle Columbus Laboratories, Columbus, Ohio, March 15, 1977.
64. W. H. Kampen, "Ethyl Alcohol–The Automobile Fuel of the Future," *Sugar y Azucar*, pp. 18-30, April, 1978.
65. C. S. Hopkinson, Jr., and D. W. Day, Jr., "Net Energy Analysis of Alcohol Production from Sugarcane," *Science*, Vol. 207, pp. 302-303, January 1980.
66. G. T. Tsao, "Sugar and Sweeteners from Sugarcane," *International Sweetener Report* No. 2, September 1979, edited by H. Ahlfeld, F. O. Licht GmbH, Ratzeburg, Fed. Rep. Germany.
67. W. K. Miller, "The Politics of Sugar," *XVII Congress of International Society of Sugar Cane Technologists*, Manila, 1980.
68. Public Law 331, 89th Congress of the United States, First Session.
69. "International Sugar Agreement and Rules," International Sugar Organization, London, 1977.
70. International Sugar Agreement, 1968.
71. "Sugar and Sweetener Report," SSR Vol. **5**, No. 2, pp. 33, February 1980, U.S. Dept. of Agriculture, Washington, D.C.

17

Industrial Gases

R. M. Neary*

The industrial gases fall into two groups; (1) the less easily liquefiable gases, hydrogen, helium, oxygen, nitrogen, argon, and carbon monoxide; and (2) the more easily compressible gases, chlorine, sulfur dioxide, ammonia, nitrous oxide, carbon dioxide, propane, and fluorocarbon refrigerant gases. At ordinary temperatures, the former do not liquefy in spite of considerable pressure whereas the latter, at ordinary temperatures, form liquids under rather moderate pressures. Hence the content by weight of a standard cylinder for the gases in the first group is small, but for those in the second group, it is considerable. It follows that gases in the first group are generally used as free gases as soon as they are generated, or they will be shipped short distances, from many plants, each serving a small territory; gases in the latter group may be economically shipped long distances, from a few central plants.

The distinction between the two groups is far less sharp today, due to many recent advancements in low-temperature science. Handling liquefied hydrogen (-253°C), liquid helium (-269°C) and other gases that liquefy below -150°C have opened a new scientific field commonly called cryogenics. However, handling cryogenic gases is not new. By 1939, oxygen in liquid form was transported by tank cars and trucks. The other atmospheric gases, nitrogen and argon, have been shipped as liquids in the same type of containers since about 1940. Liquefied hydrogen, the coldest flammable gas, has been transported since 1960 across the United States in tank cars and trucks with super (vacuum) insulation so that no hydrogen is vented en route. Liquefied natural gas (LNG) and methane are transported by boat to the United States and other countries. Thus all of the above gases are being transported in commercial quantities in the liquid state. For miscellaneous industrial uses, however, the familiar steel cylinder with its charge of compressed gas remains standard.

Acetylene (C_2H_2) lies midway between the two groups. It is in a class by itself, partly because its potential explosive decomposition requires that special precautions be taken in handling it.

*R.M. Neary Associates, Pleasantville, N.Y.

TABLE 17.1 Properties of Industrial Gases[a]

Properties of Gases		*Helium* He	*equilibrium-Hydrogen* $e\text{-}H_2$	*normal-Hydrogen* $n\text{-}H_2$	*Neon* Ne	*Nitrogen* N_2	*Air* *Air*	*Fluorine* F_2	*Argon* Ar	*Oxygen* O_2	*Methane* CH_4	*Krypton* Kr	*Xenon* Xe	*Acetylene* C_2H_2	*Carbon Dioxide* CO_2
Atomic or molecular weight		4.0026	2.0159	2.0159	20.183	28.013	28.96	37.997	39.948	31.999	16.043	83.80	131.30	26.04	44.01
Normal boiling point (nbp)	degrees F	-452.1	-423.2	-423.0	-410.9	-320.4	-317.8 to -312.4	-306.7	-302.6	-297.3	-258.7	-244.0	-162.6	-118.5[b]	-109.4[b]
Triple point (tp)	degrees F		-434.8	-434.6	-415.5	-346.0		-363.3	-308.9	-361.8	-296.5	-251.0	-169.2	-116	-69.9
	psia		1.02	1.04	6.3	1.8		0.037	10.0	0.022	1.7	10.6	11.8	17.7	75.1
Critical point	degrees F	-450.3	-400.3	-399.9	-380.0	-232.5	-221.3	-200.2	-188.1	-181.1	-115.8	-82.8	+61.9	96.8	87.8
	psia	33.2	187.7	188.1	395.	493.	547.	808.	710.	737.	673.	796.	847.	905	1072.1
Gas, NTP	lb/ft^3	0.01034	0.005209	0.005209	0.05215	0.07245	0.07493	0.0983	0.1034	0.08281	0.0416	0.2172	0.3416	0.0678	
Gas, STP	lb/ft^3	0.01114	0.005611	0.005611	0.05618	0.07805	0.08072	0.106	0.1114	0.08921	0.0448	0.2340	0.3680		0.1234
Density															
Vapor, nbp	lb/ft^3	1.04	0.0835	0.0831	0.596	0.287	0.280	0.33	0.363	0.279	0.114	0.53	0.60		
Liquid nbp	lb/ft^3	7.798	4.418	4.428	75.35	50.46	54.56	94.1	86.98	71.27	26.5	150.6	190.8		
Specific Heat, Cp, gas, NTP	Btu/lb, degree F	1.25	3.56	3.42	0.246	0.247	0.240	0.197	0.125	0.220	0.533	0.0593	0.0383	0.383	0.202
Specific heat ratio, Cp/Cv, Gas, NTP		1.66	1.38	1.41	1.66	1.41	1.40		1.67	1.40	1.31	1.68	1.66	1.26	1.307
Heat of vaporization, nbp	Btu/lb	8.72	193.	192.	37.0	85.7	88.2	74.0	70.2	91.7	219.	46.4	41.4		
Heat of fusion, T_p	Btu/lb		25.0	25.0	7.1	11.1		5.8	12.7	6.0	25.2	8.4	7.5	356[b]	246.3[c]

[a]Compiled from "Cryogenic Data Reference," Union Carbide Corporation, Linde Division.
[b]Sublimation temperatures, because at atmospheric pressure acetylene and carbon dioxide go from gaseous to solid phase without entering the liquid state.
[c]Heat of sublimation.

In addition to a variety of technical uses, some members of the second group serve as ordinary refrigerants–ammonia, carbon dioxide, dichlorodifluoromethane (Refrigerant 12) and other fluorocarbons. Members of the first group are used as extraordinary refrigerants; for example, nitrogen is used in the liquefaction of hydrogen.

OXYGEN

The production and distribution of cryogenic liquefied gases, including liquid oxygen, is a major industry which has been undergoing very rapid growth in size and technology in recent years.

Refrigeration

The basic refrigeration systems used in the liquefaction cycle for most cryogens such as liquid oxygen, hydrogen, helium, and liquefied natural gas are essentially the same, and thus it is appropriate to discuss them first. The two basic systems employed include the Joule-Thompson expansion cycle and the expansion engine cycle. A third, the cascade system which involves the throttle-expansion of a liquid, is used in LNG plants.

All plants cool the incoming compressed gas by transferring heat to the cooler outgoing plant waste, recycle gas, and product gas in countercurrent heat exchangers. After preliminary cooling and purification, if necessary, the gas is refrigerated by expansion which may partially liquefy the gas.

Figure 17.1 illustrates the fundamental cycles used in liquefying cryogens today.[2] The left side of Fig. 17.1 shows the expansion of cold gas through a throttle valve or nozzle which produces a drop in pressure with a corresponding drop in temperature. This is a constant enthalpy process called a Joule-Thompson expansion. As it passes through the expansion valve some of the gas is liquefied. The liquid is then transferred to a separator, distillation column, or storage tank; the vapor passes through the heat exchanger. This simple liquefaction pro-

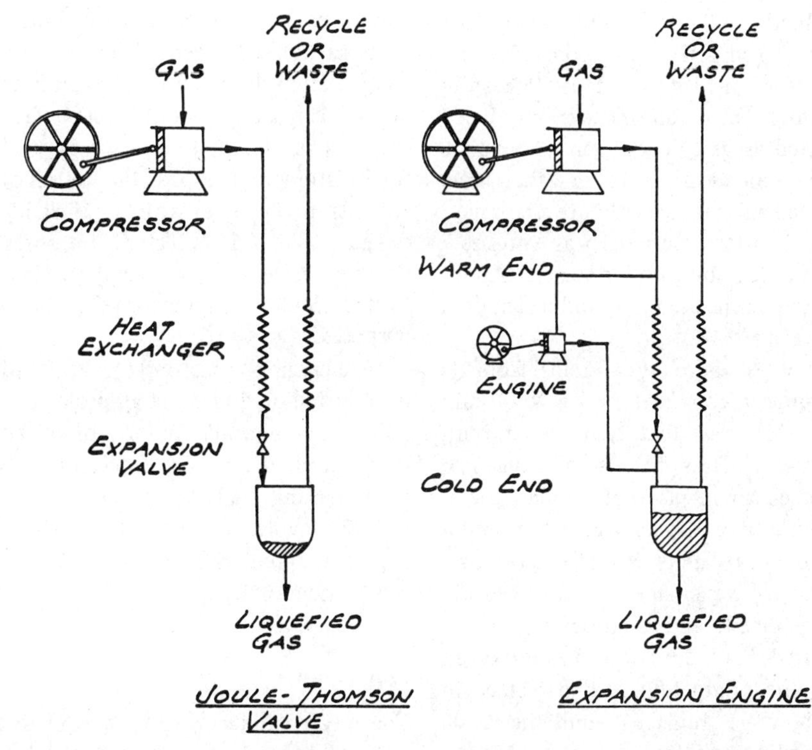

Fig. 17.1 Fundamental refrigeration cycles used in air separation plants. (Courtesy Linde Division, Union Carbide Corp.)

cess was utilized in Dr. Karl von Linde's first cycle for the commercial separation of air.

A second method of producing refrigeration in a gas, as shown at right side of Fig. 17.1, is to expand it in an engine, doing useful work; the temperature is reduced because of the removal of energy. This cycle approaches a constant entropy expansion. The temperature reduction is greater than is possible by expanding gas through a valve, making the system more efficient than the Joule-Thompson cycle and better adapted to large-scale plants.

This expansion engine cycle, known as the Claude principle, does not produce liquid during the expansion phase, as liquid would damage the machine. This cycle is normally used in conjunction with condensers and expansion valves that produce liquid.

Uses

Oxygen is vitally important to industry, medicine, and to explorers of outer space because it is one of the primary tools of their trade. It is well established as a basic industrial chemical with a production of 18,900,000 tons (456.6 billion cubic ft.) in 1979,[3] up from 12.2 and 3.3 million tons in 1969 and 1959 respectively. About 13.5 million tons were produced and distributed as gaseous oxygen via pipeline to steel mills, chemical plants, or other users that are located adjacent or within several miles. This oxygen is often referred to as "on-site," "tonnage," or "combustion" oxygen. About 2.7 million tons were produced and distributed as liquid oxygen at −183°C.

The primary use of oxygen stems from its ability to support combustion and to sustain animal life. Materials that burn in air burn faster in oxygen. Thus, the use of oxygen or oxygen-enriched air in place of ordinary air in many metallurgical and chemical processes increases the intensity and speed of reaction, resulting in shorter cycle time, greater yield per volume of equipment, and lower costs.

It is estimated that 8.5 million tons of oxygen are used in the production of ingot steel—in open hearth, blast furnaces, and the basic process. Another 2 million tons are used in the processing and fabrication of steel mostly in conjunction with the oxygen-acetylene flame, (see Fig. 17.2). Oxygen is used for the "cracking" of methane or natural gas by partial oxidation to produce acetylene, an important basic chemical. It serves as a raw material for synthesizing oxygen compounds (ethylene oxide, sodium peroxide).

Pollution-abatement tests show that oxygen provides greater efficiency and lower costs than air as the oxidizing agent for biological treatment of a sludge waste-water stream to remove its organic matter and to lower its biological oxygen demand (BOD) rating. This applies to both commercial waste streams and community waste treatment.

The missile industry consumes large volumes of liquid oxygen in the testing and firing of rocket motors. Oxygen, stored as a liquid and converted to gas as used, provides a dry supply for aviator's breathing apparatus at high altitudes. Medicinal and breathing oxygen is piped from a central supply to rooms in most hospitals where it is administered to patients by tent, mask, and catheter.

Oxygen is also delivered to the homes of elderly people with respiratory ailments, much of it as liquid oxygen. The patients can use oxygen from the 5-gallon reservoir in the home, or they can carry a portable one-quart unit that converts the liquid to gaseous oxygen warmed for breathing. The portable unit allows many patients to be ambulatory. Most high-purity oxygen sold for industrial purposes is sufficiently pure to meet the requirements of United States Pharmacopoeia for breathing purposes.

An emerging use of large volumes of oxygen is in the billion dollar energy projects that could reduce our nation's imports of oil and petroleum products. The processes for deriving clean-burning fuels from coal and shale frequently involve the use of oxygen, so many people are predicting substantial growth in oxygen production by the year 2,000.

Production

The oxygen industry is unique in that its raw material is available in abundance everywhere, for over 99 percent of it is obtained from at-

Fig. 17.2 Mechanized "hot" scarfing machine, utilizing many oxyacetylene flames, removes surface scale and impurities prior to finish rolling. (Courtesy Union Carbide Corp.)

mospheric air, the remainder from the electrolysis of water. Generally, oxygen is produced by liquefying air and then separating it into its components (oxygen, nitrogen, and argon) by fractional distillation.

There are three fundamental steps in the production of oxygen: purification, refrigeration, and rectification. Purification is the removal of dust, water vapor, carbon dioxide, and hydrocarbon contaminants. Refrigeration in an oxygen plant means cooling the compressed air until it becomes a liquid at about -190°C. Rectification is the separation of liquid air into its components, oxygen and nitrogen, by repeated distillation. In actual practice, these three steps overlap or are performed at the same time.

Purification. The atmosphere holds dirt, water vapor, and carbon dioxide which must be removed from the compressed air stream as these would "plug up" the rectification column. Purification can be accomplished in three ways: by a chemical method, a mechanical method, or a combination of both. Mechanical filters are usually employed ahead of the compressor to remove atmospheric dirt. An example of chemical purification is the removal of carbon dioxide by passing air up through towers filled with coke, down which a solution of caustic soda or caustic potash travels; also a sodium hydroxide solution will remove carbon dioxide from the air.

Removal of impurities from a large volume of air by chemical methods is expensive. Thus, mechanical methods are employed in modern plants. Water vapor is removed in traps. Ice and solid carbon dioxide are removed in cold heat exchangers, reversing heat exchangers, and various types of filters.

The compressed air is cooled and refrigerated by the basic expansion cycles described earlier. In most liquid plants, the incoming compressed air is also cooled by a conventional refrigerant, such as Refrigerant 22 or ammonia. This is necessary to make up some of the refrigeration removed with the liquid products.

The cycles used in many air-separation plants

today employ both of the refrigerating techniques shown in Fig. 17.1. About 40 percent of the air is expanded in a turbo or reciprocating expander. Since some of the air is expanded in the engine, a smaller portion is left to be cooled in the countercurrent heat exchanger. Consequently, the incoming air can be cooled to a lower temperature, with the net result that a much greater fraction of the air is liquefied by the expansion valve.

Rectification. For practical purposes, air may be considered a binary mixture of oxygen and nitrogen. As illustrated in Table 17.2, nitrogen boils at -195.8°C and oxygen at -183°C, so the difference in boiling points is almost 13°C. It is upon this fact that fractionation is based.

If a body of liquid air is warmed, the initial gas released is 93 percent nitrogen and 7 percent oxygen, leaving a liquid that becomes richer in oxygen. The last liquid to be evaporated would be about 45 percent oxygen. Also when air is condensed, the first droplets in equilibrium with the air are about 45 percent oxygen.

The distillation column consists essentially of a cylindrical shell that contains trays spaced at regular intervals. These trays are perforated metal, permitting the liquid to pass transversely over the trays while the gas rises through the perforations and bubbles through the liquid. This brings the liquid and gas into intimate contact. The lower-boiling constituents, particularly nitrogen, are boiled off at each tray so that the gas going up the column becomes richer in nitrogen. At the same time the higher-boiling constituents, particularly oxygen, are condensed at each tray so that the liquid becomes richer in oxygen. With a single column, 99.5 percent oxygen accumulates at the bottom, but the waste gas at the top is 93 percent nitrogen and 7 percent oxygen, representing an efficiency of only 66 percent.

All commercial plants employ a double rectification column separated by a boiler-condenser, commonly called a "reboiler." The lower or high-pressure column operates at 75 to 90 psi, while the upper column operates at 10 to 12 psi. The physical principle for operation of the double column is that increased pressure on the liquid increases its boiling point. The temperature of the 10-psi liquid oxygen on one side of the condenser is lower than the boiling (or condensing) point of nitrogen at 75 psi. Thus the oxygen boils producing vapor for the upper column while the nitrogen gas in the high pressure column that reaches the reboiler is condensed. It flows down over the trays in the lower column becoming increasingly rich in oxygen. The distillation in the lower column produces high-purity nitrogen at the condenser and oxygen-enriched liquid at the bottom. Each of these liquids is transferred to the upper column after being refrigerated through an expansion valve.

The high-purity-nitrogen liquid is introduced at the top of the low-pressure column where it acts as a reflux to strip out oxygen. As a result, relatively high-purity-nitrogen gas is vented from the top of the upper column so that the oxygen-recovery efficiency for a double rectification column is about 90 percent.

Liquid Oxygen

All liquid oxygen is produced as high purity—99.5 percent or higher. When the oxygen is produced as a liquid, a large quantity of refrigeration is removed from the cycle with the product and therefore, is not available to cool the incoming air. Thus, cycles for liquid oxygen must develop large amounts of refrigeration.

TABLE 17.2 Composition of Air

Atmospheric air contains:	*Percent by Volume*	*Boiling Point °C*
Nitrogen	78.14	-195.8
Oxygen	20.93	-183.0
Argon	0.93	-186.0
Carbon dioxide	0.03	- 73.3 (Solidifies)
Water vapor (not included)	amounts variable	0° (Solidifies)

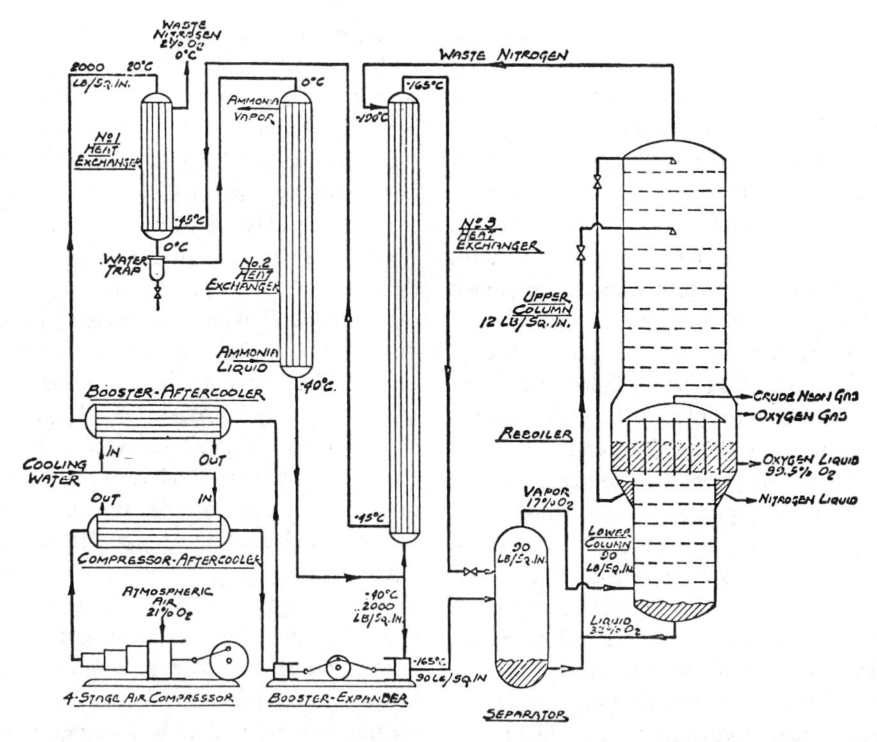

Fig. 17.3 Air separation plant for liquid oxygen production. (Courtesy Linde Division, Union Carbide Corp.)

A common way to accomplish this is by the Heylandt Cycle which includes compression of the air to a very high pressure and use of external refrigeration.

It will be noted that the Heylandt Cycle makes use of both the Linde principle of cooling by expansion from a high to a low pressure through a valve (Joule-Thompson effect), and the Claude principle of the conversion of energy (heat) into work by the expansion of a gas in an engine (or turbine). Figure 17.3 gives details of the cycle as used in most "high pressure" liquid plants.

The air is compressed to about 1500 psig in a 4-stage compressor and cooled in conventional water-cooled intercoolers and aftercoolers. The cooled air is compressed to 2000 psi in fifth stage in a booster-expander, which will be explained later. The air is cooled to room temperature in an aftercooler and delivered to the heat exchanger section, consisting of three units.

In unit No. 1, the air is not permitted to cool below the freezing point of water. This permits most of the water to be condensed out and drained off through the water trap at the bottom of the unit. The cooling agent here is waste nitrogen. It enters the heat exchanger at about -45°C and leaves at about zero. This heat exchanger operates with a relatively large temperature difference.

The high-pressure air then enters the No. 2 heat exchanger, often called a forecooler, and is cooled by the evaporation of liquid ammonia or other refrigerant that is processed in an ordinary refrigeration system. This cools the air to about -40°C and freezes out any remaining water. Eventually this heat exchanger becomes plugged with ice and must be thawed out. Consequently, the No. 2 heat exchanger is provided in duplicate.

The air stream leaving the No. 2 heat exchanger is divided, 60 percent passing through the No. 3 heat exchanger where it is cooled to an extremely low temperature, about -165°C. The air does not liquefy under these conditions because it is at a pressure of 2,000 psig, well in excess of the critical pressure of air (530 psig). This extremely cold air fluid is expanded through a valve to about 90 psig, which liquefies a sizable fraction of the air.

The 40 percent portion of air that does not

go through the No. 3 heat exchanger is expanded in an engine to about the same temperature and pressure as the vapor from the expansion valve. The air expanded in the engine is still gaseous. The work done by the engine is absorbed by direct coupling with the compression cylinder of the booster-compressor mentioned before. This expanded air stream enters the separator along with the throttled stream from the heat exchanger and the two are mixed together.

The vapor which leaves this separator is only about 17 percent oxygen. Since the liquid formed is in equilibrium with the vapor in contact with it, it is about 23 percent oxygen, much richer in oxygen than the entering air.

The liquid and vapor products of the separation are dealt with individually. The vapor stream enters at the bottom of the lower rectifying column at 90 psi. The liquid stream is combined with a similar liquid stream leaving the lower column, and the two are throttled to the pressure of the upper column (about 12 psig). The liquid stream then enters the middle of the upper column. The double rectification column previously described is used to separate the gases. The liquid oxygen accumulates in the main condenser and is piped through filters to an insulated storage tank.

Since liquid oxygen is continuously fed from the main condenser, the traces of hydrocarbon contaminants such as acetylene and methane are drained off with the oxygen. Thus, elaborate hydrocarbon removal systems are not needed in liquid-producing plants.

Gaseous Oxygen

Low-cost gaseous oxygen is produced in large quantities. This has made it economically feasible to use oxygen extensively in many steelmaking, copper-refining, and chemical processes. Oxygen-enriched air can be made economically by mixing the low-cost oxygen with appropriate volumes of air.

Gaseous oxygen is expensive to transport, therefore, these plants are built adjacent to the consuming plant, and are commonly referred to as "on-site" plants. In the 1960's, the oxygen-producing facilities of the nation were more than tripled, and most of the expansion was by "on-site" plants. These plants are equipped with modern instruments to improve their operating efficiency. Many facilities are completely automatic and operate unattended. In such cases, signal lights that indicate any operating difficulty are installed in the customers plant where they are continuously observed.

These new large-scale oxygen plants are being designed to produce a number of products to meet customer demands. These include high-purity oxygen, 99.5 percent; high-purity nitrogen, 99.85 percent for ammonia synthesis; ultrahigh-purity nitrogen, 99.99^{+} percent; and argon. Figure 17.4 illustrates the fundamental cycle used in many modern plants. Air is compressed to about 75 psig in a centrifugal compressor, cooled in an aftercooler and freed of liquid water. It then enters a regenerator or heat exchanger (installed in pairs) where it is cooled to the saturation temperature. The air is then processed to remove small amounts of contaminating solids as well as the vapor-phase carbon dioxide and some hydrocarbons. The air then enters the lower column without any expansion.

To maintain self-cleaning conditions in the regenerators and to provide a preheat stream to the turbine, some air is withdrawn at the -100 to -120°C level. This air is cleaned of carbon dioxide and its temperature is adjusted so that the turbine discharge temperature remains just above the liquid region. The turbine discharge goes to the middle of the low pressure column. It should be noted that the refrigeration for this cycle employs primarily the Claude principle of conversion of energy (heat) into work by the expansion of gas in a turbine.

The conventional double column with condenser-reboiler is used to separate the gases. The oxygen-rich liquid air and the liquid nitrogen from the lower column are subcooled by waste nitrogen prior to transfer to the upper column. From this heat exchanger the waste nitrogen goes through the regenerators where it picks up the heat, water vapor, and carbon dioxide deposited by the incoming air.

The gaseous oxygen product in high-purity plants passes through coils or heat exchangers heated by the incoming air. In low-purity plants, the oxygen passes through a pair of

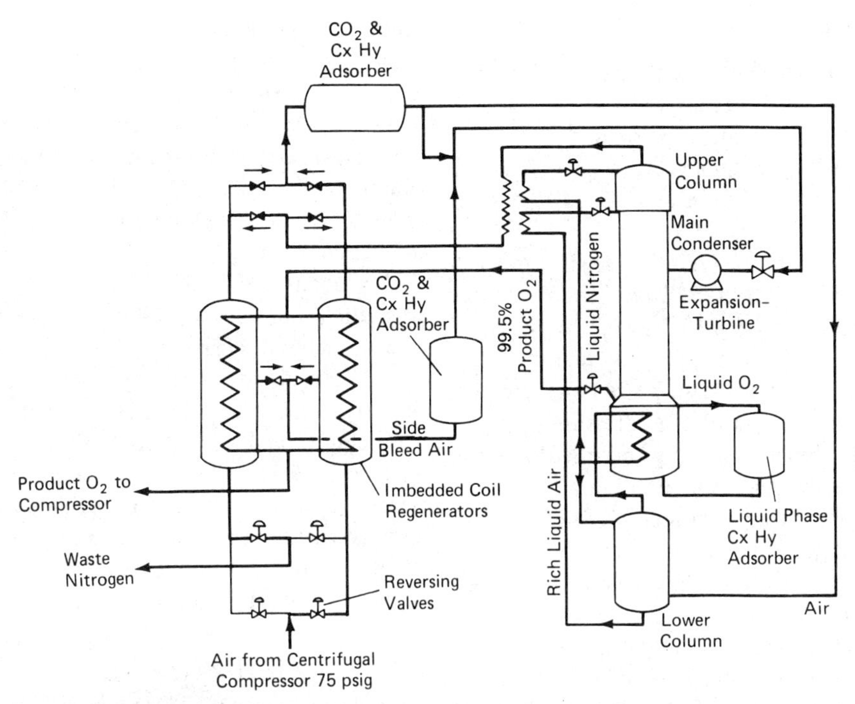

Fig. 17.4 Process diagram for an air separation unit producing 80 million cu ft of oxygen per month. (Courtesy Linde Division, Union Carbide Corp.)

reversing regenerators similar to those used for nitrogen.

The temperature of the waste nitrogen and product oxygen leaving the regenerators is a few degrees below that of the incoming air so that in this type of plant most of the refrigeration is recovered. However, since no liquid oxygen is withdrawn from the reboiler, hydrocarbon contaminants tend to concentrate in the liquid oxygen. Thus, it is necessary to remove all traces of hydrocarbons from the reboiler liquid. One method of achieving this is by the appropriate placement of silica gel adsorption traps which remove most of the hydrocarbons from the process air before they reach the liquid oxygen in the reboiler. As a further safeguard, the liquid oxygen in the reboiler is continually in contact with or circulated through another molecular sieve or silica gel adsorption trap to remove possible residual traces of those hydrocarbons that might concentrate to a hazardous level. Certain amounts of methane and ethane are normally found in the liquid oxygen in the reboiler and are nonhazardous if maintained below safe concentration levels in all parts of the reboiler.

Gaseous oxygen is also produced in a pressure-swing adsorption plant using several parallel adsorbent vessels to provide continuous flow of oxygen. As feed air, compressed to 30-60 psi passes through one of the vessels, the absorbent (molecular sieve) traps (absorbs) the water, carbon dioxide and nitrogen gas, producing a relatively high-purity oxygen product. The absorbent is regenerated by blow down to low pressure during other phases of the cycle.

In pressure-swing adsorption process, hydrocarbons which are present in the feed air are safely adsorbed on the adsorbent bed and do not pass through the system with the oxygen product. The hydrocarbons are, in turn, desorbed into a waste stream which is depleted in oxygen to the extent that the oxygen concentration is about half that which exists in the atmosphere. In such an environment, one that probably would not support combustion, the dilute concentration of hydrocarbons in the stream are far below the lower flammable limit.

Distribution

Most people are familar with the steel cylinders in industry and hospitals where the oxygen is contained at very high pressure (1800-2640 psi). The large cylinders which have a water volume of about 1.5 cu ft hold 244 cu ft of oxygen measured at atmospheric temperature and pressure. These cylinders weigh about 125 lbs and contain about 15 lbs of product. Thus, the ratio of the weight of lading to weight of container is about 1 to 8, an uneconomical ratio. Oxygen is also transported at these high pressures in clusters of high pressure cylinders and long tubes permanently mounted on semi-trailers. Although the production costs of gaseous oxygen are considerably less than for liquid oxygen, the transportation costs are higher because of the heavy containers employed.

The lowest-cost oxygen available today is produced in large-volume gaseous oxygen plants. As mentioned in the previous section, most of these are installed at the consumer's plant. However, pipelines are now employed to transport gaseous oxygen and nitrogen at moderate pressure for many miles from central large-capacity plants to several industrial users, such as in the Gary, Indiana and Houston areas. This type of installation incorporates low-cost production with low-cost distribution, and improves reliability over a single, on-site plant.

Oxygen, nitrogen, and argon are transported commercially as pressurized liquids in (Department of Transportation) DOT 4L cylinders. The cylinders have a built-in or attached vaporizer that automatically converts liquid to gas as the products are withdrawn.

One cubic foot of liquid oxygen when evaporated and warmed to atmospheric temperature and pressure will produce 862 cubic feet of gaseous oxygen. Thus, the current DOT 4L cylinder, similar to the one shown in Fig. 17.5, weighs about 250 pounds and contains about 360 pounds of lading. It has a weight-of-lading to weight-of-container ratio of 1.4 to 1 as compared to 1 to 8 for the high pressure cylinder. This vessel contains the equivalent of 18 high pressure cylinders while occupying about $\frac{1}{4}$ of the floor area, and weighing $\frac{1}{4}$ as much (a weight saving of about $\frac{1}{2}$ ton). It is about the same size as a 55-gallon drum and can be moved with the same ease.

The vessel will stand several days with no withdrawal and no release of oxygen because of the unusually good properties of the fiberglass-aluminum-laminate insulation, the conductivity of which is about 5×10^{-4} Btu/hr/sq ft/°F. This is about $\frac{1}{4}$ the value for 6 inches of powder-vacuum insulation or about 150 times better than 4 inches of cork.

Actually, liquid oxygen is pumped and handled in much the same manner as water except that the pumps and containers are well insulated (see Fig. 17.6). This liquid is normally transferred from the transport equipment to the storage tank by connecting a single pipe of hose to the liquid phase of the transport container. Pumps are usually used to transfer the liquid if

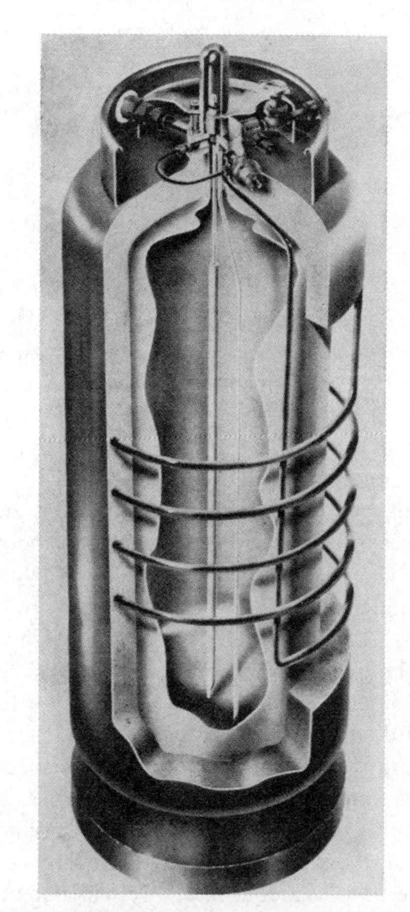

Fig. 17.5 DOT 4L cylinder cutaway view. (Courtesy Linde Division, Union Carbide Corp.)

Fig. 17.6 Containers for liquid oxygen. (Courtesy Linde Division, Union Carbide Corp.)

the pressure in the receiving container is much above atmospheric pressure. If the storage tank operates at essentially atmospheric pressure, the transfer of liquid is often made by building a small amount of pressure in the transport vessel. Some pressure transfers are made up to 200 psi.

Actually liquid oxygen has been transported by tank truck since 1932 and by tank car since 1939. It is estimated that in 1970 there were about 1200 tank trucks and 525 tank cars in liquid oxygen-nitrogen-argon service.

Customer storage tanks range in capacity from 25,000 cu ft (220 gal) to 10 million cu ft (86,700 gal). (The capacity of a tank is usually expressed as equivalent to cubic feet of gaseous product at 70°F and one atmosphere.)

NITROGEN

The most economical way to utilize liquefied air is to make use of both the oxygen and the nitrogen. This is not always practical, since there is approximately four times as much nitrogen as oxygen in the air. In general, commercial oxygen plants recover the oxygen and discharge much of the nitrogen back to the atmosphere. However the demand for nitrogen has increased to the point where the current production by volume of nitrogen and oxygen is about equal.

Nitrogen production in 1979 was 15.48 million tons (427 billion cu. ft.) up from 4.53 million tons in 1969. This does not include captive on-site production for ammonia.

The growth in nitrogen seems to be led by its uses as a blanketing atmosphere, food freezing agent, and in the oil fields.

About half of the above nitrogen production was delivered by pipeline for inert blanketing and other uses in the petrochemical and oil refinery industries. Thus large nitrogen pipeline installations are being utilized in Texas. Nitrogen-blanketing atmospheres are also extensively employed in the electronics and metals industries.

The rising value of oil and natural gas has contributed to the increased use of nitrogen to maintain pressure in underground crude oil reservoirs to increase production. Natural gas was used for this in the past. Also, nitrogen is finding application as the carrier in other processes designed to enhance the recovery of oil from low producing wells.

Although the potential use of nitrogen in oil fields is being realized, some methods remain to be proven.

Liquid nitrogen, which boils at -195.8°C at atmospheric pressure has many refrigeration applications, one being the fast freezing of foods such as hamburger patties and other items for fast-food restaurants, as well as shrimp, fish, and other frozen foods. Liquid nitrogen is also employed for refrigeration of frozen foods during transportation and storage. On trucks, the liquid nitrogen is held at 15-20 psig in a DOT 4L type container, which is not equipped with a vaporizer so that liquid is withdrawn. When the thermostat calls for refrigeration, a control valve is opened and liquid nitrogen is sprayed into the refrigerated van as shown in Figure 17.7 until the desired temperature is reached. Since the control valve is the only moving part, the reliability of this system is excellent. Another advantage in buildings is that the foods are held under an inert atmosphere, thereby preventing damage by rodents and fire.

Liquid nitrogen has important applications in the field of health. Whole blood, which used to be limited to 21 days in storage can be suitable for transfusions, can now be frozen by a liquid-nitrogen process and stored for months and probably years. Bone marrow cells and other parts of the body can be preserved in liquid-nitrogen refrigerators. And cryosurgery has been used for Parkinson's disease and removal of other cell tissue by freezing.

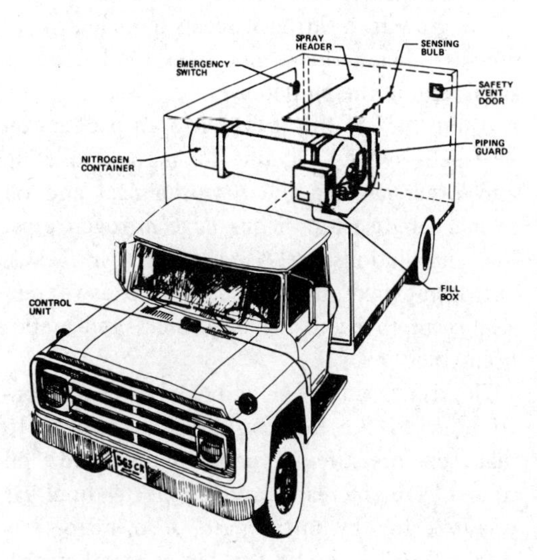

Fig. 17.7 Liquid nitrogen (or air) refrigerates the van of a truck. (Courtesy Linde Division, Union Carbide Corp.)

Liquid-nitrogen cooled refrigerators preserve bull semen used for the artificial insemination of cattle. This has reduced costs and markedly improved the quality of cattle in this country.

Other applications for liquid nitrogen, at -195°C, are as a coolant and purge in the space program, for the shrink fit of parts, and in the deflashing of molded rubber parts.

Liquid nitrogen is transported in insulated containers, which are basically of the same construction as those used for liquid oxygen and argon.

ARGON

Mass markets for argon have developed, resulting chiefly from the development of gas-shielded arc-welding processes. These are used to join hard-to-weld metals such as aluminum, bronze, copper, Monel, and stainless steels. Once considered a "rare" gas, argon is available in tonnage quantities. Production in 1979 was 0.416 million tons (8 billion cu. ft.), a 280 percent increase in ten years. Titanium and zirconium, which have found wide application in our nuclear-space technology, depend heavily on argon or helium for their production. An inert gas envelops the manufacturing process for these metals from start to finish. These metals must also be welded under an inert atmosphere.

Argon is also used in incandescent lamp bulbs, in fluorescent luminous tubes, and in the manufacture of various semiconductor devices.

Production

Argon is relatively scarce; it represents only 0.93 percent by volume of the earth's atmosphere. However, the advanced technology developed in air-separation plants plus the large air volumes handled in today's plants permits the recovery of this "rare" gas by the ton. The boiling point of argon is -187°C, under conditions used in the fractional distillation of liquid air, so it lies between the boiling points of oxygen and nitrogen. Therefore, argon tends to accumulate in the middle of the low pressure (upper) column, readily building up to 10-15 percent by volume. It is drawn off in a fraction that also contains oxygen and nitrogen and is further refined in other distillation columns that remove the nitrogen. The residual oxygen is removed by catalytic combination with hydrogen. Welding-grade argon is produced at a purity of about 99.995 percent.

RARE GASES

The rare or "noble" gases include krypton, xenon, neon, helium, and argon. The availability of helium (other than atmospheric) and argon in large quantities makes the designation "rare" a misnomer insofar as these two are concerned. However, atmospheric helium is rare, as is argon specially purified for particular applications; the terminology for argon and helium appears to be a matter of philosophy. All these gases are characterized by extreme chemical inertness and some of them, primarily neon, krypton, and xeon, ionize or become electrically conducting at a substantially lower voltage than other gases. While passing current, they also emit a brilliant colored light used to advantage in the tubular display signs so prominent in what has been termed the "neon jungle."

Present commercial applications for the rare gases rely principally on their inertness. They are used singly or in mixtures by the electronics industry in gas-filled electronic tubes. The lamp industry uses all the rare gases, including atmospheric helium and specially purified argon as fill gas in specialty lamps, neon and argon glow lamps, high output lamps, and others. The gases are also used as fill gas for ionization chambers, bubble chambers, and related devices.

Production

Partial separation of the rare gases is accomplished by the same liquefaction-fractional distillation process used to produce most other atmospheric gases, using liquid nitrogen as a refrigerant. Final purification requires the use of special processes and equipment.

HELIUM

Most of the helium produced for commercial use is obtained from certain natural gases. About 85% of the known U.S. helium reserves are contained in five helium-bearing gas fields: the Hugoton field in Kansas, Oklahoma, and Texas; the Keyes field in Oklahoma; the Panhandle and Cliffside fields in Texas; and the Tip Top field in Wyoming, Fig. 17.8.

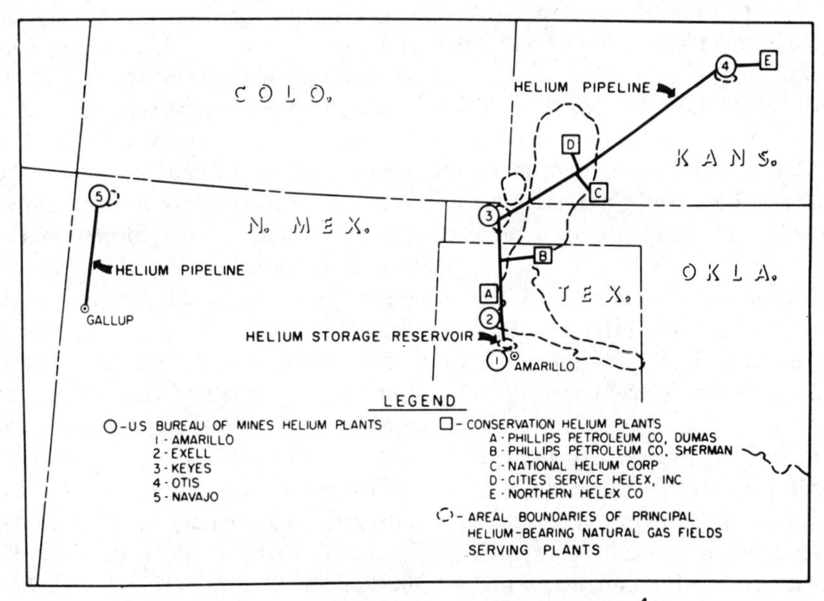

Fig. 17.8 Five major helium producing gasfields.[4]

The rapidly increasing use of helium in the 1950's heavily taxed the production facilities which were operated almost exclusively by the Bureau of Mines under the Helium Act. The quantity of helium sold by the Bureau of Mines increased from about 50 million cu ft in 1949 to 477 million cu ft in 1959. Thus, in 1960 Congress amended the Act, to provide for long-term purchase contracts for helium by the Bureau of Mines from private industry as part of an extensive helium conservation program. In 1977, the Bureau held about 37.8 billion cubic feet in storage under this conservation program.[5] Also private producers held 1.7 billion cubic feet in conservation storage for future redelivery under contract with the Bureau.

Nine plants were operational in 1977 to extract helium from natural gas being piped to market. Two of the plants were owned by the U.S. Government and operated by the Bureau of Mines; the other seven were owned and operated by private industry.

Total helium extracted from natural gas was 1.67 billion cubic feet in 1979,[3] up from 1.5 billion cubic feet in 1977. Bureau of Mines produced 22% of the crude helium in 1977, private industry the remainder. Crude (unrefined) helium is extracted by a cryogenic process that separates out a helium-nitrogen stream containing 50-85 percent helium. Most of the crude is stored in an underground reservoir near Amarillo, Texas.

There was 957 million cubic feet of high-purity helium produced in 1977, 23 percent of this by the Bureau of Mines, the remainder by private industry. Bureau of the Census shows high purity shipments to be 817 million cubic feet in 1979.

Prior to 1946, about 99 percent of the helium was used for dirigibles and balloons. Helium purity was gradually improved from 98.3 to 99.5 and then to 99.995^+ percent, which greatly expanded its use.

During the 1960's, a major user was NASA and its missile contractors. Currently, helium finds its primary applications in cryogenics, welding, purging and pressurizing, and synthetic breathing mixtures for deep underwater diving. Other uses include chromatography, leak detection, as a lifting gas, and in heat transfer systems.

Production

Bureau of Mines and private plants take gas from natural gas pipelines, remove the helium and return the stripped gas to the pipeline. Plant capacities vary over a wide range, but are usually sufficient to handle the entire flow in a pipeline, except for a few cold winter days. Also the helium content in gas varies widely from plant to plant, ranging from 0.9 to 5.8 percent. In general, helium production involves two phases: (1) production of crude helium, and (2) purification to 99.99^+ purity.

An article[1] describes the process used at the Bureau of Mines Keyes plant, which is typical of all plants and includes the following: (1) scrubbing the natural gas to remove water and condensed hydrocarbons; (2) dust removal; (3) CO_2 removal in a scrubbing tower; (4) drying; (5) chilling to liquefy and remove the natural gas; and (6) rectification of the gaseous materials from step 5 to produce the crude (75 percent He, 25 percent N_2, 0.1 percent each of H_2 and CH_4) helium.

The pure helium is obtained from the crude by: (1) oxidizing the hydrogen with air; (2) removing the water formed by the oxidation; and (3) removing the last traces of hydrogen and nitrogen with activated charcoal. The final product is 99.99 plus percent pure.

Many of the private plants and the Otis BM plant contain helium liquefiers that have sufficient capacity to liquefy most of the plant's output. The refrigeration techniques described earlier in Fig. 17.1 are used, plus precooling with liquid nitrogen as shown in Fig. 17.9.

The following cycle has been used in a 100 l/hr. helium liquefier. Pure helium gas is compressed 270 psig, passed through an aftercooler and into the cold box where it is cooled to 80°K against the exiting recycle helium and nitrogen. The helium is then refrigerated by liquid nitrogen, purified in a gel trap, and further cooled to 24.5°K by recycle helium gas. The stream then splits with part of the cold

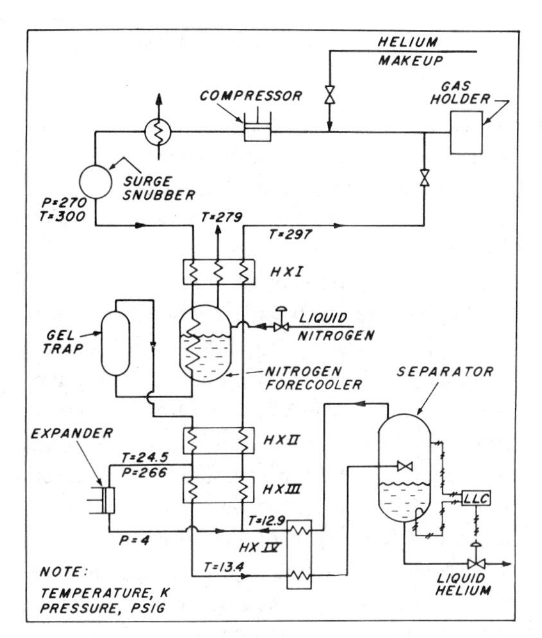

Fig. 17.9 Helium is liquefied by cryogenic refrigeration techniques, plus precooling with liquid nitrogen.[4]

helium going through an expansion turbine, while the remainder goes through a heat exchanger and expansion valve.

Liquid helium is transported in vacuum-super-insulated tank trucks, portable tanks, and cylinders without nitrogen shielding. Many liquid-helium Dewars, 100 liter and less, that operate at less than 15 psi, use nitrogen shielding. Since liquid helium at 4.25°K will readily solidify air, caution must be exercised to be sure that the Dewar neck tube does not become plugged with solid air, particularly during air transportation where it is subjected to wide changes in atmospheric pressure.[6]

Liquid helium boils at 4.25°K and this ultra-low temperature can liquefy any other gas. LHe is being used increasingly in low temperature research investigating the properties of matter, particularly electrical properties. LHe is also used in cryogenic chemistry, bubble chambers, and for cryopumping.

HYDROGEN

Large quantities of hydrogen are used in the chemical syntheses of ammonia, hydrogen chloride, and methanol. Other alcohols are produced by the hydrogenation of the corresponding acids and aldehydes.

During recent years, hydrogen has become industrially important in the hydrogenation of the edible oil in corn, cotton seed, soy bean, and other vegetables for the production of shortening and other foods, many of which are low in cholesterol. Its other uses include the oxygen-hydrogen flame that has a temperature of about 2200°C (4000°F) and which is well suited for low-temperature welding and brazing of thin metals, undersea cutting-welding where over 15 psi is involved, and in the fabrication of quartz and glass. Atomic-hydrogen welding, a form of shielded-arc welding, produces temperatures of about 6100°C (11,000°F). Hydrogen serves as de-oxidizing fuel in many applications such as annealing furnaces and fabrication of electronic components; it is used in large electric generators to reduce windage losses and heat.

Liquefied hydrogen is used as a rocket fuel by explorers of outer space; and in recent years it is the commercial source of ultrahigh purity hydrogen (99.999$^+$ percent) required in parts of the electronic industry.

Production of hydrogen, liquid and gas, was about 99 billion cubic feet in 1979, excluding amounts produced and consumed in the manufacture of synthetic ammonia and methonal.[3]

Hydrogen is obtained almost exclusively from water or hydrocarbons. From water gas, it is obtained by the removal of the nonhydrogen constituents. It is made by the catalytic action of steam on oil refinery gases and natural gas, by steam on heated iron, by the electrolysis of water, or by other miscellaneous processes. It is a by-product in the electrolytic cell for caustic,* in several fermentation processes, and in other types of processes. The choice of a process will be decided by the resources at hand, and by the degree of purity required. Should rapid generation with a minimum of apparatus in an isolated place be demanded, the steam-methanol or the ferrosilicon process would serve.

*Hydrogen from mercury cells for caustic has a slight contamination of mercury, which for certain uses, must be removed.

The Water Gas and Steam Process (Continuous Catalytic Process)

Water gas with steam in excess is passed over an iron oxide catalyst, just as is done in the Bosch process, except that since no producer gas is added, the amount of nitrogen is small. The converter has several trays, on which the catalyst rests. The reaction ($CO + H_2O \longrightarrow CO_2 + H_2$) is exothermic; as the temperature must be maintained at 450°C (842°F), the converters are insulated and the incoming gases heated in exchangers. Once the reaction has begun, no outside fuel is required. Three volumes of steam to one of gas are used; the great excess of steam drives the reaction to the right. After passing the exchangers the reacted gas is freed from its stream by water-cooling. The carbon dioxide formed, as well as the small amount which entered with the water gas (4 percent), is removed by scrubbing with cold water while under pressures of 25 to 30 atmospheres, in tall steel towers; under such pressures carbon dioxide is freely soluble in water. The gas leaving the last scrubber has the composition:

	Percent
Hydrogen	92-94
Nitrogen	1-4
Methane	0.5
Carbon monoxide	2-4
Carbon dioxide	small
Moisture	small

The crude hydrogen may be further purified from carbon monoxide by scrubbing in ammoniacal cuprous chloride solution. The nitrogen impurity may be lowered by careful operation of the water-gas plant. The methane is not wanted, and may be almost avoided by using well-burned coke.

A similar process in which the catalytic agent is lime at the temperature of 450°C (842°F) instead of iron oxide has been proposed; its great advantage is that the carbon dioxide is simultaneously removed. Unfortunately this absorption is accompanied by a powdering of the lime granules, as the carbonate forms, and the powder tends to clog the lime towers.

Water Gas Process with Liquefaction of the Carbon Monoxide

There are two processes in which the carbon monoxide in water gas is liquefied by cold and pressure and removed in that state, leaving the hydrogen gas comparatively pure. The Linde-Fränkl-Caro process uses liquid air boiling under a few millimeters of pressure for the final cooling of the water gas already pre-cooled by three steps, first by an ammonia refrigerating system (-35°C or -31°F), then in an exchanger wherein the uncondensed hydrogen takes up heat and finally by the liquid carbon monoxide separated in the final cooling. The carbon monoxide at the same time boils, and is used as a gaseous fuel. The other process is Claude's, which uses no liquid air, but obtains the necessary final cooling by the expansion of hydrogen from a pressure of 20 atmospheres to a lower pressure while doing work against a piston. The hydrogen is pre-cooled in exchangers by outgoing gases, and by the evaporation of the carbon monoxide which has been previously liquefied.

Steam on Heated Iron

The interaction of steam and iron takes place at an elevated temperature, such as 650°C (1202°F) in a multiplicity of relatively small, upright steel cylindrical retorts. The iron packing selected should have a porous structure and little tendency to disintegrate; a calcined iron carbonate (spathic ore) has been found suitable. A plant for the production of 3500 cu ft (about 100 cu m) per hr would consist of three sets of 12 retorts each, each retort being 9 inches in diameter and 12 ft high. The action is intermittent. The steaming period (hydrogen production) lasts 10 min (upward travel of steam).

$$3Fe + 4H_2O \longrightarrow Fe_3O_4 + 4H_2$$

The iron oxide formed is reduced by water gas, for example, and the water period lasts 20 min (downward travel), because the reduction of the oxide is slower than the oxidation of the iron. A brief purging with steam sends the first hydrogen to the water gas holder. By the stepwise operation of such a plant a continuous flow of hydrogen is obtained. The spent water gas is cooled and burned (for it still contains combustible gases) around the retorts to maintain the reaction temperature. The steam reaction is exothermic; the over-all reduction by the water gas is endothermic.

The hydrogen passes out with the great excess of steam which is employed to drive the reaction to the right; it is cooled to remove the steam, and freed from carbon dioxide and hydrogen sulfide by lime purifiers. The gas obtained is 98.5 to 99 percent pure; by careful purging, using closed condensers instead of scrubbing towers, and other modifications, the purity may be raised to 99.94 percent. The steam-iron process is used chiefly in connection with the hydrogenation of fatty oils. The iron mass lasts six months, the retorts one year.

For the production of 1 volume of hydrogen, the continuous catalytic process requires 1.25 volumes of water gas, the liquefaction process (CO liquefied), 2.5 volumes, and the steam-on-heated-iron process, also 2.5 volumes.

Electrolysis of Water

In commercial cells, direct current is passed between iron electrodes, which may be nickel-plated, suspended in a bath consisting of a 10 to 25 percent caustic soda or potassium hydroxide solution. Only distilled water is added to the cells, for the electrolyte is not consumed. Cells differ in the method of gathering the hydrogen (at cathode) and the oxygen (at anode), in size, and in details of construction. The efficiency is close to 7.5 cu ft of hydrogen per kw-h, and 3.8 cu ft of oxygen.

The decomposition voltage for the reaction $H_2O \longrightarrow H_2 + \frac{1}{2}O_2$ is 1.48; the operating voltage about 2. The amperage varies with the size and intensity of operation; with maximum load, it may be as much as 1000 amp for electrodes 40 in wide and 60 in high, with a number of electrodes in each cell. The electrodes are separated by an asbestos diaphragm, and suspended in cast-iron containers.

Steam Reformer Process

The most widely used process for producing hydrogen in large volume is by steam reformer; about 80 percent of the hydrogen used for ammonia production is derived by that process.

Heated natural gas (or other hydrocarbon fuel) and steam are mixed and fed into a reformer furnace, consisting of large diameter, thick walled vertical tubes manifolded together and packed with a nickel catalyst. Heat input, 900°C (1600°F), for this step is from natural gas-fired burners. The reactions in the reformer are:

$$CH_4 + 2H_2O \longrightarrow 4H_2 + CO_2$$

$$CH_4 + H_2O \longrightarrow 3H_2 + CO$$

The crude hydrogen stream, containing 8.5 percent CO and excess steam, is cooled in a steam generator to 400°C (750°F) and fed to a shift converter that reacts the CO to CO_2 while producing hydrogen. This reaction is slightly exothermic.

$$CO + H_2O \longrightarrow CO_2 + H_2$$

Liquefied Hydrogen

The above steam-reformer process is the source of hydrogen in a majority of liquefied hydrogen plants, which have four basic sections: steam reformer, purification, liquefaction, and storage as shown in Fig. 17.10.

Crude hydrogen from the reformer-shift converter contains about 20 percent carbon dioxide and is saturated with water. The CO_2 is usually removed in a monoethanolamine (MEA) scrubber with the exiting crude being about 96 percent hydrogen, 1.7 percent CO, 2.2 percent CH_4, and small amounts of nitrogen, CO_2, and moisture.

The moisture is removed to a very low dew point in a molecular sieve bed to prevent freeze-up of the low-temperature purification equipment shown in Fig. 17.11.[7] This equipment removes impurities in hydrogen to about 5 ppm, of which not more than 1 ppm is oxygen, in order to meet customer requirements and prevent freeze-up of the liquefaction section. Partial condensation is used to remove most of the methane and some of the nitrogen, and carbon monoxide is removed in propane and methane wash columns and other steps, using liquid nitrogen as the final cooling step. Then residual impurities are removed by low-tem-

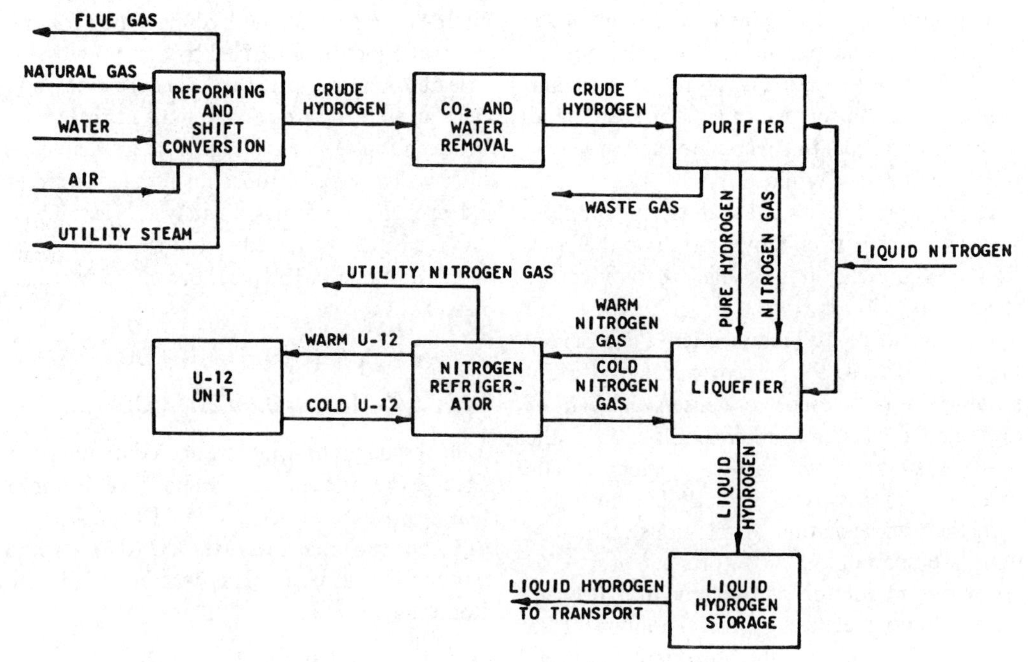

Fig. 17.10 Process block diagram of large liquid hydrogen plant.[7]

perature adsorption to obtain the desired hydrogen purity.

Pertinent properties of hydrogen are given here before discussing the liquefaction process. The boiling point of liquefied hydrogen is -252.9°C (-423.2°F), which is lower than the freezing temperature of all other gases except helium. Its density is only 4.418 lb/cu ft, about $\frac{1}{14}$ that of water. Hydrogen can exist in two forms depending upon the rotation of the two atoms in the hydrogen molecule. In *ortho*-hydrogen, the atoms spin in the same direction; in *para*-hydrogen they spin in opposite directions. Also, hydrogen has a very low inversion

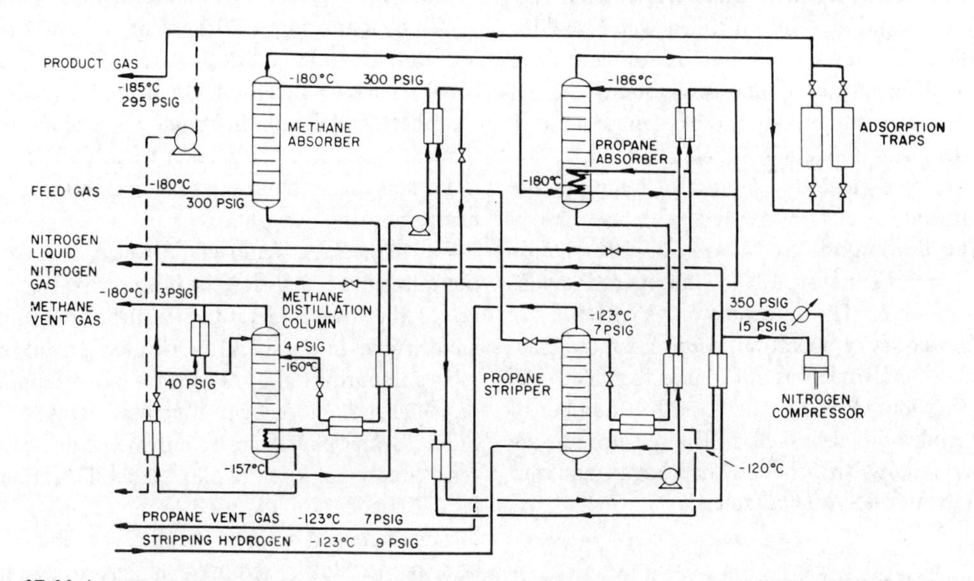

Fig. 17.11 Low-temperature purification equipment removes impurities from hydrogen. (Chem. Engr. Prog., *59*, No. 8, 61 (1963); copyright 1963 by The American Institute of Chemical Engineers, reprinted by permission.)

temperature, −70°C, above which Joule-Thompson expansion produces a rise in temperature, instead of lowering it.

The liquefier section produces liquefied *para*-hydrogen by removing heat from the purified hydrogen, and by converting the *ortho*-hydrogen to *para*-hydrogen. Conventional cryogenic refrigeration cycles described in Fig. 17.1 are used to liquefy the hydrogen. Liquid nitrogen is used to precool the hydrogen; then the hydrogen stream is split in two with portion one being cooled to near liquefaction temperature in a high speed expansion turbine. This cold, expanded hydrogen is used to cool portion two before it is partially liquefied through a (Joule-Thompson) expansion valve, Fig. 17.12.[8]

Ortho-to-*para* conversion is an important factor in the design of the liquefier cycle. Purified hydrogen, which is normal hydrogen, is only 25 percent *para*-hydrogen at atmospheric temperature. Stable liquefied hydrogen is 99.7 percent *para*-hydrogen, and the heat generated by conversion is about 11.5 percent of the total heat removed during liquefaction. If normal (liquid) hydrogen were transferred to storage, up to two-thirds of the liquid would be vaporized as conversion took place. So the liquefier cycle includes rapid conversion over ferric iron hydroxide catalyst, in two steps, at liquid nitrogen and at liquid hydrogen temperatures in the presence of liquid refrigeration to remove heat. Thus, liquefied hydrogen, which is over 95 percent *para*, is produced by the liquefier and transferred to insulated storage tank. Purity exceeds 99.999 percent.

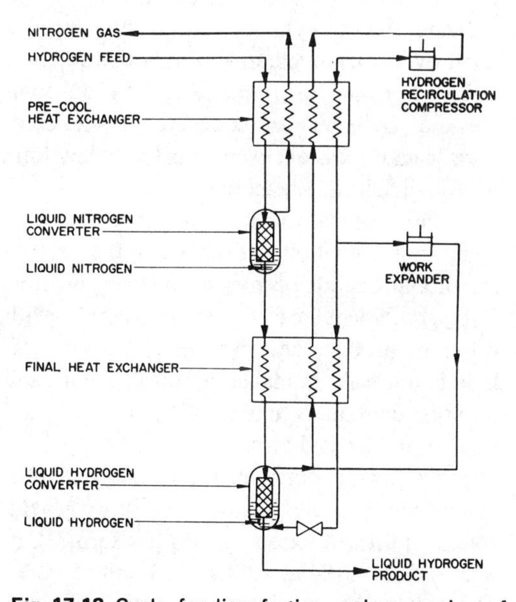

Fig. 17.12 Cycle for liquefaction and conversion of liquid hydrogen. (Baker, C. R. and Matsch, L. S., "Advances in Petroleum Chemistry and Refining", Vol. X, Interscience Publishers; by permission of John Wiley and Sons, Inc.)

Tank trucks and tanks cars with laminar and high-vacuum insulation are used to transport liquefied hydrogen from large production plants to any part of the country without loss of product enroute. Thus liquefied hydrogen is delivered commercially in containers and by procedures similar to those used for liquid oxygen, which were described earlier.

LIQUEFIED NATURAL GAS (LNG)

Natural gas as a fuel is discussed in Chapter 3. Since liquefied natural gas is a cryogen (boiling point about −162°C) and the refrigeration, handling, and transportation techniques discussed in this chapter for other cryogens is used for LNG, it seems appropriate to include a brief discussion of LNG.

The cryogenic phases used to produce LNG are purification and liquefaction, the same as for liquefied hydrogen. Natural gas contains some water vapor and carbon dioxide as it is withdrawn from the pipeline for liquefaction. Concentrations of these impurities are reduced to 1 ppm and 50 ppm respectively, to prevent plugging of the liquefying equipment. One purification method is to pass the gas through adsorption towers packed with molecular sieves, with an activated alumina bed at the inlet to remove oil vapor.

There are several liquefaction cycles used for LNG but the "cascade" concept in which several change-of-phase systems are combined is the most popular. Each system involves a fluid suitable for use at a given temperature level and is used to precool the fluid in the next change-of-phase system operating at a lower temperature level. Several levels are "cascaded" together to match the cooling curve of the natural gas, with throttling or Joule-Thompson expansion of the refrigerant between each level. An advantage of this cycle for LNG is that the various temperature levels can be adjusted to provide the most efficient cooling with changes in composition of the natural gas to be liquefied.

The composition of the refrigerant is usually a mixture of hydrocarbons and nitrogen. The equipment consists of several heat exchangers and separators in series with throttle valves between them as shown in Fig. 17.13.[9] Thus, only one refrigeration compressor is needed and a typical cycle would include change-of-phase (system) of propane-, ethane- and methane-enriched streams in succeeding heat exchangers.

LNG is stored and handled in insulated containers, including seagoing vessels. Domestic transportation is by tank trucks which are similar to those used for liquid nitrogen. Most tank trucks built today for LNG are suitable for liquid nitrogen, and vice versa, to provide greater flexibility in a carrier's fleets and reduce fabrication costs.[10]

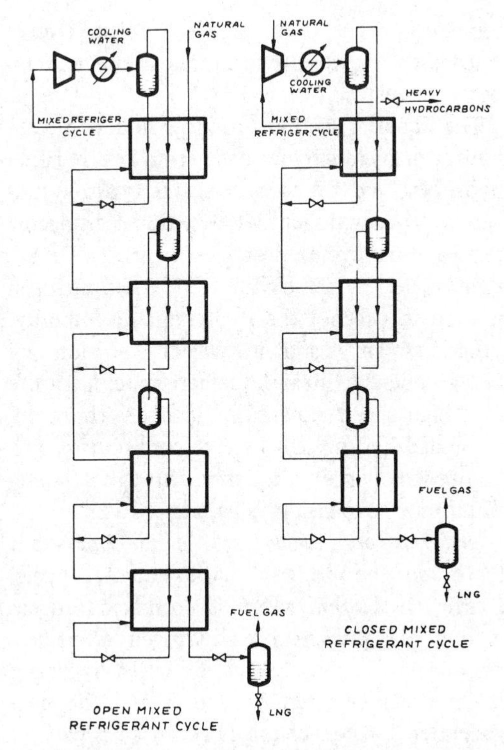

Fig. 17.13 Cycles commonly used in liquefying natural gas.[9]

ACETYLENE

Acetylene is an important industrial gas. The oxygen-acetylene flame produces the highest temperature of any combustible gas, hence its great value in welding and cutting of steel and other metals. A large volume of acetylene is being used at present as a basic raw material in the chemical industry for the production of synthetic rubber-like materials, flexible vinyl plastics, rigid plastics, paints, and textile finishes, to name a few. The original and still frequently used method of making acetylene is by the action of calcium carbide, a product of the electric furnace.

$$\underset{}{CaC_2} + 2H_2O \longrightarrow \underset{\substack{\text{Acetylene}\\\text{(gas)}}}{C_2H_2} + \underset{\substack{\text{Hydrated lime}\\\text{(by-product)}}}{Ca(OH)_2}$$

The only method of distribution, other than pipeline, is by means of portable steel tanks, containing a porous solid filler saturated with acetone, or other suitable solvent, in which the acetylene is dissolved by pressure. Acetylene alone is not handled at pressures higher than 2 atmospheres because of its tendency to decompose explosively into carbon and hydrogen; dissolved in acetone it may be under pressures of 10 to 15 atmospheres with safety. In order to fill the cylinders, the acetylene gas is dried over calcium chloride or by other method and compressed in a specially designed compressor. Cylinders are connected to the compressor and occasionally shut off to allow time for the dissolution in acetone.

In cases where the rate of consumption is high, it is sometimes feasible for the user to install and operate his own acetylene generator. This equipment reacts calcium carbide with water to produce acetylene gas. Calcium carbide is normally available in 100 lb drums, and in larger containers holding 250, 300, 500, and 600 lbs and $2\frac{1}{2}$ and 5 tons.

In the low-pressure generator, the pressure is below 6 lbs; in the "medium-pressure" generators, it must not exceed 15 lbs per sq in. Carbide is fed a little at a time to a body of water. The volume of gas generated may be as low as 1 cubic foot an hour.

There are two kinds of welding torches, high pressure and low pressure, and they differ in important details in the internal mixer. Cutting torches differ from welding torches in the following way: in addition to the oxygen and acetylene conduits, which both torches have, the cutting torch receives a stream of oxygen around the flame; it is this oxygen which cuts the steel, by oxidizing it and forcing away the particles of oxide formed; the flame serves merely to attain the oxidizing temperature. The flame in each has an inner brilliant part, whose temperature is estimated to 3300°C; in the welding torch this is surrounded by a larger envelope into which the air penetrates. The inner portion, which does the welding, is sometimes called the neutral part. The use of acetylene is not without danger; the directions and cautions of the manufacturer must be observed.

The oxyacetylene torch has many uses besides the welding and cutting of steel. It serves for metal cladding, in certain special circumstances; and for steel conditioning, pressure welding, flame spinning, and flame hardening. In shaping synthetic sapphire and ruby (hexagonal crystals of alumina) the torch is in constant application. Synthetic rubies, for example, are obtainable in the form of slim rods and boules (balls); thc rods when heated in the torch may be bent, in the form of a thread guide, let us say, which is a complete loop. At the same time, the material acquires a flame finish of extreme smoothness. Ruby in rod form is made into precision gauges. It is also made into extrusion dies, phonograph needles, and knife edges on balances.

Economic petrochemical processes have been developed for the manufacture of acetylene and now account for an important percentage of the total acetylene production. (See Chapters 14, 25.)

Acetylene undergoes a number of reactions which are the bases of several large industries. Acetylene as a chemical substances is presented in Chapter 25.

CARBON DIOXIDE

Sources

Carbon dioxide is used commercially as a gas (soda ash manufacture), compressed as a liquid in steel cylinders (for soda fountains, for refrigeration, and as convenient source of the gas), and a solid. Production in 1979 totaled about 3.5 million tons.

It is obtained from (1) the combustion of coke; (2) the calcination of limestone; (3) as a by-product in syntheses involving carbon monoxide; (4) as a by-product in fermentations; (5) by the action of sulfuric acid on dolomite; (6) and from wells. Gas from any one of these sources may be made into the gaseous, liquid, or solid form of carbon dioxide.

The production of carbon dioxide from the combustion gases of coke involves the alternate formation and decomposition of alkali bicarbonates in solution. Hard coke is burned under boilers and the fuel gases are regulated in such a way that a maximum content of carbon dioxide, 16 to 17 percent, is obtained. The gases enter a scrubber (tower) packed with limestone, to remove sulfur compounds, and fed with water, to cool the gas and arrest the dust. The cold gases enter the absorber, a tower packed with coke down which a solution of potassium carbonate passes; carbon dioxide is absorbed, and the saturated solution is run to a boiler where the absorbed gas is liberated by heat. It is under the boiler that the coke is burned. The operation is continuous; charged solution flows in constantly, the spent liquor is run off constantly. By means of an intercharger, the outgoing liquor heats the incoming liquor to some extent. The outgoing liquor, cold, returns to the absorber. The gas from the boiler is very pure. It is dried in a calcium chloride tower, and compressed to 100 atmospheres, at which pressure it liquefies at ordinary temperatures (see Figure 17.14)

In addition to the uses which have been mentioned, carbon dioxide serves as a chemical in the manufacture of salicylic acid, white lead, and other products, as well as in fire-fighting devices of various kinds. Fire extinguishers of the wall type, also called the soda-acid type, contain $2\frac{1}{2}$ gal of saturated sodium bicarbonate solution and 4 oz of concentrated sulfuric acid in a small bottle. On inversion, the acid reaches the solution and liberates carbon dioxide; the pressure developed expels the liquid through a nozzle to a distance of 30 to 40 ft. The main

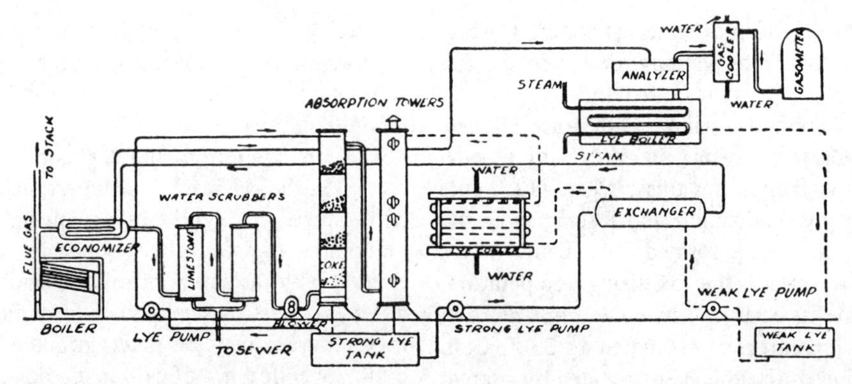

Fig. 17.14 Diagrammatic flow sheet for the abosprtion of carbon dioxide in the fire gases from burning coke. The gas dissolves in a strong lye solution from which it is driven out by heat, giving up 100% CO_2 gas. (Courtesy Frick Co, Waynesboro, Pa.)

extinguishing agent is the water. Liquid carbon dioxide under pressure in steel cylinders may be released so that a carbon dioxide snow forms which may be directed into the gaseous blanket over the fire. The "firefoam" extinguisher system relies upon the smothering action of a foam blanket produced by the interaction of a sodium bicarbonate solution with an alum solution, in the presence of a foam stabilizer.

Solid carbon dioxide is obtainable in commercial quantities. It is supplied in block form resembling the familiar artificial ice cake. Its uses are similar to the uses of ice, but it functions without melting, and without producing drips; it vaporizes, and leaves only a gas, which may be easily vented, so that it has received the rather apt name of "dry ice." Its manufacture will be described in terms of a particular plant.

Pure, liquid carbon dioxide under a pressure of 1000 lb and at a temperature of 70°F (21°C), is delivered to the plant by a pipe system. It is sent to the "evaporator" (Figure 17.15), where its pressure is reduced to 500 lb with a simultaneous drop in temperature to 32°F (0°C). With the pressure set at 500 pounds, the liquid maintains itself at that temperature; as this is lower by several degrees than the room temperature, heat flows in and causes the liquid to simmer quietly. About 25 percent of it boils away. The vaporized portion is sent to a special compressor which delivers it as gas to the main compressor gas line, at a pressure of 1000 lbs.

The 32°F liquid from the evaporator is admitted to the press chambers; these have movable tops and bottoms, worked by hydraulic pressure. The chamber is 20 by 20 in, and 24 to 30 in, deep. The liquid enters through an ordinary nozzle; part of it expands to gas and draws its heat largely from the incoming liquid, which is thus solidified to a fluffy snow. The gas formed is drawn off constantly by the suction line of the main compressors and recompressed. By operating the top and bottom walls, the snow is compacted to a solid block 20 by 20 by 10 in. Each press makes 6 to 8 cakes per hr. The density of the resulting cake is controlled by the amount of snow pressed into the 10-in space. After discharge to a conveyor, the block reaches band saws, which cut it into four smaller blocks, each a 10-in cube, weighing about 20 lbs. This is wrapped in brown paper and stacked in a specially insulated railway car for transportation to distant points, or into trucks for local delivery.

Of the liquid delivered to the press, 20 to 45 percent is solidified; the rest turns to gas and

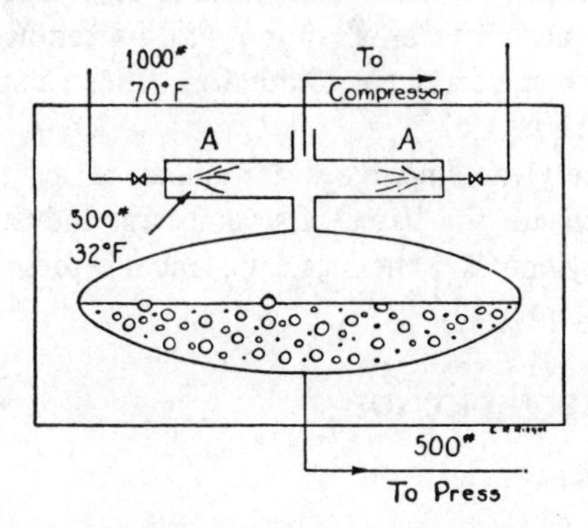

Fig. 17.15 The "evaporator" in which liquid carbon dioxide is formed and stored.

must be reliquefied. The colder the temperature of the liquid CO_2 and the colder the press chest, the higher the percentage frozen. Based on heat content, it is found that it takes 3.75 pounds of liquid to produce 1 pound of solid. The expansion in the chest is due to atmospheric pressure.

The critical temperature of carbon dioxide is 88°F (31.1°C), the critical pressure 1073 psi. At 70°F (21°C), it is considerably below the critical temperature, so that a pressure of 1000 to 1100 psi suffices to keep it in the liquid state.

It will be clear that much of the expense in the plant will be that for recirculating the carbon dioxide gasified at the presses. The compressors are four-stage machines: 0 to 5 lb, 65 to 70 lb, 300 to 325 lb, and 1000 to 1100 psi. From the last stage the gas enters oil-removing filters, then a condenser cooled with tap water, which reduces its temperature to about 70°F (12°C). In the condenser, the carbon dioxide liquefies, and enters the "evaporator" with the new liquid, at the same temperature and pressure. Carbon dioxide from any source may be made into the solid form.

The uses of carbon dioxide are as a refrigerant for the frozen food, dairy product, and meat packing industries; grinding of dyes and pigments; and in the manufacture of certain pharmaceuticals and chemicals.

SULFUR DIOXIDE

Of the more compressible gases, chlorine and ammonia are discussed elsewhere (Chapters 6 and 5, respectively). Sulfur dioxide, SO_2 anhydrous, liquefied under a moderate pressure (2 to 3 atm) at room temperature, is shipped in steel cylinders of 50- or 100-lb capacity, in 1-ton containers, and in single-unit 15-ton car tanks. It is used from such cylinders and tanks in preparing hydroxylamine sulfate, which in turn serves in making dimethylglyoxime, the nickel reagent; for refrigeration, for bleaching, and, increasingly, in petroleum refining. The boiling point is–10°C.

The burner gas from sulfur (or pyrite), freed from dust and cooled, is dissolved in water in two towers used in series; the solution from the second tower is elevated to the top of the first tower, where it meets the rich gas. Burner gas with 8 to 10 percent sulfur dioxide yields a 1 percent solution. In a third tower this solution is sprayed at the top, and flows down, while steam is injected at the base of the tower; previously the 1 percent liquor was heated in a closed coal laid in the spent liquor from the base of the still. The packing in all the towers may be coke, or special earthenware cylinders. The gas issuing from the third tower is cooled to remove most of its moisture, and is passed up a fourth tower down which concentrated sulfuric acid flows. The dried gas is compressed in a bronze pump to $2\frac{1}{2}$ atm, which suffices to liquefy it.

NITROUS OXIDE

Nitrous oxide (N_2O) is made by heating ammonium nitrate to 200°C in small lots (50 lbs) in aluminum retorts. The gas is cooled in a condenser, washed in a solution of sodium dichromate to remove nitric oxide, in caustic to absorb nitric acid, and in water. Under a pressure of 100 atm it liquefies, in small shipping cylinders, for instance; or it may be stored in a gas holder. The reaction is

$$NH_4NO_3 \longrightarrow N_2O + 2H_2O$$

Nitrous oxide is used as general anesthetic, usually mixed with oxygen, and sometimes with ether vapor.

REFERENCES

1. *Chem. Eng.*, 67, No. 15, 96–99 (1960).
2. Shaner, R. L., "Production of Industrial Gases From Air," Linde Company.
3. "Industrial Gases-Series MA28C(79)-1", U.S. Department of Commerce, Bureau of Census, Washington, D.C., 1980.
4. Kropschat, R. H., Birmingham, B. W., and Mann, D. B., "Technology of Liquid Helium," *Nat. Bur. Stand. U.S., Monograph*, III (1968).

5. Foster, Russel J., "Helium", *Minerals Yearbook 1977, Vol.* I, *Metals and Minerals*, pp. 469–476, U.S. Deparment of Interior, Bureau of Mines, U.S. Government Printing Office, Washington, D.C., 1980.
6. Neary, R. M., "Air Solidifying Cryogenics," American Society of Mechanical Engineers.
7. Baker, C. R. and Paul, R. S., "Purification of Liquefaction Grade Hydrogen," *Chem. Eng. Progr.*, 59, No. 8, 61 (1963), and U.S. Patent 3,073,093 (to Union Carbide).
8. Baker, C.R. and Matsch, L. S., "Production and Distribution of Liquid Hydrogen," in *Advances in Petroleum Chemistry and Refining*, Vol. X, New York, John Wiley & Sons.
9. "LNG Liquefaction Cycles," *Cryogenic and Industrial Gases*, 5, No. 8 (1970).
10. Prater, P. G., "Design Considerations of an LNG Peak Shaving Facility," *Cryogenic and Industrial Gases*, 5, No. 5 (1970).

Selected References

Handbook of Compressed Gases, Compressed Gas Association, New York, Van Nostrand Reinhold, 1981.
Booth, Norman, *Industrial Gases*, Pergamon Press, 1st edition, 1973.
Braker, William, and Mossman, Allen L., *Matheson Gas Data Book*, 5th edition, 1971, East Rutherford, NJ, Matheson Co.
Gibbs, Chas. W., *Compressed Air and Gas Data*, 2nd edition, Ingersoll-Rand, 1971, Woodcliff Lake, N.J.
Haselden, G. G., "Safety", in *Cryogenic Fundamentals*, Academic Press, London, 1971.
Neary, R. M., *Handling Cryogenic Fluids*, Nat. Fire Protect. Assoc. Quart. (Jul. 1960).
Ruhemann, M., "Separation of Gases," *International Series of Monographs of Physics*, Oxford University Press, 1945.
Schmidt, H. W. and Forney, D. E., "Oxygen Systems Engineering Review", in *ASRDI Oxygen Technology Survey, Vol. IX*, National Aeronautics and Space Administration, Washington, D.C., 1975.
Standard Density Data-Atmospheric Gases and Hydrogen, Compressed Gas Association, New York.

18

Industrial Fermentation

Arthur E. Humphrey* and S. Edward Lee**

EVOLUTION OF MODERN FERMENTATION PROCESSES

Man was well aware of fermentations, even though he had little knowledge of what caused them, long before he was able to record such an awareness. Sometime in pre-recorded history, man discovered that meat allowed to stand a few days was more pleasing to the taste than meat eaten soon after the kill. He also was aware that intoxicating drinks could be made from grains and fruits. The aging of meat and the manufacture of alcoholic beverages were man's first uses of fermentation.

Without even knowing that microorganisms existed, ancient man learned to put them to work. The ancient art of cheese-making involves fermentation of milk. For thousands of years, the soy sauces of China and Japan have been made from fermented beans. For centuries, the Balkan people have enjoyed fermented milk, or yogurt, and Central Asian tribesmen have found equal pleasure in sour camel's milk, or kumiss. Bread, which has been known almost as long as agriculture itself, involves a yeast fermentation. Loaves of bread have been found in Egyptian pyramids built 6,000 years ago.

The discovery of fruit fermentation was made so long ago that the ancient Greeks believed wine had been invented by the god Dionysus. The manufacture of beer is only slightly less ancient than that of wine. A Mesopotamian clay tablet written in Sumerian-Akkadian about 500 years before Christ tells us that brewing was a well-established profession 1,500 years earlier. An Assyrian tablet of 2000 B.C. lists beer among the commodities that Noah took aboard his ark. Egyptian documents dating back to the Fourth Dynasty, about 2500 B.C., describe the malting of barley and the fermentation of beer. Kui, a Chinese rice beer, has been traced back to 2300 B.C.

During the Middle Ages, experimenters learned how to improve the taste of wine, bread, beer, and cheese. Yet, after thousands of years of experience man still did not realize that in fermentations he was dealing with microorganisms. It wasn't until 1857 that Pasteur proved alcoholic fermentation was brought about by yeasts and that yeasts were living cells.

*Lehigh University
**American Cyanamid (Presently with Hoffmann LaRoche)

This discovery was a turning point in medical history and the birth of microbiology.

The inheritors of Pasteur's knowledge sought to use microbes as production workers in industry. The production of bakers' yeast in deep, aerated tanks was developed towards the end of the nineteenth century. During World War I, Chaim Weizmann used a bacterial cousin of the gas gangrene microbe to convert maize mash into acetone, which is essential in the manufacture of the explosive cordite. In 1923, Pfizer opened the world's first successful plant for citric acid fermentation. The process involved a fermentation utilizing the mold *Aspergillus niger* whereby ordinary sugar was transformed into citric acid.

Other industrial chemicals produced by fermentation were found subsequently, and the processes were reduced to commercial practice. These processes included butanol, acetic acid, oxalic acid, gluconic acid, fumaric acid, and many more.

In 1928 Alexander Fleming, working with *Staphylococcus aureus*, a bacterium that causes boils, observed a strange fact. A mold of the *Penicillium* family grew as a contaminant on a Petri dish inoculated with *Staphylococcus aureus*; a clear zone was observed where the *Staphylococcus* organism in the vicinity of the contaminating mold had been killed. Fleming nurtured the mold and then extracted a chemical from it which killed the bacteria. He named the extracted material penicillin.

Fleming's discovery received little notice until two Oxford University experimenters, under the stress of World War II, resolved to find an antibacterial agent of wider activity than the sulfa drugs. These two British workers, Dr. Howard Florey and Dr. Ernst Chain, set out to produce a therapeutic agent capable of saving the lives of war casualties. Their first candidate was the *Penicillium notatum* mold preserved from Fleming's studies. Penicillin turned out to be exactly what they were looking for. The American pharmaceutical industry helped them solve their difficulties in mass-producing the antibiotic. Three American companies led the way - Merck, Pfizer, and Squibb.

Initially, *Penicillium* was surface cultured in flasks. A chance discovery in a Peoria market provided the major breakthrough. There, a government worker found a moldy cantaloupe on which was growing a new strain of penicillium, *Penicillium chrysogenum*, which would thrive when cultured in deep, aerated tanks and which gave 200 times more penicillin than did Fleming's mold.

Other antibiotics were quick to appear. Streptomycin was next. This antibiotic was particularly effective against the causative organism of tuberculosis. The search was now on. Antibiotic prospectors combed the earth for organisms that produced different and more useful antibiotics. The list of these antibiotics is long today and include such important antibiotics as chloramphenicol, the tetracyclines, bacitracin, erythromycin, novobiocin, nystatin, kanamycin, and many others.

Future Developments

Progress in fermentation is continuing at an ever-increasing pace. Each year, new products are added to the list of compounds derived from fermentation. Several vitamins are now produced routinely, employing fermentation steps in their synthesis. Outstanding examples are B-2 (riboflavin) B-12 (cyanocobalamin), and C (ascorbic acid). Some of the more interesting fermentation processes are the specific dehydrogenations and hydroxylations of the steroid nucleus. These chemical transformations are economical short cuts used in the manufacture of the anti-arthritic cortisone and its derivatives. Fermentative syntheses of the amino acids L-lysine and L-glutamic acid are also being carried out commercially. Important agricultural uses are found for the new fermentation product gibberellin, a plant-growth regulator; and crystalline inclusions for a species of *Bacillus* are being used as specific insecticides in another agricultural application. Microbial hydrolysis of cellulose and subsequent fermentation of the resultant sugar to alcohol is a promising source of liquid fuel. Research is in progress on chemical transformations utilizing fermentation techniques, new fermentative biosyntheses, continuous algal culture, and submerged mammalian-tissue culture. Fermentation processes may not only be tomorrow's source of chemotherapeutical agents, but may very well be the manner in which food and liquid fuel is produced. The future for fermentation is indeed bright.

FERMENTATION FUNDAMENTALS

Introduction

The engineer's contribution to the development of the penicillin fermentation was a very important one. The outgrowth of the undertaking was the pure-culture technique, carried out in aerated and agitated deep-tank fermentors. This technique, similar to its antecedent used for yeast propagation, introduced to the biochemical-process-industry refined fermentation equipment capable of being maintained under aseptic conditions even when vigorously aerated. The technique has now been applied widely with minor modifications to the production of other antibiotics, amino acids, steroids, and enzymes.

Of the operations auxiliary to those in the fermentor, engineers have made a major contribution in establishing the theoretical bases for the design of equipment to provide large volumes of sterile medium and air. Close cooperation between biologists and engineers is needed to devise logical methods for screening large numbers of strains, and translating the results of shake-flask and pilot-plant experiments to production vessels.

Fermentation Unit Operations

Fermentation processes have in common many of the familiar chemical engineering unit operations. For example, aerobic fermentations involve the "mixing" of three heterogeneous phases - microorganisms, medium and air. Other unit operations include "mass transfer" of oxygen from the air to the organisms and "heat transfer" from the fermentation medium.

Analysis of fermentations by the unit-operation technique has added greatly to the understanding of their behavior. This understanding, however, is far from complete. The scale-up of fermentations for instance, is still rather empirical although the sensitive oxygen probes and sensitive gas analysis techniques now available have enabled a more rational approach to scaling-up aerated, non-Newtonian fermentations.

Fermentation Unit Processes

Analysis of the many industrial fermentation processes shows that there are common reactions from a chemical as well as a physical viewpoint. Fermentation processes can be classified by the reaction mechanisms involved in converting the raw materials into products; these include reductions, simple and complex oxidations, substrate conversions, transformations, hydrolyses, polymerizations, complex biosyntheses, and the formation of cells.

Unit-process classification provides a ready catalogue of the chemical activities and abilities of microorganisms for the biochemist. More importantly, it offers a logical approach to an examination of fermentation reaction mechanisms.

Microorganisms

Microorganisms are chemically similar to higher plant and animal cells. They perform many of the same biochemical reactions. Generally, microorganisms exist as single cells. They have much simpler nutrient requirements than the higher life forms. Their requirements for growth are usually limited to air, carbon dioxide and inorganic salts or inorganic salts supplemented by simple sugars. They occur in four main groups:

1. Bacteria
2. Viruses*
3. Fungi, including yeast and actinomyces
4. Protozoa and algae

Typically, *bacteria* are single cells, either cocci, rods, or spirals which are 0.5 to 20 μm in size and capable of independent growth. They multiply by simple division. A typical bacterial cell is illustrated in Figure 18.1.

Viruses are the smallest microbes. They are

*Not a microbial cell by strict definition.

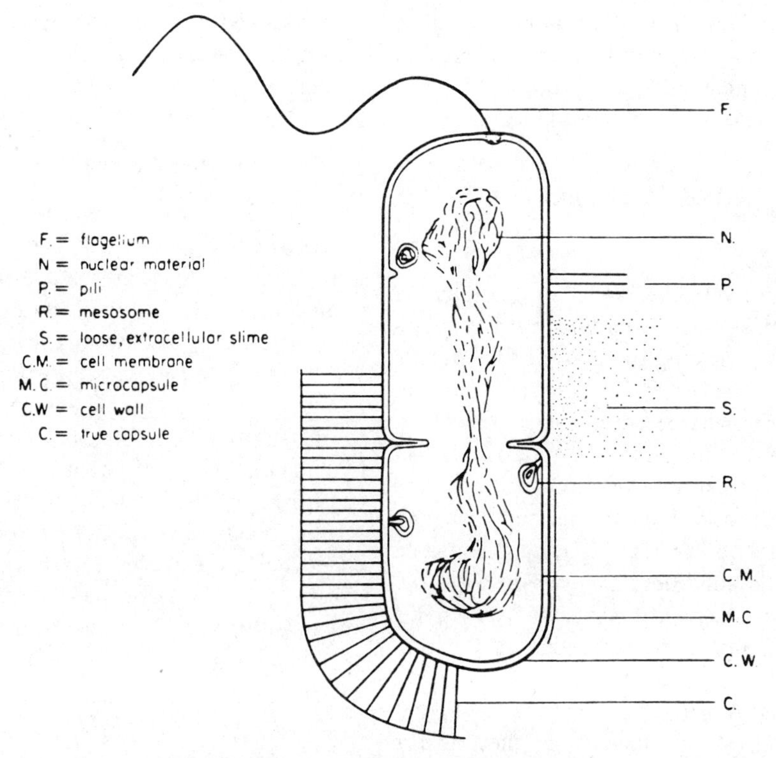

Fig. 18.1 Detailed structure of a bacterial cell.

obligate intracellular parasites of animals, plants, insects, fungi, algae, or bacteria. Bacterial viruses are called bacteriophage. Growth and multiplication take place intracellularly when the virus invades the host cell, taking over the genetic mechanism to direct the host cell to synthesize new viruses. The viruses consist primarily of genetic material, either ribonucleic acid (RNA) or deoxyribonucleic acid (DNA) and a protein coating. Because of the purity of the protein coat, it is possible to crystallize viruses and obtain X-ray diffraction patterns.

The *fungi* are widely spread in nature in environments of lower relative humidity than those which favor bacteria. The metabolism of the fungi is essentially aerobic. They generally form long filamentous nucleated cells 4 to 20 μm wide which are highly branched. Many species have complicated life cycles during which both sexual and asexual spores are formed. These spores are usually quite resistant to environmental changes.

Generally, fungi are free-living saprophytes, but a few are parasites to animal cells and many are serious pathogens of plants. They have very wide degradation and synthetic capabilities and have proved a fruitful source of industrially important organic acids, antibiotics, and enzymes.

Among the fungi are yeasts which are elliptical cells 3 to 5 μm by 8 to 15 μm. They differ from most fungi in that growth is by budding. Growth of yeast on solid medium resembles a bacterial colony. Yeasts are industrially important in making alcohol and bakers' yeast.

Actinomycetes are a group of organisms intermediate in properties between bacteria and true fungi. Industrially, the group is extremely important as a source of powerful antibiotics for the control of microbial infections of man, animals, and plants. Their growth in fermentation tanks is very similar to that of fungal fermentations; for this reason, they have been grouped with the fungi in this discussion.

The *protozoa* are widely distributed in fresh and salt water, and in lesser frequency in soil and animals. They may be unicellular or multicellular and exhibit a wide variety of morphological forms. Usually, they are divided in two main groups - the *algae* which are all capable of

TABLE 18.1 Chemical Analyses, Dry Weights, and the Populations of Different Microorganisms Obtained in Culture*

Organism	Composition (% dry weight) Protein	Nucleic Acid	Lipid	Polulation in Culture (numbers/ml)	Dry Weight of This Culture (g/100 ml)	Comments
Viruses	50–90	5–50	<1	10^8–10^9	0.0005[a]	Viruses with a lipoprotein sheath may contain 25% lipid
Bacteria	40–70	13–34	10–15	2×10^8–2×10^{11}	0.02–2.9	*Mycobacterium* may contain 30% lipid
Filamentous fungi	10–25	1–3	2–7		3–5	Some *Aspergillus* and *Penicillium* sp. contain 50% lipid
Yeast	40–50	4–10	1–6	$1\text{–}4 \times 10^8$	1–5	Some *Rhodotorula* and *Candida* sp. contain 50% lipid
Small unicellular algae	10–60 (50)	1–5 (3)	4–80 (10)	$4\text{–}8 \times 10^7$	0.4–0.9	Figure ( ) is a commonly found value but the composition varies with the growth conditions

*S. Aiba, A. E. Humphrey and N. Millis, *Biochemical Engineering*, 2nd ed., p. 26, University of Tokyo Press, Tokyo (1973)
[a]For a virus of 200 μm diameter.

photosynthesis and contain chlorophyll, and the protozoa which are not photosynthetic and resemble primitive animal cells.

The chemical composition of microorganisms can be quite varied depending upon such factors as the composition of growth medium, the age of the culture, and the cell growth rate. Table 18.1 lists some of the compositions of different microbial forms.

Microbial Activity

All organisms contain the genetic information to produce a wide variety of enzymes and hence produce a great number of chemicals. However, only some enzymes are produced at all times, while others are greatly influenced by the substrate. Certain compounds interact with the substrate to repress the translation of genetic information for synthesis. (This is called repression.) Also, the substrate sometimes reacts with a compound that is a genetic-mechanism repressor and removes its action. (This is referred to as derepression). This control is depicted in Fig. 18.2.

These processes allow cells to regulate their enzyme content in direct response to the environment. They prevent the formation of excess end product and superfluous enzymes.

Mutants can be genetically engineered that lack these controls. Such mutants have been changed so that the genetic mechanism no

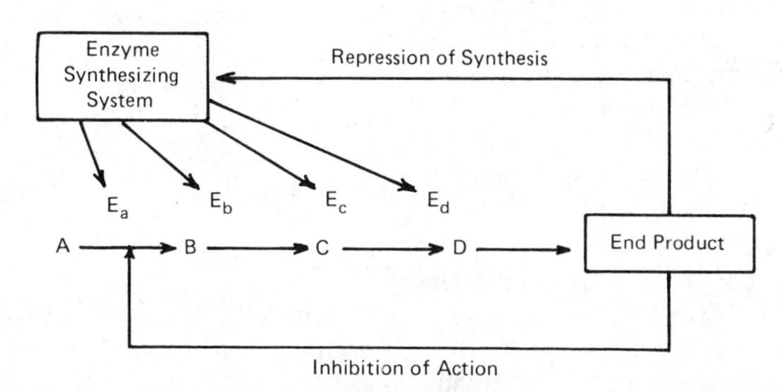

Fig. 18.2 System for the control of enzyme formation and action in biosynthesis.

longer is sensitive to a particular controlling metabolite. For industrial processes, such strains with faulty regulation, altered permeability, or metabolic deficiencies may be used to accumulate products.

Strain Development

Although in a normal bacterial population about one mutant arises in 10^6 cells, the rate of mutation is far higher than this since many mutations are lethal and so are not detected, while others involve base changes that do not change the amino acid read from the m-RNA codon. This low rate of spontaneous mutation is unsatisfactory when attempting strain improvement. However, mutagens are available, that markedly increase the rate of mutation. Recently, tremendous strides have been made in genetic engineering which allow the creation of mutants through such techniques as

1. transduction
2. transformation
3. protoplast fusion

We now know that exchange of genetic information between cells is much more common than originally thought. The trick is not only to inject genetic material with the desired characteristics into a cell but to insure that it is "expressed" and that an environment is maintained that permits these "new species" to dominate.

Although it is relatively easy to produce mutants, strain improvement requires painstaking effort, ingenuity in devising screening tests, and luck. The difficult problem is to select desirable mutants and this is accentuated when the parent strain is already a high producer. In this situation, any improvement is likely to be minor and quantitative rather than major and qualitative; major changes are most useful in the very first stages of a development program and in basic research.

Improvement in product formation can be achieved in a number of ways. Mutants can be selected which have the following properties:

1. Defective enzymes in a biosynthetic pathway so that useful intermediates accumulate.
2. Lack of feedback control by the end product itself or by a key intermediate on the pathway to the end product.
3. Defective cell membranes so that intracellular metabolites are excreted rapidly.
4. Lack of some undesirable by-product, such as a pigment.
5. Reduced toxicity to a precursor molecule such as phenyl acetate for penicillin production.

Stock Cultures

In all the industrially important bacteria and fungi, except some yeasts, the main vegetative phase of growth is haploid, so that variation in genetic material arises only from mutations, provided transduction does not occur in bacteria, and provided sexual spores are not formed in fungi. In maintaining stock cultures, genetic change must be minimized; this is best achieved by preventing nuclear divisions, since most mutations occur as errors in DNA replication.

The method of choice is to store cells or spores (if these are produced) in sealed ampoules at very low temperatures (-200°C) in liquid nitrogen. This method has the great advantage that the culture can be stored almost indefinitely, thawed and used immediately as an inoculum without loss of viability or diminution in metabolic rate. Cultures kept at -20° to -60°C are satisfactory but less active than those kept in liquid nitrogen. Although storage at 0° to 4°C allows some growth, this is better than storage at room temperature. Lyophilization (freeze-drying) is widely used and is very convenient since freeze-dried cultures retain viability without genetic change for years when stored at room temperature. It may be noted that all of these methods are, in effect, techniques to immobilize intracellular water and yet retain viability.

Microbial Kinetics

The kinetics of microbial systems may be expressed at four different system levels. These include

1. molecular or enzyme level
2. macromolecular or cellular component level
3. cellular level
4. population level

Each level of expression has a unique characteristic that leads to a rather specific kinetic treatment. For example, biological reactions at the molecular level invariabily involve enzyme-catalyzed reactions. These reactions, when they occur in solution, behave in a manner similar to homogeneous catalyzed chemical reactions. However, enzymes can be attached to inert solid supports or contained within a solid cell structure. In this case, the kinetics are similar to those for heterogeneous catalyzed chemical reactions.

Enzyme Kinetics. In their simplest form, enzyme-catalyzed reactions, occurring in a well-mixed solution, are characterized by the well-known Michaelis-Menten kinetic expression. This relationship depicts the substrate, S, combining reversibly with the enzyme, E, to form an enzyme-substrate complex, ES, that can irreversibly decompose to the product and the enzyme, i.e.

$$E + S \underset{k_{-1}}{\overset{k_{+1}}{\rightleftharpoons}} ES \xrightarrow{k_{+2}} E + P \tag{18.1}$$

This leads to a kinetic expression for the velocity of the reaction, ν, of the following form:

$$\nu = \frac{\nu_{max} S}{K_M + S} = \frac{k_{+2} E_0 S}{\dfrac{k_{-1} + k_{+2}}{k_{+1}} + S} \tag{18.2}$$

where $\nu_{max} = k_{+2} E_0$ is the maximum observable reaction rate at high substrate concentration and, hence, is only limited by the initial enzyme concentration, E_0. K_M is the dissociation constant. When

$$k_{+2} << k_{-1}, \text{ then } K_M = K_s \tag{18.3}$$

and

$$ES \xrightarrow{k_2} E + P \text{ is limiting} \tag{18.4}$$

The saturation constant, $K_s = k_{-1}/k_{+1}$, is an indication of the affinity of the enzyme active site for the substrate.

Basically two kinds of catalytic poisoning or inhibition are considered. These include inhibition by competition for the active site by a nonreactive substrate and inhibition by a substance that modifies the enzyme activity but does not compete for the active site. The competitive inhibitor adversely affects the binding of substrate and enzyme, and thus has an effect in increasing K_M. The noncompetitive inhibitor, on the other hand, only affects k_{+2} and ν_{max}.

Cell Growth Kinetics. Enzyme kinetic concepts have been utilized to express the kinetic behavior of cell growth on a single limiting substrate. Monod postulated that the growth of cells by binary fission on a single limiting substrate probably had a single limiting reaction step and therefore behaved in a manner analogous to the Michaelis-Menten enzyme kinetics, i.e.

$$\frac{dX}{dt} = \mu_{max} X \frac{S}{K_s + S} \tag{18.5}$$

where X = the cell concentration, S = the growth limiting substrate concentration, t = time, and μ_{max} = the maximum growth rate. Since the growth rate of cells, increasing by binary fission, is defined by

$$\mu = \frac{1}{X} \frac{dX}{dt} \tag{18.6}$$

The growth kinetics can be expressed as

$$\mu = \mu_{max} \left(\frac{S}{K_s + S} \right) \tag{18.7}$$

For a complete theory of cell growth and substrate utilization, it is necessary to know the relationship between the growth of cells and the utilization of substrate.

This relationship is expressed as a yield, Y, defined as

$$Y \equiv \frac{dX}{dS} \tag{18.8}$$

The simplest assumption is that Y is constant. That this is essentially true only at high growth rates will be shown later. From the kinetic equation and the yield expression one obtains the following relationship for substrate utilization:

$$\frac{dS}{dt} = \frac{1}{Y_G}\frac{dX}{dt} + mX \tag{18.9}$$

or

$$\frac{1}{X}\frac{dS}{dt} = \frac{1}{Y_G}\mu + m \tag{18.10}$$

where m is the maintenance requirement for substrate per unit of cell biomass per unit of time and Y_G is a true yield constant representing the substrate utilized only for growth.

Recently it has been suggested that the rate of cell increase should be expressed as a net growth rate which involves both growth, μ, and death, δ, i.e.

$$\frac{dx}{dt} = \mu X - \delta X \tag{18.11}$$

where

$$\mu = \mu_{max}\left(\frac{S}{K_s + S}\right) \tag{18.12}$$

and

$$\delta = \delta_{max}\left(1 - \frac{S}{K_s' + S}\right) \tag{18.13}$$

Here, cell death is depicted as having a first-order kinetic behavior and as maximal, i.e. δ_{max}, when the growth limiting substrate is zero, and minimal when the substrate is in excess.

Over the years, numerous models for depicting cell growth have evolved. Several will be discussed here. One such model is that for growth under multiple substrate limitation. It can be expressed as

$$\mu = \mu_{max}\left(\frac{S_1}{K_{S_1} + S_1}\right)\left(\frac{S_2}{K_{S_2} + S_2}\right)\cdots \tag{18.14}$$

Multiple substrate limitations frequently occurs in batch growth systems.

Another model for cell growth is that for situations of extremely low substrate concentrations, i.e. $S < K_s$. In this case, the growth kinetic equation reduces to

$$\mu = KS \tag{18.15}$$

where K is constant and approximately equal to μ_{max}/K_s. This kinetic behavior is frequently observed in large single reactor waste treatment systems.

Another situation that commonly occurs in waste treatment is growth under conditions of "shock loading," i.e. conditions in which the substrate rises to a level in which it becomes inhibitory to growth. This situation is frequently modelled by waste treatment designers by the following kind of kinetic expression,

$$\mu = \frac{\mu_{max}}{1 + \frac{K_s}{S} + \frac{S}{K_I}} \tag{18.16}$$

The three basic growth relationships are depicted in Figure 18.3. In passing, it is interesting to note that the observed μ_{max} under conditions of substrate inhibition is a fraction of the true μ_{max} if there were no inhibition, i.e.

$$\frac{\mu \text{ at } d\mu/dS = 0}{\mu_{max}} = \frac{1}{1 + 2\sqrt{K_s/K_I}} \tag{18.17}$$

Another common kinetic expression for growth is one which takes into account a substrate diffusional limitation. It has been observed in various waste treatment systems that the K_s for "floc" or sludge growth was greater than that for single cell systems. Further, the growth rate of flocs or sludges appears to be a function of their size. This situation is depicted in Figure 18.4. The growth kinetics in this case

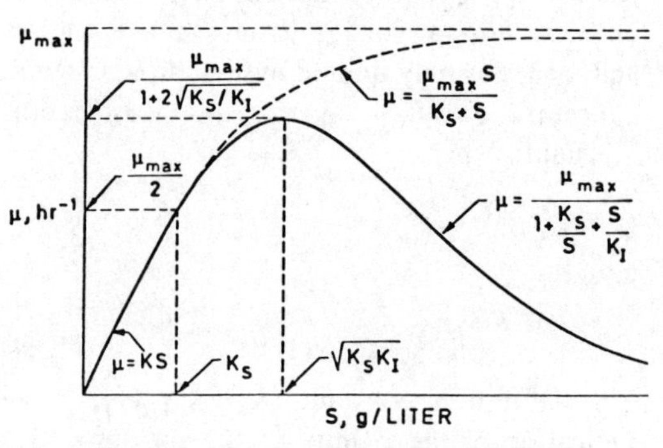

Fig. 18.3 Kinetic expressions for growth.

can be expressed as

$$\mu = \frac{\mu_{max} S}{K_s^{app} + S} \quad (18.18)$$

and

$$K_s^{app} = K_s + K_D \quad (18.19)$$

where $K_D = f$ (sludge or cell size, membrane permeability, diffusion coefficient, etc.). When $K_D = 0$ there is no diffusional limitation to substrate uptake.

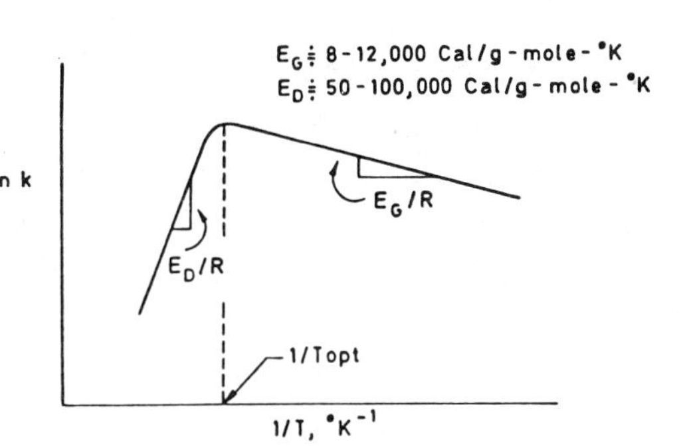

Fig. 18.5 Effect of temperature on growth.

Temperature Effects. Absolute reaction rate theory has been found applicable to both cell growth and death, i.e.

$$\frac{d\ell nk}{dT} = \frac{E_a}{RT^2} \quad (18.20)$$

or

$$d\ell nk = -(E_a/R)d(1/T) \quad (18.21)$$

where k is the rate constant for growth, i.e. μ_{max} or death, i.e. δ_{max}, T is the absolute temperature, E_a is the activation energy for the process, and R is the gas law constant. Generally speaking, E_a for growth, E_G, is the order of 8-12,000 cal/g-mole, °K and E_a for death, E_D, is the order of 50-100,000 cal/g-mole, °K. This behavior leads to an optimal temperature for growth for a given cell species. This is depicted in Figure 18.5.

pH Effects. The model for pH effect on growth has been based on the behavior of enzymes. Since enzymes are composed of amino acids, they exhibit "zwitterion" behavior, i.e. they have an acid, base, and neutral form. Cells have been depicted to have an active or neutral form and an inactive base or acid form, i.e.

$$X^+ \underset{K_1}{\rightleftharpoons} X \underset{K_2}{\rightleftharpoons} X^- \quad (18.22)$$

where K_1 and K_2 are the equilibrium constants of the above reaction. If one defines the growing or active cell fractions as y, i.e.

$$y \equiv \text{active cell fraction} = \frac{X}{X_0} \quad (18.23)$$

where

$$X_0 = X^+ + X + X^- \quad (18.24)$$

By analogy to enzyme behavior

$$y = \frac{1}{1 + \frac{[H^+]}{K_1} + \frac{K_2}{[H^+]}} \quad (18.25)$$

where $[H^+]$ is hydrogen ion concentration. Figure 18.6 depicts the effect of pH on the active cell fraction.

For most cell systems, $pH_{opt} = 6.5 \pm 1$ and $pK_2 - pK_1 = 2 \pm 1$.

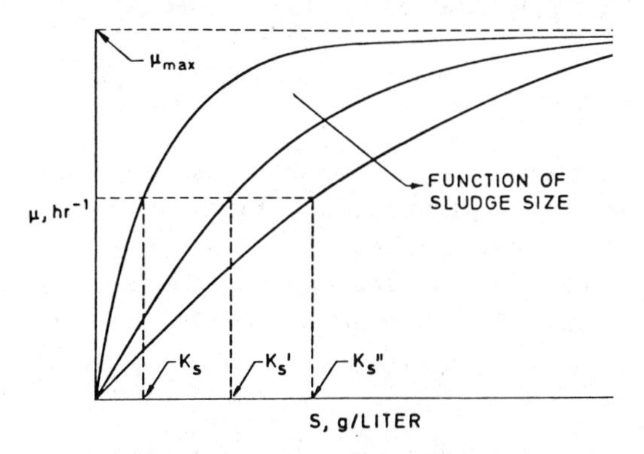

Fig. 18.4 Effect of sludge size on growth rate.

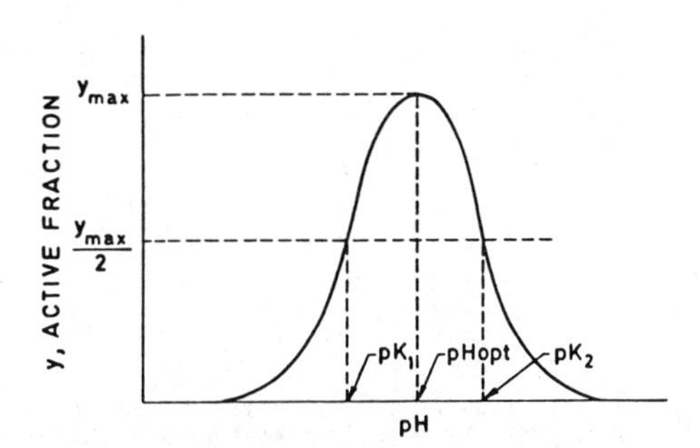

Fig. 18.6 Effect of pH on growth.

Fermentor Configurations

Numerous fermentor configurations may be found in microbial processes. These include:

- Batch
- Batch-fed
- Repeated draw-off
- Continuous
  - single stage
  - multiple stage
- Continuous with recycle
- Continuous with step feeding

For the most part, the antibiotic industry uses batch-type processes. The reason for this stems from the fact that most efficient antibiotic producing organisms are highly mutated and are readily replaced by fast growing, less efficient antibiotic producers in a continuous culture. In order to avoid substrate repression or inhibition, some batch processes are continuously fed concentrated substrate on demand during the course of the batch cycle. This is referred to as a "batch-fed" fermentation. The production of bakers' yeast is an example of a batch-fed process. In some highly mycelial antibiotic fermentation, 20 to 40 percent draw-off followed by fresh media make-up is practiced. In the trade, this is referred to as a "repeated draw-off" process. Strict continuous processes are only practiced in processes for the production of biomass for feed or food and the treatment of wastes. Continuous biomass producing systems are usually single-stage reactors without recycle. Waste treatment systems usually use multistage systems with recycle. They frequently use step feeding of several stages to improve system stability.

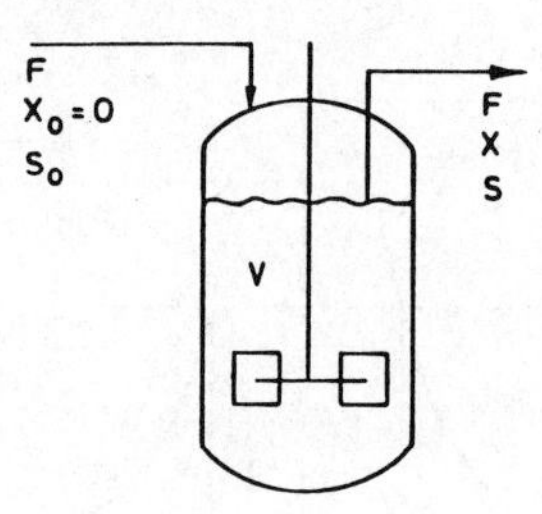

F = flow rate, liters/hr
V = volume, liters
D ≡ F/V = dilution rate, hr^{-1}
X = cell conc., g/liter
S = substrate conc., g/liter

Fig. 18.7 Typical single stage continuous fermentor.

A typical single-stage continuous fermentor is depicted in Figure 18.7.

A biomass material balance around the reactor yields the following relationship. At steady state

$$\frac{dX}{dt} = 0 \text{ (steady state)} = \frac{F}{V}(0 - X) + \mu X \tag{18.26}$$

from which

$$\mu = \frac{F}{V} \equiv D \tag{18.27}$$

This means that the dilution rate or the nominal residence time of the fermentor sets the growth rate of the biomass in the fermentor. A change in the dilution rate causes a change in the growth rate.

A similar balance around the fermentor at steady state for the growth limiting substrate, assuming the overall yield, Y, is constant, gives

$$\frac{dS}{dt} = 0 \text{ (steady state)} = \frac{F}{V}(S_0 - S) - \mu\frac{X}{Y} \tag{18.28}$$

from which

$$(S_0 - S) = \frac{X}{Y} \tag{18.29}$$

Substituting D for μ in Monod growth kinetics yields

$$S = \frac{K_s D}{\mu_{max} - D} \tag{18.30}$$

and substituting $S_0 - X/Y$ for S gives

$$X = Y\left(S_0 - \frac{K_s D}{\mu_{max} - D}\right) \tag{18.31}$$

Strictly speaking, Y is not a constant, particularly at conditions of low growth rate because the maintenance requirement, m, of the biomass utilizes substrate without producing biomass. The steady-state substrate balance equation for a single stage fermenter can be restated as

$$D(S_0 - S) = \frac{\mu X}{Y_G} + mX \tag{18.32}$$

TABLE 18.2 Effect of Maintenance on Overall Cell Yield in Continuous Culture

		Cell Yield, Y, g/g	
D,hr^{-1}	*m=0.01*	*m=0.02*	*m=0.05*
0.01	0.333	0.250	0.143
0.02	0.400	0.333	0.222
0.05	0.455	0.416	0.333
0.10	0.476	0.455	0.400
0.20	0.488	0.476	0.444
0.50	0.495	0.490	0.476

Note: $Y_G = 0.5$

Table 18.2 illustrates the effect of maintenance and cell growth or reactor dilution rate on the overall cell yield.

Since most single stage continuous fermentors are used to produce biomass, they are usually operated to optimize the biomass productivity. The unit volume biomass productivity of such a reactor is defined as DX. This unit volume productivity can be expressed as

$$DX = DY\left(S_0 - \frac{K_s D}{\mu_{max} - D}\right) \quad (18.33)$$

By taking the first derivative of the productivity expression with respect to the dilution rate and setting is equal to zero, the dilution rate of maximum productivity, D_m, can be found as

$$D_m = \mu_{max}[1 - \sqrt{K_s/(K_s + S_0)}] \quad (18.34)$$

and the maximum productivity, D_mX, as

$$D_mX = \mu_{max}[1 - \sqrt{K_s/(K_s + S_0)}] \cdot Y[S_0 - K_sD/(\mu_{max} - D)] \quad (18.35)$$

The behavior of these relationships is depicted in Figure 18.8. Note that for this system "wash out" of the biomass occurs when D approaches the maximum growth rate, i.e. $D \rightarrow \mu_{max}$.

Biomass recycle is frequently used in fermentors as a way of increasing the unit productivity. In order for the system to operate successfully, a biomass concentrator must be connected to the unit (See Figure 18.9).

Usually a centrifuge or settling tank is used as the concentrator. The increased productivity achieved with a recycle fermentor depends upon the recycle ratio, r, and the cell concentration factor, $C = X_r/X$, achieved in the concentrator. Equations expressing the recycle system behavior are derived from material balances around the reactor, i.e. for the cell biomass balance at steady state

$$\frac{dX}{dt} = 0 = \frac{F}{V}(0) + \frac{rF}{V}X_r - \frac{F}{V}(1 + r)X + \mu X \quad (18.36)$$

from which

$$\mu = D\left(1 + r - r\frac{X_r}{X}\right) = D(1 + r - rC) \quad (18.37)$$

Since $(1 + r - rC) < 1$, it is possible to operate the system at dilution rates greater than the maximum growth rate. Utilizing the Monod equation, the following expressions for the growth limiting substrate, S, system effluent biomass concentration, X_e, and biomass con-

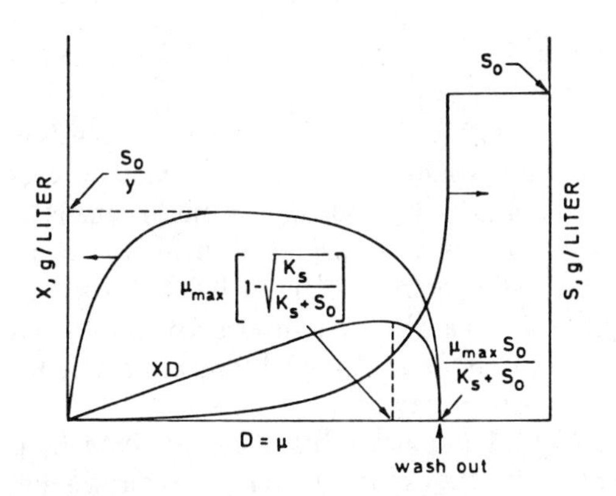

Fig. 18.8 Relationships for a single stage continuous fermentor.

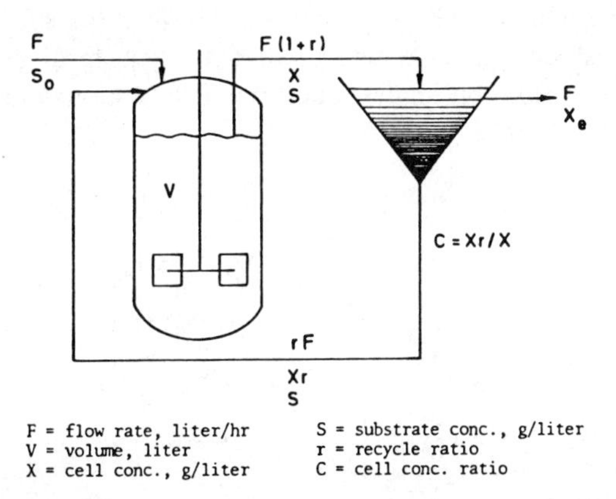

Fig. 18.9 Single stage continuous fermentor with recycle.

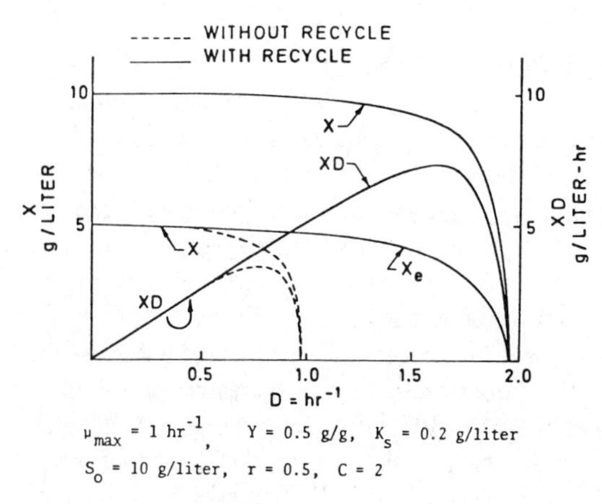

Fig. 18.10 Relationships for a single stage continuous fermentor with recycle.

centration in the exit stream from the bioreactor, X, are obtained:

$$S = \frac{K_s D(1 + r - rC)}{\mu_{max} - D(1 + r - rC)} \qquad (18.38)$$

$$X_e = Y(S_0 - S) \qquad (18.39)$$

$$X = \frac{Y(S_0 - S)}{(1 + r - rC)} \qquad (18.40)$$

These relationships are illustrated in Figure 18.10 for a particular set of conditions.

Another continuous fermentor system encountered in the microbial world is the multistage system with step feeding. This configuration is depicted in Figure 18.11.

In this system, fresh feed is step fed, i.e. fed to all reactor stages. This configuration is used in waste treatment systems to provide greater stability and to minimize the effects of substrate "shock loading" or inhibition through distribution of the concentrated feed along the

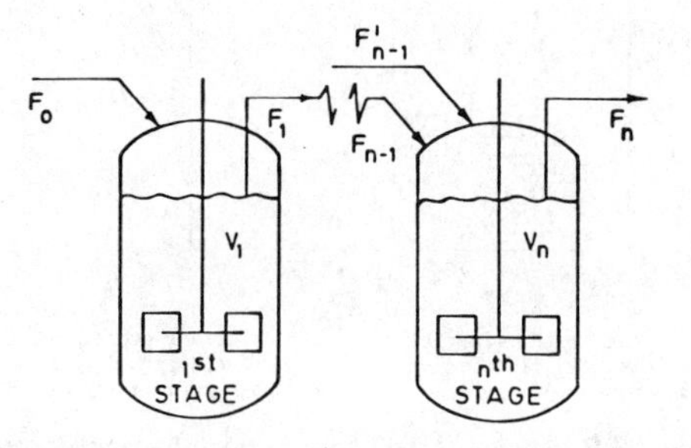

Fig. 18.11 Multistage continuous fermentor with step feeding.

system. A two-stage step feeding system is also used as a research tool to look at the inhibitory effects of high substrate loading which can't be achieved in the single stage without wash out. The material balance relationships for step feeding systems are straightforward. What is usually troublesome is the selection of the proper kinetic expression for high substrate conditions.

Oxygen Transfer in Fermentation Systems

In aerobic fermentations, oxygen is a basic substrate that must be supplied for growth. As in enzymatic reactions, the relationship between oxygen concentration and growth is of a Michaelis-Menton type. The specific rate at which cells respire (Q_{O_2}) increases rapidly with an increase in the dissolved oxygen concentrations (C) up to C_{crit}, which typically ranges from 0.5 to 2.0 ppm for well-dispersed bacteria, yeast, and fungi growing at 20 to 30°C. Beyond C_{crit}, the specific oxygen uptake increases only slightly with increasing oxygen concentrations.

If Q_{O_2} is plotted against the specific growth rate (μ) of the microbe, a linear correlation is obtained. An intersection of this straight line with the ordinate is designed as $(Q_{O_2})_m$, which represents the oxygen required for cellular maintenance.

In mathematical terms, this relation is given by

$$Q_{O_2}X = (Q_{O_2})_m X + (1/Y_{X/O_2})(dX/dt) \qquad (18.41)$$

where Y_{X/O_2} is the yield based on cell mass and t is the time.

This type of correlation can be applied to almost any substrate involved in cellular energy metabolism. It is justified from an energetic point of view and is supported by experimental steady-state data. Care must be used in applying this equation to transient conditions. It can only be used for situations at or near the steady-state equilibrium conditions.

When a product other than cell biomass is involved, the situation can be considerably more complex. Oxygen can be utilized for maintenance and growth as well as product formation. The simplest possible expression to represent the rate of oxygen utilization in this case is

$$Q_{O_2}X = (Q_{O_2})_m X + (1/Y_{X/O_2})(dX/dt) + (1/Y_{P/O_2})(dP/dt) \quad (18.42)$$

where P is the product concentration.

Of course, the oxygen transfer rate under steady-state conditions must be equal to the oxygen uptake rate, i.e.,

$$Q_{O_2}X = k_L a(C^* - C)_{mean} \quad (18.43)$$

where C^* = concentration of oxygen in the liquid that would be in equilibrium with the gas-bubble concentration and $k_L a$ = the volumetric oxygen transfer rate.

Utilizing this relationship, the volumetric oxygen transfer coefficient can be estimated, i.e.,

$$k_L a = (Q_{O_2}X)/(C^* - C)_{mean} \quad (18.44)$$

In practice, the dissolved oxygen concentration (C) is monitored by a membrane-covered probe. When the fermentation is equipped with a fast-responding probe (i.e., the response time is less then 6 s), then the fermentor can be used as a respirameter by making dynamic measurements of the oxygen concentration under aeration and nonaeration conditions. This is illustrated in Fig. 18.12.

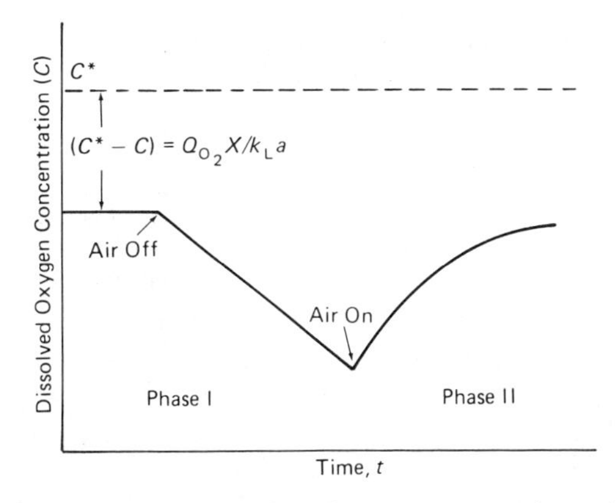

Fig. 18.12 Determination of oxygen consumption and mass transfer rate by dynamic techniques.

The following equations represent these conditions. Air off, Phase I:

$$dC/dt = -Q_{O_2}X \quad (18.45a)$$

Air on, Phase II;

$$dC/dt = K_L a(C^* - C)_{mean} - Q_{O_2}X \quad (18.45b)$$

Scale-Up

Many fermentation companies have both the problem of scaling-up new fermentations as well as the translation of process-improvement data for well-established fermentations from laboratory operations to existing plant equipment. In general, fermentations are scaled-up on the basis of achieving similar oxygen transfer capabilities in the plant equipment that proved to be optimal on the bench scale.

Fermentation biomass productivities usually range from 2 to 5 g/L/hr. This represents an oxygen demand in the range of 1.5 to 4 g O_2/L/hr. In a 500 m^3 fermentor this means achievement of a volumetric oxygen transfer coefficient in the range of 250 to 400 hr^{-1}. Such O_2 transfer capabilities can be achieved with aeration rates of the order of 0.5 VVM and mechanical agitation power inputs of 3.3 to 4.4 hp/m^3 (1.2 to 1.6 hp/100 gal).

On scale-up, however, it is usually heat removal that causes design problems. With the above mechanical agitation power inputs, between 40 and 50 Btu/gal/hr of turbulent heat is generated. This, coupled with a peak metabolic heat release from the growing biomass of 120 to 180 Btu/gal/hr means that the fermentor must be capable of removing up to 160 to 230 Btu/gal/hr. If the fermentation is a penicillin fermentation operated at 25°C, and if the best cooling water temperature that can be achieved is around 18°C, and if the overall heat transfer coefficient is in the range of 100 Btu/hr/°F/ft^2, there is no way that the heat will be removed in large fermentors without external heat exchange or extensive cooling coils in the fermentor. In highly viscous fermentations, internal cooling coils are usually not desirable because of interference with mixing patterns.

As a consequence, numerous schemes exist for heat removal in large fermentors including half coil baffles and draft tubes. No census of design seems to have emerged. Each manufacturer has his own design scheme that he promotes.

Power input to the liquid phase in a sparged system (P_g) can be calculated from

$$P_g/V = \frac{Q_g P_g}{V}\frac{\eta v_o^2}{2} + \frac{RT}{M}\left(ln\,\frac{\pi_o}{\pi}\right) \tag{18.46}$$

The first term represents the energy dissipated at the sparger holes. The second term represents the energy involved to move the gas through the static liquid head. For well-designed spargers the first term can usually be neglected.

Unaerated mechanical power input (P_{mo}) can be estimated from

$$P_{mo}/V = \frac{f(\rho N^3 D^5)}{V} \tag{18.47}$$

The power factor in this correlation can be obtained from the paper of Rushton et al.*

For estimation of mechanical power input to a sparged agitated system (P_m), the correlations of Michel and Miller are widely used:

$$P_m/V = \frac{0.706}{V}\left[\frac{P_{mo}^2 N D^3}{Q_g^{0.56}}\right]^{0.45} \tag{18.48}$$

Although the separate effect of sparged and of mechanically agitated power input can be estimated, their effect is not simply additive. In general, a correlation of the following type is used to estimate their combined effects:

$$\frac{P_e}{V} = \frac{P_m}{V} + \frac{CP_g}{V} \tag{18.49}$$

The C correlation term usually varies from 0.05 to 0.4 in value. The data of Miller** can be used to estimate the value of the constant.

There are various correlations between $k_L a$ and power inputs. Some design engineers prefer to scale-up on the following basis

$$k_L a = k(P_m/V)^{\alpha}(v_s)^{\beta} \tag{18.50}$$

where $\alpha = 0.95$ and $\beta = 0.67$ for pilot plant equipment and $\alpha = 0.4$ and $\beta = 0.5$ for large-scale plant equipment. When such a basis is used for scale-up, the designer attempts to also maintain constant tip speed on scale-up. Generally, these run from 600 to 1200 ft/min in most fermentors, with 1000 ft/min a typical value.

Once a plant is built, the conditions of agitation, aeration, mass (oxygen) transfer, and heat transfer are more or less set. Therefore, it has been suggested that the problem of translating process improvements is not one of scale-up but rather of scale-down. Those environmental conditions achievable in plant-scale equipment should be scaled down to the pilot plant and screen-size equipment (shake flask) to insure that the studies are carried out under conditions that can be duplicated.

Chem Eng Prog., *46*, 467 (1959).
A.I.Ch.E.J.*, *20*, **445 (1975).

Air Sterilization

Submerged aerobic fermentation processes require a continuous supply of large quantities of air. Stereilization of this air is mandatory in many fermentations. For pure culture operation, incomplete destruction or inadequate removal of the microorganisms carried in the air may preclude successful operation. Many ways have been suggested for sterilizing air. Only filtration through beds of fibrous and membrane-type materials have found widespread usage on an industrial scale. Examples would be filtration through beds of fibrous materials such as glass wool or plate-type filters made of polyvinyl alcohol several centimeters thick. In recent years sufficient research has been carried out to permit the design of these fibrous filters on a rational basis. Major requirements which every air sterilization system must satisfy are:

1. The system should be simple in design.
2. It should not be inordinately costly to operate.
3. It should remove or destroy air-borne contamination to the extent necessary for satisfactory fermentation performance.
4. It should be stable to repeated steam or chemical vapor sterilization.
5. It should condition the air.
6. Its ability to maintain a sterile air supply should not be jeopardized by power failure or compressor surges.

This latter requirement is frequently overlooked. Its consideration is paramount in the design of filters compounded from fibrous materials. For a particular filter there is an intermediate air velocity at which filtration efficiency is a minimum. If the filter design is based upon a performance observed at an operating velocity other than that at which minimum efficiency occurs, surges or brief power failures could create periods of operation at lower than designed for efficiencies. Hence the filter design should be based on satisfactory operation at this point of minimal efficiency.

Minimal efficiency at an intermediate air velocity occurs because different forces act to collect air-borne particles at different velocities. At low velocities, gravitational, diffusional, and electrostatic forces act on the particle. Their effect is inversely proportional to air velocity. At high velocities, inertia forces come into play, and they are directly proportional to air velocity. The nature of inertial effects is such that below a certain air velocity, collection due to inertial forces is zero.

This velocity can be estimated by the following relation:

$$V_{\text{min eff}} = \frac{1.125\,\mu d_f}{C \rho_p d_p^2} \tag{18.51}$$

where

μ = air viscosity
d_f = fiber diameter
C = Cunningham correction factor
ρ_p = particle density
d_p = particle diameter

For the collection of unit density, 1 micron-sized bacterial particles from air streams at room temperature and pressure, this velocity is equal to

$$V_{\text{min eff}} = 0.066\, d_f \tag{18.52}$$

where velocity V is in feet per second and the fiber diameter, d_f, is in micrometers.

Regardless of air velocity, some collection always occurs because air-borne particles possess a finite size and will be intercepted by some fiber blocking an air stream along which a particle moves.

Collection must always be greater than that due to interception because it represents the minimum collection physically possible. In the absence of experimental data, a reasonable estimate of the minimal collection efficiency of a particular filter is that estimated solely from interception effects calculated at a velocity just below which inertial effects are zero.

The theoretical expression for the collection efficiency of a single isolated fiber, due solely to interception effects, can be estimated from

$$n_o = 0.5[1/(2 - ln\, N_{Re})]\,[2(1+R)ln(1+R) - (1+R) + 1/(1+R)] \tag{18.53}$$

where

n_o = collection efficiency of a single isolated fiber
$R = d_p/d_f$
N_{Re} = Reynolds number

From experimentation with aerosols it has been found that the effectiveness of a filter can be expressed by

$$ln\frac{N_1}{N_2} = \frac{1.27 n_o (1 + 4.5\alpha)\alpha L}{(1 - \alpha) d_f} \tag{18.55}$$

where

N_1 = total number of particles entering the filter
N_2 = total number of particles penetrating the filter
L = filter thickness
α = volume fraction of fibers in the filter

This equation is a reasonable basis for air sterilizing filter design. The single fiber efficiency, n_o, to use in this equation when experimental data are lacking is that based on collection due solely to interception effects estimated at an air velocity where minimal filtration efficiency is expected.

Medium Sterilization

Microorganisms can be removed from fluids by mechanical methods, for example, by filtration, centrifugation, floatation, or electrostatic attraction. They may also be destroyed by heat, chemical agents, or electromagnetic irradiation. Although cells may be disrupted or killed by mechanical abrasion on a small scale, this method is not satisfactory industrially. Similarly, X rays, β rays, ultraviolet light, and sonic irradiations, while useful in the laboratory, are not applicable to the sterilization of large volumes of fluids.

While antibacterial agents have an important place in the fermentation industry, particularly for the production of a pure-water supply, they have little application for the sterilization of fermentation media. Despite the fact that heat sterilization of media is the most common method, little attention has been paid until recently to the engineering aspects of heat sterilization.

Interest in continuous methods of sterilizing media is increasing, but for the successful operation of a continuous sterilizer, foaming of the media must be carefully controlled and the viscosity of the media must be relatively low. Figure 18-13 illustrates two types of continuous media sterilizers that have been utilized in the fermentation industry. The advantages of continuous sterilization of media are as follows:

1. Increase of productivity since the short period of exposure to heat minimizes damage to media constituents
2. Better control of quality
3. Leveling of the demand for process steam
4. Suitability for automatic control

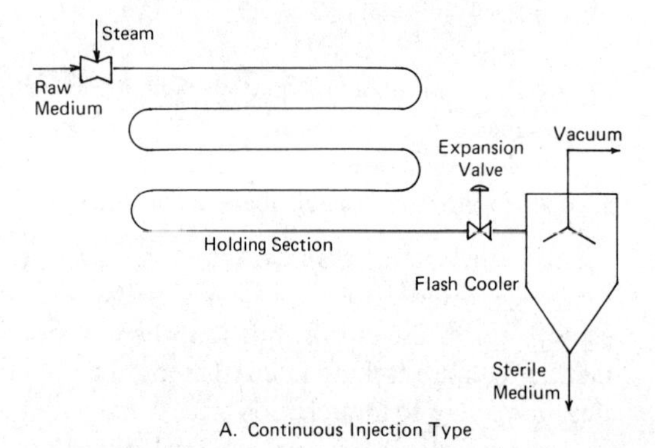

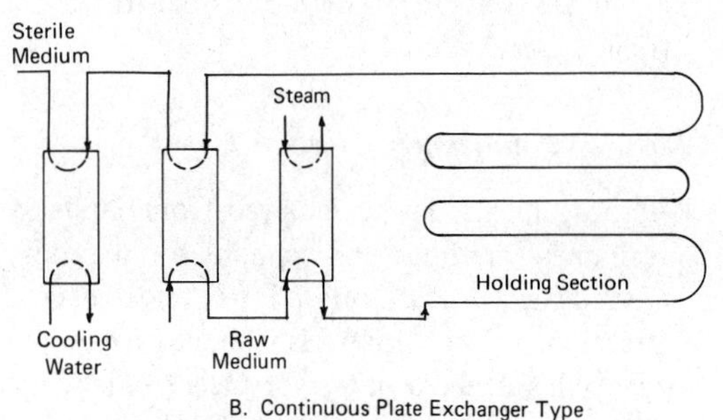

Fig. 18.13 Two types of continuous sterilizers.

At present, most media in the fermentation industry are sterilized by batch methods. Overexposure of the medium to heat is inherent in batch sterilization processes. But continuous sterilization, when properly operated, can minimize damage to the medium.

Design and operation of equipment for sterilizing media is based on the concept of thermal death of microorganisms. Consequently, an understanding of the kinetics of the death of microorganisms is important to the rational design of sterilizers.

The destruction of microorganisms by heat implies loss of viability, not destruction in the physical sense. The destruction of organisms by heat at a specific temperature follows a monomolecular rate of reaction:

$$\frac{dN}{dt} = -KN = -(Ae^{-E/RT})N \quad (18.56)$$

where

k = reaction rate constant, time^{-1}
N = number of viable organisms/unit volume
t = time
T = absolute temperature
E = energy of activation for death
R = gas law constant
A = constant

This equation can be integrated to yield the design equation

$$ln \frac{N_o}{N_f} = A \int_o^{t_s} e^{-E/RT}\, dt \quad (18.57)$$

where

N_o = number of contaminating organisms in the total fermentation medium to be sterilized
N_f = level of contamination that must be achieved to produce the desired degree of apparent sterility
t_s = sterilization time

In estimating the medium sterilization time, one must define the contamination, the desired degree of apparent sterility, and the time-temperature profile of the medium, i.e., $T = f(t)$. For typical bacterial spore contaminants, the constants used in most designs have the following values:

$$E = 68{,}700 \text{ cal/g-mol}$$

$$R = 1.987 \text{ cal/g-mol, } ^\circ\text{K}$$

$$A = 4e^{+87.82}, \text{ min}^{-1}$$

Instrumentation and Control

In scaling-up any successful fermentation, the molecular biologist would suggest the necessity for controlling the environment and hence regulating the fermentation. However, two problems arise. First, knowledge may not exist of the regulation mechanism of metabolic pathways which produce the desired product. In fact, the metabolic pathways may not be fully known or understood. Second, even if the pathways and the regulatory mechanisms are known, the necessary instrumentation to detect regulatory metabolites may not exist. The design engineer is, therefore, presented with a dilemma in developing a new fermentation or improving an existing process.

Should he focus his efforts on researching the mechanisms of product regulation and control, developing needed analysis and sensing instrumentation to provide this control, or should he use trial-and-error development procedures for strain and medium selection to evolve an apparent optimal environment for plant-scale production? The answer is that he must do both.

In the past, metabolic controls were simply bred into or out of fermentation organisms through mutation and strain selection. However, tremendous strides have been, and are being made, in sensor development. The design engineer will soon be able to rely more on environmental control than before in order to gain economical fermentation results. Figure 18.14 diagrams how a highly instrumented fermentor, which is designed to secure basic information on environmental control, can be coupled to a computer for data analysis and control.

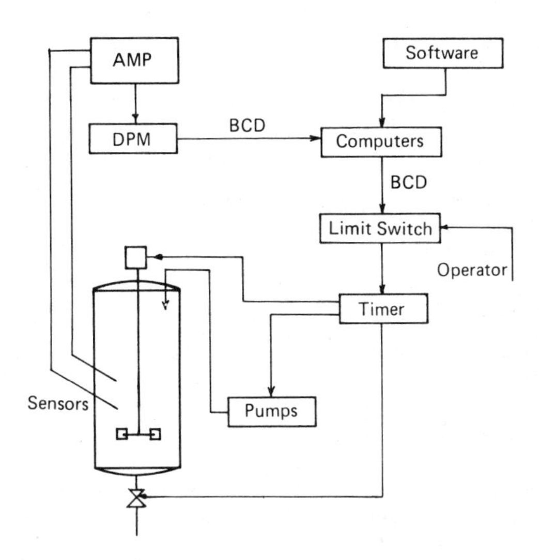

Fig. 18.14 General layout of a highly instrumented, computer-coupled fermentor.

To achieve meaningful fermentation control, the following three steps are necessary:

1. Carry out fermentation research on fully-monitored environmental systems.
2. Correlate the environmental observations with existing knowledge of cellular control mechanisms.
3. Reproduce the desired environmental control conditions through continuous computer monitoring, analysis, and feedback control of the fermentation environment.

Until recently, fermentation control was limited to that of temperature, pH, and aeration. With the development of numerous sensors and inexpensive minicomputers, the engineer can think in terms of sophisticated control systems for fermentation processes.

Indeed, he can even think in terms of utilizing computers to perform on-line dynamic optimization.

New instruments are still needed. The most important sensor needed is a reliable biomass monitoring device that can be sterilized. Also, there is need for a glucose, a nitrogen substrate, and a phosphate sensor that can withstand repeated system sterilization.

Indirect measurement via computers appears to be a viable alternative to measurement involving sampling. Certain sensor information

TABLE 18.3 Gateway Measurements*

Measurement	*Result*
pH	Acid product formation rate
Air flow rate In and out O_2 concentration	O_2 uptake rate
Air flow rate Out CO_2 concentration	CO_2 evolution rate
CO_2 evolution rate O_2 uptake rate	Respiratory quotient
Power input Air flow rate	O_2 transfer rate

*S. Aiba, A. E. Humphrey and N. Millis, *Biochemical Engineering*, 2nd ed., p. 332, University of Tokyo Press, Tokyo (1973).

can be combined to give additional information such as oxygen-uptake rate, carbon-dioxide evolution rate, and respiratory quotient. These measurements can be thought of as "gateway" measurements because they make possible the calculation of additional information (Table 18.3).

Further, the indirect measurement of a given component can be achieved by material balancing that component around the fermentor. If a model for utilization of that component for biomass or product formation is known, then either the biomass or product level can be estimated by computer summation or integration of the data using the model.

Besides these uses, the computer has application in fermentation processes for continuous nonprejudicial monitoring and (most important) continuous feedback control and dynamic optimization of the process.

Recovery of Fermentation Products

From the amount of space devoted to fermentor design and scale-up, one might gather that the recovery processes of fermentation products are rather straightforward and relatively simple. Nothing could be further from the truth. In one case of an antibiotic production plant, the investment for the recovery facilities is claimed to be about four times greater than that for the fermentor vessels and their auxiliary equipment. As much as 60% of the fixed costs of fermentation plants are attributable to the recovery portion in organic acid and amino acid fermentations.

Figure 18-15 shows a typical recovery process for antibiotics, while Fig. 18.16 presents another flow sheet for an enzyme plant. It is apparent from these diagrams that most recovery processes involve combinations of the following procedures:

1. Mechanical separations of cells from fermentation broth
2. Disruption of cells
3. Extraction
4. Preliminary fractionation procedures
5. High resolution steps
6. Concentration
7. Drying

In general, the most troublesome recovery steps are those involving cell disruption and high resolution. New developments in these areas are occurring very rapidly, and the next decade should see significant progress in these processing procedures. Particularly promising are gel fractionation techniques.

Immobilized Enzymes and Cells

Most of the soluble hydrolytic enzymes in common use in industry are formed extracellularly; the wide range of intracellular microbial enzymes are virtually unexplored commercially. In order to exploit these enzymes, it is necessary to develop economic methods of purification. Reports of continuous methods for harvesting, breaking cells and fractionating their protein are major advances towards this end. Recently, enzymes as well as whole cells have been immobilized by adsorption, encapsulation, or inclusion in gels. They may also be covalently bound with a bifunctional linking agent to an insoluble polymer, covalently cross-linked with themselves or bound directly in an enzyme-polymer complex. The preparation can be packed into a column or complexed onto porous sheets and the substrate can then react with the enzyme in a batch or continuous process. Enzymes can also be contained within an ultra-filtration membrane. For instance,

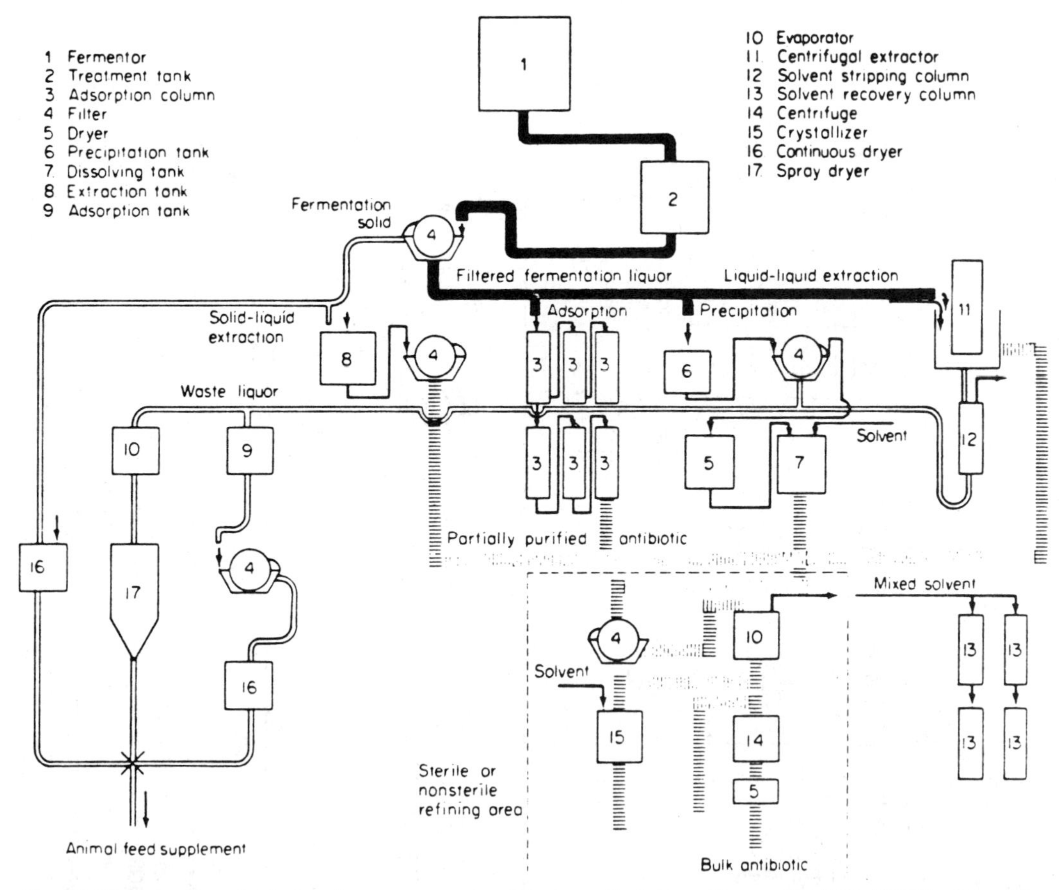

Fig. 18.15 Basic flow sheet for the recovery of antibiotics. (Ind. Eng. Chem., *49*, 1494 (1957). Copyright American Chemical Society.)

starch can be fed continuously into a membrane structure containing α-amylase and the products of hydrolysis collected outside the membrane. Whole cells or crude extracts can also be placed in capsules that allow substrate and products to diffuse to the enzymes.

Some properties of the enzyme, such as the pH optimum, Michaelis constant or stability may change when they are immobilized but the economies achieved by retaining the enzyme for reuse and obtaining products free of contaminating enzyme greatly outweigh any loss of stability or reduction in the rate of reaction. In many cases the life of the immobilized enzyme is greater than that of the soluble enzyme. If substrates or products are sensitive to pH, it may be possible to select a polymer which changes the optimum pH of the immobilized enzyme to a range more favorable to the stability of the reactants.

WASTE TREATMENT

In principle, it is widely accepted that industry must take responsibility for the wastes it produces, but in practice, wastes are frequently dumped into a convenient municipal system or discharged with minimum treatment (or none at all) into the nearby body of water. For the most past industry tends to adopt the cheapest expedient the law and the public will tolerate.

In the fermentation industry, the treatment of wastes should be regarded as an integral part of plant design. It is illogical and uneconomical to dilute a waste that has a high available BOD to a point where the substrate concentration falls to levels that limits microbial growth. Such wastes should, ideally, be treated when the high BOD allows dense cell populations and rapid rates of growth. Since the effluent from a single factory is far more uniform than that

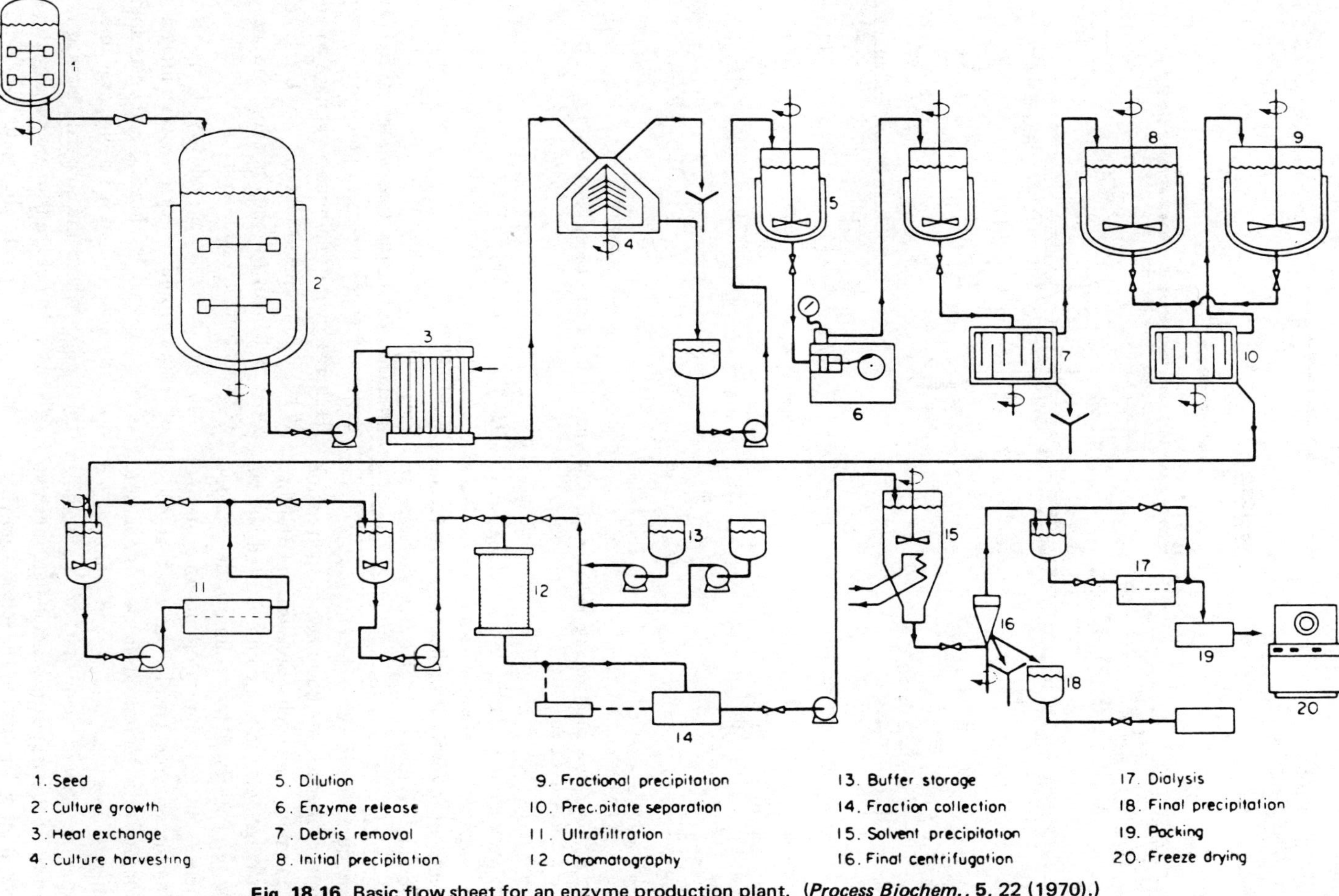

Fig. 18.16 Basic flow sheet for an enzyme production plant. (*Process Biochem.*, **5**, 22 (1970).)

reaching a sewage plant, the microbial population established on this effluent is likely to be more efficient and more readily controlled than that in a general sewage plant. Effluent treatment under these circumstances is just as amenable to optimization as any other fermentation.

BOD Removal

In recent years, the so-called activated sludge process has become a very popular way of efficiently removing biological oxygen demand (BOD) from wastewaters. This process uses a continuous culture system with cell or sludge recycle. The purpose of the recycle is to build up the active biomass concentration levels very high so that efficient volumetric BOD removal can be obtained. In a typical multistage system (see Fig. 18.17), the influent BOD is around 100 to 200 milligrams of oxygen demand per liter. This is reduced by 90 to 95% in the system. The recycle rate is usually 25% of the influent rate. The clarifier operates to concentrate the sludge by a factor of approximately 4. The typical sludge concentration in the aeration section is from 2000 to 5000 milligrams of volatile suspended solids (VSS) per liter.

Initially, a single-stage (completely mixed) aeration basin was utilized. Such a system, while not optimal in volumetric removal levels, is much less susceptible to shock loading or inhibition of BOD removal due to high levels of toxic substrates. One of the problems with such a system is that it tends to evolve a biomass which is filamentous and does not settle too well. Approximately ten years ago, it was found that by using enriched oxygen and staging, a much better settling sludge was created, allowing the system to operate at higher biomass concentrations and, hence, higher volumetric BOD removal rates. In the initial patent by McWhirter [U.S. Patent 3,547,814 (1970)], the credit for this behavior was given to high dissolved oxygen concentration. While this was correct in part, it was later discovered by Casey, et al [U.S. Patent 3,864,246 (1975)], that it was not the high dissolved oxygen that was creating the good settling sludge but rather the staging. In particular, a small first-stage, operated at very high throughout, allowed the cells to be exposed to high substrate concentrations. This minimized the competitive advantage for substrate uptake by the higher surface to volume mycelial organisms over the lower surface to volume nonmycelial organisms, the latter having better settling characteristics. As a consequence, more and more activated-sludge systems have been designed as multistage units. They typically have from 5 to 10 stages with the first stage usually comprising less than 10% of the total system volume. This practice minimizes the clarified volume required for a given BOD load and ensures much better system performance. The problem with the multistage system is that it is susceptible to what the waste industry calls shock loading or the fermentation industry would call high substrate inhibition. When this is critical, the design may involve a holding basin plus the appropriate monitoring equipment in order to dilute out the shock.

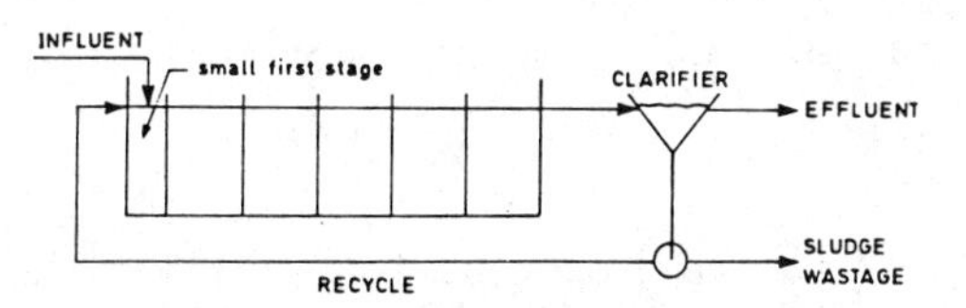

Fig. 18.17 Multistage activated sludge.

A variant of the multistage unit has evolved which is called "contact stabilization." The recycle in this system, i.e., the concentrated sludge, is aerated following clarification. The reason for this is two-fold: 1) Recycle sludge has been exposed to an anaerobic environment in the clarifier where cell autolysis and BOD bleed-back occurred. Reaeration is believed to help activate the sludge prior to mixing with fresh influent, and 2) reaeration minimizes sludge make by oxidizing some of the bleed-back BOD. In systems with a primary clarifier, it has been possible to biologically oxidize virtually all of the biological oxygen demand with negligible sludge make. However, as energy becomes more and more expensive, the cost of doing this becomes more prohibitive. Hence reaeration systems are becoming less attractive.

Denitrification

In recent years, there has been pressure to remove the nitrogenous BOD demand as well as the carbonaceous BOD from waste waters by extending the aeration time in the systems. In a properly designed waste treatment system, it is possible to remove greater than 90% of the carbonaceous BOD within two hours of residence time. Six or more hours of residence time are necessary to remove the nitrogenous BOD demand. In areas where there are particularly high nitrogenous BOD loading on the local surface waters, BOD requirements are extended to include nitrogenous oxidation, i.e., nitrification. However, because of the worry that high nitrate/nitrite levels are injurious to health, there has been an increasing demand to also perform denitrification. This must be done in an anoxic environment in which nitrate and nitrite become a source of oxygen for satisfying the biological oxygen demand. Initially, BOD removal coupled with denitrification was performed in a double system with intermediate clarification and two different biomasses.

The first system was used to nitrify, then clarify, and the second system was used to remove the nitrates and nitrites in an anoxic environment. This was believed necessary because ammonia is a repressor of denitrification at levels of 15 milligrams of ammonia-equivalent nitrogen. This level is exceeded in most waste treatment systems.

The difficulty with two separate systems is their instability and the ease of wash out if wide variations in flow cause the nitrification system to spill over into the denitrification system. Also, the denitrification system does require a carbonaceous energy source to drive the biological system.

An ingenious continuous culture system has evolved from the work of Barnard of South Africa. He has proposed reversing and combining the system, i.e., placing the anoxic system in front of the aerobic system with a high internal recycle on the order of 3 to 1 in order to dilute out the ammonia repression effects (see Fig. 18.18). This system has proved very effective and is now marketed by several companies in the United States.

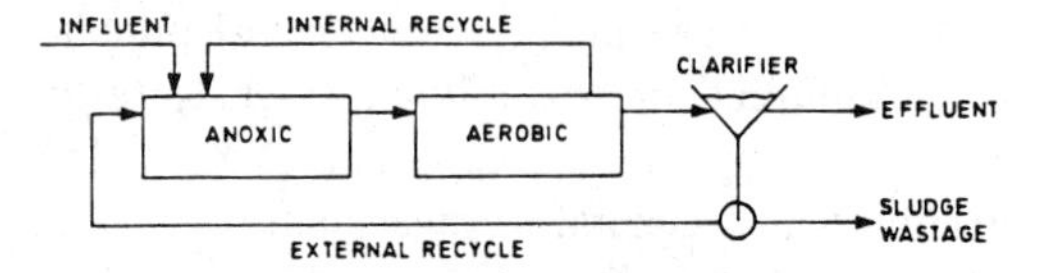

Fig. 18.18 Denitrification system.

This continuous culture system is very interesting from a control point of view for there are both internal and external recycle; it may be a multistage system; and it has both aerobic and anaerobic sections. In terms of stability and dynamics, there are a number of unsolved challenges in this system for the fermentation technologists interested in continuous culture behavior.

Phosphate Removal

One of the most recent innovations in continuous culture systems is the combined anaerobic-aerobic multistage waste treatment systems which have been marketed in just the past year or so (see Fig. 18.19). The objective of these systems is not only BOD removal but also effective phosphate removal. They operate by an environmental selection principle. Such systems allow those organisms that can accumulate polymetaphosphate to have a selective advantage over organisms that have ATP-driven substrate uptake. The polymetaphosphate accumulating organisms can store large quantities of this material which can substitute for ATP in the active transport of glucose and related carbohydrates. Hence, in the anaerobic portion of such a system, they can take up large quantities of glucose, releasing inorganic phosphate. Then, in the aerobic section, they oxidize the glucose and regenerate the polymetaphosphate. Phosphate removal occurs because these particular organisms can accumulate as much as 15 to 20% of their biomass as polymetaphosphate. Hence, the phosphate in the wastewater is removed via sludge disposal.

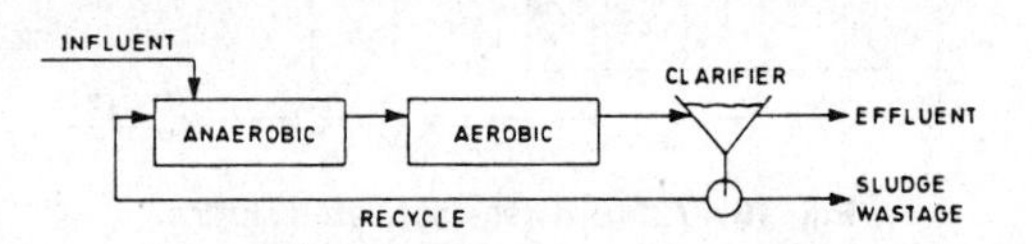

Fig. 18.19 Phosphate removal system.

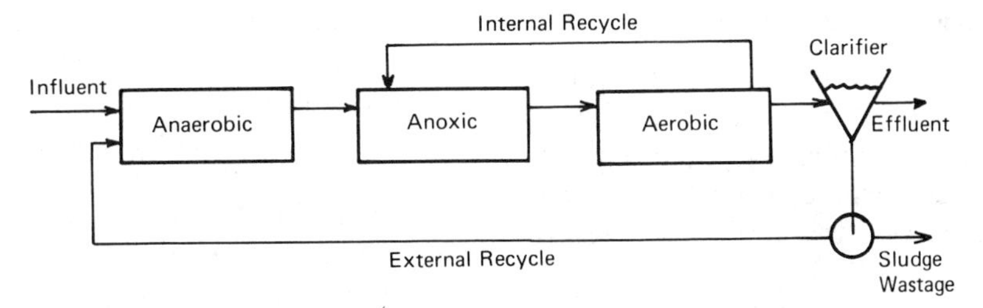

Fig. 18.20 Denitrification and phosphate removal system.

In order to make this system effective, one has to carefully control the sludge make relative to the demand for phosphate removal.

These systems are now being designed and installed in various communities in the United States and elsewhere. They are very challenging continuous-culture devices to understand and design. By controlling the aeration rate and the relative size of the anaerobic to aerobic sections, one can, in effect, control both BOD and phosphate removal. The design of these systems is difficult because they must operate under conditions of varying influent flow rate and BOD loading. An even more challenging continuous-culture system is to add denitrification to phosphate removal. One such scheme is shown in Fig. 18.20. The kinetic behavior and dynamics of this system are virtually unknown and represent a tremendous challenge to the industrial fermentation technologists.

ORGANIC ACIDS

There are many organic acids that can be produced by microbial or biochemical means. The major ones are listed in Table 18.4. However, only acetic acid (as vinegar), citric acid, itaconic acid, gluconic acid, and 2-ketogluconic acid are produced industrially by fermentation now. Other organic acids, such as fumaric, gallic, lactic, malic, and tartaric acids, once produced by fermentation or enzyme processes, are now commercially produced predominantly by the more economic means of chemical synthesis.

TABLE 18.4 Major Organic Acids that Can Be Produced by Fermentation

Organic Acids	Producing Microorganisms	*Substrates*	*Yields %*
Acetic	*Acetobacter acetic*	Ethanol	95
	Clostridium thermoaceticum	Glucose	90
Araboascorbic	*Penicillium notatum*	Glucose	45
Citric	*Aspergillus niger*	Sucrose	85
	Candida lipolytica	n-Paraffin	140
Fumaric	*Rhizopus delemar*	Glucose	58
Gluconic	*Aspergillus niger*	Glucose	95
Isocitric	*Candida brumptii*	Glucose	28
Itaconic	*Aspergillus terreus*	Glucose	60
2-ketogluconic	*Pseudomonas fluorescens*	Glucose	90
5-ketogluconic	*Gluconobacter suboxydans*	Glucose	90
α-ketoglutaric	*Candida hydrocarbofumarica*	n-Paraffin	84
Kojic	*Aspergillus oxyzae*	Glucose	50
Lactic	*Lactobacillus delbriickii*	Glucose	90
Malic	*Lactobacillus brevis*	Glucose	100
Propionic	*Propionibacter shermanii*	Glucose	60
Pyruvic	*Pseudomonas aeruginosa*	Glucose	50
Salicylic	*Pseudomonas aeruginosa*	Naphthalene	94
Succinic	*Bacterium succinicum*	Malic Acid	57
Tartaric	*Gluconobacter suboxydans*	Glucose	27
Xylonic	*Enterobacter cloacae*	Xylose	90

Acitic Acid and Vinegar

Acetic acid as a chemical feedstock is manufactured by chemical synthesis. Acetic acid in the form of vinegar (at least four percent acetic acid by law) is produced largely via the oxidation of ethanol by bacteria of the *Acetobacter* genus.

Synthetic acetic acid is presently produced by several different routes.

1. Wood distillation (being phased out),
2. Oxidation of acetaldehyde (Wacker Process),
3. Liquid-phase oxidation of n-butane, and
4. Carbonylation of methanol (Monsanto Process).

The last method is the most economical among the four, and is employed in most new acetic-acid manufacturing facilities throughout the world. Monsanto was awarded the Kirkpatrick Chemical Engineering Merit Award in 1976 for the development of this process.

The U.S. annual production of acetic acid amounts to 3.3 billion pounds, and the selling price is around \$0.23 per pound. The major producers includes Celanese, Monsanto, USI, Union Carbide, Eastman, and Borden.

High-tonnage production of acetic acid for industrial purposes was earlier done through fermentation. In 1915, USI of Baltimore, Maryland, constructed and operated two of the largest acetic-acid fermentation plants ever built, with a combined tank capacity of about 20-million gallons. The acetic acid was converted with lime to calcium acetate, and the latter pyrolyzed to acetone. The plant operated for a number of years until competition from acetone produced by direct fermentation made the process uneconomical. Today, fermentative acetic acid is limited to vinegar production only.

Vinegar is one of the oldest fermentation products known to man, predated only by wine and possibly by certain foods from milk. First derived from the spoilage of wine, vinegar has been used as a condiment, food preservative, medicinal agent, primitive antibiotic, and even today a household cleansing agent. The present production of vinegar is used almost entirely in foods.

Vinegar may be defined as the product of a double fermentation: an alcoholic fermentation of a sugary mash by a suitable yeast (usually a selected strain of *Saccharomyces cerevisiae* or *ellipsoidens*) and a second fermentation to oxidize the alcohol to acetic acid by a suitable culture of *Acetobacter* organisms.

The theoretical maximum of acetic acid yield on glucose is 67% (two moles of acetic acid produced from every mole of glucose consumed) by this route. A homofermentative culture, *Clostridium thermoaceticum*, is known to be capable of fixing CO_2 and yielding three moles of acetic acid from one mole of glucose under anaerobic conditions. The technology for this process has not been commercialized, however.

Vinegar can be made from a variety of fermentable substances, the essential requirements being that they are satisfactory and economical sources of ethanol. The commonly used substances include fruits and their juices, cereals, sugar syrups and synthetic ethanol. There are four major types of vinegar: white distilled, cider, wine, and malt vinegar. Almost the entire production of white distilled vinegar in the U.S. is derived from synthetic ethanol. Cider vinegar is made by the alcoholic and subsequent acetous fermentation of the juice of apples or its concentrate; wine vinegar is made of grape wine, and malt vinegar made of a mash containing malt, corn and/or barley. The annual U.S. production of all types of vinegar amounts to about 150-million gallons, approximately 77% being white distilled, 17% cider, 4% wine and 1% malt vinegar. The major producers include Heinz U.S.A., Hunt Foods, National Vinegar, Speas, and Standard Brands.

Several vinegar manufacturing processes are commercially used. These include the following:

1. Circulating Generators (Trickling Generators),
2. Frings Acetators (Submerged Culture Generators),
3. Yeomans Cavitators,
4. Tower Fermentors (Column Fermentors).

The circulating, trickling generator is most

widely used. It is a large tank constructed in a variety of dimensions, generally of wood, including redwood and fir but preferably cypress. The vertical timbers are held in place with steel hoops. A false bottom supports curled beachwood shavings above the lower one-fifth of the tank, which serves as a collection reservoir. Air is supplied by a simple fan-type blower, and is distributed to the generator by a number of equally spaced inlets just beneath the false bottom. An air flow rate of about 0.015 vvm (volume of air per volume of packing per minute) is adequate. A pump circulates the ethanol-water-acetic acid mixture from the collection reservoir up through a cooler to a distributing sparger arm in the top of the tank. The liquid trickles down through the packing and returns to the bottom reservoir. Cooling water to the cooler is regulated to maintain the temperature of the generator around 29°C at the top and below 35°C at the bottom. A portion of the finished vinegar is periodically withdrawn from the reservoir, and replaced with the ethanol-containing charge. The ethanol concentration in the generator should not exceed 5% or fall below 0.2%. If ethanol is depleted in the generator, the *Acetobacter* will die and the generator becomes inactive.

The Frings Acetator (produced by the Heinrich Frings Comany of Bonn, Germany) consists of a stainless steel tank with internal cooling cools, a high-speed, bottom-entering agitator and a centrifugal foam breaker. The unique feature of this Acetator is its highly efficient method of supplying air. This is accomplished by means of a high-velocity self-aspirating rotor that pulls air in from the room to the bottom of the tank. The equipment is operated batchwise. When the ethanol content falls to 0.2% by volume, about 35-40% of the finished product is then removed. Fresh feed is pumped in to restore the original level, and the cycle starts again. Cycle time for 12% vinegar is about 35 hours. The rate of production can be as much as ten times as great as that obtained in a trickling generator of equivalent size. The yield on ethanol is higher. Values of 94% and 85% have been reported respectively for the Acetator and the trickling generator. But much more extensive refining equipments are necessary for filtering vinegar produced by the submerged process, since the mash contains the bacteria that produced it.

The Yeomans Cavitator (no longer being manufactured) is a submerged culture system somewhat similar to the Acetator but differing in the way in which air is supplied. This generator has a top-driven rotor-agitator. It withdraws liquid and air from a centrally located draft tube. The system can be easily installed in existing wooden vats and adapted to continuous production. Technical difficulties have forced its abandonment. Some units are still in use to produce vinegar in continuous mode. A 98-percent efficiency of ethanol to acetic acid has been achieved in commercial operations.

The tower fermentor is a relatively new aeration system applied to vinegar production. The fermentor is constructed of polypropylene reinforced with fiber glass. Aeration is accomplished through a plastic perforated plate covering the cross section of the tower and holding up the liquid. Cost of the tower fermentor is said to be approximately half that of a Frings Acetator of equivalent productive capacity. It has been reported that the tower fermentor is satisfactory for producing all types of vinegar.

Vinegar clarification is accomplished by filtration, usually with the use of filter aids and/or fining agents such as diatomaceous earth or bentonite respectively. After clarification, vinegar is bottled, sealed tightly, pasturized at 60-65°C for 30 minutes, and then cooled to 22°C. Vinegar can be concentrated by a freezing process. Vinegar of 200-grain strength is readily obtainable from 120-grain raw vinegar. There are a number of advantages of 200-grain vinegar. The acid strength of brine solutions which would have had to be discarded because of dilution by pickle juice can easily be increased by pickle processors. Transportation costs are substantially reduced, as well.

Citric Acid

Citric acid, whose structure is shown below, is the most important organic acid produced by fermentation means.

$$\begin{array}{c} CH_2COOH \\ | \\ HO-C-COOH \\ | \\ CH_2COOH \end{array}$$

In the food and beverage industries, citric acid is used in soft drink mixes, in carbonated and still beverages, in candies, wines, desserts, jellies and jams, as an antioxidant in frozen fruits and vegetables, and as an emulsifier in cheese. These represent 70% of the use of citric acid.

In terms of pharmaceutical and cosmetic applications, citric acid provides effervescence in oral dosage forms and in products for external use by combining the citric acid with a biocarbonate/carbonate source to form carbon dioxide. Citric acid and its salts are also used in blood anticoagulants to chelate calcium, block blood clotting, and buffer the blood. Citric acid is contained in various cosmetic products such as hair shampoos, rinses, lotions, creams, and toothpastes. These areas of application account for 20% of citric acid use.

More recently, citric acid is being used for metal cleaning, in detergents substituting phosphate, as plasticizers in its ester forms, for secondary oil recovery, and as a buffer/absorber in stack gas desulfurization.

Citric acid was first isolated from lemon juice and crystallized as a solid in 1784 by Scheele. In 1893, Wehmet first described citric acid as a product of mold fermentation. In 1919, fermentation processes based on sucrose were developed commercially. Many organisms have been shown to produce citric acid from carbohydrates as well as from n-paraffins; however, *Asperigillus niger* has always given the best results in industrial production of citric acid.

The generally accepted pathway from sugar to citric acid is assumed to follow the glycolysis pathway to pyruvate. Pyruvate then enters the TCA cycle. It is decarboxylated to acetate as acetyl-CoA, or adds on CO_2 to form oxalacetate. Acetyl-CoA and oxalacetate then react to form citrate. It is generally considered that aconitase, the enzyme responsible for the conversion of citrate to isocitrate in the TCA cycle, is inhibited by depriving it of the necessary metal (iron) co-enzymes. Therefore, the TCA cycle does not operate and citric acid accumulates. Figure 18.21 summarizes the reactions leading to citric acid from glucose. It is worthwhile to note that one mole of glucose yields one mole of citric acid with no consumption of oxygen. The overall reaction is actually energy yielding. It yields one mole of ATP and two moles of $NADH_2$ per mole of citric acid produced. Both ATP and $NADH_2$ are high energy containing molecules. This aspect makes the process a good candidate for the process of cells immobilization.

Microbiological production of citric acid can be implemented by three techniques:

1. Solid state fermentation.
2. Liquid surface fermentation.
3. Submerged culture fermentation.

In solid state or Koji fermentation, *Aspergillus niger* grows on moist wheat bran (70-80% water) and produces citric acid in 5-8 days. This process is only practiced in Japan and accounts for about one fifth of Japanese citric acid production.

In liquid surface or shallow tray fermentation, beet molasses (containing 48 to 52% sugar) or cane molasses of blackstrap (containing 52 to 57% sugar) or high test (containing 70 to 80% sugar) is introduced into a mixer. Dilute sulfuric acid is added to adjust pH to about 6.0. Phosphorus, potassium, and nitrogen in the form of acids or salts are added as nutrients for proper mold growth and optimal citric production. The mix is then sterilized and finally diluted with water to a 15 to 20% sugar concentration. The medium flows by gravity into shallow aluminum pans or trays arranged in tiers in sterile chambers. Most chambers have provisions for regulation and control of temperature, relative humidity and air circulation. One plant has 80 trays per chamber. Each tray holds about 400 liters of solution at a depth of 76 mm. When the medium has cooled to about 30°C, it is inoculated with spores of *Aspergillus niger*. The tray fermentation requires 8 to 12 days. The pH drops to about 2 at the end of the fermentation, and the acid content varies from 10 to 20%. Some oxalic and gluconic acids are also formed. Temperature is main-

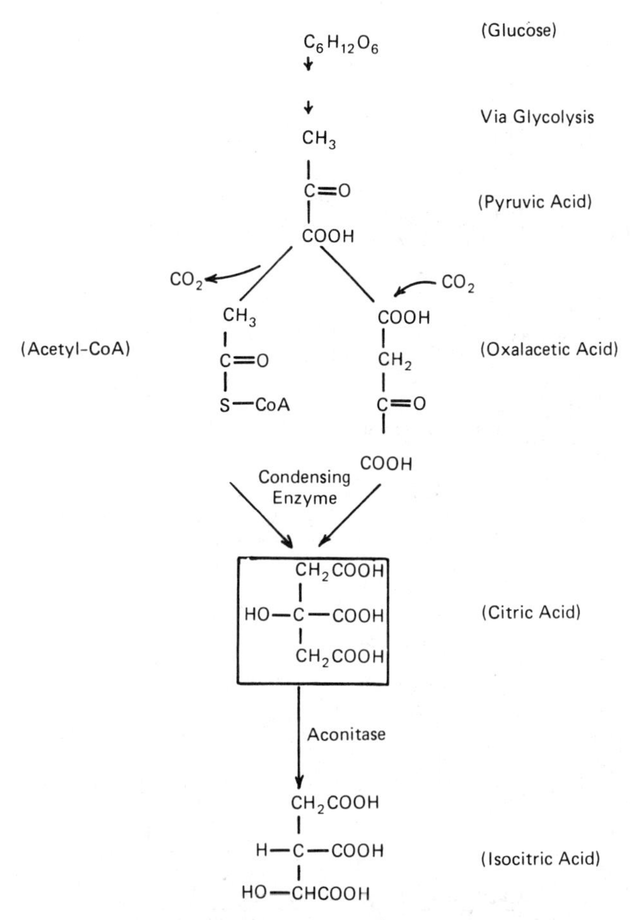

Fig. 18.21 Pathway leading to citric acid from glucose.

tained at 28 to 32°C during the fermentation. Sterile air is circulated through the chambers, and the relative humidity is controlled between 40 and 60%.

The submerged culture or deep fermentation process has been adopted by most newly constructed plants. The fermentation medium consists of sucrose (around 200 g/liter) and mineral salts to provide a balanced supply of iron, zinc, copper, magnesium, manganese, and phosphate. The provision of a suitable culture medium is the most critical factor in obtaining a high yield of citric acid. The fermentation is carried out at 25-27°C. Continuous aeration is provided by bubbling air at a rate of 0.5-1.5 volumes of air/volume of solution/minute. Mechanical stirring is not necessary. It is generally accepted that the formation of pellets between 1 and 2 mm in diameter in the fermentation mash is most desirable. Pelleting reduces broth viscosity, increases oxygen transfer and simplifies mycelium separation in the recovery scheme. The submerged fermentation has a time cycle of 6 to 9 days.

The yield of citric acid on sugar varies from process to process and from manufacturer to manufacturer. The theoretical maximum is 112% on sucrose. The liquid surface fermentation has a yield of 90-95%, and the submerged culture fermentation 75-85%. Improvements have been made by reducing the formation of by-products, mainly oxalic acid, and yields of the submerged culture process are reaching those of the surface culture process.

The fermentation broth from solid state, surface culture, or submerged culture process is treated similarly for recovering and refining citric acid. Two recovery methods are being

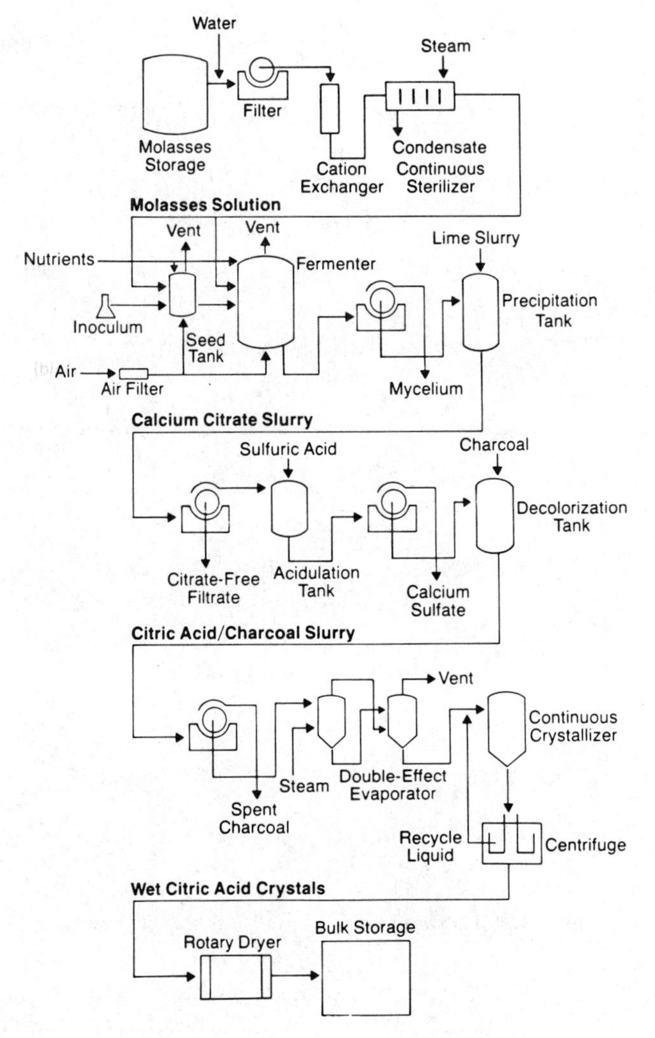

Fig. 18.22 Citric acid process flowsheet.

used: (1) Precipitation and filtration, and (2) Extraction.

A process flowsheet including the fermentation section and the refining section using the first method is shown in Figure 18.22. The mycelium is filtered out of the fermentation liquor first. The mycelium may be used as fertilizer after proper weathering and processing. The clarified liquor flows to precipitating tanks fitted with stirrers, where it is heated to a temperature of 80 to 90°C. The oxalic acid present is separated by preferential precipitation through the addition of a small amount of hydrated lime. The resulting calcium oxalate is worked up separately in a manner similar to the following process described for citric acid recovery. Approximately 1 part of hydrated lime for every 2 parts of liquor is slowly added over a one-hour period, while the temperature is raised to about 95°C. The precipitated calcium citrate is filtered on a vacuum filter, and the filtrate free of citrate is run to waste. The calcium citrate cake is run to acidulation tanks, where it is acidified with dilute sulfuric acid. It is then filtered, and the citric acid mother liquor is decolorized by a charcoal treatment. The purified liquor is then concentrated in a vacuum evaporator. It is then run into a crystallizer where, on cooling, citric acid crystallizes, generally in the form of monohydrate. The resulting acid is of USP grade.

The extraction method treats the filtered

fermentation liquor with a highly selective solvent, tri-n-butyl phosphate, then recovers free citric acid by counter-extraction with water. The aqueous solution, which is further concentrated and crystallized, yields 92% of citric acid with 8% of soluble impurities.

The major citric acid producers include La Citrigue Belge (Belgium), Sturge/Boehringer (W. Germany), Rhone-Poulenc (France), and Miles and Pfizer (U.S.). Citric acid sells for about \$1.50 per kg. The U.S. International Trade Commission does not release production and sales statistics since there are only two producers of citric acid in the U.S. Up to 1967 there was a third producer, the original Bzura Chemical at Fieldsboro, N.J. The plant was sold to Stepan Fermentation Chemicals, and was subsequently closed down because of operating difficulties. The world production of citric acid is estimated at 150,000 tons a year with about two-thirds produced in the U.S. The demand for citric acid is expected to grow continuously. To successfully produce citric acid and compete with the established producers requires extensive process know-how.

Itaconic Acid

Itaconic acid (methylene succinic acid) is an unsaturated dibasic acid.

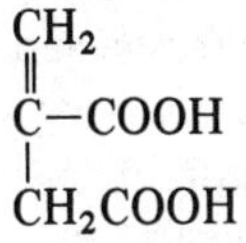

$$\begin{array}{l} CH_2 \\ \| \\ C-COOH \\ | \\ CH_2COOH \end{array}$$

It is a structurally substituted methacrylic acid. Consequently, its principal use is as a copolymer in acrylic or methacrylic resins. Acrylic fibers, by definition, contain at least 85% acrylonitrile. Since pure acrylic fibers are dye-resistant, it is necessary to include other components to make the fibers susceptible to dyes. An acrylic resin containing 5% itaconic acid offers superior properties in taking and holding printing inks and in bonding. In addition to its main application as a component of acrylic fibers, itaconic acid is also used in detergents, food ingredients, and food shortenings.

Previously, itaconic acid was isolated from pyrolytic products of citric acid or produced by converting aconitic acid present in sugar cane juice.

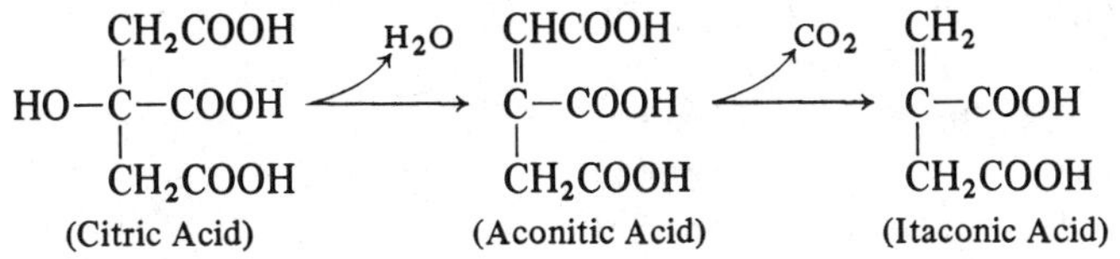

$$\begin{array}{ccccc} CH_2COOH & & CHCOOH & & CH_2 \\ | & & \| & & \| \\ HO-C-COOH & \xrightarrow{-H_2O} & C-COOH & \xrightarrow{-CO_2} & C-COOH \\ | & & | & & | \\ CH_2COOH & & CH_2COOH & & CH_2COOH \\ \text{(Citric Acid)} & & \text{(Aconitic Acid)} & & \text{(Itaconic Acid)} \end{array}$$

It is now produced on a commercial basis predominantly by direct fermentation of molasses.

The biosynthesis of itaconic acid was formally believed to follow the decarboxylation of aconitic acid of the citric acid cycle. It is

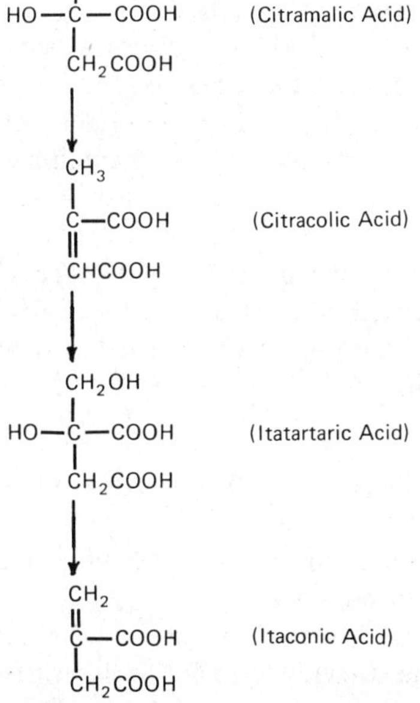

$$\begin{array}{ll} C_6H_{12}O_6 & \text{(Glucose)} \\ \downarrow \text{Via Glycolysis} & \\ CH_3COCOOH & \text{(Pyruvic Acid)} \\ \downarrow \text{Acetyl CoA} & \\ CH_3 & \\ | & \\ HO-C-COOH & \text{(Citramalic Acid)} \\ | & \\ CH_2COOH & \\ \downarrow & \\ CH_3 & \\ | & \\ C-COOH & \text{(Citracolic Acid)} \\ \| & \\ CHCOOH & \\ \downarrow & \\ CH_2OH & \\ | & \\ HO-C-COOH & \text{(Itatartaric Acid)} \\ | & \\ CH_2COOH & \\ \downarrow & \\ CH_2 & \\ \| & \\ C-COOH & \text{(Itaconic Acid)} \\ | & \\ CH_2COOH & \end{array}$$

Fig. 18.23 Proposed metabolic sequence for biosynthesis of itaconic acid.

presently believed to follow the metabolic sequence shown in Figure 18.23.

While both *Aspergillus itaconicus* and *Aspergillus terreus* are known producers of itaconic acid, the latter is superior to the former and is believed to be used industrially. Either surface (shallow-pan) or submerged (deep-tank) fermentation can be used. The medium contains molasses, cornsteep liquor, ammonium sulfate and mineral salts of calcium, zinc, magnesium, and copper. The fermentation, similar to that of citric acid, is very sensitive to concentrations of copper and iron. Copper ion favorably restricts growth and product destruction, but excessive concentration of iron results in reduction of product accumulation. Ten to twenty percent (by volume) inoculum is used. The fermentation is carried out at around 40°C and a pH of 2.0-4.0. Vigorous agitation is employed. Moderate, but continuous, aeration is required. Air failure of very brief duration is enough to damage the fermentation. The batch cycle is 3-6 days. The highest known product concentration is 180-200 g/liter from a medium containing 30% sugar. The yield of itaconic acid on sugar is typically 50-70%. The itaconic acid recovery scheme involves the following:

1. Acidification of itaconic precipitates, if present.
2. Filtration to remove mycelium and other suspended solids.
3. Activated carbon treatment (This step and the next can be omitted for industrial grade product).
4. Filtration to remove carbon.
5. Evaporation and crystallization.

If a high purity acid is desired, further purification steps such as solvent extraction, ion exchange, and carbon decolorization are required.

The current price of itaconic acid is around \$1.80 per kg. The world production is estimated to be about 6,000 tons per year. The major producers are Pfizer in the U.S., Iwati in Japan, and Pfizer and Melle Bezous in Europe.

Gluconic Acid

Gluconic acid is produced by the oxidation of the aldehyde group of glucose.

$$
\begin{array}{c}
\text{H} \\ | \\ \text{C=O} \\ | \\ \text{H—C—OH} \\ | \\ \text{HO—C—H} \\ | \\ \text{H—C—OH} \\ | \\ \text{H—C—OH} \\ | \\ \text{CH}_2\text{OH} \\ \text{(Glucose)}
\end{array}
\xrightarrow{-H_2}
\begin{array}{c}
\text{C=O} \\ | \\ \text{H—C—OH} \\ | \\ \text{HO—C—H} \quad \text{O} \\ | \\ \text{H—C—OH} \\ | \\ \text{H—C} \\ | \\ \text{CH}_2\text{OH} \\ \text{(Glucono-}\delta\text{-lactone)}
\end{array}
\xrightarrow{+H_2O}
\begin{array}{c}
\text{OH} \\ | \\ \text{C=O} \\ | \\ \text{H—C—OH} \\ | \\ \text{HO—C—H} \\ | \\ \text{H—C—OH} \\ | \\ \text{H—C—OH} \\ | \\ \text{CH}_2\text{OH} \\ \text{(Gluconic Acid)}
\end{array}
$$

Gluconic acid may be prepared from glucose by oxidation with a hypochlorite solution, by electrolysis of a solution of sugar containing a measured amount of bromine, or by fermentation of glucose by fungi or bacteria. The latter method is now perferred from an economic standpoint.

Gluconic acid is marketed in the form of 50% aqueous solution, calcium gluconate, sodium gluconate, and glucono-δ-lactone. Gluconic acid finds use in metal pickling, in foods as an acidulant, in tofu (soybean curd) manufacture as a protein coagulant, in detergent formulations as a calcium sequestrant, in the pharmaceutical area as mineral (calcium and iron) supplements, and in the construction area as cement viscosity modifier. Calcium gluconate is widely used for oral and intravenous therapy. Sodium gluconate, a sequestering agent in neutral or alkaline solutions, finds use in cleansing of glassware. Glucono-δ-lactone is used as a food flavor and an acidulant in baking powders and effervescent products.

The organism commonly used in gluconic

acid fermentation is *Aspergillus niger* or *Gluconobacter suboxydans.* The larger volume production uses the fungal process. Most of the *Gluconobacter* production is marketed as glucono-δ-lactone.

During gluconic acid fermentation, glucose is first oxidized (or more correctly, dehydrogenated) to glucono-δ-lactone. This is carried out by glucose oxidase. Hydrogen peroxide is also produced in this step, but is decomposed by catalase.

The fermentation can be by either surface or submerged culture, the latter being the most generally practiced in industry. Horizontally rotating fermentors have also been used.

Calcium gluconate fermentation, in which calcium carbonate is used for neutralization of the product, is limited to an initial glucose concentration of approximately 15% because of the low solubility of calcium gluconate in water (4% at 30°C). The addition of borate or boric acid allows the use of up to 35% glucose in the medium. However, borogluconate was found deleterious to blood vessels of animals, and the product was withdrawn from the market.

The recovery of calcium gluconate from fermentation broth involves the following:

1. Filtration to remove mycelium and other suspended solids.
2. Carbon treatment for decolorization.
3. Filtration to remove carbon.
4. Evaporation to obtain a 15-20% calcium gluconate solution.
5. Crystallization at a temperature just above 0°C.

In sodium gluconate fermentation, sodium hydroxide is used to control pH. Sodium gluconate is much more soluble than calcium gluconate. The addition of sodium hydroxide provides an easy and precise means of neutralizing the acid as it is produced. Much higher concentration of glucose (up to 35%) can be used in this fermentation. The medium also contains cornsteep liquor, urea, magnesium sulfate, and some phosphates. The pH is controlled above 6.5 by addition of sodium hydroxide. One to 1.5 volumes of air per volume of solution per minute (vvm) is supplied for efficient oxygenation. High back pressure (up to 30 psig) is desirable. The fermentation cycle is 2-3 days. Continuous fermentation is used in Japan to convert 35% glucose solution to sodium gluconate with a yield higher than 95%. The continuous process doubles the productivity of the usual batch system.

Sodium gluconate can be recovered from fermentation filtrate by concentrating to 42-45% solids, adjusting to pH 7.5 with sodium hydroxide, and drum-drying.

In glucono-δ-lactone fermentation, *Gluconobacter suboxydans* converts a 10% glucose solution to glucono-δ-lactone and free gluconic acid in about 3 days. Approximately 40% of the gluconic acid is in the form of glucono-δ-lactone. Aqueous solutions of gluconic acid are in equilibrium with glucono-δ-lactone and glucono-γ-lactone. Crystals separating out of a supersaturated solution below 30°C will be predominantly free gluconic acid; from 30° to 70°C the crystals will be principally glucono-δ-lactone, and above 70°C they will be mainly the γ-lactone.

Prices of gluconates vary. Sodium gluconate (technical grade) is about $1.20 a kg, and calcium gluconate (USP grade) is $2.60 a kg. The world production of gluconic acid is estimated around 35,000 tons a year. The major producers include Bristol-Meyers, Pfizer, Premier Malt Products (U.S.), Benckhiser, Diosyth, Merck, Orsan, Pfizer (Europe), Fujisawa Pharmaceutical, and Kyowa Hakko (Japan).

2-Ketogluconic Acid

2-Ketogluconic acid may be produced by a bacterial fermentation involving various strains of *Gluconobacter* or *Pseudomonas.* Selected strains of *Pseudomonas fluorescens* have been reported as giving the highest yield (up to 90%)

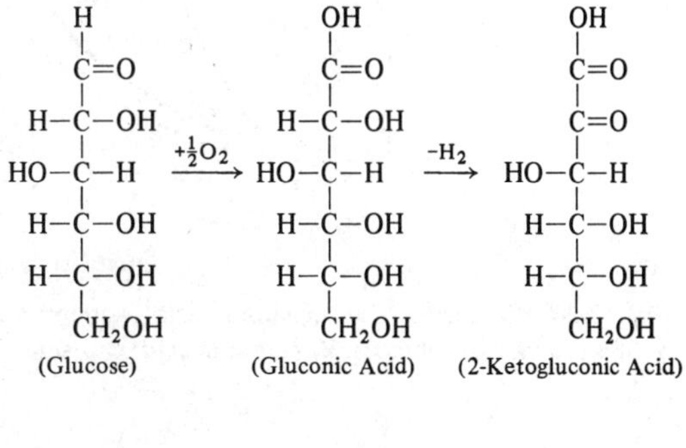

$$
\begin{array}{c} \mathrm{H} \\ | \\ \mathrm{C{=}O} \\ | \\ \mathrm{H{-}C{-}OH} \\ | \\ \mathrm{HO{-}C{-}H} \\ | \\ \mathrm{H{-}C{-}OH} \\ | \\ \mathrm{H{-}C{-}OH} \\ | \\ \mathrm{CH_2OH} \\ \text{(Glucose)} \end{array}
\xrightarrow{+\frac{1}{2}O_2}
\begin{array}{c} \mathrm{OH} \\ | \\ \mathrm{C{=}O} \\ | \\ \mathrm{H{-}C{-}OH} \\ | \\ \mathrm{HO{-}C{-}H} \\ | \\ \mathrm{H{-}C{-}OH} \\ | \\ \mathrm{H{-}C{-}OH} \\ | \\ \mathrm{CH_2OH} \\ \text{(Gluconic Acid)} \end{array}
\xrightarrow{-H_2}
\begin{array}{c} \mathrm{OH} \\ | \\ \mathrm{C{=}O} \\ | \\ \mathrm{C{=}O} \\ | \\ \mathrm{HO{-}C{-}H} \\ | \\ \mathrm{H{-}C{-}OH} \\ | \\ \mathrm{H{-}C{-}OH} \\ | \\ \mathrm{CH_2OH} \\ \text{(2-Ketogluconic Acid)} \end{array}
$$

when glucose or gluconate is used in the medium in highly aerated processes. Gluconic acid is an intermediate in the process.

Figure 18-24 illustrates the kinetic pattern for the fermentation of 10% glucose medium in the presence of an excess of calcium carbonate. 20% glucose medium is used commercially.

2-Ketogluconic acid is usually recovered and shipped in the free acid state after centrifugation or filtration to remove cells. Calcium is removed by precipitation with sulfuric acid. The filtered acid may be shipped as a syrup or as a crystalline material after evaporation under reduced pressure and below 50°C.

The principal use of 2-ketogluconic acid is as an intermediate in the preparation of isoascorbic acid; now known to the trade as erythorbic acid. Erythorbic acid, its esters, and its salts are used as water- or fat-soluble antioxidants to retain color, flavor, and nutritive values in canned fruits and vegetables and in meats and meat products.

Fujisawa Pharmaceutical (Japan) is a major producer of 2-ketogluconic acid. Its yearly production is about 2,000 tons of 2-ketogluconic acid, with most of that exported to the United States, Canada, and European countries.

ORGANIC SOLVENTS

The organic chemicals which fall into this category and can be produced by fermentation include ethanol, butanol, acetone, 2,3-butanediol, and glycerol. 2,3-Butanediol and glycerol fermentations have been developed at laboratory and pilot-plant scales, but have not been commercialized. Ethanol, butanol, and acetone have been produced industrially by fermentation, but chemical synthesis is the manufacturing practice of choice for economic reasons. However, as price and availability of ethylene and propylene as feedstocks for the synthetic processes become subjects of concern, there is renewed interest in examining the fermentation processes as means of producing all or a portion of our future needs of ethanol, butanol, and acetone.

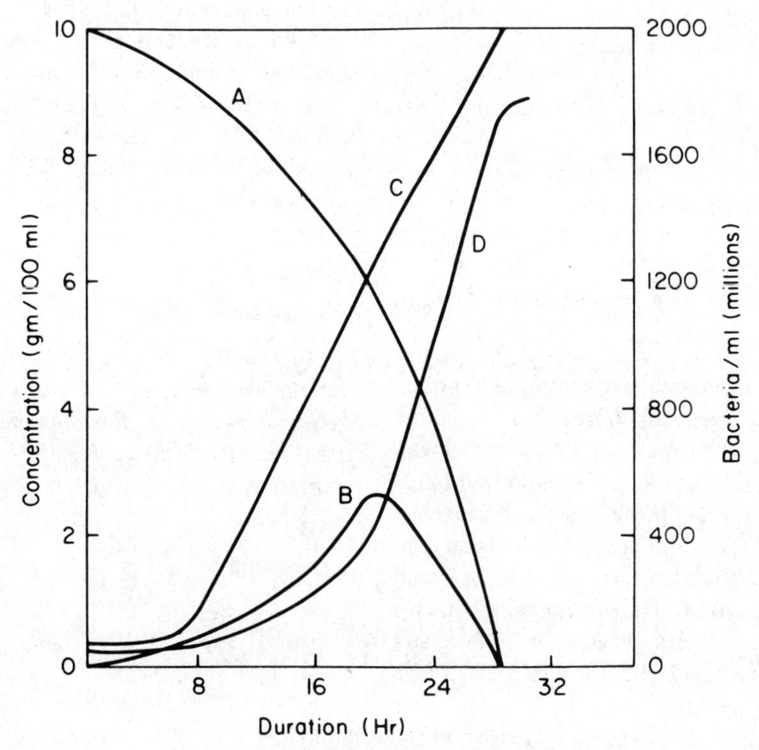

Fig. 18.24 Kinetics of 2-ketogluconic acid fermentation from glucose. (*Microbial Technology*, 2nd ed., Vol. 1, p. 384 (1979)) (A: glucose; B: gluconic acid; C: bacteria; D: 2-ketogluconic acid)

Ethanol

Ethanol is used in industrial solvents, thinners, detergents, toiletries, cosmetics and pharmaceuticals, and, most importantly, as an intermediate for manufacturing other chemicals such as glycol ethers, ethyl chloride, amines, ethyl acetate, vinegar, and acetaldehyde. With the ever-increasing price and dwindling supply of crude oil, ethanol fermented from grains and other renewable organic resources is in close competition synthetic ethanol produced from ethylene.

In 1979, 200 million gallons of ethanol were produced synthetically in the U.S. and around 80 million gallons of ethanol were produced by fermentation. Synthetic ethanol is produced from ethylene via catalytic hydration. Its major U.S. producers include Shell Chemical, Union Carbide, USI, and Texas Eastern. Fermentation ethanol is produced from sugar-containing materials such as grain products, fruits, molasses, whey, and sulfite waste liquor utilizing yeasts. ADM (Archer Daniels Midland Co.), Grain Processing, Midwest Solvents, Georgia Pacific, Milbrew, and Publicker Industries are among the major producers of fermentation ethanol for industrial and fuel consumptions. ADM with an annual production of approximately 50 million gallons is the biggest among them all. Yeasts, particularly strains of *Saccharomyces cerevisiae*, are almost exclusively used in industrial ethanol fermentation. *Saccharomyces cerevisiae* tolerates ethanol concentrations up to about 15% (by volume), and has relatively fast fermentation rates. It converts over 85% of the available carbohydrates to ethanol and carbon dioxide under anaerobic conditions. Air or oxygen suppresses the formation of ethanol (the Pasteur effect). Under aerobic conditions, a major portion of the carbohydrates goes to cell growth.

Ethanol is formed via glycolysis (the Embden-Meyerhof-Parnes pathway). The overall reaction starting from glucose can be written as follows,

$$C_6H_{12}O_6 \longrightarrow 2C_2H_5OH + 2CO_2 + 31{,}200 \text{ cal}$$

The ethanol yield from the above equation is 51% by weight. Since carbohydrate is used also for cell growth and respiration, the overall yield of ethanol from total carbohydrate consumed is typically between 42 and 46%.

Ethanol fermentation can be conducted on nearly any carbohydrate-rich substrate. Molasses, which is the waste mother liquor remaining after the crystallization of sucrose in sugar mill operations, is widely used. Blackstrap molasses contains 35-40% sucroses and 15-20% invert sugars (glucose and fructose). High-test molasses contains 22-27% sucrose and 50-55% invert sugars. Most of the blackstrap molasses does not require the addition of other nutrients for ethanol fermentation. However, high-test molasses requires considerable quantities of ammonium sulfate and other salts such as phosphates. The nonsugar solids nutrient content of high-test molasses is about 7%, compared to 28-35% found in blackstrap molasses.

In the molasses process, Blackstrap or high-test molasses is charged into a mixing tank, where it is diluted with warm water to give a sugar concentration of 15 to 20%. Mineral acid is added to adjust the pH to between 4.0 and 5.0. The diluted and acidified molasses, called "mash," is then pasteurized, cooled, and charged into fermentor tanks, where about 5% yeast inoculum is added. The fermentation is carried out nonaseptically at 23-32°C. Antibiotics may be added to control possible contaminations. Since the overall reaction is exothermic, cooling is required. The feremementation takes 28 to 72 hours (averaging arout 44 hours) to produce an ethanol concentration of 8 to 10%. Carbon dioxide is normally vented. If it is to be collected and recovered, the vent gas is scrubbed with water to remove entrained ethanol and then purified using activated carbon. The activated carbon is reactivated periodically using hot air or steam. The outgassing of the carbon dioxide from the fermentors provides sufficient agitation for small tanks. Mechnical agitation may be added for large ferementors. The fermentation may be conducted batch-wise or continuously, with or without recycling yeast. While continuous fermentation and/or cell recycle can significantly improve productivity and thus reduce required capital investment, it may have only a limited impact on lowering product costs. A major portion of the costs comes from raw materials.

After the fermentation is completed, the liquor, known as "beer," is withdrawn from the fermentors, passed through heat exchangers and pumped to the upper section of a beer still (or "whisky column"), where the alcohol and other volatiles such as aldehydes are distilled off as the overhead. The bottoms, known as "slop" or "stillage," are processed into animal feeds, known as "distillers' dried grains," which contain residual sugars, proteins, and vitamins.

The overhead from the beer column is passed through a heat exchanger and condensed. This condensate, known as "high wines," contains 50-70% alcohol. It is charged into an aldehyde column (or "head column"), where aldehydes, esters, and other low-boiling impurities are separated as overhead. The stream from about the middle of the column is run into a refining column (or "rectifying column"). The tails from the aldehyde column and the weak fractions from the beer column, called "low wines," are rerun with subsequent batches.

In the rectifying column, the heads containing a trace of aldehydes are returned to the aldehyde column. Near the top of the column, the azeotropic alcohol-water mixture of 95% alcohol is taken off, condensed, and run to storage. The higher-boiling alcohols, known as "fusel oils" containing amyl, butyl, isobutyl, propyl and hexyl alcohols, are withdrawn further down the column. The fusel oils amount to 0.5% of total carbohydrates consumed. Water is discharged from the bottom of the column. In order to prevent the diversion of industrial alcohol to potable uses, it is "denatured" by the addition of some material which renders the alcohol so treated unfit for use as a beverage. The 95% alcohol from the rectifying column is stored in Government-bonded warehouses. The alcohol is either denatured, dehydrated, or sold (tax-free or tax-paid). Anhydrous or absolute alcohol is produced by several methods. A third component such as benzene may be added, and the mixture distilled. The ternary azeotrope thus formed carries over the water, leaving behind anhydrous alcohol. Another method uses countercurrent extraction with a third component such as glycerine or ethylene glycol. The added component depresses the vapor pressure of the water and allows anhydrous alcohol to be distilled from the top of the extraction column. Both these methods are run using continuous columns. A process flow diagram of ethanol fermentation using molasses as substrate is shown in Figure 18.25.

Ethanol can also be produced by fermentation of starch, whey, and sulfite waste liquor. Grain fermentations require additional pretreatment since yeast cannot metabolize starch directly. The grain (usually corn) is ground and heated in an aqueous slurry to gelatinize or solubilize the starch. Some starch-liquifying enzymes may be added in this step at lower

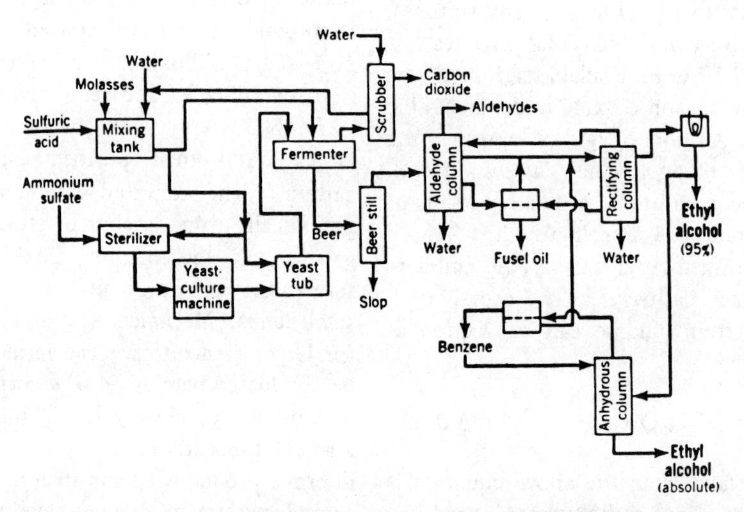

Fig. 18.25 Ethanol fermentation using molasses as substrate. (*Industrial Chemicals*, 4th ed., p. 357 (1975))

temperatures. The liquified starch is then cooled to about 65°C, and treated with malt (germinated, dried barley which is rich in starch-hydrolyzing enzymes), or fungal amylase (produced by *Aspergillus niger*) to convert starch to oligosaccharides. Yeast is then added along with amyloglucosidase (or glucoamylase) which breaks down oligosaccharides into glucose. The subsequent fermentation and refining procedures are the same as those using molasses as raw materials.

Gasohol

Gasohol, a fuel mixture of 10% alcohol and 90% gasoline, is a going business in the U.S.– and a growing one. It may evolve to be the most important use of fermentation ethanol. The 1979 U.S. production of fermentation ethanol was around 80-million gallons. The 1980 production approached 300-million gallons. This increase was largely due to the reactivation of distilleries which were shut down when synthetic ethanol became favored. The U.S. Government's objective is 1.5-billion gallons by 1985 which would require 600-million bushels of corn and 2.8-million tons of coal if coal is exclusively used as the energy source for grinding, heating, fermenting, and distilling corn into ethanol. The 1979 U.S. production of corn was 7.6-billion bushels, 5-billion bushels for domestic consumptions and 2.5-billion bushels for exports. The domestic uses of corn can be roughly broken down as follows,

Uses	*Million Bushels of Corn*
Hog Feed	1,750
Cattle Feed	1,000
Dairy Feed	650
Laying Hen Feed	380
Broiler Chicken Feed	360
Turkey Feed	90
Sheep and Horse Feed	80
Regular Corn Syrup	150
Starch	130
High Fructose Corn Syrup	120
Dextrose	40

The 1985 goal of 1.5-billion-gallon ethanol production seems feasible from the raw material's point of view. However, about 100-billion gallons of gasoline are consumed every year in the U.S. To make a real impact on this nation's energy crisis, 10-billion gallons or more of ethanol need to be produced, and 4-billion bushels of corn would be diverted to this cause if corn is to be used exclusively. Other agricultural crops and organic wastes, thus, would have to be considered for the production of ethanol. Conceptually, the most ideal resource would be cellulose for its abundance and renewability. However, at the present time cellulose saccharification and fermentation is still under development and is not economically competitive with the starch-based fermentations for ethanol production.

Ethanol from grain fermentations has been made competitive as liquid fuel in U.S. largely due to government subsidies in attempts to reduce this nation's dependence on imported foreign oil. The various government incentives include the following:

1. An extension through 1992 of the 4-cent Federal tax exemption for every gallon of gasohol sold as fuel.
2. In more than 20 states, the elimination of state gasoline taxes for gasohol.
3. Tax credits, loans and loan guarantees for biomass-based alcohol plant constructions.

These subsidies may become unnecessary in the future if the price of crude oil continues to inflate at a rate at least 5% higher than corn and other agricultural crops.

Fermentation ethanol also has the potential to become a major chemical feedstock. The U.S. production of ethylene was 29.2 billion pounds in 1979. It would take an equivalent of 3-billion bushels of corn to produce the same quantity of ethylene via ethanol fermentation and subsequent dehydration. The use of other substrates, such as wheat, sugar cane, sweet sorghum, whey, waste paper, and fast-growing trees, has to be explored.

To realize and expand the uses of ethanol as chemical feedstock and liquid fuel, fermentation research and development in the following areas would be helpful:

1. Strains which tolerate higher ethanol concentrations.

2. Strains which ferment optimally at higher temperatures.
3. Strains capable of fixing CO_2 to give higher yields.
4. Strains capable of utilizing a wider range of carbohydrates.
5. Utilization of cellulosic or fibrous components of corn and other crops.
6. Fermentations under vacuum conditions.
7. Continuous fermentations with or without recycling of yeast.
8. Simultaneous and continuous product removal by physical or chemical means such as extraction or ultrafiltration.
9. More energy-efficient ways of recovering and dehydrating ethanol.

Butanol/Acetone

The original observation of the butanol production by *Clostridia* was made by Pasteur, and acetone formation by Schardinger. Interest in commercializing butanol-actone fermentation occurred in 1909 primarily as a means of obtaining butadiene as raw material for synthetic rubber. In 1914 Weizmann established a working process to ferment starchy grains such as maize or corn to produce butanol, acetone, and ethanol using *Clostridium acetobutylicum.* With the outbreak of World War I, the production of acetone was of great interest for the manufacture of cordite. Large-scale operations were established in Canada, U.S., India, and elsewhere during the war period. Shortly after the war, DuPont developed fast-drying nitrocellulose lacquers for the automobile industry, and butyl acetate was the solvent of choice for coating. Large quantities of butanol esters were needed as solvents, and butanol became the principal product of the butanol-acetone fermentation. By the 1930's, some butanol and acetone were being produced by chemical synthesis, and the butanol-actone fermentation industry faced economic difficulties. This problem was then solved by the discovery of strains *Clostridium saccharo-acetobutylicum* and *Clostridium saccharo-butylacetonicum-liquefaciens* that would ferment molasses, a cheaper raw material than starchy grains, and the industry thrived until the end of World War II. The major U.S. producers at that time were Commercial Solvents and Publicker Industries. They operated plants at Terre Haute, Indiana, Baltimore, Maryland, and Philadelphia, Pennsylvania. The butanol-acetone fermentations were conducted in large-scale equipment; fermentors of 50,000- to 500,000-gallon capacity were commonly used. At the present time, the petrochemical processes dominate, and fermentation processes are closed down everywhere except in South Africa, where they are operated by National Chemical Products under rather special conditions.

In the grain fermentation process, 8-10 percent corn mashes were fermented (corn contains 70-72 percent starch on dry basis). Fermentation yields were on the order of 29-32 grams mixed solvent per 100 grams starch used, with solvent ratio of approximately 60-30-10 (butanol-actone-ethanol, respectively). The organisms possessed good diastatic activity, thus malting was not required. The cooked sterile cornmeal suspension was aseptically transferred to sterile fermentors, inoculated, and incubated for about 65 hours at 37°C, after which the solvents were recovered by distillation. The aqueous residue (slop or stillage) was concentrated in multiple-effect evaporators and drum dried for use in animal and poultry feeds.

With the advent of molasses-fermenting strains, more rapid fermentations were attained (40-48 hours) and the solvents produced contained as much as 65-75 percent butanol, principally at the expense of ethanol. One-hundred pounds of blackstrap molasses (containing 57 lbs of sugar) was fermented into the following spectrum of products,

Product	*Yield*
Butanol	11.5 lbs (Solvent Ratio = 68%) (Yield on Sugar = 20%)
Acetone	4.9 lbs (Solvent Ratio = 29%) (Yield on Sugar = 9%)
Ethanol	0.5 lbs (Solvent Ratio = 3%) (Yield on Sugar = 1%)
Total Solvents	16.9 lbs (Yield on sugar = 30%)
Carbon Dioxide	32.1 lbs (Yield on sugar = 56%)
Hydrogen	0.8 lbs (Yield on sugar = 1%)
Dried stillage	28.6 lbs

The anaerobic *Clostridium* yields energy by

converting glucose to acetyl-CoA, formate, CO_2 and H_2. The reducing power is then used to produce butanol and acetone via acetoacetyl-CoA. The overall reaction for butanol and acetone production can be pictured as follows,

$$2C_6H_{12}O_6 \longrightarrow C_4H_9OH + CH_3COCH_3 + 5CO_2 + 4H_2$$

According to the equation the yields on sugar for butanol, acetone, carbon dioxide and hydrogen are 21%, 16%, 60% and 2% respectively.

Continuous fermentations were described by Russian workers. A plant in Dokshukin was operated in three batteries of 7-8 fermentors of 60,000-70,000 gallons. The continuous cycle was 40-90 hours, and the flow rate through the battery was 5,000-10,000 gallons per hour. The feed contained 4-6% of carbohydrates. A combination of raw material was used: molasses, flour, and hydrolysate containing pentoses. The continuous process gave a 20% productivity increase, and saved 142 pounds of starch for every ton of solvents produced.

A considerable amount of care must be exercised in carrying out butanol-acetone fermentation. It is biologically unstable and may fail completely when contaminated. Numerous instances of contamination by bacteriophage were encountered commercially, and on several occasions plants had to suspend operation until the entire plant could be decontaminated. Accordingly, absolute cleanliness, experienced personnel familiar with phage symptoms, and the maintenance of a vigorous rapid fermentation are necessities in butanol-acetone fermentation.

The butanol-acetone-ethanol industrial fermentation of the 1950's made some remarkable advances in distillation efficiencies. Highlights included concentrating upon beer (fermentation mixture) stripping, early removal of water from solvents, thermo-compression on beer still, continuous refining of butanol, and recovery of heat. A process flowsheet is shown in Figure 18.26. When the fermentation is completed, the fermentor broth containing around 2% mixed solvents is pumped to a beer column where a 50% solvent mixture is taken off overhead and distiller's slop is removed as bottoms. The slop may be dried and sold as animal feed. Another by-product is a mixture of carbon dioxide and hydrogen. The mixed-solvent vapors from the beer column are led to a batch fractionating column from which three fractions (acetone, with a B.P. 56.1°C, ethanol with a B.P. 78.3°C, and butanol with a B.P. 117.3°C) are removed overhead, leaving water as bottoms. The acetone and ethanol fractions are purified by conventional fractionation. The butanol fractions containing 70% butanol and 30% water are removed overhead. On condensation, two layers are formed. The top layer (80% butanol and 20% water) is returned to the butanol column, and the bottom layer (4% butanol and 96% water) is returned to the beer column. Approximately 35,000 pounds of steam is consumed for every ton of solvents produced.

The U.S. production of butanol and acetone was 732- and 2,502-million pounds, respectively in 1979. Butanol costs 29 cents a pound, and acetone 27 cents a pound. The fermentation

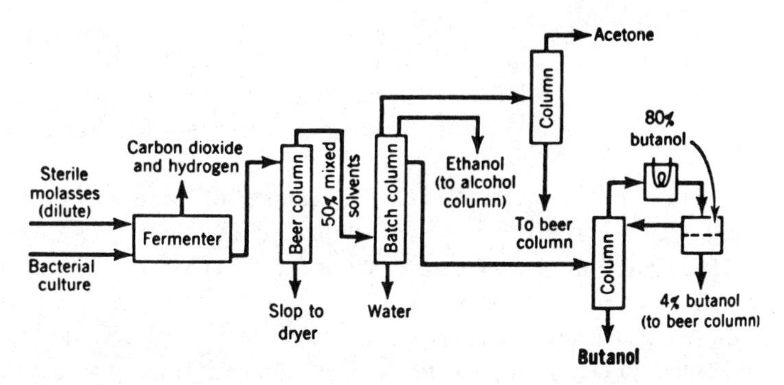

Fig. 18.26 Process flowsheet of butanol-acetone-ethanol fermentation. (*Industrial Chemicals*, 4th ed., p. 178 (1975))

industry has the potential to capture all or at least a portion of the market by concentrating on research and development along the following lines:

1. Develop strains to tolerate high concentrations of butanol and acetone.
2. Develop homo-fermentive strains to improve product yields.
3. Utilize cheaper raw materials such as waste carbohydrates.
4. Develop continuous processes to improve productivities.
5. Improve fermentation stability through both strain and equipment enhancements.
6. Develop more energy-efficient separation methods such as extraction or reverse osmosis.

BREWING

Beer is defined, in United States Federal Regulations, as a malt beverage resulting from an alcoholic fermentation of the aqueous extract of malted barley with hops. It may also include other sources of carbohydrates which are called "adjuncts." American beers are made from malted barley and hops, with corn grits, corn syrup, or rice grits as the adjunct carbohydrate source.

In 1977, breweries in U.S. sold around 170 million barrels or 5.2 billion gallons of beer. The five major brewers (all combined having two thirds of the market share) and their sale volumes are shown below,

Brewers	*Millions of Barrels*
Anheuser-Busch	36.6
Miller	24.2
Schlitz	22.1
Pabst	16.3
Coors	12.8

Carling, Falstaff, Heileman, and Schaefer are among other regional brewers with sales of more than 5 million barrels. Equipment obsolecence and sharp competition have forced the shutdown or take-over of many small and old breweries. There were about 500 brewers in the U.S. in 1930, about 100 in 1975, and only fewer than 50 in 1978.

The brewing process consists of the following steps:

1. Malting.
2. Malt and adjunct milling.
3. Adjunct cooking.
4. Mashing of malt and adjunct.
5. Wort preparation.
6. Fermentation.
7. Lagering, aging and finishing.

A process flow chart of the brewing process is given in Figure 18.27.

Before malting, barley is stored at 10-12% moisture and low temperatures for several weeks to overcome dormancy. During malting, barley is intentionally germinated under controlled temperature and humidity. The enzyme systems necessary for the reproduction of the barley plant are fully developed. As a result, proteins and carbohydrates of the barley kernel are modified and made soluble for subsequent hydrolysis in the mashing step. Growth of the germinated kernel is stopped by drying or kilning in hot air. During the kilning process, color and flavor ingredients are developed. The dried, malted barley is then stored for subsequent malt preparations.

Malt is milled, most often, in roller mills to provide ground malt with a particle size distribution which is optimal for mashing and wort preparation.

The adjunct, after passing cleaners and crushing rolls, is made up with water, mixed with a small percentage (about 10%) of malt, and heated in a pressure cooker. The weight ratio of water to adjunct is about 3 to 1. The addition of malt is to aid liquefaction and decrease viscosity. The cooker is steam-heated, equipped with an agitator, and operated under five pounds of pressure. The total time of heating and boiling is about one hour. It brings about the liquefaction of adjunct starch.

The major portion of malt, made up with brewing water, is placed in a mash tub, thoroughly mixed at a temperature between 35 and 50°C, and then allowed to rest. During the "protein-rest" period of about an hour, the

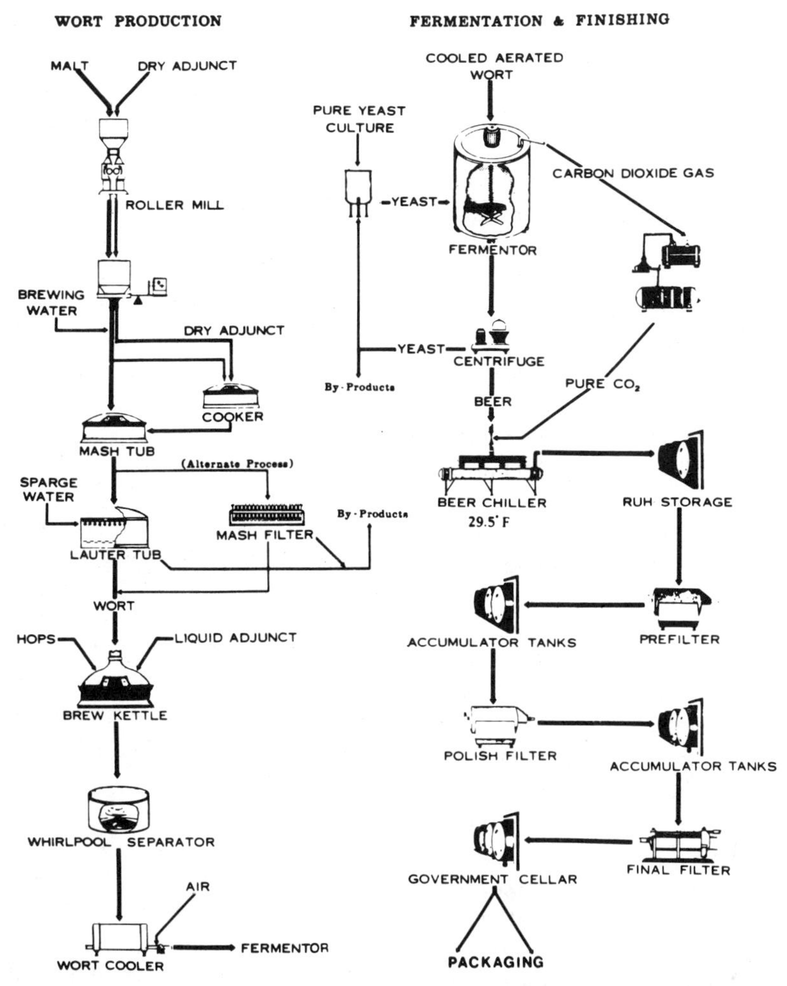

Fig. 18.27 Flowsheet of the brewing process. (*Microbial Technology,* 2nd ed., Vol. 2, p. 20 (1979))

proteolytic enzymes begin to degrade the malt and other proteins into soluble forms. After the resting period, the contents of the completed adjunct cooker are added to raise the temperature of the combined mashes to 65 to 70°C. The mash temperature is kept constant for 20-30 minutes to allow completion of starch hydrolysis. The temperature is then raised to 75-80°C, high enough to arrest the action of the conversion. This completes the mashing step.

After the mashing step, the material passes to a Lauter tub, the bottom of which has slots, through which the clear liquor, known as "wort," passes while the arms of the agitator turn slowly. "Sparge water" in spray form is introduced to extract more dissolved materials from within the solids remaining on the screen. The wort is then run into the beer kettle for boiling. During the approximately 2-hour period of boiling, the hops are added at fixed intervals. While in the boiling kettle, enzymes are destroyed and inactivated, the hop "bitter" (from the lupulin glands of hop cones) is extracted, flavor and color are developed, undesirable proteins are coagulated. Other proteins are precipitated by the tannin in the hops, hop oils are distilled off, and the whole is sterilized by the heat. The material is concentrated by about 10% during the boiling period. Solids, especially the spent hops, are removed with a hop strainer and/or a whirlpool separator. Due to centrifugal force in the sedimentation tank, the solids settle out in the middle and form a trub cone. Wort is drawn off from the outer periphery of the bottom. Trub solids are flushed to sewer, recovered as by-products, or recycled back into the process. After separation of the trub, wort is cooled in a plate heat exchanger to the starting temperature between 9 and 10°C required for the subsequent fermentation. Wort flowing to the fermentor is aerated to bring the oxygen concentration up to 8-10 ppm. After cooling and aeration, wort is inoculated with yeast usually obtained from a previous fermentation. Oxygen is necessary in the initial growth phase of the fermentation. The dissolved oxygen originally present in wort is sufficient to give a four- to five-fold increase in yeast cell population in 3-4 days.

The fermentable carbohydrates furnish energy and carbon to produce ethanol, carbon dioxide, and other minor by-products, resulting in a decrease in specific gravity. Carbon dioxide is recovered, purified and returned to the finished products. Since the reaction is exothermic, the fermentation temperature first rises. When a desired maximum temperature around 13°C is reached, it is kept constant by cooling. The bubbling of the carbon dioxide produced during the fermentation agitates the wort, and causes the yeast to be suspended. Agitation is sometimes aided by mechanical agitators. The so-called primary fermentation is complete after 4-5 days, but complete utilization of fermentable material may take about 10 days. After fermentation, yeast is removed by slurry centrifugation, decantation, or a combination of both.

Lagering or ruh storage and aging are terms referring to different periods for maturing beer. It involves a secondary fermentation to metabolize the small amount of residual carbohydrate and to complete diacetyl assimilation. This is accomplished by the action of 6-10 million cells per milliliter of residual yeast transferred from the primary fermentor. During the 2- to 6-week process, the temperature may start at 3-6°C and gradually be decreased to -1°C. After lagering and aging, the beer is recarbonated by the natural buildup of CO_2 during the secondary fermentation and/or by sparging with CO_2 in storage tanks; it is then clarified by centrifugation or filtration to remove yeast, proteinaceous precipitates, and colloidal matters, and finally bottled or barreled. The bottled beer is pasteurized at 60°C after capping. This completes the brewing process.

AMINO ACIDS

General

Amino acids in general can be represented by the following formula,

$$\begin{array}{c} \mathrm{H} \\ | \\ \mathrm{R{-}C{-}COOH} \\ | \\ \mathrm{NH_2} \end{array}$$

Since the amino group is on the α-carbon, the amino acids with this general formula are known as α amino acids. The α carbon atom becomes asymmetric when R is not an H atom. Naturally occurring amino acids have an L-configuration. Amino acids are the building blocks of proteins, and the elementary composition of most proteins are similar; the approximate precentages are,

C = 50-55 N = 15-18
H = 6-8 S = 0-4
O = 20-23

Table 18.5 gives the structure of R, molecular weight, and elementary composition for each of the twenty amino acids commonly obtained on hydrolysis of proteins.

Amino acids can be obtained from purified proteins by chemical or enzymatic hydrolysis. They can also be isolated from industrial by-products, extracted from plant or animal tissues, or synthesized by organic, enzymatic, or microbiological means. Amino acids that are produced industrially by fermentation include arginine, citrulline, glutamic acid, glutamine, histidine, isoleucine, leucine, lysine, ornithine, phenylalanine, proline, threonine, tryptophan, tyrosine, and valine.

Alanine and aspartic acid are produced commercially utilizing enzymes. In the case of alanine, the decarboxylation of aspartic acid by the aspartate decarboxylase of *Pseudomonas dacunhae* is commercialized. Aspartic acid, on

TABLE 18.5 Twenty Common Amino Acids

Amino Acids	*R–*	*M.W.*	Elemental Composition (% wt) *C*	*H*	*O*	*N*	*S*
Alanine	CH_3-	89	40	8	36	16	0
Arginine	$H_2N-C(=NH)-NH-CH_2-CH_2-CH_2-$	174	41	8	18	32	0
Aspargine	$H_2N-C(=O)-CH_2-$	132	36	6	36	21	0
Aspartic Acid	$HO-C(=O)-CH_2-$	133	36	5	48	11	0
Cysteine	$HS-CH_2-$	121	30	6	26	12	26
Glutamic Acid	$HO-C(=O)-CH_2-CH_2-$	147	41	6	44	10	0
Glutamine	$H_2N-C(=O)-CH_2-CH_2-$	146	41	7	33	19	0
Glycine	$H-$	75	32	7	43	19	0
Histidine	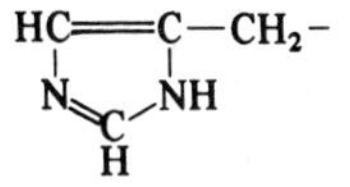 imidazole ring (HC=C–, N, NH, CH) $-CH_2-$	155	46	6	21	27	0
Isoleucine	$CH_3-CH_2-CH(CH_3)-$	131	55	10	24	11	0
Leucine	$CH_3-CH(CH_3)-CH_2-$	131	55	10	24	11	0
Lysine	$H_2N-CH_2-CH_2-CH_2-$	146	49	10	22	19	0
Methionine	$CH_3-S-CH_2-CH_2-$	149	40	7	22	9	2
Phenylalanine	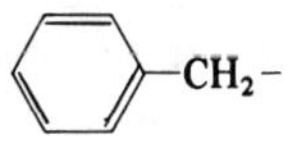 $C_6H_5-CH_2-$	165	66	7	19	8	0
Proline	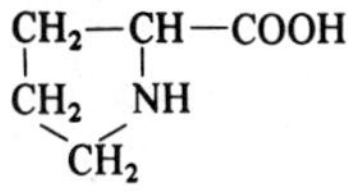 $CH_2-CH-COOH$ / CH_2 NH / CH_2 (ring)	115	52	8	27	12	0
Serine	$HOCH_2-$	105	34	7	46	13	0
Threonine	$CH_3-CH(OH)-$	119	40	8	40	12	0
Tryptophan	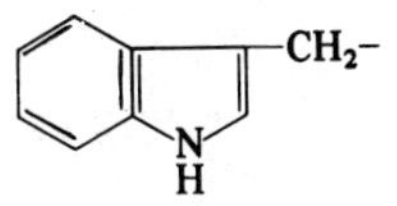 indole ring (NH) $-CH_2-$	204	65	6	16	14	0
Tyrosine	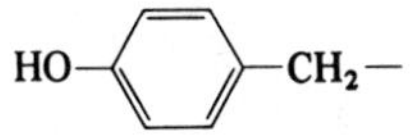 $HO-C_6H_4-CH_2-$	181	60	6	26	8	0
Valine	$CH_3-CH(CH_3)-$	117	51	9	12	27	0

the other hand, is produced commercially by condensing fumarate and ammonia using aspartase from *Escherchia coli.* This process has been made more convenient with enzyme immobilization technique. Tyrosine, cysteine, tryptophan, phenylalanine, and lysine can also be produced by enzymatic methods. Japanese companies, like Ajinomoto, Kyowa Hakko, and Tanabe Seiaka, and their foreign subsidiaries account for more than two thirds of world production of amino acids. The world production of amino acids is estimated at approximately 300,000 tons a year with a sales value of about five billion dollars.

Amino acids play important roles in many areas related to nutrition and medicine. Most amino acids are produced for medicinal purposes. Methionine, lysine, and tryptophan are three of the most important essential amino acids in the body. Both methionine and lysine are produced in large quantities (more than 10,000 tons a year) and have moderate prices ($5-10 per kilogram). Methionine is synthesized chemically, and is used exclusively as a nutritional supplement. Lysine is mainly produced by fermentation, and is used largely as an animal feed additive. Tryptophan, synthesized chemically, has a relatively small market (20-50 tones a year) because of its high selling price (about $150 a kilogram). However, its market can be potentially large when production cost is reduced.

Besides lysine and methionine, the only amino acid produced in large quantities is glutamic acid. It is produced exclusively by fermentation, with an estimated world production of 200,000 tons a year. Monosodium glutamate (MSG) is widely used for food seasoning.

The industrial fermentations of glutamic acid and lysine will be discussed in more detail.

Glutamic Acid

MSG was first produced using acid hydrolysis of wheat gluten or soybean protein in 1909 by Ajinomoto. In 1957, a glutamate-producing bacterium was isolated and subsequent research and development brought about the economic production of glutamic acid by fermentation.

A large number of glutamic acid-producing microorganisms are known. A partial list is given below.

Bacteria	*Corynebacterium glutamicum* (Synonym *Micrococcus glutamicus*)
	Brevibacterium flavum
	Brevibacterium divaricatum
Fungi	*Aspergillus terreus*
	Ustilago maydis

Among them, *Corynebacterium glutamicum* is used most commonly in industry.

The fermentation medium contains a carbon source (glucose, molasses, or acetic acid), a nitrogen source (urea, ammonium sulfate, cornsteep liquor, or casein hydrolyzate), small amounts of mineral salts which supply potassium, phosphorus, magnesium, iron and manganese, and a few (less than five) micrograms of biotin per liter. The biotin requirement is the major controlling factor in the fermentation. When too much biotin is supplied for optimal growth, the organism produces lactic acid. Under conditions of suboptimal growth, glutamic acid is excreted. The metabolic pathway involved in the biosynthesis of glutamic acid from glucose is shown in Figure 18.28.

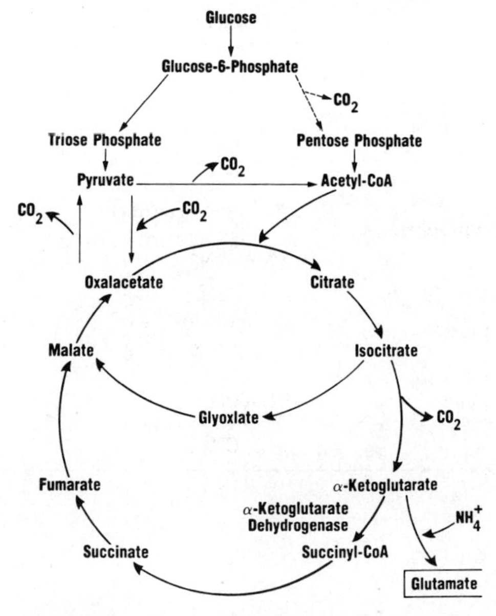

Fig. 18.28 Pathway for glutamic acid formation from glucose.

The lack, or very low content of α-ketoglutarate dehydrogenase is a special characteristic of glutamic acid-producing microorganisms.

The fermentation is conducted aerobically in tanks with k_La (volumetric oxygen transfer coefficient) values around 300 millimoles O_2/liter/hour/atm. If aeration is not adequate, lactic acid is produced and the yield of glutamic acid is poor. Too much air is no good either; it gives even more lactic acid, plus some α-ketoglutaric acid. The fermentation temperature is usually 28 to 33°C, and the pH optimal for growth and glutamic acid production is 7.0–8.0. Continuous feeding of liquid ammonium hydroxide or gaseous ammonia controls pH and supplies ammonium ions to the fermentation. Modern digital computers are in some cases used for event sequencing, control, and optimization of the fermentation operation.

The fermentation cycle is 24 to 48 hours. The final concentration of glutamic acid is about 150 grams/liter if glucose is used, and about 120 grams/liter if molasses is used. The overall yield of glutamic acid on sugar is about 65% on a weight basis. A portion of sugar is used for cellular growth with about 50% yield, while the sugar used for glutamic acid production has a yield of 86% according to the following equations,

$$C_{12}H_{22}O_{11} + 2NH_3 + 3O_2 \longrightarrow 2\,C_5H_9O_4N + 2\,CO_2 + 5\,H_2O$$

The recovery of glutamic acid from fermentation broth is relatively easy. The broth is clarified by adding acid to pH 3.4, heating to 87°C, holding for a sufficient time to coagulate suspended solids, and filtering the coagulated broth to yield a clarified broth. The clarified broth is concentrated by evaporation, and then crystallized to yield glutamic acid. Other recovery schemes exist.

MSG is an important flavor enhancer for natural and processed foods. It is also good for protecting the flavor and color of preserved foods and suppressing off-flavors. MSG is selling for about $2 a kilogram, and glutamic acid (99.5%) about $4 a kilogram. Glutamic acid is not an essential amino acid. However, it has some pharmaceutical uses. It also improves the growth of pigs.

The major manufacturers of glutamic acid include Ajinomoto, Asahi Chemical, Kyowa Hakko (Japan), Stauffer, Commercial Solvents (U.S.), Orsan (France), Eridamia (Italy), Mi-Won (S. Korea), and Wei Wang (Taiwan). The estimated world production is about 200,000 tons a year. Over half this quantitiy is produced in the Orient, where MSG is used widely for food seasoning.

Lysine

Lysine, biologically active in its L-configuration, is an essential amino acid in human and animal nutrition. It is contained in good measure in fishmeal and in soybean meal. Plant products, particularly cereal grains like corn, wheat, and rice, are usually low in lysine. Because of lysine deficiency, these grain proteins are of poorer quality than animal-derived proteins which contain higher levels of lysine and command much higher prices. For applications in human food, lysine in its salt forms can be added to cookies and bread, and in solution can be used to soak rice. As most animal feed rations are based on maize and other grains, supplementing the feedstuffs with lysine (plus methionine) significantly improves their nutritional value for breeding poultry and pigs.

Lysine can be obtained by several different processes:

1. Isolation from natural sources.
2. Chemical synthesis/enzymatic racemization.
3. Enzymatic conversion.
4. Fermentation.

Chemical pathway to lysine, although perfectly practical and giving high yields, results in a DL-racemic mixture. The resolution of the mixture to L-configuration is costly.

The enzymatic process uses DL-aminolactam as starting material. It consists of the racemization of D-aminolactam and the hydrolyzation of L-aminolactam to form lysine. The reactions are shown in Figure 18.29. One-hundred grams/liter of DL-aminolactam can be con-

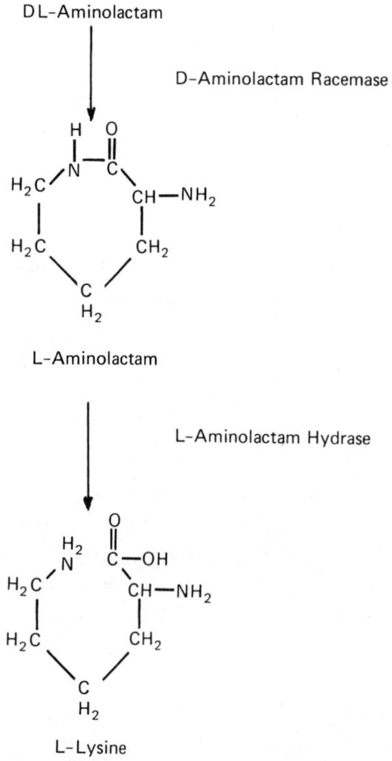

DL-Aminolactam

D-Aminolactam Racemase

L-Aminolactam

L-Aminolactam Hydrase

L-Lysine

Fig. 18.29 Enzymatic steps to form lysine from DL-aminolactam.

verted into lysine in 25 hours with 100% molar yield.

The fermentation process to produce lysine was first developed by Pfizer. *Escherichia coli* synthesized its own lysine requirements by converting carbohydrate and ammonia to α, ϵ- diaminopimelic acid (DAP). Its decarboxylation resulted in lysine. The industrial fermentation employed an *E. coli* mutant devoid of DAP-decarboxylase. It accumulated substantial DAP in the culture medium. After maximum DAP yields were attained, a second organism, another *E. coli* strain or *Aerobacter aerogenis*, was used as a source of DAP-decarboxylase to produce lysine. This second organism was devoid of lysine decarboxylase. Lysine, thus, accumulated in the broth. The reactions are shown in Figure 18.30.

A single-stage or directed fermentation process for producing lysine was later made possible by using mutants of *Micrococcus glutamicus (Corynebacterium glutamicum)* or *Brevibacterium flavum*.

Both auxotrophic and regulatory mutants have been obtained for overproduction of lysine. Figure 18.31 shows the biosynthetic pathway of lysine and its metabolic controls in *Corynebacterium glutamicum*. The key enzyme, aspartate kinase, is under the influence of the concerted feedback inhibition by lysine and threonine. Threonine, alone, also inhibits the activity of homoserine dehydrogenase. Excess homoserine increases the threonine pool within the cells, and aspartate kinase is then inhibited. An auxotrophic mutant which requires homoserine (or threonine plus methionine) and lacks of homoserine dehydrogenase, when grown under suboptimal homoserine concentrations, alleviates the feedback inhibition on aspartate kinase and shunts ASA to form lysine.

In the case of *Brevibacterium flavum*, only

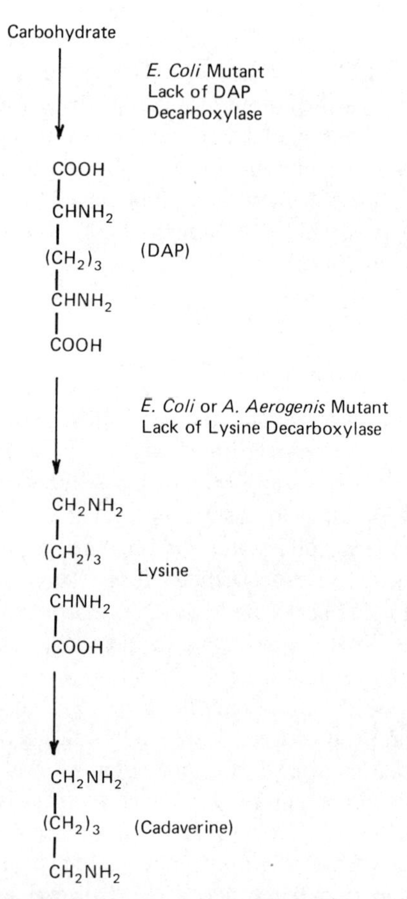

Fig. 18.30 Two-stage fermentation for lysine production.

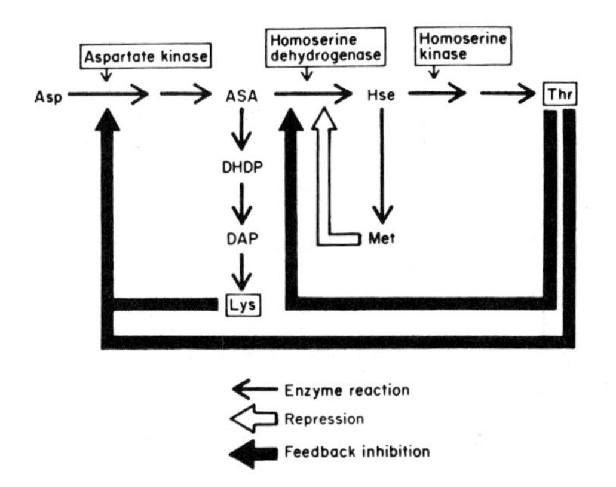

Fig. 18.31 Regulation in lysine biosynthesis in *Brevibacterium flavum* and *Corynrbacterium glutamicum.* ASA. aspartate semialdehyde; DADP. dihydrodipicolinate; Hse. hemoserine; DAP. diaminopimelate. (*Microbial Technology*, 2nd ed., Vol. 1, p. 220 (1979))

aspartate kinase is sensitive to the feedback inhibition.

Regulatory mutants, whose aspartate kinase is desensitized to the concerted feedback inhibition, have been obtained by isolation of lysine analog-resistant mutants, whose growth was not inhibited by a lysine analog, such as amino ethyl cysteine (AEC). The lysine analog behaved as a false feedback inhibitor on aspartate kinase and did not allow growth of the parent strain. Only the mutants desensitized to the feedback inhibition grew.

The *Corynebacterium glutamicum* mutant is generally used industrially in direct fermentation of lysine. Molasses is the most common carbon source. Sufficient amounts (over 30 micrograms/liter) of biotin must be included in the medium to prevent the excretion of glutamic acid. This biotin requirement is usually met by using molasses as the carbon sources. The fermentation runs at temperature about 28-33°C, and pH 6-8. High aeration is desirable. The final product concentration is around 60 gm/liter, and the fermentation cycle is 48-72 hours. The yield of lysine on carbohydrate is around 40%. The formation of lysine from molasses can be represented as follows,

$$C_{12}H_{22}O_{11} + 5\,O_2 + 2\,NH_3 \longrightarrow C_6H_{14}O_2N_2 + 6\,CO_2 + 7\,H_2O$$

Ion-exchange resins are used for isolation of lysine from fermentation broths. The eluted lysine is then crystallized from water. The most common commercial form of lysine used in animal feed is 98% lysine monohydrochloride. It is equivalent to 78.4% of the amino acid lysine, which can be metabolized by animals to body proteins. The supplementation level is about 0.5% lysine in feed. When the *Brevibacterium* mutant is used to produce lysine, the entire fermentation broth may be evaporated and dried, and the dried product used as animal feed supplement.

Food-grade lysine (as lysine monohydrochloride) is about $10/kilogram, and lysine as feed supplement is about $4-7/kilogram. Its price varies and ties in closely with the cost and availability of fishmeal and soybean cake. The annual world production of lysine is about 30,000 tons with Ajinomoto, Kyowa Hokko, Toray (Japan), Mi-won (S. Korea), Rhone-Poulenc, and Orsan (France) being the major producers. More than 80% of lysine is produced by fermentation, and the remaining by chemical or enzymatic synthesis.

The world demand for lysine is expected to continually increase. The increasing ethanol fermentation capacity for gasohol production provides an opportunity for increased lysine production. Distillers' dried grains (DDG), a by-product of ethanol fermentation and used as a protein source in animal feeds, is deficient in lysine and other essential amino acids. It needs to be supplemented with lysine for full value use.

VITAMINS

Many vitamins of medical importance can be synthesized by microorganisms. However, only "two and a half" vitamins are now produced by fermentation. Vitamin B_2 (riboflavin) and vitamin B_{12} (cyanocobalamin) are products of fermentation. The remaining half vitamin produced by fermentation is vitamin C. L-Sorbose is the precursor of vitamin C (ascorbic acid), and L-sorbose is produced microbiologically from sorbitol.

Vitamin C (Ascorbic Acid)

Figure 18.32 gives the reactions involved in the combined microbiological and chemical conver-

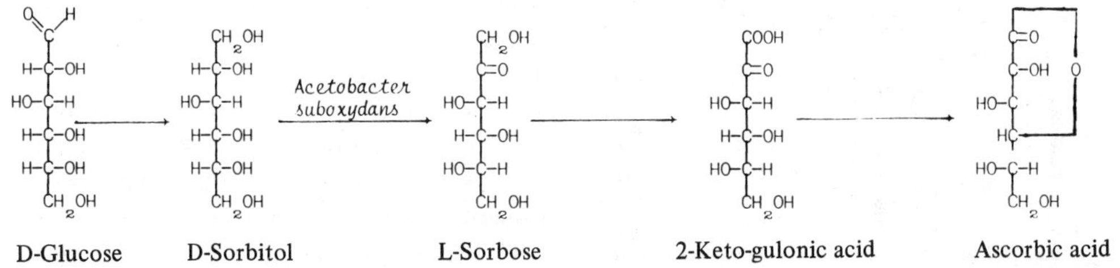

Acetobacter suboxydans

D-Glucose D-Sorbitol L-Sorbose 2-Keto-gulonic acid Ascorbic acid

Fig. 18.32 Reactions leading to vitamin C from glucose.

sions of glucose to vitamin C. Sorbitol is made by catalytic hydrogenation of glucose. L-Sorbose is produced from sorbitol by the action of several species of bacteria of the genus *Acetobacter*. The most commonly used microorganism is *Acetobacter suboxydans*. Since this organism is very sensitive to nickel ions, it is important that the medium and fermentor be free of nickel. The medium normally consists of 100–200 grams/liter sorbitol, 2.5 grams/liter cornsteep liquor and antifoam such as soybean oil. The medium is sterilized and cooled to 30–35°C, and about 2.5% inoculum is added. The tank is aerated and sometimes stirred. Yields of 80–90% of the sugar used are commonly obtained in 20–30 hours.

The chemical steps in the conversion of sorbose to ascorbic acid involve the preparation of the diacetone derivative which is then oxidized; the acetone groups are removed and the resultant 2-keto-L-gulonic acid is isomerized to the enediol with ring closure.

Sorbitol, sorbose and ascorbic acid are selling for about \$1.30, \$3 and \$10 per kilogram respectively. The estimated world production of sorbose is around 25,000 tons a year, with about half of it produced in the U.S.

Vitamin B_2 (Riboflavin)

Microbiologically produced riboflavin (structure shown in Figure 18.33) has long been available in yeast and related preparations in association with many other vitamins of the B-complex. Aside from yeast, the first organism employed primarily for riboflavin production was *Clostridium acetobutylicum*, the anaerobic bacterium used for the microbial production of acetone and butanol. Riboflavin was purely a by-product and was found in the dried stillage residues in amounts ranging from 40–70 micrograms per gram of dried fermentation solids. Further research developed improvements, adaptable only to the fermentation of cereal, grains and milk products by *Clostridium acetobutylicum* to yield residue containing as high as 7000 micrograms of vitamin B_2 per gram of dried solids. This was effected principally by reducing the iron content of the medium to 1–3 ppm, and fermenting in stainless steel, aluminum, or other iron-free tanks.

Later investigations disclosed that riboflavin could be produced by species of a yeast, *Candida flareri* or *C. guilliermondi*, when grown under aerobic conditions in a medium containing a fermentable sugar, an assimilable source of nitrogen, biotin, and less than 100 micrograms of iron per liter of medium. Yields as high as 200 mgs/liter were obtained.

Other studies on a fungus, *Eremothecium ashbyii*, and a closely related organism known

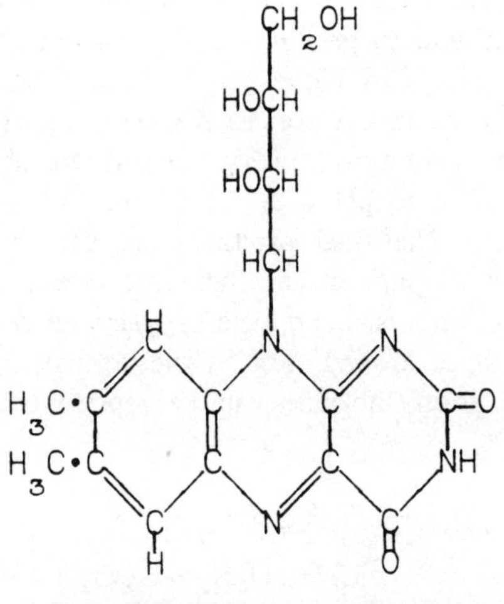

Fig. 18.33 Riboflavin: Vitamin B_2.

as *Ashbya gossypii* resulted in the production of much larger amounts of riboflavin. An aerobic process was used in which the iron content was not critical. Riboflavin was produced in large amounts by the fermentation industry using either the *Eremothecium* or *Ashbya* strains. Yields as high as 10–15 grams/liter were possible.

Late in the 60's, concurrently with the development of the riboflavin fermentation process, a synthetic means of producing riboflavin was discovered. This synthetic process dominated the production of riboflavin until 1972 when some major fermentation strain and process improvements were made with the *Ashbya gossypii* strain. Since then yields have been significantly improved. The fermentation method now accounts for essentially all the riboflavin produced.

The fermentation lasts 8 to 10 days. Cell growth occurs in the first two days, and enzymes catalyzing riboflavin synthesis are formed during the growth period. Glycine and edible oil stimulate the formation of riboflavin, but they are not its precusors. The additions of carbohydrate and oil permit the overproduction of riboflavin. The C/C yield is about 50% on carbohydrate, and about 100% on oil.

Upon completion of the fermentation, the solids are dried to a crude product for animal feed supplement or processed to a USP-grade product. In either case, the pH value of the fermented medium is adjusted to pH 4.5. For the feed-grade product, the broth is concentrated to about 30% solids and dried on double-drum driers.

When a crystalline product is required, the fermented broth is heated for one hour at 121°C to solubilize the riboflavin. Insoluble matter is removed by centrifugation, and riboflavin recovered by conversion to the less soluble form. Both chemical and microbiological methods of conversion have been used. The precipitated riboflavin is then dissolved in water, polar solvents, or an alkaline solution, oxidized by aeration, and recovered by recrystallization from the aqueous or polar solvent solution or by acidification of the alkaline solution.

The major producers of riboflavin include Merck, Hoffmann-LaRoche, and Pfizer in the U.S. with an estimated annual production of 1.3-million kilograms. The feedgrade product sells for about $35 a kilogram and the USP-grade product $55 a kilogram.

Vitamin B_{12} (Cyanocobalamin)

Vitamin B_{12}, cyanocobalamin, is an important biologically active compound. It serves as an hematopoietic factor in mammals and as a growth factor for many microbial and animal species. Its markets are divided into pharmaceutical (96–98% pure) and animal feed (80% pure) applications. All vitamin B_{12} is now made by fermentation commercially.

The microbiological production of vitamin B_{12} arose from an interesting sequence of events. For many years, liver extract was used to check cases of pernicious anemia. Investigators at Merck discovered that crystalline extracts made from liver tissue contained the highly active compound responsible for the therapeutic action. Identity with the antianemia factor in liver was then established, and the compound was called vitamin B_{12}. Later it was found that spent liquors from streptomycin and other antibiotic fermentations contained appreciable amounts of vitamin B_{12}. Soon vitamin B_{12} derived from cultures of these fermentations supplanted beef liver as a practical source of the vitamin. Around 1950, materials rich in biomass such as *Actinomycetes* or bacteria broths of antibiotic fermentations or dried sewage residues of activated sludge processes were used for isolating vitamin B_{12} either in a crude form for animal feeding or in a pure state for medicinal uses. Later, high-producing bacterial strains were specially selected for commercial production. Today vitamin B_{12} is obtained from fermentations using selected strains of *Propionibacterium* or *Pseudomonas* cultures. A full chemical synthesis of vitamin B_{12} is known. However, it requires some 70 steps and is of little value for all practical purposes.

The *Pseudomonas denitrificans* strain is most often used for commercial production of vitamin B_{12}. It only requires traditional components in the growth medium, such as sucrose, yeast extract, and several metallic salts. Dimethylbenzimidazole (10-25 mg per liter) and cobaltous nitrate (40-200 mg per liter) must be supplemented at the start of the culture in order to enhance vitamin production. Betaine stimulates the biosynthesis of vitamin B_{12}, even though it need not be metabolized by the organism. Choline also has favorable effects in activating some biosynthesis steps or altering the membrane permeability. Glutamic acid, on the other hand, stimulates cellular growth. Because of its relative cheapness and high betaine and glutamic acid contents, beet molasses (60-120 gm per liter) is preferentially used in industrial fermentations of vitamin B_{12}. The fermentation is conducted with aeration and agitation. Temperature optimal is around 28°C, and pH optimal around 7.0. The yield reported in the literature was 59 mg/liter in 1971, using a *Pseudomonas* strain. It is believed that yields of vitamin B_{12} have been significantly improved since then. A yield of 200 mg/liter was reported for vitamin B_{12} fermentations using *Propionibacteria* in 1974.

The isolation of vitamin B_{12} from fermentation broth, where it is normally present in parts per million, is a brilliant achievement on the part of the chemist and chemical engineer. About 80% of the vitamin produced is outside the cells, and 20% inside the cells. The whole broth is heated at 80-120°C for 10-30 minutes at pH 6.5-8.5. The heated broth is treated with cyanide or thiocynate to obtain cyanocobalamin. The separation can then be accomplished by adsorption on a cation-exchange resin, such as Amberlite IRC 50. Extraction can also be done by using phenol or cresol alone or in mixture with benzene, butanol, carbon tetrachloride or chloroform; or it can be done by precipitation or crystallization upon evaporation with appropriate diluents such as cresol or tannic acid. Using the extraction method, 98% pure cyanocobalamin can be obtained with a 75% yield.

The total world market for cyanocobalamin is estimated about 600 kilograms a year. The USP-grade vitamin B_{12} is selling for $10-15 per gram. The major producers include Merck (U.S.), Rhone-Poulenc, Roussel (France), Glaxo (England), Farmitalia (Italy), Chinoin and Richter (Hungary).

ENZYMES

All fermentation processes are the result of the enzyme activity of microorganisms. In fact, life itself, whether plant or animal, involves a complex myriad of enzyme reactions.

An enzyme is a protein that is synthesized in a living cell. It catalyzes or speeds up a thermodynamically possible reaction so that the rate of the reaction is compatible with the numerous biochemical processes essential for the growth and maintenance of a cell. An enzyme, like chemical catalysts, in no way modifies the equilibrium constant or the free energy change of a reaction. The synthesis of an enzyme is, thus, under tight metabolic regulations and controls which can sometimes be genetically or environmentally manipulated to cause the overproduction of an enzyme by the cell.

Being a protein, an enzyme loses its catalytic properties when subjected to agents like heat, strong acids or bases, organic solvents, or other materials that denature the protein. Each enzyme catalyzes a specific reaction or a group of reactions with certain common characteristics. The high specificity of the catalytic function of an enzyme is due to its protein nature; that is, the highly complex structure of a protein can provide both the environment for a particular reaction mechanism and the template function to recognize a limited set of substrates.

Enzymes are used quite extensively now as industrial catalysts. They offer the following advantages in comparison with chemical catalysts:

1. They are specific on action, thus minimize the occurrence of undesirable sidereactions.
2. They are relatively cheap when used in crude form.
3. They are effective for chemical conversions within the physiological range of pH and at low temperatures and pressures.
4. They are relatively nontoxic and, thus, acceptable for applications in food processes and medicinal treatments.
5. They are effective within a wide range of substrate concentrations.

The enzymes of most economical importance, such as amylases, glucose isomerase and proteases, will be discussed individually. Altogether

TABLE 18.6 Some Industrially Important Enzymes Other Than Amylases, Proteases and Glucose Isomerase.

Enzyme	*Microbial Sources*	*Applications*
Amino acylase	Bacteria	*L-Amino acid production.
Asparginase	*Escherichia coli*	*Treatment of acute lymphatic anaemia. *Remission of lymphatic leukemia. *Anticancer therapy.
Catalase	*Aspergillus niger*	*Removal of trace of hydrogen peroxide. *Used with glucose oxidase.
Cellulase and Hemicellulase	*Trichoderma reesei* *Aspergillus niger*	*As a digestive aid. *Production of sugar syrup.
Dextranase	*Penicillium funiculosum* *Trichoderma sp.*	*Prevention of dental plague. *Removal of dextran impurities in sugar refining.
Glucose oxidase (Notatin)	*Aspergillus niger* *Penicillium notatum*	*As oxygen scavenger in food industry. *Combined with catalase for removal of glucose or oxygen. *Combined with peroxidase for quantitative determination of glucose.
Invertase	*Saccharomyces cerevisiae*	*Sucrose hydrolysis to form glucose and fructose which are sweeter and have lower crystallinity. *Used in jam making and chocolate manufacturing.
Lactase	*Kluyveromyces fragilis* *Kluyveromyces lactis* *Lactobacillus sp.* *Saccharomyces lactis*	*Lactose hydrolysis to form glucose and galactose which are sweeter and more soluble.
Lipase	*Aspergillus niger* *Candida lipolytica* *Mucor javanicus* *Rhizopus arrhizus*	*As a digestive aid. *Used in waste treatment. *Used in tanning. *Used to improve flavor.
Pectinase	*Aspergillus niger* *Aspergillus sp.*	*Clarification of fruit juice. *Increased juice extraction. *Pectin decomposition and viscosity reduction.
Penicillin acylase (Penicillin amidase)	*Bacillus megaterium* *Escherichia coli*	*Formation of 6-APA from penicillin for production of semi-synthetic pencillins.
Pullulanase	*Aerobacter aerogenes* *Klebsiella aerogenes*	*Increase of fermentability of starch worts or syrups. *Structural determination of polysaccharides.
Rennets	*Bacillus polymyxa* *Bacillus subtilis* *Endothia parasitica* *Mucor miehei* *Mucor pusillus*	*As coagulants in cheese curd making.

they account for almost 90% of the total sales of enzymes. Other industrially important enzymes with relative small sales volumes are listed in Table 18.6 along with their microbial sources and commercial applications. In addition, there are many microbial enzymes that are used for analytical, clinical, and research purposes; they include hexokinases, pyruvate kinase, uricase, glucose-6-phosphate dehydrogenase, amino acid oxidase, aminopeptidase, and others.

The world sales of enzymes is estimated to be about 200-million dollars a year. Forty-eight percent of the sales is generated from proteases and microbial rennet, 30 percent from amylases, and 12 percent from glucose isomerase. Novo and Gist-Brocades have the largest share of the enzyme market.

Amylases

The commercial importance of amylolytic enzymes is rapidly increasing. These enzymes catalyze the hydrolytic reactions of amylose (the unbranched starch) and amylopectin (the branched starch). Amylases, according to their difference in modes of action, can be divided into:

1. α-Amylase which hydrolyzes α-1,4-linkages randomly to yield a mixture of oligosaccharides, maltose, and glucose.
2. β-Amylase which cleaves away successive maltose units from the nonreducing end of starch to yield maltose quantitatively.
3. Glucoamylase (or amyloglucosidase) which chops off glucose successively to yield glucose.
4. The debranching enzyme such as pullulanase which attacks the α-1,6-linkages at the branching point of amylopectin.

Amylases are used extensively in the following ways:

- for production of grain syrup, glucose syrup, liquid glucose and crystalline glucose,
- for production of high fructose corn syrup in connection with glucose isomerase,
- to solubilize and saccharify starch for alcohol fermentation in brewing, distilling, and fuel industries,
- to modify the viscosity of starch used in coating printing papers,
- to remove starch sizes applied to cotton thread before weaving in the textile industry,
- for production of maltose-containing syrups in brewing and baking industries,
- to reduce viscosity of sugar syrups used in various food and sugar products, and
- as a component in digestive aids.

α-Amylase is produced commercially using both fungal and bacterial species. The fungal amylase has a relatively low-heat stability, and its major application is in the baking industry to supplement the variable activity of the amylase present in wheat flour. The bacterial amylase is much more heat-stable, and it is used in brewing, starch degradation, alcohol, and textile industries. The organisms commonly used for the commercial production of α-amylase include:

Fungi	*Aspergillus oryzae*
Bacteria	*Bacillus subtilis*
	Bacillus licheniformis
	Bacillus amyloliquefaciens

Aspergillus oryzae (the green fungus) can be grown in either semi-solid or submerged culture. In semi-solid culture it produces several enzymes, primarily α-amylase, glucoamylase, lactase, and protease. In submerged culture the production of α-amylase is increased, and the formation of other enzymes becomes minimal. The use of this fungal amylase in the baking industry speeds up the yeast (*Saccharomyces carlsbergenis*) fermentation, produces stiffer, more stable doughs, and improves the texture, porosity, digestability, and shelf life of bread. The fungal α-amylase delivers its optimal activity at 5-7 pH and at 50-55°C.

Different amylase-producing organisms may require different fermentation conditions for optimal enzyme production. When *Bacillus subtilis* is used, the fermentation medium may contain starch, cornsteep liquor, yeast, phosphate, and some mineral salts. The amylase treatment on starch is often short to prevent the significant accumulation of glucose which is inhibitory to the *Bacillus* amylase fermentation. The fermentation is run at neutral pH and at around 35°C. Care must be taken to prevent contamination. The time cycle is about 48 hours. Whole mash may be used directly for starch liquefaction and saccharification, or the mash may be processed to produce liquid or crystal enzyme preparation with high purity. The processing, involving filtration or centrifugation, of the bacterial fermentation broth, presents real problems to the recovery plant. Pretreatment with a coagulating or floccculating agent is often needed. The amylase produced by this *Bacillus* strain is relatively unstable, but the addition of calcium chloride improves the stability.

Following the solubilization of starch by α-amylase (often of a bacterial origin), further degradation is achieved using fungal glucoamylase. *Aspergillus niger* (the black fungus) is commonly used for the production of glucoamylase. The fungal fermentation starts with a medium containing 25-30% starch and around 10% cornsteep liquor. Incremental or continuous feeding of concentrated nutrients may be used to circumvent the problems caused by a concentrated initial medium. The fermentation pH is about 4.0 and temperature around 28°C. The fermentation has a high oxygen demand.

High oxygen tension, however, inhibits enzyme production. Zero D.O. is not atypical in this fermentation. After the completion of the fermentation in 4-5 days, the fermentor mash is cooled and filtered to remove cells and insoluble matters. Transglucosidase may be removed using clay, destroyed preferentially using proteases at certain pH's and temperatures, or inactivated by magnesium oxide. Contamination of glucoamylase by the transglucosidase activity may result in loss of 5-10% of glucose to form isomaltose and panose by a reversion process. These reversion products also impede the crystallization of glucose.

Fungal glucoamylase in combination with bacterial α-amylase make a complete enzymatic mixture for hydrolysis of starch to glucose. Prior to liquefaction, starch is gelatinized by heat treatment at temperatures above 100°C. The liquefaction to form maltodextrins is aided by the action of bacterial α-amylase. α-Amylase from *Bacillus amyloliquefaciens* functions at 5.5-7 pH and 90°C, while α-amylase from *Bacillus licheniformis* functions at 5.5-9 pH and temperatures as high as 110°C. In the subsequent saccharification process, an appropriate amount of glucoamylase from *Aspergillus niger* is added to the thinned starch (30-50% dry substance) with stirring at 55-60°C and 4-5 pH for 48-72 hours. This achieves a final D.E. (dextrose equivalence) of about 97, with about 94% of the dry weight glucose. The equilibrium concentrations of the saccharides formed by resynthesis limit the maximum degree of hydrolysis obtainable. Since the activity of glucoamylase toward the branching points (the α-1,6-linkages) is low, it may be advantageous to use a debranching enzyme such as pollulanase, early in the hydrolysis process.

The enzymatic hydrolysis of starch to glucose is commercially preferred to the acid hydrolysis route using hydrochloric acid. The enzymatic process produces fewer side products, does not involves a corrosive acid, and allows the use of less pure starch products whose protein contaminants would, upon acid hydrolysis, give amino acids and browning reactions.

The world production of bacterial α-amylase, fungal α-amylase, and fungal glucoamylase is estimated to be about 300, 15, and 300 tons a year, respectively.

Glucose Isomerase and HFCS (High Fructose Corn Syrup)

Starch degradation using α-amylase and glucoamylase produces corn syrup with glucose concentration up to 94% on a dry weight basis. The glucose can then be isomerized to a mixture of glucose and fructose by glucose isomerase which is present in many microorganisms. The product, commonly known as "high fructose corn syrup" (HFCS), typically contains 42% fructose, 50% glucose, 6% maltose, and 2% maltotriose. HFCS is also available in the form of 55% syrup, 90% syrup, or 99% crystal. Figure 18.34 gives a block diagram which shows how a corn wet-milling processor produces its starch, corn syrup, glucose, and fructose syrups.

Fructose, the monosaccharide commonly called fruit sugar, is about 50% sweeter than sucrose, the disaccharide familiarly known as table sugar. Sucrose can be hydrolyzed to invert sugar which is a mixture of fructose and glucose. Liquid invert sugar has been the major sweetener used in soft drinks and as a food ingredient.

The 42% HFCS sells for about 20 cents per pound which is 15 to 20 percent cheaper than liquid invert sugar on a dry weight basis. The soft-drink industry, shifting away from liquid invert sugar, is the major user of 42% HFCS. The baking industry ranks as the second largest user.

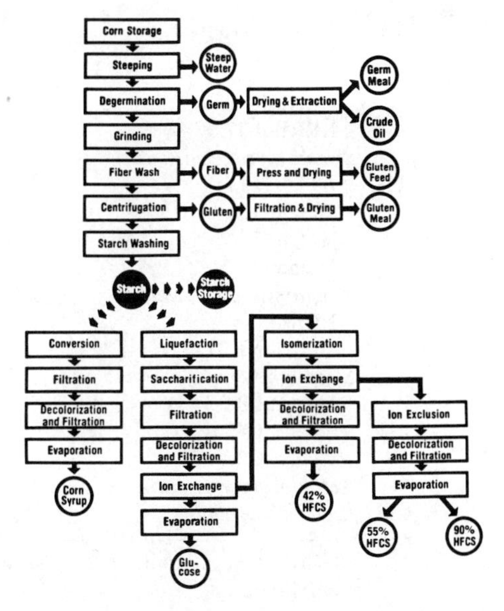

Fig. 18.34 Flowsheet of a corn wet milling process.

The 55% HFCS sells for about 25 cents per pound which is 5 to 10 percent cheaper than liquid invert sugar on a dry weight basis. Fifty-five percent HFCS, made by combining the 42% HFCS with 90% HFCS, has about the same degree of sweetness as sucrose. It is used as a sweetener and flavor-enhancer in fruit-flavored soft drinks. Fructose enhances flavors, while sucrose masks them.

The 90% HFCS, 40 to 50 percent sweeter than sucrose, sells for about 45 cents per pound which is 40 to 50 percent more expensive than liquid invert sugar on a dry weight basis. Since it is much sweeter than sucrose, less amount is needed to sweeten a product to desired levels, and sweetener calories in the product can be reduced by about one third. The major use of 90% HFCS is in dietetic foods and drinks.

The 99% fructose in crystal form is obtained by drying the 90% HFCS. It is about 70 percent sweeter than sucrose. Being an essentially pure sweetener, it allows the full taste of product flavors to develop. It is finding uses in diabetic and low calorie foods and drinks.

The HFCS industry in U.S. has a total production capacity of 3.4 billion pounds per year, which is about 1.2 billion pounds short of the demand. Among the major HFCS producers are ADM, Car-Mi (a joint venture of Cargill and Miles), A. E. Staley, Clinton/Standard Brands, CPC and Hubinger/Heinz.

Many organisms are glucose isomerase producers. Most of them produce xylose isomerase with low specificity, and glucose can be an alternative substrate for the enzyme. Table 18.7 lists microorganisms, believed to be used industrially for glucose isomerase production, along with their reported yields.

The desirable application conditions of glucose isomerase are 7.0-8.0 in pH and around 65°C in temperature.

Glucose isomerase fermentation typically has a cycle time of about two days. The fermentation conditions vary from producer to producer. Many glucose isomerase-producing organisms require xylose and cobalt for full enzyme induction. Xylose is too rare and expensive to be used in commercial fermentation processes, while cobalt ions remaining in the spent fermentation medium constitute a serious environmental hazard. However, mutants that do not require xylose and cobalt have been obtained for commercial production of glucose isomerase.

Almost every known glucose isomerase is an intracellular enzyme, and can only be extracted from the cells in relatively low concentrations. It is, thus, expensive to use it as a soluble and "once only" enzyme. Immobilized enzyme/cell

TABLE 18.7 Glucose Isomerase (GI) Producing Microorganisms

Microorganism	*Patent Assignee*	*Yield (GIU*/Liter)*
Arthrobacter sp.		
NRRL-B-3726	R. J. Reynolds	4,720
NRRL-B-3728		4,440
Streptomyces welmorensis		
ATCC-21175	Standard Brands	4,640/g
Mutant-1		7,540/g
Mutant-2		6,680/g
Mutant-3		6,000/g
Streptomyces olivaceus		
NRRL-3583	Miles	2,560
Mutant NRRL-3916		2,960
Streptomyces olivochromogenes		
Mutant CPC-3	CPC	4,800-11,440
CPC-4		5,700-9,680
CPC-8		3,960-4,440
Actinoplanes missouriensis	Anheuser-Busch	2,500-35,200
Bacillus coagulans	Novo	not known

*1 GIU = The Amount of Enzyme Which Converts 1 μmole Glucose to Fructose/Minute.

technology is the key scientific advancement which has made the use of glucose isomerase for HFCS production a commercial success. In many industrial cases, whole microbial cells are immobilized by physical means such as entrapment or encapsulation in polymeric materials or by chemical means such as intermolecular cross-linking with glutaraldehyde or covalent binding with diazotized diamino compounds. Commercially, soluble glucose isomerase is also immobilized on DEAE-cellulose. One industrial immobilization process goes as follows: The enzyme-containing cells are homogenized and mixed with glutaraldehyde and inert protein. The resulting gel is shaped into suitable granules which are then washed and dried.

The immobilized glucose isomerase can usually be used for over 1,000 hours at a temperature around 65°C. When the column enzyme activity decreases, the flow rate of the incoming glucose syrup can be adjusted so that the conversion of glucose to fructose is maintained constant. The world production of glucose isomerase is estimated to be around 50 tons a year, and it is expected to increase rapidly in the next few years due to the rising demand for HFCS.

Proteases

Proteolytic enzymes are by far the most important of the commercially available industrial enzymes. These enzymes, being essential parts of the metabolic system of most living organisms, can be isolated from innumerable sources. Microbial proteases with widely different properties are commercially produced. *Bacillus* protease, however, represents more than 95% of the sale of all proteases.

The most important use of *Bacillus* proteases is in detergents. Proteinaceous dirt often precipitates on clothes, and it coagulates during the normal washing process. The addition of proteolytic enzymes to the detergent can easily dissolve proteinaceous stains, which are otherwise difficult to remove.

The alkaline serine protease of *Bacillus licheniformis*, also known as "Subtilisin Carlsberg," is the preferred protease in most nonionic and anionic detergents. It attacks many peptide bonds and easily dissolves proteins. It may be used at temperatures up to 65°C, and its pH optimum is close to 9.0, the pH normally used in washing fluids.

A few other proteases are also used in detergents. The serine protease of *Bacillus amyloliquefaciens* has found some applications, presumably because of its substantial content of α-amylase. This may be an advantage for some applications.

Sales of proteases were small and relatively unimportant till about 1965. Then, the use of proteases in detergents created an explosion in the enzyme industry. But in 1971 the backlash came. Allergic symptoms were discovered in some workers handling enzymes in detergent factories. The public, particularly in the U.S., was scared, and proteases were taken out of most detergents. It was later found that with proper precautions in handling every risk can be eliminated by using proteases in liquid form or by encapsulating the enzymes. In the wake of the 1973 oil embargo and energy crisis, the increasing use of cold water laundry resulted in a rapid regrowth of the use of alkaline proteases in detergents. A steady growth in this area is expected.

In the tanning industry, alkaline protease from *Bacillus amyloliquefaciens* is used in combination with sulfite for hide treatment and dewooling. A protease from an alkalophilic *Bacillus sp.* is successfully used for dehairing of ox hides in combination with lime apparently because of the enzyme's stability at pH's as high as 12.

In the brewing industry, there is a development towards substitution of malt with unmalted barley and amylase, glucanase, and protease of microbial origin. The neutral protease from *Bacillus amyloliquefaciens* and the thermostable neutral protease *Bacillus subtilis var thermoproteolyticus* have been used by brewers successfully to hydrolyze barley proteins into amino acids and peptides.

Furthermore, *Bacillus* proteases are used in the preparation of protein hydrolyzates. The limited hydrolysis of soybean protein is an especially promising application.

The protease fermentation of the *Bacillus* bacteria takes place under strictly aseptic conditions in conventional equipment for submerged fermentations. The aeration rate is about 1 vvm (volume of air per volume of medium per minute). Vigorous agitation is used to improve air distribution and oxygen transfer. The fermentation temperature is around 37°C, and time cycle is 2 to 4 days.

The composition of the fermentation medium is important to the yields of protease. Proteins of many different sources are used in commercial media. Carbohydrates are used as an energy source. The C/N ratio is important to the success of the process. Protein should be present in high concentration, and carbohydrate must not be in excess. A convenient way to obtaining this is to conduct fed-batch fermentation feeding carbohydrate during the run and maintaining the carbohydrate concentration below 1%. Continuous fermentation of protease in commercial scale is not yet known.

The recovery and finishing of *Bacillus* protease involves the following steps:

1. Cooling to about 4°C to prevent microbial spoilage,
2. Precipitating undesirable salts using flocculents or filter aids,
3. Removing all particles by centrifugation or filtration,
4. Removing pigments and odors with activated carbon treatment,
5. Removing bacterial contaminants by filtration,
6. Concentrating at low temperature either by reverse osmosis or by vaccum evaporation.
7. Recovering protease using precipitation by salts (ammonium or sodium sulfate) or solvents (acetone, ethanol or iso-propanol),
8. Recovering the precipitate by filtration,
9. Drying at low temperature, and
10. Encapsulating granules of enzymes in a non-ionic sufactant.

The world production of *Bacillus* protease is estimated to be around 500 tons, with sales close to 100 million dollars.

Proteases are also produced for special applications by many different organisms such as *Streptomyces griseus, Aspergillus niger, Aspergillus flavusoryzae, Mucor miehei, Mucor pusillus* and *Endothia parasitica*. The applications include protein hydrolysis to amino acids, milk coagulation, cheese manufacturing, and preparation of digestive aids.

SCP (SINGLE CELL PROTEIN)

The term "single cell protein" (SCP) was coined at MIT by Professor C.L. Wilson in 1966 to represent the cells of algae, bacteria, yeast, and fungi grown for their protein contents. It should be noted that these microbial cells contain, in addition to proteins, carbohydrates, lipids, nucleic acids, vitamins, and minerals.

In the late 60's and early 70's, considerable research, development and commercial interests from all over the world were directed toward SCP production against a background of an increasing deficit in protein supplies, an increasing output, at the time, of cheap petrochemicals, increasing interest in converting waste materials into profitable products, and relatively stable supplies and prices of agricultural products.

Most of this research and industrial effort did not achieve commercialization of SCP, largely due to political, socialogical, and psychological factors. However, the SCP era witnessed several important advances in fermentation technology, such as continuous culture, improved bioreactor design, and computer control of fermentation processes. The following paragraphs summarize, by substrates, the status of major SCP projects.

Carbon Dioxide for SCP

The photosynthetic *Chlorella* (the green algae) has been grown on carbon dioxide mostly in the oriental countries. Its market is largely in Japan. The estimated annual consumption is around 20,000 tons.

Methane for SCP

Methane has been of interest as a substrate for SCP production because it is available in large quantity and high purity and, in many cases, is wasted at oil producing wells. Many bacteria and some fungi utilize methane as a carbon and energy source for growth. A bacterial SCP continuous process was developed to the pilot scale by Shell in England using *Methanomonas capsulatus* and mixed cultures with good productivities and yields. The following factors forced the operation to be discontinued:

- Limited transfer of methane and oxygen from the gas phase to the bacterial cells.
- Explosive hazards.
- High cooling demands because of excessive heat production.
- Presence of inhibitory products.
- High capital requirements because of problems stated above.

N-Paraffins for SCP

The *Candida* yeast grown on n-paraffins was widely investigated for production of SCP. Continuous processes, using air-lift fermentors and operating nonaseptically in most cases, were developed by British Petroleum (using *Candida lipolytica*), Gulf (using *Candida tropicalis*), Kanegafuchi (using *Candida sp.*), Liquichemica Biosintesi (using *Candida maltosa*), and others.

Italproteine, a joint venture of BP and Anic (the petrochemical arm of the Italian state-owned Ente Nazionale Idrocarburi), constructed a 100,000 tons/year plant at Sarroch, Sardinia. Italproteine tradenamed its product "Toprina," made by BP-developed know-how. Liquichimica Biosintesi, owned by Liquichimica (a subsidiary of Italy's Liquigas), built its 100,00 tons/year plant at Saline di Montebello, Calabria. The product was tradenamed "Liquipron," made by the Kanegafuchi process. The operation of these two SCP plants has been blocked by the Italian government's health authorities on questions of levels of hydrocarbon residues.

Plans to construct large-scale yeast SCP plants (60–120,000 tons/year each) in Japan by Kanegafuchi, Kyowa Hakko, Asahi, Mitsui Toatsu, and Dainippon were cancelled because of public protests over questions on safety of the products and subsequent Japanese government regulatory action banning the use of these products.

However, yeast SCP plants are believed to be in operation in the U.S.S.R. using purified n-alkanes as substrates. The overall capacity is estimated above 300,000 tons/year.

Methanol for SCP

The events in Italy and Japan, as well as the dramatic increases in petroleum prices in the mid-70's, stimulated much increased effort in selecting substrates other than n-paraffins for SCP production. Methanol has received perhaps the greatest attention. Use of methanol as a substrate for SCP production has a number advantages, principally the following:

- High solubility in water.
- Low explosive hazards of methanoloxygen mixtures.
- Moderate heat liberation (3.3–5.7 kcal/g of cells vs 4.4–8.0 kcal/g of cells grown on n-paraffins).
- Freedom from traces of aromatic hydrocarbons.
- Ease of removal from the microbial cell product.

Organizations in many parts of the world have worked on methanol processes using both yeasts and bacteria. They included Shell and ICI in England, Societa' Italiane Resine in Italy, Hoechst AG and Gelsenberg in West Germany, Norsk Hydro in Norway, AB Marabou in Sweden, Yissum Research Development in Israel, Phillips Petroleum and Tenneco in U.S., and Mitsubishi in Japan. In addition, much work has been done in academic institutions. Among the industrial organizations, ICI and Hoechst are the ones that remain most active. Both use bacterial processes which have the advantages over yeast processes of higher growth rates, productivities and yields.

Hoechst operates a 1000-ton/year pilot plant near Frankfurt. *Methylomonas clara* is grown on a mixture of air and methanol containing ammonia, water, and other essential mineral nutrients. The dried product, tradenamed "Probion," contains some 70% protein, 10% nucleic acids, 8% fats, 7% mineral "ash," and 5% water. A purified version, with essentially all nucleic acids and fats removed, has a protein content of more than 90%, with minerals and water making up the balance. Called "Probion-S," Hoechst intends to market this purified product for human consumption.

ICI has developed the most advanced process for SCP production from methanol. *Methylophilus methylotrophus* is the organism used. A novel type of air-lift bioreactor, called the "pressure cycle fermentor," has been developed for use in the process. It is designed to maintain high oxygen transfer rate and homogeneous liquid phase without excessive shear, to remove the heat liberated during the growth phase at high productivities, and to avoid contamination problems encountered with conventional stirred fermentors because of leaks through drive shafts and mechanical seals. Air and fresh medium are introduced at the base of a tall vertical column—the "riser." The riser is connected horizontally to another smaller diameter column, the "downcomer," at both the top and the base. A driving force for the movement of broth at high velocities is provided by the difference in air hold-up between the riser and the downcomer. The high hydrostatic head and the level of turbulence in the riser provide excellent oxygen transfer because of the increased pressure, the small bubble size and the mixing; up to 50% of the incoming oxygen is transferred. Spent air and CO_2 (stripped by the nitrogen content of the air) are disengaged in the upper horizontal section. The broth is cooled in a simple heat exchanger set in the downcomer. ICI has also developed a proprietary agglomeration process for the initial separation of bacterial cells from the growth medium, which permits the final centrifugation of a much higher solid slurry than is possible otherwise. A 1000-ton/year pilot plant has been operated for three years. A 70,000-ton/year plant is being started up now. A schematic diagram of the ICI protein process is shown in Figure 18.35. The final product, tradenamed "Pruteen," contains more than 70% protein and is to be sold as animal feed. The ICI pressure cycle fermentor at Billingham, England, measuring 200 ft high at 25 ft in diameter, is claimed to be the largest such vessel in the world.

The naturally occurring *Methylophilus methylotrophus* used in the ICI-SCP process is relatively inefficient in assimilating nitrogen. The inefficiency is caused by the absence of the gene that codes for glutamate dehydrogenase, the enzyme essential for incorporating hydrogen and ammonia into glutamic acid, the major pathway for the formation of α-amino groups directly from ammonia. Instead, *Methylophilus methylotrophus* posseses the gene for producing glutamate synthase, which leads to a less efficient nitrogen assimilation pathway. It is reported that ICI has successfully overcome the problem by first inactivating the glutamate synthase enzyme using classical mutation techniques and then inserting into *Methylophilus*

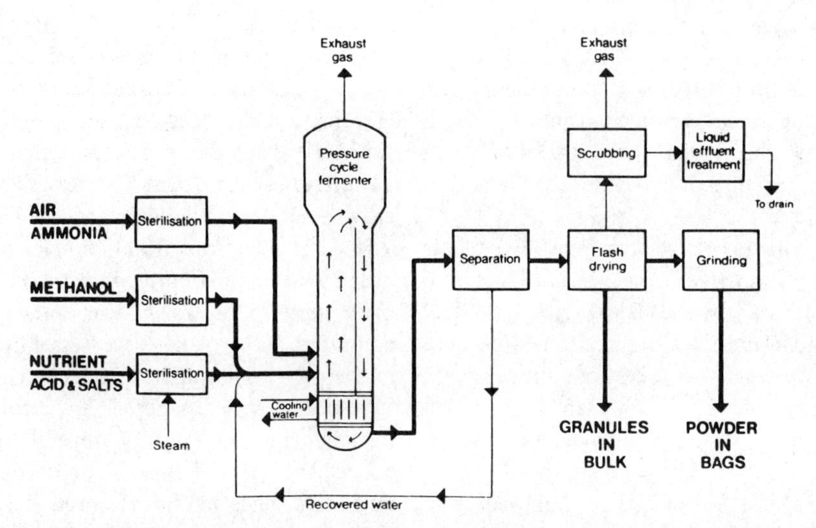

Fig. 18.35 Flowsheet of the ICI SCP process. (*Process Biochem.*, **12**, No. 1, p. 30 (1977)).

methylotrophus the glutamate dehydrogenase gene from *Escherichia coli*. The genetically modified bacterium may be used in the existing ICI-SCP plant after further study. If so, this will mark the first large scale application of genetic engineering in Western Europe.

Ethanol for SCP

Ethanol can be used by certain bacteria and yeasts as a substrate for SCP production. It is completely water-soluble, available in large quantity and high purity, and easy to store and transport. In addition, it has the advantage over methanol of low toxicity and good acceptability to regulatory agencies as a raw material for producing a food-grade end product. Organisms of interest that utilize ethanol in SCP production include:

1. *Candida utilis* (used by Pure Culture Products, the previous Amoco Foods),
2. *Acinetobacter calcoaceticus* (used by Exxon and Nestle), and
3. *Candida acidothermophilum* and *Candida ethanothermophilum* (used by Mitsubishi Petrochemical).

The yeast-based SCP processes using ethanol as a substrate have been developed to a greater extent than the bacterial processes.

The Pure Culture Products, Inc. in the U.S. has taken the approach of using torula yeast, which is well known and has long been established as a food ingredient. As such, they have been able to obtain clearance for their ethanol-grown product as a food ingredient, and they are already marketing material from their plant in Hutchinson, Minnesota, which is ultimately designed to produce up to 15 million pounds of product per year. The material, tradenamed "Torutein," is being sold in the U.S. as a nutritional supplement and flavor enhancer for such processed foods as meat patties, pasta, baked goods, frozen pizzas, and sauces.

In the Pure Culture Product's continuous process, ethanol concentration is maintained at approximately 200 ppm. The nitrogen source, either aqueous or anhydrous ammonia, is supplied continuously to maintain pH in the desired range. Other macro and micro elements are also supplied continuously. All liquid streams except ammonia are sterilized at 149°C. Ammonia is sterilized by filtration, and air is sterilized by compression and filtration. Typical operating conditions are 30°C, pH 4.6, and aeration and agitation to give oxygen absorption rates in the range of 100 to 140 millimoles of O_2 per liter per hour. Cell concentrations of 6-7 gm/liter on a dry weight basis are obtained at a dilution rate of about 0.3 hr^{-1}. The Hutchinson plant produced the pure culture food yeast continuously without interruption for 75 days from February 18 to May 4, 1980. It was then shut down for routine maintenance.

Carbohydrates for SCP

The raw materials used for SCP production in this category are carbohydrate-containing wastes or by-products of various manufacturing processes. These carbohydrate-containing materials include molasses, sulfite liquor waste, whey, and cellulosic wastes.

Yeasts grown on molasses are produced for both food and feed uses by Standard Brands, Universal Food, and Yeast Products in the U.S., and by many other processors abroad.

Sulfite liquor waste is used for feed yeast production by Boise Cascade and Lake States Yeast (a division of St. Regis Paper) in the U.S., by Ontario Paper in Canada, by Attisholz in Switzerland, and by Aanekoski in Finland. Yeast plants utilizing sulfite liquor or wood hydrolyzate are also in operation in Austria, Poland, East Germany, Japan, and the U.S.S.R.

One well-publicized SCP process using spent sulfite liquor as a substrate is the Pekilo process developed by Tampella in Finland. The Pekilo process is said to be the first commercial continuously operating process in which filamentous fungi are cultivated for SCP production. A plant, based on the Pekilo process, was built jointly by United Paper Mills and Tampella at Jamsankoski in Finland with a capacity of 10,000 tons/year using a microfungus *Paecilomyces varioti*. Its product, the Pekilo protein, is sold for animal feed.

Whey, a waste of cheese manufacturing, is

fermented with *Saccharomyces fragilis* by Amber Laboratories, Stauffer Chemical, and Kraft in the U.S. for both food and feed uses. For the production of food yeast, the yeast cream obtained by centrifugation is spray dried. For feed yeast, the whole fermentor broth is concentrated by evaporation and then spray dried.

Cellulose and cellulosic wastes have received a great deal of attention in recent years as potential substrates for SCP production. Delignification, decrystallization, hydrolysis, and microbial growth are the important steps involved in SCP production from cellulosic materials. At the present time, no commercial production of SCP from cellulose is known. However, the utilization of cellulose, the most abundant renewable organic resource on earth, is of great importance from both technical and commercial points of view, not only for SCP production, but also for chemicals and fuels.

ANTIBIOTICS

The science of microbiology was established by the French chemist, Louis Pasteur. Among his many contributions, the discovery of the existence of microscopic forms of life should be especially pointed out: yeasts were established as the cause of the alcohol fermentation of wine. He later established pathogenic bacteria as the cause of many infectious diseases and even recognized viruses as causative agents of other infections.

Shortly before World War II, another significant role of microorganisms began to emerge. Alexander Fleming, an English bacteriologist, is credited with first suggesting that the product of one microorganism might be used to inhibit the growth of another. Fleming observed that a chance contaminant (a *Penicillium* mold) clearly prevented the growth of a pathogenic *Staphylococcus* he was culturing in a Petri dish. He succeeded in establishing some of the simple properties of the mold product penicillin and published his results in 1929. Nearly a decade later, another group of English biochemists undertook to further examine the phenomenon in the course of a broad research program for better chemotherapeutic agents. Florey, Chain, Heatley, and Abraham, of Oxford University, succeeded in isolating a small quantity of penicillin concentrate and by 1941 unequivocally demonstrated its potential usefulness in *Staphylococcus* septicemia. British government officials recognized its possible usefulness in military medicine, but could not further develop the discovery because of the serious war conditions then prevailing. The research was brought to the attention of the government and industy in the United States, with the result that an international government-industry research program was established to produce this remarkable chemotherapeutic agent. Success of this program established a new class of powerful therapeutic agents, the antibiotics, which have revolutionized medical practice. Thus the product of one microorganism is used to combat an infection caused by another.

To supply the huge amounts of antibiotics needed in modern medicine, the fermentation industry, too, has undergone a virtual revolution.

Since the early 1940's, an intensive search for new and useful antibiotics has been in progress throughout the world. In a period of a little more than 35 years, over 4300 antibiotics have been discovered from microbial origins. More than 30,000 semisynthetic antibiotics have been prepared. Of the thousands of antibiotics known, about 100 are produced on a commercial scale, with approximately one half of them prepared by a combination of microbial synthesis and chemical modification.

Penicillins, cephalosporins, and tetracyclines are the most important antibiotic groups in terms of production tonnages and dollar sales.

Penicillins

The original mold observed and preserved by Fleming was a strain of *Penicillium notatum*, a common laboratory contaminant. Later cultures of *Penicillium chrysogenum* were found to be better producers of penicillin, and the present industrial strains have been derived from this species. The original strains produced the antibiotic only by surface fermentation methods and in very low yields, a few ppm. Gradually, improved media and the eventual discovery of strains productive under submerged aerobic fermentation conditions led to dramatic yield increases which made commercial production a reality. Subsequent improvements, principally in culture selection and mutation to productive strains, further improved yields, until today broths often contain 20-30 gm of penicillin per liter. Dramatic price reduction has come with improved production. For instance, a million-unit vial of penicillin (1667 units = 1 mg of potassium penicillin G) had a wholesale price of $200 in 1943. In 1952, the same vial had a wholesale value of only $1.30. Today, a million units of potassium penicillin G is selling for as little as $0.05 or approximately $20 per kilogram of free acid.

The original *P. chrysogenum* strains produced large amounts of unwanted yellow pigments which were difficult to remove from the recovered penicillin. Today, nonpigmented mutants, a strain known in the industry as Wisconsin 49-133 (or progeny therefrom), are universally employed. The desired culture is propagaged from a laboratory stock in small flasks and transferred to plant inoculum tanks. After 24 hours these are used to inoculate larger fermentors which contain a typical production medium such as that shown below.

Components	*Grams/Liter*
Corn steep liquor	30
Lactose	30
Glucose	5.0
$NaNO_3$	3.0
$MgSO_4$	0.25
$ZnSO_4$	0.044
Phenyl acetamide (precursor)	0.05
$CaCO_3$	3.0

The medium is usually sterilized batch-wise, cooled to 24°C, and inoculated. The time of fermentation may vary from 60-200 hours. Sterile air is blown through the tank, usually at a rate of one volume per volume per minute.

When penicillin concentration reaches its peak potency, as determined by microbiological or chemical assays, the broth is clarified by means of rotary vacuum filters. The penicillin, being acidic, is extracted from the aqueous phase into a solvent, such as methyl isobutyl ketone or amyl acetate, at a pH of 2.5 by means of a continuous countercurrent extractor, such as a Podbielniak. The penicillin extract is then re-extracted with an aqueous alkaline solution or a buffer at a pH of 6.5-7.0. A 90 percent recovery is made at this step. The aqueous solution is chilled, acidified, and extracted again with a solvent, such as ether or chloroform. The solvent extract is then re-extracted into water at a pH of 6.5-7.0 by titration with a solution of base. The base used depends on which salt of penicillin is desired. The popular forms are sodium or potassium salts. A typical flow cheet for antibiotic recovery is shown in Figure 18.15.

Table 18.8 gives the structural formulae of the "natural penicillins," comprising several closely related structures with aliphatic and aromatic substitutions to the common nucleus. The early impure product contained mixtures of these types. For several reasons penicillin G became the preferred type and the crystalline product of commerce. Phenylacetic acid or its derivatives are used as precursors in the fermentation medium to enhance penicillin biosynthesis and suppress the production of the less desirable types.

The fact that proper selection of precursors could lead to new variations in the penicillin side chain offered the first source of synthetic penicillins. Penicillin V, derived from a phenoxyacetic acid precursor attracted clinical use because of its greater acid tolerance, which made it more useful in oral administration.

The widespread use of penicillin eventually led to a clinical problem of penicillin-resistant staphylococci and streptococci. Resistance for the most part involved the penicillin-destroying enzyme, penicillinase, which attacked the beta-lactam structure of the 6-aminopenicellanic acid nucleus (6-APA).

In 1959, Batchelor and co-workers in the Beecham Research Laboratories in England discovered that the penicillin nucleus, 6-APA, accumulated during fermentation when side chain precursors were omitted. This 6-APA could be used for the chemical synthesis of entirely new types of penicillin by coupling with new side-chains. Shortly thereafter, several sources of penicillin amidase were found which would cleave the phenylacetyl side chain from

TABLE 18.8 Structural Formula of Natural Penicillins

Formula: O=C(R)–HN–CH–CH(–S–)C$(CH_3)_2$, with O=C——N——CHCOOH ring closure

Type of Penicillin	*Side Chain R Substitutions*
(G) Benzyl	C_6H_5–CH_2–
(X) *p*-Hydroxybenzyl	HO–C_6H_4–CH_2
(F) 2-Pentenyl	$CH_3-CH_2-CH=CH-CH_2-$
(Dihydro F) *n*-Pentyl	$CH_3-CH_2-CH_2-CH_2-CH_2-$
(K) *n*-Heptyl	$CH_3-CH_2-CH_2-CH_2-CH_2-CH_2-CH_2-$
(V) Phenoxy	C_6H_5–OCH_2–

penicillin G, thus producing a more economical source of 6-APA. A vast number of "synthetic penicillins" have been generated and a few have achieved clinical importance. Several objectives were sought:

1. To broaden the inherent utility of penicillin to include gram-negative pathogens not inhibited by the natural penicillins.
2. To improve its stability and absorption.
3. To increase its resistance to penicillinase-producing pathogens.
4. To decrease allergenicity.
5. To improve other factors pertinent to clinical use.

The broad objectives have been achieved with varying degrees of success.

Table 18.9 shows the structures of some of the semisynthetic penicillins which have become important chemotherapeutics. Semisynthetic penicillins, on the average, sell for about $150 per kilogram.

Cephalosporins

In 1948 Professor Guiseppe Brotzu isolated a *Cephalosporium* culture from seawater near the sewer discharge of Cagliarri, Sardinia. This culture produced a broth inhibiting both gram-positive and gram-negative bacteria. The team of Florey, Abraham, and Newton at Oxford University isoalted a compound identified as cephalosporin N. During the same period a group in the Michigan Department of Health isolated synnematin B from another strain of *Cephalosporium.* Synnematin B and cephalosporin N proved to be identical. Structure studies eventually proved these antibiotics to be a new type of penicillin, α-aminoadipyl-6-APA (also called penicillin N), a naturally produced penicillin with gram-negative activity.

In the course of studies on the Brotzu strain of *Cephalosporium*, Abraham and Newton detected small quantities of a second antibiotic, cephalosporin C. Painstaking work proved it to be chemically similar to penicillin N, but not a penicillin. It had pronounced gram-negative activity, was more stable to acid, and was not destroyed by penicillinase. It possessed the same α-aminoadipyl side chain as a new penicillin, but the nucleus was 7-aminocephalosporanic acid (7-ACA). 7-ACA contains a six-membered 1,3-dihydrothiazine ring instead of the five-membered thiazole ring in 6-APA. The structures of 6-APA and 7-ACA are shown in Figure 18.36.

Microbiological processes for production of cephalosporin C resemble in many respects those used for penicillin production. Special strains of *Cephalosporium* have been selected which produce more cephalosporin C and less cephalosporin N than the parent culture. The growth of these strains in certain special fer-

TABLE 18.9 Structural Formulae of Some Semi-synthetic Penicillins

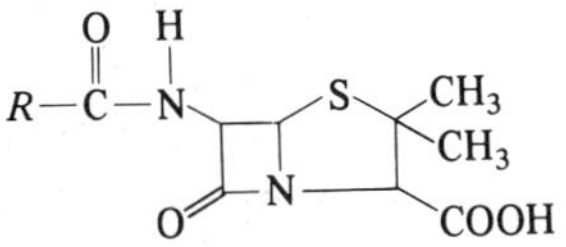

R—C(=O)—NH— (penicillin nucleus; CH_3, CH_3, COOH)

R =	R =
Carbencillin	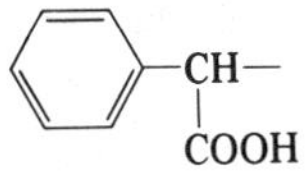 Nafcillin 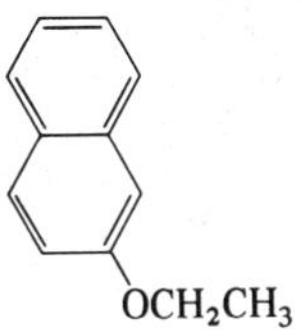
Penicillin V	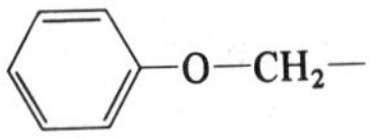 Quinacillin 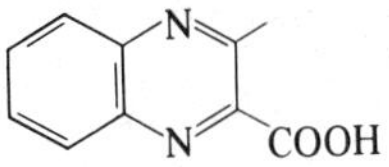
Phenethicillin	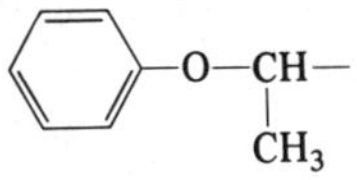 Oxacillin 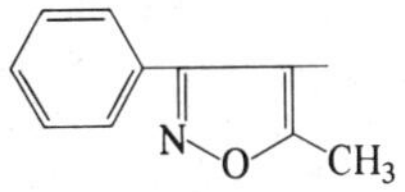
Propicillin	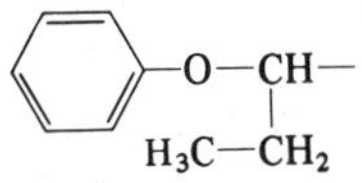 Cloxacillin 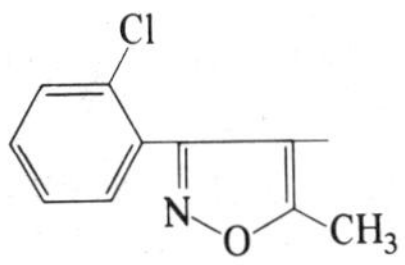
Phenbenicillin	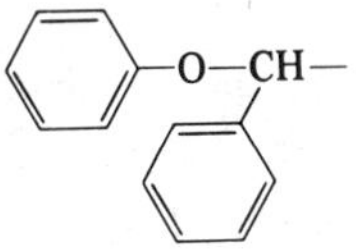 Dicloxacillin 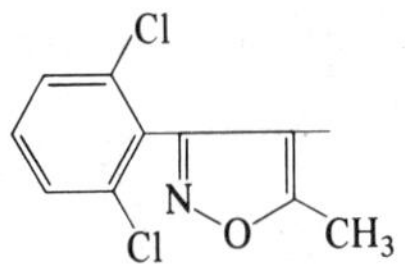

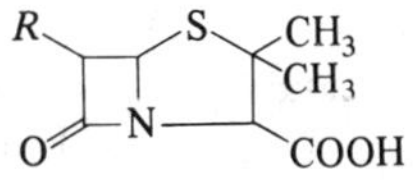

R— (penicillin nucleus; CH_3, CH_3, COOH)

R =	R =
Methicillin	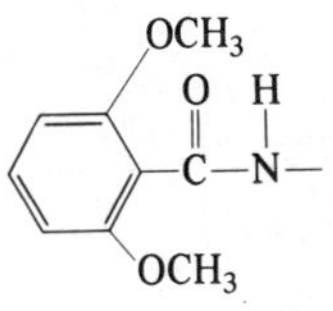 Flucloxacillin 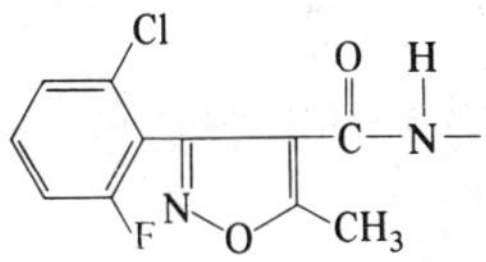
Ampicillin	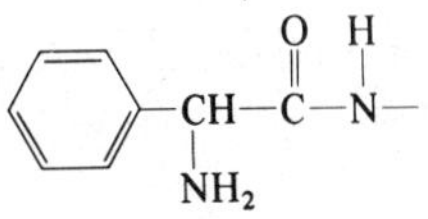 Azidocillin 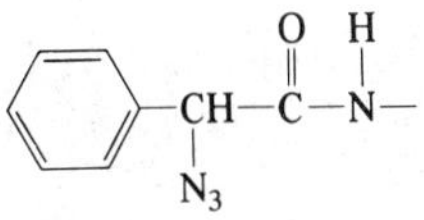
Epicillin	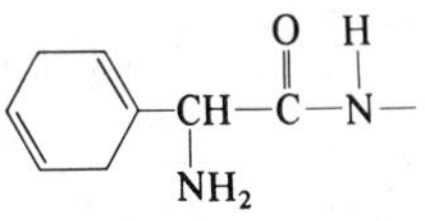 Cyclacillin 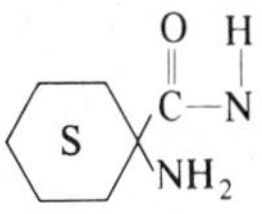
Amoxicillin	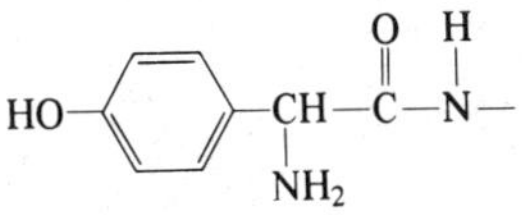 Metapilcillin 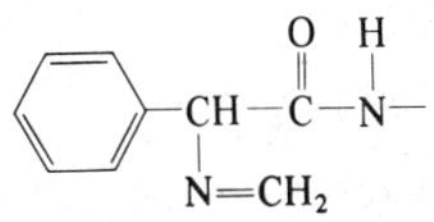
Hetacillin	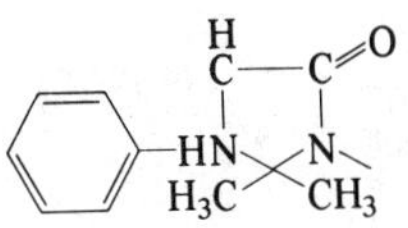 Mecillinam 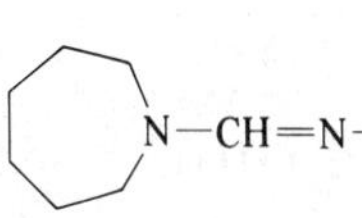

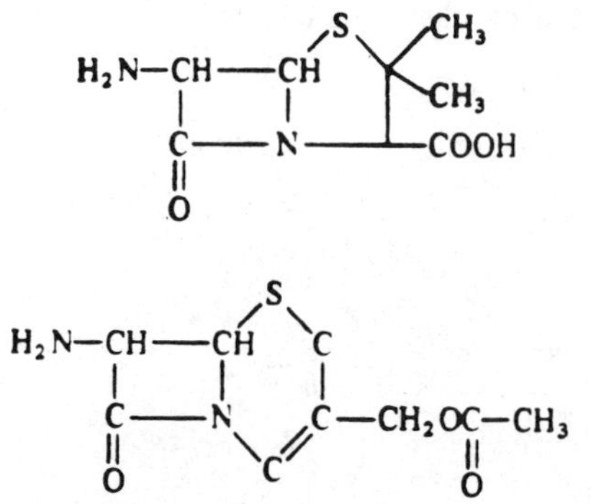

Fig. 18.36 Structures of 6-APA and 7-ACA.

mentation media has resulted in higher antibiotic titers. Even with these improvements in processing, the antibiotic yields, averaging 10-20 gm/liter, are much lower than those reported for the penicillins. Cephalosporins sell for around $250 a kilogram.

As in the penicillin studies, the possibility of further improving the chemotherapeutic properties of cephalosporin C was apparent if the 7-ACA nucleus could be obtained. Enzymatic cleavage of the side chain failed as did the use of precursors to generate new side chains; however, successful chemical methods have been found. Several semi-synthetic cephalosporins have been produced and are used clinically.

The chemical transformation of phenoxypenicillin (V) to the cephalosporin analog has been accomplished, a transformation which converts 6-APA to 7-ACA. This brilliant achievement may change the economics of all penicillin-cephalosporin chemistry.

Tetracyclines

In 1948, a broad-spectrum antibiotic, chlortetracycline ("Aureomycin"), was announced from the Lederle Laboratories, Division of American Cyanamid Company. This antibiotic is produced by *Streptomyces aureofaciens* when grown under submerged aerobic conditions on media composed of sugar, cornsteep liquor, and mineral salts. The crystalline compound has a golden yellow color, which suggested the trade name.

The following year a second related antibiotic, oxytetracycline ("Terramycin"), a product of *Streptomyces rimousus*, was announced by Pfizer Inc. It also is a yellow substance, chemically and biologically similar to chlortetracycline. Independent research by both companieis eventually led to the disclosure of the structure of these two important chemotherapeutic agents; this has been regarded as one of the brilliant achievements of modern organic chemistry.

TABLE 18.10 Structures of Clinically Important Tetracyclines

	R_1	R_2	R_3	R_4
Tetracycline	H	OH	CH_3	H
7-Chlortetracycline (Aureomycin®)	H	OH	CH_3	Cl
5-Oxytetracycline (Terramycin®)	OH	OH	CH_3	H
6-Demethyl-7-chlortetracycline (Declomycin®)	H	OH	H	Cl
6-Deoxy-5-oxytetracycline (Vibramycin®)	OH	H	CH_3	H
6-Methylene-6-deoxyl-6-demethyl-5-oxytetracycline (Rondomycin®)	OH	H	CH_2	H

TABLE 18.11 Some Antibiotics Produced on a Commercial Scale*

Antibiotic	*Microbial source*	*Antibiotic spectrum*[b]						*Chemical type*	*Therapeutic or other use*
		G+	*G–*	*My*	*AF*	*AT*	*Other*		
Amphomycin	*Streptomyces canus*	+						Peptide	Topical
Amphotericin B	*Streptomyces nodosus*				+			Polyene	Oral or parenteral
Avoparcin[a]	*Streptomyces candidus*	+	+					Glycopeptide	Animal growth promotant
Azalomycin F[a]	*Streptomyces hygroscopicus*	+	+		+				Topical (AF)
Bacitracin	*Bacillus subtilis*	+						Peptide	Topical; also animal growth promotant
Bambermycins	*Streptomyces bambergenesis*	+						Phosphoglycolipid	Animal growth promotant
Bicyclomycin[a]	*Streptomyces sapporonensis*	+	+						Topical
Blasticidin S[a]	*Streptomyces griseochromogenes*				+			Nucleoside	Agricultural (AF)
Bleomycins	*Streptomyces verticillus*	+	+			+		Peptide	Parenteral (AT)
Cactinomycin[a]	*Streptomyces chrysomallus*	+				+		Peptide	Parenteral (AT)
Candicidin B	*Streptomyces griseus*				+			Polyene	Topical
Candidin[a]	*Streptomyces viridoflavus*				+			Polyene	Topical
Capreomycin	*Streptomyces capreolus*			+				Peptide	Parenteral
Cephalosporins	Cephalosporin C is produced by *Cephalosporium acremonium* and converted to 7-ACA, which is used for prep of semisynthetic cephalosporins	+	+					Peptide	Oral and parenteral
Chloramphenicol	*Streptomyces venezuelae*; commercial manufacture is by chemical synthesis	+	+				Rickettsia		Oral or parenteral
Chromomycin A_3[a]	*Streptomyces griseus*	+				+			Parenteral (AT)
Colistin	*Bacillus colistinus*		+					Peptide	Parenteral
Cycloheximide	*Streptomyces griseus*				+				Agricultural (AF)
Cycloserine	*Streptomyces orchidaceus*			+				Amino acid	Parenteral (TB)
Dactinomycin	*Streptomyces antibioticus*					+		Peptide	Parenteral (AT)
Daunorubicin	*Streptomyces peucetius*					+			Parenteral (AT)
Doxorubicin	*Streptomyces peucetius*					+			Parenteral (AT)
Enduracidin[a]	*Streptomyces fungicidus*	+		+				Peptide	Animal growth promotant
Erythromycin	*Streptomyces erythreus*	+						Macrolide	Oral and parenteral; animal growth promotant
Fortimicins[a]	*Micromonospora olivoasterospora*	+	+					Aminoglycoside	Parenteral
Fungimycin[a]	*Streptomyces coelicolor* var *aminophilus*				+			Polyene	Topical

TABLE 18.11 (Continued)

Antibiotic	Microbial source	Antibiotic spectrum[b]						Chemical type	Therapeutic or other use
		G+	*G–*	*My*	*AF*	*AT*	*Other*		
Fusidic Acid[a]	*Fusidium coelcineum*	+						Steroid	Parenteral
Gentamicins	*Micromonospora purpurea*	+	+					Aminoglycoside	Parenteral
Gramicidin A	*Bacillus brevis*	+						Peptide	Topical
Gramicidin J[a](S)	*Bacillus brevis*	+						Peptide	Topical
Griseofulvin	*Penicillium griseofulvum*				+			Spirolactone	Oral
Hygromycin B	*Streptomyces hygroscopicus*	+	+				Helminths	Aminoglycoside	Animal feed suppl.
Josamycin[a]	*Streptomyces narbonesis*	+						Macrolide	Oral and parenteral
Kanamycins	*Streptomyces kanamycetcus*	+	+	+				Aminoglycosides	Parenteral
Kasugamycin[a]	*Streptomyces kasugaensis*	+	+					Aminoglycoside	Agricultural antibacterial
Kitasatamycin[a]	*Streptomyces kitasatoensis*	+						Macrolide	Oral and parenteral
Lasalocid	*Streptomyces hazelensis*	+					Coccidia	Polyether	Agricultural use as coccidiostat and growth promotant
Lincomycin	*Streptomyces lincolnensis*	+							Oral and parenteral
Lividomycin[a]		+	+						
Macarbomycins[a]	*Streptomyces phaechromogenes*	+						Phosphoglycolipid	Animal growth promotant
Mepartricin[a]					+			Polyene	Topical
Midecamycin[a]		+						Macrolide	Oral and topical
Mikamycins[a]	*Streptomyces mitakaensis*	+						Peptide	Animal growth promotant
Mithramycin	*Streptomyces* species	+		+		+			Parenteral (AT)
Mitomycin C	*Streptomyces caespitosus*	+	+		+				Parenteral
Mocimycin[a]		+							Animal growth promotant
Monensin	*Streptomyces cinnamonensis*	+					Coccidia	Polyether	Animal growth promotant
Myxin	*Chromobacterium iodinium* plus chemical modification				+			Phenazine	Topical in veterinary use
Neocarzinostatin	*Streptomyces carzinostaticus*	+				+		Peptide	Parenteral (AT)
Neomycins	*Streptomyces fradiae*	+	+					Aminoglycoside	Oral and topical
Nosiheptide[a]	*Streptomyces actinosus*	+						Peptide	Animal growth promotant
Nisin	*Streptococcus cremoris*	+						Peptide	Food preservative
Novobiocin	*Streptomyces niveus*	+							Oral and topical
Nystatin	*Streptomyces noursel*				+			Polyene	Oral and topical
Oleandomycin	*Streptomyces antibioticus*	+						Macrolide	Oral and parenteral
Paromomycin	*Streptomyces rimosus*	+	+				Protozoa	Aminoglycoside	Oral

Penicillin G[a]	*Penicillium chrysogenum*	+						Peptide	Oral and parenteral; also as animal growth promotant
Penicillin V[a]	*Penicillium chrysogenum*	+						Peptide	Oral
Pimaricin	*Streptomyces natalensis*					+		Polyene	Topical; also used for food preservation
Polymyxin B	*Bacillus polymyxa*				+			Peptide	Parenteral
Polyoxins[a]	*Streptomyces cacaoi* var. *asoensis*								Agriculture (AF)
Pristinamycins[a]	*Streptomyces pristinaspiralis*	+						Peptide	Parenteral
Quebemycin[a]	*Streptomyces viridans*	+						Phosphoglycolipid	Animal growth promotant
Ribostamycin[a]	*Streptomyces ribosidificus*	+	+					Aminoglycoside	Parenteral
Rifamycin SV[a]	*Nocardia mediterranei*			+				Anasamycin	Parenteral
Ristocetin	*Nocardia lurida*	+						Glycopeptide	Parenteral
Sagamycin[a]	*Micromonospora sagamiensis*	+	+					Aminoglycoside	Parenteral
Salinomycin[a]	*Streptomyces albus*	+					Coccidia	Polyether	Veterinary use
Siccanin[a]	*Helminthosporium siccans*				+				Veterinary use
Siomycin	*Streptomyces sioyaensis*	+						Peptide	Animal growth promotant
Sisomicin	*Micromonospora inyoensis*	+	+					Aminoglycoside	Parenteral
Spectinomycin	*Streptomyces spectrabilis*	+						Aminocyclitol	Parenteral
Spiramycin[a]	*Streptomyces ambofaciens*	+						Macrolide	Parenteral and oral
Streptomycin	*Streptomyces griseus*	+	+	+				Aminoglycoside	Parenteral; use in agriculture to control bacteria
Dihydrostreptomycin	*Streptomyces humidus* (Also chemical reduction of streptomycin)	+	+	+				Aminoglycoside	Parenteral
Tetracyclines									
Chlortetracycline	*Streptomyces auerofaciens*	+	+				Rickettsia	Tetracycline	Parenteral and oral; animal growth promotant
6-Demethyl-7-chlortetracycline	*Streptomcyes aureofaciens*	+	+				Rickettsia	Tetracycline	Parenteral and oral
5-Hydroxytetracycline	*Streptomyces rimosus*	+	+				Rickettsia	Tetracycline	Parenteral and oral animal growth promotant
Tetranactin[a]	*Streptomyces flaveolus*	+					Insects	Macrotetralide	Insecticide
Thiopeptin[a]	*Streptomyces tateyamensis*	+						Peptide	Animal growth promotant
Thiostrepton	*Streptomyces azureus*	+						Peptide	Animal growth promotant
Tobramycin[a]	*Streptomyces tenebrarius*	+	+					Aminoglycoside	Parenteral
Trichomycin[a]	*Streptomyces hachijoensis*				+		Trichomonas	Polyene	Topical
Tylosin	*Streptomyces fradiae*	+					PPLO	Macrolide	Veterinary; animal growth promotant
Tyrothricin	*Bacillus brevis*	+	+					Peptide	Topical
Tyrocidine	*Bacillus brevis*	+	+					Peptide	Topical

TABLE 18.11 (Continued)

Antibiotic	*Microbial source*	*Antibiotic spectrum*[b]					*Other*	*Chemical type*	*Therapeutic or other use*
		G+	*G−*	*My*	*AF*	*AT*			
Validamycin[a]	*Streptomyces hygroscopicus* var. *limonesis*	+	+					Aminoglycoside	Parenteral
Vancomycin	*Streptomyces orientalis*	+						Glycopeptide	Parenteral
Variotin[a]	*Paecilomyces varioti*				+				Topical
Viomycin	*Streptomyces floridae*	+	+	+				Peptide	Parenteral
Virginiamycin	*Streptomyces virginiae*	+						Peptide	Animal growth promotant

[a]Not distributed in the United States (1978).
[b]G+, gram-positive bacteria; G−, gram-negative bacteria; My, mycobacteria; AF, antifungal; AT, antitumor.
*Microbial Technology, 2nd ed., Vol. 1, p. 244 (1979).

Both compounds may be regarded as derivatives of a nucleus known as tetracycline. Their structures along with those of other clinically important tetracyclines are shown in Table 18.10.

Tetracycline can also be produced by *Streptomyces aureofaciens* fermentations under special conditions, i.e., chloride starvation or special strains of the organism which fail to halogenate efficiently. Tetracycline possesses many chemotherapeutic properties of chlortetracycline and oxytetracycline. It is an important broad-spectrum antibiotic.

Mutations of tetracycline-producing organisms have led to other tetracycline analogs, of which 6-demethyl-7-chlortetracycline ("Declomycin") has clinical use. Chemical modifications of oxytetracycline have generated two other useful members of the family, known as Vibramycin and Rondomycin.

Tetracyclines are active *in vivo* against numerous gram-positive and gram-negative organisms, and some of the pathogenic rickettsiae and large viruses. The systemic administration of tetracyclines may be carried out utilizing either oral or intravenous dosage forms. For veterinary use, tetracyclines are given by intravenous injection in bovines and equines, except that oral routes are used for young nonruminating calves.

In connection with work on animal protein factors, it was found that chlortetracycline fermentation mash containing some vitamin B_{12} gave growth responses well above those obtained with supraoptimal levels of vitamin B_{12} alone. At the present time, both chlortetracycline and oxytetracycline are used extensively for growth stimualtion and improvement of feed efficiency in poultry and hogs, and for the reduction of losses from certain disease conditions. Purified antibiotics as well as dried fermentation residues with the mycelium of *Streptomyces aureofaciens* are used for these purposes. Tetracyclines, used as animal feed supplements, are selling for about $60 per kilogram.

Other Antibiotics

Table 18.11 lists some antibiotics produced on a commercial scale. The importance of new antibiotics for use in clinical medicine is stressed in most discussions of antibiotics, but their importance in other areas should not be forgotten. Antibiotics such as streptomycin and tetracycline are being used against bacterial plant pathogens, while cycloheximide, blasticidin S, nystatin, and griseofulvin are being used against fungi. Antibiotics are also being used in livestock production where they improve marketable weight, and increase food utilization. Those which stimulate animal growth are bacitracin for poultry and swine; bambermycin for swine and calves; virginiamycin for poultry and swine; avoparcin for poultry and swine; chlortetracyline for poultry, swine, calves, cattle, and sheep; erythromycin for chickens; nystin, oleandomycin, and procaine penicillin for poultry and swine; streptomycin for poultry; oxytetracyline for poultry, swine, cattle, and calves; tylosin for poultry and swine; lasolocid for poultry; and monensin for poultry and cattle. Combinations of antibiotics and their mixtures with sulfa drugs are also being used.

Further research on new agents to treat both human and animal diseases is certainly in order, particularly to treat those diseases which are not successfully controlled at present.

REFERENCES

1. S. Aiba, A. E. Humphrey, and N. Millis, *Biochemical Engineering*, 2nd ed., University of Tokyo Press, Tokyo, 1973.
2. W. B. Armiger, ed., "Computer Applications in Fermentation Technology," Biotechnology and Bioengineering Symposium Series No. 9, John Wiley and Sons, New York, 1979.
3. J. E. Bailey and D. F. Ollis, *Biochemical Engineering Fundamentals*, McGraw-Hill Book Company, New York, 1977.
4. S. A. Barker, "Industrialized Enzyme Processes," *Sci. Prog. Oxf.*, 65, p. 477, 1980.

5. A. T. Bull, D. C. Ellwood and C. Ratledge, eds., *Microbial Technology*: *Current State, Future Prospects*, Cambridge University Press, Cambridge, England, 1979.
6. *Chemical Marketing Reporter*, Schnell Publishing Company, New York.
7. N. P. Cheremisinoff, *Gasohol for Energy Production*, Ann Arbor Science Publishers Inc., Ann Arbor, Michigan, 1979.
8. H. A. Conner and R. J. Allgeier, "Vinegar: Its History and Development," *Adv. Appl. Microbiol.*, **20**, p. 82, 1976.
9. A. L. Demain, "Riboflavin Oversynthesis," *Ann. Rev. Microbiol.*, **26**, p. 369, 1972.
10. A. DiMarco and P. Pennella, "The Fermentation of the Tetracyclines," *Prog. Ind. Microbiol.*, **1**, p. 45, 1959.
11. M. Dixon and E. L. Webb, *Enzymes*, 3rd ed., Academic Press, Inc., New York, 1980.
12. N. Esaka, K. Soda, H. Kumagai, and H. Yamada, "Recent Advances in Enzymatic Synthesis of Amino acids," *Biotech. Bioeng.*, **22**, Suppl. 1, p. 127, 1980.
13. W. P. K. Findlay, ed., *Modern Brewing Technology*, Macmillan, New York, 1971.
14. H. C. Friedmann and L. M. Cagen, "Microbial Biosynthesis of B_{12}-Like Compounds," *Ann. Rev. Microbiol.*, **24**, p. 159, 1970.
15. D. M. Glover, *Genetic Engineering Cloning DNA*, Chapman and Hall, New York, 1980.
16. Y. Hirose and M. Shibai, "Amino Acid Fermentation," *Biotech. Bioeng.*, **22**, Suppl. 1, p. 111, 1980.
17. A. E. Humphrey, "Biochemical Reaction Engineering," Chapter 8 in *Chemical Reaction Engineering Reviews*, ACS Symposium Series 72, 1979.
18. International Technical Information Institute, "Japan's Most Advanced Industrial Fermentation: Technology and Industry," International Technical Information Institute, Tokyo, 1977.
19. J. Kleyn and J. Hough, "The Microbiology of Brewing," *Ann. Rev. Microbial.*, **25**, p. 583, 1971.
20. A. L. Lehninger, *Biochemistry*, 2nd ed., Worth Publishers, Inc., 1975.
21. F. A. Lowenheim and M. K. Moran, eds., *Faith, Keyes, and Clark's Industrial Chemicals*, 4th ed., John Wiley and Sons, New York, 1975.
22. I. Malek and Z. Fencl, eds., *Theoretical and Methodological Basis of Continuous Culture of Microorganisms*, Publishing House of the Czechoslovak Academy of Sciences, Prague, 1966.
23. J. F. Martin and A. L. Demain, "Control of Antibiotic Biosynthesis," *Microbial. Rev.*, **44**, No. 2, p. 230, 1980.
24. Master Brewers Association of the Americas, H. M. Broderick, ed., *The Practical Brewer–A Manual for the Brewing Industry*, 2nd ed., Impressions, Inc., Madison, Wisconsin, 1977.
25. R. I. Mateles and S. Tannenbaum, eds., *Single-Cell Protein*, M.I.T. Press, Cambridge, MA, 1968.
26. C. E. Morris, "America's Gold," *Food Engineering*, **52**, No. 5, p. 95, 1980.
27. K. Mosbach, eds., "Immobilized Enzymes," *Methods in Enzymology*, vol., **XLIV**, Academic Press, New York, 1976.
28. K. Nakayama, "The Production of Amino Acids," *Process Biochemistry*, **11**, No. 2, p. 4, 1976.
29. H. J. Peppler and D. Perlman, eds., *Microbial Technology*, volumes **1** and **2**, 2nd ed., Academic Press, New York, 1979.
30. D. Perlman, ed., "Fermentation Advances," Proceedings of the 3rd International Fermentation Symposium, Academic Press, New York, 1969.
31. D. Perlman, "Influence of Penicillin Fermentation Technology on Processes for Production of Other Antibiotics," *Process Biochemistry*, **10**, No. 9, p. 23, 1975.
32. D. Perlman and G. T. Tsao, eds., *Annual Reports on Fermentation Processes*, volumes **1**, **2** and **3**, Academic Press, New York, 1977, 1978 and 1979.
33. G. Reed and H. J. Peppler, *Yeast Technology*, AVI Publishing Company, 1973.
34. J. W. Richards, *Introduction to Industrial Sterilization*, Academic Press, New York, 1968.
35. J. Schierholt, "Fermentation Processes for the Production of Citric Acid," *Process Biochemistry*, **12**, No. 9, p. 20, 1977.
36. G. L. Solomons, *Materials and Methods in Fermentation*, Academic Press, London, 1969.
37. B. Spencer, ed., *Industrial Aspects of Biochemistry*, North-Holland Publishing Company, Amsterdam, 1974.
38. R. Y. Stanier, M. Doudoroff and E. A. Aldelberg, *The Microbial World*, 4th ed., Prentice-Hall, Inc., Englewood Cliffs, NJ, 1975.
39. S. Tannenbaum and D. I. C. Wang, eds., *Single-Cell Protein II*, M.I.T. Press, Cambridge, MA, 1975.
40. G. Terui, ed., "Fermentation Technology Today," Proceedings of the 4th International Fermentation Symposium, Society of Fermentation Technology, Osaka, Japan, 1972.
41. R. W. Thoma, ed., *Industrial Microbiology*, Dowden, Hutchinson and Ross., Inc., Stroudsburg, PA, 1976.
42. D. I. C. Wang, C. L. Cooney, A. L. Demain, P. Dunnill, A. E. Humphrey and M. D. Lilly, *Fermentation and Enzyme Technology*, John Wiley and Sons, New York, 1979.
43. J. D. Watson, *Molecular Biology of the Gene*, 3rd ed., W. A. Benjamin, Inc. Menlo Park, CA, 1976.
44. M. J. Weinstein and G. H. Wagman, eds., "Antibiotics: Isolation, Separation and Purification," *Journal of*

Chromatography, Library-volume 15, Elsevier Scientific Publishing Company, Amsterdam, Netherlands, 1978.

45. A. Wiseman, ed., *Topics in Enzyme and Fermentation Biotechnology*, vols. 1, 2, 3 and 4, Ellis Horwood Limited, Chichester, England, 1977, 1978, 1979, and 1980.
46. A. Wiseman, ed., *Handbook of Enzyme Biotechnology*, Ellis Horwood Limited, Chichester, England, 1975.
47. U.S. International Trade Commission, *Synthetic Organic Chemicals*: *U.S. Production and Sales, 1979*, U.S. Government Printing Office, Washington, DC, 1980.

19

Chemical Explosives

Walter B. Sudweeks, Ray D. Larsen, Fred K. Balli*

CHEMICAL EXPLOSIVES

Explosives serve two main purposes. First, they are utilized in industry to save billions of man-hours of work each year, e.g., in mining coal and metallic and nonmetallic ores; in quarrying, clearing land, ditching, loosening formations in oil and gas wells, and in road building; for sporting ammunition; and for such important specialized applications as blind rivets and starter cartridges for aircraft and diesel engines, high-speed machining and metal forming, and perforating oil-well casings. Second, they are of major importance in the field of rockets, missiles, space vehicles, and military and civilian weapons. The manufacture of commercial explosives is a growing industry. Production has risen steadily in America from about 200 million pounds in 1920 to approximately four billion pounds in 1978. Bebie[1] lists some 135 chemicals and formulations which are useful as explosives; of these, 75 are used in the explosives industry alone, 45 are primarily military explosives, and 15 are used for both purposes.

An explosive is a substance or mixture of substances which, when raised to a sufficiently high temperature, whether by direct heating, friction, impact, shock, spark, flame, or sympathetic reaction from a primary or donor explosive, suddenly undergoes a very rapid chemical transformation with the evolution of large quantities of heat and gas, thereby exerting high pressures on surrounding media. With some explosives the rate of this transformation (or burning rate) is so great that the explosive exerts a very great shattering action (or *brisance*), while with others the reaction may take place at a much slower, controlled, but still explosive rate to give pressure-time characteristics which make them suitable for use as propellants in guns, rockets, etc., where much lower rates of pressure development and peak pressures are required. Another characteristic property of explosives is *sensitivity*, or ease of initiating the explosion, whether of the fast (shattering) type or the much slower, propellant type. *Strength*, or the maximum explosive energy available for useful work, is another

*IRECO Chemicals. This chapter is a revision of that published in the previous edition and authored by Dr. Melvin A. Cook.

important factor. It depends much less on the rate of reaction than does the brisance, peak pressure, or pressure-time curve of the explosive. The uses of explosives depend on all three of these characteristics (pressure or pressure-time curve, sensitivity, and strength); these are the bases for the groupings used in Table 19.1 and the classifications presented in the next paragraph.

The usual classification of explosives is into two general groups, *high* or *detonating* explosives, and *low* or *deflagrating*, sometimes also called *propellant* explosives. The latter have a low burning rate which permits them to have a relatively slow-rising pressure-time curve; the peak pressure seldom rises above 50,000 psi. The rate of burning of the low explosive directly into the grain never exceeds a few cm/sec, whereas in detonation the reaction rates are hundreds of times faster. The low explosives exert a powerful "heaving action" or push, and while they are used today primarily as propellants, they have a very desirable blasting action for lump coal. However, from the historical viewpoint, the most prominent of this type, namely black powder, has been used extensively in the past in borehole blasting. In high explosives the reaction takes place to a large extent in a peculiar type of shock wave known as the detonation wave. This wave propagates in accord with well-known principles of hydrodynamics at velocities ranging from one to seven miles per second, depending on the density, heat of explosion, the particle size and shape, and in gelatins, the air-bubble content and distribution. An important characteristic of detonation in condensed explosives is that the reaction zone comprises an ionized gas or plasma existing with high cohesion in a quasi-lattice structure, pictured resembling the metallic state.[2,3] This plasma causes the pressure rise in the detonation front to be much less steep than it was at first thought. That is, instead of being infinitely steep, the detonation rises to its characteristic pressure of 100,000–5,000,000 psi in a period ranging from a few tenths of a microsecond to several microseconds, depending on the explosive. Subclasses of high explosives are the primary explosives (used as detonators) and the secondary explosives. The former are characterized by the fact that even in very small quantities they develop (via the essential plasma formation) detonation waves in extremely short periods of time following simple ignition, e.g., by flames, sparks, hot wires, or friction. The secondary explosives, however, usually require detonators, and sometimes boosters also, to bring them to detonation, at least in practical applications. A booster may be one of the more sensitive secondary explosives, such as pressed tetryl, TNT, RDX, waxed RDX, or cast TNT and pentolites (TNT-PETN). The commercial detonators are the ordinary (or fuse) and electric (or EB—sometimes also called composition) caps which contain either mercury fulminate alone (fuse caps) or separate elements consisting of (1) an ignition element (e.g., lead styphnate), (2) a primary explosive (e.g., lead azide), and (3) a base charge comprising a secondary explosive (e.g., pressed tetryl, PETN, or RDX). These are the EB or composition caps. Commercial dynamites are all secondary explosives which may be detonated directly by commercial detonators or Primacord. The latter is a detonating fuse usually containing about 50 grains of PETN per foot in a special wax-impregnated cloth or plastic sheath which may or may not be reinforced by a binding wire. The least sensitive secondary explosives, of which the 94/6 prilled ammonium nitrate-fuel oil mixture is currently the most popular, require a relatively large booster.

Except for the ammonium nitrate—fuel oil mixture (ANFO) just mentioned, all of the explosive materials discussed above are "molecular" explosives, i.e., the two main components of an explosive—oxidizer and fuel—are part of the same molecule. Another significant category of secondary explosives is referred to as "composite" or explosive mixtures. With these the oxidizer and fuel components are at least two different substances which are mixed to provide the degree of intimacy required to form an explosive product. Thus, ANFO is such a composite explosive consisting of a mixture of ammonium nitrate (oxidizer) and fuel oil (fuel).

Other series of explosives of this type are the "slurry" or "watergel" explosives and "slurry" or "watergel" blasting agents. Boosters are

TABLE 19.1 Characteristics and Uses of the More Important Explosives

PRIMARY EXPLOSIVES

Name	*Composition or Chemical Formula*	*Density (g/cc)*	*Detonation Velocity*[a] *(km/sec)*	*Detonation Pressure (kilobars)*	*Detonation Temperature*[a] *(°K)*
Mercury fulminate	$Hg(ONC)_2$	3.6	4.7	220	6900
Lead azide	$Pb(N_3)_2$	4.0	5.1	250	5600
Lead styphnate	$C_6H(NO_2)_3O_2Pb$	2.5	4.8	150	–
Nitromannite (Mannitol hexanitrate)	$C_6H_8(ONO_2)_6$	1.73	8.3	300	6000
Diazodinitrophenol (DDNP)	$C_6H_2N_4O_5$	1.5	6.6	160	–

SECONDARY HIGH EXPLOSIVES

Name	*Composition or Chemical Formula*	*Density (g/cc)*	*Detonation Velocity (km/sec)*	*Available Energy (kcal/g)*
Ammonia gelatin dynamites	30–90% grades same as straight gelatins except for some NG and $NaNO_3$ replacement by NH_4NO_3	1.2–1.5	4–6.5	0.75–1.15
Semigelatin dynamite	15–20% NG, 1–2% DNT oil, AN-SN dope	1.2	3.5–5 (depends on diameter)	0.9
Prilled AN-Fuel Oil	94/6 NH_4NO_3/oil	0.8–0.9	1.5–4	0.81–0.83
Slurry explosives TNT-SE	TNT 17–40 Oxidizer* 30–65 H_2O 12–25 Al 0–20 Other 0.3–1.5	1.4–2.0	5–8	0.7–1.8
Smokeless powder SE	SP 20–40 Oxidizer* 30–60 H_2O 3–25 Al 0–20 Other 0.3–10	1.35–1.9	4–7	0.65–1.7

[a]Most important properties of detonators.
*AN, SN, perchlorates, etc.

Sensitivity	*Major Characteristics*	*Uses*
Very high	Best primary explosive for single-component (fuse) detonators; easily detonated by flame, spark, heat, or friction; easily dead-pressed.	In fuse caps (mixed with $KClO_3$); propellant primer; in fuses for shells; small arms cartridge caps.
Very high (higher than NG; less than mercury fulminate)	Powerful detonator but requires strong igniters, e.g., lead styphnate.	Primary explosive in composition (EB) caps; military fuses.
Exceedingly high	Extremely sensitive to sparks, static electricity; explodes rapidly on ignition; good thermal stability.	Igniter in composition caps, military fuses; very satisfactory detonator explosive for fast ignition.
Very high (greater than NG; less than lead azide)	Stronger and more brisant than NG, RDX, PETN.	In composition caps and fuses.
Very high (less than lead azide)	Does not dead-press. About 3/4 as strong as TNT.	In composition caps and fuses.

Sensitivity	*Major Characteristics*	*Uses*
High	More economical; only slightly less brisant than straight gelatin; exhibits low-order detonation with threshold priming and high pressures.	General small and large diameter blasting in hard rock and under water.
High	Stringy, plastic; easily loaded in "uppers;" economical; high strength; moderate brisance.	Popular small diameter metal-mining explosive.
Low (requires booster)	One of the cheapest sources of explosive energy available today; flammable and will explode when ignited under strong confinement; no water resistance; adaptable to do-it-yourself operations.	Open-pit and underground blasting where dry conditions prevail; most adaptable to soft, easy shooting.
Low (requires boosters)	Gel or thick pea-soup consistency; capable of detonation at high pressures, excellent water resistance.	Large diameter, open-pit, small diameter underground, oil well, submarine, water-filled boreholes, deep-water bombs.
Low (requires boosters)	Generally similar to TNT slurry.	Large diameters, open-pit blasting.

TABLE 19.1 (Continued)

Name	*Composition or Chemical Formula*	*Density (g/cc)*	*Detonation Velocity (km/sec)*	*Available Energy (kcal/g)*
Slurry blasting agents	Al 0–35 Oxidizer* 50–80 H_2O 4–18 Other 0.2–10	1.1–1.6	2–6	0.7–2.0
Nitrostarch powders	Nitrostarch in place of NG	1.2	4–5	0.8–1.0
Compositon B	40/59/1 TNT/RDX/wax	1.7	7.8	1.1
Composition B-3	40/60 TNT/RDX	1.73	7.9	1.15
Haleite or EDNA	$(CH_2NHNO_2)_2$	1.6 (pressed)	7.9	1.2
Ammonium picrate (Explosive D)	$(ONH_4)C_6H_2(NO_2)_3$	1.56 (pressed)	6.6	0.7
Nitrostarch	Mixtures of various nitro esters of starch	1.4 (pressed)	6.4	0.95
Tetryl	$(NO_2)_3C_6H_2N(CH_3)NO_2$	1.45 (pressed)	7.0	0.95
PETN (pentaerythritol tetranitrate)	$C(CH_2ONO_2)_4$	1.6 (pressed)	7.92	1.31
Pentolite	50/50 TNT/PETN	1.63 (cast)	7.7	1.1
Trinitrotoluene (TNT)	$CH_3C_6H_2(NO_2)_3$	1.59 (cast)	6.9	0.9
		1.45 (pressed)	6.9	–
		1.03 ("Pelletol")	5.1	–
		0.8 (grained)	4.2	0.8
Amatols	50/50 AN/TNT	1.55 (cast)	5–6.5 (depending on diameter)	0.95
	80/20 AN/TNT	1.0 (loose)	4 (large diameter)	0.93
		1.45 (pressed)	5.6 (large (diameter)	
Dinitrotolune (DNT)	$CH_3C_6H_3(NO_2)_2$	1.28 (liquid)	5	0.7
		0.8 (granular solid)	2–3.5 (depending on diameter)	
Nitromethane (NM)	CH_3NO_2	1.12	6.2	–

Sensitivity	*Major Characteristics*	*Uses*
Low (requires boosters)	Gelatin to thick or thin consistency.	Large diameter, underwater, wet- and dry-hole blasting, large bombs.
Moderately high, but less than dynamites	Good "fumes;" fair water resistance; powerful; economical.	Small diameter blasting.
Average	Very high brisance.	Bursting charge and special weapons.
High	Very high brisance.	Experimental standard.
High	High brisance; less sensitive than RDX and PETN.	In Ednatols for bursting charges.
Very low	Insensitive to shock and friction; melts with decomposition; shells filled with high-pressure pressing.	Armor-piercing shells.
High	Highly inflammable white powder.	Demolition blocks and Trojan blasting explosives.
High	Very sensitive; rapidly reacting; easily pressed with 1–2% graphite; high brisance.	Booster; base charge in caps; in tetrytols for bursting charges.
High	Very powerful and sensitive (more sensitive than RDX, less than NG).	In Primacord fuse; base charge in caps.
Moderate	High pressure or brisance; primacord sensitive	Booster and special weapons; commercial booster for prilled AN-fuel oil and slurry explosives.
Low	Easily melted and cast; suitable liquid for slurrying with other explosives; easily pressed into blocks; completely waterproof.	Military; "Nitropel" TNT used in slurry explosives and in filling annulus between charge and borehole in water-filled holes; in amatols.
Low	Insensitive; hygroscopic, not waterproof; less brisant but stronger than TNT; 50/50 can be cast; 80/20 either pressed or granulated.	Military; oil well shooting; quarrying; dry-hole booster for very low-sensitive types.
Low		
Very low	Reddish brown or yellow liquid.	Sensitizer in "Nitramons"; 60/41 NG/DNT in oil well shooting; up to 20% in TNT bursting charges; in FNH (flashless) propellant; 6% in small-arms ammunition (with guncotton).
Moderate	Clear, watery liquid	Special demolition, experimental studies of liquid explosives.

TABLE 19.1 (Continued)

Name	*Composition of Chemical Formula*	*Density (g/cc)*	*Detonation Velocity (km/sec)*	*Available Energy (kcal/g)*
Cyclonite (RDX)	$C_3H_6N_6O_6$	1.2 (loose) 1.6 (pressed)	6.8 8.0	1.32
HMX	$C_4H_8N_8O_8$	1.89	9.1	1.35
HBX	Mixtures of RDX, TNT, aluminum, and wax	1.78	7.5	1.5
Plastic Explosives (Compositions A, C, C-2, C-3, C-4)	Waxed RDX	1.45–1.6	8.0	1.1–1.3
PBX 9404	94/3/HMX/binder/ nitrocellulose	1.84	8.8	1.3
Nitroglycerin (NG)	$C_3H_5(ONO_2)_3$	1.59	7.8	1.41
Ethylene glycol dinitrate (EGDN)	$C_2H_4(ONO_2)_2$	1.48	7.4	1.43
Straight dynamites[b]	20–60% NG, in balanced SN dope 20% grade ≡ 20% NG, etc.	1.3	4-6	0.55–0.85[b]
Ammonia dynamites (and permissibles)	As above except NH_4NO_3 replaces part of NG and $NaNO_3$	0.8–1.2[b]	1.5–5.5 Depends on AN particle size, NG content.	0.7–0.9[b]
Blasting gelatin	92/8/NG/nitrocotton ("Solidified" NG contains some wood pulp to minimize low-order detonation)	1.55 (1.45)	7.5 (7.2	1.45 (1.4)

[b]Depends on grade.

usually required for these relatively insensitive explosives, although specially formulated cap-sensitive formulations are gaining in popularity. Boosters comprise 0.1 to 1.0 pounds or more of cast 50/50 pentolite, cast TNT, or other water-insoluble material, usually cast, and water-insoluble explosives of very high brisance. The first *slurry* used commercially was made essentially of ammonium nitrate, coarse TNT, and water.

Another early and still useful series of slurries substituted smokeless powder for TNT. Aluminum was added to these explosive-sensitized slurry types to increase strength and water soluble gums were added as thickeners and cross-linked to provide a water resistant gelled matrix. The most popular and economical commercial slurries in use today are the so-called *blasting agents* in which no molecular explosive per se is used. Foremost among these are the aluminum-sensitized slurries. Cap-sensitive slurries, usually sensitized with fine aluminum or methylammonium nitrate, are now widely used for small-diameter, under-

Sensitivity	*Major Characteristics*	*Uses*
High	Higher thermal stability in solid state; expressively sensitive in pure state; 1.65 times as strong as low density TNT; 1.45 times as strong as cast TNT.	Major ingredient in plastic explosives; one of the most brisant explosives in cast TNT (composition B); base charge in caps.
High	Better than RDX in all respects.	Same as RDX.
Average	Very powerful.	Underwater explosive.
Moderate	Plastic, easily molded or pressed.	Specialized military demolition.
Moderate	Plastic bonded.	Specialized military demolition.
Very high (almost a primary explosive)	Oily, toxic liquid; volatile above 50°C; gelatinized by nitrocotton; exhibits low-order detonation with threshold priming.	Shooting oil wells; main explosive in dynamites; used in dynamites; used in double-base powders.
Very high	Closely resembles NG; more volatile, toxic, slightly stronger but less brisant (owing to lower density).	Used in solution with NG as freezing point depressant.
High	Cheesy, plastic substance; packed in paper cartridges; may be slit and tamped in borehole for greatest blasting effect; fired by detonator as are all dynamites; heat, friction, shock, and flame sensitive.	Ditching, stumping, other uses where high propagation-by-influence "sensitiveness" is required.
High	Cheaper than comparable grade straight dynamites; must be waterproofed by special additives.	General small and large dynamite blasting, permissible (some grades).
High	Strongest, most brisant dynamite; completely waterproof; exhibits low-order detonation with threshold priming and under high pressures.	Oil well and submarine blasting, tunnel drilling, demolition.

ground mining. In 1978 slurries comprised roughly 12 percent of the total commercial market in the U.S.A. and 37 percent exclusive of ANFO. Explosive slurries will be discussed in more detail in a later section.

Most military explosives also require boosters; a 10-g tetryl or pentolite charge usually satisfies the booster requirements of the secondary military explosives, illustrating the level of sensitivity of the explosives used by the armed forces. Artillery rounds may contain igniters, primary explosives, boosters, and secondary explosives, all assembled so that each element serves a definite function.*

*The design of ammunition is an interesting subject. A description of the function and requirements for cartridge primers and fuse primers (which are frequently different formulations) for propellant igniters and bursting charge detonators, etc., is available in a War Department publication, T19-4-205, "Coast Artillery Ammunition." This pamphlet also provides a description of mechanical base and nose fuses of various types, and explains how they work.

Strictly speaking, a "round of ammunition" includes "everything necessary to fire the gun once."

In the development and use of explosives, it is important to be able to measure and evaluate explosive properties. Some of these measured properties include: impact sensitivity, spark sensitivity, friction sensitivity, detonation velocity, incendivity, and explosive strength. To measure these properties some very special equipment has been employed. The methods of measuring sensitivity involve dropping or sliding weights, discharging electrical sparks through explosives, and determination of how much initiating explosive is required to induce detonation.

Detonation velocities can be measured using streak cameras like the one shown in Fig. 19.1. Another method of determining detonation velocities employs wire probes or break wires placed at prescribed distances and attached to electronic counters which measure the time for the detonation front to travel between probes. Incendivity tests specifically for "permissible" explosives, i.e., those approved for use in coal mines, are conducted in large specially constructed steel cylinders filled with a flammable atmosphere. Explosive strength is measured by deformation of metal cylinders, pendulum deflection, bubble sizes created in underwater tests, and shock measurements using piezoelectric pressure transducers. Framing cameras with rates from a few hundred frames per second to two and one-half million frames per second are also used to gain information pertaining to the performance of explosives. Many of these tests have been standardized by the United States Bureau of Mines and are described in the Bureau of Mines Information Circular IC-8541.

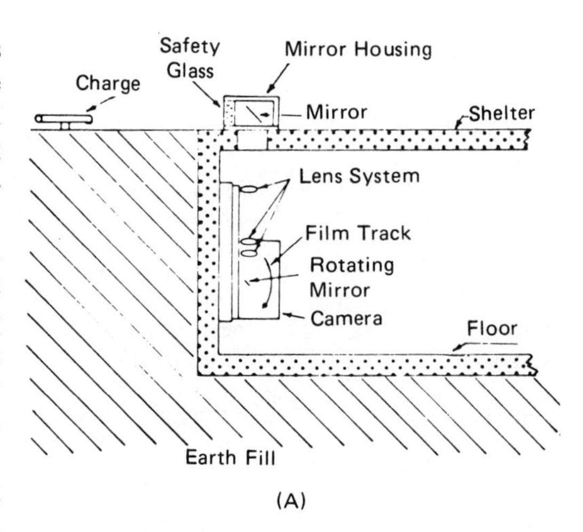

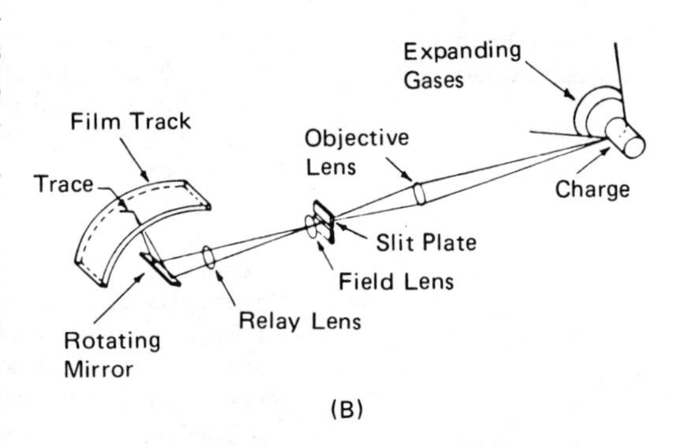

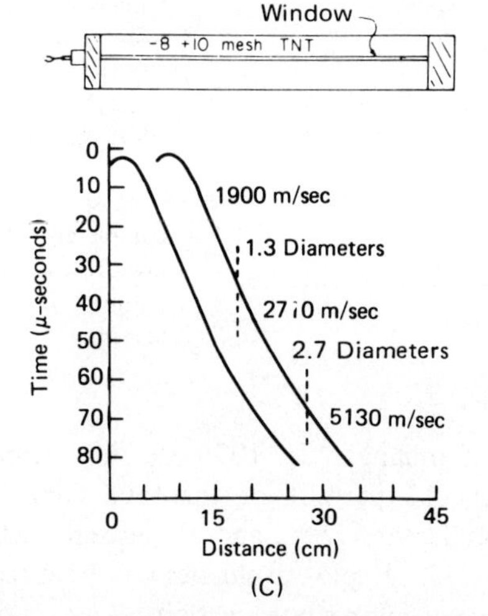

Fig. 19.1 Measurement of detonation velocity by means of the streak camera. (A) Arrangement in inside-out bombproof shelter. (B) Measurement of detonation velocity by means of the streak camera. (C) Streak camera "trace" illustrating detailed information made possible by following the detonation wave from the point of initiation until stable (steady state) detonation is established.

Some Principles of Explosives Technology

The most fundamental requirement of an explosive is that its characteristic chemical reaction be highly exothermic. The most prominent explosives exhibiting exothermic reactions are the ones which undergo oxidation-reduction reactions. These are of two general types: first, the *internal redox* compounds in which the oxidation-reduction involves oxidant and reducer radicals which are both *within* the same molecule; second, *redox mixtures*, i.e., simple mixtures in which the oxidant and reducer are separate and distinct molecules. Examples of the internal redox compounds are NG (nitroglycerin), EGDN (ethylene glycol dinitrate), PETN, RDX, TNT, tetryl, and Haleite (or EDNA). These internal redox compounds generally have a C_xH_y radical or trunk unit as the reducer part of the compound and a nitrate ($-ONO_2$), nitro ($-NO_2$), nitramine ($-NHNO_2$), or other similar oxygen-bearing radicals as the oxidant.

The practical internal redox compounds are necessarily stable compounds, otherwise they would be unduly hazardous. While nitrate, nitro, nitramine, chlorate, perchlorate, and similar groups are sometimes referred to as "explosophore" groups, this is a misnomer since they are actually relatively stable groups. For instance, there are many compounds containing nitrate and nitro groups that are nonexplosive, including many inorganic nitrates and nitro compounds. Nitric acid (HNO_3) and nitrous acid (HNO_2) are not explosive because their reducer unit (one hydrogen atom) can use up only a very small part of the oxygen of the oxidant radical. Also, ordinary metal nitrates and nitrites are nonexplosive. Ammonium nitrate is a very insensitive explosive even though its fuel radical NH_4- reacts in a redox-type reaction with two of the three oxygen atoms of the $-ONO_2$ radical. As evidence that the internal redox explosives are stable chemically, note that the heats of formation ΔH_f (positive for heat evolved and negative for heat absorbed, relative to the constituents in their standard states) are 83, 56, 123, and 13 kcal/mole for NG, EGDN, PETN, and TNT, respectively. Of the seven examples at the beginning of this discussion, only RDX and tetryl have negative heats of formation (-18.3 and -9.3 kcal/mole, respectively). The heats of explosion of RDX and tetryl are about 290 and 190 kcal/mole, respectively, showing that the "explosophore" character of the $-NHNO_2$ and $=N-NO_2$ groups is far less important than the internal redox character of these compounds.

Many internal-redox explosive compounds may be made more powerful by combining their internal redox with the redox-mixture principle. For example, TNT has appreciably more reduction than oxidation potential, i.e., it is highly oxygen-deficient. Therefore its explosion potential is enhanced by mixing it with an oxidant. Likewise, an explosive compound that has an internal redox character but is rich in oxygen may be made more powerful by mixing it with almost any fuel or combustible substance. Ammonium nitrate and ammonium perchlorate are the outstanding examples of explosives having this character. An example of an oxygen-balanced mechanical mixture of both an oxygen-rich and an oxygen-deficient explosive substance is 80/20 amatol (AN/TNT). It is nearly always advantageous to oxygen-balance an explosive by mixing it with a suitable fuel or oxidant, although the highest strength frequently occurs in mixtures having a somewhat negative oxygen balance. Aluminized and some other metallized explosives are exceptional, since the maximum strength occurs at a much more negative oxygen balance. For example, 80/20 tritonal (TNT/AL), although much more oxygen deficient, is appreciably more powerful than pure TNT.

Explosives based strictly on "explosophores," i.e., single groups or radicals whose rearrangement or decomposition generates large quantities of heat, are common among the primary explosives. For example, lead azide ($Pb(N_3)_2$) contains only an "explosophore" group ($-N{=}\overset{+}{N}{=}\overset{-}{N}$ or $-N\langle{N \atop N}\|$). All the heat of explosion, Q, of lead azide is therefore attributed to the rearrangement of this group to yield nitrogen, and in this case the heat of explosion is entirely the negative of the heat of formation ΔH_f. Another "explosophoric" group is the acetylide ($-C{\equiv}C-$) group in which $Q = -\Delta H_f$ at low pressure. Under very high pressures such as occur upon decomposition under confinement, the products from the detonation of actylene consist of CH_4 and carbon instead of H_2 and carbon; then Q is somewhat greater than $-\Delta H_f$, i.e., about 73 instead of 55 kcal/mole. The fulminate group ($-ONC$) is another "explosophore" group, however $Q \doteq -2\Delta H_f$ because of the formation of CO and CO_2 in the products of detonation at high pressures.

The heat of explosion and the nT product, i.e., maximum temperature multiplied by the expanded volume of the gaseous products, are each roughly proportional to the total available energy. But the detonation pressure, P_2, is determined by the density and velocity. This is expressed approximately by the relation $P_2 = \frac{1}{4}\rho_1 D^2$, where ρ_1 is the density and D is the detonation velocity. Furthermore, D usually increases almost linearly with ρ_1; the pressure, therefore, involves terms in the first, second, and third power of the density. Hence in order to obtain high densities, casting a molten material into a shell is preferable to loading loose grains and tamping. However, if the melting point of the explosive is excessively high, the charge may undergo decomposition or even explode upon melting. TNT, with a melting point of 80.2°C, is easily and safely cast, whereas the casting of picric acid (m.p. 123°C) is hazardous; casting substances of still higher melting points is virtually impossible. For example, PETN and tetryl (m.p. 139 and 130°C, respectively) are extremely hazardous and may explode on melting. It is therefore common practice, in obtaining high-density explosives, to make a slurry of the high-melting, more powerful explosive in TNT; or in some cases to make an eutectic solution of two explosives, e.g., tetryl and TNT, to permit casting at a lower temperature. High densities are also obtained by pressing the charges. On the other hand, the burning rate of the low explosive decreases with increasing density; this is the basis for time-delay pellets used in fuses.

Many other factors are considered in evaluating an explosive. These include the toxicity of the gases produced on explosion,* hygroscopicity or water-resistance, stability in storage, and cost of manufacture.

The high explosives used by industry are chiefly the dynamites, slurries, and the ammonium nitrate mixtures. Coal mining utilized 41.2 percent of all commercial explosives in 1950, metal mining 17.8, quarrying and nonmetal mining 19.6, railway and other construction 19.4, and all other uses, 2.0 percent. In 1978 these figures were, respectively, 55.2, 14.6, 15.4, 10.6, and 4.2 percent.† Prilled ammonium nitrate-fuel oil mixtures, which are being used extensively today, comprise roughly three fourths of the total market in the U.S.A., thus effecting considerable savings in explosive costs in open-pit and underground blasting. Slurry explosives are also gaining rapidly in commercial blasting, not only in open-pit but also in underground mining.

*Toxic gases produced in detonation are called *fumes*.

Nitroglycerin

Nitroglycerin (glyceryl trinitrate) is a colorless (when pure) oily liquid which freezes at 13°C and is very sensitive to shock, especially when it contains air bubbles or when conditions are such that air might be trapped in the shocked liquid. It is made commercially in a steel nitrator which has steel cooling coils and a mechanical agitator. In the interests of safety and stability, a very pure glycerin, above 99.9 percent pure, is required. A typical process is described below.

About 6800 lbs of mixed acid having the composition 48 percent nitric acid (HNO_3) and 52 percent oleum (H_2SO_4)‡ is charged into the nitrator with agitation and is cooled by the circulation of calcium chloride brine at -20°C through the coils. A total of about 1300 lbs of glycerin is then added slowly in a steady

†Mineral Industry Surveys, U.S. Dept. of the Interior, U.S. Bureau of Mines, Washington, D.C., Explosives Annual for 1978.
Note: The definitions of some classifications were changed in 1977.

‡A mixture of nitric and sulfuric acid is commonly used for organic nitrations because the sulfuric acid is capable of forming a hydrated molecule, thereby effectively removing one of the products of any nitration reaction, i.e., water. This shifts the equilibrium to the right, allowing more of the nitrated product to form and resulting in a higher yield than would otherwise be possible. The sulfuric acid does not otherwise take part in the reaction and normally is completely recovered for reuse, except for minor losses.

stream. The temperature of nitration is 2-3°C,* and if there is any tendency for the temperature to rise, the flow of glycerin is stopped. The addition step takes about 50-60 minutes; agitation is continued a few minutes, and then the spent acid is allowed to separate from the nitrated glycerin oil. The spent acid carries away in solution about 5 percent of the total yield. The acid is given an additional settling treatment, then is denitrated for recovery of the HNO_3 and H_2SO_4. The glycerin trinitrate is washed with warm water, then with a 2-3 percent Na_2CO_3 solution, and finally with a concentrated solution of NaCl which breaks down any emulsion present and partly dehydrates the nitroglycerin. The product, which contains about 0.5 percent water, is stored in lead-lined tanks. A process developed in Italy for continuous nitration of glycerin is rapidly being adopted throughout the world.

The Dynamites

The high sensitivity of nitroglycerin to shock was a deterrent to its use until the chance discovery by Nobel in 1862 that large quantities of the liquid explosive could be absorbed in kieselguhr (diatomaceous earth), forming a plastic cheesy substance which could then be transported and used with appreciably less hazard. This original formulation (kieselguhr dynamite) is now rarely used, however, since the presence of 25 percent of inert "guhr" diminishes the blasting effect by absorbing energy. The absorbent used today is generally a mixture of wood pulps, meals, sawdust, flour, starch, cereal products, or the like, to which is added appropriate oxidizers, e.g., ammonium nitrate and/or sodium nitrate, and a small amount of an antacid ($CaCO_3$ or ZnO). The explosively active absorbents are called "balanced dope" because they are always carefully oxygen-balanced for maximum strength and minimum fumes. Various amounts of nitroglycerin may be used in the formula, depending on the strength desired. The commercial straight dynamites contain from 20 to 60 percent, the grade being the same as the percentage of nitroglycerin present.† Ethylene glycol dinitrate ($O_2NO \cdot CH_2 \cdot CH_2 \cdot ONO_2$) or other explosive oils‡ are now added to all domestic dynamites to make them low-freezing. Nitroglycerin may be gelatinized with 7-8 percent collodion cotton to make *blasting gelatin*. Gelatin dynamites are mixtures of gelatin and balanced dope ranging from 20 to 90 percent grades, depending on the ratio of gelatin to balanced dope used. Gelatin dynamites comprise two types: the "straight" gelatins and the "ammonia" gelatins. The straight gelatins are based on sodium nitrate dopes, and the ammonia gelatins use NH_4NO_3 or mixtures of NH_4NO_3 and $NaNO_3$ dopes.

Straight dynamites have been replaced for many uses by the *ammonia dynamites*, which provide the same blasting strength but contain much less nitroglycerin for a given grade. Some of the high ammonium nitrate dynamites are really ammonium nitrate fuel or combustible explosives in which a small percentage of nitroglycerin is used to sensitize the NH_4NO_3 to detonation. Other ammonia dynamites (the highest grades) will contain appreciable quantities of nitroglycerin. Gelatin dynamites (ammonia and straight) have the greatest shattering action or brisance;§ the straight dynamites are

*This is below the freezing point of nitroglycerin but in acid solutions the oil will not freeze to the coils if there is sufficient agitation. Adapted from "Explosives Manual," Lefax, Inc., Philadelphia, Chap. XI, part 2.

†A "40% straight dynamite" contains, for example, 40% nitroglycerin, 44% $NaNO_3$, 14% carbonaceous material, 1% antacid, and 1% moisture. A "40% ammonia dynamite," with the same strength, may contain about 15% nitroglycerin, 32% NH_4NO_3, 38% $NaNO_3$, 4% sulfur, 9% carbonaceous material, 1% antacid, and 1% moisture.

‡Dinitromonochlorhydrin and tetranitrodiglycerin are also used.

§A test for comparing the brisance of explosives is the sand bomb test: 80 g of standard Ottawa sand, sized −20 to +30 mesh, is placed in a thick-walled cylinder. A cap containing 0.400 g of the explosive is then inserted and the remainder of the cavity filled with 120 g of sand. The explosive is suitably detonated. The weight of sand which then passes the 30-mesh screen is a measure of the brisance of the explosive. For further details see Olsen and Greene, "Laboratory Manual of Explosive Chemistry," New York, John Wiley and Sons, Inc., page 80, 1943.

intermediate, and the ammonia dynamites have the least brisance for a given grade of dynamite, although there is some overlapping from one series to the other if grade is not also taken into consideration. The explosives with highest brisance are preferred for "mudcapping," hole "springing," and very hard rock shooting, although brisance must also be balanced against cost. Finally, mention is made of the *semi-gelatin dynamites* which contain sufficient gelatin agent (nitrocellulose) to thicken the liquid NG appreciably, but not enough to make a stiff gel.

To make any of the above compositions, the carbonaceous material and the other "dope" ingredients are dried together in a steam-heated pan dryer, then mixed with ammonium nitrate and/or sodium nitrate in a spiral-blade mixer. For dynamite, the "balance dope" is placed in an edge runner (muller) and nitroglycerin is added carefully. For gelatin dynamite, the dope and nitrocellulose are blended with the nitroglycerin in a sigma-blade mixer. All equipment must be of nonsparking construction, and liquid nitroglycerin is handled in small lots in rubber-tired buggies and rubber pails. The blended products are packed in paper cartridges by automatic machinery and the cartridges are then paraffin-coated and boxed in sawdust.

TNT (2,4,6-trinitrotoluene)

One of the most important military explosives, TNT, is made either by continuous or by stepwise nitration of toluene. In the stepwise process, mixtures of nitric and sulfuric acid are used, and the reactions are carried out in well-agitated steel nitrators equipped with steel cooling (or heating) coils. In 1941, the accepted daily production of a TNT "line" was about 36,000 lbs. By the middle of 1945, however, the output of the same line, with but few modifications of equipment, was nearly 120,000 lbs per day. This increase was brought about by many factors, including adoption of the principle of adding the organic material to the mixed acid (indirect nitration) for all three nitrations. Previously, for mono- and trinitrations, the oil was first added to a large volume of cycle acid or oleum to "cushion" the reaction, and this was followed by slow addition of the nitrating acids. Since, in the newer method, only a small amount of nitratable material is present in the acid at any time, the hazards of the operation are appreciably reduced. Also, much less time is required for the reaction cycle.[5]

Cyclonite (RDX)

Cyclonite (RDX) (symmetrical trimethylenetrinitramine) is over 50 percent more powerful than low-density granular TNT and has good stability. Its excellent thermal stability makes it highly adaptable, especially at high temperatures. Its sensitivity and high melting point (about 200°C) were deterrents to its use until the discovery that it could be desensitized with beeswax for press-loading into shells, slurried with TNT for casting, and mixed with a special oil to make "plastic explosives" for demolition work.* The problem of its high cost was resolved first by a reduction in the cost of the raw material, hexamethylenetetramine, as it became more available, and second by the development of an ingenious method which combines two chemical reactions, one which does not require hexamethylenetetramine.

The formaldehyde and part of the ammonium nitrate required by Reaction 2 are supplied as products of Reaction 1. Two moles of cyclonite are thereby theoretically produced from each mole of hexamine. The actual yield is about 1.25 moles per mole, due to the formation of some fifty undesired compounds, but this yield is better than the total obtained if the two reactions were carried out separately. The capacity reached at one point, during

*From a paper by Ralph Conner, formerly Chief, Division 8, NDRC, presented Sept. 9, 1946, at the Chicago meeting of the American Chemical Society. See also "Chemistry, Science in World War II," edited by W. A. Noyes, Jr., Chapters 2 to 11 by G. B. Kistiakowsky and Ralph Connor, Boston, Little, Brown and Co., 1948.

Reaction 1 (nitrolysis)

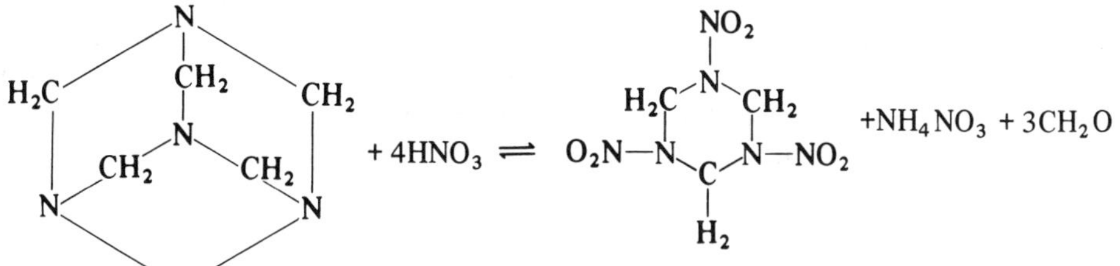

$$(CH_2)_6N_4 + 4HNO_3 \rightleftharpoons (CH_2 \cdot N \cdot NO_2)_3 + NH_4NO_3 + 3CH_2O$$

Hexamethylenetetramine $(CH_2)_6N_4$

Cyclonite (RDX) $(CH_2 \cdot N \cdot NO_2)_3$

Reaction 2

$$\underset{\text{Formaldehyde}}{3CH_2O} + \underset{\text{Ammonium nitrate}}{3NH_4NO_3} + \underset{\text{Acetic anhydride}}{6(CH_3CO)_2O} \rightleftharpoons \underset{\text{Cyclonite}}{(CH_2 \cdot N \cdot NO_2)_3} + \underset{\text{Acetic acid}}{12CH_3COOH}$$

Combined Reaction

$$(CH_2)_6N_4 + 4HNO_3 + 2NH_4NO_3 + 6(CH_3CO)_2O \rightleftharpoons 2(CH_2 \cdot N \cdot NO_2)_3 + 12CH_3COOH$$

World War II, was over a million lbs per day of RDX-TNT and waxed RDX mixtures.*

EDNA

EDNA or Haleite† (ethylenedinitramine), while not quite as powerful nor as stable as cyclonite, is somewhat less sensitive, and is appreciably more powerful than TNT. Although several reactions have been developed for its production, the most economical appears to be the synthesis of ethylene urea, starting with formaldehyde and hydrogen cyanide, and finalized by nitration of the amine groups:

$$CH_2O + HCN \longrightarrow \underset{\text{Cyanohydrin}}{HO \cdot CH_2 \cdot CN} \xrightarrow{NH_3}$$

$$H_2N \cdot CH_2 \cdot CN \xrightarrow{H_2} \underset{\text{Ethylene diamine}}{H_2N \cdot CH_2 \cdot CH_2 \cdot NH_2}$$

$$H_2N \cdot CH_2 \cdot CH_2 \cdot NH_2 + CO_2 \longrightarrow \underset{\text{Ethylene Urea}}{(CH_2-NH)_2CO} + H_2O$$

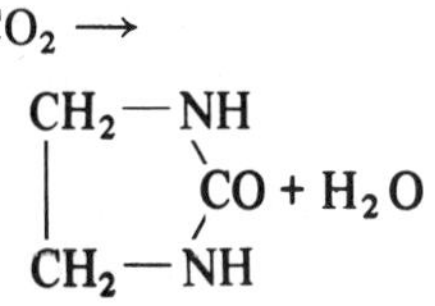

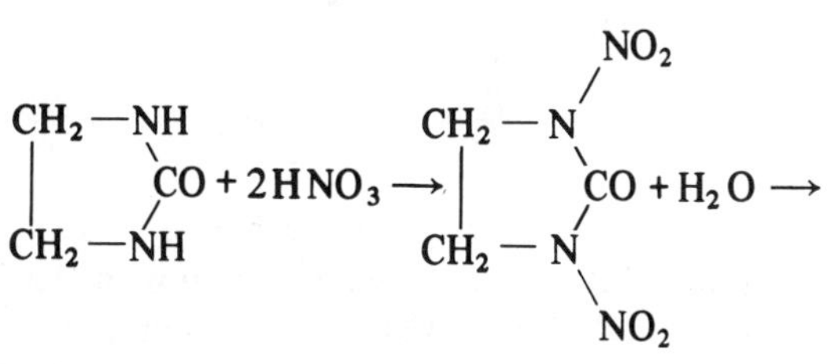

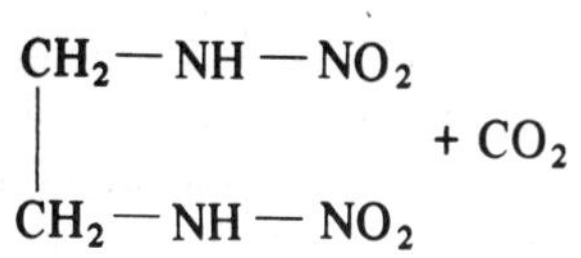

$$(CH_2-NH)_2CO + 2HNO_3 \longrightarrow (CH_2-N(NO_2))_2CO + H_2O \longrightarrow \underset{\text{EDNA}}{(CH_2-NH-NO_2)_2} + CO_2$$

Tetryl

Tetryl (2,4,6-trinitrophenylmethylnitramine), the standard booster used by the United States armed forces, may be made from dimethylani-

*RDX was used in World War II in mixtures: British and U.S. Composition "A": 91 percent RDX, 9 percent wax; Composition "B": 60 percent RDX, 40 percent TNT, 1 percent wax; Composition C: 87 percent RDX, 13 percent wax mixture; Compositions C-2, etc.; Russian: 71.9 percent RDX, 16.3 percent TNT, 11.8 percent tetryl. German compositions were similar to U.S. "A" and "B"; the Italians used an RDX-ammonium nitrate-wax mixture. Only the Japanese used RDX without a desensitizer, hence the shell loading had to be pressed to an extremely high density to avoid premature explosion.

†Named for Dr. G. C. Hale of Picatinny Arsenal.

line or from 2,4-dinitrochlorobenzene, both of which are made from benzene. Nitration of dimethylaniline is carried out by first dissolving the oil in 96 percent sulfuric acid and adding the solution slowly to a mixed acid consisting of approximately 67 percent HNO_3 and 16 percent H_2SO_4. The temperature is controlled at about 70°C by cooling coils and agitation. In the nitrator, several reactions occur in sequence: (1) *ortho-para* nitration of the benzene ring, (2) oxidation of one of the N-methyl groups to carboxyl, (3) loss of CO_2, leaving an amine, (4) introduction of the third nitro group into the benzene ring, and finally (5), nitration of the amine to nitramine. The result is tetryl $[(NO_2)_3C_6H_2 \cdot N(CH_3)NO_2]$. The solid product is purified by boiling with water, followed by filtration. It may be recrystallized from benzene. In the alternate method, the raw material is treated with methylamine to make monomethyldinitraniline, which is then nitrated with mixed acids to make tetryl directly. The yields are higher and the product does not require as many water washings. Also, the nitration is easier to control. On the other hand, the extreme toxicity of dinitrochlorobenzene is an unfavorable factor. Besides being used as a booster, where a small amount of graphite is added to facilitate pressing, tetryl is slurried with molten TNT and cast for use as a bursting charge (Tetrytol).

PETN

PETN (pentaerythritol tetranitrate) is made by nitration of the four hydroxy groups of pentaerythritol $[C(CH_2OH)_4]$ by reaction with 94 percent HNO_3 at about 50°C. Pentaerythritol may be made by the reduction of a mixture of formaldehyde and acetaldehyde, both of which are obtained from the corresponding alcohols. The pentolites, made by casting slurries of PETN with TNT, are used as boosters and in many specialized uses where relatively high detonator and Primacord sensitivity is required. In 1956, cast 50/50 pentolite was introduced by M. A. Cook, H. E. Farnam, and the Canadian Industries Limited as a booster for slurry blasting agents and prilled ammonium nitrate-fuel oil mixtures.[6] It and related boosters are currently being used extensively for that purpose throughout the world, not only as pentolite but also as the core charge along with composition B or TNT as the main charge in the popular protected core or "Procore" boosters introduced in 1958.

Picric Acid

Picric acid (trinitrophenol) may be prepared* from benzene directly by simultaneous oxidation and nitration with nitric acid in the presence of mercuric nitrate (Russian), from monochlorobenzene, or from phenol. *Ammonium picrate* is made rather easily by adding picric acid filter cake and aqua ammonia simultaneously and in small increments to a large amount of water in such a way that the batch is at all times slightly alkaline. These two explosives are of only limited importance today.

Slurry Explosives

The composite explosive products represented by aqueous-based slurries cover a wide range of formulations depending upon the intended application or the specific manufacturer involved. Basically, they consist of an aqueous solution of one or more oxidizer salts, principally ammonium nitrate (AN) alone or in combination with sodium nitrate (SN), calcium nitrate (CN), or sodium perchlorate (NaP), (specialty applications may involve sodium perchlorate alone); soluble and/or insoluble fuels such as ethylene glycol (EG), methanol, formamide, and sugar (soluble fuels) and fuel oil, powdered aluminum, coal dust, and finely divided sulfur (insoluble fuels); thickeners (water soluble polymers which also act as fuels) such as gums, starches and synthetic polymers; and trace ingredients to crosslink the thickener and to chemically foam the slurry to lower its density.[7] Some significant amount of air bubbles or "void volume" is needed to give the desired degree of sensitivity to initiation. Common gassing agents employed include

*See the Lefax "Explosives Manual" previously mentioned, or refer to the 4th Edition, page 601.

sodium nitrite, hydrogen peroxide, sodium bicarbonate, and "Unicel" (N,N'-dinitrosopentamethylene tetramine) which react under acidic conditions or with a catalyst or by heat to produce small gas bubbles. Part or all of the chemically-produced gas may be replaced by mechanically entrained air with the help of foaming agents and foam stabilizers, such as egg albumen, Aer-O-Foam, and "Duponol" (sodium laurylsulfate), or by various sources of encapsulated air, such as expanded perlite, vermiculite, or sugar cane bagasse; and glass, plastic or ceramic hollow microballoons. Some air is also incorporated into a formulation when paint grade aluminum or ammonium nitrate prills are added.

In addition to the above ingredients, many slurry formulations are "sensitized" by the addition of water soluble, highly energetic compounds such as methylammonium nitrate (MAN), hydroxyethylammonium nitrate, ethylenediammonium dinitrate, and ethyleneglycol mononitrate (EGMN). Of these, slurries containing MAN have probably been the most prominent. Slurries with relatively high levels of MAN have been formulated to produce cap sensitive explosives suitable for use in small diameter mining and construction applications as replacements for dynamite. Generally, small diameter slurries require a suitable sensitizer for cap sensitivity and high order detonation. A second category of small diameter products has been developed by sensitization with paint grade aluminum (PGAl) flakes. These are very fine particles (-325 mesh screen size) characterized by high surface area (1-4 m^2/g). They are coated with a liquid fatty acid to produce a hydrophobic surface to prevent any aluminum-water reaction and to trap small air bubbles which in turn provide sensitization.

The development of slurries for industrial use represents a significant advance in explosive technology from the standpoint of safety, performance, and cost.[8] The ability to produce an explosive formulation without using individually explosive ingredients greatly increased the safety and, in general, represents lower ingredient costs. Generally, slurries are not as susceptible to accidental initiation by fire, friction or impact, and with the development of small diameter formulations they represent the first viable substitute for cap sensitive dynamites. While ANFO is cheaper than most large diameter slurries, it has no water resistance. So the development of water-resistant slurries has resulted in improved performance of booster sensitive blasting agents. Below are shown some typical slurry formulations:

AN	57.3	36.0	62.6	43.4	53.5	53.5
MAN	–	35.2	–	20.4	21.3	–
SN	9.3	15.0	12.0	12.4	8.6	2.4
CN	–	–	–	–	–	10.7
H_2O	18.9	10.0	19.0	16.6	8.1	12.7
Thickeners	0.8	1.0	0.8	0.6	1.8	0.8
Al	10.2	–	3.4	6.6	2.5	3.0
EGMN	–	–	–	–	–	13.5
Coal	–	–	–	–	2.0	–
Sulfur	–	–	–	–	–	2.1
EG	1.9	–	–	–	–	–
Bulking Agent	1.6	2.8	2.2	–	2.2	1.3

The manufacturing process generally involves dissolving the oxidizer salts in water, along with the thickener, at an elevated temperature and then mixing in the dry ingredients and trace ingredients followed by pumping and packaging. Unless it's a special pourable formulation, explosive slurries set up upon cooling into a rubber-like gel or stiff, hard consistency due to salt crystallization. They generally have densities ranging from 1.05 to 1.30 g/cc with detonation velocities ranging from 3,000 to 5,000 meters per second. A wide range of energy levels are readily formulated by addition of aluminum or in some cases powdered silicon.

Emulsion Slurries

The most recent major development in the area of slurry or composite type explosives is the introduction of water-in-oil emulsion explosives and blasting agents.[9] In contrast to the standard slurry or water gel product which has a continuous aqueous phase (at least when first produced), the emulsion "slurries" or emulsion "gels" have an aqueous oxidizer solution dispersed in an oil continuous phase. This water-immiscible continuous phase greatly increases the water resistance of the product even above that of the crosslinked slurries. A high viscosity is obtained just by the nature of the system, i.e., high internal phase emulsion, without the need of water-soluble thickeners with their inherent problems of hydration time, thermal degradation, etc. The sensitization of emulsion slurries is provided by the intimacy between fuel and oxidizer achieved when the oxidizer solution is dispersed into very small droplets, so that the addition of aluminum, especially expensive paint grade aluminum to small diameter products, or explosive sensitizers is not needed. Therefore, emulsion formulations also offer potentially less expensive ingredient costs than most conventional slurries.

Emulsion rheology is quite different from that of aqueous phase slurries. The thickness of emulsion products can be adjusted considerably by changing the nature of the oil continuous phase. A wide variety of organic materials ranging from diesel fuel to microcrystalline or paraffin waxes may be employed with the resultant emulsion rheology ranging from that of a soft grease to that of a firm putty-like consistency. In general, however, the emulsions tend to remain relatively soft even when cold. This soft texture may be a disadvantage with regard to some handling applications, but it also makes emulsions especially suitable for other applications such as for use as a repumpable explosive product.

The widespread commercialization of the emulsion explosive products is still in its infancy, but sufficient potential has been demonstrated by products from several different manufacturers that broad applicability is anticipated. Some of the advantageous characteristics of emulsions, such as superior water resistance, high detonation velocity, rapid detonation velocity buildup, high peak pressure generation, good fume characteristics, and improved resistance to shock wave desensitization may accelerate the gradual replacement of dynamite and other molecular explosive products with inherently safer and cheaper slurry explosives. Some examples of emulsion slurry formulations are shown below.

AN	59.9	70.9	70.7	74.2	77.7
SN	18.1	11.9	10.7	–	–
H_2O	11.8	9.1	7.3	14.8	13.4
Al	0.2	–	5.7	3.0	–
Emulsifier	1.0	1.1	0.8	1.5	1.5
Oil + Wax	5.1	4.0	2.2	5.1	5.5
Microballoons	3.9	3.0	2.6	1.4	1.1

Emulsion densities tend to be higher than those observed in aqueous slurries, being generally in the range of 1.1 to 1.4 g/cc with detonation velocities of 4,000 to 6,000 meters per second and detonation pressures in the range of 100 to 120 kilobars.

Initiators, Primers, and Igniters

The common primary explosives used in detonators and fuses (military detonators) are mercury fulminate, lead azide, lead styphnate, nitromannite, and diazodinitrophenol. These are highly sensitive to shock, heat or flame, spark, etc. Igniter compositions do not always include one of the highly sensitive initiators, but may be flame- or friction-sensitive mixtures of such compounds as potassium chlorate, antimony sulfide, and an abrasive. A black-powder charge is usually used in propellant and fuse primers. In the first instance, the charge transmits the explosion to the igniter charge which may be black powder or a mixture of barium peroxide or nitrate with magnesium powder. In the second case it provides a delay action before detonating the booster charge.

Mercury Fulminate [$Hg(ONC)_2$]. This compound is made in small batches. One pound of mercury is added gradually to about 8 pounds of concentrated HNO_3 to make mercuric nitrate. This solution, which contains an excess of HNO_3, is refluxed with 10 lbs of 95 percent ethyl alcohol. The reaction is vigorous, and is moderated in the later stages by addition of dilute alcohol. The solid fulminate crystallizes out as gray crystals which are screened, washed with cold water, drained, and stored in cloth bags under water. It is commonly used with other substances, such as potassium perchlorate, in a mixture which gives a larger flash. Powdered glass, TNT, $Pb(CNS)_2$, and Sb_2S_3 may also be added.

Lead Azide. This compound, used extensively in composition caps, is made by adding a 2 percent aqueous solution of sodium azide* to a 5 percent aqueous solution of lead acetate containing a small amount of dextrin which makes the product safer to handle. When lead azide is used in a percussion primer, it must be mixed with a suitable sensitizing explosive such as lead styphnate (lead trinitroresorcinate). While lead azide is the most common primary explosive or detonator element in blasting caps and fuses (a fuse is a military detonator), it is sometimes replaced by diazodinitrophenol, nitromannite, or other primary explosives.

Lead Styphnate. This compound is a common igniter for lead azide in the composition cap and is used in direct contact with the bridgewire to obtain fast ignition of the primary explosive. It is a primary explosive, but it serves better as the igniter because of its extremely high heat sensitivity and the great rapidity with which it ignites the detonator. It is used, for example, in bead form on the bridgewire as the igniter in the du Pont seismograph cap, which is one of the fastest of the EB caps when fired by small-current blasting machines. It is extremely sensitive to spark, static electricity, and flame, and therefore requires highly specialized equipment in processing and assemblying into caps.

*Sodium azide, NaN_3, is made by heating a mixture of sodamide ($NaNH_2$) and lime or other water absorber in a stream of nitrous oxide, N_2O.

Black Powder

The many uses of this explosive are listed in Table 19.1. Since the 16th century, the formula for the standard fast-burning powder has been approximately 75 percent potassium nitrate, 15 percent charcoal, and 10 percent sulfur. A slow-burning powder contains 59 percent KNO_3. Powders suitable for time fuses are obtained by blending the two types. For blasting, a weaker powder, e.g., 40 percent nitrate, 30 percent charcoal, and 15 percent sulfur, may be used. Its manufacture consists of a series of batch operations. The finely-ground components are moistened and kneaded together in a sigma-blade mixer, then milled in an edge-runner to make a more homogeneous mixture. The mass is pressed into dense cakes which are then granulated, dried, and screened to various sizes. The dust is used in pyrotechnics and to make fuses; the oversize is recrushed. The sized grains are "polished" and provided with a coating of graphite by tumbling them in rotating cylinders or coating pans. The blasting cartridge (coal mining), which is a single cylindrical piece with one perforation in the long axis, is formed by extruding a paste made of the proper sized grains. All the operations from granulation to screening are hazardous and are performed with remotely controlled machinery located in barricated buildings. Black powder is rapidly becoming obsolete as a blasting agent, but it still remains the best fuse and igniter composition yet developed.

REFERENCES

1. Bebie, J., *Manual of Explosives, Military Pryotechnics and Chemical Warfare Agents*, New York, Macmillan, 1943.
2. Cook, M. A., *The Science of High Explosives*, New York, Van Nostrand Reinhold, 1958.
3. Cook, M. A., and McEwan, W. S., *J. App. Phys.* **29**, 1612 (1958).
4. Davis, T. L., *The Chemistry of Powders and Explosives*, New York, John Wiley and Sons, 1941.
5. Raifsnider, P. J., *Chem. Ind.* **57**, No. 7, 1054 (1945).
6. Farnam, H. E., Jr., Paper presented at Eighth Annual Drillers' and Blasters' Symposium, U. of Minn., Oct. 2, 1958; Cook, M. A., *Science*, **132**, 1105 (1960).
7. Cook, M. A., *The Science of Industrial Explosives*, Salt Lake City, Utah, IRECO Chemicals, 1974.
8. Thornley, G. M. and Funk, A. G., *Proceedings of the Seventh Conference on Explosives and Blasting Techniques*, Society of Explosives Engineers, p. 271, 1981.
9. Wade, C. G., *Proceedings of the Fourth Conference on Explosives and Blasting Techniques*, Society of Explosives Engineers, p. 222, 1978.

20

The Pharmaceutical Industry

Lincoln H. Werner* and Ellen Donoghue*

INTRODUCTION

Health care in the United States today encompasses a wide range of services. Expenditures for 1980 were 249 billion dollars or approximately 10% of the GNP, which reached 2626.1 billion dollars.[1] In 1979, there were 334,000 physicians and surgeons directly involved in patient care, for whom the pharmaceutical industry provided the necessary drugs.

The pharmaceutical industry is one of the larger industries of the United States. The total output of prescription drugs for human use, the main product of the industry, amounted to 9.34 billion dollars for 1978 in domestic sales at the manufacturing level. The total for veterinary drugs in 1977 was approximately 0.24 billion dollars. Government statistics indicate that in 1978 more than 900 companies manufactured drugs; however, 21 companies with sales over 100 million dollars in 1977 accounted for 86.7% of domestic drug sales.[2]

In 1978, the pharmaceutical industry employed 178,800 persons in the United States. Of these, 23,400 or 13.1% were employed in research and development. Expenditures for research and development for human-use drugs in the U.S. financed by the pharmaceutical industry in 1978 amounted to 1,089.2 million dollars, (11.5% of sales); 1,225.2 million dollars were budgeted for 1979.[2]

Development of the Industry

It is the responsibility of the pharmaceutical industry to provide efficacious and safe drugs for the treatment of disease, and many government controls have been imposed on drug manufacturers to ensure that these responsibilities are met.

The products of the pharmaceutical industry differ from those of other industries in that they serve the health of the nation in a direct way.

Most of the drugs which were available from prehistoric times until the beginning of the sulfa drug and antibiotic age were, in the light of today's knowledge, crude nostrums. Few of them did the patient any good, and he was probably fortunate if they did him no harm.

*Research Department, Pharmaceuticals Division CIBA-GEIGY Corporation Summit, N.J.

TABLE 20.1 Some Early Drugs

Drug	*Indication*	*Year Used in Medicine*
Laxatives		used for centuries
Quinine	Antimalarial	1639
Smallpox vaccine		approx. 1800
Morphine	Analgesic	1820
Cocaine	Local anesthetic	1879
Digitalis	Cardiac stimulant	1890
Aspirin	Analgesic (mild)	1894
Phenobarbital	Hypnotic	1912
Insulin	Hypoglycemic	1921
Arsphenamine (Salvarsan)	Antisyphilitic	1925
Salyrgan	Diuretic	1928
Sulfanilamide	Antibacterial	1932

Nevertheless, a number of drugs of proven effectiveness were known before 1940, as shown in Table 20.1.

The discovery of the antibacterial activity of sulfanilamide in 1932 and of penicillin in 1940 opened a new era in drug research and manufacture which has contributed much to the practice of medicine as we know it today.

During the period between 1940 and 1978, many new effective and potent drugs were introduced due to an intense research effort. Table 20.2 lists the therapeutic class, U.S. approved name (USAN), and year of introduction of major new drugs. This list is necessarily incomplete. The selection is based in part on the annual sales volume of the individual drugs. The USAN 1980 (list of U.S. Adopted Names) and the USP dictionary of drug names[3] lists 1,843 U.S. Adopted Names for single drug entities. The *Physicians' Desk Reference*[4] (PDR) for 1980 lists approximately 2,500 drug products by brand name.

Drug Patents and Drug Nomenclature

Recent analyses indicate that some 88% of drug innovation today comes from the pharmaceutical industry.[5] The investment in the development of a new drug is very substantial, and it is essential that it be protected by a valid patent. There are four types of pharmaceutical inventions that are patentable, i.e. a novel product, a process, a composition, or a use. Patents covering new compounds provide the best protection. U.S. patents are granted for 17 years from the date of issue.

United States law requires that all pure drug substances have a generic or nonproprietary name which is public property, apart from the trademark which is a protected name. The generic name of a drug is frequently derived from the chemical name and must be approved by the United States Adopted Names Council (USAN).

The trademark is a name selected by the manufacturer and registered with the U.S. Patent and Trademark Office. It identifies the product with the manufacturer and his reputation for quality and integrity. Some drugs are manufactured by more than one company and sold under different brand names (trademarks). Examples are given in Table 20.3.

During the life of the patent the new drug can be marketed only by the inventor(s) and the inventor's assignee and licensees. Thereafter the drug can be manufactured and marketed by any company, subject to the regulations of the Food and Drug Administration. Such drug products are generally marketed under their generic name; they are also called "generic drugs." The Drug Topics Red Book for 1980 lists no fewer than 50 companies offering "generic" hydrochlorothiazide tablets.

Regulation

Although fraud in food and medicine is an age-old problem, it did not reach alarming proportions until the industrial 19th century. In 1892, Senator Paddock of Nebraska sponsored a bill to control the adulteration of food and drugs, but it failed to pass Congress. On

TABLE 20.2 Examples of Major Drugs Introduced Between 1940 and 1980

Class	*USAN*	*Year*
Analgesics	Meperidine	1944
	Methadone	1948
Antiallergy agents	Cromolyn Sodium	1968
Antianxiety agents	Meprobamate	1955
	Chlordiazepoxide	1960
	Diazepam	1965
	Clorazepate dipotassium	1972
Antibacterial agents	Sulfisoxazole	1949
	Cotrimoxazole	1973
Antibiotics	Penicillin	1940
	Chlortetracycline	1948
	Erythromycin	1952
	Cephalexin	1971
Anticonvulsants	Carbamazepine	1968
Antidepressants	Imipramine	1959
	Amitriptyline	1962
Antihistamines	Benadryl	1946
	Chlorpheniramine	1951
Antihypertensive agents	Hydralazine	1952
	Reserpine	1953
	Guanethidine	1961
	Methyldopa	1963
	Propranolol	1968
	Clonidine	1974
	Prazosin	1976
	Metoprolol	1978
Anti-inflammatory agents (steroidal)	Cortisone acetate	1950
	Prednisolone	1955
	Dexamethasone	1960
Anti-inflammatory agents (nonsteroidal)	Phenylbutazone	1952
	Indomethacin	1965
	Ibuprofen	1974
	Piroxicam	1980
Antiulcer agents	Banthine	1950
	Propantheline bromide	1953
	Cimetidine	1977
CNS stimulants	Amphetamine	1944
	Methylphenidate	1956
Diuretics (nonmercurial)	Chlorothiazide	1958
	Hydrochlorothiazide	1959
	Spironolactone	1961
	Dyazide®	1965
	Triamterene	1965
	Furosemide	1966
	Ethacrynic acid	1967
Hypocholesterolemic agents	Clofibrate	1962
Hypoglycemic agents (oral)	Tolbutamide	1957
	Chlorpropamide	1958
Seditives and Hypnotics	Flurazepam	1970
Steroidal sex hormones	Progesterone	1936–1940
	Estradiol	1952
	Testosterone	1954
Tranquilizer, major	Chlorpromazine	1954
	Thioridazine	1960
	Haloperidol	1967
Vaccines	Poliomyelitis	1961–1962
	Measles	1964–1965
Antineoplastic agents	Cyclophosphamide	1962
	Fluorouracil	1962
	Methotrexate	

TABLE 20.3 Trademarks of Some Drugs

Generic Name	*Trademark*	*Company*
Chlorpheniramine	Chlortrimeton	Schering
	Teldrin	SKF
	Drize	B.F. Ascher & Co.
	Histaspan	U.S.V. Laboratories
	Allerbid	Amfre Grant, Inc.
	Antagonate	Dome Laboratories
Erythromycin	Illotycin	Lilly
	Erythrocin	Abbott
	Emicin	Upjohn
	Robimycin	Robins
	Kesso-Mycin	McKesson
Hydrochlorothiazide	Esidrix	CIBA
	Hydrodiuril	Merck, Sharp & Dohme
	Lexxor	Lemmon
	Oretic	Abbott
	Thiuretic	Parke Davis

February 1, 1906, the Pure Food and Drug Act was enacted. From the start, inadequacies in this act were recognized. Sometime later, after a five-year struggle, the 1938 Food, Drug, and Cosmetic Act was signed into law by President Roosevelt. The Act granted the authority needed to establish standards of identity and quality for most food products; hazardous cosmetics were outlawed, and for the first time misleading cosmetic labeling was banned. Outlawed also were drugs which were dangerous to health when used according to directions (for example, radium water). Advances were made with respect to formula disclosure, and new drugs could not be marketed until the manufacturer had shown to the satisfaction of the Food and Drug Administration (FDA) that they were safe.

Intense drug research between 1940 and the present yielded many new drugs which have revolutionized therapy. In 1962, the Kefauver-Harris Amendments to the Federal Food, Drug and Cosmetic Act were adopted by Congress. These amendments gave the FDA regulatory control over the assessment of efficacy as well as safety of new drugs. The driving force for this new legislation was the incidence of phocomelia (shortened, misformed limbs) in newborn children whose mothers had taken thalidomide, a sedative-hypnotic developed in Europe.

Precise regulations govern the investigation of new drugs. In-depth laboratory studies of the biological activity and safety in animals, a description of the structure and synthesis of the new compound as well as the composition of the formulation to be used for clinical studies, together with the protocol for the proposed clinical studies, and other required information constitute the Notice of Claimed Investigational Exemption for a new drug, the so-called IND. This document must be filed with the FDA at least 30 days prior to starting clinical studies. The 1962 regulations prescribe that clinical studies on new drugs are to be conducted in three phases. In Phase I, tolerance, absorption, excretion, and metabolism are studied. Phase II covers the initial trials in a limited number of patients for specific disease control. Phase III provides the assessment of the drug's safety and effectiveness in the treatment of a given disease in a larger number of patients. Before these studies can be started, the clinical protocols must be reviewed and approved by the FDA. When the studies are deemed adequate, all information (chemical, pharmaceutical, biological, and clinical) on the new drug is assembled in New Drug Application (NDA) and submitted to the FDA for approval.

Precise regulations govern not only the study of new drugs in humans (clinical) and the preparation of documents which must be submitted to the FDA before such studies can be started, but also the preclinical (animal safety) studies. The latter are the Good Laboratory Practice (GLP) regulations which became effective June 20, 1979 and delineate the procedures that must be followed during the conduct of animal safety studies, including a precise protocol, a study director for each study, a quality assurance unit to inspect and audit each study/facility and assure top management that the study has been conducted in compliance with the GLP's, etc. These regulations have had a perceptible impact on the quality and integrity of safety studies, although at a significant increase in expense. The Kefauver-Harris Amendments of 1962 also specifically require that the methods, facilities, and controls used by the manufacturer conform with "current good manufacturing practice." FDA issued regulations (GMP's) in 1962 spelling out exactly what constitutes current good manufacturing practice. The GMP's were revised in 1971 and underwent a second and rather massive revision which became effective March 28, 1979. The new revision includes specific requirements for the establishment of a quality control unit, written procedures (SOP's) in certain key areas, and it mandates an expiration date for just about every drug product marketed. FDA monitors the industry's compliance with these GMP regulations via periodic unannounced inspections.

Older drugs introduced between 1938 and 1962 have been reviewed by drug panels convened by the National Academy of Science-National Research Council to ascertain their efficacy.

Research

The links between chemistry and medicine are as old as the alchemical roots of the science of chemistry. Drugs derived from natural materials had their origin in folk remedies of ancient times. However, it was not until the beginning of the 20th century that chemistry began to play a significant role in health care. At this time Paul Ehrlich in Germany instituted a systematic search for synthetic chemotherapeutic agents. Chemotherapy really started to develop with the discovery of the antibacterial effects of the sulfa drugs in 1932. Dedication to the search for useful new drugs is a characteristic of the industry.

From 1940 to 1967, 848 basic new drugs were developed. Of these 525 were discovered by American companies. From 1969 to 1978, 159 new pharmaceutical products representing new chemical entities were introduced in the United States.[6] Annual research expenditures by the pharmaceutical industry have grown from 50 million dollars in 1951 to 500 million in 1968, and to 1.2 billion dollars or 11.5% of gross sales in 1979. In 1978, 13.1% of the employees in the drug industry worked in Research and Development.[2] It is estimated that of the many compounds prepared and tested in the research laboratories, only one out of 8000-10,000 ever becomes a new, marketed drug.

The development of a research compound into a drug approved by the FDA for use in patients may take from six to ten years. During the period from 1958 to 1962, it was estimated that about two years were needed to assess the safety and efficacy of a potential new drug from the time it was selected for testing and application was filed to obtain approval by the FDA for marketing. In 1962, costs for these required studies averaged about $1.2 million; between 1968 and 1972, the time for this process had lengthened to five-and-a-half to eight years, and the cost in 1972 had risen to $11.5 million on the average. In 1979, this figure had risen to approximately $30 million. This increase in the cost of drug development has not only led to a more cautious approach to drug research but also has diverted the resources otherwise available for the search for new drugs to the development of a few drug candidates. The feedback from clinical trials is extremely valuable in drug research; as fewer drug candidates go into clinical studies, this feedback decreases and may eventually result in fewer new drugs being found. This is unfortunate because a need for improved therapy continues, especially in major chronic diseases.[7]

An interesting account of the development of cimetidine, an antiulcer drug, was published by Duncan and Parsons[8] in 1980. The research program leading to this new drug was initiated at Smith, Kline, and French in the United Kingdom in 1964. In 1968, a lead was found which resulted in the discovery of cimetidine in 1973-1974. More than 700 compounds were synthesized in this research effort. The drug

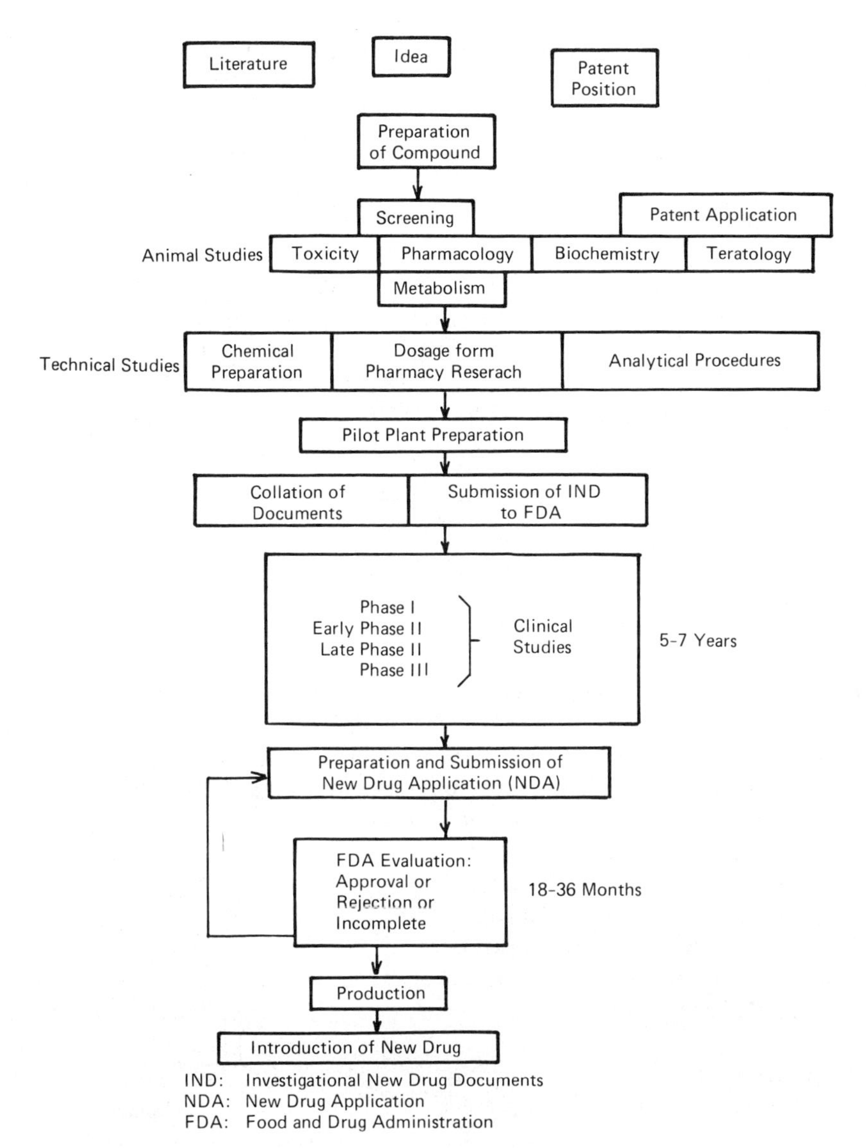

Fig. 20.1 Development of a new drug from idea to market.

was finally marketed in 1976 in the United Kingdom and in 1977 in the United States.

Figure 20.1 outlines the many steps in the development of a new drug. Such development requires the skills and experience of an organization of chemists, biochemists, biologists, microbiologists, doctors, pathologists, toxicologists, pharmacists, and engineers working together to discover and develop a new drug.

Unfavorable findings in any of the safety and efficacy studies can lead to a decision to discontinue further development and thus render useless years of careful and patient work.

Products

The drugs manufactured today can be grouped into certain main classes as shown in Table 20.4. These classes do not cover all drugs but do include those which are widely used.

In addition to manufacturing drugs which have wide application, the pharmaceutical industry, often at considerable cost, has made available certain products for the treatment of infrequent special conditions such as iron poisoning (deferoxamine), botulism poisoning (botulism antitoxin USP), snake bite (poly-

TABLE 20.4 Domestic Sales, by Product Class, Ethical Drugs for Human Use, 1976-1977[a] (percentages)

Class	*Relative share of sales* *1976*	*1977*	*% Change, Dollar Sales[b] 1976-1977*
Central Nervous System	27.4%	25.2%	-1.5%
Anti-Infectives	15.4	15.4	7.2
Neoplasms & Endocrine	11.4	11.4	6.9
Gastrointestinal & Genitourinary	10.0	10.5	12.3
Cardiovasculars	10.3	10.5	9.3
Vitamins & Nutrients	7.3	8.0	17.7
Respiratory System	6.7	6.6	6.4
Dermatologicals	3.5	3.5	6.0
Biologicals	2.0	2.2	19.0
Diagnostic Agents	1.4	1.4	9.4
Other Pharmaceutical Preparations	4.6	5.3	–
	100.0%	100.0%	7.3%

[a]Same companies for 1976–1977, representing approximately 84 percent of domestic sales.
[b]Based on actual dollar sales
Source: PMA Annual Survey Report 1977–1978

valent crotaline antivenin USP), and respiratory depression (e.g. doxapram).

Representative compounds in the categories listed in Table 20.2 are described below to illustrate the various types of chemical structures of the active ingredients in drug products.

A detailed discussion of drug products and their mode of action can be found in Goodman and Gilman, *The Pharmacological Basis of Therapeutics,*[9] Remington's *Pharmaceutical Sciences,*[10] and AMA Drug Evaluations.[11]

Nonnarcotic Analgesics. This type of analgesic is used for the relief of mild pain. Examples are aspirin, acetaminophen, and phenacetin.

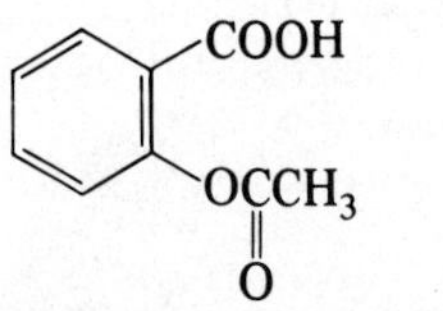

Aspirin (acetylsalicylic acid)

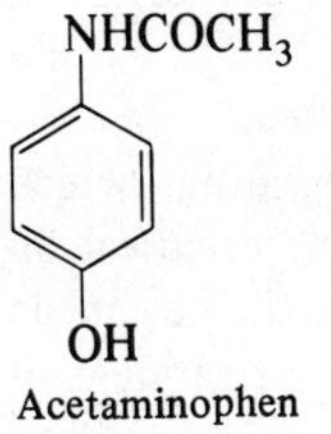

Acetaminophen

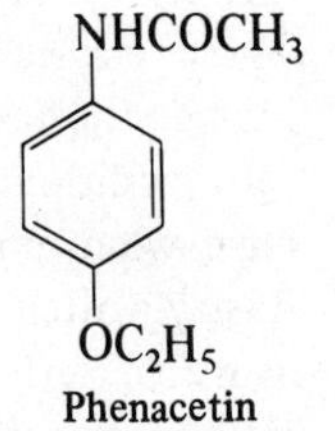

Phenacetin

Narcotic Analgesics. These substances are used primarily for the relief of severe pain; they are all addicting (i.e. producing psychological, and sometimes physical, dependence), and they all produce side effects, e.g. respiratory depression.

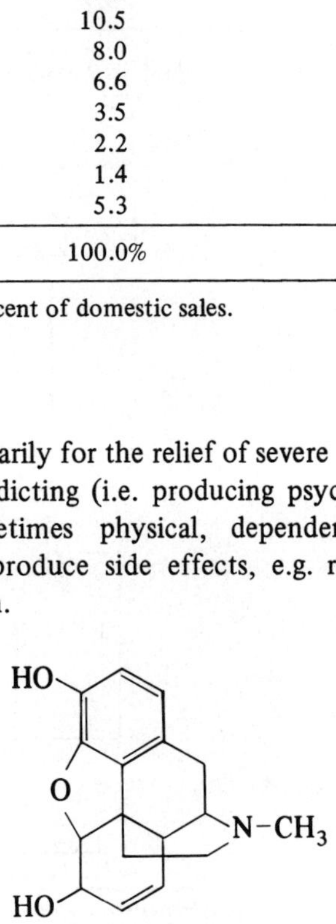

Morphine

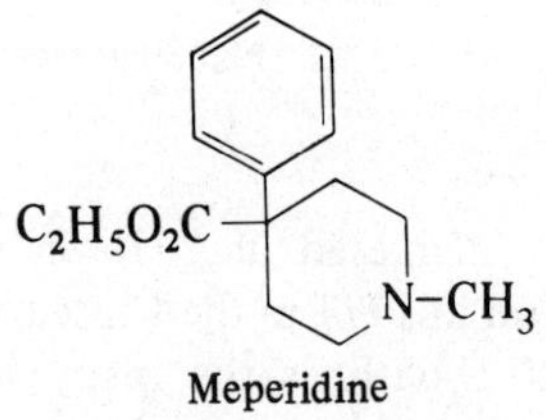

Meperidine

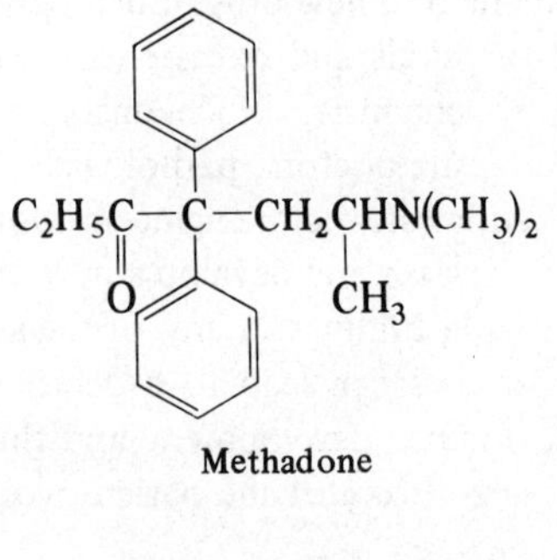

Methadone

Two more recent analgesics, propoxyphene and pentazocine, have a lower addiction potential and are more potent than aspirin, acetaminophen, and phenacetin.

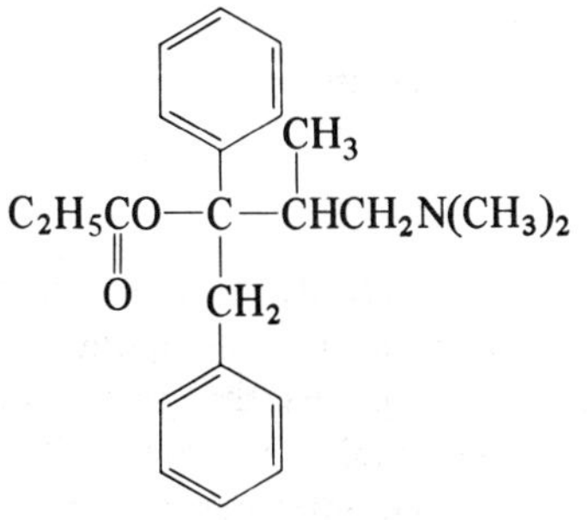

Propoxyphene

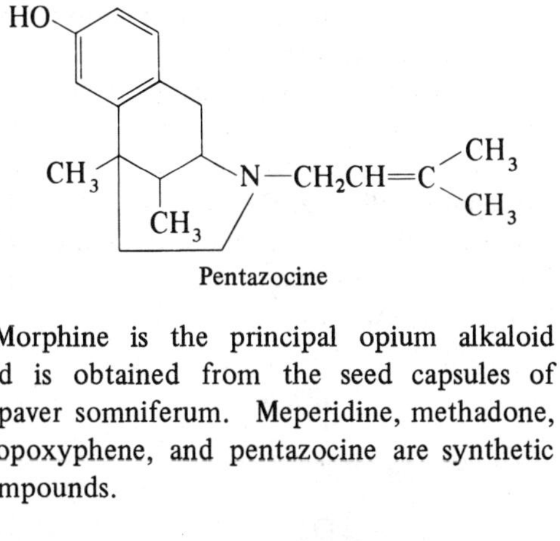

Pentazocine

Morphine is the principal opium alkaloid and is obtained from the seed capsules of Papaver somniferum. Meperidine, methadone, propoxyphene, and pentazocine are synthetic compounds.

Antacids. Gastric antacids are agents that neutralize or remove acid from the gastric contents. Generally they contain a weakly basic moiety. The following compounds are used:

Sodium bicarbonate	Magnesium oxide or Magnesium hydroxide
Magnesium carbonate	Aluminum hydroxide
	Magnesium trisilicate

Antiallergy Agents. Cromolyn sodium, introduced in 1968, has been used for the prophylaxis of severe bronchial asthma. It acts by inhibiting release of histamine and a slow-reacting substance in anaphylaxis (SRS-A) from human lung tissue during allergic responses. It represents the first drug of this type.

Antianxiety Agents. "Minor tranquilizers" are widely used for the relief of mild anxiety and tension. They are of little or no value in the treatment of psychoses. The structures of three compounds of this group are given below.

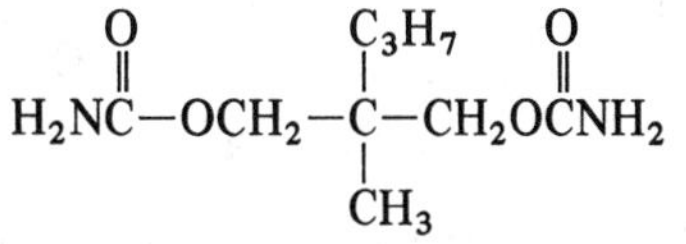

Meprobamate (1955)

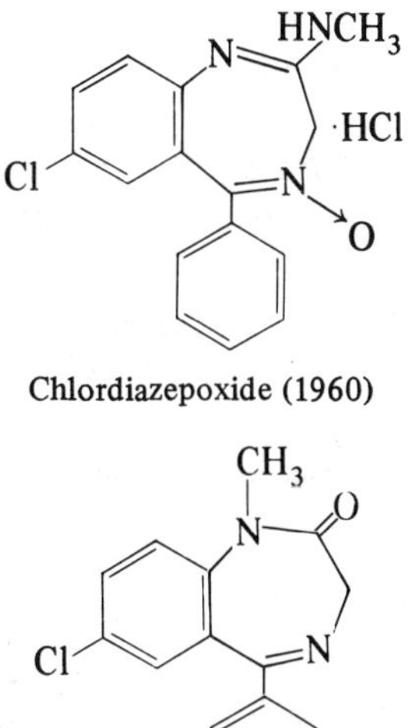

Chlordiazepoxide (1960)

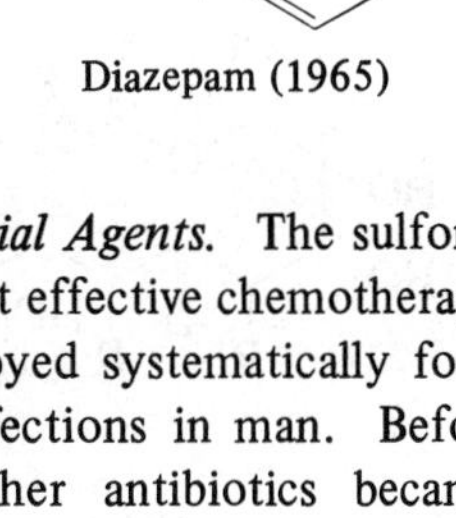

Diazepam (1965)

Antibacterial Agents. The sulfonamide drugs were the first effective chemotherapeutic agents to be employed systematically for the cure of bacterial infections in man. Before penicillin and the other antibiotics became generally

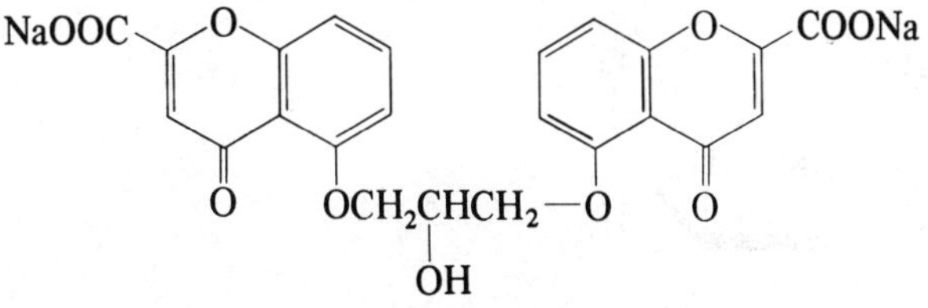

Cromolyn Sodium

available, the sulfonamides were the mainstay of antibacterial chemotherapy. They still maintain an important but smaller place in medicine. Structural formulas of two major sulfonamides are shown below.

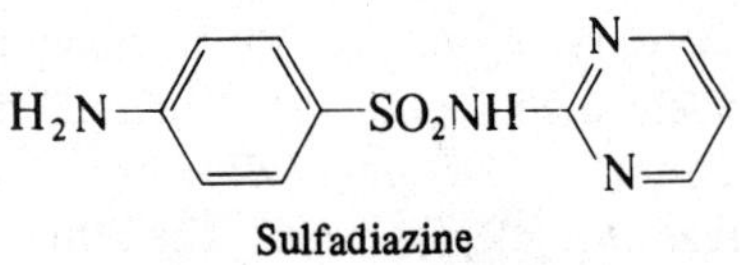

Sulfadiazine

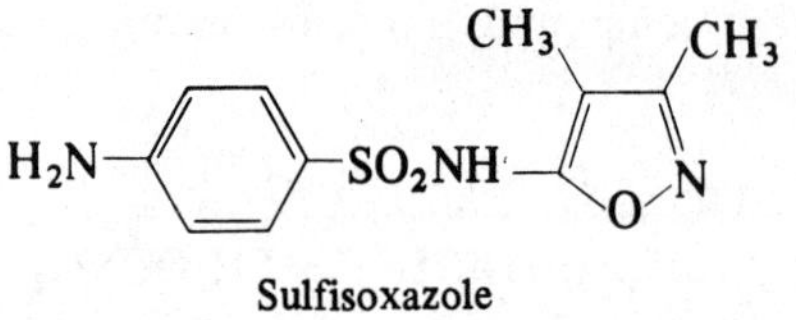

Sulfisoxazole

In 1973, a combination of trimethoprim with sulfamethoxazole (cotrimazole) was introduced. The two drugs act on sequential steps in the pathway of an enzymatic reaction in bacteria. The result of the combination is supraadditive and has led to a considerably broader therapeutic usefulness for sulfamethoxazole.

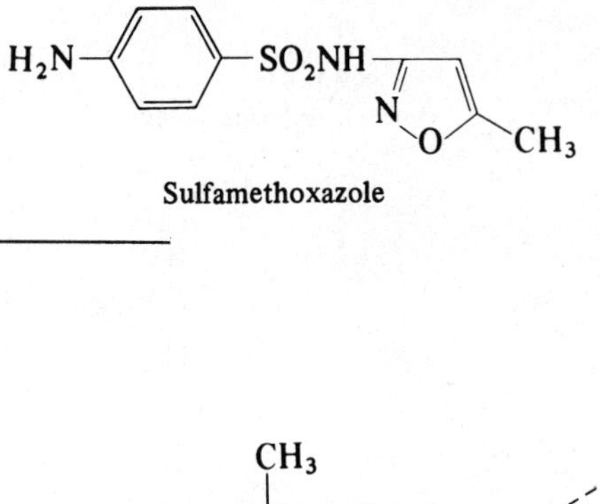

Sulfamethoxazole

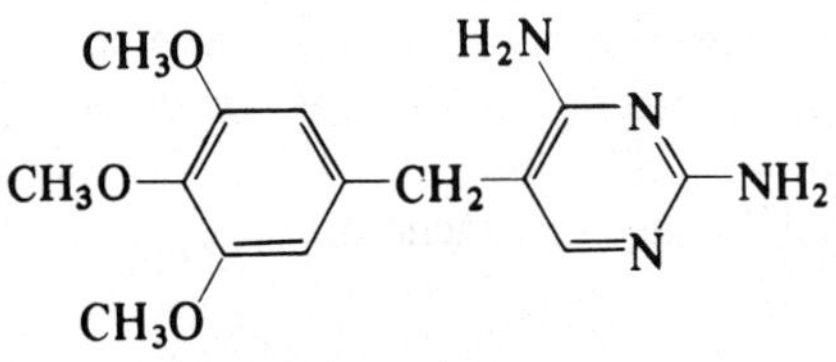

Trimethoprim

Antibiotics. Antibiotics are chemical substances produced by certain species of microorganisms that suppress the growth of other microorganisms and that may eventually destroy them. Since the introduction of penicillin (1941), over 60 antibiotics have been developed to a stage where they are of value in the therapy of infectious diseases. These substances differ markedly in their chemical and pharmacological properties and in their mechanism of action. Antibiotics are generally produced by growing the appropriate organism in a sterile medium and extracting the antibiotic from the broth. The structures of some widely used antibiotics are given below.

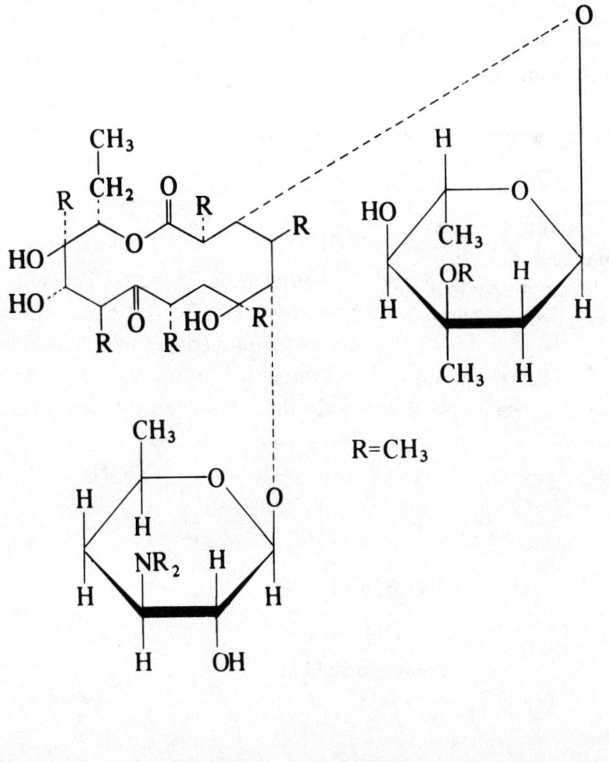

Erythromycin

$R=CH_3$

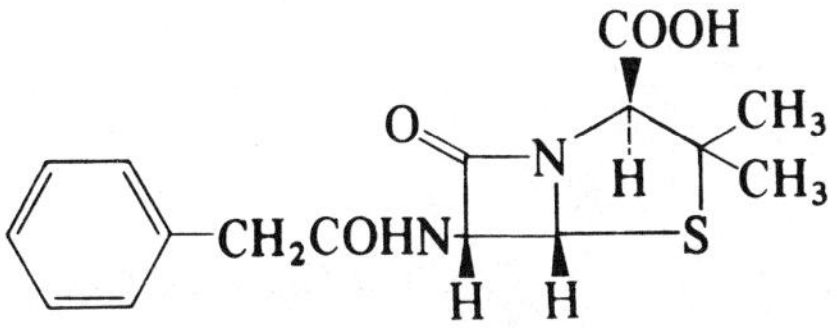

Penicillin G

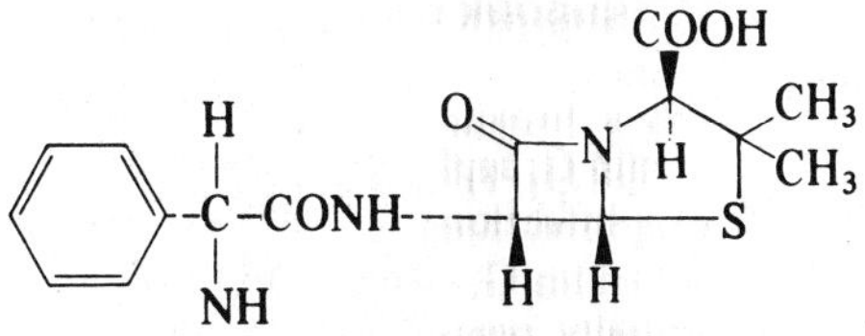

Ampicillin

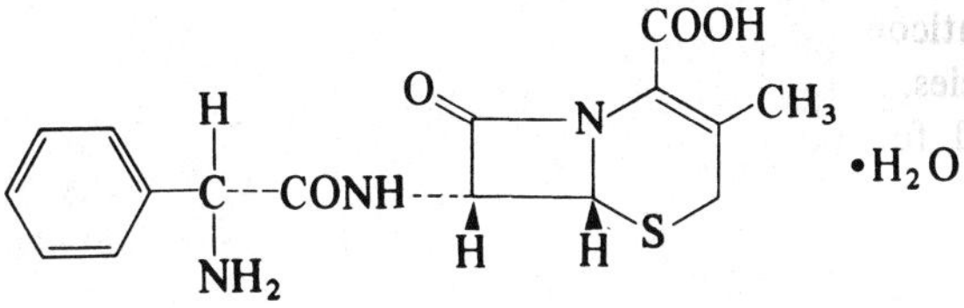

$\cdot H_2O$

Cephalexin

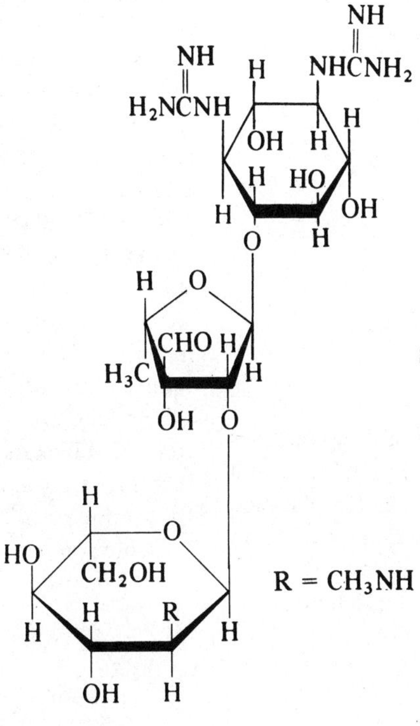

Streptomycin

R = CH_3NH

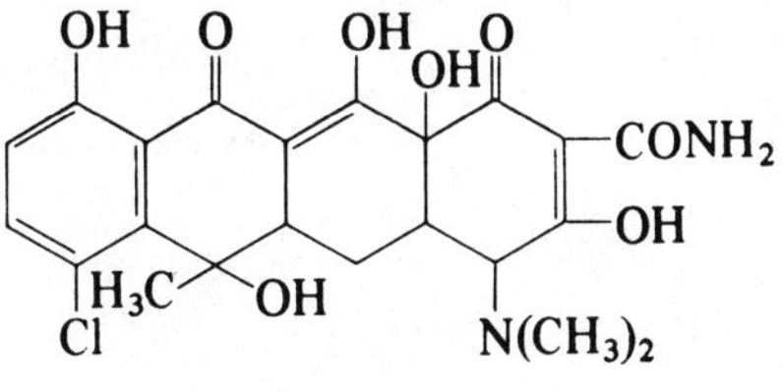

Chlortetracycline
(Aureomycin®)

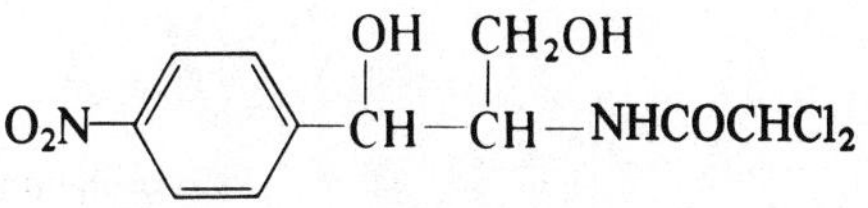

Chloramphenicol
(Chloromycetin®)

Ampicillin has a broader antibacterial spectrum than penicillin G; cephalexin is particularly useful in treating infections that have acquired resistance to penicillin G. By chemical modification of a naturally occurring antibiotic, it is possible to obtain products with desirable properties not found in the natural product.

Anticonvulsants. Anticonvulsants are used in the therapy of epilepsies. The structures of a number of drugs used for this indication are shown below.

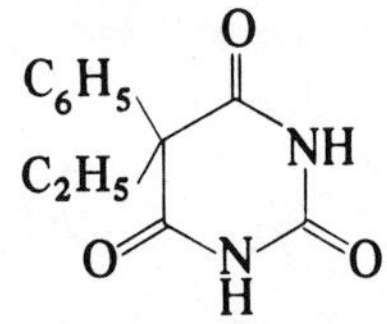

Phenobarbital

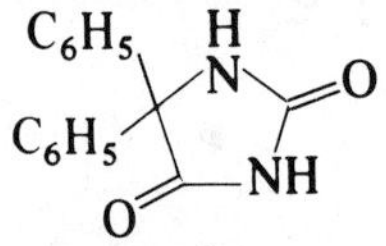

Phenytoin

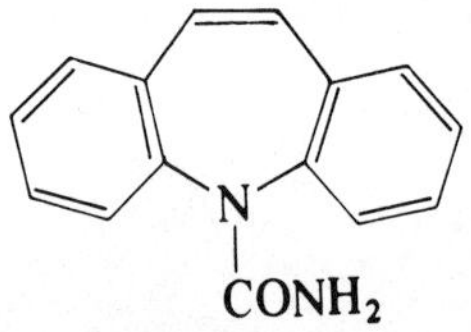

$CONH_2$

Carbamazepine

$(CH_3CH_2CH_2)_2CHCOOH$

Valproic Acid

Antidepressants. This group of drugs includes the monoamine oxidase (MAO) inhibitors and the tricyclic antidepressants. The first tricyclic antidepressant, imipramine, was introduced in 1959. Others have been developed more recently. Structures of some antidepressants are shown below.

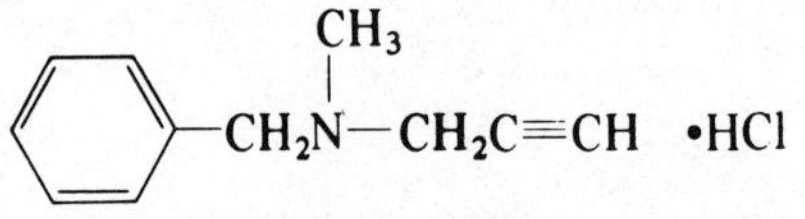

$-CH_2N(CH_3)-CH_2C{\equiv}CH \cdot HCl$

Pargyline (1963)
(MAO inhibitor)

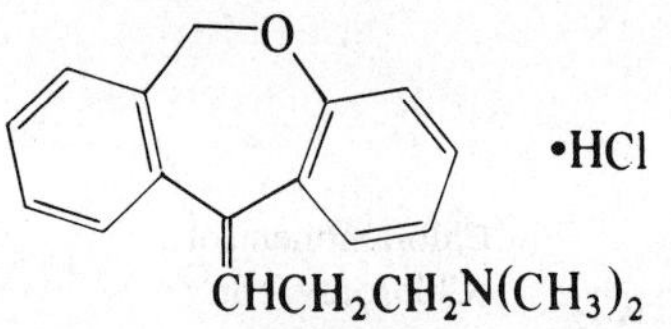

$\cdot HCl$

$CHCH_2CH_2N(CH_3)_2$

Doxepin (1969)

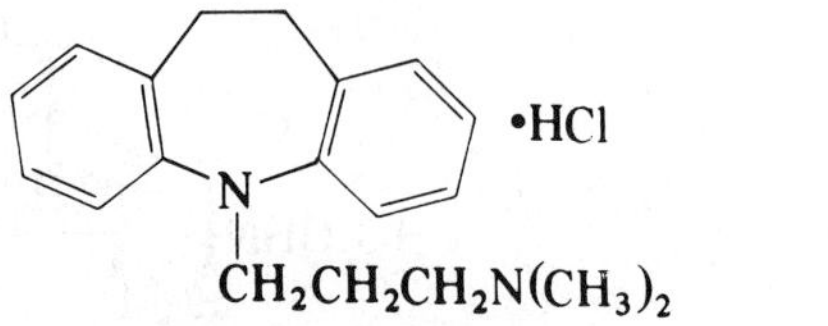

$\cdot HCl$

$CH_2CH_2CH_2N(CH_3)_2$

Imipramine (1959)

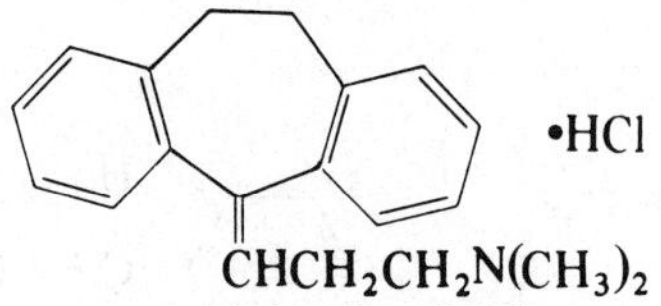

$\cdot HCl$

$CHCH_2CH_2N(CH_3)_2$

Amitriptyline (1962)

Antihistamines. The traditional antihistamines effectively antagonize many important responses to histamine and thus reduce the intensity of allergic and anaphylactic reactions; however, they uniformly fail to inhibit other responses, most notably they have no effect on gastric acid secretion.

In 1972 Black and coworkers discovered agents that preferentially blocked gastric secretion and other histamine-induced effects that were refractory to the traditional antihistamines. The traditional antihistamines are referred to as H_1-blockers; those blocking gastric secretion as H_2-blockers. Structures for some H_1-blockers are shown below. The structure of an H_2-blocker (cimetidine) is shown in the section on antiulcer agents.

H_1-blockers:

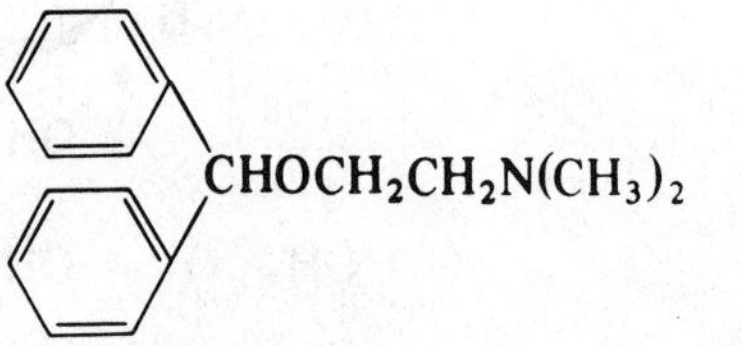

$CHOCH_2CH_2N(CH_3)_2$

Diphenhydramine

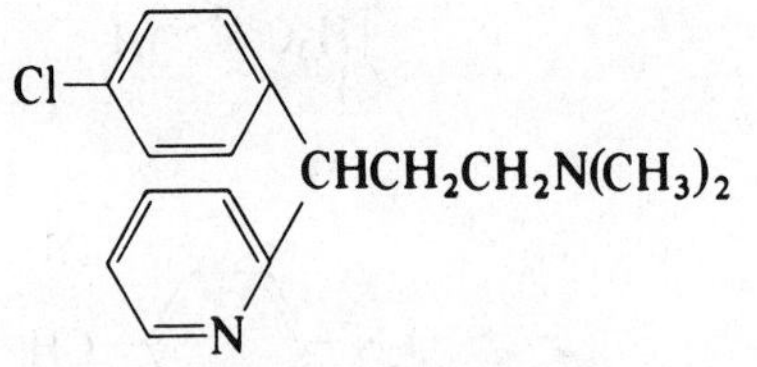

$CHCH_2CH_2N(CH_3)_2$

Chlorpheniramine

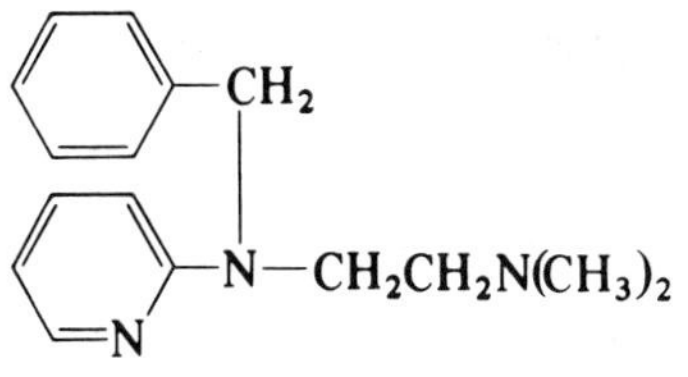

Tripelennamine

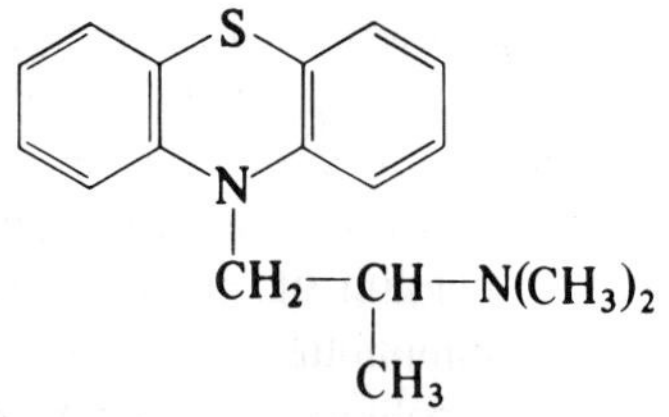

Promethazine

Antihypertensive Agents. The search for effective antihypertensive agents has led to the development of a number of useful drugs with quite diverse mechanisms of action and chemical structure. Examples of some widely used agents are shown below.

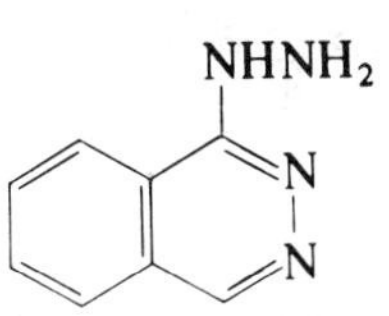

Hydralazine (1952)

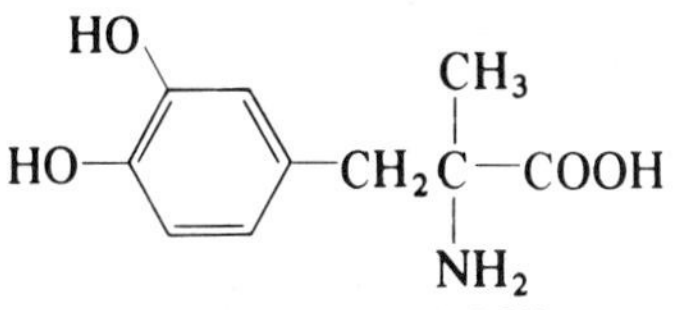

Methyldopa (1963)

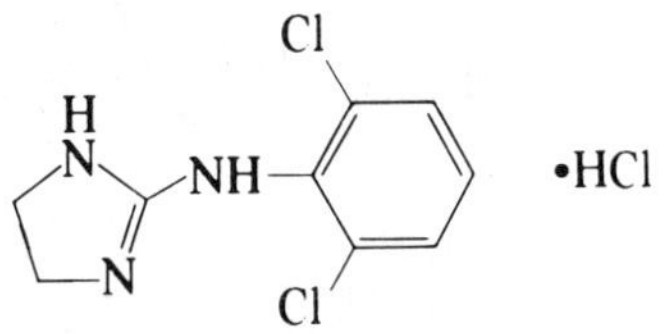

Clonidine (1974)

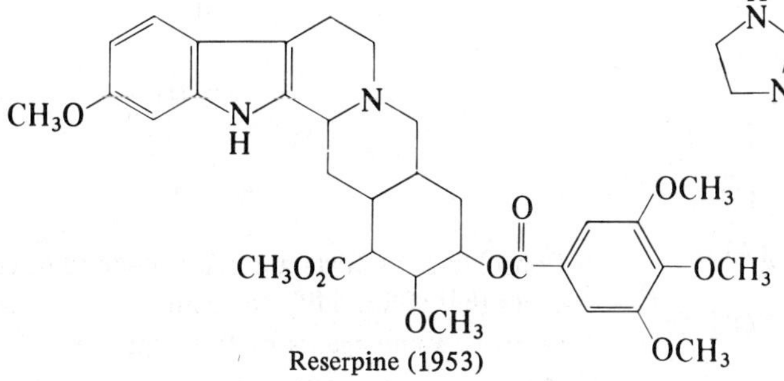

Reserpine (1953)

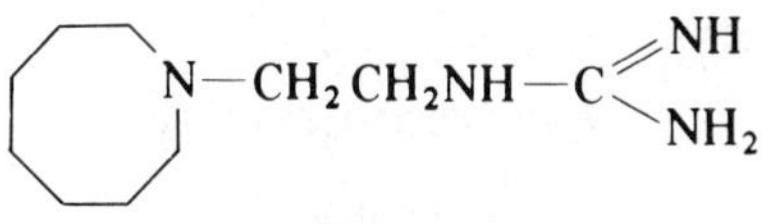

Guanethidine (1959)

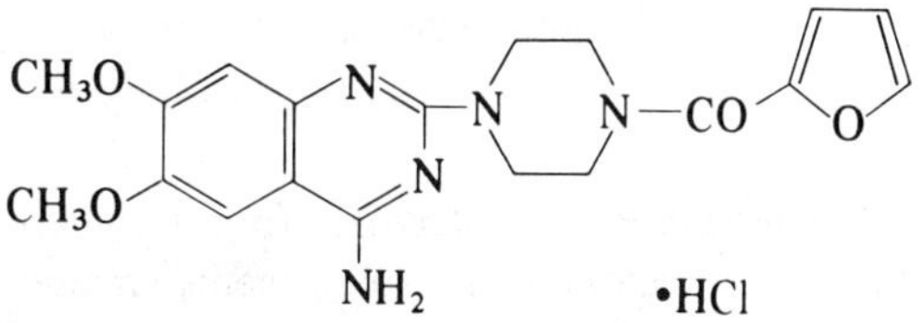

Prazosin (1976)

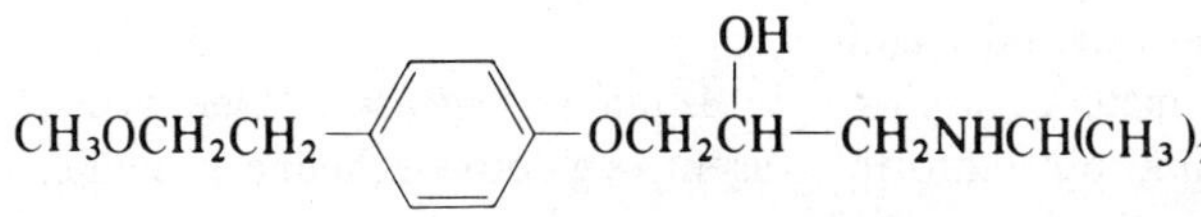

Metoprolol (1978)

Anti-inflammatory Agents, (steroidal, glucocorticoids). The most widely known application of the anti-inflammatory actions of the glucocorticoids is in the treatment of arthritis. Cortisone, introduced in 1950, was the first steroidal anti-inflammatory agent used for this purpose. Modification of the structure led to more active compounds with more specific anti-inflammatory properties. Clinically they inhibit the inflammatory response regardless of whether the inciting agent is mechanical, chemical, or immunological. The effects, however, are only palliative, and treatment with glucocorticoids does not remove the underlying disorders. Prednisolone and dexamethasone are examples of glucocorticoids widely used.

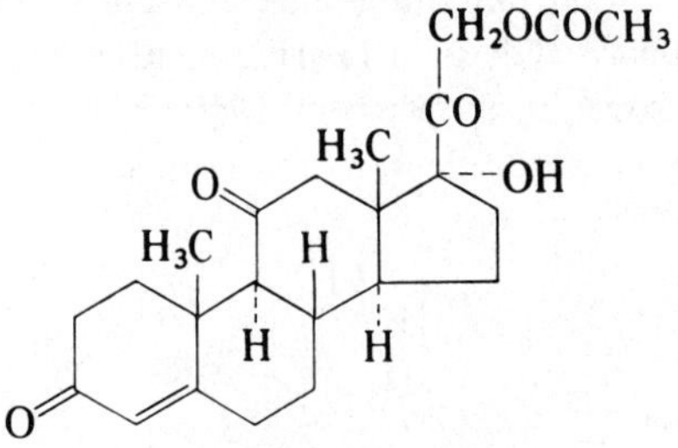

Cortisone acetate

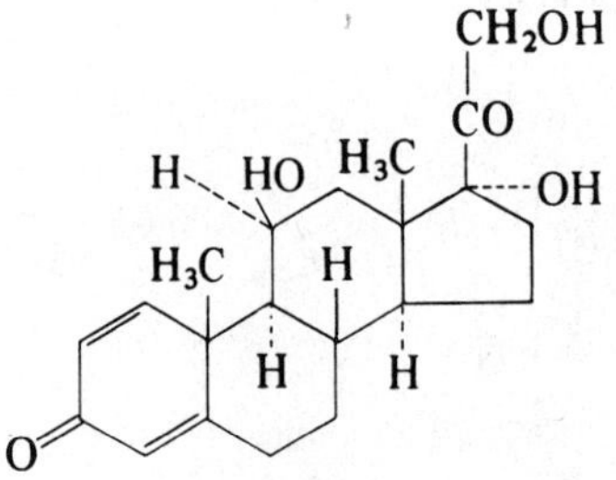

Prednisolone

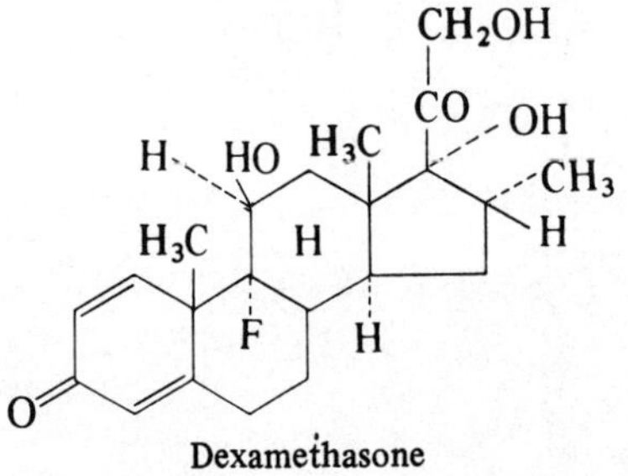

Dexamethasone

Anti-inflammatory Agents, (nonsteroidal). Nonsteroidal anti-inflammatory agents are used extensively in the treatment of rheumatoid arthritis and osteoarthrosis. The main group of these agents, the NSAIDs, (nonsteroidal anti-inflammatory drugs), which includes aspirin, reduce inflammation and pain, but unfortunately do not arrest or reverse the progress of the underlying disease. Examples of this type of agent are shown below.

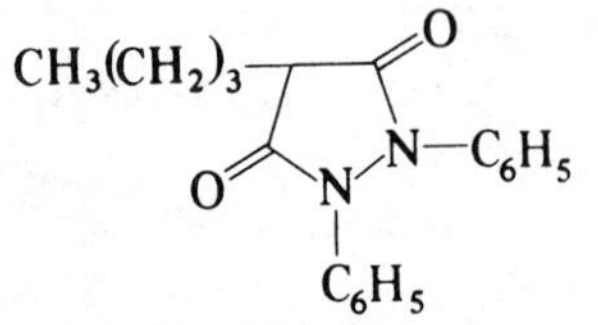

Phenylbutazone

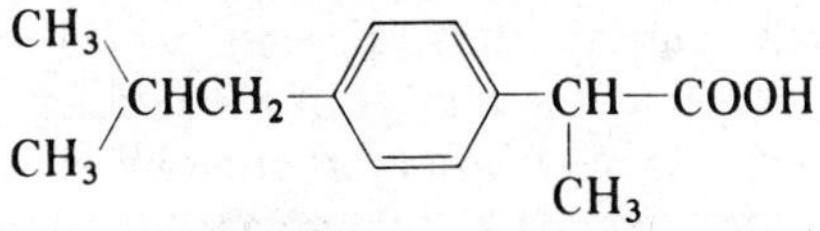

Ibuprofen

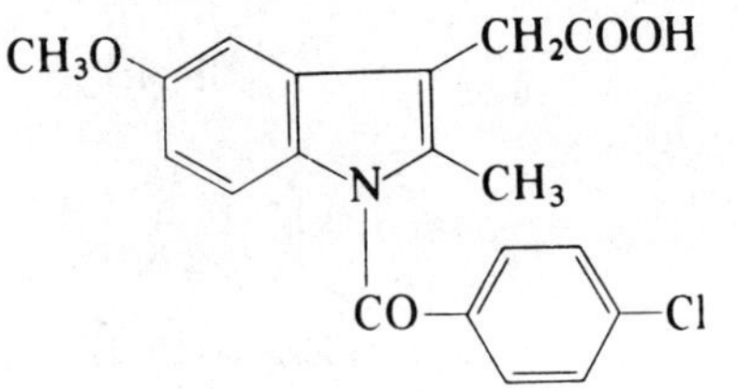

Indomethacin

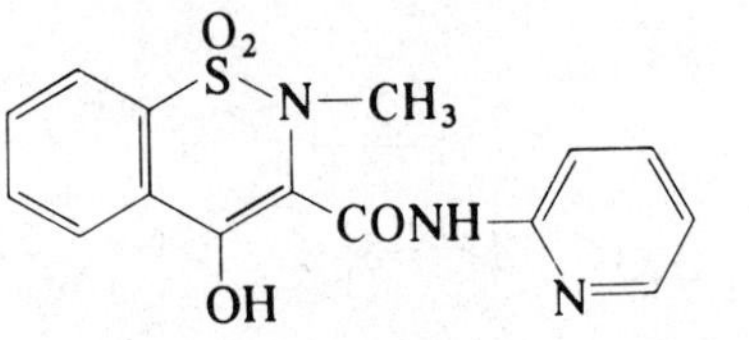

Piroxicam

Considerable efforts are being made to discover drugs that will modify the course of rheumatoid arthritis; some agents of this type are available, (e.g., gold salts, penicillamine) but toxicity and slow onset of action preclude their widespread application.

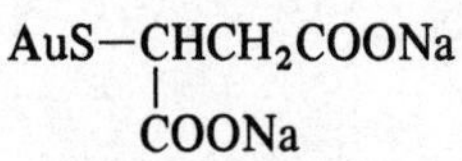

$$AuS{-}\underset{\displaystyle COONa}{\underset{|}{C}}HCH_2COONa$$

Gold sodium thiomalate

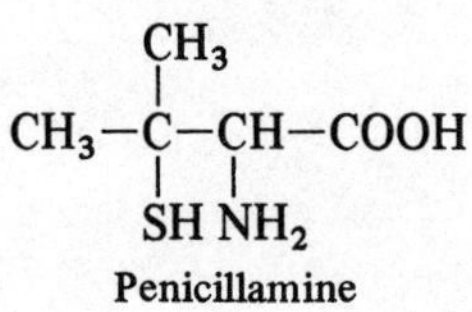

$$CH_3{-}\overset{\displaystyle CH_3}{\underset{\displaystyle SH}{C}}{-}\underset{\displaystyle NH_2}{CH}{-}COOH$$

Penicillamine

Antiulcer Agents. One type of antiulcer agent, e.g. propantheline bromide, is based on compounds with antimuscarinic activity; however, these drugs do not only block gastric acid secretion but also cause dryness of the mouth and other side effects.

A more recent development is the use of histamine H_2-blockers to reduce the secretion of gastric acid. Cimetidine, the first drug of this type, was introduced in 1977 and has found widespread use.

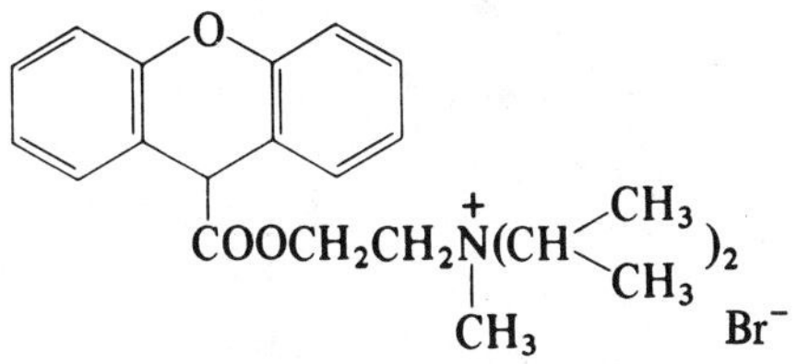

Propantheline Bromide

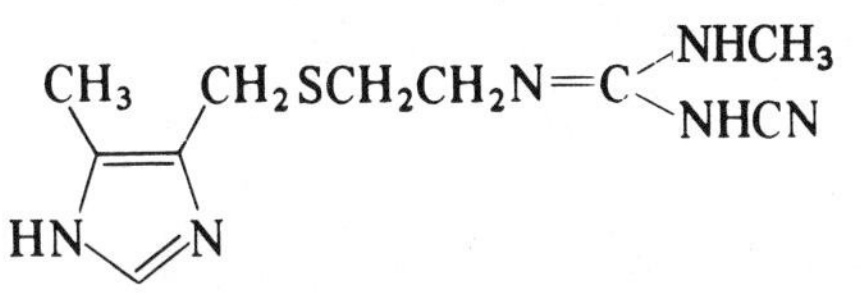

Cimetidine

Cardiovascular Drugs. A group of chemically and pharmacologically related drugs is often referred to as *cardiac glycosides.* An important and typical representative of this type is digitoxin. It is obtained by extraction of digitalis. The main pharmacological action of digitoxin is to increase the force of myocardial contraction. The treatment of congestive heart failure is the most important indication for digitoxin and other cardiac glycosides.

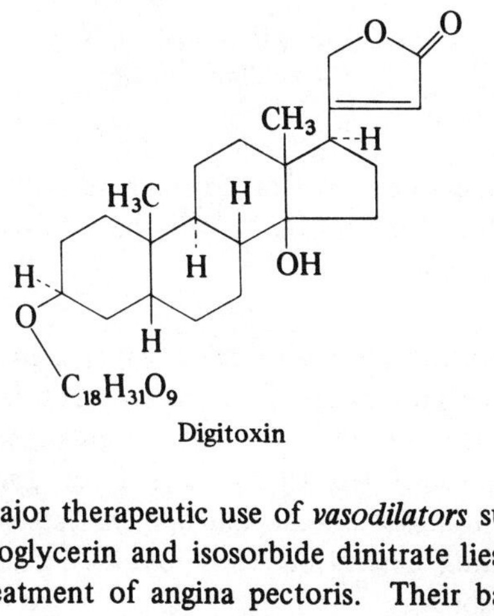

Digitoxin

The major therapeutic use of *vasodilators* such as nitroglycerin and isosorbide dinitrate lies in the treatment of angina pectoris. Their basic pharmacological action is to relax smooth muscle.

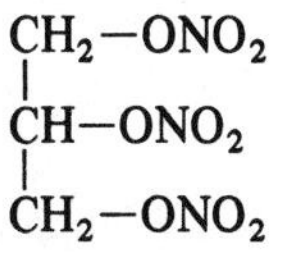

$$\begin{array}{l} CH_2-ONO_2 \\ | \\ CH-ONO_2 \\ | \\ CH_2-ONO_2 \end{array}$$

Nitroglycerin

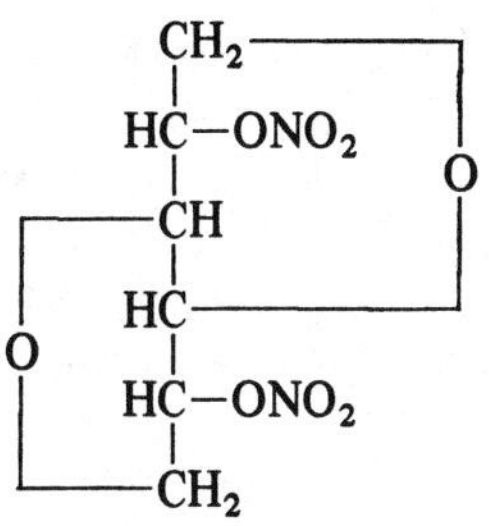

Isosorbide dinitrate

Central Nervous System Stimulants. The use of drugs as CNS stimulants has been restricted because of the abuse potential of such agents. Amphetamine and methylphenidate are examples of this type of drug.

Methylphenidate is an important adjunct in the therapy of attention deficit disorders in children. It has been clearly demonstrated that methylphenidate improves behavior and facilitates learning in these children.

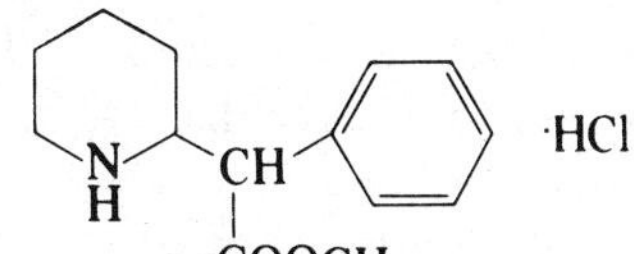

Amphetamine sulfate

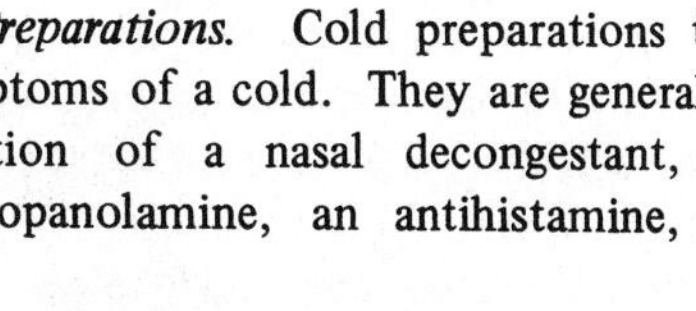

Methylphenidate

Cold Preparations. Cold preparations treat the symptoms of a cold. They are generally a combination of a nasal decongestant, e.g. phenylpropanolamine, an antihistamine, and aspirin.

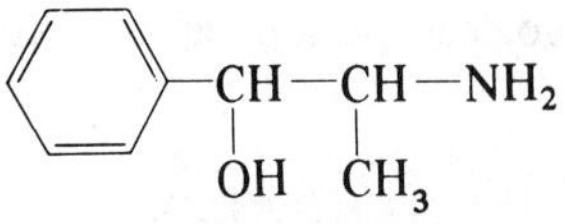

Phenylpropanolamine

Cough Preparations. Cough is a physiological mechanism to clear the respiratory passages of foreign material and excess secretions. It should not be suppressed indiscriminately.

A number of drugs are used to reduce cough, especially at night; among them are codeine and dextromethorphan. The latter, unlike the former, has no analgesic or addictive properties and acts by centrally elevating the threshold for coughing.

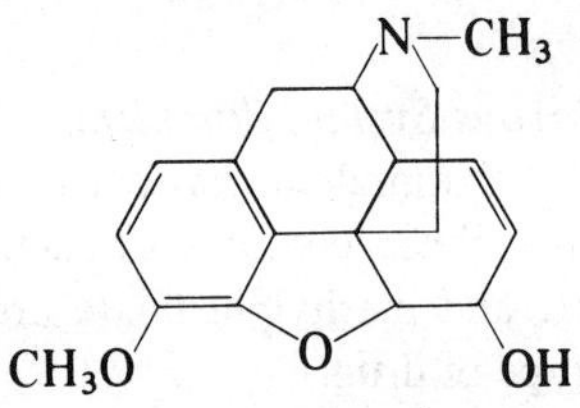

Codeine

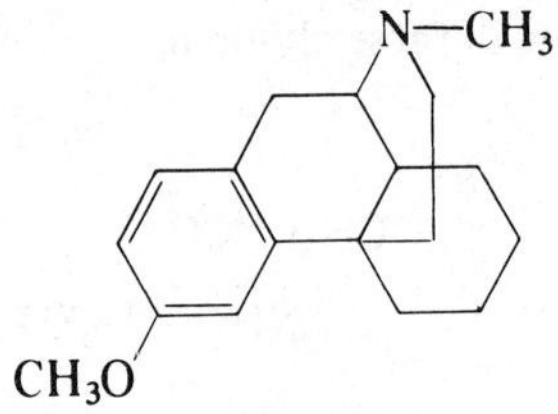

Dextromethorphan

Diuretic Agents (nonmercurial). Diuretics are drugs used to increase the volume of urine excreted by the kidneys. They are most effective in the treatment of edema associated with congestive heart failure. Some diuretics, e.g. hydrochlorothiazide, have antihypertensive activity and are used by themselves or in combination with other antihypertensive agents. There are three important groups of orally active diuretics, i.e. the thiazide diuretics, potassium-sparing diuretics, and high-ceiling diuretics. Examples are shown below.

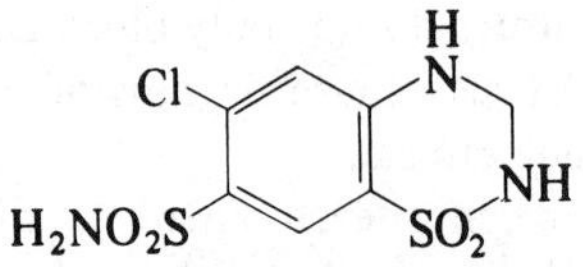

Hydrochlorothiazide
"thiazide"

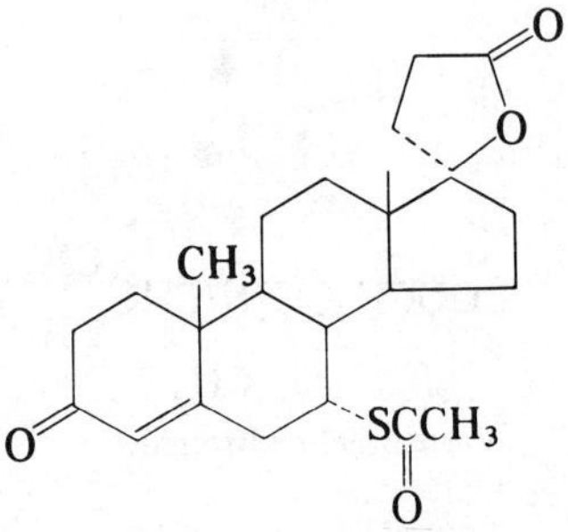

Spironolactone

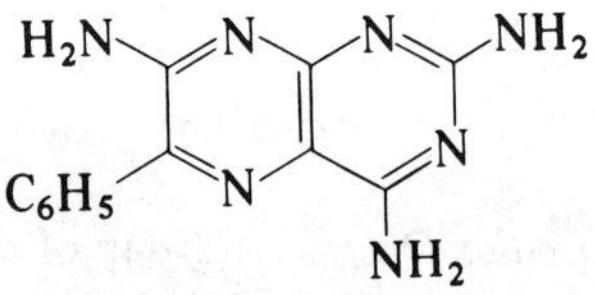

Triamterene

"Potassium-sparing diuretics"

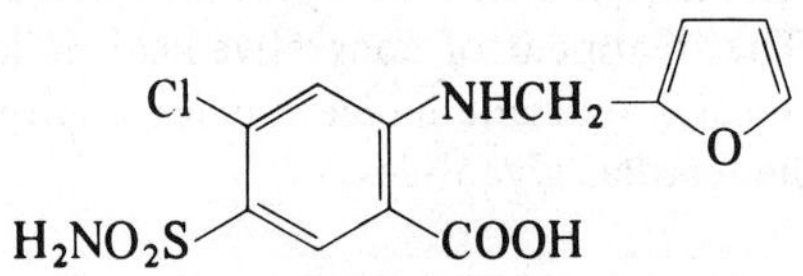

Furosemide
"high-ceiling diuretic"

A combination of triamterene with hydrochlorothiazide, Dyazide®, has found wide application.

Hypolipidemic Agents. The increased interest in the prevention of coronary heart disease and the recognition of the role of hyperlipemia as a risk factor led to the search for drugs that would lower plasma lipid and cholesterol levels. A drug extensively used for this indication is clofibrate.

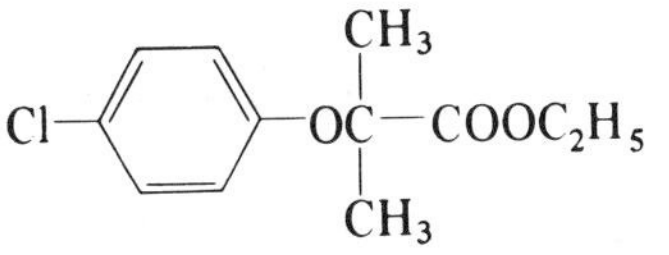

Clofibrate

Hypoglycemic Agents. These drugs are indicated in the therapy of diabetes mellitus. Insulin in its various forms is required for the treatment of juvenile onset diabetes; however, in the older patient with maturity-onset forms of the disease, oral hypoglycemic drugs can be used.

Insulin is a polypeptide with a molecular weight of approximately 6000 and is obtained by extraction of beef or swine pancreas. Several types of insulin preparations are currently in use; they differ principally in solubility, and consequently in speed of onset and duration of action. In contrast to insulin which must be administered by injection, certain sulfonylureas are orally effective hypoglycemic agents; chlorpropamide is an example of this type of drug.

$Cl-C_6H_4-SO_2NHCONHCH_2CH_2CH_3$

Chlorpropamide

Laxatives. Laxatives or cathartics act by various mechanisms and can be classified accordingly. A large number of preparations are available.

Class	*Preparation*
Stimulant	cascara, senna, castor oil phenolphthalein.
Saline	milk of magnesia, sodium sulfate, etc.
Bulk-forming	methylcellulose, plantago seed, bran, etc.
Lubricant	mineral oil.

Sedatives and Hypnotics. Sedatives and hypnotics are used mainly to produce drowsiness. They are general central nervous depressants. Derivatives of barbituric acid are widely used. Certain compounds not related to the barbiturates are also effective hypnotics.

The use of CNS depressants has been restricted because of their abuse potential.

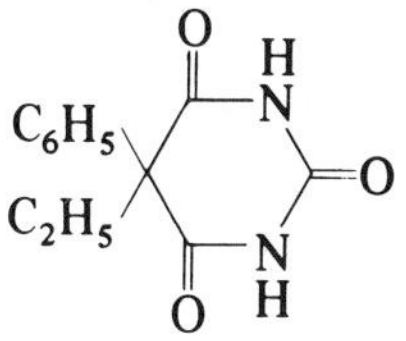

Phenobarbital

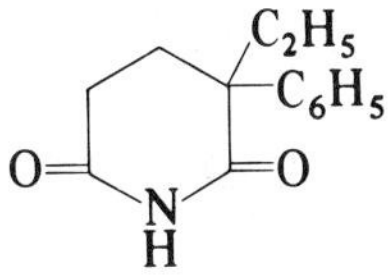

Glutethimide

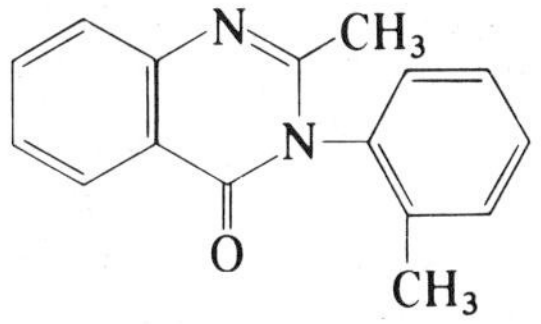

Methaqualone

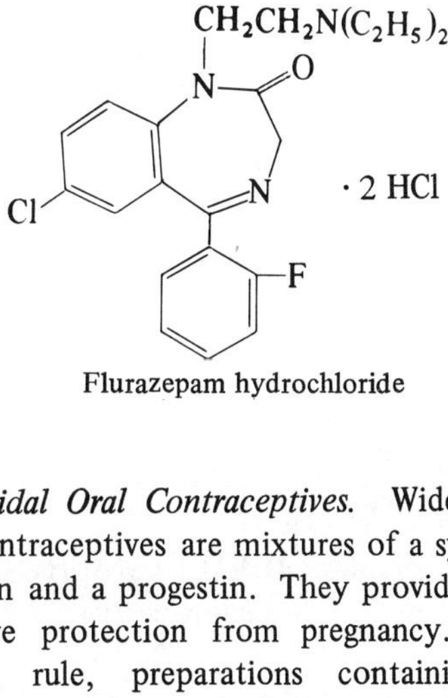

$\cdot$ 2 HCl

Flurazepam hydrochloride

Steroidal Oral Contraceptives. Widely used oral contraceptives are mixtures of a synthetic estrogen and a progestin. They provide highly effective protection from pregnancy. As a general rule, preparations containing the smallest quantity of hormones consistent with efficacy and tolerable side effects are preferred. Enovid-E® is an example of such a combination.

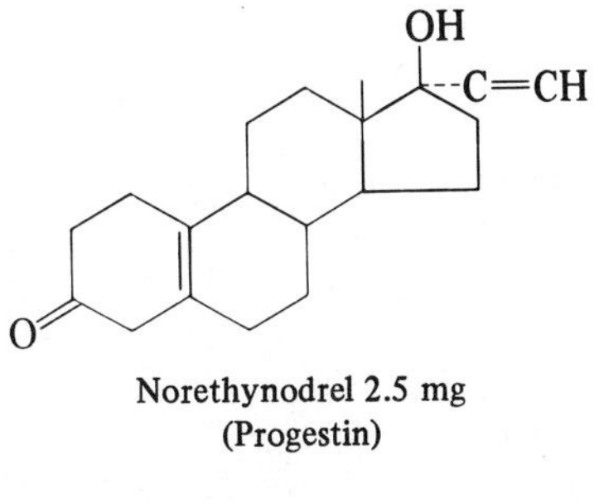

Norethynodrel 2.5 mg
(Progestin)

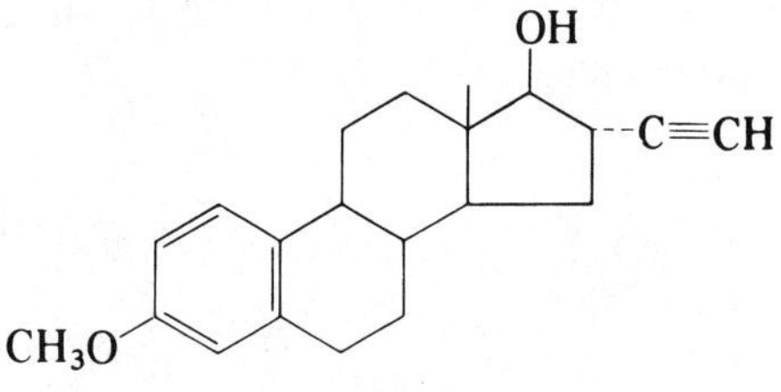

OH

$-C \equiv CH$

CH_3O

Mestranol 0.1 mg
(Estrogen)

Enovid-E®

Major Tranquilizers, Antipsychotics. Antipsychotic drugs are widely used. The prototype of this class of drugs is chlorpromazine, a phenothiazine derivative. The initial reports on the effectiveness of chlorpromazine in mental illness appeared in 1952; since then, more than twenty phenothiazine derivatives, including thioridazine, have been introduced.

The antipsychotic thioxanthene analogs and haloperidol, although structurally different from the phenothiazines, exhibit many of their pharmacological properties.

Chemotherapy of Neoplastic Diseases. In the past decade, encouraging progress has been made in the field of cancer chemotherapy. There have not been any spectacular breakthroughs, but a better understanding of cellular biology and of the mechanism of action of the available drugs has led to a more rational and efficacious therapy. To effect a cure, the entire population of cancer cells must be eradicated, as a few remaining malignant cells are capable of multiplying to an extent where they can kill the host. Drugs from several classes, e.g. alkylating agents, antimetabolites and antibiotics, are available. They are often used in combination with one another. Examples are shown below.

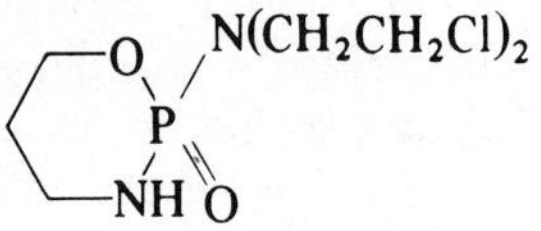

$N(CH_2CH_2Cl)_2$

Cyclophosphamide

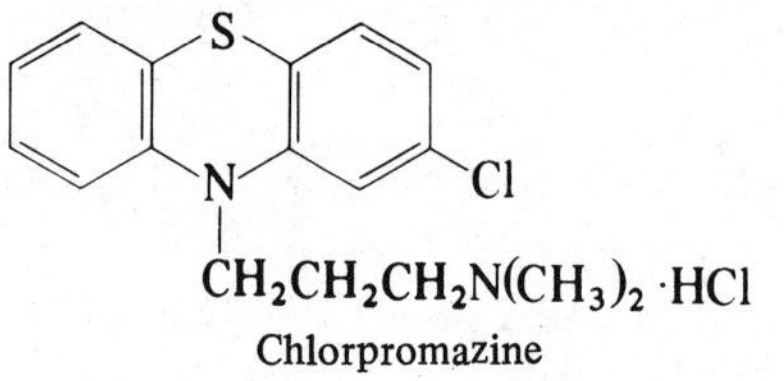

S, N, Cl

$CH_2CH_2CH_2N(CH_3)_2 \cdot HCl$

Chlorpromazine

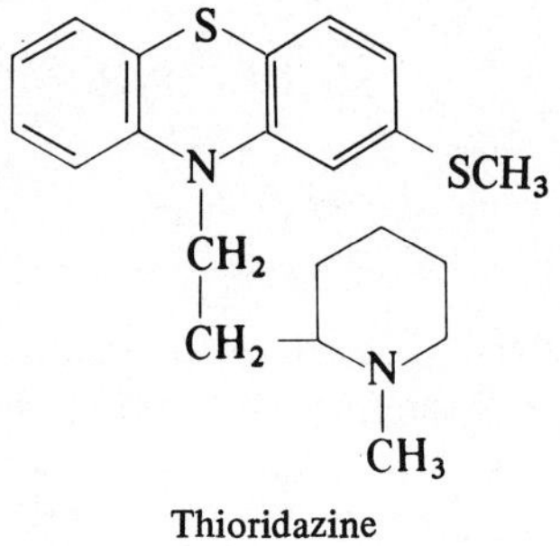

S, N, SCH_3, CH_2, CH_2, N, CH_3

Thioridazine

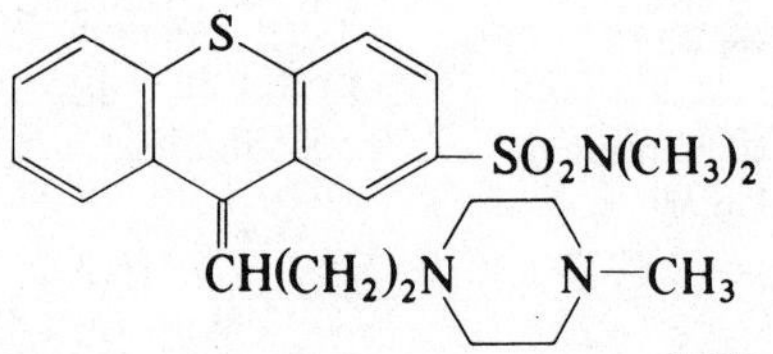

S, $SO_2N(CH_3)_2$

$CH(CH_2)_2N \quad N-CH_3$

Thiothixene 2HCl

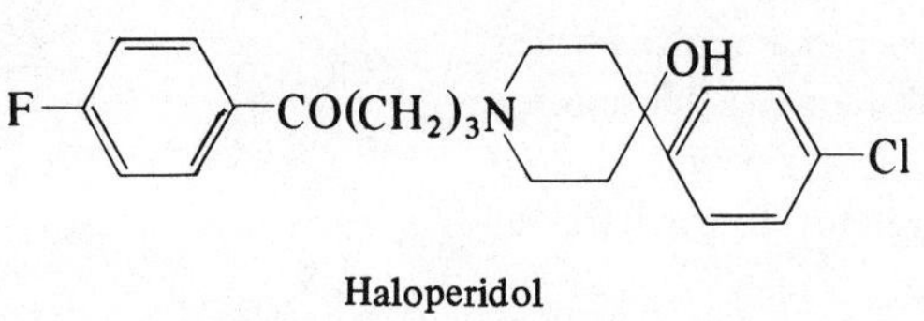

$F-C_6H_4-CO(CH_2)_3N$, OH, Cl

Haloperidol

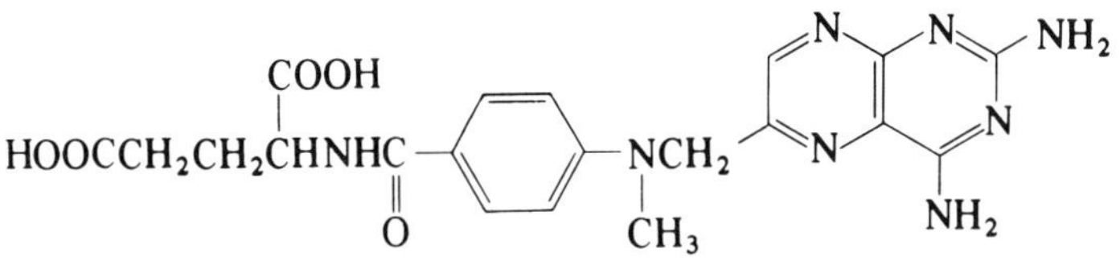

Methotrexate

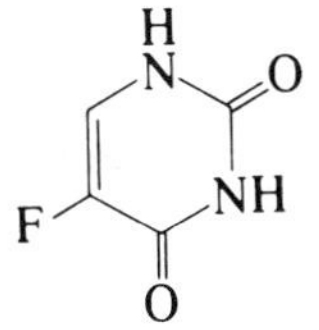

Fluorouracil

Vaccines and other Immunizing Agents. Vaccines are available which will provide an immunity against certain bacterial and viral infections. Examples are given in Table 20.5.

The group of immunizing agents also includes toxoids, e.g. diphtheria toxoid U.S.P., antitoxins, immune serums, e.g., antirabies serum U.S.P., and antivenins. As antigenic agents capable of producing active immunity, some vaccines have proved of great value, especially for prophylaxis. Outstanding examples are smallpox, poliomyelitis, typhoid, and pertussis vaccines.

Vitamins. A vitamin may be broadly defined as a substance essential for natural metabolic functions, but not synthesized in the body. It must, therefore, be furnished from an exogenous source.

A well-balanced diet should supply adequate amounts of vitamins. In certain cases, however, it is desirable to supplement the dietary intake with vitamins given in the pure form. These may be administered either as individual compounds for certain specific indications, or as a multivitamin preparation. A large number of individual vitamin and multivitamin capsules and tablets are available.

Probably no single class of drugs has been the target of as much misunderstanding and misuse as the vitamins, in spite of the fact that their mode of action and therapeutic utility has been studied in great detail. The vitamins are conveniently separated into two groups:

1. Water soluble vitamins, e.g. ascorbic acid (vitamin C); the vitamin B complex, (e.g. thiamine, riboflavin, nicotinic acid); vitamin B_{12} (cyanocobalamin); and folic acid.
2. Fat soluble vitamins, e.g. vitamins A, D, E, and K.

Most of the vitamins are obtained by synthetic procedures. Vitamins A and D are frequently added to milk and margarine; vitamin C is added to foods as an antioxidant to preserve color and flavor.

The vitamins are unlike each other in their chemical structure as well as their function in the body. The structures of vitamins A, D, and C are shown below.

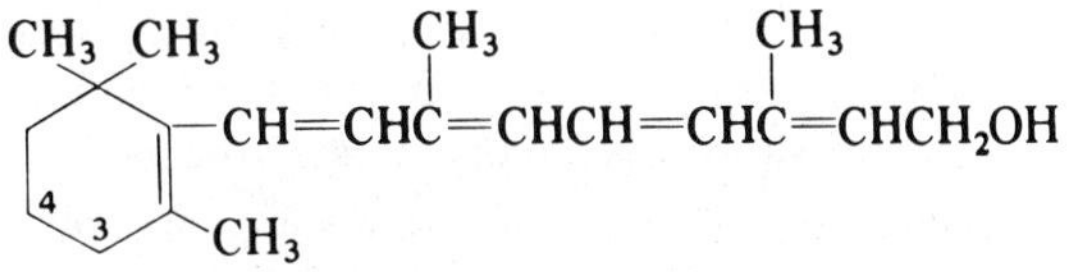

TABLE 20.5 Vaccines

Vaccine	*State of Bacterial or Virus*
Pertussis U.S.P.	Killed bacteria
Typhoid vaccine U.S.P.	Killed bacteria
Smallpox vaccine U.S.P.	Live attenuated virus
Oral poliomyelitis vaccine U.S.P.	Live attenuated virus
Mumps vaccine	Live attenuated virus
Measles vaccine U.S.P.	Live attenuated virus
Influenza vaccine U.S.P.	Killed virus
Rubella vaccine	Live attenuated virus

Vitamin A (Retinol (Vitamin A_1)

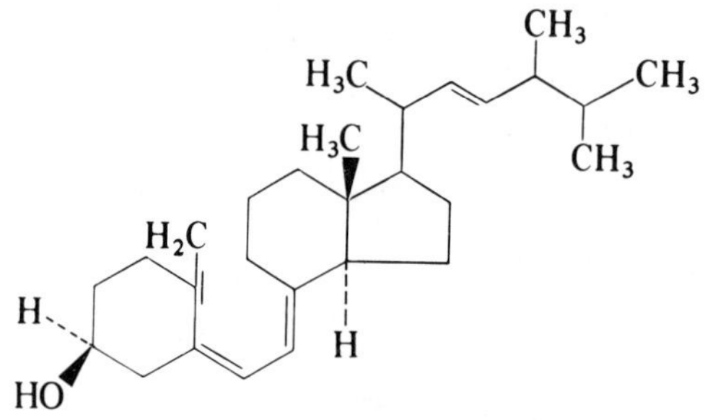

Vitamin D_2 (Ergocalciferol)

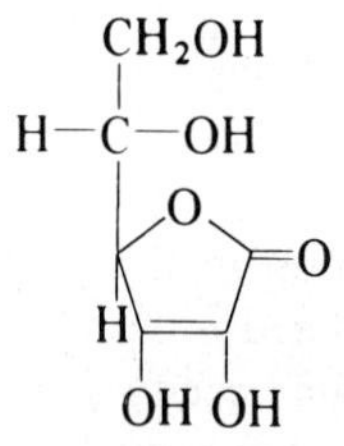

Vitamin C (L-Ascorbic Acid)

PRODUCTION

Chemical Manufacturing

Most of the compounds used as drugs today are prepared by chemical synthesis, generally by a batch process. Antibiotics are an exception and are obtained by a fermentation and extraction process.

In designing a synthesis process, consideration must be given, as in all chemical processes, to the availability of starting materials, their toxicity, as well as the toxicity of the intermediates and of possible by-products. The disposal of mother liquors, residues, and other unwanted by-products must also be carefully considered as the disposal of chemical wastes is covered by federal, state, and local regulations. Reactions are generally carried out in reactors varying in size from fifty to several thousand gallons. Depending on the reaction, either stainless steel or glass-lined steel reactors are used. Fig. 20.2 shows a drawing of a typical 100-gallon jacketed reactor equipped with agitator, various inlets and a bottom outlet. Fig. 20.3 shows a series of reactors installed in a production facility for Geopen®, a semi-synthetic penicillin derivative.

Processes for acetylsalicylic acid (aspirin), tripelennamine (an antihistamine), penicillin, and influenza vaccine are described below. They illustrate the chemical manufacturing procedures used in the pharmaceutical industry.

Acetysalicylic Acid. A commercial process engineered for good yields, high purity, and low cost is shown in Fig. 20.4. Salicylic acid powder, acetic anhydride, and mother liquor are reacted in a 500-gallon glass-lined reactor for 2 to 3 hours. The mass is then pumped through a stainless steel filter to a crystallizing kettle which may also be a 500-gallon glass-lined reactor. The temperature is reduced to 3°C, and after crystallization a portion of the mother liquor is drawn off for the next batch (step 1) in a modified nutsch-type slurry tank. The remaining slurry is centrifuged, and the crystals are dried in a rotary dryer to yield bulk aspirin. Excess mother liquor is stored and then distilled to yield acetic acid which is recycled.

Tripelennamine. A commercial process for tripelennamine (Pyribenzamine®) is outlined in the schematic flowsheet Fig. 20.5.

Step 1. Thionyl chloride and toluene are charged into the reactor R-1; a solution of dimethylaminoethanol in toluene from tank T-1 is added at a controlled rate. Following the addition, the reaction mixture is refluxed to complete the reaction. Sulfur dioxide is evolved and absorbed in the scrubber. After cooling, the crystalline product, 2-chloro-N,N-dimethylethylamine hydrochloride, is separated by centrifuging.

Step 2. 2-Aminopyridine, benzaldehyde and formic acid are charged into reactor R-2. The reaction mixture is refluxed until completion of the reaction. After cooling, the reaction mixture is acidified by addition of hydrochloric acid from tank T-2. Toluene from tank T-3 is added to extract nonbasic materials. The aqueous acidic solution is drawn from the bottom of R-2 into reactor R-2A. The product, 2-benzylaminopyridine, separates on addition of 30% sodium hydroxide solution and is isolated by centrifuging.

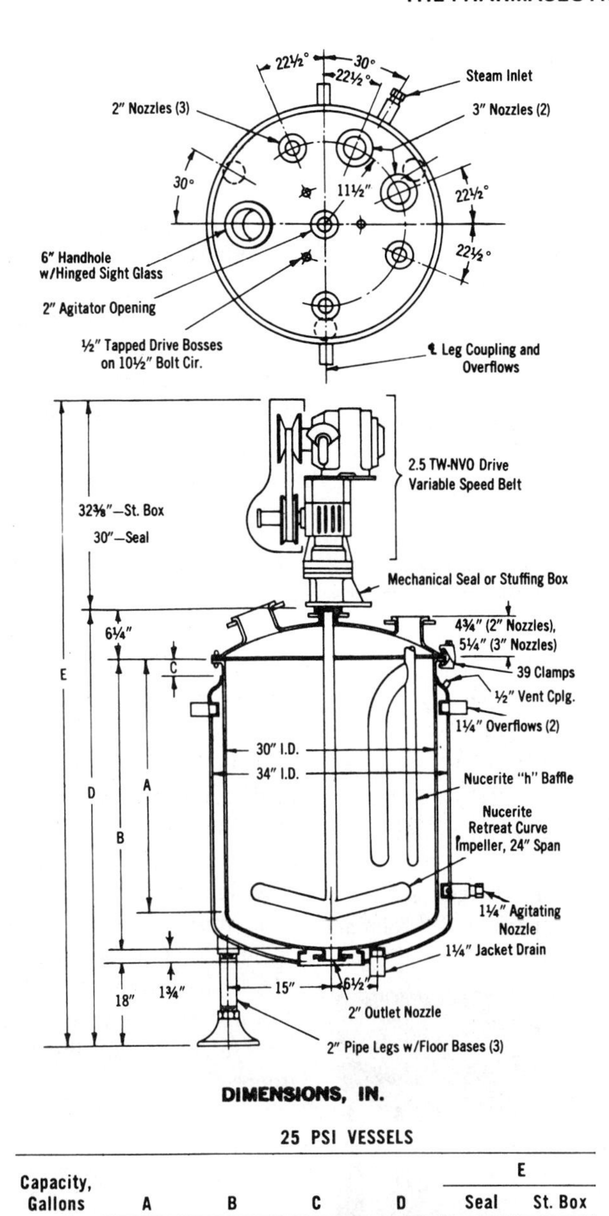

25 PSI VESSELS

Capacity, Gallons	A	B	C	D	E Seal	E St. Box
100	36	41¼	3	67½	97½	99⅞

Fig. 20.2 Plan and elevation of a 100-gallon glass-lined reactor. (*Courtesy Pfaudler Co.*)

Step 3. Water and 2-chloro-N,N-dimethylethylamine hydrochloride are charged into reactor R-3; an equivalent amount of 30% sodium hydroxide solution from T-4 is added. The free base is extracted by addition of toluene from T-5. The lower aqueous layer is drawn off and the toluene solution of the free base is dried with anhydrous sodium carbonate and sodium hydroxide flakes.

Step 1:

$$(CH_3)_2NCH_2CH_2OH \xrightarrow{SOCl_2} (CH_3)_2NCH_2CH_2Cl \cdot HCl \text{ (DMAEC)}$$

Step 2:

$$\text{2-aminopyridine } (\text{pyridine-}NH_2) + C_6H_5CHO \xrightarrow{HCOOH} \text{pyridine-}NHCH_2C_6H_5 \quad \text{(BAP)}$$

Step 3:

$$(CH_3)_2NCH_2CH_2Cl \cdot HCl \xrightarrow{NaOH} (CH_3)_2NCH_2CH_2Cl$$

Step 4:

$$\text{pyridine-}NHCH_2C_6H_5 \xrightarrow{(CH_3)_2NCH_2CH_2Cl} \text{pyridine-}N(CH_2C_6H_5)(CH_2CH_2N(CH_3)_2) \quad \text{(PBZ)}$$

Step 5: Formation of PBZ hydrochloride

Step 4. Reactor R-4 is charged with the required amount of sodium amide and toluene.

Fig. 20.3 Production facilities for Geopen. *(Courtesy Pfizer, Inc.)*

From tank T-6 a toluene solution of 2-benzylamino pyridine is added at a controlled rate. After completion of the reaction, the toluene solution of the 2-chloroethyl-N,N-dimethyl ethylamine from Step 3 is added slowly from T-7. After completion of the reaction, the toluene solution is separated from the sodium chloride formed in the reaction and concentrated *in vacuo*. The remaining oily product is distilled *in vacuo* to yield tripelennamine base. In Step 5 the base is converted to the crystalline hydrochloride, representing the final product.

Penicillin. The production of an antibiotic is typified by the manufacture of penicillin. The mold used industrially is from the *Penicillium chrysogenum* group and is particularly effective against staphylococcus, streptococcus, and pneumococcus, which are gram-positive microorganisms; generally it has little effect on gram-negative microorganisms.

The penicillin mold produces several types of penicillins; however, the only natural penicillins used commercially are penicillin G (benzylpenicillin) and penicillin V (phenoxymethyl penicillin). These compounds are carboxylic acids and since they are unstable as the free acids, the commercial products are the sodium, calcium, aluminum, potassium, or procaine salts. The formulas for penicillin G and V are shown below. Other forms of penicillin (e.g. semisynthetic penicillins) contain different R groups.

[Structure: penicillin nucleus — β-lactam (C=O, N) fused to thiazolidine ring (S, $C(CH_3)_2$, CH–COOH); R—CONH on the β-lactam carbon; H, H, H stereo-hydrogens]

R = $C_6H_5CH_2$-Penicillin G (benzylpenicillin)
R = $C_6H_5OCH_2$-Penicillin V (phenoxymethylpenicillin)

In 1943, the means of production was the surface process in which the mold grows on the surface of a shallow layer of media in trays or bottles. With the development of the commercial method of submerged fermentation in 1944, reduction in space and labor requirements resulted in a tremendous reduction in cost.

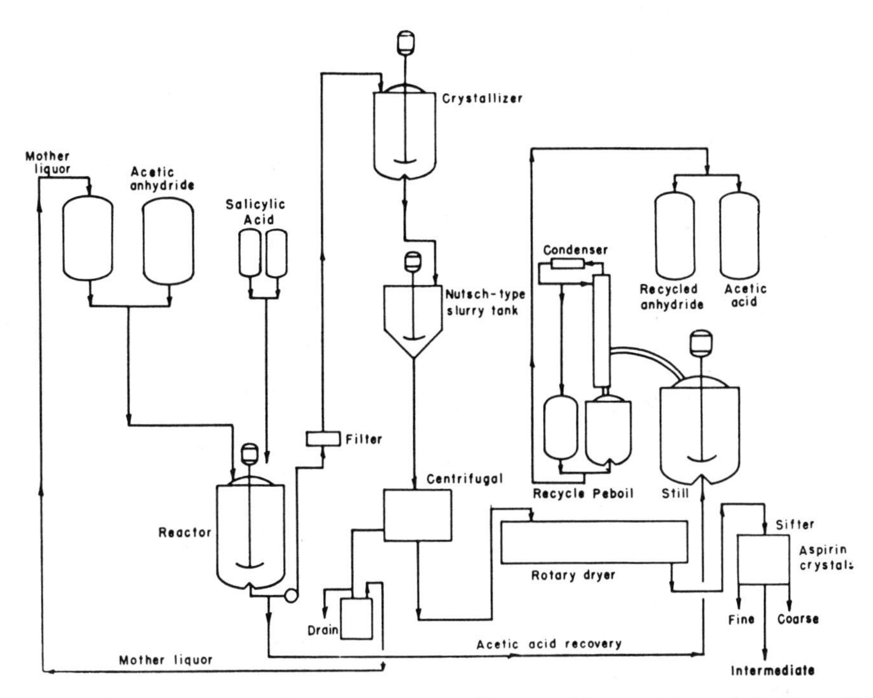

Fig. 20.4 Flow diagram for production of bulk aspirin. (Excerpted by special permission from Chem. Eng., **60**, No. 6, 116-120 (1953); copyright 1953 by McGraw-Hill, Inc.)

In producing penicillin by submerged fermentation, the inoculum or "seed" for the large fermentation tanks with a 5,000 to 30,000-gal capacity is prepared by growing a master stock culture of the mold from lyophylized spores on a nutrient agar substratum with incubation. Several gallons of culture medium, generally constituting 5 to 10 percent of the charge, are prepared in a series of seed tanks to "seed" a large tank.

The four main stages in the manufacture of bulk penicillin are:

1. Fermentation.
2. Removal of mycelium from the fermented broth and extraction of penicillin by solvents.
3. Solvent purification and formation of the penicillin sodium salts.
4. Testing, storage, and shipping.

The fermentation broth is made from corn steep liquor, to which 2 to 4 percent lactose has been added, in addition to such inorganic materials $CaCO_3$, KH_2PO_4, $MgSO_4$, and traces of iron, copper, and zinc salts. The addition of some compounds, although desirable for mold growth, must be omitted since they cannot be tolerated in the end use of the product nor economically removed. After adjusting the pH to 4.5 to 5.0, the culture medium is fed into the fermenter which is equipped with a vertical agitator, a means of introducing air which has been sterilized by filtration, and coils for controlling temperature. Following sterilization of the fermenter, the mold is introduced through sterile pipelines by air pressure. During its growth, the temperature is maintained at 23 to 25°C. Sterile air permits growth of the aerobic mold; agitation distributes it uniformly in the batch. One volume of air per minute is required for each volume of medium. Assays are made every 3 to 6 hours; after 50 to 90 hours, when the potency stops increasing, the mold is harvested. The batch is cooled to 5°C because of the instability of penicillin at room temperature, and the mycelia are removed by filtration on a rotary drum filter.

In the old process, penicillin was recovered from the filtrate by charcoal adsorption. It was eluted with amyl acetate, the eluate being concentrated, cooled at 0°C, and acidified with

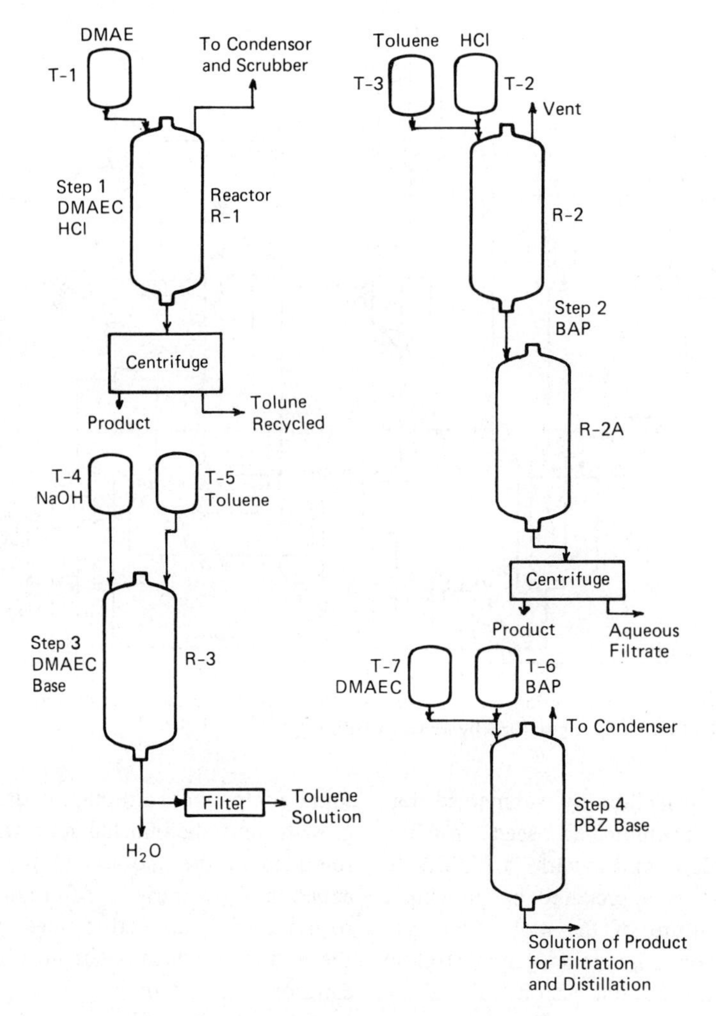

Fig. 20.5 Schematic process for tripelennamine.

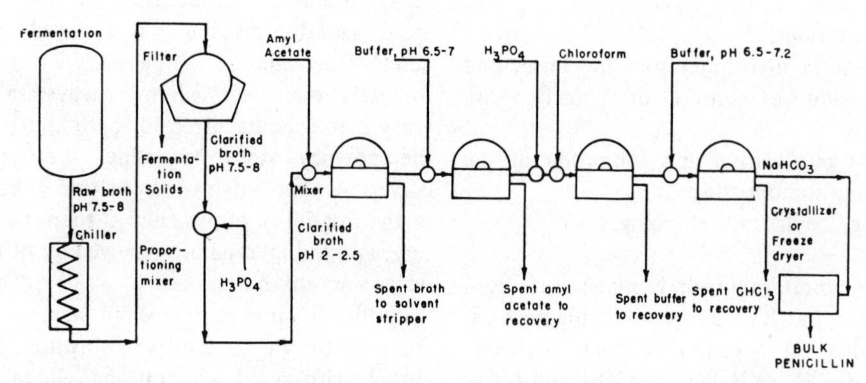

Fig. 20.6 Flow diagram for manufacture of penicillin. (Chem. Eng., **64**, No. 5, 247 (1957); copyright 1957 by McGraw-Hill, Inc.)

an organic acid to pH 2.0. In the solvent-extraction process the activated carbon step is omitted and the filtered liquor ["beer"] is adjusted to a pH of 2.5 with phosphoric acid in the line (Fig. 20.6). Continuous countercurrent extraction is carried out with amyl acetate and then chloroform, with successive concentration in Podbielniak centrifugal extractors; the final liquor is treated with buffered phosphate and sodium bicarbonate to form the sodium salt. This material is made sterile by filtration and is freed aseptically of water and other solvents by crystallization; crystalline penicillin is thereby formed which, when dried, may be packed in bulk in polyethylene bags, or stainless steel containers.

Section 507 of the 1938 Food, Drug and Cosmetic Act requires that all forms of penicillin be tested prior to sale. Specifically, potency, toxicity, pyrogens, sterility, and moisture (since this affects stability) are extremely important. Potency is determined by comparison with a standard to assure consistent clinical results; toxicity refers to a test made with mice that determines the toxic effect of the drug on humans when taken internally; pryogen testing determines the presence of fever-inducing substances which may be undesirable in treatment; sterility, defined earlier, refers to the absence of microorganisms.

A numerical lot number and code system is maintained for each lot from bulk to finished product so that in case of necessity (e.g., lapse of the "expiration" period) the material may be withdrawn from trade channels.

Penicillin is now produced in the United States, Canada, Mexico, Australia, Japan, Italy, England, and many other countries. Initial United States production was 250 billion units a year; in 1946, production was 25,000 billion units; in 1957, 525,738 billion units, and in 1979 reached 3.7 million BIU[13] (B.I.U. = billion international units).

Influenza Vaccine. Influenza vaccine is a sterile, aqueous suspension of suitably inactivated influenza virus. It usually contains two or more different virus strains. The strains used are those designated by the Division of Biologics Standards of the National Institutes of Health.

Fig. 20.7 Seeding of eggs with influenza virus. (*Courtesy of Pfizer, Inc.*)

The actual production cycle starts with fertile eggs which are first candled, then disinfected by a coating of an iodine solution, and finally punctured with a small drill (Fig. 20.7). The virus is introduced through this hole which is then sealed with collodion gelatin and incubated at 99°F for 48 hours. A circular top section of the egg is removed and the allantoic egg fluid is extracted in the harvesting operation. The live virus is separated from the egg fluid by centrifuging at 50,000 rpm (Fig. 20.8); the heavier virus particles separate and cake out on the centrifuge cartridge. The virus is collected and reconstituted in a saline solution. The live virus is inactivated with formalin under prescribed conditions. The suspension is tested for inactivation of the virus, sterility and potency (antigenicity) then filled aseptically into vials.

Pharmaceutical Production

Drug substances are seldom administered as pure chemical compounds but are almost always given in some kind of formulation. The high degree of uniformity, the physiological availability and the therapeutic quality expected of modern drug products are the result of considerable effort and research on the part of the formulation pharmacist. Careful selection and

Fig. 20.8 Separation of viruses by centrifugation. (*Courtesy of Pfizer, Inc.*)

control of the quality of the various ingredients employed, application of well-defined manufacturing processes, and adequate consideration of the many variables which may influence the stability and utility of the product are required. Prime considerations are the stability of the active ingredient in the formulation and its bioavailability. Pharmaceutical products should also appeal to the patient; they must look pleasing and, if not flavored, should be free of objectionable odor and taste.

Drugs for specific applications may be formulated into various types of dosage forms. These can be classified as follows:

- Solutions (aqueous and non-aqueous), emulsions and suspensions.
- Parenteral solutions
- Ophthalmic solutions
- Ointments, creams, and suppositories
- Tablets, capsules, and pills
- Aerosols
- Advanced drug delivery systems.

A detailed discussion of the preparation of these dosage forms and other minor classes is presented in Remington's *Pharmaceutical Sciences.*[10] A brief overview of the production of tablets, capsules, and parenteral solutions is given below.

Tablets and Capsules. Tablets and capsules are the most frequently used dosage forms for oral administration. They are greatly favored because they provide exact drug doses, protect the drug during storage, and are convenient and easy to take. However, before the drug contained in a tablet can be absorbed, the tablet must, after ingestion, disintegrate, and the drug must go into solution. Whatever hinders any step in the dissolution process can potentially slow drug absorption and thus decrease the bioavailability of the drug contained in the tablet. This requires that tablets be carefully and properly designed and produced.

Capsules present similar problems. The gelatin shell of the capsule must first dissolve in the stomach fluids to release the powder contents, which in turn must be dispersed and dissolved before absorption can occur.

Many factors influence dissolution rates such as particle size, crystalline form, and surface area of the drug; diluents, disintegrants and wetting agents in the formulation; and speed and force of compression in the processing of tablets. The dissolution rate of tablets and capsules, i.e. the time required to dissolve 90% of the drug, is determined in special equipment at 37 ± 0.5°C under standardized conditions. The results are validated by bioavailability studies in man. The relationship between time and blood levels attained following administration of the solid dosage form and an aqueous solution of the drug are compared. Ideally, identical blood levels should be reached and the areas under the curves (AUC) should be comparable.

In addition to the active ingredient, a tablet will usually contain a binder to provide cohesiveness, a filler to provide an acceptable size tablet, a disintegrant which causes the tablet to break up when exposed to an aqueous medium, and a lubricant to prevent sticking in the tableting machine.

There are three general methods of tablet preparation: wet granulation, dry granulation, and direct compression. The purpose of the granulation procedure is to prepare a homogeneous mixture of the ingredients in such a form that will flow freely into the processing machinery. A modern high-speed tableting machine capable of producing up to 10,000 tablets per minute is shown in Fig. 20.9.

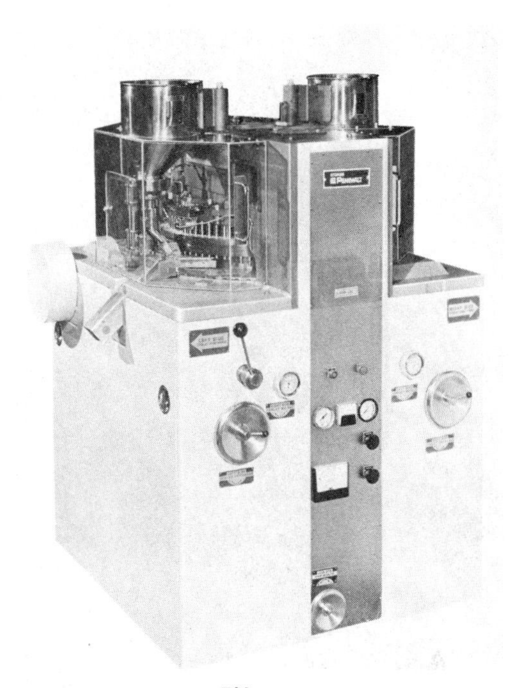

Fig. 20.9 Stokes GTP™ tabletting press. *(Courtesy of Stokes, Pennwalt Corp.)*

In addition to uncoated compound tablets, other types listed below are produced for specific purposes.

- Sugar-coated tablets.
- Film-coated tablets
- Enteric-coated tablets
- Multiple-compressed tablets, layered tablets, and press-coated tablets
- Prolonged-action tablets

Capsules are solid dosage forms in which the drug substance is enclosed in either a hard or soft soluble shell of gelatin. The hard gelatin capsule consists of two sections, one slipping over the other. They are filled with the drug, usually mixed with a diluent and lubricant to provide complete filling of the capsule and to ensure proper flow through the encapsulating machine. The filled capsules are sealed either by spotwelding or applying a band of gelatin around the joint. Figure 20.10 illustrates the filling of hard gelatin capsules.

Soft elastic gelatin capsules are prepared in a rotary die process from two continuous gelatin ribbons; the capsule is formed, filled, and sealed in delicately timed operations. The capsules can be filled with any liquid, semiliquid, paste, or even powder that will not dissolve or interact with the gelatin.

Parenteral Preparations. Parenteral preparations are administered by injection and can provide immediate physiological action. They are of special usefulness in those cases where the patient is unconscious. The therapeutic response of a drug is also more readily controlled by parenteral administration since it avoids the irregularities of intestinal absorption. However, since the protective barriers of skin and mucuous membranes are bypassed, it is imperative that parenteral solutions be as nearly perfect as possible with respect to purity and freedom from contamination by bacteria or viruses.

Single- and multiple-dose vials and ampules

Fig. 20.10 Filling of capsules. *(Courtesy of Pfizer, Inc.)*

are used in medical practice. The vehicle (solvent), which is of greatest importance for parenteral products, is specially distilled, pyrogen-free water; the container of choice is glass.

Prior to large-scale production, painstaking studies must be carried out to determine the stability of the drug in its final dosage form. That is, its stability to heat, light, and humidity over an extended period of time. Fig. 20.11 shows a step in the production of multiple dose vials under sterile conditions.

Advanced Drug Delivery Systems

The industrial pharmaceutical scientist today is not just a formulator of dosage forms but also a designer of new drug delivery systems. One example is an I.U.D. which continuously releases very low levels of progesterone; another is a transdermal therapeutic system which could be used for drugs which may be poorly absorbed from the gastrointestinal tract or rapidly inactivated in the body. Implantable drug delivery systems such as a drug pellet, a reservoir, or a pump have also been considered.[14]

Noval dosage forms and the concepts underlying their application will permit consideration of drug candidates that could not have been developed with yesterday's technology and knowledge of drug absorption, distribution, metabolism, and excretion.[12]

Quality Assurance—Quality Control

In the industrial manufacture of drugs, the variety and complexity of operations make it necessary to assign to a separate and independent group of scientists within the company the responsibility for controlling the quality of the final product. This group is referred to as Quality Control.

To ensure the required high standards of safety, purity, and effectiveness of drug products, it is necessary to carefully control each step in the manufacturing procedure. This means control of all raw materials including packaging components and labels, control of the product during manufacturing by means of in-process analysis, and control during packaging of the final product.

The Quality Assurance and Control group is also responsible for ensuring that all products are manufactured, processed, and manufactured in accordance with the current Good Manufacturing Practices (GMP) issued by the FDA and also with company policies.

As an example, tablets and capsules are examined for the following properties:

1. Identity
2. Content of active drug
3. Size
4. Physical appearance
5. Disintegration and/or dissolution time (time required for a tablet or capsule to disintegrate or dissolve in water or other specified medium at 37°C)
6. Friability (mechanical stability of tablet)
7. Weight variation
8. Content uniformity

To determine content uniformity, a representa-

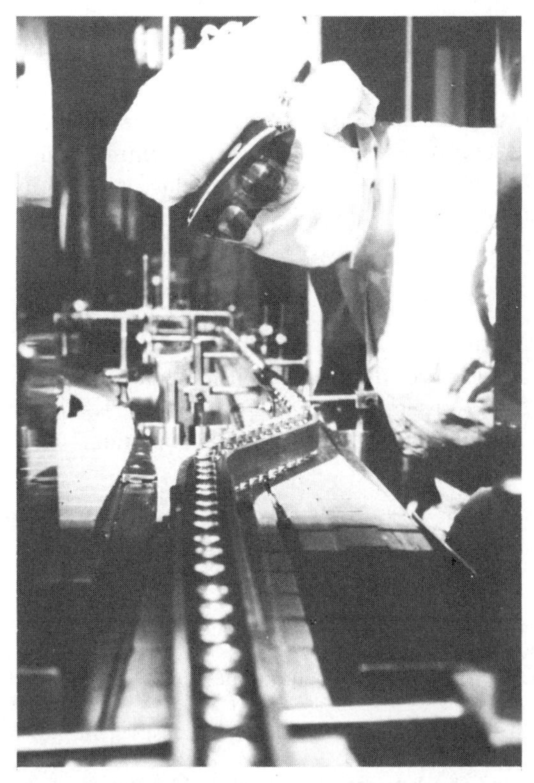

Fig. 20.11 Production of injectable dosage forms under sterile conditions. (*Courtesy of Pfizer, Inc.*)

tive sample of tablets is taken from each lot and each tablet is assayed individually for drug content. Automated equipment is generally used for this test (Fig. 20.12).

All injectable solutions are tested for sterility and the absence of pyrogens.

Spectroscopic, chromatographic, and other analytical methods are used to determine the purity and quantity of drug present in the various dosage forms.

Specifications for containers, closures, and other component parts of drug packages must

Fig. 20.12 Instrument for automatic analysis of individual tablets. (*Courtesy CIBA-GEIGY*)

be drawn up to ensure that they are suitable for their intended use. Also, packaging and labeling operations must be adequately controlled to assure that correct labeling is employed for the drug. Finished products must be identified with lot or control numbers that permit determination of the history of the manufacture and control of the batch.

CONCLUSION

The involvement of the pharmaceutical industry, from basic research to consumer education, constitutes a major contribution to health progress. Largely because of advances in drug therapy, four of the ten leading causes of death in 1900 no longer prevail. In addition, new drugs have been partially or completely successful in fighting many once feared diseases, accounting in good measure for a dramatic decline in death rates in the past 25 years. Lower death rates and less time lost from work have contributed significantly to our economic well-being. Given the past record, it is reasonable to expect that the pharmaceutical industry will continue to develop and produce new useful products for the betterment of the human condition.

REFERENCES

1. *Annual Time Series*, Predicasts, Inc., 200 University Circle, Research Center, 11001 Cedar Avenue, Cleveland, Ohio 44106.
2. *Annual Survey Report, Ethical Pharmaceutical Industry Operations 1977–1978.* Pharmaceutical Manufacturers Association, Washington, D.C.
3. *USAN and USP Dictionary of Drug Names.* United States Pharmacopeial Convention, Inc., Rockville, MD 20852.
4. *Physicians' Desk Reference*, 34th Edition, 1980. Litton Industries, Inc., Medical Economics Company, Oradell, NJ 07649.
5. Karl H. Beyer, Jr., *Discovery, Development and Delivery of New Drugs*, S. P. Medical and Scientific Books, New York, London 1978, p. 178.
6. P. deHaen, *New Product Survey*, Paul deHaen Information Systems, Micromedex, Inc., Englewood, CO 80110.
7. *Chemistry in Medicine.* A Study by the Committee on Chemistry and Public Affairs, American Chemical Society, Washington, DC 1977.
8. William H. M. Duncan and Michael E. Parsons, *Gastroenterology* 78, 620 (1980).
9. Louis S. Goodman and Alfred Gilman, Eds., *The Pharmacological Basis of Therapeutics*, 5th Edition, MacMillan Publishing Co., New York, NY 1975.
10. *Remington's Pharmaceutical Sciences*, 15th Edition, A. Osol and J. E. Hoover, Eds., Mack Publishing Co., Easton, PA 18042, 1975.
11. *AMA Drug Evaluations*, 4th Edition, American Medical Association, Chicago, IL, 1980.
12. Louis C. Schroeter in *How Modern Drugs are Developed*, F. H. Clarke, Ed., Futura Publishing Co., Inc., Mount Kisco, NY 10549, 1977, p. 85.
13. *Chemical Industry Notes*, Predicasts, Inc., 200 University Circle, Research Center, 11001 Cedar Ave., Cleveland, Ohio 44106.
14. Perry J. Blackshear, *Scientific American*, **241**, 66 (Dec. 1979).

21

The Pesticide Industry

Gustave K. Kohn*

INTRODUCTION

Scope of the Chapter

This chapter deals with the chemicals used in agriculture mainly to protect, preserve, and improve crop yields. Arbitrarily excluded are those substances that serve as fundamental nutrients, which are treated in Chapters 5 and 7 on nitrogen and phosphorus technology, respectively. Nevertheless, it is the current practice of the farmer, particularly in advanced American agriculture, to integrate nutritional and plant-protection application schedules, and even provide single formulations that include both fertilizers and pesticides. Further, plant nutrition at this stage of scientific sophistication is far more complex than the older and conventional N-P-K applications. Many chemicals that accelerate plant growth act as hormonal agents modifying plant metabolic processes at some stage of development. Since these substances are manufactured and marketed by the pesticide industry, they are included as subject matter here.

Also included in this chapter are chemicals which are significant in public health. Many organisms are vectors in the dissemination of human and animal disease. Since products of the pesticide industry control the insect or the rodent or the mollusk, etc. (the vectors), they are often the most effective and sometimes the only practical means for controlling some of the most serious health problems of mankind, especially, but not exclusively, in the underdeveloped countries. Thus, chemicals that control such vectors are also included in the subject matter of this chapter. In this area, and in the area of animal health products, Chapter 20 on the pharmaceutical industry overlaps to some extent.

Finally, the products of the pesticide industry are employed in urban, suburban, and rural areas in home and garden for a variety of useful and aesthetic purposes. Moreover, industry employs herbicides, algicides, fungicides, and bactericides, and the railroads utilize herbicides to keep rights-of-way free of vegetation.

The Language of the Pesticide Industry

There is a need for a generic term to cover the diversity of functional applications in this

*Zoecon Corporation, formerly with Chevron Chemical Co.

industry. The term "insecticide" has been commonly applied to all control agents. The Federal Insecticide Act of 1910 provided an extension in meaning to include both insecticides and fungicides. The Federal Insecticide, Fungicide, and Rodenticide Act (1947) uses "economic poison" for all these groups, and embraces herbicides as well. The term "pesticide" is now used officially to include all toxic chemicals, whether used against insects, fungi, weeds, or rodents, etc. Agricultural disinfectants and animal health products are in many instances also included under the term "pesticides."

"Plant protection agents" is another generic term. A common practice within the industry is to add the suffix "cide" to the group or biological unit under consideration. The term "insecticide" has both generic and specific meaning that depends upon context. Obvious in meaning, also, are such frequently used words as fungicide, bactericide, miticide or acaricide, ovicide, herbicide, rodenticide, muscicide, molluscicide, nematicide, and algicide. This practice of utilizing the "cide" suffix reflects also upon the state of agricultural technology as it was practiced during the first half of the twentieth century. Whereas the objective in most cases, until recently, was to kill or attempt to eradicate the offending pest, many of the newer programs reflect more complex approaches.

Hence the terms attractants, repellents, antifeeding compounds, hormones (plant and insect growth control agents), defoliants, dessicants, etc., all describe functional chemical agents that are within the purview of the pesticide or plant-protection industry. There are many who believe that these chemicals will become more significant in the agriculture of the future.

History

It is probable that man's treatment of crops with foreign substances dates back into prehistory. The Bible abounds with references to insect depredations, plant diseases, and some basic agricultural principles such as periodic withholding of land in the fallow state. Homer speaks of "pest-averting sulfur."

More recently, in the nineteenth century, a great increase in the application of foreign chemicals to agriculture ensued. Discovered or, more precisely, rediscovered was the usefulness of sulfur, lime sulfur (calcium polysulfides) and Bordeaux mixture (basic copper sulfates). With the exception of formaldehyde, inorganic chemicals provided the farmer with his major weapons.

The earliest of the organic compounds were generally chemicals derived from natural products or crude mixtures of chemicals in states of very elementary refinement. Ground-up extracts of plant tissue were useful in the control of insects. Such extracts were employed in agriculture in many cases before the chemist had elucidated the structure or synthesized the molecule responsible for the biological activity. These extracts included the pyrethroids, rotenoids, and nicotinoids, which continue to be derived in large part from plant extracts. Crude petroleum fractions were recognized for their effectiveness in the control of mites, scale, and various fungi, as well as for their phytopathological properties.

Although a few synthetic organics were already known, the great revolution in the use of organic chemicals in agriculture roughly coincides with the period of the onset of World War II. The more important of these discoveries were DDT (Mueller – 1939), 2,4-D (Jones patent–1945), benzenehexachloride (ICI and French development – circa 1940), and the organic phosphate esters (Schräder – begun in the late 30's, revealed in the forties). These new chemicals were so enormously more potent than their predecessors in their biological activity (frequently by orders of magnitude) that they very rapidly displaced almost all of the chemicals previously employed. The chemicals of today, largely discovered in the 1950's and 1960's, are predominantly extensions of this almost revolutionary transition from inorganics to synthetic organics that dates from the period of World War II.

These trends are vividly illustrated in Fig. 21.1. As late as 1945, inorganic chemicals accounted for almost 75 percent of all pesticide sales, while oil sprays and natural products accounted for most of the remainder. The rapid reduction in the sales of inorganics was

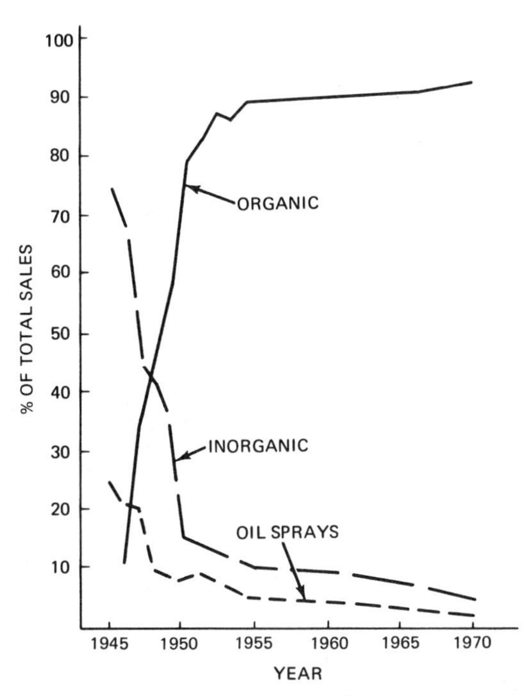

Fig. 21.1 Impact of synthetic organic chemicals on pesticide sales. (This chart is based on estimates for U.S. industry derived from data collected from one major chemical company (Chevron Chemical Co.) which is a manufacturer and marketer of organic and inorganic pesticides and oil sprays.) In the 1970's oil ceased to be a low cost commodity and this fact, combined with toxicological considerations, further reduced its percentage impact on pesticide sales. In this figure metalloorganics (maneb, zineb, etc.) are included in the organic category.

virtually complete by the early 1950's, and oil sprays and natural products reached a quantitatively low but constant level at approximately the same period.

It is fair to say that within the United States the pesticide industry is largely dominated by synthetic organic chemicals. The inorganics, oil sprays, and natural products continue to share a small portion of the total pesticide market.

Federal government statistics on the supply of pesticides in the United States emphasize the dominance of the synthetic organics. Many of the botanicals are derived from imports of plant parts or their extracts to the U.S. In contrast to the supply and sales of chemicals from the standpoint of their derivation, it is interesting to examine trends according to the biological function of the pesticide.

TABLE 21.1 U.S. Production of Synthetic Pesticides (lbs × 10⁶)*

Year	*Insecticides*	*Fungicides*	*Herbicides*
1945	43.0	–	–
1950	177.3	28.5	1.1
1955	331.1	76.8	70.6
1960	351.0	98.2	102.5
1965	421.6	123.9	262.9
1970	358.8	140.2	403.8
1975	445.0	155.3	788.0
1980 (Est)	390–540	125–170	900–1055

Source: SRI International Chemical Economics Handbook.

*Note: World production figures increasing more rapidly, but in different proportions.

The data in Table 21.1 show the enormous growth of the pesticide industy (see also Fig. 21.2). Although fungicides used in the United States are presently the smallest of the three main classifications of pesticides, it must be remembered that in tropical and high moisture areas the problem of plant disease is much more serious. High labor costs are largely responsible for the extremely rapid growth rate of herbicide production. Herbicides, which diminish the need for cultivation (hoe and more modern mechanical weeding) in advanced agricultural countries, will probably find more use in the future. On the other hand, ecological pressures and the halting of herbicide use for military purposes did lower the production figures of certain insecticides (DDT and chlorinated hydrocarbons) and certain herbicides (2,4,5-T and 2,4-D) for 1970 and succeeding years.

Although the U.S. continues to provide the largest national market for pesticides, the intensity of such usage based on either agricultural acreage or tons of product per acre is higher in certain countries of the world. (Japan is a notable example.)

If one takes U.S. consumption of pesticides as unity, then the total world consumption can roughly be estimated for the present as *two*. There is little doubt that this index (that is, world wide pesticide usage in relation to U.S. consumption) will markedly increase as will world population and its consequential requirement for greater quantities of food. Pesticide

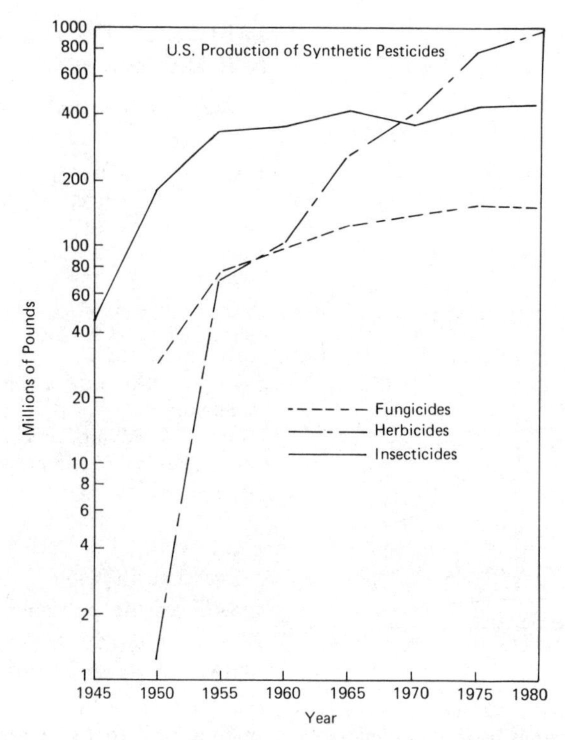

Fig. 21.2 U.S. production of synthetic pesticides.

production and the pesticide industry are inextricably related to the Malthusian paradox.

CHARACTERISTICS OF THE PESTICIDE INDUSTRY

Among the distinguishing characteristics of the pesticide industry are (1) the multiplicity of chemical agents employed, (2) a limited price range (and hence limited chemical complexity governed primarily by the economics of agricultural production), (3) a fairly rapid obsolescence for the chemicals employed, and (4) a high degree of government regulation for the production, application, shipment, and use of pesticide chemicals.

Government Regulation

The first state laws on insecticides were enacted in 1900 to establish standards of purity for the arsenical, Paris green (still employed in diminished quantity today).

$$Cu(CH_3-\overset{\overset{\displaystyle O}{\|}}{C}-O-)_2 \cdot 3Cu(AsO_2)$$

Gradually these laws were extended to cover a wide list of inorganic compounds and plant extracts, many of which are extremely toxic to man. Included in this group are such compounds as arsenic combined with copper, lead, and calcium; phosphorus pastes for ants and roaches; strychnine in rodent baits; thallium in ant and rodent baits; and selenium for plant-feeding mites. Mercury, both as corrosive sublimate and calomel, was used as an insect repellent and later as a seed disinfectant. Sodium fluoride was the common ant poison, and sodium cyanide, calcium cyanide, and HCN itself were the universal fumigants. Nicotine sulfate was used generally in the garden and on the farm. These compounds, among the most toxic of any known at that time, were widely marketed without supervision under any of the early state laws. There was no provision for public health, either in regulating the amounts applied, or the possible danger of minute amounts (residues) remaining on the marketed produce. The need to protect the applicator and the public against the dangerous qualities of the insecticides, or their residues on crops, provided the motives for all the legislation that followed.

In response to reports from England of the poisoning of people eating apples grown in the United States, the Federal Food and Drug Administration in 1927 banned any fruit involved in interstate commerce if it possessed more than 3.57 ppm equivalent of arsenic trioxide.

Government concern was first related to standardization of the manufactured chemical and protection of the farmer in relation to the product that he purchased.. This was then extended to the handling of the chemical in interstate commerce, and then to protection of the consumer of raw agricultural products (apples, corn, and lettuce, for example), and in other legislation to protection of the consumer of finished foods (canned juice, margarine, cereal food, meat, and milk, for example). Included in this legislation were provisions that protect the shipper of the chemical, the applicator of the chemical, and all personnel proximal to the application of the chemical. Legislation now regulates chemicals applied to crops or foods as protective agents – pesticides, emulsifiers, solvents, packaging materials (wax, container materials, plasticizers, antioxidants, etc.). Table 21.2 summarizes government regulation of the pesticide industry.

Toward the end of the 1960's, a new dimension of concern arose. The effect of the manufacture and application of pesticide chemicals to the environment was recognized. This culminated in 1970 with the establishment of the Environmental Protection Agency (EPA) which was given authority to regulate virtually all aspects of pesticide manufacture and use in the United States.

Since its inception, the principal objectives of the EPA's pesticide activities have been (1) to establish procedures which ensure that new pesticides will not pose unreasonable risks to human health and the environment, and (2) to terminate the use of those previously registered pesticides which exceed certain risk criteria.

The approach to the first of these objectives, that regarding new pesticide products, has been to give careful scrutiny to a comprehensive and expanding set of tests. Among the requirements called for are studies on mammalian toxicology (including lifetime feeding studies), environmental chemistry (persistence, mobility, etc.) and effects on fish and wildlife. New product registrations are granted only after EPA scientists are satisfied that use of the product does not pose unreasonable risk to humans or hazard to the environment.

As to the second objective, much the same criteria are used in judging whether or not to allow continued use of previously registered pesticides. With the 1972 Amendments to the FIFRA, EPA received authority to forbid or restrict use of pesticides which present unreasonable hazards. The mechanism used to accomplish this end is the RPAR (Rebuttable Presumption Against Registration) procedure, established by EPA in 1975. The RPAR is a stepwise process which begins with EPA's statement of the hazards supporting its contention that the use of the pesticide should be discontinued. These arguments may then be rebutted by any interested parties. Final determination is made by EPA after consideration of information on both risk and benefit. Between 1976 and 1979 certain registrations for a number of chlorinated pesticides (Dieldrin, Aldrin, heptachlor, chlordane) were cancelled through the RPAR process. In other cases, restrictions were placed on the use of pesticides to limit exposure to hazards disclosed through the RPAR process.

Residues of pesticides on food commodities are also regulated by EPA, although the statutory basis of this control lies in the Federal Food, Drug, and Cosmetic Act (most of which is administered by the Food and Drug Administration). Since the use of pesticides on food crops often leads to residues consumed by man, a complex regulatory mechanism has evolved to determine residue levels which may be presumed safe for human consumption. These levels, called "tolerances," are defined after consideration of chronic feeding studies in laboratory animals and analytical data to determine what levels may follow from approved agricultural practices.

Many agricultural chemicals and their formulations manufactured in the United States are used throughout the world. Although the

TABLE 21.2 Development of Government Regulation of the Pesticide Industry in the United States

Date	*Legislation*	*Content*
1900	State laws on product purity (Paris green)	Quality control standard for protection of user of chemical.
1910	Federal Insecticide Act	First specific pesticide regulation; standards for insecticides and fungicides moving in interstate commerce.
1938	Federal Food Drug and Cosmetic Act	Protection of public against contamination of foods, drugs, etc.
1947	Federal Insecticide Fungicide and Rodenticide Act (FIFRA)	Regulates interstate shipment of pesticides. Safety standards for the handling of chemicals in interstate commerce. Required federal registration of labeling.
1954	Miller Pesticide Amendment to FIFRA (Public Law 518)	Requires setting of residue tolerances for pesticides on raw agric. product; protects consumer.
1958	Food Additives Amendment (contains Delaney Amendment)	Protects public against contamination of processed agric. products -foods- from pesticide and other contamination; protects consumer; prohibits direct food additives that cause cancer *at any dosage level* of test protocol.
1947–66	State enactments embodying FIFRA and food additives legislation	By 1966, 49 States and Puerto Rico had supplemented protection within state boundaries in varying degree for above federal legislation.
1970–71	Environmental Protection Agency (EPA) Established by Executive fiat	EPA assumes regulatory responsibility for pesticides (previously held by USDA).
1970–80	Occupational Safety and Health Act, OSHA. Enacted into law Dec., 1970. Grew as an effective agency throughout decade.	Regulates to minimize hazards associated with manufacture of all chemicals, including all pesticides.
1972	Federal Environmental Pest Control Act of 1972 - Amendments to FIFRA	FIFRA strengthened to comprehensive regulatory statute. Grants EPA authority to conduct extensive pre-market review of pesticide safety and efficacy data. Emphasis on human health and environmental effects on pesticides. Provides statutory authority for cancellation of registration for pesticides which pose risk of "unreasonable adverse effect."
1975	EPA establishes RPAR process by Regulation	*R*ebuttable *P*resumption *A*gainst *R*egistration is established as key mechanism for study of presumed risk of old pesticides.
1975-80	EPA proposed "Guidelines for Registering Pesticides"	Detailed testing requirements are imposed on new pesticides for health and environmental effects.
1976	TSCA Congressional Act–(Toxic Substances Control Act)	Regulation of toxic substances includes intermediates, solvents, all chemical compositions. Pre-manufacturing notification required for all new substances.
1978	Federal Pesticide Act of 1978	Provides for "Conditional" registration and "Generic" registration (reregistration) of pesticides.

application of science and technology is high in the United States, and comprehensive regulatory legislation was developed first and most extensively in this country, all the technically developed nations regulate to some degree the manufacture, sale, and use of agricultural chemicals. Germany, Japan, Great Britain, France, and many smaller technically advanced nations have legislation and regulating agencies in this area. This is also true to a lesser degree in most of the smaller nations of the world and for many of the developing nations.

The World Health Organization (WHO) and the Food and Agricultural Organization (FAO) play an important role in the international regulation of pesticides. A joint undertaking of these two agencies, the Codex Alimentarius Commission, reviews data on pesticide residues and toxicity and sets tolerance limits on residues which may be considered safe for purposes of international commerce in agricultural commodities.

In the Unites States, enormous and complex questions of values and priorities–despite numerous studies and reports–remain unresolved. Many chemicals have been banned, or their use has been seriously restricted. The rate of discovery and commercialization of new pesticide products has slowed dramatically. Resolution of these matters relating to ecology, toxicology, and practical need is critical for the pesticide chemical industry, for agriculture, for the protection of the environment, and in the interests of public health.

Obsolescence of Pesticidal Compounds

Obsolescence of pesticidal chemicals derives from two main causes. The first of these is a consequence of the universal property of living things to adapt by selection to environmental changes, including foreign chemicals. From a practical standpoint, this is best exemplified by the tolerance of certain strains of the housefly to DDT. Certain resistant strains of the housefly require 2,000 times the dose of DDT used on susceptible strains to achieve an LD_{50} (the lethal dose required for a 50 percent kill of a given population). In agriculture, resistance applies as a practical consideration particularly, but not exclusively, to insecticides.

The second cause of obsolescence is the development of more useful chemicals. The term "useful" here covers a host of unrelated properties. The new chemical may be cheaper to produce, easier to apply, less hazardous (to the applicator, consumer, or the environment), possess a greater scope or more intense and extensive biological activity; or, in contrast, it may be highly specific in its biological properties. The chemicals enumerated in the charts that follow were all largely the result of technology that contributed some advantageous property. In certain cases, subsequent research or misuse may have demonstrated that these advantages were overshadowed by the discovery of undesirable consequences.

The case of DDT is an excellent example of the obsolesence of an important pesticidal chemical (Table 21.3). Many chemicals of the chlorinated hydrocarbon group were found which were more highly active–either broadly

Table 21.3 Obsolescence in the Pesticide Industry, DDT and its History

$(Cl{-}C_6H_4)_2CHCCl_3$

	Year
First synthesized, Zeidler, Germany.	1874
Laboratory curiosity until discovery of insecticidal properties, Paul Mueller, Switzerland.	1939
Nobel Prize (Medicine) Mueller.	1948
Mass introduction into agriculture and public health areas.	1949 Onward
Proliferation of modern analytical techniques. Recognition of insect resistance.	1950-1965
Recognition of residue, physiological, and ecological interrelations.	1960-1970
State and federal legislation prohibiting or severely restricting use.	1969-1970
Virtually eliminated from use in the U.S. and advanced countries, DDT continues to be used in certain "third world" and poorer nations.	1970-1980

or narrowly effective—than DDT. Indeed this compound stimulated the research and development of toxaphene, the cyclodiene insecticides DDD, dicofol, etc., and even some of the fungicides and herbicides.

In the sixties, after the development of sensitive analytical techniques (mostly chromatography or spectroscopic methods) and concern for the proliferation of environmental hazards, DDT was virtually eliminated from agriculture. It is significant that just two decades after the discoverer of the insecticidal properties of DDT received the plaudits of the scientific world through the Nobel Prize, the chemical has been almost eliminated from use in the United States and its use much reduced in the world at large.

THE CHEMICALS OF THE PESTICIDE INDUSTRY

Complex chemicals such as streptomycin (which is employed against certain plant bacterial infections), rotenone (a nonpersistent insecticide), and gibberellic acid (a plant growth regulator) all have limited practical applicability in agriculture. By far the majority of pesticides are of simpler structure. This is a consequence in part, of the realities of agricultural economics. Structures like those in Table 21.4 containing sequences of optically active sugars or peptides, *cis-trans*-isomeric multiple bonds, numerous asymmetric centers, or multiple annelated rings, present formidable obstacles to the synthetic organic chemist and well-nigh insuperable obstacles to engineers who would try to mass produce them at low to moderate cost.

Because pesticides frequently owe their activity to their "fit" on some biological surface e.g., an enzyme or cell membrane, configuration is frequently decisive for biological activity. The difference in activity resulting from isomeric variations is outlined in the following discussion.

Structural isomerism, such as *ortho, meta*, or

TABLE 21.4 Typical Complex Structures Employed in Agriculture

*Structure**	*Name, Process, and Function*
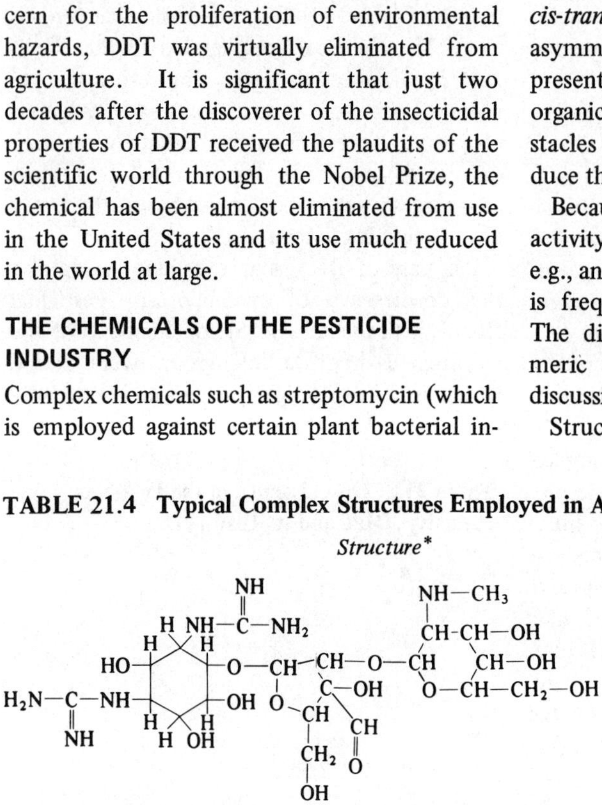	Streptomycin Produced by fermentation Plant bactericide
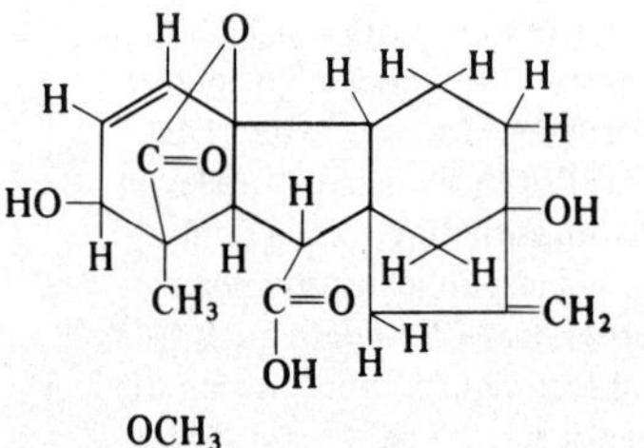	Gibberellic acid Produced by fermentation Plant-growth regulator
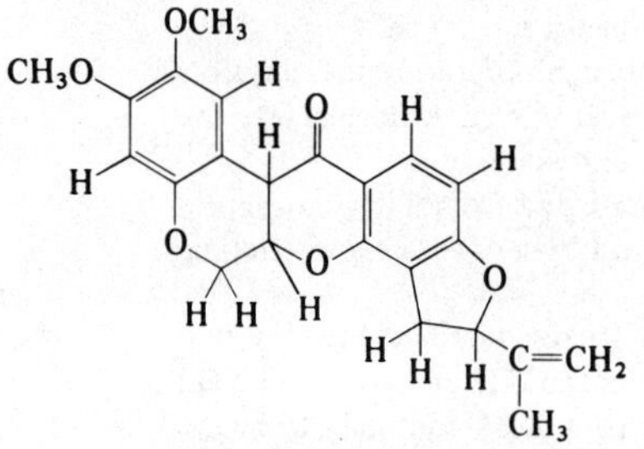	Rotenone Extracted from derris root Insecticide

*Note multiplicity of assymetric carbon atoms in all three structures.

TABLE 21.5 Structural Isomerism of DDT

Structure	*LC_{50} Drosophila Melanogaster (mg/kg wt.)*	*% in Technical DDT*
Cl–, Cl–, $CHCl_3$ *p,p*-isomer	1	65–85
Cl–, $CHCl_3$, Cl *o,p*-isomer	145	10–20

para orientation on a benzene ring can be reasonably controlled by the synthetic process. At times, the insecticide is defined by the process employed in its manufacture. Technical DDT is a pesticide which contains both isomers and impurities totaling as many as 14 distinct compounds. The technical product meets a standard defined by a minimum setting point at 89°C. The principal impurity is the *ortho-para* structure which has very little insecticidal potency (see Table 21.5).

The isomers of DDT illustrate the effects of ring substitution on biological activity. Large differences in activity also exist for *cis-trans* isomers. These are given for two organophosphate insecticides in Table 21.6. Optical isomers of pesticidal compounds also exhibit differences in biological activity. This difference is exhibited in the activity of pyrethroids, and in the isomers of benzenehexachloride of which lindane, the γ isomer, possesses almost all the useful insecticidal activity.

TABLE 21.6 *Cis-Trans* Isomerism and Biological Activity[a]

Structure[b]	*Name and Isomer*	*LD_{50} Housefly (mg/kg)*
$(CH_3O)_2P(O)–O–C(CH_3)=C(H)–C(=O)–O–CH_3$	Mevinphos, *cis.*	0.27
$(CH_3O)_2P(O)–O–C(CH_3)=C(H)–C(=O)–O–CH_3$	Mevinphos, *trans.*	23.0
$(C_2H_5O)_2P(O)–O–C(H)=C(H)(Cl)$	Diethyl β-chlorovinyl phosphate, *cis.*	1
$(C_2H_5O)_2P(O)–O–C(H)=C(Cl)(H)$	Diethyl β-chlorovinyl phosphate, *trans.*	80

[a]Based on Lichtenthaler, F. W., *Chem. Rev.*, **61**, 630 (1961).

[b]In both organophosphate insecticides, the *cis* isomer is about 80 times as active as the *trans* isomer as measured by topical application to the housefly.

In the past decade, there has been increasing interest in the ecologically "soft" synthetic pyrethroids. These molecules possess chiral carbon atoms and the configuration at these centers largely determines the degree of biological activity. Illustrative of this structural specificity is the fact that for the control of important lepidopterous cotton pests, only approximately 0.01 to 0.02 lbs/acre of decamethrin are required. This compares to about 0.1 to 0.2 lbs of other *unresolved* synthetic pyrethroids and about 1 to 2 lbs of conventional organophosphate or chlorinated hydrocarbon insecticides. Since these pyrethroids have low environmental persistence and are used at such low dosages, the environmental burden is much reduced.

It can be concluded from these examples, as well as from the structures that follow, that relatively minor changes in chemical substitution or configuration can cause enormous differences in activity. These structural characteristics do provide some information about the requirements of the receptor site of the biological system affected.

Despite some very striking correlations found for the physical-chemical parameters of organophosphate and carbamate insecticides, the guiding principle for the development of new pesticides from the fifties onward has been largely empirical.

The molecular weight of most pesticides provides a simple generalization of the size and complexity of the molecules that are commercially employed as pesticides. The common names of some extensively used pesticides are followed by their molecular weights in parentheses: DDT (354.5), 2,4-D (249), captan (300.6), parathion (291.3), atrazine (215.7), maneb (265.3), and carbaryl (201.2). These are small to medium-sized organic molecules, frequently containing halogen, nitrogen, sulfur, and phosphorus, as well as carbon, hydrogen, and oxygen.

Comprehensive descriptions of contemporary pesticides are provided in the bibliography. What follows here are groups of significant pesticides generally typical of a given class or identifying a given principle and commonly used in world agriculture.

Pesticide Nomenclature

Because no uniform practice exists for the naming of pesticides, and because commercial pesticides generally possess more than one designation, some clarification of the nomenclature adopted here is in order. In the charts that follow and in the text, the so-called generic or common name has been used most frequently.

Where a pesticide is a specific chemical composition, there is a systematic chemical name generally consistent with the rules of nomenclature in *Chemical Abstracts.* Such a name is often quite cumbersome. The systematic name for the structure of streptomycin given in the text occupies four or five lines of print and, from the standpoint of the chemist, is less informative than the two-dimensional structural formula that is provided.

Pesticides developed by private companies usually are initially given an arbitrary designation (generally containing a number) which is employed during the period of exploration and development. This name may prevail during the period that research teams, farm adivsers, and experiment stations work with the material. During the course of this period, a proprietary name is generally selected by the company responsible for the development of a potentially successful pesticide. This proprietary name falls within the purview of the U.S. trademark laws, and the similar foreign trademark names which are selected are not always the same as the U.S. name. The product is usually sold under a variety of trademark designations which denote the company of its source. A name which is the generic or common name may then be selected for the commercialized chemical. Most scientific texts involved with the discussion of structure and biological activity, metabolism, toxicology, etc., of pesticides employ this generic or common name.

Table 21.7 provides some alternate names for pesticides whose structures are given in the text. For a given pesticide, one of the alternate names may be a more common designation than the generic name. It is obvious that even for a relatively simple structure, such as endo-

TABLE 21.7 Names of Some Pesticides

Generic Name	*Systematic Name*	*Proprietary Name*[a]	*Other Names*
dimethoate	O, O-Dimethyl S-(N-methylcarbamoylmethyl) phosphorodithioate	Cygon®	Rogor® (Europe) Foslion®
carbaryl	1-Naphthyl N-methylcarbamate	Sevin®	
naled	1,2-Dibromo-2,2-dichloroethyl dimethyl phosphate	DIBROM®	RE-4355
captan	N-Trichloromethylmercapto-4-cycloexene-*cis*-1,2-dicarboximide	Orthocide®	
endosulfan	6,7,8,9,10,10-Hexachloro-1,5,5a,6,9,9a-hexahydro-6,9-methano-2,4,3-benzodioxathiepin-3-oxide	Thiodan®	Malix®

[a]Proprietary names are generally capitalized for the initial letter and followed by the symbol ®, as in Sevin®; common names are written in small letters.

sulfan, the systematic chemical name is quite inconvenient.

Chlorinated Organic Group

The halogenated organics were among the earliest synthetic organics employed as pesticides. DDT exhibited a broad spectrum of insecticidal activity and its success stimulated the research that led to many of the other synthetic organic chlorine compounds. DDT has a relatively moderate acute mammalian toxicity–about 250 mg/kg for the technical product. It is oil-soluble and very insoluble in water–about one part per billion (1 ppb).

In the fifties and sixties there was on the one hand increasing evidence of resistance to DDT by various insect species, and on the other hand a proliferation of chemical-analytical techniques enabling quantitative pesticide analyses in complex substrates down to, and in certain cases below, the ppb level. These conclusively

TABLE 21.8 Group I–Miscellaneous Chlorinated Hydrocarbons

Name	*Structure*	*Significant Intermediates*	*Uses–Properties*
DDT	$(Cl\text{–}C_6H_4)_2CH\text{–}CCl_3$	Chlorbenzene, chloral.	General insecticide (malaria control).
Toxaphene	Complex mixture	Chlorinated camphene (68–69% chlorine).	Insecticide (cotton).
BHC Lindane (99.5% pure γ-isomer)	$C_6H_6Cl_6$ (cyclohexane ring, each carbon bearing H and Cl)	Benzene, chlorine ($h\nu$).	Insecticide used heavily in Japan (rice).
Hexachlorobenzene	C_6Cl_6	Benzene, chlorine.	Fungicide (cereal-seed treater).
dicofol	$(Cl\text{–}C_6H_4)_2C(OH)\text{–}CCl_3$	DDT, chlorine (hydrolysis).	Acaricide; miticide.

showed that DDT and its chief metabolite DDE (dichlorodiphenyldichloroethylene) persisted in the environment and concentrated in fatty tissue, particularly in organisms at the top of any given ecosystem.

There is considerable, but variable, cross resistance from DDT to other chlorinated hydrocarbons such as benzene hexachloride, and the cyclodiene insecticides. However, when one substitutes a *para*-methoxy group for *para*-chlorine in the DDT molecule, the far less persistent—and somewhat less active—but useful insecticide methoxychlor is obtained. The substitution of OH for H in the trichloroethane group greatly reduces and almost eliminates the general insecticidal properties of the DDT configuration. Nevertheless, such molecules posses high acaricidal and ovicidal properties, e.g., dicofol.

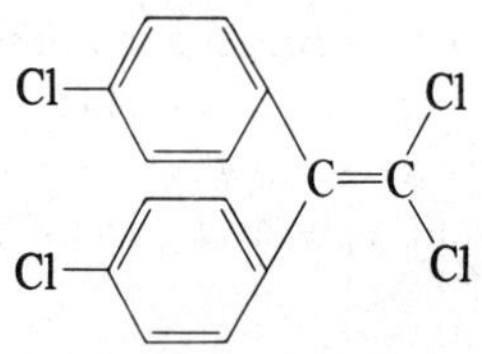

DDT metabolite, DDE

The simplicity of the chlorination of benzene led to the discovery of the usefulness of benzene hexachloride and lindane as insecticides—and of hexachlorobenzene, the true hexachlorobenzene, as a grain fungicide. (Benzene hexachloride as a chemical is a misnomer. Benzene hexachloride, originally prepared by Michael Faraday in 1825, is hexachlorocyclohexane.) Lindane has the spatial configuration that possesses the highest activity: six hexachlorocyclohexanes are found in commercial preparations, although the stereochemistry of the cyclohexane ring does provide many more theoretical isomers. Lindane has 3 axial and 3 equatorial chlorines.

The Cyclodiene Group

Hexachlorocyclopentadiene (C_5Cl_6)

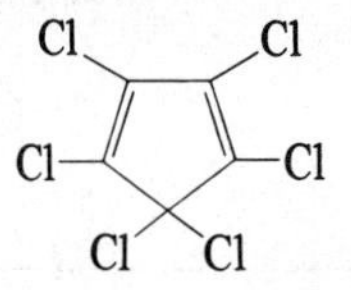

itself possesses herbicidal activity. As a diene combined with a variety of dienophiles and modified thereafter, it has given rise to a large number of very useful insecticides. The earliest of these was chlordane which was employed both in agriculture and for the control of home and garden pests. Chlordane is somewhat herbicidal as well and finds some use for crabgrass control.

Aldrin, dieldrin, heptachlor, and endrin are all very highly active toxicants with moderate to highly acute mammalian toxicity, great potency, and a wide spectrum of insecticidal activity. They and their metabolites tend to concentrate in fatty tissue and have considerable persistence in the environment. Recently, federal regulations (RPAR process; see preceeding section on government regulation) as well as state legislation have either eliminated or severely limited their agricultural application. The names, uses, and structural formulae of several of these chemicals are given in Table 21.9.

Despite the fact that this group of chemicals has had wide agricultural and other usage, little has been unequivocally established concerning the essential biochemistry involved in their activity. Whereas one of the mechanisms of resistance to DDT can be at least partially related to the enyzmatic reaction

$$\text{DDT} \xrightarrow[\text{dehydrochlorinase enzyme}]{-\text{HCl}} \text{DDE}$$

no such simple mechanism can be associated with cyclodiene resistance mechanisms. Frequently, in the field, some insect varieties exhibit a generalized chlorinated hydrocarbon resistance.

Derivatives of Organic Acids

Many important pesticides may be considered chemically as derivatives of organic acids. (Table 21.10) For many years 2,4-D (2,4-dichlorophenoxyacetic acid) and 2,4,5-T (an additional chlorine in the 5-position on the phenoxy moiety—2,4,5-trichlorophenoxyacetic acid) dominated the herbicide market. They possess potent auxin-like behavior and have found their greatest usefulness in the control of dicotyledonous (broad-leafed) weeds. 2,4,5-T

TABLE 21.9 Cyclodiene Group Toxicants

Name	*Structure*	*Significant Intermediates*	*Uses*
Chlordane	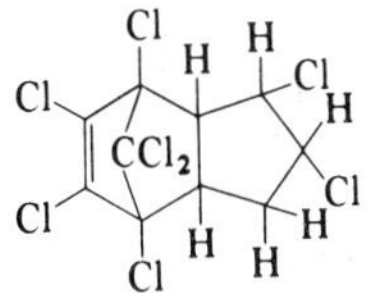	Hexachlorocyclopentadiene, cyclopentadiene, chlorine (isomeric mixture).	Insecticide; crabgrass control.
Aldrin	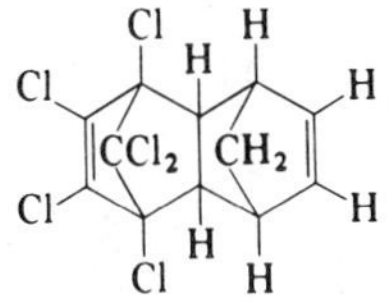	Hexachlorocyclopentadiene, acetylene, cyclopentadiene.	Broad spectrum insecticide (soil insects).
Dieldrin	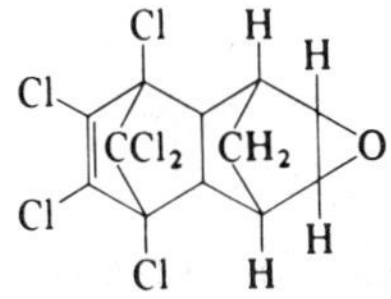	Hexachlorocyclopentadiene, acetylene, cyclopentadiene, peracetic acid.	Broad spectrum insecticide.
Mirex	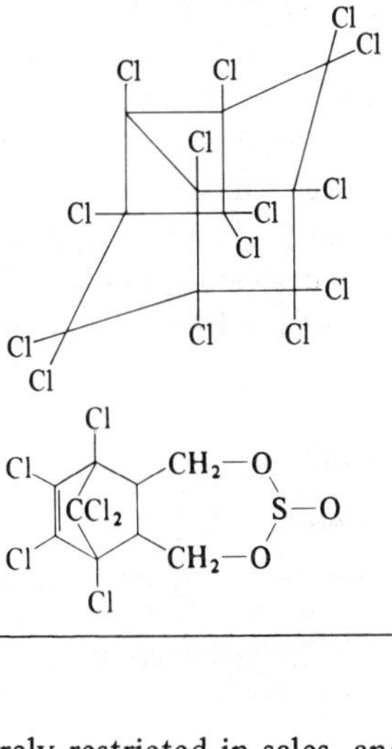	Hexachlorocyclopentadiene, chlorine.	Insecticide (fire ant control).
endosulfan	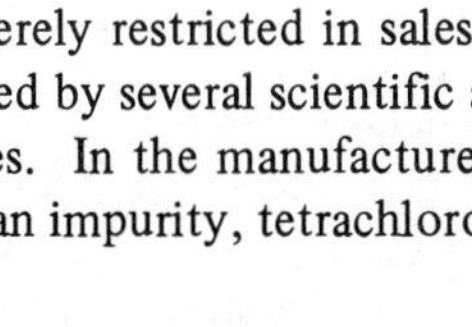	Hexachlorocyclopentadiene, 1,4-butenediol, thionyl chloride.	Insecticide; aphicide.

has been severely restricted in sales, and 2,4-D is being studied by several scientific and administrative bodies. In the manufacture of 2,4,5-T, particularly, an impurity, tetrachlorodioxin

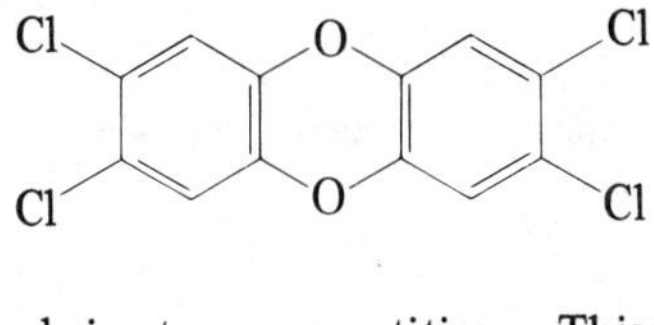

is formed in trace quantities. This 2,3,7,8-tetrachlorodibenzodioxin is the most toxic of the chlorinated dioxins and related benzofurans. It possesses carcinogenic, mutagenic, and teratogenic (birth defects) effects, as well as other toxic effects. Still not completely resolved is the question as to whether aspects of mammalian toxicity of 2,4-D and 2,4,5-T derive almost entirely or only in part from the dioxin content of these useful herbicides.

Much chemistry has also been done (see formulation section) on altering the parent acids, e.g., to form esters (both volatile and nonvolatile), salts, and nonvolatile amides. Volatility is extremely important because with the potency of these chemicals the dilute vapor as well as suspended droplets can injure broad leaf crops downwind from the point of application.

Trichlorobenzoic acid is one of a large number of related herbicides and growth regulants. (See Table 21.10.) Of this class,

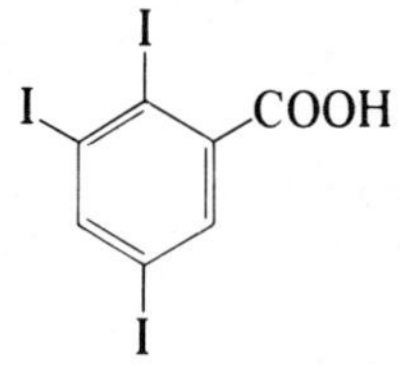

TIBA

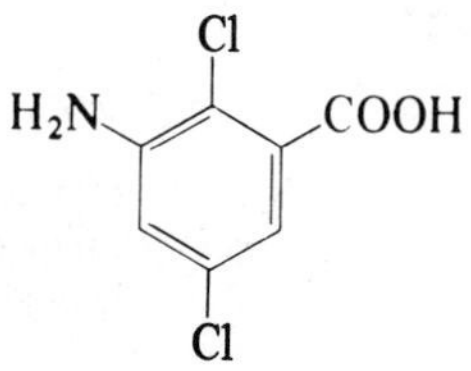

Amiben

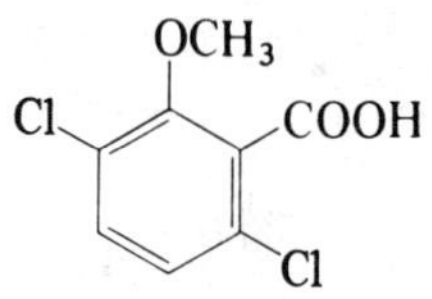

Dicamba

TIBA, amiben, and dicamba are among the more prominent. Halogenated terephthalics are also herbicidal. TCA (trichloroacetic acid) is one of the simplest herbicides, but it is biologically weaker and less specific than many of those cited above. The related dichloropropionic acid ($CH_3CCL_2-\overset{O}{\overset{\|}{C}}-OH$ dalapon) is an important postemergent herbicide.

Folpet, a derivative of *o*-phthalic acid, and captan, from tetrahydrophthalic acid, are compounds of very low acute mammalian toxicity and low persistence in the environment, and they have a wide spectrum of useful fungitoxicity. They are used on fruit crops, grapes, potatoes, tomatoes, seed dressings, and so on.

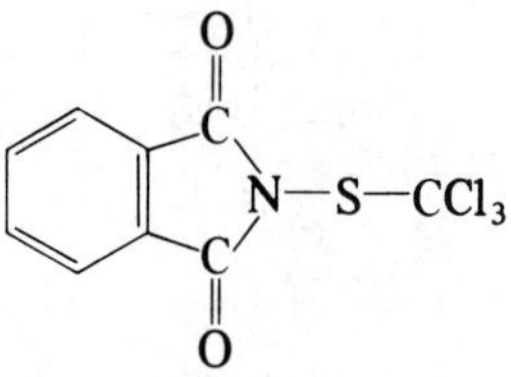

folpet

An important plant-growth regulator is alpha-naphthyl acetic acid (ANA). It stimulates root formation and affects the production of ethylene within the plant, and this compound, in turn, is implicated in many plant physiological processes. Ethylene itself is possibly the simplest hormonal substance known. It has very recently been shown to be derived in plants from methionine through the

TABLE 21.10 Pesticides Derived From Common Organic Acids

Name	*Structure*	*Significant Intermediates*	*Uses*
2, 4-D Acid salts Esters	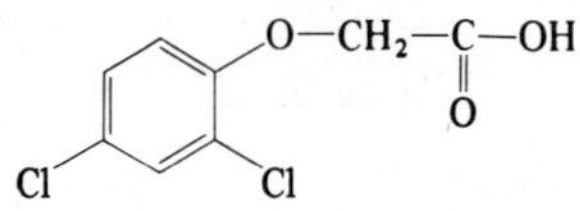 $Cl_2C_6H_3-O-CH_2-C(=O)-OH$	Phenol, chlorine, chloracetic acid.	Herbicide and growth regulator, broad-leaf control (grain and cereal).
TBA TBC	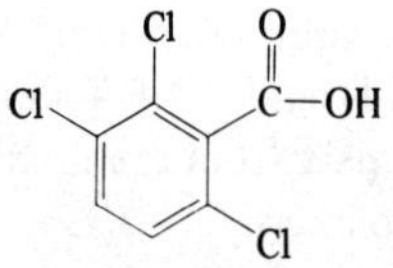 $Cl_3C_6H_2-C(=O)-OH$	Toluene, chlorine (oxidation).	Preemergence broad leaf.
TCA	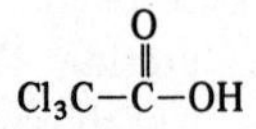 $Cl_3C-C(=O)-OH$	Acetic acid, chlorine.	Postemergence herbicide.
Captan	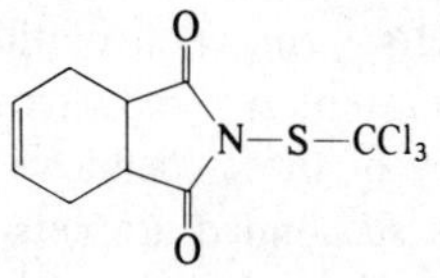 $N-S-CCl_3$ (tetrahydrophthalimide)	Maleic anhydride, butadiene, ammonia, carbon disulfide, chlorine.	General fungicide.
ANA	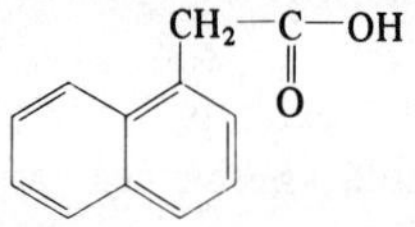 $C_{10}H_7-CH_2-C(=O)-OH$	Naphthalene, chloracetic acid; or formaldehyde, hydrochloric acid, potassium cyanide (hydrolysis).	Plant-growth regulator.

intermediate 1-aminocyclopropane carboxylic acid, H_2N–cyclopropyl–COOH. Many phenoxy, naphthyl, and naphthoxy acids, as well as more complex acids such as indoleacetic acid, influence plant physiological activity.

The Organic Phosphate Pesticides

These phosphorus esters, which have the generic structure:

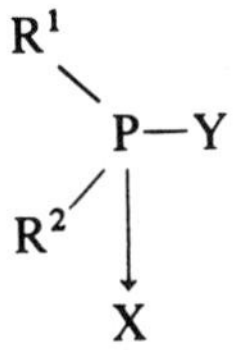

$$\begin{array}{c} R^1 \\ \quad\searrow \\ \quad\quad P-Y \\ \quad\nearrow \quad\downarrow \\ R^2 \quad\quad X \end{array}$$

owe their insecticidal activity to their capacity to phosphorylate the enzyme cholinesterase, which moderates nerve-impulse transmission. The inactivated, phosphorylated-enzyme-reaction product possesses varying degrees of persistence. The effect of this deactivation is a build-up of the chemically neuroactive acetylcholine

$$(CH_3)_3\overset{+}{N}-CH_2CH_2O-\overset{\overset{O}{\|}}{C}-CH_3$$

This results in energy transmission through the neuron and, among other effects, causes convulsive activity at some vital muscular center.

R^1 and R^2 are generally lower alkyl, alkoxy, alkylthio, or alkylamino groups usually containing fewer than four carbons; X can be O or S, but if it is S, the activity is derived by an *in situ* oxidation to the oxoderivative; Y is usually a hydrolytically unstable group which becomes detached from the molecules, thus providing a residue that can combine with the enzyme. The first six molecules in Table 21.11 all satisfy these generalizations, although changes within the plant or soil or insect may be required before the biochemically active species is generated.

These organophosphates can act as contact, stomach, respiratory, or systemic toxicants. Many of them are highly toxic to mammals as well. However, in the case of malathion, mammals possess enzymes that can saponify the succinic ester before it arrives at vital centers, thereby providing relatively low mammalian toxicity. Insects appear to be less able to accomplish this detoxification. Many of the newer (and some of the earliest) organophosphorus derivatives possess amidate structures, that is

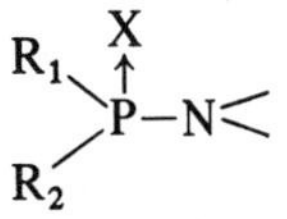

$$\begin{array}{c} R_1 \quad X \\ \searrow \uparrow \\ \quad P-N\!\!< \\ \nearrow \\ R_2 \end{array}$$

bonds. Some of these are useful animal systemics, i.e., they possess a differential toxicity for the cow, sheep, fowl, horse, dog, etc., and the insects that attack them. They are tolerated by the host animals at concentrations that kill or suppress the particular parasitic species.

Two defoliants have been included in Table 21.11. In addition, there are other organophosphorus derivatives of the amidate types, as well as other classes, which possess herbicidal and fungicidal properties. It is no wonder then that each year sees the development of many variations on organophosphorus structures, for they have large capabilities for differential toxicity, variation in persistence, and specificity. Particularly with the interdiction of the chlorinated hydrocarbons, the organophosphate and carbamate groups provide opportunities for replacement with environmentally less persistent pesticides, even though they are somewhat more expensive as practical pest-control agents.

Principles of physical organic chemistry can be successfully applied in relating the biological activity to the phosphate insecticides (also the carbamates). The Hammett σ and ρ functions for aromatic substitution provide useful correlations to anti-cholinesterase activity. Practical field-insect effectiveness involves more complex variables.

While the organo-phosphorus insecticides are generally not highly persistent in the environment, and do not concentrate in fatty tissues, evidence of *resistance to this genus of chemicals exists and is mounting.*

The Carbamate Groups

Many of the more recently developed insecticides possess the carbamate structure.

TABLE 21.11 Important Organophosphates

Name	*Structure*	*Significant Intermediates*	*Uses*
parathion	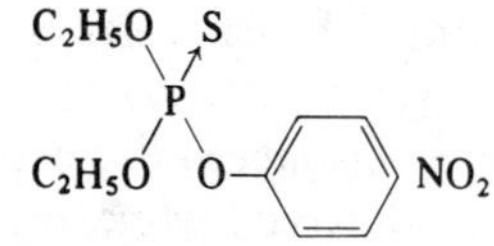 C_2H_5O S P C_2H_5O O–(benzene ring)–NO_2	*p*-Nitrophenol, ethanol, phosphorus trichloride, sulfur.	General insecticide.
Methyl parathion	Methyl homolog of parathion.		Insecticide (cotton).
Bolstar	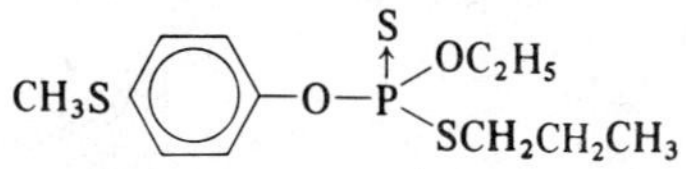 S CH_3S–(benzene ring)–O–P(↑S)(OC_2H_5)($SCH_2CH_2CH_3$)	$PSCl_3$ Ethanol, propylmercaptan, *p*-thiomethylphenol	*Lepidoptera* cotton insects
Kitazin	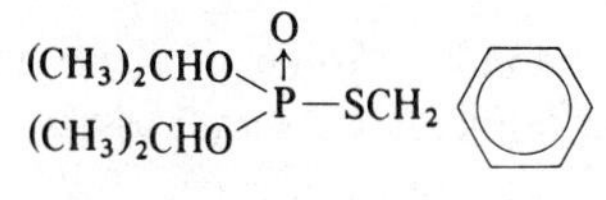 O $(CH_3)_2CHO$ ↑ P–SCH_2–(benzene ring) $(CH_3)_2CHO$	2-Propanol, benzylmercaptan, PCl_3	Rice fungicide.
Malathion	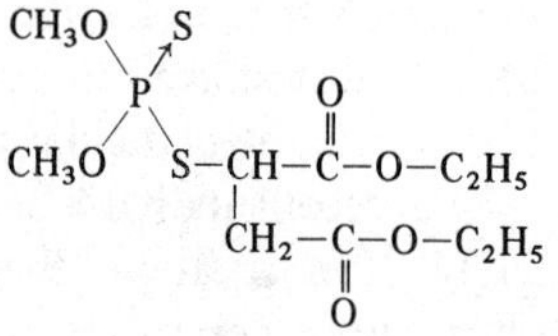 CH_3O S P O CH_3O S–CH–C(=O)–O–C_2H_5 CH_2–C(=O)–O–C_2H_5	Methanol, phosphorus pentasulfide, ethyl maleate.	General insecticide.
Dimethoate	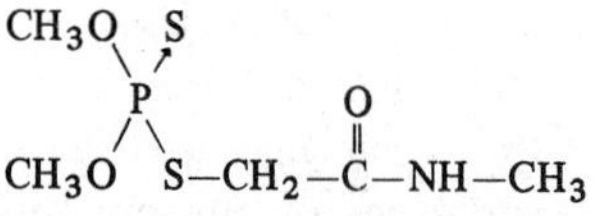 CH_3O S P O CH_3O S–CH_2–C(=O)–NH–CH_3	Methanol, phosphorus pentasulfide, chloroacetic acid, methylamine.	Systemic insecticide.
Naled	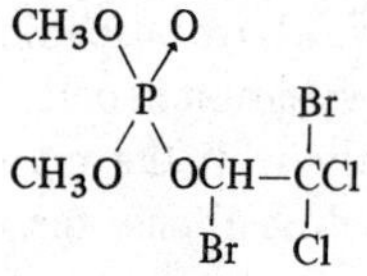 CH_3O O P Br CH_3O OCH–CCl Br Cl	Chloral, bromine, trimethyl phosphite.	Nonpersistent insecticide.
Demeton (O) (Many variations of this structure are commercial insecticides.)	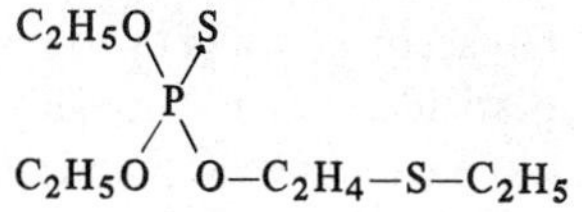 C_2H_5O S P C_2H_5O O–C_2H_4–S–C_2H_5	Ethanol, phosphorus thiochloride, ethyl thioethanol.	Systemic insecticide.
Folex®	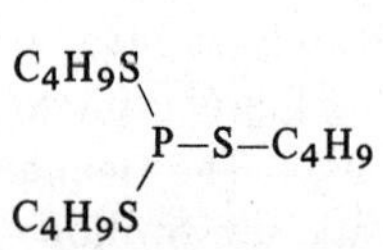 C_4H_9S P–S–C_4H_9 C_4H_9S	Phosphorus trichloride, butyl mercaptan.	Defoliant (cotton).
DEF®	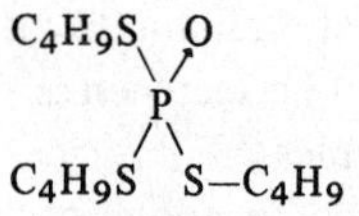 C_4H_9S O P C_4H_9S S–C_4H_9	Phosphorus oxychloride, butyl mercaptan.	Defoliant (cotton).
Orthene	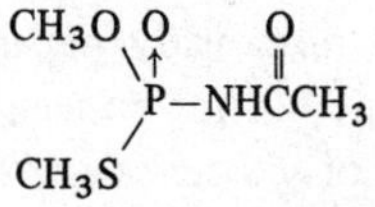 CH_3O O O ↑ ‖ P–$NHCCH_3$ CH_3S	Dimethyl thiono-phosphorochloridate, NH_3, acetic anhydride	Cabbage looper, aphids.
Ethrel	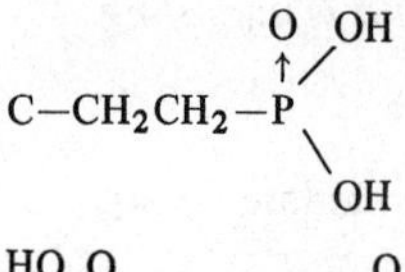 O OH ↑ C–CH_2CH_2–P OH	PCl_3 ethyleneoxide H_2O.	Plant growth regulator.
Glyphosate	 HO O O ↑ ‖ P–CH_2NHCH_2COH HO	PCl_3, CH_2O, glycine.	Herbicide post emergent.

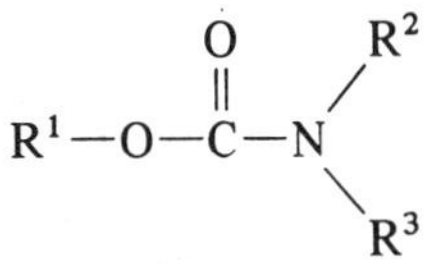

$$R^1{-}O{-}\overset{\displaystyle O}{\overset{\|}{C}}{-}N\begin{matrix}\diagup R^2\\ \diagdown R^3\end{matrix}$$

Like the organic phosphate insecticides, these compounds owe their activity to their interference with the enzyme cholinesterase which governs the concentration of acetylcholine at the synapse. Earlier it was thought that the insecticidal carbamates were competitive inhibitors competing with acetylcholine for complexation at the enzyme surface. It is now held that actual carbamoylation of the enzyme occurs. This in turn governs the transmission of impulses through the synaptic junctions of the nervous system. In this way, their mode of action resembles, but is not quite identical to that of the organic phosphate esters previously discussed. For many of these carbamates, R^1 is an aromatic moiety such as a substituted phenyl (BUX) or naphthyl group (carbaryl). Some of the more active carbamate structures contain oximino groups, e.g., methomyl and the systemic and very potent temik (note structural resemblance to acetylcholine)

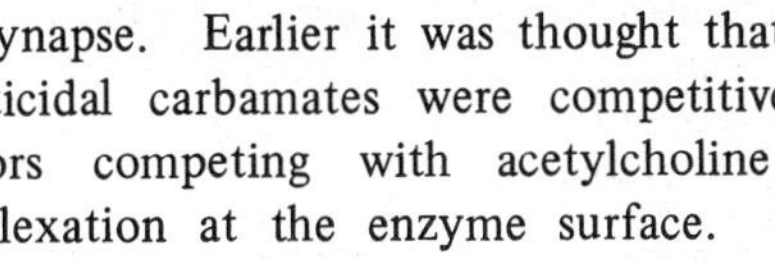

$$CH_3S{-}\underset{CH_3}{\overset{CH_3}{C}}{-}CH{=}N{-}O{-}\overset{O}{\overset{\|}{C}}{-}\underset{H}{N}{-}CH_3$$

temik

as well as the heterocyclic variants such as carbofuran.

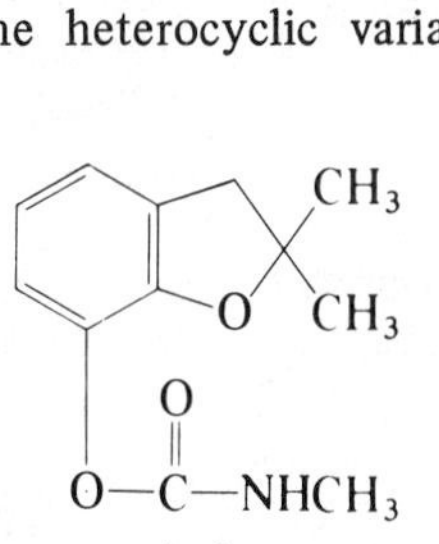

$$\text{(2,2-dimethyl-2,3-dihydrobenzofuran-7-yl)}{-}O{-}\overset{O}{\overset{\|}{C}}{-}NHCH_3$$

carbofuran

Usually R^2 is methyl and R^3 is hydrogen, although sometimes R^2 and R^3 are both methyl. In very recent pesticidal structures, the proton of R^3 is substituted by an acyl, sulfenyl, or phosphoryl moiety. The same fully oxygenated generic structure also provides important herbicidal compositions when R^2 is aryl or substituted aryl, R^3 is H and R^1 is aliphatic or substituted aliphatic. For example, there follows the substitutions in the order R^1, R^2, R^3 and the name of the herbicide, respectively.

Isopropyl, Phneyl, H, IPC; Isopropyl, *m*-chlorophenyl, H, CIP; 4-chloro-2-butynyl, *m*-chlorophenyl, H, barban.

These carbamates are used in grain production and are involved with the inhibition of mitosis in the offending weed species. These compounds are *not* effective cholinesterase inhibitors.

While many of the derivatives of the fully oxygenated carbamates are useful insecticides, the *dithiocarbamate group*

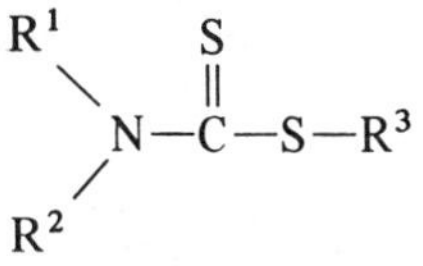

$$\begin{matrix}R^1\diagdown\\ R^2\diagup\end{matrix}N{-}\overset{S}{\overset{\|}{C}}{-}S{-}R^3$$

provides many very significant fungicides. The sodium, ammonium, zinc, manganese, and iron dithiocarbamates derived from ethylene diamine ($R^1 = -CH_2CH_2-$; $R^2 = H$) and from dimethylamine ($R^1 = R^2 = CH_3$) are standard fungicides for a wide variety of fungal plant diseases. Two of these structures (maneb and ferbam) are given in Table 21.12.

Finally, the mixed oxygen-sulfur carbamates, possessing the general structure

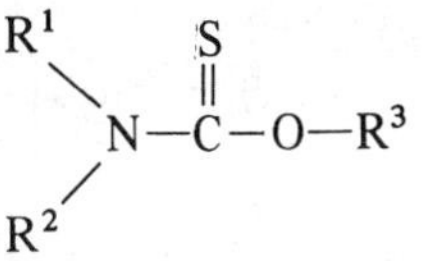

$$\begin{matrix}R^1\diagdown\\ R^2\diagup\end{matrix}N{-}\overset{S}{\overset{\|}{C}}{-}O{-}R^3$$

contribute important examples of herbicidal chemicals. The structure of eptam, an important preemergence herbicide, is given in Table 21.12. There are many homologues with variations of R^1, R^2, and R^3 of which pebulate (C_2H_5, n-C_4H_4, n-C_3H_7), vernolate ($R^1 = R^2 = R^3$, n-C_3H_7), and di-allate (i-C_3H_7, i-C_3H_7, $C_3Cl_2H_3$), are significant examples.

A new discovery in the last decade has been made in the area of *herbicidal antidotes*. These antidotes provide greater tolerance by the crop for the particular herbicide, thereby providing the farmer a degree of insurance against crop damage. Many compositions have been identi-

TABLE 21.12 Carbamate Pesticides

Name	*Structure*	*Chemical Intermediates*	*Uses*
Carbaryl	1-naphthyl–O–C(=O)–NH–CH_3	α-Naphthol, methyl isocyanate.	Broad spectrum insecticide (*Lepidopterous* larvae).
Methomyl	$CH_3-S-C(CH_3)=N-O-C(=O)-NH-CH_3$	Acetyl chloride, methyl mercaptan, hydroxylamine, methyl isocyanate.	Contact insecticide.
BUX	phenyl–O–C(=O)–NH–CH_3, with ring substituent $CH(CH_3)-CH_2-CH_2-CH_3$	Phenol, pentene, methyl isocyanate.	Soil insecticide (corn).
Maneb	$^{-}S-C(=S)-N(H)-CH_2-CH_2-N(H)-C(=S)-S^{-}$ Mn^{++}	Ethylene diamine, carbon disulfide, caustic, Mn^{++}	Fungicide.
Vapam	$CH_3-NH-C(=S)-S-Na$	Methylamine, carbon disulfide, caustic.	Soil fungicide; nematicide.
Ferbam	$[(CH_3)_2N-C(=S)-S-]_3$ Fe	Dimethylamine, carbon disulfide, Fe^{+++}	Fungicide (ornamental plants).
Eptam	$C_2H_5-S-C(=O)-N(CH_2-CH_2-CH_3)_2$	Dipropylamine, phosgene, ethylthiol.	Preemergence herbicide.
Sulfallate	$CH_2=C(Cl)-CH_2-S-C(=S)-N(CH_2-CH_3)_2$	Diethylamine, phosgene, 2, 3-dichloropropene.	Herbicide.

fied as antidotal including some carbamate compositions.

Another recent discovery in the herbicidal carbamate area is that the active species for the thiocarbamate herbicides are the oxidized metabolites, the sulfoxides and the sulfones.

$$R_1R_2N-C(=O)-S(=O)-R_3 \qquad R_1R_2N-C(=O)-S(=O)_2-R_3$$

This is another example of biochemical *intoxification*, the production, *in situ*, of a more active species.

Although the above generalizations concerning the types of carbamates and their biological activity are generally sound, there are occasional exceptions. The biological activity of carbamates is quite broad and occasionally nematicidal, acaricidal, and other properties are exhibited. The major economic usefulness of the dithiocarbamate sulfallate is in the herbicidal area.

Nonsymmetrical Ureas

The nonsymmetrical ureas (Table 21.13)

$$\begin{matrix} R^1 & & O & & R^3 \\ & \diagdown & \| & \diagup & \\ & N & -C- & N & \\ & \diagup & & \diagdown & \\ R^2 & & & & R^4 \end{matrix}$$

provide many valuable herbicidal compounds and continue to be under active development.

Generally, one nitrogen is substituted with a cyclic group, such as an aryl, substituted aryl, heterocyclic, mono- or bicyclic group, and the other nitrogen with lower-alkyl or alkoxy moieties. The compounds interfere with the photosynthetic process. The nature of the cyclic group provides interesting specificities, i.e., relative tolerance of certain crop species, and effectiveness against groups of weed species. Many of the compounds possess degrees of preemergence and postemergence activity.

A preemergence herbicide is one generally applied prior to the emergence of the weed seed from the ground. The chemical then may be applied at or near the time of crop seeding. Some preemergence herbicides must be worked into the soil for optimum activity; others are quite active when applied to the soil surface. A postemergence herbicide is one that exhibits its activity by killing or impeding the growth of the weed species from seedling stage onwards and is applied after the weed's emergence.

A sterilant is a chemical toxic to all forms of plant life, i.e., it exhibits very little useful *differential* plant toxicity.

Some Heterocyclic Pesticides

In recent years pesticides based upon heterocyclic structures (Table 21.14) have become more prominent. The corn, sorghum weed-control market has been dominated by the symmetrical triazine group, of which atrazine is a most important representative. The corn plant possesses an enzyme that degrades atrazine much more effectively than the weed species which compete with it. Many other triazine herbicides are based upon appropriate

TABLE 21.13 Urea Herbicides

Name	*Structure*	*Chemical Intermediates*	*Uses*
Monuron	$Cl-C_6H_4-NH-C(=O)-N(CH_3)_2$ (para)	*p*-Chloroaniline, dimethylamine, phosgene.	Preemergence herbicide (cotton).
Linuron	$3,4-Cl_2C_6H_3-NH-C(=O)-N(OCH_3)(CH_3)$	3, 4-Dichloroaniline, phosgene, hydroxylamine, dimethyl sulfate.	Herbicide.
Fluometuron	$m-F_3C-C_6H_4-NH-C(=O)-N(CH_3)_2$	*m*-trifluoromethylaniline, dimethylamine, phosgene.	Herbicide (cotton).
Noruron	$C_{10}H_{15}-NH-C(=O)-N(CH_3)_2$ (hexahydro-4,7-methanoindanyl)	Dicyclopentadiene (reduction), ammonia, phosgene, dimethylamine.	Herbicide.
Cycluron	$C_8H_{15}-NH-C(=O)-N(CH_3)_2$ (cyclooctyl)	Butadiene, (dimerization; reduction), ammonia, phosgene, dimethylamine.	Herbicide.

TABLE 21.14 Heterocyclic Pesticides

Name	*Structure*	*Significant Intermediates*	*Uses*
Atrazine	Cl; N, N, N ring; C_2H_5—NH; NH—CH—CH_3; CH_3	Cyanuric chloride, ethylamine, isopropylamine.	Preemergence herbicide (corn).
Paraquat dimethylsulfate	CH_3—$\overset{+}{N}$ (bipyridyl) $\overset{+}{N}$—CH_3; $(CH_3{-}O{-}SO_2{-}O^-)_2$	Pyridine, dimethyl sulfate.	Postermergence herbicide dessicant.
Picloram	NH_2; Cl, Cl, Cl; N; C—OH, O	Chlorine, ammonia, α-picoline (oxidation).	General herbicide; plant sterilant.
Benomyl	O; C—NH—C_4H_9; N; C—NH—C—OCH_3; N; O	Phenylenediamine, phosgene, methanol, butylamine.	Systematic broad spectrum fungicide.
Bromacil	H; CH_3; N; O; N—CH—C_2H_5; Br; O; CH_3	Acetoacetic ester, *sec*-butylamine, phosgene, ammonia, bromine.	General herbicide (garden and home).
Carboxin	O; CH_3; S; C—NH—(cyclohexyl); O	Aniline, ketene, mercaptoethanol.	Systematic fungicide (grain, soybeans–smut and rust).
Bentazone Basagran	O; N—CH(CH_3)CH_3; SO_2; N; H	*n*-Anthranilic acid, isopropylamine, sulfuryl chloride	Postemergence
Metribuzin Sencor	CH_3; CH_3—C—; CH_3; O; N—NH_2; N; N; —SCH_3	2-*t*-butyl isopropyl pyruvate, hydrazine, methyl thiol	Preemergence soybeans, other crops

substitution of the chlorines in cyanuric chloride:

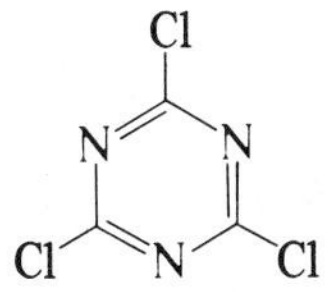

cyanuric chloride

e.g., simazine, prometrone, and propazine. Each gains an area of specificity by the substitution usually of methoxy or methylthio and lower alkylamino groups.

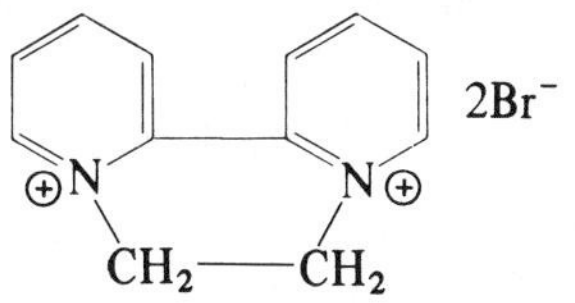

$2Br^-$

Diquat

Like paraquat, diquat is a member of a group of quaternary dipyridyls which owe their activity to their ability ro produce inside the chloroplast a stable free radical that intercepts the transfer of energy in the early stages of the photosynthetic process. These compounds are irreversibly bound to montmorillonite clays and deactivated thereby so that the surrounding clay soil becomes tolerant to plant growth. The trend to reduce labor in agriculture is exemplified by the use of paraquat (and more recently glyphosate) as a "chemical hoe." Planting is performed with minimal cultivation in certain areas by the use of paraquat before the drilling of seeds.

Picloram is an extremely effective sterilant and, from a biological view, is perhaps one of the most powerful herbicides known. Its usefulness in the U.S. is presently limited because of the combination of the properties of persistence and its nonselective phytotoxicity.

Two examples of interesting heterocyclic systemic fungicides are provided in Table 21.14. These fungicides are absorbed by the plant, and they, or their metabolities, inhibit the growth of the fungal organism. The sulfone homologue of carboxin, called oxycarboxin, is also a systemic fungicide.

There are many compounds, some of very recent origin, based upon cyclization of the phenylenediamine structure. One of these, benomyl, exhibits a broad scope of contact and systemic fungal activity. However fungal resistance to the *systemic* fungicides is becoming increasingly manifest.

Miscellaneous Chemicals

Dinitrophenols. One of the earliest groups of synthetic organic chemicals, predating World War II and still in use, are the alkyldinitrophenols. The chemicals have broad biological activity embracing the insecticide, fungicide, acaricide, and herbicide fields. This broad activity results from their capacity to block oxidative phosphorylation, hence, interfering with vital biochemical synthesis cycles found both in animal and plant life.

In addition to the example given in Table 21.15, there are many other structural variations with biological activity. Salts, esters, and free phenols are all utilized in commercial preparations.

Generic Formula	R^1	R^2	*Functions*	*Name*
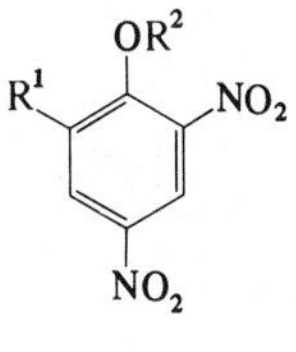 OR^2, R^1, NO_2, NO_2	$-CH_3$	H	Insecticide Herbicide	DNOC
	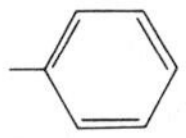 (ring)	H	Insecticide Herbicide	DNOCHP
	$-CH(CH_3)-(CH_2)_5CH_3$	$-C(=O)-CH=CH-CH_3$	Acaricide Fungicide	Dinitrocaprylphenyl crotonate

Soil Fumigants. In many areas, the population of nematodes and pathogenic fungi is so high that seeds are destroyed, or the emerging seedling and plant are similarly attacked so that little or no crop can be harvested. Such soil is treated with a fumigant, usually a volatile general toxicant which is frequently injected into the soil prior to planting. These sub-

TABLE 21.15 Miscellaneous Organic Pesticides

Name	*Structure*	*Significant Intermediates*	*Properties*
DNOBP	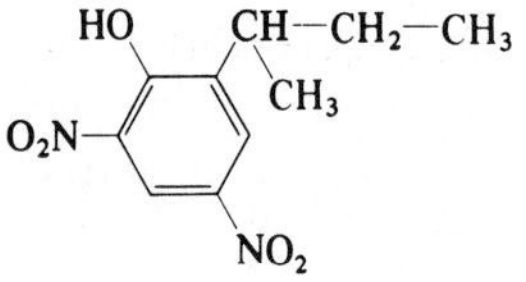 HO, $CH-CH_2-CH_3$, CH_3, O_2N, NO_2	Phenol, nitric acid, butene-2.	Insecticide; herbicide.
Nemagon	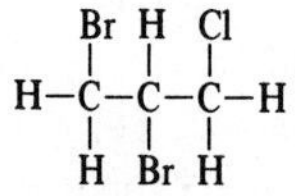 Br H Cl / H–C–C–C–H / H Br H	Propylene, bromine, chlorine.	Soil fumigant; nematocide.
EDB	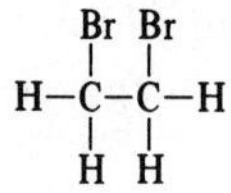 Br Br / H–C–C–H / H H	Ethylene, bromine.	Fumigant; nematocide.
PCP	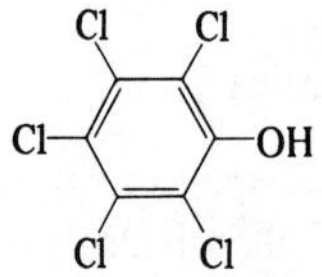 Cl, Cl, Cl, OH, Cl, Cl	Phenol, chlorine,	Wood preservative; fungicide; herbicide.
Piperonyl butoxide	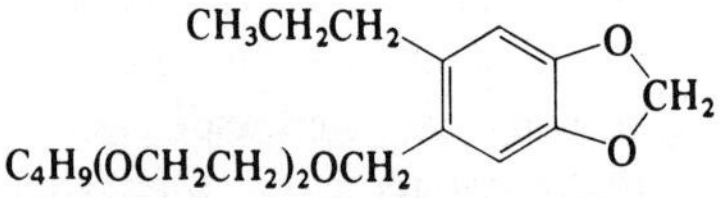 $CH_3CH_2CH_2$, $C_4H_9(OCH_2CH_2)_2OCH_2$, O, CH_2, O	Safrole, hydrogenation, formaldehyde, hydrochloric acid, ethylene oxide, butanol.	Insecticide; synergist; aerosol spray.
Trifluralin (Treflan)	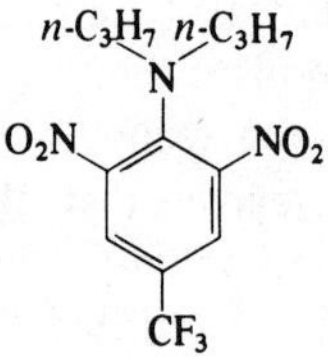 n-C_3H_7, n-C_3H_7, N, O_2N, NO_2, CF_3	4-Difluorodinitrochlorobenzene, n-propylamine.	Preemergence herbicide (cotton).
Warfarin	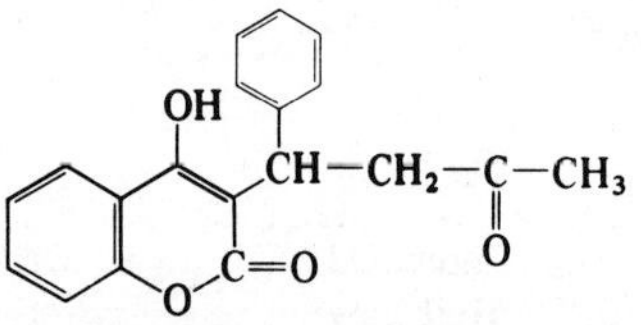 OH, $CH-CH_2-C-CH_3$, O, C=O, O	Benzalacetone, 4-hydroxycoumarin.	Rodenticide.
Plictran	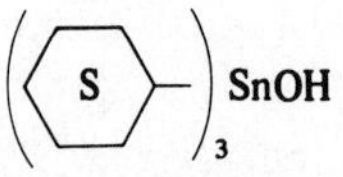 (S)$_3$ SnOH	Chlorocyclohexane $SnCl_4$, Mg or Li.	Acaricide.

stances are generally, but not exclusively, low-molecular weight halogenated compounds. In addition to those chemicals given in Table 21.15, chloropicrin (CCl_3NO_2), methyl bromide (CH_3Br), and other halogenated C_1, C_2, and C_3 hydrocarbons find extensive application as soil fumigants.

In recent years, the toxicological properties of some of these halogenated fumigants have been intensively reinvestigated. DBCP for example, has been shown to cause mammalian male sterility. Ethylene dibromide is suspected of mutagenic and carcinogenic potential. As a result, their agricultural application is either totally interdicted or very severely limited. A current need exists for *effective*, *safe* nematicides and soil pathogenic organism sterilants.

Fumigants of various structures are also employed to preserve harvested crops, corn, small grain, etc., from insect and other pests.

Wood Preservatives. The pesticide industry also provides chemicals to prevent insect and fungal attack on lumber and wood products. In some cases, conventional pesticides are employed. Formulations of halogenated phenols are among the most common wood preservatives. Mercury and tin compounds have been also used to control wood rot organisms. Lindane has wide acceptance for control of destructive wood and lumber insects.

Synergists. It has long been known in medicine and in agriculture that the use of two chemicals can result in additive, less than additive (antagonistic) or greater than additive biological effects. Where a greater-than-additive result occurs, the two chemicals are said to behave synergistically. In the pesticide industry, there are a number of moieties which have little or no activity themselves, but when used with a toxicant improve the activity of the toxicant. These synergists are most advantageously employed when the toxicant is an expensive chemical, for the combination possesses a decisive, practical economic advantage.

The group

$$R{-}CH_2{-}C_6H_3{<}(O_2CH_2)$$

(piperonyl group) where R has a multiplicity of organic functions, synergizes many insecticides. Certain alkynes, heterocycles, imido, and thiocyanato linkages also possess degrees of synergy with insecticides. Synergists find considerable use in aerosol sprays with pyrethrins or synthetic pyrethroids such as allethrin.

Organic Mercurials. For many years, organic mercurials were important fungicides and bactericides. They were applied as foliage sprays, e.g., phenyl mercuric acetate, and especially as seed dressings. Cotton and grain seed, particularly, could be treated at the rate (for grain) of 1 to 4 oz per 100 lbs, of a 2 to 4 percent solution. Enough volatility exists for many mercury compounds to sterilize not only the areas external to the seed coat, but to penetrate into the seed coat itself. In 1970, severe limitation of the use of mercurial pesticides was instituted, largely because of mercury contamination of the environment, as well as the danger of mercury-treated seed entering (illegally) man's food supplies.

The structures of some common mercurial pesticides are listed below. (NOTE: A safe, effective organic substitute for the mercurials with approximately equivalent beneficial properties does not currently exist and is an R & D objective.)

$$CH_3HgS{-}CH_2{-}CH(OH){-}CH_2(OH) \text{ and } CH_3HgOC(=O){-}CH_3$$

Ceresan L

$$C_6H_5{-}HgOC(=O){-}CH_3$$

Phenyl mercuric acetate

$$C_9H_6N{-}O{-}HgCH_3$$

LM

Fluorinated Herbicides. Included in Table 21.15 is an important preemergence cotton herbicide, Trifluralin (Treflan). It is noteworthy as one of the earliest fluorinated organic pesticides to satisfy the toxicological requirements for registration. Its effectiveness is markedly increased if it is "worked" into the soil. The CF_3 group and other fluorinated moieties are now frequently found in both medicinally and agriculturally useful compositions.

Rodenticide. Damage to agricultural crops by rodents and the capacity of this group of mammals to harbor disease vectors has stimulated research for rodenticides. Effective control of various rodent species is achieved through impregnating grain with warfarin. This coumarin compound is related to the chemicals used in medicine to prevent blood clotting, particularly in cardiac disease. As a rodenticide, it causes internal hemorrhage and eventual death.

In addition to the coumarins, certain indanediones, ureas, and other organics find application as rodenticides. Phosphine formed by the slow hydrolysis of metal (usually aluminum and zinc) phosphides continues to be employed for rodent control despite the obvious hazard. In many 'Third World' countries, rodents are a major problem, limiting both the harvest and storage of grains as well as acting as an important public health hazard.

Sulfur and Inorganic Sulfur Compounds. The oldest of the inorganics, sulfur and the water soluble polysulfides, continue to be employed, and their use may increase during the next decade in many agricultural areas. Sulfur, usually as a coarse powder or a small granule, is a soil additive. It is a required nutrient for all plant life, and the soils in many parts of this country and of the world are deficient in it. The oxidation of sulfur by soil microorganisms assists in the production of acid bodies that help solubilize other insoluble nutrients.

Sulfur is used as a wettable powder or dust, alone or in combination with other pesticides for foliar fungal control of mildews and other organisms. Calcium polysulfide (CaS_x, $x = 4.5$) is usually found in an about 30 percent aqueous solution in many commercial preparations which are used for soil conditioning and fungal, mite, and insect control. The ammonium and sodium variants are also used.

Because sulfur is a low-cost chemical in abundant supply, an essential nutrient, relatively nontoxic to man, and ecologically desirable (except in gaseous forms as an air pollutant), the agricultural usefulness of sulfur may expand. Both industry and agriculture will continue to compete for the sulfur that is removed from fossil fuels (coal, oil, gas).

Inorganic Pesticides. Though synthetic organic pesticides dominate the market in countries with advanced agricultural technology, the inorganics still find useful application in several areas. Copper salts, particularly the basic sulfate (Bordeaux mixture) as well as copper oxychloride, control many foliar fungal infections. In Europe, copper products continue to dominate in wine grape mildew-control programs, particularly for the "vin ordinaire" production in the Latin countries.

Basic sulfates of zinc and manganese are used as minor-element nutrient sprays as well as for fungicidal purposes, with or without copper. Basic phosphates are also employed. All of these salts are generally prepared under controlled pH conditions (sometimes *in situ* in the field) in such a way that a fairly insoluble salt is provided. Thus, its solubility limits the quantity of soluble metal ion available at any given time for absorption by the plant. The high quantities of Zn^{++}, Cu^{++}, Fe^{+++}, etc., which would be obtained from the normal sulfate, nitrate, or halide sprays would be too phytotoxic for practical use.

In this respect, modern chelation chemistry is also employed to provide metallic minor-element additives. The EDTA (ethylenediaminetetraacetic acid) group combined with metal salts is utilized both for soil and foliar application. Generally, however, less expensive chelants are employed. These include polyphosphates and partially refined products derived from the paper and lumber industry containing metal coordinating groups.

Calcium arsenate and lead arsenates (orthoarsenates and more complex arsenates) are still employed for certain cotton insects and codling moth control in apples, as well as for other insect-control purposes. These uses are dwindling as are the uses of arsenites for soil sterilization. At present, the residue and toxicological properties of the arsenicals in relation to human toxicity and ecological factors are under review by government, scientific, and administrative bodies. The above is true also for organic arsenicals such as cacodylic acid $[(CH_3)_2As(O)OH]$ and sodium methyl arsonate $[(NaO)_2As(O)CH_3]$ which are used as herbicides and defoliants.

NEWER PYRETHROIDS

With increasing evidence of resistance and cross resistance to chlorinated hydrocarbon, organo phosphate, and carbamate insecticides–and particularly with mediocre control in particular localities of certain of the significant cotton pests–the discovery of new, relatively photostable pyrethroid molecules was viewed with great interest. These compounds have low persistence in the soil but provide sufficient plant surface stability to control the Heliothis complex - the army worms, the cabbage looper, and to a lesser degree the boll weevil, as well as many other insects. Laboratory resistance to the pyrethroids can be easily induced and there is evidence of some cross resistance by insects previously under heavy pesticide application pressure. Nevertheless, these compounds are finding increasing usefulness in cotton, vegetable, fruit, and other crops. As the accompanying table shows, these molecules are more complex than the conventional pesticide, which makes their production cost considerably higher. But this is more than offset by the extraordinary low dosage required for acceptable field control. In addition, all of these photostable molecules have two to three asymetric carbon atoms providing opportunities for various degrees of resolution. Decamethrin is indeed a totally resolved molecule and is effective against most insects at 0.01 to 0.02 lbs per acre. This provides a very low environmental burden. Permethrin and fenvalerate are currently unresolved. Their effective field dosage is about 0.1 to 0.2 lbs/acre. Because biological activity is centered almost exclusively in a single isomer, various half-resolved and fully-resolved preparations have been tested and have shown to be field effective at $\frac{1}{2}$ and $\frac{1}{4}$ the required dosage of the currently unresolved commercial product. The 1980's will probably experience the introduction of the chirally pure or half-resolved diastereoisomer pairs.

One disadvantage possessed by all of the current pyrethroids is their toxicity to bees and beneficial insects. It is interesting to note that a new experimental pyrethroid almost equivalent to permethrin and fenvalerate in activity against the major U.S. pests is 1000 times less toxic to the honey bee, an important consideration for many crops. The structure of this experimental compound exemplifies again the enormous difference in biological specificity that can be achieved by a relatively minor structural change.

In Table 21.16 we have written racemic (totally unresolved) structures for all the pyrethroids except decamethrin. Provided below is a reaction sequence for the manufacture of fenvalerate. The pyrethroid alcohol is made by the addition of HCN to *m*-phenoxybenzaldehyde which, in turn, is derived from *m*-phenoxytoluene.

Reaction Sequence for the Preparation of Pydrin (Fenvalerate)

Step No.	*Pydrin*
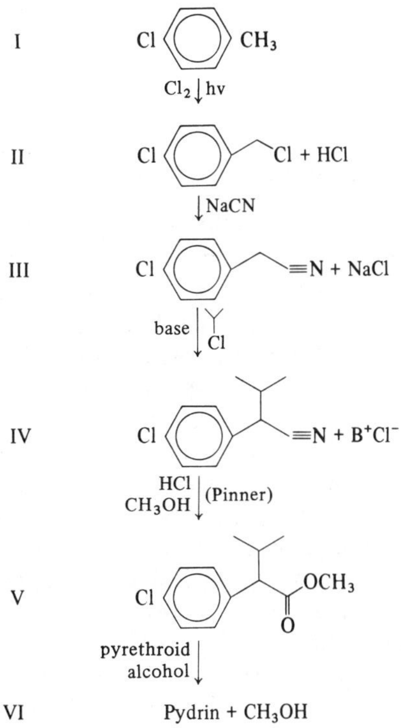 I	Cl (benzene ring) CH_3
	Cl_2 ↓ hv
II	Cl (benzene ring) Cl + HCl
	↓ NaCN
III	Cl (benzene ring) ≡N + NaCl
	base ↓ Cl
IV	Cl (benzene ring) ≡N + B^+Cl^-
	HCl, CH_3OH ↓ (Pinner)
V	Cl (benzene ring) OCH_3, O
	pyrethroid alcohol ↓
VI	Pydrin + CH_3OH

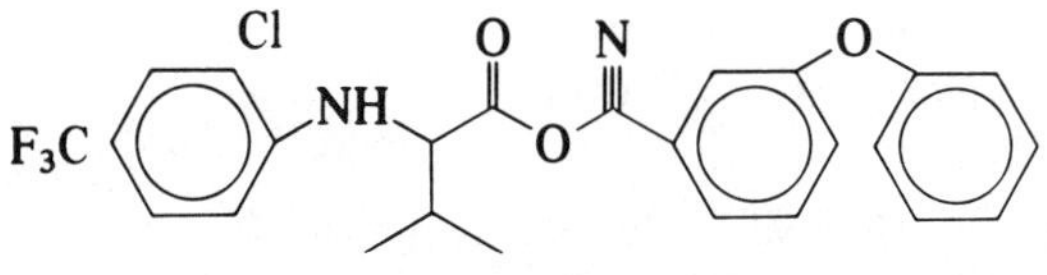

F_3C Cl NH O O N O

Fluvalinate

TABLE 21.16 Significant Synthetic Pyrethroids

Name	*Structure*	*Uses*
Allethrin	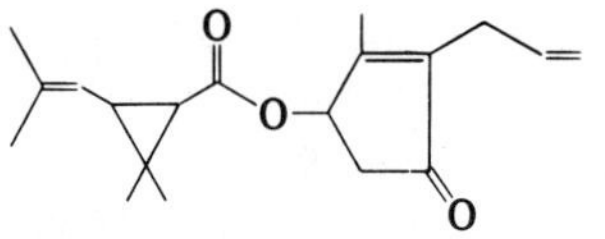	Knockdown agent flies
fenvalerate (Pydrin)	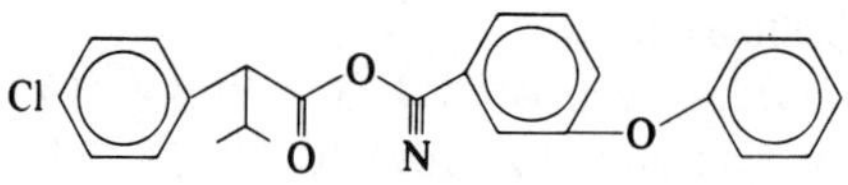	boll worms boll weevil cotton, fruit, vegetable pests
permethrin (Ambush) (Pounce)	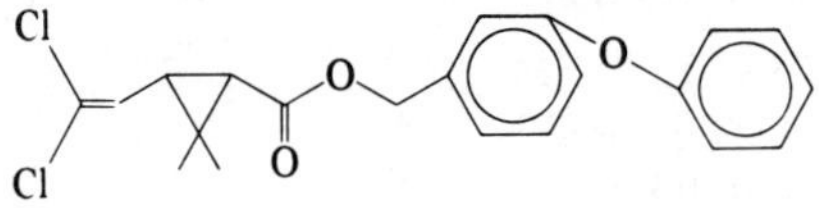	same as above
decamethrin (Decis)	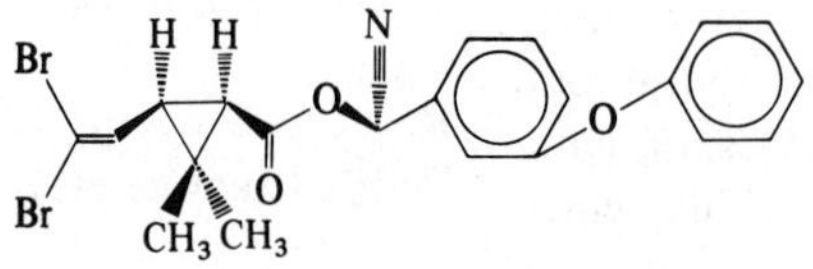	same as above
cypermethrin	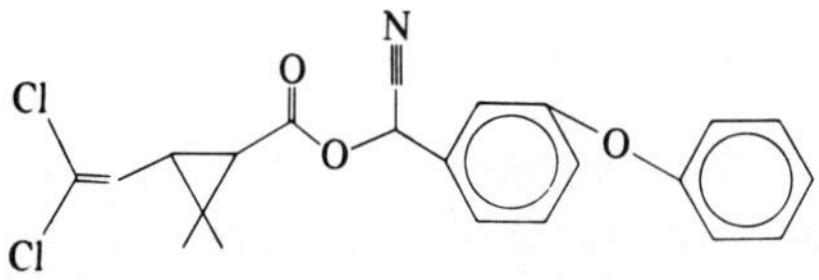	same as above

NEWER HERBICIDAL COMPOSITIONS

Consistent with the statistical data in Table 21.1, which show the tremendous increase in herbicide production, there has been a rapid proliferation of new compositions. The biological distance between plants and the higher mammals makes for a somewhat greater ease for producing active herbicides (than for insecticides for example) that are more or less innocuous to man. In the case of preemergent herbicides, residues on the harvested crop are generally nil or infinitesimal. This again results in easier registration for the herbicide than, for example, an insecticide. However, the discovery of the toxicological properties of the dioxins and the possibly mammalian toxic aspects of the highly chlorinated herbicides, especially 2,4-D and 2,4,5-T, has encouraged industrial development of safer substitutes. Moreover, the high costs of labor and energy stimulate the search for alternatives to human and mechanical weed control.

Although only treflan has been given as an example of a type of dinitro aniline herbicide, there are numerous homologs of this compound (Prowl, Surflan, etc.) currently registered. These compounds all affect the root systems of weeds (and at higher levels, of many crops) but each has some area of interesting specificity.

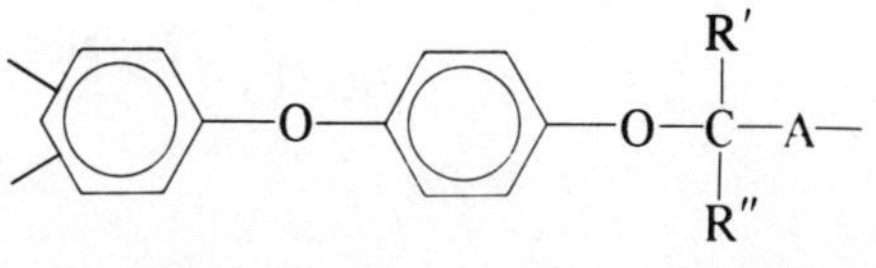

The diphenyl ethers which interfere with electron transport and phosphorylation reactions in plants are another very rapidly expanding development area. Space permits the providing of only two structures, Tok & Hoelon (see Table 12.17). Newer compositions include many variations of substitutions on the first ring and of side chain substitutions on a second ring. These changes modify greatly both pre- and postemergent activity and crop specificity. In addition, the pyridine-oxy-phenols show similar biological specificities. Tok is representative of the earliest members of this class. More recent compounds have the general structure where R′ and R″ are alkyl and alkyloxy alkylthio, moieties respectively. A also has great variability, frequently ending as an acid or ester moiety. The total side chain varies usually from 3 to 5 carbons saturated and unsaturated.

Whereas the older herbicides were generally employed at rates between two to 10 lbs/acre, the newer compounds both show greater specificity and are extraordinarily active. DPX 4189, a sterilant, is alleged to be active at 0.1 lbs/acre and even lower dosages. Alachlor is an important and selective preemergent herbicide. There has been a great deal of homolog synthesis performed, and the most interesting new developments are substitutions on the nitrogen to provide fungicidal rather than herbicidal properties. It will be noted also that there is greater molecular complexity (implying greater production costs) for the newer compositions.

Finally, the newer compositions reveal a greater overall understanding of plant biochemistry. Various sites in the photosynthetic process, in electron transport, in phosphorylation, in lipid and protein biosynthesis, in mitotic inhibition, etc., are all rather specific targets of many of the newer herbicides.

TABLE 21.17 Additional Herbicidal Compositions and Antidotes

Name	*Structure*	*Significant Intermediates*	*Properties*
TOK	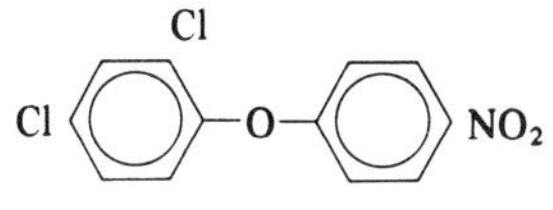 Cl, Cl, O, NO_2	2,4-dichlorophenol, *p*-chloronitrobenzene	Pre and post cereals current use outside U.S.A.
dichloro *p*-methyl Hoelon	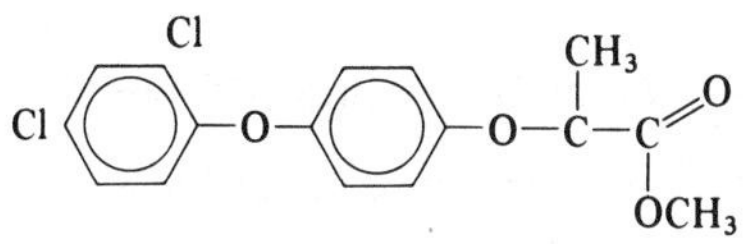 Cl, Cl, O, O—C—C(=O), CH_3, OCH_3	2,4 dichlorophenol, hydroquinone, 2-chloropropionic acid	
R 25788 Eradicane (comb. with Eptam)	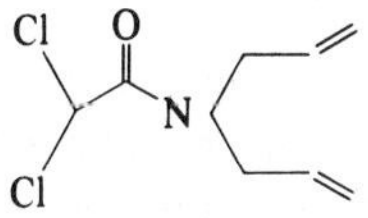 Cl, O, N, Cl	dichloroacetyl chloride, diallylamine	Herbicidal antidote for thiol carbamates
Protect	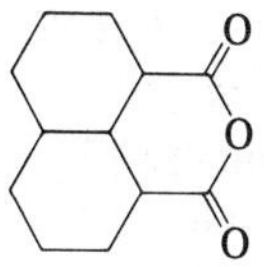 O, O, O	dimethyl naphthalene oxidation	Herbicidal antidote with thiol carbamates
NP 48 Alloxydin sodium	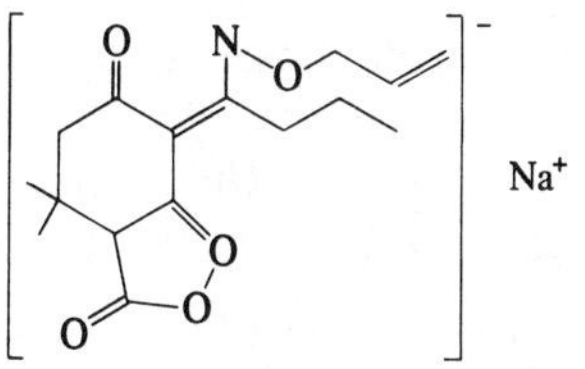 O, N, O, O, O, O, Na^+	dimedone acetyl chloride, allyl bromide, *N*-hydroxyphthalimide	Postemergence systemic
Alachlor Lasso	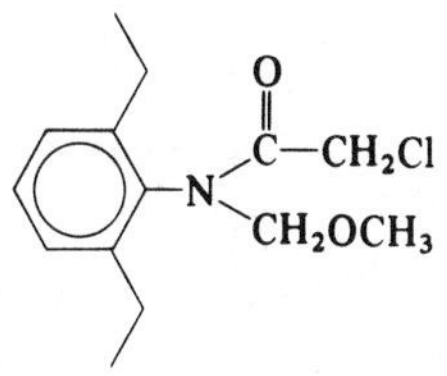 O, C—CH_2Cl, N, CH_2OCH_3	2,6 diethyl aniline, chloracetyl chloride, formaldehyde, methanol	Preemergence corn, soy, cotton
DPX 4189	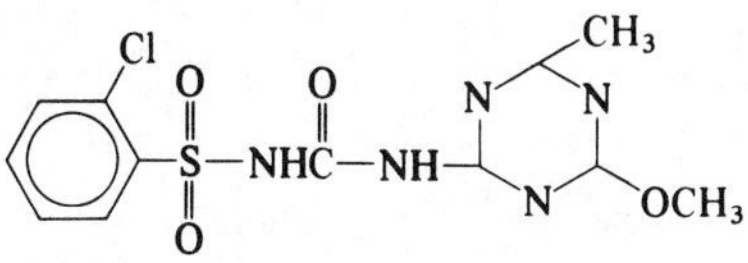 Cl, O, O, S—NHC—NH, O, N, N, N, CH_3, OCH_3	*o*-chlorobenzene, sulfonyl isocyanate, cyanuric chloride	Extremely potent sterilant. Pre and post activity at low doses

CHEMICAL PESTICIDE MANUFACTURE

The essential raw materials and intermediates and, by implication, the routes employed in the manufacture of pesticidal chemicals were given in the foregoing tables. There follows in this section a discussion of some chemical, economic, and engineering aspects of a number of typical pesticidal processes.

Cyclodiene Group of Insecticides

The Diels-Alder addition of dienophiles to hexachlorocyclopentadiene (C_5Cl_6) is the first step in the preparation of all compounds in this group of insecticides. Chlorination of C_5H_6 or other pentanes (e.g., in the presence of suitable catalysts) provides C_5Cl_6. One of the oldest and simplest examples is provided by the insecticide chlordane.

The equation illustrates the exothermic reaction which can be carried out neat or preferably with a low-boiling solvent. The second reaction is carried out with sufficient chlorine to provide a product containing 68-69 percent chlorine. The technical product is a dark brown oil. It is obviously a mixture of a number of discrete chemicals (which can be separated chromatographically), and each component possesses its individual biological activity. In the Diels-Alder addition, *exo* and *endo* configurations can result. Some of each of these configurations are under- and overchlorinated, hence the crude final product with 68 to 69 percent chlorine contains a large number of isomers and closely related compounds.

Chlorination by free radical mechanism (peroxides) yields the important insecticide heptachlor which is written below to portray spatial relationships. Such spatial differences as *exo* and *endo* attachments make for large differences in biological activity.

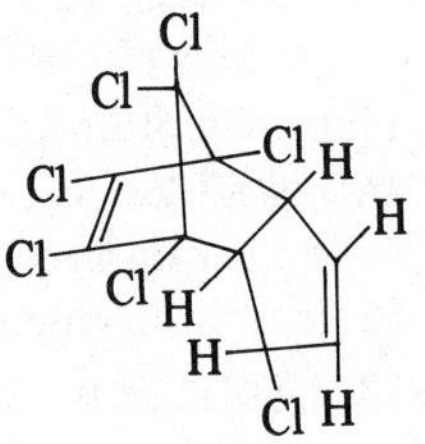

heptachlor

The preparative chemisty is very similar for aldrin, dieldrin, endrin, endosulfan, etc., the differences lying in the nature of the dienophiles and the subsequent chemistry and purifications.

Some Organophosphorous Pesticide Syntheses

A reaction distinctive to organophosphorous chemistry is the Arbusov reaction, which consists of the formation of a phosphonate by reaction of a phosphite with an organic halide. However, if a halogen is alpha to a carbonyl group, a rearrangement, known as the Perkow rearrangement, takes place to produce a vinyl phosphate. This reaction is the basis for a whole series of commercially significant organophosphate insecticides including dichlorvos, naled, mevinphos, phosphamidon, monocrotophos,* dicrotophos,* etc.

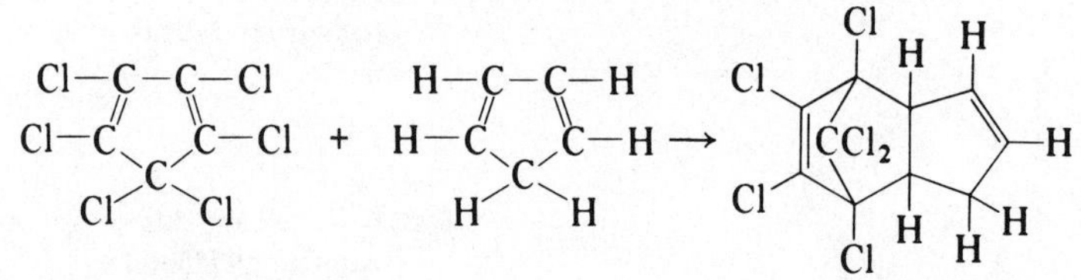

hexachlorocyclopentadiene + cyclopentadiene → chlordene (1)

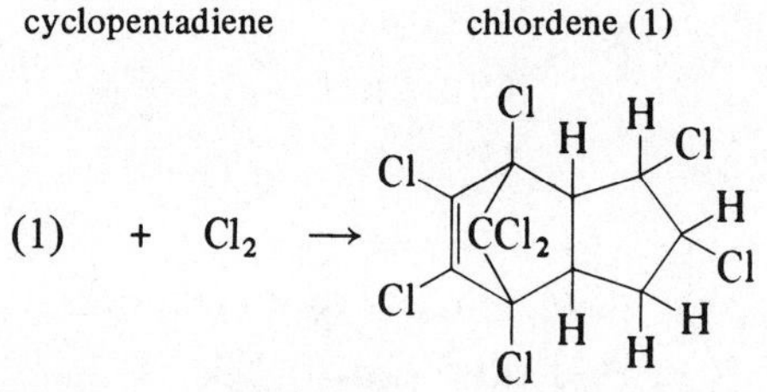

(1) + Cl_2 →

chlordene chlorine chlordane (a)

The synthesis of naled is illustrative of the route followed to obtain these compounds.

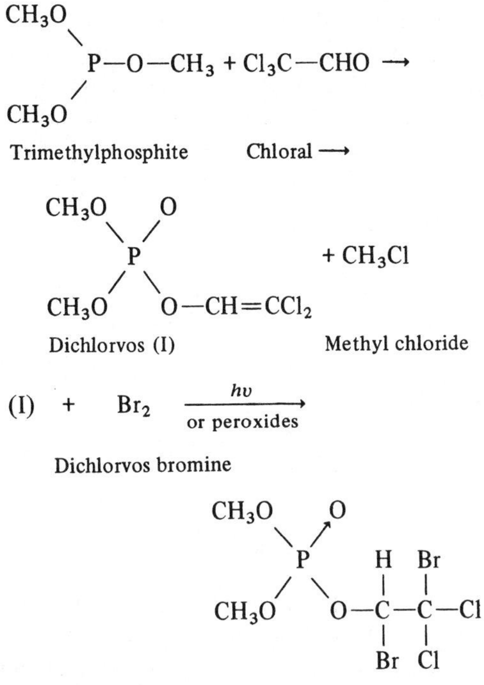

$$(CH_3O)_2P{-}O{-}CH_3 + Cl_3C{-}CHO \longrightarrow$$

Trimethylphosphite Chloral ⟶

$$(CH_3O)_2P(O){-}O{-}CH{=}CCl_2 + CH_3Cl$$

Dichlorvos (I) Methyl chloride

$$(I) + Br_2 \xrightarrow[\text{or peroxides}]{h\nu}$$

Dichlorvos bromine

$$(CH_3O)_2P(\rightarrow O){-}O{-}CH(Br){-}C(Br)(Cl){-}Cl$$

Naled

The Arbusov reaction combined with the Perkow rearrangement is generally an exothermic reaction, as in this case. Utilization of a low-boiling solvent is convenient. The bromination of the dichlorvos is carried out in the presence of light or peroxides to procure a fast rate of reaction with a minimum of destructive side reactions.

Another reaction frequently employed in organophosphate pesticide manufacture is the addition of $(RO)_2P(S)SH$ to an olefin or carbonyl, or its displacement of a reactive halide. $(CH_3O)_2$-PSSH is prepared by the reaction of methanol and P_4S_{10}.

An important commercial synthesis is the manufacture of malathion. By esterifying maleic anhydride with ethanol, the intermediate diethyl maleate may be prepared.

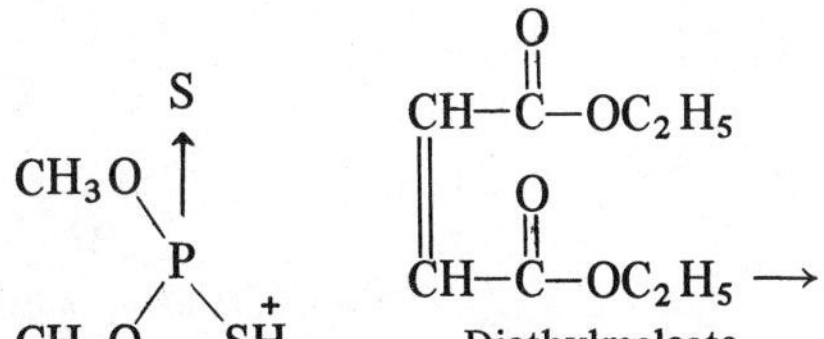

$$(CH_3O)_2P(\rightarrow S){-}\overset{+}{S}H \; + \; CH{-}C(=O){-}OC_2H_5 \,\|\, CH{-}C(=O){-}OC_2H_5 \longrightarrow$$

O, O-Dimethylphosphorodithioic acid Diethylmaleate

$$(CH_3O)_2P(\rightarrow S){-}S{-}CH(C(=O){-}OC_2H_5){-}CH_2{-}C(=O){-}OC_2H_5$$

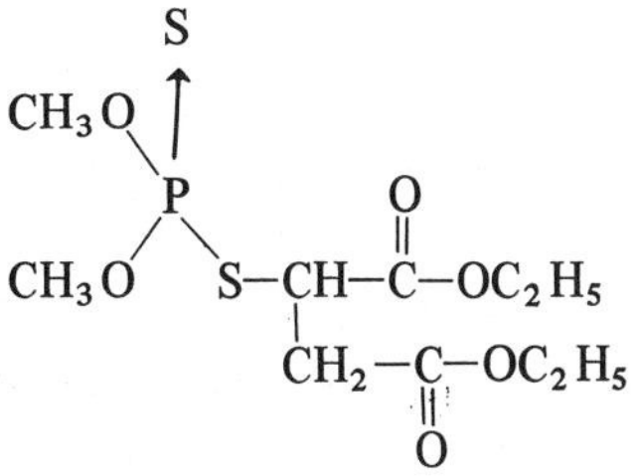

Malathion

Many commercial pesticides contain this dithioate group. Additional examples are: dimethoate, phosmet,* azinphos.*

Another frequently employed reaction is the displacement of a phosphorohalidate, usually a chloridate, by a nucleophilic reagent. A route to parathion is described. By reaction of P_4S_{10} with ethanol and subsequent chlorination the intermediate, $(C_2H_5O)_2P(S)Cl$ is formed; this is then reacted with sodium *p*-nitrophenate in a solvent which is later removed. Filtration and topping provide a crude insecticidal product that may be further purified.

Some Herbicide Manufacture

It was indicated that the triazine herbicides, including the very significant atrazine and simazine, are prepared by the appropriate nucleophilic attack of amines and other nucleophiles on cyanuric chloride. The art centers upon the means of controlling the rates of displacement of the first, second, and third chlorine atoms (see Table 21.14).

The equally important group of urea herbicides depend directly or indirectly upon phosgene chemistry, as does the preparation of the insecticidal carbamates and the herbicidal thiol carbamates. The treatment of *p*-chloroaniline (as its hydrochloride) with phosgene proceeds as follows:

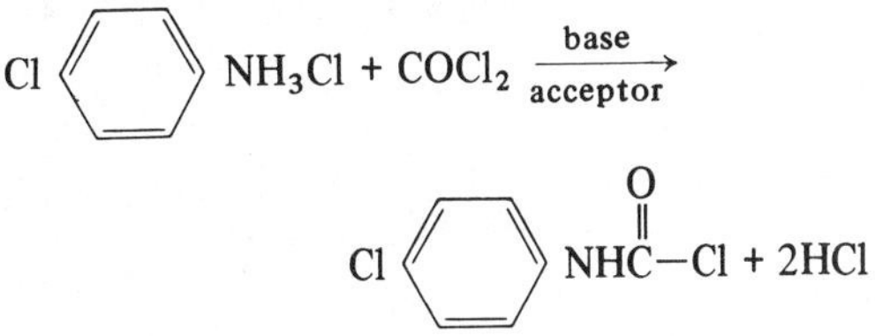

$$Cl{-}C_6H_4{-}NH_3Cl + COCl_2 \xrightarrow[\text{acceptor}]{\text{base}} Cl{-}C_6H_4{-}NHC(=O){-}Cl + 2HCl$$

*Space does not permit providing the structural formulae of all pesticides referred to in this and other sections. The reader is referred to the references in the bibliography for the given structures.

or

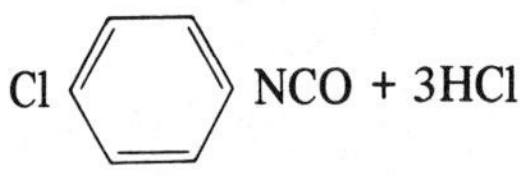

$$Cl-C_6H_4-NCO + 3HCl$$

Either the above isocyanate or the carbamoyl chloride is then reacted with an amine to provide the urea derivative.

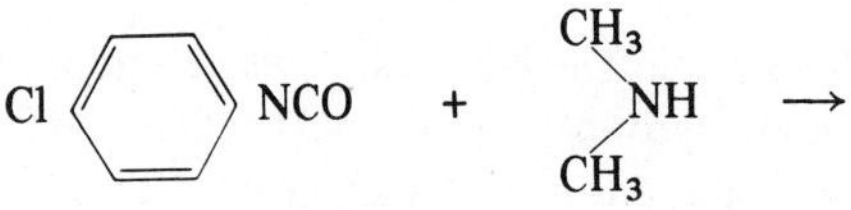

$$Cl-C_6H_4-NCO + (CH_3)_2NH \longrightarrow$$

p-Chlorophenyl isocyanate Dimethylamine

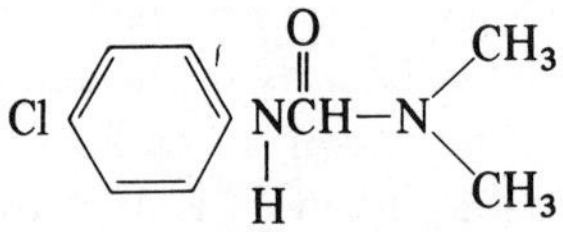

$$Cl-C_6H_4-NH-C(=O)-N(CH_3)_2$$

Monuron

These ureas are easily separated from the reaction mixtures as high-melting solids. The salt of the base-acceptor amine or alkali metal chloride can be removed by aqueous wash.

Insecticidal Carbamates

For the insecticidal carbamates, methyl isocyanate is the preferred intermediate. The corn root worm insecticide BUX is prepared by an interesting series of reactions.

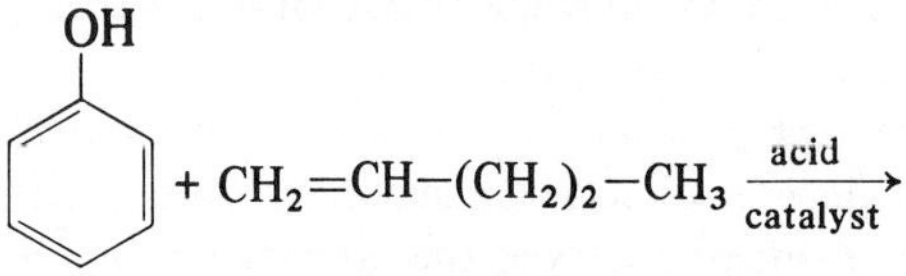

$$C_6H_5OH + CH_2{=}CH{-}(CH_2)_2{-}CH_3 \xrightarrow[\text{catalyst}]{\text{acid}}$$

Phenol 1-Pentene

$$HO-C_6H_4-CH(CH_3)-CH_2-CH_2-CH_3 + HO-C_6H_4-CH(CH_2CH_3)-CH_2-CH_3$$

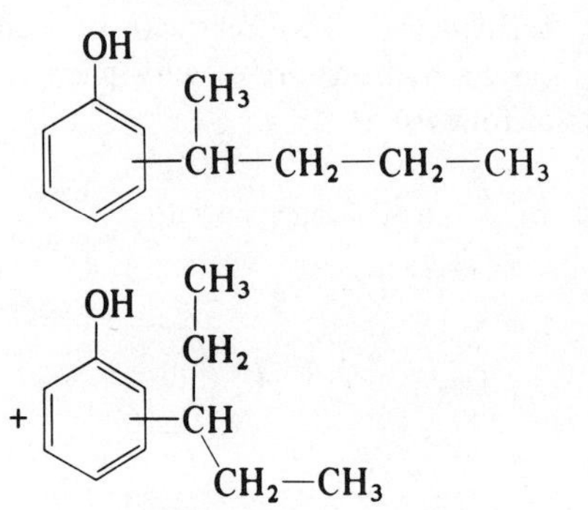

Mixed amylphenols, mostly ortho and para

This alkylation is performed at a low temperature using certain catalytically active clays. On elevating the temperature, an equilibrium mixture of amylphenols is formed in which *meta* isomers are predominant. This is a key step because the *o*- and *p*-amylphenols give rise to carbamates which are biologically less active.

The final reaction involves the condensation of the predominantly *m*-amylphenols with methyl isocyanate according to the equation:

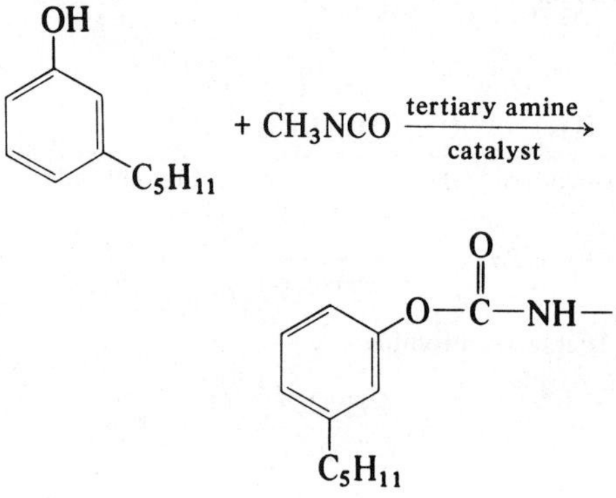

$$m\text{-}C_5H_{11}C_6H_4OH + CH_3NCO \xrightarrow[\text{catalyst}]{\text{tertiary amine}} m\text{-}C_5H_{11}C_6H_4-O-C(=O)-NH-CH_3$$

Because the alkylation step gives rise to both methylbutyl and the ethylpropyl isomers these species (both highly active carbamates) are found in the predominantly *meta* amylphenyl N-methylcarbamate which is BUX. Carbaryl is formed by the analogous reaction of alphanaphthol with methyl isocyanate.

Fungicide Manufacture

The preparation of captan begins with the Diels-Alder addition of butadiene and maleic anhydride to form *cis* - Δ^4- tetrahydrophthalic anhydride.

$$CH_2{=}CH-CH{=}CH_2 + \text{CH}{=}\text{CH}(\text{CO})_2\text{O} \longrightarrow$$

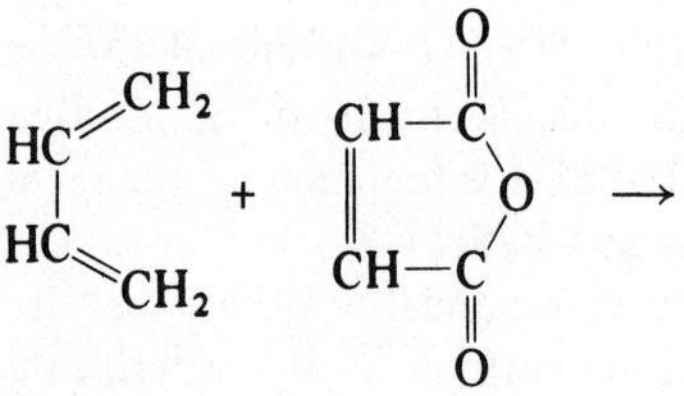

Butadiene Maleic anhydride

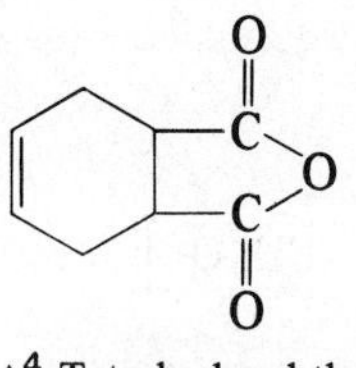

cis-Δ^4-Tetrahydrophthalic anhydride

The anhydride is then reacted with ammonia to produce the imide.

$$cis\text{-}\Delta^4\text{-Tetrahydrophthalic anhidride} + NH_3 \text{ (Ammonia)} \rightarrow cis\text{-}\Delta^4\text{-Tetrahydrophthalimide} + H_2O \text{ (Water)}$$

Carbon disulfide, which is derived from sulfur and hydrocarbons or carbon, is reacted with chlorine in the presence of catalytic quantities of iodine to provide, in an overall reaction, trichloromethanesulfenyl chloride (which is known by the common name perchloromethyl mercaptan) and sulfur monochloride.

$$\underset{\text{Carbon disulfide}}{2CS_2} + \underset{\text{Chlorine}}{5Cl_2} \xrightarrow{I_2} \underset{\text{perchloromethyl mercaptan}}{2CCl_3SCl} + \underset{\text{sulfur monochloride}}{S_2Cl_2}$$

Finally, the tetrahydrophthalimide is dissolved in cold caustic and reacted with perchloromethyl mercaptan to give captan.

$$\underset{\text{Tetrahydrophthalimide}}{C_8H_9NO_2} + \underset{\text{Perchloro methyl mercaptan}}{CCl_3SCl} + \underset{\text{Caustic}}{NaOH} \rightarrow \underset{\text{Captan}}{C_8H_8O_2NSCCl_3} + \underset{\text{salt}}{NaCl} + \underset{\text{water}}{H_2O}$$

Captan, a solid, is separated from the aqueous slurry by filtration, and air-dried. The low solubility of captan in water provides a finely divided powder which, after drying, is formulated most frequently as a wettable powder after some further grinding.

Some Special Considerations of Pesticide Manufacture

Each pesticide of course has its particular chemistry determined by the nature of the composition employed. The few examples above illustrate some simple manufacturing manipulations. Obviously, the manufacture of Decamethrin and other pyrethroids involves more complex chemistry including the separation of diastereoisomers and enantiomers. Space does not permit a thorough and critical review of the industrial manufacture of all the currently useful pesticides. However, a few general considerations particular to pesticide manufacture merit comment in this chapter.

Because of the corrosive nature of many pesticides, their intermediates, and their by-products, exotic alloys, stainless steel, monel, etc. are often employed as construction material for reactors, piping, columns, and so on. Much glass or glass-lined equipment is also used for the same reason, or because of lability of the chemical to traces of iron or other metals.

Many of the pesticides and intermediates are highly toxic to humans, and all of them, because of their biological activity, must be prevented from contaminating other manufacturing processes and the environment. In the production of the organophosphates, carbamates, and certain chlorinated hydrocarbons, both the intermediates and the final chemicals may be hazardous to man. Fume scrubbing, decontamination processes, and gas, liquid, and solid waste disposal require careful monitoring. This requires detection devices, automation of equipment, and reactors for the decontamination of by-products. Oxidation, hydrolysis, and reactions with nucleophilic reagents are all used for decontamination. The toxicity of pesticides to fish is often extremely high, and aqueous effluents are treated so that toxic components do not enter streams, rivers or bays. Experience has shown that *frequently 50 per-*

cent or more of the total capital costs of the chemical plants must be expended to protect human beings and the environment. This proportion of total capital expenditure is increasing. There is also a significant increment of the operating cost that relates to the problems of human and environmental hazard. The choice of plant site is also critical. There are an increasing number of examples in recent years where the combination of these factors (increased capital costs, increased operating costs, and plantsite location) has made a new pesticide-manufacturing project appear unfeasible. There is no doubt that the total recycling of by-products can be quite expensive, but that is the direction toward which pesticide manufacture is tending. The relatively rapid obsolescence of most pesticides and the increasing severity of regulations compounds the difficulties faced by the manufacturer.

FORMULATION OF PESTICIDES

The art and science employed in transformation of a manufactured chemical into a form that the farmer or the applicator can readily use in the field is of particular significance in the pesticide industry. This transformation process of chemical to agricultural product is the formulation of the pesticide.

The technical product, DDT, for example, is a waxy microcrystalline solid usually recovered in small lumps. The problem is to alter these lumps so that $\frac{1}{4}$ to 1 lb of product can effectively cover the foliage on an acre of farm land.

Formulations of DDT (or other pesticides) are usually of two major forms, solid or liquid. The solid formulations require aging, or hardening, of the microcrystals, followed by comminution and fracturing of the lumps into small particles–generally micron-sized (1-100μ) particles. This requires a large expenditure of energy and complex engineering equipment. Among the techniques employed are air grinding, and rarely, wet grinding.

The formulation usually requires inert additives such as clay, celite, talc, etc., to absorb sticky particles and help prevent adhesion in the final product. If the micronized solid is to be used with water–that is, if it is to be suspended in water and sprayed on the foliage as an aqueous suspension–then surfactants of various types must be added. Such a formulation of finely ground particles is called a wettable powder. Frequently a combination of surfactants is employed including those that wet the solid pesticide, maintain a stable suspension of particles in the water, and help in the adhesion of the suspended pesticide to the foliage surface. Hence the formulation art is replete with terms such as suspending agents, stickers, spreaders, and antiflocculants. Since pesticides are used on substances that will become either raw agricultural products or, ultimately, food products, the surfactants, additives, adjuvants, etc., must all be substances approved by the regulating agencies. In other words, the most efficient surfactant may or may not be acceptable for pesticide formulation. Toxicological and residue factors must be considered for the additives as well as for the pesticides themselves.

Liquid pesticides can also be transformed into wettable powders. This is generally done by dispersing the liquid on to fine particles of an absorbent clay, celite, etc. Again, surfactants and other adjuvants are added to provide the physical and chemical properties necessary for optimum dispersibility in water, wetting of foliage and insect, etc.

Although used less in recent years, "dust" formulations are still employed in significant amounts. Solids can be ground into fine particles, mixed with inerts, such as talc and clay, and sometimes with adjuvants, and dusted on the plant surface or the soil.

Many pesticides are liquids, and many of the solid ones are preferably employed as solutions. An emulsive concentrate is a formulation of a pesticide, which, when mixed with water, provides a dispersion or emulsion of a solvent containing the pesticide. Here again, choice of solvent and surfactant systems involves questions of art and knowledge of colloid and surface chemistry–as well as the toxicology and residue characteristics of the additives. In the formulation of petroleum oil sprays, heavy deposits with little runoff may be required. Formulation requires a compromise, then, between effective dispersion of the emulsive in the water during the spraying operation, and quick breaking of the emulsion upon the plant surface. In other cases, high surfactant activity is desired to perform the biological objective. Hence, for a single pesticide chemical the farmer may require many types of formulation. Each formulation requires a separate label approved by federal and state regulations.

Nonionic, anionic, and cationic surfactants are employed in pesticide formulations. Although oil in water emulsions are most commonly employed, i.e., the pesticide system is dispersed in a continuous aqueous phase, the reverse (called in invert emulsion) is also, but less frequently, used.

Another variation of liquid formulation of pesticides is called the "flowable." Certain plant parts are variably sensitive to organic solvents, particularly aromatics. The aliphatics are generally inferior solvents. Solid pesticides can be ground to micron-sized dimensions and suspended at high concentration in an aqueous medium. The adjuvants include surfactants, viscosity-controlling agents, etc., which enable the formulator to make a useful metastable flowable containing a minimum of water and a high concentration (frequently 50% or over) of solid toxicant. This "flowable" is then poured by the applicator in the field into his spray tank for dilution and direct application to the crop. There is of course an economic advantage deriving from the elimination of the solvent.

One of the newer formulation techniques developed during the sixties is the use of low-volume and ultralow-volume sprays, particularly of liquid formulations. These formulations, developed to reduce application costs, contain concentrated pesticides with little or no solvent or additives. Mechanical equipment is used to form liquid pesticide droplets of fairly uniform size (40 to 100μ). There is evidence to suggest that such formulations increase the biological effectiveness of insecticides, e.g., malathion and naled. The problem is one of providing uniform distribution; very fine, submicron droplets will drift past the objective, and large droplets are most inefficient, for many of them never reach the objective. Air is generally used for dispersion. Ground equipment, planes, and helicopters fitted with calibrated delivery systems are commonly used.

A variation of this technique, which is applicable to ground equipment, consists of the vaporization and rapid condensation of the pesticide solution to create a fog which is then carried by air currents to its objective. In this case, the insect respires submicron particles of the toxicant.

Another expanding type of formulation is the pesticide granule. One area of its usefulness is the control of disease vectors, particularly mosquito control, in swampy areas. Such targets may be practically inaccessible except by air. However, conventional formulations employing droplets or fine solids are of little use since they will deposit on the foliage surrounding the watery areas in which the mosquitoes breed. Granules of a solid (compacted clay, organic materials, etc.) impregnated with the toxicant can be dispersed by air; they will penetrate through the foliage, and many of them will drop into the surface waters. These granules can be made to release the toxicant at various rates. Granules of various types are now employed in rice culture and in soil-insecticide formulations in the corn belt. They are increasing as new applications are found.

There are other types of formulations employed. Almost all cotton, corn, and wheat seed, and many other crop seeds are treated with special preparations which generally contain pesticides to combat insect and disease attacks, but they may also contain growth stimulants, nutrients, and other components. Not only must the additives employed be effective against the intended target, but they must also be tolerated by the seed and the emerging seedling under a wide variety of external conditions including variations of soil type, temperature, pH, and moisture. Such seed must be easily distinguished from untreated seed so that the toxicants employed in the treatment are prevented from entering into the animal and human food chain. Legislation requires the coloring of the treated seed for ready identification. Such treated seed is proscribed from any but its intended use. Nevertheless, tragedies have occurred through human ingestion of seed containing organophosphates, mercurials, and other toxicants employed in the formulation of seed treaters.

Since labor in modern American agriculture provides an increasing increment of the total cost, the farmer tries to reduce the number of applications of the formulations employed. To accomplish this, he sometimes combines fertilizer and pesticide applications. This leads to other special formulation problems. Most pesticides are covalent bonded—nonionic organic entities—and the surfactants used with them function best in aqueous solutions of low ionic strength. Fertilizers, on the other hand, are generally water-soluble ionic compounds. The problem of compatibility must be solved, and novel surfactants that will function in solutions of relatively high ionic strength must be found for combined fertilizer and pesticide formulation to be successful.

Aerosol formulations, dilute pesticide formulations containing a volatile component (usually a fluoromethane, carbon dioxide, propane, or nitrous oxide) which propels the pesticide in fine droplets, are widely used in the home and garden.

No attempt is made here to cover the entire area of formulation, but rather to emphasize that formulation is frequently the key to successful application of the pesticide. The formulation process requires the application of physical-chemical principles, engineering skills and familiarity with the agricultural problem as a whole. Factors such as corrosion, solubility, surface activity, phytotoxicity, odor, vapor pressure, flammability, compatibility with additives and other formulations, stability, toxicity, and residue characteristics, are all part of this art and science.

RESEARCH IN THE PESTICIDE INDUSTRY; DIRECTIONS FOR THE FUTURE

During the past few years, certain chemical companies, some quite large, have re-evaluated their research and development programs relating to agriculture in the light of social, political, and economic trends and have either abandoned pesticide research or have contracted it markedly. The industry most certainly is in a state of transition, and it is interesting to consider some of the new directions being taken by research and development.

Resistance to Pesticides

The adaptation of organisms to the environment is a fundamental biological principle. Nevertheless, in the euphoria that followed the discovery of DDT and the early organic insecticides, insufficient consideration was given to the consequences of large scale and continued use of insecticides.

The process of natural selection takes many forms. Resistance is the common name applied to this biological response to chemotherapeutants. Resistance to chemicals that control insects, fungi, bacteria, weeds, rodents, etc., is now well acknowledged. No group of chemicals is immune to this biological adaptation. Resistance to both organic or inorganic compositions occurs; insect resistance to the arsenicals, the chlorinated hydrocarbons, to the organophosphates, the carbamates, the pyrethroids, etc., has been well documented. Resistance can involve mechanisms of detoxification in the environment or in the target organism, enzyme adaptations such as variations in the form of cholinesterase, physical factors such as penetrability, behavioral factors such as feeding adaptations, etc., or a combination of any or all of the above. Insects known to be resistant to chlorinated hydrocarbons or the organophosphates, for example, show differences in susceptibility to the pyrethroids as compared with their nonresistant close relatives, even though these insects have never previously been exposed to this new group of insecticides. This phenomenon is called cross resistance and is increasingly prevalent. Thus, the rate of development of resistance to a new chemical group of insecticides is also adversely affected by this so-called cross resistance.

Many of the pesticides listed in the various tables that precede this discussion have only limited usefulness presently because of resistance. We have included BUX in the table of carbamates only as an illustrative example because resistance in varous forms was responsible for its demise as a practical agent for the control of the corn root worm. Both the insect and the soil organisms partially responsible for its degradation all adapted to this chemical so that at present it has little or no practical usefulness in many parts of the U.S. corn belt.

In the fungicide area, resistance to varying degrees occurs primarily with the systemics, e.g., benomyl, carboxin, etc., and their homologs, the antibiotics polyoxins Kasugamycin, etc., whenever a given area is subjected to continuous high-level application of these chemicals.

Among the rodenticides, rats and mice, in certain parts of the world, have become highly resistant to the coumarins, the indanediones, etc.

In the herbicide field, resistance generally takes another form. For any given chemical, certain weeds are more tolerant than others. These tolerant species then become the prevalent weed pest, whereas they may have been only minor pests before the introduction of the pre- or postemergent herbicide. Nutsedge and Johnson grass are examples in certain areas of this type of 'resistance' or adaptation. Direct resistance of a plant species to a chemical is also possible.

Where the time for generation is brief, e.g., bacterial species, the probability for resistance to chemotherapeutants is great. For this reason, resistance is acquired earlier for insect species such as flies (about 3 weeks per generation) mosquitoes, mites, etc., than for those insect species that have but one generation per year.

The widespread prevalence of 'resistance' and the recognition of its ubiquity are major driving forces for the development of newer approaches to pest control.

Hormone Research

One of the recent, most exciting, academic triumphs has been in the understanding of the physiology, the isolation, and the synthesis of the moulting hormone Ecdysone and the juvenile hormone. Figure 21.3 gives their structures.

Most insects proceed through an elaborate metamorphosis from egg to larvae to pupa to adult, including less dramatic intermediate stages as well. The juvenile hormone, at a given titre in the insect at each stage, tends to maintain the insect at any given stage. Ecdysone, on the other hand, tends to promote moulting. Through a complex interaction of these and other endocrine secretions, the insect develops through its normal lifecycle. By interfering in the normal equilibrium, one can prevent insects from reaching the sexually mature adult stage or from producing fertile mature eggs that will develop into larvae. This is the strategy then of the hormone approach to insect control. A hormone differs from a conventional insecticide in that its aim is not to kill the insect directly but to interfere with its lifecycle.

Much attention centers about compounds that will simulate the juvenile hormone in biological activity. Both the juvenile hormone

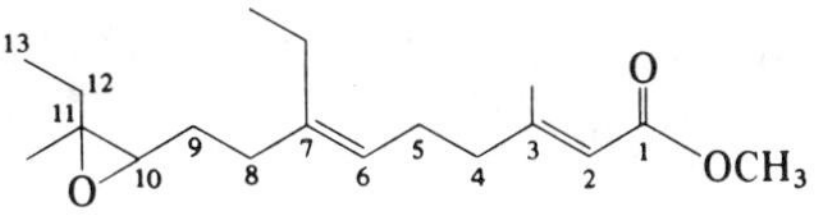

METHYL TRANS, TRANS, CIS-10-EPOXY-7-ETHYL-3,11-DIMETHYL-2,6-TRIDECADIENOATE (J.H.)

ECDYSON–MOULTING HORMONE

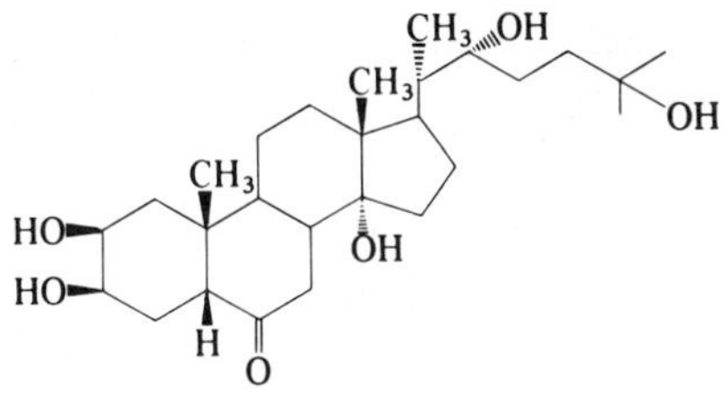

SEX PHEROMONE OF THE CABBAGE LOOPER TRICHOPLUSIA NI

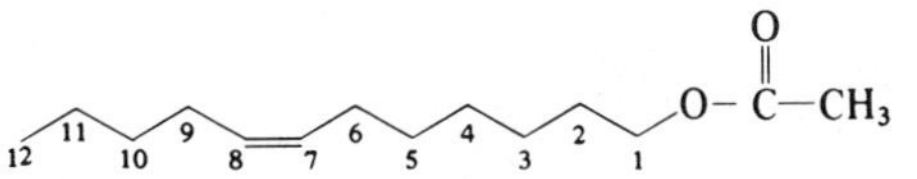

CIS-7-DODECENYLACETATE

Fig. 21.3 Juvenile Hormone (*J. H. Hyalophora Cecropia*), Ecdysone, and the Cabbage Looper sex hormone.

and Ecdysone appear to have little mammalian toxicity. At present, no successful large-scale field control of insects has been achieved, and there remain many biological, chemical, and economic problems that must be resolved.

Nevertheless, interesting applications of juvenile hormone mimics such as methoprene have been developed. These include mosquito control, particularly where the use of toxic pesticides must be avoided; 'feed-through' formulations for horn fly control in ruminants where the flies breeding in the manure fail to metamorphose into sexually mature adults; silk worm larval treatments where greater sized cocoons (more silk) are produced, the protection of stored tobacco from a variety of insect pests, and the control of certain fly species in mushroom culture.

One of the weakness in the J.H. or J.H. mimic approach to field insect control relates to the fact that the larval form is frequently the more voracious and destructive of the insect life stages. Chemicals that would induce premature pupation would conceivably be quite useful, and considerable effort has been expended toward the search for such substances. Indeed, some compositions are known, for example, precocene II, that induce premature metamorphosis, but as yet no useful agricultural application has been developed from such research. Investigation also is proceeding in relation to the hormonal control of other insect vital processes in the search for new vulnerable targets.

Nevertheless, a more easily synthesized, smaller molecule within the confines of agricultural economics that possesses one or all of the above physiological properties would indeed provide a breakthrough in insect control.

Pheromone Research

Another area of active investigation is in the use of insect signal agents (pheromones) often combined with traps, biocides, or sterilants. The chemicals responsible for the sexual stimulation of certain male or female insects have been isolated in a number of cases. These are usually fairly long-chain hydrocarbons, frequently olefinic and often with carbonyl or alcohol or ester terminal functions (Fig. 21.5). In addition to sex attractants, there are other types of signal agents. There are chemicals that simulate attractive food entities, danger, and various social insect functions. As with the hormone approach, there appears to be less ecological hazard with the use of pheromones. The strategy here is to attract the insect and then either trap, or by the use of chemicals in confined locations, sterilize or kill the specific insect attracted to the given pheromone. Some success has been achieved, usually in locations where migration of insects is difficult. As an example, an infestation of the oriental fruit fly on Rota (a 3-sq mile island) was eliminated by the combination of an attractant and the insecticide naled.

With the sex attractants, another approach called the 'confusion' technique is employed. The sex attractant saturates a given area and the opposite sex is confused by the ubiquity of the signal and fails to find its mate. In the American Southwest, this technique has been employed for the control of thc pink boll worm, (*Pectinophora gossypiella*). Both qualified successes and failures with this approach have been reported.

The main attractiveness of the pheromone approach is that the biocides need not be

TABLE 21.18 Synthetic J.H. and ant. J.H. Substances

Hormone	*Structure*	*Use*
Methoprene	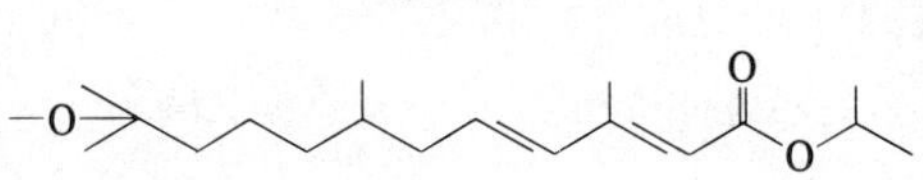 —O, O, O	See text
Precocene II	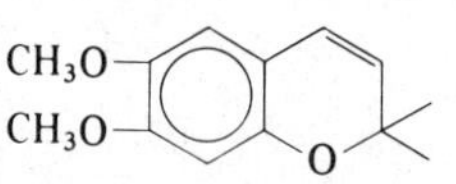 CH_3O, CH_3O, O	Laboratory use only

broadcast on the crop or soil. In present-day monoculture (i.e., where very large areas are mainly planted in one crop, such as the corn belt in the Midwest) the problem of using pheromones remains to be solved. Insect pheromones are highly specific. In agriculture, the farmer is generally confronted with a complex of insect pests—only rarely a single species. The pheromone approach then must be supplemented with other strategies in order to solve the real agricultural problem. Although the sex pheromones of many economically important insects have been isolated and their structures elaborated, recent research points to somewhat greater complexity relating to the signalling systems of the insects under study than was originally indicated. Nevertheless, the area of pheromone research provides a broad field for fundamental chemical and field biological exploration. Pheromones have found useful applications in the determination of the presence and populations of harmful insect species.

Repellents and Antifeeding Compounds

Metadiethyltoluamide

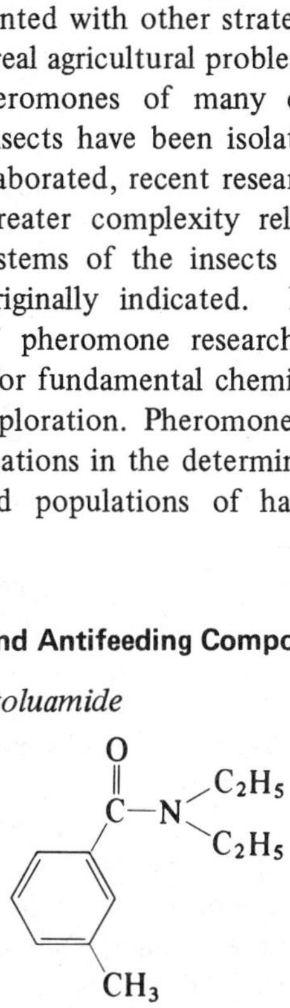

O ‖ C—N(C_2H_5)(C_2H_5), attached to a benzene ring bearing CH_3

is used as a mosquito repellent, but no repellents have been successfully employed in large-scale agriculture to repel field insects from their normal food supply. Many chemicals exhibit varying degrees of laboratory repellent activity.

Certain chemicals exhibit the property of discouraging insects from feeding without killing the undesirable species. No large scale use of such chemicals exists, although an increasing number of 'antifeedants' have been isolated and their structures elaborated.

Biological Control Methods

With the attention presently being given to ecological aspects of chemical pesticide control, increased exploration of biological methods for pest control appears timely and, indeed, various approaches are presently being explored.

One biological method involves the culturing of a bacterium. *Bacillus thuringiensis* effectively infects the larval stage of certain lepidoptera. Economic control has at times been successful with the use of preparations containing the spores of this bacillus, for example, against the larval stage of the cabbage looper, an insect that has developed resistance to many pesticides including many potent organic phosphates. *Bacillus thuringiensis* preparations are being sold commercially. As a result of successful genetic experiments, highly active strains are presently available.

Experiments are also proceeding with the cultivation of the polyhedrosis virus for control of lepidopterous larvae. Certain U.S. registrations have been granted for the application of this virus. Up to this time, there seems to be no evidence that these insect pathogenic bacteria or viruses produce effects other than those for which they were intended.

Another biological method that has been quite widely investigated and that has been utilized with varying degrees of success, is the sterile-male technique. Male insects are treated with a sterilizing agent, usually, but not exclusively, X-ray or ^{60}Co radiation. Then, large numbers of sterile males are introduced into the field. After mating, the females produce no offspring and the population of the next generation drops.

This method can be successful, but many problems arise, both mathematical and physiological, particularly where there are large populations of migrating adults. One as-yet-unsolved problem consists of the relative attractiveness and competitive vigor of the natural population of males as compared to the treated population. Under WHO auspices, the sterile-male technique has been employed against the tsetse fly (sleeping sickness) in Africa. In North America, the campaign against the screw worm (a cattle parasite) using this technique was largely successful. At present, this technique is best suited for government or international agency supervision where logistical and political considerations can be resolved.

Another aspect of biological control consists of the release of parasitic and predator insects. Such predator and parasitic species attack crop-destroying insects. Certain wasps, lace-wings, ladybugs, predator mites, as well as others, suppress, but do not eliminate the undesirable species. The practical problems, including the economics of such control, are currently being investigated by government agencies and by certain pest-control companies.

Great enthusiasm existed a few years ago for the development of chemicals which caused sterility in insects. Unfortunately, most of the active chemicals, though quite effective, were powerful alkylating agents, and as such exhibited carcinogenic potential. Some very simple phosphorus amides containing the ethyleneimine

$$\begin{array}{c} CH_2-CH_2 \\ \diagdown \;\; \diagup \\ N \\ | \end{array}$$

group were among the most active insect sterilants. $P(O)A_3$, and $P(S)A_3$, where A is the ethyleneimine group, are such agents. Even the solvent hexamethylphosphorictriamide $[[(CH_3)_2N]_3PO]$ possesses some sterilizing capacity. Newer compositions based upon hormonal control lead to sterility but have not as yet penetrated the agricultural market.

At present, an integrated approach (IPM) which will include some of the biochemical and biological methods as well as improved chemical agents in the more classic sense seems to be the direction in which current exploration is heading.

More Fundamental Investigation

Systemic agents for the control of plant disease are presently being studied and developed. These agents or their metabolic products provide a practical sort of plant immunity with little or no toxic residue on the plant surface. Some of these induce the elaboration of the plant's own defense mechanism, the phytoalexins. Progress has been achieved, and more is anticipated, in the genetic development of insect- and disease-resistant plant species.

The biochemistry of photosynthesis, of the important energy and synthesis pathways, of transpiration, of respiration, and of reproduction of plants is being studied both for fundamental knowledge and for new and better practical approaches to plant disease-control and growth-control agents, including selective herbicides. New chemicals for the control of insects are being developed which are more target-specific, and which can overcome prevailing resistance problems and yet still be non-hazardous to man and the environment.

Effort also is being expended to develop systems in agriculture that improve upon current practice. Examples include seed-treating methods which will permit fall planting, improvements in harvesting practices, and protection of harvested crops in storage. Many of these approaches employ chemicals in new and previously unforeseen ways.

Plant Molecular Biology

Molecular biology is currently the glamor science of the 1980's and is at a stage of development not too dissimilar to solid state physics and electronics some decades ago. It may provide markedly better ways for the genetic improvement of crop species. Many large agricultural and pharmaceutical companies are currently devoting considerable sums and much manpower to the exploration of molecular biological agricultural applications.

In a way, the chemistry will be done in the petri dish and the test tube rather than on the farm. The generation of whole plants from single cells (tissue culture techniques) provides massive opportunities to generate plants that not only provide higher crop yields but that also may be resistant to insects or fungi or bacteria or herbicides. Such plants could possess tolerance to salt and drought as well. The use of molecular engineering methods to provide chemicals, glycols, fructose, etc. has been projected. Bound enzymes from improved microorganisms may produce some of the intermediates used in pesticide and industrial manufacture.

None of what is written here is science fiction—neither have any commercially exploitable improved plant species or chemical processes yet resulted. The 80's will determine whether the visions outlined above are indeed achievable. Companies such as DuPont, Monsanto, Chevron,

Stauffer, Zoecon and others are involved in varying degrees in exploring the many opportunities. This is an area of research and aspiration not conceived when this same chapter was written a decade ago.

CONCLUSION

The pesticide industry is an industry that continues to be in transition. The interrelatedness of agricultural problems to environmental changes and to ecology in general has become recognized. With this recognition, there has come about the implementation of more restrictive legislation. Old and established methods are being challenged and new approaches are needed.

Agriculture in the U.S. has become energy rather than labor intensive. Only 4 to 5 percent of the total U.S. population is directly involved in the production of food, fiber, and vegetable oils. Yet agricultural products are by far the single most significant factor in U.S. exports.

It is the conviction of the author that chemicals can be developed that are effective and that possess minimal ecological hazard and that are safe to man and to animal species. These combined with some of the newer approaches will lead to integrated campaigns that can maintain an adequate food supply and an improved public health. The optimum training for such a career must include specialization in chemistry and engineering, but be broader so that biological and human consequences are fully appreciated. The methods and tools employed in 1990 will not be the same as those used in 1980. They will be better and will reflect a better understanding of man's relationship to the world in which he lives.

REFERENCES

General*

*General references relating to structure and/or mode of action, biological, chemical properties, etc.

1. Audus, L. J., Ed., *The Physiology and Biochemistry of Herbicides*, New York, Academic Press, 1964.
2. Ashton, Floyd M., *Mode of Action of Herbicides*, Wiley, N.Y., 1973.
3. Crafts, A. S., *The Chemistry and Mode of Action of Herbicides*, New York, John Wiley and Sons, 1961.
4. Frear, D. H., *Pesticide Index*, State College, Pa., College Station Pub., 1968.
5. Horsfall, J. C., "Principles of Fungicidal Action," *Chronica Botanica*, Waltham, Mass., 1956.
6. Kearney, Philip C., Eds., *Herbicides. Chemistry, Degradation and Mode of Action.* Marcel Dekker, N.Y. 1975.
7. Kohn, G. K., Ospenson, J. N., and Gardner, I. R., "Synthetic Pesticides from Petroleum," in *Advances in Petroleum Chemistry and Refining*, Vol. **VII**, p. 323–363, New York, John Wiley & Sons, 1963.
8. Metcalf, R. L. *Organic Insecticides*, New York, John Wiley & Sons, 1955.
9. Metcalf, R. L., Ed., *Advances in Pest Control Research*, Volumes **I–IX**, New York, John Wiley & Sons, 1957–196.
10. Siegel, Malcolm R., and Sisler, Hugh D., (ed.), *Antifungal Compounds* 2nd vol. Marcel Dekker, N.Y. 1977.
11. Silk, J. A. and Brown, S., *Crop Protection Chemicals Index*, 5th Ed., Plant Protection, Limited, ICI, Jealott's Hill Research Station, Bracknell, Barkshire, 1969.
12. Spencer, E. Y., "Guide to Chemicals Used in Crop Protection," Publication 1083, Canada Dept. of Agriculture, Ottawa, 1968.
13. Torgeson, D. C., Ed., *Fungicides–An Advanced Treatise*, Vols. **I** and **II**, New York, Academic Press, 1967 and 1969.

Specific

1. Hamner, C. L. and Tukey, H. B., *Science*, **100**, 154, 1944; Jones, F. D., U.S. Patent 2,390,941 (1945).
2. Mueller, P., U.S. Patent 2,329,074 (1943); Lauger, P., Martin, H., and Mueller, P., *Helv. Chim. Acta* **27**, 892 (1944).
3. Kittleson, A. R., *et al.*, U.S.P. 2,553,770–7.
4. Cupery, H. E., Searle, N. E., and Todd, C. W., U.S. Patent 2,705,195.
5. Lambrech, J. A., U.S. Patents 2,903,478 and 3,009,855.

6. Schrader, G., U.S. Patents 2,597,534 and 2,571,989.
7. Hyman, J., U.S. Patent 2,519,190.
8. Kohn, G. K., Ospenson, J. N., Moore, J. E., *J. Agri. Food Chem.* 13, 232 (1965).

Statistical*

*Statistical data except for Fig. 1 was derived from references in this section.

1. U.S. Tariff Commission, "Synthetic Organic Chemicals, U.S. Production and Sales" (yearly). U.S. Government Printing Office, Washington, D.C.
2. *Chemical Economics Handbook*, Stanford Research Institute, Palo Alto, Cal.

Journals*

*Contents of these journals devoted largely to matters pertinent to the pesticide industry were used in the preparation of this chapter.

1. *J. Food Agri. Chem.* **1-28** (ACS Publication).
2. Gunther, F. A. and Gunther, J. D., *Residue Reviews* **1-73**. (Springer-Verlag, New York.)
3. *Pesticide Science* Vol. **1** thru **11** Society of Chemical Industry, Blackwell Scientific Publications, Oxford, Great Britain.
4. *Weed Research* Vol. **1** thru **20** Society of Chemical Industry, Blackwell Scientific Publications, Oxford, Great Britain.
5. *Journal of Pesticide Science* Vol. **1** thru **5** Pesticide Science–Society of Japan, Tokyo, Japan.

22

Pigments, Paints, Varnishes, Lacquers, and Printing Inks

Charles R. Martens*

PIGMENTS

Pigments are used in paints, plastics, rubber, textiles, inks, and other materials to impart color, opaqueness, and other desirable properties to the product. In paints, pigments may also adjust the gloss, impart anticorrosive properties, and reinforce the film. Extender pigments have neither color nor hiding power under most circumstances, but they impart other desirable properties to the product at low cost.

Pigments are insoluble powders of very fine particle size, i.e., as small as 0.01 micron and usually no larger than one micron. They are both natural and synthetic in origin, and organic and inorganic in composition.

Hiding power, or the ability of paint to obscure underlying color, varies with different pigments. In general, dark pigments, since they are more opaque, are more effective than light pigments in this respect. The difference between the index of refraction of the vehicle and that of the pigment largely determines the hiding power of paint, i.e., the greater the different, the greater the hiding power. For example, white lead has a refractive index of 1.59, zinc oxide, 2.00, and titanium dioxide, 2.70. Most nonvolatile vehicles and resins have a refractive index of from 1.47 to 1.52 (see Table 22.1). Thus titanium dioxide is the most effective pigment for hiding power. The hiding power of pigments is also useful in minimizing the damaging effects of sunlight to the coating and the substrate.

Another factor affecting the hiding power is the particle size of the pigment. Within limits, the finer the pigments, the greater the hiding power. The optimum particle size for pigments to give maximum light scattering and hiding is approximately one-half the wave length of light in air or 0.2–0.4 μ. Below this size, the particle loses scattering power, above this size the number of interfaces in a given weight of pigment decreases.

Fading and change of color in paints results largely from instability of the pigment. Pigment stability is particularly important for paints that are designed for exposure to sunlight and industrial fumes. Lead pigments should not be used in many industrial areas as they darken in the presence of sulfide fumes.

*Consultant, Cleveland, O.

TABLE 22.1 Indices of Refraction of Some Common Paint Materials

Material	*Refractive Index*
Rutile titanium dioxide	2.76
Anatase titanium dioxide	2.55
Zinc sulfide	2.37
Antimony oxide	2.09
Zinc oxide	2.02
Basic lead carbonate	2.00
Basic lead sulfate	1.93
Barytes	1.64
Calcium sulfate (anhydrite)	1.59
Magnesium silicate	1.59
Calcium carbonate	1.57
China clay	1.56
Silica	1.55
Phenolic resins	1.55-1.68
Melamine resins	1.55-1.68
Urea-formaldehyde resins	1.55-1.60
Alkyd resins	1.50-1.60
Natural resins	1.50-1.55
China wood oil	1.52
Linseed oil	1.48
Soya bean oil	1.48

Fig. 22.1 Bridge painted with aluminum paint. (*Courtesy Sherwin-Williams Co.*)

An ideal pigment should be chemically inert, free of soluble salts, insoluble in all media used, and unaffected by normal temperatures. It should be easily dispersed, nontoxic, and have low oil-absorption characteristics. The trend over the last few years has been toward easy dispersing pigments for use in high-speed dispersing equipment. These pigments, because of special processing, contain very few agglomerates. Pigments can be classified as follows:

Inorganic	*Organic*
White	Colors
Extender	Black
Colors	
Black	
Metallic	

Inorganic Pigments

All white pigments are inorganic compounds of titanium, zinc, lead, or antimony. They are classified as nonreactive and reactive, depending on whether they react with the vehicle, which may be acidic.

The most important of the nonreactive white pigments is titanium dioxide. Because of its high refractive index (2.76), it gives the greatest amount of hiding per dollar (see Table 22.2). Titanium dioxide is available in two forms, depending on the crystal structure—rutile and anatase. Anatase is considered to be a chalking-type pigment.

Two other nonreactive pigments are zinc sulfide and lithopone. Lithopone is a composite pigment of zinc sulfide coprecipitated upon calcium sulfate or barium sulfate crystals.

The reactive white pigments are basic carbonate white lead, basic sulfate white lead, basic silicate white lead, dibasic lead phosphite, zinc oxide, leaded zinc oxide. and antimony oxide. Reactive pigments serve several functions in coatings; for example, zinc oxide, being basic in nature, readily forms zinc soaps with the acid constituents of the paint vehicle. This reactivity is utilized by the paint chemist to increase the hardening of the paint film, to increase the consistency of the fluid paint, to aid in mixing and grinding the pigments because of better wetting, to reduce after-yellowing, and to improve the

TABLE 22.2 Tinting Strength and Hiding Power of White Pigments

Pigment	*Tinting *Strength*	*Hiding Power (sq ft/lb)*
Rutile titanium dioxide (PSC)	1850	157
Rutile titanium dioxide (conventional)	1750	147
Anatase titanium dioxide	1250	115
50 percent Rutile calcium-base	880	82
Zinc sulfide	640	58
30 percent Rutile calcium-base	600	57
Lithopone	280	27
Antimony oxide	300	22
Dibasic lead phosphite	250	20
Zinc oxide	210	20
35 percent Leaded zinc oxide	175	20
Basic carbonate	160	18
Basic sulfate white lead	120	14
Basic silicate white lead	80	12

[a]Tinting strength is the relative capacity of a pigment to impart color to a white base.

self-cleaning and mildew-resistance of exterior house paints.

Antimony oxide is used because of its fire-retardant properties.

White extender pigments have several functions, such as controlling gloss, texture, suspension, and viscosity. Extenders have low refractive indices, i.e., 1.40 to 1.65. Common extenders are whiting (calcium carbonate), talc (magnesium silicate), clay (aluminum silicate), calcium sulfate, barytes (barium sulfate), silica, and mica. Titanium dioxide and some extender pigments are available in water slurry form for use in waterborne paints. The pigment slurry is delivered to the plant in tank cars or tank trucks, pumped to storage tanks and then to the paint production equipment. The advantages of slurries are that production is speeded up and labor costs are reduced. Slurries can be pumped into the batch much faster than can be done by adding bags of pigments to the dispersing tanks. Titanium dioxide slurries contain 64.5 to 76.5% by weight of pigment in water. Particle size and shape are the most significant properties of extenders.

Colored inorganic pigments are both natural and synthetic in origin. These are:

Iron oxides—from both mineral and synthetic sources including Spanish oxide, Persian Gulf oxide, and domestic iron oxides; and siennas, ochers, umbers, and black iron oxide.

Lead chromate yellows and oranges—$PbCrO_4$ and modifications.

Molybdate oranges and reds—lead chromate modified with $PbMoO_4$.

Zinc chromates—approximate composition: $4(ZnO \cdot K_2O) \cdot 4(CrO_3) \cdot 3(H_2O)$.

Red lead—Pb_2O_3.

Cadmium colors—yellow, orange, and red—mixtures of cadmium and zinc sulfide, cadmium sulfoselenides, and cadmium selenides.

Iron blues—potassium, sodium or ammonia coordination compounds of ferriferroxyanide.

Ultramarine blues—formula unknown, made from china clay, sodium carbonate, silica, sulfur, and a reducing agent.

chrome greens—blends of chrome yellows and iron blue.

Chrome oxide green—Cr_2O_3.

Chrome oxide green—$Cr_2O_3 \cdot 2H_2O$.

Nickel titanate—oxides of titanium, nickel and antimony.

Organic Pigments

The difference between dyes and pigments is their relative solubility; dyes are soluble, while pigments are essentially insoluble in the liquid media in which they are dispersed. In the manufacture of organic pigments, certain coloring materials become insoluble in the pure form, whereas others require a metal or an inorganic base to precipitate them. The coloring materials which are insoluble in the pure form are known as "toner pigments" and those which require a base are referred to as "lakes."

Organic pigments in general have lower hiding power but greater tinting strength than inorganic pigments. Although there are a great many organic pigments available, they may be classified into about six groups based on some general characteristic of their chemical composition. These are:

1. *Azo insoluble*—toluidine, *para*-chlorinated nitroanalines, naphthol reds, Hansa, benzidine,

dinitroanaline orange; these are all members of the azo dyestuff family, which is insoluble in water.

2. *Acid-azo*–lithol, lithol rubine, BON colors, red lake C, Persian orange, tartrazine; these are all acid-azo pigments obtained from dyestuffs which contain acid groups ($-SO_3H$, $-COOH$) in their structure, and they are insolubilized by reaction with sodium, barium, calcium, or strontium.

3. *Anthraquinone*–alizarine, madder lake, indathrene, vat colors.

4. *Indigoid*–indigo blue and maroons.

5. *Phthalocyanine*–phthalocyanine green and blue.

6. *Basic PMA*, *PTA*–PMA and PTA toners and lakes, rhodamine, malachite green methyl violet, Victoria blue.

While normally not classified as such, carbon blacks are organic in nature. The carbon blacks are channel, furnace, and lamp blacks, and graphite.

Pigment Manufacture

Titanium Dioxide. Titanium dioxide pigments are manufactured by two processes, viz., the sulfate process and the chloride process. The older process is the sulfate method, which uses ilmenite ore. The ore, previously dried and ground, is digested with sulfuric acid (85-90 percent). After digestion and solvation, the iron is reduced to the ferrous state and much of it is removed by crystallization as ferrous sulfate. The liquid is filtered, concentrated in vacuum evaporators, and boiled with sulfuric acid to precipitate the titanium dioxide pigment, which is washed, dried, and calcined at a temperature of about 1650°F. The calcining operation converts the titanium dioxide from the amorphous to the crystalline state, thereby raising the refractive index. Controlled grinding and bagging follow. After the purification steps, the process varies depending on the grade and type of product desired.

The chloride process produces pigment by the oxidation of titanium tetrachloride, itself obtained from the mineral rutile by chlorination in the presence of carbon. The process is attractive because the titanium tetrachloride is a definite chemical compound which may be produced in a high degree of purity. All of the newer titanium dioxide plants use this process.

"Extended" titanium pigments are prepared either by mixing or coprecipitating TiO_2 with cheaper pigments of low hiding power. Titanium calcium may be prepared by two methods: (1) by precipitating hydrated TiO_2 in the presence of $CaSO_4$, the coprecipitate being filtered, washed, calcined, and dry ground; or (2) TiO_2

Fig. 22.2 Florida exposure for paints. (*Courtesy Sherwin-Williams Co.*)

and $CaSO_4$ may be mixed as a wet slurry, filtered, dried, calcined, and dryground. The composite pigment contains 30 percent TiO_2 and 70 percent $CaSO_4$ and has much better hiding power than would be obtained from a simple dry mix of the two components in the same proportion.

Titanium dioxide paints which have controlled chalking are extensively used for house paints; they stay white longer because of the gradual erosion of the soiled surface. A large proportion of anatase, sometimes combined with a small amount of an oxide of antimony or aluminum, is used for this purpose.

Zinc Oxide. Zinc oxide is made in several ways. In one of these, the ore (frankenite) is mixed with coal and burned on a grate. The natural oxide is first reduced, and then reformed by the air and carbon dioxide from combustion.

Lithopone. Lithopone is formed when a solution of zinc sulfate is mixed with one of barium sulfide. Barium sulfate and zinc sulfide are formed, both of which are white. The precipitate is not suitable for use as a pigment, however, until it has been dried, heated to a high temperature, and then plunged, while still hot, into cold water. Lithopone is 30 percent zinc sulfide and 70 percent barium sulfate, with slight variations.

Red Lead. Red lead is made by calcining litharge in a muffle furnace. A current of air is admitted into the muffle, the temperature is maintained within narrow limits near 640°F, and the time required is usually 48 hours.

Carbon Black. There are several kinds of carbon black: *thermal black* produced by the thermal decomposition of natural gas; *channel black* produced by the impingement of numerous small regulated flames against a relatively cold steel surface which is constantly scraped free of the soot deposit; *furnace black* produced by the partial combustion of the gas in the furnace with recovery of the carbon product in cyclones and electrical precipitators; and *lampblack*, used mainly as a tinctorial pigment.

Iron Oxide. Iron oxide (Fe_2O_3) is made on a large scale by roasting ferrous sulfate obtained from the vats used for pickling steel. Water and sulfur oxides are driven off and led through a stack to the atmosphere. The shade may be varied by altering the firing time, the temperature, and the atmosphere. It is a relatively cheap pigment and is usually used in red barn paint and metal primers. The use of selected grades for polishing glass and lenses is determined by their resistance to grit and the hardness of the glass; such grades of iron oxides are called *rouges.*

Metallic Pigments. Although gold, zinc, and copper bronze powders are used as pigments, powdered aluminum is the most important. In addition to its principal use in organic coatings it is also used in solid rocket propellants and as a filler for thermosetting resin systems. As a leafing pigment it reflects sunlight, thus preventing degradation of the organic film. As a nonleafing pigment it is used in automobile finishes to provide a metallic sparkle.

Aluminum powder is prepared in a stamping mill. Then aluminum sheets, usually mixed with a small amount of lubricant (e.g., stearic acid), are pounded into a powder, which is then screened and polished. Much of the powder is converted to a paste by grinding in a ball mill with mineral spirits.

PAINTS

Paint is a substance composed of solid coloring matter suspended in a liquid medium and applied as a coating to various types of surfaces.

The purpose of the coating may be decorative, protective, or functional. Decorative effcts may be produced by color, gloss, or texture. A secondary decorative function is lighting, as the color of the surface affects the reflectance. The proportion of light reflected by a surface is expressed as a percentage of complete reflectance. The following are approximate reflectance values for various colors:

White	90-80%
Very light tints	80-70%
Light tints	70-60%
Medium to dark tints	60-20%
Deep colors	20-3%
Black	2-1%
Aluminum	45-35%

Because there is a multiplie reflection of light

TABLE 22.3 Pigment Production in USA in 1977*

	In millions	
	Quantity lbs.	*Value $*
Inorganic Pigments	–	960.2
Total	1510.8	627.1
Titanium Dioxide	20.8	9.4
White Lead	20.8	9.4
Zinc Oxide	396.2	148.8
Chrome Colors	131.2	104.7
Iron Oxides	214.8	84.9
Red Lead	18.6	7.8
Carbon Blacks	12.0	8.6
Cadmium sulfide	5.0	9.8

*1977 Census of Manufacturers.

from one surface to another, the effect on room lighting is more pronounced than indicated by the above figures. By the proper selection of colors on ceiling and walls, the amount of lighting can be reduced, thereby, saving energy. The protective coating may be the paint on a wooden boat which serves as a barrier against moisture and prevents rotting; the interior lining of metal cans or drums which prevents corrosion from foods or chemicals; the coating on electrical parts to exclude moisture; the fire-retardant paint which protects combustible surfaces; the coating on plaster or concrete which makes for ease of cleaning, etc. An example of the functional use of paint would be as a traffic paint which marks the center and edge of a road for safer driving.

Paints are classified by gloss. Gloss as applied to a paint film may be defined as luster, shininess or reflecting ability of a surface. The gloss of a paint is controlled by the amount of pigment or extender pigment. The greater the pigment content the lower the gloss.

Gloss Classifications

Gloss	
70	High or full gloss
70–30	Semi gloss or medium gloss
30–6	Eggshell
6–2	Flat to eggshell (intermediate)
2	Flat

There were between 1050 and 1150 million gallons of paint produced in the United States in the year 1979, with a value of about seven billion dollars.

The paint industry serves two distinct types of markets, trade sales (shelf goods) and industrial sales (chemical coatings).

Trade sales is the large consumer-oriented portion of the business. Trade-sale paints are house paints and other products marketed through wholesale and retail channels to the general public and professional painters for use on new construction or for the maintenance of old buildings. Also included in this category are the paints sold to garages and repair shops for automobile refinishing, marine finishes for boats, paints for graphic arts such as signs, paints for refinishing machinery and equipment, and paints sold to government agencies, especially the traffic paints used for road-marking.

Industrial sales comprise the coatings sold directly to the manufacturer for factory application. The products on which paints are applied include durable goods such as automobiles, appliances, and house sidings, and nondurable goods such as cans for food and beverages.

All coatings contain a resinous or resin-forming constituent called the *binder.* This can be a liquid such as a drying oil, or a resin syrup that can be converted to a solid gel by chemical reaction. In some instances, where the binder is either a solid or is too viscous to be applied as a fluid film, a volatile solvent or *thinner* is also added. This evaporates after a film is deposited, causing solidification of the film. The binder plus solvent is known as the *vehicle.* Most paints also contain *pigments* which are described in the first part of this chapter.

In addition to pigment, binders, and thinners, a paint may contain many additives, such as defoamers, thickeners, flow agents, and driers to improve specific properties.

Paints can be classified according to the binder used, e.g., alkyds, vinyls, and epoxies. Paints are also classified according to their properties or end use. Alkyd enamels, for instance, are gloss paints with good abrasion resistance and good cleanability, while alkyd flat wall paints are characterized by very low sheen and good film build.

Paints applied directly to the surface are

called undercoaters or primers. Primers are used to aid the adhesion of the top coat to a surface and to prevent absorption of the top coat into a porous surface. Primers can also be used to prevent corrosion of a metal surface. Fillers or surfacers are types of primers that fill scratches and surface imperfections to give a smooth surface.

Paints used as the final coat are referred to as finish coats or top coats. Some top coats are self-priming.

Binders

The protective properties of a coating are determined primarily by the binder. In the early days of paint technology, binders were limited to materials of natural origin, such as drying oils, congo resins, and asphalts. These still find some usage in the protective-coating industry. However, the chemical and plastic industries during the past sixty years have supplied a large number of synthetic binders which make possible paints with greatly improved protective and decorative properties.

Coatings in liquid form can be divided into two types, *solutions* and *dispersions.* In solution systems the resin binder is dissolved in a solvent. In dispersion systems the resin is in the form of tiny spheres (usually 10 microns or less in size) suspended in a volatile liquid carrier. If the liquid in a dispersion system is water, the system is called an emulsion; if it is an organic material, it is called an organosol. When the liquid evaporates, a mixture of soft resin and pigment is left behind and fuses into a continuous film.

The general types of resins used are listed below. Mixtures of two or more types may sometimes be used to improve certain properties.

Oils—easy application; soluble in aliphatic solvents.

Alkyds—all purpose; combined with other resins; most are soluble in aliphatic solvents.

Cellulosics—(nitrate and acetate)—used in lacquers; fast dry.

Acrylics—good color and durability.

Vinyls—good durability; abrasion resistant.

Phenolics—good chemical resistance; yellow color.

Epoxies—good chemical resistance.

Polyurethanes—good flexibility; abrasion resistance.

Silicones—good heat resistance.

Amino Resins (ureas and melamines)—blended with alkyds for baking finishes; tough, good color.

*Styrene-butadiene**—low cost, alkali resistant.

*Polyvinyl acetates**—low cost; good color retention.

*Acrylics**—good color and durability.

Solvents

The type of binder determines the type of solvent or thinner used in a paint formulation. The preferred type of thinner is an odorless aliphatic hydrocarbon which can be used in all areas including the home. Unfortunately, aliphatic thinners do not dissolve all resins. In such cases, strong solvents such as aromatic hydrocarbons, esters, and ketones are used.

Solvents are generally classified as low boilers, medium boilers, and high boilers depending upon the evaporation rate. Various ranges are required depending on the specific application. Examples of solvents classified by their chemical composition are as follows:

Hydrocarbons, aliphatic-VM&P naphtha and mineral spirits; and aromatics—benzene, toluene, and xylene.

Alcohols—methyl alcohol, ethyl alcohol, and butyl alcohol.

Ethers—dimethyl ether and ethylene glycol monoethyl ether.

Ketones—acetone, methyl ethyl ketone, and methyl isobutyl ketone.

Esters—ethyl acetate, butyl acetate, and butyl lactate.

Chlorinated—tetrachlorethane.

Nitrated—nitromethane, nitroethane, and 1-nitropropane.

Latex or emulsion paints use water as the volatile component so that they may be used in the home.

---

*Latex form.

Fig. 22.3 Reactor for manufacturing alkyresins. (*Courtesy Sherwin-Williams Co.*)

General Properties of the Various Types of Paints

Acrylics. Acrylic resins are used in protective and decorative lacquers for paper, fabrics, leather, plastics, wood, and metal. White baking enamels having excellent resistance to chemicals, and chemical fumes can be made from acrylic resins. Acrylic vehicles are used in luminescent paints. The present automobile finishes are based on acrylic resins.

Thermosetting types of acrylic resins have been developed with cross-link upon heating to form hard, insoluble, infusible coatings for appliance finishes that will withstand severe service on washing machines, stoves, and dishwashers.

Alkyds. Alkyds are oil-modified phthalic resins that dry by reacting with oxygen from the surrounding air. Alkyd finishes are usually of the general-purpose type and are available as clear or pigmented coatings. Paints are available in flat, semigloss, and high-gloss finishes in a wide range of colors. They are easy to apply and may be used on most surfaces with the exception of fresh concrete, masonry, and plaster which are alkaline. Alkyd finishes have good color and gloss, and retain these characteristics in normal interior and exterior environments except under corrosive conditions.

Alkyd finishes are available in odorless formulations for use in hospitals, kitchens, sleeping quarters, and other areas where odor during painting might be objectional.

In trade sales applications, alkyds are used for interior walls and woodwork in both flat and gloss, and on the exterior as trim paints.

In industrial finishes, alkyds are combined with amino resins to produce hard baking finishes for use on appliances, metal furniture, etc.

Cellulosics. The most widely used cellulosic derivatives are the esters, particularly cellulose nitrate and cellulose acetate. Cellulose nitrate has been combined with many different resins to produce useful coatings. Perhaps the most outstanding combinations have been those with alkyd and amino resins to produce tough, hard, durable lacquers capable of withstanding the severe service requirements of automotive, aircraft, and other industrial finishes.

Successful coatings have also been based on blends of nitrocellulose with natural resins and drying oils, resin derivatives, phenolic resins, acrylic resins, certain vinyl resins, and other materials. Applications include coatings for metal, wood, paper, fabrics, leather, and cellophane.

Epoxies. Epoxy binders are of two types: (1) the oil-modified compositions, which dry by oxidation; and (2) the two-component materials, which comprise epoxies and amine or polyamide hardeners and are mixed just prior to use. In the latter type, when the two ingredients are mixed they react to form the cured coating. These coatings have a limited pot-life, usually a working day. Anything left at the end of the day must be discarded.

Epoxy paints can be used on any surface and can be applied with high-solids content, thus producing high film build per coat. The cured film has outstanding hardness, adhesion, flexibility, and resistance to abrasion, alkali, and solvents, as well as being highly corrosion resistant. Their major uses are as tile-like glaze coatings for concrete and masonry and for the protection of structural steel in corrosive environments. Their cost per gallon is high, but this is offset by the higher solids content and the reduced number of coats required to provide adequate film thickness. When used on exterior surfaces, epoxy paints tend to chalk to low-gloss levels and fade. Apart from this, their durability is excellent.

Epoxy-coal tar coatings are made by adding coal tar as an ingredient to epoxy paints, thereby reducing the cost. They have outstanding corrosion resistance and are used for interior and submerged surfaces. Color choice is limited to black.

Oils. Linseed oil is the major binder used in oil house paints. These paints are the oldest type of coatings in use. They are used primarily on exterior wood and metal since they dry too slowly for interior use. They are sensitive to alkaline masonry. Oil paints are easy to use and give high film build per coat. They also wet the surface very well so that surface preparation is less critical than with other types of paints for metal. Oil paints are not particularly hard or resistant to abrasion, chemicals, or strong solvents, but they arc durable in normal environments.

Oleoresinous. These binders are made by processing drying oils with hard resins, such as natural resins, rosin esters, and hydrocarbon resins. They are generally used as spar varnishes or as mixing vehicles for aluminum paint.

Phenolics. Phenolic binders are made by processing a drying oil with a phenolic resin and, thus, are a type of oleoresinous binder. They may be used as flat (lusterless) or high gloss finishes which are clear or pigmented in a range of colors. The clear finishes may be used on exterior wood and as mixing vehicles for producing aluminum paints. The durability of the clear fninsh is very good for this class of material, i.e., one to two years; the durability of the aluminum paints is excellent.

Phenolic paints are used as topcoats on metal in extremely humid environments and as primers for fresh-water immersion. These paints require the same degree of surface preparation as alkyds, but they are slightly higher in cost than alkyds. Phenolic coatings have excellent resistance to abrasion, water, and mild chemical environments. They are not available in white or light tints because of the relatively dark color of the binder.

Phenolic and alkyd binders are often blended to combine the hardness and resistance of phenolics with the color and color-retention of the alkyds. This may be done either by blending phenolic varnish with the alkyd vehicle or by the addition of phenolic resins during the processing of the alkyd resin.

Silicones. Silicone resins are used for heat-resistant finishes. They have good water-resistance and outstanding gloss-retention. When pigmented with aluminum, heat-resistant organic finishes containing a high concentration of silicone resins have the ability to withstand temperatures up to 1200°F. A combination of silicone and alkyd resins provides some heat resistance at a lower cost.

Urethanes. Urethane binders are of three types: (1) oil-modified, which are cured by oxygen from the air; (2) moisture-curing, which are cured by moisture in the air; and (3) two-part systems which are mixed just prior to use.

Oil-modified urethanes are similar to phenolic varnishes. Although somewhat more expensive, they have better initial color and color-retention, dry more rapidly, are harder, and have better abrasion-resistance. They can be used as exterior spar varnishes or as tough floor finishes. Oil-modified urethanes can be used on all surfaces. In common with all clear finishes, they have limited durability when used on exterior surfaces.

Moisture-curing urethanes are used in a manner similar to other one-package coatings except that the containers must be full to exclude moisture during storage. They have outstanding abrasion resistance and chemical resistance.

Vinyls. Lacquers based on modified polyvinyl chloride resins are used on steel where the ultimate in durability under abnormal environments is desired. They are moderate in cost, but they have low solids and require the most extensive degree of surface preparation to secure a firm bond. Because of their low solids, vinyl finishes require numerous coats to achieve adequate dry film thickness; thus the total cost of painting is higher than with most other paints. Since vinyl coatings are lacquers, they are best applied by spray, and they dry quickly, even at low temperatures. Recoating must be done with care to avoid lifting by the strong solvents which are present. In addition, these solvents present an odor problem. Vinyls can be used on metal or masonry, but they are not recommended for use on wood. They have exceptional resistance to water, chemicals, and corrosive environments, but they are not resistant to solvents.

Vinyl-alkyd combinations offer a compromise between the excellent durability and resistance of vinyls with the lower cost, higher film build, ease of handling, and adhesion of alkyds. They can be applied by brush or spray, and they are widely used on structural steel in marine and moderately severe corrosive environments.

Rubber-Base. So-called rubber-base binders are solvent-thinned and should not be confused with latex binders which are often called rubber-base emulsions. Four types are available: chlorinated rubber, styrene-butadiene, vinyl-toluene-butadiene, and styrene-acrylic. They are lacquer-type products which dry rapidly to form finishes which are highly resistant to water and mild chemicals. Recoating must be done carefully to avoid lifting by the strong solvents used. Rubber-base paints are available in a wide range of colors and levels of gloss. They are used for exterior masonry, also for areas which are wet, humid, or subject to frequent washing, e.g., swimming pools, wash rooms, shower rooms, kitchens, and laundry rooms.

Latices. Latex paints are based on aqueous emulsions of three basic types of polymers: polyvinyl acetate, polyacrylics, and polystyrene-butadiene. They dry by evaporation of the water, followed by coalescense of the polymer particles to form tough insoluble films. They have little odor, are easy to apply, and dry very rapidly. Interior latex paints are generally used either as a primer or finish coat on interior walls and ceilings made of plaster or wall board. Exterior latex paints are used directly on exterior masonry or on primed wood. They are noninflammable, economical, and have excellent color and color-retention. Latex paint films are somewhat porous so that blistering due to moisture vapor is less of a problem than with solvent-thinned paints. They do not adhere readily to chalked or dirty surfaces nor to glossy surfaces under eaves. Therefore, careful surface preparation is required for their use.

Latex paints are very durable in normal environments, at least as durable as oil paints. The popularity of latex paints is due mainly to easy cleaning of brushes and equipment with water.

Inorganics. The major inorganic binders used in paints are sodium, potassium, lithium, and ethyl silicates. These binders are used in zinc-dust-pigmented primers in which they react with the fine zinc metal to form very hard films. These films are extremely resistant to corrosion in humid or marine environments. Many of these primers also contain substantial concentrations of lead oxides which react with the silicates in conjunction with the zinc to form an even more corrosion-resistant coating.

Another type of inorganic binder is Portland cement. The paint is supplied as a powder to which water is added before use. Cement paints are used on rough surfaces such as concrete, masonry, and stucco. They dry to form hard, flat, porous films which permit water vapor to pass through readily. Cement paints should not be used in arid regions. When properly cured, cement paints of good quality are quite durable; when improperly cured, they chalk excessively on exposure, and then they may present problems on repainting.

A typical paint formula for an outside house oil paint is shown in Table 22.4, and for an acrylic latex house paint in Table 22.5.

Manufacture of Paints

The manufacture of paint involves the following operations: mixing, grinding, thinning, adjusting, and filling.

TABLE 22.4 Composition of a White Oil House Paint

Component	*Pounds*	*Gallons*	*Percent (by wt)*
Titanium dioxide (rutile)	105	3.05	7.4
White lead–basic carbonate	160	1.88	11.3
Zinc oxide (35 percent leaded)	370	7.55	26.2
Talc	240	14.70	16.9
Linseed oil–bodied 3	85	10.54	6.0
Linseed oil alkali refined	350	45.40	24.8
Drier–manganese napthenate 6 percent	2	00.26	–
Drier–lead napthenate 24 percent	7	00.72	–
Mineral spirits	105	15.90	7.4
TOTAL	1419	100.00	100.0

Constants

Viscosity	90 KU
PVC	30 percent
Vehicle solids	80 percent
Total solids (wt.)	92.5 percent
Total solids (vol.)	84.0 percent
Wt/gallon	13.1 lbs

One of the older methods consisted of mixing all the pigment and part of the vehicle to make a paste of suitable consistency in a tub with rotating blades. This paste was then fed by means of a trough into a slow-speed stong mill which had two circular stones, one stationary and the other rotating above it. The pressure developed between the two stones was sufficient to disperse the pigments and liquids intimately, eliminating unwetted agglomerates and air pockets. Next, this paste was fed in a continuous stream into a third piece of equipment, the thinning tank, which had rotating blades. At this point the remainder of the liquids was

TABLE 22.5 Composition of a White Acrylic Latex House Paint

Component	*Pounds*	*Gallons*	*Percent (by wt.)*
Titanium Dioxide	250.0	7.60	20.8
Zinc Oxide	50.0	1.07	4.2
Talc	178.9	7.73	14.8
Dispersant (30.0%)	10.5	1.05	0.9
Potassium tripolyphosphate	1.5	0.07	0.1
Surfactant	2.5	0.28	0.2
Antifoamer	4.0	0.52	0.3
Methylcellulose 250 cp	3.0	0.26	0.2
Ethylene glycol	25.0	2.69	2.2
Propylene glycol	34.0	3.94	2.8
Ethanol ethylene glycol	11.0	1.39	0.9
Fungicide	2.0	0.23	0.2
Acrylic latex (60.5%)	380.0	42.37	31.4
Ammonium hydroxide	2.0	0.27	0.2
Water	251.3	30.53	20.8
TOTAL	1205.7	100.00	100.00

Constants

Viscoity	75-80 K.U.
PVC	40.0%
Wt. Solids	59.0%
Volume Solids	41.0%
Wt./gallon	12.05 lbs.

Fig. 22.4 A three-roll mill. *(Courtesy Sherwin-Williams Co.)*

added, and the completed product was tested for viscosity, color, and other physical properties pertinent to the formulation being prepared. The batch, having been approved, was then strained and filled into appropriate size containiners.

The term "grinding" is a misnomer, as little or no breakdown of pigment particles takes place, but this term, nevertheless, is generally used in the trade. The word dispersion would be more descriptive.

Fig. 22.5 Sandmills: one 16-gallon mill and one 8-gallon mill. *(Courtesy Glidden-Durkee Div. of SCM Corp., Carrolton, Texas)*

Dispersion by old-fashioned low-speed stone mills is extremely slow and costly and is obsolete. Various types of more efficient equipment are now in use. The general principle, however, is the same, that is, to wet each individual particle thoroughly with the vehicle and to eliminate flocculated aggregates. Mills in use today include steel roller, ball, pebble, sand, high-speed impeller, Morehouse (a high speed stone type), and Cowles disperser.

Application of Paint

Paints can be applied in many different ways. Although most architectural paints are applied with a brush or roller, much paint is now applied by professional painters with air or airless spray equipment. With airless spray equipment, the paint is atomized by forcing it through a very small orifice under very high pressure. Other types of spray application are electrostatic spraying, hot spraying, steam spraying, two-component spraying and aerosol spraying.

In electrostatic spraying, the atomized paint is attracted to the conductive object to be painted by an electrostatic potential between the two. Advantages of this process include efficient use of coating material, rapid application, and relative ease of coating irregular shapes uniformly. Two-component spray equipment consists of two lines leading to the spray gun so that two materials, for example, an epoxy and a catalyst, can be mixed in the gun just before application.

There are several other methods for industrial application of paints. Dip application of coatings is a simple method wherein objects to be coated are suspended and dipped into a large tank containing the paint. This method is often used for undercoating objects were the uniformity and appearance of the paint are not important. Electrodeposition consists of depositing paint on a conductive surface from a water bath containing the paint. The negatively charged paint-component particles are attracted to the object being coated which becomes the

Fig. 22.6 Phenolic coated ducts. These exhaust fume ducts are being coated with Bakelite® phenolic resin baking finishes. Resistance to most solvents and chemicals at elevated processing temperatures is an inherent property of this type of resin. Other uses include linings for lard pails and fish cans. (*Courtesy Sherwin-Williams Co.*)

anode when an electrical potential is applied. Paint can be applied to very irregular surfaces with very uniform thicknesses and little loss. The system is limited to one coat of limited film thicknesses and the equipment-cost is high.

Yet, another method utilizes roller-coating machines to apply paint to one or both sides of flat objects, such as metal or fiberboard. The thickness of the coating can be controlled by the clearance between a doctor blade and the applicator rolls. Decorative effects such as wood-grain patterns can be applied with these machines.

Other methods include flow coating, where the paint is allowed to flow over the object being coated, and powder coating, in which paint in dry powder form is fused on the surface of the object being coated.

Testing

Paint products undergo extensive testing to be sure that they are up to standard and to meet rigid performance requirements. Testing for paint products may be divided into the following basic groups.

1. Package properties
2. Application properties
3. Film appearance
4. Film performance
5. Durability

There are standard ASTM (American Society for Testing and Materials) and Federal tests available to test all of these properties. Package tests include viscosity, skinning, settling weight per gallon, flash point, freeze-thaw stability, and fineness of grind. Application properties include ease of brushing, spraying or rolling, leveling, sag, spatter, and drying time. Film appearance tests include gloss, color, opacity, color acceptance, and color development. Film resistance tests are for hardness, abrasion, adhesion, flexibility, impact, scrubbability, and chemical and water resistance. Durability tests include weatherometer, salt spray, humidity, and exterior exposure.

The lead content of paints used in the home, schools, etc. is now limited to 0.06% lead. This eliminates the use of lead pigments for these uses.

Performance

The average thickness of a one coat paint film of a dry paint is 1 to 2 mils. Coverage of paint is expressed in square feet per gallon and refers to the number of square feet that can be painted to give satisfactory hiding and performance. For example, a coverage of 400 square feet per gallon might be specified for application.

To give satisfactory service, a paint must adhere to the substrate during the service life. For this reason surface preparation is very important. The paint film should expand and contract with the substrate. Also, for exterior use the paint film must be resistant to breakdown from moisture and sunlight.

Solvent Limitations

Air pollution regulations limit the amount of organic solvents that can be discharged into the atmosphere. Rules will become stricter in the future. Because of this, there has been a trend to use more waterborne coatings. At the present time, 55% of all trade sales paints and 15% of all industrial paints are waterborne coatings. Moreover, the use of waterborne coatings is expected to increase in the future.

Other types of coatings used to reduce solvent release to the atmosphere are high-solids coatings and powder coatings. These are primarily for industrial use. High-solids coatings have solids contents of 70% or higher and are usually two component materials. Powder coatings contain no solvent and the particles fuse together by the application of heat.

VARNISHES

The term "varnish" is applied to clear, transparent coating materials that dry by a process comprising evaporation of the solvent, followed by oxidation and polymerization of the drying oils and resins. Varnishes are a homogeneous mixture of resins, drying oils, driers, and solvents, and they contain no pigment.

Varnishes fall into three general classes, *spar varnishes* (exterior), *floor varnishes*, and *furniture finishes.*

The types of oils and resins and the ratio of oil to resin are the principal factors which determine the properties of a varnish. The selection depends on the compatibility of different oils and resins and on the intended use of the varnish. It is generally accepted that the oils in the finished coating contribute to its elasticity, and the resin, to its hardness. In oleoresinous varnishes, the ratio of oil to resin is expressed as the number of gallons of oil that are combined with 100 pounds of resin, and it is commonly referred to as the "length" of the varnish. Thus, where 50 gallons of oil are used with 100 pounds of resin, the varnish has a 50-gallon length. Varnishes containing less than 20 gallons of oil per 100 pounds of resin are usually classed as *short-oil* varnishes. A *medium-oil* varnish contains from 20 to 30 gallons of oil, and a *long-oil* varnish is one in which 30 gallons or more are used. Short-oil varnishes dry more rapidly than long-oil varnishes. They are used primarily where hardness, a high degree of impermeability, or resistance to alcohols, alkalies, or acids is desirable and where elasticity in the film is relatively unimportant. Short-oil varnishes are especially suitable where a "rubbed" finish is desired, e.g., on furniture, but they are too brittle for floors, for which a medium-oil varnish is best suited. Varnishes for exterior exposure are usually of a long-oil formulation because of the beneficial effect of the drying oil in providing elasticity and resistance to weathering. Spar varnish, a high-grade exterior varnish formulated originally for use on the wooden spars of ships, is a varnish of this type.

Interior varnishes are often based on linseed oil in combination with ester gum or a maleic-modified-rosin resin. Rosin-modified phenolic resins form varnishes with fairly good water- and alkali-resistance, but for exterior durability a long-oil varnish is required. The most durable spar (wood) varnishes for exterior exposure are usually made from tung oil combined with 100 percent phenolic resin. The phenolics as a group yield varnishes of relatively dark color. Terpene-phenolic resins are lighter in color and lower in cost than the comparable phenolic resins and have very good water- and alkali-resistance but only fair exterior durability. The terpene resins themselves are low in cost and make good light-colored interior varnishes with good color retention and good resistance to water and alkali.

Coumarone-indene resins are thermoplastic resins available in a wide range of hardness. Although poor in color and weathering, they contribute excellent water- and alkali-resistance and good dielectric properties to varnishes.

TABLE 22.6 Composition of Phenolic Marine Spar Varnish

Lbs		*Gals*
75	*p*-Phenylphenol pure phenolic resin	7.5
30	High m.p. modified phenolic resin	3.3
180	Tung oil	23.1
125	Alkali-refined linseed oil	16.5
328	Mineral spirits	48.0
2	Cobalt napthenate 6 percent	.25
4.5	Lead napthenate 24 percent	.5
2.0	Antiskinning agent	.25
743.5	TOTAL	100.4

Cooking directions: heat resins and linseed oil to 575°F; hold for one hour; add tung oil; reheat to 450°C; hold for body, then add drier and antiskinning agent.

Oil length	37.5 gallon
Viscosity	E-F Gardner Holdt
NVM	55 percent
Wt/gal	7.40 lbs

Petroleum hydrocarbon resins are soluble in drying oils and form varnishes with good water-, alkali-, and alcohol-resistance, but they have only fair color.

Natural resins, such as, congo, kauri, pontlianak, and batu are often used in varnishes because of their low cost and their ability to impart specific properties such as hardness, gloss, and moisture-resistance.

Alkyd varnishes differ from the conventional type in that the resin is formed in the kettle and co-reacted with the oil.

The principal uses of varnishes are for interior woodwork, floors, and furniture, and outdoors for buildings, furniture, and boats.

A floor varnish should dry to a tack-free finish within four hours. It must be able to stand scuffing from shoes and moving of furniture, and have excellent adhesion, high gloss, good holdout, and good resistance to the water and alkali used in cleaning. Furniture varnishes should dry within about four hours and must not soften from body heat, otherwise clothing would tend to stick when people sit on furniture such as chairs. They should have good sanding and polishing properties, resistance to water, acids (food), alkalis, alcohol, etc.

Exterior varnishes should have good durability or weather-resistance. They must withstand the elements without failure by cracking, peeling, whitening, spotting, etc. and with a minimum loss of gloss for a maximum period of time.

LACQUERS

A lacquer is a protective coating which dries by evaporation of volatile components.

The film-forming constituent is usually a cellulosic ester, i.e., nitrate, acetate, acetate-butyrate, or other high molecular weight polymer. Other types of lacquers are based on acrylics, polyurethanes, vinyls, etc. There are several differences between varnishes and lacquers. As stated previously, lacquers dry essentially by evaporation, while varnishes dry by a combination of oxidation and polymerization. Lacquers are characterized by very rapid drying and distinctive odor. They are usually based on high molecular weight polymers which require low-boiling solvents of high solvency power such as alcohols, ketones, and esters. The film will redissolve in the original solvent. Lacquers are usually available in a solids range of 20-30 percent, while varnishes have a solids range of 45-55 percent.

Lacquers dry "tack-free" in 5 to 15 minutes and to a firm film in 30 minutes to 4 hours. They are usually applied by spraying, brush application often being impractical because of the very rapid drying. Pigmented lacquers predominate in use; clear lacquers are used where colorless, tough films are desired. Clear lacquers are not generally considered as durable as high grade varnishes on exposure to sunlight and moisture.

Nitrocellulose, which is one of the principal film-formers, will be considered in detail.

In addition to nitrated cotton, a nitrocellulose lacquer contains, for example, (1) a solvent mixture which usually contains a ketone, an alcohol, an ester, and frequently an ether-alcohol; (2) a resin such as an alkyd, a phenolic, or ester gum to increase solids and improve adhesion; (3) a plasticizer for flexibilizing the film; (4) an inexpensive volatile diluent such as toluene; and (5) a dye or pigment which is omitted if a clear lacquer is desired.

The main outlets for lacquers are automobile finishes, furniture finishes, metal finishes, and plastic, rubber, paper, and textile finishes.

PRINTING INKS

Printing inks consist essentially of coloring matter: dispersed (pigment) or dissolved (dye) in a vehicle producing a fluid or paste that can be printed on a substrate (paper, plastic, metal, ceramic etc.) and then dried (rendered non-fluid). Commercial printing inks are nearly all pigmented products.

There are thousands of ink formulations depending on color, use, and the method of application.

A conventional method of roughly classifying printing inks is in terms of intended usage (news ink) and method of application (letterpress, lithographic, gravure, flexographic etc.).

Another classification could be based on the consistency of the ink. Letterpress and lithographic inks (oil or paste inks) are quite viscous, whereas flexographic and rotogravure inks (solvent inks) are relatively fluid.

From a mechanical standpoint, there are four major printing systems:

1. Printing from raised type (relief, typographic, letterpress, flexographic).
2. Printing from a planar surface (planographic, lithographic).
3. Printing from recessed or engraved type (intaglio, rotogravure).
4. Printing through a stencil (silk or wire screen).

In addition, the printing process may be carried out directly (ink applied directly from a printing plate cylinder to an intermediate offset roll, then transferring the imprinted ink image to a receptive substrate).

The rheology of an ink imposes another variable that must be considered in the mechanical design of a printing press. Viscous inks may require several ink rolls to render the ink suitable for printing. Solvent-type inks, however, normally involve only a few rolls, since they are fluid and the reduced number of rolls minimizes difficulties due to solvent evaporation.

Direct letterpress printing is shown in Figure 22.7. In this process, viscous ink is fed from the ink fountain through a series of soft, resilient rollers to the top surface of the printing (plate) cylinder (metal or hard rubber). Since the inking rolls are arranged to contact only the surface of the raised print, no ink is deposited below the raised area.

Indirect or offset printing using raised type (letterset printing) is also shown in Figure 22.7. Here the ink that is applied to the raised type (at the top) is printed onto an offset (blanket)

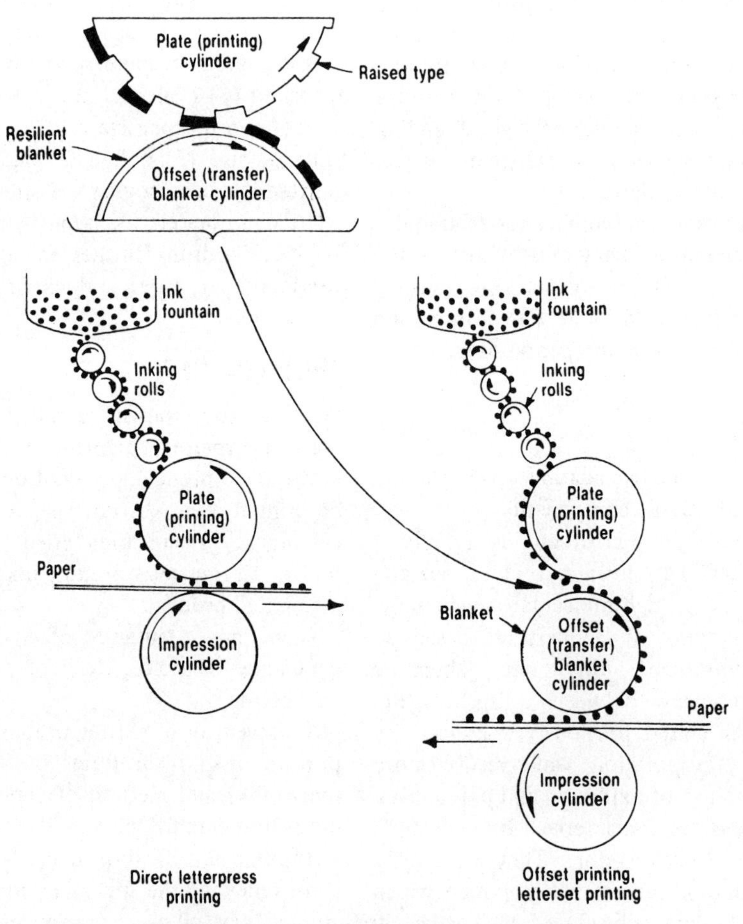

Fig. 22.7 Direct and letterset (offset) printing.

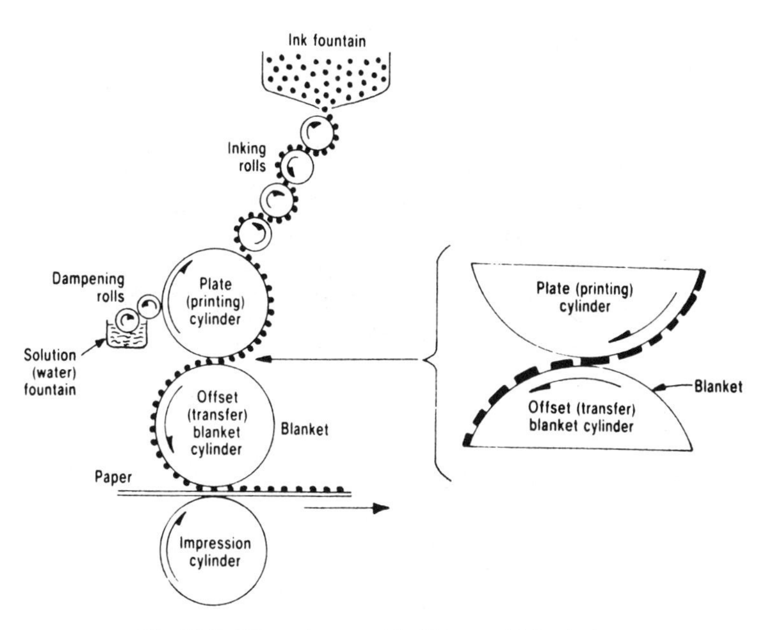

Fig. 22.8 Offset planographic (lithographic) printing.

cylinder as illustrated in the schematic detail. In turn, this imprint is transferred to the paper substrate, to provide the printed image.

Flexographic printing is a direct system of printing from raised type. However, as distinguished from letterpress printing, flexographic inks are relatively fluid and of low solids.

Lithographic printing takes places from a plane surface (planograph printing). It is based on the fact that oil and water systems tend to be incompatable and that printing surfaces can be fashioned so that certain areas attract oil systems and repel water systems and vice versa. A lithographic printing surface (substantially planar) then comprises printing areas (ink receptive) and nonprinting areas (water receptive). The printed image deposited on the inked (oil receptive) area is transferred either directly to a substrate or more usually to an offset blanket cylinder that in turn transfers this inked image to a substrate as schematically diagramed in Figure 22.8. Lithographic printing is becoming increasingly popular due to the fact that it is a realtively inexpensive system.

Rotogravure (gravure) printing is a system of printing from recessed areas that have been etched or inscribed below the surface of a printing plate or printing cylinder. The surface of a rotogravure printing cylinder consists of numerous recessed cells. During the printing operation, these cells pick up the ink directly from an ink fountain, and after the surface has been wiped clean by a doctor blade, the residual ink in the cells is transferred to the substrate to provide the printed image. A gravure press is shown in Figure 22.9. Intaglio printing is a term also used to describe printing systems based on recessed or engraved print. Intaglio printing uses deeply engraved cavities, which are copperplated to provide fairly thick quality prints required for banknotes, stock certificates, business cards, etc.

Stencil screen (silk) printing refers to a printing process wherein a thick film of viscous ink is forced (squeezed) through a stencilled fine mesh screen (silk or wire). The ink that is forced through the open areas of the meshwork deposits on the substrate to provide the required printed image. Screen printing is used mainly for outdoor posters, special textile applications, and irregular objects where the screen can be adopated to surface irregularities.

Composition

As mentioned before, printing inks may contain pigments, vehicles, solvents, and other additives so that they are in some ways similar to paint.

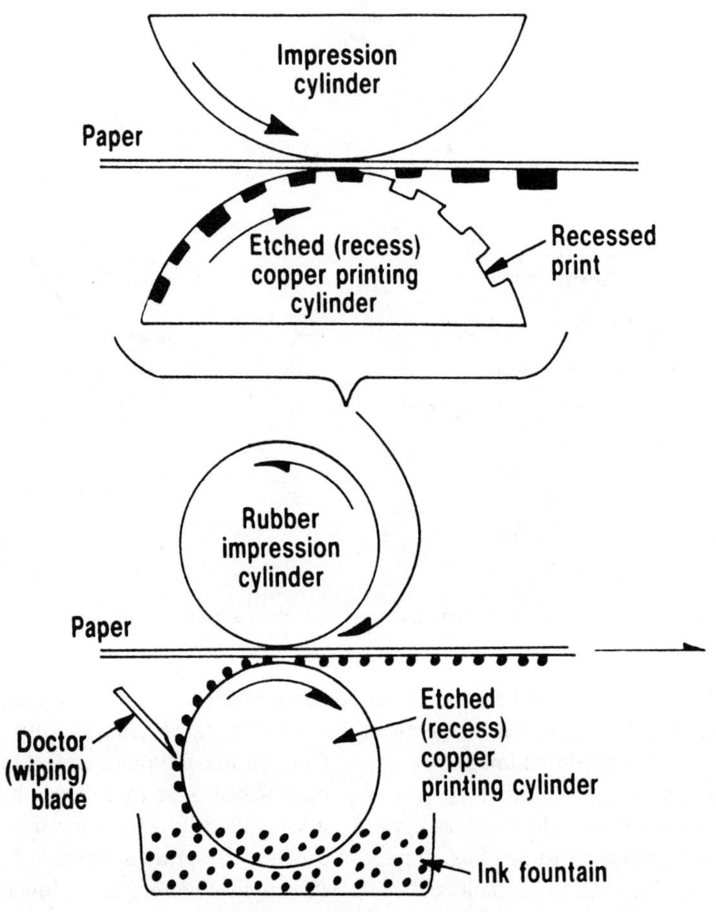

Fig. 22.9 Rotogravure printing.

Pigments. In general, pigments used in printing inks are similar to those used in paints. Pigments also contribute to the inks printing properties such as rheology, bleeding, damping, water incompatability, print appearance, color, luster, useful service life, fading, and resistance to chemical attack.

The same pigments are generally used for all types of commercial printing inks, with a few exceptions. For example, pigments used in lithographic inks must not bleed in water or aqueous solutions of weak acids, such as one percent chromic or phosphoric acid.

Vehicles. The composition of the vehicle varies depending on the types of printing ink. Compositions are nondrying oils. These are used on soft absorbent papers such as news-

TABLE 22.7 A Summary of the Types of Vehicles and Drying Systems Used in Various Printing Processes

Printing Process	*Drying System*	*Type of Vehicle*
News & Comics	Absorption	Nondrying oil
Letterpress; offset	Oxidation	Drying oil
Gravure, flexographic	Evaporation	Resin-solvent
Letterpress, letterset	Precipitation	Resin-glycol
Letterpress, offset	Quick setting	Resin-oil
Letterpress	Cold setting	Resin-wax

paper, comics, etc. which dry by absorption of the vehicle into the paper. The vehicle consists of nondrying, penetrating mineral oils, rosin oils, and so on, used in combination with various resins to import suitable tack and flow characteristics.

Drying Oil. Oxidation drying is the type used in most letterpress and offset inks today. Linseed oil or litho varnish is still the most widely used. The linseed oil is boiled by heat to the proper viscosity. Linseed oil varnishes have excellent wetting properties for pigments, have good transfer qualities, and provide good binding to the paper.

Other oils used in printing inks are tung, cottonseed, castor, perilla, soybean, petroleum, fish, rosin, etc. Many of these are combined with synthetic resins in order to obtain faster and harder drying oils.

Solvent-Resin Vehicles. These are used in inks which dry by the evaporation of solvents. They are composed of suitable resins or gum dissolved in low-boiling hydrocarbon solvents or alcohol.

Glycol. These are resins such maleic-modified rosin products, modified phenolics, and so on, which are dissolved in a glycol. These are used in moisture-set inks from which the resin precipitates out when subjected to steam.

Resin-Oil. These are carefully balanced combinations of resin, oil, and solvent. The solvent is rapidly absorbed by the paper, leaving a relatively dry ink film which is subsequently hardened by oxidation. Alkyd resins are sometimes used for this purpose.

Resin-Wax. These inks are solid at room temperature and are liquefied by heating. They are composed of resin, paraffin waxes, and solvent.

Solvents. Solvents are used to reduce the viscosity of the ink and then evaporate off after application. Depending on the type of ink, hydrocarbon solvents, alcohols, or glycols may be used.

Miscellaneous Ingredients. Other materials may be added to the printing ink for specific uses. If the ink contains a vehicle that drys by oxidation, driers (metallic soaps) will be used. Bodying materials to give the proper rheology are also sometimes used.

Manufacture

As with paint, the bulk of printing inks are made by the batch process. Only the large volume standardized inks, such as gravure and news ink are manufactured by continuous processes.

Mixing is a major step in ink manufacture since the solid ingredients such as the pigments must be wet down or dispersed into the vehicle. To accomplish this, a "change can" or dough-type mixer is used.

Fine printing inks, such as those used for letterpress and offset printing, require intimate dispersal in dough-type impeller mixers and grinding on a three-roll mill. Other types of equipment used are ball mills, colloid mills, sand mills, high-speed dispersers, and turbine mixers.

Types of Printing Inks

Different types or printing inks are required for various processes and end uses. These are:

Heat-Set Inks	Metallic Inks
Moisture-Set Inks	Wax-Set Inks
Quick-Setting Inks	Water Color Inks
High-Gloss Inks	Cold-Set Inks
News Inks	Magnetic Inks

Heat-Set Inks. The printing of publications in large runs at high speed and with good quality is made possible by the use of heat-set inks. The vehicle is composed of high melting resins with good solvent release at elevated temperatures. The solvents have boiling point ranges between temperatures of 400°-500°F.

Moisture-Set Inks. This type of ink has a vehicle composed of a water-insoluble resin dissolved in a water-miscible solvent. Upon subjecting the printing to either steam or a fine water mist, the water-miscible solvent picks up some of the water which causes the resin to precipitate out, binding the pigment to the paper.

Quick-Setting Inks. These inks dry by absorption and coagulation. The vehicles are generally special resin-oil combinations, which after the ink has been printed, separate into a solid material which remains on the surface as a dry film plus an oil material which penetrates into the paper. Oxidation hardens the film as it ages.

Water-Reducible Inks. The vehicle consists of modified rosin soap in a glycol solvent and is water reducible.

High-Gloss Inks. High gloss inks are lower in pigment content and have good holdout. In general, the more resistant the paper is to penetration of the ink the higher the gloss is.

News Inks. Black news inks consist of mineral oil and carbon black. Colored news inks are based on pigments flushed in mineral oil vehicles.

Metallic Inks. These inks consist of metallic powders such as aluminum, bronze, and copper suspended in a low-acid value vehicle.

Wax-Set Inks. These are used on bread wrappers and other waxed papers and consist of a vehicle composed of a wax-insoluble resin dissolved in a wax-soluble solvent. These inks work on the principal that as the ink is applied some of the solvent migrates to the wax paper causing drying of the ink.

Water Color Inks. These inks are based on a vehicle composed of gum arabic, dextrin, glycerine, and water.

Cold-Set Inks. Inks of this type are solid at room temperature and the vehicles are plasticized waxes having melting points ranging from 150° to 200°F.

Magnetic Inks. Magentic inks are made with pigments which can be magnetized after printing so that the printed characters can later be recognized by electronic reading equipment.

Commercial Inks

Letterpress Inks. Inks for this use must have the proper penetration, evaporation, and oxidation rate for the particular absorbent paper used. Letterpress printing ink is applied at the highest color strength possible, consistant with acceptable flow properties, to achieve the thinnest ink film possible.

Black News Inks. These inks are based on carbon black pigments, fortified with small amounts of organic toner and mineral oil with minor additions of rosin or similar additives.

Job Press Inks. These inks are standard items for use on all types of paper. They dry mainly by oxidation and are of high viscosity.

Publication Inks. Paper for books and periodicals can vary in porosity. For porous stock, inks based on mineral oil/rosin are used. For coated papers the major portion of the vehicle is a drying oil such as linseed.

Lithographic Inks. Lithographic printing is done from an essentially plane surface that contains selected (hydrophobic) areas that attract and convey ink to give a rpinted image. The nonprinting (hydrophilic) areas are conventionally rendered oil repellent by a very fine film of gum arabic deposited from water solutions.

Lithographic inks are of high viscosity so that they cannot mix with the water phase. Lithographic printing deposits relatively thin films so that high-strength colorants must be employed. The standard vehicle for lithographic ink is heat-bodied linseed oil in an aliphatic solvent. Alternative vehicles such as alkyds, rosin, and phenolic varnishes are often used.

Flexographic Inks. Flexography is a form of rotary letterpress printing that uses flexible rubber plates and fluid inks. Solvents must be used that do not affect the rubber plates.

TABLE 22.8 Printing Ink Production in U.S.A. in 1977*

	In millions quantity lbs.	*Value $*
Total	–	960.2
Letterpress Inks (black & color)	–	119.6
News Inks	335.7	58.5
Publication Inks	20.3	12.1
Packaging Inks	23.2	24.5
Others	12.1	24.5
Lithographic & Offset Inks (black & color)	–	281.2
News Inks	153.0	61.1
Publication Inks	60.9	85.0
Packaging Inks	33.7	60.2
Others	41.2	74.9
Gravure Inks	–	194.5
Packaging Inks	76.5	80.3
Publication Inks	219.5	99.4
Others	18.6	14.8
Flexographic Inks	–	194.7
Packaging Inks	118.0	159.0
Others	36.5	35.4
Printing Inks (miscellaneous)	–	61.3
Textile printing inks	4.0	26.6
Screen printing inks	18.6	20.3
Others	–	19.3

*1977 Census of Manufacturers.

Today flexography is used for printing on plastic films such as cellulose acetate, polyethylene, polystyrene, and cellophane. It prints equally well on glassine, tissue, sulfite, kraft, and other paper stocks, on aluminum foils, paperboard, corrugated liners, bags, paper labels, gift wrappings, and corrugated boxes. The early flexographic inks were colored with analine dyes. Today pigmented flexographic inks account for the major share of the market. Alcohol-type flexographic inks are generally based on nitrocellulose as the binder. Water-based flexographic inks are usually based on ammonia or amine soluble resins such as shellac and esterified fumarated rosins. Polyamide inks are most widely used for printing plastic packaging and are used in a solvent mixture of propyl alcohol and VM & P naptha. Polyamide flexographic inks have excellent adhesion to plastic surfaces, superior gloss, and good water resistance. Their principal disadvantage is poor heat resistance.

Acrylic flexographic inks have good adhesion, good abrasion resistance, oil and water resistance, and good heat stability. They are made of acrylic resins in a mixture of alcohol and ester solvents.

Rotogravure Inks. Rotogravure inks are quite similar to flexographic in that they use the same pigments, contain major percentages of solvent in their composition, possess low viscosities, and dry primarily by solvent evaporation. However, there is a major difference in that flexographic inks are limited to solvents that do not affect the flexographic rubber printing plates while rotogravure inks are not subject to such restrictions since they are printed directly from copper plates.

Binders used may be rosin soaps, Gilsonite, and hydrocarbon resins. The so-called "lacquer types" are made of nitrocellulose, ethyl cellulose, chlorinated rubber, and vinyls. Solvent contents range from 40 to 60%. Pigment contents range from 6 percent with strong colorants to 35 percent when diluted with extender pigments. Binder contents, including additives, range from 16 to 44 percent.

REFERENCES

1. Martens, C. R., *Technology of Paints, Varnishes and Lacquers*, Krieger Publishing Co. Inc., Huntington, New York, 1974.
2. Martens, C. R., *Water-borne Coatings*, New York, Van Nostrand Reinhold, 1980.
3. Sward, G. G., *Paint Testing Manual*, 13th Edition, American Society for Testing and Materials, Philadelphia, Penna., 1972.
4. Patton, T. C., *Pigment Handbooks*, Vol. **I**, **II**, **III**, New York, John Wiley & Sons.
5. *Federation Series on Coating Technology*, Units **I** to **XXV** Federation of Societies for Paint Technology, Philadelphia, Penna., 1964–1979.

23

Dye Application, Manufacture of Dye Intermediates and Dyes

C. W. Maynard, Jr.*

Dyes are intensely colored substances that can be used to produce a significant degree of coloration when dispersed in, or reacted with, other materials by a process which, at least temporarily, destroys the crystal structure of the substance. This latter point distinguishes dyes from pigments which are almost always applied in an aggregated or crystalline-insoluble form. Modern dyes are products of synthetic organic chemistry. To be of commercial interest, dyes must have high color intensity and produce dyeings of some permanance. The degree of permanence required varies with the end use of the dyed material.

All molecules absorb energy over various parts of the electromagnetic spectrum. The characteristic of dye molecules is that they absorb radiation strongly in the visible region, which extends from 4000-7000 angstroms. Only organic molecules of considerable complexity which contain extensive conjugation systems linked to electron withdrawing and attracting groups give sufficient absorption (tinctorial value) in the visible region to be useful as dyes. The shade and fastness of a given dye may vary depending on the substrate, due to different interactions of the molecular orbitals of the dye with the substrate, and the ease with which the dye may dissipate its absorbed energy to its environment without itself decomposing.

The primary use for dyes is textile coloration, although substantial quantities are consumed for coloring such diverse materials as leather, paper, plastics, petroleum products, and food.

The manufacture and use of dyes is an important part of modern technology. Because of the variety of materials that must be dyed in a complete spectrum of hues, manufacturers now offer many hundreds of distinctly different dyes. An understanding of the chemistry of these dyes requires that they be classified in some way. From the viewpoint of the dyer, they are best classified according to application method. The dye manufacturer, on the other hand, prefers to classify dyes according to chemical type.

Both the dyer and the dye manufacturer must consider the properties of dyes in relation

*Retired, formerly with E. I. duPont de Nemours & Co., Wilmington, Del.

to the properties of the materials to be dyed. In general, dyes must be selected and applied so that, color excepted, a minimum of change is produced in the properties of the substrate. It is necessary, therefore, to consider the chemistry of textile fibers as a background for an understanding of the chemistry of dyes.

The major uses of dyes are in the coloration of textile fibers and paper. The substrates can be grouped into two major classes–hydrophobic and hydrophilic. Hydrophilic substances such as cotton, wool, silk, and paper are readily swollen by water making access of the dye to the substrate relatively easy. On the other hand, the ease of penetration also allows easy removal in aqueous systems and special techniques must be used where a high degree of wet-fastness is required.

On the other hand, hydrophobic fibers, such as the synthetic polyesters, acrylics, polyamides, and polyolefin fibers, are not readily swollen by water, hence, higher application temperatures and smaller molecules are generally required.

The polymer chemist has increased the versatility of the newer fibers by incorporating dye sites of a varying nature as needed to achieve dyeability with a predetermined class of dyes. It is now possible to have polyesters, acrylics, and polyamide fibers which can be dyed with positive (basic, cationic), negative (acid, anionic), or neutral (disperse) dyes. These recent developments have allowed the fabric designer to produce materials (textiles, carpets) fabricated in patterns which can be dyed three different colors from one dyebath containing three types of dyes. This concept is called cross dyeing and is becoming increasingly popular as a low-cost method of coloration.

TEXTILE FIBERS

Cotton and Rayon

Cotton and rayon (regenerated cellulose) fibers are composed of cellulose in quite pure form. Cellulose lacks significant acidic or basic properties but has a large number of alcoholic hydroxyl groups. It is hydrolyzed by hot acid and swollen by concentrated alkali. When cotton is swollen by concentrated alkali under tension, so that the fibers cannot shrink lengthwise, it develops a silk-like luster. This process is called mercerization. The affinity of mercerized cotton for dyes is greater than that of untreated cotton.

Cotton and rayon fibers are easily wetted by water and afford ready access to dye molecules. Dyeing may take place by absorption, occlusion, or reaction with the hydroxyl groups. It is also possible to make cotton and rayon receptive to a variety of dyes by pretreatment or mordanting with a material capable of binding the dyes.

Wool and Silk

Wool and silk fibers are protein substances with both acidic and basic properties. They are destroyed by strong alkali. Strong acid causes hydrolysis, but the process may be controlled to permit dyeing from acidic solutions.

Wool and silk are wetted by water and are dyed with either acid or basic dyes through formation of salt linkages.* They may also be dyed with reactive dyes that form covalent bonds with available amino groups. Mordanting is sometimes used to alter the dyeability of wool and silk.

Cellulose Acetates

Acetylated cellulose fibers differ from cellulose fibers in that they are more hydrophobic and lack large numbers of free hydroxyl groups. The higher the degree of acetylation the more unlike cotton and rayon the acetates become. Strong acid and strong alkali degrade cellulose acetates, although the initial attack is slow under moderate conditions because of the difficulty of wetting the fiber. The triacetate is the most hydrophobic and the most stable.

Dyeing of cellulose acetates is effected with dyes of low water solubility which become dissolved in the fiber, or by occlusion of dyes formed *in situ*. Acid, basic, and reactive dyes

*This refers to ionic bonds, the forces acting between ions of opposite charges.

cannot be used because of the lack of sites for attachment.

Polyamides

Polyamide fibers (nylon) are synthetic fibers possessing properties somewhat like those of wool and silk. They are more hydrophobic, however, with only a limited number of basic or acidic groups. Polyamides are degraded by strong acid but may be dyed from acidic dye baths under controlled conditions.

Polyamide fibers are dyeable near the boiling point of water with acid dyes that form salt linkages with basic sites. Dyeing by this means is limited by the availability of these sites. Dyes like those used on cellulose acetates (i.e., that dissolve in the fiber), or reactive dyes that bond to available amino groups may also be used.

Polyesters

Polyester fibers are synthetic fibers unlike any produced in nature. They are hydrophobic and possess good stability to acid and alkali as a result of this hydrophobicity. They are hydrolyzed under sufficiently drastic conditions, however. Some polyester fibers lack functional groups; others are provided with acidic groups or otherwise modified to make them more hydrophilic.

Unmodified polyester fibers are dyed by solution of dyes in the fiber or, to a limited extent, by occlusion of dyes formed *in situ*. Modified polyester fibers may be dyed in these ways or with dyes selected according to the nature of the sites introduced by the modification. Both unmodified and modified polyester fibers must be dyed under vigorous conditions, often with the assistance of a swelling agent to open up the fiber.

Acrylics

Acrylic fibers are hydrophobic synthetic fibers with excellent chemical stability. They do not resemble any natural product. The only functional groups present are those introduced for the purpose of providing sites for dyeing.

Acrylic fibers are dyed by solution of dyes in the fiber, by occlusion of dyes formed *in situ*, and by formation of salt linkages with dyes capable of attachment to sites provided for that purpose. Basic dyes are used on acrylic fibers bearing sulfonic acid groups.

Vinyls

Vinyl polymers and copolymers make up a class of fiber-forming materials that vary greatly in properties, depending on constitution. Some vinyl fibers are very resistant to degradation by acids. Dyes are selected according to the nature of the specific polymer to be dyed.

Polyolefins

Polyolefin fibers are formed from the products of polymerization of unsaturated compounds of carbon and hydrogen, for example, propylene. They do not absorb water and are chemically quite inert. They can be dyed with special disperse dyes but are colored best by introducing a colorant into the polymer before the fibers are spun. Some types of polypropylene incorporate metal ions such as Ni^{++} to act as dye sites for chelatable dyes.

Glass Fibers

Glass fibers are used for special purposes, for example, where flammable materials cannot be tolerated. They are often colored during manufacture, but they can be dyed by special techniques which involve the use of surface coatings that have an affinity for dyes.

Paper

Paper is a nonwoven material made up primarily from cellulose of varying degrees of refining (see Chapter 15). Paper may be colored in the pulp as a watery fibrous slurry by either continuous or batch methods. The dyeing process takes place at ambient temperature and the dyes are adsorbed on the pulp by their affinity for the cellulose. Direct dyes are most commonly used. In continuous coloration, the dye solutions are metered directly into a moving stream of pulp. In batch operations, dye is added to a pulper, beater, or blending chest containing a given quantity of slurry.

Paper may also be colored on its surface after the initial sheet is formed, pressed, and partially dried. This can be done at the size press of the paper machine, or color can be carried by a calender roll for heavier sheets. A wide variety of low-cost dyes can be used for surface coloration.

THE PROPERTIES OF DYES

The properties of dyes may be classified as application properties and end-use properties. Application properties include solubility, affinity, and dyeing rate. End-use properties include hue, and fastness to degrading influences such as light, washing, heat (sublimation), and bleaching. Dyes are selected for acceptable end-use properties at minimum expense. Involved application procedures are used only when necessary to achieve unusually good results.

It has become common practice to treat dyed textiles with agents designed to improve resistance to shrinking, wrinkling, and the like. These agents frequently alter the appearance and fastness of dyes. Stability to after-treatments must therefore be considered as an important end-use property of dyes.

The amount of dye required to obtain a light shade is usually about 1 percent of the weight of the fiber; heavier shades may require as much as 8 percent. These values are very approximate, since dyes differ in color strength and are usually sold in diluted form. These amounts of dye are not sufficient, in most cases, to markedly affect the properties, other than color, of the fiber. Care must be exercised, however, to apply the dye under conditions that do not cause fiber degradation.

Color strength of a dye can be measured quantitatively as molar absorptivity which falls within the general ranges given below.

Dye Type	*Molar Absorptivity**
Anthraquinone	5,000–15,000
Azo	20,000–40,000
Basic	
cyanine	40,000–80,000
triarylmethane	40,000–160,000

It is obvious from the list above that many basic dyes have about 10–20 times the color value of the anthraquinone types. Unfortunately, light fastness is in the reverse order, the anthraquinones being used where maximum durability to light is needed. The challenge to the dye chemist or engineer is to increase the color strength of the light-fast dyes or to improve the fastness of the high absorption dyes.

CLASSIFICATION OF DYES

For the convenience of the dyer, dyes are classified according to application method. The best classification method available is that used in the Colour Index,[15] a publication sponsored by the Society of Dyers and Colourists (England) and the American Association of Textile Chemists and Colorists.

Acid Dyes

Acid dyes depend on the presence of one or more acidic groups for their attachment to textile fibers. These are usually sulfonic acid groups which serve to make the dye soluble in water. An example of this class is Acid Yellow 36† (Metalin Yellow).

*Molar absorptivity of a dye is its absorbance of visible light in a cell one centimer thick in a molar solution (one gram-mol per liter). Since a dye absorbs visible light strongly, by definition, absorptivity is normally measured at high dilution and molar absorptivity is calculated by extrapolation, assuming that Beer's Law holds.

†The dye names used in this chapter are those given in the "Colour Index."[15] Trivial names, when well established, are given in parentheses following the "Colour Index" name.

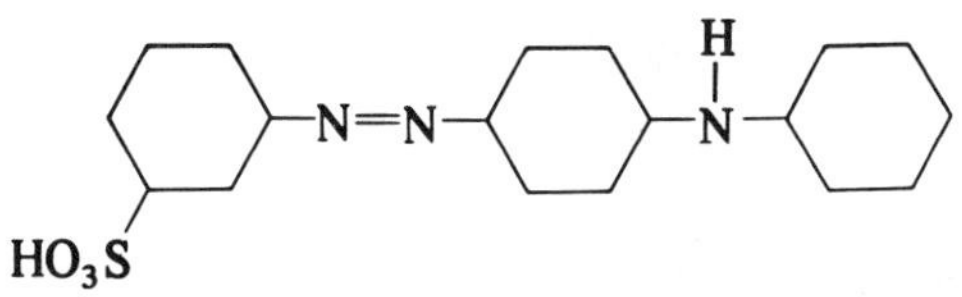

Acid Yellow 36

Acid dyes are used to dye fibers containing basic groups, such as wool, silk, and polyamides. Application is usually made under acidic conditions which cause protonation of the basic groups. The dyeing process may be described as follows:

$$Dye^{-} + H^{+} + Fiber \leftrightarrows Dye^{-}\ H^{+}\text{—}Fiber$$

It should be noted that this process is reversible. Generally, acid dyes can be removed from fibers by washing. The rate of removal depends on the rate at which the dye can diffuse through the fiber under the conditions of washing. For a given fiber, the diffusion rate is determined by temperature, size and shape of the dye molecules, and the number and kind of linkages formed with the fiber.

Chrome Dyes. A special kind of acid dyes used mainly on wool, chrome dyes possess improved fastness when converted to chromium complexes. A suitable chromium salt is applied to the fiber (1) before the dye, (2) at the same time as the dye, or (3) after the dye. All these methods are satisfactory, but more complicated than is desired. In recent years, manufacturers have made available dyes in which chromium is already a part of the molecule. These dyes are simpler to apply than the older types and, as a consequence, are increasing in importance.

Basic or Cationic Dyes

Cationic dyes become attached to fibers by formation of salt linkages with anionic or acidic groups in the fibers. Basic dyes are those which have a basic amino group which is protonated under the acid conditions of the dyebath. Cationic dyes can be divided into the three classes which are illustrated.

Basic Brown 1 (Bismark Brown) is an amino-containing dye which is readily protonated under the pH 2–5 conditions of dyeing.

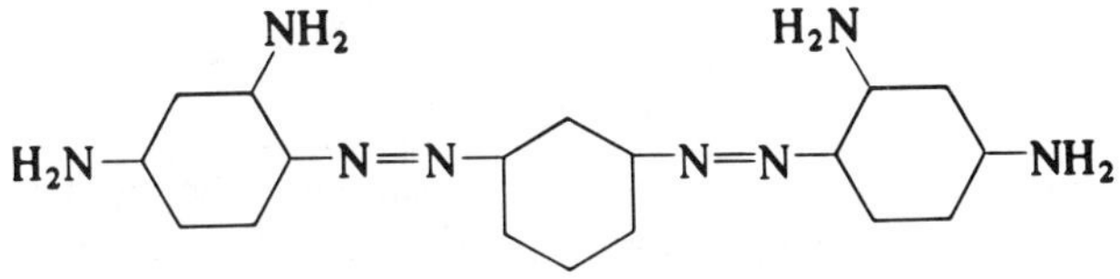

Basic Brown 1

Basic Violet 3 (Crystal Violet) is an example of a cationic dye in which the cationic charge is delocalized by resonance and may be present at any one of the basic centers at any time. These resonance forms of almost equivalent energy are one of the reasons that Crystal Violet is among the strongest dyes known. This high color value (tinctorial strength) has important commercial interest in the hectograph copying system. In this system, Crystal Violet in a wax base is transferred to the back of a typewritten copy sheet. By using paper moistened with alcohol more than 200 good copies may be made from the master.

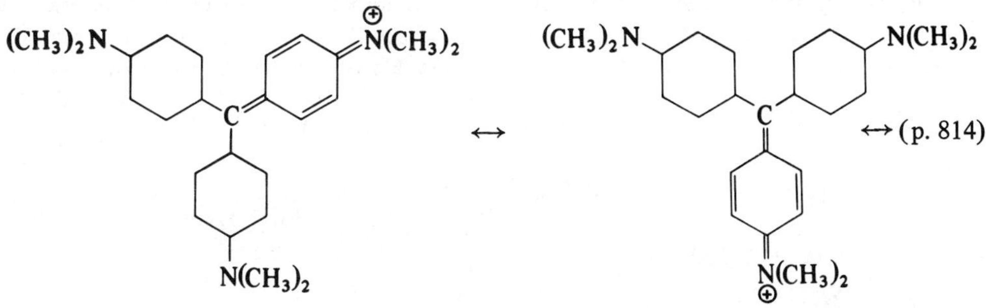

↔ (p. 814)

(p. 813) ↔

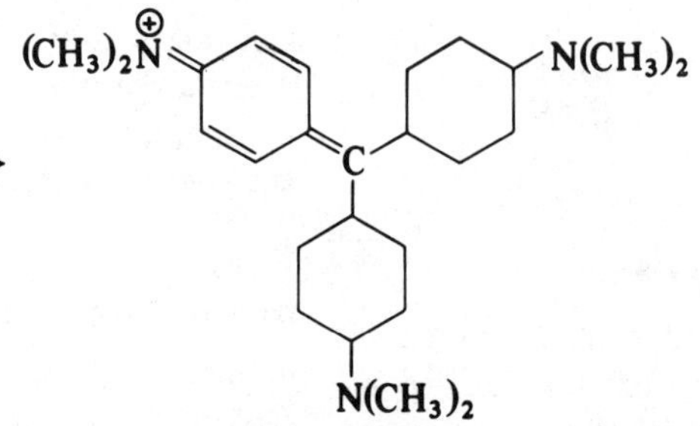

Basic Violet 3

The so-called "classical" basic dyes illustrated above are noted for their high color value, but on the average they also have very poor light fastness. When the newer, more durable synthetic fibers emerged having affinity for cationic dyes, it was necessary to develop more light-fast types. These were formed by taking known light-fast azo or anthraquinone chromophores and attaching a cationic "tail" insulated from the chromophore. C. I. Basic Red 18 is a typical example of this pendant cationic dye type.

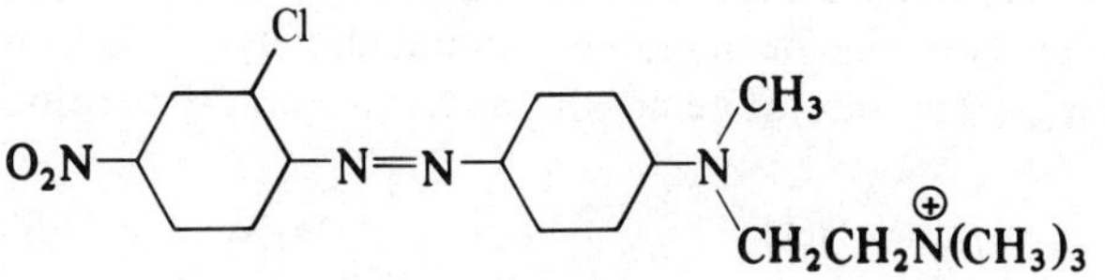

Basic Red 18

Further improvements in light-fast dyes for acrylic fibers are the diazo hemicyanines. These are unusual in that they contain a delocalized charged chromophore, which produces tinctorially strong bright shades, but they also have unusually good light-fastness. The blue dye below is an example.

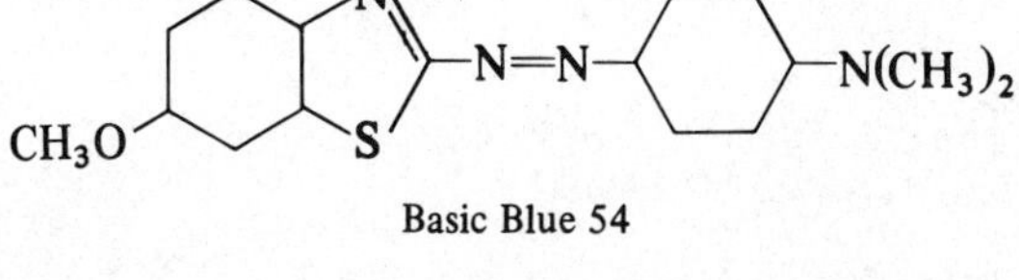

Basic Blue 54

Direct Dyes

Direct dyes are a class of dyes that become strongly adsorbed on cellulose. They usually bear sulfonic acid groups, but are not considered acid dyes since these groups are not used as a means of attachment to the fiber. Direct dyes are large, flat, linear molecules which can enter the water-swollen amorphous regions of cellulose and orient themselves along the crystalline regions. Common salt or Glauber's salt is often used to promote dyeing since the presence of excess sodium ions favors establishment of equilibrium with a minimum of dye remaining in the dye bath. Direct Orange 26 is a typical direct dye.

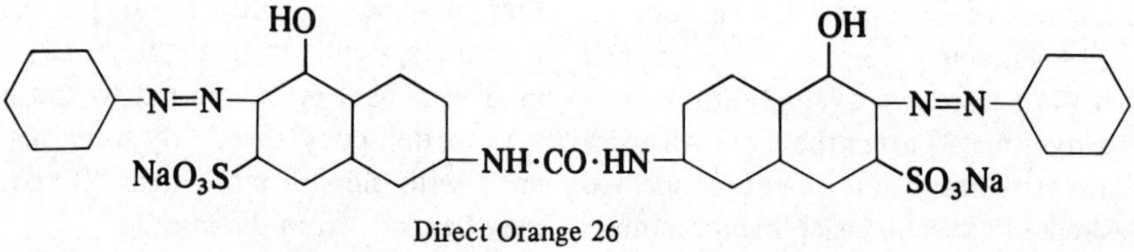

Direct Orange 26

Since direct dyes are held in cellulosic fibers by adsorption, the dyeing process is reversible. Unless after-treated with resins and dye-fixing agents, direct dyes, as a class, have poor fastness to washing. They are used mainly because they are economical and easy to apply.

A special type of direct dye having free amino groups is designed to be diazotized and coupled (developed) on the fiber. This operation increases the size of the molecules and improves their fastness to washing. An example of this type is Direct Black 17 (Zambesi Black D).

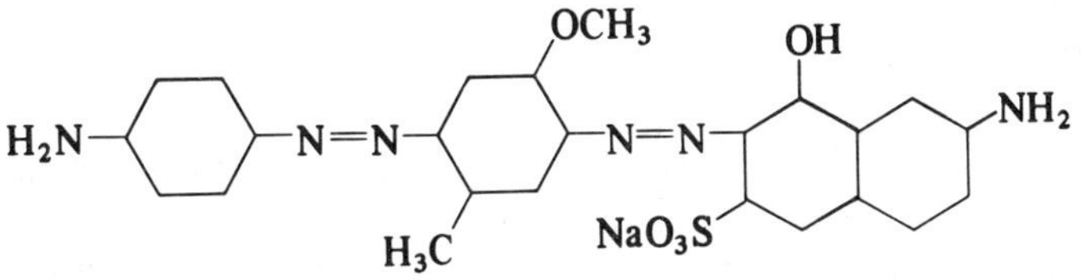

Direct Black 17

These dyes are used primarily to color plain grounds which are later to be printed in a pattern with vat dyes. The sodium hydrosulfite used to reduce the vat dyes destroys the ground color upon contact. Oxidation of the reduced vat dyes then produces two-color effects.

During the application of a "diazotized and developed" dye, the direct dye is first applied in the usual manner. The goods are then passed into a bath containing nitrous acid, where the amino groups are diazotized. Next, the goods are passed through an alkaline solution of beta-naphthol which couples with the diazonium groups to give the final dye.

Sulfur Dyes

Sulfur dyes are insoluble dyes which must be reduced with sodium sulfide before use. In the reduced form they are soluble and exhibit affinity for cellulose. They dye by adsorption, as do the direct dyes, but upon exposure to air they are oxidized to re-form the original insoluble dye inside the fiber. Thus, unlike the direct dyes, they become very resistant to removal by washing.

The exact constitutions of most sulfur dyes are unknown, although the conditions required to reproduce given types are well established. They are fairly cheap and give dyeings of good fastness to washing, as noted above. Their brightness and fastness to bleaching are often inferior, however.

Vat Dyes

Vat dyes, like sulfur dyes, are insoluble. They are reduced with sodium hydrosulfite in a strongly alkaline medium to give a soluble form that has affinity for cellulose. This reducing operation was formerly carried out in wooden vats, giving rise to the name vat dye.

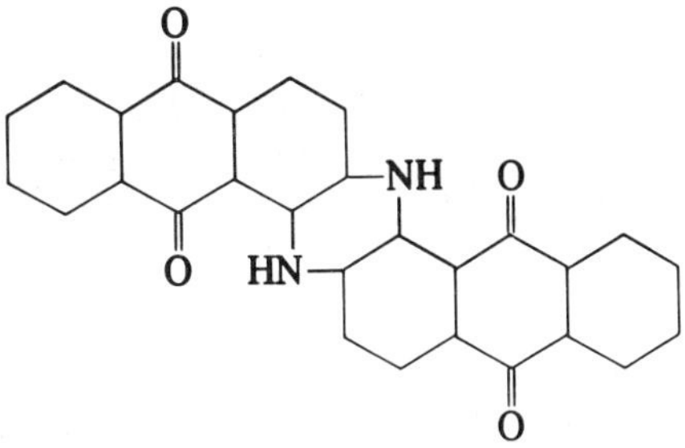

Vat Blue 4

After the reduced dye has been adsorbed on the fiber, the original insoluble dye is reformed by oxidation with air or chemicals. The dyeings produced in this way are very fast to washing, and, in most cases, the dyes are designed to be fast to light and bleaching as well. An example of a vat dye is Vat Blue 4 (Indanthrone).

Vat dyes are quite expensive and must be applied with care. They offer excellent fastness when properly selected and are the dyes most often used on cotton fabrics which are to be subjected to severe conditions of washing and bleaching.

It is sometimes impossible to tolerate the strongly alkaline conditions used to reduce vat dyes, for example, when dyeing fibers that are sensitive to alkali. For this reason, and for added convenience, some manufacturers offer soluble vat dyes. These are usually the sodium or potassium salts of the sulfuric esters of reduced vat dyes. When applied to the fiber and subjected to an acid treatment in the presence of an oxidizing agent, they hydrolyze, reverting to the original form of the dye.

Reactive Dyes

Reactive dyes are a relatively new class of dyes that form covalent bonds with fibers possessing hydroxyl or amino groups. One important type of reactive dye contains chlorine atoms that react with hydroxyl groups in cellulose when applied in the presence of alkali. It is believed that an ether linkage is established between the dye and the fiber. An example of this type is the blue anthraquinone dye shown below:

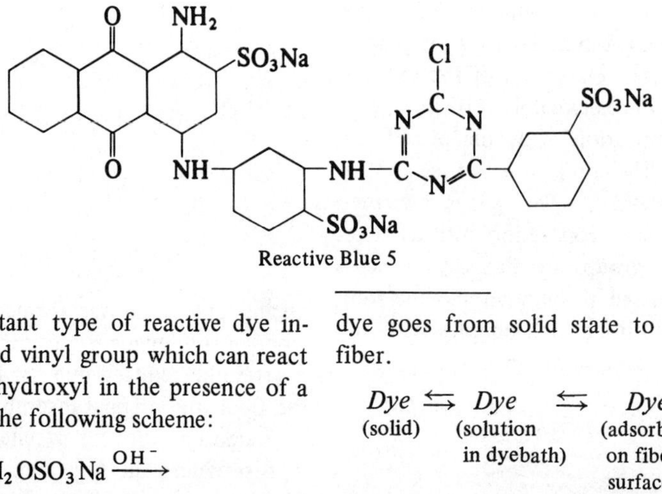

Reactive Blue 5

Another important type of reactive dye involves an activated vinyl group which can react with a cellulose hydroxyl in the presence of a base according to the following scheme:

$$\textit{Dye}-SO_2CH_2CH_2OSO_3Na \xrightarrow{OH^-} \textit{Dye}-SO_2CH{=}CH_2 + HO-\textit{Cellulose} \longrightarrow \textit{Dye}-SO_2CH_2CH_2O-\textit{Cellulose}$$

Reactive dyes offer excellent fastness to washing since the dye becomes a part of the fiber. The other properties depend on the structure of the colored part of the molecule and the means by which it is attached to the reactive part.

Disperse Dyes

Disperse dyes are nonionic dyes having low water solubility which are capable of dissolving in certain synthetic fibers. Their fiber attraction is due to the formation of a solid solution, since, being uncharged, there is no driving force to form salt linkages. Disperse dyes are now primarily used for polyester fibers although they were originally developed for cellulose acetate and polyamide fibers. In the decade 1968-1978 the production of disperse dyes doubled, reflecting the steady growth in polyesters. This growth is expected to continue as polyesters achieve a larger share of the textile markets.

Disperse dyes like all other dyes go through a monomolecular form during the dyeing process. This means that when applied from an aqueous bath they must have a finite solubility under the dyeing conditions. This limits the size and polarity of the molecules used. The dyeing mechanisms involve the equilibrium processes shown below, with the net result being that the dye goes from solid state to solution in the fiber.

$$\underset{\text{(solid)}}{\textit{Dye}} \rightleftarrows \underset{\text{(solution in dyebath)}}{\textit{Dye}} \rightleftarrows \underset{\text{(adsorbed on fiber surface)}}{\textit{Dye}} \rightleftarrows \underset{\text{(in fiber)}}{\textit{Dye}}$$

Since the dye chemists design molecules which are in the lowest energy system where the dye is in the fiber, the dye will gradually move from solid phase to solution in the fiber.

This imposes a unique technological problem for the manufacture of disperse dyes. The rate of solution of the solid particle in the dyebath is a function of its particle size. In practice, commercial disperse dyes must be ground down or milled to particle sizes of <5 microns and preferably <1 micron. The steps required to convert a disperse dye, after chemical synthesis, to a salable product may as much as double the cost of a commercial dye.

Disperse dyes may be applied by a dry heat (Thermosol) process to polyester fibers. In this case, the dye achieves molecular form by sublimation (vaporization) from the solid dye to the fiber surface. Extremely small particle size is also important for this process.

Disperse dye molecules are generally small and have some hydroxyl or amino groups to give finite water solubility at dyeing temperatures. Examples of disperse dyes for polyester are shown below.

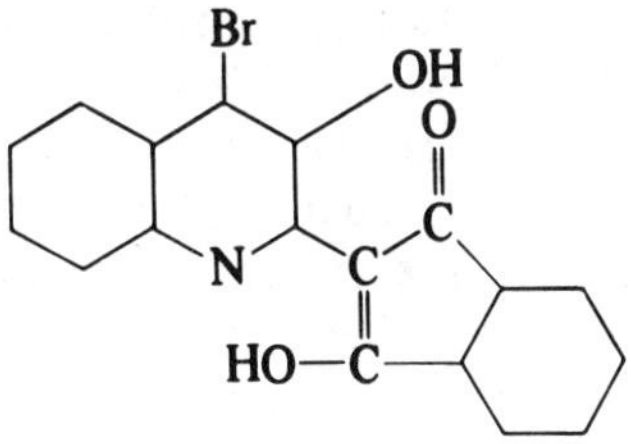

Disperse Yellow 64 (Quinophthalone)

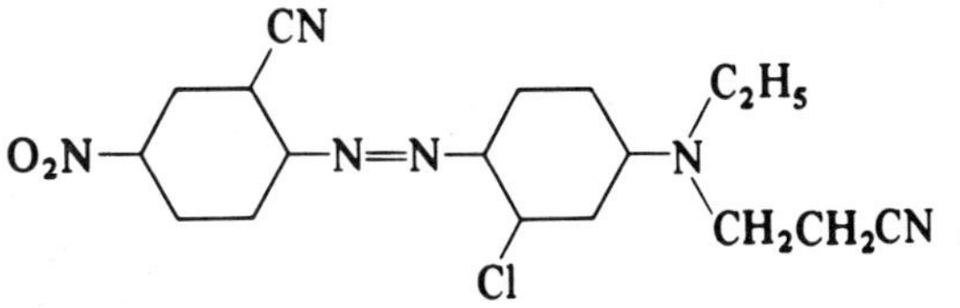

Disperse Red 73 (Azo)

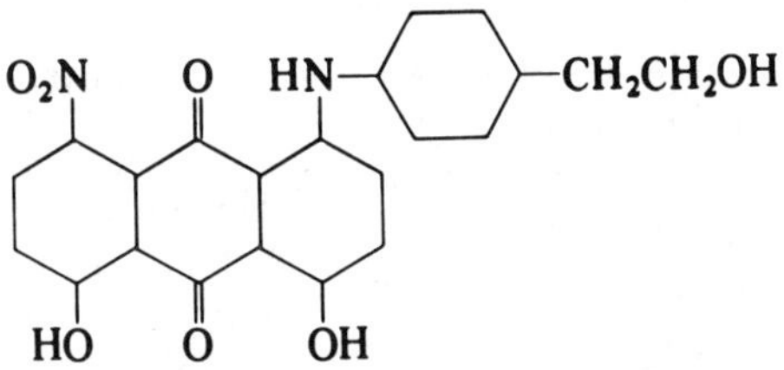

Disperse Blue 27 (Anthraquinone)

Since disperse dyes become dissolved in textile fibers, the dyeing process is reversible. Fibers that are swollen easily by water can be dyed with disperse dyes under moderate conditions, but the dyeings have only modest fastness to washing. Fibers that are more difficult to swell must be dyed under more drastic conditions, but offer the advantage that these conditions are not duplicated during normal washing procedures. For example, most polyester fibers must be dyed under pressure or with the use of organic swelling agents. The washing fastness of disperse dyes on these fibers is excellent.

Mordant Dyes

Mordant dyes require a pretreatment of the fiber with a mordant material designed to bind the dye. The mordant becomes attached to the fiber and then combines with the dye to form an insoluble complex called a "lake." An example of a mordant is aluminum hydroxide that has been precipitated in cotton fiber. This mordant is capable of binding such dyes as Mordant Red 11 (Alizarin) by formation of an aluminum "lake."

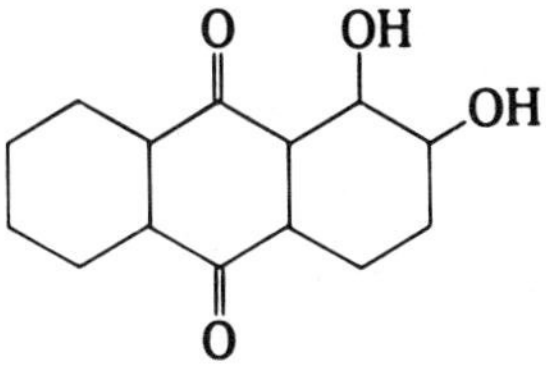

Mordant Red 11

Mordant dyes have declined in importance mainly because their use is no longer necessary. Equal or superior results can be obtained with other classes of dyes at less expense in time and labor.

Azoic Dyes

Azoic dyes are produced on textile fibers, usually cotton, by azo coupling. The dye is firmly occluded and is fast to washing. A variety of hues can be obtained by proper choice of diazo and coupling components. For example, a bluish red is produced from

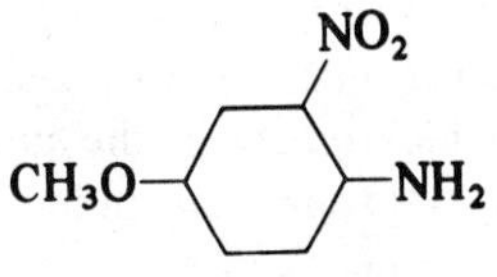

Azoic Diazo Component 1

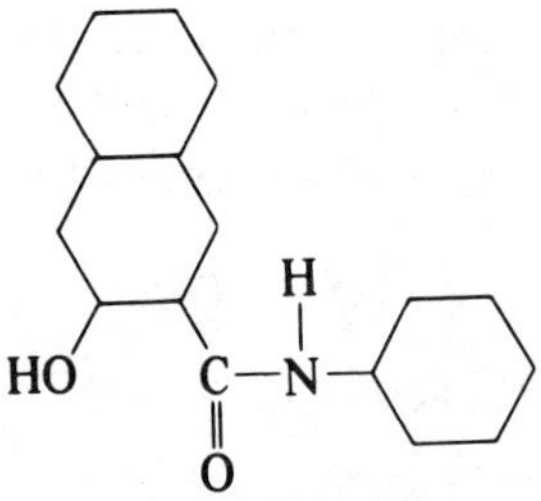

Azoic Coupling Component 2

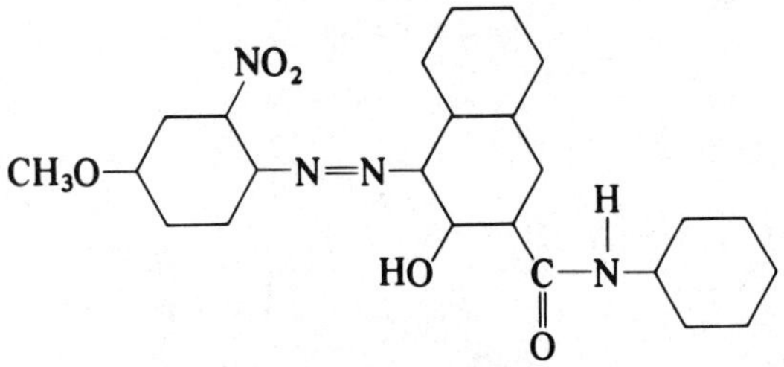

Bluish Red Azoic Dye

diazotized Azoic Diazo Component 1 and Azoic Coupling Component 2 (Naphthol AS).

In the usual procedure for the development of azoic dyes, the fiber is first impregnated with an alkaline solution of the coupling component and then treated with a solution of the diazonium compound. Finally, the dyed goods are soaped and rinsed.

The diazonium compound may be produced in the dyehouse by diazotization of the azoic diazo component, or it may be purchased as a stabilized complex ready for use. Examples of stabilized diazonium complexes are the zinc chloride double salts, nitrosamines, and diazoamino compounds.

Special techniques have been developed for the use of azoic dyes on synthetic fibers. It is sometimes possible to apply both the diazo component and the coupling component simultaneously from aqueous dispersion and then to treat the goods with nitrous acid to produce the color.

Oxidation Dyes

Oxidation dyes are produced on textile fibers by oxidation of a colorless compound. For example, aniline may be oxidized in cotton with sodium bichromate in the presence of a metal catalyst to produce an aniline black. This is an economical way to produce full black shades. The appearance and fastness of the dyeings may be varied over a wide range by the choice of oxidant, conditions, and catalyst. The exact structures of the aniline blacks are not known.

Ingrain Dyes

Ingrain dyes are produced on textile fibers. The azoic and oxidation dyes already discussed are examples. In addition, there is a small number of ingrain dyes on the market which do not fall into these classes. One such group, called precursors, is capable of generating the very bright blue dye copper phthalocyanine on cotton fibers. The dyeings are extremely fast to light and washing. The structure of copper phthalocyanine is given on page 859.

THE APPLICATION OF DYES

The process of dyeing may be carried out in batches or on a continuous basis. The fiber may be dyed as stock, yarn, or fabric. However, no matter how the dyeing is done, the process is always fundamentally the same: dye must be transferred from a bath–usually aqueous–to the fiber itself. The basic operations of dyeing include: (1) preparation of the fiber; (2) preparation of the dye bath; (3) application of the dye; and (4) finishing. There are many variations of these operations, depending on the kind of dye. The dyeing process is complicated by the fact that single dyes are seldom used. The matching of a specified shade may require from two to a dozen dyes.

Fiber Preparation

Fiber preparation ordinarily involves scouring to remove foreign materials and ensure even access to dye liquor. Some natural fibers are contaminated with fatty materials and dirt. Synthetic fibers may have been treated with spinning lubricants or sizing which must be removed. Some fibers may also require bleaching before they are ready for use.

Dye-Bath Preparation

Preparation of the dye bath may involve simply dissolving the dye in water, or it may be necessary to carry out more involved operations such as reducing the vat dyes. Wetting agents, salts, "carriers," retarders, and other dyeing assistants may also be added. "Carriers" are swelling agents which improve the dyeing rate of very hydrophobic fibers such as the polyesters. Examples are *o*-phenylphenol and biphenyl. Retarders are colorless substances that compete with dyes for dye sites or form a complex with the dye in the bath and act to slow the dyeing rate. Their use is necessary when too rapid dyeing tends to cause unevenness in the dyeings.

Dye Application

During application, dye must be transferred from the bath to the fiber and allowed to penetrate. In the simplest cases, this is done by immersing the fiber in the bath for a prescribed period of time at a suitable temperature. Unless the dye is unstable, the bath is usually heated to increase the rate of dyeing. To ensure an even uptake of dye, it is desirable to stir the bath. This is done with paddles, by pumping, or by moving the fiber. For continuous dyeing of fabrics, the dye liquor is picked up from a shallow pan, the excess is squeezed out by rollers, and penetration is assured by steaming or heating in air.

Finishing

The finishing steps for many dyes, such as the direct dyes, are very simple. The dyed material is merely rinsed and dried. Vat-dyed materials, on the other hand, must be rinsed to remove reducing agent, oxidized, rinsed again, and soaped before the final rinsing and drying steps are carried out. Generally, the finishing steps must fix the color (if this has not occurred during application) and remove any loose dye from the surface of the colored substrate. Residual dyeing assistants such as "carriers" must also be removed.

DYEING EQUIPMENT

The equipment used in modern dyehouses varies with the type of dyeing to be done. Stock dyeing is often carried out in large heated kettles made of stainless steel or other corrosion-resistant metal. These kettles can be sealed and used for dyeing at temperatures somewhat above the boiling point of water at atmospheric pressure. Yarns in packages are dyed in closed machines that circulate hot dye liquor through them. Fabrics are dyed in machines that move them through the dye liquor either under tension (jig) or relaxed (beck).

The newly developed pressure-jet dyeing machine is unique in that it has no moving parts. It is illustrated in the photograph Fig. 23.1, and it is illustrated schematically in Fig. 23.2. The cloth, in rope form, is introduced into a unidirectional liquid stream enclosed in a pipe. Liquor is pumped through a specially-designed Venturi jet imparting a driving force which moves the fabric. The two fabric ends are sewn together to form a continuous loop. The jet dyeing machine is becoming very popular for dyeing knit goods due to the absence of reels or drives which might chafe the fabric or tangle strands.

Fig. 23.1 Jet dyeing machine. *(Courtesy Gaston County Dyeing Machine Co.)*

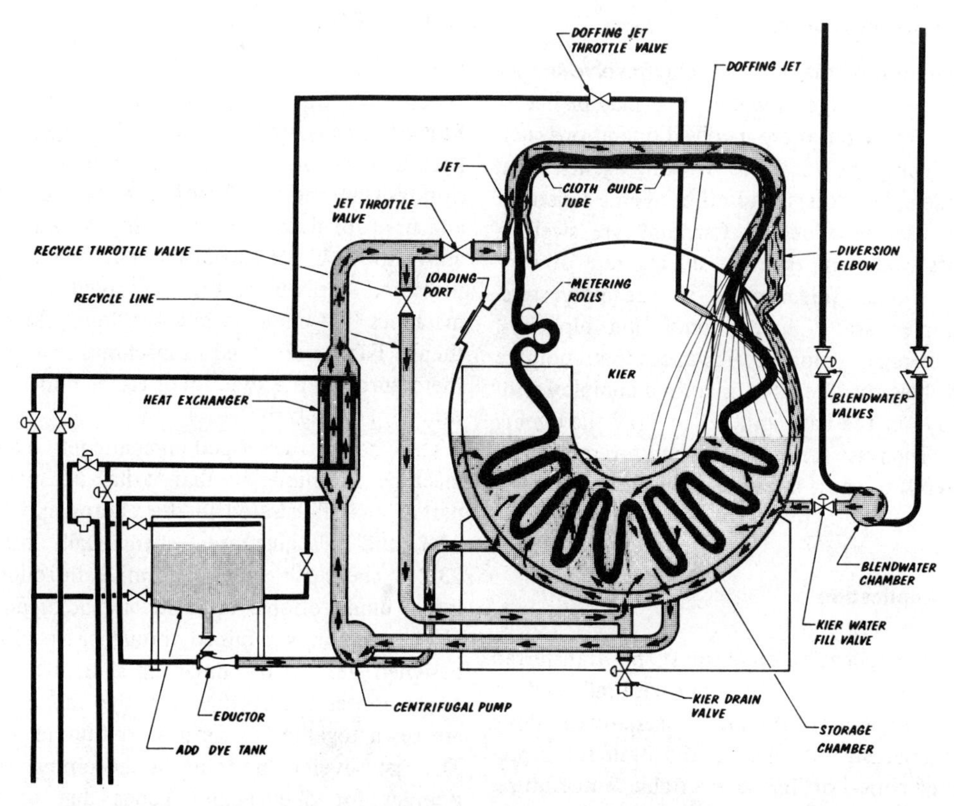

Fig. 23.2 Schematic diagram of jet dyeing machine. *(Courtesy Gaston County Dyeing Machine Co.)*

Fabrics can also be dyed in full width by winding them on a perforated beam through which hot dye liquor is pumped. This is the principle of the Burlington beam dyeing machine.

Some of the most important developments over the past 35 years have been the evolution of continuous processes having high color reproducibility, high throughput, and relatively low labor costs. Batch dyeing of standard fabrics involves lengths of about 1,000 yards depending on weight, with a dyeing cycle of 4-8 hours. This contrasts with a modern continuous dyeing machine capable of dyeing at rates up to 100 yards/minute. One such machine is capable of producing 25 million yards of dyed cloth per year, while still leaving sufficient downtime for cleanup and repair.

The first volume-yardage continuous process was the continuous pad-steam process for vat dyes on cotton.[11] The vat-dye dispersion was padded onto the cloth, dried, then passed through a reducing bath, steamed for 30 seconds, followed by an oxidizing bath, and then washing. When it was discovered that disperse dyes could be thermosoled into polyesters by dry heat for 60 seconds at 400°F, this procedure was readily adapted to continuous processing. The advent of large volumes of dyed polyester-cotton-blend fabrics in the late 1960's was made possible by combining these two processes into one thermosol pad-steam system. This is illustrated schematically in Fig. 23.3. Over one billion yards of cloth were dyed by this process in the United States in 1970.

Recently several machines have been developed in Europe which enable continuous dyeing of carpet at a rate of 10-15 yards/minute in 15-foot widths. The problems are much different from the case of textiles due to the varying pile height, heterogeneity of the substrate, and the aesthetics of the final product. Nonetheless, continuous carpet dyeing was initiated on a commercial scale in the United States in 1969. The principal processing steps

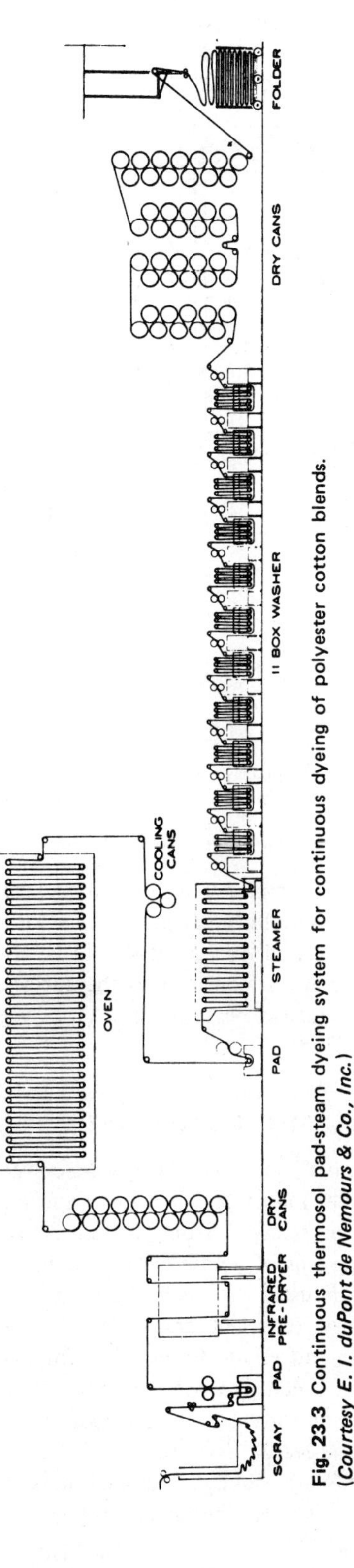

Fig. 23.3 Continuous thermosol pad-steam dyeing system for continuous dyeing of polyester cotton blends. *(Courtesy E. I. duPont de Nemours & Co., Inc.)*

are saturation with assisting chemicals, application of dyes, steaming, washing, and drying.[2] A perspective drawing is given in Figure 23.4.

In addition to the above, special machines are available to dye textile fibers at any stage during their conversion to finished goods. The principles involved are the same as those outlined in the preceding section "Application of Dyes."

PRINTING

Printing is a special kind of localized dyeing which produces patterns. Four kinds of printing have long been recognized: (1) direct, (2) dyed, (3) discharge, and (4) resist. In *direct* printing, a thickened paste of the dye is printed on the fabric to produce a pattern. The fabric is then steamed to fix the dye, and it is finished by washing and drying. *Dyed* printing requires that the pattern be printed on the fabric with a mordant. The entire piece is then placed in a dyebath containing a mordant dye, but only the mordanted areas are dyeable. Washing then clears the dye from the unmordanted areas, leaving the pattern in color.

In *discharge* printing, the cloth is dyed all over and then printed with a substance which can destroy the dye by oxidation or reduction, leaving the pattern in white. When a reducing agent such as sodium hydrosulfite is used to destroy the dye, the paste may contain a reduced vat dye. Finishing the goods by oxidation and soaping then produces the pattern in color. In *resist* printing, certain colorless substances are printed on the fabric. The whole piece is then dyed but the dye is repelled from the printed areas, thus producing a colored ground with the pattern in white.

Printing is most often done with copper rollers etched in the design to be printed. Printing paste is fed constantly to the roller from a trough. A scraper then clears the surface of the roller, leaving the dye paste only in the etched areas from which it is transferred to the fabrics.

An important recent advance in the pattern-coloring of textiles is heat-transfer printing.[5] In this process, the desired pattern is first printed on paper, using inks containing sublimable

Fig. 23.4 Basic elements of the Küsters carpet dyeing process: (1) chemical saturation, (2) extraction, (3) dyestuff application and (4) steaming. (Reprinted from *Textile Chemist and Colorist*, **1**, (2) Jan. 15, 1969.) (*By permission.*)

disperse dyes. Next, the printed paper is placed in contact with a disperse-dyeable fabric, such as a polyester knit, and heat and pressure are applied to the back of the paper. The dyes transfer by subliming from the paper, diffusing across an air gap to the surface of the fabric, condensing, and then diffusing into the interior of the fiber.

PIGMENT DYEING AND PRINTING

Pigment dyeing and printing are processes that compete with the more conventional means of dyeing and printing described above. These processes use water-insoluble dyes or pigments which are bound to the surfaces of fabrics with resins. A paste or emulsion containing pigment and resin or resin-former is applied to the fabric. The goods are then dried and cured by heat to produce the finished dyeing or print. During the heating or cure, fabric, resin, and pigment become firmly bonded together. This method of color application is economical and produces good results. It should be noted that the pigment is confined to the surface of the fabric and can be selected without regard for fiber affinity.

NONTEXTILE USES OF DYES

Substantial quantities of dye are consumed by the paper industry. The most important paper dyes are selected from the direct, acid, and basic groups of textile dyes. The needs of the paper industry are such that it is frequently necessary to select and finish the dyes in a manner different from that employed for textiles. Such dyes are now normally supplied in solution for the convenience of the user.

Natural leather is a protein substance and can be dyed with acid dyes, among others. Finished leather also can be dyed or pigmented, the choice of color type depending on the nature of the finish.

Dyes for many materials such as gasoline, lacquers, inks, and varnishes are selected on the basis of solubility, as well as hue and fastness. It is necessary to modify some dyes to achieve the desired degree of solubility. For example, the solubility in alcohol of many acid dyes is greatly increased by forming the diphenylguanidine salts.

Important and growing markets for dyes are the plastic and metal industries. Dyed anodized aluminum, in particular, is finding increasing use in automotive and architectural applications.

Dyes for foods, drugs, and cosmetics must be chosen with great care to avoid toxic effects. In the United States, dyes are certified by the government when judged safe for such uses. Only the certified dyes may legally be used. This restriction does not apply to natural colors.

PRODUCTION AND USES

Table 23.1 shows the distribution of the free world's production of dyes by country.[12] In 1975 the United States and Germany were the two largest manufacturers, with Japan and the United Kingdom next.

Table 23.2 gives the sales of synthetic dyes in the United States according to application class. These figures are taken from the report of the U.S. International Trade Commission.

TABLE 23.1 World Production of Dyes in 1975

Country	*Million Pounds*	*Million Dollars*
Belgium	3.7	10
France	43.8	123
Germany	191	694
Italy	30.8	67
Japan	125	277
Netherlands	4.8	8
Spain	24.2	40
Switzerland	44.2	297
United Kingdom	95	244
United States	206	476
TOTAL	768.5	2,236

RAW MATERIALS FOR THE MANUFACTURE OF DYES

The raw materials for the manufacture of dyes are mainly aromatic hydrocarbons, such as benzene, toluene, naphthalene, anthracene, pyrene, phenol, pyridine, and carbazole. In the past, these aromatic hydrocarbons came almost exclusively from the distillation of coal tar, but in recent years increasing quantities, especially of benzene and toluene, have become available from petroleum and natural gas. The term "coal-tar dyes," still widely considered synonymous with synthetic dyes, is no longer an entirely correct description. A great variety of inorganic materials are required by the dye industry. These include sulfuric acid, oleum, nitric acid, chlorine, bromine, caustic soda, sodium nitrite, hydrochloric acid, sodium carbonate, sodium hydrosulfite, sodium sulfide, aluminum chloride, sodium bichromate, and manganese dioxide.

DYE INTERMEDIATES

The raw materials for dyes are almost never directly useful in dye synthesis. It is necessary to convert them to a variety of derivatives which are in turn made into dyes. These derivatives are called *intermediates.* They are produced by reactions such as nitration, reduction, sulfonation, halogenation, oxidation, and condensation. Most of these reactions lead to the formation of substituted hydrocarbons which are functional in nature, that is, they bear groups capable of undergoing further chemical reaction. The number of dye intermediates actually or potentially available is very large, and the technology of their manufacture is an important part of industrial organic chemistry. Intermediates are used not only for dye manufacture but also for the manufacture of other important products such as pharmaceuticals.

The substitution reactions of aromatic hydrocarbons, in which a hydrogen atom is replaced by another group, frequently lead to the formation of position isomers. Position isomers are compounds which are alike in the groups that they contain, but different in the relative

TABLE 23.2 Benzenoid Dyes: U.S. Sales, by Class of Application, 1978

Class of Application	*Quantity (million pounds)*	*Value (million dollars)*
Acid	36.7	130.2
Azoic dyes and components	1.3	1.8
Basic	15.3	67.4
Direct	26.8	80.5
Disperse	39.7	156.8
Fiber-reactive	5.3	28.7
Fluorescent brightening agents	29.4	49.6
Food, drug, and cosmetic colors	6.0	42.1
Mordant	0.4	1.7
Solvent	10.3	32.6
Vat	36.9	102.2
All other	24.7	39.9
TOTAL	232.8	733.5

positions of these groups, e.g., the two nitrophenols shown in the section on nitration. When position isomers are formed, it is almost always necessary to effect a separation since the isomers differ in properties and lead to different dyes. One of the most difficult problems in the manufacture of dye intermediates is the efficient separation of isomers. A further problem is the control of the relative quantities in which isomers are formed, or the discovery of uses for all of them in the event that control is impractical.

The many reactions employed in the manufacture of dye intermediates, and the delicate nature of many of the operations make it mandatory that all processes be well-planned and controlled. Some of the processes, nitration for example, are inherently dangerous if not properly run. For these reasons, close supervision by technically trained men is the rule in plants where intermediates are made.

In recent years, there has been an ever increasing emphasis on high standards of uniformity and quality in chemical manufacturing. At the same time, the increasing cost of labor has discouraged the use of numerous manual controls to meet these standards. Consequently, many plants have turned to partly or fully automatic control. Such practice requires advanced instrumentation that can insure that all process variables are closely controlled.

Because of the large number of compounds that are required, often in limited amounts, most dye intermediates are manufactured in batches. Some of the more fundamental processes can be run continuously, however, with a decided economic advantage. Where continuous production is not justified, the largest practical batches are made to hold the costs of operation to a minimum.

A wide variety of batch-type equipment is used for the manufacture of dye intermediates. Reaction kettles are made from wood, cast iron, stainless steel, or steel, often lined with rubber, glass (enamel), brick, or other corrosion-resistant material. Usual production sizes are 500-10,000 gallons, and the kettles are equipped with mechanical agitators, thermometers, condensers, and cooling or heating coils or jackets, depending on the nature of the operation. Products are generally transferred by gravity flow, pumping, or blowing. Isolation is generally achieved by plate and frame filter presses, filter nutsches, centrifugation, or continuous rotary filters. Both pressure and vacuum filters are used. Drying of dyes and intermediates is done in air or vacuum tray dryers or graining bowls. For larger volumes, spray dryers are becoming increasingly important.

Nitration

The nitration of aromatic hydrocarbons is a fundamental operation in the manufacture of many dye intermediates. Nitration involves the replacement of one or more ring hydrogen atoms by the nitro ($-NO_2$) group. The nitration of benzene is an example.

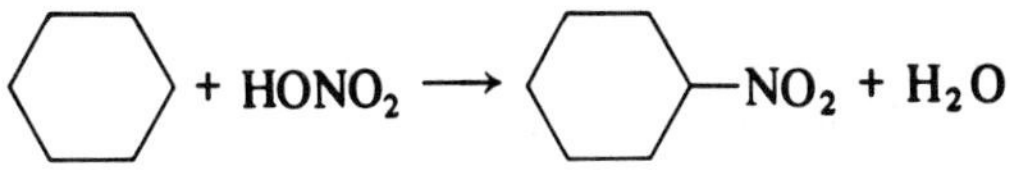

$$C_6H_6 + HONO_2 \longrightarrow C_6H_5NO_2 + H_2O$$

Nitrobenzene

Nitration of compounds of high reactivity is carried out with nitric acid in water or an organic solvent. Less reactive compounds are nitrated in a combination of nitric and sulfuric acids ("mixed acids"). The sulfuric acid serves as a solvent for the reaction and facilitates nitration by reaction with nitric acid to form the nitronium ion ($-NO_2^+$), generally believed to be the active nitrating agent. It also serves to maintain the strength of the nitrating mixture by combining with the water which is formed. A typical mixed acid consists of 33 percent nitric acid and 67 percent sulfuric acid. Since nitric acid is a strong oxidizing agent, nitrations are carried out at low temperatures to avoid destructive side reactions. Some functional groups, such as the amino group, in the compounds to be nitrated must be protected against oxidative destruction by acetylation. The acetyl group is removed by hydrolysis when nitration is complete.

When nitrating benzene to nitrobenzene, an excess of nitric acid must be avoided or dinitration will occur. The procedure is to run the "mixed acids" into the benzene, not the converse; this is a general rule when mononitration is desired. About 2,500 pounds of benzene is nitrated in one batch over a period of three to four hours. Heat is evolved during the reaction and is removed by means of cooling coils, or by means of cold water or brine circulated in a jacket surrounding the nitration vessel. Cast iron is used since it is not attacked by "mixed acids." The vessel is agitated to provide good contact between the two layers and to facilitate heat transfer. When the reaction is complete, the agitation is stopped and the nitrobenzene separates as an oil over the acid. This oil is drawn off and agitated with water or dilute alkali to remove residual acid. It may then be distilled if pure nitrobenzene is required. In recent years, continuous vapor-phase nitration methods have been devised. Aqueous nitric acid is used and the water resulting from the reaction is distilled off continuously to keep the nitric acid concentration high enough to be effective. Benzene can be nitrated continuously at about 80°C using 61 percent aqueous nitric acid.

When benzene derivatives are nitrated, isomers of the desired product are obtained in addition to the product itself. For example, nitration of phenol by nitric acid gives *o*- and *p*-nitrophenol.

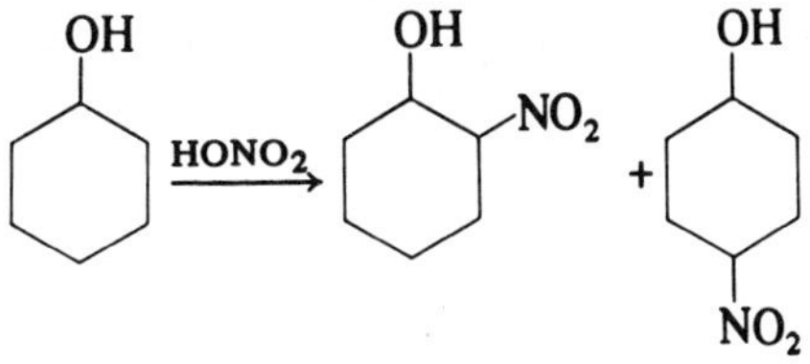

OH ——HONO₂——→ OH, NO₂ + OH, NO₂

Phenol *o*-Nitrophenol *p*-Nitrophenol

The isomeric *o*- and *p*-nitrophenols can be separated by steam distillation. The *ortho* isomer is volatile in steam, while the *para* isomer is not and, therefore, remains in the distillation vessel. Commercially, *p*- and *o*-nitrophenols are made by hydrolysis of the respective chloronitrobenzenes.

Mononitration of anthraquinone at about 50°C gives mainly 1-nitroanthraquinone. At 80-95°C, dinitration occurs to give a mixture of the 1,5- and 1,8-isomers. These isomers are important as starting materials for the preparation of other intermediates. In some cases the mixture is used; in others, isomer separation is necessary.

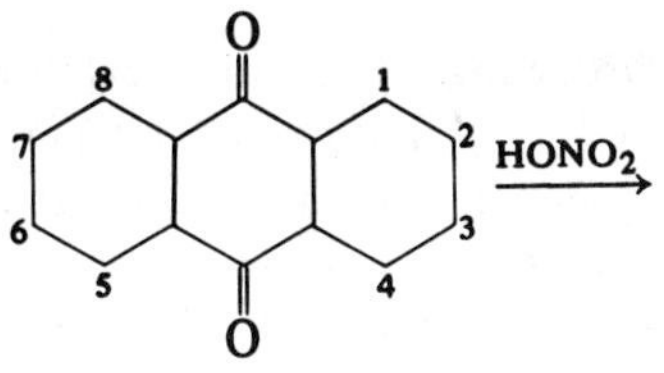

O, O, 8, 1, 7, 2, 6, 3, 5, 4 ——HONO₂——→

Anthraquinone

O, NO₂, O ——HONO₂——→

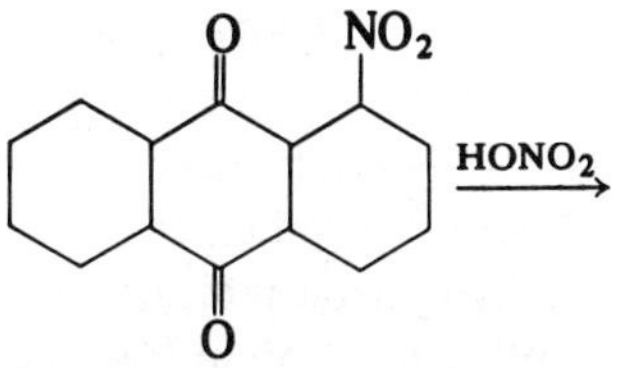

1-Nitroanthraquinone

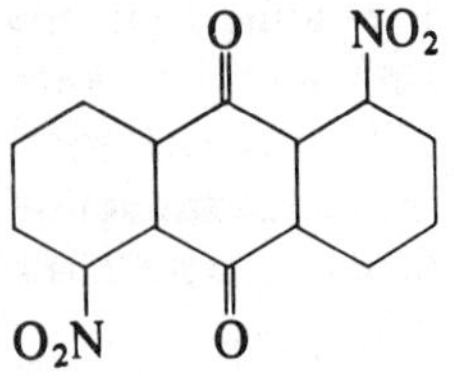

1,5-Dinitro-anthraquinone

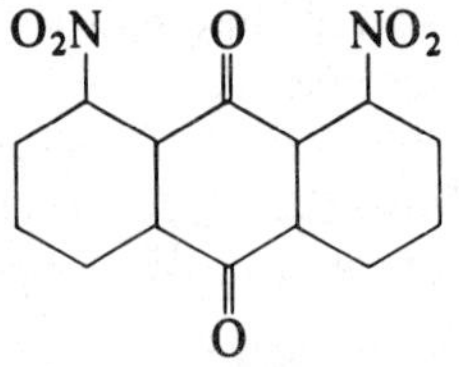

1,8-Dinitro-anthraquinone

These nitrations are performed in cast-iron or steel vessels with steel agitators. Since the starting materials are solids they are first dissolved in sulfuric acid and then treated with "mixed acids"; the products are also solids.

Reduction

The most common reduction reaction in the manufacture of dye intermediates is the conversion of a nitro compound to the corresponding amine. This reaction is illustrated by the reduction of nitrobenzene to aniline.

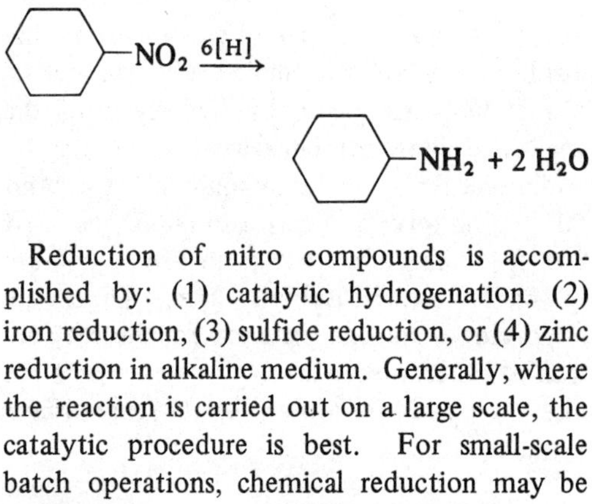

$$C_6H_5-NO_2 \xrightarrow{6[H]} C_6H_5-NH_2 + 2\,H_2O$$

Reduction of nitro compounds is accomplished by: (1) catalytic hydrogenation, (2) iron reduction, (3) sulfide reduction, or (4) zinc reduction in alkaline medium. Generally, where the reaction is carried out on a large scale, the catalytic procedure is best. For small-scale batch operations, chemical reduction may be preferred.

Catalytic hydrogenation requires a catalyst such as nickel, copper, platinum, molybdenum, or tungsten. These catalysts are usually supported on other materials and are especially prepared for the type of reduction to be carried out. Reduction conditions vary widely, depending on the nature of the nitro compound and the catalyst. Reduction may be carried out in solvent in the vapor phase or in the liquid phase. Aniline can be made by continuous vapor-phase reduction of nitrobenzene at 350–460°C at nearly atmospheric pressure. Some reductions, on the other hand, are run at 1000 to 4000 psi.

Iron reduction is employed on a large scale because of its simplicity. Iron turnings are used in an agitated aqueous system containing a small amount of acid to promote reaction. The over-all reaction is illustrated for nitrobenzene as follows:

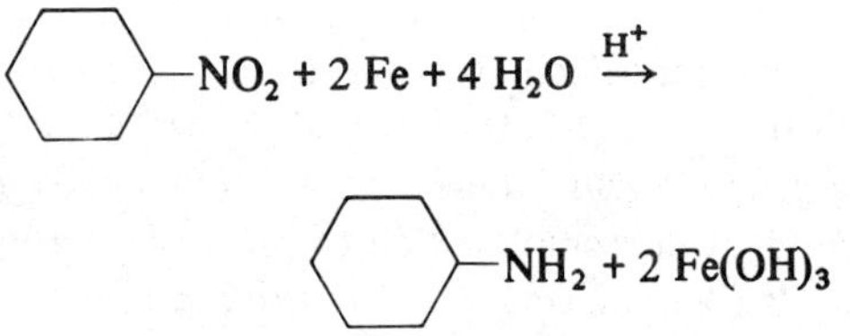

$$C_6H_5-NO_2 + 2\,Fe + 4\,H_2O \xrightarrow{H^+} C_6H_5-NH_2 + 2\,Fe(OH)_3$$

The nitrobenzene is placed in a reducer, a vertical cylindrical vessel provided with cover, steam jacket, and agitator. The iron turnings, or powder, and a small amount of hydrochloric acid are added in small portions. A brisk reaction is maintained by means of steam circulated in the jacket of the reducer or blown directly into the charge. A condenser returns to the reducer any vapors that escape. After the nitrobenzene is completely converted to aniline, a strong current of live steam is passed into the charge; a mixture of steam and aniline vapors passes to the condenser and is collected in storage tanks. The bulk of the aniline separates as a lower layer and is drawn off; the water over it still contains aniline, which must be recovered by distilling this "aniline water" again, or by extracting it with nitrobenzene. The iron sludge is washed out of the reducer through a side outlet by flushing. A reducer 6 feet in diameter and 10 feet high takes a charge of 5000 pounds of nitrobenzene in one batch and requires about 10 hours for reduction. The aniline may be redistilled, which renders it water white.

However, iron reduction of nitro compounds is being de-emphasized in favor of catalytic reduction which is more efficient in a labor intensive industry such as dyes.

Sulfide reduction employs sodium sulfide, sodium polysulfide, or sodium hydrosulfide. An important feature of this type of reducing system is its adaptability to bring about stepwise reduction of dinitro compounds. Partial reduction is illustrated with *m*-dinitrobenzene which can be reduced to *m*-nitroaniline with sodium sulfide under controlled conditions.

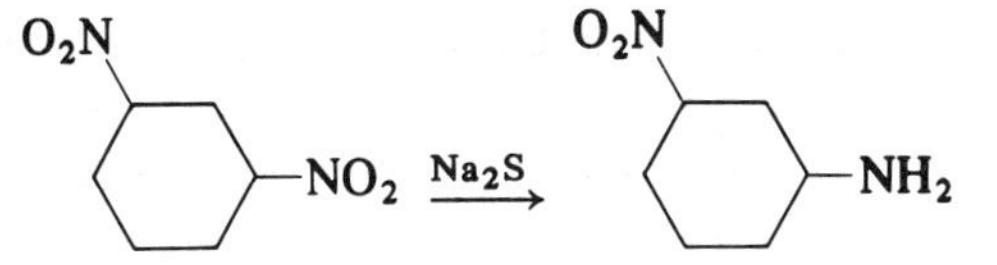

$$m\text{-Dinitrobenzene} \xrightarrow{Na_2S} m\text{-Nitroaniline}$$

m-Dinitrobenzene *m*-Nitroaniline

The sodium sulfide is dissolved in alcohol and placed in a steam-jacketed reducer; the dinitrobenzene is added either in solid form or dissolved in alcohol. The mixture is boiled for two hours; then the alcohol is distilled off and collected for re-use. The *m*-nitroaniline mixed with inorganic salt remains in the reducer. The mass is agitated with water, which dissolves the salt, and is then pumped into a filter press. The press cake of *m*-nitroaniline is washed, and then discharged and dried in a vacuum drier.

Zinc reduction in alkaline aqueous or alcoholic medium is especially useful to bring about bimolecular reduction. This kind of reaction is illustrated by the conversion of nitrobenzene to hydrazobenzene. Rearrangement of hydrazobenzene with acid gives benzidine, a formerly important intermediate for azo dyes, which is now banned due to its carcinogenic activity.

$$2\ C_6H_5NO_2 \xrightarrow{Zn,\ OH^{\ominus}} C_6H_5NH{-}NHC_6H_5 \xrightarrow{H^{\oplus}} H_2N{-}C_6H_4{-}C_6H_4{-}NH_2$$

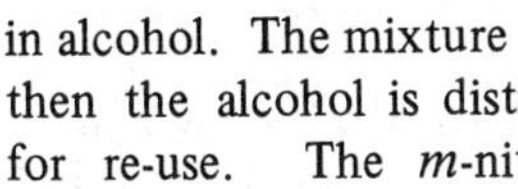

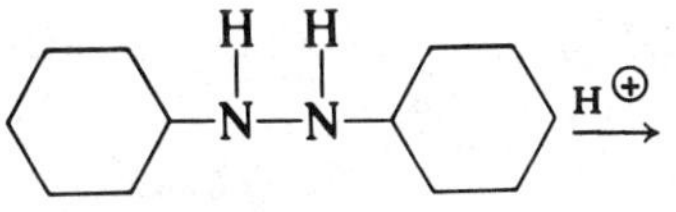

Hydrazobenzene

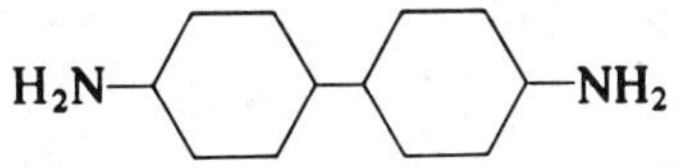

Benzidine

In a similar way, *o*-nitroanisole is converted to hydrazoanisole and then to *o*-dianisidine.

Amination

The introduction of an amino group into an aromatic nucleus by replacement of another functional group is called amination. This process is to be distinguished from reduction of a nitro group in that one group is totally displaced by another and not simply altered in character.

An example of amination in the benzene series is the conversion of *p*-nitrochlorobenzene to *p*-nitroaniline with ammonia. This reaction may be carried out continuously with 40 percent aqueous ammonia under 200 atmospheres pressure at 235–240°C.

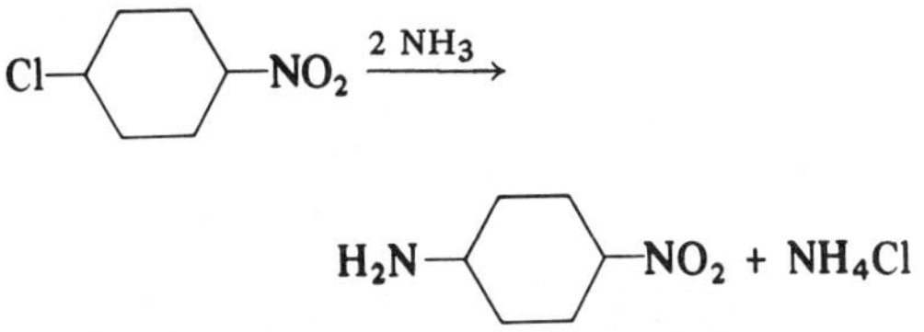

$$Cl{-}C_6H_4{-}NO_2 \xrightarrow{2\ NH_3} H_2N{-}C_6H_4{-}NO_2 + NH_4Cl$$

The Bucherer reaction illustrates amination in the naphthalene series. A naphthol is heated with ammonium sulfite or ammonia and alkali metal bisulfite. The result is replacement of the hydroxyl group by an amino group, probably by way of a bisulfite addition product of the keto form of the naphthol. However, 2-naphthylamine is no longer used as a chemical of commerce due to its carcinogenic activity.

$$\text{2-Naphthol} \underset{}{\overset{NaHSO_3}{\rightleftharpoons}} \text{Addition product (H, H, OH, } SO_3Na) \overset{NH_3}{\rightleftharpoons} \text{2-Naphthylamine } (NH_2)$$

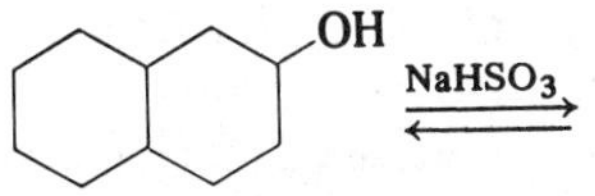

2-Naphthol

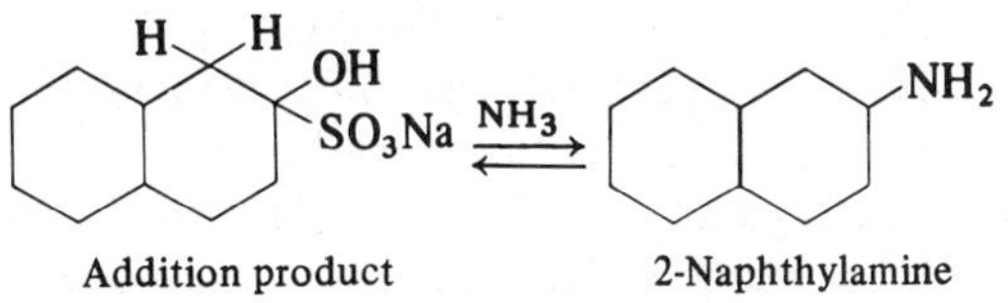

Addition product

2-Naphthylamine

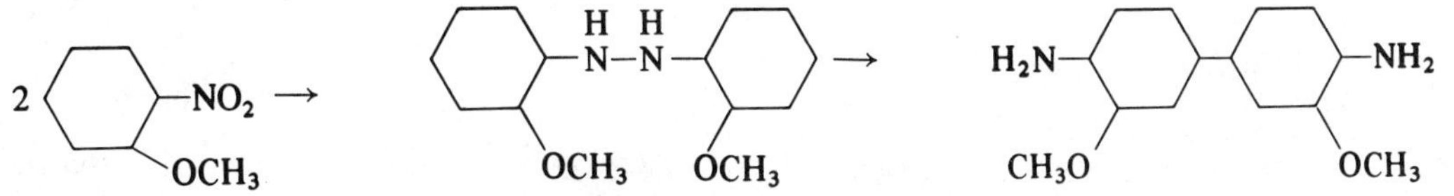

$$2\ o\text{-}C_6H_4(NO_2)(OCH_3) \rightarrow \text{Hydrazoanisole } (NH{-}NH,\ OCH_3,\ OCH_3) \rightarrow H_2N{-}C_6H_3(OCH_3){-}C_6H_3(OCH_3){-}NH_2$$

o-Nitroanisole Hydrazoanisole *o*-Dianisidine

The reaction is reversible and may be used to convert naphthylamines to naphthols.

In the anthraquinone series, amination is frequently a convenient means of preparing amines. 1-Aminoanthraquinone-2-carboxylic acid can be made by reaction of 1-nitroanthraquinone-2-carboxylic acid with 15 percent aqueous ammonia at 130°C. The nitro group is displaced, not reduced. In a similar manner, 1-anthraquinonesulfonic acid can be aminated to give 1-aminoanthraquinone or, if desired, ammonia may be replaced by methylamine to give 1-(methylamino)anthraquinone.

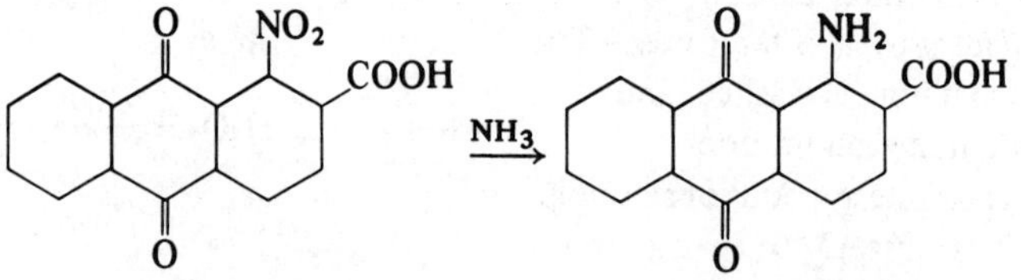

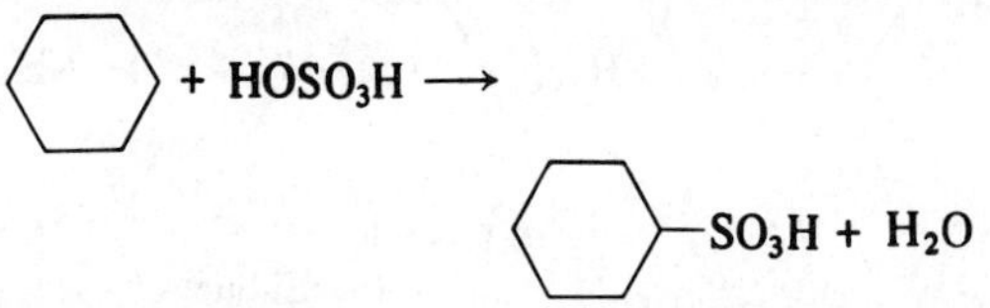

The actual sulfonating agent is believed to be the cation, $^{+}SO_3H$. In carrying out this reaction, oleum containing 8 percent free sulfur trioxide is added slowly to offset the dilution caused by the water formed in the process. The temperature is maintained at 30°C until near the end, when it is raised to 50°C. When reaction is complete, the charge is diluted by running it into water, and the product is precipitated by adding salt. It is isolated by filtration as the sodium sulfonate.

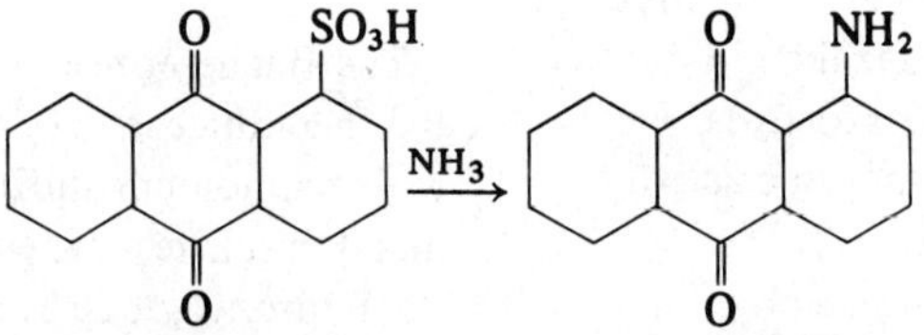

Substitution rules for sulfonation are similar to those for nitration. The groups, $-NO_2$,

1-Nitroanthraquinone-2-carboxylic acid $\xrightarrow{NH_3}$ 1-Aminoanthraquinone-2-carboxylic acid

1-Anthraquinonesulfonic acid $\xrightarrow{NH_3}$ 1-Aminoanthraquinone

Sulfonation

The sulfonic acid group ($-SO_3H$) is one of the more common substituents in dye intermediates. It is introduced to render intermediates soluble in water, or to provide a route to other substituents, such as the hydroxyl group which is obtained by subsequent alkaline fusion.

Direct sulfonation is achieved with: (1) strong sulfuric acid, (2) oleum (sulfuric acid plus sulfur trioxide), (3) sulfur trioxide in organic solvent or as a complex, or (4) chlorosulfonic acid. The sulfonation of benzene with sulfuric acid is illustrated by the following equation:

$$C_6H_6 + HOSO_3H \longrightarrow C_6H_5SO_3H + H_2O$$

$-COOH$, and $-SO_3H$ direct the entering group to the *meta* position.

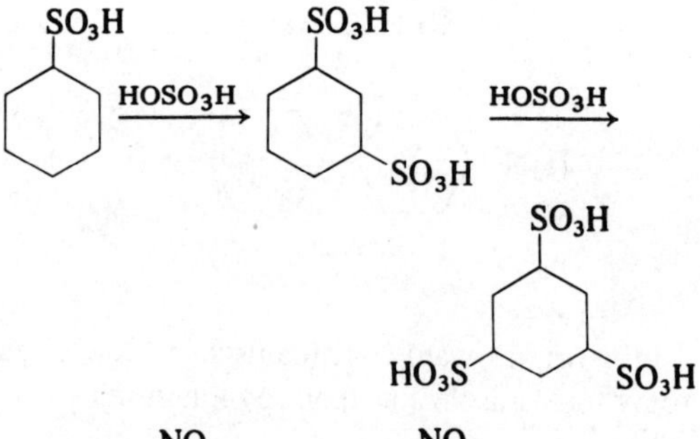

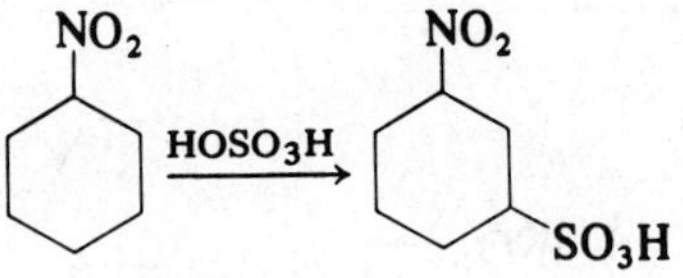

$$C_6H_5SO_3H \xrightarrow{HOSO_3H} m\text{-}C_6H_4(SO_3H)_2 \xrightarrow{HOSO_3H} 1,3,5\text{-}C_6H_3(SO_3H)_3$$

$$C_6H_5NO_2 \xrightarrow{HOSO_3H} m\text{-}NO_2C_6H_4SO_3H$$

Alkyl groups, for example, methyl, direct the entering group predominantly to the *ortho* and *para* positions. Usually, a mixture of isomers is formed.

CH_3 —$HOSO_3H$→ CH_3, SO_3H (ortho) + CH_3, SO_3H (para)

Chlorine is similar to the methyl group in its effect on orientation but gives less of the *ortho* isomer.

The directing effect of amino groups depends on their basicity. The less basic amines are *ortho-para* directing. Aniline is an example of this type. More basic amines, for example, dimethylaniline, form *meta*-directing salts in acid.

H_3C—N—CH_3 —$H^{\oplus}$→ H_3C—$\overset{H}{N}^{\oplus}$—CH_3 —$HOSO_3H$→ H_3C—N—CH_3, SO_3H (meta)

In addition to sulfonation with sulfuric acid or its equivalent, amines may be sulfonated by baking the sulfates at elevated temperatures. This procedure offers the advantage of giving fewer isomers. The baking of aniline sulfate at 260–280°C gives a high yield of the *para* sulfonic acid.

Indirect sulfonation may be achieved in a number of ways. Sodium bisulfite will often replace a labile functional group with the sulfonic acid group. An example is *o*-chlorobenzoic acid which is converted to *o*-sulfobenzoic acid by aqueous sodium bisulfite.

Cl, COOH —$NaHSO_3$→ SO_3H, COOH

The sulfonation of naphthalene yields a number of isomers. The product obtained may be controlled to some extent by the choice of agent. With any one agent, temperature and time of reaction determine the result. It is rarely possible to obtain a single isomer; but effort is directed toward forming a preponderant amount of one isomer. For example, for the monosulfonates, made by direct sulfonation with sulfuric acid, there is formed at 80°C in eight hours, 96 percent 1-naphthalenesulfonic acid. As the temperature is raised, correspondingly less of this isomer is formed and more of the 2-sulfonic acid. At 150°C, for example, 18 percent of the 1-isomer is formed and over 80 percent of the 2-isomer.

Naphthalene —$HOSO_3H$→ SO_3H (1-isomer) + SO_3H (2-isomer)

When carrying out this sulfonation, the acid is run into the melted naphthalene to avoid disulfonation. The amount of acid added is the calculated amount for one sulfonic acid group. The water formed during the reaction retards but does not prevent it. Oleum may be added toward the end to hasten the reaction.

The sulfonation of amino and hydroxy derivatives of naphthalene usually leads to a large number of isomers. To avoid isomer-formation as much as possible, naphthionic acid is often prepared by baking the sulfate of 1-naphthylamine.

$NH_3^{\oplus}\,\frac{SO_4^{\ominus}}{2}$ —heat→ NH_2, SO_3H

By direct sulfonation of 1-naphthylamine, four of the seven possible 1-naphthylamine-sulfonic acids may be formed. The main product under proper conditions is the 1,4-isomer. Direct sulfonation of 2-naphthylamine yields primarily a mixture of 2,5- and 2,8-isomers.

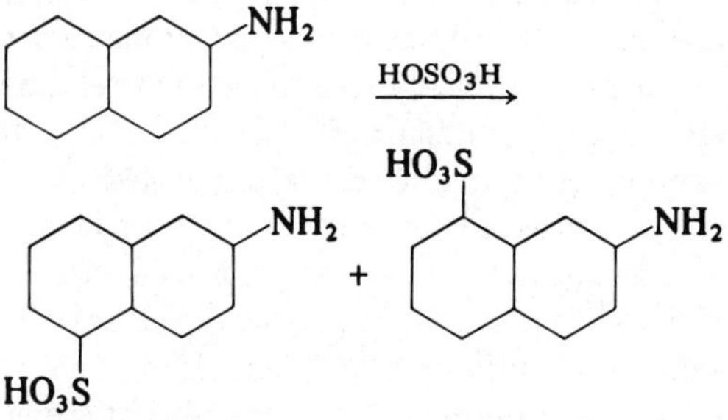

Two important monosulfonic acids resulting from the sulfonation of 2-naphthol are Schaeffer's acid and Crocein acid. At 110°C, Schaeffer's acid is preponderant; at lower temperatures more Crocein acid is formed.

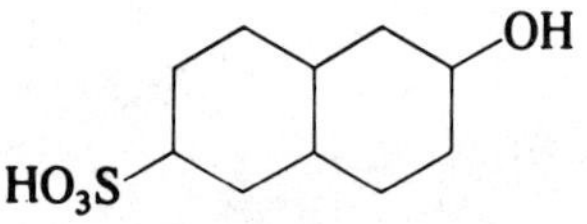

Schaeffer's acid

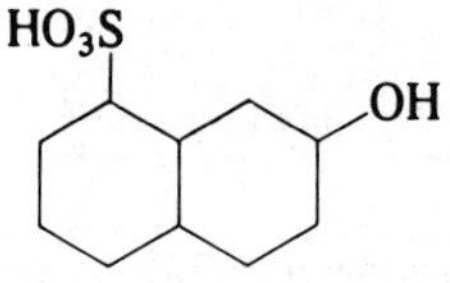

Crocein acid

By further sulfonation, two isomeric disulfonic acids are the main products. In the cold, G acid predominates, while at higher temperatures R acid is formed in greater amount. Both isomers are important intermediates for azo dyes.

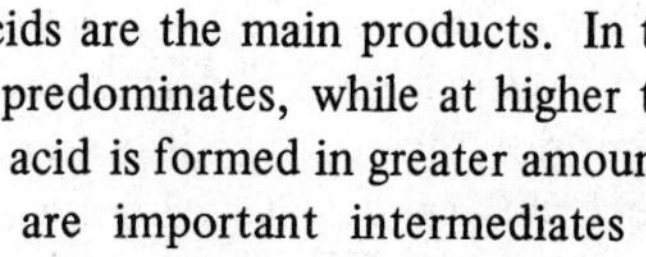

R acid

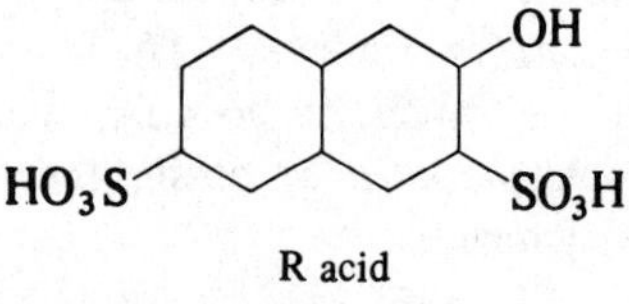

G acid

Anthraquinone is sulfonated by suspending it in oleum containing 45 percent free sulfur trioxide and heating at 150°C for one hour. The resulting melt is run into water and neutralized with sodium hydroxide while still hot. On cooling, the sodium salt of the 2-sulfonic acid separates. Further sulfonation produces a mixture of the 2,6- and 2,7-disulfonic acids.

When anthraquinone is sulfonated in the presence of mercury sulfate, the results differ from those just described. A single sulfonic acid group enters at position 1. Two groups enter to form the 1,5- and 1,8-disulfonic acids. The 1,5-isomer is salted out from the more soluble 1,8-isomer after dilution of the sulfonation mass. The 1,5- and 1,8-disulfonic acids are of great importance for the manufacture of other derivatives, which can be made by replacement of the sulfonic acid groups. Examples are the chloro- and hydroxyanthraquinones.

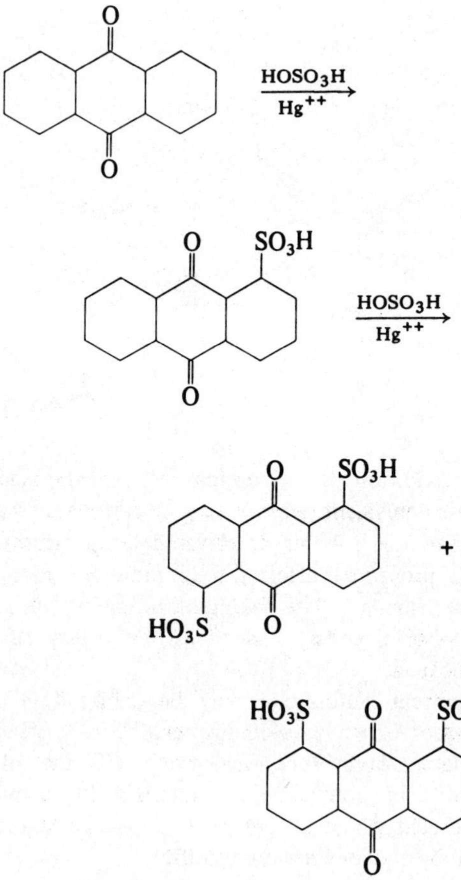

Halogenation

Chlorine is the most widely used of the halogens because it is comparatively economical. Most often, chlorinations are performed by dried chlorine gas, that is, by direct chlorination with or without a catalyst. An alternate procedure consists of generating nascent chlorine *in situ* by the oxidation of a chlorine-containing compound. In a few cases, chlorination may be achieved with reagents such as thionyl chloride, phosphorus oxychloride, phosphorus pentachloride, or sulfuryl chloride.

Chlorination of alkylated aromatic compounds, for example, toluene, can occur either in the aromatic ring or in the side chain. The use of an iron catalyst directs the chlorine to the aromatic ring, probably by inducing formation of the Cl^+ cation as the active agent. Without catalyst, chlorination takes place in the side chain, especially under ultraviolet light; the active agent in this case is believed to be the Cl · radical.

Chlorobenzene is made by passing a stream of dried chlorine into benzene in the presence of ferric chloride; some *p*-dichlorobenzene is formed at the same time. To produce monochlorobenzene as the sole product, excess benzene is only partially converted and unreacted benzene is recycled. Even so, some polychlorobenzene is obtained.

$$C_6H_6 + Cl_2 \xrightarrow{FeCl_3} C_6H_5Cl + HCl$$

Chlorination of toluene in the complete absence of iron produces side-chain chlorination products; these are benzyl chloride, benzal chloride, and benzotrichloride. All these compounds are valuable, although the reaction is difficult to stop to produce pure intermediate products. Benzal chloride is converted to benzaldehyde with calcium carbonate in water, while benzotrichloride gives benzoic acid under the same conditions.

$$C_6H_5CH_3 \xrightarrow{Cl_2} C_6H_5CH_2Cl \xrightarrow{Cl_2} C_6H_5CHCl_2 \xrightarrow{Cl_2} C_6H_5CCl_3$$

Toluene, Benzyl chloride, Benzal chloride, Benzotrichloride

$$C_6H_5CHCl_2 \xrightarrow[CaCO_3]{H_2O} C_6H_5CHO \qquad C_6H_5CCl_3 \xrightarrow[CaCO_3]{H_2O} C_6H_5COOH$$

Benzaldehyde, Benzoic acid

In the naphthalene series, direct chlorination is seldom used. The reaction takes place readily but leads to numerous isomers. In the anthraquinone series, both direct and indirect chlorination are employed. An example of indirect chlorination is the conversion of 1-anthraquinonesulfonic acid to the corresponding chloro compound. This reaction is carried out at approximately 100°C with sodium or potassium chlorate in hydrochloric acid.

$$\text{1-anthraquinone-}SO_3H \xrightarrow[HCl]{NaClO_3} \text{1-chloroanthraquinone}$$

Chlorination of aliphatic compounds is illustrated by the chlorination of acetic acid. One, two, or three of the hydrogen atoms on carbon can be replaced by direct chlorination of the warm liquid in the presence of sulfur.

$$CH_3COOH \xrightarrow{Cl_2} ClCH_2COOH \xrightarrow{Cl_2}$$

Monochloroacetic acid

$$Cl_2CHCOOH \xrightarrow{Cl_2} Cl_3CCOOH$$

Dichloroacetic acid   Trichloroacetic acid

Both fluorine and bromine find some use in the manufacture of dye intermediates, but high cost restricts them to applications where they offer some unique advantage over chlorine.

Alkaline Fusion

Alkaline fusion is an important procedure for the hydroxylation of aromatic compounds. In alkaline fusion, a sulfonic acid group is replaced by a hydroxyl group. This reaction cannot be used when nitro or chloro groups are present but is applicable to amino compounds.

Alkaline fusion is usually carried out with a concentrated solution of sodium hydroxide in a cast-iron pot which is equipped with a scraping agitator and is heated externally. The water is evaporated; then the mass fuses. The reaction temperature is between 190 and 350°C, depending on the reactivity of the sulfonic acid.

Phenol can be made by fusing benzenesulfonic acid according to the following equations:

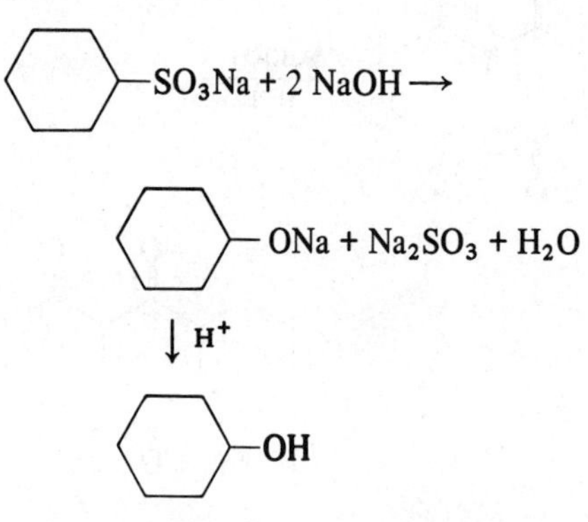

$$C_6H_5SO_3Na + 2\,NaOH \longrightarrow C_6H_5ONa + Na_2SO_3 + H_2O$$

$$\downarrow H^+$$

$$C_6H_5OH$$

Phenol

In a similar way, resorcinol is made by fusion of *m*-benzenedisulfonic acid.

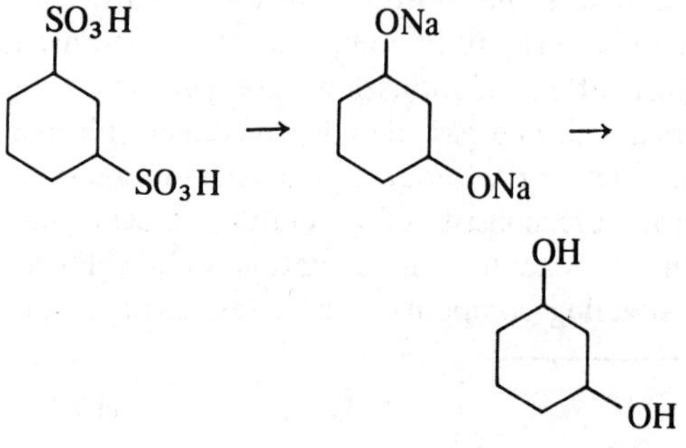

$$C_6H_4(SO_3H)_2 \longrightarrow C_6H_4(ONa)_2 \longrightarrow C_6H_4(OH)_2$$

Resorcinol

2-Naphthol is made by fusion of the corresponding sulfonic acid with sodium hydroxide. The naphtholate is treated with carbon dioxide to precipitate the naphthol which is then purified by vacuum distillation.

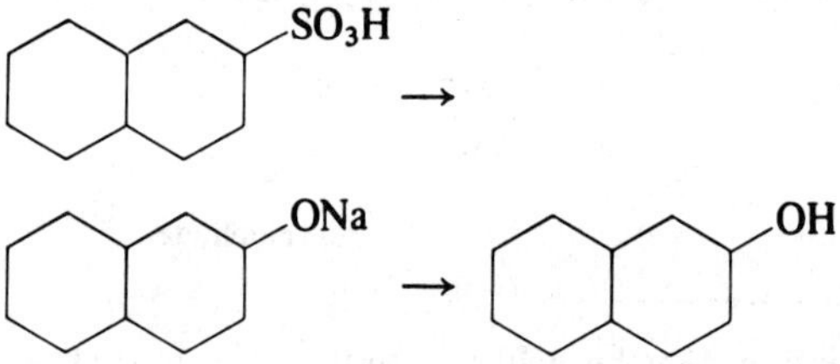

$$C_{10}H_7SO_3H \longrightarrow C_{10}H_7ONa \longrightarrow C_{10}H_7OH$$

2-Naphthol

1-Naphthol is not made from the sulfonic acid. Instead, 1-naphthylamine sulfate is heated with water at 200°C in a closed vessel. On cooling, 1-naphthol crystallizes out.

For some compounds, the conditions of alkaline fusion are too severe. In such cases the reaction may often be effected in water solution under pressure. Chromotropic acid is prepared from 1-naphthol-3,6,8-trisulfonic acid in 60 percent sodium hydroxide solution under pressure.

$$C_{10}H_4(OH)(SO_3H)_3 \longrightarrow C_{10}H_4(OH)_2(SO_3H)_2$$

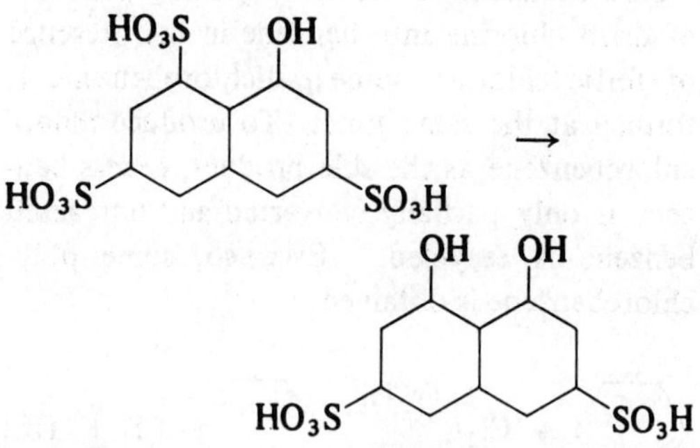

Chromotropic acid

Sulfonated naphthylamines may be fused without destruction of the amino group. The important azo dye intermediates, H acid and J acid, are made in this way.

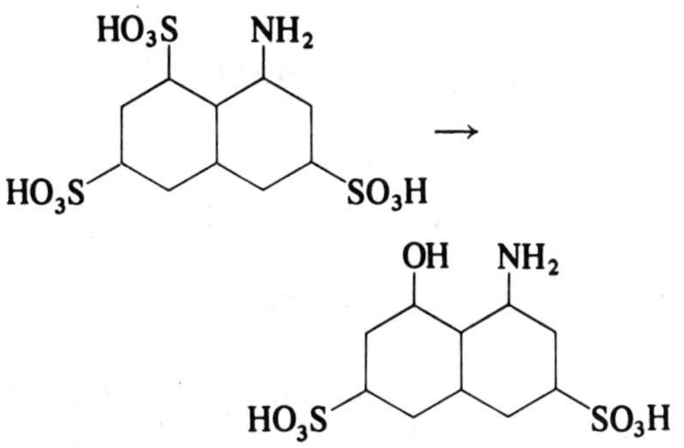

HO_3S NH_2 HO_3S SO_3H →

OH NH_2 HO_3S SO_3H

H acid

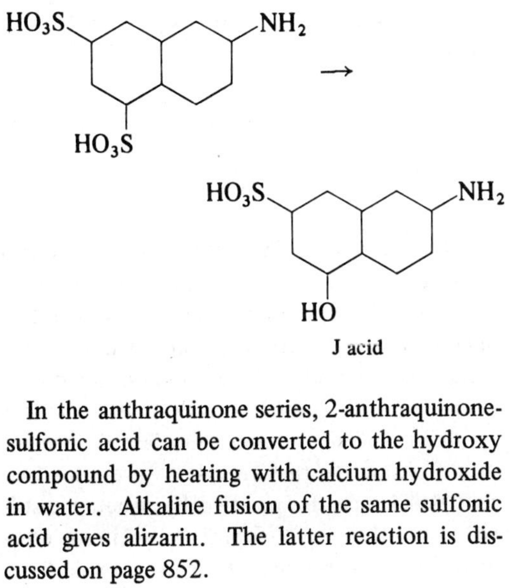

HO_3S NH_2 HO_3S →

HO_3S NH_2 HO

J acid

In the anthraquinone series, 2-anthraquinonesulfonic acid can be converted to the hydroxy compound by heating with calcium hydroxide in water. Alkaline fusion of the same sulfonic acid gives alizarin. The latter reaction is discussed on page 852.

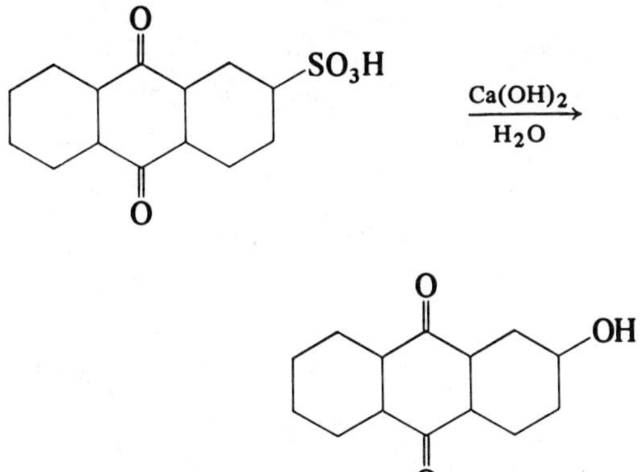

O O SO_3H $\xrightarrow[H_2O]{Ca(OH)_2}$

O O OH

2-Hydroxyanthraquinone

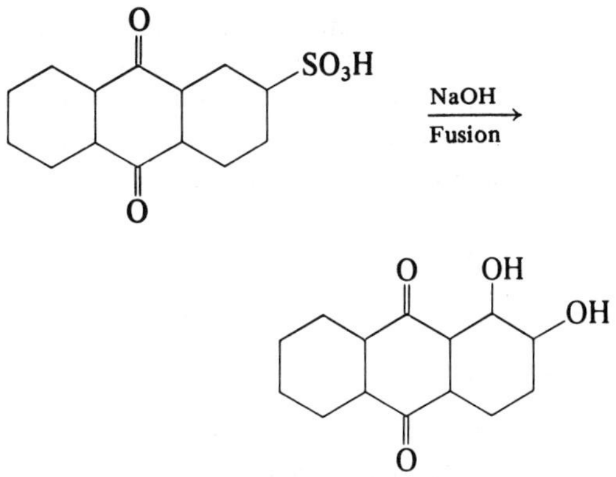

O O SO_3H $\xrightarrow[\text{Fusion}]{NaOH}$

O O OH OH

Alizarin

Oxidation

Oxidation may be effected by air in the presence of a catalyst or by a variety of chemical oxidants, such as manganese dioxide and potassium permanganate

Catalytic vapor-phase oxidation is illustrated by the conversion of naphthalene to phthalic anhydride. This reaction is carried out over a vanadium pentoxide catalyst at 450°C.

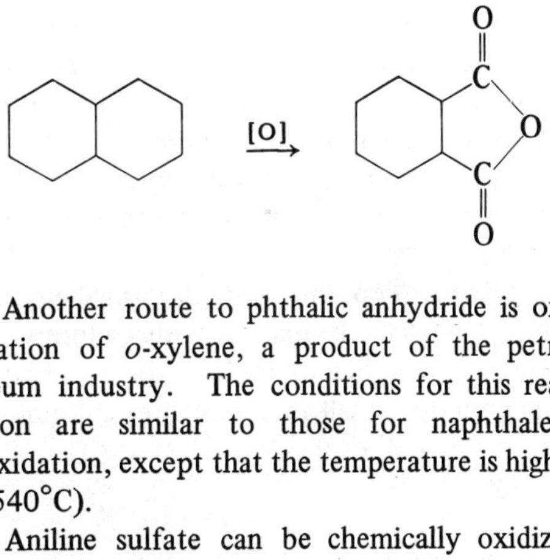

$\xrightarrow{[O]}$ O C O C O

Another route to phthalic anhydride is oxidation of *o*-xylene, a product of the petroleum industry. The conditions for this reaction are similar to those for naphthalene oxidation, except that the temperature is higher (540°C).

Aniline sulfate can be chemically oxidized with manganese dioxide in sulfuric acid. The product is *p*-benzoquinone.

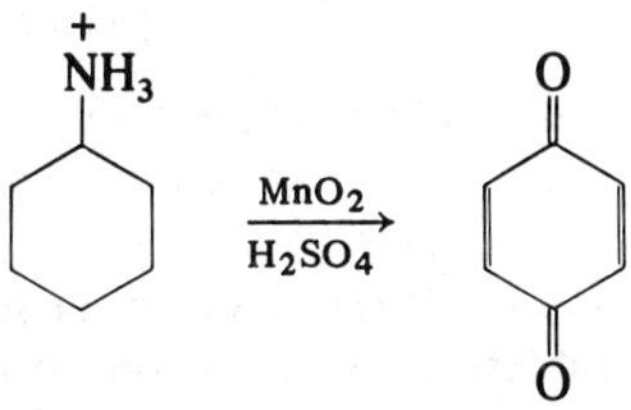

$\overset{+}{N}H_3$ $\xrightarrow[H_2SO_4]{MnO_2}$ O O

p-Benzoquinone

The use of potassium bichromate in sulfuric acid effects the oxidation of 1-nitro-2-methylanthraquinone to the corresponding carboxylic acid. This reaction illustrates side-chain oxidation, an important route to carboxylic acids.

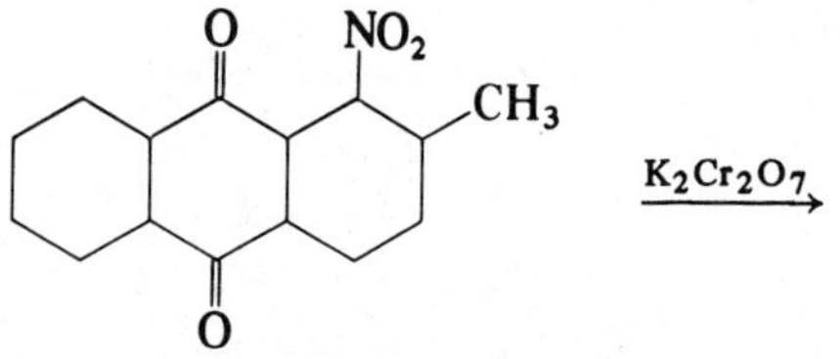

O NO$_2$ CH$_3$ O $\xrightarrow{K_2Cr_2O_7}$

1-Nitro-2-methylanthraquinone

O NO$_2$ COOH O

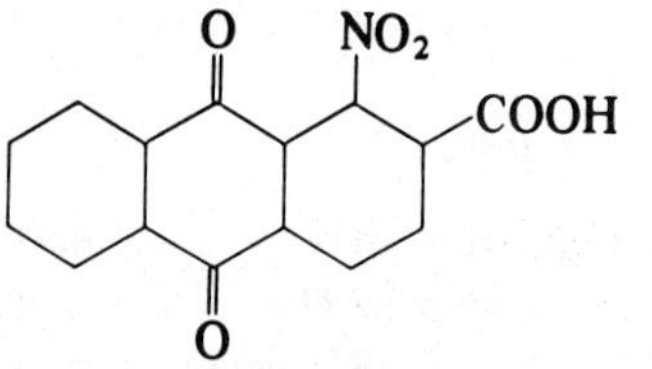

1-Nitro-2-anthraquinonecarboxylic acid

Other Important Reactions

Condensation. The term condensation describes a variety of reactions which join molecules or parts of the same molecule with elimination of a molecule of water or other low molecular weight substance. An example is the conversion of benzanthrone to dibenzanthronyl, described in detail on pages 854 and 855.

Addition. An important intermediate, cyanuric chloride, is made by the addition of cyanogen chloride to itself. The reaction is catalyzed by a small amount of free chlorine.

3 ClCN → Cyanuric chloride

Cyanogen chloride

Cl, N, N, Cl—C, C—Cl, N

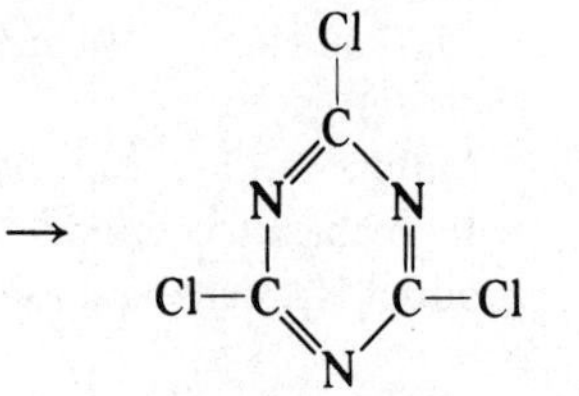

Cyanuric chloride

Alkylation. Alkylation refers to the introduction of an aliphatic group, such as methyl, into an organic molecule. Alkylation may occur on carbon, nitrogen, oxygen, or sulfur. Of these possibilities, alkylation on nitrogen and oxygen are most important in the manufacture of dye intermediates. A methyl group may be introduced into the amino group of aniline by heating with methyl alcohol under pressure in the presence of a mineral acid.

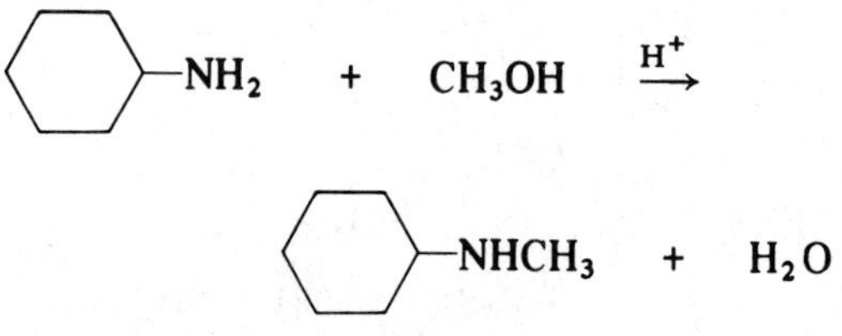

$$C_6H_5{-}NH_2 + CH_3OH \xrightarrow{H^+} C_6H_5{-}NHCH_3 + H_2O$$

N-Methylaniline

Phenol may be methylated with methyl sulfate in cold alkaline medium to give anisole.

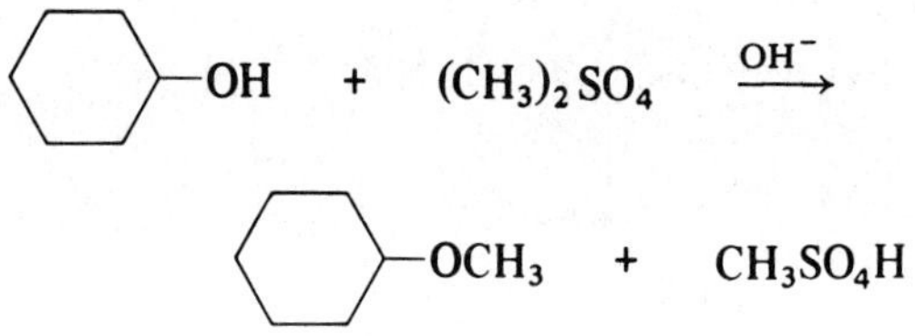

$$C_6H_5{-}OH + (CH_3)_2SO_4 \xrightarrow{OH^-} C_6H_5{-}OCH_3 + CH_3SO_4H$$

Anisole

Carboxylation. The carboxylic acid group may be introduced by side-chain oxidation as described above. In addition, it may be introduced by direct action of carbon dioxide on certain compounds. When sodium phenolate is treated with carbon dioxide under pressure at about 150°C, sodium salicylate is formed. Acidification of the salicylate gives the free acid which may be purified by vacuum distillation. 3-Hydroxy-2-naphthoic acid, an intermediate for developed azo dyes, is made from sodium 2-naphtholate.

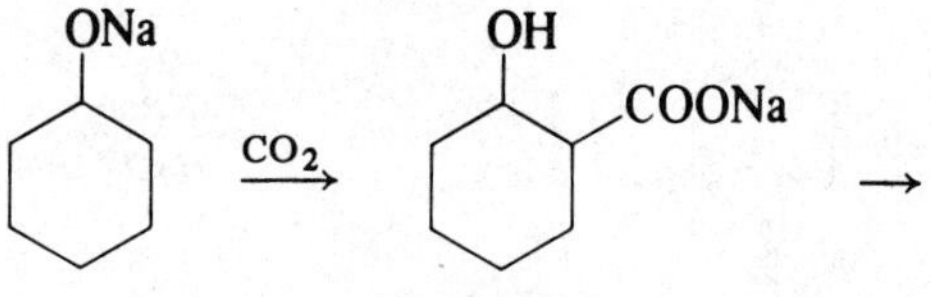

ONa $\xrightarrow{CO_2}$ OH COONa →

Sodium salicylate

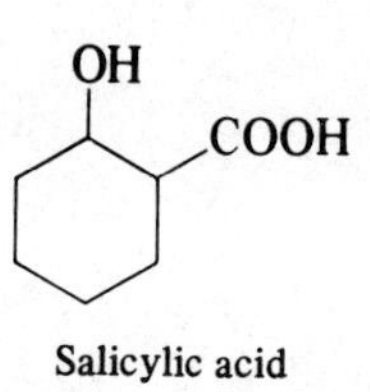

OH COOH

Salicylic acid

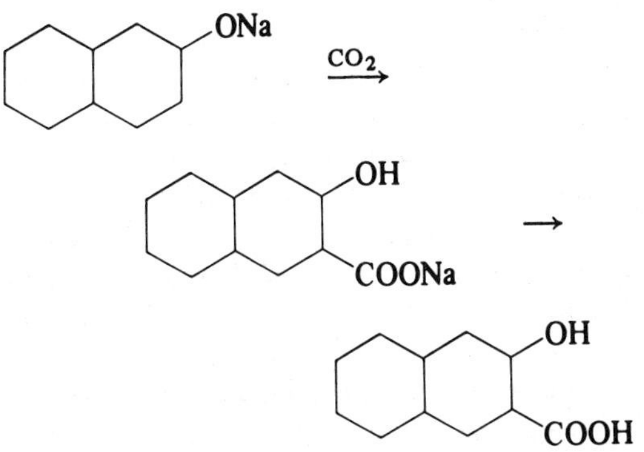

ONa $\xrightarrow{CO_2}$ OH, COONa → OH, COOH

3-Hydroxy-2-naphthoic acid

Sandmeyer Reaction. The replacement of a diazonium group by halogen, nitrile, nitro, sulfhydryl, and other groups in the presence of a cuprous salt, is known as the Sandmeyer reaction. An illustration of the use of this reaction is the preparation of 2-chloro-5-nitrophenol from the corresponding aminophenol.

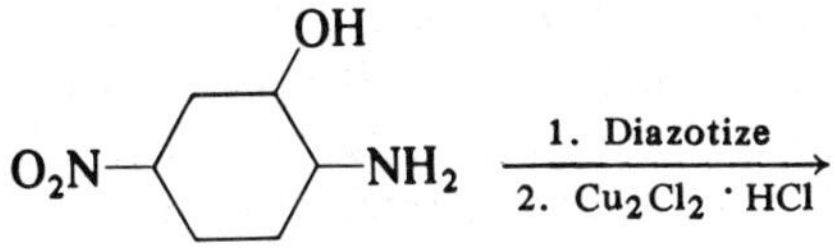

OH, O_2N—, —NH_2 $\xrightarrow[\text{2. } Cu_2Cl_2 \cdot HCl]{\text{1. Diazotize}}$

2-Amino-5-nitrophenol

OH, O_2N—, —Cl

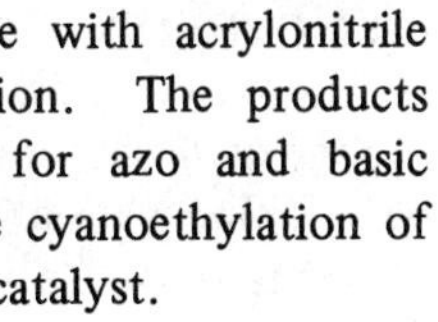

2-Chloro-5-nitrophenol

Cyanoethylation. Reaction of a primary or secondary aromatic amine with acrylonitrile results in N-cyanoethylation. The products are useful intermediates for azo and basic dyes. An example is the cyanoethylation of aniline with cupric sulfate catalyst.

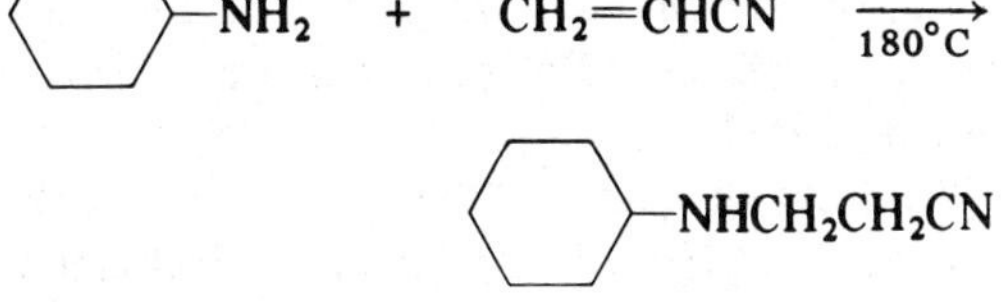

—NH_2 + $CH_2{=}CHCN$ $\xrightarrow[180°C]{Cu^{++}}$ —$NHCH_2CH_2CN$

N-(Cyanoethyl)aniline

PRODUCTION OF DYE INTERMEDIATES

Table 23.3 gives the production figures for some important raw materials and intermediates in the United States for 1978. These figures are taken from the report of the U.S. International Trade Commission. It should be remembered that dye intermediates are often used for end-products other than dyes, and that the figures in Table 23.3 do not necessarily correlate with figures on dye production.

TABLE 23.3 United States Production of Some Raw Materials and Dye Intermediates, 1978 (In Millions of Pounds)

Aniline	606
o-Dichlorobenzene	41
1,4-Dihydroxyanthraquinone	1.3
N,N-Dimethylaniline	11
o-Nitroaniline	16
Nitrobenzene	576
Phthalic anhydride	978
Toluene-2,4-diamine	139

THE MANUFACTURE OF DYES

Dyes owe their color to their ability to absorb light in the visible region of the spectrum, between 400 and 800 nanometers. Absorption is caused by electronic transitions in the molecules and can occur in the visible region only when the electrons are reasonably mobile. Mobility is encouraged by unsaturation and resonance. The main structural unit of a dye, which is always unsaturated, is called the *chromophore*, and a compound containing a chromophore is called a *chromogen*. Any substituent atom or group that increases the intensity of the color is called an *auxochrome*. An auxochrome may also serve to shift the absorption band of a chromophore to a longer wavelength, or may play a part in solubilizing the dye and attaching it to fibers.

Dyes are classified according to the type of chromophore that they contain. An example is the azo class, which is characterized by the presence of the —N=N— linkage. A typical chromogen is azobenzene.

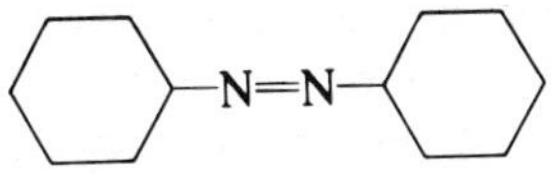

—N=N—

Common auxochromes are hydroxyl, amino, and carboxyl groups.

The hue, strength, and brightness of a dye depend on the entire light-absorbing system, consisting of the chromophore and auxochromes acting together. The nature of these groups and their relative positions in the molecule must be worked out correctly to produce a dye of desired appearance.

In general, for a given type of dye, extension of the unsaturated system and increased opportunities for resonance shifts the absorption of light toward longer wave lengths. Assuming a single main absorption band, the color absorbed progresses across the visible spectrum from violet to red. As this progression occurs, the light that is not absorbed is reflected and seen by the human eye as the color complementary to that absorbed. The wave lengths absorbed, the corresponding colors, and the observed complementary colors are given below for several major hues.

Wave Length Absorbed nm	*Corresponding Color Absorbed*	*Observed Color*
430	Violet	Yellow
470	Blue	Orange
530	Green	Red
570	Yellow	Violet
600	Orange	Blue
660	Red	Green

In addition to securing the desired appearance, it is necessary in dye synthesis to provide the dye with any groups necessary to confer solubility and affinity for textile fibers. It is also important to use only color systems that have the required fastness to light and other degrading influences. Considering that all these properties must be exhibited by one compound, it can readily be understood that the development and manufacture of dyes require a high degree of technical competence in the laboratory and in the plant.

The "Colour Index"[15] lists 30 chemical classes of dyes. Many of these classes are closely related; others are of relatively minor importance. Only several of the more important classes will be discussed in the following sections of this chapter.

Azo Dyes

The azo dye class, one of the most important, includes many hundreds of commercial dyes of various application types. Azo dyes are characterized by the presence of one or more azo (—N=N—) groups.

The principal method of forming azo dyes involves *diazotization* of primary aromatic amines, followed by *coupling* with hydroxy or amino derivatives of aromatic hydrocarbons or with certain aliphatic keto compounds. Both the aromatic amine, which is diazotized, and the compound to which it is coupled may bear a variety of substituents, such as alkyl, alkoxyl, halogen, and sulfonic acid. Because of the large number of compounds that can be combined, often in more than one sequence, the number of possible azo dyes is almost infinite.

Diazotization takes place when nitrous acid (HNO_2) reacts with a primary aromatic amino group in acid medium. Usually sulfuric or hydrochloric acid is used and the nitrous acid is generated from sodium nitrite. The equation for diazotization, using aniline in hydrochloric acid, is as follows:

$$C_6H_5{-}NH_2 + HNO_2 + H^+Cl^- \longrightarrow C_6H_5{-}N_2{}^+Cl^- + 2\,H_2O$$

Diazotization is usually carried out with excess acid to prevent partial diazotization and to inhibit secondary reactions. If the reaction is to proceed easily, the amine must be in solution, or its hydrochloride, if insoluble, must be in a fine state of subdivision. Temperatures of from 0 to 5°C are usually employed for diazotization since diazonium salts are generally unstable. There are exceptions to this, and temperatures of 20°C or higher are occasionally preferred.

The ease of diazotization depends markedly on the basicity of the amine. Extremely weakly basic amines are diazotizable only by special methods.

The coupling reaction with aromatic hydroxy compounds is illustrated with benzene diazonium chloride and phenol.

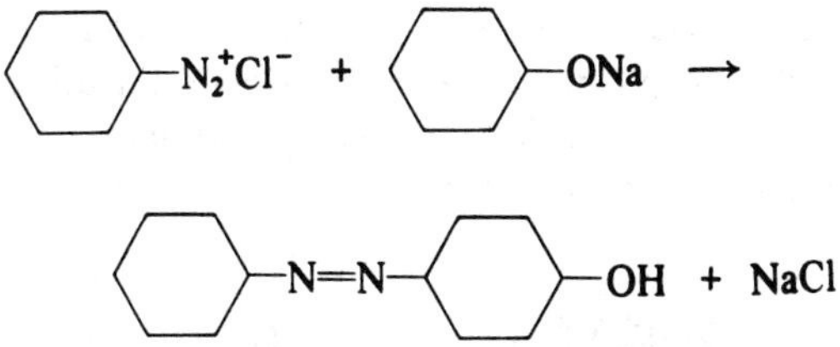

$$C_6H_5{-}N_2^+Cl^- + C_6H_5{-}ONa \longrightarrow C_6H_5{-}N{=}N{-}C_6H_4{-}OH + NaCl$$

Coupling to phenols, naphthols, and related hydroxyl compounds is carried out in alkaline solution. Under alkaline conditions the hydroxy compound is soluble, and coupling is usually rapid. Ordinarily the coupling must be carried out in the cold to prevent decomposition of the diazonium salt.

An example of coupling to an aromatic amine is the reaction of *p*-nitrobenzenediazonium chloride with *m*-toluidine.

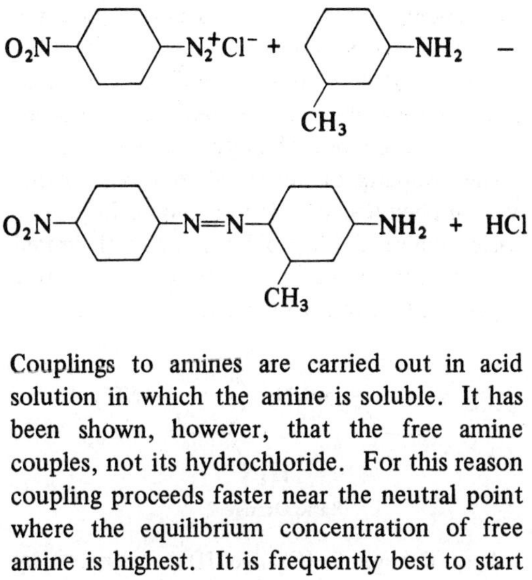

$$O_2N{-}C_6H_4{-}N_2^+Cl^- + C_6H_4(CH_3){-}NH_2 \longrightarrow O_2N{-}C_6H_4{-}N{=}N{-}C_6H_3(CH_3){-}NH_2 + HCl$$

Couplings to amines are carried out in acid solution in which the amine is soluble. It has been shown, however, that the free amine couples, not its hydrochloride. For this reason coupling proceeds faster near the neutral point where the equilibrium concentration of free amine is highest. It is frequently best to start the reaction at a low pH and then to raise the alkalinity slowly with sodium acetate or soda ash as the reaction proceeds.

It should be noted that in the above examples of coupling the diazonium group is shown entering the position *para* to the hydroxyl and amino groups. This specificity is characteristic of azo coupling. The diazo group does not enter at random but in certain definite positions. For the benzene derivatives, the attack is always on the position *para* to the activating group. If this position is blocked, coupling will sometimes occur in one of the *ortho* positions, but at a much slower rate.

In naphthalene derivatives, orientation of the entering diazo group is somewhat different. In *alpha*-naphthol, the attack is at position 4. If 4 is blocked, the diazo group enters at 2. In *beta*-naphthol, coupling takes place at 1, never at 3 or 4. The same rules apply to the corresponding naphthylamines.

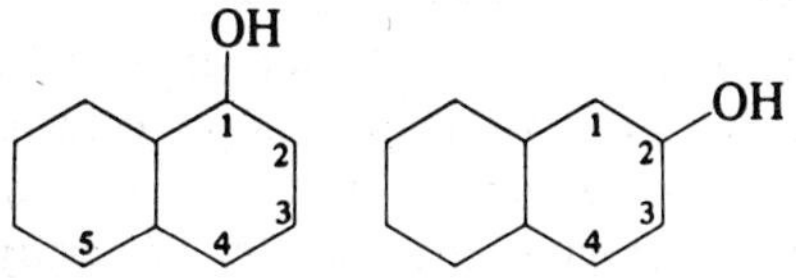

alpha-Naphthol

beta-Naphthol

In the naphthalene series, it is found that the presence of certain substituents, especially the sulfonic acid group, can influence the position of coupling even when not in a directly blocking position. For example, a sulfonic acid group in position 5 of *alpha*-naphthol causes the coupling to occur predominantly at position 2, rather than 4.

Certain aminonaphthol sulfonic acids couple twice. In this case, the place of entry depends on whether the coupling is performed in acid or alkaline solution. If acid, the coupling is *ortho* to the amino group; if alkaline, it is *ortho* to the hydroxyl. In the threc examples following, the place of entry for acid coupling is marked X, for alkaline coupling, Z.

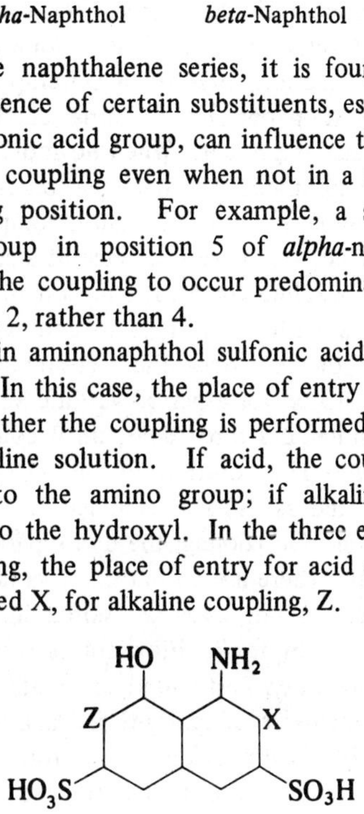

H acid

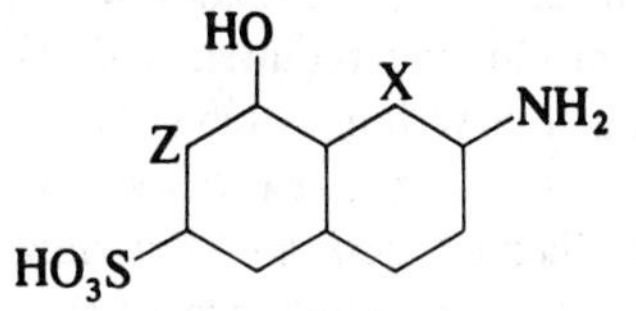

Gamma acid

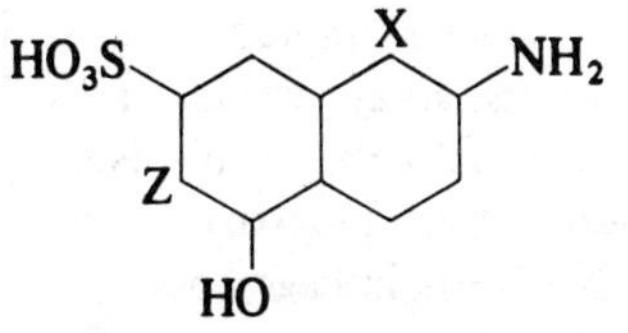

J acid

Ortho and *para* diamines in both the benzene and naphthalene series do not couple at all; only the *meta* diamines do so.

A generalized procedure for diazotization and coupling as practiced in both laboratory and plant is as follows: Sodium nitrite is added slowly with stirring to an acid solution of the amine in a wooden or brick-lined tub. The tub is not externally cooled, but ice is added directly to the amine solution to control the temperature between 0 and 5°C. The progress of the reaction is followed by observing the disappearance of the nitrous acid generated from the sodium nitrite. This is done by spotting the reaction mass on a starch-potassium iodide test paper which turns black when nitrous acid is present. When no more nitrous acid is consumed, the diazotization is judged to be complete. The diazonium compound is not isolated, but is run at once at a slow rate into the alkaline (or acid) solution of the intermediate with which it is to couple. After addition of the diazonium compound, the batch is stirred for a period varying between a few minutes and three days, until coupling is complete. The solution of the dye is then warmed, and salt (NaCl) is added and allowed to dissolve. On cooling, the dye separates and is filtered. There are a number of variations in procedure. Some dyes are salted hot, some cold, some not at all. Filtration is carried out at a temperature best suited to isolate the dye without inclusion of impurities. As a result of salting, dyes with sulfonic acid groups are always isolated as the sodium sulfonates.

Diazotization and coupling are usually carried out as batch processes because of the limited production required and the necessity of testing the various batches of intermediates that are used. In a few cases, however, continuous diazotization and coupling are practical. When continuous processes are used, it becomes necessary to devise a means of isolating the product as rapidly as it is made. Centrifugal filters, spray dryers, and other special apparatus are suitable for this use.

Monoazo Dyes. Monoazo dyes contain the azo (—N=N—) linkage once; the simplest example is aminoazobenzene. This compound is used as a dye for oils, lacquers, and stains under the name Solvent Yellow 1 (Aniline Yellow), and is also an intermediate for other dyes. It is made by coupling benzenediazonium chloride with aniline in acid medium:

$$C_6H_5-N_2^{\oplus}Cl^{\ominus} + C_6H_5-NH_2 \rightarrow C_6H_5-N{=}N-C_6H_4^{*}-NH_2 + HCl$$

Solvent Yellow 1

To simplify the description of other azo dyes to follow, only the formulas of the dyes will be given, but the place of attack of the diazonium group will be indicated by an asterisk on the active carbon of the coupling component. It will be understood that the diazonium compound was prepared by diazotization of the corresponding amine and that coupling took place on the coupling component with the elimination of a hydrogen ion.

The coupling of benzenediazonium chloride with *m*-phenylenediamine in acid medium gives Basic Orange 2 (Chrysoidine), a dye of limited use on textiles but of importance for coloring paper, leather, and woodstains.

$$C_6H_5-N{=}N-C_6H_3^{*}(NH_2)-NH_2$$

Basic Orange 2

Acid Orange 7 (Orange II) is prepared by coupling diazotized sulfanilic acid with *beta*-naphthol in basic medium. Acid Orange 7 is useful on textile fibers, such as wool, silk, and nylon, as well as on paper and leather.

$$NaO_3S-C_6H_4-N{=}N-C_{10}H_6^{*}-OH$$

Acid Orange 7

Other acid monoazo dyes of importance are derived from a variety of sulfonated intermediates. Acid Red 26 (Ponceau R) is made from diazotized 2,4-xylidine and R-acid. Acid Red 14 (Azo Rubine) is made from diazotized naphthionic acid and Nevile and Winther's acid. Acid Violet 6 is *p*-aminoacetanilide coupled with Chromotropic acid.

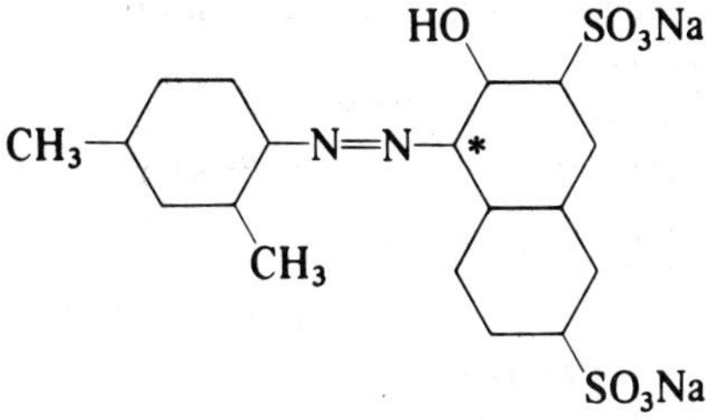

Acid Red 26

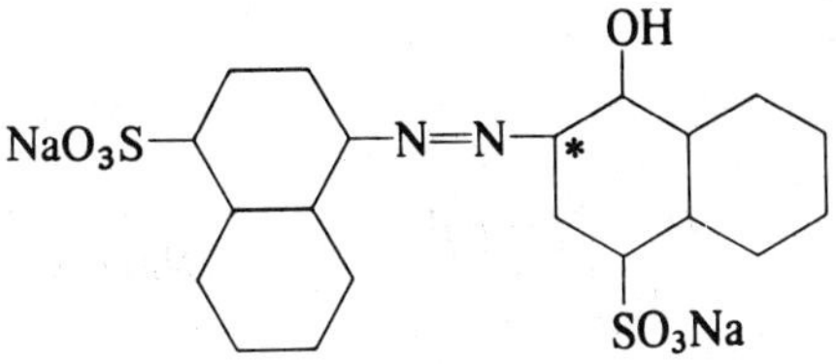

Acid Red 14

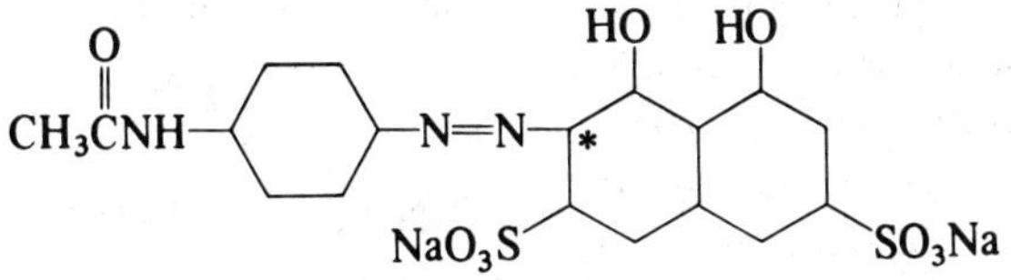

Acid Violet 6

Mordant Black 17 is an example of a type of dye that shows improved fastness on wool when treated on the fiber with a soluble chromium salt. Reaction occurs between the dye and the chromium ion to form a complex of enhanced fastness to washing and light. The component parts of Mordant Black 17 will be recognized as 1-amino-2-naphthol-4-sulfonic acid and *beta*-naphthol. Another dye of this type is Mordant Brown 4, which is made by diazotizing picramic acid and coupling it in acid medium with toluene-2,4-diamine.

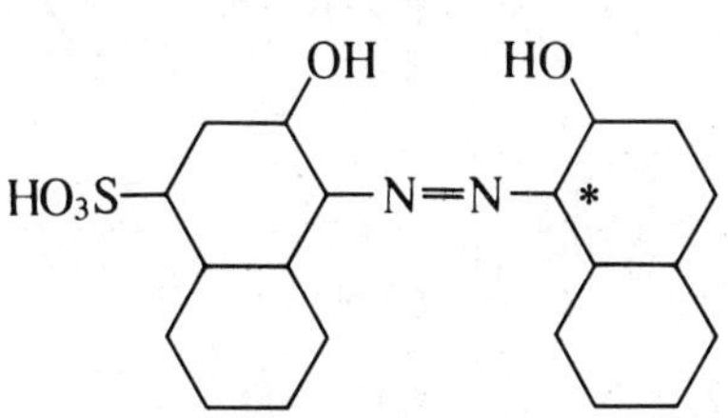

Mordant Black 17

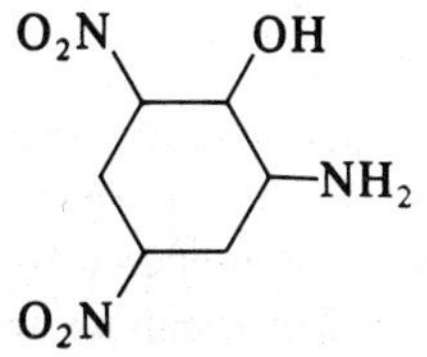

Picramic acid

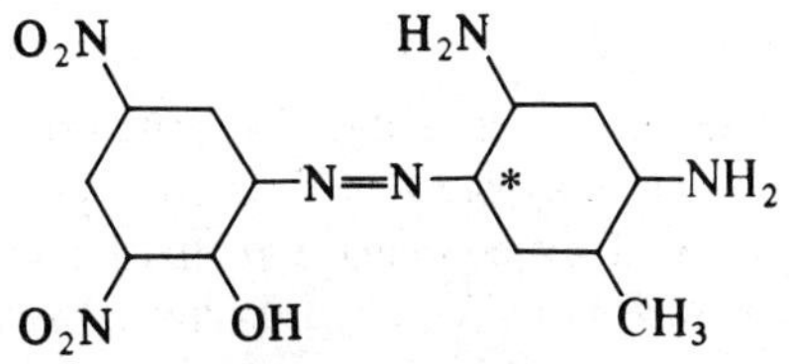

Mordant Brown 4

Acid Yellow 23 (Tartrazine) contains the pyrazolone nucleus. It is prepared by coupling diazotized sulfanilic acid with 5-oxo-1-(4-sulfophenyl)-2-pyrazoline-3-carboxylic acid or by heating oxidized tartaric acid (1 mole) with phenylhydrazinesulfonic acid (2 moles). Acid Yellow 23 is useful as a dye for basic fibers, paper, and leather. It is also used as a filter dye in photography.

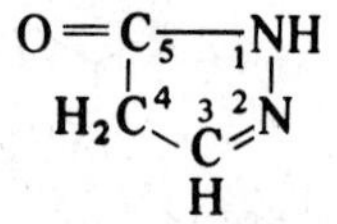

Pyrazolone

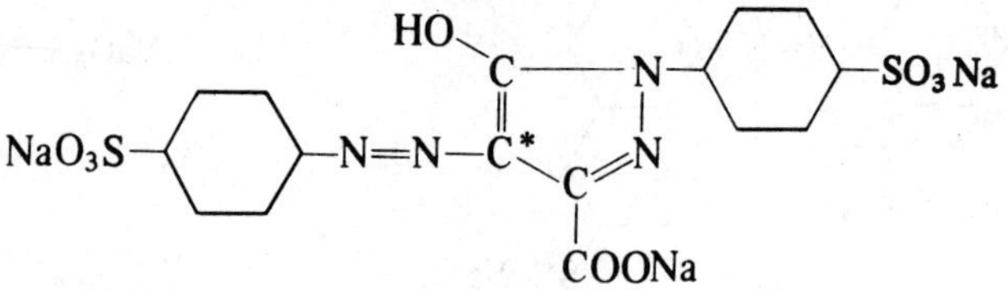

Acid Yellow 23

A further example of a pyrazolone dye is Acid Red 38. This dye is the chromium complex of the azo compound made by coupling diazotized 2-amino-5-nitrophenol with *meta*-(3-methyl-5-oxo-2-pyrazolin-1-yl)-benzenesulfonamide. It is used as a dye for wool and nylon. Acid Red 38 has one advantage over dyes such as Mordant Black 17 in that the dyer does not have to form the chromium complex by a separate operation during the dyeing process.

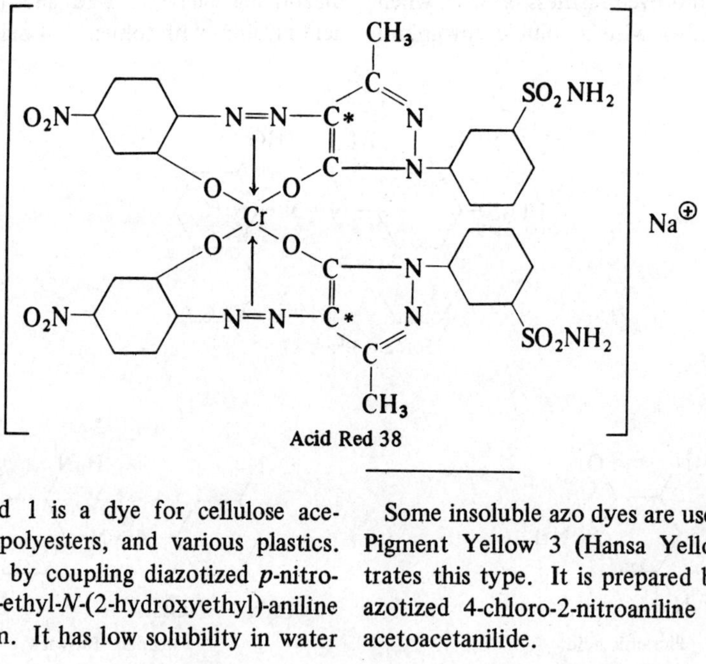

Acid Red 38

Disperse Red 1 is a dye for cellulose acetates, nylon, polyesters, and various plastics. It is prepared by coupling diazotized *p*-nitroaniline with *N*-ethyl-*N*-(2-hydroxyethyl)-aniline in acid medium. It has low solubility in water and is supplied in finely divided form mixed with a dispersing agent.

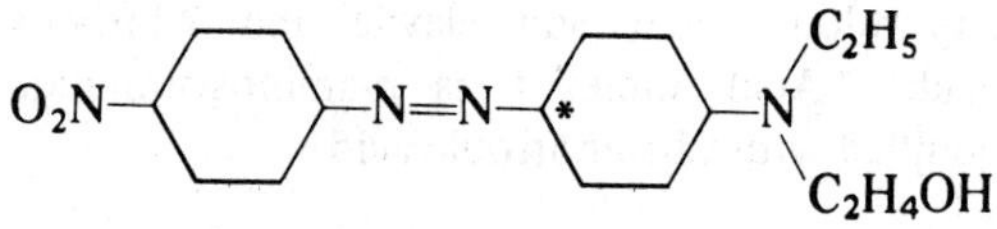

Disperse Red 1

The hue can be deepened into a useful blue by adding electronegative auxochromes in the diazo component, as in Disperse Blue 165.

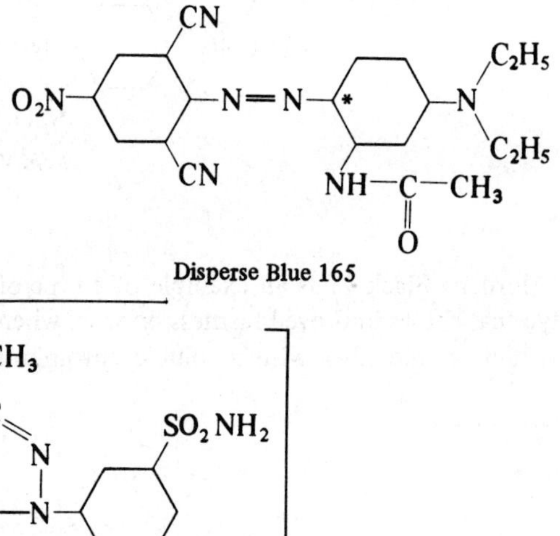

Disperse Blue 165

Some insoluble azo dyes are used as pigments. Pigment Yellow 3 (Hansa Yellow 10G) illustrates this type. It is prepared by coupling diazotized 4-chloro-2-nitroaniline with *o*-chloroacetoacetanilide.

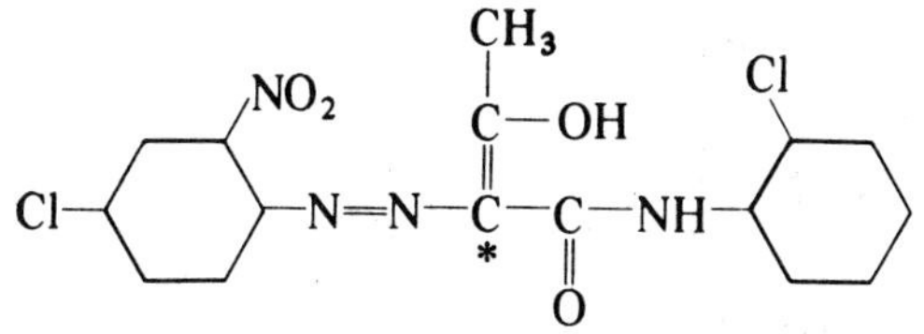

Pigment Yellow 3

A final example of a monoazo color is an orange fiber-reactive dye. It is prepared by diazotizing 2-amino-1,5-naphthalenedisulfonic acid, coupling it with *N*-Methyl-J-acid, condensing with cyanuric chloride, and then with ammonia. One chlorine atom remains for reaction with textile fibers bearing hydroxyl or amino groups.

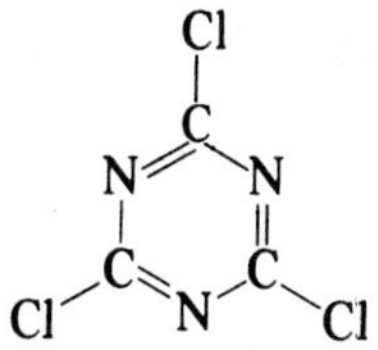

Cyanuric chloride

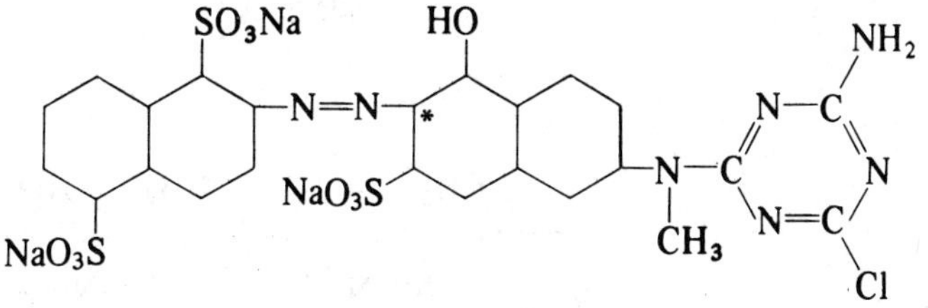

Reactive Orange 13

Disazo Dyes. Disazo dyes contain two azo linkages. They are formed: (1) by the further diazotization and coupling of monoazo dyes containing free amino groups; (2) by the coupling of two moles of diazonium compound to a coupling component that can couple twice; or (3) by the tetrazotization of a diamine followed by coupling to two moles of coupling component. When two moles of diazonium compound or coupling component are used, they may be the same, or they may be different in structure.

Acid Red 150 (Cloth Red 2R) is a disazo dye prepared by diazotizing aminoazobenzene and coupling to R acid.

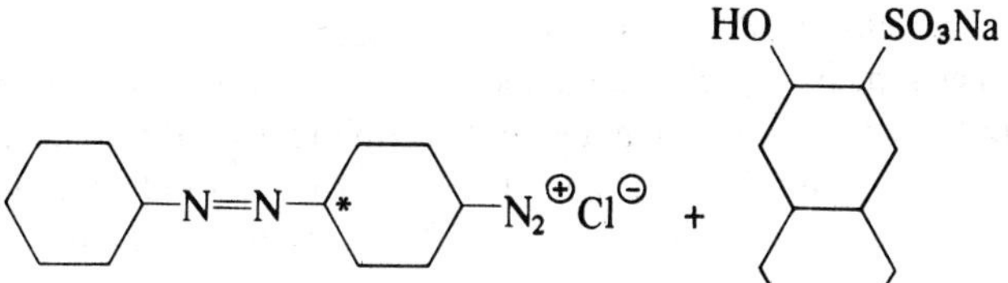

Diazotized aminoazobenzene

R acid

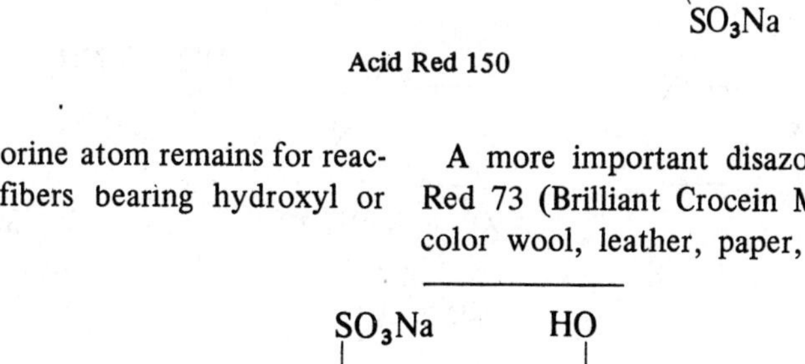

Acid Red 150

A more important disazo acid dye is Acid Red 73 (Brilliant Crocein M) which is used to color wool, leather, paper, and anodized aluminum. It is prepared from diazotized aminoazobenzene and G acid.

HO
—N=N— —N=N—
NaO_3S—
SO_3Na

Acid Red 73

Other disazo dyes of this kind are Acid Red 115 (Cloth Red B), which is made by coupling diazotized aminoazotoluene with R acid, and Acid Black 18. Acid Black 18 will be recognized as having been made by the combination, through diazotization and coupling, of sulfanilic acid, *alpha*-naphthylamine, and H acid.

Solvent Red 27 is a disazo dye used to color gasoline. Its formula indicates that it is made by diazotizing aminoazoxylene and coupling to *beta*-naphthol.

An example of a disazo dye prepared by coupling two different diazonium compounds to a single coupling component is Acid Black 1. This dye is made by coupling diazotized

HO SO_3Na
—N=N— —N=N—
CH_3 CH_3
SO_3Na

Acid Red 115

HO NH_2
NaO_3S— —N=N— —N=N—
NaO_3S SO_3Na

Acid Black 18

CH_3 CH_3 HO
—N=N— —N=N—
H_3C CH_3

Solvent Red 27

p-nitroaniline (acid) and aniline (alkaline) to H acid.

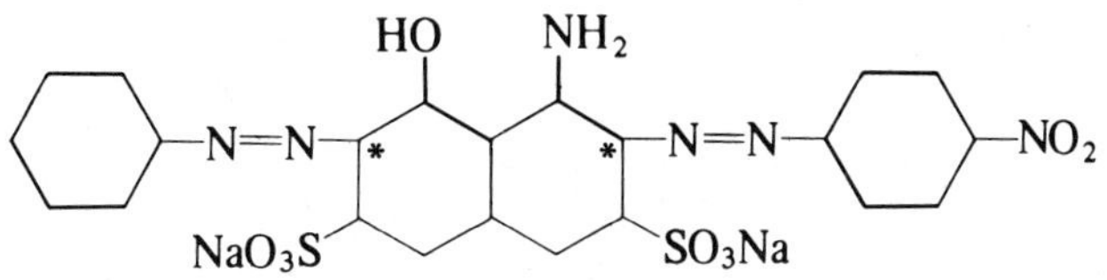

Acid Black 1

A direct dye of importance is Direct Blue 1 (Sky Blue FF). It is made by coupling tetrazotized *o*-dianisidine twice with Chicago acid. Direct Blue 1 is used as a dye for cellulosic textiles, for paper, and for nylon. It is applied to cellulosic textiles and paper under neutral conditions; it dyes them by adsorption. It is used as an acid dye on nylon and must be applied under acidic conditions.

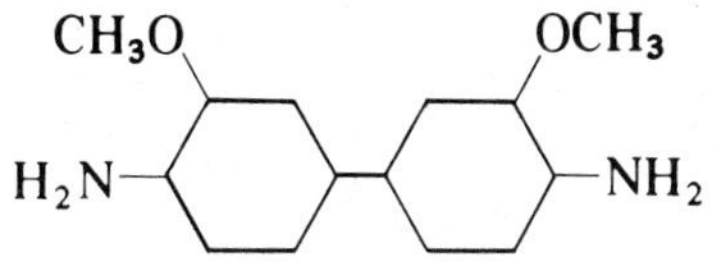

o-Dianisidine

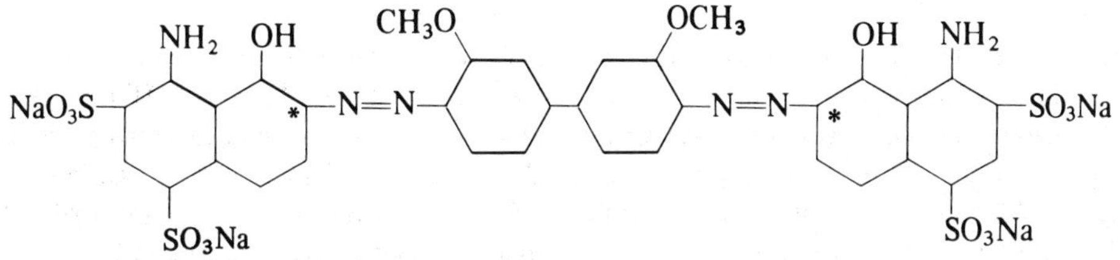

Direct Blue 1

As mentioned above, it is not necessary to couple both diazonium groups of a tetrazotized diamine to the same coupling component. Acid Red 114 is made by coupling tetrazotized *o*-tolidine with G acid and with phenol, followed by formation of an ester of the phenol with *p*-toluenesulfonyl chloride. This dye is used on wool, silk, and leather.

The fastness to light of direct dyes is often improved by the formation of copper chelates. Direct Red 83 is the copper complex of the

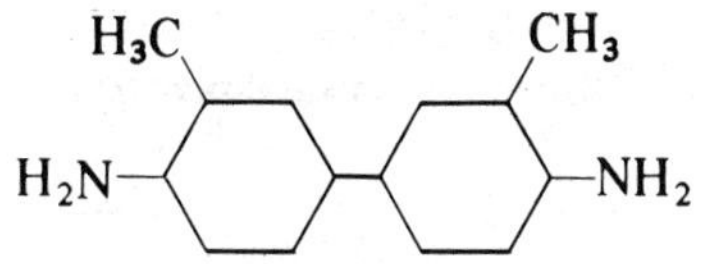

o-Tolidine

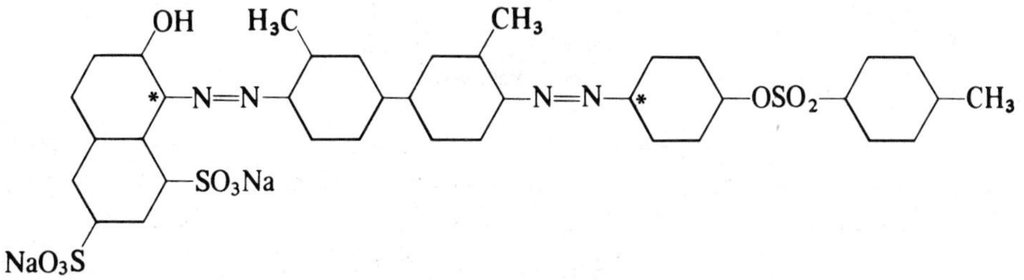

Acid Red 114

disazo dye prepared by coupling two moles of diazotized 2-aminophenol-4-sulfonic acid to one mole of J acid urea.

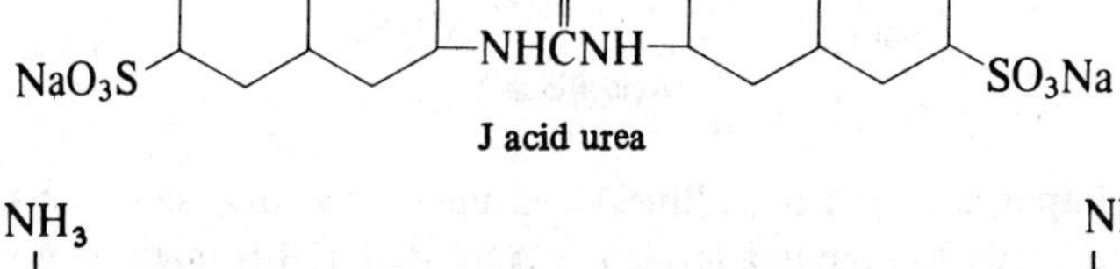

J acid urea

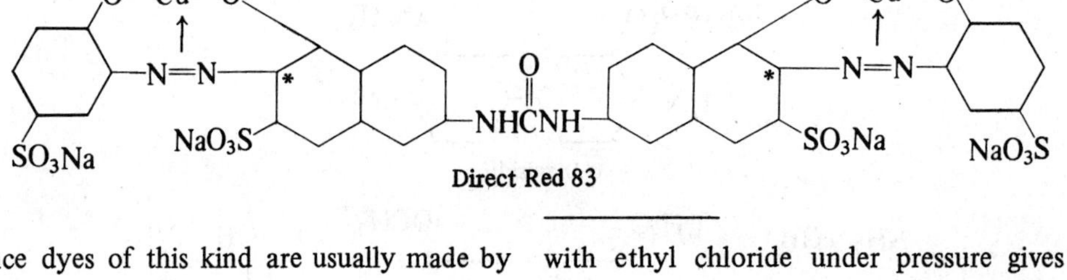

Direct Red 83

Since dyes of this kind are usually made by reacting the coupling product with a cupric salt in aqueous ammonia, ammonia molecules are shown coordinated with the copper in the complex. This position may sometimes be filled by water or by various amines.

A number of important dyes are based on diaminostilbenedisulfonic acid. An example is Direct Yellow 4 (Brilliant Yellow), which is made by coupling tetrazotized diaminostilbenedisulfonic acid twice to phenol. Direct Yellow 4 is an important dye for paper; it is also used as an indicator since it turns red in strong alkali. Reaction of Direct Yellow 4 with ethyl chloride under pressure gives the ethyl ether, Direct Yellow 12 (Chrysophenine). This is an important dye for textiles and paper. Unlike Direct Yellow 4, it is not an indicator.

Some stillbene azo dyes are prepared by methods other than diazotization and coupling. Direct Orange 61, for example, is prepared by the condensation in alkaline solution of aminoazobenzene-*m*-sulfonic acid and dinitrostilbenedisulfonic acid. The exact structure of Direct Orange 61 is not known, but the pres-

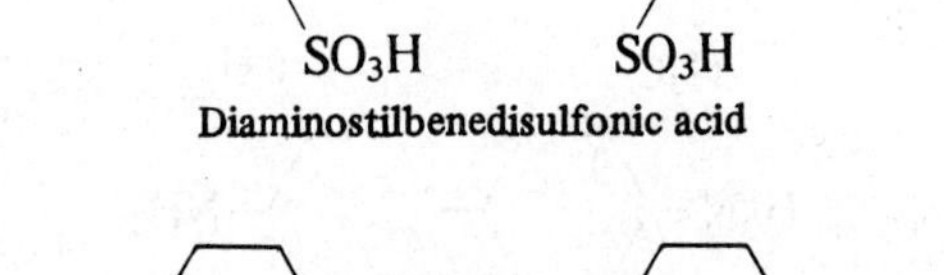

Diaminostilbenedisulfonic acid

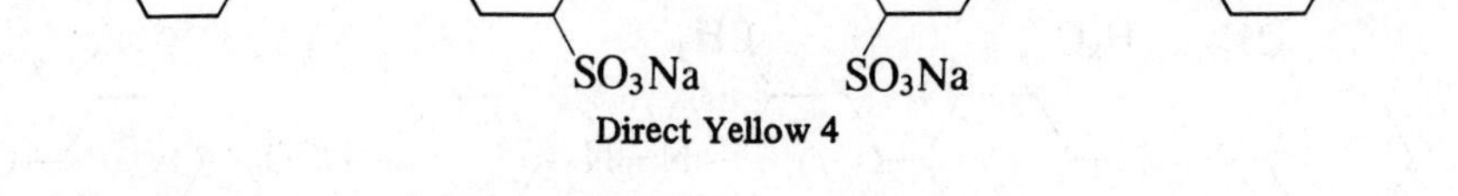

Direct Yellow 4

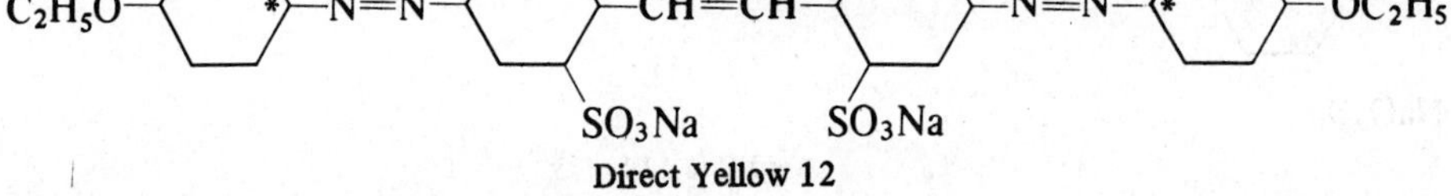

Direct Yellow 12

ence of an azo linkage formed by reaction of an amine and a nitro group can be demonstrated.

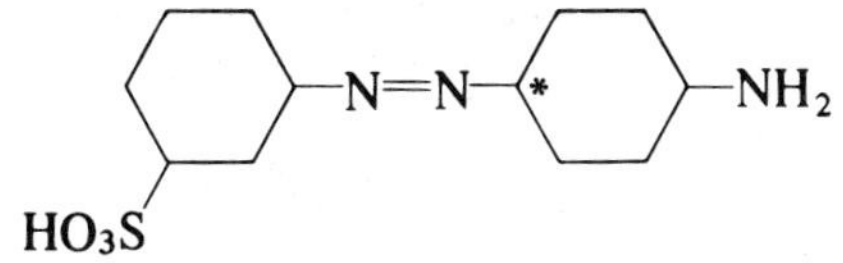

$-N{=}N-$ * $-NH_2$

HO_3S

Aminoazobenzene-*m*-sulfonic acid

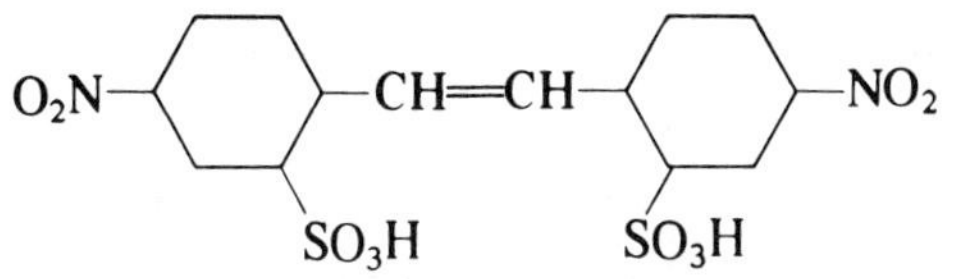

O_2N- $-CH{=}CH-$ $-NO_2$

SO_3H SO_3H

Dinitrostilbenedisulfonic acid

Polyazo Dyes. Azo dyes can be made with three, four, or more azo linkages. Those with three (trisazo) and four (tetrakisazo) are quite common. An example of an important trisazo dye is Direct Blue 110, a dye for cellulose, wool, and silk. Examination of the formula shows that this dye is made from five intermediates: metanilic acid, 1,6 and 1,7-Cleve's acid, 1-naphthylamine, and J acid.

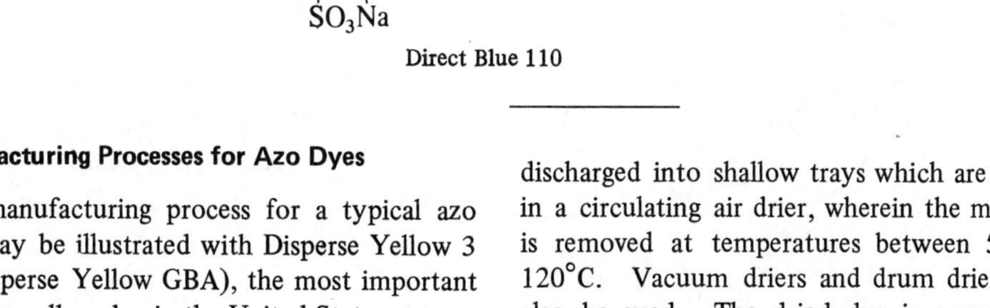

The apparatus required for the manufacture of a typical azo dye is indicated in Fig. 23.5. (Some typical equipment used in dye manufacture is shown in Figs. 23.6 and 23.7.)

p-Aminoacetanilide is suspended in water in a wooden or brick-lined tub (Tub 2) equipped with an agitator. Two moles of hydrochloric acid per mole of amine are added, and the whole is brought to a boil by passing in live steam. As soon as the *p*-aminoacetanilide has dissolved, ice is added. One mole of sodium nitrite, in solution in Tank 1, is run in slowly for about two hours. In the meantime, two moles of *p*-cresol are dissolved in an excess of soda ash in water (Tub 3), and to this the contents of Tub 2 are added with stirring. Enough ice is added to keep the temperature at about 5°C. The dye separates in part as the coupling proceeds. When the coupling is complete, the contents of Tub 3 are warmed with steam.

The suspension of dye is run into a wooden plate-and-frame filter press. The filter cake is washed and freed from most of the adhering mother liquor by blowing with compressed air while still in the press. The moist cake is discharged into shallow trays which are placed in a circulating air drier, wherein the moisture is removed at temperatures between 50 and 120°C. Vacuum driers and drum driers may also be used. The dried dye is ground and mixed with a diluent and dispersant to make it equal in color strength to a predetermined standard. Dilution is necessary because batches differ in their content of pure dye; if sold "as is," the user would have to adjust his dyeing recipes for each batch. Therefore, uniformity is assured by dilution to a standard strength.

NaO_3S $-N{=}N-$ * $-N{=}N-$ * $-N{=}N-$ HO * NaO_3S NH_2

SO_3Na

Direct Blue 110

Manufacturing Processes for Azo Dyes

The manufacturing process for a typical azo dye may be illustrated with Disperse Yellow 3 (Intrasperse Yellow GBA), the most important disperse yellow dye in the United States.

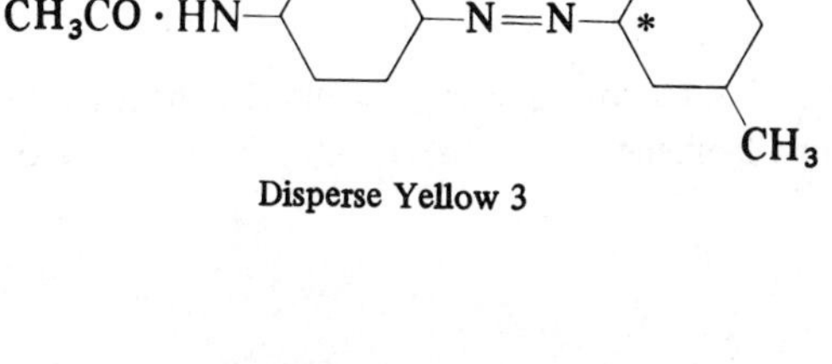

$CH_3CO \cdot HN-$ $-N{=}N-$ * HO CH_3

Disperse Yellow 3

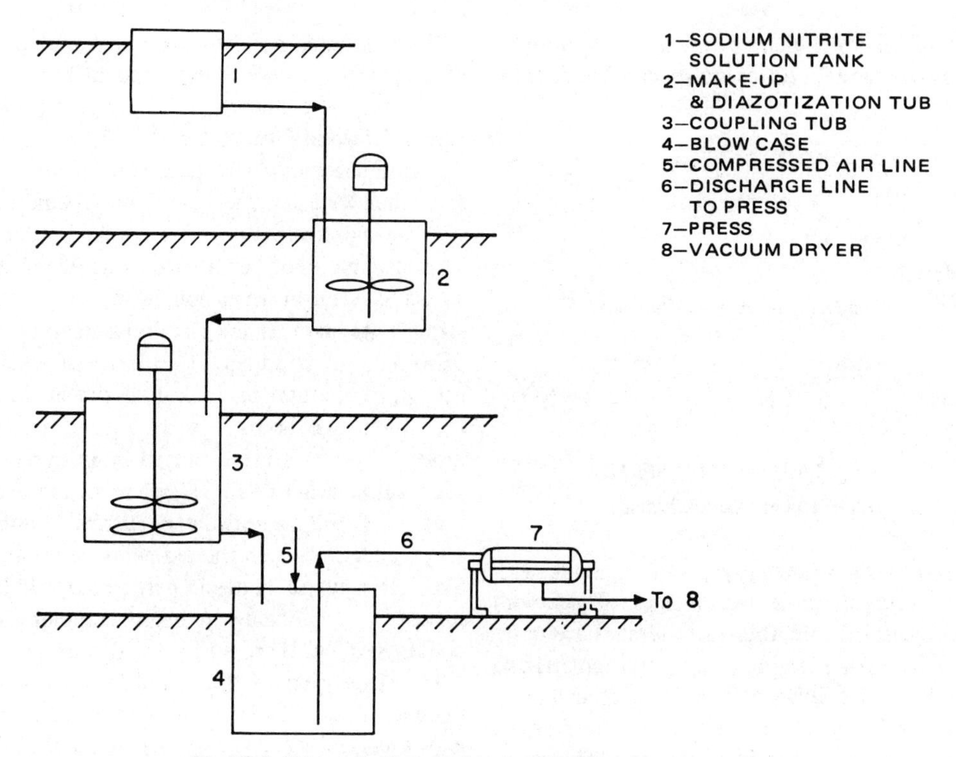

Fig. 23.5 Schematic diagram for manufacturing a monoazo dye.

Triphenylmethane Dyes

Triphenylmethane dyes are characterized by the highly resonance-stabilized chromophore shown below. They are among the strongest and brightest of synthetic dyes, but usually do not exhibit good fastness to light. An exception is found with acrylic fibers, which are dyed in bright light-fast hues by selected triphenylmethane dyes.

The triphenylmethane chromophore

Basic Red 9 (Pararosaniline) is the simplest triphenylmethane dye. It is prepared by the condensation of one mole of *p*-toluidine and two moles of aniline, usually by heating in nitrobenzene which serves as solvent and oxi-

H_2N—C(OH)—NH_2 $\xrightarrow{HCl}$

NH_2

Colorless carbinol

H_2N—C= $\overset{\oplus}{N}H_2$ $Cl^{\ominus}$

NH_2

Basic Red 9

Fig. 23.6 (a) Filter press room; (b) internals of a plate and frame filter. (*Courtesy Crompton and Knowles Corp.*)

dant. The first product is a colorless carbinol which is converted to the dye with hydrochloric acid.

Closely related to Basic Red 9 is Basic Violet 14 (Fuchsine). This dye is made by condensing one mole each of aniline, *p*-toluidine, and *o*-toluidine in nitrobenzene. The dye is formed on neutralizing the carbinol with hydrochloric acid.

A second method for the synthesis of triphenylmethane dyes is illustrated with Basic Green 4 (Malachite Green). Benzaldehyde (one mole) is condensed with two moles of *N,N*-dimethylaniline to form a leuco base which is oxidized with lead peroxide and treated with hydrochloric acid to give the dye.

Basic Green 4 is used as a dye for bast fibers, acrylic fibers, leather, paper, and lacquers. When *o*-chlorobenzaldehyde is used in place of benzaldehyde, the product is Basic Blue 1, a more alkali-resistant dye which finds use on tannin-mordanted cotton, wool, leather, and paper.

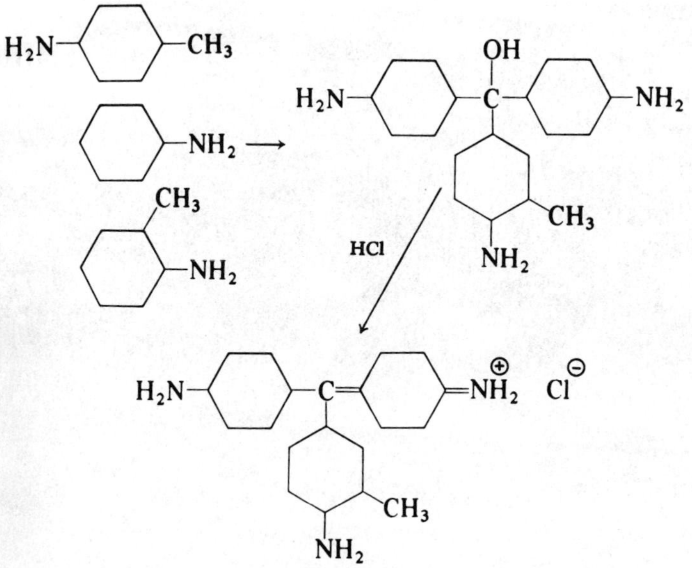

H_2N–C6H4–CH_3 + C6H5–NH_2 + CH_3–C6H4–NH_2 ⟶ H_2N–C6H4–C(OH)(C6H3(CH_3)NH_2)–C6H4–NH_2

HCl

H_2N–C6H4–C(C6H3(CH_3)NH_2)=C6H4=$\overset{\oplus}{N}H_2$ $Cl^{\ominus}$

Basic Violet 14

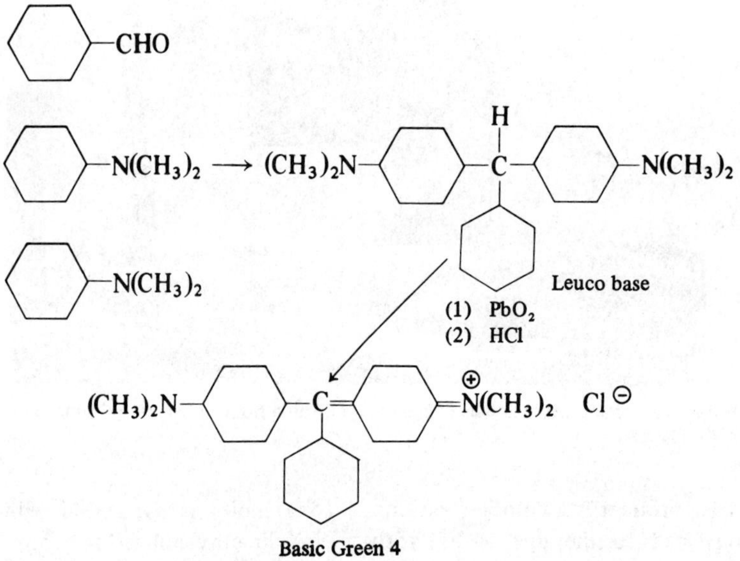

C6H5–CHO + C6H5–$N(CH_3)_2$ + C6H5–$N(CH_3)_2$ ⟶ $(CH_3)_2N$–C6H4–CH(C6H5)–C6H4–$N(CH_3)_2$

Leuco base

(1) PbO_2
(2) HCl

$(CH_3)_2N$–C6H4–C(C6H5)=C6H4=$\overset{\oplus}{N}(CH_3)_2$ $Cl^{\ominus}$

Basic Green 4

$(CH_3)_2N$–C6H4–C(C6H4Cl)=C6H4=$\overset{\oplus}{N}(CH_3)_2$ $Cl^{\ominus}$

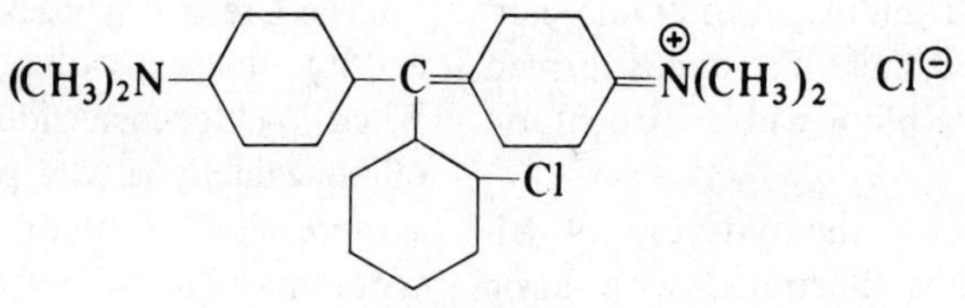

Basic Blue 1

Ketone condensation, a third method of manufacture of triphenylmethane dyes, is illustrated with Basic Violet 3 (Crystal Violet). Michler's Ketone, made by passing phosgene into *N,N*-dimethylaniline, is condensed with *N,N*-dimethylaniline in a solvent using phosphorus oxychloride. A carbinol is formed which is converted to the dye with hydrochloric acid.

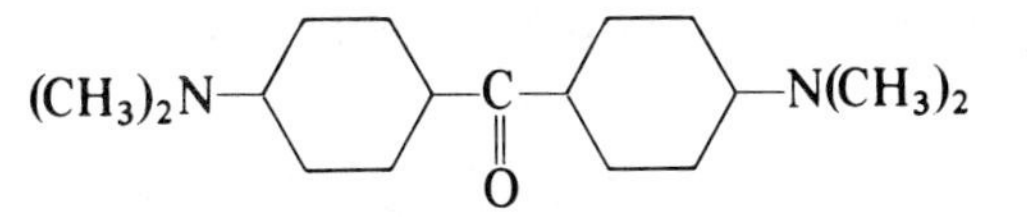

Michler's Ketone

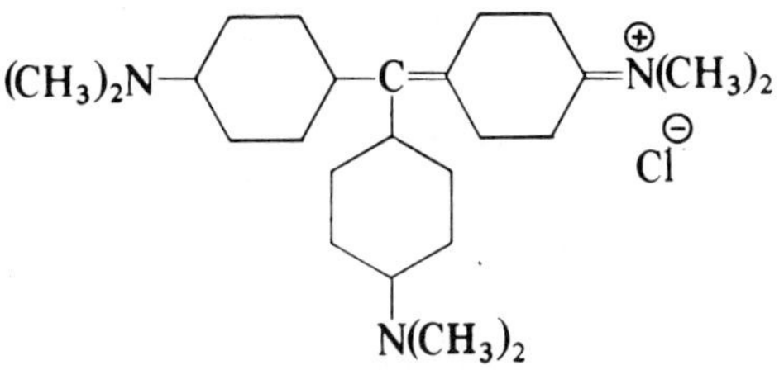

Basic Violet 3

Fig. 23.7 Glass-lined steel reactor. *(Courtesy Crompton and Knowles Corp.)*

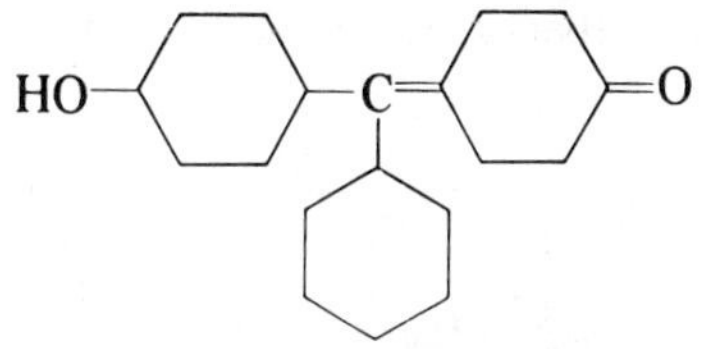

Not all triphenylmethane dyes are basic. Acid Green 3, for example, is a dye for wool, silk, and leather. There are two sulfonate groups in the molecule, one of which forms an inner salt with the positive charge on the chromophore.

Aurin is a triphenylmethane dye containing no nitrogen. It is weakly colored, lacks solubility in water, and has no affinity for fibers. Some of its derivatives, however, with a carboxyl group *ortho* to one of the hydroxyl groups, are useful as mordant dyes.

Aurin

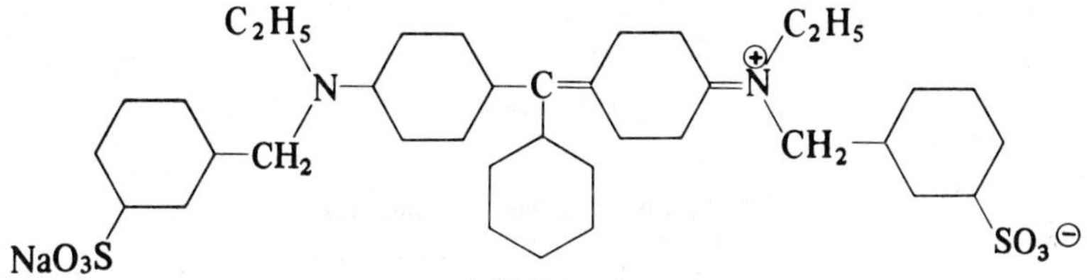

Acid Green 3

Closely related to the triphenylmethane dyes are the diphenylmethane dyes. Only one of these, Basic Yellow 2 (Auramine), is important. It is useful on cotton, nylon, silk, wool, leather, and paper. It is sometimes classified as a ketonimine dye since it is the ketonimine hydrochloride of Michler's Ketone.

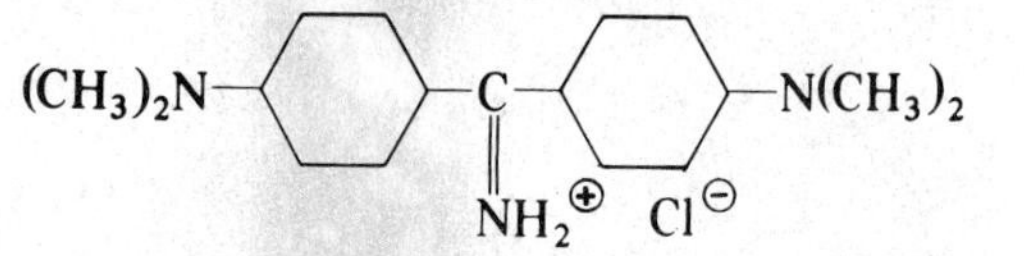

$(CH_3)_2N$—C—$N(CH_3)_2$ with $NH_2^{\oplus}$ $Cl^{\ominus}$

Basic Yellow 2

in an iron kettle. The temperature is regulated by means of an oil or metal bath and kept at 220°C for seven hours. The melt is dissolved in caustic soda, and the product is precipitated by acidifying. It is a yellow-red powder that fluoresces green-yellow in alkaline solution. The sodium salt is called Uranine and is used to trace underground flow of water.

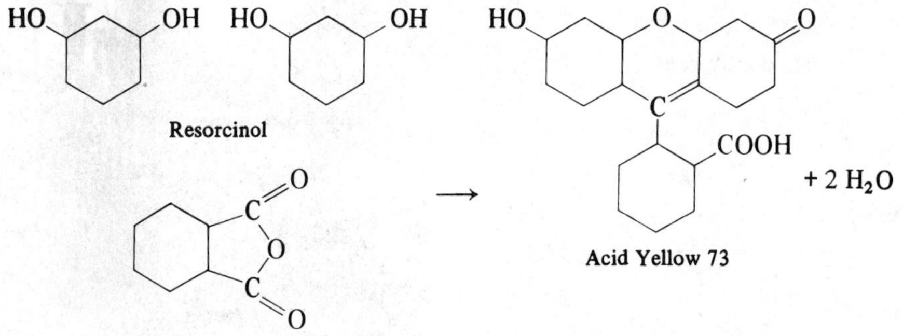

HO OH HO OH

Resorcinol

Phthalic anhydride

→ HO O O C COOH + 2 H_2O

Acid Yellow 73

Xanthene Dyes

The chromophore of the aminoxanthene dyes is the resonance-stabilized structure shown below, where R is H, or alkyl, or aryl. The hydroxyxanthenes can be stabilized by the loss

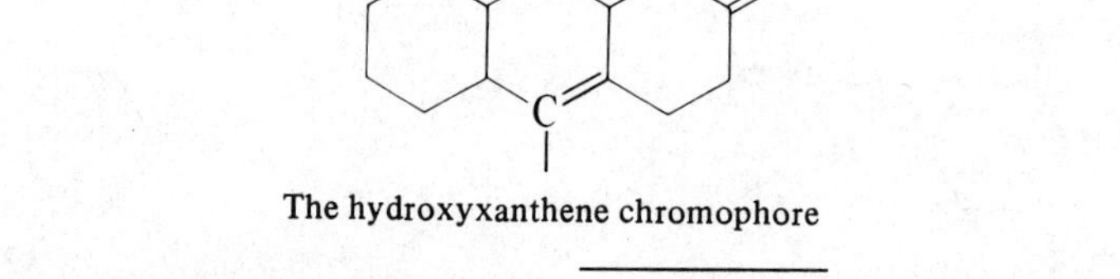

When Acid Yellow 73 is dissolved in alcohol and treated, while warm, with bromine, four equivalents are absorbed to form Acid Red 87 (Eosine). This dye is used to a limited extent on wool; its primary uses are for paper and inks.

The aminoxanthene chromophore

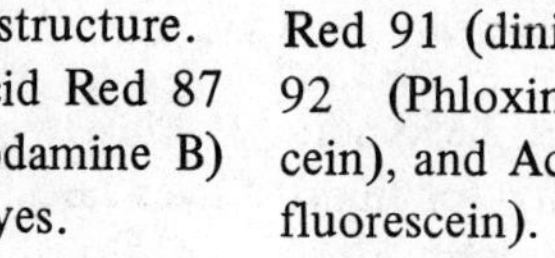

The hydroxyxanthene chromophore

of a proton, forming an uncharged system in which the chromophore is a quinoid structure.

Acid Yellow 73 (Fluorescein), Acid Red 87 (Eosine), and Basic Violet 10 (Rhodamine B) each represent a series of xanthene dyes.

Acid Yellow 73 is made by heating phthalic anhydride (1 mole) and resorcinol (2 moles)

Other dyes related to Acid Red 87 are Acid Red 91 (dinitrodibromofluorescein), Acid Red 92 (Phloxine; tetrabromotetrachlorofluorescein), and Acid Red 51 (Erythrosine; tetraiodofluorescein).

Basic Violet 10 (Rhodamine B) is a dye for bast fibers, mordanted cotton, leather, and

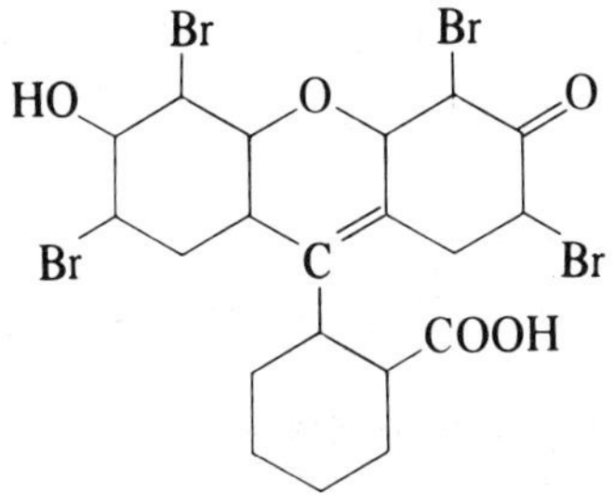

Acid Red 87

paper. It is prepared by condensing phthalic anhydride with *m*-diethylaminophenol and treating with hydrochloric acid. This is the general preparative route for rhodamines.

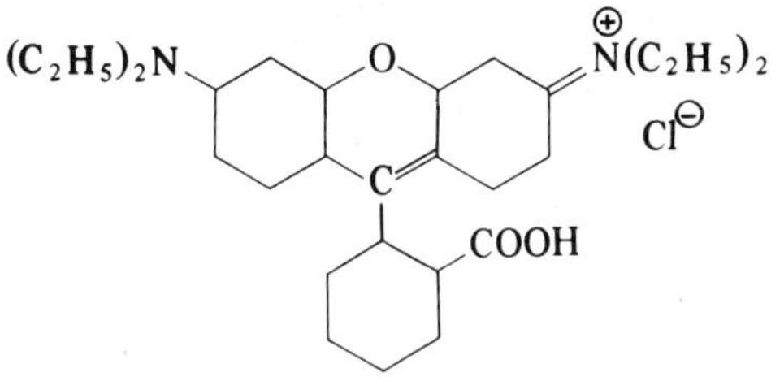

Basic Violet 10

Related to the xanthene dyes are the *acridines*, *azines*, *oxazines*, and *thiazines.* Acridine Yellow results from the fusion of toluene-2,4-diamine with glycerin and oxalic acid, followed by oxidation with ferric chloride. Basic Orange 15 (Phosphine) is a by-product of the manufacture of Basic Violet 14 (Fuchsine), and is an acridine dye.

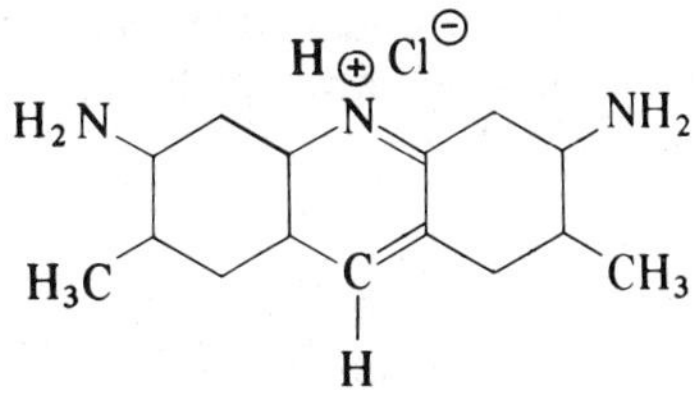

Acridine Yellow

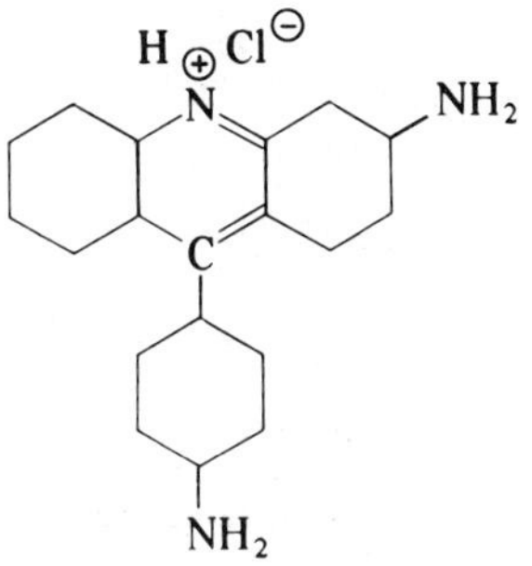

Basic Orange 15

The preparation of azine dyes is illustrated with Safranine B. *p*-Phenylenediamine and aniline are reacted to form indamine. Further reaction of indamine with aniline under oxidizing conditions, followed by treatment with hydrochloric acid, gives Safranine B.

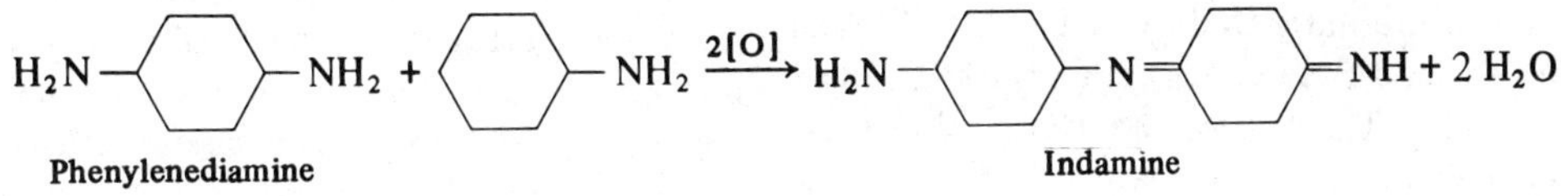

Phenylenediamine

Indamine

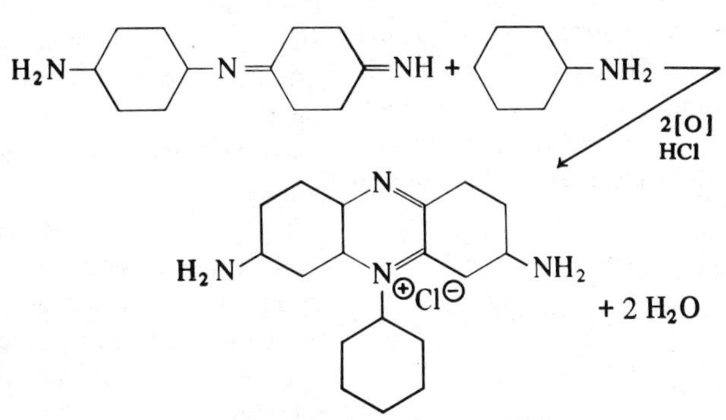

Safranine B

Indamine (above) is an example of a *quinone-imine* dye. Dyes of the indamine type are used in color photography; the dye is formed during development of the film.

Basic Blue 6 (Meldola's Blue) is an oxazine dye. Basic Blue 9 (Methylene Blue) illustrates the thiazine dye class.

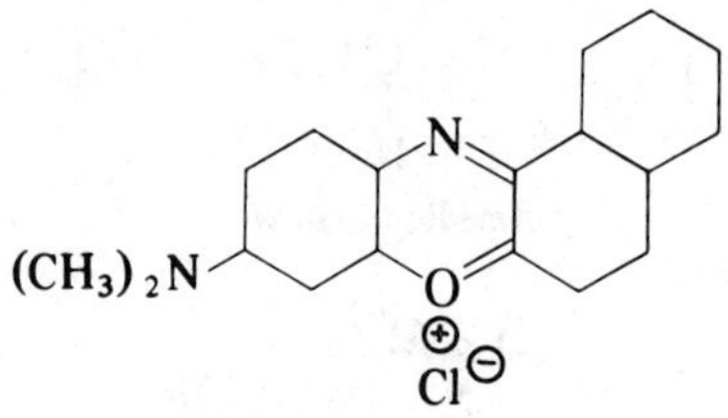

Basic Blue 6

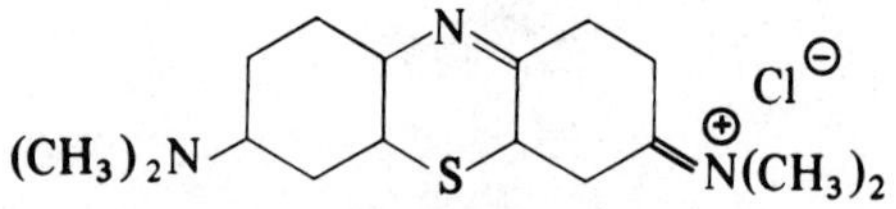

Basic Blue 9

Anthraquinone and Related Dyes

Dyes based on anthraquinone and related polycyclic aromatic quinones are of great importance. Many of the most light-fast acid, mordant, disperse, and vat dyes are of this kind. The chromophore is the carbonyl group, >C=O, which may be present once or several times.

Anthraquinone Acid Dyes. Anthraquinone acid dyes are illustrated by Acid Blue 25, Acid Blue 45, and Acid Green 25 (Alizarin Cyanine Green G). These dyes are water-soluble anthraquinone derivatives which are used for dyeing wool, silk, nylon, leather, and paper. The chemistry of this type of dye is shown with Blue 25 as an example. The starting material is 1-aminoanthraquinone which is sulfonated and brominated to give Bromamine acid. Condensation with aniline then yields Acid Blue 25 which is isolated as the sodium salt.

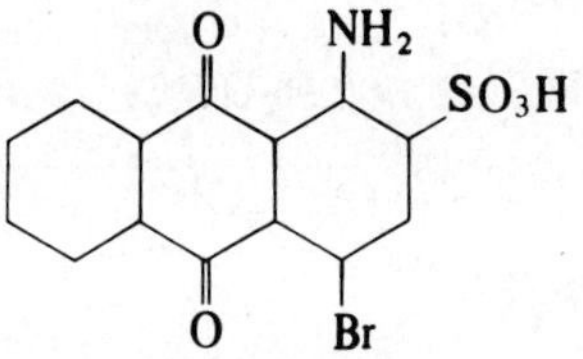

Bromamine acid

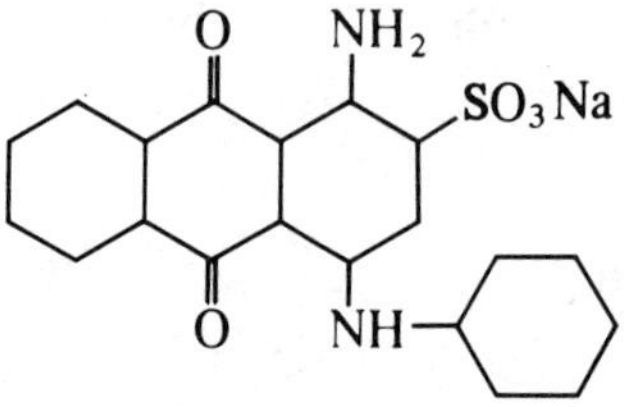

Acid Blue 25

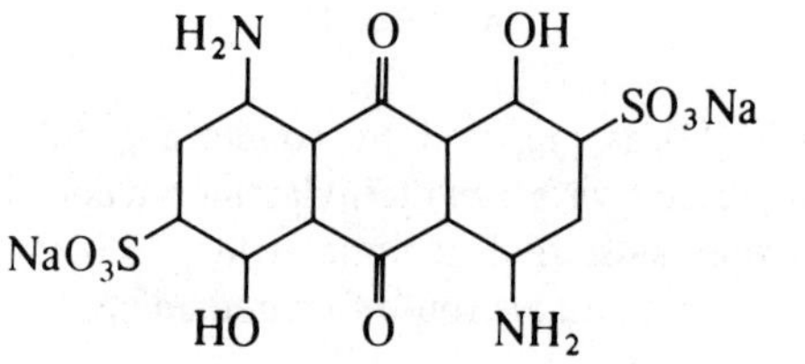

Acid Blue 45

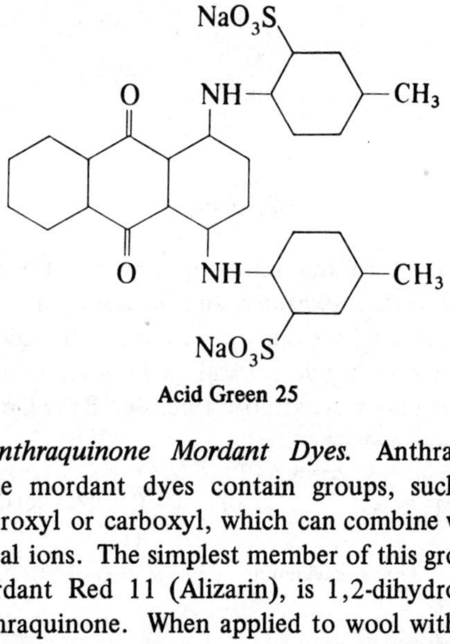

Acid Green 25

Anthraquinone Mordant Dyes. Anthraquinone mordant dyes contain groups, such as hydroxyl or carboxyl, which can combine with metal ions. The simplest member of this group, Mordant Red 11 (Alizarin), is 1,2-dihydroxyanthraquinone. When applied to wool with an aluminum mordant it gives the well-known Turkey Red, and when converted to its calcium salt forms a bluish red powder useful as a pigment.

Mordant Red 11 is made by heating, under pressure, Silver Salt (sodium anthraquinone-2-sulfonate, so called because of its silvery crystals), caustic soda, potassium chlorate, and water. A steel autoclave is ordinarily used and the temperature is maintained at about 180°C. The resulting melt is blown into water and acidified to decompose the sodium derivative of alizarin; the precipitated alizarin is filtered, washed, and standardized as a 20 percent paste.

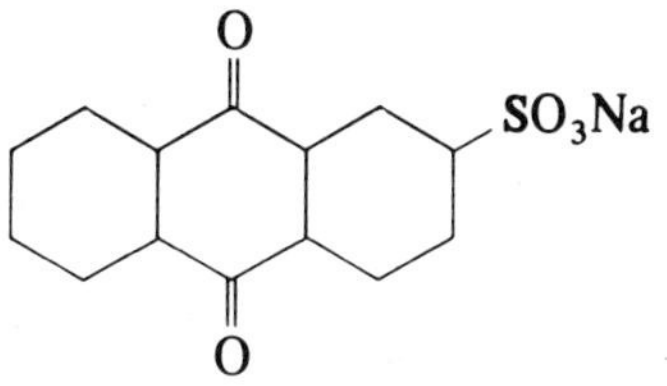

Silver Salt

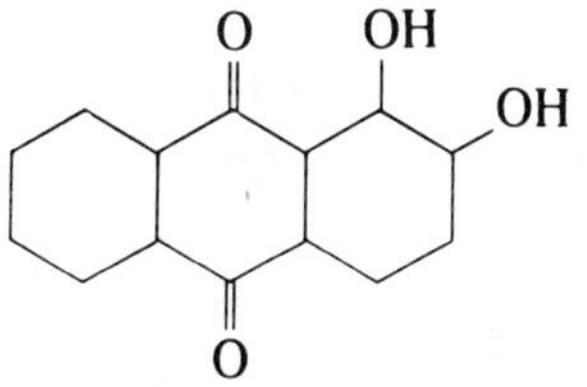

Mordant Red 11

While Mordant Red 11 is very sparingly soluble in water, the introduction of sulfonic acid groups gives soluble derivatives. The most important water-soluble dye of the alizarin mordant class is Mordant Black 13, a dye primarily for wool. It is made by condensing aniline with 1,2,4-trihydroxyanthraquinone and sulfonating the resulting base. It is applied to wool with a chromium mordant and is quite fast to light and washing.

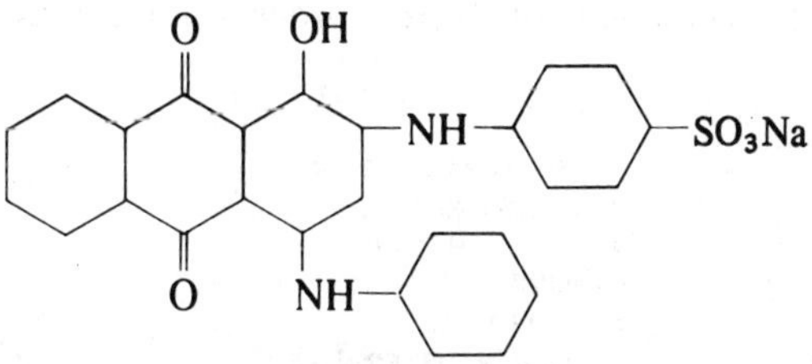

Mordant Black 13

Anthraquinone Vat Dyes. Many of the best vat dyes are derivatives of anthraquinone or related compounds. A relatively simple dye of this class is Vat Yellow 3, which is made by benzoylation of 1,5-diaminoanthraquinone. It is a yellow pigment that is used as such when properly ground and dried. As a vat dye, it is usually supplied as an aqueous paste. Reduction with sodium hydrosulfite in caustic soda solution gives the alkali-soluble anthrahydroquinone, which has affinity for cellulosic fibers. After application to the fiber, the insoluble dye is reformed by oxidation, as described earlier on page 815.

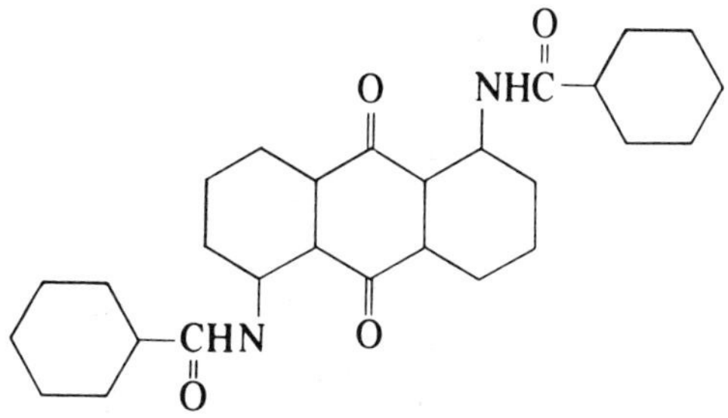

Vat Yellow 3

Perhaps the best known of the anthraquinone vat dyes is Vat Blue 4 (Indanthrone). This attractive dye is made by the fusion of 2-aminoanthraquinone with caustic potash. The fastness to chlorine bleaching of Vat Blue 4 is improved by chlorination. A number of chlorinated indanthrones are in commercial use and make up the most important group of fast vat blues. The preparative route for Vat Blue 4 is outlined below:

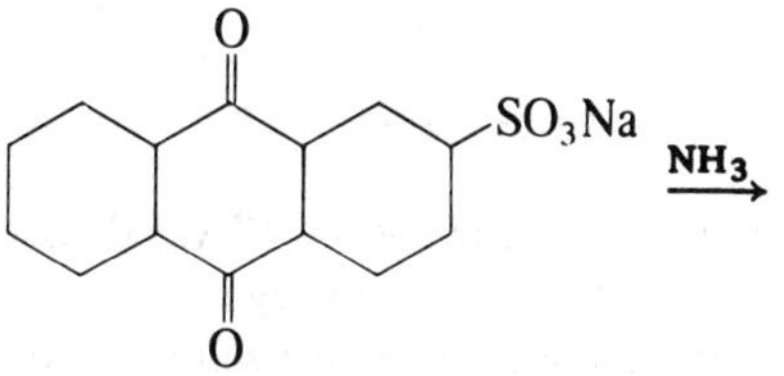

Silver Salt $\xrightarrow{NH_3}$ **2-Aminoanthraquinone** $\xrightarrow{[O],\ KOH}$ **Vat Blue 4**

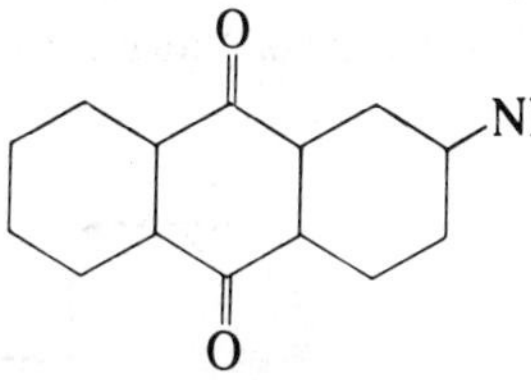

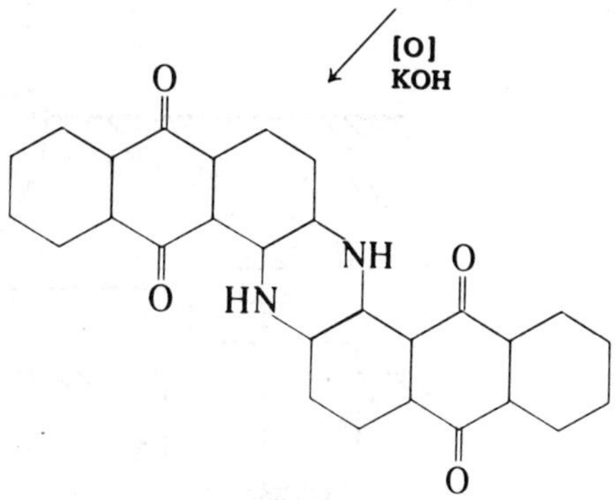

Related to Vat Blue 4 are Vat Yellow 1 (Flavanthrone) and Vat Orange 9 (Pyranthrone), as well as other important dyes.

Vat Yellow 1

Vat Orange 9

Vat Blue 20

Vat Green 1

One of the most attractive of all vat dyes is Vat Green 1 (Jade Green). This dye is a derivative of dibenzanthrone, which is itself a vat dye (Vat Blue 20 or Violanthrone).

Anthraquinone-vat-dye manufacture is illustrated by the preparation of Vat Blue 20. Phthalic anhydride is condensed with benzene in aluminum chloride to give *o*-benzoylbenzoic acid which is ring-closed to anthraquinone with sulfuric acid. Treatment of anthraquinone in 82 percent sulfuric acid with metal powder (iron, copper) and glycerin gives benzanthrone, probably by the series of steps shown below. Condensation of benzanthrone with caustic

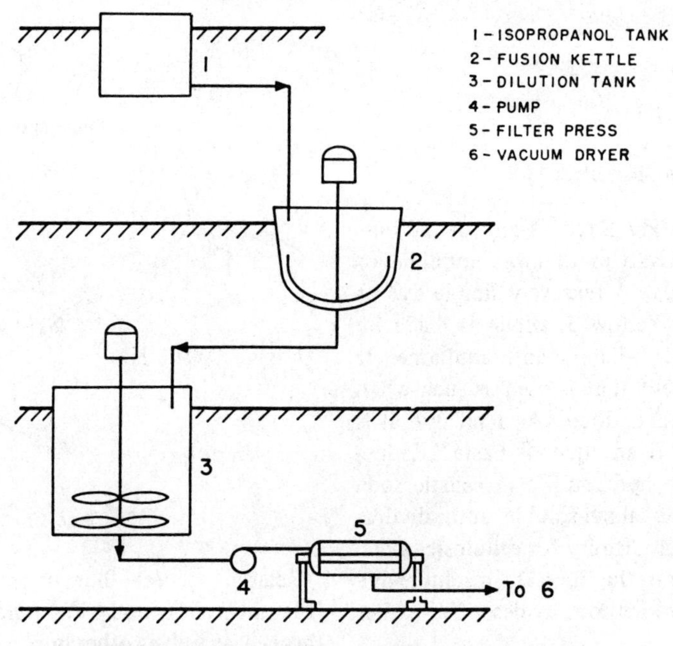

Fig. 23.8 Schematic diagram for manufacturing dibenzanthrone.

potash in isopropanol and sodium acetate gives dibenzanthronyl, which undergoes ring-closure to dibenzanthrone upon further heating. The dibenzanthrone usually is dissolved in sulfuric acid and reprecipitated as a paste of fine particle size by dilution with water. The paste is washed free of acid and standardized as Vat Blue 20. The equipment for this series of reactions must be able to resist the strongly corrosive conditions and must be externally heated and cooled. Cast iron, stainless steel, and enamel-lined closed vessels are used. (Fig. 23.8)

Indigoid and Thioindigoid Dyes

The parent compound of the indigoid dye class, indigo, has been in use as a vat dye since ancient times. Natural indigo is obtained from a species of plant, *Indigofera.* It can be reduced by fermentation as well as by the modern method using sodium hydrosulfite and caustic soda. Once reduced, it can be applied to cellulosic fibers and then oxidized by air to produce rich blue dyeings. Natural indigo is no longer important, but large quantities of the synthetic dye are used to produce low-cost

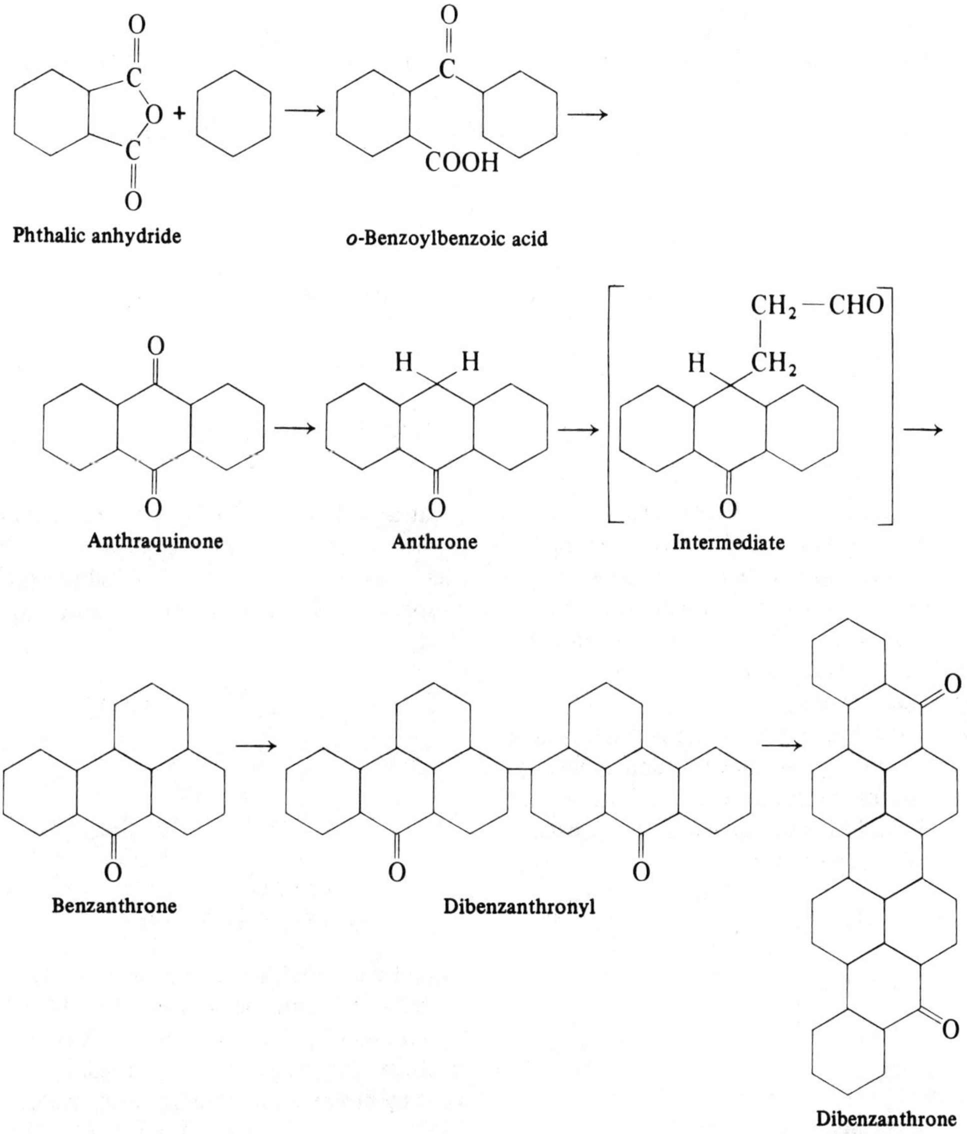

Phthalic anhydride → ***o*-Benzoylbenzoic acid** →

Anthraquinone → **Anthrone** → **Intermediate** (CH_2-CHO, CH_2) →

Benzanthrone → **Dibenzanthronyl** → **Dibenzanthrone Vat Blue 20**

blues on cotton. The oxidized form is also used as a pigment. The "Colour Index" designation for indigo is Vat blue 1.[15]

A modern synthesis of Vat Blue 1 involves first the preparation of phenylglycine. To a water solution of sodium bisulfite are added formaldehyde, aniline, and sodium cyanide. The nitrile of phenylglycine which is formed is isolated and washed free of sulfite. Warming of the nitrile in an alkaline slurry then gives a solution of phenylglycine as its sodium salt. Ammonia is evolved during this step.

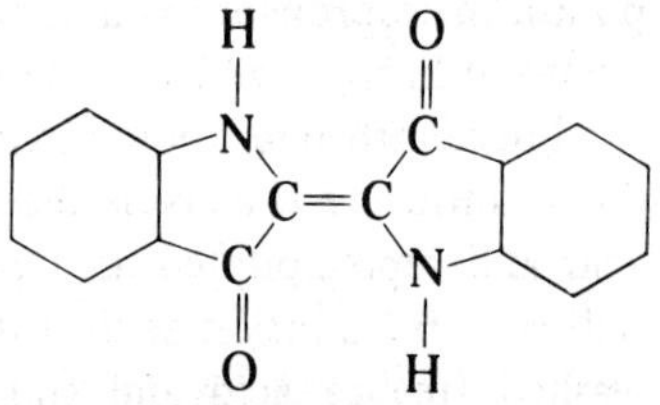

Vat Blue 1

Since very large volumes of indigo are produced, it has been worthwhile to devise continuous processes. Such processes were pioneered by the former I. G. Farbenindustrie in Germany.

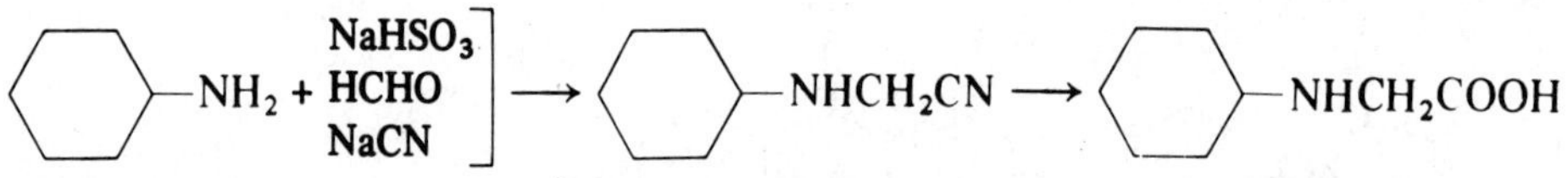

$$C_6H_5{-}NH_2 + \left.\begin{matrix} NaHSO_3 \\ HCHO \\ NaCN \end{matrix}\right] \longrightarrow C_6H_5{-}NHCH_2CN \longrightarrow C_6H_5{-}NHCH_2COOH$$

Nitrile of phenylglycine | Phenylglycine

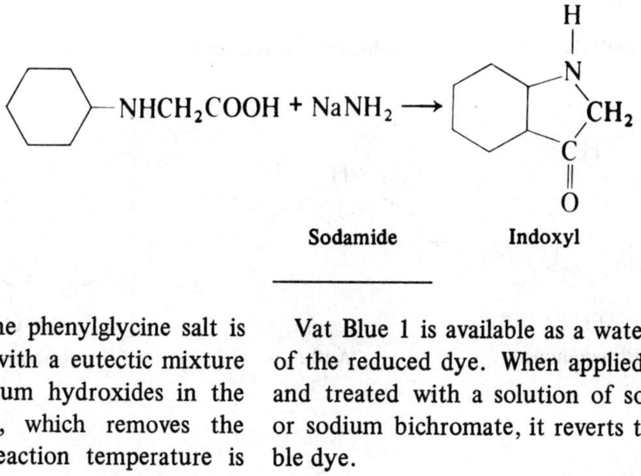

$$C_6H_5{-}NHCH_2COOH + NaNH_2 \longrightarrow \text{Indoxyl}$$

Sodamide | Indoxyl

In the second step the phenylglycine salt is condensed by heating with a eutectic mixture of potassium and sodium hydroxides in the presence of sodamide, which removes the water formed. The reaction temperature is 200-220°C. The product, which is formed in high yield, is indoxyl.

After the fusion step, water is added and air is blown into the alkaline solution of indoxyl. Indigo forms and separates from solution. It is filtered, washed, and standardized as a paste of about 20 percent solids.

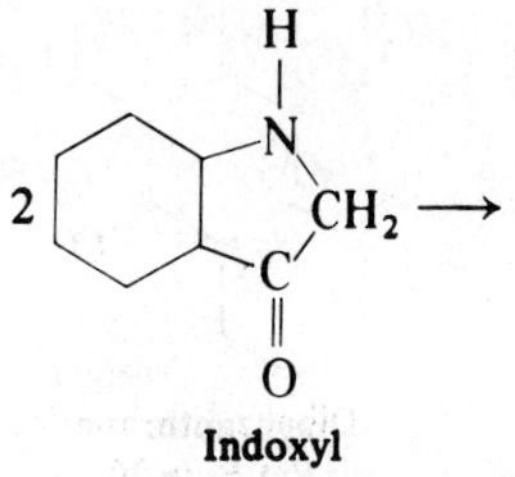

$$2\ \text{Indoxyl} \longrightarrow$$

Indoxyl

Vat Blue 1 is available as a water-soluble salt of the reduced dye. When applied to the fiber and treated with a solution of sodium nitrite or sodium bichromate, it reverts to the insoluble dye.

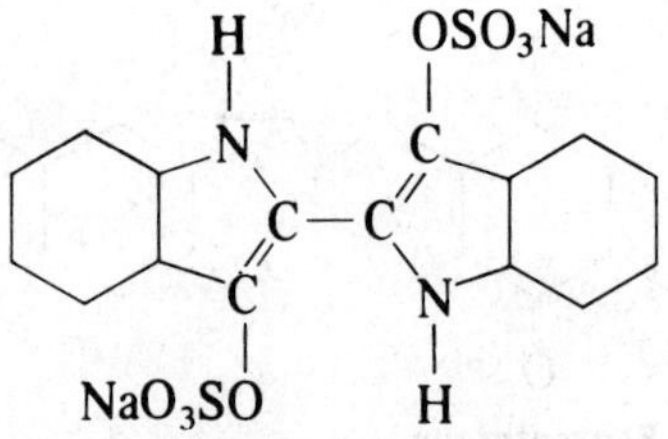

Solubilized Vat Blue 1 (OSO_3Na / NaO_3SO)

Important derivatives of indigo are the chloro and bromo derivatives. Vat Blue 5 is tetrabromoindigo. It is similar to Vat Blue 1 in shade but is considerably brighter. It is made by bromination of indigo in acetic acid.

Vat Red 41 (Thioindigo) is made by the reaction of thiosalicylic acid with chloroacetic acid, followed by fusion with caustic soda and oxidation by air. Thiosalicylic acid is made from anthranilic acid by diazotization and reaction with hydrogen sulfide.

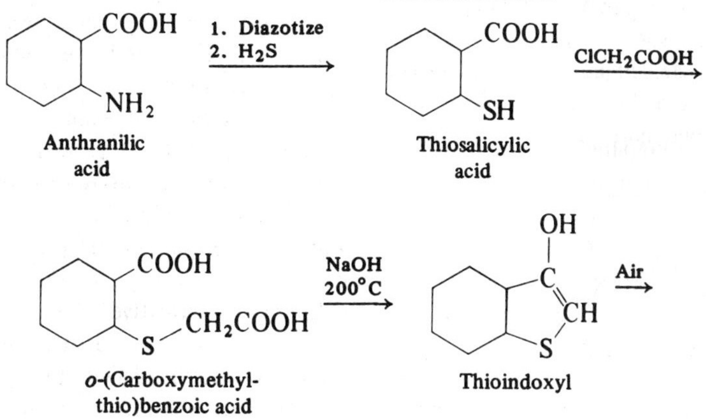

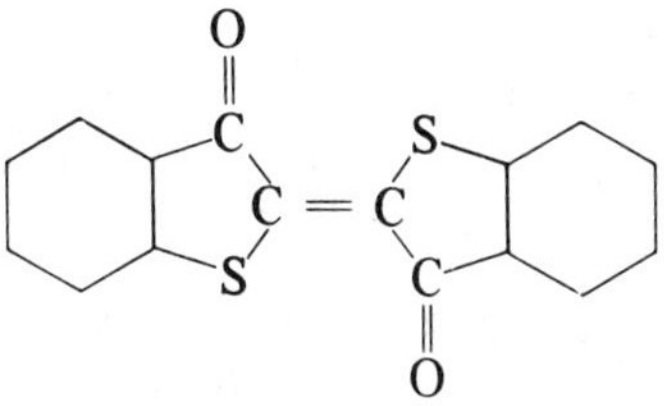

Anthranilic acid → (1. Diazotize; 2. H_2S) → Thiosalicylic acid → ($ClCH_2COOH$) → *o*-(Carboxymethylthio)benzoic acid → (NaOH, 200°C) → Thioindoxyl → (Air) →

Vat Red 41

Vat Red 41 is a dull bluish red on cotton. On polyester fibers it produces bright pink shades. Many of its derivatives however are of greater commercial importance on cotton. An example is Vat Orange 5.

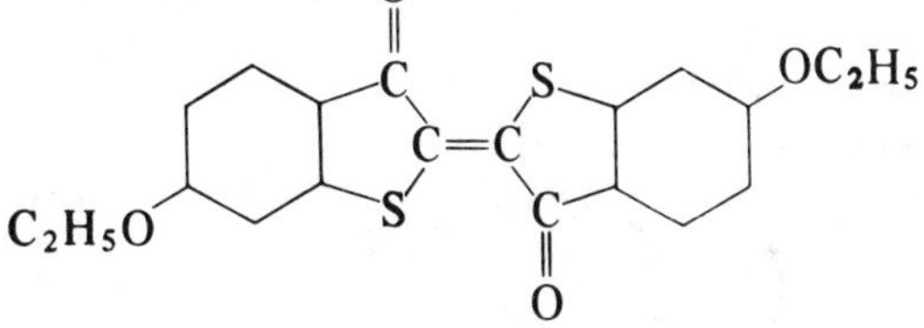

Vat Orange 5

Sulfur Dyes

Sulfur dyes are dyes of unknown constitution which can be applied to fibers when reduced with sodium sulfide. Most of them are insoluble in water before reduction. After reduction they are soluble and can be adsorbed by fibers and then oxidized to an insoluble form with air. Although structures cannot be written for the sulfur dyes, the methods for reproducing individual types are well established.

m-Diamines (toluene-2,4-diamine for example) fused with sulfur cause the evolution of hydrogen sulfide and produce a melt which yields brown sulfur dyes. Redder shades are obtained by using derivatives of phenazine, such as the compound below which produces Sulfur Red 6.

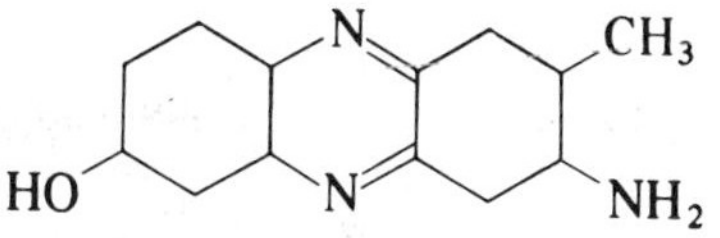

Intermediate for Sulfur Red 6

Blue sulfur dyes are obtained by treating diphenylamine derivatives with sodium polysulfide. Sulfur Blue 7, Sulfur Blue 9, and Sulfur Blue 13 are made from the intermediates below:

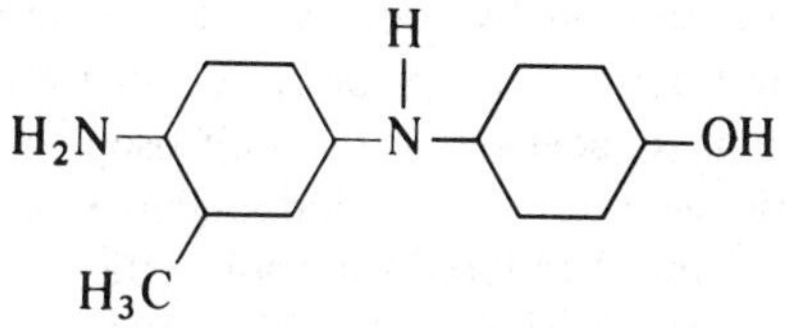

Intermediate for Sulfur Blue 7

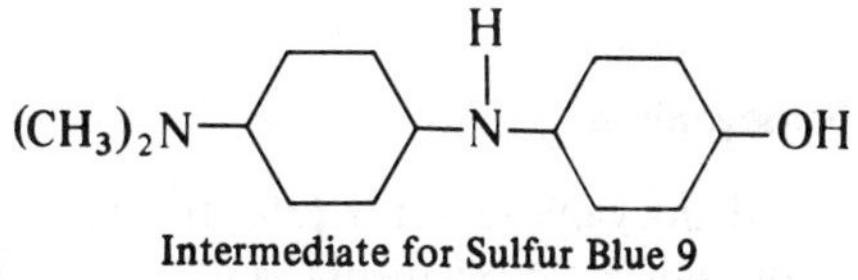

Intermediate for Sulfur Blue 9

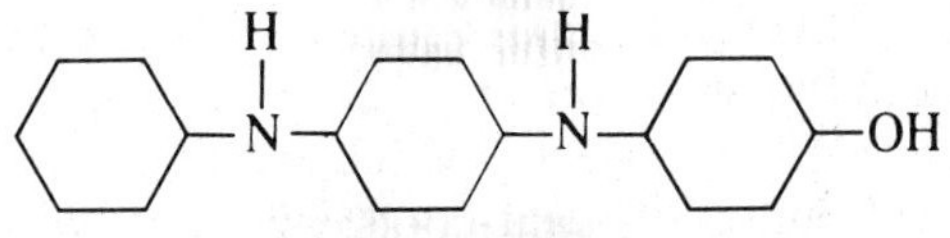

Intermediate for Sulfur Blue 13

Sulfur Black 1 is prepared by heating 2,4-dinitrophenol with sodium polysulfide. The fusion mass is dissolved in water and blown with air until all the dye has separated. It is then filtered, washed, and dried.

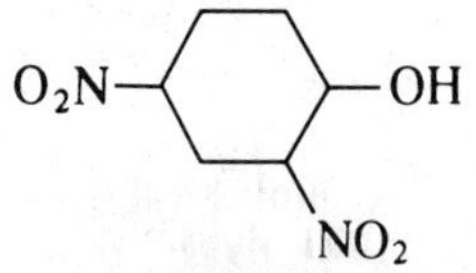

Intermediate for Sulfur Black 1

Vat Blue 43 (Hydron Blue R) is a kind of sulfur dye which, unlike typical sulfur dyes, may be reduced by sodium hydrosulfite without destruction. To prepare Vat Blue 43, carbazole is condensed with *p*-nitrosophenol in sulfuric acid to give an indophenol which is reduced and then fused with polysulfide.

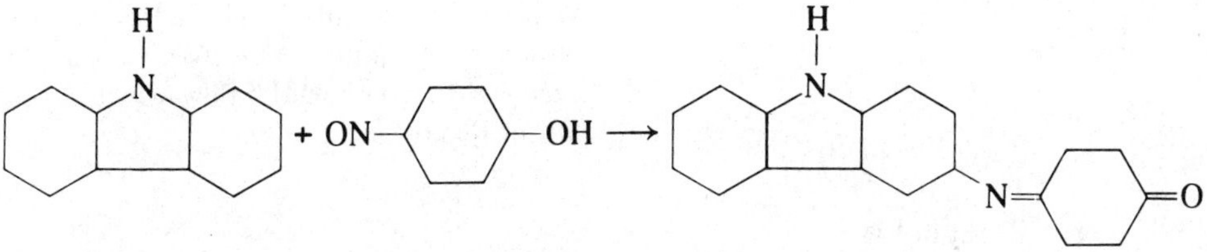

Carbazole

p-Nitrosophenol

Condensation product

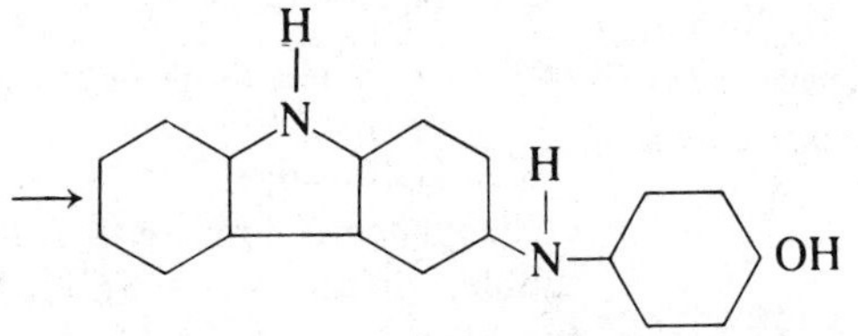

→ Vat Blue 43

Reduced condensation product

Vat Blue 42 (Hydron Blue G) gives a greener shade than Vat Blue 43. It is prepared by using *N*-ethylcarbazole in place of carbazole.

Phthalocyanines

The phthalocyanines constitute an important class of synthetic pigments and dyes. The parent compound is Pigment Blue 16 (Phthalocyanine). One method of preparation involves the fusion of phthalonitrile with cyclohexylamine in an inert solvent. The two central hydrogen atoms can be replaced by metals such as copper, nickel, iron, and cobalt. In actual practice, the metal derivatives are not made from the metal-free compound, but are synthesized directly. Pigment Blue 15 (Copper Phthalocyanine), for example, is made by the fusion of phthalonitrile with copper metal or a copper salt. An alternative method involves the fusion of phthalic anhydride with urea and a copper salt in the presence of a molybdenum catalyst. Pigment Blue 15 and its polychlorinated derivative, Pigment Green 7, are the most important of the phthalocyanines. They are distinguished by great brilliance, strength and stability, and are used in every field in which colored pigments are used.

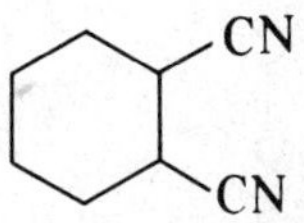

Phthalonitrile

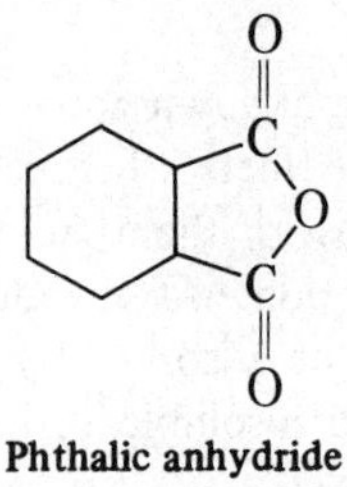

Phthalic anhydride

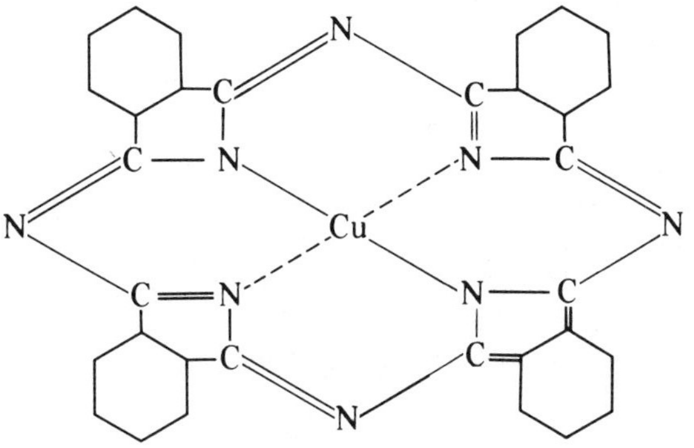

Pigment Blue 15

A number of phthalocyanine derivatives can be made by replacing the hydrogen atoms on the benzene rings. Pigment Green 7, mentioned above, is an example. Vat Blue 29 is a partially sulfonated cobalt phthalocyanine. Reduction with sodium hydrosulfite in caustic soda solution converts it to a soluble form that can be applied to textiles and reoxidized to the pigment. The structure of the reduced form is not known.

Direct Blue 86 is copper phthalocyanine which has been sulfonated to the extent of between two and three sulfonic acid groups per molecule. It produces very bright blue-green shades on cotton, viscose, and paper. Solvent Blue 25 is a spirit-soluble derivative of copper phthalocyanine, useful for inks, lacquers, and stains. It is the reaction product of the tetrasulfonyl chloride of copper phthalocyanine with 4-methylpentylamine.

Fluorescent Brightening Agents

The fluorescent brightening agents, also known as fluorescent whiteners or optical bleaches, comprise a valuable class of dyes. Although they were not commercialized until just prior to World War II, their use has grown at an exceedingly rapid rate. They are now one of the major dye classes produced in the United States, having a value of over $49 million in 1978 (see Table 23.2).

Fluorescent brighteners derive their value from the fact that although they are colorless, they strongly absorb light of shorter wave length in the ultraviolet region and fluoresce or re-emit light of longer wave length in the visible region of the spectrum. Thus, they produce a brightening or whitening effect which is useful in making yellowed products appear whiter and white materials brighter. The major use of optical brighteners is in detergents. However, considerable amounts are used in the paper and textile industries.

A number of molecules have the ability to fluoresce, but relatively few have achieved commercial importance. The organic chemist can tailor fluorescent dyes to be useful on almost any substrate; that is, they can be made into direct, cationic, or disperse dyes. By far the major volume is in the direct-dye type for cotton and paper. The major fluorophore is the diaminostilbene nucleus which is then made substantive by incorporation of solubilizing and affinity-conferring groups. A typical commercial fluorescent brightener is made by condensing diaminostilbenedisulfonic acid with cyanuric chloride, and then further substituting the cyanuric chloride residue with colorless amines to give a product such as the one shown on page 860.

PRODUCTION STATISTICS

Table 23.4 gives the production of synthetic dyes in the United States for the year 1972, the last year for which such data are available arranged according to chemical class. These figures are taken from the report of the United States Tariff Commission. As in Table 23.2, it will be noted that the classifications used are

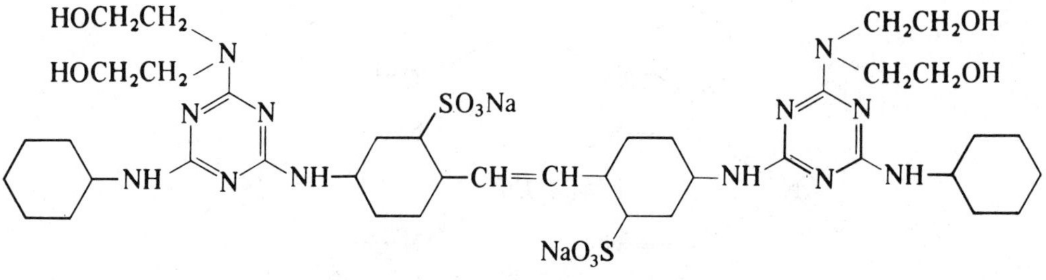

C.I. Fluorescent Brightening Agent 28

not exactly the same as those given in the "Colour Index."

TABLE 23.4 Synthetic Dyes: U.S. Production and Sales, by Chemical Class, 1972

Chemical Class	*Production (Million pounds)*
Anthraquinone	46.6
Azo	92.0
Azoic	10.3
Cyanine	0.9
Methine	5.6
Nitro	1.4
Oxazine	0.5
Phthalocyanine	1.4
Quinoline	2.7
Stilbene	30.9
Thiazole	0.4
Triarylmethane	8.9
Xanthene	1.2
All Other	60.2
TOTAL	263.0

NEW DEVELOPMENTS IN DYES

The most significant development influencing the dye industry during the past decade involved the rapid growth of synthetic fibers, which, in the late 1960's, surpassed natural fibers in over-all poundage. Commensurate with the rise of the synthetic fibers was the development of new dyes and dyeing methods which were needed to make these new fibers economically attractive. Higher labor costs and the need for greatly expanded output required the development of new low-cost continuous dyeing processes. Advanced instrumental techniques, with the aid of computers, are making exact color matching possible from lot to lot.

The future will continue to see improvements and refinements in these techniques. Dye manufacture will be automated from batch to batch to produce spectrally reproducible products. New fiber variations will require continual development of new dyes to optimize fastness, application simplicity, and hue. Pollution abatement will be a major concern in the dye manufacturing as well as the dye application industries. Equipment and processes must be designed to minimize noxious fumes and polluted waste streams. This will require new designs for reactors and scrubbers, and efficient cleanup of effluents of both the colored and the not-so-obvious soluble colorless salts. Water, which is a major raw material for the manufacture of dyes and in dye application, must be conserved and left essentially free of chemical or thermal contamination. There is much activity in the direction of solvent processing of textiles as a means of water conservation. Dyeing processes using recoverable organic solvents in place of water are attractive because of their reduced energy requirements for drying, and their reduced water requirements. As yet, no large-scale applications of solvent dyeing have been commercialized. The decade of the 1980's represents a real challenge to the engineer to demonstrate that man can freely use colored materials as a means of self-expression, motivation, and beautification, while maintaining a sound ecological environment.

REFERENCES

1. American Association of Textile Chemists and Colorists, *Technical Manual* (an annual publication) vol. 55, AATCC, Research Triangle Park, North Carolina, 1979.
2. Amidon, C. H., *Textile Chemist and Colorist*, **1**, 40 (1969).
3. Fierz-David, H. E., and L. Blangey, *Fundamental Processes of Dye Chemistry*, trans. by P. W. Vittum, John Wiley & Sons, New York, 1949.
4. Friedlaender, P., ed., vols. **1-13**; H. E. Fierz-David, ed., vols. **14-25**, *Fortschritte der Teerfarbenfabrikation*, Julius Springer, Berlin, 1888-1942.
5. Gorondy, E. J., *Textile Chemist and Colorist*, **10**, 105 (1978).
6. Groggins, P. H., ed., *Unit Processes in Organic Synthesis*, 5th Ed., McGraw-Hill, New York, 1958.
7. Houben, J., *Das Anthracen und die Anthrachinone*, Thieme, Leipzig, 1929; Photo-lithoprint by Edwards Bros., Ann Arbor, 1944.
8. I. G. Farbenindustrie, FIAT 764 (PB 60946), "Dyestuffs Manufacturing Processes of I. G. Farbenindustrie A. G.," 1947.
9. I. G. Farbenindustrie, FIAT 1313 (PB 85172), "German Dyestuffs and Dyestuff Intermediates, including Manufacturing Processes, Plant Design, and Research Data," 1948.
10. Lubs, H. A., ed., *The Chemistry of Synthetic Dyes and Pigments*, Reinhold, New York, 1955; Hafner reprint, New York, 1965.
11. Meunier, P. L., *Textile Chemist and Colorist*, **2**, 386 (1970).
12. Organisation for Economic Co-operation and Development, "The Chemical Industry, 1975," Paris, 1977.
13. Peters, R. H., *Textile Chemistry*, vol. **III**, "The Physical Chemistry of Dyeing," Elsevier, New York, 1975.
14. Rys, P. and H. Zollinger, *Fundamentals of the Chemistry and Application of Dyes*, Wiley-Interscience, New York, 1972.
15. Society of Dyers and Colourists and American Association of Textile Chemists and Colorists, *Colour Index*, 3rd ed., 5 volumes, 1971; 1st Revision of vol. 5, 1975; vol. 6 (1st Supplement to vols. 1-4), 1975.
16. Trotman, E. R., *Dyeing and Chemical Technology of Textile Fibres*, 5th ed., Griffin, London, 1975.
17. Venkataraman, K., ed., *The Chemistry of Synthetic Dyes*, Academic Press, New York, 8 volumes, 1952-1978.
18. Vickerstaff, T., *The Physical Chemistry of Dyeing*, 2nd ed., Oliver and Boyd, London, 1954.
19. Zollinger, H., *Azo and Diazo Chemistry*, Wiley, New York, 1961.

24

The Nuclear Industry

Warren K. Eister*

INTRODUCTION

The production of thermal energy is the objective of the nuclear industry. To achieve this objective, it was necessary to develop a unique family of materials to meet specifications for the nuclear properties of materials as well as the physical and chemical properties. Therefore for many materials, interest was centered in the atomic nuclide as well as in the chemical element. Furthermore, because of the radiation characteristics of the nuclear fuels, uranium and thorium, along with those of the materials produced, the actinides and fission-products, a tremendous effort was expended to evaluate their biological significance and to control their processing and use.

In this chapter, attention will be given to the nuclear technology related to the products from this industry (Table 24.1).[1] This will include 1) the nuclear fission and fusion processes, 2) the nuclear explosives, 3) the fission-type power reactors, 4) the related materials processes, 5) waste management, and 6) isotope products. However, first the status and outlook of the nuclear industry will be reviewed, including the Earth's energy balance and energy economics.

Safety will be the final topic to be considered. The chemical industry, with its considerable experience in the safe, controlled processing of toxic materials, was involved in the pioneering development of the nuclear fission reactor and related chemical process technology for plutonium production and for uranium isotopic separation. At the start of the nuclear industry in the early 1940's, safety was and has continued to be the prime responsibility of the people in the research and development of nuclear technology along with those involved in the construction and operation of the nuclear facilities.

Status and Outlook

In 1979, 12 percent of electric power generating capacity in the USA was nuclear, and the large scale demonstration of sodium-cooled breeder reactors (LMFBR's) was underway in France, Japan, Russia, and the United King-

*Division of Waste Isolation, United States Department of Energy, Washington, D.C.

TABLE 24.1 U.S.A. Nuclear Industry Products (1978)

	Number of Units	
Industry Product	*Present*	*Additional Planned*
Commercial		
Electric generators	77	124
Radioisotope applications	millions	a
Defense		
Weapons	~10,000	b
Submarines	115	35
Cruisers	7	2
Aircraft carriers	4	–

[a] Increasing use of radioisotopes for medical diagnosis, industrial radiography and instrumentation, chemical polymerization and power sources for space exploration.
[b] Sufficient to maintain viable defense.

dom. The light water reactors (LWR's) developed in the USA were producing the majority of the electricity from nuclear power throughout the world. (Fig. 24.1) In 1980, the 575 MWe LWR, Connecticut Yankee unit at Haddan Neck, Conn., produced its 50th million MWh. This was shortly after its 12th anniversary in commercial operation. The USA development of LMFBR's was at the test reactor stage and being delayed pending the resolution of proliferation questions.

The fusion process for the production of thermal energy has thus far been limited to use in nuclear explosives, the hydrogen bomb. There has been continuing progress since the early 1960's in the development of controlled thermonuclear reactors employing the fusion process. However, the successes of recent experiments still do not assure the ultimate success of a controlled fusion reactor, and a

Fig. 24.1 The Phoenix liquid metal fast breeder reactor located on the Rhone River in France started operation in 1973. It uses the pool-type technology and generates 230 MW of electricity.

tremendous engineering development effort spanning decades will be required to produce commercial power.

The nuclear industry has made both stable and radioactive nuclides common products of commerce. Uranium is enriched in the isotope 235 to optimize the performance of LWR type reactors. Deuterium is separated from hydrogen as a more efficient moderator for natural uranium fueled reactors. Cobalt-60 is being produced for use in process irradiation and medical therapy. Iodine-131 and technetium-99m are in routine use in medical diagnosis. Plutonium-238 has provided thermal energy to power instrument systems in the space programs. In a decade, the supply of materials went from less than 92 chemical elements to more than 2000 chemical nuclides and isomers.

Perhaps most significant, the nuclear industry has created and has been thrust into the forefront of serious public concerns regarding technology's impact on worldwide social institutions. In the 1960's, the development of nuclear power was confidently projected to provide a significant part of the future USA electric energy requirement, roughly equivalent to the oil imports today. However, by the early 1970's there developed a concern regarding reactor safety, radioactive waste management, and the possible diversion of nuclear fuel for the proliferation of nuclear weapons. In the late 1970's, the projections of the 1960's were, at the very least, significantly delayed.

Several important studies were performed in the 1970's that related to the major nuclear issues: reactor safety, waste management, and weapons proliferation. The reactor safety study concluded that the risk from reactor accidents was a small fraction of risks from other human activities. (Table 24.2)[1] The first major accident involving a power reactor occurred at Three Mile Island in March 1979, and it confirmed these conclusions. No one received radiation in excess of the licensed limits. In spite of numerous management, equipment, and operator malfunctions the engineered safety features worked.

The presidential-appointed Interagency Review Group on Radioactive Waste Management concluded that mined geologic vaults would be a safe place for terminal storage of high-level wastes.[2] However, the Review Group recommended a conservative, step-wise approach that would delay the operation of the waste repositories until the late 1990's. This delay had no significant effect on nuclear power costs or safety.

The study of proliferation control called the "International Fuel Cycle Evaluation" was an international analysis, with participation of fifty countries.[3] This study concluded that administrative control was the only effective

TABLE 24.2 Average Risk of Fatality by Various Causes in the United States

Accident Type	*Annual Total Number*	*Individual Chance per Year*
Motor Vehicle	55,791	1 in 4,000
Falls	17,827	1 in 10,000
Fires and Hot Substances	7,451	1 in 25,000
Drowning	6,181	1 in 30,000
Firearms	2,309	1 in 100,000
Air Travel	1,778	1 in 100,000
Falling Objects	1,271	1 in 160,000
Electrocution	1,148	1 in 160,000
Lightning	160	1 in 2,000,000
Tornadoes	91	1 in 2,500,000
Hurricanes	93	1 in 2,500,000
All Accidents	111,992	1 in 1,600
Nuclear Reactor Accidents (100 plants)	–	1 in 5,000,000,000

method to prevent nations from diverting nuclear power fuel to use for military weapons.

THE EARTH'S ENERGY SUPPLY AND USE

The earth offers sources of both income and capital energy for the benefit of man (see Table 24.3[5] and Figure 24.2[4]). The income energy, equivalent to 5.3 Q/yr ($Q = 10^{18}$ Btu), is constantly replenished and includes solar, gravitational, and subterranean sources. The capital energy, equivalent to 22 to 340 Q cannot be replenished and includes fossil and nuclear forms. The proven U.S. and world supplies of nuclear energy are presently estimated to be 21 and 73 Q, respectively. The fossil fuel deposits of coal, petroleum, natural gas, tar sands, and oil shale resulted from the interaction of the sun with the earth's environment. The development of the deposits was a very low yield process, equivalent to about 40 solar years* of the five billion years the earth is believed to have existed. The fission fuels, uranium and thorium, were deposited on the earth at the time of its consolidation.

*A solar year represents the amount of solar energy reaching the earth in one year.

Little use was made of the capital energy supply until the last century, and now about 5 percent of the fossil fuel supply has been consumed. Yet even though 95 percent of the fossil fuel remains, there are increasingly serious problems today concerned with its extraction, distribution, and use. Petroleum production may have reached its peak in the United States in the late 1970's and may peak within the next 20 years on a world-wide basis. While the peak production rate for coal may not be reached for a hundred years, the high-grade deposits in the industrialized regions of the world are being rapidly depleted. In addition, there are increasing concerns regarding sulfur dioxide and carbon dioxide, as well as radioactive pollution of the atmosphere resulting from coal-fired power plants.

The solar, gravitational, and subterranean energy sources are projected to supply about 20 percent of U.S. energy needs in the year

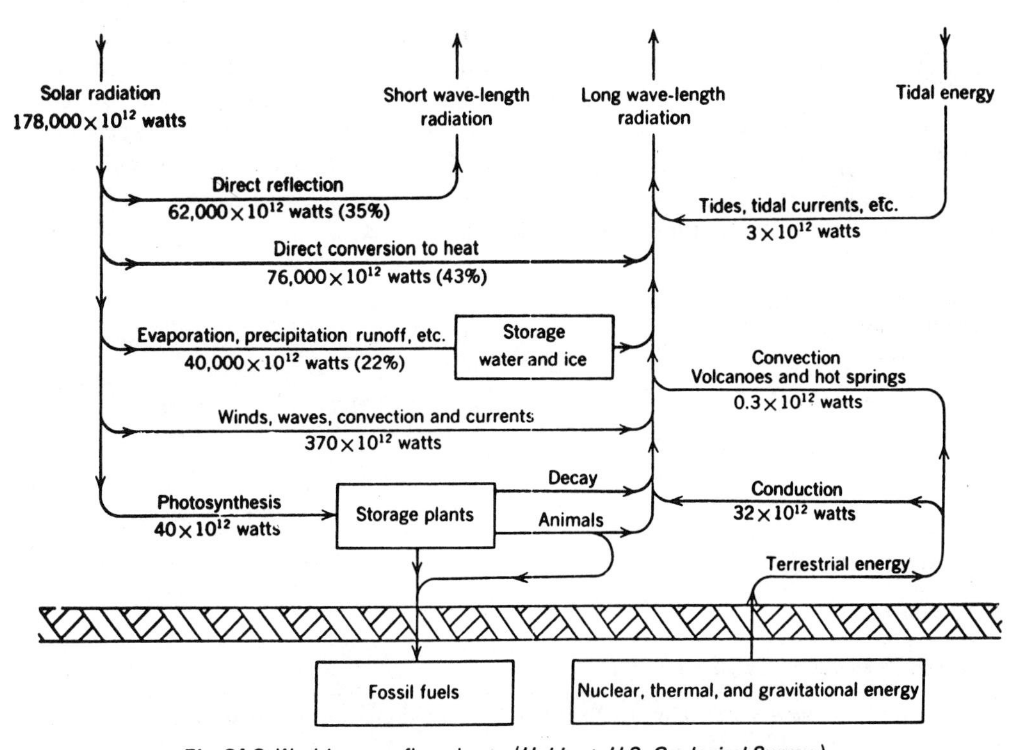

Fig. 24.2 World energy flow sheet. (*Hubbert, U.S. Geological Survey.*)

2000. This will include energy generated by water, wind, and biomass (including wood, geothermal, ocean thermal, solar thermal, and solar voltaic methods). Today these sources, principally water power, provide about 10 percent of the U.S. electrical energy needs.

The demand for energy has increased as the world has continued to industrialize, with the present total demand for all types of energy for all uses about 0.19 Q per year. The annual per capita consumption ranges from 180 million Btu in the United States to 70, 10 and 5 million Btu, respectively, in Russia, Brazil, and India. Of course, only a fraction of this energy is consumed in the form of electricity–about 34 percent in the United States in 1978.

The fossil fuels are adequate to meet world and U.S. requirements for the next 110 years. The use of nuclear breeders could more than double the energy supply in both cases, but based on the present knowledge of nuclear fuel resources, without breeders, nuclear power would have very limited benefit. However, it is very important to recognize that continuing geologic exploration may greatly improve this resource picture in a few decades.

The increasing demand for energy can be largely attributed to the growth of population, to the rise in standard of living, to the decrease in resource availability, and to the increased need to treat the wastes resulting from these human activities in order to maintain a hospitable environment. In the 1960's, nuclear energy was being considered as a means of providing water, fertilizer, and power for colonizing several arid coastal regions of the world in order to provide viable habitats for the world's expanding population. These arid regions have the necessary soil and climate. The study of these nuclear agro-industrial projects is no longer active, but such projects could serve a future social need.

NUCLEAR ENERGY ECONOMICS

In the free world in 1979 approximately 600 million MWh of electricity was generated in nuclear stations, with half being generated in the USA. The total invested cost in the U.S. nuclear industry was about $60 billion in 1980 for 52 GWe capacity, along with the supporting fuel processing and reactor fabrication facilities. The cost of electricity from nuclear reactors was competitive with coal and oil, and considerably cheaper than solar and geothermal (see Table 24.4).

The Electric Power Research Institute estimates that the projected annual growth for energy requirements from all sources in the USA will be in the range of 1.5 to 2.8% per year, leading to an energy requirement in 2000 of 0.1 to 0.15 Q ($Q = 10^{18}$ Btu). However,

TABLE 24.3 Energy Status of the Earth ($Q = 10^{18}$ Btu)

Income Energy Supply				
Solar	5.3			
Terrestrial core	0.001			
Gravitational	0.0001			
	Proved		*Estimated*	
Capital Energy Supply	*World*	*USA*	*World*	*USA*
Gas	2.4	0.2	10	1
Oil	3.1	0.16	8.7	0.79
Shale Oil	1.6	0.44	14.	6.0
Coal	14.	4.8	110	21
Uranium				
Burner	1.0	0.28	2.6	1.1
Breeder	73.	21.	190	80
1978 Consumption	0.19	0.07		
*Fossil Fuel Life**	2015	2009	2090	2085
(Based on 3% annual consumption growth)				

Source: Parent, J., *A Survey of the U.S. and Total World Fossil and Uranium Fuel Resources–1977*, Institute of Gas Technology, Chicago, Ill., 1979.

*Year when Reserve drops to 10 years based on 3% annual consumption growth.

TABLE 24.4 Cost Comparison of Alternative Electricity Generation Systems—1979[7]

Electricity Generation System	*Output MW*	*Capital ($/kWe)*	*Operating (Mills/kWh) Fixed*	*Operating (Mills/kWh) Variable*
Nuclear–LWR	1000	800	0.4	1.5
Nuclear–LMFBR	1000	1200	0.4	1.5
Coal	1000	750	1.5	3.5
Oil	500	425	0.2	1.5
Geothermal	50	720	3.2	1.6
Wind	2.5	860	0.8	–
Solar/Oil	100	1520	2.4	0.3

Note: A 1000-MWe power plant requires each day 45,000 barrels of oil, 12,000 tons of coal, 0.3 trillion cubic feet of natural gas or 0.6 ton of uranium.

electricity use is estimated to increase at approximately 5% per year, requiring 1300 GWe capacity in 2000 along with 44,000 miles of new high voltage transmission lines. Coal and uranium will have to fuel these stations. Gas will be phased out by 1990 and oil by 2000. Geothermal sources may provide 20 GWe, and solar, including wind and biomass, 5 GWe. Annual construction costs for the power generation system will increase from $30 billion in 1977 to $65 billion in 2000. The overall trend will be towards higher construction and fuel costs, increasing environmental requirements and longer construction times. Design, licensing, and construction now take about 7 years for a coal-fired plant and 10 years for a nuclear plant.

The nuclear industry includes three major categories of activity (see Fig. 24.3). The first category includes uranium and thorium mining, ore milling, feed materials processing, and uranium-isotope separation. The second includes reactor-fuel-element fabrication, nuclear reactor operation (including electric power generation, and coolant treatment), and effluent treatment. The third category includes recovery of uranium, plutonium, thorium, actinide products, and fission products from the spent reactor fuel, together with radioactive waste management. Because of inflation, cost

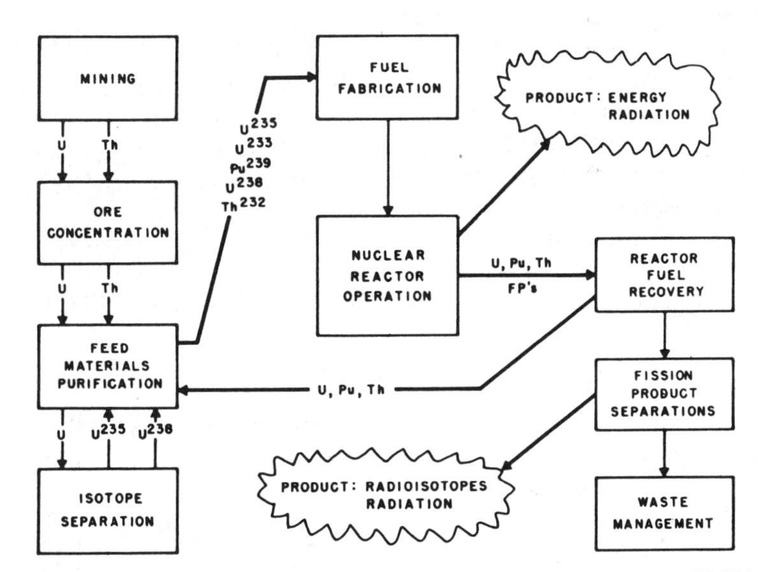

Fig. 24.3 Nuclear industry materials flowsheet; represented are 115 organizations with 50,000 employees and product shipments valued at $562 million in 1969.

TABLE 24.5 Nuclear Industry Capital Cost Ratios

Light-water reactors		0.75
Reactor component fabrication plants		
Turbine generators	0.05	
Pressure vessels	0.02	
Steam generators	0.03	
Subtotal		0.1
Fuel-Cycle plants		
Uranium ore mills	0.04	
Uranium refineries	0.01	
Uranium enrichment	0.09	
Fuel fabrication	0.02	
Spent-fuel recovery	0.01	
Subtotal		0.15
Total		1.00

TABLE 24.6 Typical Fuel-Cycle Cost Ratio Breakdown for First 5 Years of Operation of a 1000 MWe Light Water Reactor Plant

Fuel-Cycle Cost Component	
Uranium hexafluoride	0.24
Isotopic separative work	0.28
Fuel element fabrication	0.30
Spent fuel recovery	0.1
Plutonium credit	(0.13)*
Working capital	0.21
TOTAL	1.00

*(0.13) Credit for plutonium and uranium recovery and use

ratios, which are generally constant, are given rather than costs for these various activities in Table 24.5, 24.6, and 24.7.

In the technical evaluation of alternative nuclear-power generating systems, there are two bases: fuel efficiency and thermal efficiency. The fuel efficiency represents the most significant factor, since the availability of fuel is limited. Furthermore, uranium is the essential fuel material since it contains the only naturally occurring fissionable nuclide, uranium-235. With the once-through LWR system, approximately 0.005 GWy is produced for each ton of uranium mined and extracted from the ore (see Table 24.8).

The fuel efficiency of the nuclear reactor can be imporved by several routes: by using a more efficient neutron moderator, by recycling the spent fuel, and by improving the yield of fissionable material by neutron capture. The HWR and HTGR concepts use more efficient moderators, heavy water and carbon. Greater gain is achieved by going to the recycle of the spent fuel as in the U-Pu and Th-U cycles. It is to be noted that in the Th-U cycle only a small amount of thorium is required. Finally the LMFBR promises the ultimate performance—approximately 170 times more efficient than the LWR once-through system. The LMFBR promises 0.85 GWy of electricity per metric ton of uranium ore mined.

The thermal efficiency affects both the fuel requirement and the environmental impact. Thermal efficiencies range from about 30% for LWR and HWR reactors to 43% for HTGR with LMFBR's at about 41%. This is principally a function of the reactor coolant and vessel materials. While the fuel efficiency ranged over a factor of 170, the thermal efficiency represents only a factor of 1.4.

THE NUCLEAR REACTIONS

The reactions on which the nuclear industry is based involve the nucleus of the atom. These nuclear reactions result in the conversion of mass, m, to energy, E, as defined by Einstein's equation

$$E = mc^2$$

where c is the velocity of light. Based on this

TABLE 24.7 Waste Management Costs (Ratios*)

	Once–Through	*U–Pu Cycle*
Mining and Milling	0.03	0.02
Conversion and Enrichment	0.06	0.04
Fuel Fabrication	0.03	0.04
Nuclear Power Plant	0.48	0.62
Spent Fuel	0.40	0.28
*% of electricity cost	2	1.6

TABLE 24.8 Comparison of Alternative Nuclear Reactor Systems

	Once–Through		*U–Pu Cycle*			*U–Th Cycle*	
	LWR	*HWR*	*LWR*	*HWR*	*LMFBR*	*HTGR*	*HWR*
Fuel Efficiency[a]							
U	1.0	1.1	1.7	2.9	170	3.7	33
Th	–	–	–	–	–	(240)	(100)
Thermal Efficiency[b]	1.0	0.91	1.0	0.91	1.4	1.25	0.91

[a]1.0 = 0.005 GWye/metric ton of U or Th from ore
[b]1.0 = 32% conversion of thermal energy to electric energy

relationship, one atomic mass unit is equivalent to 931 Mev, or 4.2×10^{-17} kw h. As noted earlier, there are two types of nuclear reactions that lead to the production of energy: fission of heavy nuclei and fusion of light nuclei. In addition, nuclear reactions involving the capture of neutrons are important in the production of additional fission and fusion fuels. The energy evolved from grams of nuclear fuel is equivalent to that evolved from tons of chemical fuels.[2]

Fission

Controlled nuclear fission was first achieved on December 2, 1942, at the University of Chicago by a group of scientists under the guidance of Enrico Fermi. The device in which it was accomplished, then called an "atomic pile," consisted of an ordered pile of graphite blocks along with natural uranium cubes and cylinders arranged in such a way that a sustained chain reaction could be obtained and controlled (Fig. 24.4). The fission reaction depended on uranium-235, the only naturally occurring nuclide that is fissionable with thermal neutrons. However, other fissionable nuclides are readily produced in the nuclear reactor, the most important being plutonium-239 and uranium-233.

There is a high probability of fission when a neutron is captured in the nucleus of a fissionable nuclide. A typical fission reaction is as follows:

$$ {}^{235}_{92}\text{U} + {}^{1}_{0}\text{n} \longrightarrow {}^{236}_{92}\text{U} \xrightarrow{\text{fission}} {}^{89}_{35}\text{Br} + {}^{145}_{57}\text{La} + 2.3\text{n} + 192\ \text{Mev} \quad (1) $$

At the time of fission, 178 Mev is released, and additional energy, about 23 Mev, is subsequently released by the decay of the radioactive fission products. While the nuclear reaction is readily initiated at room temperature

Fig. 24.4 Uranium metal fuel from the first atomic pile. This reactor contained 40 tons of uranium oxide along ith 6.2 tons of uranium metal. (*ORNL Newa, 1-01-076.*)

and pressure, nuclear reactor operation at high temperatures is required for efficient electric power generation. The temperature, along with the associated pressure, is limited by the properties of the materials of construction as in the case of fossil fuel systems.

Plutonium-239 and uranium-233 are produced by the capture of a neutron by the fertile nuclides uranium-238 and thorium-232, respectively.

$$^{238}_{92}\mathrm{U} \xrightarrow{n} {}^{239}_{92}\mathrm{U} \xrightarrow[23.5\ \mathrm{min}]{\beta} {}^{239}_{93}\mathrm{Np} \xrightarrow[23.3\ \mathrm{days}]{\beta} {}^{239}_{94}\mathrm{Pu} \quad (2)$$

$$^{232}_{90}\mathrm{Th} \xrightarrow{n} {}^{233}_{90}\mathrm{Th} \xrightarrow[23.3\ \mathrm{min}]{\beta} {}^{233}_{91}\mathrm{Pa} \xrightarrow[27.4\ \mathrm{days}]{\beta} {}^{233}_{92}\mathrm{U} \quad (3)$$

Neutrons are produced in fission (Equation 1), and the nuclear reaction can be made self-sustaining. In addition, since more than two neutrons are produced in the fission reaction, Reactions 2 and 3 may be used to produce more fissionable material than is consumed.

In the present generation of nuclear reactors, fission reactions are initiated by thermal neutrons, neutrons with an energy of 0.025 to 0.1 ev (electron volt), equivalent to temperatures from 22 to 870°C, which is the temperature of the environment in which the neutron is present. The 5-Mev kinetic energy of the fission neutron must therefore be reduced to thermal energy before it escapes from the nuclear reaction zone or is captured by other materials in this zone (Fig. 24.5)[9]. Even when it is captured by fissionable material, fission does not always occur. The probability of a neutron reacting with materials is expressed in terms of their so-called cross sections (see Tables 24.9 and 24.10).

The fission reaction can also employ fast neutrons, as in the plutonium cycle which gives a high breeding ratio. The neutron cross sections of most materials are significantly less for fast neutrons, thus permitting greater freedom of choice in materials of construction and coolants. The neutron multiplication factor η for plutonium at 1 Mev is 2.99 compared with 2.28 and 2.07, respectively, for uranium-233 and -235 under thermal conditions, about 0.1 ev (Table 24.11). In addition, with fast neutrons uranium-238 has a significant fission cross section, thus adding to the interest in the fast-flux cycle. The thorium cycle provides the only possibility of breeding when using a thermal reactor.

Fusion

Nuclear fusion, the stellar energy process, was first achieved in 1932 by the accelerator-produced collisions of deuterium nuclei. In 1952, the spontaneous release of fusion energy was demonstrated with the first test of a thermonuclear device, the so-called hydrogen bomb. Since that time, research has been carried on to devise a method for carrying out controlled nuclear fusion. The fuels for the fusion reaction are hydrogen-2 and hydrogen-3, more commonly called deuterium and tritium. Deuterium is a naturally occurring isotope of hydrogen with an abundance of

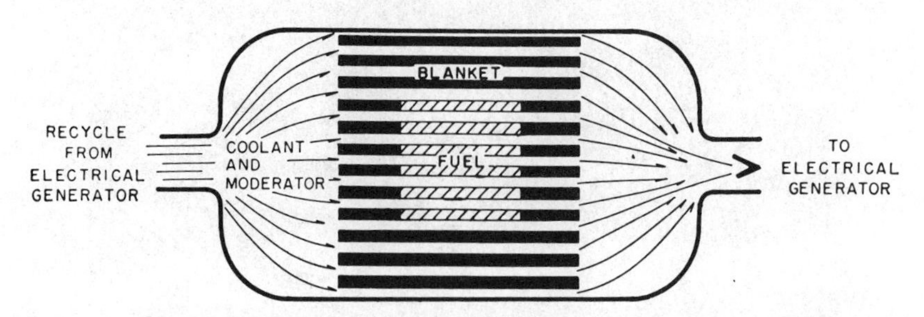

Fig. 24.5 Nuclear reactor core; fissionable material is principally in the blanket for greater neutron economy; but economics for converter type power reactor require reversal of this pattern to achieve high fuel burn-up and uniform power distribution across the core.

TABLE 24.9 Nuclear Properties of Fissionable and Fertile Materials at Thermal Energy[a]

	σ_{abs}	σ_F	$1+\alpha$	η*	ν†
^{233}U	581 ± 7	527 ± 4	1.102 ± 0.005	2.28 ± 0.02	2.51 ± 0.03
^{235}U	694 ± 8	582 ± 6	1.19 ± 0.01	2.07 ± 0.02	2.47 ± 0.03
^{239}Pu	1026 ± 13	746 ± 8	1.38 ± 0.02	2.10 ± 0.02	2.90 ± 0.04
^{238}U	2.71 ± 0.02				
^{232}Th	7.56 ± 0.11				

σ_{abs} = cross section for neutron absorption in barns at a neutron velocity of 220 meters/sec; one barn = 10^{-24} cm^2.

σ_F = cross section for fission in barns at a neutron velocity of 220 meters/sec.

$$1+\alpha = \frac{\sigma_{abs}}{\sigma_F}$$

[a]Neutron cross sections, D. J. Hughes and R. B. Schwartz, BNL-325, July 1, 1958.
*η = Neutrons emitted/neutrons adsorbed.
†ν = Neutrons per atom fissioned.

0.015 percent. Tritium is the neutron-capture product of lithium-6, readily produced in either a nuclear fission or fusion reactor. A third possible fuel for fusion is helium-3, the decay product of radioactive tritium. The fusion reactions of practical interest which have been identified so far are as follows.[10]

Reaction	Ignition Temperature in Millions of °C
$^2_1H + ^2_1H \longrightarrow ^3_1H + ^1_1H + 4$ Mev	100
$\longrightarrow ^3_2He + ^1_0n + 3.25$ Mev	
$^2_1H + ^3_1H \longrightarrow ^4_2He + ^1_0n + 17.6$ Mev	45
$^2_1H + ^3_2He \longrightarrow ^4_2He + ^1_1H + 18.3$ Mev	

In addition, the success of the fusion reaction depends on the capture of the fusion-produced neutrons in the lithium blanket of the reactor to produce both tritium and energy:

$$^6_3Li + n \longrightarrow ^3_1H + ^4_2He + 4.6 \text{ Mev}$$

While the fission reaction, as noted earlier, can be initiated under conditions of room temper-

TABLE 24.10 Thermal Neutron Cross Sections of Typical Nuclear Reactor Materials (Barns, σ)

Chemical Element	σ_{abs}	*Chemical Element*	σ_{abs}
C	0.0032	B	755
Be	0.010	Cd	3300
H	0.33	Hf	105
D	0.00057	Al	0.023
Na	0.53	Zr	0.18
K	2.0	Fe	2.5
He	0.007	Ni	4.6
N	1.9	Cr	3.1
O	0.0002	Xe	35

TABLE 24.11 Fission Cross Sections and Neutron Multiplication Factors for Fissile and Fertile Nuclei at Higher Energy

Neutron Energy (kev)	*10*	*100*	*1000*
Fission Cross Sections (Barns, σ_F)			
U-233	4.5	2.5	2.0
U-235	3.3	1.7	1.2
Pu-239	1.9	1.8	1.7
Th-232	–	–	0.076
U-238	–	–	0.411
Neutrons Produced per Neutron Absorbed (ν)			
U-233	2.24	2.26	2.40
U-235	1.77	1.90	2.32
Pu-239	2.00	2.41	2.99

ature and pressure, the fusion reaction requires very high temperatures and pressure.

Fusion technology studies have two major objectives: 1) breakeven, and 2) practical power production. To achieve the breakeven objective as much energy from fusion must be produced as was required to initiate the fusion reaction. This is known as the "gain" and at breakeven the gain is equal to one.[11] For a feasible power generating station, the gain would have to exceed 10. There are several potential approaches to achieving this, including continuous magnetic torodial, pulsed magnetic pinch, and pulsed inertial devices. All three could be employed as the reactor "driver" for the engineered electric power generation system shown in Figure 24.6. They all, however, are faced with similar technological barriers.[12]

The torodial type of fusion reactor is one of the most intensely studied systems today and gives promise of providing for continuous operation.[11a] For this reactor the favored fuel

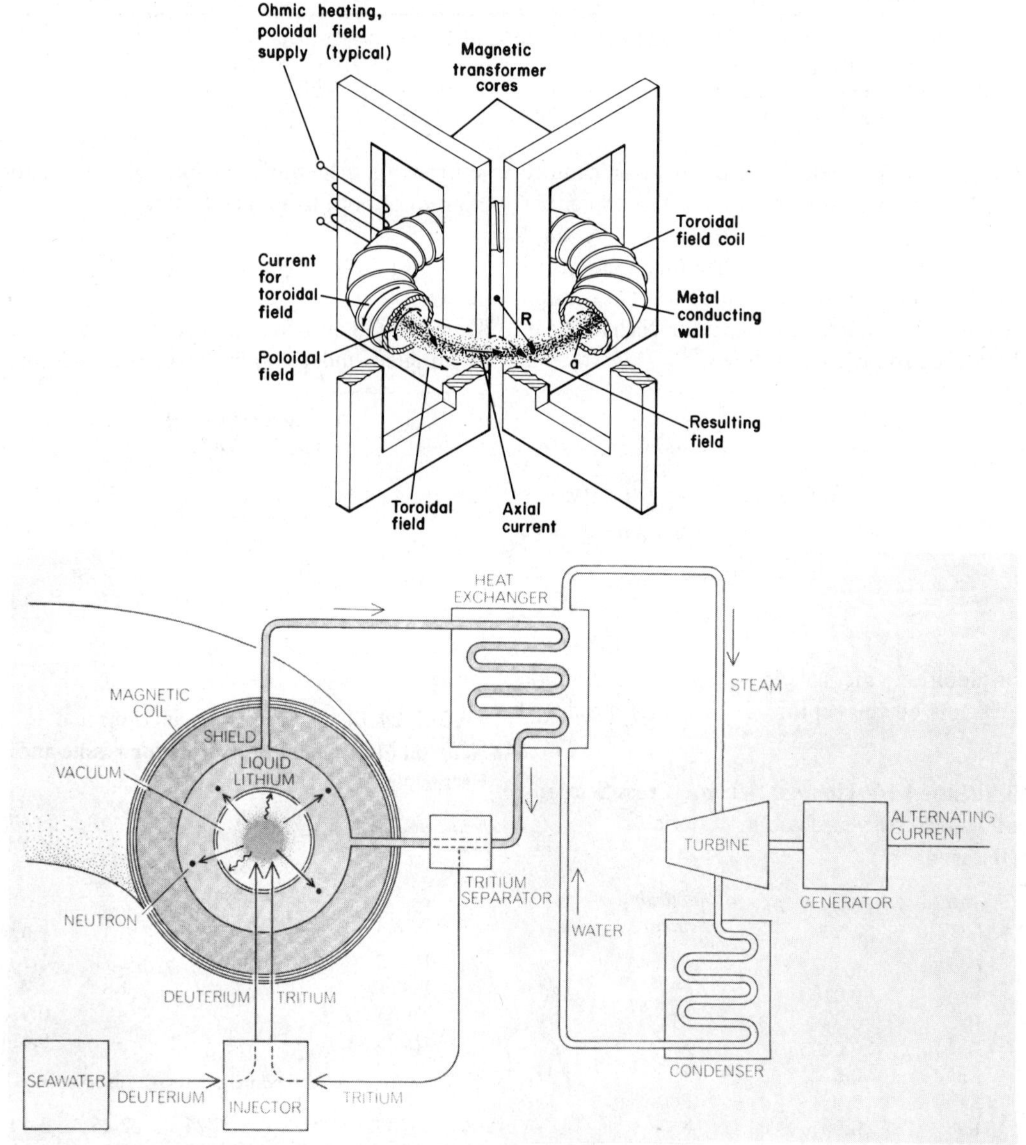

Fig. 24.6 Fusion-energy conversion system using the molten lithium blanket loop to extract the heat and the tritium; the plasma contained in the toroid is shown in the inset. (*USAEC.*)

TABLE 24.12 Status of Tokamak Technology[12]

	Achieved	*Breakeven*	*Reactor*
Confinement (cm^3.sec)	3×10^3	1×10^{13}	3×10^{14}
Temperature (keV)	1.0	5	7–10
B (plasma pressure/magnetic field)	0.02	0.02	0.06–0.10
Gain (energy output/input)	0.01	1.0	10–20

is the deuterium-tritium system. In one version, an induction heater is used to bring the ionized reactants to temperature with the resulting plasma being magnetically retained in a torodial orbit.

There are two products from this reaction, helium-4 and neutrons. The helium-4 is retained within the plasma and its associated energy, 3.5 Mev, maintains the plasma temperature. The neutrons escape and are captured in a lithium blanket. The energy of the neutrons, 14.1 Mev, and that from the neutron-lithium reaction, 4.6 Mev, is extracted and converted into electricity (Fig. 24.6).

The conditions required for a productive reaction are about 10^{15} ions per cm^3 and a temperature of 10 kev for 700 milliseconds. In more familiar terms, this corresponds to a pressure of about 100 atmospheres at 10^8 °C.

The timetable for the achievement of economic electric power from fusion is difficult to predict. Even in view of continuing progress, many more years and billions of dollars will be required to demonstrate engineering feasibility. See Table 24.12.

EXPLOSIVES

When used as an explosive, a nuclear device creates cataclysmic physical, thermal, and nuclear effects of significance for military applications, and in the past was considered for civil applications. The fission of about 56 grams of uranium or the fusion of 6.8 grams of tritium is equivalent to the physical force of about one thousand tons of trinitrotoluene (TNT) (Table 24.13). The force of nuclear explosives ranges from fractions of kilotons to tens of megatons of TNT, whereas the cost is estimated to be about 1 percent the cost of TNT (Table 24.14).

To generate a nuclear fission explosion, it is necessary to bring together a supercritical mass of fissile material, and then to hold it together long enough to produce the desired energy. In the fission-type explosive, the critcality event is initiated with chemical explosives, which either drive together two subcritical masses of fissile metal, or decrease the volume of a subcritical assembly. The duration of the fission reaction is less than one millionth of a second.

A fusion explosion using the deuterium-tritium reaction is initiated by the fission reaction, which provides the necessary conditions of temperature and pressure. The fusion-type explosive produces a lesser quantity of radioactivity per unit of explosive force.[13]

Plowshare

The Atomic Energy Commission's Plowshare program investigated the use of nuclear explo-

TABLE 24.13 Energy Equivalents of One Thousand Tons of TNT

Fission of 1.45×10^{23} atomic nuclei
Fission of 56 g Pu-239
Fusion of 16.5×10^{23} atomic nuclei
Fusion of 6.8 g TD
10^{12} calories
4.2×10^{19} ergs
1.2×10^6 kilowatt hours
4.0×10^9 British thermal units (Btu)

TABLE 24.14 Comparative Explosives Cost

Chemical	
Ammonium nitrate	1*
Trinitrotoluene	60
Nuclear	
Fission	2
Fusion	0.02

*Cost of ammonium nitrate is estimated to be $4.50 per million Btu sives for civil purposes. Underground explosions were carried out to create underground cavities and porosity, the surface craters. An underground explosion initially forms a bubble of vaporized material with a temperature of about one million degrees Celsius and a pressure of several million atmospheres. The high pressures cause the cavity, filled with hot gases, to expand rapidly and initiate an outward-moving spherical shock wave which crushes the surrounding medium. In deeply buried explosions in competent rock, a chimney of broken rock is formed when the cavity gases cool and the cavity ceiling collapses (Figure 24.7). The size of the chimney is a function of the magnitude of the explosion, the depth of burial, and the type of rock.

The resulting interstitial volume within the chimney and the surrounding fractures provides the basis for several potential applications classified as underground engineering. In the "Gasbuggy" shot, a chimney with a volume of 220,000 cubic yards was formed; about 10^9 ft^3 of natural gas may be recovered from it, at least 5 times more than normally expected. In view of the anticipated shortage of natural gas within the next ten years, this method of gas extraction from low-porosity deposits had considerable interest. Since the solid fission products remain deeply buried underground, the only radioactive species of consequence in natural gas from nuclear-stimulated wells are tritium and, to a lesser extent, Kr-85. Other applications under consideration included the *in situ* extraction of copper from low-grade ores by continuous leaching, and the extraction of geothermal energy.

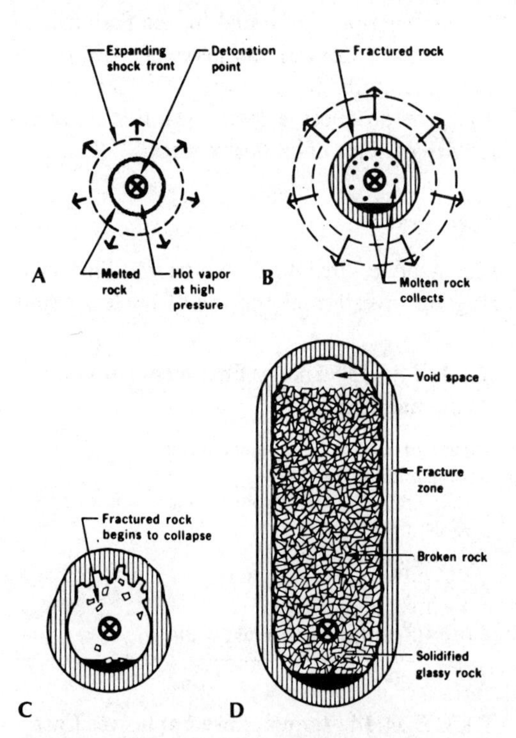

Fig. 24.7 An underground nuclear explosion (A) initially vaporizes, melts, and fractures the adjacent rock; (B) it then sends out a shock wave; (C) the zone of thermal energy rises toward the surface, continuing to fracture rock; (D) as the cavity cools, the fractured rock is imploded into the cavity, resulting in a column of broken rock. If the cavity reaches the surface, the explosion lifts the rock and dirt and forms a crater. (*USAEC.*)

The excavation of a crater 1280 feet in diameter by 320 feet deep has been demonstrated using a 100 kiloton thermonuclear explosive. Also, a ditch 850 feet long, 250 feet wide, and 65 feet deep was excavated by using a row of five simultaneously detonated 1.1 kiloton nuclear explosives (Fig. 24.8). The development and use of very low-fission thermonuclear explosives would reduce the production and release of radioactivity from cratering explosions. This, in turn, would reduce the waiting time for reentry into the crater area to a few weeks. These and other nuclear cratering experiments demonstrated the potential usefulness of nuclear explosives in such large excavation projects as canals, harbors, dams, and transits through mountainous terrain.

However, largely because of problems of residual radioactivity, the development of civil applications of nuclear explosives has been terminated.

Military Applications

During World War II, nuclear energy for military applications was an important development. The subsequent nuclear-weapons development has been concerned with increasing their efficiency, evaluating their effects, and establishing protective measures. In military

Fig. 24.8 The first nuclear row-charge experiment created a channel 855 ft long, 255 ft wide, and 65 ft deep. (*USAEC.*)

applications, nuclear explosions may inflict severe damage from blast, heat, and radiation effects at distances greater than 30 miles. Moreover, the radioactive debris from the explosion may extend the radiation effects from fall-out several hundred miles downwind. The radiation from stratospheric explosions may also affect radio and radar performance for several hours following the event.

Studies of protective measures have been concerned with two aspects: the immediate effects of blast and neutron radiation and the long term fall-out effects of the radioactive bomb debris. Blast-resistant structures, important in urban areas, generally provide adequate protection from all radiation effects, while fall-out shelters, adequate for rural areas, provide only radiation protection. In the design of certain new structures, blast or fall-out protection can be provided at nominal cost.[13]

THE NUCLEAR REACTOR

The development of a commercial nuclear power industry has required about 30 years. While the first commercial electric power was produced using a pressurized water reactor (PWR) at Shippingport in 1957, only in the early 1970s did large numbers of power reactors come into operation. Nuclear-reactor development required the evaluation of many reactor concepts (Fig. 24.9) and involved two fuel cycles, the uranium-plutonium and the thorium-uranium. The light water reactors (LWRs) were the first to achieve widespread acceptance, with the high temperature gas cooled reactor (HTGR) being the only other thermal-converter type extensively developed in the U.S.

In England and France, a family of gas-cooled power reactors (GCR) were developed, and several were built in other countries.

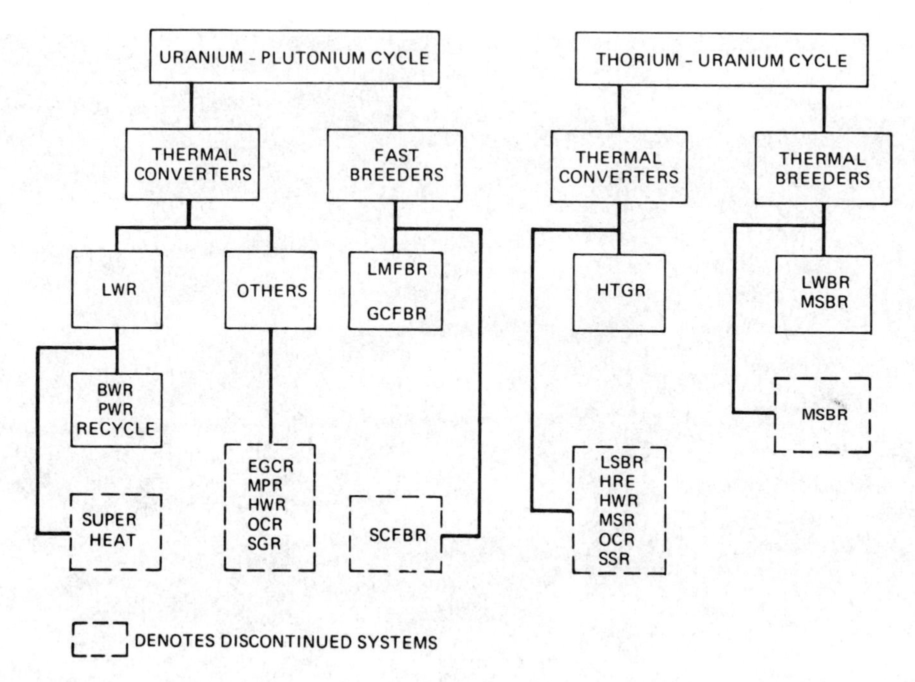

Fig. 24.9 U.S. civilian power reactors: *B*oiling *W*ater, *P*ressurized *W*ater, *L*iquid *M*etal, *G*as *C*ooled, *H*igh *T*emperature *G*raphite, *L*ight *W*ater, *M*olten *S*alt.—Discontinued systems: *E*nriched *G*as *C*ooled, *M*olten *P*lutonium, *H*eavy *W*ater, *O*rganic *C*ooled, *S*odium *G*raphite, *S*team *C*ooled, *L*arge *S*eed *B*lanket, *H*omogeneous *R*eactor *E*xperiment, *M*olten *S*alt, *S*pectral *S*hift.

TABLE 24.15 Characteristics of Nuclear Power Reactors

	PWR[a]	*BWR*[b]	*Candu*[c]	*LMFBR*[d]	*HTGR*[e]	*AGR*[f]
Electric Power (GWe)	1.1	1.06	0.74	0.23	0.33	0.63
Plant Efficiency (%)	33	33	30	40	42	42
Fuel	UO_2	UO_2	UO_2	UO_2	UO_2 ThO_2	UO_2
Clad	Zr	Zr	Zr	SS	C	Mg
Uranium Loading (t)	99	155	130		95	
U-235(5)	3.0	2.9	0.81	12*	40	0.75
Burnup (GWd/t)	30	28	7.3	100	110	
SWU(t/GWey)	109	114	0	0.03	130	
Fueling Internal (y)	1.1	1	0.003	0.5	1	
Fueling Fraction	0.3	0.2	0.003	0.3	0.3	
Power density (kW/kgU)	38	25	26	160	115	38
Coolant/Moderator	H_2O	H_2O	D_2O	Sodium	He/C	CO_2
Pressure (bars)	155	73	100	2	45	40
Temp (°C)	327	–	310	560	750	670
Flowrate (t/s)	21	15	12	3	0.4	
Conversion Ratio	0.6	0.6	0.6	1.2	0.8	0.6

*Including Pu - 229

[a]Zion 1 1973, [b] Peach Bottom 2 1974, [c] Bruce 1 1977, [d] Phenix 1973, [e] Fort St. Varain 1979, [f] Hinkley Point Bl 1976

Candu heavy water reactors (HWR) were developed by Canada and are in effective commercial operation.[16] They utilize natural uranium as the fuel and heavy water as both the coolant and modulator. To achieve high burn-up, somewhat better than the LWR on a mined ore basis, on-line refueling is required daily. With the average life of the fuel being about one year with approximately 0.003% replaced each day, the reactor requires refueling while it is in operation.

The current major development work is concerned with the liquid-metal fast breeder reactor (LMFBR), with more limited efforts being expended on the gas-cooled fast breader (GCFBR) and the light water breeder (LWBR). (Table 24.15.)

Light Water Reactors

There are two types of LWRs, the PWR and the boiling-water reactor (BWR). In the PWR, the reactor water coolant transfers its energy through an intermediate heat exchanger to generate steam in a second cycle for electric power generation. In the BWR, the steam is generated in the reactor. Although the BWR eliminates the intermediate heat exchanger, it places the steam turbine in a radioactive environment. These two types have similar costs, but the PWR has more suppliers throughout the world.

In the following sections, three important aspects of the LWRs will be discussed: the fuel elements, the reactor vessels, and the reactor-containment system. (Figures 24.10, 24.11.)

Fuel Elements. The fuel element is designed to provide primary containment of the radioactive fuel and fission products over the three-year operating life of the PWR and BWR fuel (Fig. 24.12 and Fig. 24.13). This depends on the integrity of about 80 miles of 24 and 32 mil Zircaloy-4 tubing at temperatures up to 660°F. Zircaloy-4 has replaced stainless steel to increase neutron economy. The economics of the fuel cycle also favor low enrichment, long operating life and high burn-up.

The 1000 Mwe BWR core contains about 165

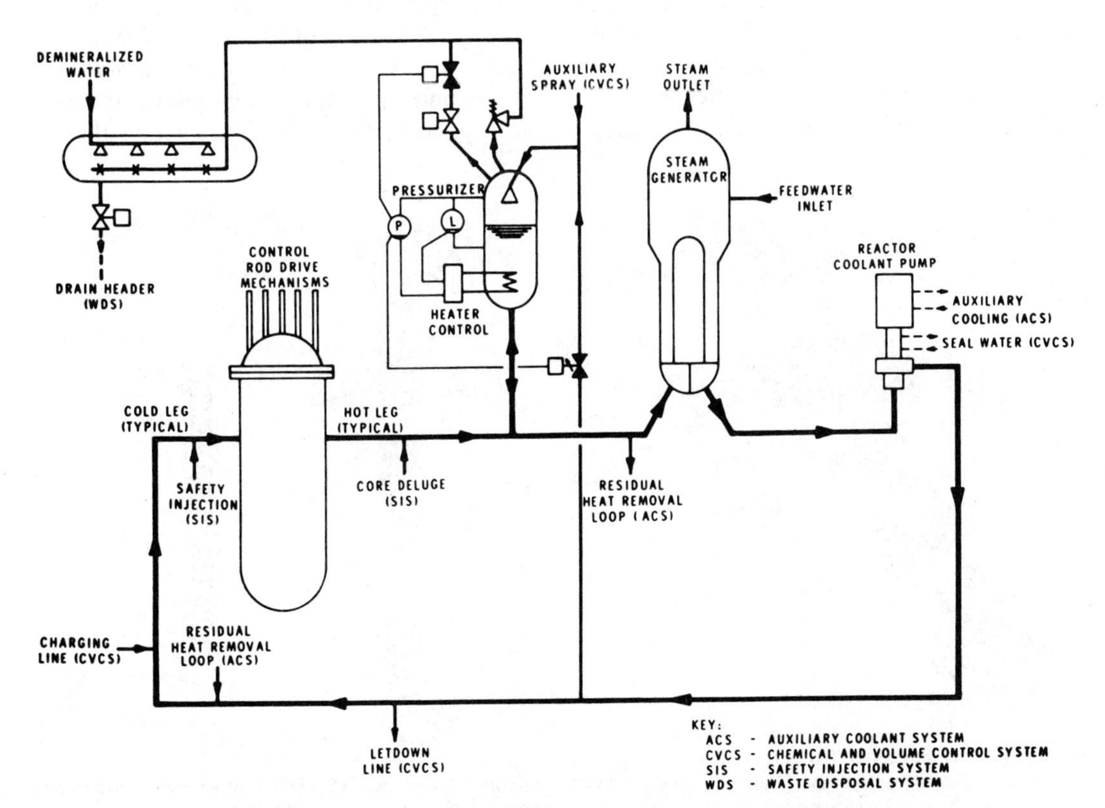

Fig. 24.10 Pressurized-water reactor flow sheet for power generation. (*USAEC.*)

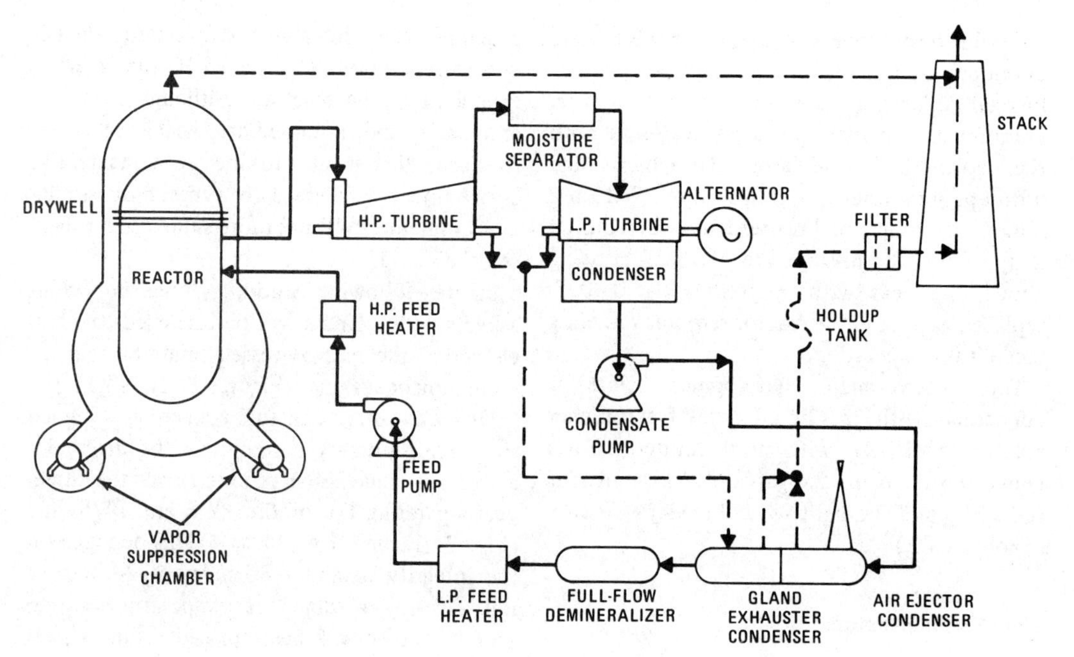

Fig. 24.11 Boiling water reactor flow sheet for power generation. (*USAEC.*)

tons of UO_2 which is charged with 2.5 percent uranium-235, the PWR, about 110 tons charged with 3.3 percent uranium-235. Since these power stations represent a significant part of the power system capacity, frequent shutdowns are not desirable. Fuel is charged to the reactor about once a year, with the new fuel being placed in the outer regions of the reactor. This serves to flatten the power distribution in the reactor while at the same time greater burn-up is achieved by placing the partially spent fuel towards the center of the re-

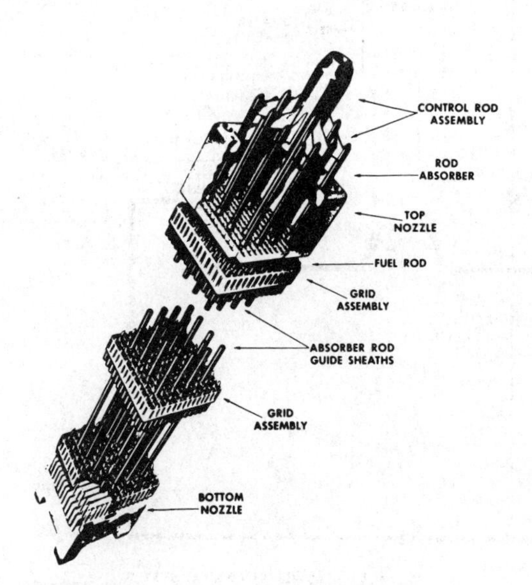

Fig. 24.12 Cutaway of PWR fuel element with the control-rod-cluster assembly. Element contains 1140 lb UO_2 in 193 rods. (*USAEC.*)

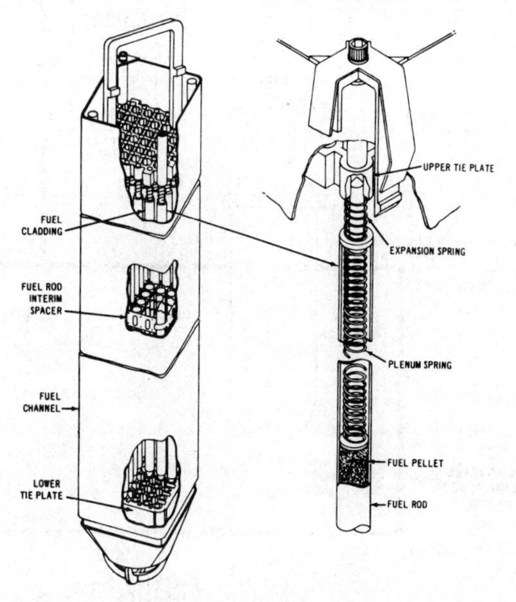

Fig. 24.13 Cutaway of BWR fuel element containing 488 lb UO_2 in 49 rods. (*USAEC.*)

TABLE 24.16 PWR and BWR Fuel Burn-Up

	Fuel Burn-up (Mwd/MT)	*Fuel Composition (g/MT)* U-235	*Pu-239*	*Pu-241*
	PWR–Plutonium Withdrawal			
Charge	–	32,500	0	0
Discharge	32,000	8,359	5,327	1,213
	PWR–Plutonium Recycle			
Charge	–	6,869	19,849	4,082
Discharge	32,000	2,919	8,765	4,647
	BWR–Plutonium Withdrawal			
Charge	–	25,000	0	0
Discharge	27,000	6,403	4,808	1,034

actor where the neutron flux is higher (Table 24.16). Significant improvements continue to be made in the design of the fuel elements.[14]

Two fuel-related projects were performed to further extend the value of the LWRs: (1) recycle of the plutonium to reduce the enriched uranium-235 requirements, or (2) conversion to the thorium-uranium cycle to achieve thermal breeding. The demonstration phase of the plutonium recycle development was carried out in several power reactors. Several LWRs were originally started up on the thorium-uranium-cycle but were changed over to the uranium-plutonium cycle to reduce fuel-cycle costs. A LWBR core using the thorium cycle was tested in the Shippingport reactor to evaluate the feasibility of converting existing and future pressurized water reactors to self-sustaining breeders. In principle, such reactors would make possible the conversion of 50 percent of the mined thorium into energy sufficient for hundreds of years. This improvement could be realized without the major development required for other systems. The advantages of the other breeders are greater thermal and fuel cycle efficiency.

Reactor Vessels. While the nuclear industry has posed challenges on most frontiers of engineering, the fabrication of the reactor vessel is one of the more significant. The design criteria for optimum performance requires operating pressures of 1050 psia for BWRs and 2250 psia for PWRs. Typical dimensions for vessels serving 1000-MWE stations range from 21 ft diameter by 70 ft height for BWRs to 14.4 ft by 42.8 ft for PWRs (Fig. 24.14) with vessel weights of 782 and 459 tons, respectively. These are shop-fabricated (Fig. 24.15). The use of a prestressed-concrete vessel combines the shielding with the pressure containment for the nuclear reactor. With a gas-cooled reactor, where prestressed concrete vessels are now used, the thermal insulation is placed inside the pressure vessel. (Fig. 24.16).

Coolant Technology. The coolant technology for nuclear reactors is principally concerned with inhibiting corrosion; however, there are many other considerations. Two aspects are discussed below as they uniquely apply to the pressurized water reactors (PWR) and the boiling water reactors (BWR), respectively.

In the PWR, there are three coolant systems: (1) the primary that extracts the heat from the reactor and partially controls nuclear criticality, (2) the secondary that transfers the heat from the primary to a turbine-driven electric generator, and (3) the service water system that dumps the residual coolant energy from the turbine condenser to the environment. The service water is recirculated from a river, lake, ocean, or a cooling tower. In the primary system, boron is present to control nuclear criticality. Ion exchange in fixed-bed units is used to maintain the water quality in both the primary and secondary systems. In addition, the chemical and volume control system also reduces boron concentration to compensate for the fissile material burn-up. These operations are continuously carried out through bypass systems.

In a BWR, steam is generated in the reactor and goes directly to the turbine. Here, there are full-flow ion exchange units treating about 30,000 gallons of water per minute. This serves two objectives: radioactivity removal and reactor protection from a possible leakage of service water to the primary coolant from the turbine condenser (Table 24.17).[15]

Containment. Under all conditions of normal operation and maximum credible accidents, any release of radioactivity from the nuclear reactor to the environment must be below acceptable limits. The maximum credible accident in the LWR is postulated to be the loss of coolant. This could result in melt-down of the fuel with release of the radioactive fission products. The

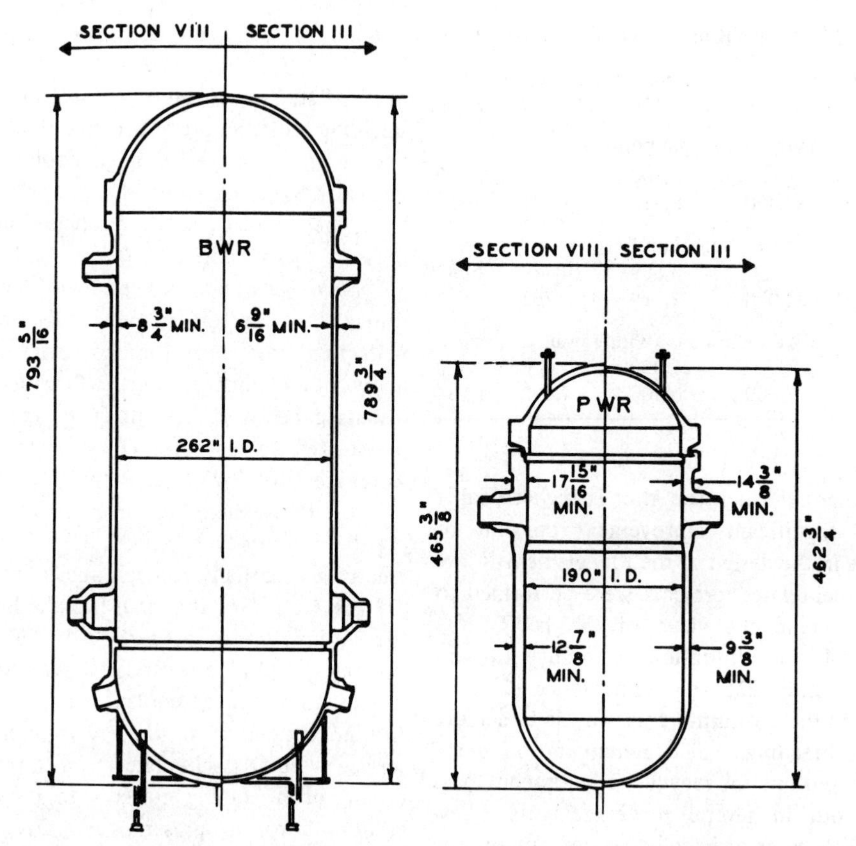

Fig. 24.14 BWR and PWR reactor vessels for 1000 Mwe plants; comparison of ASME Code Sections III and VIII. (*USAEC.*)

fission products in turn could be transported out of the core in the escaping vaporized coolant.

To prevent the escape of this radioactive coolant from the facility, a containment vessel is built around the reactor. Provision is made to rapidly shut down the nuclear reaction, condense the coolant, and scrub out and concentrate the radioactivity. A PWR approach to accomplishing this uses a large-surface ice condenser. The ice is always present during reactor operation. In the BWR, the escaping steam is vented into a large water pool where it is condensed. In both reactor systems, there are redundant systems for cooling the reactor and treating the radioactive effluents and an inert atmosphere is established to eliminate the possibility of metal combustion. Decontamination of the environment within the containment vessel is started within minutes.

As anticipated, there have been occasional equipment failures involving reactors, but the safety systems have been sufficiently redundant that one or more have always worked. Even in the Three Mile Island accident in 1979 the safety systems worked. The damage to the fuel resulted from operator actions to override the safety systems.[16] As concluded in the Rassmussen Report, the limitations of the plant operator created and seriously aggravated the Three Mile Island incident. In spite of these limitations the reactor design features prevented radiation exposure to the plant personnel or any off-site individual.[17]

LWR tests on failure are in progress in order to evaluate accident scenarios involving loss of coolant events, such as occurred in the Three Mile Island incident. The Power Burst tests in a 20 MWT PWR have created fuel failures and defined the initiating conditions. The Loss-of-Coolant tests with a 50 MWT PWR have

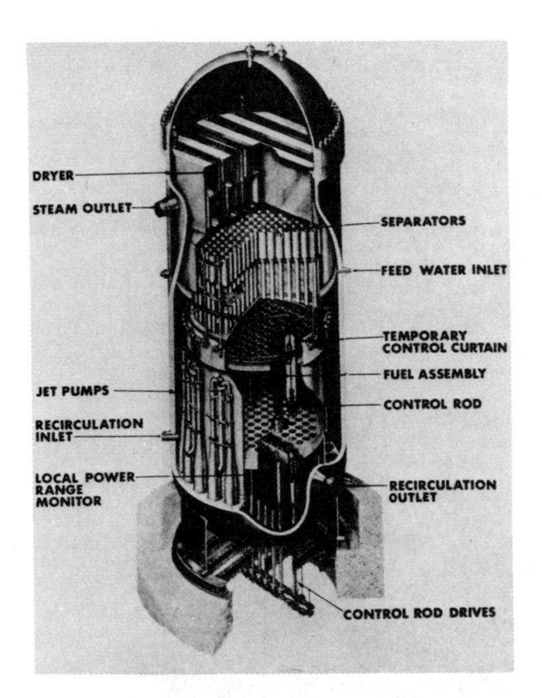

Fig. 24.15 BWR reactor vessel contains jet pumps to recirculate water in order to increase heat transfer; steam conditioner in the top of the reactor requires bottom control rod drives. (*USAEC.*)

demonstrated recovery from catastrophic major feed-water and steam line breaks without fuel damage.

Liquid Metal Fast Breeder Reactor

The salient features of the liquid-metal fast-breeder reactor (LMFBR) include a fuel-doubling time of 7 to 10 years along with a

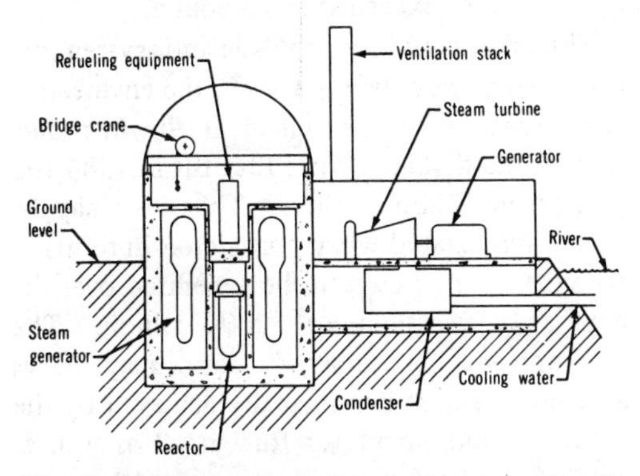

Fig. 24.16 Model of dual-cycle pressurized water reactor. (*Combustion Engineering.*)

TABLE 24.17 Process Technology for Nuclear Reactor Coolants[15]

Consideration	*Process*
Corrosion–Fuel clad	pH/Eh control
–Components	
Errosion–Pumps	Filtration
–Turbine blades	Demisters
Scaling–Boiler tubes	Inhibitor addition
–Reactors	Clarification
Radioactivity Transport	Ion exchange
Radiolysis	H_2/O_2 Recombiners
Radioactive Effluent Control	Gas Filtration
	Iodine Sorption
	Kr/Xe Cryogenic Traps
Criticality Control	Boron ion exchange
Radioactivity Decontamination	Inhibited acid/base flush

high coolant temperature for more efficient energy conversion and low operating pressure through the use of metallic sodium as the coolant (Fig. 24.18). Operation of the reactor with fast neutrons increases the neutron multiplication factor and reduces the parasitic capture of neutrons in the coolant and structural materials. This enhances the fuel-conversion ratio while permitting the use of high-temperature cladding material for the mixed uranium-plutonium oxide fuel. The power density in the LMFBR core is about 5 times greater than that in the LWR cores and therefore the former is much smaller in size. This core is surrounded with radial and axial blankets of fertile material. These blanket regions which absorb the leakage neutrons must be used to achieve breeding. Finally, the use of the uranium-plutonium fuel cycle makes optimum use of the plutonium produced in the LWRs.

All the major countries are participating in LMFBR demonstration plants ranging from about 200 to 500 MWE. In the U.S., the Enriched Breeder Reactor-II (EBR-2) has operated at power up to 62.5 Mwt and demonstrated plutonium-uranium oxide fuel burn-ups in excess of the design goal of 100,000 megawatt-days per metric ton. In January 1971, the Southwest Experimental Fast Oxide Reactor (SEFOR) was operated at its licensed power of

Fig. 24.17 Oyster Creek Nuclear Power Station utilizing 640 Mwe boiling water reactor. *(Courtesy Jersey Central Power and Light Co.)*

20 Mwt, and tests demonstrated the inherent shutdown capabilities of the LMFBR type reactor. The 400 Mwt Fast Flux Test Facility started operation in 1980 to test fuels and materials at the flux of 7×10^{15} and a power density of 300 to 700 kw per liter. EBR-1, the starting point of this program, went critical in 1951.

The liquid metal fast breeder reactors (LMF-BR) are also operating in the USA, England, France, Japan, and Russia. In France, the Phoenix, a 590 Mwt/230 Mwe reactor, has operated satisfactorily since 1976, and the Super Phoenix a 1200 Mwe reactor is being constructed.[17] In Russia, the BR-350 Mwe reactor has been producing power and steam heat for Shevchenko on the shore of the Caspian Sea, and they have the BN-600 nearing operation. The development of these reactors involves several unique problems because of the sodium coolant. It is chemically and neutron reactive and opaque.[18] The opaque property prohibits the visual inspection that is easily accomplished in water and gas cooled reactors. While other countries are proceeding rapidly through the demonstration stage, the USA does not see the LMFBR to be economical until the uranium price exceeds $100 a pound.

While the feasibility of this reactor system has been extensively demonstrated, the engineering development program aimed at the introduction of the first large scale LMFBR in 1984 will cost several billion dollars. This cost may be better appreciated when considered in terms of the cost of the power to be generated over the next fifty years, about $2000 billion. The development cost-to-benefit ratio may be as great as 8, and it is principally achieved by the reduced requirement for foreign oil as well as reduced need for uranium ore and enrichment.

An alternate fast breeder reactor, the gas-cooled fast breeder, uses helium instead of

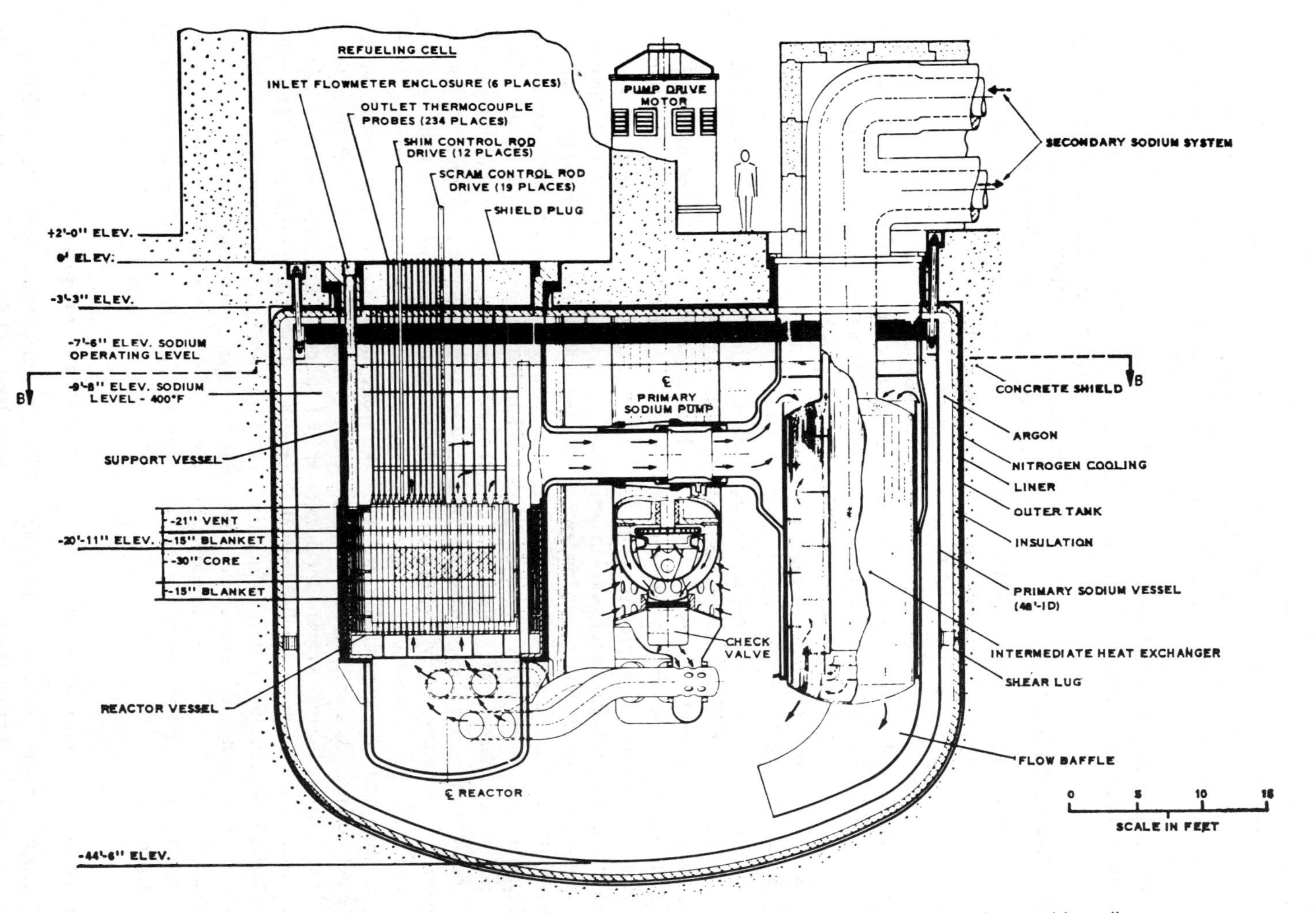

Fig. 24.18 Pool-type LMFBR for 1000 Mwe plant; both reactor and heat exchanger are submerged in sodium coolant to avoid loss-of-coolant accident with prestressed concrete reactor vessel. (*General Electric.*)

sodium as the coolant. It incorporates important components of the HTGR including the prestressed-concrete reactor vessel and the turbine-driven helium circulators. While the fuel-doubling time for this reactor could be only half that of the LMFBR, a counter balancing factor is the higher capital costs attributable to the high-pressure-gas coolant system. Development of the gas-cooled breeder is unlikely.

Molten Salt Breeder Reactor

The molten-salt breeder reactor (MSBR) was interesting from a chemical systems point of view, since it employed uranium and thorium fluoride in a solution of $^7LiF\text{-}BeF_2$ molten salt as the fuel.[21] This molten-salt homogenous reactor provided high thermal efficiency with low fissile material inventory and on-site fuel recycle. A 7.3 Mwt experimental reactor at Oak Ridge National Laboratory was operated for about five years with good results. During this period, uranium-235, uranium-233, and plutonium-239 were used as the fuel.

Frequent processing would have been required to achieve breeding. These processing operations would have included continuous removal of the xenon, protactinium-extraction every two or three days, and rare earth-fission-products removal about every 50 days. The xenon is readily removed from the recirculating molten-salt stream using a gas separator. The uranium may be recovered by fluoridation to the volatile UF, but the separation methods for protactinium and rare earth-fission products were not defined. As a breeder, only thorium would be added to the fuel cycle and the power level would be largely controlled by the uranium-233 concentration in the molten salt.

While the feasibility of this concept was established, much engineering development remained to be completed before commercial acceptance.

High Temperature Graphite Reactor

High-temperature gas-cooled reactors are under development in the United States, the United Kingdom, and Germany.[8] The Fort St. Varain HTGR, a 300 Mwe reactor started operation in 1979. However, escalating costs and reduced government support make further development doubtful.

The concepts for these reactors include: (1) only fissile, fertile, and modulator ceramic materials in the core for high temperature stability, (2) the good neutron economy of the thorium-uranium fuel cycle, and (3) high-temperature helium coolant coupled to a closed-cycle trubine for high-efficiency power conversion. The HTGR core has a large heat capacity that provides more time to provide alternate cooling if a loss-of-coolant accident occurred. Other features of this reactor design include:

1. Field-constructed prestressed concrete reactor vessels that also serve as primary radiation shields, subject to in-service inspection to eliminate possibility of vessel failure.
2. Steam-driven helium circulators and control rods.
3. A fissile-fuel requirement about half that of the LWR cycle.

The core of the Gulf General Atomic HTGR consists of graphite-coated uranium and thorium particles embedded in a hexagonal graphite block (Fig. 24.19). The initial operation employs uranium-235 which must be rejected after one cycle to avoid neutron losses in the uranium-236 formed. Therefore, the size of the uranium and thorium particles differ so that they can be mechanically separated before processing the spent fuel. The uranium-233 is recycled with the thorium. The reprocessing of the spent fuel from this reactor could use the existing plants which treat LWR fuels, but a special head-end treatment facility would be required. The tentative process would involve crushing, burning in a fluidized bed, and leaching to prepare the uranium and thorium solutions for solvent extraction recovery of the uranium-235, uranium-233, and the thorium as separate products.

Other Nuclear Reactors

There are many other nuclear reactors that have been developed for research, engineering development, nuclide production, and mobile power. Most noteworthy are the nuclear propulsion

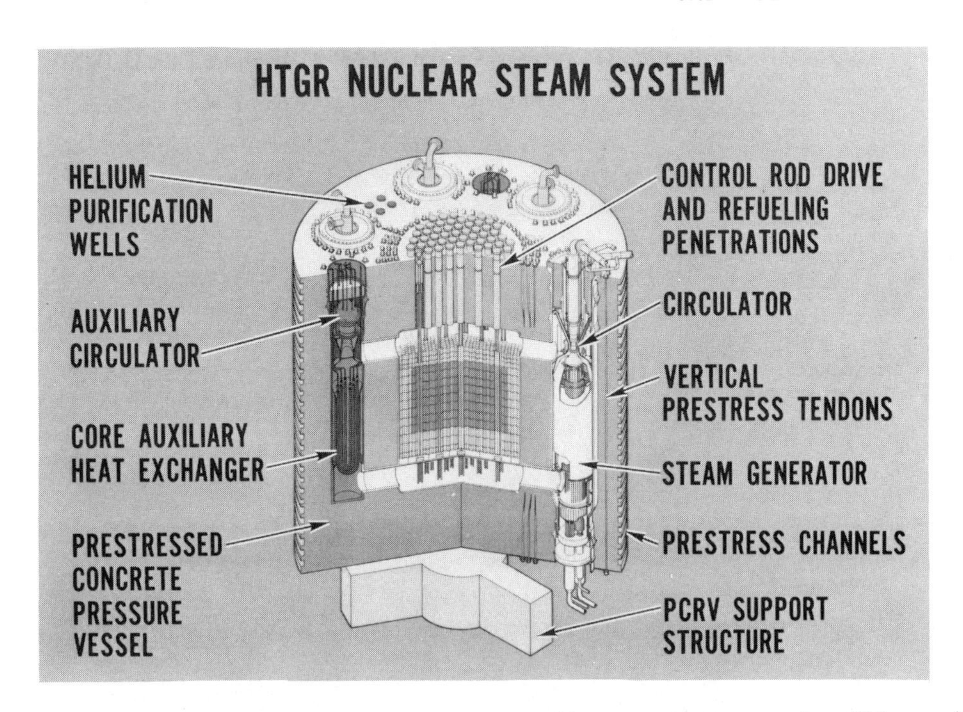

Fig. 24.19 High-temperature-graphite reactor provides about 44 percent energy conversion efficiency. (*Gulf General Atomics.*)

systems for naval applications. The first nuclear submarine, the Nautilus, was commissioned in 1954. Today there are more than a hundred such submarines, along with a number of surface vessels.[20] In the United States, the PWR technology for electric-power generation was an outgrowth of the nuclear naval-development program.

Other nuclear propulsion systems studied for arctic tractor trains, aircraft, and interplanetary rockets were terminated before completion due to insufficient need. The nuclear rocket provides a significant extension of interplanetary transport capability, with final development dependent on a national commitment to space exploration.

The remaining classes of nuclear reactors range from zero-power, subcritical neutron sources for university training to large-scale reactor systems for plutonium-239 production. Portable reactors provided heat, power, and water to U.S. bases in Alaska and Anarctica, and another provided electric power to the base in Panama. Private industry operated test reactors for development of reactor components and for commercial production of a wide variety of radioisotopes.[21]

REACTOR MATERIALS PROCESSING

Reactor-materials processing is concerned with preparation of the special materials used for nuclear-reactor fuels, coolants, moderators, plumbing, heat exchangers, and other components. Attention in this chapter will be limited to uranium and thorium. However, complete coverage would require the consideration of structural materials including zirconium, stainless steel, and aluminum; coolants including water, sodium, and helium; moderators including deuterium, graphite, and beryllium; and control rod materials including boron, cadmium, and hafnium.

Uranium Raw Materials

Uranium is the essential material in the nuclear industry. In 1979, it was estimated that the total, reasonably assured resources of uranium outside the Communist bloc in the production

TABLE 24.18 Estimated Resources of Uranium, 1979 (10^3 Short Tons U_3O_8)

Country	*Cost Range < $30/lb* Reasonably Assured	*Cost Range < $30/lb* Additional Estimated	*Cost Range $30–50/lb* Reasonably Assured	*Cost Range $30–50/lb* Additional Estimated
Algeria	36	–	–	7
Argentina	30	5	6	7
Australia	380	60	10	10
Brazil	96	117	–	–
Canada	280	480	25	465
Central African Republic	23	–	–	–
Denmark (Greenland)	0	35	–	21
Finland	–	4	–	–
France	51	34	21	26
Gabon	48	–	–	–
India	39	–	1	30
Italy	–	2	–	3
Japan	10	–	–	–
Mexico	6	–	44	–
Namibia	152	39	21	30
Niger	210	69	–	–
Portugal	9	3	1	–
South Africa	320	70	188	110
Spain	13	–	11	–
Sweden	–	–	390	4
USA	690	1010	230	495
Others	15	25	19	44
TOTAL (approximate)	2400	2000	1000	1300

Source: Nuclear Energy Agency.

cost category of under $30 per pound U_3O_8 amounted to 2,400,000 tons U_3O_8 (Table 24.18). Substantial discoveries of uranium made in the late 1960's and 1970's in Australia, Brazil, Canada, Namibia (Southwest Africa), Niger, and the U.S.A. have increased known reserves by several hundred thousand tons. Other significant discoveries were made in Algeria, Argentina, Gabon, and India. World demand for uranium was projected to be about 85,000 tons U_3O_8 in 1995 with cumulative requirements from 1979-1995 of 1.0 million tons U_3O_8 (Table 24.19). Estimates of uranium demand through 1995, however, have been progressively decreased since the mid-1970's due to problems licencing with nuclear power plants, longer lead times for construction, lower energy demand projections, and opposition to nuclear power by various environmental groups. Introduction of the breeder reactor is not contemplated before the year 2000; after that, uranium supply will be adequate to meet anticipated requirements for several centuries.

TABLE 24.19 Estimated World Uranium Demand Tons U_3O_8 ($\times 10^3$)

	Annual	*Cumulative*
1978	31	31
1979	34	65
1980	36	101
1981	39	140
1982	42	182
1983	46	228
1984	49	277
1985	52	329
1986	54	383
1987	57	440
1988	62	502
1989	66	568
1990	68	636
1991	71	707
1992	74	781
1993	78	859
1994	81	940
1995	85	1025

Source: R. Gene Clark, Director, Nuclear Energy Analysis Division, at Uranium Industry Seminar, Grand Junction, Colorado, October 16, 1979.

As of January 1, 1979, the U.S. reserves of uranium ore from which U_3O_8 was recoverable at $30 per pound were 714,000,000 tons of ore at 0.10 percent U_3O_8. This is equivalent to a total of 690,000 tons of U_3O_8. There was no net change in the $30 reserves total from the previous year. The total surface exploration and development drilling for 1978 was at an all-time high of 47 million feet, and the total exploration expenditures for the year were $273.6 million, excluding land acquisitions.[22]

Uranium Milling

In 1979, there were 21 uranium ore processing mills in operation in the U.S., and nine new plants were under construction or planned for start-up by 1982. In addition to the conventional ore-processing plants, there were 9 in situ solution mining operations recovering uranium and 5 plants recovering uranium as a by-product as the result of another mining activity. The capacity of individual ore-processing plants varied from 600 to 7,000 tons of ore per day.

The uranium-bearing ores processed commercially in the United States generally contain 0.05 to 0.2 percent U_3O_8. The average concentration is about 0.1 percent. In addition, some of the ores contain vanadium which can be recovered as a by-product.[20] Because the ores are quite low in uranium-content, the uranium-recovery plants or mills must be located near the mines to minimize transportation costs. The final product is a uranium concentrate containing 70 to 95 percent U_3O_8. This material can be delivered to refineries located a long distance away at comparatively lost cost (Fig. 24.20).[24]

Physical Methods of Concentration. Most uranium minerals are friable, and they become so finely pulverized and mixed with clay slimes in the grinding operation that they cannot be

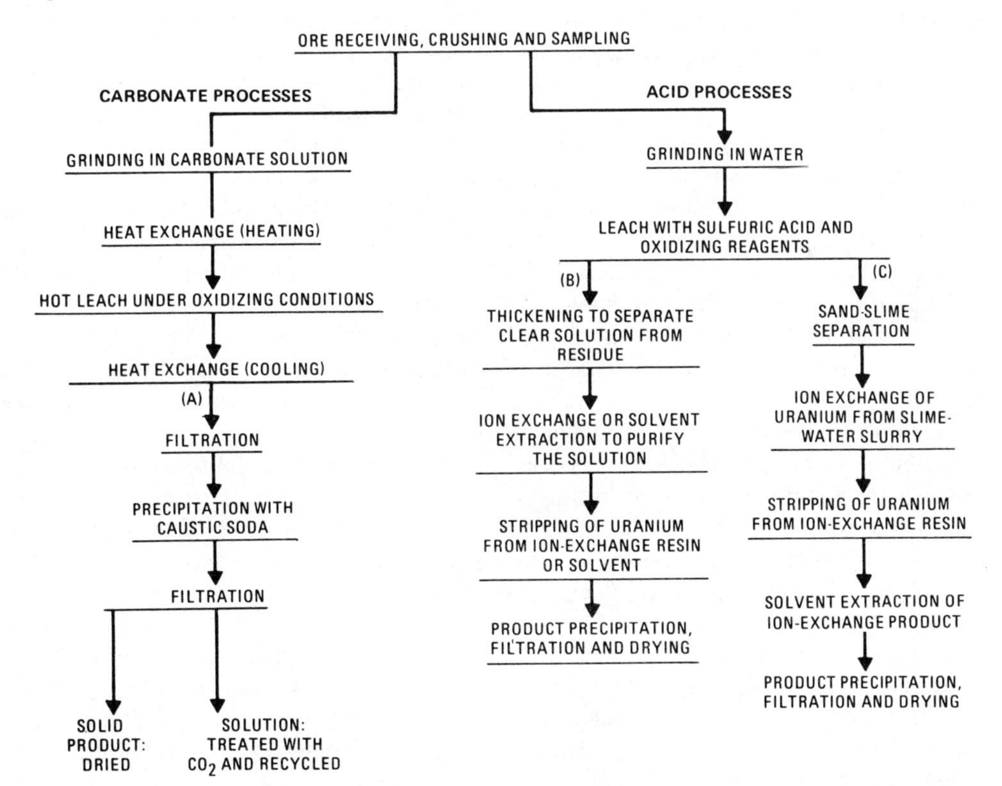

Fig. 24.20 Uranium ore mill processes; (A) carbonate leach, caustic soda precipitation process; (B) acid leach, ion exchange and acid leach, solvent extraction processes; and (C) acid leach, resin-in-pulp process.

effectively separated from the gangue or waste constituents by physical methods. Flotation methods are successful only in the few cases in which the uranium is closely associated with another mineral that can be floated, such as pyrite.

Uranium Dissolution. Uranium can be dissolved from its ores by any mineral acid or by sodium carbonate solutions. If uranium is present in the ore in its chemically reduced state, as is generally the case, an oxidizing agent in conjunction with either acid or sodium carbonate must be used to dissolve it from the ore.

Carbonate Leaching. In this process, the ore is first crushed and ground to very fine particle sizes in an aqueous slurry, commonly called a pulp, containing sodium carbonate and bicarbonate. Then the pulp is heated to temperature near boiling and agitated for approximately 12 to 24 hours with the continuous addition of air to supply oxygen, and in some cases with the addition of chemical oxidizing agents as well. Next the slurry is cooled in heat exchangers and filtered on rotary drum filters to separate the uranium-bearing solution from the residual ore solids. Sodium hydroxide is added to the filtrate to precipitate the uranium as semi-refined sodium diuranate.

As the carbonate leach is quite selective for uranium, the product, after filtration and drying, may be sufficiently pure to meet specifications without further treatment. Additional steps are often required, however, to remove such impurities as vanadium or sodium, depending on the needs of the refinery to which the product is to be sent for subsequent treatment. The solution from which the uranium has been precipitated is regenerated with carbon dioxide, usually from boiler flue gas, and returned to the process.

Acid Leaching. In this process, after the ore is crushed and ground in water, the slurry is leached for about 8 to 12 hours in a dilute acid solution containing an oxidizing reagent. Because of its relatively low cost, sulfuric acid is universally employed for leaching. Heat is often used to improve dissolution, but its effect is not as important as in the carbonate leach process. The two most commonly used oxidizing reagents are sodium chlorate and manganese dioxide. The manganese dioxide is often in the form of a crude pyrolusite ore containing roughly 30 to 40 percent MnO_2. After the uranium is dissolved, the uranium-bearing solution is usually separated from the residual ore solids in a series of large settling tanks called thickeners. The solids are transferred from one thickener to the next countercurrent to the flow of wash water.

Many elements in the ore other than uranium are soluble in mineral acids. Thus, after separation from the ore pulp, the uranium-bearing solutions require further purification. Ion-exchange or solvent-extraction methods are employed for this purpose. If the uranium were to be precipitated directly from the leach solution with sodium hydroxide, as is done in the carbonate process, the resultant product would contain about 2 to 10 percent U_3O_8, and would be contaminated with large quantities of iron, aluminum, silica, and many other impurities. The ion-exchange and solvent-extraction procedures make it possible to separate the uranium effectively from most of the impurities in the leach solution, and to obtain a product of satisfactory purity in a single unit operation.

In Situ Solution Mining. Where uranium ore occurs in a permeable aquifer with impermeable formations above and below the aquifer, uranium may be leached from the ore in place by circulating a leaching solution. The solutions used for leaching are sulfuric acid, ammonium carbonate, or sodium carbonate-bicarbonate, each containing an oxidant such as air, hydrogen peroxide, or sodium chlorate.

Vertical holes for injection of leach solutions are drilled to the bottom of the ore zone on a grid pattern, with a spacing of 50 to 200 feet between holes. Each hole is cased above the ore zone and fitted with a well screen in the ore zone. The leach solutions are pulled through the ore zone by pumping solution from production wells each located in the center of a group of four injection wells, and uranium is recovered from the loaded solutions by ion exchange.[23]

Ion Exchange. There are two anion exchange processes for the acid leach liquors, fixed-bed and resin-in-pulp. In the fixed-bed process, the resin is held in cylindrical tanks, and the clear uranium-bearing leach solution is passed through a bed of resin several feet deep. The anion-exchange resins that are used for uranium recovery are small spherical beads ranging generally from $\frac{1}{10}$ to $\frac{1}{60}$ of an inch in diameter. A uranium sulfate complex is preferentially retained by the resin, while the other soluble metals, which do not form such complexes, pass through the resin bed and are rejected. The uranium usually is removed from the resin by elution with an acidified chloride or nitrate solution. Partial neutralization precipitates the iron and aluminum, removing them from the solution. Then, the uranium is precipitated from the eluate with additional alkali, filtered, and dried.

The resin-in-pulp process is employed as an alternative to recover uranium from unfiltered solutions. After the leach, the pulp is diluted so that the coarser sand particles can be settled out, washed, and discarded. The remaining solution containing the very fine slime particles is then mixed with the ion-exchange resin. The presence of the slime does not interfere with uranium absorption from solution by the resin. Because the slime particles are very fine compared to the resin beads, screens can be used to separate the resin from the pulp after the absorption step. Uranium is then recovered from the resin in the manner previously described.

Solvent Extraction. In some plants, solvent extraction is the primary method for recovery of uranium from the acid leach liquors. However, several plants using resin-in-pulp as the primary separation method elute the resin with a strong sulfate solution and further purify the uranium-bearing eluate by a solvent-extraction step before precipitation. The final product of this process generally contains over 90 percent U_3O_8.

The active solvents used for uranium recovery are either organic amines or phosphates in which each organic group generally contains 8 to 14 carbon atoms. The amines are more selective solvents for uranium than the phosphates. However, the latter are sometimes preferred in plants which recover and purify vanadium as well as uranium. Due to excessive losses of expensive active solvent, the solvent extraction process is not satisfactory for recovery of uranium from carbonate solutions, nor can it be employed to recover uranium from ore pulps. Thus, the procedure is limited to recovery of uranium from clear solutions produced by the acid leach process.

The active solvent, which is a liquid ion exchange material, is diluted to a concentration of about 5 percent in kerosene. The diluted solvent is contacted with the clear leach solution in three or four countercurrent steps. Mixer-settler units are commonly used. In each stage, the two immiscible solutions are agitated in a small mixing compartment from which they overflow into a settling tank. The organic solvent rises, while the aqueous solution sinks to the bottom. After four contact stages, the organic stream containing the uranium is transferred to a separate circuit where it is stripped with very dilute acid. The product is precipitated from the strip solution, then filtered and dried.[24]

Uranium Product. The end product of all the milling procedures now being used is a crude U_3O_8 concentrate, generally referred to as yellow cake. While some study is being given to the potential production of refined uranium compounds in the ore-processing mills, no mill is now producing such materials.

Thorium Raw Materials

Thorium, which is several times more abundant in the earth's crust than uranium, is a fertile material which can be converted to the fissionable isotope U-233 by bombardment with neutrons in a reactor.[25] Most of the presently known resources of low-cost thorium (500,000 tons ThO_2) are in placer deposits containing monazite sands in India and elsewhere. Thorium is also found in large quantities in vein deposits in the Lemhi Pass area of Idaho and Montana. It is present in significant concentrations in the uranium ores in the Elliot Lake district in Canada, and has been produced as a by-product of uranium in this area. These known reserves are adequate to support the present need. The future requirement for thorium depends on the acceptance of the Candu and LWBR reactor programs.[32]

FEED MATERIALS PROCESSING

Uranium Feed Materials

In uranium feed materials processing, the crude uranium product from the ore mills—or the recycled product from the reactor-spent-fuel processing—is converted into the enriched uranium fuel required for the nuclear reactor (Fig. 24.21). For the light water reactor, the fuel is slightly enriched uranium oxide, usually 2 percent to 4 percent uranium-235, contained in zirconium tubing. In the past, the standard LWR fuel was stainless steel-clad uranium oxide, but zirconium provides greater neutron economy.

Using high-grade ore concentrate, it is possible to produce the uranium hexafluoride used for isotopic enrichment by hydrogen reduction followed by hydrofluorination and fluorination. Fluid-bed reactors are used for these conversion steps which are followed by fractional distillation of the UF_6.

In another process, ore concentrates are first purified by tributyl phosphate solvent extraction to provide a product that requires no further purification and greatly reduces the operating problems in subsequent conversion to UF_6 in fluid-bed reactors. In addition, when uranium metal is to be the end product, the intermediate uranium tetrafluoride product from the solvent-extraction route meets metal specifications.

The UF_6 product goes to the gaseous diffusion plant for enrichment followed by hydrogen reduction to UF_4 and, then, ammonia precipitation and calcining to UO_2. In the final step, the UO_2 is pelleted, sintered in a hydrogen furnace, machined to size and loaded into zircaloy tubing on its way to becoming a fuel element for the nuclear reactor.

Uranium Solvent Extraction. The ore concentrate is dissolved in nitric acid and solvent-extracted with tributyl phosphate (TBP) in a kerosene or hexane diluent. The diluent reduces the density of the TBP, enhancing the aqueous/solvent phase separation.

$$UO_2^{++}{}_{(aq)} + 2NO_3^-{}_{(aq)} + 2TBP_{org} \longrightarrow UO_2(NO_3)_2 \cdot 2TBP_{(org)}$$

This is an easily reversible reaction, directly dependent on the salting ion (NO_3^-) concentration in the aqueous phase. The salting agent is nitric acid. The extraction is very specific for uranium, providing decontamination factors of 10^3 to 10^5. The uranium is extracted into the solvent, leaving most other ions in the aqueous phase; the extract is scrubbed with a small amount of water for further purification; and the uranium is then stripped from the solvent with water (Fig. 24.22). The product is an aqueous uranyl nitrate solution.[25]

Uranium Denitration. Denitration of the uranyl nitrate product from solvent extraction to UO_3 involves evaporation followed by calcination. The evaporation is carried out to a

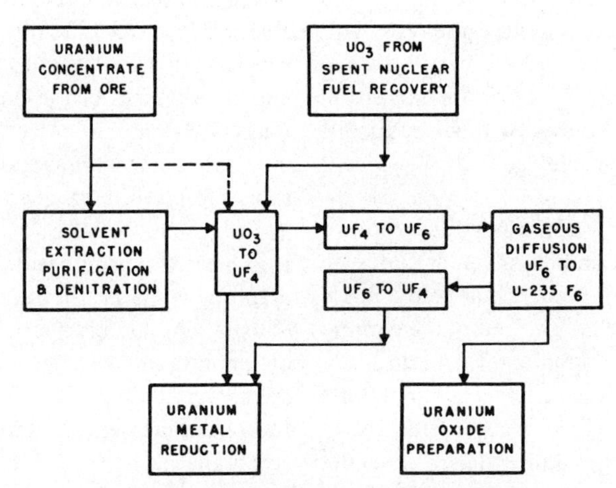

Fig. 24.21 Uranium feed materials flow sheet.

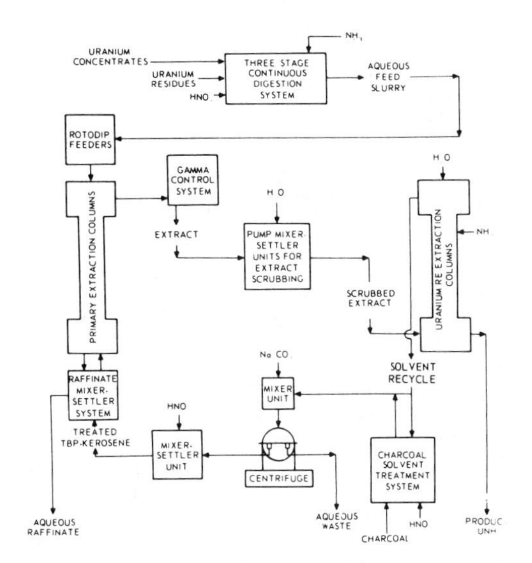

Fig. 24.22 Uranium solvent-extraction process for purification of ore concentrate and scrap; slurry feed eliminates clarification cost and losses. (*USAEC.*)

final boiling point of 120 to 143°C (9.8 to 12.5 lb U/gal) and the residue is transferred to a stirred-pot calciner where it is heated to 621°C.

$$UO_2(NO_3)_2 \cdot 6H_2O \xrightarrow[621°C]{\Delta} UO_3 + NO_2 + NO + O_2 + 6H_2O$$

Other calciner types in use include the stirred-trough and the fluidized-bed types. The type of equipment and the processing conditions significantly affect the physical properties of the UO_3, and hence the kinetics of the subsequent reactions in the production of UO_2, UF_4, and the metal (Figure 24.23).[25]

Uranium Oxide Conversion to Uranium Hexafluoride. At Allied's Metrolopis, Illinois plant using U_3O_8 as the feed, the conversion of uranium oxide to UF_6 is accomplished in a series of three fluid-bed reactors; first, for the hydrogen reduction of the uranium oxide to UO_2; second, for hydrofluorination to $UF_4\cdot$ and third, for fluorination to UF_6.[26]

$$U_3O_8 + 2H_2 \xrightarrow{650°C} 3UO_2 + 2H_2O$$

$$UO_2 + 4HF \xrightarrow{400°C} UF_4 + 2H_2O$$

$$UF_4 + F_2 \xrightarrow{500°C} UF_6$$

However, a second plant built by Kerr-McGee uses solvent extraction and denitration procedures, similar to those employed for metal production, providing chemically pure UO_3 before the reduction to UO_2 operation. In either case, the feed for UO_2 reduction is fluidized by the countercurrent flow of hydrogen at about 0.5 ft/sec with the temperature

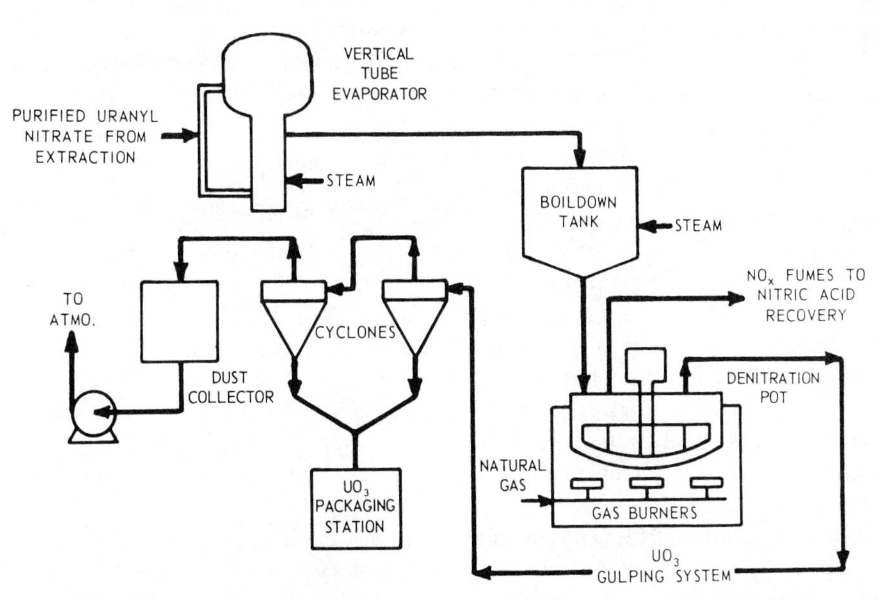

Fig. 24.23 Uranium denitration process prepares UO_3 for conversion to metal and UF_6. Continuous fluid-bed process was developed for large-scale use. (*USAEC.*)

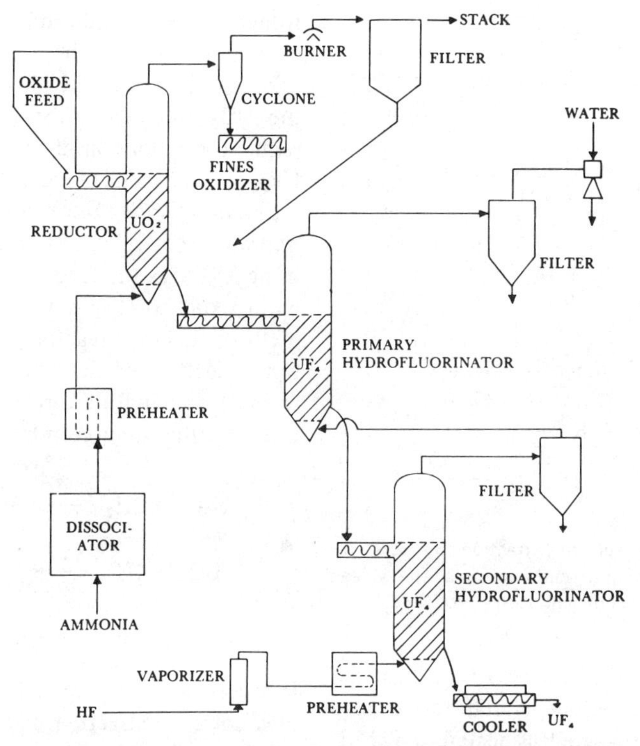

Fig. 24.24 Fluid-bed system for UO_2 conversion to UF_4. *(Allied Chemical Company process.)*

maintained at 600°C. The hydrofluorination step employs a two-stage fluid bed to achieve economic use of the hydrogen fluoride and maximum conversion (Fig. 24.24). The UF_4 product then goes either to metal production or on to UF_6 preparation.

In the fluorination step, complete use of the high cost fluorine and complete recovery of the uranium as UF_6 are important. Here the fluorine feed first contacts the fluorinator ash for maximum uranium recovery, then goes to the fluorinator where most of the fluorine and uranium are reacted, and finally, the last traces of unreacted fluorine are stripped by passing through the UF_4 in a screw reactor that is feeding the UF_4 to the fluorinator.

Since there is no solvent extraction purification step in the Allied process, the UF_6 product still contains vanadium, molybdenum, and other impurities. These are removed by distillation, first in a 120-foot, 100-plate column at 200°F and 85 psia to remove the high-volatile impurities, then in a second 45-plate column operating at 240°F and 95 psia to remove the low-volatile impurities. The final UF_6 product is condensed and packaged in 10-ton cylinders, each holding about $200,000 worth of product for delivery to the gaseous diffusion plant for isotopic enrichment.

Reduction to Uranium Metal. Uranium metal is produced from UF_4 by reduction with magnesium.

$$UF_4 + 2Mg \longrightarrow U\,(\text{metal}) + 2MgF_2 \quad (\Delta H_{298} = -83.5 \text{ kcal})$$

This is a batch process, carried out in a steel reactor lined with the reaction by-product, magnesium fluoride. The resulting 350-lb uranium regulus, called the "derby," is remelted, held at 1454°C in a vacuum furnace to vaporize and remove impurities, and then recast in a graphite mold to produce the ingot. The ingot is fabricated into reactor fuel by extrusion, rolling-mill operation, and machining. Uranium metal is primarily used for defense production reactors. However, depleted metal also finds some limited use as radiation shielding for spent-fuel shipping casks, for military

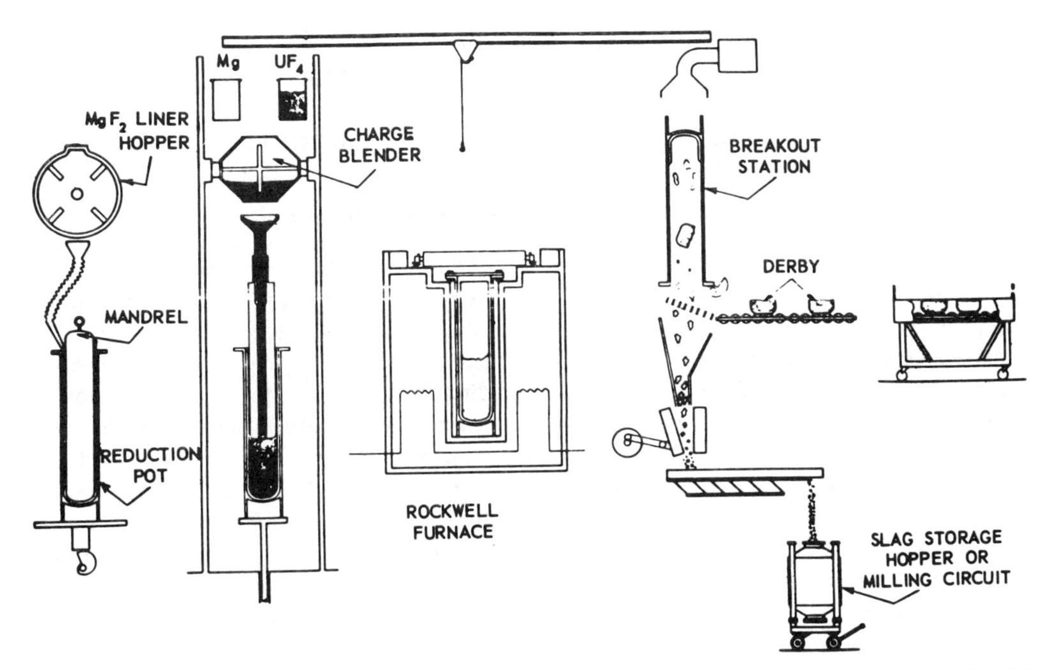

Fig. 24.25 Uranium metal reduction process; similar process is used for plutonium metal preparation. (*USAEC.*)

projectiles, and in counter weights (Figure 24.25).[27]

Uranium Enrichment. This phase of the nuclear industry represents about 4% of the cost of nuclear power. Following early development work, gaseous diffusion was selected over electromagnetic separation and thermal diffusion to provide enriched uranium-235. While gaseous diffusion today is still the established large-scale method of isotopic separation, centrifugation plants are now coming into operation in Europe and in the USA. Centrifugation may provide a 50% reduction in cost, but operating reliability is still to be established. The following discussion will be limited to gaseous diffusion.[28]

In gaseous diffusion, the porous membrane or barrier, as it is usually called, and the temperature and pressure are adjusted to optimize the separation of gaseous molecules of differing mass. The theoretical separation factor for uranium-235 is:

$$\alpha = \sqrt{\frac{^{238}UF_6\ (352)}{^{235}UF_6\ (349)}} = 1.0043$$

The opimum conditions require the use of elevated temperature and reduced pressure with a pressure differential across the barrier sufficient to transport about half of the UF_6 while flowing through that stage. By varying this flow ratio it is possible to taper the stages in the plant. This is very important since with the small separation factor a large number of stages are required to achieve the required enrichment. The large number of stages in turn require a large in-plant inventory and the large inventory results in a long operating time for the plant to reach equilibrium. By tapering the plant, the volume of the plant stages is decreased as the isotopic concentration of the uranium-235 increases. This volume reduction greatly reduces the time it takes to bring the plant to equilibrium, along with reducing the cost of the diffusion plant equipment, and plant operation.

The AEC gaseous diffusion plants contain 10,812 stages and consume up to 6000 Mw of power and 1350 million gallons of water per day. The power and water requirements would meet the needs of a city of a million people. The buildings housing the system have a combined floor area of one square mile.

The DOE gaseous diffusion plants now have an annual base capacity of 17.2 million separation work units (SWU), and this is being increased to 27.3 million SWU. The SWU is the work required to separate uranium of a given U-235 content into two components having higher and lower U-235 content. Axial flow compressors powered by electric motors rated up to 3300 horsepower transport the UF_6 gas. In the large scale components, over 640,000 kilograms of UF_6 are circulated to produce one SWU. About 150 metric tons of UF_6 are fed to the plants each day.

Other isotopic separations processes also being investigated to achieve further reductions in cost include separation nozzle and laser excitation devices. The separation nozzle depends on centrifugal force as produced in the centrifuge. In the laser process, by tuning the energy band of the laser beam, the uranium-235 is selectively excited and precipitated from either a molecular or an ionic state. Each of these enrichment processes while quite simple in concept challenge technologic frontiers to make them work. The feasibility of gas centrifuges was established in the 1940's with the first practical use achieved only in the 1980's, despite obvious advantages. However each of these techniques have processing considerations similar to gaseous diffusion, namely, tapered cascades involving large numbers of process units (Figure 24.26, Table 24.20).[29,30,31]

Uranium Hexafluoride Conversion to Uranium (IV) Oxide. The production of uranium (IV) oxide (UO_2) for nuclear fuel is particularly important to power reactors which operate at high temperatures. The oxide has greater chemical and physical stability than uranium metal, and these factors override the disadvantage of lower neutron economy. To compensate for the loss of neutron economy, the usual practice is to use uranium with a slightly higher U-235 content. The uranium hexafluoride (UF_6) is hydrolyzed in an aqueous ammonia solution to precipitate ammonium diuranate, reduced with hydrogen, and then calcined to UO_2 (Fig. 24.27).

$$2UF_6 + 14NH_3 + 7H_2O \longrightarrow (NH_4)_2U_2O_7 + 12NH_4F$$

$$(NH_4)_2U_2O_7 + 2H_2 \longrightarrow 2UO_2 + 2NH_3 + 3H_2O$$

The oxide is ground to provide the optimum material for preparation of high-density oxide pellets. Close control of many empirical factors in this operation is necessary to produce a material having the desired properties.

Thorium Feed Material

Thorium Feed Material Processing. Thorium is at present essentially a by-product of uranium ore milling and of rare earth recovery from monazite ore. The monazite process involves grinding the ore to -200 to -325 mesh, followed by leaching with sulfuric acid. The thorium, rare earths, and uranium are recovered by precipitation (Fig. 24.28). The crude thorium product is purified by solvent extraction with tributyl phosphate from a nitric acid solution using a process quite similar to that used for crude uranium mill product.

SPENT FUEL REPROCESSING

A broad range of processes for spent nuclear fuel have been developed and many operated on a production plant scale. The products from these processes have included:

- plutonium for military weapons
- plutonium, uranium-233, uranium, and thorium for reactor fuel
- neptunium for production of plutonium-238 thermal heat sources
- americium and curium for production of californium neutron sources
- americium , strontium, cesium, promethium, and krypton for radiation sources
- strontium for thermal heat sources
- iodine, actinides, strontium, cesium, krypton, xenon, tritium, and technetium for waste management

The principal process involved acid dissolution of the spent fuel followed by solvent extraction in an aqueous system. However other processes have included precipitation, ion exchange, volatilization, distillation, and sorption in aqueous, gaseous, molten salt, and molten metal systems.

Following India's secret development of an

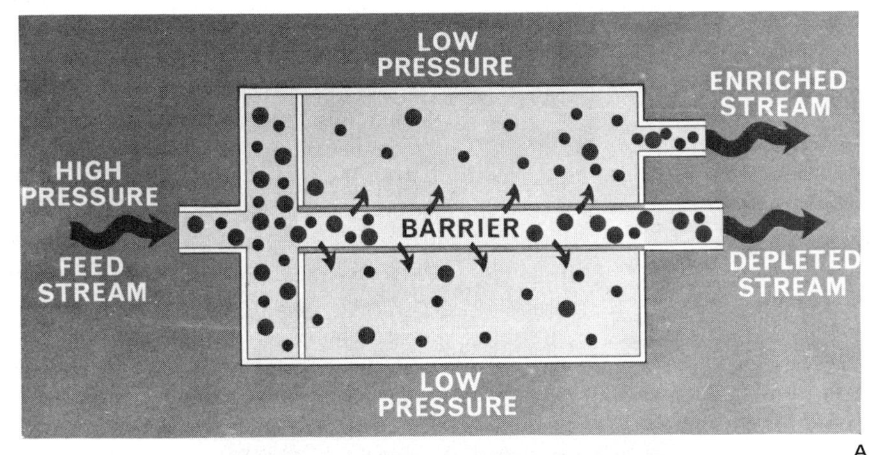

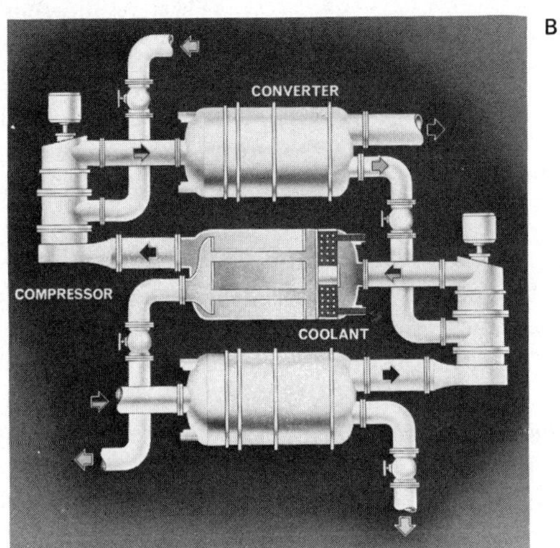

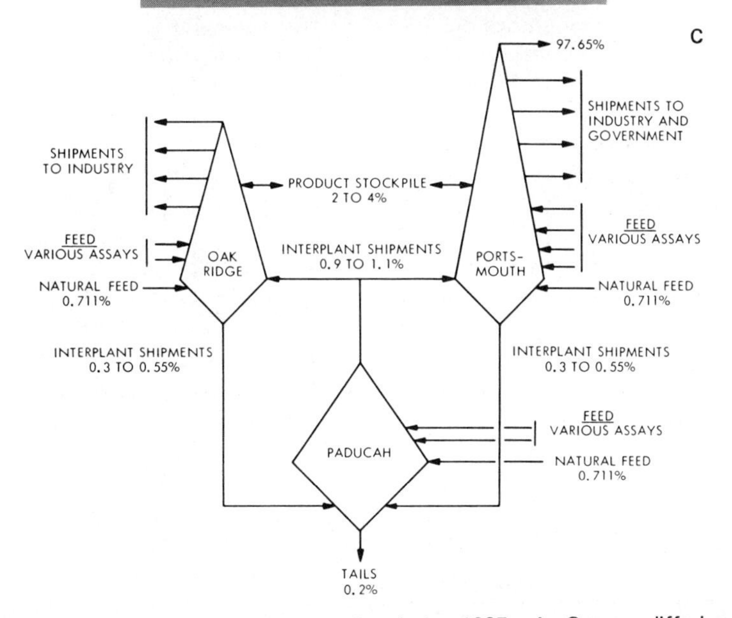

Fig. 24.26 U.S. AEC gaseous diffusion plant complex during 1967. A. Gaseous diffusion stage. B. Stage arrangement. C. Mode of operation for gaseous diffusion complex during last half of 1970. (% values are weight % U-235.) (*USAEC.*)

TABLE 24.20 Separative Work of Uranium Enrichment

Assay (wt.% U-235)	*Separative Work Unit*[a]
0.20	0
0.50	-0.173
0.60	-0.107
0.711 (natural	0.000
1.000	0.380
2.00	2.194
2.60	3.441
3.00	4.306
3.60	5.638
4.00	6.544
5.00	8.851
10.00	20.863
20.00	45.747
50.00	122.344
98.00	269.982

[a]Negative units represent additional charges for depleted feed. Unit per kg uranium product.

atomic bomb in 1978, the U.S. government declared a moratorium on commercial spent fuel processing. Although it achieves the benefits of nuclear power, spent fuel processing, along with uranium enrichment, are sensitive operations, because of the associated problems of nuclear arms proliferation. In 1981, the moratorium on spent fuel reprocessing was terminated in the U.S. with a commerical plant expected by 1990. Other countries are proceeding to establish these plants with the ultimate objective of a breeder-type nuclear power system. This is considered essential to reduce to a minimum the requirements for uranium.

The following discussion of the chemical process technology is generic to all the countries. This technology is in routine plant-scale operation to produce weapons-grade plutonium. However, plants for commercial spent fuel processing including those in the U.S. have experience numerous difficulties, resulting primarily from innovations to achieve marginal economic improvements.

The steps in spent fuel reprocessing are dissolution, separation and purification, and waste disposal. The unique feature that makes spent-fuel recovery different from other industrial operations is the radiation from the radioactive materials produced in the nuclear fission reaction. Because many of the fission products, which form a major part of the radioactive contaminants, have sufficiently short half-lives, they may be avoided by allowing time for them to decay to their stable end-products.

The radionuclides in the spent fuel that have a controlling influence are as follows:

Fe 55 & Co 60–limiting the reuse of the stainless steel and Inconel fittings of the fuel assembly

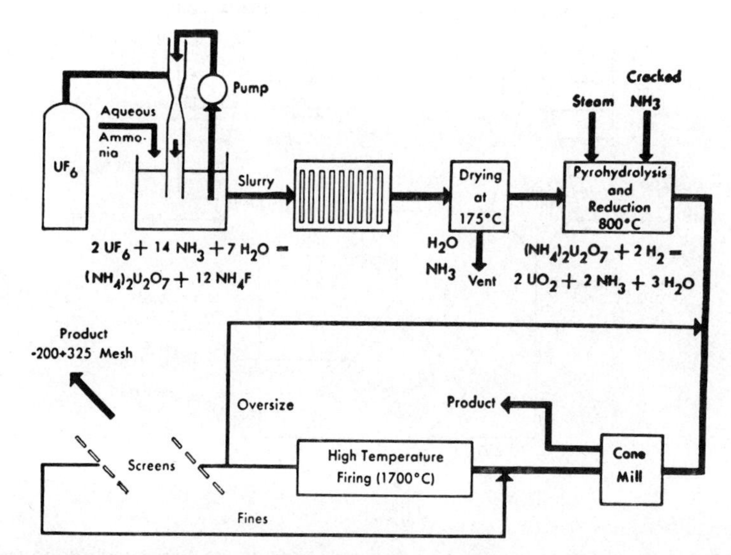

Fig. 24.27 UF_6 conversion to UO_2 by ammonium diuranate precipitation. (*USAEC.*)

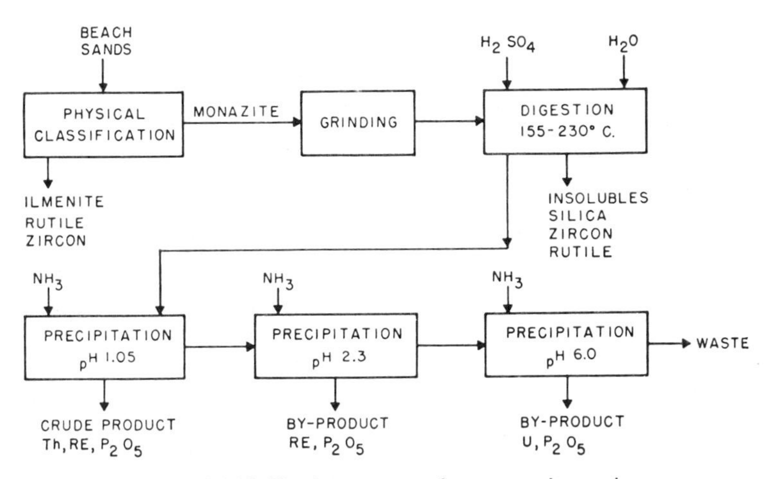

Fig. 24.28 Thorium recovery from monazite sands.

Zr 93—limiting the reuse of the zirconium clad on the fuel

I 129 & Tc 99—migrate most rapidly if waste release occurs from the geologic host rock

Ra 226—decay product of the very long lived uranium 238 to be eventually released to the biosphere

Sr/Y 90 & Cs/Ba 137m—source of short term (300 years) thermal energy

Pu 240 & Am 241—source of long term (10,000 years) thermal energy

The most significant delay-and-decay contaminant is iodine-131 with a half-life of 8 days. The properties of iodine-131, including short half-life, high fission yield, biochemical action, and volatility make it normal practice to delay the processing of spent power-reactor fuel for about 150 days. But even after 150 days, it is still necessary to process and trap the off-gas in order to limit the iodine release to about 0.1 percent and allow additional decay of the trapped iodine to provide another 10^3 reduction. When the LMFBRs come into operation, economics could require processing after only 30-days decay, and under these conditions the release may be limited to only 10^{-6} percent. Decay cooling time also reduces the irradiation and thermal effects on the process materials, and eases the requirements for chemical purification.

Radioactive-waste treatment is primarily concerned with reducing the volume of the radioactive material contained in the water and air associated with the spent-fuel processing operations. The guide lines recommended by international commissions limit the discharge of radioactivity to an amount that is a fraction of natural radiation background; therefore, essentially all of the radio-activity from the spent fuel must be retained during reprocessing and placed in long-term storage (Figure 24.29, Table 24.21).[33,34,35]

Dissolution

While the uranium oxide in the spent nuclear fuel from LWRs is readily dissolved in nitric acid, the jacket material and associated fittings of the fuel element are not. Mechanical shearing of the fuel elements to expose the nuclear fuel material to permit efficient dissolution has been demonstrated in the West Valley Plant of the Nuclear Fuels Services Corp. The typical fuel element is a 12-ft long by 5- to 8-in square bundle of from 50 to 225 half-inch diameter fuel rods, and contains 500 to 1100 pounds of uranium oxide jacketed in zirconium or stainless steel. At West Valley, after the end fittings have been sawed off, the active tube bundle was placed in a 300-ton shear where the rods are cut in $\frac{1}{2}$- to 2-inch lengths; the cut pieces are dropped into a boron-stainless steel cannister. The cannister is transferred to the dissolver where the nuclear fuel material is dissolved in nitric acid, leaving behind the jackets and associated fittings (Fig. 24.30). While there are chemical and electrochemical

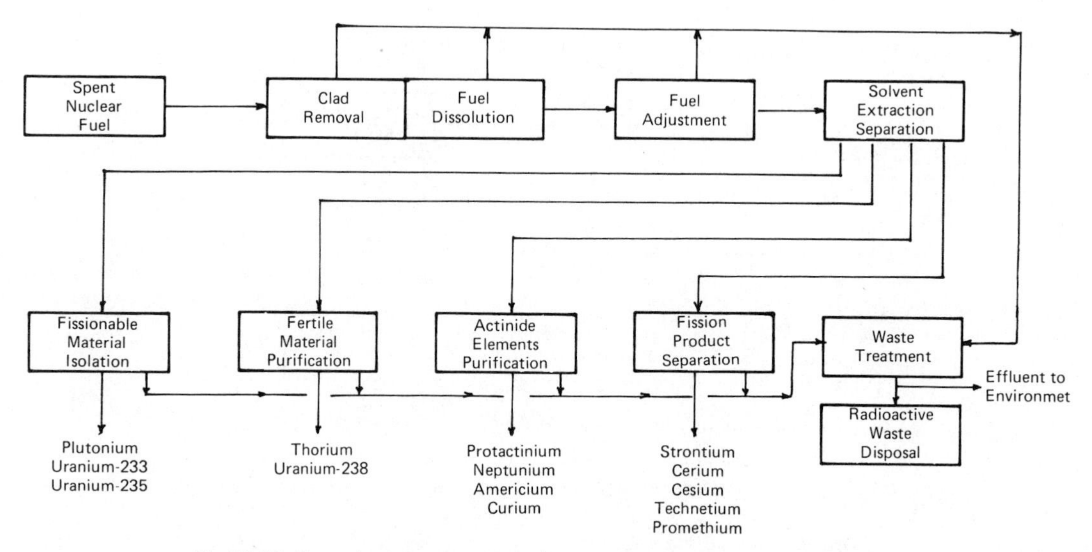

Fig. 24.29 General solvent-extraction process for spent nuclear fuel recovery.

TABLE 24.21 Radionuclides in a PWR Spent Fuel Assembly. Characteristics: 33000 MWd/t burnup, 14.7 kg U^{235}, 446 kg U^{238}, 108 kg Zr, 10 kg steel and Inconel

			(*Curies*: *years after discharge*)			
Nuclide	*Half life* (*y*)	*mpc* (*Ci*/*l*)[a]	*10*	*100*	*1000*	*10,000**
			Zr Clad			
Fe 55	2.7	8E-7	180	10^{-8}	–	–
Co 60	5.3	5E-8	970	0.01	–	–
ZR 93	9E5	6E-8	0.10	0.10	0.10	0.10
Total			1600	150	3.7	2.8
			UO_2 Fuel–fission products			
Kr 85	10.4	3E-10	2300	6.8	10^{-25}	–
Sr/Y 90	2E5	3E-7	6.1	6.1	6.1	5.9
Rh 106	2.2 hr		270	10^{-25}	–	–
Pd 107	7E6		0.05	0.05	0.05	0.05
I 129	2E7	6E-11	0.01	0.01	0.01	0.01
Xe 133	0.014	3E-10	–	–	–	–
Cs/Ba 137m	30	2E-8	74000	9300	10^{-5}	–
Total			14,000	15,000	10	10
			UO_2 Fuel–transuranium products			
Ra 226	1620	3E-11	10^{-7}	10^{-5}	10^{-3}	10^{-1}
Np 237	2E6	3E-9	7.6	7.6	7.0	3.1
Pu 238	89	5E-8	1000	500	0.5	0.06
Pu 239	24,360	5E-8	142	142	139	108
Pu 240	6580	5E-8	236	235	214	85
Pu 241	13	2E-7	36000	500	0.01	0.003
Am 241	458	4E-9	770	1700	410	0.005
Cm 244	17.6	7E-9	480	15	10^{-14}	–
Total			38,000	3,100	780	200

[a]Normal human consumption in water would be equivalent to a radiation dose of 500 mrem/year, approximately 2 to 5 times the natural radiation background.

*Note: After 10,000 years the radioactive toxicity is approximately constant for several million years.

procedures capable of concurrently dissolving the zirconium and stainless steel components, the shearing procedure and preferential dissolution of only the nuclear fuel materials significantly reduces the volume of materials in solution to be processed and disposed of as radiochemical waste.

The West Valley Plant was shut down in the mid 1970s when increasing regulatory requirements made it uneconomic.

A critical aspect of the dissolution step is treatment of the off-gas. Nitrogen oxides are evolved from the reaction of nitric acid with the fuel material along with the release of volatile fission products: iodine, krypton, xenon, ad tritium. Most of the tritium stays in the dissolver or is trapped in the down-draft condenser. Some of the iodine leaves the dissolver, and a significant fraction is scrubbed out in the condenser with the remainder being sorbed on a silver nitrate bed. The krypton and xenon could be removed by low-temperature charcoal adsorption or by freon scrubbing. When required for large-scale operations, the release of these radioactive gases from the dissolver may be greatly reduced by an additional head-end procedure.

Separation and Purification

Uranium, plutonium and the other transuranium products, and the fission products are separated and purified by solvent extraction followed by anion exchange and precipitation. The Purex Process employs tributyl phosphate as the solvent for this separation of products from the nitric acid solution of the spent fuel and is the only process now in commercial use for this purpose. About 99 percent of the uranium and plutonium is recovered in separate product streams, and decontamination from fission products by a factor greater than 10^7 is effected. The solids content of the final radioactive waste is only about 150 pounds per ton of fuel processed, an important consideration. This is a great advantage over the earlier Bismuth Phosphate and Redox processes, with quantities of waste equivalent to the mass of the fuel.

In the standard Purex Process, in a first cycle, the uranium, plutonium, and residual radioactivity are separated into three separate streams. The residual stream is the principal source of high-level waste, and it contains the fission products along with the transplutonium products. Second cycles of solvent extraction are used for both the uranium and plutonium streams (Fig. 24.31). The wastes from the second cycles may be recycled back to the first cycle, both to reduce product losses and to minimize the quantity of radioactive waste.

There are two interesting modifications of the solvent-extraction process. The first uses a co-extraction of the uranium and plutonium in the first cycle, thereby eliminating one set of solvent-extraction columns. A 5000 ton per year plant was built by Allied General Nuclear Systems using this modification but operation has been delayed by the proliferation morato-

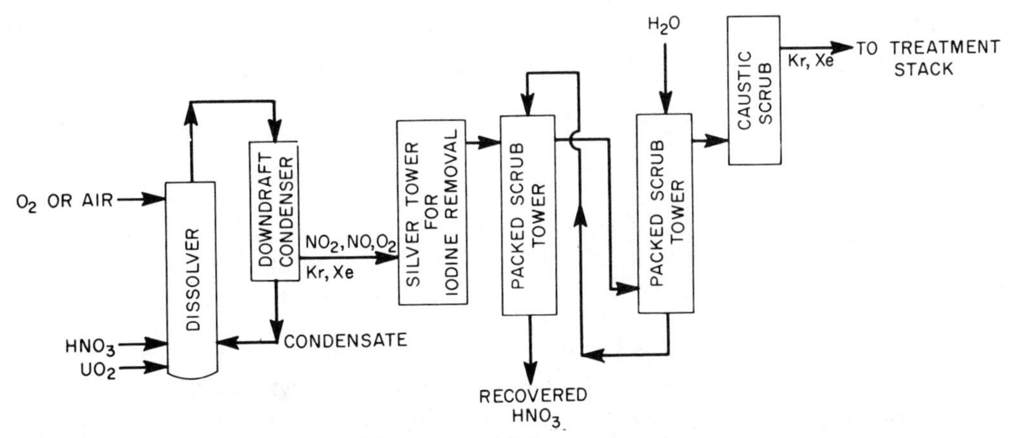

Fig. 24.30 Spent-fuel dissolution flow sheet.

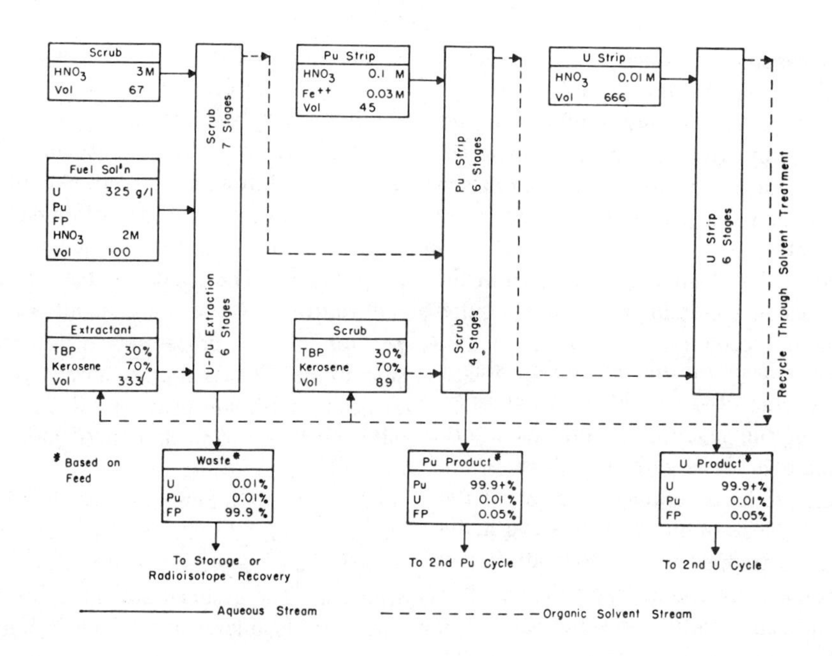

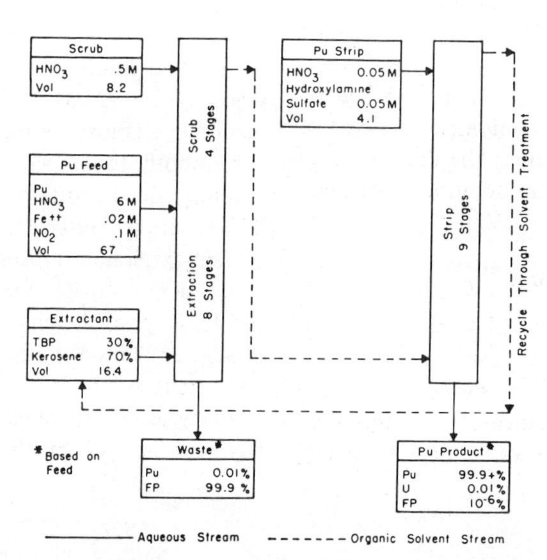

Fig. 24.31 Purex process showing (top) first solvent-extraction cycle, and (bottom) second plutonium solvent-extraction cycle.

rium. In the second moficiation (in the Aquaflor Process developed by General Electric for their Midwest Fuel Recovery Plant), the uranium stream, following the normal Purex first cycle, goes to a fluidized-bed operation for conversion to a purified UF_6. The plutonium stream goes to anion exchange as discussed next. However, solids handling problems related to the fluidized bed led to a GE decision to not operate the plant.

The uranium and plutonium product streams from the standard process go through sorption cycles for final purification, and then they can be recycled to the feed materials operation. The uranium product passes through a silica gel bed, with corrosion products being adsorbed on the silica gel along with final traces of plutonium and fission-product zirconium and niobium. The plutonium product is sorbed on an anion exchange bed, while corrosion products and traces of uranium pass through (Fig. 24.32). The plutonium is then eluted with dilute nitric acid, precipitated, and converted to the oxide.

The basic Purex process will be used to process the LMFBR fuel. But there are a number of conditions that will require several significant modifications to the flow sheet for LWR fuels as follows: the shearing and dissolution steps will have to accommodate the metallic sodium in failed fuel elements; the sodium, tritium, and possibly iodine will have to be removed by a high-temperature treatment before dissolution; the capability of the dissolver-off-gas treatment system must be increased for better iodine-removal efficiency (the iodine concentration will be much greater because fuel-inventory costs may justify rapid recovery and reuse of the plutonium); the solvent-extraction operation will have to be modified to minimize solvent exposure to the much higher fission-product radiation; and continuous equipment for plutonium purification will probably need to replace the present fixed-bed anion-exchange procedure.

RADIOACTIVE WASTE MANAGEMENT

Radioactive-waste management includes the treatment, storage, and disposal of liquid, airborne, and solid effluents from nuclear reactor and fuel cycle operations. It employs three philosophies: delay-and-decay, concentrate-and-contain, and dilute-and-disperse.[25] Combinations of all three are usually employed to solve each waste problem. Concentrate-and-contain is the major procedure since far more radioactive waste is handled by the nuclear industry than may be dispersed into the environment under federal regulations. However, delay-and-decay is an inherent part of both the containment and dispersal technologies.

The general strategy for waste management is to convert the liquid, airborne, and solid wastes to solid packaged forms for shipment to, and emplacement into, geologic repositories. The liquid and airborne effluents from the effluent-treatment and waste-conversion processes contain some residual radioactivity. However, the

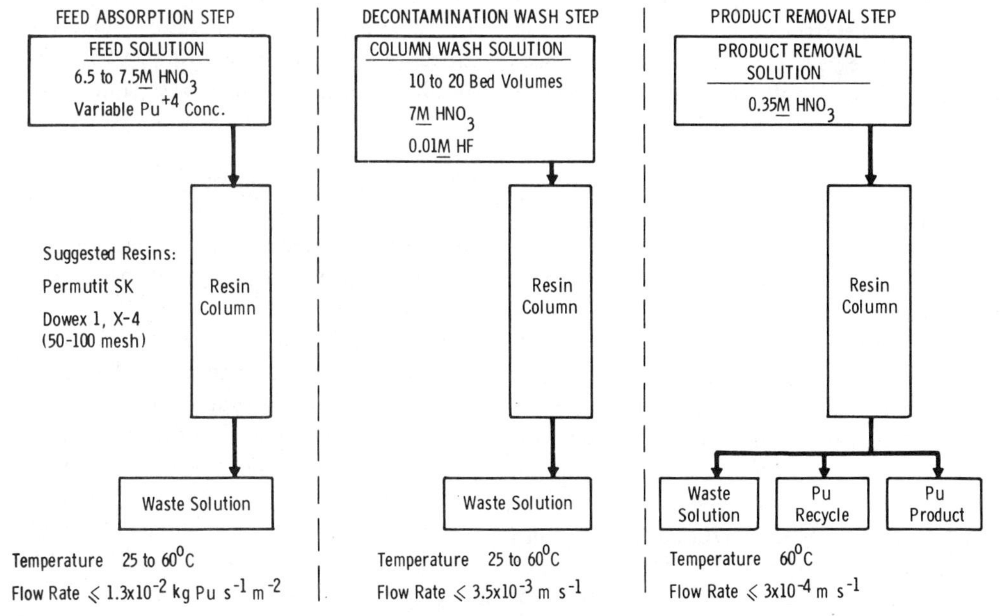

Fig. 24.32 Plutonium ion exchange flow sheet.

TABLE 24.22 Estimated Cumulative Waste Quantities and Land Requirements–USA[a] (148 GWe in Yr. 2000)

	Volume (10^6 ft^3)			*Land (acres)*		
	1980	*1990*	*2000*	*1980*	*1990*	*2000*
Low Level						
Defense	60	80	90	540	650	680
Commerical	15	45	83	150	300	420
TRU[b]						
Defense	0.05	0.05	0.05	20	100	200
Commerical	0.01	0.01	0.1	nil	nil	nil
High Level						
Defense (Sludge & Calcine)[c]			0.004	0	0	900
Defense (LLW Salt Cake)			0.008	nil	nil	15
Commerical Spent Fuel[d]	0.014	0.7	1.4	20	50	2200
Disposal Facilities						
Near Surface Burial Grounds[e]				690	950	1100
Deep Geologic Vaults				nil	nil	2700

[a]IRG Appendix D, assuming 148 GWe in 2000 and commerical fuel not processed.
[b]Defense TRU in interim surface storage until a repository for disposal is available in late 1980's.
[c]Packaging of Defense waste started in early 1990's.
[d]Commerical spent fuel in interim water basin storage until one or two repositories for disposal are available in the late 1990's.
[e]Includes decommissioned facilities.

radiation level of this residual radioactivity in the effluents released to the environment is limited to a few percent of natural background radiation. Short-lived, low-level wastes are placed in near surface burial grounds. The long-lived, high-level wastes are planned to be placed in deep mined vaults in coherent geologic formations (Table 24.22).[36,37]

Liquid Waste Treatment

The most effective concentration procedure for liquid wastes is evaporation. The decontamination achieved by this approach is 10^3 to 10^5 per cycle for the water removed. Other more economical but less effective procedures include ion exchange and coprecipitation, with the choice of procedures depending on the degree of decontamination required. Tritium, the heaviest isotope of hydrogen, is the only radionuclide not removed.

Procedures have been developed to solidify the high-level liquid wastes from fuel reprocessing. The anticipated procedure includes evaporation, fixation in silicate glass, and encapsulation in welded steel pots. The pot, approximately 1 foot in diameter and 10 feet long, provides the containment for handling, transport, and interim storage. The evaporator sludges and ion exchange resin slurries are frequently solidified by adding cement, asphalt, or organic polymers.

Airborne Waste Treatment

Air contaminated with radioactive iodine, xenon, krypton, tritium, and particulates is the major problem. The iodine is removed by sorption on a silver catalyst bed followed by nitric acid scrubbers, and eventually it decays to stable xenon. Krypton can be recovered along with the xenon from the dissolver off-gas by low-temperature sorption or freon scrubbing, or it can be released to the environment. The krypton, when recovered, will require long-term storage because of the 10-year half-life of krypton-85. Since xenon represents about 8 times the volume of krypton and the half-life of the longest-lined xenon fission product is

12 days, it may be economically attractive after adequate decay to separate it from krypton-85 to reduce the volume of tankage required for the long-term krypton storage. Tritium may be removed from the spent fuel before dissolution by high temperature retorting.

Dry filtration with fibre filters is the standard final treatment for decontaminating radioactive-particulate-contaminated air. The paper filter developed for this purpose has a 99.97 percent efficiency for removal of 0.3 micron particles. When necessary, these filters are preceded by scrubbing, sorption, and roughing filters.

Solid Waste Treatment

The nuclear operations produce solid wastes directly in the form of paper, clothing, laboratory glassware, and failed equipment, as well as the solidified waste from liquid and air waste treatment. Some of the solid wastes may be compacted or incinerated to reduce the volume. Incineration is not yet well established. While incineration provides the opportunity to decrease volume and increase the waste stability, it does not offer significant benefit for transportation or for geologic containment and does create additional airborne and liquid wastes.

Waste Disposal

The terminal storage of the radioactive wastes from the nuclear industry involves the wastes from mining, milling, enrichment, fuel fabrication, reactor operations, and spent fuel. The mining and milling wastes are placed in terminal storage by covering the surface storage piles with approximately 10 ft. of earth. This controls errosion and delays for decay the radon release to the atmosphere. The wastes from the reactor plants, enrichment, and uranium fuel fabrication go to shallow land burial. There have been some difficulties with shallow land burial due to surface water leaching; however, the resulting releases have been lower than the licensed limits. The use of deep-mined repositories in geologic media is planned for spent fuel if not reprocessed and for the wastes from the spent fuel processing and refabrication. While the technology for these operations is established, there is aggressive effort to improve the understanding regarding the long term thermal effects on the geologic storage method. Time (thousands of years) precludes experimental verification, but historic and exploratory geologic information indicate the risk of future health effects would be less than 1% of those from natural radiation.

The land requirements, population dose, and costs for waste management are modest. Depending on the fuel cycle and reactor type, permanent land requirements range from approximately 1 to 4 acres/GWey for near surface burial with 1 to 2 acres/GWey for mined repositories. Within the U.S. the land requirements through 2000 will be approximately 1100 acres for shallow land burial and decommissioned nuclear facilities, and approximately 2700 acres for subterranean acres for deep geologic vaults. The deep vault area will be decreased to about 1000 acres if the spent fuel is reprocessed and the uranium and plutonium recycled.

In the event of leakage from both the waste containers and the host rock, the radiation dose to the regional population would be in the range of 0.1 to 0.8% of natural background radiation. The costs for waste management including treatment, interim storage, and final disposal would be on the order of 2% of the consumers' cost of the electricity generated. It is to be noted that the costs are primarily related to the nuclear power plant and the spent fuel.

Many alternatives have been studied for waste disposal. Only the shallow land burial and deep-mined geological vaults, along with emplacement within the thick sediments on the floor of the ocean (sub sea bed), could handle all types of waste. The other methods would only reduce the volume going to deep geological vaults. The results of the continuing study of sub sea bed emplacement are encouraging. (Tables 24.23 and 24.24).

The low level, less toxic waste is buried in privately operated burial grounds, and requires about an acre of land for each 400,000 ft^3 of licensed-for-burial-type waste. At present, about a million cubic feet of this waste is generated and buried each year.

TABLE 24.23 Conceptual Methods for Radioactive Waste Disposal

Terrestrial	Shallow land burial Deep geological vaults – continental sites – island sites Caves Deep hole Sub sea bed Ice cap Hydrofracturing Insitu rock melting
Extraterrestrial	Solar orbit – space shuttle Moon crater – Rocket/soft lander Solar escape – Electric cannon
Transmutation	Fission reactors Fusion reactors Electromagnetic accelerators

ISOTOPE PRODUCTS

There are more than 2000 isotopes, or nuclides and isomers as they are more properly called. Isotopes are atomic species with the same atomic number but different mass numbers. Nuclides are characterized by a particular number of protons and neutrons in the nucleus of an atomic species, and isomers represent different energy states of the same nuclide. The stable nuclides number 276 with over 250 radioactive isomers, and there are approximately 1500 radioactive nuclides.

The nuclides are very significant by-products of the nuclear industry. Both the stable and the radioactive nuclides find considerable use as diagnostic agents in research, medicine, and industry, where they serve as tracers for physical, chemical, and biological processes. In addition, the radiation from radioactive nuclides is used to detect changes in density and other characteristics of a material, to promote physical and chemical changes in materials, and to provide a source of thermal energy. Making these nuclides available has been a noteworthy contribution of the nuclear industry.[27]

Nuclide Conversion

The widely reported dream of the alchemist was to convert base metals into gold. In 1934, this dream was realized in principle when I. and F. Joliot-Curie announced that boron and

TABLE 24.24 Effect of Reactor Cycle on Nuclear Waste Volumes (per GWye)

	Once - Through		*U - Pu Recycle*			*Th - U Recycle*
	LWR	*HWR*	*LWR*	*HWR*	*FBR*	*HWR*
Mill Tailings (m^3)	51500	38600	27600	17500	313	2100
TRU[a] Drums	–	–	120	640	619	1920
TRU Drums (shielded)	45	63	310	1200	157	480
HLW[b] Canisters	50	123	29	33	23	42
ILW[c] Canisters	8	2	157	171	189	183
Gas Cylinders	–	–	18	41	18	27
LLW[d] Drums	4220	3950	4180	3350	7570	12600
LLW Drums (shielded)	800	400	400	430	415	428

[a]Transuranic wastes
[b]High level wastes
[c]Intermediate level wastes
[d]Low level wastes

aluminum could be made radioactive by bombardment with the alpha rays from polonium. This demonstrated that atomic particles could be captured by the nucleus to convert one nuclide into another. The Curies had produced phosphorus by the A1-27 (α, n) P-30 reaction. Unfortunately for the modern alchemist, the cost of converting a base metal to gold would be several orders of magnitude greater than the present value of gold.[46] These reactions are a function of both the type and energy of the bombarding particles. In turn, the compound nucleus formed by particle-capture goes through an instantaneous rearrangement, emitting other particles and energy. The resulting nucleus may be stable or may continue to decay, emitting particles and energy. In some cases, it decays through other unstable states before reaching a stable nuclide state. The time for half of the unstable nuclei to decay is a constant, the so-called half-life, which ranges from milliseconds to millions of years.

Nuclide-conversion is primarily concerned with the preparation of the unstable (radioactive) nuclides called radioisotopes. Nuclear reactors and charged-particle accelerators are employed for this purpose. In the nuclear reactor, the most common reactions are neutron-capture and fission, while in the accelerators the reaction is charged-particle capture. The quantities produced are usually in the millicurie to curie range, this being sufficient to satisfy the tracer, diagnostic, and analytical applications. However, nuclear reactors have also been used to produce megacurie quantities of fissile and fusion nuclides—plutonium and tritium—for energy production as discussed in earlier sections of this chapter. A final point to be made is that most of the neutron-capture products are the same chemical species as the target material, while most of the charged-particle products are a different chemical atom.

Radioisotopes

Of the several hundred radioisotopes prepared and used in research, about twenty-five have been involved in extensive development efforts. The fission products and trans-uranium (neutron-capture) products have received the most development attention since their production results directly from the operation of nuclear reactors. As the use of nuclear-power reactors increases so will the availability of these isotopes. The processes to recover these products involve precipitation, solvent extraction, and ion exchange. Relatively large-scale production is required to achieve economic costs. This is required even when the spent fuel is being processed for plutonium-239 recovery, and the rest of the products need only cover the incremental added cost.

In comparison with the fission and transuranium products, the production of the other reactor-neutron-capture and accelerator products is relatively simple. Most reactor products such as cobalt-60 are activation products in which no chemical separations are involved. For example, when cobalt-59 is irradiated with neutrons, chemical separation of the cobalt-60 product is not possible. Most of the accelerator products and the other reactor products are transmuted to different chemical elements, for example the reactor-neutron irradiation of sulfur to give phosphorous-32, or the cyclotron-proton irradiation of iron to produce cobalt-57. In this case, chemical separations yield pure nuclide products, and the form of the target material may be selected to minimize the separation problem (Table 24.25).[40]

Fission Products. In the fission of uranium and plutonium, the nuclides produced range in mass from about 72 to 162 with maximum yields occurring in two broad peaks in the regions of 95 and 138. In addition, some tritium results from triple fission. The maximum yields in the 95 and 138 regions are about 6 percent.

The major products recovered in the past from spent nuclear fuel include strontium-90, cesium-137, promethium-147, and krypton-85. The strontium-90 was recovered from Purex waste by precipitation followed by solvent extraction, and the product is then converted to strontium titanate or strontium flouride for heat-source applications. The promethium was separated by ion exchange and converted to the oxide for thermal and radiation applications.

TABLE 24.25 Major Radioisotopes and Applications

Am-241	A	I-131	M
Cf-252	AM	Ir-192	A
C-14	A	Kr-85	AL
Cs-137	AMP	Tc-99m	M
Cr-51	M	Ni-63	A
Co-57	AM	P-32	M
Co-60	AMPT	Po-210	A
Cm-244	T	Pu-238	T
F-18	M	Pu-239	T
H-3	AL	Pm-147	AT
I-125	M	Sr-90	AT
Xe-133	M	Sr-85	M

A–Analysis and control.
M–Medical.
P–Process radiation.
T–Thermal.
L–Luminescent.

The cesium has been extracted by ion exchange and subsequently converted to the chloride for radiation-source applications. Krypton-recovery is part of the dissolver off-gas treatment system.[37]

To reduce the heat load on the waste storage tanks at Hanford, 100 megacuric quantities of strontium-90 and cesium-137 have been extracted from the waste. This provides for about 30 kW of electric power using strontium-90 in thermoelectric generators and approximately 100 process irradiators using the cesium-137 (see Figure 24.33).

A promethium-147 capability was established in 1964 at the Pacific Northwest Laboratory and about 4 kilowatts (2750 curies/watt) processed for thermal and self-luminescent applications. The 5 percent krypton-85 product from fission has been enriched by thermal diffusion to a 45 percent product for luminescent applications.

The stable fission products are receiving increased attention. Rhodium and xenon are probably the most important materials because of their limited availability in nature and short-lived radioisotopic contamination. Developmental quantities of these materials along with technetium and palladium have been prepared.

The short-lived fission products represent a broad spectrum of nuclides. In this group, xenon-133 and technetium 99m are the most widely used products, especially for medical diagnostic purposes. These are produced by special irradiation of gram-size targets of uranium-235.

Reactor Products. The reactor products are made by neutron capture in target materials. There are three general groups of reactor products: (1) the transuranium actinide products resulting from the transmuting capture of neutrons in the reactor fuel and fertile materials; (2) the transmutation products made by irradiation of special targe materials; and (3) the activation products also made by irradiation of special target materials. Most radioisotopes are produced in research and test reactors such as those described in Table 24.26. The cost of these products ranges from a few cents per curie for thulium-170, and 50 cents per curie for cobalt to ten dollars a microcurie for californium.

Since most of the reactor-produced radioisotopes are the activation products of the same chemical element, very little chemical processing is required. In the case of cobalt-60, the naturally occurring cobalt-59 metal is irradiated and then encapsultated for use as a gamma-radiation or heat source. Megacurie quantities of cobalt-60 are produced for use in process irradiators (Figure 24.34). Cobalt-60 is also an effective thermal energy source similar to strontium-90.

For the products involving transmutation, the form of the target is selected to keep the processing to simple batch precipitation, distillation, and solvent extraction or ion exchange. The quantities involved are grams and millicuries. The irradiation is carried out in small quartz flame-sealed vials and the processing is performed in laboratory glassware.

The production of transuranium nuclides in the Savannah River reactor is a much more complex and challenging operation. The principal transuranium product other than plutonium 239 and tritium is plutonium-238 for use as a thermal energy source for space applications. Curium-242 and -244, americium-241, and californium-252 were produced in the past for research purposes. The other nuclides in this production chain have little usefulness because of half-life or other limiting nuclear characteristics. Although the production of plutonium-239 involves the nuclear capture of

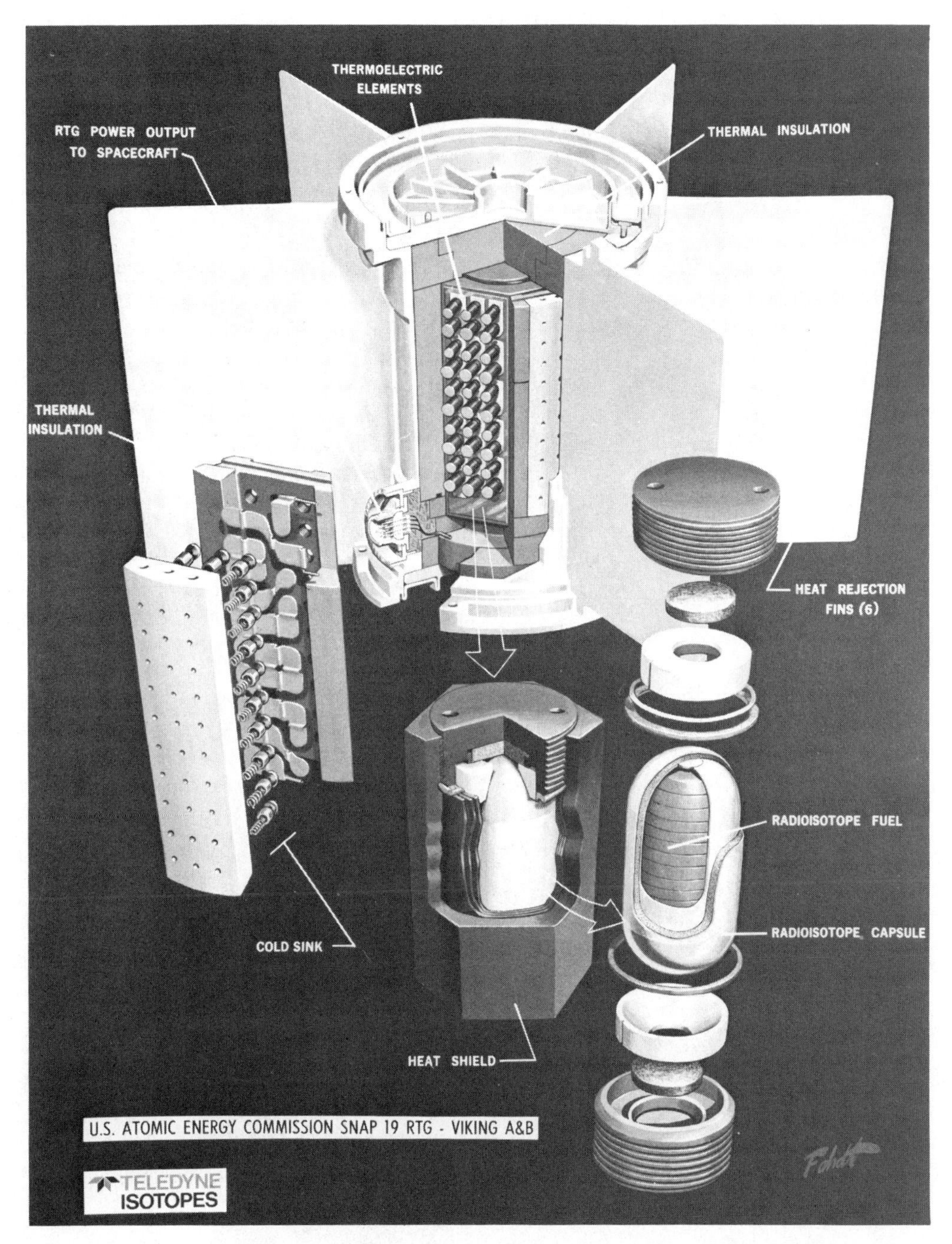

Fig. 24.33 The multi-hundred watt electric generator. Three units on Voyager generated 475 watts from the decay heat of plutonium-238 acting on thermoelectric couples. (*Courtesy U.S. Department of Energy.*)

only one neutron in uranium-238, the production of either plutonium-238 or americium-241 involves the capture of 3 neutrons, while curium-242 requires 4, curium-244 requires 6, and californium-252 requires the capture of 14 neutrons. In each case, there are intermediate products that require processing either to achieve more efficient conversion by recycling the target material or to recover a decay product from a long-lived nuclide parent.[41, 47, 48]

Solvent extraction and ion exchange compete as the principal techniques for use in these

TABLE 24.26 Characteristics of Various Nuclear Reactors for Radioisotope Production

Reactor	Power (Mw)	Neutron Flux (10^{14}) Thermal	Fast
ORR			
(HT)	30	2.5	1.5
(S-1)		1.8	0.83
(S-2)		4.8	0.65
HFIR			
(T)	100	28	7.5
(RB)		15	2.6
SR[a]	700	60	20
PWR	3300	0.5	3.0
BWR	3300	0.4	1.9
LMFBR			
(core)	2600	nil	51
(blanket)		nil	9.7

[a]Savannah River C Reactor; flux demonstration.

transuranium product separations. The chemical oxidation effects from the intense radiation (about 1 watt per liter) is one of the factors influencing the choice of technique. In solvent extraction, these effects tend to neutralize the acid, while in ion exchange they cause gas-binding of the bed. The addition of methanol reduces the oxidation rate, and rapid high-pressure processing prevents gas-binding of the ion exchange columns. Other unique problems include the need for both neutron and gamma shielding, and the presence of considerable thermal energy, 0.5 w/g for plutonium-238 to 160 w/g for curium-242. (Figures 24.33, 24.34; Tables 24.26, 24.27).[41,42,43]

Accelerator Products. The accelerator products are prepared by capturing in special target materials the high-energy isotopes of hydrogen or helium: proton, deuteron, triton, alpha or helium-3 ions. The ion energy is usually in the range of 10 to 40 Mev with beam currents up

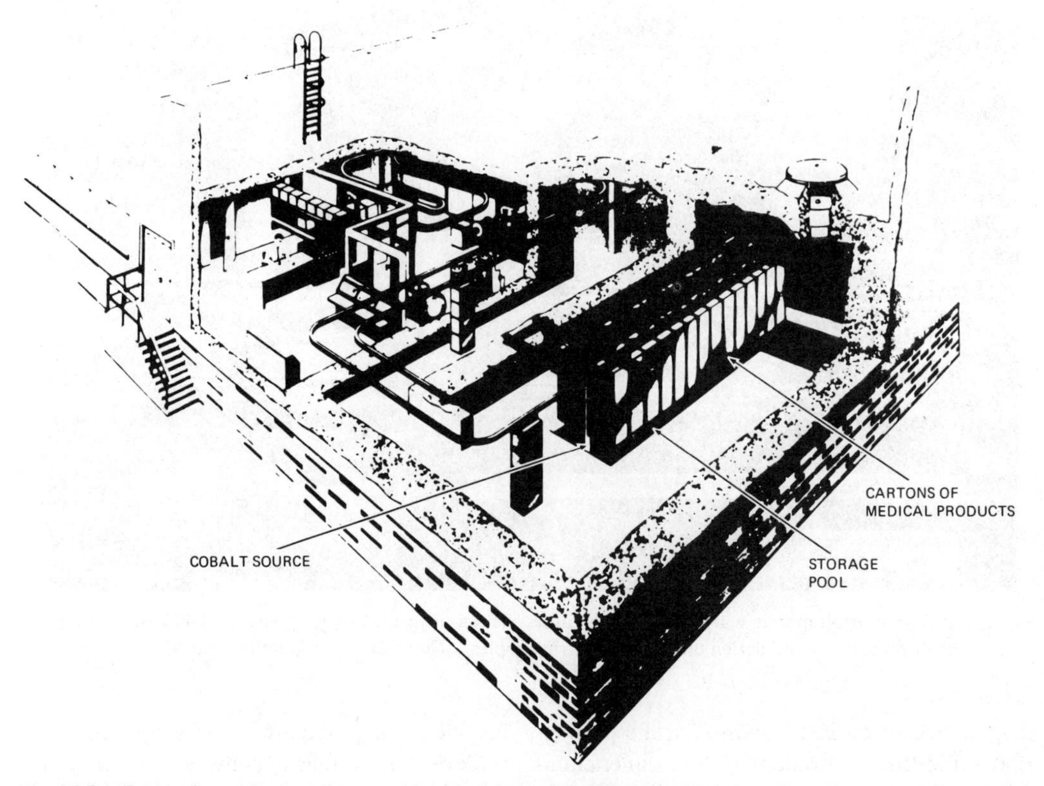

Fig. 24.34 Ethicon's medical-products irradiator, Somerville, N.J., sterilizes prepackaged sutures and syringes. Approximately 2.5 megarads is delivered to 140 cu ft of product per hour using 950,000 curies of cobalt-60. The cobalt-60 is installed in a slab geometry with the product cartons conveyed in two roes on either side. The slab is lowered into the water-filled pool below the conveyor when maintenance is required.

TABLE 24.27 Reactor Products

Activation Series		*Transmutation Series*	
Nuclide	*Half-Life*	*Nuclide*	*Half-Life*
Mo-99	2.75 d	F-18	110 m
Cr-51	27.8 d	K-43	22 h
Fe-59	45 d	Cu-67	61.6 h
Ir-192	74 d	I-131	8.06 d
Sn-113	118 d	P-32	14.3 d
Tm-170	130 d	P-33	25.2 d
Co-60	5.26 y	Po-210	138 d
Ni-63	125 y	Gd-153	242 d
Cl-36	3.1×10^5 y	H-3	12.5 y
		C-14	5730 y

to several milliamperes. For many years, the ORNL 86-inch cyclotron has been the principal source of accelerator products and this machine still has a greater isotope-production capability than any other accelerator employing low-energy protons. The BNL 60-inch and other cyclotrons provide the broader capability to use deuteron, alpha and helium-3 ions. In addition, BNL Van de Graaffs are being used to develop triton and deuteron-capture products. The characteristics of several machines are given in Table 24.28.

While the particle flux in an accelerator is very modest, the machine yield is adequate for most tracer-type needs. The irradiation cost is on the order of $200 per hour, as compared with $350 per week in a nuclear reactor, because accelerators are usually single-target machines. Accelerator products are somewhat more difficult to process than reactor products due to the differences in target-irradiation conditions. The charged particle beam in the accelerator is very hot, 10 to 1000 kw/cm^2 with the optimum target being a high-temperature metal plated on a water-cooled copper heat exchanger. When this situation is impossible, water-cooled capsules are used in many cases to contain the target material.

A new family of cyclotrons located at large hospitals are producing short lived oxygen-15, nitrogen-13 and carbon-11 tracers for a broad range of diagnostic medical procedures. Similar cyclotrons are being operated by industry for the longer life products.

Several accelerator products are listed in Table 24.29.

Stable Isotopes

The separation techniques for stable isotopes are based on differences in their physical or chemical properties. For the lighter weight nuclides, chemical exchange processes are employed because the differences in chemical equilibrium constants are sufficient to allow separation. For the heavier nuclides, the major

TABLE 24.28 Characteristics of DOE Accelerators for Radioisotope Production

Machine	*Particle*	*Energy (Mev)*	*Current (μAmp)[a]* *Internal*	*External*
		Cyclotrons		
ORNL 86-inch	p	17.5	3,000	–
		21	200–1,500	–
		22.4	–	15
BNL 60-inch	p	40	–	100
	d	20	–	
	α	40	–	50
	He-3	60	–	
		Van de Graaff		
BNL Physics		3.5	–	50
BNL Radiobiology	d	4.0	–	10
		Linear Accelerators		
BLIP	p	200	–	180
LAMPF	p	800	–	750

[a]μAmp (microamperes) = 2.3×10^{13} particles per hour.

TABLE 24.29 Accelerator Products

Nuclide	*Half-Life*	*Nuclide*	*Half-Life*
O-15	2 m	V-48	16 d
N-13	10 m	Cr-51	28 d
C-11	20 m	Rb-84	33 d
		Be-7	53 d
F-18	110 m	Sr-85	65 d
Fe-52	8.3 h	Co-56	77 d
I-123	13 h	W-181	130 d
Mg-28	21.3 h	Ce-139	137 d
Cs-129	31 h	Co-57	271 d
Au-194	39 h	Ge-68	287 d
Ga-67	78 h	Na-22	2.5 y
Y-87	80 h	Fe-55	2.7 y

techniques depend on differences in mass. They are electromagnetic, thermal diffusion, and gaseous diffusion (Table 24.30). Both the chemical-exchange and the mass-separation techniques are multiple-stage diffusion processes which, with the exception of the electromagnetic process, have very small separation coefficients, on the order of 1.1 to 1.001. Also characteristic of these processes are their relatively low yields and the long time required to reach equilibrium. For example, in the case of uranium enrichment by gaseous diffusion, the tails contain 0.2 mole percent uranium-235 which represents 28 percent of the uranium-235 entering the process in the natural uranium feed. This process requires months to reach equilibrium.[45]

In the electromagnetic process, the separation factor may be on the order of 10 to 100. In this process, the feed is ionized and then accelerated through a bending magnetic field. The differences in mass cause the heavier particles to pass through larger arcs with the products captured in slit-like pockets located 180° from the feed point (Fig. 24.35). The accelerating voltage and the magnetic field strength may be adjusted to give good separation although the yield per pass is only 1 to 10 percent. The low yields result from inefficiences in the ionizer, the accelerator, and the catchers. When the feed has value, the system is scrubbed and flushed to recover the material which is recycled. Although the cost of this process is high, it provides the most economical method for producing gram quantities, which are sufficient to meet most research and development needs.

TABLE 24.30 Isotope Separation Processes

Process	*Applicable Isotopes*
Electromagnetic	All
Chemical exchange	H, Li, B, C, N
Thermal diffusion	H, He, N, O, Ne, A, Kr, Xe, Cl, C
Gaseous diffusion	U
Fractional distillation	H, O, N, C
Gas centrifuge	Ge, Xe, U
Ion migration	U, K, Cl, Cu, Mg, Li, Na
Molecular distillation	Li, K, Cl

Thermal diffusion is a second general purpose method for isotope separation and is applicable to all gases. The gas is circulated in a small column, about 1 inch in diameter. Chilling the wall and heating the center with a very hot wire, 700 to 800°C, causes the light isotopes to migrate to the center and move upward, while the heavier isotopes move to the wall and move downward. These columns usually operate on a batch basis with several months being required to reach equilibrium.

There are several types of chemical exchange systems involving gas-liquid, liquid-liquid, and liquid-solid equilibria operating in multistage fashion. Nitrogen-15 is usually enriched by stripping from nitrogen-14 using the gaseous-nitrous-oxide/aqueous-nitric-acid system. The nitrogen-15 in this equilibrium favors the aqueous phase and therefore concentrates at the bottom of the plant. To achieve maximum concentration, the nitrogen is refluxed by decomposing the nitric acid in the bottom to nitrous oxide with sulfur dioxide. The nitrogen reflux at the other end of the plant reacts the nitrous oxide with oxygen and water to produce nitric acid.

The heavy water plants producing hundreds of tons per year represent the largest chemical exchange operations to date. Heavy water is a very efficient moderator for nuclear reactors and was used in the AEC's Savannah River reactors for nuclide conversion producing plutonium-239, plutonium-238, tritium, cobalt-60, and the transuranium elements through californium-252. It uses a dual-temperature

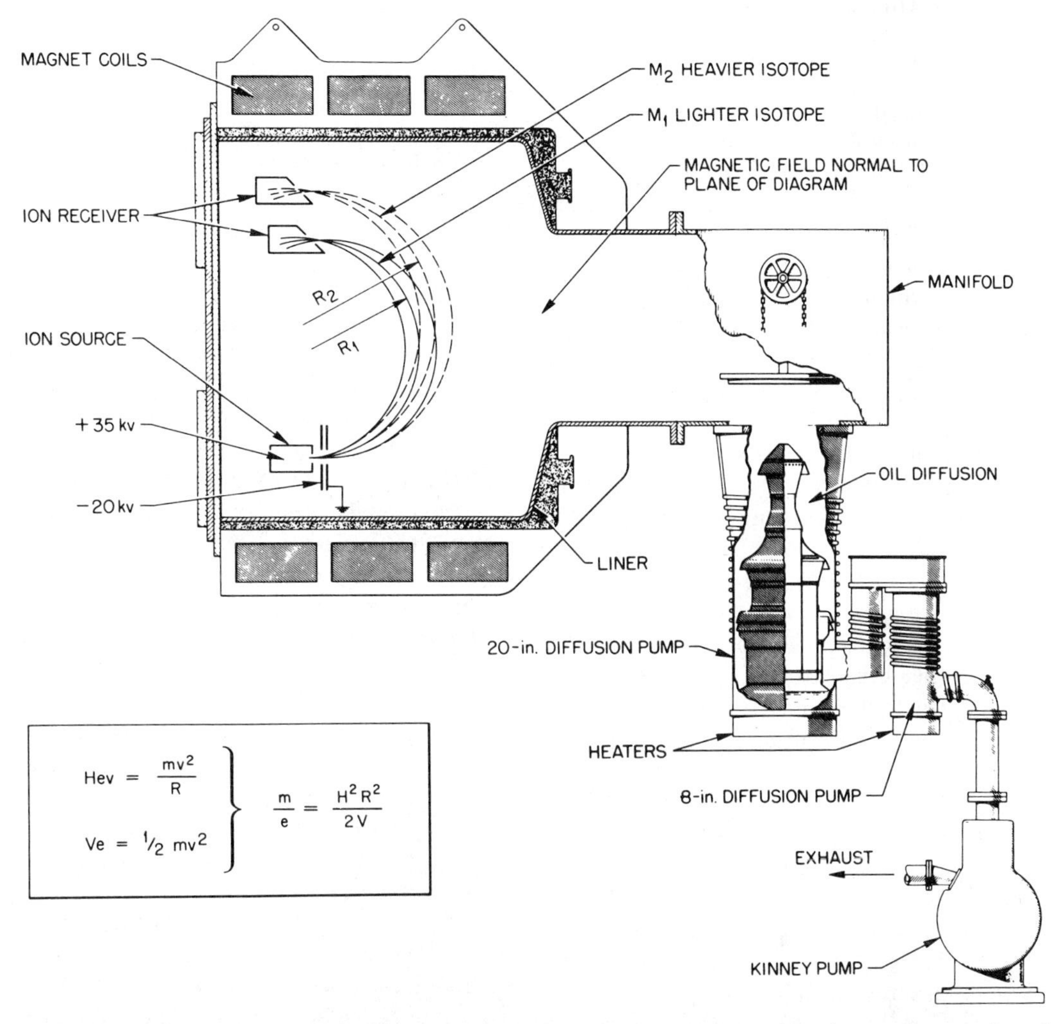

Fig. 24.35 Oak Ridge Calutron for separation of isotopes. First method to achieve large-scale separation of U-235, and today the source of research quantities of most stable and a few radioactive nuclides. (*USAEC.*)

process which is based on the fact that the separation factor for the exchange of deuterium between H_2S and water varies with temperature. Isotopic separation is effected by running a pair of columns, one at about 27°C and the other at about 220°C. Each column may have about two hundred plates. In the hot column, the deuterium is stripped from the water, enriching the H_2S in deuterium. The enriched H_2S then passes to the bottom of the cold column where the equilibrium now favors the water. The H_2S is on internal recycle, stripping the deuterium from the water. The fresh water feed, containing about 0.016 weight percent of deuterium, passes through first the cold, then the hot columns. The deuterium concentration builds up at the bottom of the cold column and at the top of the hot column. After days of operation when equilibrium is reached, the gas and/or the liquid phase can be withdrawn and passed on to a subsequent cycle. In the AEC plant at Savannah River a 10 to 20 percent deuterium product is achieved in two cycles. This is further enriched to 99.8 percent by distillation.

The economics of the process depend on efficient exchange of heat from the hot to the cold stream to minimize the quantity of energy that must be added to the dual-temperature flow control, pumping energy, corrosion, product-extraction methods, and the number of cycles.

SAFETY

Safety has always been and will continue to be the single most essential feature of the nuclear industry. Although radiation has always been part of our environment (Fig. 24.36), the nuclear industry has brought with it potential increases from man-made radioactive materials. Since the beginning of this industry, there have been no accidents, with either reactors or radioisotopes, that have resulted in any significant exposure of the public. A few employee fatalities have resulted from nuclear criticality accidents.

The excellent safety record is due in large part to development of both national and international agreements governing all aspects of the nuclear industry. Government and industry have cooperated with various standards and regulatory organizations to establish required performance criteria and to define the design specifications needed to mett these criteria.

The U.S. Atomic Energy Commission and later the Nuclear Regulatory Commission have been responsible for establishing and controlling the safety of the nuclear industry.[44] The Three Mile Island accident in 1979 led the industry to recognize its dependence on the NRC for operational safety. This dependence, however, has been greatly reduced by two industry organizations: (1) the Institute of Nuclear Operators to coordinate management organization, execution, and control functions and (2) the Nuclear Safety Accident Center to collect and evaluate abnormal event information and to provide ready resources to deal promptly and methodically with accidents.

In 1975, the Reactor Safety Study was issued, the most thorough hazard evaluation of any technology ever performed. It concluded that on the basis of past licensing review practices the hazard from LWR type nuclear power reactors was orders of magnitude less than other commonly accepted hazards (Figure 24.37). Again, the Three Mile Island accident confirmed this result; while there was very expensive damage to the reactor fuel core, there was no injury to either the public or the plant personnel.[1,45]

The International Commission for Radiation Protection (ICRP) has recommended that the radiation-exposure limit for workers be less than five roentgen per year after age 18. This limit is based on thorough and careful consideration of life-shortening and genetic damage that could be caused by radiation. The maximum allowable radiation exposure to the general public has been set at 0.5 roentgen per year.

Nuclear facilities are now designed to limit the radiation exposure from normal operations to less than 5 milliroentgens per year to any individual outside the facility property. This radiation level is a few percent of the natural radiation background. In addition, under the ALARA philosophy ("as low as reasonably achievable"), the need for additional effluent treatment is evaluated on the basis of an added cost of $10,000 per annual roentgen reduction to the total regional population.

Through the cooperative efforts of industry and government, systems have been developed and demonstrated that can limit the release of radioactivity from the nuclear power reactors to a few percent of natural background. Similar performance can be expected for the fuel cycle and waste disposal operations that support the operations of the nuclear power reactors.

The future well-being of our environment has been further assured by the National Environmental Protection Act of 1969. The National environmental goals as expressed by the National Environmental Protection Act are as follows:

> . . . it is the continuing responsibility of the Federal Government to use all practical means, consistent with other essential considerations of national policy, to improve and coordinate Federal plans, functions, programs, and resources to the end that the Nation may
>
> (1) fulfill the responsibilities of each generation as trustee of the environment for succeeding generations;
> (2) assure for all Americans safe, healthful, productive and esthetically and culturally pleasing surroundings;
> (3) attain the widest range of beneficial use of the environment without degradation, risk to health or safety, or other undesirable and unintended consequences;

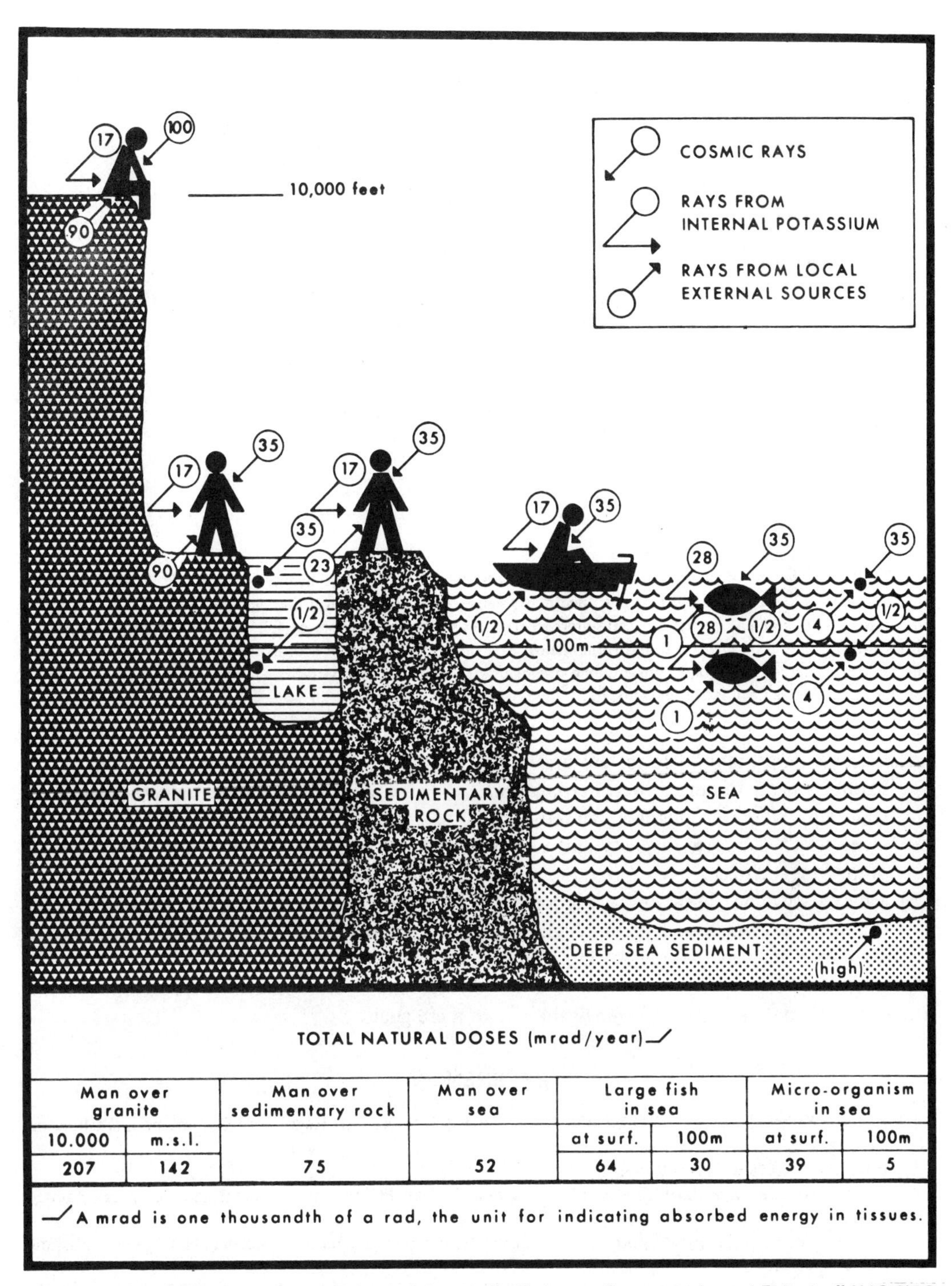

Man over granite		Man over sedimentary rock	Man over sea	Large fish in sea		Micro-organism in sea	
10.000	m.s.l.			at surf.	100m	at surf.	100m
207	142	75	52	64	30	39	5

Fig. 24.36 Natural radiation. (*"The Effects of Atomic Radiation on Oceanography and Fisheries," NAS-NRC Pub. 551/1957.*)

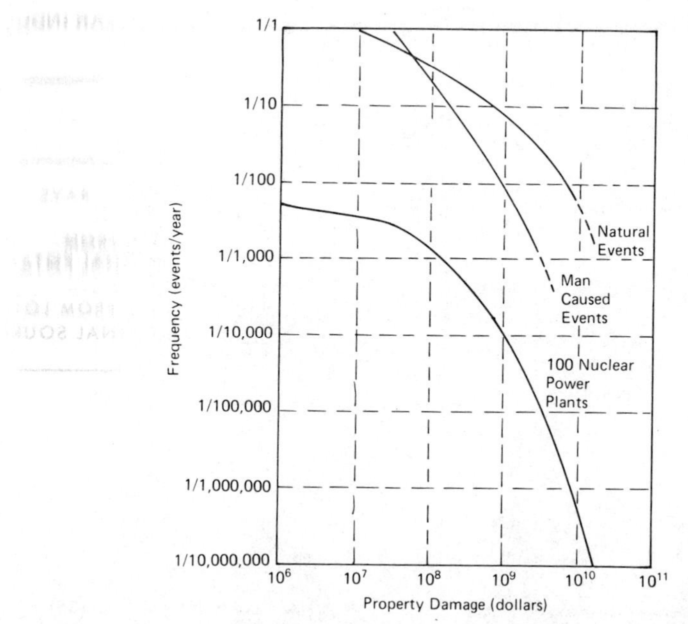

Fig. 24.37 Frequency of property damage due to natural and man-caused events. (Notes: 1. Property damage due to auto accidents not included; auto accidents cause about $15 billion damage each year; 2. Approximate uncertainties for nuclear events are estimated to be represented by factors of 1/5 and 2 on consequence magnitudes and by factors of 1/5 and 5 on probabilities; 3. For natural and man-caused occurrences the uncertainty in probability of largest recorded consequence magnitude is estimated to be represented by factors of 1/20 and 5. Smaller magnitudes have less uncertainty.)

(4) Preserve important historic, cultural, and natural aspects of our national heritage and maintain, wherever possible, an environment which supports diversity and variety of individual choice;

(5) achieve a balance between population and resource use which will permit high standards of living and a wide sharing of life's amenities; and

(6) enhance the quality of renewable resources and approach the maximum attainable recycling of depletable resources.

These goals bring to focus interfaces between private and public needs, technology and quality-of-life concepts, industrial and ecological systems, and national and international economics. The nuclear industry is in the forefront of the studies in progress to assure these goals.

REFERENCES

1. *Reactor Safety Study*, WASH1400 (NUREG 75/014) U.S. Nuclear Regulatory Commission, Washington, DC, 1975.
2. Interagency Review Group Report.
3. *Overview of the Internation Fuel Cycle Evaluation*, INFCE/PC/2/9, International Atomic Energy Agency, 1980.
4. Hubbert, M. K., "Energy Resources," NAS/NRC Report 1000-D, National Academy of Science, Washington, DC, 1962.
5. Parent, J., *A Survey of the U.S. and Total World Fossil and Uranium Fuel Resources 1977*, Institute of Gas Technology, Chicago, ILL, 1979.
6. *Monthly Energy Review*, DOE/EIA 0035/12(79) Department of Energy, Washington, DC, 1979.
7. *Technical Assessment Guide*, EPRI PS-1201-SR, Electric Power Institute, Palo Alto, CA, 7/79.
8. Glasstone, S., *Sourcebook on Atomic Energy*, 2nd ed., New York, Van Nostrand Reinhold, 1958.
9. Glasstone, S. and Edlund, M. C., *The Elements of Nuclear Theory*, New York, Van Nostrand Reinhold, 1952.

10. Glasstone, S. and Lovberg, R. H., *Controlled Thermonuclear Reactions*, New York, Van Nostrand Reinhold, 1960.
11. Ng. R. - Personal communication.
11a. Gough, W. C. and Eastland, B. J., *Sci. Am.*, **224** no. 2, 50–64, Feb., 1971.
12. Rawls, J. Ed., *Status of Tokamak Research*, DOE/ER-0034, Department of Energy, Washington, DC, October, 1979.
13. Glasstone, S., *The Effects of Nuclear Weapons*, U.S. Government Printing Office, Washington, DC, 1977.
14. Weber, C. E. et al, "Fuel Element Design" section in *Reactor Handbook*, New York, John Wiley and Sons, 1964.
15. Cohen, P., *Water Coolant Technology of Power Reactors*, Gordon and Breach, 1969.
16. Blake, M., "Three Mile Island One Year Later," *Nuclear News*, March, 1980, vol. **23**, no. 3, p. 51–54.
17. Aubert, M., "Phenix Fast Breeder Reactor Power Plant," *Nuclear Engineering, Int.* v. **19**, p. 563.
18. Rippon, S., "Prototype Fast Breeder Reactor Operating in Europe and the U.S.S.R," *Nuclear Engineering International*, v. **20**, p. 545.
19. Remshaw, R. and Smith, E., "The Standard Candu 600 MWe Nuclear Plant," *Nuclear Engineering International*, June, 1977.
20. *Jane's Fighting Ships*, Moore, J., ed., *Jane's Yearbooks*, Paulton House, London Nl 7LW, England, 1978.
21. Hewlett, R., Duncan, F., *Nuclear Navy*, University of Chicago Press, 1974.
22. Everhart, D., Personal Communication, U.S. Department of Energy, Grand Junction, Co., 1979.
23. Larson, W., "Uranium In Situ Leach Mining in the United States," U.S. Department of Interior, Bureau of Mines Circular 8777.
24. Clegg, J. W. and Foley, D. D., *Uranium Ore Processing*, Addison-Wesley, Reading, Mass., 1958.
25. Leist, N. B., "The Beneficiation and Refining of Uranium Concentrates," USAEC Report NLCO 1067, U.S. Atomic Energy Comm., Washington, DC, 1970.
26. Ruch, W. C., "Production of Pure Uranium Hexafluoride From Ore Concentrates," Allied Chemical Corp., Engineers Joint Council, New York, April, 1959.
27. Mantz, E. W., "Production of Uranium Tetrafluoride and Uranium Metal,"USAEC Report NLCO 1068, U.S. Atomic Energy Comm., Washington, DC, 1970.
28. "AEC Gaseous Diffusion Plant Operations," USAEC Report ORO 658, U.S. Atomic Energy Comm., Washington, DC, 1968.
29. Olander, D., "The Gas Centrifuge," *Scientific American*, v. **239**, No. 2, August, 1978.
30. Bechner, E. et al, "Technological Aspects of the Separation Nozzle Process," *AIChE Symposium Series*, v. 72, no. 169, 1976.
31. Littman, A. and Thomasson, N., "Environmental Development Plan for Advanced Isotope Separation," U.S. Department of Energy, Washington, DC, 1979.
32. "The Use of Thorium in Nuclear Power Reactors," USAEC Report WASH 1097, U.S. Atomic Energy Comm., Washington, DC, 1969.
33. Eister, W., Stoughton, R., Sullivan, W., "Processing of Nuclear Reactor Fuel, in *Principles of Nuclear Reactor Engineering*, ed., Glasstone, S., Van Nostrand, 1955.
34. Long, J. T., *Engineering for Nuclear Fuel Reprocessing*, New York, Gordon and Breach.
35. "Aqueous Processing for LMFBR Fuels-Technical Assessment and Experimental Program Definition," USAEC Report ORNL 4436, U.S. Atomic Energy Comm., Washington, DC, 1970.
36. Eister, W., "Material Considerations in Radioactive Waste Storage," p. 6, *Nuclear Technology*, vol. **32**, no. 1, January, 1977. Table of Isotopes: Heath, R., B-4 to 96.
37. "International Nuclear Fuel Cycle Evaluation," Waste Management INFCE/PC/2/7, Internation Atomic Energy Commission, January, 1980, Vienna.
38. Heath, R., "Table of Isotopes," *Handbook of Chemistry and Physics*, 45th edition, The Chemical Rubber Co., Cleveland, 1964.
39. Friedlander, G., Kennedy, J. W. and Miller, J. N., *Nuclear and Radiochemistry*, New York, John Wiley and Sons, 1955.
40. Eister, W. et al, "Radioisotope Production in the U.S.," *Radioisotope Production Study*, Sao Paulo, Brazil, LAEA-124, International Atomic Energy Agency, Vienna, 1970.
41. Crandall, J. L., "The Savannah River High Flux Demonstration," USAEC Report DP-999, U.S. Atomic Energy Comm., Washington, DC.
42. Eister, W. et al, "Radioisotope Generators in America" (see above).
43. Eister, W. et al, "Radiopharmaceuticals and Short-lived Radioisotopes" (see above).
44. Code of Federal Regulations, Title 10, Energy, Chapter 1-Nuclear Regulatory Commission, Washington, DC, January 1, 1977.
    Part 20: Standards for Protection Against Radiation
    Part 40: Licensing of Source Material

Part 31: General Licenses for Byproduct Material
Part 50: Licensing of Production and Utilization Facilities
Part 60: Licensing of Repositories for High Level Radioactive Waste
Part 70: Special Nuclear Material
Part 71: Packaging of Radioactive Material for Transport and Transportation of Radioactive Material Under Certain Conditions
Part 73: Physical Protection of Plants and Materials
Part 100: Reactor Site Criteria.

45. Lewis, H., "The Safety of Fission Reactors," *Scientific American*, v. **242**, no. 3, March, 1980.
46. Barbler, M., "Induced Radioactivity," CERN, Geneva, New York, John Wiley & Sons, 1969.
47. Groh, H. J. and Schlea, C. S., *Progress in Nuclear Energy, Process Chemistry*, vol. **4**, Oxford, Pergamon Press, 1970.
48. Leuze, R. E. and Lloyd, M. H., *Progress in Nuclear Energy, Process Chemistry*, vol. **4**, Oxford, Pergamon Press, 1970.

25

Synthetic Organic Chemicals

William H. Haberstroh* and Daniel E. Collins*

Synthetic organic chemicals can be defined as derivative products of naturally occurring materials (petroleum, natural gas, and coal) which have undergone at least one chemical reaction, such as oxidation, hydrogenation, halogenation, sulfonation, or alkylation.

The volume of synthetic organic chemicals increased from 17 billion pounds in 1949 to more than 186 billion pounds in 1978. The production for the past three decades is shown in Fig. 25.1. It can be seen that the effect of the 1974-75 recession on output of synthetic organic chemicals is reflected very clearly. Much of the phenomenal growth overall reflects replacement of "natural" organic chemicals up to the mid-1960's. Since that time, growth for synthetic materials has been dictated by the expansion of present markets and development of new organic chemical end uses. However, the future will show a slow return to renewable raw material sources such as ethanol from sugar sources and corn, methanol from waste paper, the recycling of garbage, and so on. These materials will become the chemical building blocks of the future.

More than 2700 organic chemical products are derived principally from petro-chemical sources. These are commercially produced from five logical starting points. Consequently, this chapter has been subdivided into five major raw material classifications: methane, ethylene, propylene, C_4 and higher aliphatics, and aromatics.

CHEMICALS DERIVED FROM METHANE

It has been stated that every synthetic organic chemical listed in Beilstein can be made in some way or other starting with methane. This section, however, deals only with the relatively small number which can be made economically and which are useful enough to warrant large volume production. A diagram of the principal materials covered is shown in Fig. 25.2.

Synthesis Gas

The most important route for the conversion of methane to petrochemicals is via either

*Dow Chemical Co., Midland, Michigan

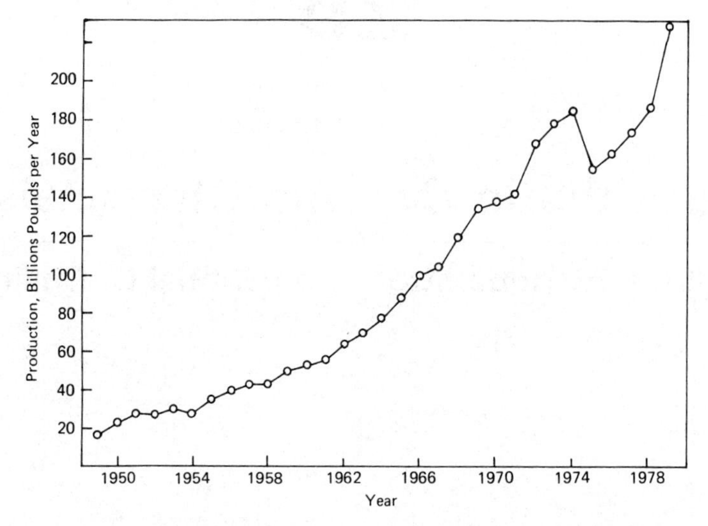

Fig. 25.1 Total production of synthetic organic chemicals. (*U.S. International Trade Commission.*)

hydrogen, or a mixture of hydrogen and carbon monoxide. This latter material is known as 'synthesis gas." The manufacture of carbon monoxide-hydrogen mixtures from coal was first established industrially by the well-known water-gas reaction

$$C + H_2O \longrightarrow CO + H_2$$

Two important methods are presently used to produce the gas mixture from methane. The first is the methane-steam reaction, where methane and steam at about 900°C are passed

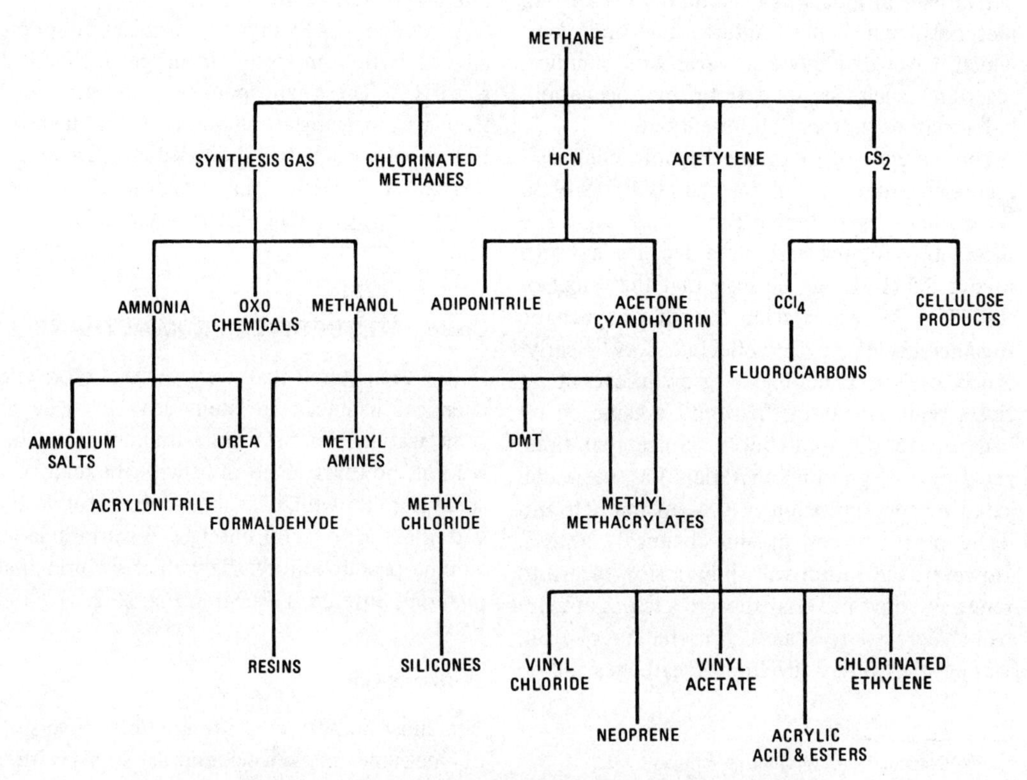

Fig. 25.2 Some important synthetic chemicals derived from methane.

through a tubular reactor packed with a promoted iron oxide catalyst. Two reactions are possible, depending on the conditions:

$$CH_4 + H_2O \longrightarrow CO + 3H_2$$

$$CO + H_2O \longrightarrow CO_2 + H_2$$

The second commercial method involves the partial combustion of methane to provide the heat and steam needed for the conversion. Thus the reaction can be considered to take place in at least two steps. The combustion step

$$CH_4 + 2O_2 \longrightarrow CO_2 + 2H_2O$$

followed by the reaction step

$$CH_4 + CO_2 \longrightarrow 2CO + 2H_2$$

$$CH_4 + H_2O \longrightarrow CO + 3H_2$$

The process is usually run with nickel catalysts in the temperature range of 800–1000°C. Steam reforming is usually used on the lighter feed stocks, while partial oxidation is used for the heavier fractions.

Synthesis gas is the starting material for the manufacture of ammonia and its derivatives, and also for methanol, as well as for other oxo-synthesis processes. It is also a source of carbon monoxide in the manufacture of such chemicals as acetic acid and acrylates, and a source of hydrogen for various petroleum refining processes. However, the shortages and corresponding high prices of natural gas and naphtha have generated interest in reconsidering other synthesis gas feedstocks, such as coal and residual oil.

Ammonia. Ammonia derived from petroleum and natural gas sources accounts for almost all of the 35 billion pounds produced annually. Consequently, it can be termed the number one petrochemical in volume. About one-third goes directly into fertilizers, and the rest is used to produce such chemicals as ammonium nitrate, urea, nitric acid, acrylonitrile, ammonium sulfates and phosphates, caprolactam, adipic acid, and organic amines. A detailed description of the ammonia processes is given in Chapter 5.

Methanol. Before 1926, all American methanol was obtained commercially as a by-product of the wood distillation process (wood alcohol). That year, however, marked the first appearance of German synthetic methanol. Presently, only a negligible amount of the seven billion pounds produced comes from wood.

The methyl alcohol synthesis is well known. It resembles the synthesis of ammonia in that the catalysts operate only at high-temperature levels, and conversion and equilibrium are greatly assisted by high-pressure operation. The industrial reaction conditions are pressures of 250–350 atmospheres and temperatures in the range of 300–400°C. The catalysts employed are based on zinc oxide, which is mixed with other oxides to provide temperature resistance. Variations between synthetic methanol plants are quite similar to those between synthetic ammonia plants. In fact, many ammonia operations are designed so that methanol could also be produced in them.

Recently there has been increased interest in the U.S. and other parts of the world in the use of medium and low pressure processing, both for new plants, and for conversion of the older ones.

Methanol can be readily made from any carbon containing compound, including coal, lignite, natural gas, heavier oils, wood, and agricultural wastes. Since methanol is easily handled, it could potentially be produced on a large scale from surplus natural gas in areas such as the Middle East, and shipped to the industrialized nations for conversion. Such a scheme would involve extremely large methanol plants. If a low-cost process for converting coal or cellulose is commercialized, it would also give impetus to its use as a starting material for synthetic organic chemicals.

Methanol is used as a solvent, antifreeze, refrigerant, and chemical intermediate. Significant derivatives include formaldehyde, methyl methacrylate, methyl amines, methyl halides, acetic acid, and methyl *t*-butyl ether (MTBE). Formaldehyde accounts for almost one-half of the methanol usage. Use in methyl chloride and methyl esters is declining, while its use in acetic acid and methyl methacrylate is increasing. The biggest growth in the near future should occur in the amount used to produce the fuel-additive MTBE.

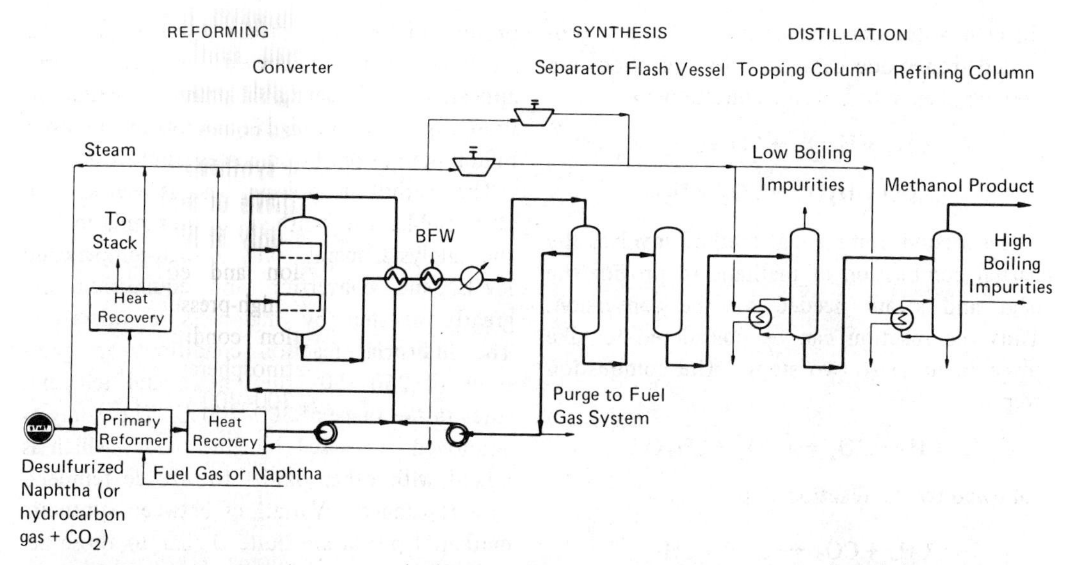

Fig. 25.3 Methanol process (low pressure). (*Hydrocarbon Processing, 58, no. 11, 191 (1979)*; *(copyright 1979 by the Gulf Publishing Co.)*

Formaldehyde. Formaldehyde may be made from methanol either by catalytic vapor-phase oxidation,

$$CH_3OH + \tfrac{1}{2}O_2\ (\text{air}) \longrightarrow CH_2O + H_2O$$

or by a combination oxidation-dehydrogenation process

$$CH_3OH \longrightarrow CH_2O + H_2$$

It can also be produced directly from natural gas, methane, and other aliphatic hydrocarbons, but this process yields mixtures of various oxygenated materials.

Since both gaseous and liquid formaldehyde readily polymerize at room temperature, it is not available in the pure form. It is sold instead as a 37 percent solution in water, or in the polymeric form as paraformaldehyde $[HO(CH_2O)_nH$, where n is between 8 and 50] or as trioxane $[(CH_2O)_3]$. The largest end use of formaldehyde is in the field of synthetic resins, either as a homopolymer or as a copolymer with phenol, urea, or melamine. It is also reacted with acetaldehyde to produce pentaerythritol $[C(CH_2OH)_4]$ which finds use in polyester resins. Two smaller-volume uses are in urea-formaldehyde fertilizers and hexamethylenetetramine, the latter being formed by condensation with ammonia.

Methyl Methacrylate. Methyl methacrylate is formed in a three-step process from methanol, acetone, and HCN.

$$CH_3\overset{\overset{O}{\|}}{C}CH_3 \xrightarrow{HCN} CH_3\underset{\underset{OH}{|}}{\overset{\overset{CH_3}{|}}{C}}CN \xrightarrow{H_2SO_4}$$

$$CH_2{=}\underset{\underset{CH_3}{|}}{C}CONH_2 \cdot H_2SO_4 \xrightarrow{CH_3OH}$$

$$CH_2{=}\underset{\underset{CH_3}{|}}{C}COOCH_3 + NH_4HSO_4$$

While this is the major process in operation, there have been reports of a route involving the oxidation of isobutylene to methacrylic acid.

Production of methyl methacrylate now totals more than 800 million pounds. The largest part of this goes into cast sheet, where the clarity and resistance of poly(methyl methacrylate) are desirable. Other uses are in surface-coating resins and molding powders.

Acetic Acid. The production of acetic acid from methanol is a relatively new process. One method involves reacting methanol first with hydrogen iodide and then with a complex containing cobalt or rhodium. This product is then reacted with CO and then with water. The final reaction produces acetic acid and regenerates the HI and metal complex. About one-third of the acetic acid capacity is based on this route, and it is the basis for almost all new plants.

Methyl t-Butyl Ether. MTBE is by far the fastest growing chemical derivative of methanol. Starting with essentially zero production in 1978, it will consume almost 10% of the available methanol, or 120 million gallons, in 1980. This promising compound is an antiknock additive or octane improver for unleaded gasolines. It is easily produced by passing methanol and isobutylene over an ion exchange catalyst under mild operating conditions. Since saturated hydrocarbons, butylenes, and butadiene are inert under these conditions, a mixed C_4 feed stream can be used and the isobutylene selectively removed by the reaction.

Other Compounds From Methanol. Other new or potential uses of methanol are in making chlorinated methanes, acetic anhydride, bioprotein, and ethylene. A large part of the chlorinated methane production is now based on methanol rather than on methane. The process is described later in this chapter. A new plant has been announced which will produce acetic anhydride from methanol and CO. It is a several-step process and has the advantage of being based on coal instead of petroleum. Single cell protein can be produced by converting methanol and ammonia in the presence of selected micro-organisms and mineral salts. The product can be used directly for fodder or further enriched. A process has been proposed for converting methanol to ethylene and thereby permit a large number of chemicals to become methanol-based. It involves making dimethyl ether from methanol over a fixed-bed catalyst and then converting the ether to ethylene.

Oxo Chemicals. The oxo chemicals are compounds–primarily C_4 and higher alcohols–made by the so-called "oxo" process. This process is a method of reacting olefins with carbon monoxide and hydrogen to produce aldehydes containing one more carbon atom than the olefin; these in turn are converted into alcohols. The earliest reaction studied used ethylene to produce both an aldehyde and a ketone. Thus the name "oxo," which was adapted from the German *oxierung*, meaning ketonization. However, even though other names such as hydroformylation would much more accurately describe the process, the term oxo appears too deeply entrenched to be replaced.

A flow sheet of a typical process is shown in Figure 25.4. The steps involved in the reaction are

$$RCH{=}CH_2 + CO + H_2 \xrightarrow[250^\circ C]{Co \; 3000 \text{ psi}} \begin{Bmatrix} RCH_2CH_2CHO \\ RCH(CHO)CH_3 \end{Bmatrix}$$

$$\begin{Bmatrix} RCH_2CH_2CHO \\ RCH(CHO)CH_3 \end{Bmatrix} + H_2 \xrightarrow[200^\circ C]{3000 \text{ psi}} \begin{Bmatrix} RCH_2CH_2CH_2OH \\ RCH(CH_2OH)CH_3 \end{Bmatrix}$$

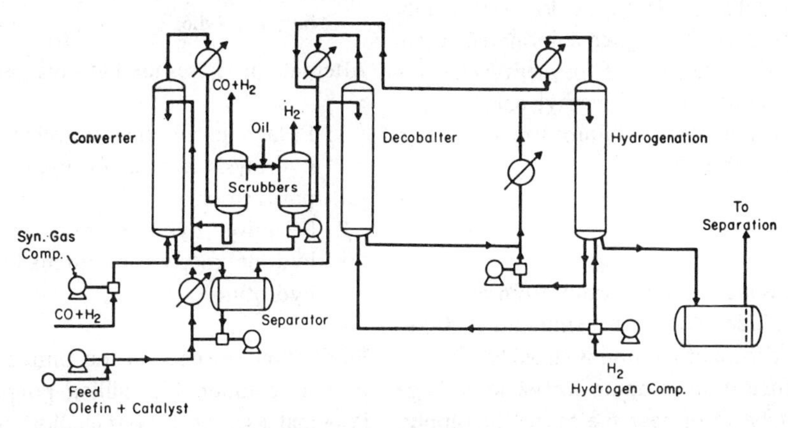

Fig. 25.4 The Oxo process. *(Pet. Ref., 38, no. 11, 280 (1959); copyright by the Gulf Publishing Co.)*

The cobalt catalyst used under these conditions is in the form of dicobalt octacarbonyl and cobalt hydrocarbonyl.

At the present time, the plant capacity in the United States amounts to more than 1.1 billion pounds per year of oxo chemicals. These include such products as *n*-butanol, isobutanol, propionaldehyde, butyraldehydes, butyronitriles, isooctyl alcohol, decyl alcohol, and tridecyl alcohol.

Chlorinated Methanes

The chlorination of methane can be carried out thermally (450°C) to produce methyl chloride (CH_3Cl), methylene chloride (CH_2Cl_2), chloroform ($CHCl_3$), and carbon tetrachloride (CCl_4). This route accounts for about one-third of the present chlorinated methanes capacity. Most of the methyl chloride, methylene chloride, and chloroform, however, is based on the reaction of methanol with hydrogen chloride. Most of the carbon tetrachloride is a by-product in the manufacture of perchloroethylene. Some carbon tetrachloride is also produced from carbon disulfide.

About one billion pounds of chlorinated methanes are now produced. In order of importance they are carbon tetrachloride, methylene chloride, methyl chloride, and chloroform. Carbon tetrachloride is used mainly for fluorocarbons 11 and 12. However, this demand is expected to decline because of air pollution controls. Methylene chloride goes into paint removers, blowing agents, and aerosol propellants. Methyl chloride is mainly used to produce the higher chlorinated methanes. Another growing use for methyl chloride is in the manufacture of methyl chlorosilanes for silicone polymers. Chloroform is used to make fluorocarbon 22.

Acetylene

Acetylene is made commercially from methane, calcium carbide, or in the manufacture of olefins. The choice of method is strongly affected by the difficulty in shipping acetylene, so large users must be at or near the source of supply. The calcium carbide process, which requires a supply of cheap electric power, was the sole source of acetylene until about 1950. Since then its use has dropped drastically, and now almost 85% of the acetylene produced for chemical use is derived from hydrocarbons. The manufacture of acetylene is described in Chapters 14 and 17.

Acetylene has long been a valuable building block in the chemical industry. But its use as a chemical feed has decreased over the last 15 years to a present consumption of only 300 million pounds. The decline is due mainly to a switch from acetylene-based to ethylene-based routes for vinyl chloride and vinyl acetate monomers. Propylene has replaced it in the manufacture of acrylonitrile and acrylic acid and esters, and neoprene is now made from butadiene. The major remaining use is in the production of acetylinic chemicals such as butanediol, vinyl ethers, and propargyl alcohol. Smaller users are vinylidene fluoride and acetylene black.

1,4-Butanediol. 1,4-Butanediol is currently produced in the United States using the Reppe process based on acetylene and formaldehyde as shown below. This reaction is carried out under high pressure using a copper-bismuth catalyst. The resulting butynediol is hydrogenated using a Raney nickel catalyst.

$$\underset{\text{acetylene}}{HC{\equiv}CH} + \underset{\text{formaldehyde}}{2HCHO} \longrightarrow \underset{\text{2-butyne-1,4-diol}}{HOCH_2C{\equiv}CCH_2OH}$$

$$\underset{\text{2-butyne-1,4-diol}}{HOCH_2C{\equiv}CCH_2OH} + 2H_2 \longrightarrow \underset{\text{1,4-butanediol}}{HO(CH_2)_4OH}$$

Alternate processes used in Japan are:

1. Butadiene chlorination followed by caustic hydrolysis and finally hydrogenation and purification.
2. Selective hydrogenation of maleic anhydride to give 1,4-butanediol and tetrahydrofuran.

Production capacity in the United States is reported at about 300 million pounds per year. Principal uses of 1,4-butanediol are the manufacture of tetrahydrofuran, 54%; the manufacture of butyrolactone, 23%; polybutylene terephthalate resins, 10.5%; polyurethanes, 10.5%; and 2% to other uses. It is estimated that production is at approximately 80% of the above reported capacity.

Hydrogen Cyanide

Hydrogen cyanide is prepared, as shown in Fig. 25.5, by passing a mixture of air, ammonia, and natural gas over a platinum catalyst. The converter is operated at a temperature of about 1800°F, and care must be taken to minimize the decomposition of the ammonia and methane, as well as the oxidation of methane to carbon monoxide and hydrogen. The effluent gases are cooled, washed with dilute sulfuric acid, and then passed through a column where the hydrogen cyanide is absorbed in water. This is concentrated by distillation, and an inhibitor is added to prevent polymerization. Although all new plants follow the methane-ammonia route, HCN can also be produced from coke-oven gas, from sodium and calcium cyanides, and by the decomposition of formamide.

Because of the safety problems, most production is captive to avoid the need for shipment. About one-third of the HCN goes into the production of acetone cyanohydrin, while almost 45 percent is used to make adiponitrile. Other uses are for chelating agents, sodium cyanide, and cyanuric acid.

The advent of the propylene-ammonia process for acrylonitrile has had an interesting effect on this material. Several years ago acrylonitrile manufacture was a major consumer of HCN. Now almost 25 percent of our HCN is produced as a by-product in acrylonitrile manufacture.

Carbon Disulfide

Carbon disulfide is made by the catalytic reaction of methane and sulfur vapor. Production is about 200 million pounds, with the largest portion going to the manufacture of rayon and cellophane. The other major use is production of carbon tetrachloride.

CHEMICALS DERIVED FROM ETHYLENE

Ethylene continues to far surpass all other hydrocarbons both in volume and in diversity of commercial use. In the whole field of petrochemicals, it is exceeded in tonnage only by synthetic ammonia. Consumption of ethylene has grown remarkably in the last 40 years. In 1940, 300 million pounds were produced, mostly for ethanol and ethylene oxide. The war time demand for styrene and the postwar impact of polyethylene aided in causing this figure to swell to almost 5 billion pounds in 1960. A boom in polyethylene use and strong growth in ethylene dichloride and ethylene oxide expanded ethylene production to over 16 billion pounds in 1968 and over 18 billion pounds in 1970. This volume expanded to 27 billion pounds in 1978.

The major consumers of ethylene in 1978 are

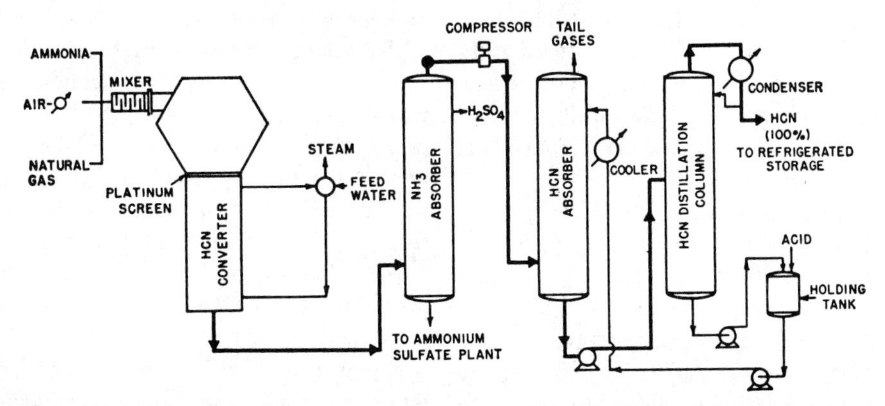

Fig. 25.5 The hydrogen cyanide process. (*Reprinted from Ind. Eng. Chem.,* **51**, *no. 10, 1235 (1959); copyright 1959 by the American Chemical Society and reprinted by permission of the copyright owner.*)

TABLE 25.1 Ethylene Derivatives

	Ethylene Consumed	
	(million lbs)	*(% of total)*
Low density polyethylene	7,078	26.2
Ethylene oxide	5,012	18.5
High density polyethylene	4,849	17.9
Ethylene dichloride	3,945	14.6
Ethylbenzene	2,264	8.4
Ethylene oligomers	992	3.7
Ethanol	798	3.0
Acetaldehyde	675	2.5
Vinyl acetate	563	2.1
Ethyl chloride	260	0.9
Ethylene-propylene elastomers	224	0.8
Propionaldehyde	135	0.5
Ethylene dibromide	45	0.2
Other	200	0.7
	27,040	

[a]U.S Intl. Trade Comm, 1978.
[b]*Chemical Economics Handbook*, S.R.I. Menlo Park, Calif.

shown in Table 25.1. A chart showing major ethylene derivatives is shown in Figure 25.6.

Polyethylene

Polyethylene has shown a spectacular growth, accounting for only 4 percent of total ethylene consumption in 1950, and almost 25 percent ten years later. In 1961, polyethylene surpassed ethylene oxide as the principle ethylene consumer. In 1978, polyethylene and ethylene copolymer manufacture consumed nearly 45 percent of all ethylene produced that year.

Low Density Polyethylene. Low density polyethylene (LDPE) is the most widely used thermoplastic in the world. It is the largest consumer of ethylene, consuming 1.08 billion pounds of ethylene in 1978. Low density polyethylene products are used for packaging, communication, transportation, electric transmission, and many applications around the home.

The largest use is in films, where low cost and favorable properties lead to packaging of produce, frozen foods, meats, baked goods, and many others. These LDPE films find large industrial uses for stretch and shrink wrapping of shipping trays and pallets. Other uses are for garment bags, soft goods containers, trash bags, and for moisture, dirt or light barriers, or coverings in building and pipeline construction, as well as in agriculture.

LDPE products made by injection-molding consume the second largest quantity of polyethylene polymer. These items include toys, housewares, molded furniture and furniture parts, and many other such applications.

Extrusion coatings of LDPE are used in paper coating and paperboard, particularly where heat sealing properties of LDPE can be used to advantage in consumer packaging and similar uses.

Cable and wire insulation consume a considerable quantity of LDPE. It is now the main insulation for communication cables and finds extensive application in power-cable and building-wire insulation.

LDPE is produced primarily under high pressure by free-radical polymerization of high purity ethylene. Although LDPE is sold primarily as a commodity homopolymer, copolymers with ethyl acrylate, vinyl acetate, or acrylic acid yield products with increased flexibility, clarity, and impact strength.

High Density Polyethylene. High density polyethylene (HDPE) is a stronger, tougher, more rigid plastic material than LDPE. Thus, it is chosen over LDPE in applications requiring these properties. The volume of ethylene consumed for HDPE is surpassing that used for

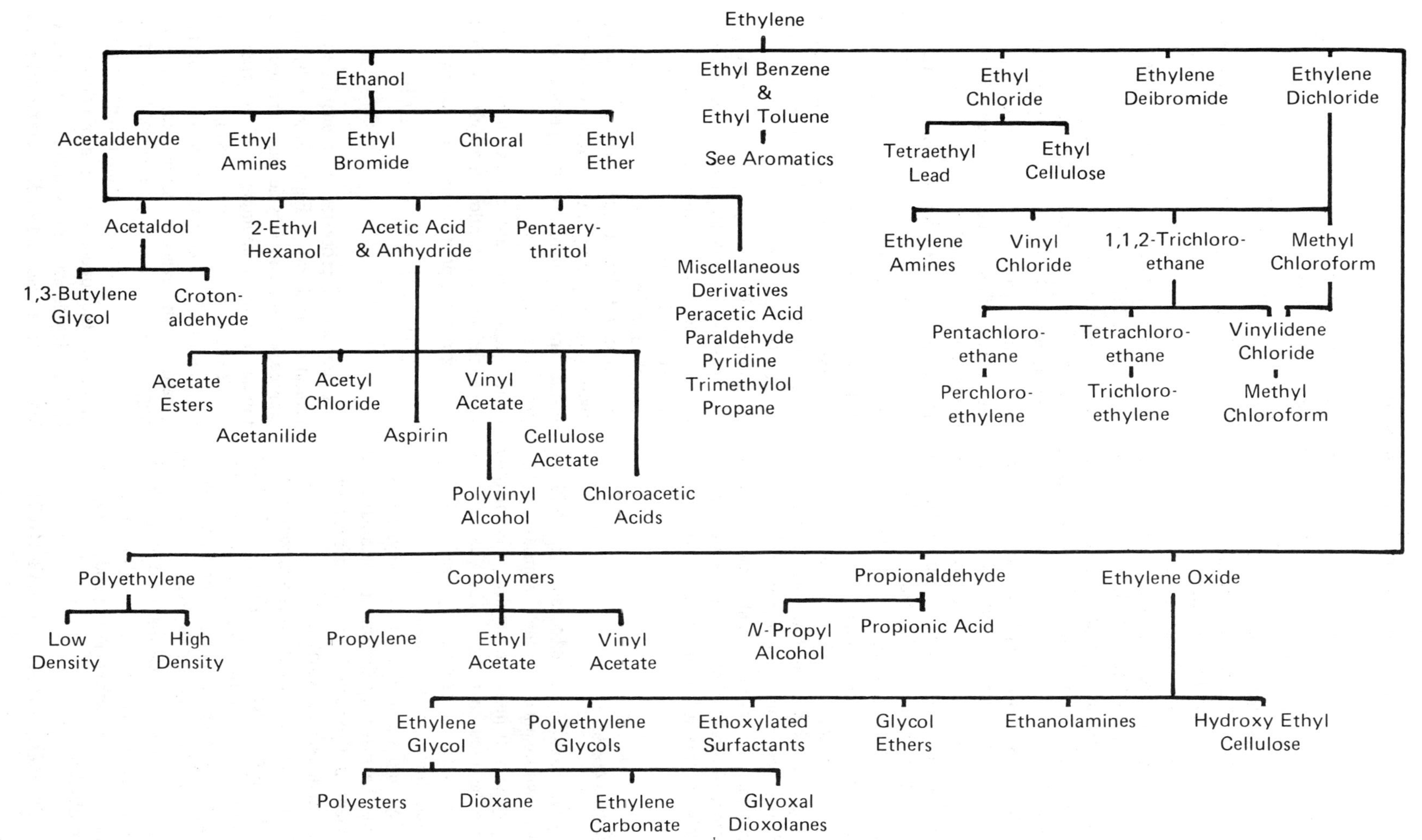

Fig. 25.6 Major ethylene derivatives.

ethylene oxide—the second largest consumer of ethylene after LDPE. Ethylene consumed for HDPE in 1978 was 4.85 billion pounds.

The blow-molded plastics market is dominated by HDPE due to its favorable properties of toughness, rigidity, and chemical resistance, as well as low cost. Bottles or cans for holding milk, bleach, detergents, and antifreeze consume about half of the HDPE going into containers. Significant growth is likely to occur in drain pipes, drums, and fuel tanks for autos and trucks.

HDPE is used in injection molding in applications where heat resistance, rigidity, and toughness are required. Such end uses as pallets, cases, packaging containers of many sorts, tote bins, and so on, comprise these applications. Major growth will occur in packaging film as well as wire and cable insulation.

HDPE is manufactured by polymerization of high purity ethylene in low-pressure processes utilizing transition metal catalysts. Specialty copolymers containing small amounts of 1-hexene, 1-butene, or propylene are utilized.

Polyethylene is covered in detail in Chapter 10.

Ethylene Oxide

Ethylene oxide was discovered in 1859 by Wurtz, who named it as such because of certain analogies with inorganic oxides. The method which he used to make it was what is today known as the chlorohydrin process. He considered that direct oxidation was an impossibility, and stated flatly that "ethylene oxide cannot be made by the direct combination of ethylene and oxygen." It took almost eighty years to disprove this statement.

There are two basic processes presently used in the production of ethylene oxide from ethylene; the chlorohydrin process and the catalytic oxidation process. The chlorohydrin process is the older of the two. It was first commercialized in Germany during World War I. It is based on the addition of hypochlorous acid to ethylene to produce ethylene chlorohydrin,

$$CH_2{=}CH_2 + HOCl \longrightarrow CH_2ClCH_2OH$$

which in turn is dehydrochlorinated to produce ethylene oxide.

$$2CH_2ClCH_2OH + Ca(OH)_2 \longrightarrow CaCl_2 + 2H_2O + 2\underset{\diagdown\,O\,\diagup}{CH_2{-}CH_2}$$

A flow sheet of this process is shown in the first part of Fig. 25.7. Ethylene, chlorine, and water are fed into the bottom of a large acid-proof brick-lined tower at somewhere below 50°C. The water and chlorine form hypochlorous acid, which then reacts rapidly with the ethylene. The dilute solution emerging from the tower contains about 5 percent chlorohydrin. The major side reaction is the formation of ethylene dichloride. The solution next passes to a hydrolyzer where the chlorohydrin is treated with either slaked lime or caustic soda to produce the oxide. The crude ethylene oxide contains about 10 percent ethylene dichloride which is removed by distillation. Ethylene oxide molar yield is at least 80 percent based on ethylene. Most of the chlorine used ends up as an aqueous salt brine that must be decontaminated for disposal or recycle. The chlorohydrin process accounts for less than 5 percent of ethylene oxide capacity today, but many former ethylene oxide-chlorohydrin plants are now used for the production of propylene oxide.

The most important process now used involves the direct oxidation of ethylene with air or oxygen in the presence of a silver catalyst.

$$2CH_2{=}CH_2 + O_2 \longrightarrow 2\underset{\diagdown\,O\,\diagup}{CH_2{-}CH_2}$$

A number of processes are available, and a typical one is shown in Figure 25.8. Ethylene compressed air, and recycle gases are fed to a tubular reactor containing a silver catalyst. The oxygen and ethylene concentrations are maintained at a low level to avoid explosion hazards. The reaction temperature is 250-300°C with a pressure of 120-300 psi. Two competing side reactions which must be minimized are the total combustion of ethylene to carbon dioxide and the isomerization of ethylene oxide to acetaldehyde. Some ethylene

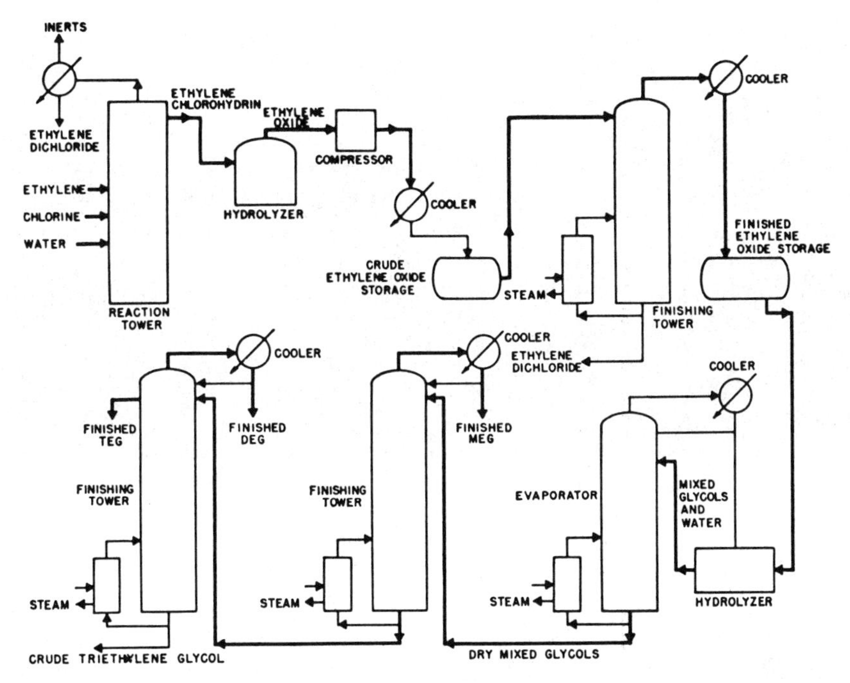

Fig. 25.7 Ethylene glycols and ethylene oxide by the chlorohydrin process. *(Ind. Eng. Chem., **51**, no. 8, 896 (1959); copyright 1959 by the American Chemical Society and reprinted by permission of the copyright owner.)*

oxide direct-oxidation plants use purified oxygen instead of air as the oxidizing agent.

Plants utilizing air require additional investment for the purge reactors and associated absorbers and for energy recovery from the vent gas. This capital investment is offset by the need, in the oxygen process, for an oxygen production unit and a carbon dioxide removal system. In general, it can be said* that the oxygen-based unit is lower in capital for smaller plants up to 100 million pounds per year even including the air-separation unit. But for plants larger than 150 million pounds per year, the capital for the air process is less unless oxygen is available from a very large air separation unit serving other uses as well. A flow sheet of an oxygen-based ethylene oxide process is shown in Figure 25.9. Major uses of ethylene oxide are given in Table 25.2, and several of these are discussed in the following paragraphs.

Ethylene Glycols. Ethylene glycol can be

*Kirk-Othmer, Encyclopedia of Chemical Technology, 3rd ed., Vol. 9, pp. 432–471 (John Wiley & Sons) (1980).

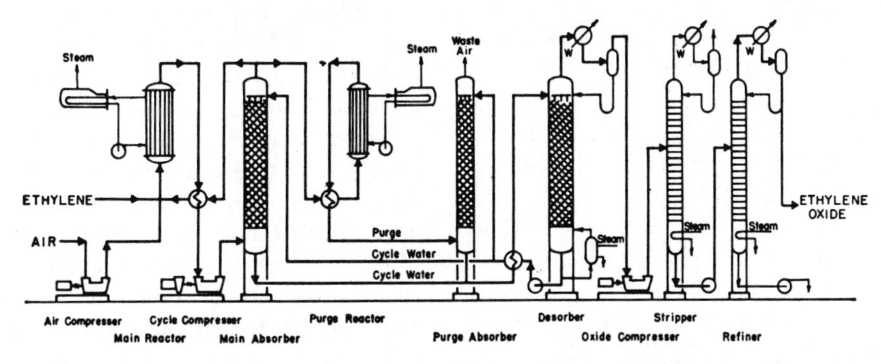

Fig. 25.8 Ethylene oxide by direct oxidation. *(Pet. Ref., **38**, no. 11, 248 (1959); copyright 1959 by Gulf Publishing Co.)*

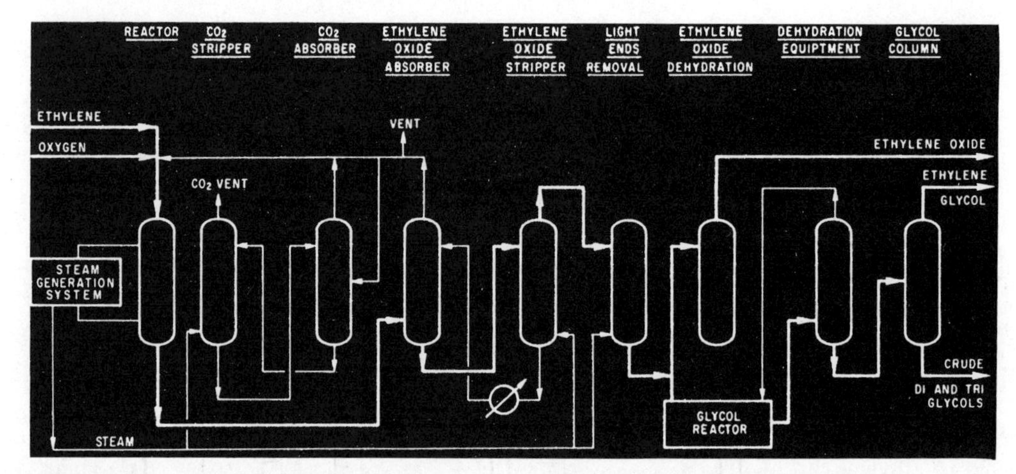

Fig. 25.9 Ethylene glycols and ethylene oxide by direct oxidation. (*Hydrocarbon Processing,* **50**, *no. 11, 158 (1971); copyright 1971 by Gulf Publishing Co.*)

prepared directly by the hydrolysis of ethylene chlorohydrin, but the indirect hydrolysis via ethylene oxide is the preferred method. This is shown in the second part of Figure 25.7. The feed stream consists of ethylene oxide and water. This mixture is fed into the reactor vessel at about 100°C.

$$\overset{O}{\overbrace{CH_2\,CH_2}} + H_2O \longrightarrow HOCH_2CH_2OH$$

By the end of the reaction, the temperature will rise depending on reactor design and configuration. Some diethylene glycol and triethylene glycol are produced by the reaction of ethylene glycol with the oxide. The crude glycol solution is then concentrated in a multiple-effect evaporator. Final separation of the mono, di, and triethylene glycols is accomplished by vacuum distillation.

In order to make a marked improvement in the yield of ethylene glycol from ethylene, a new process has been developed. This is a liquid-phase oxidation of ethylene in the presence of acetic acid using a tellurium-bromide-lithium catalyst. Ethylene glycol acetate esters are formed, then in a separate reaction step the esters are hydrolyzed to ethylene glycol. A large (800 $\overline{M}$ lb/yr) plant utilizing this technology was started up in 1978. It was reported however in late 1979* that the plant was shut down indefinitely due to very difficult corrosion problems. It thus appears that for some time to come ethylene glycol will be derived from ethylene oxide.

Use of ethylene glycol for polyethylene terephthalate has helped obtain a continued healthy growth rate of over 7 percent annually from 1968 to 1978. Polyethylene terephthalate is converted into polyester fibers or films. Polyester films are used primarily for photographic film, magnetic tape, and wherever else dimensionally stable tapes and labels can justify the premium price. As a result, the use of ethylene glycol in polyethylene terephthalate will soon exceed its use in antifreeze. In 1978, antifreeze consumed 48 percent of the

TABLE 25.2 Production of Chemicals from Ethylene Oxide[a,b]

Product	*Output (million lbs)*	*Oxide Requirement (million lbs)*
Ethylene glycol	3866	2899
Surfactants	1087	573
Glycol Ethers	743	384
Ethanolamines	391	313
Diethylene glycol	276	229
Polyglycols	–	132
Triethylene glycol	134	121
Others		265
TOTAL		4936

[a]*Chemical Economics Handbook*, Stanford Research Institute, 1978.
[b]U.S. International Trade Commission, 1978.

***Chemical Marketing Reporter*, August 25, 1980, page 42.

ethylene glycol used; polyester fibers and film consumed 44 percent, while 8 percent went to such uses as alkyd and polyester resins for surface coatings, polyester resins for glass fiber laminates, heat transfer agents in refrigeration and electron tubes, in tracing systems, as an ingredient in de-icing fluid for airport runways, and in solvent uses.

Growth of ethylene glycol uses will probably average at slightly below 5 percent in the next ten years. This slower rate is due to plateauing of antifreeze use due to smaller volumes used in smaller automobiles.

Diethylene glycol has properties similar to those of ethylene glycol. It is mainly produced as a by-product of ethylene glycol manufacture. Its main uses are in unsaturated polyester resins, as a textile lubricant and conditioner, in solvent extraction, and for back blending into antifreeze when the supply is long and specifications will allow.

Triethylene glycol (either a by-product of ethylene glycol manufacture or made by adding ethylene oxide to diethylene glycol), due to its hydroscopicity finds large use as a drying agent for natural gas, as a humectant for tobacco and other materials, and as an intermediate in the manufacture of vinyl plasticizers. Other uses are for polyester resins, polyester polyols for polyurethanes, and as a Udex solvent for extracting aromatic hydrocarbons from mixed hydrocarbons.

Polyethylene glycols are produced by passing ethylene oxide into a small amount of a low molecular weight glycol using a sodium or caustic soda catalyst. The molecular weight of liquid polyglycol products ranges from 200 to 1000. These liquid polyethylene glycols are used as plasticizers, dispersants, lubricants, and humectants. Above a molecular weight of about 1000, the polyglycols are waxy solids, suitable for use as softening agents in ointments, and cosmetics, and/or lubricants.

Surfactants. Ethylene oxide consuming surfactants are primarily non-ionic cyclic and acyclic types. It is estimated that 593 million pounds ethylene oxide were consumed in 1978 to produce1.087 billion pounds of surfactants.

Cyclic nonionic surfactants are primarily those derived from alkyl phenols. These products consumed roughly 25 percent of the ethylene oxide used in surfactants in 1978 and appear to have plateaued or are dropping slowly. Most of the uses are in industrial processing and appear to be well established. Replacement, if accomplished, will likely come from the more biodegradable ethoxymers of mixed linear alcohols.

Acyclic surfactants consuming ethylene oxide are comprised of polyethylene glycol esters, ethoxylated alcohols, polyether polyols, ethoxylated fats and oils, and miscellaneous other ethoxylated products. Most of the growth in recent years has been accounted for by the trend to heavy duty liquid laundry detergents. The linear alcohol ethoxylates are the products with properties to satisfy this use, such as detergent action to effectively remove soil from polyester fabrics, excellent detergent action at cooler wash temperatures, and the ability to dissolve instantly. These products provide the detergency properties without phosphate builders.

Ethanolamines. Ethanolamines are manufactured by reacting ethylene oxide and ammonia. The relative amounts of the three amines will depend primarily on the ammonia to oxide feed ratio.

$$NH_3 \xrightarrow{C_2H_4O} (HOC_2H_4)NH_2 \xrightarrow{C_2H_4O} (HOC_2H_4)_2NH \xrightarrow{C_2H_4O} (HOC_2H_4)_3N$$

The products from the reaction are separated by distillation. During the last few years, each of the amines has in turn been in the greatest demand, so processing flexibility must be maintained.

Monoethanolamine is used primarily in detergents and as an absorbent for acid-gas (H_2S, CO_2) removal. It is to a lesser extent, a chemical intermediate for compounds such as ethylene imine. Diethanolamine's major end use is in detergents, but it is also utilized in textiles and as a gas-purification agent. Most of the triethanolamine goes into the production of cosmetics and textile specialities.

The greatest potential for growth is in heavy duty liquid detergents which are still gaining on the solid heavy duty detergents.

Isopropanolamines, derived from propylene oxide and ammonia, are competitive with the ethanolamines, and both are unique in that they are organic compounds and yet strongly alkaline.

Glycol Ethers. In the same way that water reacts with one or more molecules of ethylene oxide, alcohols react to give monoethers of ethylene glycol, producing monoethers of diethylene glycol, triethylene glycol, etc., as by-products.

$$ROH + CH_2{-}CH_2 \text{ (with bridging O)} \longrightarrow ROC_2H_4OH$$

$$+ CH_2{-}CH_2 \text{ (with bridging O)} \longrightarrow ROC_2H_4OC_2H_4OH$$

$$+ CH_2{-}CH_2 \text{ (with bridging O)} \longrightarrow ROC_2H_4OC_2H_4OC_2H_4OH$$

Since their commercial introduction in 1926, glycol ethers have become valuable as industrial solvents and chemical intermediates. Because glycol monoethers contain a $-OCH_2CH_2OH$ group, they resemble a combination of ether and ethyl alcohol in solvent properties. The most common alcohols used are methanol, ethanol, and butanol. Principal uses for the glycol ethers are as solvents for paints and lacquers, as intermediates in the production of plasticizers, and as ingredients in brake fluid formulations. The most common trade names are Dowanol®, Cellosolve®, and Polysolve®. Condensation of the monoethers produces glycol diethers which are also useful as solvents.

$$2ROCH_2CH_2OH \xrightarrow{H_2SO_4} ROC_2H_4OC_2H_4OR + H_2O$$

Solvent characteristics of glycol ether esters are enhanced by esterifying to make the acetate ester. This ester is used extensively in paints and lacquers.

Alcohol-ethylene oxide products have also been developed in which the number of oxide units is considerably higher in order to improve water solubility. Long-chain fatty alcohols are condensed with 10–40 molecules of ethylene oxide to produce detergents for the textile industry. Also important are the water-soluble alkyl phenyl ethers of the higher polyethylene glycols. Phenols react in the same way as alcohols to give polyglycol ethers. The reaction is rapid, and essentially quantitative. These products are detergents of the same general class as the long-chain alcohol-ethylene oxide condensates.

Of recent origin is the interest in the extremely high molecular weight homopolymers of ethylene oxide. These resins, trade-marked Polyox®, have good water and organic solvent solubility and thus are used for thickening agents and water-soluble films.

Other Ethylene Oxide Uses. Some 265 million pounds of ethylene oxide were consumed by this category of uses in 1978. The major use is in polyether polyols for flexible polyurethane foams, accounting for approximately 55 percent of the ethylene oxide used. Other ethylene oxide uses in this category include 10 percent to hydroxyethyl cellulose, 9 percent to acetal copolymer resins, 8 percent each to choline and ethylene chlorohydrin, 4 percent to hydroxyethyl starch, and 2 percent to each of aryl ethanolamines, cationic surface-active agents and fumigant, food sterilant, and hospital sterilization uses.

Chlorinated Hydrocarbons

The manufacture of chlorinated hydrocarbons forms an important part of industrial chemistry today. The products are useful as solvents, chemical intermediates, pesticides, monomers, and in many other ways. Table 25.3 shows some of the large-volume materials.

Chlorinated derivatives of aliphatic hydrocarbons are usually prepared by one of three general methods: (1) addition of hydrogen chloride to unsaturated hydrocarbons, (2) addition of chlorine to unsaturated hydrocarbons, or (3) substitution of chlorine for hydrogen in either saturated or unsaturated hydrocarbons. In the last case, hydrogen chloride is a by-product. Examples of the first method are the addition of HCl to ethylene to form ethyl

TABLE 25.3 Production and Sales of Some Halogenated Aliphatic Hydrocarbons[a,b]

Products	*Production (1000 lbs)*	*Sales (1000 lbs)*	*Sales Value ($1000)*
Ethylene dichloride	12,877,000	1,033,313	82,645
Vinyl chloride	6,941,123	4,885,688	618,407
Carbon tetrachloride	737,030	363,406	41,698
Perchloroethylene	725,457	549,111	53,589
Methyl chloroform	644,475	631,243	135,388
Methylene chloride	570,098	490,678	114,342
Ethyl chloride	539,743	159,079	23,428
Methyl chloride	453,810	200,797	28,977
Chloroform	349,169	302,114	53,423
Dichlorodifluoromethane	327,097	316,864	134,743
Trichloroethylene	298,986	298,557	46,588
Ethylene dibromide	229,913	–	–
Chlorodifluoromethane	205,612	139,797	106,236
Trichlorofluoromethane	193,735	166,898	57,341
Chlorinated paraffins	99,896	95,200	28,245
Other brominated hydrocarbons	76,800	–	–
Propylene dichloride	74,112	33,382	2,120
Tetrafluoroethylene	27,733	–	–
Iodinated hydrocarbons	50	–	–
All others	1,369,293	303,461	148,875
TOTAL HALOGENATED HYDROCARBONS	26,751,112	9,969,588	1,676,045

[a]U.S. International Trade Commission, 1978.
[b]*Chemical Economics Handbook*, SRI, Menlo Park, California (1978).

chloride and the addition of HCl to acetylene to form vinyl chloride. Typical of the second method is the addition of chlorine to ethylene to form dichloroethane. The third method, the direct substitution of chlorine for hydrogen, usually involves a free-radical mechanism. The formation of chlorine free radicals occurs spontaneously at temperatures above 250°C and increases with temperature. The formation may also be brought about by the action of actinic light at lower temperatures. These light-activated, or photochemical, chlorinations may be carried out in either gas or liquid phase.

The principal uses for chlorinated hydrocarbons are as solvents, chemical intermediates, and insecticides. Their value as solvents derives from a combination of good solvent power, low flammability, and high vapor density. The latter property is particularly important in vapor degreasing of metal parts (a major use for trichloroethylene, perchloroethylene, and methyl chloroform). Certain of the compounds are of value as chemical intermediates. Thus, ethyl chloride is used to make tetraethyl lead; and ethylene dichloride is used to make vinyl chloride, which in turn is polymerized to polyvinyl chloride.

Ethylene Dichloride (EDC). While some ethylene dichloride (1,2-dichloroethane) occurs as a by-product of chlorohydrin and ethyl chloride processes, the bulk comes from the chlorination of ethylene. About 87 percent of the EDC consumed in 1978 went into the manufacture of vinyl chloride monomer. Figure 25.10 shows a vinyl chloride monomer and ethylene dichloride (EDC) process using ethylene, chlorine, and air, thus including oxychlorination.

In this process, vinyl chloride is produced by thermal cracking of EDC. The feed, EDC, is supplied from two sources. In the first source, ethylene and chlorine are reacted in essentially stoichiometric proportions to produce EDC by direct addition. In the second source, ethylene is reacted with the HCl produced from the thermal cracking operation to produce EDC by oxychlorination. Chemical reactions are as follows:

$$\textit{Cracking EDC:}\quad C_2H_4Cl_2 \xrightarrow{\Delta} HCl + C_2H_3Cl$$

Direct chlorination: $Cl_2 + C_2H_4 \longrightarrow C_2H_4Cl_2$

Oxychlorination: $C_2H_4 + 2HCl + \frac{1}{2}O_2 \longrightarrow C_2H_4Cl_2 + H_2O$

The oxychlorination reaction is carried out in a fluid bed of copper chloride-impregnated catalyst. In direct chlorination, ethylene and chlorine gas are charged to a reactor containing catalyst and excess ethylene dichloride as solvent. The reaction is carried out at 50°C and 20 psig. Purified ethylene dichloride from both sources is cracked in a furnace at about 400°C and elevated pressure. The hot gases are quenched and distilled to remove HCl, then the vinyl chloride. Unconverted EDC is returned to the EDC purification train.

Profitable disposal or utilization of HCl was once one of the major restrictions in the growth of ethylene dichloride as the source of vinyl chloride and other chlorinated hydrocarbon derivatives. The advent of oxychlorination which uses hydrochloric acid and molecular oxygen in contact with ethylene has opened the door for very rapid replacement of the acetylene process. The oxygen reacts with HCl in the oxychlorination step and generates chlorine *in situ* which reacts with the ethylene, forming the EDC.

Since oxychlorination has greatly improved the process economics of EDC, the trend now is to make many of the chlorinated hydrocarbons previously derived from acetylene by this route. With proper conditions, EDC can be chlorinated to mainly tetrachloroethane, then catalytic dehydrochlorination of the tetra gives trichloroethylene. With different chlorination conditions, mainly pentachloroethane can be formed, and with dehydrochlorination, perchloroethylene is formed.

Another modification of EDC chlorination is to adjust the conditions to maximize 1,1,2-trichloroethane as the product. This product when dehydrochlorinated gives vinylidene chloride (1,1-dichloroethylene), a monomer used in a growing number of plastic polymers. Vinylidene chloride can, however, be hydrochlorinated to methyl chloroform, a very rapidly growing solvent, mainly by virtue of its low toxicity.

Less than 1.5 percent of EDC produced in

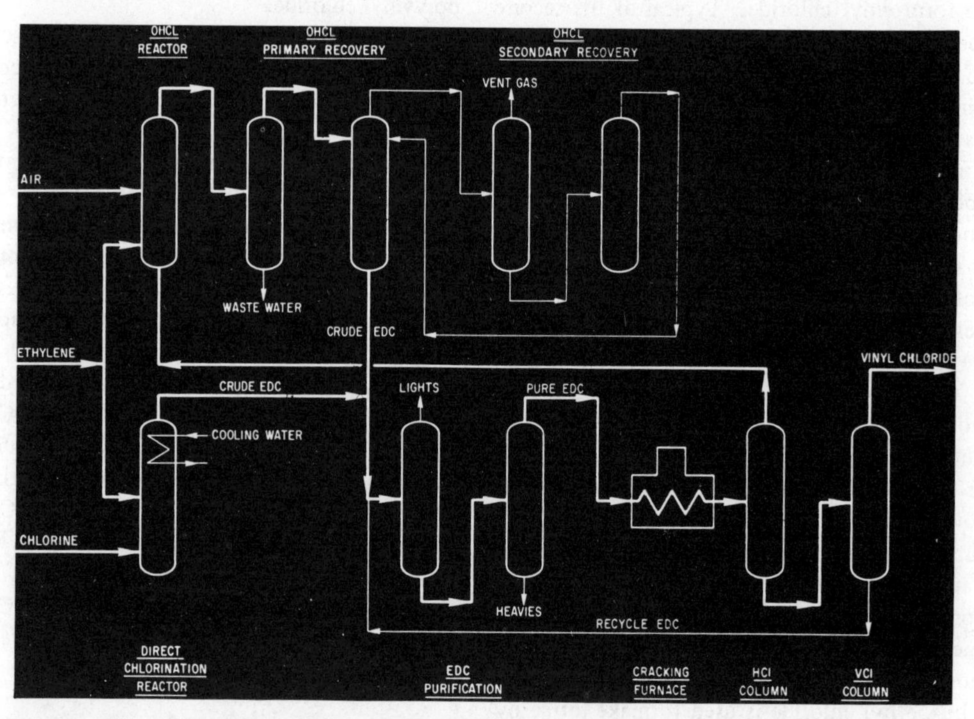

Fig. 25.10 Vinyl chloride and ethylene dichloride manufacture. *(Hydrocarbon Processing,* **50,** *no. 11, 220 (1971); copyright 1971 by Gulf Publishing Co.)*

1978 was used as a lead scavenger, while about 2.5 percent was used as an intermediate to produce about 70 million pounds of ethylene amines.

Ethyl Chloride. About 90 percent of the ethyl chloride is produced by the addition of hydrogen chloride to ethylene in the presence of an aluminum chloride catalyst.

$$CH_2{=}CH_2 + HCl \xrightarrow{AlCl_3} CH_3CH_2Cl$$

This is a liquid phase reaction, carried out at about 40°C. The remaining production comes from the chlorination of ethane. This latter route made up about 8 percent of U.S. capacity in 1979.

The production of ethyl chloride is closely tied in with that of tetraethyl lead, which consumes over 90 percent of the former. It is also used in the production of ethyl cellulose, as a refrigerant, and as an anesthetic, and it is actively exported.

Tetraethyl lead is made by reacting ethyl chloride with a lead-sodium alloy (about 9 parts lead and 1 part sodium.)

$$4NaPb + 4C_2H_5Cl \longrightarrow Pb(C_2H_5)_4 + 3Pb + 4NaCl$$

The crude product is purified by vacuum distillation and water washing. It is then blended with ethylene dichloride and ethylene dibromide to make the gasoline antiknock compound. At present, the outlook for tetraethyl lead is dim due to serious concern for the environmental effects of lead compounds in automobile exhausts. This has led to lead-free gasolines.

However, many possibilities similar to those of ethylene dichloride exist; ethyl chloride with properly controlled chlorination conditions yields methyl chloroform and 1,1,2-trichloroethane. The 1,1,2-trichloroethane can be chlorinated to tetrachloroethane, which in turn can be dehydrochlorinated to trichloroethylene. This use for ethyl chloride may utilize some of the ethyl chloride capacity idled by sharp cutbacks in the production of lead tetraethyl for gasoline.

Thus, starting with ethylene, a valuable large volume of chlorinated hydrocarbons can be produced with economics such that acetylene is rapidly being replaced. The new processes are also very flexible in the product mix they can produce. Prior to 1969, about 85 percent of trichloroethylene was made from acetylene. In 1969, this proportion dropped to about 55 percent, and the replacement of acetylene by ethylene for manufacture of trichloroethylene was complete when the last plant was shut down in early 1978.

A similar shift has taken place with vinyl chloride, where in 1962 about 50 percent of vinyl chloride was based on acetylene and 50 percent on ethylene via EDC. By 1969 ethylene-based vinyl chloride accounted for about 85 percent of production, and by the end of 1978 less than 4 percent of vinyl chloride was still produced from acetylene.

Chlorinated Solvents. This group of compounds consists mainly of carbon tetrachloride, trichloroethylene, perchloroethylene, methyl chloroform, chloroform, and methylene chloride. Some of the intermediates to these products such as 1,1,2-trichloroethane, tetrachloroethane, and pentachloroethane find special solvent uses. Since the manufacturing processes are difficult to untangle, the sketches that follow are added for clarity, even though some start with methane, propane, or propylene.

The preparation of trichloroethylene is accomplished wherein tetrachloroethane is dehydrochlorinated either thermally, catalytically, or by reaction with lime. The catalytic method is the most common commercially, and it involves a vapor-phase reaction over a bed of barium chloride at about 400-700°F.

$$CHCl_2CHCl_2 \longrightarrow CHCl{=}CCl_2 + HCl$$

More than 90 percent of the trichloroethylene produced is for vapor degreasing of metals, a field in which it has almost completely supplanted carbon tetrachloride, but it in turn is now being rapidly replaced by methyl chloroform and perchloroethylene because trichloroethylene has been found to be a smog producer and its toxicity has been found to be much higher than was formerly thought.

Perchloroethylene is made by the pyrolysis of carbon tetrachloride, although it may also be produced by the dehydrochlorination of pentachloroethane. At a temperature of 800–900°C, carbon tetrachloride readily decomposes into perchloroethylene and hexachloroethane. The latter is recycled to produce more perchloroethylene. A modification of this process has been reported in which chlorine and a light hydrocarbon (such as natural gas or LPG) are fed into a chlorination furnace at 900–1200°F. Chlorination takes place readily, producing carbon tetrachloride and perchloroethylene. Undoubtedly, the latter is again formed by the pyrolysis of CCl_4. Most perchloroethylene is consumed by the dry cleaning industry. Some is used as a vapor degreasing solvent.

Other Halogenated Hydrocarbons. About 230 million pounds of ethylene dibromide is made annually by the addition of bromine to ethylene. It is used as a lead scavenger in antiknock fluids and as an agricultural fumigant.

Smaller uses are as a solvent and in the synthesis of pharmaceuticals and dye intermediates.

Fluorocarbons made their first impact on the chemical industry in 1931 with the introduction of *Freon* refrigerants (Freon® and Genetron® are the trade-marks used by the two largest manufacturers). The next major advance came during World War II with the development of fluorocarbon polymers and nonfood aerosol propellents. By 1969, production was estimated

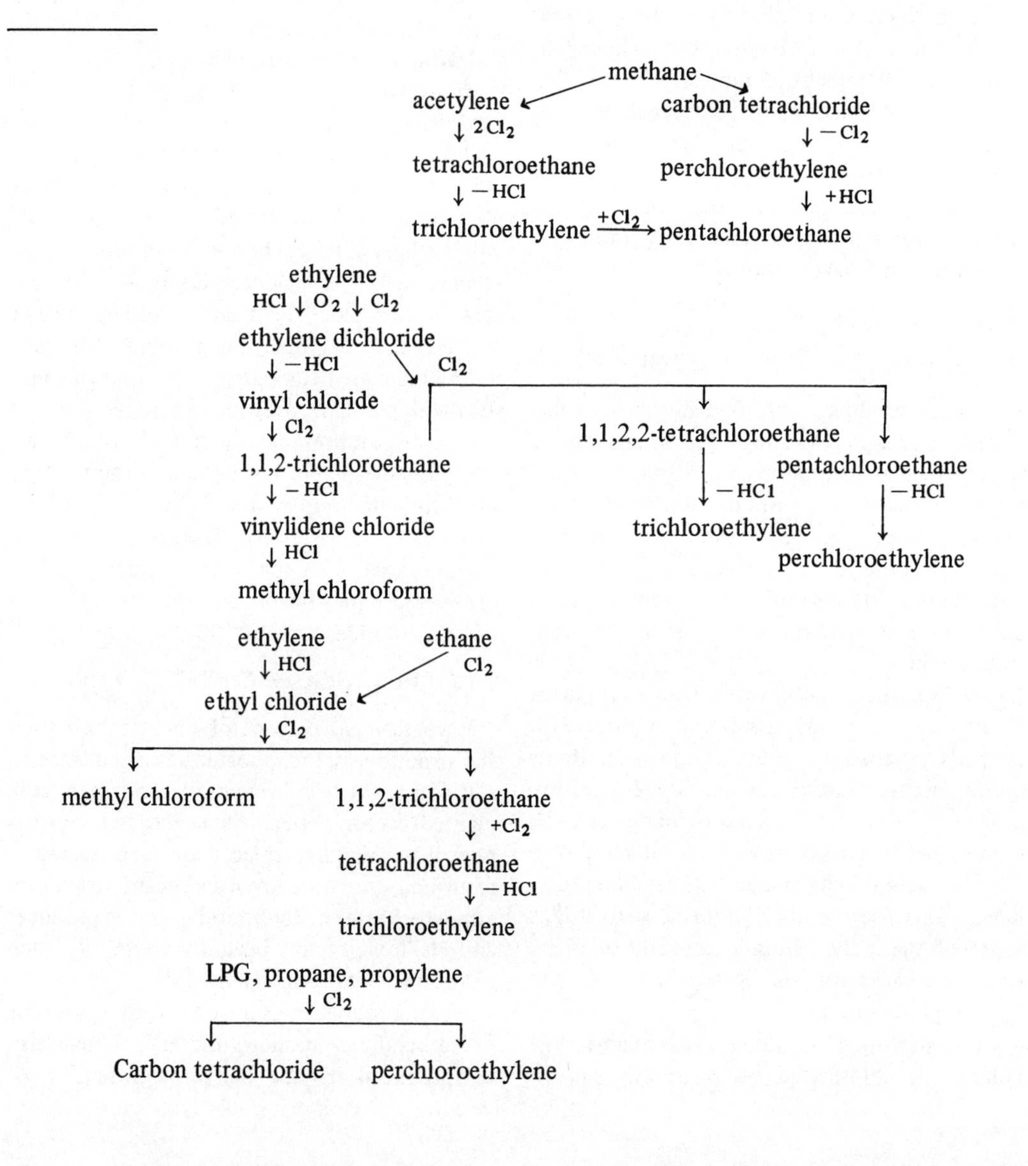

at 700 million pounds, of which dichlorodifluoromethane accounted for more than half. In recent years, fluorocarbons have plateaued because of their alleged impact on the earth's ozone layer. Propellant use is now essentially zero.

The five main compounds today are as follows:

- *Fluorocarbon-12* (CCl_2F_2) is the most widely used fluorocarbon. It is an excellent refrigerant.
- *Fluorocarbon-11* (CCl_3F) is used with fluorocarbon-12 propellent to raise the boiling point. The blend is widely employed in air conditioning and process-water cooling.
- *Fluorocarbon-22* ($CHClF_2$) is used in small-scale refrigeration and air conditioning units. Tetrafluoroethylene can be produced from this compound.
- *Fluorocarbon-113* (CCl_2-$FCClF_2$) is used to improve the solvent properties of fluorocarbon-12 propellents. It can be dechlorinated to produce chlorotrifluoroethylene.
- *Fluorocarbon-114* ($CClF_2CClF_2$) is also used with fluorocarbon-12 propellent. The latter two materials are used as expanding agents for plastic foams.

In addition to refrigerants, fluorocarbons are finding use in plastics, in particular the homopolymers of tetrafluoroethylene and chlorotrifluorethylene (Teflon® and Kel-F®, respectively). These polymers are specially noted for their temperature-resistance and chemical inertness. Also of importance is the elastomer (Viton®) produced by the copolymerization of vinylidene fluoride and hexafluoropropylene.

Fluorocarbons find smaller markets as fire extinguishers (bromotrifluoromethane), as blowing agents for urethane foams, as solvents, and as specialty lubricants.

Ethanol

Less than thirty years ago, synthetic ethanol was the largest consumer of ethylene. Today it is in seventh place. Just prior to World War II, the fermentation of molasses accounted for about 72 percent of the ethanol production. Today less than 10 percent of the ethanol manufactured is made by this route; over 90 percent is synthesized from ethylene by the direct catalytic hydration. In 1978, over 1.27 billion pounds of synthetic ethanol was produced.

The direct hydration process involves a water-ethylene reaction, over a phosphoric acid catalyst, at about 400°C and 1000 psi.

$$C_2H_4 + H_2O \xrightarrow{H_3PO_4} C_2H_5OH$$

This method has the advantage of producing less by-product diethyl ether. Over-all yield of ethanol is reported to be better than 97 percent. A flow sheet of the direct hydration process is shown in Fig. 25.11. Forty-one percent of the synthetic ethanol produced in 1969 was by direct hydration; in 1970, this had increased to 48 percent and reached essentially 100% in 1976. Direct hydration has replaced the esterification-hydrolysis process in the last ten years because of higher yields, lower plant maintenance costs, and elimination of waste and pollution problems with the sulfuric acid route.

The production of acetaldehyde was the principal use of industrial ethyl alcohol until 1968 when solvent uses surpassed it. Acetaldehyde-use is showing a steady decline, while solvent-use continues to grow at a rate somewhat above 5 percent per year. Other uses for ethyl alcohol are in synthetic rubber, drugs, and in the synthesis of various chemicals such as acetic acid and ethyl acetate.

Fermentation ethyl alcohol can be manufactured from sugar, starch, or cellulosic raw materials. The attractiveness of these materials is that they are renewable agricultural or forestry products. In the past, these materials fluctuated too much in value to compete with hydrocarbon sources. Looking into the future, however, there is renewed interest in producing ethyl alcohol first from waste materials or by-product molasses. Government-subsidized energy programs are leading to the development of large projects based on corn, wherein waste cornstalks, cobs, etc. are used for fuel to give more favorable overall net energy yield. Utilization of by-product protein solids in cattle feed is an important factor in making such schemes economically feasible.

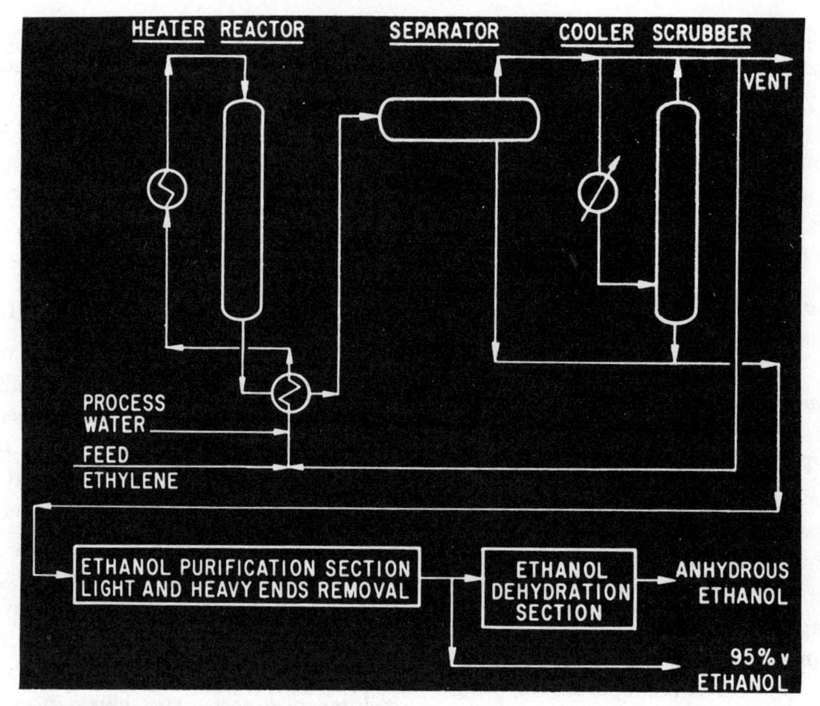

Fig. 25.11 Vapor-phase direct hydration of ethylene to ethanol. (*Hydrocarbon Processing,* **50**, *no. 11, 151 (1971); copyright 1971 by Gulf Publishing Co.*)

Ethylbenzene

Ethylbenzene has but one major end-use and that is the manufacture of styrene. In 1978, ethylbenzene was the fifth largest consumer of ethylene, accounting for about 2.26 billion pounds. This represents about a 7 percent average annual growth in the last ten years. In addition to being produced by the alkylation of benzene with ethylene, ethylbenzene is also isolated from mixed xylene streams or as a by-product of cumene manufacture. These latter sources account for less than 10 percent of the ethylbenzene used today.

Ethylbenzene synthesis by reaction of ethylene with benzene is carried out either in the liquid phase using aluminum chloride as the catalyst, or in the vapor phase with a phosphoric acid or alumina-silica catalyst. Yields about 95 percent are obtained in either case. Figure 25.12 shows a flow diagram of an ethylbenzene process.

The principal uses for styrene, made by catalytic cracking of ethylbenzene, are in polystyrene, styrene-butadiene latex and plastics, SBR rubber, and in the cross-linking of unsaturated polyester resins. For more details, see Chapter 10.

Acetaldehyde, Acetic Acid, Vinyl Acetate

Acetaldehyde. Acetaldehyde production peaked in 1969 at 1.65 billion pounds. Of this, 685 million or 42 percent was made by direct oxidation of ethylene. In 1978, production was one billion pounds. Over 97 percent of this was by ethylene oxidation. The direct oxidation process uses cupric chloride and a small amount of palladium chloride in aqueous solution as the catalyst. The reaction is exothermic and is controlled by evaporation. The gaseous reaction mixture is scrubbed to remove acetaldehyde and overhead gases are recycled. The liquid from the reaction mixture is regenerated with oxygen to return the copper to the cupric state and palladium metal back to palladium chloride. See Figure 25.13.

The alternate processes consist of oxidation of ethyl alcohol and glycerin by-product.

About 70 percent of acetaldehyde consumed is used for the production of acetic acid and

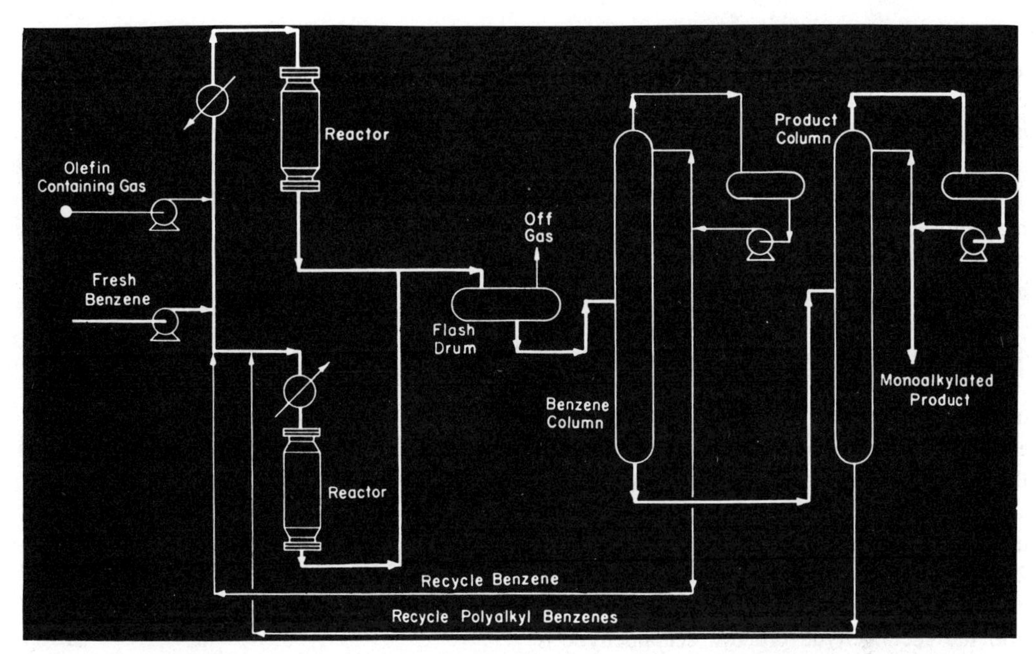

Fig. 25.12 Alkar ethylbenzene process. (*Hydrocarbon Processing,* **50**, *no. 11, 125 (1971); copyright 1971 by Gulf Publishing Co.*)

anhydride. Pentaerythritol, peracetic acid, pyridine, glyoxal, 1,3-butylene glycol, and chloral all combined consume the remaining 30 percent of acetaldehyde. Since the fate of acetaldehyde is determined by its use as an acetic acid-acetic anhydride intermediate and future acetic acid technology appears to be firmly based on methanol, acetaldehyde consumption will drop over the next few years. High unit-value products are derived from pyridine compounds; thus, although mass consumption of acetaldehyde will not look exciting, growth in value of products will be attractive.

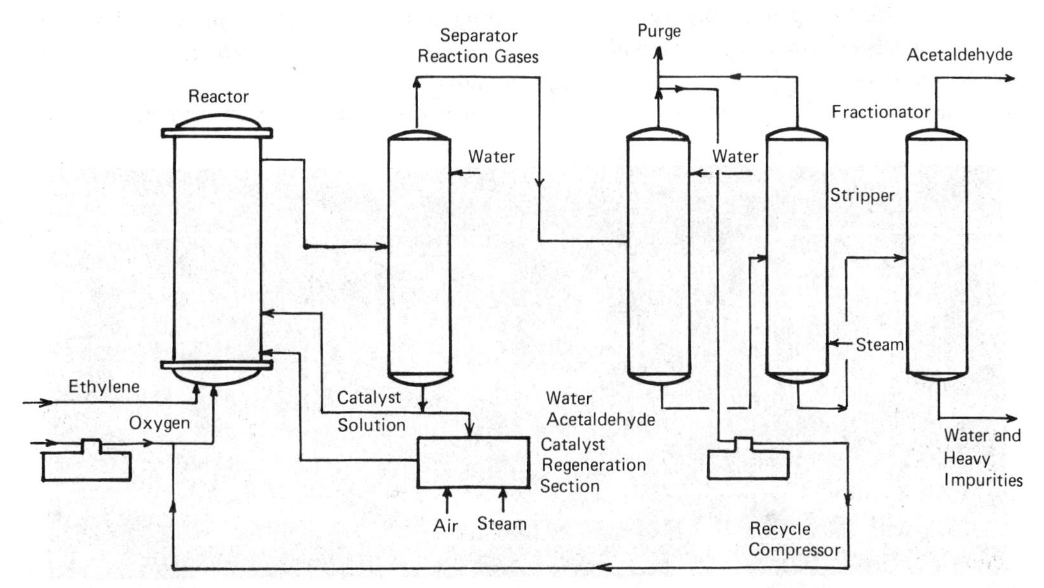

Fig. 25.13 Acetaldehyde by single-stage Wacker process. (*Hahn, A. V., "The Petrochemical Industry," p. 243, McGraw-Hill, 1970.*)

Acetic Acid. Acetic acid production is reported at 2,776 million pounds in 1978. Domestic capacity in 1978 was almost equal between *n*-butane oxidation, methanol carbonylation, and acetaldehyde oxidation, with a minor amount from glycerin by-product. By 1980, methanol carbonylation exceeded 40 percent of the capacity and will continue to increase in share.

Growth of acetic acid consumption was about 4 percent annually in the last ten years and will maintain this growth, while acetate, the major user, will show over 5 percent growth. Acetic esters and terephthalic acid/dimethyl terephthalate will each consume about 12 percent of acetic acid used. Cellulose acetate consumption has essentially plateaued and consumes about 20 percent of acetic acid. Other uses are chloroacetic acid, textiles, and others.

In the acetaldehyde process, acetaldehyde is fed to a reactor where air is passed through the liquid at 55 to 65°C and 70 to 75 psi. Manganous acetate, 0.1 to 0.5 percent, is used in the liquid to control the formation of the explosive intermediate peracetic acid. Acetaldehyde is scrubbed from the off-gases with water and recovered for recycle, while the liquid reactor mass is distilled to 99 percent glacial acetic acid. Continuous modern plants use cobalt acetate dissolved in acetic acid as the catalyst. Conversions of 20-30 percent per pass are used and oxygen allows lower pressures and minimizes the off-gas.

The methanol carbonylation process uses carbon monoxide and methanol as raw materials. Either the older high pressure process or the newer low pressure process can be used. Yields range from 90 to 99.5 percent based on methanol. Using M to symbolize the metal complex catalyst either rhodium or cobalt, the chemical steps can be shown as follows:

1. $CH_3OH + HI \longrightarrow CH_3I + H_2O$

2. $CH_3I + [\text{M complex}] \longrightarrow \begin{bmatrix} CH_3 \\ \text{M complex} \\ I \end{bmatrix}$

3. $\begin{bmatrix} CH_3 \\ \text{M complex} \\ I \end{bmatrix} + CO \longrightarrow \begin{bmatrix} CH_3 \\ C{=}O \\ \text{M complex} \\ I \end{bmatrix}$

4. $\begin{bmatrix} CH_3 \\ C{=}O \\ \text{M complex} \\ I \end{bmatrix} + H_2O \longrightarrow CH_3\overset{O}{\overset{\|}{C}}OH + HI + [\text{M complex}]$

The economics of the methanol carbonylation process are reported to have marked advantage over the acetaldehyde route, and it appears destined to take over as the manufacturing route. Fig. 25.14 shows a flow sheet of acetic acid from methanol and carbon monoxide.

Large quantities of acetic acid are recovered

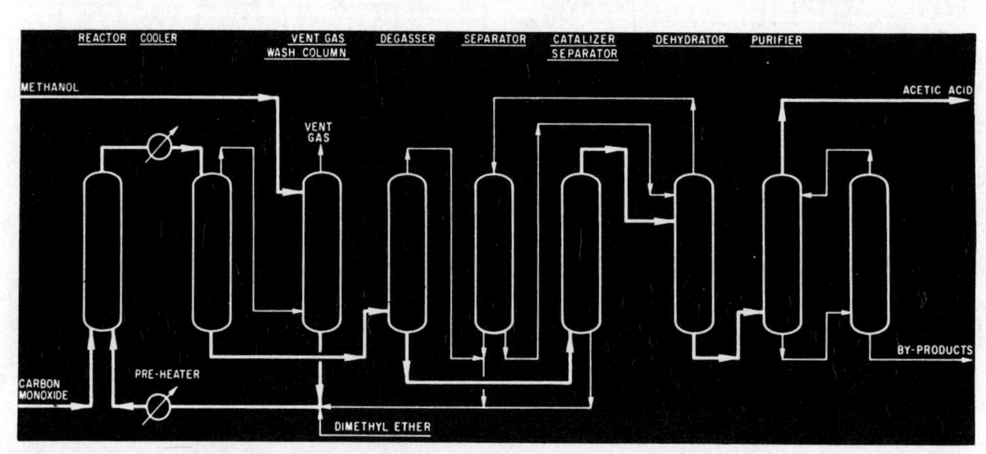

Fig. 25.14 High-pressure acetic acid from methanol and carbon monoxide. (*Hydrocarbon Processing,* **50,** *no. 11, 115 (1971); copyright 1971 by Gulf Publishing Co.*)

and reconverted to acetic anhydride for cellulose acetate production. This acid is not reported in the consumption figures reported above.

Acetic Anhydride. Acetic anhydride can be produced by processes using acetaldehyde and acetic acid. This process is reported to be highly corrosive; thus, essentially all acetic anhydride is produced by the reaction of ketene and acetic acid as shown in Fig. 25.15. New process technology* is reported for producing acetic anhydride from carbon monoxide and methanol. A plant using this technology is scheduled to come on stream in 1983 and reportedly will be based on coal rather than petroleum.

It is estimated that 1.5-1.6 billion pounds of acetic anhydride was produced in 1978, with 90 percent of this used in cellulose acetate, 1.5 percent in aspirin, and the remainder in other miscellaneous esters and other uses.

Vinyl acetate. Vinyl acetate is produced today primarily from ethylene in a vapor-phase process wherein ethylene, acetic acid, and oxygen are reacted using a palladium chloride catalyst to form vinyl acetate and water. This process yields acetaldehyde as a by-product in adequate quantity to be converted to the acetic acid needed. This means that the only net feed to the complex is ethylene. A flow sheet of this process is shown in Figure 25.16.

Vinyl acetate was first produced by vapor phase catalytic reaction of acetylene and acetic acid. Only a minor amount is still produced by this route today.

Major uses of the 1.69 billion pounds of vinyl acetate reportedly produced in 1978 comprise polyvinyl acetate emulsions and resins 55 percent; polyvinyl alcohol 21 percent, polyvinyl butyral 7 percent, vinyl chloride copolymer 6 percent, ethylene-vinyl acetate resins and emulsions 5 percent, and others 6 percent.

Polyvinyl acetate emulsions and resins are used mainly in paints, paper coatings, and adhesives. Major uses of polyvinyl alcohol are in the warp size market and as a component in industrial adhesives. Ethylene-vinyl acetate copolymer resins are used in hot-melt adhesives. Polyvinyl butyral is used as a laminating adhesive for automobile windshields. Vinyl chloride copolymers are used mainly in floor tile and in phonograph records.

Good growth of 5-7 percent annually is predicted for most vinyl acetate derivatives.

*Chemical Processing, Chem. Trends, July 1980.

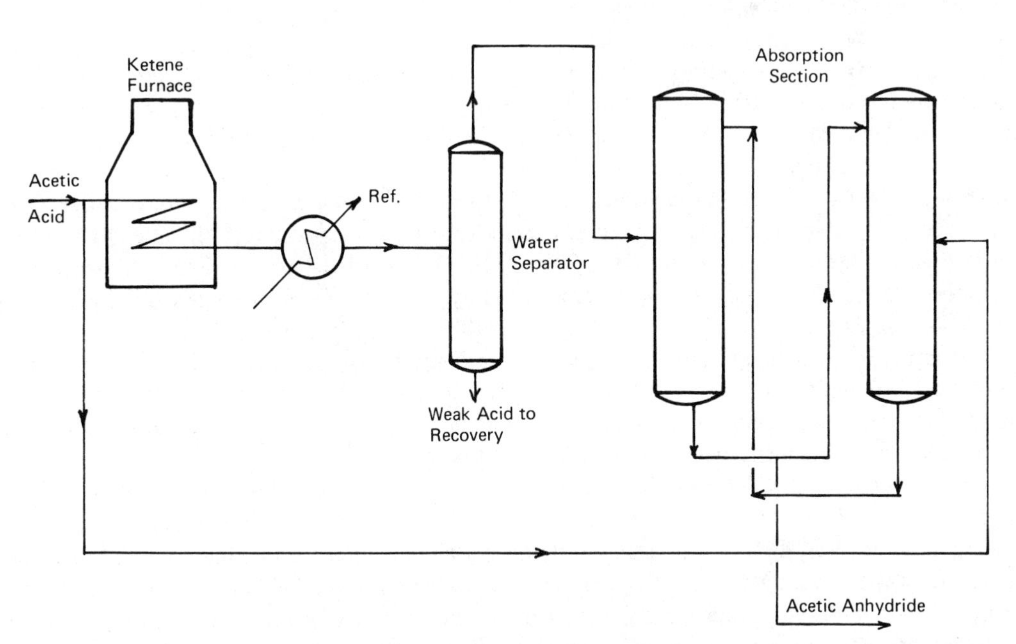

Fig. 25.15 Acetic anhydride (*Hahn, A. V., "The Petrochemical Industry," p. 253, McGraw-Hill, 1970.*)

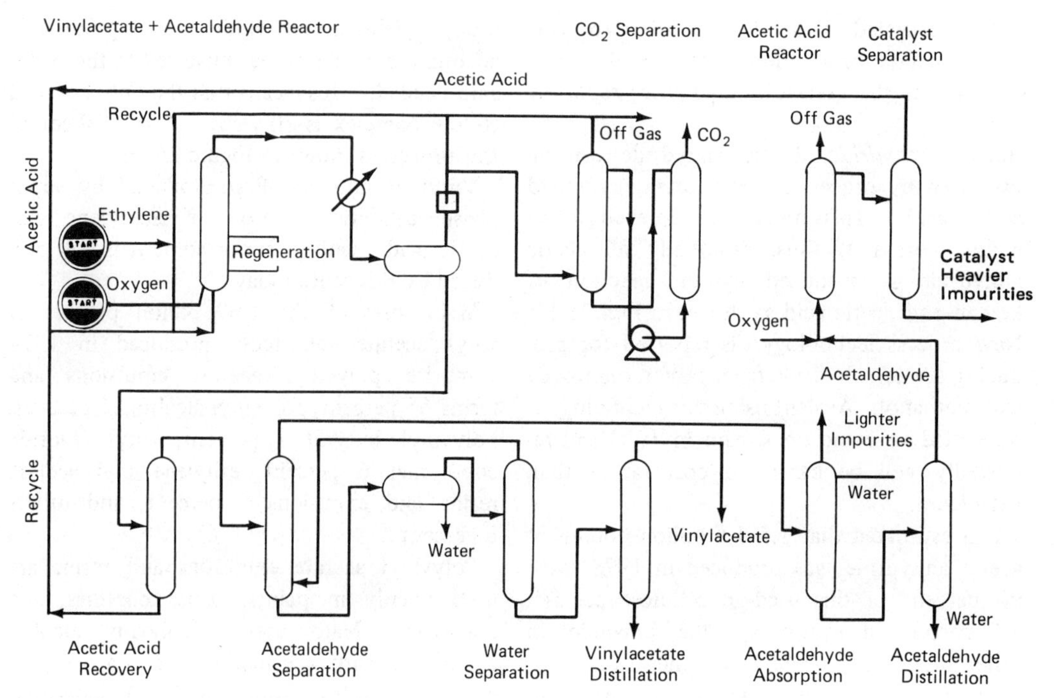

Fig. 25.16 Vinyl acetate by oxyacetylation process. (Hydrocarbon Processing, *50*, no. 11, 219 (1971); copyright 1971 by Gulf Publishing Co.)

Ethylene Oligomers

Linear primary alcohols and α-olefins in the C_6 to C_{10} range and the C_{12} to C_{18} range have become important industrial chemicals in the last 15 years. The linear alcohols in the C_6 to C_{10} range are used to make plasticizers for flexible polyvinyl chloride. These plasticizers give more desirable properties than phthalates, adipates, or sebacates made from conventional alcohols. The C_{12} to C_{18} linear primary alcohols are used to produce highly biogradable surface active agents, which in the final forms are ethoxylates, alcohol sulfates, or sulfates of alcohol ethoxylates.

Production of linear primary alcohols approached 700 million pounds in 1978 and is increasing at a 10 percent rate. These compounds are replacing many of the natural fatty alcohols now used for detergents. The production of α-olefin compounds as such is taking hold as uses develop for things other than the alcohols. Production in 1978 was over 400 million pounds and projections for growth are still optimistic. Other low molecular weight oligomers also appear to have considerable potential to become volume chemical intermediates.

Figure 25.17 is a flow diagram for the manufacture of Alfol® α-alcohols. In this process, aluminum triethyl is prepared by hydrogenation of aluminum powder in a solvent followed by ethylation with ethylene. The mixture must be kept scrupulously dry. Aluminum triethyl is then reacted with ethylene under pressure to form higher alkyls. A spectrum of alkyls from C_2 to C_{22} are obtained. The aluminum alkyls are then oxidized to aluminum alkoxides with very dry air. The alkoxides are hydrolyzed with sulfuric acid, after which the residual acid is neutralized with caustic and washed free of sodium sulfate with water. The alcohols are then dried and fractionated into the desired cuts.

Figure 25.18 is a flow diagram of an α-olefin process. In this process, ethylene and a solvent containing an alkyl aluminum are fed to a reactor where the alkyl aluminum adds one or more ethylene molecules in sequence, forming linear alkyl groups averaging 10 to 20 carbon atoms.

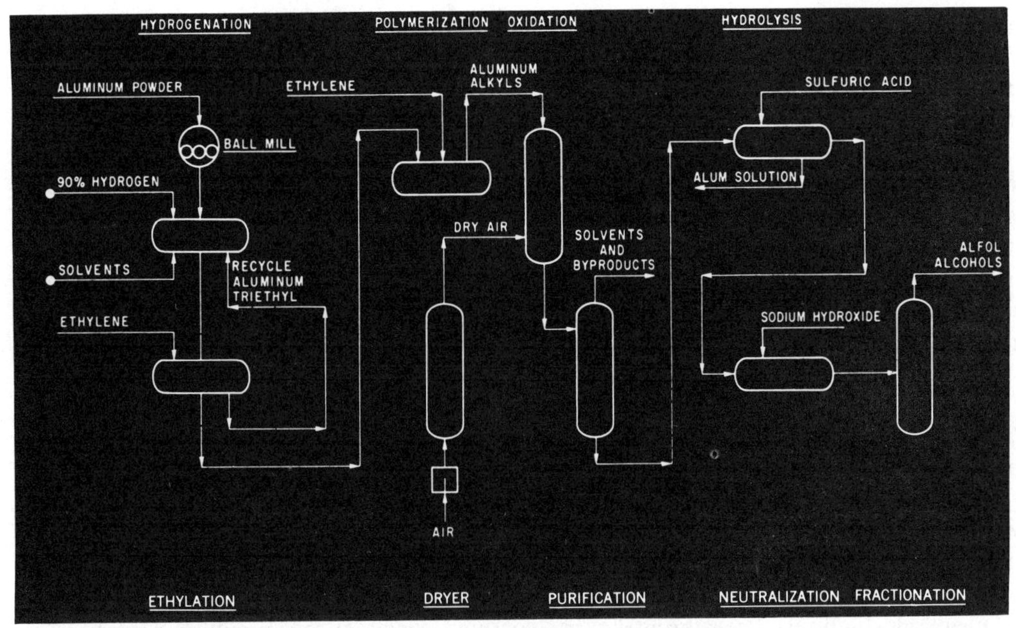

Fig. 25.17 Alpha alcohols by Alfol® process. *(Hydrocarbon Processing,* **50,** *no. 11, 126 (1971); copyright 1971 by Gulf Publishing Co.)*

Eventually, ethylene displaces the alkyl groups to form α-olefins plus triethyl aluminum that is re-used. The product is gas-liquid separated to recover ethylene for recycle, then washed and distilled into various fractions.

Ethylene-Propylene Elastomers

Most polymers in this family of products are produced by polymerization of ethylene, propylene, and a small amount (5 percent) of a nonconjugated diolefin such as ethylidene norbornene, 1,4-hexadiene, or dicyclopentadiene. These polymers are readily oil extended and are used with 20 to 50 weight percent oil in many applications.

About 80 percent of ethylene-propylene elastomers are used in non-tire applications. Some of these uses include radiator and other water and steam hoses, weather stripping, appliance parts, car bumpers, and wire and cable insulation. Other uses are as components

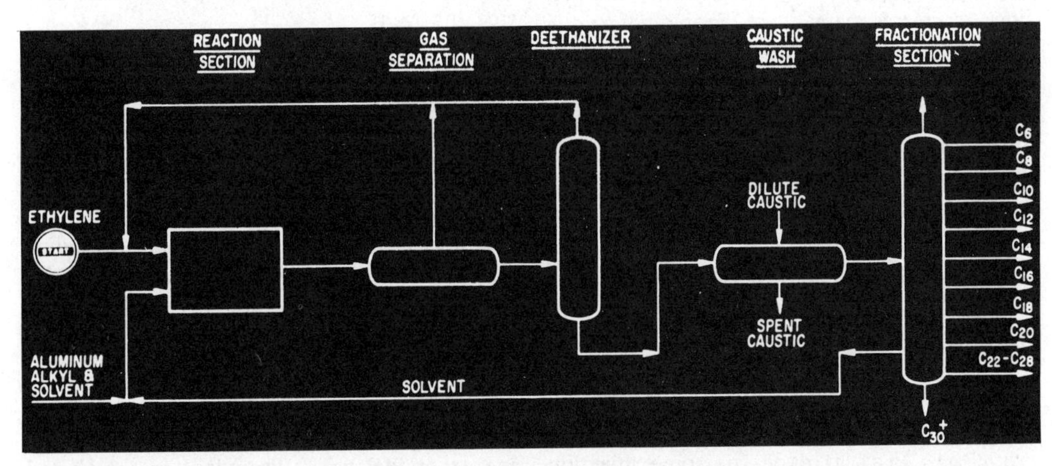

Fig. 25.18 Alpha-olefins process. *(Hydrocarbon Processing,* **50,** *no. 11, 127 (1971); copyright 1971 by Gulf Publishing Co.)*

in olefinic thermoplastic elastomers, as impact modifiers for plastics, and as viscosity index improvers for lube oils. Major tire uses are in white sidewall compounds, cover strips, and specialty tires and tubes.

Ethylene-propylene elastomer manufacture consumed 224 million pounds of ethylene in 1978. This produced approximately 350 million pounds of polymer. A growth rate of some 6–8 percent is anticipated for several years.

Propionaldehyde

Propionaldehyde is produced by the oxo process reaction of ethylene and carbon monoxide. A yield of approximately 80 percent is obtained. Primary uses are for the manufacture of *n*-propyl alcohol and propionic acid. Both of these chemicals find their uses primarily in agricultural products. *n*-Propanol is used in the manufacture of thiocarbamate and dinitroaniline herbicides and propionic acid is used to make carboxylic acid and amide herbicides. *n*-Propanol also finds some solvent uses in coatings and inks. Propionic acid finds its major uses as a grain preservative, while its sodium and calcium salts are used as food preservatives.

n-Propanol is now all derived from ethylene-based acetaldehyde, and it is enjoying about a 5 percent growth rate. Propionic acid use is growing at about the same rate, but ethylene consumption is growing about twice this fast since a greater percent of propionic acid is produced from propionaldehyde.

Propionic acid helps preserve wet corn. Expansion of this use, possibly forced by high drying costs offers a good growth potential. Penetration of this market into other grains has heretofore been blocked by the fact that propionic acid leaves a highly acidic taste, leaving the grain unpalatable to humans.

Other Ethylene Uses

Some ethylene is *copolymerized with propylene* to improve the impact resistance of polypropylene. This is done for some injection molding applications, wire and cable applications, and film.

Ethylene is used in varying degrees with *vinyl ester adhesive* resins to impart special properties.

Vinyl ester co-and terpolymer coatings for special uses incorporate ethylene to give the desired properties.

Vinyl toluene is prepared by alkylation of toluene with ethylene, then thermal cat cracking to vinyl toluene. Uses for vinyl toluene are believed to be in unsaturated polyester resins, and it replaces sytrene in some polymers where solvency is a factor. Some 15 to 20 million pounds of ethylene are used annually to produce vinyl toluene.

Aluminum alkyls used primarily as polymerization catalysts for alpha-olefins, high density polyethylene, polybutadiene, and others consume some 10 to 15 million pounds per year of ethylene.

Ethylanilines are intermediates used in the production of dyes, pesticides, and pharmaceuticals. Ethylene is used to alkylate aniline in these compounds.

Diethyl sulfate is used as an ethylating agent in the synthesis of dyes, textile finishes, pharmaceuticals, and pesticides.

1,4-Hexadiene is produced by the reaction of ethylene and butadiene. This is used as the third monomer in one company's ethylene-propylene terpolymer elastomer.

Agricultural uses of ethylene include use directly on fruits and vegetables to serve as a ripening agent.

Ethyl bromide is produced from ethylene and hydrogen bromide. It is used as an intermediate in chemical synthesis and as a solvent.

CHEMICALS FROM PROPYLENE

The use of propylene as a chemical building block has been increasing at a rapid rate. Its current consumption by the chemical industry is almost one-half as great as that of ethylene. In 1977, about 12 billion pounds were used to make chemicals, but even more propylene (15 billion pounds) went into gasoline alkylate and other fuel uses. Propylene in the United

States is currently derived about 70 percent from petroleum refining and 30 percent as a coproduct in the manufacture of ethylene. Because of its purity and convenient location, virtually all ethylene-coproduct propylene is used as chemical plant feed.

The most important derivatives of propylene are polypropylene, acrylonitrile, propylene oxide, and isopropanol. The break-down according to propylene consumption is;

Polypropylene	27.5%
Acrylonitrile	15.6
Propylene Oxide	13.8
Isopropanol	10.9
Cumene	10.2
Oxo Chemicals	8.1
Oligomers	6.2
Others	6.8

Chemical uses of propylene are expected to grow at about 5–6% per year, with the greatest growth being in acrylic acid, cumene, and polypropylene.

Polypropylene

In the twenty-odd years since its first commercial production, polypropylene has become by far the largest chemical use of propylene in the United States. It is primarily produced in a slurry-type reactor by polymerization of high-purity propylene in the presence of a catalyst such as $TiCl_3$-aluminum alkyl. A similar process can also be used to make high-density polyethylene. A new gas-phase polymerization has been introduced which shows promise of reducing the catalyst residue problem and making a purer product.

The major uses of polypropylene are as an injection-molding polymer and in fibers for carpet and industrial applications. Other uses are in wire insulation, film, and blow-molding. It is also copolymerized with ethylene to give a polymer with improved impact strength.

Polypropylene is described in more detail in Chapter 10.

Acrylonitrile

One of the biggest "success stories" in the last decades in the field of synthetic chemicals resulted from the development of a process for making acrylonitrile directly from propylene. At the same time, there was a huge growth in demand for the material. In 1960, almost all of the 260 million pounds of acrylonitrile produced came from acetylene. Ten years later, more than 1100 million pounds were produced, and the industry had converted almost entirely to the propylene-ammonia process. Today the conversion is complete and more than 2 billion pounds per year are made this way.

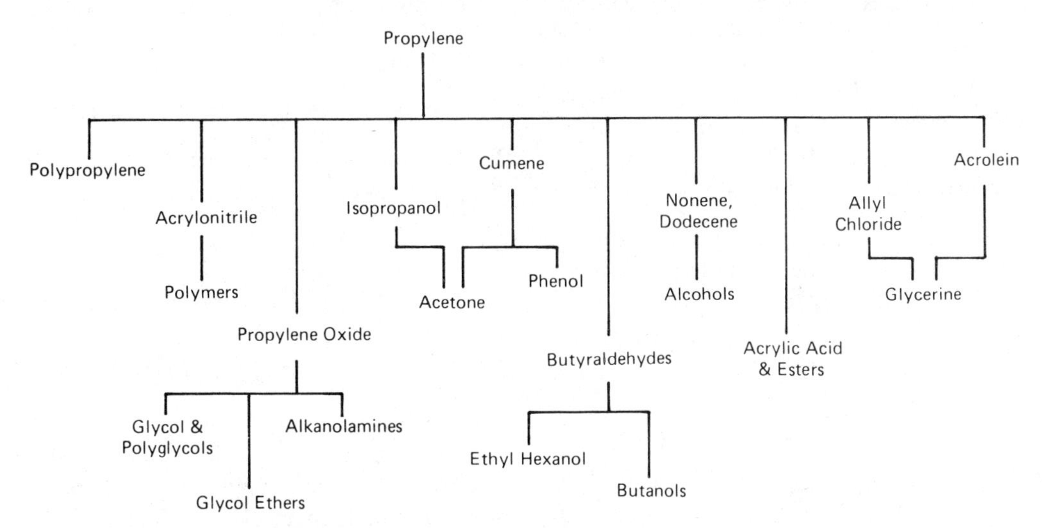

Fig. 25.19 Some chemicals derived from propylene.

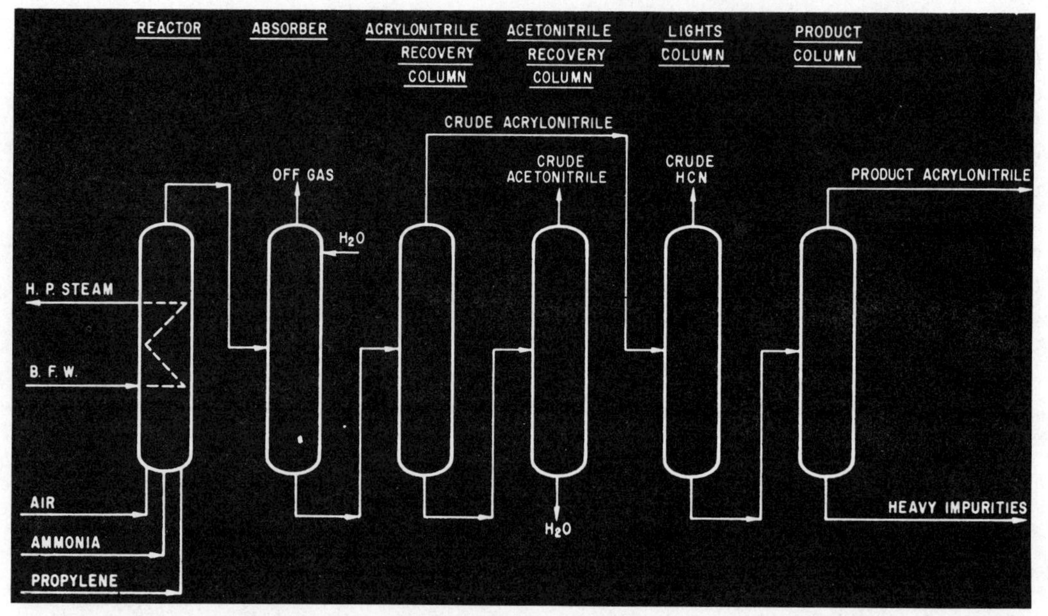

Fig. 25.20 Manufacture of acrylonitrile from propylene, ammonia and air. *(Hydrocarbon Processing,* **58,** *no. 11, 124 (1979); copyright 1979 by Gulf Publishing Co.)*

This process is shown in Fig. 25.20. Propylene, ammonia, and air are fed to a fluid-bed catalytic reactor operating at 800-900°F. The reactor effluent is scrubbed, and the organic materials are recovered by distillation. Hydrogen cyanide, water, and organic impurities are removed by fractionation. A high conversion to acrylonitrile is obtained with acetonitrile and HCN as by-products.

The major growth for acrylonitrile has been in the area of acrylic fibers. In 1977, more than 52% went into this use. Other uses are for ABS and SAN resins, adiponitrile, and nitrile rubber.

Propylene Oxide

Until recently, almost all propylene oxide was manufactured by the chlorohydrin process. This involves two steps:

1. Formation of the propylene chlorohydrin

$$H_2C{=}\overset{\displaystyle CH_3}{\overset{|}{CH}} + HOCl \longrightarrow \overset{\displaystyle Cl}{\overset{|}{CH_2}}-\overset{\displaystyle OH}{\overset{|}{CH}}-CH_3$$

2. Reaction with a base

$$\overset{\displaystyle Cl}{\overset{|}{CH_2}}-\overset{\displaystyle OH}{\overset{|}{CH}}-CH_3 + NAOH \longrightarrow \overset{\displaystyle O}{\diagup\,\diagdown}\!\!\!\!\!\!\!\!\!\!\!\!\!\!\quad CH_2CHCH_3$$

The process generates a mole of salt in the form of a dilute brine for every mole of product. Depending on reaction conditions, appreciable amounts of propylene dichloride and propylene glycol can also be produced.

In 1969, the first major plant using a non-chlorohydrin route was brought on stream. It uses a two-step process involving a hydroperoxide intermediate. Starting with isobutane, the steps are:

1. Formation of the hydroperoxide

$$(CH_3)_3CH + O_2 \longrightarrow (CH_3)_3COOH$$

2. Reaction with the propylene

$$(CH_3)_3COOH + CH_3CH{=}CH_3 \longrightarrow$$

$$CH_2\overset{O}{-}CHCH_3 + (CH_3)_3COH$$

The *t*-butanol can then be dehydrated to isobutylene if desired.

In some plants, ethylbenzene is used as the starting material which is dehydrated to make sytrene. Presently, almost one-half of the propylene oxide in the United States is made by the hydroperoxide route. It should be noted that the oxide is really the coproduct in this process, since almost twice as much styrene as oxide is produced.

A new variation of the chlorohydrin process has been described wherein *t*-butyl hypochlorite is the chlorinating agent, and the waste brine stream is converted back to chlorine and caustic in a special electrolytic cell.

The major uses for propylene oxide are propylene glycols and polyglycols and polypropylene glycols. Other uses are in isopropanolamines, glycol ethers, surfactants, and demulsifiers.

Glycols and Polyglycols. Propylene glycol manufacture is carried out with the same processes as those used for ethylene glycol, i.e., oxide hydrolysis, water removal, and product purification. The major uses are in resins and cellophane, as hydraulic fluids, as a tobacco humectant, and in cosmetics. Production in 1977 was 485 million pounds.

Polypropylene glycols are prepared commercially by the base-catalyzed addition of propylene oxide to propylene glycol. Propylene oxide can also be added to such starting materials as glycerine, pentaerythritol, sucrose, and sorbitol, depending on the type of product desired.

Polypropylene glycols and the polyglycols made by cocondensing ethylene and propylene oxide are used as lubricants, hydraulic fluids, mold release agents, and as the polyether portion in polyurethane foam. Approximately 1240 million pounds of propylene oxide is used to produce polypropylene glycols. Most of these polyols were used in the fast-growing polyurethane industry.

Isopropyl Alcohol

Isopropyl alcohol is claimed by some to be the first petrochemical. During the latter part of World War I, the first isopropanol was manufactured by the Ellis process, which was quite similar to the process used today. The flow sheet in Fig. 25.21 shows a typical isopropanol process scheme.

The dilute liquid-propylene feed stock is combined with recycled hydrocarbons and the mixture sulfated with H_2SO_4 at 300-400 psig. The sulfated mixture is then hydrolyzed to the alcohol, stripped from the acid, neutralized, degassed, and distilled to approach the binary azeotrope; then, it is finished in an azeotropic finishing column with an entrainer such as ethyl ether. The anhydrous product is taken from the bottom of the azeotropic still.

New plants have been installed in Europe

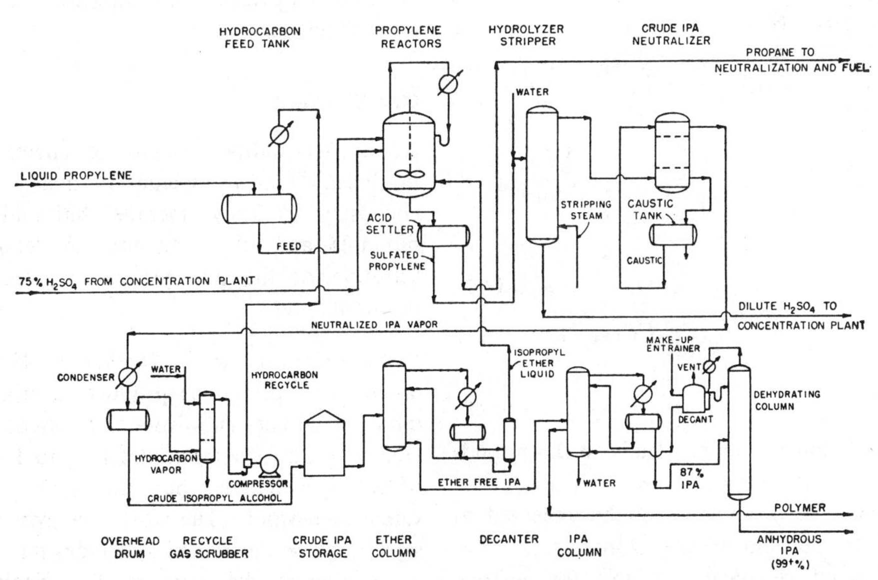

Fig. 25.21 Isopropyl alcohol manufacture. (Pet. Ref., *38*, No. 11, 264 (1959); copyright 1959 by Gulf Publishing Co.)

utilizing a direct hydration process in which propylene and water are reacted in the presence of a catalyst such as phosphoric acid on bentonite. The elimination of sulfuric acid should decrease processing and maintenance costs. The drawback of this new route is that it requires a highly concentrated propylene feed rather than a dilute refinery stream.

During 1977, some 1900 million pounds of isopropanol were produced. The major use was in the manufacture of acetone, which accounted for about 47 percent of the total. Other chemical uses were in the manufacture of acetates and xanthates.

Acetone. Acetone can be made from isopropanol by several routes, but catalytic dehydrogenation is the main one.

$$CH_3CHOHCH_3 \longrightarrow CH_3\overset{O}{\overset{\|}{C}}CH_3 + H_2$$

Copper, brass, or supported zinc is used as the catalyst. Operation at a high temperature (400-500°C) and moderate pressure (45 psig) gives about a 90 percent yield of acetone.

Less than 34 percent of the current acetone production comes from isopropanol as opposed to more than 80 percent in 1959. The decline in popularity of acetone from the isopropanol process can be attributed to the increasing number of phenol plants using the cumene route.

$$\underset{\text{Cumene}}{C_6H_5CH(CH_3)_2} + \text{air} \longrightarrow C_6H_5C(OOH)(CH_3)_2 \xrightarrow{H^{\oplus}} \underset{\text{Acetone}}{CH_3\overset{O}{\overset{\|}{C}}CH_3} + \underset{\text{Phenol}}{C_6H_5OH}$$

The oxidation of cumene leads to 0.6 pounds of acetone per pound of phenol produced. The importance of this source can be expected to grow as the demand for phenol increases.

The major chemical end uses for acetone are in the production of methyl methacrylate and methyl isobutyl ketone (MIBK). Methyl methacrylate is made by the acetone-cyanohydrin process, and MIBK by condensing acetone to form mesityl oxide. Other uses are in making bisphenol A, methyl isobutyl carbinol, pharmaceuticals, and as a solvent.

Isopropyl Alcohol Solvent. A significant amount of isopropanol is used as a solvent for essential and other oils, gums, shellac, rosins, and synthetic resins. Isopropanol is an excellent solvent for these materials and thus finds extensive use for compounding or blending numerous incompatible substances. As a component of nitrocellulose lacquer solutions, isopropyl alcohol improves blush resistance and increases solvency in esters and ketones.

Cumene

Cumene accounts for about 10% of the chemical usage of propylene. Almost all of the cumene is used to produce phenol and acetone by the cumene hydroperoxide process. Smaller amounts go into α-methylstyrene, acetophenone, and various solvents.

Although some cumene is present in petroleum refinery streams, most is made from benzene. The benzene is reacted with propylene in the presence of an alkylation catalyst at elevated temperature and pressure. The yields range above 90%.

Oxo Chemicals

About 1.1 billion pounds of chemicals are produced from propylene by means of oxo chemistry. They include butyraldehydes, butanols, and ethyl hexanol. A detailed description of the oxo process is given earlier in this chapter.

Butyl Alcohols and Aldehydes. Hydroformylation of propylene produces a mixture of *n*-butyraldehyde and isobutyraldehyde. The aldehyde ratio is approximately 2 to 1 in favor of *n*-butyraldehyde, but this ratio may be varied somewhat. The aldehydes may be used separately or the mixed aldehydes may be hydrogenated to the corresponding alcohols (*n*-butanol and isobutanol), which are then separated. The demand for n-butyraldehyde is much greater than that for the isobutyraldehyde.

The three major products of the butyraldehydes are 2-ethylhexanol, n-butanol, and isobutanol. The ethylhexanol is used mainly to make phthalate-type plasticizers for vinyl surface coatings. The butanols are used as solvents and as intermediates for plasticizers and resins.

Recent advances in catalysis have made it possible to also produce the butanols and ethylhexanol directly without isolating the butyraldehyde intermediate.

Other products which can be made by the oxo process and start originally with propylene are decyl alcohol, tridecyl alcohol, trimethylolpropane, and butyric acid.

Dodecene, Nonene

The manufacturing processes for these materials are very similar to cumene. The flow sheet in Fig. 25.22 specifically illustrates the manufacture of dodecene (propylene tetramer). When nonene is the desired product, additional fractionation is required, the extent of which is determined by product specifications.

In the reactor portion of this process, the olefin stock is mixed with benzene (for cumene) or recycle lights (for tetramer). The resulting charge is pumped to the reaction chamber. The catalyst, solid phosphoric acid, is maintained in separate beds in the reactor. Suitable propane quench is provided between beds for temperature control purposes since the reaction is exothermic.

Dodecene is an intermediate for surfactants, going mainly through two routes. One, the largest user, produces dodecylbenzene sulfonate for anionic detergents. The other goes through the oxo process to tridecyl alcohol, which is then converted into a nonionic detergent by the addition of alkylene oxides.

Nonene has two major outlets. The larger one is oxo production of decyl alcohol, which is used in the manufacture of esters, etc. for plasticizers. The other significant nonene use is in the manufacture of nonylphenol, an intermediate for the important series of ethoxylated nonylphenol nonionic surfactants.

Acrylic Acid and Esters

The first production of acrylic acid and esters from propylene began in 1970, and in just a few years it has replaced essentially all of the acetylene-based processes. The total propylene consumption for these products is estimated at almost 400 million pounds.

Propylene is first oxidized to acrolein, which can then be further oxidized to give acrylic acid. The esters are made by reacting the acid with the appropriate alcohol. Although both reactions can be carried out simultaneously, they are usually run separately in order to have the optimum catalyst and reaction conditions.

About 85% of the acrylic acid produced is

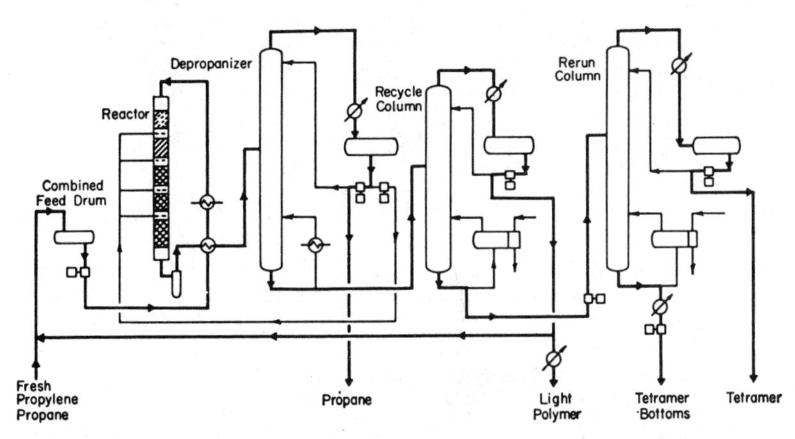

Fig. 25.22 Propylene tetramer process. (*Pet. Ref.*, **36**, *no. 11, 278 (1957); copyright 1957 by Gulf Publishing Co.*)

consumed in the manufacture of esters, particularly the methyl, ethyl, butyl, and ethylhexyl acrylates. The greater proportion of these esters goes to make resin emulsion polymers for surface coatings, textiles, and polishes. Other uses for the acrylic acid are in specialty acrylates, water-soluble resins, and salts.

Glycerine

An estimated 330 million pounds of glycerine was produced in 1979. Of this amount, less than 50% was made synthetically; the rest was obtained from natural sources, primarily as a by-product in the manufacture of soaps and fatty acids. All of the synthetic glycerine was derived from propylene, either by way of allyl chloride, acrolein, or propylene oxide.

Glycerine by the Epichlorohydrin Process. In the epichlorohydrin process, synthetic glycerine is produced in three successive operations; the end products of these are allyl chloride, epichlorohydrin, and finished glycerine, respectively. A flow sheet for this process is shown in Fig. 25.23. A portion of the allyl chloride is used to manufacture allyl alcohol, and a portion of the epichlorohydrin is used in the manufacture of epoxy resins.

The key reaction in this process is the hot chlorination of propylene which fairly selectively gives substitution rather than the addition reactions. In this chlorination step, fresh propylene is first mixed with recycle propylene. This mixture is dried over a desiccant, heated to 650-700°F, and then mixed rapdily with chlorine (C_3H_6 to Cl_2 ratio is 4:1) and

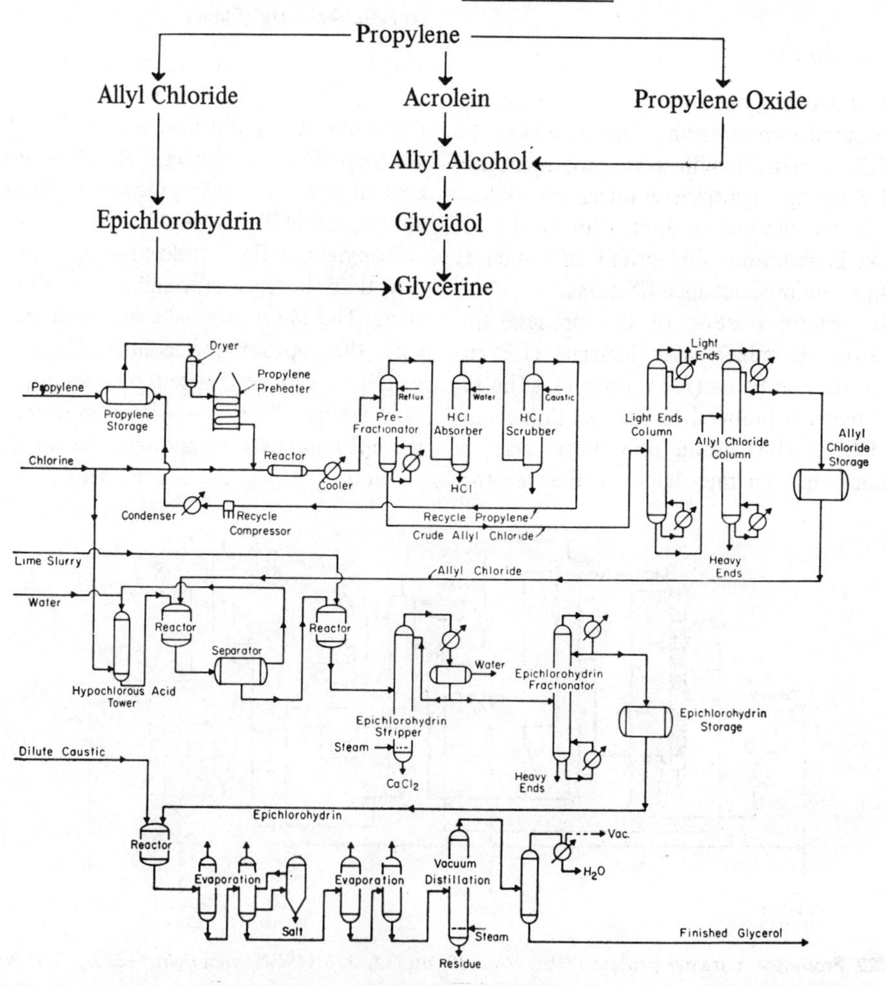

Fig. 25.23 Manufacture of glycerol. (*Pet Ref.*, **38**, *no. 11, 252 (1959); copyright 1959 by Gulf Publishing Co.*)

fed to a simple steel-tube adiabatic reactor. The effluent gases (950°F) are cooled quickly to 120°F and fractionated. Yield of allyl chloride is 80–85 percent.

Hypochlorous acid is then reacted with the allyl chloride at 85 to 100°F to form a mixture of dichlorohydrins. The reactor effluent is separated, the aqueous phase is returned to make-up hypochlorous acid, while the non-aqueous phase containing the dichlorohydrins is reacted with a lime slurry to form epichlorohydrin. The epichlorohydrin is steam-distilled out and given a finishing distillation.

Glycerine is formed by the hydrolysis of epichlorohydrin with 10 percent caustic. Crude glycerine is separated from this reaction mass by multiple-effect evaporation to remove salt and most of the water. A final vacuum distillation yields a 99^+ percent product.

A large amount of the epichlorohydrin intermediate is purified and used in the manufacture of unmodified epoxy resins.

Glycerine by Acrolein and Hydrogen Peroxide Process. This process is used to manufacture glycerine, but large amounts of acetone are obtained as a coproduct. Basic raw materials consumed are propylene and oxygen. Glycerine is synthesized in this process by hydroxylation of allyl alcohol with hydrogen peroxide.

The major end uses for glycerine are in drugs and cosmetics, alkyd resins, tobacco, food, cellophane, polyols, and explosives.

BUTANES, BUTYLENES, AND BUTADIENE

Saturated four-carbon hydrocarbons occur in natural petroleum products. They are found as heavy vapors in wet natural gas and in crude oil. The C_4's are also produced from other hydrocarbons during the various petroleum refining processes. The butylenes and butadiene—unsaturated C_4's—do not occur in nature, but are derived from saturated C_4's or from other hydrocarbons either as prime products or as by-products. The main sources of the C_4's are as shown in Table 25.4.

Wet natural gas, petroleum refining, and the by-product of ethylene manufacture are primary sources of C_4 hydrocarbons. A great majority of the C_4's come from wet natural gas and petroleum refining. About 59 percent of butadiene was obtained as a by-product of ethylene manufacture in 1978.

The complex interrelation between C_4 hydrocarbons showing how they are produced and used is shown most descriptively in Figure 25.24.

Chemical uses of C_4 hydrocarbons in 1978 show butanes consuming the most at about 5.5 billion pounds, with *n*-butenes second at approximately 1.85 billion pounds, and isobutylene estimated at 1.23 billion pounds. The chemical uses of the butanes, butenes, butadiene, and some of the processes used in making the derivatives are described in succeeding pages.

To place the volume of C_4's used in chemical manufacture in perspective with refinery uses for gasoline and other fuels, one finds that approximately 6 percent of the butanes and about 10.5 percent of the butylenes were used as chemical raw materials in 1978. Conversion of *n*-butane to gasoline and isobutane to gasoline in 1978 was estimated at 8 billion gallons (36 billion pounds) and 9 billion gallons (39 billion pounds), respectively. In the same year, about 5 billion gallons (27 billion pounds) of butenes were used in gasoline.

As shown in the source chart in Figure 25.25, the trends that effect availability of C_4 hydrocarbons for chemical and energy end-uses are determined by the natural gas processors, petroleum refiners, and, to a growing extent, ethylene manufacture. Changes in technology and in the availability of optimum feed stocks have far-reaching effects on the entire product mix. For example, as the availability of LPG and ethane for ethylene manufacture has decreased sharply, *n*-butane and the higher cuts of crudes have been used, and the proportion of by-product butediene produced has increased from the higher molecular weight feed stock. This by-product, butadiene, has moved into the market place first.

n-Butane

Oxidation of Butane. The old noncatalytic process for oxidation of LPG gases is essentially extinct today; however, it may still be used in

TABLE 25.4 Main Sources of C_4 Hydrocarbons[a]

	n-*BUTANE*	*ISOBUTANE*	n-*BUTENES*	*ISOBUTYLENE*	*BUTADIENE*
Wet Natural Gas	X	X			
Petroleum Refining					
Crude oil distillation	X	X			
Catalytic cracking	X	X	X	X	
Hydrocracking	X	X			
Catalytic reforming	X	X			
Thermal operations	X	X	X	X	
Isomerization		X			
By-Product of Ethylene Manufacture			X	X	X
Dehydrogenation of n-Butane			X		X
Dehydrogenation of n-Butenes					X
Dehydration of Tertiary-Butyl Alcohol Coproduct from Propylene Oxide Manufacture				X	
By-Product from Alpha-Olefin Production			X		

[a]*Chemical Economics Handbook*, Sept. 1976, p. 620.5010A, Stanford Research Insititue, Menlo Park, Ca. 1976.

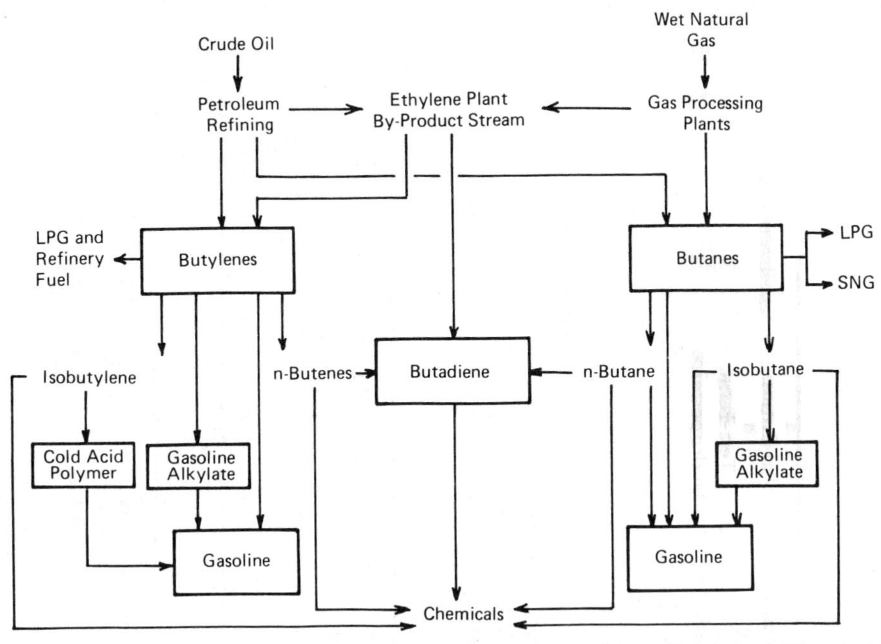

Fig. 25.24 Origins, interrelationships, and end uses of C_4 hydrocarbons. (*Reproduced from "Chemical Economics Handbook," p. 300.200A, March 1980; Stanford Research Institute, Menlo Park, Ca., 1980.*)

some foreign countries. The change in available feedstocks, as well as the need for more selectivity of the product mix, along with the complex separations necessary to recover all of the products, have all led to the replacement of this process.

Of more recent origin is the catalytic oxidation of butane, which incidentally emphasizes the trend toward increased product selectivity in oxidation processes. This employs a liquid-phase, high-pressure (850 psi) and 160-180°C oxidation using acetic acid as a diluent, and a metal acetate catalyst (cobalt or manganese). Figure 25.26 shows the flow sheet for this process. From the reactor, the product mixture is passed through coolers, and then through separators where the dissolved gases are released. The major components of the oxidized crude are acetic acid, methyl ethyl ketone, ethyl alcohol, methyl alcohol, propionic and butyric acids, and acetone.

It should be noted that unlike the propane-butane oxidation, this process produces no formaldehyde. The separation and purification procedure is similar to that described previously. Part of the acetic acid, which is the major product, is converted in a separate unit at high efficiencies to acetic anhydride.

Most acetic anhydride is made by pyrolysis of acetic acid to ketene, then reacting a mol of ketene with a mol of acetic acid to form the anhydride as shown below.

$$CH_3COOH \longrightarrow CH_2{=}C{=}O + H_2O$$

$$CH_3COOH + CH_2{=}C{=}O \longrightarrow (CH_3CO)_2O$$

This reaction takes place at high temperatures, and uses a triethyl phosphate catalyst. It is also possible to produce ketene from acetone at high temperatures.

$$CH_3COCH_3 \longrightarrow CH_2{=}C{=}O + CH_4$$

It is expected that the future of acetic anhydride will continue to be closely allied with that of its principal user, cellulose acetate.

Also connected with this plant are units for producing vinyl acetate, methyl acrylate, and ethyl acrylate.

The two large butane-oxidation plants in the U.S. together produced approximately one billion pounds of acetic acid and about 160 million pounds of methyl ethyl ketone in 1978.

Reaction conditions in this process can be changed to produce more methyl ethyl ketone at the expense of some acetic acid.

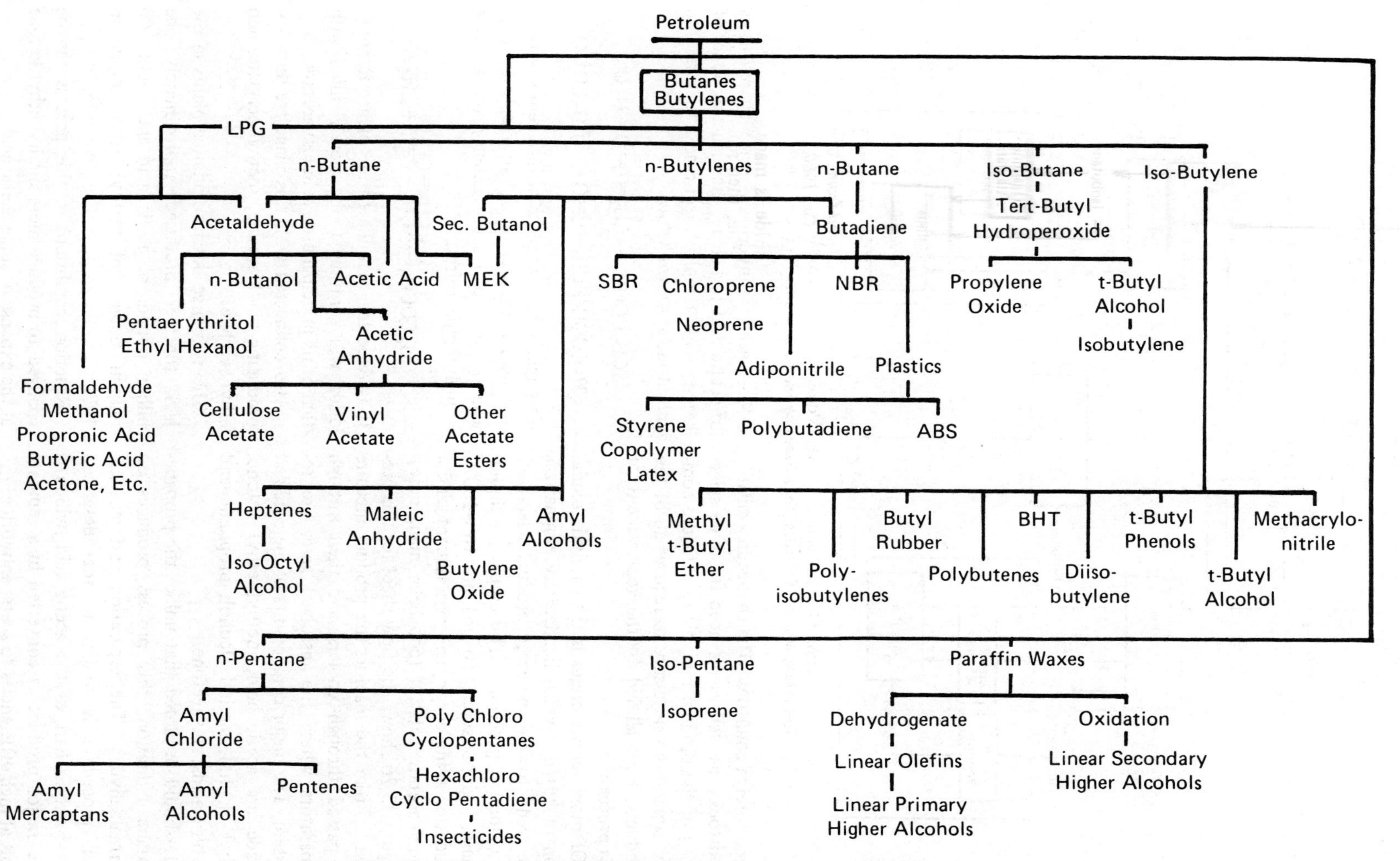

Fig. 25.25 Chemicals from butanes, butylenes, LPG, and higher aliphatic hydrocarbons.

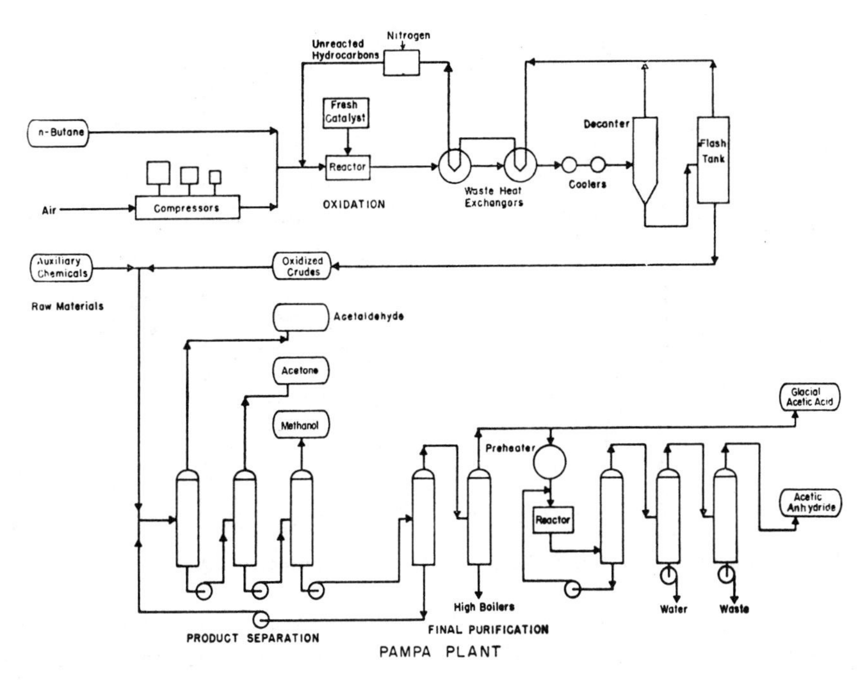

Fig. 25.26 Oxidation of butane. (*Pet. Ref.,* **38,** *no. 11, 234* (*1959*); *copyright 1959 by Gulf Publishing Co.*)

Maximum acetic acid/MEK ratio is 6.5 to 7 on weight basis. If ethyl acetate is also formed, the ratio can go to acetic/ethyl acetate of 3.6–4.0 with MEK at 55–57 percent of the by-product.

In addition to being used to make acetic acid and anhydride, some of the acetaldehyde is up-graded to *n*-butanol by the aldol process. The steps in the reaction are: (1) the acetaldehyde is condensed to aldol; (2) the aldol is dehydrated to crotonaldehyde; and (3) the crotonaldehyde is hydrogenated (180°C, nickel-chrome catalyst) to yield *n*-butanol.

$$2CH_3CHO \longrightarrow CH_3CH(OH)CH_2CHO \longrightarrow$$
$$CH_3CH{=}CHCHO \longrightarrow CH_3CH_2CH_2CH_2OH$$

This route to *n*-butanol is currently losing out to the oxo process which uses propylene and carbon monoxide.

n-Butane Dehydrogenation. The Phillips and Houdry processes are used for dehydrogenation of *n*-butane. In the Phillips process, high purity (98^+%) *n*-butane is first dehydrogenated to *n*-butenes, and then further dehydrogenated to butadiene. The Houdry or one-step process is fed 95^+% *n*-butane and conditions can be varied so that either butadiene or *n*-butenes are produced. Figure 25.27 shows a flow sheet of such a process. This is an adiabatic fixed-bed catalytic process. The catalyst is compressed cylindrical pellets consisting of activated alumina impregnated with chromic oxide. The reactor will normally operate in the range of 1000–1200°F, at a pressure of $\frac{1}{6}$ atmosphere or higher. The feed is preheated to reaction temperature before contact with the catalyst in the reactors. The hot effluent from the reactors is cooled in a quench tower by direct contact with circulating oil, then goes to compression, cooling, and conventional absorption and stabilization. The stabilized product stream is then extracted for recovery of high-purity product. Monoolefins, when not taken as product, are returned with the paraffin stream for recycle.

This process is suitable not only for dehydrogenation of *n*-butane to butadiene or *n*-butenes, but it can also be used for the production of propylene from propane; C_4 mono-olefins from *n*- or isobutane; butadiene from *n*-butenes; C_5 mono-olefins from *n*- or isopentane; isoprene from isopentane, isopentene, or mixtures thereof; or piperylene from *n*-pentane, *n*-pentenes, or mixtures thereof.

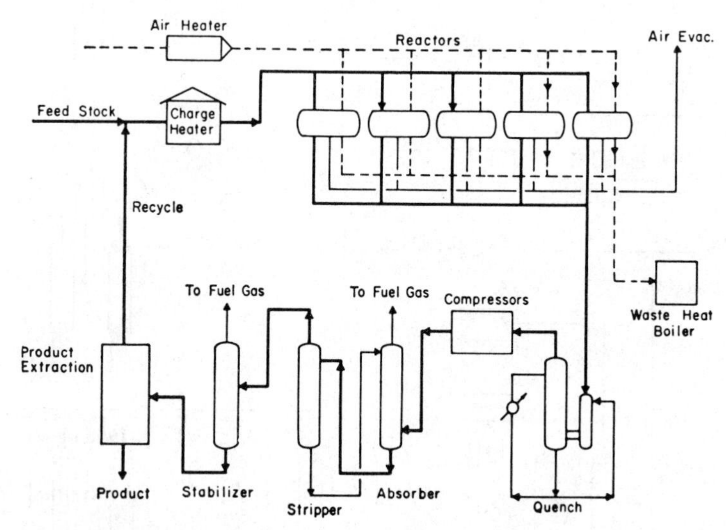

Fig. 25.27 Dehydrogenation. (*Pet. Ref.*, **38**, *no. 11, 233 (1959); copyright 1959 by Gulf Publishing Co.*)

Typical ultimate yields of butadiene from an *n*-butane feed stream are shown in the table below.

In 1978, 390 million pounds of butadiene was produced by the dehydrogenation of *n*-butane. This is out of a total of 3,515 million pounds produced by all manufacturing routes.

The trends in butadiene capacity, by the various manufacturing routes from 1968–1978 inclusive, are shown in Table 25.5

Since the Houdry dehydrogenation process is energy intensive, butane-based capacity is the first idled when butadiene demand decreases.

Ethylene Cracking. *n*-Butane as a feedstock for ethylene steam-cracking plants has accounted for only 2–4 percent of ethylene production since 1970. *n*-Butane is not used as a pure feedstock. However, it can be used to 10–15 percent of total feed in ethane/propane crackers with no major modifications. *n*-Butane can also be used as a supplemental feed to as high as 20–30 percent to heavy naphtha crackers. Relative prices and availability of alternate feedstocks will determine the use level of *n*-butane. As imported NGLs become available, and since the decline of the use of *n*-butane dehydrogenation as a source of butadiene, it is likely *n*-butane may become a more important factor in ethylene cracking feedstocks in the near future.

Maleic Anhydride. For many years, vapor-phase catalytic oxidation of benzene was essentially the exclusive process for the manufacture of maleic anhydride. In 1974, the first

TABLE 25.5 Butadiene Production Source Trends[a]

Year	Total Capacity Million Pounds	Percentage Production Based On: n-*butane*	n-*butene*	By-product from Ethylene Cracking and Coking
1968	3,008	43	38	19
1973	3,800	42	22	36
1975	4,220	37	20	43
1977	4,100	11	33	55
1978	4,430	11	30	59

[a]*Chemical Economics Handbook*, October, 1979, p. 300.350IV, Stanford Research Institute, Menlo Park, Ca. 1979.

plant utilizing the oxidation of *n*-butane came on stream. Since then, two additional plants utilizing *n*-butane oxidation were put into operation. By 1983, it is expected that over 40 percent of maleic anhydride capacity will be based on this technology.

The oxidation reaction to produce maleic acid from *n*-butane is as follows:

$$2C_4H_{10} + 7O_2 \longrightarrow 2 \begin{array}{c} C-C{=}O \\ \| \quad \backslash \\ \| \quad O \\ \| \quad / \\ C-C{=}O \end{array} + 8H_2O$$

n-Butane is oxidized by air over a zinc-vanadium-phosphorous catalyst in a fixed reactor. Yield is in the range of 41-45 percent based on *n*-butane.

Maleic anhydride is used primarily in the production of unsaturated polyester resins, agricultural chemicals, fumaric acid, and in lubricating additives.

Isobutane

Propylene Oxide/t-Butyl Alcohol. Peroxidation of propylene with *t*-butyl hydroperoxide is the basis of this new technology. This process involves oxidation of propylene with a molybdenum catalyst at 110°C. with *t*-butyl hydroperoxide, which is a product of reaction of isobutane and molecular oxygen in the liquid phase at 135-140°C. *t*-Butyl alcohol is a coproduct.

$$\underset{\text{isobutane}}{2(CH_3)_3CH} + 3/2\,O_2 \longrightarrow \underset{t\text{-butyl hydroperoxide}}{(CH_3)_3C{-}O{-}OH} + \underset{t\text{-butyl alcohol}}{(CH_3)_3COH}$$

$$\underset{\text{propylene}}{CH_2{=}CHCH_3} + \underset{t\text{-butyl hydroperoxide}}{(CH_3)_3C{-}O{-}OH} \longrightarrow \underset{\text{propylene oxide}}{H_2C\overset{O}{\text{—}}CHCH_3} + \underset{t\text{-butyl alcohol}}{(CH_3)_3COH}$$

t-Butyl alcohol is produced at a ratio of 2.0-2.5/1 to propylene oxide at nameplate capacity of 920 million pounds per year of propylene oxide; this plant has a coproduct capacity of 1.8-2.3 billion pounds per year of *t*-butyl alcohol.

t-Butyl alcohol can be used in gasoline as an effective octane improvement additive. If isobutylene is in high demand, the *t*-butyl alcohol can be readily dehydrated to high purity isobutylene.

Production of propylene oxide in 1978 was 635 million pounds with *t*-butyl alcohol coproduct production of 1.27-1.59 billion pounds, assuming an operating rate of 80 percent. Consumption of isobutane is in the range of 2.0-2.5 pounds per pound propylene oxide produced. Extensive expansions using this unique technology are reported for the next five-year period.

Other Isobutane-Based Chemicals. A new process for dehydration of isobutane to isobutylene by the Houdry technology is reported. This could provide an excellent base for the production of the gasoline octane-improver blending compound MTBE (methyl *t*-butyl ether). Isobutylene is reacted with methanol in the presence of an appropriate acid catalyst in large volumes to produce the MTBE. The market for this material is in the multiple billions of pounds annually.

Blends of purified *n*-butane and isobutane have essentially replaced fluorocarbons as aerosol propellants. This market is 50 million pounds annually.

n-Butylenes

Over 90 percent of C_4 olefins come from refinery streams. This represents only about 10-15 percent of the total available in these streams. The remainder comes from dehydrogenation of butane and as by-products of ethylene manufacture.

The basic problems of obtaining C_4 olefins is that of separation. Isobutylene is removed by absorption in 65 percent sulfuric acid. With isobutylene removed, isobutane and 1-butene are separated from *n*-butane and the 2-butenes by fractionation. The olefins can then be separated from the paraffin hydrocarbons by extractive distillation (with furfural, acetone, etc.). Available today are commercial quantities of 1-butene and 2-butene, 95 percent pure or higher. The 2-butene is a mixture of the *cis* and *trans* isomers.

The major derivatives of *n*-butenes are butadiene, *sec*-butyl alcohol, heptenes, butylene oxide, and crack-resistant high-density polyethylene.

Butadiene. Butadiene is produced by the dehydrogenation of *n*-butenes, or it may be produced as a coproduct with butenes in the dehydrogenation of butane. A flow sheet of the latter processing scheme is shown in Fig. 25.26. Both are fixed-bed catalytic dehydrogenation processes. Slightly higher yields of butadiene may be obtained from butene, but overall economics, utilization of the resulting product mixture, and readily available feed streams determine the route.

Butadiene production in 1978 was approximately 3.5 billion pounds. 30 percent of this was produced by *n*-butylene dehydrogenation. Some 44 percent of butadiene goes into styrene-butadiene rubber. Other uses are polybutadiene, 19 percent; high-impact polystyrene and paint latices, 13 percent; adiponitrile, 11 percent; polychloroprene, 7 percent; nitrile rubber, 3 percent; and miscellaneous uses, 3 percent.

The more promising growth areas for butadiene use are Nylon resins, 7–8 percent; high-impact polystyrene, 6–8 percent; ABS resins, 5–6 percent; and adiponitrile, 4–5 percent.

For more details on polymerization, manufacture, and uses of butadiene in rubber and plastics, see Chapters 9 and 10.

sec-Butanol and Methyl Ethyl Ketone. The largest end-product chemical use for *n*-butenes is in the manufacture of *sec*-butanol. A mixed feed containing butane and *n*-butenes is contacted with 80 percent sulfuric acid. Dilution and steam-stripping then produce the alcohol. The only important use of *sec*-butanol is the production of methyl ethyl ketone (MEK).

MEK is obtained by catalytic dehydrogenation of *sec*-butanol. A yield of approximately 75 percent is obtained in this step. A flow sheet, Fig. 25.28, shows an integrated process for producing *sec*-butanol and MEK.

Approximately 85 percent of MEK is manufactured by the *sec*-butanol route. MEK is also obtained as a by-product from the butane-oxidation process for acetic acid.

MEK is used almost entirely as a solvent in such applications as vinyl-resin lacquers, nitrocellulose, lacquers, lube oil dewaxing, paint removers, rubber cement, and adhesives.

U.S. production of MEK in 1978 was reported as 598 million pounds.

Heptenes and Isooctyl Alcohol. The heptenes comprise a C_7 olefin cut fractionated from polymer gasoline that is produced by polymerizing C_3-C_4 refinery gases. The C_4 used is generally *n*-butene. Heptenes are used mainly to feed oxo units for the manufacture of isooctyl alcohol. Small quantities of hep-

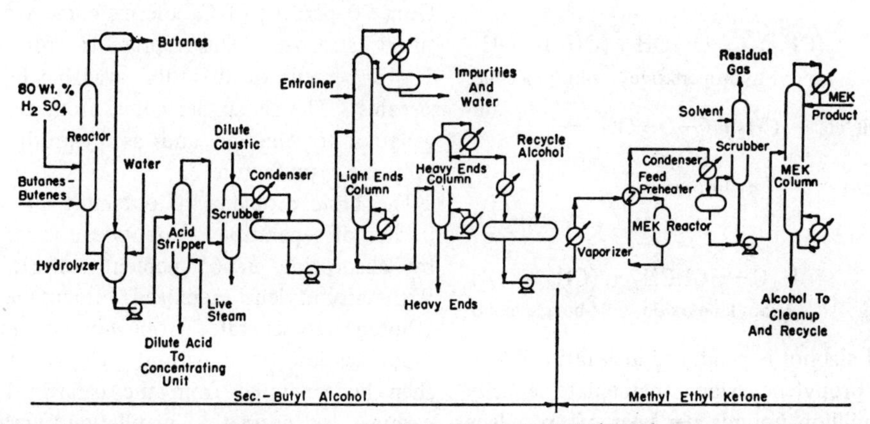

Fig. 25.28 Methyl ethyl ketone or acetone. (*Pet. Ref.*, **38**, *no. 11, 272* (*1959*); *copyright 1959 by Gulf Publishing Co.*)

tenes are used to make isohexadecyl alcohol, heptylphenol, and heptylbenzene.

In 1978, isooctyl alcohol equivalent of heptenes produced was about 135 million pounds. In the same year, 130 million pounds of isononyl alcohol was produced.

Butylene Oxide. Approximately 6 million pounds of butylene oxide was produced in 1978. The feed used was 1-butene to give 1,2 butylene oxide. Manufacture is carried out by the chlorohydrin process. Butylene oxide is used as a corrosion inhibitor in chlorinated solvents such as methyl chloroform and trichloroethylene.

Amyl Alcohols. Primary amyl alcohols are produced by the oxo reaction of *n*-butenes. The resulting alcohol mixture is about 60-percent *n*-amyl alcohol, 35 percent 2-methyl-1-butanol and 5 percent 3-methyl-1-butanol. This alcohol may be fractionated or used as the mix for solvent uses, for acetate esters, and esters of dithiophosphates which are used as lube-oil and hydraulic-fluid additives. Production in 1978 was estimated at 60 million pounds.

High-Density Polyethylene Copolymers. High-density polyethylene made with 3-5 percent high-purity 1-butene ($98^{+}\%$) gives a plastic with improved crack-resistance under stress. This property has led to extensive use in blow-molding applications where this resistance is important. In 1978, some 45 million pounds of butene-1 were used in stress resistant copolymer. New low-density polyethylene polymers are soon to enter the market-place containing 4-10 percent butene-1 as comonomers. Use here will approach that used in high-density materials noted above.

Isobutylene

Isobutylene is more reactive than the *n*-butenes, but many of the compounds formed are quite readily reversible under less than extreme conditions.

Over 95 percent of isobutylene used in the chemical industry goes into di- and trisobutylenes, butyl rubber, and other polymers. Total isobutylene consumption for chemicals was estimated at nearly 1,233 million pounds in 1978.

Dimers and Trimers of Isobutylene. A mixture of dimers and trimers of isobutylene is produced by absorbing isobutylene in 60-65 percent sulfuric acid at 10-20°C, then reacting at 80-100°C for $\frac{1}{2}$ hour.

The main chemical use of the dimer mixture is in the alkylation of phenols to octylphenols for detergent use. Nonyl alcohol is also produced from this dimer by the oxo process.

Use of diisobutylene is limited by its tendency to thermally depolymerize to isobutylene. Thus, only low-temperature reaction conditions can be utilized.

Approximately 60 million pounds of diisobutylene are produced annually for chemical uses.

Polybutenes and Polyisobutylenes. Polymerization of isobutylene can be carried out with such catalysts as boron trifluoride and aluminum chloride. This gives a wide range of polymers—from viscous liquids (called polybutenes) to semisolid and solid polymers (called polyisobutylenes).

Polybutenes are very stable materials with good oxygen or ozone resistance. These are highly saturated materials which do not set or "dry" on storage or use. Important industrial applications are in caulking and sealing compounds, adhesives, surgical tapes, vibration dampers, electrical insulation, and special lubricants. Output of polybutenes in 1978 was estimated at 465 million pounds.

Polyisobutylenes vary from soft, sticky gums to tough, elastic materials. Stability and resistance to chemical attack are excellent. Polyisobutylenes are used as tacky agents for oils to avoid dripping and splattering from bearings and rotating shafts, etc. These materials are also used as a viscosity-index-improver for oil and hydraulic fluids. Estimated 1978 output of polyisobutylenes was 60 million pounds.

A flow sheet showing a polybutene process is shown in Fig. 25.29.

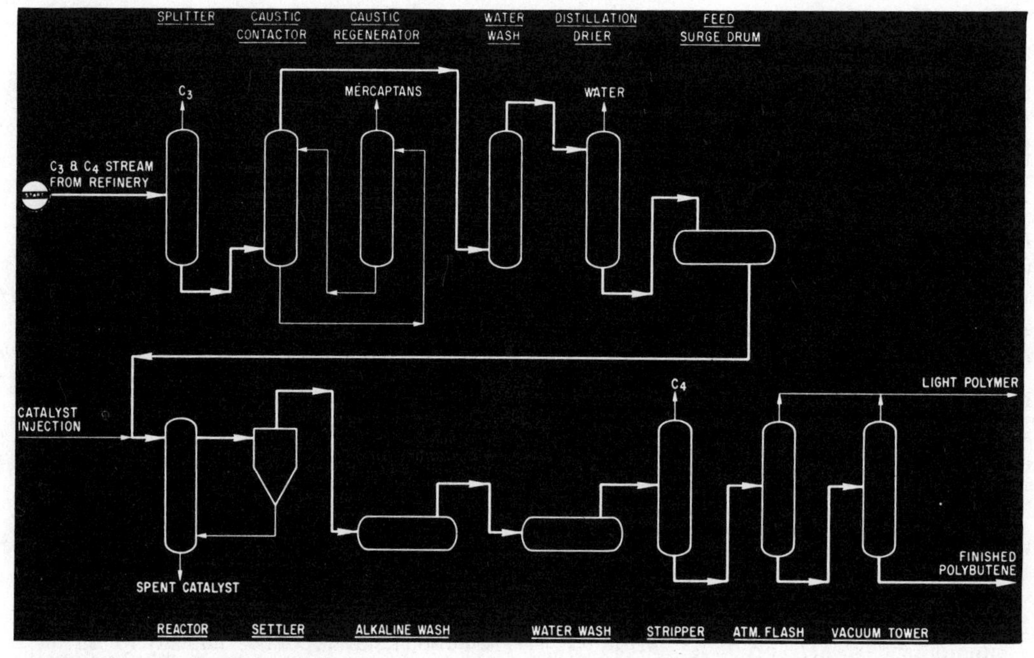

Fig. 25.29 Liquid polybutene from mixed C_4 hydrocarbons. (*Hydrocarbon Processing*, **50**, *no. 11, 194 (1971); copyright 1971 by Gulf Publishing Co.*)

Butyl Rubber. Low-temperature (-100°C) copolymerization of isobutylene (98 percent) and isoprene (2 percent) produces a solid, rubber-like, vulcanizable polymer. Butyl rubber has unusually low gas permeability and thus has found extensive use in tire inner liners and tubes and pneumatic bags. Chemical variations such as chlorine-containing butyl rubbers have made butyl rubber compatible with SBR and natural rubber for blending purposes. See Chapter 9 for more details.

Output of some 337 million pounds of butyl rubber was estimated for 1978, and this volume is essentially static.

Butyl Hydroxy Toluene (*BHT*). The more proper chemical name for BHT is 4-methyl-2,6-di-*t*-butylphenol. The material is produced by alkylating *p*-cresol with high-purity isobutylene. BHT is an antioxidant and is sold in food grade for use in fats, oils, and fat-containing foods. The technical-grade product is used chiefly as a gum inhibitor in gasoline. Some 23 million pounds of BHT were produced in 1978.

t-*Butylphenols*. *t*-Butylphenols are formed by the reaction of phenolic compounds with isobutylene using sulfuric acid as catalyst.

These compounds are used as intermediates for bactericides, oil soluble phenol-formaldehyde resins, and antioxidants. Some 56 million pounds of compounds in this class are used annually, requiring about 30 million pounds of isobutylene.

t-*Butyl Alcohol*. The hydration of isobutylene to *t*-butyl alcohol goes readily under mild acid conditions. The alcohol is easily dehydrated under acidic conditions, thus it must be isolated from a dilute acid or neutral system. This property limits the use of this alcohol. Also limiting its use is its relatively high (25.6°C) freezing point. Production in 1978 was approximately 12 million pounds.

Isoprene. The Prins reaction between isobutylene and formaldehyde has been used somewhat to produce 4,4-dimethyl-1,3-dioxane as an intermediate. The second step catalytically converts this intermediate to isoprene and formaldehyde. Yields of approximately 76 percent on formaldehyde and 83 percent on isobutylene have been reported. Commercial use is still restricted by cost, since such materials as butadiene and styrene which are made on a mammoth scale make competition stiff. A more typical route to isoprene is shown in the C_5 section of this chapter. In the United States today, isoprene from heavy-liquid crackers is satisfying the market.

Methacrylonitrile. This is produced by the ammoxidation of isobutylene in a process similar to that used to manufacture acrylonitrile from propylene, ammonia, and air.

t-*Butyl Mercaptan.* This product is made from isobutylene. 1978 production was 13 million pounds. Its uses are as an odorant in natural gas and as an intermediate in an organophosphate corn insecticide.

t-*Butylamine.* *t*-Butylamine is formed by the reaction of isobutylene with HCN in the presence of strong sulfuric acid. The intermediate *t*-butyl formamide is then hydrolized to form *t*-butylamine. This amine is used mainly to synthesize sulfenamide rubber accelerator compounds. 18 million pounds of the amine were produced in 1978.

HIGHER ALIPHATIC HYDROCARBONS

n-Pentane and Cyclopentane

Derivatives of *n*-pentane are used for small specialty type products. Perhaps the most important are: (1) amyl chlorides, which in turn are converted to amyl mercaptan for use as a warning agent in natural gas; and (2) amyl alcohols, which are converted to sweet-smelling acetate esters for flavoring and photographic uses. Amyl chlorides can be dehydrohalogenated to specialty pentenes.

Cyclopentane finds its major outlet as an intermediate in the production of some Diels-Alder reactions with cyclopentadiene, in which the next step leads to such insecticide products as aldrin, chlordane, dieldrin, and heptachlor. Because these insecticides are very stable and difficult to biodegrade, they are losing their strength in the market place.

Isopentane

Of major importance is the use of isopentane as a base stock for dehydrogenation to isopentene and to isoprene. A flow sheet of a process for converting isopentane and tertiary amylenes to isoprene is shown in Fig. 25.30. This process exists essentially in two sections. The first part is a cleanup of cracked stock to obtain some tertiary amylenes of reasonable quality to be used in the final dehydrogenation to isoprene. The remainder of the process is a cleanup and purification, where the final purification uses the Shell ACN (acetonitrile) extractive distillation to give high-purity isoprene.

Isoprene is used mainly in polyisoprene elastomers which are currently growing rapidly. A much smaller use is in the manufacture of butyl rubber, a copolymer of isobutylene and isoprene. Isoprene production in the United States in 1978 is reported as 184 million pounds.

n-Paraffins, Monoolefins, Primary and Secondary Higher Alcohols

n-*Paraffins.* A process for separation of high purity normal paraffins in the C_{10} to C_{24} range from branched-chain and cyclic hydrocarbons is shown in Fig. 25.31. The principle employs the highly selective absorption action of molecular sieves in a multibed system. The beds are loaded with normal paraffins successively and desorption of normal paraffins is carried out with a purge stream. The purge material is then separated from the *n*-paraffin by distillation.

Monoolefins. Linear internal monoolefins are produced from the appropriate-carbon-number linear paraffins. Products are produced with a linear-monoolefin purity of about 94 wt percent. A flow sheet of the processing scheme is shown in Fig. 25.32. In this process, the linear paraffin is catalytically dehydrogenated

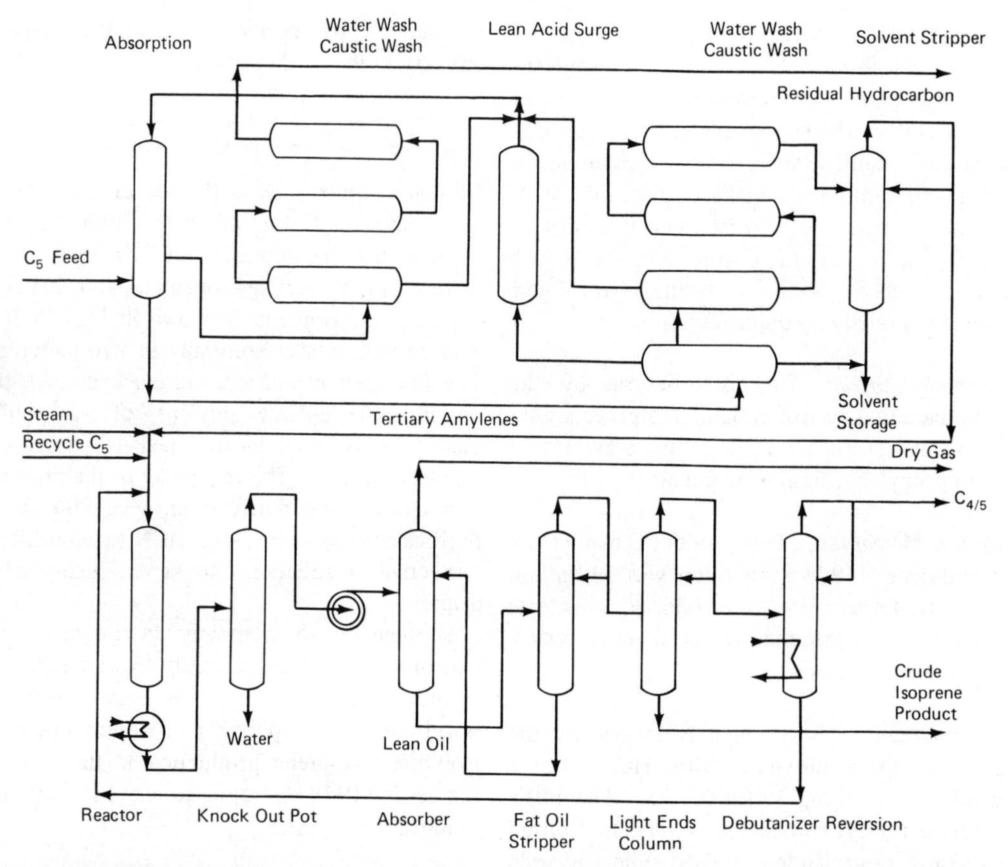

Fig. 25.30 High purity isoprene monomer—USSR process. (*Hydrocarbon Processing,* **50,** *no. 11, 168 (1971); copyright 1971 by Gulf Publishing Co.*)

to the monoolefin. The olefin and paraffin are then separated by using a fixed-bed absorbent-extraction system. A desorbent is added to each stream, i.e., the olefin extract and the paraffin raffinate, and is separated from each in a distillation column. The desorbent is a low-boiling hydrocarbon, chosen so that it will separate easily overhead in each of the final distillation columns.

Primary and Secondary Higher Alcohols. Essentially three methods are used to arrive at the higher alcohols. In one case *n*-dodecyl (lauryl) alcohol (C_{12}) is made by hydrogenation of coconut oil to give the linear primary alcohol.

In another case, the linear olefin is subjected to a modified oxo reaction to give a primary alcohol having some alpha-methyl branching.

In the third case, the linear paraffin is oxidized giving a linear secondary alcohol.

The major use for these higher alcohols, of course, is in the synthesis of biodegradable synthetic detergents. Depending on the final functionality desired, the alcohols are ethoxylated, sulfated, phosphated, etc. to add the appropriate hydrophilic functional groups. These materials generally compete with the alpha-alcohols which are made directly or from alpha-olefins as described in the *Ethylene* section of this chapter.

AROMATIC CHEMICALS

Until World War II, coal tar was the only source of aromatic raw materials. Since then, petroleum has overtaken it to the point where now about 92% of the benzene, 97% of the toluene,

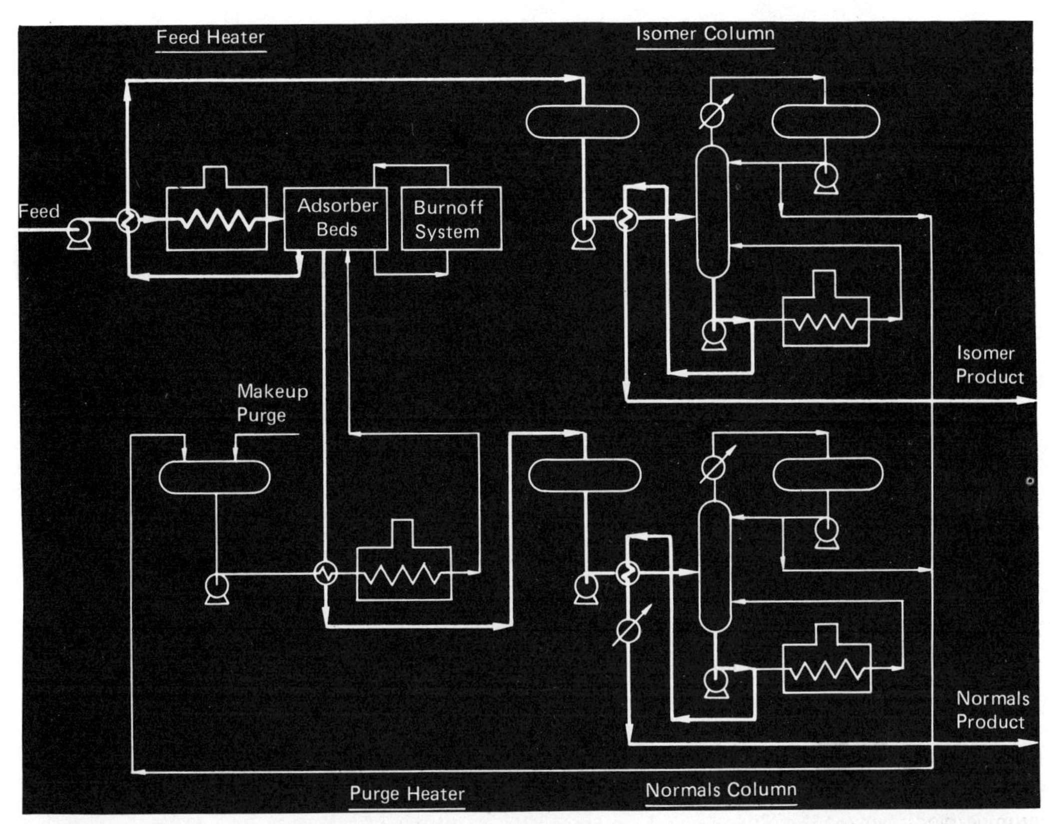

Fig. 25.31 Separation of high purity C_{10}–C_{24} normal paraffins from kerosene. (*Hydrocarbon Processing,* **50,** *no. 11, 184* (*1971*)*; copyright 1971 by Gulf Publishing Co.*)

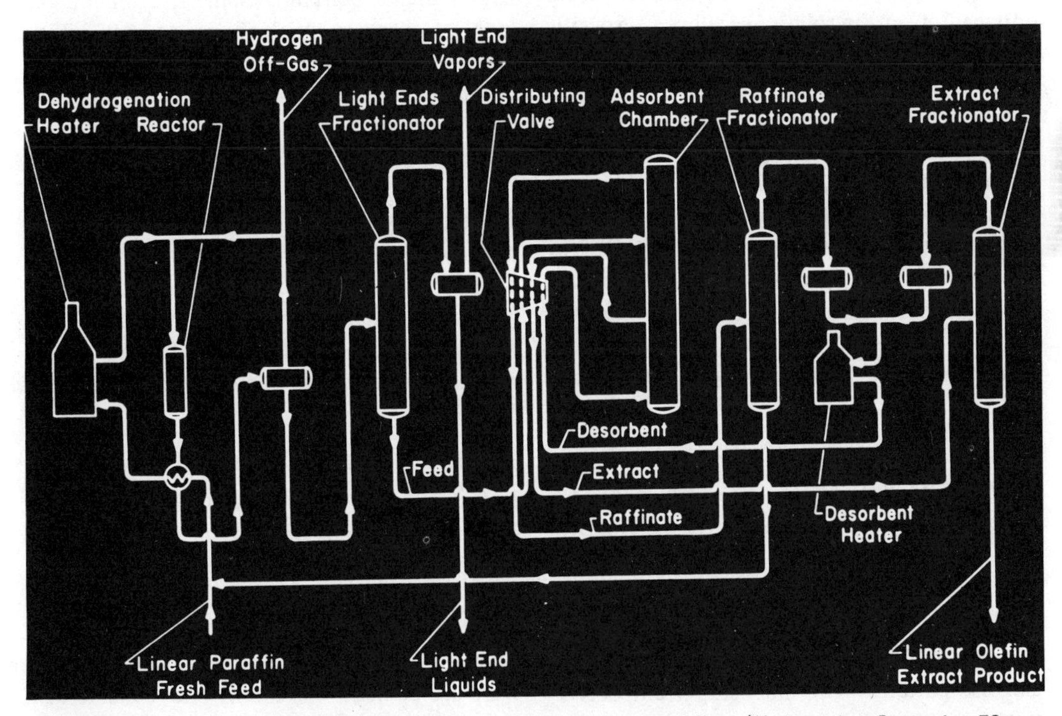

Fig. 25.32 Catalytic dehydrogenation process for linear internal monoolefins. (*Hydrocarbon Processing,* **50,** *no. 11, 174* (*1971*)*; copyright 1971 by Gulf Publishing Co.*)

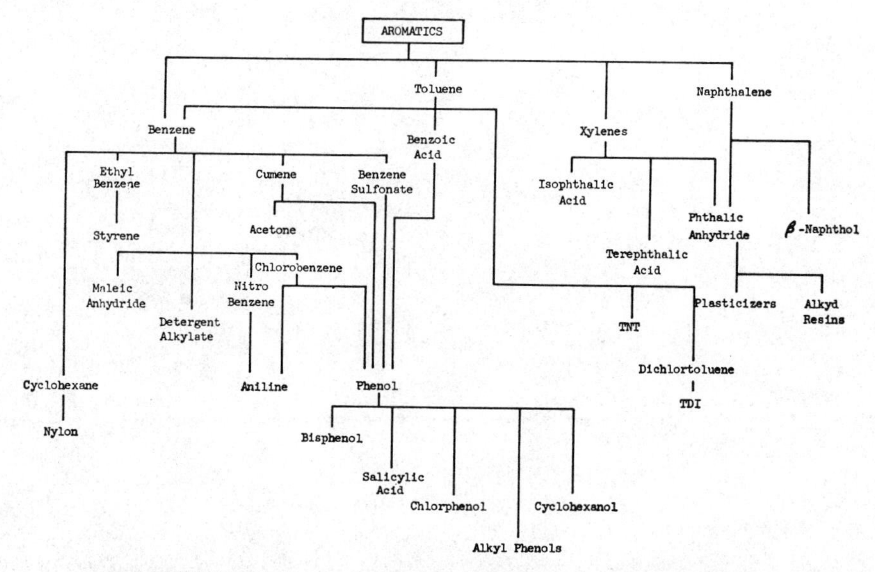

Fig. 25.33 Synthetic organic chemicals derived from aromatic compounds.

and 99% of the xylenes are petroleum-based. The first petroleum-derived naphthalene plant was put on stream in early 1961. Most petroleum-derived benzene, toluene, and xylenes (BTX) are obtained from the catalytic reformate streams in refineries. A smaller amount is obtained from pyrolysis gasoline. Specific processing methods for producing these materials from petroleum are described in Chapter 14.

The volume of aromatic chemicals used in chemical processing is fairly small compared to the amount that goes into fuel. About 40% of the benzene, 91% of the toluene, and 92% of the xylenes find their way into the gasoline pool. With the demise of leaded gasoline, the demand for aromatics in fuel is expected to remain high unless a better octane-improver can be found.

Benzene Products

Benzene is by far the most important aromatic raw material. During 1978, some 1.6 billion gallons were consumed, which ranks it about even with propylene as a synthetic organic chemical building block. Benzene has a broad end use pattern. The most important derivatives are ethylbenzene (styrene), cumene (phenol), cyclohexane, nitrobenzene (aniline), chlorobenzene, maleic anhydride, and detergent alkylate. Most of these products are expected to show continued growth.

Styrene. Styrene manufacture is by far the largest use of benzene. Some 815 million gallons of benzene yielded about 6.9 billion pounds of styrene monomer in 1979. Styrene monomer is made by cracking (dehydrogenating) ethylbenzene (see Chapter 9 for details).

The major uses for styrene are in plastics, rubber latex paints and coatings, synthetic rubber, polyesters, and styrene-alkyd protective coatings. In these uses styrene is polymerized to a homopolymer or in copolymers with such materials as acrylonitrile, butadiene, maleic anhydride, and glycols. Further details may be found in the chapters on plastics, rubber, and paints.

Phenol. Synthetic phenol is now the second largest market for benzene. Approximately 305 million gallons of benzene were consumed to make 2.4 billion pounds of phenol in 1978. Before 1970, there were five different processes used to make phenol in the United States. They were the sulfonation, chlorobenzene, Raschig, cumene oxidation, and benzoic acid processes. By 1978, the first three had disappeared entirely, and 98% of the remaining

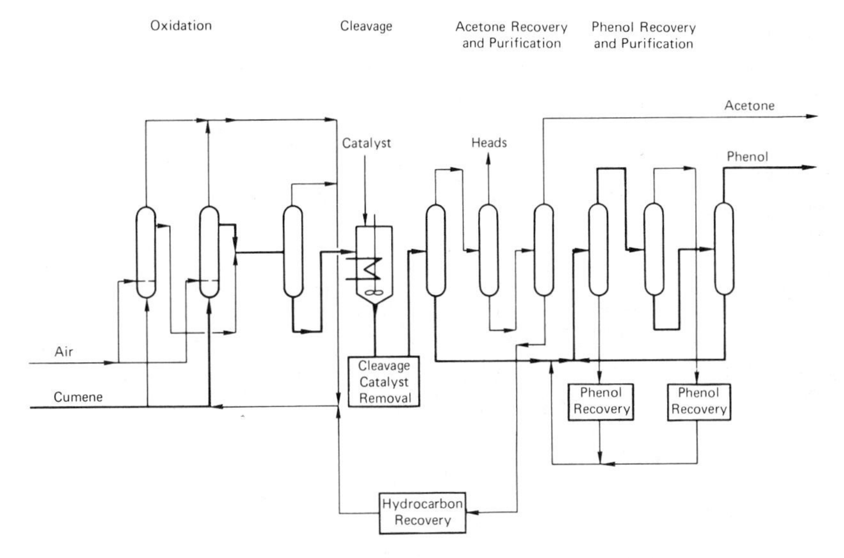

Fig. 25.34 Manufacture of phenol and acetone by oxidation of cumene. (*Hydrocarbon Processing, Nov., 1979; copyright 1979 by Gulf Publishing Co.*)

plant capacity was based on the cumene oxidation route. In the process shown in Fig. 25.34, cumene, which is made by reacting benzene and propylene, is oxidized with air to cumene hydroperoxide. This is then concentrated and cleaved to phenol and acetone by acid catalysis. It is reported that the total capacity of licensee's plants built and under construction in 1979 amounted to 4 billion pounds per year. The benzoic acid route to phenol involves the oxidation of toluene. The two major uses of phenol are for phenolic resins and bisphenol A.

Phenolic Resins, such as those made from phenol and formaldehyde, are by far the major user of phenol and should maintain this position. They are used in such applications as plywood adhesives, insulation binders, molding compounds and laminating resins. A more detailed description can be found in Chapter 10.

Bisphenol A is used in the production of polycarbonates and epoxy resins, which are growing plastics. Bisphenol A is in turn obtained from phenol by the following reaction:

$$2\,C_6H_5OH + CH_3\overset{\overset{\displaystyle O}{\|}}{C}CH_3 \xrightarrow{H^+} HO\text{-}C_6H_4\text{-}\underset{CH_3}{\overset{CH_3}{C}}\text{-}C_6H_4\text{-}OH + H_2O$$

Bisphenol A

A flow sheet for bisphenol A manufacture is shown in Fig. 25.35.

Phenol and acetone in a molar ratio of 3:1 or 4:1 are charged to an acid-resistant stirred bisphenol A reactor. Glass-lined equipment is used ordinarily. A sulfur-containing catalyst is added; then dry HCl gas is bubbled into the reaction mass. The temperature is maintained at 30 to 40°C for 8-12 hours. The product crystallizes from the reaction mixture to form a slurry.

At the end of the reaction, the mixture is washed with water and treated with just enough lime to neutralize the free acid. Vacuum and heat are applied to the reaction kettle, and water and phenol are distilled separately from the mixture. The batch is finished by blowing the molten product with steam under vacuum at 150°C to remove the odor of the sulfur-containing catalyst.

The molten bisphenol A product is quenched in a large volume of water, filtered, and dried. It is a light tan powder, which may be further purified by recrystallization from solvents.

The patent literature indicates that ion-exchange resins promoted with various mercaptans or amines may be used in some of the newer plants in operation today.

U.S. consumption of bisphenol A during 1977 was mainly in epoxy resins, and in polycarbonate resins. Total consumption was over 450 million pounds.

Other products from phenol include aspirin, alkylated phenols, chlorinated phenols (to 2,4-D), and caprolactam. Chemical reactions for some of these materials are shown below:

Important by-products from the chlorobenzene process which are now substantial volume chemicals are phenylphenols, numerous chlorophenols, and Dowtherm® A (a eutectic mixture of diphenyl and diphenyl oxide sold as a heat transfer medium).

Cyclohexane. Cyclohexane is the third largest consumer of benzene. During 1978, some 250 million gallons were used to make it. Almost all of the cyclohexane goes to make nylon, either through adipic acid, hexamethylene-diamine, or caprolactam.

1. Phenol (OH) $\xrightarrow{NaOH}$ ONa $\xrightarrow[100°C,\ 5\text{-}6\ atm.]{CO_2}$ OH, COONa $\xrightarrow{H^+}$ Salicylic Acid (OH, COOH) $\xrightarrow[CH_3COOH,\ H_2SO_4,\ 90°C]{acetic\ anhydride}$ Aspirin ($OC(=O)-CH_3$, COOH)

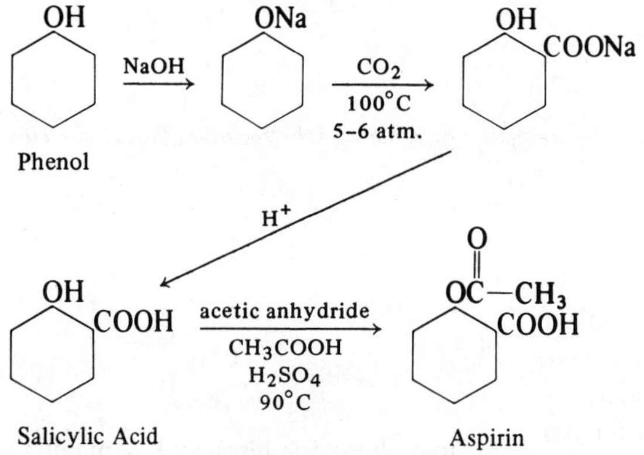

2. OH + $RC(H)=C(H)H$ (alkene) $\xrightarrow[metal\ halide]{H^+}$ OH, CH_2CH_2R (*o*-Alkyl phenol) + OH, CH_2CH_2R (*p*-Alkyl phenol)

3. OH + Cl_2 $\xrightarrow{90°C}$ OH, Cl, Cl $\xrightarrow{ClCH_2C(=O)OH}$ $OCH_2C(=O)OH$, Cl, Cl (2,4-D)

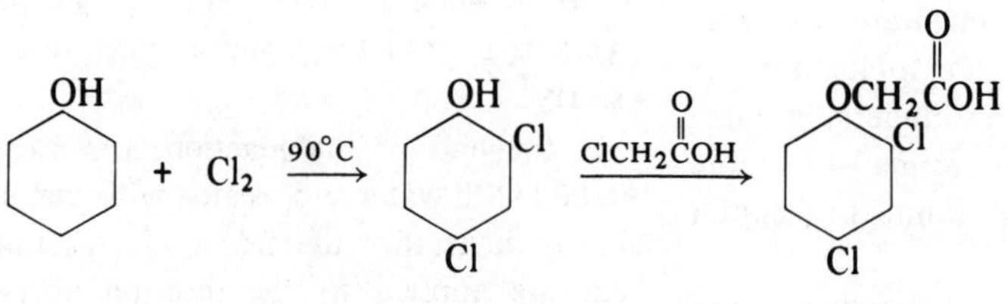

4. OH $\xrightarrow[Ni]{H_2}$ OH (S) $\xrightarrow{(O)}$ O (S) $\xrightarrow{NH_2OH}$ NOH (S) $\xrightarrow[rearrangement]{H^+\ Beckmann}$ Caprolactam (H_2C, CH_2, C=O, NH, CH_2, CH_2, H_2C ring)

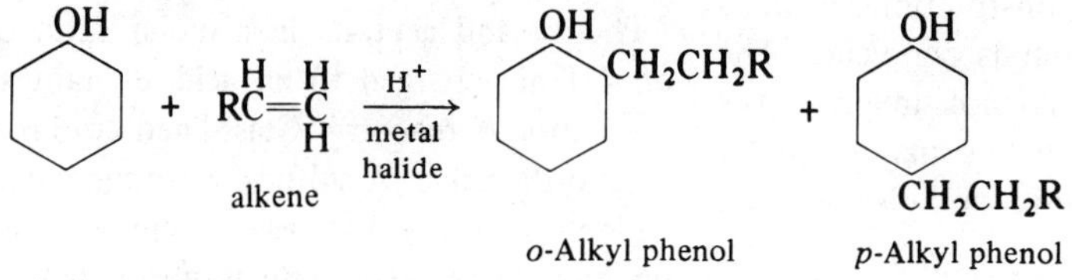

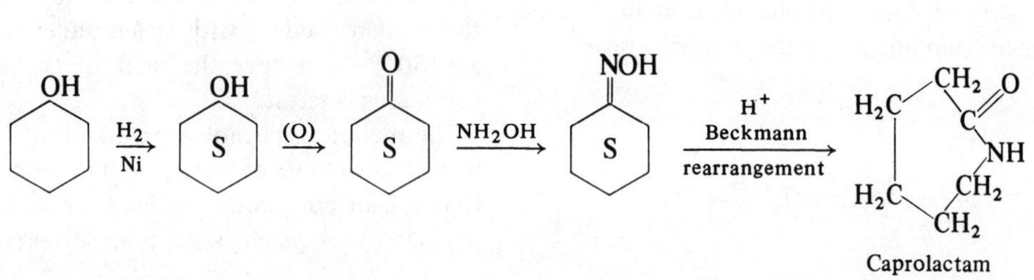

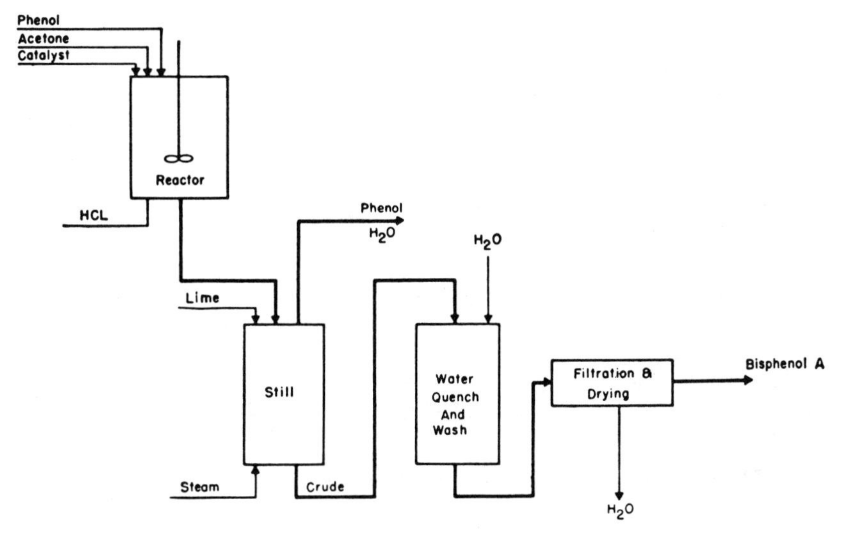

Fig. 25.35 Manufacture of bisphenol A. *(Pet. Ref., 38, no. 11, 225 (1959); copyright 1959 by Gulf Publishing Co.)*

Nylon 66, produced from adipic acid and hexamethylenediamine, is the major current domestic nylon. Nylon 6, derived from caprolactam, is growing more rapidly. Although these nylon intermediates are mainly derived from cyclohexane, there are various other routes. For example, adipic acid can be made by butylene oxidation as well as by cyclohexane oxidation; hexamethylenediamine can be made from butadiene, adipic acid, or acrylonitrile, as well as from cyclohexane; and caprolactam can be made by starting with phenol as well as with cyclohexane. See Chapter 11 for more details.

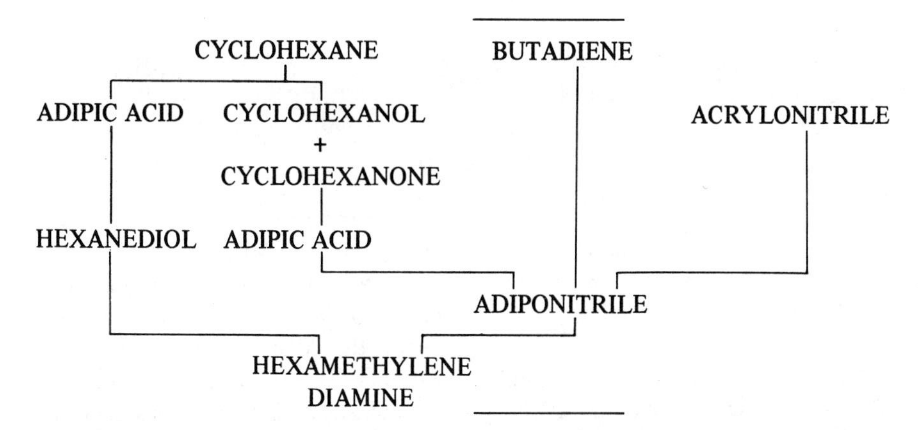

Maleic Anhydride. Maleic anhydride produced during 1978 reached almost 330 million pounds. Some 83% of this was derived from benzene by the process shown in Figure 25.36. The remainder came from the oxidation of butylene, which is described in the C_4 section of this chapter.

Maleic anhydride is used mainly for making polyester resins, agricultural chemicals, lubricating oil additives, and fumaric and malic acids.

Detergent Alkylate. Alkylbenzene is an intermediate in the manufacture of synthetic detergents. Production during 1977 was approximately 700 million pounds from about 60 million gallons of benzene.

Alkylbenzenes are mostly dodecyl- and tridecylbenzene. To make synthetic detergents, α-olefins or internal linear olefins are reacted with benzene to form the alkylbenzene; this is

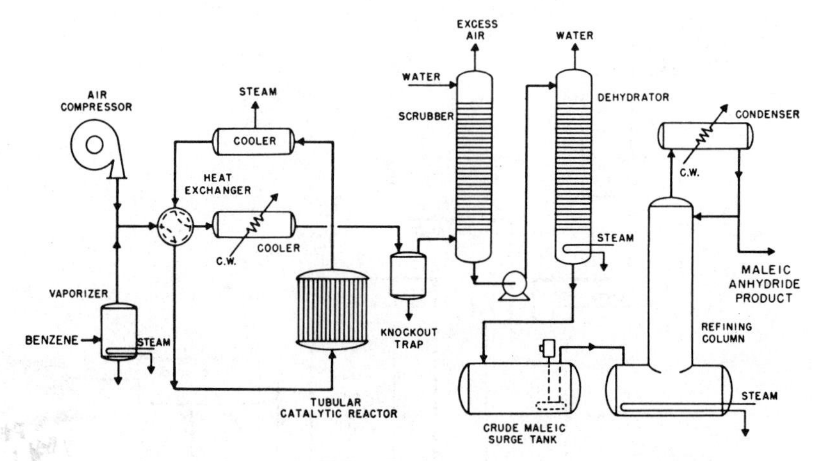

Fig. 25.36 Maleic anhydride manufacture. (Pet. Ref., *38*, no. 11, 265 (1959); copyright 1959 by Gulf Publishing Co.)

then sulfonated, neutralized, blended with chemical "builders," and flake dried. The following chemical reactions are carried out in the processing:

$$C_6H_6 + C_{10}H_{21}\overset{H}{C}{=}CH_2 \xrightarrow[115^\circ F\ max.]{AlCl_3} C_6H_5C_{12}H_{25}$$

Dodecylbenzene

$$\downarrow \; H_2SO_4\ (100\%),\ 20\%\ oleum\ or\ SO_3,\ 100\text{-}400^\circ F$$

$$C_{12}H_{25}C_6H_4SO_3H \xrightarrow[NaOH]{125^\circ F} C_{12}H_{25}C_6H_4SO_3Na$$

The manufacture of detergent alkylate is described in Chapter 13, "Soaps and Synthetic Detergents."

Aniline. Some 500 million pounds of aniline was produced in the United States in 1978. It is produced by one of the two routes shown below, of which the first process is illustrated by the flow sheet in Figure 25.37.

About 52% of the aniline is used to manufacture *p,p'*-methylene diphenyl diisocyanate (MDI) by the reaction with formaldehyde and phosgene. The MDI is then reacted further to make rigid urethane foams for insulation. A new process has been announced which produces the MDI directly from nitrobenzene. Another 30% of the aniline goes into the manufacture of rubber chemicals such as thiazole derivatives. Other aniline uses are in dyes, drugs, veterinary medicines, and photographic chemicals.

$$1.\quad C_6H_5NO_2 + 3H_2 \xrightarrow[\Delta]{\text{Cat.}} C_6H_5NH_2 + 2H_2O$$

Nitrobenzene Aniline

$$2.\quad C_6H_5NO_2 \xrightarrow[\text{HCl}]{\text{iron filings}} C_6H_5NH_2 + 2H_2O$$

Some Other Benzene-Derived Chemicals. Other significant chemicals derived from benzene are mono- and dichlorobenzene, nitrobenzene, and resorcinol.

Nitrobenzene is manufactured from benzene and mixed nitrating acids as shown below:

$$C_6H_6 \xrightarrow[45\text{-}60^\circ C]{HNO_3,\ H_2SO_4} C_6H_5NO_2 + H_2O$$

Nitrobenzene, is used mainly to produce aniline. Of the 900 million pounds produced in 1979, over 95 percent was used in aniline production.

Mono- and dichlorobenzene are made by two routes—direct chlorination or oxychlorination with HCl and air. Figure 25.38 shows a flow sheet of the Dow process for direct chlorination.

Monochlorobenzene is used for the manufacture of sulfur dyes such as sulfur black, drugs, perfumes, and numerous solvent applications, *o*-Dichlorobenzene is used mostly as a solvent or cleaner, particularly for metal degreasing. It has utility, when purified and stabilized, as a heat-transfer fluid in the temperature range of 150–260°C. *p*-Dichlorobenzene is used extensively in moth protection for wool. Small cakes of *p*-dichlorobenzene are used in the sanitary field. Its vapor pressure and pleasant odor make it highly suitable for this purpose.

Resorcinol (m-dihydroxybenzene) production for 1977 is reported at approximately 35 million pounds. Current production is reported to be mainly by the sulfonation process where the sodium salt of *m*-disulfonic acid is hydrolyzed with caustic, then neutralized with a strong acid. A new plant started up in 1971 using a new process wherein *m*-diisopropylbenzene is oxidized to *m*-diisopropylbenzene dihydroperoxide which in turn is cleaved to yield resorcinol and acetone. The reactions involved are similar to those used in producing phenol from cumene.

Resorcinol is used by the tire industry in a resorcinol-formaldehyde resin form to bond the tire cord to rubber. This is a particularly

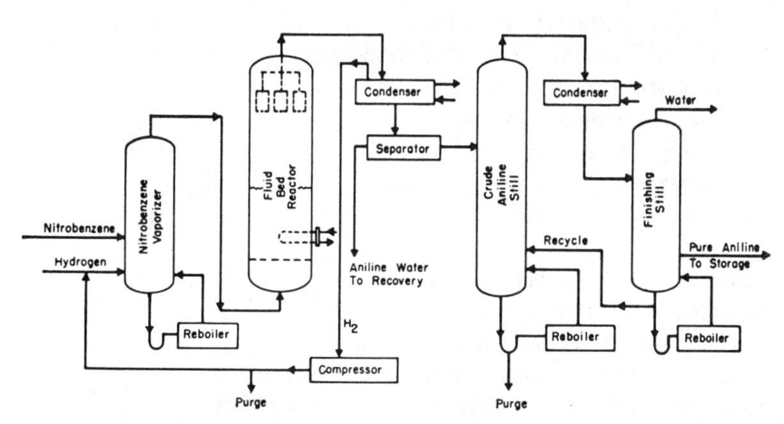

Fig. 25.37 Manufacture of aniline. (*Pet. Ref.,* **38,** *no. 11, 224 (1959); copyright 1959 by Gulf Publishing Co.*)

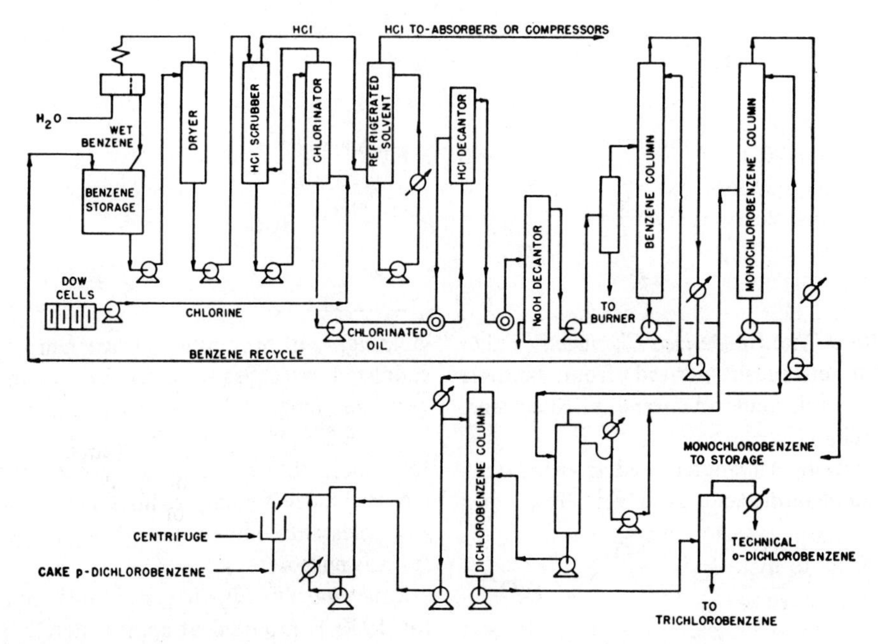

Fig. 25.38 Chlorobenzene manufacture. *(Ind. Eng. Chem., 52, no. 11, 895 (1960); copyright 1960 by American Chemical Society and reprinted by permission of the copyright owner.)*

effective adhesive for polyester and fiber glass cords.

Other uses are as a wood adhesive in the resorcinol-formaldehyde resin form–serving as a fast setting resin at low temperature–used mainly for laminated beam applications. Some resorcinol is used in the synthesis of ultraviolet absorbers, dyes, and for some pharmaceutical uses.

Toluene Products

Approximately 5.5 billion pounds of toluene were consumed in the United States in 1978 for chemical manufacture and solvent uses. This amounts to something less than 10% of total toluene consumption, the remainder going into gasoline. The breakdown into major users is; benzene (66%), solvents (16%), toluene diisocyanate (8%), benzoic acid (3%), and benzyl chloride (2%).

Toluene is converted into benzene by a hydrodealkylation process. This use of toluene is very sensitive to the overall price and demand for benzene, since the benzene produced this way is more costly than that isolated directly from refinery reformate streams. About one-fourth of the benzene produced in 1978 was derived from toluene.

The future use of toluene as a solvent is expected to show a steady decline because of government regulations controlling air pollution.

Toluene Diisocyanate (TDI). Toluene diisocyanate can be manufactured from toluene by the route indicated in the following equations.

$$\text{Toluene } (C_6H_5CH_3) \xrightarrow[80^\circ C]{H_3SO_4,\ HNO_3} CH_3C_6H_3(NO_2)_2 \xrightarrow[HCl]{Fe}$$

$$CH_3C_6H_3(NH_2)_2 \xrightarrow{COCl_2}$$

$$CH_3C_6H_3(NHCOCl)_2 \xrightarrow[\text{80\% yield}]{\Delta} CH_3C_6H_3(NCO)_2$$

Toluene diisocyanate

Toluenediamine is dissolved in a high-boiling compatible solvent. The solvent mixture (10 percent amine) is then mixed with phosgene (1.5 to 3.5 lb phosgene per lb diamine) in a reactor held at 20 to 50°C. In this temperature range, the first phosgene unit is reacting.

The resulting slurry is pumped to a second reactor where further reaction with gaseous phosgene is carried out at 185°C. Unreacted phosgene and HCl are vented from the top of the reactor. The final conversion to the diisocyanate occurs in a third vessel where an inert gas is blown through the solution at 110 to 115°C. The resulting solution of crude TDI is distilled to 2 to 5 mm Hg pressure. The diisocyanate product distills overhead.

Toluene diisocyanate is reacted with polyols or polyesters to make polyurethanes. Flexible polyurethane forms are used for cushioning and padding in automobiles, furniture, carpeting, etc. Semirigid urethane foams are used for padding such as crash pads on automobiles. Rigid urethane foams possess excellent thermal insulation properties; thus, they are used for plastic panels for home construction and in insulation.

Benzoic Acid. Benzoic acid is produced by the oxidation of toluene with air in the presence of a catalyst. Almost one-half of the benzoic acid thus produced is reacted further to make phenol. The other main uses are in manufacture of plasticizers, benzoyl chloride, and sodium benzoate.

Benzyl Chloride. The principal method for producing benzyl chloride involves the dark-phase chlorination of toluene, followed by mild neutralization and distillation. Almost two-thirds of the benzyl chloride is used to manufacture the diester, butyl benzyl phthalate, which is widely used as a plasticizer in PVC resins. The second important use is to make benzyl alcohol. Benzyl alcohol finds use in photographic chemicals, perfumes, and cosmetics.

Vinyltoluene. Vinyltoluene is produced by a process similar to the one used for styrene, i.e., alkylation to ethyltoluene by Friedel-Crafts reaction between toluene and ethylene, then catalytic dehydrogenation to vinyltoluene. Processing is complicated by the existence of 3 isomers; also, *o*-vinyltoluene undergoes side reactions in the dehydrogenation, and therefore, it is separated before dehydrogenation.

Vinyltoluene is in many uses a replacement for styrene, in some cases producing a superior product, as in certain types of coatings.

Chemicals from Xylene

Xylenes are obtained from petroleum reformate in the form of a "mixed xylenes" stream. A typical composition of this stream is 20 percent ethylbenzene, 18 percent paraxylene, 40 percent *m*-xylene, and 22 percent *o*-xylene. The major chemical uses of xylene require the pure isomers. *o*-Xylene can be separated by distillation, while most *p*-xylene is separated by low-temperature crystallization. About 400 million gallons of xylene go into the manufacture of chemicals. In terms of the specific isomers this amounts to 1040 million pounds of *o*-xylene, 90 million pounds of *m*-xylene, and more than 1000 million pounds of *p*-xylene. Their major uses are phthalic anhydride, isophthalic acid, and terephthalates, respectively.

Phthalic Anhydride from o*-Xylene.* The flow sheet, Fig. 25.39, shows the process for making phthalic anhydride from *o*-xylene. The xylene is vaporized by injection into the hot stream and then passes through a catalyst-filled 10,000-tube reactor. The crude phthalic anhydride desublimes in a switch condenser and any phthalic acid present is dehydrated in the predecomposer vessel. The crude is finally purified in two distillations. Although the fixed-bed process is currently the most important, there are a number of plants in operation which use a fluidized-bed reactor.

About 960 million pounds of phthalic anhydride are produced annually in the United States. More than 72 percent of the present plant capacity is based on *o*-xylene, and it is expected that the trend to this process instead of the naphthalene-based route will continue.

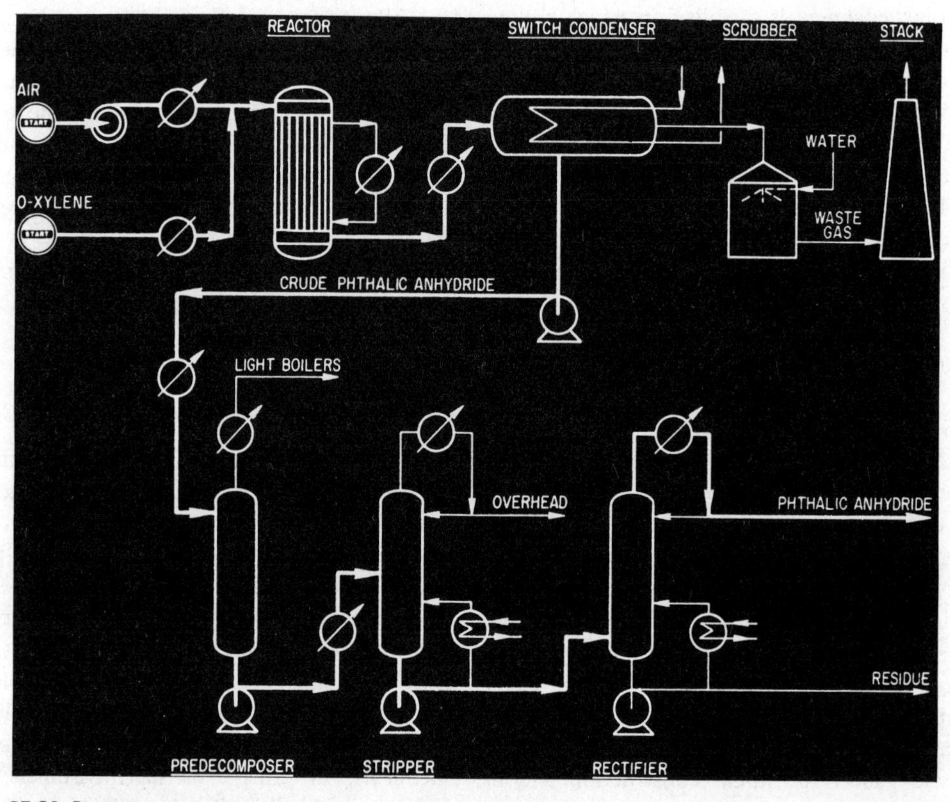

Fig. 25.39 Phthalic anhydride from *o*-xylene and air. (*Hydrocarbon Processing,* **58**, *no. 11, 207* (*1979*); *copyright 1979 by Gulf Publishing Co.*)

The major uses of phthalic anhydride are in plasticizers, 50 percent; alkyd resins, 26 percent; and unsaturated polyester resins, 13 percent. The plasticizers are esters made by reacting two moles of an alcohol, such as 2-ethylhexanol, with one mole of phthalic anhydride. By far the largest use is in plasticizing vinyl chloride polymer and copolymers. Alkyd resins are a type of polyester resin used in surface coatings. The most rapidly growing end use is in unsaturated polyester resins for reinforced plastics. Here the phthalic anhydride is used to modify the resin by replacing a portion of the unsaturated acid. Other outlets for phthalic anhydride are in dyes, agricultural chemicals, and pharmaceuticals.

Isophthalic Acid. Although *m*-xylene is the most abundant xylene isomer, it is in least demand as a chemical raw material. The only major outlet is in the manufacture of isophthalic acid, which is made mainly by liquid-phase oxidation, using heavy metal catalysts.

Terephthalic Acid. p-Xylene is in strong demand as the major raw material for the manufacture of terephthalic acid and dimethyl terephthalates, intermediates used in the production of polyester fibers and film. Processing techniques are very similar to those used for converting *m*-xylene. The crude terephthalic acid obtained after oxidation is either purified as the acid, or it is reacted with methanol to give dimethyl terephthalate. Most of the present production of polyethylene terephthalate comes from the methyl ester, but there are indications that the acid will be the preferred route in the future.

More than a billion pounds of *p*-xylene are consumed in America alone, with an almost equivalent amount in the rest of the world. Better than 90 percent of this goes into fibers, and the remainder is used to make films.

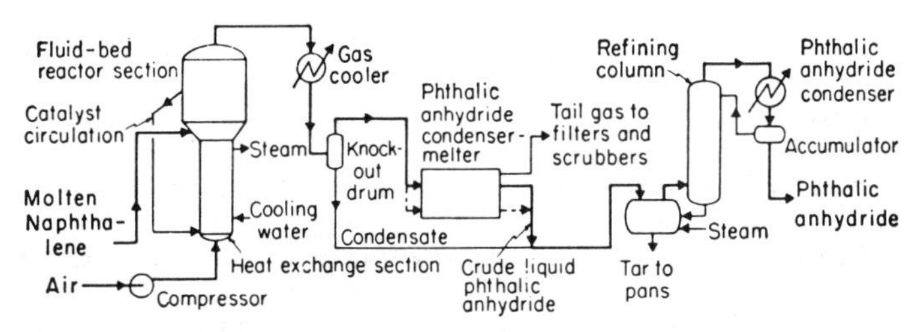

Fig. 25.40 Phthalic anhydride by Badger fluid-bed process. *(Reprinted with permission from Chem. Eng. (Dec. 14, 1959); copyright 1959 by McGraw-Hill, Inc.)*

Naphthalene Chemicals

The U.S. supply of naphthalene was approximately 157 million pounds in 1978. Prior to 1960, all naphthalene came from coke, but now almost half comes from petroleum sources. About 50 percent is consumed in the manufacture of phthalic anhydride. The remainder goes into insecticides, β-naphthol, mothballs, tanning agents, and surfactants.

Phthalic Anhydride. Phthalic anhydride is produced by catalytic oxidation of naphthalene. Two major types of processes are available today. One utilizes a fluid-bed oxidation reactor, while the other utilizes a fixed-bed reactor. Both processes utilize naphthalene and air as raw materials, carry out the oxidation under the influence of catalysts, and then finish the product by distillation. The fluid-bed process is depicted in Fig. 25.40.

A process for phthalic anhydride derived from *o*-xylene is described in an earlier section.

Other Naphthalene Uses. The second largest use for naphthalene is in the manufacture of carbaryl (1-naphthyl N-methylcarbamate) for insecticides. About 45 million pounds went to this use in 1971. The naphthalene is first made into 1-naphthol and then converted to carbaryl by reaction with methyl isocyanate.

β-Naphthol is the most significant other use for naphthalene taking some 30 million pounds in 1978. It is manufactured commercially by first making β-naphthalenesulfonic acid. The acid is then fused with caustic, acidified, washed, and vacuum distilled. β-Naphthol has numerous uses in dye, rubber, perfume, and pharmaceutical industries.

Sulfonated naphthalenes are available commercially as surface active agents of various sorts.

Other Polymethylbenzenes

Among the newer aromatic raw materials are the higher polymethylbenzenes. The most important of these is pseudocumene (1,2,4-trimethylbenzene) which is used in the manufacture of trimellitates for vinyl plasticizers. Other materials in this area are mesitylene and durene.

REFERENCES

1. Lowenheim, F. A. and Moran, M. K., *Faith, Keyes and Clark, Industrial Chemicals*, Wiley Interscience Publication (1975).
2. *Chemical Marketing Reporter* (May 12 and November 24, 1980).
3. Hahn, A. V., *The Petrochemical Industry, Maket and Economics*, New York, McGraw-Hill (1970).
4. Hancack, E. G., ed., *Propylene and Its Industrial Derivatives*, New York, John Wiley & Sons (1973).
5. Shreve, R. H. and Brink, J. A., Jr., *Chemical Process Industries*, New York, McGraw-Hill Book Co. (1977).
6. Goldstein, R. F. and Waddams, A. L. *The Petroleum Chemicals Industry*, London, E. & F.N. Span Ltd., (1967).
7. *Chemical Economics Handbook*, Menlo Park, CA., Stanford Research Insitute.
8. *World Petrochemicals*, Stanford Research Institute International, Menlo Park, CA.
9. *Synthetic Organic Chemicals*, U.S. International Trade Commission Reports, 1950–1978 incl.
10. Kent, J. A. Ed., *Riegel's Industrial Chemistry*, 7th ed., Chapter 25, New York, Van Nostrand Reinhold.

11. "Petrochemical Handbook", *Hydrocarbon Processing*, **50** No. 11, 113 (1971); **56** No. 11, 115 (1977); **59** No. 11, 115 (1979).
12. Landaw, R. and Schaffel, G. S., "Recent Developments in Ethylene Chemistry", in *Origin and Refining of Petroleum*, Part 8, Washington, American Chemical Society, 1971.
13. Kehde, H. *C_4 Hydrocarbon Production and Distribution*, New York, American Institute of Chemical Engineers, 1970.
14. Ockerboom, N. E., "Xylene and Higher Aromatics", *Hydrocarbon Processing* Series, beginning with **50**, No. 7, 112 (1971).
15. *Encyclopedia of Chemical Technology*, Vol. 9, pp. 432–47, Kirk and Othmer, 3rd ed., John Wiley & Sons, (1980).

INDEX